Get Connected to

ConnectED

connectED.mcgraw-hill.com

- 100% Online
- One-Stop Shop, One Personalized Password
- Easy Intuitive Navigation
- Resources, Resources, Resources

For Students

Leave your books at school. Now you can go online and interact with your **StudentWorks™ Plus** digital Student Edition from any place, any time!

For Teachers

ConnectED is your one-stop online center for everything you need to teach using **Glencoe Precalculus**, including: **TeacherWorks™ Plus** digital Teacher Edition, lesson planning and scheduling tools, pacing, and assessment.

For Parents

Get homework help, help your student prepare for testing, and review math topics.

Get Connected to the way students learn and teachers teach TODAY!

Glencoe McGraw-Hill

Precalculus

Authors
Carter · Cuevas · Day · Malloy · Bryan · Holliday · Hovsepian

Glencoe

About the Cover

To throw a discus, an athlete spins around one and a half times to build momentum before releasing the discus. When viewed from above, a right-handed thrower spins clockwise, and a left-handed thrower spins counterclockwise. The path of the discus through the spin and throw can be graphed on a polar coordinate system. You will study polar coordinates in Chapter 9.

Photo Credits:

Endsheet (t) Richard Hutchings, (c) Getty Images, (b) Jupiter Images/Think stock/Alamy; **iv–v** The McGraw-Hill Companies; **vi–vii** Bruce Heinemann/Getty Images; **viii** Anderson Ross/Blend Images/ Getty Images; **ix** Brand X/JupiterImages; **x** Inti St. Clair/Digital Vision/Getty Images; **xi** Daniel J Cox/Getty Images; **xii** PatitucciPhoto/Getty Images; **xiii** PM Images/The Image Bank/Getty Images; **xiv** Image Source; **xv** Mike Powell/Allsport Concepts/Getty Images; **xvi** Ariel Skelley/ Blend Images/Getty Images; **xvii** Jessica Peterson/Rubberball Productions/Getty Images; **xviii** Huntstock, Inc/Alamy; **xix** Noel Hendrickson/Digital Vision/Getty Images; **xx** Steven Wooster/ Dorling Kindersley/Getty Images.

TI-Nspire is a trademark of Texas Instruments.
TI-Nspire images used by permission.

The McGraw·Hill Companies

 Glencoe

Send all inquiries to:
Glencoe/McGraw-Hill
8787 Orion Place
Columbus, OH 43240-4027

ISBN: 978-0-07-880273-7
MHID: 0-07-880273-3

Printed in the United States of America.

3 4 5 6 7 8 9 10 11 12 WDQ/LEH 19 18 17 16 15 14 13 12 11 10

Designed for Success

Contents in Brief

Authors

O ur lead authors ensure that the Macmillan/McGraw-Hill and Glencoe/McGraw-Hill mathematics programs are truly vertically aligned by beginning with the end in mind—success in Algebra 1 and beyond. By "backmapping" the content from the high school programs, all of our mathematics programs are well articulated in their scope and sequence.

Lead Authors

John A. Carter, Ph.D.
Assistant Principal for Teaching and Learning
Adlai E. Stevenson High School
Lincolnshire, Illinois
Areas of Expertise: Using technology and
   manipulatives to visualize concepts;
   Mathematics Achievement of
   English-Language Learners

Gilbert J. Cuevas, Ph.D.
Professor of Mathematics Education
Texas State University–San Marcos
San Marcos, Texas
Areas of Expertise: Applying concepts and
   skills in mathematically rich contexts;
   Mathematical Representations

Roger Day, Ph.D., NBCT
Mathematics Department Chairperson
Pontiac Township High School
Pontiac, Illinois
Areas of Expertise: Understanding and applying
   probability and statistics; Mathematics Teacher
   Education

Carol Malloy, Ph.D.
Associate Professor
University of North Carolina at Chapel Hill
Chapel Hill, North Carolina
Areas of Expertise: Representations and critical
   thinking; Student Success in Algebra 1

Program Authors

Luajean Bryan
Mathematics Teacher
2009 Tennessee Teacher of the Year
Walker Valley High School
Cleveland, Tennessee
Areas of Expertise: Meaningful projects
that make precalculus and calculus
real to students

Dr. Berchie Holliday, Ed.D.
National Mathematics Consultant
Silver Spring, Maryland
Areas of Expertise: Using mathematics
to model and understand real-world
data; the effect of graphics on
mathematical understanding

Viken Hovsepian
Professor of Mathematics
Rio Hondo College
Whittier, California
Areas of Expertise: Polynomials
and Field Theory

Consultants and Reviewers

These professionals were instrumental in providing valuable input and suggestions for improving the effectiveness of the mathematics instruction.

Consultants

Preparation for Advanced Placement

Elizabeth W. Black
Mathematics Teacher
Greenwich High School
Greenwich, Connecticut

Dixie Ross
Lead Teacher for Advanced Placement
  Mathematics
Pflugerville High School
Pflugerville, Texas

Test Preparation

Christopher F. Black
Director
College Hill Coaching
Greenwich, Connecticut

Graphing Calculator

Ruth M. Casey
T^3 National Instructor
(Teachers Teaching with Technology)
Frankfort, Kentucky

Jerry J. Cummins
Former President, National Council
  of Supervisors of Mathematics
Hinsdale, Illinois

Science/Physics

Jane Bray Nelson
Instructor
Santa Fe College
Gainesville, Florida

Jim Nelson
Instructor
Santa Fe College
Gainesville, Florida

Pre-Engineering

Celeste Baine
Consultant
Springfield, Oregon

Alfred C. Soldavini, P.E.
Adjunct Professor
Devry University
Alpharetta, Georgia

Mt. Rainier National Park, Washington

Reviewers

Cliff Allred
Mathematics Department Chair
C. F. Vigor High School
Mobile, Alabama

Fernando Antunez
Mathematics Teacher
Stranahan High School
Fort Lauderdale, Florida

Mark B. Baetz
Mathematics Teacher/School Test
  Coordinator
Salem High School
Salem, Virginia

Jamie Buchanan
Mathematics Department Chair
Petoskey High School
Petoskey, Michigan

Douglas Caringer
Mathematics Teacher
S. F. Austin High School
Houston, Texas

Debby A. Churella
Mathematics Teacher
Oak Glen High School
New Cumberland, West Virginia

Franco A. DiPasqua
Director of K-12 Mathematics
West Seneca Central
West Seneca, New York

Robert Dubesa
Mathematics Teacher
Pioneer High School
Whittier, California

Kimberly Garcia
Mathematics Teacher
Leon High School
Tallahassee, Florida

Courtney Glass
Mathematics Teacher
Monrovia High School
Monrovia, California

Jo Grayson
Mathematics Teacher, Dept. Chairperson
Martinsville High School
Martinsville, Virginia

Marie H. Hanes
Mathematics Teacher
Carroll High School
Fort Wayne, Indiana

Leah B. Hudgens
Mathematics Teacher
Deane Bozeman School
Panama City, Florida

Wanda Jeans LeBert
Mathematics Teacher, Masters in Science
A. M. Barbe High School
Lake Charles, Louisiana

Grey Nakayama
Mathematics Teacher
Spain Park High School
Birmingham, Alabama

Steven Parisi
Mathematics Teacher
Williamsville North High School
Williamsville, New York

Paul E. Peelle
Mathematics Teacher
Easthampton High School
Easthampton, Massachusetts

Penelope Sather
Mathematics Teacher
R. L. Turner High School
Carrollton, Texas

Julie Skalka Martin
District Mathematics Curriculum
  Coordinator
Metro Nashville Public Schools
Nashville, Tennessee

Richard R. Soendlin
Former Mathematics Department Head
Arsenal Technical High School
Indianapolis, Indiana

Alex Taylor
AP Calculus AB Teacher
Z. B. Vance High School
Charlotte, North Carolina

Vicalyn Tyson
Mathematics Teacher
Pickerington Local School District
Pickerington, Ohio

Anne Uma
Mathematics Teacher
Kingstree Senior High School
Kingstree, South Carolina

Fred Vogt
Mathematics Department Chair
Pekin Community High School
Pekin, Illinois

Julie Williams
Mathematics Instructor
Lawton High School
Lawton, Oklahoma

Kathleen Williams
Mathematics Teacher
Glendale Senior High School
Glendale, California

Paula S. Wyatt
Instructor and Head of Mathematics
  Department
East Gaston High School
Mt. Holly, North Carolina

Marianne D. Zamor
Mathematics Department Chair
Owasso High School
Owasso, Oklahoma

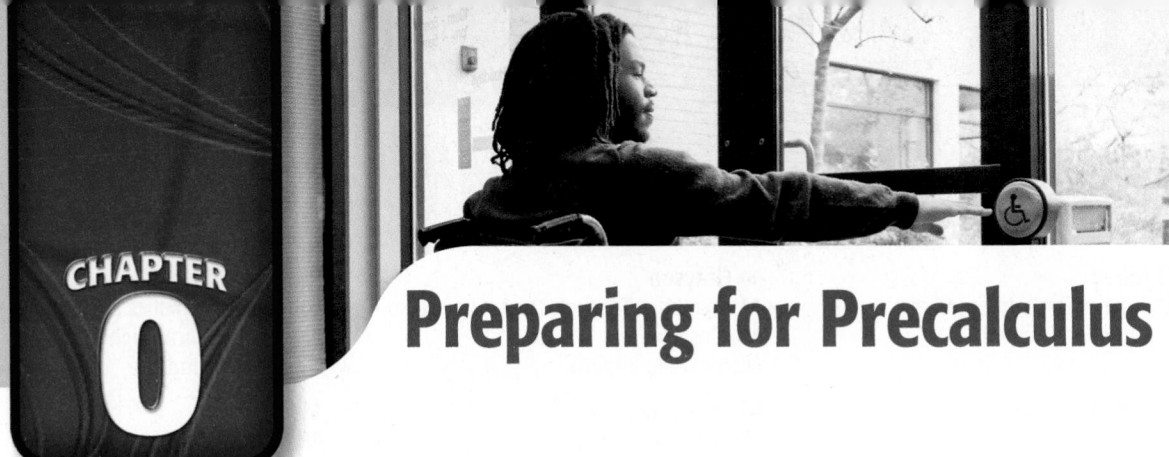

Preparing for Precalculus

Chapter 0 Support

📖 Helping You Learn

- **New Vocabulary** P3, P6, P9, P14, P18, P23, P28, P32
- **Key Concepts** P6, P9, P10, P11, P12, P14, P15, P16, P18, P19, P21, P24, P25, P28, P29, P30, P32
- **Exercises** P5, P8, P13, P17, P22, P27, P31, P37

Math Online ▷

- **Animations** P9, P21
- **Self-Check Quizzes** P3, P6, P9, P14, P18, P23, P28, P32

Functions from a Calculus Perspective

Chapter 1 Support

Helping You Learn
- **New Vocabulary** 4, 13, 24, 34, 45, 57, 65
- **Key Concepts** 4, 5, 6, 16, 18, 24, 27, 34, 36, 38, 45, 46, 47, 48, 49, 51, 57, 58, 65, 66, 68
- **Concept Summary** 25
- **Exercises** 9, 19, 30, 40, 52, 61, 70
- **Multiple Representations** 11, 22, 32, 54, 63, 72
- **H.O.T. Problems** 11, 22, 32, 42, 54, 63, 72

Math Online
- **Animations** 6, 49
- **Self-Check Quizzes** 4, 13, 24, 34, 45, 57, 65

Preparing for Testing
- **Multiple Choice** 12, 23, 33, 43, 55, 64, 73
- **Free Response** 64

CHAPTER 2

Power, Polynomial, and Rational Functions

Chapter 2 Support

Helping You Learn

- **New Vocabulary** 86, 97, 109, 119, 130, 141
- **Key Concepts** 86, 90, 98, 100, 102, 110, 111, 113, 119, 121, 123, 124, 131, 132, 134
- **Concept Summary** 114
- **Exercises** 92, 104, 115, 127, 138, 145
- **Multiple Representations** 94, 106, 116, 128, 139, 146
- **H.O.T. Problems** 94, 106, 116, 128, 139, 146

 Math Online

- **Animations** 98, 123
- **Self-Check Quizzes** 86, 97, 109, 119, 130, 141

Preparing for Testing

- **Multiple Choice** 95, 107, 117, 129, 140, 147
- **Free Response** 147

Exponential and Logarithmic Functions

Chapter 3 Support

📖 **Helping You Learn**
- **New Vocabulary** 158, 172, 181, 190, 200
- **Key Concepts** 158, 159, 162, 163, 164, 172, 173, 174, 176, 181, 183, 190, 191, 202, 204
- **Exercises** 166, 178, 185, 196, 207
- **Multiple Representations** 168, 179
- **H.O.T. Problems** 168, 179, 187, 198, 209

▶ **Math Online**
- **Animations** 159, 172
- **Self-Check Quizzes** 158, 172, 181, 190, 200

✏️ **Preparing for Testing**
- **Multiple Choice** 169, 180, 188, 199, 210
- **Free Response** 210

CHAPTER 4

Trigonometric Functions

Chapter 4 Support

📖 Helping You Learn

- **New Vocabulary** 220, 231, 242, 256, 269, 280, 291
- **Key Concepts** 220, 222, 223, 232, 233, 234, 235, 237, 242, 243, 244, 245, 248, 250, 256, 257, 259, 260, 261, 269, 270, 272, 273, 276, 284, 286, 291, 293, 295, 296, 297
- **Concept Summary** 262
- **Exercises** 227, 238, 251, 264, 277, 288, 298
- **Multiple Representations** 229, 252, 265, 289
- **H.O.T. Problems** 229, 240, 252, 265, 278, 289, 300

Math Online ▷

- **Animations** 244, 259
- **Self-Check Quizzes** 220, 231, 242, 256, 269, 280, 291

✏️ Preparing for Testing

- **Multiple Choice** 230, 241, 253, 266, 279, 290, 301
- **Free Response** 301

CHAPTER 5

Trigonometric Identities and Equations

Chapter 5 Support

CHAPTER 6

Systems of Equations and Matrices

Chapter 6 Support

📖 Helping You Learn

- **New Vocabulary** 364, 375, 388, 398, 405
- **Key Concepts** 364, 366, 367, 375, 377, 378, 379, 381, 382, 388, 390, 405, 406
- **Concept Summary** 380, 401
- **Exercises** 372, 383, 392, 402, 410
- **Multiple Representations** 385, 403
- **H.O.T. Problems** 373, 385, 393, 403, 411

Math Online ▷

- **Animations** 375, 406
- **Self-Check Quizzes** 364, 375, 388, 398, 405

✏️ Preparing for Testing

- **Multiple Choice** 374, 386, 394, 404, 412
- **Free Response** 412

CHAPTER 7

Conic Sections and Parametric Equations

Chapter 7 Support

Helping You Learn
- **New Vocabulary** 422, 432, 442, 464
- **Key Concepts** 423, 427, 433, 435, 437, 443, 447, 454, 455, 457, 464, 467
- **Exercises** 428, 438, 449, 459, 469
- **Multiple Representations** 430, 451, 460, 470
- **H.O.T. Problems** 430, 440, 451, 460, 470

Math Online
- **Animations** 433, 454
- **Self-Check Quizzes** 422, 432, 442, 454, 464

Preparing for Testing
- **Multiple Choice** 431, 441, 452, 461, 471
- **Free Response** 471

Vectors

Chapter 8 Support

CHAPTER 9

Polar Coordinates and Complex Numbers

Chapter 9 Support

CHAPTER 10

Sequences and Series

Chapter 10 Support

📖 Helping You Learn

- **New Vocabulary** 590, 599, 608, 621, 628, 636
- **Key Concepts** 594, 599, 603, 609, 612, 613, 621, 630, 632, 636, 638, 639, 640
- **Exercises** 595, 605, 615, 625, 633, 642
- **Multiple Representations** 597, 617, 634
- **H.O.T. Problems** 597, 606, 617, 626, 634, 643

Math Online ➤

- **Animations** 594, 629
- **Self-Check Quizzes** 590, 599, 608, 621, 628, 636

✏️ Preparing for Testing

- **Multiple Choice** 598, 607, 618, 627, 635, 644
- **Free Response** 644

CHAPTER 11

Inferential Statistics

Chapter 11 Support

 Helping You Learn

- **New Vocabulary** 654, 664, 674, 685, 696, 705, 713
- **Key Concepts** 654, 657, 664, 665, 666, 667, 668, 669, 670, 674, 675, 676, 677, 686, 689, 690, 697, 699, 701, 707, 713, 715, 716
- **Concept Summary** 691, 719
- **Exercises** 660, 671, 681, 692, 702, 710, 720
- **Multiple Representations** 662, 672, 682, 721
- **H.O.T. Problems** 662, 672, 682, 693, 703, 711, 721

Math Online

- **Animations** 654, 675
- **Self-Check Quizzes** 654, 664, 674, 685, 696, 705, 713

Preparing for Testing

- **Multiple Choice** 663, 673, 683, 694, 704, 712, 722
- **Free Response** 722

CHAPTER 12

Limits and Derivatives

Chapter 12 Support

Helping You Learn

- **New Vocabulary** 736, 746, 758, 766, 775, 784
- **Key Concepts** 737, 738, 740, 746, 750, 751, 758, 760, 767, 769, 770, 771, 777, 785, 786
- **Concept Summary** 740
- **Exercises** 743, 754, 762, 772, 781, 789
- **Multiple Representations** 763, 773, 782, 790
- **H.O.T. Problems** 744, 755, 763, 773, 782, 790

Math Online

- **Animations** 758, 775
- **Self-Check Quizzes** 736, 746, 758, 766, 775, 784

Preparing for Testing

- **Multiple Choice** 745, 756, 764, 774, 783, 791
- **Free Response** 791

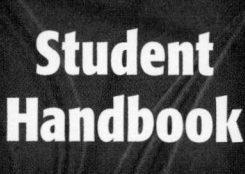

Student Handbook

Contents

Get Started on Chapter 0

Chapter 0 contains lessons on topics from previous courses. You can use this chapter in various ways.

- Begin the school year by taking the Pretest. If you need additional review, complete the lessons in this chapter. To verify that you have successfully reviewed the topics, take the Posttest.

- As you work through the text, you may find that there are topics you need to review. When this happens, complete the individual lessons that you need.

- Use this chapter for reference. When you have questions about any of these topics, flip back to this chapter to review definitions or key concepts.

Chapter 0 Contents

Math Online > glencoe.com

- Study the chapter online
- Explore **Get Animated**
- Get extra help from your own **Personal Tutor**
- Use **Extra Examples** for additional help
- Take a **Self-Check Quiz**
- **Review Vocabulary** in fun ways

Review Vocabulary

English		Español
set	• P3 •	conjunto
element	• P3 •	elemento
subset	• P3 •	subconjunto
universal set	• P3 •	conjunto universal
complement	• P3 •	complemento
union	• P4 •	unión
intersection	• P4 •	intersección
empty set	• P4 •	conjunto vacío
imaginary unit	• P6 •	unidad imaginario
complex number	• P6 •	número complejo
standard form	• P6 •	forma estándar
imaginary number	• P6 •	número imaginario
complex conjugates	• P7 •	conjugados complejos
parabola	• P9 •	parábola
axis of symmetry	• P9 •	eje de simetría
vertex	• P9 •	vértice
completing the square	• P12 •	completer el cuadrado
*n*th root	• P14 •	raíz enésima
principal root	• P14 •	raíz principal
system of equations	• P18 •	sistema de ecuaciones
substitution method	• P18 •	método de sustitución
elimination method	• P19 •	método de eliminación
matrix	• P23 •	matriz
element	• P23 •	elemento
dimension	• P23 •	dimensión
experiment	• P28 •	experimento
factorial	• P28 •	factorial
permutation	• P29 •	permutación
combination	• P30 •	combinación
measure of central tendency	• P32 •	medida de tendencia central
measures of spread	• P32 •	medidas de propagación
frequency distribution	• P34 •	distribución de frecuencias

> **Multilingual eGlossary** glencoe.com

Use set notation to write the elements of each set. Then determine whether the statement about the set is *true* or *false*.

1. L is the set of whole number multiples of 2 that are less than 22. $18 \in L$

2. S is the set of integers that are less than 5 but greater than -6. $-8 \in S$

Let $C = \{0, 1, 2, 3, 4\}$, $D = \{3, 5, 7, 8\}$, $E = \{0, 1, 2\}$, and $F = \{0, 8\}$. Find each of the following.

3. $D \cap E$

4. $C \cap E$

5. $C \cap F$

6. $D \cup E$

Simplify.

7. $(6 + 5i) + (-3 + 2i)$

8. $(-3 + 4i) - (4 - 5i)$

9. $(1 + 8i)(6 + 2i)$

10. $(-3 + 3i)(-2 + 2i)$

11. $\dfrac{3 + i}{5 - 2i}$

12. $\dfrac{-3 + i}{4 - 3i}$

Determine whether each function has a *maximum* or *minimum* value. Then find the value of the maximum or minimum, and state the domain and range of the function.

13.

14.

Solve each equation.

15. $x^2 - x - 20 = 0$

16. $x^2 - 3x + 5 = 0$

17. $x^2 + 2x - 1 = 0$

18. $x^2 + 11x + 24 = 0$

19. **CARS** The current value V and the original value v of a car are related by $V = v(1 - r)^n$, where r is the rate of depreciation per year and n is the number of years. If the original value of a car is $10,000, what would be the current value of the car after 30 months at an annual depreciation rate of 10%?

Simplify each expression.

20. $\sqrt[6]{x^{12}y^{15}}$

21. $\sqrt[3]{8a^9b^7}$

22. $\sqrt{25r^5t^4u^2}$

23. $\sqrt[5]{32x^{11}y^{20}z^5}$

Simplify.

24. $\dfrac{x^{\frac{1}{2}}}{x^{\frac{1}{4}}}$

25. $\sqrt[4]{81x^8y^{14}}$

26. $\sqrt[7]{x^{15}y^{23}}$

27. $\dfrac{\sqrt[8]{49}}{\sqrt[6]{7}}$

28. **JOBS** Destiny mows lawns for $8 per lawn and weeds gardens for $10 per garden. If she had 8 jobs and made $72, how many of the jobs were mowing? How many were weeding?

Solve each system of equations. State whether the system is *consistent and independent*, *consistent and dependent*, or *inconsistent*.

29. $3x + y = 4$
    $x - y = 12$

30. $2x - y = 2$
    $-4x + 2y = -4$

31. $2x + 4y - z = -1$
    $2x - 3y + 2z = 6$
    $-x - 5y + z = -2$

32. $-3x + 9y - 3z = -12$
    $-3x + y - z = -1$
    $2x - 6y + 2z = 9$

Solve each system of inequalities. If the system has no solution, state *no solution*.

33. $y \geq x + 5$
    $y \leq 2x + 2$

34. $y + x < 3$
    $y > -2x - 4$

35. $4x - 3y < 7$
    $2y - x < -6$

36. $3y \leq 2x - 8$
    $y \geq \dfrac{2}{3}x - 1$

Find each of the following for $D = \begin{bmatrix} -2 & 4 \\ 0 & 1 \\ 4 & -3 \end{bmatrix}$, $E = \begin{bmatrix} 3 & 5 \\ -2 & -1 \\ 0 & -1 \end{bmatrix}$, and $F = \begin{bmatrix} 8 & 2 \\ -3 & -5 \\ 2 & 2 \end{bmatrix}$.

37. $D - F$

38. $D + 2F$

39. $2D - E$

40. $D + E + F$

41. $3D - 2E$

42. $D - 3E + 3F$

Find each permutation or combination.

43. $_9C_5$

44. $_9P_5$

45. $_5P_5$

46. $_5C_5$

47. $_4P_2$

48. $_4C_2$

49. **CARDS** Three cards are randomly drawn from a standard deck of 52 cards. Find each probability.

  a. P(all even)

  b. P(two clubs and one heart)

Find the mean, median, and mode for each set of data. Then find the range, variance, and standard deviation for each population.

50. $\{7, 7, 8, 10, 10, 10\}$

51. $\{0.5, 0.4, 0.2, 0.5, 0.2\}$

0-1 Sets

Objective

- Use set notation to denote elements, subsets, and complements.
- Find intersections and unions of sets.

Vocabulary

set
element
subset
universal set
complement
union
intersection
empty set

Math Online

glencoe.com

Set Notation A **set** is a collection of objects. Each object in a set is called an **element**. A set is named using a capital letter and is written with its elements listed within braces { }.

Set Name	Description of Set	Set Notation
C	pages in a chapter of a book	$C = \{35, 36, 37, 38, 39, 40\}$
A	students who made an A on the test	$A = \{\text{Olinda, Mario, Karen}\}$
L	the letters from A to H	$L = \{A, B, C, D, E, F, G, H\}$
N	positive odd numbers	$N = \{1, 3, 5, 7, 9, 11, 13, \ldots\}$

To write that Olinda is *an element* of set A, write Olinda $\in A$.

EXAMPLE 1 | **Use Set Notation**

Use set notation to write the elements of each set. Then determine whether the statement about the set is *true* or *false*.

a. N is the set of whole numbers greater than 12 and less than 16.

$15 \in N$

The elements in this set are 13, 14, and 15, so $N = \{13, 14, 15\}$. Because 15 is an element of N, $15 \in N$ is a true statement.

b. V is the set of vowels.

$g \in V$

The elements in this set are the letters a, e, i, o, and u, so $V = \{a, e, i, o, u\}$. Because the letter g is not an element of V, a correct statement is $g \notin V$. Therefore, $g \in V$ is a false statement.

c. M is the set of months that begin with J.

April $\in M$

The elements in this set are the months January, June, and July, so $M = \{\text{January, June, July}\}$. Because the month of April is not an element of this set, a correct statement is April $\notin M$. Therefore, April $\in M$ is a false statement.

d. X is the set of numbers on a die.

$4 \in X$

The elements in this set are 1, 2, 3, 4, 5, and 6, so $X = \{1, 2, 3, 4, 5, 6\}$. Because 4 is an element of X, $4 \in X$ is a true statement.

If every element of set B is also contained in set A, then B is called a **subset** of A, and is written as $B \subset A$. The **universal set** U is the set of all possible elements for a situation. All other sets in this situation are subsets of the universal set.

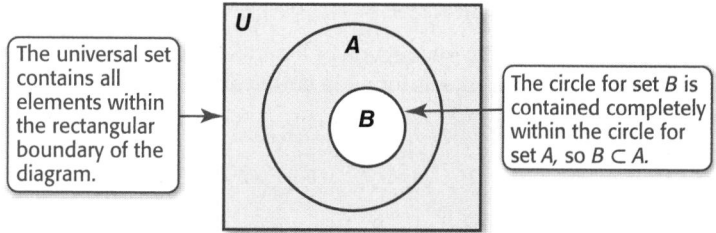

The universal set contains all elements within the rectangular boundary of the diagram.

The circle for set B is contained completely within the circle for set A, so $B \subset A$.

Suppose $A = \{0, 1, 2, 3, 4, 5, 6, 7, 8\}$ and $B = \{1, 2, 3\}$. Because $1 \in \{0, 1, 2, 3, 4, 5, 6, 7, 8\}$, $2 \in \{0, 1, 2, 3, 4, 5, 6, 7, 8\}$, and $3 \in \{0, 1, 2, 3, 4, 5, 6, 7, 8\}$, $B \subset A$.

The set of elements in U that are not elements of set B is called the **complement** of B, and is written as B'. In the Venn diagram, the complement of B is all of the shaded regions.

EXAMPLE 2 **Identify Subsets and Complements of Sets**

Let $U = \{0, 1, 2, 3, 4, 5, 6, 7, 8, 9\}$, $A = \{1, 4, 5, 7, 8, 9\}$, $B = \{5, 7\}$, $C = \{1, 5, 7, 8\}$, $D = \{2, 3\}$, and $E = \{6, 3\}$.

a. State whether $B \subset A$ is *true* or *false*.

$B = \{5, 7\}$ $\qquad\qquad\qquad$ $A = \{1, 4, 5, 7, 8, 9\}$

True; $5 \in A$ and $7 \in A$, so all of the elements of B are also elements of A. Therefore, B is a subset of A.

b. State whether $E \subset D$ is *true* or *false*.

$E = \{6, 3\}$ $\qquad\qquad\qquad$ $D = \{2, 3\}$

False; $6 \notin D$, so not all of the elements of E are in D. Therefore, E is not a subset of D.

c. Find A'.

Identify the elements of U that are not in A.

$A = \{1, 4, 5, 7, 8, 9\}$ $\qquad$ $U = \{0, 1, 2, 3, 4, 5, 6, 7, 8, 9\}$

So, $A' = \{0, 2, 3, 6\}$.

d. Find D'.

Identify the elements of U that are not in D.

$D = \{2, 3\}$ $\qquad\qquad\qquad$ $U = \{0, 1, 2, 3, 4, 5, 6, 7, 8, 9\}$

So, $D' = \{0, 1, 4, 5, 6, 7, 8, 9\}$.

StudyTip

Math Symbols

$\in$ $\quad$ is an element of

$\notin$ $\quad$ is not an element of

$\subset$ $\quad$ is a subset of

A' $\quad$ the complement of set A

$A \cap B$ $\quad$ the intersection of set A and set B

$A \cup B$ $\quad$ the union of set A and set B

Unions and Intersections The **union** of sets A and B, written $A \cup B$, is a new set consisting of all of the elements that are in either A or B. The **intersection** of sets A and B, written $A \cap B$, is a new set consisting of elements found in A and B. If two sets have no elements in common, their intersection is called the **empty set**, and is written as $\varnothing$ or $\{\ \}$.

EXAMPLE 3 **Find the Union and Intersection of Two Sets**

Let $U = \{1, 2, 3, 4, 5, 6, 7, 8\}$, $R = \{2, 4, 6\}$, $S = \{4, 5, 6, 7\}$, and $T = \{1\}$.

a. Find $R \cup S$.

The union of R and S is the set of all elements that belong to R, S, or to both sets.

So, $R \cup S = \{2, 4, 5, 6, 7\}$.

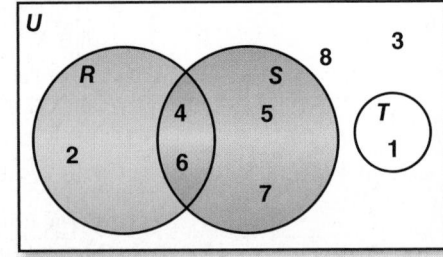

b. Find $R \cap S$.

The intersection of R and S is the set of all elements found in both R and S.

So, $R \cap S = \{4, 6\}$.

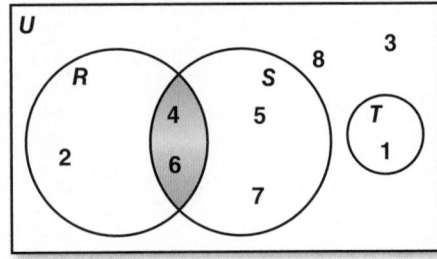

Watch Out!

Disjoint Sets Sets A and B are said to be *disjoint* if they have no elements in common. The intersection of two disjoint sets is the empty set, while the union of two disjoint sets includes all of the elements from each set.

---

c. Find $T \cap S$.

Because there are no elements that belong to both T and S, the intersection of T and S is the empty set. So, $T \cap S = \varnothing$.

Exercises

● = **Step-by-Step Solutions** begin on page R29.

Use set notation to write the elements of each set. Then determine whether the statement about the set is *true* or *false*. (Example 1)

1. *J* is the set of whole number multiples of 3 that are less than 15. $15 \in J$

2. *K* is the set of consonant letters in the English alphabet. $h \in K$

3. *L* is the set of the first six prime numbers. $9 \in L$

4. *V* is the set of states in the U.S. that border Georgia. Alabama $\notin V$

5. *N* is the set of natural numbers less than 12. $0 \in N$

6. *D* is the set of days that start with S. Sunday $\in D$

7. *A* is the set of girls names that start with A. Ashley $\in A$

8. *S* is the set of the 48 continental states in the U.S. Hawaii $\notin S$

For Exercises 9–24, use the following information.
Let $U = \{0, 1, 2, 3, 4, 5, 6, 7, 8, 9, 10, 11, 12\}$,
$A = \{1, 2, 6, 9, 10, 12\}$, $B = \{2, 9, 10\}$, $C = \{0, 1, 6, 9, 11\}$,
$D = \{4, 5, 10\}$, $E = \{2, 3, 6\}$, and $F = \{2, 9\}$.

Determine whether each statement is *true* or *false*. Explain your reasoning. (Examples 1 and 2)

9. $3 \in D$
10. $8 \notin A$
11. $B \subset A$
12. $U \subset A$
13. $5 \notin D$
14. $2 \in E$
15. $0 \in F$
16. $6 \notin F$

Find each of the following. (Examples 2 and 3)

17. C'
18. U'
19. A'
20. $D \cap E$
21. $C \cap E$
22. $E \cup F$
23. $A \cup B$
24. $A \cap B$

Use the Venn diagram to find each of the following. (Examples 2 and 3)

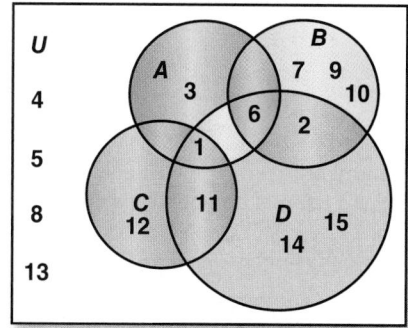

25. $A \cup B$
26. $A \cap D$
27. $C \cup D$
28. A'
29. $A \cap B \cap D$
30. $(A \cup B) \cup C$

31. **SPORTS** Sixteen students in Mr. Frank's gym class each participate in one or more sports as shown in the table. (Examples 2 and 3)

Mr. Frank's Gym Class		
Basketball	**Soccer**	**Volleyball**
Ayanna	Lisa	Pam
Pam	Ayanna	Lisa
Sue	Ron	Shiv
Lisa	Tyron	Max
Ron	Max	Aida
Max	Aida	Juan
Ito	Evita	Tino
Juan	Nelia	Kai
Nelia	Percy	Percy

a. Let *B* represent the set of basketball players, *S* represent the set of soccer players, and *V* represent the set of volleyball players. Draw a Venn diagram of this situation.

b. Find $S \cap V$. What does this set represent?

c. Find S'. What does this set represent?

d. Find $B \cup V$. What does this set represent?

32. **ACADEMICS** There are 26 students at West High School who take either calculus or physics or both. Each student is represented by a letter of the alphabet below. Draw a Venn diagram of this situation. (Examples 2 and 3)

Calculus	A, D, F, I, J, K, L, M, P, R, T, V, X, Y, Z
Physics	B, C, D, E, F, G, H, I, J, K, L, N, O, Q, S, U, W

33. **BEVERAGES** Suppose you can select a juice from three possible kinds: apple, orange, or grape, or you can select a soda from two possible kinds, Brand A or Brand B. If you can choose a juice *or* a soda to drink, according to the Addition Principle, you have 3 + 2 or 5 possible choices. Using notation that you have learned in this lesson, justify this result. In what situation could this principle not be applied?

GEOMETRY Use the figure to find the simplest name for each of the following.

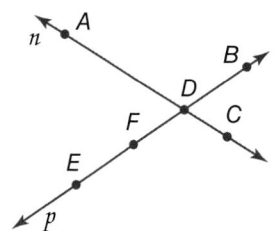

34. $\overline{DE} \cap \overline{BF}$
35. $\overline{AD} \cup \overline{DC}$
36. $\overrightarrow{DE} \cup \overrightarrow{DC}$
37. line *n* $\cap$ line *p*
38. $\overleftrightarrow{AC} \cap \overline{EF}$
39. $\overline{FB} \cup \overrightarrow{EB}$

Operations with Complex Numbers

Imaginary and Complex Numbers The **imaginary unit** i is defined as the principal square root of -1 and can be written as $i = \sqrt{-1}$. The first eight powers of i are shown below.

$$i^1 = i \qquad\qquad i^5 = i^4 \cdot i = i$$
$$i^2 = -1 \qquad\qquad i^6 = i^4 \cdot i^2 = -1$$
$$i^3 = i^2 \cdot i = -i \qquad\qquad i^7 = i^4 \cdot i^3 = -i$$
$$i^4 = i^2 \cdot i^2 = 1 \qquad\qquad i^8 = i^4 \cdot i^4 = 1$$

Notice that the pattern $i, -1, -i, 1, \ldots$ repeats after the first four results. In general, the value of i^n, where n is a whole number, can be found by dividing n by 4 and examining the remainder.

Key Concept   **Powers of i**

To find the value of i^n, let R be the remainder when n is divided by 4.

$n < 0$	$n > 0$
$R = -3 \rightarrow i^n = i$	$R = 1 \rightarrow i^n = i$
$R = -2 \rightarrow i^n = -1$	$R = 2 \rightarrow i^n = 1$
$R = -1 \rightarrow i^n = -i$	$R = 3 \rightarrow i^n = -i$
$R = 0 \rightarrow i^n = 1$	$R = 0 \rightarrow i^n = 1$

EXAMPLE 1   **Powers of i**

Simplify.

a. i^{53}

Method 1

$53 \div 4 = 13 \text{ R } 1$

If $R = 1$, $i^n = i$.

$i^{53} = i$

Method 2

$i^{53} = (i^4)^{13} \cdot i$

$= (1)^{13} \cdot i$

$= i$

b. i^{-18}

Method 1

$-18 \div 4 = -4 \text{ R } -2$

If $R = -2$, $i^n = -1$.

$i^{-18} = -1$

Method 2

$i^{-18} = (i^4)^{-5} \cdot i^2$

$= (1)^{-5} \cdot (-1)$

$= -1$

Sums of real numbers and real number multiples of i belong to an extended system of numbers, known as *complex numbers*. A **complex number** is a number that can be written in the **standard form** $a + bi$, where the real number a is the **real part** and the real number b is the **imaginary part**.

If $a \neq 0$ and $b = 0$, the complex number is $a + 0i$, or the real number a. Therefore, all real numbers are also complex numbers. If $b \neq 0$ the complex number is known as an **imaginary number**. If $a = 0$ and $b \neq 0$, such as $4i$ or $-9i$, the complex number is a **pure imaginary number**.

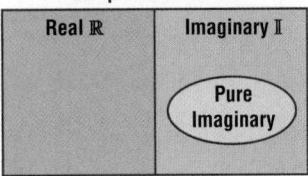

Complex Numbers $\mathbb{C}$

Real $\mathbb{R}$	Imaginary $\mathbb{I}$
	Pure Imaginary

Complex numbers can be added and subtracted by performing the chosen operation on the real and imaginary parts separately.

EXAMPLE 2   Add and Subtract Complex Numbers

Simplify.

a. $(5 - 3i) + (-2 + 4i)$

$$(5 - 3i) + (-2 + 4i) = [5 + (-2)] + (-3i + 4i)$$ Group the real and imaginary parts together.
$$= 3 + i$$ Simplify.

b. $(10 - 2i) - (14 - 6i)$

$$(10 - 2i) - (14 - 6i) = 10 - 2i - 14 + 6i$$ Distributive Property
$$= -4 + 4i$$ Simplify.

Many properties of real numbers, such as the Distribution Property, are also valid for complex numbers. Because of this, complex numbers can be multiplied using similar techniques to those that are used when multiplying binomials.

EXAMPLE 3   Multiply Complex Numbers

Simplify.

a. $(2 - 3i)(7 - 4i)$

$$(2 - 3i)(7 - 4i) = 14 - 8i - 21i + 12i^2$$ FOIL method
$$= 14 - 8i - 21i + 12(-1)$$ $i^2 = -1$
$$= 2 - 29i$$ Simplify.

b. $(4 + 5i)(4 - 5i)$

$$(4 + 5i)(4 - 5i) = 16 - 20i + 20i - 25i^2$$ FOIL method
$$= 16 - 25(-1)$$ $i^2 = -1$
$$= 41$$ Simplify.

Use Complex Conjugates Two complex numbers of the form $a + bi$ and $a - bi$ are called **complex conjugates**. Complex conjugates can be used to rationalize an imaginary denominator. Multiply the numerator and the denominator by the complex conjugate of the denominator.

EXAMPLE 4   Rationalize a Complex Expression

Simplify $(5 - 3i) \div (1 - 2i)$.

$$(5 - 3i) \div (1 - 2i) = \frac{5 - 3i}{1 - 2i}$$ Rewrite as a fraction.

$$= \frac{5 - 3i}{1 - 2i} \cdot \frac{1 + 2i}{1 + 2i}$$ Multiply the numerator and denominator by the conjugate of the denominator, $1 + 2i$.

$$= \frac{5 + 10i - 3i - 6i^2}{1 - 4i^2}$$ Multiply.

$$= \frac{5 + 7i - 6(-1)}{1 - 4(-1)}$$ $i^2 = -1$

$$= \frac{11 + 7i}{5}$$ Simplify.

$$= \frac{11}{5} + \frac{7}{5}i$$ Write the answer in the form $a + bi$.

Exercises

Simplify. (Example 1)

1. i^{-10}

2. $i^2 + i^8$

3. $i^3 + i^{20}$

4. i^{100}

5. i^{77}

6. $i^4 + i^{-12}$

7. $i^5 + i^9$

8. i^{18}

Simplify. (Example 2)

9. $(3 + 2i) + (-4 + 6i)$

10. $(7 - 4i) + (2 - 3i)$

11. $(0.5 + i) - (2 - i)$

12. $(-3 - i) - (4 - 5i)$

13. $(2 + 4.1i) - (-1 - 6.3i)$

14. $(2 + 3i) + (-6 + i)$

15. $(-2 + 4i) + (5 - 4i)$

16. $(5 + 7i) - (-5 + i)$

17. ELECTRICITY Engineers use imaginary numbers to express the two-dimensional quantity of alternating current, which involves both amplitude and angle. In these imaginary numbers, i is replaced with j because engineers use I as a variable for the entire quantity of current. *Impedance* is the measure of how much hinderance there is to the flow of the charge in a circuit with alternating current. The impedance in one part of a series circuit is $2 + 5j$ ohms and the impedance in another part of the circuit is $7 - 3j$ ohms. Add these complex numbers to find the total impedance in the circuit. (Example 2)

Simplify. (Example 3)

18. $(-2 - i)^2$

19. $(1 + 4i)^2$

20. $(5 + 2i)^2$

21. $(3 + i)^2$

22. $(2 + i)(4 + 3i)$

23. $(3 + 5i)(3 - 5i)$

24. $(5 + 3i)(2 + 6i)$

25. $(6 + 7i)(6 - 7i)$

Simplify. (Example 4)

26. $\dfrac{5 + i}{6 + i}$

27. $\dfrac{i}{1 + 2i}$

28. $\dfrac{5 - i}{5 + i}$

29 $\dfrac{3 - 2i}{-4 - i}$

30. $\dfrac{1 + 2i}{2 - 3i}$

31. $\dfrac{3 + 4i}{1 + 5i}$

32. $\dfrac{2 - \sqrt{2}i}{3 + \sqrt{6}i}$

33. $\dfrac{1 + \sqrt{3}i}{1 - \sqrt{2}i}$

ELECTRICITY The voltage E, current I, and impedance Z in a circuit are related by $E = I \cdot Z$. Find the voltage (in volts) in each of the following circuits given the current and impedance.

34. $I = 1 + 3j$ amps, $Z = 7 - 5j$ ohms

35. $I = 2 - 7j$ amps, $Z = 4 - 3j$ ohms

36. $I = 5 - 4j$ amps, $Z = 3 + 2j$ ohms

37. $I = 3 + 10j$ amps, $Z = 6 - j$ ohms

Solve each equation.

38. $5x^2 + 5 = 0$

39. $4x^2 + 64 = 0$

40. $2x^2 + 12 = 0$

41. $6x^2 + 72 = 0$

42. $8x^2 + 120 = 0$

43. $3x^2 + 507 = 0$

44. ELECTRICITY The impedance Z of a circuit depends on the resistance R, the reactance due to capacitance X_C, and the reactance due to inductance X_L, and can be written as a complex number $R + (X_L - X_C)j$. The values (in ohms) for R, X_C, and X_L in the first and second parts of a particular series circuit are shown.

Series Circuit

R	X_C	X_L
10	2	1
3	1	1

a. Write complex numbers that represent the impedances in the two parts of the circuit.

b. Add your answers from part **a** to find the total impedance in the circuit.

c. The *admittance* S of a circuit is the measure of how easily the circuit allows current to flow and is the reciprocal of impedance. Find the admittance (in siemens) in a circuit with an impedance of $6 + 3j$ ohms.

Find values of x and y to make each equation true.

45. $3x + 2iy = 6 + 10i$

46. $5x + 3iy = 5 - 6i$

47. $x - iy = 3 + 4i$

48. $-5x + 3iy = 10 - 9i$

49. $2x + 3iy = 12 + 12i$

50. $4x - iy = 8 + 7i$

Simplify.

51. $(2 - i)(3 + 2i)(1 - 4i)$

52. $(-1 - 3i)(2 + 2i)(1 - 2i)$

53. $(2 + i)(1 + 2i)(3 - 4i)$

54. $(-5 - i)(6i + 1)(7 - i)$

0-3 Quadratic Functions and Equations

Graph Quadratic Functions The graph of a quadratic function is called a **parabola**. To graph a quadratic function, graph ordered pairs that satisfy the function.

EXAMPLE 1 | Graph a Quadratic Function Using a Table

Graph $f(x) = 2x^2 - 8x + 4$ by making a table of values.

Step 1 Choose integer values for x and evaluate the function for each value.

x	$f(x) = 2x^2 - 8x + 4$	$f(x)$	$(x, f(x))$
0	$f(0) = 2(0)^2 - 8(0) + 4$	4	$(0, 4)$
1	$f(1) = 2(1)^2 - 8(1) + 4$	-2	$(1, -2)$
2	$f(2) = 2(2)^2 - 8(2) + 4$	-4	$(2, -4)$
3	$f(3) = 2(3)^2 - 8(3) + 4$	-2	$(3, -2)$
4	$f(4) = 2(4)^2 - 8(4) + 4$	4	$(4, 4)$

Step 2 Graph the resulting coordinate pairs, and connect the points with a smooth curve.

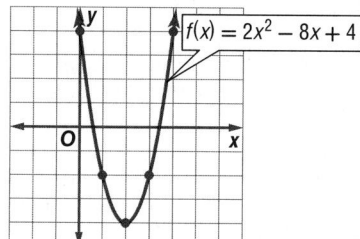

Parabolas exhibit symmetry. The **axis of symmetry** is a line through a parabola that divides the graph into two congruent halves, where each side of the parabola is a reflection of the other side.

The axis of symmetry intersects a parabola at a point called the **vertex**. The vertex of the graph at the right is $A(3, -3)$.

Notice that the x-coordinates of points B and C are both 4 units away from the x-coordinate of the vertex, and they have the same y-coordinate. This is due to the symmetrical nature of the graph.

Key Concept | Graph of a Quadratic Function

Consider the graph of $y = ax^2 + bx + c$, where $a \neq 0$.

- The y-intercept is $a(0)^2 + b(0) + c$ or c.

- The equation of the axis of symmetry is $x = -\dfrac{b}{2a}$.

- The x-coordinate of the vertex is $-\dfrac{b}{2a}$.

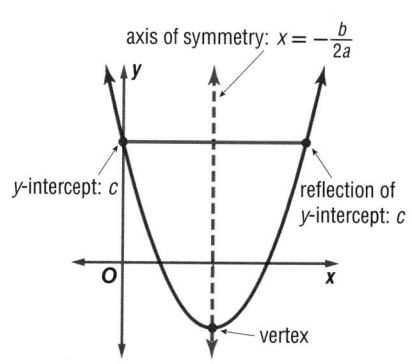

EXAMPLE 2 | **Axis of Symmetry, *y*-intercept, and Vertex**

Use the axis of symmetry, *y*-intercept, and vertex to graph $f(x) = x^2 + 2x - 3$.

Step 1 Determine *a*, *b*, and *c*.

$$f(x) = ax^2 + bx + c$$

$\longrightarrow$ *a* = 1, *b* = 2, and *c* = −3

$$f(x) = 1x^2 + 2x - 3$$

Step 2 Use *a* and *b* to find the equation of the axis of symmetry.

$x = -\dfrac{b}{2a}$ **Equation of the axis of symmetry**

$= -\dfrac{2}{2(1)}$ or −1   ***a* = 1 and *b* = 2**

Step 3 Find the coordinates of the vertex.

Because the equation of the axis of symmetry is $x = -1$, the *x*-coordinate of the vertex is −1.

$f(x) = x^2 + 2x - 3$ **Original equation**

$f(-1) = (-1)^2 + 2(-1) - 3$ **Evaluate *f*(x) for *x* = −1.**

$f(-1) = -4$ **Simplify.**

The vertex is at (−1, −4).

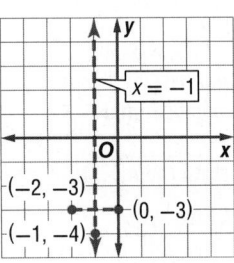

Step 4 Find the *y*-intercept and its reflection.

Because $c = -3$, the coordinates of the *y*-intercept are (0, −3). The axis of symmetry is $x = -1$, so the reflection of the *y*-intercept is (−2, −3).

Step 5 Graph the axis of symmetry, vertex, *y*-intercept and its reflection. Find and graph one or more additional points and their reflections. Then connect the points with a smooth curve.

StudyTip

Graphing Quadratic Functions You can always use a table of values to generate more points for the graph of a quadratic function.

The *y*-coordinate of the vertex of a quadratic function is the *maximum* or *minimum value* of the function. These values represent the greatest or least possible value that the function can reach.

Key Concept | **Maximum and Minimum Values**

The graph of $f(x) = ax^2 + bx + c$, where $a \neq 0$,
- opens up and has a minimum value when $a > 0$, and
- opens down and has a maximum value when $a < 0$.

a > 0

a < 0

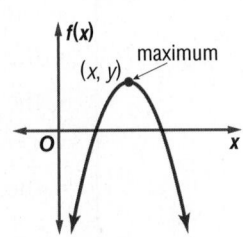

The *y*-coordinate of the vertex is the minimum value.

The *y*-coordinate of the vertex is the maximum value.

The domain of a quadratic function is all real numbers. The range will either be all real numbers less than or equal to the maximum value or all real numbers greater than or equal to the minimum value.

EXAMPLE 3 **Maximum and Minimum Values**

Consider $f(x) = -3x^2 + 12x + 11$.

a. Determine whether the function has a *maximum* or *minimum* value.

For this function, $a = -3$. Because a is negative, the graph opens down and the function has a maximum value.

b. Find the maximum or minimum value of the function.

The maximum value of the function is the y-coordinate of the vertex. To find the coordinates of the vertex, first find the equation of the axis of symmetry.

$$x = -\frac{b}{2a} \qquad \text{Equation of the axis of symmetry}$$

$$= -\frac{12}{2(-3)} \qquad a = -3 \text{ and } b = 12$$

$$= 2 \qquad \text{Simplify.}$$

Because the equation of the axis of symmetry is $x = 2$, the x-coordinate of the vertex is 2. You can now find the y-coordinate of the vertex, or maximum value of the function, by evaluating the original function for $x = 2$.

$$f(x) = -3x^2 + 12x + 11 \qquad \text{Original function}$$

$$f(2) = -3(2)^2 + 12(2) + 11 \qquad x = 2$$

$$= -12 + 24 + 11 \qquad \text{Multiply.}$$

$$= 23 \qquad \text{Simplify.}$$

Therefore, $f(x)$ has a maximum value of 23 at $x = 2$.

c. State the domain and range of the function.

The domain of the function is all real numbers or $\mathbb{R}$. The range of the function is all real numbers less than or equal to the maximum value 23 or $y \leq 23$.

Solve Quadratic Equations A **quadratic equation** is a polynomial equation of degree 2. The standard form of a quadratic equation is $ax^2 + bx + c = 0$, where $a \neq 0$. The factors of a quadratic polynomial can be used to solve the related quadratic equation. Solving quadratic equations by factoring is an application of the Zero Product Property.

Key Concept **Zero Product Property**

For any real numbers a and b, if $ab = 0$, then either $a = 0$, $b = 0$, or both a and b equal zero.

EXAMPLE 4 **Solve by Factoring**

Solve $x^2 - 8x + 12 = 0$ by factoring.

$$x^2 - 8x + 12 = 0 \qquad \text{Original equation}$$

$$(x - 2)(x - 6) = 0 \qquad \text{Factor.}$$

$$x - 2 = 0 \quad \text{or} \quad x - 6 = 0 \qquad \text{Zero Product Property}$$

$$x = 2 \qquad\qquad x = 6 \qquad \text{Simplify.}$$

The solutions of the equation are 2 and 6.

Another method for solving quadratic equations is to **complete the square**.

Key Concept   **Completing the Square**

To complete the square for any quadratic expression of the form $x^2 + bx$, follow the steps below.

Step 1   Find one half of b, the coefficient of x.

Step 2   Square the result in Step 1.

Step 3   Add the result of Step 2 to $x^2 + bx$.

$$x^2 + bx + \left(\frac{b}{2}\right)^2 = \left(x + \frac{b}{2}\right)^2$$

EXAMPLE 5   **Solve by Completing the Square**

Solve $x^2 - 4x + 1 = 0$ by completing the square.

$x^2 - 4x + 1 = 0$	Original equation
$x^2 - 4x = -1$	Rewrite so that the left side is of the form $x^2 + bx$.
$x^2 - 4x + 4 = -1 + 4$	Because $\left(\frac{-4}{2}\right)^2 = 4$, add 4 to each side.
$(x - 2)^2 = 3$	Write the left side as a perfect square.
$x - 2 = \pm\sqrt{3}$	Take the square root of each side.
$x = 2 \pm \sqrt{3}$	Add 2 to each side.
$x = 2 + \sqrt{3}$ or $x = 2 - \sqrt{3}$	Write as two equations.
≈ 3.73 ≈ 0.27	Use a calculator.

The solutions of the equation are approximately 0.27 and 3.73.

Completing the square can be used to develop a general formula that can be used to solve any quadratic equation of the form $ax^2 + bx + c = 0$, known as the **Quadratic Formula**.

Key Concept   **Quadratic Formula**

The solutions of a quadratic equation of the form $ax^2 + bx + c = 0$, where $a \neq 0$, are given by the following formula.

$$x = \frac{-b \pm \sqrt{b^2 - 4ac}}{2a}$$

EXAMPLE 6   **Solve by Using the Quadratic Formula**

Solve $x^2 - 4x + 15 = 0$ by using the Quadratic Formula.

$x = \dfrac{-b \pm \sqrt{b^2 - 4ac}}{2a}$	Quadratic Formula
$= \dfrac{-(-4) \pm \sqrt{(-4)^2 - 4(1)(15)}}{2(1)}$	Replace a with 1, b with -4, and c with 15.
$= \dfrac{4 \pm \sqrt{16 - 60}}{2}$	Multiply.
$= \dfrac{4 \pm \sqrt{-44}}{2}$	Simplify.
$= \dfrac{4 \pm 2i\sqrt{11}}{2}$	$\sqrt{-44} = \sqrt{4}\,\sqrt{-1}\,\sqrt{11}$ or $2i\sqrt{11}$
$= 2 \pm \sqrt{11}i$	Simplify.

The solutions are $2 + \sqrt{11}i$ and $2 - \sqrt{11}i$.

Graph each equation by making a table of values. (Example 1)

1. $f(x) = x^2 + 5x + 6$

2. $f(x) = x^2 - x - 2$

3. $f(x) = 2x^2 + x - 3$

4. $f(x) = 3x^2 + 4x - 5$

5. $f(x) = x^2 - x - 6$

6. $f(x) = -x^2 - 3x - 1$

7. BASEBALL A batter hits a baseball with an initial speed of 80 feet per second. If the initial height of the ball is 3.5 feet above the ground, the function $d(t) = 80t - 16t^2 + 3.5$ models the ball's height above the ground in feet as a function of time in seconds. Graph the function using a table of values. (Example 1)

Use the axis of symmetry, y-intercept, and vertex to graph each function. (Example 2)

8. $f(x) = x^2 + 3x + 2$

9. $f(x) = x^2 - 9x + 8$

10. $f(x) = x^2 - 2x + 1$

11. $f(x) = x^2 - 6x - 16$

12. $f(x) = 2x^2 - 8x - 5$

13. $f(x) = 3x^2 + 12x - 4$

14. HEALTH The normal systolic pressure P in millimeters of mercury (mm Hg) for a woman can be modeled by $P = 0.01x^2 + 0.05x + 107$, where x is age in years. (Example 2)

a. Find the axis of symmetry, y-intercept, and vertex for the graph of P.

b. Graph P using the values you found in part **a**.

Determine whether each function has a *maximum* or *minimum* value. Then find the value of the maximum or minimum, and state the domain and range of the function. (Example 3)

15.

$f(x) = -2x^2 + 3x - 5$

16.

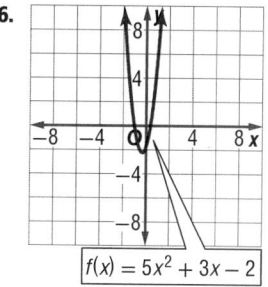

$f(x) = 5x^2 + 3x - 2$

17. $f(x) = -x^2 + 3x - 5$

18. $f(x) = x^2 - 5x + 6$

19. $f(x) = 2x^2 + 4x + 7$

20. $f(x) = 6x^2 + 3x - 1$

21. $f(x) = -3x^2 - 2x - 1$

22. $f(x) = -5x^2 + 10x - 6$

Solve each equation by factoring. (Example 4)

23. $x^2 - 10x + 21 = 0$

24. $p^2 - 6p + 5 = 0$

25. $x^2 - 3x - 28 = 0$

26. $4w^2 + 19w - 5 = 0$

27. $4r^2 - r = 5$

28. $g^2 + 6g - 16 = 0$

Solve each equation by completing the square. (Example 5)

29. $x^2 + 8x - 20 = 0$

30. $2a^2 + 11a - 21 = 0$

31. $z^2 - 2z - 24 = 0$

32. $p^2 - 3p - 88 = 0$

33 $t^2 - 3t - 7 = 0$

34. $3g^2 - 12g = -4$

Solve each equation by using the Quadratic Formula. (Example 6)

35. $m^2 + 12m + 36 = 0$

36. $t^2 - 6t + 13 = 0$

37. $6m^2 + 7m - 3 = 0$

38. $c^2 - 5c + 9 = 0$

39. $4x^2 - 2x + 9 = 0$

40. $3p^2 + 4p = 8$

41. PHOTOGRAPHY Jocelyn wants to frame a photograph that has an area of 20 square inches with a uniform width of matting between the photograph and the edge of the frame as shown.

a. Write an equation to model the situation if the length and width of the matting must be 8 inches by 10 inches, respectively, to fit in the frame.

b. Graph the related function.

c. What is the width of the exposed part of the matting x?

Solve each equation.

42. $x^2 + 5x - 6 = 0$

43. $a^2 - 13a + 40 = 0$

44. $x^2 - 11x + 24 = 0$

45. $q^2 - 12q + 36 = 0$

46. $-x^2 + 4x - 6 = 0$

47. $7x^2 + 3 = 0$

48. $x^2 - 4x + 7 = 0$

49. $2x^2 + 6x - 3 = 0$

50. PETS A rectangular turtle pen is 6 feet long by 4 feet wide. The pen is enlarged by increasing the length and width by an equal amount in order to double its area. What are the dimensions of the new pen?

NUMBER THEORY Use a quadratic equation to find two real numbers that satisfy each situation, or show that no such numbers exist.

51. Their sum is -17 and their product is 72.

52. Their sum is 7 and their product is 14.

53. Their sum is -9 and their product is 24.

54. Their sum is 12 and their product is -28.

0-4

nth Roots and Real Exponents

Objective

- Simplify expressions in radical form.
- Simplify expressions in exponential form.

Vocabulary

nth root
principal nth root

Math Online

glencoe.com

Simplify Radicals A square root of a number is one of two equal factors of that number. For example, 4 is a square root of 16 because $4 \cdot 4$ or $4^2 = 16$. In general, if a and b are real numbers and n is a positive integer greater than 1, if $b^n = a$, then b is an **nth root** of a.

Key Concept   nth Root of a Number

Let a and b be real numbers and let n be any positive integer greater than 1.

- If $a = b^n$, then b is an nth root of a.
- If a has an nth root, the **principal nth root** of a is the root having the same sign as a.

The principal nth root of a is denoted by the radical expression $\sqrt[n]{a}$, where n is the index of the radical and a is the radicand.

Some examples of nth roots are listed below.

$\sqrt[4]{81} = 3$ $\sqrt[4]{81}$ indicates the principal fourth root of 81.

$-\sqrt[4]{81} = -3$ $-\sqrt[4]{81}$ indicates the opposite of the principal fourth root of 81.

$\pm\sqrt[4]{81} = \pm 3$ $\pm\sqrt[4]{81}$ indicates both real fourth roots of 81.

Whether a radical expression has positive and/or negative roots is dependent upon the value of the radicand and whether the index is even or odd.

Real nth Roots of Real Numbers		
Suppose n is an integer greater than 1, and a is a real number.		
a	n is even.	n is odd.
$a > 0$	1 positive and 1 negative real root: $\pm\sqrt[n]{a}$	1 positive and 0 negative real root: $\sqrt[n]{a}$
$a < 0$	0 real roots	0 positive and 1 negative real root: $\sqrt[n]{a}$
$a = 0$	1 real root: $\sqrt[n]{0} = 0$	1 real root: $\sqrt[n]{0} = 0$

EXAMPLE 1   Find nth Roots

Evaluate.

a. $-\sqrt{49}$

$-\sqrt{49} = -(\sqrt{49})$ or -7 **Simplify.**

b. $\sqrt[3]{\dfrac{64}{27}}$

$\sqrt[3]{\dfrac{64}{27}} = \dfrac{\sqrt[3]{64}}{\sqrt[3]{27}}$ or $\dfrac{4}{3}$ **Simplify.**

c. $\sqrt[4]{-121}$

Because there is no real number that can be raised to the fouth power to produce -121, $\sqrt[4]{-121}$ is not a real number.

When you find an even root of an even power and the result is an odd power, you must use the absolute value of the result to ensure that the answer is nonnegative.

EXAMPLE 2   Simplify Using Absolute Value

Simplify.

a. $\sqrt[6]{n^{18}}$

$$\sqrt[6]{n^{18}} = \sqrt[6]{(n^3)^6}$$
$$= |n^3|$$

Notice that n^3 is a sixth root of n^{18}. Because the index is even and the exponent is odd, you must use the absolute value of n^3.

b. $\sqrt[4]{81(a+1)^{12}}$

$$\sqrt[4]{81(a+1)^{12}} = \sqrt[4]{[3(a+1)^3]^4}$$
$$= 3|(a+1)^3|$$

Because the index is even and the exponent is odd, you must use the absolute value of $(a+1)^3$.

c. $\sqrt{63y^3}$

$$\sqrt{63y^3} = \sqrt{9y^2 \cdot 7y}$$
$$= \sqrt{(3y)^2} \cdot \sqrt{7y}$$
$$= 3y\sqrt{7y}$$

Notice that $\sqrt{63y^3}$ is only a real number when y is nonnegative. Therefore, it is not necessary to use absolute value.

d. $\sqrt[5]{-p^{10}q^7}$

$$\sqrt[5]{-p^{10}q^7} = \sqrt[5]{-1p^{10}q^5 \cdot q^2}$$
$$= \sqrt[5]{(-1p^2q)^5} \cdot \sqrt[5]{q^2}$$
$$= -p^2q\sqrt[5]{q^2}$$

Because the index is odd, it is not necessary to use absolute value.

StudyTip

Odd Index If n is odd, there is only one real root. Therefore, absolute value symbols are never needed.

Rational Exponents Squaring a number and taking the square root of a number are inverse operations. This relationship can be used to express radicals in exponential form.

Let $\sqrt{b} = b^n$.

$$\sqrt{b} = b^n \qquad \textbf{Given}$$

$$(\sqrt{b})^2 = (b^n)^2 \qquad \textbf{Square each side.}$$

$$b = b^{2n} \qquad \textbf{Simplify.}$$

$$1 = 2n \qquad \textbf{If } \boldsymbol{a^m = a^n}\textbf{, then } \boldsymbol{m = n.}$$

$$\frac{1}{2} = n \qquad \textbf{Divide each side by 2.}$$

So, $\sqrt{b} = b^{\frac{1}{2}}$. This process can be used to determine the exponential form for any nth root. You can determine the exponential form for any nth root using the properties shown below.

Key Concept **Rational Exponents**

If b is a real number, variable, or algebraic expression and m and n are positive integers greater than 1, then

- $b^{\frac{1}{n}} = \sqrt[n]{b}$, if the principal nth root of b exists, and

- $b^{\frac{m}{n}} = \left(\sqrt[n]{b}\right)^m$ or $\sqrt[n]{b^m}$, if $\frac{m}{n}$ is in reduced form.

Watch Out!

Common Error Be careful not to confuse rational and negative exponents.

$b^{-n} \neq b^{\frac{1}{n}}$

$b^{-n} = \frac{1}{b^n}$

$b^{\frac{1}{n}} = \sqrt[n]{b}$

EXAMPLE 3 **Simplify Expressions with Rational Exponents**

Simplify.

a. $\dfrac{\sqrt[10]{32}}{\sqrt[6]{2}}$

$$\frac{\sqrt[10]{32}}{\sqrt[6]{2}} = \frac{32^{\frac{1}{10}}}{2^{\frac{1}{6}}} \qquad \sqrt[n]{b} = b^{\frac{1}{n}}$$

$$= \frac{(2^5)^{\frac{1}{10}}}{2^{\frac{1}{6}}} \qquad 32 = 2^5$$

$$= \frac{2^{\frac{1}{2}}}{2^{\frac{1}{6}}} \qquad \textbf{Power of a Power}$$

$$= 2^{\frac{1}{2} - \frac{1}{6}} \qquad \textbf{Quotient of Powers}$$

$$= 2^{\frac{1}{3}} \text{ or } \sqrt[3]{2} \qquad \textbf{Simplify.}$$

b. $\sqrt[4]{x^{21}y^{15}}$

$$\sqrt[4]{x^{21}y^{15}} = \sqrt[4]{x^{21}} \cdot \sqrt[4]{y^{15}} \qquad \textbf{Product of Roots}$$

$$= x^{\frac{21}{4}} \cdot y^{\frac{15}{4}} \qquad \sqrt[n]{b^m} = b^{\frac{m}{n}}$$

$$= x^{\frac{20}{4}} \cdot x^{\frac{1}{4}} \cdot y^{\frac{12}{4}} \cdot y^{\frac{3}{4}} \qquad \textbf{Product of Powers}$$

$$= x^5 y^3 \sqrt[4]{xy^3} \qquad \textbf{Simplify.}$$

c. $\dfrac{16^{\frac{3}{4}}}{16^{\frac{1}{4}}}$

$$\frac{16^{\frac{3}{4}}}{16^{\frac{1}{4}}} = 16^{\frac{3}{4} - \frac{1}{4}} \qquad \textbf{Quotient of Powers}$$

$$= 16^{\frac{1}{2}} \qquad \textbf{Simplify.}$$

$$= \sqrt{16} \text{ or } 4 \qquad b^{\frac{1}{n}} = \sqrt[n]{b}$$

Evaluate. (Example 1)

1. $-\sqrt{169}$

2. $\sqrt{-100}$

3. $\sqrt[3]{\dfrac{216}{125}}$

4. $\sqrt[3]{-\dfrac{64}{343}}$

5. $\sqrt[4]{-81}$

6. $\sqrt[4]{625}$

7. $\sqrt[5]{243}$

8. $\sqrt[5]{-1024}$

Simplify. (Example 2)

9. $\sqrt[3]{-27x^9}$

10. $\sqrt[4]{16a^{20}}$

11. $\sqrt[8]{8y^{16}}$

12. $\sqrt[3]{54x^{17}}$

13. $\sqrt{20x^{16}}$

14. $\sqrt{121(z-2)^{14}}$

15. $\sqrt[4]{a^{12}b^9}$

16. $\sqrt[7]{-q^{13}r^{16}}$

Simplify. (Example 3)

17. $\dfrac{b^{\frac{5}{4}} \cdot b^{\frac{3}{4}}}{b^{\frac{1}{4}}}$

18. $\left(2x^{\frac{1}{4}}y^{\frac{1}{3}}\right)\left(3x^{\frac{1}{4}}y^{\frac{2}{3}}\right)$

19. $\sqrt[6]{640a^3}$

20. $\sqrt[6]{128b^4}$

21. $\dfrac{\sqrt[3]{16}}{\sqrt[5]{4}}$

22. $\dfrac{\sqrt[4]{27}}{\sqrt[3]{81}}$

23. **BOATING** The motion comfort ratio M of a boat is given by

$$M = \dfrac{D}{0.65(B)^{\frac{4}{3}}(0.7W + 0.3A)},$$

where D is the water displacement of the boat in pounds, B is the boat's beam or width in feet, W is the boat's length in feet at the waterline, and A is the boat's overall length in feet. The higher the ratio, the greater the level of comfort experienced by those on board as the boat encounters waves. (Example 3)

a. Find the motion comfort ratio of the boat shown below.

Overall length: 28.1 ft
Beam: 10 ft
Water displacement: 13,800 lb
Waterline length: 26.2 ft

b. Find the beam of a boat to the nearest foot with a comfort ratio of 27 that displaces 15,000 pounds of water, has a waterline length of 30.4 feet, and an overall length of 32.3 feet.

24. **CARS** The value of a car depreciates or declines over the course of its useful life. The new value V and the original value v of a car are related by the formula $V = v(1 - r)^n$, where r is the rate of depreciation per year and n is the number of years. Suppose the current value of a used car is \$12,000. What would be the value of the car after 18 months at an annual depreciation rate of 20%? (Example 3)

Evaluate.

25. $216^{\frac{1}{3}}$

26. $4096^{\frac{1}{4}}$

27. $49^{-\frac{1}{2}}$

28. $27^{-\frac{1}{3}}$

29. **MUSIC** The note progression of the twelve tone scale is comprised of a series of half tones. In order for an instrument to be "in tune," the frequency of each note has an optimum ratio with the frequency of middle C, called the perfect 1st.

The optimum frequency ratio r can be calculated using $r = \left(\sqrt[12]{2}\right)^n$, where n is the number of half tones the note is above the perfect 1st, including the note itself. (Example 1)

a. Approximate the optimum frequency ratio of the middle 3rd with the perfect 1st.

b. Without the use of a calculator, approximate the optimum frequency ratio of the perfect 8th and the perfect 1st. Justify your answer.

Simplify.

30. $\sqrt[3]{-250r^{11}t^6u^5}$

31. $\sqrt[3]{128a^9b^7c^4}$

32. $\sqrt[4]{96a^8b^6c^{20}}$

33. $\sqrt[6]{64x^7y^6z^{18}}$

34. $\sqrt[4]{a^2b^3c^4d^5}$

35. $\sqrt[5]{w^6x^8y^{10}z^{13}}$

Systems of Linear Equations and Inequalities

Objective

- Use various techniques to solve systems of equations.
- Solve systems of inequalities by graphing.

Vocabulary

system of equations
substitution method
elimination method
consistent
independent
dependent
inconsistent
system of inequalities

Math Online

glencoe.com

Systems of Equations A **system of equations** is a set of two or more equations. To *solve* a system of equations means to find values for the variables in the equations that make all of the equations true. One way to solve a system of equations is by graphing the equations on the same coordinate plane. The point of intersection of the graphs of the equations represents the solution of the system.

EXAMPLE 1 **Solve by Graphing**

Solve the system of equations by graphing.

$$3x - 2y = -6$$
$$x + y = -2$$

Solve each equation for y. Then graph each equation.

$$y = \frac{3}{2}x + 3 \qquad y = -x - 2$$

The lines intersect at the point $(-2, 0)$. This ordered pair is the solution of the system.

CHECK	$3x - 2y = -6$	**Original equations**	$x + y = -2$
	$3(-2) - 2(0) \overset{?}{=} -6$	$x = -2$ and $y = 0$	$-2 + 0 \overset{?}{=} -2$
	$-6 = -6 ✔$	**Simplify.**	$-2 = -2 ✔$

Algebraic methods are used to find exact solutions of systems of equations. One algebraic method is called the **substitution method**.

Key Concept **Substitution Method**

Step 1 Solve one equation for one of the variables in terms of the other.

Step 2 Substitute the expression found in Step 1 into the other equation to obtain an equation in one variable. Then solve the equation.

Step 3 Substitute to solve for the other variable.

EXAMPLE 2 **Solve by Substitution**

Use substitution to solve the system of equations.

$$2x + 3y = 9$$
$$5x - y = 14$$

Step 1 $5x - y = 14$ ➡ $y = 5x - 14$ **Solve for y.**

Step 2
	$2x + 3y = 9$	**First equation**
	$2x + 3(5x - 14) = 9$	**Substitute $5x - 14$ for y.**
	$17x - 42 = 9$	**Simplify.**
	$x = 3$	**Solve for x.**

Step 3 $y = 5x - 14$ **Step 1 equation**

$= 5(3) - 14$ or 1 The solution is $(3, 1)$.

CHECK From the graph in Figure 0.5.1, you can see that the lines intersect at the point $(3, 1)$.

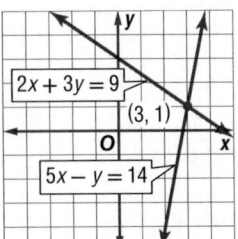

Figure 0.5.1

You can use the **elimination method** to solve a system when one of the variables has the same coefficient in both equations.

Key Concept — Elimination Method

Step 1 Multiply one or both equations by a number to result in two equations that contain opposite terms.

Step 2 Add the equations, eliminating one variable. Then solve the equation.

Step 3 Substitute to solve for the other variable.

EXAMPLE 3 Solve by Elimination

Use elimination to solve the system of equations.

$1. 5x + 2y = 20$
$2. 5x - 5y = -25$

$5x + 2y = 20$ → **Multiply by 5.** → $7.5x + 10y = 100$
$5x - 5y = -25$ → **Multiply by 2.** → $(+)\ 5x - 10y = -50$

$$12.5x \qquad = 50 \qquad \text{Add.}$$
$$x = 4 \qquad \text{Divide each side by 12.5.}$$

$1.5x + 2y = 20$	Equation 1
$1.5(4) + 2y = 20$	$x = 4$
$y = 7$	Solve for y.

The solution is $(4, 7)$.

You can use any of the methods or a combination of the methods for solving systems of equations in two variables to solve systems of equations in three variables.

EXAMPLE 4 Systems of Equations in Three Variables

Solve the system of equations.
$x - 2y + z = 15$
$2x + 3y - z = 7$
$4x + 10y - 5z = -3$

Step 1 Eliminate one variable by using two pairs of equations.

$x - 2y + z = 15$	Equation 1		$5x - 10y + 5z = 75$	5 × Equation 1
$(+)\ 2x + 3y - z = 7$	Equation 2		$(+)\ 4x + 10y - 5z = -3$	Equation 3
$3x + y = 22$	Add.		$9x \qquad = 72$	Add.
			$x = 8$	Divide.

Step 2 Solve the system of two equations.

$3x + y = 22$	Equation in two variables
$3(8) + y = 22$	$x = 8$
$y = -2$	Solve for y.

Step 3 Substitute the two values into one of the original equations to find z.

$x - 2y + z = 15$	Equation 1
$8 - 2(-2) + z = 15$	$x = 8$ and $y = -2$
$z = 3$	Solve for z.

The solution is $(8, -2, 3)$.

The graphs of two equations do not always intersect at one point. For example, a system of linear equations could contain parallel lines or the same line. In these cases, the system of equations may have no solution or infinitely many solutions. A **consistent** system has at least one solution. If there is exactly one solution, the system is **independent**. If there are infinitely many solutions, the system is **dependent**. If there is no solution, the system is **inconsistent**.

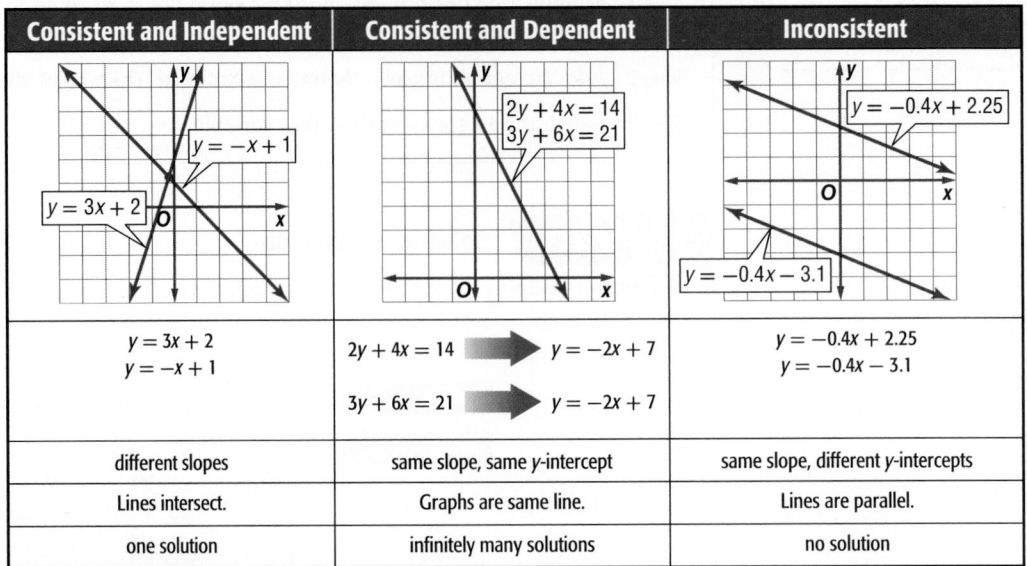

Consistent and Independent	Consistent and Dependent	Inconsistent
$y = 3x + 2$ $y = -x + 1$	$2y + 4x = 14 \implies y = -2x + 7$ $3y + 6x = 21 \implies y = -2x + 7$	$y = -0.4x + 2.25$ $y = -0.4x - 3.1$
different slopes	same slope, same y-intercept	same slope, different y-intercepts
Lines intersect.	Graphs are same line.	Lines are parallel.
one solution	infinitely many solutions	no solution

EXAMPLE 5   No Solution and Infinitely Many Solutions

Solve each system of equations. State whether the system is *consistent and independent*, *consistent and dependent*, or *inconsistent*.

a. $-7x + 3y = 21$
$7x - 3y = 17$

$$\begin{array}{ll} -7x + 3y = 21 & \textbf{Add.} \\ \underline{(+)\; 7x - 3y = 17} & \\ 0 = 4 & \end{array}$$

Because $0 = 55$ is not true, this system has no solutions. Therefore, the system is inconsistent, and the equations in the system are parallel lines, as shown in Figure 0.5.2.

b. $5x + 2y = 12$
$20x + 8y = 48$

$$\begin{array}{ll} 20x + 8y = 48 & \textbf{4} \times \textbf{Equation 1} \\ \underline{(-)\; 20x + 8y = 48} & \\ 0 = 0 & \end{array}$$

Because $0 = 0$ is always true, there are an infinite number of solutions. Therefore, the system is consistent and dependent, and the equations in the system have the same graph, as shown in Figure 0.5.3.

Figure 0.5.2

Figure 0.5.3

Systems of Inequalities Solving a **system of inequalities** means finding all of the ordered pairs that satisfy the inequalities in the system.

> ### Key Concept — Solving Systems of Inequalities
>
> **Step 1** Graph each inequality, shading the correct area. Use a solid line to graph inequalities that contain $\geq$ or $\leq$. Use a dashed line to graph inequalities that contain $>$ or $<$.
>
> **Step 2** Identify the region that is shaded for all of the inequalities. This is the solution of the system.
>
> **Step 3** Check the solution using a test point within the solution region.

EXAMPLE 6 — Intersecting Regions

Solve the system of inequalities.

$y \geq 0.5x - 3$
$y \leq -2x + 7$

Step 1 Graph each inequality.

Use a solid line to graph each inequality, since each contains either $\geq$ or $\leq$. Then shade the region either above or below the line that contains the coordinates that make each inequality true.

Step 2 Identify the region that is shaded for all of the inequalities.

The solution of $y \geq 0.5x - 3$ is Regions 1 and 3.

The solution of $y \leq -2x + 7$ is Regions 2 and 3.

Region 3 is part of the solution of both inequalities, so it is the solution of the system.

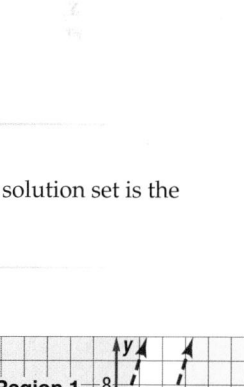

CHECK You can use a test point from the solution region to check your solution. Substitute the x- and y-values of the test point into the inequalities.

$y \geq 0.5x - 3$	$y \leq -2x + 7$
$0 \overset{?}{\geq} 0.5(0) - 3$	$0 \overset{?}{\leq} -2(0) + 7$
$0 \geq -3$ ✔	$0 \leq 7$ ✔

When the regions do not intersect, the system has no solution. That is, the solution set is the empty set.

EXAMPLE 7 — System of Inequalities with Separate Regions

Solve the system of inequalities.

$y < 3x - 8$
$y > 3x + 4$

Graph each line and shade the region either above or below the line that makes the inequality true.

The solution of $y < 3x - 8$ is Region 1.

The solution of $y > 3x + 4$ is Region 2.

Because the graphs of the inequalities do not overlap, there are no points in common and there is no solution to the system.

Watch Out!

▶ **Parallel Inequalities**
Not all systems of inequalities with boundaries having the same slope have no solution. For example, if the system in Example 7 had been
$y > 3x - 8$
$y > 3x + 4$,
the solution would have been the region to the right of the line $y = 3x + 4$.

Exercises

Solve each system of equations by graphing. (Example 1)

1. $y = 5x - 2$
$y = -2x + 5$

2. $y = 2x - 5$
$y = 0.5x + 1$

3. $x + y = -2$
$3x - y = 10$

4. $y = -3$
$2x = 8$

5. $3y = 4x + 6$
$2y = x - 1$

6. $x = 5$
$4x + 5y = 20$

Use substitution to solve each system of equations.
(Example 2)

7. $5x - y = 16$
$2x + 3y = 3$

8. $3x - 5y = -8$
$x + 2y = 1$

9. $y = 6 - x$
$x = 4.5 + y$

10. $x = 2y - 8$
$2x - y = -7$

11. $4x - 5y = 6$
$x + 3 = 2y$

12. $x - 3y = 6$
$2x + 4y = -2$

13. JOBS Connor works at a movie rental store earning $8 per hour. He also walks dogs for $10 per hour on the weekends. Connor worked 13 hours this week and made $110. How many hours did he work at the movie rental store? How many hours did he walk dogs over the weekend?

Use elimination to solve each system of equations.
(Example 3)

14. $7x + y = 9$
$5x - y = 15$

15. $2x - 3y = 1$
$4x - 5y = 7$

16. $-3x + 10y = 5$
$2x + 7y = 24$

17. $2x + 3y = 3$
$12x - 15y = -4$

18. $3x + 4y = -1$
$6x - 2y = 3$

19. $5x - 6y = 10$
$-2x + 3y = -7$

Solve each system of equations. (Example 4)

20. $x + 2y + 3z = 5$
$3x + 2y - 2z = -13$
$5x + 3y - z = -11$

21. $x - y - z = 7$
$-x + 2y - 3z = -12$
$3x - 2y + 7z = 30$

22. $7x + 5y + z = 0$
$-x + 3y + 2z = 16$
$x - 6y - z = -18$

23. $3x - 5y + z = 9$
$x - 3y - 2z = -8$
$5x - 6y + 3z = 15$

24. $4x + 2y + z = 7$
$2x + 2y - 4z = -4$
$x + 3y - 2z = -8$

25. $x - 3z = 7$
$2x + y - 2z = 11$
$-x - 2y + 2z = 6$

26. $8x - z = 4$
$y + z = 5$
$11x + y = 15$

27 $4x - 2y + z = -5$
$5x + y + 3z = 6$
$-2x + 3y + 2z = -4$

Solve each system of equations. State whether the system is *consistent and independent*, *consistent and dependent*, **or** *inconsistent*. (Example 5)

28. $8x - 5y = -11$
$-8x + 9y = 7$

29. $x - y = 2$
$2x = 2y + 10$

30. $5x + 4y = 2$
$6x + 5y = 4$

31. $12x - 9y = 3$
$4x - 3y = 1$

32. $1.5x + y = 3.5$
$3x + 2y = 7$

33. $10x - 3y = -4$
$-8x + 5y = 11$

34. $2x - 2y + 3z = 2$
$2x - 3y + 7z = -1$
$4x - 3y + 2z = 0$

35. $-3x + 2y + z = -23$
$4x + 2y + z = 5$
$5x + 3y + 3z = 11$

36. CAMPING The Mountaineers Club held two camping trips during the summer. The club rented 5 tents and 1 cabin for the 30 members who went on the first trip. The club rented 4 tents and 2 cabins for the 36 members who went on the second trip. If the tents and cabins were filled to capacity on both trips, how many people can each tent and each cabin accommodate? (Example 5)

Solve each system of inequalities. If the system has no solution, state *no solution*. (Examples 6 and 7)

37. $y \geq x - 3$
$y \leq 2x + 1$

38. $y + x < 1$
$y > -x - 1$

39. $x + 2y \geq 12$
$x - y \geq 3$

40. $y \leq \frac{1}{3}x - 7$
$3y \geq x + 6$

41. $y + 5 < 4x$
$2y > -2x + 10$

42. $y \leq -x + 8$
$y \geq 0.5x - 4$

43. $8y \leq -2x - 1$
$4y + x \geq 3$

44. $y + 7 < 3x$
$2y + 5x > 8$

45. $-6y \geq -5x + 6$
$y \leq -3x - 1$

46. $y + 4 \leq \frac{4}{3}x$
$3y \geq 4x + 9$

47. $y \leq 2x + 1$
$y \geq 2x - 2$
$3x + y \leq 9$

48. $x - 3y > 2$
$2x - y < 4$
$2x + 4y \geq -7$

49. ART Charlie can spend no more than $225 on the art club's supply of brushes and paint. He needs at least 20 brushes and 56 tubes of paint. Graph the region that shows how many packages of each item can be purchased. (Example 6)

SALE!
3 brushes for $7.50

10 colors
10 Paint Tubes
$21.45

0-6 Matrix Operations

Objective

- Use characteristics to describe matrices.
- Add, subtract, and multiply matrices by a scalar.

Vocabulary

matrix
element
dimensions
row matrix
column matrix
square matrix
zero matrix
equal matrices
scalar

Math Online

glencoe.com

Describe and Analyze Matrices A **matrix** is a rectangular array of variables or constants in horizontal rows and vertical columns, usually enclosed in brackets. In a matrix, the numbers or data are organized so that each position in the matrix has a purpose. Each value in the matrix is called an **element**. A matrix is usually named using an uppercase letter.

$$A = \begin{bmatrix} 8 & -2 & 5 & 6 \\ -1 & 3 & -3 & 6 \\ 7 & -8 & 1 & 4 \end{bmatrix} \} \text{ 3 rows}$$

The element -1 is in Row 2, Column 1, depicted by a_{21}.

The element -8 is in Row 3, Column 2, depicted by a_{32}.

4 columns

A matrix can be described by its **dimensions**. A matrix with m rows and n columns is an $m \times n$ matrix, which is read m by n. Matrix A above is a 3×4 matrix because it has 3 rows and 4 columns.

EXAMPLE 1   Dimensions and Elements of a Matrix

Use $A = \begin{bmatrix} 4 & 9 & -18 \\ -2 & 11 & 3 \end{bmatrix}$ to answer the following.

a. State the dimensions of A.

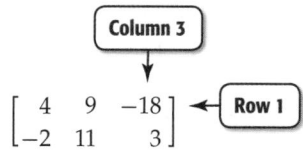

$\begin{bmatrix} 4 & 9 & -18 \\ -2 & 11 & 3 \end{bmatrix} \} \text{ 2 rows}$

3 columns

Because A has 2 rows and 3 columns, the dimensions of A are 2×3.

b. Find the value of a_{13}.

Column 3

$\begin{bmatrix} 4 & 9 & -18 \\ -2 & 11 & 3 \end{bmatrix}$ ← **Row 1**

Because a_{13} is the element in row 1, column 3, the value of a_{13} is -18.

Certain matrices have special names. For example, a matrix that has one row is called a **row matrix**, and a matrix with one column is a **column matrix**. A matrix that has the same number of rows and columns is known as a **square matrix**, and a matrix in which every element is zero is called a **zero matrix**.

Row Matrix

$[8 \quad -5 \quad 2 \quad 4]$

Column Matrix

$\begin{bmatrix} 8 \\ -1 \end{bmatrix}$

Square Matrix

$\begin{bmatrix} -4 & 2 \\ 3 & 9 \end{bmatrix}$

Zero Matrix

$\begin{bmatrix} 0 & 0 & 0 \\ 0 & 0 & 0 \end{bmatrix}$

Two matrices are **equal matrices** if and only if each element of one matrix is equal to the corresponding element in the other matrix. So, matrix A and B shown below are equal matrices.

$$A = \begin{bmatrix} 2 & 4 \\ -1 & 3 \end{bmatrix} \qquad B = \begin{bmatrix} 2 & 4 \\ -1 & 3 \end{bmatrix}$$

Notice that for two matrices to be equal, they must have the same number of rows and columns.

Matrix Operations Matrices can be added or subtracted if and only if they have the same dimensions.

Key Concept — Adding and Subtracting Matrices

To add or subtract two matrices with the same dimensions, add or subtract their corresponding elements.

$$A \quad + \quad B \quad = \quad A + B$$

$$\begin{bmatrix} a & b \\ c & d \end{bmatrix} + \begin{bmatrix} e & f \\ g & h \end{bmatrix} = \begin{bmatrix} a+e & b+f \\ c+g & d+h \end{bmatrix}$$

$$A \quad - \quad B \quad = \quad A - B$$

$$\begin{bmatrix} a & b \\ c & d \end{bmatrix} - \begin{bmatrix} e & f \\ g & h \end{bmatrix} = \begin{bmatrix} a-e & b-f \\ c-g & d-h \end{bmatrix}$$

StudyTip

Corresponding Elements *Corresponding* refers to elements that are in the exact same position in each matrix.

EXAMPLE 2 — Add and Subtract Matrices

Find each of the following for $A = \begin{bmatrix} 8 & 3 \\ -5 & 14 \end{bmatrix}$, $B = \begin{bmatrix} 12 & -7 \\ 6 & -23 \end{bmatrix}$, and $C = \begin{bmatrix} 2 \\ 9 \end{bmatrix}$.

a. $A + B$

$$A + B = \begin{bmatrix} 8 & 3 \\ -5 & 14 \end{bmatrix} + \begin{bmatrix} 12 & -7 \\ 6 & -23 \end{bmatrix} \qquad \text{Substitution}$$

$$= \begin{bmatrix} 8+12 & 3+(-7) \\ -5+6 & 14+(-23) \end{bmatrix} \text{ or } \begin{bmatrix} 20 & -4 \\ 1 & -9 \end{bmatrix} \qquad \text{Add corresponding elements.}$$

b. $B - C$

$$B - C = \begin{bmatrix} 12 & -7 \\ 6 & -23 \end{bmatrix} - \begin{bmatrix} 2 \\ 9 \end{bmatrix} \qquad \text{Substitution}$$

B is a 2×2 matrix and C is a 2×1 matrix. Since these dimensions are not the same, you cannot subtract the matrices.

You can multiply any matrix by a constant called a **scalar**. When you do this, you multiply each individual element by the value of the scalar.

EXAMPLE 3 — Scalar Multiplication

Find each product.

a. $3\begin{bmatrix} -6 & -3 & 7 \\ 10 & 2 & -15 \end{bmatrix}$

$$3\begin{bmatrix} -6 & -3 & 7 \\ 10 & 2 & -15 \end{bmatrix} = \begin{bmatrix} 3(-6) & 3(-3) & 3(7) \\ 3(10) & 3(2) & 3(-15) \end{bmatrix} \text{ or } \begin{bmatrix} -18 & -9 & 21 \\ 30 & 6 & -45 \end{bmatrix}$$

b. $-4\begin{bmatrix} 2 & -9 \\ 7 & 3 \\ -11 & 4 \end{bmatrix}$

$$-4\begin{bmatrix} 2 & -9 \\ 7 & 3 \\ -11 & 4 \end{bmatrix} = \begin{bmatrix} -4(2) & -4(-9) \\ -4(7) & -4(3) \\ -4(-11) & -4(4) \end{bmatrix} \text{ or } \begin{bmatrix} -8 & 36 \\ -28 & -12 \\ 44 & -16 \end{bmatrix}$$

StudyTip

Scalar Multiplication Matrix brackets behave like other grouping symbols. So when multiplying by a scalar, distribute the same way as with a grouping symbol.

Many properties of real numbers also hold true for matrices. A summary of these properties is listed below.

Key Concept — Properties of Matrix Operations

For any matrices A, B, and C for which the matrix sum and product are defined and any scalar k, the following properties are true.

Commutative Property of Addition	$A + B = B + A$
Associative Property of Addition	$(A + B) + C = A + (B + C)$
Left Scalar Distributive Property	$k(A + B) = kA + kB$
Right Scalar Distributive Property	$(A + B)k = kA + kB$

Multi-step operations can be performed on matrices. The order of these operations is the same as with real numbers.

EXAMPLE 4 — Multi-Step Operations

Find $4(P + Q)$. If $P = \begin{bmatrix} 3 & 8 & -2 \\ -5 & 5 & -4 \end{bmatrix}$ and $Q = \begin{bmatrix} -4 & 5 & 7 \\ 3 & -10 & -6 \end{bmatrix}$.

$4(P + Q) = 4\left(\begin{bmatrix} 3 & 8 & -2 \\ -5 & 5 & -4 \end{bmatrix} + \begin{bmatrix} -4 & 5 & 7 \\ 3 & -10 & -6 \end{bmatrix} \right)$ **Substitution**

$= 4 \begin{bmatrix} 3 & 8 & -2 \\ -5 & 5 & -4 \end{bmatrix} + 4 \begin{bmatrix} -4 & 5 & 7 \\ 3 & -10 & -6 \end{bmatrix}$ **Distributive Property**

$= \begin{bmatrix} 12 & 32 & -8 \\ -20 & 20 & -16 \end{bmatrix} + \begin{bmatrix} -16 & 20 & 28 \\ 12 & -40 & -24 \end{bmatrix}$ **Multiply by the scalar.**

$= \begin{bmatrix} 12 + (-16) & 32 + 20 & -8 + 28 \\ -20 + 12 & 20 + (-40) & -16 + (-24) \end{bmatrix}$ **Add.**

$= \begin{bmatrix} -4 & 52 & 20 \\ -8 & -20 & -40 \end{bmatrix}$ **Simplify.**

You can use the same algebraic methods for solving equations with real numbers to solve equations with matrices.

EXAMPLE 5 — Solving a Matrix Equation

Given $A = \begin{bmatrix} -9 & 15 & 4 \\ 2 & -10 & -5 \end{bmatrix}$ and $B = \begin{bmatrix} 5 & -7 & 8 \\ 14 & 10 & -3 \end{bmatrix}$, solve $4X - B = A$ for X.

$4X - B = A$ **Original equation**

$4X = A + B$ **Add B to each side.**

$X = \frac{1}{4}(A + B)$ **Divide each side by 4.**

$X = \frac{1}{4}\left(\begin{bmatrix} -9 & 15 & 4 \\ 2 & -10 & -5 \end{bmatrix} + \begin{bmatrix} 5 & -7 & 8 \\ 14 & 10 & -3 \end{bmatrix} \right)$ **Substitution**

$X = \frac{1}{4} \begin{bmatrix} -4 & 8 & 12 \\ 16 & 0 & -8 \end{bmatrix}$ **Add.**

$X = \begin{bmatrix} -1 & 2 & 3 \\ 4 & 0 & -2 \end{bmatrix}$ **Multiply by the scalar.**

Watch Out!

Matrix Equations
Remember that the variable X in matrix equations stands for a matrix, while the variable x in algebraic equations stands for a number.

Matrix equations can be used in real-world situations.

Real-World Link

On average, 13- to 17-year-olds text more than they talk, sending and receiving over 1,700 text messages a month but only making and receiving about 230 calls per month.

Source: Nielsen Mobile

EXAMPLE 6 | **Use a Matrix Equation**

CELL PHONES Allison took a survey of her high school to see which class sent the most text messages, pictures, and talked for the most minutes on their cell phones each week. The averages for the freshmen, sophomores, juniors, and seniors are shown.

Class	Texts	Pictures	Calls
freshman	20	3	163
sophomore	25	4	170
junior	15	7	178
senior	22	3	190

a. If each text message costs $0.10, each picture costs $0.75, and each minute on the phone costs $0.05, find the average weekly cell phone costs for each class. Express your answer as a matrix.

Step 1 Write a matrix equation for the total cost X. Let T represent the number of texts for all classes, P represent the number of pictures, and C represent the number of call minutes.

$$X = 0.10T + 0.75P + 0.05C$$

Step 2 Solve the equation.

$X = 0.10T + 0.75P + 0.05C$ **Original equation**

$$= 0.10 \begin{bmatrix} 20 \\ 25 \\ 15 \\ 22 \end{bmatrix} + 0.75 \begin{bmatrix} 3 \\ 4 \\ 7 \\ 3 \end{bmatrix} + 0.05 \begin{bmatrix} 163 \\ 170 \\ 178 \\ 190 \end{bmatrix}$$ **Substitution**

$$= \begin{bmatrix} 2.00 \\ 2.50 \\ 1.50 \\ 2.20 \end{bmatrix} + \begin{bmatrix} 2.25 \\ 3.00 \\ 5.25 \\ 2.25 \end{bmatrix} + \begin{bmatrix} 8.15 \\ 8.50 \\ 8.90 \\ 9.50 \end{bmatrix} \text{ or } \begin{bmatrix} 12.40 \\ 14.00 \\ 15.65 \\ 13.95 \end{bmatrix}$$ **Multiply by the scalars.**

The final matrix indicates average weekly cell phone costs for each class. Therefore, on average, each freshman spent $12.40, each sophomore spent $14.00, each junior spent $15.65, and each senior spent $13.95.

b. If there are 100 freshmen, 180 sophomores, 250 juniors, and 300 seniors that use cell phones at Allison's school, use her survey results to estimate the total number of text messages sent, pictures sent, and minutes used on the cell phone each week by these students. Express your answer as a matrix.

Step 1 Write a matrix equation for the total usage X. Let F represent freshmen, S represent sophomores, J represent juniors, and N represent seniors.

$$X = 100F + 180S + 250J + 300N$$

Step 2 Solve the equation.

$$X = 100F + 180S + 250J + 300N$$

$$= 100[20 \quad 3 \quad 163] + 180[25 \quad 4 \quad 170] + 250[15 \quad 7 \quad 178] + 300[22 \quad 3 \quad 190]$$

$$= [16{,}850 \quad 3670 \quad 148{,}400]$$

The final matrix indicates the average weekly totals for each type of cell phone use. Therefore, there were 16,850 texts, 3670 pictures, and 148,400 minutes used by these students.

State the dimensions of each matrix. (Example 1)

1. $\begin{bmatrix} 1 & -8 \\ 6 & -2 \end{bmatrix}$

2. $\begin{bmatrix} -9 & -8 \\ 2 & 17 \\ 11 & -6 \end{bmatrix}$

3. $\begin{bmatrix} 10 & 12 & 25 & 48 \\ 53 & 62 & 74 & 89 \end{bmatrix}$

4. $\begin{bmatrix} -5 & -9 & 4 \\ -7 & 12 & 1 \\ 14 & 6 & -8 \end{bmatrix}$

Find the value of each element in

$A = \begin{bmatrix} -3 & 45 & 28 & -19 \\ 24 & 36 & -22 & 5 \\ 8 & -11 & 54 & 17 \\ -15 & 4 & 29 & -9 \end{bmatrix}$. (Example 1)

5. a_{22} **6.** a_{21} **7.** a_{43}

8. a_{13} **9.** a_{32} **10.** a_{34}

Find each of the following for $W = \begin{bmatrix} 13 & -6 \\ 2 & -10 \\ -4 & 8 \end{bmatrix}$,

$X = \begin{bmatrix} 1 & -3 \\ -5 & 9 \\ 12 & 7 \end{bmatrix}$, $Y = \begin{bmatrix} 5 & -2 & 1 \\ -6 & 14 & 8 \end{bmatrix}$, and $Z = \begin{bmatrix} -11 & 3 & 7 \\ 4 & -9 & 16 \end{bmatrix}$.

If the matrix does not exist, write *impossible.* (Example 2)

11. $W + X$ **12.** $Z - X$ **13.** $Z - Y$

14. $X + Y$ **15.** $W - X$ **16.** $Y + Z$

17. BUSINESS Two car companies are planning a merger. The numbers in thousands of car models i of each color j sold by each company are represented in the matrices. After the merger, how many of each car model will be sold in each color? (Example 2)

$$[a_{ij}] = \begin{bmatrix} 42 & 56 & 85 \\ 41 & 57 & 89 \\ 45 & 53 & 84 \end{bmatrix} \quad [b_{ij}] = \begin{bmatrix} 51 & 45 & 79 \\ 53 & 48 & 81 \\ 56 & 46 & 83 \end{bmatrix}$$

Find each product. (Example 3)

18. $2\begin{bmatrix} 6 & -18 & 7 \\ 3 & 4 & 11 \end{bmatrix}$ **19.** $9\begin{bmatrix} -1 & -5 \\ 8 & 4 \end{bmatrix}$

20. $3\begin{bmatrix} 2 & 8 \\ -7 & 15 \\ 12 & -6 \end{bmatrix}$ **21.** $6[-3 \quad 10 \quad -5 \quad 9]$

22. $7\begin{bmatrix} 20 & -9 & 4 \\ -1 & 5 & 11 \end{bmatrix}$ **23.** $4\begin{bmatrix} -4 & 6 \\ -12 & 5 \\ 3 & 4 \end{bmatrix}$

24. CHILD CARE An adult is responsible for taking two children to the community swimming pool once a week for six weeks. The daily admission fees are $4.50 for a child and $6.75 for an adult. Write 1×3 matrix with a scalar multiple that represents the total cost of admission. What is the total cost? (Example 3)

Find each of the following if $D = \begin{bmatrix} -2 & 5 \\ 9 & -11 \\ 4 & -7 \end{bmatrix}$,

$E = \begin{bmatrix} 8 & 10 \\ -5 & 5 \\ 1 & -12 \end{bmatrix}$, and $F = \begin{bmatrix} 5 & -1 \\ -4 & 2 \\ 6 & 10 \end{bmatrix}$. (Example 4)

25. $2D + E$ **26.** $3(E - F)$

27. $\frac{1}{2}(D + F)$ **28.** $3D - 2E$

29. $D + E - F$ **30.** $2(D + F) - E$

Given $J = \begin{bmatrix} 8 & -10 & 3 \\ -4 & 1 & 12 \end{bmatrix}$, $K = \begin{bmatrix} 2 & 5 & -9 \\ -6 & 7 & -3 \end{bmatrix}$ **and**

$L = \begin{bmatrix} 4 & 1 & -8 \\ 11 & -7 & 6 \end{bmatrix}$, **solve each equation for X.** (Examples 5 and 6)

31. $2X = J + K$ **32.** $L - K = \frac{1}{3}X$

33 $2J - L = 3X$ **34.** $3K - X = J$

35. $3L - 2K = X$ **36.** $2(J - X) = -L$

Use matrices A, B, C, D, E **and** F **to solve for X. If the matrix does not exist, write** *impossible.*

$A = \begin{bmatrix} 5 & 7 \\ -6 & 1 \end{bmatrix}$ $B = \begin{bmatrix} 3 & 5 \\ -1 & 8 \end{bmatrix}$ $C = \begin{bmatrix} 4 & -2 & 3 \\ 5 & 0 & -1 \\ 9 & 0 & 1 \end{bmatrix}$

$D = \begin{bmatrix} 0 & 1 & 2 \\ -2 & 3 & 0 \\ 4 & 4 & -2 \end{bmatrix}$ $E = \begin{bmatrix} 8 & -4 & 2 \\ 3 & 1 & -5 \end{bmatrix}$ $F = \begin{bmatrix} -6 & -1 & 0 \\ 1 & 4 & 0 \end{bmatrix}$

37. $A + B = X$ **38.** $X = -2F$

39. $C - D = X$ **40.** $X = D + B$

41. $X = 4D$ **42.** $X = 3B - A$

43. $F - 2(E + C) = X$ **44.** $2X = 3(E + F)$

45. SWIMMING The table shows some of the women's freestyle swimming records.

Distance (meters)	World	Olympic	American
50	23.96 s	24.06 s	24.07 s
100	52.88 s	53.12 s	53.39 s
200	1 min 54.47 s	1 min 54.82 s	1 min 55.78 s
800	8 min 14.10 s	8 min 14.10 s	8 min 16.22 s

Source: Fédération Internationale de Natation

a. Find the difference between American and World records expressed as a column matrix.

b. If all the data in the table were expressed in seconds and represented by a matrix A, what matrix expression could be used to convert all the data to minutes?

Probability with Permutations and Combinations

Objective

- Find the number of possible outcomes of an experiment.
- Use permutations and combinations with probability.

Vocabulary

experiment
sample space
independent events
dependent events
factorial
permutation
combination

Math Online

glencoe.com

Sample Space An **experiment** is a situation involving chance or probability that leads to specific outcomes. The set of all possible outcomes is called the **sample space**. One method that can be used to determine the *number* of possible outcomes of an experiment is the Fundamental Counting Principle.

> **Key Concept**   **Fundamental Counting Principle**
>
> Let A and B be two events. If event A has n_1 possible outcomes and is followed by event B that has n_2 possible outcomes, then event A followed by event B has $n_1 \cdot n_2$ possible outcomes.

The Fundamental Counting Principle can also be used to find the number of possible outcomes for three or more events. For example, the number of ways that k events can occur is given by $n_1 \cdot n_2 \cdot n_3 \cdot \cdots \cdot n_k$.

Events with outcomes that do not affect each other are called **independent events**, and events with outcomes that do affect each other are called **dependent events.**

> **EXAMPLE 1**   **Fundamental Counting Principle**
>
> **a. A restaurant offers a dinner special in which a customer can select from one of 6 appetizers, a soup or salad, one of 12 entrees, and one of 8 desserts. How many different dinner specials are possible?**
>
> Because the selection of one menu item does not affect the selection of any other item, each selection is independent. To determine the number of possible dinner specials, multiply the number of ways each item can be selected.
>
> $6 \cdot 2 \cdot 12 \cdot 8 = 1152$
>
> Therefore, there are 1152 different dinner specials.
>
> **b. Garrett works for a bookstore. He is arranging the five best-sellers for a shelf display. If he can place the books in any order, how many different ways can Garrett arrange the books?**
>
> The selection of the book for the first position affects the books available for the second position, the selection for the second position affects the books available for the third position, and so on. So, the selections of books are dependent events.
>
> There are 5 books from which to choose for the first position, 4 books for the second, 3 for the third, 2 for the fourth, and 1 for the fifth. To determine the total number of ways that the books can be arranged, multiply by the number of ways that the books can be chosen for each position.
>
> $5 \cdot 4 \cdot 3 \cdot 2 \cdot 1 = 120$
>
> Therefore, there are 120 possible ways for Garrett to arrange the books.

The expression used in Example 1b to calculate the number of arrangements of books, $5 \cdot 4 \cdot 3 \cdot 2 \cdot 1$, can be written as 5!, which is read *5 factorial*. The **factorial** of a positive integer n is the product of the positive integers less than or equal to n, and is given by

$$n! = n \cdot (n-1) \cdot (n-2) \cdot \cdots \cdot 1, \text{ where } 0! = 1.$$

Permutations and Combinations The Fundamental Counting Principle can also be used to determine the number of ways that n objects can be arranged in a certain order. An arrangement of n objects is called a **permutation** of the objects.

Key Concept   **Permutations**

The number of permutations of n objects taken n at a time is	The number of permutations of n objects taken r at a time is
$$_nP_n = n!.$$	$$_nP_r = \frac{n!}{(n-r)!}.$$

EXAMPLE 2   **Permutations with Probability**

An alarm system requires a 7-digit code using the digits 0 through 9. Each digit may be used only once.

a. How many different codes are possible?

The order of the numbers in the code is important, so this situation is a permutation of 10 digits taken 7 at a time.

$$_nP_r = \frac{n!}{(n-r)!} \qquad \text{Definition of a permutation}$$

$$_{10}P_7 = \frac{10!}{(10-7)!} \qquad n = 10 \text{ and } r = 7$$

$$= \frac{10 \cdot 9 \cdot 8 \cdot 7 \cdot 6 \cdot 5 \cdot 4 \cdot \cancel{3!}}{\cancel{3!}} \qquad \text{Expand 10! and divide out common factorials.}$$

$$= 10 \cdot 9 \cdot 8 \cdot 7 \cdot 6 \cdot 5 \cdot 4 \qquad \text{Simplify.}$$

$$= 604{,}800 \qquad \text{Multiply.}$$

So, 604,800 codes are possible.

b. If a code is randomly generated, what is the probability that the first three digits are odd?

To find the probability of the first three digits being odd, find the number of ways to select three odd digits and multiply by the number of ways to select the remaining digits and then divide by the total possible codes.

$$P(\text{1st three digits are odd}) = \frac{\text{ways to select 3 odd digits} \cdot \text{ways to select last 4 digits}}{\text{total possible codes}}$$

$$= \frac{_5P_3 \cdot _7P_4}{_{10}P_7}$$

$$= \frac{\dfrac{5!}{(5-3)!} \cdot \dfrac{7!}{(7-4)!}}{\dfrac{10!}{(10-7)!}} \qquad _nP_r = \frac{n!}{(n-r)!}$$

$$= \frac{\dfrac{5!}{2!} \cdot \dfrac{7!}{3!}}{\dfrac{10!}{3!}} \qquad \text{Subtract.}$$

$$= \frac{\dfrac{5 \cdot 4 \cdot 3 \cdot \cancel{2!}}{\cancel{2!}} \cdot \dfrac{7 \cdot 6 \cdot 5 \cdot 4 \cdot \cancel{3!}}{\cancel{3!}}}{\dfrac{10 \cdot 9 \cdot 8 \cdot 7 \cdot 6 \cdot 5 \cdot 4 \cdot \cancel{3!}}{\cancel{3!}}} \qquad \begin{array}{l}\text{Expand 5!, 7!, and 10!, and divide out} \\ \text{common factorials.}\end{array}$$

$$= \frac{5 \cdot 4 \cdot 3 \cdot 7 \cdot 6 \cdot 5 \cdot 4}{10 \cdot 9 \cdot 8 \cdot 7 \cdot 6 \cdot 5 \cdot 4} \qquad \text{Simplify.}$$

$$= \frac{50{,}400}{604{,}800} \text{ or } \frac{1}{12} \qquad \text{Multiply.}$$

Therefore, the probability is $\frac{1}{12}$ or about 0.08.

In a combination, order is *not* important. A **combination** of n objects taken r at a time is calculated by dividing the number of permutations by the number of arrangements containing the same elements and is denoted by ${}_nC_r$.

Key Concept   Combinations

The number of combinations of n objects taken r at a time is

$$_nC_r = \frac{n!}{(n-r)!\, r!}.$$

The main difference between a permutation and a combination is whether order is considered (as in permutation) or not (as in combination). For example, for objects E, F, G, and H taken two at a time, the permutations and combinations are listed below.

Permutations			
EF	FE	GE	HE
EG	FG	GF	HF
EH	FH	GH	HG

Combinations	
EF	FG
EG	FH
EH	GH

In permutations, EF is different from FE. But in combinations, EF is the same as FE.

EXAMPLE 3   Combinations with Probability

There are 7 seniors, 5 juniors, and 4 sophomores on the pep squad. Mr. Rinehart needs to choose 12 students out of the group to sell spirit buttons during lunch.

a. How many ways can the 12 students be chosen?

$$_nC_r = \frac{n!}{(n-r)!\, r!} \qquad \text{Definition of combination}$$

$$_{16}C_{12} = \frac{16!}{(16-12)!\, 12!} \qquad n = 16 \text{ and } r = 12$$

$$= \frac{16!}{4!\, 12!} \qquad \text{Subtract.}$$

$$= \frac{16 \cdot 15 \cdot 14 \cdot 13 \cdot \cancel{12!}}{4 \cdot 3 \cdot 2 \cdot 1 \cdot \cancel{12!}} \qquad \text{Expand 16! and 4!, and divide out common factorials.}$$

$$= \frac{43{,}680}{24} \qquad \text{Multiply.}$$

$$= 1820 \qquad \text{Simplify.}$$

So, there are 1820 ways that the 12 students can be chosen.

b. If the students are randomly chosen, what is the probability that 4 seniors, 4 juniors, and 4 sophomores will be chosen?

ways to choose 4 seniors out of 7: ${}_7C_4 = \dfrac{7!}{(7-4)!\, 4!} = \dfrac{7!}{3!\, 4!}$

ways to choose 4 juniors out of 5: ${}_5C_4 = \dfrac{5!}{(5-4)!\, 4!} = \dfrac{5!}{1!\, 4!}$

ways to choose 4 sophomores out of 4: ${}_4C_4 = \dfrac{4!}{(4-4)!\, 4!} = \dfrac{4!}{0!\, 4!}$

There are $\dfrac{7!}{3!\, 4!} \cdot \dfrac{5!}{1!\, 4!} \cdot \dfrac{4!}{0!\, 4!}$ or 175 ways to choose 4 seniors, 4 juniors, and 4 sophomores.

Therefore, the probability is $\dfrac{175}{1820}$ or $\dfrac{5}{52}$.

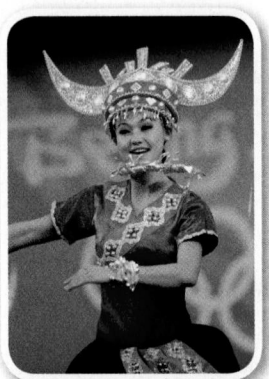

Real-World Link

Two hundred thousand people from around the world volunteered to be part of the pep squad for the 2008 Summer Olympics. Of this number, about 600 made the team.

Source: NFL Fanhouse

Use the Fundamental Counting Principle to determine the number of outcomes for each event. (Example 1)

1. How many different T-shirts are available?

Size	Colors
XS, S, M, L, XL, XXL	blue, red, green, gray, black

2. For a particular model of car, a dealer offers 3 sizes of engines, 2 types of stereos, 18 body colors, and 7 upholstery colors. How many different possibilities are available for that model?

3. If you toss a coin, roll a die, and then spin a 4-colored spinner with equal sections, how many outcomes are possible?

4. If a deli offers 12 different meats, 5 different cheeses, and 6 different breads, how many different sandwiches can be made from 1 meat, 1 cheese, and 1 bread?

5. An ice cream shop offers 20 flavors of ice cream, 5 different toppings, and 3 different sizes. How many different sundaes are available?

6. How many different 1-topping pizzas are available?

7. How many ways can six different books be arranged on a shelf if the books can be arranged in any order?

8. How many ways can eight actors be listed in the opening credits of a movie if the leading actor must be listed first?

Find each value. (Examples 2 and 3)

9. $_6P_6$ **10.** $_5P_3$ **11.** $_7C_4$

12. $_{20}C_{15}$ **13.** $_8P_1$ **14.** $_6P_1$

15. $_6P_3$ **16.** $_7P_4$ **17.** $_9P_5$

18. $_4C_2$ **19.** $_{12}C_4$ **20.** $_9C_9$

21. CLASS OFFICERS At Grant Senior High School, there are 15 names on the ballot for junior class officers. Five will be selected to form a class committee. (Examples 2 and 3)

  a. How many different committees can be formed?

  b. In how many ways can the committee be formed if each student has a different responsibility?

  c. If there are 8 girls and 7 boys on the ballot, what is the probability that a committee of 2 boys and 3 girls is formed?

22. ART An art gallery curator wants to select four paintings out of twenty to put on display. How many groups of four paintings can be chosen?

23. PHONE NUMBERS In the United States, standard local telephone numbers consist of 7 digits, where the first digit cannot be 1 or 0.

  a. Find the number of possibilities for telephone numbers.

  b. Find the probability of randomly selecting a given telephone number from all the possible numbers.

  c. How many different telephone numbers are possible if only even digits are used?

  d. Find the probability of choosing a telephone number in which only even digits are used.

  e. Find the number of possibilities for telephone numbers if the first three digits are 593. What is the probability of randomly choosing a telephone number in which the first three digits are 593?

24. CARDS Five cards are drawn from a standard deck of 52 cards.

  a. Determine the number of possible five-card selections.

  b. Find the probability of an arrangement containing 3 hearts and 2 clubs.

  c. Find the probability of an arrangement containing all face cards.

  d. Find the probability of an arrangement containing 1 ace, 2 jacks, and 2 kings.

A gumball machine contains 7 red (R), 8 orange (N), 9 purple (P), 7 white (W), and 5 yellow (Y) gumballs. Tyson buys 3 gumballs. Find each probability, assuming that the machine dispenses 3 gumballs at random all at once.

25. $P(3\ R)$ **26.** $P(2\ W\ \text{and}\ 1\ P)$

27 $P(1\ R\ \text{and}\ 2\ N)$ **28.** $P(1\ N\ \text{and}\ 2\ Y)$

29. $P(2\ R\ \text{and}\ 1\ Y)$ **30.** $P(1\ P, 1\ W, \text{and}\ 1\ R)$

31. COMPUTERS A circuit board with 20 computer chips contains 4 chips that are defective. If 3 chips are selected at random, what is the probability that all 3 are defective?

32. BOOKS Dan has twelve books on his shelf that he has not read, including seven novels and five biographies. If he wants to take four books with him on vacation, what is the probability that he randomly selects two novels and two biographies?

33. SCHOLARSHIPS Twelve male and 16 female students have been selected as equal qualifiers for 6 college scholarships. If the awarded recipients are to be chosen at random, what is the probability that 3 will be male and 3 will be female?

Statistics

Measures of Center and Spread **Statistics** is the science of collecting, analyzing, interpreting, and presenting data. The branch of statistics that focuses on collecting, summarizing, and displaying data is known as *descriptive statistics*. Data in one *variable*, or data type, are called **univariate data**. These data can be described by a **measure of central tendency** which represents the center or middle of the data. The three most common measures of central tendency are mean, median, and mode.

A **population** is the entire membership of people, objects, or events of interest to be analyzed. A **sample** is a subset of a population. The formulas for mean use x to represent the data values in a sample or population, Σx to represent the sum of all x-values, n to represent the number of x-values, μ to represent the population mean, $\bar{x}$ to represent the sample mean.

Objective

- Find measures of center and spread.
- Organize statistical data.

Vocabulary

statistics
univariate data
measure of central tendency
population
sample
mean, median, mode
measures of spread
   (or variation)
range
variance
standard deviation
frequency distribution
class (or interval)
relative frequency
class width
cumulative frequency
cumulative relative
   frequency
quartiles
five-number summary
interquartile range
outliers

Math Online

glencoe.com

Key Concept   **Measures of Central Tendency**

| **Mean** | the sum of the numbers in a set of data divided by the number of items |

Population Mean   **Sample Mean**

$$\mu = \frac{\Sigma x}{n} \qquad\qquad \bar{x} = \frac{\Sigma x}{n}$$

| **Median** | the middle number in a set of data when the data are arranged in numerical order or the mean of the middle two values |
| **Mode** | the number or numbers that appear most often in a set of data |

EXAMPLE 1   **Find Measures of Central Tendency**

Find the mean, median, and mode for the data 14, 7, 12, 4, 13, 20, 2, 3, 5, 15, 10, 4.

Mean $\dfrac{\Sigma x}{n} = \dfrac{14 + 7 + 12 + 4 + 13 + 20 + 2 + 3 + 5 + 15 + 10 + 4}{12} \approx 9.08$

Median 2 3 4 4 5 7 10 12 13 14 15 20

$\dfrac{7 + 10}{2}$ or 8.5

Mode The value that occurs most often in the set is 4, so the mode is 4.

Measures of spread or **variation** describe the distribution of a set of data. Three measures of spread are range, variance, and standard deviation. The formulas for population variance σ^2 and standard deviation σ use $x - \mu$ to represent the deviation or *difference* an x-value is from the population mean and $\Sigma(x - \mu)^2$ to represent the sum of the squares of these deviations. Similar notation is used for sample variance s^2 and standard deviation s.

StudyTip

Population vs. Sample
Statisticians have found that using sample data to approximate measures of spread for a population consistently underestimates these measures. To counteract this error, the formulas for sample variance and standard deviation use division by $n - 1$ instead of n.

Key Concept   **Measures of Spread**

| **Range** | the difference between the greatest and least values in a set of data |
| **Variance** | the mean of the squares of the deviations from the mean |

Population Variance   **Sample Variance**

$$\sigma^2 = \frac{\Sigma(x - \mu)^2}{n} \qquad\qquad s^2 = \frac{\Sigma(x - \bar{x})^2}{n - 1}$$

| **Standard Deviation** | the average amount by which individual items deviate from the mean of all the data |

Population Standard Deviation   **Sample Standard Deviation**

$$\sigma = \sqrt{\frac{\Sigma(x - \mu)^2}{n}} \qquad\qquad s = \sqrt{\frac{\Sigma(x - \bar{x})^2}{n - 1}}$$

| EXAMPLE 2 | Find Measures of Spread |

The quiz scores for a class of 25 students are shown.

Quiz Scores				
7	8	9	9	9
10	5	7	5	6
10	9	7	8	2
8	9	9	7	5
3	6	8	10	10

a. Find the measures of spread for the entire class.

Range   maximum − minimum

$$= 10 - 2 \text{ or } 8$$

Variance   Find the mean of the data.

$$\mu = \frac{\Sigma x}{n} \qquad \text{Mean of a population}$$

$$= \frac{7 + 8 + \cdots + 10 + 10}{25} \qquad \Sigma x \text{ is the sum of the data values and } n = 25.$$

$$= 7.44 \text{ or about } 7.4 \qquad \text{Simplify.}$$

Use the unrounded mean to find the variance.

$$\sigma^2 = \frac{\Sigma(x - \mu)^2}{n} \qquad \text{Variance of a population}$$

$$= \frac{\Sigma(x - 7.44)^2}{25} \qquad \mu = 7.4 \text{ and } n = 25$$

$$= \frac{(7 - 7.44)^2 + (8 - 7.44)^2 + \cdots + (10 - 7.44)^2}{25} \qquad \text{Substitution}$$

$$= 4.5664 \text{ or about } 4.6 \qquad \text{Simplify.}$$

Standard Deviation   Take the square root of the variance.

$$\sigma = \sqrt{4.5664}$$

$$\approx 2.1$$

b. Use the last column of the quiz scores to find the measures of spread for a sample of the class.

Range   The sample is 9, 6, 2, 5, and 10. The range of the sample is $10 - 2$ or 8.

Variance   The sample mean is $\bar{x} = \dfrac{9 + 6 + 2 + 5 + 10}{5}$ or 6.4.

$$s^2 = \frac{\Sigma(x - \bar{x})^2}{n - 1} \qquad \text{Variance of a sample}$$

$$= \frac{\Sigma(x - 6.4)^2}{5 - 1} \qquad \bar{x} = 6.4 \text{ and } n = 5$$

$$= \frac{(9 - 6.4)^2 + (6 - 6.4)^2 + (2 - 6.4)^2 + (5 - 6.4)^2 + (10 - 6.4)^2}{4} \qquad \text{Substitution}$$

$$= 10.3 \qquad \text{Simplify.}$$

Standard Deviation $\sigma = \sqrt{10.3}$

$$\approx 3.2$$

StudyTip

Rounding The mean and standard deviation should be rounded to one more decimal place than the original data. To avoid rounding error, round only final answers. Avoid using a round value to do further calculations.

In a given set of data, the majority of the values fall within one standard deviation of the mean, and almost all of the values will fall within 2 standard deviations. The quiz scores in Example 2a had a mean of about 7.4 and a standard deviation of about 2.1. This can be illustrated graphically.

If the quiz scores were compared with other scores throughout the country on a national test, this class would be considered a sample of all of the students who took the test. A sample mean $\bar{x}$ and a sample standard deviation s need to be calculated.

When comparing data sets, it is important to analyze the center *and* spread of each distribution. This is important because two sets of data can have the same mean but different spreads.

EXAMPLE 3	Compare Data Sets Using Measures of Spread

HEALTH *Metabolic rate* is the rate at which the body consumes energy, measured in Calories per 24 hours. During a study on diet and exercise, the metabolic rates for two different groups of men were observed. Which group has a greater variation in metabolic rates?

Group 1				
1507	1619	1731	1468	1533
1744	1588	1675	1552	1475
1593	1745	1523	1590	1764
1429	1604	1574	1708	1656

Group 2				
1498	1589	1634	1702	1629
1621	1629	1589	1592	1603
1573	1476	1613	1585	1582
1723	1619	1615	1601	1607

Enter the data into **L1** and **L2** on a graphing calculator. Press STAT and select **1-Var Stats** from the **CALC** menu, press 2nd [L1] or 2nd [L2] to select the Group 1 or Group 2 data, and press ENTER. Record the values for the sample mean $\bar{x}$, median **Med**, standard deviation **Sx**, and use **maxX − minX** to calculate the range.

Group 1 mean = 1603.9
  range = 335
  standard deviation = 100.2
  median = 1691.5 − 1429 or 1591.5

Group 2 mean = 1604
  range = 247
  standard deviation = 54.6
  median = 1723 − 1476 or 1605

Although the measures of center are reasonably close, the standard deviation of 100.2 for Group 1 is much larger than Group 2's value of 54.6. The range for Group 1 is also much larger than that of Group 2. Therefore, there is a greater variation in metabolic rates in Group 1.

Organize Data Data can be organized into a table called a **frequency distribution** to show how often each data value or group of data values, called a **class** or **interval**, appears in a data set. The **relative frequency** of a class is the ratio of data within the class to all the data. The **cumulative frequency** for a class is the sum of its frequency and all previous classes. The **cumulative relative frequency** for a class is the ratio of the cumulative frequency of the class to all the data.

Each class can be described in several ways. The **class width** is the range of values for each class. A *lower class limit* is the least value that can belong to a specific class, and an *upper class limit* is the greatest value that can belong to a specific class.

Real-World **EXAMPLE 4**	Frequency Distribution

FOOTBALL The winning scores for the first 42 Super Bowls are shown below.
35 33 16 23 16 24 14 24 16 21 32 27 35 31 27 26 27 38 38 46 39
42 20 55 20 37 52 30 49 27 35 31 34 23 34 20 48 32 24 21 29 17

a. Make a distribution table that shows the frequency and relative frequency of the data.

Step 1 Determine the number of classes and an appropriate class interval. The scores range from 14 to 55, so use 5 classes with a class interval of 10 points. Make a table listing the class limits. Begin with 10 points and end with 59 points.

Step 2 Tally the data. Then calculate the relative frequencies.

Winning Score	Tallies	Frequency	Relative Frequency
10–19	卌	5	$\frac{5}{42}$ or about 0.12
20–29	卌 卌 卌 l	16	$\frac{16}{42}$ or about 0.38
30–39	卌 卌 卌	15	$\frac{15}{42}$ or about 0.36
40–49	llll	4	$\frac{4}{42}$ or about 0.10
50–59	ll	2	$\frac{2}{42}$ or about 0.05
		42	

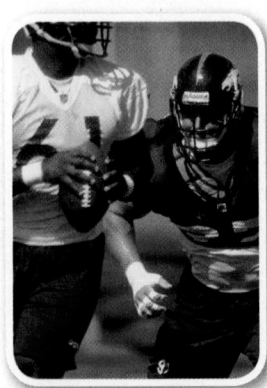

◈ Real-World Link

Mike Lodish has played in more Super Bowls than anyone else, four times with the Buffalo Bills and twice with the Denver Broncos.

Source: About: Football

b. Construct a histogram for both the frequency distribution and the relative frequency distribution. Then compare the graphs.

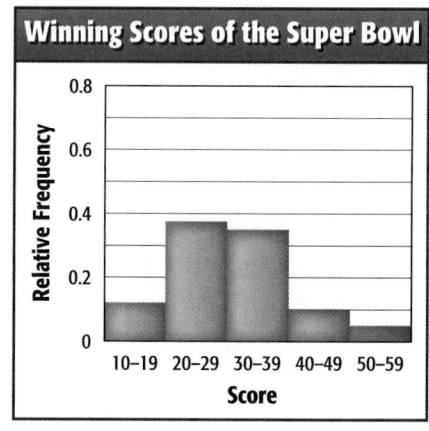

The overall shapes of the histograms are the same. The only difference is the vertical scale.

c. Make a cumulative frequency distribution for the data. Then determine the cumulative relative frequency distribution.

Winning Score	Frequency	Cumulative Frequency	Cumulative Relative Frequency
10–19	5	5	$\frac{5}{42}$ or about 0.12
20–29	16	16 + 5 or 21	$\frac{21}{42}$ or 0.50
30–39	15	15 + 21 or 36	$\frac{36}{42}$ or about 0.86
40–49	4	4 + 36 or 40	$\frac{40}{42}$ or about 0.95
50–59	2	2 + 40 or 42	$\frac{42}{42}$ or 1

d. Construct a histogram for the cumulative frequency distribution. Then compare the graph to the graph of the frequency distribution.

The shape of the cumulative frequency histogram shows an increasing pattern. There is a large increase in the number of teams scoring 10 to 19 points to the number of teams scoring 20 to 39 points. Then there is very little change in the number of teams scoring 40 or more points.

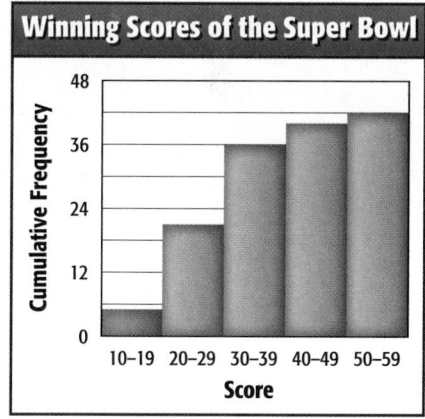

In a set of data, **quartiles** are values that divide the data into four equal parts.

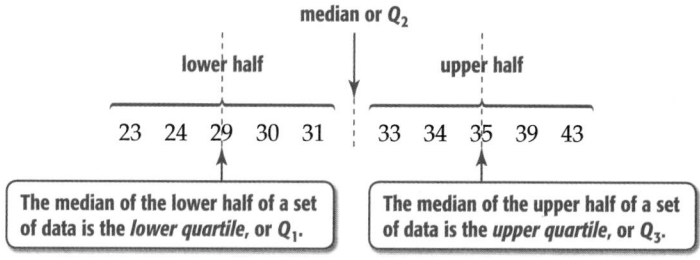

This **five-number summary**, which includes the minimum value, lower quartile, median, upper quartile, and the maximum value of a data set, provides another numerical way of characterizing a set of data. The five-number summary can be described visually with a box-and-whisker plot, as shown.

The difference between the upper quartile and lower quartile is called the **interquartile range**. **Outliers** are data that are more than 1.5 times the interquartile range beyond the upper or lower quartiles.

Real-World EXAMPLE 5 | Box-and-Whisker Plots

EDUCATION The enrollment for state universities in Ohio is shown. Display the data using a box-and-whisker plot.

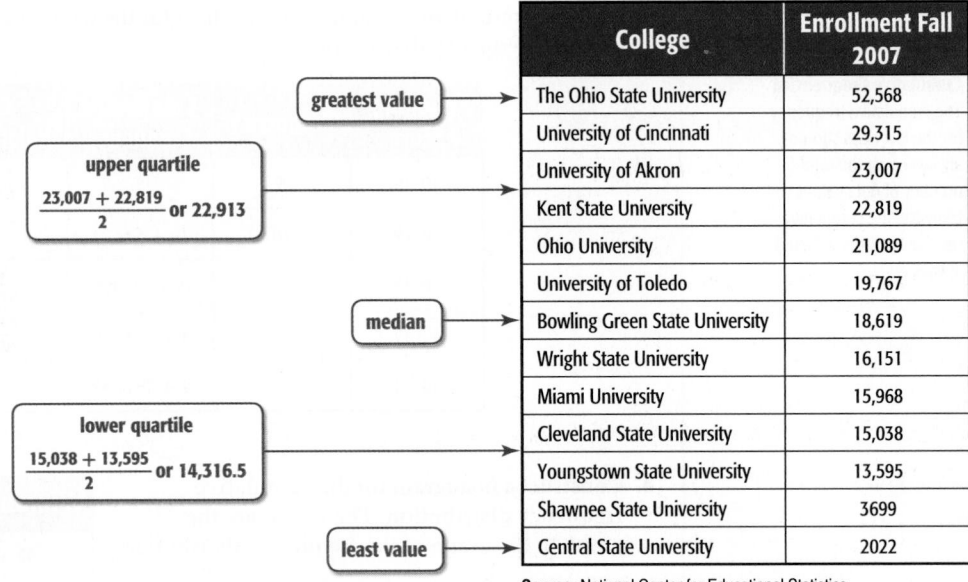

College	Enrollment Fall 2007
The Ohio State University	52,568
University of Cincinnati	29,315
University of Akron	23,007
Kent State University	22,819
Ohio University	21,089
University of Toledo	19,767
Bowling Green State University	18,619
Wright State University	16,151
Miami University	15,968
Cleveland State University	15,038
Youngstown State University	13,595
Shawnee State University	3699
Central State University	2022

Source: National Center for Educational Statistics

greatest value

upper quartile $\dfrac{23{,}007 + 22{,}819}{2}$ or 22,913

median

lower quartile $\dfrac{15{,}038 + 13{,}595}{2}$ or 14,316.5

least value

Step 1 Find the maximum and minimum values, and draw a number line that covers the range of the data. Then find the median and the upper and lower quartiles. Mark these points and the extreme values above the number line.

Step 2 Identify any outliers.

The interquartile range is 22,913 − 14,316.5 or 8596.5. There are no data values less than 14,316.5 − 1.5(8596.5) or 1421.75. There is one data value greater than 22,913 + 1.5(8596.5) or 35,807.75. The enrollment at The Ohio State University, 52,568, is an outlier.

Step 3 Draw a box around the upper and lower quartiles, a vertical line through the median, and use horizontal lines or *whiskers* to connect the lower value and the greatest value that is not an outlier. The greatest value that is not an outlier is 29,315.

Exercises

Find the mean, median, and mode of each set of data. (Example 1)

1. {24, 28, 21, 37, 31, 29, 23, 22, 34, 31}

2. {64, 87, 62, 87, 63, 98, 76, 54, 87, 58, 70, 76}

3. {6, 9, 11, 11, 12, 7, 6, 11, 5, 8, 10, 6}

4. **PACKING** Crates of books are being stored. The weights of the crates in pounds are shown in the table. (Example 1)

Weights in Pounds				
142.6	160.8	151.3	139.1	145.2
117.9	172.4	155.7	124.5	126.4
133.8	141.6	119.4	121.2	157.0

   a. What is the mean of the weights?

   b. Find the median of the weights.

   c. If 5 pounds is added to each crate, how will the mean and median be affected?

Find the range, variance, and standard deviation for each set of data. Use the formula for a population. (Example 2)

5. {$4.45, $5.50, $5.50, $6.30, $7.80, $11.00, $12.20, $17.20}

6. {200, 476, 721, 579, 152, 158}

7. {5.7, 5.7, 5.6, 5.5, 5.3, 4.9, 4.4, 4.0, 4.0, 3.8}

8. {369, 398, 381, 392, 406, 413, 376, 454, 420, 385, 402, 446}

9. **HEIGHTS** The heights in inches of Ms. Turner's astronomy students are listed below. (Example 2)

Heights in Inches									
66	72	70	74	64	65	60	62	66	67
68	71	70	72	73	65	63	62	62	61

   a. Find the measures of spread for the heights of the astronomy students.

   b. Using the second row as a sample, find the measures of spread for the sample of the astronomy students.

10. **MANUFACTURING** Sample lifetimes, measured in number of charging cycles, for two brands of rechargeable batteries are shown. Which brand has a greater variation in lifetimes? (Example 3)

Brand A				
998	950	1020	1003	990
942	1115	973	1018	981
1047	1002	997	1110	1003

Brand B				
892	1044	1001	999	903
950	998	993	1002	995
990	1000	1005	997	1004

11. **GRAPHIC NOVELS** The prices of 18 randomly selected graphic novels at two stores are shown. Which store has a greater variation in graphic novel prices? (Example 3)

Store 1 ($)					
18.99	12.99	15.95	12.99	12.95	29.95
24.99	39.99	9.95	14.99	24.95	9.99
17.99	13.99	4.99	29.95	9.99	12.95

Store 2 ($)					
19.99	7.95	7.95	4.99	12.99	7.95
25.65	7.95	9.99	14.99	9.95	14.99
8.99	9.99	7.95	12.95	14.95	29.95

12. **HISTORY** The ages of the first 44 presidents upon taking office are listed below. (Example 4)

Ages of U.S. Presidents										
57	61	57	57	58	57	61	54	68	51	49
64	50	48	65	52	56	46	54	49	50	47
55	55	54	42	51	56	55	51	54	51	60
62	43	55	56	61	52	69	64	46	54	47

   a. Make a distribution table of the data. Show the frequency, relative frequency, cumulative frequency, and cumulative relative frequency.

   b. Construct histograms for the frequency, relative frequency, and cumulative relative frequency distributions.

   c. Name the interval or intervals that describe the age of most presidents upon taking office.

13. **SPORTS DRINKS** Carlos surveyed his friends to find the number of bottles of sports drinks that they consume in an average week. Display the data using a box-and-whisker plot. (Example 5)

Sports Drinks Consumed Weekly										
0	0	0	1	1	1	2	2	3	4	4
5	5	7	10	10	10	11	11			

14. **DRIVING** Kara surveyed 20 randomly selected students at her school about how many miles they drive in an average day. The results are shown. (Example 5)

   a. What percent of the students drive more than 30 miles in a day?

   b. What is the interquartile range of the box-and-whisker plot shown?

   c. Does a student at Kara's school have a better chance of meeting someone who drives about the same mileage that he or she does if 15 miles or 50 miles are driven in a day? Why?

Use set notation to write the elements of each set. Then determine whether the statement about the set is *true* or *false*.

1. M is the set of natural number multiples of 5 that are less than 50. $12 \in M$

2. S is the set of integers that are less than -40 but greater than -50. $-49 \in S$

Let $B = \{0, 1, 2, 3\}$, $C = \{0, 1, 2, 3, 4, 5, 6\}$, $D = \{1, 3, 5, 7, 9\}$, $E = \{0, 2, 4, 6, 8, 10\}$, and $F = \{0, 10\}$. Find each of the following.

3. $D \cap C$

4. $D \cap F$

5. $E \cup B$

6. $D \cup F$

Simplify.

7. $(1 + 4i) + (-2 - 3i)$

8. $(2 + 4i) - (-1 + 5i)$

9. $(6 + 7i)(-5 + 3i)$

10. $(-1 + i)(-6 + 2i)$

11. $\dfrac{2 + 3i}{1 - 3i}$

12. $\dfrac{1 + 2i}{1 - 2i}$

Determine whether each function has a *maximum* or *minimum* value. Then find the value of the maximum or minimum, and state the domain and range of the function.

13.

$f(x) = -x^2 + 5x - 1$

14.

$f(x) = 3x^2 + 4x + 3$

Solve each equation.

15. $x^2 - x - 72 = 0$

16. $x^2 - 6x + 4 = 0$

17. $2x^2 - 5x + 4 = 0$

18. $2x^2 - x - 3 = 0$

19. **RECREATION** The current value C and the original value v of a recreational vehicle are related by $C = v(1 - r)^n$, where r is the rate of depreciation per year and n is the number of years. If the current value of a recreational vehicle is $47,500, what would be the value of the vehicle after 75 months at an annual depreciation rate of 15%?

Simplify each expression.

20. $\sqrt[6]{x^{18}y^{20}}$

21. $\sqrt[5]{a^{10}b^7}$

22. $\sqrt{16t^8u^{16}}$

23. $\sqrt[5]{243x^{10}y^{25}z^6}$

Simplify.

24. $\dfrac{y^{\frac{3}{4}} \cdot y^{\frac{2}{3}}}{y^{\frac{5}{12}}}$

25. $\sqrt[9]{512x^{10}y^{28}}$

26. $\sqrt[4]{m^{21}n^{18}}$

27. $\dfrac{\sqrt[12]{25}}{\sqrt[9]{125}}$

28. **JOBS** Leah babysits during the day for $3 per hour and at night for $5 per hour. If she worked 5 hours and earned $19, how many hours did she babysit during the day? How many at night?

Solve each system of equations. State whether the system is *consistent and independent, consistent and dependent,* or *inconsistent.*

29. $15x - 4.5y = 15$
    $6x - 3y = 10$

30. $5x + y = 2$
    $x - y = 22$

31. $9x - 3y + 12z = 39$
    $12x - 4y + 16z = 54$
    $3x - 8y + 12z = 23$

32. $6x + 2y + 4z = 2$
    $3x + 4y - 8z = -3$
    $-3x - 6y + 12z = 5$

Solve each system of inequalities. If the system has no solution, state *no solution.*

33. $y \geq x - 3$
    $y \leq 3x + 1$

34. $y + x < 6$
    $y > -3x + 2$

35. $3x + 2y \geq 6$
    $4x - y \geq 2$

36. $2x + 5y \leq -15$
    $y > -\dfrac{2}{5}x + 2$

Find each of the following for $A = \begin{bmatrix} 3 & 0 \\ 2 & -1 \\ -6 & -5 \end{bmatrix}$, $B = \begin{bmatrix} 1 & -7 \\ -8 & 4 \\ 10 & 2 \end{bmatrix}$, and $C = \begin{bmatrix} 2 & 2 \\ -7 & 8 \\ -1 & -3 \end{bmatrix}$.

37. $A + B + C$

38. $B - C$

39. $2A - B$

Find each permutation or combination.

40. $_{10}C_3$

41. $_{10}P_3$

42. $_6P_6$

43. $_6C_6$

44. $_8P_4$

45. $_8C_4$

46. **CARDS** Four cards are randomly drawn from a standard deck of 52 cards. Find each probability.

   a. $P(1 \text{ ace and } 3 \text{ kings})$

   b. $P(2 \text{ odd and } 2 \text{ face cards})$

Find the mean, median, and mode for each set of data. Then find the range, variance, and standard deviation for each population.

47. $\{1, 1, 1, 2, 2, 3\}$

48. $\{0.8, 0.9, 0.4, 0.8, 0.6, 0.8, 0.6\}$

Functions from a Calculus Perspective

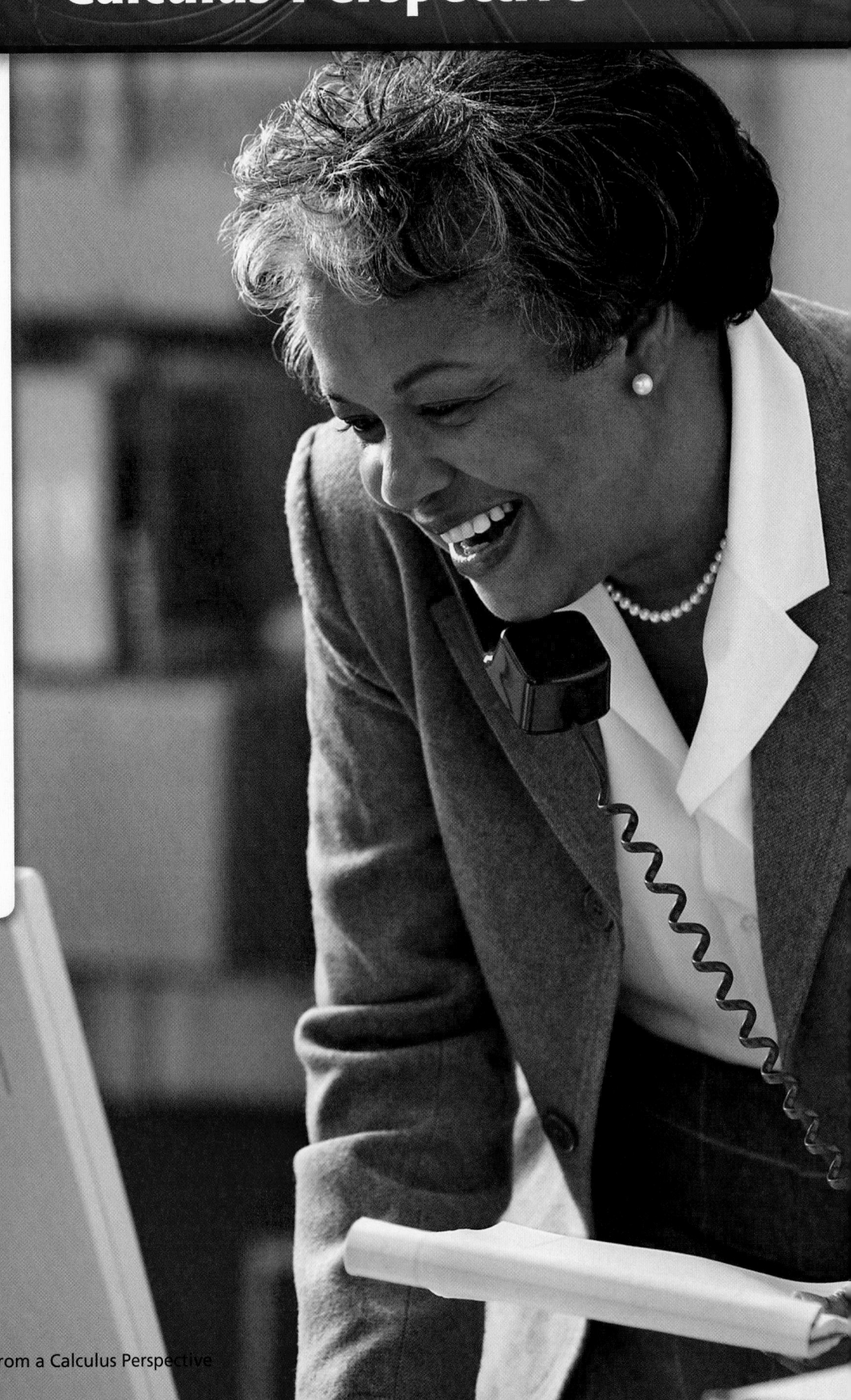

Then

In Algebra 2, you analyzed functions from a graphical perspective.

Now

In Chapter 1, you will:

- Explore symmetries of graphs.
- Determine continuity and average rates of change of functions.
- Use limits to describe end behavior.
- Find inverse functions algebraically and graphically.

Why?

BUSINESS Functions are often used throughout the business world. Some of the uses of functions are to analyze costs, predict sales, calculate profit, forecast future costs and revenue, estimate depreciation, and determine the proper labor force.

PREREAD Create a list of two or three things that you already know about functions. Then make a prediction of what you will learn in Chapter 1.

Get Ready for Chapter 1

Diagnose Readiness You have two options for checking Prerequisite Skills.

Text Option

Take the Quick Check below.

*Quick*Check

Graph each inequality on a number line.
(Prerequisite Skill)

1. $x > -3$
2. $x \le -2$
3. $x \le -5$
4. $x > 1$
5. $7 \ge x$
6. $-4 < x$

Solve each equation for y. (Prerequisite Skill)

7. $y - 3x = 2$
8. $y + 4x = -5$
9. $2x - y^2 = 7$
10. $y^2 + 5 = -3x$
11. $9 + y^3 = -x$
12. $y^3 - 9 = 11x$

13. **DONUTS** A bakery uses the formula $12D = n$, where D is the number of dozens of donuts and n is the total number of donuts sold to determine how many dozens of donuts were sold. Solve the equation for D, and determine how many dozens of donuts were sold if 306 donuts were sold. (Prerequisite Skill)

Evaluate each expression given the value of the variable. (Prerequisite Skill)

14. $3y - 4, y = 2$
15. $2b + 7, b = -3$
16. $x^2 + 2x - 3, x = -4a$
17. $5z - 2z^2 + 1, z = 5x$
18. $-4c^2 + 7, c = 7a^2$
19. $2 + 3p^2, p = -5 + 2n$

20. **TEMPERATURE** The formula $C = \frac{5}{9}(F - 32)$, where C represents a temperature in degrees Celsius and F in degrees Fahrenheit, can be used to convert between the two measures. If the temperature on a thermometer reads 73°F, what is the temperature in degrees Celsius rounded to the nearest tenth? (Prerequisite Skill)

Online Option

Math Online Take a self-check Chapter Readiness Quiz at **glencoe.com**.

New Vocabulary

English	• Page •	Español
interval notation	• 5 •	notación del intervalo
function	• 5 •	función
function notation	• 7 •	notación de la función
implied domain	• 7 •	dominio implicado
zeros	• 15 •	ceros
roots	• 15 •	raíz
even function	• 18 •	función uniforme
odd function	• 18 •	función impar
limit	• 24 •	límite
end behavior	• 28 •	comportamiento final
increasing	• 34 •	aumento
decreasing	• 34 •	el disminuir
constant	• 34 •	constante
maximum	• 36 •	máximo
minimum	• 36 •	mínimo
extrema	• 36 •	extrema
secant line	• 38 •	línea secante
parent function	• 45 •	función del padre
transformation	• 46 •	transformación
reflection	• 48 •	reflexión
dilation	• 49 •	dilatación
composition	• 58 •	composición

Review Vocabulary

parabola • p. P9 • parábola the graph of a quadratic function

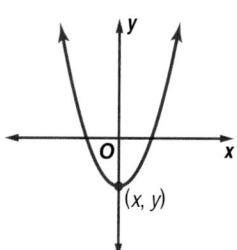

slope • **Prerequisite Skill** • línea pendiente the ratio of the change in y-coordinates to the change in x-coordinates

Functions

Then
You used set notation to denote elements, subsets, and complements.
(Lesson 0-1)

Now
- Describe subsets of real numbers.
- Identify and evaluate functions and state their domains.

New Vocabulary
set-builder notation
interval notation
function
function notation
independent variable
dependent variable
implied domain
piecewise-defined function
relevant domain

Math Online
glencoe.com

Why?

Many events that occur in everyday life involve two related quantities. For example, to operate a vending machine, you insert money and make a selection. The machine gives you the selected item and any change due. Once your selection is made, the amount of change you receive *depends* on the amount of money you put into the machine.

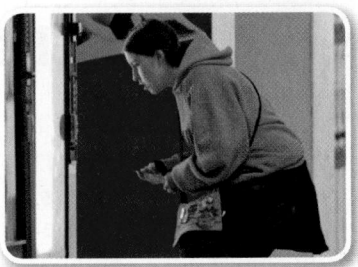

Describe Subsets of Real Numbers Real numbers are used to describe quantities such as money, time, and distance. The set of real numbers $\mathbb{R}$ includes the following subsets of numbers.

Key Concept	Real Numbers

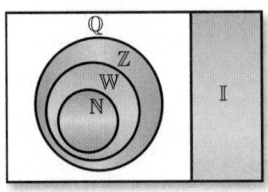

Real Numbers ($\mathbb{R}$)

Letter	Set	Examples
$\mathbb{Q}$	rationals	$0.125, -\frac{7}{8}, \frac{2}{3} = 0.666\ldots$
$\mathbb{I}$	irrationals	$\sqrt{3} = 1.73205\ldots$
$\mathbb{Z}$	integers	$-5, 17, -23, 8$
$\mathbb{W}$	wholes	$0, 1, 2, 3\ldots$
$\mathbb{N}$	naturals	$1, 2, 3, 4\ldots$

These and other sets of real numbers can be described using set-builder notation. **Set-builder notation** uses the properties of the numbers in the set to define the set.

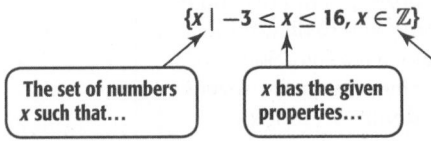

$$\{x \mid -3 \leq x \leq 16, x \in \mathbb{Z}\}$$

| The set of numbers x such that... | x has the given properties... | and x is an element of the given set of numbers. |

EXAMPLE 1   Use Set-Builder Notation

Describe the set of numbers using set-builder notation.

a. {8, 9, 10, 11, ...}

The set includes all whole numbers greater than or equal to 8.

$\{x \mid x \geq 8, x \in \mathbb{W}\}$ **Read as *the set of all x such that x is greater than or equal to 8 and x is an element of the set of whole numbers.***

b. $x < 7$

Unless otherwise stated, you should assume that a given set consists of real numbers. Therefore, the set includes all real numbers less than 7.

$\{x \mid x < 7, x \in \mathbb{R}\}$

c. all multiples of three

The set includes all integers that are multiples of three.

$\{x \mid x = 3n, n \in \mathbb{Z}\}$

✔ Guided Practice

1A. {1, 2, 3, 4, 5, ...}   **1B.** $x \leq -3$ **1C.** all multiples of π

StudyTip

Look Back You can review set notation, including unions and intersections of sets, in Lesson 0-1.

Interval notation uses inequalities to describe subsets of real numbers. The symbols [or] are used to indicate that an endpoint is included in the interval, while the symbols (or) are used to indicate that an endpoint is not included in the interval. The symbols ∞, positive infinity, and $-\infty$, negative infinity, are used to describe the unboundedness of an interval. An interval is *unbounded* if it goes on indefinitely.

Bounded Intervals		Unbounded Intervals	
Inequality	**Interval Notation**	**Inequality**	**Interval Notation**
$a \leq x \leq b$	$[a, b]$	$x \geq a$	$[a, \infty)$
$a < x < b$	(a, b)	$x \leq a$	$(-\infty, a]$
$a \leq x < b$	$[a, b)$	$x > a$	(a, ∞)
$a < x \leq b$	$(a, b]$	$x < a$	$(-\infty, a)$
		$-\infty < x < \infty$	$(-\infty, \infty)$

EXAMPLE 2 **Use Interval Notation**

Write each set of numbers using interval notation.

a. $-8 < x \leq 16$ $(-8, 16]$

b. $x < 11$ $(-\infty, 11)$

c. $x \leq -16$ or $x > 5$ $(-\infty, -16] \cup (5, \infty)$ $\cup$ read as *union*

☑ **Guided Practice**

2A. $-4 \leq y < -1$ **2B.** $a \geq -3$ **2C.** $x > 9$ or $x < -2$

Identify Functions Recall that a *relation* is a rule that relates two quantities. Such a rule pairs the elements in a set A with elements in a set B. The set A of all inputs is the *domain* of the relation, and set B contains all outputs or the *range*.

Relations are commonly represented in four ways.

1. Verbally A sentence describes how the inputs and outputs are related.

The output value is 2 more than the input value.

2. Numerically A table of values or a set of ordered pairs relates each input (x-value) with an output value (y-value).

{(0, 2), (1, 3), (2, 4), (3, 5)}

3. Graphically Points on a graph in the coordinate plane represent the ordered pairs.

4. Algebraically An equation relates the x- and y-coordinates of each ordered pair.

$y = x + 2$

A **function** is a special type of relation.

StudyTip

Domain and Range In this text, the notation for domain and range will be D = and R =, respectively.

Key Concept **Function**

Words   A function f from set A to set B is a relation that assigns to each element x in set A *exactly one* element y in set B.

Example   The relation from set A to set B is a function.

Set A is the domain.
D = {1, 2, 3, 4}

Set B contains the range.
R = {6, 8, 9}

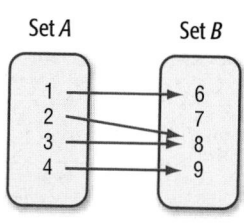

An alternate definition of a function is a set of ordered pairs in which no two different pairs have the same x-value. Interpreted graphically, this means that no two points on the graph of a function in the coordinate plane can lie on the same vertical line.

Key Concept **Vertical Line Test**

Words

A set of points in the coordinate plane is the graph of a function if each possible vertical line intersects the graph in at most one point.

Model

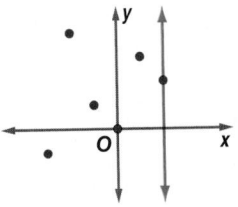

EXAMPLE 3 **Identify Relations that are Functions**

Determine whether each relation represents y as a function of x.

a. **The input value x is a student's ID number, and the output value y is that student's score on a physics exam.**

Each value of x cannot be assigned to more than one y-value. A student cannot receive two different scores on an exam. Therefore, the sentence describes y as a function of x.

b.

x	y
−8	−5
−5	−4
0	−3
3	−2
6	−3

Each x-value is assigned to exactly one y-value. Therefore, the table represents y as a function of x.

c.

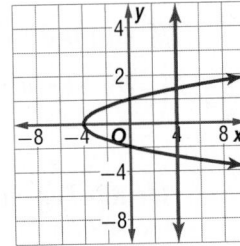

A vertical line at $x = 4$ intersects the graph at more than one point. Therefore, the graph does not represent y as a function of x.

d. $y^2 - 2x = 5$

To determine whether this equation represents y as a function of x, solve the equation for y.

$y^2 - 2x = 5$	Original equation
$y^2 = 5 + 2x$	Add $2x$ to each side.
$y = \pm\sqrt{5 + 2x}$	Take the square root of each side.

This equation does not represent y as a function of x because there will be two corresponding y-values, one positive and one negative, for any x-value greater than −2.5.

☑ **Guided Practice**

3A. The input value x is the area code, and the output value y is a phone number in that area code.

3B.

x	y
−6	−7
2	3
5	8
5	9
9	22

3C.

3D. $3y + 6x = 18$

In **function notation**, the symbol $f(x)$ is read *f of x* and interpreted as *the value of the function f at x*. Because $f(x)$ corresponds to the *y*-value of *f* for a given *x*-value, you can write $y = f(x)$.

Equation	Related Function
$y = -6x$	$f(x) = -6x$

Because it can represent any value in the function's domain, *x* is called the **independent variable**. A value in the range of *f* is represented by the **dependent variable**, *y*.

EXAMPLE 4   Find Function Values

If $g(x) = x^2 + 8x - 24$, find each function value.

a. $g(6)$

To find $g(6)$, replace *x* with 6 in $g(x) = x^2 + 8x - 24$.

$g(x) = x^2 + 8x - 24$ **Original function**

$g(6) = (6)^2 + 8(6) - 24$ **Substitute 6 for x.**

$\quad\ = 36 + 48 - 24$ **Simplify.**

$\quad\ = 60$ **Simplify.**

b. $g(-4x)$

$g(x) = x^2 + 8x - 24$ **Original function**

$g(-4x) = (-4x)^2 + 8(-4x) - 24$ **Substitute −4x for x.**

$\quad\ = 16x^2 - 32x - 24$ **Simplify.**

c. $g(5c + 4)$

$g(x) = x^2 + 8x - 24$ **Original function**

$g(5c + 4) = (5c + 4)^2 + 8(5c + 4) - 24$ **Substitute 5c + 4 for x.**

$\quad\ = 25c^2 + 40c + 16 + 40c + 32 - 24$ **Expand $(5c + 4)^2$ and $8(5c + 4)$.**

$\quad\ = 25c^2 + 80c + 24$ **Simplify.**

✔ Guided Practice

If $f(x) = \dfrac{2x + 3}{x^2 - 2x + 1}$, find each function value.

4A. $f(12)$ **4B.** $f(6x)$ **4C.** $f(-3a + 8)$

When you are given a function with an unspecified domain, the **implied domain** is the set of all real numbers for which the expression used to define the function is real. In general, you must exclude values from the domain of a function that would result in division by zero or taking the even root of a negative number.

EXAMPLE 5   Find Domains Algebraically

State the domain of each function.

a. $f(x) = \dfrac{2 + x}{x^2 - 7x}$

When the denominator of $\dfrac{2 + x}{x^2 - 7x}$ is zero, the expression is undefined. Solving $x^2 - 7x = 0$, the excluded values for the domain of this function are $x = 0$ and $x = 7$. The domain of this function is all real numbers except $x = 0$ and $x = 7$, or $\{x \mid x \neq 0, x \neq 7, x \in \mathbb{R}\}$.

b. $g(t) = \sqrt{t - 5}$

Because the square root of a negative number cannot be real, $t - 5 \geq 0$. Therefore, the domain of $g(t)$ is all real numbers *t* such that $t \geq 5$ or $[5, \infty)$.

---

c. $h(x) = \dfrac{1}{\sqrt{x^2 - 9}}$

This function is defined only when $x^2 - 9 > 0$. Therefore, the domain of $h(x)$ is $(-\infty, -3) \cup (3, \infty)$.

✓ Guided Practice

State the domain of each function.

5A. $f(x) = \dfrac{5x - 2}{x^2 + 7x + 12}$ **5B.** $h(a) = \sqrt{a^2 - 4}$ **5C.** $g(x) = \dfrac{8x}{\sqrt{2x + 6}}$

A function that is defined using two or more equations for different intervals of the domain is called a **piecewise-defined function**.

🌐 Real-World EXAMPLE 6   Evaluate a Piecewise-Defined Function

HEIGHT The average maximum height of children in inches as a function of their parents' maximum height in inches can be modeled by the following piecewise function. Find the average maximum heights of children whose parents have the given maximum heights. Use $h(x)$, where x is the independent variable representing the parents' height and $h(x)$ is the dependent variable representing the child's height.

$$h(x) = \begin{cases} 1.6x - 41.6 & \text{if } 63 < x < 66 \\ 3x - 132 & \text{if } 66 \le x \le 68 \\ 2x - 66 & \text{if } x > 68 \end{cases}$$

a. $h(67)$

Because 67 is between 66 and 68, use $h(x) = 3x - 132$ to find $h(67)$.

$h(67) = 3x - 132$ **Function for 66 ≤ x ≤ 68**

$ = 3(67) - 132$ **Substitute 67 for x.**

$ = 201 - 132 \text{ or } 69$ **Simplify.**

According to this model, children whose parents have a maximum height of 67 inches will attain an average maximum height of 69 inches.

b. $h(72)$

Because 72 is greater than 68, use $h(x) = 2x - 66$.

$h(72) = 2x - 66$ **Function for x > 68**

$ = 2(72) - 66$ **Substitute 72 for x.**

$ = 144 - 66 \text{ or } 78$ **Simplify.**

According to this model, children whose parents have a maximum height of 72 inches will attain an average maximum height of 78 inches.

✓ Guided Practice

6. SPEED The speed v of a vehicle in miles per hour can be represented by the following piecewise function when t is the time in seconds. Find the speed of the vehicle at each indicated time.

$$v(t) = \begin{cases} 4t & \text{if } 0 \le t \le 15 \\ 60 & \text{if } 15 < t < 240 \\ -6t + 1500 & \text{if } 240 \le t \le 250 \end{cases}$$

A. $v(5)$ **B.** $v(15)$ **C.** $v(245)$

Write each set of numbers in set-builder and interval notation, if possible. (Examples 1 and 2)

1. $x > 50$

2. $x < -13$

3. $x \le -4$

4. $\{-4, -3, -2, -1, \ldots\}$

5. $8 < x < 99$

6. $-31 < x \le 64$

7. $x < -19$ or $x > 21$

8. $x < 0$ or $x \ge 100$

9. $\{-0.25, 0, 0.25, 0.50, \ldots\}$

10. $x \le 61$ or $x \ge 67$

11. $x \le -45$ or $x > 86$

12. all multiples of 8

13. all multiples of 5

14. $x \ge 32$

Determine whether each relation represents y as a function of x. (Example 3)

15. The input value x is a bank account number and the output value y is the account balance.

16. The input value x is the year and the output value y is the day of the week.

17.

x	y
−50	2.11
−40	2.14
−30	2.16
−20	2.17
−10	2.17
0	2.18

18.

x	y
0.01	423
0.04	449
0.04	451
0.07	466
0.08	478
0.09	482

19. $\frac{1}{x} = y$

20. $x^2 = y + 2$

21. $3y + 4x = 11$

22. $4y^2 + 18 = 96x$

23. $\sqrt{48y} = x$

24. $\frac{x}{y} = y - 6$

25.

26.

27.

28.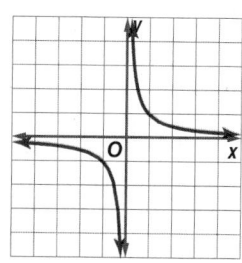

29. **METEOROLOGY** The five-day forecast for a city is shown. (Example 3)

	1	2	3	4	5
Hi	70°F	75°F	70°F	62°F	65°F
Lo	49°F	53°F	51°F	57°F	56°F

a. Represent the relation between the day of the week and the estimated high temperature as a set of ordered pairs.

b. Is the estimated high temperature a function of the day of the week? the low temperature? Explain your reasoning.

Find each function value. (Example 4)

30. $g(x) = 2x^2 + 18x - 14$
   a. $g(9)$
   b. $g(3x)$
   c. $g(1 + 5m)$

31. $h(y) = -3y^3 - 6y + 9$
   a. $h(4)$
   b. $h(-2y)$
   c. $h(5b + 3)$

32. $f(t) = \dfrac{4t + 11}{3t^2 + 5t + 1}$
   a. $f(-6)$
   b. $f(4t)$
   c. $f(3 - 2a)$

33 $g(x) = \dfrac{3x^3}{x^2 + x - 4}$
   a. $g(-2)$
   b. $g(5x)$
   c. $g(8 - 4b)$

34. $h(x) = 16 - \dfrac{12}{2x + 3}$
   a. $h(-3)$
   b. $h(6x)$
   c. $h(10 - 2c)$

35. $f(x) = -7 + \dfrac{6x + 1}{x}$
   a. $f(5)$
   b. $f(-8x)$
   c. $f(6y + 4)$

36. $g(m) = 3 + \sqrt{m^2 - 4}$
   a. $g(-2)$
   b. $g(3m)$
   c. $g(4m - 2)$

37. $t(x) = 5\sqrt{6x^2}$
   a. $t(-4)$
   b. $t(2x)$
   c. $t(7 + n)$

38. **DIGITAL AUDIO PLAYERS** The sales of digital audio players in millions of dollars for a five-year period can be modeled using $f(t) = 24t^2 - 93t + 78$, where t is the year. The actual sales data are shown in the table. (Example 4)

Year	Sales ($)
1	1 million
2	3 million
3	14 million
4	74 million
5	219 million

a. Find $f(1)$ and $f(5)$.

b. Do you think that the model is more accurate for the earlier years or the later years? Explain your reasoning.

State the domain of each function. (Example 5)

39. $f(x) = \dfrac{8x + 12}{x^2 + 5x + 4}$

40. $g(x) = \dfrac{x + 1}{x^2 - 3x - 40}$

41. $g(a) = \sqrt{1 + a^2}$

42. $h(x) = \sqrt{6 - x^2}$

43. $f(a) = \dfrac{5a}{\sqrt{4a - 1}}$

44. $g(x) = \dfrac{3}{\sqrt{x^2 - 16}}$

45. $f(x) = \dfrac{2}{x} + \dfrac{4}{x + 1}$

46. $g(x) = \dfrac{6}{x + 3} + \dfrac{2}{x - 4}$

47. PHYSICS The period T of a pendulum is the time for one cycle and can be calculated using the formula $T = 2\pi\sqrt{\dfrac{\ell}{9.8}}$, where ℓ is the length of the pendulum and 9.8 is the gravitational acceleration due to gravity in meters per second squared. Is this formula a function of ℓ? If so, determine the domain. If not, explain why not. (Example 5)

Length ℓ

Period T

Find $f(-5)$ and $f(12)$ for each piecewise function. (Example 6)

48. $f(x) = \begin{cases} -4x + 3 & \text{if } x < 3 \\ -x^3 & \text{if } 3 \le x \le 8 \\ 3x^2 + 1 & \text{if } x > 8 \end{cases}$

49. $f(x) = \begin{cases} -5x^2 & \text{if } x < -6 \\ x^2 + x + 1 & \text{if } -6 \le x \le 12 \\ 0.5x^3 - 4 & \text{if } x > 12 \end{cases}$

50. $f(x) = \begin{cases} 2x^2 + 6x + 4 & \text{if } x < -4 \\ 6 - x^2 & \text{if } -4 \le x < 12 \\ 14 & \text{if } x \ge 12 \end{cases}$

51. $f(x) = \begin{cases} -15 & \text{if } x < -5 \\ \sqrt{x + 6} & \text{if } -5 \le x \le 10 \\ \dfrac{2}{x} + 8 & \text{if } x > 10 \end{cases}$

52. INCOME TAX Federal income tax for a person filing single in the United States in a recent year can be modeled using the following function, where x represents income and $T(x)$ represents total tax. (Example 6)

$$T(x) = \begin{cases} 0.10x & \text{if } 0 \le x \le 7285 \\ 782.5 + 0.15x & \text{if } 7285 < x \le 31,850 \\ 4386.25 + 0.25x & \text{if } 31,850 < x \le 77,100 \end{cases}$$

a. Find $T(7000)$, $T(10,000)$, and $T(50,000)$.

b. If a person's annual income were $7285, what would his or her income tax be?

53. PUBLIC TRANSPORTATION The nationwide use of public transportation can be modeled using the following function. The year 1996 is represented by $t = 0$, and $P(t)$ represents passenger trips in millions. (Example 6)

$$P(t) = \begin{cases} 0.35t + 7.6 & \text{if } 0 \le t \le 5 \\ 0.04t^2 - 0.6t + 11.6 & \text{if } 5 < t \le 10 \end{cases}$$

a. Approximately how many passenger trips were there in 1999? in 2004?

b. State the domain of the function.

Use the vertical line test to determine whether each graph represents a function. Write *yes* or *no*. Explain your reasoning.

54.

55.

56.

57.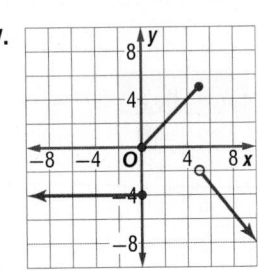

58. TRIATHLON In a triathlon, athletes swim 2.4 miles, then bike 112 miles, and finally run 26.2 miles. Jesse's average rates for each leg of a triathlon are shown in the table.

Leg	Rate
swim	4 mph
bike	20 mph
run	6 mph

a. Write a piecewise function to describe the distance D that Jesse has traveled in terms of time t. Round t to the nearest tenth, if necessary.

b. State the domain of the function.

59. ELECTIONS Describe the set of presidential election years beginning in 1792 in interval notation or in set-builder notation. Explain your reasoning.

60. CONCESSIONS The number of students working the concession stands at a football game can be represented by $f(x) = \dfrac{x}{50}$, where x is the number of tickets sold. Describe the relevant domain of the function.

61. ATTENDANCE The Chicago Cubs franchise has been in existence since 1874. The total season attendance for its home games can be modeled by $f(x) = 70{,}050x - 137{,}400{,}000$, where x represents the year. Describe the relevant domain of the function.

62. ACCOUNTING A business' assets, such as equipment, wear out or depreciate over time. One way to calculate depreciation is the straight-line method, using the value of the estimated life of the asset. Suppose $v(t) = 10{,}440 - 290t$ describes the value $v(t)$ of a copy machine after t months. Describe the relevant domain of the function.

Find $f(a)$, $f(a + h)$, and $\dfrac{f(a + h) - f(a)}{h}$ if $h \neq 0$.

63. $f(x) = -5$ **64.** $f(x) = \sqrt{x}$

65. $f(x) = \dfrac{1}{x + 4}$ **66.** $f(x) = \dfrac{2}{5 - x}$

67. $f(x) = x^2 - 6x + 8$ **68.** $f(x) = -\dfrac{1}{4}x + 6$

69. $f(x) = -x^5$ **70.** $f(x) = x^3 + 9$

71. $f(x) = 7x - 3$ **72.** $f(x) = 5x^2$

73. $f(x) = x^3$ **74.** $f(x) = 11$

75. MAIL The U.S. Postal Service requires that envelopes have an aspect ratio (length divided by height) of 1.3 to 2.5, inclusive. The minimum allowable length is 5 inches and the maximum allowable length is $11\frac{1}{2}$ inches.

a. Write the area of the envelope A as a function of length ℓ if the aspect ratio is 1.8. State the domain of the function.

b. Write the area of the envelope A as a function of height h if the aspect ratio is 2.1. State the domain of the function.

c. Find the area of an envelope with the maximum height at the maximum aspect ratio.

76. GEOMETRY Consider the circle below with area A and circumference C.

a. Represent the area of the circle as a function of its circumference.

b. Find $A(0.5)$ and $A(4)$.

c. What do you notice about the area as the circumference increases?

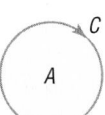

Determine whether each equation is a function of x. Explain.

77. $x = |y|$ **78.** $x = y^3$

79. ⟳ **MULTIPLE REPRESENTATIONS** In this problem, you will investigate the range of a function.

a. **GRAPHICAL** Use a graphing calculator to graph $f(x) = x^n$ for whole-number values of n from 1 to 6, inclusive.

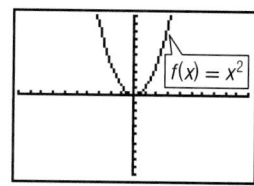

[−10, 10] scl: 1 by [−10, 10] scl: 1

b. **TABULAR** Predict the range of each function based on the graph, and tabulate each value of n and the corresponding range.

c. **VERBAL** Make a conjecture about the range of $f(x)$ when n is even.

d. **VERBAL** Make a conjecture about the range of $f(x)$ when n is odd.

H.O.T. Problems   Use **H**igher-**O**rder **T**hinking Skills

80. ERROR ANALYSIS Ana and Mason are evaluating $f(x) = \dfrac{2}{x^2 - 4}$. Ana thinks that the domain of the function is $(-\infty, -2)$ or $(1, 1)$ or $(2, \infty)$. Mason thinks that the domain is $\{x \mid x \neq -2, x \neq 2, x \in \mathbb{R}\}$. Is either of them correct? Explain.

81 **WRITING IN MATH** Write the domain of $f(x) = \dfrac{1}{(x + 3)(x + 1)(x - 5)}$ in interval notation and in set-builder notation. Which notation do you prefer? Explain.

82. CHALLENGE $G(x)$ is a function for which $G(1) = 1$, $G(2) = 2$, $G(3) = 3$, and $G(x + 1) = \dfrac{G(x - 2)\,G(x - 1) + 1}{G(x)}$ for $x \geq 3$. Find $G(6)$.

REASONING Determine whether each statement is *true* or *false* given a function from set X to set Y. If a statement is false, rewrite it to make a true statement.

83. Every element in X must be matched with only one element in Y.

84. Every element in Y must be matched with an element in X.

85. Two or more elements in X may not be matched with the same element in Y.

86. Two or more elements in Y may not be matched with the same element in X.

WRITING IN MATH Explain how you can identify a function described as each of the following.

87. a verbal description of inputs and outputs

88. a set of ordered pairs

89. a table of values

90. a graph

91. an equation

Find the standard deviation of each population of data. (Lesson 0-8)

92. {200, 476, 721, 579, 152, 158}

93. {5.7, 5.7, 5.6, 5.5, 5.3, 4.9, 4.4, 4.0, 4.0, 3.8}

94. {369, 398, 381, 392, 406, 413, 376, 454, 420, 385, 402, 446}

95. BASEBALL How many different 9-player teams can be made if there are 3 players who can only play catcher, 4 players who can only play first base, 6 players who can only pitch, and 14 players who can play in any of the remaining 6 positions? (Lesson 0-7)

Find the values for x and y that make each matrix equation true. (Lesson 0-6)

96. $\begin{bmatrix} y \\ x \end{bmatrix} = \begin{bmatrix} 4x - 3 \\ y - 2 \end{bmatrix}$

97. $\begin{bmatrix} 3y \\ 10 \end{bmatrix} = \begin{bmatrix} 27 + 6x \\ 5y \end{bmatrix}$

98. $[9 \quad 11] = [3x + 3y \quad 2x + 1]$

Use any method to solve the system of equations. State whether the system is *consistent*, *dependent*, *independent*, or *inconsistent*. (Lesson 0-5)

99. $2x + 3y = 36$
$4x + 2y = 48$

100. $5x + y = 25$
$10x + 2y = 50$

101. $7x + 8y = 30$
$7x + 16y = 46$

102. BUSINESS A used book store sells 1400 paperback books per week at $2.25 per book. The owner estimates that he will sell 100 fewer books for each $0.25 increase in price. What price will maximize the income of the store? (Lesson 0-3)

Use the Venn diagram to find each of the following. (Lesson 0-1)

103. A'

104. $A \cup B$

105. $B \cap C$

106. $A \cap B$

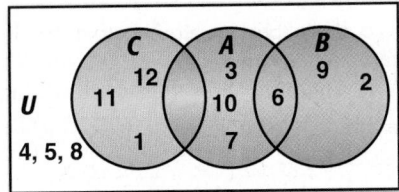

107. SAT/ACT A circular cone with a base of radius 5 has been cut as shown in the figure.

What is the height of the smaller top cone?

A $\frac{8}{13}$

C $\frac{96}{12}$

E $\frac{104}{5}$

B $\frac{96}{13}$

D $\frac{96}{5}$

108. REVIEW Which function is linear?

F $f(x) = x^2$

H $f(x) = \sqrt{9 - x^2}$

G $g(x) = 2.7$

J $g(x) = \sqrt{x - 1}$

109. Louis is flying from Denver to Dallas for a convention. He can park his car in the Denver airport long-term lot or in the nearby shuttle parking facility. The long-term lot costs $1 per hour or any fraction thereof with a maximum charge of $6 per day. In the shuttle facility, he has to pay $4 for each day or part of a day. Which parking lot is less expensive if Louis returns after 2 days and 3 hours?

A shuttle facility

B airport lot

C They will both cost the same.

D cannot be determined with the information given

110. REVIEW Given $y = 2.24x + 16.45$, which statement best describes the effect of moving the graph down two units?

F The y-intercept increases.

G The x-intercept remains the same.

H The x-intercept increases.

J The y-intercept remains the same.

Analyzing Graphs of Functions and Relations

Then
You identified functions.
(Lesson 1-1)

Now
- Use graphs of functions to estimate function values and find domains, ranges, *y*-intercepts, and zeros of functions.
- Explore symmetries of graphs, and identify even and odd functions.

New Vocabulary

zeros
roots
line symmetry
point symmetry
even function
odd function

Math Online >
glencoe.com

Why?

With more people turning to the Internet for news and entertainment, Internet advertising is big business. The total revenue R in millions of dollars earned by U.S. companies from Internet advertising from 1999 to 2008 can be approximated by $R(t) = 17.7t^3 - 269t^2 + 1458t - 910$, $1 \leq t \leq 10$, where t represents the number of years since 1998. Graphs of functions like this can help you visualize relationships between real-world quantities.

Analyzing Function Graphs The graph of a function f is the set of ordered pairs $(x, f(x))$ such that x is in the domain of f. In other words, the graph of f is the graph of the equation $y = f(x)$. So, the value of the function is the directed distance y of the graph from the point x on the x-axis as shown.

You can use the graph of a function to estimate function values.

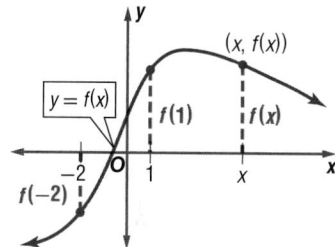

🌐 Real-World EXAMPLE 1   Estimate Function Values

INTERNET Consider the graph of function R shown.

a. Use the graph to estimate total Internet advertising revenue in 2007. Confirm the estimate algebraically.

The year 2007 is 9 years after 1998. The function value at $x = 9$ appears to be about $3300 million, so the total Internet advertising revenue in 2007 was about $3.3 billion.

To confirm this estimate algebraically, find $f(9)$.

$f(9) = 17.7(9)^3 - 269(9)^2 + 1458(9) - 910$

≈ 3326.3 million or 3.326 billion

Therefore, the graphical estimate of $3.3 billion is reasonable.

b. Use the graph to estimate the year in which total Internet advertising revenue reached $2 billion. Confirm the estimate algebraically.

The value of the function appears to reach $2 billion or $2000 million for x-values between 6 and 7. So, the total revenue was nearly $2 billion in 1998 + 6 or 2004 but had exceeded $2 billion by the end of 1998 + 7 or 2005.

To confirm algebraically, find $f(6)$ and $f(7)$.

$f(6) = 17.7(6)^3 - 269(6)^2 + 1458(6) - 910$ or about 1977 million

$f(7) = 17.7(7)^3 - 269(7)^2 + 1458(7) - 910$ or about 2186 million

In billions, $f(6) \approx 1.977$ billion and $f(7) \approx 2.186$ billion. Therefore, the graphical estimate that total Internet advertising revenue reached $2 billion in 2005 is reasonable.

☑ Guided Practice

1. **STOCKS** An investor assessed the average daily value of a share of a certain stock over a 20-day period. The value of the stock can be approximated by $v(d) = 0.002d^4 - 0.11d^3 + 1.77d^2 - 8.6d + 31$, $0 \leq d \leq 20$, where d represents the day of the assessment.

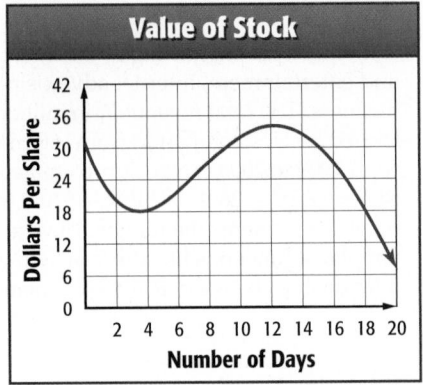

A. Use the graph to estimate the value of the stock on the 10th day. Confirm your estimate algebraically.

B. Use the graph to estimate the days during which the stock was valued at $30 per share. Confirm your estimate algebraically.

You can also use a graph to find the domain and range of a function. Unless the graph of a function is bounded on the left by a circle or a dot, you can assume that the function extends beyond the edges of the graph.

TechnologyTip

Choosing an Appropriate Window The viewing window of a graph is a picture of the graph for a specific domain and range. This may not represent the entire graph. Notice the difference in the graphs of $f(x) = x^4 - 20x^3$ shown below.

[−10, 10] scl: 1 by
[−10, 10] scl: 1

[−15, 25] scl: 4 by
[−20,000, 20,000] scl: 4000

EXAMPLE 2   Find Domain and Range

Use the graph of f to find the domain and range of the function.

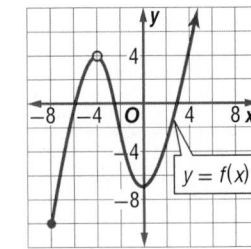

Domain

- The dot at $(-8, -10)$ indicates that the domain of f starts at and includes -8.

- The circle at $(-4, 4)$ indicates that -4 is not part of the domain.

- The arrow on the right side indicates that the graph will continue without bound.

The domain of f is $[-8, -4) \cup (-4, \infty)$. In set-builder notation, the domain is $\{x \mid -8 \leq x, x \neq -4, x \in \mathbb{R}\}$.

Range

The graph does not extend below $f(-8)$ or -10, but $f(x)$ increases without bound for greater and greater values of x. So, the range of f is $[-10, \infty)$.

☑ Guided Practice

Use the graph of g to find the domain and range of each function.

2A.

2B.

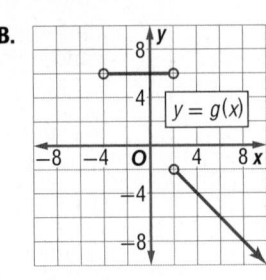

A point where a graph intersects or meets the x- or y-axis is called an intercept. An x-intercept of a graph occurs where $y = 0$. A y-intercept of a graph occurs where $x = 0$. The graph of a function can have 0, 1, or more x-intercepts, but at most one y-intercept.

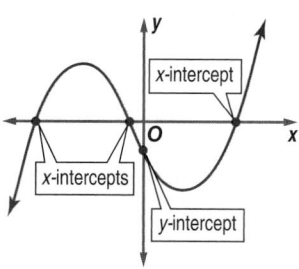

To find the y-intercept of a graph of a function f algebraically, find $f(0)$.

EXAMPLE 3   Find y-Intercepts

Use the graph of each function to approximate its y-intercept. Then find the y-intercept algebraically.

a.
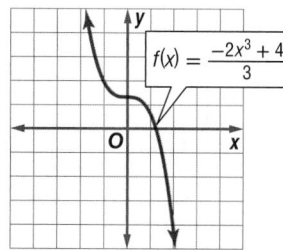

$$f(x) = \frac{-2x^3 + 4}{3}$$

b.
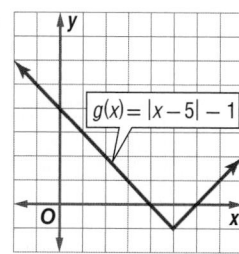

$$g(x) = |x - 5| - 1$$

Estimate Graphically

It appears that $f(x)$ intersects the y-axis at approximately $\left(0, 1\frac{1}{3}\right)$, so the y-intercept is about $1\frac{1}{3}$.

Solve Algebraically

Find $f(0)$.

$$f(0) = \frac{-2(0)^3 + 4}{3} \text{ or } \frac{4}{3}$$

The y-intercept is $\frac{4}{3}$ or $1\frac{1}{3}$.

Estimate Graphically

It appears that $g(x)$ intersects the y-axis at $(0, 4)$, so the y-intercept is 4.

Solve Algebraically

Find $g(0)$.

$$g(0) = |0 - 5| - 1 \text{ or } 4$$

The y-intercept is 4.

✔ Guided Practice

3A.
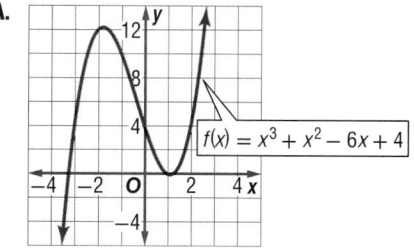

$$f(x) = x^3 + x^2 - 6x + 4$$

3B.
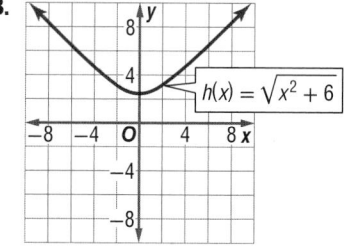

$$h(x) = \sqrt{x^2 + 6}$$

The x-intercepts of the graph of a function are also called the **zeros** of a function. The solutions of the corresponding equation are called the **roots** of the equation. To find the zeros of a function f, set the function equal to 0 and solve for the independent variable.

EXAMPLE 4   Find Zeros

Use the graph of $f(x) = 2x^2 + x - 15$ to approximate its zero(s). Then find its zero(s) algebraically.

Estimate Graphically

The x-intercepts appear to be at about -3 and 2.5.

Solve Algebraically

$2x^2 + x - 15 = 0$	Let $f(x) = 0$.
$(2x - 5)(x + 3) = 0$	Factor.
$2x - 5 = 0$ or $x + 3 = 0$	Zero Product Property
$x = 2.5$ $x = -3$	Solve for x.

The zeros of f are -3 and 2.5.

☑ Guided Practice

Use the graph of each function to approximate its zero(s). Then find its zero(s) algebraically.

4A.

4B.
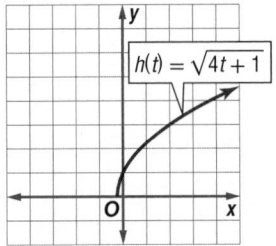

Symmetry of Graphs Graphs of relations can have two different types of symmetry. Graphs with **line symmetry** can be folded along a line so that the two halves match exactly. Graphs that have **point symmetry** can be rotated 180° with respect to a point and appear unchanged. The three most common types of symmetry are shown below.

StudyTip

Symmetry, Relations, and Functions There are numerous *relations* that have x-axis, y-axis, and origin symmetry. However, the only *function* that has all three types of symmetry is the zero function, $f(x) = 0$.

Key Concept   Tests for Symmetry

Graphical Test	Model	Algebraic Test
The graph of a relation is *symmetric with respect to the x-axis* if and only if for every point (x, y) on the graph, the point $(x, -y)$ is also on the graph.		Replacing y with $-y$ produces an equivalent equation.
The graph of a relation is *symmetric with respect to the y-axis* if and only if for every point (x, y) on the graph, the point $(-x, y)$ is also on the graph.		Replacing x with $-x$ produces an equivalent equation.
The graph of a relation is *symmetric with respect to the origin* if and only if for every point (x, y) on the graph, the point $(-x, -y)$ is also on the graph.		Replacing x with $-x$ and y with $-y$ produces an equivalent equation.

EXAMPLE 5 Test for Symmetry

Use the graph of each equation to test for symmetry with respect to the x-axis, y-axis, and the origin. Support the answer numerically. Then confirm algebraically.

a. $x - y^2 = 1$

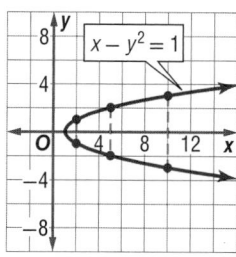

Analyze Graphically

The graph appears to be symmetric with respect to the x-axis because for every point (x, y) on the graph, there is a point $(x, -y)$.

Support Numerically

A table of values supports this conjecture.

x	2	2	5	5	10	10
y	1	−1	2	−2	3	−3
(x, y)	(2, 1)	(2, −1)	(5, 2)	(5, −2)	(10, 3)	(10, −3)

Confirm Algebraically

Because $x - (-y)^2 = 1$ is equivalent to $x - y^2 = 1$, the graph is symmetric with respect to the x-axis.

b. $xy = 4$

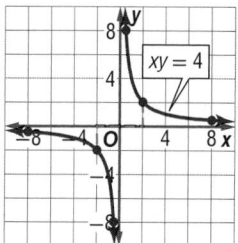

Analyze Graphically

The graph appears to be symmetric with respect to the origin because for every point (x, y) on the graph, there is a point $(-x, -y)$.

Support Numerically

A table of values supports this conjecture.

x	−8	−2	−0.5	0.5	2	8
y	−0.5	−2	−8	8	2	0.5
(x, y)	(−8, −0.5)	(−2, −2)	(−0.5, −8)	(0.5, 8)	(2, 2)	(8, 0.5)

Confirm Algebraically

Because $(-x)(-y) = 4$ is equivalent to $xy = 4$, the graph is symmetric with respect to the origin.

✔️ **Guided Practice**

5A.

5B.
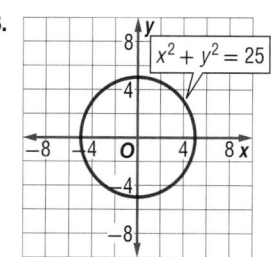

Graphs of functions can have y-axis or origin symmetry. Functions with these types of symmetry have special names.

Key Concept	Even and Odd Functions

Type of Function	Algebraic Test
Functions that are symmetric with respect to the y-axis are called **even functions**.	For every x in the domain of f, $f(-x) = f(x)$.
Functions that are symmetric with respect to the origin are called **odd functions**.	For every x in the domain of f, $f(-x) = -f(x)$.

EXAMPLE 6 · Identify Even and Odd Functions

GRAPHING CALCULATOR Graph each function. Analyze the graph to determine whether each function is *even*, *odd*, or *neither*. Confirm algebraically. If odd or even, describe the symmetry of the graph of the function.

a. $f(x) = x^3 - 2x$

It appears that the graph of the function is symmetric with respect to the origin. Test this conjecture.

$$f(-x) = (-x)^3 - 2(-x) \qquad \text{Substitute } -x \text{ for } x.$$
$$= -x^3 + 2x \qquad \text{Simplify.}$$
$$= -(x^3 - 2x) \qquad \text{Distributive Property}$$
$$= -f(x) \qquad \text{Original function } f(x) = x^3 - 2x$$

$[-10, 10]$ scl: 1 by $[-10, 10]$ scl:

The function is odd because $f(-x) = -f(x)$. Therefore, the function is symmetric with respect to the origin.

b. $g(x) = x^4 + 2$

It appears that the graph of the function is symmetric with respect to the y-axis. Test this conjecture.

$$g(-x) = (-x)^4 + 2 \qquad \text{Substitute } -x \text{ for } x.$$
$$= x^4 + 2 \qquad \text{Simplify.}$$
$$= g(x) \qquad \text{Original function } g(x) = x^4 + 2$$

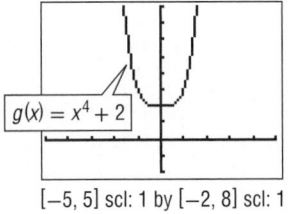

$[-5, 5]$ scl: 1 by $[-2, 8]$ scl: 1

The function is even because $g(-x) = g(x)$. Therefore, the function is symmetric with respect to the y-axis.

c. $h(x) = x^3 - 0.5x^2 - 3x$

It appears that the graph of the function may be symmetric with respect to the origin. Test this conjecture algebraically.

$$h(-x) = (-x)^3 - 0.5(-x)^2 - 3(-x) \qquad \text{Substitute } -x \text{ for } x.$$
$$= -x^3 - 0.5x^2 + 3x \qquad \text{Simplify.}$$

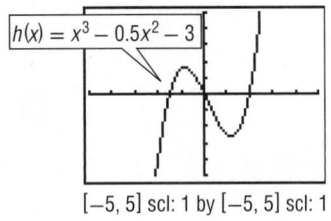

$[-5, 5]$ scl: 1 by $[-5, 5]$ scl: 1

Because $-h(x) = -x^3 + 0.5x^2 + 3x$, the function is neither even nor odd because $h(-x) \neq h(x)$ and $h(-x) \neq -h(x)$.

✓ Guided Practice

6A. $f(x) = \dfrac{2}{x^2}$ **6B.** $g(x) = 4\sqrt{x}$ **6C.** $h(x) = x^5 - 2x^3 + x$

Use the graph of each function to estimate the indicated function values. Then confirm the estimate algebraically. Round to the nearest hundredth, if necessary. (Example 1)

1.

$g(x) = -5\sqrt{x} + 50$

a. $g(6)$ **b.** $g(12)$ **c.** $g(19)$

2.

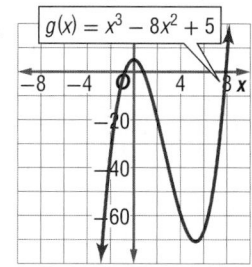

$g(x) = x^3 - 8x^2 + 5$

a. $g(-2)$ **b.** $g(1)$ **c.** $g(8)$

3.

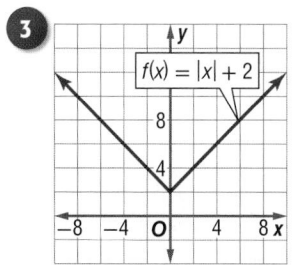

$f(x) = |x| + 2$

a. $f(-8)$ **b.** $f(-3)$ **c.** $f(0)$

4.

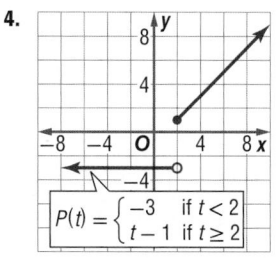

$P(t) = \begin{cases} -3 & \text{if } t < 2 \\ t - 1 & \text{if } t \geq 2 \end{cases}$

a. $P(-6)$ **b.** $P(2)$ **c.** $P(9)$

5.

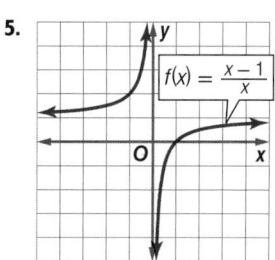

$f(x) = \dfrac{x-1}{x}$

a. $f(-3)$ **b.** $f(0.5)$ **c.** $f(0)$

6.

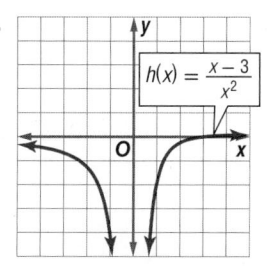

$h(x) = \dfrac{x-3}{x^2}$

a. $h(-1)$ **b.** $h(1.5)$ **c.** $h(2)$

7. RECYLING The quantity of paper recycled in the United States in thousands of tons from 1993 to 2007 can be modeled by $p(x) = -0.0013x^4 + 0.0513x^3 - 0.662x^2 + 4.128x + 35.75$, where x is the number of years since 1993. (Example 1)

U. S. Paper Recycling

a. Use the graph to estimate the amount of paper recycled in 1993, 1999, and 2006. Then find each value algebraically.

b. Use the graph to estimate the year in which the quantity of paper recycled reached 50,000 tons. Confirm algebraically.

8. WATER Bottled water consumption from 1977 to 2006 can be modeled using $f(x) = 9.35x^2 - 12.7x + 541.7$, where x represents the number of years since 1977. (Example 1)

Bottled Water Consumption

a. Use the graph to estimate the amount of bottled water consumed in 1994.

b. Find the 1994 consumption algebraically. Round to the nearest ten million gallons.

c. Use the graph to estimate when water consumption was 6 billion gallons. Confirm algebraically.

Use the graph of h to find the domain and range of each function. (Example 2)

9.

$y = h(x)$

10.

$y = h(x)$

11.

$y = h(x)$

12.

$y = h(x)$

13.

$y = h(x)$

14.

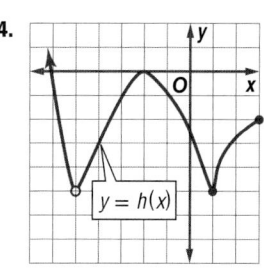

$y = h(x)$

15. **ENGINEERING** Tests on the physical behavior of four metal specimens are performed at various temperatures in degrees Celsius. The impact energy, or energy absorbed by the sample during the test, is measured in joules. The test results are shown. (Example 2)

Impact Test Results

a. State the domain and range of each function.

b. Use the graph to estimate the impact energy of each metal at 0°C.

Use the graph of each function to find its *y*-intercept and zero(s). Then find these values algebraically. (Examples 3 and 4)

16.
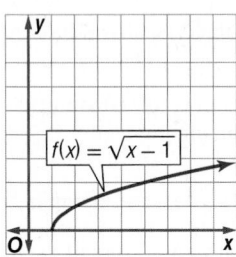
$f(x) = \sqrt{x} - 1$

17.

$f(x) = 2x^3 - x^2 - 3x$

18.

$f(x) = \sqrt[3]{x}$

19.
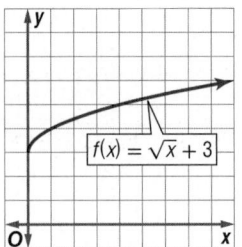
$f(x) = \sqrt{x} + 3$

20.
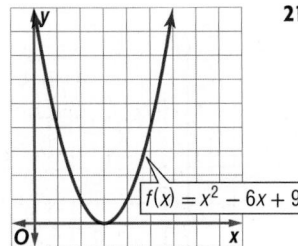
$f(x) = x^2 - 6x + 9$

21.

$f(x) = 6x^2 - x - 2$

22.

$f(x) = x^3 + 6x^2 + 12x + 8$

23.
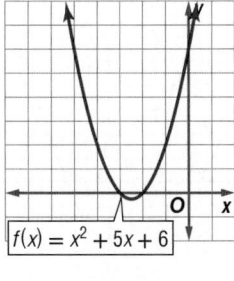
$f(x) = x^2 + 5x + 6$

Use the graph of each equation to test for symmetry with respect to the *x*-axis, *y*-axis, and the origin. Support the answer numerically. Then confirm algebraically. (Example 5)

24.
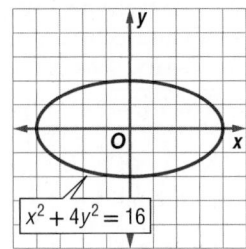
$x^2 + 4y^2 = 16$

25.

$x = y^2 - 3$

26.

$x = -y$

27.
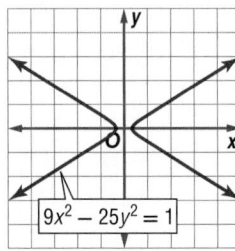
$9x^2 - 25y^2 = 1$

28.

$y = \dfrac{x^3}{4}$

29.

$y = -\dfrac{10}{x}$

30.
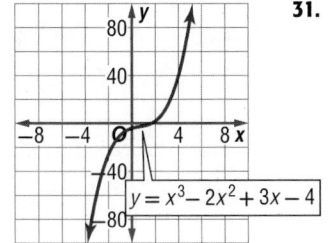
$y = x^3 - 2x^2 + 3x - 4$

31.
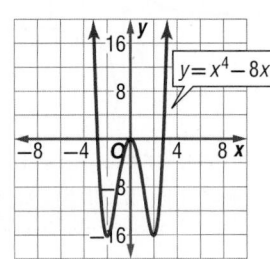
$y = x^4 - 8x^2$

32.
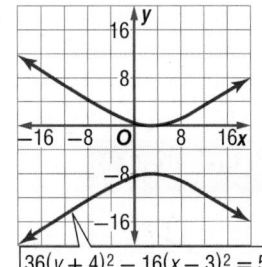
$36(y + 4)^2 - 16(x - 3)^2 = 576$

33.
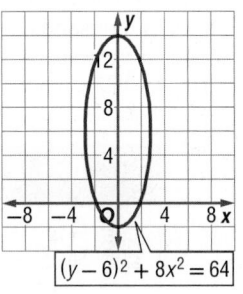
$(y - 6)^2 + 8x^2 = 64$

GRAPHING CALCULATOR Graph each function. Analyze the graph to determine whether each function is *even*, *odd*, or *neither*. Confirm algebraically. If odd or even, describe the symmetry of the graph of the function. (Example 6)

34. $f(x) = x^2 + 6x + 10$

35. $f(x) = -2x^3 + 5x - 4$

36. $g(x) = \sqrt{x + 6}$

37. $h(x) = \sqrt{x^2 - 9}$

38. $h(x) = |8 - 2x|$

39. $f(x) = |x^3|$

40. $f(x) = \dfrac{x + 4}{x - 2}$

41. $g(x) = \dfrac{x^2}{x + 1}$

Use the graph of each function to estimate the indicated function values.

42.

43.

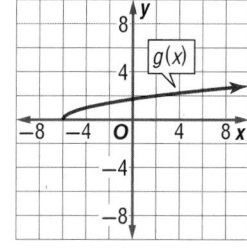

42. a. $f(-2)$ b. $f(-6)$ c. $f(0)$

43. a. $g(-8)$ b. $g(-6)$ c. $g(-2)$

44. FOOTBALL A running back's rushing yards for each game in a season are shown.

a. State the domain and range of the relation.

b. In what game did the player rush for no yards?

45 PHONES The number of households h in millions with only wireless phone service from 2001 to 2005 can be modeled by $h(x) = 0.5x^2 + 0.5x + 1.2$, where x represents the number of years after 2001.

a. State the relevant domain and approximate the range.

b. Use the graph to estimate the number of households with only wireless phone service in 2003. Then find it algebraically.

c. Use the graph to approximate the y-intercept of the function. Then find it algebraically. What does the y-intercept represent?

d. Does this function have any zeros? If so, estimate them and explain their meaning. If not, explain why.

46. FUNCTIONS Consider $f(x) = x^n$.

a. Use a graphing calculator to graph $f(x)$ for values of n in the range $1 \le n \le 6$, where $n \in \mathbb{N}$.

b. Describe the domain and range of each function.

c. Describe the symmetry of each function.

d. Predict the domain, range, and symmetry for $f(x) = x^{35}$. Explain your reasoning.

47. PHARMACOLOGY Suppose the number of milligrams of a pain reliever in the bloodstream x hours after taking a dose is modeled by $f(x) = 0.5x^4 + 3.45x^3 - 96.65x^2 + 347.7x$.

a. Use a graphing calculator to graph the function.

b. State the relevant domain. Explain your reasoning.

c. What was the approximate maximum amount of pain reliever, in milligrams, that entered the bloodstream?

GRAPHING CALCULATOR Graph and locate the zeros for each function. Confirm your answers algebraically.

48. $f(x) = \dfrac{4x - 1}{x}$

49. $f(x) = \dfrac{x^2 + 9}{x + 3}$

50. $h(x) = \sqrt{x^2 + 4x + 3}$

51. $h(x) = 2\sqrt{x + 12} - 8$

52. $g(x) = -12 + \dfrac{4}{x}$

53. $g(x) = \dfrac{6}{x} + 3$

54. TELEVISION The percent of households h with basic cable for the years 1980 through 2006 can be modeled using $h(x) = -0.115x^2 + 4.43x + 25.6$, $1980 \le x \le 2006$, where x represents the number of years after 1980.

a. Use a graphing calculator to graph the function.

b. What percent of households had basic cable in 1999? Round to the nearest percent.

c. For what years was the percent of subscribers greater than 65%?

Use the graph of f to find the domain and range of each function.

55.

56.

57.

58.

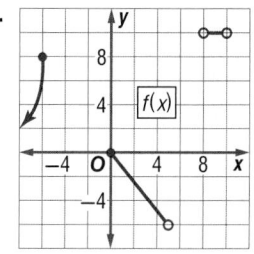

59. POPULATION The percent population change from 1930 to 1940, 1940 to 1950, and so on, for a certain U.S. city from 1930 to 2000 can be modeled by $f(x) = 0.0001x^3 - 0.001x^2 - 0.825x + 12.58$, where x is the number of years since 1930.

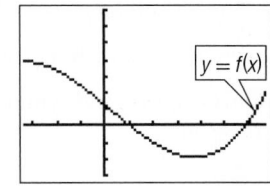

[−50, 100] scl: 15 by [−30, 70] scl: 10

a. State the relevant domain and estimate the range for this domain.

b. Use the graph to approximate the y-intercept. Then find the y-intercept algebraically. What does the y-intercept represent?

c. Find and interpret the zeros of the function.

d. Use the model to determine what the percent population change will be in 2080. Does this value seem realistic? Explain your reasoning.

60. STOCK MARKET The percent p a stock price has fluctuated in one year can be modeled by $p(x) = 0.0005x^4 - 0.0193x^3 + 0.243x^2 - 1.014x + 1.04$, where x is the number of months since January.

a. Use a graphing calculator to graph the function.

b. State the relevant domain and estimate the range.

c. Use the graph to approximate the y-intercept. Then find the y-intercept algebraically. What does the y-intercept represent?

d. Find and interpret any zeros of the function.

61. ⬛ **MULTIPLE REPRESENTATIONS** In this problem, you will investigate the range values of $f(x) = \dfrac{1}{x-2}$ as x approaches 2.

a. TABULAR Copy and complete the table below. Add an additional value to the left and right of 2.

x	1.99	1.999	2	2.001	2.01
f(x)					

b. ANALYTICAL Use the table from part **a** to describe the behavior of the function as x approaches 2.

c. GRAPHICAL Graph the function. Does the graph support your conjecture from part **b**? Explain.

d. VERBAL Make a conjecture as to why the graph of the function approaches the value(s) found in part **c**, and explain any inconsistencies present in the graph.

GRAPHING CALCULATOR Graph each function. Analyze the graph to determine whether each function is *even*, *odd*, or *neither*. Confirm algebraically. If odd or even, describe the symmetry of the graph of the function.

62. $f(x) = x^2 - x - 6$

63. $g(n) = n^2 - 37$

64. $h(x) = x^6 + 4$

65. $f(g) = g^9$

66. $g(y) = y^4 + 8y^2 + 81$

67. $h(y) = y^5 - 17y^3 + 16y$

68. $h(b) = b^4 - 2b^3 - 13b^2 + 14b + 24$

OPEN ENDED Sketch a graph that matches each description.

69. passes through $(-3, 8)$, $(-4, 4)$, $(-5, 2)$, and $(-8, 1)$ and is symmetric with respect to the y-axis

70. passes through $(0, 0)$, $(2, 6)$, $(3, 12)$, and $(4, 24)$ and is symmetric with respect to the x-axis

71 passes through $(-3, -18)$, $(-2, -9)$, and $(-1, -3)$ and is symmetric with respect to the origin

72. passes through $(4, -16)$, $(6, -12)$, and $(8, -8)$ and represents an even function

73. WRITING IN MATH Explain why a function can have 0, 1, or more x-intercepts but only one y-intercept.

74. CHALLENGE Use a graphing calculator to graph $f(x) = \dfrac{2x^2 + 3x - 2}{x^3 - 4x^2 - 12x}$, and predict its domain. Then confirm the domain algebraically. Explain your reasoning.

REASONING Determine whether each statement is *true* or *false*. Explain your reasoning.

75. The range of $f(x) = nx^2$, where n is any integer, is $\{y \mid y \geq 0, y \in \mathbb{R}\}$.

76. The range of $f(x) = \sqrt{nx}$, where n is any integer, is $\{y \mid y \geq 0, y \in \mathbb{R}\}$.

77. All odd functions are also symmetric with respect to the line $y = -x$.

78. An even function rotated $180n°$ about the origin, where n is any integer, remains an even function.

REASONING If $a(x)$ is an odd function, determine whether $b(x)$ is *odd*, *even*, *neither*, or *cannot be determined*. Explain your reasoning.

79. $b(x) = a(-x)$

80. $b(x) = -a(x)$

81. $b(x) = [a(x)]^2$

82. $b(x) = a(|x|)$

83. $b(x) = [a(x)]^3$

REASONING State whether a graph with each type of symmetry *always*, *sometimes*, or *never* represents a function. Explain your reasoning.

84. symmetric with respect to the line $x = 4$

85. symmetric with respect to the line $y = 2$

86. symmetric with respect to the line $y = x$

87. symmetric with respect to both the x- and y-axes

88. WRITING IN MATH Can a function be both even and odd? Explain your reasoning.

Find each function value. (Lesson 1-1)

89. $g(x) = x^2 - 10x + 3$
   a. $g(2)$
   b. $g(-4x)$
   c. $g(1 + 3n)$

90. $h(x) = 2x^2 + 4x - 7$
   a. $h(-9)$
   b. $h(3x)$
   c. $h(2 + m)$

91. $p(x) = \dfrac{2x^3 + 2}{x^2 - 2}$
   a. $p(3)$
   b. $p(x^2)$
   c. $p(x + 1)$

92. GRADES The midterm grades for a Chemistry class of 25 students are shown. Find the measures of spread for the data set. (Lesson 0-8)

Midterm Grades				
89	76	91	72	81
81	65	74	80	74
73	92	76	83	96
66	61	80	74	70
97	78	73	62	72

93. PLAYING CARDS From a standard 52-card deck, find how many 5-card hands are possible that fit each description. (Lesson 0-7)
   a. 3 hearts and 2 clubs
   b. 1 ace, 2 jacks, and 2 kings
   c. all face cards

Find the following for $A = \begin{bmatrix} -6 & 3 \\ -5 & 11 \end{bmatrix}$, $B = \begin{bmatrix} 3 & -7 \\ 2 & -3 \end{bmatrix}$, and $C = \begin{bmatrix} 2 \\ 4 \end{bmatrix}$. (Lesson 0-6)

94. $4A - 2B$

95. $3C + 2A$

96. $-2(B - 3A)$

Evaluate each expression. (Lesson 0-4)

97. $27^{\frac{1}{3}}$

98. $64^{\frac{5}{6}}$

99. $49^{-\frac{1}{2}}$

100. $16^{-\frac{3}{4}}$

101. $25^{\frac{3}{2}}$

102. $36^{-\frac{3}{2}}$

103. GENETICS Suppose R and W represent two genes that a plant can inherit from its parents. The terms of the expansion of $(R + W)^2$ represent the possible pairings of the genes in the offspring. Write $(R + W)^2$ as a polynomial. (Lesson 0-3)

Simplify. (Lesson 0-2)

104. $(2 + i)(4 + 3i)$

105. $(1 + 4i)^2$

106. $(2 - i)(3 + 2i)(1 - 4i)$

107. SAT/ACT In the figure, if n is a real number greater than 1, what is the value of x in terms of n?

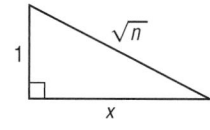

   A $\sqrt{n^2 - 1}$ **C** $\sqrt{n + 1}$ **E** $n + 1$
   B $\sqrt{n - 1}$ **D** $n - 1$

108. REVIEW Which inequality describes the range of $f(x) = x^2 + 1$ over the domain $-2 < x < 3$?
   F $5 \leq y < 9$ **H** $1 < y < 9$
   G $2 < y < 10$ **J** $1 \leq y < 10$

109. Which of the following is an even function?
   A $f(x) = 2x^4 + 6x^3 - 5x^2 - 8$
   B $g(x) = 3x^6 + x^4 - 5x^2 + 15$
   C $m(x) = x^4 + 3x^3 + x^2 + 35x$
   D $h(x) = 4x^6 + 2x^4 + 6x - 4$

110. Which of the following is the domain of $g(x) = \dfrac{1 + x}{x^2 - 16x}$?
   F $(-\infty, 0) \cup (0, 16) \cup (16, \infty)$
   G $(-\infty, 0] \cup [16, \infty)$
   H $(-\infty, -1) \cup (-1, \infty)$
   J $(-\infty, -4) \cup (-4, 4) \cup (4, \infty)$

Continuity, End Behavior, and Limits

Then
You found domain and range using the graph of a function. (Lesson 1-2)

Now
- Use limits to determine the continuity of a function, and apply the Intermediate Value Theorem to continuous functions.
- Use limits to describe end behavior of functions.

New Vocabulary
continuous function
limit
discontinuous function
infinite discontinuity
jump discontinuity
removable discontinuity
nonremovable discontinuity
end behavior

Math Online

glencoe.com

Why?
The minimum wage has increased significantly since the early 1980s. Rather than increasing steadily, the current minimum wage is the result of several jumps. The graph of the minimum wage as a function of time shows these jumps as breaks in the graph, such as those at $x = 1990$, $x = 1996$, and $x = 2008$.

Continuity The graph of a **continuous function** has no breaks, holes, or gaps. You can trace the graph of a continuous function without lifting your pencil.

One condition for a function $f(x)$ to be continuous at $x = c$ is that the function must approach a unique function value as x-values approach c from the left and right sides. The concept of approaching a value without necessarily ever reaching it is called a **limit**.

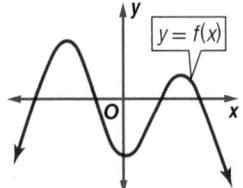

$f(x)$ is continuous for all x.

Key Concept · Limits

Words If the value of $f(x)$ approaches a unique value L as x approaches c from each side, then the limit of $f(x)$ as x approaches c is L.

Symbols $\lim\limits_{x \to c} f(x) = L$, which is read *The limit of $f(x)$ as x approaches c is L.*

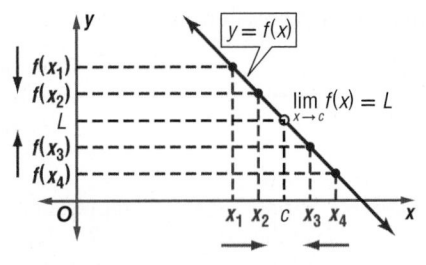

To understand what it means for a function to be continuous from an algebraic perspective, it helps to examine the graphs of **discontinuous functions**, or functions that are not continuous. Functions can have many different types of discontinuity.

Key Concept · Types of Discontinuity

A function has an **infinite discontinuity** at $x = c$ if the function value increases or decreases indefinitely as x approaches c from the left and right.

Example

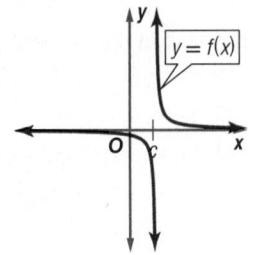

A function has a **jump discontinuity** at $x = c$ if the limits of the function as x approaches c from the left and right exist but have two distinct values.

Example

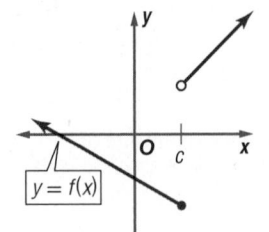

A function has a **removable discontinuity** if the function is continuous everywhere except for a hole at $x = c$.

Example

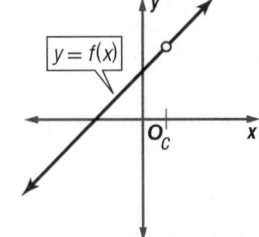

Notice that for graphs of functions with a removable discontinuity, the limit of $f(x)$ at point c exists, but either the value of the function at c is undefined, or, as with the graph shown, the value of $f(c)$ is not the same as the value of the limit at point c.

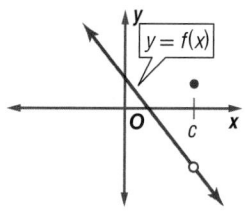

Infinite and jump discontinuities are classified as **nonremovable discontinuities**. A nonremovable discontinuity cannot be eliminated by redefining the function at that point, since the function approaches different values from the left and right sides at that point or does not approach a single value at all. Instead it is increasing or decreasing indefinitely.

These observations lead to the following test for the continuity of a function.

Concept Summary   **Continuity Test**

A function $f(x)$ is continuous at $x = c$ if it satisfies the following conditions.

- $f(x)$ is defined at c. That is, $f(c)$ exists.

- $f(x)$ approaches the same value from either side of c. That is, $\lim\limits_{x \to c} f(x)$ exists.

- The value that $f(x)$ approaches from each side of c is $f(c)$. That is, $\lim\limits_{x \to c} f(x) = f(c)$.

EXAMPLE 1   **Identify a Point of Continuity**

Determine whether $f(x) = 2x^2 - 3x - 1$ is continuous at $x = 2$. Justify using the continuity test.

Check the three conditions in the continuity test.

1. Does $f(2)$ exist?

Because $f(2) = 1$, the function is defined at $x = 2$.

2. Does $\lim\limits_{x \to 2} f(x)$ exist?

Construct a table that shows values of $f(x)$ for x-values approaching 2 from the left and from the right.

	— x approaches 2 →				← x approaches 2 —		
x	1.9	1.99	1.999	**2.0**	2.001	2.01	2.1
$f(x)$	0.52	0.95	0.995		1.005	1.05	1.52

The pattern of outputs suggests that as the value of x gets closer to 2 from the left and from the right, $f(x)$ gets closer to 1. So, we estimate that $\lim\limits_{x \to 2} f(x) = 1$.

3. Does $\lim\limits_{x \to 2} f(x) = f(2)$?

Because $\lim\limits_{x \to 2} (2x^2 - 3x - 1)$ is estimated to be 1 and $f(2) = 1$, we conclude that $f(x)$ is continuous at $x = 2$. The graph of $f(x)$ shown in Figure 1.3.1 supports this conclusion.

✓ Guided Practice

Determine whether each function is continuous at $x = 0$. Justify using the continuity test.

1A. $f(x) = x^3$

1B. $f(x) = \begin{cases} \frac{1}{x} & \text{if } x < 0 \\ x & \text{if } x \geq 0 \end{cases}$

(2, 1)

$f(x) = 2x^2 - 3x - 1$

Figure 1.3.1

If just one of the conditions for continuity is not satisfied, the function is discontinuous at $x = c$. Examining a function can help you identify the type of discontinuity at that point.

EXAMPLE 2   Identify a Point of Discontinuity

Determine whether each function is continuous at the given x-value(s). Justify using the continuity test. If discontinuous, identify the type of discontinuity as *infinite*, *jump*, or *removable*.

a. $f(x) = \begin{cases} 3x - 2 & \text{if } x > -3 \\ 2 - x & \text{if } x \leq -3 \end{cases}$; at $x = -3$

    1. Because $f(-3) = 5$, $f(-3)$ exists.

    2. Investigate function values close to $f(-3)$.

	← x approaches −3 →				← x approaches −3 →		
x	−3.1	−3.01	−3.001	**−3.0**	−2.999	−2.99	−2.9
f(x)	5.1	5.01	5.001		−10.997	−10.97	10.7

The pattern of outputs suggests that $f(x)$ approaches 5 as x approaches -3 from the left and -11 as $f(x)$ approaches -3 from the right. Because these values are not the same, $\lim\limits_{x \to -3} f(x)$ does not exist. Therefore, $f(x)$ is discontinuous at $x = -3$. Because $f(x)$ approaches two different values when $x = -3$, $f(x)$ has a jump discontinuity at $x = -3$. The graph of $f(x)$ in Figure 1.3.2 supports this conclusion.

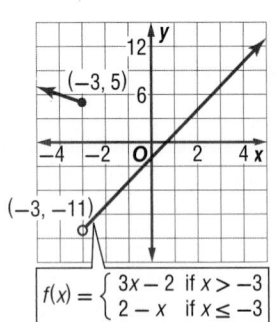

$f(x) = \begin{cases} 3x - 2 & \text{if } x > -3 \\ 2 - x & \text{if } x \leq -3 \end{cases}$

Figure 1.3.2

b. $f(x) = \dfrac{x + 3}{x^2 - 9}$; at $x = -3$ and $x = 3$

    1. Because $f(-3) = \frac{0}{0}$ and $f(3) = \frac{6}{0}$, both of which are undefined, $f(-3)$ and $f(3)$ do not exist. Therefore, $f(x)$ is discontinuous at both $x = -3$ and at $x = 3$.

    2. Investigate function values close to $f(-3)$.

	← x approaches −3 →				← x approaches −3 →		
x	−3.1	−3.01	−3.001	**−3.0**	−2.999	−2.99	−2.9
f(x)	−0.164	−0.166	−0.167		−0.167	−0.167	−0.169

The pattern of outputs suggests that $f(x)$ approaches a limit close to -0.167 as x approaches -3 from each side, so $\lim\limits_{x \to -3} f(x) \approx -0.167$ or $-\frac{1}{6}$.

Investigate function values close to $f(3)$.

	← x approaches 3 →				← x approaches 3 →		
x	2.9	2.99	2.999	**3.0**	3.001	3.01	3.1
f(x)	−10	−100	−1000		1000	100	10

The pattern of outputs suggests that for values of x approaching 3 from the left, $f(x)$ becomes increasingly more negative. For values of x approaching 3 from the right, $f(x)$ becomes increasingly more positive. Therefore, $\lim\limits_{x \to 3} f(x)$ does not exist.

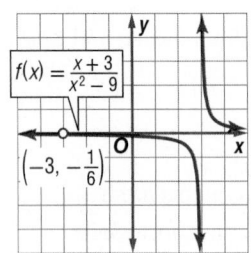

$f(x) = \dfrac{x + 3}{x^2 - 9}$

$\left(-3, -\frac{1}{6}\right)$

Figure 1.3.3

    3. Because $\lim\limits_{x \to -3} f(x)$ exists, but $f(-3)$ is undefined, $f(x)$ has a removable discontinuity at $x = -3$. Because $f(x)$ decreases without bound as x approaches 3 from the left and increases without bound as x approaches 3 from the right, $f(x)$ has an infinite discontinuity at $x = 3$. The graph of $f(x)$ in Figure 1.3.3 supports these conclusions.

✔ Guided Practice

2A. $f(x) = \dfrac{1}{x^2}$; at $x = 0$

2B. $f(x) = \begin{cases} 5x + 4 & \text{if } x > 2 \\ 2 - x & \text{if } x \leq 2 \end{cases}$; at $x = 2$

If a function is continuous, you can approximate the location of its zeros by using the Intermediate Value Theorem and its corollary The Location Principle.

Key Concept   **Intermediate Value Theorem**

If $f(x)$ is a continuous function and $a < b$ and there is a value n such that n is between $f(a)$ and $f(b)$, then there is a number c, such that $a < c < b$ and $f(c) = n$.

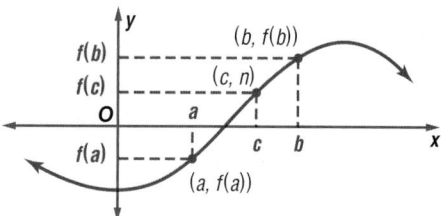

Corollary: The Location Principle If $f(x)$ is a continuous function and $f(a)$ and $f(b)$ have opposite signs, then there exists at least one value c, such that $a < c < b$ and $f(c) = 0$. That is, there is a zero between a and b.

EXAMPLE 3   **Approximate Zeros**

Determine between which consecutive integers the real zeros of each function are located on the given interval.

a. $f(x) = x^3 - 4x + 2$; $[-4, 4]$

x	−4	−3	−2	−1	0	1	2	3	4
f(x)	−46	−13	2	5	2	−1	2	17	50

Because $f(-3)$ is negative and $f(-2)$ is positive, by the Location Principle, $f(x)$ has a zero between −3 and −2. The value of $f(x)$ also changes sign for $0 \le x \le 1$ and $1 \le x \le 2$. This indicates the existence of real zeros in each of these intervals.

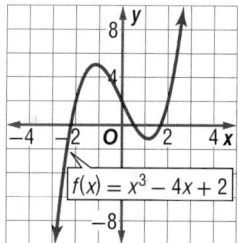

The graph of $f(x)$ shown at the right supports the conclusion that there are real zeros between −3 and −2, 0 and 1, and 1 and 2.

b. $f(x) = x^2 + x + 0.16$; $[-3, 3]$

x	−3	−2	−1	0	1	2	3
f(x)	6.16	2.16	0.16	0.16	2.16	6.16	12.16

The values of $f(x)$ do not change sign for the x-values used. However, as the x-values approach −1 from the left, $f(x)$ decreases, then begins increasing at $x = 0$. So, there may be real zeros between consecutive integers −1 and 0. Graph the function to verify.

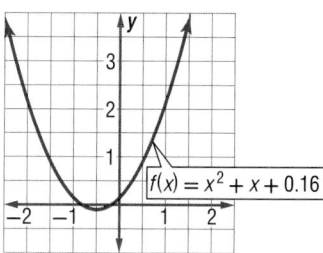

The graph of $f(x)$ crosses the x-axis twice on the interval $[-1, 0]$, so there are real zeros between −1 and 0.

StudyTip

Approximating Zeros with No Sign Changes While a sign change on an interval *does* indicate the location of a real zero, the absence of a sign change *does not* indicate that there are no real zeros on that interval. The best method of checking this is to graph the function.

✔ **Guided Practice**

3A. $f(x) = \dfrac{x^2 - 6}{x + 4}$; $[-3, 4]$

3B. $f(x) = 8x^3 - 2x^2 - 5x - 1$; $[-5, 0]$

End Behavior The **end behavior** of a function describes how a function *behaves* at either *end* of the graph. That is, end behavior is what happens to the value of $f(x)$ as x increases or decreases without bound—becoming greater and greater or more and more negative. To describe the end behavior of a graph, you can use the concept of a limit.

Left-End Behavior

$$\lim_{x \to -\infty} f(x)$$

Right-End Behavior

$$\lim_{x \to \infty} f(x)$$

One possibility for the end behavior of the graph of a function is for the value of $f(x)$ to increase or decrease without bound. This end behavior is described by saying that $f(x)$ approaches positive or negative infinity.

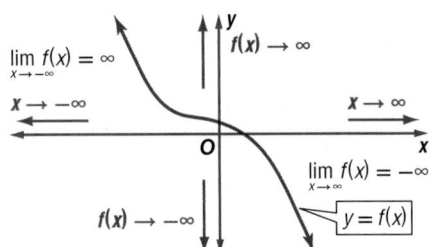

EXAMPLE 4 Graphs that Approach Infinity

Use the graph of $f(x) = -x^4 + 8x^3 + 3x^2 + 6x - 80$ to describe its end behavior. Support the conjecture numerically.

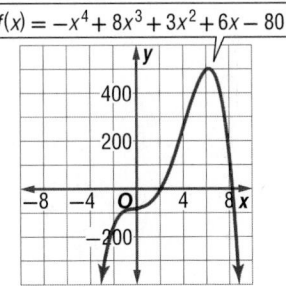

Analyze Graphically

In the graph of $f(x)$, it appears that $\lim_{x \to -\infty} f(x) = -\infty$ and $\lim_{x \to \infty} f(x) = -\infty$.

Support Numerically

Construct a table of values to investigate function values as $|x|$ increases. That is, investigate the value of $f(x)$ as the value of x becomes greater and greater or more and more negative.

	← x approaches −∞				x approaches ∞ →		
x	−10,000	−1000	−100	0	100	1000	10,000
f(x)	$-1 \cdot 10^{16}$	$-1 \cdot 10^{12}$	$-1 \cdot 10^{8}$	−80	$-1 \cdot 10^{8}$	$-1 \cdot 10^{12}$	$-1 \cdot 10^{16}$

The pattern of outputs suggests that as x approaches $-\infty$, $f(x)$ approaches $-\infty$ and as x approaches ∞, $f(x)$ approaches $-\infty$. This supports the conjecture.

✓ Guided Practice

Use the graph of each function to describe its end behavior. Support the conjecture numerically.

4A.

4B.

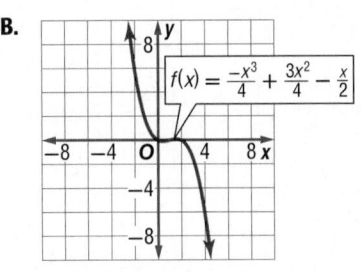

Instead of $f(x)$ being unbounded, approaching ∞ or $-\infty$ as $|x|$ increases, some functions approach, but never reach, a fixed value.

EXAMPLE 5 Graphs that Approach a Specific Value

Use the graph of $f(x) = \dfrac{x}{x^2 - 2x + 8}$ to describe its end behavior. Support the conjecture numerically.

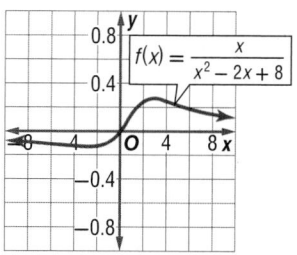

Analyze Graphically

In the graph of $f(x)$, it appears that $\lim\limits_{x \to -\infty} f(x) = 0$ and $\lim\limits_{x \to \infty} f(x) = 0$.

Support Numerically

	← x approaches $-\infty$					x approaches ∞ →	
x	−10,000	−1000	−100	0	100	1000	10,000
f(x)	$-1 \cdot 10^{-4}$	−0.001	−0.01	0	0.01	0.001	$1 \cdot 10^{-4}$

The pattern of outputs suggests that as x approaches $-\infty$, $f(x)$ approaches 0 and as x approaches ∞, $f(x)$ approaches 0. This supports the conjecture.

✓ Guided Practice

Use the graph of each function to describe its end behavior. Support the conjecture numerically.

5A.
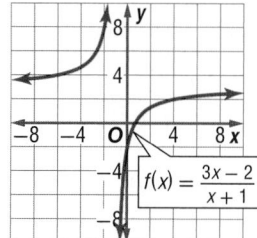
$f(x) = \dfrac{3x - 2}{x + 1}$

5B.
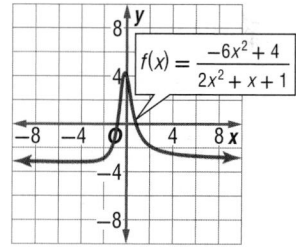
$f(x) = \dfrac{-6x^2 + 4}{2x^2 + x + 1}$

Knowing the end behavior of a function can help you solve real-world problems.

🌐 Real-World EXAMPLE 6 Apply End Behavior

PHYSICS Gravitational potential energy of an object is given by $U(r) = -\dfrac{GmM_e}{r}$, where G is Newton's gravitational constant, m is the mass of the object, M_e is the mass of Earth, and r is the distance from the object to the center of Earth as shown. What happens to the gravitational potential energy of the object as it moves farther and farther from Earth?

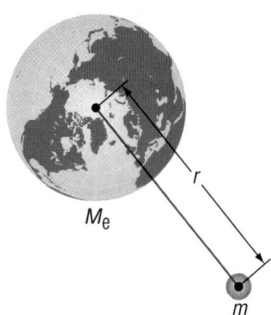

We are asked to describe the end behavior $U(r)$ for large values of r. That is, we are asked to find $\lim\limits_{r \to \infty} U(r)$. Because G, m, and M_e are constant values, the product GmM_e is also a constant value. For increasing values of r, the fraction $-\dfrac{GmM_e}{r}$ will approach 0, so $\lim\limits_{r \to \infty} U(r) = 0$. Therefore, as an object moves farther from Earth, its gravitational potential energy approaches 0.

✓ Guided Practice

6. PHYSICS Dynamic pressure is the pressure generated by the velocity of the moving fluid and is given by $q(v) = \dfrac{\rho v^2}{2}$, where ρ is the density of the fluid and v is the velocity of the fluid. What would happen to the dynamic pressure of a fluid if the velocity were to continuously increase?

Exercises

Determine whether each function is continuous at the given *x*-value(s). Justify using the continuity test. If discontinuous, identify the type of discontinuity as *infinite*, *jump*, or *removable*. (Examples 1 and 2)

1. $f(x) = \sqrt{x^2 - 4}$; at $x = -5$

2. $f(x) = \sqrt{x + 5}$; at $x = 8$

3. $h(x) = \dfrac{x^2 - 36}{x + 6}$; at $x = -6$ and $x = 6$

4. $h(x) = \dfrac{x^2 - 25}{x + 5}$; at $x = -5$ and $x = 5$

5. $g(x) = \dfrac{x}{x - 1}$; at $x = 1$

6. $g(x) = \dfrac{2 - x}{2 + x}$; at $x = -2$ and $x = 2$

7. $h(x) = \dfrac{x - 4}{x^2 - 5x + 4}$; at $x = 1$ and $x = 4$

8. $h(x) = \dfrac{x(x - 6)}{x^3}$; at $x = 0$ and $x = 6$

9. $f(x) = \begin{cases} 4x - 1 & \text{if } x \le -6 \\ -x + 2 & \text{if } x > -6 \end{cases}$; at $x = -6$

10. $f(x) = \begin{cases} x^2 - 1 & \text{if } x > -2 \\ x - 5 & \text{if } x \le -2 \end{cases}$; at $x = -2$

11. PHYSICS A wall separates two rooms with different temperatures. The heat transfer in watts between the two rooms can be modeled by $f(w) = \dfrac{7.4}{w}$, where w is the wall thickness in meters. (Examples 1 and 2)

Cool Room 60°F   Warm Room 80°F

Wall

a. Determine whether the function is continuous at $w = 0.4$. Justify your answer using the continuity test.

b. Is the function continuous? Justify your answer using the continuity test. If discontinuous, identify the type of discontinuity as *infinite*, *jump*, or *removable*.

c. Graph the function to verify your conclusion from part **b**.

12. CHEMISTRY A solution must be diluted so it can be used in an experiment. Adding a 4-molar NaCl solution to a 10-molar solution will decrease the concentration. The concentration C of the mixture can be modeled by $C(x) = \dfrac{500 + 4x}{50 + x}$, where x is the number of liters of 4-molar solution added. (Examples 1 and 2)

a. Determine whether the function is continuous at $x = 10$. Justify the answer using the continuity test.

b. Is the function continuous? Justify your answer using the continuity test. If discontinuous, identify the type of discontinuity as *infinite*, *jump*, or *removable* and describe what affect, if any, the discontinuity has on the concentration of the mixture.

c. Graph the function to verify your conclusion from part **b**.

Determine between which consecutive integers the real zeros of each function are located on the given interval. (Example 3)

13. $f(x) = x^3 - x^2 - 3$; $[-2, 4]$

14. $g(x) = -x^3 + 6x + 2$; $[-4, 4]$

15. $f(x) = 2x^4 - 3x^3 + x^2 - 3$; $[-3, 3]$

16. $h(x) = -x^4 + 4x^3 - 5x - 6$; $[3, 5]$

17. $f(x) = 3x^3 - 6x^2 - 2x + 2$; $[-2, 4]$

18. $g(x) = \dfrac{x^2 + 3x - 3}{x^2 + 1}$; $[-4, 3]$

19. $h(x) = \dfrac{x^2 + 4}{x - 5}$; $[-2, 4]$

20. $f(x) = \sqrt{x^2 - 6} - 6$; $[3, 8]$

21. $g(x) = \sqrt{x^3 + 1} - 5$; $[0, 5]$

Use the graph of each function to describe its end behavior. Support the conjecture numerically. (Examples 4 and 5)

22.

$f(x) = 4x^4 - 6x^3 + 3x$

23.

$f(x) = -5x^3 + 7x - 1$

24.

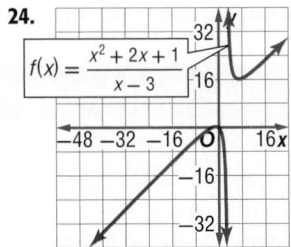

$f(x) = \dfrac{x^2 + 2x + 1}{x - 3}$

25.

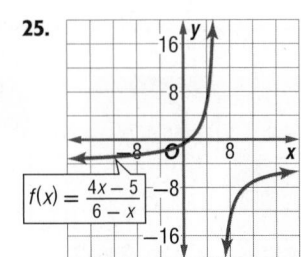

$f(x) = \dfrac{4x - 5}{6 - x}$

26.

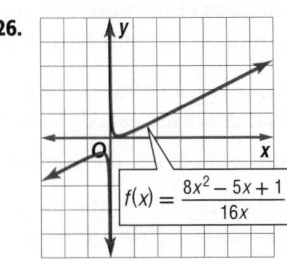

$f(x) = \dfrac{8x^2 - 5x + 1}{16x}$

27.

$f(x) = \dfrac{16x^2}{2x^3 + 5x + 2}$

28.

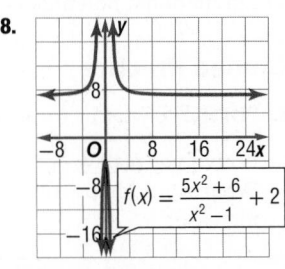

$f(x) = \dfrac{5x^2 + 6}{x^2 - 1} + 2$

29.

$f(x) = \dfrac{12x^3 + 4x}{4x^3 - 9} - 5$

30. POPULATION The U.S. population from 1790 to 1990 can be modeled by $p(x) = 0.0057x^3 + 0.4895x^2 + 0.3236x + 3.8431$, where x is the number of decades after 1790. Use the end behavior of the graph to describe the population trend. Support the conjecture numerically. Does this trend seem realistic? Explain your reasoning. (Example 4)

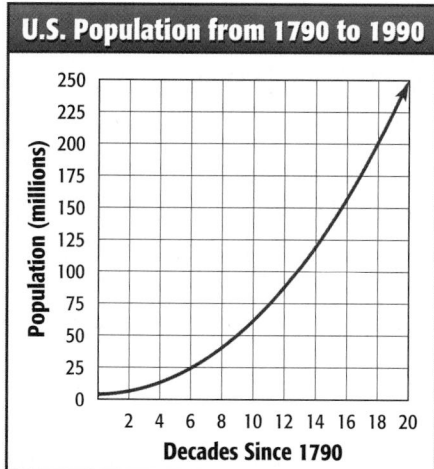

U.S. Population from 1790 to 1990

31. CHEMISTRY A catalyst is used to increase the rate of a chemical reaction. The reaction rate R, or the speed at which the reaction is occurring, is given by $R(x) = \frac{0.5x}{x + 12}$, where x is the concentration of the solution in milligrams of solute per liter of solution. (Example 5)

a. Graph the function using a graphing calculator.

b. What does the end behavior of the graph mean in the context of this experiment? Support the conjecture numerically.

32. ROLLER COASTERS The speed of a roller coaster after it drops from a height A to a height B is given by $f(h_A) = \sqrt{2g(h_A - h_B)}$, where h_A is the height at point A, h_B is the height at point B, and g is the acceleration due to gravity. What happens to $f(h_A)$ as h_B decreases to 0? (Example 6)

Use logical reasoning to determine the end behavior or limit of the function as x approaches infinity. Explain your reasoning. (Example 6)

33. $q(x) = -\frac{24}{x}$

34. $f(x) = \frac{0.8}{x^2}$

35. $p(x) = \frac{x + 1}{x - 2}$

36. $m(x) = \frac{4 + x}{2x + 6}$

37. $c(x) = \frac{5x^2}{x^3 + 2x + 1}$

38. $k(x) = \frac{4x^2 - 3x - 1}{11x}$

39. $h(x) = 2x^5 + 7x^3 + 5$

40. $g(x) = x^4 - 9x^2 + \frac{x}{4}$

41. PHYSICS The kinetic energy of an object in motion can be expressed as $E(m) = \frac{p^2}{2m}$, where p is the momentum and m is the mass of the object. If sand is added to a moving railway car, what would happen as m continues to increase? (Example 6)

Use each graph to determine the x-value(s) at which each function is discontinuous. Identify the type of discontinuity. Then use the graph to describe its end behavior. Justify your answers.

42.

43

44. PHYSICS The wavelength λ of a periodic wave is the distance between consecutive corresponding points on the wave, such as two crests or troughs.

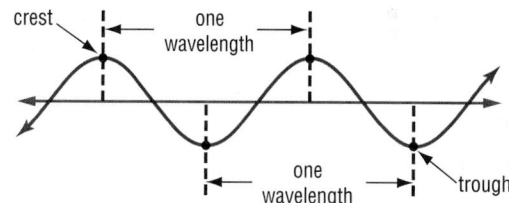

The frequency f, or number of wave crests that pass any given point during a given period of time, is given by $f(\lambda) = \frac{c}{\lambda}$, where c is the speed of light or $2.99 \cdot 10^8$ meters per second.

a. Graph the function using a graphing calculator.

b. Use the graph to describe the end behavior of the function. Support your conjecture numerically.

c. Is the function continuous? If not, identify and describe any points of discontinuity.

GRAPHING CALCULATOR Graph each function and determine whether it is continuous. If discontinuous, identify and describe any points of discontinuity. Then describe its end behavior and locate any zeros.

45. $f(x) = \frac{x^2}{x^3 - 4x^2 + x + 6}$

46. $g(x) = \frac{x^2 - 9}{x^3 - 5x^2 - 18x + 72}$

47. $h(x) = \frac{4x^2 + 11x - 3}{x^2 + 3x - 18}$

48. $h(x) = \frac{x^3 - 4x^2 - 29x - 24}{x^2 - 2x - 15}$

49. $h(x) = \frac{x^3 - 5x^2 - 26x + 120}{x^2 + x - 12}$

50. VEHICLES The number A of alternative-fueled vehicles in use in the United States from 1995 to 2004 can be approximated by $f(t) = 2044t^2 - 3388t + 206{,}808$, where t represents the year and $t = 5$ corresponds to 1995.

  a. Graph the function.

  b. About how many alternative-fueled vehicles were there in the United States in 1998?

  c. As time goes by, what will the number of alternative-fueled vehicles approach, according to the model? Do you think that the model is valid after 2004? Explain.

GRAPHING CALCULATOR Graph each function, and describe its end behavior. Support the conjecture numerically, and provide an effective viewing window for each graph.

51. $f(x) = -x^4 + 12x^3 + 4x^2 - 4$

52. $g(x) = x^5 - 20x^4 + 2x^3 - 5$

53. $f(x) = \dfrac{16x^2}{x^2 + 15x}$

54. $g(x) = \dfrac{8x - 24x^3}{14 + 2x^3}$

55. BUSINESS Gabriel is starting a small business screen-printing and selling T-shirts. Each shirt costs $3 to produce. He initially invested $4000 for a screen printer and other business needs.

  a. Write a function to represent the average cost per shirt as a function of the number of shirts sold n.

  b. Use a graphing calculator to graph the function.

  c. As the number of shirts sold increases, what value does the average cost approach?

56. 🔲 **MULTIPLE REPRESENTATIONS** In this problem, you will investigate limits. Consider $f(x) = \dfrac{ax^3 + b}{cx^3 + d}$, where a and c are nonzero integers, and b and d are integers.

  a. TABULAR Let $c = 1$, and choose three different sets of values for a, b, and d. Write the function with each set of values. Copy and complete the table below.

c = 1				
a	b	d	$\lim\limits_{x \to \infty} f(x)$	$\lim\limits_{x \to -\infty} f(x)$

  b. TABULAR Choose three different sets of values for each variable: one set with $a > c$, one set with $a < c$, and one set with $a = c$. Write each function, and create a table as you did in part **a**.

  c. ANALYTICAL Make a conjecture about the limit of $f(x) = \dfrac{ax^3 + b}{cx^3 + d}$ as x approaches positive and negative infinity.

57. GRAPHING CALCULATOR Graph several different functions of the form $f(x) = x^n + ax^{n-1} + bx^{n-2}$, where n, a, and b are nonnegative integers.

  a. Make a conjecture about the end behavior of the function when n is positive and even. Include a graph to support your conjecture.

  b. Make a conjecture about the end behavior of the function when n is positive and odd. Include a graph to support your conjecture.

REASONING Determine whether each function has an *infinite*, *jump*, or *removable* discontinuity at $x = 0$. Explain.

58. $f(x) = \dfrac{x^5 + x^6}{x^5}$ **59.** $f(x) = \dfrac{x^4}{x^5}$

60. ERROR ANALYSIS Keenan and George are determining whether the relation graphed below is continuous at point c. Keenan thinks that it is the graph of a function $f(x)$ that is discontinuous at point c because $\lim\limits_{x \to c} f(x) = f(c)$ from only one side of c. George thinks that the graph is not a function because when $x = c$, the relation has two different y-values. Is either of them correct? Explain your reasoning.

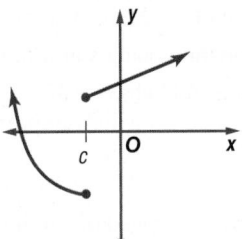

61 CHALLENGE Determine the values of a and b so that f is continuous.

$$f(x) = \begin{cases} x^2 + a & \text{if } x \geq 3 \\ bx + a & \text{if } -3 < x < 3 \\ \sqrt{-b - x} & \text{if } x \leq -3 \end{cases}$$

REASONING Find $\lim\limits_{x \to -\infty} f(x)$ for each of the following. Explain your reasoning.

62. $\lim\limits_{x \to \infty} f(x) = -\infty$ and f is an even function.

63. $\lim\limits_{x \to \infty} f(x) = -\infty$ and f is an odd function.

64. $\lim\limits_{x \to \infty} f(x) = \infty$ and the graph of f is symmetric with respect to the origin.

65. $\lim\limits_{x \to \infty} f(x) = \infty$ and the graph of f is symmetric with respect to the y-axis.

66. WRITING IN MATH Provide an example of a function with a removable discontinuity. Explain how this discontinuity can be eliminated. How does eliminating the discontinuity affect the function?

GRAPHING CALCULATOR Graph each function. Analyze the graph to determine whether each function is *even*, *odd*, or *neither*. Confirm algebraically. If odd or even, describe the symmetry of the graph of the function. (Lesson 1-2)

67. $h(x) = \sqrt{x^2 - 16}$

68. $f(x) = \dfrac{2x + 1}{x}$

69. $g(x) = x^5 - 5x^3 + x$

State the domain of each function. (Lesson 1-1)

70. $f(x) = \dfrac{4x + 6}{x^2 + 3x + 2}$

71. $g(x) = \dfrac{x + 3}{x^2 - 2x - 10}$

72. $g(a) = \sqrt{2 - a^2}$

73. POSTAL SERVICE The U.S. Postal Service uses five-digit ZIP codes to route letters and packages to their destinations. (Lesson 0-7)

   a. How many ZIP codes are possible if the numbers 0 through 9 are used for each of the five digits?

   b. Suppose that when the first digit is 0, the second, third, and fourth digits cannot be 0. How many five-digit ZIP codes are possible if the first digit is 0?

   c. In 1983, the U.S. Postal Service introduced the ZIP + 4, which added four more digits to the existing five-digit ZIP codes. Using the numbers 0 through 9, how many additional ZIP codes were possible?

Given $A = \begin{bmatrix} -4 & 10 & -2 \\ 3 & -3 & 1 \end{bmatrix}$ and $B = \begin{bmatrix} 8 & -5 & 4 \\ 4 & 9 & -3 \end{bmatrix}$, solve each equation for X. (Lesson 0-6)

74. $3X - B = A$

75. $2B + X = 4A$

76. $A - 5X = B$

Solve each system of equations. (Lesson 0-5)

77. $4x - 6y + 4z = 12$
$6x - 9y + 6z = 18$
$5x - 8y + 10z = 20$

78. $x + 2y + z = 10$
$2x - y + 3z = -5$
$2x - 3y - 5z = 27$

79. $2x - y + 3z = -2$
$x + 4y - 2z = 16$
$5x + y - z = 14$

80. SAT/ACT At Lincoln County High School, 36 students are taking either calculus or physics or both, and 10 students are taking both calculus and physics. If there are 31 students in the calculus class, how many students are there in the physics class?

   A 5   C 11   E 21
   B 8   D 15

81. Which of the following statements could be used to describe the end behavior of $f(x)$?

   F $\lim\limits_{x \to -\infty} f(x) = -\infty$ and $\lim\limits_{x \to \infty} f(x) = -\infty$

   G $\lim\limits_{x \to -\infty} f(x) = -\infty$ and $\lim\limits_{x \to \infty} f(x) = \infty$

   H $\lim\limits_{x \to -\infty} f(x) = \infty$ and $\lim\limits_{x \to \infty} f(x) = -\infty$

   J $\lim\limits_{x \to -\infty} f(x) = \infty$ and $\lim\limits_{x \to \infty} f(x) = \infty$

$y = f(x)$

82. REVIEW Amy's locker code includes three numbers between 1 and 45, inclusive. None of the numbers can repeat. How many possible locker permutations are there?

83. REVIEW Suppose a figure consists of three concentric circles with radii of 1 foot, 2 feet, and 3 feet. Find the probability that a point chosen at random lies in the outermost region (between the second and third circles).

   A $\dfrac{1}{3}$ C $\dfrac{4}{9}$

   B $\dfrac{\pi}{9}$ D $\dfrac{5}{9}$

1-4

Extrema and Average Rates of Change

Then
You found function values.
(Lesson 1-1)

Now
- Determine intervals on which functions are increasing, constant, or decreasing, and determine maxima and minima of functions.
- Determine the average rate of change of a function.

New Vocabulary
increasing
decreasing
constant
maximum
minimum
extrema
average rate of change
secant line

Math Online

glencoe.com

Why?
The graph shows that the 2007 U.S. average price of regular-grade gasoline increased from January to May, decreased between May and August, remained fairly constant from August to October, increased again from October to November, and then declined slightly from November to December.

The highest average price was about $3.15 per gallon in May.

The slopes of the red and blue dashed lines show that the price of gasoline changed more rapidly in the first half of 2007 than in the second half.

Increasing and Decreasing Behavior An analysis of a function can also include a description of the intervals on which the function is increasing, decreasing, or constant.

Consider the graph of $f(x)$ shown. As you move from *left to right*, $f(x)$ is

- increasing or *rising* on $(-\infty, -5)$,
- constant or *flat* on $(-5, 0)$, and
- decreasing or *falling* on $(0, \infty)$.

These graphical interpretations can also be described algebraically.

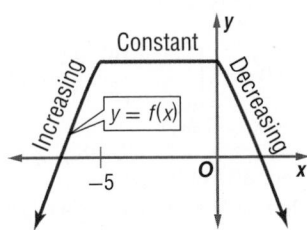

Key Concept   Increasing, Decreasing, and Constant Functions

Words	A function f is **increasing** on an interval I if and only if for any two points in I, a positive change in x results in a positive change in $f(x)$.	**Example** 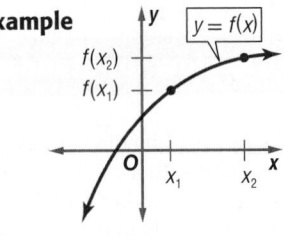
Symbols	For every x_1 and x_2 in an interval I, $f(x_1) < f(x_2)$ when $x_1 < x_2$.	Interval: $(-\infty, \infty)$

Words	A function f is **decreasing** on an interval I if and only if for any two points in I, a positive change in x results in a negative change in $f(x)$.	**Example**
Symbols	For every x_1 and x_2 in an interval I, $f(x_1) > f(x_2)$ when $x_1 < x_2$.	Interval: $(-\infty, \infty)$

Words	A function f is **constant** on an interval I if and only if for any two points in I, a positive change in x results in a zero change in $f(x)$.	**Example**
Symbols	For every x_1 and x_2 in an interval I, $f(x_1) = f(x_2)$ when $x_1 < x_2$.	Interval: (a, b)

EXAMPLE 1   Analyze Increasing and Decreasing Behavior

Use the graph of each function to estimate intervals to the nearest 0.5 unit on which the function is increasing, decreasing, or constant. Support the answer numerically.

a. $f(x) = -2x^3$

Analyze Graphically

When viewed from left to right, the graph of f falls for all real values of x. Therefore, we can conjecture that f is decreasing on $(-\infty, \infty)$.

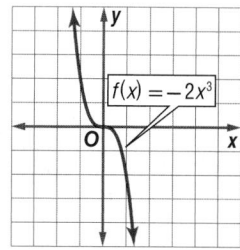

Support Numerically

Create a table using values in the interval.

x	−8	−6	−4	−2	0	2	4	6	8
f(x)	1024	432	128	16	0	−16	−128	−432	−1024

The table shows that as x increases, $f(x)$ decreases. This supports the conjecture.

b. $g(x) = x^3 - 3x$

Analyze Graphically

From the graph, we can estimate that f is increasing on $(-\infty, -1)$, decreasing on $(-1, 1)$, and increasing on $(1, \infty)$.

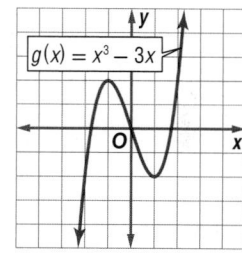

Support Numerically

Create a table of values using x-values in each interval.

$(-\infty, -1)$:

x	−13	−11	−9	−7	−5	−3
f(x)	−2158	−1298	−702	−322	−110	−18

$(-1, 1)$:

x	−0.75	−0.5	0	0.5	0.75
f(x)	1.828	1.375	0	−1.375	−1.828

$(1, \infty)$:

x	3	5	7	9	11	13
f(x)	18	110	322	702	1298	2158

The tables show that as x increases to -1, $f(x)$ increases; as x increases from -1 to 1, $f(x)$ decreases; as x increases from 1, $f(x)$ increases. This supports the conjecture.

✔ **Guided Practice**

1A.

1B.

While a graphical approach to identify the intervals on which a function is increasing, decreasing, or constant can be supported numerically, calculus is often needed to confirm this behavior and to confirm that a function does not change its behavior beyond the domain shown.

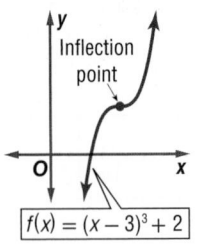
The points at which a function changes its increasing or decreasing behavior are seen as peaks or valleys on its graph. At these critical points, called **extrema**, the function has a **maximum** or a **minimum** value, either relative or absolute.

Key Concept   **Relative and Absolute Extrema**

Words	A *relative maximum* of a function f is the greatest value $f(x)$ can attain on some interval of the domain.	**Model**
Symbols	$f(a)$ is a relative maximum of f if there exists an interval (x_1, x_2) containing a such that $f(a) > f(x)$ for every $x \neq a$ in (x_1, x_2).	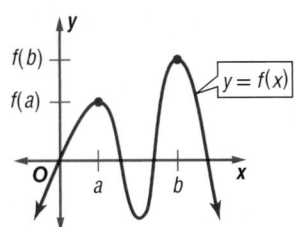
Words	If a relative maximum is the greatest value a function f can attain over its entire domain, then it is the *absolute maximum*.	$f(a)$ is a relative maximum of f.
Symbols	$f(b)$ is the absolute maximum of f if $f(b) > f(x)$ for every $x \neq b$, in the domain of f.	$f(b)$ is the absolute maximum of f.

Words	A *relative minimum* of a function f is the least value $f(x)$ can attain on some interval of the domain.	**Model**
Symbols	$f(a)$ is a relative minimum of f if there exists an interval (x_1, x_2) containing a such that $f(a) < f(x)$ for every $x \neq a$ in (x_1, x_2).	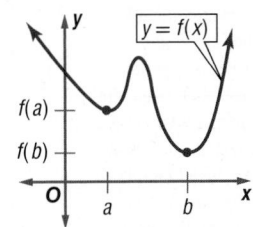
Words	If a relative minimum is the least value a function f can attain over its entire domain, then it is the *absolute minimum*.	$f(a)$ is a relative minimum of f.
Symbols	$f(b)$ is the absolute minimum of f if $f(b) < f(x)$ for every $x \neq b$, in the domain of f.	$f(b)$ is the absolute minimum of f.

EXAMPLE 2   **Estimate and Identify Extrema of a Function**

Estimate and classify the extrema for the graph of $f(x)$. Support the answers numerically.

Analyze Graphically

It appears that $f(x)$ has a relative maximum at $x = -0.5$ and a relative minimum at $x = 1$. It also appears that $\lim\limits_{x \to -\infty} f(x) = -\infty$ and $\lim\limits_{x \to \infty} f(x) = \infty$, so we conjecture that this function has no absolute extrema.

$f(x) = x^3 - x^2 - x$

Support Numerically

Choose x-values in half unit intervals on either side of the estimated x-value for each extremum, as well as one very large and one very small value for x.

x	-100	-1	-0.5	0	0.5	1	1.5	100
$f(x)$	$-1.0 \cdot 10^6$	-1.00	0.125	0	-0.63	-1	-0.38	$9.9 \cdot 10^5$

Because $f(-0.5) > f(-1)$ and $f(-0.5) > f(0)$, there is a relative maximum in the interval $(-1, 0)$ near -0.5. The approximate value of this relative maximum is $f(-0.5)$ or about 0.13.

Likewise, because $f(1) < f(0.5)$ and $f(1) < f(1.5)$, there is a relative minimum in the interval $(0.5, 1.5)$ near 1. The approximate value of this relative maximum is $f(1)$ or -1.

$f(100) > f(-0.5)$ and $f(-100) < f(1)$, which supports our conjecture that f has no absolute extrema.

✔ **Guided Practice**

Estimate and classify the extrema for the graph of each function. Support the answers numerically.

2A.

$f(x) = -x^4 - x^3 + 3x^2 + 2x$

2B.

$g(x) = x^5 - 2x^4 - 2x^3 + 3x^2$

Because calculus is needed to confirm the increasing and decreasing behavior of a function, calculus is also needed to confirm the relative and absolute extrema of a function. For now, however, you can use a graphing calculator to help you better approximate the location and function value of extrema.

EXAMPLE 3   Use a Graphing Calculator to Approximate Extrema

GRAPHING CALCULATOR Approximate to the nearest hundredth the relative or absolute extrema of $f(x) = -4x^3 - 8x^2 + 9x - 4$. State the x-value(s) where they occur.

Graph the function and adjust the window as needed so that all of the graph's behavior is visible.

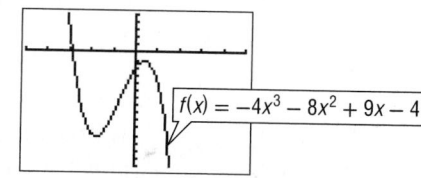

$f(x) = -4x^3 - 8x^2 + 9x - 4$

[−5, 5] scl: 1 by [−30, 10] scl: 4

From the graph of f, it appears that the function has one relative minimum in the interval $(-2, -1)$ and one relative maximum in the interval $(0, 1)$ of the domain. The end behavior of the graph suggests that this function has no absolute extrema.

Using the **minimum** and **maximum** selections from the **CALC** menu of your graphing calculator, you can estimate that $f(x)$ has a relative minimum of -22.81 at $x \approx -1.76$ and a relative maximum of -1.93 at $x \approx 0.43$.

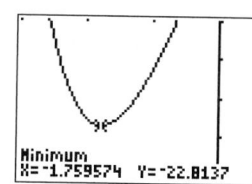

[−3, 0.5] scl: 1 by [−28, 12] scl: 4

[−0.9, 1.6] scl: 1 by [−7.3, 2.7] scl: 4

✔ **Guided Practice**

GRAPHING CALCULATOR Approximate to the nearest hundredth the relative or absolute extrema of each function. State the x-value(s) where they occur.

3A. $h(x) = 7 - 5x - 6x^2$

3B. $g(x) = 2x^3 - 4x^2 - x + 5$

Optimization is an application of mathematics where one searches for a maximum or a minimum quantity given a set of constraints. If a set of real-world quantities can be modeled by a function, the extrema of the function will indicate these optimal values.

🌐 Real-World EXAMPLE 4　Use Extrema for Optimization

AGRICULTURE Suppose each of the 75 orange trees in a Florida grove produces 400 oranges per season. Also suppose that for each additional tree planted in the orchard, the yield per tree decreases by 2 oranges. How many additional trees should be planted to achieve the greatest total yield?

Write a function $P(x)$ to describe the orchard yield as a function of x, the number of additional trees planted in the existing orchard.

orchard yield	=	number of trees in orchard	·	number of oranges produced per tree
$P(x)$	=	$(75 + x)$	·	$(400 - 2x)$

We want to maximize the orchard yield or $P(x)$. Graph this function using a graphing calculator. Then use the **maximum** selection from the **CALC** menu to approximate the x-value that will produce the greatest value for $P(x)$.

The graph has a maximum of 37,812.5 for $x \approx 62.5$. So by planting an additional 62 trees, the orchard can produce a maximum yield of 37,812 oranges.

[−100, 221.3] scl: 1 by [−12270.5, 87900] scl: 5000

☑ Guided Practice

4. **CRAFTS** A glass candle holder is in the shape of a right circular cylinder that has a bottom and no top and has a total surface area of 10π square inches. Determine the radius and the height of the candle holder that will allow the maximum volume.

Average Rate of Change　In algebra, you learned that the slope between any two points on the graph of a linear function represents a *constant* rate of change. For a nonlinear function, the slope changes between different pairs of points, so we can only talk about the *average* rate of change between any two points.

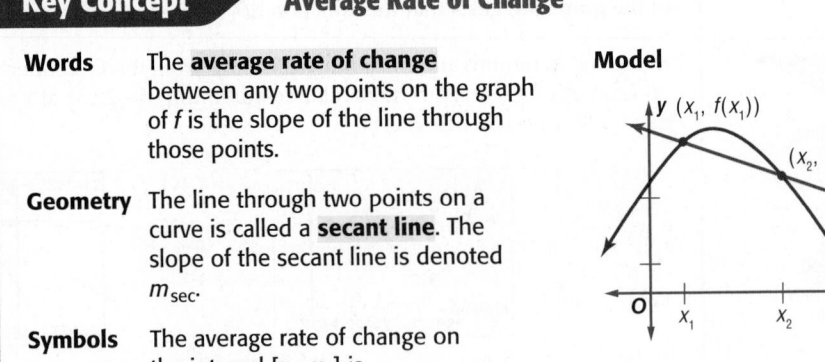

Key Concept　　**Average Rate of Change**

Words　　The **average rate of change** between any two points on the graph of f is the slope of the line through those points.

Model

Geometry　　The line through two points on a curve is called a **secant line**. The slope of the secant line is denoted m_{sec}.

Symbols　　The average rate of change on the interval $[x_1, x_2]$ is

$$m_{\text{sec}} = \frac{f(x_2) - f(x_1)}{x_2 - x_1}.$$

When the average rate of change over an interval is positive, the function increases on average over that interval. When the average rate of change is negative, the function decreases on average over that interval.

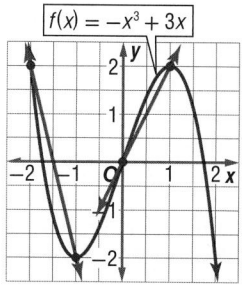

$f(x) = -x^3 + 3x$

Figure 1.4.1

EXAMPLE 5   **Find Average Rates of Change**

Find the average rate of change of $f(x) = -x^3 + 3x$ on each interval.

a. $[-2, -1]$

Use the Slope Formula to find the average rate of change of f on the interval $[-2, -1]$.

$\dfrac{f(x_2) - f(x_1)}{x_2 - x_1} = \dfrac{f(-1) - f(-2)}{-1 - (-2)}$ Substitute −1 for x_2 and −2 for x_1.

$= \dfrac{[-(-1)^3 + 3(-1)] - [-(-2)^3 + 3(-2)]}{-1 - (-2)}$ Evaluate $f(-1)$ and $f(-2)$

$= \dfrac{-2 - 2}{-1 - (-2)}$ or -4 Simplify.

The average rate of change on the interval $[-2, -1]$ is −4. Figure 1.4.1 supports this conclusion.

b. $[0, 1]$

$\dfrac{f(x_2) - f(x_1)}{x_2 - x_1} = \dfrac{f(1) - f(0)}{1 - 0}$ Substitute 1 for x_2 and 0 for x_1.

$= \dfrac{2 - 0}{1 - 0}$ or 2 Evaluate $f(1)$ and $f(0)$ and simplify.

The average rate of change on the interval $[0, 1]$ is 2. Figure 1.4.1 supports this conclusion.

✅ Guided Practice

Find the average rate of change of each function on the given interval.

5A. $f(x) = x^3 - 2x^2 - 3x + 2; [2, 3]$ **5B.** $f(x) = x^4 - 6x^2 + 4x; [-5, -3]$

Average rate of change has many real-world applications. One common application involves the average speed of an object traveling over a distance d or from a height h in a given period of time t. Because speed is distance traveled per unit time, the average speed of an object cannot be negative.

🌐 **Real-World EXAMPLE 6**   **Find Average Speed**

PHYSICS The height of an object that is thrown straight up from a height of 4 feet above ground is given by $h(t) = -16t^2 + 30t + 4$, where t is the time in seconds after the object is thrown. Find and interpret the average speed of the object from 1.25 to 1.75 seconds.

$\dfrac{h(t_2) - h(t_1)}{t_2 - t_1} = \dfrac{h(1.75) - h(1.25)}{1.75 - 1.25}$ Substitute 1.75 for t_2 and 1.25 for t_1.

$= \dfrac{[-16(1.75)^2 + 30(1.75) + 4] - [-16(1.25)^2 + 30(1.25) + 4]}{0.5}$ Evaluate $h(1.75)$ and $h(1.25)$.

$= \dfrac{7.5 - 16.5}{0.5}$ or -18 Simplify.

The average rate of change on the interval is −18. Therefore, the average *speed* of the object from 1.25 to 1.75 seconds is 18 feet per second, and the distance the object is from the ground is decreasing on average over that interval, as shown in the figure at the right.

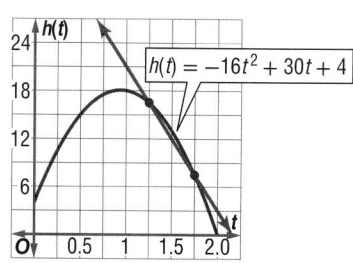

$h(t) = -16t^2 + 30t + 4$

✅ Guided Practice

6. PHYSICS If wind resistance is ignored, the distance $d(t)$ in feet an object travels when dropped from a high place is given by $d(t) = 16t^2$, where t is the time is seconds after the object is dropped. Find and interpret the average speed of the object from 2 to 4 seconds.

🌐 **Real-World Link**

Due to air resistance, a falling object will eventually reach a constant velocity known as *terminal velocity*. A skydiver with a closed parachute typically reaches terminal velocity of 120 to 150 miles per hour.

Source: *MSN Encarta*

= **Step-by-Step Solutions** begin on page R29.

Exercises

Use the graph of each function to estimate intervals to the nearest 0.5 unit on which the function is increasing, decreasing, or constant. Support the answer numerically. (Example 1)

1.

$f(x) = x^3 - x^2 - 2x + 3$

2.

$f(x) = -x^3 - 4x^2 + 2$

3.
$f(x) = x^4 - 3x^3 - x + 1$

4.

$f(x) = -x^5 + 3x^3$

5.

$f(x) = \dfrac{x - 2}{x}$

6.

$f(x) = \dfrac{x^2}{x + 3}$

7.

$f(x) = \begin{cases} 2.5x + 11 & \text{if } x \le -2 \\ 0.5x^2 - 4x + 2 & \text{if } x > 0 \end{cases}$

8.

$f(x) = \begin{cases} -0.5x^2 - 4x & \text{if } x \le -4 \\ -0.5x & \text{if } -4 < x \le 4 \\ -8x^2 + 80x - 190 & \text{if } x > 4 \end{cases}$

9.

$f(x) = \begin{cases} -4 & \text{if } x \le -5 \\ -x^2 - 7x & \text{if } x > -5 \end{cases}$

10.

$f(x) = \begin{cases} \sqrt{x} & \text{if } x \ge 0 \\ -\sqrt{-x} & \text{if } x < 0 \end{cases}$

11. BASKETBALL The height of a free-throw attempt can be modeled by $f(t) = -16t^2 + 23.8t + 5$, where t is time in seconds and $f(t)$ is the height in feet. (Example 2)

a. Graph the height of the ball.

b. Estimate the greatest height reached by the ball. Support the answer numerically.

Estimate and classify the extrema for the graph of each function. Support the answers numerically. (Example 2)

12.
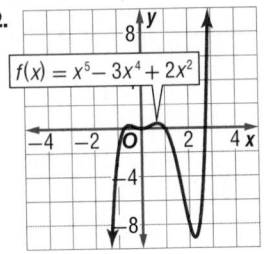
$f(x) = x^5 - 3x^4 + 2x^2$

13

$f(x) = -x^4 + 4x^2 - 1$

14.

$f(x) = -x^5 + 10x^3$

15.

$f(x) = x^6 - 20x^4 + 3x^3$

16.

$f(x) = 0.5x^4 - 4x^3 - 3x^2 + 10x$

17.
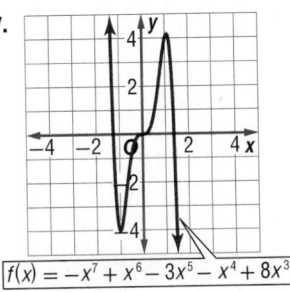
$f(x) = -x^7 + x^6 - 3x^5 - x^4 + 8x^3$

18.

$f(x) = x^7 - 3x^5$

19.

$f(x) = -x^5 + 4x^4 - 4x^3$

20.

$f(x) = -0.5x^4 + 2.5x^3 + x^2 - 6.5x$

21.

$f(x) = x^7 - 4x^3 - 5$

GRAPHING CALCULATOR Approximate to the nearest hundredth the relative or absolute extrema of each function. State the x-value(s) where they occur. (Example 3)

22. $f(x) = 3x^3 - 6x^2 + 8$

23. $g(x) = -2x^3 + 7x - 5$

24. $f(x) = -x^4 + 3x^3 - 2$

25. $f(x) = x^4 - 2x^2 + 5x$

26. $f(x) = x^5 - 2x^3 - 6x - 2$

27. $f(x) = -x^5 + 3x^2 + x - 1$

28. $g(x) = x^6 - 4x^4 + x$

29. $g(x) = x^7 + 6x^2 - 4$

30. $f(x) = 0.008x^5 - 0.05x^4 - 0.2x^3 + 1.2x^2 - 0.7x$

31. $f(x) = 0.025x^5 - 0.1x^4 + 0.57x^3 + 1.2x^2 - 3.5x - 2$

32. GRAPHIC DESIGN A graphic designer wants to create a rectangular graphic that has a 2-inch margin on each side and a 4-inch margin on the top and the bottom. The design, including the margins, should have an area of 392 square inches. What overall dimensions will maximize the size of the design, excluding the margins? (*Hint*: If one side of the design is x, then the other side is 392 divided by x.) (Example 4)

33. GEOMETRY Determine the radius and height that will maximize the volume of the drinking glass shown. Round to the nearest hundredth of an inch, if necessary. (Example 4)

$SA = 20.5\pi$ in²

Find the average rate of change of each function on the given interval. (Example 5)

34. $g(x) = -4x^2 + 3x - 4; [-1, 3]$

35. $g(x) = 3x^2 - 8x + 2; [4, 8]$

36. $f(x) = 3x^3 - 2x^2 + 6; [2, 6]$

37. $f(x) = -2x^3 - 4x^2 + 2x - 8; [-2, 3]$

38. $f(x) = 3x^4 - 2x^2 + 6x - 1; [5, 9]$

39. $f(x) = -2x^4 - 5x^3 + 4x - 6; [-1, 5]$

40. $h(x) = -x^5 - 5x^2 + 6x - 9; [3, 6]$

41. $h(x) = x^5 + 2x^4 + 3x - 12; [-5, -1]$

42. $f(x) = \frac{x - 3}{x}; [5, 12]$

43. $f(x) = \frac{x + 5}{x - 4}; [-6, 2]$

44. $f(x) = \sqrt{x + 8}; [-4, 4]$

45. $f(x) = \sqrt{x - 6}; [8, 16]$

46. WEATHER The average high temperature by month in Pensacola, Florida, can be modeled by $f(x) = -0.9x^2 + 13x + 43$, where x is the month and $x = 1$ represents January. Find the average rate of change for each time interval, and explain what this rate represents. (Example 6)

  a. April to May   **b.** July to November

47. COFFEE The world coffee consumption from 1990 to 2000 can be modeled by $f(x) = -0.004x^4 + 0.077x^3 - 0.38x^2 + 0.46x + 12$, where x is the year, $x = 0$ corresponds with 1990, and the consumption is measured in millions of pounds. Find the average rate of change for each time interval. (Example 6)

  a. 1990 to 2000   **b.** 1995 to 2000

48. TOURISM Tourism in Hawaii for a given year can be modeled using $f(x) = 0.0635x^6 - 2.49x^5 + 37.67x^4 - 275.3x^3 + 986.6x^2 - 1547.1x + 1390.5$, where $1 \le x \le 12$, x represents the month, $x = 1$ corresponds with May 1st, and $f(x)$ represents the number of tourists in thousands.

  a. Graph the equation.

  b. During which month did the number of tourists reach its absolute maximum?

  c. During which month did the number of tourists reach a relative maximum?

49. Use the graph to complete the following.

  a. Find the average rate of change for [5, 15], [15, 20], and [25, 45].

  b. Compare and contrast the nature of the speed of the object over these time intervals.

  c. What conclusions can you make about the magnitude of the rate of change, the steepness of the graph, and the nature of the function?

50. TECHNOLOGY A computer company's research team determined that the profit per chip for a new processor chip can be modeled by $P(x) = -x^3 + 5x^2 + 8x$, where x is the sales price of the chip in hundreds of dollars.

  a. Graph the function.

  b. What is the optimum price per chip?

  c. What is the profit per chip at the optimum price?

51. INCOME The average U.S. net personal income from 1997 to 2007 can be modeled by $I(x) = -1.465x^5 + 35.51x^4 - 277.99x^3 + 741.06x^2 + 847.8x + 25362$, $0 \le x \le 10$, where x is the number of years since 1997.

  a. Graph the equation.

  b. What was the average rate of change from 2000 to 2007? What does this value represent?

  c. In what 4-year period was the average rate of change highest? lowest?

52. BUSINESS A company manufactures rectangular aquariums that have a capacity of 12 cubic feet. The glass used for the base of each aquarium is $1 per square foot. The glass used for the sides is $1.75 per square foot.

  a. If the height and width of the aquarium are equal, find the dimensions that will minimize the cost to build an aquarium.

  b. What is the minimum cost?

  c. If the company also manufactures a cube-shaped aquarium with the same capacity, what is the difference in manufacturing costs?

53. PACKAGING Kali needs to design an enclosed box with a square base and a volume of 3024 cubic inches. What dimensions minimize the surface area of the box? Support your reasoning.

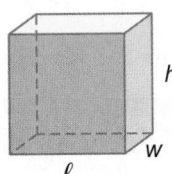

Sketch a graph of a function with each set of characteristics.

54. $f(x)$ is continuous and always increasing.

55. $f(x)$ is continuous and always decreasing.

56. $f(x)$ is continuous, always increasing, and $f(x) > 0$ for all values of x.

57. $f(x)$ is continuous, always decreasing, and $f(x) > 0$ for all values of x.

58. $f(x)$ is continuous, increasing for $x < -2$ and decreasing for $x > -2$.

59. $f(x)$ is continuous, decreasing for $x < 0$ and increasing for $x > 0$.

Determine the coordinates of the absolute extrema of each function. State whether each extremum is a *maximum* or *minimum* value.

60. $f(x) = 2(x - 3)^2 + 5$

61. $f(x) = -0.5(x + 5)^2 - 1$

62. $f(x) = -4|x - 22| + 65$

63. $f(x) = 4(3x - 7)^4 + 8$

64. $f(x) = (36 - x^2)^{0.5}$

65. $f(x) = -(25 - x^2)^{0.5}$

66. $f(x) = x^3 + x$

67. TRAVEL Each hour, Simeon recorded and graphed the total distance in miles his family drove during a trip. Give some reasons as to why the average rate of change varies and even appears constant during two intervals.

Total Distance Traveled

H.O.T. Problems Use **H**igher-**O**rder **T**hinking Skills

OPEN ENDED Sketch a graph of a function with each set of characteristics.

68. continuous
increasing on $(-\infty, 4)$
constant on $[4, 8]$
decreasing on $(8, \infty)$
$f(5) = 3$

69. infinite discontinuity at $x = -2$
increasing on $(-\infty, -2)$
increasing on $(-2, \infty)$
$f(-6) = -6$

70. continuous
average rate of change for $[3, 8]$ is 4.
decreasing on $(8, \infty)$
$f(-4) = 2$

71. REASONING What is the slope of the secant line from $(a, f(a))$ to $(b, f(b))$ when $f(x)$ is constant for the interval $[a, b]$? Explain your reasoning.

72. REASONING If the average rate of change of $f(x)$ on the interval (a, b) is positive, is $f(x)$ *sometimes*, *always*, or *never* increasing on (a, b)? Explain your reasoning.

73. CHALLENGE Use a calculator to graph $f(x) = \sin x$ in degree mode. Describe the relative extrema of the function and the window used for your graph.

74. REASONING A continuous function f has a relative minimum at c and is increasing as x increases from c. Describe the behavior of the function as x increases to c. Explain your reasoning.

75. WRITING IN MATH Describe how the average rate of change of a function relates to a function when it is increasing, decreasing, and constant on an interval.

Determine whether each function is continuous at the given *x*-value(s). Justify using the continuity test. If discontinuous, identify the type of discontinuity as *infinite, jump,* or *removable.* (Lesson 1-3)

76. $f(x) = \sqrt{x^2 - 2}; x = -3$

77. $f(x) = \sqrt{x + 1}; x = 3$

78. $h(x) = \dfrac{x^2 - 25}{x + 5}; x = -5$ and $x = 5$

GRAPHING CALCULATOR Graph each function. Analyze the graph to determine whether each function is *even, odd,* or *neither.* Confirm algebraically. If odd or even, describe the symmetry of the graph of the function. (Lesson 1-2)

79. $f(x) = |x^5|$

80. $f(x) = \dfrac{x + 8}{x - 4}$

81. $g(x) = \dfrac{x^2}{x + 3}$

State the domain of each function. (Lesson 1-1)

82. $f(x) = \dfrac{3x}{x^2 - 5}$

83. $g(x) = \sqrt{x^2 - 9}$

84. $h(x) = \dfrac{x + 2}{\sqrt{x^2 - 7}}$

85. Find the values of x, y, and z for $3\begin{bmatrix} x & y - 1 \\ 4 & 3z \end{bmatrix} = \begin{bmatrix} 15 & 6 \\ 6z & 3x + y \end{bmatrix}$. (Lesson 0-6)

86. If possible, find the solution of $y = x + 2z$, $z = -1 - 2x$, and $x = y - 14$. (Lesson 0-5)

Solve each equation. (Lesson 0-3)

87. $x^2 + 3x - 18 = 0$

88. $2a^2 + 11a - 21 = 0$

89. $z^2 - 4z - 21 = 0$

Simplify. (Lesson 0-2)

90. i^{19}

91. $(7 - 4i) + (2 - 3i)$

92. $\left(\dfrac{1}{2} + i\right) - (2 - i)$

93. ELECTRICITY On a cold day, a 12-volt car battery has a resistance of 0.02 ohm. The power available to start the motor is modeled by the equation $P = 12I - 0.02I^2$, where I is the current in amperes. What current is needed to produce 1600 watts of power to start the motor? (Lesson 0-2)

94. SAT/ACT In the figure, if $q \neq n$, what is the slope of the line segment?

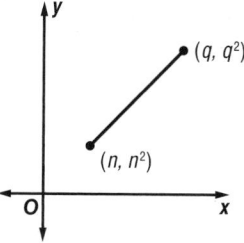

(q, q^2)

(n, n^2)

A $q + n$

C $\dfrac{q^2 + q}{n^2 - n}$

E $\dfrac{1}{q - n}$

B $q - n$

D $\dfrac{1}{q + n}$

95. REVIEW When the number of a year is divisible by 4, then a leap year occurs. However, when the year is divisible by 100, then a leap year does not occur unless the year is divisible by 400. Which is *not* an example of a leap year?

F 1884

H 1904

G 1900

J 1940

96. The function $f(x) = x^3 + 2x^2 - 4x - 6$ has a relative maximum and relative minimum located at which of the following *x*-values?

A relative maximum at $x \approx -0.7$, relative minimum at $x \approx 2$

B relative maximum at $x \approx -0.7$, relative minimum at $x \approx -2$

C relative maximum at $x \approx -2$, relative minimum at $x \approx 0.7$

D relative maximum at $x \approx 2$, relative minimum at $x \approx 0.7$

97. REVIEW A window is in the shape of an equilateral triangle. Each side of the triangle is 8 feet long. The window is divided in half by a support from one vertex to the midpoint of the side of the triangle opposite the vertex. Approximately how long is the support?

F 5.7 ft

G 6.9 ft

H 11.3 ft

J 13.9 ft

Determine whether each relation represents y as a function of x. (Lesson 1-1)

1. $3x + 7y = 21$

2.

x	−1	1	3	5	7
y	−1	3	7	11	15

3.

4.

5. Evaluate $f(2)$ for $f(x) = \begin{cases} x^2 + 3x & \text{if } x < 2 \\ x + 10 & \text{if } x \geq 2 \end{cases}$. (Lesson 1-1)

6. SPORTS During a baseball game, a batter pops up the ball to the infield. After t seconds the height of the ball in feet can be modeled by $h(t) = -16t^2 + 50t + 5$. (Lesson 1-1)

  a. What is the baseball's height after 3 seconds?

  b. What is the relevant domain of this function? Explain your reasoning.

Use the graph of each function to find its y-intercept and zero(s). Then find these values algebraically. (Lesson 1-2)

7.

8.

Use the graph of h to find the domain and range of each function. (Lesson 1-2)

9.

10.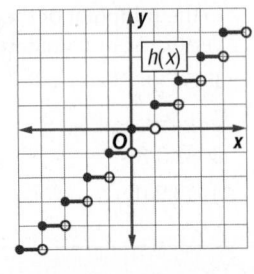

Determine whether each function is continuous at $x = 5$. Justify your answer using the continuity test. (Lesson 1-3)

11. $f(x) = \sqrt{x^2 - 36}$

12. $f(x) = \dfrac{x^2}{x + 5}$

Use the graph of each function to describe its end behavior. (Lesson 1-3)

13.

14.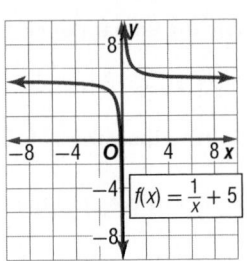

15. MULTIPLE CHOICE The graph of $f(x)$ contains a(n) _____ discontinuity at $x = 3$. (Lesson 1-3)

$f(x) = \begin{cases} \dfrac{x^2 - 9}{x - 3} & \text{if } x \neq 3 \\ 3 & \text{if } x = 3 \end{cases}$

  A undefined

  B infinite

  C jump

  D removable

Use the graph of each function to estimate intervals to the nearest 0.5 unit on which the function is increasing, decreasing, or constant. (Lesson 1-4)

16.

17.

18. PHYSICS The height of an object dropped from 80 feet above the ground after t seconds is $f(t) = -16t^2 + 80$. What is the average speed for the object during the first 2 seconds after it is dropped? (Lesson 1-4)

1-5

Parent Functions and Transformations

Why?

The path of a 60-yard punt can be modeled by the function at the right. This function is related to the basic quadratic function $f(x) = x^2$.

Parent Functions A *family of functions* is a group of functions with graphs that display one or more similar characteristics. A **parent function** is the simplest of the functions in a family. This is the function that is transformed to create other members in a family of functions.

In this lesson, you will study eight of the most commonly used parent functions. You should already be familiar with the graphs of the following linear and polynomial parent functions.

Then
You analyzed graphs of functions. (Lessons 1-2 through 1-4)

Now
- Identify, graph, and describe parent functions.
- Identify and graph transformations of parent functions.

New Vocabulary
parent function
constant function
zero function
identity function
quadratic function
cubic function
square root function
reciprocal function
absolute value function
step function
greatest integer function
transformation
translation
reflection
dilation

Math Online
glencoe.com

Punted Football

$h(x) = -\frac{1}{15}x^2 + 4x + 1$

Vertical Distance (yd)

Horizontal Distance (yd)

Key Concept   **Linear and Polynomial Parent Functions**

A **constant function** has the form $f(x) = c$, where c is any real number. Its graph is a horizontal line. When $c = 0$, $f(x)$ is the **zero function**.

The **identity function** $f(x) = x$ passes through all points with coordinates (a, a).

The **quadratic function** $f(x) = x^2$ has a U-shaped graph.

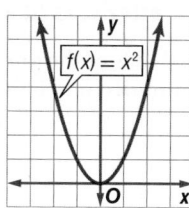

The **cubic function** $f(x) = x^3$ is symmetric about the origin.

You should also be familiar with the graphs of both the square root and reciprocal functions.

Key Concept   **Square Root and Reciprocal Parent Functions**

The **square root function** has the form $f(x) = \sqrt{x}$.

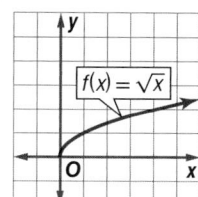

The **reciprocal function** has the form $f(x) = \frac{1}{x}$.

Another parent function is the piecewise-defined absolute value function.

Key Concept   Absolute Value Parent Function

Words   The **absolute value function**, denoted
$f(x) = |x|$, is a V-shaped function defined as
$$f(x) = \begin{cases} -x & \text{if } x < 0 \\ x & \text{if } x \geq 0 \end{cases}.$$

Model

Examples $|-5| = 5$, $|0| = 0$, $|4| = 4$

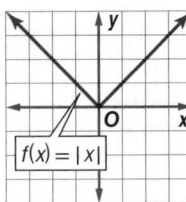

A piecewise-defined function in which the graph resembles a set of stairs is called a **step function**. The most well-known step function is the greatest integer function.

StudyTip

Floor Function The greatest integer function is also known as the *floor function*.

Key Concept   Greatest Integer Parent Function

Words   The **greatest integer function**, denoted
$f(x) = [\![x]\!]$, is defined as the greatest integer
less than or equal to x.

Model

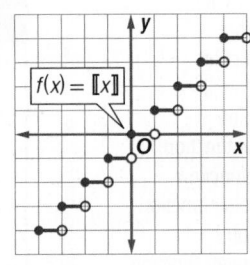

Examples $[\![-4]\!] = -4$, $[\![-1.5]\!] = -2$, $\left[\!\left[\frac{1}{3}\right]\!\right] = 0$

Using the tools you learned in Lessons 1-1 through 1-4, you can describe characteristics of each parent function. Knowing the characteristics of a parent function can help you analyze the shapes of more complicated graphs in that family.

EXAMPLE 1   Describe Characteristics of a Parent Function

Describe the following characteristics of the graph of the parent function $f(x) = \sqrt{x}$: domain, range, intercepts, symmetry, continuity, end behavior, and intervals on which the graph is increasing/decreasing.

The graph of the square root function (Figure 1.5.1) has the following characteristics.

- The domain of the function is $[0, \infty)$, and the range is $[0, \infty)$.
- The graph has one intercept at $(0, 0)$.
- The graph has no symmetry. Therefore, $f(x)$ is neither odd nor even.
- The graph is continuous for all values in its domain.
- The graph begins at $x = 0$ and $\lim\limits_{x \to \infty} f(x) = \infty$.
- The graph is increasing on the interval $(0, \infty)$.

Figure 1.5.1

☑ Guided Practice

1. Describe the following characteristics of the graph of the parent function $f(x) = |x|$: domain, range, intercepts, symmetry, continuity, end behavior, and intervals on which the graph is increasing/decreasing.

Transformations **Transformations** of a parent function can affect the appearance of the parent graph. *Rigid transformations* change only the position of the graph, leaving the size and shape unchanged. *Nonrigid transformations* distort the shape of the graph.

A **translation** is a rigid transformation that has the effect of shifting the graph of a function. A *vertical translation* of a function f shifts the graph of f up or down, while a *horizontal translation* shifts the graph left or right. Horizontal and vertical translations are examples of rigid transformations.

Key Concept   **Vertical and Horizontal Translations**

Vertical Translations

The graph of $g(x) = f(x) + k$ is the graph of $f(x)$ translated

• **k units up** when $k > 0$, and
• **k units down** when $k < 0$.

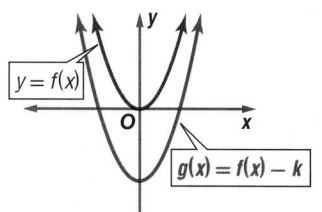

Horizontal Translations

The graph of $g(x) = f(x - h)$ is the graph of $f(x)$ translated

• **h units right** when $h > 0$, and
• **h units left** when $h < 0$.

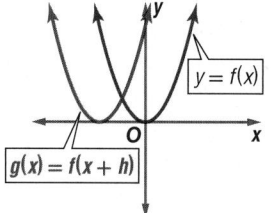

EXAMPLE 2   **Graph Translations**

Use the graph of $f(x) = |x|$ to graph each function.

a. $g(x) = |x| + 4$

This function is of the form $g(x) = f(x) + 4$. So, the graph of $g(x)$ is the graph of $f(x) = |x|$ translated 4 units up, as shown in Figure 1.5.2.

b. $g(x) = |x + 3|$

This function is of the form $g(x) = f(x + 3)$ or $g(x) = f[x - (-3)]$. So, the graph of $g(x)$ is the graph of $f(x) = |x|$ translated 3 units left, as shown in Figure 1.5.3.

c. $g(x) = |x - 2| - 1$

This function is of the form $g(x) = f(x - 2) - 1$. So, the graph of $g(x)$ is the graph of $f(x) = |x|$ translated 2 units right and 1 unit down, as shown in Figure 1.5.4.

TechnologyTip

Translations You can translate a graph using a graphing calculator. Under [Y=], place an equation in Y1. Move to the Y2 line, and then press [VARS] [▶] [ENTER] [ENTER]. This will place Y1 in the Y2 line. Enter a number to translate the function.
Press [Graph]. The two equations will be graphed in the same window.

Figure 1.5.2

Figure 1.5.3

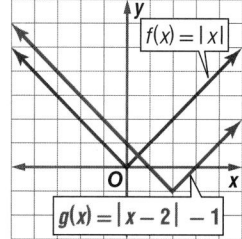

Figure 1.5.4

✓ **Guided Practice**   Use the graph of $f(x) = x^3$ to graph each function.

2A. $h(x) = x^3 - 5$ **2B.** $h(x) = (x - 3)^3$ **2C.** $h(x) = (x + 2)^3 + 4$

Another type of rigid transformation is a **reflection**, which produces a mirror image of the graph of a function with respect to a specific line.

Key Concept — Reflections in the Coordinate Axes

Reflection in *x*-axis	Reflection in *y*-axis
$g(x) = -f(x)$ is the graph of $f(x)$ **reflected in the *x*-axis**.	$g(x) = f(-x)$ is the graph of $f(x)$ **reflected in the *y*-axis**.

 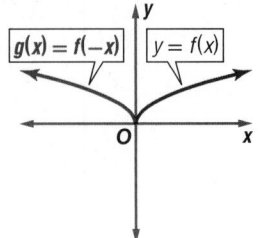

When writing an equation for a transformed function, be careful to indicate the transformations correctly. The graph of $g(x) = -\sqrt{x-1} + 2$ is different from the graph of $g(x) = -\left(\sqrt{x-1} + 2\right)$.

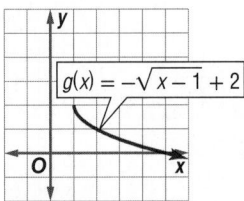

reflection of $f(x) = \sqrt{x}$ in the *x*-axis, then translated 1 unit to the right and 2 units up

translation of $f(x) = \sqrt{x}$ 1 unit to the right and 2 units up, then reflected in the *x*-axis

EXAMPLE 3   Write Equations for Transformations

Describe how the graphs of $f(x) = x^2$ and $g(x)$ are related. Then write an equation for $g(x)$.

Figure 1.5.5

a.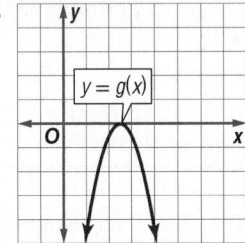

The graph of $g(x)$ is the graph of $f(x) = x^2$ translated 5 units to the right and reflected in the *x*-axis. So, $g(x) = -(x - 5)^2$.

b.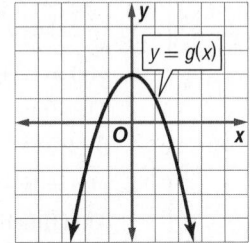

The graph of $g(x)$ is the graph of $f(x) = x^2$ reflected in the *x*-axis and translated 2 units up. So, $g(x) = -x^2 + 2$.

✓ Guided Practice

Describe how the graphs of $f(x) = \frac{1}{x}$ and $g(x)$ are related. Then write an equation for $g(x)$.

3A.

3B.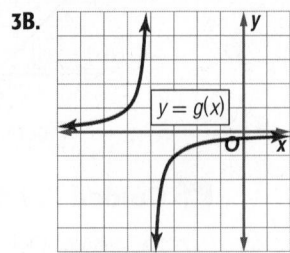

A **dilation** is a nonrigid transformation that has the effect of compressing (shrinking) or expanding (enlarging) the graph of a function vertically or horizontally.

Key Concept   Vertical and Horizontal Dilations

Vertical Dilations

If a is a positive real number, then $g(x) = a \cdot f(x)$, is

- the graph of $f(x)$ **expanded vertically**, if $a > 1$.
- the graph of $f(x)$ **compressed vertically**, if $0 < a < 1$.

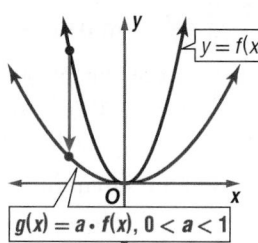

Horizontal Dilations

If a is a positive real number, then $g(x) = f(ax)$, is

- the graph of $f(x)$ **compressed horizontally**, if $a > 1$.
- the graph of $f(x)$ **expanded horizontally**, if $0 < a < 1$.

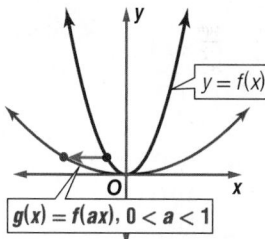

StudyTip

Dilations Sometimes pairs of dilations look similar such as a vertical expansion and a horizontal compression. It is not possible to tell which dilation a transformation is from the graph. You must compare the equation of the transformed function to the parent function.

EXAMPLE 4   Describe and Graph Transformations

Identify the parent function $f(x)$ of $g(x)$, and describe how the graphs of $g(x)$ and $f(x)$ are related. Then graph $f(x)$ and $g(x)$ on the same axes.

a. $g(x) = \frac{1}{4}x^3$

The graph of $g(x)$ is the graph of $f(x) = x^3$ compressed vertically because $g(x) = \frac{1}{4}x^3 = \frac{1}{4}f(x)$ and $0 < \frac{1}{4} < 1$.

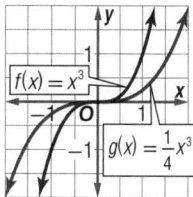

b. $g(x) = -(0.2x)^2$

The graph of $g(x)$ is the graph of $f(x) = x^2$ expanded horizontally and then reflected in the x-axis because $g(x) = -(0.2x)^2 = -f(0.2x)$ and $0 < 0.2 < 1$.

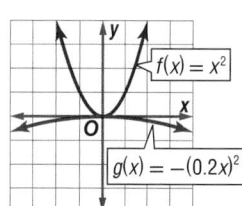

✔ Guided Practice

4A. $g(x) = [\![x]\!] - 4$

4B. $g(x) = \frac{15}{x} + 3$

You can use what you have learned about transformations of functions to graph a piecewise-defined function.

EXAMPLE 5   Graph a Piecewise-Defined Function

$$\text{Graph } f(x) = \begin{cases} 3x^2 & \text{if } x < -1 \\ -1 & \text{if } -1 \le x < 4. \\ (x-5)^3 + 2 & \text{if } x \ge 4 \end{cases}$$

On the interval $(-\infty, -1)$, graph $y = 3x^2$.
On the interval $[-1, 4)$, graph the constant function $y = -1$.
On the interval $[4, \infty)$, graph $y = (x-5)^3 + 2$.

Draw circles at $(-1, 3)$ and $(4, -1)$ and dots at $(-1, -1)$ and $(4, 1)$ because $f(-1) = -1$ and $f(4) = 1$.

✔ Guided Practice

Graph each function.

5A. $g(x) = \begin{cases} x - 5 & \text{if } x \le 0 \\ x^3 & \text{if } 0 < x \le 2 \\ \frac{2}{x} & \text{if } x > 2 \end{cases}$

5B. $h(x) = \begin{cases} (x+6)^2 & \text{if } x < -5 \\ 7 & \text{if } -5 \le x \le 2 \\ |4 - x| & \text{if } x > 2 \end{cases}$

You can also use what you have learned about transformations to transform functions that model real-world data or phenomena.

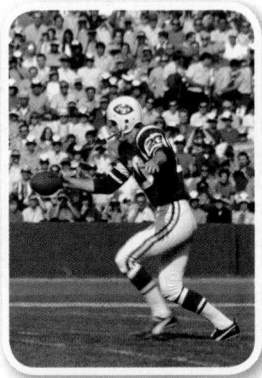

🌐 Real-World EXAMPLE 6   Transformations of Functions

FOOTBALL The path of a 60-yard punt can be modeled by $g(x) = -\frac{1}{15}x^2 + 4x + 1$, where $g(x)$ is the vertical distance in yards of the football from the ground and x is the horizontal distance in yards such that $x = 0$ corresponds to the kicking team's 20-yard line.

a. Describe the transformations of the parent function $f(x) = x^2$ used to graph $g(x)$.

Rewrite the function so that it is in the form $g(x) = a(x-h)^2 + k$ by completing the square.

$g(x) = -\frac{1}{15}x^2 + 4x + 1$	**Original function**
$= -\frac{1}{15}(x^2 - 60x) + 1$	**Factor $-\frac{1}{15}x^2 + 4x$.**
$= -\frac{1}{15}(x^2 - 60x + 900) + 1 + \frac{1}{15}(900)$	**Complete the square.**
$= -\frac{1}{15}(x-30)^2 + 61$	**Write $x^2 - 60x + 900$ as a perfect square and simplify.**

So, $g(x)$ is the graph of $f(x)$ translated 30 units right, compressed vertically, reflected in the x-axis, and then translated 61 units up.

b. Suppose the punt was from the kicking team's 30-yard line. Rewrite $g(x)$ to reflect this change. Graph both functions on the same graphing calculator screen.

A change of position from the kicking team's 20- to 30-yard line is a horizontal translation of 10 yards to the right, so subtract an additional 10 yards from inside the squared expression.

[0, 100] scl: 1 by [0, 100] scl: 20

$g(x) = -\frac{1}{15}(x - 30 - 10)^2 + 61$ or $g(x) = -\frac{1}{15}(x - 40)^2 + 61$

✔ Guided Practice

6. ELECTRICITY The current in amps flowing through a DVD player is described by $I(x) = \sqrt{\frac{x}{11}}$, where x is the power in watts and 11 is the resistance in ohms.

A. Describe the transformations of the parent function $f(x) = \sqrt{x}$ used to graph $I(x)$.

B. The resistance of a lamp is 15 ohms. Write a function to describe the current flowing through the lamp.

C. Graph the resistance for the DVD player and the lamp on the same graphing calculator screen.

Another nonridgid transformation involves absolute value.

Key Concept   **Transformations with Absolute Value**

$g(x) = |f(x)|$	$g(x) = f(|x|)$
This transformation reflects any portion of the graph of $f(x)$ that is below the x-axis so that it is above the x-axis.	This transformation results in the portion of the graph of $f(x)$ that is to the left of the y-axis being replaced by a reflection of the portion to the right of the y-axis.

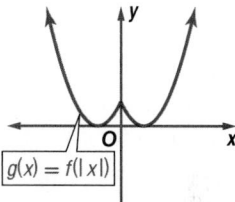

EXAMPLE 7   **Describe and Graph Transformations**

Use the graph of $f(x) = x^3 - 4x$ in Figure 1.5.6 to graph each function.

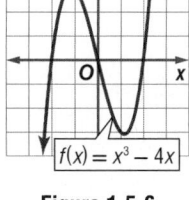

Figure 1.5.6

a. $g(x) = |f(x)|$

The graph of $f(x)$ is below the x-axis on the intervals $(-\infty, -2)$ and $(0, 2)$, so reflect those portions of the graph in the x-axis and leave the rest unchanged.

b. $h(x) = f(|x|)$

Replace the graph of $f(x)$ to the left of the y-axis with a reflection of the graph to the right of the y-axis.

✔ **Guided Practice**

Use the graph of $f(x)$ shown to graph $g(x) = |f(x)|$ and $h(x) = f(|x|)$.

7A.

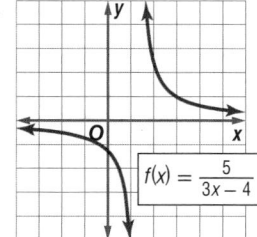

$f(x) = \dfrac{5}{3x - 4}$

7B.

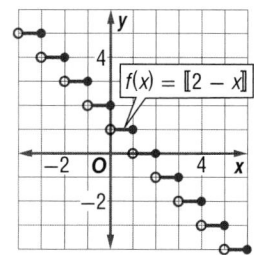

$f(x) = [\![2 - x]\!]$

Describe the following characteristics of the graph of each parent function: domain, range, intercepts, symmetry, continuity, end behavior, and intervals on which the graph is increasing/decreasing. (Example 1)

1. $f(x) = [\![x]\!]$ **2.** $f(x) = \frac{1}{x}$ **3.** $f(x) = x^3$

4. $f(x) = x^4$ **5.** $f(x) = c$ **6.** $f(x) = x$

Use the graph of $f(x) = \sqrt{x}$ to graph each function. (Example 2)

7. $g(x) = \sqrt{x - 4}$ **8.** $g(x) = \sqrt{x + 3}$

9. $g(x) = \sqrt{x + 6} - 4$ **10.** $g(x) = \sqrt{x - 7} + 3$

Use the graph of $f(x) = \frac{1}{x}$ to graph each function. (Example 2)

11. $g(x) = \frac{1}{x} + 4$ **12.** $g(x) = \frac{1}{x} - 6$

13. $g(x) = \frac{1}{x - 6} + 8$ **14.** $g(x) = \frac{1}{x + 7} - 4$

Describe how the graphs of $f(x) = [\![x]\!]$ and $g(x)$ are related. Then write an equation for $g(x)$. (Example 3)

15.   **16.**

17.   **18.**

 19. PROFIT An automobile company experienced an unexpected two-month delay on manufacturing of a new car. The projected profit of the car sales before the delay $p(x)$ is shown below. Describe how the graph of $p(x)$ and the graph of a projection including the delay $d(x)$ are related. Then write an equation for $d(x)$. (Example 3)

Projected First-Quarter Profit

$p(x) = 10x^3 - 70x^2 + 150x - 2$

Months After January

Describe how the graphs of $f(x) = |x|$ and $g(x)$ are related. Then write an equation for $g(x)$. (Example 3)

20.   **21.**

22.   **23.**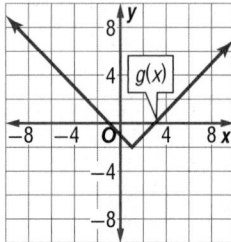

Identify the parent function $f(x)$ of $g(x)$, and describe how the graphs of $g(x)$ and $f(x)$ are related. Then graph $f(x)$ and $g(x)$ on the same axes. (Example 4)

24. $g(x) = 3|x| - 4$ **25.** $g(x) = 3\sqrt{x + 8}$

26. $g(x) = \frac{4}{x + 1}$ **27.** $g(x) = 2[\![x - 6]\!]$

28. $g(x) = -5[\![x - 2]\!]$ **29.** $g(x) = -2|x + 5|$

30. $g(x) = \frac{1}{6x} + 7$ **31.** $g(x) = \frac{\sqrt{x + 3}}{4}$

Graph each function. (Example 5)

32. $f(x) = \begin{cases} -x^2 & \text{if } x < -2 \\ 3 & \text{if } -2 \le x < 7 \\ (x - 5)^2 + 2 & \text{if } x \ge 7 \end{cases}$

33. $g(x) = \begin{cases} x + 4 & \text{if } x < -6 \\ \frac{1}{x} & \text{if } -6 \le x < 4 \\ 6 & \text{if } x \ge 4 \end{cases}$

34. $f(x) = \begin{cases} 4 & \text{if } x < -5 \\ x^3 & \text{if } -2 \le x \le 2 \\ \sqrt{x + 3} & \text{if } x > 3 \end{cases}$

35. $h(x) = \begin{cases} |x - 5| & \text{if } x < -3 \\ 4x - 3 & \text{if } -1 \le x < 3 \\ \sqrt{x} & \text{if } x \ge 4 \end{cases}$

36. $g(x) = \begin{cases} 2 & \text{if } x < -4 \\ x^4 - 3x^3 + 5 & \text{if } -1 \le x < 1 \\ [\![x]\!] + 1 & \text{if } x \ge 3 \end{cases}$

37. $f(x) = \begin{cases} -3x - 1 & \text{if } x \le -1 \\ 0.5x + 5 & \text{if } -1 < x \le 3 \\ -|x - 5| + 3 & \text{if } x > 3 \end{cases}$

38. POSTAGE The cost of a first-class postage stamp in the U.S. from 1988 to 2008 is shown in the table below. Use the data to graph a step function. (Example 5)

Year	Price (¢)
1988	25
1991	29
1995	32
1999	33
2001	34
2002	37
2006	39
2007	41
2008	42

39. BUSINESS A no-contract cell phone company charges a flat rate for daily access and $0.10 for each minute. The cost of the plan can be modeled by $c(x) = 1.99 + 0.1[\![x]\!]$, where x is the number of minutes used. (Example 6)

a. Describe the transformation(s) of the parent function $f(x) = [\![x]\!]$ used to graph $c(x)$.

b. The company offers another plan in which the daily access rate is $2.49, and the per-minute rate is $0.05. What function $c(x)$ can be used to describe the second plan?

c. Graph both functions on the same graphing calculator screen.

d. Would the cost of the plans ever equal each other? If so, at how many minutes?

40. GOLF The path of a drive can be modeled by the function shown, where $g(x)$ is the vertical distance in feet of the ball from the ground and x is the horizontal distance in feet such that $x = 0$ corresponds to the initial point. (Example 6)

Drive Path

$g(x) = 0.176x - 0.0004x^2$

a. Describe the transformation(s) of the parent function $f(x) = x^2$ used to graph $g(x)$.

b. If a second golfer hits a similar shot 30 feet farther down the fairway from the first player, what function $h(x)$ can be used to describe the second golfer's shot?

c. Graph both golfers' shots on the same graphing calculator screen.

d. If both golfers hit their shots at the same time, at what horizontal and vertical distances will the shots cross paths?

Use the graph of $f(x)$ to graph $g(x) = |f(x)|$ and $h(x) = f(|x|)$. (Example 7)

41. $f(x) = \dfrac{2}{x}$ **42.** $f(x) = \sqrt{x - 4}$

43. $f(x) = x^4 - x^3 - 4x^2$ **44.** $f(x) = \dfrac{1}{2}x^3 + 2x^2 - 8x - 2$

45. $f(x) = \dfrac{1}{x - 3} + 5$ **46.** $f(x) = \sqrt{x + 2} - 6$

47. TRANSPORTATION In New York City, the standard cost for taxi fare is shown. One unit is equal to a distance of 0.2 mile or a time of 60 seconds when the car is not in motion.

CAB SERVICE

$2.50 per trip plus
$0.40 per unit

a. Write a greatest integer function $f(x)$ that would represent the cost for units of cab fare, where $x > 0$. Round to the nearest unit.

b. Graph the function.

c. How would the graph of $f(x)$ change if the fare for the first unit increased to $3.70? Graph the new function.

48. PHYSICS The potential energy in joules of a spring that has been stretched or compressed is given by $p(x) = \dfrac{cx^2}{2}$, where c is the spring constant and x is the distance from the equilibrium position. When x is negative, the spring is compressed, and when x is positive, the spring is stretched.

Compressed   Equilibrium   Stretched

a. Describe the transformation(s) of the parent function $f(x) = x^2$ used to graph $p(x)$.

b. The graph of the potential energy for a second spring passes through the point (3, 315). Find the spring constant for the spring and write the function for the potential energy.

Write and graph the function with the given parent function and characteristics.

49. $f(x) = \dfrac{1}{x}$; expanded vertically by a factor of 2; translated 7 units to the left and 5 units up

50. $f(x) = [\![x]\!]$; expanded vertically by a factor of 3; reflected in the x-axis; translated 4 units down

PHYSICS The distance an object travels as a function of time is given by $f(t) = \dfrac{1}{2}at^2 + v_0t + x_0$, where a is the acceleration, v_0 is the initial velocity, and x_0 is the initial position of the object. Describe the transformations of the parent function $f(t) = t^2$ used to graph $f(t)$ for each of the following.

51. $a = 2, v_0 = 2, x_0 = 0$ **52.** $a = 2, v_0 = 0, x_0 = 10$

53. $a = 4, v_0 = 8, x_0 = 1$ **54.** $a = 3, v_0 = 5, x_0 = 3$

Write an equation for each $g(x)$.

55.

56.

57.

58.
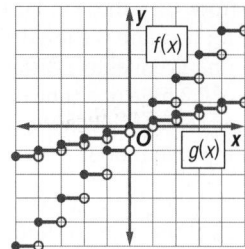

59. SHOPPING The management of a new shopping mall originally predicted that attendance in thousands would follow $f(x) = \sqrt{7x}$ for the first 60 days of operation, where x is the number of days after opening and $x = 1$ corresponds with opening day. Write $g(x)$ in terms of $f(x)$ for each situation below.

   a. Attendance was consistently 12% higher than expected.

   b. The opening was delayed 30 days due to construction.

   c. Attendance was consistently 450 less than expected.

Identify the parent function $f(x)$ of $g(x)$, and describe the transformation of $f(x)$ used to graph $g(x)$.

60.

61.

62.

63.
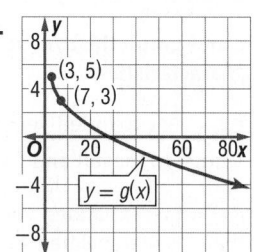

Use $f(x)$ to graph $g(x)$.

64. $g(x) = 0.25f(x) + 4$

65. $g(x) = 3f(x) - 6$

66. $g(x) = f(x - 5) + 3$

67. $g(x) = -2f(x) + 1$

Use $f(x) = \dfrac{8}{\sqrt{x+6}} - 4$ to graph each function.

68. $g(x) = 2f(x) + 5$

69. $g(x) = -3f(x) + 6$

70. $g(x) = f(4x) - 5$

71. $g(x) = f(2x + 1) + 8$

72. 🔁 MULTIPLE REPRESENTATIONS In this problem, you will investigate operations with functions. Consider

   • $f(x) = x^2 + 2x + 7$,

   • $g(x) = 4x + 3$, and

   • $h(x) = x^2 + 6x + 10$.

   a. TABULAR Copy and complete the table below for three values for a.

a	$f(a)$	$g(a)$	$f(a) + g(a)$	$h(a)$

   b. VERBAL How are $f(x)$, $g(x)$, and $h(x)$ related?

   c. ALGEBRAIC Prove the relationship from part **b** algebraically.

H.O.T. Problems   Use **H**igher-**O**rder **T**hinking Skills

73. ERROR ANALYSIS Danielle and Miranda are describing the transformation $g(x) = [\![x + 4]\!]$. Danielle says that the graph is shifted 4 units to the left, while Miranda says that the graph is shifted 4 units up. Is either of them correct? Explain.

74. REASONING Let $f(x)$ be an odd function. If $g(x)$ is a reflection of $f(x)$ in the x-axis and $h(x)$ is a reflection of $g(x)$ in the y-axis, what is the relationship between $f(x)$ and $h(x)$? Explain.

75. WRITING IN MATH Explain why order is important when transforming a function with reflections and translations.

REASONING Determine whether the following statements are *sometimes*, *always*, or *never* true. Explain your reasoning.

76. If $f(x)$ is an even function, then $f(x) = |f(x)|$.

77. If $f(x)$ is an odd function, then $f(-x) = -|f(x)|$.

78. If $f(x)$ is an even function, then $f(-x) = -|f(x)|$.

79 CHALLENGE Describe the transformation of $f(x) = \sqrt{x}$ if $(-2, -6)$ lies on the curve.

80. REASONING Suppose (a, b) is a point on the graph of $f(x)$. Describe the difference between the transformations of (a, b) when the graph of $f(x)$ is expanded vertically by a factor of 4 and when the graph of $f(x)$ is compressed horizontally by a factor of 4.

81. WRITING IN MATH Use words, graphs, tables, and equations to relate parent functions and transformations. Show this relationship through a specific example.

Find the average rate of change of each function on the given interval. (Lesson 1-4)

82. $g(x) = -2x^2 + x - 3; [-1, 3]$ **83.** $g(x) = x^2 - 6x + 1; [4, 8]$ **84.** $f(x) = -2x^3 - x^2 + x - 4; [-2, 3]$

Use the graph of each function to describe its end behavior. Support the conjecture numerically. (Lesson 1-3)

85. $q(x) = -\dfrac{12}{x}$ **86.** $f(x) = \dfrac{0.5}{x^2}$ **87.** $p(x) = \dfrac{x + 2}{x - 3}$

Use the graph of each function to estimate its y-intercept and zero(s). Then find these values algebraically. (Lesson 1-2)

88.

89.

90.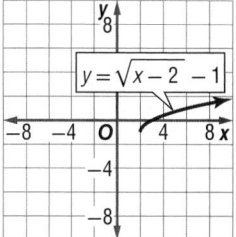

91. GOVERNMENT The number of times each of the first 42 presidents vetoed bills are listed below. What is the standard deviation of the data? (Lesson 0-8)

2,   0,   0,   7,   1,   0,   12,   1,   0,   10,   3,   0,   0,   9,
7,   6,   29,   93,   13,   0,   12,   414,   44,   170,   42,   82,   39,   44,
6,   50,   37,   635,   250,   181,   21,   30,   43,   66,   31,   78,   44,   25

92. LOTTERIES In a multi-state lottery, the player must guess which five of the white balls numbered from 1 to 49 will be drawn. The order in which the balls are drawn does not matter. The player must also guess which one of the red balls numbered from 1 to 42 will be drawn. How many ways can the player complete a lottery ticket? (Lesson 0-7)

93. SAT/ACT The figure shows the graph of $y = g(x)$, which has a minimum located at $(1, -2)$. What is the maximum value of $h(x) = -3g(x) - 1$?

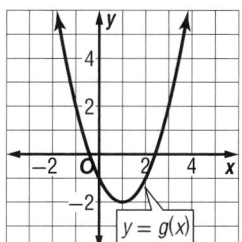

  A   0   D   3

  B   1   E   It cannot be determined from the

  C   2   information given.

94. REVIEW What is the simplified form of $\dfrac{4x^3y^2z^{-1}}{\left(x^{-2}y^3z^2\right)^2}$?

95. What is the range of $y = \dfrac{x^2 + 8}{2}$?

  F $\{y \mid y \neq \pm 2\sqrt{2}\}$

  G $\{y \mid y \geq 4\}$

  H $\{y \mid y \geq 0\}$

  J $\{y \mid y \leq 0\}$

96. REVIEW What is the effect on the graph of $y = kx^2$ as k decreases from 3 to 2?

  A   The graph of $y = 2x^2$ is a reflection of the graph of $y = 3x^2$ across the y-axis.

  B   The graph is rotated 90° about the origin.

  C   The graph becomes narrower.

  D   The graph becomes wider.

Objective
Use a graphing calculator to solve nonlinear inequalities.

A nonlinear inequality in one variable can be solved graphically by converting it into two inequalities in two variables and finding the intersection. You can use a graphing calculator to find this intersection.

ACTIVITY | **Solve an Inequality by Graphing**

Solve $2|x - 4| + 3 < 15$.

Step 1 Separate this inequality into two inequalities, one for each side of the inequality symbol. Replace each side with y to form the new inequalities.

$$2|x - 4| + 3 < Y_1; Y_2 < 15$$

Step 2 Graph each inequality. Make each inequality symbol by scrolling to the left of the equals sign and selecting ENTER until the shaded triangles are flashing. The triangle above represents *greater than* and the triangle below represents *less than*.

To access **abs(**, press MATH ▶ 1.

Step 3 Graph the inequalities in the appropriate window. Either use the **zoom** feature or adjust the window manually to display both graphs. Any window that shows the two intersection points will work.

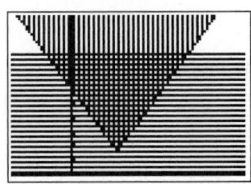

[−5, 15] scl: 1 by [0, 20] scl: 1

> **StudyTip**
>
> **Adjusting the Window**
> You can use the ZoomFit or ZoomOut options or manually adjust the window to include both graphs.

Step 4 The darkly shaded area indicates the intersection of the graphs and the solution of the system of inequalities. The two graphs intersect at two different points. Use the **intersection** feature to locate the first point. Move the cursor over the intersection and select ENTER 3 times to get $(-2, 15)$. The other intersection is at $(10, 15)$.

[−5, 15] scl: 1 by [0, 20] scl: 1

Step 5 The solution of the system of inequalities occurs in the region of the graph where $-2 < x < 10$. Thus, the solution to $2|x - 4| + 3 < 15$ is the set of x-values such that $-2 < x < 10$. Check your solution algebraically by confirming that an x-value in this interval is a solution of the inequality.

Exercises

Solve each inequality by graphing.

1. $3|x + 2| - 4 > 8$

2. $-2|x + 4| + 6 \leq 2$

3. $5|2x + 1| > 15$

4. $-3|2x - 3| + 1 \leq 10$

5. $|x - 6| > x + 2$

6. $|2x + 1| \geq 4x - 3$

Extension

7. **REASONING** Describe the appearance of the graph for an inequality with no solution.

8. **CHALLENGE** Solve $-10x - 32 < |x + 3| - 2 < -|x + 4| + 8$ by graphing.

1-6

Function Operations and Composition of Functions

Then
You evaluated functions.
(Lesson 1-1)

Now
- Perform operations with functions.
- Find compositions of functions.

New Vocabulary
composition

Math Online
glencoe.com

Why?
In April 2008, the top social networking site, founded by Chris DeWolfe and Tom Anderson, had over 60.4 million unique visitors. By comparison, the number two site at that time had 24.9 million unique visitors, $60.4 - 24.9$ million or 35.5 million fewer visitors.

Suppose $A(t)$ and $B(t)$ model the number of unique visitors to the number one and two social networking sites, respectively, t years since 2000. The difference in the number of unique visitors between the two sites for any year t could then be represented by the function $A(t) - B(t)$.

Operations with Functions Just as you can combine two real numbers using the operations of addition, subtraction, multiplication, and division, you can combine two functions.

Key Concept | **Operations with Functions**

Let f and g be two functions with intersecting domains. Then for all x-values in the intersection, the sum, product, difference, and quotient of f and g are new functions defined as follows.

Sum $\quad (f + g)(x) = f(x) + g(x)$ $\qquad$ **Product** $\quad (f \cdot g)(x) = f(x) \cdot g(x)$

Difference $(f - g)(x) = f(x) - g(x)$ $\qquad$ **Quotient** $\left(\dfrac{f}{g}\right)(x) = \dfrac{f(x)}{g(x)}, g(x) \neq 0$

For each new function, the domain consists of those values of x common to the domains of f and g. The domain of the quotient function is further restricted by excluding any values that make the denominator, $g(x)$, zero.

EXAMPLE 1 | **Operations with Functions**

Given $f(x) = x^2 + 4x$, $g(x) = \sqrt{x + 2}$, and $h(x) = 3x - 5$, find each function and its domain.

a. $(f + g)(x)$

$(f + g)(x) = f(x) + g(x)$
$\qquad = (x^2 + 4x) + \left(\sqrt{x + 2}\right)$
$\qquad = x^2 + 4x + \sqrt{x + 2}$

The domain of f is $(-\infty, \infty)$, and the domain of g is $[-2, \infty)$. So, the domain of $(f + g)$ is the intersection of these domains or $[-2, \infty)$.

b. $(f - h)(x)$

$(f - h)(x) = f(x) - h(x)$
$\qquad = (x^2 + 4x) - (3x - 5)$
$\qquad = x^2 + 4x - 3x + 5$
$\qquad = x^2 + x + 5$

The domains of f and h are both $(-\infty, \infty)$, so the domain of $(f - h)$ is $(-\infty, \infty)$.

c. $(f \cdot h)(x)$

$(f \cdot h)(x) = f(x) \cdot h(x)$
$\qquad = (x^2 + 4x)(3x - 5)$
$\qquad = 3x^3 - 5x^2 + 12x^2 - 20x$
$\qquad = 3x^3 + 7x^2 - 20x$

The domains of f and h are both $(-\infty, \infty)$, so the domain of $(f \cdot h)$ is $(-\infty, \infty)$.

d. $\left(\dfrac{h}{f}\right)(x)$

$\left(\dfrac{h}{f}\right)(x) = \dfrac{h(x)}{f(x)}$ or $\dfrac{3x - 5}{x^2 + 4x}$

The domain of h and f are both $(-\infty, \infty)$, but $x = 0$ or $x = -4$ yields a zero in the denominator of $\left(\dfrac{h}{f}\right)$. So, the domain of $\left(\dfrac{h}{f}\right)$ is $(-\infty, -4) \cup (-4, 0) \cup (0, \infty)$.

Guided Practice

Find $(f + g)(x)$, $(f - g)(x)$, $(f \cdot g)(x)$, and $\left(\dfrac{f}{g}\right)(x)$ for each $f(x)$ and $g(x)$. State the domain of each new function.

1A. $f(x) = x - 4$, $g(x) = \sqrt{9 - x^2}$ **1B.** $f(x) = x^2 - 6x - 8$, $g(x) = \sqrt{x}$

Composition of Functions The function $y = (x - 3)^2$ combines the linear function $y = x - 3$ with the squaring function $y = x^2$, but the combination does not involve addition, subtraction, multiplication, or division. This combining of functions, called *composition*, is the result of one function being used to evaluate a second function.

Key Concept   **Composition of Functions**

The **composition** of function f with function g is defined by

$$[f \circ g](x) = f[g(x)].$$

The domain of $f \circ g$ includes all x-values in the domain of g that map to $g(x)$-values in the domain of f as shown.

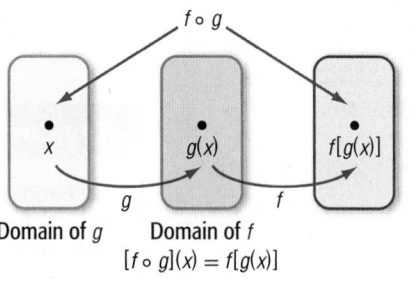

Domain of g Domain of f
$[f \circ g](x) = f[g(x)]$

In the composition $f \circ g$, which is read as *f composition g* or *f of g*, the function g is applied first and then f.

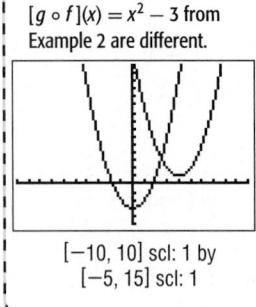
EXAMPLE 2   **Compose Two Functions**

Given $f(x) = x^2 + 1$ and $g(x) = x - 4$, find each of the following.

a. $[f \circ g](x)$

$\begin{aligned}[f \circ g](x) &= f[g(x)] &&\text{Definition of } f \circ g\\ &= f(x - 4) &&\text{Replace } g(x) \text{ with } x - 4.\\ &= (x - 4)^2 + 1 &&\text{Substitute } x - 4 \text{ for } x \text{ in } f(x).\\ &= x^2 - 8x + 16 + 1 \text{ or } x^2 - 8x + 17 &&\text{Simplify.}\end{aligned}$

b. $[g \circ f](x)$

$\begin{aligned}[g \circ f](x) &= g[f(x)] &&\text{Definition of } g \circ f\\ &= g(x^2 + 1) &&\text{Replace } f(x) \text{ with } x^2 + 1.\\ &= (x^2 + 1) - 4 \text{ or } x^2 - 3 &&\text{Substitute } x^2 + 1 \text{ for } x \text{ in } g(x).\end{aligned}$

c. $[f \circ g](2)$

Evaluate the expression $[f \circ g](x)$ you wrote in part **a** for $x = 2$.

$[f \circ g](2) = (2)^2 - 8(2) + 17 \text{ or } 5$ Substitute 2 for x in $x^2 - 8x + 17$.

Guided Practice

For each pair of functions, find $[f \circ g](x)$, $[g \circ f](x)$, and $[f \circ g](3)$.

2A. $f(x) = 3x + 1$, $g(x) = 5 - x^2$ **2B.** $f(x) = 6x^2 - 4$, $g(x) = x + 2$

Because the domains of f and g in Example 2 include all real numbers, the domain of $f \circ g$ is all real numbers, $\mathbb{R}$.

When the domains of f or g are restricted, the domain of $f \circ g$ is restricted to all x-values in the domain of g whose range values, $g(x)$, are in the domain of f.

EXAMPLE 3 **Find a Composite Function with a Restricted Domain**

Find $f \circ g$.

a. $f(x) = \dfrac{1}{x+1}, g(x) = x^2 - 9$

To find $f \circ g$, you must first be able to find $g(x) = x^2 - 9$, which can be done for all real numbers. Then you must be able to evaluate $f(x) = \dfrac{1}{x+1}$ for each of these $g(x)$-values, which can only be done when $g(x) \neq -1$. Excluding from the domain those values for which $x^2 - 9 = -1$, namely when $x = \pm\sqrt{8}$ or $\pm 2\sqrt{2}$, the domain of $f \circ g$ is $\{x \mid x \neq \pm 2\sqrt{2}, x \in \mathbb{R}\}$.

Now find $[f \circ g](x)$.

$$[f \circ g](x) = f[g(x)] \qquad \text{Definition of } f \circ g$$
$$= f(x^2 - 9) \qquad \text{Replace } g(x) \text{ with } x^2 - 9.$$
$$= \frac{1}{x^2 - 9 + 1} \text{ or } \frac{1}{x^2 - 8} \qquad \text{Substitute } x^2 - 9 \text{ for } x \text{ in } f(x).$$

Notice that $\dfrac{1}{x^2 - 8}$ is undefined when $x^2 - 8 = 0$, which is when $x = \pm 2\sqrt{2}$. Because the implied domain is the same as the domain determined by considering the domains of f and g, the composition can be written as $[f \circ g](x) = \dfrac{1}{x^2 - 8}$ for $x \neq \pm 2\sqrt{2}$.

b. $f(x) = x^2 - 2, g(x) = \sqrt{x - 3}$

To find $f \circ g$, you must first be able to find $g(x)$, which can only be done for $x \geq 3$. Then you must be able to square each of these $g(x)$-values and subtract 2, which can be done for all real numbers. Therefore, the domain of $f \circ g$ is $\{x \mid x \geq 3, x \in \mathbb{R}\}$. Now find $[f \circ g](x)$.

$$[f \circ g](x) = f[g(x)] \qquad \text{Definition of } f \circ g$$
$$= f(\sqrt{x - 3}) \qquad \text{Replace } g(x) \text{ with } \sqrt{x - 3}.$$
$$= (\sqrt{x - 3})^2 - 2 \qquad \text{Substitute } \sqrt{x - 3} \text{ for } x \text{ in } f(x).$$
$$= x - 3 - 2 \text{ or } x - 5 \qquad \text{Simplify.}$$

Once the composition is simplified, it appears that the function is defined for all reals, which is known to be untrue. Therefore, write the composition as $[f \circ g](x) = x - 5$ for $x \geq 3$.

CHECK Use a graphing calculator to check this result. Enter the function as $y = (\sqrt{x - 3})^2 - 2$. The graph appears to be part of the line $y = x - 5$. Then use the **TRACE** feature to help determine that the domain of the composite function begins at $x = 3$ and extends to ∞.

$[-1, 9]$ scl: 1 by $[-5, 5]$ scl: 1

✔ **Guided Practice**

3A. $f(x) = \sqrt{x + 1}, g(x) = x^2 - 1$

3B. $f(x) = \dfrac{5}{x}, g(x) = x^2 + x$

An important skill in calculus is to be able to *decompose* a function into two simpler functions. To decompose a function h, find two functions with a composition of h.

StudyTip

Domains of Composite Functions It is very important to complete the domain analysis before performing the composition. Domain restrictions may not be evident after the composition is simplified.

StudyTip

Using Absolute Value Recall from Lesson 0-4 that when you find an even root of an even power and the result is an odd power, you must use the absolute value of the result to ensure that the answer is nonnegative. For example, $\sqrt{x^2} = |x|$.

EXAMPLE 4 | Decompose a Composite Function

Find two functions f and g such that $h(x) = [f \circ g](x)$. Neither function may be the identity function $f(x) = x$.

a. $h(x) = \sqrt{x^3 - 4}$

Observe that h is defined using the square root of $x^3 - 4$. So one way to write h as a composition of two functions is to let $g(x) = x^3 - 4$ and $f(x) = \sqrt{x}$. Then

$$h(x) = \sqrt{x^3 - 4} = \sqrt{g(x)} = f[g(x)] \text{ or } [f \circ g](x).$$

b. $h(x) = 2x^2 + 20x + 50$

$h(x) = 2x^2 + 20x + 50$	Notice that $h(x)$ is factorable.
$= 2(x^2 + 10x + 25)$ or $2(x + 5)^2$	Factor.

One way to write $h(x)$ as a composition is to let $f(x) = 2x^2$ and $g(x) = x + 5$.

$$h(x) = 2(x + 5)^2 = 2[g(x)]^2 = f[g(x)] \text{ or } [f \circ g](x).$$

☑ Guided Practice

4A. $h(x) = x^2 - 2x + 1$ **4B.** $h(x) = \dfrac{1}{x + 7}$

You can use the composition of functions to solve real-world problems.

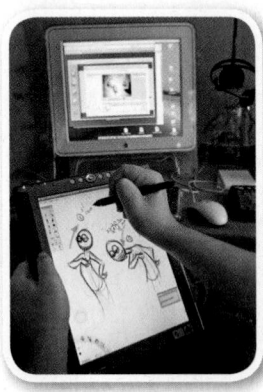

🌐 Real-World Career

Computer Animator
Animators work in many industries to create the animated images used in movies, television, and video games. Computer animators must be artistic, and most have received post-secondary training at specialized schools.

🌐 Real-World EXAMPLE 5 | Compose Real-World Functions

COMPUTER ANIMATION To animate the approach of an opponent directly in front of a player, a computer game animator starts with an image of a 20-pixel by 60-pixel rectangle. The animator then increases each dimension of the rectangle by 15 pixels per second.

a. Find functions to model the data.

The length L of the rectangle increases at a rate of 15 pixels per second, so $L(t) = 20 + 15t$, where t is the time in seconds and $t \geq 0$. The area of the rectangle is its length L times its width. The width is 40 pixels more than its length or $L + 40$. So, the area of the rectangle is $A(L) = L(L + 40)$ or $L^2 + 40L$ and $L \geq 20$.

b. Find $A \circ L$. What does this function represent?

$A \circ L = A[L(t)]$	Definition of $A \circ L$
$= A(20 + 15t)$	Replace $L(t)$ with $20 + 15t$.
$= (20 + 15t)^2 + 40(20 + 15t)$	Substitute $20 + 15t$ for L in $A(L)$.
$= 225t^2 + 1200t + 1200$	Simplify.

This composite function models the area of the rectangle as a function of time.

c. How long does it take for the rectangle to triple its original size?

The initial area of the rectangle is $20 \cdot 60$ or 1200 pixels. The rectangle will be three times its original size when $[A \circ L](t) = 225t^2 + 1200t + 1200 = 3600$. Solve for t to find that $t \approx 1.55$ or -6.88. Because a negative t-value is not part of the domain of $L(t)$, it is also not part of the domain of the composite function. The area will triple after about 1.55 seconds.

☑ Guided Practice

5. BUSINESS A computer store offers a 15% discount to college students on the purchase of any notebook computer. The store also advertises $100 coupons.

    A. Find functions to model the data.

    B. Find $[c \circ d](x)$ and $[d \circ c](x)$. What does each composite function represent?

    C. Which composition of the coupon and discount results in the lower price? Explain.

Find $(f + g)(x)$, $(f - g)(x)$, $(f \cdot g)(x)$, and $\left(\dfrac{f}{g}\right)(x)$ for each $f(x)$ and $g(x)$. State the domain of each new function. (Example 1)

1. $f(x) = x^2 + 4$
$g(x) = \sqrt{x}$

2. $f(x) = 8 - x^3$
$g(x) = x - 3$

3. $f(x) = x^2 + 5x + 6$
$g(x) = x + 2$

4. $f(x) = x - 9$
$g(x) = x + 5$

5. $f(x) = x^2 + x$
$g(x) = 9x$

6. $f(x) = x - 7$
$g(x) = x + 7$

7. $f(x) = \dfrac{6}{x}$
$g(x) = x^3 + x$

8. $f(x) = \dfrac{x}{4}$
$g(x) = \dfrac{3}{x}$

9. $f(x) = \dfrac{1}{\sqrt{x}}$
$g(x) = 4\sqrt{x}$

10. $f(x) = \dfrac{3}{x}$
$g(x) = x^4$

11. $f(x) = \sqrt{x + 8}$
$g(x) = \sqrt{x + 5} - 3$

12. $f(x) = \sqrt{x + 6}$
$g(x) = \sqrt{x - 4}$

13. BUDGETING Suppose a budget in dollars for one person for one month is approximated by $f(x) = 25x + 350$ and $g(x) = 15x + 200$, where f is the cost of rent and groceries, g is the cost of gas and all other expenses, and $x = 1$ represents the total cost at the end of the first week. (Example 1)

a. Find $(f + g)(x)$ and the relevant domain.

b. What does $(f + g)(x)$ represent?

c. Find $(f + g)(4)$. What does this value represent?

14. PHYSICS Two different forces act on an object being pushed across a floor: the force of the person pushing the object and the force of friction. If W is work in joules, F is force in newtons, and d is displacement of the object in meters, $W_p(d) = F_p d$ describes the work of the person and $W_f(d) = F_f d$ describes the work done by friction. The increase in kinetic energy of the object is the difference between the work done by the person W_p and the work done by friction W_f. (Example 1)

a. Find $(W_p - W_f)(d)$.

b. Determine the net work expended when a person pushes a box 50 meters with a force of 95 newtons and friction exerts a force of 55 newtons.

For each pair of functions, find $[f \circ g](x)$, $[g \circ f](x)$, and $[f \circ g](6)$. (Example 2)

15. $f(x) = 2x - 3$
$g(x) = 4x - 8$

16. $f(x) = -2x^2 - 5x + 1$
$g(x) = -5x + 6$

17. $f(x) = 8 - x^2$
$g(x) = x^2 + x + 1$

18. $f(x) = x^2 - 16$
$g(x) = x^2 + 7x + 11$

19. $f(x) = 3 - x^2$
$g(x) = x^3 + 1$

20. $f(x) = 2 + x^4$
$g(x) = -x^2$

Find $f \circ g$. (Example 3)

21. $f(x) = \dfrac{1}{x + 1}$
$g(x) = x^2 - 4$

22. $f(x) = \dfrac{2}{x - 3}$
$g(x) = x^2 + 6$

23. $f(x) = \sqrt{x + 4}$
$g(x) = x^2 - 4$

24. $f(x) = x^2 - 9$
$g(x) = \sqrt{x + 3}$

25. $f(x) = \dfrac{5}{x}$
$g(x) = \sqrt{6 - x}$

26. $f(x) = -\dfrac{4}{x}$
$g(x) = \sqrt{x + 8}$

27. $f(x) = \sqrt{x + 5}$
$g(x) = x^2 + 4x - 1$

28. $f(x) = \sqrt{x - 2}$
$g(x) = x^2 + 8$

29 RELATIVITY In the theory of relativity, $m(v) = \dfrac{100}{\sqrt{1 - \dfrac{v^2}{c^2}}}$, where c is the speed of light, 300 million meters per second, and m is the mass of a 100-kilogram object at speed v in meters per second. (Example 4)

a. Are there any restrictions on the domain of the function? Explain their meaning.

b. Find $m(10)$, $m(10,000)$, and $m(1,000,000)$.

c. Describe the behavior of $m(v)$ as v approaches c.

d. Decompose the function into two separate functions.

Find two functions f and g such that $h(x) = [f \circ g](x)$. Neither function may be the identity function $f(x) = x$. (Example 4)

30. $h(x) = \sqrt{4x + 2} + 7$

31. $h(x) = \dfrac{6}{x + 5} - 8$

32. $h(x) = |4x + 8| - 9$

33. $h(x) = [\![-3(x - 9)]\!]$

34. $h(x) = \sqrt{\dfrac{5 - x}{x + 2}}$

35. $h(x) = (\sqrt{x} + 4)^3$

36. $h(x) = \dfrac{6}{(x + 2)^2}$

37. $h(x) = \dfrac{8}{(x - 5)^2}$

38. $h(x) = \dfrac{\sqrt{4 + x}}{x - 2}$

39. $h(x) = \dfrac{x + 5}{\sqrt{x - 1}}$

40. QUANTUM MECHANICS The wavelength λ of a particle with mass m kilograms moving at v meters per second is represented by $\lambda = \dfrac{h}{mv}$, where h is a constant equal to $6.626 \cdot 10^{-34}$.

a. Find a function to represent the wavelength of a 25-kilogram object as a function of its speed.

b. Are there any restrictions on the domain of the function? Explain their meaning.

c. If the object is traveling 8 meters per second, find the wavelength in terms of h.

d. Decompose the function into two separate functions.

41. JOBS A salesperson for an insurance agency is paid an annual salary plus a bonus of 4% of sales made over 300,000. Let $f(x) = x - 300,000$ and $h(x) = 0.04x$, where x is total sales. (Example 5)

   a. If x is greater than 300,000, is the bonus represented by $f[h(x)]$ or by $h[f(x)]$? Explain your reasoning.

   b. Determine the amount of bonus for one year with sales of 450,000.

42. TRAVEL An airplane flying above a landing strip at 275 miles per hour passes a control tower 0.5 mile below at time $t = 0$ hours. (Example 5)

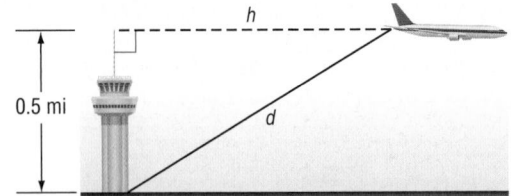

   a. Find the distance d between the airplane and the control tower as a function of the horizontal distance h from the control tower to the plane.

   b. Find h as a function of time t.

   c. Find $d \circ h$. What does this function represent?

   d. If the plane continued to fly the same distance from the ground, how far would the plane be from the control tower after 10 minutes?

Find two functions f and g such that $h(x) = [f \circ g](x)$. Neither function may be the identity function $f(x) = x$.

43. $h(x) = \sqrt{x - 1} - \dfrac{4}{x}$

44. $h(x) = \sqrt{2x + 6} + \dfrac{6}{x}$

45. $h(x) = \dfrac{8}{x^2 + 2} + 5|x|$

46. $h(x) = \sqrt{-7x} + 9x$

47. $h(x) = \dfrac{x}{2x - 1} + \sqrt{\dfrac{4}{x}}$

48. $h(x) = \dfrac{x^2 - 4}{x} + \dfrac{3x - 5}{5x}$

Use the given information to find $f(0.5)$, $f(-6)$, and $f(x + 1)$. Round to the nearest tenth if necessary.

49. $f(x) - g(x) = x^2 + x - 6,\ g(x) = x + 4$

50. $f(x) + g(x) = \dfrac{2}{x^2} + \dfrac{1}{x} - \dfrac{1}{3},\ g(x) = 2x$

51. $g(x) - f(x) + \dfrac{3}{5} = 9x^2 + 4x,\ g(x) = \dfrac{x}{10}$

52. $g(x) = f(x) - 18x^2 + \dfrac{\sqrt{2}}{x},\ g(x) = \sqrt{1 - x}$

Find $[f \circ g \circ h](x)$.

53 $f(x) = x + 8$
   $g(x) = x^2 - 6$
   $h(x) = \sqrt{x} + 3$

54. $f(x) = x^2 - 2$
   $g(x) = 5x + 12$
   $h(x) = \sqrt{x} - 4$

55. $f(x) = \sqrt{x + 5}$
   $g(x) = x^2 - 3$
   $h(x) = \dfrac{1}{x}$

56. $f(x) = \dfrac{3}{x}$
   $g(x) = x^2 - 4x + 1$
   $h(x) = x + 2$

57. If $f(x) = x + 2$, find $g(x)$ such that:

   a. $(f + g)(x) = x^2 + x + 6$

   b. $\left(\dfrac{f}{g}\right)(x) = \dfrac{1}{4}$

58. If $f(x) = \sqrt{4x}$, find $g(x)$ such that:

   a. $[f \circ g](x) = |6x|$

   b. $[g \circ f](x) = 200x + 25$

59. If $f(x) = 4x^2$, find $g(x)$ such that:

   a. $(f \cdot g)(x) = x$

   b. $(f \cdot g)(x) = \dfrac{1}{8}x^{\frac{7}{3}}$

60. INTEREST An investment account earns interest compounded quarterly. If x dollars are invested in an account, the investment $I(x)$ after one quarter is the initial investment plus accrued interest or $I(x) = 1.016x$.

   a. Find $[I \circ I](x)$, $[I \circ I \circ I](x)$, and $[I \circ I \circ I \circ I](x)$.

   b. What do the compositions represent?

   c. What is the account's annual percentage yield?

Use the graphs of $f(x)$ and $g(x)$ to find each function value.

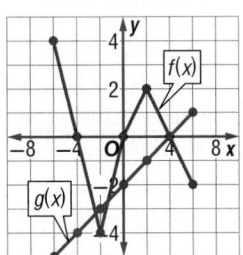

61. $(f + g)(2)$

62. $(f - g)(-6)$

63. $(f \cdot g)(4)$

64. $\left(\dfrac{f}{g}\right)(-2)$

65. $[f \circ g](-4)$

66. $[g \circ f](6)$

67. CHEMISTRY The average speed $v(m)$ of gas molecules at 30°C in meters per second can be represented by

$$v(m) = \sqrt{\dfrac{(24.9435)(303)}{m}},\text{ where } m \text{ is the molar mass of}$$

the gas in kilograms per mole.

   a. Are there any restrictions on the domain of the function? Explain their meaning.

   b. Find the average speed of 145 kilograms per mole gas molecules at 30°C.

   c. How will the average speed change as the molar mass of gas increases?

   d. Decompose the function into two separate functions.

Find functions f, g, and h such that $a(x) = [f \circ g \circ h](x)$.

68. $a(x) = (\sqrt{x - 7} + 4)^2$

69. $a(x) = \sqrt{(x - 5)^2 + 8}$

70. $a(x) = \dfrac{3}{(x - 3)^2 + 4}$

71. $a(x) = \dfrac{4}{(\sqrt{x} + 3)^2 + 1}$

For each pair of functions, find $f \circ g$ and $g \circ f$.

72. $f(x) = x^2 - 6x + 5$
$g(x) = \sqrt{x + 4} + 3$

73. $f(x) = x^2 + 8x - 3$
$g(x) = \sqrt{x + 19} - 4$

74. $f(x) = \sqrt{x + 6}$
$g(x) = \sqrt{16 + x^2}$

75. $f(x) = \sqrt{x}$
$g(x) = \sqrt{9 - x^2}$

76. $f(x) = -\dfrac{8}{5 - 4x}$
$g(x) = \dfrac{2}{3 + x}$

77. $f(x) = \dfrac{6}{2x + 1}$
$g(x) = \dfrac{4}{4 - x}$

Graph each of the following.

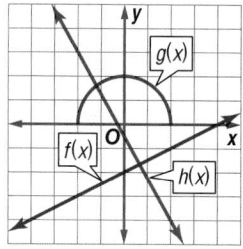

78. $(f + h)(x)$

79. $(h - f)(x)$

80. $(f + g)(x)$

81. $(h + g)(x)$

82. ⚙ **MULTIPLE REPRESENTATIONS** In this problem, you will investigate inverses of functions.

a. ALGEBRAIC Find the composition of f with g and of g with f for each pair of functions.

$f(x)$	$g(x)$
$x + 3$	$x - 3$
$4x$	$\dfrac{x}{4}$
x^3	$\sqrt[3]{x}$

b. VERBAL Describe the relationship between the composition of each pair of functions.

c. GRAPHICAL Graph each pair of functions on the coordinate plane. Graph the line of reflection by finding the midpoint of the segment between corresponding points.

d. VERBAL Make a conjecture about the line of reflection between the functions.

e. ANALYTICAL The compositions $[f \circ g](x)$ and $[g \circ f](x)$ are equivalent to which parent function?

f. ANALYTICAL Find $g(x)$ for each $f(x)$ such that $[f \circ g](x) = [g \circ f](x) = x$.

 i. $f(x) = x - 6$

 ii. $f(x) = \dfrac{x}{3}$

 iii. $f(x) = x^5$

 iv. $f(x) = 2x - 3$

 v. $f(x) = x^3 + 1$

State the domain of each composite function.

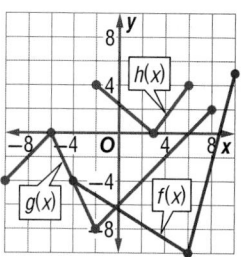

83. $[f \circ g](x)$

84. $[g \circ f](x)$

85. $[h \circ f](x)$

86. $[h \circ g](x)$

H.O.T. Problems Use **H**igher-**O**rder **T**hinking Skills

REASONING Determine whether $[f \circ g](x)$ is *even, odd, neither,* or *not enough information* for each of the following.

87. f and g are odd.

88. f and g are even.

89. f is even and g is odd.

90. f is odd and g is even.

CHALLENGE Find a function f other than $f(x) = x$ such that the following are true.

91. $[f \circ f](x) = x$

92. $[f \circ f \circ f](x) = f(x)$

93. WRITING IN MATH Explain how $f(x)$ might have a domain restriction while $[f \circ g](x)$ might not. Provide an example to justify your reasoning.

94. REASONING Determine whether the following statement is *true* or *false*. Explain your reasoning.
If f is a square root function and g is a quadratic function, then $f \circ g$ is always a linear function.

95 **CHALLENGE** State the domain of $[f \circ g \circ h](x)$ for $f(x) = \dfrac{1}{x - 2}$, $g(x) = \sqrt{x + 1}$, and $h(x) = \dfrac{4}{x}$.

96. WRITING IN MATH Describe how you would find the domain of $[f \circ g](x)$.

97. WRITING IN MATH Explain why order is important when finding the composition of two functions.

98. FINANCIAL LITERACY The cost of labor for servicing cars at B & B Automotive is displayed in the advertisement. (Lesson 1-5)

B & B
AUTOMOTIVE SERVICE
$50 per hour
Each fraction of an hour is considered a full hour

a. Graph the function that describes the cost for x hours of labor.

b. Graph the function that would show a $25 additional charge if you decide to also get the oil changed and fluids checked.

c. What would be the cost of servicing a car that required 3.45 hours of labor if the owner requested to have the oil changed and the fluids checked?

Approximate to the nearest hundredth the relative or absolute extrema of each function. State the *x*-values where they occur. (Lesson 1-4)

99. $f(x) = 2x^3 - 3x^2 + 4$

100. $g(x) = -x^3 + 5x - 3$

101. $f(x) = x^4 + x^3 - 2$

Approximate the real zeros of each function for the given interval. (Lesson 1-3)

102. $g(x) = 2x^5 - 2x^4 - 4x^2 - 1; [-1, 3]$

103. $f(x) = \dfrac{x^2 - 3}{x - 4}; [-3, 3]$

104. $g(x) = \dfrac{x^2 - 2x - 1}{x^2 + 3x}; [1, 5]$

105. SPORTS The table shows the leading home run and runs batted in totals in the American League for 2004–2008. (Lesson 1-1)

Year	2004	2005	2006	2007	2008
HR	43	48	54	54	48
RBI	150	148	137	156	146

Source: *World Almanac*

a. Make a graph of the data with home runs on the horizontal axis and runs batted in on the vertical axis.

b. Identify the domain and range.

c. Does the graph represent a function? Explain your reasoning.

106. SAT/ACT A jar contains only red, green, and blue marbles. It is three times as likely that you randomly pick a red marble as a green marble, and five times as likely that you pick a green one as a blue one. Which could be the number of marbles in the jar?

A 39   C 41   E 63

B 40   D 42

107. If $g(x) = x^2 + 9x + 21$ and $h(x) = 2(x - 5)^2$, then $h[g(x)] =$

F $x^4 + 18x^3 + 113x^2 + 288x + 256$

G $2x^4 + 36x^3 + 226x^2 + 576x + 512$

H $3x^4 + 54x^3 + 339x^2 + 864x + 768$

J $4x^4 + 72x^3 + 452x^2 + 1152x + 1024$

108. FREE RESPONSE The change in temperature of a substance in degrees Celsius as a function of time for $0 \le t \le 8$ is shown in the graph.

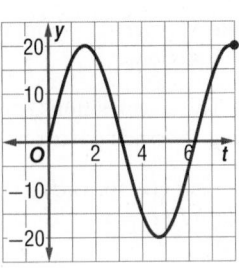

a. This graph represents a function. Explain why.

b. State the domain and range.

c. If the initial temperature is 25°C, what is the approximate temperature of the substance at $t = 7$?

d. Analyze the graph for symmetry and zeros. Determine if the function is *even*, *odd*, or *neither*.

e. Is the function continuous at $t = 2$? Explain.

f. Determine the intervals on which the function is increasing or decreasing.

g. Estimate the average rate of change for [2, 5].

h. What is the significance of your answers to parts **f** and **g** in the context of the situation?

Inverse Relations and Functions

Then
You found the composition of two functions.
(Lesson 1-6)

Now
- Use the horizontal line test to determine whether a function has an inverse function.
- Find inverse functions algebraically and graphically.

New Vocabulary
inverse relation
inverse function
one-to-one

Math Online
glencoe.com

Why?

The Band Boosters at Julia's high school are selling raffle tickets. Table A relates the cost in dollars to the number of tickets purchased. Table B relates the number of tickets that can be purchased to the number of dollars spent. By interchanging the input and output from Table A, Julia obtains Table B.

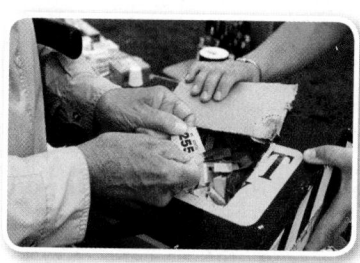

Table A

Tickets	1	2	3	4	6
Cost ($)	2	4	6	8	10

Table B

Money Spent ($)	2	4	6	8	10
Tickets	1	2	3	4	6

Inverse Functions In the example above, the relation shown in Table A is said to be the *inverse relation* of the relation shown in Table B. Two relations are **inverse relations** if and only if one relation contains (b, a) whenever the other relation contains (a, b). When a relation is expressed as an equation, its inverse relation can be found by interchanging the independent and dependent variables. Consider the following inverse relations.

Relation

$y = x^2 - 4$

x	y
−3	5
−2	0
−1	−3
0	−4
1	−3
2	0
3	5

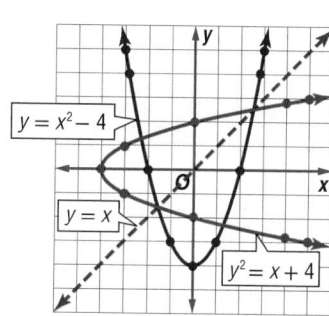

Inverse Relation

$x = y^2 - 4$ or $y^2 = x + 4$

x	y
5	−3
0	−2
−3	−1
−4	0
−3	1
0	2
5	3

Notice that these inverse relations are reflections of each other in the line $y = x$. This relationship is true for the graphs of all relations and their inverse relations.

While every relation has an inverse relation, we are most interested in *functions* with inverse relations that are also *functions*. If the inverse relation of a function f is also a function, then it is called the **inverse function** of f and is denoted f^{-1}, read *f inverse*. Not all functions have inverse functions. In the graph above, notice that the original relation is a function because it passes the vertical line test. But its inverse relation fails this test, so it is not a function.

The reflective relationship between the graph of a function and its inverse relation leads us to the following graphical test for determining whether the inverse function of a function exists.

Key Concept	**Horizontal Line Test**	
Words	A function f has an inverse function f^{-1} if and only if each horizontal line intersects the graph of the function in at most one point.	**Model** 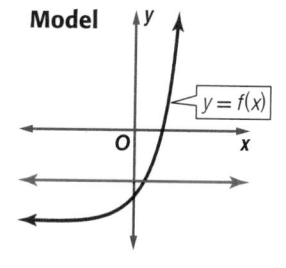
Example	Since no horizontal line intersects the graph of f more than once, the inverse function f^{-1} exists.	

EXAMPLE 1 **Apply the Horizontal Line Test**

Graph each function using a graphing calculator, and apply the horizontal line test to determine whether its inverse function exists. Write *yes* or *no*.

a. $f(x) = |x - 1|$

The graph of $f(x)$ in Figure 1.7.1 shows that it is possible to find a horizontal line that intersects the graph of $f(x)$ more than once. Therefore, you can conclude that f^{-1} does not exist.

b. $g(x) = x^3 - 6x^2 + 12x - 8$

The graph of $g(x)$ in Figure 1.7.2 shows that it is not possible to find a horizontal line that intersects the graph of $g(x)$ in more than one point. Therefore, you can conclude that g^{-1} exists.

$[-4, 6]$ scl: 1 by $[-2, 8]$ scl: 1

Figure 1.7.1

$[-4, 6]$ scl: 1 by $[-5, 5]$ scl: 1

Figure 1.7.2

✓ **Guided Practice**

1A. $h(x) = \dfrac{4}{x}$

1B. $f(x) = x^2 + 5x - 7$

Find Inverse Functions If a function passes the horizontal line test, then it is said to be **one-to-one**, because no *x*-value is matched with more than one *y*-value and no *y*-value is matched with more than one *x*-value.

If a function f is one-to-one, it has an inverse function f^{-1} such that the domain of f is equal to the range of f^{-1}, and the range of f is equal to the domain of f^{-1}.

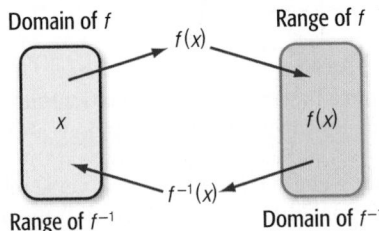

To find an inverse function algebraically, follow the steps below.

Key Concept **Finding an Inverse Function**

Step 1 Determine whether the function has an inverse by checking to see if it is one-to-one using the horizontal line test.

Step 2 In the equation for $f(x)$, replace $f(x)$ with y and then interchange x and y.

Step 3 Solve for y and then replace y with $f^{-1}(x)$ in the new equation.

Step 4 State any restrictions on the domain of f^{-1}. Then show that the domain of f is equal to the range of f^{-1} and the range of f is equal the domain of f^{-1}.

The last step implies that only part of the function you find algebraically may be the inverse function of f. Therefore, be sure to analyze the domain of f when finding f^{-1}.

EXAMPLE 2 **Find Inverse Functions Algebraically**

Determine whether f has an inverse function. If it does, find the inverse function and state any restrictions on its domain.

a. $f(x) = \dfrac{x-1}{x+2}$

The graph of f shown passes the horizontal line test. Therefore, f is a one-to-one function and has an inverse function. From the graph, you can see that f has domain $(-\infty, -2) \cup (-2, \infty)$ and range $(-\infty, 1) \cup (1, \infty)$. Now find f^{-1}.

[−10, 10] scl: 1 by [−10, 10] scl: 1

$f(x) = \dfrac{x-1}{x+2}$ **Original function**

$y = \dfrac{x-1}{x+2}$ **Replace $f(x)$ with y.**

$x = \dfrac{y-1}{y+2}$ **Interchange x and y.**

$xy + 2x = y - 1$ **Multiply each side by $y + 2$. Then apply the Distributive Property.**

$xy - y = -2x - 1$ **Isolate the y-terms.**

$y(x - 1) = -2x - 1$ **Distributive Property**

$y = \dfrac{-2x - 1}{x - 1}$ **Solve for y.**

$f^{-1}(x) = \dfrac{-2x - 1}{x - 1}$ **Replace y with $f^{-1}(x)$. Note that $x \neq 1$.**

From the graph at the right, you can see that f^{-1} has domain $(-\infty, 1) \cup (1, \infty)$ and range $(-\infty, -2) \cup (-2, \infty)$. The domain and range of f are equal to the range and domain of f^{-1}, respectively. So, $f^{-1}(x) = \dfrac{-2x - 1}{x - 1}$ for $x \neq 1$.

[−10, 10] scl: 1 by [−10, 10] scl: 1

b. $f(x) = \sqrt{x - 4}$

The graph of f shown passes the horizontal line test. Therefore, f is a one-to-one function and has an inverse function. From the graph, you can see that f has domain $[4, \infty)$ and range $[0, \infty)$. Now find f^{-1}.

[−5, 15] scl: 1 by [−10, 10] scl: 1

$f(x) = \sqrt{x - 4}$ **Original function**

$y = \sqrt{x - 4}$ **Replace $f(x)$ with y.**

$x = \sqrt{y - 4}$ **Interchange x and y.**

$x^2 = y - 4$ **Square each side.**

$y = x^2 + 4$ **Solve for y.**

$f^{-1}(x) = x^2 + 4$ **Replace y with $f^{-1}(x)$.**

From the graph of $y = x^2 + 4$ shown, you can see that the inverse relation has domain $(-\infty, \infty)$ and range $[4, \infty)$. By restricting the domain of the inverse relation to $[0, \infty)$, the domain and range of f are equal to the range and domain of f^{-1}, respectively. So, $f^{-1}(x) = x^2 + 4$, for $x \geq 0$.

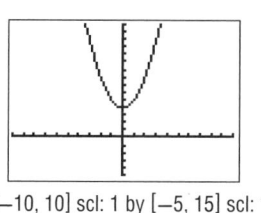

[−10, 10] scl: 1 by [−5, 15] scl: 1

✓ **Guided Practice**

2A. $f(x) = -16 + x^3$

2B. $f(x) = \dfrac{x + 7}{x}$

2C. $f(x) = \sqrt{x^2 - 20}$

An inverse function f^{-1} has the effect of "undoing" the action of a function f. For this reason, inverse functions can also be defined in terms of their composition with each other.

Key Concept　　**Compositions of Inverse Functions**

Two functions, f and g, are inverse functions if and only if
- $f[g(x)] = x$ for every x in the domain of $g(x)$ and
- $g[f(x)] = x$ for every x in the domain of $f(x)$.

Notice that the composition of a function with its inverse function is always the identity function. You can use this fact to verify that two functions are inverse functions of each other.

EXAMPLE 3　**Verify Inverse Functions**

Show that $f(x) = \dfrac{6}{x-4}$ and $g(x) = \dfrac{6}{x} + 4$ are inverse functions.

Show that $f[g(x)] = x$ and that $g[f(x)] = x$.

$$f[g(x)] = f\left(\frac{6}{x} + 4\right) \qquad\qquad g[f(x)] = g\left(\frac{6}{x-4} + 4\right)$$

$$= \frac{6}{\left(\frac{6}{x} + 4\right) - 4} \qquad\qquad = \frac{6}{\left(\frac{6}{x-4}\right)} + 4$$

$$= \frac{6}{\left(\frac{6}{x}\right)} \text{ or } x \qquad\qquad = x - 4 + 4 \text{ or } x$$

Because $f[g(x)] = g[f(x)] = x$, $f(x)$ and $g(x)$ are inverse functions. This is supported graphically because $f(x)$ and $g(x)$ appear to be reflections of each other in the line $y = x$.

[−15.16, 15.16] scl: 1 by [−10, 10] scl: 1

✓ **Guided Practice**

Show that f and g are inverse functions.

3A. $f(x) = 18 - 3x, g(x) = 6 - \dfrac{x}{3}$ 　　　**3B.** $f(x) = x^2 + 10, x \geq 0; g(x) = \sqrt{x - 10}$

The inverse functions of most one-to-one functions are often difficult to find algebraically. However, it is possible to graph the inverse function by reflecting the graph of the original function in the line $y = x$.

EXAMPLE 4　**Find Inverse Functions Graphically**

Use the graph of $f(x)$ in Figure 1.7.3 to graph $f^{-1}(x)$.

Graph the line $y = x$. Locate a few points on the graph of $f(x)$. Reflect these points in $y = x$. Then connect them with a smooth curve that mirrors the curvature of $f(x)$ in line $y = x$ (Figure 1.7.4).

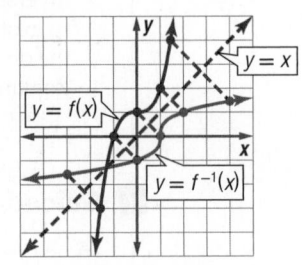

Figure 1.7.3　　　　　　　　　**Figure 1.7.4**

✓ Guided Practice

Use the graph of each function to graph its inverse function.

4A.

4B.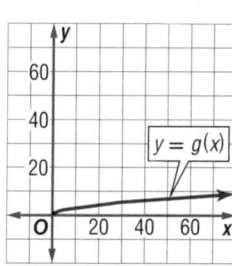

🌐 Real-World EXAMPLE 5 / Use an Inverse Function

SUMMER EARNINGS Kendra earns $8 an hour, works at least 40 hours per week, and receives overtime pay at 1.5 times her regular hourly rate for any time over 40 hours. Her total earnings $f(x)$ for a week in which she worked x hours is given by $f(x) = 320 + 12(x - 40)$.

a. Explain why the inverse function $f^{-1}(x)$ exists. Then find $f^{-1}(x)$.

The function simplifies to $f(x) = 320 + 12x - 480$ or $12x - 160$. The graph of $f(x)$ passes the horizontal line test. Therefore, $f(x)$ is a one-to-one function and has an inverse function. Find $f^{-1}(x)$.

$f(x) = 12x - 160$	**Original function**
$y = 12x - 160$	**Replace $f(x)$ with y.**
$x = 12y - 160$	**Interchange x and y.**
$x + 160 = 12y$	**Add 160 to each side.**
$y = \dfrac{x + 160}{12}$	**Solve for y.**
$f^{-1}(x) = \dfrac{x + 160}{12}$	**Replace y with $f^{-1}(x)$.**

Kendra's Summer Job Earnings

b. What do $f^{-1}(x)$ and x represent in the inverse function?

In the inverse function, x represents Kendra's earnings for a particular week and $f^{-1}(x)$ represents the number of hours Kendra worked that week.

c. What restrictions, if any, should be placed on the domain of $f(x)$ and $f^{-1}(x)$? Explain.

The function $f(x)$ assumes that Kendra works at least 40 hours in a week. There are $7 \cdot 24$ or 168 hours in a week, so the domain of $f(x)$ is [40, 168]. Because $f(40) = 320$ and $f(168) = 1856$, the range of $f(x)$ is [320, 1856]. Because the range of $f(x)$ must equal the domain of $f^{-1}(x)$, the domain of $f^{-1}(x)$ is [320, 1856].

d. Find the number of hours Kendra worked last week if her earnings were $380.

Because $f^{-1}(380) = \dfrac{380 + 160}{12}$ or 45, Kendra worked 45 hours last week.

✓ Guided Practice

5. SAVINGS Solada's net pay is 65% of her gross pay, and she budgets $600 per month for living expenses. She estimates that she can save 20% of her remaining money, so her one-month savings $f(x)$ for a gross pay of x dollars is given by $f(x) = 0.2(0.65x - 600)$.

A. Explain why the inverse function $f^{-1}(x)$ exists. Then find $f^{-1}(x)$.

B. What do $f^{-1}(x)$ and x represent in the inverse function?

C. What restrictions, if any, should be placed on the domains of $f(x)$ and $f^{-1}(x)$? Explain.

D. Determine Solada's gross pay for one month if her savings for that month were $120.

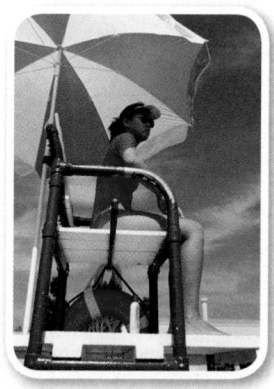

🌐 Real-World Link

From 1999 to 2006, the number of 16- to 19-year-olds in the United States who had summer jobs decreased from 48% to 37%.

Source: U.S. Bureau of Labor Statistics

Exercises

Graph each function using a graphing calculator, and apply the horizontal line test to determine whether its inverse function exists. Write *yes* or *no*. (Example 1)

1. $f(x) = x^2 + 6x + 9$ **2.** $f(x) = x^2 - 16x + 64$

3. $f(x) = x^2 - 10x + 25$ **4.** $f(x) = 3x - 8$

5. $f(x) = \sqrt{2x}$ **6.** $f(x) = 4$

7. $f(x) = \sqrt{x + 4}$ **8.** $f(x) = -4x^2 + 8$

9. $f(x) = \dfrac{5}{x - 6}$ **10.** $f(x) = \dfrac{8}{x + 2}$

11. $f(x) = x^3 - 9$ **12.** $f(x) = \dfrac{1}{4}x^3$

Determine whether each function has an inverse function. If it does, find the inverse function and state any restrictions on its domain. (Example 2)

13. $g(x) = -3x^4 + 6x^2 - x$ **14.** $f(x) = 4x^5 - 8x^4$

15. $h(x) = x^7 + 2x^3 - 10x^2$ **16.** $f(x) = \sqrt{x + 8}$

17. $f(x) = \sqrt{6 - x^2}$ **18.** $f(x) = |x - 6|$

19. $f(x) = \dfrac{4 - x}{x}$ **20.** $g(x) = \dfrac{x - 6}{x}$

21. $f(x) = \dfrac{6}{\sqrt{8 - x}}$ **22.** $g(x) = \dfrac{7}{\sqrt{x + 3}}$

23. $f(x) = \dfrac{6x + 3}{x - 8}$ **24.** $h(x) = \dfrac{x + 4}{3x - 5}$

25. $g(x) = |x + 1| + |x - 4|$

26. SPEED The speed of an object in kilometers per hour y is $y = 1.6x$, where x is the speed of the object in miles per hour. (Example 2)

   a. Find an equation for the inverse of the function. What does each variable represent?

   b. Graph each equation on the same coordinate plane.

Show algebraically that *f* and *g* are inverse functions. (Example 3)

27. $f(x) = -6x + 3$
     $g(x) = \dfrac{3 - x}{6}$

28. $f(x) = 4x + 9$
     $g(x) = \dfrac{x - 9}{4}$

29. $f(x) = -3x^2 + 5, x \geq 0$
     $g(x) = \sqrt{\dfrac{5 - x}{3}}$

30. $f(x) = \dfrac{x^2}{4} + 8, x \geq 0$
     $g(x) = \sqrt{4x - 32}$

31. $f(x) = 2x^3 - 6$
     $g(x) = \sqrt[3]{\dfrac{x + 6}{2}}$

32. $f(x) = (x + 8)^{\frac{3}{2}}$
     $g(x) = x^{\frac{2}{3}} - 8, x \geq 0$

33. $g(x) = \sqrt{x + 8} - 4$
     $f(x) = x^2 + 8x + 8, x \geq -4$

34. $g(x) = \sqrt{x - 8} + 5$
     $f(x) = x^2 - 10x + 33, x \geq 5$

35. $f(x) = \dfrac{x + 4}{x}$
     $g(x) = \dfrac{4}{x - 1}$

36. $f(x) = \dfrac{x - 6}{x + 2}$
     $g(x) = \dfrac{2x + 6}{1 - x}$

37. PHYSICS The kinetic energy of an object in motion in joules can be described by $f(x) = 0.5mx^2$, where m is the mass of the object in kilograms and x is the speed of the object in meters per second. (Example 3)

   a. Find the inverse of the function. What does each variable represent?

   b. Show that $f(x)$ and the function you found in part **a** are inverses.

   c. Graph $f(x)$ and $f^{-1}(x)$ on the same graphing calculator screen if the mass of the object is 1 kilogram.

Use the graph of each function to graph its inverse function. (Example 4)

38.

39.

40.

41.

42.

43.
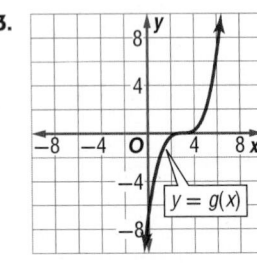

44. JOBS Jamie sells shoes at a department store after school. Her base salary each week is $140, and she earns a 10% commission on each pair of shoes that she sells. Her total earnings $f(x)$ for a week in which she sold x dollars worth of shoes is $f(x) = 140 + 0.1x$. (Example 5)

   a. Explain why the inverse function $f^{-1}(x)$ exists. Then find $f^{-1}(x)$.

   b. What do $f^{-1}(x)$ and x represent in the inverse function?

   c. What restrictions, if any, should be placed on the domains of $f(x)$ and $f^{-1}(x)$? Explain.

   d. Find Jamie's total sales last week if her earnings for that week were $220.

45. CURRENCY The average exchange rate from Euros to U.S. dollars for a recent four-month period can be described by $f(x) = 0.66x$, where x is the currency value in Euros. (Example 5)

   a. Explain why the inverse function $f^{-1}(x)$ exists. Then find $f^{-1}(x)$.

   b. What do $f^{-1}(x)$ and x represent in the inverse function?

   c. What restrictions, if any, should be placed on the domains of $f(x)$ and $f^{-1}(x)$? Explain.

   d. What is the value in Euros of 100 U.S. dollars?

Determine whether each function has an inverse function.

46.

47.

48.

49.
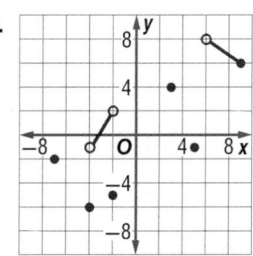

Determine if f^{-1} exists. If so, complete a table for f^{-1}.

50.

x	−6	−4	−1	3	6	10
f(x)	−4	0	3	5	9	13

51.

x	−3	−2	−1	0	1	2
f(x)	14	11	8	10	11	16

52.

x	1	2	3	4	5	6
f(x)	2	8	16	54	27	16

53.

x	−10	−9	−8	−7	−6	−5
f(x)	8	7	6	5	4	3

54. TEMPERATURE The formula $f(x) = \frac{9}{5}x + 32$ is used to convert x degrees Celsius to degrees Fahrenheit. To convert x degrees Fahrenheit to Kelvin, the formula $k(x) = \frac{5}{9}(x + 459.67)$ is used.

   a. Find f^{-1}. What does this function represent?

   b. Show that f and f^{-1} are inverse functions. Graph each function on the same graphing calculator screen.

   c. Find $[k \circ f](x)$. What does this function represent?

   d. If the temperature is 60°C, what would the temperature be in Kelvin?

Restrict the domain of each function so that the resulting function is one-to-one. Then determine the inverse of the function.

55

56.

57.

58.
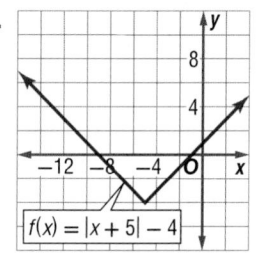

State the domain and range of f and f^{-1}, if f^{-1} exists.

59. $f(x) = \sqrt{x - 6}$

60. $f(x) = x^2 + 9$

61. $f(x) = \dfrac{3x + 1}{x - 4}$

62. $f(x) = \dfrac{8x + 3}{2x - 6}$

63. ENVIRONMENT Once an endangered species, the bald eagle was downlisted to threatened status in 1995. The table shows the number of nesting pairs each year.

Year	Nesting Pairs
1984	1757
1990	3035
1994	4449
1998	5748
2000	6471
2005	7066

   a. Use the table to approximate a linear function that relates the number of nesting pairs to the year. Let 0 represent 1984.

   b. Find the inverse of the function you generated in part **a**. What does each variable represent?

   c. Using the inverse function, in approximately what year was the number of nesting pairs 5094?

64. FLOWERS Bonny needs to purchase 75 flower stems for banquet decorations. She can choose between lilies and hydrangea, which cost $5.00 per stem and $3.50 per stem, respectively.

   a. Write a function for the total cost of the flowers.

   b. Find the inverse of the cost function. What does each variable represent?

   c. Find the domain of the cost function and its inverse.

   d. If the total cost for the flowers was $307.50, how many lilies did Bonny purchase?

Find an equation for the inverse of each function, if it exists. Then graph the equations on the same coordinate plane. Include any domain restrictions.

65. $f(x) = \begin{cases} x^2 & \text{if } -4 \geq x \\ -2x + 5 & \text{if } -4 < x \end{cases}$

66. $f(x) = \begin{cases} -4x + 6 & \text{if } -5 \geq x \\ 2x - 8 & \text{if } -5 < x \end{cases}$

67. FLOW RATE The flow rate of a gas is the volume of gas that passes through an area during a given period of time. The speed v of air flowing through a vent can be found using $v(r) = \dfrac{r}{A}$, where r is the flow rate in cubic feet per second and A is the cross-sectional area of the vent in square feet.

2 ft
1.5 ft

a. Find v^{-1} of the vent shown. What does this function represent?

b. Determine the speed of air flowing through the vent in feet per second if the flow rate is 15,000 feet cubed per second.

c. Determine the gas flow rate of a circular vent that has a diameter of 5 feet with a gas stream that is moving at 1.8 feet per second.

68. COMMUNICATION A cellular phone company is having a sale as shown. Assume that the $50 rebate is given only after the 10% discount is given.

> **Central Mobile**
> **Cell Phone Sale**
> 10% off new phones
> $50 rebate with
> proof of purchase

a. Write a function r for the price of the phone as a function of the original price if only the rebate applies.

b. Write a function d for the price of the phone as a function of the original price if only the discount applies.

c. Find a formula for $T(x) = [r \circ d](x)$ if both the discount and the rebate apply.

d. Find T^{-1} and explain what the inverse represents.

e. If the total cost of the phone after the discount and the rebate was $49, what was the original price of the phone?

Use $f(x) = 8x - 4$ and $g(x) = 2x + 6$ to find each of the following.

69. $[f^{-1} \circ g^{-1}](x)$

70. $[g^{-1} \circ f^{-1}](x)$

71. $[f \circ g]^{-1}(x)$

72. $[g \circ f]^{-1}(x)$

73. $(f \cdot g)^{-1}(x)$

74. $(f^{-1} \cdot g^{-1})(x)$

Use $f(x) = x^2 + 1$ with domain $[0, \infty)$ and $g(x) = \sqrt{x - 4}$ to find each of the following.

75. $[f^{-1} \circ g^{-1}](x)$

76. $[g^{-1} \circ f^{-1}](x)$

77. $[f \circ g]^{-1}(x)$

78. $[g \circ f]^{-1}(x)$

79. $(f \cdot g^{-1})(x)$

80. $(f^{-1} \cdot g)(x)$

81. COPIES Karen's Copies charges users $0.40 for every minute or part of a minute to use their computer scanner. Suppose you use the scanner for x minutes, where x is any real number greater than 0.

a. Sketch the graph of the function, $C(x)$, that gives the cost of using the scanner for x minutes.

b. What are the domain and range of $C(x)$?

c. Sketch the graph of the inverse of $C(x)$.

d. What are the domain and range of the inverse?

e. What real-world situation is modeled by the inverse?

82. ⬡ MULTIPLE REPRESENTATIONS In this problem, you will investigate inverses of even and odd functions.

a. GRAPHICAL Sketch the graphs of three different even functions. Do the graphs pass the horizontal line test?

b. ANALYTICAL What pattern can you discern regarding the inverses of even functions? Confirm or deny the pattern algebraically.

c. GRAPHICAL Sketch the graphs of three different odd functions. Do the graphs pass the horizontal line test?

d. ANALYTICAL What pattern can you discern regarding the inverses of odd functions? Confirm or deny the pattern algebraically.

83 REASONING If f has an inverse and a zero at 6, what can you determine about the graph of f^{-1}?

84. WRITING IN MATH Explain what type of restriction on the domain is needed to determine the inverse of a quadratic function and why a restriction is needed. Provide an example.

85. REASONING *True* or *False*. Explain your reasoning. *All linear functions have inverse functions.*

86. CHALLENGE If $f(x) = x^3 - ax + 8$ and $f^{-1}(23) = 3$, find the value of a.

87. REASONING Can $f(x)$ pass the horizontal line test when $\lim\limits_{x \to \infty} f(x) = 0$ and $\lim\limits_{x \to -\infty} f(x) = 0$? Explain.

88. REASONING Why is $\pm$ not used when finding the inverse function of $f(x) = \sqrt{x + 4}$?

89. WRITING IN MATH Explain how an inverse of f can exist. Give an example provided that the domain of f is restricted and f does not have an inverse when the domain is unrestricted.

For each pair of functions, find $f \circ g$ and $g \circ f$. Then state the domain of each composite function. (Lesson 1-6)

90. $f(x) = x^2 - 9$
$g(x) = x + 4$

91. $f(x) = \frac{1}{2}x - 7$
$g(x) = x + 6$

92. $f(x) = x - 4$
$g(x) = 3x^2$

Use the graph of the given parent function to describe the graph of each related function. (Lesson 1-5)

93. $f(x) = x^2$
 a. $g(x) = (0.2x)^2$
 b. $h(x) = (x - 5)^2 - 2$
 c. $m(x) = 3x^2 + 6$

94. $f(x) = x^3$
 a. $g(x) = |x^3 + 3|$
 b. $h(x) = -(2x)^3$
 c. $m(x) = 0.75(x + 1)^3$

95. $f(x) = |x|$
 a. $g(x) = |2x|$
 b. $h(x) = |x - 5|$
 c. $m(x) = |3x| - 4$

96. ADVERTISING A newspaper surveyed companies on the annual amount of money spent on television commercials and the estimated number of people who remember seeing those commercials each week. A soft-drink manufacturer spends $40.1 million a year and estimates 78.6 million people remember the commercials. For a package-delivery service, the budget is $22.9 million for 21.9 million people. A telecommunications company reaches 88.9 million people by spending $154.9 million. Use a matrix to represent these data. (Lesson 0-6)

Solve each system of equations. (Lesson 0-5)

97. $x + 2y + 3z = 5$
$3x + 2y - 2z = -13$
$5x + 3y - z = -11$

98. $7x + 5y + z = 0$
$-x + 3y + 2z = 16$
$x - 6y - z = -18$

99. $x - 3z = 7$
$2x + y - 2z = 11$
$-x - 2y + 9z = 13$

100. BASEBALL A batter pops up the ball. Suppose the ball was 3.5 feet above the ground when he hit it straight up with an initial velocity of 80 feet per second. The function $d(t) = 80t - 16t^2 + 3.5$ gives the ball's height above the ground in feet as a function of time t in seconds. How long did the catcher have to get into position to catch the ball after it was hit? (Lesson 0-3)

101. SAT/ACT What is the probability that the spinner will land on a number that is either even or greater than 5?

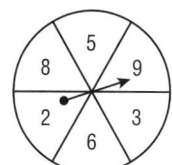

 A $\frac{1}{6}$ **C** $\frac{1}{2}$ **E** $\frac{5}{6}$
 B $\frac{1}{3}$ **D** $\frac{2}{3}$

102. REVIEW If m and n are both odd natural numbers, which of the following must be true?

 I. $m^2 + n^2$ is even.
 II. $m^2 + n^2$ is divisible by 4.
 III. $(m + n)^2$ is divisible by 4.

 F none **H** I and II only
 G I only **J** I and III only

103. Which of the following is the inverse of $f(x) = \frac{3x - 5}{2}$?

 A $g(x) = \frac{2x + 5}{3}$

 B $g(x) = \frac{3x + 5}{2}$

 C $g(x) = 2x + 5$

 D $g(x) = \frac{2x - 5}{3}$

104. REVIEW A train travels d miles in t hours and arrives at its destination 3 hours late. At what average speed, in miles per hour, should the train have gone in order to have arrived on time?

 F $t - 3$

 G $\frac{t - 3}{d}$

 H $\frac{d}{t - 3}$

 J $\frac{d}{t} - 3$

Objective
Use a graphing calculator and parametric equations to graph inverses on the calculator.

Parametric equations are equations that can express the position of an object as a function of time. The basic premise of parametric equations is the introduction of an extra variable t, called a *parameter*. For example, $y = x + 4$ can be expressed parametrically using $x = t$ and $y = t + 4$. Parametric equations will be studied in depth in Lesson 7-5.

ACTIVITY 1 | **Parametric Graph**

Graph $x = t, y = 0.1t^2 - 4$.

Step 1 Set the mode. In the [MODE] menu, select **par** and **simul**. This allows the equations to be graphed simultaneously.

StudyTip

Standard Window You can use zoomstandard to set the window in standard form.

Step 2 Enter the parametric equations. In parametric form, [X,T,θ,n] will use t instead of x.

Step 3 Set the window as shown.

Step 4 Graph the equations. Notice that the graph looks like $y = 0.1x^2 - 4$ but is traced from $t = -10$ to $t = 10$.

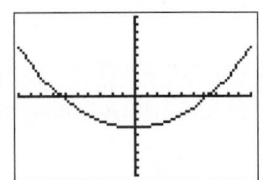

[−10, 10] scl: 1 by [−10, 10] scl: 1
t: [−10, 10]; tstep: 1

Exercises

1. **REASONING** Graph the equations using Tstep = 10, 5, 0.5, and 0.1. How does this affect the way the graph is shown?

2. In this problem, you will investigate the relationship between x, y, and t.

   a. Graph $X_{1T} = t - 3, Y_{1T} = t + 4$ in the standard viewing window.

   b. Replace the equations in part **a** with $X_{1T} = t, Y_{1T} = t + 7$ and graph.

   c. What do you notice about the two graphs?

   d. **REASONING** What conclusions can you make about the relationship between x, y, and t? In other words, how do you think the second set of parametric equations was formed using the first set?

One benefit of parametric equations is the ability to graph inverses without determining them.

ACTIVITY 2 | **Graph an Inverse**

Graph the inverse of $x = t$, $y = 0.1t^2 - 4$.

Step 1 Inverses are found by switching x and y. Enter the given equations as X_{1T} and Y_{1T}. To graph the inverse, set $X_{2T} = Y_{1T}$ and $Y_{2T} = X_{1T}$. These are found in the VARS menu. Select **Y-Vars, parametric, X_{1T}**.

Step 2 Graph the relation and the inverse. Notice how the x- and y-coordinates are reversed. You can use **ZSquare** to see the symmetry of the two graphs more clearly.

[−15, 15] scl: 1 by [−10, 10] scl: 1
t: [−10, 10]; $t$step: 1

Exercises

3. REASONING What needs to be true about the ordered pairs of each graph in Activity 2?

4. REASONING Does the graph of $x = t$, $y = 0.1t^2 - 4$ represent a one-to-one function? Explain.

ACTIVITY 3 | **Domains and One-to-One Functions**

Limit the domain of $x = t$, $y = 0.1t^2 - 4$ in order to make it one-to-one.

StudyTip

Symmetry You may need to use the trace feature and adjust Tstep in order to locate the axis of symmetry.

Step 1 A graph must pass the horizontal line test in order to be one-to-one. The graph of $x = t$, $y = 0.1t^2 - 4$ is symmetric with respect to the y-axis. We can produce a one-to-one function for $0 \le x \le 10$. By substituting this range into $x = t$, we can determine the corresponding t values for this range of x-values, which is $0 \le t \le 10$.

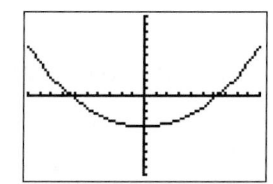

[−10, 10] scl: 1 by [−10, 10] scl: 1
t: [−10, 10]; $t$step: 1

Step 2 Change **Tmin** from −10 to 0 in order to limit the domain. Graph the one-to-one function and its inverse. Again, you can use **ZSquare** to see the symmetry of the two graphs more clearly.

[−15, 15] scl: 1 by [−10, 10] scl: 1
t: [0, 10]; $t$step: 1

Exercises

Graph each function. Then graph the inverse function and indicate the limited domain if necessary.

5. $x = t - 6$, $y = t^2 + 2$

6. $x = 3t - 1$, $y = t^2 + t$

7. $x = 3 - 2t$, $y = t^2 - 2t + 1$

8. $x = 2t^2 + 3$, $y = \sqrt{t}$

9. $x = 4t$, $y = \sqrt{t + 2}$

10. $x = t - 8$, $y = t^3$

11. CHALLENGE Consider a quintic function with two relative maxima and two relative minima. Into how many different one-to-one functions can this function be separated if each separation uses the largest interval possible?

Chapter Summary

Key Concepts

Functions (Lesson 1-1)

- Common subsets of the real numbers are integers, rational numbers, irrational numbers, whole numbers, and natural numbers.
- A function f is a relation that assigns each element in the domain exactly one element in the range.
- The graph of a function passes the vertical line test.

Analyzing Graphs of Functions and Relations (Lesson 1-2)

- Graphs may be symmetric with respect to the y-axis, the x-axis, and the origin.
- An even function is symmetric with respect to the y-axis. An odd function is symmetric with respect to the origin.

Continuity, End Behavior, and Limits (Lesson 1-3)

- If the value of $f(x)$ approaches a unique value L as x approaches c from either side, then the limit of $f(x)$ as x approaches c is L. It is written $\lim_{x \to c} f(x) = L$.
- A function may be discontinuous because of infinite discontinuity, jump discontinuity, or removable discontinuity.

Extrema and Average Rate of Change (Lesson 1-4)

- A function can be described as increasing, decreasing, or constant.
- Extrema of a function include relative maxima and minima and absolute maxima and minima.
- The average rate of change between two points can be represented by $m_{\text{sec}} = \dfrac{f(x_2) - f(x_1)}{x_2 - x_1}$.

Parent Functions and Transformations (Lesson 1-5)

Transformations of parent functions include translations, reflections, and dilations.

Operations with and Composition of Functions (Lesson 1-6)

The sum, difference, product, quotient, and composition of two functions form new functions.

Inverse Relations and Functions (Lesson 1-7)

- Two relations are inverse relations if and only if one relation contains the element (b, a) whenever the other relation contains the element (a, b).
- Two functions, f and f^{-1}, are inverse functions if and only if $f[f^{-1}(x)] = x$ and $f^{-1}[f(x)] = x$.

Key Vocabulary

composition (p. 58)
constant (p. 34)
continuous function (p.24)
decreasing function (p. 34)
dilation (p. 49)
discontinuous function (p. 24)
end behavior (p. 28)
even function (p. 18)
extrema (p. 36)
function (p. 5)
increasing (p. 34)
interval notation (p. 5)
inverse function (p. 65)
inverse relation (p. 65)

limit (p. 24)
line symmetry (p. 16)
maximum (p. 36)
minimum (p. 36)
nonremovable discontinuity (p. 25)
odd function (p. 18)
one-to-one (p. 66)
parent function (p. 45)
piecewise-defined function (p. 8)
point symmetry (p. 16)
reflection (p. 48)
roots (p. 15)
translation (p. 47)
zero function (p. 45)
zeros (p. 15)

Vocabulary Check

State whether each sentence is *true* or *false*. If *false*, replace the underlined term to make a true sentence.

1. A <u>function</u> assigns every element of its domain to exactly one element of its range.

2. Graphs that have <u>point symmetry</u> can be rotated 180° with respect to a point and appear unchanged.

3. An <u>odd function</u> has a point of symmetry.

4. The graph of a <u>continuous function</u> has no breaks or holes.

5. The <u>limit</u> of a graph describes approaching a value without necessarily ever reaching it.

6. A function $f(x)$ with values that decrease as x increases is a <u>decreasing</u> function.

7. The <u>extrema</u> of a function can include relative maxima or minima.

8. The <u>translation</u> of a graph produces a mirror image of the graph with respect to a line.

9. A <u>one-to-one function</u> passes the horizontal line test.

10. One-to-one functions have <u>line symmetry</u>.

Lesson-by-Lesson Review

1-1   Functions (pp. 4–12)

Determine whether each relation represents y as a function of x.

11. $3x - 2y = 18$ **12.** $y^3 - x = 4$

13.

x	y
5	7
7	9
9	11
11	13

14.

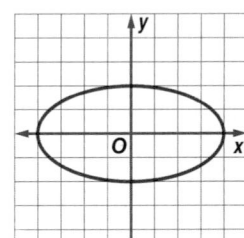

Let $f(x) = x^2 - 3x + 4$. Find each function value.

15. $f(5)$ **16.** $f(-3x)$

State the domain of each function.

17. $f(x) = 5x^2 - 17x + 1$ **18.** $g(x) = \sqrt{6x - 3}$

19. $h(a) = \dfrac{5}{a + 5}$ **20.** $v(x) = \dfrac{x}{x^2 - 4}$

EXAMPLE 1

Determine whether $y^2 - 8 = x$ represents y as a function of x.

Solve for y.

$y^2 - 8 = x$	**Original equation**
$y^2 = x + 8$	**Add 8 to each side.**
$y = \pm\sqrt{x + 8}$	**Take the square root of each side.**

This equation does not represent y as a function of x because for any x-value greater than -8, there will be two corresponding y-values.

EXAMPLE 2

Let $g(x) = -3x^2 + x - 6$. Find $g(2)$.

Substitute 2 for x in the expression $-3x^2 + x - 6$.

$g(2) = -3(2)^2 + 2 - 6$	$x = 2$
$= -12 + 2 - 6 \text{ or } -16$	**Simplify.**

1-2   Analyzing Graphs of Functions and Relations (pp. 13–23)

Use the graph of g to find the domain and range of each function.

21.

22.

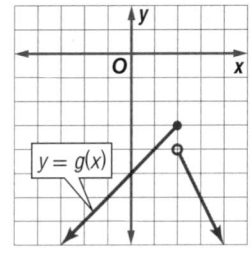

Find the y-intercept(s) and zeros for each function.

23. $f(x) = 4x - 9$

24. $f(x) = x^2 - 6x - 27$

25. $f(x) = x^3 - 16x$

26. $f(x) = \sqrt{x + 2} - 1$

EXAMPLE 3

Use the graph of $f(x) = x^3 - 8x^2 + 12x$ to find its y-intercept and zeros. Then find these values algebraically.

Estimate Graphically

It appears that $f(x)$ intersects the y-axis at $(0, 0)$, so the y-intercept is 0.

The x-intercepts appear to be at about 0, 2, and 6.

Solve Algebraically

Find $f(0)$.

$f(0) = (0)3 - 8(0)2 + 12(0)$ or 0

The y-intercept is 0.

Factor the related equation.

$x(x^2 - 8x + 12) = 0$

$x(x - 6)(x - 2) =$

The zeros of f are 0, 6, and 2.

Lesson-by-Lesson Review

1-3   Continuity, End Behavior, and Limits (pp. 24–33)

Determine whether each function is continuous at the given *x*-value(s). Justify using the continuity test. If discontinuous, identify the type of discontinuity as *infinite, jump,* or *removable.*

27. $f(x) = x^2 - 3x; x = 4.$

28. $f(x) = \sqrt{2x - 4}; x = 10$

29. $f(x) = \dfrac{x}{x + 7}; x = 0$ and $x = 7$

30. $f(x) = \dfrac{x}{x^2 - 4}; x = 2$ and $x = 4$

31. $f(x) = \begin{cases} 3x - 1 & \text{if } x < 1 \\ 2x & \text{if } x \geq 1 \end{cases}; x = 1$

Use the graph of each function to describe its end behavior.

32.   **33.**

EXAMPLE 4

Determine whether $f(x) = \dfrac{1}{x - 4}$ is continuous at $x = 0$ and $x = 4$. Justify your answer using the continuity test. If discontinuous, identify the type of discontinuity as *infinite, jump,* or *removable.*

$f(0) = -0.25$, so f is defined at 0. The function values suggest that as f gets closer to -0.25 x gets closer to 0.

x	−0.1	−0.01	0	0.01	0.1
f(x)	−0.244	−0.249	−0.25	−0.251	−0.256

Because $\lim\limits_{x \to 0} f(x)$ is estimated to be -0.25 and $f(0) = -0.25$, we can conclude that $f(x)$ is continuous at $x = 0$. Because f is not defined at 4, f is not continuous at 4.

EXAMPLE 5

Use the graph of $f(x) = -2x^4 - 5x + 1$ to describe its end behavior.

Examine the graph of $f(x)$.
As $x \to \infty, f(x) \to -\infty$.
As $x \to -\infty, f(x) \to -\infty$.

1-4   Extrema and Average Rate of Change (pp. 34–43)

Use the graph of each function to estimate intervals to the nearest 0.5 unit on which the function is increasing, decreasing, or constant. Then estimate to the nearest 0.5 unit, and classify the extrema for the graph of each function.

34.   **35.**

Find the average rate of change of each function on the given interval.

36. $f(x) = -x^3 + 3x + 1; [0, 2]$

37. $f(x) = x^2 + 2x + 5; [-5, 3]$

EXAMPLE 6

Use the graph of $f(x) = x^3 - 4x$ to estimate intervals to the nearest 0.5 unit on which the function is increasing, decreasing, or constant. Then estimate to the nearest 0.5 unit and classify the extrema for the graph of each function.

From the graph, we can estimate that f is increasing on $(-\infty, -1)$, decreasing on $(-1, 1)$, and increasing on $(1, \infty)$.

We can estimate that f has a relative maximum at $(-1, 3)$ and a relative minimum at $(1, -3)$.

Lesson-by-Lesson Review

1-5 Parent Functions and Transformations (pp. 45–55)

Identify the parent function $f(x)$ of $g(x)$, and describe how the graphs of $g(x)$ and $f(x)$ are related. Then graph $f(x)$ and $g(x)$ on the same axes.

38. $g(x) = \sqrt{x-3} + 2$ **39.** $g(x) = -(x-6)^2 - 5$

40. $g(x) = \dfrac{1}{2(x+7)}$ **41.** $g(x) = \frac{1}{4}[\![x]\!] + 3$

Describe how the graphs of $f(x) = \sqrt{x}$ and $g(x)$ are related. Then write an equation for $g(x)$.

42.

43.

EXAMPLE 7

Identify the parent function $f(x)$ of $g(x) = -|x-3| + 7$, and describe how the graphs of $g(x)$ and $f(x)$ are related. Then graph $f(x)$ and $g(x)$ on the same axes.

The parent function for $g(x)$ is $f(x) = |x|$. The graph of g will be the same as the graph of f reflected in the x-axis, translated 3 units to the right, and translated 7 units up.

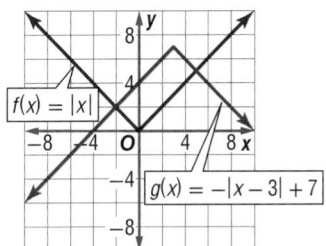

1-6 Function Operations and Composition of Functions (pp. 57–64)

Find $(f+g)(x)$, $(f-g)(x)$, $(f \cdot g)(x)$, and $\left(\dfrac{f}{g}\right)(x)$ for each $f(x)$ and $g(x)$. State the domain of each new function.

44. $f(x) = x + 3$
$g(x) = 2x^2 + 4x - 6$

45. $f(x) = 4x^2 - 1$
$g(x) = 5x - 1$

46. $f(x) = x^3 - 2x^2 + 5$
$g(x) = 4x^2 - 3$

47. $f(x) = \dfrac{1}{x}$
$g(x) = \dfrac{1}{x^2}$

For each pair of functions, find $[f \circ g](x)$, $[g \circ f](x)$, and $[f \circ g](2)$.

48. $f(x) = 4x - 11$; $g(x) = 2x^2 - 8$

49. $f(x) = x^2 + 2x + 8$; $g(x) = x - 5$

50. $f(x) = x^2 - 3x + 4$; $g(x) = x^2$

Find $f \circ g$.

51. $f(x) = \dfrac{1}{x-3}$
$g(x) = 2x - 6$

52. $f(x) = \sqrt{x-2}$
$g(x) = 6x - 7$

EXAMPLE 8

Given $f(x) = x^3 - 1$ and $g(x) = x + 7$, find $(f+g)(x)$, $(f-g)(x)$, $(f \cdot g)(x)$, and $\left(\dfrac{f}{g}\right)(x)$. State the domain of each new function.

$(f+g)(x) = f(x) + g(x)$
$= (x^3 - 1) + (x + 7)$
$= x^3 + x + 6$

The domain of $(f+g)(x)$ is $(-\infty, \infty)$.

$(f-g)(x) = f(x) - g(x)$
$= (x^3 - 1) - (x + 7)$
$= x^3 - x - 8$

The domain of $(f-g)(x)$ is $(-\infty, \infty)$.

$(f \cdot g)(x) = f(x) \cdot g(x)$
$= (x^3 - 1)(x + 7)$
$= x^4 + 7x^3 - x - 7$

The domain of $(f \cdot g)(x)$ is $(-\infty, \infty)$.

$\left(\dfrac{f}{g}\right)(x) = \dfrac{f(x)}{g(x)}$ or $\dfrac{x^3 - 1}{x + 7}$

The domain of $\left(\dfrac{f}{g}\right)(x)$ is $D = (-\infty, -7) \cup (-7, \infty)$.

Lesson-by-Lesson Review

1-7 | **Inverse Relations and Functions** (pp. 65–73)

Graph each function using a graphing calculator, and apply the horizontal line test to determine whether its inverse function exists. Write *yes* or *no*.

53. $f(x) = |x| + 6$

54. $f(x) = x^3$

55. $f(x) = -\dfrac{3}{x + 6}$

56. $f(x) = x^3 - 4x^2$

Find the inverse function and state any restrictions on the domain.

57. $f(x) = x^3 - 2$

58. $g(x) = -4x + 8$

59. $h(x) = 2\sqrt{x + 3}$

60. $f(x) = \dfrac{x}{x + 2}$

EXAMPLE 9

Find the inverse function of $f(x) = \sqrt{x} - 3$ and state any restrictions on its domain.

Note that f has domain $[0, \infty)$ and range $[-3, \infty)$. Now find the inverse relation of f.

$\begin{aligned}
y &= \sqrt{x} - 3 & &\text{Replace } f(x) \text{ with } y. \\
x &= \sqrt{y} - 3 & &\text{Interchange } x \text{ and } y. \\
x + 3 &= \sqrt{y} & &\text{Add 3 to each side.} \\
(x + 3)^2 &= y & &\text{Square each side. Note that } D = (-\infty, \infty) \\
& & &\text{and } R = [0, \infty).
\end{aligned}$

The domain of $y = (x + 3)^2$ does not equal the range of f unless restricted to $[-3, \infty)$. So, $f^{-1}(x) = (x + 3)^2$ for $x \geq -3$.

Applications and Problem Solving

61. CELL PHONES Basic Mobile offers a cell phone plan that charges $39.99 per month. Included in the plan are 500 daytime minutes that can be used Monday through Friday between 7 A.M. and 7 P.M. Users are charged $0.20 per minute for every daytime minute over 500 used. (Lesson 1-1)

   a. Write a function $p(x)$ for the cost of a month of service during which you use x daytime minutes.

   b. How much will you be charged if you use 450 daytime minutes? 550 daytime minutes?

   c. Graph $p(x)$.

62. AUTOMOBILES The fuel economy for a hybrid car at various highway speeds is shown. (Lesson 1-2)

   a. Approximately what is the fuel economy for the car when traveling 50 miles per hour?

   b. At approximately what speed will the car's fuel economy be less than 40 miles per gallon?

63. SALARIES After working for a company for five years, Ms. Washer was given a promotion. She is now earning $1500 per month more than her previous salary. Will a function modeling her monthly income be a continuous function? Explain. (Lesson 1-3)

64. BASEBALL The table shows the number of home runs by a baseball player in each of the first 5 years he played professionally. (Lesson 1-4)

Year	2004	2005	2006	2007	2008
Number of Home Runs	5	36	23	42	42

   a. Explain why 2006 represents a relative minimum.

   b. Suppose the average rate of change of home runs between 2008 and 2011 is 5 home runs per year. How many home runs were there in 2011?

   c. Suppose the average rate of change of home runs between 2007 and 2012 is negative. Compare the number of home runs in 2007 and 2012.

65. PHYSICS A stone is thrown horizontally from the top of a cliff. The velocity of the stone measured in meters per second after t seconds can be modeled by $v(t) = -\sqrt{(9.8t)^2 + 49}$. The speed of the stone is the absolute value of its velocity. Draw a graph of the stone's speed during the first 6 seconds. (Lesson 1-5)

66. FINANCIAL LITERACY A department store advertises $10 off the price of any pair of jeans. How much will a pair of jeans cost if the original price is $55 and there is 8.5% sales tax? (Lesson 1-6)

67. MEASUREMENT One inch is approximately equal to 2.54 centimeters. (Lesson 1-7)

   a. Write a function $A(x)$ that will convert the area x of a rectangle from square inches to square centimeters.

   b. Write a function $A^{-1}(x)$ that will convert the area x of a rectangle from square centimeters to square inches.

Determine whether the given relation represents y as a function of x.

1. $x = y^2 - 5$

3. $y = \sqrt{x^2 + 3}$

2.

4. PARKING The cost of parking a car downtown is $0.75 per 30 minutes for a maximum of $4.50. Parking is charged per second.

   a. Write a function for $c(x)$, the cost of parking a car for x hours.

   b. Find $c(2.5)$.

   c. What is the domain for $c(x)$? Explain your reasoning.

State the domain and range of each function.

5.

6.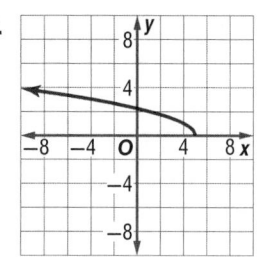

Find the y-intercept(s) and zeros for each function.

7. $f(x) = 4x^2 - 8x - 12$

8. $f(x) = x^3 + 4x^2 + 3x$

9. MULTIPLE CHOICE Which relation is symmetric with respect to the x-axis?

   A $-x^2 - yx = 2$

   B $x^3 y = 8$

   C $y = |x|$

   D $-y^2 = -4x$

Determine whether each function is continuous at $x = 3$. If discontinuous, identify the type of discontinuity as *infinite*, *jump*, or *removable*.

10. $f(x) = \begin{cases} 2x & \text{if } x < 3 \\ 9 - x & \text{if } x \geq 3 \end{cases}$

11. $f(x) = \dfrac{x - 3}{x^2 - 9}$

Find the average rate of change for each function on the interval $[-2, 6]$.

12. $f(x) = -x^4 + 3x$

13. $f(x) = \sqrt{x + 3}$

Use the graph of each function to estimate intervals to the nearest 0.5 unit on which the function is increasing or decreasing.

14.

15.

16. MULTIPLE CHOICE Which function is shown in the graph?

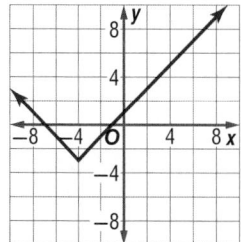

   F $f(x) = |x - 4| - 3$

   G $f(x) = |x - 4| + 3$

   H $f(x) = |x + 4| - 3$

   J $f(x) = |x + 4| + 3$

Identify the parent function $f(x)$ of $g(x)$. Then sketch the graph of $g(x)$.

17. $g(x) = -(x + 3)^3$

18. $g(x) = |x^2 - 4|$

Given $f(x) = x - 6$ and $g(x) = x^2 - 36$, find each function and its domain.

19. $\left(\dfrac{f}{g}\right)(x)$

20. $[g \circ f](x)$

21. TEMPERATURE In most countries, temperature is measured in degrees Celsius. The equation that relates degrees Fahrenheit with degrees Celsius is $F = \dfrac{9}{5}C + 32$.

   a. Write C as a function of F.

   b. Find two functions f and g such that $C = [f \circ g](F)$.

Determine whether f has an inverse function. If it does, find the inverse function and state any restrictions on its domain.

22. $f(x) = (x - 2)^3$

23. $f(x) = \dfrac{x + 3}{x - 8}$

24. $f(x) = \sqrt{4 - x}$

25. $f(x) = x^2 - 16$

Objective
Approximate the rate of change of a function at a point.

Differential calculus is a branch of calculus that focuses on the rates of change of functions at individual points. You have learned to calculate the constant rate of change, or slope, for linear functions and the average rate of change for nonlinear functions. Using differential calculus, you can determine the *exact* rate of change of any function at a single point, as represented by the slopes of the tangent lines in the figure at the right.

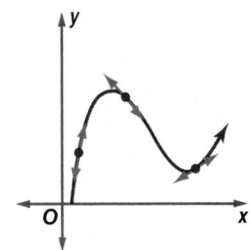

The constant rate of change for a linear function not only represents the slope of the graph between two points, but also the *exact* rate of change of the function at each of its points. For example, notice in the figure at the right that the slope m of the function is 1. This value also refers to the *exact* rate at which this function is changing at any point in its domain. This is the focus of differential calculus.

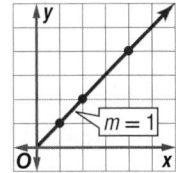

The average rate of change for nonlinear functions is represented by the slope of a line created by any two points on the graph of the function. This line is called a *secant line*. This slope is not the *exact* rate of change of the function at any one point. We can, however, use this process to give us an approximation for that instantaneous rate of change.

ACTIVITY 1 Approximate Rate of Change

Approximate the rate of change of $f(x) = 2x^2 - 3x$ at $x = 2$.

Step 1 Graph $f(x) = 2x^2 - 3x$, and plot the point $P = (2, f(2))$.

Step 2 Draw a secant line through $P = (2, f(2))$ and $Q = (4, f(4))$.

Step 3 Calculate the average rate of change m for $f(x)$ using P and Q, as shown in the figure.

Step 4 Repeat Steps 1–3 four more times. Use $Q = (3, f(3))$, $Q = (2.5, f(2.5))$, $Q = (2.25, f(2.25))$, and $Q = (2.1, f(2.1))$.

▶ Analyze the Results

1. As Q approaches P, what does the average rate of change m appear to approach?

2. Using a secant line to approximate rate of change at a point can produce varying results. Make a conjecture as to when this process will produce accurate approximations.

In differential calculus, we express the formula for average rate of change in terms of x and the horizontal distance h between the two points that determine the secant line.

ACTIVITY 2 Approximate Rate of Change

Write a general formula for finding the slope m of any secant line. Use the formula to approximate the rate of change of $f(x) = 2x^2 - 3x$ for $x = 2$.

Step 1 Generate an expression for finding the average rate of change for the figure at the right. This expression is called the *difference quotient*.

Step 2 Use your difference quotient to approximate the rate of change of $f(x)$ at $x = 2$. Let $h = 0.4, 0.25,$ and 0.1.

Analyze the Results

3. As the value of h gets closer and closer to 0, what does the average rate of change appear to approach?

4. Graph $f(x)$ and the secant line created when $h = 0.1$.

5. What does the secant line appear to become as h approaches 0?

6. Make a conjecture about the rate of change of a function at a point as it relates to your answer to the previous question.

We can use the difference quotient to find the *exact* rate of change of a function at a single point.

ACTIVITY 3 | **Calculate Rate of Change**

Use the difference quotient to calculate the *exact* rate of change of $f(x) = 2x^2 - 3x$ at the point $x = 2$.

Step 1 Substitute $x = 2$ into the difference quotient, as shown.

$$m = \frac{f(2 + h) - f(2)}{h}$$

Step 2 Expand the difference quotient by evaluating for $f(2 + h)$ and $f(2)$.

$$m = \frac{[2(2 + h)^2 - 3(2 + h)] - [2(2)^2 - 3(2)]}{h}$$

Step 3 Simplify the expression. At some point, you will need to factor h from the numerator and then reduce.

Step 4 Find the *exact* rate of change of $f(x)$ at $x = 2$ by substituting $h = 0$ into your expression.

Analyze the Results

7. Compare the *exact* rate of change found in Step 4 to the previous rates of change that you found.

8. What happens to the secant line for $f(x)$ at $x = 2$ when $h = 0$?

9. Explain the process for calculating the exact rate of change of a function at a point using the difference quotient.

Model and Apply

10. In this problem, you will approximate the rate of change for, and calculate the *exact* rate for, $f(x) = x^2 + 1$ at $x = 1$.

a. Approximate the rate of change of $f(x)$ at $x = 1$ by calculating the average rates of change of the three secant lines through $f(3)$, $f(2)$, and $f(1.5)$. Graph $f(x)$ and the three secant lines on the same coordinate plane.

b. Approximate the rate of change of $f(x)$ at $x = 1$ by using the difference quotient and three different values for h. Let $h = 0.4, 0.25$, and 0.1.

c. Calculate the exact rate of change of $f(x)$ at $x = 1$ by first evaluating the difference quotient for $f(1 + h)$ and $f(1)$ and then substituting $h = 0$. Follow the steps in Activity 3.

CHAPTER 2

Power, Polynomial, and Rational Functions

Then

In Chapter 1, you analyzed functions and their graphs and determined whether inverse functions existed.

Now

In Chapter 2, you will:

- Model real-world data with polynomial functions.
- Use the Remainder and Factor Theorems.
- Find real and complex zeros of polynomial functions.
- Analyze and graph rational functions.
- Solve polynomial and rational inequalities.

Why?

BUSINESS Polynomial functions are often used throughout the business world. Some of the uses of functions are to analyze costs, predict sales, calculate profit, forecast future costs and revenue, estimate depreciation, and determine the proper labor force.

PREREAD Scan the lessons of Chapter 2, and use what you already know about functions to make a prediction of the purpose of this chapter.

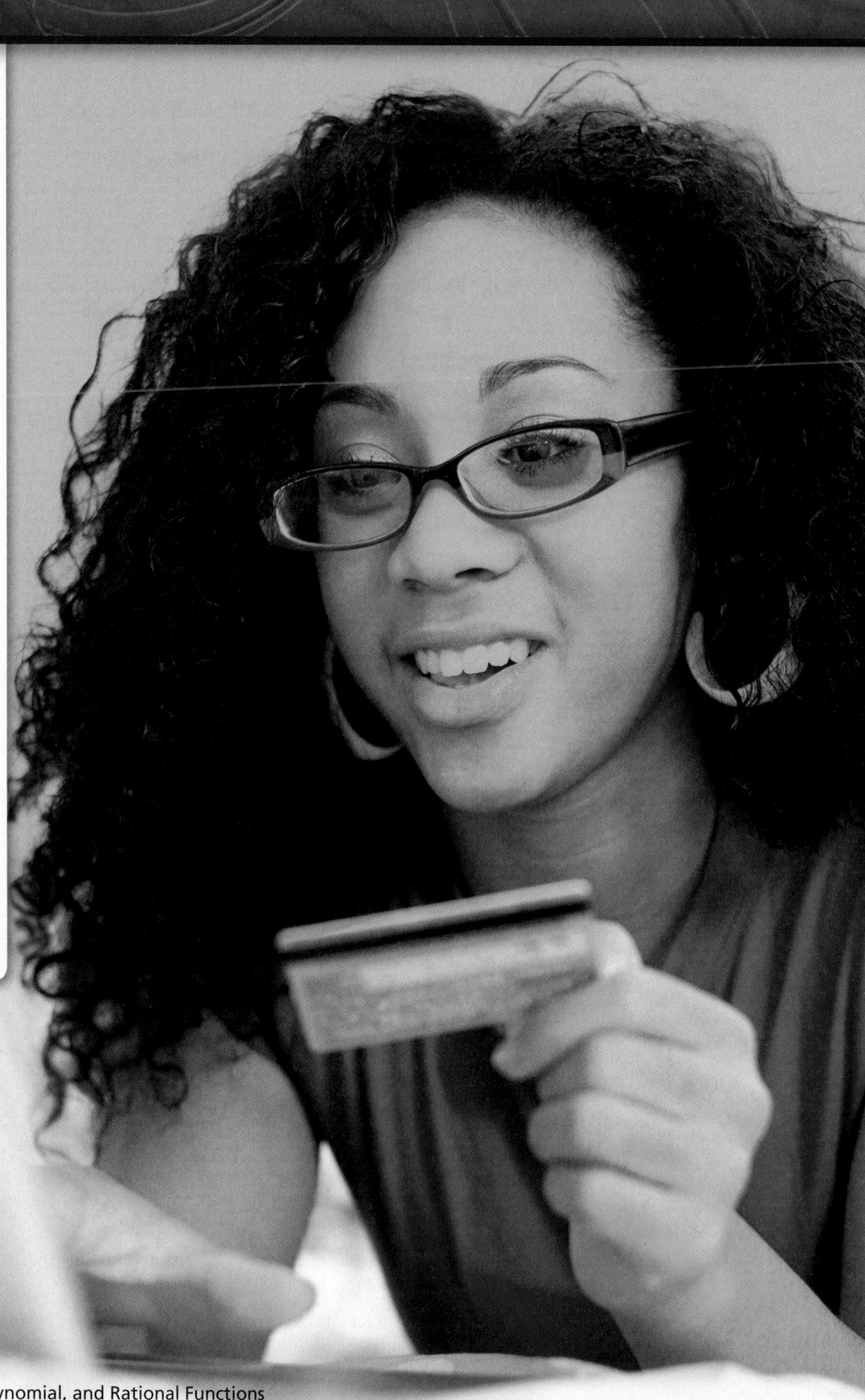

Get Ready for Chapter 2

Diagnose Readiness You have two options for checking Prerequisite Skills.

Text Option

Take the Quick Check below.

*Quick*Check

Factor each polynomial. (Lesson 0-3)

1. $x^2 + x - 20$
2. $x^2 + 5x - 24$
3. $2x^2 - 17x + 21$
4. $3x^2 - 5x - 12$
5. $12x^2 + 13x - 35$
6. $8x^2 - 42x + 27$

7. GEOMETRY The area of a square can be represented by $16x^2 + 56x + 49$. Determine the expression that represents the width of the square.

Use a table to graph each function. (Lesson 0-3)

8. $f(x) = \frac{1}{2}x$
9. $f(x) = -2$
10. $f(x) = x^2 + 3$
11. $f(x) = -x^2 + x - 6$
12. $f(x) = 2x^2 - 5x - 3$
13. $f(x) = 3x^2 - x - 2$

14. TELEVISIONS An electronics magazine estimates that the total number of plasma televisions sold worldwide can be represented by $f(x) = 2t + 0.5t^2$, where t is the number of days after their release date. Graph this function for $0 \le t \le 40$.

Write each set of numbers in set-builder and interval notation, if possible. (Lesson 1-1)

15. $x \le 6$
16. $\{-2, -1, 0, \ldots\}$
17. $-2 < x < 9$
18. $1 < x \le 4$
19. $x < -4$ or $x > 5$
20. $x < -1$ or $x \ge 7$

21. MUSIC At a music store, all of the compact discs are between $9.99 and $19.99. Describe the prices in set-builder and interval notation.

Online Option

Math Online Take a self-check Chapter Readiness Quiz at glencoe.com.

New Vocabulary

English	•Page•	Español
power function	•86•	función potencia
monomial function	•86•	función monomio
radical function	•89•	función radical
extraneous solutions	•91•	solución extrana
polynomial function	•97•	función polinomial
leading coefficient	•97•	coeficiente líder
leading-term test	•98•	conducción de prueba de término
quartic function	•99•	función quartic
quadratic form	•100•	forma de ecuación cuadrática
repeated zero	•101•	cero repetido
lower bound	•121•	más abajo ligado
upper bound	•121•	superior ligado
rational function	•130•	función racional
asymptotes	•130•	asíntota
vertical asymptote	•131•	asíntota vertical
horizontal asymptote	•131•	asíntota horizontal
polynomial inequality	•141•	desigualdad de polinomio
sign chart	•141•	carta de signo
rational inequality	•143•	desigualdad racional

Review Vocabulary

complex conjugates • p. P7 • conjugados complejos   a pair of complex numbers in the form $a + bi$ and $a - bi$

reciprocal functions • p. 45 • funciones recíprocas   functions of the form $f(x) = \frac{a}{x}$

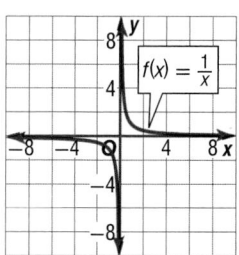

Power and Radical Functions

Then
You analyzed parent functions and their families of graphs. (Lesson 1-5)

Now
- Graph and analyze power functions.
- Graph and analyze radical functions, and solve radical equations.

New Vocabulary
power function
monomial function
radical function
extraneous solution

> **Math Online**

glencoe.com

Why?

Suspension bridges are used to span long distances by hanging, or suspending, the main deck using steel cables. The amount of weight that a steel cable can support is a function of the cable's diameter and can be modeled by a power function.

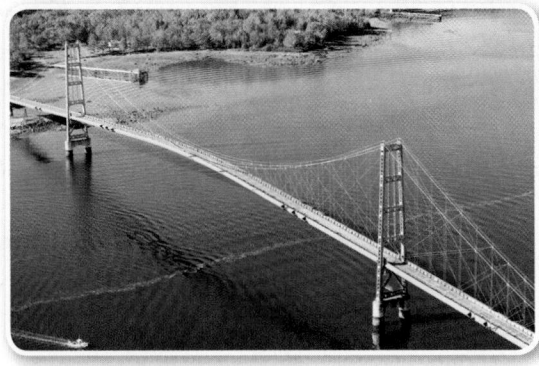

Power Functions In Lesson 1-5, you studied several parent functions that can be classified as power functions. A **power function** is any function of the form $f(x) = ax^n$, where a and n are nonzero constant real numbers.

A power function is also a type of monomial function. A **monomial function** is any function that can be written as $f(x) = a$ or $f(x) = ax^n$, where a and n are nonzero constant real numbers.

Key Concept — Monomial Functions

Let f be the power function $f(x) = ax^n$, where n is a positive integer.

n Even, a Positive

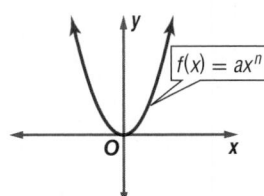

Domain: $(-\infty, \infty)$ **Range:** $[0, \infty)$
x- and y-Intercept: 0
Continuity: continuous for $x \in \mathbb{R}$
Symmetry: y-axis   **Minimum:** $(0, 0)$
Decreasing: $(-\infty, 0)$ **Increasing:** $(0, \infty)$
End behavior: $\lim\limits_{x \to -\infty} f(x) = \infty$ and
$\lim\limits_{x \to \infty} f(x) = \infty$

n Even, a Negative

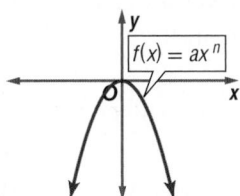

Domain: $(-\infty, \infty)$ **Range:** $(-\infty, 0]$
x- and y-Intercept: 0
Continuity: continuous for $x \in \mathbb{R}$
Symmetry: y-axis   **Maximum:** $(0, 0)$
Decreasing: $(0, \infty)$ **Increasing:** $(-\infty, 0)$
End behavior: $\lim\limits_{x \to -\infty} f(x) = -\infty$ and
$\lim\limits_{x \to \infty} f(x) = -\infty$

n Odd, a Positive

Domain and Range: $(-\infty, \infty)$
x- and y-Intercept: 0
Continuity: continuous on $(-\infty, \infty)$
Symmetry: origin
Extrema: none   **Increasing:** $(-\infty, \infty)$
End Behavior: $\lim\limits_{x \to -\infty} f(x) = -\infty$ and
$\lim\limits_{x \to \infty} f(x) = \infty$

n Odd, a Negative

Domain and Range: $(-\infty, \infty)$
x- and y-Intercept: 0
Continuity: continuous for $x \in \mathbb{R}$
Symmetry: origin
Extrema: none   **Decreasing:** $(-\infty, \infty)$
End Behavior: $\lim\limits_{x \to -\infty} f(x) = \infty$ and
$\lim\limits_{x \to \infty} f(x) = -\infty$

Monomial functions with an even degree are also *even* in the sense that $f(-x) = f(x)$. Likewise, monomial functions with an odd degree are also *odd*, or $f(-x) = -f(x)$.

EXAMPLE 1 | **Analyze Monomial Functions**

Graph and analyze each function. Describe the domain, range, intercepts, end behavior, continuity, and where the function is increasing or decreasing.

a. $f(x) = \frac{1}{2}x^4$

Evaluate the function for several x-values in its domain. Then use a smooth curve to connect each of these points to complete the graph.

x	−3	−2	−1	0	1	2	3
f(x)	40.5	8	0.5	0	0.5	8	40.5

Domain: $(-\infty, \infty)$ Range: $[0, \infty)$

Intercept: 0

End behavior: $\lim\limits_{x \to -\infty} f(x) = \infty$ and $\lim\limits_{x \to \infty} f(x) = \infty$

Continuity: continuous on $(-\infty, \infty)$

Decreasing: $(-\infty, 0)$ Increasing: $(0, \infty)$

b. $f(x) = -x^7$

x	−3	−2	−1	0	1	2	3
f(x)	2187	128	1	0	−1	−128	−2187

Domain: $(-\infty, \infty)$ Range: $(-\infty, \infty)$

Intercept: 0

End behavior: $\lim\limits_{x \to -\infty} f(x) = \infty$ and $\lim\limits_{x \to \infty} f(x) = -\infty$

Continuity: continuous on $(-\infty, \infty)$

Decreasing: $(-\infty, \infty)$

✓ Guided Practice

1A. $f(x) = 3x^6$ **1B.** $f(x) = -\frac{2}{3}x^5$

Recall that $f(x) = \frac{1}{x}$ or x^{-1} is undefined at $x = 0$. Similarly, $f(x) = x^{-2}$ and $f(x) = x^{-3}$ are undefined at $x = 0$. Because power functions can be undefined when $n < 0$, the graphs of these functions will contain discontinuities.

EXAMPLE 2 | **Functions with Negative Exponents**

Graph and analyze each function. Describe the domain, range, intercepts, end behavior, continuity, and where the function is increasing or decreasing.

a. $f(x) = 3x^{-2}$

x	−3	−2	−1	0	1	2	3
f(x)	$0.\overline{3}$	0.75	3	undefined	3	0.75	$0.\overline{3}$

Domain: $(-\infty, 0) \cup (0, \infty)$ Range: $(0, \infty)$

Intercepts: none

End behavior: $\lim\limits_{x \to -\infty} f(x) = 0$ and $\lim\limits_{x \to \infty} f(x) = 0$

Continuity: infinite discontinuity at $x = 0$

Increasing: $(-\infty, 0)$ Decreasing: $(0, \infty)$

Review Vocabulary

Degree of a Monomial The sum of the exponents of the variables of a monomial.

Review Vocabulary

Reciprocal Functions Reciprocal functions have the form $f(x) = \frac{a}{x}$. (Lesson 1-5)

b. $f(x) = -\frac{3}{4}x^{-5}$

x	−3	−2	−1	0	1	2	3
f(x)	0.0031	0.0234	0.75	undefined	−0.75	−0.0234	−0.0031

Domain: $(-\infty, 0) \cup (0, \infty)$ Range: $(-\infty, 0) \cup (0, \infty)$

Intercepts: none

End behavior: $\lim\limits_{x \to -\infty} f(x) = 0$ and $\lim\limits_{x \to \infty} f(x) = 0$

Continuity: infinite discontinuity at $x = 0$

Increasing: $(-\infty, 0)$ and $(0, \infty)$

✔ Guided Practice

2A. $f(x) = -\frac{1}{2}x^{-4}$

2B. $f(x) = 4x^{-3}$

Recall that $x^{\frac{1}{n}}$ indicates the nth root of x, and $x^{\frac{p}{n}}$, where $\frac{p}{n}$ is in simplest form, indicates the nth root of x^p. If n is an even integer, then the domain must be restricted to nonnegative values.

Review Vocabulary

Rational Exponents exponents written as fractions in simplest form. (Lesson 0-4)

EXAMPLE 3   Rational Exponents

Graph and analyze each function. Describe the domain, range, intercepts, end behavior, continuity, and where the function is increasing or decreasing.

a. $f(x) = x^{\frac{5}{2}}$

x	0	1	2	3	4	5	6
f(x)	0	1	5.657	15.588	32	55.902	88.182

Domain: $[0, \infty)$ Range: $[0, \infty)$

x- and y-Intercepts: 0

End behavior: $\lim\limits_{x \to \infty} f(x) = \infty$

Continuity: continuous on $[0, \infty)$

Increasing: $(0, \infty)$

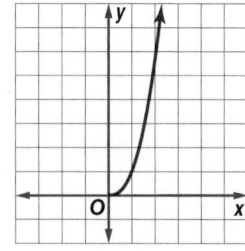

b. $f(x) = 6x^{-\frac{2}{3}}$

x	−3	−2	−1	0	1	2	3
f(x)	2.884	3.780	6	undefined	6	3.780	2.884

Domain: $(-\infty, 0) \cup (0, \infty)$ Range: $(0, \infty)$

Intercepts: none

End behavior: $\lim\limits_{x \to -\infty} f(x) = 0$ and $\lim\limits_{x \to \infty} f(x) = 0$

Continuity: infinite discontinuity at $x = 0$

Increasing: $(-\infty, 0)$ Decreasing: $(0, \infty)$

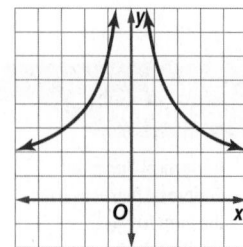

✔ Guided Practice

3A. $f(x) = 2x^{\frac{3}{4}}$

3B. $f(x) = 10x^{\frac{5}{3}}$

StudyTip

Regression Model A polynomial function with rounded coefficients will produce estimates different from values calculated using the unrounded regression equation. From this point forward, you can assume that when asked to use a model to estimate a value, you are to use the unrounded regression equation.

EXAMPLE 4 Power Regression

BIOLOGICAL SCIENCE The following data represents the resting metabolic rate R in kilocalories per day for the mass m in kilograms of several selected animals.

m	0.3	0.4	0.7	0.8	0.85	2.4	2.6	5.5	6.4	6
R	28	35	54	66	46	135	143	331	293	292
m	7	7.9	8.41	8.5	13	29.3	29.8	39.5	83.6	
R	265	327	346	363	520	956	839	1036	1948	

Source: *American Journal of Physical Anthropology*

a. Create a scatter plot of the data.

The scatter plot appears to resemble the square root function, which is a power function. Therefore, test a power regression model.

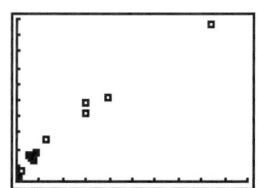

[0, 100] scl: 10 by [0, 2000] scl: 200

b. Write a polynomial function to model the data set. Round each coefficient to the nearest thousandth, and state the correlation coefficient.

Using the **PwrReg** tool on a graphing calculator and rounding each coefficient to the nearest thousandth yields $f(x) = 69.582x^{0.759}$. The correlation coefficient r for the data, 0.995, suggests that a power regression may accurately reflect the data.

We can graph the complete (unrounded) regression by sending it to the Y= menu. In the Y= menu, pick up this regression equation by entering VARS, **Statistics**, **EQ**. Graph this function and the scatter plot in the same viewing window. The function appears to fit the data reasonably well.

[0, 100] scl: 10 by [0, 2000] scl: 200

c. Use the equation to predict the resting metabolic rate for a 60-kilogram animal.

Use the **CALC** feature on the calculator to find $f(60)$. The value of $f(60)$ is about 1554, so the resting metabolic rate for a 60-kilogram animal is about 1554 kilocalories.

✓ Guided Practice

4. CARS The table shows the braking distance in feet at several speeds in miles per hour for a specific car on a dry, well-paved roadway.

Speed	10	20	30	40	50	60	70
Distance	4.2	16.7	37.6	66.9	104.5	150.5	204.9

A. Create a scatter plot of the data.

B. Determine a power function to model the data.

C. Predict the braking distance of a car going 80 miles per hour.

Radical Functions An expression with rational exponents can be written in radical form.

Exponential Form		Radical Form
$x^{\frac{p}{n}}$	=	$\sqrt[n]{x^p}$

Power functions with rational exponents represent the most basic of radical functions. A **radical function** is a function that can be written as $f(x) = \sqrt[n]{x^p}$, where n and p are positive integers greater than 1 that have no common factors. Some examples of radical functions are shown below.

$$f(x) = 3\sqrt{5x^3} \qquad f(x) = -5\sqrt[3]{x^4 + 3x^2 - 1} \qquad f(x) = \sqrt[4]{x + 12} + \frac{1}{2}x - 7$$

It is important to understand the characteristics of the graphs of radical functions as well.

Key Concept | **Radical Functions**

Let f be the radical function $f(x) = \sqrt[n]{x}$ where n is a positive integer.

n Even

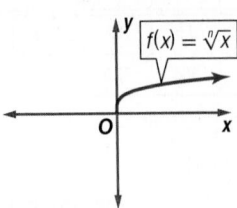

Domain and Range: $[0, \infty)$
x- and y-Intercept: 0
Continuity: continuous on $[0, \infty)$
Symmetry: none   **Increasing:** $(0, \infty)$
Extrema: absolute minimum at $(0, 0)$
End Behavior: $\lim\limits_{x \to \infty} f(x) = \infty$

n Odd

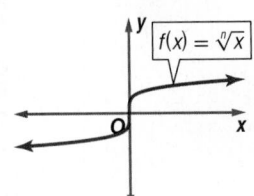

Domain and Range: $(-\infty, \infty)$
x- and y-Intercept: 0
Continuity: continuous on $(-\infty, \infty)$
Symmetry: origin   **Increasing:** $(-\infty, \infty)$
Extrema: none
End Behavior: $\lim\limits_{x \to -\infty} f(x) = -\infty$ and
$\lim\limits_{x \to \infty} f(x) = \infty$

EXAMPLE 5 | **Graph Radical Functions**

Graph and analyze each function. Describe the domain, range, intercepts, end behavior, continuity, and where the function is increasing or decreasing.

a. $f(x) = 2\sqrt[4]{5x^3}$

x	0	1	2	3	4	5
$f(x)$	0	2.99	5.03	6.82	8.46	10

Domain and Range: $[0, \infty)$
x- and y-Intercepts: 0
End behavior: $\lim\limits_{x \to \infty} f(x) = \infty$

Continuity: continuous on $[0, \infty)$
Increasing: $(0, \infty)$

b. $f(x) = \frac{1}{4}\sqrt[5]{6x - 8}$

x	-3	-2	-1	0	1	2	3
$f(x)$	-0.48	-0.46	-0.42	-0.38	-0.29	0.33	0.40

Domain and Range: $(-\infty, \infty)$

x-Intercept: $\frac{4}{3}$ y-Intercept: about -0.38

End behavior: $\lim\limits_{x \to -\infty} f(x) = -\infty$ and $\lim\limits_{x \to \infty} f(x) = \infty$

Continuity: continuous on $(-\infty, \infty)$
Increasing: $(-\infty, \infty)$

Watch Out!

Radical Functions
Remember that when n is even, the domain and range will have restrictions.

✓ **Guided Practice**

5A. $f(x) = -\sqrt[3]{12x^2 - 5}$

5B. $f(x) = \frac{1}{2}\sqrt[4]{2x^3 - 16}$

Like radical functions, a radical equation is any equation in which a variable is in the radicand. To solve a radical equation, first isolate the radical expression. Then raise each side of the equation to a power equal to the index of the radical to eliminate the radical.

Raising each side of an equation to a power sometimes produces **extraneous solutions**, or solutions that do not satisfy the original equation. It is important to check for extraneous solutions.

EXAMPLE 6   Solve Radical Equations

Solve each equation.

a. $2x = \sqrt{100 - 12x} - 2$

$2x = \sqrt{100 - 12x} - 2$	Original equation
$2x + 2 = \sqrt{100 - 12x}$	Isolate the radical.
$4x^2 + 8x + 4 = 100 - 12x$	Square each side to eliminate the radical.
$4x^2 + 20x - 96 = 0$	Subtract $100 - 12x$ from each side.
$4(x^2 + 5x - 24) = 0$	Factor.
$4(x + 8)(x - 3) = 0$	Factor.
$x + 8 = 0$ or $x - 3 = 0$	Zero Product Property
$x = -8$ $x = 3$	Solve.

CHECK $x = -8$

$2x = \sqrt{100 - 12x} - 2$

$-16 \stackrel{?}{=} \sqrt{100 - 12(-8)} - 2$

$-16 \stackrel{?}{=} \sqrt{196} - 2$

$-16 \neq 12$ ✗

CHECK $x = 3$

$2x = \sqrt{100 - 12x} - 2$

$6 \stackrel{?}{=} \sqrt{100 - 12(3)} - 2$

$6 \stackrel{?}{=} \sqrt{64} - 2$

$6 = 6$ ✔

One solution checks and the other solution does not. Therefore, the solution is 3.

b. $\sqrt[3]{(x - 5)^2} + 14 = 50$

$\sqrt[3]{(x - 5)^2} + 14 = 50$	Original equation
$\sqrt[3]{(x - 5)^2} = 36$	Isolate the radical.
$(x - 5)^2 = 46{,}656$	Raise each side to the third power. (The index is 3.)
$x - 5 = \pm 216$	Take the square root of each side.
$x = 221 \text{ or } -211$	Add 5 to each side.

A check of the solution in the original equation confirms that the solutions are valid.

c. $\sqrt{x - 2} = 5 - \sqrt{15 - x}$

$\sqrt{x - 2} = 5 - \sqrt{15 - x}$	Original equation
$x - 2 = 25 - 10\sqrt{15 - x} + (15 - x)$	Square each side.
$2x - 42 = -10\sqrt{15 - x}$	Isolate the radical.
$4x^2 - 168x + 1764 = 100(15 - x)$	Square each side.
$4x^2 - 168x + 1764 = 1500 - 100x$	Distributive Property
$4x^2 - 68x + 264 = 0$	Combine like terms.
$4(x^2 - 17x + 66) = 0$	Factor.
$4(x - 6)(x - 11) = 0$	Factor.
$x - 6 = 0$ or $x - 11 = 0$	Zero Product Property
$x = 6$ $x = 11$	Solve.

A check of the solutions in the original equation confirms that both solutions are valid.

✔ Guided Practice

6A. $3x = 3 + \sqrt{18x - 18}$ **6B.** $\sqrt[3]{4x + 8} + 3 = 7$ **6C.** $\sqrt{x + 7} = 3 + \sqrt{2 - x}$

● = **Step-by-Step Solutions** begin on page R29.

Exercises

Graph and analyze each function. Describe the domain, range, intercepts, end behavior, continuity, and where the function is increasing or decreasing. (Examples 1 and 2)

1. $f(x) = 5x^2$

2. $g(x) = 8x^5$

3. $h(x) = -x^3$

4. $f(x) = -4x^4$

5. $g(x) = \frac{1}{3}x^9$

6. $f(x) = \frac{5}{8}x^8$

7. $f(x) = -\frac{1}{2}x^7$

8. $g(x) = -\frac{1}{4}x^6$

9. $f(x) = 2x^{-4}$

10. $h(x) = -3x^{-7}$

11. $f(x) = -8x^{-5}$

12. $g(x) = 7x^{-2}$

13. $f(x) = -\frac{2}{5}x^{-9}$

14. $h(x) = \frac{1}{6}x^{-6}$

15. $h(x) = \frac{3}{4}x^{-3}$

16. $f(x) = -\frac{7}{10}x^{-8}$

17. GEOMETRY The volume of a sphere is given by $V(r) = \frac{4}{3}\pi r^3$, where r is the radius. (Example 1)

    a. State the domain and range of the function.

    b. Graph the function.

Graph and analyze each function. Describe the domain, range, intercepts, end behavior, continuity, and where the function is increasing or decreasing. (Example 3)

18. $f(x) = 8x^{\frac{1}{4}}$

19. $f(x) = -6x^{\frac{1}{5}}$

20. $g(x) = -\frac{1}{5}x^{-\frac{1}{3}}$

21. $f(x) = 10x^{-\frac{1}{6}}$

22. $g(x) = -3x^{\frac{5}{8}}$

23. $h(x) = \frac{3}{4}x^{\frac{3}{5}}$

24. $f(x) = -\frac{1}{2}x^{-\frac{3}{4}}$

25. $f(x) = x^{-\frac{2}{3}}$

26. $h(x) = 7x^{\frac{5}{3}}$

27. $h(x) = -4x^{\frac{7}{4}}$

28. $h(x) = -5x^{-\frac{3}{2}}$

29. $h(x) = \frac{2}{3}x^{-\frac{8}{5}}$

Complete each step.

    a. Create a scatter plot of the data.

    b. Determine a power function to model the data.

    c. Calculate the value of each model at $x = 30$. (Example 4)

30.

x	y
1	4
2	22
3	85
4	190
5	370
6	650
7	1000
8	1500

31.

x	y
1	1
2	32
3	360
4	2000
5	7800
6	25,000
7	60,000
8	130,000

32. CLIFF DIVING In the sport of cliff diving, competitors perform three dives from a height of 28 meters. Judges award divers a score from 0 to 10 points based on degree of difficulty, take-off, positions, and water entrance. The table shows the speed of a diver at various distances in the dive. (Example 4)

Distance (m)	Speed (m/s)
4	8.85
8	12.52
12	15.34
16	17.71
20	19.80
24	21.69
28	23.43

    a. Create a scatter plot of the data.

    b. Determine a power function to model the data.

    c. Use the function to predict the speed at which a diver would enter the water from a cliff dive of 30 meters.

33. WEATHER The wind chill temperature is the apparent temperature felt on exposed skin, taking into account the effect of the wind. The table shows the wind chill temperature produced at winds of various speeds when the actual temperature is 50°F. (Example 4)

Wind Speed (mph)	Wind Chill (°F)
5	48.22
10	46.04
15	44.64
20	43.60
25	42.76
30	42.04
35	41.43
40	40.88

    a. Create a scatter plot of the data.

    b. Determine a power function to model the data.

    c. Use the function to predict the wind chill temperature when the wind speed is 65 miles per hour.

Graph and analyze each function. Describe the domain, range, intercepts, end behavior, continuity, and where the function is increasing or decreasing. (Example 5)

34. $f(x) = 3\sqrt{6 + 3x}$

35. $g(x) = -2\sqrt[5]{1024 + 8x}$

36. $f(x) = -\frac{3}{8}\sqrt[6]{16x + 48} - 3$

37 $h(x) = 4 + \sqrt{7x - 12}$

38. $g(x) = \sqrt{(1 - 4x)^3} - 16$

39. $f(x) = -\sqrt[3]{(25x - 7)^2} - 49$

40. $h(x) = \frac{1}{2}\sqrt[3]{27 - 2x} - 8$

41. $g(x) = \sqrt{22 - x} - \sqrt{3x - 3}$

42. FLUID MECHANICS The velocity of the water flowing through a hose with a nozzle can be modeled using $V(P) = 12.1\sqrt{P}$, where V is the velocity in feet per second and P is the pressure in pounds per square inch. (Example 5)

    a. Graph the velocity through a nozzle as a function of pressure.

    b. Describe the domain, range, end behavior, and continuity of the function and determine where it is increasing or decreasing.

43. AGRICULTURAL SCIENCE The net energy NE_m required to maintain the body weight of beef cattle, in megacalories (Mcal) per day, is estimated by the formula $NE_m = 0.077 \sqrt[4]{m^3}$, where m is the animal's mass in kilograms. One megacalorie is equal to one million calories. (Example 6)

   a. Find the net energy per day required to maintain a 400-kilogram steer.

   b. If 0.96 megacalorie of energy is provided per pound of whole grain corn, how much corn does a 400-kilogram steer need to consume daily to maintain its body weight?

Solve each equation. (Example 6)

44. $4 = \sqrt{-6 - 2x} + \sqrt{31 - 3x}$ **45.** $0.5x = \sqrt{4 - 3x} + 2$

46. $-3 = \sqrt{22 - x} - \sqrt{3x - 3}$ **47.** $\sqrt{(2x - 5)^3} - 10 = 17$

48. $\sqrt[4]{(4x + 164)^3} + 36 = 100$ **49.** $x = \sqrt{2x - 4} + 2$

50. $7 + \sqrt{(-36 - 5x)^5} = 250$ **51.** $x = 5 + \sqrt{x + 1}$

52. $\sqrt{6x - 11} + 4 = \sqrt{12x + 1}$ **53.** $\sqrt{4x - 40} = -20$

54. $\sqrt{x + 2} - 1 = \sqrt{-2 - 2x}$ **55.** $7 + \sqrt[5]{1054 - 3x} = 11$

Determine whether each function is a monomial function given that a and b are positive integers. Explain your reasoning.

56. $y = \dfrac{5}{b}x^{4a}$ **57.** $G(x) = -2ax^4$

58. $F(b) = 3ab^{5x}$ **59.** $y = \dfrac{7}{3}t^{ab}$

60. $H(t) = \dfrac{1}{ab}t^{\frac{4b}{2}}$ **61.** $y = 4abx^{-2}$

62. CHEMISTRY The function $r = R_0(A)^{\frac{1}{3}}$ can be used to approximate the nuclear radius of an element based on its molecular mass, where r is length of the radius in meters, R_0 is a constant (about 1.2×10^{-15} meter), and A is the molecular mass.

Element	Molecular Mass
Carbon (C)	12.0
Helium (H)	4.0
Iodine (I)	126.9
Lead (Pb)	207.2
Sodium (Na)	?
Sulfur (S)	32.1

   a. If the nuclear radius of sodium is about 3.412×10^{-15} meter, what is its molecular mass?

   b. The approximate nuclear radius of an element is 6.030×10^{-15} meter. Identify the element.

   c. The ratio of the molecular masses of two elements is 27:8. What is the ratio of their nuclear radii?

Solve each inequality.

63. $\sqrt[5]{1040 + 8x} \geq 4$ **64.** $\sqrt[13]{41 - 7x} \geq -1$

65. $(1 - 4x)^{\frac{3}{2}} \geq 125$ **66.** $\sqrt{6 + 3x} \leq 9$

67. $(19 - 4x)^{\frac{5}{3}} - 12 \leq -13$ **68.** $(2x - 68)^{\frac{2}{3}} \geq 64$

69 CHEMISTRY Boyle's Law states that, at constant temperature, the pressure of a gas is inversely proportional to its volume. The results of an experiment to explore Boyle's Law are shown.

Volume (liters)	Pressure (atmospheres)
1.0	3.65
1.5	2.41
2.0	1.79
2.5	1.46
3.0	1.21
3.5	1.02
4.0	0.92

   a. Create a scatter plot of the data.

   b. Determine a power function to model the pressure P as a function of volume v.

   c. Based on the information provided in the problem statement, does the function you determined in part **b** make sense? Explain.

   d. Use the model to predict the pressure of the gas if the volume is 3.25 liters.

   e. Use the model to predict the pressure of the gas if the volume is 6 liters.

Without using a calculator, match each graph with the appropriate function.

70.   **71.**

72.   **73.**

   a. $f(x) = \dfrac{1}{2}\sqrt[4]{3x^5}$ **b.** $g(x) = \dfrac{2}{3}x^6$

   c. $h(x) = 4x^{-3}$ **d.** $p(x) = 5\sqrt[3]{2x + 1}$

74. ELECTRICITY The voltage used by an electrical device such as a DVD player can be calculated using $V = \sqrt{PR}$, where V is the voltage in volts, P is the power in watts, and R is the resistance in ohms. The function $I = \sqrt{\dfrac{P}{R}}$ can be used to calculate the current, where I is the current in amps.

   a. If a lamp uses 120 volts and has a resistance of 11 ohms, what is the power consumption of the lamp?

   b. If a DVD player has a current of 10 amps and consumes 1200 watts of power, what is the resistance of the DVD player?

   c. Ohm's Law expresses voltage in terms of current and resistance. Use the equations given above to write Ohm's Law using voltage, resistance, and amperage.

Use the points provided to determine the power function represented by the graph.

75

76.

77.

78.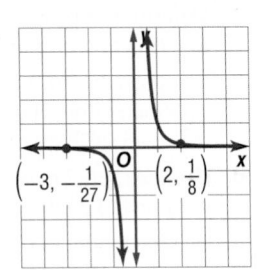

79. OPTICS A contact lens with the appropriate depth ensures proper fit and oxygen permeation. The depth of a lens can be calculated using the formula $S = r - \sqrt{r^2 - \left(\dfrac{d}{2}\right)^2}$, where S is the depth, r is the radius of curvature, and d is the diameter, with all units in millimeters.

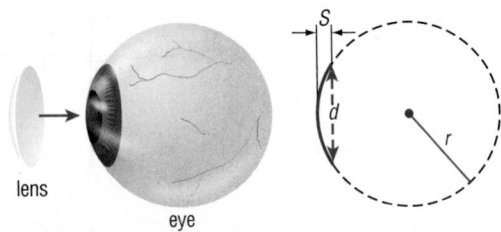

   a. If the depth of the contact lens is 1.15 millimeters and the radius of curvature is 7.50 millimeters, what is the diameter of the contact lens?

   b. If the depth of the contact lens is increased by 0.1 millimeter and the diameter of the lens is 8.2 millimeters, what radius of curvature would be required?

   c. If the radius of curvature remains constant, does the depth of the contact lens increase or decrease as the diameter increases?

80. MULTIPLE REPRESENTATIONS In this problem, you will investigate the average rates of change of power functions.

   a. GRAPHICAL For power functions of the form $f(x) = x^n$, graph a function for two values of n such that $0 < n < 1$, $n = 1$, and two values of n such that $n > 1$.

   b. TABULAR Copy and complete the table, using your graphs from part **a** to analyze the average rates of change of the functions as x approaches infinity. Describe this rate as *increasing*, *constant*, or *decreasing*.

n	$f(x)$	Average Rate of Change as $x \longrightarrow \infty$
$0 < n < 1$		
$n = 1$		
$n > 1$		

   c. VERBAL Make a conjecture about the average rate of change of a power function as x approaches infinity for the intervals $0 < n < 1$, $n = 1$, and $n > 1$.

H.O.T. Problems   Use **H**igher-**O**rder **T**hinking Skills

81. CHALLENGE Show that $\sqrt{\dfrac{8^n \cdot 2^7}{4^{-n}}} = 2^{2n+3}\sqrt{2^{n+1}}$.

82. REASONING Consider $y = 2^x$.

   a. Describe the value of y if $x < 0$.

   b. Describe the value of y if $0 < x < 1$.

   c. Describe the value of y if $x > 1$.

   d. Write a conjecture about the relationship between the value of the base and the value of the power if the exponent is greater than or less than 1. Justify your answer.

83. PREWRITE Your senior project is to tutor an underclassman for four sessions on power and radical functions. Make a plan for writing that addresses purpose and audience, and has a controlling idea, logical sequence, and time frame for completion.

84. REASONING Given $f(x) = x^{\frac{a}{b}}$, where a and b are integers with no common factors, determine whether each statement is *true* or *false*. Explain.

   a. If the value of b is even and the value of a is odd, then the function is undefined for $x < 0$.

   b. If the value of a is even and the value of b is odd, then the function is undefined for $x < 0$.

   c. If the value of a is 1, then the function is defined for all x.

85. REASONING Consider $f(x) = x^{\frac{1}{n}} + 5$. How would you expect the graph of the function to change as n increases if n is odd and greater than or equal to 3?

86. WRITING IN MATH Use words, graphs, tables, and equations to show the relationship between functions in exponential form and in radical form.

87. FINANCE If you deposit $1000 in a savings account with an interest rate of r compounded annually, then the balance in the account after 3 years is given by $B(r) = 1000(1 + r)^3$, where r is written as a decimal. (Lesson 1-7)

    a. Find a formula for the interest rate r required to achieve a balance of B in the account after 3 years.

    b. What interest rate will yield a balance of $1100 after 3 years?

Find $(f + g)(x)$, $(f - g)(x)$, $(f \cdot g)(x)$, and $\left(\dfrac{f}{g}\right)(x)$ for each $f(x)$ and $g(x)$. State the domain of each new function. (Lesson 1-6)

88. $f(x) = x^2 - 2x$

    $g(x) = x + 9$

89. $f(x) = \dfrac{x}{x + 1}$

    $g(x) = x^2 - 1$

90. $f(x) = \dfrac{3}{x - 7}$

    $g(x) = x^2 + 5x$

Use the graph of $f(x)$ to graph $g(x) = |f(x)|$ and $h(x) = f(|x|)$. (Lesson 1-5)

91. $f(x) = -4x + 2$

92. $f(x) = \sqrt{x + 3} - 6$

93. $f(x) = x^2 - 3x - 10$

Use the graph of each function to estimate intervals to the nearest 0.5 unit on which the function is increasing, decreasing, or constant. Support the answer numerically. (Lesson 1-4)

94.

$f(x) = 0.5(x + 4)(x + 1)(x - 2)$

95.

$f(x) = \dfrac{x - 3}{x + 4}$

96.

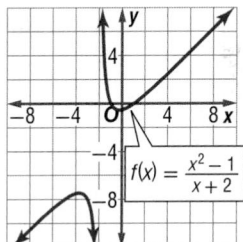

$f(x) = \dfrac{x^2 - 1}{x + 2}$

Simplify. (Lesson 0-2)

97. $\dfrac{\frac{1}{2} + \sqrt{3}i}{1 - \sqrt{2}i}$

98. $\dfrac{2 - \sqrt{2}i}{3 + \sqrt{6}i}$

99. $\dfrac{(1 + i)^2}{(-3 + 2i)^2}$

100. SAT/ACT If m and n are both positive, then which of the following is equivalent to $\dfrac{2m\sqrt{18n}}{m\sqrt{2}}$?

    A $3m\sqrt{n}$ **D** $6\sqrt{n}$

    B $6m\sqrt{n}$ **E** $8\sqrt{n}$

    C $4\sqrt{n}$

101. REVIEW If $f(x, y) = x^2 y^3$ and $f(a, b) = 10$, what is the value of $f(2a, 2b)$?

    F 50   **J** 320

    G 100   **K** 640

    H 160

102. REVIEW The number of minutes m it takes c children to eat p pieces of pizza varies directly as the number of pieces of pizza and inversely as the number of children. If it takes 5 children 30 minutes to eat 10 pieces of pizza, how many minutes should it take 15 children to eat 50 pieces of pizza?

    A 30   **C** 50

    B 40   **D** 60

103. If $\sqrt[3]{5m + 2} = 3$, then $m = $?

    F 3   **H** 5

    G 4   **J** 6

Objective
Graph and analyze the behavior of polynomial functions.

In Lesson 1-3, you analyzed the end behavior of functions by making a table of values and graphing them. For a *polynomial function*, the behavior of the graph can be determined by analyzing specific terms of the function.

ACTIVITY   **Graph Polynomial Functions**

Sketch each graph, and identify the end behavior of the function.

a. $f(x) = x^3 + 6x^2 - 4x + 2$

Use a table of values to sketch the graph.

x	−10	−5	−2	0	2	5	10
f(x)	−358	47	26	2	26	257	1562

In the graph of $f(x)$, it appears that $\lim\limits_{x \to -\infty} f(x) = -\infty$ and $\lim\limits_{x \to \infty} f(x) = \infty$.

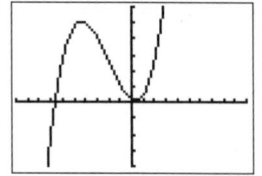

[−10, 10] scl: 1 by [−40, 60] scl: 10

StudyTip

Table of Values Be sure to use enough points to get the overall shape of the graph.

b. $g(x) = -2x^3 + 6x^2 - 4x + 2$

x	−8	−5	−2	0	2	5	8
g(x)	1442	422	50	2	2	−118	−670

In the graph of $g(x)$, it appears that $\lim\limits_{x \to -\infty} g(x) = \infty$ and $\lim\limits_{x \to \infty} g(x) = -\infty$.

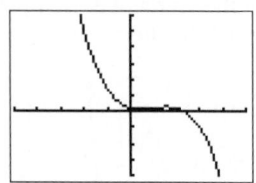

[−5, 5] scl: 1 by [−40, 60] scl: 10

c. $h(x) = -x^4 + x^3 + 6x^2 - 4x + 2$

x	−8	−5	−2	0	2	5	8
h(x)	−4190	−578	10	2	10	−368	−3230

In the graph of $h(x)$, it appears that $\lim\limits_{x \to -\infty} h(x) = -\infty$ and $\lim\limits_{x \to \infty} h(x) = -\infty$.

[−5, 5] scl: 1 by [−20, 20] scl: 4

Analyze the Results

1. Look at the terms of each function above. What differences do you see?

2. How is the end behavior of the graphs of each function affected by these differences?

3. Develop a pattern for every possible type of end behavior of a polynomial function.

4. Give an example of a polynomial function with a graph that approaches positive infinity when x approaches both negative infinity and positive infinity.

Exercises

Describe the end behavior of each function without making a table of values or graphing.

5. $f(x) = -2x^3 + 4x$ **6.** $f(x) = 5x^4 + 3$ **7.** $f(x) = -x^5 + 2x - 4$

8. $g(x) = 6x^6 - 2x^2 + 10x$ **9.** $g(x) = 3x - 4x^4$ **10.** $h(x) = 6x^2 - 3x^3 - 2x^6$

Polynomial Functions

Then
You analyzed graphs of functions. (Lesson 1-2)

Now
- Graph polynomial functions.
- Model real-world data with polynomial functions.

New Vocabulary
polynomial function
polynomial function of degree n
leading coefficient
leading-term test
quartic function
turning point
quadratic form
repeated zero
multiplicity

Math Online >
glencoe.com

Why?
The scatter plot shows total personal savings as a percent of disposable income in the United States. Often data with multiple relative extrema are best modeled by a polynomial function.

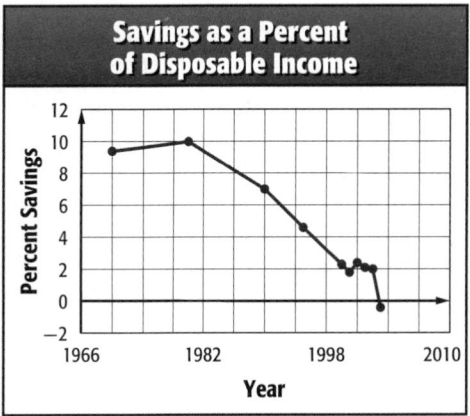

Savings as a Percent of Disposable Income

Graph Polynomial Functions In Lesson 2-1, you learned about the basic characteristics of monomial functions. Monomial functions are the most basic polynomial functions. The sums and differences of monomial functions form other types of **polynomial functions**.

Let n be a nonnegative integer and let $a_0, a_1, a_2, \ldots, a_{n-1}, a_n$ be real numbers with $a_n \neq 0$. Then the function given by

$$f(x) = a_n x^n + a_{n-1} x^{n-1} + \cdots + a_2 x^2 + a_1 x + a_0$$

is called a **polynomial function of degree n**. The **leading coefficient** of a polynomial function is the coefficient of the variable with the greatest exponent. The leading coefficient of $f(x)$ is a_n.

You are already familiar with the following polynomial functions.

Constant Functions

Degree: 0

Linear Functions

Degree: 1

Quadratic Functions

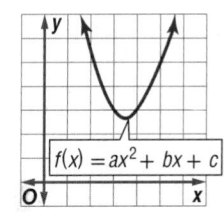

Degree: 2

The zero function is a constant function with no degree. The graphs of polynomial functions share certain characteristics.

Graphs of Polynomial Functions	
Example	**Nonexamples**
Polynomial functions are defined and continuous for all real numbers and have smooth, rounded turns.	Graphs of polynomial functions do not have breaks, holes, gaps, or sharp corners.

Recall that the graphs of even-degree, non-constant monomial functions resemble the graph of $f(x) = x^2$, while the graphs of odd-degree monomial functions resemble the graph of $f(x) = x^3$. You can use the basic shapes and characteristics of even- and odd-degree monomial functions and what you learned in Lesson 1-5 about transformations to transform graphs of monomial functions.

EXAMPLE 1 **Graph Transformations of Monomial Functions**

Graph each function.

a. $f(x) = (x - 2)^5$

This is an odd-degree function, so its graph is similar to the graph of $y = x^3$. The graph of $f(x) = (x - 2)^5$ is the graph of $y = x^5$ translated 2 units to the right.

b. $g(x) = -x^4 + 1$

This is an even-degree function, so its graph is similar to the graph of $y = x^2$. The graph of $g(x) = -x^4 + 1$ is the graph of $y = x^4$ reflected in the x-axis and translated 1 unit up.

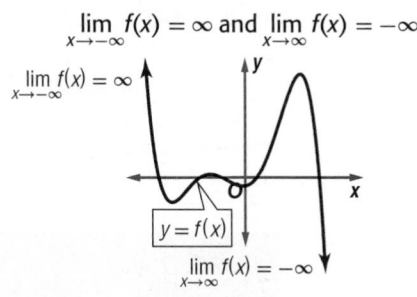

☑ **Guided Practice**

1A. $f(x) = 4 - x^3$

1B. $g(x) = (x + 7)^4$

In Lesson 1-3, you learned that the end behavior of a function describes how the function behaves, rising or falling, at either end of its graph. As $x \to -\infty$ and $x \to \infty$, the end behavior of any polynomial function is determined by its leading term. The **leading term test** uses the power and coefficient of this term to determine polynomial end behavior.

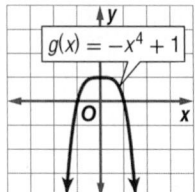

Key Concept **Leading Term Test for Polynomial End Behavior**

The end behavior of any non-constant polynomial function $f(x) = a_n x^n + \cdots + a_1 x + a_0$ can be described in one of the following four ways, as determined by the degree n of the polynomial and its leading coefficient a_n.

n odd, a_n positive

$\lim\limits_{x \to -\infty} f(x) = -\infty$ and $\lim\limits_{x \to \infty} f(x) = \infty$

n odd, a_n negative

$\lim\limits_{x \to -\infty} f(x) = \infty$ and $\lim\limits_{x \to \infty} f(x) = -\infty$

n even, a_n positive

$\lim\limits_{x \to -\infty} f(x) = \infty$ and $\lim\limits_{x \to \infty} f(x) = \infty$

n even, a_n negative

$\lim\limits_{x \to -\infty} f(x) = -\infty$ and $\lim\limits_{x \to \infty} f(x) = -\infty$

EXAMPLE 2   Apply the Leading Term Test

Describe the end behavior of the graph of each polynomial function using limits. Explain your reasoning using the leading term test.

a. $f(x) = 3x^4 - 5x^2 - 1$

The degree is 4, and the leading coefficient is 3. Because the degree is even and the leading coefficient is positive, $\lim\limits_{x \to -\infty} f(x) = \infty$ and $\lim\limits_{x \to \infty} f(x) = \infty$.

b. $g(x) = -3x^2 - 2x^7 + 4x^4$

Write in standard form as $g(x) = -2x^7 + 4x^4 - 3x^2$. The degree is 7, and the leading coefficient is -2. Because the degree is odd and the leading coefficient is negative, $\lim\limits_{x \to -\infty} f(x) = \infty$ and $\lim\limits_{x \to \infty} f(x) = -\infty$.

c. $h(x) = x^3 - 2x^2$

The degree is 3, and the leading coefficient is 1. Because the degree is odd and the leading coefficient is positive, $\lim\limits_{x \to -\infty} f(x) = -\infty$ and $\lim\limits_{x \to \infty} f(x) = \infty$.

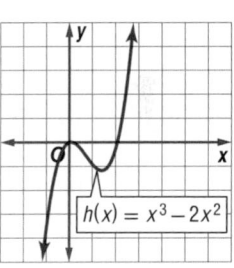

✓ **Guided Practice**

2A. $g(x) = 4x^5 - 8x^3 + 20$ **2B.** $h(x) = -2x^6 + 11x^4 + 2x^2$

Consider the shapes of a few typical third-degree polynomial or cubic functions and fourth-degree polynomial or **quartic functions** shown.

Typical Cubic Functions   **Typical Quartic Functions**

Observe the number of x-intercepts for each graph. Because an x-intercept corresponds to a real zero of the function, you can see that cubic functions have at most 3 zeros and quartic functions have at most 4 zeros.

Turning points indicate where the graph of a function changes from increasing to decreasing, and vice versa. Maxima and minima are also located at turning points. Notice that cubic functions have at most 2 turning points, and quartic functions have at most 3 turning points. These observations can be generalized as follows and shown to be true for any polynomial function.

Key Concept   **Zeros and Turning Points of Polynomial Functions**

A polynomial function f of degree $n \geq 1$ has at most n distinct real zeros and at most $n - 1$ turning points.

Example   Let $f(x) = 3x^6 - 10x^4 - 15x^2$. Then f has at most 6 distinct real zeros and at most 5 turning points. The graph of f suggests that the function has 3 real zeros and 3 turning points.

$f(x) = 3x^6 - 10x^4 - 15x^2$

Recall that if f is a polynomial function and c is an x-intercept of the graph of f, then it is equivalent to say that:

- c is a zero of f,
- $x = c$ is a solution of the equation $f(x) = 0$, and
- $(x - c)$ is a factor of the polynomial $f(x)$.

You can find the zeros of some polynomial functions using the same factoring techniques you used to solve quadratic equations.

EXAMPLE 3   **Zeros of a Polynomial Function**

State the number of possible real zeros and turning points of $f(x) = x^3 - 5x^2 + 6x$. Then determine all of the real zeros by factoring.

The degree of the function is 3, so f has at most 3 distinct real zeros and at most $3 - 1$ or 2 turning points. To find the real zeros, solve the related equation $f(x) = 0$ by factoring.

$x^3 - 5x^2 + 6x = 0$	Set $f(x)$ equal to 0.
$x(x^2 - 5x + 6) = 0$	Factor the greatest common factor, x.
$x(x - 2)(x - 3) = 0$	Factor completely.

So, f has three distinct real zeros, 0, 2, and 3. This is consistent with a cubic function having at most 3 distinct real zeros.

CHECK   You can use a graphing calculator to graph $f(x) = x^3 - 5x^2 + 6x$ and confirm these zeros. Additionally, you can see that the graph has 2 turning points, which is consistent with cubic functions having at most 2 turning points.

[−5, 5] scl: 1 by [−5, 5] scl: 1

✓ **Guided Practice**

State the number of possible real zeros and turning points of each function. Then determine all of the real zeros by factoring.

3A. $f(x) = x^3 - 6x^2 - 27x$ **3B.** $f(x) = x^4 - 8x^2 + 15$

In some cases, a polynomial function can be factored using quadratic techniques if it has *quadratic form*.

Key Concept   **Quadratic Form**

Words   A polynomial expression in x is in **quadratic form** if it is written as $au^2 + bu + c$ for any numbers a, b, and c, $a \neq 0$, where u is some expression in x.

Example $x^4 - 5x^2 - 14$ is in quadratic form because the expression can be written as $(x^2)^2 - 5(x^2) - 14$. If $u = x^2$, then the expression becomes $u^2 - 5u - 14$.

EXAMPLE 4 Zeros of a Polynomial Function in Quadratic Form

State the number of possible real zeros and turning points for $g(x) = x^4 - 3x^2 - 4$. Then determine all of the real zeros by factoring.

The degree of the function is 4, so g has at most 4 distinct real zeros and at most $4 - 1$ or 3 turning points. This function is in quadratic form because $x^4 - 3x^2 - 4 = (x^2)^2 - 3(x^2) - 4$. Let $u = x^2$.

$$(x^2)^2 - 3(x^2) - 4 = 0 \qquad \text{Set } g(x) \text{ equal to 0.}$$

$$u^2 - 3u - 4 = 0 \qquad \text{Substitute } u \text{ for } x^2.$$

$$(u + 1)(u - 4) = 0 \qquad \text{Factor the quadratic expression.}$$

$$(x^2 + 1)(x^2 - 4) = 0 \qquad \text{Substitute } x^2 \text{ for } u.$$

$$(x^2 + 1)(x + 2)(x - 2) = 0 \qquad \text{Factor completely.}$$

$x^2 + 1 = 0$ or	$x + 2 = 0$ or	$x - 2 = 0$ **Zero Product Property**
$x = \pm\sqrt{-1}$	$x = -2$	$x = 2$ **Solve for x.**

Because $\pm\sqrt{-1}$ are not real zeros, g has two distinct real zeros, -2 and 2. This is consistent with a quartic function. The graph of $g(x) = x^4 - 3x^2 - 4$ in Figure 2.2.1 confirms this. Notice that there are 3 turning points, which is also consistent with a quartic function.

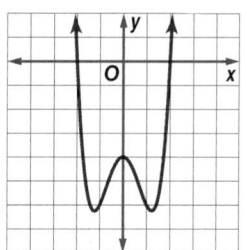

Figure 2.2.1

✔ Guided Practice

State the number of possible real zeros and turning points of each function. Then determine all of the real zeros by factoring.

4A. $g(x) = x^4 - 9x^2 + 18$

4B. $h(x) = x^5 - 6x^3 - 16x$

If a factor $(x - c)$ occurs more than once in the completely factored form of $f(x)$, then its related zero c is called a **repeated zero**. When the zero occurs an even number of times, the graph will be tangent to the x-axis at that point. When the zero occurs an odd number of times, the graph will cross the x-axis at that point. A graph is tangent to an axis when it touches the axis at that point, but does not cross it.

EXAMPLE 5 Polynomial Function with Repeated Zeros

State the number of possible real zeros and turning points of $h(x) = -x^4 - x^3 + 2x^2$. Then determine all of the real zeros by factoring.

The degree of the function is 4, so h has at most 4 distinct real zeros and at most $4 - 1$ or 3 turning points. Find the real zeros.

$$-x^4 - x^3 + 2x^2 = 0 \qquad \text{Set } h(x) \text{ equal to 0.}$$

$$-x^2(x^2 + x - 2) = 0 \qquad \text{Factor the greatest common factor, } -x^2.$$

$$-x^2(x - 1)(x + 2) = 0 \qquad \text{Factor completely.}$$

The expression above has 4 factors, but solving for x yields only 3 distinct real zeros, 0, 1, and -2. Of the zeros, 0 occurs twice.

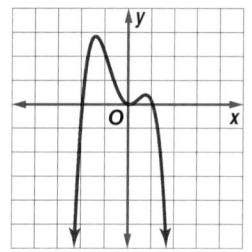

Figure 2.2.2

The graph of $h(x) = -x^4 - x^3 + 2x^2$ shown in Figure 2.2.2 confirms these zeros and shows that h has three turning points. Notice that at $x = 1$ and $x = -2$, the graph crosses the x-axis, but at $x = 0$, the graph is tangent to the x-axis.

✔ Guided Practice

State the number of possible real zeros and turning points of each function. Then determine all of the real zeros by factoring.

5A. $g(x) = -2x^3 - 4x^2 + 16x$

5B. $f(x) = 3x^5 - 18x^4 + 27x^3$

In $h(x) = -x^2(x - 1)(x + 2)$ from Example 5, the zero $x = 0$ occurs 2 times. In $k(x) = (x - 1)^3(x + 2)^4$, the zero $x = 1$ occurs 3 times, while $x = -2$ occurs 4 times. Notice that in the graph of k shown, the curve crosses the x-axis at $x = 1$ but not at $x = -2$. These observations can be generalized as follows and shown to be true for all polynomial functions.

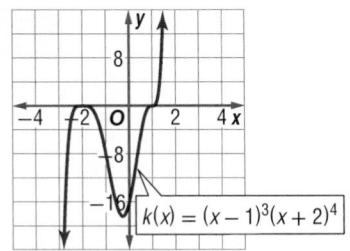

$$k(x) = (x - 1)^3(x + 2)^4$$

Key Concept   **Repeated Zeros of Polynomial Functions**

If $(x - c)^m$ is the highest power of $(x - c)$ that is a factor of polynomial function f, then c is a zero of **multiplicity** m of f, where m is a natural number.

- If a zero c has odd multiplicity, then the graph of f crosses the x-axis at $x = c$ and the value of $f(x)$ changes signs at $x = c$.
- If a zero c has even multiplicity, then the graph of f is tangent to the x-axis at $x = c$ and the value of $f(x)$ does not change signs at $x = c$.

You now have several tests and tools to aid you in graphing polynomial functions.

EXAMPLE 6   **Graph a Polynomial Function**

For $f(x) = x(2x + 3)(x - 1)^2$, (a) apply the leading-term test, (b) determine the zeros and state the multiplicity of any repeated zeros, (c) find a few additional points, and then (d) graph the function.

a. The product $x(2x + 3)(x - 1)^2$ has a leading term of $x(2x)(x)^2$ or $2x^4$, so f has degree 4 and leading coefficient 2. Because the degree is even and the leading coefficient is positive, $\lim\limits_{x \to -\infty} f(x) = \infty$ and $\lim\limits_{x \to \infty} f(x) = \infty$.

b. The distinct real zeros are 0, −1.5, and 1. The zero at 1 has multiplicity 2.

c. Choose x-values that fall in the intervals determined by the zeros of the function.

Interval	x-value in Interval	$f(x)$	$(x, f(x))$
$(-\infty, -1.5)$	-2	$f(-2) = 18$	$(-2, 18)$
$(-1.5, 0)$	-1	$f(-1) = -4$	$(-1, -4)$
$(0, 1)$	0.5	$f(0.5) = 0.5$	$(0.5, 0.5)$
$(1, \infty)$	1.5	$f(1.5) = 2.25$	$(1.5, 2.25)$

d. Plot the points you found (Figure 2.2.3). The end behavior of the function tells you that the graph eventually rises to the left and to the right. You also know that the graph crosses the x-axis at nonrepeated zeros −1.5 and 0, but does not cross the x-axis at repeated zero 1, because its multiplicity is even. Draw a continuous curve through the points as shown in Figure 2.2.4.

Figure 2.2.3

Figure 2.2.4

✔ **Guided Practice**

For each function, (a) apply the leading-term test, (b) determine the zeros and state the multiplicity of any repeated zeros, (c) find a few additional points, and then (d) graph the function.

6A. $f(x) = -2x(x - 4)(3x - 1)^3$ **6B.** $h(x) = -x^3 + 2x^2 + 8x$

Model Data You can use a graphing calculator to model data that exhibit linear, quadratic, cubic, and quartic behavior by first examining the number of turning points suggested by a scatter plot of the data.

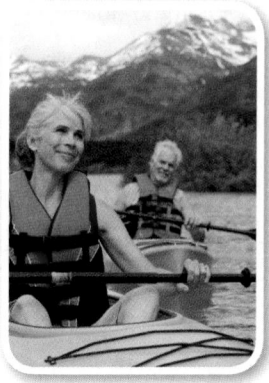

Real-World Link

A college graduate planning to retire at 65 needs to save an average of $10,000 per year toward retirement.

Source: Monroe Bank

Figure 2.2.5

🌐 Real-World EXAMPLE 7 | Model Data Using Polynomial Functions

SAVINGS Refer to the beginning of the lesson. The average personal savings as a percent of disposable income in the United States is given in the table.

Year	1970	1980	1990	1995	2000	2001	2002	2003	2004	2005
% Savings	9.4	10.0	7.0	4.6	2.3	1.8	2.4	2.1	2.0	−0.4

Source: U.S. Department of Commerce

a. Create a scatter plot of the data and determine the type of polynomial function that could be used to represent the data.

Enter the data using the list feature of a graphing calculator. Let L1 be the number of years since 1970. Then create a scatter plot of the data. The curve of the scatter plot resembles the graph of a quadratic equation, so we will use a quadratic regression.

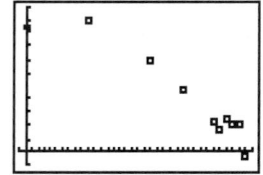

[−1, 36] scl: 1 by [−1, 11] scl: 1

b. Write a polynomial function to model the data set. Round each coefficient to the nearest thousandth, and state the correlation coefficient.

Using the **QuadReg** tool on a graphing calculator and rounding each coefficient to the nearest thousandth yields $f(x) = -0.009x^2 + 0.033x + 9.744$. The correlation coefficient r^2 for the data is 0.96, which is close to 1, so the model is a good fit.

We can graph the complete (unrounded) regression by sending it to the Y= menu. If you enter L_1, L_2, and Y_1 after **QuadReg**, as shown in Figure 2.2.5, the regression equation will be entered into Y_1. Graph this function and the scatter plot in the same viewing window. The function appears to fit the data reasonably well.

[−1, 36] scl: 1 by [−1, 11] scl: 1

c. Use the model to estimate the percent savings in 1993.

Because 1993 is 23 years after 1970, use the **CALC** feature on a calculator to find $f(23)$. The value of $f(23)$ is 5.94, so the percent savings in 1993 was about 5.94%.

d. Use the model to determine the approximate year in which the percent savings reached 6.5%.

Graph the line $y = 6.5$ for Y_2. Then use **5: intersect** on the **CALC** menu to find the point of intersection of $y = 6.5$ with $f(x)$. The intersection occurs when $x \approx 21$, so the approximate year in which the percent savings reached 6.5% was about 1970 + 21 or 1991.

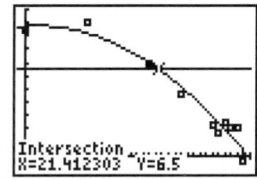

[−1, 36] scl: 1 by [−1, 11] scl: 1

✓ Guided Practice

7. POPULATION The median age of the U.S. population by year predicted through 2080 is shown.

Year	1900	1930	1960	1990	2020	2050	2080
Median Age	22.9	26.5	29.5	33.0	40.2	42.7	43.9

Source: U.S. Census Bureau

a. Write a polynomial function to model the data. Let L1 be the number of years since 1900.

b. Estimate the median age of the population in 2005.

c. According to your model, in what year did the median age of the population reach 30?

Exercises

Graph each function. (Example 1)

1. $f(x) = (x + 5)^2$

2. $f(x) = (x - 6)^3$

3. $f(x) = x^4 - 6$

4. $f(x) = x^5 + 7$

5. $f(x) = (2x)^4$

6. $f(x) = (2x)^5 - 16$

7. $f(x) = (x - 3)^4 + 6$

8. $f(x) = (x + 4)^3 - 3$

9. $f(x) = \frac{1}{3}(x - 9)^5$

10. $f(x) = \left(\frac{1}{2}x\right)^3 + 8$

11. WATER If it takes exactly one minute to drain a 10-gallon tank of water, the volume of water remaining in the tank can be approximated by $v(t) = 10(1 - t)^2$, where t is time in minutes, $0 \le t \le 1$. Graph the function. (Example 1)

Describe the end behavior of the graph of each polynomial function using limits. Explain your reasoning using the leading term test. (Example 2)

12. $f(x) = -5x^7 + 6x^4 + 8$

13. $f(x) = 2x^6 + 4x^5 + 9x^2$

14. $g(x) = 5x^4 + 7x^5 - 9$

15. $g(x) = -7x^3 + 8x^4 - 6x^6$

16. $h(x) = 8x^2 + 5 - 4x^3$

17. $h(x) = 4x^2 + 5x^3 - 2x^5$

18. $f(x) = x(x + 1)(x - 3)$

19. $g(x) = x^2(x + 4)(-2x + 1)$

20. $f(x) = -x(x - 4)(x + 5)$

21. $g(x) = x^3(x + 1)(x^2 - 4)$

22. ORGANIC FOOD The number of acres in the United States used for organic apple production from 2000 to 2005 can be modeled by $a(x) = 43.77x^4 - 498.76x^3 + 1310.2x^2 + 1626.2x + 6821.5$, where $x = 0$ is 2000. (Example 2)

    a. Graph the function using a graphing calculator.

    b. Describe the end behavior of the graph of the function using limits. Explain using the leading term test.

State the number of possible real zeros and turning points of each function. Then determine all of the real zeros by factoring. (Examples 3–5)

23 $f(x) = x^5 + 3x^4 + 2x^3$

24. $f(x) = x^6 - 8x^5 + 12x^4$

25. $f(x) = x^4 + 4x^2 - 21$

26. $f(x) = x^4 - 4x^3 - 32x^2$

27. $f(x) = x^6 - 6x^3 - 16$

28. $f(x) = 4x^8 + 16x^4 + 12$

29. $f(x) = 9x^6 - 36x^4$

30. $f(x) = 6x^5 - 150x^3$

31. $f(x) = 4x^4 - 4x^3 - 3x^2$

32. $f(x) = 3x^5 + 11x^4 - 20x^3$

For each function, (a) apply the leading-term test, (b) determine the zeros and state the multiplicity of any repeated zeros, (c) find a few additional points, and then (d) graph the function. (Example 6)

33. $f(x) = x(x + 4)(x - 1)^2$

34. $f(x) = x^2(x - 4)(x + 2)$

35. $f(x) = -x(x + 3)^2(x - 5)$

36. $f(x) = 2x(x + 5)^2(x - 3)$

37. $f(x) = -x(x - 3)(x + 2)^3$

38. $f(x) = -(x + 2)^2(x - 4)^2$

39. $f(x) = 3x^3 - 3x^2 - 36x$

40. $f(x) = -2x^3 - 4x^2 + 6x$

41. $f(x) = x^4 + x^3 - 20x^2$

42. $f(x) = x^5 + 3x^4 - 10x^3$

43. RESERVOIRS The number of feet below the maximum water level in Wisconsin's Rainbow Reservoir during ten months in 2007 is shown. (Example 7)

Month	Level	Month	Level
January	4	July	9
February	5.5	August	11
March	10	September	16.5
April	9	November	11.5
May	7.5	December	8.5

Source: Wisconsin Valley Improvement Company

    a. Write a model that best relates the water level as a function of the number of months since January.

    b. Use the model to estimate the water level in the reservoir in October.

Use a graphing calculator to write a polynomial function to model each set of data. (Example 7)

44.

x	−3	−2	−1	0	1	2	3
f(x)	8.75	7.5	6.25	5	3.75	2.5	1.25

45.

x	5	7	8	10	11	12	15	16
f(x)	2	5	6	4	−1	−3	5	9

46.

x	−2.53	−2	−1.5	−1	−0.5	0	0.5	1	1.5
f(x)	23	11	7	6	6	5	3	2	4

47.

x	30	35	40	45	50	55	60	65	70	75
f(x)	52	41	32	44	61	88	72	59	66	93

48. ELECTRICITY The average retail electricity prices in the U.S. from 1970 to 2005 are shown. Projected prices for 2010 and 2020 are also shown. (Example 7)

Year	Price (¢ / kWh)	Year	Price (¢ / kWh)
1970	6.125	1995	7.5
1974	7	2000	6.625
1980	7.25	2005	6.25
1982	9.625	2010	6.25
1990	8	2020	6.375

Source: Energy Information Administration

    a. Write a model that relates the price as a function of the number of years since 1970.

    b. Use the model to predict the average price of electricity in 2015.

    c. According to the model, during which year was the price 7¢ for the second time?

49. COMPUTERS The numbers of laptops sold each quarter from 2005 to 2007 are shown. Let the first quarter of 2005 be 1, and the fourth quarter of 2007 be 12.

Quarters	Sale (Thousands)
1	423
2	462
3	495
4	634
5	587
6	498
7	798
8	986
9	969
10	891
11	1130
12	1347

a. Predict the end behavior of a graph of the data as x approaches infinity.

b. Use a graphing calculator to graph and model the data. Is the model a good fit? Explain your reasoning.

c. Describe the end behavior of the graph using limits. Was your prediction accurate? Explain your reasoning.

Determine whether each graph could show a polynomial function. Write *yes* or *no*. If not, explain why not.

50.

51.

52.

53.
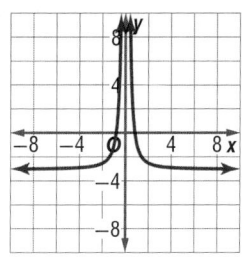

Find a polynomial function of degree n with only the following real zeros. More than one answer is possible.

54. $-1; n = 3$

55. $3; n = 3$

56. $6, -3; n = 4$

57. $-5, 4; n = 4$

58. $7; n = 4$

59. $0, -4; n = 5$

60. $2, 1, 4; n = 5$

61. $0, 3, -2; n = 5$

62. no real zeros; $n = 4$

63. no real zeros; $n = 6$

Determine whether the degree n of the polynomial for each graph is *even* or *odd* and whether its leading coefficient a_n is *positive* or *negative*.

64.

65.

66.

67.
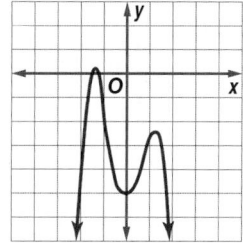

68. MANUFACTURING A company manufactures aluminum containers for energy drinks.

a. Write an equation V that represents the volume of the container.

b. Write a function A in terms of r that represents the surface area of a container with a volume of 15 cubic inches.

c. Use a graphing calculator to determine the minimum possible surface area of the can.

Determine a polynomial function that has each set of zeros. More than one answer is possible.

69. $5, -3, 6$

70. $4, -8, -2$

71. $3, 0, 4, -1, 3$

72. $1, 1, -4, 6, 0$

73. $\frac{3}{4}, -3, -4, -\frac{2}{3}$

74. $-1, -1, 5, 0, \frac{5}{6}$

75 POPULATION The percent of the United States population living in metropolitan areas has increased.

Year	Percent of Population
1950	56.1
1960	63
1970	68.6
1980	74.8
1990	74.8
2000	79.2

Source: U.S. Census Bureau

a. Write a model that relates the percent as a function of the number of years since 1950.

b. Use the model to predict the percent of the population that will be living in metropolitan areas in 2015.

c. Use the model to predict the year in which 85% of the population will live in metropolitan areas.

Create a function with the following characteristics. Graph the function.

76. degree = 5, 3 real zeros, $\lim_{x\to\infty} = \infty$

77. degree = 6, 4 real zeros, $\lim_{x\to\infty} = -\infty$

78. degree = 5, 2 distinct real zeros, 1 of which has a multiplicity of 2, $\lim_{x\to\infty} = \infty$

79. degree = 6, 3 distinct real zeros, 1 of which has a multiplicity of 2, $\lim_{x\to\infty} = -\infty$

80. WEATHER The temperatures in degrees Celsius from 10 A.M. to 7 P.M. during one day for a city are shown where x is the number of hours since 10 A.M.

Time	Temp.	Time	Temp.
0	4.1	5	10
1	5.7	6	7
2	7.2	7	4.6
3	7.3	8	2.3
4	9.4	9	−0.4

   a. Graph the data.

   b. Use a graphing calculator to model the data using a polynomial function with a degree of 3.

   c. Repeat part **b** using a function with a degree of 4.

   d. Which function is a better model? Explain.

For each of the following graphs:

   a. Determine the degree and end behavior.

   b. Locate the zeros and their multiplicity. Assume all of the zeros are integral values.

   c. Use the given point to determine a function that fits the graph.

81.

82.

83.

84.

State the number of possible real zeros and turning points of each function. Then find all of the real zeros by factoring.

85. $f(x) = 16x^4 + 72x^2 + 80$ **86.** $f(x) = -12x^3 - 44x^2 - 40x$

87 $f(x) = -24x^4 + 24x^3 - 6x^2$ **88.** $f(x) = x^3 + 6x^2 - 4x - 24$

89. **MULTIPLE REPRESENTATIONS** In this problem, you will investigate the behavior of combinations of polynomial functions.

   a. GRAPHICAL Graph $f(x)$, $g(x)$, and $h(x)$ in each row on the same graphing calculator screen. For each graph, modify the window to observe the behavior both on a large scale and very close to the origin.

$f(x) =$	$g(x) =$	$h(x) =$
$x^2 + x$	x^2	x
$x^3 - x$	x^3	$-x$
$x^3 + x^2$	x^3	x^2

   b. ANALYTICAL Describe the behavior of each graph of $f(x)$ in terms of $g(x)$ or $h(x)$ near the origin.

   c. ANALYTICAL Describe the behavior of each graph of $f(x)$ in terms of $g(x)$ or $h(x)$ as x approaches ∞ and $-\infty$.

   d. VERBAL Predict the behavior of a function that is a combination of two functions a and b such that $f(x) = a + b$, where a is the term of higher degree.

H.O.T. Problems   Use Higher-Order Thinking Skills

90. ERROR ANALYSIS Colleen and Martin are modeling the data shown. Colleen thinks the model should be $f(x) = 5.754x^3 + 2.912x^2 - 7.516x + 0.349$. Martin thinks it should be $f(x) = 3.697x^2 + 11.734x - 2.476$. Is either of them correct? Explain your reasoning.

x	$f(x)$	x	$f(x)$
−2	−19	0.5	−2
−1	5	1	1.5
0	0.4	2	43

91. REASONING Can a polynomial function have both an absolute maximum and an absolute minimum? Explain your reasoning.

92. REASONING Explain why the constant function $f(x) = c$, $c \neq 0$, has degree 0, but the zero function $f(x) = 0$ has no degree.

93. CHALLENGE Use factoring by grouping to determine the zeros of $f(x) = x^3 + 5x^2 - x^2 - 5x - 12x - 60$. Explain each step.

94. REASONING How is it possible for more than one function to be represented by the same degree, end behavior, and distinct real zeros? Provide an example to explain your reasoning.

95. REASONING What is the minimum degree of a polynomial function that has an absolute maximum, a relative maximum, and a relative minimum? Explain your reasoning.

96. WRITING IN MATH Explain how you determine the best polynomial function to use when modeling data.

Solve each equation. (Lesson 2-1)

97. $\sqrt{z+3} = 7$

98. $d + \sqrt{d^2 - 8} = 4$

99. $\sqrt{x-8} = \sqrt{13+x}$

100. REMODELING An installer is replacing the carpet in a 12-foot by 15-foot living room. The new carpet costs $13.99 per square yard. The formula $f(x) = 9x$ converts square yards to square feet. (Lesson 1-7)

  a. Find the inverse $f^{-1}(x)$. What is the significance of $f^{-1}(x)$?

  b. How much will the new carpet cost?

Given $f(x) = 2x^2 - 5x + 3$ and $g(x) = 6x + 4$, find each function. (Lesson 1-6)

101. $(f + g)(x)$

102. $[f \circ g](x)$

103. $[g \circ f](x)$

Describe how the graphs of $f(x) = x^2$ and $g(x)$ are related. Then write an equation for $g(x)$. (Lesson 1-5)

104.

105.

106.
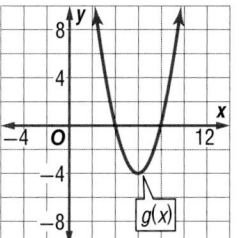

107. BUSINESS A company creates a new product that costs $25 per item to produce. They hire a marketing analyst to help determine a selling price. After collecting and analyzing data relating selling price s to yearly consumer demand d, the analyst estimates demand for the product using $d = -200s + 15{,}000$. (Lesson 1-4)

  a. If yearly profit is the difference between total revenue and production costs, determine a selling price $s \geq 25$, that will maximize the company's yearly profit P. (Hint: $P = sd - 25d$)

  b. What are the risks of determining a selling price using this method?

The scores for an exam given in physics class are given. (Lesson 0-8)

  82, 77, 84, 98, 93, 71, 76, 64, 89, 95, 78, 89, 65, 88, 54,
  96, 87, 92, 80, 85, 93, 89, 55, 62, 79, 90, 86, 75, 99, 62

108. Make a box-and-whisker plot of the test.

109. What is the standard deviation of the test scores?

110. SAT/ACT The figure shows the intersection of three lines. The figure is not drawn to scale.

  $x =$

  A 16
  B 20
  C 30
  D 60
  E 90

111. Over the domain $2 < x \leq 3$, which of the following functions contains the greatest values of y?

  F $y = \dfrac{x+3}{x-2}$
  H $y = x^2 - 3$

  G $y = \dfrac{x-5}{x+1}$
  J $y = 2x$

112. MULTIPLE CHOICE Which of the following equations represents the result of shifting the parent function $y = x^3$ up 4 units and right 5 units?

  A $y + 4 = (x + 5)^3$
  C $y + 4 = (x - 5)^3$

  B $y - 4 = (x + 5)^3$
  D $y - 4 = (x - 5)^3$

113. REVIEW Which of the following describes the numbers in the domain of $h(x) = \dfrac{\sqrt{2x-3}}{x-5}$?

  F $x \neq 5$
  H $x \geq \dfrac{3}{2}, x \neq 5$

  G $x \geq \dfrac{3}{2}$
  J $x \neq \dfrac{3}{2}$

Objective
Use TI-Nspire to explore the hidden behavior of graphs.

Using graphing technologies such as computers and calculators is an efficient way to be able to graph and evaluate functions. It is important, however, to consider the limitations of graphing technology when interpreting graphs.

ACTIVITY   **Hidden Behavior of Graphs**

Determine the zeros of $f(x) = x^3 - x^2 - 60.7x + 204$ graphically.

Step 1   Open a new **Graphs and Geometry** page, and graph the function.

In the default window, it appears that the function has two zeroes, one between -10 and -8 and one between 4 and 6.

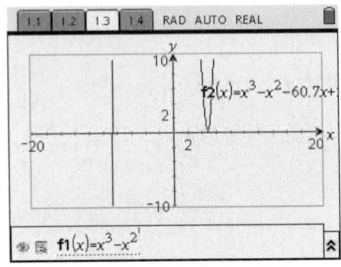

StudyTip

Window Settings You can choose values for the window based on observation of your graph, or you can use one of the zoom tools such as the box zoom that allows you to zoom in on a certain area of a graph.

Step 2   From the **Window** menu, choose **Window Settings**. Change the dimensions of the window as shown.

The behavior of the graph is much clearer in the larger window. It still appears that the function has two zeros, one between -8 and -10 and one between 4 and 6.

Step 3   From the **Window** menu, choose **Window Settings**. Change the window to [2, 8] by [−2, 2].

By enlarging the graph in the area where it appears that the zero occurs, it is clear that there is no zero between the values of 4 and 6. Therefore, the graph only has one zero.

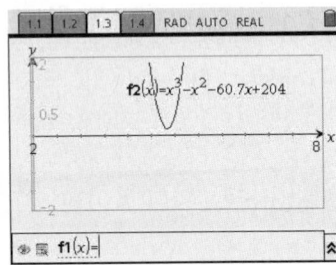

Analyze the Results

1. In addition to the limitation discovered in the previous steps, how can graphing calculators limit your ability to interpret graphs?

2. What are some ways to avoid these limitations?

Exercises

Determine the zeros of each polynomial graphically. Watch for hidden behavior.

3. $x^3 + 6.5x^2 - 46.5x + 60$

4. $x^4 - 3x^3 + 12x^2 + 6x - 7$

5. $x^5 + 7x^3 + 4x^2 - x + 10.9$

6. $x^4 - 19x^3 + 107.2x^2 - 162x + 73$

The Remainder and Factor Theorems

Then
You factored quadratic expressions to solve equations. (Lesson 0–3)

Now
- Divide polynomials using long division and synthetic division.
- Use the Remainder and Factor Theorems.

New Vocabulary
synthetic division
depressed polynomial
synthetic substitution

Math Online
glencoe.com

Why?

The redwood trees of Redwood National Park in California are the oldest living species in the world. The trees can grow up to 350 feet and can live up to 2000 years. Synthetic division can be used to determine the height of one of the trees during a particular year.

Divide Polynomials Consider the polynomial function $f(x) = 6x^3 - 25x^2 + 18x + 9$. If you know that f has a zero at $x = 3$, then you also know that $(x - 3)$ is a factor of $f(x)$. Because $f(x)$ is a third-degree polynomial, you know that there exists a second-degree polynomial $q(x)$ such that

$$f(x) = (x - 3) \cdot q(x).$$

This implies that $q(x)$ can be found by dividing $6x^3 - 25x^2 + 18x + 9$ by $(x - 3)$ because

$$q(x) = \frac{f(x)}{x - 3}, \text{ if } x \neq 3.$$

To divide polynomials, we can use an algorithm similar to that of long division with integers.

EXAMPLE 1 **Use Long Division to Factor Polynomials**

Factor $6x^3 - 25x^2 + 18x + 9$ completely using long division if $(x - 3)$ is a factor.

$$
\begin{array}{r}
6x^2 - 7x - 3 \\
x - 3 \overline{)6x^3 - 25x^2 + 18x + 9} \\
(-)\underline{6x^3 - 18x^2} \\
-7x^2 + 18x \\
(-)\underline{-7x^2 + 21x} \\
-3x + 9 \\
(-)\underline{-3x + 9} \\
0
\end{array}
$$

← Multiply divisor by $6x^2$ because $\frac{6x^3}{x} = 6x^2$.

← Subtract and bring down next term.

← Multiply divisor by $-7x$ because $-\frac{7x^2}{x} = -7x$.

← Subtract and bring down next term.

← Multiply divisor by -3 because $\frac{-3x}{x} = -3$.

← Subtract. Notice that the remainder is 0.

From this division, you can write $6x^3 - 25x^2 + 18x + 9 = (x - 3)(6x^2 - 7x - 3)$.

Factoring the quadratic expression yields $6x^3 - 25x^2 + 18x + 9 = (x - 3)(2x - 3)(3x + 1)$.

So, the zeros of the polynomial function $f(x) = 6x^3 - 25x^2 + 18x + 9$ are 3, $\frac{3}{2}$, and $-\frac{1}{3}$. The x-intercepts of the graph of $f(x)$ shown support this conclusion.

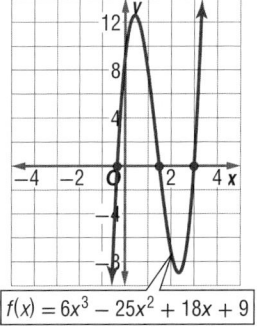

$f(x) = 6x^3 - 25x^2 + 18x + 9$

✔ Guided Practice

Factor each polynomial completely using the given factor and long division.

1A. $x^3 + 7x^2 + 4x - 12;\ x + 6$

1B. $6x^3 - 2x^2 - 16x - 8;\ 2x - 4$

Long division of polynomials can result in a zero remainder, as in Example 1, or a nonzero remainder, as in the example below. Notice that just as with integer long division, the result of polynomial division is expressed using the quotient, remainder, and divisor.

$$
\begin{array}{r}
x + 3 \quad \longleftarrow \text{Quotient} \\
x + 2 \overline{)\, x^2 + 5x - 4} \quad \longleftarrow \text{Dividend} \\
(-)\ \underline{x^2 + 2x} \\
3x - 4 \\
(-)\ \underline{3x + 6} \\
\text{Remainder} \longrightarrow -10
\end{array}
$$

Dividend ⟶   Divisor ⟶

$$\frac{x^2 + 5x - 4}{x + 2} = x + 3 + \frac{-10}{x + 2}, \; x \neq -2$$

Quotient — Remainder — Divisor — Excluded value

Recall that a dividend can be expressed in terms of the divisor, quotient, and remainder.

divisor	·	quotient	+	remainder	=	dividend
$(x + 2)$	·	$(x + 3)$	+	(-10)	=	$x^2 + 5x - 4$

This leads to a definition for polynomial division.

Key Concept   **Polynomial Division**

Let $f(x)$ and $d(x)$ be polynomials such that the degree of $d(x)$ is less than or equal to the degree of $f(x)$ and $d(x) \neq 0$. Then there exist unique polynomials $q(x)$ and $r(x)$ such that

$$\frac{f(x)}{d(x)} = q(x) + \frac{r(x)}{d(x)} \qquad \text{or} \qquad f(x) = d(x) \cdot q(x) + r(x),$$

where $r(x) = 0$ or the degree of $r(x)$ is less than the degree of $d(x)$. If $r(x) = 0$, then $d(x)$ *divides evenly* into $f(x)$.

Before dividing, be sure that each polynomial is written in standard form and that placeholders with zero coefficients are inserted where needed for missing powers of the variable.

EXAMPLE 2   **Long Division with Nonzero Remainder**

Divide $9x^3 - x - 3$ by $3x + 2$.

First rewrite $9x^3 - x - 3$ as $9x^2 + 0x^2 - x - 3$. Then divide.

$$
\begin{array}{r}
3x^2 - 2x + 1 \\
3x + 2 \overline{)\, 9x^3 + 0x^2 - x - 3} \\
(-)\ \underline{9x^3 + 6x^2} \\
-6x^2 - x \\
(-)\ \underline{-6x^2 - 4x} \\
3x - 3 \\
(-)\ \underline{3x + 2} \\
-5
\end{array}
$$

You can write this result as

$$\frac{9x^3 - x - 3}{3x + 2} = 3x^2 - 2x + 1 + \frac{-5}{3x + 2}, \; x \neq -\frac{2}{3}$$

$$= 3x^2 - 2x + 1 - \frac{5}{3x + 2}, \; x \neq -\frac{2}{3}.$$

CHECK Multiply to check this result.

$$(3x + 2)(3x^2 - 2x + 1) + (-5) \stackrel{?}{=} 9x^3 - x - 3$$
$$9x^3 - 6x^2 + 3x + 6x^2 - 4x + 2 - 5 \stackrel{?}{=} 9x^3 - x - 3$$
$$9x^3 - x - 3 = 9x^3 - x - 3 \;\checkmark$$

✓ Guided Practice

Divide using long division.

2A. $(8x^3 - 18x^2 + 21x - 20) \div (2x - 3)$ **2B.** $(-3x^3 + x^2 + 4x - 66) \div (x - 5)$

When dividing polynomials, the divisor can have a degree higher than 1. This can sometimes result in a quotient with missing terms.

StudyTip

Division by Zero In Example 3, this division is not defined for $x^2 - 2x + 7 = 0$. From this point forward in this lesson, you can assume that x cannot take on values for which the indicated division is undefined.

EXAMPLE 3 **Division by Polynomial of Degree 2 or Higher**

Divide $2x^4 - 4x^3 + 13x^2 + 3x - 11$ by $x^2 - 2x + 7$.

$$
\begin{array}{r}
2x^2 -1 \\
x^2 - 2x + 7 \overline{\smash{)}\, 2x^4 - 4x^3 + 13x^2 + 3x - 11} \\
(-)\ \underline{2x^4 - 4x^3 + 14x^2 } \\
-x^2 + 3x - 11 \\
(-)\ \underline{-x^2 + 2x - 7} \\
x - 4
\end{array}
$$

You can write this result as $\dfrac{2x^4 - 4x^3 + 13x^2 + 3x - 11}{x^2 - 2x + 7} = 2x^2 - 1 + \dfrac{x - 4}{x^2 - 2x + 7}$.

☑ **Guided Practice**

Divide using long division.

3A. $(2x^3 + 5x^2 - 7x + 6) \div (x^2 + 3x - 4)$ **3B.** $(6x^5 - x^4 + 12x^2 + 15x) \div (3x^3 - 2x^2 + x)$

Synthetic division is a shortcut for dividing a polynomial by a linear factor of the form $x - c$. Consider the long division from Example 1.

Long Division	**Suppress Variables**	**Collapse Vertically**	**Synthetic Division**
Notice the coefficients highlighted in colored text.	Suppress x and powers of x.	Collapse the long division vertically, eliminating duplications.	Change the signs of the divisor and the numbers on the second line.

Long Division:
$$
\begin{array}{r}
6x^2 - 7x - 3 \\
x - 3 \overline{\smash{)}\, 6x^3 - 25x^2 + 18x + 9} \\
(-)\ \underline{6x^3 - 18x^2} \\
-7x^2 + 18x \\
(-)\ \underline{-7x^2 + 21x} \\
-3x + 9 \\
(-)\ \underline{-3x + 9} \\
0
\end{array}
$$

Suppress Variables:
$$
\begin{array}{r}
6 \quad -7 \quad -3 \\
-3 \overline{\smash{)}\, 6 \quad -25 \quad +18 \quad +9} \\
(-)\ \underline{6 \quad -18} \\
-7 \quad +18 \\
(-)\ \underline{-7 \quad +21} \\
-3 \quad +9 \\
(-)\ \underline{-3 \quad +9} \\
0
\end{array}
$$

Collapse Vertically:
$$
\begin{array}{r|rrrr}
-3 & 6 & -25 & 18 & 9 \\
& & -18 & 21 & 9 \\
\hline
& 6 & -7 & -3 & 0
\end{array}
$$

Synthetic Division:
$$
\begin{array}{r|rrrr}
3 & 6 & -25 & 18 & 9 \\
& & 18 & -21 & -9 \\
\hline
& 6 & -7 & -3 & 0
\end{array}
$$

The number now representing the divisor is the related zero of the binomial $x - c$. Also, by changing the signs on the second line, we are now adding instead of subtracting.

We can use the synthetic division shown in the example above to outline a procedure for synthetic division of any polynomial by a binomial.

Key Concept **Synthetic Division Algorithm**

To divide a polynomial by the factor $x - c$, complete each step.

Step 1 Write the coefficients of the dividend in standard form. Write the related zero c of the divisor $x - c$ in the box. Bring down the first coefficient.

Step 2 Multiply the first coefficient by c. Write the product under the second coefficient.

Step 3 Add the product and the second coefficient.

Step 4 Repeat Steps 2 and 3 until you reach a sum in the last column. The numbers along the bottom row are the coefficients of the quotient. The power of the first term is one less than the degree of the dividend. The final number is the remainder.

Example

Divide $6x^3 - 25x^2 + 18x + 9$ by $x - 3$.

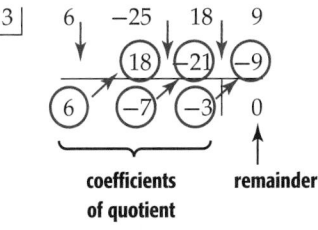

coefficients of quotient   remainder

↓ = Add terms.   ↗ = Multiply by c, and write the product.

As with division of polynomials by long division, remember to use zeros as placeholders for any missing terms in the dividend. When a polynomial is divided by one of its binomial factors $x - c$, the quotient is called a **depressed polynomial**.

EXAMPLE 4   Synthetic Division

Divide using synthetic division.

a. $(2x^4 - 5x^2 + 5x - 2) \div (x + 2)$

Because $x + 2 = x - (-2)$, $c = -2$. Set up the synthetic division as follows, using zero as a placeholder for the missing x^3-term in the dividend. Then follow the synthetic division procedure.

TechnologyTip

Using Graphs To check your division, you can graph the polynomial division expression and the depressed polynomial with the remainder. The graphs should coincide.

The quotient has degree one less than that of the dividend, so

$$\frac{2x^4 - 5x^2 + 5x - 2}{x + 2} = 2x^3 - 4x^2 + 3x - 1.$$ **Check this result.**

b. $(10x^3 - 13x^2 + 5x - 14) \div (2x - 3)$

Rewrite the division expression so that the divisor is of the form $x - c$.

$$\frac{10x^3 - 13x^2 + 5x - 14}{2x - 3} = \frac{(10x^3 - 13x^2 + 5x - 14) \div 2}{(2x - 3) \div 2} \text{ or } \frac{5x^3 - \frac{13}{2}x^2 + \frac{5}{2}x - 7}{x - \frac{3}{2}}.$$

So, $c = \frac{3}{2}$. Perform the synthetic division.

$$
\begin{array}{r|rrrr}
\frac{3}{2} & 5 & -\frac{13}{2} & \frac{5}{2} & -7 \\
\end{array}
$$

$$
\begin{array}{r|rrrr}
\frac{3}{2} & 5 & -\frac{13}{2} & \frac{5}{2} & -7 \\
& & \frac{15}{2} & \frac{3}{2} & 6 \\
\hline
& 5 & 1 & 4 & -1 \\
\end{array}
$$

So, $\dfrac{10x^3 - 13x^2 + 5x - 14}{2x - 3} = 5x^2 + x + 4 - \dfrac{1}{x - \frac{3}{2}}$ or $5x^2 + x + 4 - \dfrac{2}{2x - 3}$.   **Check this result.**

✔ Guided Practice

4A. $(4x^3 + 3x^2 - x + 8) \div (x - 3)$ **4B.** $(6x^4 + 11x^3 - 15x^2 - 12x + 7) \div (3x + 1)$

The Remainder and Factor Theorems When $d(x)$ is the divisor $(x - c)$ with degree 1, the remainder is the real number r. So, the division algorithm simplifies to

$$f(x) = (x - c) \cdot q(x) + r.$$

Evaluating $f(x)$ for $x = c$, we find that

$$f(c) = (c - c) \cdot q(c) + r = 0 \cdot q(c) + r \text{ or } r.$$

So, $f(c) = r$, which is the remainder. This leads us to the following theorem.

Key Concept   Remainder Theorem

If a polynomial $f(x)$ is divided by $x - c$, the remainder is $r = f(c)$.

The Remainder Theorem indicates that to evaluate a polynomial function $f(x)$ for $x = c$, you can divide $f(x)$ by $x - c$ using synthetic division. The remainder will be $f(c)$. Using synthetic division to evaluate a function is called **synthetic substitution**.

⊕ Real-World EXAMPLE 5 — Use the Remainder Theorem

FOOTBALL The number of tickets sold during the Northside High School football season can be modeled by $t(x) = x^3 - 12x^2 + 48x + 74$, where x is the number of games played. Use the Remainder Theorem to find the number of tickets sold during the twelfth game of the Northside High School football season.

To find the number of tickets sold during the twelfth game, use synthetic substitution to evaluate $t(x)$ for $x = 12$.

```
12 | 1   -12    48    74
   |       12     0   576
   ----------------------
     1     0     48 | 650
```

The remainder is 650, so $t(12) = 650$. Therefore, 650 tickets were sold during the twelfth game of the season.

CHECK You can check your answer using direct substitution.

$t(x) = x^3 - 12x^2 + 48x + 74$ **Original function**

$t(12) = (12)^3 - 12(12)^2 + 48(12) + 74$ or 650 ✔   **Substitute 12 for x and simplify.**

✅ Guided Practice

5. **FOOTBALL** Use the Remainder Theorem to determine the number of tickets sold during the thirteenth game of the season.

⊕ Real-World Link

High school football rules are similar to most college and professional football rules. Two major differences are that the quarters are 12 minutes as opposed to 15 minutes and kick-offs take place at the 40-yard line instead of the 30-yard line.

Source: National Federation of State High School Associations

If you use the Remainder Theorem to evaluate $f(x)$ at $x = c$ and the result is $f(c) = 0$, then you know that c is a zero of the function and $(x - c)$ is a factor. This leads us to another useful theorem that provides a test to determine whether $(x - c)$ is a factor of $f(x)$.

Key Concept — Factor Theorem

A polynomial $f(x)$ has a factor $(x - c)$ if and only if $f(c) = 0$.

You can use synthetic division to perform this test.

EXAMPLE 6 — Use the Factor Theorem

Use the Factor Theorem to determine if the binomials given are factors of $f(x)$. Use the binomials that are factors to write a factored form of $f(x)$.

a. $f(x) = 4x^4 + 21x^3 + 25x^2 - 5x + 3$; $(x - 1)$, $(x + 3)$

Use synthetic division to test each factor, $(x - 1)$ and $(x + 3)$.

```
 1 | 4    21    25    -5    3            -3 | 4    21    25    -5    3
   |       4    25    50   45               |     -12   -27     6   -3
   ------------------------------          --------------------------------
     4    25    50    45 | 48                 4     9    -2     1 | 0
```

Because the remainder when $f(x)$ is divided by $(x - 1)$ is 48, $f(1) = 48$ and $(x - 1)$ is not a factor.

Because the remainder when $f(x)$ is divided by $(x + 3)$ is 0, $f(-3) = 0$ and $(x + 3)$ is a factor.

Because $(x + 3)$ is a factor of $f(x)$, we can use the quotient of $f(x) \div (x + 3)$ to write a factored form of $f(x)$.

$$f(x) = (x + 3)(4x^3 + 9x^2 - 2x + 1)$$

(continued on the next page)

CHECK If $(x + 3)$ is a factor of $f(x) = 4x^4 + 21x^3 + 25x^2 - 5x + 3$, then -3 is a zero of the function and $(-3, 0)$ is an x-intercept of the graph. Graph $f(x)$ using a graphing calculator and confirm that $(-3, 0)$ is a point on the graph. ✔

$[-10, 10]$ scl: 1 by $[-10, 30]$ scl: 2

b. $f(x) = 2x^3 - x^2 - 41x - 20;\ (x + 4),\ (x - 5)$

Use synthetic division to test the factor $(x + 4)$.

```
-4 | 2   -1   -41   -20
   |      -8    36    20
   ─────────────────────
     2   -9    -5  |  0
```

Because the remainder when $f(x)$ is divided by $(x + 4)$ is 0, $f(-4) = 0$ and $(x + 4)$ is a factor of $f(x)$.

Next, test the second factor, $(x - 5)$, with the depressed polynomial $2x^2 - 9x - 5$.

```
5 | 2   -9   -5
  |      10    5
  ──────────────
    2    1  |  0
```

Because the remainder when the quotient of $f(x) \div (x + 4)$ is divided by $(x - 5)$ is 0, $f(5) = 0$ and $(x - 5)$ is a factor of $f(x)$.

Because $(x + 4)$ and $(x - 5)$ are factors of $f(x)$, we can use the final quotient to write a factored form of $f(x)$.

$$f(x) = (x + 4)(x - 5)(2x + 1)$$

CHECK The graph of $f(x) = 2x^3 - x^2 - 41x - 20$ confirms that $x = -4$, $x = 5$, and $x = -\frac{1}{2}$ are zeros of the function. ✔

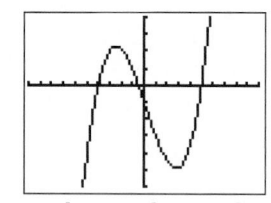

$[-10, 10]$ scl: 1 by $[-120, 80]$ scl: 20

✓ **Guided Practice**

Use the Factor Theorem to determine if the binomials given are factors of $f(x)$. Use the binomials that are factors to write a factored form of $f(x)$.

6A. $f(x) = 3x^3 - x^2 - 22x + 24;\ (x - 2),\ (x + 5)$

6B. $f(x) = 4x^3 - 34x^2 + 54x + 36;\ (x - 6),\ (x - 3)$

You can see that synthetic division is a useful tool for factoring and finding the zeros of polynomial functions.

Concept Summary   **Synthetic Division and Remainders**

If r is the remainder obtained after a synthetic division of $f(x)$ by $(x - c)$, then the following statements are true.

- r is the value of $f(c)$.
- If $r = 0$, then $(x - c)$ is a factor of $f(x)$.
- If $r = 0$, then c is an x-intercept of the graph of f.
- If $r = 0$, then $x = c$ is a solution of $f(x) = 0$.

Exercises

Factor each polynomial completely using the given factor and long division. (Example 1)

1. $x^3 + 2x^2 - 23x - 60; x + 4$

2. $x^3 + 2x^2 - 21x + 18; x - 3$

3. $x^3 + 3x^2 - 18x - 40; x - 4$

4. $4x^3 + 20x^2 - 8x - 96; x + 3$

5. $-3x^3 + 15x^2 + 108x - 540; x - 6$

6. $6x^3 - 7x^2 - 29x - 12; 3x + 4$

7. $x^4 + 12x^3 + 38x^2 + 12x - 63; x^2 + 6x + 9$

8. $x^4 - 3x^3 - 36x^2 + 68x + 240; x^2 - 4x - 12$

Divide using long division. (Examples 2 and 3)

9. $(5x^4 - 3x^3 + 6x^2 - x + 12) \div (x - 4)$

10. $(x^6 - 2x^5 + x^4 - x^3 + 3x^2 - x + 24) \div (x + 2)$

11. $(4x^4 - 8x^3 + 12x^2 - 6x + 12) \div (2x + 4)$

12. $(2x^4 - 7x^3 - 38x^2 + 103x + 60) \div (x - 3)$

13. $(6x^6 - 3x^5 + 6x^4 - 15x^3 + 2x^2 + 10x - 6) \div (2x - 1)$

14. $(108x^5 - 36x^4 + 75x^2 + 36x + 24) \div (3x + 2)$

15. $(x^4 + x^3 + 6x^2 + 18x - 216) \div (x^3 - 3x^2 + 18x - 54)$

16. $(4x^4 - 14x^3 - 14x^2 + 110x - 84) \div (2x^2 + x - 12)$

17. $\dfrac{6x^5 - 12x^4 + 10x^3 - 2x^2 - 8x + 8}{3x^3 + 2x + 3}$

18. $\dfrac{12x^5 + 5x^4 - 15x^3 + 19x^2 - 4x - 28}{3x^3 + 2x^2 - x + 6}$

Divide using synthetic division. (Example 4)

19. $(x^4 - x^3 + 3x^2 - 6x - 6) \div (x - 2)$

20. $(2x^4 + 4x^3 - 2x^2 + 8x - 4) \div (x + 3)$

21. $(3x^4 - 9x^3 - 24x - 48) \div (x - 4)$

22. $(x^5 - 3x^3 + 6x^2 + 9x + 6) \div (x + 2)$

23. $(12x^5 + 10x^4 - 18x^3 - 12x^2 - 8) \div (2x - 3)$

24. $(36x^4 - 6x^3 + 12x^2 - 30x - 12) \div (3x + 1)$

25. $(45x^5 + 6x^4 + 3x^3 + 8x + 12) \div (3x - 2)$

26. $(48x^5 + 28x^4 + 68x^3 + 11x + 6) \div (4x + 1)$

27. $(60x^6 + 78x^5 + 9x^4 - 12x^3 - 25x - 20) \div (5x + 4)$

28. $\dfrac{16x^6 - 56x^5 - 24x^4 + 96x^3 - 42x^2 - 30x + 105}{2x - 7}$

29. **EDUCATION** The number of U.S. students, in thousands, that graduated with a bachelor's degree from 1970 to 2006 can be modeled by $g(x) = 0.0002x^5 - 0.016x^4 + 0.512x^3 - 7.15x^2 + 47.52x + 800.27$, where x is the number of years since 1970. Use synthetic substitution to find the number of students that graduated in 2005. Round to the nearest thousand. (Example 5)

30. **SKIING** The distance in meters that a person travels on skis can be modeled by $d(t) = 0.2t^2 + 3t$, where t is the time in seconds. Use the Remainder Theorem to find the distance traveled after 45 seconds. (Example 5)

Find each $f(c)$ using synthetic substitution. (Example 5)

31. $f(x) = 4x^5 - 3x^4 + x^3 - 6x^2 + 8x - 15; c = 3$

32. $f(x) = 3x^6 - 2x^5 + 4x^4 - 2x^3 + 8x - 3; c = 4$

33. $f(x) = 2x^6 + 5x^5 - 3x^4 + 6x^3 - 9x^2 + 3x - 4; c = 5$

34. $f(x) = 4x^6 + 8x^5 - 6x^3 - 5x^2 + 6x - 4; c = 6$

35. $f(x) = 10x^5 + 6x^4 - 8x^3 + 7x^2 - 3x + 8; c = -6$

36. $f(x) = -6x^7 + 4x^5 - 8x^4 + 12x^3 - 15x^2 - 9x + 64; c = 2$

37. $f(x) = -2x^8 + 6x^5 - 4x^4 + 12x^3 - 6x + 24; c = 4$

Use the Factor Theorem to determine if the binomials given are factors of $f(x)$. Use the binomials that are factors to write a factored form of $f(x)$. (Example 6)

38. $f(x) = x^4 - 2x^3 - 9x^2 + x + 6; (x + 2), (x - 1)$

39. $f(x) = x^4 + 2x^3 - 5x^2 + 8x + 12; (x - 1), (x + 3)$

40. $f(x) = x^4 - 2x^3 + 24x^2 + 18x + 135; (x - 5), (x + 5)$

41. $f(x) = 3x^4 - 22x^3 + 13x^2 + 118x - 40; (3x - 1), (x - 5)$

42. $f(x) = 4x^4 - x^3 - 36x^2 - 111x + 30; (4x - 1), (x - 6)$

43. $f(x) = 3x^4 - 35x^3 + 38x^2 + 56x + 64; (3x - 2), (x + 2)$

44. $f(x) = 5x^5 + 38x^4 - 68x^2 + 59x + 30; (5x - 2), (x + 8)$

45. $f(x) = 4x^5 - 9x^4 + 39x^3 + 24x^2 + 75x + 63; (4x + 3), (x - 1)$

46. **TREES** The height of a tree in feet at various ages in years is given in the table.

Age	Height	Age	Height
2	3.3	24	73.8
6	13.8	26	82.0
10	23.0	28	91.9
14	42.7	30	101.7
20	60.7	36	111.5

   a. Use a graphing calculator to write a quadratic equation to model the growth of the tree.

   b. Use synthetic division to evaluate the height of the tree at 15 years.

47. **BICYCLING** Patrick is cycling at an initial speed v_0 of 4 meters per second. When he rides downhill, the bike accelerates at a rate a of 0.4 meter per second squared. The vertical distance from the top of the hill to the bottom of the hill is 25 meters. Use $d(t) = v_0 t + \frac{1}{2}at^2$ to find how long it will take Patrick to ride down the hill, where $d(t)$ is distance traveled and t is given in seconds..

Factor each polynomial using the given factor and long division. Assume $n > 0$.

48. $x^{3n} + x^{2n} - 14x^n - 24; x^n + 2$

49. $x^{3n} + x^{2n} - 12x^n + 10; x^n - 1$

50. $4x^{3n} + 2x^{2n} - 10x^n + 4; 2x^n + 4$

51. $9x^{3n} + 24x^{2n} - 171x^n + 54; 3x^n - 1$

52. MANUFACTURING An 18-inch by 20-inch sheet of cardboard is cut and folded into a bakery box.

a. Write a polynomial function to model the volume of the box.

b. Graph the function.

c. The company wants the box to have a volume of 196 cubic inches. Write an equation to model this situation.

d. Find a positive integer for x that satisfies the equation found in part **c**.

Find the value of k so that each remainder is zero.

53. $\dfrac{x^3 - kx^2 + 2x - 4}{x - 2}$

54. $\dfrac{x^3 + 18x^2 + kx + 4}{x + 2}$

55. $\dfrac{x^3 + 4x^2 - kx + 1}{x + 1}$

56. $\dfrac{2x^3 - x^2 + x + k}{x - 1}$

57. SCULPTING Esteban will use a block of clay that is 3 feet by 4 feet by 5 feet to make a sculpture. He wants to reduce the volume of the clay by removing the same amount from the length, the width, and the height.

a. Write a polynomial function to model the situation.

b. Graph the function.

c. He wants to reduce the volume of the clay to $\dfrac{3}{5}$ of the original volume. Write an equation to model the situation.

d. How much should he take from each dimension?

Use the graphs and synthetic division to completely factor each polynomial.

58. $f(x) = 8x^4 + 26x^3 - 103x^2 - 156x + 45$ (Figure 2.3.1)

59. $f(x) = 6x^5 + 13x^4 - 153x^3 + 54x^2 + 724x - 840$ (Figure 2.3.2)

Figure 2.3.1

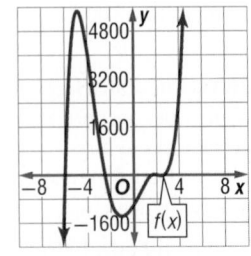

Figure 2.3.2

60. MULTIPLE REPRESENTATIONS In this problem, you will explore the upper and lower bounds of a function.

a. GRAPHICAL Graph each related polynomial function, and determine the greatest and least zeros. Then copy and complete the table.

Polynomial	Greatest Zero	Least Zero
$x^3 - 2x^2 - 11x + 12$		
$x^4 + 6x^3 + 3x^2 - 10x$		
$x^5 - x^4 - 2x^3$		

b. NUMERICAL Use synthetic division to evaluate each function in part **a** for three integer values greater than the greatest zero.

c. VERBAL Make a conjecture about the characteristics of the last row when synthetic division is used to evaluate a function for an integer greater than its greatest zero.

d. NUMERICAL Use synthetic division to evaluate each function in part **a** for three integer values less than the least zero.

e. VERBAL Make a conjecture about the characteristics of the last row when synthetic division is used to evaluate a function for a number less than its least zero.

H.O.T. Problems   Use Higher-Order Thinking Skills

61. CHALLENGE Is $(x - 1)$ a factor of $18x^{165} - 15x^{135} + 8x^{105} - 15x^{55} + 4$? Explain your reasoning.

62. WRITING IN MATH Explain how you can use a graphing calculator, synthetic division, and factoring to completely factor a fifth-degree polynomial with rational coefficients, three integral zeros, and two non-integral, rational zeros.

63. REASONING Determine whether the statement below is *true* or *false*. Explain.
If $h(y) = (y + 2)(3y^2 + 11y - 4) - 1$, then the remainder of $\dfrac{h(y)}{y + 2}$ is -1.

CHALLENGE Find k so that the quotient has a 0 remainder.

64. $\dfrac{x^3 + kx^2 - 34x + 56}{x + 7}$

65. $\dfrac{x^6 + kx^4 - 8x^3 + 173x^2 - 16x - 120}{x - 1}$

66. $\dfrac{kx^3 + 2x^2 - 22x - 4}{x - 2}$

67. CHALLENGE If $2x^2 - dx + (31 - d^2)x + 5$ has a factor $x - d$, what is the value of d if d is an integer?

68. WRITING IN MATH Compare and contrast polynomial division using long division and using synthetic division.

Determine whether the degree n of the polynomial for each graph is *even* or *odd* and whether its leading coefficient a_n is *positive* or *negative*. (Lesson 2–2)

69.

70.

71.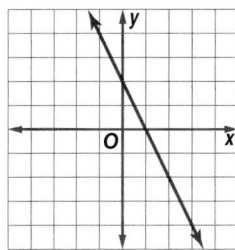

72. SKYDIVING The approximate time t in seconds that it takes an object to fall a distance of d feet is given by $t = \sqrt{\dfrac{d}{16}}$. Suppose a skydiver falls 11 seconds before the parachute opens. How far does the skydiver fall during this time period? (Lesson 2–1)

73. FIRE FIGHTING The velocity v and maximum height h of water being pumped into the air are related by $v = \sqrt{2gh}$, where g is the acceleration due to gravity (32 feet/second^2). (Lesson 1–7)

   a. Determine an equation that will give the maximum height of the water as a function of its velocity.

   b. The Mayfield Fire Department must purchase a pump that is powerful enough to propel water 80 feet into the air. Will a pump that is advertised to project water with a velocity of 75 feet/second meet the fire department's needs? Explain.

Solve each system of equations algebraically. (Lesson 0–5)

74. $5x - y = 16$
$2x + 3y = 3$

75. $3x - 5y = -8$
$x + 2y = 1$

76. $y = 6 - x$
$x = 4.5 + y$

77. $2x + 5y = 4$
$3x + 6y = 5$

78. $7x + 12y = 16$
$5y - 4x = -21$

79. $4x + 5y = -8$
$3x - 7y = 10$

80. SAT/ACT In the figure, an equilateral triangle is drawn with an altitude that is also the diameter of the circle. If the perimeter of the triangle is 36, what is the circumference of the circle?

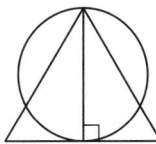

  A $6\sqrt{2}\pi$ **C** $12\sqrt{2}\pi$ **E** 36π
  B $6\sqrt{3}\pi$ **D** $12\sqrt{3}\pi$

81. REVIEW If $(3, -7)$ is the center of a circle and $(8, 5)$ is on the circle, what is the circumference of the circle?

  F 13π **H** 18π **K** 26π
  G 15π **J** 25π

82. REVIEW The first term in a sequence is x. Each subsequent term is three less than twice the preceding term. What is the 5th term in the sequence?

  A $8x - 21$ **C** $16x - 39$ **E** $32x - 43$
  B $8x - 15$ **D** $16x - 45$

83. Use the graph of the polynomial function. Which is not a factor of $x^5 + x^4 - 3x^3 - 3x^2 - 4x - 4$?

  F $(x - 2)$
  G $(x + 2)$
  H $(x - 1)$
  J $(x + 1)$

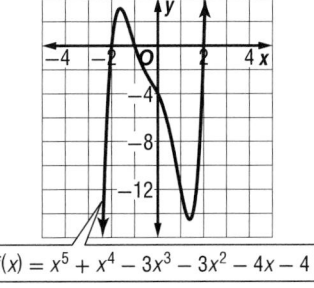

$f(x) = x^5 + x^4 - 3x^3 - 3x^2 - 4x - 4$

Graph and analyze each function. Describe its domain, range, intercepts, end behavior, continuity, and where the function is increasing or decreasing. (Lesson 2-1)

1. $f(x) = 2x^3$

2. $f(x) = -\frac{2}{3}x^4$

3. $f(x) = 3x^{-8}$

4. $f(x) = 4x^{\frac{2}{5}}$

5. TREES The heights of several fir trees and the areas under their branches are shown in the table. (Lesson 2-1)

Height (m)	Area (m²)
4.2	37.95
2.1	7.44
3.4	23.54
1.7	4.75
4.6	46.48

a. Create a scatter plot of the data.

b. Determine a power function to model the data.

c. Predict the area under the branches of a fir tree that is 7.6 meters high.

Solve each equation. (Lesson 2-1)

6. $\sqrt{5x + 7} = 13$

7. $\sqrt{2x - 2} + 1 = x$

8. $\sqrt{3x + 10} + 1 = \sqrt{x + 11}$

9. $-5 = \sqrt[4]{(6x + 3)^3} - 32$

State the number of possible real zeros and turning points of each function. Then find all of the real zeros by factoring. (Lesson 2-2)

10. $f(x) = x^2 - 11x - 26$

11. $f(x) = 3x^5 + 2x^4 - x^3$

12. $f(x) = x^4 + 9x^2 - 10$

13. MULTIPLE CHOICE Which of the following describes the possible end behavior of a polynomial of odd degree? (Lesson 2-2)

A $\lim\limits_{x \to \infty} f(x) = 5; \lim\limits_{x \to -\infty} f(x) = 5$

B $\lim\limits_{x \to \infty} f(x) = -\infty; \lim\limits_{x \to -\infty} f(x) = -\infty$

C $\lim\limits_{x \to \infty} f(x) = \infty; \lim\limits_{x \to -\infty} f(x) = \infty$

D $\lim\limits_{x \to \infty} f(x) = -\infty; \lim\limits_{x \to -\infty} f(x) = \infty$

Describe the end behavior of the graph of each polynomial function using limits. Explain your reasoning using the leading term test. (Lesson 2-2)

14. $f(x) = -7x^4 - 3x^3 - 8x^2 + 23x + 7$

15. $f(x) = -5x^5 + 4x^4 + 12x^2 - 8$

16. ENERGY Crystal's electricity consumption measured in kilowatt hours (kWh) for the past 12 months is shown below. (Lesson 2-2)

Month	Consumption (kWh)	Month	Consumption (kWh)
January	240	July	300
February	135	August	335
March	98	September	390
April	110	October	345
May	160	November	230
June	230	December	100

a. Determine a model for the number of kilowatt hours Crystal used as a function of the number of months since January.

b. Use the model to predict how many kilowatt hours Crystal will use the following January. Does this answer make sense? Explain your reasoning.

Divide using synthetic division. (Lesson 2-3)

17. $(5x^3 - 7x^2 + 8x - 13) \div (x - 1)$

18. $(x^4 - x^3 - 9x + 18) \div (x - 2)$

19. $(2x^3 - 11x^2 + 9x - 6) \div (2x - 1)$

Determine each $f(c)$ using synthetic substitution. (Lesson 2-3)

20. $f(x) = 9x^5 + 4x^4 - 3x^3 + 18x^2 - 16x + 8; c = 2$

21. $f(x) = 6x^6 - 3x^5 + 8x^4 + 12x^2 - 6x + 4; c = -3$

22. $f(x) = -2x^6 + 8x^5 - 12x^4 + 9x^3 - 8x^2 + 6x - 3; c = -2$

Use the Factor Theorem to determine if the binomials given are factors of $f(x)$. Use the binomials that are factors to write a factored form of $f(x)$. (Lesson 2-3)

23. $f(x) = x^3 + 2x^2 - 25x - 50; (x + 5)$

24. $f(x) = x^4 - 6x^3 + 7x^2 + 6x - 8; (x - 1), (x - 2)$

25. MULTIPLE CHOICE Find the remainder when $f(x) = x^3 - 4x + 5$ is divided by $x + 3$. (Lesson 2-3)

F -10

G 8

H 20

J 26

Zeros of Polynomial Functions

Then

You learned that a polynomial function of degree *n* can have at most *n* real zeros. (Lesson 2-1)

Now

- Find real zeros of polynomial functions.
- Find complex zeros of polynomial functions.

New Vocabulary

Rational Zero Theorem
lower bound
upper bound
Descartes' Rule of Signs
Fundamental Theorem of Algebra
Linear Factorization Theorem
Conjugate Root Theorem
complex conjugates
irreducible over the reals

Math Online

glencoe.com

Why?

A company estimates that the profit *P* in thousands of dollars from a certain model of video game controller is given by $P(x) = -0.0007x^2 + 2.45x$, where *x* is the number of thousands of dollars spent marketing the controller. To find the number of advertising dollars the company should spend to make a profit of $1,500,000, you can use techniques presented in this lesson to solve the polynomial equation $P(x) = 1500$.

Real Zeros Recall that a polynomial function of degree *n* can have at most *n* real zeros. These real zeros are either rational or irrational.

Rational Zeros	Irrational Zeros
$f(x) = 3x^2 + 7x - 6$ or $f(x) = (x + 3)(3x - 2)$	$g(x) = x^2 - 5$ or $g(x) = (x + \sqrt{5})(x - \sqrt{5})$
There are two rational zeros, -3 or $\frac{2}{3}$.	There are two irrational zeros, $\pm\sqrt{5}$.

The **Rational Zero Theorem** describes how the leading coefficient and constant term of a polynomial function with integer coefficients can be used to determine a list of all possible rational zeros.

Key Concept · Rational Zero Theorem

If *f* is a polynomial function of the form $f(x) = a_n x^n + a_{n-1}x^{n-1} + \ldots + a_2 x^2 + a_1 x + a_0$, with degree $n \geq 1$, integer coefficients, and $a_0 \neq 0$, then every rational zero of *f* has the form $\frac{p}{q}$, where

- *p* and *q* have no common factors other than ± 1,
- *p* is an integer factor of the constant term a_0, and
- *q* is an integer factor of the leading coefficient a_n.

Corollary If the leading coefficient a_n is 1, then any rational zeros of *f* are integer factors of the constant term a_0.

Once you know all of the *possible* rational zeros of a polynomial function, you can then use direct or synthetic substitution to determine which, if any, are actual zeros of the polynomial.

EXAMPLE 1 · Leading Coefficient Equal to 1

List all possible rational zeros of each function. Then determine which, if any, are zeros.

a. $f(x) = x^3 + 2x + 1$

   Step 1 Identify possible rational zeros.

   Because the leading coefficient is 1, the possible rational zeros are the integer factors of the constant term 1. Therefore, the possible rational zeros of *f* are 1 and -1.

   Step 2 Use direct substitution to test each possible zero.

   $f(1) = (1)^3 + 2(1) + 1$ or 4

   $f(-1) = (-1)^3 + 2(-1) + 1$ or -2

   Because $f(1) \neq 0$ and $f(-1) \neq 0$, you can conclude that *f* has no rational zeros. From the graph of *f* you can see that *f* has one real zero. Applying the Rational Zeros Theorem shows that this zero is irrational.

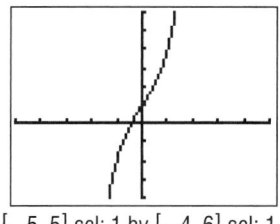

$[-5, 5]$ scl: 1 by $[-4, 6]$ scl: 1

b. $g(x) = x^4 + 4x^3 - 12x - 9$

Step 1 Because the leading coefficient is 1, the possible rational zeros are the integer factors of the constant term -9. Therefore, the possible rational zeros of g are ± 1, ± 3, and ± 9.

Step 2 Begin by testing 1 and -1 using synthetic substitution.

$$
\begin{array}{r|rrrrr}
1 & 1 & 4 & 0 & -12 & -9 \\
  &   & 1 & 5 & 5 & -7 \\
\hline
  & 1 & 5 & 5 & -7 & -16 \\
\end{array}
\qquad
\begin{array}{r|rrrrr}
-1 & 1 & 4 & 0 & -12 & -9 \\
   &   & -1 & -3 & 3 & 9 \\
\hline
   & 1 & 3 & -3 & -9 & 0 \\
\end{array}
$$

Because $g(-1) = 0$, you can conclude that -1 is a zero of g. Testing -3 on the depressed polynomial shows that -3 is another rational zero.

$$
\begin{array}{r|rrrr}
-3 & 1 & 3 & -3 & -9 \\
   &   & -3 & 0 & 9 \\
\hline
   & 1 & 0 & -3 & 0 \\
\end{array}
$$

Thus, $g(x) = (x + 1)(x + 3)(x^2 - 3)$. Because the factor $(x^2 - 3)$ yields no rational zeros, we can conclude that g has only two rational zeros, -1 and -3.

CHECK The graph of $g(x) = x^4 + 4x^3 - 12x - 9$ in Figure 2.4.1 has x-intercepts at -1 and -3, and close to $(2, 0)$ and $(-2, 0)$. By the Rational Zeros Theorem, we know that these last two zeros must be irrational. In fact, the factor $(x^2 - 3)$ yields two irrational zeros, $\sqrt{3}$ and $-\sqrt{3}$. ✔

[-5, 5] scl: 1 by [-20, 10] scl: 3

Figure 2.4.1

☑ Guided Practice

List all possible rational zeros of each function. Then determine which, if any, are zeros.

1A. $f(x) = x^3 + 5x^2 - 4x - 2$ **1B.** $h(x) = x^4 + 3x^3 - 7x^2 + 9x - 30$

When the leading coefficient of a polynomial function is not 1, the list of possible rational zeros can increase significantly.

EXAMPLE 2   **Leading Coefficient not Equal to 1**

List all possible rational zeros of $h(x) = 3x^3 - 7x^2 - 22x + 8$. Then determine which, if any, are zeros.

Step 1 The leading coefficient is 3 and the constant term is 8.

$$\text{Possible rational zeros: } \frac{\text{Factors of 8}}{\text{Factors of 3}} = \frac{\pm 1, \pm 2, \pm 4, \pm 8}{\pm 1, \pm 3} \text{ or } \pm 1, \pm 2, \pm 4, \pm 8, \pm\frac{1}{3}, \pm\frac{2}{3}, \pm\frac{4}{3}, \pm\frac{8}{3}$$

Step 2 By synthetic substitution, you can determine that -2 is a rational zero.

$$
\begin{array}{r|rrrr}
-2 & 3 & -7 & -22 & 8 \\
   &   & -6 & 26 & -8 \\
\hline
   & 3 & -13 & 4 & 0 \\
\end{array}
$$

By the division algorithm, $h(x) = (x + 2)(3x^2 - 13x + 4)$. Once $3x^2 - 13x + 4$ is factored, the polynomial becomes $h(x) = (x + 2)(3x - 1)(x - 4)$, and you can conclude that the rational zeros of h are -2, $\frac{1}{3}$, and 4. Check this result by graphing.

☑ Guided Practice

List all possible rational zeros of each function. Then determine which, if any, are zeros.

2A. $g(x) = 2x^3 - 4x^2 + 18x - 36$ **2B.** $f(x) = 3x^4 - 18x^3 + 2x - 21$

Real-World EXAMPLE 3   Solve a Polynomial Equation

BUSINESS After the first half-hour, the number of video games that were sold by a company on their release date can be modeled by $g(x) = 2x^3 + 4x^2 - 2x$, where $g(x)$ is the number of games sold in hundreds and x is the number of hours after the release. How long did it take to sell 400 games?

Because $g(x)$ represents the number of games sold in hundreds, you need to solve $g(x) = 4$ to determine how long it will take to sell 400 games.

$$g(x) = 4 \qquad \text{Write the equation.}$$
$$2x^3 + 4x^2 - 2x = 4 \qquad \text{Substitute } 2x^3 + 4x^2 - 2x \text{ for } g(x).$$
$$2x^3 + 4x^2 - 2x - 4 = 0 \qquad \text{Subtract 4 from each side.}$$

Apply the Rational Zeros Theorem to this new polynomial function, $f(x) = 2x^3 + 4x^2 - 2x - 4$.

Step 1   Possible rational zeros: $\dfrac{\text{Factors of 4}}{\text{Factors of 2}} = \dfrac{\pm 1, \pm 2, \pm 4}{\pm 1, \pm 2}$

$$= \pm 1, \pm 2, \pm 4, \pm \tfrac{1}{2}$$

Step 2   By synthetic substitution, you can determine that 1 is a rational zero.

$$
\begin{array}{r|rrrr}
1 & 2 & 4 & -2 & -4 \\
  &   & 2 & 6 & 4 \\
\hline
  & 2 & 6 & 4 & 0 \\
\end{array}
$$

Because 1 is a zero of f, $x = 1$ is a solution of $f(x) = 0$. The depressed polynomial $2x^2 + 6x + 4$ can be written as $2(x + 2)(x + 1)$. The zeros of this polynomial are -2 and -1. Because time cannot be negative, the solution is $x = 1$. So, it took 1 hour to sell 400 games.

☑ Guided Practice

3. **VOLLEYBALL** A volleyball that is returned after a serve with an initial speed of 40 feet per second at a height of 4 feet is given by $f(t) = 4 + 40t - 16t^2$, where $f(t)$ is the height the ball reaches in feet and t is time in seconds. At what time(s) will the ball reach a height of 20 feet?

One way to narrow the search for real zeros is to determine an interval within which all real zeros of a function are located. A real number a is a **lower bound** for the real zeros of f if $f(x) \neq 0$ for $x < a$. Similarly, b is an **upper bound** for the real zeros of f if $f(x) \neq 0$ for $x > b$.

The real zeros of f are in the interval $[a, b]$.

You can test whether a given interval contains all real zeros of a function by using the following upper and lower bound tests.

Key Concept   Upper and Lower Bound Tests

Let f be a polynomial function of degree $n \geq 1$, real coefficients, and a positive leading coefficient. Suppose $f(x)$ is divided by $x - c$ using synthetic division.

- If $c \leq 0$ and every number in the last line of the division is alternately nonnegative and nonpositive, then c is a *lower bound* for the real zeros of f.
- If $c \geq 0$ and every number in the last line of the division is nonnegative, then c is an *upper bound* for the real zeros of f.

To make use of the upper and lower bound tests, follow these steps.

Step 1 Graph the function to determine an interval in which the zeros lie.

Step 2 Using synthetic substitution, confirm that the upper and lower bounds of your interval are in fact upper and lower bounds of the function by applying the upper and lower bound tests.

Step 3 Use the Rational Zero Theorem to help find all the real zeros.

EXAMPLE 4 Use the Upper and Lower Bound Tests

Determine an interval in which all real zeros of $h(x) = 2x^4 - 11x^3 + 2x^2 - 44x - 24$ must lie. Explain your reasoning using the upper and lower bound tests. Then find all the real zeros.

Step 1 Graph $h(x)$ using a graphing calculator. From this graph, it appears that the real zeros of this function lie in the interval $[-1, 7]$.

[−3, 7] scl: 1 by [−400, 200] scl: 60

Step 2 Test a lower bound of $c = -1$ and an upper bound of $c = 7$.

$$
\begin{array}{r|rrrrr}
-1 & 2 & -11 & 2 & -44 & -24 \\
   &   & -2 & 13 & -15 & 59 \\
\hline
   & 2 & -13 & 15 & -59 & 35 \\
\end{array}
$$

Values alternate signs in the last line, so −1 is a lower bound.

$$
\begin{array}{r|rrrrr}
7 & 2 & -11 & 2 & -44 & -24 \\
  &   & 14 & 21 & 161 & 819 \\
\hline
  & 2 & 3 & 23 & 117 & 795 \\
\end{array}
$$

Values are all nonnegative in last line, so 7 is an upper bound.

Step 3 Use the Rational Zero Theorem.

Possible rational zeros: $\dfrac{\text{Factors of } 24}{\text{Factors of } 2} = \dfrac{\pm 1, \pm 2, \pm 3, \pm 4, \pm 6, \pm 8, \pm 12, \pm 24}{\pm 1, \pm 2}$

$$= \pm 1, \pm 2, \pm 4, \pm 6, \pm 8, \pm 12, \pm 24, \pm\tfrac{1}{2}, \pm\tfrac{3}{2}$$

Because the real zeros are in the interval $[-1, 7]$, you can narrow this list to just $\pm 1, \pm\tfrac{1}{2}, \pm\tfrac{3}{2}, 2, 4,$ or 6. From the graph, it appears that only 6 and $-\tfrac{1}{2}$ are reasonable.

Begin by testing 6.

$$
\begin{array}{r|rrrrr}
6 & 2 & -11 & 2 & -44 & -24 \\
  &   & 12 & 6 & 48 & 24 \\
\hline
  & 2 & 1 & 8 & 4 & 0 \\
\end{array}
$$

Now test $-\tfrac{1}{2}$ in the depressed polynomial.

$$
\begin{array}{r|rrrr}
-\tfrac{1}{2} & 2 & 1 & 8 & 4 \\
              &   & -1 & 0 & -4 \\
\hline
              & 2 & 0 & 8 & 0 \\
\end{array}
$$

By the division algorithm, $h(x) = 2(x - 6)\left(x + \tfrac{1}{2}\right)(x^2 + 4)$. Notice that the factor $(x^2 + 4)$ has no real zeros associated with it because $x^2 + 4 = 0$ has no real solutions. So, f has two real solutions that are both rational, 6 and $-\tfrac{1}{2}$. The graph of $h(x) = 2x^4 - 11x^3 + 2x^2 - 44x - 24$ supports this conclusion.

☑ **Guided Practice**

Determine an interval in which all real zeros of the given function must lie. Explain your reasoning using the upper and lower bound tests. Then find all the real zeros.

4A. $g(x) = 6x^4 + 70x^3 - 21x^2 + 35x - 12$ **4B.** $f(x) = 10x^5 - 50x^4 - 3x^3 + 22x^2 - 41x + 30$

ReadingMath

Variation in Sign
A *variation in sign* occurs in a polynomial written in standard form when consecutive coefficients have opposite signs.

Another way to narrow the search for real zeros is to use **Descartes' Rule of Signs**. This rule gives us information about the number of positive and negative real zeros of a polynomial function by looking at a polynomial's variations in sign.

Key Concept — Descartes' Rule of Signs

If $f(x) = a_n x^n + a_{n-1} x^{n-1} + \ldots + a_1 x + a_0$ is a polynomial function with real coefficients, then

- the number of *positive* real zeros of f is equal to the number of variations in sign of $f(x)$ or less than that number by some even number and

- the number of *negative* real zeros of f is the same as the number of variations in sign of $f(-x)$ or less than that number by some even number.

EXAMPLE 5 — Use Descartes' Rule of Signs

Describe the possible real zeros of $g(x) = -3x^3 + 2x^2 - x - 1$.

Examine the variations in sign for $g(x)$ and for $g(-x)$.

$$g(x) = -3x^3 + 2x^2 - x - 1$$
(+ to −, − to +)

$$g(-x) = -3(-x)^3 + 2(-x)^2 - (-x) - 1$$
$$= 3x^3 + 2x^2 + x - 1$$
(+ to −)

The original function $g(x)$ has *two* variations in sign, while $g(-x)$ has *one* variation in sign. By Descartes' Rule of Signs, you know that $g(x)$ has either 2 or 0 positive real zeros and 1 negative real zero.

From the graph of $g(x)$ shown, you can see that the function has one negative real zero close to $x = -0.5$ and no positive real zeros.

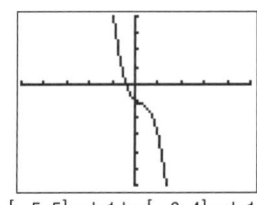
$[-5, 5]$ scl: 1 by $[-6, 4]$ scl: 1

✓ Guided Practice

Describe the possible real zeros of each function.

5A. $h(x) = 6x^5 + 8x^2 - 10x - 15$ **5B.** $f(x) = -11x^4 + 20x^3 + 3x^2 - x + 18$

When using Descartes' Rule of Signs, the number of real zeros indicated includes any repeated zeros. Therefore, a zero with multiplicity m should be counted as m zeros.

Complex Zeros Just as quadratic functions can have real or imaginary zeros, polynomials of higher degree can also have zeros in the complex number system. This fact, combined with the **Fundamental Theorem of Algebra**, allows us to improve our statement about the number of zeros for any nth-degree polynomial.

Math History Link

**René Descartes
(1596–1650)**
A French mathematician, scientist, and philosopher, Descartes wrote many philosophical works such as *Discourse on Method* and mathematical works such as *Geometry*.

Key Concept — Fundamental Theorem of Algebra

A polynomial function of degree n, where $n > 0$, has at least one zero (real or imaginary) in the complex number system.

Corollary A polynomial function of degree n has *exactly* n zeros, including repeated zeros, in the complex number system.

By extending the Factor Theorem to include both real and imaginary zeros and applying the Fundamental Theorem of Algebra, we obtain the **Linear Factorization Theorem**.

Key Concept — Linear Factorization Theorem

If $f(x)$ is a polynomial function of degree $n > 0$, then f has exactly n linear factors and

$$f(x) = a_n(x - c_1)(x - c_2) \cdots (x - c_n)$$

where a_n is some nonzero real number and $c_1, c_2, \ldots, c_n$ are the complex zeros (including repeated zeros) of f.

According to the **Conjugate Root Theorem**, when a polynomial equation in one variable with real coefficients has a root of the form $a + bi$, where $b \neq 0$, then its **complex conjugate**, $a - bi$, is also a root. You can use this theorem to write a polynomial function given its complex zeros.

EXAMPLE 6 — Find a Polynomial Function Given Its Zeros

Write a polynomial function of least degree with real coefficients in standard form that has $-2, 4$, and $3 - i$ as zeros.

Because $3 - i$ is a zero and the polynomial is to have real coefficients, you know that $3 + i$ must also be a zero. Using the Linear Factorization Theorem and the zeros $-2, 4, 3 - i$, and $3 + i$, you can write $f(x)$ as follows.

$$f(x) = a[x - (-2)](x - 4)[x - (3 - i)][x - (3 + i)]$$

While a can be any nonzero real number, it is simplest to let $a = 1$. Then write the function in standard form.

$$\begin{aligned} f(x) &= (1)(x + 2)(x - 4)[x - (3 - i)][x - (3 + i)] & \text{Let } a = 1. \\ &= (x^2 - 2x - 8)(x^2 - 6x + 10) & \text{Multiply.} \\ &= x^4 - 8x^3 + 14x^2 + 28x - 80 & \text{Multiply.} \end{aligned}$$

Therefore, a function of least degree that has $-2, 4, 3 - i$, and $3 + i$ as zeros is $f(x) = x^4 - 8x^3 + 14x^2 + 28x - 80$ or any nonzero multiple of $f(x)$.

StudyTip

Infinite Polynomials
Because a can be any nonzero real number, there are an infinite number of polynomial functions that can be written for a given set of zeros.

✓ Guided Practice

Write a polynomial function of least degree with real coefficients in standard form with the given zeros.

6A. $-3, 1$ (multiplicity: 2), $4i$

6B. $2\sqrt{3}, -2\sqrt{3}, 1 + i$

StudyTip

Prime Polynomials
Note the difference between expressions which are irreducible over the reals and expressions which are *prime*. The expression $x^2 - 8$ is prime because it cannot be factored into expressions with integral coefficients. However, $x^2 - 8$ is *not* irreducible over the reals because there are real zeros associated with it, $\sqrt{8}$ and $-\sqrt{8}$.

In Example 6, you wrote a function with real and complex zeros. A function has complex zeros when its factored form contains a quadratic factor which is irreducible over the reals. A quadratic expression is **irreducible over the reals** when it has real coefficients but no real zeros associated with it. This example illustrates the following theorem.

Key Concept — Factoring Polynomial Functions Over the Reals

Every polynomial function of degree $n > 0$ with real coefficients can be written as the product of linear factors and irreducible quadratic factors, each with real coefficients.

As indicated by the Linear Factorization Theorem, when factoring a polynomial function over the complex number system, we can write the function as the product of only linear factors.

EXAMPLE 7 **Factor and Find the Zeros of a Polynomial Function**

Consider $k(x) = x^5 - 18x^3 + 30x^2 - 19x + 30$.

a. Write $k(x)$ as the product of linear and irreducible quadratic factors.

The possible rational zeros are $\pm 1, \pm 2, \pm 3, \pm 5, \pm 6, \pm 10, \pm 15, \pm 30$. The original polynomial has 4 sign variations.

$$k(-x) = (-x)^5 - 18(-x)^3 + 30(-x)^2 - 19(-x) + 30$$
$$= -x^5 + 18x^3 + 30x^2 + 19x + 30$$

$k(-x)$ has 1 sign variation, so $k(x)$ has 4, 2, or 0 positive real zeros and 1 negative real zero.

The graph shown suggests -5 as one real zero of $k(x)$. Use synthetic substitution to test this possibility.

```
-5 | 1    0   -18    30   -19    30
   |     -5    25   -35    25   -30
   --------------------------------
     1   -5     7    -5     6 |   0
```

$[-8, 8]$ scl: 1 by $[-100, 800]$ scl: 50

Because $k(x)$ has only 1 negative real zero, you do not need to test any other possible negative rational zeros. Zooming in on the positive real zeros in the graph suggests 2 and 3 as other rational zeros. Test these possibilities successively in the depressed quartic and then cubic polynomials.

```
2 | 1   -5    7   -5    6      Begin by
  |       2   -6    2   -6     testing 2.
  --------------------------
    1   -3    1   -3 |  0
```

$[-8, 8]$ scl: 1 by $[-20, 20]$ scl: 4

```
3 | 1   -3    1   -3          Now test 3 on the
  |       3    0    3          depressed polynomial.
  -----------------------
    1    0    1 |  0
```

The remaining quadratic factor $(x^2 + 1)$ yields no real zeros and is therefore irreducible over the reals. So, $k(x)$ written as a product of linear and irreducible quadratic factors is $k(x) = (x + 5)(x - 2)(x - 3)(x^2 + 1)$.

Quadratic Formula
You could also use the Quadratic Formula to find the zeros of $x^2 + 1$ in order to factor the expression.

$$x = \frac{-0 \pm \sqrt{0^2 - 4(1)(1)}}{2(1)}$$

$$= \pm \frac{\sqrt{-4}}{2}$$

$$= \pm \frac{2i}{2} \text{ or } \pm i$$

So, i and $-i$ are zeros and $(x - i)$ and $(x + i)$ are factors.

b. Write $k(x)$ as the product of linear factors.

You can factor $x^2 + 1$ by writing the expression first as a difference of squares $x^2 - (\sqrt{-1})^2$ or $x^2 - i^2$. Then factor this difference of squares as $(x - i)(x + i)$. So, $k(x)$ written as a product of linear factors is as follows.

$$k(x) = (x + 5)(x - 2)(x - 3)(x - i)(x + i)$$

c. List all the zeros of $k(x)$.

Because the function has degree 5, by the corollary to the Fundamental Theorem of Algebra $k(x)$ has exactly five zeros, including any that may be repeated. The linear factorization gives us these five zeros: $-5, 2, 3, i,$ and $-i$.

✓ Guided Practice

Write each function as **(a)** the product of linear and irreducible quadratic factors and **(b)** the product of linear factors. Then **(c)** list all of its zeros.

7A. $f(x) = x^4 + x^3 - 26x^2 + 4x - 120$

7B. $f(x) = x^5 - 2x^4 - 2x^3 - 6x^2 - 99x + 108$

You can use synthetic substitution with complex numbers in the same way you use it with real numbers. Doing so can help you factor a polynomial in order to find all of its zeros.

Watch Out!

Complex Numbers Recall from Lesson 0-2 that all real numbers are also complex numbers.

EXAMPLE 8 Find the Zeros of a Polynomial When One is Known

Find all complex zeros of $p(x) = x^4 - 6x^3 + 20x^2 - 22x - 13$ given that $2 - 3i$ is a zero of p. Then write the linear factorization of $p(x)$.

Use synthetic substitution to verify that $2 - 3i$ is a zero of $p(x)$.

$$
\begin{array}{r|rrrrr}
2-3i & 1 & -6 & 20 & -22 & -13 \\
      &   & 2-3i & -17+6i & & \\
\hline
      & 1 & -4-3i & & &
\end{array}
$$

$$
\begin{aligned}
(2-3i)(-4-3i) &= -8 + 6i + 9i^2 \\
&= -8 + 6i + 9(-1) \\
&= -17 + 6i
\end{aligned}
$$

$$
\begin{array}{r|rrrrr}
2-3i & 1 & -6 & 20 & -22 & -13 \\
      &   & 2-3i & -17+6i & 24+3i & \\
\hline
      & 1 & -4-3i & 3+6i & & 0
\end{array}
$$

$$
\begin{aligned}
(2-3i)(3+6i) &= 6 + 3i - 18i^2 \\
&= 6 + 3i - 18(-1) \\
&= 24 + 3i
\end{aligned}
$$

$$
\begin{array}{r|rrrrr}
2-3i & 1 & -6 & 20 & -22 & -13 \\
      &   & 2-3i & -17+6i & 24+3i & 13 \\
\hline
      & 1 & -4-3i & 3+6i & 2+3i & 0
\end{array}
$$

$$
\begin{aligned}
(2-3i)(2+3i) &= 4 - 9i^2 \\
&= 4 - 9(-1) \\
&= 4 + 9 \text{ or } 13
\end{aligned}
$$

Because $2 - 3i$ is a zero of p, you know that $2 + 3i$ is also a zero of p. Divide the depressed polynomial by $2 + 3i$.

$$
\begin{array}{r|rrrr}
2+3i & 1 & -4-3i & 3+6i & 2+3i \\
      &   & 2+3i & -4-6i & -2-3i \\
\hline
      & 1 & -2 & -1 & 0
\end{array}
$$

Using these two zeros and the depressed polynomial from this last division, you can write $p(x) = [x - (2 - 3i)][x - (2 + 3i)](x^2 - 2x - 1)$.

Because $p(x)$ is a quartic polynomial, you know that it has exactly 4 zeros. Having found 2, you know that 2 more remain. Find the zeros of $x^2 - 2x - 1$ by using the Quadratic Formula.

$$x = \frac{-b \pm \sqrt{b^2 - 4ac}}{2a} \qquad \text{Quadratic Formula}$$

$$= \frac{-(-2) \pm \sqrt{(-2)^2 - 4(1)(-1)}}{2(1)} \qquad a = 1, b = -2, \text{ and } c = -1$$

$$= \frac{2 \pm \sqrt{8}}{2} \qquad \text{Simplify.}$$

$$= 1 \pm \sqrt{2} \qquad \text{Simplify.}$$

StudyTip

Dividing Out Common Factors Before applying any of the methods in this lesson, remember to factor out any common monomial factors. For example, $g(x) = -2x^4 + 6x^3 - 4x^2 - 8x$ should first be factored as $g(x) = -2x(x^3 - 3x^2 + 2x + 4)$, which implies that 0 is a zero of g.

Therefore, the four zeros of $p(x)$ are $2 - 3i$, $2 + 3i$, $1 + \sqrt{2}$, and $1 - \sqrt{2}$. The linear factorization of $p(x)$ is $[x - (2 - 3i)] \cdot [x - (2 + 3i)][x - (1 + \sqrt{2})][x - (1 - \sqrt{2})]$.

Using the graph of $p(x)$, you can verify that the function has two real zeros at $1 + \sqrt{2}$ or about 2.41 and $1 - \sqrt{2}$ or about -0.41.

$[-4, 6]$ scl: 1 by $[-40, 40]$ scl: 8

✓ **Guided Practice**

For each function, use the given zero to find all the complex zeros of the function. Then write the linear factorization of the function.

8A. $g(x) = x^4 - 10x^3 + 35x^2 - 46x + 10; 2 + \sqrt{3}$

8B. $h(x) = x^4 - 8x^3 + 26x^2 - 8x - 95; 1 - \sqrt{6}$

List all possible rational zeros of each function. Then determine which, if any, are zeros. (Examples 1 and 2)

1. $g(x) = x^4 - 6x^3 - 31x^2 + 216x - 180$

2. $f(x) = 4x^3 - 24x^2 - x + 6$

3. $g(x) = x^4 - x^3 - 31x^2 + x + 30$

4. $g(x) = -4x^4 + 35x^3 - 87x^2 + 56x + 20$

5. $h(x) = 6x^4 + 13x^3 - 67x^2 - 156x - 60$

6. $f(x) = 18x^4 + 12x^3 + 56x^2 + 48x - 64$

7. $h(x) = x^5 - 11x^4 + 49x^3 - 147x^2 + 360x - 432$

8. $g(x) = 8x^5 + 18x^4 - 5x^3 - 72x^2 - 162x + 45$

9. MANUFACTURING The specifications for the dimensions of a new cardboard container are shown. If the volume of the container is modeled by $V(h) = 2h^3 - 9h^2 + 4h$ and it will hold 45 cubic inches of merchandise, what are the container's dimensions? (Example 3)

Solve each equation. (Example 3)

10. $x^4 + 2x^3 - 7x^2 - 20x - 12 = 0$

11. $x^4 + 9x^3 + 23x^2 + 3x - 36 = 0$

12. $x^4 - 2x^3 - 7x^2 + 8x + 12 = 0$

13. $x^4 - 3x^3 - 20x^2 + 84x - 80 = 0$

14. $x^4 + 34x = 6x^3 + 21x^2 - 48$

15. $6x^4 + 41x^3 + 42x^2 - 96x + 6 = -26$

16. $-12x^4 + 77x^3 = 136x^2 - 33x - 18$

17. SALES The sales $S(x)$ in thousands of dollars that a store makes during one month can be approximated by $S(x) = 2x^3 - 2x^2 + 4x$, where x is the number of days after the first day of the month. How many days will it take the store to make $16,000? (Example 3)

Determine an interval in which all real zeros of each function must lie. Explain your reasoning using the upper and lower bound tests. Then find all the real zeros. (Example 4)

18. $f(x) = x^4 - 9x^3 + 12x^2 + 44x - 48$

19. $f(x) = 2x^4 - x^3 - 29x^2 + 34x + 24$

20. $g(x) = 2x^4 + 4x^3 - 18x^2 - 4x + 16$

21. $g(x) = 6x^4 - 33x^3 - 6x^2 + 123x - 90$

22. $f(x) = 2x^4 - 17x^3 + 39x^2 - 16x - 20$

23. $f(x) = 2x^4 - 13x^3 + 21x^2 + 9x - 27$

24. $h(x) = x^5 - x^4 - 9x^3 + 5x^2 + 16x - 12$

25. $h(x) = 4x^5 - 20x^4 + 5x^3 + 80x^2 - 75x + 18$

Describe the possible real zeros of each function. (Example 5)

26. $f(x) = -2x^3 - 3x^2 + 4x + 7$

27. $f(x) = 10x^4 - 3x^3 + 8x^2 - 4x - 8$

28. $f(x) = -3x^4 - 5x^3 + 4x^2 + 2x - 6$

29. $f(x) = 12x^4 + 6x^3 + 3x^2 - 2x + 12$

30. $g(x) = 4x^5 + 3x^4 + 9x^3 - 8x^2 + 16x - 24$

31. $h(x) = -4x^5 + x^4 - 8x^3 - 24x^2 + 64x - 124$

Write a polynomial function of least degree with real coefficients in standard form that has the given zeros. (Example 6)

32. $3, -4, 6, -1$

33. $-2, -4, -3, 5$

34. $-5, 3, 4 + i$

35. $-1, 8, 6 - i$

36. $2\sqrt{5}, -2\sqrt{5}, -3, 7$

37. $-5, 2, 4 - \sqrt{3}, 4 + \sqrt{3}$

38. $\sqrt{7}, -\sqrt{7}, 4i$

39. $\sqrt{6}, -\sqrt{6}, 3 - 4i$

40. $2 + \sqrt{3}, 2 - \sqrt{3}, 4 + 5i$

41. $6 - \sqrt{5}, 6 + \sqrt{5}, 8 - 3i$

Write each function as (a) the product of linear and irreducible quadratic factors and (b) the product of linear factors. Then (c) list all of its zeros. (Example 7)

42. $g(x) = x^4 - 3x^3 - 12x^2 + 20x + 48$

43 $g(x) = x^4 - 3x^3 - 12x^2 + 8$

44. $h(x) = x^4 + 2x^3 - 15x^2 + 18x - 216$

45. $f(x) = 4x^4 - 35x^3 + 140x^2 - 295x + 156$

46. $f(x) = 4x^4 - 15x^3 + 43x^2 + 577x + 615$

47. $h(x) = x^4 - 2x^3 - 17x^2 + 4x + 30$

48. $g(x) = x^4 + 31x^2 - 180$

Use the given zero to find all complex zeros of each function. Then write the linear factorization of the function. (Example 8)

49. $h(x) = 2x^5 + x^4 - 7x^3 + 21x^2 - 225x + 108; 3i$

50. $h(x) = 3x^5 - 5x^4 - 13x^3 - 65x^2 - 2200x + 1500; -5i$

51. $g(x) = x^5 - 2x^4 - 13x^3 + 28x^2 + 46x - 60; 3 - i$

52. $g(x) = 4x^5 - 57x^4 + 287x^3 - 547x^2 + 83x + 510; 4 + i$

53. $f(x) = x^5 - 3x^4 - 4x^3 + 12x^2 - 32x + 96; -2i$

54. $g(x) = x^4 - 10x^3 + 35x^2 - 46x + 10; 3 + i$

55. ARCHITECTURE An architect is constructing a scale model of a building that is in the shape of a pyramid.

a. If the height of the scale model is 9 inches less than its length and its base is a square, write a polynomial function that describes the volume of the model in terms of its length.

b. If the volume of the model is 6300 cubic inches, write an equation describing the situation.

c. What are the dimensions of the scale model?

56. CONSTRUCTION The height of a tunnel that is under construction is 1 foot more than half its width and its length is 32 feet more than 324 times its width. If the volume of the tunnel is 62,231,040 cubic feet and it is a rectangular prism, find the length, width, and height.

Write a polynomial function of least degree with integer coefficients that has the given number as a zero.

57. $\sqrt[3]{6}$

58. $\sqrt[3]{5}$

59. $-\sqrt[3]{2}$

60. $-\sqrt[3]{7}$

Use each graph to write g as the product of linear factors. Then list all of its zeros.

61. $g(x) = 3x^4 - 15x^3 + 87x^2 - 375x + 300$

[−2, 8] scl: 1 by [−300, 200] scl: 50

62. $g(x) = 2x^5 + 2x^4 + 28x^3 + 32x^2 - 64x$

[−4, 4] scl: 1 by [−40, 80] scl: 12

Determine all rational zeros of the function.

63. $h(x) = 6x^3 - 6x^2 + 12$

64. $f(y) = \frac{1}{4}y^4 + \frac{1}{2}y^3 - y^2 + 2y - 8$

65. $w(z) = z^4 - 10z^3 + 30z^2 - 10z + 29$

66. $b(a) = a^5 - \frac{5}{6}a^4 + \frac{2}{3}a^3 - \frac{2}{3}a^2 - \frac{1}{3}a + \frac{1}{6}$

67. ENGINEERING A steel beam is supported by two pilings 200 feet apart. If a weight is placed x feet from the piling on the left, a vertical deflection represented by $d = 0.0000008x^2(200 - x)$ occurs. How far is the weight from the piling if the vertical deflection is 0.8 feet?

Write each polynomial as the product of linear and irreducible quadratic factors.

68. $x^3 - 3$

69. $x^3 + 16$

70. $8x^3 + 9$

71. $27x^6 + 4$

72. MULTIPLE REPRESENTATIONS In this problem, you will explore even- and odd-degree polynomial functions.

a. ANALYTICAL Identify the degree and number of zeros of each polynomial function.

   i. $f(x) = x^3 - x^2 + 9x - 9$

   ii. $g(x) = 2x^5 + x^4 - 32x - 16$

   iii. $h(x) = 5x^3 + 2x^2 - 13x + 6$

   iv. $f(x) = x^4 + 25x^2 + 144$

   v. $h(x) = 3x^6 + 5x^5 + 46x^4 + 80x^3 - 32x^2$

   vi. $g(x) = 4x^4 - 11x^3 + 10x^2 - 11x + 6$

b. NUMERICAL Find the zeros of each function.

c. VERBAL Does an odd-degree function have to have a minimum number of real zeros? Explain.

H.O.T. Problems   Use Higher-Order Thinking Skills

73. ERROR ANALYSIS Angie and Julius are using the Rational Zeros Theorem to find all the possible rational zeros of $f(x) = 7x^2 + 2x^3 - 5x - 3$. Angie thinks the possible zeros are $\pm\frac{1}{7}$, $\pm\frac{3}{7}$, ±1, ±3. Julius thinks they are $\pm\frac{1}{2}$, $\pm\frac{3}{2}$, ±1, ±3. Is either of them correct? Explain your reasoning.

74. REASONING Explain why $g(x) = x^9 - x^8 + x^5 + x^3 - x^2 + 2$ must have a root between $x = -1$ and $x = 0$.

75. CHALLENGE Use $f(x) = x^2 + x - 6$, $f(x) = x^3 + 8x^2 + 19x + 12$, and $f(x) = x^4 - 2x^3 - 21x^2 + 22x + 40$ to make a conjecture about the relationship between the graphs and zeros of $f(x)$ and the graphs and zeros of each of the following.

a. $-f(x)$

b. $f(-x)$

76. OPEN ENDED Write a function of 4th degree with an imaginary zero and an irrational zero.

77. REASONING Determine whether the statement is *true* or *false*. If false, provide a counterexample.
A third-degree polynomial with real coefficients has at least one nonreal zero.

CHALLENGE Find the zeros of each function if $h(x)$ has zeros at x_1, x_2, and x_3.

78. $c(x) = 7h(x)$

79. $k(x) = h(3x)$

80. $g(x) = h(x - 2)$

81. $f(x) = h(-x)$

82. REASONING If $x - c$ is a factor of $f(x) = a_1x^5 - a_2x^4 + \ldots$, what value must c be greater than or equal to in order to be an upper bound for the zeros of $f(x)$? Assume $a \neq 0$. Explain your reasoning.

83. WRITING IN MATH Explain why a polynomial with real coefficients and one imaginary zero must have at least two imaginary zeros.

Divide using synthetic division. (Lesson 2-3)

84. $(x^3 - 9x^2 + 27x - 28) \div (x - 3)$

85. $(x^4 + x^3 - 1) \div (x - 2)$

86. $(3x^4 - 2x^3 + 5x^2 - 4x - 2) \div (x + 1)$

87. $(2x^3 - 2x - 3) \div (x - 1)$

Describe the end behavior of the graph of each polynomial function using limits. Explain your reasoning using the leading term test. (Lesson 2-2)

88. $f(x) = -4x^7 + 3x^4 + 6$

89. $f(x) = 4x^6 + 2x^5 + 7x^2$

90. $g(x) = 3x^4 + 5x^5 - 11$

Estimate to the nearest 0.5 unit and classify the extrema for the graph of each function. Support the answers numerically. (Lesson 1-4)

91.

$f(x) = (x - 1)(x + 1)(x + 3)$

92.

$f(x) = x(x - 2)(x + 2)(x - 4)$

93.

$f(x) = x^2(x - 3)(x + 4)^2$

94. FINANCE Investors choose different stocks to comprise a balanced portfolio. The matrices show the prices of one share of each of several stocks on the first business day of July, August, and September. (Lesson 0-6)

	July	August	September
Stock A	[33.81	30.94	27.25]
Stock B	[15.06	13.25	8.75]
Stock C	[54	54	46.44]
Stock D	[52.06	44.69	34.38]

a. Mrs. Rivera owns 42 shares of stock A, 59 shares of stock B, 21 shares of stock C, and 18 shares of stock D. Write a row matrix to represent Mrs. Rivera's portfolio.

b. Use matrix multiplication to find the total value of Mrs. Rivera's portfolio for each month to the nearest cent.

95. SAT/ACT A circle is inscribed in a square and intersects the square at points A, B, C, and D. If $AC = 12$, what is the total area of the shaded regions?

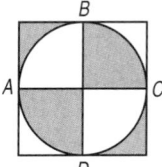

   A 18   **D** 24π

   B 36   **E** 72

   C 18π

96. REVIEW $f(x) = x^2 - 4x + 3$ has a relative minimum located at which of the following x-values?

   F -2 **H** 3

   G 2   **J** 4

97. Find all of the zeros of $p(x) = x^3 + 2x^2 - 3x + 20$.

   A $-4, 1 + 2i, 1 - 2i$ **C** $-1, 1, 4 + i, 4 - i$

   B $1, 4 + i, 4 - i$ **D** $4, 1 + i, 1 - i$

98. REVIEW Which expression is equivalent to $(t^2 + 3t - 9)(5 - t)^{-1}$?

   F $t + 8 - \dfrac{31}{5 - t}$

   G $-t - 8$

   H $-t - 8 + \dfrac{31}{5 - t}$

   J $-t - 8 - \dfrac{31}{5 - t}$

Rational Functions

Why?

Water desalination, or removing the salt from sea water, is currently in use in areas of the world with limited water availability and on many ships and submarines. It is also being considered as an alternative for providing water in the future. The cost for various extents of desalination can be modeled using rational functions.

5. Freshwater storage
4. Post-treatment
3. Reverse osmosis process
Concentrated sea-water disposal
Porous layer
Treated water
2. Pretreatment system
Mesh spacer
Semi-permeable membrane
1. Seawater supply
Salt water

Rational Functions A **rational function** $f(x)$ is the quotient of two polynomial functions $a(x)$ and $b(x)$, where b is nonzero.

$$f(x) = \frac{a(x)}{b(x)}, \; b(x) \neq 0$$

The domain of a rational function is the set of all real numbers excluding those values for which $b(x) = 0$ or the zeros of $b(x)$.

One of the simplest rational functions is the reciprocal function $f(x) = \frac{1}{x}$. The graph of the reciprocal function, like many rational functions, has branches that approach specific x- and y-values. The lines representing these values are called **asymptotes**.

The reciprocal function is undefined when $x = 0$, so its domain is $(-\infty, 0)$ or $(0, \infty)$. The behavior of $f(x) = \frac{1}{x}$ to the left (0^-) and right (0^+) of $x = 0$ can be described using limits.

$$\lim_{x \to 0^-} f(x) = -\infty \qquad\qquad \lim_{x \to 0^+} f(x) = \infty$$

From Lesson 1-3, you should recognize 0 as a point of infinite discontinuity in the domain of f. The line $x = 0$ in Figure 2.5.1 is called a *vertical asymptote* of the graph of f. The end behavior of f can be also be described using limits.

$$\lim_{x \to -\infty} f(x) = 0 \qquad\qquad \lim_{x \to +\infty} f(x) = 0$$

The line $y = 0$ in Figure 2.5.2 is called a *horizontal asymptote* of the graph of f.

Figure 2.5.1

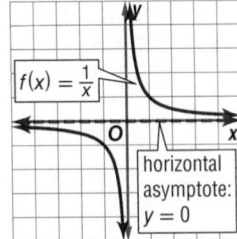

Figure 2.5.2

These definitions of vertical and horizontal asymptotes can be generalized.

You can use your knowledge of limits, discontinuity, and end behavior to determine the vertical and horizontal asymptotes, if any, of a rational function.

Key Concept — Vertical and Horizontal Asymptotes

Words The line $x = c$ is a **vertical asymptote** of the graph of f if $\lim\limits_{x \to c^-} f(x) = \pm\infty$ or $\lim\limits_{x \to c^+} f(x) = \pm\infty$.	**Words** The line $y = c$ is a **horizontal asymptote** of the graph of f if $\lim\limits_{x \to -\infty} f(x) = c$ or $\lim\limits_{x \to \infty} f(x) = c$.
Example	**Example** 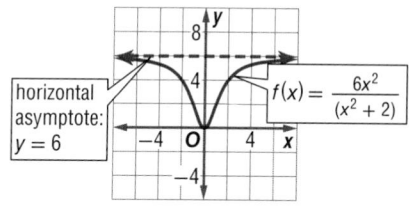

ReadingMath

Limit Notation The expression $\lim\limits_{x \to c^-} f(x)$ is read as the *limit of f of x as x approaches c from the left* and the expression $\lim\limits_{x \to c^+} f(x)$ is read as the *limit of f of x as x approaches c from the right*.

EXAMPLE 1 — Find Vertical and Horizontal Asymptotes

Find the domain of each function and the equations of the vertical or horizontal asymptotes, if any.

a. $f(x) = \dfrac{x + 4}{x - 3}$

Step 1 Find the domain.

The function is undefined at the real zero of the denominator $b(x) = x - 3$. The real zero of $b(x)$ is 3. Therefore, the domain of f is all real numbers except $x = 3$.

Step 2 Find the asymptotes, if any.

Check for vertical asymptotes.

Determine whether $x = 3$ is a point of infinite discontinuity. Find the limit as x approaches 3 from the left and the right.

x	2.9	2.99	2.999	3	3.001	3.01	3.1
f(x)	−69	−699	−6999	undefined	7001	701	71

Because $\lim\limits_{x \to 3^-} f(x) = -\infty$ and $\lim\limits_{x \to 3^+} f(x) = \infty$, you know that $x = 3$ is a vertical asymptote of f.

Check for horizontal asymptotes.

Use a table to examine the end behavior of $f(x)$.

x	−10,000	−1000	−100	0	100	1000	10,000
f(x)	0.9993	0.9930	0.9320	−1.33	1.0722	1.0070	1.0007

The table suggests that $\lim\limits_{x \to -\infty} f(x) = \lim\limits_{x \to \infty} f(x) = 1$. Therefore, you know that $y = 1$ is a horizontal asymptote of f.

CHECK The graph of $f(x) = \dfrac{x + 4}{x - 3}$ shown supports each of these findings. ✔

b. $g(x) = \dfrac{8x^2 + 5}{4x^2 + 1}$

Step 1 The zeros of the denominator $b(x) = 4x^2 + 1$ are imaginary, so the domain of g is all real numbers.

Step 2 Because the domain of g is all real numbers, the function has no vertical asymptotes.

Using division, you can determine that

$$g(x) = \frac{8x^2 + 5}{4x^2 + 1} = 2 + \frac{3}{4x^2 + 1}.$$

As the value of $|x|$ increases, $4x^2 + 1$ becomes an increasing large positive number and $\dfrac{3}{4x^2 + 1}$ decreases, approaching 0. Therefore,

$$\lim_{x \to -\infty} g(x) = \lim_{x \to \infty} g(x) = 2 + 0 \text{ or } 2.$$

CHECK You can use a table of values to support this reasoning. The graph of $g(x) = \dfrac{8x^2 + 5}{4x^2 + 1}$ shown also supports each of these findings. ✔

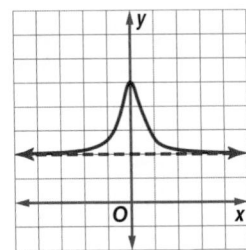

☑ Guided Practice

Find the domain of each function and the equations of the vertical or horizontal asymptotes, if any.

1A. $m(x) = \dfrac{15x + 3}{x + 5}$

1B. $h(x) = \dfrac{x^2 - x - 6}{x + 4}$

The analysis in Example 1 suggests a connection between the end behavior of a function and its horizontal asymptote. This relationship, along with other features of the graphs of rational functions, is summarized below.

Key Concept   **Graphs of Rational Functions**

If f is the rational function given by

$$f(x) = \frac{a(x)}{b(x)} = \frac{a_n x^n + a_{n-1}x^{n-1} + \ldots + a_1 x + a_0}{b_m x^m + b_{m-1}x^{m-1} + \ldots + b_1 x + b_0},$$

where $b(x) \neq 0$ and $a(x)$ and $b(x)$ have no common factors other than ± 1, then the graph of f has the following characteristics.

Vertical Asymptotes Vertical asymptotes may occur at the real zeros of $b(x)$.

Horizontal Asymptote The graph has either one or no horizontal asymptotes as determined by comparing the degree n of $a(x)$ to the degree m of $b(x)$.

- If $n < m$, the horizontal asymptote is $y = 0$.
- If $n = m$, the horizontal asymptote is $y = \dfrac{a_n}{b_m}$.
- If $n > m$, there is no horizontal asymptote.

Intercepts The x-intercepts, if any, occur at the real zeros of $a(x)$. The y-intercept, if it exists, is the value of f when $x = 0$.

StudyTip

Poles A vertical asymptote in the graph of a rational function is also called a *pole* of the function.

StudyTip

Test Intervals A rational function can change sign at its zeros and its undefined values, so when these x-values are ordered, they divide the domain of the function into test intervals that can help you determine if the graph lies above or below the x-axis.

To graph a rational function, simplify f, if possible, and then follow these steps.

Step 1 Find the domain.

Step 2 Find and sketch the asymptotes, if any.

Step 3 Find and plot the x-intercepts and y-intercept, if any.

Step 4 Find and plot at least one point in the *test intervals* determined by any x-intercepts and vertical asymptotes.

EXAMPLE 2   Graph Rational Functions: $n < m$ and $n > m$

For each function, determine any vertical and horizontal asymptotes and intercepts. Then graph the function, and state its domain.

a. $g(x) = \dfrac{6}{x + 3}$

Step 1 The function is undefined at $b(x) = 0$, so the domain is $\{x \mid x \neq -3, x \in \mathbb{R}\}$.

Step 2 There is a vertical asymptote at $x = -3$.

  The degree of the polynomial in the numerator is 0, and the degree of the polynomial in the denominator is 1. Because $0 < 1$, the graph of g has a horizontal asymptote at $y = 0$.

Step 3 The polynomial in the numerator has no real zeros, so g has no x-intercepts. Because $g(0) = 2$, the y-intercept is 2.

Step 4 Graph the asymptotes and intercepts. Then choose x-values that fall in the test intervals determined by the vertical asymptote to find additional points to plot on the graph. Use smooth curves to complete the graph.

StudyTip

Hyperbola The graphs of the reciprocal functions $f(x) = \dfrac{1}{x}$ and $g(x) = \dfrac{6}{x+3}$ are called *hyperbolas*. You will learn more about hyperbolas in Chapter 7.

Interval	x	(x, g(x))
$(-\infty, -3)$	-8	$(-8, -1.2)$
	-6	$(-6, -2)$
	-4	$(-4, -6)$
$(-3, \infty)$	-2	$(-2, 6)$
	2	$(2, 1.2)$

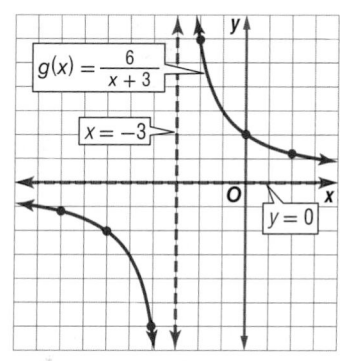

b. $k(x) = \dfrac{x^2 - 7x + 10}{x - 3}$

Factoring the numerator yields $k(x) = \dfrac{(x - 2)(x - 5)}{x - 3}$. Notice that the numerator and denominator have no common factors, so the expression is in simplest form.

Step 1 The function is undefined at $b(x) = 0$, so the domain is $\{x \mid x \neq 3, x \in \mathbb{R}\}$.

Step 2 There is a vertical asymptote at $x = 3$.

  Compare the degrees of the numerator and denominator. Because $2 > 1$, there is no horizontal asymptote.

Step 3 The numerator has zeros at $x = 2$ and $x = 5$, so the x-intercepts are 2 and 5. $k(0) = -\dfrac{10}{3}$, so the y-intercept is at about -3.3.

Step 4 Graph the asymptotes and intercepts. Then find and plot points in the test intervals determined by the intercepts and vertical asymptotes: $(-\infty, 0)$, $(0, 3)$, $(3, \infty)$. Use smooth curves to complete the graph.

✔ **Guided Practice**

2A. $h(x) = \dfrac{2}{x^2 + 2x - 3}$

2B. $n(x) = \dfrac{x}{x^2 + x - 2}$

In Example 3, the degree of the numerator is *equal to* the degree of the denominator.

Graph a Rational Function: $n = m$

Determine any vertical and horizontal asymptotes and intercepts for $f(x) = \dfrac{3x^2 - 3}{x^2 - 9}$. Then graph the function, and state its domain.

Factoring both numerator and denominator yields $f(x) = \dfrac{3(x-1)(x+1)}{(x-3)(x+3)}$ with no common factors.

Step 1 The function is undefined at $b(x) = 0$, so the domain is $\{x \mid x \neq -3, 3, x \in \mathbb{R}\}$.

Step 2 There are vertical asymptotes at $x = 3$ and $x = -3$.

There is a horizontal asymptote at $y = \dfrac{3}{1}$ or $y = 3$, the ratio of the leading coefficients of the numerator and denominator, because the degrees of the polynomials are equal.

Step 3 The x-intercepts are 1 and -1, the zeros of the numerator. The y-intercept is $\dfrac{1}{3}$ because $f(0) = \dfrac{1}{3}$.

Step 4 Graph the asymptotes and intercepts. Then find and plot points in the test intervals $(-\infty, -3)$, $(-3, -1)$, $(-1, 1)$, $(1, 3)$, and $(3, \infty)$.

✔ Guided Practice

For each function, determine any vertical and horizontal asymptotes and intercepts. Then graph the function and state its domain.

3A. $h(x) = \dfrac{x - 6}{x + 2}$

3B. $h(x) = \dfrac{x^2 - 4}{5x^2 - 5}$

When the degree of the numerator is *exactly one more* than the degree of the denominator, the graph has a *slant* or **oblique asymptote**.

Nonlinear Asymptotes
Horizontal, vertical, and oblique asymptotes are all linear. A rational function can also have a nonlinear asymptote. For example, the graph of $f(x) = \dfrac{x^3}{x-1}$ has a quadratic asymptote.

Key Concept **Oblique Asymptotes**

If f is the rational function given by

$$f(x) = \frac{a(x)}{b(x)} = \frac{a_n x^n + a_{n-1}x^{n-1} + \ldots + a_1 x + a_0}{b_m x^m + b_{m-1}x^{m-1} + \ldots + b_1 x + b_0},$$

where $b(x)$ has a degree greater than 0 and $a(x)$ and $b(x)$ have no common factors other than 1, then the graph of f has an oblique asymptote if $n = m + 1$. The function for the oblique asymptote is the quotient polynomial $q(x)$ resulting from the division of $a(x)$ by $b(x)$.

$$f(x) = \frac{a(x)}{b(x)} = \underbrace{q(x)}_{\text{function for oblique asymptote}} + \frac{r(x)}{b(x)},$$

Example

EXAMPLE 4 Graph a Rational Function: $n = m + 1$

Determine any asymptotes and intercepts for $f(x) = \dfrac{2x^3}{x^2 + x - 12}$. Then graph the function, and state its domain.

Factoring the denominator yields $f(x) = \dfrac{2x^3}{(x + 4)(x - 3)}$.

Step 1 The function is undefined at $b(x) = 0$, so the domain is $\{x \mid x \neq -4, 3, x \in \mathbb{R}\}$.

Step 2 There are vertical asymptotes at $x = -4$ and $x = 3$.

The degree of the numerator is greater than the degree of the denominator, so there is no horizontal asymptote.

Because the degree of the numerator is exactly one more than the degree of the denominator, f has a slant asymptote. Using polynomial division, you can write the following.

$$f(x) = \frac{2x^3}{x^2 + x - 12}$$
$$= 2x - 2 + \frac{26x - 24}{x^2 + x - 12}$$

Therefore, the equation of the slant asymptote is $y = 2x - 2$.

Step 3 The x- and y-intercepts are 0 because 0 is the zero of the numerator and $f(0) = 0$.

Step 4 Graph the asymptotes and intercepts. Then find and plot points in the test intervals $(-\infty, -4)$, $(-4, 0)$, $(0, 3)$, and $(3, \infty)$.

StudyTip

End-Behavior Asymptote In Example 4, the graph of f *approaches* the slant asymptote $y = 2x - 2$ as $x \to \pm\infty$. Between the vertical asymptotes $x = -4$ and $x = 3$, however, the graph *crosses* the line $y = 2x - 2$. For this reason, a slant or horizontal asymptote is sometimes referred to as an *end-behavior asymptote*.

✔ **Guided Practice**

For each function, determine any asymptotes and intercepts. Then graph the function and state its domain.

4A. $h(x) = \dfrac{x^2 + 3x - 3}{x + 4}$

4B. $p(x) = \dfrac{x^2 - 4x + 1}{2x - 3}$

When the numerator and denominator of a rational function have common factors, the graph of the function has removable discontinuities called **holes**, at the zeros of the common factors. Be sure to indicate these points of discontinuity when you graph the function.

StudyTip

Removable and Nonremovable Discontinuities If the function is not continuous at $x = a$, but could be made continuous at that point by simplifying, then the function has a *removable discontinuity* at $x = a$. Otherwise, it has a *nonremovable discontinuity* $x = a$.

$$f(x) = \frac{\cancel{(x - a)}(x - b)}{\cancel{(x - a)}(x - c)}$$

Divide out the common factor in the numerator and denominator. The zero of $x - a$ is a.

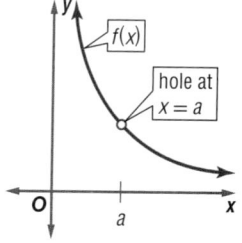

EXAMPLE 5　Graph a Rational Function with Common Factors

Determine any vertical and horizontal asymptotes, holes, and intercepts for $h(x) = \frac{x^2 - 4}{x^2 - 2x - 8}$.
Then graph the function, and state its domain.

Factoring both the numerator and denominator yields $h(x) = \frac{(x - 2)(x + 2)}{(x - 4)(x + 2)}$ or $\frac{x - 2}{x - 4}$, $x \neq -2$.

Step 1　The function is undefined at $b(x) = 0$, so the domain is $\{x \mid x \neq -2, 4, x \in \mathbb{R}\}$.

Step 2　There is a vertical asymptote at $x = 4$, the real zero of the simplified denominator.
There is a horizontal asymptote at $y = \frac{1}{1}$ or 1, the ratio of the leading coefficients of the
numerator and denominator, because the degrees of the polynomials are equal.

Step 3　The x-intercept is 2, the zero of the simplified numerator. The y-intercept is
$\frac{1}{2}$ because $h(0) = \frac{1}{2}$.

Step 4　Graph the asymptotes and intercepts.
Then find and plot points in the test
intervals $(-\infty, 2)$, $(2, 4)$, and $(4, \infty)$.
There is a hole at $\left(-2, \frac{2}{3}\right)$ because
the original function is undefined
when $x = -2$.

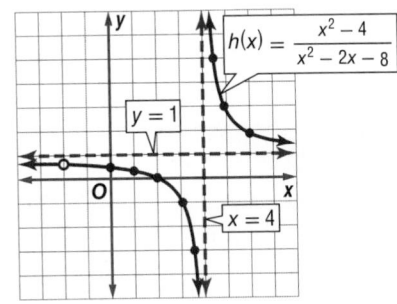

StudyTip

Hole　For Example 5,
$x + 2$ was divided out of
the original expression.
Substitute -2 into the
new expression.

$h(-2) = \frac{(-2) - 2}{(-2) - 4}$

$\quad\quad = \frac{-4}{-6}$ or $\frac{2}{3}$

There is a hole at $\left(-2, \frac{2}{3}\right)$.

✓ Guided Practice

For each function, determine any vertical and horizontal asymptotes, holes, and intercepts.
Then graph the function and state its domain.

5A.　$g(x) = \frac{x^2 + 10x + 24}{x^2 + x - 12}$

5B.　$c(x) = \frac{x^2 - 2x - 3}{x^2 - 4x - 5}$

Rational Equations　Rational equations involving fractions can be solved by multiplying each
term in the equation by the least common denominator (LCD) of all the terms of the equation.

StudyTip

Check for Reasonableness
You can also check the
result in Example 6 by
using a graphing calculator
to graph $y = x + \frac{6}{x - 8}$.
Use the CALC menu to
locate the zeros. Because
the zeros of the graph
appear to be at about
$x = 7.16$ and $x = 0.84$, the
solution is reasonable.

[−20, 20] scl: 2 by
[−20, 20] scl: 2

EXAMPLE 6　Solve a Rational Equation

Solve $x + \frac{6}{x - 8} = 0$.

$x + \frac{6}{x - 8} = 0$　　　　Original equation

$x(x - 8) + \frac{6}{x - 8}(x - 8) = 0(x - 8)$　　　Multiply by the LCD, $x - 8$.

$x^2 - 8x + 6 = 0$　　　Distributive Property

$x = \frac{8 \pm \sqrt{(-8)^2 - 4(1)(6)}}{2(1)}$　　　Quadratic Formula

$x = \frac{8 \pm 2\sqrt{10}}{2}$ or $4 \pm \sqrt{10}$　　Simplify.

✓ Guided Practice

Solve each equation.

6A.　$\frac{20}{x + 3} - 4 = 0$

6B.　$\frac{9x}{x - 2} = 6$

Solving a rational equation can produce extraneous solutions. Always check your answers in the original equation.

EXAMPLE 7 Solve a Rational Equation with Extraneous Solutions

Solve $\dfrac{4}{x^2 - 6x + 8} = \dfrac{3x}{x - 2} + \dfrac{2}{x - 4}$.

The LCD of the expressions is $(x - 2)(x - 4)$, which are the factors of $x^2 - 6x + 8$.

$$\dfrac{4}{x^2 - 6x + 8} = \dfrac{3x}{x - 2} + \dfrac{2}{x - 4}$$ Original equation

$$(x - 2)(x - 4)\,\dfrac{4}{x^2 - 6x + 8} = (x - 2)(x - 4)\left(\dfrac{3x}{x - 2} + \dfrac{2}{x - 4}\right)$$ Multiply by the LCD.

$$4 = 3x(x - 4) + 2(x - 2)$$ Distributive Property

$$4 = 3x^2 - 10x - 4$$ Distributive Property

$$0 = 3x^2 - 10x - 8$$ Subtract 4 from each side.

$$0 = (3x + 2)(x - 4)$$ Factor.

$$x = -\dfrac{2}{3} \text{ or } x = 4$$ Solve.

Because the original equation is not defined when $x = 4$, you can eliminate this extraneous solution. So, the only solution is $-\dfrac{2}{3}$.

✔ **Guided Practice**

Solve each equation.

7A. $\dfrac{2x}{x + 3} + \dfrac{3}{x - 6} = \dfrac{27}{x^2 - 3x - 18}$

7B. $-\dfrac{12}{x^2 + 6x} = \dfrac{2}{x + 6} + \dfrac{x - 2}{x}$

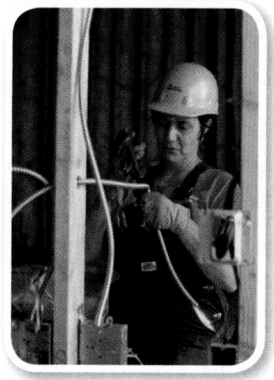

Real-World Career

Electrician Electricians install and maintain various components of electricity, such as wiring and fuses. They must maintain compliance with national, state, and local codes. Most electricians complete an apprenticeship program that includes both classroom instruction and on-the-job training.

🌐 **Real-World EXAMPLE 8** Solve a Rational Equation

ELECTRICITY The diagram of an electric circuit shows three parallel resistors. If R is the equivalent resistance of the three resistors, then $\dfrac{1}{R} = \dfrac{1}{R_1} + \dfrac{1}{R_2} + \dfrac{1}{R_3}$. In this circuit, R_1 is twice the resistance of R_2, and R_3 equals 20 ohms. Suppose the equivalent resistance is equal to 10 ohms. Find R_1 and R_2.

$$\dfrac{1}{R} = \dfrac{1}{R_1} + \dfrac{1}{R_2} + \dfrac{1}{R_3}$$ Original equation

$$\dfrac{1}{10} = \dfrac{1}{2R_2} + \dfrac{1}{R_2} + \dfrac{1}{20}$$ $R = 10$, $R_1 = 2R_2$, and $R_3 = 20$

$$\dfrac{1}{20} = \dfrac{1}{2R_2} + \dfrac{1}{R_2}$$ Subtract $\dfrac{1}{20}$ from each side.

$$(20R_2)\,\dfrac{1}{20} = (20R_2)\left(\dfrac{1}{2R_2} + \dfrac{1}{R_2}\right)$$ Multiply each side by the LCD, $20R_2$.

$$R_2 = 10 + 20 \text{ or } 30$$ Simplify.

R_2 is 30 ohms and $R_1 = 2R_2$ or 60 ohms.

✔ **Guided Practice**

8. ELECTRONICS Suppose the current I, in amps, in an electric circuit is given by the formula $I = t + \dfrac{1}{10 - t}$, where t is time in seconds. At what time is the current 1 amp?

Exercises

Find the domain of each function and the equations of the vertical or horizontal asymptotes, if any. (Example 1)

1. $f(x) = \dfrac{x^2 - 2}{x^2 - 4}$

2. $h(x) = \dfrac{x^3 - 8}{x + 4}$

3. $f(x) = \dfrac{x(x-1)(x+2)^2}{(x+3)(x-4)}$

4. $g(x) = \dfrac{x - 6}{(x+3)(x+5)}$

5. $h(x) = \dfrac{2x^2 - 4x + 1}{x^2 + 2x}$

6. $f(x) = \dfrac{x^2 + 9x + 20}{x - 4}$

7. $h(x) = \dfrac{(x-1)(x+1)}{(x-2)^2(x+4)^2}$

8. $g(x) = \dfrac{(x-4)(x+2)}{(x+1)(x-3)}$

For each function, determine any asymptotes and intercepts. Then graph the function, and state its domain. (Examples 2–5)

9. $f(x) = \dfrac{(x+2)(x-3)}{(x+4)(x-5)}$

10. $g(x) = \dfrac{(2x+3)(x-6)}{(x+2)(x-1)}$

11. $f(x) = \dfrac{8}{(x-2)(x+2)}$

12. $f(x) = \dfrac{x+2}{x(x-6)}$

13. $g(x) = \dfrac{(x+2)(x+5)}{(x+5)^2(x-6)}$

14. $h(x) = \dfrac{(x+6)(x+4)}{x(x-5)(x+2)}$

15. $h(x) = \dfrac{x^2(x-2)(x+5)}{x^2 + 4x + 3}$

16. $f(x) = \dfrac{x(x+6)^2(x-4)}{x^2 - 5x - 24}$

17. $f(x) = \dfrac{x-8}{x^2 + 4x + 5}$

18. $g(x) = \dfrac{-4}{x^2 + 6}$

19. SALES The business plan for a new car wash projects that profits in thousands of dollars will be modeled by the function $p(z) = \dfrac{3z^2 - 3}{2z^2 + 7z + 5}$, where z is the week of operation and $z = 0$ represents opening. (Example 4)

    a. State the domain of the function.

    b. Determine any vertical and horizontal asymptotes and intercepts for $p(z)$.

    c. Graph the function.

For each function, determine any asymptotes, holes, and intercepts. Then graph the function and state its domain. (Examples 2–5)

20. $h(x) = \dfrac{3x - 4}{x^3}$

21. $h(x) = \dfrac{4x^2 - 2x + 1}{3x^3 + 4}$

22. $f(x) = \dfrac{x^2 + 2x - 15}{x^2 + 4x + 3}$

23. $g(x) = \dfrac{x + 7}{x - 4}$

24. $h(x) = \dfrac{x^3}{x + 3}$

25. $g(x) = \dfrac{x^3 + 3x^2 + 2x}{x - 4}$

26. $f(x) = \dfrac{x^2 - 4x - 21}{x^3 + 2x^2 - 5x - 6}$

(27) $g(x) = \dfrac{x^2 - 4}{x^3 + x^2 - 4x - 4}$

28. $f(x) = \dfrac{(x+4)(x-1)}{(x-1)(x+3)}$

29. $g(x) = \dfrac{(2x+1)(x-5)}{(x-5)(x+4)^2}$

30. STATISTICS A number x is said to be the *harmonic mean* of y and z if $\frac{1}{x}$ is the average of $\frac{1}{y}$ and $\frac{1}{z}$. (Example 7)

    a. Write an equation for which the solution is the harmonic mean of 30 and 45.

    b. Find the harmonic mean of 30 and 45.

31. OPTICS The lens equation is $\dfrac{1}{f} = \dfrac{1}{d_i} + \dfrac{1}{d_o}$, where f is the focal length, d_i is the distance from the lens to the image, and d_o is the distance from the lens to the object. Suppose the object is 32 centimeters from the lens and the focal length is 8 centimeters. (Example 7)

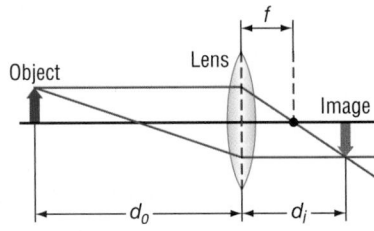

    a. Write a rational equation to model the situation.

    b. Find the distance from the lens to the image.

Solve each equation. (Examples 6–8)

32. $y + \dfrac{6}{y} = 5$

33. $\dfrac{8}{z} - z = 4$

34. $\dfrac{x-1}{2x-4} + \dfrac{x+2}{3x} = 1$

35. $\dfrac{2}{y+2} - \dfrac{y}{2-y} = \dfrac{y^2 + 4}{y^2 - 4}$

36. $\dfrac{3}{x} + \dfrac{2}{x+1} = \dfrac{23}{x^2 + x}$

37. $\dfrac{4}{x-2} - \dfrac{2}{x} = \dfrac{14}{x^2 - 2x}$

38. $\dfrac{x}{x+1} - \dfrac{x-1}{x} = \dfrac{1}{20}$

39. $\dfrac{6}{x-3} - \dfrac{4}{x+2} = \dfrac{12}{x^2 - x - 6}$

40. $\dfrac{x-1}{x-2} + \dfrac{3x+6}{2x+1} = 3$

41. $\dfrac{2}{a+3} - \dfrac{3}{4-a} = \dfrac{2a-2}{a^2 - a - 12}$

42. WATER The cost per day to remove x percent of the salt from seawater at a desalination plant is $c(x) = \dfrac{994x}{100 - x}$, where $0 \le x < 100$.

    a. Graph the function using a graphing calculator.

    b. Graph the line $y = 8000$ and find the intersection with the graph of $c(x)$ to determine what percent of salt can be removed for $8000 per day.

    c. According to the model, is it feasible for the plant to remove 100% of the salt? Explain your reasoning.

Write a rational function for each set of characteristics.

43. x-intercepts at $x = 0$ and $x = 4$, vertical asymptotes at $x = 1$ and $x = 6$, and a horizontal asymptote at $y = 0$

44. x-intercepts at $x = 2$ and $x = -3$, vertical asymptote at $x = 4$, and point discontinuity at $(-5, 0)$

45 TRAVEL When distance and time are held constant, the average rates, in miles per hour, during a round trip can be modeled by $r_2 = \dfrac{30r_1}{r_1 - 30}$, where r_1 represents the average rate during the first leg of the trip and r_2 represents the average rate during the return trip.

a. Find the vertical and horizontal asymptotes of the function, if any. Verify your answer graphically.

b. Copy and complete the table shown.

r_1	45	50	55	60	65	70
r_2						

c. Is a domain of $r_1 > 30$ reasonable for this situation? Explain your reasoning.

Use your knowledge of asymptotes and the provided points to express the function represented by each graph.

46.

47.

Use the intersection feature of a graphing calculator to solve each equation.

48. $\dfrac{x^4 - 2x^3 + 1}{x^3 + 6} = 8$

49. $\dfrac{2x^4 - 5x^2 + 3}{x^4 + 3x^2 - 4} = 1$

50. $\dfrac{3x^3 - 4x^2 + 8}{4x^4 + 2x - 1} = 2$

51. $\dfrac{2x^5 - 3x^3 + 5x}{4x^3 + 5x - 12} = 6$

52. CHEMISTRY When a 60% acetic acid solution is added to 10 liters of a 20% acetic acid solution in a 100-liter tank, the concentration of the total solution changes.

$V = 100$ L
a L, 60% acetic
10 L, 20% acetic acid

a. Show that the concentration of the solution is $f(a) = \dfrac{3a + 10}{5a + 50}$, where a is the volume of the 60% solution.

b. Find the relevant domain of $f(a)$ and the vertical or horizontal asymptotes, if any.

c. Explain the significance of any domain restrictions or asymptotes.

d. Disregarding domain restrictions, are there any additional asymptotes of the function? Explain.

53. **MULTIPLE REPRESENTATIONS** In this problem, you will investigate asymptotes of rational functions.

a. **TABULAR** Copy and complete the table. Determine the horizontal asymptote of each function algebraically.

Function	Horizontal Asymptote
$f(x) = \dfrac{x^2 - 5x + 4}{x^3 + 2}$	
$h(x) = \dfrac{x^3 - 3x^2 + 4x - 12}{x^4 - 4}$	
$g(x) = \dfrac{x^4 - 1}{x^5 + 3}$	

b. **GRAPHICAL** Graph each function and its horizontal asymptote from part **a**.

c. **TABULAR** Copy and complete the table below. Use the Rational Zero Theorem to help you find the real zeros of the numerator of each function.

Function	Real Zeros of Numerator
$f(x) = \dfrac{x^2 - 5x + 4}{x^3 + 2}$	
$h(x) = \dfrac{x^3 - 3x^2 + 4x - 12}{x^4 - 4}$	
$g(x) = \dfrac{x^4 - 1}{x^5 + 3}$	

d. **VERBAL** Make a conjecture about the behavior of the graph of a rational function when the degree of the denominator is greater than the degree of the numerator and the numerator has at least one real zero.

H.O.T. Problems   Use Higher-Order Thinking Skills

54. REASONING Given $f(x) = \dfrac{ax^3 + bx^2 + c}{dx^3 + ex^2 + f}$, will $f(x)$ *sometimes*, *always*, or *never* have a horizontal asymptote at $y = 1$ if a, b, c, d, e, and f are constants with $a \neq 0$ and $d \neq 0$. Explain.

55. PREWRITE Design a lesson plan to teach the graphing rational functions topics in this lesson. Make a plan that addresses purpose, audience, a controlling idea, logical sequence, and time frame for completion.

56. CHALLENGE Write a rational function that has vertical asymptotes at $x = -2$ and $x = 3$ and an oblique asymptote $y = 3x$.

57. WRITING IN MATH Use words, graphs, tables, and equations to show how to graph a rational function.

58. CHALLENGE Solve for k so that the rational equation has exactly one extraneous solution and one real solution.

$$\frac{2}{x^2 - 4x + k} = \frac{2x}{x - 1} + \frac{1}{x - 3}$$

59. WRITING IN MATH Explain why all of the test intervals must be used in order to get an accurate graph of a rational function.

List all the possible rational zeros of each function. Then determine which, if any, are zeros. (Lesson 2-4)

60. $f(x) = x^3 + 2x^2 - 5x - 6$ **61.** $f(x) = x^3 - 2x^2 + x + 18$ **62.** $f(x) = x^4 - 5x^3 + 9x^2 - 7x + 2$

Use the Factor Theorem to determine if the binomials given are factors of $f(x)$. Use the binomials that are factors to write a factored form of $f(x)$. (Lesson 2-3)

63. $f(x) = x^4 - 2x^3 - 13x^2 + 14x + 24; x - 3, x - 2$

64. $f(x) = 2x^4 - 5x^3 - 11x^2 - 4x; x - 4, 2x - 1$

65. $f(x) = 6x^4 + 59x^3 + 138x^2 - 45x - 50; 3x - 2, x - 5$

66. $f(x) = 4x^4 - 3x^3 - 12x^2 + 17x - 6; 4x - 3; x - 1$

67. $f(x) = 4x^5 + 15x^4 + 12x^3 - 4x^2; x + 2, 4x + 1$

68. $f(x) = 4x^5 - 8x^4 - 5x^3 + 10x^2 + x - 2; x + 1, x - 1$

Graph each function. (Lesson 2-2)

69. $f(x) = (x + 7)^2$ **70.** $f(x) = (x - 4)^3$ **71.** $f(x) = x^4 - 5$

72. RETAIL Sara is shopping at a store that offers $10 cash back for every $50 spent. Let $f(x) = \left\lfloor \dfrac{x}{50} \right\rfloor$ and $h(x) = 10x$, where x is the amount of money Sara spends. (Lesson 1-6)

   a. If Sara spends money at the store, is the cash back bonus represented by $f[h(x)]$ or $h[f(x)]$? Explain your reasoning.

   b. Determine the cash back bonus if Sara spends $312.68 at the store.

73. INTERIOR DESIGN Adrienne Herr is an interior designer. She has been asked to locate an oriental rug for a new corporate office. The rug should cover half of the total floor area with a uniform width surrounding the rug. (Lesson 0-3)

   a. If the dimensions of the room are 12 feet by 16 feet, write an equation to model the area of the rug in terms of x.

   b. Graph the related function.

   c. What are the dimensions of the rug?

Simplify. (Lesson 0-2)

74. $i^{10} + i^2$ **75.** $(2 + 3i) + (-6 + i)$ **76.** $(2.3 + 4.1i) - (-1.2 - 6.3i)$

77. SAT/ACT A company sells ground coffee in two sizes of cylindrical containers. The smaller container holds 10 ounces of coffee. If the larger container has twice the radius of the smaller container and 1.5 times the height, how many ounces of coffee does the larger container hold? (The volume of a cylinder is given by the formula $V = \pi r^2 h$.)

   A 30   C 60   E 90
   B 45   D 75

78. What are the solutions of $1 = \dfrac{2}{x^2} + \dfrac{2}{x}$?

   F $x = 1, x = -2$ H $x = 1 + \sqrt{3}, x = 1 - \sqrt{3}$

   G $x = -2, x = 1$ J $x = \dfrac{1 + \sqrt{3}}{2}, x = \dfrac{1 - \sqrt{3}}{2}$

79. REVIEW Alex wanted to determine the average of his 6 test scores. He added the scores correctly to get T but divided by 7 instead of 6. The result was 12 less than his actual average. Which equation could be used to determine the value of T?

   A $6T + 12 = 7T$ C $\dfrac{T}{7} + 12 = \dfrac{T}{6}$

   B $\dfrac{T}{7} = \dfrac{T - 12}{6}$ D $\dfrac{T}{6} = \dfrac{T - 12}{7}$

80. Diana can put a puzzle together in three hours. Ella can put the same puzzle together in five hours. How long will it take them if they work together?

   F $1\dfrac{3}{8}$ hours   H $1\dfrac{3}{4}$ hours

   G $1\dfrac{5}{8}$ hours   J $1\dfrac{7}{8}$ hours

2-6 Nonlinear Inequalities

Then
You solved polynomial and rational equations. (Lessons 2-3 and 2-4)

Now
- Solve polynomial inequalities.
- Solve rational inequalities.

New Vocabulary
polynomial inequality
sign chart
rational inequality

Math Online
glencoe.com

Why?

Many factors are involved when starting a new business, including the amount of the initial investment, maintenance and labor costs, the cost of manufacturing the product being sold, and the actual selling price of the product. Nonlinear inequalities can be used to determine the price at which to sell a product in order to make a specific profit.

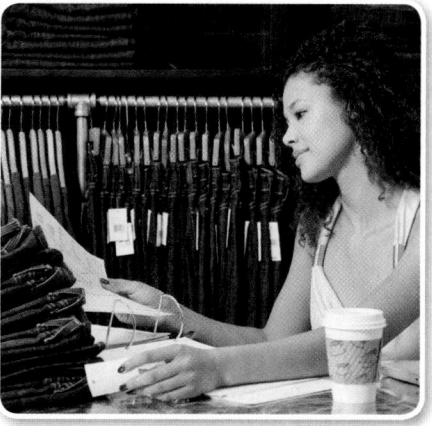

Polynomial Inequalities If $f(x)$ is a polynomial function, then a **polynomial inequality** has the general form $f(x) \leq 0$, $f(x) < 0$, $f(x) \neq 0$, $f(x) > 0$, or $f(x) \geq 0$. The inequality $f(x) < 0$ is true when $f(x)$ is *negative*, while $f(x) > 0$ is true when $f(x)$ is *positive*.

In Lesson 1-2, you learned that the x-intercepts of a polynomial function are the real zeros of the function. When ordered, these zeros divide the x-axis into intervals for which the value of $f(x)$ is either entirely positive (above the x-axis) or entirely negative (below the x-axis).

By finding the sign of $f(x)$ for just one x-value in each interval, you can determine on which intervals the function is positive or negative. From the test intervals represented by the **sign chart** at the right, you know that:

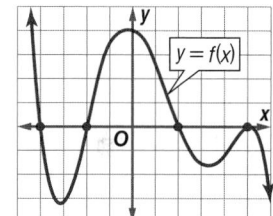

- $f(x) < 0$ on $(-4, -2) \cup (2, 5) \cup (5, \infty)$,
- $f(x) \leq 0$ on $[-4, -2] \cup [2, \infty)$,
- $f(x) = 0$ at $x = -4, -2, 2, 5$,
- $f(x) > 0$ on $(-\infty, -4) \cup (-2, 2)$, and
- $f(x) \geq 0$ on $(-\infty, -4] \cup [-2, 2] \cup [5, 5]$.

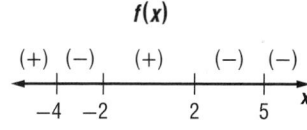

$f(x)$

EXAMPLE 1 | Solve a Polynomial Inequality

Solve $x^2 - 6x - 30 > -3$.

Adding 3 to each side, you get $x^2 - 6x - 27 > 0$. Let $f(x) = x^2 - 6x - 27$. Factoring yields $f(x) = (x + 3)(x - 9)$, so $f(x)$ has real zeros at -3 and 9. Create a sign chart using these zeros. Then substitute an x-value in each test interval into the factored form of the polynomial to determine if $f(x)$ is positive or negative at that point.

$f(x) = (x + 3)(x - 9)$ $f(x) = (x + 3)(x - 9)$

Think: $(x + 3)$ and $(x - 9)$ are both negative when $x = -4$.

Test $x = -4$. Test $x = 0$. Test $x = 10$.

$(-)(-) \quad (+)(-) \quad (+)(+)$
$ -3 9 x$

$(+) \quad\quad (-) \quad\quad (+)$
$ -3 9 x$

Because $f(x)$ is positive on the first and last intervals, the solution set of $x^2 - 6x - 30 > -3$ is $(-\infty, -3) \cup (9, \infty)$. The graph of $f(x)$ supports this conclusion, because $f(x)$ is above the x-axis on these same intervals.

$f(x) = x^2 - 6x - 30$

✔ Guided Practice

Solve each inequality.

1A. $x^2 + 5x + 6 < 20$

1B. $(x - 4)^2 > 4$

If you know the real zeros of a function, including their multiplicity, and the function's end behavior, you can create a sign chart without testing values.

EXAMPLE 2 Solve a Polynomial Inequality Using End Behavior

Solve $3x^3 - 4x^2 - 13x - 6 \le 0$.

Step 1 Let $f(x) = 3x^3 - 4x^2 - 13x - 6$. Use the techniques from Lesson 2-4 to determine that f has real zeros with multiplicity 1 at -1, $-\frac{2}{3}$, and 3. Set up a sign chart.

Step 2 Determine the end behavior of $f(x)$. Because the degree of f is odd and its leading coefficient is positive, you know $\lim\limits_{x \to -\infty} f(x) = -\infty$ and $\lim\limits_{x \to \infty} f(x) = \infty$. This means that the function starts off negative at the left and ends positive at the right.

Step 3 Because each zero listed is the location of a sign change, you can complete the sign chart.

The solutions of $3x^3 - 4x^2 - 13x - 6 \le 0$ are x-values such that $f(x)$ is negative or equal to 0. From the sign chart, you can see that the solution set is $(-\infty, -1] \cup \left[-\frac{2}{3}, 3\right]$.

CHECK The graph of $f(x) = 3x^3 - 4x^2 - 13x - 6$ is on or below the x-axis on $(-\infty, -1] \cup \left[-\frac{2}{3}, 3\right]$. ✔

$[-4, 6]$ scl: 1 by $[-25, 5]$ scl: 3

✔ **Guided Practice**

Solve each inequality.

2A. $2x^2 - 10x \le 2x - 16$

2B. $2x^3 + 7x^2 - 12x - 45 \ge 0$

When a polynomial function does not intersect the x-axis, the related inequalities have unusual solutions.

EXAMPLE 3 Polynomial Inequalities with Unusual Solution Sets

Solve each inequality.

a. $x^2 + 5x + 8 < 0$

The related function $f(x) = x^2 + 5x + 8$ has no real zeros, so there are no sign changes. This function is positive for all real values of x. Therefore, $x^2 + 5x + 8 < 0$ has no solution. The graph of $f(x)$ supports this conclusion, because the graph is never on or below the x-axis. The solution set is $\varnothing$.

$f(x) = x^2 + 5x + 8$

$[-12, 8]$ scl: 1 by $[-5, 10]$ scl: 1

b. $x^2 + 5x + 8 \ge 0$

Because the related function $f(x) = x^2 + 5x + 8$ is positive for all real values of x, the solution set of $x^2 + 5x + 8 \ge 0$ is all real numbers or $(-\infty, \infty)$.

c. $x^2 - 10x + 25 > 0$

The related function $f(x) = x^2 - 10x + 25$ has one real zero, 5, with multiplicity 2, so the value of $f(x)$ does not change signs. This function is positive for all real values of x except $x = 5$. Therefore, the solution set of $x^2 - 10x + 25 > 0$ is $(-\infty, 5) \cup (5, \infty)$. The graph of $f(x)$ supports this conclusion.

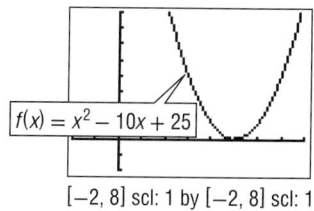

[−2, 8] scl: 1 by [−2, 8] scl: 1

d. $x^2 - 10x + 25 \le 0$

The related function $f(x) = x^2 - 10x + 25$ has a zero at 5. For all other values of x, the function is positive. Therefore, the solution set of $x^2 - 10x + 25 \le 0$ is {5}.

✔ Guided Practice

Solve each inequality.

3A. $x^2 + 2x + 5 > 0$ **3B.** $x^2 + 2x + 5 \le 0$

3C. $x^2 - 2x - 15 \le -16$ **3D.** $x^2 - 2x - 15 > -16$

Rational Inequalities Consider the rational function at the right. Notice the intervals on which $f(x)$ is positive and negative. While a polynomial function can change signs only at its real zeros, a rational function can change signs at its real zeros or at its points of discontinuity. For this reason, when solving a **rational inequality**, you must include the zeros of both the numerator and the denominator in your sign chart.

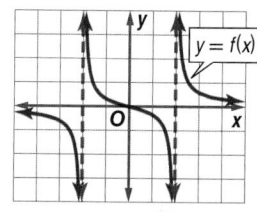

You can begin solving a rational inequality by first writing the inequality in general form with a single rational expression on the left and a zero on the right.

EXAMPLE 4 | **Solve a Rational Inequality**

Solve $\dfrac{4}{x - 6} + \dfrac{2}{x + 1} > 0$.

$\dfrac{4}{x - 6} + \dfrac{2}{x + 1} > 0$ **Original inequality**

$\dfrac{4x + 4 + 2x - 12}{(x - 6)(x + 1)} > 0$ **Use the LCD, $(x - 6)(x + 1)$, to rewrite each fraction. Then add.**

$\dfrac{6x - 8}{(x - 6)(x + 1)} > 0$ **Simplify.**

Let $f(x) = \dfrac{6x - 8}{(x - 6)(x + 1)}$. The zeros and undefined points of the inequality are the zeros of the numerator, $\frac{4}{3}$, and denominator, 6 and −1. Create a sign chart using these numbers. Then choose and test x-values in each interval to determine if $f(x)$ is positive or negative.

$$f(x) = \frac{6x - 8}{(x - 6)(x + 1)} \qquad\qquad f(x) = \frac{6x - 8}{(x - 6)(x + 1)}$$

Test $x = -2$. Test $x = 0$. Test $x = 2$. Test $x = 7$.

Figure 2.6.1

The solution set of the original inequality is the union of those intervals for which $f(x)$ is positive, $\left(-1, \frac{4}{3}\right) \cup (6, \infty)$. The graph of $f(x) = \dfrac{4}{x - 6} + \dfrac{2}{x + 1}$ in Figure 2.6.1 supports this conclusion.

✓ Guided Practice

Solve each inequality.

4A. $\dfrac{x+6}{4x-3} \geq 1$ 　　　　 **4B.** $\dfrac{x^2-x-11}{x-2} \leq 3$ 　　　　 **4C.** $\dfrac{1}{x} > \dfrac{1}{x+5}$

You can use nonlinear inequalities to solve real-world problems.

🌐 Real-World EXAMPLE 5 　　 Solve a Rational Inequality

AMUSEMENT PARKS A group of high school students is renting a bus for $600 to take to an amusement park the day after prom. The tickets to the amusement park are $60 less an extra $0.50 group discount per person in the group. Write and solve an inequality that can be used to determine how many students must go on the trip for the total cost to be less than $40 per student.

Let x represent the number of students.

Ticket cost per student	+	bus cost per student	must be less than	$40.
$60 - 0.5x$	+	$\dfrac{600}{x}$	$<$	40

$60 - 0.5x + \dfrac{600}{x} < 40$ 　 **Write the inequality.**

$60 - 0.5x + \dfrac{600}{x} - 40 < 0$ 　 **Subtract 40 from each side.**

$\dfrac{60x - 0.5x^2 + 600 - 40x}{x} < 0$ 　 **Use the LCD, x, to rewrite each fraction. Then add.**

$\dfrac{-0.5x^2 + 20x + 600}{x} < 0$ 　 **Simplify.**

$\dfrac{x^2 - 40x - 1200}{x} > 0$ 　 **Multiply each side by −2. Reverse the inequality sign.**

$\dfrac{(x+20)(x-60)}{x} > 0$ 　 **Factor.**

Let $f(x) = \dfrac{(x+20)(x-60)}{x}$. The zeros of this inequality are −20, 60, and 0. Use these numbers to create and complete a sign chart for this function.

$$f(x) = \frac{(x+20)(x-60)}{x} \qquad\qquad f(x) = \frac{(x+20)(x-60)}{x}$$

Test $x = -30$.　Test $x = -10$.　Test $x = 10$.　Test $x = 70$.

So, the solution set of $60 - 0.5x + \dfrac{600}{x} < 40$ is $(-20, 0) \cup (60, \infty)$.

Because there cannot be a negative number of students, more than 60 students must go to the amusement park for the total cost to be less than $40 per student.

✓ Guided Practice

5. **LANDSCAPING** A landscape architect is designing a fence that will enclose a rectangular garden that has a perimeter of 250 feet. If the area of the garden is to be at least 1000 square feet, write and solve an inequality to find the possible lengths of the fence.

● = **Step-by-Step Solutions** begin on page R29.

Solve each inequality. (Examples 1–3)

1. $(x + 4)(x - 2) \le 0$

2. $(x - 6)(x + 1) > 0$

3. $(3x + 1)(x - 8) \ge 0$

4. $(x - 4)(-2x + 5) < 0$

5. $(4 - 6y)(2y + 1) < 0$

6. $2x^3 - 9x^2 - 20x + 12 \le 0$

7. $-8x^3 - 30x^2 - 18x < 0$

8. $5x^3 - 43x^2 + 72x + 36 > 0$

9. $x^2 + 6x > -10$

10. $2x^2 \le -x - 4$

11. $4x^2 + 8 \le 5 - 2x$

12. $2x^2 + 8x \ge 4x - 8$

13. $2b^2 + 16 \le b^2 + 8b$

14. $c^2 + 12 \le 3 - 6c$

15. $-a^2 \ge 4a + 4$

16. $3d^2 + 16 \ge -d^2 + 16d$

17. BUSINESS A new company projects that its first-year revenue will be $r(x) = 120x - 0.0004x^2$ and the start-up cost will be $c(x) = 40x + 1,000,000$, where x is the number of products sold. The net profit p that they will make the first year is equal to $p = r - c$. Write and solve an inequality to determine how many products the company must sell to make a profit of at least \$2,000,000. (Example 1)

Solve each inequality. (Example 4)

18. $\dfrac{x - 3}{x + 4} > 3$

19. $\dfrac{x + 6}{x - 5} \le 1$

20. $\dfrac{2x + 1}{x - 6} \ge 4$

21 $\dfrac{3x - 2}{x + 3} < 6$

22. $\dfrac{3 - 2x}{5x + 2} < 5$

23. $\dfrac{4x + 1}{3x - 5} \ge -3$

24. $\dfrac{(x + 2)(2x - 3)}{(x - 3)(x + 1)} \le 6$

25. $\dfrac{(4x + 1)(x - 2)}{(x + 3)(x - 1)} \le 4$

26. $\dfrac{12x + 65}{(x + 4)^2} \ge 5$

27. $\dfrac{2x + 4}{(x - 3)^2} < 12$

28. CHARITY The Key Club at a high school is having a dinner as a fundraiser for charity. A dining hall that can accommodate 80 people will cost \$1000 to rent. If each ticket costs \$20 in advance or \$22 the day of the dinner, and the same number of people bought tickets in advance as bought the day of the dinner, write and solve an inequality to determine the minimum number of people that must attend for the club to make a profit of at least \$500. (Example 5)

29. PROM A group of friends decides to share a limo for prom. The cost of rental is \$750 plus a \$25 fee for each occupant. There is a minimum of two passengers, and the limo can hold up to 14 people. Write and solve an inequality to determine how many people can share the limo for less than \$120 per person. (Example 5)

Find the domain of each expression.

30. $\sqrt{x^2 + 5x + 6}$

31. $\sqrt{x^2 - 3x - 40}$

32. $\sqrt{16 - x^2}$

33. $\sqrt{x^2 - 9}$

34. $\sqrt{\dfrac{x}{x^2 - 25}}$

35. $\sqrt[3]{\dfrac{x}{36 - x^2}}$

Find the solution set of $f(x) - g(x) \ge 0$.

36.

37.

38. SALES A vendor sells hot dogs at each school sporting event. The cost of each hot dog is \$0.38 and the cost of each bun is \$0.12. The vendor rents the hot dog cart that he uses for \$1000. If he wants his costs to be less than his profits after selling 400 hot dogs, what should the vendor charge for each hot dog?

39. PARKS AND RECREATION A rectangular playing field for a community park is to have a perimeter of 112 feet and an area of at least 588 square feet.

   a. Write an inequality that could be used to find the possible lengths to which the field can be constructed.

   b. Solve the inequality you wrote in part **a** and interpret the solution.

   c. How does the inequality and solution change if the area of the field is to be no more than 588 square feet? Interpret the solution in the context of the situation.

Solve each inequality. (*Hint:* **Test every possible solution interval that lies within the domain using the original inequality.**)

40. $\sqrt{9y + 19} - \sqrt{6y - 5} > 3$

41. $\sqrt{4x + 4} - \sqrt{x - 4} \le 4$

42. $\sqrt{12y + 72} - \sqrt{6y - 11} \ge 7$

43. $\sqrt{25 - 12x} - \sqrt{16 - 4x} < 5$

Determine the inequality shown in each graph.

44.

45.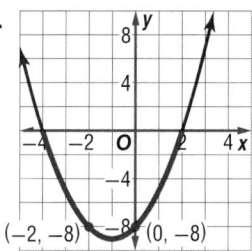

Solve each inequality.

46. $2y^4 - 9y^3 - 29y^2 + 60y + 36 > 0$

47. $3a^4 + 7a^3 - 56a^2 - 80a < 0$

48. $c^5 + 6c^4 - 12c^3 - 56c^2 + 96c \ge 0$

49. $3x^5 + 13x^4 - 137x^3 - 353x^2 + 330x + 144 \le 0$

50. PACKAGING A company sells cylindrical oil containers like the one shown.

2 L

a. Use the volume of the container to express its surface area as a function of its radius in centimeters. (*Hint*: 1 liter = 1000 cubic centimeters)

b. The company wants the surface area of the container to be less than 2400 square centimeters. Write an inequality that could be used to find the possible radii to meet this requirement.

c. Use a graphing calculator to solve the inequality you wrote in part **b** and interpret the solution.

Solve each inequality.

51. $(x + 3)^2(x - 4)^3(2x + 1)^2 < 0$

52. $(y - 5)^2(y + 1)(4y - 3)^4 \geq 0$

53. $(a - 3)^3(a + 2)^3(a - 6)^2 > 0$

54. $c^2(c + 6)^3(3c - 4)^5(c - 3) \leq 0$

55. STUDY TIME Jarrick determines that with the information that he currently knows, he can achieve a score of a 75% on his test. Jarrick believes that for every 5 complete minutes he spends studying, he will raise his score by 1%.

a. If Jarrick wants to obtain a score of at least 89.5%, write an inequality that could be used to find the time t that he will have to spend studying.

b. Solve the inequality that you wrote in part **a** and interpret the solution.

56. GAMES A skee ball machine pays out 3 tickets each time a person plays and then 2 additional tickets for every 80 points the player scores.

a. Write a nonlinear function to model the amount of tickets received for an x-point score.

b. Write an inequality that could be used to find the score a player would need in order to receive at least 11 tickets.

c. Solve the inequality in part **b** and interpret your solution.

57. The area of a region bounded by a parabola and a horizontal line is $A = \frac{2}{3}bh$, where b represents the base of the region along the horizontal line and h represents the height of the region. Find the area bounded by f and g.

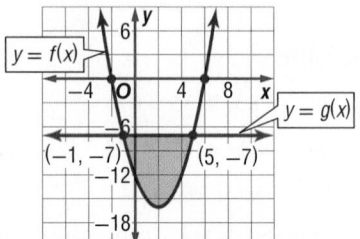

If k is nonnegative, find the interval for x for which each inequality is true.

58. $x^2 + kx + c \geq c$

59. $(x + k)(x - k) < 0$

60. $x^3 - kx^2 - k^2x + k^3 > 0$

61. $x^4 - 8k^2x^2 + 16k^4 \geq 0$

62. 🔧 **MULTIPLE REPRESENTATIONS** In this problem, you will investigate absolute value nonlinear inequalities.

a. **TABULAR** Copy and complete the table below.

Function	Zeros	Undefined Points
$f(x) = \dfrac{x - 1}{\lvert x + 2\rvert}$		
$g(x) = \dfrac{\lvert 2x - 5\rvert}{x - 3}$		
$h(x) = \dfrac{\lvert x + 4\rvert}{\lvert 3x - 1\rvert}$		

b. **GRAPHICAL** Graph each function in part **a**.

c. **SYMBOLIC** Create a sign chart for each inequality. Include zeros and undefined points and evaluate the sign of the numerators and denominators separately.

   i. $\dfrac{x - 1}{\lvert x + 2\rvert} < 0$

   ii. $\dfrac{\lvert 2x - 5\rvert}{x - 3} \geq 0$

   iii. $\dfrac{\lvert x + 4\rvert}{\lvert 3x - 1\rvert} > 0$

d. **NUMERICAL** Write the solution for each inequality in part **c**.

H.O.T. Problems Use **H**igher-**O**rder **T**hinking Skills

63. ERROR ANALYSIS Ajay and Mae are solving $\dfrac{x^2}{(3 - x)^2} \geq 0$. Ajay thinks that the solution is $(-\infty, 0]$ or $[0, \infty)$, and Mae thinks that the solution is $(-\infty, \infty)$. Is either of them correct? Explain your reasoning.

64. REASONING If the solution set of a polynomial inequality is $(-3, 3)$, what will be the solution set if the inequality symbol is reversed? Explain your reasoning.

65. CHALLENGE Determine the values for which $(a + b)^2 > (c + d)^2$ if $a < b < c < d$.

66. REASONING If $0 < c < d$, find the interval on which $(x - c)(x - d) \leq 0$ is true. Explain your reasoning.

67 CHALLENGE What is the solution set of $(x - a)^{2n} > 0$ if n is a natural number?

68. REASONING What happens to the solution set of $(x + a)(x - b) < 0$ if the expression is changed to $-(x + a)(x - b) < 0$, where a and $b > 0$? Explain your reasoning.

69. WRITING IN MATH Explain why you cannot solve $\dfrac{3x + 1}{x - 2} < 6$ by multiplying each side by $x - 2$.

Find the domain of each function and the equations of the vertical or horizontal asymptotes, if any. (Lesson 2-5)

70. $f(x) = \dfrac{2x}{x + 4}$

71. $h(x) = \dfrac{x^2}{x + 6}$

72. $f(x) = \dfrac{x - 1}{(2x + 1)(x - 5)}$

73. GEOMETRY A cone is inscribed in a sphere with a radius of 15 centimeters. If the volume of the cone is 1152π cubic centimeters, find the length represented by x. (Lesson 2-4)

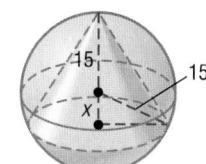

Divide using long division. (Lesson 2-3)

74. $(x^2 - 10x - 24) \div (x + 2)$

75. $(3a^4 - 6a^3 - 2a^2 + a - 6) \div (a + 1)$

76. $(z^5 - 3z^2 - 20) \div (z - 2)$

77. $(x^3 + y^3) \div (x + y)$

78. FINANCE The closing prices in dollars for a share of stock during a one-month period are shown. (Lesson 2-2)

  a. Graph the data.

  b. Use a graphing calculator to model the data using a polynomial function with a degree of 3.

  c. Use the model to estimate the closing price of the stock on day 25.

Day	Price($)	Day	Price($)
1	30.15	15	15.64
5	27.91	20	10.38
7	26.10	21	9.56
10	22.37	28	9.95
12	19.61	30	12.25

79. HOME SECURITY A company offers a home-security system that uses the numbers 0 through 9, inclusive, for a 5-digit security code. (Lesson 0-7)

  a. How many different security codes are possible?

  b. If no digits can be repeated, how many security codes are available?

  c. Suppose the homeowner does not want to use 0 or 9 as the first digit and wants the last digit to be 1. How many codes can be formed if the digits can be repeated? If no repetitions are allowed, how many codes are available?

80. SAT/ACT Two circles, A and B, lie in the same plane. If the center of circle B lies on circle A, then in how many points could circle A and circle B intersect?

$\quad$ **I.** 0 $\qquad$ **II.** 1 $\qquad$ **III.** 2

  A I only $\qquad$ **C** I and III only $\quad$ **E** I, II, and III

  B III only $\qquad$ **D** II and III only

81. A rectangle is 6 centimeters longer than it is wide. Find the possible widths if the area of the rectangle is more than 216 square centimeters.

  F $w > 12$ $\qquad\qquad$ **H** $w > 18$

  G $w < 12$ $\qquad\qquad$ **J** $w < 18$

82. FREE RESPONSE The amount of drinking water reserves in millions of gallons available for a town is modeled by $f(t) = 80 + 10t - 4t^2$. The minimum amount of water needed by the residents is modeled by $g(t) = (2t)^{\frac{4}{3}}$, where t is the time in years.

  a. Identify the types of functions represented by $f(t)$ and $g(t)$.

  b. What is the relevant domain and range for $f(t)$ and $g(t)$? Explain.

  c. What is the end behavior of $f(t)$ and $g(t)$?

  d. Sketch $f(t)$ and $g(t)$ for $0 \le t \le 6$ on the same graph.

  e. Explain why there must be a value c for $[0, 6]$ such that $f(c) = 50$.

  f. For what value in the relevant domain does f have a zero? What is the significance of the zero in this situation?

  g. If this were a true situation and these projections were accurate, when would the residents be expected to need more water than they have in reserves?

Chapter Summary

Key Concepts

Power and Radical Functions (Lesson 2-1)

- A *power function* is any function of the form $f(x) = ax^n$, where a and n are nonzero real numbers.

- A *monomial function* is any function that can be written as $f(x) = a$ or $f(x) = ax^n$, where a and n are nonzero constant real numbers.

- A *radical function* is a function that can be written as $f(x) = \sqrt[n]{n^p}$, where n and p are positive integers greater than 1 that have no common factors.

Polynomial Functions (Lesson 2-2)

- A polynomial function is any function of the form $f(x) = a_n x^n + a_{n-1} x^{n-1} + \ldots + a_1 x + a_0$, where $a_n \neq 0$. The degree is n.

- The graph of a polynomial function has at most n distinct real zeros and at most $n - 1$ turning points.

- The behavior of a polynomial graph at its zero c depends on the multiplicity of the factor $(x - c)$.

The Remainder and Factor Theorems (Lesson 2-3)

- Synthetic division is a shortcut for dividing a polynomial by a linear factor of the form $x - c$.

- If a polynomial f is divided by $x - c$, the remainder is equal to $f(c)$.

- $x - c$ is a factor of a polynomial f if and only if $f(c) = 0$.

Zeros of Polynomial Functions (Lesson 2-4)

- If $f(x) = a_n x^n + \ldots + a_1 x + a_0$ with integer coefficients, then any rational zero of $f(x)$ is of the form $\frac{p}{q}$, where p and q have no common factors, p is a factor of a_0, and q is a factor of a_n.

- A polynomial of degree n has n zeros, including repeated zeros, in the complex system. It also has n factors:
$$f(x) = a_n(x - c_1)(x - c_2) \ldots (x - c_n).$$

Rational Functions (Lesson 2-5)

- The graph of f has a vertical asymptote $x = c$ if
$\lim\limits_{x \to c^-} f(x) = \pm\infty$ or $\lim\limits_{x \to c^+} f(x) = \pm\infty$.

- The graph of f has a horizontal asymptote $y = c$ if
$\lim\limits_{x \to \infty} f(x) = c$ or $\lim\limits_{x \to -\infty} f(x) = c$.

- A rational function $f(x) = \frac{a(x)}{b(x)}$ may have vertical asymptotes, horizontal asymptotes, or oblique asymptotes, x-intercepts, and y-intercepts. They can all be determined algebraically.

Nonlinear Inequalities (Lesson 2-6)

- The sign chart for a rational inequality must include zeros and undefined points.

Key Vocabulary

complex conjugates (p. 124)
extraneous solution (p. 91)
horizontal asymptote (p. 131)
irreducible over the reals (p. 124)
leading coefficient (p. 97)
leading-term test (p. 98)
lower bound (p. 121)
multiplicity (p. 102)
oblique asymptote (p. 134)
polynomial function (p. 97)

power function (p. 86)
quartic function (p. 99)
rational function (p. 130)
repeated zero (p. 101)
sign chart (p. 141)
synthetic division (p. 111)
synthetic substitution (p. 113)
turning point (p. 99)
upper bound (p. 121)
vertical asymptote (p. 131)

Vocabulary Check

Identify the word or phrase that best completes each sentence.

1. The coefficient of the term with the greatest exponent of the variable is the (leading coefficient, degree) of the polynomial.

2. A (polynomial function, power function) is a function of the form $f(x) = a_n x^n + a_{n-1} x^{n-1} + \ldots + a_1 x + a_0$, where $a_1, a_2, \ldots, a_n$ are real numbers and n is a natural number.

3. A function that has multiple factors of $(x - c)$ has (repeated zeros, turning points).

4. (Polynomial division, Synthetic division) is a short way to divide polynomials by linear factors.

5. The (Remainder Theorem, Factor Theorem) relates the linear factors of a polynomial with the zeros of its related function.

6. Some of the possible zeros for a polynomial function can be listed using the (Factor, Rational Zeros) Theorem.

7. (Vertical, Horizontal) asymptotes are determined by the zeros of the denominator of a rational function.

8. The zeros of the (denominator, numerator) determine the x-intercepts of the graph of a rational function.

9. (Horizontal, Oblique) asymptotes occur when a rational function has a denominator with a degree greater than 0 and a numerator with degree one greater than its denominator.

10. A (quartic function, power function) is a function of the form $f(x) = ax^n$, where a and n are nonzero constant real numbers.

Lesson-by-Lesson Review

2-1 Power and Radical Functions (pp. 86–95)

Graph and analyze each function. Describe the domain, range, intercepts, end behavior, continuity, and where the function is increasing or decreasing.

11. $f(x) = 5x^6$

12. $f(x) = -8x^3$

13. $f(x) = x^{-9}$

14. $f(x) = \frac{1}{3}x^{-4}$

15. $f(x) = \sqrt{5x - 6} - 11$

16. $f(x) = -\frac{3}{4}\sqrt[3]{6x^2 - 1} + 2$

Solve each equation.

17. $2x = 4 + \sqrt{7x - 12}$

18. $\sqrt{4x + 5} + 1 = 4x$

19. $4 = \sqrt{6x + 1} - \sqrt{17 - 4x}$

20. $\sqrt[4]{x^2 + 31} - 1 = 3$

EXAMPLE 1

Graph and analyze $f(x) = -4x^{-5}$. Describe the domain, range, intercepts, end behavior, continuity, and where the function is increasing or decreasing.

x	f(x)
−3	0.016
−2	0.125
−1	4
0	undefined
1	−4
2	−0.125
3	−0.016

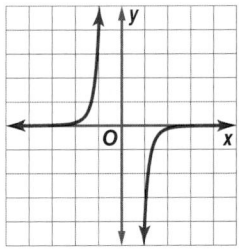

Domain: $(-\infty, 0) \cup (0, \infty)$ Range: $(-\infty, 0) \cup (0, \infty)$

Intercepts:: none

End behavior: $\lim\limits_{x \to -\infty} f(x) = 0$ and $\lim\limits_{x \to \infty} f(x) = 0$

Continuity: infinite discontinuity at $x = 0$

Increasing: $(-\infty, 0)$ Increasing: $(0, \infty)$

2-2 Polynomial Functions (pp. 97–107)

Describe the end behavior of the graph of each polynomial function using limits. Explain your reasoning using the leading term test.

21. $f(x) = -4x^4 + 7x^3 - 8x^2 + 12x - 6$

22. $f(x) = -3x^5 + 7x^4 + 3x^3 - 11x - 5$

23. $f(x) = \frac{2}{3}x^2 - 8x - 3$

24. $f(x) = x^3(x - 5)(x + 7)$

State the number of possible real zeros and turning points of each function. Then determine all of the real zeros by factoring.

25. $f(x) = x^3 - 7x^2 + 12x$ 26. $f(x) = x^5 + 8x^4 - 20x^3$

27. $f(x) = x^4 - 10x^2 + 9$ 28. $f(x) = x^4 - 25$

For each function, (a) apply the leading term test, (b) find the zeros and state the multiplicity of any repeated zeros, (c) find a few additional points, and then (d) graph the function.

29. $f(x) = x^3(x - 3)(x + 4)^2$ 30. $f(x) = (x - 5)^2(x - 1)^2$

EXAMPLE 2

Describe the end behavior of the graph of $f(x) = -2x^5 + 3x^3 - 8x^2 - 6$ using limits. Explain your reasoning using the leading term test.

The degree is 5 and the leading coefficient is −2. Because the degree is odd and the leading coefficient is negative, $\lim\limits_{x \to -\infty} f(x) = \infty$ and $\lim\limits_{x \to \infty} f(x) = -\infty$.

EXAMPLE 3

State the number of possible real zeros and turning points for $f(x) = x^3 + 6x^2 + 9x$. Then find all the real zeros by factoring.

The degree of f is 3, so f has at most 3 distinct real zeros and at most $3 - 1$ or 2 turning points. To find the real zeros, solve the related equation $f(x) = 0$ by factoring.

$$x^3 + 6x^2 + 9x = x(x^2 + 6x + 9)$$
$$= x(x + 3)(x + 3) \text{ or } x(x + 3)^2$$

The expression has 3 factors but only 2 distinct real zeros, 0 and −3.

Lesson-by-Lesson Review

2-3 The Remainder and Factor Theorems (pp. 109–117)

Divide using long division.

31. $(x^3 + 8x^2 - 5) \div (x - 2)$

32. $(-3x^3 + 5x^2 - 22x + 5) \div (x^2 + 4)$

33. $(2x^5 + 5x^4 - 5x^3 + x^2 - 18x + 10) \div (2x - 1)$

Divide using synthetic division.

34. $(x^3 - 8x^2 + 7x - 15) \div (x - 1)$

35. $(x^4 - x^3 + 7x^2 - 9x - 18) \div (x - 2)$

36. $(2x^4 + 3x^3 - 10x^2 + 16x - 6) \div (2x - 1)$

Use the Factor Theorem to determine if the binomials given are factors of $f(x)$. Use the binomials that are factors to write a factored form of $f(x)$.

37. $f(x) = x^3 + 3x^2 - 8x - 24;\ (x + 3)$

38. $f(x) = 2x^4 - 9x^3 + 2x^2 + 9x - 4;\ (x - 1),\ (x + 1)$

39. $f(x) = x^4 - 2x^3 - 3x^2 + 4x + 4;\ (x + 1),\ (x - 2)$

EXAMPLE 4

Divide $(2x^3 - 3x^2 + 5x - 4) \div (2x - 1)$ using synthetic division.

Rewrite the division expression $\dfrac{2x^3 - 3x^2 + 5x - 4}{2x - 1}$ so that the denominator is of the form $x - c$.

$$\frac{2x^3 - 3x^2 + 5x - 4}{2x - 1} = \frac{(2x^3 - 3x^2 + 5x - 4) \div 2}{(2x - 1) \div 2}$$

$$= \frac{x^3 - \frac{3}{2}x^2 + \frac{5}{2}x - 2}{x - \frac{1}{2}}$$

Therefore, $c = \dfrac{1}{2}$. Perform synthetic division.

$$\begin{array}{r|rrrr} \frac{1}{2} & 1 & -\frac{3}{2} & \frac{5}{2} & -2 \\ & & \frac{1}{2} & -\frac{1}{2} & 1 \\ \hline & 1 & -1 & 2 & -1 \end{array}$$

$$\frac{(2x^3 - 3x^2 + 5x - 4)}{(2x - 1)} = x^2 - x + 2 - \frac{2}{(2x - 1)}$$

2-4 Zeros of Polynomial Functions (pp. 119–129)

List all possible rational zeros of each function. Then determine which, if any, are zeros.

40. $f(x) = x^3 - x^2 - x + 1$

41. $f(x) = x^3 - 14x - 15$

42. $f(x) = x^4 + 5x^2 + 4$

43. $f(x) = 3x^4 - 14x^3 - 2x^2 + 31x + 10$

Solve each equation.

44. $x^4 - 9x^3 + 29x^2 - 39x + 18 = 0$

45. $6x^3 - 23x^2 + 26x - 8 = 0$

46. $x^4 - 7x^3 + 8x^2 + 28x = 48$

47. $2x^4 - 11x^3 + 44x = -4x^2 + 48$

Use the given zero to find all complex zeros of each function. Then write the linear factorization of the function.

48. $f(x) = x^4 + x^3 - 41x^2 + x - 42;\ i$

49. $f(x) = x^4 - 4x^3 + 7x^2 + 16x + 12;\ -2i$

EXAMPLE 5

Solve $x^3 + 2x^2 - 16x - 32 = 0$.

Because the leading coefficient is 1, the possible rational zeros are the factors of -32. So the possible rational zeros are ±1, ±2, ±4, ±8, ±16, and ±32. Using synthetic substitution, you can determine that -2 is a rational zero.

$$\begin{array}{r|rrrr} -2 & 1 & 2 & -16 & -32 \\ & & -2 & 0 & 32 \\ \hline & 1 & 0 & -16 & 0 \end{array}$$

Therefore, $f(x) = (x + 2)(x^2 - 16)$. This polynomial can be written $f(x) = (x + 2)(x - 4)(x + 4)$. The rational zeros of f are -2, 4, and -4.

Lesson-by-Lesson Review

2-5 Rational Functions (pp. 130–140)

Find the domain of each function and the equations of the vertical or horizontal asymptotes, if any.

50. $f(x) = \dfrac{x^2 - 1}{x + 4}$

51. $f(x) = \dfrac{x^2}{x^2 - 25}$

52. $f(x) = \dfrac{x(x - 3)}{(x - 5)^2(x + 3)^2}$

53. $f(x) = \dfrac{(x - 5)(x - 2)}{(x + 3)(x + 9)}$

For each function, determine any asymptotes and intercepts. Then graph the function, and state its domain.

54. $f(x) = \dfrac{x}{x - 5}$

55. $f(x) = \dfrac{x - 2}{x + 4}$

56. $f(x) = \dfrac{(x + 3)(x - 4)}{(x + 5)(x - 6)}$

57. $f(x) = \dfrac{x(x + 7)}{(x + 6)(x - 3)}$

58. $f(x) = \dfrac{x + 2}{x^2 - 1}$

59. $f(x) = \dfrac{x^2 - 16}{x^3 - 6x^2 + 5x}$

Solve each equation.

60. $\dfrac{12}{x} + x - 8 = 1$

61. $\dfrac{2}{x + 2} + \dfrac{3}{x} = -\dfrac{x}{x + 2}$

62. $\dfrac{1}{d + 4} = \dfrac{2}{d^2 + 3d - 4} - \dfrac{1}{1 - d}$

63. $\dfrac{1}{n - 2} = \dfrac{2n + 1}{n^2 + 2n - 8} + \dfrac{2}{n + 4}$

EXAMPLE 6

Find the domain of $f(x) = \dfrac{x + 7}{x + 1}$ and any vertical or horizontal asymptotes.

Step 1 Find the domain.

The function is undefined at the zero of the denominator $h(x) = x + 1$, which is -1. The domain of f is all real numbers except $x = -1$.

Step 2 Find the asymptotes, if any.

Check for vertical asymptotes.

The zero of the denominator is -1, so there is a vertical asymptote at $x = -1$.

Check for horizontal asymptotes.

The degree of the numerator is equal to the degree of the denominator. The ratio of the leading coefficient is $\dfrac{1}{1} = 1$. Therefore, $y = 1$ is a horizontal asymptote.

2-6 Nonlinear Inequalities (pp. 141–147)

Solve each inequality.

64. $(x + 5)(x - 3) \leq 0$

65. $x^2 - 6x - 16 > 0$

66. $x^3 + 5x^2 \leq 0$

67. $2x^2 + 13x + 15 < 0$

68. $x^2 + 12x + 36 \leq 0$

69. $x^2 + 4 < 0$

70. $x^2 + 4x + 4 > 0$

71. $\dfrac{x - 5}{x} < 0$

72. $\dfrac{x + 1}{(12x + 6)(3x + 4)} \geq 0$

73. $\dfrac{5}{x - 3} + \dfrac{2}{x - 4} > 0$

EXAMPLE 7

Solve $x^3 + 5x^2 - 36x \leq 0$.

Factoring the polynomial $f(x) = x^3 + 5x^2 - 36x$ yields $f(x) = x(x + 9)(x - 4)$, so $f(x)$ has real zeros at 0, -9, and 4.

Create a sign chart using these zeros. Then substitute an x-value from each test interval into the function to determine whether $f(x)$ is positive or negative at that point.

Because $f(x)$ is negative on the first and third intervals, the solution of $x^3 + 5x^2 - 36x \leq 0$ is $(-\infty, -9] \cup [0, 4]$.

Applications and Problem Solving

74. PHYSICS Kepler's Third Law of Planetary Motion implies that the time T it takes for a planet to complete one revolution in its orbit about the Sun is given by $T = R^{\frac{3}{2}}$, where R is the planet's mean distance from the Sun. Time is measured in Earth years, and distance is measured in astronomical units. (Lesson 2-1)

  a. State the relevant domain and range of the function.

  b. Graph the function.

  c. The time for Mars to orbit the Sun is observed to be 1.88 Earth years. Determine Mars' average distance from the Sun in miles, given that one astronomical unit equals 93 million miles.

75. PUMPKIN LAUNCH Mr. Roberts' technology class constructed a catapult to compete in the county's annual pumpkin launch. The speed v in miles per hour of a launched pumpkin after t seconds is given. (Lesson 2-1)

t	0.5	1.0	1.5	2.0	2.5	3.0
v	85	50	30	20	15	12

  a. Create a scatter plot of the data.

  b. Determine a power function to model the data.

  c. Use the function to predict the speed at which a pumpkin is traveling after 1.2 seconds.

  d. Use the function to predict the time at which the pumpkin's speed is 47 miles per hour.

76. AMUSEMENT PARKS The elevation above the ground for a rider on the Big Monster roller coaster is given in the table. (Lesson 2-2)

Time (seconds)	5	10	15	20	25
Elevation (feet)	85	62	22	4	17

  a. Create a scatter plot of the data and determine the type of polynomial function that could be used to represent the data.

  b. Write a polynomial function to model the data set. Round each coefficient to the nearest thousandth and state the correlation coefficient.

  c. Use the model to estimate a rider's elevation at 17 seconds.

  d. Use the model to determine approximately the first time a rider is 50 feet above the ground.

77. GARDENING Mark's parents seeded their new lawn in 2001. From 2001 until 2011, the amount of crab grass increased following the model $f(x) = 0.021x^3 - 0.336x^2 + 1.945x - 0.720$, where x is the number of years since 2001 and $f(x)$ is the number of square feet each year. Use synthetic division to find the number of square feet of crab grass in the lawn in 2011. Round to the nearest thousandth. (Lesson 2-3)

78. BUSINESS A used bookstore sells an average of 1000 books each month at an average price of $10 per book. Due to rising costs the owner wants to raise the price of all books. She figures she will sell 50 fewer books for every $1 she raises the prices. (Lesson 2-4)

  a. Write a function for her total sales after raising the price of her books x dollars.

  b. How many dollars does she need to raise the price of her books so that the total amount of sales is $11,250?

  c. What is the maximum amount that she can increase prices and still achieve $10,000 in total sales? Explain.

79. AGRICULTURE A farmer wants to make a rectangular enclosure using one side of her barn and 80 meters of fence material. Determine the dimensions of the enclosure. Assume that the width of the enclosure w will not be greater than the side of the barn. (Lesson 2-4)

80. ENVIRONMENT A pond is known to contain 0.40% acid. The pond contains 50,000 gallons of water. (Lesson 2-5)

  a. How many gallons of acid are in the pond?

  b. Suppose x gallons of pure water was added to the pond. Write $p(x)$, the percentage of acid in the pond after x gallons of pure water are added.

  c. Find the horizontal asymptote of $p(x)$.

  d. Does the function have any vertical asymptotes? Explain.

81. BUSINESS For selling x cakes, a baker will make $b(x) = x^2 - 5x - 150$ hundreds of dollars in revenue. Determine the minimum number of cakes the baker needs to sell in order to make a profit. (Lesson 2-6)

82. DANCE The junior class would like to organize a school dance as a fundraiser. A hall that the class wants to rent costs $3000 plus an additional charge of $5 per person. (Lesson 2-6)

  a. Write and solve an inequality to determine how many people need to attend the dance if the junior class would like to keep the cost per person under $10.

  b. The hall will provide a DJ for an extra $1000. How many people would have to attend the dance to keep the cost under $10 per person?

Graph and analyze each function. Describe the domain, range, intercepts, end behavior, continuity, and where the function is increasing or decreasing.

1. $f(x) = 0.25x^{-3}$

2. $f(x) = 8x^{\frac{4}{3}}$

Solve each equation.

3. $x = \sqrt{4 - x} - 8$

4. $\sqrt{5x + 4} = \sqrt{9 - x} + 7$

5. $-2 + \sqrt{3x + 2} = x$

6. $56 - \sqrt[8]{7x^2 + 4} = 54$

7. $x^4 - 5x^3 - 14x^2 = 0$

8. $x^3 - 3x^2 - 10x = -24$

Describe the end behavior of the graph of each polynomial function using limits. Explain your reasoning using the leading term test.

9. $f(x) = 5x^4 - 3x^3 - 7x^2 + 11x - 8$

10. $f(x) = -3x^5 - 8x^4 + 7x^2 + 5$

State the number of possible real zeros and turning points of each function. Then find all of the real zeros by factoring.

11. $f(x) = 4x^3 + 8x^2 - 60x$

12. $f(x) = x^5 - 16x$

13. **MULTIPLE CHOICE** Which function has 3 turning points?

A $f(x) = x^4 - 4$

C $f(x) = x^3 + 9x^2 + 20x$

B $f(x) = x^4 - 11x^3$

D $f(x) = x^4 - 5x^2 + 4$

14. **BASEBALL** The height h in feet of a baseball after being struck by a batter is given by $h(t) = -32t^2 + 128t + 4$, where t is the time in seconds after the ball is hit. Describe the end behavior of the graph of the function using limits. Explain using the leading term test.

For each function, (a) apply the leading term test, (b) find the zeros and state the multiplicity of any repeated zeros, (c) find a few additional points, and then (d) graph the function.

15. $f(x) = x(x - 1)(x + 3)$

16. $f(x) = x^4 - 9x^2$

Use the Factor Theorem to determine if the binomials given are factors of $f(x)$. Use the binomials that are factors to write a factored form of $f(x)$.

17. $f(x) = x^3 - 3x^2 - 13x + 15; (x + 3)$

18. $f(x) = x^4 - x^3 - 34x^2 + 4x + 120; (x + 5), (x - 2)$

19. **WEATHER** The table shows the average high temperature in Bay Town each month.

Jan	Feb	Mar	Apr	May	Jun
62.3°	66.5°	73.3°	79.1°	85.5°	90.7°

Jul	Aug	Sep	Oct	Nov	Dec
93.6°	93.5°	89.3°	82.0°	72.0°	64.6°

a. Make a scatter plot for the data.

b. Use a graphing calculator to model the data using a polynomial function with a degree of 3. Use $x = 1$ for January and round each coefficient to the nearest thousandth.

c. Use the model to predict the average high temperature for the following January. Let $x = 13$.

Write a polynomial function of least degree with real coefficients in standard form that has the given zeros.

20. $-1, 4, -\sqrt{3}$

21. $5, -5, 1 - i$

22. **MULTIPLE CHOICE** Which function graphed below must have imaginary zeros?

F

H

G

J

Divide using synthetic division.

23. $f(x) = (x^3 - 7x^2 + 13) \div (x - 2)$

24. $f(x) = (x^4 + x^3 - 2x^2 + 3x + 8) \div (x + 3)$

Determine any asymptotes and intercepts. Then graph the function and state its domain.

25. $f(x) = \dfrac{2x - 6}{x + 5}$

26. $f(x) = \dfrac{x^2 + x - 6}{x - 4}$

Solve each inequality.

27. $x^2 - 5x - 14 < 0$

28. $\dfrac{x^2}{x - 6} \geq 0$

Area Under a Curve

Objective
Approximate the area between a curve and the x-axis.

Integral calculus is a branch of calculus that focuses on the processes of finding areas, volumes, and lengths. In geometry, you learned how to calculate the perimeters, areas, and volumes of polygons, polyhedrons, and composite figures by using your knowledge of basic shapes, such as triangles, pyramids, and cones. The perimeters, areas, and volumes of irregular shapes, objects that are not a combination of basic shapes, can be found in a similar manner. Calculating the area between the curve and the x-axis, as shown to the right, is an application of integral calculus.

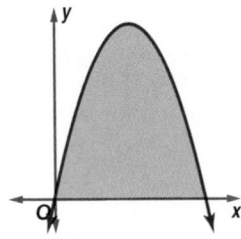

ACTIVITY 1 **Approximate Area Under a Curve**

Approximate the area between the curve $f(x) = \sqrt{-x^2 + 8x}$ and the x-axis using rectangles.

Step 1 Draw 4 rectangles with a width of 2 units between $f(x)$ and the x-axis. The height of the rectangle should be determined when the left endpoint of the rectangle intersects $f(x)$, as shown in the figure. Notice that the first rectangle will have a height of $f(0)$ or 0.

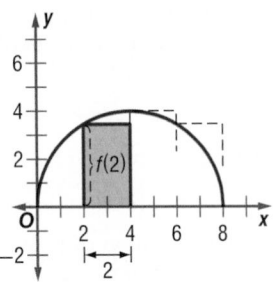

Step 2 Calculate the area of each rectangle.

Step 3 Approximate the area of the region by taking the sum of the areas of the rectangles.

▶ Analyze the Results

1. What is the approximation for the area?

2. How does the area of a rectangle that lies outside the graph affect the approximation?

3. Calculate the actual area of the semicircle. How does the approximation compare to the actual area?

4. How can rectangles be used to find a more accurate approximation? Explain your reasoning.

Using relatively large rectangles to estimate the area under a curve may not produce an approximation that is as accurate as desired. Significant sections of area under the curve may go unaccounted for. Similarly, if the rectangles extend beyond the curve, substantial amounts of areas that lie above a curve may be included in the approximation.

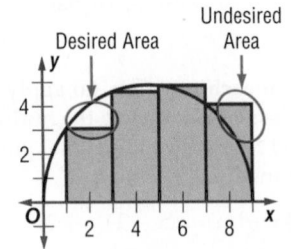

In addition, regions are also not always bound by a curve intersecting the x-axis. You have studied many functions with graphs that have different end behaviors. These graphs do not necessarily have two x-intercepts that allow for obvious start and finish points. In those cases, we often estimate the area under the curve for an x-axis interval.

Approximate the area between the curve $f(x) = x^2 + 2$ and the x-axis on the interval $[1, 5]$ using rectangles.

Step 1   Draw 4 rectangles with a width of 1 unit between $f(x)$ and the x-axis on the interval $[1, 5]$, as shown in the figure. Use the left endpoint of each sub interval to determine the height of each rectangle.

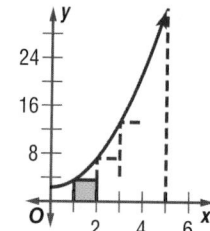

Step 2   Calculate the area of each rectangle.

Step 3   Approximate the area of the region by determining the sum of the areas of the rectangles.

Step 4   Repeat Steps 1–3 using 8 rectangles, each with a width of 0.5 units, and 16 rectangles, each with a width of 0.25 units.

▶ **Analyze the Results**

5. What value for total area are the approximations approaching?

6. Using left endpoints, all of the rectangles completely lie under the curve. How does this affect the approximation for the area of the region?

7. Would the approximations differ if each rectangle's height was determined by its right endpoint? Is this always true? Explain your reasoning.

8. What would happen to the approximations if we continued to increase the number of rectangles being used? Explain your reasoning.

9. Make a conjecture about the relationship between the area under a curve and the number of rectangles used to find the approximation. Explain your answer.

Model and Apply

10. In this problem, you will approximate the area between the curve $f(x) = -x^2 + 12x$ and the x-axis.

    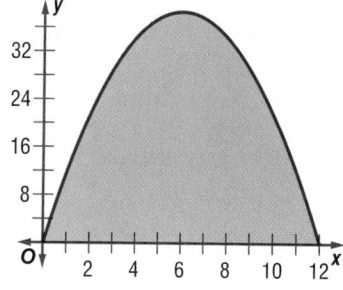

    a. Approximate the area by using 6 rectangles, 12 rectangles, and 24 rectangles. Determine the height of each rectangle using the left endpoints.

    b. What value for total area are the approximations approaching?

    c. Does using right endpoints opposed to left endpoints for the rectangles' heights produce a different approximation? Explain your reasoning.

11. In this problem, you will approximate the area between the curve $f(x) = \frac{1}{2}x^3 - 3x^2 + 3x + 6$ and the x-axis on the interval $[1, 5]$.

    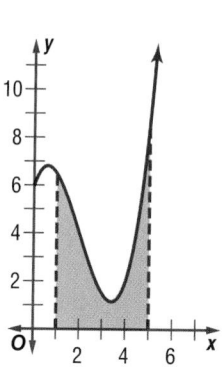

    a. Approximate the area by first using 4 rectangles and then using 8 rectangles. Determine the height of each rectangle using left endpoints.

    b. Does estimating the area by using 4 or 8 rectangles give sufficient approximations? Explain your reasoning.

    c. Does using right endpoints opposed to left endpoints for the rectangles' heights produce a different approximation? Explain your reasoning.

CHAPTER 3

Exponential and Logarithmic Functions

Then

In Chapter 2, you graphed and analyzed power, polynomial, and rational functions.

Now

In Chapter 3, you will:

- Evaluate, analyze, and graph exponential and logarithmic functions.
- Apply properties of logarithms.
- Solve exponential and logarithmic equations.
- Model data using exponential, logarithmic, and logistic functions.

Why?

ENDANGERED SPECIES Exponential functions are often used to model the growth and decline of populations of endangered species. For example, an exponential function can be used to model the population of the North American gray wolf since it became an endangered species in 1974.

PREREAD Use the Concept Summary Boxes in the chapter to predict the organization of Chapter 3.

Get Ready for Chapter 3

Diagnose Readiness You have two options for checking Prerequisite Skills.

Text Option

Take the Quick Check below.

*Quick*Check

Simplify. (Lesson 0-4)

1. $(3x^2)^4 \cdot 2x^3$

2. $(3b^3)(2b^4)$

3. $\dfrac{y^7}{y^4}$

4. $\left(\dfrac{1}{2a}\right)^3$

5. $\dfrac{c^4d}{cd}$

6. $\dfrac{(2n^2)^4}{4n}$

7. CARPET The length of a bedroom carpet can be represented by $2a^2$ feet and the width by $5a^3$ feet. Determine the area of the carpet.

Use a graphing calculator to graph each function. Determine whether the inverse of the function is a function. (Lesson 1-7)

8. $f(x) = \sqrt{4 - x^2}$

9. $f(x) = \sqrt{x + 2}$

10. $f(x) = \dfrac{8 - x}{x}$

11. $g(x) = \dfrac{x - 3}{x}$

12. $f(x) = \dfrac{2}{\sqrt{1 - x}}$

13. $g(x) = \dfrac{5}{\sqrt{x + 7}}$

14. STAMPS The function $v(t) = 200(1.6)^t$ can be used to predict the value v of a rare stamp after t years. Graph the function, and determine whether the inverse of the function is a function.

Graph and analyze each function. Describe the domain, range, intercepts, end behavior, continuity, and where the function is increasing or decreasing. (Lesson 2-1)

15. $f(x) = 2x^2$

16. $g(x) = 4x^3$

17. $h(x) = -3x^3$

18. $f(x) = -x^5$

Online Option

Math Online Take a self-check Chapter Readiness Quiz at **glencoe.com**.

New Vocabulary

English	• Page •	Español
algebraic functions	• 158 •	funciones algebraicas
transcendental functions	• 158 •	funciones transcendentales
exponential functions	• 158 •	funciones exponenciales
natural base	• 160 •	base natural
continuous compound interest	• 163 •	interés compuesto continuo
logarithmic function with base b	• 172 •	función logarítmica con base b
logarithm	• 172 •	logaritmo
common logarithm	• 173 •	logaritmo común
natural logarithm	• 174 •	logaritmo natural
logistic growth function	• 202 •	función de crecimiento logística
linearize	• 204 •	linearize

Review Vocabulary

one-to-one • p. 66 • de uno a uno   a function that passes the horizontal line test, and no y-value is matched with more than one x-value

inverse functions • p. 65 • funciones inversas   Two functions, f and f^{-1}, are inverse functions if and only if $f[f^{-1}(x)] = x$ and $f^{-1}[f(x)] = x$.

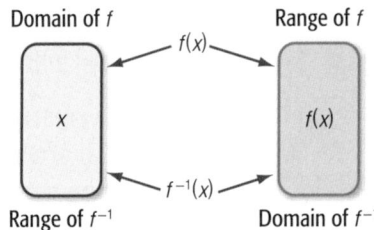

end behavior • p. 159 • comportamiento de final   describes the behavior of $f(x)$ as x increases or decreases without bound—becoming greater and greater or more and more negative

continuous function • p. 159 • función continua   a function with a graph that has no breaks, holes or gaps

Exponential Functions

Why?

Worldwide water consumption has increased rapidly over the last several decades. Most of the world's water is used for agriculture, and increasing population has resulted in an increasing agricultural demand. The increase in water consumption can be modeled using an exponential function.

Then

You identified, graphed, and described several parent functions. (Lesson 1-5)

Now

- Evaluate, analyze, and graph exponential functions.
- Solve problems involving exponential growth and decay.

New Vocabulary

algebraic function
transcendental function
exponential function
natural base
continuous compound
  interest

Math Online

glencoe.com

Exponential Functions In Chapter 2, you studied power, radical, polynomial, and rational functions. These are examples of **algebraic functions**—functions with values that are obtained by adding, subtracting, multiplying, or dividing constants and the independent variable or raising the independent variable to a rational power. In this chapter, we will explore exponential and logarithmic functions. These are considered to be **transcendental functions** because they cannot be expressed in terms of algebraic operations. In effect, they *transcend* algebra.

Consider functions $f(x) = x^3$ and $g(x) = 3^x$. Both involve a base raised to a power; however, in $f(x)$, a power function, the base is a variable and the exponent is a constant. In $g(x)$, the base is a constant and the exponent is a variable. Functions of a form similar to $g(x)$ are called **exponential functions**.

Key Concept | Exponential Function

An exponential function with base b has the form $f(x) = ab^x$, where x is any real number and a and b are real number constants such that $a \neq 0$, b is positive, and $b \neq 1$.

Examples	**Nonexamples**
$f(x) = 4^x$, $f(x) = \left(\frac{1}{3}\right)^x$, $f(x) = 7^{-x}$	$f(x) = 2x^{-3}$, $f(x) = 5^{\pi}$, $f(x) = 1^x$

When the inputs are rational numbers, exponential functions can be evaluated using the properties of exponents. For example, if $f(x) = 4^x$, then

$$f(2) = 4^2 \qquad f\left(\frac{1}{3}\right) = 4^{\frac{1}{3}} \qquad f(-3) = 4^{-3}$$

$$= 16 \qquad\qquad = \sqrt[3]{4} \qquad\qquad = \frac{1}{4^3}$$

$$= \frac{1}{64}$$

Since exponential functions are defined for all real numbers, you must also be able to evaluate an exponential function for *irrational* values of x, such as $\sqrt{2}$. But what does the expression $4^{\sqrt{2}}$ mean?

The value of this expression can be approximated using successively closer rational approximations of $\sqrt{2}$ as shown below.

x	1	1.4	1.41	1.414	1.4142	1.41421	...
$f(x) = 4^x$	4	7.0	7.06	7.101	7.1029	7.10296	...

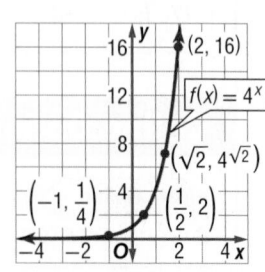

From this table, we can conclude that $4^{\sqrt{2}}$ is a real number approximately equal to 7.10. Since $f(x) = 4^x$ has real number values for every x-value in its domain, this function is continuous and can be graphed as a smooth curve as shown.

EXAMPLE 1 | Sketch and Analyze Graphs of Exponential Functions

Sketch and analyze the graph of each function. Describe its domain, range, intercepts, asymptotes, end behavior, and where the function is increasing or decreasing.

a. $f(x) = 3^x$

Evaluate the function for several x-values in its domain. Then use a smooth curve to connect each of these ordered pairs.

x	−4	−2	−1	0	2	4	6
f(x)	0.01	0.11	0.33	1	9	81	729

Domain: $(-\infty, \infty)$ Range $(0, \infty)$

y-Intercept: 1   Asymptote: x-axis

End behavior: $\lim\limits_{x \to -\infty} f(x) = 0$ and $\lim\limits_{x \to \infty} f(x) = \infty$

Increasing: $(-\infty, \infty)$

b. $g(x) = 2^{-x}$

x	−6	−4	−2	0	2	4	6
f(x)	64	16	4	1	0.25	0.06	0.02

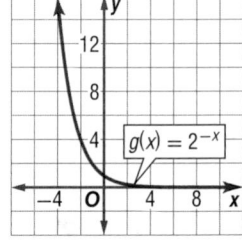

Domain: $(-\infty, \infty)$ Range $(0, \infty)$

y-Intercept: 1   Asymptote: x-axis

End behavior: $\lim\limits_{x \to -\infty} g(x) = \infty$ and $\lim\limits_{x \to \infty} g(x) = 0$

Decreasing: $(-\infty, \infty)$

> **StudyTip**
>
> **Negative Exponents**
> Notice that $f(x) = \left(\frac{1}{b}\right)^x$ and $g(x) = b^{-x}$ are equivalent because $\frac{1}{b^x} = (b^{-1})^x$ or b^{-x}.

✔ Guided Practice

1A. $f(x) = 6^{-x}$ **1B.** $g(x) = 5^x$ **1C.** $h(x) = \left(\frac{1}{4}\right)^x + 1$

The increasing and decreasing graphs in Example 1 are typical of the two basic types of exponential functions: *exponential growth* and *exponential decay*.

Key Concept | Properties of Exponential Functions

Exponential Growth

Exponential Decay

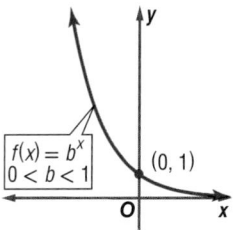

Domain: $(-\infty, \infty)$	**Range:** $(0, \infty)$
y-Intercept: 1	**x-Intercept:** none
Extrema: none	**Asymptote:** x-axis

End Behavior: $\lim\limits_{x \to -\infty} f(x) = 0$
and $\lim\limits_{x \to \infty} f(x) = \infty$

Continuity: continuous on $(-\infty, \infty)$

Domain: $(-\infty, \infty)$	**Range:** $(0, \infty)$
y-Intercept: 1	**x-Intercept:** none
Extrema: none	**Asymptote:** x-axis

End Behavior: $\lim\limits_{x \to -\infty} f(x) = \infty$
and $\lim\limits_{x \to \infty} f(x) = 0$

Continuity: continuous on $(-\infty, \infty)$

> **StudyTip**
>
> **Coefficient a** An exponential function of the form $f(x) = ab^x$ has a y-intercept at $(0, a)$.

The same techniques that you used to transform graphs of algebraic functions can be applied to graphs of exponential functions.

EXAMPLE 2 **Graph Transformations of Exponential Functions**

Use the graph of $f(x) = 2^x$ to describe the transformation that results in each function. Then sketch the graphs of the functions.

a. $g(x) = 2^{x+1}$

This function is of the form $g(x) = f(x + 1)$. Therefore, the graph of $g(x)$ is the graph of $f(x) = 2^x$ translated 1 unit to the left (Figure 3.1.1).

b. $h(x) = 2^{-x}$

This function is of the form $h(x) = f(-x)$. Therefore, the graph of $h(x)$ is the graph of $f(x) = 2^x$ reflected in the y-axis (Figure 3.1.2).

c. $j(x) = -3(2^x)$

This function is of the form $j(x) = -3f(x)$. Therefore, the graph of $j(x)$ is the graph of $f(x) = 2^x$ reflected across the x-axis and expanded vertically by a factor of 3 (Figure 3.1.3).

Figure 3.1.1

Figure 3.1.2

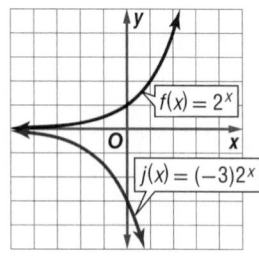

Figure 3.1.3

☑ **Guided Practice**

Use the graph of $f(x) = 4^x$ to describe the transformation that results in each function. Then sketch the graphs of the functions.

2A. $k(x) = 4^x - 2$ **2B.** $m(x) = -4^{x+2}$ **2C.** $p(x) = 2(4^{-x})$

It may surprise you to learn that for most real-world applications involving exponential functions, the most commonly used base is not 2 or 10 but an irrational number e called the **natural base**, where

$$e = \lim_{x \to \infty} \left(1 + \frac{1}{x}\right)^x.$$

By calculating the value of $\left(1 + \frac{1}{x}\right)^x$ for greater and greater values of x, we can estimate that the value of this expression approaches a number close to 2.7183. In fact, using calculus, it can be shown that this value approaches the irrational number we call e, named after the Swiss mathematician Leonhard Euler who computed e to 23 decimal places.

$$e = 2.718281828\ldots$$

The number e can also be defined as $\lim_{x \to 0}(1 + x)^{\frac{1}{x}}$, since for fractional values of x closer and closer to 0, $(1 + x)^{\frac{1}{x}} = 2.718281828\ldots$ or e.

x	$\left(1 + \frac{1}{x}\right)^x$
1	2
10	2.59374...
100	2.70481...
1000	2.71692...
10,000	2.71814...
100,000	2.71827...
1,000,000	2.71828...

Figure 3.1.4

The function given by $f(x) = e^x$, is called the *natural base exponential function* (Figure 3.1.4) and has the same properties as those of other exponential functions.

EXAMPLE 3 | **Graph Natural Base Exponential Functions**

Use the graph of $f(x) = e^x$ to describe the transformation that results in the graph of each function. Then sketch the graphs of the functions.

a. $a(x) = e^{4x}$

This function is of the form $a(x) = f(4x)$. Therefore, the graph of $a(x)$ is the graph of $f(x) = e^x$ compressed horizontally by a factor of 4 (Figure 3.1.5).

b. $b(x) = e^{-x} + 3$

This function is of the form $b(x) = f(-x) + 3$. Therefore, the graph of $b(x)$ is the graph of $f(x) = e^x$ reflected in the y-axis and translated 3 units up (Figure 3.1.6).

c. $c(x) = \frac{1}{2}e^x$

This function is of the form $c(x) = \frac{1}{2}f(x)$. Therefore, the graph of $c(x)$ is the graph of $f(x) = e^x$ compressed vertically by a factor of $\frac{1}{2}$ (Figure 3.1.7).

Figure 3.1.5

Figure 3.1.6

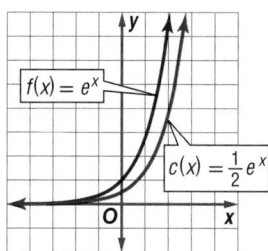

Figure 3.1.7

✓ **Guided Practice**

3A. $q(x) = e^{-2x}$ **3B.** $r(x) = e^x - 5$ **3C.** $t(x) = 3e^x$

Exponential Growth and Decay A common application of exponential growth is compound interest. Suppose an initial principal P is invested into an account with an annual interest rate r, and the interest is *compounded* or reinvested annually. At the end of each year, the interest earned is added to the account balance. This sum becomes the new principal for the next year.

Year	Account Balance After Each Compounding	
0	$A_0 = P$	P = original investment or principal
1	$A_1 = A_0 + A_0 r$	Interest from year 0, $A_0 r$, is added.
	$= A_0(1 + r)$	Distributive Property
	$= P(1 + r)$	$A_0 = P$
2	$A_2 = A_1(1 + r)$	Interest from year 1 is added.
	$= P(1 + r)(1 + r)$	$A_1 = P(1 + r)$
	$= P(1 + r)^2$	Simplify.
3	$A_3 = A_2(1 + r)$	Interest from year 2 is added.
	$= P(1 + r)^2(1 + r)$	$A_2 = P(1 + r)^2$
	$= P(1 + r)^3$	Simplify.
4	$A_4 = A_3(1 + r)$	Interest from year 3 is added.
	$= P(1 + r)^3(1 + r)$	$A_3 = P(1 + r)^3$
	$= P(1 + r)^3$	Simplify.

The pattern that develops leads to the following exponential function with base $(1 + r)$.

$$A(t) = P(1 + r)^t$$ **Account balance after t years**

To allow for quarterly, monthly, or even daily compoundings, let n be the number of times the interest is compounded each year. Then

- the rate per compounding $\frac{r}{n}$ is a fraction of the annual rate r, and

- the number of compoundings after t years is nt.

Replacing r with $\frac{r}{n}$ and t with nt in the formula $A(t) = P(1 + r)^t$, we obtain a general formula for compound interest.

Key Concept   Compound Interest Formula

If a principal P is invested at an annual interest rate r (in decimal form) compounded n times a year, then the balance A in the account after t years is given by

$$A = P\left(1 + \frac{r}{n}\right)^{nt}.$$

EXAMPLE 4   Use Compound Interest

FINANCIAL LITERACY Krysti invests \$300 in an account with a 6% interest rate, making no other deposits or withdrawals. What will Krysti's account balance be after 20 years if the interest is compounded:

a. semiannually?

For semiannually compounding, $n = 2$.

$A = P\left(1 + \frac{r}{n}\right)^{nt}$ **Compound Interest Formula**

$= 300\left(1 + \frac{0.06}{2}\right)^{2(20)}$ **$P = 300$, $r = 0.06$, $n = 2$, and $t = 20$**

≈ 978.61 **Simplify.**

With semiannual compounding, Krysti's account balance after 20 years will be \$978.61.

b. monthly?

For monthly compounding, $n = 12$, since there are 12 months in a year.

$A = P\left(1 + \frac{r}{n}\right)^{nt}$ **Compound Interest Formula**

$= 300\left(1 + \frac{0.06}{12}\right)^{12(20)}$ **$P = 300$, $r = 0.06$, $n = 12$, and $t = 20$**

≈ 993.06 **Simplify.**

With monthly compounding, Krysti's account balance after 20 years will be \$993.06.

c. daily?

For daily compounding, $n = 365$.

$A = P\left(1 + \frac{r}{n}\right)^{nt}$ **Compound Interest Formula**

$= 300\left(1 + \frac{0.06}{365}\right)^{365(20)}$ **$P = 300$, $r = 0.06$, $t = 20$, and $n = 365$**

≈ 995.94 **Simplify.**

With daily compounding, Krysti's account balance after 20 years will be \$995.94.

✓ Guided Practice

4. FINANCIAL LITERACY If \$1000 is invested in an online savings account earning 8% per year, how much will be in the account at the end of 10 years if there are no other deposits or withdrawals and interest is compounded:

   A. semiannually?   **B.** quarterly?   **C.** daily?

Notice that as the number of compoundings increases in Example 4, the account balance also increases. However, the increase is relatively small, only \$995.94 − \$993.06 or \$2.88.

The table below shows the amount A computed for several values of n. Notice that while the account balance is increasing, the amount of increase slows down as n increases. In fact, it appears that the amount tends towards a value close to $996.03.

Compounding	n	$A = 300\left(1 + \dfrac{0.06}{n}\right)^{20n}$
annually	1	$962.14
semiannually	2	$978.61
quarterly	4	$987.20
monthly	12	$993.06
daily	365	$995.94
hourly	8760	$996.03

Suppose the interest were compounded *continuously* so that there was no waiting period between interest payments. We can derive a formula for **continuous compound interest** by first using algebra to manipulate the regular compound interest formula.

$$A = P\left(1 + \frac{1}{\frac{n}{r}}\right)^{nt}$$ Compound interest formula with $\frac{r}{n}$ written as $\frac{1}{\frac{n}{r}}$.

$$= P\left(1 + \frac{1}{x}\right)^{xrt}$$ Let $x = \frac{n}{r}$ and $n = xr$.

$$= P\left[\left(1 + \frac{1}{x}\right)^{x}\right]^{rt}$$ Power Property of Exponents

The expression in brackets should look familiar. Recall from page 160 that $\lim\limits_{x \to \infty}\left(1 + \frac{1}{x}\right)^{x} = e$. Since r is a fixed value and $x = \frac{n}{r}$, $x \to \infty$ as $n \to \infty$. Thus,

$$\lim_{n \to \infty} P\left(1 + \frac{r}{n}\right)^{nt} = \lim_{x \to \infty} P\left[\left(1 + \frac{1}{x}\right)^{x}\right]^{rt} = Pe^{rt}.$$

This leads us to the formula for calculating continuous compounded interest shown below.

> **Key Concept**   **Continuous Compound Interest Formula**
>
> If a principal P is invested at an annual interest rate r (in decimal form) compounded continuously, then the balance A in the account after t years is given by
>
> $$A = Pe^{rt}.$$

EXAMPLE 5   **Use Continuous Compound Interest**

FINANCIAL LITERACY Suppose Krysti finds an account that will allow her to invest her $300 at a 6% interest rate compounded continuously. If there are no other deposits or withdrawals, what will Krysti's account balance be after 20 years?

$A = Pe^{rt}$ Continuous Compound Interest Formula

$= 300e^{(0.06)(20)}$ $P = 300$, $r = 0.06$, and $t = 20$

≈ 996.04 Simplify.

With continuous compounding, Krysti's account balance after 20 years will be $996.04.

✓ **Guided Practice**

5. **ONLINE BANKING** If $1000 is invested in an online savings account earning 8% per year compounded continuously, how much will be in the account at the end of 10 years if there are no other deposits or withdrawals?

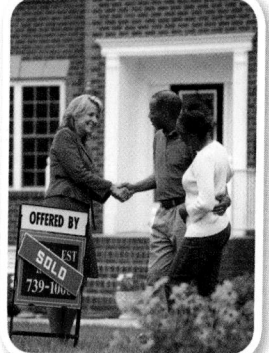

Real-World Link

The prime rate is the interest rate that banks charge their most credit-worthy borrowers. Changes in this rate can influence other rates, including mortgage interest rates.

Source: Federal Reserve System

In addition to investments, populations of people, animals, bacteria, and amounts of radioactive material can also change at an exponential rate. Exponential growth and decay models apply to any situation where growth is proportional to the initial size of the quantity being considered.

Key Concept **Exponential Growth or Decay Formulas**

If an initial quantity N_0 grows or decays at an exponential rate r or k (as a decimal), then the final amount N after a time t is given by the following formulas.

Exponential Growth or Decay

$$N = N_0(1 + r)^t$$

If r is a *growth rate*, then $r > 0$.

If r is a *decay rate*, then $r < 0$.

Continuous Exponential Growth or Decay

$$N = N_0e^{kt}$$

If k is a *continuous growth rate*, then $k > 0$.

If k is a *continuous decay rate*, then $k < 0$.

Continuous growth or decay is similar to continuous compound interest. The growth or decay is compounded continuously rather than just yearly, monthly, hourly, or at some other time interval. Population growth can be modeled exponentially, continuously, and by other models.

⊕ Real-World Link

In 2008, the population of Mexico was estimated to be growing at a rate of about 1.42% annually.

Source: *CIA–The World Fact Book*

⊕ Real-World **EXAMPLE 6** Model Using Exponential Growth or Decay

POPULATION Mexico has a population of approximately 110 million. If Mexico's population continues to grow at the described rate, predict the population of Mexico in 10 and 20 years.

a. 1.42% annually

Use the exponential growth formula to write an equation that models this situation.

$N = N_0(1 + r)^t$ **Exponential Growth Formula**

$\quad = 110,000,000(1 + 0.0142)^t$ $N_0 = 110,000,000$ and $r = 0.0142$

$\quad = 110,000,000(1.0142)^t$ **Simplify.**

Use this equation to find N when $t = 10$ and $t = 20$.

$N = 110,000,000(1.0142)^t$ **Modeling equation** $N = 110,000,000(1.0142)^t$

$\quad = 110,000,000(1.0142)^{10}$ $t = 10$ or $t = 20$ $\quad = 110,000,000(1.0142)^{20}$

$\quad \approx 126,656,869$ **Simplify.** $\quad = 145,836,022$

If the population of Mexico continues to grow at an annual rate of 1.42%, its population in 10 years will be about 126,656,869; and in 20 years, it will be about 145,836,022.

b. 1.42% continuously

Use the continuous exponential growth formula to write a modeling equation.

$N = N_0e^{kt}$ **Continuous Exponential Growth Formula**

$\quad = 110,000,000e^{0.0142t}$ $N_0 = 110,000,000$ and $k = 0.0142$

Use this equation to find N when $t = 10$ and $t = 20$.

$N = 110,000,000e^{0.0142t}$ **Modeling equation** $N = 110,000,000e^{1.0142t}$

$\quad = 110,000,000e^{0.0142(10)}$ $t = 10$ and $t = 20$ $\quad = 110,000,000e^{0.0142(20)}$

$\quad \approx 126,783,431$ **Simplify.** $\quad \approx 146,127,622$

If the population of Mexico continues to grow at a continuous rate of 1.42%, its population in 10 years will be about 126,783,431; in 20 years, it will be about 146,127,622.

☑ Guided Practice

6. POPULATION The population of a town is declining at a rate of 6%. If the current population is 12,426 people, predict the population in 5 and 10 years using each model.

    A. annually   **B.** continuously

Watch Out!

Using Rates of Decay
Remember to write rates of decay as negative values.

After finding a model for a situation, you can use the graph of the model to solve problems.

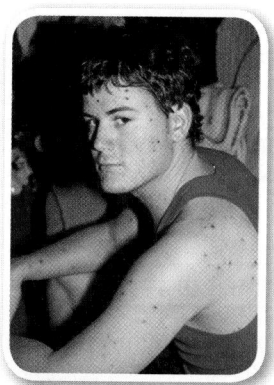
Real-World EXAMPLE 7 — Use the Graph of an Exponential Model

DISEASE The table shows the number of reported cases of chicken pox in the United States in 1980 and 2005.

U.S. Reported Cases of Chicken Pox	
Year	Cases (thousands)
1980	190.9
2005	32.2

Source: U.S. Centers for Disease Control and Prevention

a. If the number of reported cases of chicken pox is decreasing at an exponential rate, identify the rate of decline and write an exponential equation to model this situation.

If we let $N(t)$ represent the number of cases t years after 1980 and assume exponential decay, then the initial number of cases $N_0 = 190.9$ and at time $t = 2005 - 1980$ or 25, the number of reported cases $N(25) = 32.2$. Use this information to find the rate of decay r.

$N(t) = N_0(1 + r)^t$ **Exponential Decay Formula**

$32.2 = 190.9(1 + r)^{25}$ **$N(25) = 32.2$, $N_0 = 190.9$, and $t = 25$**

$\dfrac{32.2}{190.9} = (1 + r)^{25}$ **Divide each side by 190.9.**

$\sqrt[25]{\dfrac{32.2}{190.9}} = 1 + r$ **Take the positive 25th root of each side.**

$\sqrt[25]{\dfrac{32.2}{190.9}} - 1 = r$ **Subtract 1 from each side.**

$-0.069 \approx r$ **Simplify.**

The number of reported cases is decreasing at a rate of approximately 6.9% per year. Therefore, an equation modeling this situation is $N(t) = 190.9[1 + (-0.069)]^t$ or $N(t) = 190.9(0.931)^t$.

b. Use your model to predict when the number of cases will drop below 20,000.

To find when the number of cases will drop below 20,000, find the intersection of the graph of $N(t) = 190.9(0.931)^t$ and the line $N(t) = 20$. A graphing calculator shows that value of t for which $190.9(0.931)^t = 20$ is about 32.

Since t is the number of years after 1980, this model suggests that after the year $1980 + 32$ or 2012, the number of cases will drop below 20,000 if this rate of decline continues.

Intersection X=31.554456 Y=20
[−5, 50] scl: 5 by [−25, 200] scl: 25

✓ Guided Practice

7. POPULATION Use the data in the table and assume that the population of Miami-Dade County is growing exponentially.

Estimated Population of Miami-Dade County, Florida	
Year	Population (million)
1990	1.94
2000	2.25

Source: U.S. Census Bureau

A. Identify the rate of growth and write an exponential equation to model this growth.

B. Use your model to predict in which year the population of Miami-Dade County will surpass 2.7 million.

Exercises

Sketch and analyze the graph of each function. Describe its domain, range, intercepts, asymptotes, end behavior, and where the function is increasing or decreasing. (Example 1)

1. $f(x) = 2^{-x}$
2. $r(x) = 5^x$
3. $h(x) = 0.2^{x+2}$
4. $k(x) = 6^x$
5. $m(x) = -(0.25)^x$
6. $p(x) = 0.1^{-x}$
7. $q(x) = \left(\frac{1}{6}\right)^x$
8. $g(x) = \left(\frac{1}{3}\right)^x$
9. $c(x) = 2^x - 3$
10. $d(x) = 5^{-x} + 2$

Use the graph of $f(x)$ to describe the transformation that results in the graph of $g(x)$. Then sketch the graphs of $f(x)$ and $g(x)$. (Examples 2 and 3)

11. $f(x) = 4^x; g(x) = 4^x - 3$
12. $f(x) = \left(\frac{1}{2}\right)^x; g(x) = \left(\frac{1}{2}\right)^{x+4}$
13. $f(x) = 3^x; g(x) = -2(3^x)$
14. $f(x) = 2^x; g(x) = 2^{x-2} + 5$
15. $f(x) = 10^x; g(x) = 10^{-x+3}$
16. $f(x) = e^x; g(x) = e^{2x}$
17. $f(x) = e^x; g(x) = e^{x+2} - 1$
18. $f(x) = e^x; g(x) = e^{-x+1}$
19. $f(x) = e^x; g(x) = 3e^x$
20. $f(x) = e^x; g(x) = -(e^x) + 4$

FINANCIAL LITERACY Copy and complete the table below to find the value of an investment A for the given principal P, rate r, and time t if the interest is compounded n times annually. (Examples 4 and 5)

n	1	4	12	365	continuously
A					

21. $P = \$500, r = 3\%, t = 5$ years
22. $P = \$1000, r = 4.5\%, t = 10$ years
23. $P = \$1000, r = 5\%, t = 20$ years
24. $P = \$5000, r = 6\%, t = 30$ years

25. **FINANCIAL LITERACY** Brady acquired an inheritance of $20,000 at age 8, but he will not have access to it until he turns 18. (Examples 4 and 5)
    a. If his inheritance is placed in a savings account earning 4.6% interest compounded monthly, how much will Brady's inheritance be worth on his 18th birthday?
    b. How much will Brady's inheritance be worth if it is placed in an account earning 4.2% interest compounded continuously?

26. **FINANCIAL LITERACY** Katrina invests $1200 in a certificate of deposit (CD). The table shows the interest rates offered by the bank on 3- and 5-year CDs. (Examples 4 and 5)

CD Offers		
Years	3	5
Interest	3.45%	4.75%
Compounded	continuously	monthly

   a. How much would her investment be worth with each option?
   b. How much would her investment be worth if the 5-year CD was compounded continuously?

POPULATION Copy and complete the table to find the population N of an endangered species after a time t given its initial population N_0 and annual rate r or continuous rate k of increase or decline. (Example 6)

t	5	10	15	20	50
N					

27. $N_0 = 15{,}831, r = -4.2\%$
28. $N_0 = 23{,}112, r = 0.8\%$
29. $N_0 = 17{,}692, k = 2.02\%$
30. $N_0 = 9689, k = -3.7\%$

31. **WATER** Worldwide water usage in 1950 was about 294.2 million gallons. If water usage has grown at the described rate, estimate the amount of water used in 2000 and predict the amount in 2050. (Example 6)
    a. 3% annually
    b. 3.05% continuously

32. **WAGES** Jasmine receives a 3.5% raise at the end of each year from her employer to account for inflation. When she started working for the company in 1994, she was earning a salary of $31,000. (Example 6)
    a. What was Jasmine's salary in 2000 and 2004?
    b. If Jasmine continues to receive a raise at the end of each year, how much money will she earn during her final year if she plans on retiring in 2024?

33. **PEST CONTROL** Consider the termite guarantee made by Exterm-inc in their ad below.

If the first statement in this claim is true, assess the validity of the second statement. Explain your reasoning. (Example 6)

34. INFLATION The Consumer Price Index (CPI) is an index number that measures the average price of consumer goods and services. A change in the CPI indicates the growth rate of inflation. In 1958 the CPI was 28.6, and in 2008 the CPI was 211.08. (Example 7)

   a. Determine the growth rate of inflation between 1958 and 2008. Use this rate to write an exponential equation to model this situation.

   b. What will be the CPI in 2015? At this rate, when will the CPI exceed 350?

35 GASOLINE Jordan wrote an exponential equation to model the cost of gasoline. He found the average cost per gallon of gasoline for two years and used these data for his model. (Example 7)

Average Cost per Gallon of Gasoline	
Year	**Cost($)**
1990	1.19
2007	3.86

   a. If the average cost of gasoline increased at an exponential rate, identify the rate of increase. Write an exponential equation to model this situation.

   b. Use your model to predict the average cost of a gallon of gasoline in 2011 and 2013.

   c. When will the average cost per gallon of gasoline exceed $7?

   d. Why might an exponential model not be an accurate representation of average gasoline prices?

36. PHYSICS The pressure of the atmosphere at sea level is 15.191 pounds per square inch (psi). It decreases continuously at a rate of 0.004% as altitude increases by x feet.

   a. Write a modeling function for the continuous exponential decay representing the atmospheric pressure $a(x)$.

   b. Use the model to approximate the atmospheric pressure at the top of Mount Everest.

Highest point above sea level—peak of Mount Everest
29,035 ft
Sea Level: 15.191 psi

   c. If a certain rescue helicopter can fly only in atmospheric pressures greater than 5.5 pounds per square inch, how high can it fly up Mount Everest?

37. RADIOACTIVITY The half-life of a radioactive substance is the amount of time it takes for half of the atoms of the substance to disintegrate. Uranium-235 is used to fuel a commercial power plant. It has a half-life of 704 million years.

   a. How many grams of uranium-235 will remain after 1 million years if you start with 200 grams?

   b. How many grams of uranium-235 will remain after 4540 million years if you start with 200 grams?

38. BOTANY Under the right growing conditions, a particular species of plant has a doubling time of 12 days. Suppose a pasture contains 46 plants of this species. How many plants will there be after 20, 65, and x days?

39. RADIOACTIVITY Radiocarbon dating uses carbon-14 to estimate the age of organic materials found commonly at archaeological sites. The half-life of carbon-14 is approximately 5.73 thousand years.

   a. Write a modeling equation for the exponential decay.

   b. How many grams of carbon-14 will remain after 12.82 thousand years if you start with 7 grams?

   c. Use your model to estimate when only 1 gram of the original 7 grams of carbon-14 will remain.

40. MICROBIOLOGY A certain bacterium used to treat oil spills has a doubling time of 15 minutes. Suppose a colony begins with a population of one bacterium.

   a. Write a modeling equation for this exponential growth.

   b. About how many bacteria will be present after 55 minutes?

   c. A population of 8192 bacteria is sufficient to clean a small oil spill. Use your model to predict how long it will take for the colony to grow to this size.

41. ENCYCLOPEDIA The number of articles making up an online open-content encyclopedia increased exponentially during its first few years. The number of articles, $A(t)$, t years after 2001 can be modeled by $A(t) = 16{,}198 \cdot 2.13^t$.

   a. According to this model, how many articles made up the encyclopedia in 2001? At what percentage rate is the number of articles increasing?

   b. During which year did the encyclopedia reach 1 million articles?

   c. Predict the number of articles there will be at the beginning of 2012.

42. RISK The chance of having an automobile accident increases exponentially if the driver has consumed alcohol. The relationship can be modeled by $A(c) = 6e^{12.8c}$, where A is the percent chance of an accident and c is the driver's blood alcohol concentration (BAC).

   a. The legal BAC is 0.08. What is the percent chance of having a car accident at this concentration?

   b. What BAC would correspond to a 50% chance of having a car accident?

43. GRAPHING CALCULATOR The table shows the number of blogs in millions semiannually from September 2003 to March 2006.

Months	1	7	13	19	25	31
Blogs	0.7	2	4	8	16	31

  a. Using the calculator's exponential regression tool, find a function that models the data.

  b. After how many months did the number of blogs reach 20 million?

  c. Predict the number of blogs after 48 months.

44. LANGUAGES *Glottochronology* is an area of linguistics that studies the divergence of languages. The equation $c = e^{-Lt}$, where c is the proportion of words that remain unchanged, t is the time since two languages diverged, and L is the rate of replacement, models this divergence.

  a. If two languages diverged 5 years ago and the rate of replacement is 43.13%, what proportion of words remains unchanged?

  b. After how many years will only 1% of the words remain unchanged?

45. FINANCIAL LITERACY A couple just had a child and wants to immediately start a college fund. Use the information below to determine how much money they should invest.

$60,000

Interest Rate: 9%
Compounding: daily

0 years old   18 years old

GRAPHING CALCULATOR Determine the value(s) of x that makes each equation or inequality below true. Round to the nearest hundredth, if necessary.

46. $2^x < 4$

47. $e^{2x} = 3$

48. $-e^x > -2$

49. $2^{-4x} \leq 8$

Describe the domain, range, continuity, and increasing/decreasing behavior for an exponential function with the given intercept and end behavior. Then graph the function.

50. $f(0) = -1,\ \lim\limits_{x \to -\infty} f(x) = 0,\ \lim\limits_{x \to \infty} f(x) = -\infty$

51. $f(0) = 4,\ \lim\limits_{x \to -\infty} f(x) = \infty,\ \lim\limits_{x \to \infty} f(x) = 3$

52. $f(0) = 3,\ \lim\limits_{x \to -\infty} f(x) = 2,\ \lim\limits_{x \to \infty} f(x) = \infty$

Determine the equation of each function after the given transformation of the parent function.

53. $f(x) = 5^x$ translated 3 units left and 4 units down

54. $f(x) = 0.25^x$ compressed vertically by a factor of 3 and translated 9 units left and 12 units up

55. $f(x) = 4^x$ reflected across the x-axis and translated 1 unit left and 6 units up

Determine the transformations of the given parent function that produce each graph.

56. $f(x) = \left(\dfrac{1}{2}\right)^x$

57. $f(x) = 3^x$

58. ⟳ **MULTIPLE REPRESENTATIONS** In this problem, you will investigate the average rate of change for exponential functions.

  a. GRAPHICAL Graph $f(x) = b^x$ for $b = 2, 3, 4,$ or 5.

  b. ANALYTICAL Find the average rate of change of each function on the interval $[0, 2]$.

  c. VERBAL What can you conclude about the average rate of change of $f(x) = b^x$ as b increases? How is this shown in the graphs in part **a**?

  d. GRAPHICAL Graph $f(x) = b^{-x}$ for $b = 2, 3, 4,$ or 5.

  e. ANALYTICAL Find the average rate of change of each function on the interval $[0, 2]$.

  f. VERBAL What can you conclude about the average rate of change of $f(x) = b^{-x}$ as b increases. How is this shown in the graphs in part **d**?

H.O.T. Problems Use **H**igher-**O**rder **T**hinking Skills

59 ERROR ANALYSIS Eric and Sonja are determining the worth of a $550 investment after 12 years in a savings account earning 3.5% interest compounded monthly. Eric thinks the investment is worth $837.08, while Sonja thinks it is worth $836.57. Is either of them correct? Explain.

REASONING State whether each statement is *true* or *false*. Explain your reasoning.

60. Exponential functions can never have restrictions on the domain.

61. Exponential functions always have restrictions on the range.

62. Graphs of exponential functions always have an asymptote.

63. OPEN ENDED Write an example of an increasing exponential function with a negative y-intercept.

64. CHALLENGE Trina invests $1275 in an account that compounds quarterly at 8%, but at the end of each year she takes 100 out. How much is the account worth at the end of the fifth year?

65. REASONING Two functions of the form $f(x) = b^x$ *sometimes*, *always*, or *never* have at least one ordered pair in common.

66. WRITING IN MATH Compare and contrast the domain, range, intercepts, symmetry, continuity, increasing/decreasing behavior, and end behavior of exponential and power parent functions.

Solve each inequality. (Lesson 2-6)

67. $(x - 3)(x + 2) \leq 0$

68. $x^2 + 6x \leq -x - 4$

69. $3x^2 + 15 \geq x^2 + 15x$

Find the domain of each function and the equations of any vertical or horizontal asymptotes, noting any holes. (Lesson 2-5)

70. $f(x) = \dfrac{3}{x^2 - 4x + 4}$

71. $f(x) = \dfrac{x - 1}{x^2 + 4x - 5}$

72. $f(x) = \dfrac{x^2 - 8x + 16}{x - 4}$

73. TEMPERATURE A formula for converting degrees Celsius to Fahrenheit is $F(x) = \dfrac{9}{5}x + 32$. (Lesson 1-7)

  a. Find the inverse $F^{-1}(x)$. Show that $F(x)$ and $F^{-1}(x)$ are inverses.

  b. Explain what purpose $F^{-1}(x)$ serves.

74. SHOPPING Lily wants to buy a pair of inline skates that are on sale for 30% off the original price of $149. The sales tax is 5.75%. (Lesson 1-6)

  a. Express the price of the inline skates after the discount and the price of the inline skates after the sales tax using function notation. Let x represent the price of the inline skates, $p(x)$ represent the price after the 30% discount, and $s(x)$ represent the price after the sales tax.

  b. Which composition of functions represents the price of the inline skates, $p[s(x)]$ or $s[p(x)]$? Explain your reasoning.

  c. How much will Lily pay for the inline skates?

75. EDUCATION The table shows the number of freshmen who applied to and the number of freshmen attending selected universities in a certain year. (Lesson 1-1)

  a. State the relation as a set of ordered pairs.

  b. State the domain and range of the relation.

  c. Determine whether the relation is a function. Explain.

  d. Assuming the relation is a function, is it reasonable to determine a prediction equation for this situation? Explain.

University	Applied	Attending
Auburn University	13,264	4184
University of California-Davis	27,954	4412
University of Illinois-Urbana-Champaign	21,484	6366
Florida State University	13,423	4851
State University of New York-Stony Brook	16,849	2415
The Ohio State University	19,563	5982
Texas A&M University	17,284	6949

Source: *How to Get Into College*

76. SAT/ACT A set of n numbers has an average (arithmetic mean) of $3k$ and a sum of $12m$, where k and m are positive. What is the value of n in terms of k and m?

  A $\dfrac{4m}{k}$ **C** $\dfrac{4k}{m}$ **E** $\dfrac{k}{4m}$

  B $36km$ **D** $\dfrac{m}{4k}$

77. The number of bacteria in a colony were growing exponentially. Approximately how many bacteria were there at 7 P.M.?

  F 15,700

  G 159,540

  H 1,011,929

  J 6,372,392

Time	Number of Bacteria
2 P.M.	100
4 P.M.	4000

78. REVIEW If $4^{x + 2} = 48$, then $4^x = ?$

  A 3.0   **C** 6.9

  B 6.4   **D** 12.0

79. REVIEW What is the equation of the function?

  F $y = 2(3)^x$

  G $y = 2\left(\dfrac{1}{3}\right)^x$

  H $y = 3\left(\dfrac{1}{2}\right)^x$

  J $y = 3(2)^x$

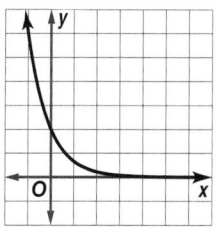

Objective
Calculate future values of annuities and monthly payments.

In Lesson 3-1, you used exponential functions to calculate compounded interest. In the compounding formula, you assume that an initial deposit is made and the investor never deposits nor withdrawals any money. Other types of investments do not follow this simple compounding rule.

When an investor takes out an *annuity*, he or she makes identical deposits into the account at regular intervals or periods. The compounding interest is calculated at the time of each deposit. We can determine the *future value* of an annuity, or its value at the end of a period, using the formula below.

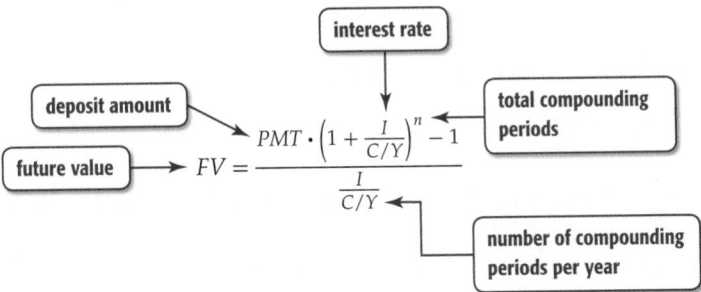

StudyTip

Future Value Formula
The payments must be periodic and of equal value in order for the formula to be accurate.

Because solving this equation by hand can be tedious, you can use the finance application on a TI-84. The *time value of money* solver can be used to find any unknown value in this formula. The known variables are all entered and zeros are entered for the unknown variables.

ACTIVITY 1   **Find a Future Value of an Annuity**

An investor pays $600 quarterly into an annuity. The annuity earns 7.24% annual interest. What will be the value of the annuity after 15 years?

Step 1   Select Finance in the APPS Menu. Then select CALC, TVM Solver.

Step 2   **Enter the data.**

Payments are made quarterly over 15 years, so there are 4 · 15 or 60 payments. The present value, or amount at the beginning, is $0. The future value is unknown, 0 is used as a placeholder. Interest is compounded quarterly, so **P/Y** and **C/Y** are 4. (**C/Y** and **P/Y** are identical.) Payment is made at the end of each month, so select **end**.

Step 3   **Calculate.**

Quit the screen then go back into the **Finance** application. Select **tvm_FV** to calculate the future value. Then press ENTER. The result is the future value subtracted from the present value, so the negative sign is ignored.

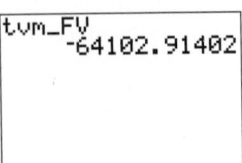

After 15 years, the value of the annuity will be about $64,103.

When taking out a loan for a large purchase like a home or car, consumers are typically concerned with how much their monthly payment will be. While the exponential function below can be used to determine the monthly payment, it can also be calculated using the finance application in the TI-84.

$$PMT = \frac{PV \cdot \dfrac{I}{C/Y}}{1 - \left(1 + \dfrac{I}{C/Y}\right)^{-n}}$$

present value

ACTIVITY 2 — Calculate Monthly Payment

You would like to borrow $170,000 from the bank in order to purchase a home. The loan would be for 30 years and have an annual interest rate of 4.5%. Calculate your monthly payment. What is the amount you will have paid after 30 years?

Step 1 Select Finance in the APPS Menu. Then select CALC, TVM Solver.

Step 2 Enter the data.

The number of payments is $N = 30 \cdot 12$ or 360.
The interest rate I is 4.5%.
The present value of the loan PV is $170,000.
The monthly payment and future value are unknown, so use zeros for each.
The number of payments per year P/Y and C/Y is 12.
Payment is made at the end of month, so select **end**.

Step 3 Calculate.

Select **tvm_Pmt** to calculate the monthly payment. Then press ENTER. There will be 360 payments, so multiply the monthly payments by 360.

Your monthly payment will be $861.37 and the total that will be repaid is $310,091.41.

Exercises

Calculate the future value of each annuity.

1. $800 semiannually, 12 years, 4%

2. $400 monthly, 6 years, 5.5%

3. $200 monthly, 3 years, 7%

4. $1,000 annually, 14 years, 6.25%

5. $450, quarterly, 8 years, 5.5%

6. $300 bimonthly, 18 years, 4.35%

Calculate the monthly payment and the total amount to be repaid for each loan.

7. $220,000, 30 years, 5.5%
8. $140,000, 20 years, 6.75%
9. $20,000, 5 years, 8.5%
10. $5,000, 5 years, 4.25%
11. $45,000, 10 years, 3.5%
12. $180,000, 30 years, 6.5%

13. **CHANGING VALUES** Changing a value of any of the variables may dramatically affect the loan payments. The monthly payment for a 30-year loan for $150,000 at 6% interest is $899.33, with a total payment amount of $323,757.28. Calculate the monthly payment and the total amount of the loan for each scenario.

   a. Putting down $20,000 on the purchase.

   b. Paying 4% interest instead of 6%.

   c. Paying the loan off in 20 years instead of 30.

   d. Making 13 payments per year.

   e. Which saved the most money? Which had the lowest monthly payment?

Logarithmic Functions

Why?

The intensity level of sound is measured in decibels. A whisper measures 20 decibels, a normal conversation 60 decibels, and a vacuum cleaner at 80 decibels. The music playing in headphones maximizes at 100 decibels. The decibel scale is an example of a logarithmic scale.

Then
You graphed and analyzed exponential functions. (Lesson 3-1)

Now
- Evaluate expressions involving logarithms.
- Sketch and analyze graphs of logarithmic functions.

New Vocabulary
logarithmic function with base b
logarithm
common logarithm
natural logarithm

Math Online

glencoe.com

Logarithmic Functions and Expressions Recall from Lesson 1-7 that graphs of functions that pass the horizontal line test are said to be *one-to-one* and have inverses that are also functions. Looking back at the graphs on page 159, you can see that exponential functions of the form $f(x) = b^x$ pass the horizontal line test and are therefore one-to-one with inverses that are functions.

The inverse of $f(x) = b^x$ is called **a logarithmic function with base b**, denoted $\log_b x$ and read *log base b of x*. This means that if $f(x) = b^x$, $b > 0$ and $b \neq 1$, then $f^{-1}(x) = \log_b x$, as shown in the graph of these two functions. Notice that the graphs are reflections of each other in the line $y = x$.

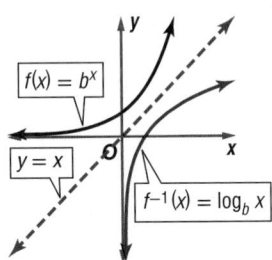

This inverse definition provides a useful connection between exponential and logarithmic equations.

Key Concept — Relating Logarithmic and Exponential Forms

If $b > 0$, $b \neq 1$, and $x > 0$, then

Logarithmic Form

$\log_b x = y$

base ↑ ↑ exponent

if and only if

Exponential Form

$b^y = x.$

base ↑ ↑ exponent

The statement above indicates that $\log_b x$ is the exponent to which b must be raised in order to obtain x. Therefore, when evaluating **logarithms**, remember that a logarithm is an exponent.

EXAMPLE 1 Evaluate Logarithms

Evaluate each logarithm.

a. $\log_3 81$

$\log_3 81 = y$	Let $\log_3 81 = y$.
$3^y = 81$	Write in exponential form.
$3^y = 3^4$	$81 = 3^4$
$y = 4$	Equality Prop. of Exponents

Therefore, $\log_3 81 = 4$, because $3^4 = 81$.

b. $\log_5 \sqrt{5}$

$\log_5 \sqrt{5} = y$	Let $\log_5 \sqrt{5} = y$.
$5^y = \sqrt{5}$	Write in exponential form.
$5^y = 5^{\frac{1}{2}}$	$5^{\frac{1}{2}} = \sqrt{5}$
$y = \frac{1}{2}$	Equality Prop. of Exponents

Therefore, $\log_5 \sqrt{5} = \frac{1}{2}$, because $5^{\frac{1}{2}} = \sqrt{5}$.

c. $\log_7 \frac{1}{49}$

$\log_7 \frac{1}{49} = -2$, because $7^{-2} = \frac{1}{7^2}$ or $\frac{1}{49}$.

d. $\log_2 2$

$\log_2 2 = 1$, because $2^1 = 2$.

✅ Guided Practice

1A. $\log_8 512$ **1B.** $\log_4 4^{3.2}$ **1C.** $\log_2 \frac{1}{32}$ **1D.** $\log_{16} \sqrt{2}$

Example 1 and other examples suggest the following basic properties of logarithms.

Key Concept   **Basic Properties of Logarithms**

If $b > 0$, $b \neq 1$, and x is a real number, then the following statements are true.
- $\log_b 1 = 0$
- $\log_b b = 1$
- $\log_b b^x = x$ } Inverse Properties
- $b^{\log_b x} = x$, $x > 0$

These properties follow directly from the statement relating the logarithmic and exponential forms of equations.

$\log_b 1 = 0$, because $b^0 = 1$. $\log_b b^y = y$, because $b^y = b^y$.

$\log_b b = 1$, because $b^1 = b$. $b^{\log_b x} = x$, because $\log_b x = \log_b x$.

You can use these basic properties to evaluate logarithmic and exponential expressions.

EXAMPLE 2   **Apply Properties of Logarithms**

Evaluate each expression.

a. $\log_5 125$

$\log_5 125 = \log_5 5^3$ $5^3 = 125$

         $= 3$ $\log_b b^x = x$

b. $12^{\log_{12} 4.7}$

$12^{\log_{12} 4.7} = 4.7$ $b^{\log_b x} = x$

✅ Guided Practice

2A. $\log_9 81$

2B. $3^{\log_3 1}$

A logarithm with base 10 or $\log_{10}$ is called a **common logarithm** and is often written without the base. The common logarithm function $y = \log x$ is the inverse of the exponential function $y = 10^x$. Therefore,

$y = \log x$ if and only if $10^y = x$, for all $x > 0$.

The properties for logarithms also hold true for common logarithms.

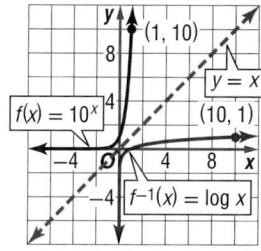

Key Concept   **Basic Properties of Common Logarithms**

If x is a real number, then the following statements are true.
- $\log 1 = 0$
- $\log 10 = 1$
- $\log 10^x = x$ } Inverse Properties
- $10^{\log x} = x$, $x > 0$

Common logarithms can be evaluated using the basic properties described above. Approximations of common logarithms of positive real numbers can be found by using LOG on a calculator.

TechnologyTip

▶ **Error Message** If you try to take the common logarithm of a negative number, your calculator will display either the error message ERR: NONREAL ANS or an imaginary number.

EXAMPLE 3 | **Common Logarithms**

Evaluate each expression.

a. log 0.001

$\log 0.001 = \log 10^{-3}$ **0.001** $= \frac{1}{10^3}$ or 10^{-3}

$ = -3$ **log $10^x = x$**

b. $10^{\log 5}$

$10^{\log 5} = 5$ **$10^{\log x} = x$**

c. log 26

$\log 26 \approx 1.42$ **Use a calculator.**

CHECK Since 26 is between 10 and 100, log 26 is between log 10 and log 100. Since log 10 = 1 and log 100 = 2, log 26 has a value between 1 and 2. ✔

d. log (−5)

Since $f(x) = \log_b x$ is only defined when $x > 0$, log (−5) is undefined on the set of real numbers.

☑ **Guided Practice**

3A. log 10,000   **3B.** log 0.081   **3C.** log −0   **3D.** $10^{\log 3}$

A logarithm with base e or $\log_e$ is called a **natural logarithm** and is denoted ln. The natural logarithmic function $y = \ln x$ is the inverse of the exponential function $y = e^x$. Therefore,

$$y = \ln x \qquad \text{if and only if} \qquad e^y = x, \text{ for all } x > 0.$$

The properties for logarithms also hold true for natural logarithms.

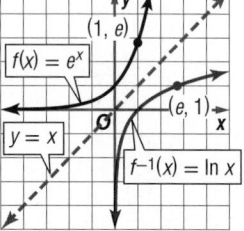

Key Concept | **Basic Properties of Natural Logarithms**

If x is a real number, then the following statements are true.
- $\ln 1 = 0$
- $\ln e = 1$
- $\ln e^x = x$ ⎫ **Inverse Properties**
- $e^{\ln x} = x, x > 0$ ⎭

Natural logarithms can be evaluated using the basic properties described above. Approximations of natural logarithms of positive real numbers can be found by using LN on a calculator.

EXAMPLE 4 | **Natural Logarithms**

Evaluate each expression.

a. $\ln e^{0.73}$

$\ln e^{0.73} = 0.73$ **$\ln e^x = x$**

b. ln (−5)

ln (−5) is undefined.

c. $e^{\ln 6}$

$e^{\ln 6} = 6$ **$e^{\ln x} = x$**

d. ln 4

$\ln 4 \approx 1.39$ **Use a calculator.**

☑ **Guided Practice**

4A. ln 32   **4B.** $e^{\ln 4}$ **4C.** $\ln\left(\frac{1}{e^3}\right)$ **4D.** $-\ln 9$

Graphs of Logarithmic Functions You can use the inverse relationship between exponential and logarithmic functions to graph functions of the form $y = \log_b x$.

EXAMPLE 5   Graphs of Logarithmic Functions

Sketch and analyze the graph of each function. Describe its domain, range, intercepts, asymptotes, end behavior, and where the function is increasing or decreasing.

a. $f(x) = \log_3 x$

Construct a table of values and graph the inverse of this logarithmic function, the exponential function $f^{-1}(x) = 3^x$.

x	−4	−2	−1	0	1	2
$f^{-1}(x)$	0.01	0.11	0.33	1	3	9

Since $f(x) = \log_3 x$ and $f^{-1}(x) = 3^x$ are inverses, you can obtain the graph of $f(x)$ by plotting the points $(f^{-1}(x), x)$.

$f^{-1}(x)$	0.01	0.11	0.33	1	3	9
x	−4	−2	−1	0	1	2

The graph of $f(x) = \log_3 x$ has the following characteristics.

Domain: $(0, \infty)$ Range: $(-\infty, \infty)$

x-intercept: 1   Asymptote: y-axis

End behavior: $\lim\limits_{x \to 0^+} f(x) = -\infty$ and $\lim\limits_{x \to \infty} f(x) = \infty$

Increasing: $(0, \infty)$

b. $g(x) = \log_{\frac{1}{2}} x$

Construct a table of values and graph the inverse of this logarithmic function, the exponential function $g^{-1}(x) = \left(\dfrac{1}{2}\right)^x$.

x	−4	−2	0	1	2	4
$g^{-1}(x)$	16	4	1	0.5	0.25	0.06

Graph $g(x)$ by plotting the points $(g^{-1}(x), x)$.

$g^{-1}(x)$	16	4	1	0.5	0.25	0.06
x	−4	−2	0	1	2	4

The graph of $g(x) = \log_{\frac{1}{2}} x$ has the following characteristics.

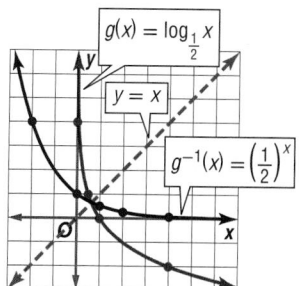

Domain: $(0, \infty)$ Range: $(-\infty, \infty)$

x-intercept: 1   Asymptote: y-axis

End behavior: $\lim\limits_{x \to 0^+} g(x) = \infty$ and $\lim\limits_{x \to \infty} g(x) = -\infty$

Decreasing: $(0, \infty)$

✔ Guided Practice

5A. $h(x) = \log_2 x$

5B. $j(x) = \log_{\frac{1}{3}} x$

The characteristics of typical *logarithmic growth*, or increasing logarithmic functions, and *logarithmic decay*, or decreasing logarithmic functions, are summarized below.

Key Concept — **Properties of Logarithmic Functions**

Logarithmic Growth

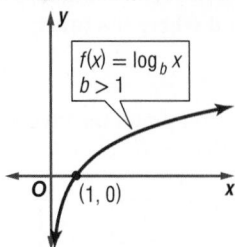

$f(x) = \log_b x$
$b > 1$

(1, 0)

Domain: $(0, \infty)$ **Range:** $(-\infty, \infty)$
y-Intercept: none   **x-Intercept:** 1
Extrema: none   **Asymptote:** y-axis
End Behavior: $\lim\limits_{x \to 0^+} f(x) = -\infty$ and
           $\lim\limits_{x \to \infty} f(x) = \infty$
Continuity: continuous on $(0, \infty)$

Logarithmic Decay

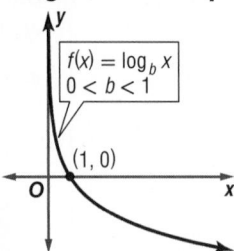

$f(x) = \log_b x$
$0 < b < 1$

(1, 0)

Domain: $(0, \infty)$ **Range:** $(-\infty, \infty)$
y-Intercept: none   **x-Intercept:** 1
Extrema: none   **Asymptote:** y-axis
End Behavior: $\lim\limits_{x \to 0^+} f(x) = \infty$ and
           $\lim\limits_{x \to \infty} f(x) = -\infty$
Continuity: continuous on $(0, \infty)$

The same techniques used to transform the graphs of exponential functions can be applied to the graphs of logarithmic functions.

Watch Out!

Transformations
Remember that horizontal translations are dependent on the constant *inside* the parentheses, and vertical translations are dependent on the constant *outside* of the parentheses.

EXAMPLE 6   **Graph Transformations of Logarithmic Functions**

Use the graph of $f(x) = \log x$ to describe the transformation that results in each function. Then sketch the graphs of the functions.

a. $k(x) = \log(x + 4)$

This function is of the form $k(x) = f(x + 4)$. Therefore, the graph of $k(x)$ is the graph of $f(x)$ translated 4 units to the left (Figure 3.2.1).

b. $m(x) = -\log x - 5$

The function is of the form $m(x) = -f(x) - 5$. Therefore, the graph of $m(x)$ is the graph of $f(x)$ reflected in the x-axis and then translated 5 units down (Figure 3.2.2).

c. $p(x) = 3 \log(x + 2)$

The function is of the form $p(x) = 3f(x + 2)$. Therefore, the graph of $p(x)$ is the graph of $f(x)$ expanded vertically by a factor of 3 and then translated 2 units to the left. (Figure 3.2.3).

$k(x) = \log(x + 4)$

$f(x) = \log x$

Figure 3.2.1

$f(x) = \log x$

$m(x) = -\log x - 5$

Figure 3.2.2

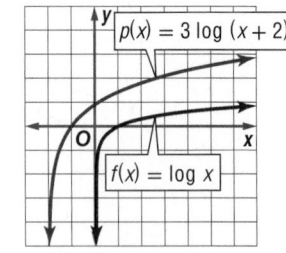

$p(x) = 3 \log(x + 2)$

$f(x) = \log x$

Figure 3.2.3

✓ **Guided Practice**

Use the graph of $f(x) = \ln x$ to describe the transformation that results in each function. Then sketch the graphs of the functions.

6A. $a(x) = \ln(x - 6)$ **6B.** $b(x) = 0.5 \ln x - 2$ **6C.** $c(x) = \ln(x + 4) + 3$

Logarithms can be used in scientific calculations, such as with pH acidity levels and the intensity level of sound.

Real-World Career

Sound Engineer Sound engineers operate and maintain sound recording equipment. They also regulate the signal strength, clarity, and range of sounds of recordings or broadcasts. To become a sound engineer, you should take high school courses in math, physics, and electronics.

⊕ Real-World EXAMPLE 7 — Use Logarithmic Functions

SOUND The intensity level of a sound, measured in decibels, can be modeled by $d(w) = 10 \log \frac{w}{w_0}$, where w is the intensity of the sound in watts per square meter and w_0 is the constant 1.0×10^{-12} watts per square meter.

a. If the intensity of the sound of a person talking loudly is 3.16×10^{-8} watts per square meter, what is the intensity level of the sound in decibels?

Evaluate $d(w)$ when $w = 3.16 \times 10^{-8}$.

$d(w) = 10 \log \frac{w}{w_0}$ **Original function**

$= 10 \log \frac{3.16 \times 10^{-8}}{1.0 \times 10^{-12}}$ **$w = 3.16 \times 10^{-8}$ and $w_0 = 1.0 \times 10^{-12}$**

≈ 45 **Use a calculator.**

The intensity level of the sound is 45 decibels.

b. If the threshold of hearing for a certain person with hearing loss is 5 decibels, will a sound with an intensity level of 2.1×10^{-12} watts per square meter be audible to that person?

Evaluate $d(w)$ when $w = 2.1 \times 10^{-12}$.

$d(w) = 10 \log \frac{w}{w_0}$ **Original function**

$= 10 \log \frac{2.1 \times 10^{-12}}{1.0 \times 10^{-12}}$ **$w = 2.1 \times 10^{-12}$ and $w_0 = 1.0 \times 10^{-12}$**

≈ 3.22 **Use a calculator.**

Because the person can only hear sounds that are 5 decibels or higher, he or she would not be able to hear a sound with an intensity level of 3.22 decibels.

c. Sounds in excess of 85 decibels can cause hearing damage. Determine the intensity of a sound with an intensity level of 85 decibels.

Use a graphing calculator to graph $d(w) = 10 \log \frac{w}{1 \times 10^{-12}}$ and $d(w) = 85$ on the same screen and find the point of intersection.

When the intensity level of the sound is 85 decibels, the intensity of the sound is 3.1623×10^{-4} watts per square meter.

✓ Guided Practice

7. TECHNOLOGY The number of machines infected by a specific computer virus can be modeled by $c(d) = 6.8 + 20.1 \ln d$, where d is the number of days since the first machine was infected.

  A. About how many machines were infected on day 12?

  B. How many more machines were infected on day 30 than on day 12?

  C. On about what day will the number of infected machines reach 75?

Exercises

Evaluate each expression. (Examples 1–4)

1. $\log_2 8$

2. $\log_{10} 10$

3. $\log_6 \frac{1}{36}$

4. $4^{\log_4 1}$

5. $\log_{11} 121$

6. $\log_2 2^3$

7. $\log_{\sqrt{9}} 81$

8. $\log 0.01$

9. $\log 42$

10. $\log_x x^2$

11. $\log 5275$

12. $\ln e^{-14}$

13. $3 \ln e^4$

14. $\ln (5 - \sqrt{6})$

15. $\log_{36} \sqrt[5]{6}$

16. $4 \ln (7 - \sqrt{2})$

17. $\log 635$

18. $\frac{\ln 2}{\ln 7}$

19. $\ln (-6)$

20. $\ln \left(\frac{1}{e^{12}}\right)$

21. $\ln 8$

22. $\log_{\sqrt[7]{4}} 64$

23. $\frac{7}{\ln e}$

24. $\log 1000$

25. LIGHT The amount of light A absorbed by a sample solution is given by $A = 2 - \log 100T$, where T is the fraction of the light transmitted through the solution as shown in the diagram below. (Example 3)

Percent of light transmitted $T = \frac{I_1}{I_0}$

I_0
Intensity of light entering solution

I_1
Intensity of light leaving solution

Sample solution

In an experiment, a student shines light through two sample solutions containing different concentrations of a certain dye.

a. If the percent of light transmitted through the first sample solution is 72%, how much light does the sample solution absorb to the nearest hundredth?

b. If the absorption of the second sample solution is 0.174, what percent of the light entering the solution is transmitted?

26. SOUND While testing the speakers for a concert, an audio engineer notices that the sound level reached a relative intensity of 2.1×10^8 watts per square meter. The equation $D = \log I$ represents the loudness in decibels D given the relative intensity I. What is the level of the loudness of this sound in decibels? Round to the nearest thousandth if necessary. (Example 3)

27. MEMORY The students in Mrs. Ross' class were tested on exponential functions at the end of the chapter and then were retested monthly to determine the amount of information they retained. The average exam scores can be modeled by $f(x) = 85.9 - 9 \ln x$, where x is the number of months since the initial exam. What was the average exam score after 3 months? (Example 4)

Sketch and analyze the graph of each function. Describe its domain, range, intercepts, asymptotes, end behavior, and where the function is increasing or decreasing. (Example 5)

28. $f(x) = \log_4 x$

29. $g(x) = \log_5 x$

30. $h(x) = \log_8 x$

31. $j(x) = \log_{\frac{1}{4}} x$

32. $m(x) = \log_{\frac{1}{5}} x$

33. $n(x) = \log_{\frac{1}{8}} x$

Use the graph of $f(x)$ to describe the transformation that results in the graph of $g(x)$. Then sketch the graphs of $f(x)$ and $g(x)$. (Example 6)

34. $f(x) = \log_2 x; g(x) = \log_2 (x + 4)$

35. $f(x) = \log_3 x; g(x) = \log_3 (x - 1)$

36. $f(x) = \log x; g(x) = \log 2x$

37. $f(x) = \ln x; g(x) = 0.5 \ln x$

38. $f(x) = \log x; g(x) = -\log (x - 2)$

39. $f(x) = \ln x; g(x) = 3 \ln (x) + 1$

40. $f(x) = \log x; g(x) = -2 \log x + 5$

41. $f(x) = \ln x; g(x) = \ln (-x)$

42. INVESTING The annual growth rate for an investment can be found using $r = \frac{1}{t} \ln \frac{P}{P_0}$, where r is the annual growth rate, t is time in years, P is the present value, and P_0 is the original investment. An investment of \$10,000 was made in 2002 and had a value of \$15,000 in 2009. What was the average annual growth rate of the investment? (Example 7)

Determine the domain, range, x-intercept, and vertical asymptote of each function.

43. $y = \log (x + 7)$

44. $y = \log x - 1$

45. $y = \ln (x - 3)$

46. $y = \ln \left(x + \frac{1}{4}\right) - 3$

Find the inverse of each equation.

47. $y = e^{3x}$

48. $y = \log 2x$

49 $y = 4e^{2x}$

50. $y = 6 \log 0.5x$

51. $y = 20^x$

52. $y = 4(2^x)$

Determine the domain and range of the inverse of each function.

53. $y = \log x - 6$

54. $y = 0.25e^{x + 2}$

55. COMPUTERS Gordon Moore, the cofounder of Intel, made a prediction in 1975 that is now known as Moore's Law. He predicted that the number of transistors on a computer processor at a given price point would double every two years.

    a. Write Moore's Law for the predicted number of transmitters P in terms of time in years t and the initial number of transistors.

    b. In October 1985, a specific processor had 275,000 transistors. About how many years later would you expect the processor at the same price to have about 4.4 million transistors?

Describe the domain, range, symmetry, continuity, and increasing/decreasing behavior for each logarithmic function with the given intercept and end behavior. Then sketch a graph of the function.

56. $f(1) = 0$; $\lim\limits_{x \to 0} f(x) = -\infty$; $\lim\limits_{x \to \infty} f(x) = \infty$

57. $g(-2) = 0$; $\lim\limits_{x \to -3} g(x) = -\infty$; $\lim\limits_{x \to \infty} g(x) = \infty$

58. $h(-1) = 0$; $\lim\limits_{x \to -\infty} h(x) = \infty$; $\lim\limits_{x \to 0} h(x) = -\infty$

59. $j(1) = 0$; $\lim\limits_{x \to 0} j(x) = \infty$; $\lim\limits_{x \to \infty} j(x) = -\infty$

Use the parent graph of $f(x) = \log x$ to find the equation of each function.

60.

61.

62.

63
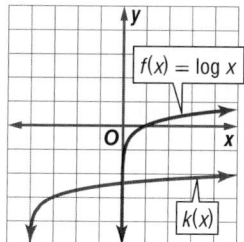

GRAPHING CALCULATOR Create a scatter plot of the values shown in the table. Then use the graph to determine whether each statement is *true* or *false*.

x	1	3	9	27
y	0	1	2	3

64. y is an exponential function of x.

65. x is an exponential function of y.

66. y is a logarithmic function of x.

67. y is inversely proportional to x.

68. BACTERIA The function $t = \dfrac{\ln B - \ln A}{2}$ models the amount of time t in hours for a specific bacteria to reach amount B from the initial amount A.

    a. If the initial number of bacterial present is 750, how many hours would it take for the number of bacteria to reach 300,000? Round to the nearest hour.

    b. Determine the average rate of change in bacteria per hour for the bacterial amounts in part **a**.

69. ⚙ **MULTIPLE REPRESENTATIONS** In this problem, you will compare the average rates of change for an exponential, a power, and a radical function.

    a. GRAPHICAL Graph $f(x) = 2^x$ and $g(x) = x^2$ for $0 \le x \le 8$.

    b. ANALYTICAL Find the average rate of change of each function from part **a** on the interval $[4, 6]$.

    c. VERBAL Compare the growth rates of the functions from part **a** as x increases.

    d. GRAPHICAL Graph $f(x) = \ln x$ and $g(x) = \sqrt{x}$.

    e. ANALYTICAL Find the average rate of change of each function from part **d** on the interval $[4, 6]$.

    f. VERBAL Compare the growth rates of the functions from part **d** as x increases.

H.O.T. Problems   Use Higher-Order Thinking Skills

70. WRITING IN MATH Compare and contrast the domain, range, intercepts, symmetry, continuity, increasing/decreasing behavior and end behavior of logarithmic functions with $a(x) = x^n$, $b(x) = x^{-1}$, $c(x) = a^x$, and $d(x) = e^x$.

71 REASONING Explain why b cannot be negative in $f(x) = \log_b x$.

72. CHALLENGE For $f(x) = \log_{10}(x - k)$, where k is a constant, what are the coordinates of the x-intercept?

73. WRITING IN MATH Compare the large-scale behavior of exponential and logarithmic functions with base b for $b = 2, 6,$ and 10.

REASONING Determine whether each statement is *true* or *false*.

74. Logarithmic functions will always have a restriction on the domain.

75. Logarithmic functions will never have a restriction on the range.

76. Graphs of logarithmic functions always have an asymptote.

77. WRITING IN MATH Use words, graphs, tables, and equations to compare logarithmic and exponential functions.

78. AVIATION When kerosene is purified to make jet fuel, pollutants are removed by passing the kerosene through a special clay filter. Suppose a filter is fitted in a pipe so that 15% of the impurities are removed for every foot that the kerosene travels. (Lesson 3-1)

   a. Write an exponential function to model the percent of impurity left after the kerosene travels x feet.

   b. Graph the function.

   c. About what percent of the impurity remains after the kerosene travels 12 feet?

   d. Will the impurities ever be completely removed? Explain.

Solve each inequality. (Lesson 2-6)

79. $x^2 - 3x - 2 > 8$

80. $4 \geq -(x - 2)^3 + 3$

81. $\frac{2}{x} + 3 > \frac{29}{x}$

82. $\dfrac{(x - 3)(x - 4)}{(x - 5)(x - 6)^2} \leq 0$

83. $\sqrt{2x + 3} - 4 \leq 5$

84. $\sqrt{x - 5} + \sqrt{x + 7} \leq 4$

Solve each equation. (Lesson 2-5)

85. $\dfrac{2a - 5}{a - 9} + \dfrac{a}{a + 9} = \dfrac{-6}{a^2 - 81}$

86. $\dfrac{2q}{2q + 3} - \dfrac{2q}{2q - 3} = 1$

87. $\dfrac{4}{z - 2} - \dfrac{z + 6}{z + 1} = 1$

Graph and analyze each function. Describe its domain, range, intercepts, end behavior, continuity, and where the function is increasing or decreasing. (Lesson 2-1)

88. $f(x) = -\frac{1}{2}x^7$

89. $g(x) = 3x^{-6}$

90. $h(x) = 2x^{-\frac{3}{4}}$

91. MICROBIOLOGY One model for the population P of bacteria in a sample after t days is given by $P(t) = 1000 - 19.75t + 20t^2 - \frac{1}{3}t^3$. (Lesson 1-2)

   a. What type of function is $P(t)$?

   b. When is the bacteria population increasing?

   c. When is it decreasing?

Skills Review for Standardized Tests

92. SAT/ACT The table below shows the per unit revenue and cost of three products at a sports equipment factory.

Product	Revenue per Unit ($)	Cost per Unit ($)
football	f	4
baseball	b	3
soccer ball	6	y

If profit equals revenue minus cost, how much profit do they make if they produce and sell two of each item?

A $2f + 2b - 2y - 2$ **C** $f + b - y - 1$

B $2y - 2b - 2f - 2$ **D** $b + 2f + y - 7$

93. What is the value of n if $\log_3 3^{4n - 1} = 11$?

   F 3   **H** 6

   G 4   **J** 12

94. REVIEW The curve represents a portion of the graph of which function?

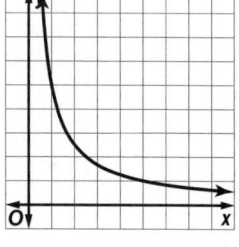

A $y = 50 - x$

B $y = \log x$

C $y = e^{-x}$

D $y = \frac{5}{x}$

95. REVIEW A radioactive element decays over time according to

$$y = x\left(\frac{1}{4}\right)^{\frac{t}{200}},$$

where x = the number of grams present initially and t = time in years. If 500 grams were present initially, how many grams will remain after 400 years?

F 12.5 grams   **H** 62.5 grams

G 31.25 grams   **J** 125 grams

3-3 Properties of Logarithms

Then
You evaluated logarithmic expressions with different bases. (Lesson 3–2)

Now
- Apply properties of logarithms.
- Apply the Change of Base Formula.

Math Online
glencoe.com

Why?
Plants take in carbon-14 through photosynthesis, and animals and humans take in carbon-14 by ingesting plant material. When an organism dies, it stops taking in new carbon, and the carbon-14 already in its system starts to decay. Scientists can calculate the age of organic materials using a logarithmic function that estimates the decay of carbon-14. Properties of logarithms can be used to analyze this function.

Properties of Logarithms In Lesson 0-4, you reviewed the following properties of exponents, where b, x, and y are positive real numbers.

Product Property	**Quotient Property**	**Power Property**
$b^x \cdot b^y = b^{x+y}$	$\dfrac{b^x}{b^y} = b^{x-y}$	$(b^x)^y = b^{xy}$

Since logarithms and exponents have an inverse relationship, these properties of exponents imply these corresponding properties of logarithms.

> ### Key Concept — Properties of Logarithms
>
> If b, x, and y are positive real numbers, $b \neq 1$, and p is a real number, then the following statements are true.
>
> | **Product Property** | $\log_b xy = \log_b x + \log_b y$ |
> |---|---|
> | **Quotient Property** | $\log_b \dfrac{x}{y} = \log_b x - \log_b y$ |
> | **Power Property** | $\log_b x^p = p \log_b x$ |

You will prove the Quotient and Power Properties in Exercises 113 and 114.

To show that the Product Property of Logarithms is true, let $m = \log_b x$ and $n = \log_b y$. Then, using the definition of logarithm, $b^m = x$ and $b^n = y$.

$$
\begin{aligned}
\log_b xy &= \log_b b^m b^n &&x = b^m \text{ and } y = b^n \\
&= \log_b b^{m+n} &&\textbf{Product Property of Exponents} \\
&= m + n &&\textbf{Inverse Property of Logarithms} \\
&= \log_b x + \log_b y &&m = \log_b x \text{ and } n = \log_b y
\end{aligned}
$$

The Product, Quotient, and Power Properties can be used to express logarithms in terms of other logarithms.

> ### EXAMPLE 1 — Use the Properties of Logarithms
>
> Express each logarithm in terms of ln 2 and ln 3.
>
> **a.** ln 54
>
> $$
> \begin{aligned}
> \ln 54 &= \ln (2 \cdot 3^3) &&54 = 2 \cdot 3^3 \\
> &= \ln 2 + \ln 3^3 &&\textbf{Product Property} \\
> &= \ln 2 + 3 \ln 3 &&\textbf{Power Property}
> \end{aligned}
> $$
>
> **b.** $\ln \dfrac{9}{8}$
>
> $$
> \begin{aligned}
> \ln \frac{9}{8} &= \ln 9 - \ln 8 &&\textbf{Quotient Property} \\
> &= \ln 3^2 - \ln 2^3 &&3^2 = 9 \text{ and } 2^3 = 8 \\
> &= 2 \ln 3 - 3 \ln 2 &&\textbf{Power Property}
> \end{aligned}
> $$

✔ **Guided Practice**   Express each logarithm in terms of log 5 and log 3.

1A. log 75   **1B.** log 5.4

The Product, Quotient, and Power Properties can also be used to simplify logarithms.

Math History Link

**Joost Burgi
(1550–1617)**
A Swiss mathematician, Burgi was a renowned clockmaker who also created and designed astronomical instruments. His greatest works in mathematics came when he discovered logarithms independently from John Napier.

EXAMPLE 2 **Simplify Logarithms**

Evaluate each logarithm.

a. $\log_4 \sqrt[5]{64}$

Since the base of the logarithm is 4, express $\sqrt[5]{64}$ as a power of 4.

$$\log_4 \sqrt[5]{64} = \log_4 64^{\frac{1}{5}} \qquad \text{Rewrite using rational exponents.}$$
$$= \log_4 (4^3)^{\frac{1}{5}} \qquad 4^3 = 64$$
$$= \log_4 4^{\frac{3}{5}} \qquad \text{Power Property of Exponents}$$
$$= \frac{3}{5} \log_4 4 \qquad \text{Power Property of Logarithms}$$
$$= \frac{3}{5}(1) \text{ or } \frac{3}{5} \qquad \log_x x = 1$$

b. $5 \ln e^2 - \ln e^3$

$$5 \ln e^2 - \ln e^3 = 5(2 \ln e) - 3 \ln e \qquad \text{Power Property of Logarithms}$$
$$= 10 \ln e - 3 \ln e \qquad \text{Multiply.}$$
$$= 10(1) - 3(1) \text{ or } 7 \qquad \ln e = 1$$

✔ **Guided Practice**

2A. $\log_6 \sqrt[3]{36}$

2B. $\ln e^9 + 4 \ln e^3$

The properties of logarithms provide a way of expressing logarithmic expressions in forms that use simpler operations, converting multiplication into addition, division into subtraction, and powers and roots into multiplication.

EXAMPLE 3 **Expand Logarithmic Expressions**

Expand each expression.

a. $\log 12x^5 y^{-2}$

The expression is the logarithm of the product of 12, x^5, and y^2.

$$\log 12x^5 y^{-2} = \log 12 + \log x^5 + \log y^{-2} \qquad \text{Product Property}$$
$$= \log 12 + 5 \log x - 2 \log y \qquad \text{Power Property}$$

b. $\ln \dfrac{x^2}{\sqrt{4x + 1}}$

The expression is the logarithm of the quotient of x^2 and $\sqrt{4x + 1}$.

$$\ln \frac{x^2}{\sqrt{4x + 1}} = \ln x^2 - \ln \sqrt{4x + 1} \qquad \text{Quotient Property}$$
$$= \ln x^2 - \ln (4x + 1)^{\frac{1}{2}} \qquad \sqrt{4x + 1} = (4x + 1)^{\frac{1}{2}}$$
$$= 2 \ln x - \frac{1}{2} \ln (4x + 1) \qquad \text{Power Property}$$

✔ **Guided Practice**

3A. $\log_{13} 6a^3 bc^4$

3B. $\ln \dfrac{3y + 2}{4\sqrt[3]{y}}$

The same methods used to expand logarithmic expressions can be used to condense them.

Watch Out!

Logarithm of a Sum
The logarithm of a sum or difference does not equal the sum or difference of logarithms. For example, $\ln (x \pm 4) \neq \ln x \pm \ln 4$.

EXAMPLE 4 **Condense Logarithmic Expressions**

Condense each expression.

a. $4 \log_3 x - \frac{1}{3} \log_3 (x + 6)$

$$4 \log_3 x - \frac{1}{3} \log_3 (x + 6) = \log_3 x^4 - \log_3 (x + 6)^{\frac{1}{3}} \qquad \text{Power Property}$$

$$= \log_3 x^4 - \log_3 \sqrt[3]{x + 6} \qquad (x + 6)^{\frac{1}{3}} = \sqrt[3]{x + 6}$$

$$= \log_3 \frac{x^4}{\sqrt[3]{x + 6}} \qquad \text{Quotient Property}$$

$$= \log_3 \frac{x^4 \sqrt[3]{(x + 6)^2}}{x + 6} \qquad \text{Rationalize the denominator.}$$

b. $6 \ln (x - 4) + 3 \ln x$

$$6 \ln (x - 4) + 3 \ln x = \ln (x - 4)^6 + \ln x^3 \qquad \text{Power Property}$$

$$= \ln x^3 (x - 4)^6 \qquad \text{Product Property}$$

✔ **Guided Practice**

4A. $-5 \log_2 (x + 1) + 3 \log_2 (6x)$

4B. $\ln (3x + 5) - 4 \ln x - \ln (x - 1)$

Change of Base Formula Sometimes you may need to work with a logarithm that has an inconvenient base. For example, evaluating $\log_3 5$ presents a challenge because calculators have no key for evaluating base 3 logarithms. The Change of Base Formula provides a way of changing such an expression into a quotient of logarithms with a different base.

Key Concept **Change of Base Formula**

For any positive real numbers a, b, and x, $a \neq 1$, $b \neq 1$,

$$\log_b x = \frac{\log_a x}{\log_a b}.$$

You will prove the Change of Base Formula in Exercise 115.

Most calculators have only two log keys, [LOG] for base 10 logarithms and [LN] for base e logarithms. Therefore, you will often use the Change of Base Formula in one of the following two forms. Either method will provide the correct answer.

$$\log_b x = \frac{\log x}{\log b} \qquad \qquad \log_b x = \frac{\ln x}{\ln b}$$

StudyTip

Check for Reasonableness
You can check your answer in Example 5a by evaluating $3^{1.47}$. Because $3^{1.47} \approx 5$, the answer is reasonable.

EXAMPLE 5 **Use the Change of Base Formula**

Evaluate each logarithm.

a. $\log_3 5$

$$\log_3 5 = \frac{\ln 5}{\ln 3} \qquad \text{Change of Base Formula}$$

$$\approx 1.47 \qquad \text{Use a calculator.}$$

b. $\log_{\frac{1}{2}} 6$

$$\log_{\frac{1}{2}} 6 = \frac{\log 6}{\log \frac{1}{2}} \qquad \text{Change of Base Formula}$$

$$\approx -2.58 \qquad \text{Use a calculator.}$$

✔ **Guided Practice**

5A. $\log_{78} 4212$

5B. $\log_{15} 33$

5C. $\log_{\frac{1}{3}} 10$

You can use properties of logarithms to solve real-world problems. For example, the ratio of the frequencies of a note in one octave and the same note in the next octave is $2:1$. Therefore, further octaves will occur at 2^n times the frequency of that note, where n is an integer. This relationship can be used to find the difference in pitch between any two notes.

EXAMPLE 6 | **Use the Change of Base Formula**

MUSIC The musical *cent* (¢) is a unit of relative pitch. One octave consists of 1200 cents.

Musical Cents

semitone — 100¢
whole tone — 200¢
minor third — 300¢
400¢ major third
500¢ fourth
700¢ fifth
1200¢ octave

The formula to determine the difference in cents between two notes with beginning frequency a and ending frequency b is $n = 1200\left(\log_2 \frac{a}{b}\right)$. Find the difference in pitch between each of the following pairs of notes.

a. 493.9 Hz, 293.7 Hz

Let $a = 493.9$ and $b = 293.7$. Substitute for the values of a and b and solve.

$n = 1200\left(\log_2 \dfrac{a}{b}\right)$ **Original equation**

$\quad = 1200\left(\log_2 \dfrac{493.9}{293.7}\right)$ $a = 493.9$ and $b = 293.7$

$\quad = 1200\left(\dfrac{\log \frac{493.9}{293.7}}{\log 2}\right)$ **Change of Base Formula**

$\quad \approx 899.85$ **Simplify.**

The difference in pitch between the notes is approximately 899.85 cents.

b. 3135.9 Hz, 2637 Hz

Let $a = 3135.9$ and $b = 2637$. Substitute for the values of a and b and solve.

$n = 1200\left(\log_2 \dfrac{a}{b}\right)$ **Original equation**

$\quad = 1200\left(\log_2 \dfrac{3135.9}{2637}\right)$ $a = 3135.9$ and $b = 2637$

$\quad = 1200\left(\dfrac{\log \frac{3135.9}{2637}}{\log 2}\right)$ **Change of Base Formula**

$\quad \approx 299.98$ **Simplify.**

The difference in pitch between the notes is approximately 299.98 cents.

☑ **Guided Practice**

6. **PHOTOGRAPHY** In photography, exposure is the amount of light allowed to strike the film. Exposure can be adjusted by the number of stops used to take a photograph. The change in the number of stops n needed is related to the change in exposure c by $n = \log_2 c$.

   A. How many stops would a photographer use to triple the exposure?

   B. How many stops would a photographer use to get $\frac{1}{5}$ the exposure?

Express each logarithm in terms of ln 2 and ln 5. (Example 1)

1. $\ln \frac{4}{5}$

2. $\ln 200$

3. $\ln 80$

4. $\ln 12.5$

5. $\ln \frac{0.8}{2}$

6. $\ln \frac{2}{5}$

7. $\ln 2000$

8. $\ln 1.6$

Express each logarithm in terms of ln 3 and ln 7. (Example 1)

9. $\ln 63$

10. $\ln \frac{49}{81}$

11. $\ln \frac{7}{9}$

12. $\ln 147$

13. $\ln 1323$

14. $\ln \frac{343}{729}$

15. $\ln \frac{2401}{81}$

16. $\ln 1701$

17. CHEMISTRY The ionization constant of water K_w is the product of the concentrations of hydrogen (H^+) and hydroxide (OH^-) ions.

Nonionized Water Molecule

After Ionization

The formula for the ionization constant of water is $K_w = [H^+][OH^-]$, where the brackets denote concentration in moles per liter. (Example 1)

a. Express $\log K_w$ in terms of $\log [H^+]$ and $\log [OH^-]$.

b. The value of the constant K_w is 1×10^{-14}. Simplify your equation from part **a** to reflect the numerical value of K_w.

c. If the concentration of hydrogen ions in a sample of water is 1×10^{-9} moles per liter, what is the concentration of hydroxide ions?

18. TORNADOES The distance d in miles that a tornado travels is $d = 10^{\frac{w-65}{93}}$, where w is the wind speed in miles per hour of the tornado. (Example 1)

a. Express w in terms of $\log d$.

b. If a tornado travels 100 miles, estimate the wind speed.

Evaluate each logarithm. (Example 2)

19. $\log_5 \sqrt[4]{25}$

20. $8 \ln e^2 - \ln e^{12}$

21. $9 \ln e^3 + 4 \ln e^5$

22. $\log_2 \sqrt[5]{32}$

23. $2 \log_3 \sqrt{27}$

24. $3 \log_7 \sqrt[6]{49}$

25. $4 \log_2 \sqrt{8}$

26. $50 \log_5 \sqrt{125}$

27. $\log_3 \sqrt[6]{243}$

28. $36 \ln e^{0.5} - 4 \ln e^5$

Expand each expression. (Example 3)

29. $\log_9 6x^3y^5z$

30. $\ln \dfrac{x^7}{\sqrt[3]{x+2}}$

31. $\log_3 \dfrac{p^2q}{\sqrt[5]{3q-1}}$

32. $\ln \dfrac{4df^5}{\sqrt[8]{1-3d}}$

33. $\log_{11} ab^{-4}c^{12}d^7$

34. $\log_7 h^2j^{11}k^{-5}$

35. $\log_4 10t^2uv^{-3}$

36. $\log_5 a^6b^{-3}c^4$

37. $\ln \dfrac{3a^4b^7c}{\sqrt[4]{b-9}}$

38. $\log_2 \dfrac{3x+2}{\sqrt[7]{1-5x}}$

Condense each expression. (Example 4)

39. $3 \log_5 x - \frac{1}{2} \log_5 (6-x)$

40. $5 \log_7 (2x) - \frac{1}{3} \log_7 (5x+1)$

41. $7 \log_3 a + \log_3 b - 2 \log_3 (8c)$

42. $4 \ln (x+3) - \frac{1}{5} \ln (4x+7)$

43. $2 \log_8 (9x) - \log_8 (2x-5)$

44. $\ln 13 + 7 \ln a - 11 \ln b + \ln c$

45. $2 \log_6 (5a) + \log_6 b + 7 \log_6 c$

46. $\log_2 x - \log_2 y - 3 \log_2 z$

47. $\frac{1}{4} \ln (2a-b) - \frac{1}{5} \ln (3b+c)$

48. $\log_3 4 - \frac{1}{2} \log_3 (6x-5)$

Evaluate each logarithm. (Example 5)

49. $\log_6 14$

50. $\log_3 10$

51. $\log_7 5$

52. $\log_{128} 2$

53. $\log_{12} 145$

54. $\log_{22} 400$

55. $\log_{100} 101$

56. $\log_{\frac{1}{2}} 3$

57. $\log_{-2} 8$

58. $\log_{13,000} 13$

59 COMPUTERS Computer programs are written in sets of instructions called *algorithms*. To execute a task in a computer program, the algorithm coding in the program must be analyzed. The running time in seconds R that it takes to analyze an algorithm of n steps can be modeled by $R = \log_2 n$. (Example 6)

a. Determine the running time to analyze an algorithm of 240 steps.

b. To the nearest step, how many steps are in an algorithm with a running time of 8.45 seconds?

60. TRUCKING Bill's Trucking Service purchased a new delivery truck for \$56,000. Suppose $t = \log_{(1-r)} \frac{V}{P}$ represents the time t in years that has passed since the purchase given its initial price P, present value V, and annual rate of depreciation r. (Example 6)

a. If the truck's present value is \$40,000 and it has depreciated at a rate of 15% per year, how much time has passed since its purchase to the nearest year?

b. If the truck's present value is \$34,000 and it has depreciated at a rate of 10% per year, how much time has passed since its purchase to the nearest year?

Estimate each logarithm to the nearest whole number.

61. $\log_4 5$

62. $\log_2 13$

63. $\log_3 10$

64. $\log_7 400$

65. $\log_5 \frac{1}{124}$

66. $\log_{12} 177$

67. $\log_{\frac{1}{5}} \frac{1}{6}$

68. $\log_4 \frac{1}{165}$

Expand each expression.

69. $\ln \sqrt[5]{x^3(x+3)}$

70. $\log_5 \frac{x^2 y^5}{\sqrt[3]{4x-y}}$

71 $\log_{14} \frac{11}{\sqrt[4]{x^5(8x-1)}}$

72. $\ln \frac{9x^2 y z^3}{(y-5)^4}$

73. $\log_8 \sqrt[7]{x^3 y^2(z-1)}$

74. $\log_{12} \frac{5x}{\sqrt[6]{x^7(x+13)}}$

75. EARTHQUAKES The Richter scale measures the intensity of an earthquake. The magnitude M of the seismic energy in joules E released by an earthquake can be calculated by $M = \frac{2}{3} \log \frac{E}{10^{4.4}}$.

The Richter Scale

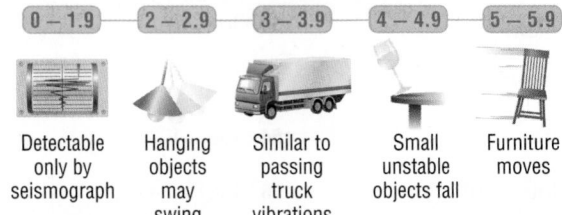

| 0 – 1.9 | 2 – 2.9 | 3 – 3.9 | 4 – 4.9 | 5 – 5.9 |

Detectable only by seismograph | Hanging objects may swing | Similar to passing truck vibrations | Small unstable objects fall | Furniture moves

a. Use the properties of logarithms to expand the equation.

b. What magnitude would an earthquake releasing 7.94×10^{11} joules have?

c. The 2007 Alum Rock earthquake in California released 4.47×10^{12} joules of energy. The 1964 Anchorage earthquake in Alaska released 1.58×10^{18} joules of energy. How many times as great was the magnitude of the Anchorage earthquake as the magnitude of the Alum Rock earthquake?

d. Generally, earthquakes cannot be felt until they reach a magnitude of 3 on the Richter scale. How many joules of energy does an earthquake of this magnitude release?

Condense each expression.

76. $\frac{3}{4} \ln x + \frac{7}{4} \ln y + \frac{5}{4} \ln z$

77. $\log_2 15 + 6 \log_2 x - \frac{4}{3} \log_2 x - \frac{1}{3} \log_2 (x+3)$

78. $\ln 14 - \frac{2}{3} \ln 3x - \frac{4}{3} \ln (4-3x)$

79. $3 \log_6 2x + 9 \log_6 y - \frac{4}{5} \log_6 x - \frac{8}{5} \log_6 y - \frac{1}{5} \log_6 z$

80. $\log_4 25 - \frac{5}{2} \log_4 x - \frac{7}{2} \log_4 y - \frac{3}{2} \log_4 (z+9)$

81. $\frac{5}{2} \ln x + \frac{1}{2} \ln (y+8) - 3 \ln y - \ln (10-x)$

Use the properties of logarithms to rewrite each logarithm below in the form $a \ln 2 + b \ln 3$, where a and b are constants. Then approximate the value of each logarithm given that $\ln 2 \approx 0.69$ and $\ln 3 \approx 1.10$.

82. $\ln 4$

83. $\ln 48$

84. $\ln 162$

85. $\ln 216$

86. $\ln \frac{3}{2}$

87. $\ln \frac{4}{9}$

88. $\ln \frac{4}{27}$

89. $\ln \frac{32}{9}$

Determine the graph that corresponds to each equation.

a.

b.

c.

d.

e.

f.
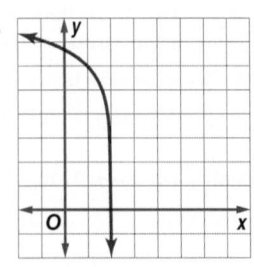

90. $f(x) = \ln x + \ln (x+3)$

91. $f(x) = \ln x - \ln (x+5)$

92. $f(x) = 2 \ln (x+1)$

93. $f(x) = 0.5 \ln (x-2)$

94. $f(x) = \ln (2-x) + 6$

95. $f(x) = \ln 2x - 4 \ln x$

Write each set of logarithmic expressions in increasing order.

96. $\log_3 \frac{12}{4}$, $\log_3 \frac{36}{3} + \log_3 4$, $\log_3 12 - 2\log_3 4$

97. $\log_5 55$, $\log_5 \sqrt{100}$, $3\log_5 \sqrt[3]{75}$

98. BIOLOGY The generation time for bacteria is the time that it takes for the population to double. The generation time G can be found using $G = \frac{t}{3.3 \log_b f}$, where t is the time period, b is the number of bacteria at the beginning of the experiment, and f is the number of bacteria at the end of the experiment. The generation time for mycobacterium tuberculosis is 16 hours. How long will it take 4 of these bacteria to multiply into 1024 bacteria?

Write an equation for each graph.

99.

100.

101.

102.

103. CHEMISTRY pK_a is the logarithmic acid dissociation constant for the acid HF, which is composed of ions H^+ and F^-. The pK_a can be calculated by $pK_a = -\log \frac{[H^+][F^-]}{[HF]}$, where $[H^+]$ is the concentration of H^+ ions, $[F^-]$ is the concentration of F^- ions, and $[HF]$ is the concentration of the acid solution. All of the concentrations are measured in moles per liter.

a. Use the properties of logs to expand the equation for pK_a.

b. What is the pK_a of a reaction in which $[H^+] = 0.01$ moles per liter, $[F^-] = 0.01$ moles per liter, and $[HF] = 2$ moles per liter?

c. The acid dissociation constant K_a of a substance can be calculated by $K_a = \frac{[H^+][F^-]}{[HF]}$. If a substance has a $pK_a = 25$, what is its K_a?

d. Aldehydes are a common functional group in organic molecules. Aldehydes have a pK_a around 17. To what K_a does this correspond?

Evaluate each expression.

104. $\ln\left[\ln\left(e^{e^6}\right)\right]$

105. $10^{\log e^{\ln 4}}$

106. $4\log_{17} 17^{\log_{10} 100}$

107. $e^{\log_4 4^{\ln 2}}$

Simplify each expression.

108. $(\log_3 6)(\log_6 13)$

109 $(\log_2 7)(\log_5 2)$

110. $(\log_4 9) \div (\log_4 2)$

111. $(\log_5 12) \div (\log_8 12)$

112. MOVIES Traditional movies are a sequence of still pictures which, if shown fast enough, give the viewer the impression of motion. If the frequency of the stills shown is too small, the moviegoer notices a flicker between each picture. Suppose the minimum frequency f at which the flicker first disappears is given by $f = K \log I$, where I is the intensity of the light from the screen that reaches the viewer and K is the constant of proportionality.

a. The intensity of the light perceived by a moviegoer who sits at a distance d from the screen is given by $I = \frac{k}{d^2}$, where k is a constant of proportionality. Show that $f = K(\log k - 2 \log d)$.

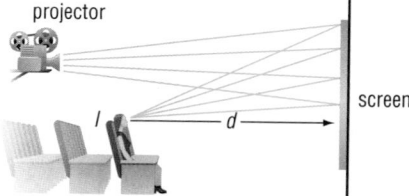

b. Suppose you notice the flicker from a movie projection and move to double your distance from the screen. In terms of K, how does this move affect the value of f? Explain your reasoning.

H.O.T. Problems Use Higher-Order Thinking Skills

PROOF Investigate graphically and then prove each of the following properties of logarithms.

113. Quotient Property

114. Power Property

115. PROOF Prove that $\log_b x = \frac{\log_a x}{\log_a b}$.

116. REASONING How can the graph of $g(x) = \log_4 x$ be obtained using a transformation of the graph of $f(x) = \ln x$?

117. CHALLENGE If $x \in \mathbb{N}$, for what values of x can $\ln x$ not be simplified?

118. ERROR ANALYSIS Omar and Nate expanded $\log_2 \left(\frac{xy}{z}\right)^4$ using the properties of logarithms. Is either of them correct? Explain.

Omar: $4\log_2 x + 4\log_2 y - 4\log_2 z$
Nate: $2\log_4 x + 2\log_4 y - 2\log_4 z$

119 **PROOF** Use logarithmic properties to prove
$$\frac{\log_5 (nt)^2}{\log_4 \frac{t}{r}} = \frac{2\log n \log 4 + 2\log t \log 4}{\log 5 \log t - \log 5 \log r}.$$

120. WRITING IN MATH The graph of $g(x) = \log_b x$ is actually a transformation of $f(x) = \log x$. Use the Change of Base Formula to find the transformation that relates these two graphs. Then explain the effect that different values of b have on the common logarithm graph.

Sketch and analyze each function. Describe its domain, range, intercepts, asymptotes, end behavior, and where the function is increasing or decreasing. (Lesson 3-2)

121. $f(x) = \log_6 x$

122. $g(x) = \log_{\frac{1}{3}} x$

123. $h(x) = \log_5 x - 2$

Use the graph of $f(x)$ to describe the transformation that yields the graph of $g(x)$. Then sketch the graphs of $f(x)$ and $g(x)$. (Lesson 3-1)

124. $f(x) = 2^x; g(x) = -2^x$

125. $f(x) = 5^x; g(x) = 5^{x+3}$

126. $f(x) = \left(\frac{1}{4}\right)^x; g(x) = \left(\frac{1}{4}\right)^x - 2$

127. GEOMETRY The volume of a rectangular prism with a square base is fixed at 120 cubic feet. (Lesson 2-5)

  a. Write the surface area of the prism as a function $A(x)$ of the length of the side of the square x.

  b. Graph the surface area function.

  c. What happens to the surface area of the prism as the length of the side of the square approaches 0?

Divide using synthetic division. (Lesson 2-3)

128. $(x^2 - x + 4) \div (x - 2)$

129. $(x^3 + x^2 - 17x + 15) \div (x + 5)$

130. $(x^3 - x^2 + 2) \div (x + 1)$

Show that f and g are inverse functions. Then graph each function on the same graphing calculator screen. (Lesson 1-7)

131. $f(x) = -\frac{2}{3}x + \frac{1}{6}$

  $g(x) = -\frac{3}{2}x + \frac{1}{4}$

132. $f(x) = \frac{1}{x+2}$

  $g(x) = \frac{1}{x} - 2$

133. $f(x) = (x-3)^3 + 4$

  $g(x) = \sqrt[3]{x - 4} + 3$

134. SCIENCE Specific heat is the amount of energy per unit of mass required to raise the temperature of a substance by one degree Celsius. The table lists the specific heat in joules per gram for certain substances. The amount of energy transferred is given by $Q = cmT$, where c is the specific heat for a substance, m is its mass, and T is the change in temperature. (Lesson 1-5)

  a. Find the function for the change in temperature.

  b. What is the parent graph of this function?

  c. What is the relevant domain of this function?

Substance	Specific Heat (j/g)
aluminum	0.902
gold	0.129
mercury	0.140
iron	0.45
ice	2.03
water	4.179
air	1.01

135. SAT/ACT If $b \neq 0$, let $a \triangle b = \frac{a^2}{b^2}$. If $x \triangle y = 1$, then which statement must be true?

  A $x = y$

  B $x = -y$

  C $x^2 - y^2 = 0$

  D $x > 0$ and $y > 0$

  E $x = |y|$

136. REVIEW Find the value of x for $\log_2 (9x + 5) = 2 + \log_2 (x^2 - 1)$.

  F -0.4

  G 0

  H 1

  J 3

137. To what is $2 \log_5 12 - \log_5 8 - 2 \log_5 3$ equal?

  A $\log_5 2$

  B $\log_5 3$

  C $\log_5 0.5$

  D 1

138. REVIEW The weight of a bar of soap decreases by 2.5% each time it is used. If the bar of soap weighs 95 grams when it is new, what is its weight to the nearest gram after 15 uses?

  F 58 g

  G 59 g

  H 65 g

  J 93 g

Sketch and analyze the graph of each function. Describe its domain, range, intercepts, asymptotes, end behavior, and where the function is increasing or decreasing. (Lesson 3-1)

1. $f(x) = 5^{-x}$

2. $f(x) = \left(\frac{2}{3}\right)^x + 3$

Use the graph of $f(x)$ to describe the transformation that results in the graph of $g(x)$. Then sketch the graphs of $f(x)$ and $g(x)$. (Lesson 3-1)

3. $f(x) = \left(\frac{3}{2}\right)^x$; $g(x) = \left(\frac{3}{2}\right)^{-x}$

4. $f(x) = 3^x$; $g(x) = 2 \cdot 3^{x-2}$

5. $f(x) = e^x$; $g(x) = -e^x - 6$

6. $f(x) = 10^x$; $g(x) = 10^{2x}$

7. **MULTIPLE CHOICE** In the formula for compound interest $A = P\left(1 + \frac{r}{n}\right)^{nt}$, which variable has NO effect on the amount of time it takes an investment to double? (Lesson 3-1)

A P C n

B r D t

8. **FINANCIAL LITERACY** Clarissa has saved $1200 from working summer jobs and would like to invest it so that she has some extra money when she graduates from college in 5 years. (Lesson 3-1)

a. How much money will Clarissa have if she invests at an annual rate of 7.2% compounded monthly?

b. How much money will Clarissa have if she invests at an annual rate of 7.2% compounded continuously?

9. **MULTIPLE CHOICE** The parent function for the graph shown is $f(x) = \log_2 x$.

The graph contains the given point and has the vertical asymptote shown. Which of the following is the function for the graph? (Lesson 3-2)

F $f(x) = \log_2 (x + 3) + 1$

G $f(x) = \log_2 (x - 4) + 1$

H $f(x) = -\log_2 (x - 3) + 1$

J $f(x) = \log_2 (x - 3) + 1$

Evaluate each expression. (Lesson 3-2)

10. $\log_2 64$

11. $\log_5 \frac{1}{125}$

12. $\ln e^{23}$

13. $\log 0.001$

14. **TECHNOLOGY** The number of children infected by a virus can be modeled by $c(d) = 4.9 + 11.2 \ln d$, where d is the number of days since the first child was infected. About how many children are infected on day 8? (Lesson 3-2)

Evaluate each function for the given value. (Lesson 3-2)

15. $T(x) = 2 \ln (x + 3); x = 18$

16. $H(a) = 4 \log \frac{2a}{5} - 8; a = 25$

Express each logarithm in terms of ln 3 and ln 4. (Lesson 3-3)

17. $\ln 48$

18. $\ln 2.25$

19. $\ln \frac{64}{27}$

20. $\ln \frac{9}{16}$

21. **CHEMISTRY** The half-life of a radioactive isotope is 7 years. (Lesson 3-3)

a. If there is initially 75 grams of the substance, how much of the substance will remain after 14 years?

b. After how many years will there be $\frac{1}{16}$ of the original amount remaining?

c. The time it takes for a substance to decay from N_0 to N can be modeled by $t = 7 \log_{0.5} \frac{N}{N_0}$. Approximately how many years will it take for any amount of the radioactive substance to decay to $\frac{1}{3}$ its original amount?

Expand each expression. (Lesson 3-3)

22. $\log_3 \sqrt[4]{x^2 y^3 z^5}$ 23. $\log_9 \frac{3x^3}{y}$

Condense each expression. (Lesson 3-3)

24. $5 \log_4 a + 6 \log_4 b - \frac{1}{3} \log_4 7c$

25. $2 \log (x + 1) - \log (x^2 - 1)$

Exponential and Logarithmic Equations

Why?

The intensity of an earthquake can be calculated using $R = \log \frac{a}{T} + B$, where R is the Richter scale number, a is the amplitude of the vertical ground motion, T is the period of the seismic wave in seconds, and B is a factor that accounts for the weakening of the seismic waves.

One-to-One Property of Exponential Functions In Lesson 3-2, exponential functions were shown to be one-to-one. Recall from Lesson 1-7 that if a function f is one-to-one, no y-value is matched with more than one x-value. That is, $f(a) = f(b)$ if and only if $a = b$. This leads us to the following One-to-One Property of Exponential Functions.

Key Concept   **One-to-One Property of Exponential Functions**

Words   For $b > 0$ and $b \neq 1$, $b^x = b^y$ if and only if $x = y$.

Examples   If $3^x = 3^5$, then $x = 5$.   If $\log x = 3$, then $10^{\log x} = 10^3$.

This is also known as the Property of Equality for exponential functions.

The *if and only if* wording in this property implies two separate statements. One of them, $b^x = b^y$ if $x = y$, can be used to solve some simple exponential equations by first expressing both sides of the equation in terms of a common base.

EXAMPLE 1   **Solve Exponential Equations Using One-to-One Property**

Solve each equation.

a. $36^{x+1} = 6^{x+6}$

$36^{x+1} = 6^{x+6}$ Original equation

$(6^2)^{x+1} = 6^{x+6}$ $6^2 = 36$

$6^{2x+2} = 6^{x+6}$ Power of a Power

$2x + 2 = x + 6$ One-to-One Property

$x + 2 = 6$ Subtract x from each side.

$x = 4$ Subtract 2 from each side. Check this solution in the original equation.

b. $\left(\frac{1}{2}\right)^c = 64^{\frac{1}{2}}$

$\left(\frac{1}{2}\right)^c = 64^{\frac{1}{2}}$ Original equation

$2^{-c} = (2^6)^{\frac{1}{2}}$ $2^{-1} = \frac{1}{2}$, $2^6 = 64$

$2^{-c} = 2^3$ Power of a Power

$-c = 3$ One-to-One Property

$c = -3$ Solve for c. Check this solution in the original equation.

✓ **Guided Practice**

1A. $16^{x+3} = 4^{4x+7}$ **1B.** $\left(\frac{2}{3}\right)^{x-5} = \left(\frac{9}{4}\right)^{\frac{3x}{4}}$

Another statement that follows from the One-to-One Property of Exponential Functions, if $x = y$, then $b^x = b^y$, can be used to solve *logarithmic equations* such as $\log_2 x = 3$.

$$\log_2 x = 3 \qquad \text{Original equation}$$
$$2^{\log_2 x} = 2^3 \qquad \text{One-to-One Property}$$
$$x = 2^3 \qquad \text{Inverse Property}$$

This application of the One-to-One Property is called *exponentiating* each side of an equation. Notice that the effect of exponentiating each side of $\log_2 x = 3$ is to convert the equation from logarithmic to exponential form.

<table>
<tr><td>StudyTip</td></tr>
</table>

Solutions to Logarithmic Equations While it is always a good idea to check your solutions to equations, this is especially true of logarithmic equations, since logarithmic functions are only defined on the set of positive real numbers.

EXAMPLE 2 | Solve Logarithmic Equations Using One-to-One Property

Solve each logarithmic equation. Round to the nearest hundredth.

a. $\ln x = 6$

Method 1 Use exponentiation.

$\ln x = 6$	**Original equation**
$e^{\ln x} = e^6$	**Exponentiate each side.**
$x = e^6$	**Inverse Property**
$x \approx 403.43$	**Use a calculator.**

CHECK $\ln 403.43 \approx 6$ ✔

Method 2 Write in exponential form.

$\ln x = 6$	**Original equation**
$x = e^6$	**Write in exponential form.**
$x \approx 403.43$	**Use a calculator.**

b. $6 + 2 \log 5x = 18$

$6 + 2 \log 5x = 18$	**Original equation**
$2 \log 5x = 12$	**Subtract 6 from each side.**
$\log 5x = 6$	**Divide each side by 2.**
$5x = 10^6$	**Write in exponential form.**
$x = \dfrac{10^6}{5}$	**Divide each side by 5.**
$x = 200{,}000$	**Simplify. Check this solution in the original equation.**

c. $\log_8 x^3 = 12$

$\log_8 x^3 = 12$	**Original equation**
$3 \log_8 x = 12$	**Power Property**
$\log_8 x = 4$	**Divide each side by 3.**
$x = 8^4$ or 4096	**Write in exponential form and simplify. Check this solution.**

☑ **Guided Practice**

2A. $-3 \ln x = -24$ **2B.** $4 - 3 \log (5x) = 16$ **2C.** $\log_3 (x^2 - 1) = 4$

One-to-One Property of Logarithmic Functions Logarithmic functions are also one-to-one. Therefore, we can state the following One-to-One Property of Logarithmic Functions.

StudyTip

Property of Equality The One-to-One Property of Logarithmic Functions is also known as the Property of Equality for Logarithmic Functions.

Key Concept | **One-to-One Property of Logarithmic Functions**

Words For $b > 0$ and $b \neq 1$, $\log_b x = \log_b y$ if and only if $x = y$.

Examples If $\log_2 x = \log_2 6$, then $x = 6$.   If $e^y = 2$, then $\ln e^y = \ln 2$.

One statement implied by this property is that $\log_b x = \log_b y$ if $x = y$. You can use this statement to solve some simple logarithmic equations by first condensing each side of an equation into logarithms with the same base.

Solve each equation.

a. $\log_4 x = \log_4 3 + \log_4 (x - 2)$

$\log_4 x = \log_4 3 + \log_4 (x - 2)$	Original equation
$\log_4 x = \log_4 3(x - 2)$	Product Property
$\log_4 x = \log_4 (3x - 6)$	Distributive Property
$x = 3x - 6$	One-to-One Property
$-2x = -6$	Subtract 3x from each side.
$x = 3$	Divide each side by -2. Check this solution.

b. $\log_3 (x^2 + 3) = \log_3 52$

$\log_3 (x^2 + 3) = \log_3 52$	Original equation
$x^2 + 3 = 52$	One-to-One Property
$x^2 = 49$	Subtract 3 from each side.
$x = \pm 7$	Take the square root of each side. Check this solution.

✔ **Guided Practice**

3A. $\log_6 2x = \log_6 (x^2 - x + 2)$ **3B.** $\log_{12} (x + 3) = \log_{12} x + \log_{12} 4$

Another statement that follows from the One-to-One Property of Logarithmic Functions, if $x = y$, then $\log_b x = \log_b y$, can be used to solve exponential equations such as $e^x = 3$.

$e^x = 3$	Original equation
$\ln e^x = \ln 3$	One-to-One Property
$x = \ln 3$	Inverse Property

This application of the One-to-One Property is called *taking the logarithm of each side* of an equation. While natural logarithms are more convenient to use when the base of the exponential expression is e, you can use logarithms to any base to help solve exponential equations.

StudyTip

Alternate Solution The problem in Example 4a could also have been solved by taking the $\log_4$ of each side. The result would be $x = \log_4 13$. Notice that when the Change of Base Formula is applied, this is equivalent to the solution $x = \dfrac{\log 13}{\log 4}$.

EXAMPLE 4   **Solve Exponential Equations**

Solve each equation. Round to the nearest hundredth.

a. $4^x = 13$

$4^x = 13$	Original equation
$\log 4^x = \log 13$	Take the common logarithm of each side.
$x \log 4 = \log 13$	Power Property
$x = \dfrac{\log 13}{\log 4}$ or about 1.85	Divide each side by log 4 and use a calculator.

b. $e^{4 - 3x} = 6$

$e^{4 - 3x} = 6$	Original equation
$\ln e^{4 - 3x} = \ln 6$	Take the natural logarithm of each side.
$4 - 3x = \ln 6$	Inverse Property
$x = \dfrac{\ln 6 - 4}{-3}$ or about 0.74	Solve for x and use a calculator.

✔ **Guided Practice**

4A. $8^y = 0.165$ **4B.** $1.43^a + 3.1 = 8.48$ **4C.** $e^{2 + 5w} = 12$

EXAMPLE 5   Solve in Logarithmic Terms

Solve $4^{3x-1} = 3^{2-x}$. Round to the nearest hundredth.

Solve Algebraically

$4^{3x-1} = 3^{2-x}$	Original equation
$\ln 4^{3x-1} = \ln 3^{2-x}$	Take the natural logarithm of each side.
$(3x-1)\ln 4 = (2-x)\ln 3$	Power Property
$3x \ln 4 - \ln 4 = 2\ln 3 - x\ln 3$	Distributive Property
$3x \ln 4 + x\ln 3 = 2\ln 3 + \ln 4$	Isolate the variables on the left side of the equation.
$x(3\ln 4 + \ln 3) = 2\ln 3 + \ln 4$	Distributive Property
$x(\ln 4^3 + \ln 3) = \ln 3^2 + \ln 4$	Power Property
$x \ln [3(4^3)] = \ln 36$	Product Property
$x \ln 192 = \ln 36$	$3(4^3) = 192$
$x = \dfrac{\ln 36}{\ln 192}$ or about 0.68	Divide each side by ln 192.

> **Watch Out!**
>
> **Simplifying** Notice that the Quotient Property cannot be used to further simplify $\dfrac{\ln 36}{\ln 192}$.

Confirm Graphically

Graph $y = 4^{3x-1}$ and $y = 3^{2-x}$. The point of intersection of these two graphs given by the calculator is approximately 0.68, which is consistent with our algebraic solution.

[−10, 10] scl: 1 by [−10, 10] scl: 1

✔ Guided Practice

Solve each equation. Round to the nearest hundredth.

5A. $6^{2x+4} = 5^{-x+1}$

5B. $4^{3x+2} = 6^{2x-1}$

Equations involving multiple exponential expressions can be solved by applying quadratic techniques, such as factoring or the Quadratic Formula. Be sure to check for extraneous solutions.

EXAMPLE 6   Solve Exponential Equations in Quadratic Form

Solve $e^{2x} + 6e^x - 16 = 0$.

$e^{2x} + 6e^x - 16 = 0$	Original equation
$u^2 + 6u - 16 = 0$	Write in quadratic form by letting $u = e^x$.
$(u+8)(u-2) = 0$	Factor.

$u = -8$ or	$u = 2$	Zero Product Property
$e^x = -8$	$e^x = 2$	Replace u with e^x.
$\ln e^x = \ln(-8)$	$\ln e^x = \ln 2$	Take the natural logarithm of each side.
$x = \ln(-8)$	$x = \ln 2$ or about 0.69	Inverse Property

The only solution is $x = \ln 2$ because $\ln(-8)$ is extraneous. Check this solution.

CHECK		
	$e^{2x} + 6e^x - 16 = 0$	Original equation
	$e^{2(\ln 2)} + 6e^{\ln 2} - 16 \stackrel{?}{=} 0$	Replace x with ln 2.
	$e^{\ln 2^2} + 6e^{\ln 2} - 16 \stackrel{?}{=} 0$	Power Property
	$2^2 + 6(2) - 16 = 0$ ✔	Inverse Property

> **TechnologyTip**
>
> **Finding Zeros** You can confirm the solution of $e^{2x} + 6e^x - 16 = 0$ graphically by using a graphing calculator to locate the zero of $y = e^{2x} + 6e^x - 16$. The graphical solution of about 0.69 is consistent with the algebraic solution of $\ln 2 \approx 0.69$.
>
>
>
> [−5, 5] scl: 1 by [−40, 40] scl: 5

✔ Guided Practice   Solve each equation.

6A. $e^{2x} + 2e^x = 8$

6B. $4e^{4x} + 8e^{2x} = 5$

Equations having multiple logarithmic expressions may be solved by first condensing expressions using the Power, Product, and Quotient Properties, and then applying the One-to-One Property.

EXAMPLE 7 | Solve Logarithmic Equations

Solve $\ln (x + 2) + \ln (3x - 2) = 2 \ln 2x$.

$\ln (x + 2) + \ln (3x - 2) = 2 \ln 2x$	Original equation
$\ln (x + 2)(3x - 2) = \ln (2x)^2$	Product and Power Property
$\ln (3x^2 + 4x - 4) = \ln 4x^2$	Simplify.
$3x^2 + 4x - 4 = 4x^2$	One-to-One Property
$0 = x^2 - 4x + 4$	Simplify.
$0 = (x - 2)(x - 2)$	Factor.
$x = 2$	Zero Product Property

CHECK You can check this solution in the original equation, or confirm graphically by locating the intersection of the graphs of $y = \ln (x + 2) + \ln (3x - 2)$ and $y = 2 \ln 2x$.

Intersection
X=1.9999977 Y=2.7725864
[−3, 3] scl: 1 by [−3, 3] scl: 1

✅ Check Your Progress

Solve each equation.

7A. $\ln (7x + 3) - \ln (x + 1) = \ln (2x)$

7B. $\ln (2x + 1) + \ln (2x - 3) = 2 \ln (2x - 2)$

It may not be obvious that a solution of a logarithmic equation is extraneous until you check it in the original equation.

StudyTip

Identify the Domain of an Equation Another way to check for extraneous solutions is to identify the domain of the equation. In Example 8, the domain of $\log_{12} 12x$ is $x > 0$ while the domain of $\log_{12} (x - 1)$ is $x > 1$. Therefore, the domain of the equation is $x > 1$. Since $-3 \not> 1$, -3 cannot be a solution of the equation.

EXAMPLE 8 | Check for Extraneous Solutions

Solve $\log_{12} 12x + \log_{12} (x - 1) = 2$.

$\log_{12} 12x + \log_{12} (x - 1) = 2$	Original equation
$\log_{12} 12x(x - 1) = 2$	Product Property
$\log_{12} (12x^2 - 12x) = 2$	Distributive Property
$\log_{12} (12x^2 - 12x) = \log_{12} 12^2$	Inverse Property
$\log_{12} (12x^2 - 12x) = \log_{12} 144$	$12^2 = 144$
$12x^2 - 12x = 144$	One-to-One Property
$12x^2 - 12x - 144 = 0$	Subtract 144 from each side.
$12(x - 4)(x + 3) = 0$	Factor.
$x = 4 \text{ or } x = -3$	Zero Product Property

CHECK

$\log_{12} 12x + \log_{12} (x - 1) = 2$

$\log_{12} 12(4) + \log_{12} (4 - 1) \overset{?}{=} 2$

$\log_{12} 48 + \log_{12} 3 \overset{?}{=} 2$

$\log_{12} 48 \cdot 3 \overset{?}{=} 2$

$\log_{12} 144 = 2 ✔$

$\log_{12} 12x + \log_{12} (x - 1) = 2$

$\log_{12} 12(-3) + \log_{12} (-3 - 1) \overset{?}{=} 2$

$\log_{12} (-36) + \log_{12} (-4) \overset{?}{=} 2$

Since neither $\log_{12} (-36)$ nor $\log_{12} (-4)$ is defined, $x = -3$ is an extraneous solution.

✅ Guided Practice

Solve each equation.

8A. $\ln (6y + 2) - \ln (y + 1) = \ln (2y - 1)$

8B. $\log (x - 12) = 2 + \log (x - 2)$

You can use information about growth or decay to write the equation of an exponential function.

🌐 Real-World EXAMPLE 9 — Model Exponential Growth

INTERNET The table shows the number of hits a new Web site received by the end of January and the end of April of the same year.

Web Site Traffic	
Month	**Number of Hits**
January	125
April	2000

a. If the number of hits is increasing at an exponential rate, identify the continuous rate of growth. Then write an exponential equation to model this situation.

Let $N(t)$ represent the number of hits at the end of t months and assume continuous exponential growth. Then the initial number N_0 is 125 hits and the number of hits N after a time of 3 months, the number of months from January to April, is 2000. Use this information to find the continuous growth rate k.

$N(t) = N_0 e^{kt}$	**Exponential Growth Formula**
$2000 = 125 e^{k(3)}$	$N(3) = 2000$, $N_0 = 125$, and $t = 3$
$16 = e^{3k}$	**Divide each side by 125.**
$\ln 16 = \ln e^{3k}$	**Take the natural logarithm of each side.**
$\ln 16 = 3k$	**Inverse Property**
$\dfrac{\ln 16}{3} = k$	**Divide each side by 3.**
$0.924 \approx k$	**Use a calculator.**

The number of hits is increasing at a continuous rate of approximately 92.4% per month. Therefore, an equation modeling this situation is $N(t) = 125 e^{0.924t}$.

b. Use your model to predict the number of months it will take for the Web site to receive 2 million hits.

$N(t) = 125 e^{0.924t}$	**Exponential growth model**
$2{,}000{,}000 = 125 e^{0.924t}$	$N(t) = 2{,}000{,}000$
$16{,}000 = e^{0.924t}$	**Divide each side by 125.**
$\ln 16{,}000 = \ln e^{0.924t}$	**Take the natural logarithm of each side.**
$\ln 16{,}000 = 0.924t$	**Inverse Property**
$\dfrac{\ln 16{,}000}{0.924} = t$	**Divide each side by 0.924.**
$10.48 \approx t$	**Use a calculator.**

According to this model, the Web site will receive 2,000,000 hits in about 10.48 months.

✔ Guided Practice

9. MEMORABILIA The table shows revenue from sales of T-shirts and other memorabilia sold by two different vendors during and one week after the World Series.

World Series Memorabilia Sales		
Days after Series	**Vendor A Sales ($)**	**Vendor B Sales ($)**
0	300,000	200,000
7	37,000	49,000

A. If the sales are decreasing at an exponential rate, identify the continuous rate of decline for each vendor's sales. Then write an exponential equation to model each situation.

B. Use your models to predict the World Series memorabilia sales by each vendor 4 weeks after the series ended.

C. Will the two vendors' sales ever be the same? If so, at what point in time?

🌐 Real-World Link

Championship hats and shirts are printed for both teams before a major athletic contest like the Bowl Championship Series. The losing team's merchandise is often donated to nonprofit organizations that distribute it to families in need in other countries. In 2007, an estimated $2.5 million of unusable sports clothing was donated.

Source: World Vision

Exercises

Solve each equation. (Example 1)

1. $4^{x+7} = 8^{x+3}$

2. $8^{x+4} = 32^{3x}$

3. $49^{x+4} = 7^{18-x}$

4. $32^{x-1} = 4^{x+5}$

5. $\left(\frac{9}{16}\right)^{3x-2} = \left(\frac{3}{4}\right)^{5x+4}$

6. $12^{3x+11} = 144^{2x+7}$

7. $25^{\frac{x}{3}} = 5^{x-4}$

8. $\left(\frac{5}{6}\right)^{4x} = \left(\frac{36}{25}\right)^{9-x}$

9. INTERNET The number of people P in millions using two different search engines to surf the Internet t weeks after the creation of the search engine can be modeled by $P_1(t) = 1.5^{t+4}$ and $P_2(t) = 2.25^{t-3.5}$, respectively. During which week did the same number of people use each search engine? (Example 1)

10. FINANCIAL LITERACY Brandy is planning on investing $5000 and is considering two savings accounts. The first account is continuously compounded and offers a 3% interest rate. The second account is annually compounded and also offers a 3% interest rate, but the bank will match 4% of the initial investment. (Example 1)

a. Write an equation for the balance of each savings account at time t years.

b. How many years will it take for the continuously compounded account to catch up with the annually compounded savings account?

c. If Brandy plans on leaving the money in the account for 30 years, which account should she choose?

Solve each logarithmic equation. (Example 2)

11. $\ln a = 4$

12. $-8 \log b = -64$

13. $\ln (-2) = c$

14. $2 + 3 \log 3d = 5$

15. $14 + 20 \ln 7x = 54$

16. $100 + 500 \log_1 g = 1100$

17. $7000 \ln h = -21{,}000$

18. $-18 \log_0 j = -126$

19. $12{,}000 \log_2 k = 192{,}000$

20. $\log_2 m^4 = 32$

21. CARS If all other factors are equal, the higher the displacement D in liters of the air/fuel mixture of an engine, the more horsepower H it will produce. The horsepower of a naturally aspirated engine can be modeled by $H = \log_{1.003} \frac{D}{1.394}$. Find the displacement when horsepower is 200. (Example 2)

Solve each equation. (Example 3)

22. $\log_6 (x^2 + 5) = \log_6 41$

23. $\log_8 (x^2 + 11) = \log_8 92$

24. $\log_9 (x^4 - 3) = \log_9 13$

25. $\log_7 6x = \log_7 9 + \log_7 (x - 4)$

26. $\log_5 x = \log_5 (x + 6) - \log_5 4$

27. $\log_{11} 3x = \log_{11} (x + 5) - \log_{11} 2$

Solve each equation. Round to the nearest hundredth. (Example 4)

28. $6^x = 28$

29. $1.8^x = 9.6$

30. $3e^{4x} = 45$

31. $e^{3x+1} = 51$

32. $8^x - 1 = 3.4$

33. $2e^{7x} = 84$

34. $8.3e^{9x} = 24.9$

35. $e^{2x} + 5 = 16$

36. $2.5e^{x+4} = 14$

37. $0.75e^{3.4x} - 0.3 = 80.1$

38. GENETICS PCR (Polymerase Chain Reaction) is a technique commonly used in forensic labs to amplify DNA. PCR uses an enzyme to cut a designated nucleotide sequence from the DNA and then replicates the sequence. The number of identical nucleotide sequences N after t minutes can be modeled by $N(t) = 100 \cdot 1.17^t$. (Example 4)

a. At what time will there be 1×10^4 sequences?

b. At what time will the DNA have been amplified to 1 million sequences?

Solve each equation. (Example 5)

39. $7^{2x+1} = 3^{x+3}$

40. $11^{x+1} = 7^{x-1}$

41. $9^{x+2} = 2^{5x-4}$

42. $4^{x-3} = 6^{2x-1}$

43. $3^{4x+3} = 8^{-x+2}$

44. $5^{3x-1} = 4^{x+1}$

45. $6^{x-2} = 5^{2x+3}$

46. $8^{-2x-1} = 5^{-x+2}$

47. $2^{5x+6} = 4^{2x+1}$

48. $6^{-x-2} = 9^{-x-1}$

49. ASTRONOMY The brightness of two celestial bodies as seen from Earth can be compared by determining the variation in brightness between the two bodies. The variation in brightness V can be calculated by $V = 2.512^{m_f - m_b}$, where m_f is the magnitude of brightness of the fainter body and m_b is the magnitude of brightness of the brighter body. (Example 5)

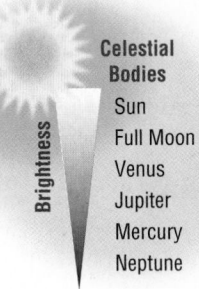

Celestial Bodies

Sun
Full Moon
Venus
Jupiter
Mercury
Neptune

Brightness

a. The Sun has $m = -26.73$, and the full Moon has $m = -12.6$. Determine the variation in brightness between the Sun and the full Moon.

b. The variation in brightness between Mercury and Venus is 5.25. Venus has a magnitude of brightness of -3.7. Determine the magnitude of brightness of Mercury.

c. Neptune has a magnitude of brightness of 7.7, and the variation in brightness of Neptune and Jupiter is 15,856. What is the magnitude of brightness of Jupiter?

Solve each equation. (Example 6)

50. $e^{2x} + 3e^x - 130 = 0$ **51.** $e^{2x} - 15e^x + 56 = 0$

52. $e^{2x} + 3e^x = -2$ **53.** $6e^{2x} - 5e^x = 6$

54. $9e^{2x} - 3e^x = 6$ **55.** $8e^{4x} - 15e^{2x} + 7 = 0$

56. $2e^{8x} + e^{4x} - 1 = 0$ **57.** $2e^{5x} - 7e^{2x} - 15e^{-x} = 0$

58. $10e^x - 15 - 45e^{-x} = 0$ **59.** $11e^x - 51 - 20e^{-x} = 0$

Solve each logarithmic equation. (Example 7)

60. $\ln x + \ln (x + 2) = \ln 63$

61. $\ln x + \ln (x + 7) = \ln 18$

62. $\ln (3x + 1) + \ln (2x - 3) = \ln 10$

63. $\ln (x - 3) + \ln (2x + 3) = \ln (-4x^2)$

64. $\log (5x^2 + 4) = 2 \log 3x^2 - \log (2x^2 - 1)$

65. $\log (x + 6) = \log (8x) - \log (3x + 2)$

66. $\ln (4x^2 - 3x) = \ln (16x - 12) - \ln x$

67. $\ln (3x^2 - 4) + \ln (x^2 + 1) = \ln (2 - x^2)$

68. SOUND Noise-induced hearing loss (NIHL) accounts for 25% of hearing loss in the United States. Exposure to sounds of 85 decibels or higher for an extended period can cause NIHL. Recall that the decibels (dB) produced by a sound of intensity I can be calculated by

$$dB = 10 \log \left(\frac{I}{1 \times 10^{-12}} \right). \quad \text{(Example 7)}$$

Intensity (W/m³)	Sound
316.227	fireworks
31.623	jet plane
3.162	ambulance
0.316	rock concert
0.032	headphones
0.003	hair dryer

Source: Dangerous Decibels

a. Which of the sounds listed in the table produce enough decibels to cause NIHL?

b. Determine the number of hair dryers that would produce the same number of decibels produced by a rock concert. Round to the nearest whole number.

c. How many jet planes would it take to produce the same number of decibels as a firework display? Round to the nearest whole number.

Solve each logarithmic equation. (Example 8)

69. $\log_2 (2x - 6) = 3 + \log_2 x$

70. $\log (3x + 2) = 1 + \log 2x$

71. $\log x = 1 - \log (x - 3)$

72. $\log 50x = 2 + \log (2x - 3)$

73. $\log_9 9x - 2 = -\log_9 x$

74. $\log (x - 10) = 3 + \log (x - 3)$

Solve each logarithmic equation. (Example 8)

75. $\log (29{,}995x + 40{,}225) = 4 + \log (3x + 4)$

76. $\log_{\frac{1}{4}} \left(\frac{1}{4}x \right) = -\log_{\frac{1}{4}} (x + 8) - \frac{5}{2}$

77. $\log x = 3 - \log (100x + 900)$

78. $\log_5 \frac{x^2}{8} - 3 = \log_5 \frac{x}{40}$

79. $\log 2x + \log \left(4 - \frac{16}{x} \right) = 2 \log (x - 2)$

80. TECHNOLOGY A chain of retail computer stores opened 2 stores in its first year of operation. After 8 years of operation, the chain consisted of 206 stores. (Example 9)

a. Write a continuous exponential equation to model the number of stores N as a function of year of operation t. Round k to the nearest hundredth.

b. Use the model you found in part **a** to predict the number of stores in the 12th year of operation.

81. STOCK The price per share of a coffee chain's stock was $0.93 in a month during its first year of trading. During its fifth year of trading, the price per share of stock was $3.52 during the same month. (Example 9)

a. Write a continuous exponential equation to model the price of stock P as a function of year of trading t. Round k to the nearest ten-thousandth.

b. Use the model you found in part **a** to predict the price of the stock during the ninth year of trading.

Solve each logarithmic equation.

82. $5 + 5 \log_{100} x = 20$ **83.** $6 + 2 \log_{e^2} x = 30$

84. $5 - 4 \log_{\frac{1}{2}} x = -19$ **85.** $36 + 3 \log_3 x = 60$

86. ACIDITY The acidity of a substance is determined by its concentration of H^+ ions. Because the H^+ concentration of substances can vary by several orders of magnitude, the logarithmic pH scale is used to indicate acidity. pH can be calculated by $pH = -\log [H^+]$, where $[H^+]$ is the concentration of H^+ ions in moles per liter.

Item	pH
ammonia	11.0
baking soda	8.3
human blood	7.4
water	7.0
milk	6.6
apples	3.0
lemon juice	2.0

a. Determine the H^+ concentration of baking soda.

b. How many times as acidic is milk than human blood?

c. By how many orders of magnitude is the $[H^+]$ of lemon juice greater than $[H^+]$ of ammonia?

d. How many moles of H^+ ions are in 1500 liters of human blood?

GRAPHING CALCULATOR Solve each equation algebraically, if possible. If not possible, approximate the solution to the nearest hundredth using a graphing calculator.

87. $x^3 = 2^x$

88. $\log_2 x = \log_8 x$

89. $3^x = x(5^x)$

90. $\log_x 5 = \log_5 x$

91. RADIOACTIVITY The isotopes phosphorous-32 and sulfur-35 both exhibit radioactive decay. The half-life of phosphorous-32 is 14.282 days. The half-life of sulfur-35 is 87.51 days.

  a. Write equations to express the radioactive decay of phosphorous-32 and sulfur-35 in terms of time t in days and ratio R of remaining isotope using the general equation for radioactive decay, $A = t \cdot \dfrac{\ln R}{-0.693}$, where A is the number of days the isotope has decayed and t is the half-life in days.

  b. At what value of R will sulfur-35 have been decaying 5 days longer than phosphorous-32?

Solve each exponential inequality.

92. $2 \le 2^x \le 32$

93. $9 < 3^y < 27$

94. $\dfrac{1}{4096} \le 8^p \le \dfrac{1}{64}$

95. $\dfrac{1}{2197} < 13^f \le \dfrac{1}{13}$

96. $10 < 10^d < 100{,}000$

97. $4000 > 5^q > 125$

98. $49 < 7^z < 1000$

99. $10{,}000 < 10^a < 275{,}000$

100. $\dfrac{1}{15} \ge 4^b \ge \dfrac{1}{64}$

101. $\dfrac{1}{2} \ge e^c \ge \dfrac{1}{100}$

102. FORENSICS Forensic pathologists perform autopsies to determine time and cause of death. The time t in hours since death can be calculated by $t = -10 \ln\left(\dfrac{T - R_t}{98.6 - R_t}\right)$, where T is the temperature of the body and R_t is the room temperature.

  a. A forensic pathologist measures the body temperature to be 93°F in a room that is 72°F. What is the time of death?

  b. A hospital patient passed away 4 hours ago. If the hospital has an average temperature of 75°F, what is the body temperature?

  c. A patient's temperature was 89°F 3.5 hours after the patient passed away. Determine the room temperature.

103. MEDICINE Fifty people were treated for a virus on the same day. The virus is highly contagious, and the individuals must stay in the hospital until they have no symptoms. The number of people p who show symptoms after t days can be modeled by $p = \dfrac{52.76}{1 + 0.03e^{0.75t}}$.

  a. How many show symptoms after 5 days?

  b. Solve the equation for t.

  c. How many days will it take until only one person shows symptoms?

Solve each equation.

104. $27 = \dfrac{12}{1 - \frac{1}{2}e^{-x}}$

105. $22 = \dfrac{L}{1 + \frac{L-3}{3}e^{-15}}$

106. $1000 = \dfrac{10{,}000}{1 + 19e^{-t}}$

107. $300 = \dfrac{400}{1 + 3e^{-2k}}$

108. $16^x + 4^x - 6 = 0$

109. $\dfrac{e^x + e^{-x}}{e^x - e^{-x}} = 6$

110. $\dfrac{\ln(4x + 2)}{\ln(4x - 2)} = 3$

111. $\dfrac{e^x - e^{-x}}{e^x + e^{-x}} = \dfrac{1}{2}$

112. POLLUTION Some factories have added filtering systems called *scrubbers* to their smokestacks in order to reduce pollution emissions. The percent of pollution P removed after f feet of length of a particular scrubber can be modeled by $P = \dfrac{0.9}{1 + 70e^{-0.28f}}$.

Scrubber   Chimney stack

  a. Graph the percent of pollution removed as a function of scrubber length.

  b. Determine the maximum percent of pollution that can be removed by the scrubber. Explain your reasoning.

  c. Approximate the maximum length of scrubber that a factory should choose to use. Explain.

H.O.T. Problems   Use **Higher-Order Thinking Skills**

113. REASONING What is the maximum number of extraneous solutions that a logarithmic equation can have? Explain your reasoning.

114. OPEN ENDED Give an example of a logarithmic equation with infinite solutions.

115 CHALLENGE If an investment is made with an interest rate r compounded monthly, how long will it take for the investment to triple?

116. REASONING How can you solve an equation involving logarithmic expressions with three different bases?

117. CHALLENGE For what x values do the domains of $f(x) = \log(x^4 - x^2)$ and $g(x) = \log x + \log x + \log(x - 1) + \log(x + 1)$ differ?

118. WRITING IN MATH Explain how to algebraically solve for t in $P = \dfrac{L}{1 + \left(\frac{L-I}{I}\right)e^{-kt}}$.

Evaluate each logarithm. (Lesson 3-3)

119. $\log_8 15$

120. $\log_2 8$

121. $\log_5 625$

122. SOUND An equation for loudness L, in decibels, is $L = 10 \log_{10} R$, where R is the relative intensity of the sound. (Lesson 3-2)

a. Solve $130 = 10 \log_{10} R$ to find the relative intensity of a fireworks display with a loudness of 130 decibels.

b. Solve $75 = 10 \log_{10} R$ to find the relative intensity of a concert with a loudness of 75 decibels.

c. How many times as intense is the fireworks display as the concert? In other words, find the ratio of their intensities.

Sound	Decibels
fireworks	130–190
car racing	100–130
parades	80–120
yard work	95–115
movies	90–110
concerts	75–110

For each function, (a) apply the leading term test, (b) determine the zeros and state the multiplicity of any repeated zeros, (c) find a few additional points, and then (d) graph the function. (Lesson 2-2)

123. $f(x) = x^3 - 8x^2 + 7x$

124. $f(x) = x^3 + 6x^2 + 8x$

125. $f(x) = -x^4 + 6x^3 - 32x$

Solve each equation. (Lesson 2-1)

126. $\frac{1}{6}(12a)^{\frac{1}{3}} = 1$

127. $\sqrt[3]{x-4} = 3$

128. $(3y)^{\frac{1}{3}} + 2 = 5$

Use logical reasoning to determine the end behavior or limit of the function as x approaches infinity. Explain your reasoning. (Lesson 1-3)

129. $f(x) = x^{10} - x^9 + 5x^8$

130. $g(x) = \dfrac{x^2 + 5}{7 - 2x^2}$

131. $h(x) = |(x-3)^2 - 1|$

Find the variance and standard deviation of each population to the nearest tenth. (Lesson 0-8)

132. $\{48, 36, 40, 29, 45, 51, 38, 47, 39, 37\}$

133. $\{321, 322, 323, 324, 325, 326, 327, 328, 329, 330\}$

134. $\{43, 56, 78, 81, 47, 42, 34, 22, 78, 98, 38, 46, 54, 67, 58, 92, 55\}$

Skills Review for Standardized Tests

135. SAT/ACT In a movie theater, 2 boys and 3 girls are randomly seated together in a row. What is the probability that the 2 boys are seated next to each other?

A $\dfrac{1}{5}$

C $\dfrac{1}{2}$

E $\dfrac{2}{5}$

B $\dfrac{3}{5}$

D $\dfrac{2}{3}$

136. REVIEW Which equation is equivalent to $\log_4 \frac{1}{16} = x$?

F $\dfrac{1^4}{16} = x^4$

G $\left(\dfrac{1}{16}\right)^4 = x$

H $4^x = \dfrac{1}{16}$

J $4^{\frac{1}{16}} = x$

137. If $2^4 = 3^x$, then what is the approximate value of x?

A 0.63

C 2.52

B 2.34

D 2.84

138. REVIEW The pH of a person's blood is given by pH $= 6.1 + \log_{10} B - \log_{10} C$, where B is the concentration base of bicarbonate in the blood and C is the concentration of carbonic acid in the blood. Determine which substance has a pH closest to a person's blood if their ratio of bicarbonate to carbonic acid is $17.5 : 2.25$.

F lemon juice

G baking soda

H milk

J ammonia

Substance	pH
lemon juice	2.3
milk	6.4
baking soda	8.4
ammonia	11.9

Modeling with Nonlinear Regression

Why?

While exponential growth is not a perfect model for the growth of a human population, government agencies can use estimates from such models to make strategic plans that ensure they will be prepared to meet the future needs of their people.

Then
You modeled data using polynomial functions.
(Lesson 2-1)

Now
- Model data using exponential, logarithmic, and logistic functions.
- Linearize and analyze data.

New Vocabulary
logistic growth function
linearize

Math Online
glencoe.com

Exponential, Logarithmic, and Logistic Modeling In this lesson, we will use the exponential regression features on a graphing calculator, rather than algebraic techniques, to model data exhibiting exponential or logarithmic growth or decay.

EXAMPLE 1 | Exponential Regression

POPULATION Mesa, Arizona, is one of the fastest-growing cities in the United States. Use exponential regression to model the Mesa population data. Then use your model to estimate the population of Mesa in 2020.

Population of Mesa, Arizona (thousands)										
Year	1900	1910	1920	1930	1940	1950	1960	1970	1980	1990	2000
Population	0.7	1.6	3.0	3.7	7.2	16.8	33.8	63	152	288	396

Step 1 Make a scatter plot.

Let $P(t)$ represent the population in thousands of Mesa t years after 1900. Enter and graph the data on a graphing calculator to create the scatter plot (Figure 3.5.1). Notice that the plot very closely resembles the graph of an exponential growth function.

Step 2 Find an exponential function to model the data.

With the diagnostic feature turned on and using **ExpReg** from the list of regression models, we get the values shown in Figure 3.5.2. The population in 1900 is represented by a and the growth rate, 6.7% per year, is represented by b. Notice that the correlation coefficient $r \approx 0.9968$ is close to 1, indicating a close fit to the data. In the $\boxed{Y=}$ menu, pick up this regression equation by entering $\boxed{VARS}$, Statistics, EQ, RegEQ.

Step 3 Graph the regression equation and scatter plot on the same screen.

Notice that the graph of the regression fits the data fairly well. (Figure 3.5.3).

[0, 130] scl: 10 by [−50, 500] scl: 50

Figure 3.5.1

Figure 3.5.2

[0, 130] scl: 10 by [−50, 500] scl: 50

Figure 3.5.3

Step 4 Use the model to make a prediction.

To predict the population of Mesa in 2020, 120 years after 1900, use the **CALC** feature to evaluate the function for $P(120)$ as shown. Based on the model, Mesa will have about 1675 thousand or 1.675 million people in 2020.

✔ Guided Practice

1. **INTERNET** The Internet experienced rapid growth in the 1990s. The table shows the number of users in millions for each year during the decade. Use exponential regression to model the data. Then use your model to predict the number of users in 2008. Let x be the number of years after 1990.

Year	1991	1992	1993	1994	1995	1996	1997	1998	1999	2000
Internet Users	1	1.142	1.429	4.286	5.714	10	21.429	34.286	59.143	70.314

While data exhibiting rapid growth or decay tend to suggest an exponential model, data that grow or decay rapidly at first and then more slowly over time tend to suggest a logarithmic model calculated using natural logarithmic regression.

EXAMPLE 2 Logarithmic Regression

BIRTHS Use logarithmic regression to model the data in the table about twin births in the United States. Then use your model to predict when the number of twin births in the U.S. will reach 150,000.

Number of Twin Births in the United States						
Year	1995	1997	1998	2000	2002	2004	2005
Births	96,736	104,137	110,670	118,916	125,134	132,219	133,122

Step 1 Let $B(t)$ represent the number of twin births t years after 1990. Then create a scatter plot (Figure 3.5.5). The plot resembles the graph of a logarithmic growth function.

Step 2 Calculate the regression equation using **LnReg**. The correlation coefficient $r \approx 0.9949$ indicates a close fit to the data. Rounding each value to three decimal places, a natural logarithm function that models the data is $B(t) = 38,428.963 + 35,000.168 \ln x$.

Step 3 In the Y= menu, pick up this regression equation. Figure 3.5.4 shows the results of the regression $B(t)$. The number of twin births in 1990 is represented by a. The graph of $B(t)$ fits the data fairly well (Figure 3.5.6).

Step 4 To find when the number of twin births will reach 150,000, graph the line $y = 150,000$ and the modeling equation on the same screen. Calculating the point of intersection (Figure 3.5.7), we find that according to this model, the number of twin births will reach 150,000 when $t \approx 24$, which is in 1990 + 24 or 2014.

```
LnReg
 y=a+blnx
 a=38428.96308
 b=35000.1679
 r²=.9897472487
 r=.9948604167
```

Figure 3.5.4

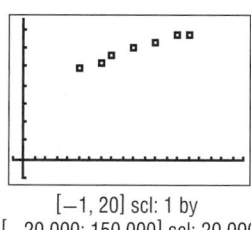

[−1, 20] scl: 1 by
[−20,000; 150,000] scl: 20,000
Figure 3.5.5

[−1, 20] scl: 1 by
[−20,000; 150,000] scl: 20,000
Figure 3.5.6

[−1, 30] scl: 2 by
[−20,000; 200,000] scl: 20,000
Figure 3.5.7

✔ Guided Practice

2. **LIFE EXPECTANCY** The table shows average U.S. life expectancies according to birth year. Use logarithmic regression to model the data. Then use the function to predict the life expectancy of a person born in 2020. Let x be the number of years after 1900.

Birth Year	1950	1960	1970	1980	1990	1995	2000	2005
Life Expectancy	68.2	69.7	70.8	73.7	75.4	75.8	77.0	77.8

Exponential and logarithmic growth is unrestricted, increasing at an ever-increasing rate with no upper bound. In many growth situations, however, the amount of growth is limited by factors that sustain the population, such as space, food, and water. Such factors cause growth that was initially exponential to slow down and level out, approaching a horizontal asymptote. A **logistic growth function** models such resource-limited exponential growth.

Key Concept — Logistic Growth Function

A logistic growth function has the form

$$f(t) = \frac{c}{1 + ae^{-bt}},$$

where t is any real number, a, b, and c are positive constants, and c is the limit to growth.

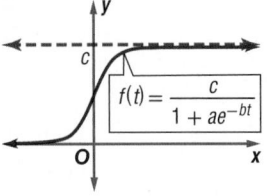

Logistic growth functions are bounded by two horizontal asymptotes, $y = 0$ and $y = c$. The limit to growth c is also called the carrying capacity of the function.

StudyTip

Logistic Decay If $b < 0$, then $f(t) = \frac{c}{1 + ae^{-bt}}$ would represent logistic decay. Unless otherwise stated, all logistic models in this text will represent logistic growth.

Real-World EXAMPLE 3 — Logistic Regression

BIOLOGY Use logistic regression to find a logistic growth function to model the data in the table about the number of yeast growing in a culture. Then use your model to predict the limit to the growth of the yeast in the culture.

Yeast Population in a Culture																	
Time (h)	1	2	3	4	5	6	7	8	9	10	11	12	13	14	15	16	17	18
Yeast	10	19	31	45	68	120	172	255	353	445	512	561	597	629	641	653	654	658

Step 1 Let $Y(t)$ represent the number of yeast in the culture after t hours. Then create a scatter plot (Figure 3.5.8). The plot resembles the graph of a logistic growth function.

Step 2 Calculate the regression equation using **Logistic** (Figure 3.5.9). Rounding each value to three decimal places, a logistic function that models the data is

$$Y(t) = \frac{661.565}{1 + 131.178e^{-0.555t}}.$$

Step 3 The graph of $Y(t) = \frac{661.565}{1 + 131.178e^{-0.555t}}$ fits the data fairly well (Figure 3.5.10).

Step 4 The limit to growth in the modeling equation is the numerator of the fraction or 661.565. Therefore, according to this model, the population of yeast in the culture will approach, but never reach, 662.

StudyTip

Correlation Coefficients Logistic regressions do not have a corresponding correlation coefficient due to the nature of the models.

[−2, 22] scl: 2 by [−50, 700] scl: 50

Figure 3.5.8

Logistic
y=c/(1+ae^(-bx))
a=131.1781232
b=.5545874884
c=661.5649456

Figure 3.5.9

[−2, 22] scl: 2 by [−50, 700] scl: 50

Figure 3.5.10

✓ Guided Practice

3. FISH Use logistic regression to model the data in the table about a lake's fish population. Then use your model to predict the limit to the growth of the fish population.

Time (mo)	0	4	8	12	16	20	24
Fish	125	580	2200	5300	7540	8280	8450

While you can use a calculator to find a linear, quadratic, power, exponential, logarithmic, or logistic regression equation for a set of data, it is up to you to determine which model *best* fits the data by looking at the graph and/or by examining the correlation coefficient of the regression. Consider the graphs of each regression model and its correlation coefficient using the same set of data below.

Over the domain displayed, the exponential and logistic regression models appear to most accurately fit the data, with the exponential model having the strongest correlation coefficient.

EXAMPLE 4   Choose a Regression

EARTHQUAKES Use the data below to determine a regression equation that best relates the distance a seismic wave can travel from an earthquake's epicenter to the time since the earthquake. Then determine how far from the epicenter the wave will be felt 8.5 minutes after the earthquake.

Travel Time (min)	1	2	5	7	10	12	13
Distance (km)	400	800	2500	3900	6250	8400	10,000

Step 1 From the shape of the scatter plot, it appears that these data could best be modeled by any of the regression models above except logarithmic. (Figure 3.5.10)

Step 2 Use the LinReg(ax+b), QuadReg, CubicReg, QuartReg, LnReg, ExpReg, PwrReg, and Logistic regression features to find regression equations to fit the data, noting the corresponding correlation coefficients. The regression equation with a correlation coefficient closest to 1 is the quartic regression with equation rounded to $y = 0.702x^4 - 16.961x^3 + 160.826x^2 - 21.045x + 293.022$. Remember to use [VARS] to transfer the entire equation to the graph.

Step 3 The quartic regression equation does indeed fit the data very well. (Figure 3.5.11)

Step 4 Use the CALC feature to evaluate this regression equation for $x = 8.5$. (Figure 3.5.12) Since $y \approx 4981$ when $x = 8.5$, you would expect the wave to be felt approximately 4981 kilometers away after 8.5 minutes.

[−1, 15] scl: 1 by [0, 12,000] scl: 1000   [−1, 15] scl: 1 by [0, 12,000] scl: 1000   [−1, 15] scl: 1 by [0, 12,000] scl: 1000

Figure 3.5.10   **Figure 3.5.11**   **Figure 3.5.12**

4. INTERNET Use the data in the table to determine a regression equation that best relates the cumulative number of domain names that were purchased from an Internet provider each month. Then predict how many domain names will be purchased during the 18th month.

Time (mo)	1	2	3	4	5	6	7	8
Domain Names	211	346	422	468	491	506	522	531

Time (mo)	9	10	11	12	13	14	15	16
Domain Names	540	538	551	542	565	571	588	593

Linearizing Data The correlation coefficient is a measure calculated from a *linear* regression. How then do graphing calculators provide correlation coefficients for *nonlinear* regression? The answer is that the calculators have **linearized** the data, transforming it so that it appears to cluster about a line. The process of transforming nonlinear data so that it appears to be linear is called *linearization*.

To linearize data, a function is applied to one or both of the variables in the data set as shown in the example below.

Original Data

x	y
0	1
1	1.4
2	1.9
3	2.7
4	3.7
5	5.2
6	7.2
7	10.0

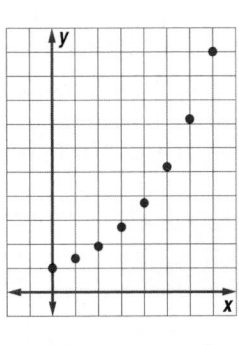

Linearized Data

x	$\ln y$
0	0
1	0.3
2	0.6
3	1.0
4	1.3
5	1.6
6	2.0
7	2.3

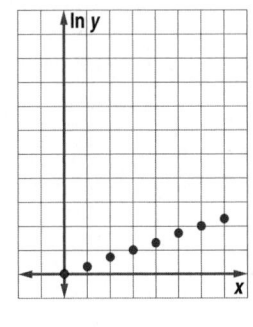

By calculating the equation of the line that best fits the linearized data and then applying inverse functions, a calculator can provide you with an equation that models the original data. The correlation coefficient for this nonlinear regression is actually a measure of how well the calculator was able to fit the *linearized data*.

Data modeled by a quadratic function are linearized by applying a square root function to the y-variable, while data modeled by exponential, power, or logarithmic functions are linearized by applying a logarithmic function to one or both variables.

Key Concept **Transformations for Linearizing Data**

To linearize data modeled by:

• a quadratic function $y = ax^2 + bx + c$, graph $(x, \sqrt{y})$.

• an exponential function $y = ab^x$, graph $(x, \ln y)$.

• a logarithmic function $y = a \ln x + b$, graph $(\ln x, y)$.

• a power function $y = ax^b$, graph $(\ln x, \ln y)$.

You will justify two of these linear transformations algebraically in Exercises 34 and 35.

EXAMPLE 5 Linearizing Data

A graph of the data below is shown at the right. Linearize the data assuming a power model. Graph the linearized data, and find the linear regression equation. Then use this linear model to find a model for the original data.

x	0.5	1	1.5	2	2.5	3	3.5	4
y	0.13	2	10.1	32	78.1	162	300.1	512

[0, 5] scl: 0.5 by [0, 1000] scl: 100

Step 1 Linearize the data.

To linearize data that can be modeled by a power function, take the natural log of both the x- and y-values.

ln x	−0.7	0	0.4	0.7	0.9	1.1	1.3	1.4
ln y	−2	0.7	2.3	3.5	4.4	5.1	5.7	6.2

Step 2 Graph the linearized data and find the linear regression equation.

The graph of $(\ln x, \ln y)$ appears to cluster about a line. Let $\hat{x} = \ln x$ and $\hat{y} = \ln y$. Using linear regression, the approximate equation modeling the linearized data is $\hat{y} = 4\hat{x} + 0.7$.

[0, 5] scl: 0.5 by [0, 10] scl: 1

Step 3 Use the model for the linearized data to find a model for the original data.

Replace $\hat{x}$ with $\ln x$ and $\hat{y}$ with $\ln y$, and solve for y.

$\hat{y} = 4\hat{x} + 0.7$	**Equation for linearized data**
$\ln y = 4 \ln x + 0.7$	$\hat{x} = \ln x$ and $\hat{y} = \ln y$
$e^{\ln y} = e^{4 \ln x + 0.7}$	**Exponentiate each side.**
$y = e^{4 \ln x + 0.7}$	**Inverse Property of Logarithms**
$y = e^{4 \ln x} e^{0.7}$	**Product Property of Exponents**
$y = e^{\ln x^4} e^{0.7}$	**Power Property of Exponents**
$y = x^4 e^{0.7}$	**Inverse Property of Logarithms**
$y = 2x^4$	$e^{0.7} \approx 2$

Therefore, a power function that models these data is $y = 2x^4$. The graph of this function with the scatter plot of the original data shows that this model is a good fit for the data.

[0, 5] scl: 1 by [0, 1000] scl: 100

☑ **Guided Practice**

Make a scatter plot of each set of data, and linearize the data according to the given model. Graph the linearized data, and find the linear regression equation. Then use this linear model for the transformed data to find a model for the original data.

5A. quadratic model

x	0	1	2	3	4	5	6	7
y	0	1	2	9	20	35	54	77

5B. logarithmic model

x	1	2	3	4	5	6	7	8
y	5	7.1	8.3	9.5	9.8	10.4	10.8	11.2

You can linearize data to find models for real-world data without the use of a calculator.

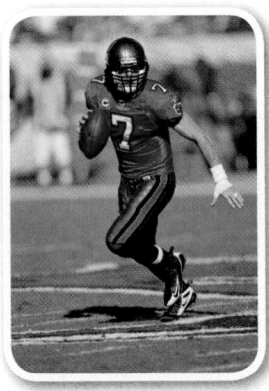

Real-World EXAMPLE 6 | Use Linearization

SPORTS The table shows the average professional football player's salary for several years. Find an exponential model relating these data by linearizing the data and finding the linear regression equation. Then use your model to predict the average salary in 2012.

Year	1990	1995	2000	2002	2003	2004	2005	2006
Average Salary ($1000)	354	584	787	1180	1259	1331	1400	1700

Source: NFL Players Association

Step 1 **Make a scatter plot, and linearize the data.**

Let x represent the number of years after 1900 and y the average salary in thousands.

x	90	95	100	102	103	104	105	106
y	354	584	787	1180	1259	1331	1400	1700

The plot is nonlinear and its shape suggests that the data could be modeled by an exponential function. Linearize the data by finding $(x, \ln y)$.

[80, 120] scl: 5 by [0, 2000] scl: 200

x	90	95	100	102	103	104	105	106
ln y	5.9	6.4	6.7	7.1	7.1	7.2	7.2	7.4

Step 2 **Graph the linearized data, and find a linear regression equation.**

A plot of the linearized data appears to form a straight line. Letting $\hat{y} = \ln y$, the rounded regression equation is about $\hat{y} = 0.096x - 2.754$.

[80, 120] scl: 5 by [0, 10] scl: 1

Step 3 **Use the model for the linearized data to find a model for the original data.**

Replace $\hat{y}$ with $\ln y$, and solve for y.

$\hat{y} = 0.096x - 2.754$	**Equation for linearized data**
$\ln y = 0.096x - 2.754$	$\hat{y} = \ln y$
$e^{\ln y} = e^{0.096x - 2.754}$	**Exponentiate each side.**
$y = e^{0.096x - 2.754}$	**Inverse Property of Logarithms**
$y = e^{0.096x} e^{-2.754}$	**Product Property of Exponents**
$y = 0.06e^{0.096x}$	$e^{-2.754} \approx 0.06$

Therefore, an exponential equation that models these data is $y = 0.06e^{0.096x}$.

Step 4 **Use the equation that models the original data to solve the problem.**

To find the average salary in 2012, find y when $x = 2012 - 1900$ or 112. According to this model, the average professional football player's salary in 2012 will be $0.06e^{0.096(112)} \approx \$2803$ thousand or about $2.8 million.

✓ Guided Practice

6. **FALLING OBJECT** Roger drops one of his shoes out of a hovering helicopter. The distance d in feet the shoe has dropped after t seconds is shown in the table.

t	0	1	1.5	2	2.5	3	4	5
d	0	15.7	35.4	63.8	101.4	144.5	258.1	404.8

Find a quadratic model relating these data by linearizing the data and finding the linear regression equation. Then use your model to predict the distance the shoe has traveled after 7 seconds.

For Exercises 1–3, complete each step.
a. Find an exponential function to model the data.
b. Find the value of each model at $x = 20$. (Example 1)

1.

x	y
1	7
2	11
3	25
4	47
5	96
6	193
7	380

2.

x	y
0	1
1	6
2	23
3	124
4	620
5	3130
6	15,600

3.

x	y
0	25
1	6
2	1.6
3	0.4
4	0.09
5	0.023
6	0.006

4. GENETICS *Drosophila melanogaster*, a species of fruit fly, are a common specimen in genetics labs because they reproduce about every 8.5 days, allowing researchers to study several generations. The table shows the population of *drosophila* over a period of days. (Example 1)

Generation	Drosophila	Generation	Drosophila
1	80	5	1180
2	156	6	2314
3	307	7	4512
4	593	8	8843

a. Find an exponential function to model the data.

b. Use the function to predict the population of *drosophila* after 93.5 days.

5 SHARKS Sharks have numerous rows of teeth embedded directly into their gums and not connected to their jaws. As a shark loses its teeth, teeth from the next row move forward. The rate of replacement of a row of teeth in days per row increases with the water temperature. (Example 1)

Temp. (°C)	20	21	22	23	24	25	26	27
Days per Row	66	54	44	35	28	22	18	16

a. Find an exponential function to model the data.

b. Use the function to predict the temperature at which sharks lose a row of teeth in 12 days.

6. WORDS A word family consists of a base word and all of its derivations. The table shows the percentage of words in an average English text comprised of the most common word families. (Example 2)

Word Families	1000	2000	3000	4000	5000
Percentage of Words	73.1	79.7	84.0	86.7	88.6

a. Find a logarithmic function to model the data.

b. Predict the number of word families that make up 95% of the words in an average English text.

For Exercises 7–9, complete each step.
a. Find a logarithmic function to model the data.
b. Find the value of each model at $x = 15$. (Example 2)

7.

x	y
1	50
2	42
3	37
4	33
5	31
6	28
7	27

8.

x	y
2	8.6
4	7.2
6	6.4
8	5.8
10	5.4
12	5.0
14	4.7

9.

x	y
1	40
2	49.9
3	55.8
4	59.9
5	63.2
6	65.8
7	68.1

10. CHEMISTRY A lab received a sample of an isotope of cobalt in 1999. The amount of cobalt in grams remaining per year is shown in the table below. (Example 2)

Year	2000	2001	2002	2003	2004	2005	2006	2007
Cobalt (g)	877	769	674	591	518	454	398	349

a. Find a logarithmic function to model the data. Let $x = 1$ represent 2000.

b. Predict the amount of cobalt remaining in 2020.

For Exercises 11–13, complete each step.
a. Find a logistic function to model the data.
b. Find the value of each model at $x = 25$. (Example 3)

11.

x	y
0	50
2	67
4	80
6	89
8	94
10	97
12	98
14	99

12.

x	y
1	3
2	5
3	7
4	8
5	13
6	16
7	19
8	20

13.

x	y
3	21
6	25
9	28
12	31
15	33
18	34
21	35
24	35

14. CHEMISTRY A student is performing a titration in lab. To perform the titration, she uses a burette to add a basic solution of NaOH to a neutral solution. The table shows the pH of the solution as the NaOH is added. (Example 3)

NaOH (mL)	0	1	2	3	5	7.5	10
pH	10	10.4	10.6	11.0	11.3	11.5	11.5

a. Find a logistic function to model the data.

b. Use the model to predict the pH of the solution after 12 milliliters of NaOH have been added.

15 CENSUS The table shows the projected population of Maine from the 2000 census. Let x be the number of years after 2000. (Example 3)

Year	Population (millions)
2000	1.275
2005	1.319
2010	1.357
2015	1.389
2020	1.409
2025	1.414
2030	1.411

a. Find a logistic function to model the data.

b. Based on the model, at what population does the 2000 census predict Maine's growth to level off?

c. Discuss the effectiveness of the model to predict the population as time increases significantly beyond the domain of the data.

16. **SCUBA DIVING** Scuba divers search for dive locations with good visibility, which can be affected by the murkiness of the water and the penetration of surface light. The table shows the percent of surface light reaching a diver at different depths as the diver descends. (Example 4)

Depth (ft)	Light (%)
15	89.2
30	79.6
45	71.0
60	63.3
75	56.5
90	50.4
105	44.9
120	40.1

a. Use the regression features on a calculator to determine the regression equation that best relates the data.

b. Use the graph of your regression equation to approximate the percent of surface light that reaches the diver at a depth of 83 feet.

17. **EELS** The table shows the average length of female king snake eels at various ages. (Example 4)

Age (yr)	Length (in.)	Age (yr)	Length (in.)
4	8	14	17
6	11	16	18
8	13	18	18
10	15	20	19
12	16		

a. Use the regression features on a calculator to determine if a logarithmic regression is better than a logistic regression. Explain.

b. Use the graph of your regression equation to approximate the length of an eel at 19 years.

For Exercises 18–21, complete each step.
a. Linearize the data according to the given model.
b. Graph the linearized data, and find the linear regression equation.
c. Use the linear model to find a model for the original data. Check its accuracy by graphing. (Examples 5 and 6)

18. exponential

x	y
0	11
1	32
2	91
3	268
4	808
5	2400
6	7000
7	22,000

19. quadratic

x	y
0	1.0
1	6.6
2	17.0
3	32.2
4	52.2
5	77.0
6	106.6
7	141.0

20. logarithmic

x	y
2	80.0
4	83.5
6	85.5
8	87.0
10	88.1
12	89.0
14	90.0
16	90.5

21. power

x	y
1	5
2	21
3	44
4	79
5	120
6	180
7	250
8	320

22. **TORNADOES** A tornado with a greater wind speed near the center of its funnel can travel greater distances. The table shows the wind speeds near the centers of tornadoes that have traveled various distances. (Example 6)

Distance (mi)	Wind Speed (mph)
0.50	37
0.75	53
1.00	65
1.25	74
1.50	81
1.75	88
2.00	93
2.25	98
2.50	102
2.75	106

a. Linearize the data assuming a logarithmic model.

b. Graph the linearized data, and find the linear regression equation.

c. Use the linear model to find a model for the original data, and approximate the wind speed of a funnel that has traveled 3.7 miles.

23. **HOUSING** The table shows the appreciation in the value of a house every 3 years since the house was purchased. (Example 6)

Years	0	3	6	9	12	15
Value ($)	78,000	81,576	85,992	90,791	95,859	101,135

a. Linearize the data assuming an exponential model.

b. Graph the linearized data, and find the linear regression equation.

c. Use the linear model to find a model for the original data, and approximate the value of the house 24 years after it is purchased.

24. **COOKING** Cooking times, temperatures, and recipes are often different at high altitudes than at sea level. This is due to the difference in atmospheric pressure, which causes boiling points for varius substances, such as water, to be lower at higher altitudes. The table shows the boiling point of water at different elevations above sea level.

Elevation (m)	Boiling Point (°C)
0	100
1000	99.29
2000	98.81
3000	98.43
4000	98.10
5000	97.80
6000	97.53
7000	97.28
8000	97.05
9000	96.83
10,000	96.62

a. Linearize the data for exponential, power, and logarithmic models.

b. Graph the linearized data, and determine which model best represents the data.

c. Write an equation to model the data based on your analysis of the linearizations.

Determine the model most appropriate for each scatter plot. Explain your reasoning.

25.

26.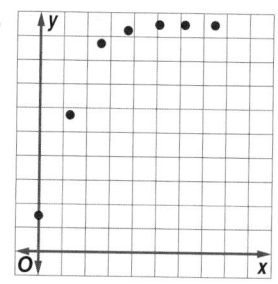

Linearize the data in each table. Then determine the most appropriate model.

27.

x	y
2	2.5
4	7.3
6	13.7
8	21.3
10	30.2
12	40.0
14	50.8
16	62.5

28.

x	y
1	6
2	29
3	42
4	52
5	59
6	65
7	70
8	75

29.

x	y
1	37.8
2	17.0
3	7.7
4	3.4
5	1.6
6	0.7
7	0.3
8	0.1

30. **FISH** Several ichthyologists are studying the smallmouth bass population in a lake. The table shows the smallmouth bass population of the lake over time.

Year	2001	2002	2003	2004	2005	2006	2007	2008
Bass	673	891	1453	1889	2542	2967	3018	3011

a. Determine the most appropriate model for the data. Explain your reasoning.

b. Find a function to model the data.

c. Use the function to predict the smallmouth bass population in 2012.

d. Discuss the effectiveness of the model to predict the population of the bass as time increases significantly beyond the domain of the data.

H.O.T. Problems   Use **H**igher-**O**rder **T**hinking Skills

31. **REASONING** Why are logarithmic regressions invalid when the domain is 0?

32. **CHALLENGE** Show that $y = ab^x$ can be converted to $y = ae^{kx}$.

33. **REASONING** Can the graph of a logistic function ever have any intercepts? Explain your reasoning.

PROOF Use algebra to verify that data modeled by each type of function can be linearized, or expressed as a function $y = mx + b$ for some values m and b, by replacing (x, y) with the indicated coordinates.

34. exponential, $(x, \ln y)$

35 power, $(\ln x, \ln y)$

36. **REASONING** How is the graph of $g(x) = \dfrac{5}{1 + e^{-x}} + a$ related to the graph of $f(x) = \dfrac{5}{1 + e^{-x}}$? Explain.

37. **WRITING IN MATH** Explain how the parameters of an exponential or logarithmic model relate to the data set or situation being modeled.

Lesson 3-5 Modeling with Nonlinear Regression   **209**

Solve each equation. (Lesson 3-4)

38. $3^{4x} = 3^{3-x}$

39. $3^{5x} \cdot 81^{1-x} = 9^{x-3}$

40. $49^x = 7^{x^2 - 15}$

41. $\log_5 (4x - 1) = \log_5 (3x + 2)$

42. $\log_{10} z + \log_{10} (z + 3) = 1$

43. $\log_6 (a^2 + 2) + \log_6 2 = 2$

44. ENERGY The energy E, in kilocalories per gram molecule, needed to transport a substance from the outside to the inside of a living cell is given by $E = 1.4(\log_{10} C_2 - \log_{10} C_1)$, where C_1 and C_2 are the concentrations of the substance inside and outside the cell, respectively. (Lesson 3-3)

   a. Express the value of E as one logarithm.

   b. Suppose the concentration of a substance inside the cell is twice the concentration outside the cell. How much energy is needed to transport the substance on the outside of the cell to the inside? (Use $\log_{10} 2 \approx 0.3010$.)

   c. Suppose the concentration of a substance inside the cell is four times the concentration outside the cell. How much energy is needed to transport the substance from the outside of the cell to the inside?

45. FINANCIAL LITERACY In 2003, Maya inherited $1,000,000 from her grandmother. She invested all of the money and by 2015, the amount will grow to $1,678,000. (Lesson 3-1)

   a. Write an exponential function that could be used to model the amount of money y. Write the function in terms of x, the number of years since 2003.

   b. Assume that the amount of money continues to grow at the same rate. Estimate the amount of money in 2025. Is this estimate reasonable? Explain your reasoning.

Simplify. (Lesson 0-2)

46. $(-2i)(-6i)(4i)$

47. $3i(-5i)^2$

48. i^{13}

49. $(1 - 4i)(2 + i)$

50. $\dfrac{4i}{3 + i}$

51. $\dfrac{4}{5 + 3i}$

Skills Review for Standardized Tests

52. SAT/ACT A recent study showed that the number of Australian homes with a computer doubles every 8 months. Assuming that the number is increasing continuously, at approximately what monthly rate must the number of Australian computer owners be increasing for this to be true?

   A 6.8%   **C** 12.5%   **E** 2%

   B 8.66%   **D** 8.0%

53. The data below gives the number of bacteria found in a certain culture. The bacteria are growing exponentially.

Hours	0	1	2	3	4
Bacteria	5	8	15	26	48

Approximately how much time will it take the culture to double after hour 4?

   F 1.26 hours   **H** 1.68 hours

   G 1.35 hours   **J** 1.76 hours

54. FREE RESPONSE The speed in miles per hour at which a car travels is represented by $v(t) = 60\left(1 - e^{-t^2}\right)$ where t is the time in seconds. Assume the car never needs to stop.

   a. Graph $v(t)$ for $0 \le t \le 10$.

   b. Describe the domain and range of $v(t)$. Explain your reasoning.

   c. What type of function is $v(t)$?

   d. What is the end behavior of $v(t)$? What does this mean in the context of the situation?

   e. Let $d(t)$ represent the total distance traveled by the car. What type of function does $d(t)$ represent as t approaches infinity? Explain.

   f. Let $a(t)$ represent the acceleration of the car. What is the end behavior of $a(t)$? Explain.

Chapter Summary

Key Concepts

Exponential Functions (Lesson 3-1)

- Exponential functions are of the form $f(x) = ab^x$, where $a \neq 0$, b is positive and $b \neq 1$. For natural base exponential functions, the base is the constant e.

- If a principal P is invested at an annual interest rate r (in decimal form), then the balance A in the account after t years is given by

  $A = P\left(1 + \frac{r}{n}\right)^{nt}$, if compounded n times a year or

  $A = Pe^{rt}$, if compounded continuously.

- If an initial quantity N_0 grows or decays at an exponential rate r or k (as a decimal), then the final amount N after a time t is given by $N = N_0(1 + r)^t$ or $N = N_0 e^{kt}$, where r is the rate of growth per time t and k is the continuous rate of growth at any time t.

Logarithmic Functions (Lesson 3-2)

- The inverse of $f(x) = b^x$, where $b > 0$ and $b \neq 1$, is the logarithmic function with base b, denoted $f^{-1}(x) = \log_b x$.

- If $b > 0$, $b \neq 1$, and $x > 0$, then the exponential form of $\log_b x = y$ is $b^y = x$ and the logarithmic form of $b^y = x$ is $\log_b x = y$. A logarithm is an exponent.

- Common logarithms: $\log_{10} x$ or $\log x$

- Natural logarithms: $\log_e x$ or $\ln x$

Properties of Logarithms (Lesson 3-3)

- Product Property: $\log_b xy = \log_b x + \log_b y$

- Quotient Property: $\log_b \frac{x}{y} = \log_b x - \log_b y$

- Power Property: $\log_b x^p = p \cdot \log_b x$

- Change of Base Formula: $\log_b x = \dfrac{\log_a x}{\log_a b}$

Exponential and Logarithmic Equations (Lesson 3-4)

- One-to-One Property of Exponents: For $b > 0$ and $b \neq 1$, $b^x = b^y$ if and only if $x = y$.

- One-to-One Property of Logarithms: For $b > 0$ and $b \neq 1$, $\log_b x = \log_b y$ if and only if $x = y$.

Modeling with Nonlinear Regression (Lesson 3-5)

To linearize data modeled by:

- a quadratic function $y = ax^2 + bx + c$, graph $(x, \sqrt{y})$.

- an exponential function $y = ab^x$, graph $(x, \ln y)$.

- a logarithmic function $y = a \ln x + b$, graph $(\ln x, y)$.

- a power function $y = ax^b$, graph $(\ln x, \ln y)$.

Key Vocabulary

algebraic function (p. 158)

common logarithm (p. 173)

continuous compound interest (p. 163)

exponential function (p. 158)

linearize (p. 204)

logarithm (p. 172)

logarithmic function with base *b* (p. 172)

logistic growth function (p. 202)

natural base (p. 160)

natural logarithm (p. 174)

transcendental function (p. 158)

Vocabulary Check

Choose the correct term from the list above to complete each sentence.

1. A logarithmic expression in which no base is indicated uses the _____.

2. _____ are functions in which the variable is the exponent.

3. Two examples of _____ are exponential functions and logarithmic functions.

4. The inverse of $f(x) = b^x$ is called a(n) _____.

5. The graph of a(n) _____ contains two horizontal asymptotes. Such a function is used for growth that has a limiting factor.

6. Many real-world applications use the _____ e, which, like π or $\sqrt{5}$, is an irrational number that requires a decimal approximation.

7. To _____ data, a function is applied to one or both of the variables in the data set, transforming the data so that it appears to cluster about a line.

8. Power, radical, polynomial, and rational functions are examples of _____.

9. _____ occurs when there is no waiting period between interest payments.

10. The _____ is denoted by ln.

Lesson-by-Lesson Review

3-1 Exponential Functions (pp. 158–169)

Sketch and analyze the graph of each function. Describe its domain, range, intercepts, asymptotes, end behavior, and where the function is increasing or decreasing.

11. $f(x) = 3^x$

12. $f(x) = 0.4^x$

13. $f(x) = \left(\frac{3}{2}\right)^x$

14. $f(x) = \left(\frac{1}{3}\right)^x$

Use the graph of $f(x)$ to describe the transformation that results in the graph of $g(x)$. Then sketch the graphs of $f(x)$ and $g(x)$.

15. $f(x) = 4^x;\ g(x) = 4^x + 2$

16. $f(x) = 0.1^x;\ g(x) = 0.1^{x-3}$

17. $f(x) = 3^x;\ g(x) = 2 \cdot 3^x - 5$

18. $f(x) = \left(\frac{1}{2}\right)^x;\ g(x) = \left(\frac{1}{2}\right)^{x+4} + 2$

Copy and complete the table below to find the value of an investment A for the given principal P, rate r, and time t if the interest is compounded n times annually.

n	1	4	12	365	continuously
A					

19. $P = \$250,\ r = 7\%,\ t = 6$ years

20. $P = \$1000,\ r = 4.5\%,\ t = 3$ years

EXAMPLE 1

Use the graph of $f(x) = 2^x$ to describe the transformation that results in the graph of $g(x) = -2^{x-5}$. Then sketch the graphs of g and f.

This function is of the form $g(x) = -f(x - 5)$.

So, $g(x)$ is the graph of $f(x) = 2^x$ translated 5 units to the right and reflected in the x-axis.

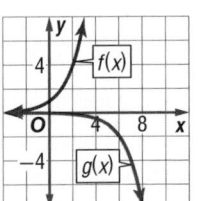

EXAMPLE 2

What is the value of $2000 invested at 6.5% after 12 years if the interest is compounded quarterly? continuously?

$A = P\left(1 + \frac{r}{n}\right)^{nt}$ **Compound Interest Formula**

$= 2000\left(1 + \frac{0.065}{4}\right)^{4(12)}$ $P = 2000, r = 0.065, n = 4, t = 12$

$= \$4335.68$ **Simplify.**

$A = Pe^{rt}$ **Continuous Interest Formula**

$= 2000e^{0.065(12)}$ $P = 2000, r = 0.065, t = 12$

$= \$4362.94$ **Simplify.**

3-2 Logarithmic Funtions (pp. 172–180)

Evaluate each expression.

21. $\log_2 32$

22. $\log_3 \frac{1}{81}$

23. $\log_{25} 5$

24. $\log_{13} 1$

25. $\ln e^{11}$

26. $3^{\log_3 9}$

27. $\log 80$

28. $e^{\ln 12}$

Use the graph of $f(x)$ to describe the transformation that results in the graph of $g(x)$. Then sketch the graphs of $f(x)$ and $g(x)$.

29. $f(x) = \log x;\ g(x) = -\log(x + 4)$

30. $f(x) = \log_2 x;\ g(x) = \log_2 x + 3$

31. $f(x) = \ln x;\ g(x) = \frac{1}{4}\ln x - 2$

EXAMPLE 3

Use the graph of $f(x) = \ln x$ to describe the transformation that results in the graph of $g(x) = -\ln(x - 3)$. Then sketch the graphs of $g(x)$ and $f(x)$.

This function is of the form $g(x) = -f(x - 3)$. So, $g(x)$ is the graph of $f(x)$ reflected in the x-axis translated 3 units to the right.

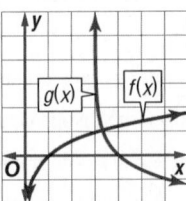

Lesson-by-Lesson Review

3-3 **Properties of Logarithms** (pp. 181–188)

Expand each expression.

32. $\log_3 9x^3y^3z^6$

33. $\log_5 x^2 a^7 \sqrt{b}$

34. $\ln \dfrac{e}{x^2 y^3 z}$

35. $\log \dfrac{\sqrt{gj^5k}}{100}$

Condense each expression.

36. $3 \log_3 x - 2 \log_3 y$

37. $\dfrac{1}{3} \log_2 a + \log_2 (b + 1)$

38. $5 \ln (x + 3) + 3 \ln 2x - 4 \ln (x - 1)$

EXAMPLE 4

Condense $3 \log_3 x + \log_3 7 - \dfrac{1}{2} \log_3 x$.

$3 \log_3 x + \log_3 7 - \dfrac{1}{2} \log_3 x$

$= \log_3 x^3 + \log_3 7 - \log_3 \sqrt{x}$ **Power Property**

$= \log_3 7x^3 - \log_3 \sqrt{x}$ **Product Property**

$= \log_3 \dfrac{7x^3}{\sqrt{x}}$ **Quotient Property**

3-4 **Exponential and Logarithmic Equations** (pp. 190–199)

Solve each equation.

39. $3^{x + 3} = 27^{x - 2}$

40. $25^{3x + 2} = 125$

41. $e^{2x} - 8e^x + 15 = 0$

42. $e^x - 4e^{-x} = 0$

43. $\log_2 x + \log_2 3 = \log_2 18$

44. $\log_6 x + \log_6 (x - 5) = 2$

EXAMPLE 5

Solve $7 \ln 2x = 28$.

$7 \ln 2x = 28$	**Original equation**
$\ln 2x = 4$	**Divide each side by 7.**
$e^{\ln 2x} = e^4$	**Exponentiate each side.**
$2x = e^4$	**Inverse Property**
$x = 0.5e^4$ or about 27.299	**Solve and simplify.**

3-5 **Modeling With Nonlinear Regression** (pp. 200–210)

Complete each step.
  a. Linearize the data according to the given model.
  b. Graph the linearized data, and find the linear regression equation.
  c. Use the linear model to find a model for the original data and graph it.

45. exponential

x	0	1	2	3	4	5	6
y	2	5	17	53	166	517	1614

46. logarithmic

x	1	2	3	4	5	6	7
y	−3	4	8	10	12	14	15

EXAMPLE 6

Linearize the data shown assuming a logarithmic model, and calculate the equation for the line of best fit. Use this equation to find a logarithmic model for the original data.

x	1	3	5	7	9	10
y	12	−7	−15	−21	−25	−27

Step 1   To linearize $y = a \ln x + b$, graph $(\ln x, y)$.

ln x	0	1.1	1.6	1.9	2.2	2.3
y	12	−7	−15	−21	−25	−27

Step 2   The line of best fit is $y = -16.94x + 11.86$.

Step 3 $y = -16.94 \ln x + 11.86$ $x = \ln x$

Applications and Problem Solving

47. INFLATION Prices of consumer goods generally increase each year due to inflation. From 2000 to 2008, the average rate of inflation in the United States was 4.5%. At this rate, the price of milk t years after January 2000 can be modeled with $M(t) = 2.75(1.045)^t$. (Lesson 3-1)

  a. What was the price of milk in 2000? 2005?

  b. If inflation continues at 4.5%, approximately what will the price of milk be in 2015?

  c. In what year will the price of milk reach $4?

48. CARS The value of a new vehicle depreciates the moment the car is driven off the dealership's lot. The value of the car will continue to depreciate every year. The value of one car t years after being bought is $f(x) = 18,000(0.8)^t$. (Lesson 3-1)

  a. What is the rate of depreciation for the car?

  b. How many years after the car is bought will it be worth half of its original value?

49. CHEMISTRY A radioactive substance has a half-life of 16 years. The number of years t it takes to decay from an initial amount N_0 to N can be determined using $t = \dfrac{16 \log \frac{N}{N_0}}{\log \frac{1}{2}}$. (Lesson 3-2)

  a. Approximately how many years will it take 100 grams to decay to 30 grams?

  b. Approximately what percentage of 100 grams will there be in 40 years?

50. EARTHQUAKES The Richter scale is a number system for determining the strength of earthquakes. The number R is dependent on energy E released by the earthquake in kilowatt-hours. The value of R is determined by $R = 0.67 \cdot \log (0.37E) + 1.46$. (Lesson 3-2)

  a. Find R for an earthquake that releases 1,000,000 kilowatt-hours.

  b. Estimate the energy released by an earthquake that registers 7.5 on the Richter scale.

51. BIOLOGY The time it takes for a species of animal to double is defined as its *generation time* and is given by $G = \dfrac{t}{2.5 \log_b d}$, where b is the initial number of animals, d is the final number of animals, t is the time period, and G is the generation time. If the generation time G of a species is 6 years, how much time t will it take for 5 animals to grow into a population of 3125 animals? (Lesson 3-3)

52. SOUND The intensity level of a sound, measured in decibels, can be modeled by $d(w) = 10 \log \frac{w}{w_0}$, where w is the intensity of the sound in watts per square meter and w_0 is the constant 1×10^{-12} watts per square meter. (Lesson 3-4)

  a. Determine the intensity of the sound at a concert that reaches 100 decibels.

  b. Tory compares the concert with the music she plays at home. She plays her music at 50 decibels, so the intensity of her music is half the intensity of the concert. Is her reasoning correct? Justify your answer mathematically.

  c. Soft music is playing with an intensity of 1×10^{-8} watts per square meters. By how much do the decimals increase if the intensity is doubled?

53. FINANCIAL LITERACY Delsin has $8000 and wants to put it into an interest-bearing account that compounds continuously. His goal is to have $12,000 in 5 years. (Lesson 3-4)

  a. Delsin found a bank that is offering 6% on investments. How long would it take his investment to reach $12,000 at 6%?

  b. What rate does Delsin need to invest at to reach his goal of $12,000 after 5 years?

54. INTERNET The number of people to visit a popular Web site is given below. (Lesson 3-5)

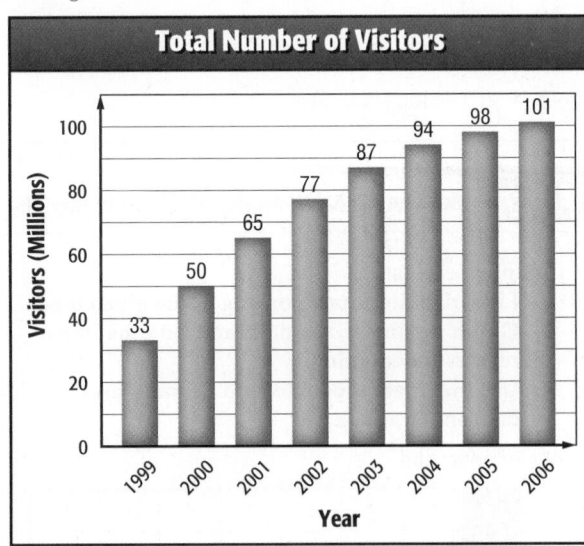

Total Number of Visitors

  a. Make a scatterplot of the data. Let 1990 = 0.

  b. Linearize the data with a logarithmic model.

  c. Graph the linearized data, and find the linear regression equation.

  d. Use the linear model to find a model for the original data and graph it.

Sketch and analyze the graph of each function. Describe its domain, range, intercepts, asymptotes, end behavior, and where the function is increasing or decreasing.

1. $f(x) = -e^{x+7}$

2. $f(x) = 2\left(\frac{3}{5}\right)^{-x} - 4$

Use the graph of $f(x)$ to describe the transformation that results in the graph of $g(x)$. Then sketch the graphs of $f(x)$ and $g(x)$.

3. $f(x) = \left(\frac{1}{2}\right)^x$ $g(x) = \left(\frac{1}{2}\right)^{x-3} + 4$

4. $f(x) = 5^x$ $g(x) = -5^{-x} - 2$

5. MULTIPLE CHOICE For which function is $\lim\limits_{x \to \infty} f(x) = -\infty$?

A $f(x) = -2 \cdot 3^{-x}$

B $f(x) = -\left(\frac{1}{10}\right)^x$

C $f(x) = -\log_8(x - 5)$

D $f(x) = \log_3(-x) - 6$

Evaluate each expression.

6. $\log_3 \dfrac{1}{81}$

7. $\log_{32} 2$

8. $\log 10^{12}$

9. $9^{\log_9 5.3}$

Sketch the graph of each function.

10. $f(x) = -\log_4(x + 3)$

11. $g(x) = \log(-x) + 5$

12. FINANCIAL LITERACY You invest $1500 in an account with an interest rate of 8% for 12 years, making no other deposits or withdrawals.

  a. What will be your account balance if the interest is compounded monthly?

  b. What will be your account balance if the interest is compounded continuously?

  c. If your investment is compounded daily, about how long will it take for it to be worth double the initial amount?

Expand each expression.

13. $\log_6 36xy^2$

14. $\log_3 \dfrac{a\sqrt{b}}{12}$

15. GEOLOGY Richter scale magnitude of an earthquake can be calculated using $R = \frac{2}{3}\log \frac{E}{E_0}$, where E is the energy produced and E_0 is a constant.

  a. An earthquake with a magnitude of 7.1 hit San Francisco in 1989. Find the scale of an earthquake that produces 10 times the energy of the 1989 earthquake.

  b. In 1906, San Francisco had an earthquake registering 8.25. How many times as much energy did the 1906 earthquake produce as the 1989 earthquake?

Condense each expression.

16. $2\log_4 m + 6\log_4 n - 3(\log_4 3 + \log_4 j)$

17. $1 + \ln 3 - 4\ln x$

Solve each equation.

18. $3^{x+8} = 9^{2x}$

19. $e^{2x} - 3e^x + 2 = 0$

20. $\log x + \log(x - 3) = 1$

21. $\log_2(x - 1) + 1 = \log_2(x + 3)$

22. MULTIPLE CHOICE Which equation has no solution?

F $e^x = e^{-x}$

G $2^{x-1} = 3^{x+1}$

H $\log_5 x = \log_9 x$

J $\log_2(x + 1) = \log_2 x$

For Exercises 23 and 24, complete each step.

  a. Find an exponential or logarithmic function to model the data.

  b. Find the value of each model at $x = 20$.

23.

x	1	3	5	7	9	11	13
y	8	3	0	-2	-3	-4	-5

24.

x	1	3	5	7	9	11	13
y	3	4	5	6	7	9	10

25. CENSUS The table gives the U.S. population between 1790 and 1940. Let 1780 = 0.

Year	Population (millions)
1790	4
1820	10
1850	23
1880	50
1910	92
1940	132

  a. Linearize the data, assuming a quadratic model. Graph the data, and write an equation for a line of best fit.

  b. Use the linear model to find a model for the original data. Is a quadratic model a good representation of population growth? Explain.

Connect to AP Calculus
Approximating Rates of Change

Objective
Use secant lines and the difference quotient to approximate rates of change.

In Chapter 1, we explored the rate of change of a function at a point using secant lines and the difference quotient. You learned that the rate of change of a function at a point can be represented by the slope of the line tangent to the function at that point. This is called the *instantaneous rate of change* at that point.

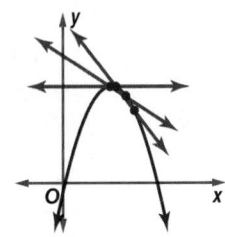

The constant e is used in applications of continuous growth and decay. This constant also has many applications in differential and integral calculus. The rate of change of $f(x) = e^x$ at any of its points is unique, which makes it a useful function for exploration and application in calculus.

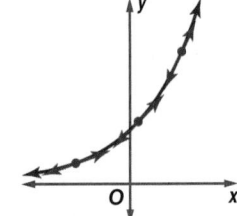

ACTIVITY 1 | **Approximate Rate of Change**

Approximate the rate of change of $f(x) = e^x$ at $x = 1$.

Step 1 Graph $f(x) = e^x$, and plot the points $P(1, f(1))$ and $Q(2, f(2))$.

Step 2 Draw a secant line of $f(x)$ through P and Q.

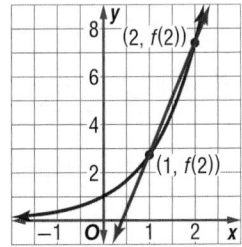

Step 3 Use $m = \dfrac{f(x_2) - f(x_1)}{x_2 - x_1}$ to calculate the average rate of change m for $f(x)$ using P and Q.

Step 4 Repeat Steps 1–3 two more times. First use $P(1, f(1))$ and $Q(1.5, f(1.5))$ and then use $P(1, f(1))$ and $Q(1.25, f(1.25))$.

▶ Analyze the Results

1. As the x-coordinate of Q approaches 1, what does the average rate of change m appear to approach?

2. Evaluate and describe the overall efficiency and the overall effectiveness of using secant lines to approximate the instantaneous rate of change of a function at a given point.

In Chapter 1, you developed an expression, the *difference quotient*, to calculate the slope of secant lines for different values of h.

Difference Quotient

$$m = \frac{f(x + h) - f(x)}{h}$$

As h decreases, the secant line moves closer and closer to a line tangent to the function. Substituting decreasing values for h into the difference quotient produces secant-line slopes that approach a limit. This limit represents the slope of the tangent line and the instantaneous rate of change of the function at that point.

ACTIVITY 2 — Approximate Rate of Change

Approximate the rate of change of $f(x) = e^x$ at several points.

Step 1 Substitute $f(x) = e^x$ into the difference quotient.

$$m = \frac{f(x + h) - f(x)}{h}$$

Step 2 Approximate the rate of change of $f(x)$ at $x = 1$ using values of h that approach 0. Let $h = 0.1, 0.01, 0.001$, and 0.0001.

$$m = \frac{e^{x + h} - e^x}{h}$$

Step 3 Repeat Steps 1 and 2 for $x = 2$ and for $x = 3$.

▶ Analyze the Results

3. As $h \to 0$, what does the average rate of change appear to approach for each value of x?

4. Write an expression for the rate of change of $f(x) = e^x$ at any point x.

5. Find the rate of change of $g(x) = 3e^x$ at $x = 1$. How did multiplying e^x by a constant affect the rate of change at $x = 1$?

6. Write an expression for the rate of change of $g(x) = ae^x$ at any point x.

In this chapter, you learned that $f(x) = \ln x$ is the inverse of $g(x) = e^x$, and you also learned about some of its uses in exponential growth and decay applications. Similar to e, the rate of change of $f(x) = \ln x$ at any of its points is unique, thus also making it another useful function for calculus applications.

ACTIVITY 3 — Approximate Rate of Change

Approximate the rate of change of $f(x) = \ln x$ at several points.

Step 1 Substitute $f(x) = \ln x$ into the difference quotient.

$$m = \frac{f(x + h) - f(x)}{h}$$

Step 2 Approximate the rate of change of $f(x)$ at $x = 2$ using values of h that approach 0. Let $h = 0.1, 0.01, 0.001$, and 0.0001.

$$m = \frac{\ln (x + h) - \ln x}{h}$$

Step 3 Repeat Steps 1 and 2 for $x = 3$ and for $x = 4$.

▶ Analyze the Results

7. As $h \to 0$, what does the average rate of change appear to approach for each value of x?

8. Write an expression for the rate of change of the function $f(x) = \ln x$ at any point x.

Model and Apply

9. In this problem, you will investigate the rate of change of the function $g(x) = -3 \ln x$ at any point x.

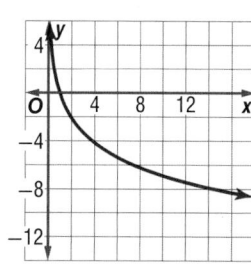

   a. Approximate the rates of change of $g(x)$ at $x = 2$ and then at $x = 3$.

   b. How do these rates of change compare to the rates of change for $f(x) = \ln x$ at these points?

   c. Write an expression for the rate of change of the function $g(x) = a \ln x$ for any point x.

Connect to AP Calculus: Approximating Rates of Change **217**

CHAPTER 4

Trigonometric Functions

Then

- In Chapter 3, you studied exponential and logarithmic functions, which are two types of transcendental functions.

Now

In Chapter 4, you will:

- Use trigonometric functions to solve right triangles.
- Find values of trigonometric functions for any angle.
- Graph trigonometric and inverse trigonometric functions.

Why?

SATELLITE NAVIGATION Satellite navigation systems operate by receiving signals from satellites in orbit, determining the distance to each of the satellites, and then using trigonometry to establish the location on Earth's surface. These techniques are also used when navigating cars, planes, ships, and spacecraft.

PREREAD Use the prereading strategy of previewing to make two or three predictions of what Chapter 4 is about.

Get Ready for Chapter 4

Diagnose Readiness You have two options for checking Prerequisite Skills.

Text Option

Take the Quick Quiz below.

*Quick*Check

Find the missing value in each figure.
(Prerequisite Skill)

1.

2.

3.

4.
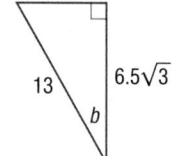

Determine whether each of the following could represent the measures of three sides of a triangle. Write *yes* or *no*. (Prerequisite Skill)

5. 4, 8, 12

6. 12, 15, 18

7. ALGEBRA The sides of a triangle have lengths x, $x + 17$, and 25. If the length of the longest side is 25, what value of x makes the triangle a right triangle? (Prerequisite Skill)

Find the equations of any vertical or horizontal asymptotes. (Lesson 2-5)

8. $f(x) = \dfrac{x^2 - 4}{x^2 + 8}$

9. $h(x) = \dfrac{x^3 - 27}{x + 5}$

10. $f(x) = \dfrac{x(x - 1)^2}{(x - 2)(x + 4)}$

11. $g(x) = \dfrac{x + 5}{(x - 3)(x - 5)}$

12. $h(x) = \dfrac{x^2 + x - 20}{x + 5}$

13. $f(x) = \dfrac{2x^2 + 5x - 12}{2x - 3}$

Online Option

Math Online Take a self-check Chapter Readiness Quiz at glencoe.com.

New Vocabulary

English	• Page •	Español
trignometric functions	• 220 •	funciones trigonométricas
sine	• 220 •	seno
cosine	• 220 •	coseno
tangent	• 220 •	tangente
cosecant	• 220 •	cosecant
secant	• 220 •	secant
cotangent	• 220 •	función recíproca
reciprocal function	• 220 •	cotangente
inverse sine	• 223 •	seno inverso
inverse cosine	• 223 •	coseno inverso
inverse tangent	• 223 •	tangente inversa
radian	• 232 •	radian
coterminal angles	• 234 •	ángulos coterminales
reference angle	• 244 •	ángulo de referencia
unit circle	• 247 •	círculo de unidad
circular function	• 248 •	función circular
period	• 250 •	período
sinusoid	• 256 •	sinusoid
amplitude	• 257 •	amplitud
frequency	• 260 •	frecuencia
phase shift	• 261 •	cambio de fase
Law of Sines	• 291 •	ley de senos
Law of Cosines	• 295 •	ley de cosenos

Review Vocabulary

reflection • p. 48 • reflexíon the mirror image of the graph of a function with respect to a specific line

dilation • p. 49 • homotecia a nonrigid transformation that has the effect of compressing (shrinking) or expanding (enlarging) the graph of a function vertically or horizontally

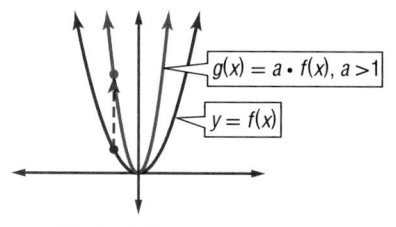

Vertical dilation

Right Triangle Trigonometry

Then
You evaluated functions.
(Lesson 1-1)

Now
- Find values of trigonometric functions for acute angles of right triangles.
- Solve right triangles.

New Vocabulary
trigonometric ratios
trigonometric functions
sine
cosine
tangent
cosecant
secant
cotangent
reciprocal function
inverse trigonometric function
inverse sine
inverse cosine
inverse tangent
angle of elevation
angle of depression
solve a right triangle

Math Online

glencoe.com

Why?

Large helium-filled balloons are a tradition of many holiday parades. Long cables attached to the balloon are used by volunteers to lead the balloon along the parade route.

Suppose two of these cables are attached to a balloon at the same point, and the volunteers holding these cables stand so that the ends of the cables lie in the same vertical plane. If you know the measure of the angle that each cable makes with the ground and the distance between the volunteers, you can use right triangle trigonometry to find the height of the balloon above the ground.

Values of Trigonometric Ratios The word *trigonometry* means *triangle measure*. In this chapter, you will study trigonometry as the relationships among the sides and angles of triangles and as a set of functions defined on the real number system. In this lesson, you will study *right triangle trigonometry*.

Using the side measures of a right triangle and a reference angle labeled θ (the Greek letter theta), we can form the six **trigonometric ratios** that define six **trigonometric functions**.

Key Concept   Trigonometric Functions

Let θ be an acute angle in a right triangle and the abbreviations opp, adj, and hyp refer to the length of the side opposite θ, the length of the side adjacent to θ, and the length of the hypotenuse, respectively.

Then the six trigonometric functions of θ are defined as follows.

sine $(\theta) = \sin \theta = \dfrac{\text{opp}}{\text{hyp}}$ **cosecant** $(\theta) = \csc \theta = \dfrac{\text{hyp}}{\text{opp}}$

cosine $(\theta) = \cos \theta = \dfrac{\text{adj}}{\text{hyp}}$ **secant** $(\theta) = \sec \theta = \dfrac{\text{hyp}}{\text{adj}}$

tangent $(\theta) = \tan \theta = \dfrac{\text{opp}}{\text{adj}}$ **cotangent** $(\theta) = \cot \theta = \dfrac{\text{adj}}{\text{opp}}$

The cosecant, secant, and cotangent functions are called **reciprocal functions** because their ratios are reciprocals of the sine, cosine, and tangent ratios, respectively. Therefore, the following statements are true.

$$\csc \theta = \frac{1}{\sin \theta} \qquad \sec \theta = \frac{1}{\cos \theta} \qquad \cot \theta = \frac{1}{\tan \theta}$$

From the definitions of the sine, cosine, tangent, and cotangent functions, you can also derive the following relationships. **You will prove these relationships in Exercise 83.**

$$\tan \theta = \frac{\sin \theta}{\cos \theta} \qquad \text{and} \qquad \cot \theta = \frac{\cos \theta}{\sin \theta}$$

EXAMPLE 1 Find Values of Trigonometric Ratios

Find the exact values of the six trigonometric functions of θ.

The length of the side opposite θ is 8, the length of the side adjacent to θ is 15, and the length of the hypotenuse is 17.

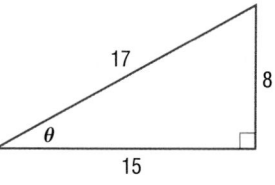

$\sin \theta = \dfrac{opp}{hyp}$ or $\dfrac{8}{17}$ **opp = 8 and hyp = 17** $\csc \theta = \dfrac{hyp}{opp}$ or $\dfrac{17}{8}$

$\cos \theta = \dfrac{adj}{hyp}$ or $\dfrac{15}{17}$ **adj = 15 and hyp = 17** $\sec \theta = \dfrac{hyp}{adj}$ or $\dfrac{17}{15}$

$\tan \theta = \dfrac{opp}{adj}$ or $\dfrac{8}{15}$ **opp = 8 and adj = 15** $\cot \theta = \dfrac{adj}{opp}$ or $\dfrac{15}{8}$

✔ **Guided Practice**

1A.

1B.
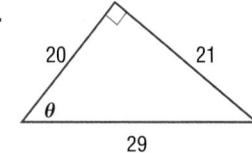

Consider $\sin \theta$ in the figure.

Using $\triangle ABC$: $\sin \theta = \dfrac{BC}{AB}$ Using $\triangle AB'C'$: $\sin \theta = \dfrac{B'C'}{AB'}$

Notice that the triangles are similar because they are two right triangles that share a common angle, θ. Because the triangles are similar, the ratios of the corresponding sides are equal. So, $\dfrac{BC}{AB} = \dfrac{B'C'}{AB'}$.

Therefore, $\sin \theta$ has the same value regardless of the triangle used. The values of the functions are constant for a given angle measure. They do not depend on the size of the right triangle.

EXAMPLE 2 Use One Trigonometric Value to Find Others

If $\cos \theta = \dfrac{2}{5}$, find the exact values of the five remaining trigonometric functions for the acute angle θ.

Begin by drawing a right triangle and labeling one acute angle θ.

Because $\cos \theta = \dfrac{adj}{hyp} = \dfrac{2}{5}$, label the adjacent side 2 and the hypotenuse 5.

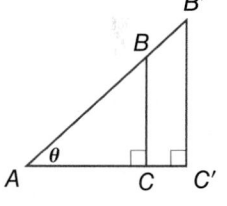

By the Pythagorean Theorem, the length of the leg opposite θ is $\sqrt{5^2 - 2^2}$ or $\sqrt{21}$.

$\sin \theta = \dfrac{opp}{hyp} = \dfrac{\sqrt{21}}{5}$ $\tan \theta = \dfrac{opp}{adj} = \dfrac{\sqrt{21}}{2}$ $\sec \theta = \dfrac{hyp}{adj} = \dfrac{5}{2}$

$\csc \theta = \dfrac{hyp}{opp} = \dfrac{5}{\sqrt{21}}$ or $\dfrac{5\sqrt{21}}{21}$ $\cot \theta = \dfrac{adj}{opp} = \dfrac{2}{\sqrt{21}}$ or $\dfrac{2\sqrt{21}}{21}$

✔ **Guided Practice**

2. If $\tan \theta = \dfrac{1}{2}$, find the exact values of the five remaining trigonometric functions for the acute angle θ.

You will often be asked to find the trigonometric functions of specific acute angle measures. The table below gives the values of the six trigonometric functions for three common angle measures: 30°, 45°, and 60°. To remember these values, you can use the properties of 30°-60°-90° and 45°-45°-90° triangles.

Key Concept **Trigonometric Values of Special Angles**

30°-60°-90° Triangle

45°-45°-90° Triangle

θ	30°	45°	60°
$\sin \theta$	$\frac{1}{2}$	$\frac{\sqrt{2}}{2}$	$\frac{\sqrt{3}}{2}$
$\cos \theta$	$\frac{\sqrt{3}}{2}$	$\frac{\sqrt{2}}{2}$	$\frac{1}{2}$
$\tan \theta$	$\frac{\sqrt{3}}{3}$	1	$\sqrt{3}$
$\csc \theta$	2	$\sqrt{2}$	$\frac{2\sqrt{3}}{3}$
$\sec \theta$	$\frac{2\sqrt{3}}{3}$	$\sqrt{2}$	2
$\cot \theta$	$\sqrt{3}$	1	$\frac{\sqrt{3}}{3}$

You will verify some of these values in Exercises 57–62.

Solving Right Triangles Trigonometric functions can be used to find missing side lengths and angle measures of right triangles.

EXAMPLE 3 **Find a Missing Side Length**

Find the value of x. Round to the nearest tenth, if necessary.

Because you are given an acute angle measure and the length of the hypotenuse of the triangle, use the cosine function to find the length of the side adjacent to the given angle.

$\cos \theta = \dfrac{\text{adj}}{\text{hyp}}$ Cosine function

$\cos 42° = \dfrac{x}{18}$ $\theta = 42°$, adj = x, and hyp = 18

$18 \cos 42° = x$ Multiply each side by 18.

$13.4 \approx x$ Use a calculator.

Therefore, x is about 13.4.

CHECK You can check your answer by substituting $x = 13.4$ into $\cos 42° = \dfrac{x}{18}$.

$\cos 42° = \dfrac{x}{18}$

$\cos 42° = \dfrac{13.4}{18}$ $x = 13.4$

$0.74 = 0.74$ ✓ Simplify.

☑ **Guided Practice**

3A.

3B.

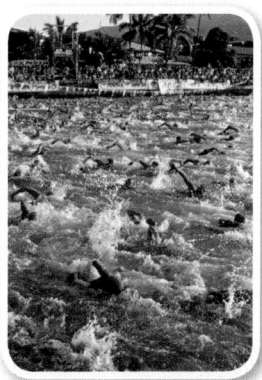

Real-World Link

The Ironman Triathlon held in Kailua-Kona Bay, Hawaii, consists of three endurance events, including a 2.4-mile swim, a 112-mile bike ride, and a 26.2-mile marathon.

Source: World Triathlon Corporation

Real-World EXAMPLE 4   Find a Missing Side Length

TRIATHLONS A competitor in a triathlon is running along the course shown. Determine the length in feet that the runner must cover to reach the finish line.

An acute angle measure and the opposite side length are given, so the sine function can be used to find the hypotenuse.

$\sin \theta = \dfrac{\text{opp}}{\text{hyp}}$ **Sine function**

$\sin 63° = \dfrac{200}{x}$ $\theta = 63°$, opp = 200, and hyp = x

$x \sin 63° = 200$ **Multiply each side by x.**

$x = \dfrac{200}{\sin 63°}$ or about 224.47   **Divide each side by sin 63°.**

So, the competitor must run about 224.5 feet to finish the triathlon.

✔ Guided Practice

4. **TRIATHLONS** Suppose a competitor in the swimming portion of the race is swimming along the course shown. Find the distance the competitor must swim to reach the shore.

When a trigonometric value of an acute angle is known, the corresponding **inverse trigonometric function** can be used to find the measure of the angle.

ReadingMath

Inverse Trigonometric Ratios The expression $\sin^{-1} x$ is read *the inverse sine of x*. Be careful not to confuse this notation with the notation for negative exponents: $\sin^{-1} x \neq \dfrac{1}{\sin x}$. Instead, this notation is similar to the notation for an inverse function, $f^{-1}(x)$.

Key Concept	**Inverse Trigonometric Functions**
Inverse Sine	If θ is an acute angle and the sine of θ is x, then the **inverse sine** of x is the measure of angle θ. That is, if $\sin \theta = x$, then $\sin^{-1} x = \theta$.
Inverse Cosine	If θ is an acute angle and the cosine of θ is x, then the **inverse cosine** of x is the measure of angle θ. That is, if $\cos \theta = x$, then $\cos^{-1} x = \theta$.
Inverse Tangent	If θ is an acute angle and the tangent of θ is x, then the **inverse tangent** of x is the measure of angle θ. That is, if $\tan \theta = x$, then $\tan^{-1} x = \theta$.

EXAMPLE 5   Find a Missing Angle Measure

Use a trigonometric function to find the measure of θ. Round to the nearest degree, if necessary.

Because the measures of the sides opposite and adjacent to θ are given, use the tangent function.

$\tan \theta = \dfrac{\text{opp}}{\text{adj}}$ **Tangent function**

$\tan \theta = \dfrac{26}{11}$ **opp = 26 and adj = 11**

$\theta = \tan^{-1} \dfrac{26}{11}$ or about 67°   **Definition of inverse tangent**

✔ Guided Practice

5A.

5B.

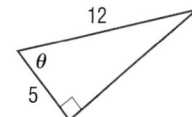

Some applications of trigonometry use an angle of elevation or depression. An **angle of elevation** is the angle formed by a horizontal line and an observer's line of sight to an object above. An **angle of depression** is the angle formed by a horizontal line and an observer's line of sight to an object below.

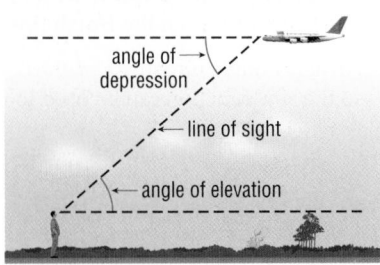

In the figure, the angles of elevation and depression are congruent because they are alternate interior angles of parallel lines.

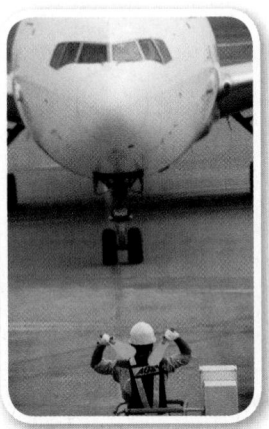
🌐 Real-World EXAMPLE 6   Use an Angle of Elevation

AIRPLANES A ground crew worker who is 6 feet tall is directing a plane on a runway. If the worker sights the plane at an angle of elevation of 32°, what is the horizontal distance from the worker to the plane?

Not drawn to scale

Because the worker is 6 feet tall, the vertical distance from the worker to the plane is $150 - 6$, or 144 feet. Because the measures of an angle and opposite side are given in the problem, you can use the tangent function to find x.

$$\tan \theta = \frac{\text{opp}}{\text{adj}}$$ **Tangent function**

$$\tan 32° = \frac{144}{x}$$ $\theta = 32°$, opp = 144, and adj = x

$$x \tan 32° = 144$$ **Multiply each side by x.**

$$x = \frac{144}{\tan 32°}$$ **Divide each side by tan 32°.**

$$x \approx 230.4$$ **Use a calculator.**

So, the horizontal distance from the worker to the plane is approximately 230.4 feet.

✔ Guided Practice

6. **CAMPING** A group of hikers on a camping trip climb to the top of a 1500-foot mountain. When the hikers look down at an angle of depression of 36°, they can see the campsite in the distance. What is the horizontal distance between the campsite and the group to the nearest foot?

Angles of elevation or depression can be used to estimate the distance between two objects, as well as the height of an object when given two angles from two different positions of observation.

StudyTip

Indirect Measurement When calculating the distance between two objects using angles of depression, it is important to remember that the objects must lie in the same horizontal plane.

🌐 Real-World EXAMPLE 7 | Use Two Angles of Elevation or Depression

BALLOONING A hot air balloon that is moving above a neighborhood has an angle of depression of 28° to one house and 52° to a house down the street. If the height of the balloon is 650 feet, estimate the distance between the two houses.

Draw a diagram to model this situation. Because the angle of elevation from a house to the balloon is congruent to the angle of depression from the balloon to that house, you can label the angles of elevation as shown. Label the horizontal distance from the balloon to the first house x and the distance between the two houses y.

From the smaller right triangle, you can use the tangent function to find x.

$\tan \theta = \dfrac{\text{opp}}{\text{adj}}$ **Tangent function**

$\tan 52° = \dfrac{650}{x}$ $\theta = 52°$, opp = 650, and adj = x

$x \tan 52° = 650$ **Multiply each side by x.**

$x = \dfrac{650}{\tan 52°}$ **Divide each side by tan 52°.**

TechnologyTip

Using Parentheses When evaluating a trigonometric expression using a graphing calculator, be careful to close parentheses. While a calculator returns the same value for the expressions tan(26 and tan(26), it does not for expressions tan(26 + 50 and tan(26) + 50.

From the larger triangle, you can use the tangent function to find y.

$\tan \theta = \dfrac{\text{opp}}{\text{adj}}$ **Tangent function**

$\tan 28° = \dfrac{650}{x + y}$ $\theta = 28°$, opp = 650, and adj = $x + y$

$(x + y) \tan 28° = 650$ **Multiply each side by $x + y$.**

$x + y = \dfrac{650}{\tan 28°}$ **Divide each side by tan 28°.**

$\dfrac{650}{\tan 52°} + y = \dfrac{650}{\tan 28°}$ **Substitute $x = \dfrac{650}{\tan 52°}$.**

$y = \dfrac{650}{\tan 28°} - \dfrac{650}{\tan 52°}$ **Subtract $\dfrac{650}{\tan 52°}$ from each side.**

$y \approx 714.6$ **Use a calculator.**

Therefore, the houses are about 714.6 feet apart.

✔ Guided Practice

7. **BUILDINGS** The angle of elevation from a car to the top of an apartment building is 48°. If the angle of elevation from another car that is 22 feet directly in front of the first car is 64°, how tall is the building?

Trigonometric functions and inverse relations can be used to **solve a right triangle**, which means to find the measures of all of the sides and angles of the triangle.

ReadingMath

Labeling Triangles
Throughout this chapter, a capital letter will be used to represent both a vertex of a triangle and the measure of the angle at that vertex. The same letter in lowercase will be used to represent both the side opposite that angle and the length of that side.

EXAMPLE 8   Solve a Right Triangle

Solve each triangle. Round side lengths to the nearest tenth and angle measures to the nearest degree.

a.

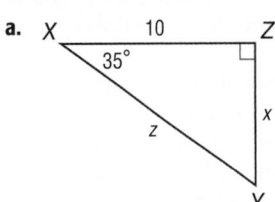

Find x and z using trigonometric functions.

$\tan 35° = \dfrac{x}{10}$	**Substitute.**	$\cos 35° = \dfrac{10}{z}$	**Substitute.**
$10 \tan 35° = x$	**Multiply.**	$z \cos 35° = 10$	**Multiply.**
$7.0 \approx x$	**Use a calculator.**	$z = \dfrac{10}{\cos 35°}$	**Divide.**
		$z \approx 12.2$	**Use a calculator.**

Because the measures of two angles are given, Y can be found by subtracting X from 90°.

$35° + Y = 90°$ **Angles X and Y are complementary.**

$\quad\;\; Y = 55°$ **Subtract.**

Therefore, $Y = 55°$, $x \approx 7.0$, and $z \approx 12.2$.

b.

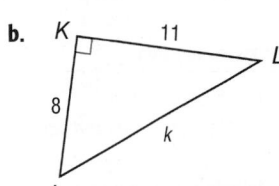

Because two side lengths are given, you can use the Pythagorean Theorem to find that $k = \sqrt{185}$ or about 13.6. You can find J by using any of the trigonometric functions.

$\tan J = \dfrac{11}{8}$ **Substitute.**

$J = \tan^{-1} \dfrac{11}{8}$ **Definition of inverse tangent**

$J \approx 53.97°$ **Use a calculator.**

Because J is now known, you can find L by subtracting J from 90°.

$53.97° + L \approx 90°$ **Angles J and L are complementary.**

$\qquad\;\; L \approx 36.03°$ **Subtract.**

Therefore, $J \approx 54°$, $L \approx 36°$, and $k \approx 13.6$.

✔ **Guided Practice**

8A.

8B.

Find the exact values of the six trigonometric functions of θ.
(Example 1)

1.

2.

3.

4.

5.

6.

7.

8.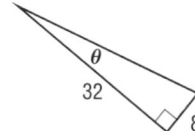

Use the given trigonometric function value of the acute angle θ to find the exact values of the five remaining trigonometric function values of θ. (Example 2)

9. $\sin \theta = \frac{4}{5}$

10. $\cos \theta = \frac{6}{7}$

11. $\tan \theta = 3$

12. $\sec \theta = 8$

13. $\cos \theta = \frac{5}{9}$

14. $\tan \theta = \frac{1}{4}$

15. $\cot \theta = 5$

16. $\csc \theta = 6$

17. $\sec \theta = \frac{9}{2}$

18. $\sin \theta = \frac{8}{13}$

Find the value of x. Round to the nearest tenth, if necessary.
(Example 3)

19.

20.

21.

22.

23.

24.

25.

26.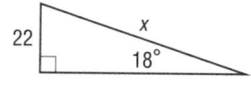

27 **MOUNTAIN CLIMBING** A team of climbers must determine the width of a ravine in order to set up equipment to cross it. If the climbers walk 25 feet along the ravine from their crossing point, and sight the crossing point on the far side of the ravine to be at a 35° angle, how wide is the ravine? (Example 4)

28. SNOWBOARDING Brad built a snowboarding ramp with a height of 3.5 feet and an 18° incline. (Example 4)

   a. Draw a diagram to represent the situation.

   b. Determine the length of the ramp.

29. DETOUR Traffic is detoured from Elwood Ave., left 0.8 mile on Maple St., and then right on Oak St., which intersects Elwood Ave. at a 32° angle. (Example 4)

   a. Draw a diagram to represent the situation.

   b. Determine the length of Elwood Ave. that is detoured.

30. PARACHUTING A paratrooper encounters stronger winds than anticipated while parachuting from 1350 feet, causing him to drift at an 8° angle. How far from the drop zone will the paratrooper land? (Example 4)

Find the measure of angle θ. Round to the nearest degree, if necessary. (Example 5)

31.

32.

33.

34.

35.

36.

37.

38.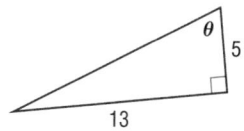

39. PARASAILING Kayla decided to try parasailing. She was strapped into a parachute towed by a boat. An 800-foot line connected her parachute to the boat, which was at a 32° angle of depression below her. How high above the water was Kayla? (Example 6)

40. OBSERVATION WHEEL The London Eye is a 135-meter-tall observation wheel. If a passenger at the top of the wheel sights the London Aquarium at a 58° angle of depression, what is the distance between the aquarium and the London Eye? (Example 6)

41. ROLLER COASTER On a roller coaster, 375 feet of track ascend at a 55° angle of elevation to the top before the first and highest drop. (Example 6)

   a. Draw a diagram to represent the situation.

   b. Determine the height of the roller coaster.

42. SKI LIFT A company is installing a new ski lift on a 225-meter-high mountain that will ascend at a 48° angle of elevation. (Example 6)

   a. Draw a diagram to represent the situation.

   b. Determine the length of cable the lift requires to extend from the base to the peak of the mountain.

43. BASKETBALL Both Derek and Sam are 5 feet 10 inches tall. Derek looks at a 10-foot basketball goal with an angle of elevation of 29°, and Sam looks at the goal with an angle of elevation of 43°. If Sam is directly in front of Derek, how far apart are the boys standing? (Example 7)

44. PARIS A tourist on the first observation level of the Eiffel Tower sights the Musée D'Orsay at a 1.4° angle of depression. A tourist on the third observation level located 219 meters directly above the first, sights the Musée D'Orsay at a 6.8° angle of depression. (Example 7)

   a. Draw a diagram to represent the situation.

   b. Determine the distance between the Eiffel Tower and the Musée D'Orsay.

45 LIGHTHOUSE Two ships are spotted from the top of a 156-foot lighthouse. The first ship is at a 27° angle of depression, and the second ship is directly behind the first at a 7° angle of depression. (Example 7)

   a. Draw a diagram to represent the situation.

   b. Determine the distance between the two ships.

46. MOUNT RUSHMORE The faces of the presidents at Mount Rushmore are 60 feet tall. A visitor sees the top of George Washington's head at a 48° angle of elevation and his chin at a 44.76° angle of elevation. Find the height of Mount Rushmore. (Example 7)

Solve each triangle. Round side lengths to the nearest tenth and angle measures to the nearest degree. (Example 8)

47.

48.

49.

50.

51.

52.

53.

54.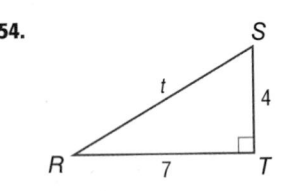

55. BASEBALL Michael's seat at a game is 65 feet behind home plate. His line of vision is 10 feet above the field.

   a. Draw a diagram to represent the situation.

   b. What is the angle of depression to home plate?

56. HIKING Jessica is standing 2 miles from the center of the base of Pikes Peak and looking at the summit of the mountain, which is 1.4 miles from the base.

   a. Draw a diagram to represent the situation.

   b. With what angle of elevation is Jessica looking at the summit of the mountain?

Find the exact value of each expression without using a calculator.

57. sin 60° **58.** cot 30° **59.** sec 30°

60. cos 45° **61.** tan 60° **62.** csc 45°

Without using a calculator, find the measure of the acute angle θ in a right triangle that satisfies each equation.

63. $\tan \theta = 1$ **64.** $\cos \theta = \dfrac{\sqrt{3}}{2}$

65. $\cot \theta = \dfrac{\sqrt{3}}{3}$ **66.** $\sin \theta = \dfrac{\sqrt{2}}{2}$

67. $\csc \theta = 2$ **68.** $\sec \theta = 2$

Without using a calculator, determine the value of x.

69.

70.
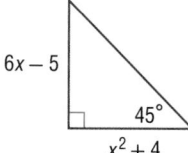

71. SCUBA DIVING A scuba diver located 20 feet below the surface of the water spots a shipwreck at a 70° angle of depression. After descending to a point 45 feet above the ocean floor, the diver sees the shipwreck at a 57° angle of depression. Draw a diagram to represent the situation, and determine the depth of the shipwreck.

Find the value of cos θ if θ is the measure of the smallest angle in each type of right triangle.

72. 3-4-5 **73.** 5-12-13

74. SOLAR POWER Find the total area of the solar panel shown below.

Without using a calculator, insert the appropriate symbol >, <, or = to complete each equation.

75. sin 45° ⬤ cot 60° **76.** tan 60° ⬤ cot 30°

77. cos 30° ⬤ csc 45° **78.** cos 30° ⬤ sin 60°

79. sec 45° ⬤ csc 60° **80.** tan 45° ⬤ sec 30°

81. ENGINEERING Determine the depth of the shaft at the large end d of the air duct shown below if the taper of the duct is 3.5°.

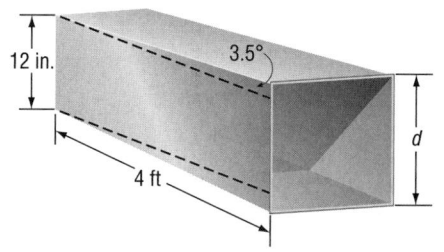

82. 🔲 **MULTIPLE REPRESENTATIONS** In this problem, you will investigate trigonometric functions of acute angles and their relationship to points on the coordinate plane.

  a. GRAPHICAL Let $P(x, y)$ be a point in the first quadrant. Graph the line through point P and the origin. Form a right triangle by connecting the points P, $(x, 0)$, and the origin. Label the lengths of the legs of the triangle in terms of x or y. Label the length of the hypotenuse as r and the angle the line makes with the x-axis θ.

  b. ANALYTICAL Express the value of r in terms of x and y.

  c. ANALYTICAL Express sin θ, cos θ, and tan θ in terms of x, y, and/or r.

  d. VERBAL Under what condition can the coordinates of point P be expressed as (cos θ, sin θ)?

  e. ANALYTICAL Which trigonometric ratio involving θ corresponds to the slope of the line?

  f. ANALYTICAL Find an expression for the slope of the line perpendicular to the line in part **a** in terms of θ.

H.O.T. Problems   Use Higher-Order Thinking Skills

83. PROOF Prove that if θ is an acute angle of a right triangle, then $\tan \theta = \dfrac{\sin \theta}{\cos \theta}$ and $\cot \theta = \dfrac{\cos \theta}{\sin \theta}$.

84. ERROR ANALYSIS Jason and Nadina know the value of sin $\theta = a$ and are asked to find csc θ. Jason says that this is not possible, but Nadina disagrees. Is either of them correct? Explain your reasoning.

85. WRITING IN MATH Explain why the six trigonometric functions are transcendental functions.

86. CHALLENGE Write an expression in terms of θ for the area of the scalene triangle shown. Explain.

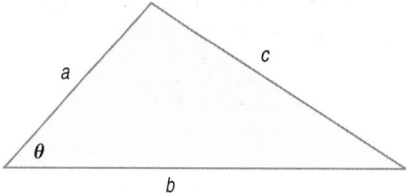

87. PROOF Prove that if θ is an acute angle of a right triangle, then $(\sin \theta)^2 + (\cos \theta)^2 = 1$.

REASONING If A and B are the acute angles of a right triangle and $m\angle A < m\angle B$, determine whether each statement is true or false. If false, give a counterexample.

88. $\sin A < \sin B$

89. $\cos A < \cos B$

90. $\tan A < \tan B$

91. WRITING IN MATH Notice on a graphing calculator that there is no key for finding the secant, cosecant, or cotangent of an angle measure. Explain why you think this might be so.

92. ECONOMICS The Consumer Price Index (CPI) measures inflation. It is based on the average prices of goods and services in the United States, with the annual average for the years 1982–1984 set at an index of 100. The table shown gives some annual average CPI values from 1955 to 2005. Find an exponential model relating this data (year, CPI) by linearizing the data. Let $x = 0$ represent 1955. Then use your model to predict the CPI for 2025. (Lesson 3-5)

Year	CPI
1955	26.8
1965	31.5
1975	53.8
1985	107.6
1995	152.4
2005	195.3

Source: Bureau of Labor Statistics

Solve each equation. Round to the nearest hundredth. (Lesson 3-4)

93. $e^{5x} = 24$

94. $2e^{x-7} - 6 = 0$

Sketch and analyze the graph of each function. Describe its domain, range, intercepts, asymptotes, end behavior, and where the function is increasing or decreasing. (Lesson 3-1)

95. $f(x) = -3^{x-2}$

96. $f(x) = 2^{3x-4} + 1$

97. $f(x) = -4^{-x+6}$

Solve each equation. (Lesson 2-5)

98. $\dfrac{x^2 - 16}{(x+4)(2x-1)} = \dfrac{4}{x+4} - \dfrac{1}{2x-1}$

99. $\dfrac{x^2 - 7}{(x+1)(x-5)} = \dfrac{6}{x+1} + \dfrac{3}{x-5}$

100. $\dfrac{2x^2 + 3}{3x^2 + 5x + 2} = \dfrac{5}{3x+2} - \dfrac{1}{x+1}$

101. NEWSPAPERS The circulation in thousands of papers of a national newspaper is shown. (Lesson 2-1)

Year	2002	2003	2004	2005	2006	2007	2008
Circulation (in thousands)	904.3	814.7	773.9	725.5	716.2	699.1	673.0

a. Let x equal the number of years after 2001. Create a scatter plot of the data.

b. Determine a power function to model the data.

c. Use the function to predict the circulation of the newspaper in 2015.

102. SAT/ACT In the figure, what is the value of z?

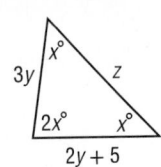

Note: Figure not drawn to scale.

A 15

B $15\sqrt{2}$

C $15\sqrt{3}$

D $30\sqrt{2}$

E $30\sqrt{3}$

103. REVIEW Joseph uses a ladder to reach a window 10 feet above the ground. If the ladder is 3 feet away from the wall, how long should the ladder be?

F 9.39 ft

G 10.44 ft

H 11.23 ft

J 12.05 ft

104. A person holds one end of a rope that runs through a pulley and has a weight attached to the other end. Assume that the weight is at the same height as the person's hand. What is the distance from the person's hand to the weight?

A 7.8 ft

B 10.5 ft

C 12.9 ft

D 14.3 ft

105. REVIEW A kite is being flown at a 45° angle. The string of the kite is 120 feet long. How high is the kite above the point at which the string is held?

F 60 ft

G $60\sqrt{2}$ ft

H $60\sqrt{3}$ ft

J 120 ft

Degrees and Radians

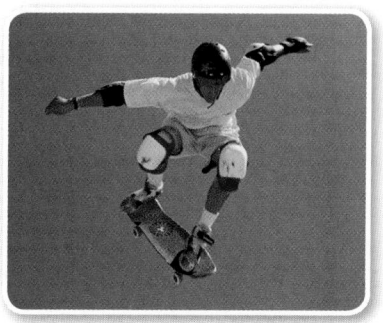

Why?

In Lesson 4-1, you worked only with acute angles, but angles can have *any* real number measurement. For example, in skateboarding, a 540 is an aerial trick in which a skateboarder and the board rotate through an angle of 540°, or one and a half complete turns, in midair.

Then
You used the measures of acute angles in triangles given in degrees. (Lesson 4-1)

Now
- Convert degree measures of angles to radian measures, and vice versa.
- Use angle measures to solve real-world problems.

New Vocabulary
vertex
initial side
terminal side
standard position
radian
coterminal angles
linear speed
angular speed
sector

Math Online
glencoe.com

Angles and Their Measures From geometry, you may recall an angle being defined as two noncollinear rays that share a common endpoint known as a **vertex**. An angle can also be thought of as being formed by the action of rotating a ray about its endpoint. From this dynamic perspective, the starting position of the ray forms the **initial side** of the angle, while the ray's position after rotation forms the angle's **terminal side**. In the coordinate plane, an angle with its vertex at the origin and its initial side along the positive *x*-axis is said to be in **standard position**.

Angle

Angle in Standard Position

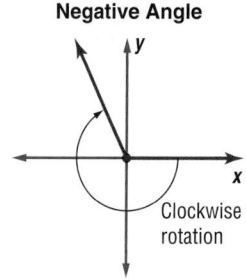

The measure of an angle describes the amount and direction of rotation necessary to move from the initial side to the terminal side of the angle. A *positive angle* is generated by a counterclockwise rotation and a *negative angle* by a clockwise rotation.

Positive Angle

Negative Angle

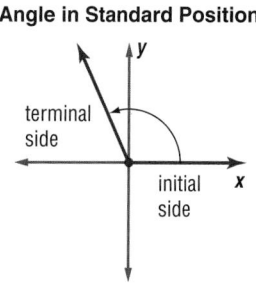

The most common angular unit of measure is the *degree* (°), which is equivalent to $\frac{1}{360}$ of a full rotation (counterclockwise) about the vertex. From the diagram shown, you can see that 360° corresponds to 1 complete rotation, 180° to a $\frac{1}{2}$ rotation, 90° to a $\frac{1}{4}$ rotation, and so on, as marked along the circumference of the circle.

Degree measures can also be expressed using a decimal degree form or a degree-minute-second (DMS) form where each degree is subdivided into 60 minutes (′) and each minute is subdivided into 60 seconds (″).

EXAMPLE 1   Convert Between DMS and Decimal Degree Form

Write each decimal degree measure in DMS form and each DMS measure in decimal degree form to the nearest thousandth.

a. 56.735°

First, convert 0.735° into minutes and seconds.

$$56.735° = 56° + 0.735° \left(\frac{60'}{1°}\right) \qquad\qquad 1° = 60'$$
$$= 56° + 44.1' \qquad\qquad\qquad\quad \text{Simplify.}$$

Next, convert 0.1′ into seconds.

$$56.735° = 56° + 44' + 0.1' \left(\frac{60''}{1'}\right) \qquad 1' = 60''$$
$$= 56° + 44' + 6'' \qquad\qquad\qquad \text{Simplify.}$$

Therefore, 56.735° can be written as 56° 44′ 6″.

b. 32° 5′ 28″

Each minute is $\frac{1}{60}$ of a degree and each second is $\frac{1}{60}$ of a minute, so each second is $\frac{1}{3600}$ of a degree.

$$32° 5' 28'' = 32° + 5'\left(\frac{1°}{60'}\right) + 28''\left(\frac{1°}{3600''}\right) \qquad 1' = \frac{1}{60}(1°) \text{ and } 1'' = \frac{1}{3600}(1°)$$
$$\approx 32° + 0.083 + 0.008 \qquad\qquad\qquad\qquad \text{Simplify.}$$
$$\approx 32.091° \qquad\qquad\qquad\qquad\qquad\qquad\quad \text{Add.}$$

Therefore, 32° 5′ 28″ can be written as about 32.091°.

✓ Guided Practice

1A. 213.875°

1B. 89° 56′ 7″

Measuring angles in degrees is appropriate when applying trigonometry to solve many real-world problems, such as those in surveying and navigation. For other applications with trigonometric functions, using an angle measured in degrees poses a significant problem. A degree has no relationship to any linear measure; inch-degrees or $\frac{\text{inch}}{\text{degree}}$ has no meaning. Measuring angles in **radians** provides a solution to this problem.

Key Concept   Radian Measure

Words   The measure θ in radians of a central angle of a circle is equal to the ratio of the length of the intercepted arc s to the radius r of the circle.

Symbols $\theta = \frac{s}{r}$, where θ is measured in radians (rad)

Example   A central angle has a measure of 1 radian if it intercepts an arc with the same length as the radius of the circle.

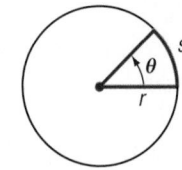

$\theta = 1$ radian when $s = r$.

Notice that as long as the arc length s and radius r are measured using the same linear units, the ratio $\frac{s}{r}$ is unitless. For this reason, the word *radian* or its abbreviation *rad* is usually omitted when writing the radian measure of an angle.

The central angle representing one full rotation counterclockwise about a vertex corresponds to an arc length equivalent to the circumference of the circle, $2\pi r$. From this, you can obtain the following radian measures.

$$1 \text{ rotation} = \frac{2\pi r}{r}$$
$$= 2\pi \text{ rad}$$

$$\frac{1}{2} \text{ rotation} = \frac{1}{2} \cdot 2\pi$$
$$= \pi \text{ rad}$$

$$\frac{1}{4} \text{ rotation} = \frac{1}{4} \cdot 2\pi$$
$$= \frac{\pi}{2} \text{ rad}$$

$$\frac{1}{6} \text{ rotation} = \frac{1}{6} \cdot 2\pi$$
$$= \frac{\pi}{3} \text{ rad}$$

   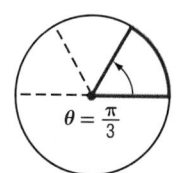

Because 2π radians and 360° both correspond to one complete revolution, you can write $360° = 2\pi$ radians or $180° = \pi$ radians. This last equation leads to the following equivalence statements.

$$1° = \frac{\pi}{180} \text{ radians} \qquad \text{and} \qquad 1 \text{ radian} = \left(\frac{180}{\pi}\right)°$$

Using these statements, we obtain the following conversion rules.

Key Concept   **Degree/Radian Conversion Rules**

1. To convert a degree measure to radians, multiply by $\frac{\pi \text{ radians}}{180°}$.

2. To convert a radian measure to degrees, multiply by $\frac{180°}{\pi \text{ radians}}$.

EXAMPLE 2   **Convert Between Degree and Radian Measure**

Write each degree measure in radians as a multiple of π and each radian measure in degrees.

a. 120°

$120° = 120° \left(\frac{\pi \text{ radians}}{180°}\right)$ Multiply by $\frac{\pi \text{ radians}}{180°}$.

$= \frac{2\pi}{3} \text{ radians or } \frac{2\pi}{3}$ Simplify.

b. −45°

$-45° = -45° \left(\frac{\pi \text{ radians}}{180°}\right)$ Multiply by $\frac{\pi \text{ radians}}{180°}$.

$= -\frac{\pi}{4} \text{ radians or } -\frac{\pi}{4}$ Simplify.

c. $\frac{5\pi}{6}$

$\frac{5\pi}{6} = \frac{5\pi}{6} \text{ radians}$ Multiply by $\frac{180°}{\pi \text{ radians}}$.

$= \frac{5\pi}{6} \text{ radians} \left(\frac{180°}{\pi \text{ radians}}\right) \text{ or } 150°$ Simplify.

d. $-\frac{3\pi}{2}$

$-\frac{3\pi}{2} = -\frac{3\pi}{2} \text{ radians}$ Multiply by $\frac{180°}{\pi \text{ radians}}$.

$= -\frac{3\pi}{2} \text{ radians} \left(\frac{180°}{\pi \text{ radians}}\right) \text{ or } -270°$ Simplify.

✓ **Guided Practice**

2A. 210°   **2B.** −60°   **2C.** $\frac{4\pi}{3}$ **2D.** $-\frac{\pi}{6}$

By defining angles in terms of their rotation about a vertex, two angles can have the same initial and terminal sides but different measures. Such angles are called **coterminal angles**. In the figures below, angles α and β are coterminal.

Positive and Negative Coterminal Angles

Positive Coterminal Angles

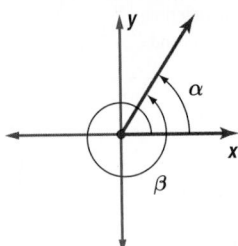

The two positive coterminal angles shown differ by one full rotation. A given angle has infinitely many coterminal angles found by adding or subtracting integer multiples of 360° or 2π radians.

Key Concept   **Coterminal Angles**

Degrees	**Radians**
If α is the degree measure of an angle, then all angles measuring $\alpha + 360n°$, where n is an integer, are coterminal with α.	If α is the radian measure of an angle, then all angles measuring $\alpha + 2n\pi$, where n is an integer, are coterminal with α.

EXAMPLE 3   **Find and Draw Coterminal Angles**

Identify all angles that are coterminal with the given angle. Then find and draw one positive and one negative angle coterminal with the given angle.

a. 45°

All angles measuring $45° + 360n°$ are coterminal with a 45° angle. Let $n = 1$ and -1.

$$45° + 360(1)° = 45° + 360° \text{ or } 405°$$

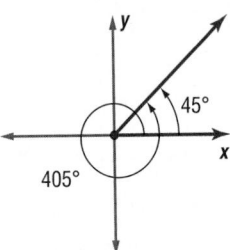

$$45° + 360(-1)° = 45° - 360° \text{ or } -315°$$

b. $-\dfrac{\pi}{3}$

All angles measuring $-\dfrac{\pi}{3} + 2n\pi$ are coterminal with a $-\dfrac{\pi}{3}$ angle. Let $n = 1$ and -1.

$$-\frac{\pi}{3} + 2(1)\pi = -\frac{\pi}{3} + 2\pi \text{ or } \frac{5\pi}{3}$$

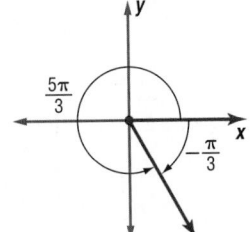

$$-\frac{\pi}{3} + 2(-1)\pi = -\frac{\pi}{3} - 2\pi \text{ or } -\frac{7\pi}{3}$$

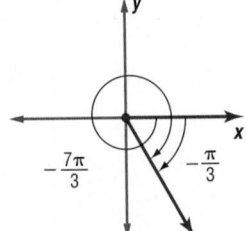

✓ **Guided Practice**

3A. $-30°$

3B. $\dfrac{3\pi}{4}$

Applications with Angle Measure Solving $\theta = \frac{s}{r}$ for the arc length s yields a convenient formula for finding the length of an arc of a circle.

Key Concept   **Arc Length**

If θ is a central angle in a circle of radius r, then the length of the intercepted arc s is given by

$$s = r\theta,$$

where θ is measured in radians.

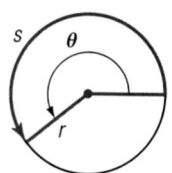

When θ is measured in degrees, you could also use the equation $s = \frac{\pi r \theta}{180}$, which already incorporates the degree-radian conversion.

EXAMPLE 4   **Find Arc Length**

Find the length of the intercepted arc in each circle with the given central angle measure and radius. Round to the nearest tenth.

a. $\frac{\pi}{4}$, $r = 5$ cm

$$\begin{aligned}
s &= r\theta & &\text{Arc length} \\
&= 5\left(\frac{\pi}{4}\right) & &r = 5 \text{ and } \theta = \frac{\pi}{4} \\
&= \frac{5\pi}{4} & &\text{Simplify.}
\end{aligned}$$

The length of the intercepted arc is $\frac{5\pi}{4}$ or about 3.9 centimeters.

b. $60°$, $r = 2$ in.

Method 1   Convert $60°$ to radian measure, and then use $s = r\theta$ to find the arc length.

$$\begin{aligned}
60° &= 60°\left(\frac{\pi \text{ radians}}{180°}\right) & &\text{Multiply by } \frac{\pi \text{ radians}}{180°}. \\
&= \frac{\pi}{3} & &\text{Simplify.}
\end{aligned}$$

Substitute $r = 2$ and $\theta = \frac{\pi}{3}$.

$$\begin{aligned}
s &= r\theta & &\text{Arc length} \\
&= 2\left(\frac{\pi}{3}\right) & &r = 2 \text{ and } \theta = \frac{\pi}{3} \\
&= \frac{2\pi}{3} & &\text{Simplify.}
\end{aligned}$$

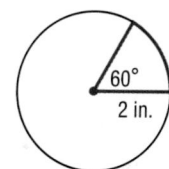

Method 2   Use $s = \frac{\pi r \theta}{180°}$ to find the arc length.

$$\begin{aligned}
s &= \frac{\pi r \theta}{180°} & &\text{Arc length} \\
&= \frac{\pi(2)(60°)}{180°} & &r = 2 \text{ and } \theta = 60° \\
&= \frac{2\pi}{3} & &\text{Simplify.}
\end{aligned}$$

The length of the intercepted arc is $\frac{2\pi}{3}$ or about 2.1 inches.

✓ **Guided Practice**

4A. $\frac{2\pi}{3}$, $r = 2$ m   **4B.** $135°$, $r = 0.5$ ft

---

The formula for arc length can be used to analyze circular motion. The rate at which an object moves along a circular path is called its **linear speed**. The rate at which the object *rotates* about a fixed point is called its **angular speed**. Linear speed is measured in units like miles per hour, while angular speed is measured in units like revolutions per minute.

Key Concept   **Linear and Angular Speed**

Suppose an object moves at a constant speed along a circular path of radius r.

If s is the arc length traveled by the object during time t, then the object's *linear speed* v is given by $v = \frac{s}{t}$.

If θ is the angle of rotation (in radians) through which the object moves during time t, then the *angular speed* ω of the object is given by $\omega = \frac{\theta}{t}$.

Real-World Link

In some U.S. cities, it is possible for bicycle messengers to ride an average of 30 to 35 miles a day while making 30 to 45 deliveries.

Source: New York Bicycle Messenger Association

Real-World EXAMPLE 5   **Find Angular and Linear Speeds**

BICYCLING A bicycle messenger rides the bicycle shown.

a. During one delivery, the tires rotate at a rate of 140 revolutions per minute. Find the angular speed of the tire in radians per minute.

Because each rotation measures 2π radians, 140 revolutions correspond to an angle of rotation θ of $140 \times 2\pi$ or 280π radians.

$\omega = \frac{\theta}{t}$ **Angular speed**

$= \frac{280\pi \text{ radians}}{1 \text{ minute}}$ $\theta = 280\pi$ **radians and** $t = 1$ **minute**

30 in.

Therefore, the angular speed of the tire is 280π or about 879.6 radians per minute.

b. On part of the trip to the next delivery, the tire turns at a constant rate of 2.5 revolutions per second. Find the linear speed of the tire in miles per hour.

A rotation of 2.5 revolutions corresponds to an angle of rotation θ of $2.5 \times 2\pi$ or 5π.

$v = \frac{s}{t}$ **Linear speed**

$= \frac{r\theta}{t}$ $s = r\theta$

$= \frac{15(5\pi) \text{ inches}}{1 \text{ second}}$ or $\frac{75\pi \text{ inches}}{1 \text{ second}}$ $r = 15$ **inches,** $\theta = 5\pi$ **radians, and** $t = 1$ **second**

Use dimensional analysis to convert this speed from inches per second to miles per hour.

$$\frac{75\pi \text{ inches}}{1 \text{ second}} \times \frac{60 \text{ seconds}}{1 \text{ minute}} \times \frac{60 \text{ minutes}}{1 \text{ hour}} \times \frac{1 \text{ foot}}{12 \text{ inches}} \times \frac{1 \text{ mile}}{5280 \text{ feet}} \approx \frac{13.4 \text{ miles}}{\text{hour}}$$

Therefore, the linear speed of the tire is about 13.4 miles per hour.

✓ Guided Practice

MEDIA Consider the DVD shown.

5A. Find the angular speed of the DVD in radians per second if the disc rotates at a rate of 3.5 revolutions per second.

5B. If the DVD player overheats and the disc begins to rotate at a slower rate of 3 revolutions per second, find the disc's linear speed in meters per minute.

120 mm

Recall from geometry that a **sector** of a circle is a region bounded by a central angle and its intercepted arc. For example, the shaded portion in the figure is a sector of circle P. The ratio of the area of a sector to the area of a whole circle is equal to the ratio of the corresponding arc length to the circumference of the circle. Let A represent the area of the sector.

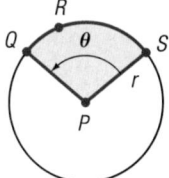

$$\frac{A}{\pi r^2} = \frac{\text{length of } \overset{\frown}{QRS}}{2\pi r} \qquad \frac{\text{area of sector}}{\text{area of circle}} = \frac{\text{arc length}}{\text{circumference of circle}}$$

$$\frac{A}{\pi r^2} = \frac{r\theta}{2\pi r} \qquad \text{The length of } \overset{\frown}{QRS} \text{ is } r\theta.$$

$$A = \frac{1}{2}r^2\theta \qquad \text{Solve for } A.$$

Key Concept   Area of a Sector

The area A of a sector of a circle with radius r and central angle θ is

$$A = \frac{1}{2}r^2\theta,$$

where θ is measured in radians.

EXAMPLE 6   Find Areas of Sectors

a. Find the area of the sector of the circle.

The measure of the sector's central angle θ is $\frac{7\pi}{8}$, and the radius is 3 centimeters.

$$A = \frac{1}{2}r^2\theta \qquad \text{Area of a sector}$$

$$= \frac{1}{2}(3)^2\left(\frac{7\pi}{8}\right) \text{ or } \frac{63\pi}{16} \qquad r = 3 \text{ and } \theta = \frac{7\pi}{8}$$

Therefore, the area of the sector is $\frac{63\pi}{16}$ or about 12.4 square centimeters.

b. WIPERS Find the approximate area swept by the wiper blade shown, if the total length of the windshield wiper mechanism is 26 inches.

The area swept by the wiper blade is the difference between the areas of the sectors with radii 26 inches and $26 - 16$ or 10 inches.

Convert the central angle measure to radians.

$$130° = 130°\left(\frac{\pi \text{ radians}}{180°}\right) = \frac{13\pi}{18}$$

Then use the radius of each sector to find the area swept. Let A_1 = the area of the sector with a 26-inch radius, and let A_2 = the area of the sector with a 10-inch radius.

$$A = A_1 - A_2 \qquad \text{Swept area}$$

$$= \frac{1}{2}(26)^2\left(\frac{13\pi}{18}\right) - \frac{1}{2}(10)^2\left(\frac{13\pi}{18}\right) \qquad \text{Area of a sector}$$

$$= \frac{2197\pi}{9} - \frac{325\pi}{9} \qquad \text{Simplify.}$$

$$= 208\pi \text{ or about } 653.5 \qquad \text{Simplify.}$$

Therefore, the swept area is about 653.5 square inches.

✓ Guided Practice

Find the area of the sector of a circle with the given central angle θ and radius r.

6A. $\theta = \frac{3\pi}{4}$, $r = 1.5$ ft

6B. $\theta = 50°$, $r = 6$ m

● = **Step-by-Step Solutions** begin on page R29.

Exercises

Write each decimal degree measure in DMS form and each DMS measure in decimal degree form to the nearest thousandth. (Example 1)

1. 11.773°

2. 58.244°

3. 141.549°

4. 273.396°

5. 87° 53′ 10″

6. 126° 6′ 34″

7. 45° 21′ 25″

8. 301° 42′ 8″

9. NAVIGATION A sailing enthusiast uses a sextant, an instrument that can measure the angle between two objects with a precision to the nearest 10 seconds, to measure the angle between his sailboat and a lighthouse. If his reading is 17° 37′ 50″, what is the measure in decimal degree form to the nearest hundredth? (Example 1)

17° 37′ 50″

Write each degree measure in radians as a multiple of π and each radian measure in degrees. (Example 2)

10. 30°

11. 225°

12. −165°

13. −45°

14. $\frac{2\pi}{3}$

15. $\frac{5\pi}{2}$

16. $-\frac{\pi}{4}$

17. $-\frac{7\pi}{6}$

Identify all angles that are coterminal with the given angle. Then find and draw one positive and one negative angle coterminal with the given angle. (Example 3)

18. 120°

19. −75°

20. 225°

21. −150°

22. $\frac{\pi}{3}$

23. $-\frac{3\pi}{4}$

24. $-\frac{\pi}{12}$

25. $\frac{3\pi}{2}$

26. GAME SHOW Sofia is spinning a wheel on a game show. There are 20 values in equal-sized spaces around the circumference of the wheel. The value that Sofia needs to win is two spaces above the space where she starts her spin, and the wheel must make at least one full rotation for the spin to count. Describe a spin rotation in degrees that will give Sofia a winning result. (Example 3)

Space Sofia needs to land on →

Start of → Sofia's spin

Find the length of the intercepted arc with the given central angle measure in a circle with the given radius. Round to the nearest tenth. (Example 4)

27. $\frac{\pi}{6}$, $r = 2.5$ m

28. $\frac{2\pi}{3}$, $r = 3$ in.

29. $\frac{5\pi}{12}$, $r = 4$ yd

30. 105°, $r = 18.2$ cm

31. 45°, $r = 5$ mi

32. 150°, $r = 79$ mm

33 AMUSEMENT PARK A carousel at an amusement park rotates 3024° per ride. (Example 4)

  a. How far would a rider seated 13 feet from the center of the carousel travel during the ride?

  b. How much farther would a second rider seated 18 feet from the center of the carousel travel during the ride than the rider in part **a**?

Find the rotation in revolutions per minute given the angular speed and the radius given the linear speed and the rate of rotation. (Example 5)

34. $\omega = \frac{2}{3}\pi \frac{\text{rad}}{\text{s}}$

35. $\omega = 135\pi \frac{\text{rad}}{\text{h}}$

36. $\omega = 104\pi \frac{\text{rad}}{\text{min}}$

37. $v = 82.3 \frac{\text{m}}{\text{s}}$, $131 \frac{\text{rev}}{\text{min}}$

38. $v = 144.2 \frac{\text{ft}}{\text{min}}$, $10.9 \frac{\text{rev}}{\text{min}}$

39. $v = 553 \frac{\text{in.}}{\text{h}}$, $0.09 \frac{\text{rev}}{\text{min}}$

40. MANUFACTURING A company manufactures several circular saws with the blade diameters and motor speeds shown below. (Example 5)

Blade Diameter (in.)	Motor Speed (rps)
3	2800
5	5500
$5\frac{1}{2}$	4500
$6\frac{1}{8}$	5500
$7\frac{1}{4}$	5000

  a. Determine the angular and linear speeds of the blades in each saw. Round to the nearest tenth.

  b. How much faster is the linear speed of the $6\frac{1}{8}$-inch saw compared to the 3-inch saw?

41. CARS On a stretch of interstate, a vehicle's tires range between 646 and 840 revolutions per minute. The diameter of each tire is 26 inches. (Example 5)

  a. Find the range of values for the angular speeds of the tires in radians per minute.

  b. Find the range of values for the linear speeds of the tires in miles per hour.

42. TIME A wall clock has a face diameter of $8\frac{1}{2}$ inches. The length of the hour hand is 2.4 inches, the length of the minute hand is 3.2 inches, and the length of the second hand is 3.4 inches. (Example 5)

$8\frac{1}{2}$ in.

a. Determine the angular speed in radians per hour and the linear speed in inches per hour for each hand.

b. If the linear speed of the second hand is 20 inches per minute, is the clock running fast or slow? How much time would it gain or lose per day?

GEOMETRY Find the area of each sector. (Example 6)

43.

1.5 in.
102°

44.

$\frac{4\pi}{3}$
3.4 m

45.

$\frac{2\pi}{5}$ 12 yd

46.

146° 21.4 km

47.

18 ft
177°

48.

43.5 cm
$\frac{\pi}{12}$

49. GAMES The dart board shown is divided into twenty equal sectors. If the diameter of the board is 18 inches, what area of the board does each sector cover? (Example 6)

50. LAWN CARE A sprinkler waters an area that forms one third of a circle. If the stream from the sprinkler extends 6 feet, what area of the grass does the sprinkler water? (Example 6)

The area of a sector of a circle and the measure of its central angle are given. Find the radius of the circle.

51 $A = 29$ ft^2, $\theta = 68°$

52. $A = 808$ cm^2, $\theta = 210°$

53. $A = 377$ in^2, $\theta = \frac{5\pi}{3}$

54. $A = 75$ m^2, $\theta = \frac{3\pi}{4}$

55. Describe the radian measure between 0 and 2π of an angle θ that is in standard position with a terminal side that lies in:
a. Quadrant I   **c.** Quadrant III
b. Quadrant II   **d.** Quadrant IV

56. If the terminal side of an angle that is in standard position lies on one of the axes, it is called a *quadrantal angle*. Give the radian measures of four quadrantal angles.

57. GEOGRAPHY Phoenix, Arizona, and Ogden, Utah, are located on the same line of longitude, which means that Ogden is directly north of Phoenix. The latitude of Phoenix is 33° 26′ N, and the latitude of Ogden is 41° 12′ N. If Earth's radius is approximately 3963 miles, about how far apart are the two cities?

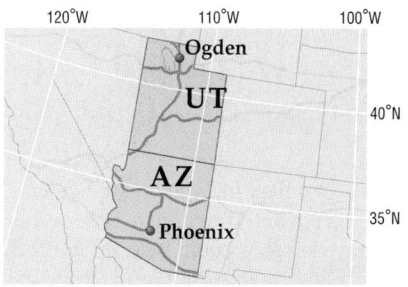

Find the measure of angle θ in radians and degrees.

58.

9 in.
θ
5 in.

59.

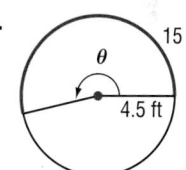

15
θ
4.5 ft

60.

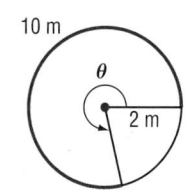

10 m
θ
2 m

61.

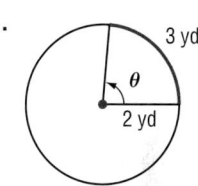

3 yd
θ
2 yd

62. TRACK The curve of a standard 8-lane track is semicircular as shown.

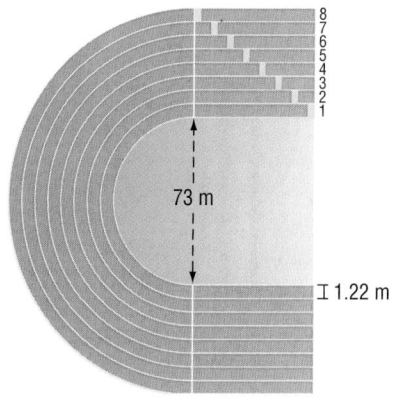

73 m
1.22 m

a. What is the length of the outside edge of Lane 4 in the curve?

b. How much longer is the inside edge of Lane 7 than the inside edge of Lane 3 in the curve?

63. DRAMA A pulley with radius r is being used to remove part of the set of a play during intermission. The height of the pulley is 12 feet.

   a. If the radius of the pulley is 6 inches and it rotates 180°, how high will the object be lifted?

   b. If the radius of the pulley is 4 inches and it rotates 900°, how high will the object be lifted?

12 ft

64. ENGINEERING A pulley like the one in Exercise 63 is being used to lift a crate in a warehouse. Determine which of the following scenarios could be used to lift the crate a distance of 15 feet the fastest. Explain how you reached your conclusion.

   I. The radius of the pulley is 5 inches rotating at 65 revolutions per minute.

   II. The radius of the pulley is 4.5 inches rotating at 70 revolutions per minute.

   III. The radius of the pulley is 6 inches rotating at 60 revolutions per minute.

GEOMETRY Find the area of each shaded region.

65

53°
72°
11 in.

66.

88°
104°
8 cm
12 cm

67. CARS The speedometer shown measures the speed of a car in miles per hour.

mph

   a. If the angle between 25 mi/h and 60 mi/h is 81.1°, about how many miles per hour are represented by each degree?

   b. If the angle of the speedometer changes by 95°, how much did the speed of the car increase?

Find the complement and supplement of each angle, if possible. If not possible, explain your reasoning.

68. $\frac{2\pi}{5}$ **69.** $\frac{5\pi}{6}$ **70.** $\frac{3\pi}{8}$ **71.** $-\frac{\pi}{3}$

72. SKATEBOARDING A physics class conducted an experiment to test three different wheel sizes on a skateboard with constant angular speed.

   a. Write an equation for the linear speed of the skateboard in terms of the radius and angular speed. Explain your reasoning.

   b. Using the equation you wrote in part **a**, predict the linear speed in meters per second of a skateboard with an angular speed of 3 revolutions per second for wheel diameters of 52, 56, and 60 millimeters.

   c. Based on your results in part **b**, how do you think wheel size affects linear speed?

H.O.T. Problems   Use **H**igher-**O**rder **T**hinking Skills

73. ERROR ANALYSIS Sarah and Mateo are told that the perimeter of a sector of a circle is 10 times the length of the circle's radius. Sarah thinks that the radian measure of the sector's central angle is 8 radians. Mateo thinks that there is not enough information given to solve the problem. Is either of them correct? Explain your reasoning.

74. CHALLENGE The two circles shown are concentric. If the length of the arc from A to B measures 8π inches and $DB = 2$ inches, find the arc length from C to D in terms of π.

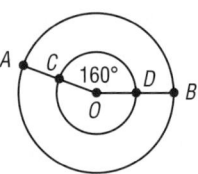

A C 160° D B O

REASONING Describe how the linear speed would change for each parameter below. Explain.

75. a decrease in the radius

76. a decrease in the unit of time

77. an increase in the angular speed

78. PROOF If $\dfrac{s_1}{r_1} = \dfrac{s_2}{r_2}$, prove that $\theta_1 = \theta_2$.

79. REASONING What effect does doubling the radius of a circle have on each of the following measures? Explain your reasoning.

   a. the perimeter of the sector of the circle with a central angle that measures θ radians

   b. the area of a sector of the circle with a central angle that measures θ radians

80. WRITING IN MATH Compare and contrast degree and radian measures. Then create a diagram similar to the one on page 231. Label the diagram using degree measures on the inside and radian measures on the outside of the circle.

Use the given trigonometric function value of the acute angle θ to find the exact values of the five remaining trigonometric function values of θ. (Lesson 4-1)

81. $\sin \theta = \dfrac{8}{15}$

82. $\sec \theta = \dfrac{4\sqrt{7}}{10}$

83. $\cot \theta = \dfrac{17}{19}$

84. BANKING An account that Hally's grandmother opened in 1955 earned continuously compounded interest. The table shows the balances of the account from 1955 to 1959. (Lesson 3-5)

a. Use regression to find a function that models the amount in the account. Use the number of years after Jan. 1, 1955, as the independent variable.

b. Write the equation from part **a** in terms of base e.

c. What was the interest rate on the account if no deposits or withdrawals were made during the period in question?

	Date	Balance
1	Jan. 1, 1955	$2137.52
2	Jan. 1, 1956	$2251.61
3	Jan. 1, 1957	$2371.79
4	Jan. 1, 1958	$2498.39
5	Jan. 1, 1959	$2631.74

Express each logarithm in terms of ln 2 and ln 5. (Lesson 3-2)

85. $\ln \dfrac{25}{16}$

86. $\ln 250$

87. $\ln \dfrac{10}{25}$

List all possible rational zeros of each function. Then determine which, if any, are zeros. (Lesson 2-4)

88. $f(x) = x^4 - x^3 - 12x - 144$

89. $g(x) = x^3 - 5x^2 - 4x + 20$

90. $g(x) = 6x^4 + 35x^3 - x^2 - 7x - 1$

Write each set of numbers in set-builder and interval notation if possible. (Lesson 1-3)

91. $f(x) = 4x^5 + 2x^4 - 3x - 1$

92. $g(x) = -x^6 + x^4 - 5x^2 + 4$

93. $h(x) = -\dfrac{1}{x^3} + 2$

Describe each set using interval notation. (Lesson 0-1)

94. $n > -7$

95. $-4 \le x < 10$

96. $y < 1$ or $y \ge 11$

97. SAT/ACT In the figure, C and D are the centers of the two circles with radii of 3 and 2, respectively. If the larger shaded region has an area of 9, what is the area of the smaller shaded region?

 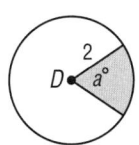

Note: Figure not drawn to scale.

A 3   C 5   E 8
B 4   D 7

98. REVIEW If $\cot \theta = 1$, then $\tan \theta =$

F -1 H 1
G 0   J 3

99. REVIEW If $\sec \theta = \dfrac{25}{7}$, then $\sin \theta =$

A $\dfrac{7}{25}$

B $\dfrac{24}{25}$

C $\dfrac{24}{25}$ or $-\dfrac{24}{25}$

D $\dfrac{25}{7}$

100. Which of the following radian measures is equal to 56°?

F $\dfrac{\pi}{15}$ H $\dfrac{14\pi}{45}$

G $\dfrac{7\pi}{45}$ J $\dfrac{\pi}{3}$

Trigonometric Functions on the Unit Circle

Then
You found values of trigonometric functions for acute angles using ratios in right triangles. (Lesson 4-1)

Now
- Find values of trigonometric functions for any angle.
- Find values of trigonometric functions using the unit circle.

New Vocabulary
quadrantal angle
reference angle
unit circle
circular function
periodic function
period

Math Online
glencoe.com

Why?

A blood pressure of 120 over 80, measured in millimeters of mercury, means that a person's blood pressure oscillates or cycles between 20 millimeters above and below a pressure of 100 millimeters of mercury for a given time t in seconds. A complete cycle of this oscillation takes about 1 second.

If the pressure exerted by the blood at time $t = 0.25$ second is 120 millimeters of mercury, then at time $t = 1.25$ seconds the pressure is also 120 millimeters of mercury.

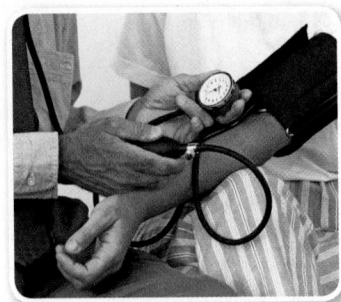

Trigonometric Functions of Any Angle In Lesson 4-1, the definitions of the six trigonometric functions were restricted to positive acute angles. In this lesson, these definitions are extended to include *any* angle.

Key Concept   **Trigonometric Functions of Any Angle**

Let θ be any angle in standard position and point $P(x, y)$ be a point on the terminal side of θ. Let r represent the nonzero distance from P to the origin. That is, let $r = \sqrt{x^2 + y^2} \neq 0$. Then the trigonometric functions of θ are as follows.

$\sin \theta = \dfrac{y}{r}$ $\csc \theta = \dfrac{r}{y}, y \neq 0$

$\cos \theta = \dfrac{x}{r}$ $\sec \theta = \dfrac{r}{x}, x \neq 0$

$\tan \theta = \dfrac{y}{x}, x \neq 0$ $\cot \theta = \dfrac{x}{y}, y \neq 0$

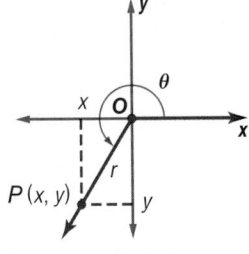

EXAMPLE 1   **Evaluate Trigonometric Functions Given a Point**

Let $(8, -6)$ be a point on the terminal side of an angle θ in standard position. Find the exact values of the six trigonometric functions of θ.

Use the values of x and y to find r.

$r = \sqrt{x^2 + y^2}$ **Pythagorean Theorem**

$ = \sqrt{8^2 + (-6)^2}$ $x = 8$ and $y = -6$

$ = \sqrt{100}$ or 10   **Take the positive square root.**

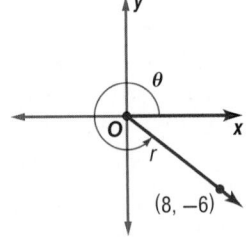

Use $x = 8$, $y = -6$, and $r = 10$ to write the six trigonometric ratios.

$\sin \theta = \dfrac{y}{r} = \dfrac{-6}{10}$ or $-\dfrac{3}{5}$ $\cos \theta = \dfrac{x}{r} = \dfrac{8}{10}$ or $\dfrac{4}{5}$ $\tan \theta = \dfrac{y}{x} = \dfrac{-6}{8}$ or $-\dfrac{3}{4}$

$\csc \theta = \dfrac{r}{y} = \dfrac{10}{-6}$ or $-\dfrac{5}{3}$ $\sec \theta = \dfrac{r}{x} = \dfrac{10}{8}$ or $\dfrac{5}{4}$ $\cot \theta = \dfrac{x}{y} = \dfrac{8}{-6}$ or $-\dfrac{4}{3}$

☑ **Guided Practice**

The given point lies on the terminal side of an angle θ in standard position. Find the values of the six trigonometric functions of θ.

1A. $(4, 3)$ **1B.** $(-2, -1)$

In Example 1, you found the trigonometric values of θ without knowing the measure of θ. Now we will discuss methods for finding these function values when only θ is known. Consider trigonometric functions of quadrantal angles. When the terminal side of an angle θ that is in standard position lies on one of the coordinate axes, the angle is called a **quadrantal angle**.

Key Concept **Common Quadrantal Angles**

| $\theta = 0°$ or 0 radians | $\theta = 90°$ or $\frac{\pi}{2}$ radians | $\theta = 180°$ or π radians | $\theta = 270°$ or $\frac{3\pi}{2}$ radians |

You can find the values of the trigonometric functions of quadrantal angles by choosing a point on the terminal side of the angle and evaluating the function at that point. Any point can be chosen. However, to simplify calculations, pick a point for which r equals 1.

EXAMPLE 2   **Evaluate Trigonometric Functions of Quadrantal Angles**

Find the exact value of each trigonometric function, if defined. If not defined, write *undefined*.

a. $\sin(-180°)$

The terminal side of $-180°$ in standard position lies on the negative x-axis. Choose a point P on the terminal side of the angle. A convenient point is $(-1, 0)$ because $r = 1$.

$\sin(-180°) = \frac{y}{r}$ **Sine function**

$= \frac{0}{1}$ or 0 **$y = 0$ and $r = 1$**

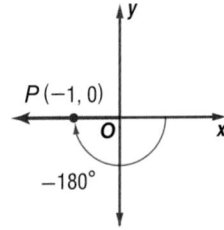

b. $\tan \frac{3\pi}{2}$

The terminal side of $\frac{3\pi}{2}$ in standard position lies on the negative y-axis. Choose a point $P(0, -1)$ on the terminal side of the angle because $r = 1$.

$\tan \frac{3\pi}{2} = \frac{y}{x}$ **Tangent function**

$= \frac{-1}{0}$ or undefined **$y = -1$ and $x = 0$**

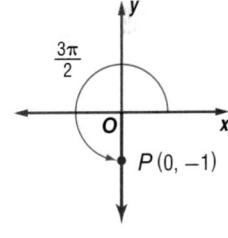

c. $\sec 4\pi$

The terminal side of 4π in standard position lies on the positive x-axis. The point $(1, 0)$ is convenient because $r = 1$.

$\sec 4\pi = \frac{r}{x}$ **Secant function**

$= \frac{1}{1}$ or 1 **$r = 1$ and $x = 1$**

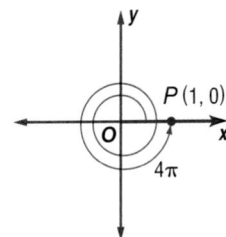

✓ **Guided Practice**

2A. $\cos 270°$ **2B.** $\csc \frac{\pi}{2}$ **2C.** $\cot(-90°)$

To find the values of the trigonometric functions of angles that are neither acute nor quadrantal, consider the three cases shown below in which a and b are positive real numbers. Compare the values of sine, cosine, and tangent of θ and θ'.

Quadrant II		Quadrant III		Quadrant IV	
$\sin \theta = \dfrac{b}{r}$	$\sin \theta' = \dfrac{b}{r}$	$\sin \theta = -\dfrac{b}{r}$	$\sin \theta' = \dfrac{b}{r}$	$\sin \theta = -\dfrac{b}{r}$	$\sin \theta' = \dfrac{b}{r}$
$\cos \theta = -\dfrac{a}{r}$	$\cos \theta' = \dfrac{a}{r}$	$\cos \theta = -\dfrac{a}{r}$	$\cos \theta' = \dfrac{a}{r}$	$\cos \theta = \dfrac{a}{r}$	$\cos \theta' = \dfrac{a}{r}$
$\tan \theta = -\dfrac{b}{a}$	$\tan \theta' = \dfrac{b}{a}$	$\tan \theta = \dfrac{b}{a}$	$\tan \theta' = \dfrac{b}{a}$	$\tan \theta = -\dfrac{b}{a}$	$\tan \theta' = \dfrac{b}{a}$

This angle θ', called a **reference angle**, can be used to find the trigonometric values of any angle θ.

Key Concept **Reference Angle Rules**

If θ is an angle in standard position, its reference angle θ' is the acute angle formed by the terminal side of θ and the x-axis. The reference angle θ' for any angle θ, $0° < \theta < 360°$ or $0 < \theta < 2\pi$, is defined as follows.

Quadrant I

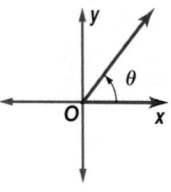

$\theta' = \theta$

Quadrant II

$\theta' = 180° - \theta$
$\theta' = \pi - \theta$

Quadrant III

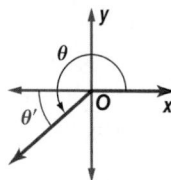

$\theta' = \theta - 180°$
$\theta' = \theta - \pi$

Quadrant IV

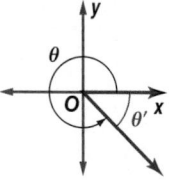

$\theta' = 360° - \theta$
$\theta' = 2\pi - \theta$

To find a reference angle for angles outside the interval $0° < \theta < 360°$ or $0 < \theta < 2\pi$, first find a corresponding coterminal angle in this interval.

EXAMPLE 3 **Find Reference Angles**

Sketch each angle. Then find its reference angle.

a. 300°

The terminal side of 300° lies in Quadrant IV. Therefore, its reference angle is $\theta' = 360° - 300°$ or 60°.

b. $-\dfrac{2\pi}{3}$

A coterminal angle is $2\pi - \dfrac{2\pi}{3}$ or $\dfrac{4\pi}{3}$. The terminal side of $\dfrac{4\pi}{3}$ lies in Quadrant III, so its reference angle is $\dfrac{4\pi}{3} - \pi$ or $\dfrac{\pi}{3}$.

✔ **Guided Practice**

3A. $\dfrac{5\pi}{4}$ **3B.** −240°   **3C.** 390°

Because the trigonometric values of an angle and its reference angle are equal or differ only in sign, you can use the following steps to find the value of a trigonometric function of any angle θ.

Key Concept   **Evaluating Trigonometric Functions of Any Angle**

Step 1   Find the reference angle θ'.

Step 2   Find the value of the trigonometric function for θ'.

Step 3   Using the quadrant in which the terminal side of θ lies, determine the sign of the trigonometric function value of θ.

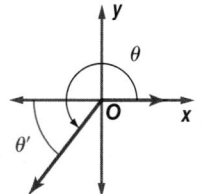

StudyTip

Memorizing Trigonometric Values
To memorize the exact values of sine for 0°, 30°, 45°, 60°, and 90°, consider the following pattern.

$\sin 0° = \dfrac{\sqrt{0}}{2}$, or 0

$\sin 30° = \dfrac{\sqrt{1}}{2}$, or $\dfrac{1}{2}$

$\sin 45° = \dfrac{\sqrt{2}}{2}$

$\sin 60° = \dfrac{\sqrt{3}}{2}$

$\sin 90° = \dfrac{\sqrt{4}}{2}$, or 1

A similar pattern exists for the cosine function, except the values are given in reverse order.

The signs of the trigonometric functions in each quadrant can be determined using the function definitions given on page 242. For example, because $\sin \theta = \dfrac{y}{r}$, it follows that $\sin \theta$ is negative when $y < 0$, which occurs in Quadrants III and IV. Using this same logic, you can verify each of the signs for $\sin \theta$, $\cos \theta$, and $\tan \theta$ shown in the diagram. Notice that these values depend only on x and y because r is always positive.

Quadrant II	Quadrant I
sin θ: +	sin θ: +
cos θ: −	cos θ: +
tan θ: −	tan θ: +
Quadrant III	**Quadrant IV**
sin θ: −	sin θ: −
cos θ: −	cos θ: +
tan θ: +	tan θ: −

Because you know the exact trigonometric values of 30°, 45°, and 60° angles, you can find the exact trigonometric values of *all* angles for which these angles are reference angles. The table lists these values for θ in both degrees and radians.

θ	30° or $\dfrac{\pi}{6}$	45° or $\dfrac{\pi}{4}$	60° or $\dfrac{\pi}{3}$
$\sin \theta$	$\dfrac{1}{2}$	$\dfrac{\sqrt{2}}{2}$	$\dfrac{\sqrt{3}}{2}$
$\cos \theta$	$\dfrac{\sqrt{3}}{2}$	$\dfrac{\sqrt{2}}{2}$	$\dfrac{1}{2}$
$\tan \theta$	$\dfrac{\sqrt{3}}{3}$	1	$\sqrt{3}$

EXAMPLE 4   **Use Reference Angles to Find Trigonometric Values**

Find the exact value of each expression.

a. $\cos 120°$

Because the terminal side of θ lies in Quadrant II, the reference angle θ' is $180° - 120°$ or $60°$.

$\cos 120° = -\cos 60°$ **In Quadrant II, cos θ is negative.**

$\qquad\quad = -\dfrac{1}{2}$ $\cos 60° = \dfrac{1}{2}$

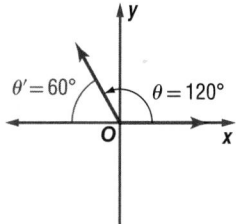

b. $\tan \dfrac{7\pi}{6}$

Because the terminal side of θ lies in Quadrant III, the reference angle θ' is $\dfrac{7\pi}{6} - \pi$ or $\dfrac{\pi}{6}$.

$\tan \dfrac{7\pi}{6} = \tan \dfrac{\pi}{6}$ **In Quadrant III, tan θ is positive.**

$\qquad\quad = \dfrac{\sqrt{3}}{3}$ $\tan \dfrac{\pi}{6} = \dfrac{\sqrt{3}}{3}$

c. $\csc \dfrac{15\pi}{4}$

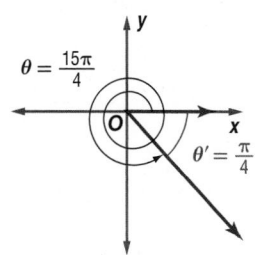

A coterminal angle of θ is $\dfrac{15\pi}{4} - 2\pi$ or $\dfrac{7\pi}{4}$, which lies in Quadrant IV. So, the reference angle θ' is $2\pi - \dfrac{7\pi}{4}$ or $\dfrac{\pi}{4}$. Because sine and cosecant are reciprocal functions and $\sin \theta$ is negative in Quadrant IV, it follows that $\csc \theta$ is also negative in Quadrant IV.

$\csc \dfrac{15\pi}{4} = -\csc \dfrac{\pi}{4}$ **In Quadrant IV, csc θ is negative.**

$= -\dfrac{1}{\sin \dfrac{\pi}{4}}$ $\csc \theta = \dfrac{1}{\sin \theta}$

$= -\dfrac{1}{\dfrac{\sqrt{2}}{2}}$ or $-\sqrt{2}$ $\sin \dfrac{\pi}{4} = \dfrac{\sqrt{2}}{2}$

CHECK You can check your answer by using a graphing calculator.

$\csc \dfrac{15\pi}{4} \approx -1.414$ ✔

$-\sqrt{2} \approx -1.414$ ✔

✔ Guided Practice

Find the exact value of each expression.

4A. $\tan \dfrac{5\pi}{3}$ **4B.** $\sin \dfrac{5\pi}{6}$ **4C.** $\sec(-135°)$

If the value of one or more of the trigonometric functions and the quadrant in which the terminal side of θ lies is known, the remaining function values can be found.

EXAMPLE 5   Use One Trigonometric Value to Find Others

Let $\tan \theta = \dfrac{5}{12}$, where $\sin \theta < 0$. Find the exact values of the five remaining trigonometric functions of θ.

To find the other function values, you must find the coordinates of a point on the terminal side of θ. You know that $\tan \theta$ is positive and $\sin \theta$ is negative, so θ must lie in Quadrant III. This means that both x and y are negative.

Because $\tan \theta = \dfrac{y}{x}$ or $\dfrac{5}{12}$, use the point $(-12, -5)$ to find r.

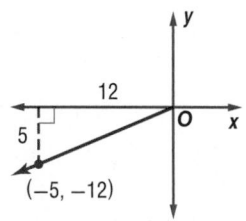

$r = \sqrt{x^2 + y^2}$ **Pythagorean Theorem**

$= \sqrt{(-12)^2 + (-5)^2}$ $x = -12$ **and** $y = -5$

$= \sqrt{169}$ or 13 **Take the positive square root.**

Use $x = -12$, $y = -5$, and $r = 13$ to write the five remaining trigonometric ratios.

$\sin \theta = \dfrac{y}{r}$ or $-\dfrac{5}{13}$ $\cos \theta = \dfrac{x}{r}$ or $-\dfrac{12}{13}$ $\cot \theta = \dfrac{x}{y}$ or $\dfrac{12}{5}$

$\csc \theta = \dfrac{r}{y}$ or $-\dfrac{13}{5}$ $\sec \theta = \dfrac{r}{x}$ or $-\dfrac{13}{12}$

Watch Out!

Rationalizing the Denominator Be sure to rationalize the denominator, if necessary.

✔ Guided Practice

Find the exact values of the five remaining trigonometric functions of θ.

5A. $\sec \theta = \sqrt{3}$, $\tan \theta < 0$ **5B.** $\sin \theta = \dfrac{5}{7}$, $\cot \theta > 0$

🌐 Real-World EXAMPLE 6   Find Coordinates Given a Radius and an Angle

ROBOTICS As part of the range of motion category in a high school robotics competition, a student programmed a 20-centimeter long robotic arm to pick up an object at point C and rotate through an angle of exactly 225° in order to release it into a container at point D. Find the position of the object at point D, relative to the pivot point O.

With the pivot point at the origin and the angle through which the arm rotates in standard position, point C has coordinates $(20, 0)$. The reference angle θ' for 225° is $225° - 180°$ or 45°.

Let the position of point D have coordinates (x, y). The definitions of sine and cosine can then be used to find the values of x and y. The value of r, 20 centimeters, is the length of the robotic arm. Since D is in Quadrant III, the sine and cosine of 225° are negative.

$\cos \theta = \dfrac{x}{r}$	**Cosine ratio**	$\sin \theta = \dfrac{y}{r}$	**Sine ratio**
$\cos 225° = \dfrac{x}{20}$	$\theta = 225°$ **and** $r = 20$	$\sin 225° = \dfrac{y}{20}$	$\theta = 225°$ **and** $r = 20$
$-\cos 45° = \dfrac{x}{20}$	$\cos 225° = -\cos 45°$	$-\sin 45° = \dfrac{y}{20}$	$\sin 225° = -\sin 45°$
$-\dfrac{\sqrt{2}}{2} = \dfrac{x}{20}$	$\cos 45° = \dfrac{\sqrt{2}}{2}$	$-\dfrac{\sqrt{2}}{2} = \dfrac{y}{20}$	$\sin 45° = \dfrac{\sqrt{2}}{2}$
$-10\sqrt{2} = x$	**Solve for x.**	$-10\sqrt{2} = y$	**Solve for y.**

The exact coordinates of D are $\left(-10\sqrt{2}, -10\sqrt{2}\right)$. Since $10\sqrt{2}$ is about 14.14, the object is about 14.14 centimeters to the left of the pivot point and about 14.14 centimeters below the pivot point.

✓ Guided Practice

6. CLOCKWORK A 3-inch-long minute hand on a clock shows a time of 45 minutes past the hour. What is the new position of the end of the minute hand relative to the pivot point at 10 minutes past the next hour?

Trigonometric Functions on the Unit Circle A **unit circle** is a circle of radius 1 centered at the origin. Notice that on a unit circle, the radian measure of a central angle $\theta = \dfrac{s}{1}$ or s, so the arc length intercepted by θ corresponds exactly to the angle's radian measure. This provides a way of mapping a real number input value for a trigonometric function to a real number output value.

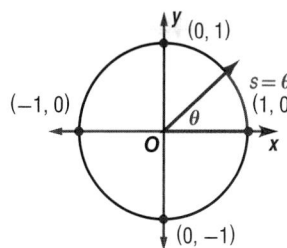

Consider the real number line placed vertically tangent to the unit circle at $(1, 0)$ as shown below. If this line were wrapped about the circle in both the positive (counterclockwise) and negative (clockwise) direction, each point t on the line would map to a unique point $P(x, y)$ on the circle. Because $r = 1$, we can define the trigonometric ratios of angle t in terms of just x and y.

Positive Values of _t_

Negative Values of _t_

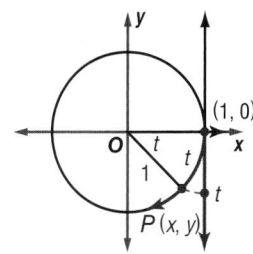

Key Concept | Trigonometric Functions on the Unit Circle

Let t be any real number on a number line and let $P(x, y)$ be the point on t when the number line is wrapped onto the unit circle. Then the trigonometric functions of t are as follows.

$\sin t = y$ $\qquad$ $\cos t = x$ $\qquad$ $\tan t = \dfrac{y}{x}, x \neq 0$

$\csc t = \dfrac{1}{y}, y \neq 0$ $\qquad$ $\sec t = \dfrac{1}{x}, x \neq 0$ $\qquad$ $\cot t = \dfrac{x}{y}, y \neq 0$

Therefore, the coordinates of P corresponding to the angle t can be written as $P(\cos t, \sin t)$.

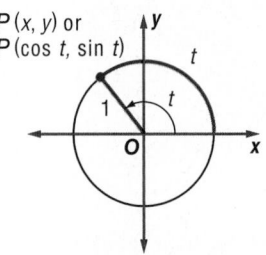

Notice that the input value in each of the definitions above can be thought of as an angle measure or as a real number t. When defined as functions of the real number system using the unit circle, the trigonometric functions are often called **circular functions**.

Using reference angles or quadrantal angles, you should now be able to find the trigonometric function values for all integer multiples of 30°, or $\dfrac{\pi}{6}$ radians, and 45°, or $\dfrac{\pi}{4}$ radians. These special values wrap to 16 special points on the unit circle, as shown below.

StudyTip

16-Point Unit Circle You have already memorized these values in the first quadrant. The remaining values can be determined using the x-axis, y-axis, and origin symmetry of the unit circle along with the signs of x and y in each quadrant.

16-Point Unit Circle

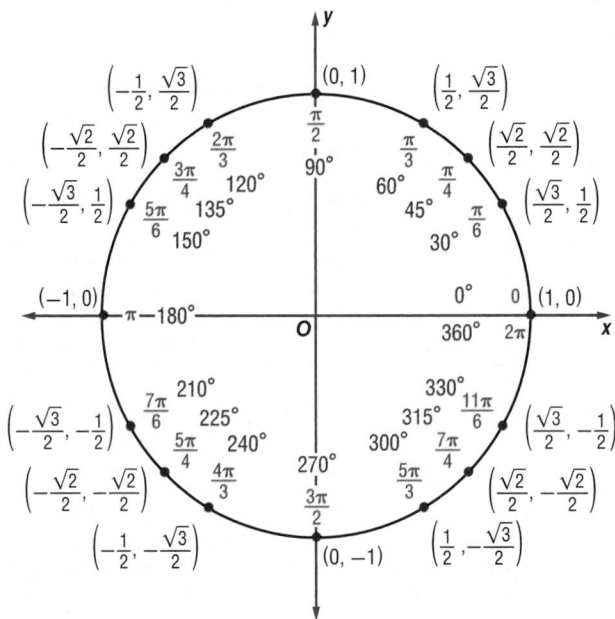

Using the (x, y) coordinates in the 16-point unit circle and the definitions in the Key Concept Box at the top of the page, you can find the values of the trigonometric functions for common angle measures. It is helpful to memorize these exact function values so you can quickly perform calculations involving them.

EXAMPLE 7 | Find Trignometric Values Using the Unit Circle

Find the exact value of each expression. If undefined, write *undefined*.

a. $\sin \dfrac{\pi}{3}$

$\dfrac{\pi}{3}$ corresponds to the point $(x, y) = \left(\dfrac{1}{2}, \dfrac{\sqrt{3}}{2} \right)$ on the unit circle.

$\sin t = y$ $\qquad$ **Definition of sin t**

$\sin \dfrac{\pi}{3} = \dfrac{\sqrt{3}}{2}$ $\qquad$ $y = \dfrac{\sqrt{3}}{2}$ when $t = \dfrac{\pi}{3}$.

b. cos 135°

135° corresponds to the point $(x, y) = \left(-\frac{\sqrt{2}}{2}, \frac{\sqrt{2}}{2}\right)$ on the unit circle.

$\cos t = x$ **Definition of cos t**

$\cos 135° = -\frac{\sqrt{2}}{2}$ $x = -\frac{\sqrt{2}}{2}$ **when t = 135°.**

c. tan 270°

270° corresponds to the point $(x, y) = (0, -1)$ on the unit circle.

$\tan t = \frac{y}{x}$ **Definition of tan t**

$\tan 270° = \frac{-1}{0}$ **x = 0 and y = −1, when t = 270°.**

Therefore, tan 270° is undefined.

d. csc $\frac{11\pi}{6}$

$\frac{11\pi}{6}$ corresponds to the point $(x, y) = \left(\frac{\sqrt{3}}{2}, -\frac{1}{2}\right)$ on the unit circle.

$\csc t = \frac{1}{y}$ **Definition of csc t**

$\csc \frac{11\pi}{6} = \frac{1}{-\frac{1}{2}}$ $y = -\frac{1}{2}$ **when t = $\frac{11\pi}{6}$.**

$= -2$ **Simplify.**

✓ Guided Practice

7A. $\cos \frac{\pi}{4}$ **7B.** $\sin 120°$ **7C.** $\cot 210°$ **7D.** $\sec \frac{7\pi}{4}$

As defined by wrapping the number line around the unit circle, the domain of both the sine and cosine functions is the set of all real numbers $(-\infty, \infty)$. Extending infinitely in either direction, the number line can be wrapped multiple times around the unit circle, mapping more than one t-value to the same point $P(x, y)$ with each wrapping, positive or negative.

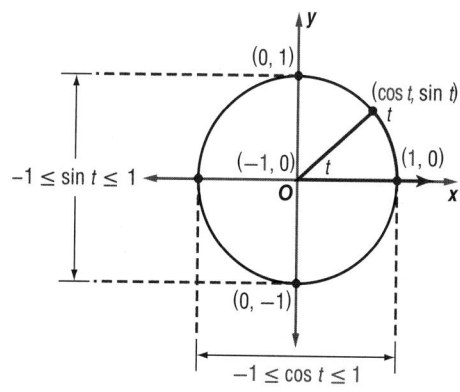

Because $\cos t = x$, $\sin t = y$, and one wrapping corresponds to a distance of 2π,

$$\cos (t + 2n\pi) = \cos t \quad \text{and} \quad \sin (t + 2n\pi) = \sin t,$$

for any integer n and real number t.

StudyTip

Periodic Functions
The other three circular functions are also periodic. The periods of these functions will be discussed in Lesson 4-5.

The values for the sine and cosine function therefore lie in the interval $[-1, 1]$ and repeat for every integer multiple of 2π on the number line.

Functions with values that repeat at regular intervals are called **periodic functions**.

Key Concept   **Periodic Functions**

A function $y = f(t)$ is periodic if there exists a positive real number c such that $f(t + c) = f(t)$ for all values of t in the domain of f.

The smallest number c for which f is periodic is called the **period** of f.

The sine and cosine functions are periodic, repeating values after 2π, so these functions have a period of 2π. It can be shown that the values of the tangent function repeat after a distance of π on the number line, so the tangent function has a period of π and

$$\tan t = \tan (t + n\pi),$$

for any integer n and real number t, unless both $\tan t$ and $\tan (t + n\pi)$ are undefined. You can use the periodic nature of the sine, cosine, and tangent functions to evaluate these functions.

EXAMPLE 8   **Use the Periodic Nature of Circular Functions**

Find the exact value of each expression.

a. $\cos \dfrac{11\pi}{4}$

$\cos \dfrac{11\pi}{4} = \cos \left(\dfrac{3\pi}{4} + 2\pi\right)$ Rewrite $\dfrac{11\pi}{4}$ as the sum of a number and 2π.

$\qquad = \cos \dfrac{3\pi}{4}$ $\dfrac{3\pi}{4}$ and $\dfrac{3\pi}{4} + 2\pi$ map to the same point $(x, y) = \left(-\dfrac{\sqrt{2}}{2}, \dfrac{\sqrt{2}}{2}\right)$ on the unit circle.

$\qquad = -\dfrac{\sqrt{2}}{2}$ $\cos t = x$ and $x = -\dfrac{\sqrt{2}}{2}$ when $t = \dfrac{3\pi}{4}$.

b. $\sin \left(-\dfrac{2\pi}{3}\right)$

$\sin \left(-\dfrac{2\pi}{3}\right) = \sin \left(\dfrac{4\pi}{3} + 2(-1)\pi\right)$ Rewrite $-\dfrac{2\pi}{3}$ as the sum of a number and an integer multiple of 2π.

$\qquad = \sin \dfrac{4\pi}{3}$ $\dfrac{4\pi}{3}$ and $\dfrac{4\pi}{3} - 2(-1)\pi$ map to the same point $(x, y) = \left(-\dfrac{1}{2}, -\dfrac{\sqrt{3}}{2}\right)$ on the unit circle.

$\qquad = -\dfrac{\sqrt{3}}{2}$ $\sin t = y$ and $y = -\dfrac{\sqrt{3}}{2}$ when $t = \dfrac{4\pi}{3}$.

c. $\tan \dfrac{19\pi}{6}$

$\tan \dfrac{19\pi}{6} = \tan \left(\dfrac{\pi}{6} + 3\pi\right)$ Rewrite $\dfrac{19\pi}{6}$ as the sum of a number and an integer multiple of π.

$\qquad = \tan \dfrac{\pi}{6}$ $\dfrac{\pi}{6}$ and $\dfrac{\pi}{6} + 3\pi$ map to points on the unit circle with the same tangent values.

$\qquad = \dfrac{\frac{1}{2}}{\frac{\sqrt{3}}{2}}$ or $\dfrac{\sqrt{3}}{3}$ $\tan t = \dfrac{y}{x}$; $x = \dfrac{\sqrt{3}}{2}$ and $y = \dfrac{1}{2}$ when $t = \dfrac{\pi}{6}$.

✓ **Guided Practice**

8A. $\sin \dfrac{13\pi}{4}$ **8B.** $\cos \left(-\dfrac{4\pi}{3}\right)$ **8C.** $\tan \dfrac{15\pi}{6}$

Recall from Lesson 1-2 that a function f is *even* if for every x in the domain of f, $f(-x) = f(x)$ and *odd* if for every x in the domain of f, $f(-x) = -f(x)$. You can use the unit circle to verify that the cosine function is even and that the sine and tangent functions are odd. That is,

$$\cos (-t) = \cos t \qquad \sin (-t) = -\sin t \qquad \tan (-t) = -\tan t.$$

Exercises

The given point lies on the terminal side of an angle θ in standard position. Find the values of the six trigonometric functions of θ. (Example 1)

1. $(3, 4)$ **2.** $(-6, 6)$

3. $(-4, -3)$ **4.** $(2, 0)$

5. $(1, -8)$ **6.** $(5, -3)$

7. $(-8, 15)$ **8.** $(-1, -2)$

Find the exact value of each trigonometric function, if defined. If not defined, write _undefined_. (Example 2)

9. $\sin \dfrac{\pi}{2}$ **10.** $\tan 2\pi$

11. $\cot (-180°)$ **12.** $\csc 270°$

13. $\cos (-270°)$ **14.** $\sec 180°$

15. $\tan \pi$ **16.** $\sec \left(-\dfrac{\pi}{2}\right)$

Sketch each angle. Then find its reference angle. (Example 3)

17. $135°$ **18.** $210°$

19. $\dfrac{7\pi}{12}$ **20.** $\dfrac{11\pi}{3}$

21. $-405°$ **22.** $-75°$

23. $\dfrac{5\pi}{6}$ **24.** $\dfrac{13\pi}{6}$

Find the exact value of each expression. (Example 4)

25. $\cos \dfrac{4\pi}{3}$ **26.** $\tan \dfrac{7\pi}{6}$

27. $\sin \dfrac{3\pi}{4}$ **28.** $\cot (-45°)$

29. $\csc 390°$ **30.** $\sec (-150°)$

31. $\tan \dfrac{11\pi}{6}$ **32.** $\sin 300°$

Find the exact values of the five remaining trigonometric functions of θ. (Example 5)

33. $\tan \theta = 2$, where $\sin \theta > 0$ and $\cos \theta > 0$

34. $\csc \theta = 2$, where $\sin \theta > 0$ and $\cos \theta < 0$

35. $\sin \theta = -\dfrac{1}{5}$, where $\cos \theta > 0$

36. $\cos \theta = -\dfrac{12}{13}$, where $\sin \theta < 0$

37. $\sec \theta = \sqrt{3}$, where $\sin \theta < 0$ and $\cos \theta > 0$

38. $\cot \theta = 1$, where $\sin \theta < 0$ and $\cos \theta < 0$

39. $\tan \theta = -1$, where $\sin \theta < 0$

40. $\cos \theta = -\dfrac{1}{2}$, where $\sin \theta > 0$

41 **CAROUSEL** Zoe is on a carousel at the carnival. The diameter of the carousel is 80 feet. Find the position of her seat from the center of the carousel after a rotation of 210°. (Example 6)

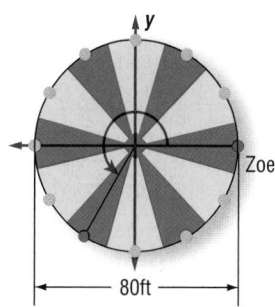

42. **COIN FUNNEL** A coin is dropped into a funnel where it spins in smaller circles until it drops into the bottom of the bank. The diameter of the first circle the coin makes is 24 centimeters. Before completing one full circle, the coin travels 150° and falls over. What is the new position of the coin relative to the center of the funnel? (Example 6)

Find the exact value of each expression. If undefined, write _undefined_. (Examples 7 and 8)

43. $\sec 120°$ **44.** $\sin 315°$

45. $\cos \dfrac{11\pi}{3}$ **46.** $\tan \left(-\dfrac{5\pi}{4}\right)$

47. $\csc 390°$ **48.** $\cot 510°$

49. $\csc 5400°$ **50.** $\sec \dfrac{3\pi}{2}$

51. $\cot \left(-\dfrac{5\pi}{6}\right)$ **52.** $\csc \dfrac{17\pi}{6}$

53. $\tan \dfrac{5\pi}{3}$ **54.** $\sec \dfrac{7\pi}{6}$

55. $\sin \left(-\dfrac{5\pi}{3}\right)$ **56.** $\cos \dfrac{7\pi}{4}$

57. $\tan \dfrac{14\pi}{3}$ **58.** $\cos \left(-\dfrac{19\pi}{6}\right)$

59. **RIDES** Jae and Anya are on a ride at an amusement park. After the first several swings, the angle the ride makes with the vertical is modeled by $\theta = 22 \cos \pi t$, with θ measured in radians and t measured in seconds. Determine the measure of the angle in radians for $t = 0$, 0.5, 1, 1.5, 2, and 2.5. (Example 8)

Complete each trigonometric expression.

60. $\cos 60° = \sin$ ____

61. $\tan \frac{\pi}{4} = \sin$ ____

62. $\sin \frac{2\pi}{3} = \cos$ ____

63. $\cos \frac{7\pi}{6} = \sin$ ____

64. $\sin(-45°) = \cos$ ____

65. $\cos \frac{5\pi}{3} = \sin$ ____

66. ICE CREAM The monthly sales in thousands of dollars for Fiona's Fine Ice Cream shop can be modeled by $y = 71.3 + 59.6 \sin \frac{\pi(t-4)}{6}$, where $t = 1$ represents January, $t = 2$ represents February, and so on.

   a. Estimate the sales for January, March, July, and October.

   b. Describe why the ice cream shop's sales can be represented by a trigonometric function.

Use the given values to evaluate the trigonometric functions.

67. $\cos(-\theta) = \frac{8}{11}$; $\cos \theta = ?$; $\sec \theta = ?$

68. $\sin(-\theta) = \frac{5}{9}$; $\sin \theta = ?$; $\csc \theta = ?$

69. $\sec \theta = \frac{13}{12}$; $\cos \theta = ?$; $\cos(-\theta) = ?$

70. $\csc \theta = \frac{19}{17}$; $\sin \theta = ?$; $\sin(-\theta) = ?$

71. GRAPHS Suppose the terminal side of an angle θ in standard position coincides with the graph of $y = 2x$ in Quadrant III. Find the six trigonometric functions of θ.

Find the coordinates of P for each circle with the given radius and angle measure.

72.

73.

74.

75.

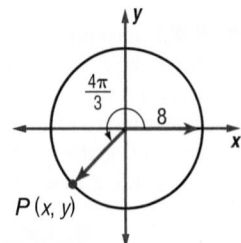

76. COMPARISON Suppose the terminal side of an angle θ_1 in standard position contains the point $(7, -8)$, and the terminal side of a second angle θ_2 in standard position contains the point $(-7, 8)$. Compare the sines of θ_1 and θ_2.

77. TIDES The depth y in meters of the tide on a beach varies as a sine function of x, the hour of the day. On a certain day, that function was $y = 3 \sin \left[\frac{\pi}{6}(x - 4)\right] + 8$, where $x = 0, 1, 2, \ldots, 24$ corresponds to 12:00 midnight, 1:00 A.M., 2:00 A.M., ..., 12:00 midnight the next night.

   a. What is the maximum depth, or high tide, that day?

   b. At what time(s) does the high tide occur?

78. **MULTIPLE REPRESENTATIONS** In this problem, you will investigate the period of the sine function.

   a. TABULAR Copy and complete a table similar to the one below that includes all 16 angle measures from the unit circle.

θ	0	$\frac{\pi}{6}$	$\frac{\pi}{4}$	$\frac{\pi}{3}$	...	2π
$\sin \theta$						
$\sin 2\theta$						
$\sin 4\theta$						

   b. VERBAL After what values of θ do $\sin \theta$, $\sin 2\theta$, and $\sin 4\theta$, repeat their range values? In other words, what are the periods of these functions?

   c. VERBAL Make a conjecture as to how the period of $y = \sin n\theta$ is affected for different values of n.

H.O.T. Problems   Use **H**igher-**O**rder **T**hinking Skills

79 **CHALLENGE** For each statement, describe n.

   a. $\cos \left(n \cdot \frac{\pi}{2}\right) = 0$

   b. $\csc \left(n \cdot \frac{\pi}{2}\right)$ is undefined.

REASONING Determine whether each statement is *true* or *false*. Explain your reasoning.

80. If $\cos \theta = 0.8$, $\sec \theta - \cos(-\theta) = 0.45$.

81. Since $\tan(-t) = -\tan t$, the tangent of a negative angle is a negative number.

82. WRITING IN MATH Explain why the attendance at a year-round theme park could be modeled by a periodic function. What issues or events could occur over time to alter this periodic depiction?

REASONING Use the unit circle to verify each relationship.

83. $\sin(-t) = -\sin t$

84. $\cos(-t) = \cos t$

85. $\tan(-t) = -\tan t$

86. WRITING IN MATH Make a conjecture as to the periods of the secant, cosecant, and cotangent functions. Explain your reasoning.

Write each decimal degree measure in DMS form and each DMS measure in decimal degree form to the nearest thousandth. (Lesson 4-2)

87. $168.35°$

88. $27.465°$

89. $14° 5' 20''$

90. $173° 24' 35''$

91. EXERCISE A preprogrammed workout on a treadmill consists of intervals walking at various rates and angles of incline. A 1% incline means 1 unit of vertical rise for every 100 units of horizontal run. (Lesson 4-1)

a. At what angle, with respect to the horizontal, is the treadmill bed when set at a 10% incline? Round to the nearest degree.

b. If the treadmill bed is 40 inches long, what is the vertical rise when set at an 8% incline?

Evaluate each logarithm. (Lesson 3-3)

92. $\log_8 64$

93. $\log_{125} 5$

94. $\log_2 32$

95. $\log_4 128$

List all possible rational zeros of each function. Then determine which, if any, are zeros. (Lesson 2-4)

96. $f(x) = x^3 - 4x^2 + x + 2$

97. $g(x) = x^3 + 6x^2 + 10x + 3$

98. $h(x) = x^4 - x^2 + x - 1$

99. $h(x) = 2x^3 + 3x^2 - 8x + 3$

100. $f(x) = 2x^4 + 3x^3 - 6x^2 - 11x - 3$

101. $g(x) = 4x^3 + x^2 + 8x + 2$

102. NAVIGATION A global positioning system (GPS) uses satellites to allow a user to determine his or her position on Earth. The system depends on satellite signals that are reflected to and from a hand-held transmitter. The time that the signal takes to reflect is used to determine the transmitter's position. Radio waves travel through air at a speed of 299,792,458 meters per second. Thus, $d(t) = 299{,}792{,}458t$ relates the time t in seconds to the distance traveled $d(t)$ in meters. (Lesson 1-1)

a. Find the distance a radio wave will travel in 0.05, 0.2, 1.4, and 5.9 seconds.

b. If a signal from a GPS satellite is received at a transmitter in 0.08 second, how far from the transmitter is the satellite?

103. SAT/ACT In the figure, $\overline{AB}$ and $\overline{AD}$ are tangents to circle C. What is the value of m?

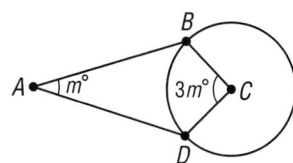

Note: Figure not drawn to scale.

104. Suppose θ is an angle in standard position with $\sin \theta > 0$. In which quadrant(s) could the terminal side of θ lie?

A I only

B I and II

C I and III

D I and IV

105. REVIEW Find the angular speed in radians per second of a point on a bicycle tire if it completes 2 revolutions in 3 seconds.

F $\dfrac{\pi}{3}$

G $\dfrac{\pi}{2}$

H $\dfrac{2\pi}{3}$

J $\dfrac{4\pi}{3}$

106. REVIEW Which angle has a tangent and cosine that are both negative?

A $110°$

B $180°$

C $210°$

D $340°$

Objective
Use a graphing calculator and parametric equations to graph the sine function and its inverse.

As functions of the real number system, you can graph trigonometric functions on the coordinate plane and apply the same graphical analysis that you did to functions in Chapter 1. As was done in Extend 1-7, parametric equations will be used to graph the sine function.

ACTIVITY 1 | Parametric Graph of $y = \sin x$

Graph $x = t$, $y = \sin t$.

Step 1 Set the mode. In the MODE menu, select RADIAN, PAR, and SIMUL. This allows the equations to be graphed simultaneously. Next, enter the parametric equations. In parametric form, X,T,θ,n will use t instead of x.

Step 2 Set the x- and t-values to range from 0 to 2π. Set Tstep and x-scale to $\frac{\pi}{12}$. Set y to $[-1, 1]$ scl: 0.1. The calculator automatically converts to decimal form.

```
WINDOW
 Tmin=0
 Tmax=6.2831853…
 Tstep=.2617993…
 Xmin=0
 Xmax=6.2831853…
 Xscl=.26179938…
↓Ymin=⁻1
```

Step 3 Graph the equations. Trace the function to identify points along the graph. Select Trace and use the right arrow to move along the curve. Record the corresponding x- and y-values.

$[0, 2\pi]$ scl: $\frac{\pi}{12}$ by $[-1, 1]$ scl: 0.1
t: $[0, 2\pi]$; tstep $\frac{\pi}{12}$

Step 4 The table shows angle measures from 0° to 180°, or 0 to π, and the corresponding values for sin t on the unit circle. The figures below illustrate the relationship between the graph and the unit circle.

Degrees	0	30	45	60	90	120	135	150	180
Radians	0	0.52	0.79	1.05	1.571	2.094	2.356	2.618	3.14
$y = \sin t$	0	0.5	0.707	0.866	1	0.866	0.707	0.5	0

StudyTip

Decimal Equivalents
Below are the decimal equivalents of common trigonometric values.

$\dfrac{\sqrt{3}}{2} \approx 0.866$

$\dfrac{\sqrt{2}}{2} \approx 0.707$

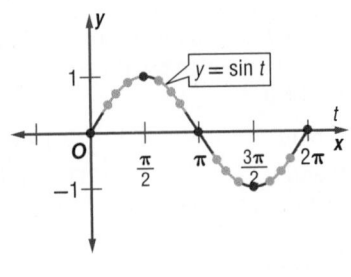

Exercises

Graph each function on [0, 2π].

1. $x = t, y = \cos t$

2. $x = t, y = \sin 2t$

3. $x = t, y = 3 \cos t$

4. $x = t, y = 4 \sin t$

5. $x = t, y = \cos(t + \pi)$

6. $x = t, y = 2 \sin\left(t - \dfrac{\pi}{4}\right)$

By definition, $\sin t$ is the y-coordinate of the point $P(x, y)$ on the unit circle to which the real number t on the number line gets wrapped. As shown in the diagram on the previous page, the graph of $y = \sin t$ follows the y-coordinate of the point determined by t as it moves counterclockwise around the unit circle.

The graph of the sine function is called a *sine curve*. From Lesson 4-3, you know that the sine function is periodic with a period of 2π. That is, the sine curve graphed from 0 to 2π would repeat every distance of 2π in either direction, positive or negative. Parametric equations can be used to graph the inverse of the sine function.

ACTIVITY 2 Graph an Inverse

Graph $x + t$, $y + \sin t$ and its inverse. Then determine a domain for which $y = \sin t$ is one-to-one.

Step 1 Inverses are found by switching x and y. Enter the given equations as X1T and Y1T. To graph the inverse, set X2T = Y1T and Y2T = X1T. These are found in the ⌊ VARS ⌋ menu. Select Y-VARS, parametric, Y1T. Repeat for X1T.

```
Plot1 Plot2 Plot3
\X1T=T
 Y1T=sin(T)
\X2T=Y1T
 Y2T=X1T
\X3T=
 Y3T=
\X4T=
```

Step 2 Graph the equations. Adjust the window so that both of the graphs can be seen, as shown. You may need to set the tstep to a smaller value in order to get a smooth curve.

$[-3\pi, 3\pi]$ scl: $\dfrac{\pi}{12}$ by $[-10, 10]$ scl: 2

t: $[-3\pi, 3\pi]$; $t$step $\dfrac{\pi}{12}$

Step 3 Because the sine curve is periodic, there are an infinite number of domains for which the curve will pass the horizontal line test and be one-to-one. One such domain is $\left[\dfrac{\pi}{2}, \dfrac{3\pi}{2}\right]$.

> **StudyTip**
>
> **Tstep** If your graph appears to be pointed, you can change the tstep to a smaller value in order to get a smoother curve.

Exercises

Graph each function and its inverse. Then determine a domain for which each function is one-to-one.

7. $x = t, y = \cos 2t$

8. $x = t, y = -\sin t$

9. $x = t, y = 2 \cos t$

10. $x = t + \dfrac{\pi}{4}, y = \sin t$

11. $x = t, y = 2 \cos(t - \pi)$

12. $x = t - \dfrac{\pi}{6}, y = \sin t$

4-4 Graphing Sine and Cosine Functions

Then
You analyzed graphs of functions. (Lesson 1-5)

Now
- Graph transformations of the sine and cosine functions.
- Use sinusoidal functions to solve problems.

New Vocabulary
sinusoid
amplitude
frequency
phase shift
vertical shift
midline

Math Online

glencoe.com

Why?
As you ride a Ferris wheel, the height that you are above the ground varies periodically as a function of time. You can model this behavior using a *sinusoidal function*.

Transformations of Sine and Cosine Functions As shown in Explore 4-4, the graph $y = \sin t$ follows the y-coordinate of the point determined by t as it moves around the unit circle. Similarly, the graph of $y = \cos t$ follows the x-coordinate of this point. The graphs of these functions are periodic, repeating after a period of 2π. The properties of the sine and cosine functions are summarized below.

Key Concept | **Properties of the Sine and Cosine Functions**

Sine Function

Domain: $(-\infty, \infty)$ **Range:** $[-1, 1]$

y-intercept: 0

x-intercepts: $n\pi, n \in \mathbb{Z}$

Continuity: continuous on $(-\infty, \infty)$

Symmetry: origin (odd function)

Extrema: maximum of 1 at
$x = \dfrac{\pi}{2} + 2n\pi, n \in \mathbb{Z}$
minimum of -1 at
$x = \dfrac{3\pi}{2} + 2n\pi, n \in \mathbb{Z}$

End Behavior: $\lim\limits_{x \to -\infty} \sin x$ and $\lim\limits_{x \to \infty} \sin x$ do not exist.

Oscillation: between -1 and 1

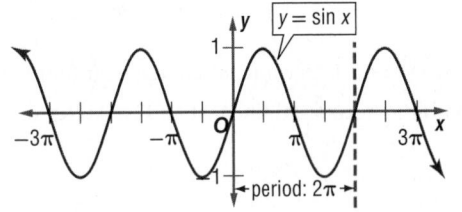

Cosine Function

Domain: $(-\infty, \infty)$ **Range:** $[-1, 1]$

y-intercept: 1

x-intercepts: $\dfrac{\pi}{2}n, n \in \mathbb{Z}$

Continuity: continuous on $(-\infty, \infty)$

Symmetry: y-axis (even function)

Extrema: maximum of 1 at $x = 2n\pi$, $n \in \mathbb{Z}$

minimum of -1 at $x = \pi + 2n\pi$, $n \in \mathbb{Z}$

End Behavior: $\lim\limits_{x \to -\infty} \cos x$ and $\lim\limits_{x \to \infty} \cos x$ do not exist.

Oscillation: between -1 and 1

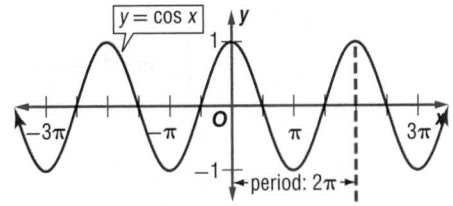

The portion of each graph on $[0, 2\pi]$ represents one period or *cycle* of the function. Notice that the cosine graph is a horizontal translation of the sine graph. Any transformation of a sine function is called a **sinusoid**. The general form of the sinusoidal functions sine and cosine are

$$y = a \sin (bx + c) + d \qquad \text{and} \qquad y = a \cos (bx + c) + d$$

where $a, b, c,$ and d are constants and neither a nor b is 0.

Notice that the constant factor a in $y = a \sin x$ and $y = a \cos x$ expands the graphs of $y = \sin x$ and $y = \cos x$ vertically if $|a| > 1$ and compresses them vertically if $|a| < 1$.

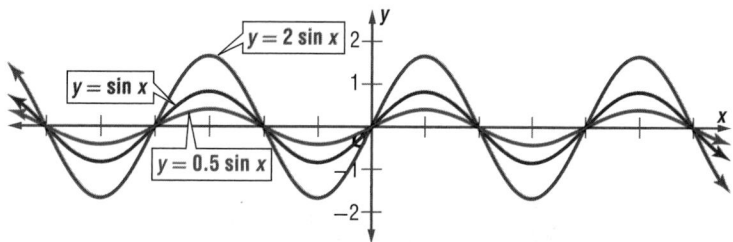

StudyTip

Dilations and x-intercepts
Notice that a dilation of a sinusoidal function does not affect where the curve crosses the x-axis, at its x-intercepts.

Vertical dilations affect the *amplitude* of sinusoidal functions.

Key Concept   **Amplitudes of Sine and Cosine Functions**

Words   The **amplitude** of a sinusoidal function is half the distance between the maximum and minimum values of the function or half the height of the wave.

Model

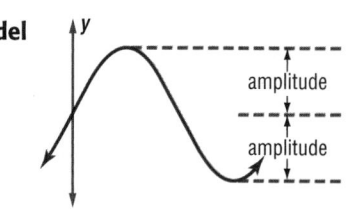

Symbols   For $y = a \sin (bx + c) + d$ and $y = a \cos (bx + c) + d$, amplitude $= |a|$.

To graph a sinusoidal function of the form $y = a \sin x$ or $y = a \cos x$, plot the x-intercepts of the parent sine or cosine function and use the amplitude $|a|$ to plot the new maximum and minimum points. Then sketch the sine wave through these points.

EXAMPLE 1   **Graph Vertical Dilations of Sinusoidal Functions**

Describe how the graphs of $f(x) = \sin x$ and $g(x) = \frac{1}{4} \sin x$ are related. Then find the amplitude of $g(x)$, and sketch two periods of both functions on the same coordinate axes.

The graph of $g(x)$ is the graph of $f(x)$ compressed vertically. The amplitude of $g(x)$ is $\left|\frac{1}{4}\right|$ or $\frac{1}{4}$.

Create a table listing the coordinates of the x-intercepts and extrema for $f(x) = \sin x$ for one period on $[0, 2\pi]$. Then use the amplitude of $g(x)$ to find corresponding points on its graph.

Function	x-intercept	Maximum	x-intercept	Minimum	x-intercept
$f(x) = \sin x$	$(0, 0)$	$\left(\frac{\pi}{2}, 1\right)$	$(\pi, 0)$	$\left(\frac{3\pi}{2}, -1\right)$	$(2\pi, 0)$
$g(x) = \frac{1}{4} \sin x$	$(0, 0)$	$\left(\frac{\pi}{2}, \frac{1}{4}\right)$	$(\pi, 0)$	$\left(\frac{3\pi}{2}, -\frac{1}{4}\right)$	$(2\pi, 0)$

Sketch the curve through the indicated points for each function. Then repeat the pattern suggested by one period of each graph to complete a second period on $[2\pi, 4\pi]$. Extend each curve to the left and right to indicate that the curve continues in both directions.

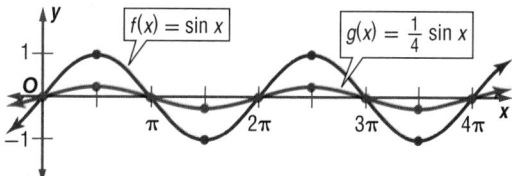

StudyTip

Radians Versus Degrees
You could rescale the x-axis in terms of degrees and produce sinusoidal graphs that look similar to those produced using radian measure. In calculus, however, you will encounter rules that depend on radian measure. So, in this book, we will graph all trigonometric functions in terms of radians.

☑ **Guided Practice**

Describe how the graphs of $f(x)$ and $g(x)$ are related. Then find the amplitude of $g(x)$, and sketch two periods of both functions on the same coordinate axes.

1A. $f(x) = \cos x$
     $g(x) = \frac{1}{3} \cos x$

1B. $f(x) = \sin x$
     $g(x) = 5 \sin x$

1C. $f(x) = \cos x$
     $g(x) = 2 \cos x$

If $a < 0$, the graph of the sinusoidal function is reflected in the x-axis.

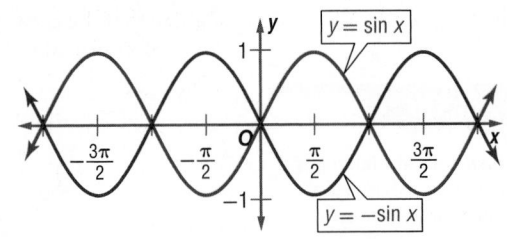

EXAMPLE 2 Graph Reflections of Sinusoidal Functions

Describe how the graphs of $f(x) = \cos x$ and $g(x) = -3 \cos x$ are related. Then find the amplitude of $g(x)$, and sketch two periods of both functions on the same coordinate axes.

The graph of $g(x)$ is the graph of $f(x)$ expanded vertically and then reflected in the x-axis. The amplitude of $g(x)$ is $|-3|$ or 3.

Create a table listing the coordinates of key points of $f(x) = \cos x$ for one period on $[0, 2\pi]$. Use the amplitude of $g(x)$ to find corresponding points on the graph of $y = 3 \cos x$. Then reflect these points in the x-axis to find corresponding points on the graph of $g(x)$.

Function	Extremum	x-intercept	Extremum	x-intercept	Extremum
$f(x) = \cos x$	$(0, 1)$	$\left(\frac{\pi}{2}, 0\right)$	$(\pi, -1)$	$\left(\frac{3\pi}{2}, 0\right)$	$(2\pi, 1)$
$y = 3 \cos x$	$(0, 3)$	$\left(\frac{\pi}{2}, 0\right)$	$(\pi, -3)$	$\left(\frac{3\pi}{2}, 0\right)$	$(2\pi, 3)$
$g(x) = -3 \cos x$	$(0, -3)$	$\left(\frac{\pi}{2}, 0\right)$	$(\pi, 3)$	$\left(\frac{3\pi}{2}, 0\right)$	$(2\pi, -3)$

Sketch the curve through the indicated points for each function. Then repeat the pattern suggested by one period of each graph to complete a second period on $[2\pi, 4\pi]$. Extend each curve to the left and right to indicate that the curve continues in both directions.

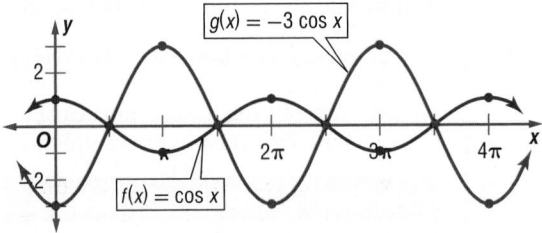

Watch Out!

Amplitude Notice that Example 2 does not state that the amplitude of $g(x) = -3 \cos x$ is -3. Amplitude is a height and is not directional.

✔ Guided Practice

Describe how the graphs of $f(x)$ and $g(x)$ are related. Then find the amplitude of $g(x)$, and sketch two periods of both functions on the same coordinate axes.

2A. $f(x) = \cos x$
$g(x) = -\frac{1}{5} \cos x$

2B. $f(x) = \sin x$
$g(x) = -4 \sin x$

In Lesson 1-5, you learned that if $g(x) = f(bx)$, then $g(x)$ is the graph of $f(x)$ compressed horizontally if $|b| > 1$ and expanded horizontally if $|b| < 1$. Horizontal dilations affect the *period* of a sinusoidal function—the length of one full cycle.

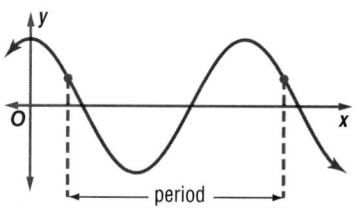

Key Concept — Periods of Sine and Cosine Functions

Words
The period of a sinusoidal function is the distance between any two sets of repeating points on the graph of the function.

Symbols
For $y = a \sin (bx + c) + d$ and $y = a \cos (bx + c) + d$, where $b \neq 0$, period $= \dfrac{2\pi}{|b|}$.

Model

To graph a sinusoidal function of the form $y = \sin bx$ or $y = \cos bx$, find the period of the function and successively add $\dfrac{\text{period}}{4}$ to the left endpoint of an interval with that length. Then use these values as the x-values for the key points on the graph.

EXAMPLE 3 — Graph Horizontal Dilations of Sinusoidal Functions

Describe how the graphs of $f(x) = \cos x$ and $g(x) = \cos \dfrac{x}{3}$ are related. Then find the period of $g(x)$, and sketch at least one period of both functions on the same coordinate axes.

Because $\cos \dfrac{x}{3} = \cos \dfrac{1}{3}x$, the graph of $g(x)$ is the graph of $f(x)$ expanded horizontally. The period of $g(x)$ is $\dfrac{2\pi}{\left|\frac{1}{3}\right|}$ or 6π.

Because the period of $g(x)$ is 6π, to find corresponding points on the graph of $g(x)$, change the x-coordinates of those key points on $f(x)$ so that they range from 0 to 6π, increasing by increments of $\dfrac{6\pi}{4}$ or $\dfrac{3\pi}{2}$.

Function	Maximum	x-intercept	Minimum	x-intercept	Maximum
$f(x) = \cos x$	$(0, 1)$	$\left(\dfrac{\pi}{2}, 0\right)$	$(\pi, -1)$	$\left(\dfrac{3\pi}{2}, 0\right)$	$(2\pi, 1)$
$g(x) = \cos \dfrac{x}{3}$	$(0, 1)$	$\left(\dfrac{3\pi}{2}, 0\right)$	$(3\pi, -1)$	$\left(\dfrac{9\pi}{2}, 0\right)$	$(6\pi, 1)$

Sketch the curve through the indicated points for each function, continuing the patterns to complete one full cycle of each.

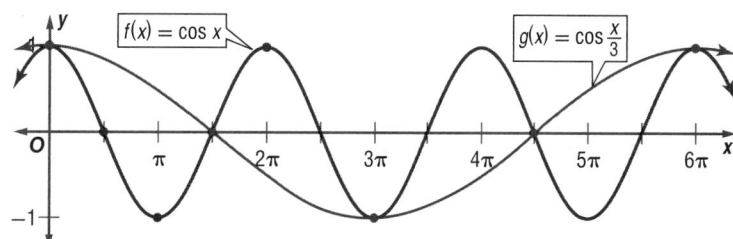

✓ Guided Practice

Describe how the graphs of $f(x)$ and $g(x)$ are related. Then find the period of $g(x)$, and sketch at least one period of each function on the same coordinate axes.

3A. $f(x) = \cos x$
$g(x) = \cos \dfrac{x}{2}$

3B. $f(x) = \sin x$
$g(x) = \sin 3x$

3C. $f(x) = \cos x$
$g(x) = \cos \dfrac{1}{4}x$

Horizontal dilations also affect the *frequency* of sinusoidal functions.

Key Concept   **Frequency of Sine and Cosine Functions**

Words   The **frequency** of a sinusoidal function is the number of cycles the function completes in a one unit interval. The frequency is the reciprocal of the period.

Model

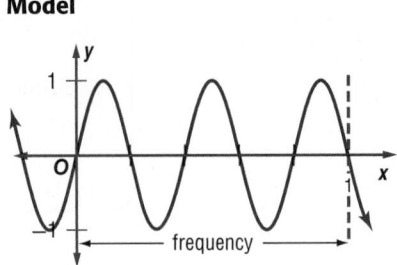

Symbols   For $y = a \sin (bx + c) + d$ and $y = a \cos (bx + c) + d$,

$$\text{frequency} = \frac{1}{\text{period}} \text{ or } \frac{|b|}{2\pi}.$$

Because the frequency of a sinusoidal function is the reciprocal of the period, it follows that the period of the function is the reciprocal of its frequency.

Real-World Link

In physics, frequency is measured in *hertz* or oscillations per second. For example, the number of sound waves passing a point *A* in one second would be the wave's frequency.

Source: *Science World*

Real-World EXAMPLE 4   **Use Frequency to Write a Sinusoidal Function**

MUSIC Musical notes are classified by frequency. In the equal tempered scale, middle C has a frequency of 262 hertz. Use this information and the information at the left to write an equation for a sine function that can be used to model the initial behavior of the sound wave associated with middle C having an amplitude of 0.2.

The general form of the equation will be $y = a \sin bt$, where t is the time in seconds. Because the amplitude is 0.2, $|a| = 0.2$. This means that $a = \pm 0.2$.

The period is the reciprocal of the frequency or $\frac{1}{262}$. Use this value to find b.

$\text{period} = \dfrac{2\pi}{|b|}$ **Period formula**

$\dfrac{1}{262} = \dfrac{2\pi}{|b|}$ $\text{period} = \frac{1}{262}$

$|b| = 2\pi(262) \text{ or } 524\pi$ **Solve for** $|b|$.

$b = \pm 524\pi$ **Solve for** b.

By arbitrarily choosing the positive values of a and b, one sine function that models the initial behavior is $y = 0.2 \sin 524\pi t$.

☑ Guided Practice

4. MUSIC In the same scale, the C above middle C has a frequency of 524 hertz. Write an equation for a sine function that can be used to model the initial behavior of the sound wave associated with this C having an amplitude of 0.1.

A *phase* of a sinusoid is the position of a wave relative to some reference point. A horizontal translation of a sinusoidal function results in a *phase shift*. Recall from Lesson 1-5 that the graph of $y = f(x + c)$ is the graph of $y = f(x)$ translated or *shifted* $|c|$ units left if $c > 0$ and $|c|$ units right if $c < 0$.

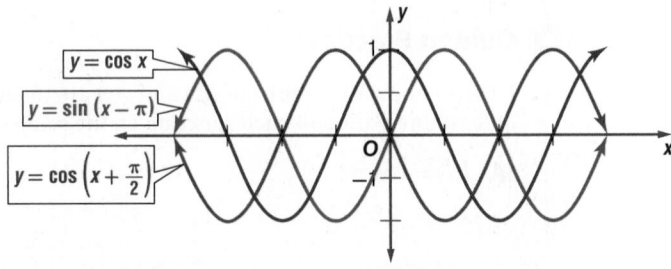

$y = \cos x$

$y = \sin (x - \pi)$

$y = \cos \left(x + \dfrac{\pi}{2}\right)$

Key Concept — Phase Shift of Sine and Cosine Functions

Words   The **phase shift** of a sinusoidal function is the difference between the horizontal position of the function and that of an otherwise similar sinusoidal function.

Model

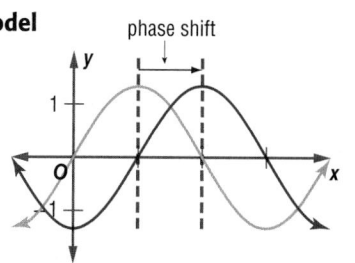

Symbols   For $y = a \sin (bx + c) + d$ and $y = a \cos (bx + c) + d$, where $b \neq 0$,

phase shift $= -\dfrac{c}{|b|}$.

You will verify the formula for phase shift in Exercise 44.

To graph the phase shift of a sinusoidal function of the form $y = a \sin (bx + c) + d$ or $y = a \cos (bx + c) + d$, first determine the endpoints of an interval that corresponds to one cycle of the graph by adding $-\dfrac{c}{b}$ to each endpoint on the interval $[0, 2\pi]$ of the parent function.

EXAMPLE 5   Graph Horizontal Translations of Sinusoidal Functions

State the amplitude, period, frequency, and phase shift of $y = \sin \left(3x - \dfrac{\pi}{2}\right)$. Then graph two periods of the function.

In this function, $a = 1$, $b = 3$, and $c = -\dfrac{\pi}{2}$.

Amplitude: $|a| = |1|$ or 1

Period: $\dfrac{2\pi}{|b|} = \dfrac{2\pi}{|3|}$ or $\dfrac{2\pi}{3}$

Frequency: $\dfrac{|b|}{2\pi} = \dfrac{|3|}{2\pi}$ or $\dfrac{3}{2\pi}$

Phase shift: $-\dfrac{c}{|b|} = -\dfrac{-\dfrac{\pi}{2}}{|3|}$ or $\dfrac{\pi}{6}$

To graph $y = \sin \left(3x - \dfrac{\pi}{2}\right)$, consider the graph of $y = \sin 3x$. The period of this function is $\dfrac{2\pi}{3}$. Create a table listing the coordinates of key points of $y = \sin 3x$ on the interval $\left[0, \dfrac{2\pi}{3}\right]$. To account for a phase shift of $\dfrac{\pi}{6}$, add $\dfrac{\pi}{6}$ to the x-values of each of the key points for the graph of $y = \sin 3x$.

Function	x-intercept	Maximum	x-intercept	Minimum	x-intercept
$y = \sin 3x$	$(0, 0)$	$\left(\dfrac{\pi}{6}, 1\right)$	$\left(\dfrac{\pi}{3}, 0\right)$	$\left(\dfrac{\pi}{2}, -1\right)$	$\left(\dfrac{2\pi}{3}, 0\right)$
$y = \sin \left(3x - \dfrac{\pi}{2}\right)$	$\left(\dfrac{\pi}{6}, 0\right)$	$\left(\dfrac{\pi}{3}, 1\right)$	$\left(\dfrac{\pi}{2}, 0\right)$	$\left(\dfrac{2\pi}{3}, -1\right)$	$\left(\dfrac{5\pi}{6}, 0\right)$

Sketch the graph of $y = \sin \left(3x - \dfrac{\pi}{2}\right)$ through these points, continuing the pattern to complete two cycles.

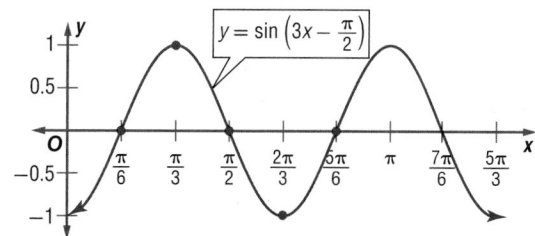

✔ Guided Practice

State the amplitude, period, frequency, and phase shift of each function. Then graph two periods of the function.

5A. $y = \cos \left(\dfrac{x}{2} + \dfrac{\pi}{4}\right)$

5B. $y = 3 \sin \left(2x - \dfrac{\pi}{3}\right)$

The final way to transform the graph of a sinusoidal function is through a vertical translation or **vertical shift**. Recall from Lesson 1-5 that the graph of $y = f(x) + d$ is the graph of $y = f(x)$ translated or *shifted* $|d|$ units up if $d > 0$ and $|d|$ units down if $d < 0$. The vertical shift is the average of the maximum and minimum values of the function.

The parent functions $y = \sin x$ and $y = \cos x$ oscillate about the x-axis. After a vertical shift, a new horizontal axis known as the **midline** becomes the reference line or equilibrium point about which the graph oscillates. For example, the midline of $y = \sin x + 1$ is $y = 1$, as shown.

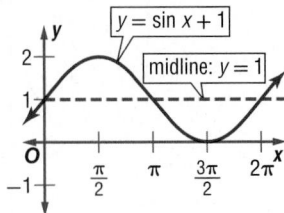

In general, the midline for the graphs of $y = a \sin(bx + c) + d$ and $y = a \cos(bx + c) + d$ is $y = d$.

EXAMPLE 6 **Graph Vertical Translations of Sinusoidal Functions**

State the amplitude, period, frequency, phase shift, and vertical shift of $y = \sin(x + 2\pi) - 1$. Then graph two periods of the function.

In this function, $a = 1$, $b = 1$, $c = 2\pi$, and $d = -1$.

Amplitude: $|a| = |1|$ or 1   Period: $\dfrac{2\pi}{|b|} = \dfrac{2\pi}{|1|}$ or 2π Frequency: $\dfrac{|b|}{2\pi} = \dfrac{|1|}{2\pi}$ or $\dfrac{1}{2\pi}$

Phase shift: $-\dfrac{c}{|b|} = -\dfrac{2\pi}{|1|} = -2\pi$ Vertical shift: d or -1 Midline: $y = d$ or $y = -1$

First, graph the midline $y = -1$. Then graph $y = \sin x$ shifted 2π units to the left and 1 unit down.

Notice that this transformation is equivalent to a translation 1 unit down because the phase shift was one period to the left.

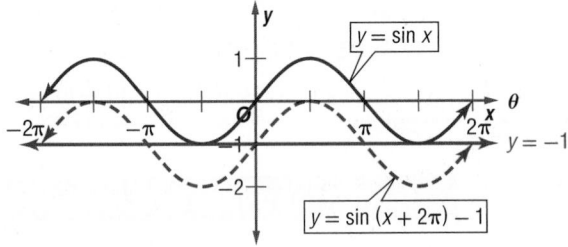

☑ **Guided Practice**

State the amplitude, period, frequency, phase shift, and vertical shift of each function. Then graph two periods of the function.

6A. $y = 2 \cos x + 1$

6B. $y = \dfrac{1}{2} \sin\left(\dfrac{x}{4} - \dfrac{\pi}{2}\right) - 3$

The characteristics of transformations of the parent functions $y = \sin x$ and $y = \cos x$ are summarized below.

Concept Summary **Graphs of Sinusoidal Functions**

The graphs of $y = a \sin(bx + c) + d$ and $y = a \cos(bx + c) + d$, where $a \neq 0$ and $b \neq 0$, have the following characteristics.

Amplitude: $|a|$ Period: $\dfrac{2\pi}{|b|}$ Frequency: $\dfrac{|b|}{2\pi}$ or $\dfrac{1}{\text{Period}}$

Phase shift: $-\dfrac{c}{|b|}$ Vertical shift: d Midline: $y = d$

Applications of Sinusoidal Functions Many real-world situations that exhibit periodic behavior over time can be modeled by transformations of $y = \sin x$ or $y = \cos x$.

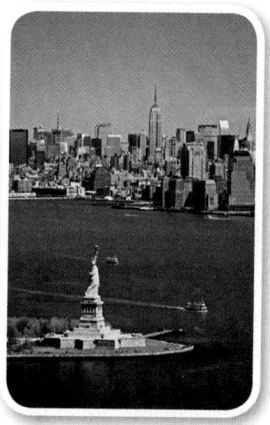

Real-World EXAMPLE 7 **Modeling Data Using a Sinusoidal Function**

METEOROLOGY Use the information at the left to write a sinusoidal function that models the number of hours of daylight for New York City as a function of time x, where $x = 1$ represents January 15, $x = 2$ represents February 15, and so on. Then use your model to estimate the number of hours of daylight on September 30 in New York City.

Step 1 Make a scatter plot of the data and choose a model.

The graph appears wave-like, so you can use a sinusoidal function of the form $y = a \sin (bx + c) + d$ or $y = a \cos (bx + c) + d$ to model the data. We will choose to use $y = a \cos (bx + c) + d$ to model the data.

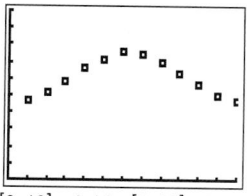

[0, 12] scl: 1 by [0, 20] scl: 2

Step 2 Find the maximum M and minimum m values of the data, and use these values to find a, b, c, and d.

The maximum and minimum hours of daylight are 15.07 and 9.27, respectively. The amplitude a is half of the distance between the extrema.

$$a = \frac{1}{2}(M - m) = \frac{1}{2}(15.07 - 9.27) \text{ or } 2.9$$

The vertical shift d is the average of the maximum and minimum data values.

$$d = \frac{1}{2}(M + m) = \frac{1}{2}(15.07 + 9.27) \text{ or } 12.17$$

A sinusoid completes half of a period in the time it takes to go from its maximum to its minimum value. One period is twice this time.

Period $= 2(x_{max} - x_{min}) = 2(12 - 6)$ or 12 $x_{max} =$ **December 15 or month 12 and** $x_{min} =$ **June 15 or month 6**

Because the period equals $\frac{2\pi}{|b|}$, you can write $|b| = \frac{2\pi}{\text{Period}}$. Therefore, $|b| = \frac{2\pi}{12}$, or $\frac{\pi}{6}$.

The maximum data value occurs when $x = 6$. Since $y = \cos x$ attains its first maximum when $x = 0$, we must apply a phase shift of $6 - 0$ or 6 units. Use this value to find c.

Phase shift $= -\dfrac{c}{|b|}$ **Phase shift formula**

$6 = -\dfrac{c}{\frac{\pi}{6}}$ **Phase shift = 6 and** $|b| = \frac{\pi}{6}$

$c = -\pi$ **Solve for c.**

Step 3 Write the function using the values for a, b, c, and d. Use $b = \frac{\pi}{6}$.

$y = 2.9 \cos \left(\frac{\pi}{6}x - \pi\right) + 12.17$ is one model for the hours of daylight

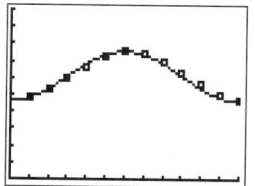

[0, 12] scl: 1 by [0, 20] scl: 2

Figure 4.4.1

Graph the function and scatter plot in the same viewing window, as in Figure 4.4.1. To find the number of hours of daylight on September 30, evaluate the model for $x = 9.5$.

$y = 2.9 \cos \left(\frac{\pi}{6}(9.5) - \pi\right) + 12.17$ or about 11.42 hours of daylight

✔ Guided Practice

METEOROLOGY The average monthly temperatures for Seattle, Washington, are shown.

Month	Jan	Feb	Mar	Apr	May	Jun	Jul	Aug	Sept	Oct	Nov	Dec
Temp. (°)	41	44	47	50	56	61	65	66	61	54	46	42

7A. Write a function that models the monthly temperatures, using $x = 1$ to represent January.

7B. According to your model, what is Seattle's average monthly temperature in February?

Describe how the graphs of $f(x)$ and $g(x)$ are related. Then find the amplitude of $g(x)$, and sketch two periods of both functions on the same coordinate axes. (Examples 1 and 2)

1. $f(x) = \sin x$
$g(x) = \frac{1}{2} \sin x$

2. $f(x) = \cos x$
$g(x) = -\frac{1}{3} \cos x$

3. $f(x) = \cos x$
$g(x) = 6 \cos x$

4. $f(x) = \sin x$
$g(x) = -8 \sin x$

Describe how the graphs of $f(x)$ and $g(x)$ are related. Then find the period of $g(x)$, and sketch at least one period of both functions on the same coordinate axes. (Example 3)

5. $f(x) = \sin x$
$g(x) = \sin 4x$

6. $f(x) = \cos x$
$g(x) = \cos 2x$

7. $f(x) = \cos x$
$g(x) = \cos 0.2x$

8. $f(x) = \sin x$
$g(x) = \sin 0.25x$

9. VOICES The contralto vocal type includes the deepest female singing voice. Some contraltos can sing as low as the E below middle C (E3), which has a frequency of 165 hertz. Write an equation for a sine function that models the initial behavior of the sound wave associated with E3 having an amplitude of 0.15. (Example 4)

Write a sine function that can be used to model the initial behavior of a sound wave with the frequency and amplitude given. (Example 4)

10. $f = 440, a = 0.3$

11. $f = 932, a = 0.25$

12. $f = 1245, a = 0.12$

13. $f = 623, a = 0.2$

State the amplitude, period, frequency, phase shift, and vertical shift of each function. Then graph two periods of the function. (Examples 5 and 6)

14. $y = 3 \sin \left(x - \frac{\pi}{4} \right)$

15. $y = \cos \left(\frac{x}{3} + \frac{\pi}{2} \right)$

16. $y = 0.25 \cos x + 3$

17. $y = \sin 3x - 2$

18. $y = \cos \left(x - \frac{3\pi}{2} \right) - 1$

19. $y = \sin \left(x + \frac{5\pi}{6} \right) + 4$

20. VACATIONS The average number of reservations R that a vacation resort has at the beginning of each month is shown. (Example 7)

Month	R	Month	R
Jan	200	May	121
Feb	173	Jun	175
Mar	113	Jul	198
Apr	87	Aug	168

a. Write an equation of a sinusoidal function that models the average number of reservations using $x = 1$ to represent January.

b. According to your model, approximately how many reservations can the resort anticipate in November?

21. TIDES The table shown below provides data for the first high and low tides of the day for a certain bay during one day in June. (Example 7)

Tide	Height (ft)	Time
first high tide	12.95	4:25 A.M.
first low tide	2.02	10:55 A.M.

a. Determine the amplitude, period, phase shift, and vertical shift of a sinusoidal function that models the height of the tide. Let x represent the number of hours that the high or low tide occurred after midnight.

b. Write a sinusoidal function that models the data.

c. According to your model, what was the height of the tide at 8:45 P.M. that night?

22. METEOROLOGY The average monthly temperatures for Boston, Massachusetts are shown. (Example 7)

Month	Temp. (°F)	Month	Temp. (°F)
Jan	29	Jul	74
Feb	30	Aug	72
Mar	39	Sept	65
Apr	48	Oct	55
May	58	Nov	45
Jun	68	Dec	34

a. Determine the amplitude, period, phase shift, and vertical shift of a sinusoidal function that models the monthly temperatures using $x = 1$ to represent January.

b. Write an equation of a sinusoidal function that models the monthly temperatures.

c. According to your model, what is Boston's average temperature in August?

GRAPHING CALCULATOR Find the values of x in the interval $-\pi < x < \pi$ that make each equation or inequality true. (*Hint*: Use the intersection function.)

23. $-\sin x = \cos x$

24. $\sin x - \cos x = 1$

25. $\sin x + \cos x = 0$

26. $\cos x \leq \sin x$

27. $\sin x \cos x > 1$

28. $\sin x \cos x \leq 0$

29 CAROUSELS A wooden horse on a carousel moves up and down as the carousel spins. When the ride ends, the horse usually stops in a vertical position different from where it started. The position y of the horse after t seconds can be modeled by $y = 1.5 \sin (2t + c)$, where the phase shift must be continuously adjusted to compensate for the different starting positions. If during one ride the horse reached a maximum height after $\frac{7\pi}{12}$ seconds, find the equation that models the horse's position.

30. AMUSEMENT PARKS The position y in feet of a passenger cart relative to the center of a Ferris wheel over t seconds is shown below.

Side view of Ferris wheel over time interval [0, 5.5]

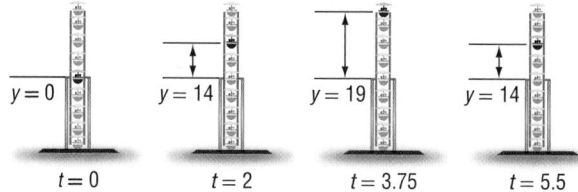

$y = 0$ $y = 14$ $y = 19$ $y = 14$

$t = 0$ $t = 2$ $t = 3.75$ $t = 5.5$

a. Find the time t that it takes for the cart to return to $y = 0$ during its initial spin.

b. Find the period of the Ferris wheel.

c. Sketch the graph representing the position of the passenger cart over one period.

d. Write a sinusoidal function that models the position of the passenger cart as a function of time t.

Write an equation that corresponds to each graph.

31.

32.

33.

34.

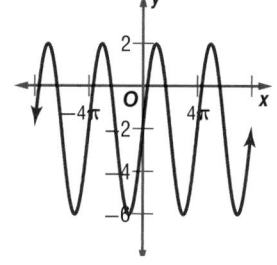

Write a sinusoidal function with the given period and amplitude that passes through the given point.

35. period: π; amplitude: 5; point: $\left(\dfrac{\pi}{6}, \dfrac{5}{2}\right)$

36. period: 4π; amplitude: 2; point: $(\pi, 2)$

37. period: $\dfrac{\pi}{2}$; amplitude: 1.5; point: $\left(\dfrac{\pi}{2}, \dfrac{3}{2}\right)$

38. period: 3π; amplitude: 0.5; point: $\left(\pi, \dfrac{\sqrt{3}}{4}\right)$

39. ⚡ **MULTIPLE REPRESENTATIONS** In this problem, you will investigate the change in the graph of a sinusoidal function of the form $y = \sin x$ or $y = \cos x$ when multiplied by a polynomial function.

a. **GRAPHICAL** Use a graphing calculator to sketch the graphs of $y = 2x$, $y = -2x$, and $y = 2x \cos x$ on the same coordinate plane, on the interval $[-20, 20]$.

b. **VERBAL** Describe the behavior of the graph of $y = 2x \cos x$ in relation to the graphs of $y = 2x$ and $y = -2x$.

c. **GRAPHICAL** Use a graphing calculator to sketch the graphs of $y = x^2$, $y = -x^2$, and $y = x^2 \sin x$ on the same coordinate plane, on the interval $[-20, 20]$.

d. **VERBAL** Describe the behavior of the graph of $y = x^2 \sin x$ in relation to the graphs of $y = x^2$ and $y = -x^2$.

e. **ANALYTICAL** Make a conjecture as to the behavior of the graph of a sinusoidal function of the form $y = \sin x$ or $y = \cos x$ when multiplied by polynomial function of the form $y = f(x)$.

H.O.T. Problems   Use Higher-Order Thinking Skills

40. CHALLENGE Without graphing, find the exact coordinates of the first maximum point to the right of the y-axis for $y = 4 \sin\left(\dfrac{2}{3}x - \dfrac{\pi}{9}\right)$.

REASONING Determine whether each statement is *true* or *false*. Explain your reasoning.

41. Every sine function of the form $y = a \sin(bx + c) + d$ can also be written as a cosine function of the form $y = a \cos(bx + c) + d$.

42. The period of $f(x) = \cos 8x$ is equal to four times the period of $g(x) = \cos 2x$.

43. CHALLENGE How many zeros does $y = \cos 1500x$ have on the interval $0 \le x \le 2\pi$?

44. PROOF Prove the phase shift formula.

45. WRITING IN MATH The Power Tower ride in Sandusky, Ohio, is shown below. Along the side of each tower is a string of lights that send a continuous pulse of light up and down each tower at a constant rate. Explain why the distance d of this light from the ground over time t cannot be represented by a sinusoidal function.

300 ft

The given point lies on the terminal side of an angle θ in standard position. Find the values of the six trigonometric functions of θ. (Lesson 4-3)

46. $(-4, 4)$ **47.** $(8, -2)$ **48.** $(-5, -9)$ **49.** $(4, 5)$

Write each degree measure in radians as a multiple of π and each radian measure in degrees. (Lesson 4-2)

50. $25°$ **51.** $-420°$ **52.** $-\dfrac{\pi}{4}$ **53.** $\dfrac{8\pi}{3}$

54. SCIENCE Radiocarbon dating is a method of estimating the age of an organic material by calculating the amount of carbon-14 present in the material. The age of a material can be calculated using $A = t \cdot \dfrac{\ln R}{-0.693}$, where A is the age of the object in years, t is the half-life of carbon-14 or 5700 years, and R is the ratio of the amount of carbon-14 in the sample to the amount of carbon-14 in living tissue. (Lesson 3-4)

  a. A sample of organic material contains 0.000076 gram of carbon-14. A living sample of the same material contains 0.00038 gram. About how old is the sample?

  b. A specific sample is at least 20,000 years old. What is the maximum percent of carbon-14 remaining in the sample?

State the number of possible real zeros and turning points of each function. Then determine all of the real zeros by factoring. (Lesson 2-2)

55. $f(x) = x^3 + 2x^2 - 8x$ **56.** $f(x) = x^4 - 10x^2 + 9$

57. $f(x) = x^5 + 2x^4 - 4x^3 - 8x^2$ **58.** $f(x) = x^4 - 1$

Determine whether f has an inverse function. If it does, find the inverse function and state any restrictions on its domain. (Lesson 1-7)

59. $f(x) = -x - 2$ **60.** $f(x) = \dfrac{1}{x + 4}$ **61.** $f(x) = (x - 3)^2 - 7$ **62.** $f(x) = \dfrac{1}{(x - 1)^2}$

63. SAT/ACT If $x + y = 90°$ and x and y are both nonnegative angles, what is equal to $\dfrac{\cos x}{\sin y}$?

  A 0   **D** 1.5

  B $\dfrac{1}{2}$ **E** Cannot be determined from the information given.

  C 1

64. REVIEW If $\tan x = \dfrac{10}{24}$ in the figure below, what are $\sin x$ and $\cos x$?

  F $\sin x = \dfrac{26}{10}$ and $\cos x = \dfrac{24}{26}$

  G $\sin x = \dfrac{10}{26}$ and $\cos x = \dfrac{24}{26}$

  H $\sin x = \dfrac{26}{10}$ and $\cos x = \dfrac{26}{24}$

  J $\sin x = \dfrac{10}{26}$ and $\cos x = \dfrac{26}{24}$

65. Identify the equation represented by the graph.

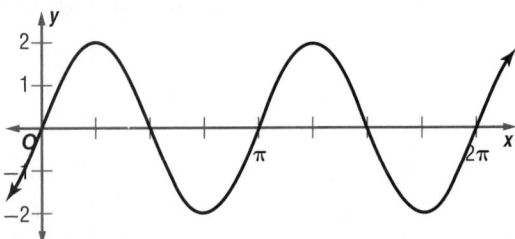

  A $y = \dfrac{1}{2}\sin 4x$

  B $y = \dfrac{1}{4}\sin 2x$

  C $y = 2\sin 2x$

  D $y = 4\sin \dfrac{1}{2}x$

66. REVIEW If $\cos \theta = \dfrac{8}{17}$ and the terminal side of the angle is in Quadrant IV, what is the exact value of $\sin \theta$?

  F $-\dfrac{15}{8}$ **H** $-\dfrac{15}{17}$

  G $-\dfrac{17}{15}$ **J** $-\dfrac{8}{15}$

Objective
Graph and examine the periods of sums and differences of sinusoids.

The graphs of the sums and differences of two sinusoids will often have different periods than the graphs of the original functions.

| **ACTIVITY** | **Sum of Sinusoids** |

Determine a common interval on which both $f(x) = 2 \sin 3x$ and $g(x) = 4 \cos \frac{x}{2}$ complete a whole number of cycles. Then graph $h(x) = f(x) + g(x)$, and identify the period of the function.

Step 1 Enter $f(x)$ for Y1 and $g(x)$ for Y2. Then adjust the window until each graph completes one or more whole cycles on the same interval. One interval on which this occurs is $[0, 4\pi]$. On this interval, $g(x)$ completes one whole cycle and $f(x)$ completes six whole cycles.

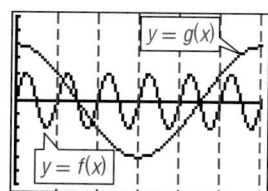

$[0, 4\pi]$ scl: π by $[-6, 6]$ scl: 1

Step 2 To graph $h(x)$ as Y3, under the ⬚VARS menu, select **Y-VARS**, **function**, and Y1 to enter Y1. Then press ⬚+ and select **Y-VARS**, **function**, and Y2 to enter Y2.

Step 3 Graph f(x), g(x) and h(x) on the same screen. To make the graph of $h(x)$ stand out, scroll to the left of the equals sign next to Y3, and press ⬚ENTER. Then graph the functions using the same window as above.

$[0, 4\pi]$ scl: π by $[-6, 6]$ scl: 1

TechnologyTip

Hiding Graphs Scroll to the equals sign and select enter to make a graph disappear.

Step 4 By adjusting the x-axis from $[0, 4\pi]$ to $[0, 8\pi]$ to observe the full pattern of $h(x)$, we can see that the period of the sum of the two sinusoids is 4π.

$[0, 8\pi]$ scl: 2π by $[-6, 6]$ scl: 1

Exercises

Determine a common interval on which both $f(x)$ and $g(x)$ complete a whole number of cycles. Then graph $a(x) = f(x) + g(x)$ and $b(x) = f(x) - g(x)$, and identify the period of the function.

1. $f(x) = 4 \sin 2x$
$g(x) = -2 \cos 3x$

2. $f(x) = \sin 8x$
$g(x) = \cos 6x$

3. $f(x) = 3 \sin (x - \pi)$
$g(x) = -2 \cos 2x$

4. $f(x) = \frac{1}{2} \sin 4x$
$g(x) = 2 \sin \left(x - \frac{\pi}{2}\right)$

5. $f(x) = \frac{1}{4} \cos \frac{x}{2}$
$g(x) = -2 \cos \left(x - \frac{\pi}{2}\right)$

6. $f(x) = -\frac{1}{2} \sin 2x$
$g(x) = 3 \cos 2x$

7. MAKE A CONJECTURE Explain how you can use the periods of two sinusoids to find the period of the sum or difference of the two sinusoids.

Find the exact values of the six trigonometric functions of θ. (Lesson 4-1)

1.

2.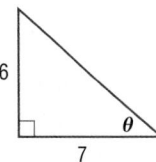

Find the value of *x*. Round to the nearest tenth if necessary. (Lesson. 4-1)

3.

4.

5. SHADOWS A pine tree casts a shadow that is 7.9 feet long when the Sun is 80° above the horizon. (Lesson 4-1)

   a. Find the height of the tree.

   b. Later that same day, a person 6 feet tall casts a shadow 6.7 feet long. At what angle is the Sun above the horizon?

Find the measure of angle θ. Round to the nearest degree if necessary. (Lesson 4-1)

6.

7.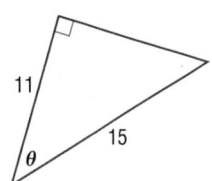

8. Write $\frac{2\pi}{9}$ in degrees. (Lesson 4-2)

Identify all angles that are coterminal with the given angle. Then find and draw one positive and one negative angle coterminal with the given angle. (Lesson 4-2)

9. $\frac{3\pi}{10}$

10. −22°

11. MULTIPLE CHOICE Find the approximate area of the shaded region. (Lesson 4-2)

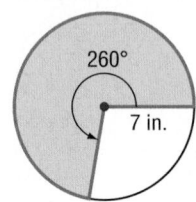

260°

7 in.

   A 12.2 in²
   B 42.8 in²
   C 85.5 in²
   D 111.2 in²

12. TRAVEL A car is traveling at a speed of 55 miles per hour on tires that measure 2.6 feet in diameter. Find the approximate angular speed of the tires in radians per minute. (Lesson 4-2)

Sketch each angle. Then find its reference angle. (Lesson 4-3)

13. 175°

14. $\frac{21\pi}{13}$

Find the exact value of each expression. If undefined, write *undefined*. (Lesson 4-3)

15. cos 315°

16. $\sec \frac{3\pi}{2}$

17. $\sin \frac{5\pi}{3}$

18. $\tan \frac{5\pi}{6}$

Find the exact values of the five remaining trigonometric functions of θ. (Lesson 4-3)

19. $\cos \theta = -\frac{2}{5}$, where $\sin \theta < 0$ and $\tan \theta > 0$

20. $\cot \theta = \frac{4}{3}$, where $\cos \theta > 0$ and $\sin \theta > 0$

State the amplitude, period, frequency, phase shift, and vertical shift of each function. Then graph two full periods of the function. (Lesson 4-4)

21. $y = -3 \sin \left(x - \frac{3\pi}{2}\right)$

22. $y = 5 \cos 2x - 2$

23. MULTIPLE CHOICE Which of the functions has the same graph as $y = 3 \sin (x - \pi)$? (Lesson 4-4)

   F $y = 3 \sin (x + \pi)$
   H $y = -3 \sin (x - \pi)$
   G $y = 3 \cos \left(x - \frac{\pi}{2}\right)$
   J $y = -3 \cos \left(x + \frac{\pi}{2}\right)$

24. SPRING The motion of an object attached to a spring oscillating across its original position of rest can be modeled by $x(t) = A \cos \omega t$, where *A* is the initial displacement of the object from its resting position, ω is a constant dependent on the spring and the mass of the object attached to the spring, and *t* is time measured in seconds. (Lesson 4-4)

   a. Draw a graph for the motion of an object attached to a spring and displaced 4 centimeters where ω = 3.

   b. How long will it take for the object to return to its initial position for the first time?

   c. The constant ω is equal to $\sqrt{\frac{k}{m}}$, where *k* is the spring constant, and *m* is the mass of the object. How would increasing the mass of an object affect the period of its oscillations? Explain your reasoning.

25. BUOY The height above sea level in feet of a signal buoy's transmitter is modeled by $h = a \sin bt + \frac{11}{2}$. In rough waters, the height cycles between 1 and 10 feet, with 4 seconds between cycles. Find the values of *a* and *b*.

Graphing Other Trigonometric Functions

Why?

There are two types of radio transmissions known as amplitude modulation (AM) and frequency modulation (FM). When sound is transmitted by an AM radio station, the amplitude of a sinusoidal wave called the *carrier wave* is changed to produce sound. The transmission of an FM signal results in a change in the frequency of the carrier wave. You will learn more about the graphs of these waves, known as *damped waves*, in this lesson.

AM signal

FM signal

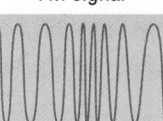

Then
You analyzed graphs of trigonometric functions. (Lesson 4-4)

Now
- Graph tangent and reciprocal trigonometric functions.
- Graph damped trigonometric functions.

New Vocabulary
damped trigonometric function
damping factor
damped oscillation
damped wave
damped harmonic motion

Math Online

glencoe.com

Tangent and Reciprocal Functions In Lesson 4-4, you graphed the sine and cosine functions on the coordinate plane. You can use the same techniques to graph the tangent function and the reciprocal trigonometric functions—cotangent, secant, and cosecant.

Since $\tan x = \frac{\sin x}{\cos x}$, the tangent function is undefined when $\cos x = 0$. Therefore, the tangent function has a *vertical asymptote* whenever $\cos x = 0$. Similarly, the tangent and sine functions each have zeros at integer multiples of π because $\tan x = 0$ when $\sin x = 0$.

The properties of the tangent function are summarized below.

Key Concept	Properties of the Tangent Function

Domain: $x \in \mathbb{R}, x \neq \frac{\pi}{2} + n\pi, n \in \mathbb{Z}$

Range: $(-\infty, \infty)$

x-intercepts: $n\pi, n \in \mathbb{Z}$

y-intercept: 0

Continuity: infinite discontinuity at $x = \frac{\pi}{2} + n\pi, n \in \mathbb{Z}$

Asymptotes: $x = \frac{\pi}{2} + n\pi, n \in \mathbb{Z}$

Symmetry: origin (odd function)

Extrema: none

End Behavior: $\lim\limits_{x \to -\infty} \tan x$ and $\lim\limits_{x \to \infty} \tan x$ do not exist. The function oscillates between $-\infty$ and ∞.

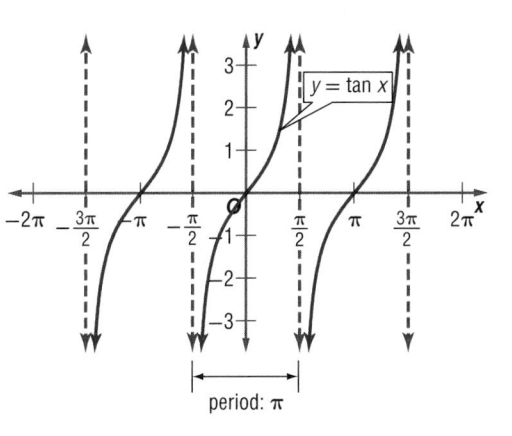

The general form of the tangent function, which is similar to that of the sinusoidal functions, is $y = a \tan (bx + c) + d$, where a produces a vertical stretch or compression, b affects the period, c produces a phase shift, d produces a vertical shift and neither a or b are 0.

Key Concept — Period of the Tangent Function

Words The *period* of a tangent function is the distance between any two consecutive vertical asymptotes.

Model

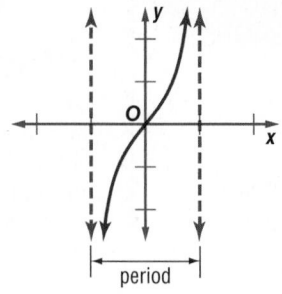

Symbols For $y = a \tan (bx + c)$, where $b \neq 0$, period $= \dfrac{\pi}{|b|}$.

StudyTip

Amplitude The term *amplitude* does not apply to the tangent or cotangent functions because the heights of these functions are infinite.

Two consecutive vertical asymptotes for $y = \tan x$ are $x = -\dfrac{\pi}{2}$ and $x = \dfrac{\pi}{2}$. You can find two consecutive vertical asymptotes for any tangent function of the form $y = a \tan (bx + c) + d$ by solving the equations $bx + c = -\dfrac{\pi}{2}$ and $bx + c = \dfrac{\pi}{2}$.

You can sketch the graph of a tangent function by plotting the vertical asymptotes, x-intercepts, and points between the asymptotes and x-intercepts.

EXAMPLE 1 Graph Horizontal Dilations of the Tangent Function

Locate the vertical asymptotes, and sketch the graph of $y = \tan 2x$.

The graph of $y = \tan 2x$ is the graph of $y = \tan x$ compressed horizontally. The period is $\dfrac{\pi}{|2|}$ or $\dfrac{\pi}{2}$. Find two consecutive vertical asymptotes.

$bx + c = -\dfrac{\pi}{2}$	**Tangent asymptote equations**	$bx + c = \dfrac{\pi}{2}$
$2x + 0 = -\dfrac{\pi}{2}$	$b = 2, c = 0$	$2x + 0 = \dfrac{\pi}{2}$
$x = -\dfrac{\pi}{4}$	**Simplify.**	$x = \dfrac{\pi}{4}$

Create a table listing key points, including the x-intercept, that are located between the two vertical asymptotes at $x = -\dfrac{\pi}{4}$ and $x = \dfrac{\pi}{4}$.

Function	Vertical Asymptote	Intermediate Point	x-intercept	Intermediate Point	Vertical Asymptote
$y = \tan x$	$x = -\dfrac{\pi}{2}$	$\left(-\dfrac{\pi}{4}, -1\right)$	$(0, 0)$	$\left(\dfrac{\pi}{4}, 1\right)$	$x = \dfrac{\pi}{2}$
$y = \tan 2x$	$x = -\dfrac{\pi}{4}$	$\left(-\dfrac{\pi}{8}, -1\right)$	$(0, 0)$	$\left(\dfrac{\pi}{8}, 1\right)$	$x = \dfrac{\pi}{4}$

Sketch the curve through the indicated key points for the function. Then sketch one cycle to the left on the interval $\left(-\dfrac{3\pi}{4}, -\dfrac{\pi}{4}\right)$ and one cycle to the right on the interval $\left(\dfrac{\pi}{4}, \dfrac{3\pi}{4}\right)$.

✔ Guided Practice

Locate the vertical asymptotes, and sketch the graph of each function.

1A. $y = \tan 4x$

1B. $y = \tan \dfrac{x}{2}$

EXAMPLE 2	Graph Reflections and Translations of the Tangent Function

Locate the vertical asymptotes, and sketch the graph of each function.

a. $y = -\tan \frac{x}{2}$

The graph of $y = -\tan \frac{x}{2}$ is the graph of $y = \tan x$ expanded horizontally and then reflected in the x-axis. The period is $\frac{\pi}{\left|\frac{1}{2}\right|}$ or 2π. Find two consecutive vertical asymptotes.

$\frac{x}{2} + 0 = -\frac{\pi}{2}$ $b = \frac{1}{2}, c = 0$ $\frac{x}{2} + 0 = \frac{\pi}{2}$

$x = 2\left(-\frac{\pi}{2}\right)$ or $-\pi$ **Simplify.** $x = 2\left(\frac{\pi}{2}\right)$ or π

Create a table listing key points, including the x-intercept, that are located between the two vertical asymptotes at $x = -\pi$ and $x = \pi$.

Function	Vertical Asymptote	Intermediate Point	x-intercept	Intermediate Point	Vertical Asymptote
$y = \tan x$	$x = -\frac{\pi}{2}$	$\left(-\frac{\pi}{4}, -1\right)$	$(0, 0)$	$\left(\frac{\pi}{4}, 1\right)$	$x = \frac{\pi}{2}$
$y = -\tan \frac{x}{2}$	$x = -\pi$	$\left(-\frac{\pi}{2}, 1\right)$	$(0, 0)$	$\left(\frac{\pi}{2}, -1\right)$	$x = \pi$

Sketch the curve through the indicated key points for the function. Then repeat the pattern for one cycle to the left and right of the first curve.

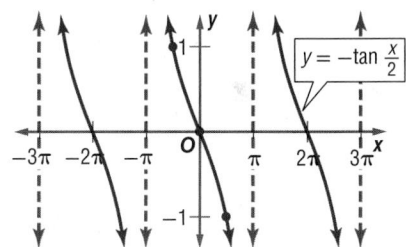

b. $y = \tan\left(x - \frac{3\pi}{2}\right)$

The graph of $y = \tan\left(x - \frac{3\pi}{2}\right)$ is the graph of $y = \tan x$ shifted $\frac{3\pi}{2}$ units to the right. The period is $\frac{\pi}{|1|}$ or π. Find two consecutive vertical asymptotes.

$x - \frac{3\pi}{2} = -\frac{\pi}{2}$ $b = 1, c = -\frac{3\pi}{2}$ $x - \frac{3\pi}{2} = \frac{\pi}{2}$

$x = -\frac{\pi}{2} + \frac{3\pi}{2}$ or π **Simplify.** $x = \frac{\pi}{2} + \frac{3\pi}{2}$ or 2π

StudyTip

Alternate Method When graphing a function with only a horizontal translation c, you can find the key points by adding c to each of the x-coordinates of the key points of the parent function.

Function	Vertical Asymptote	Intermediate Point	x-intercept	Intermediate Point	Vertical Asymptote
$y = \tan x$	$x = -\frac{\pi}{2}$	$\left(-\frac{\pi}{4}, -1\right)$	$(0, 0)$	$\left(\frac{\pi}{4}, 1\right)$	$x = \frac{\pi}{2}$
$y = \tan\left(x - \frac{3\pi}{2}\right)$	$x = \pi$	$\left(\frac{5\pi}{4}, -1\right)$	$\left(\frac{3\pi}{2}, 0\right)$	$\left(\frac{7\pi}{4}, 1\right)$	$x = 2\pi$

Sketch the curve through the indicated key points for the function. Then sketch one cycle to the left and right of the first curve.

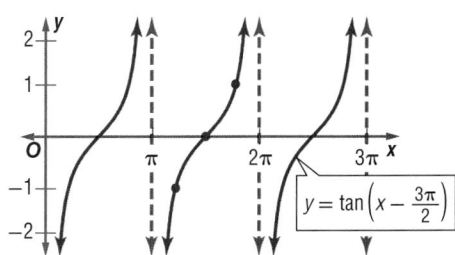

✔ **Guided Practice**

2A. $y = \tan\left(2x + \frac{\pi}{2}\right)$ **2B.** $y = -\tan\left(x - \frac{\pi}{6}\right)$

The cotangent function is the reciprocal of the tangent function, and is defined as $\cot x = \dfrac{\cos x}{\sin x}$. Like the tangent function, the period of a cotangent function of the form $y = a \cot (bx + c) + d$ can be found by calculating $\dfrac{\pi}{|b|}$. Two consecutive vertical asymptotes can be found by solving the equations $bx + c = 0$ and $bx + c = \pi$. The properties of the cotangent function are summarized below.

TechnologyTip

Graphing a Cotangent Function When using a calculator to graph a cotangent function, enter the reciprocal of tangent, $y = \dfrac{1}{\tan x}$. Graphing calculators may produce solid lines where the asymptotes occur. Setting the mode to DOT will eliminate the line.

Key Concept **Properties of the Cotangent Function**

Domain:	$x \in \mathbb{R}, x \neq n\pi, n \in \mathbb{Z}$
Range:	$(-\infty, \infty)$
x-intercepts:	$\dfrac{\pi}{2} + n\pi, n \in \mathbb{Z}$
y-intercept:	none
Continuity:	infinite discontinuity at $x = n\pi, n \in \mathbb{Z}$
Asymptotes:	$x = n\pi, n \in \mathbb{Z}$
Symmetry:	origin (odd function)
Extrema:	none
End Behavior:	$\lim\limits_{x \to -\infty} \cot x$ and $\lim\limits_{x \to \infty} \cot x$ do not exist. The function oscillates between $-\infty$ and ∞.

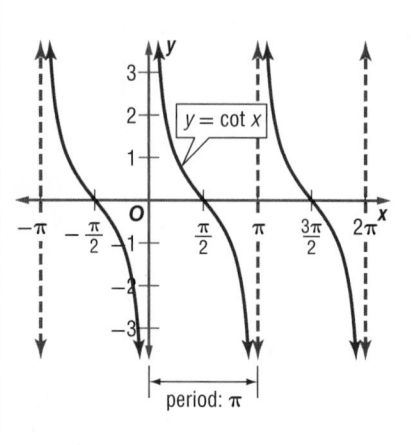

You can sketch the graph of a cotangent function using the same techniques that you used to sketch the graph of a tangent function.

EXAMPLE 3 **Sketch the Graph of a Cotangent Function**

Locate the vertical asymptotes, and sketch the graph of $y = \cot \dfrac{x}{3}$.

The graph of $y = \cot \dfrac{x}{3}$ is the graph of $y = \cot x$ expanded horizontally. The period is $\dfrac{\pi}{\left|\frac{1}{3}\right|}$ or 3π. Find two consecutive vertical asymptotes by solving $bx + c = 0$ and $bx + c = \pi$.

$$\dfrac{x}{3} + 0 = 0 \qquad\qquad b = \tfrac{1}{3}, c = 0 \qquad \dfrac{x}{3} + 0 = \pi$$

$$x = 3(0) \text{ or } 0 \qquad \textbf{Simplify.} \qquad x = 3(\pi) \text{ or } 3\pi$$

Create a table listing key points, including the x-intercept, that are located between the two vertical asymptotes at $x = 0$ and $x = 3\pi$.

Function	Vertical Asymptote	Intermediate Point	x-intercept	Intermediate Point	Vertical Asymptote
$y = \cot x$	$x = 0$	$\left(\dfrac{\pi}{4}, 1\right)$	$\left(\dfrac{\pi}{2}, 0\right)$	$\left(\dfrac{3\pi}{4}, -1\right)$	$x = \pi$
$y = \cot \dfrac{x}{3}$	$x = 0$	$\left(\dfrac{3\pi}{4}, 1\right)$	$\left(\dfrac{3\pi}{2}, 0\right)$	$\left(\dfrac{9\pi}{4}, -1\right)$	$x = 3\pi$

Following the same guidelines that you used for the tangent function, sketch the curve through the indicated key points that you found. Then sketch one cycle to the left and right of the first curve.

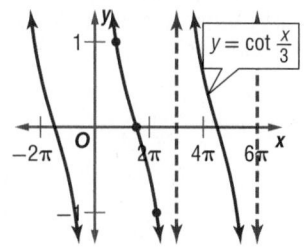

☑ **Guided Practice**

Locate the vertical asymptotes, and sketch the graph of each function.

3A. $y = -\cot 3x$

3B. $y = 3 \cot \dfrac{x}{2}$

The reciprocals of the sine and cosine functions are defined as $\csc x = \dfrac{1}{\sin x}$ and $\sec x = \dfrac{1}{\cos x}$, as shown below.

 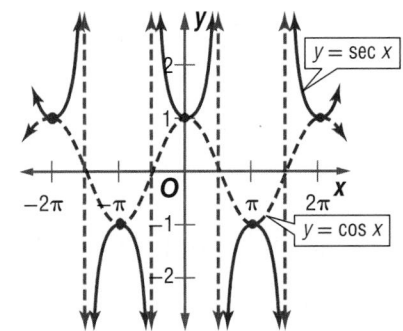

The cosecant function has asymptotes when $\sin x = 0$, which occurs at integer multiples of π. Likewise, the secant function has asymptotes when $\cos x = 0$, located at odd multiples of $\dfrac{\pi}{2}$.

Notice also that the graph of $y = \csc x$ has a relative minimum at each maximum point on the sine curve, and a relative maximum at each minimum point on the sine curve. The same is true for the graphs of $y = \sec x$ and $y = \cos x$.

The properties of the cosecant and secant functions are summarized below.

Key Concept | **Properties of the Cosecant and Secant Functions**

Cosecant Function	**Secant Function**
Domain: $x \in \mathbb{R}, x \neq n\pi, n \in \mathbb{Z}$	**Domain:** $x \in \mathbb{R}, x \neq \dfrac{\pi}{2} + n\pi, n \in \mathbb{Z}$
Range: $(-\infty, -1]$ and $[1, \infty)$	**Range:** $(-\infty, -1]$ and $[1, \infty)$
x-intercepts: none	**x-intercepts:** none
y-intercept: none	**y-intercept:** 1
Continuity: infinite discontinuity at $x = n\pi, n \in \mathbb{Z}$	**Continuity:** infinite discontinuity at $x = \dfrac{\pi}{2} + n\pi, n \in \mathbb{Z}$
Asymptotes: $x = n\pi, n \in \mathbb{Z}$	**Asymptotes:** $x = \dfrac{\pi}{2} + n\pi, n \in \mathbb{Z}$
Symmetry: origin (odd function)	**Symmetry:** y-axis (even function)
End Behavior: $\displaystyle\lim_{x \to -\infty} \csc x$ and $\displaystyle\lim_{x \to \infty} \csc x$ do not exist. The function oscillates between $-\infty$ and ∞.	**Behavior:** $\displaystyle\lim_{x \to -\infty} \sec x$ and $\displaystyle\lim_{x \to \infty} \sec x$ do not exist. The function oscillates between $-\infty$ and ∞.

 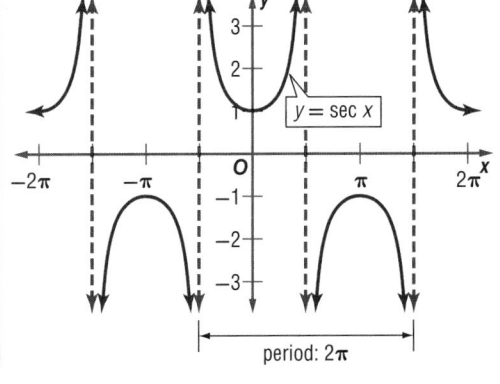

Like the sinusoidal functions, the period of a secant function of the form $y = a \sec (bx + c) + d$ or cosecant function of the form $y = a \csc (bx + c) + d$ can be found by calculating $\dfrac{2\pi}{|b|}$. Two vertical asymptotes for the secant function can be found by solving the equations $bx + c = -\dfrac{\pi}{2}$ and $bx + c = \dfrac{3\pi}{2}$ and two vertical asymptotes for the cosecant function can be found by solving $bx + c = -\pi$ and $bx + c = \pi$.

To sketch the graph of a cosecant or secant function, locate the asymptotes of the function and find the corresponding relative maximum and minimums points.

EXAMPLE 4   **Sketch Graphs of Cosecant and Secant Functions**

Locate the vertical asymptotes, and sketch the graph of each function.

a. $y = \csc\left(x + \frac{\pi}{2}\right)$

The graph of $y = \csc\left(x + \frac{\pi}{2}\right)$ is the graph of $y = \csc x$ shifted $\frac{\pi}{2}$ units to the left. The period is $\frac{2\pi}{|1|}$ or 2π. Two vertical asymptotes occur when $bx + c = -\pi$ and $bx + c = \pi$. Therefore, two asymptotes are $x + \frac{\pi}{2} = -\pi$ or $x = -\frac{3\pi}{2}$ and $x + \frac{\pi}{2} = \pi$ or $x = \frac{\pi}{2}$.

Create a table listing key points, including the relative maximum and minimum, that are located between the two vertical asymptotes at $x = -\frac{3\pi}{2}$ and $x = \frac{\pi}{2}$.

StudyTip

Finding Asymptotes and Key Points You can use the periodic nature of trigonometric graphs to help find asymptotes and key points. In Example 4a, notice that the vertical asymptote $x = -\frac{\pi}{2}$ is equidistant from the calculated asymptotes, $x = -\frac{3\pi}{2}$ and $x = \frac{\pi}{2}$.

Function	Vertical Asymptote	Relative Maximum	Vertical Asymptote	Relative Minimum	Vertical Asymptote
$y = \csc x$	$x = -\pi$	$\left(-\frac{\pi}{2}, -1\right)$	$x = 0$	$\left(\frac{\pi}{2}, 1\right)$	$x = \pi$
$y = \csc\left(x + \frac{\pi}{2}\right)$	$x = -\frac{3\pi}{2}$	$(-\pi, -1)$	$x = -\frac{\pi}{2}$	$(0, 1)$	$x = \frac{\pi}{2}$

Sketch the curve through the indicated key points for the function. Then sketch one cycle to the left and right. The graph is shown in Figure 4.5.1 below.

b. $y = \sec\frac{x}{4}$

The graph of $y = \sec\frac{x}{4}$ is the graph of $y = \sec x$ expanded horizontally. The period is $\frac{2\pi}{\left|\frac{1}{4}\right|}$ or 8π. Two vertical asymptotes occur when $bx + c = -\frac{\pi}{2}$ and $bx + c = \frac{3\pi}{2}$. Therefore, two asymptotes are $\frac{x}{4} + 0 = -\frac{\pi}{2}$ or $x = -2\pi$ and $\frac{x}{4} + 0 = \frac{3\pi}{2}$ or $x = 6\pi$.

Create a table listing key points that are located between the asymptotes at $x = -2\pi$ and $x = 6\pi$.

Function	Vertical Asymptote	Relative Minimum	Vertical Asymptote	Relative Maximum	Vertical Asymptote
$y = \sec x$	$x = -\frac{\pi}{2}$	$(0, 1)$	$x = \frac{\pi}{2}$	$(\pi, -1)$	$x = \frac{3\pi}{2}$
$y = \sec\frac{x}{4}$	$x = -2\pi$	$(0, 1)$	$x = 2\pi$	$(4\pi, -1)$	$x = 6\pi$

Sketch the curve through the indicated key points for the function. Then sketch one cycle to the left and right. The graph is shown in Figure 4.5.2 below.

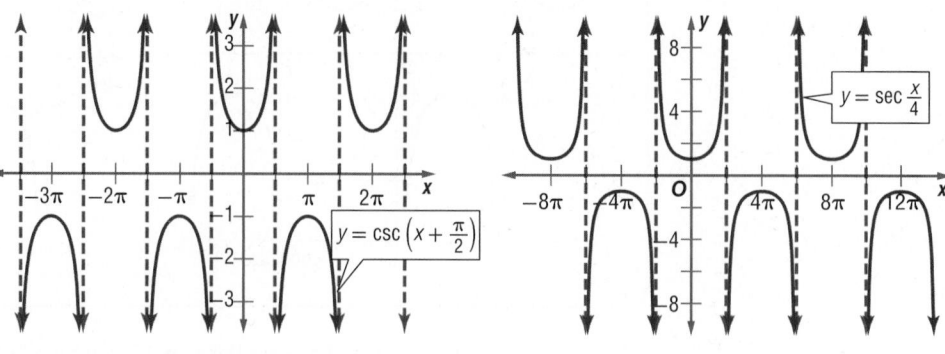

Figure 4.5.1   Figure 4.5.2

✓ **Guided Practice**

4A. $y = \csc 2x$ **4B.** $y = \sec(x + \pi)$

Damped Trigonometric Functions When a sinusoidal function is multiplied by another function $f(x)$, the graph of their product oscillates between the graphs of $y = f(x)$ and $y = -f(x)$. When this product reduces the amplitude of the wave of the original sinusoid, it is called **damped oscillation**, and the product of the two functions is known as a **damped trigonometric function**. This change in oscillation can be seen in Figures 4.5.3 and 4.5.4 for the graphs of $y = \sin x$ and $y = 2x \sin x$.

Figure 4.5.3

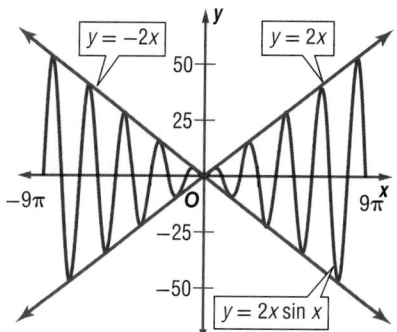

Figure 4.5.4

A damped trigonometric function is of the form $y = f(x) \sin bx$ or $y = f(x) \cos bx$, where $f(x)$ is the **damping factor**.

Damped oscillation occurs as x approaches $\pm\infty$ or as x approaches 0 from both directions.

EXAMPLE 5 **Sketch Damped Trigonometric Functions**

Identify the damping factor $f(x)$ of each function. Then use a graphing calculator to sketch the graphs of $f(x)$, $-f(x)$, and the given function in the same viewing window. Describe the behavior of the graph.

a. $y = -3x \cos x$

The function $y = -3x \cos x$ is the product of the functions $y = -3x$ and $y = \cos x$, so $f(x) = -3x$.

The amplitude of the function is decreasing as x approaches 0 from both directions.

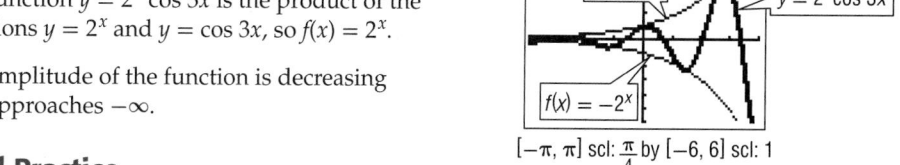

$[-4\pi, 4\pi]$ scl: π by $[-40, 40]$ scl: 5

b. $y = x^2 \sin x$

The function $y = x^2 \sin x$ is the product of the functions $y = x^2$ and $y = \sin x$. Therefore, the damping factor is $f(x) = x^2$.

The amplitude of the function is decreasing as x approaches 0 from both directions.

$[-4\pi, 4\pi]$ scl: π by $[-100, 100]$ scl: 10

c. $y = 2^x \cos 3x$

The function $y = 2^x \cos 3x$ is the product of the functions $y = 2^x$ and $y = \cos 3x$, so $f(x) = 2^x$.

The amplitude of the function is decreasing as x approaches $-\infty$.

$[-\pi, \pi]$ scl: $\frac{\pi}{4}$ by $[-6, 6]$ scl: 1

✔️ **Guided Practice**

5A. $y = 5x \sin x$ **5B.** $y = \frac{1}{x} \cos x$ **5C.** $y = 3^x \sin x$

When the amplitude of the motion of an object decreases with time due to friction, the motion is called *damped harmonic motion.*

Key Concept | **Damped Harmonic Motion**

Words An object is in **damped harmonic motion** when the amplitude is determined by the function $a(t) = ke^{-ct}$.

Symbols For $y = ke^{-ct} \sin \omega t$ and $y = ke^{-ct} \cos \omega t$, where $c > 0$, k is the displacement, c is the damping constant, t is time, and ω is the period.

Model

The greater the damping constant c, the faster the amplitude approaches 0. The magnitude of c depends on the size of the object and the material of which it is composed.

🌐 Real-World **EXAMPLE 6** | **Damped Harmonic Motion**

MUSIC A guitar string is plucked at a distance of 0.8 centimeter above its rest position and then released, causing a vibration. The damping constant for the string is 2.1, and the note produced has a frequency of 175 cycles per second.

a. Write a trigonometric function that models the motion of the string.

The maximum displacement of the string occurs when $t = 0$, so $y = ke^{-ct}\cos \omega t$ can be used to model the motion of the string because the graph of $y = \cos t$ has a y-intercept other than 0.

The maximum displacement occurs when the string is plucked 0.8 centimeter. The total displacement is the maximum displacement M minus the minimum displacement m, so
$$k = M - m = 0.8 - 0 \text{ or } 0.8 \text{ cm.}$$

You can use the value of the frequency to find ω.

$\dfrac{|\omega|}{2\pi} = 175$ $\qquad$ $\dfrac{|\omega|}{2\pi}$ = frequency

$|\omega| = 350\pi$ $\qquad$ **Multiply each side by 2π.**

Write a function using the values of k, ω, and c.

$y = 0.8e^{-2.1t}\cos 350\pi t$ is one model that describes the motion of the string.

b. Determine the amount of time t that it takes the string to be damped so that $-0.28 \le y \le 0.28$.

Use a graphing calculator to determine the value of t when the graph of $y = 0.8e^{-2.1t}\cos 350\pi t$ is oscillating between $y = -0.28$ and $y = 0.28$.

From the graph, you can see that it takes approximately 0.5 second for the graph of $y = 0.8e^{-2.1t}\cos 350\pi t$ to oscillate within the interval $-0.28 \le y \le 0.28$.

[0, 1] scl: 0.5 by [−0.75, 0.75] scl: 0.25

✓ Guided Practice

6. MUSIC Suppose another string on the guitar was plucked 0.5 centimeter above its rest position with a frequency of 98 cycles per second and a damping constant of 1.7.

A. Write a trigonometric function that models the motion of the string y as a function of time t.

B. Determine the time t that it takes the string to be damped so that $-0.15 \le y \le 0.15$.

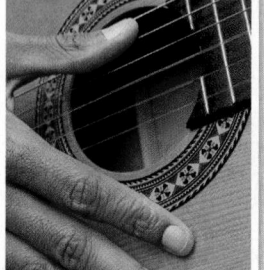

Locate the vertical asymptotes, and sketch the graph of each function. (Examples 1–4)

1. $y = 2 \tan x$

2. $y = \tan\left(x + \frac{\pi}{4}\right)$

3. $y = \cot\left(x - \frac{\pi}{6}\right)$

4. $y = -3 \tan \frac{x}{3}$

5. $y = -\frac{1}{4} \cot x$

6. $y = -\tan 3x$

7. $y = -2 \tan (6x - \pi)$

8. $y = \cot \frac{x}{2}$

9. $y = \frac{1}{5} \csc 2x$

10. $y = \csc\left(4x + \frac{7\pi}{6}\right)$

11. $y = \sec (x + \pi)$

12. $y = -2 \csc 3x$

13. $y = 4 \sec\left(x - \frac{3\pi}{4}\right)$

14. $y = \sec\left(\frac{x}{5} + \frac{\pi}{5}\right)$

15. $y = \frac{3}{2} \csc\left(x - \frac{2\pi}{3}\right)$

16. $y = -\sec \frac{x}{8}$

Identify the damping factor $f(x)$ of each function. Then use a graphing calculator to sketch the graphs of $f(x)$, $-f(x)$, and the given function in the same viewing window. Describe the behavior of the graph. (Example 5)

17. $y = \frac{3}{5}x \sin x$

18. $y = 4x \cos x$

19. $y = 2x^2 \cos x$

20. $y = \frac{x^3}{2} \sin x$

21. $y = \frac{1}{3}x \sin 2x$

22. $y = (x - 2)^2 \sin x$

23. $y = e^{0.5x} \cos x$

24. $y = 3^x \sin x$

25. $y = |x| \cos 3x$

26. $y = \ln x \cos x$

27. MECHANICS When the car shown below hit a bump in the road, the shock absorber was compressed 8 inches, released, and then began to vibrate in damped harmonic motion with a frequency of 2.5 cycles per second. The damping constant for the shock absorber is 3. (Example 6)

Rest Position
8 in.

a. Write a trigonometric function that models the displacement of the shock absorber y as a function of time t. Let $t = 0$ be the instant the shock absorber is released.

b. Determine the amount of time t that it takes for the amplitude of the vibration to decrease to 4 inches.

28. DIVING The end of a diving board is 20.3 centimeters above its resting position at the moment a diver leaves the board. Two seconds later, the board has moved down and up 12 times. The damping constant for the board is 0.901. (Example 6)

20.3

resting position

a. Write a trigonometric function that models the motion of the diving board y as a function of time t.

b. Determine the amount of time t that it takes the diving board to be damped so that $-0.5 \le y \le 0.5$.

Locate the vertical asymptotes, and sketch the graph of each function.

29. $y = \sec x + 3$

30. $y = \sec\left(x - \frac{\pi}{2}\right) + 4$

31. $y = \csc \frac{x}{3} - 2$

32. $y = \csc\left(3x + \frac{\pi}{6}\right) + 3$

33. $y = \cot (2x + \pi) - 3$

34. $y = \cot\left(\frac{x}{2} + \frac{\pi}{2}\right) - 1$

35 PHOTOGRAPHY Jeff is taking pictures of a hawk that is flying 150 feet above him. The hawk will eventually fly directly over Jeff. Let d be the distance Jeff is from the hawk and θ be the angle of elevation to the hawk from Jeff's camera.

150 ft

d

θ

a. Write d as a function of θ.

b. Graph the function on the interval $0 < \theta < \pi$.

c. Approximately how far away is the hawk from Jeff when the angle of elevation is 45°?

36. DISTANCE A spider is slowly climbing up a wall. Brianna is standing 6 feet away from the wall watching the spider. Let d be the distance Brianna is from the spider and θ be the angle of elevation to the spider from Brianna.

a. Write d as a function of θ.

b. Graph the function on the interval $0 < \theta < \frac{\pi}{2}$.

c. Approximately how far away is the spider from Brianna when the angle of elevation is 32°?

GRAPHING CALCULATOR Find the values of θ on the interval −π < θ < π that make each equation true.

37. $\cot \theta = 2 \sec \theta$

38. $\sin \theta = \cot \theta$

39. $4 \cos \theta = \csc \theta$

40. $\tan \frac{\theta}{2} = \sin \theta$

41. $\csc \theta = \sec \theta$

42. $\tan \theta = \sec \frac{\theta}{2}$

43. **TENSION** A helicopter is delivering a large mural that is to be displayed in the center of town. Two ropes are used to attach the mural to the helicopter, as shown. The tension T on each rope is equal to half the downward force times $\sec \frac{\theta}{2}$.

a. The downward force in newtons equals the mass of the mural times gravity, which is 9.8 newtons per kilogram. If the mass of the mural is 544 kilograms, find the downward force.

b. Write an equation that represents the tension T on each rope.

c. Graph the equation from part **b** on the interval [0, 180°].

d. Suppose the mural is 9.14 meters long and the ideal angle θ for tension purposes is a right angle. Determine how much rope is needed to transport the mural and the tension that is being applied to each rope.

e. Suppose you have 12.2 meters of rope to use to transport the mural. Find θ and the tension that is being applied to each rope.

Match each function with its graph.

a.

b.

c.

d.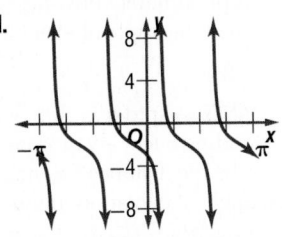

44. $y = \csc \left(\frac{x}{3} + \frac{\pi}{4} \right) - 2$

45. $y = \sec \left(\frac{x}{3} + \frac{\pi}{4} \right) - 2$

46. $y = \cot \left(2x - \frac{\pi}{4} \right) - 2$

47. $y = \tan \left(2x - \frac{\pi}{4} \right) - 2$

GRAPHING CALCULATOR Graph each pair of functions on the same screen and make a conjecture as to whether they are equivalent for all real numbers. Then use the properties of the functions to verify each conjecture.

48. $f(x) = \sec x \cos x; g(x) = 1$

49. $f(x) = \sec^2 x; g(x) = \tan^2 x + 1$

50. $f(x) = \cos x \csc x; g(x) = \cot x$

51. $f(x) = \dfrac{1}{\sec \left(x - \frac{\pi}{2} \right)}; g(x) = \sin x$

Write an equation for the given function given the period, phase shift (ps), and vertical shift (vs).

52. function: sec; period: 3π; ps: 0; vs: 2

53. function: tan; period: $\frac{\pi}{2}$; ps: $\frac{\pi}{4}$; vs: −1

54. function: csc; period: $\frac{\pi}{4}$; ps: −π; vs: 0

55. function: cot; period: 3π; ps: $\frac{\pi}{2}$; vs: 4

56. function: csc; period: $\frac{\pi}{3}$; ps: $-\frac{\pi}{2}$; vs: −3

H.O.T. Problems   Use **H**igher-**O**rder **T**hinking Skills

57. **PROOF** Verify that the y-intercept for the graph of any function of the form $y = ke^{-ct} \cos \omega t$ is k.

REASONING Determine whether each statement is *true* or *false*. Explain your reasoning.

58. If $b \neq 0$, then $y = a + b \sec x$ has extrema of $\pm(a + b)$.

59. If $x = \theta$ is an asymptote of $y = \csc x$, then $x = \theta$ is also an asymptote of $y = \cot x$.

60. **ERROR ANALYSIS** Mira and Arturo are studying the graph shown. Mira thinks that it is the graph of $y = -\frac{1}{3} \tan 2x$, and Arturo thinks that it is the graph of $y = \frac{1}{3} \cot 2x$. Is either of them correct? Explain your reasoning.

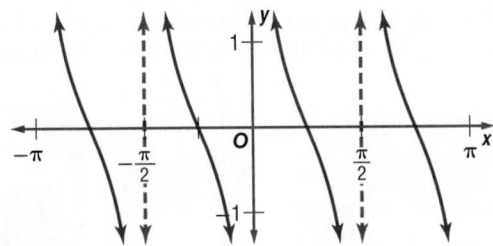

61. **CHALLENGE** Write a cosecant function and a cotangent function that have the same graphs as $y = \sec x$ and $y = \tan x$ respectively. Check your answers by graphing.

62. **WRITING IN MATH** A damped trigonometric function oscillates between the positive and negative graphs of the damping factor. Explain why a damped trigonometric function oscillates between the positive and negative graphs of the damping factor and why the amplitude of the function depends on the damping factor.

State the amplitude, period, frequency, phase shift, and vertical shift of each function. Then graph two periods of the function. (Lesson 4-4)

63. $y = 3 \sin \left(2x - \frac{\pi}{3}\right) + 10$

64. $y = 2 \cos \left(3x + \frac{3\pi}{4}\right) - 6$

65. $y = \frac{1}{2} \cos (4x - \pi) + 1$

Find the exact values of the five remaining trigonometric functions of θ. (Lesson 4-3)

66. $\sin \theta = \frac{4}{5}, \cos \theta > 0$

67. $\cos \theta = \frac{6\sqrt{37}}{37}, \sin \theta > 0$

68. $\tan \theta = \frac{24}{7}, \sin \theta > 0$

69. POPULATION The population of a city 10 years ago was 45,600. Since then, the population has increased at a steady rate each year. If the population is currently 64,800, find the annual rate of growth for this city. (Lesson 3-5)

70. MEDICINE The half-life of a radioactive substance is the amount of time it takes for half of the atoms of the substance to disintegrate. Nuclear medicine technologists use the iodine isotope I-131, with a half-life of 8 days, to check a patient's thyroid function. After ingesting a tablet containing the iodine, the isotopes collect in the patient's thyroid, and a special camera is used to view its function. Suppose a patient ingests a tablet containing 9 microcuries of I-131. To the nearest hour, how long will it be until there are only 2.8 microcuries in the patient's thyroid? (Lesson 3-2)

Factor each polynomial completely using the given factor and long division. (Lesson 2-3)

71. $x^3 + 2x^2 - x - 2; x - 1$

72. $x^3 + x^2 - 16x - 16; x + 4$

73. $x^3 - x^2 - 10x - 8; x + 1$

74. EXERCISE The American College of Sports Medicine recommends that healthy adults exercise at a target level of 60% to 90% of their maximum heart rates. You can estimate your maximum heart rate by subtracting your age from 220. Write a compound inequality that models age a and target heart rate r. (Lesson 0-5)

75. SAT/ACT In the figure, A and D are the centers of the two circles, which intersect at points C and E. $\overline{CE}$ is a diameter of circle D. If $AB = CE = 10$, what is AD?

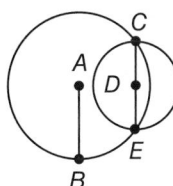

A 5

B $5\sqrt{2}$

C $5\sqrt{3}$

D $10\sqrt{2}$

E $10\sqrt{3}$

76. REVIEW Refer to the figure below. If $c = 14$, find the value of b.

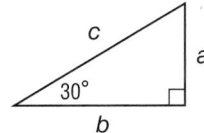

F $\frac{\sqrt{3}}{2}$

G $14\sqrt{3}$

H 7

J $7\sqrt{3}$

77. Which equation is represented by the graph?

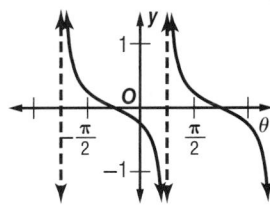

A $y = \cot \left(\theta + \frac{\pi}{4}\right)$

B $y = \cot \left(\theta - \frac{\pi}{4}\right)$

C $y = \tan \left(\theta + \frac{\pi}{4}\right)$

D $y = \tan \left(\theta - \frac{\pi}{4}\right)$

78. REVIEW If $\sin \theta = -\frac{1}{2}$ and $\pi < \theta < \frac{3\pi}{2}$, then $\theta = ?$

F $\frac{13\pi}{12}$

G $\frac{7\pi}{6}$

H $\frac{5\pi}{4}$

J $\frac{4\pi}{3}$

Inverse Trigonometric Functions

Why?

Inverse trigonometric functions can be used to model the changing horizontal angle of rotation needed for a television camera to follow the motion of a drag-racing vehicle.

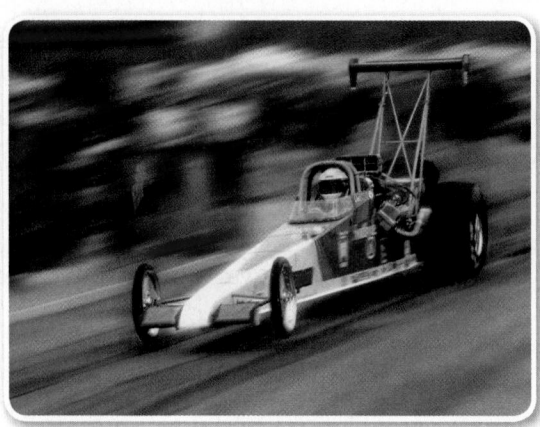

Then
You found and graphed the inverses of relations and functions. (Lesson 1-7)

Now
- Evaluate and graph inverse trigonometric functions.
- Find compositions of trigonometric functions.

New Vocabulary
arcsine function
arccosine function
arctangent function

Math Online
glencoe.com

Inverse Trigonometric Functions In Lesson 1-7, you learned that a function has an inverse function if and only if it is *one-to-one*, meaning that each y-value of the function can be matched with no more than one x-value. Because the sine function fails the horizontal line test, it is not one-to-one.

If, however, we restrict the domain of the sine function to the interval $\left[-\frac{\pi}{2}, \frac{\pi}{2}\right]$, the restricted function *is* one-to-one and takes on all possible range values $[-1, 1]$ of the unrestricted function. It is on this restricted domain that $y = \sin x$ has an inverse function called the *inverse sine function* $y = \sin^{-1} x$. The graph of $y = \sin^{-1} x$ is found by reflecting the graph of the restricted sine function in the line $y = x$.

Restricted Sine Function

Inverse Sine Function

Notice that the domain of $y = \sin^{-1} x$ is $[-1, 1]$, and its range is $\left[-\frac{\pi}{2}, \frac{\pi}{2}\right]$. Because angles and arcs given on the unit circle have equivalent radian measures, the inverse sine function is sometimes referred to as the **arcsine function** $y = \arcsin x$.

In Lesson 4-1, you used the inverse relationship between the sine and inverse sine functions to find acute angle measures. From the graphs above, you can see that in general,

$$y = \sin^{-1} x \text{ or } y = \arcsin x \text{ iff } \sin y = x, \text{ when } -1 \le x \le 1 \text{ and } -\frac{\pi}{2} \le y \le \frac{\pi}{2}. \qquad \textbf{iff means if and only if.}$$

This means that $\sin^{-1} x$ or $\arcsin x$ can be interpreted as *the angle (or arc) between* $-\frac{\pi}{2}$ *and* $\frac{\pi}{2}$ *with a sine of x.* For example, $\sin^{-1} 0.5$ is the angle with a sine of 0.5.

Recall that sin t is the y-coordinate of the point on the unit circle corresponding to the angle or arc length t. Because the range of the inverse sine function is restricted to $\left[-\frac{\pi}{2}, \frac{\pi}{2}\right]$, the possible angle measures of the inverse sine function are located on the right half of the unit circle, as shown.

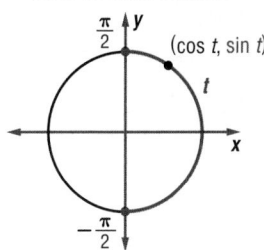

Inverse Sine Values

You can use the unit circle to find the exact value of some expressions involving $\sin^{-1} x$ or arcsin x.

TechnologyTip

▶ **Evaluate sin⁻¹** You can also use a graphing calculator to find the angle that has a sine of $\frac{1}{2}$.

```
sin⁻¹(0.5)
         .5235987756
π/6
         .5235987756
```

Make sure you select RADIAN on the MODE feature of your graphing calculator.

EXAMPLE 1 **Evaluate Inverse Sine Functions**

Find the exact value of each expression, if it exists.

a. $\sin^{-1} \frac{1}{2}$

Find a point on the unit circle on the interval $\left[-\frac{\pi}{2}, \frac{\pi}{2}\right]$ with a y-coordinate of $\frac{1}{2}$. When $t = \frac{\pi}{6}$, sin $t = \frac{1}{2}$.

Therefore, $\sin^{-1} \frac{1}{2} = \frac{\pi}{6}$.

CHECK If $\sin^{-1} \frac{1}{2} = \frac{\pi}{6}$, then $\sin \frac{\pi}{6} = \frac{1}{2}$. ✔

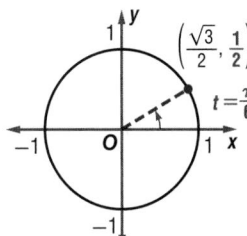

b. $\arcsin\left(-\frac{\sqrt{2}}{2}\right)$

Find a point on the unit circle on the interval $\left[-\frac{\pi}{2}, \frac{\pi}{2}\right]$ with a y-coordinate of $-\frac{\sqrt{2}}{2}$. When $t = -\frac{\pi}{4}$, sin $t = -\frac{\sqrt{2}}{2}$.

Therefore, $\arcsin\left(-\frac{\sqrt{2}}{2}\right) = -\frac{\pi}{4}$.

CHECK If $\arcsin\left(-\frac{\sqrt{2}}{2}\right) = \frac{\pi}{4}$, then $\sin\left(-\frac{\pi}{4}\right) = -\frac{\sqrt{2}}{2}$. ✔

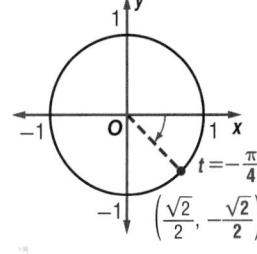

c. $\sin^{-1} 3$

Because the domain of the inverse sine function is $[-1, 1]$ and $3 > 1$, there is no angle with a sine of 3. Therefore, the value of $\sin^{-1} 3$ does not exist.

☑ **Guided Practice**

1A. $\arcsin\left(\frac{\sqrt{3}}{2}\right)$ **1B.** $\sin^{-1}(-2\pi)$ **1C.** $\arcsin(-1)$

Notice in Example 1a that while sin $\frac{5\pi}{6}$ is also $\frac{1}{2}$, $\frac{5\pi}{6}$ is not in the interval $\left[-\frac{\pi}{2}, \frac{\pi}{2}\right]$. Therefore, $\sin^{-1} \frac{1}{2} \neq \frac{5\pi}{6}$.

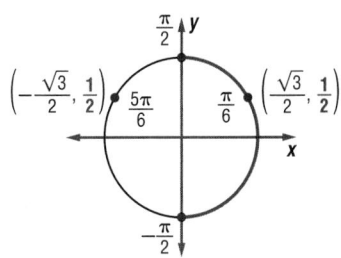

StudyTip

Principal Values
Trigonometric functions with restricted domains are sometimes indicated with capital letters. For example, $y = \text{Sin } x$ represents the function $y = \sin x$, where $-\frac{\pi}{2} \le x \le \frac{\pi}{2}$. The values in these restricted domains are often called *principal values*.

When restricted to a domain of $[0, \pi]$, the cosine function is one-to-one and takes on all of its possible range values on $[-1, 1]$. It is on this restricted domain that the cosine function has an inverse function, called the *inverse cosine function* $y = \cos^{-1} x$ or **arccosine function** $y = \arccos x$. The graph of $y = \cos^{-1} x$ is found by reflecting the graph of the restricted cosine function in the line $y = x$.

Restricted Cosine Function

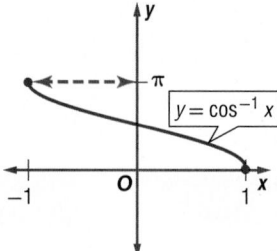

Inverse Cosine Function

Recall that $\cos t$ is the x-coordinate of the point on the unit circle corresponding to the angle or arc length t. Because the range of $y = \cos^{-1} x$ is restricted to $[0, \pi]$, the values of an inverse cosine function are located on the upper half of the unit circle.

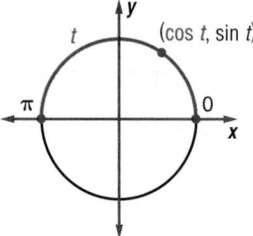

Inverse Cosine Values

EXAMPLE 2 **Evaluate Inverse Cosine Functions**

Find the exact value of each expression, if it exists.

a. $\cos^{-1}\left(-\frac{\sqrt{2}}{2}\right)$

Find a point on the unit circle in the interval $[0, \pi]$ with an x-coordinate of $-\frac{\sqrt{2}}{2}$. When $t = \frac{3\pi}{4}$, $\cos t = -\frac{\sqrt{2}}{2}$.

Therefore, $\cos^{-1}\left(-\frac{\sqrt{2}}{2}\right) = \frac{3\pi}{4}$.

CHECK If $\cos^{-1}\left(-\frac{\sqrt{2}}{2}\right) = \frac{3\pi}{4}$, then $\cos \frac{3\pi}{4} = -\frac{\sqrt{2}}{2}$. ✔

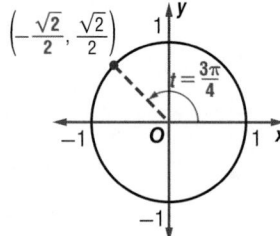

b. $\arccos (-2)$

Since the domain of the cosine function is $[-1, 1]$ and $-2 < -1$, there is no angle with a cosine of -2. Therefore, the value of $\arccos (-2)$ does not exist.

c. $\cos^{-1} 0$

Find a point on the unit circle in the interval $[0, \pi]$ with an x-coordinate of 0. When $t = \frac{\pi}{2}$, $\cos t = 0$.

Therefore, $\cos^{-1} 0 = \frac{\pi}{2}$.

CHECK If $\cos^{-1} 0 = \frac{\pi}{2}$, then $\cos \frac{\pi}{2} = 0$. ✔

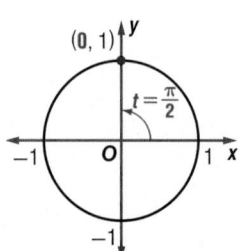

✅ **Guided Practice**

2A. $\cos^{-1}\left(-\frac{\sqrt{3}}{2}\right)$ **2B.** $\arccos 2.5$ **2C.** $\cos^{-1}\left(-\frac{1}{2}\right)$

When restricted to a domain of $\left(-\frac{\pi}{2}, \frac{\pi}{2}\right)$, the tangent function is one-to-one. It is on this restricted domain that the tangent function has an inverse function called the *inverse tangent function* $y = \tan^{-1} x$ or **arctangent function** $y = \arctan x$. The graph of $y = \tan^{-1} x$ is found by reflecting the graph of the restricted tangent function in the line $y = x$. Notice that unlike the sine and cosine functions, the domain of the inverse tangent function is $(-\infty, \infty)$.

Restricted Tangent Function

Inverse Tangent Function

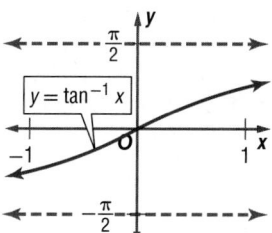

You can also use the unit circle to find the value of an inverse tangent expression. On the unit circle, $\tan t = \frac{\sin t}{\cos t}$ or $\frac{y}{x}$. The values of $y = \tan^{-1} x$ will be located on the right half of the unit circle, not including $-\frac{\pi}{2}$ and $\frac{\pi}{2}$, because the tangent function is undefined at those points.

Inverse Tangent Values

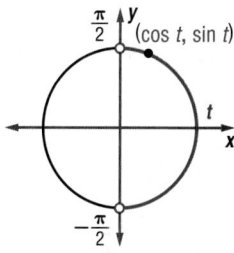

EXAMPLE 3   **Evaluate Inverse Tangent Functions**

Find the exact value of each expression, if it exists.

a. $\tan^{-1}\sqrt{3}$

Find a point (x, y) on the unit circle in the interval $\left(-\frac{\pi}{2}, \frac{\pi}{2}\right)$ such that $\frac{y}{x} = \sqrt{3}$. When $t = \frac{\pi}{3}$, $\tan t = \frac{\frac{\sqrt{3}}{2}}{\frac{1}{2}}$ or $\sqrt{3}$. Therefore, $\tan^{-1} \sqrt{3} = \frac{\pi}{3}$.

CHECK If $\tan^{-1} \sqrt{3} = \frac{\pi}{3}$, then $\tan \frac{\pi}{3} = \sqrt{3}$. ✔

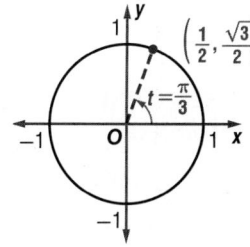

b. $\arctan 0$

Find a point (x, y) on the unit circle in the interval $\left(-\frac{\pi}{2}, \frac{\pi}{2}\right)$ such that $\frac{y}{x} = 0$. When $t = 0$, $\tan t = \frac{0}{1}$ or 0.

Therefore, $\arctan 0 = 0$.

CHECK If $\arctan 0 = 0$, then $\tan 0 = 0$. ✔

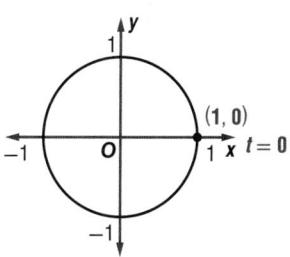

✔ **Guided Practice**

3A. $\arctan\left(-\frac{\sqrt{3}}{3}\right)$ **3B.** $\tan^{-1}(-1)$

While inverse functions for secant, cosecant, and cotangent do exist, these functions are rarely used in computations because the inverse functions for their reciprocals exist. Also, deciding how to restrict the domains of secant, cosecant, and cotangent to obtain arcsecant, arccosecant, and arccotangent is not as apparent. You will explore these functions in Exercise 66.

The three most common inverse trigonometric functions are summarized below.

Key Concept · Inverse Trigonometric Functions

	Inverse Sine of x	**Inverse Cosine of x**	**Inverse Tangent of x**
Words	The angle (or arc) between $-\frac{\pi}{2}$ and $\frac{\pi}{2}$ with a sine of x.	The angle (or arc) between 0 and π with a cosine of x.	The angle (or arc) between $-\frac{\pi}{2}$ and $\frac{\pi}{2}$ with a tangent of x.
Symbols	$y = \sin^{-1} x$ if and only if $\sin y = x$, for $-1 \le x \le 1$ and $-\frac{\pi}{2} \le y \le \frac{\pi}{2}$.	$y = \cos^{-1} x$ if and only if $\cos y = x$, for $-1 \le x \le 1$ and $0 \le y \le \pi$.	$y = \tan^{-1} x$ if and only if $\tan y = x$, for $-\infty < x < \infty$ and $-\frac{\pi}{2} < y < \frac{\pi}{2}$.
Domain:	$[-1, 1]$	$[-1, 1]$	$(-\infty, \infty)$
Range:	$\left[-\frac{\pi}{2}, \frac{\pi}{2}\right]$	$[0, \pi]$	$\left(-\frac{\pi}{2}, \frac{\pi}{2}\right)$

  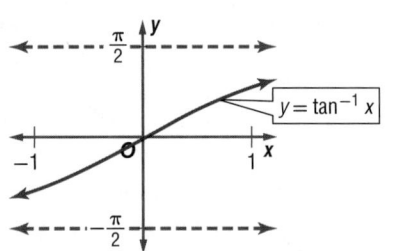

You can sketch the graph of one of the inverse trigonometric functions shown above by rewriting the function in the form $\sin y = x$, $\cos y = x$, or $\tan y = x$, assigning values to y and making a table of values, and then plotting the points and connecting the points with a smooth curve.

Watch Out!

Remember that $\pi = 3.14$ radians or 180°.

EXAMPLE 4 · Sketch Graphs of Inverse Trigonometric Functions

Sketch the graph of $y = \arccos 2x$.

By definition, $y = \arccos 2x$ and $\cos y = 2x$ are equivalent on $0 \le y \le \pi$, so their graphs are the same. Rewrite $\cos y = 2x$ as $x = \frac{1}{2} \cos y$ and assign values to y on the interval $[0, \pi]$ to make a table of values.

y	0	$\frac{\pi}{4}$	$\frac{\pi}{6}$	$\frac{\pi}{2}$	$\frac{5\pi}{6}$	$\frac{3\pi}{4}$	π
$x = \frac{1}{2} \cos y$	$\frac{1}{2}$	$\frac{\sqrt{2}}{4}$	$\frac{\sqrt{3}}{4}$	0	$-\frac{\sqrt{3}}{4}$	$-\frac{\sqrt{2}}{4}$	$-\frac{1}{2}$

Then plot the points (x, y) and connect them with a smooth curve. Notice that this curve has endpoints at $\left(-\frac{1}{2}, \pi\right)$ and $\left(\frac{1}{2}, 0\right)$, indicating that the entire graph of $y = \arccos 2x$ is shown.

✓ Guided Practice

Sketch the graph of each function.

4A. $y = \arcsin 3x$

4B. $y = \tan^{-1} 2x$

🌐 **Real-World EXAMPLE 5**   **Use an Inverse Trigonometric Function**

MOVIES In a movie theater, a person's viewing angle for watching a movie changes depending on where he or she sits in the theater.

a. Write a function modeling the viewing angle θ for a person in the theater whose eye-level when sitting is 4 feet above ground.

Draw a diagram to find the measure of the viewing angle. Let θ_1 represent the angle formed from eye-level to the bottom of the screen, and let θ_2 represent the angle formed from eye-level to the top of the screen.

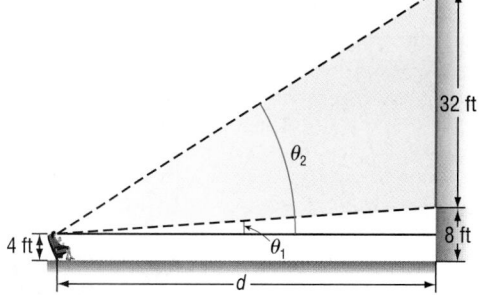

So, the viewing angle is $\theta = \theta_2 - \theta_1$. You can use the tangent function to find θ_1 and θ_2. Because the eye-level of the person when seated is 4 feet above the floor, the distance opposite θ_1 is $8 - 4$ feet or 4 feet long.

$\tan \theta_1 = \dfrac{4}{d}$ opp = 4 and adj = d

$\theta_1 = \tan^{-1} \dfrac{4}{d}$ Inverse tangent function

The distance opposite θ_2 is $(32 + 8) - 4$ feet or 36 feet.

$\tan \theta_2 = \dfrac{36}{d}$ opp = 36 and adj = d

$\theta_2 = \tan^{-1} \dfrac{36}{d}$ Inverse tangent function

So, the viewing angle can be modeled by $\theta = \tan^{-1} \dfrac{36}{d} - \tan^{-1} \dfrac{4}{d}$.

b. Determine the distance that corresponds to the maximum viewing angle.

The distance at which the maximum viewing angle occurs is the maximum point on the graph. You can use a graphing calculator to find this point.

From the graph, you can see that the maximum viewing angle occurs approximately 12 feet from the screen.

[0, 100] scl: 10 by [0, 60] scl: 5

✔️ **Guided Practice**

5. TELEVISION Lucas has purchased a new flat-screen television. So that his family will be able to see, he has decided to hang the television on the wall as shown.

A. Write a function modeling the distance d of the maximum viewing angle θ for Lucas if his eye level when sitting is 3 feet above ground.

B. Determine the distance that corresponds to the maximum viewing angle.

Compositions of Trigonometric Functions In Lesson 1-7, you learned that if x is in the domain of $f(x)$ and $f^{-1}(x)$, then

$$f[f^{-1}(x)] = x \qquad \text{and} \qquad f^{-1}[f(x)] = x.$$

Because the domains of the trigonometric functions are restricted to obtain the inverse trigonometric functions, the properties do not apply for all values of x.

For example, while $\sin x$ is defined for all x, the domain of $\sin^{-1} x$ is $[-1, 1]$. Therefore, $\sin(\sin^{-1} x) = x$ is only true when $-1 \le x \le 1$. A different restriction applies for the composition $\sin^{-1}(\sin x)$. Because the domain of $\sin x$ is restricted to the interval $\left[-\frac{\pi}{2}, \frac{\pi}{2}\right]$, $\sin^{-1}(\sin x) = x$ is only true when $-\frac{\pi}{2} \le x \le \frac{\pi}{2}$.

These domain restrictions are summarized below.

Key Concept · **Domain of Compositions of Trigonometric Functions**

$f[f^{-1}(x)] = x$

If $-1 \le x \le 1$, then $\sin(\sin^{-1} x) = x$.

If $-1 \le x \le 1$, then $\cos(\cos^{-1} x) = x$.

If $-\infty < x < \infty$, then $\tan(\tan^{-1} x) = x$.

$f^{-1}[f(x)] = x$

If $-\frac{\pi}{2} \le x \le \frac{\pi}{2}$, then $\sin^{-1}(\sin x) = x$.

If $0 \le x \le \pi$, then $\cos^{-1}(\cos x) = x$.

If $-\frac{\pi}{2} < x < \frac{\pi}{2}$, then $\tan^{-1}(\tan x) = x$.

EXAMPLE 6 · **Use Inverse Trigonometric Properties**

Find the exact value of each expression, if it exists.

a. $\sin\left[\sin^{-1}\left(-\frac{1}{4}\right)\right]$

The inverse property applies because $-\frac{1}{4}$ lies on the interval $[-1, 1]$.

Therefore, $\sin\left[\sin^{-1}\left(-\frac{1}{4}\right)\right] = -\frac{1}{4}$.

b. $\arctan\left(\tan\frac{\pi}{2}\right)$

Because $\tan x$ is not defined when $x = \frac{\pi}{2}$, $\arctan\left(\tan\frac{\pi}{2}\right)$ does not exist.

c. $\arcsin\left(\sin\frac{7\pi}{4}\right)$

Notice that the angle $\frac{7\pi}{4}$ does not lie on the interval $\left[-\frac{\pi}{2}, \frac{\pi}{2}\right]$. However, $\frac{7\pi}{4}$ is coterminal with $\frac{7\pi}{4} - 2\pi$ or $-\frac{\pi}{4}$, which is on the interval $\left[-\frac{\pi}{2}, \frac{\pi}{2}\right]$.

$\arcsin\left(\sin\frac{7\pi}{4}\right) = \arcsin\left[\sin\left(-\frac{\pi}{4}\right)\right]$ $\sin\frac{7\pi}{4} = \sin\left(-\frac{\pi}{4}\right)$

$\qquad\qquad\qquad\quad = -\frac{\pi}{4}$ Since $-\frac{\pi}{2} \le -\frac{\pi}{4} \le \frac{\pi}{2}$, $\arcsin(\sin x) = x$.

Therefore, $\arcsin\left(\sin\frac{7\pi}{4}\right) = -\frac{\pi}{4}$.

✔ **Guided Practice**

6A. $\tan\left(\tan^{-1}\frac{\pi}{3}\right)$ **6B.** $\cos^{-1}\left(\cos\frac{3\pi}{4}\right)$ **6C.** $\arcsin\left(\sin\frac{2\pi}{3}\right)$

Watch Out!

Compositions and Inverses When computing $f^{-1}[f(x)]$ with trigonometric functions, the domain appears to be $(-\infty, \infty)$. However, because the ranges of the inverse functions are restricted, coterminal angles must sometimes be found.

You can also evaluate the composition of two different inverse trigonometric functions.

EXAMPLE 7 Evaluate Compositions of Trigonometric Functions

Find the exact value of $\cos\left[\tan^{-1}\left(-\frac{3}{4}\right)\right]$.

To simplify the expression, let $u = \tan^{-1}\left(-\frac{3}{4}\right)$, so $\tan u = -\frac{3}{4}$.

Because the tangent function is negative in Quadrants II and IV, and the domain of the inverse tangent function is restricted to Quadrants I and IV, u must lie in Quadrant IV.

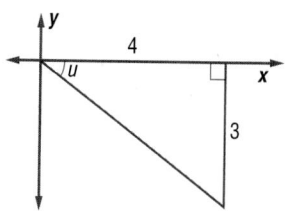

Using the Pythagorean Theorem, you can find that the length of the hypotenuse is 5. Now, solve for $\cos u$.

$\cos u = \dfrac{\text{adj}}{\text{hyp}}$ **Cosine function**

$\quad\ = \dfrac{4}{5}$ **adj = 4 and hyp = 5**

So, $\cos\left[\tan^{-1}\left(-\frac{3}{4}\right)\right] = \frac{4}{5}$.

✔ Guided Practice

Find the exact value of each expression.

7A. $\cos^{-1}\left(\sin\frac{\pi}{3}\right)$

7B. $\sin\left(\arctan\frac{5}{12}\right)$

Sometimes the composition of two trigonometric functions reduces to an algebraic expression that does not involve *any* trigonometric expressions.

EXAMPLE 8 Evaluate Compositions of Trigonometric Functions

Write **tan (arcsin *a*)** as an algebraic expression of *a* that does not involve trigonometric functions.

Let $u = \arcsin a$, so $\sin u = a$.

Because the domain of the inverse sine function is restricted to Quadrants I and IV, u must lie in Quadrant I or IV. The solution is similar for each quadrant, so we will solve for Quadrant I.

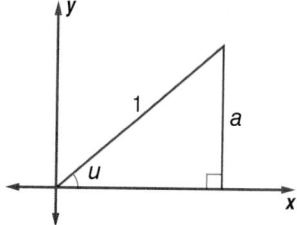

From the Pythagorean Theorem, you can find that the length of the side adjacent to u is $\sqrt{1 - a^2}$. Now, solve for $\tan u$.

$\tan u = \dfrac{\text{opp}}{\text{adj}}$ **Tangent function**

$\quad\ = \dfrac{a}{\sqrt{1 - a^2}} \text{ or } \dfrac{a\sqrt{1 - a^2}}{1 - a^2}$ **opp = *a* and adj = $\sqrt{1 - a^2}$**

So, $\tan(\arcsin a) = \dfrac{a\sqrt{1 - a^2}}{1 - a^2}$.

StudyTip

Decomposing Algebraic Functions The technique used to convert a trigonometric expression into an algebraic expression can be reversed. Decomposing an algebraic function as the composition of two trigonometric functions is a technique used frequently in calculus.

✔ Guided Practice

Write each expression as an algebraic expression of *x* that does not involve trigonometric functions.

8A. $\sin(\arccos x)$

8B. $\cot[\sin^{-1} x]$

Exercises

= **Step-by-Step Solutions** begin on page R29.

Find the exact value of each expression, if it exists.
(Examples 1–3)

1. $\sin^{-1} 0$

2. $\arcsin \dfrac{\sqrt{3}}{2}$

3. $\arcsin \dfrac{\sqrt{2}}{2}$

4. $\sin^{-1} \dfrac{1}{2}$

5. $\sin^{-1} \left(-\dfrac{\sqrt{2}}{2}\right)$

6. $\arccos 0$

7. $\cos^{-1} \dfrac{\sqrt{2}}{2}$

8. $\arccos(-1)$

9. $\arccos \dfrac{\sqrt{3}}{2}$

10. $\cos^{-1} \dfrac{1}{2}$

11. $\arctan 1$

12. $\arctan(-\sqrt{3})$

13. $\tan^{-1} \dfrac{\sqrt{3}}{3}$

14. $\tan^{-1} 0$

15. ARCHITECTURE The support for a roof is shaped like two right triangles, as shown below. Find θ. (Example 3)

16. RESCUE A cruise ship sailed due west 24 miles before turning south. When the cruise ship became disabled and the crew radioed for help, the rescue boat found that the fastest route covered a distance of 48 miles. Find the angle θ at which the rescue boat should travel to aid the cruise ship. (Example 3)

Sketch the graph of each function. (Example 4)

17. $y = \arcsin x$

18. $y = \sin^{-1} 2x$

19. $y = \sin^{-1}(x + 3)$

20. $y = \arcsin x - 3$

21. $y = \arccos x$

22. $y = \cos^{-1} 3x$

23. $y = \arctan x$

24. $y = \tan^{-1} 3x$

25. $y = \tan^{-1}(x + 1)$

26. $y = \arctan x - 1$

27 DRAG RACE A television camera is filming a drag race. The camera rotates as the vehicles move past it. The camera is 30 meters away from the track. Consider θ and x as shown in the figure. (Example 5)

a. Write θ as a function of x.

b. Find θ when $x = 6$ meters and $x = 14$ meters.

28. SPORTS Steve and Ravi want to project a pro soccer game on the side of their apartment building. They have placed a projector on a table that stands 5 feet above the ground and have hung a 12-foot-tall screen that is 10 feet above the ground. (Example 5)

a. Write a function expressing θ in terms of distance d.

b. Use a graphing calculator to determine the distance for the maximum projecting angle.

Find the exact value of each expression, if it exists.
(Examples 6 and 7)

29. $\sin\left(\sin^{-1} \dfrac{3}{4}\right)$

30. $\sin^{-1}\left(\sin \dfrac{\pi}{2}\right)$

31. $\cos\left(\cos^{-1} \dfrac{2}{9}\right)$

32. $\cos^{-1}(\cos \pi)$

33. $\tan\left(\tan^{-1} \dfrac{\pi}{4}\right)$

34. $\tan^{-1}\left(\tan \dfrac{\pi}{3}\right)$

35. $\cos(\tan^{-1} 1)$

36. $\sin^{-1}\left(\cos \dfrac{\pi}{2}\right)$

37. $\sin\left(2\cos^{-1} \dfrac{\sqrt{2}}{2}\right)$

38. $\sin(\tan^{-1} 1 - \sin^{-1} 1)$

39. $\cos(\tan^{-1} 1 - \sin^{-1} 1)$

40. $\cos\left(\cos^{-1} 0 + \sin^{-1} \dfrac{1}{2}\right)$

Write each trigonometric expression as an algebraic expression of *x*. (Example 8)

41 $\tan(\arccos x)$

42. $\csc(\cos^{-1} x)$

43. $\sin(\cos^{-1} x)$

44. $\cos(\arcsin x)$

45. $\csc(\sin^{-1} x)$

46. $\sec(\arcsin x)$

47. $\cot(\arccos x)$

48. $\cot(\arcsin x)$

Describe how the graphs of *g*(*x*) and *f*(*x*) are related.

49. $f(x) = \sin^{-1} x$ and $g(x) = \sin^{-1}(x-1) - 2$

50. $f(x) = \arctan x$ and $g(x) = \arctan 0.5x - 3$

51. $f(x) = \cos^{-1} x$ and $g(x) = 3(\cos^{-1} x - 2)$

52. $f(x) = \arcsin x$ and $g(x) = \frac{1}{2}\arcsin(x+2)$

53. $f(x) = \arccos x$ and $g(x) = 5 + \arccos 2x$

54. $f(x) = \tan^{-1} x$ and $g(x) = \tan^{-1} 3x - 4$

55. SAND When piling sand, the angle formed between the pile and the ground remains fairly consistent and is called the *angle of repose*. Suppose Jade creates a pile of sand at the beach that is 3 feet in diameter and 1.1 feet high.

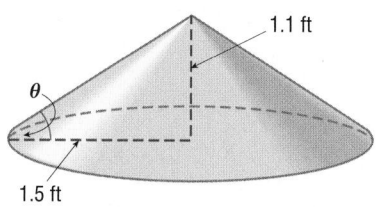

1.1 ft

θ

1.5 ft

a. What is the angle of repose?

b. If the angle of repose remains constant, how many feet in diameter would a pile need to be to reach a height of 4 feet?

Give the domain and range of each composite function. Then use your graphing calculator to sketch its graph.

56. $y = \cos(\tan^{-1} x)$

57. $y = \sin(\cos^{-1} x)$

58. $y = \arctan(\sin x)$

59. $y = \sin^{-1}(\cos x)$

60. $y = \cos(\arcsin x)$

61. $y = \tan(\arccos x)$

62. INVERSES The arcsecant function is graphed by restricting the domain of the secant function to the intervals $\left[0, \frac{\pi}{2}\right)$ and $\left(\frac{\pi}{2}, \pi\right]$, and the arccosecant function is graphed by restricting the domain of the cosecant function to the intervals $\left[-\frac{\pi}{2}, 0\right)$ and $\left(0, \frac{\pi}{2}\right]$.

a. State the domain and range of each function.

b. Sketch the graph of each function.

c. Explain why a restriction on the domain of the secant and cosecant functions is necessary in order to graph the inverse functions.

Write each algebraic expression as a trigonometric function of an inverse trigonometric function of *x*.

63. $\dfrac{x}{\sqrt{1-x^2}}$

64. $\dfrac{\sqrt{1-x^2}}{x}$

65. ⬚ **MULTIPLE REPRESENTATIONS** In this problem, you will explore the graphs of compositions of trigonometric functions.

a. ANALYTICAL Consider $f(x) = \sin x$ and $f^{-1}(x) = \arcsin x$. Describe the domain and range of $f \circ f^{-1}$ and $f^{-1} \circ f$.

b. GRAPHICAL Create a table of several values for each composite function on the interval $[-2, 2]$. Then use the table to sketch the graphs of $f \circ f^{-1}$ and $f^{-1} \circ f$. Use a graphing calculator to check your graphs.

c. ANALYTICAL Consider $g(x) = \cos x$ and $g^{-1}(x) = \arccos x$. Describe the domain and range of $g \circ g^{-1}$ and $g^{-1} \circ g$ and make a conjecture as to what the graphs of $g \circ g^{-1}$ and $g^{-1} \circ g$ will look like. Explain your reasoning.

d. GRAPHICAL Sketch the graphs of $g \circ g^{-1}$ and $g^{-1} \circ g$. Use a graphing calculator to check your graphs.

e. VERBAL Make a conjecture as to what the graphs of the two possible compositions of the tangent and arctangent functions will look like. Explain your reasoning. Then check your conjecture using a graphing calculator.

H.O.T. Problems *Use Higher-Order Thinking Skills*

66. ERROR ANALYSIS Alisa and Trey are discussing inverse trigonometric functions. Because $\tan x = \frac{\sin x}{\cos x}$, Alisa conjectures that $\tan^{-1} x = \frac{\sin^{-1} x}{\cos^{-1} x}$. Trey disagrees. Is either of them correct? Explain.

67. CHALLENGE Use the graphs of $y = \sin^{-1} x$ and $y = \cos^{-1} x$ to find the value of $\sin^{-1} x + \cos^{-1} x$ on the interval $[-1, 1]$. Explain your reasoning.

68. REASONING Determine whether the following statement is *true* or *false*: If $\cos \frac{7\pi}{4} = \frac{\sqrt{2}}{2}$, then $\cos^{-1}\frac{\sqrt{2}}{2} = \frac{7\pi}{4}$. Explain your reasoning.

REASONING Determine whether each function is *odd*, *even*, or *neither*. Justify your answer.

69. $y = \sin^{-1} x$

70. $y = \cos^{-1} x$

71. $y = \tan^{-1} x$

72. WRITING IN MATH Explain how the restrictions on the sine, cosine, and tangent functions dictate the domain and range of their inverse functions.

Locate the vertical asymptotes, and sketch the graph of each function. (Lesson 4-5)

73. $y = 3 \tan \theta$

74. $y = \cot 5\theta$

75. $y = 3 \csc \frac{1}{2}\theta$

76. WAVES A leaf floats on the water bobbing up and down. The distance between its highest and lowest points is 4 centimeters. It moves from its highest point down to its lowest point and back to its highest point every 10 seconds. Write a cosine function that models the movement of the leaf in relationship to the equilibrium point. (Lesson 4-4)

Find the value of x. Round to the nearest tenth, if necessary. (Lesson 4-1)

77.

78.

79.

For each pair of functions, find $[f \circ g](x)$, $[g \circ f](x)$, and $[f \circ g](4)$. (Lesson 1-6)

80. $f(x) = x^2 + 3x - 6$
$g(x) = 4x + 1$

81. $f(x) = 6 - 5x$
$g(x) = \frac{1}{x}$

82. $f(x) = \sqrt{x + 3}$
$g(x) = x^2 + 1$

83. EDUCATION Todd has answered 11 of his last 20 daily quiz questions correctly. His baseball coach told him that he must raise his average to at least 70% if he wants to play in the season opener. Todd vows to study diligently and answer all of the daily quiz questions correctly in the future. How many consecutive daily quiz questions must he answer correctly to raise his average to 70%? (Lesson 0-8)

Skills Review for Standardized Tests

84. SAT/ACT To the nearest degree, what is the angle of depression θ between the shallow end and the deep end of the swimming pool?

Side View of Swimming Pool

A 25° **C** 41° **E** 73°

B 37° **D** 53°

85. Which of the following represents the exact value of $\sin\left(\tan^{-1}\frac{1}{2}\right)$?

F $-\dfrac{2\sqrt{5}}{5}$

H $\dfrac{\sqrt{5}}{5}$

G $-\dfrac{\sqrt{5}}{5}$

J $\dfrac{2\sqrt{5}}{5}$

86. REVIEW The hypotenuse of a right triangle is 67 inches. If one of the angles has a measure of 47°, what is the length of the shortest leg of the triangle?

A 45.7 in. **C** 62.5 in.

B 49.0 in. **D** 71.8 in.

87. REVIEW Two trucks, A and B, start from the intersection C of two straight roads at the same time. Truck A is traveling twice as fast as truck B and after 4 hours, the two trucks are 350 miles apart. Find the approximate speed of truck B in miles per hour.

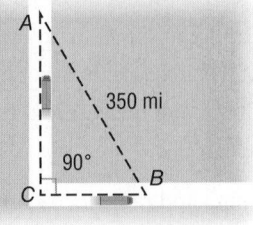

F 39 **H** 51

G 44 **J** 78

4-7

The Law of Sines and the Law of Cosines

Then

You solved right triangles using trigonometric functions. (Lesson 4-1)

Now

- Solve oblique triangles by using the Law of Sines or the Law of Cosines.
- Find areas of oblique triangles.

New Vocabulary

oblique triangles
Law of Sines
ambiguous case
Law of Cosines
Heron's Formula

Math Online >

glencoe.com

Why?

Triangulation is the process of finding the coordinates of a point and the distance to that point by calculating the length of one side of a triangle, given the measurements of the angles and sides of the triangle formed by that point and two other known reference points. Weather spotters can use triangulation to determine the location of a tornado.

Solve Oblique Triangles In Lesson 4-1, you used trigonometric functions to solve *right* triangles. In this lesson, you will solve **oblique triangles**—triangles that are not right triangles.

You can apply the **Law of Sines** to solve an oblique triangle if you know the measures of two angles and a nonincluded side (AAS), two angles and the included side (ASA), or two sides and a nonincluded angle (SSA).

Key Concept | **Law of Sines**

If $\triangle ABC$ has side lengths a, b, and c representing the lengths of the sides opposite the angles with measures A, B, and C, then $\dfrac{\sin A}{a} = \dfrac{\sin B}{b} = \dfrac{\sin C}{c}$.

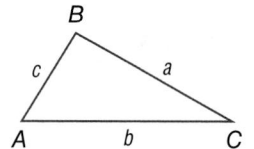

You will derive the Law of Sines in Exercise 69.

EXAMPLE 1 | **Apply the Law of Sines (AAS)**

Solve $\triangle ABC$. Round side lengths to the nearest tenth and angle measures to the nearest degree.

Because two angles are given, $C = 180° - (103° + 35°)$ or $42°$.

Use the Law of Sines to find a and c.

$\dfrac{\sin B}{b} = \dfrac{\sin A}{a}$	Law of Sines	$\dfrac{\sin B}{b} = \dfrac{\sin C}{c}$	
$\dfrac{\sin 103°}{20} = \dfrac{\sin 35°}{a}$	Substitution	$\dfrac{\sin 103°}{20} = \dfrac{\sin 42°}{c}$	
$a \sin 103° = 20 \sin 35°$	Multiply.	$c \sin 103° = 20 \sin 42°$	
$a = \dfrac{20 \sin 35°}{\sin 103°}$	Divide.	$c = \dfrac{20 \sin 42°}{\sin 103°}$	
$a \approx 11.8$	Use a calculator.	$c \approx 13.7$	

Therefore, $a \approx 11.8$, $c \approx 13.7$, and $\angle C = 42°$.

✓ Guided Practice

Solve each triangle. Round side lengths to the nearest tenth and angle measures to the nearest degree.

1A.

1B.

Real-World EXAMPLE 2 **Apply the Law of Sines (ASA)**

SATELLITES An Earth-orbiting satellite is passing between the Oak Ridge Laboratory in Tennessee and the Langley Research Center in Virginia, which are 446 miles apart. If the angles of elevation to the satellite from the Oak Ridge and Langley facilities are 58° and 72°, respectively, how far is the satellite from each station?

Because two angles are given, $C = 180° - (58° + 72°)$ or 50°. Use the Law of Sines to find the distance to the satellite from each station.

$\dfrac{\sin C}{c} = \dfrac{\sin B}{b}$	**Law of Sines**	$\dfrac{\sin C}{c} = \dfrac{\sin A}{a}$
$\dfrac{\sin 50°}{446} = \dfrac{\sin 72°}{b}$	**Substitution**	$\dfrac{\sin 50°}{446} = \dfrac{\sin 58°}{a}$
$b \sin 50° = 446 \sin 72°$	**Multiply.**	$a \sin 50° = 446 \sin 58°$
$b = \dfrac{446 \sin 72°}{\sin 50°}$	**Divide.**	$a = \dfrac{446 \sin 58°}{\sin 50°}$
$b \approx 553.72$	**Use a calculator.**	$a \approx 493.74$

So, the satellite is about 554 miles from Oak Ridge and about 494 miles from Langley.

Guided Practice

2. **SHIPPING** Two ships are 250 feet apart and traveling to the same port as shown. Find the distance from the port to each ship.

From geometry, you know that the measures of two sides and a nonincluded angle (SSA) do not necessarily define a unique triangle. Consider the angle and side measures given in the figures below.

In general, given the measures of two sides and a nonincluded angle, one of the following will be true: (1) no triangle exists, (2) exactly one triangle exists, or (3) two triangles exist. In other words, when solving an oblique triangle for this **ambiguous case**, there may be no solution, one solution, or two solutions.

Key Concept — The Ambiguous Case (SSA)

Consider a triangle in which a, b, and A are given. For the acute case, $\sin A = \frac{h}{b}$, so $h = b \sin A$.

A is Acute.
($A < 90°$)

$a < b$ and $a < h$
no solution

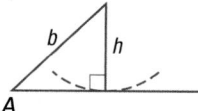

$a < b$ and $a = h$
one solution

$a < b$ and $a > h$
two solutions

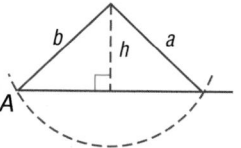

$a \geq b$
one solution

A is Right or Obtuse.
($A \geq 90°$)

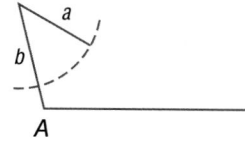

$a \leq b$, no solution

$a > b$, one solution

To solve an ambiguous case oblique triangle, first determine the number of possible solutions. If the triangle has one or two solutions, use the Law of Sines to find them.

EXAMPLE 3 The Ambiguous Case—One or No Solution

Find all solutions for the given triangle, if possible. If no solution exists, write *no solution*. Round side lengths to the nearest tenth and angle measures to the nearest degree.

a. $a = 15, c = 12, A = 94°$

Notice that A is obtuse and $a > c$ because $15 > 12$. Therefore, one solution exists. Apply the Law of Sines to find C.

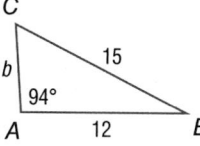

$$\frac{\sin C}{12} = \frac{\sin 94°}{15} \qquad \text{Law of Sines}$$

$$\sin C = \frac{12 \sin 94°}{15} \qquad \text{Multiply each side by 12.}$$

$$C = \sin^{-1}\left(\frac{12 \sin 94°}{15}\right) \text{ or about } 53° \qquad \text{Definition of } \sin^{-1}$$

Because two angles are now known, $B \approx 180° - (94° + 53°)$ or about $33°$. Apply the Law of Sines to find b. Choose the ratios with the fewest calculated values to ensure greater accuracy.

$$\frac{\sin 94°}{15} \approx \frac{\sin 33°}{b} \qquad \text{Law of Sines}$$

$$b \approx \frac{15 \sin 33°}{\sin 94°} \text{ or about } 8.2 \qquad \text{Solve for } b.$$

Therefore, the remaining measures of $\triangle ABC$ are $B \approx 33°$, $C \approx 53°$, and $b \approx 8.2$.

b. $a = 9, b = 11, A = 61°$

Notice that A is acute and $a < b$ because $9 < 11$. Find h.

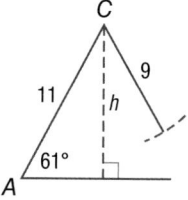

$$\sin 61° = \frac{h}{11} \qquad \text{Definition of sine}$$

$$h = 11 \sin 61° \text{ or about } 9.6 \qquad h = b \sin A$$

Because $a < h$, no triangle can be formed with sides $a = 9$, $b = 11$, and $A = 61°$. Therefore, this problem has no solution.

✓ Guided Practice

3A. $a = 12, b = 8, B = 61°$

3B. $a = 13, c = 26, A = 30°$

EXAMPLE 4   **The Ambiguous Case-Two Solutions**

Find two triangles for which $A = 43°$, $a = 25$, and $b = 28$. Round side lengths to the nearest tenth and angle measures to the nearest degree.

A is acute, and $h = 28 \sin 43°$ or about 19.1. Notice that $a < b$ because $25 < 28$, and $a > h$ because $25 > 19.1$. Therefore, two different triangles are possible with the given angle and side measures. Angle B will be acute, while angle B' will be obtuse.

Make a reasonable sketch of each triangle and apply the Law of Sines to find each solution. Start with the case in which B is acute.

Solution 1 $\angle B$ is acute.

Find B.

$\dfrac{\sin B}{28} = \dfrac{\sin 43°}{25}$ **Law of Sines**

$\sin B = \dfrac{28 \sin 43°}{25}$ **Solve for sin B.**

$\sin B \approx 0.7638$ **Use a calculator.**

$B \approx \sin^{-1} 0.7638$ or about 50°   **Definition of $\sin^{-1}$**

Find C.

$C \approx 180° - (43° + 50°) \approx 87°$

Apply the Law of Sines to find c.

$\dfrac{\sin 87°}{c} \approx \dfrac{\sin 43°}{25}$ **Law of Sines**

$c \approx \dfrac{25 \sin 87°}{\sin 43°}$ or about 36.6   **Solve for c.**

Solution 2 $\angle B'$ is obtuse.

Note that $m\angle CB'B \cong m\angle CBB'$. To find B', you need to find an obtuse angle with a sine that is also 0.7638. To do this, subtract the measure given by your calculator to the nearest degree, 50°, from 180°. Therefore, B' is approximately $180° - 50°$ or 130°.

Find C.

$C \approx 180° - (43° + 130°)$ or 7°

Apply the Law of Sines to find c.

$\dfrac{\sin 7°}{c} \approx \dfrac{\sin 43°}{25}$ **Law of Sines**

$c \approx \dfrac{25 \sin 7°}{\sin 43°}$ or about 4.5   **Solve for c.**

Therefore, the missing measures for acute $\triangle ABC$ are $B \approx 50°$, $C \approx 87°$, and $c \approx 36.6$, while the missing measures for obtuse $\triangle AB'C$ are $B' \approx 130°$, $C \approx 7°$, and $c \approx 4.5$.

TechnologyTip

Using $\sin^{-1}$ Notice that when calculating $\sin^{-1}$ of a ratio, your calculator will never return two possible angle measures because $\sin^{-1}$ is a *function*. Also, your calculator will never return an obtuse angle measure for $\sin^{-1}$ because the inverse sine function has a range of $-\dfrac{\pi}{2}$ to $\dfrac{\pi}{2}$ or $-90°$ to $90°$.

☑ **Guided Practice**

Find two triangles with the given angle measure and side lengths. Round side lengths to the nearest tenth and angle measures to the nearest degree.

4A. $A = 38°$, $a = 8$, $b = 10$ **4B.** $A = 65°$, $a = 55$, $b = 57$

You can use the **Law of Cosines** to solve an oblique triangle for the remaining two cases: when you are given the measures of three sides (SSS) or the measures of two sides and their included angle (SAS).

Key Concept   **Law of Cosines**

In △ABC, if sides with lengths a, b, and c are opposite angles with measures A, B, and C, respectively, then the following are true.

$$a^2 = b^2 + c^2 - 2bc \cos A$$
$$b^2 = a^2 + c^2 - 2ac \cos B$$
$$c^2 = a^2 + b^2 - 2ab \cos C$$

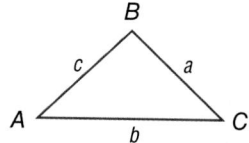

You will derive the first formula for the Law of Cosines in Exercise 70.

Real-World EXAMPLE 5   **Apply the Law of Cosines (SSS)**

HOCKEY When a hockey player attempts a shot, he is 20 feet from the left post of the goal and 24 feet from the right post, as shown. If a regulation hockey goal is 6 feet wide, what is the player's shot angle to the nearest degree?

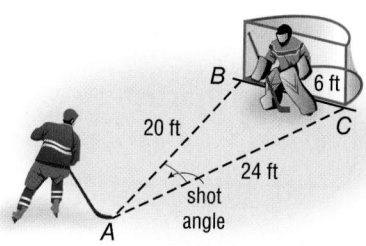

Since three side lenghts are given, you can use the Law of Cosines to find the player's shot angle, A.

$a^2 = b^2 + c^2 - 2bc \cos A$	**Law of Cosines**
$6^2 = 24^2 + 20^2 - 2(24)(20) \cos A$	$a = 6$, $b = 24$, and $c = 20$
$36 = 576 + 400 - 960 \cos A$	**Simplify.**
$36 = 976 - 960 \cos A$	**Add.**
$-940 = -960 \cos A$	**Subtract 976 from each side.**
$\dfrac{940}{960} = \cos A$	**Divide each side by −960.**
$\cos^{-1}\left(\dfrac{940}{960}\right) = A$	**Use the $\cos^{-1}$ function.**
$11.7° \approx A$	**Use a calculator.**

So, the player's shot angle is about 12°.

✓ **Guided Practice**

5. HIKING A group of friends who are on a camping trip decide to go on a hike. According to the map shown, what is the angle that is formed by the two trails that lead to the camp?

EXAMPLE 6 **Apply the Law of Cosines (SAS)**

Solve $\triangle ABC$. Round side lengths to the nearest tenth and angle measures to the nearest degree.

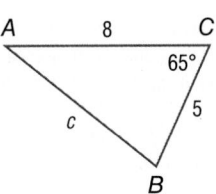

Step 1 Use the Law of Cosines to find the missing side measure.

$c^2 = a^2 + b^2 - 2ab \cos C$	Law of Cosines
$c^2 = 5^2 + 8^2 - 2(5)(8) \cos 65°$	$a = 5, b = 8,$ and $C = 65°$
$c^2 \approx 55.19$	Use a calculator.
$c \approx 7.4$	Take the positive square root of each side.

Step 2 Use the Law of Sines to find a missing angle measure.

$\dfrac{\sin A}{5} = \dfrac{\sin 65°}{7.4}$	$\dfrac{\sin A}{a} = \dfrac{\sin C}{c}$
$\sin A = \dfrac{5 \sin 65°}{7.4}$	Multiply each side by 5.
$A \approx 38°$	Definition of $\sin^{-1}$

Step 3 Find the measure of the remaining angle.

$B \approx 180° - (65° + 38°)$ or $77°$

Therefore, $c \approx 7.4$, $A \approx 38°$, and $B \approx 77°$.

✓ **Guided Practice**

6. Solve $\triangle HJK$ if $H = 34°$, $j = 7$, and $k = 10$.

Find Areas of Oblique Triangles When the measures of all three sides of a triangle are known, the Law of Cosines can be used to prove **Heron's Formula** for the area of the triangle.

Key Concept **Heron's Formula**

If the measures of the sides of $\triangle ABC$ are a, b, and c, then the area of the triangle is

$$\text{Area} = \sqrt{s(s - a)(s - b)(s - c)},$$

where $s = \frac{1}{2}(a + b + c)$.

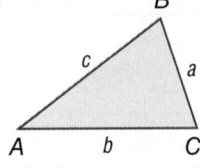

You will prove this formula in Lesson 5-1.

EXAMPLE 7 **Heron's Formula**

Find the area of $\triangle XYZ$.

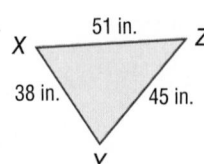

The value of s is $\frac{1}{2}(45 + 51 + 38)$ or 67.

$\text{Area} = \sqrt{s(s - x)(s - y)(s - z)}$	Heron's Formula
$= \sqrt{67(67 - 45)(67 - 51)(67 - 38)}$	$s = 67, x = 45, y = 51,$ and $z = 38$
$= \sqrt{683{,}936}$	Simplify.
$\approx 827 \text{ in}^2$	Use a calculator.

✓ **Guided Practice**

7A. $x = 24$ cm, $y = 53$ cm, $z = 39$ cm   **7B.** $x = 61$ ft, $y = 70$ ft, $z = 88$ ft

In the ambiguous case of the Law of Sines, you compared the length of a to the value $h = b \sin A$. In the triangle shown, h represents the length of the altitude to side c in $\triangle ABC$. You can use this expression for the height of the triangle to develop a formula for the area of the triangle.

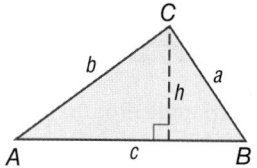

$$\text{Area} = \tfrac{1}{2}ch \qquad \text{Formula for area of a triangle}$$

$$\phantom{\text{Area}} = \tfrac{1}{2}c(b \sin A) \qquad \text{Replace } h \text{ with } b \sin A.$$

$$\phantom{\text{Area}} = \tfrac{1}{2}bc \sin A \qquad \text{Simplify.}$$

By a similar argument, you can develop the formulas

$$\text{Area} = \tfrac{1}{2}ab \sin C \qquad \text{and} \qquad \text{Area} = \tfrac{1}{2}ac \sin B.$$

Notice that in each of these formulas, the information needed to find the area of the triangle is the measures of two sides and the included angle.

Key Concept **Area of a Triangle Given SAS**

Words The area of a triangle is one half the product of the lengths of two sides and the sine of their included angle.

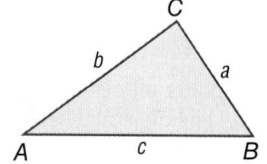

Symbols $\text{Area} = \dfrac{1}{2}bc \sin A$

$\text{Area} = \dfrac{1}{2}ac \sin B$

$\text{Area} = \dfrac{1}{2}ab \sin C$

Because the area of a triangle is constant, the formulas above can be written as one formula.

$$\text{Area} = \tfrac{1}{2}bc \sin A = \tfrac{1}{2}ab \sin C = \tfrac{1}{2}ac \sin B$$

If the included angle measures 90°, notice that each formula simplifies to the formula for the area of a right triangle, $\tfrac{1}{2}(\text{base})(\text{height})$, because $\sin 90° = 1$.

EXAMPLE 8 **Find the Area of a Triangle Given SAS**

Find the area of $\triangle GHJ$ to the nearest tenth.

In $\triangle GHJ$, $g = 7$, $h = 10$, and $J = 108°$.

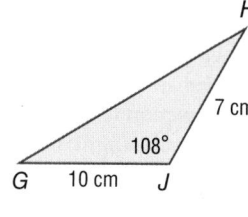

$$\text{Area} = \tfrac{1}{2}gh \sin J \qquad \text{Area of a triangle using SAS}$$

$$\phantom{\text{Area}} = \tfrac{1}{2}(7)(10) \sin 108° \qquad \text{Substitution}$$

$$\phantom{\text{Area}} \approx 33.3 \qquad \text{Simplify.}$$

So, the area is about 33.3 square centimeters.

✓ **Guided Practice**

Find the area of each triangle to the nearest tenth.

8A.

8B.

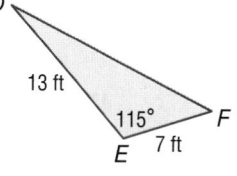

Exercises

Solve each triangle. Round to the nearest tenth, if necessary. (Examples 1 and 2)

1.

2.

3.

4.

5.

6.

7. GOLF A golfer misses a 12-foot putt by putting 3° off course. The hole now lies at a 129° angle between the ball and its spot before the putt. What distance does the golfer need to putt in order to make the shot? (Examples 1 and 2)

8. ARCHITECTURE An architect's client wants to build a home based on the architect Jon Lautner's Sheats-Goldstein House. The length of the patio will be 60 feet. The left side of the roof will be at a 49° angle of elevation, and the right side will be at an 18° angle of elevation. Determine the lengths of the left and right sides of the roof and the angle at which they will meet. (Examples 1 and 2)

9 TRAVEL For the initial 90 miles of a flight, the pilot heads 8° off course in order to avoid a storm. The pilot then changes direction to head toward the destination for the remainder of the flight, making a 157° angle to the first flight course. (Examples 1 and 2)

a. Determine the total distance of the flight.

b. Determine the distance of a direct flight to the destination.

Find all solutions for the given triangle, if possible. If no solution exists, write _no solution_. Round side lengths to the nearest tenth and angle measures to the nearest degree. (Example 3)

10. $a = 9, b = 7, A = 108°$

11. $a = 14, b = 15, A = 117°$

12. $a = 18, b = 12, A = 27°$

13. $a = 35, b = 24, A = 92°$

14. $a = 14, b = 6, A = 145°$

15. $a = 19, b = 38, A = 30°$

16. $a = 5, b = 6, A = 63°$

17. $a = 10, b = \sqrt{200}, A = 45°$

18. SKIING A ski lift rises at a 28° angle during the first 20 feet up a mountain to achieve a height of 25 feet, which is the height maintained during the remainder of the ride up the mountain. Determine the length of cable needed for this initial rise. (Example 3)

Find two triangles with the given angle measure and side lengths. Round side lengths to the nearest tenth and angle measures to the nearest degree. (Example 4)

19. $A = 39°, a = 12, b = 17$

20. $A = 26°, a = 5, b = 9$

21. $A = 61°, a = 14, b = 15$

22. $A = 47°, a = 25, b = 34$

23. $A = 54°, a = 31, b = 36$

24. $A = 18°, a = 8, b = 13$

25. BROADCASTING A radio tower located 38 miles along Industrial Parkway transmits radio broadcasts over a 30-mile radius. Industrial Parkway intersects the interstate at a 41° angle. How far along the interstate can vehicles pick up the broadcasting signal? (Example 4)

26. BOATING The light from a lighthouse can be seen from an 18-mile radius. A boat is anchored so that it can just see the light from the lighthouse. A second boat is located 25 miles from the lighthouse and is headed straight toward it, making a 44° angle with the lighthouse and the first boat. Find the distance between the two boats when the second boat enters the radius of the lighthouse light. (Example 4)

Solve each triangle. Round side lengths to the nearest tenth and angle measures to the nearest degree. (Examples 5 and 6)

27. $\triangle ABC$, if $A = 42°, b = 12$, and $c = 19$

28. $\triangle XYZ$, if $x = 5, y = 18$, and $z = 14$

29. $\triangle PQR$, if $P = 73°, q = 7$, and $r = 15$

30. $\triangle JKL$, if $J = 125°, k = 24$, and $l = 33$

31. $\triangle RST$, if $r = 35, s = 22$, and $t = 25$

32. $\triangle FGH$, if $f = 39, g = 50$, and $h = 64$

33. $\triangle BCD$, if $B = 16°, c = 27$, and $d = 3$

34. $\triangle LMN$, if $\ell = 12, m = 4$, and $n = 9$

35. AIRPLANES During her shift, a pilot flies from Columbus to Atlanta, a distance of 448 miles, and then on to Phoenix, a distance of 1583 miles. From Phoenix, she returns home to Columbus, a distance of 1667 miles. Determine the angles of the triangle created by her flight path. (Examples 5 and 6)

36. CATCH Lola rolls a ball on the ground at an angle of 23° to the right of her dog Buttons. If the ball rolls a total distance of 48 feet, and she is standing 30 feet away, how far will Buttons have to run to retrieve the ball? (Examples 5 and 6)

Use Heron's Formula to find the area of each triangle. Round to the nearest tenth. (Example 7)

37. $x = 9$ cm, $y = 11$ cm, $z = 16$ cm

38. $x = 29$ in., $y = 25$ in., $z = 27$ in.

39. $x = 58$ ft, $y = 40$ ft, $z = 63$ ft

40. $x = 37$ mm, $y = 10$ mm, $z = 34$ mm

41. $x = 8$ yd, $y = 15$ yd, $z = 8$ yd

42. $x = 133$ mi, $y = 82$ mi, $z = 77$ mi

43. LANDSCAPING The Steele family wants to expand their backyard by purchasing a vacant lot adjacent to their property. To get a rough measurement of the area of the lot, Mr. Steele counted the steps needed to walk around the border and diagonal of the lot. (Example 7)

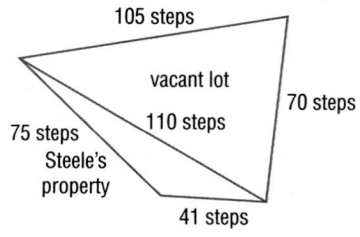

a. Estimate the area of the lot in steps.

b. If Mr. Steele measured his step to be 1.8 feet, determine the area of the lot in square feet.

44. DANCE During a performance, a dancer remained within a triangular area of the stage. (Example 7)

a. Find the area of stage used in the performance.

b. If the stage is 250 square feet, determine the percentage of the stage used in the performance.

Find the area of each triangle to the nearest tenth. (Example 8)

45. $\triangle ABC$, if $A = 98°$, $b = 13$ mm, and $c = 8$ mm

46. $\triangle JKL$, if $L = 67°$, $j = 11$ yd, and $k = 24$ yd

47. $\triangle RST$, if $R = 35°$, $s = 42$ ft, and $t = 26$ ft

48. $\triangle XYZ$, if $Y = 124°$, $x = 16$ m, and $z = 18$ m

49. $\triangle FGH$, if $F = 41°$, $g = 22$ in., and $h = 36$ in.

50. $\triangle PQR$, if $Q = 153°$, $p = 27$ cm, and $r = 21$ cm

51. DESIGN A free-standing art project requires a triangular support piece for stability. Two sides of the triangle must measure 18 and 15 feet in length and a nonincluded angle must measure 42°. If support purposes require the triangle to have an area of at least 75 square feet, what is the measure of the third side? (Example 8)

Use Heron's Formula to find the area of each figure. Round answers to the nearest tenth.

52.

53.

54.

55.

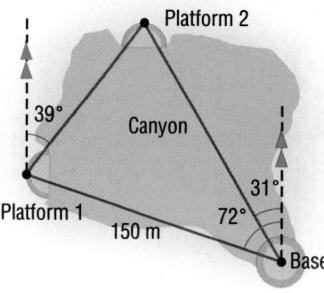

56. ZIP LINES A tourist attraction currently has its base connected to a tree platform 150 meters away by a zip line. The owners now want to connect the base to a second platform located across a canyon and then connect the platforms to each other. The bearings from the base to each platform and from platform 1 to platform 2 are given. Find the distances from the base to platform 2 and from platform 1 to platform 2.

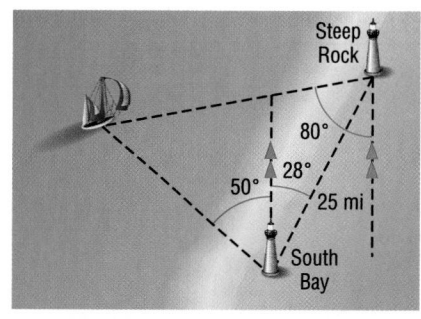

57. LIGHTHOUSES The bearing from the South Bay lighthouse to the Steep Rock lighthouse 25 miles away is N 28° E. A small boat in distress spotted off the coast by each lighthouse has a bearing of N 50° W from South Bay and S 80° W from Steep Rock. How far is each tower from the boat?

Find the area of each figure. Round answers to the nearest tenth.

58.

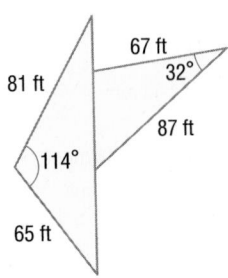

43 cm 32 cm
30° 35°
51 cm

59.

36 mm 42 mm
26°

60.

67 ft
32°
81 ft
87 ft
114°
65 ft

61.

37.5 in.
38°
50° 32 in.
42 in.

62. BRIDGE DESIGN In the figure below, $\angle FDE = 45°$, $\angle CED = 55°$, $\angle FDE \cong \angle FGE$, B is the midpoint of AC, and $DE \cong EG$. If $AD = 4$ feet, $DE = 12$ feet, and $CE = 14$ feet, find BF.

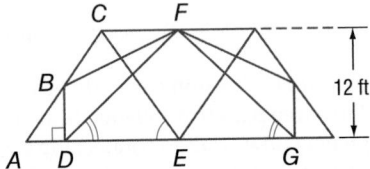

C   F
B   12 ft
A D   E   G

63. BUILDINGS Barbara wants to know the distance between the tops of two buildings R and S. On the top of her building, she measures the distance between the points T and U and finds the given angle measures. Find the distance between the two buildings.

S
R
53°
65° 54°
38°
T 20 m U

64. DRIVING After a high school football game, Della left the parking lot traveling at 35 miles per hour in the direction N 55° E. If Devon left 20 minutes after Della at 45 miles per hour in the direction S 10° W, how far apart are Devon and Della an hour and a half after Della left?

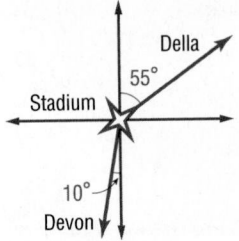

Della
55°
Stadium
10°
Devon

65. ERROR ANALYSIS Monique and Rogelio are solving an acute triangle in which $\angle A = 34°$, $a = 16$, and $b = 21$. Monique thinks that the triangle has one solution, while Rogelio thinks that the triangle has no solution. Is either of them correct? Explain your reasoning.

66. WRITING IN MATH Explain the different circumstances in which you would use the Law of Cosines, the Law of Sines, the Pythagorean Theorem, and the trigonometric ratios to solve a triangle.

67. REASONING Why does an obtuse measurement appear on the graphing calculator for inverse cosine while negative measures appear for inverse sine?

68. PROOF Show for a given rhombus with a side length of s and an included angle of θ that the area can be found with the formula $A = s^2 \sin \theta$.

69. PROOF Derive the Law of Sines.

70. PROOF Consider the figure below.

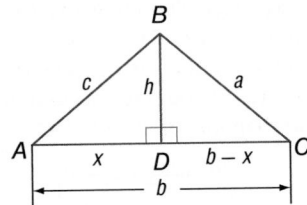

B
c   h   a
A   x   D b − x   C
b

a. Use the figure and hints below to derive the first formula $a^2 = b^2 + c^2 - 2bc \cos A$ in the Law of Cosines.

- Use the Pythagorean Theorem for $\triangle DBC$.
- In $\triangle ADB$, $c^2 = x^2 + h^2$.
- $\cos A = \dfrac{x}{c}$

b. Explain how you would go about deriving the other two formulas in the Law of Cosines.

71 CHALLENGE A satellite is orbiting 850 miles above Mars and is now positioned directly above one of the poles. The radius of Mars is 2110 miles. If the satellite was positioned at point X 14 minutes ago, approximately how many hours does it take for the satellite to complete a full orbit, assuming that it travels at a constant rate around a circular orbit?

X
35°

72. WRITING IN MATH Describe why solving a triangle in which $h < a < b$ using the Law of Sines results in two solutions. Is this also true when using the Law of Cosines? Explain your reasoning.

Find the exact value of each expression, if it exists. (Lesson 4-6)

73. $\cos^{-1} -\frac{1}{2}$

74. $\sin^{-1} \frac{\sqrt{2}}{2}$

75. arctan 1

76. $\sin^{-1} \frac{\sqrt{3}}{2}$

Identify the damping factor $f(x)$ of each function. Then use a graphing calculator to sketch the graphs of $f(x)$, $-f(x)$, and the given function in the same viewing window. Describe the behavior of the graph. (Lesson 4-5)

77. $y = -2x \sin x$

78. $y = \frac{3}{5}x \cos x$

79. $y = (x - 1)^2 \sin x$

80. $y = -4x^2 \cos x$

81. CARTOGRAPHY The distance around Earth along a given latitude can be found using $C = 2\pi r \cos L$, where r is the radius of Earth and L is the latitude. The radius of Earth is approximately 3960 miles. Make a table of values for the latitude and corresponding distance around Earth that includes $L = 0°, 30°, 45°, 60°$, and $90°$. Use the table to describe the distances along the latitudes as you go from $0°$ at the equator to $90°$ at a pole. (Lesson 4-3)

82. RADIOACTIVITY A scientist starts with a 1-gram sample of lead-211. The amount of the sample remaining after various times is shown in the table below. (Lesson 3-5)

Time (min)	10	20	30	40
Pb-211 present (g)	0.83	0.68	0.56	0.46

    a. Find an exponential regression equation for the amount y of lead as a function of time x.

    b. Write the regression equation in terms of base e.

    c. Use the equation from part **b** to estimate when there will be 0.01 gram of lead-211 present.

Write a polynomial function of least degree with real coefficients in standard form that has the given zeros. (Lesson 2-4)

83. $-1, 1, 5$

84. $-2, -0.5, 4$

85. $-3, -2i, 2i$

86. $-5i, -i, i, 5i$

87. SAT/ACT Which of the following is the perimeter of the triangle shown?

    A 49.0 cm

    B 66.0 cm

    C 71.2 cm

    D 91.4 cm

    E 93.2 cm

36°
22 cm

88. In $\triangle DEF$, what is the value of θ to the nearest degree?

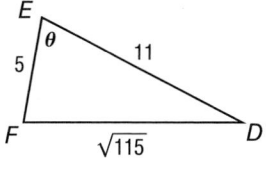

    F 26°

    G 74°

    H 80°

    J 141°

89. FREE RESPONSE The pendulum at the right moves according to $\theta = \frac{1}{4} \cos 12t$, where θ is the angular displacement in radians and t is the time in seconds.

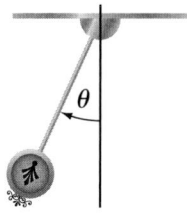

    a. Set the mode to radians and graph the function for $0 \leq t \leq 2$.

    b. What are the period, amplitude, and frequency of the function? What do they mean in the context of this situation?

    c. What is the maximum angular displacement of the pendulum in degrees?

    d. What does the midline of the graph represent?

    e. At what times is the pendulum displaced 5 degrees?

Chapter Summary

Key Concepts

Right Triangle Trigonometry (Lesson 4-1)

$$\sin \theta = \frac{\text{opp}}{\text{hyp}} \qquad \cos \theta = \frac{\text{adj}}{\text{hyp}} \qquad \tan \theta = \frac{\text{opp}}{\text{adj}}$$

$$\csc \theta = \frac{\text{hyp}}{\text{opp}} \qquad \sec \theta = \frac{\text{hyp}}{\text{adj}} \qquad \cot \theta = \frac{\text{adj}}{\text{opp}}$$

Degrees and Radians (Lesson 4-2)

• To convert from degrees to radians, multiply by $\frac{\pi \text{ radians}}{180°}$.

• To convert from radians to degrees, multiply by $\frac{180°}{\pi \text{ radians}}$.

• Linear speed: $v = \frac{s}{t}$, where s is the arc length traveled during time t

• Angular speed: $\omega = \frac{\theta}{t}$, where θ is the angle of rotation (in radians) moved during time t

Trigonometric Functions on the Unit Circle (Lesson 4-3)

• For an angle θ in radians containing (x, y), $\cos \theta = \frac{x}{r}$, $\sin \theta = \frac{y}{r}$, and $\tan \theta = \frac{y}{x}$, where $r = \sqrt{x^2 + y^2}$.

• For an angle t containing (x, y) on the unit circle, $\cos \theta = x$, $\sin \theta = y$, and $\tan \theta = \frac{y}{x}$.

Graphing Sine and Cosine Functions (Lesson 4-4)

• A sinusoidal function is of the form $y = a \sin (bx + c) + d$ or $y = a \cos (bx + c) + d$, where amplitude $= |a|$, period $= \frac{2\pi}{|b|}$, frequency $= \frac{|b|}{2\pi}$, phase shift $= -\frac{c}{|b|}$, and vertical shift $= d$.

Graphing Other Trigonometric Functions (Lesson 4-5)

• A damped trigonometric function is of the form $y = f(x) \sin bx$ or $y = f(x) \cos bx$, where $f(x)$ is the damping factor.

Inverse Trigonometric Functions (Lesson 4-6)

• $y = \sin^{-1} x$ iff $\sin y = x$, for $-1 \le x \le 1$ and $-\frac{\pi}{2} \le y \le \frac{\pi}{2}$.

• $y = \cos^{-1} x$ iff $\cos y = x$, for $-1 \le x \le 1$ and $0 \le y \le \pi$.

• $y = \tan^{-1} x$ iff $\tan y = x$, for $-\infty < x < \infty$ and $-\frac{\pi}{2} < y < \frac{\pi}{2}$.

The Law of Sines and the Law of Cosines (Lesson 4-7)

Let $\triangle ABC$ be any triangle.

• The Law of Sines: $\frac{\sin A}{a} = \frac{\sin B}{b} = \frac{\sin C}{c}$

• The Law of Cosines: $a^2 = b^2 + c^2 - 2bc \cos A$

$$b^2 = a^2 + c^2 - 2ac \cos B$$

$$c^2 = a^2 + b^2 - 2ab \cos C$$

Key Vocabulary

amplitude (p. 257)
angle of depression (p. 224)
angle of elevation (p. 224)
angular speed (p. 236)
circular function (p. 248)
cosecant (p. 220)
cosine (p. 220)
cotangent (p. 220)
coterminal angles (p. 234)
damped trigonometric function (p. 275)
damped wave (p. 275)
damping factor (p. 275)
frequency (p. 260)
initial side (p. 231)
inverse trigonometric function (p. 223)
Law of Cosines (p. 295)
Law of Sines (p. 291)
linear speed (p. 236)
midline (p. 262)

oblique triangles (p. 291)
period (p. 250)
periodic function (p. 250)
phase shift (p. 261)
quadrantal angle (p. 243)
radian (p. 232)
reciprocal function (p. 220)
reference angle (p. 244)
secant (p. 220)
sector (p. 237)
sine (p. 220)
sinusoid (p. 256)
standard position (p. 231)
tangent (p. 220)
terminal side (p. 231)
trigonometric functions (p. 220)
trigonometric ratios (p. 220)
unit circle (p. 247)
vertical shift (p. 262)

Vocabulary Check

State whether each sentence is *true* or *false*. If *false*, replace the underlined term to make a true sentence.

1. The <u>sine</u> of an acute angle in a right triangle is the ratio of the lengths of its opposite leg to the hypotenuse.

2. The <u>secant</u> ratio is the reciprocal of the sine ratio.

3. An <u>angle of elevation</u> is the angle formed by a horizontal line and an observer's line of sight to an object below the line.

4. The <u>radian</u> measure of an angle is equal to the ratio of the length of its intercepted arc to the radius.

5. The rate at which an object moves along a circular path is called its <u>linear speed</u>.

6. $0°$, π, and $-\frac{\pi}{2}$ are examples of <u>reference angles</u>.

7. The <u>period</u> of the graph of $y = 4 \sin 3x$ is 4.

8. For $f(x) = \cos bx$, as b increases, the <u>frequency</u> decreases.

9. The range of the <u>arcsine</u> function is $[0, \pi]$.

10. The <u>Law of Sines</u> can be used to determine unknown side lengths or angle measures of some triangles.

Lesson-by-Lesson Review

4-1 **Right Triangle Trigonometry** (pp. 220–230)

Find the exact values of the six trigonometric functions of θ.

11.

12.

Find the value of x. Round to the nearest tenth, if necessary.

13.

14.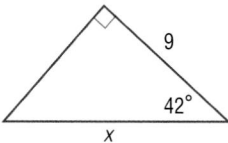

Find the measure of angle θ. Round to the nearest degree, if necessary.

15.

16.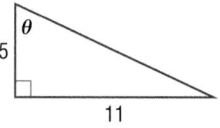

EXAMPLE 1

Find the value of x. Round to the nearest tenth, if necessary.

$\tan \theta = \dfrac{\text{opp}}{\text{adj}}$ **Tangent function**

$\tan 38° = \dfrac{10}{x}$ $\theta = 38°$, opp $= 10$, and adj $= x$

$x \tan 38° = 10$ **Multiply each side by x.**

$x = \dfrac{10}{\tan 38°}$ **Divide each side by tan 38°.**

$x \approx 12.8$ **Use a calculator.**

4-2 **Degrees and Radians** (pp. 231–241)

Write each degree measure in radians as a multiple of π and each radian measure in degrees.

17. $135°$

18. $450°$

19. $\dfrac{7\pi}{4}$

20. $\dfrac{13\pi}{10}$

Identify all angles coterminal with the given angle. Then find and draw one positive and one negative angle coterminal with the given angle.

21. $342°$

22. $-\dfrac{\pi}{6}$

Find the area of each sector.

23.

24.

EXAMPLE 2

Identify all angles coterminal with $\dfrac{5\pi}{12}$. Then find and draw one positive and one negative coterminal angle.

All angles measuring $\dfrac{5\pi}{12} + 2n\pi$ are coterminal with a $\dfrac{5\pi}{12}$ angle. Let $n = 1$ and -1.

$\dfrac{5\pi}{6} + 2\pi(1) = \dfrac{17\pi}{6}$ $\dfrac{5\pi}{6} - 2\pi(-1) = -\dfrac{7\pi}{6}$

 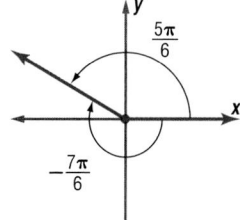

Lesson-by-Lesson Review

4-3 Trigonometric Functions on the Unit Circle (pp. 242–253)

Sketch each angle. Then find its reference angle.

25. $240°$ **26.** $75°$

27. $-\dfrac{3\pi}{4}$ **28.** $\dfrac{11\pi}{18}$

Find the exact values of the five remaining trigonometric functions of θ.

29. $\cos \theta = \dfrac{2}{5}$, where $\sin \theta > 0$ and $\tan \theta > 0$

30. $\tan \theta = -\dfrac{3}{4}$, where $\sin \theta > 0$ and $\cos \theta < 0$

31. $\sin \theta = -\dfrac{5}{13}$, where $\cos \theta > 0$ and $\cot \theta < 0$

32. $\cot \theta = \dfrac{2}{3}$, where $\sin \theta < 0$ and $\tan \theta > 0$

Find the exact value of each expression. If undefined, write *undefined*.

33. $\sin 180°$ **34.** $\cot \dfrac{11\pi}{6}$

35. $\sec 450°$ **36.** $\cos \left(-\dfrac{19\pi}{6}\right)$

EXAMPLE 3

Let $\cos \theta = \dfrac{5}{13}$, where $\sin \theta < 0$. Find the exact values of the five remaining trigonometric functions of θ.

Since $\cos \theta$ is positive and $\sin \theta$ is negative, θ lies in Quadrant IV. This means that the x-coordinate of a point on the terminal side of θ is positive and the y-coordinate is negative.

Since $\cos \theta = \dfrac{x}{r} = \dfrac{5}{13}$, use $x = 5$ and $r = 13$ to find y.

$y = \sqrt{r^2 - x^2}$ **Pythagorean Theorem**

$\quad = \sqrt{169 - 25}$ or 12 $r = 13$ and $x = 5$

$\sin \theta = \dfrac{y}{r}$ or $\dfrac{12}{13}$ $\tan \theta = \dfrac{y}{x}$ or $\dfrac{12}{5}$ $\sec \theta = \dfrac{r}{x}$ or $\dfrac{13}{5}$

$\csc \theta = \dfrac{r}{y}$ or $\dfrac{13}{12}$ $\cot \theta = \dfrac{x}{y}$ or $\dfrac{5}{12}$

4-4 Graphing Sine and Cosine Functions (pp. 256–266)

Describe how the graphs of $f(x)$ and $g(x)$ are related. Then find the amplitude and period of $g(x)$, and sketch at least one period of both functions on the same coordinate axes.

37. $f(x) = \sin x$ **38.** $f(x) = \cos x$
    $g(x) = 5 \sin x$ $g(x) = \cos 2x$

39. $f(x) = \sin x$ **40.** $f(x) = \cos x$
    $g(x) = \dfrac{1}{2} \sin x$ $g(x) = -\cos \dfrac{1}{3}x$

State the amplitude, period, frequency, phase shift, and vertical shift of each function. Then graph two periods of the function.

41. $y = 2 \cos (x - \pi)$ **42.** $y = -\sin 2x + 1$

43. $y = \dfrac{1}{2} \cos \left(x + \dfrac{\pi}{2}\right)$ **44.** $y = 3 \sin \left(x + \dfrac{2\pi}{3}\right)$

EXAMPLE 4

State the amplitude, period, frequency, phase shift, and vertical shift of $y = 4 \sin \left(x - \dfrac{\pi}{2}\right) - 4$. Then graph two periods of the function.

In this function, $a = 4$, $b = 1$, $c = -\dfrac{\pi}{2}$, and $d = -4$.

Amplitude: $|a| = |4|$ or 4 Period: $\dfrac{2\pi}{|b|} = \dfrac{2\pi}{|1|}$ or 2π

Frequency: $\dfrac{|b|}{2\pi} = \dfrac{|1|}{2\pi}$ or $\dfrac{1}{2\pi}$ Vertical shift: d or -4

Phase shift: $-\dfrac{c}{|b|} = -\dfrac{-\frac{\pi}{2}}{|1|}$ or $\dfrac{\pi}{2}$

First, graph the midline $y = -4$. Then graph $y = 4 \sin x$ shifted $\dfrac{\pi}{2}$ units to the right and 4 units down.

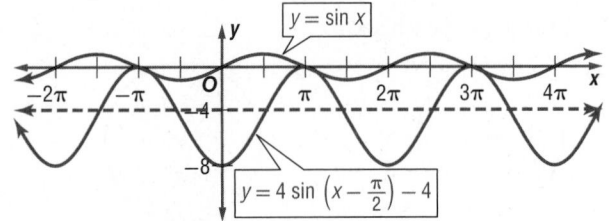

Lesson-by-Lesson Review

4-5 **Graphing Other Trigonometric Functions** (pp. 269–279)

Locate the vertical asymptotes, and sketch the graph of each function.

45. $y = 3 \tan x$

46. $y = \frac{1}{2}\tan\left(x - \frac{\pi}{2}\right)$

47. $y = \cot\left(x + \frac{\pi}{3}\right)$

48. $y = -\cot(x - \pi)$

49. $y = 2 \sec\left(\frac{x}{2}\right)$

50. $y = -\csc(2x)$

51. $y = \sec(x - \pi)$

52. $y = \frac{2}{3}\csc\left(x + \frac{\pi}{2}\right)$

EXAMPLE 5

Locate the vertical asymptotes, and sketch the graph of $y = 2 \sec\left(x + \frac{\pi}{4}\right)$.

Because the graph of $y = 2 \sec\left(x + \frac{\pi}{4}\right)$ is the graph of $y = 2 \sec x$ shifted to the left $\frac{\pi}{4}$ units, the vertical asymptotes for one period are located at $-\frac{3\pi}{4}$, $\frac{\pi}{4}$, and $\frac{5\pi}{4}$. Graph two cycles on the interval $\left[-\frac{3\pi}{4}, \frac{13\pi}{4}\right]$.

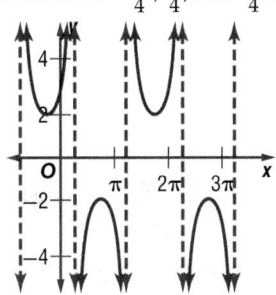

4-6 **Inverse Trigonometric Functions** (pp. 280–290)

Find the exact value of each expression, if it exists.

53. $\sin^{-1}(-1)$

54. $\cos^{-1}\frac{\sqrt{3}}{2}$

55. $\tan^{-1}\left(-\frac{\sqrt{3}}{3}\right)$

56. $\arcsin 0$

57. $\arctan(-1)$

58. $\arccos\frac{\sqrt{2}}{2}$

59. $\sin^{-1}\left[\sin\left(-\frac{\pi}{3}\right)\right]$

60. $\cos^{-1}[\cos(-3\pi)]$

EXAMPLE 6

Find the exact value of $\arctan -\sqrt{3}$.

Find a point on the unit circle in the interval $\left(-\frac{\pi}{2}, \frac{\pi}{2}\right)$ with a tangent of $-\sqrt{3}$. When $t = -\frac{\pi}{3}$, $\tan t = -\sqrt{3}$.

Therefore, $\arctan -\sqrt{3} = -\frac{\pi}{3}$.

4-7 **The Law of Sines and the Law of Cosines** (pp. 291–301)

Find all solutions for the given triangle, if possible. If no solution exists, write *no solution*. Round side lengths to the nearest tenth and angle measurements to the nearest degree.

61. $a = 11, b = 6, A = 22°$

62. $a = 9, b = 10, A = 42°$

63. $a = 20, b = 10, A = 78°$

64. $a = 2, b = 9, A = 88°$

Solve each triangle. Round side lengths to the nearest tenth and angle measures to the nearest degree.

65. $a = 13, b = 12, c = 8$

66. $a = 4, b = 5, C = 96°$

EXAMPLE 7

Solve the triangle if $a = 3$, $b = 4$, and $A = 71°$.

In the figure, $h = 4 \sin 71°$ or about 3.8

Because $a \leq h$, there is no triangle that can be formed with sides $a = 3$, $b = 4$, and $A = 71°$. Therefore, this problem has no solution.

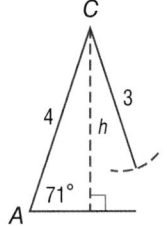

Applications and Problem Solving

67. CONSTRUCTION A construction company is installing a three-foot-high wheelchair ramp onto a landing outside of an office. The angle of the ramp must be 4°. (Lesson 4-1)

 a. What is the length of the ramp?

 b. What is the slope of the ramp?

68. NATURE For a photography project, Maria is photographing deer from a tree stand. From her sight 30 feet above the ground, she spots two deer in a straight line, as shown below. How much farther away is the second deer than the first? (Lesson 4-1)

69. FIGURE SKATING An Olympic ice skater performs a routine in which she jumps in the air for 2.4 seconds while spinning 3 full revolutions. (Lesson 4-2)

 a. Find the angular speed of the figure skater.

 b. Express the angular speed of the figure skater in degrees per minute.

70. TIMEPIECES The length of the minute hand of a pocket watch is 1.5 inches. What is the area swept by the minute hand in 40 minutes? (Lesson 4-2)

1.5 in.

71. WORLD'S FAIR The first Ferris wheel had a diameter of 250 feet and took 10 minutes to complete one full revolution. (Lesson 4-3)

 a. How many degrees would the Ferris wheel rotate in 100 seconds?

 b. How far has a person traveled if he or she has been on the Ferris wheel for 7 minutes?

 c. How long would it take for a person to travel 200 feet?

72. AIR CONDITIONING An air-conditioning unit turns on and off to maintain the desired temperature. On one summer day, the air conditioner turns on at 8:30 A.M. when the temperature is 80° Fahrenheit and turns off at 8:55 A.M. when the temperature is 74°. (Lesson 4-4)

 a. Find the amplitude and period if you were going to use a trigonometric function to model this change in temperature, assuming that the temperature cycle will continue.

 b. Is it appropriate to model this situation with a trigonometric function? Explain your reasoning.

73. TIDES In Lewis Bay, the low tide is recorded as 2 feet at 4:30 A.M., and the high tide is recorded as 5.5 feet at 10:45 A.M. (Lesson 4-4)

 a. Find the period for the trigonometric model.

 b. At what time will the next high tide occur?

74. MUSIC When plucked, a bass string is displaced 1.5 inches, and its damping factor is 1.9. It produces a note with a frequency of 90 cycles per second. Determine the amount of time it takes the string's motion to be dampened so that $-0.1 < y < 0.1$. (Lesson 4-5)

75. PAINTING A painter is using a 15-foot ladder to paint the side of a house. If the angle the ladder makes with the ground is less than 65°, it will slide out from under him. What is the greatest distance that the bottom of the ladder can be from the side of the house and still be safe for the painter? (Lesson 4-6)

15 ft

76. NAVIGATION A boat is 20 nautical miles from a port at a bearing 30° north of east. The captain sees a second boat and reports to the port that his boat is 15 nautical miles from the second boat, which is located due east of the port. Can port personnel be sure of the second boat's position? Justify your answer. (Lesson 4-7)

77. GEOMETRY Consider quadrilateral ABCD. (Lesson 4-7)

 a. Find C.

 b. Find the area of ABCD.

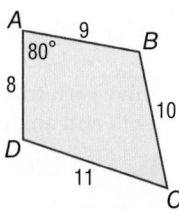

Find the value of x. Round to the nearest tenth, if necessary.

1.

2.

Find the measure of angle θ. Round to the nearest degree, if necessary.

3.

4.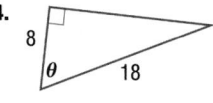

5. MULTIPLE CHOICE What is the linear speed of a point rotating at an angular speed of 36 radians per second at a distance of 12 inches from the center of the rotation?

A 420 in./s   C 439 in./s

B 432 in./s   D 444 in./s

Write each degree measure in radians as a multiple of π and each radian measure in degrees.

6. 200°

7. $-\dfrac{8\pi}{3}$

8. Find the area of the sector of the circle shown.

Sketch each angle. Then find its reference angle.

9. 165°

10. $\dfrac{21\pi}{13}$

Find the exact value of each expression.

11. $\sec \dfrac{7\pi}{6}$

12. $\cos(-240°)$

13. MULTIPLE CHOICE An angle θ satisfies the following inequalities: $\csc \theta < 0$, $\cot \theta > 0$, and $\sec \theta < 0$. In which quadrant does θ lie?

F I   H III

G II   J IV

State the amplitude, period, frequency, phase shift, and vertical shift of each function. Then graph two periods of the function.

14. $y = 4 \cos \dfrac{x}{2} - 5$

15. $y = -\sin\left(x + \dfrac{\pi}{2}\right)$

16. TIDES The table gives the approximate times that the high and low tides occurred in San Azalea Bay over a 2-day period.

Tide	High 1	Low 1	High 2	Low 2
Day 1	2:35 A.M.	8:51 A.M.	3:04 P.M.	9:19 P.M.
Day 2	3:30 A.M.	9:48 A.M.	3:55 P.M.	10:20 P.M.

a. The tides can be modeled with a trigonometric function. Approximately what is the period of this function?

b. The difference in height between the high and low tides is 7 feet. What is the amplitude of this function?

c. Write a function that models the tides where t is measured in hours. Assume the function has no phase shift or vertical shift.

Locate the vertical asymptotes, and sketch the graph of each function.

17. $y = \tan\left(x + \dfrac{\pi}{4}\right)$

18. $y = \dfrac{1}{2}\sec 4x$

Find all solutions for the given triangle, if possible. If no solution exists, write *no solution*. Round side lengths to the nearest tenth and angle measurements to the nearest degree.

19. $a = 8, b = 16, A = 22°$

20. $a = 9, b = 7, A = 84°$

21. $a = 3, b = 5, c = 7$

22. $a = 8, b = 10, C = 46°$

Find the exact value of each expression, if it exists.

23. $\cos^{-1}\left(-\dfrac{\sqrt{3}}{2}\right)$

24. $\sin^{-1}\left(-\dfrac{1}{2}\right)$

25. NAVIGATION A boat leaves a dock and travels 45° north of west averaging 30 knots for 2 hours. The boat then travels directly west averaging 40 knots for 3 hours.

a. How many nautical miles is the boat from the dock after 5 hours?

b. How many degrees south of east is the dock from the boat's present position?

Objective
Model and solve related rates problems.

If air is being pumped into a balloon at a given rate, can we find the rate at which the volume of the balloon is expanding? How does the rate a company spends money on advertising affect the rate of its sales? *Related rates* problems occur when the rate of change for one variable can be found by *relating* that to rates of change for other variables.

Suppose two cars leave a point at the same time. One car is traveling 40 miles per hour due north, while the second car is traveling 30 miles per hour due east. How far apart are the two cars after 1 hour? 2 hours? 3 hours? We can use the formula $d = rt$ and the Pythagorean Theorem to solve for these values.

In this situation, we know the rates of change for each car. What if we want to know the rate at which the distance between the two cars is changing?

ACTIVITY 1   Rate of Change

Two cars leave a house at the same time. One car travels due north at 35 miles per hour, while the second car travels due east at 55 miles per hour. Approximate the rate at which the distance between the two cars is changing.

Step 1 Make a sketch of the situation.

Step 2 Write equations for the distance traveled by each car after t hours.

Step 3 Find the distance traveled by each car after 1, 2, 3, and 4 hours.

Step 4 Use the Pythagorean Theorem to find the distance between the two cars at each point in time.

Step 5 Find the average rate of change of the distance between the two cars for $1 \le t \le 2$, $2 \le t \le 3$, and $3 \le t \le 4$.

▶ Analyze the Results

1. Make a scatter plot displaying the total distance between the two cars. Let time t be the independent variable and total distance d be the dependent variable. Draw a line through the points.

2. What type of function does the graph seem to model? How is your conjecture supported by the values found in Step 5?

3. What would happen to the average rate of change of the distance between the two cars if one of the cars slowed down? sped up? Explain your reasoning.

The rate that the distance between the two cars is changing is *related* to the rates of the two cars. In calculus, problems involving related rates can be solved using *implicit differentiation*. However, before we can use advanced techniques of differentiation, we need to understand how the rates involved relate to one another. Therefore, the first step to solving any related rates problem should always be to model the situation with a sketch or graph and to write equations using the relevant values and variables.

ACTIVITY 2   Model Related Rates

When a rock is tossed into a still body of water, it creates a circular ripple in the water. Suppose the radius of the circle grows at a rate of 5 centimeters per second. Find the area of the circle after 3 seconds if the radius of the circle is 5 centimeters at $t = 1$.

Step 1 Make a sketch of the situation.

Step 2 Write an equation for the radius r of the circle after t seconds.

Step 3 Find the radius for $t = 3$, and then find the area.

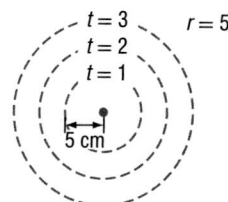

▶ Analyze the Results

4. Find an equation for the area A of the circle in terms of t by substituting the equation found in Step 2 into the equation for the area of a circle.

5. Find the area of the circle for $t = 1, 2, 3, 4$, and 5 seconds.

6. Make a graph of the values. What type of function does the graph seem to model?

For Activity 2, the graph of the area of the circle does not appear to be linear. Calculating the rate of change for the area of the circle at a certain point in time requires using the difference quotient.

ACTIVITY 3   Approximate Related Rate

Approximate the rate of change for the area of the circle in Activity 2 by using the difference quotient.

Step 1 Substitute the expression for the area of the circle into the difference quotient.

$$m = \frac{\pi[5(t + h)]^2 - \pi(5t)^2}{h}$$

Step 2 Approximate the rate of change of the circle at 2 seconds. Let $h = 0.1, 0.01$, and 0.001.

Step 3 Repeat Steps 1 and 2 for $t = 3$ seconds and $t = 4$ seconds.

▶ Analyze the Results

7. What do the rates of change appear to approach for each value of t?

8. What is happening to the rate of change of the area of the circle as the radius is increasing? Explain your reasoning.

9. How does this approach differ from the approach you used in Activity 1 to find the rate of change for the distance between the two cars? Explain why this was necessary.

Model and Apply

10. A 13-foot ladder is leaning against a wall so that the base of the ladder is exactly 5 feet from the base of the wall. If the bottom of the ladder starts to slide away from the wall at a rate of 2 feet per second, how fast is the top of the ladder sliding down the wall?

   a. Sketch a model of the situation. Let d be the distance from the top of the ladder to the ground and m be the rate at which the top of the ladder is sliding down the wall.

   b. Write an expression for the distance from the base of the ladder to the wall after t seconds.

   c. Find an equation for the distance d from the top of the ladder to the ground in terms of t by substituting the expression found in part **b** into the Pythagorean Theorem.

   d. Use the Pythagorean Theorem to find the distance d from the top of the ladder to the ground for $t = 0, 1, 2, 3, 3.5$, and 3.75.

   e. Make a graph of the values. What type of function does the graph seem to model?

   f. Use the difference quotient to approximate the rate of change m for the distance from the top of the ladder to the ground at $t = 2$. Let $h = 0.1, 0.01$, and 0.001. As h approaches 0, what do the values for m appear to approach?

Then

In Chapter 4, you learned to graph trigonometric functions and to solve right and oblique triangles.

Now

In Chapter 5, you will:

- Use and verify trigonometric identities.
- Solve trigonometric equations.
- Use sum and difference identities to evaluate trigonometric expressions and solve equations.
- Use double-angle, power-reducing, half-angle, and product-sum identities to evaluate trigonometric expressions and solve equations.

Why?

MUSIC Musicians tune their instruments by listening for a *beat*, which is an interference between two sound waves with slightly different frequencies. The sum of the sound waves can be represented using a trigonometric equation.

PREREAD Using what you know about trigonometric functions, make a prediction of what you will learn in Chapter 5.

Get Ready for Chapter 5

Diagnose Readiness You have two options for checking Prerequisite Skills.

Text Option

Take the Quick Quiz below.

*Quick*Check

Solve each equation by factoring. (Lesson 0-3)

1. $x^2 + 5x - 24 = 0$ **2.** $x^2 - 11x + 28 = 0$

3. $2x^2 - 9x - 5 = 0$ **4.** $15x^2 + 26x + 8 = 0$

5. $2x^3 - 2x^2 - 12x = 0$ **6.** $12x^3 + 78x^2 - 42x = 0$

7. ROCKETS A rocket is projected vertically into the air. Its distance in feet after t seconds is represented by $s(t) = -16t^2 + 192t$. Find the amount of time that the rocket is in the air. (Lesson 0-3)

Find the missing side lengths and angle measures of each triangle. (Lessons 4-1 and 4-7)

8.   **9.**

10.   **11.**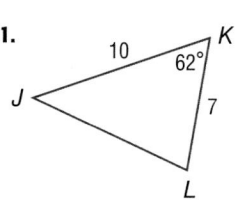

Find the exact value of each expression. (Lesson 4-3)

12. $\cot 420°$ **13.** $\cos \dfrac{7\pi}{4}$

14. $\sec \dfrac{10\pi}{3}$ **15.** $\tan 480°$

16. $\csc \dfrac{2\pi}{3}$ **17.** $\sin 510°$

Online Option

Math Online Take a self-check Chapter Readiness Quiz at **glencoe.com**.

New Vocabulary

English	• Page •	Español
trigonometric identity	• 312 •	identidad trigonométrica
reciprocal identity	• 312 •	identidad recíproca
quotient identity	• 312 •	cociente de identidad
Pythagorean identity	• 313 •	Pitágoras identidad
odd-even-identity	• 314 •	impar-incluso-de identidad
cofunction	• 314 •	co función
verify an identity	• 320 •	verificar una identidad
sum identity	• 337 •	suma de identidad
reduction identity	• 340 •	identidad de reducción
double-angle identity	• 346 •	doble-ángulo de la identidad
power-reducing identity	• 347 •	poder-reducir la identidad
half-angle identity	• 348 •	medio ángulo identidad

Review Vocabulary

extraneous solution • p. 91 • solución extraña a solution that does not satisfy the original equation

quadrantal angle • p. 243 • ángulo cuadranta If the terminal side of an angle θ that is in standard position lies on one of the coordinate axes, it forms a quadrantal angle.

unit circle • p. 247 • círculo unitario a circle of radius 1 centered at the origin

periodic function • p. 250 • función periódica a function with range values that repeat at regular intervals

trigonometric functions • p. 220 • funciones trigonométricas Let θ be any angle in standard position and point $P(x, y)$ be a point on the terminal side of θ. Let r represent the nonzero distance from P to the origin or $|r| = \sqrt{x^2 + y^2} \neq 0$. Then the trigonometric functions of θ are as follows.

$$\sin \theta = \frac{y}{r} \qquad\qquad \csc \theta = \frac{r}{y}, y \neq 0$$

$$\cos \theta = \frac{x}{r} \qquad\qquad \sec \theta = \frac{r}{x}, x \neq 0$$

$$\tan \theta = \frac{y}{x}, x \neq 0 \qquad\qquad \cot \theta = \frac{x}{y}, y \neq 0$$

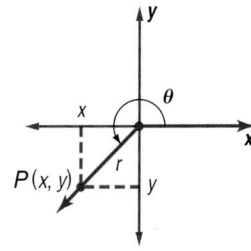

Trigonometric Identities

Why?

Many physics and engineering applications, such as determining the path of an aircraft, involve trigonometric functions. These functions are made more flexible if you can change the trigonometric expressions involved from one form to an equivalent but more convenient form. You can do this by using trigonometric identities.

Then
You found trigonometric values using the unit circle. (Lesson 4-3)

Now
- Identify and use basic trigonometric identities to find trigonometric values.
- Use basic trigonometric identities to simplify and rewrite trigonometric expressions.

New Vocabulary
identity
trigonometric identity
cofunction

Math Online

glencoe.com

Basic Trigonometric Identities An equation is an **identity** if the left side is equal to the right side for all values of the variable for which both sides are defined. Consider the equations below.

$\dfrac{x^2 - 9}{x - 3} = x + 3$ **This *is* an identity since both sides of the equation are defined and equal for all x such that $x \neq 3$.**

$\sin x = 1 - \cos x$ **This *is not* an identity. Both sides of this equation are defined and equal for certain values, such as when $x = 0$, but not for other values for which both sides are defined, such as when $x = \dfrac{\pi}{4}$.**

Trigonometric identities are identities that involve trigonometric functions. You already know a few basic trigonometric identities. The reciprocal and quotient identities below follow directly from the definitions of the six trigonometric functions introduced in Lesson 4-1.

Key Concept **Reciprocal and Quotient Identities**

Reciprocal Identities

$\sin \theta = \dfrac{1}{\csc \theta}$ $\cos \theta = \dfrac{1}{\sec \theta}$ $\tan \theta = \dfrac{1}{\cot \theta}$

$\csc \theta = \dfrac{1}{\sin \theta}$ $\sec \theta = \dfrac{1}{\cos \theta}$ $\cot \theta = \dfrac{1}{\tan \theta}$

Quotient Identities

$\tan \theta = \dfrac{\sin \theta}{\cos \theta}$

$\cot \theta = \dfrac{\cos \theta}{\sin \theta}$

You can use these basic trigonometric identities to find trigonometric values. As with any fraction, the denominator cannot equal zero.

EXAMPLE 1 **Use Reciprocal and Quotient Identities**

a. If $\csc \theta = \dfrac{7}{4}$, find $\sin \theta$.

$\sin \theta = \dfrac{1}{\csc \theta}$ **Reciprocal Identity**

$= \dfrac{1}{\frac{7}{4}}$ $\csc \theta = \dfrac{7}{4}$

$= \dfrac{4}{7}$ **Divide.**

b. If $\cot x = \dfrac{2}{5\sqrt{5}}$ and $\sin x = \dfrac{\sqrt{5}}{3}$, find $\cos x$.

$\cot x = \dfrac{\cos x}{\sin x}$ **Quotient Identity**

$\dfrac{2}{5\sqrt{5}} = \dfrac{\cos x}{\frac{\sqrt{5}}{3}}$ $\cot x = \dfrac{2}{5\sqrt{5}}$; $\sin x = \dfrac{\sqrt{5}}{3}$

$\dfrac{2}{5\sqrt{5}} \cdot \dfrac{\sqrt{5}}{3} = \cos x$ **Multiply each side by $\dfrac{\sqrt{5}}{3}$.**

$\dfrac{2}{15} = \cos x$ **Simplify.**

☑ **Guided Practice**

1A. If $\sec x = \dfrac{5}{3}$, find $\cos x$.

1B. If $\csc \beta = \dfrac{25}{7}$ and $\sec \beta = \dfrac{25}{24}$, find $\tan \beta$.

Recall from Lesson 4-3 that trigonometric functions can be defined on a unit circle as shown. Notice that for any angle θ, sine and cosine are the directed lengths of the legs of a right triangle with hypotenuse 1. We can apply the Pythagorean Theorem to this right triangle to establish another basic trigonometric identity.

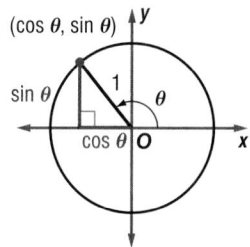

$(\sin \theta)^2 + (\cos \theta)^2 = 1^2$ **Pythagorean Theorem**

$\sin^2 \theta + \cos^2 \theta = 1$ **Simplify.**

While the signs of these directed lengths may change depending on the quadrant in which the triangle lies, notice that because these lengths are squared, the equation above holds true for any value of θ. This equation is one of three **Pythagorean identities**.

ReadingMath

Powers of Trigonometric Functions $\sin^2 \theta$ is read as *sine squared theta* and interpreted as the square of the quantity $\sin \theta$.

Key Concept   **Pythagorean Identities**

$$\sin^2 \theta + \cos^2 \theta = 1 \qquad \tan^2 \theta + 1 = \sec^2 \theta \qquad \cot^2 \theta + 1 = \csc^2 \theta$$

You will prove the remaining two Pythagorean Identities in Exercises 69 and 70.

Notice the shorthand notation used to represent powers of trigonometric functions: $\sin^2 \theta = (\sin \theta)^2$, $\cos^2 \theta = (\cos \theta)^2$, $\tan^2 \theta = (\tan \theta)^2$, and so on.

EXAMPLE 2   **Use Pythagorean Identities**

If $\tan \theta = -8$ and $\sin \theta > 0$, find $\sin \theta$ and $\cos \theta$.

Use the Pythagorean Identity that involves $\tan \theta$.

$\tan^2 \theta + 1 = \sec^2 \theta$ **Pythagorean Identity**

$(-8)^2 + 1 = \sec^2 \theta$ $\tan \theta = -8$

$65 = \sec^2 \theta$ **Simplify.**

$\pm\sqrt{65} = \sec \theta$ **Take the square root of each side.**

$\pm\sqrt{65} = \dfrac{1}{\cos \theta}$ **Reciprocal Identity**

$\pm\dfrac{\sqrt{65}}{65} = \cos \theta$ **Solve for $\cos \theta$.**

Since $\tan \theta = \dfrac{\sin \theta}{\cos \theta}$ is negative and $\sin \theta$ is positive, $\cos \theta$ must be negative. So, $\cos \theta = -\dfrac{\sqrt{65}}{65}$. You can then use this quotient identity again to find $\sin \theta$.

$\tan \theta = \dfrac{\sin \theta}{\cos \theta}$ **Quotient Identity**

$-8 = \dfrac{\sin \theta}{-\dfrac{\sqrt{65}}{65}}$ $\tan \theta = -8$ and $\cos \theta = -\dfrac{\sqrt{65}}{65}$

$\dfrac{8\sqrt{65}}{65} = \sin \theta$ **Multiply each side by $-\dfrac{\sqrt{65}}{65}$.**

CHECK $\sin^2 \theta + \cos^2 \theta = 1$ **Pythagorean Identity**

$\left(\dfrac{8\sqrt{65}}{65}\right)^2 + \left(-\dfrac{\sqrt{65}}{65}\right)^2 \stackrel{?}{=} 1$ $\sin \theta = \dfrac{8\sqrt{65}}{65}$ and $\cos \theta = -\dfrac{\sqrt{65}}{65}$

$\dfrac{64}{65} + \dfrac{1}{65} = 1$ ✔   **Simplify.**

StudyTip

Checking Answers It is beneficial to confirm your answers using a different identity than the ones you used to solve the problem, as in Example 2, so that you do not make the same mistake twice.

☑ **Guided Practice**

Find the value of each expression using the given information.

2A. $\csc \theta$ and $\tan \theta$; $\cot \theta = -3$, $\cos \theta < 0$ **2B.** $\cot x$ and $\sec x$; $\sin x = \dfrac{1}{6}$, $\cos x > 0$

Another set of basic trigonometric identities involve cofunctions. A trigonometric function f is a **cofunction** of another trigonometric function g if $f(\alpha) = g(\beta)$ when α and β are complementary angles. In the right triangle shown, angles α and β are complementary angles. Using the right triangle ratios, you can show that the following statements are true.

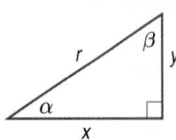

$$\sin \alpha = \cos \beta = \cos (90° - \alpha) = \frac{y}{r}$$

$$\tan \alpha = \cot \beta = \cot (90° - \alpha) = \frac{y}{x}$$

$$\sec \alpha = \csc \beta = \csc (90° - \alpha) = \frac{r}{y}$$

From these statements, we can write the following cofunction identities, which are valid for all real numbers, not just acute angle measures.

Key Concept   **Cofunction Identities**

$$\sin \theta = \cos \left(\frac{\pi}{2} - \theta\right) \qquad \tan \theta = \cot \left(\frac{\pi}{2} - \theta\right) \qquad \sec \theta = \csc \left(\frac{\pi}{2} - \theta\right)$$

$$\cos \theta = \sin \left(\frac{\pi}{2} - \theta\right) \qquad \cot \theta = \tan \left(\frac{\pi}{2} - \theta\right) \qquad \csc \theta = \sec \left(\frac{\pi}{2} - \theta\right)$$

You will prove these identities for any angle in Lesson 5-3.

You have also seen that each of the basic trigonometric functions—sine, cosine, tangent, cosecant, secant, and cotangent—is either odd or even. Using the unit circle, you can show that the following statements are true.

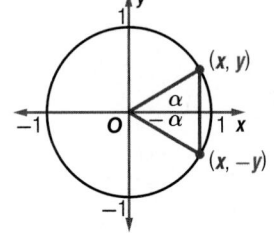

$$\sin \alpha = y \qquad \sin (-\alpha) = -y$$
$$\cos \alpha = x \qquad \cos (-\alpha) = x$$

Recall from Lesson 1-2 that a function f is even if for every x in the domain of f, $f(-x) = f(x)$ and odd if for every x in the domain of f, $f(-x) = -f(x)$. These relationships lead to the following odd-even identities.

Key Concept   **Odd-Even Identities**

$$\sin (-\theta) = -\sin \theta \qquad\qquad \cos (-\theta) = \cos \theta \qquad\qquad \tan (-\theta) = -\tan \theta$$

$$\csc (-\theta) = -\csc \theta \qquad\qquad \sec (-\theta) = \sec \theta \qquad\qquad \cot (-\theta) = -\cot \theta$$

You can use cofunction and odd-even identities to find trigonometric values.

EXAMPLE 3   **Use Cofunction and Odd-Even Identities**

If $\tan \theta = 1.28$, find $\cot \left(\theta - \frac{\pi}{2}\right)$.

$$\cot \left(\theta - \frac{\pi}{2}\right) = \cot \left[-\left(\frac{\pi}{2} - \theta\right)\right] \qquad \text{Factor.}$$

$$= -\cot \left(\frac{\pi}{2} - \theta\right) \qquad \text{Odd-Even Identity}$$

$$= -\tan \theta \qquad\qquad \text{Cofunction Identity}$$

$$= -1.28 \qquad\qquad \tan \theta = 1.28$$

✔ **Guided Practice**

3. If $\sin x = -0.37$, find $\cos \left(x - \frac{\pi}{2}\right)$.

Simplify and Rewrite Trigonometric Expressions To simplify a trigonometric expression, start by rewriting it in terms of one trigonometric function or in terms of sine and cosine only.

EXAMPLE 4 | **Simplify by Rewriting Using Only Sine and Cosine**

Simplify $\csc \theta \sec \theta - \cot \theta$.

Solve Algebraically

$$\csc \theta \sec \theta - \cot \theta = \frac{1}{\sin \theta} \cdot \frac{1}{\cos \theta} - \frac{\cos \theta}{\sin \theta} \qquad \text{Rewrite in terms of sine and cosine using Reciprocal and Quotient Identities.}$$

$$= \frac{1}{\sin \theta \cos \theta} - \frac{\cos \theta}{\sin \theta} \qquad \text{Multiply.}$$

$$= \frac{1}{\sin \theta \cos \theta} - \frac{\cos^2 \theta}{\sin \theta \cos \theta} \qquad \text{Rewrite fractions using a common denominator.}$$

$$= \frac{1 - \cos^2 \theta}{\sin \theta \cos \theta} \qquad \text{Subtract.}$$

$$= \frac{\sin^2 \theta}{\sin \theta \cos \theta} \qquad \text{Pythagorean Identity}$$

$$= \frac{\sin \theta}{\cos \theta} \text{ or } \tan \theta \qquad \text{Divide the numerator and denominator by } \sin \theta.$$

Support Graphically The graphs of $y = \csc \theta \sec \theta - \cot \theta$ and $y = \tan \theta$ appear to be identical.

$y = \csc \theta \sec \theta - \cot \theta$
$[-2\pi, 2\pi]$ scl: $\frac{\pi}{2}$ by $[-2, 2]$ scl: 0.5

$y = \tan \theta$
$[-2\pi, 2\pi]$ scl: $\frac{\pi}{2}$ by $[-2, 2]$ scl: 0.5

☑ **Guided Practice**

4. Simplify $\sec x - \tan x \sin x$.

TechnologyTip

Graphing Reciprocal Functions When using a calculator to graph a reciprocal function, such as $y = \csc x$, you can enter the reciprocal of the function.

```
Plot1 Plot2 Plot3
\Y1◻1/sin(X)
\Y2=
\Y3=
\Y4=
\Y5=
\Y6=
\Y7=
```

Some trigonometric expressions can be simplified by applying identities and factoring.

EXAMPLE 5 | **Simplify by Factoring**

Simplify $\sin^2 x \cos x - \sin\left(\frac{\pi}{2} - x\right)$.

Solve Algebraically

$$\sin^2 x \cos x - \sin\left(\frac{\pi}{2} - x\right) = \sin^2 x \cos x - \cos x \qquad \text{Cofunction Identity}$$

$$= -\cos x \, (-\sin^2 x + 1) \qquad \text{Factor } -\cos x \text{ from each term.}$$

$$= -\cos x \, (1 - \sin^2 x) \qquad \text{Commutative Property}$$

$$= -\cos x \, (\cos^2 x) \text{ or } -\cos^3 x \qquad \text{Pythagorean Identity}$$

Support Graphically The graphs below appear to be identical.

$y = \sin^2 x \cos x - \sin\left(\frac{\pi}{2} - x\right)$
$[-2\pi, 2\pi]$ scl: $\frac{\pi}{2}$ by $[-2, 2]$ scl: 0.5

$y = -\cos^3 x$
$[-2\pi, 2\pi]$ scl: $\frac{\pi}{2}$ by $[-2, 2]$ scl: 0.5

☑ **Guided Practice**

5. Simplify $-\csc\left(\frac{\pi}{2} - x\right) - \tan^2 x \sec x$.

Watch Out!

Graphing While the graphical approach shown in Examples 4 and 5 can lend support to the equality of two expressions, it cannot be used to prove that two expressions are equal. It is impossible to show that the graphs are identical over their entire domain using only the portion of the graph shown on your calculator.

You can simplify some trigonometric expressions by combining fractions.

EXAMPLE 6 Simplify by Combining Fractions

Simplify $\dfrac{\sin x \cos x}{1 - \sin x} - \dfrac{1 + \sin x}{\cos x}$.

$\dfrac{\sin x \cos x}{1 - \sin x} - \dfrac{1 + \sin x}{\cos x} = \dfrac{\sin x \cos x\,(\cos x)}{(1 - \sin x)(\cos x)} - \dfrac{(1 + \sin x)(1 - \sin x)}{(\cos x)(1 - \sin x)}$ **Common denominator**

$= \dfrac{\sin x \cos^2 x}{\cos x - \sin x \cos x} - \dfrac{1 - \sin^2 x}{\cos x - \sin x \cos x}$ **Multiply.**

$= \dfrac{\sin x \cos^2 x}{\cos x - \sin x \cos x} - \dfrac{\cos^2 x}{\cos x - \sin x \cos x}$ **Pythagorean Identity**

$= \dfrac{\sin x \cos^2 x - \cos^2 x}{\cos x - \sin x \cos x}$ **Subtract.**

$= \dfrac{(\cos^2 x)(\sin x - 1)}{(-\cos x)(\sin x - 1)}$ **Factor the numerator and denominator.**

$= -\cos x$ **Divide out common factors.**

✓ Guided Practice

Simplify each expression.

6A. $\dfrac{\cos x}{1 + \sin x} + \dfrac{1 + \sin x}{\cos x}$

6B. $\dfrac{\csc x}{1 + \sec x} + \dfrac{\csc x}{1 - \sec x}$

In calculus, you will sometimes need to rewrite a trigonometric expression so it does not involve a fraction. When the denominator is of the form $1 \pm u$ or $u \pm 1$, you can sometimes do so by multiplying the numerator and denominator by the conjugate of the denominator and applying a Pythagorean identity.

Review Vocabulary

conjugate a binomial factor which when multiplied by the original binomial factor has a product that is the difference of two squares (Lesson 0-3)

EXAMPLE 7 Rewrite to Eliminate Fractions

Rewrite $\dfrac{1}{1 + \cos x}$ as an expression that does not involve a fraction.

$\dfrac{1}{1 + \cos x} = \dfrac{1}{1 + \cos x} \cdot \dfrac{1 - \cos x}{1 - \cos x}$ **Multiply numerator and denominator by the conjugate of $1 + \cos x$, which is $1 - \cos x$.**

$= \dfrac{1 - \cos x}{1 - \cos^2 x}$ **Multiply.**

$= \dfrac{1 - \cos x}{\sin^2 x}$ **Pythagorean Identity**

$= \dfrac{1}{\sin^2 x} - \dfrac{\cos x}{\sin^2 x}$ **Write as the difference of two fractions.**

$= \dfrac{1}{\sin^2 x} - \dfrac{\cos x}{\sin x} \cdot \dfrac{1}{\sin x}$ **Factor.**

$= \csc^2 x - \cot x \csc x$ **Reciprocal and Quotient Identities**

✓ Guided Practice

Rewrite as an expression that does not involve a fraction.

7A. $\dfrac{\cos^2 x}{1 - \sin x}$

7B. $\dfrac{4}{\sec x + \tan x}$

Exercises

Find the value of each expression using the given information. (Example 1)

1. If $\cot \theta = \frac{5}{7}$, find $\tan \theta$.

2. If $\cos x = \frac{2}{3}$, find $\sec x$.

3. If $\tan \alpha = \frac{1}{5}$, find $\cot \alpha$.

4. If $\sin \beta = -\frac{5}{6}$, find $\csc \beta$.

5. If $\cos x = \frac{1}{6}$ and $\sin x = \frac{\sqrt{35}}{6}$, find $\cot x$.

6. If $\sec \varphi = 2$ and $\tan \varphi = \sqrt{3}$, find $\sin \varphi$.

7. If $\csc \alpha = \frac{7}{3}$ and $\cot \alpha = \frac{2\sqrt{10}}{3}$, find $\sec \alpha$.

8. If $\sec \theta = 8$ and $\tan \theta = 3\sqrt{7}$, find $\csc \theta$.

Find the value of each expression using the given information. (Example 2)

9. $\sec \theta$ and $\cos \theta$; $\tan \theta = -5$, $\cos \theta > 0$

10. $\cot \theta$ and $\sec \theta$; $\sin \theta = \frac{1}{3}$, $\tan \theta < 0$

11. $\tan \theta$ and $\sin \theta$; $\sec \theta = 4$, $\sin \theta > 0$

12. $\sin \theta$ and $\cot \theta$; $\cos \theta = \frac{2}{5}$, $\sin \theta < 0$

13. $\cos \theta$ and $\tan \theta$; $\csc \theta = \frac{8}{3}$, $\tan \theta > 0$

14. $\sin \theta$ and $\cos \theta$; $\cot \theta = 8$, $\csc \theta < 0$

15. $\cot \theta$ and $\sin \theta$; $\sec \theta = -\frac{9}{2}$, $\sin \theta > 0$

16. $\tan \theta$ and $\csc \theta$; $\cos \theta = -\frac{1}{4}$, $\sin \theta < 0$

Find the value of each expression using the given information. (Example 3)

17. If $\csc \theta = -1.24$, find $\sec \left(\theta - \frac{\pi}{2}\right)$.

18. If $\cos x = 0.61$, find $\sin \left(x - \frac{\pi}{2}\right)$.

19. If $\tan \theta = -1.52$, find $\cot \left(\theta - \frac{\pi}{2}\right)$.

20. If $\sin \theta = 0.18$, find $\cos \left(\theta - \frac{\pi}{2}\right)$.

21. If $\cot x = 1.35$, find $\tan \left(x - \frac{\pi}{2}\right)$.

Simplify each expression. (Examples 4 and 5)

22. $\csc x \sec x - \tan x$

23. $\csc x - \cos x \cot x$

24. $\sec x \cot x - \sin x$

25. $\dfrac{\tan x + \sin x \sec x}{\csc x \tan x}$

26. $\dfrac{1 - \sin^2 x}{\csc^2 x - 1}$

27. $\dfrac{\csc x \cos x + \cot x}{\sec x \cot x}$

28. $\dfrac{\sec x \csc x - \tan x}{\sec x \csc x}$

29. $\dfrac{\sec^2 x}{\cot^2 x + 1}$

30. $\cot x - \csc^2 x \cot x$

31. $\cot x - \cos^3 x \csc x$

Simplify each expression. (Example 6)

32. $\dfrac{\cos x}{\sec x + 1} + \dfrac{\cos x}{\sec x - 1}$

33. $\dfrac{1 - \cos x}{\tan x} + \dfrac{\sin x}{1 + \cos x}$

34. $\dfrac{1}{\sec x + 1} + \dfrac{1}{\sec x - 1}$

35. $\dfrac{\cos x \cot x}{\sec x + \tan x} + \dfrac{\sin x}{\sec x - \tan x}$

36. $\dfrac{\sin x}{\csc x + 1} + \dfrac{\sin x}{\csc x - 1}$

37. **SUNGLASSES** Many sunglasses are made with polarized lenses, which reduce the intensity of light. The intensity of light emerging from a system of two polarizing lenses I can be calculated by $I = I_0 - \dfrac{I_0}{\csc^2 \theta}$, where I_0 is the intensity of light entering the system of lenses and θ is the angle of the axis of the second lens in relation to that of the first lens. (Example 6)

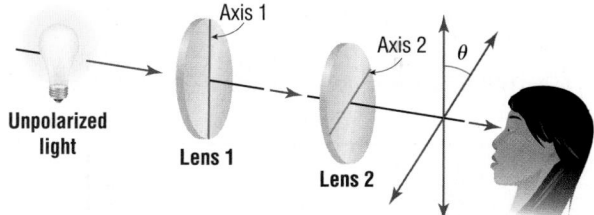

a. Simplify the formula for the intensity of light emerging from a system of two polarized lenses.

b. If a pair of sunglasses contains a system of two polarizing lenses with axes at 30° to one another, what proportion of the intensity of light entering the sunglasses emerges?

Rewrite as an expression that does not involve a fraction. (Example 7)

38. $\dfrac{\sin x}{\csc x - \cot x}$

39. $\dfrac{\csc x}{1 - \sin x}$

40. $\dfrac{\cot x}{\sec x - \tan x}$

41. $\dfrac{\cot x}{1 + \sin x}$

42. $\dfrac{3 \tan x}{1 - \cos x}$

43. $\dfrac{2 \sin x}{\cot x + \csc x}$

44. $\dfrac{\sin x}{1 - \sec x}$

45. $\dfrac{\cot^2 x \cos x}{\csc x - 1}$

46. $\dfrac{5}{\sec x + 1}$

47. $\dfrac{\sin x \tan x}{\cos x + 1}$

Determine whether each parent trigonometric function shown is odd or even. Explain your reasoning.

48.

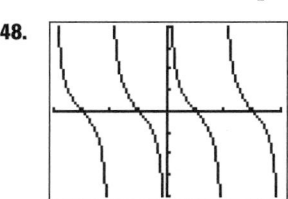

$[-2\pi, 2\pi]$ scl: $\frac{\pi}{2}$ by $[-4, 4]$ scl: 1

49.

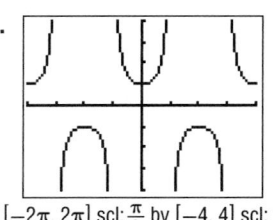

$[-2\pi, 2\pi]$ scl: $\frac{\pi}{2}$ by $[-4, 4]$ scl: 1

50. SOCCER When a soccer ball is kicked from the ground, its height y and horizontal displacement x are related by $y = \dfrac{-gx^2}{2v_0^2 \cos^2 \theta} + \dfrac{x \sin \theta}{\cos \theta}$, where v_0 is the initial velocity of the ball, θ is the angle at which it was kicked, and g is the acceleration due to gravity. Rewrite this equation so that $\tan \theta$ is the only trigonometric function that appears in the equation.

Write each expression in terms of a single trigonometric function.

51. $\tan x - \csc x \sec x$

52. $\cos x + \tan x \sin x$

53. $\csc x \tan^2 x - \sec^2 x \csc x$

54. $\sec x \csc x - \cos x \csc x$

55. 🔧 **MULTIPLE REPRESENTATIONS** In this problem, you will investigate the verification of trigonometric identities. Consider the functions shown.

i. $y_1 = \tan x + 1$
$y_2 = \sec x \cos x - \sin x \sec x$

ii. $y_3 = \tan x \sec x - \sin x$
$y_4 = \sin x \tan^2 x$

a. TABULAR Copy and complete the table below, without graphing the functions.

x	-2π	$-\pi$	0	π	2π
y_1					
y_2					
y_3					
y_4					

b. GRAPHICAL Graph each function on a graphing calculator.

c. VERBAL Make a conjecture about the relationship between y_1 and y_2. Repeat for y_3 and y_4.

d. ANALYTICAL Are the conjectures that you made in part **c** valid for the entire domain of each function? Explain your reasoning.

Rewrite each expression as a single logarithm and simplify the answer.

56. $\ln |\sin x| - \ln |\cos x|$

57. $\ln |\sec x| - \ln |\cos x|$

58. $\ln (\cot^2 x + 1) + \ln |\sec x|$

59. $\ln (\sec^2 x - \tan^2 x) - \ln (1 - \cos^2 x)$

60. ELECTRICITY A current in a wire in a magnetic field causes a force to act on the wire. The strength of the magnetic field can be determined using the formula $B = \dfrac{F \csc \theta}{I\ell}$, where F is the force on the wire, I is the current in the wire, ℓ is the length of the wire, and θ is the angle the wire makes with the magnetic field. Some physics books give the formula as $F = I\ell B \sin \theta$. Show that the two formulas are equivalent.

61. LIGHT WAVES When light shines through two narrow slits, a series of light and dark fringes appear. The angle θ, in radians, locating the mth fringe can be calculated by $\sin \theta = \dfrac{m\lambda}{d}$, where d is the distance between the two slits, and λ is the wavelength of light.

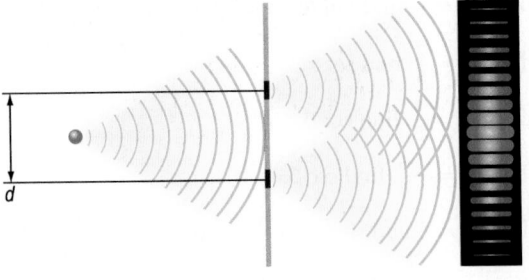

a. Rewrite the formula in terms of $\csc \theta$.

b. Determine the angle locating the 100th fringe when light having a wavelength of 550 nanometers is shined through double slits spaced 0.5 millimeters apart.

H.O.T. Problems   Use **H**igher-**O**rder **T**hinking Skills

62. PROOF Prove that the area of the triangle is $A = \sqrt{s(s-a)(s-b)(s-c)}$ where $s = \frac{1}{2}(a + b + c)$. (*Hint:* The area of an oblique triangle is $A = \frac{1}{2} bc \sin A$.)

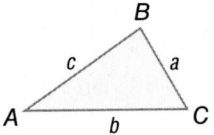

63. ERROR ANALYSIS Jenelle and Chloe are simplifying $\dfrac{1 - \sin^2 x}{\sin^2 x - \cos^2 x}$. Jenelle thinks that the expression simplifies to $\dfrac{\cos^2 x}{1 - 2\cos^2 x}$, and Chloe thinks that it simplifies to $\csc^2 x - \tan^2 x$. Is either of them correct? Explain your reasoning.

CHALLENGE Write each of the basic trigonometric functions in terms of the following functions.

64. $\sin x$ **65.** $\cos x$ **66.** $\tan x$

REASONING Determine whether each statement is *true* or *false*. Explain your reasoning.

67. $\csc^2 x \tan x = \csc x \sec x$ is true for all real numbers.

68. The odd-even identities can be used to prove that the graphs of $y = \cos x$ and $y = \sec x$ are symmetric with respect to the y-axis.

PROOF Prove each Pythagorean identity.

69 $\tan^2 \theta + 1 = \sec^2 \theta$ **70.** $\cot^2 \theta + 1 = \csc^2 \theta$

71. PREWRITE Use a chart or a table to help you organize the major trigonometric identities found in Lesson 5-1.

Solve each triangle. Round to the nearest tenth, if necessary. (Lesson 4-7)

72.

73.

74.

75.

76.

77.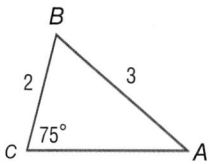

Find the exact value of each expression, if it exists. (Lesson 4-6)

78. $\cot\left(\sin^{-1}\frac{7}{9}\right)$

79. $\tan(\arctan 3)$

80. $\cos\left[\arccos\left(-\frac{1}{2}\right)\right]$

81. $\cos\left(\frac{\pi}{2} - \cos^{-1}\frac{\sqrt{2}}{2}\right)$

82. $\cos^{-1}\left(\sin^{-1}\frac{\pi}{2}\right)$

83. $\sin\left(\cos^{-1}\frac{3}{5}\right)$

84. ANTHROPOLOGY *Allometry* is the study of the relationship between the size of an organism and the size of any of its parts. A researcher decided to test for an allometry between the size of the human head compared to the human body as a person ages. The data in the table represent the average American male. (Lesson 3-5)

  a. Find a quadratic model relating these data by linearizing the data and finding the linear regression equation.

  b. Use the model for the linearized data to find a model for the original data.

  c. Use your model to predict the height of an American male whose head circumference is 24 inches.

Growth of the Average American Male (0-3 years of age)	
Head Circumference (in.)	Height (in.)
14.1	19.5
18.0	26.4
18.3	29.7
18.7	32.3
19.1	34.4
19.4	36.2
19.6	37.7

Source: National Center for Health Statistics

Let $U = \{0, 1, 2, 3, 4, 5\}$, $A = \{6, 9\}$, $B = \{6, 9, 10\}$, $C = \{0, 1, 6, 9, 11\}$, $D = \{2, 5, 11\}$. **Determine whether each statement is *true* or *false*. Explain your reasoning.** (Lesson 0-1)

85. $A \subset B$

86. $D \subset U$

87. SAT/ACT If $x > 0$, then
$$\frac{x^2 - 1}{x + 1} + \frac{(x + 1)^2 - 1}{x + 2} + \frac{(x + 2)^2 - 1}{x + 3} =$$

  A $(x + 1)^2$

  B $(x - 1)^2$

  C $3x - 1$

  D $3x$

  E $3(x - 1)^2$

88. REVIEW If $\sin x = m$ and $0 < x < 90°$, then $\tan x =$

  F $\frac{1}{m^2}$

  G $\frac{1 - m^2}{m}$

  H $\frac{m\sqrt{1 - m^2}}{1 - m^2}$

  J $\frac{m}{1 - m^2}$

89. Which of the following is equivalent to
$$\frac{1 - \sin^2\theta}{1 - \cos^2\theta} \cdot \tan\theta?$$

  A $\tan\theta$

  C $\sin\theta$

  B $\cot\theta$

  D $\cos\theta$

90. REVIEW Refer to the figure. If $\cos D = 0.8$, what is the length of $\overline{DF}$?

  F 5

  G 4

  H 3.2

  J $\frac{4}{5}$

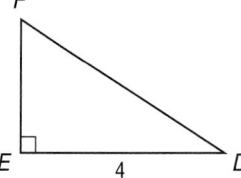

Verifying Trigonometric Identities

Then
You simplified trigonometric expressions. (Lesson 5-1)

Now
- Verify trigonometric identities.
- Determine whether equations are identities.

New Vocabulary
verify an identity

Math Online
glencoe.com

Why?

Two fireworks travel at the same speed v. The fireworks technician wants to explode one firework higher than another by adjusting the angle θ of the path each rocket makes with the ground. To calculate the maximum height h of each rocket, the formula $h = \dfrac{v^2 \tan^2 \theta}{2g \sec^2 \theta}$ could be used, but would $h = \dfrac{v^2 \sin^2 \theta}{2g}$ give the same result?

Verify Trigonometric Identities In Lesson 5-1, you used trigonometric identities to rewrite expressions in equivalent and sometimes more useful forms. Once verified, these *new* identities can also be used to solve problems or to rewrite other trigonometric expressions.

To **verify an identity** means to *prove* that both sides of the equation are equal for all values of the variable for which both sides are defined. This is done by transforming the expression on one side of the identity into the expression on the other side through a sequence of intermediate expressions that are each equivalent to the first. As with other types of proofs, each step is justified by a reason, usually another verified trigonometric identity or an algebraic operation.

You will find that it is often easier to start the verification of a trigonometric identity by beginning on the side with the more complicated expression and working toward the less complicated expression.

EXAMPLE 1 Verify a Trigonometric Identity

Verify that $\dfrac{\csc^2 x - 1}{\csc^2 x} = \cos^2 x.$

The left-hand side of this identity is more complicated, so start with that expression first.

$$\frac{\csc^2 x - 1}{\csc^2 x} = \frac{\cot^2 x}{\csc^2 x} \qquad \textbf{Pythagorean Identity}$$

$$= \cot^2 x \sin^2 x \qquad \textbf{Reciprocal Identity}$$

$$= \left(\frac{\cos^2 x}{\sin^2 x}\right)\sin^2 x \qquad \textbf{Quotient Identity}$$

$$= \cos^2 x \; \checkmark \qquad \textbf{Pythagorean Identity}$$

Notice that the verification ends with the expression on the other side of the identity.

✓ Guided Practice   Verify each identity.

1A. $\sec^2 \theta \cot^2 \theta - 1 = \cot^2 \theta$ **1B.** $\tan^2 \alpha = \sec \alpha \csc \alpha \tan \alpha - 1$

There is usually more than one way to verify an identity. For example, the identity in Example 1 can also be verified as follows.

$$\frac{\csc^2 x - 1}{\csc^2 x} = \frac{\csc^2 x}{\csc^2 x} - \frac{1}{\csc^2 x} \qquad \textbf{Write as the difference of two fractions.}$$

$$= 1 - \sin^2 x \qquad \textbf{Simplify and apply a Reciprocal Identity.}$$

$$= \cos^2 x \qquad \textbf{Pythagorean Identity}$$

When there are multiple fractions with different denominators in an expression, you can find a common denominator to reduce the expression to one fraction.

EXAMPLE 2 Verify a Trigonometric Identity by Combining Fractions

Verify that $2 \csc x = \dfrac{1}{\csc x + \cot x} + \dfrac{1}{\csc x - \cot x}$.

The right-hand side of this identity is more complicated, so start there, rewriting each fraction using the common denominator $(\csc x + \cot x)(\csc x - \cot x)$.

$$\dfrac{1}{\csc x + \cot x} + \dfrac{1}{\csc x - \cot x}$$ Start with the right-hand side of the identity.

$$= \dfrac{\csc x - \cot x}{(\csc x + \cot x)(\csc x - \cot x)} + \dfrac{\csc x + \cot x}{(\csc x + \cot x)(\csc x - \cot x)}$$ Common denominator

$$= \dfrac{2 \csc x}{(\csc x + \cot x)(\csc x - \cot x)}$$ Add.

$$= \dfrac{2 \csc x}{\csc^2 x - \cot^2 x}$$ Multiply.

$$= 2 \csc x \checkmark$$ Pythagorean Identity

☑ **Guided Practice**

2. Verify that $\dfrac{\cos \alpha}{1 + \sin \alpha} + \dfrac{\cos \alpha}{1 - \sin \alpha} = 2 \sec \alpha$.

To eliminate a fraction in which the denominator is of the form $1 \pm u$ or $u \pm 1$, remember to try multiplying the numerator and denominator by the conjugate of the denominator. Then you can potentially apply a Pythagorean Identity.

StudyTip

Alternate Method You do not always have to start with the more complicated side of the equation. If you start with the right-hand side in Example 3, you can still prove the identity.

$\csc \alpha + \cot \alpha$

$= \dfrac{1}{\sin \alpha} + \dfrac{\cos \alpha}{\sin \alpha}$

$= \dfrac{1 + \cos \alpha}{\sin \alpha}$

$= \dfrac{1 + \cos \alpha}{\sin \alpha} \cdot \dfrac{1 - \cos \alpha}{1 - \cos \alpha}$

$= \dfrac{\sin \alpha}{1 - \cos \alpha} \checkmark$

EXAMPLE 3 Verify a Trigonometric Identity by Multiplying

Verify that $\dfrac{\sin \alpha}{1 - \cos \alpha} = \csc \alpha + \cot \alpha$.

Because the left-hand side of this identity involves a fraction, it is slightly more complicated than the right side. So, start with the left side.

$$\dfrac{\sin \alpha}{1 - \cos \alpha} = \dfrac{\sin \alpha}{1 - \cos \alpha} \cdot \dfrac{1 + \cos \alpha}{1 + \cos \alpha}$$ Multiply numerator and denominator by the conjugate of $1 - \cos \alpha$, which is $1 + \cos \alpha$.

$$= \dfrac{\sin \alpha (1 + \cos \alpha)}{1 - \cos^2 \alpha}$$ Multiply.

$$= \dfrac{\sin \alpha (1 + \cos \alpha)}{\sin^2 \alpha}$$ Pythagorean Identity

$$= \dfrac{1 + \cos \alpha}{\sin \alpha}$$ Divide out the common factor of $\sin \alpha$.

$$= \dfrac{1}{\sin \alpha} + \dfrac{\cos \alpha}{\sin \alpha}$$ Write as the sum of two fractions.

$$= \csc \alpha + \cot \alpha \checkmark$$ Reciprocal and Quotient Identities

☑ **Guided Practice**

3. Verify that $\dfrac{\tan x}{\sec x + 1} = \csc x - \cot x$.

Until an identity has been verified, you cannot assume that both sides of the equation are equal. Therefore, you cannot use the properties of equality to perform algebraic operations on each side of an identity, such as adding the same quantity to each side of the equation.

When the more complicated expression in an identity involves powers, try factoring.

EXAMPLE 4 | **Verify a Trigonometric Identity by Factoring**

Verify that $\cot \theta \sec \theta \csc^2 \theta - \cot^3 \theta \sec \theta = \csc \theta$.

$\cot \theta \sec \theta \csc^2 \theta - \cot^3 \theta \sec \theta$	Start with the left-hand side of the identity.
$= \cot \theta \sec \theta (\csc^2 \theta - \cot^2 \theta)$	Factor.
$= \cot \theta \sec \theta$	Pythagorean Identity
$= \dfrac{\cos \theta}{\sin \theta} \cdot \dfrac{1}{\cos \theta}$	Reciprocal and Quotient Identities
$= \dfrac{1}{\sin \theta}$	Multiply.
$= \csc \theta \checkmark$	Reciprocal Identity

☑ **Guided Practice**

4. Verify that $\sin^2 x \tan^2 x \csc^2 x + \cos^2 x \tan^2 x \csc^2 x = \sec^2 x$.

It is sometimes helpful to work each side of an identity separately to obtain a common intermediate expression.

StudyTip

Additional Steps When verifying an identity, the number of steps that are needed to justify the verification may be obvious. However, if it is unclear, it is usually safer to include too many steps, rather than too few.

EXAMPLE 5 | **Verify an Identity by Working Each Side Separately**

Verify that $\dfrac{\tan^2 x}{1 + \sec x} = \dfrac{1 - \cos x}{\cos x}$.

Both sides look complicated, but the left-hand side is slightly more complicated since its denominator involves two terms. So, start with the expression on the left.

$\dfrac{\tan^2 x}{1 + \sec x} = \dfrac{\sec^2 x - 1}{1 + \sec x}$	Pythagorean Identity
$= \dfrac{(\sec x - 1)(\sec x + 1)}{1 + \sec x}$	Factor.
$= \sec x - 1$	Divide out common factor of sec $x + 1$.

From here, it is unclear how to transform $\sec x - 1$ into $\dfrac{1 - \cos x}{\cos x}$, so start with the right-hand side and work to transform it into the intermediate form $\sec x - 1$.

$\dfrac{1 - \cos x}{\cos x} = \dfrac{1}{\cos x} - \dfrac{\cos x}{\cos x}$	Write as the difference of two fractions.
$= \sec x - 1$	Use the Quotient Identity and simplify.

To complete the proof, work backward to connect the two parts of the proof.

$\dfrac{\tan^2 x}{1 + \sec x} = \dfrac{\sec^2 x - 1}{1 + \sec x}$	Pythagorean Identity
$= \dfrac{(\sec x - 1)(\sec x + 1)}{1 + \sec x}$	Factor.
$= \sec x - 1$	Divide out common factor of sec $x + 1$.
$= \dfrac{1}{\cos x} - \dfrac{\cos x}{\cos x}$	Use the Quotient Identity and write 1 as $\frac{\cos x}{\cos x}$.
$= \dfrac{1 - \cos x}{\cos x} \checkmark$	Combine fractions.

☑ **Guided Practice**

5. Verify that $\sec^4 x - \sec^2 x = \tan^4 x + \tan^2 x$.

Concept Summary — Strategies for Verifying Trigonometric Identities

- Start with the more complicated side of the identity and work to transform it into the simpler side, keeping the other side of the identity in mind as your goal.

- Use reciprocal, quotient, Pythagorean, and other basic trigonometric identities.

- Use algebraic operations such as combining fractions, rewriting fractions as sums or differences, multiplying expressions, or factoring expressions.

- Convert a denominator of the form $1 \pm u$ or $u \pm 1$ to a single term using its conjugate and a Pythagorean Identity.

- Work each side separately to reach a common intermediate expression.

- If no other strategy presents itself, try converting the entire expression to one involving only sines and cosines.

Identifying Identities and Nonidentities You can use a graphing calculator to investigate whether an equation may be an identity by graphing the functions related to each side of the equation.

EXAMPLE 6 Determine Whether an Equation is an Identity

Use a graphing calculator to test whether each equation is an identity. If it appears to be an identity, verify it. If not, find a value for which both sides are defined but not equal.

a. $\dfrac{\cos \beta + 1}{\tan^2 \beta} = \dfrac{\cos \beta}{\sec \beta + 1}$

The graphs of the related functions do not coincide for all values of x for which the both functions are defined. When $x = \dfrac{\pi}{4}$, $Y_1 \approx 1.7$ but $Y_2 \approx 0.3$. The equation is not an identity.

$[-2\pi, 2\pi]$ scl: π by $[-1, 3]$ scl: 1

$[-2\pi, 2\pi]$ scl: π by $[-1, 3]$ scl: 1

b. $\dfrac{\cos \beta + 1}{\tan^2 \beta} = \dfrac{\cos \beta}{\sec \beta - 1}$

The equation *appears* to be an identity because the graphs of the related functions coincide. Verify this algebraically.

$[-2\pi, 2\pi]$ scl: π by $[-1, 3]$ scl: 1

$\dfrac{\cos \beta}{\sec \beta - 1} = \dfrac{\cos \beta}{\sec \beta - 1} \cdot \dfrac{\sec \beta + 1}{\sec \beta + 1}$ **Multiply numerator and denominator by the conjugate of $\sec \beta - 1$.**

$= \dfrac{\cos \beta \sec \beta + \cos \beta}{\sec^2 \beta - 1}$ **Multiply.**

$= \dfrac{\cos \beta \left(\dfrac{1}{\cos \beta}\right) + \cos \beta}{\sec^2 \beta - 1}$ **Reciprocal Identity**

$= \dfrac{1 + \cos \beta}{\sec^2 \beta - 1}$ **Simplify.**

$= \dfrac{\cos \beta + 1}{\tan^2 \beta}$ ✓   **Commutative Property and Pythagorean Identity**

✓ Guided Practice

6A. $\csc \theta = \dfrac{\cot \theta \tan^2 \theta + \cot \theta}{\sec \theta}$

6B. $\dfrac{\cos x + 1}{\sec^2 x} = \dfrac{\cos x}{\sec x - 1}$

Exercises

Verify each identity. (Examples 1–3)

1. $(\sec^2 \theta - 1) \cos^2 \theta = \sin^2 \theta$

2. $\sec^2 \theta (1 - \cos^2 \theta) = \tan^2 \theta$

3. $\sin \theta - \sin \theta \cos^2 \theta = \sin^3 \theta$

4. $\csc \theta - \cos \theta \cot \theta = \sin \theta$

5. $\cot^2 \theta \csc^2 \theta - \cot^2 \theta = \cot^4 \theta$

6. $\tan \theta \csc^2 \theta - \tan \theta = \cot \theta$

7. $\dfrac{\sec \theta}{\sin \theta} - \dfrac{\sin \theta}{\cos \theta} = \cot \theta$

8. $\dfrac{\sin \theta}{1 - \cos \theta} + \dfrac{1 - \cos \theta}{\sin \theta} = 2 \csc \theta$

9. $\dfrac{\cos \theta}{1 + \sin \theta} + \tan \theta = \sec \theta$

10. $\dfrac{\sin \theta}{1 - \cot \theta} + \dfrac{\cos \theta}{1 - \tan \theta} = \sin \theta + \cos \theta$

11. $\dfrac{1}{1 - \tan^2 \theta} + \dfrac{1}{1 - \cot^2 \theta} = 1$

12. $\dfrac{1}{\csc \theta + 1} + \dfrac{1}{\csc \theta - 1} = 2 \sec^2 \theta \sin \theta$

13. $(\csc \theta - \cot \theta)(\csc \theta + \cot \theta) = 1$

14. $\cos^4 \theta - \sin^4 \theta = \cos^2 \theta - \sin^2 \theta$

15. $\dfrac{1}{1 - \sin \theta} + \dfrac{1}{1 + \sin \theta} = 2 \sec^2 \theta$

16. $\dfrac{\cos \theta}{1 + \sin \theta} + \dfrac{\cos \theta}{1 - \sin \theta} = 2 \sec \theta$

17. $\csc^4 \theta - \cot^4 \theta = 2 \cot^2 \theta + 1$

18. $\dfrac{\csc^2 \theta + 2 \csc \theta - 3}{\csc^2 \theta - 1} = \dfrac{\csc \theta + 3}{\csc \theta + 1}$

19. **FIREWORKS** If a rocket is launched from ground level, the maximum height that it reaches is given by $h = \dfrac{v^2 \sin^2 \theta}{2g}$, where θ is the angle between the ground and the initial path of the rocket, v is the rocket's initial speed, and g is the acceleration due to gravity, 9.8 meters per second squared. (Example 3)

a. Verify that $\dfrac{v^2 \sin^2 \theta}{2g} = \dfrac{v^2 \tan^2 \theta}{2g \sec^2 \theta}$.

b. Suppose a second rocket is fired at an angle of 80° from the ground with an initial speed of 110 meters per second. Find the maximum height of the rocket.

Verify each identity. (Examples 4 and 5)

20. $(\csc \theta + \cot \theta)(1 - \cos \theta) = \sin \theta$

21. $\sin^2 \theta \tan^2 \theta = \tan^2 \theta - \sin^2 \theta$

22. $\dfrac{1 - \tan^2 \theta}{1 - \cot^2 \theta} = \dfrac{\cos^2 \theta - 1}{\cos^2 \theta}$

23. $\dfrac{1 + \csc \theta}{\sec \theta} = \cos \theta + \cot \theta$

24. $(\csc \theta - \cot \theta)^2 = \dfrac{1 - \cos \theta}{1 + \cos \theta}$

25. $\dfrac{1 + \tan^2 \theta}{1 - \tan^2 \theta} = \dfrac{1}{2 \cos^2 \theta - 1}$

26. $\tan^2 \theta \cos^2 \theta = 1 - \cos^2 \theta$

27. $\sec \theta - \cos \theta = \tan \theta \sin \theta$

28. $1 - \tan^4 \theta = 2 \sec^2 \theta - \sec^4 \theta$

29. $(\csc \theta - \cot \theta)^2 = \dfrac{1 - \cos \theta}{1 + \cos \theta}$

30. $\dfrac{1 + \tan \theta}{\sin \theta + \cos \theta} = \sec \theta$

31. $\dfrac{2 + \csc \theta \sec \theta}{\csc \theta \sec \theta} = (\sin \theta + \cos \theta)^2$

32. **OPTICS** If two prisms of the same power are placed next to each other, their total power can be determined using $z = 2p \cos \theta$, where z is the combined power of the prisms, p is the power of the individual prisms, and θ is the angle between the two prisms. Verify that $2p \cos \theta = 2p(1 - \sin^2 \theta) \sec \theta$. (Example 4)

33. **PHOTOGRAPHY** The amount of light passing through a polarization filter can be modeled using $I = I_m \cos^2 \theta$, where I is the amount of light passing through the filter, I_m is the amount of light shined on the filter, and θ is the angle of rotation between the light source and the filter. Verify that $I_m \cos^2 \theta = I_m - \dfrac{I_m}{\cot^2 \theta + 1}$. (Example 4)

GRAPHING CALCULATOR Test whether each equation is an identity by graphing. If it appears to be an identity, verify it. If not, find a value for which both sides are defined but not equal. (Example 6)

34. $\dfrac{\tan x + 1}{\tan x - 1} = \dfrac{1 + \cot x}{1 - \cot x}$

35. $\sec x + \tan x = \dfrac{1}{\sec x - \tan x}$

36. $\sec^2 x - 2 \sec x \tan x + \tan^2 x = \dfrac{1 - \cos x}{1 + \cos x}$

37. $\dfrac{\cot^2 x - 1}{1 + \cot^2 x} = 1 - 2 \sin^2 x$

38. $\dfrac{\tan x - \sec x}{\tan x + \sec x} = \dfrac{\tan^2 x - 1}{\sec^2 x}$

39. $\cos^2 x - \sin^2 x = \dfrac{\cot x - \tan x}{\tan x + \cot x}$

Verify each identity.

40. $\sqrt{\dfrac{\sin x \tan x}{\sec x}} = |\sin x|$

41. $\sqrt{\dfrac{\sec x - 1}{\sec x + 1}} = \left|\dfrac{\sec x - 1}{\tan x}\right|$

42. $\ln |\csc x + \cot x| + \ln |\csc x - \cot x| = 0$

43. $\ln |\cot x| + \ln |\tan x \cos x| = \ln |\cos x|$

Verify each identity.

44. $\sec^2 \theta + \tan^2 \theta = \sec^4 \theta - \tan^4 \theta$

45. $-2 \cos^2 \theta = \sin^4 \theta - \cos^4 \theta - 1$

46. $\sec^2 \theta \sin^2 \theta = \sec^4 - (\tan^4 \theta + \sec^2 \theta)$

47. $3 \sec^2 \theta \tan^2 \theta + 1 = \sec^6 \theta - \tan^6 \theta$

48. $\sec^4 x = 1 + 2 \tan^2 x + \tan^4 x$

49. $\sec^2 x \csc^2 x = \sec^2 x + \csc^2 x$

50. ENVIRONMENT A biologist studying pollution situates a net across a river and positions instruments at two different stations on the river bank to collect samples. In the diagram shown, d is the distance between the stations and w is width of the river.

a. Determine an equation in terms of tangent α that can be used to find the distance between the stations.

b. Verify that $d = \dfrac{w \cos (90° - \alpha)}{\cos \alpha}$.

c. Complete the table shown for $d = 40$ feet.

w	20	40	60	80	100	120
α						

d. If $\alpha > 60°$ or $\alpha < 20°$, the instruments will not function properly. Use the table from part **c** to determine whether sites in which the width of the river is 5, 35, or 140 feet could be used for the experiment.

HYPERBOLIC FUNCTIONS The *hyperbolic trigonometric functions* are defined in the following ways.

$\sinh x = \frac{1}{2}(e^x - e^{-x})$ $\qquad$ $\operatorname{csch} x = \dfrac{1}{\sinh x}, x \neq 0$

$\cosh x = \frac{1}{2}(e^x + e^{-x})$ $\qquad$ $\operatorname{sech} x = \dfrac{1}{\cosh x}$

$\tanh x = \dfrac{\sinh x}{\cosh x}$ $\qquad$ $\coth x = \dfrac{1}{\tanh x}, x \neq 0$

Verify each identity using the functions shown above.

51. $\cosh^2 x - \sinh^2 x = 1$ $\qquad$ **52.** $\sinh(-x) = -\sinh x$

53. $\operatorname{sech}^2 x = 1 - \tanh^2 x$ $\qquad$ **54.** $\cosh(-x) = \cosh x$

GRAPHING CALCULATOR Graph each side of each equation. If the equation appears to be an identity, verify it algebraically.

55. $\dfrac{\sec x}{\cos x} - \dfrac{\tan x \sec x}{\csc x} = 1$

56. $\sec x - \cos^2 x \csc x = \tan x \sec x$

57. $(\tan x + \sec x)(1 - \sin x) = \cos x$

58. $\dfrac{\sec x \cos x}{\cot^2 x} - \dfrac{1}{\tan^2 x - \sin^2 x \tan^2 x} = -1$

59. **MULTIPLE REPRESENTATIONS** In this problem, you will investigate methods used to solve trigonometric equations. Consider $1 = 2 \sin x$.

a. NUMERICAL Isolate the trigonometric function in the equation so that $\sin x$ is the only expression on one side of the equation.

b. GRAPHICAL Graph the left and right sides of the equation you found in part **a** on the same graph over $[0, 2\pi)$. Locate any points of intersection and express the values in terms of radians.

c. GEOMETRIC Use the unit circle to verify the answers you found in part **b**.

d. GRAPHICAL Graph the left and right sides of the equation you found in part **a** on the same graph over $-2\pi < x < 2\pi$. Locate any points of intersection and express the values in terms of radians.

e. VERBAL Make a conjecture as to the solutions of $1 = 2 \sin x$. Explain your reasoning.

H.O.T. Problems Use **H**igher-**O**rder **T**hinking Skills

60. REASONING Can substitution be used to determine whether an equation is an identity? Explain your reasoning.

61. CHALLENGE Verify that the area A of a triangle is given by
$$A = \dfrac{a^2 \sin \beta \sin \gamma}{2 \sin (\beta + \gamma)}$$
where a, b, and c represent the sides of the triangle and α, β, and γ are the respective opposite angles.

62. WRITING IN MATH Use the properties of logarithms to explain why the sum of the natural logarithms of the six basic trigonometric functions for any angle θ is 0.

63. OPEN ENDED Create identities for $\sec x$ and $\csc x$ in terms of two or more of the other basic trigonometric functions.

64. REASONING If two angles α and β are complementary, is $\cos^2 \alpha + \cos^2 \beta = 1$? Explain your reasoning. Justify your answers.

65. WRITING IN MATH Explain how you would verify a trigonometric identity in which both sides of the equation are equally complex.

Simplify each expression. (Lesson 5-1)

66. $\cos \theta \csc \theta$

67. $\tan \theta \cot \theta$

68. $\sin \theta \cot \theta$

69. $\dfrac{\cos \theta \csc \theta}{\tan \theta}$

70. $\dfrac{\sin \theta \csc \theta}{\cot \theta}$

71. $\dfrac{1 - \cos^2 \theta}{\sin^2 \theta}$

72. BALLOONING As a hot-air balloon crosses over a straight portion of interstate highway, its pilot eyes two consecutive mileposts on the same side of the balloon. When viewing the mileposts, the angles of depression are 64° and 7°. How high is the balloon to the nearest foot? (Lesson 4-7)

1 mi = 5280 ft

Locate the vertical asymptotes, and sketch the graph of each function. (Lesson 4-5)

73. $y = \dfrac{1}{4} \tan x$

74. $y = \csc 2x$

75. $y = \dfrac{1}{2} \sec 3x$

Write each degree measure in radians as a multiple of π and each radian measure in degrees. (Lesson 4-2)

76. 660°

77. 570°

78. 158°

79. $\dfrac{29\pi}{4}$

80. $\dfrac{17\pi}{6}$

81. 9

Solve each inequality. (Lesson 2-6)

82. $x^2 - 3x - 18 > 0$

83. $x^2 + 3x - 28 < 0$

84. $x^2 - 4x \le 5$

85. $x^2 + 2x \ge 24$

86. $-x^2 - x + 12 \ge 0$

87. $-x^2 - 6x + 7 \le 0$

88. FOOD The manager of a bakery is randomly checking slices of cake prepared by employees to ensure that the correct amount of flavor is in each slice. Each 12-ounce slice should contain half chocolate and half vanilla flavored cream. The amount of chocolate by which each slice varies can be represented by $g(x) = \frac{1}{2}|x - 12|$. Describe the transformations in the function. Then graph the function. (Lesson 1-5)

Solving Trigonometric Equations

Why?

A baseball leaves a bat at a launch angle θ and returns to its initial batted height after a distance of d meters. To find the velocity v_0 of the ball as it leaves the bat, you can solve the trigonometric equation $d = \dfrac{2v_0^2 \sin\theta\cos\theta}{9.8}$.

Then
You verified trigonometric identities. (Lesson 5-2)

Now
- Solve trigonometric equations using algebraic techniques.
- Solve trigonometric equations using basic identities.

Math Online
glencoe.com

Use Algebraic Techniques to Solve In Lesson 5-2, you verified trigonometric equations called identities that are true for all values of the variable for which both sides are defined. In this lesson we will consider *conditional* trigonometric equations, which may be true for certain values of the variable but false for others.

Consider the graphs of both sides of the conditional trigonometric equation $\cos x = \dfrac{1}{2}$.

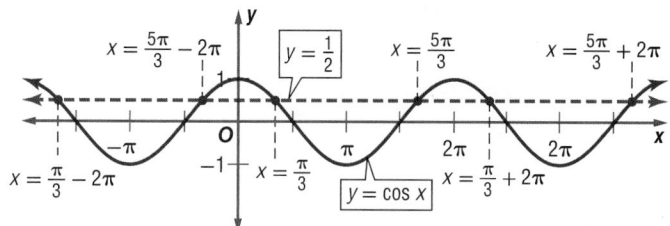

The graph shows that $\cos x = \dfrac{1}{2}$ has two solutions on the interval $[0, 2\pi)$, $x = \dfrac{\pi}{3}$ and $x = \dfrac{5\pi}{3}$. Since $y = \cos x$ has a period of 2π, $\cos x = \dfrac{1}{2}$ has infinitely many solutions on the interval $(-\infty, \infty)$. Additional solutions are found by adding integer multiples of the period, so we express all solutions by writing

$$x = \frac{\pi}{3} + 2n\pi \qquad \text{and} \qquad x = \frac{5\pi}{3} + 2n\pi, \qquad \text{where } n \text{ is an integer.}$$

To solve a trigonometric equation that involves only one trigonometric expression, begin by isolating this expression.

EXAMPLE 1 **Solve by Isolating Trigonometric Expressions**

Solve $2\tan x - \sqrt{3} = \tan x$.

$2\tan x - \sqrt{3} = \tan x$	**Original equation**
$\tan x - \sqrt{3} = 0$	**Subtract tan x from each side to isolate the trigonometric expression.**
$\tan x = \sqrt{3}$	**Add $\sqrt{3}$ to each side.**

The period of tangent is π, so you only need to find solutions on the interval $[0, \pi)$. The only solution on this interval is $x = \dfrac{\pi}{3}$. The solutions on the interval $(-\infty, \infty)$ are then found by adding integer multiples π. Therefore, the general form of the solutions is

$$x = \frac{\pi}{3} + n\pi, \qquad \text{where } n \text{ is an integer.}$$

✓ **Guided Practice**

1. Solve $4\sin x = 2\sin x + \sqrt{2}$.

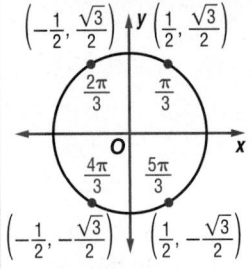
EXAMPLE 2   **Solve by Taking the Square Root of Each Side**

Solve $4 \sin^2 x + 1 = 4$.

$4 \sin^2 x + 1 = 4$	**Original equation**
$4 \sin^2 x = 3$	**Subtract 1 from each side.**
$\sin^2 x = \dfrac{3}{4}$	**Divide each side by 4.**
$\sin x = \pm\dfrac{\sqrt{3}}{2}$	**Take the square root of each side.**

On the interval $[0, 2\pi)$, $\sin x = \dfrac{\sqrt{3}}{2}$ when $x = \dfrac{\pi}{3}$ and $x = \dfrac{2\pi}{3}$ and $\sin x = -\dfrac{\sqrt{3}}{2}$ when $x = \dfrac{4\pi}{3}$ and $x = \dfrac{5\pi}{3}$. Since sine has a period of 2π, the solutions on the interval $(-\infty, \infty)$ have the general form $x = \dfrac{\pi}{3} + 2n\pi$, $x = \dfrac{2\pi}{3} + 2n\pi$, $x = \dfrac{4\pi}{3} + 2n\pi$, and $x = \dfrac{5\pi}{3} + 2n\pi$, where n is an integer.

☑ **Guided Practice**

2. Solve $3 \cot^2 x + 4 = 7$.

When trigonometric functions cannot be combined on one side of an equation, try factoring and applying the Zero Product Property. If the equation has quadratic form, factor if possible. If not possible, apply the Quadratic Formula.

EXAMPLE 3   **Solve by Factoring**

Find all solutions of each equation on the interval $[0, 2\pi)$.

a. $\cos x \sin x = 3 \cos x$

$\cos x \sin \theta = 3 \cos x$	**Original equation**
$\cos x \sin x - 3 \cos x = 0$	**Isolate the trigonometric expression.**
$\cos x(\sin x - 3) = 0$	**Factor.**
$\cos x = 0$ or $\sin x - 3 = 0$	**Zero Product Property**
$x = \dfrac{\pi}{2}$ and $\dfrac{3\pi}{2}$ $\sin x = 3$	**Solve for x on $[0, 2\pi)$.**

The equation $\sin x = 3$ has no solution since the maximum value the sine function can attain is 1. Therefore, on the interval $[0, 2\pi)$, the solutions of the original equation are $\dfrac{\pi}{2}$ and $\dfrac{3\pi}{2}$.

b. $\cos^4 x + \cos^2 x - 2 = 0$

$\cos^4 x + \cos^2 x - 2 = 0$	**Original equation**
$(\cos^2 x)^2 + \cos^2 x - 2 = 0$	**Write in quadratic form.**
$(\cos^2 x + 2)(\cos^2 x - 1) = 0$	**Factor.**
$\cos^2 x + 2 = 0$ or $\cos^2 x - 1 = 0$	**Zero Product Property**
$\cos^2 x = -2$ $\cos^2 x = 1$	**Solve for $\cos^2 x$.**
$\cos x = \pm\sqrt{-2}$ $\cos x = \pm\sqrt{1}$ or ± 1	**Take the square root of each side.**

The equation $\cos x = \pm\sqrt{-2}$ has no real solutions. On the interval $[0, 2\pi)$, the equation $\cos x = \pm 1$ has solutions 0 and π.

☑ **Guided Practice**

3A. $2 \sin x \cos x = \sqrt{2} \cos x$ **3B.** $4 \cos^2 x + 2 \cos x - 2\sqrt{2} \cos x = \sqrt{2}$

Some trigonometric equations involve functions of multiple angles, such as $\cos 2x = \frac{1}{2}$. To solve these equations, first solve for the multiple angle.

🌐 Real-World EXAMPLE 4 | Trigonometric Functions of Multiple Angles

BASEBALL A baseball leaves a bat with an initial speed of 30 meters per second and clears a fence 90.5 meters away. The height of the fence is the same height as the initial height of the batted ball. If the distance the ball traveled is given by $d = \frac{v_0^2 \sin 2\theta}{9.8}$, where 9.8 is in meters per second squared, find the interval of possible launch angles of the ball.

90.5 m

$d = \dfrac{v_0^2 \sin 2\theta}{9.8}$	**Original formula**
$90.5 = \dfrac{30^2 \sin 2\theta}{9.8}$	$d = 90.5$ and $v_0 = 30$
$90.5 = \dfrac{900 \sin 2\theta}{9.8}$	**Simplify.**
$886.9 = 900 \sin 2\theta$	**Multiply each side by 9.8.**
$\dfrac{886.9}{900} = \sin 2\theta$	**Divide each side by 900.**
$\sin^{-1} \dfrac{886.9}{900} = 2\theta$	**Definition of inverse sine**

Recall from Lesson 4-6 that the range of the inverse sine function is restricted to acute angles of θ in the interval $[-90°, 90°]$. Since we are finding the inverse sine of 2θ instead of θ, we need to consider angles in the interval $[-2(90°), 2(90°)]$ or $[-180°, 180°]$. Use your calculator to find the acute angle and the reference angle relationship $\sin(180° - \theta) = \sin \theta$ to find the obtuse angle.

$\sin^{-1} \dfrac{886.9}{900} = 2\theta$	**Definition of inverse sine**
$80.2°$ or $99.8° = 2\theta$	$\sin^{-1} \frac{886.9}{900} \approx 80.2°$ and $\sin(180° - 80.2°) = \sin 99.8°$
$40.1°$ or $49.9° = \theta$	**Divide by 2.**

The interval is $[40.1°, 49.9°]$. The ball will clear the fence if the angle is between 40.1° and 49.9°.

CHECK Substitute the angle measures into the original equation to confirm the solution.

$d = \dfrac{v_0^2 \sin 2\theta}{9.8}$	**Original formula**	$d = \dfrac{v_0^2 \sin 2\theta}{9.8}$
$90.5 \overset{?}{=} \dfrac{30^2 \sin(2 \cdot 40.1°)}{9.8}$	$\theta = 40.1°$ or $\theta = 49.9°$	$90.5 \overset{?}{=} \dfrac{30^2 \sin(2 \cdot 49.9°)}{9.8}$
$90.5 \approx 90.497$ ✔	**Use a calculator.**	$90.5 \approx 90.497$ ✔

✅ Guided Practice

4. BASEBALL Find the interval of possible launch angles required to clear the fence if:

   A. the initial speed was increased to 35 meters per second.

   B. the initial speed remained the same, but the distance to the fence was 80 meters.

StudyTip

Exact Versus Approximate Solutions When solving trigonometric equations that are not in a real-world context, write your answers using exact values rather than decimal approximations. For example, the general solutions of the equation $\tan x = 2$ should be expressed as $x = \tan^{-1} 2 + n\pi$ or $x = \arctan 2 + n\pi$.

StudyTip

Optimal Angle Ignoring wind resistance and other factors, a baseball will travel the greatest distance when it is hit at a 45° angle. This is because $\sin 2(45) = 1$, which maximizes the distance formula in the example.

Use Trigonometric Identities to Solve You can use trigonometric identities along with algebraic methods to solve trigonometric equations.

EXAMPLE 5   Solve by Rewriting Using a Single Trigonometric Function

Find all solutions of $2 \cos^2 x - \sin x - 1 = 0$ on the interval $[0, 2\pi)$.

$2 \cos^2 x - \sin x - 1 = 0$	**Original equation**
$2(1 - \sin^2 x) - \sin x - 1 = 0$	**Pythagorean Identity**
$2 - 2 \sin^2 x - \sin x - 1 = 0$	**Multiply.**
$-2 \sin^2 x - \sin x + 1 = 0$	**Simplify.**
$-1(2 \sin x - 1)(\sin x + 1) = 0$	**Factor.**
$2 \sin x - 1 = 0$ $\quad$ or $\quad$ $\sin x + 1 = 0$	**Zero Product Property**
$\sin x = \dfrac{1}{2}$ $\qquad\qquad$ $\sin x = -1$	**Solve for sin x.**
$x = \dfrac{\pi}{6}$ or $\dfrac{5\pi}{6}$ $\qquad$ $x = \dfrac{3\pi}{2}$	**Solve for x on [0, 2π].**

CHECK The graphs of Y1 = $2 \cos^2 x - \sin x$ and Y2 = 1 intersect at $\dfrac{\pi}{6}, \dfrac{5\pi}{6}$, and $\dfrac{3\pi}{2}$ on the interval $[0, 2\pi)$ as shown ✔

$[0, 2\pi]$ scl: $\dfrac{\pi}{2}$ by $[-2, 4]$ scl: 1

✓ Guided Practice

Find all solutions of each equation on the interval $[0, 2\pi)$.

5A. $1 - \cos x = 2 \sin^2 x$

5B. $\cot^2 x \csc^2 x + 2 \csc^2 x - \cot^2 x = 2$

Sometimes you can obtain an equation in one trigonometric function by squaring each side, but this technique may produce extraneous solutions.

EXAMPLE 6   Solve by Squaring

Find all solutions of $\csc x - \cot x = 1$ on the interval $[0, 2\pi)$.

$\csc x - \cot x = 1$	**Original equation**
$\csc x = 1 + \cot x$	**Add cot x to each side.**
$(\csc x)^2 = (1 + \cot x)^2$	**Square each side.**
$\csc^2 x = 1 + 2 \cot x + \cot^2 x$	**Multiply.**
$1 + \cot^2 x = 1 + 2 \cot x + \cot^2 x$	**Pythagorean Identity**
$0 = 2 \cot x$	**Subtract 1 + cot² x from each side.**
$0 = \cot x$	**Divide each side by 2.**
$x = \dfrac{\pi}{2}$ or $\dfrac{3\pi}{2}$	**Solve for x on [0, 2π].**

CHECK $\quad$ $\csc x - \cot x = 1$ $\qquad$ **Original equation** $\qquad\qquad$ $\csc x - \cot x = 1$

$\csc \dfrac{\pi}{2} - \cot \dfrac{\pi}{2} \overset{?}{=} 1$ $\qquad$ **Substitute.** $\qquad\qquad$ $\csc \dfrac{3\pi}{2} - \cot \dfrac{3\pi}{2} \overset{?}{=} 1$

$1 - 0 = 1$ ✔ $\qquad\qquad$ **Simplify.** $\qquad\qquad\qquad$ $-1 - 0 \neq 1$ ✗

Therefore, the only valid solution is $\dfrac{\pi}{2}$ on the interval $[0, 2\pi)$.

✓ Guided Practice

6A. $\sec x + 1 = \tan x$

6B. $\cos x = \sin x - 1$

Exercises

Solve each equation for all values of *x*. (Examples 1 and 2)

1. $5 \sin x + 2 = \sin x$

2. $5 = \sec^2 x + 3$

3. $2 = 4 \cos^2 x + 1$

4. $4 \tan x - 7 = 3 \tan x - 6$

5. $9 + \cot^2 x = 12$

6. $2 - 10 \sec x = 4 - 9 \sec x$

7. $3 \csc x = 2 \csc x + \sqrt{2}$

8. $11 = 3 \csc^2 x + 7$

9. $6 \tan^2 x - 2 = 4$

10. $9 + \sin^2 x = 10$

11. $7 \cot x - \sqrt{3} = 4 \cot x$

12. $7 \cos x = 5 \cos x + \sqrt{3}$

Find all solutions of each equation on $[0, 2\pi)$. (Example 3)

13. $\sin^4 x + 2 \sin^2 x - 3 = 0$

14. $-2 \sin x = -\sin x \cos x$

15. $4 \cot x = \cot x \sin^2 x$

16. $\csc^2 x - \csc x + 9 = 11$

17. $\cos^3 x + \cos^2 x - \cos x = 1$

18. $2 \sin^2 x = \sin x + 1$

19. **TENNIS** A tennis ball leaves a racket and heads toward a net 40 feet away. The height of the net is the same height as the initial height of the tennis ball. (Example 4)

$d = 40$ ft

a. If the ball is hit at 50 feet per second, neglecting air resistance, use $d = \frac{1}{32}{v_0}^2 \sin 2\theta$ to find the interval of possible angles of the ball needed to clear the net.

b. Find θ if the initial velocity remained the same but the distance to the net was 50 feet.

20. **SKIING** In the Olympic aerial skiing competition, skiers speed down a slope that launches them into the air, as shown. The maximum height a skier can reach is given by $h_{\text{peak}} = \frac{{v_0}^2 \sin^2 \theta}{2g}$, where g is the acceleration due to gravity or 9.8 meters per second squared. (Example 4)

58°

a. If a skier obtains a height of 5 meters above the end of the ramp, what was the skier's initial speed?

b. Use your answer from part **a** to determine how long it took the skier to reach the maximum height if $t_{\text{peak}} = \frac{v_0 \sin \theta}{g}$.

Find all solutions of each equation on the interval $[0, 2\pi)$. (Examples 5 and 6)

21. $1 = \cot^2 x + \csc x$

22. $\sec x = \tan x + 1$

23. $\tan^2 x = 1 - \sec x$

24. $\csc x + \cot x = 1$

25. $2 - 2 \cos^2 x = \sin x + 1$

26. $\cos x - 4 = \sin x - 4$

27. $3 \sin x = 3 - 3 \cos x$

28. $\cot^2 x \csc^2 x - \cot^2 x = 9$

29. $\sec^2 x - 1 + \tan x - \sqrt{3} \tan x = \sqrt{3}$

30. $\sec^2 x \tan^2 x + 3 \sec^2 x - 2 \tan^2 x = 3$

31. **OPTOMETRY** Optometrists sometimes join two oblique or tilted prisms to correct vision. The resultant refractive power P_R of joining two oblique prisms can be calculated by first resolving each prism into its horizontal and vertical components, P_H and P_V.

Ophthalmic Prism

Base

Position of Base

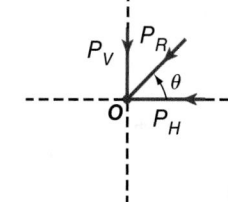

Resolving Prism Power
$P_V = P_R \sin \theta$
$P_H = P_R \cos \theta$

Using the equations above, determine for what values of θ P_V and P_H are equivalent.

Find all solutions of each equation on the interval $[0, 2\pi)$.

32. $\dfrac{\tan^2 x}{\sec x} + \cos x = 2$

33. $\dfrac{\cos x}{1 + \sin x} + \dfrac{1 + \sin x}{\cos x} = -4$

34. $\dfrac{\sin x + \cos x}{\tan x} + \dfrac{1 - \sin x}{\sin x} = \cos x$

35. $\cot x \cos x + 1 = \dfrac{1}{\sec x - 1} + \dfrac{\sin x}{\tan^2 x}$

GRAPHING CALCULATOR Solve each equation on the interval $[0, 2\pi)$ by graphing. Round to the nearest hundredth.

36. $3 \cos 2x = e^x + 1$

37. $\sin \pi x + \cos \pi x = 3x$

38. $x^2 = 2 \cos x + x$

39. $x \log x + 5x \cos x = -2$

40. METEOROLOGY The average daily temperature in degrees Fahrenheit for a city can be modeled by $t = 8.05 \cos\left(\frac{\pi}{6}x - \pi\right) + 66.95$, where x is a function of time, $x = 1$ represents January 15, $x = 2$ represents February 15, and so on.

a. Use a graphing calculator to estimate the temperature on January 31.

b. Approximate the number of months that the average daily temperature is greater than 70° throughout the entire month.

c. Estimate the highest temperature of the year and the month in which it occurs.

Find the x-intercepts of each graph on the interval $[0, 2\pi)$.

41.

42.

43.

44.

Find all solutions of each equation on the interval $[0, 4\pi)$.

45. $4 \tan x = 2 \sec^2 x$

46. $2 \sin^2 x + 1 = -3 \sin x$

47. $\csc x \cot^2 x = \csc x$

48. $\sec x + 5 = 2 \sec x + 3$

49. GEOMETRY Consider the circle below.

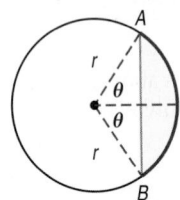

a. The length s of arc AB is given by $s = r(2\theta)$ where $0 \le \theta \le \pi$. When $s = 18$ and $AB = 14$, the radius is $r = \frac{7}{\sin\theta}$. Use a graphing calculator to find the measure of 2θ in radians.

b. The area of the shaded region is given by $A = \frac{r^2(\theta - \sin\theta)}{2}$. Use a graphing calculator to find the radian measure of θ if the radius is 5 inches and the area is 36 square inches. Round to the nearest hundredth.

Solve each inequality on the interval $[0, 2\pi)$.

50. $1 > 2 \sin x$

51. $0 < 2 \cos x - \sqrt{2}$

52. $\cos\left(\frac{\pi}{2} - x\right) \ge \frac{\sqrt{3}}{2}$

53. $\sin\left(x - \frac{\pi}{2}\right) \le \tan x \cot x$

54. $\cos x \le -\frac{\sqrt{3}}{2}$

55. $\sqrt{2} \sin x - 1 < 0$

56. REFRACTION When light travels from one transparent medium to another it bends or *refracts*, as shown.

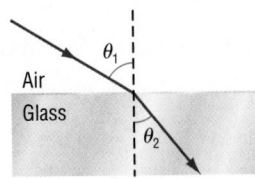

Refraction is described by $n_2 \sin\theta_1 = n_1 \sin\theta_2$, where n_1 is the index of refraction of the medium the light is entering, n_2 is the index of refraction of the medium the light is exiting, θ_1 is the angle of incidence, and θ_2 is the angle of refraction.

a. Find θ_2 for each material shown if the angle of incidence is 40° and the index of refraction for air is 1.00.

b. If the angle of incidence is doubled to 80°, will the resulting angles of refraction be twice as large as those found in part **a**?

Material	Index of Refraction
glass	1.52
ice	1.31
plastic	1.50
water	1.33

H.O.T. Problems Use **H**igher-**O**rder **T**hinking Skills

57. ERROR ANALYSIS Vijay and Alicia are solving $\tan^2 x - \tan x + \sqrt{3} = \sqrt{3} \tan x$. Vijay thinks that the solutions are $x = \frac{\pi}{4} + n\pi$, $x = \frac{5\pi}{4} + n\pi$, $x = \frac{\pi}{3} + n\pi$, and $x = \frac{4\pi}{3} + n\pi$. Alicia thinks that the solutions are $x = \frac{\pi}{4} + n\pi$ and $x = \frac{\pi}{3} + n\pi$. Is either of them correct? Explain your reasoning.

CHALLENGE Solve each equation on the interval $[0, 2\pi)$.

58. $16 \sin^5 x + 2 \sin x = 12 \sin^3 x$

59 $4 \cos^2 x - 4 \sin^2 x \cos^2 x + 3 \sin^2 x = 3$

60. REASONING Are the solutions of $\csc x = \sqrt{2}$ and $\cot^2 x + 1 = 2$ equivalent? If so, verify your answer algebraically. If not, explain your reasoning.

OPEN ENDED Write a trigonometric equation that has each of the following solutions.

61.

62.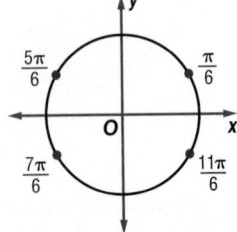

63. WRITING IN MATH Explain the difference in the techniques that are used when solving equations and verifying identities.

Verify each identity. (Lesson 5-2)

64. $\dfrac{1 + \sin\theta}{\sin\theta} = \dfrac{\cot^2\theta}{\csc\theta - 1}$

65. $\dfrac{1 + \tan\theta}{1 + \cot\theta} = \dfrac{\sin\theta}{\cos\theta}$

66. $\dfrac{1}{\sec^2\theta} + \dfrac{1}{\csc^2\theta} = 1$

Find the value of each expression using the given information. (Lesson 5-1)

67. $\tan\theta$; $\sin\theta = \dfrac{1}{2}$, $\tan\theta > 0$

68. $\csc\theta$, $\cos\theta = -\dfrac{3}{5}$, $\csc\theta < 0$

69. $\sec\theta$; $\tan\theta = -1$, $\sin\theta < 0$

70. POPULATION The population of a certain species of deer can be modeled by $p = 30{,}000 + 20{,}000\cos\left(\dfrac{\pi}{10}t\right)$, where p is the population and t is the time in years. (Lesson 4-4)

  a. What is the amplitude of the function? What does it represent?

  b. What is the period of the function? What does it represent?

  c. Graph the function.

Given $f(x) = 2x^2 - 5x + 3$ and $g(x) = 6x + 4$, find each. (Lesson 1-6)

71. $(f - g)(x)$

72. $(f \cdot g)(x)$

73. $\left(\dfrac{f}{g}\right)(x)$

74. BUSINESS A small business owner must hire seasonal workers as the need arises. The following list shows the number of employees hired monthly for a 5-month period.

If the mean of these data is 9, what is the population standard deviation for these data? Round to the nearest tenth. (Lesson 0-8)

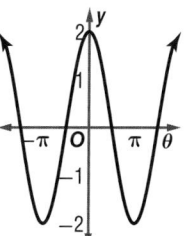

Month	Number of Additional Employees Needed
August	5
September	14
October	6
November	8
December	12

75. SAT/ACT For all positive values of m and n, if $\dfrac{3x}{m - nx} = 2$, then $x =$

  A $\dfrac{2m - 2n}{3}$

  B $\dfrac{3 + 2n}{2m}$

  C $\dfrac{2m - 3}{2n}$

  D $\dfrac{2m}{3 + 2n}$

  E $\dfrac{3}{2m - 2n}$

76. If $\cos x = -0.45$, what is $\sin\left(x - \dfrac{\pi}{2}\right)$?

  F -0.55

  G -0.45

  H 0.45

  J 0.55

77. Which of the following is *not* a solution of $0 = \sin\theta + \cos\theta\tan^2\theta$?

  A $\dfrac{3\pi}{4}$

  B $\dfrac{7\pi}{4}$

  C 2π

  D $\dfrac{5\pi}{2}$

78. REVIEW The graph of $y = 2\cos\theta$ is shown. Which is a solution for $2\cos\theta = 1$?

  F $\dfrac{8\pi}{3}$

  G $\dfrac{10\pi}{3}$

  H $\dfrac{13\pi}{3}$

  J $\dfrac{15\pi}{3}$

Objective
Use a graphing calculator to solve trigonometric inequalities.

You can use a graphing calculator to solve trigonometric inequalities. Graph each inequality. Then locate the end points of each intersection in the graph in order to determine the intervals within which the inequality is true.

ACTIVITY   Graph a Trigonometric Inequality

Graph and solve $\sin 2x \geq \cos x$.

Step 1 Separate this inequality into two inequalities, one for each side of the inequality symbol. Replace each side with y to form the new inequalities.

$$\sin 2x \geq Y_1 \text{ and } Y_2 \geq \cos x$$

Step 2 Graph each inequality. Make each inequality symbol by scrolling to the left of the equal sign and selecting ENTER until the shaded triangles are flashing. The triangle above represents *greater than*, and the triangle below represents *less than* (Figure 5.3.1). In the MODE menu, select **RADIAN**.

Step 3 Graph the equations in the appropriate window. Use the domain and range of each trigonometric function as a guide (Figure 5.3.2).

$[-2\pi, 2\pi]$ scl: $\frac{\pi}{4}$ by $[-1, 1]$ scl: 0.1

Figure 5.3.1   **Figure 5.3.2**

Step 4 The darkly shaded area indicates the intersection of the graphs and the solution of the system of inequalities. The two graphs intersect at an infinite number of points. Use **CALC: intersect** to locate these intersections. Move the cursor over the intersection and select ENTER 3 times.

$[-2\pi, 2\pi]$ scl: $\frac{\pi}{4}$ by $[-1, 1]$ scl: 0.1

Step 5 The first intersection is when $y = 0.866$ or $\frac{\sqrt{3}}{2}$. Since $\cos \frac{\pi}{6} = \frac{\sqrt{3}}{2}$, the intersection is at $x = \frac{\pi}{6}$. The next intersection is at $x = \frac{\pi}{2}$. Therefore, one of the solution intervals is $\left[\frac{\pi}{6}, \frac{\pi}{2}\right]$. Another interval is $\left[\frac{5\pi}{6}, \frac{3\pi}{2}\right]$. There are an infinite number of intervals, so the solutions for all values of x are $\left[\frac{\pi}{6} + 2n\pi, \frac{\pi}{2} + 2n\pi\right]$ and $\left[\frac{5\pi}{6} + 2n\pi, \frac{3\pi}{2} + 2n\pi\right]$.

Exercises

Graph and solve each inequality.

1. $\sin 3x < 2 \cos x$ **2.** $3 \cos x \geq 0.5 \sin 2x$ **3.** $\sec x < 2 \cos x$

4. $\csc 2x > \sin 8x$ **5.** $2 \tan 2x < 3 \sin 2x$ **6.** $\tan x \geq \cos x$

Find the value of each expression using the given information. (Lesson 5-1)

1. $\sin \theta$ and $\cos \theta$, $\cot \theta = 4$, $\cos \theta > 0$

2. $\sec \theta$ and $\sin \theta$, $\tan \theta = -\frac{2}{3}$, $\csc \theta > 0$

3. $\tan \theta$ and $\csc \theta$, $\cos \theta = \frac{1}{4}$, $\sin \theta > 0$

Simplify each expression. (Lesson 5-1)

4. $\dfrac{\sin (-x)}{\tan (-x)}$

5. $\dfrac{\sec^2 x}{\cot^2 x + 1}$

6. $\dfrac{\sin (90° - x)}{\cot^2 (90° - x) + 1}$

7. $\dfrac{\sin x}{1 + \sec x}$

8. **ANGLE OF DEPRESSION** From his apartment window, Tim can see the top of the bank building across the street at an angle of elevation of θ, as shown below. (Lesson 5-1)

a. If a bank employee looks down at Tim's apartment from the top of the bank, what identity could be used to conclude that $\sin \theta = \cos \theta'$?

b. If Tim looks down at a lower window of the bank with an angle of depression of 35°, how far below his apartment is the bank window?

9. **MULTIPLE CHOICE** Which of the following is not equal to $\csc \theta$? (Lesson 5-1)

A $\sec (90° - \theta)$

B $\sqrt{\cot^2 \theta + 1}$

C $\dfrac{1}{\sin \theta}$

D $\dfrac{1}{\sin (90° - \theta)}$

Verify each identity. (Lesson 5-2)

10. $\dfrac{\cos \theta}{1 + \sin \theta} - \dfrac{\cos \theta}{1 - \sin \theta} = -2 \tan \theta$

11. $\csc^2 \theta - \sin^2 \theta - \cos^2 \theta - \cot^2 \theta = 0$

12. $\sin \theta + \dfrac{\cos \theta}{\tan \theta} = \csc \theta$

13. $\dfrac{\cos \theta}{1 + \sin \theta} = \sec \theta - \tan \theta$

14. $\dfrac{\csc \theta}{\sin \theta} + \dfrac{\cot \theta}{\cos \theta} = \cot^2 \theta + \csc \theta + 1$

15. $\dfrac{1 + \sin \theta}{\sin \theta} + \dfrac{\sin \theta}{1 - \sin \theta} = \dfrac{\csc \theta}{1 - \sin \theta}$

Find all solutions of each equation on the interval $[0, 2\pi)$. (Lesson 5-3)

16. $4 \sec \theta + 2\sqrt{3} = \sec \theta$

17. $2 \tan \theta + 4 = \tan \theta + 5$

18. $4 \cos^2 \theta + 2 = 3$

19. $\cos \theta - 1 = \sin \theta$

20. **MULTIPLE CHOICE** Which of the following is the solution set for $\cos \theta \tan \theta - \sin^2 \theta = 0$? (Lesson 5-3)

F $\dfrac{\pi}{2}n$, where n is an integer

G $\dfrac{\pi}{2} + n\pi$, where n is an integer

H $\pi + 2n\pi$, where n is an integer

J $n\pi$, where n is an integer

Solve each equation for all values of θ. (Lesson 5-3)

21. $3 \sin^2 \theta + 6 = 2 \sin^2 \theta + 7$

22. $\sin \theta + \cos \theta = 0$

23. $\sec \theta + \tan \theta = 0$

24. $2 \tan^2 \theta = 1$

25. **PROJECTILE MOTION** The distance d that a kick ball travels (in feet) is given by $d = \dfrac{v_0^2 \sin 2\theta}{32}$, where v_0 is the object's initial speed, θ is the angle at which the object is launched, and 32 is in feet per second squared. If the ball is kicked with an initial speed of 82 feet per second, and the ball travels 185 feet, what is the launch angle of the ball? (Lesson 5-3)

26. **FERRIS WHEEL** The height h of a rider in feet on a Ferris wheel after t seconds is shown below. (Lesson 5-3)

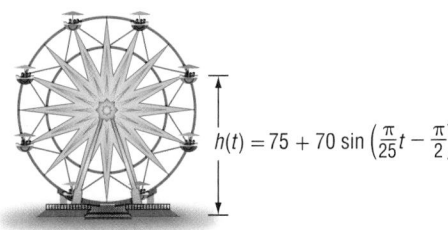

$h(t) = 75 + 70 \sin \left(\dfrac{\pi}{25}t - \dfrac{\pi}{2}\right)$

a. If the Ferris wheel begins at $t = 0$, what is the initial height of a rider?

b. When will the rider first reach the maximum height of 145 feet?

Sum and Difference Identities

Why?

When the picture on a television screen is blurry or a radio station will not tune in properly, the problem is too often due to *interference*. Interference results when waves pass through the same space at the same time. You can use trigonometric identities to determine the type of interference that is taking place.

Evaluate Trigonometric Functions In Lesson 5-1, you used basic identities involving only one variable. In this lesson, we will consider identities involving two variables. One of these is the *cosine of a difference* identity.

Let points A, B, C, and D be located on the unit circle, α and β be angles on the interval $[0, 2\pi]$, and $\alpha > \beta$ as shown in Figure 5.4.1. Because each point is located on the unit circle, $x_1^2 + y_1^2 = 1$, $x_2^2 + y_2^2 = 1$, and $x_3^2 + y_3^2 = 1$. Notice also that the measure of arc $CD = \alpha - (\alpha - \beta)$ or β and the measure of arc $AB = \beta$.

Figure 5.4.1

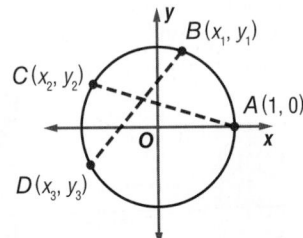

Figure 5.4.2

Since arcs AB and CD have the same measure, chords AC and BD shown in Figure 5.4.2 are congruent.

$$AC = BD \qquad \text{Chords } \textbf{AC} \text{ and } \textbf{BD} \text{ are congruent.}$$

$$\sqrt{(x_2 - 1)^2 + (y_2 - 0)^2} = \sqrt{(x_3 - x_1)^2 + (y_3 - y_1)^2} \qquad \text{Distance Formula}$$

$$(x_2 - 1)^2 + (y_2 - 0)^2 = (x_3 - x_1)^2 + (y_3 - y_1)^2 \qquad \text{Square each side.}$$

$$x_2^2 - 2x_2 + 1 + y_2^2 = x_3^2 - 2x_3 x_1 + x_1^2 + y_3^2 - 2y_3 y_1 + y_1^2 \qquad \text{Square each binomial.}$$

$$(x_2^2 + y_2^2) - 2x_2 + 1 = (x_1^2 + y_1^2) + (x_3^2 + y_3^2) - 2x_3 x_1 - 2y_3 y_1 \qquad \text{Group similar squared terms.}$$

$$1 - 2x_2 + 1 = 1 + 1 - 2x_3 x_1 - 2y_3 y_1 \qquad \text{Substitution}$$

$$2 - 2x_2 = 2 - 2x_3 x_1 - 2y_3 y_1 \qquad \text{Add.}$$

$$-2x_2 = -2x_3 x_1 - 2y_3 y_1 \qquad \text{Subtract 2 from each side.}$$

$$x_2 = x_3 x_1 + y_3 y_1 \qquad \text{Divide each side by } -2.$$

In Figure 5.4.1, notice that by the unit circle definitions for cosine and sine, $x_1 = \cos \beta$, $x_2 = \cos (\alpha - \beta)$, $x_3 = \cos \alpha$, $y_1 = \sin \beta$, and $y_3 = \sin \alpha$. Substituting, $x_2 = x_3 x_1 + y_3 y_1$ becomes

$$\cos (\alpha - \beta) = \cos \alpha \cos \beta + \sin \alpha \sin \beta. \qquad \text{Cosine Difference Identity}$$

By rewriting $\alpha - \beta$ as $\alpha + (-\beta)$ and applying the above identity, we obtain the identity for the cosine of a sum.

$$\cos [\alpha + (-\beta)] = \cos \alpha \cos (-\beta) + \sin \alpha \sin (-\beta) \qquad \text{Cosine Difference Identity}$$

$$= \cos \alpha \cos \beta - \sin \alpha \sin \beta \qquad \text{Even-Odd Identity}$$

These two cosine identities can be used to establish each of the other sum and difference identities listed below.

Key Concept Sum and Difference Identities

Sum Identities	Difference Identities
$\cos(\alpha + \beta) = \cos \alpha \cos \beta - \sin \alpha \sin \beta$	$\cos(\alpha - \beta) = \cos \alpha \cos \beta + \sin \alpha \sin \beta$
$\sin(\alpha + \beta) = \sin \alpha \cos \beta + \cos \alpha \sin \beta$	$\sin(\alpha - \beta) = \sin \alpha \cos \beta - \cos \alpha \sin \beta$
$\tan(\alpha + \beta) = \dfrac{\tan \alpha + \tan \beta}{1 - \tan \alpha \tan \beta}$	$\tan(\alpha - \beta) = \dfrac{\tan \alpha - \tan \beta}{1 + \tan \alpha \tan \beta}$

You will prove the sine and tangent sum and difference identities in Exercises 57–60.

By writing angle measures as the sums or differences of special angle measures, you can use these sum and difference identities to find exact values of trigonometric functions of angles that are less common.

EXAMPLE 1 Evaluate a Trigonometric Expression

Find the exact value of each trigonometric expression.

a. $\sin 15°$

Write 15° as the sum or difference of angle measures with sines that you know.

$\sin 15° = \sin(45° - 30°)$ .... **$45° - 30° = 15°$**

$\quad = \sin 45° \cos 30° - \sin 30° \cos 45°$ .... **Sine Difference Identity**

$\quad = \dfrac{\sqrt{2}}{2} \cdot \dfrac{\sqrt{3}}{2} - \dfrac{1}{2} \cdot \dfrac{\sqrt{2}}{2}$ .... **$\sin 30° = \dfrac{1}{2}$, $\sin 45° = \cos 45° = \dfrac{\sqrt{2}}{2}$, $\cos 30° = \dfrac{\sqrt{3}}{2}$**

$\quad = \dfrac{\sqrt{6}}{4} - \dfrac{\sqrt{2}}{4}$ .... **Multiply.**

$\quad = \dfrac{\sqrt{6} - \sqrt{2}}{4}$ .... **Combine the fractions.**

b. $\tan \dfrac{7\pi}{12}$

$\tan \dfrac{7\pi}{12} = \tan\left(\dfrac{\pi}{3} + \dfrac{\pi}{4}\right)$ .... **$\dfrac{\pi}{3} + \dfrac{\pi}{4} = \dfrac{7\pi}{12}$**

$\quad = \dfrac{\tan \dfrac{\pi}{3} + \tan \dfrac{\pi}{4}}{1 - \tan \dfrac{\pi}{3} \tan \dfrac{\pi}{4}}$ .... **Tangent Sum Identity**

$\quad = \dfrac{\sqrt{3} + 1}{1 - \sqrt{3}(1)}$ .... **$\tan \dfrac{\pi}{3} = \sqrt{3}$ and $\tan \dfrac{\pi}{4} = 1$**

$\quad = \dfrac{\sqrt{3} + 1}{1 - \sqrt{3}}$ .... **Simplify.**

$\quad = \dfrac{\sqrt{3} + 1}{1 - \sqrt{3}} \cdot \dfrac{1 + \sqrt{3}}{1 + \sqrt{3}}$ .... **Rationalize the denominator.**

$\quad = \dfrac{\sqrt{3} + 1 + 3 + \sqrt{3}}{1 - \sqrt{3} + \sqrt{3} - 3}$ .... **Multiply.**

$\quad = \dfrac{4 + 2\sqrt{3}}{-2}$ .... **Simplify.**

$\quad = -2 - \sqrt{3}$ .... **Simplify.**

StudyTip

Check Your Answer You can check your answers by using a graphing calculator. In Example 1a, sin 15° ≈ 0.259 and $\dfrac{\sqrt{6} - \sqrt{2}}{4} \approx$ 0.259. Make sure that the calculator is in the correct mode.

✔ Guided Practice

1A. $\cos 15°$

1B. $\sin \dfrac{5\pi}{12}$

Sum and difference identities are often used to solve real-world problems.

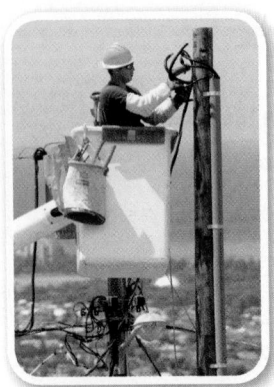

🌐 **Real-World EXAMPLE 2** | **Use a Sum or Difference Identity**

ELECTRICITY An alternating current i in amperes in a certain circuit can be found after t seconds using $i = 3 (\sin 165)t$, where 165 is a degree measure.

a. Rewrite the formula in terms of the sum of two angle measures.

$i = 3 (\sin 165)t$	Original equation
$\ \ = 3 \sin (120 + 45)t$	$120 + 45 = 165$

b. Use a sine sum identity to find the exact current after 1 second.

$i = 3 \sin (120 + 45)t$	Rewritten equation
$\ \ = 3 \sin (120 + 45)$	$t = 1$
$\ \ = 3[(\sin 120)(\cos 45) + (\cos 120)(\sin 45)]$	Sine Sum Identity
$\ \ = 3\left[\left(\dfrac{\sqrt{3}}{2}\right)\left(\dfrac{\sqrt{2}}{2}\right) + \left(-\dfrac{1}{2}\right)\left(\dfrac{\sqrt{2}}{2}\right)\right]$	Substitute.
$\ \ = 3\left(\dfrac{\sqrt{6}}{4} - \dfrac{\sqrt{2}}{4}\right)$	Multiply.
$\ \ = \dfrac{3\sqrt{6} - 3\sqrt{2}}{4}$	Simplify.

The exact current after 1 second is $\dfrac{3\sqrt{6} - 3\sqrt{2}}{4}$ amperes.

✓ **Guided Practice**

2. **ELECTRICITY** An alternating current i in amperes in another circuit can be found after t seconds using $i = 2 (\sin 285)t$, where 285 is a degree measure.

A. Rewrite the formula in terms of the difference of two angle measures.

B. Use a sine difference identity to find the exact current after 1 second.

If a trigonometric expression has the form of a sum or difference identity, you can use the identity to find an exact value or to simplify an expression by rewriting the expression as a function of a single angle.

EXAMPLE 3 | **Rewrite as a Single Trigonometric Expression**

a. Find the exact value of $\dfrac{\tan 32° + \tan 13°}{1 - \tan 32° \tan 13°}$.

$\dfrac{\tan 32° + \tan 13°}{1 - \tan 32° \tan 13°} = \tan (32° + 13°)$	Tangent Sum Identity
$\qquad\qquad = \tan 45°$ or 1	Simplify.

b. Simplify $\sin x \sin 3x - \cos x \cos 3x$.

$\sin x \sin 3x - \cos x \cos 3x = -(\cos x \cos 3x - \sin x \sin 3x)$	Distributive and Commutative Properties
$\qquad\qquad = -\cos (x + 3x)$ or $-\cos 4x$	Tangent Sum Identity

✓ **Guided Practice**

3A. Find the exact value of $\cos \dfrac{7\pi}{8} \cos \dfrac{5\pi}{24} + \sin \dfrac{7\pi}{8} \sin \dfrac{5\pi}{24}$.

3B. Simplify $\dfrac{\tan 6x - \tan 7x}{1 + \tan 6x \tan 7x}$.

Sum and difference identities can be used to rewrite trigonometric expressions as algebraic expressions.

EXAMPLE 4 **Write as an Algebraic Expression**

Write sin (arctan $\sqrt{3}$ + arcsin x) as an algebraic expression of x that does not involve trigonometric functions.

Applying the Sine of a Sum Identity, we find that

$$\sin (\arctan \sqrt{3} + \arcsin x) = \sin (\arctan \sqrt{3}) \cos (\arcsin x) + \cos (\arctan \sqrt{3}) \sin (\arcsin x).$$

If we let $\alpha = \arctan \sqrt{3}$ and $\beta = \arcsin x$, then $\tan \alpha = \sqrt{3}$ and $\sin \beta = x$. Sketch one right triangle with an acute angle α and another with an acute angle β. Label the sides such that $\tan \alpha = \sqrt{3}$ and the $\sin \beta = x$. Then use the Pythagorean Theorem to express the length of each third side.

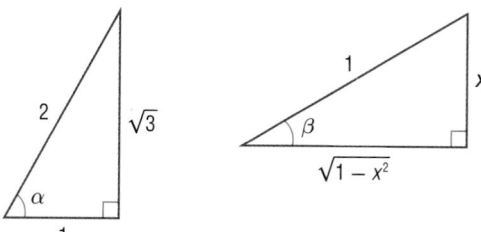

Using these triangles, we find that $\sin (\arctan \sqrt{3}) = \sin \alpha$ or $\frac{\sqrt{3}}{2}$, $\cos (\arctan \sqrt{3}) = \cos \alpha$ or $\frac{1}{2}$, $\cos (\arcsin x) = \cos \beta$ or $\sqrt{1 - x^2}$, and $\sin (\arcsin x) = \sin \beta$ or x.

Now apply substitution and simplify.

$$\sin (\arctan \sqrt{3} + \arcsin x) = \sin (\arctan \sqrt{3}) \cos (\arcsin x) + \cos (\arctan \sqrt{3}) \sin (\arcsin x)$$

$$= \frac{\sqrt{3}}{2} \left(\sqrt{1 - x^2} \right) + \frac{1}{2} x$$

$$= \frac{\sqrt{3 - 3x^2}}{2} + \frac{x}{2} \text{ or } \frac{x + \sqrt{3 - 3x^2}}{2}$$

✔️ **Guided Practice**

Write each trigonometric expression as an algebraic expression.

4A. cos (arcsin $2x$ + arccos x)

4B. $\sin \left(\arctan x - \arccos \frac{1}{2} \right)$

Sum and difference identities can be used to verify other identities.

ReadingMath

Cofunction Identities
The "co" in cofunction stands for "complement." Therefore, sine and cosine, tangent and cotangent, and secant and cosecant are all complementary functions or cofunctions.

EXAMPLE 5 **Verify Cofunction Identities**

Verify $\sin \left(\frac{\pi}{2} - x \right) = \cos x$.

$$\sin \left(\frac{\pi}{2} - x \right) = \sin \frac{\pi}{2} \cos x - \cos \frac{\pi}{2} \sin x \qquad \text{Sine Difference Identity}$$

$$= 1(\cos x) - 0(\sin x) \qquad \sin \frac{\pi}{2} = 1 \text{ and } \cos \frac{\pi}{2} = 0$$

$$= \cos x \checkmark \qquad \text{Multiply.}$$

✔️ **Guided Practice**

Verify each cofunction identity using a difference identity.

5A. $\cos \left(\frac{\pi}{2} - x \right) = \sin x$

5A. $\csc \left(\frac{\pi}{2} - \theta \right) = \sec \theta$

Sum and difference identities can be used to rewrite trigonometric expressions in which one of the angles is a multiple of 90° or $\frac{\pi}{2}$ radians. The resulting identity is called a **reduction identity** because it *reduces* the complexity of the expression.

EXAMPLE 6   **Verify Reduction Identities**

Verify each reduction identity.

a. $\sin\left(\theta + \frac{3\pi}{2}\right) = -\cos\theta$

$\sin\left(\theta + \frac{3\pi}{2}\right) = \sin\theta\cos\frac{3\pi}{2} + \cos\theta\sin\frac{3\pi}{2}$ Sine Sum Formula

$= \sin\theta(0) + \cos\theta(-1)$ $\cos\frac{3\pi}{2} = 0$ and $\sin\frac{3\pi}{2} = -1$

$= -\cos\theta$ ✔   Simplify.

b. $\tan(x - 180°) = \tan x$

$\tan(x - 180)° = \dfrac{\tan x - \tan 180°}{1 + \tan x \tan 180°}$ Tangent Sum Formula

$= \dfrac{\tan x - 0}{1 + \tan x(0)}$ $\tan 180° = 0$

$= \tan x$ ✔   Simplify.

☑ **Guided Practice**

Verify each cofunction identity.

6A. $\cos(360° - \theta) = \cos\theta$ **6B.** $\sin\left(\frac{\pi}{2} + x\right) = \cos x$

Solve Trigonometric Equations You can solve trigonometric equations using the sum and difference identities and the same techniques that you used in Lesson 5-3.

EXAMPLE 7   **Solve a Trigonometric Equation**

Find the solutions of $\cos\left(\frac{\pi}{3} + x\right) + \cos\left(\frac{\pi}{3} - x\right) = \frac{1}{2}$ on the interval $[0, 2\pi)$.

$\cos\left(\frac{\pi}{3} + x\right) + \cos\left(\frac{\pi}{3} - x\right) = \frac{1}{2}$ Original equation

$\cos\frac{\pi}{3}\cos x - \sin\frac{\pi}{3}\sin x + \cos\frac{\pi}{3}\cos x + \sin\frac{\pi}{3}\sin x = \frac{1}{2}$ Cosine Sum Identities

$\frac{1}{2}(\cos x) - \frac{\sqrt{3}}{2}(\sin x) + \frac{1}{2}(\cos x) + \frac{\sqrt{3}}{2}(\sin x) = \frac{1}{2}$ Substitute.

$\cos x = \frac{1}{2}$ Simplify.

On the interval $[0, 2\pi)$, $\cos x = \frac{1}{2}$ when $x = \frac{\pi}{3}$ and $x = \frac{5\pi}{3}$.

CHECK The graph of $y = \cos\left(\frac{\pi}{3} + x\right) + \cos\left(\frac{\pi}{3} - x\right) - \frac{1}{2}$ has zeros at $\frac{\pi}{3}$ and $\frac{5\pi}{3}$ on the interval $[0, 2\pi)$. ✔

$[0, 2\pi]$ scl: $\frac{\pi}{3}$ by $[-2, 2]$ scl: 1

☑ **Guided Practice**

7. Find the solutions of $\cos(x + \pi) - \sin(x - \pi) = 0$ on the interval $[0, 2\pi)$.

TechnologyTip

Viewing Window When checking your answer on a graphing calculator, remember that one period for $y = \sin x$ or $y = \cos x$ is 2π and the amplitude is 1. This will help you define the viewing window.

69. PHYSICS According to Snell's law, the angle at which light enters water α is related to the angle at which light travels in water β by $\sin \alpha = 1.33 \sin \beta$. At what angle does a beam of light enter the water if the beam travels at an angle of 23° through the water? **(Lesson 5-3)**

Verify each identity. (Lesson 5-2)

70. $\dfrac{\cos \theta}{1 - \sin^2 \theta} = \sec \theta$

71. $\dfrac{\sec \theta}{\sin \theta} - \dfrac{\sin \theta}{\cos \theta} = \cot \theta$

Find the exact value of each expression, if it exists. (Lesson 4-6)

72. $\sin^{-1}(-1)$

73. $\tan^{-1} \sqrt{3}$

74. $\tan\left(\arcsin \dfrac{3}{5}\right)$

75. MONEY Suppose you deposit a principal amount of P dollars in a bank account that pays compound interest. If the annual interest rate is r (expressed as a decimal) and the bank makes interest payments n times every year, the amount of money A you would have after t years is given by $A(t) = P\left(1 + \dfrac{r}{n}\right)^{nt}$. **(Lesson 3-1)**

  a. If the principal, interest rate, and number of interest payments are known, what type of function is $A(t) = P\left(1 + \dfrac{r}{n}\right)^{nt}$? Explain your reasoning.

  b. Write an equation giving the amount of money you would have after t years if you deposit $1000 into an account paying 4% annual interest compounded quarterly (four times per year).

  c. Find the account balance after 20 years.

List all possible rational zeros of each function. Then determine which, if any, are zeros. (Lesson 2-4)

76. $p(x) = x^4 + x^3 - 11x - 5x + 30$

77. $d(x) = 2x^4 - x^3 - 6x^2 + 5x - 1$

78. $f(x) = x^3 - 2x^2 - 5x - 6$

79. SAT/ACT There are 16 green marbles, 2 red marbles, and 6 yellow marbles in a jar. How many yellow marbles need to be added to the jar in order to double the probability of selecting a yellow marble?

  A 4   **C** 8   **E** 16

  B 6   **D** 12

80. REVIEW Refer to the figure below. Which equation could be used to find $m\angle G$?

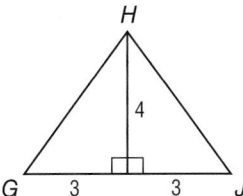

  F $\sin G = \dfrac{3}{4}$ **H** $\cot G = \dfrac{3}{4}$

  G $\cos G = \dfrac{3}{4}$ **I** $\tan G = \dfrac{3}{4}$

81. Find the exact value of $\sin \theta$.

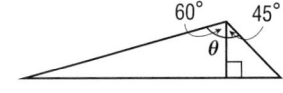

  A $\dfrac{\sqrt{2} + \sqrt{6}}{4}$

  B $\dfrac{\sqrt{2} - \sqrt{6}}{4}$

  C $\dfrac{2 + \sqrt{3}}{4}$

  D $\dfrac{2 - \sqrt{3}}{4}$

82. REVIEW Which of the following is equivalent to $\dfrac{\cos \theta (\cot^2 \theta + 1)}{\csc \theta}$?

  F $\tan \theta$

  G $\cot \theta$

  H $\sec \theta$

  J $\csc \theta$

Objective
Use TI-Nspire and quadrantal angles to reduce identities.

Another reduction identity involves the sum or difference of the measures of an angle and a quadrantal angle. For example, $\sin (x + \pi) = -\sin x$. This can be illustrated by comparing graphs of functions on the unit circle with TI-Nspire.

ACTIVITY 1   Use the Unit Circle

Use the unit circle to explore a reduction identity graphically.

Step 1 Open a new **Graphs & Geometry** page. Select **Zoom-Trig** from the **Window** menu, and select **Show Grid** from the **View** menu. From the **File** menu under **Tools**, choose **Document Settings**, set the **Display Digits** to **Float 2**, and confirm that the angle measure is in radians.

Step 2 Select **Points & Lines** and then **Point** from the menu. Place a point at $(1, 0)$. Next, select **Shapes**, and then **Circle** from the menu. To construct a circle centered at the origin through $(1, 0)$, click on the screen and define the center point at the origin. Move the cursor away from the center, and the circle will appear. Stop when you get to a radius of 1 and $(1, 0)$ lies on the circle.

Step 3 Place a point to the right of the circle on the x-axis, and label it A. Choose **Actions** and then **Coordinates and Equations** from the menu, and then double-click the point to display its coordinates. From the **Construction** menu, choose **Measurement transfer**. Select the x-coordinate of A, the circle, and the point at $(1, 0)$. Label the point created on the circle as B, and display its coordinates.

Step 4 With O as the origin, calculate and label the measure of $\angle AOB$. Select **Text** from the **Actions** menu to write the expression $a - \pi$. Then select **Calculate** from the **Actions** menu to calculate the difference of the x-coordinate of A and π.

StudyTip

Angle Measure The TI-Nspire only measures angles between 0 and π.

Step 5 Move A along the x-axis, and observe the effect on the measure of $\angle AOB$.

Location of A	$m\angle AOB$ (radians)
$(4, 0)$	2.7124
$(3, 0)$	3.0708
$(2, 0)$	2.5708
$(5, 0)$	2.2123
$(-2, 0)$	0.5708

Step 6 From the **Construction** menu, choose **Measurement transfer**. Select the x-axis and the value of $a - \pi$. Label the point as C and display its coordinates. Using **Measurement transfer** again, select the x-coordinate of C, the circle, and the point at $(1, 0)$. Label the point as D and display its coordinates.

Analyze the Results

1A. In Step 5, how are the location of A and the measure of $\angle AOB$ related?

1B. Consider the locations of points B and D. What reduction identity or identities does this relationship suggest are true?

1C. **MAKE A CONJECTURE** If you change the expression $a - \pi$ to $a + \pi$, what reduction identities do you think would result?

Use graphs to identify equal trigonometric functions.

Step 1 Open a new **Graphs & Geometry** page. Select **Zoom-Trig** from the Window menu.

Step 2 Graph $f(x) = \cos x$, $f(x) = \cos(x - \pi)$, and $f(x) = \sin x$. Using the **Attributes** feature from the **Actions** menu, make the line weight of $f(x) = \cos(x - \pi)$ medium and the line style of $f(x) = \sin x$ dotted.

Step 3 Use translations, reflections, or dilations to transform $f(x) = \sin x$ so that the graph coincides with $f(x) = \cos x$. Select the graph and drag it over $f(x) = \cos x$. As you move the graph, its function will change on the screen.

Step 4 Use translations, reflections, or dilations to transform $f(x) = \cos(x - \pi)$ so that the graph coincides with the other two graphs. Again, as you move the graph, its function will change on the screen.

Analyze the Results

2A. Write the identity that results from your transformation of $f(x) = \sin x$ in Step 3. Graph the functions to confirm your identity.

2B. Write the identity that results from your alteration of $f(x) = \cos(x - \pi)$ in Step 3. Graph the functions to confirm your identity.

2C. MAKE A CONJECTURE What does the reflection of a graph suggest for the purpose of developing an identity? a translation?

Exercises

Use the unit circle to write an identity relating the given expressions. Verify your identity by graphing.

1. $\cos(90° - x)$, $\sin x$

2. $\cos\left(\dfrac{3\pi}{2} - x\right)$, $\sin x$

Insert the trigonometric function that completes each identity.

3. $\cos x = $ _____ $\left(x - \dfrac{3\pi}{2}\right)$

4. $\cot x = $ _____ $(x + 90°)$

5. $\sec x = $ _____ $(x - 180°)$

6. $\csc x = $ _____ $\left(x + \dfrac{\pi}{2}\right)$

Use transformations to find the value of a for each expression.

7. $\sin ax = 2 \sin x \cos x$

8. $\cos 4ax = \cos^2 x - \sin^2 x$

9. $a \sin^2 x = 1 - \cos 2x$

10. $1 + \cos 6ax = 2\cos^2 x$

5-5 Multiple-Angle and Product-to-Sum Identities

Then
You proved and used sum and difference identities. (Lesson 5-4)

Now
- Use double-angle, power-reducing, and half-angle identities to evaluate trigonometric expressions and solve trigonometric equations.
- Use product-to-sum identities to evaluate trigonometric expressions and solve trigonometric equations.

Math Online

glencoe.com

Why?

The speed at which a plane travels can be described by a *mach number*, a ratio of the plane's speed to the speed of sound. Exceeding the speed of sound produces a shock wave in the shape of a cone behind the plane. The angle θ at the vertex of this cone is related to the mach number M describing the plane's speed by the half-angle equation $\sin \frac{\theta}{2} = \frac{1}{M}$.

Use Multiple-Angle Identities By letting α and β both equal θ in each of the angle sum identities you learned in the previous lesson, you can derive the following double-angle identities.

Key Concept — Double-Angle Identities

$\sin 2\theta = 2 \sin \theta \cos \theta$

$\tan 2\theta = \dfrac{2 \tan \theta}{1 - \tan^2 \theta}$

$\cos 2\theta = \cos^2 \theta - \sin^2 \theta$

$\cos 2\theta = 2 \cos^2 \theta - 1$

$\cos 2\theta = 1 - 2 \sin^2 \theta$

Proof — Double-Angle Identity for Sine

$\sin 2\theta = \sin(\theta + \theta)$ $2\theta = \theta + \theta$

$\qquad = \sin \theta \cos \theta + \cos \theta \sin \theta$ Sine Sum Identity where $\alpha = \beta = \theta$

$\qquad = 2 \sin \theta \cos \theta$ Simplify.

You will prove the double-angle identities for cosine and tangent in Exercises 63–65.

EXAMPLE 1 — Evaluate Expressions Involving Double Angles

If $\sin \theta = -\dfrac{7}{25}$ on the interval $\left(\pi, \dfrac{3\pi}{2}\right)$, find $\sin 2\theta$, $\cos 2\theta$, and $\tan 2\theta$.

Since $\sin \theta = \frac{y}{r} = -\frac{7}{25}$ on the interval $\left(\pi, \frac{3\pi}{2}\right)$, one point on the terminal side of θ has y-coordinate -7 and a distance of 25 units from the origin, as shown. The x-coordinate of this point is therefore $-\sqrt{25^2 - 7^2}$ or -24. Using this point, we find that

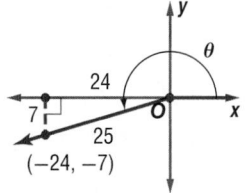

$$\cos \theta = \frac{x}{r} \text{ or } -\frac{24}{25} \quad \text{and} \quad \tan \theta = \frac{y}{x} \text{ or } \frac{7}{25}.$$

Use these values and the double-angle identities for sine and cosine to find $\sin 2\theta$ and $\cos 2\theta$. Then find $\tan 2\theta$ using either the tangent double-angle identity or the definition of tangent.

$\sin 2\theta = 2 \sin \theta \cos \theta$

$\qquad = 2\left(-\dfrac{7}{25}\right)\left(-\dfrac{24}{25}\right) \text{ or } \dfrac{336}{625}$

$\cos 2\theta = 2 \cos^2 \theta - 1$

$\qquad = 2\left(-\dfrac{24}{25}\right)^2 - 1 \text{ or } \dfrac{527}{625}$

Method 1 $\tan 2\theta = \dfrac{2 \tan \theta}{1 - \tan^2 \theta}$

$\qquad = \dfrac{2\left(\frac{7}{24}\right)}{1 - \left(\frac{7}{24}\right)^2} \text{ or } \dfrac{336}{527}$

Method 2 $\tan 2\theta = \dfrac{\sin 2\theta}{\cos 2\theta}$

$\qquad = \dfrac{\frac{336}{625}}{\frac{527}{625}} \text{ or } \dfrac{336}{527}$

✓ Guided Practice

1. If $\cos \theta = \dfrac{3}{5}$ on the interval $\left(0, \dfrac{\pi}{2}\right)$, find $\sin 2\theta$, $\cos 2\theta$, and $\tan 2\theta$.

EXAMPLE 2   Solve an Equation Using a Double-Angle Identity

Solve $\sin 2\theta - \sin\theta = 0$ on the interval $[0, 2\pi)$.

Use the sine double-angle identity to rewrite the equation as a function of a single angle.

$\sin 2\theta - \sin\theta = 0$	Original equation
$2\sin\theta\cos\theta - \sin\theta = 0$	Sine Double-Angle Identity
$\sin\theta\,(2\cos\theta - 1) = 0$	Factor.
$\sin\theta = 0 \qquad$ or $\qquad 2\cos\theta - 1 = 0$	Zero Product Property
$\theta = 0$ or $\pi \qquad\qquad \cos\theta = \dfrac{1}{2}$	Therefore, $\theta = \dfrac{\pi}{3}$ or $\dfrac{5\pi}{3}$.

The solutions on the interval $[0, 2\pi)$ are $\theta = 0$, $\dfrac{\pi}{3}$, π, or $\dfrac{5\pi}{3}$.

✓ **Guided Practice**   Solve each equation on the interval $[0, 2\pi)$.

2A. $\cos 2\alpha = -\sin^2\alpha$

2B. $\tan 2\beta = 2\tan\beta$

The double angle identities can be used to derive the power-reducing identities below. These identities make calculus-related manipulations of functions like $y = \cos^2 x$ much easier.

Key Concept   Power-Reducing Identities

$$\sin^2\theta = \frac{1 - \cos 2\theta}{2} \qquad \cos^2\theta = \frac{1 + \cos 2\theta}{2} \qquad \tan^2\theta = \frac{1 - \cos 2\theta}{1 + \cos 2\theta}$$

Proof   Power-Reducing Identity for Sine

$\dfrac{1 - \cos 2\theta}{2} = \dfrac{1 - (1 - 2\sin^2\theta)}{2}$	Cosine Double-Angle Identity
$= \dfrac{2\sin^2\theta}{2}$	Subtract.
$= \sin^2\theta$	Simplify.

You will prove the power-reducing identities for cosine and tangent in Exercises 82 and 83.

EXAMPLE 3   Use an Identity to Reduce a Power

Rewrite $\sin^4 x$ in terms with no power greater than 1.

$\sin^4 x = (\sin^2 x)^2$	$(\sin^2 x)^2 = \sin^4 x$
$= \left(\dfrac{1 - \cos 2x}{2}\right)^2$	Sine Power-Reducing Identity
$= \dfrac{1 - 2\cos 2x + \cos^2 2x}{4}$	Multiply.
$= \dfrac{1 - 2\cos 2x + \dfrac{1 + \cos 4x}{2}}{4}$	Cosine Power-Reducing Identity
$= \dfrac{2 - 4\cos 2x + 1 + \cos 4x}{8}$	Common denominator
$= \dfrac{1}{8}(3 - 4\cos 2x + \cos 4x)$	Factor.

✓ **Guided Practice**

Rewrite each expression in terms with no power greater than 1.

3A. $\cos^4 x$

3B. $\sin^3\theta$

EXAMPLE 4   Solve an Equation Using a Power-Reducing Identity

Solve $\cos^2 x - \cos 2x = \frac{1}{2}$.

Solve Algebraically

$\cos^2 x - \cos 2x = \frac{1}{2}$	Original equation
$\frac{1 + \cos 2x}{2} - \cos 2x = \frac{1}{2}$	Cosine Power-Reducing Identity
$1 + \cos 2x - 2\cos 2x = 1$	Multiply each side by 2.
$\cos 2x - 2\cos 2x = 0$	Subtract 1 from each side.
$-\cos 2x = 0$	Subtract like terms.
$\cos 2x = 0$	Multiply each side by -1.
$2x = \frac{\pi}{2}$ or $\frac{3\pi}{2}$	Solutions for double angle in $[0, 2\pi)$
$x = \frac{\pi}{4}$ or $\frac{3\pi}{4}$	Divide each solution by 2.

The graph of $y = \cos 2x$ has a period of π, so the solutions are $x = \frac{\pi}{4} + n\pi$ or $\frac{3\pi}{4} + n\pi, n \in \mathbb{Z}$.

Support Graphically

The graph of $y = \cos^2 x - \cos 2x - \frac{1}{2}$ has zeros at $\frac{\pi}{4}$ and $\frac{3\pi}{4}$ on the interval $[0, \pi)$. ✔

Zero
X=.78539816 Y=-1.1E-13

$[0, \pi]$ scl: $\frac{\pi}{4}$ by $[-1.5, 1.5]$ scl: 1

☑ **Guided Practice**   Solve each equation.

4A. $\cos^4 \alpha - \sin^4 \alpha = \frac{1}{2}$ **4B.** $\sin^2 3\beta = \sin^2 \beta$

By replacing θ with $\frac{\theta}{2}$ in each of the power-reducing identities, you can derive each of the following half-angle identities. The sign of each identity that involves the $\pm$ symbol is determined by checking the quadrant in which the terminal side of $\frac{\theta}{2}$ lies.

Watch Out!

Determining Signs To determine which sign is appropriate when using a half-angle identity, check the quadrant in which $\frac{\theta}{2}$ lies, *not* the quadrant in which θ lies.

Key Concept   Half-Angle Identities

$\sin \frac{\theta}{2} = \pm\sqrt{\frac{1 - \cos \theta}{2}}$ $\tan \frac{\theta}{2} = \pm\sqrt{\frac{1 - \cos \theta}{1 + \cos \theta}}$

$\cos \frac{\theta}{2} = \pm\sqrt{\frac{1 + \cos \theta}{2}}$ $\tan \frac{\theta}{2} = \frac{1 - \cos \theta}{\sin \theta}$

                                            $\tan \frac{\theta}{2} = \frac{\sin \theta}{1 + \cos \theta}$

Proof   Half-Angle Identity for Cosine

$\pm\sqrt{\frac{1 + \cos\theta}{2}} = \pm\sqrt{\frac{1 + \cos\left(2 \cdot \frac{\theta}{2}\right)}{2}}$	Rewrite θ as $2 \cdot \frac{\theta}{2}$.
$= \pm\sqrt{\frac{1 + \cos 2x}{2}}$	Substitute $x = \frac{\theta}{2}$.
$= \pm\sqrt{\cos^2 x}$	Cosine Power-Reducing Identity
$= \cos x$	Simplify.
$= \cos \frac{\theta}{2}$	Substitute.

You will prove the half-angle identities for sine and tangent in Exercises 66–68.

EXAMPLE 5 Evaluate an Expression Involving a Half Angle

Find the exact value of cos 112.5°.

Notice that 112.5° is half of 225°. Therefore, apply the half-angle identity for cosine, noting that since 112.5° lies in Quadrant II, its cosine is negative.

$$\cos 112.5° = \cos\frac{225°}{2} \qquad\qquad 112.5° = \frac{225°}{2}$$

$$= -\sqrt{\frac{1 + \cos 225°}{2}} \qquad \text{Cosine Half-Angle Identity (Quadrant II angle)}$$

$$= -\sqrt{\frac{1 - \frac{\sqrt{2}}{2}}{2}} \qquad\qquad \cos 225° = -\frac{\sqrt{2}}{2}$$

$$= -\sqrt{\frac{2 - \sqrt{2}}{4}} \qquad\qquad \text{Subtract and then divide.}$$

$$= -\frac{\sqrt{2 - \sqrt{2}}}{\sqrt{4}} \text{ or } -\frac{\sqrt{2 - \sqrt{2}}}{2} \qquad \text{Quotient Property of Square Roots}$$

CHECK Use a calculator to support your assertion that $\cos 112.5° = -\frac{\sqrt{2 - \sqrt{2}}}{2}$.

$$\cos 112.5° \approx -0.3826834324 \qquad \text{and} \qquad -\frac{\sqrt{2 - \sqrt{2}}}{2} \approx -0.3826834324 \checkmark$$

☑ **Guided Practice**

Find the exact value of each expression.

5A. $\sin 75°$

5B. $\tan\frac{7\pi}{12}$

Recall that you can use sum and difference identities to solve equations. Half-angle identities can also be used to solve equations.

EXAMPLE 6 Solve an Equation Using a Half-Angle Identity

Solve $\sin^2 x = 2\cos^2\frac{x}{2}$ on the interval $[0, 2\pi)$.

$$\sin^2 x = 2\cos^2\frac{x}{2} \qquad \text{Original equation}$$

$$\sin^2 x = 2\left(\pm\sqrt{\frac{1 + \cos x}{2}}\right)^2 \qquad \text{Cosine Half-Angle Identity}$$

$$\sin^2 x = 2\left(\frac{1 + \cos x}{2}\right) \qquad \text{Simplify.}$$

$$\sin^2 x = 1 + \cos x \qquad \text{Multiply.}$$

$$1 - \cos^2 x = 1 + \cos x \qquad \text{Pythagorean Identity}$$

$$-\cos^2 x - \cos x = 0 \qquad \text{Subtract 1 from each side.}$$

$$\cos x\,(-\cos x - 1) = 0 \qquad \text{Factor.}$$

$$\cos x = 0 \qquad\qquad \text{or} \qquad -\cos x - 1 = 0 \qquad \text{Zero Product Property}$$

$$x = \frac{\pi}{2} \text{ or } \frac{3\pi}{2} \qquad\qquad \cos x = -1; \text{ therefore, } x = \pi. \qquad \text{Solutions in } [0, 2\pi)$$

The solutions on the interval $[0, 2\pi)$ are $x = \frac{\pi}{2}, \frac{3\pi}{2}$, or π.

☑ **Guided Practice**

Solve each equation on the interval $[0, 2\pi)$.

6A. $2\sin^2\frac{x}{2} + \cos x = 1 + \sin x$

6B. $8\tan\frac{x}{2} + 8\cos x \tan\frac{x}{2} = 1$

Use Product-to-Sum Identities To work with functions such as $y = \cos 5x \sin 3x$ in calculus, you will need to apply one of the following product-to-sum identities.

Key Concept   **Product-to-Sum Identities**

$$\sin \alpha \sin \beta = \frac{1}{2}[\cos(\alpha - \beta) - \cos(\alpha + \beta)] \qquad \sin \alpha \cos \beta = \frac{1}{2}[\sin(\alpha + \beta) + \sin(\alpha - \beta)]$$

$$\cos \alpha \cos \beta = \frac{1}{2}[\cos(\alpha - \beta) + \cos(\alpha + \beta)] \qquad \cos \alpha \sin \beta = \frac{1}{2}[\sin(\alpha + \beta) - \sin(\alpha - \beta)]$$

Proof   **Product-to-Sum Identity for $\sin \alpha \cos \beta$**

$\frac{1}{2}[\sin (\alpha + \beta) + \sin (\alpha - \beta)]$ — **More complicated side of identity**

$= \frac{1}{2}(\sin \alpha \cos \beta + \cos \alpha \sin \beta + \sin \alpha \cos \beta - \cos \alpha \sin \beta)$ — **Sum and Difference Identities**

$= \frac{1}{2}(2 \sin \alpha \cos \beta)$ — **Combine like terms.**

$= \sin \alpha \cos \beta$ — **Multiply.**

You will prove the remaining three product-to-sum identities in Exercises 84–86.

> **StudyTip**
>
> **Proofs** Remember to work the more complicated side first when proving these identities.

EXAMPLE 7   **Use an Identity to Write a Product as a Sum or Difference**

Rewrite $\cos 5x \sin 3x$ as a sum or difference.

$\cos 5x \sin 3x = \frac{1}{2}[\sin (5x + 3x) - \sin (5x - 3x)]$ — **Product-to-Sum Identity**

$= \frac{1}{2}(\sin 8x - \sin 2x)$ — **Simplify.**

$= \frac{1}{2} \sin 8x - \frac{1}{2} \sin 2x$ — **Distributive Property**

✅ **Guided Practice**   Rewrite each product as a sum or difference.

7A. $\sin 4\theta \cos \theta$ **7B.** $\sin 7x \sin 6x$

These product-to-sum identities have corresponding sum-to-product identities.

Key Concept   **Sum-to-Product Identities**

$$\sin \alpha + \sin \beta = 2 \sin \left(\frac{\alpha + \beta}{2}\right) \cos \left(\frac{\alpha - \beta}{2}\right) \qquad \cos \alpha + \cos \beta = 2 \cos \left(\frac{\alpha + \beta}{2}\right) \cos \left(\frac{\alpha - \beta}{2}\right)$$

$$\sin \alpha - \sin \beta = 2 \cos \left(\frac{\alpha + \beta}{2}\right) \sin \left(\frac{\alpha - \beta}{2}\right) \qquad \cos \alpha - \cos \beta = -2 \sin \left(\frac{\alpha + \beta}{2}\right) \sin \left(\frac{\alpha - \beta}{2}\right)$$

Proof   **Sum-to-Product Identity for $\sin \alpha + \sin \beta$**

$2 \sin \left(\frac{\alpha + \beta}{2}\right) \cos \left(\frac{\alpha - \beta}{2}\right)$

$= 2 \sin x \cos y$ — **Substitute $x = \frac{\alpha + \beta}{2}$ and $y = \frac{\alpha - \beta}{2}$.**

$= 2 \left\{\frac{1}{2}[\sin (x + y) + \sin (x - y)]\right\}$ — **Product-to-Sum Identity**

$= \sin \left(\frac{\alpha + \beta}{2} + \frac{\alpha - \beta}{2}\right) + \sin \left(\frac{\alpha + \beta}{2} - \frac{\alpha - \beta}{2}\right)$ — **Substitute and simplify.**

$= \sin \left(\frac{2\alpha}{2}\right) + \sin \left(\frac{2\beta}{2}\right)$ — **Combine fractions.**

$= \sin \alpha + \sin \beta$ — **Simplify.**

You will prove the remaining three sum-to-product identities in Exercises 87–89.

EXAMPLE 8

EXAMPLE 8   Use a Product-to-Sum or Sum-to-Product Identity

Find the exact value of $\sin \frac{5\pi}{12} + \sin \frac{\pi}{12}$.

$$\sin \frac{5\pi}{12} + \sin \frac{\pi}{12} = 2 \sin \left(\frac{\frac{5\pi}{12} + \frac{\pi}{12}}{2} \right) \cos \left(\frac{\frac{5\pi}{12} - \frac{\pi}{12}}{2} \right) \quad \text{**Sum-to-Product Identity**}$$

$$= 2 \sin \frac{\pi}{4} \cos \frac{\pi}{6} \quad \text{**Simplify.**}$$

$$= 2 \left(\frac{\sqrt{2}}{2} \right) \left(\frac{\sqrt{3}}{2} \right) \quad \sin \frac{\pi}{4} = \frac{\sqrt{2}}{2} \text{ and } \cos \frac{\pi}{6} = \frac{\sqrt{3}}{2}$$

$$= \frac{\sqrt{6}}{2} \quad \text{**Simplify.**}$$

✓ Guided Practice

Find the exact value of each expression.

8A. $3 \cos 37.5° \cos 187.5°$ **8B.** $\cos \frac{7\pi}{12} - \cos \frac{\pi}{12}$

You can also use sum-to-product identities to solve some trigonometric equations.

EXAMPLE 9   Solve an Equation Using a Sum-to-Product Identity

Solve $\cos 4x + \cos 2x = 0$.

Solve Algebraically

$$\cos 4x + \cos 2x = 0 \quad \text{**Original equation**}$$

$$2 \cos \left(\frac{4x + 2x}{2} \right) \cos \left(\frac{4x - 2x}{2} \right) = 0 \quad \text{**Cosine Sum-to-Product Identity**}$$

$$(2 \cos 3x)(\cos x) = 0 \quad \text{**Simplify.**}$$

Set each factor equal to zero and find solutions on the interval $[0, 2\pi)$.

$2 \cos 3x = 0$	**First factor set equal to 0**	$\cos x = 0$	**Second factor set equal to 0**
$\cos 3x = 0$	**Divide each side by 2.**	$x = \frac{\pi}{2}$ or $\frac{3\pi}{2}$	**Solutions in $[0, 2\pi)$**
$3x = \frac{\pi}{2}$ or $\frac{3\pi}{2}$	**Multiple angle solutions in $[0, 2\pi)$**		
$x = \frac{\pi}{6}$ or $\frac{\pi}{2}$	**Divide each solution by 3.**		

The period of $y = \cos 3x$ is $\frac{2\pi}{3}$, so the solutions are

$$x = \frac{\pi}{6} + \frac{2\pi}{3}n, \quad \frac{\pi}{2} + \frac{2\pi}{3}n, \quad \frac{\pi}{2} + 2\pi n, \text{ or } \quad \frac{3\pi}{2} + 2\pi n, n \in \mathbb{Z}.$$

Support Graphically

The graph of $y = \cos 4x + \cos 2x$ has zeros at $\frac{\pi}{6}, \frac{\pi}{2}, \frac{5\pi}{6}$, $\frac{7\pi}{6}, \frac{3\pi}{2}$, and $\frac{11\pi}{6}$ on the interval $[0, 2\pi)$. ✔

$[0, 2\pi]$ scl: $\frac{\pi}{6}$ by $[-3, 3]$ scl: 1

✓ Guided Practice

Solve each equation.

9A. $\sin x + \sin 5x = 0$ **9B.** $\cos 3x - \cos 5x = 0$

> ### Watch Out!
>
> **Periods for Multiple Angle Trigonometric Functions** Recall from Lesson 4-4 that the periods of $y = \sin kx$ and $y = \cos kx$ are $\frac{2\pi}{k}$, not 2π.

Exercises

Find the values of sin 2θ, cos 2θ, and tan 2θ for the given value and interval. (Example 1)

1. $\cos \theta = \frac{3}{5}, (270°, 360°)$

2. $\tan \theta = \frac{8}{15}, (180°, 270°)$

3. $\cos \theta = -\frac{9}{41}, (90°, 180°)$

4. $\sin \theta = -\frac{7}{12}, \left(\frac{3\pi}{2}, 2\pi\right)$

5. $\tan \theta = -\frac{1}{2}, \left(\frac{3\pi}{2}, 2\pi\right)$

6. $\tan \theta = \sqrt{3}, \left(0, \frac{\pi}{2}\right)$

7. $\sin \theta = \frac{4}{5}, \left(\frac{\pi}{2}, \pi\right)$

8. $\cos \theta = -\frac{5}{13}, \left(\pi, \frac{3\pi}{2}\right)$

Solve each equation on the interval [0, 2π). (Example 2)

9. $\sin 2\theta = \cos \theta$

10. $\cos 2\theta = \cos \theta$

11. $\cos 2\theta - \sin \theta = 0$

12. $\tan 2\theta - \tan 2\theta \tan^2 \theta = 2$

13. $\sin 2\theta \csc \theta = 1$

14. $\cos 2\theta + 4 \cos \theta = -3$

15 **GOLF** A golf ball is hit with an initial velocity of 88 feet per second. The distance the ball travels is found by $d = \frac{v_0^2 \sin 2\theta}{32}$, where v_0 is the initial velocity, θ is the angle that the path of the ball makes with the ground, and 32 is in feet per second squared. (Example 2)

a. If the ball travels 242 feet, what is θ to the nearest degree?

b. Use a double-angle identity to rewrite the equation for d.

Rewrite each expression in terms with no power greater than 1. (Example 3)

16. $\cos^3 \theta$

17. $\tan^3 \theta$

18. $\sec^4 \theta$

19. $\cot^3 \theta$

20. $\cos^4 \theta - \sin^4 \theta$

21. $\sin^2 \theta \cos^3 \theta$

22. $\sin^2 \theta - \cos^2 \theta$

23. $\frac{\sin^4 \theta}{\cos^2 \theta}$

Solve each equation. (Example 4)

24. $1 - \sin^2 \theta - \cos 2\theta = \frac{1}{2}$

25. $\cos^2 \theta - \frac{3}{2} \cos 2\theta = 0$

26. $\sin^2 \theta - 1 = \cos^2 \theta$

27. $\cos^2 \theta - \sin \theta = 1$

28. **MACH NUMBER** The angle θ at the vertex of the cone-shaped shock wave produced by a plane breaking the sound barrier is related to the mach number M describing the plane's speed by the half-angle equation $\sin \frac{\theta}{2} = \frac{1}{M}$. (Example 5)

a. Express the mach number of the plane in terms of cosine.

b. Use the expression found in part **a** to find the mach number of a plane if $\cos \theta = \frac{4}{5}$.

Find the exact value of each expression.

29. $\sin 67.5°$

30. $\cos \frac{\pi}{12}$

31. $\tan 157.5°$

32. $\sin \frac{11\pi}{12}$

Solve each equation on the interval [0, 2π). (Example 6)

33. $\sin \frac{\theta}{2} + \cos \theta = 1$

34. $\tan \frac{\theta}{2} = \sin \frac{\theta}{2}$

35. $2 \sin \frac{\theta}{2} = \sin \theta$

36. $1 - \sin^2 \frac{\theta}{2} - \cos \frac{\theta}{2} = \frac{3}{4}$

Rewrite each product as a sum or difference. (Example 7)

37. $\cos 3\theta \cos \theta$

38. $\cos 12x \sin 5x$

39. $\sin 3x \cos 2x$

40. $\sin 8\theta \sin \theta$

Find the exact value of each expression. (Example 8)

41. $2 \sin 135° \sin 75°$

42. $\cos \frac{7\pi}{12} + \cos \frac{\pi}{12}$

43. $\frac{2}{3} \sin 172.5° \sin 127.5°$

44. $\sin 142.5° \cos 352.5°$

45. $\sin 75° + \sin 195°$

46. $2 \cos 105° + 2 \cos 195°$

47. $3 \sin \frac{17\pi}{12} - 3 \sin \frac{\pi}{12}$

48. $\cos \frac{13\pi}{12} + \cos \frac{5\pi}{12}$

Solve each equation. (Example 9)

49. $\cos \theta - \cos 3\theta = 0$

50. $2 \cos 4\theta + 2 \cos 2\theta = 0$

51. $\sin 3\theta + \sin 5\theta = 0$

52. $\sin 2\theta - \sin \theta = 0$

53. $3 \cos 6\theta - 3 \cos 4\theta = 0$

54. $4 \sin \theta + 4 \sin 3\theta = 0$

Simplify each expression.

55. $\sqrt{\dfrac{1 + \cos 6x}{2}}$

56. $\sqrt{\dfrac{1 - \cos 16\theta}{2}}$

Write each expression as a sum or difference.

57. $\cos (a + b) \cos (a - b)$

58. $\sin (\theta - \pi) \sin (\theta + \pi)$

59. $\sin (b + \theta) \cos (b + \pi)$

60. $\cos (a - b) \sin (b - a)$

61. **MAPS** A Mercator projection is a flat projection of the globe in which the distance between the lines of latitude increases with their distance from the equator.

The calculation of a point on a Mercator projection contains the expression $\tan \left(45° + \dfrac{\ell}{2}\right)$, where ℓ is the latitude of the point.

a. Write the expression in terms of $\sin \ell$ and $\cos \ell$.

b. Find the value of this expression if $\ell = 60°$.

62. **BEAN BAG TOSS** Ivan constructed a bean bag tossing game as shown in the figure below.

a. Exactly how far will the back edge of the board be from the ground?

b. Exactly how long is the entire setup?

PROOF Prove each identity.

63. $\cos 2\theta = \cos^2 \theta - \sin^2 \theta$

64. $\cos 2\theta = 2 \cos^2 \theta - 1$

65. $\tan 2\theta = \dfrac{2 \tan \theta}{1 - \tan^2 \theta}$

66. $\sin \dfrac{\theta}{2} = \pm\sqrt{\dfrac{1 - \cos \theta}{2}}$

67. $\tan \dfrac{\theta}{2} = \pm\sqrt{\dfrac{1 - \cos \theta}{1 + \cos \theta}}$

68. $\tan \dfrac{\theta}{2} = \dfrac{\sin \theta}{1 + \cos \theta}$

Verify each identity by using the power-reducing identities and then again by using the product-to-sum identities.

69. $2 \cos^2 5\theta - 1 = \cos 10\theta$

70. $\cos^2 2\theta - \sin^2 2\theta = \cos 4\theta$

Rewrite each expression in terms of cosines of multiple angles with no power greater than 1.

71. $\sin^6 \theta$

72. $\sin^8 \theta$

73. $\cos^7 \theta$

74. $\sin^4 \theta \cos^4 \theta$

75. **MULTIPLE REPRESENTATIONS** In this problem, you will investigate how graphs of functions can be used to find identities.

a. **GRAPHICAL** Use a graphing calculator to graph $f(x) = 4\left(\sin \theta \cos \dfrac{\pi}{4} - \cos \theta \sin \dfrac{\pi}{4}\right)$ on the interval $[-2\pi, 2\pi]$.

b. **ANALYTICAL** Write a sine function $h(x)$ that models the graph of $f(x)$. Then verify that $f(x) = h(x)$ algebraically.

c. **GRAPHICAL** Use a graphing calculator to graph $g(x) = \cos^2 \left(\theta - \dfrac{\pi}{3}\right) - \sin^2 \left(\theta - \dfrac{\pi}{3}\right)$ on the interval $[-2\pi, 2\pi]$.

d. **ANALYTICAL** Write a cosine function $k(x)$ that models the graph of $g(x)$. Then verify that $g(x) = k(x)$ algebraically.

H.O.T. Problems Use **H**igher-**O**rder **T**hinking Skills

76. **CHALLENGE** Verify the following identity.

$$\sin 2\theta \cos \theta - \cos 2\theta \sin \theta = \sin \theta$$

REASONING Consider an angle in the unit circle. Determine what quadrant a double angle and half angle would lie in if the terminal side of the angle is in each quadrant.

77. I **78.** II **79.** III

CHALLENGE Verify each identity.

80. $\sin 4\theta = 4 \sin \theta \cos \theta - 8 \sin^3 \theta \cos \theta$

81 $\cos 4\theta = 1 - 8 \sin^2 \theta \cos^2 \theta$

PROOF Prove each identity.

82. $\cos^2 \theta = \dfrac{1 + \cos 2\theta}{2}$

83. $\tan^2 \theta = \dfrac{1 - \cos 2\theta}{1 + \cos 2\theta}$

84. $\cos \alpha \cos \beta = \dfrac{1}{2}[\cos (\alpha - \beta) + \cos (\alpha + \beta)]$

85. $\sin \alpha \cos \beta = \dfrac{1}{2}[\sin (\alpha + \beta) + \sin (\alpha - \beta)]$

86. $\cos \alpha \sin \beta = \dfrac{1}{2}[\sin (\alpha + \beta) - \sin (\alpha - \beta)]$

87. $\cos \alpha + \cos \beta = 2 \cos \left(\dfrac{\alpha + \beta}{2}\right) \cos \left(\dfrac{\alpha - \beta}{2}\right)$

88. $\sin \alpha - \sin \beta = 2 \cos \left(\dfrac{\alpha + \beta}{2}\right) \sin \left(\dfrac{\alpha - \beta}{2}\right)$

89. $\cos \alpha - \cos \beta = -2 \sin \left(\dfrac{\alpha + \beta}{2}\right) \sin \left(\dfrac{\alpha - \beta}{2}\right)$

90. **WRITING IN MATH** Describe the steps that you would use to find the exact value of $\cos 8\theta$ if $\cos \theta = \dfrac{\sqrt{2}}{5}$.

Find the exact value of each trigonometric expression. (Lesson 5-4)

91. $\cos \dfrac{\pi}{12}$

92. $\cos \dfrac{19\pi}{12}$

93. $\sin \dfrac{5\pi}{6}$

94. $\sin \dfrac{13\pi}{12}$

95. $\cos \left(-\dfrac{7\pi}{6}\right)$

96. $\sin \left(-\dfrac{7\pi}{12}\right)$

97. GARDENING Eliza is waiting for the first day of spring in which there will be 14 hours of daylight to start a flower garden. The number of hours of daylight H in her town can be modeled by $H = 11.45 + 6.5 \sin(0.0168d - 1.333)$, where d is the day of the year, $d = 1$ represents January 1, $d = 2$ represents January 2, and so on. On what day will Eliza begin gardening? (Lesson 5-3)

Find the exact value of each expression. If undefined, write *undefined*. (Lesson 4-3)

98. $\csc \left(-\dfrac{\pi}{3}\right)$

99. $\tan 210°$

100. $\sin \dfrac{19\pi}{4}$

101. $\cos(-3780°)$

Graph and analyze each function. Describe the domain, range, intercepts, end behavior, continuity, and where the function is increasing or decreasing. (Lesson 2-1)

102. $f(x) = -\dfrac{1}{5}x^{\frac{2}{3}}$

103. $f(x) = 4x^{\frac{5}{4}}$

104. $f(x) = -3x^6$

105. $f(x) = 4x^5$

106. REVIEW Identify the equation for the graph.

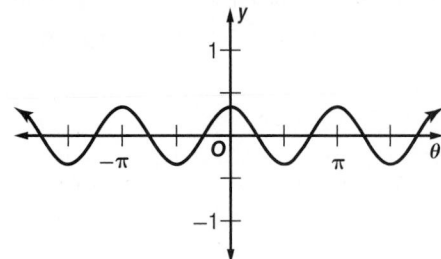

A $y = 3 \cos 2\theta$

B $y = \dfrac{1}{3} \cos 2\theta$

C $y = 3 \cos \dfrac{1}{2}\theta$

D $y = \dfrac{1}{3} \cos \dfrac{1}{2}\theta$

107. REVIEW From a lookout point on a cliff above a lake, the angle of depression to a boat on the water is 12°. The boat is 3 kilometers from the shore just below the cliff. What is the height of the cliff from the surface of the water to the lookout point?

F $\dfrac{3}{\sin 12°}$

G $\dfrac{3}{\tan 12°}$

H $\dfrac{3}{\cos 12°}$

J $3 \tan 12°$

108. FREE RESPONSE Use the graph to answer each of the following.

a. Write a function of the form $f(x) = a \cos(bx + c) + d$ that corresponds to the graph.

b. Rewrite $f(x)$ as a sine function.

c. Rewrite $f(x)$ as a cosine function of a single angle.

d. Find all solutions of $f(x) = 0$.

e. How do the solutions that you found in part **d** relate to the graph of $f(x)$?

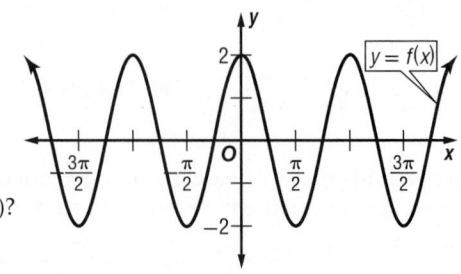

Chapter Summary

Key Concepts

Trigonometric Identities (Lesson 5-1)

• Trigonometric identities are identities that involve trigonometric functions and can be used to find trigonometric values.

• Trigonometric expressions can be simplified by writing the expression in terms of one trigonometric function or in terms of sine and cosine only.

• The most common trigonometric identity is the Pythagorean identity $\sin^2 \theta + \cos^2 \theta = 1$.

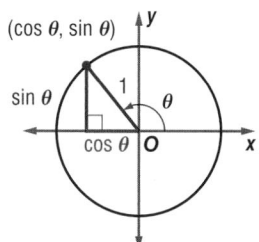

Verifying Trigonometric Identities (Lesson 5-2)

• Start with the more complicated side of the identity and work to transform it into the simpler side.

• Use reciprocal, quotient, Pythagorean, and other basic trigonometric identities.

• Use algebraic operations such as combining fractions, rewriting fractions as sums or differences, multiplying expressions, or factoring expressions.

• Convert a denominator of the form $1 \pm u$ or $u \pm 1$ to a single term using its conjugate and a Pythagorean Identity.

• Work each side separately to reach a common expression.

Solving Trigonometric Equations (Lesson 5-3)

• Algebraic techniques that can be used to solve trigonometric equations include isolating the trigonometric expression, taking the square root of each side, and factoring.

• Trigonometric identities can be used to solve trigonometric equations by rewriting the equation using a single trigonometric function or by squaring each side to obtain an identity.

Sum and Difference Identities (Lesson 5-4)

• Sum and difference identities can be used to find exact values of trigonometric functions of uncommon angles.

• Sum and difference identities can also be used to rewrite a trigonometric expression as an algebraic expression.

Multiple-Angle and Product-Sum Identities (Lesson 5-5)

• Trigonometric identities can be used to find the values of expressions that otherwise could not be evaluated.

Key Vocabulary

cofunction (p. 314)

difference identity (p. 337)

double-angle identity (p. 346)

half-angle identity (p. 348)

identity (p. 312)

odd-even identity (p. 314)

power-reducing identity (p. 347)

product-to-sum identity (p. 350)

Pythagorean identity (p. 313)

quotient identity (p. 312)

reciprocal identity (p. 312)

reduction identity (p. 340)

sum identity (p. 337)

trigonometric identity (p. 312)

verify an identity (p. 320)

Vocabulary Check

Complete each identity by filling in the blank. Then name the identity.

1. $\sec \theta = $ _____

2. _____ $= \dfrac{\sin \theta}{\cos \theta}$

3. _____ $+ 1 = \sec^2 \theta$

4. $\cos (90° - \theta) = $ _____

5. $\tan (-\theta) = $ _____

6. $\sin (\alpha + \beta) = \sin \alpha$ _____ $+ \cos \alpha$ _____

7. _____ $= \cos^2 \alpha - \sin^2 \alpha$

8. _____ $= \pm \sqrt{\dfrac{1 + \cos \theta}{2}}$

9. $\dfrac{1 - \cos 2\theta}{2} = $ _____

10. _____ $= \dfrac{1}{2}[\cos (\alpha - \beta) + \cos (\alpha + \beta)]$

Lesson-by-Lesson Review

Find the value of each expression using the given information.

11. $\sec \theta$ and $\cos \theta$; $\tan \theta = 3$, $\cos \theta > 0$

12. $\cot \theta$ and $\sin \theta$; $\cos \theta = -\frac{1}{5}$, $\tan \theta < 0$

13. $\csc \theta$ and $\tan \theta$; $\cos \theta = \frac{3}{5}$, $\sin \theta < 0$

14. $\cot \theta$ and $\cos \theta$; $\tan \theta = \frac{2}{7}$, $\csc \theta > 0$

15. $\sec \theta$ and $\sin \theta$; $\cot \theta = -2$, $\csc \theta < 0$

16. $\cos \theta$ and $\sin \theta$; $\cot \theta = \frac{3}{8}$, $\sec \theta < 0$

Simplify each expression.

17. $\sin^2(-x) + \cos^2(-x)$ **18.** $\sin^2 x + \cos^2 x + \cot^2 x$

19. $\dfrac{\sec^2 x - \tan^2 x}{\cos(-x)}$ **20.** $\dfrac{\sec^2 x}{\tan^2 x + 1}$

21. $\dfrac{1}{1 - \sin x}$ **22.** $\dfrac{\cos x}{1 + \sec x}$

EXAMPLE 1

If $\sec \theta = -3$ and $\sin \theta > 0$, find $\sin \theta$.

Since $\sin \theta > 0$ and $\sec \theta < 0$, θ must be in Quadrant II. To find $\sin \theta$, first find $\cos \theta$ using the Reciprocal Identity for $\sec \theta$ and $\cos \theta$.

$\cos \theta = \dfrac{1}{\sec \theta}$ **Reciprocal Identity**

$\quad = -\dfrac{1}{3}$ **$\sec \theta = -3$**

Now you can use the Pythagorean identity that includes $\sin \theta$ and $\cos \theta$ to find $\sin \theta$.

$\sin^2 \theta + \cos^2 \theta = 1$ **Pythagorean Identity**

$\sin^2 \theta + \left(-\dfrac{1}{3}\right)^2 = 1$ **$\cos \theta = -\dfrac{1}{3}$**

$\sin^2 \theta + \dfrac{1}{9} = 1$ **Multiply.**

$\sin^2 \theta = \dfrac{8}{9}$ **Subtract.**

$\sin \theta = \dfrac{\sqrt{8}}{3}$ or $\dfrac{2\sqrt{2}}{3}$ **Simplify.**

Verify each identity.

23. $\dfrac{\sin \theta}{1 - \cos \theta} + \dfrac{\sin \theta}{1 + \cos \theta} = 2 \csc \theta$

24. $\dfrac{\cos \theta}{\sec \theta} + \dfrac{\sin \theta}{\csc \theta} = 1$

25. $\dfrac{\cot \theta}{1 + \csc \theta} + \dfrac{1 + \csc \theta}{\cot \theta} = 2 \sec \theta$

26. $\dfrac{\cos \theta}{1 - \sin \theta} = \dfrac{1 + \sin \theta}{\cos \theta}$

27. $\dfrac{\cot^2 \theta}{1 + \csc \theta} = \csc \theta - 1$

28. $\dfrac{\sec \theta}{\tan \theta} + \dfrac{\csc \theta}{\cot \theta} = \sec \theta + \csc \theta$

29. $\dfrac{\sec \theta + \csc \theta}{1 + \tan \theta} = \csc \theta$

30. $\cot \theta \csc \theta + \sec \theta = \csc^2 \theta \sec \theta$

31. $\dfrac{\sin \theta}{\sin \theta + \cos \theta} = \dfrac{\tan \theta}{1 + \tan \theta}$

32. $\cos^4 \theta - \sin^4 \theta = \dfrac{1 - \tan^2 \theta}{\sec^2 \theta}$

EXAMPLE 2

Verify that $\dfrac{\sin \theta}{1 + \cos \theta} + \dfrac{1 + \cos \theta}{\sin \theta} = 2 \csc \theta$.

The left-hand side of this identity is more complicated, so start with that expression.

$\dfrac{\sin \theta}{1 + \cos \theta} + \dfrac{1 + \cos \theta}{\sin \theta} = \dfrac{\sin^2 \theta + (1 + \cos \theta)^2}{\sin \theta(1 + \cos \theta)}$

$= \dfrac{\sin^2 \theta + 1 + 2\cos \theta + \cos^2 \theta}{\sin \theta(1 + \cos \theta)}$

$= \dfrac{\sin^2 \theta + \cos^2 \theta + 1 + 2\cos \theta}{\sin \theta(1 + \cos \theta)}$

$= \dfrac{1 + 1 + 2\cos \theta}{\sin \theta(1 + \cos \theta)}$

$= \dfrac{2 + 2\cos \theta}{\sin \theta(1 + \cos \theta)}$

$= \dfrac{2(1 + \cos \theta)}{\sin \theta(1 + \cos \theta)}$

$= \dfrac{2}{\sin \theta}$

$= 2 \csc \theta$

Lesson-by-Lesson Review

Find all solutions of each equation on the interval $[0, 2\pi)$.

33. $2 \sin x = \sqrt{2}$ **34.** $4 \cos^2 x = 3$

35. $\tan^2 x - 3 = 0$ **36.** $9 + \cot^2 x = 12$

37. $2 \sin^2 x = \sin x$ **38.** $3 \cos x + 3 = \sin^2 x$

Solve each equation for all values of x.

39. $\sin^2 x - \sin x = 0$

40. $\tan^2 x = \tan x$

41. $3 \cos x = \cos x - 1$

42. $\sin^2 x = \sin x + 2$

43. $\sin^2 x = 1 - \cos x$

44. $\sin x = \cos x + 1$

EXAMPLE 3

Solve the equation $\sin \theta = 1 - \cos \theta$ for all values of θ.

$\sin \theta = 1 - \cos \theta$	Original equation.
$\sin^2 \theta = (1 - \cos \theta)^2$	Square each side.
$\sin^2 \theta = 1 - 2 \cos \theta + \cos^2 \theta$	Expand.
$1 - \cos^2 \theta = 1 - 2 \cos \theta + \cos^2 \theta$	Pythagorean Identity
$0 = 2 \cos^2 \theta - 2 \cos \theta$	Subtract.
$0 = 2 \cos \theta(\cos \theta - 1)$	Factor.

Solve for x on $[0, 2\pi)$.

$\cos \theta = 0$ or $\cos \theta = 1$

$\theta = \cos^{-1} 0$ $\theta = \cos^{-1} 1$

$\theta = \dfrac{\pi}{2}$ or $\dfrac{3\pi}{2}$ $\theta = 0$

A check shows that $\dfrac{3\pi}{2}$ is an extraneous solution. So the solutions are $\theta = \dfrac{\pi}{2} + 2n\pi$ or $\theta = 0 + 2n\pi$.

Find the exact value of each trigonometric expression.

45. $\cos 15°$ **46.** $\sin 345°$ **47.** $\tan \dfrac{13\pi}{12}$

48. $\sin \dfrac{7\pi}{12}$ **49.** $\cos -\dfrac{11\pi}{12}$ **50.** $\tan \dfrac{5\pi}{12}$

Simplify each expression.

51. $\dfrac{\tan \dfrac{\pi}{9} + \tan \dfrac{8\pi}{9}}{1 - \tan \dfrac{\pi}{9} \tan \dfrac{8\pi}{9}}$

52. $\cos 24° \cos 36° - \sin 24° \sin 36°$

53. $\sin 95° \cos 50° - \cos 95° \sin 50°$

54. $\cos \dfrac{2\pi}{9} \cos \dfrac{\pi}{18} + \sin \dfrac{2\pi}{9} \sin \dfrac{\pi}{18}$

Verify each identity.

55. $\cos (\theta + 30°) - \sin (\theta + 60°) = -\sin \theta$

56. $\cos \left(\theta + \dfrac{\pi}{4}\right) = \dfrac{\sqrt{2}}{2} (\cos \theta - \sin \theta)$

57. $\cos \left(\theta - \dfrac{\pi}{3}\right) + \cos \left(\theta + \dfrac{\pi}{3}\right) = \cos \theta$

58. $\tan \left(\theta + \dfrac{3\pi}{4}\right) = \dfrac{\tan \theta - 1}{\tan \theta + 1}$

EXAMPLE 4

Find the exact value of $\tan \dfrac{23\pi}{12}$.

$\tan \dfrac{23\pi}{12} = \tan \left(\dfrac{5\pi}{4} + \dfrac{2\pi}{3}\right)$	$\dfrac{23\pi}{12} = \dfrac{5\pi}{4} + \dfrac{2\pi}{3}$
$= \dfrac{\tan \dfrac{5\pi}{4} + \tan \dfrac{2\pi}{3}}{1 - \tan \dfrac{5\pi}{4} \tan \dfrac{2\pi}{3}}$	Sum Identity
$= \dfrac{1 - \sqrt{3}}{1 - (-\sqrt{3})}$	Evaluate for tangent.
$= \dfrac{1 - \sqrt{3}}{1 + \sqrt{3}}$	Simplify.
$= \dfrac{1 - \sqrt{3}}{1 + \sqrt{3}} \cdot \dfrac{1 - \sqrt{3}}{1 - \sqrt{3}}$	Rationalize the denominator.
$= \dfrac{4 - 2\sqrt{3}}{1 - 3}$	Multiply.
$= \dfrac{4 - 2\sqrt{3}}{-2}$ or $-2 + \sqrt{3}$	Simplify.

Lesson-by-Lesson Review

Find the values of sin 2θ, cos 2θ, and tan 2θ for the given value and interval.

59. $\cos \theta = \frac{1}{3}$, $(0°, 90°)$ **60.** $\tan \theta = 2$, $(180°, 270°)$

61. $\sin \theta = \frac{4}{5}$, $\left(\frac{\pi}{2}, \pi\right)$ **62.** $\sec \theta = \frac{13}{5}$, $\left(\frac{3\pi}{2}, 2\pi\right)$

Find the exact value of each expression.

63. $\sin 75°$ **64.** $\cos \frac{11\pi}{12}$

65. $\tan 67.5°$ **66.** $\cos \frac{3\pi}{8}$

67. $\sin \frac{15\pi}{8}$ **68.** $\tan \frac{13\pi}{12}$

EXAMPLE 5

Find the values of sin 2θ, cos 2θ, and tan 2θ if θ is in the fourth quadrant and $\tan \theta = -\frac{24}{7}$.

θ is in the fourth quadrant, so $\cos \theta = \frac{7}{25}$ and $\sin \theta = -\frac{24}{25}$.

$\sin 2\theta = 2 \sin \theta \cos \theta$ $\cos 2\theta = 2 \cos^2 \theta - 1$

$= 2\left(-\frac{24}{25}\right)\frac{7}{25}$ or $-\frac{336}{625}$ $= 2\left(\frac{7}{25}\right)^2 - 1$ or $-\frac{527}{625}$

$\tan 2\theta = \frac{2 \tan \theta}{1 - \tan^2 \theta} = \frac{2\left(-\frac{24}{7}\right)}{1 - \left(-\frac{24}{7}\right)^2} = \frac{-\frac{48}{7}}{-\frac{527}{49}}$ or $\frac{336}{527}$

Applications and Problem Solving

69. CONSTRUCTION Find the tangent of the angle that the ramp makes with the building if $\sin \theta = \frac{\sqrt{145}}{145}$ and $\cos \theta = \frac{12\sqrt{145}}{145}$. (Lesson 5-1)

70. LIGHT The intensity of light that emerges from a system of two polarizing lenses can be calculated by $I = I_0 - \frac{I_0}{\csc^2 \theta}$, where I_0 is the intensity of light entering the system and θ is the angle of the axis of the second lens with the first lens. Write the equation for the light intensity using only tan θ. (Lesson 5-1)

71. MAP PROJECTIONS Stereographic projection is used to project the contours of a three-dimensional sphere onto a two-dimensional map. Points on the sphere are related to points on the map using $r = \frac{\sin \alpha}{1 - \cos \alpha}$. Verify that $r = \frac{1 + \cos \alpha}{\sin \alpha}$. (Lesson 5-2)

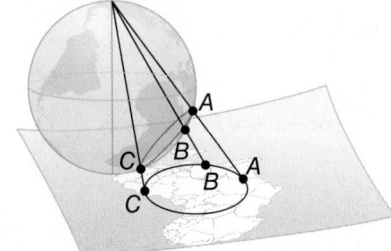

72. PROJECTILE MOTION A ball thrown with an initial speed v_0 at an angle θ that travels a horizontal distance d will remain in the air t seconds, where $t = \frac{d}{v_0 \cos \theta}$. Suppose a ball is thrown with an initial speed of 50 feet per second, travels 100 feet, and is in the air for 4 seconds. Find the angle at which the ball was thrown. (Lesson 5-3)

73. BROADCASTING Interference occurs when two waves pass through the same space at the same time. It is destructive if the amplitude of the sum of the waves is less than the amplitudes of the individual waves. Determine whether the interference is destructive when signals modeled by $y = 20 \sin (3t + 45°)$ and $y = 20 \sin (3t + 225°)$ are combined. (Lesson 5-4)

74. TRIANGULATION *Triangulation* is the process of measuring a distance d using the angles α and β and the distance ℓ using $\ell = \frac{d}{\tan \alpha} + \frac{d}{\tan \beta}$. (Lesson 5-5)

a. Solve the formula for d.

b. Verify that $d = \frac{\ell \sin \alpha \sin \beta}{\sin \alpha \cos \beta + \cos \alpha \sin \beta}$.

c. Verify that $d = \frac{\ell \sin \alpha \sin \beta}{\sin (\alpha + \beta)}$.

d. Show that if $\alpha = \beta$, then $d = 0.5\ell \tan \alpha$.

Find the value of each expression using the given information.

1. $\sin \theta$ and $\cos \theta$, $\csc \theta = -4$, $\cos \theta < 0$

2. $\csc \theta$ and $\sec \theta$, $\tan \theta = \frac{2}{5}$, $\csc \theta < 0$

Simplify each expression.

3. $\dfrac{\sin (90° - x)}{\tan (90° - x)}$

4. $\dfrac{\sec^2 x - 1}{\tan^2 x + 1}$

5. $\sin \theta (1 + \cot^2 \theta)$

Verify each identity.

6. $\dfrac{\csc^2 \theta - 1}{\csc^2 \theta} + \dfrac{\sec^2 \theta - 1}{\sec^2 \theta} = 1$

7. $\dfrac{\cos \theta}{1 + \sin \theta} + \dfrac{1 - \sin \theta}{\cos \theta} = \dfrac{2 \cos \theta}{1 + \sin \theta}$

8. $\dfrac{1}{1 + \cos \theta} + \dfrac{1}{1 - \cos \theta} = 2 \csc^2 \theta$

9. $-\sec^2 \theta \sin^2 \theta = \dfrac{\cos^2 \theta - 1}{\cos^2 \theta}$

10. $\sin^4 x - \cos^4 x = 2 \sin^2 x - 1$

11. **MULTIPLE CHOICE** Which expression is *not* true?

A $\tan (-\theta) = -\tan \theta$

B $\tan (-\theta) = \dfrac{1}{\cot (-\theta)}$

C $\tan (-\theta) = \dfrac{\sin (-\theta)}{\cos (-\theta)}$

D $\tan (-\theta) + 1 = \sec (-\theta)$

Find all solutions of each equation on the interval $[0, 2\pi)$.

12. $\sqrt{2} \sin \theta + 1 = 0$

13. $\sec^2 \theta = \dfrac{4}{3}$

Solve each equation for all values of θ.

14. $\tan^2 \theta - \tan \theta = 0$

15. $\dfrac{1 - \sin \theta}{\cos \theta} = \cos \theta$

16. $\dfrac{1}{\sec \theta - 1} - \dfrac{1}{\sec \theta + 1} = 2$

17. $\sec \theta - 2 \tan \theta = 0$

18. **CURRENT** The current produced by an alternator is given by $I = 40 \sin 135\pi t$, where I is the current in amperes and t is the time in seconds. At what time t does the current first reach 20 amperes? Round to the nearest ten-thousandths.

Find the exact value of each trigonometric expression.

19. $\tan 165°$

20. $\cos -\dfrac{\pi}{12}$

21. $\sin 75°$

22. $\cos 465° - \cos 15°$

23. $6 \sin 675° - 6 \sin 45°$

24. **MULTIPLE CHOICE** Which identity is true?

F $\cos (\theta + \pi) = -\sin \pi$

G $\cos (\pi - \theta) = \cos \theta$

H $\sin \left(\theta - \dfrac{3\pi}{2}\right) = \cos \theta$

J $\sin (\pi + \theta) = \sin \theta$

Simplify each expression.

25. $\cos \dfrac{\pi}{8} \cos \dfrac{3\pi}{8} - \sin \dfrac{\pi}{8} \sin \dfrac{3\pi}{8}$

26. $\dfrac{\tan 135° - \tan 15°}{1 + \tan 135° \tan 15°}$

27. **PHYSICS** A soccer ball is kicked from ground level with an initial speed of v at an angle of elevation θ.

a. The horizontal distance d the ball will travel can be determined using $d = \dfrac{v^2 \sin 2\theta}{g}$, where g is the acceleration due to gravity. Verify that this expression is the same as $\dfrac{2}{g} v^2 (\tan \theta - \tan \theta \sin^2 \theta)$.

b. The maximum height h the object will reach can be determined using $h = \dfrac{v^2 \sin^2 \theta}{2g}$. Find the ratio of the maximum height attained to the horizontal distance traveled.

Find the values of $\sin 2\theta$, $\cos 2\theta$, and $\tan 2\theta$ for the given value and interval.

28. $\tan \theta = -3, \left(\dfrac{3\pi}{2}, 2\pi\right)$

29. $\cos \theta = \dfrac{1}{5}, (0°, 90°)$

30. $\cos \theta = \dfrac{5}{9}, \left(0, \dfrac{\pi}{2}\right)$

Objective
Approximate rates of change for sine and cosine functions using the difference quotient.

In Chapter 4, you learned that many real-world situations exhibit periodic behavior over time and thus, can be modeled by sinusoidal functions. Using transformations of the parent functions $\sin x$ and $\cos x$, trigonometric models can be used to represent data, analyze trends, and predict future values.

Hours of Daylight

While you are able to model real-world situations using graphs of sine and cosine, differential calculus can be used to determine the rate that the model is changing at any point in time. Your knowledge of the difference quotient, the sum identities for sine and cosine, and the evaluation of limits now makes it possible to discover the rates of change for these functions at any point in time.

$f(x) = \sin x$

ACTIVITY 1   **Approximate Rate of Change**

Approximate the rate of change of $f(x) = \sin x$ at several points.

Step 1   Substitute $f(x) = \sin x$ into the difference quotient.

$$m = \frac{f(x+h) - f(x)}{h} \qquad\longrightarrow\qquad m = \frac{\sin(x+h) - \sin x}{h}$$

Step 2   Approximate the rate of change of $f(x)$ at $x = \frac{\pi}{2}$. Let $h = 0.1, 0.01, 0.001,$ and 0.0001.

Step 3   Repeat Steps 1 and 2 for $x = 0$ and for $x = \pi$.

▶ **Analyze the Results**

1. Use tangent lines and the graph of $f(x) = \sin x$ to interpret the values found in Steps 2 and 3.

2. What will happen to the rate of change of $f(x)$ as x increases?

Unlike the natural base exponential function $g(x) = e^x$ and the natural logarithmic function $h(x) = \ln x$, an expression to represent the rate of change of $f(x) = \sin x$ at any point is not as apparent. However, we can substitute $f(x)$ into the difference quotient and then simplify the expression.

$$m = \frac{f(x+h) - f(x)}{h} \qquad\qquad \textbf{Difference quotient}$$

$$= \frac{\sin(x+h) - \sin x}{h} \qquad\qquad \textbf{\textit{f}(x) = sin \textit{x}}$$

$$= \frac{(\sin x \cos h + \cos x \sin h) - \sin x}{h} \qquad\qquad \textbf{Sine Sum Identity}$$

$$= \frac{\sin x \cos h - \sin x}{h} + \frac{\cos x \sin h}{h} \qquad\qquad \textbf{Group terms with sin \textit{x} and cos \textit{x}.}$$

$$= \sin x \left(\frac{\cos h - 1}{h}\right) + \cos x \left(\frac{\sin h}{h}\right) \qquad\qquad \textbf{Factor sin \textit{x} and cos \textit{x}.}$$

We now have two expressions that involve h, $\sin x\left(\dfrac{\cos h - 1}{h}\right)$ and $\cos x\left(\dfrac{\sin h}{h}\right)$. To obtain an accurate approximation of the rate of change of $f(x)$ at a point, we want h to be as close to 0 as possible. Recall that in Chapter 1, we were able to substitute $h = 0$ into an expression to find the exact slope of a function at a point. However, both of the fractional expressions are undefined at $h = 0$.

$\sin x\left(\dfrac{\cos h - 1}{h}\right)$	**Original expressions**	$\cos x\left(\dfrac{\sin h}{h}\right)$
$= \sin x\left(\dfrac{\cos 0 - 1}{0}\right)$	$h = 0$	$= \cos x\left(\dfrac{\sin 0}{0}\right)$
undefined		undefined

We can approximate values for the two expressions by finding the limit of each as h approaches 0 using techniques discussed in Lesson 1-3.

ACTIVITY 2   Calculate Rate of Change

Find an expression for the rate of change of $f(x) = \sin x$.

Step 1   Use a graphing calculator to estimate $\displaystyle\lim_{h \to 0} \dfrac{\cos h - 1}{h}$.

Step 2   Verify the value found in Step 1 by using the **TABLE** function of your calculator.

Step 3   Repeat Steps 1 and 2 to estimate $\displaystyle\lim_{h \to 0} \dfrac{\sin h}{h}$.

Step 4   Substitute the values found in Step 2 and Step 3 into the slope equation

$$m = \sin x\left(\dfrac{\cos h - 1}{h}\right) + \cos x\left(\dfrac{\sin h}{h}\right).$$

Step 5   Simplify the expression in Step 4.

$[-\pi, \pi]$ scl: $\dfrac{\pi}{4}$ by $[-1.5, 1.5]$ scl: 1

▶ Analyze the Results

3. Find the rate of change of $f(x) = \sin x$ at $x = \dfrac{3\pi}{2}$, 2, and $\dfrac{5\pi}{2}$.

4. Make a conjecture as to why the rates of change for all trigonometric functions must be modeled by other trigonometric functions.

Model and Apply

5. In this problem, you will find an expression for the rate of change of $f(x) = \cos x$ at any point x.

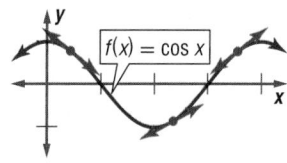

a. Substitute $f(x) = \cos x$ into the difference quotient.

b. Simplify the expression from part **a**.

c. Use a graphing calculator to find the limit of the two fractional expressions as h approaches 0.

d. Substitute the values found in part **c** into the slope equation found in part **b**.

e. Simplify the slope equation in part **d**.

f. Find the rate of change of $f(x) = \cos x$ at $x = 0$, $\dfrac{\pi}{2}$, π, and $\dfrac{3\pi}{2}$.

CHAPTER 6

Systems of Equations and Matrices

Then

In Chapter 0, you solved systems of equations and performed matrix operations.

Now

In Chapter 6, you will:

- Multiply matrices, and find determinants and inverses of matrices.
- Solve systems of linear equations.
- Write partial fraction decompositions of rational expressions.
- Use linear programming to solve applications.

Why?

Linear programming has become a standard tool for many businesses, like farming. Farmers must take into account many constraints in order to maximize profits resulting from the sale of crops or livestock, including the cost of labor, land, and feed.

PREREAD Discuss what you already know about solving equations with a classmate. Then scan the lesson titles and write two or three predictions about what you will learn in this chapter.

Get Ready for Chapter 6

Diagnose Readiness You have two options for checking Prerequisite Skills.

Text Option

Take the Quick Quiz below.

QuickCheck

Use any method to solve each system of equations. (Lesson 0-5)

1. $2x - y = 7$
   $3x + 2y = 14$

2. $3x + y = 14$
   $2x - 2y = -4$

3. $x + 3y = 10$
   $-2x + 3y = 16$

4. $4x + 2y = -34$
   $-3x - y = 24$

5. $2x + 5y = -16$
   $3x + 4y = -17$

6. $-5x + 2y = -33$
   $6x - 3y = 42$

7. **VETERINARY** A veterinarian charges different amounts to trim the nails of dogs and cats. On Monday, she made $96 trimming 4 dogs and 3 cats. On Tuesday, she made $126 trimming 6 dogs and 3 cats. What is the charge to trim the nails of each animal? (Lesson 0-5)

Find each of the following for
$A = \begin{bmatrix} 3 & -5 & 1 \\ -7 & 6 & 4 \end{bmatrix}$, $B = \begin{bmatrix} -2 & -9 & 1 \\ 10 & 8 & -1 \end{bmatrix}$, and
$C = \begin{bmatrix} 0 & 11 & -3 \\ 9 & -3 & 5 \end{bmatrix}$. (Lesson 0-6)

8. $A + 3C$

9. $2(B - A)$

10. $2A - 3B$

11. $3C + 2A$

12. $A + B - C$

13. $2(B + C) - A$

Divide. (Lesson 2-3)

14. $(x^4 - 2x^3 + 4x^2 - 5x - 5) \div (x - 1)$

15. $(2x^4 + 4x^3 - x^2 + 2x - 4) \div (x + 2)$

16. $(3x^4 - 6x^3 - 12x - 36) \div (x - 4)$

17. $(2x^5 - x^3 + 2x^2 + 9x + 6) \div (x - 2)$

18. $(2x^6 + 2x^5 + 3x^3 + x^2 - 8x - 6) \div (x + 1)$

Online Option

Math Online > Take a self-check Chapter Readiness Quiz at **glencoe.com**.

New Vocabulary

English	• Page •	Español
multivariable linear system	• 364 •	sistema lineal multivariable
Gaussian elimination	• 364 •	Eliminación de Gaussian
augmented matrix	• 366 •	matriz aumentada
coefficient matrix	• 366 •	matriz de coeffcient
row-echelon form	• 364 •	forma de grado de fila
reduced row-echelon form	• 369 •	reducir fila escalón forma
Gauss-Jordan elimination	• 369 •	Eliminación de Gauss-Jordania
identity matrix	• 378 •	matriz de identidad
inverse matrix	• 379 •	matriz inversa
inverse	• 379 •	inverso
invertible	• 379 •	invertible
singular matrix	• 379 •	matriz singular
determinant	• 381 •	determinante
square system	• 388 •	sistema cuadrado
Cramer's Rule	• 390 •	La Regla de Cramer
partial fraction	• 398 •	fracción parcial
optimization	• 405 •	optimización
linear programming	• 405 •	programación lineal
objective function	• 405 •	función objetivo
feasible solutions	• 405 •	soluciones viables
constraint	• 405 •	coacción
multiple optimal solutions	• 408 •	múltiples soluciones óptimas
unbounded	• 408 •	no acotado

Review Vocabulary

system of equations • p. P19 • sistema de ecuaciones a set of two or more equations

matrix • p. P24 • matriz a rectangular array of elements

square matrix • p. P24 • matriz cuadrada a matrix that has the same number of rows as columns

scalar • p. P25 • escalar a constant that a matrix is multiplied by

Square Matrix
$$\begin{bmatrix} 2 & -5 & 3 & 1 \\ 7 & 4 & 0 & 6 \\ 2 & 0 & -1 & 2 \\ 0 & 3 & 2 & -6 \end{bmatrix}$$

Multivariable Linear Systems and Row Operations

Why?

Metal alloys are often developed in the automotive industry to improve the performance of cars. You can solve a system of equations to determine what percent of each metal is needed for a specific alloy.

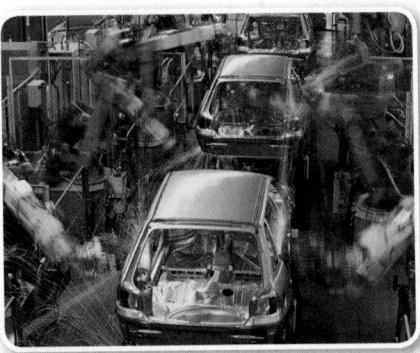

Then

You solved systems of equations algebraically and represented data using matrices. (Lessons 0-5 and 0-6)

Now

- Solve systems of linear equations using matrices and Gaussian elimination.
- Solve systems of linear equations using matrices and Gauss-Jordan elimination.

New Vocabulary

multivariable linear system
row-echelon form
Gaussian elimination
augmented matrix
coefficient matrix
reduced row-echelon form
Gauss-Jordan elimination

Math Online

glencoe.com

Gaussian Elimination A **multivariable linear system**, or *multivariate* linear system, is a system of linear equations in two or more variables. In previous courses, you may have used the *elimination method* to solve such systems. One elimination method begins by rewriting a system using an inverted triangular shape in which the leading coefficients are 1.

The substitution and elimination methods you have previously learned can be used to convert a multivariable linear system into an equivalent system in *triangular* or **row-echelon form**.

System in Row-Echelon Form

$$
\begin{aligned}
x - y - 2z &= 5 \\
y + 4z &= -5 \\
z &= -2
\end{aligned}
$$

Notice that the left side of the system forms a triangle in which the leading coefficients are 1. The last equation contains only one variable, and each equation above it contains the variables from the equation immediately below it.

Once a system is in this form, the solutions can be found by substitution. The final equation determines the final variable. In the example above, the final equation determines that $z = -2$.

Substitute the value for z in the second equation to find y.

$$
\begin{aligned}
y + 4z &= -5 && \text{Second equation} \\
y + 4(-2) &= -5 && z = -2 \\
y &= 3 && \text{Solve for } y.
\end{aligned}
$$

Substitute the values for y and z in the first equation to find x.

$$
\begin{aligned}
x - y - 2z &= 5 && \text{First equation} \\
x - 3 - 2(-2) &= 5 && y = 3 \text{ and } z = -2 \\
x &= 4 && \text{Solve for } x.
\end{aligned}
$$

So, the solution of the system is $x = 4$, $y = 3$, and $z = -2$.

The algorithm used to transform a system of linear equations into an equivalent system in row-echelon form is called **Gaussian elimination**, named after the German mathematician Carl Friedrich Gauss. The operations used to produce equivalent systems are given below.

Key Concept | Operations that Produce Equivalent Systems

Each of the following operations produces an equivalent system of linear equations.

- Interchange any two equations.
- Multiply one of the equations by a nonzero real number.
- Add a multiple of one equation to another equation.

EXAMPLE 1 **Gaussian Elimination with a System**

Write the system of equations in triangular form using Gaussian elimination. Then solve the system.

$$5x - 5y - 5z = 35 \qquad \text{Equation 1}$$
$$-x + 2y - 3z = -12 \qquad \text{Equation 2}$$
$$3x - 2y + 7z = 30 \qquad \text{Equation 3}$$

Step 1 The leading coefficient in Equation 1 is not 1, so multiply this equation by the reciprocal of its leading coefficient.

$$x - y - z = 7 \qquad \longleftarrow \tfrac{1}{5}(5x - 5y - 5z = 35)$$
$$-x + 2y - 3z = -12$$
$$3x - 2y + 7z = 30$$

Step 2 Eliminate the x-term in Equation 2. To do this, replace Equation 2 with (Equation 1 + Equation 2).

$$x - y - z = 7$$
$$y - 4z = -5 \qquad \longleftarrow \begin{cases} x - y - z = 7 \\ (+)\ -x + 2y - 3z = -12 \\ \hline y - 4z = -5 \end{cases}$$
$$3x - 2y + 7z = 30$$

Step 3 Eliminate the x-term in Equation 3 by replacing Equation 3 with $[-3(\text{Equation 1}) + \text{Equation 3}]$.

$$x - y - z = 7$$
$$y - 4z = -5$$
$$y + 10z = 9 \qquad \longleftarrow \begin{cases} -3x + 3y + 3z = -21 \\ (+)\ 3x - 2y + 7z = 30 \\ \hline y + 10z = 9 \end{cases}$$

Step 4 The leading coefficient in Equation 2 is 1, so next eliminate the y-term from Equation 3 by replacing Equation 3 with $[-1(\text{Equation 2}) + \text{Equation 3}]$.

$$x - y - z = 7$$
$$y - 4z = -5$$
$$14z = 14 \qquad \longleftarrow \begin{cases} -y + 4z = 5 \\ (+)\ y + 10z = 9 \\ \hline 14z = 14 \end{cases}$$

Step 5 The leading coefficient in Equation 3 is not 1, so multiply this equation by the reciprocal of its leading coefficient.

$$x - y - z = 7$$
$$y - 4z = -5$$
$$z = 1 \qquad \longleftarrow \tfrac{1}{14}(14z = 14)$$

You can use substitution to find that $y = -1$ and $x = 7$. So, the solution of the system is $x = 7$, $y = -1$, and $z = 1$, or the ordered triple $(7, -1, 1)$.

☑ **Guided Practice**

Write each system of equations in triangular form using Gaussian elimination. Then solve the system.

1A.
$$x + 2y - 3z = -28$$
$$3x - y + 2z = 3$$
$$-x + y - z = -5$$

1B.
$$3x + 5y + 8z = -20$$
$$-x + 2y - 4z = 18$$
$$-6x + 4z = 0$$

Solving a system of linear equations using Gaussian elimination only affects the coefficients of the variables to the left and the constants to the right of the equals sign, so it is often easier to keep track of just these numbers using a matrix.

The **augmented matrix** of a system is derived from the coefficients and constant terms of the linear equations, each written in standard form with the constant terms to the right of the equals sign. If the column of constant terms is not included, the matrix reduces to that of the **coefficient matrix** of the system. You will use this type of matrix in Lesson 6-3.

System of Equations	Augmented Matrix	Coefficient Matrix
$5x - 5y - 5z = 35$	$\begin{bmatrix} 5 & -5 & -5 & \vdots & 35 \\ -1 & 2 & -3 & \vdots & -12 \\ 3 & -2 & 7 & \vdots & 30 \end{bmatrix}$	$\begin{bmatrix} 5 & -5 & -5 \\ -1 & 2 & -3 \\ 3 & -2 & 7 \end{bmatrix}$
$-x + 2y - 3z = -12$		
$3x - 2y + 7z = 30$		

EXAMPLE 2 Write an Augmented Matrix

Write the augmented matrix for the system of linear equations.

$w + 4x + z = 2$
$x + 2y - 3z = 0$
$w - 3y - 8z = -1$
$3w + 2x + 3y = 9$

While each linear equation is in standard form, not all of the four variables of the system are represented in each equation, so the like terms do not align. Rewrite the system, using the coefficient 0 for missing terms. Then write the augmented matrix.

System of Equations	Augmented Matrix
$w + 4x + 0y + z = 2$	$\begin{bmatrix} 1 & 4 & 0 & 1 & \vdots & 2 \\ 0 & 1 & 2 & -3 & \vdots & 0 \\ 1 & 0 & -3 & -8 & \vdots & -1 \\ 3 & 2 & 3 & 0 & \vdots & 9 \end{bmatrix}$
$0w + x + 2y - 3z = 0$	
$w + 0x - 3y - 8z = -1$	
$3w + 2x + 3y + 0z = 9$	

☑ **Guided Practice**

Write the augmented matrix for each system of linear equations.

2A. $4w - 5x + 7z = -11$
$-w + 8x + 3y = 6$
$15x - 2y + 10z = 9$

2B. $-3w + 7x + y = 21$
$4w - 12y + 8z = 5$
$16w - 14y + z = -2$
$w + x + 2y = 7$

The three operations used to produce equivalent systems have corresponding matrix operations that can be used to produce an equivalent augmented matrix. Each row in an augmented matrix corresponds to an equation of the original system, so these operations are called *elementary row operations*.

Key Concept Elementary Row Operations

Each of the following row operations produces an equivalent augmented matrix.

- Interchange any two rows.
- Multiply one row by a nonzero real number.
- Add a multiple of one row to another row.

Row operations are termed *elementary* because they are simple to perform. However, it is easy to make a mistake, so you should record each step using the notation illustrated below.

① Row 1, Row 2, Row 3

$\begin{matrix} R_1 \\ R_2 \\ R_3 \end{matrix} \begin{bmatrix} 5 & -5 & -5 & \vdots & 35 \\ -1 & 2 & -3 & \vdots & -12 \\ 3 & -2 & 7 & \vdots & 30 \end{bmatrix}$

② Interchange Rows 2 and 3.

$\begin{matrix} R_3 \\ R_2 \end{matrix} \begin{bmatrix} 5 & -5 & -5 & \vdots & 35 \\ 3 & -2 & 7 & \vdots & 30 \\ -1 & 2 & -3 & \vdots & -12 \end{bmatrix}$

③ Multiply Row 1 by $\frac{1}{5}$.

$\frac{1}{5}R_1 \rightarrow \begin{bmatrix} 1 & -1 & -1 & \vdots & 7 \\ 3 & -2 & 7 & \vdots & 30 \\ -1 & 2 & -3 & \vdots & -12 \end{bmatrix}$

④ Add −3 times Row 1 to Row 2.

$-3R_1 + R_2 \rightarrow \begin{bmatrix} 1 & -1 & -1 & \vdots & 7 \\ 0 & 1 & 10 & \vdots & 9 \\ -1 & 2 & -3 & \vdots & -12 \end{bmatrix}$

Compare the Gaussian elimination from Example 1 to its matrix version using row operations.

System of Equations	**Augmented Matrix**

$\frac{1}{5}$(Eqn. 1) $\longrightarrow$
$$\begin{aligned} x - y - z &= 7 \\ -x + 2y - 3z &= -12 \\ 3x - 2y + 7z &= 30 \end{aligned}$$

$\frac{1}{5}R_1 \longrightarrow \begin{bmatrix} 1 & -1 & -1 & \vdots & 7 \\ -1 & 2 & -3 & \vdots & -12 \\ 3 & -2 & 7 & \vdots & 30 \end{bmatrix}$

Eqn. 1 + Eqn. 2 $\longrightarrow$
$$\begin{aligned} x - y - z &= 7 \\ y - 4z &= -5 \\ 3x - 2y + 7z &= 30 \end{aligned}$$

$R_1 + R_2 \longrightarrow \begin{bmatrix} 1 & -1 & -1 & \vdots & 7 \\ 0 & 1 & -4 & \vdots & -5 \\ 3 & -2 & 7 & \vdots & 30 \end{bmatrix}$

-3(Eqn. 1) + Eqn. 3 $\longrightarrow$
$$\begin{aligned} x - y - z &= 7 \\ y - 4z &= -5 \\ y + 10z &= 9 \end{aligned}$$

$-3R_1 + R_3 \longrightarrow \begin{bmatrix} 1 & -1 & -1 & \vdots & 7 \\ 0 & 1 & -4 & \vdots & -5 \\ 0 & 1 & 10 & \vdots & 9 \end{bmatrix}$

$-$(Eqn. 2) + Eqn. 3 $\longrightarrow$
$$\begin{aligned} x - y - z &= 7 \\ y - 4z &= -5 \\ 14z &= 14 \end{aligned}$$

$-R_2 + R_3 \longrightarrow \begin{bmatrix} 1 & -1 & -1 & \vdots & 7 \\ 0 & 1 & -4 & \vdots & -5 \\ 0 & 0 & 14 & \vdots & 14 \end{bmatrix}$

$\frac{1}{14}$(Eqn. 3) $\longrightarrow$
$$\begin{aligned} x - y - z &= 7 \\ y - 4z &= -5 \\ z &= 1 \end{aligned}$$

$\frac{1}{14}R_3 \longrightarrow \begin{bmatrix} 1 & -1 & -1 & \vdots & 7 \\ 0 & 1 & -4 & \vdots & -5 \\ 0 & 0 & 1 & \vdots & 1 \end{bmatrix}$

The augmented matrix that corresponds to the row-echelon form of the original system of equations is also said to be in row-echelon form.

Key Concept   **Row-Echelon Form**

A matrix is in row-echelon form if the following conditions are met.
- Rows consisting entirely of zeros (if any) appear at the bottom of the matrix.
- The first nonzero entry in a row is 1, called a *leading 1*.
- For two successive rows with nonzero entries, the leading 1 in the higher row is farther to the left than the leading 1 in the lower row.

$$\begin{bmatrix} 1 & a & b & \vdots & c \\ 0 & 1 & d & \vdots & e \\ 0 & 0 & 1 & \vdots & f \\ 0 & 0 & 0 & \vdots & 0 \end{bmatrix}$$

EXAMPLE 3   **Identify an Augmented Matrix in Row-Echelon Form**

Determine whether each matrix is in row-echelon form.

a. $\begin{bmatrix} 1 & 2 & 0 & \vdots & -1 \\ 0 & 1 & 4 & \vdots & 2 \end{bmatrix}$

b. $\begin{bmatrix} 1 & 6 & 2 & -11 & \vdots & 10 \\ 0 & 1 & -5 & 8 & \vdots & -7 \\ 0 & 0 & 0 & 1 & \vdots & 14 \\ 0 & 0 & 0 & 0 & \vdots & 0 \end{bmatrix}$

c. $\begin{bmatrix} 1 & 5 & -6 & \vdots & 10 \\ 0 & 1 & 9 & \vdots & -3 \\ 0 & 1 & 0 & \vdots & 14 \end{bmatrix}$

There is a zero below the leading one in the first row. The matrix is in row-echelon form.

There is a zero below each of the leading ones in each row. The matrix is in row-echelon form.

There is not a zero below the leading one in Row 2. The matrix is not in row-echelon form.

✔ **Guided Practice**

3A. $\begin{bmatrix} 1 & -6 & 2 & \vdots & -1 \\ 0 & 0 & 0 & \vdots & 0 \\ 0 & 1 & 3 & \vdots & 9 \end{bmatrix}$

3B. $\begin{bmatrix} 1 & 0 & 0 & \vdots & 19 \\ 0 & 1 & 0 & \vdots & 4 \\ 0 & 0 & 1 & \vdots & -20 \end{bmatrix}$

3C. $\begin{bmatrix} 0 & 1 & 0 & 4 & \vdots & 10 \\ 1 & 0 & -3 & 10 & \vdots & -7 \\ 0 & 1 & 6 & 0 & \vdots & 8 \\ 0 & 0 & 1 & -2 & \vdots & -4 \end{bmatrix}$

To solve a system of equations using an augmented matrix and Gaussian elimination, use row operations to transform the matrix so that it is in row-echelon form. Then write the corresponding system of equations and use substitution to finish solving the system. Remember, if you encounter a false equation, this means that the system has no solution.

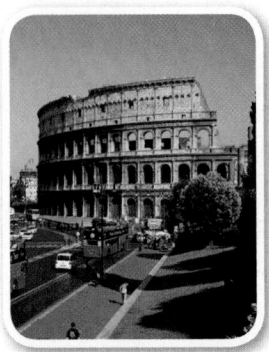

Real-World Link

In a recent year, Italy was the fourth most-visited country in the world, with over 40 million tourists.

Source: World Tourism Organization

Real-World EXAMPLE 4 — Gaussian Elimination with a Matrix

TRAVEL Manuel went to Italy during spring break. The average daily hotel, food, and transportation costs for each city he visited are shown. Write and solve a system of equations to determine how many days Manuel spent in each city. Interpret your solution.

Expense	Venice	Rome	Naples	Total
hotels	$60	$120	$60	$720
food	$40	$90	$30	$490
transportation	$15	$10	$20	$130

Write the information as a system of equations. Let x, y, and z represent the number of days Manuel spent in Venice, Rome, and Naples, respectively.

$$60x + 120y + 60z = 720$$
$$40x + 90y + 30z = 490$$
$$15x + 10y + 20z = 130$$

Next, write the augmented matrix and apply elementary row operations to obtain a row-echelon form of the matrix.

Augmented matrix $\begin{bmatrix} 60 & 120 & 60 & \vdots & 720 \\ 40 & 90 & 30 & \vdots & 490 \\ 15 & 10 & 20 & \vdots & 130 \end{bmatrix}$

$\frac{1}{60}R_1 \longrightarrow \begin{bmatrix} 1 & 2 & 1 & \vdots & 12 \\ 40 & 90 & 30 & \vdots & 490 \\ 15 & 10 & 20 & \vdots & 130 \end{bmatrix}$

$-40R_1 + R_2 \longrightarrow \begin{bmatrix} 1 & 2 & 1 & \vdots & 12 \\ 0 & 10 & -10 & \vdots & 10 \\ 15 & 10 & 20 & \vdots & 130 \end{bmatrix}$

$\frac{1}{10}R_2 \longrightarrow \begin{bmatrix} 1 & 2 & 1 & \vdots & 12 \\ 0 & 1 & -1 & \vdots & 1 \\ 15 & 10 & 20 & \vdots & 130 \end{bmatrix}$

$-15R_1 + R_3 \longrightarrow \begin{bmatrix} 1 & 2 & 1 & \vdots & 12 \\ 0 & 1 & -1 & \vdots & 1 \\ 0 & -20 & 5 & \vdots & -50 \end{bmatrix}$

$20R_2 + R_3 \longrightarrow \begin{bmatrix} 1 & 2 & 1 & \vdots & 12 \\ 0 & 1 & -1 & \vdots & 1 \\ 0 & 0 & -15 & \vdots & -30 \end{bmatrix}$

$-\frac{1}{15}R_3 \longrightarrow \begin{bmatrix} 1 & 2 & 1 & \vdots & 12 \\ 0 & 1 & -1 & \vdots & 1 \\ 0 & 0 & 1 & \vdots & 2 \end{bmatrix}$

You can use substitution to find that $y = 3$ and $x = 4$. Therefore, the solution of the system is $x = 4$, $y = 3$, and $z = 2$, or the ordered triple $(4, 3, 2)$.

Manuel spent 4 days in Venice, 3 days in Rome, and 2 days in Naples.

StudyTip

Types of Solutions Recall that a system of equations can have one solution, no solution, or infinitely many solutions.

✔ Guided Practice

4. **TRAVEL** The following year, Manuel traveled to France for spring break. The average daily hotel, food, and transportation costs for each city in France that he visited are shown. Write and solve a system of equations to determine how many days Manuel spent in each city. Interpret your solution.

Expense	Paris	Lyon	Marseille	Total
hotels	$80	$70	$80	$500
food	$50	$40	$50	$330
transportation	$10	$10	$10	$70

Gauss-Jordan Elimination If you continue to apply elementary row operations to the row-echelon form of an augmented matrix, you can obtain a matrix in which the first nonzero element of each row is 1 and the rest of the elements in the same column as this element are 0. This is called the **reduced row-echelon form** of the matrix and is shown at the right. The reduced row-echelon form of a matrix is always unique, regardless of the order of operations that were performed.

Reduced Row-Echelon Form

$$\begin{bmatrix} 1 & 0 & 0 & | & a \\ 0 & 1 & 0 & | & b \\ 0 & 0 & 1 & | & c \\ 0 & 0 & 0 & | & 0 \end{bmatrix}$$

Solving a system by transforming an augmented matrix so that it is in reduced row-echelon form is called **Gauss-Jordan elimination**, named after Carl Friedrich Gauss and Wilhelm Jordan.

StudyTip

Patterns While different elementary row operations can be used to solve the same system of equations, a general pattern can be used as a guide to help avoid wasteful operations. For the system at the right, begin by producing a 0 in the first term of the second row and work your way around the matrix in the order shown, producing 0s and 1s. Once this is completed, the terms in the first row can be converted to 0s and 1s as well.

$$\begin{bmatrix} 1 & 0 & 0 & | & a \\ 0 & 1 \leftarrow 0 & | & b \\ 0 \rightarrow 0 \rightarrow 1 & | & c \end{bmatrix}$$

EXAMPLE 5 Use Gauss-Jordan Elimination

Solve the system of equations.

$$x - y + z = 0$$
$$-x + 2y - 3z = -5$$
$$2x - 3y + 5z = 8$$

Write the augmented matrix. Apply elementary row operations to obtain a reduced row-echelon form.

Augmented matrix $\begin{bmatrix} 1 & -1 & 1 & | & 0 \\ -1 & 2 & -3 & | & -5 \\ 2 & -3 & 5 & | & 8 \end{bmatrix}$

$R_1 + R_2 \longrightarrow \begin{bmatrix} 1 & -1 & 1 & | & 0 \\ 0 & 1 & -2 & | & -5 \\ 2 & -3 & 5 & | & 8 \end{bmatrix}$

$-2R_1 + R_3 \longrightarrow \begin{bmatrix} 1 & -1 & 1 & | & 0 \\ 0 & 1 & -2 & | & -5 \\ 0 & -1 & 3 & | & 8 \end{bmatrix}$

$R_2 + R_3 \longrightarrow \begin{bmatrix} 1 & -1 & 1 & | & 0 \\ 0 & 1 & -2 & | & -5 \\ 0 & 0 & 1 & | & 3 \end{bmatrix}$ **Row-echelon form**

$R_2 + R_1 \longrightarrow \begin{bmatrix} 1 & 0 & -1 & | & -5 \\ 0 & 1 & -2 & | & -5 \\ 0 & 0 & 1 & | & 3 \end{bmatrix}$

$R_3 + R_1 \longrightarrow \begin{bmatrix} 1 & 0 & 0 & | & -2 \\ 0 & 1 & -2 & | & -5 \\ 0 & 0 & 1 & | & 3 \end{bmatrix}$

$2R_3 + R_2 \longrightarrow \begin{bmatrix} 1 & 0 & 0 & | & -2 \\ 0 & 1 & 0 & | & 1 \\ 0 & 0 & 1 & | & 3 \end{bmatrix}$ **Reduced row-echelon form**

TechnologyTip

You can check the reduced row-echelon form of a matrix by using the rref(feature on a graphing calculator.

```
rref([A])
    [[1 0 0 -2]
     [0 1 0 1 ]
     [0 0 1 3 ]]
```

The solution of the system is $x = -2$, $y = 1$, and $z = 3$ or the ordered triple $(-2, 1, 3)$. Check this solution in the original system of equations.

✓ **Guided Practice**

Solve each system of equations.

5A. $x + 2y - 3z = 7$
$\quad\ -3x - 7y + 9z = -12$
$\quad\ 2x + y - 5z = 8$

5B. $4x + 9y + 16z = 2$
$\quad\ -x - 2y - 4z = -1$
$\quad\ 2x + 4y + 9z = -5$

When solving a system of equations, if a matrix cannot be written in reduced row-echelon form, then the system either has no solution or infinitely many solutions.

EXAMPLE 6   **No Solution and Infinitely Many Solutions**

Solve each system of equations.

a. $-5x - 2y + z = 2$
$4x - y - 6z = 2$
$-3x - y + z = 1$

Write the augmented matrix. Then apply elementary row operations to obtain a reduced row-echelon matrix.

Augmented matrix
$\begin{bmatrix} -5 & -2 & 1 & \vdots & 2 \\ 4 & -1 & -6 & \vdots & 2 \\ -3 & -1 & 1 & \vdots & 1 \end{bmatrix}$

$3R_1 + R_3 \longrightarrow \begin{bmatrix} 1 & 0 & -1 & \vdots & 0 \\ 0 & -1 & -2 & \vdots & 2 \\ 0 & -1 & -2 & \vdots & 1 \end{bmatrix}$

$-2R_3 + R_1 \longrightarrow \begin{bmatrix} 1 & 0 & -1 & \vdots & 0 \\ 4 & -1 & -6 & \vdots & 2 \\ -3 & -1 & 1 & \vdots & 1 \end{bmatrix}$

$-2R_3 + R_2 \longrightarrow \begin{bmatrix} 1 & 0 & -1 & \vdots & 0 \\ 0 & 1 & 2 & \vdots & 0 \\ 0 & -1 & -2 & \vdots & 1 \end{bmatrix}$

$-4R_1 + R_2 \longrightarrow \begin{bmatrix} 1 & 0 & -1 & \vdots & 0 \\ 0 & -1 & -2 & \vdots & 2 \\ -3 & -1 & 1 & \vdots & 1 \end{bmatrix}$

$R_2 + R_3 \longrightarrow \begin{bmatrix} 1 & 0 & -1 & \vdots & 0 \\ 0 & 1 & 2 & \vdots & 0 \\ 0 & 0 & 0 & \vdots & 1 \end{bmatrix}$

According to the last row, $0x + 0y + 0z = 1$. This is impossible, so the system has no solution.

b. $3x + 5y - 8z = -3$
$2x + 5y - 2z = -7$
$-x - y + 4z = -1$

Write the augmented matrix. Then apply elementary row operations to obtain a reduced row-echelon matrix.

Augmented matrix
$\begin{bmatrix} 3 & 5 & -8 & \vdots & -3 \\ 2 & 5 & -2 & \vdots & -7 \\ -1 & -1 & 4 & \vdots & -1 \end{bmatrix}$

$R_1 + R_3 \longrightarrow \begin{bmatrix} 1 & 0 & -6 & \vdots & 4 \\ 0 & 3 & 6 & \vdots & -9 \\ 0 & -1 & -2 & \vdots & 3 \end{bmatrix}$

$R_1 - R_2 \longrightarrow \begin{bmatrix} 1 & 0 & -6 & \vdots & 4 \\ 2 & 5 & -2 & \vdots & -7 \\ -1 & -1 & 4 & \vdots & -1 \end{bmatrix}$

$2R_3 + R_2 \longrightarrow \begin{bmatrix} 1 & 0 & -6 & \vdots & 4 \\ 0 & 1 & 2 & \vdots & -3 \\ 0 & -1 & -2 & \vdots & 3 \end{bmatrix}$

$2R_3 + R_2 \longrightarrow \begin{bmatrix} 1 & 0 & -6 & \vdots & 4 \\ 0 & 3 & 6 & \vdots & -9 \\ -1 & -1 & 4 & \vdots & -1 \end{bmatrix}$

$R_2 + R_3 \longrightarrow \begin{bmatrix} 1 & 0 & -6 & \vdots & 4 \\ 0 & 1 & 2 & \vdots & -3 \\ 0 & 0 & 0 & \vdots & 0 \end{bmatrix}$

StudyTip

Infinitely Many Solutions
The solution of the system in Example 6b is not a unique answer because the solution could be expressed in terms of any of the variables in the system.

Write the corresponding system of linear equations for the reduced row-echelon form of the augmented matrix.

$x - 6z = 4$
$ y + 2z = -3$

Because the value of z is not determined, this system has infinitely many solutions. Solving for x and y in terms of z, you have $x = 6z + 4$ and $y = -2z - 3$.

So, a solution of the system can be expressed as $(6z + 4, -2z - 3, z)$, where z is any real number.

✔ **Guided Practice**

Solve each system of equations.

6A. $3x - y - 5z = 9$
$4x + 2y - 3z = 6$
$-7x - 11y - 3z = 3$

6B. $x + 3y + 4z = 8$
$4x - 2y - z = 6$
$8x - 18y - 19z = -2$

When a system has fewer equations than variables, the system either has no solution or infinitely many solutions. When solving a system of equations with three or more variables, it is important to check your answer using all of the original equations. This is necessary because it is possible for an incorrect solution to work for some of the equations but not the others.

EXAMPLE 7 | **Infinitely Many Solutions**

Solve the system of equations.

$3x - 8y + 19z - 12w = 6$

$2x - 4y + 10z = -8$

$x - 3y + 5z - 2w = -1$

Write the augmented matrix. Then apply elementary row operations to obtain leading 1s in each row and zeros below these 1s in each column.

$$\begin{bmatrix} 3 & -8 & 19 & -12 & | & 6 \\ 2 & -4 & 10 & 0 & | & -8 \\ 1 & -3 & 5 & -2 & | & -1 \end{bmatrix}$$

$$\begin{matrix} R_3 \\ \\ R_1 \end{matrix} \begin{bmatrix} 1 & -3 & 5 & -2 & | & -1 \\ 2 & -4 & 10 & 0 & | & -8 \\ 3 & -8 & 19 & -12 & | & 6 \end{bmatrix}$$

$$-2R_1 + R_2 \longrightarrow \begin{bmatrix} 1 & -3 & 5 & -2 & | & -1 \\ 0 & 2 & 0 & 4 & | & -6 \\ 3 & -8 & 19 & -12 & | & 6 \end{bmatrix}$$

$$-3R_1 + R_3 \longrightarrow \begin{bmatrix} 1 & -3 & 5 & -2 & | & -1 \\ 0 & 2 & 0 & 4 & | & -6 \\ 0 & 1 & 4 & -6 & | & 9 \end{bmatrix}$$

$$\tfrac{1}{2}R_2 \longrightarrow \begin{bmatrix} 1 & -3 & 5 & -2 & | & -1 \\ 0 & 1 & 0 & 2 & | & -3 \\ 0 & 1 & 4 & -6 & | & 9 \end{bmatrix}$$

$$-R_2 + R_3 \longrightarrow \begin{bmatrix} 1 & -3 & 5 & -2 & | & -1 \\ 0 & 1 & 0 & 2 & | & -3 \\ 0 & 0 & 4 & -8 & | & 12 \end{bmatrix}$$

$$3R_2 + R_1 \longrightarrow \begin{bmatrix} 1 & 0 & 5 & 4 & | & -10 \\ 0 & 1 & 0 & 2 & | & -3 \\ 0 & 0 & 4 & -8 & | & 12 \end{bmatrix}$$

$$\tfrac{1}{4}R_3 \longrightarrow \begin{bmatrix} 1 & 0 & 5 & 4 & | & -10 \\ 0 & 1 & 0 & 2 & | & -3 \\ 0 & 0 & 1 & -2 & | & 3 \end{bmatrix}$$

$$-5R_3 + R_1 \longrightarrow \begin{bmatrix} 1 & 0 & 0 & 14 & | & -25 \\ 0 & 1 & 0 & 2 & | & -3 \\ 0 & 0 & 1 & -2 & | & 3 \end{bmatrix}$$

Write the corresponding system of linear equations for the reduced row-echelon form of the augmented matrix.

$$x + \qquad 14w = -25$$
$$y + \qquad 2w = -3$$
$$z - 2w = 3$$

This system of equations has infinitely many solutions because for every value of w there are three equations that can be used to find the corresponding values of x, y, and z. Solving for x, y, and z in terms of w, you have $x = -14w - 25$, $y = -2w - 3$, and $z = 2w + 3$.

So, a solution of the system can be expressed as $(-14w - 25, -2w - 3, 2w + 3, w)$, where w is any real number.

CHECK Using different values for w, calculate a few solutions and check them in the original system of equations. For example, if $w = 1$, a solution of the system is $(-39, -5, 5, 1)$. This solution checks in each equation of the original system.

$3(-39) - 8(-5) + 19(5) - 12(1) = 6$ ✔

$2(-39) - 4(-5) + 10(5) = -8$ ✔

$(-39) - 3(-5) + 5(5) - 2(1) = -1$ ✔

☑ **Guided Practice**

Solve each system of equations.

7A. $-5w + 10x + 4y + 54z = 15$
$w - 2x - y - 9z = -1$
$-2w + 3x + y + 19z = 9$

7B. $3w + x - 2y - 3z = 14$
$-w + x - 10y + z = -11$
$-2w - x + 4y + 2z = -9$

Write each system of equations in triangular form using Gaussian elimination. Then solve the system. (Example 1)

1. $5x = -3y - 31$
$2y = -4x - 22$

2. $4y + 17 = -7x$
$8x + 5y = -19$

3. $12x = 21 - 3y$
$2y = 6x + 7$

4. $4y = 12x - 3$
$9x = 20y - 2$

5. $-3x + y + 6z = 15$
$2x + 2y - 5z = 9$
$4x - 5y + 2z = -3$

6. $8x - 24y + 16z = -7$
$40x - 9y + 2z = 10$
$32x + 8y - z = -2$

7. $3x + 9y - 6z = 17$
$-2x - y + 24z = 12$
$2x - 5y + 12z = -30$

8. $5x - 50y + z = 24$
$2x + 10y + 3z = 23$
$-5x - 20y + 10z = 13$

Write the augmented matrix for each system of linear equations. (Example 2)

9. $12x - 5y = -9$
$-3x + 8y = 10$

10. $-4x - 6y = 25$
$7x + 2y = 16$

11. $3x - 5y + 7z = 9$
$-10x + y + 8z = 6$
$4x - 15z = -8$

12. $4x - z = 27$
$-8x + 7y - 6z = -35$
$12x - 3y + 5z = 20$

13. $w - 8x + 5y = 11$
$7w + 2x - 3y + 9z = -5$
$6w + 12y - 15z = 4$
$3x + 4y - 8z = -13$

14. $14x - 2y + 3z = -22$
$5w - 4x + 11z = -8$
$2w - 6y + 3z = 15$
$3w + 7x - y = 1$

15 BAKE SALE Members of a youth group held a bake sale to raise money for summer trip. They sold 30 cakes, 40 pies, and 200 giant cookies and raised $684.50. A pie cost $2 less than a cake and cake cost 5 times as much as a giant cookie. (Example 2)

a. Let c = number of cakes, p = number of pies, and g = number of giant cookies. Write a system of three linear equations to represent the problem.

b. Write the augmented matrix for the system of linear equations that you wrote in part **a**.

c. Solve the system of equations. Interpret your solution.

Determine whether each matrix is in row-echelon form. (Example 3)

16. $\begin{bmatrix} 1 & 8 & \vdots & 7 \\ 0 & 1 & \vdots & 3 \end{bmatrix}$

17. $\begin{bmatrix} 1 & -2 & \vdots & 9 \\ 0 & 0 & \vdots & 1 \end{bmatrix}$

18. $\begin{bmatrix} 1 & -8 & \vdots & 12 \\ 1 & 3 & \vdots & -7 \\ 0 & 1 & \vdots & 4 \end{bmatrix}$

19. $\begin{bmatrix} 1 & -4 & \vdots & 10 \\ 0 & 1 & \vdots & -6 \\ 0 & 0 & \vdots & 1 \end{bmatrix}$

20. $\begin{bmatrix} 0 & 1 & -7 & \vdots & -3 \\ 0 & 0 & 1 & \vdots & 5 \\ 0 & 0 & 1 & \vdots & 8 \\ 0 & 0 & 0 & \vdots & 1 \end{bmatrix}$

21. $\begin{bmatrix} 0 & 1 & -8 & \vdots & 5 \\ 0 & 0 & 1 & \vdots & 13 \\ 0 & 0 & 0 & \vdots & 1 \\ 0 & 0 & 0 & \vdots & 0 \end{bmatrix}$

Solve each system of equations using Gaussian or Gauss-Jordan elimination. (Examples 4 and 5)

22. $2x = -10y + 11$
$-8y = -9x + 23$

23. $4y + 17 = -7x$
$8x + 5y = -19$

24. $x + 7y = 10$
$3x + 9y = -6$

25. $7y = 9 - 5x$
$8x = 2 - 5y$

26. $3x - 4y + 8z = 27$
$9x - y - z = 3$
$x + 8y - 2z = 9$

27. $x + 9y + 8z = 0$
$5x + 8y + z = 35$
$x - 4y - z = 17$

28. $4x + 8y - z = 10$
$3x - 8y + 9z = 14$
$7x + 6y + 5z = 0$

29. $2x - 10y + z = 28$
$-5x + 11y + 7z = 18$
$6x - y - 12z = 14$

30. COFFEE A local coffee shop specializes in espresso drinks. The table shows the cups of each drink sold throughout the day. Write and solve a system of equations to determine the price of each espresso drink. Interpret your solution. (Example 4)

Hours	Cappuccino	Latte	Macchiato	Earnings ($)
8–11	103	86	79	1040.25
11–2	48	32	26	406.50
2–5	45	25	18	334.00

31. FLORIST An advertisement for a floral shop shows the price of several flower arrangements and a list of the flowers included in each arrangement as shown below. Write and solve a system of equations to determine the price of each type of flower. Interpret your solution. (Example 6)

Birthday bouquet 4 roses, 12 lilies, and 5 irises		**$35.00**
Sunny Garden 6 roses, 9 lilies, and 12 irises		**$50.25**
Summer Expressions 10 roses, 15 lilies, and 20 irises		**$83.75**

Solve each system of equations. (Examples 6 and 7)

32. $-2x + y - 3z = 0$
$3x - 4y + 10z = -7$
$5x + 2y + 8z = 23$

33. $4x - 5y - 9z = -25$
$-6x + y + 7z = -21$
$7x - 3y - 10z = 8$

34. $-x + 3y + 10z = 8$
$4x - 9y - 34z = -17$
$3x + 5y - 2z = 46$

35. $5x - 4y - 7z = -31$
$2x + y - 8z = 11$
$-4x + 3y + 6z = 23$

36. $-3x + 4y - z = -10$
$6x - y - 5z = -29$
$4x - 5y + z = 11$

37. $8x - 9y - 4z = -33$
$-2x + 3y - 2z = 9$
$-7x + 6y + 11z = 27$

38. $2x - 5y + 4z + 4w = 2$
$-3x + 6y - 2z - 7w = 11$
$5x - 4y + 8z - 5w = 29$

39. $x - 4y + 4z + 3w = 2$
$-2x - 3y + 7z - 3w = -9$
$3x - 5y + z + 10w = 15$

GRAPHING CALCULATOR Find the row-echelon form and reduced row-echelon form of each system. Round to the nearest tenth, if necessary.

40. $3x + 2.5y = 18$
$6.8x - 4y = 29.2$

41. $\frac{2}{5}x - \frac{1}{2}y = 8$
$\frac{3}{4}x + \frac{5}{8}y = \frac{5}{2}$

42. $7x + \frac{2}{3}y + \frac{1}{6}z = -\frac{13}{3}$
$-\frac{3}{5}x + y - \frac{1}{3}z = \frac{11}{10}$
$2x - \frac{2}{5}y - \frac{1}{2}z = -6$

43. $15.9x - y + 4.3z = 14.8$
$-8.2x + 14y = 14.6$
$-11x + 0.5y - 1.6z = -20.4$

44. FINANCIAL LITERACY A sports equipment company took out three different loans from a bank to buy treadmills. The bank statement after the first year is shown below. The amount borrowed at the 6.5% rate was $50,000 less than the amounts borrowed at the other two rates combined.

Bay Bank Co.
STATEMENT SUMMARY
Amount Borrowed..$350,000

Loan 1 ....6.5% Interest Rate
Loan 2 ....7% Interest Rate
Loan 3 ....9% Interest Rate
Interest Paid...$24,950

a. Write a system of three linear equations to represent this situation.

b. Use a graphing calculator to solve the system of equations. Interpret the solution.

Determine the row operation performed to obtain each matrix.

45. $\begin{bmatrix} 1 & 5 & -6 & | & 3 \\ 0 & 1 & -3 & | & -2 \\ 0 & -1 & 2 & | & 1 \end{bmatrix} \longrightarrow \begin{bmatrix} 1 & 5 & -6 & | & 3 \\ 0 & 1 & -3 & | & -2 \\ 0 & 0 & -1 & | & -1 \end{bmatrix}$

46. $\begin{bmatrix} 3 & 1 & -5 & | & 4 \\ 9 & -1 & 4 & | & -2 \\ 8 & 4 & -3 & | & 1 \end{bmatrix} \longrightarrow \begin{bmatrix} 3 & 1 & -5 & | & 4 \\ 9 & -1 & 4 & | & -2 \\ 2 & 2 & 7 & | & -7 \end{bmatrix}$

47 $\begin{bmatrix} 1 & 15 & 2 & 4 & | & 14 \\ 0 & 8 & 5 & -5 & | & 15 \\ 2 & 1 & 0 & 16 & | & 20 \\ -3 & -11 & -1 & 6 & | & -4 \end{bmatrix} \longrightarrow \begin{bmatrix} 1 & 15 & 2 & 4 & | & 14 \\ 0 & 8 & 5 & -5 & | & 15 \\ 2 & 1 & 0 & 16 & | & 20 \\ 0 & 34 & 5 & 18 & | & 38 \end{bmatrix}$

48. $\begin{bmatrix} 8 & -2 & 0 & 2 & 12 & | & -2 \\ 8 & 5 & -7 & 1 & 6 & | & 9 \\ -1 & 0 & 9 & 3 & 3 & | & 2 \end{bmatrix} \rightarrow \begin{bmatrix} 8 & -2 & 0 & 2 & 12 & | & -2 \\ 0 & 7 & -7 & -1 & -6 & | & 11 \\ -1 & 0 & 9 & 3 & 3 & | & 2 \end{bmatrix}$

49. MEDICINE A diluted saline solution is needed for routine procedures in a hospital. The supply room has a large quantity of 20% saline solution and 40% saline solution, but needs 10 liters of 25% saline solution.

a. Write a system of equations to represent this situation.

b. Solve the system of equations. Interpret the solution.

H.O.T. Problems Use **H**igher-**O**rder **T**hinking Skills

50. OPEN ENDED Create a system of 3 variable equations that has infinitely many solutions. Explain your reasoning.

51. CHALLENGE Consider the following system of equations. What value of k would make the system consistent and independent?

$$3x + 2y = 5$$
$$5y - kz = -22$$
$$2x + 5z = 26$$
$$-2x + ky + z = -8$$

52. ERROR ANALYSIS Ken and Sari are writing the augmented matrix of the system below in row-echelon form.

$$2x - y + z = 0$$
$$x + y - 2z = -7$$
$$x - 3y + 4z = 9$$

Ken
$\begin{bmatrix} 1 & 1 & -2 & | & -7 \\ 0 & 1 & -1 & | & -2 \\ 0 & 0 & 1 & | & 4 \end{bmatrix}$

Sari
$\begin{bmatrix} 1 & 3 & -4 & | & -11 \\ 0 & 1 & -1 & | & -2 \\ 0 & 0 & 1 & | & 4 \end{bmatrix}$

Is either of them correct? Explain your reasoning.

53. REASONING *True* or *false*: If an augmented square matrix in row-echelon form has a row of zeros as its last row, then the corresponding system of equations has no solution. Explain your reasoning.

54. CHALLENGE A parabola passes through the three points shown on the graph below.

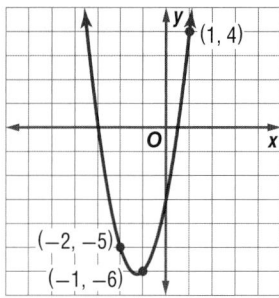

a. Write a system of equations that can be used to find the equation of the parabola in the form $f(x) = ax^2 + bx + c$.

b. Use matrices to solve the system of equations that you wrote in part **a.**

c. Use the solution you found in part **b** to write an equation of the parabola. Verify your results using a graphing calculator.

55. WRITING IN MATH Compare and contrast Gaussian elimination and Gauss-Jordan elimination.

Verify each identity. (Lesson 5-5)

56. $2 \cos^2 \frac{x}{2} = 1 + \cos x$

57. $\tan^2 \frac{x}{2} = \frac{1 - \cos x}{1 + \cos x}$

58. $\frac{1}{\sin x \cos x} - \frac{\cos x}{\sin x} = \tan x$

Find the exact value of each trigonometric expression. (Lesson 5-4)

59. $\cos 105°$

60. $\sin 165°$

61. $\cos \frac{7\pi}{12}$

62. $\sin \frac{\pi}{12}$

63. $\cot \frac{113\pi}{12}$

64. $\sec 1275°$

65. SOFTBALL In slow-pitch softball, the diamond is a square that is 65 feet on each side. The distance between the pitcher's mound and home plate is 50 feet. How far does the pitcher have to throw the softball from the pitcher's mound to third base to stop a player who is trying to steal third base? (Lesson 4-7)

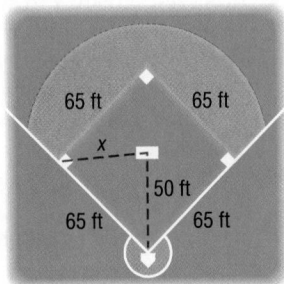

66. TRAVEL In a sightseeing boat near the base of the Horseshoe Falls at Niagara Falls, a passenger estimates the angle of elevation to the top of the falls to be 30°. If the Horseshoe Falls are 173 feet high, what is the distance from the boat to the base of the falls? (Lesson 4-1)

67. RABBITS Rabbits reproduce at a tremendous rate and their population increases exponentially in the absence of natural enemies. Suppose there were originally 65,000 rabbits in a region, and two years later there were 2,500,000. (Lesson 3-1)

   a. Write an exponential function that could be used to model the rabbit population y in that region. Write the function in terms of x, the number of years since the original year.

   b. Assume that the rabbit population continued to grow at that rate. Estimate the rabbit population in that region seven years after the initial year.

Solve each equation. (Lesson 2-5)

68. $\frac{3}{x} + \frac{2}{x-1} = \frac{17}{12}$

69. $\frac{4}{x+3} - \frac{2}{x+1} = \frac{2}{15}$

70. $\frac{4}{3} - \frac{1}{x-2} = \frac{13}{2x}$

71. SAT/ACT $\triangle ACF$ is equilateral with sides of length 4. If B, D, and E are the midpoints of their respective sides, what is the sum of the areas of the shaded regions?

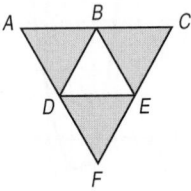

   A $3\sqrt{2}$ **C** $4\sqrt{2}$ **E** $6\sqrt{3}$

   B $3\sqrt{3}$ **D** $4\sqrt{3}$

72. REVIEW The caterer for a lunch bought several pounds of chicken and tuna salad. The chicken salad cost $9 per pound, and the tuna salad cost $6 per pound. He bought a total of 14 pounds of salad and paid a total of $111. How much chicken salad did the caterer buy?

   F 6 pounds   **H** 8 pounds

   G 7 pounds   **J** 9 pounds

73. The Yogurt Shoppe sells small cones for $0.89, medium cones for $1.19, and large cones for $1.39. One day, Scott sold 52 cones. He sold seven more medium cones than small cones. If he sold $58.98 in cones, how many medium cones did he sell?

   A 11   **C** 24

   B 17   **D** 36

74. REVIEW To practice at home, Tate purchased a basketball and a volleyball that cost a total of $67, not including tax. If the cost of the basketball b was $4 more than twice the cost of the volleyball v, which system of linear equations could be used to determine the cost of each ball?

   F $b + v = 67$
      $b = 2v - 4$

   H $b + v = 4$
      $b = 2v - 67$

   G $b + v = 67$
      $b = 2v + 4$

   J $b + v = 4$
      $b = 2v + 67$

Matrix Multiplication, Inverses, and Determinants

Then
You performed operations on matrices. (Lesson 0-5)

Now
- Multiply matrices.
- Find determinants and inverses of 2×2 and 3×3 matrices.

New Vocabulary
identity matrix
inverse matrix
inverse
invertible
singular matrix
determinant

Math Online
glencoe.com

Why?

Matrices are used in many industries as a convenient method to store data. In the restaurant business, matrix multiplication can be used to determine the amount of raw materials that are necessary to produce the desired final product, or items on the menu.

Multiply Matrices The three basic matrix operations are matrix addition, scalar multiplication, and matrix multiplication. You have seen that adding matrices is similar to adding real numbers, and multiplying a matrix by a scalar is similar to multiplying real numbers.

Matrix Addition

$$\begin{bmatrix} a_{11} & a_{12} & a_{13} \\ a_{21} & a_{22} & a_{23} \end{bmatrix} + \begin{bmatrix} b_{11} & b_{12} & b_{13} \\ b_{21} & b_{22} & b_{23} \end{bmatrix} = \begin{bmatrix} a_{11} + b_{11} & a_{12} + b_{12} & a_{13} + b_{13} \\ a_{21} + b_{21} & a_{22} + b_{22} & a_{23} + b_{23} \end{bmatrix}$$

Scalar Multiplication

$$k \begin{bmatrix} a_{11} & a_{12} & a_{13} \\ a_{21} & a_{22} & a_{23} \end{bmatrix} = \begin{bmatrix} ka_{11} & ka_{12} & ka_{13} \\ ka_{21} & ka_{22} & ka_{23} \end{bmatrix}$$

Matrix multiplication has no operational counterpart in the real number system. To multiply matrix A by matrix B, the number of columns in A *must* be equal to the number of rows in B. This can be determined by examining the dimensions of A and B. If it exists, product matrix AB has the same number of rows as A and the same number of columns as B.

$$\begin{array}{ccccc} \text{matrix } A & \cdot & \text{matrix } B & = & AB \\ 3 \times 2 & & 2 \times 4 & & 3 \times 4 \end{array}$$

Equal

Dimensions of *AB*

Key Concept   **Matrix Multiplication**

Words   If A is an $m \times r$ matrix and B is an $r \times n$ matrix, then the product AB is an $m \times n$ matrix obtained by adding the products of the entries of a row in A to the corresponding entries of a column in B.

Symbols   If A is an $m \times r$ matrix and B is an $r \times n$ matrix, then the product AB is an $m \times n$ matrix in which

$$c_{ij} = a_{i1}b_{1j} + a_{i2}b_{2j} + \cdots + a_{ir}b_{rj}.$$

$$\begin{bmatrix} a_{11} & a_{12} & \cdots & a_{1r} \\ a_{21} & a_{22} & \cdots & a_{2r} \\ \vdots & \vdots & & \vdots \\ a_{i1} & a_{i2} & \cdots & a_{ir} \\ \vdots & \vdots & & \vdots \\ a_{m1} & a_{m2} & \cdots & a_{mr} \end{bmatrix} \cdot \begin{bmatrix} b_{11} & b_{12} & \cdots & b_{1j} & \cdots & b_{1n} \\ b_{21} & b_{22} & \cdots & b_{2j} & \cdots & b_{2n} \\ \vdots & \vdots & & \vdots & & \vdots \\ b_{r1} & b_{r2} & \cdots & b_{rj} & \cdots & b_{rn} \end{bmatrix} = \begin{bmatrix} c_{11} & c_{12} & \cdots & c_{1j} & \cdots & c_{1n} \\ c_{21} & c_{22} & \cdots & c_{2j} & \cdots & c_{2n} \\ \vdots & \vdots & & \vdots & & \vdots \\ c_{i1} & c_{i2} & \cdots & c_{ij} & \cdots & c_{in} \\ \vdots & \vdots & & \vdots & & \\ c_{m1} & c_{m2} & \cdots & c_{mj} & \cdots & c_{mn} \end{bmatrix}$$

Each entry in the product of two matrices can be thought of as the product of a $1 \times r$ row matrix and an $r \times 1$ column matrix. Consider the product of the 1×3 row matrix and 3×1 column matrix shown.

$$\begin{bmatrix} -2 & 1 & 3 \end{bmatrix} \cdot \begin{bmatrix} 4 \\ -6 \\ 5 \end{bmatrix} = \begin{bmatrix} -2(4) + 1(-6) + 3(5) \end{bmatrix} \text{ or } \begin{bmatrix} 1 \end{bmatrix}$$

EXAMPLE 1   Multiply Matrices

Use matrices $A = \begin{bmatrix} 3 & -1 \\ 4 & 0 \end{bmatrix}$ and $B = \begin{bmatrix} -2 & 0 & 6 \\ 3 & 5 & 1 \end{bmatrix}$ to find each product, if possible.

a. AB

$$AB = \begin{bmatrix} 3 & -1 \\ 4 & 0 \end{bmatrix} \cdot \begin{bmatrix} -2 & 0 & 6 \\ 3 & 5 & 1 \end{bmatrix} \qquad \text{Dimensions of } A: 2 \times 2, \text{ Dimensions of } B: 2 \times 3$$

A is a 2×2 matrix and B is a 2×3 matrix. Because the number of columns for A is equal to the number of rows for B, the product AB exists.

To find the first entry in AB, write the sum of the products of the entries in row 1 of A and in column 1 of B.

$$\begin{bmatrix} 3 & -1 \\ 4 & 0 \end{bmatrix} \cdot \begin{bmatrix} -2 & 0 & 6 \\ 3 & 5 & 1 \end{bmatrix} = \begin{bmatrix} 3(-2) + (-1)(3) & & \\ & & \end{bmatrix}$$

Follow this same procedure to find the entry for row 1, column 2 of AB.

$$\begin{bmatrix} 3 & -1 \\ 4 & 0 \end{bmatrix} \cdot \begin{bmatrix} -2 & 0 & 6 \\ 3 & 5 & 1 \end{bmatrix} = \begin{bmatrix} 3(-2) + (-1)(3) & 3(0) + (-1)(5) & \\ & & \end{bmatrix}$$

Continue multiplying each row by each column to find the sum for each entry.

$$\begin{bmatrix} 3 & -1 \\ 4 & 0 \end{bmatrix} \cdot \begin{bmatrix} -2 & 0 & 6 \\ 3 & 5 & 1 \end{bmatrix} = \begin{bmatrix} 3(-2) + (-1)(3) & 3(0) + (-1)(5) & 3(6) + (-1)(1) \\ 4(-2) + 0(3) & 4(0) + 0(5) & 4(6) + 0(1) \end{bmatrix}$$

Finally, simplify each sum.

$$\begin{bmatrix} 3 & -1 \\ 4 & 0 \end{bmatrix} \cdot \begin{bmatrix} -2 & 0 & 6 \\ 3 & 5 & 1 \end{bmatrix} = \begin{bmatrix} -9 & -5 & 17 \\ -8 & 0 & 24 \end{bmatrix}$$

b. BA

$$BA = \begin{bmatrix} -2 & 0 & 6 \\ 3 & 5 & 1 \end{bmatrix} \cdot \begin{bmatrix} 3 & -1 \\ 4 & 0 \end{bmatrix} \qquad \text{Dimensions of } B: 2 \times 3, \text{ Dimensions of } A: 2 \times 2$$

Because the number of columns for B is not the same as the number of rows for A, the product BA does *not* exist. BA is undefined.

✓ Guided Practice

Find AB and BA, if possible.

1A. $A = \begin{bmatrix} 3 & 1 & -5 \\ -2 & 0 & 4 \end{bmatrix}$

$B = \begin{bmatrix} -6 & 1 & 7 \\ 4 & -5 & 3 \end{bmatrix}$

1B. $A = \begin{bmatrix} -2 & 0 & 3 \\ 5 & -7 & 1 \end{bmatrix}$

$B = \begin{bmatrix} 2 & 0 \\ -1 & 0 \\ 9 & 3 \end{bmatrix}$

Notice in Example 1 that the two products AB and BA are different. In most cases, even when both products are defined, $AB \neq BA$. This means that the Commutative Property does not hold for matrix multiplication. However, some of the properties of real numbers do hold for matrix multiplication.

Key Concept — Properties of Matrix Multiplication

For any matrices A, B, and C for which the matrix product is defined and any scalar k, the following properties are true.

Associative Property of Matrix Multiplication	$(AB)C = A(BC)$
Associative Property of Scalar Multiplication	$k(AB) = (kA)B = A(kB)$
Left Distributive Property	$C(A + B) = CA + CB$
Right Distributive Property	$(A + B)C = AC + BC$

You will prove these properties in Exercises 72–75.

Matrix multiplication can be used to solve real-world problems.

🌐 Real-World EXAMPLE 2 — Multiply Matrices

VOTING The percent of voters of different ages who were registered as Democrats, Republicans, or Independents in a recent city election are shown. Use this information to determine whether there were more male voters registered as Democrats than there were female voters registered as Republicans.

Distribution by Party and Age (%)

Party	18–25	26–40	41–50	50+
Democrat	0.55	0.50	0.35	0.40
Republican	0.30	0.40	0.45	0.55
Independent	0.15	0.10	0.20	0.05

Distribution by Age and Gender

Age	Female	Male
18–25	18,500	16,000
26–40	20,000	24,000
41–50	24,500	22,500
50+	16,500	14,000

🌐 Real-World Link

In the 2008 election, Barack Obama received 66,882,230 votes or 53% of the popular vote.

Source: CNN

Let matrix X represent the distribution by party and age, and let matrix Y represent the distribution by age and gender. Then find the product XY.

$$XY = \begin{bmatrix} 0.55 & 0.50 & 0.35 & 0.40 \\ 0.30 & 0.40 & 0.45 & 0.55 \\ 0.15 & 0.10 & 0.20 & 0.05 \end{bmatrix} \cdot \begin{bmatrix} 18,500 & 16,000 \\ 20,000 & 24,000 \\ 24,500 & 22,500 \\ 16,500 & 14,000 \end{bmatrix} = \begin{bmatrix} 35,350 & 34,275 \\ 33,650 & 32,225 \\ 10,500 & 10,000 \end{bmatrix}$$

The product XY represents the distribution of male and female voters that were registered in each party. You can use the product matrix to find the number of male voters that were registered as Democrat and the number of female voters registered as Republican.

$$\begin{array}{c} \\ \text{Democrat} \\ \text{Republican} \\ \text{Independent} \end{array} \begin{array}{cc} \text{Female} & \text{Male} \\ \begin{bmatrix} 35,350 & 34,275 \\ 33,650 & 32,225 \\ 10,500 & 10,000 \end{bmatrix} \end{array}$$

More male voters were registered as Democrat than female voters registered as Republican because 34,275 > 33,650.

✔ Guided Practice

2. SALES The number of laptops that a company sold in the first three months of the year is shown, as well as the price per model during those months. Use this information to determine which model generated the most income for the first three months.

Month	Model 1	Model 2	Model 3
Jan.	150	250	550
Feb.	200	625	100
Mar.	600	100	350

Model	Jan.	Feb.	Mar.
1	$650	$575	$485
2	$800	$700	$775
3	$900	$1050	$925

You know that the *multiplicative identity* for real numbers is 1, because for any real number a, $a \cdot 1 = a$. The multiplicative identity $n \times n$ square matrices is called the **identity matrix**.

ReadingMath

Identity Matrix The notation I_n is used to represent the identity for $n \times n$ matrices. The notation I is used instead of I_2, I_3, I_4, etc., when the order of the identity is known.

Key Concept **Identity Matrix**

Words The identity matrix of order n, denoted I_n, is an $n \times n$ matrix consisting of all 1s on its main diagonal, from upper left to lower right, and 0s for all other elements.

Symbols
$$I_n = \begin{bmatrix} 1 & 0 & 0 & \cdots & 0 \\ 0 & 1 & 0 & \cdots & 0 \\ 0 & 0 & 1 & \cdots & 0 \\ \vdots & \vdots & \vdots & \ddots & \vdots \\ 0 & 0 & 0 & \cdots & 1 \end{bmatrix}$$

So, if A is an $n \times n$ matrix, then $AI_n = I_nA = A$. You may find an identity matrix as the left side of many augmented matrices in reduced row-echelon form. In general, if A is the coefficient matrix of a system of equations, X is the column matrix of variables, and B is the column matrix of constants, then you can write the system of equations as an equation of matrices.

System of Equations

$$a_{11}x_1 + a_{12}x_2 + a_{13}x_3 = b_1$$
$$a_{21}x_1 + a_{22}x_2 + a_{23}x_3 = b_2$$
$$a_{31}x_1 + a_{32}x_2 + a_{33}x_3 = b_3$$

Matrix Equation

$$\underbrace{\begin{bmatrix} a_{11} & a_{12} & a_{13} \\ a_{21} & a_{22} & a_{23} \\ a_{31} & a_{32} & a_{33} \end{bmatrix}}_{A} \cdot \underbrace{\begin{bmatrix} x_1 \\ x_2 \\ x_3 \end{bmatrix}}_{X} = \underbrace{\begin{bmatrix} b_1 \\ b_2 \\ b_3 \end{bmatrix}}_{B}$$

EXAMPLE 3 **Solve a System of Linear Equations**

Write the system of equations as a matrix equation, $AX = B$. Then use Gauss-Jordan elimination on the augmented matrix to solve the system.

$$-x_1 + x_2 - 2x_3 = 2$$
$$-2x_1 + 3x_2 - 4x_3 = 5$$
$$3x_1 - 4x_2 + 7x_3 = -1$$

Write the system matrix in form, $AX = B$.

$$\begin{bmatrix} -1 & 1 & -2 \\ -2 & 3 & -4 \\ 3 & -4 & 7 \end{bmatrix} \cdot \begin{bmatrix} x_1 \\ x_2 \\ x_3 \end{bmatrix} = \begin{bmatrix} 2 \\ 5 \\ -1 \end{bmatrix} \qquad A \cdot X = B$$

ReadingMath

Augmented Matrices The notation $[A \vdots B]$, read *A augmented with B*, represents the augmented matrix that results when matrix B is attached to matrix A.

Write the augmented matrix $[A \vdots B]$. Use Gauss-Jordan elimination to solve the system.

$$[A \vdots B] = \left[\begin{array}{ccc:c} -1 & 1 & -2 & 2 \\ -2 & 3 & -4 & 5 \\ 3 & -4 & 7 & -1 \end{array}\right] \qquad \text{Augmented matrix}$$

$$[I \vdots X] = \left[\begin{array}{ccc:c} 1 & 0 & 0 & -13 \\ 0 & 1 & 0 & 1 \\ 0 & 0 & 1 & 6 \end{array}\right] \qquad \text{Use elementary row operations to transform } A \text{ into } I.$$

$$X = \begin{bmatrix} x_1 \\ x_2 \\ x_3 \end{bmatrix} = \begin{bmatrix} -13 \\ 1 \\ 6 \end{bmatrix} \qquad \text{The solution of the equation is given by } X.$$

Therefore, the solution of the system of equations is $(-13, 1, 6)$.

✔ **Guided Practice**

Write each system of equations as a matrix equation, $AX = B$. Then use Gauss-Jordan elimination on the augmented matrix $[A \vdots B]$ to solve the system.

3A. $x_1 - 2x_2 - 3x_3 = 9$
$-4x_1 + x_2 + 8x_3 = -16$
$2x_1 + 3x_2 + 2x_3 = 6$

3B. $x_1 + x_2 + x_3 = 2$
$2x_1 - x_2 + 2x_3 = 4$
$-x_1 + 4x_2 + x_3 = 3$

Inverses and Determinants You know that if a is a nonzero real number, then $\frac{1}{a}$ or a^{-1} is the multiplicative inverse of a because $a\left(\frac{1}{a}\right) = a \cdot a^{-1} = 1$. The multiplicative inverse of a square matrix is called its **inverse matrix**.

Key Concept   Inverse of a Square Matrix

Let A be an $n \times n$ matrix. If there exists a matrix B such that $AB = BA = I_n$, then B is called the **inverse** of A and is written as A^{-1}. So, $AA^{-1} = A^{-1}A = I_n$.

EXAMPLE 4   Verify an Inverse Matrix

Determine whether $A = \begin{bmatrix} -3 & 2 \\ -2 & 1 \end{bmatrix}$ and $B = \begin{bmatrix} 1 & -2 \\ 2 & -3 \end{bmatrix}$ are inverse matrices.

If A and B are inverse matrices, then $AB = BA = I$.

$$AB = \begin{bmatrix} -3 & 2 \\ -2 & 1 \end{bmatrix} \cdot \begin{bmatrix} 1 & -2 \\ 2 & -3 \end{bmatrix} = \begin{bmatrix} -3+4 & 6+(-6) \\ -2+2 & 4+(-3) \end{bmatrix} \text{ or } \begin{bmatrix} 1 & 0 \\ 0 & 1 \end{bmatrix}$$

$$BA = \begin{bmatrix} 1 & -2 \\ 2 & -3 \end{bmatrix} \cdot \begin{bmatrix} -3 & 2 \\ -2 & 1 \end{bmatrix} = \begin{bmatrix} -3+4 & 2+(-2) \\ -6+6 & 4+(-3) \end{bmatrix} \text{ or } \begin{bmatrix} 1 & 0 \\ 0 & 1 \end{bmatrix}$$

Because $AB = BA = I$, it follows that $B = A^{-1}$ and $A = B^{-1}$.

✓ Guided Practice

Determine whether A and B are inverse matrices.

4A. $A = \begin{bmatrix} -4 & 3 \\ 3 & -2 \end{bmatrix}$, $B = \begin{bmatrix} 2 & 3 \\ 3 & 4 \end{bmatrix}$

4B. $A = \begin{bmatrix} 6 & 2 \\ 2 & 1 \end{bmatrix}$, $B = \begin{bmatrix} 1 & -2 \\ -2 & 6 \end{bmatrix}$

If a matrix A has an inverse, then A is said to be **invertible** or *nonsingular*. A **singular matrix** does not have an inverse. Not all square matrices are invertible. To find the inverse of a square matrix A, you need to find a matrix A^{-1}, assuming A^{-1} exists, such that the product of A and A^{-1} is the identity matrix. In other words, you need to solve the matrix equation $AA^{-1} = I_n$ for B. Once B is determined, you will then need to confirm that $AA^{-1} = A^{-1}A = I_n$.

One method for finding the inverse of a square matrix is to use a system of equations. Let $A = \begin{bmatrix} 8 & -5 \\ -3 & 2 \end{bmatrix}$, and suppose A^{-1} exists. Write the matrix equation $AA^{-1} = I_2$, where $A^{-1} = \begin{bmatrix} a & b \\ c & d \end{bmatrix}$.

$$\begin{bmatrix} 8 & -5 \\ -3 & 2 \end{bmatrix}\begin{bmatrix} a & b \\ c & d \end{bmatrix} = \begin{bmatrix} 1 & 0 \\ 0 & 1 \end{bmatrix} \qquad AA^{-1} = I_2$$

$$\begin{bmatrix} 8a - 5c & 8b - 5d \\ -3a + 2c & -3b + 2d \end{bmatrix} = \begin{bmatrix} 1 & 0 \\ 0 & 1 \end{bmatrix} \qquad \text{Matrix multiplication}$$

$$\begin{array}{ll} 8a - 5c = 1 & 8b - 5d = 0 \\ -3a + 2c = 0 & -3b + 2d = 1 \end{array} \qquad \text{Equate corresponding elements.}$$

From this set of four equations, you can see that there are two systems of equations that each have two unknowns. Write the corresponding augmented matrices.

$$\begin{bmatrix} 8 & -5 & \vdots & 1 \\ -3 & 2 & \vdots & 0 \end{bmatrix}\begin{bmatrix} 8 & -5 & \vdots & 0 \\ -3 & 2 & \vdots & 1 \end{bmatrix}$$

Notice that the augmented matrix of each system has the same coefficient matrix, $\begin{bmatrix} 8 & -5 \\ -3 & 2 \end{bmatrix}$.

Because the coefficient matrix of the systems is the same, we can perform row reductions on the two augmented matrices simultaneously by creating a *doubly augmented matrix*, $[A \vdots I]$. To find A^{-1}, use the doubly augmented matrix $\begin{bmatrix} 8 & -5 & \vdots & 1 & 0 \\ -3 & 2 & \vdots & 0 & 1 \end{bmatrix}$.

EXAMPLE 5 · Inverse of a Matrix

Find A^{-1}, if it exists. If A^{-1} does not exist, write *singular*.

a. $A = \begin{bmatrix} 8 & -5 \\ -3 & 2 \end{bmatrix}$

Step 1 Create the doubly augmented matrix $[A \vdots I]$.

$$[A \vdots I] = \begin{bmatrix} 8 & -5 & \vdots & 1 & 0 \\ -3 & 2 & \vdots & 0 & 1 \end{bmatrix} \qquad \textbf{Doubly augmented matrix}$$

Step 2 Apply elementary row operations to write the matrix in reduced row-echelon form.

$$\begin{bmatrix} 8 & -5 & \vdots & 1 & 0 \\ -3 & 2 & \vdots & 0 & 1 \end{bmatrix} \quad \overset{R_1 + 5R_2}{\longrightarrow} \quad \begin{bmatrix} 8 & 0 & \vdots & 16 & 40 \\ 0 & 1 & \vdots & 3 & 8 \end{bmatrix}$$

$$\overset{3R_1 + 8R_2}{\longrightarrow} \begin{bmatrix} 8 & -5 & \vdots & 1 & 0 \\ 0 & 1 & \vdots & 3 & 8 \end{bmatrix} \quad \bigg| \quad \overset{\frac{1}{8}R_1}{\longrightarrow} \begin{bmatrix} 1 & 0 & \vdots & 2 & 5 \\ 0 & 1 & \vdots & 3 & 8 \end{bmatrix} = [I \vdots A^{-1}]$$

The first two columns are the identity matrix. Therefore, A is invertible and $A^{-1} = \begin{bmatrix} 2 & 5 \\ 3 & 8 \end{bmatrix}$.

CHECK Confirm that $AA^{-1} = A^{-1}A = I$.

$$AA^{-1} = \begin{bmatrix} 8 & -5 \\ -3 & 2 \end{bmatrix}\begin{bmatrix} 2 & 5 \\ 3 & 8 \end{bmatrix} \qquad\qquad A^{-1}A = \begin{bmatrix} 2 & 5 \\ 3 & 8 \end{bmatrix}\begin{bmatrix} 8 & -5 \\ -3 & 2 \end{bmatrix}$$

$$= \begin{bmatrix} 1 & 0 \\ 0 & 1 \end{bmatrix} \text{ or } I ✔ \qquad\qquad\qquad = \begin{bmatrix} 1 & 0 \\ 0 & 1 \end{bmatrix} \text{ or } I ✔$$

TechnologyTip

Inverse You can use $\boxed{x^{-1}}$ on your graphing calculator to find the inverse of a square matrix.

```
[A]
        [[8  -5]
         [-3 2 ]]
[A]⁻¹
        [[2 5]
         [3 8]]
```

b. $A = \begin{bmatrix} 2 & 4 \\ -3 & -6 \end{bmatrix}$

Step 1 $[A \vdots I] = \begin{bmatrix} 2 & 4 & \vdots & 1 & 0 \\ -3 & -6 & \vdots & 0 & 1 \end{bmatrix}$ **Doubly augmented matrix**

Step 2 $\overset{\frac{1}{2}R_1}{\longrightarrow} \begin{bmatrix} 1 & 2 & \vdots & \frac{1}{2} & 0 \\ -3 & -6 & \vdots & 0 & 1 \end{bmatrix}$ Apply elementary row operations to write the matrix in reduced row-echelon form.

$$\overset{3R_1 + R_2}{\longrightarrow} \begin{bmatrix} 1 & 2 & \vdots & \frac{1}{2} & 0 \\ 0 & 0 & \vdots & \frac{3}{2} & 1 \end{bmatrix}$$ Notice that it is impossible to obtain the identity matrix *I* on the left-side of the doubly augmented matrix

Therefore, A is singular.

✓ Guided Practice

5A. $\begin{bmatrix} -1 & 2 \\ 3 & -6 \end{bmatrix}$ 　　 **5B.** $\begin{bmatrix} -1 & -2 & -2 \\ 3 & 7 & 9 \\ 1 & 4 & 7 \end{bmatrix}$ 　　 **5C.** $\begin{bmatrix} 4 & 2 \\ -6 & -3 \end{bmatrix}$

The process used to find the inverse of a square matrix is summarized below.

TechnologyTip

Singular Matrices If a matrix is singular, your graphing calculator will display the following error message.
ERR: SINGULAR MAT

Concept Summary · Finding the Inverse of a Square Matrix

Let A be an $n \times n$ matrix.

1. Write the augmented matrix $[A \vdots I_n]$.

2. Perform elementary row operations on the augmented matrix to reduce A to its reduced row-echelon form.

3. Decide whether A is invertible.
   - If A can be reduced to the identity matrix I_n, then A^{-1} is the matrix on the right of the transformed augmented matrix, $[I_n \vdots A^{-1}]$.
   - If A cannot be reduced to the identity matrix I_n, then A is singular.

While the method of finding an inverse matrix used in Example 5 works well for any square matrix, you may find the following formula helpful when finding the inverse of a 2 × 2 matrix.

Key Concept | **Inverse and Determinant of a 2 × 2 Matrix**

Let $A = \begin{bmatrix} a & b \\ c & d \end{bmatrix}$. A is invertible if and only if $ad - cb \neq 0$.

If A is invertible, then $A^{-1} = \dfrac{1}{ad - cb} \begin{bmatrix} d & -b \\ -c & a \end{bmatrix}$.

The number $ad - cb$ is called the **determinant** of the 2 × 2 matrix and is denoted

$$\det(A) = |A| = \begin{vmatrix} a & b \\ c & d \end{vmatrix} = ad - cb.$$

You will prove this Theorem in Exercise 66.

Therefore, the determinant of a 2 × 2 matrix provides a test for determining if the matrix is invertible.

Notice that the determinant of a 2 × 2 matrix is the difference of the product of the two diagonals of the matrix.

$$\det(A) = \begin{vmatrix} a & b \\ c & d \end{vmatrix} = ad - cb.$$

StudyTip

Inverse of a 2 × 2 Matrix The formula for the inverse of a 2 × 2 matrix is sometimes written as

$A^{-1} = \dfrac{1}{\det(A)} \begin{bmatrix} d & -b \\ -c & a \end{bmatrix}$.

EXAMPLE 6 | **Determinant and Inverse of a 2 × 2 Matrix**

Find the determinant of each matrix. Then find the inverse of the matrix, if it exists.

a. $A = \begin{bmatrix} 2 & -3 \\ 4 & 4 \end{bmatrix}$

$\det(A) = \begin{vmatrix} 2 & -3 \\ 4 & 4 \end{vmatrix}$ $a = 2, b = -3, c = 4$, and $d = 4$

$= 2(4) - 4(-3)$ or 20 $ad - cb$

Because $\det(A) \neq 0$, A is invertible. Apply the formula for the inverse of a 2 × 2 matrix.

$A^{-1} = \dfrac{1}{ad - cb} \begin{bmatrix} d & -b \\ -c & a \end{bmatrix}$ **Inverse of 2 × 2 matrix**

$= \dfrac{1}{20} \begin{bmatrix} 4 & 3 \\ -4 & 2 \end{bmatrix}$ $a = 2, b = -3, c = 4, d = 4$, and $ad - cb = 20$

$= \begin{bmatrix} \dfrac{1}{5} & \dfrac{3}{20} \\ -\dfrac{1}{5} & \dfrac{1}{10} \end{bmatrix}$ **Scalar multiplication**

CHECK $AA^{-1} = A^{-1}A = \begin{bmatrix} 1 & 0 \\ 0 & 1 \end{bmatrix}$ ✔

b. $B = \begin{bmatrix} 6 & 4 \\ 9 & 6 \end{bmatrix}$

$\det(B) = \begin{vmatrix} 6 & 4 \\ 9 & 6 \end{vmatrix} = 6(6) - 9(4)$ or 0

Because $\det(B) = 0$, B is not invertible. Therefore, B^{-1} does not exist.

✓ **Guided Practice**

6A. $\begin{bmatrix} -4 & 6 \\ 8 & -12 \end{bmatrix}$

6B. $\begin{bmatrix} 2 & -3 \\ -2 & -2 \end{bmatrix}$

The determinant for a 3 × 3 matrix is defined using 2 × 2 determinants as shown.

TechnologyTip

Determinants You can use the det(function on a graphing calculator to find the determinant of a square matrix. If you try to find the determinant of a matrix with dimensions other than $n \times n$, your calculator will display the following error message.

ERR:INVALID DIM

Key Concept   **Determinant of a 3 × 3 Matrix**

Let $A = \begin{bmatrix} a & b & c \\ d & e & f \\ g & h & i \end{bmatrix}$. Then $\det(A) = |A| = a\begin{vmatrix} e & f \\ h & i \end{vmatrix} - b\begin{vmatrix} d & f \\ g & i \end{vmatrix} + c\begin{vmatrix} d & e \\ g & h \end{vmatrix}$.

As with 2 × 2 matrices, a 3 × 3 matrix A has an inverse if and only if $\det(A) \neq 0$. A formula for calculating the inverse of 3 × 3 and higher order matrices exists. However, due to the complexity of this formula, we will use a graphing calculator to calculate the inverse of 3 × 3 and higher-order square matrices.

EXAMPLE 7   **Determinant and Inverse of a 3 × 3 Matrix**

Find the determinant of $C = \begin{bmatrix} -3 & 2 & 4 \\ 1 & -1 & 2 \\ -1 & 4 & 0 \end{bmatrix}$. Then find C^{-1}, if it exists.

$\det(C) = \begin{vmatrix} -3 & 2 & 4 \\ 1 & -1 & 2 \\ -1 & 4 & 0 \end{vmatrix}$

$= -3\begin{vmatrix} -1 & 2 \\ 4 & 0 \end{vmatrix} - 2\begin{vmatrix} 1 & 2 \\ -1 & 0 \end{vmatrix} + 4\begin{vmatrix} 1 & -1 \\ -1 & 4 \end{vmatrix}$

$= -3[(-1)(0) - 4(2)] - 2[1(0) - (-1)(2)] + 4[1(4) - (-1)(-1)]$

$= -3(-8) - 2(2) + 4(3) \text{ or } 32$

Because $\det(A)$ does not equal zero, C^{-1} exists. Use a graphing calculator to find C^{-1}.

```
[C]⁻¹
[[-.25   .5    …
 [-.0625 .125  …
 [.09375 .3125 …
```

```
[C]⁻¹
…  .5    .25  ]
…  .125  .3125]
…  .3125 .03125]]
```

You can use the ▶Frac feature under the **MATH** menu to write the inverse using fractions, as shown below.

```
…  .5    .25  ]
…  .125  .3125]
…  .3125 .03125]]
Ans▶Frac
[[-1/4  1/2   1/…
 [-1/16 1/8   5/…
 [3/32  5/16  1/…
```

```
…  .5    .25  ]
…  .125  .3125]
…  .3125 .03125]]
Ans▶Frac
…/4   1/2   1/4 ]
…/16  1/8   5/16]
…32   5/16  1/32]]
```

Therefore, $C^{-1} = \begin{bmatrix} -\dfrac{1}{4} & \dfrac{1}{2} & \dfrac{1}{4} \\[6pt] -\dfrac{1}{16} & \dfrac{1}{8} & \dfrac{5}{16} \\[6pt] \dfrac{3}{32} & \dfrac{5}{16} & \dfrac{1}{32} \end{bmatrix}$.

✔ **Guided Practice**

Find the determinant of each matrix. Then find its inverse, if it exists.

7A. $\begin{bmatrix} 3 & 1 & 2 \\ 1 & 2 & -1 \\ 2 & -1 & 3 \end{bmatrix}$ **7B.** $\begin{bmatrix} -1 & -2 & 1 \\ 4 & 0 & 3 \\ -3 & 1 & -2 \end{bmatrix}$

Find *AB* and *BA*, if possible. (Example 1)

1. $A = \begin{bmatrix} 8 & 1 \end{bmatrix}$

$B = \begin{bmatrix} 3 & -7 \\ -5 & 2 \end{bmatrix}$

2. $A = \begin{bmatrix} 2 & 9 \\ -7 & 3 \end{bmatrix}$

$B = \begin{bmatrix} 6 & -4 \\ 0 & 3 \end{bmatrix}$

3. $A = \begin{bmatrix} 3 & -5 \end{bmatrix}$

$B = \begin{bmatrix} 4 & 0 & -2 \\ 1 & -3 & 2 \end{bmatrix}$

4. $A = \begin{bmatrix} 4 \\ 5 \end{bmatrix}$

$B = \begin{bmatrix} 6 & 1 & -10 & 9 \end{bmatrix}$

5. $A = \begin{bmatrix} 2 \\ 5 \\ -6 \end{bmatrix}$

$B = \begin{bmatrix} 6 & 0 & -1 \\ -4 & 9 & 8 \end{bmatrix}$

6. $A = \begin{bmatrix} 2 & 0 \\ -4 & -3 \\ 1 & -2 \end{bmatrix}$

$B = \begin{bmatrix} 0 & 6 & -5 \\ 2 & -7 & 1 \end{bmatrix}$

7. $A = \begin{bmatrix} 3 & 4 \\ -7 & 1 \end{bmatrix}$

$B = \begin{bmatrix} 5 & 2 & -8 \\ -6 & 0 & 9 \end{bmatrix}$

8. $A = \begin{bmatrix} 6 & -9 & 10 \\ 4 & 3 & 8 \end{bmatrix}$

$B = \begin{bmatrix} 6 & -8 \\ 3 & -9 \\ -2 & 5 \\ 4 & 1 \end{bmatrix}$

9 **BASKETBALL** Different point values are awarded for different shots in basketball. Use the information to determine the total amount of points scored by each player. (Example 2)

Player	FT	2-pointer	3-pointer
Rey	44	32	25
Chris	37	24	31
Jerry	35	39	29

Shots	Points
free throw	1
2-pointer	2
3-pointer	3

10. CARS The number of vehicles that a company manufactures each day from two different factories is shown, as well as the price of the vehicle during each sales quarter of the year. Use this information to determine which factory produced the highest sales in the 4th quarter. (Example 2)

Factory	Model			
	Coupe	Sedan	SUV	Mini Van
1	500	600	150	250
2	250	350	250	400

Model	Quarter			
	1st ($)	2nd ($)	3rd ($)	4th ($)
Coupe	18,700	17,100	16,200	15,600
Sedan	25,400	24,600	23,900	23,400
SUV	36,300	35,500	34,900	34,500
Mini Van	38,600	37,900	37,400	36,900

Write each system of equations as a matrix equation, *AX = B*. Then use Gauss-Jordan elimination on the augmented matrix to solve the system. (Example 3)

11. $\begin{aligned} 2x_1 - 5x_2 + 3x_3 &= 9 \\ 4x_1 + x_2 - 6x_3 &= 35 \\ -3x_1 + 9x_2 - 7x_3 &= -6 \end{aligned}$

12. $\begin{aligned} 3x_1 - 10x_2 - x_3 &= 6 \\ -5x_1 + 12x_2 + 2x_3 &= -5 \\ -4x_1 - 8x_2 + 3x_3 &= 16 \end{aligned}$

13. $\begin{aligned} 2x_1 - 10x_2 + 7x_3 &= 7 \\ 6x_1 - x_2 + 5x_3 &= -2 \\ -4x_1 + 8x_2 - 3x_3 &= -22 \end{aligned}$

14. $\begin{aligned} x_1 + 5x_2 + 5x_3 &= -18 \\ -7x_1 - 3x_2 + 2x_3 &= -3 \\ 6x_1 + 7x_2 - x_3 &= 42 \end{aligned}$

15. $\begin{aligned} 2x_1 + 6x_2 - 5x_3 &= -20 \\ 8x_1 - 12x_2 + 7x_3 &= 28 \\ -4x_1 + 10x_2 - x_3 &= 7 \end{aligned}$

16. $\begin{aligned} 3x_1 - 5x_2 + 12x_3 &= 9 \\ 2x_1 + 4x_2 - 11x_3 &= 1 \\ -5x_1 + 7x_2 - 15x_3 &= -28 \end{aligned}$

17. $\begin{aligned} -x_1 - 3x_2 + 9x_3 &= 25 \\ -5x_1 + 11x_2 + 8x_3 &= 33 \\ 2x_1 + x_2 - 13x_3 &= -45 \end{aligned}$

18. $\begin{aligned} x_1 - 8x_2 - 3x_3 &= -4 \\ -3x_1 + 10x_2 + 5x_3 &= -42 \\ 2x_1 + 7x_2 + 3x_3 &= 20 \end{aligned}$

Determine whether *A* and *B* are inverse matrices. (Example 4)

19. $A = \begin{bmatrix} 12 & -7 \\ -5 & 3 \end{bmatrix}$

$B = \begin{bmatrix} 3 & 7 \\ 5 & 12 \end{bmatrix}$

20. $A = \begin{bmatrix} 4 & -5 \\ 5 & -6 \end{bmatrix}$

$B = \begin{bmatrix} -6 & 5 \\ -5 & 4 \end{bmatrix}$

21. $A = \begin{bmatrix} -5 & 3 \\ 6 & -4 \end{bmatrix}$

$B = \begin{bmatrix} 4 & 3 \\ 6 & 5 \end{bmatrix}$

22. $A = \begin{bmatrix} -8 & 4 \\ 6 & -3 \end{bmatrix}$

$B = \begin{bmatrix} 3 & 4 \\ 6 & 8 \end{bmatrix}$

23. $A = \begin{bmatrix} 9 & 2 \\ 5 & 1 \end{bmatrix}$

$B = \begin{bmatrix} -1 & 2 \\ 5 & -9 \end{bmatrix}$

24. $A = \begin{bmatrix} 7 & 5 \\ -6 & -4 \end{bmatrix}$

$B = \begin{bmatrix} -4 & -5 \\ 6 & 7 \end{bmatrix}$

25. $A = \begin{bmatrix} 2 & -3 \\ -3 & 4 \end{bmatrix}$

$B = \begin{bmatrix} -4 & -3 \\ -3 & -2 \end{bmatrix}$

26. $A = \begin{bmatrix} 9 & -7 \\ 8 & -5 \end{bmatrix}$

$B = \begin{bmatrix} 1 & -6 \\ 4 & 10 \end{bmatrix}$

Find A^{-1}, if it exists. If A^{-1} does not exist, write *singular*. (Example 5)

27. $A = \begin{bmatrix} -4 & 2 \\ -6 & 3 \end{bmatrix}$

28. $A = \begin{bmatrix} -4 & 8 \\ 1 & -2 \end{bmatrix}$

29. $A = \begin{bmatrix} 3 & 5 \\ -2 & -3 \end{bmatrix}$

30. $A = \begin{bmatrix} 8 & 5 \\ 6 & 4 \end{bmatrix}$

31. $A = \begin{bmatrix} -1 & -1 & -3 \\ 3 & 6 & 4 \\ 2 & 1 & 8 \end{bmatrix}$

32. $A = \begin{bmatrix} 4 & 2 & 1 \\ -2 & 3 & 5 \\ 6 & -1 & -4 \end{bmatrix}$

33. $A = \begin{bmatrix} 5 & 2 & -1 \\ 4 & 7 & -3 \\ 1 & -5 & 2 \end{bmatrix}$

34. $A = \begin{bmatrix} 2 & 3 & -4 \\ 3 & 6 & -5 \\ -2 & -8 & 1 \end{bmatrix}$

Find the determinant of each matrix. Then find the inverse of the matrix, if it exists. (Examples 6 and 7)

35. $\begin{bmatrix} 6 & -5 \\ 3 & -2 \end{bmatrix}$

36. $\begin{bmatrix} -2 & 7 \\ 1 & 8 \end{bmatrix}$

37. $\begin{bmatrix} -4 & -7 \\ 6 & 9 \end{bmatrix}$

38. $\begin{bmatrix} 12 & -9 \\ -4 & 3 \end{bmatrix}$

39. $\begin{bmatrix} 3 & 1 & -2 \\ 8 & -5 & 2 \\ -4 & 3 & -1 \end{bmatrix}$

40. $\begin{bmatrix} 1 & -1 & -2 \\ 5 & 9 & 3 \\ 2 & 7 & 4 \end{bmatrix}$

41. $\begin{bmatrix} 9 & 3 & 7 \\ -6 & -2 & -5 \\ 3 & 1 & 4 \end{bmatrix}$

42. $\begin{bmatrix} 2 & 3 & -1 \\ -4 & -5 & 2 \\ 6 & 1 & 3 \end{bmatrix}$

43. $\begin{bmatrix} -1 & 3 & 2 \\ 3 & -5 & -3 \\ 4 & 2 & 6 \end{bmatrix}$

44. $\begin{bmatrix} 6 & -1 & 2 \\ 1 & -2 & -4 \\ -3 & 1 & -5 \end{bmatrix}$

Find the area A of each triangle with vertices (x_1, y_1), (x_2, y_2), and (x_3, y_3), by using $A = \frac{1}{2} |\det(X)|$,

where X is $\begin{bmatrix} x_1 & y_1 & 1 \\ x_2 & y_2 & 1 \\ x_3 & y_3 & 1 \end{bmatrix}$.

45

46.

47.

48.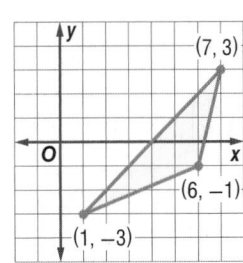

Given A and AB, find B.

49. $A = \begin{bmatrix} 8 & -4 \\ 3 & 6 \end{bmatrix}$, $AB = \begin{bmatrix} 36 & 48 \\ -24 & 48 \end{bmatrix}$

50. $A = \begin{bmatrix} 5 & 0 & 1 \\ 2 & -3 & 2 \\ 1 & -1 & 4 \end{bmatrix}$, $AB = \begin{bmatrix} 1 & 4 \\ -16 & -6 \\ -2 & -5 \end{bmatrix}$

Find x and y.

51. $A = \begin{bmatrix} 2x & -y \\ -3y & 5x \end{bmatrix}$, $B = \begin{bmatrix} 4 \\ -2 \end{bmatrix}$, and $AB = \begin{bmatrix} -2 \\ 31 \end{bmatrix}$

Find the determinant of each matrix.

52. $\begin{bmatrix} r & 0 & 0 \\ 0 & s & 0 \\ 0 & 0 & t \end{bmatrix}$

53. $\begin{bmatrix} c & c & c \\ 0 & c & c \\ 0 & 0 & c \end{bmatrix}$

54. FUNDRAISER Hawthorne High School had a fundraiser selling popcorn. The school bought the four flavors of popcorn by the case. The prices paid for the different types of popcorn and the selling prices are shown.

Class	Cases of Popcorn			
	Butter	Kettle	Cheese	Caramel
freshman	152	80	125	136
sophomore	112	92	112	150
junior	176	90	118	122
senior	140	102	106	143

Flavor	Price Paid per Case ($)	Selling Price per Case ($)	Profit per Case ($)
butter	18.90	42.00	
kettle	21.00	45.00	
cheese	23.10	48.00	
caramel	25.20	51.00	

a. Complete the last column of the second table.

b. Which class had the highest total sales?

c. How much more profit did the seniors earn than the sophomores?

55. Consider $A = \begin{bmatrix} 1 & 1 \\ 0 & 1 \end{bmatrix}$, $B = \begin{bmatrix} 1 & 1 \\ 1 & 1 \end{bmatrix}$, and $C = \begin{bmatrix} 2 & 0 \\ 2 & 2 \end{bmatrix}$.

a. Find A^2, A^3, and A^4. Then use the pattern to write a matrix for A^n.

b. Find B^2, B^3, B^4, and B^5. Then use the pattern to write a general formula for B^n.

c. Find C^2, C^3, C^4, C^5, ... until you notice a pattern. Then use the pattern to write a general formula for C^n.

d. Use the formula that you wrote in part **c** to find C^7.

56. HORSES The owner of each horse stable listed below buys bales of hay and bags of feed each month. In May, hay cost $2.50 per bale and feed cost $7.95 per bag. In June, the cost per bale of hay was $3.00 and the cost per bag of feed was $6.75.

Stables	Bales of Hay	Bags of Feed
Galloping Hills	45	5
Amazing Acres	75	9
Fairwind Farms	135	16
Saddle-Up Stables	90	11

a. Write a matrix X to represent the bales of hay i and bags of feed j that are bought monthly by each stable.

b. Write a matrix Y to represent the costs per bale of hay and bags of feed for May and June.

c. Find the product YX, and label its rows and columns.

d. How much more were the total costs in June for Fairwind Farms than the total costs in May for Galloping Hills?

Evaluate each expression.

$$A = \begin{bmatrix} 4 & 1 & -3 \\ 0 & 2 & 8 \end{bmatrix} \qquad C = \begin{bmatrix} -1 & 9 & -6 \\ 7 & 5 & 0 \end{bmatrix}$$

$$B = \begin{bmatrix} 2 & 1 \\ 0 & 1 \\ 3 & -2 \end{bmatrix} \qquad D = \begin{bmatrix} 7 & 2 \\ -4 & -1 \end{bmatrix}$$

57. $BD + B$

58. $DC - A$

59. $B(A + C)$

60. $AB + CB$

Solve each equation for X, if possible.

$$A = \begin{bmatrix} 1 & 7 \\ 2 & 0 \end{bmatrix} \qquad C = \begin{bmatrix} 5 & 3 \\ -6 & -2 \end{bmatrix}$$

$$B = \begin{bmatrix} 8 & 4 \\ 1 & -3 \\ 6 & 7 \end{bmatrix} \qquad D = \begin{bmatrix} 2 & 1 \\ 0 & 1 \\ 3 & -2 \end{bmatrix}$$

61. $A + C = 2X$

62. $4X + A = C$

63. $B - 3X = D$

64. $DA = 7X$

65. 3 × 3 DETERMINANTS In this problem, you will investigate an alternative method for calculating the determinant of a 3 × 3 matrix.

a. Calculate $\det(A) = \begin{vmatrix} -2 & 3 & 1 \\ 4 & 6 & 5 \\ 0 & 2 & 1 \end{vmatrix}$ using the method shown in this lesson.

b. Adjoin the first two columns to the right of $\det(A)$ as shown. Then find the difference between the sum of the products along the indicated downward diagonals and the sum of the products along the indicated upward diagonals.

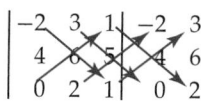

c. Compare your answers in parts **a** and **b**.

d. Show that, in general, the determinant of a 3 × 3 matrix can be found using the procedure described above.

e. Does this method work for a 4 × 4 matrix? If so, explain your reasoning. If not, provide a counterexample.

66. PROOF Suppose $A = \begin{bmatrix} a & b \\ c & d \end{bmatrix}$ and $A^{-1} = \begin{bmatrix} x_1 & y_1 \\ x_2 & y_2 \end{bmatrix}$.

Use the matrix equation $AA^{-1} = I_2$ to derive the formula for the inverse of a 2 × 2 matrix.

67. PROOF Write a paragraph proof to show that if a square matrix has an inverse, that inverse is unique. (*Hint:* Assume that a square matrix A has inverses B and C. Then show that $B = C$.)

68. **MULTIPLE REPRESENTATIONS** In this problem, you will explore square matrices. A square matrix is called *upper triangular* if all elements below the main diagonal are 0, and *lower triangular* if all elements above the main diagonal are 0. If all elements not on the diagonal of a matrix are 0, then the matrix is called *diagonal*. In this problem, you will investigate the determinants of 3 × 3 upper triangular, lower triangular, and diagonal matrices.

a. **ANALYTICAL** Write one upper triangular, one lower triangular, and one diagonal 2 × 2 matrix. Then find the determinant of each matrix.

b. **ANALYTICAL** Write one upper triangular, one lower triangular, and one diagonal 3 × 3 matrix. Then find the determinant of each matrix.

c. **VERBAL** Make a conjecture as to the value of the determinant of any 3 × 3 upper triangular, lower triangular, or diagonal matrix.

d. **ANALYTICAL** Find the inverse of each of the diagonal matrices you wrote in part **a** and **b**.

e. **VERBAL** Make a conjecture about the inverse of any 3 × 3 diagonal matrix.

H.O.T. Problems   Use Higher-Order Thinking Skills

69 CHALLENGE Given A and AB, find B.

$$A = \begin{bmatrix} 3 & -1 & 5 \\ 1 & 0 & 2 \\ 6 & 4 & 1 \end{bmatrix}, AB = \begin{bmatrix} 14 & 6 & 33 \\ 4 & 4 & 13 \\ 1 & 18 & 12 \end{bmatrix}$$

70. REASONING Explain why a nonsquare matrix cannot have an inverse.

71. OPEN ENDED Write two matrices A and B such that $AB = BA$, but neither A nor B is the identity matrix.

PROOF Show that each property is true for all 2 × 2 matrices.

72. Right Distributive Property

73. Left Distributive Property

74. Associative Property of Matrix Multiplication

75. Associative Property of Scalar Multiplication

76. ERROR ANALYSIS Alexis and Paul are discussing determinants. Alexis theorizes that the determinant of a 2 × 2 matrix A remains unchanged if two rows of the matrix are interchanged. Paul theorizes that the determinant of the new matrix will have the same absolute value but will be different in sign. Is either of them correct? Explain your reasoning.

77. REASONING If AB has dimensions 5 × 8, with A having dimensions 5 × 6, what are the dimensions of B?

78. WRITING IN MATH Explain why order is important when finding the product of two matrices A and B. Give some general examples to support your answer.

Write the augmented matrix for each system of linear equations. (Lesson 6-1)

79. $10x - 3y = -12$
$-6x + 4y = 20$

80. $15x + 7y - 2z = 41$
$9x - 8y + z = 32$
$5x + y - 11z = 36$

81. $w - 6x + 14y = 19$
$3w + 2x - 4y + 8z = -2$
$9w + 18y - 12z = 3$
$5x + 10y - 16z = -26$

82. PHYSICS The work done to move an object is given by $W = Fd \cos \theta$, where θ is the angle between the displacement d and the force exerted F. If Lisa does 2400 joules of work while exerting a force of 120 newtons over 40 meters, at what angle was she exerting the force? (Lesson 5-5)

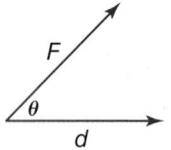

Write each degree measure in radians as a multiple of π and each radian measure in degrees. (Lesson 4-2)

83. $-10°$

84. $485°$

85. $\dfrac{3\pi}{4}$

Solve each equation. (Lesson 3-4)

86. $\log_{10} \sqrt[3]{10} = x$

87. $2 \log_5 (x - 2) = \log_5 36$

88. $\log_5 (x + 4) + \log_5 8 = \log_5 64$

89. $\log_4 (x - 3) + \log_4 (x + 3) = 2$

90. $\dfrac{1}{2}(\log_7 x + \log_7 8) = \log_7 16$

91. $\log_{12} x = \dfrac{1}{2} \log_{12} 9 + \dfrac{1}{3} \log_{12} 27$

92. AERONAUTICS The data below represent the lift of a jet model's wing in a wind tunnel at certain angles of attack. The angle of attack α of the wing is the angle between the wing and the flow of the wind. (Lesson 2-1)

Angle of Attack α	0.1	0.5	1.0	1.5	2.0	3.0	5.0	10.0
Lift (lbs)	11.7	236.0	476.2	711.6	950.3	1782.6	2384.4	4049.3

a. Determine a power function to model the data.

b. Use the function to predict the lift of the wing at 4.0 degrees.

Skills Review for Standardized Tests

93. SAT/ACT In the figure, $\ell_1 \parallel \ell_2$. If $EF = x$, and $EG = y$, which of the following represents the ratio of CD to BC?

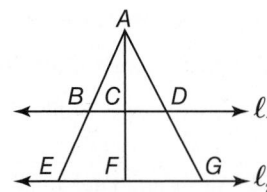

A $1 - \dfrac{y}{x}$ **C** $\dfrac{y}{x} - 1$ **E** $1 + \dfrac{x}{y}$

B $1 + \dfrac{y}{x}$ **D** $1 - \dfrac{x}{y}$

94. What are the dimensions of the matrix that results from the multiplication shown?

$$\begin{bmatrix} a & b & c \\ d & e & f \\ g & h & i \end{bmatrix} \cdot \begin{bmatrix} j \\ k \\ l \end{bmatrix}$$

F 1×3 **H** 3×3

G 3×1 **J** 4×3

95. REVIEW Shenae spent $42 on 1 can of primer and 2 cans of paint for her room. If the price of one can of paint p is 150% of the price of one can of primer r, which system of equations can be used to find the price of paint and primer?

A $p = r + \dfrac{1}{2}r, r + 2p = 42$

B $p = r + 2r, r + \dfrac{1}{2}p = 42$

C $r = p + \dfrac{1}{2}p, r + 2r = 42$

D $r = p + 2p, r + \dfrac{1}{2} = 42$

96. REVIEW To join the football team, a student must have a GPA of at least 2.0 and must have attended at least five after-school practices. Which system of inequalities best represents this situation if x represents a student's GPA, and y represents the number of after-school practices the student attended?

F $x \geq 2, y \geq 5$ **H** $x < 2, y < 5$

G $x \leq 2, y \leq 5$ **J** $x > 2, y > 5$

Objective
Use a graphing calculator to find areas of polygons using determinants.

In Lesson 6-2, you learned that the area of a triangle X with vertices (x_1, y_1), (x_2, y_2), and (x_3, y_3) can be found by calculating $\frac{1}{2}|\det(X)|$. This process can be used to find the area of any polygon.

ACTIVITY **Area of a Quadrilateral**

a. Find the area of the quadrilateral with vertices (1, 1), (2, 6), (8, 5), and (7, 2).

Step 1 Sketch the quadrilateral, and divide it into two triangles.

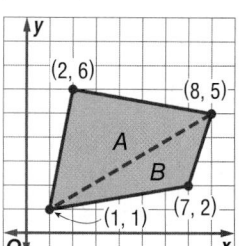

Step 2 Create a matrix for each triangle.

$$A = \begin{bmatrix} 1 & 1 & 1 \\ 2 & 6 & 1 \\ 8 & 5 & 1 \end{bmatrix} \qquad B = \begin{bmatrix} 1 & 1 & 1 \\ 8 & 5 & 1 \\ 7 & 2 & 1 \end{bmatrix}$$

Step 3 Enter each matrix into your graphing calculator, and find $\det(A)$ and $\det(B)$.

Step 4 Multiply the absolute value of each determinant by $\frac{1}{2}$, and find the sum.
The area is $\frac{1}{2}|-31| + \frac{1}{2}|-17|$ or 24 square units.

---

b. Find the area of the polygon with vertices (1, 5), (4, 8), (8, 5), (6, 2), and (2, 1).

Step 1 Sketch the pentagon, and divide it into three triangles.

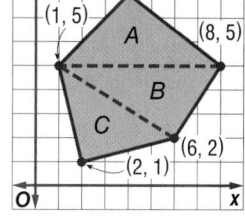

Step 2 Create a matrix for each triangle.

$$A = \begin{bmatrix} 1 & 5 & 1 \\ 4 & 8 & 1 \\ 8 & 5 & 1 \end{bmatrix} \qquad B = \begin{bmatrix} 1 & 5 & 1 \\ 8 & 5 & 1 \\ 6 & 2 & 1 \end{bmatrix}$$

$$C = \begin{bmatrix} 1 & 5 & 1 \\ 6 & 2 & 1 \\ 2 & 1 & 1 \end{bmatrix}$$

Step 3 Enter each matrix into your graphing calculator, and find the determinants.
The determinants are −21, −21, and −17.

Step 4 Multiply the absolute value of each determinant by $\frac{1}{2}$ and find the sum.
The area is $\frac{1}{2}|-21| + \frac{1}{2}|-21| + \frac{1}{2}|-17|$ or 29.5 square units.

StudyTip

Dividing Polygons There may be various ways to divide a given polygon into triangles. For instance, the quadrilateral in Example 2 could have also been divided as shown below.

(4, 8)

(1, 5) — A — (8, 5)

B

C (6, 2)

(2, 1)

Exercises

Find the area of the polygon with the given vertices.

1. (3, 2), (1, 9), (10, 12), (8, 3)

2. (−2, −4), (−11, −1), (−9, −8), (−1, −12)

3. (1, 3), (2, 9), (10, 11), (13, 7), (6, 2)

4. (−7, −6), (−10, 2), (−9, 8), (−5, 10), (8, 6), (13, 2)

Solving Linear Systems using Inverses and Cramer's Rule

Then
You found determinants and inverses of 2 × 2 and 3 × 3 matrices. (Lesson 6-2)

Now
- Solve systems of linear equations using inverse matrices.
- Solve systems of linear equations using Cramer's Rule.

New Vocabulary
square system
Cramer's Rule

Math Online
glencoe.com

Why?

Marcela downloads her favorite shows to her portable media player. A nature show requires twice as much memory as a sitcom, and a movie requires twice as much memory as a nature show. When given the amount of memory that has been used, you can use an inverse matrix to solve a system of equations to find the number of each type of show that Marcella downloaded.

Use Inverse Matrices If a system of linear equations has the same number of equations as variables, then its coefficient matrix is square and the system is said to be a **square system**. If this square coefficient matrix is invertible, then the system has a unique solution.

Key Concept — Invertible Square Linear Systems

Let A be the coefficient matrix of a system of n linear equations in n variables given by $AX = B$, where X is the matrix of variables and B is the matrix of constants. If A is invertible, then the system of equations has a unique solution given by $X = A^{-1}B$.

EXAMPLE 1 — Solve a 2 × 2 System Using an Inverse Matrix

Use an inverse matrix to solve the system of equations, if possible.

$2x - 3y = -1$
$-3x + 5y = 3$

Write the system in matrix form $AX = B$.

$$\begin{bmatrix} 2 & -3 \\ -3 & 5 \end{bmatrix} \cdot \begin{bmatrix} x \\ y \end{bmatrix} = \begin{bmatrix} -1 \\ 3 \end{bmatrix} \qquad AX = B$$

Use the formula for the inverse of a 2 × 2 matrix to find the inverse A^{-1}.

$$A^{-1} = \frac{1}{ad - cb} \begin{bmatrix} d & -b \\ -c & a \end{bmatrix} \qquad \text{Formula for the inverse of a 2 × 2 matrix } \begin{bmatrix} a & b \\ c & d \end{bmatrix}$$

$$= \frac{1}{2(5) - (-3)(-3)} \begin{bmatrix} 5 & 3 \\ 3 & 2 \end{bmatrix} \qquad a = 2, b = -3, c = -3, \text{ and } d = 5$$

$$= \begin{bmatrix} 5 & 3 \\ 3 & 2 \end{bmatrix} \qquad \text{Simplify.}$$

Multiply A^{-1} by B to solve the system.

$$X = \begin{bmatrix} 5 & 3 \\ 3 & 2 \end{bmatrix} \cdot \begin{bmatrix} -1 \\ 3 \end{bmatrix} = \begin{bmatrix} 4 \\ 3 \end{bmatrix} \qquad X = A^{-1}B$$

Therefore, the solution of the system is (4, 3).

✔ Guided Practice

Use an inverse matrix to solve the system of equations, if possible.

1A. $6x + y = -8$
$-4x - 5y = -12$

1B. $-3x + 9y = 36$
$7x - 8y = -19$

To solve a 3 × 3 system of equations using an inverse matrix, use a calculator.

Real-World Link

A bond is essentially an IOU issued by a company or government to fund its day-to-day operations or a specific project. If you invest in bonds, you are loaning your money for a certain period of time to the issuer. In return, you will receive your money back plus interest.

Source: CNN

Real-World EXAMPLE 2 — Solve a 3 × 3 System Using an Inverse Matrix

FINANCIAL LITERACY Belinda is investing $20,000 by purchasing three bonds with expected annual returns of 10%, 8%, and 6%. Investments with a higher expected return are often riskier than other investments. She wants an average annual return of $1340. If she wants to invest three times as much money in the bond with a 6% return than the other two combined, how much money should she invest in each bond?

Her investment can be represented by

$x + y + z = 20{,}000$

$3x + 3y - z = 0$

$0.10x + 0.08y + 0.06z = 1340,$

where x, y, and z represent the amounts invested in the bonds with 10%, 8%, and 6% annual returns, respectively.

Write the system in matrix form $AX = B$.

$$
\underset{A}{\begin{bmatrix} 1 & 1 & 1 \\ 3 & 3 & -1 \\ 0.10 & 0.08 & 0.06 \end{bmatrix}} \cdot \underset{X}{\begin{bmatrix} x \\ y \\ z \end{bmatrix}} = \underset{B}{\begin{bmatrix} 20{,}000 \\ 0 \\ 1340 \end{bmatrix}}
$$

Use a graphing calculator to find A^{-1}.

```
[A]⁻¹
[[-3.25 -.25 50...
 [3.5    .5   -5...
 [.75   -.25 0 ...
```

$$
A^{-1} = \begin{bmatrix} -3.25 & -0.25 & 50 \\ 3.5 & 0.5 & -50 \\ 0.75 & -0.25 & 0 \end{bmatrix}
$$

Multiply A^{-1} by B to solve the system.

$$X = A^{-1}B$$

$$
= \begin{bmatrix} -3.25 & -0.25 & 50 \\ 3.5 & 0.5 & -50 \\ 0.75 & -0.25 & 0 \end{bmatrix} \cdot \begin{bmatrix} 20{,}000 \\ 0 \\ 1340 \end{bmatrix}
$$

$$
= \begin{bmatrix} 2000 \\ 3000 \\ 15{,}000 \end{bmatrix}
$$

The solution of the system is (2000, 3000, 15,000). Therefore, Belinda invested $2000 in the bond with a 10% annual return, $3000 in the bond with an 8% annual return, and $15,000 in the bond with a 6% annual return.

CHECK You can check the solution by substituting back into the original system.

$$2000 + 3000 + 15{,}000 = 20{,}000$$
$$20{,}000 = 20{,}000 \checkmark$$

$$3(2000) + 3(3000) - 15{,}000 = 0$$
$$0 = 0 \checkmark$$

$$0.10(2000) + 0.08(3000) + 0.06(15{,}000) = 1340$$
$$1340 = 1340 \checkmark$$

✓ Guided Practice

2. **INDUSTRY** During three consecutive years, an auto assembly plant produced a total of 720,000 cars. If 50,000 more cars were made in the second year than the first year, and 80,000 more cars were made in the third year than the second year, how many cars were made in each year?

Use Cramer's Rule Another method for solving square systems, known as **Cramer's Rule**, uses determinants instead of row reduction or inverse matrices.

Consider the following 2 × 2 system.

$ax + by = e$
$cx + dy = f$

Use the elimination method to solve for x.

Multiply by d. $\longrightarrow$ $adx + bdy = ed$

Multiply by −b. $\longrightarrow$ $\underline{(+) \quad -bcx - bdy = -fb}$ So, $x = \dfrac{ed - fb}{ad - bc}$.

$\qquad\qquad\qquad\qquad (ad - bc)x \qquad = ed - fb$

Similarly, it can be shown that $y = \dfrac{af - ce}{ad - bc}$. You should recognize the denominator of each fraction as the determinant of the system's coefficient matrix $A = \begin{bmatrix} a & b \\ c & d \end{bmatrix}$. Both the numerator and denominator of each solution can be expressed using determinants.

$$x = \frac{ed - fb}{ad - bc} = \frac{\begin{vmatrix} e & b \\ f & d \end{vmatrix}}{\begin{vmatrix} a & b \\ c & d \end{vmatrix}} = \frac{|A_x|}{|A|} \qquad\qquad y = \frac{af - ce}{ad - bc} = \frac{\begin{vmatrix} a & e \\ c & f \end{vmatrix}}{\begin{vmatrix} a & b \\ c & d \end{vmatrix}} = \frac{|A_y|}{|A|}$$

Notice that numerators $|A_x|$ and $|A_y|$ are the determinants of the matrices formed by replacing the coefficients of x or y, respectively, in the coefficient matrix with the column of constant terms $\begin{smallmatrix} e \\ f \end{smallmatrix}$ from the original system $\begin{bmatrix} a & b & | & e \\ c & d & | & f \end{bmatrix}$.

Cramer's Rule can be generalized to systems of n equations in n variables.

Key Concept   **Cramer's Rule**

Let A be the coefficient matrix of a system of n linear equations in n variables given by $AX = B$. If $\det(A) \neq 0$, then the unique solution of the system is given by

$$x_1 = \frac{|A_1|}{|A|}, x_2 = \frac{|A_2|}{|A|}, x_3 = \frac{|A_3|}{|A|}, \dots, x_n = \frac{|A_n|}{|A|},$$

where A_i is obtained by replacing the ith column of A with the column of constant terms B. If $\det(A) = 0$, then $AX = B$ has either no solution or infinitely many solutions.

EXAMPLE 3   **Use Cramer's Rule to Solve a 2 × 2 System**

Use Cramer's Rule to find the solution of the system of linear equations, if a unique solution exists.

$3x_1 + 2x_2 = 6$
$-4x_1 - x_2 = -13$

The coefficient matrix is $A = \begin{bmatrix} 3 & 2 \\ -4 & -1 \end{bmatrix}$. Calculate the determinant of A.

$A = \begin{vmatrix} 3 & 2 \\ -4 & -1 \end{vmatrix} = 3(-1) - (-4)(2)$ or 5

Because the determinant of A does not equal zero, you can apply Cramer's Rule.

$$x_1 = \frac{|A_1|}{|A|} = \frac{\begin{vmatrix} 6 & 2 \\ -13 & -1 \end{vmatrix}}{5} = \frac{6(-1) - (-13)(2)}{5} = \frac{20}{5} \text{ or } 4$$

$$x_2 = \frac{|A_2|}{|A|} = \frac{\begin{vmatrix} 3 & 6 \\ -4 & -13 \end{vmatrix}}{5} = \frac{3(-13) - (-4)(6)}{5} = \frac{-15}{5} \text{ or } -3$$

So, the solution is $x_1 = 4$ and $x_2 = -3$ or $(4, -3)$. Check your answer in the original system.

Watch Out!

Division by Zero
Remember that Cramer's Rule does not apply when the determinant of the coefficient matrix is 0, because this would introduce division by zero, which is undefined.

☑ **Guided Practice**

Use Cramer's Rule to find the solution of each system of linear equations, if a unique solution exists.

3A. $2x - y = 4$
$5x - 3y = -6$

3B. $-9x + 3y = 8$
$2x - y = -3$

3C. $12x - 9y = -5$
$4x - 3y = 11$

EXAMPLE 4　**Use Cramer's Rule to Solve a 3 × 3 System**

Use Cramer's Rule to find the solution of the system of linear equations, if a unique solution exists.

$-x - 2y = -4z + 12$
$3x - 6y + z = 15$
$2x + 5y + 1 = 0$

The coefficient matrix is $A = \begin{bmatrix} -1 & -2 & 4 \\ 3 & -6 & 1 \\ 2 & 5 & 0 \end{bmatrix}$. Calculate the determinant of A.

$$|A| = \begin{vmatrix} -1 & -2 & 4 \\ 3 & -6 & 1 \\ 2 & 5 & 0 \end{vmatrix} = -1 \begin{vmatrix} -6 & 1 \\ 5 & 0 \end{vmatrix} - (-2) \begin{vmatrix} 3 & 1 \\ 2 & 0 \end{vmatrix} + 4 \begin{vmatrix} 3 & -6 \\ 2 & 5 \end{vmatrix}$$

Formula for the determinant of a 3 × 3 matrix

$$= -1[-6(0) - 5(1)] - (-2)[3(0) - 1(2)] + 4[3(5) - 2(-6)]$$　Simplify.

$$= -1(-5) + 2(-2) + 4(27) \text{ or } 109$$　Simplify.

Because the determinant of A does not equal zero, you can apply Cramer's Rule.

ReadingMath

Replacing Columns The notation $|A_x|$ is read as *the determinant of the coefficient matrix A with the column of x-coefficients replaced with the column of constants.*

$$x = \frac{|A_x|}{|A|} = \frac{\begin{vmatrix} 12 & -2 & 4 \\ 15 & -6 & 1 \\ -1 & 5 & 0 \end{vmatrix}}{109} = \frac{12[(-6)(0) - 5(1)] - (-2)[15(0) - (-1)(1)] + 4[15(5) - (-1)(-6)]}{109} = \frac{218}{109} \text{ or } 2$$

$$y = \frac{|A_y|}{|A|} = \frac{\begin{vmatrix} -1 & 12 & 4 \\ 3 & 15 & 1 \\ 2 & -1 & 0 \end{vmatrix}}{109} = \frac{(-1)[15(0) - 1(-1)] - 12[3(0) - 2(1)] + 4[3(-1) - 2(15)]}{109} = \frac{-109}{109} \text{ or } -1$$

$$z = \frac{|A_z|}{|A|} = \frac{\begin{vmatrix} -1 & -2 & 12 \\ 3 & -6 & 15 \\ 2 & 5 & -1 \end{vmatrix}}{109} = \frac{(-1)[(-6)(-1) - 5(15)] - (-2)[3(-1) - 2(15)] + 12[3(5) - 2(-6)]}{109} = \frac{327}{109} \text{ or } 3$$

Therefore, the solution is $x = 2$, $y = -1$, and $z = 3$ or $(2, -1, 3)$.

CHECK　Check the solution by substituting back into the original system.

$-(2) - 2(-1) = -4(3) + 12$
$0 = 0$ ✔

$3(2) - 6(-1) + 3 = 15$
$15 = 15$ ✔

$2(2) + 5(-1) + 1 = 0$
$0 = 0$ ✔

☑ **Guided Practice**

Use Cramer's Rule to find the solution of each system of linear equations, if a unique solution exists.

4A. $8x + 12y - 24z = -40$
$3x - 8y + 12z = 23$
$2x + 3y - 6z = -10$

4B. $-2x + 4y - z = -3$
$3x + y + 2z = 6$
$x - 3y = 1$

● = **Step-by-Step Solutions** begin on page R29.

Use an inverse matrix to solve each system of equations, if possible. (Examples 1 and 2)

1. $5x - 2y = 11$
$-4x + 7y = 2$

2. $2x + 3y = 2$
$x - 4y = -21$

3. $-3x + 5y = 33$
$2x - 4y = -26$

4. $-4x + y = 19$
$3x - 2y = -18$

5. $2x + y - z = -13$
$3x + 2y - 4z = -36$
$x + 6y - 3z = 12$

6. $3x - 2y + 8z = 38$
$6x + 3y - 9z = -12$
$4x + 4y + 20z = 0$

7. $x + 2y - z = 2$
$2x - y + 3z = 4$
$3x + y + 2z = 6$

8. $4x + 6y + z = -1$
$-x - y + 8z = 8$
$6x - 4y + 11z = 21$

9. DOWNLOADING Marcela downloaded some programs on her portable media player. In general, a 30-minute sitcom uses 0.3 gigabyte of memory, a 1-hour talk show uses 0.6 gigabyte, and a 2-hour movie uses 1.2 gigabytes. She downloaded 9 programs totaling 5.4 gigabytes. If she downloaded two more sitcoms than movies, what number of each type of show did Marcela download? (Example 2)

10. BASKETBALL Trevor knows that he has scored 37 times for a total of 70 points thus far this basketball season. He wants to know how many free throws, 2-point and 3-point field goals he has made. The sum of his 2- and 3-point field goals equals twice the number of free throws minus two. How many free throws, 2-point field goals, and 3-point field goals has Trevor made? (Example 2)

Use Cramer's Rule to find the solution of each system of linear equations, if a unique solution exists. (Examples 3 and 4)

11. $-3x + y = 4$
$2x + y = -6$

12. $2x + 3y = 4$
$5x + 6y = 5$

13. $5x + 4y = 7$
$-x - 4y = -3$

14. $4x + \frac{1}{3}y = 8$
$3x + y = 6$

15. $2x - y + z = 1$
$x + 2y - 4z = 3$
$4x + 3y - 7z = -8$

16. $x + y + z = 12$
$6x - 2y - z = 16$
$3x + 4y + 2z = 28$

17. $x + 2y = 12$
$3y - 4z = 25$
$x + 6y + z = 20$

18. $9x + 7y = -30$
$8y + 5z = 11$
$-3x + 10z = 73$

19. ROAD TRIP Dena stopped for gasoline twice during a road trip. The price of gasoline at each station is shown below. She bought a total of 33.5 gallons and spent $134.28. Use Cramer's Rule to determine the number of gallons of gasoline Dena bought for $3.96 a gallon. (Example 3)

20. GROUP PLANNING A class reunion committee is planning for 400 guests for its 10-year reunion. The guests can choose one of the three options for dessert that are shown below. The chef preparing the desserts must spend 5 minutes on each pie, 8 minutes on each trifle, and 12 minutes on each cheesecake. The total cost of the desserts was $1170, and the chef spends exactly 45 hours preparing them. Use Cramer's Rule to determine how many servings of each dessert were prepared. (Example 4)

21 PHONES Megan, Emma, and Mora all went over their allotted phone plans. For an extra 30 minutes of gaming, 12 minutes of calls, and 40 text messages, Megan paid $52.90. Emma paid $48.07 for 18 minutes of gaming, 15 minutes of calls, and 55 text messages. Mora only paid $13.64 for 6 minutes of gaming and 7 minutes of calls. If they all have the same plan, find the cost of each service. (Example 4)

Find the solution to each matrix equation.

22. $\begin{bmatrix} -2 & 1 \\ 3 & -1 \end{bmatrix} \begin{bmatrix} x_1 & y_1 \\ x_2 & y_2 \end{bmatrix} = \begin{bmatrix} 4 & 0 \\ -2 & 3 \end{bmatrix}$

23. $\begin{bmatrix} 5 & 2 \\ -2 & 1 \end{bmatrix} \begin{bmatrix} x_1 & y_1 \\ x_2 & y_2 \end{bmatrix} = \begin{bmatrix} 9 & -8 \\ 0 & 5 \end{bmatrix}$

24. $\begin{bmatrix} 6 & 4 \\ 0 & 3 \end{bmatrix} \begin{bmatrix} x_1 & y_1 \\ x_2 & y_2 \end{bmatrix} = \begin{bmatrix} 10 & 8 \\ 12 & 6 \end{bmatrix}$

25. $\begin{bmatrix} -2 & 1 \\ 3 & -5 \end{bmatrix} \begin{bmatrix} x_1 & y_1 \\ x_2 & y_2 \end{bmatrix} = \begin{bmatrix} -2 & -1 \\ 17 & -9 \end{bmatrix}$

26. FITNESS Eva is training for a half-marathon and consumes energy gels, bars, and drinks every week. This week, she consumed 12 energy items for a total of 1450 Calories and 310 grams of carbohydrates. The nutritional content of each item is shown.

Energy Item	gel	bar	drink
Calories	100	250	50
Carbohydrates (g)	25	43	14

How many energy gels, bars, and drinks did Eva consume this week?

GRAPHING CALCULATOR Solve each system of equations using inverse matrices.

27. $2a - b + 4c = 6$
$a + 5b - 2c = -6$
$3a - 2b + 6c = 8$

28. $3x - 5y + 2z = 22$
$2x + 3y - z = -9$
$4x + 3y + 3z = 1$

29. $r + 5s - 2t = 16$
$-2r - s + 3t = 3$
$3r + 2s - 4t = -2$

30. $-4m + n + 6p = 17$
$3m - n - p = 5$
$-5m - 2n + 3p = 2$

Find the values of n such that the system represented by the given augmented matrix cannot be solved using an inverse matrix.

31. $\begin{bmatrix} n & -8 & \vdots & 6 \\ 1 & 2 & \vdots & 3 \end{bmatrix}$

32. $\begin{bmatrix} 3 & n & \vdots & 4 \\ n & 2 & \vdots & -5 \end{bmatrix}$

33. $\begin{bmatrix} -5 & -9 & \vdots & 3 \\ n & n & \vdots & 11 \end{bmatrix}$

34. $\begin{bmatrix} n & -n & \vdots & 0 \\ 7 & n & \vdots & -8 \end{bmatrix}$

35. CHEMICALS Three alloys of copper and silver contain 35% pure silver, 55% pure silver, and 60% pure silver, respectively. How much of each type should be mixed to produce 2.5 kilograms of an alloy containing 54.4% silver if there is to be 0.5 kilogram more of the 60% alloy than the 55% alloy?

36. DELI A Greek deli sells the gyros shown below. During one lunch, the deli sold a total of 74 gyros and earned $320.50. The total amount of meat used for the small, large, and jumbo gyros was 274 ounces. The number of large gyros sold was one more than twice the number of jumbo gyros sold. How many of each type of gyro did the deli sell during lunch?

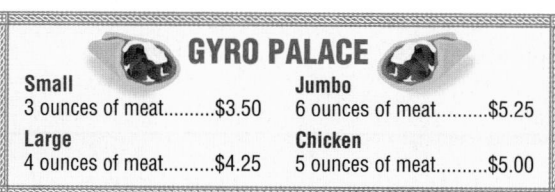

GYRO PALACE
Small
3 ounces of meat..........$3.50
Large
4 ounces of meat..........$4.25
Jumbo
6 ounces of meat..........$5.25
Chicken
5 ounces of meat..........$5.00

37. GEOMETRY The perimeter of $\triangle ABC$ is 89 millimeters. The length of $\overline{AC}$ is 47 millimeters less than the sum of the lengths of the other two sides. The length of $\overline{BC}$ is 20 millimeters more than half the length of $\overline{AB}$. Use a system of equations to find the length of each side.

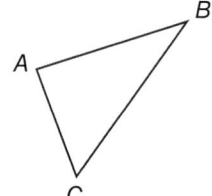

Find the inverse of each matrix, if possible.

38. $\begin{bmatrix} e^{2x} & e^{-x} \\ e^x & e^{-3x} \end{bmatrix}$

39. $\begin{bmatrix} \frac{1}{x} & \frac{3}{x} \\ x & 2 \end{bmatrix}$

40. $\begin{bmatrix} \pi^x & 1 \\ 0 & \pi^{-2x} \end{bmatrix}$

41. $\begin{bmatrix} i & -3 \\ i^2 & 2i \end{bmatrix}$

Let A and B be $n \times n$ matrices and let C, D, and X be $n \times 1$ matrices. Solve each equation for X. Assume that all inverses exist.

42. $AX = BX - C$

43. $D = AX + BX$

44. $AX + BX = 2C - X$

45. $X + C = AX - D$

46. $3X - D = C - BX$

47. $BX = AD + AX$

48. CALCULUS In calculus, systems of equations can be obtained using *partial derivatives*. These equations contain λ, which is called a *Lagrange multiplier*. Find values of x and y that satisfy $x + \lambda + 1 = 0$; $2y + \lambda = 0$; $x + y + 7 = 0$.

H.O.T. Problems   Use Higher-Order Thinking Skills

49. ERROR ANALYSIS Trent and Kate are trying to solve the system below using Cramer's Rule. Is either of them correct? Explain your reasoning.

$2x + 7y = 10$
$6x + 21y = 30$

Trent	Kate
The system has no solution because the determinant of the coefficient matrix is 0.	The system has one solution but cannot be found by using Cramer's Rule.

50. CHALLENGE The graph shown below goes through points at $(-2, -1)$, $(-1, 7)$, $(1, 5)$, and $(2, 19)$. The equation of the graph is of the form $f(x) = ax^3 + bx^2 + cx + d$.

Find the equation of the graph by solving a system of equations using an inverse matrix.

51. REASONING If $A = \begin{bmatrix} a & b \\ c & d \end{bmatrix}$ and A is nonsingular, does $(A^2)^{-1} = (A^{-1})^2$? Explain your reasoning.

52. OPEN ENDED Give an example of a system of equations in two variables that does not have a unique solution, and demonstrate how the system expressed as a matrix equation would have no solution.

53. WRITING IN MATH Describe what types of systems can be solved using each method. Explain your reasoning.

   a. Gauss-Jordan elimination

   b. inverse matrices

   c. Cramer's Rule

Find AB and BA, if possible. (Lesson 6-2)

54. $A = \begin{bmatrix} 3 & -2 \\ 5 & 1 \end{bmatrix}, B = \begin{bmatrix} 4 & 1 \\ 2 & 7 \end{bmatrix}$

55. $A = \begin{bmatrix} 4 & -2 & 6 \\ 7 & 8 & 3 \\ 11 & -5 & -1 \end{bmatrix}, B = \begin{bmatrix} 17 & 2 & -4 \\ 10 & -9 & 6 \\ 1 & 0 & -8 \end{bmatrix}$

Determine whether each matrix is in row-echelon form. (Lesson 6-1)

56. $\begin{bmatrix} 0 & -3 & -6 & | & 4 \\ 9 & -1 & -2 & | & -1 \\ 3 & 1 & -2 & | & -3 \\ 0 & 3 & -1 & | & 1 \end{bmatrix}$

57. $\begin{bmatrix} 1 & 0 & -2 & 3 & 0 & | & -24 \\ 0 & 1 & -2 & 2 & 0 & | & -7 \\ 0 & 0 & 0 & 0 & 1 & | & 4 \end{bmatrix}$

58. TRACK AND FIELD A shot put must land in a 40° sector. The vertex of the sector is at the origin, and one side lies along the x-axis. If an athlete puts the shot at a point with coordinates (18, 17), will the shot land in the required region? Explain your reasoning. (Lesson 4-6)

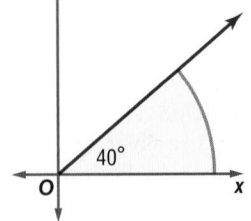

59. STARS Some stars appear bright only because they are very close to us. Absolute magnitude M is a measure of how bright a star would appear if it were 10 parsecs, or about 32 light years, away from Earth. A lower magnitude indicates a brighter star. Absolute magnitude is given by $M = m + 5 - 5 \log d$, where d is the star's distance from Earth measured in parsecs and m is its apparent magnitude. (Lesson 3-3)

Star	Apparent Magnitude	Distance (parsecs)
Sirius	−1.44	2.64
Vega	0.03	7.76

a. Sirius and Vega are two of the brightest stars. Which star appears brighter?

b. Find the absolute magnitudes of Sirius and Vega.

c. Which star is actually brighter? That is, which has a lower absolute magnitude?

Skills Review for Standardized Tests

60. SAT/ACT Point C is the center of the circle in the figure below. The shaded region has an area of 3π square centimeters. What is the perimeter of the shaded region in centimeters?

A $2\pi + 6$

B $2\pi + 9$

C $2\pi + 12$

D $3\pi + 6$

E $3\pi + 12$

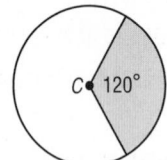

61. In March, Claudia bought 2 standard and 2 premium ring tones from her cell phone provider for $8.96. In May, she paid $9.46 for 1 standard and 3 premium ring tones. What are the prices for standard and premium ring tones?

F $1.99, $2.49   H $1.99, $2.79

G $2.29, $2.79   J $2.49, $2.99

62. REVIEW Each year, the students at Capital High School vote for a homecoming dance theme. The theme "A Night Under the Stars" received 225 votes. "The Time of My Life" received 480 votes. If 40% of girls voted for the star theme, 75% of boys voted for the life theme and all of the students voted, how many girls and boys are there at Capital High School?

A 854 boys and 176 girls

B 705 boys and 325 girls

C 395 boys and 310 girls

D 380 boys and 325 girls

63. REVIEW What is the solution of $\frac{1}{8}x - \frac{2}{3}y + \frac{5}{6}z = -8$, $\frac{3}{4}x + \frac{1}{6}y - \frac{1}{3}z = -12$, and $\frac{3}{16}x - \frac{5}{8}y - \frac{7}{12}z = -25$?

F $(-4, 6, 3)$ H $(-16, 24, 12)$

G $(-8, 12, 6)$ J no solution

Objective
Use a graphing calculator and matrices to encode and decode messages.

Cryptography is the study of coded messages. Matrices can be used to code messages so that they can only be read after being deciphered by a *key*.

The first step is to decide on a key that can be used to encode the matrix. The key must be an $n \times n$ invertible matrix. The next step is to convert the message to numbers and write it as a matrix. Each letter of the alphabet is represented by a number. The number 0 is used to represent a blank space.

A	B	C	D	E	F	G	H	I	J	K	L	M
1	2	3	4	5	6	7	8	9	10	11	12	13

N	O	P	Q	R	S	T	U	V	W	X	Y	Z
14	15	16	17	18	19	20	21	22	23	24	25	26

Finally, the message is encoded by multiplying it by the key.

ACTIVITY 1 **Encode a Message**

Use $\begin{bmatrix} 1 & -2 \\ -1 & 3 \end{bmatrix}$ to encode the message SATURDAY AT NOON.

Step 1 Convert the message to numbers and write it as a matrix.

S	A	T	U	R	D	A	Y	_	A	T	_	N	O	O	N
19	1	20	21	18	4	1	25	0	1	20	0	14	15	15	14

StudyTip

Conversion Add zeros to the end of a message if additional entries are needed to fill a matrix.

The key is a 2×2 matrix. To make the matrix multiplication possible, write the message as an 8×2 matrix.

$$\begin{bmatrix} 19 & 1 \\ 20 & 21 \\ 18 & 4 \\ 1 & 25 \\ 0 & 1 \\ 20 & 0 \\ 14 & 15 \\ 15 & 14 \end{bmatrix}$$

Step 2 Multiply the converted message by the key using a graphing calculator.

```
[[18   -35]
 [-1    23 ]
 [14   -24]
 [-24   73 ]
 [-1    3  ]
 [20   -40]
 [-1    17 ]↓
```

Step 3 Remove the matrix notation to reveal the encoded message.

18 −35 −1 23 14 −24 −24 73 −1 3 20 −40 −1 17 1 12

Exercises

Use $\begin{bmatrix} 3 & 5 \\ -2 & -3 \end{bmatrix}$ to encode each message.

1. CALL ME

2. SEE YOU LATER

3. ORDER PIZZA

4. CHALLENGE Use $\begin{bmatrix} 2 & -4 & 3 \\ -1 & 5 & 1 \\ 6 & -2 & 4 \end{bmatrix}$ to encode the message MEET ME AT FIVE.

To decode a message, the inverse of the key must be found. The coded message is then written in matrix form to make the multiplication possible. For instance, if the key is an $n \times n$ matrix, the message is written as a $k \times n$ matrix, where k is the number of rows necessary to include each number in the matrix. If there are not enough characters to fill a row, insert "0"s as spaces. Finally, the coded matrix is multiplied by the inverse of the key.

ACTIVITY 2 Decode a Message

Use the inverse of $\begin{bmatrix} -1 & -1 & -3 \\ 3 & 6 & 4 \\ 2 & 1 & 8 \end{bmatrix}$ to decode the message 38 83 39 77 99 202.

Step 1 Use a graphing calculator to find the inverse of the key.

Step 2 Write the coded message as a matrix. The coded matrix will have 3 columns because the key is a 3×3 matrix. It is a 2×3 matrix because there are enough numbers to fill two rows. Enter it into your graphing calculator.

Step 3 Use a graphing calculator to multiply the coded matrix by the inverse of the key.

Step 4 Remove the matrix notation and convert the numbers to letters.

$$\begin{array}{cccccc} 7 & 15 & 0 & 14 & 15 & 23 \\ G & O & \_ & N & O & W \end{array}$$

Exercises

Use the inverse of $\begin{bmatrix} 12 & -7 \\ -5 & 3 \end{bmatrix}$ to decode each message.

5. 128 −73 232 −135 300 −175 99 −56 83 −48 180 −104 300 −175

6. −27 17 38 −21 84 −49 21 −11 131 −76 201 −116 161 −93

7. 151 −88 150 −86 93 −54 −35 22 −5 3 191 −111 −30 18 182 −105

8. 102 −58 45 −26 −48 29 −69 42 39 −21 228 −133 141 −81 −19 12 228 −133

9. CHALLENGE Use the inverse of $\begin{bmatrix} 2 & 4 & 6 & 0 \\ 1 & 8 & -4 & -6 \\ 7 & 6 & -5 & 3 \\ 1 & 7 & 9 & 2 \end{bmatrix}$ to decode

126 265 −49 −34 198 347 193 96 174 239 49 72 177 286 −61 −27 48 200 70 −76 122 162 −21 35 81 190 −37 −63 130 331 214 17 67 267 94 −25 93 161 120 25.

Write each system of equations in triangular form using Gaussian elimination. Then solve the system. (Lesson 6-1)

1. $2x - y = 13$
    $2x + y = 23$

2. $x + y + z = 6$
    $2x - y - z = -3$
    $3x - 5y + 7z = 14$

Solve each system of equations. (Lesson 6-1)

3. $3x + 3y = -8$
    $6x - 5y = 28$

4. $-x + 8y - 2z = -37$
    $2x + 5y - 11z = -7$
    $4x - 7y + 6z = 4$

5. $-2x + 2y + z = 5$
    $3x - 2y + 2z = 7$
    $5x - y + 4z = 8$

6. $x - 5y + 8z = 7$
    $-8x + 3y + 12z = -9$
    $5x - 4y - 3z = 9$

7. PET CARE Amelia purchased 25 total pounds of dog food, bird seed, and cat food for $100. She purchased 10 pounds more dog food than bird seed. The cost per pound for each type of food is shown. (Lesson 6-1)

$4.00/lb $7.00/lb $3.00/lb

a. Write a set of linear equations for this situation.

b. Determine the number of pounds of each type of food Amelia purchased.

8. MULTIPLE CHOICE Which matrix is nonsingular? (Lesson 6-2)

A $\begin{bmatrix} 2 & 3 & -5 & 4 \\ 0 & 0 & 0 & 0 \\ 1 & 6 & -5 & 0 \\ 4 & 4 & 3 & 4 \end{bmatrix}$

C $\begin{bmatrix} 3 & 2 & 1 & 4 \\ 2 & 0 & 0 & 5 \\ 2 & 0 & 0 & 5 \\ 5 & 1 & -7 & 8 \end{bmatrix}$

B $\begin{bmatrix} 1 & 2 & 3 & 4 \\ 0 & 1 & 2 & 3 \\ 0 & 0 & 1 & 2 \\ 0 & 0 & 0 & 1 \end{bmatrix}$

D $\begin{bmatrix} 5 & 3 & 1 & 0 \\ 10 & 6 & 2 & 0 \\ 4 & 3 & -1 & 5 \\ 7 & 7 & 3 & 9 \end{bmatrix}$

Find AB and BA, if possible. (Lesson 6-2)

9. $A = \begin{bmatrix} 1 & -3 & 4 \\ -2 & 5 & 1 \\ 0 & -4 & -6 \end{bmatrix}$

$B = \begin{bmatrix} 1 & 1 & 0 & 1 \\ 1 & 0 & 1 & 1 \\ 1 & 1 & -1 & 0 \end{bmatrix}$

10. $A = \begin{bmatrix} 3 \\ 8 \\ 9 \end{bmatrix}$

$B = \begin{bmatrix} 2 & -3 & 4 \\ 1 & 1 & 0 \\ 5 & -8 & 2 \end{bmatrix}$

Find the determinant of each matrix. Then find the inverse of the matrix, if it exists. (Lesson 6-2)

11. $\begin{bmatrix} 3 & 8 \\ -1 & -2 \end{bmatrix}$

12. $\begin{bmatrix} -9 & -5 \\ -7 & -4 \end{bmatrix}$

13. $\begin{bmatrix} -4 & 3 \\ 7 & -5 \end{bmatrix}$

14. $\begin{bmatrix} 5 & -10 \\ 4 & -6 \end{bmatrix}$

15. NURSING Troy is an Emergency Room nurse. He earns $24 per hour during regular shifts and $30 per hour when working overtime. The table shows the hours Troy worked during the past three weeks. (Lesson 6-2)

Week	Regular Hours	Overtime Hours
1	35	7
2	38	0
3	40	9

a. Use matrices to determine how much Troy earned during each week.

b. During week 4, Troy worked four times more regular hours than overtime hours. Determine the number of hours he worked if he earned $1008.

Use an inverse matrix to solve each system of equations, if possible. (Lesson 6-3)

16. $2x - y = 6$
    $3x + 2y = 37$

17. $2x + y + z = 19$
    $3x - 2y + 3z = 2$
    $4x - 6y + 5z = -26$

18. MULTIPLE CHOICE Which of the augmented matrices represents the solutions of the system of equations? (Lesson 6-3)

$$x + y = 13$$
$$2x - 3y = -9$$

F $\begin{bmatrix} 1 & 0 & \vdots & \frac{3}{5} & \frac{1}{5} \\ 0 & 1 & \vdots & \frac{2}{5} & -\frac{1}{5} \end{bmatrix}$

H $\begin{bmatrix} 1 & 0 & \vdots & 6 \\ 0 & 1 & \vdots & 7 \end{bmatrix}$

G $\begin{bmatrix} 1 & 0 & \vdots & 1 & 1 \\ 0 & 1 & \vdots & 2 & -3 \end{bmatrix}$

J $\begin{bmatrix} 1 & 0 & \vdots & 8 \\ 0 & 1 & \vdots & 5 \end{bmatrix}$

Use Cramer's Rule to find the solution of each system of linear equations, if a unique solution exists. (Lesson 6-3)

19. $2x - y = 6$
    $4x - 2y = 12$

20. $3x - y - z = 13$
    $3x - 2y + 3z = 16$

6-4 Partial Fractions

Why?

In calculus, you will learn how to find the area under the graph of a function over a specified interval using various techniques. To find the area under the curve of a rational function such as $f(x) = \dfrac{x + 13}{x^2 - x - 20}$, you will first need to *decompose* the rational expression or rewrite it as the sum of two simpler rational expressions.

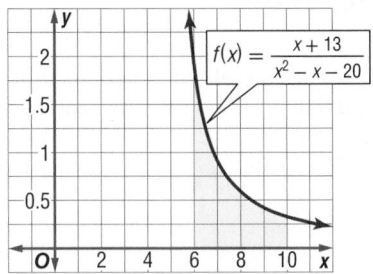

Then
You graphed rational functions. (Lesson 2-4)

Now
- Write partial fraction decompositions of rational expressions with linear factors in the denominator.
- Write partial fraction decompositions of rational expressions with prime quadratic factors in the denominator.

New Vocabulary
partial fraction
partial fraction
    decomposition

Math Online

glencoe.com

Linear Factors In Lesson 2-3, you learned that many polynomial functions with real coefficients can be expressed as the *product* of linear factors and irreducible quadratic factors, each with real coefficients. Similarly, many rational functions can be expressed as the *sum* of two or more simpler rational functions with numerators that are real constants and with denominators that are a power of a linear factor or an irreducible quadratic factor.

For example, the rational function $f(x)$ below can be written as the sum of two fractions with denominators that are linear factors of the original denominator.

$$f(x) = \frac{x + 13}{x^2 - x - 20} = \frac{2}{x - 5} + \frac{-1}{x + 4}$$

Each fraction in the sum is a **partial fraction**. The sum of these partial fractions makes up the **partial fraction decomposition** of the original rational function.

EXAMPLE 1 Denominator with Nonrepeated Linear Factors

Find the partial fraction decomposition of $\dfrac{x + 13}{x^2 - x - 20}$.

Rewrite the expression as partial fractions with constant numerators, A and B, and denominators that are the linear factors of the original denominator.

$\dfrac{x + 13}{x^2 - x - 20} = \dfrac{A}{x - 5} + \dfrac{B}{x + 4}$ **Form of partial fraction decomposition**

$x + 13 = A(x + 4) + B(x - 5)$ **Multiply each side by the LCD, $x^2 - x - 20$.**

$x + 13 = Ax + 4A + Bx - 5B$ **Distributive Property**

$1x + 13 = (A + B)x + (4A - 5B)$ **Group like terms.**

Equate the coefficients on the left and right side of the equation to obtain a system of two equations. To solve the system, you can write it in matrix form $CX = D$ and solve for X.

$$
\begin{array}{l}
A + B = 1 \\
4A - 5B = 13
\end{array}
\longrightarrow
\overset{\displaystyle C \quad\quad \cdot \quad X \;=\; D}{\begin{bmatrix} 1 & 1 \\ 4 & -5 \end{bmatrix} \cdot \begin{bmatrix} A \\ B \end{bmatrix} = \begin{bmatrix} 1 \\ 13 \end{bmatrix}}
$$

You can use a graphing calculator to find $X = C^{-1}D$. So, $A = 2$ and $B = -1$. Use substitution to find the partial fraction decomposition.

```
[C]⁻¹*[D]
          [[2 ]
           [-1]]
```

$\dfrac{x + 13}{x^2 - x - 20} = \dfrac{A}{x - 5} + \dfrac{B}{x + 4}$ **Form of partial fraction decomposition**

$\dfrac{x + 13}{x^2 - x - 20} = \dfrac{2}{x - 5} + \dfrac{-1}{x + 4}$ **$A = 2$ and $B = -1$**

✓ Guided Practice

Find the partial fraction decomposition of each rational expression.

1A. $\dfrac{2x + 5}{x^2 - x - 2}$

1B. $\dfrac{x + 11}{2x^2 - 5x - 3}$

If a rational expression $\frac{f(x)}{d(x)}$ is improper, with the degree of $f(x)$ greater than or equal to the degree of $d(x)$, you must first use the division algorithm $\frac{f(x)}{d(x)} = q(x) + \frac{r(x)}{d(x)}$ to rewrite the expression as the sum of a polynomial and a proper rational expression. Then decompose the remaining rational expression.

EXAMPLE 2   Improper Rational Expression

Find the partial fraction decomposition of $\frac{2x^2 + 5x - 4}{x^2 - x}$.

Because the degree of the numerator is greater than or equal to the degree of the denominator, the rational expression is improper. To rewrite the expression, divide the numerator by the denominator using polynomial division.

$$
\begin{array}{r}
2 \\
x^2 - x \overline{\smash{)}\, 2x^2 + 5x - 4} \\
(-)\ \underline{2x^2 - 2x} \\
7x - 4
\end{array}
$$

← Multiply the divisor by 2 because $\frac{2x^2}{x^2} = 2$.

← Subtract and bring down next term.

So, the original expression is equal to $2 + \frac{7x - 4}{x^2 - x}$.

Because the remaining rational expression is now proper, you can factor its denominator as $x(x - 1)$ and rewrite the expression using partial fractions.

$\dfrac{7x - 4}{x^2 - x} = \dfrac{A}{x} + \dfrac{B}{x - 1}$	**Form of decomposition**
$7x - 4 = A(x - 1) + B(x)$	**Multiply by the LCD, $x^2 - x$.**
$7x - 4 = Ax - A + Bx$	**Distributive Property**
$7x - 4 = (A + B)x - A$	**Group like terms.**

Write and solve the system of equations obtained by equating coefficients.

$$
\begin{aligned}
A + B &= 7 \\
-A &= -4
\end{aligned}
\quad\Longrightarrow\quad
\begin{aligned}
A &= 4 \\
B &= 3
\end{aligned}
$$

Therefore, $\dfrac{2x^2 + 5x - 4}{x^2 - x} = 2 + \dfrac{7x - 4}{x^2 - x}$ or $2 + \dfrac{4}{x} + \dfrac{3}{x - 1}$.

CHECK   You can check your answer by simplifying the expression on the right side of the equation.

$\dfrac{2x^2 + 5x - 4}{x^2 - x} = 2 + \dfrac{4}{x} + \dfrac{3}{x - 1}$	**Partial fraction decomposition**
$= \dfrac{2x(x - 1)}{x(x - 1)} + \dfrac{4(x - 1)}{x(x - 1)} + \dfrac{3x}{x(x - 1)}$	**Rewrite using LCD, $x(x - 1)$.**
$= \dfrac{2(x^2 - x) + 4(x - 1) + 3x}{x(x - 1)}$	**Add.**
$= \dfrac{2x^2 - 2x + 4x - 4 + 3x}{x^2 - x}$	**Multiply.**
$= \dfrac{2x^2 + 5x - 4}{x^2 - x}$ ✔	**Simplify.**

StudyTip

▶ **Alternate Method** By design, the equation $7x - 4 = A(x - 1) + B(x)$ obtained after clearing the fractions in Example 2 is true for all x. Therefore, you can substitute any *convenient values* of x to find the values of A and B. Convenient values are those that are zeros of the original denominator. If $x = 0$, $A = 4$. If $x = 1$, $B = 3$.

✔ Guided Practice

Find the partial fraction decomposition of each rational expression.

2A. $\dfrac{3x^2 + 12x + 4}{x^2 + 2x}$

2B. $\dfrac{x^4 - 3x^3 + x^2 - 9x + 4}{x^2 - 4x}$

If the denominator of a rational expression has a linear factor that is repeated n times, the partial fraction decomposition must include a partial fraction with its own constant numerator for each power 1 to n of the linear factor. For example, to find the partial fraction decomposition of $\dfrac{5x-1}{x^3(x-1)^2}$, you would write

$$\frac{5x-1}{x^3(x-1)^2}=\frac{A}{x}+\frac{B}{x^2}+\frac{C}{x^3}+\frac{D}{x-1}+\frac{E}{(x-1)^2}.$$

EXAMPLE 3   Denominator with Repeated Linear Factors

Find the partial fraction decomposition of $\dfrac{-x^2-3x-8}{x^3+4x^2+4x}$.

This rational expression is proper, so begin by factoring the denominator as $x(x^2+4x+4)$ or $x(x+2)^2$. Because the factor $(x+2)$ has multiplicity 2, include partial fractions with denominators of x, $(x+2)$, and $(x+2)^2$.

$\dfrac{-x^2-3x-8}{x^3+4x^2+4x}=\dfrac{A}{x}+\dfrac{B}{x+2}+\dfrac{C}{(x+2)^2}$	**Form of partial fraction decomposition**
$-x^2-3x-8=A(x+2)^2+Bx(x+2)+Cx$	**Multiply each side by the LCD, $x(x+2)^2$.**
$-x^2-3x-8=Ax^2+4Ax+4A+Bx^2+2Bx+Cx$	**Distributive Property**
$-1x^2-3x-8=(A+B)x^2+(4A+2B+C)x+4A$	**Group like terms.**

Once the system of equations obtained by equating coefficients is found, there are two methods that can be used to find the values of A, B, and C.

Method 1   You can write and solve the system of equations using the same method as Example 2.

$$A+B=-1 \qquad\qquad\qquad A=-2$$
$$4A+2B+C=-3 \quad\longrightarrow\quad B=1$$
$$4A=-8 \qquad\qquad\qquad C=3$$

Method 2   Another way to solve this system is to let x equal a convenient value to eliminate a variable in the equation found by multiplying each side by the LCD.

$-x^2-3x-8=A(x+2)^2+Bx(x+2)+Cx$	**Original equation**
$-(0)^2-3(0)-8=A(0+2)^2+B(0)(0+2)+C(0)$	**Let $x=0$ to eliminate B and C.**
$-8=4A$	
$-2=A$	

$-x^2-3x-8=A(x+2)^2+Bx(x+2)+Cx$	**Original equation**
$-(-2)^2-3(-2)-8=A(-2+2)^2+B(-2)(-2+2)+C(-2)$	**Let $x=-2$ to eliminate A and B.**
$-6=-2C$	
$3=C$	

Substitute these values for A and C and any value for x into the equation to solve for B.

$-x^2-3x-8=A(x+2)^2+Bx(x+2)+Cx$	**Original equation**
$-1^2-3(1)-8=-2(1+2)^2+B(1)(1+2)+3(1)$	**Let $x=1$, $A=-2$, and $C=3$.**
$-12=-15+3B$	**Simplify.**
$B=1$	**Solve for B.**

Therefore, $\dfrac{-x^2-3x-8}{x^3+4x^2+4x}=\dfrac{-2}{x}+\dfrac{1}{x+2}+\dfrac{3}{(x+2)^2}$.

✓ Guided Practice

Find the partial fraction decomposition of each rational expression.

3A. $\dfrac{x+2}{x^3-2x^2+x}$ **3B.** $\dfrac{x+18}{x^3-6x^2+9x}$

Prime Quadratic Factors If the denominator of a rational expression contains a prime quadratic factor, the partial fraction decomposition must include a partial fraction with a *linear numerator* of the form $Bx + C$ for each power of this factor.

EXAMPLE 4 Denominator with Prime Quadratic Factors

Find the partial fraction decomposition of $\dfrac{x^4 - 2x^3 + 8x^2 - 5x + 16}{x(x^2 + 4)^2}$.

This expression is proper. The denominator has one linear factor and one prime quadratic factor of multiplicity 2.

$$\frac{x^4 - 2x^3 + 8x^2 - 5x + 16}{x(x^2 + 4)^2} = \frac{A}{x} + \frac{Bx + C}{x^2 + 4} + \frac{Dx + E}{(x^2 + 4)^2}$$

$$x^4 - 2x^3 + 8x^2 - 5x + 16 = A(x^2 + 4)^2 + (Bx + C)x(x^2 + 4) + (Dx + E)x$$

$$x^4 - 2x^3 + 8x^2 - 5x + 16 = Ax^4 + 8Ax^2 + 16A + Bx^4 + Cx^3 + 4Bx^2 + 4Cx + Dx^2 + Ex$$

$$1x^4 - 2x^3 + 8x^2 - 5x + 16 = (A + B)x^4 + Cx^3 + (8A + 4B + D)x^2 + (4C + E)x + 16A$$

Write and solve the system of equations obtained by equating coefficients.

$$A + B = 1 \qquad\qquad\qquad A = 1$$
$$C = -2 \qquad\qquad\qquad\quad B = 0$$
$$8A + 4B + D = 8 \qquad\Longrightarrow\qquad C = -2$$
$$4C + E = -5 \qquad\qquad\quad D = 0$$
$$16A = 16 \qquad\qquad\qquad E = 3$$

Therefore, $\dfrac{x^4 - 2x^3 + 8x^2 - 5x + 16}{x(x^2 + 4)^2} = \dfrac{1}{x} - \dfrac{2}{x^2 + 4} + \dfrac{3}{(x^2 + 4)^2}$.

☑ **Guided Practice**

Find the partial fraction decomposition of each rational expression.

4A. $\dfrac{x^3 + 2x}{(x^2 + 1)^2}$

4B. $\dfrac{4x^3 - 7x}{(x^2 + x + 1)^2}$

Concept Summary Partial Fraction Decomposition of $f(x)/d(x)$

1. If the degree of $f(x) \geq$ the degree of $d(x)$, use polynomial long division and the division algorithm to write $\dfrac{f(x)}{d(x)} = q(x) + \dfrac{r(x)}{d(x)}$. Then apply partial fraction decomposition to $\dfrac{r(x)}{d(x)}$.

2. If $\dfrac{f(x)}{d(x)}$ is proper, factor $d(x)$ into a product of linear and/or prime quadratic factors.

3. For each factor of the form $(ax + b)^n$ in the denominator, the partial fraction decomposition must include the sum of n fractions

$$\frac{A_1}{ax + b} + \frac{A_2}{(ax + b)^2} + \frac{A_3}{(ax + b)^3} + \cdots + \frac{A_n}{(ax + b)^n},$$

where $A_1, A_2, A_3, \ldots, A_n$ are real numbers.

4. For each prime quadratic factor that occurs n times in the denominator, the partial fraction decomposition must include the sum of n fractions

$$\frac{B_1x + C_1}{ax^2 + bx + c} + \frac{B_2x + C_2}{(ax^2 + bx + c)^2} + \frac{B_3x + C_3}{(ax^2 + bx + c)^3} + \cdots + \frac{B_nx + C_n}{(ax^2 + bx + c)^n},$$

where $B_1, B_2, B_3, \ldots, B_n$ and $C_1, C_2, C_3, \ldots, C_n$ are real numbers.

5. The partial fraction decomposition of the original function is the sum of $q(x)$ from part 1 and the fractions in parts 3 and 4.

Exercises

= Step-by-Step Solutions begin on page R29.

Find the partial fraction decomposition of each rational expression. (Example 1)

1. $\dfrac{x+1}{x^2+5x+6}$

2. $\dfrac{x-18}{x^2-13x+42}$

3. $\dfrac{x+13}{x^2+7x+12}$

4. $\dfrac{x+12}{x^2+14x+48}$

5. $\dfrac{x+6}{-2x^2-19x-45}$

6. $\dfrac{x+7}{2x^2+15x+28}$

Find the partial fraction decomposition of each improper rational expression. (Example 2)

7. $\dfrac{3x^2+x-4}{x^2-2x}$

8. $\dfrac{-5x^2-30x-21}{x^2+7x}$

9. $\dfrac{-2x^3+4x^2+22x-32}{x^3+2x^2-8x}$

10. $\dfrac{x^4-2x^3-2x^2+8x-6}{x^2-2x}$

11. $\dfrac{x^3+12x^2+33x+2}{x^2+8x+15}$

12. $\dfrac{x^4-9x^3+24x^2-4x-12}{x^3-6x^2+8x}$

Find the partial fraction decomposition of each rational expression with repeated factors in the denominator. (Example 3)

13. $\dfrac{x^2-3}{x^3+2x^2+x}$

14. $\dfrac{5x^2-18x+24}{x^3-4x^2+4x}$

15. $\dfrac{-x^2-22x-50}{x^3+10x^2+25x}$

16. $\dfrac{-5x^2-6x+16}{x^3+8x^2+16x}$

17. $\dfrac{17x+256}{x^3-16x^2+64x}$

18. $\dfrac{-10x-108}{x^3+12x^2+36x}$

19. **ENGINEERING** The sum of the average tensile and shear stresses in the bar shown below can be approximated by $s(x)=\dfrac{20x+10\pi x+20}{\pi x^3+\pi x^2}$, where x is the diameter of the pin. (Example 3)

a. Find the partial fraction decomposition.

b. Graph $s(x)$ and the answer to part **a** in the same viewing window.

Find the partial fraction decomposition of each rational expression with prime quadratic factors in the denominator. (Example 4)

20. $\dfrac{x^3+5x-5}{(x^2+4)^2}$

21. $\dfrac{3x^4+4x^2+8x+18}{x(x^2+3)^2}$

22. $\dfrac{4x^4+x^2-25x+32}{x^5-4x^3+4x}$

23. $\dfrac{8x^3-48x+7}{(x^2-6)^2}$

24. $\dfrac{-5x^3-10x^2-6x+4}{(x^2+2x+3)^2}$

25. $\dfrac{4x^3-12x^2-5x+20}{(x^2-3x+3)^2}$

CALCULUS In calculus, you can find the area of the region between the graph of a rational function and the x-axis on a restricted domain. The first step in this process is to write the partial fraction decomposition of the rational expression. Find the partial fraction decomposition of each rational expression.

26.

27.
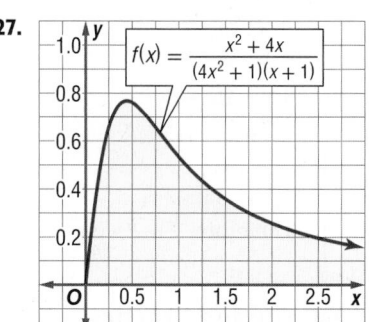

Find the partial fraction decomposition of each rational expression. Then use a graphing calculator to check your answer.

28. $\dfrac{x+4}{3x^2-x-2}$

29. $\dfrac{5x^2-2x+8}{x^3-4x}$

30. $\dfrac{4x^2-3x+3}{4x(x-1)^2}$

31. $\dfrac{x^2+x+5}{(x^2+3)^2}$

32. $\dfrac{2x^3}{(x-1)^2(x+1)^2}$

33. $\dfrac{2x^3+12x^2-3x+3}{x^2+6x+5}$

34. Find two rational expressions that have the sum $\dfrac{x+4}{3x^2-x-2}$.

35. Find three rational expressions that have the sum $\dfrac{6-x}{x^3+2x^2+x}$.

Find A, B, C, and D in terms of r and t.

36. $\dfrac{rx-t}{x^2-x-2}=\dfrac{A}{x-2}+\dfrac{B}{x+1}$

37. $\dfrac{4x^2+rx+2t}{x^2+3x}=4+\dfrac{A}{x}+\dfrac{B}{x+3}$

38. $\dfrac{rx+t}{x^3+x^2}=\dfrac{A}{x}+\dfrac{B}{x^2}+\dfrac{C}{x+1}$

39. $\dfrac{3x^3+5rx^2-16tx+32}{x^2(x^2+16)}=\dfrac{A}{x}+\dfrac{B}{x^2}+\dfrac{Cx+D}{x^2+16}$

Find the partial fraction decomposition of each rational expression.

40. $\dfrac{x^3 + 2x - 1}{(x^2 - x - 2)^2}$

41 $\dfrac{x^3 + 4}{(x^2 - 1)(x^2 + 3x + 2)}$

42. $\dfrac{4x^3 + x^2 - 3x + 3}{x(x - 1)^2}$

43. $\dfrac{7x^7 + 2x^6 - 13x^5 + 32x^4 - 19x^3 + 8x^2 - 7x + 2}{x(x - 1)^2(x + 2)(x^2 + 1)}$

44. ⬡ **MULTIPLE REPRESENTATIONS** In this problem, you will discover the relationship between the partial fraction decomposition of a rational function and its graph. Consider the rational function shown below.

$f(x) = \dfrac{3x^2 - x - 5}{x^2 - 4}$

a. VERBAL Describe the end behavior and vertical and horizontal asymptotes of the function.

b. ANALYTICAL Write the partial fraction decomposition of $f(x)$.

c. GRAPHICAL Graph each addend of the partial fraction decomposition you wrote in part **b** as a separate function.

d. VERBAL Compare the graphs from part **c** with the graph of $f(x)$ and the analysis you wrote in part **b**.

e. ANALYTICAL Make a conjecture as to how the partial fraction decomposition of a function can be used to graph a rational function.

45. GRAPH ANALYSIS The rational functions shown make up the partial fraction decomposition of $f(x)$.

 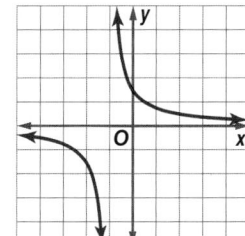

Determine which of the four functions listed below could be the original function $f(x)$.

I. $f(x) = \dfrac{6}{x^2 - 2x - 3}$ **II.** $f(x) = \dfrac{6}{x^2 + 2x - 3}$

III. $f(x) = \dfrac{6}{x^2 + 4x + 3}$ **IV.** $f(x) = \dfrac{6}{x^2 - 4x + 3}$

H.O.T. Problems Use **H**igher-**O**rder **T**hinking Skills

REASONING Use the partial fraction decompositions of $f(x)$ to explain each of the following.

46. If $f(x) = \dfrac{-2x^3 - 7x^2 + 13x + 43}{(x - 2)(x + 3)^2}$, explain why $\lim\limits_{x \to \infty} f(x) = -2$.

47. If $f(x) = \dfrac{x^2 - x + 1}{(x + 1)^3}$, then what is $\lim\limits_{x \to \infty} f(x)$?

CHALLENGE Match the graph of each rational function with its equation.

48.   **49.**

50.   **51.**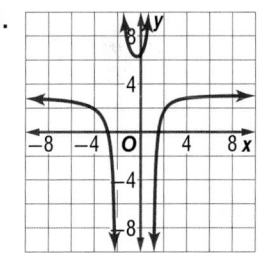

a. $y = x + 2 + \dfrac{1}{x - 1} + \dfrac{-3}{x + 2}$

b. $y = x + 2 + \dfrac{-2}{x - 1} + \dfrac{3}{x + 2}$

c. $y = 3 + \dfrac{1}{x - 1} + \dfrac{-3}{x + 2}$

d. $y = 3 + \dfrac{-2}{x - 1} + \dfrac{3}{x + 2}$

REASONING Determine whether each of the following statements is *true* or *false*. Explain your reasoning.

52. If $f(x) = \dfrac{x^3 + 8}{(x^2 - 1)(x - 2)}$, then $\lim\limits_{x \to \infty} f(x) = 8$.

53. The partial fraction decomposition of
$$f(x) = \dfrac{-4x^4 + 5x^3 + 27x^2 - 11x - 45}{x(x^2 - 3)^2} \text{ is }$$
$$\dfrac{-5}{x} + \dfrac{4 + x}{(x^2 - 3)} + \dfrac{x^2 + 1}{(x^2 - 3)^2}.$$

54. OPEN ENDED Write a rational expression of the form $\dfrac{P(x)}{Q(x)}$ in which the partial fraction decomposition contains each of the following in the denominator.

a. nonrepeated linear factors only

b. at least one repeated linear factor

55. WRITING IN MATH Describe the steps used to obtain the partial fraction decomposition of a rational expression.

Use Cramer's Rule to find the solution of each system of linear equations, if a unique solution exists. (Lesson 6-3)

56. $x + y + z = 6$
$2x + y - 4z = -15$
$5x - 3y + z = -10$

57. $a - 2b + c = 7$
$6a + 2b - 2c = 4$
$4a + 6b + 4c = 14$

58. $p - 2r - 5t = -1$
$p + 2r - 2t = 5$
$4p + r + t = -1$

59. FINANCE For a class project, Jane "bought" shares of stock in three companies. She bought 150 shares of a utility company, 100 shares of a computer company, and 200 shares of a food company. At the end of the project, she "sold" all of her stock. (Lesson 6-2)

Company	Purchase Price per share ($)	Selling Price per share ($)
utility	54.00	55.20
computer	48.00	58.60
food	60.00	61.10

a. Organize the data in two matrices and use matrix multiplication to find the total amount that Jane spent for the stock.

b. Write two matrices and use matrix multiplication to find the total amount she received for selling the stock.

c. How much money did Jane "make" or "lose" in her project?

Simplify each expression. (Lesson 5-1)

60. $\csc \theta \cos \theta \tan \theta$

61. $\sec^2 \theta - 1$

62. $\dfrac{\tan \theta}{\sin \theta}$

63. MEDICINE Doctors may use a tuning fork that resonates at a given frequency as an aid to diagnose hearing problems. The sound wave produced by a tuning fork can be modeled using a sine function. (Lesson 4-4)

a. If the amplitude of the sine function is 0.25, write the equations for tuning forks that resonate with a frequency of 64, 256, and 512 Hertz.

b. How do the periods of the tuning forks compare?

64. SAT/ACT In the figure, what is the value of x?

A 40   C 60   E 90

B 45   D 75

65. Decompose $\dfrac{3p - 1}{p^2 - 1}$ into partial fractions.

F $\dfrac{2}{p + 1} + \dfrac{1}{p - 1}$ H $\dfrac{2}{p + 1} - \dfrac{1}{p - 1}$

G $\dfrac{2}{p - 1} + \dfrac{1}{p + 1}$ J $\dfrac{2}{p - 1} - \dfrac{1}{p + 1}$

66. REVIEW A sprinkler waters a circular section of lawn about 20 feet in diameter. The homeowner decides that placing the sprinkler at $(7, 5)$ will maximize the area of grass being watered. Which equation represents the boundary of the area that the sprinkler waters?

A $(x - 7)^2 + (y - 5)^2 = 100$

B $(x + 7)^2 - (y + 5)^2 = 100$

C $(x - 7)^2 - (y + 5)^2 = 100$

D $(x + 7)^2 + (y - 5)^2 = 100$

67. REVIEW Which of the following is the sum of $\dfrac{x + 2}{x + 3}$ and $\dfrac{4}{x^2 + x - 6}$?

F $\dfrac{-3x - 9}{x^2 + x - 6}$ H $\dfrac{x^2}{x^2 + x - 6}$

G $\dfrac{x^2 - 3x - 24}{x^2 + x - 6}$ J $\dfrac{x^2 + x - 1}{x^2 + x - 6}$

Linear Optimization

Then

You solved systems of linear inequalities. (Lesson 0-4)

Now

- Use linear programming to solve applications.
- Recognize situations in which there are no solutions or more than one solution of a linear programming application.

New Vocabulary

optimization
linear programming
objective function
constraints
feasible solutions
multiple optimal solutions
unbounded

Math Online

glencoe.com

Why?

In general, businesses strive to minimize costs in order to maximize profits. Factors that create or increase business costs and limit or decrease profits are called *business constraints*.

For a shipping company, one constraint might be the number of hours per day that a trucker can safely drive. For a daycare center, one constraint might be a state regulation restricting the number of children per caregiver for certain age groups.

Linear Programming Many applications in business and economics involve **optimization**—the process of finding a minimum value or a maximum value for a specific quantity. When the quantity to be optimized is represented by a linear function, this process is called **linear programming**.

A two-dimensional linear programming problem consists of a linear function to be optimized, called the **objective function**, of the form $f(x, y) = ax + by + c$ and a system of linear inequalities called **constraints**. The solution set of the system of inequalities is the set of possible or **feasible solutions**, which are points of the form (x, y).

Recall from Lesson 0-4 that the solution of a system of linear inequalities is the set of ordered pairs that satisfy each inequality. Graphically, the solution is the intersection of the regions representing the solution sets of the inequalities in the system.

For example, the solution of the system below is the shaded region shown in the graph.

$$y \geq \frac{1}{2}x - 1$$

$$y \leq 3$$

$$x \geq 0$$

$$y \geq 0$$

Suppose you were asked to find the maximum value of $f(x, y) = 3x + 5y$ subject to the constraints given by the system above. Because the shaded region representing the set of feasible solutions contains infinitely many points, it would be impossible to evaluate $f(x, y)$ for all of them. Fortunately, the Vertex Theorem provides a strategy for finding the solution, if it exists.

Key Concept   **Vertex Theorem for Optimization**

Words   If a linear programming problem can be optimized, an optimal value will occur at one of the vertices of the region representing the set of feasible solutions.

Example   The maximum or minimum value of $f(x, y) = ax + by + c$ over the set of feasible solutions graphed occurs at point *A, B, C, D, E,* or *F.*

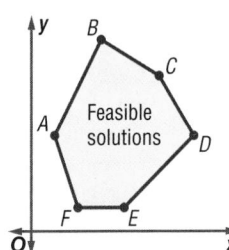

To solve a linear programming problem, follow these steps.

Step 1 Graph the region corresponding to the solution of the system of constraints.

Step 2 Find the coordinates of the vertices of the region formed.

Step 3 Evaluate the objective function at each vertex to determine which x- and y-values, if any, maximize or minimize the function.

EXAMPLE 1 **Maximize and Minimize an Objective Function**

Find the maximum and minimum values of the objective function $f(x, y) = x + 3y$ and for what values of x and y they occur, subject to the following constraints.

$x + y \leq 8$
$2x - y \leq 5$
$x \geq 0$
$y \geq 0$

Begin by graphing the given system of four inequalities. The solution of the system, which makes up the set of feasible solutions for the objective function, is the shaded region, including its boundary segments.

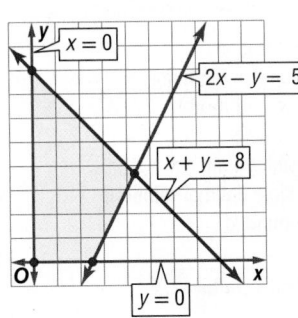

The polygonal region of feasible solutions has four vertices. One vertex is located at $(0, 0)$.

Solve each of the three systems below to find the coordinates of the remaining vertices.

System of Boundary Equations	$x + y = 8$ $2x - y = 5$	$2x - y = 5$ $y = 0$	$x + y = 8$ $x = 0$
Solution (Vertex Point)	$\left(\dfrac{13}{3}, \dfrac{11}{3}\right)$	$\left(\dfrac{5}{2}, 0\right)$	$(0, 8)$

Find the value of the objective function $f(x, y) = x + 3y$ at each of the four vertices.

$f(0, 0) = 0 + 3(0)$ or 0 ← **Minimum value of $f(x, y)$**

$f\left(\dfrac{5}{2}, 0\right) = \dfrac{5}{2} + 3(0) = \dfrac{5}{2}$ or $2\dfrac{1}{2}$

$f\left(\dfrac{13}{3}, \dfrac{11}{3}\right) = \dfrac{13}{3} + 3\left(\dfrac{11}{3}\right) = \dfrac{46}{3}$ or $15\dfrac{1}{3}$

$f(0, 8) = 0 + 3(8)$ or 24 ← **Maximum value of $f(x, y)$**

So, the maximum value of f is 24 when $x = 0$ and $y = 8$. The minimum value of f is 0 when $x = 0$ and $y = 0$.

✓ Guided Practice

Find the maximum and minimum values of the objective function $f(x, y)$ and for what values of x and y they occur, subject to the given constraints.

1A. $f(x, y) = 2x + 5y$
  $x + y \geq -3$
  $6x + 3y \leq 24$
  $x \geq 0$
  $y \geq 0$

1B. $f(x, y) = 5x - 6y$
  $y \leq 6$
  $y \geq 2x - 2$
  $y \geq -3x - 12$

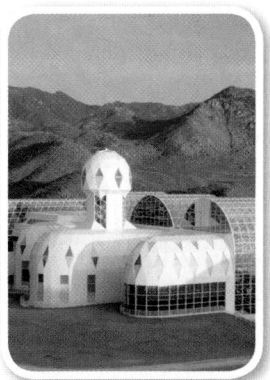

● Real-World Link

Biosphere 2 in Oracle, Arizona, is a center for research and development of self-sustaining space-colonization technology. The greenhouse consists of 7,200,000 cubic feet of sealed glass, 6500 windows, with a high point of 91 feet.

Source: The University of Arizona

● Real-World EXAMPLE 2 / Maximize Profit

BUSINESS A garden center grows only junipers and azaleas in a greenhouse that holds up to 3000 shrubs. Due to labor costs, the number of azaleas grown must be less than or equal to 1200 plus three times the number of junipers. The market demand for azaleas is at least twice that of junipers. The center makes a profit of $2 per juniper and $1.50 per azalea.

a. Write an objective function and a list of constraints that model the given situation.

Let x represent the number of junipers produced and y the number of azaleas. The objective function is then given by $f(x, y) = 2x + 1.5y$.

The constraints are given by the following.

$y \geq 2x$	**Market demand constraint**
$y \leq 3x + 1200$	**Production constraint**
$x + y \leq 3000$	**Greenhouse capacity constraint**

Because x and y cannot be negative, additional constraints are that $x \geq 0$ and $y \geq 0$.

b. Sketch a graph of the region determined by the constraints from part a to find how many of each plant the company should grow to maximize profit.

The shaded polygonal region has four vertex points at $(0, 0)$, $(0, 1200)$, $(450, 2550)$, and $(1000, 2000)$. Find the value of $f(x, y) = 2x + 1.5y$ at each of the four vertices.

$f(0, 0) = 2(0) + 1.5(0)$ or 0

$f(0, 1200) = 2(0) + 1.5(1200)$ or 1800

$f(450, 2550) = 2(450) + 1.5(2550)$ or 4725

$f(1000, 2000) = 2(1000) + 1.5(2000)$ or 5000 ⟵ **Maximum value of f(x, y)**

Because f is greatest at $(1000, 2000)$, the garden center should grow 1000 junipers and 2000 azaleas to earn a maximum profit of $5000.

✔ Guided Practice

2. MANUFACTURING A lumber mill can produce up to 600 units of product each week. To meet the needs of its regular customers, the mill must produce at least 150 units of lumber and at least 225 units of plywood. The lumber mill makes a profit of $30 for each unit of lumber and $45 for each unit of plywood.

A. Write an objective function and a list of constraints that model the given situation.

B. Sketch a graph of the region determined by the constraints to find how many units of each type of wood product the mill should produce to maximize profit.

To better understand why the maximum value of $f(x, y) = 2x + 1.5y$ must occur at a vertex in Example 2, assign f different positive values from 0 to 5000 and then graph the corresponding family of parallel lines.

Notice that the distance of a line in this family from the origin increases as f increases, sweeping across the region of feasible solutions.

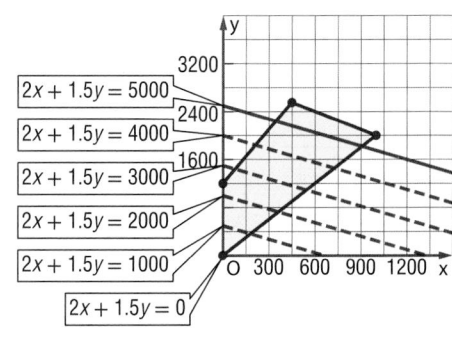

Geometrically, to maximize f over the set of feasible solutions, you want the line with the greatest f-value that still intersects the shaded region. From the graph you can see that such a line will intersect the shaded region at one point, the vertex at $(1000, 2000)$.

No or Multiple Optimal Solutions As with systems of linear equations, linear programming problems can have one, multiple, or no optimal solutions. If the graph of the equation related to the objective function f to be optimized is coincident with one side of the region of feasible solutions, f has **multiple optimal solutions**. In Figure 6.5.1, any point on the segment connecting vertices at (3, 7) and (7, 3) is an optimal solution of f. If the region does not form a polygon, but is instead **unbounded**, f may have no minimum value or no maximum value. In Figure 6.5.2, f has no maximum value.

Figure 6.5.1

Figure 6.5.2

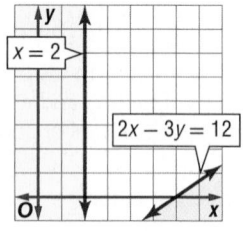
EXAMPLE 3 **Optimization at Multiple Points**

Find the maximum value of the objective function $f(x, y) = 4x + 2y$ and for what values of x and y it occurs, subject to the following constraints.

$y + 2x \leq 18$

$y \leq 6$

$x \leq 8$

$x \geq 0$

$y \geq 0$

Graph the region bounded by the given constraints. The polygon region of feasible solutions has five vertices at (0, 0), (8, 2), (0, 6), (8, 0), and (6, 6). Find the value of the objective function $f(x, y) = 4x + 2y$ at each vertex.

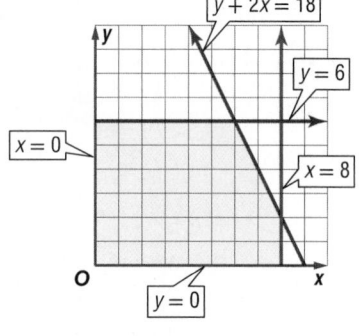

$f(0, 0) = 4(0) + 2(0)$ or 0

$f(8, 2) = 4(8) + 2(2)$ or 36

$f(0, 6) = 4(0) + 2(6)$ or 12

$f(8, 0) = 4(8) + 2(0)$ or 32

$f(6, 6) = 4(6) + 2(6)$ or 36

Because $f(x, y) = 36$ at (6, 6) and (8, 2), there are multiple points at which f is optimized. An equation of the line through these two vertices is $y = -2x + 18$. Therefore, f has a maximum value of 36 at every point on $y = -2x + 18$ for $6 \leq x \leq 8$.

✓ **Guided Practice**

Find the maximum and minimum values of the objective function $f(x, y)$ and for what values of x and y it occurs, subject to the given constraints.

3A. $f(x, y) = 3x + 3y$
$4x + 3y \geq 12$
$y \leq 3$
$y \geq 0$
$x \leq 4$
$x \geq 0$

3B. $f(x, y) = 4x + 8y$
$x + 2y \leq 16$
$y \geq 2$
$x \geq 3$

Real-World EXAMPLE 4   Unbounded Feasible Region

VETERINARY MEDICINE A veterinarian recommends that a new puppy eat a diet that includes at least 1.54 ounces of protein and 0.56 ounce of fat each day. Use the table to determine how much of each dog food should be used in order to satisfy the dietary requirements at the minimum cost.

Dog Food Brand	Protein (oz/cup)	Fat (oz/cup)	Cost per cup ($)
Good Start	0.84	0.21	0.36
Sirius	0.56	0.49	0.22

a. Write an objective function, and list the constraints that model the given situation.

Let x represent the number of cups of *Good Start* eaten and y represent the number of cups of *Sirius* eaten. The objective function is then given by $f(x, y) = 0.36x + 0.22y$.

The constraints on required fat and protein are given by

$0.84x + 0.56y \geq 1.54$ **Protein constraint**

$0.21x + 0.49y \geq 0.56$ **Fat constraint**

Because x and y cannot be negative, there are also constraints of $x \geq 0$ and $y \geq 0$.

b. Sketch a graph of the region determined by the constraints from part a to find how many cups of each dog food should be used in order to satisfy the dietary requirements at the optimal cost.

The shaded polygonal region has three vertex points at (0, 2.75), (1.5, 0.5), and (2.67, 0).

The optimal cost would be the minimum value of $f(x, y) = 0.36x + 0.22y$. Find the value of the objective function at each vertex.

$f(\mathbf{0, 2.75}) = 0.36(\mathbf{0}) + 0.22(\mathbf{2.75})$, or 0.605   ← **Minimum value of f(x, y)**

$f(\mathbf{1.5, 0.5}) = 0.36(\mathbf{1.5}) + 0.22(\mathbf{0.5})$, or 0.65

$f(\mathbf{2.67, 0}) = 0.36(\mathbf{2.67}) + 0.22(\mathbf{0})$, or 0.9612

Therefore, to meet the veterinarian's requirements at a minimum cost of about $0.61 per cup, the puppy should eat 2.75 cups of only the *Sirius* brand.

✓ Guided Practice

4. MANAGEMENT According to the manager of a pizza shop, the productivity in worker-hours of her workers is related to their positions. One worker-hour is the amount of work done by an average employee in one hour. For the next 8-hour shift, she will need two shift leaders, at least two associates, and at least 10 total workers. She will also need to schedule at least 120 worker-hours to meet customer demand during that shift.

Employees Working	Productivity (in worker-hours)	Wage ($)
associate	1.5	7.50
employee	1.0	6.50
shift leader	2.0	9.00

A. Assuming that each employee works an entire 8-hour shift, write an objective function and list the constraints that model the given situation.

B. Sketch a graph of the region determined by the constraints to find how many workers should be scheduled to optimize labor costs.

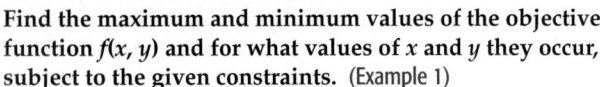
Find the maximum and minimum values of the objective function $f(x, y)$ **and for what values of** x **and** y **they occur, subject to the given constraints.** (Example 1)

1. $f(x, y) = 3x + y$
$y \le 2x + 1$
$x + 2y \le 12$
$1 \le y \le 3$

2. $f(x, y) = -x + 4y$
$y \le x + 4$
$y \ge -x + 3$
$1 \le x \le 4$

3. $f(x, y) = x - y$
$x + 2y \le 6$
$2x - y \le 7$
$x \ge -2$
$y \ge -3$

4. $f(x, y) = 3x - 5y$
$x \ge 0, y \ge 0$
$x + 2y \le 6$
$2y - x \le 2$
$x + y \le 5$

5. $f(x, y) = 3x - 2y$
$y \le x + 3$
$1 \le x \le 5$
$y \ge 2$

6. $f(x, y) = 3y + x$
$4y \le x + 8$
$2y \ge 3x - 6$
$2x + 2y \ge 4$

7. $f(x, y) = x - 4y$
$x \ge 2, y \ge 1$
$x - 2y \ge -4$
$2x - y \le 7$
$x + y \le 8$

8. $f(x, y) = x - y$
$3x - 2y \ge -7$
$x + 6y \ge -9$
$5x + y \le 13, x - 3y \ge -7$

9 **MEDICAL OFFICE** Olivia is a receptionist for a medical clinic. One of her tasks is to schedule appointments. She allots 20 minutes for a checkup and 40 minutes for a physical. The doctor can do no more than 6 physicals per day, and the clinic has 7 hours available for appointments. A checkup costs $55, and a physical costs $125. (Example 2)

 a. Write an objective function and list the constraints that model the given situation.

 b. Sketch a graph of the region determined by the constraints from part **a** to find the set of feasible solutions for the objective function.

 c. How many of each appointment should Olivia make to maximize income? What is the maximum income?

10. **INCOME** Josh is working part-time to pay for some of his college expenses. Josh delivers pizza for $5 per hour plus tips, which run about $8 per hour, and he also tutors in the math lab for $15 per hour. The math lab is open only 2 hours daily, Monday through Friday, when Josh is available to tutor. Josh can work no more than 20 hours per week due to his class schedule. (Example 2)

 a. Write an objective function and list the constraints that model the given situation.

 b. Sketch a graph of the region determined by the constraints from part **a** to find the set of feasible solutions for the objective function.

 c. How can Josh make the most money, and how much is it?

11. **SMALL BUSINESS** A design company creates Web sites and E-albums. Each Web site requires 10 hours of planning and 4.5 hours of page design. Each family E-album requires 15 hours of planning and 9 hours of page design. There are 70 hours available each week for the staff to plan and 36 hours for page design. (Example 2)

 a. If the profit is $600 for each Web site and $700 for each family E-album, write an objective function and list the constraints that model the given situation.

 b. Sketch a graph of the region determined by the constraints from part **a** to find the set of feasible solutions for the objective function.

 c. How much of each product should be produced to achieve maximum profit? What is the profit?

Find the maximum and minimum values of the objective function $f(x, y)$ **and for what values of** x **and** y **they occur, subject to the given constraints.** (Examples 3 and 4)

12. $f(x, y) = 4x - 3y$
$2x + y \ge -7$
$y \le x + 2$
$y \le 11 - 2x$

13. $f(x, y) = 3x + 6y$
$y \le -\frac{1}{2}x + \frac{5}{2}$
$y \le 2, y \ge 0$
$x \le 3, x \ge 0$

14. $f(x, y) = -3x - 6y$
$y \le -\frac{1}{2}x + 5$
$y \le 4, y \ge 0$
$x \le 6, x \ge 0$

15. $f(x, y) = 6x - 4y$
$2x + 3y \ge 6$
$3x - 2y \ge -4$
$5x + y \ge 15$

16. $f(x, y) = 3x + 4y$
$y \le x - 3$
$y \le 6 - 2x$
$2x + y \ge -3$

17. $f(x, y) = 8x + 10y$
$y \le -\frac{4}{5}x + 4$
$y \le 4, y \ge 0$
$x \le 5, x \ge 0$

18. **NUTRITION** Michelle wants to consume more nutrients. She wants to receive at least 40 milligrams of calcium, 600 milligrams of potassium, and 50 milligrams of vitamin C. Michelle's two favorite fruits are apples and bananas. The average nutritional content of both are given. (Example 4)

Fruit	Calcium	Potassium	Vitamin C
apple	9.5 mg	158 mg	9 mg
banana	7.0 mg	467 mg	11 mg

 a. If each apple costs $0.55 and each banana costs $0.35, write an objective function. List the constraints that model the given situation.

 b. Sketch a graph of the region determined by the constraints from part **a** to find the set of feasible solutions for the objective function.

 c. Determine the number of each type of fruit that Michelle should eat to minimize her cost while still obtaining her desired nutritional intake.

Find a value of *a* so that each objective function has maximum values at the indicated vertex, subject to the following constraints.

$$x \geq 0$$
$$y \geq 2$$
$$x + y \leq 9$$
$$-4x + 3y \leq 6$$

19. $f(x, y) = -4x + ay$, $(0, 2)$ **20.** $f(x, y) = -4x + ay$, $(3, 6)$

21. $f(x, y) = x - ay$, $(3, 6)$ **22.** $f(x, y) = x - ay$, $(7, 2)$

23. $f(x, y) = ax + 4y$, $(7, 2)$ **24.** $f(x, y) = ax - 3y$, $(0, 2)$

Find an objective function that has a maximum or minimum value at each indicated vertex.

25. minimum at *A*

26. maximum at *C*

27. maximum at *B*

28. minimum at *D*

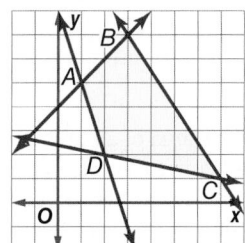

29. BUSINESS A batch of Mango Sunrise uses 3 liters of mango juice and 1 liter of strawberry juice. A batch of Oasis Dream uses 2 liters of mango juice and 1 liter of strawberry juice. The store has 40 liters of mango juice and 15 liters of strawberry juice that it wants to use up before the end of the day. The profit on Mango Sunrise is $16 per batch, and the profit on the Oasis Dream is $12 per batch.

  a. Write an objective function, and list the constraints that model the given situation.

  b. In order to maximize profits, how many batches of each drink should the juice store make?

Find the maximum and minimum values of the objective function *f(x, y)* and for what values of *x* and *y* they occur, subject to the given constraints.

30. $f(x, y) = 4x - 8y$
$$y \geq x^2 - 8x + 18$$
$$y \leq -x^2 + 8x - 10$$
$$y \leq 8 - x$$

31. $f(x, y) = -2x + 5y$
$$y \geq x^2 + 6x + 3$$
$$y \leq -x^2 - 4x + 15$$
$$y \leq x + 9$$

Find the area enclosed by the polygonal convex set defined by each system of inequalities.

32. $x \geq 0$
$$x \leq 12$$
$$2x + 6y \leq 84$$
$$2x - 3y \leq -3$$
$$8x + 3y \geq 33$$

33. $y \leq 9$
$$x \geq 2$$
$$x - y \leq 5$$
$$2x + y \leq 25$$
$$3x + 2y \leq 20$$

34. $x \geq 0$
$$y \geq \frac{1}{2}x$$
$$3x + 2y \geq 8$$
$$-3x + 4y \leq 28$$
$$x + 2y \leq 24$$

35. $x \leq 10$
$$x + y \leq 14$$
$$x + 3y \geq 13$$
$$-x + 5y \leq 40$$
$$4x + y \geq 8$$

36. CHALLENGE Provide a system of inequalities that forms the polygonal convex set shown below.

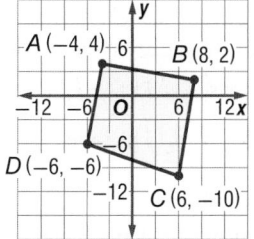

37. REASONING Consider the profit function $P(x, y) = ax + by$. Will *P* always have a positive maximum value if the feasible region lies entirely within the first quadrant? Explain your reasoning.

38. CHALLENGE Find the maximum and minimum values of the objective function $f(x, y) = -6x + 3y$ and for what values of *x* and *y* they occur, subject to the given constraints.

$$y \geq 4$$
$$3x + 2y \geq 14$$
$$-2x + 5y \leq 60$$
$$-x + y \geq -3$$
$$-7x + 5y \leq 35$$
$$2x + y \leq 36$$

39. OPEN ENDED Linear programming has numerous real-world applications.

  a. Write a real-world problem that could be solved using linear programming.

  b. Using at least 4 constraints, write an objective function to be maximized or minimized.

  c. Sketch a graph of the region determined by the constraints from part **a** to find the set of feasible solutions.

  d. Find the solution to the problem.

40. WRITING IN MATH Is it possible for a linear programming problem to have no maximum solution and no minimum solution? Explain your reasoning.

Find the partial fraction decomposition of each rational expression. (Lesson 6-4)

41. $\dfrac{8y + 7}{y^2 + y - 2}$

42. $\dfrac{x - 6}{x^2 - 2x}$

43. $\dfrac{5m - 4}{m^2 - 4}$

44. $\dfrac{-4y}{3y^2 - 4y + 1}$

45. ARCADE GAMES Marcus and Cody purchased game cards to play virtual games at the arcade. Marcus used 47 points from his game card to drive the race car and snowboard simulators four times each. Cody used 48.25 points from his game card to drive the race car simulator five times and the snowboard simulator three times. How many points did each game require per play? (Lesson 6-3)

46. WAVES After a wave is created by a boat, the height of the wave can be modeled using $y = \frac{1}{2}h + \frac{1}{2}h \sin \frac{2\pi t}{P}$, where h is the maximum height of the wave in feet, P is the period in seconds, and t is the propagation of the wave in seconds. (Lesson 5-3)

   a. If $h = 3$ and $P = 2$, write the equation for the wave. Draw its graph over a 10-second interval.

   b. How many times over the first 10 seconds does the graph predict the wave to be one foot high?

47. Verify that $\tan \theta \sin \theta \cos \theta \csc^2 \theta = 1$ is an identity. (Lesson 5-2)

Find the area of each triangle to the nearest tenth. (Lesson 4-7)

48. $\triangle ABC$, if $A = 127°$, $b = 12$ m, and $c = 9$ m

49. $\triangle ABC$, if $a = 7$ yd, $b = 8$ yd, and $C = 44°$

50. $\triangle ABC$, if $A = 50°$, $b = 15$ in., and $c = 10$ in.

51. $\triangle ABC$, if $a = 6$ cm, $B = 135°$, and $c = 3$ cm

52. SAT/ACT In the figure below, lines l, m, and n intersect in a single point. What is the value of $x + y$?

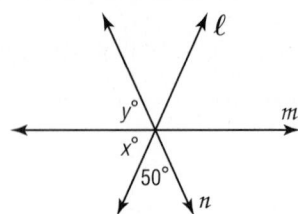

   A 40   **C** 90   **E** 260

   B 70   **D** 130

53. The area of a parking lot is 600 square meters. A car requires 6 square meters of space, and a bus requires 30 square meters of space. The attendant can handle no more than 60 vehicles. If it cost $3 to park a car and $8 to park a bus, how many of each should the attendant accept to maximize income?

   F 20 buses and 0 cars

   G 10 buses and 50 cars

   H 5 buses and 55 cars

   J 0 buses and 60 cars

54. FREE RESPONSE Use the two systems of equations to answer each of the following.

$$\begin{array}{ll}
\quad\quad\quad A & \quad\quad\quad B \\
-5x + 2y + 11z = 31 & x + 2y + 2z = 3 \\
2y + 6z = 26 & 3x + 7y + 9z = 30 \\
2x - y - 5z = -15 & -x - 4y - 7z = -37
\end{array}$$

   a. Write the coefficient matrix for each system. Label the matrices A and B.

   b. Find AB and BA if possible.

   c. Write the augmented matrix for system A in reduced row-echelon form.

   d. Find the determinant of each coefficient matrix. Which matrices are invertible? Explain your reasoning.

   e. Find the inverse of matrix B.

   f. Use the inverse of B to solve the system.

   g. Which systems could you use Cramer's Rule to solve? Explain your reasoning.

Chapter Summary

Key Concepts

Multivariable Linear Systems and Row Operations
(Lesson 6-1)

- Each of these row operations produces an equivalent augmented matrix.
  - Interchange any two rows.
  - Multiply one row by a nonzero real number.
  - Add a multiple of one row to another row.

Multiplying Matrices (Lesson 6-2)

- If A is an $m \times r$ matrix and B is an $r \times n$ matrix, then the product AB is an $m \times n$ matrix in which
$$c_{ij} = a_{i1}b_{1j} + a_{i2}b_{2j} + \ldots + a_{ir}b_{rj}.$$

- I_n is an $n \times n$ matrix consisting of all 1s on its main diagonal and 0s for all other elements.

- The inverse of A is A^{-1} where $AA^{-1} = A^{-1}A = I_n$.

- If $A = \begin{bmatrix} a & b \\ c & d \end{bmatrix}$ and $ad - cb \neq 0$, then $A^{-1} = \dfrac{1}{ad - cb}$
$\begin{bmatrix} d & -b \\ -c & a \end{bmatrix}$. The number $ad - cb$ is called the *determinant* of the 2×2 matrix and is denoted by $\det(A) = |A| = \begin{vmatrix} a & b \\ c & d \end{vmatrix}$.

Solving Systems (Lesson 6-3)

- Suppose $AX = B$, where A is the matrix of coefficients of a linear system, X is the matrix of variables, and B is the matrix of constant terms. If A is invertible, then $AX = B$ has a unique solution given by $X = A^{-1}B$.

- If $\det(A) \neq 0$, then the unique solution of a system is given by
$x_1 = \dfrac{|A_1|}{|A|}, x_2 = \dfrac{|A_2|}{|A|}, x_3 = \dfrac{|A_3|}{|A|}, \ldots, x_n = \dfrac{|A_n|}{|A|}$. If $\det(A) = 0$, then $AX = B$ has no solution or infinitely many solutions.

Partial Fractions (Lesson 6-4)

- If the degree of $f(x)$ is greater than or equal to the degree of $d(x)$, use polynomial long division and the division algorithm to write $\dfrac{f(x)}{d(x)} = q(x) + \dfrac{r(x)}{d(x)}$. Then apply partial fraction decomposition to $\dfrac{r(x)}{d(x)}$.

Linear Optimization (Lesson 6-5)

- The maximum and minimum values of a linear function in x and y are determined by linear programming techniques.
  Step 1. Graph the solution of the system of constraints.
  Step 2. Find the coordinates of the vertices of the region.
  Step 3. Evaluate the objective function at each vertex to find which values maximize or minimize the function.

Key Vocabulary

augmented matrix (p. 366)
coefficient matrix (p. 366)
constraint (p. 405)
Cramer's Rule (p. 390)
determinant (p. 381)
feasible solution (p. 405)
Gaussian elimination (p. 364)
identity matrix (p. 378)
inverse (p. 379)
inverse matrix (p. 379)
invertible (p. 379)
linear programming (p. 405)

multiple optimal solutions (p. 408)
multivariable linear system (p. 364)
objective function (p. 405)
optimization (p. 405)
partial fraction (p. 398)
partial fraction decomposition (p. 398)
reduced row-echelon form (p. 369)
row-echelon form (p. 364)
singular matrix (p. 379)
square system (p. 388)
unbounded (p. 408)

Vocabulary Check

Choose the word or phrase that best completes each sentence.

1. A(n) (augmented matrix, coefficient matrix) is a matrix made up of all the coefficients and constant terms of a linear system.

2. (Gaussian elimination, Elementary row operations) reduce(s) a system of equations to an equivalent, simpler system, making it easier to solve.

3. The result of Gauss-Jordan elimination is a matrix that is in (reduced row-echelon, invertible) form.

4. The product of an $n \times n$ matrix A with the (inverse matrix, identity matrix) is A.

5. The identity matrix I is its own (augmented matrix, inverse matrix).

6. A square matrix that has no inverse is (nonsingular, singular).

7. The (determinant, square system) of $A = \begin{bmatrix} a & b \\ c & d \end{bmatrix}$ is $ad - bc$.

8. When solving a square linear system, an alternative to Gaussian elimination is (Cramer's Rule, partial fraction decomposition).

9. A two-dimensional linear programming problem contains (constraints, feasible solutions), which are linear inequalities.

10. If the graph of the objective function to be optimized, f, is coincident with one side of the region of feasible solutions, then there may be (multiple optimal, unbounded) solutions.

Lesson-by-Lesson Review

6-1   Multivariable Linear Systems and Row Operations (pp. 364–374)

Write each system of equations in triangular form using Gaussian elimination. Then solve the system.

11. $3x + 4y = 7$
$2y = -5x + 7$

12. $5x - 3y = 16$
$x + 3y = -4$

13. $x + y + z = 4$
$2x - y - 3z = 4$
$-3x - 4y - 5z = -13$

14. $x + y - z = 5$
$2x - 3y + 5z = -1$
$3x - y + 2z = 10$

15. $2x - 5y = 2z + 11$
$3y + 4z = x - 28$
$3z - x = -18 - 3y$

16. $2x - 3z = y - 1$
$5z - 8 = 3x + 4y$
$x + y + z = 3$

Solve each system of equations.

17. $2x + 2y = 8$
$3x - 8y = -21$

18. $x - 2y = 13$
$-5x - 6y = 15$

19. $x + y = 4$
$x + y + z = 7$
$x - z = -1$

20. $x + y = 1$
$3x - 7y + z = -7$
$4x + 8y + 3z = -9$

21. $3x - y + z = 8$
$2x - 3y = 3z - 13$
$x + z = 6 - y$

22. $x + y = z - 1$
$2x + 2y + z = 13$
$3x - 5y + 4z = 8$

EXAMPLE 1

Solve the system of equations using Gaussian elimination.
$x + 2y + 3z = 8$
$2x - 4y + z = 2$
$-3x - 6y + 7z = 8$

Write the augmented matrix. Then apply elementary row operations to obtain a row echelon form.

Augmented matrix
$$\begin{bmatrix} 1 & 2 & 3 & | & 8 \\ 2 & -4 & 1 & | & 2 \\ -3 & -6 & 7 & | & 8 \end{bmatrix}$$

$$\begin{array}{c} -2R_1 + R_2 \rightarrow \\ 3R_1 + R_3 \rightarrow \end{array} \begin{bmatrix} 1 & 2 & 3 & | & 8 \\ 0 & -8 & -5 & | & -14 \\ 0 & 0 & 16 & | & 32 \end{bmatrix}$$

$$\frac{1}{16}R_3 \rightarrow \begin{bmatrix} 1 & 2 & 3 & | & 8 \\ 0 & -8 & -5 & | & -14 \\ 0 & 0 & 1 & | & 2 \end{bmatrix}$$

You can use substitution to find that $y = 0.5$ and $x = 1$. Therefore, the solution of the system is $x = 1$, $y = 0.5$, and $z = 2$, or the ordered triple $(1, 0.5, 2)$.

6-2   Matrix Multiplication, Inverses, and Determinants (pp. 375–386)

Find AB and BA, if possible.

23. $A = \begin{bmatrix} 1 & -3 & 7 \\ 2 & 0 & 1 \end{bmatrix}$
$B = \begin{bmatrix} -5 & -4 \\ 1 & 7 \end{bmatrix}$

24. $A = \begin{bmatrix} -2 & 3 \\ -4 & 7 \end{bmatrix}$
$B = \begin{bmatrix} 1 & 1 \\ 0 & 1 \end{bmatrix}$

25. $A = \begin{bmatrix} 4 & -3 \end{bmatrix}$
$B = \begin{bmatrix} -8 & 5 \\ -7 & -1 \end{bmatrix}$

26. $A = \begin{bmatrix} 1 & 3 \end{bmatrix}$
$B = \begin{bmatrix} 5 \\ -4 \end{bmatrix}$

Find A^{-1}, if it exists. If A^{-1} does not exist, write *singular*.

27. $A = \begin{bmatrix} 2 & 1 \\ -3 & 8 \end{bmatrix}$

28. $A = \begin{bmatrix} 1 & 2 & 3 \\ 0 & 0 & 1 \\ 2 & -1 & 3 \end{bmatrix}$

29. $A = \begin{bmatrix} 2 & 3 \\ -8 & -12 \end{bmatrix}$

30. $A = \begin{bmatrix} -5 & 2 \\ 3 & 8 \end{bmatrix}$

EXAMPLE 2

Let $A = \begin{bmatrix} 1 & 4 \\ 2 & 9 \end{bmatrix}$. Find A^{-1}, if it exists. If A^{-1} does not exist, write *singular*.

First, write a doubly augmented matrix. Then apply elementary row operations to write the matrix in reduced row-echelon form.

Augmented matrix $\rightarrow \begin{bmatrix} 1 & 4 & | & 1 & 0 \\ 2 & 9 & | & 0 & 1 \end{bmatrix}$

$-2R_1 + R_2 \rightarrow \begin{bmatrix} 1 & 4 & | & 1 & 0 \\ 0 & 1 & | & -2 & 1 \end{bmatrix}$

$-4R_2 + R_1 \rightarrow \begin{bmatrix} 1 & 0 & | & 9 & -4 \\ 0 & 1 & | & -2 & 1 \end{bmatrix}$

Because the system has a solution, $a = 9$, $b = -4$, $c = -2$, and $d = 1$, A is invertible and $A^{-1} = \begin{bmatrix} 9 & -4 \\ -2 & 1 \end{bmatrix}$.

Lesson-by-Lesson Review

6-3 Solving Linear Systems Using Inverses and Cramer's Rule (pp. 388–394)

Use an inverse matrix to solve each system of equations, if possible.

31. $2x - 3y = -23$
$3x + 7y = 23$

32. $3x - 6y = 9$
$-5x - 8y = -6$

33. $2x + y = 1$
$x - 3y + z = -4$
$y + 8z = -7$

34. $x + y + z = 1$
$x + y - z = -7$
$y + z = -1$

35. $3y + 5z = 25$
$2x - 7y - 3z = 15$
$x + y - z = -11$

36. $x - 2y - 3z = 0$
$2x - 3y + 4z = 11$
$x - 8y + 2z = -1$

Use Cramer's Rule to find the solution of each system of linear equations, if a unique solution exists.

37. $2x - 4y = 30$
$3x + 5y = 12$

38. $2x + 6y = 14$
$x - 3y = 1$

39. $2x + 3y - z = 1$
$x + y - 3z = 12$
$5x - 7y + 2z = 28$

40. $x + 2y + z = -2$
$2x + 2y - 5z = -19$
$3x - 4y + 8z = -1$

41. $-3x - 4y + z = 15$
$x - 5y - z = 3$
$4x - 3y - 2z = -8$

42. $2x + 3y + 4z = 29$
$x - 8y - z = -3$
$2x + y + z = 4$

EXAMPLE 3

Use an inverse matrix to solve the system of equations, if possible.
$x - y + z = -5$
$2x + 2y - 3z = -27$
$-3x - y + z = 17$

Write the system in matrix form.

$$\underset{A}{\begin{bmatrix} 1 & -1 & 1 \\ 2 & 2 & -3 \\ -3 & -1 & 1 \end{bmatrix}} \cdot \underset{X}{\begin{bmatrix} x \\ y \\ z \end{bmatrix}} = \underset{B}{\begin{bmatrix} -5 \\ -27 \\ 17 \end{bmatrix}}$$

Use a graphing calculator to find A^{-1}.

$$A^{-1} = \begin{bmatrix} 0.25 & 0 & -0.25 \\ -1.75 & -1 & -1.25 \\ -1 & -1 & -1 \end{bmatrix}$$

Multiply A^{-1} by B to solve the system.

$$X = \begin{bmatrix} 0.25 & 0 & -0.25 \\ -1.75 & -1 & -1.25 \\ -1 & -1 & -1 \end{bmatrix} \cdot \begin{bmatrix} -5 \\ -27 \\ 17 \end{bmatrix} = \begin{bmatrix} -5.5 \\ 14.5 \\ 15 \end{bmatrix}$$

Therefore, the solution is $(-5.5, 14.5, 15)$.

6-4 Partial Fractions (pp. 398–404)

Find the partial fraction decomposition of each rational expression.

43. $\dfrac{2x}{x^2 - 4}$

44. $\dfrac{7x - 6}{x^2 - x - 6}$

45. $\dfrac{-2x + 9}{x^2 - 11x + 30}$

46. $\dfrac{6x^2 - 4x - 6}{x^3 - 2x^2 - 3x}$

47. $\dfrac{2x^3 - 14x^2 + 2x + 7}{x^2 - 7x}$

48. $\dfrac{2x^4 + 3x^3 + 5x^2 + 3x + 2}{x(x^2 + 1)^2}$

49. $\dfrac{2x^2 + 4}{x^2 - 2x}$

50. $\dfrac{2x^2 - 12x - 20}{x^2 + 4x}$

51. $\dfrac{x^2 + x - 6}{2x^2 - 3x}$

52. $\dfrac{3x^2 - 10x - 20}{2x^2 + 5x}$

EXAMPLE 4

Find the partial fraction decomposition of $\dfrac{x - 8}{x^2 - 11x + 18}$.

Rewrite the equation as partial fractions with constant numerators, A and B.

$$\frac{x + 12}{x^2 - 11x + 18} = \frac{A}{x - 9} + \frac{B}{x - 2}$$

$$x + 12 = A(x - 2) + B(x - 9)$$

$$x + 12 = Ax - 2A + Bx - 9B$$

$$x + 12 = (A + B)x + (-2A - 9B)$$

Equate the coefficients on the left and right sides of the equation to obtain a system of two equations.

$$A + B = 1$$

$$-2A - 9B = 12$$

The solution of the system is $A = 3$ and $B = -2$.

Therefore, $\dfrac{x + 12}{x^2 - 11x + 18} = \dfrac{3}{x - 9} + \dfrac{-2}{x - 2}$.

Lesson-by-Lesson Review

6-5   Linear Optimization   (pp. 405–412)

Find the maximum and minimum values of the objective function $f(x, y)$ and for what values of x and y they occur, subject to the given constraints.

53. $f(x, y) = 2x - y$
$x \geq 0$
$y \leq -x + 7$
$y \geq x + 1$

54. $f(x, y) = 3x + y$
$2x - y \leq 1$
$1 \leq y \leq 9$
$x \geq 1$

55. $f(x, y) = x + y$
$0 \leq x \leq 10$
$x + 2y \geq 8$
$0 \leq y \leq 8$

56. $f(x, y) = 2x - 4y$
$x \geq 3$
$y \geq 3$
$4x + 5y \leq 47$

57. $f(x, y) = 4x + 3y$
$3x + y \geq 8$
$2x + y \leq 12$
$y \geq x$

58. $f(x, y) = 2y - 5x$
$2x + y \geq 0$
$x - 5y \leq 0$
$3x + 7y \leq 22$

EXAMPLE 5

Find the maximum value of the objective function $f(x, y) = 2x + 6y$ and for what values of x and y they occur, subject to the following constraints.
$y \geq 0, x \geq 0, y \leq 6, 3x + 2y \leq 18$

Graph the region bounded by the given constraints. Find the value of the objective function $f(x, y) = 2x + 6y$ at each vertex.

$f(0, 0) = 0 + 0$ or 0
$f(0, 6) = 0 + 36$ or 36
$f(6, 0) = 12 + 0$ or 12
$f(2, 6) = 4 + 36$ or 40

So, the maximum value of f is 40 when $x = 2$ and $y = 6$.

Applications and Problem Solving

59. HAMBURGERS The table shows the number of hamburgers, cheeseburgers, and veggie burgers sold at a diner over a 3-hour lunch period. Find the price for each type of burger. (Lesson 6-1)

Hours	Plain	Cheese	Veggie	Total Sales ($)
11 A.M.–12 P.M.	2	8	2	53
12–1 P.M.	7	12	8	119
1–2 P.M.	1	5	7	64

60. GRADING Ms. Hebert decides to base grades on tests, homework, projects and class participation. She assigns a different percentage weight for each category, as shown. Find the final grade for each student to the nearest percent. (Lesson 6-2)

Category	tests	HW	projects	participation
Weight	40%	30%	20%	10%

Category	Serena	Andrew	Corey	Shannon
tests	88	72	78	91
HW	95	90	68	71
projects	80	73	75	85
participation	100	95	100	80

61. SHAVED ICE A shaved ice stand sells 3 flavors: strawberry, pineapple, and cherry. Each flavor sells for $1.25. One day, the stand had $60 in total sales. The stand made $13.75 more in cherry sales than pineapple sales and $16.25 more than strawberry sales. Use Cramer's Rule to determine how many of each flavor was sold. (Lesson 6-3)

62. BIKING On a biking trip, a couple traveled 240 miles on Day 1 and 270 miles on Day 2. The average rate traveled during Day 1 is 5 miles per hour faster than the average rate traveled during Day 2. The total number of hours spent biking is $T = \dfrac{510r - 1200}{r(r - 5)}$. (Lesson 6-4)

  a. Find the partial fraction decomposition of T.

  b. Each fraction represents the time spent biking each day. If the couple biked 6 hours longer on Day 2, how many total hours did they bike?

63. RECYCLING A recycling company will collect from private sites if the site produces at least 60 pounds of recycling a week. The company can collect at most 50 pounds of paper and 30 pounds of glass from each site. The company earns $20 for each pound of glass and $25 for each pound of paper. (Lesson 6-5)

  a. Write an objective function, and list the constraints that model the given situation.

  b. Determine the number of pounds of glass and paper needed to produce the maximum profit.

  c. What is the maximum profit?

Write each system of equations in triangular form using Gaussian elimination. Then solve the system.

1. $-3x + y = 4$
   $5x - 7y = 20$

2. $x + 4y - 3z = -8$
   $5x - 7y + 3z = -4$
   $3x - 2y + 4z = 24$

Solve the system of equations.

3. $5x - 6y = 28$
   $6x + 5y = -3$

4. $2x - 4y + z = 8$
   $3x + 3y + 4z = 20$
   $5x + y - 3z = -13$

5. **LIBRARY** Kristen checked out books, CDs, and DVDs from the library. She checked out a total of 16 items. The total number of CDs and DVDs equaled the number of books. She checked out two more CDs than DVDs.

   a. Let b = number of books, c = number of CDs, and d = number of DVDs. Write a system of three linear equations to represent the problem.

   b. Solve the system of equations. Interpret your solution.

Find AB and BA, if possible.

6. $A = \begin{bmatrix} 1 & 0 & 0 & 0 \\ 0 & 1 & 3 & 3 \\ 0 & 0 & 1 & 2 \\ 0 & 0 & 0 & 6 \end{bmatrix}$, $B = \begin{bmatrix} 1 & 1 \\ 2 & 3 \\ 5 & -1 \\ -1 & 6 \end{bmatrix}$

7. $A = \begin{bmatrix} 1 & -5 & 4 \\ -2 & 3 & 5 \\ 6 & -3 & 1 \end{bmatrix}$, $B = [2 \quad -1 \quad -8]$

8. **GEOMETRY** The coordinates of a point (x, y) can be written as a 2×1 matrix $\begin{bmatrix} x \\ y \end{bmatrix}$. Let $A = \begin{bmatrix} 0 & -1 \\ 1 & 0 \end{bmatrix}$.

   a. Let P be the point $(-3, 4)$. Discuss what effect multiplying A by P has on P.

   b. A triangle contains vertices $(0, 0)$, $(2, 6)$, and $(8, 3)$. Create B, a 2×3 matrix to represent the triangle. Find AB. What is the effect on the triangle? Does it agree with your answer to part **a**?

Find A^{-1}, if it exists. If A^{-1} does not exist, write *singular*.

9. $A = \begin{bmatrix} 1 & 2 \\ -3 & -4 \end{bmatrix}$

10. $A = \begin{bmatrix} -3 & -5 \\ -6 & 8 \end{bmatrix}$

Use an inverse matrix to solve each system of equations, if possible.

11. $2x - 3y = -7$
    $5x + 2y = 11$

12. $2x + 2y + 5z = -6$
    $2x - 3y + 7z = -7$
    $x - 5y + 9z = 4$

Use Cramer's Rule to find the solution of each system of linear equations, if a unique solution exists.

13. $3x - 2y = -2$
    $4x - 2y = 2$

14. $3x - 2y - 3z = -24$
    $3x + 5y + 2z = 7$
    $-x + 5y + 3z = 25$

Find the partial fraction decomposition of each rational expression.

15. $\dfrac{4x}{x^2 - 9}$

16. $\dfrac{2x + 10}{x^2 - 4x + 3}$

Find the maximum and minimum values of the objective function $f(x, y)$ and for what values of x and y they occur, subject to the given constraints.

17. $f(x, y) = 2x - y$
    $x \geq 0$
    $y \geq 0$
    $y \geq -2x + 8$

18. $f(x, y) = -x + 2y$
    $x - 3y \leq 0$
    $x \geq 0$
    $y \leq 9$

19. **PRICING** The Harvest Nut Company sells create-your-own trail mixes where customers can choose whatever combinations they want. Colin's favorite mix contains peanuts, dried cranberries, and carob-coated pretzels. The prices for each are shown below. If Colin bought a 5-pound mixture for $16.80 that contained twice as many pounds of carob-coated pretzels as cranberries, how many pounds of each item did he buy?

Peanuts	Cranberries	Pretzels
$3.20/lb	$2.40/lb	$4.00/lb

20. **MULTIPLE CHOICE** The graph displays the constraints for an objective function. Which of the following CANNOT be one of the constraints?

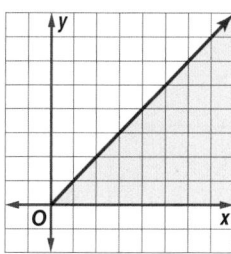

A $y \geq 0$

B $x \geq 0$

C $x - y \leq 0$

D $x - y \geq 0$

Objective
Approximate solutions to nonlinear optimization problems.

In Lesson 6-5, you learned how to solve optimization problems by using linear programming. The objective function and the system of constraints were represented by linear functions. Unfortunately, not all situations that require optimization can be defined by linear functions.

Advanced optimization problems involving quadratic, cubic, and other nonlinear functions require calculus to find exact solutions. However, we can find good approximations using graphing calculators.

ACTIVITY 1   Maximum Volume

A 16-inch × 20-inch piece of cardboard is made into a box with no top by cutting congruent squares from each corner and folding the sides up. What are the dimensions of the box with the largest possible volume? What is the maximum volume?

Step 1   Sketch a diagram of the situation.

Step 2   Let x represent the side length of one of the squares that is to be removed. Write expressions for the length, width, and height of the box in terms of x.

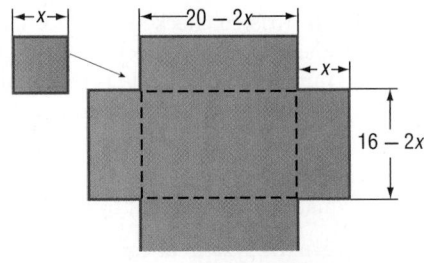

Step 3   Find an equation for the volume of the box V in terms of x using the dimensions found in Step 2.

Step 4   Use a calculator to graph the equation from Step 3.

▶ Analyze the Results

1. Describe the domain of x. Explain your reasoning.

2. Use your calculator to find the coordinates of the maximum point on your graph. Interpret the meaning of these coordinates.

3. What are the dimensions of the box with the largest possible volume? What is the maximum volume?

The desired outcome and complexity of each optimization problem differs. You can use the following steps to analyze and solve each problem.

Key Concept   Optimization

To solve an optimization problem, review these steps.

Step 1   Sketch a diagram of the situation and label all known and unknown quantities.

Step 2   Determine the quantity that needs to be maximized or minimized. Decide on the values necessary to find the desired quantity and represent each value with a number, a variable, or an expression.

Step 3   Write an equation for the quantity that is to be optimized in terms of one variable.

Step 4   Graph the equation and find either the maximum or minimum value. Determine the allowable domain of the variable.

ACTIVITY 2   Minimum Surface Area

A typical soda can is about 2.5 inches wide and 4.75 inches tall yielding a volume of about 23.32 cubic inches. What would be the dimensions of a soda can if you kept the volume constant but minimized the amount of material used to construct the can?

Step 1   Sketch a diagram of the situation.

Step 2   The quantity to be minimized is surface area. Values for the radius and height of the can are needed. Find an expression for the height h of the can in terms of the radius r using the given volume.

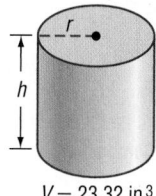

$V = 23.32$ in^3

Step 3   Using the expression found in Step 2, write an equation for surface area SA.

Step 4   Use a calculator to graph the equation from Step 3. State the domain of r.

▶ **Analyze the Results**

4. Find the coordinates of the minimum point. Interpret the meaning of these coordinates.

5. What are the dimensions and surface area of the can with the smallest possible surface area?

6. A right cylinder with no top is to be constructed with a surface area of 6π square inches. What height and radius will maximize the volume of the cylinder? What is the maximum volume?

Minimizing materials is not the only application of optimization.

ACTIVITY 3   Quickest Path

A race requires participants to travel on foot over a sandy beach to the entrance of a pier as shown. Racers can take any path they choose. There is a sidewalk to the pier that contestants can run on as well. If a racer can run 6 miles per hour on the sand and 7.5 miles per hour on the sidewalk, what path will require the shortest amount of time?

Step 1   Sketch a diagram of the situation.

Step 2   To minimize time, write expressions for the distances traveled on each surface and the respective rates. Let x represent the distance the runner does not run on the sidewalk as shown. Find expressions for the distances traveled on each surface in terms of x.

Step 3   Using the expressions found in Step 2, write an equation for time t.

Step 4   Use a calculator to graph the equation from Step 3. State the domain of x.

▶ **Analyze the Results**

7. Find the coordinates of the minimum point. Interpret the meaning of these coordinates.

8. What path will require the shortest amount of time? How long will it take?

9. Find the average rate of change m at the minimum point of your graph using the difference quotient. What does this value suggest about the line tangent to the graph at this point?

10. Make a conjecture about the rates of change and the tangent lines of graphs at minimum and maximum points. Does your conjecture hold true for the first two activities? Explain.

CHAPTER 7

Conic Sections and Parametric Equations

Then

In Chapter 6, you learned how to solve systems of linear equations using matrices.

Now

In Chapter 7, you will:

- Analyze and graph equations of parabolas, circles, ellipses, and hyperbolas.
- Write equations of parabolas, circles, ellipses, and hyperbolas.
- Use equations to identify types of conic sections.
- Graph rotated conic sections.
- Solve problems related to the motion of projectiles.

Why?

BASEBALL When a baseball is hit, the path of the ball can be represented and traced by parametric equations.

PREREAD Scan the Study Guide and Review and use it to make two or three predictions about what you will learn in Chapter 7.

Get Ready for Chapter 7

Diagnose Readiness You have two options for checking Prerequisite Skills.

Text Option

Take the Quick Quiz below.

*Quick*Check

For each function, find the axis of symmetry, the y-intercept, and the vertex. (Lesson 0-3)

1. $f(x) = x^2 - 2x - 12$ **2.** $f(x) = x^2 + 2x + 6$

3. $f(x) = 2x^2 + 4x - 8$ **4.** $f(x) = 2x^2 - 12x + 3$

5. $f(x) = 3x^2 - 12x - 4$ **6.** $f(x) = 4x^2 + 8x - 1$

7. BUSINESS The cost of producing x bicycles can be represented by $C(x) = 0.01x^2 - 0.5x + 550$. Find the axis of symmetry, the y-intercept, and the vertex of the function. (Lesson 0-3)

Find the discriminant of each quadratic function. (Lesson 0-3)

8. $f(x) = 2x^2 - 5x + 3$ **9.** $f(x) = 2x^2 + 6x - 9$

10. $f(x) = 3x^2 + 2x + 1$ **11.** $f(x) = 3x^2 - 8x - 3$

12. $f(x) = 4x^2 - 3x - 7$ **13.** $f(x) = 4x^2 - 2x + 11$

Find the equations of any vertical or horizontal asymptotes. (Lesson 2-5)

14. $f(x) = \dfrac{x - 2}{x + 4}$ **15.** $h(x) = \dfrac{x^2 - 4}{x + 5}$

16. $f(x) = \dfrac{x(x - 1)}{(x + 2)(x - 3)}$ **17.** $g(x) = \dfrac{x + 3}{(x - 1)(x + 5)}$

18. $h(x) = \dfrac{2x^2 - 5x - 12}{x^2 + 4x}$ **19.** $f(x) = \dfrac{2x^2 - 13x + 6}{x - 4}$

20. WILDLIFE The number of deer $D(x)$ after x years living on a wildlife preserve can be represented by $D(x) = \dfrac{12x + 50}{0.02x + 4}$. Determine the maximum number of deer that can live in the preserve. (Lesson 2-5)

Online Option

Math Online   Take a self-check Chapter Readiness Quiz at glencoe.com.

New Vocabulary

English	• Page •	Español
conic section	• 422 •	sección cónica
degenerate conic	• 422 •	degenerado cónico
locus	• 422 •	lugar geométrico
parabola	• 422 •	parábola
focus	• 422 •	foco
directrix	• 422 •	directrix
axis of symmetry	• 422 •	eje de simetría
vertex	• 422 •	vértice
latus rectum	• 430 •	recto de latus
ellipse	• 432 •	elipse
foci	• 432 •	focos
major axis	• 432 •	eje principal
center	• 432 •	centro
minor axis	• 432 •	eje menor
vertices	• 432 •	vértices
co-vertices	• 432 •	co-vértices
eccentricity	• 435 •	excentricidad
hyperbola	• 442 •	hipérbola
transverse axis	• 442 •	eje transversal
conjugate axis	• 442 •	eje conjugado
parametric equation	• 464 •	ecuación paramétrica
parameter	• 464 •	parámetro
orientation	• 464 •	orientación

Review Vocabulary

transformations • p. 48 • transformaciones **changes that affect the appearance of a parent function**

asymptotes • p. 46 • asíntotas **lines or curves that graphs approach**

Parabolas

Why?

Trough solar collectors use the properties of parabolas to focus radiation onto a receiver and generate solar power.

Then
You identified, analyzed, and graphed quadratic functions. (Lesson 1-5)

Now
- Analyze and graph equations of parabolas.
- Write equations of parabolas.

New Vocabulary
conic section
degenerate conic
locus
parabola
focus
directrix
axis of symmetry
vertex
latus rectum

Math Online
glencoe.com

Analyze and Graph Parabolas **Conic sections**, or *conics*, are the figures formed when a plane intersects a double-napped right cone. A double-napped cone is two cones opposite each other and extending infinitely upward and downward. The four common conic sections that will be covered in this chapter are the *parabola*, the *ellipse*, the *circle*, and the *hyperbola*.

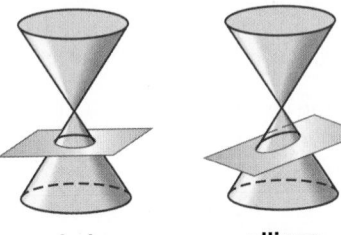

| circle | ellipse | parabola | hyperbola |

When the plane intersects the vertex of the cone, the figures formed are **degenerate conics**. A degenerate conic may be a point, a line, or two intersecting lines.

  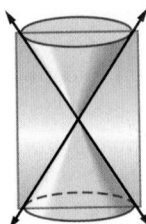

| point (degenerate ellipse) | line (degenerate parabola) | intersecting lines (degenerate hyperbola) |

The general form of the equations for conic sections is $Ax^2 + Bxy + Cy^2 + Dx + Ey + F = 0$, where A, B, and C cannot all be zero. More specific algebraic forms for each type of conic will be addressed as they are introduced.

A **locus** is a set of all points that fulfill a geometric property. A **parabola** represents the locus of points in a plane that are equidistant from a fixed point, called the **focus**, and a specific line, called the **directrix**.

A parabola is symmetric about the line perpendicular to the directrix through the focus called the **axis of symmetry**. The **vertex** is the intersection of the parabola and the axis of symmetry.

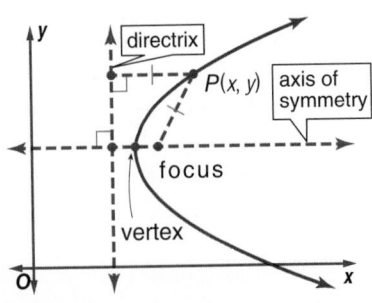

Previously, you learned the quadratic function $f(x) = ax^2 + bx + c$, where $a \neq 0$, represents a parabola that opens either up or down. The definition of a parabola as a locus can be used to derive a general equation of a parabola that opens up, down, left, or right.

Let $P(x, y)$ be any point on the parabola with vertex $V(h, k)$ where $p = FV$, the distance from the vertex to the focus. By the definition of a parabola, the distance from any point on the parabola to the focus must equal the distance from that point to the directrix. So, if $FV = p$, then $VT = p$. From the definition of a parabola, you know that $PF = PM$. Because M lies on the directrix, the coordinates of M are $(h - p, y)$.

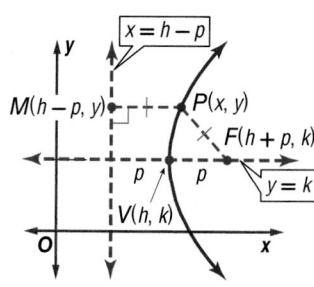

You can use the Distance Formula to determine the equation for the parabola.

$$PF = PM$$

$$\sqrt{[x - (h + p)]^2 + (y - k)^2} = \sqrt{[x - (h - p)]^2 + (y - y)^2} \qquad \textbf{Distance Formula}$$

$$[x - (h + p)]^2 + (y - k)^2 = [x - (h - p)]^2 + 0^2 \qquad \textbf{Square each side.}$$

$$x^2 - 2x(h + p) + (h + p)^2 + (y - k)^2 = x^2 - 2x(h - p) + (h - p)^2 \qquad \textbf{Multiply.}$$

$$x^2 - 2xh - 2xp + h^2 + 2hp + p^2 + (y - k)^2 = x^2 - 2xh + 2xp + h^2 - 2hp + p^2 \qquad \textbf{Multiply.}$$

$$(y - k)^2 = 4xp - 4hp \qquad \textbf{Simplify.}$$

$$(y - k)^2 = 4p(x - h) \qquad \textbf{Factor.}$$

ReadingMath

Concavity In this lesson, you will refer to parabolas as curves that open up, down, to the right, or to the left. In calculus, you will learn and use the term *concavity*. In this case, the curves are *concave up*, *concave* down, *concave right*, or *concave left*, respectively.

An equation for a parabola that opens horizontally is $(y - k)^2 = 4p(x - h)$. Similarly, for parabolas that open vertically, you can derive the equation $(x - h)^2 = 4p(y - k)$.

These represent the standard equations for parabolas, where $p \neq 0$. The values of the constants h, k, and p determine characteristics of parabolas such as the coordinates of the vertex and the direction of the parabola.

Key Concept **Standard Form of Equations for Parabolas**

$(x - h)^2 = 4p(y - k)$

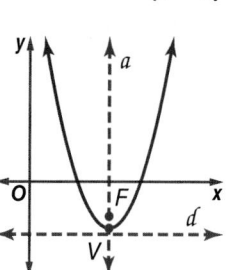

$p > 0$ \qquad $p < 0$

Orientation: opens vertically

Vertex: (h, k)

Focus: $(h, k + p)$

Axis of Symmetry a: $x = h$

Directrix d: $y = k - p$

$(y - k)^2 = 4p(x - h)$

 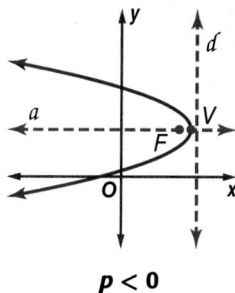

$p > 0$ \qquad $p < 0$

Orientation: opens horizontally

Vertex: (h, k)

Focus: $(h + p, k)$

Axis of Symmetry a: $y = k$

Directrix d: $x = h - p$

You can use the standard form of the equation for a parabola to determine characteristics of the parabola such as the vertex, focus, and directrix.

| EXAMPLE 1 | Determine Characteristics and Graph |

For $(y + 5)^2 = -12(x - 2)$, identify the vertex, focus, axis of symmetry, and directrix. Then graph the parabola.

The equation is in standard form and the squared term is y, which means that the parabola opens horizontally. The equation is in the form $(y - k)^2 = 4p(x - h)$, so $h = 2$ and $k = -5$. Because $4p = -12$, $p = -3$ and the graph opens to the left. Use the values of h, k, and p to determine the characteristics of the parabola.

vertex: $(2, -5)$ **(h, k)**

focus: $(-1, -5)$ **(h + p, k)**

directrix: $x = 5$ **x = h − p**

axis of symmetry: $y = -5$ **y = k**

Graph the vertex, focus, axis, and directrix of the parabola. Then make a table of values to graph the general shape of the curve.

x	y
0	−0.1, −9.9
−2	1.9, −11.9
−4	3.5, −13.5
−6	4.8, −14.8

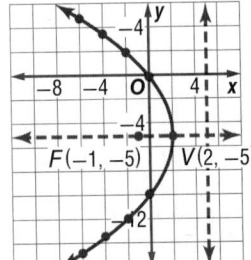

✓ Guided Practice

For each equation, identify the vertex, focus, axis of symmetry, and directrix. Then graph the parabola.

1A. $8(y + 3) = (x - 4)^2$

1B. $2(x + 6) = (y + 1)^2$

| 🌐 Real-World EXAMPLE 2 | Characteristics of Parabolas |

SOLAR ENERGY A trough solar collector is a length of mirror in a parabolic shape that focuses radiation from the Sun onto a linear receiver located at the focus of the parabola. The cross section of a single trough can be modeled using $x^2 = 3.04y$, where x and y are measured in meters. Where is the linear receiver located in this cross section?

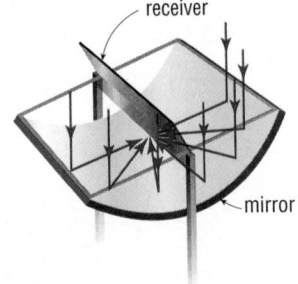

The linear receiver is located at the focus of the parabola. Because the x-term is squared and p is positive, the parabola opens up and the focus is located at $(h, k + p)$.

The equation is provided in standard form, and h and k are both zero. Because $4p = 3.04$, p is 0.76. So, the location of the focus is $(0, 0 + 0.76)$ or $(0, 0.76)$.

The location of the focus for the cross-section of the given parabola is $(0, 0.76)$. Therefore, the linear receiver is 0.76 meter above the vertex of the parabola.

✓ Guided Practice

2. ASTRONOMY Liquid-mirror telescopes consist of a thin layer of liquid metal in the shape of a parabola with a camera located at the focal point. Suppose a liquid mirror telescope can be modeled using the equation $x^2 = 44.8y - 268.8$ when $-5 \leq x \leq 5$. If x and y are measured in feet, what is the shortest distance between the surface of the liquid mirror and the camera?

To determine the characteristics of a parabola, you may sometimes need to write an equation in standard form. In some cases, you can simply rearrange the equation, but other times it may be necessary to use mathematical skills such as completing the square.

EXAMPLE 3 **Write in Standard Form**

Write $y = -\frac{1}{4}x^2 + 3x + 6$ in standard form. Identify the vertex, focus, axis of symmetry, and directrix. Then graph the parabola.

$y = -\frac{1}{4}x^2 + 3x + 6$ **Original equation**

$y = -\frac{1}{4}(x^2 - 12x) + 6$ **Factor $-\frac{1}{4}$ from x-terms.**

$y = -\frac{1}{4}(x^2 - 12x + 36 - 36) + 6$ **Complete the square.**

$y = -\frac{1}{4}(x^2 - 12x + 36) + 9 + 6$ $-\frac{1}{4}(-36) = 9$

$y = -\frac{1}{4}(x - 6)^2 + 15$ **Factor.**

$-4(y - 15) = (x - 6)^2$ **Standard form of a parabola**

Because the x-term is squared and $p = -1$, the graph opens down. Use the standard form of the equation to determine the characteristics of the parabola.

vertex: $(6, 15)$ **(h, k)**   directrix: $y = 16$ **$y = k - p$**

focus: $(6, 14)$ **$(h, k + p)$**   axis of symmetry: $x = 6$ **$x = h$**

Graph the vertex, focus, axis, and directrix of the parabola. Then make a table of values to graph the curve. The curve should be symmetric about the axis of symmetry.

x	y
0	6
4	14
8	14
12	6

✔ Guided Practice

Write each equation in standard form. Identify the vertex, focus, axis of symmetry, and directrix. Then graph each parabola.

3A. $x^2 - 4y + 3 = 7$ **3B.** $3y^2 + 6y + 15 = 12x$

Equations of Parabolas Specific characteristics can be used to determine the equation of a parabola.

StudyTip

Orientation If the vertex and focus share a common x-coordinate, then the parabola opens up or down. If the vertex and focus share a common y-coordinate, then the parabola opens to the right or left.

EXAMPLE 4 **Write Equations Given Characteristics**

Write an equation for and graph a parabola with the given characteristics.

a. focus $(3, -4)$ and vertex $(1, -4)$

Because the focus and vertex share the same y-coordinate, the graph is horizontal. The focus is $(h + p, k)$, so the value of p is $3 - 1$ or 2. Because p is positive, the graph opens to the right.

Write the equation for the parabola in standard form using the values of h, p, and k.

$(y - k)^2 = 4p(x - h)$ **Standard form**

$[y - (-4)]^2 = 4(2)(x - 1)$ **$k = -4$, $p = 2$, and $h = 1$**

$(y + 4)^2 = 8(x - 1)$ **Simplify.**

The standard form of the equation is $(y + 4)^2 = 8(x - 1)$.

Graph the vertex and focus. Then graph the parabola.

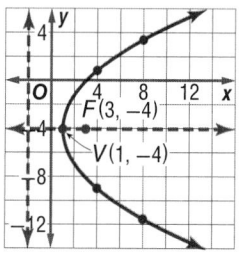

b. vertex $(-2, 4)$, directrix $y = 1$

The directrix is a horizontal line, so the parabola opens vertically. Because the directrix lies below the vertex, the parabola opens up.

Use the equation of the directrix to find p.

$y = k - p$ **Equation of directrix**

$1 = 4 - p$ **$y = 1$ and $k = 4$**

$-3 = -p$ **Subtract 4 from each side.**

$3 = p$ **Divide each side by -1.**

Substitute the values for h, k, and p into the standard form equation for a parabola opening vertically.

$(x - h)^2 = 4p(y - k)$ **Standard form**

$[x - (-2)]^2 = 4(3)(y - 4)$ **$h = -2$, $k = 4$, and $p = 3$**

$(x + 2)^2 = 12(y - 4)$ **Simplify.**

Graph the parabola.

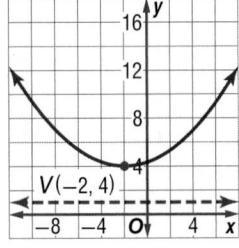

c. focus $(2, 1)$, opens left, contains $(2, 5)$

Because the parabola opens to the left, the vertex is $(2 - p, 1)$. Use the standard form of the equation of a horizontal parabola and the point $(2, 5)$ to find the equation.

$(y - k)^2 = 4p(x - h)$ **Standard form**

$(5 - 1)^2 = 4p[2 - (2 - p)]$ **$h = 2 - p$, $k = 1$, $x = 2$, and $y = 5$**

$16 = 4p(p)$ **Simplify.**

$4 = p^2$ **Multiply; then divide each side by 4.**

$\pm 2 = p$ **Take the square root of each side.**

Because the parabola opens to the left, the value of p must be negative. Therefore, $p = -2$. The vertex is $(4, 1)$, and the standard form of the equation is $(y - 1)^2 = -8(x - 4)$.

Graph the parabola.

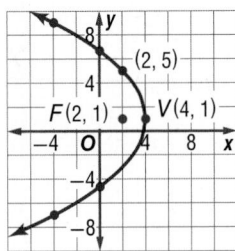

✔ Guided Practice

Write an equation for and graph a parabola with the given characteristics.

4A. focus $(-6, 2)$, vertex $(-6, -1)$

4B. focus $(5, -2)$, vertex $(9, -2)$

4C. focus $(-3, -4)$, opens down, contains $(5, -10)$

4D. focus $(-1, 5)$, opens right, contains $(8, -7)$

In calculus, you will often be asked to determine equations of lines that are tangent to curves. Equations of lines that are tangent to parabolas can be found without using calculus.

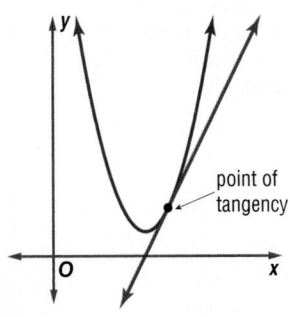

point of tangency

Key Concept — Line Tangent to a Parabola

A line ℓ that is tangent to a parabola at a point P forms an isosceles triangle such that:

- The segment from P to the focus forms one leg of the triangle.
- The segment along the axis of symmetry from the focus to another point on the tangent line forms the other leg.

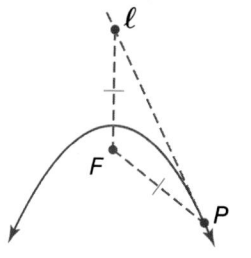

EXAMPLE 5   Find a Tangent Line at a Point

Write an equation for the line tangent to $x = y^2 + 3$ at $C(7, 2)$.

The graph opens horizontally. Determine the vertex and focus.

$$x = y^2 + 3 \qquad \text{Original equation}$$
$$1(x - 3) = (y - 0)^2 \qquad \text{Write in standard form.}$$

Because $4p = 1$, $p = 0.25$. The vertex is $(3, 0)$ and the focus is $(3.25, 0)$. As shown in the two figures, we need to determine d, the distance between the focus and the point of tangency, C.

StudyTip

Normals To Conics A *normal* to a conic at a point is the line perpendicular to the tangent line to the conic at that point. In Example 5, since the equation of the tangent line to the graph of $x = y^2 + 3$ at $(7, 2)$ is $y - 2 = \frac{1}{4}(x - 7)$, the equation of the normal line to the parabola at that same point is $y - 2 = -4(x - 7)$.

This is one leg of the isosceles triangle.

$$d = \sqrt{(x_2 - x_1)^2 + (y_2 - y_1)^2} \qquad \text{Distance Formula}$$
$$= \sqrt{(7 - 3.25)^2 + (2 - 0)^2} \qquad (x_1, y_1) = (3.25, 0) \text{ and } (x_2, y_2) = (7, 2)$$
$$= 4.25 \qquad \text{Simplify.}$$

Use d to find A, the endpoint of the other leg of the isosceles triangle.
$A(3.25 - 4.25, 0)$ or $A(-1, 0)$

Points A and C both lie on the line tangent to the parabola. Find an equation of this line.

$$m = \frac{2 - 0}{7 - (-1)} \text{ or } \frac{1}{4} \qquad \text{Slope Formula}$$
$$y - y_1 = m(x - x_1) \qquad \text{Point-slope form}$$
$$y - 2 = \frac{1}{4}(x - 7) \qquad m = \frac{1}{4}, y_1 = 2, \text{ and } x_1 = 7$$
$$y - 2 = \frac{x}{4} - \frac{7}{4} \qquad \text{Distributive Property}$$
$$y = \frac{x}{4} + \frac{1}{4} \qquad \text{Add 2 to each side.}$$

As shown in Figure 7.1.1, the equation for the line tangent to $x = y^2 + 3$ at $(7, 2)$ is $y = \frac{x}{4} + \frac{1}{4}$.

✔ Guided Practice

Write an equation for the line tangent to each parabola at each given point.

5A. $y = 4x^2 + 4$; $(-1, 8)$

5B. $x = 5 - \frac{y^2}{4}$; $(1, -4)$

Figure 7.1.1

Exercises

For each equation, identify the vertex, focus, axis of symmetry, and directrix. Then graph the parabola. (Example 1)

1. $(x - 3)^2 = 12(y - 7)$

2. $(x + 1)^2 = -12(y - 6)$

3. $(y - 4)^2 = 20(x + 2)$

4. $-1(x + 7) = (y + 5)^2$

5. $(x + 8)^2 = 8(y - 3)$

6. $-40(x + 4) = (y - 9)^2$

7. $(y + 5)^2 = 24(x - 1)$

8. $2(y + 12) = (x - 6)^2$

9. $-4(y + 2) = (x + 8)^2$

10. $10(x + 11) = (y + 3)^2$

11. SKATEBOARDING A group of high school students designing a half-pipe have decided that the ramps, or transitions, could be obtained by splitting a parabola in half. A parabolic cross section of the ramps can be modeled by $x^2 = 8(y - 2)$, where the values of x and y are measured in feet. Where is the focus of the parabola in relation to the ground if the ground represents the directrix? (Example 2)

12. COMMUNICATION The cross section of a satellite television dish has a parabolic shape that focuses the satellite signals onto a receiver located at the focus of the parabola. The parabolic cross section can be modeled by $(x - 6)^2 = 12(y - 10)$, where the values of x and y are measured in inches. Where is the receiver located in relation to this particular cross section? (Example 2)

13 BOATING As a speed boat glides through the water, it creates a wake in the shape of a parabola. The vertex of this parabola meets with the stern of the boat. A swimmer on a wakeboard, attached by a piece of rope, is being pulled by the boat. When he is directly behind the boat, he is positioned at the focus of the parabola. The parabola formed by the wake can be modeled using $y^2 - 180x + 10y + 565 = 0$, where x and y are measured in feet. (Example 3)

a. Write the equation in standard form.

b. How long is the length of rope attaching the swimmer to the stern of the boat?

14. BASEBALL During Philadelphia Phillies baseball games, the team's mascot, The Phanatic, launches hot dogs into the stands. The launching device propels the hot dogs into the air at an initial velocity of 64 feet per second. A hot dog's distance y above ground after x seconds can be illustrated by $y = -16x^2 + 64x + 6$. (Example 3)

a. Write the equation in standard form.

b. What is the maximum height that a hot dog can reach?

Write each equation in standard form. Identify the vertex, focus, axis of symmetry, and directrix. Then graph the parabola. (Example 3)

15. $x^2 - 17 = 8y + 39$

16. $y^2 + 33 = -8x - 23$

17. $3x^2 + 72 = -72y$

18. $-12y + 10 = x^2 - 4x + 14$

19. $60x - 80 = 3y^2 + 100$

20. $-33 = x^2 - 12y - 6x$

21. $-72 = 2y^2 - 16y - 20x$

22. $y^2 + 21 = -20x - 6y - 68$

23. $x^2 - 18y + 12x = 126$

24. $-34 = 2x^2 + 20x + 8y$

25. LIGHTING Stadium lights at an athletic field need to reflect light at maximum intensity. The bulb should be placed at the focal point of the parabolic globe surrounding it. If a globe is given by $x^2 = 36y$, where x and y are in inches, how far from the shell of the globe should the bulb be placed for maximum light? (Example 3)

Write an equation for and graph a parabola with the given focus F and vertex V. (Example 4)

26. $F(-9, -7), V(-9, -4)$

27. $F(2, -1), V(-4, -1)$

28. $F(-3, -2), V(1, -2)$

29. $F(-3, 4), V(-3, 2)$

30. $F(-2, -4), V(-2, -5)$

31. $F(-1, 4), V(7, 4)$

32. $F(14, -8), V(7, -8)$

33. $F(1, 3), V(1, 6)$

34. $F(-4, 9), V(-2, 9)$

35. $F(8, -3), V(8, -7)$

Write an equation for and graph each parabola with focus F and the given characteristics. (Example 4)

36. $F(3, 3)$; opens up; contains $(23, 18)$

37. $F(1, 2)$; opens down; contains $(7, 2)$

38. $F(11, 4)$; opens right; contains $(20, 16)$

39. $F(-4, 0)$; opens down; contains $(4, -15)$

40. $F(1, 3)$; opens left; contains $(-14, 11)$

41. $F(-5, -9)$; opens right; contains $(10, -1)$

42. $F(-7, 6)$; opens left; contains $(-4, 10)$

43. $F(-5, -2)$; opens up; contains $(-13, -2)$

44. ARCHITECTURE The entrance to an open-air plaza has a parabolic arch above two columns. The light in the center is located at the focus of the parabola. (Example 4)

a. Write an equation that models the parabola.

b. Graph the equation.

Write an equation for the line tangent to each parabola at each given point. (Example 5)

45. $(x + 7)^2 = -\frac{1}{2}(y - 3)$, $(-5, -5)$

46. $y^2 = \frac{1}{5}(x - 4)$, $(24, 2)$

47. $(x + 6)^2 = 3(y - 2)$, $(0, 14)$

48. $(x - 3)^2 = y + 4$, $(-1, 12)$

49 $-0.25(x - 6)^2 = y - 9$, $(10, 5)$

50. $-4x = (y + 5)^2$, $(0, -5)$

Determine the orientation of each parabola.

51. directrix $y = 4$, $p = -2$

52. $y^2 = -8(x - 6)$

53. vertex $(-5, 1)$, focus $(-5, 3)$

54. focus $(7, 10)$, directrix $x = 1$

Write an equation for each parabola.

55.

56.

57.

58.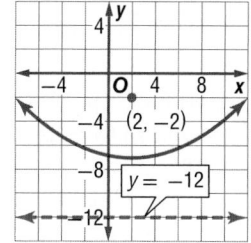

59. BRIDGES The arch of the railroad track bridge below is in the shape of a parabola. The two main support towers are 208 meters apart and 80 meters tall. The distance from the top of the parabola to the water below is 60 meters.

  a. Write an equation that models the shape of the arch. Let the railroad track represent the x-axis.

  b. Two vertical supports attached to the arch are equidistant from the center of the parabola as shown in the diagram. Find their lengths if they are 86.4 meters apart.

Write an equation for and graph a parabola with each set of characteristics.

60. vertex $(1, 8)$, contains $(11, 13)$, opens vertically

61. vertex $(-6, 4)$, contains $(-10, 8)$, opens horizontally

62. opens vertically, passes through points $(-12, -14)$, $(0, -2)$, and $(6, -5)$

63. opens horizontally, passes through points $(-1, -1)$, $(5, 3)$, and $(15, 7)$

64. SOUND Parabolic reflectors with microphones located at the focus are used to capture sounds from a distance. The sound waves enter the reflector and are concentrated toward the microphone.

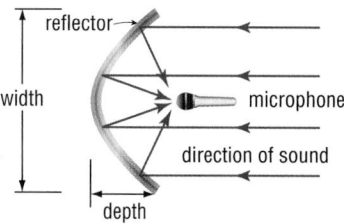

  a. How far from the reflector should a microphone be placed if the reflector has a width of 3 feet and a depth of 1.25 feet?

  b. Write an equation to model a different parabolic reflector that is 4 feet wide and 2 feet deep, if the vertex of the reflector is located at $(3, 5)$ and the parabola opens to the left.

  c. Graph the equation. Specify the domain and range.

Determine the point of tangency for each equation and line.

65. $(x + 2)^2 = 2y$
$y = 4x$

66. $(y - 8)^2 = 12x$
$y = x + 11$

67. $(y + 3)^2 = -x + 4$
$y = -\frac{1}{4}x - 1$

68. $(x + 5)^2 = -4(y + 1)$
$y = 2x + 13$

69. ILLUMINATION In a searchlight, the bulb is placed at the focus of a parabolic mirror 1.5 feet from the vertex. This causes the light rays from the bulb to bounce off the mirror as parallel rays, thus providing a concentrated beam of light.

  a. Write an equation for the parabola if the focal diameter of the bulb is 2 feet, as shown in the diagram.

  b. Suppose the focal diameter is increased to 3 feet. If the depth of both searchlights is 3.5 feet, how much greater is the width of the opening of the larger light? Round to the nearest hundredth.

70. PROOF Use the standard form of the equation of a parabola to prove the general form of the equation.

71. The **latus rectum** of a parabola is the line segment that passes through the focus, is perpendicular to the axis of the parabola, and has endpoints on the parabola. The length of the latus rectum is $|4p|$ units, where p is the distance from the vertex to the focus.

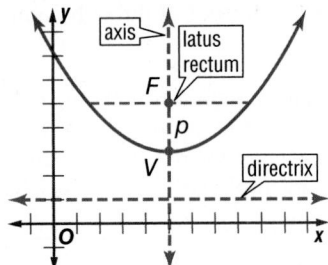

a. Write an equation for the parabola with vertex at $(-3, 2)$, axis $y = 2$, and latus rectum 8 units long.

b. Prove that the endpoints of the latus rectum and point of intersection of the axis and directrix are the vertices of a right isosceles triangle.

72. SOLAR ENERGY A solar furnace in France's Eastern Pyrenees uses a parabolic mirror that is illuminated by sunlight reflected off a field of heliostats, which are devices that track and redirect sunlight. Experiments in solar research are performed in the *focal zone* part of a tower. If the parabolic mirror is 6.25 meters deep, how many meters in front of the parabola is the focal zone?

Write a possible equation for a parabola with focus F such that the line given is tangent to the parabola.

73.

74.

75.

76.

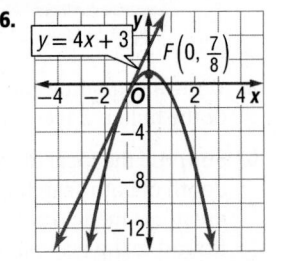

77. ⛭ **MULTIPLE REPRESENTATIONS** In this problem, you will investigate how the shape of a parabola changes as the position of the focus changes.

a. GEOMETRIC Find the distance between the vertex and the focus of each parabola.

   i. $y^2 = 4(x - 2)$ **ii.** $y^2 = 8(x - 2)$ **iii.** $y^2 = 16(x - 2)$

b. GRAPHICAL Graph the parabolas in part **a** using a different color for each. Label each focus.

c. VERBAL Describe the relationship between a parabola's shape and the distance between its vertex and focus.

d. ANALYTICAL Write an equation for a parabola that has the same vertex as $(x + 1)^2 = 20(y + 7)$ but is more narrow.

e. ANALYTICAL Make a conjecture about the graphs of $x^2 = -2(y + 1)$, $x^2 = -12(y + 1)$, and $x^2 = -5(y + 1)$. Check your conjecture by graphing the parabolas.

H.O.T. Problems   Use Higher-Order Thinking Skills

78. ERROR ANALYSIS Abigail and Jaden are graphing $x^2 + 6x - 4y + 9 = 0$. Is either of them correct? Explain your reasoning.

Abigail

Jaden

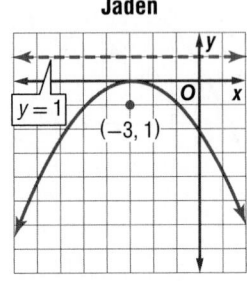

79. CHALLENGE The area of the *parabolic sector* shaded in the graph at the right is given by $A = \frac{4}{3}xy$. Find the equation of the parabola if the sector area is 2.4 square units, and the width of the sector is 3 units.

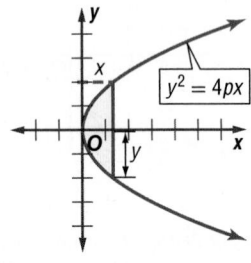

80. REASONING Which point on a parabola is closest to the focus? Explain your reasoning.

81. REASONING Without graphing, determine the quadrants in which the graph of $(y - 5)^2 = -8(x + 2)$ will have *no* points. Explain your reasoning.

82. WRITING IN MATH The *concavity* of a function's graph describes whether the graph opens upward (concave up) or downward (concave down). Explain how you can determine the concavity of a parabola given its focus and vertex.

83. PREWRITE Write a letter outlining and explaining the content you have learned in this lesson. Make an outline that addresses purpose, audience, a controlling idea, logical sequence, and time frame for completion.

Find the maximum and minimum values of the objective function $f(x, y)$ and for what values of x and y they occur, subject to the given constraints. (Lesson 6-5)

84. $x \leq 5$
$y \geq -2$
$y \leq x - 1$
$f(x, y) = x - 2y$

85. $y \geq -x + 2$
$2 \leq x \leq 7$
$y \leq \frac{1}{2}x + 5$
$f(x, y) = 8x + 3y$

86. $x \geq -3$
$y \geq 1$
$3x + y \leq 6$
$f(x, y) = 5x - 2y$

87. Find the partial fraction decomposition of $\dfrac{2y + 5}{y^2 + 3y + 2}$. (Lesson 6-4)

88. SURVEYING Talia is surveying a rectangular lot for a new office building. She measures the angle between one side of the lot and the line from her position to the opposite corner of the lot as 30°. She then measures the angle between that line and the line to a telephone pole on the edge of the lot as 45°. If Talia is 100 yards from the opposite corner of the lot, how far is she from the telephone pole? (Lesson 5-4)

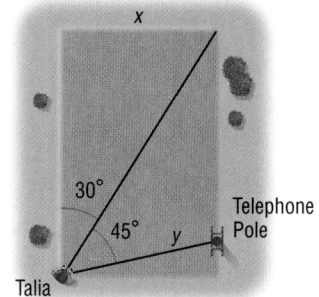

Find the value of each expression using the given information. (Lesson 5-1)

89. $\cot \theta$ and $\csc \theta$;
$\tan \theta = \frac{6}{7}, \sec \theta > 0$

90. $\cos \theta$ and $\tan \theta$;
$\csc \theta = -2, \cos \theta < 0$

Locate the vertical asymptotes, and sketch the graph of each function. (Lesson 4-5)

91. $y = \tan x + 4$

92. $y = \sec x + 2$

93. $y = \csc x - \frac{3}{4}$

Evaluate each expression. (Lesson 3-2)

94. $\log_{16} 4$

95. $\log_4 16^x$

96. $\log_3 27^x$

Graph each function. (Lesson 2-2)

97. $f(x) = x^3 - x^2 - 4x + 4$

98. $g(x) = x^4 - 7x^2 + x + 5$

99. $h(x) = x^4 - 4x^2 + 2$

100. REVIEW What is the solution set for $3(4x + 1)^2 = 48$?

A $\left\{\frac{5}{4}, -\frac{3}{4}\right\}$

B $\left\{-\frac{5}{4}, \frac{3}{4}\right\}$

C $\left\{\frac{15}{4}, -\frac{17}{4}\right\}$

D $\left\{\frac{1}{3}, -\frac{4}{3}\right\}$

E $\left\{\frac{7}{4}, -\frac{9}{4}\right\}$

101. SAT/ACT If x is a positive number, then $\dfrac{x^{\frac{3}{2}} \cdot x^{\frac{1}{2}}}{x^{\frac{1}{2}}} = ?$

F $x^{-\frac{1}{4}}$

G $\sqrt{x^3}$

H $x^{\frac{3}{4}}$

J $\sqrt{x^5}$

102. Which is the parent function of the graph shown below?

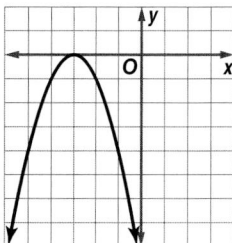

A $y = x$

B $y = \sqrt{x}$

C $y = |x|$

D $y = x^2$

103. REVIEW What are the x-intercepts of the graph of $y = -2x^2 - 5x + 12$?

F $-\frac{3}{2}, 4$

G $-4, \frac{3}{2}$

H $-2, \frac{1}{2}$

J $-\frac{1}{2}, 2$

Ellipses and Circles

Then
You analyzed and graphed parabolas. (Lesson 7-1)

Now
- Analyze and graph equations of ellipses and circles.
- Use equations to identify ellipses and circles.

New Vocabulary
ellipse
foci
major axis
center
minor axis
vertices
co-vertices
eccentricity

Math Online
glencoe.com

Why?

Due to acceleration and inertia, the safest shape for a roller coaster loop can be approximated using an ellipse rather than a circle. The elliptical shape helps to minimize force on the riders' heads and necks.

Analyze and Graph Ellipses and Circles An **ellipse** is the locus of points in a plane such that the sum of the distances from two fixed points, called **foci**, is constant. To visualize this concept, consider a length of string tacked at the foci of an ellipse. You can draw an ellipse by using a pencil to trace a curve with the string pulled tight. For any two points on the ellipse, the sum of the lengths of the segments to each focus is constant. In other words, $d_1 + d_2 = d_3 + d_4$, and this sum is constant.

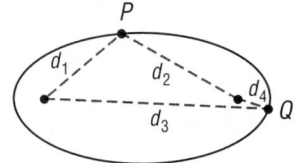

The segment that contains the foci of an ellipse and has endpoints on the ellipse is called the **major axis,** and the midpoint of the major axis is the **center**. The segment through the center with endpoints on the ellipse and perpendicular to the major axis is the **minor axis**. The two endpoints of the major axis are the **vertices,** and the endpoints of the minor axis are the **co-vertices**.

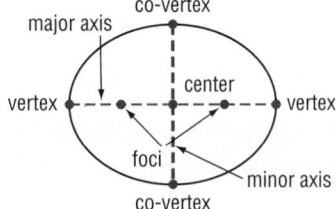

The center of the ellipse is the midpoint of both the major and minor axes. So, the segments from the center to each vertex are congruent, and the segments from the center to each co-vertex are congruent. The distance from each vertex to the center is a units, and the distance from the center to each co-vertex is b units. The distance from the center to each focus is c units.

Consider $\overline{V_1F_1}$ and $\overline{V_1F_2}$. Because $\triangle F_1V_1C \cong \triangle F_2V_1C$ by the Leg-Leg Theorem, $V_1F_1 = V_1F_2$. We can use the definition of an ellipse to find the lengths V_1F_1 and V_1F_2 in terms of the lengths given.

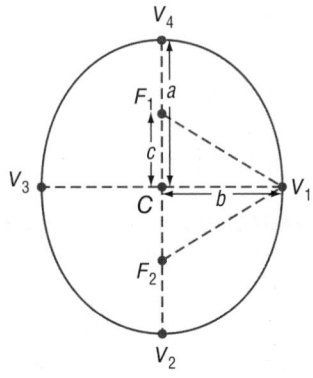

$V_1F_1 + V_1F_2 = V_2F_1 + V_2F_2$	**Definition of an ellipse**
$V_1F_1 + V_1F_2 = V_2F_1 + V_4F_1$	$V_2F_2 = V_4F_1$
$V_1F_1 + V_1F_2 = V_2V_4$	$V_2F_1 + V_4F_1 = V_2V_4$
$V_1F_1 + V_1F_2 = 2a$	$V_2V_4 = 2a$
$V_1F_1 + V_1F_1 = 2a$	$V_1F_1 = V_1F_2$
$2(V_1F_1) = 2a$	**Simplify.**
$V_1F_1 = a$	**Divide.**

Because $V_1F_1 = a$ and $\triangle F_1V_1C$ is a right triangle, $b^2 + c^2 = a^2$ by the Pythagorean Theorem.

Let $P(x, y)$ be any point on the ellipse with center $C(h, k)$. The coordinates of the foci, vertices, and co-vertices are shown at the right. By the definition of an ellipse, the sum of distances from any point on the ellipse to the foci is constant. Thus, $PF_1 + PF_2 = 2a$.

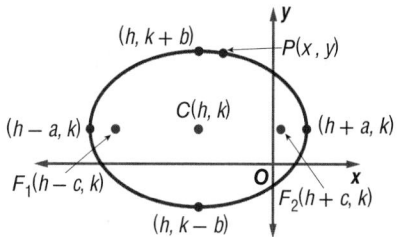

$PF_1 + PF_2 = 2a$	**Definition of ellipse**
$\sqrt{[x - (h - c)]^2 + (y - k)^2} + \sqrt{[x - (h + c)]^2 + (y - k)^2} = 2a$	**Distance Formula**
$\sqrt{(x - h + c)^2 + (y - k)^2} = 2a - \sqrt{(x - h - c)^2 + (y - k)^2}$	**Distributive and Subtraction Properties**
$\sqrt{[(x - h) + c]^2 + (y - k)^2} = 2a - \sqrt{[(x - h) - c]^2 + (y - k)^2}$	**Associative Property**
$[(x - h) + c]^2 + (y - k)^2 = 4a^2 - 4a\sqrt{[(x - h) - c]^2 + (y - k)^2} + [(x - h) - c]^2 + (y - k)^2$	
$(x - h)^2 + 2c(x - h) + c^2 + (y - k)^2 = 4a^2 - 4a\sqrt{[(x - h) - c]^2 + (y - k)^2} + (x - h)^2 - 2c(x - h) + c^2 + (y - k)^2$	
$4a\sqrt{[(x - h) - c]^2 + (y - k)^2} = 4a^2 - 4c(x - h)$	**Subtraction and Addition Properties**
$a\sqrt{[(x - h) - c]^2 + (y - k)^2} = a^2 - c(x - h)$	**Divide each side by 4.**
$a^2[(x - h)^2 - 2c(x - h) + c^2 + (y - k)^2] = a^4 - 2a^2c(x - h) + c^2(x - h)^2$	**Square each side.**
$a^2(x - h)^2 - 2a^2c(x - h) + a^2c^2 + a^2(y - k)^2 = a^4 - 2a^2c(x - h) + c^2(x - h)^2$	**Distributive Property**
$a^2(x - h)^2 - c^2(x - h)^2 + a^2(y - k)^2 = a^4 - a^2c^2$	**Subtraction Property**
$(a^2 - c^2)(x - h)^2 + a^2(y - k)^2 = a^2(a^2 - c^2)$	**Factor.**
$b^2(x - h)^2 + a^2(y - k)^2 = a^2b^2$	$a^2 - c^2 = b^2$
$\dfrac{(x - h)^2}{a^2} + \dfrac{(y - k)^2}{b^2} = 1$	**Divide each side by a^2b^2.**

The standard form for an ellipse centered at (h, k), where $a > b$, is given below.

Key Concept   **Standard Forms of Equations for Ellipses**

$$\frac{(x - h)^2}{a^2} + \frac{(y - k)^2}{b^2} = 1$$

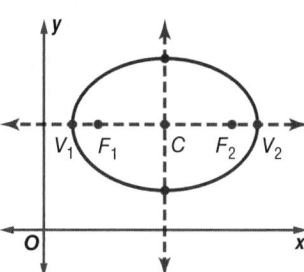

$$\frac{(x - h)^2}{b^2} + \frac{(y - k)^2}{a^2} = 1$$

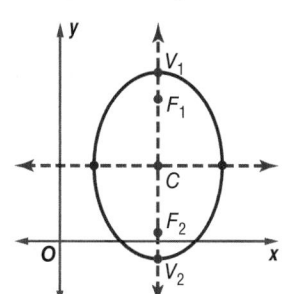

Orientation: horizontal major axis
Center: (h, k)
Foci: $(h \pm c, k)$
Vertices: $(h \pm a, k)$
Co-vertices: $(h, k \pm b)$
Major axis: $y = k$
Minor axis: $x = h$
a, b, c relationship: $c^2 = a^2 - b^2$ or $c = \sqrt{a^2 - b^2}$

Orientation: vertical major axis
Center: (h, k)
Foci: $(h, k \pm c)$
Vertices: $(h, k \pm a)$
Co-vertices: $(h \pm b, k)$
Major axis: $x = h$
Minor axis: $y = k$
a, b, c relationship: $c^2 = a^2 - b^2$ or $c = \sqrt{a^2 - b^2}$

EXAMPLE 1 | **Graph Ellipses**

Graph the ellipse given by each equation.

a. $\dfrac{(x-3)^2}{36} + \dfrac{(y+1)^2}{9} = 1$

The equation is in standard form with $h = 3$, $k = -1$, $a = \sqrt{36}$ or 6, $b = \sqrt{9}$ or 3, and $c = \sqrt{36 - 9}$ or $3\sqrt{3}$. Use these values to determine the characteristics of the ellipse.

orientation:	horizontal	When the equation is in standard form, the x^2-term contains a^2
center:	$(3, -1)$	(h, k)
foci:	$(3 \pm 3\sqrt{3}, -1)$	$(h \pm c, k)$
vertices:	$(-3, -1)$ and $(9, -1)$	$(h \pm a, k)$
co-vertices:	$(3, -4)$ and $(3, 2)$	$(h, k \pm b)$
major axis:	$y = -1$	$y = k$
minor axis:	$x = 3$	$x = h$

Graph the center, vertices, foci, and axes. Then make a table of values to sketch the ellipse.

x	y
0	1.60, −3.60
6	1.60, −3.60

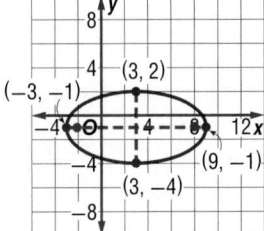

StudyTip

Orientation If the y-coordinate is the same for both vertices of an ellipse, then the major axis is horizontal. If the x-coordinate is the same for both vertices of an ellipse, then the major axis is vertical.

b. $4x^2 + y^2 - 24x + 4y + 24 = 0$

First, write the equation in standard form.

$4x^2 + y^2 - 24x + 4y + 24 = 0$	Original equation
$(4x^2 - 24x) + (y^2 + 4y) = -24$	Isolate and group like terms.
$4(x^2 - 6x) + (y^2 + 4y) = -24$	Factor.
$4(x^2 - 6x + 9) + (y^2 + 4y + 4) = -24 + 4(9) + 4$	Complete the squares.
$4(x - 3)^2 + (y + 2)^2 = 16$	Factor and simplify.
$\dfrac{(x-3)^2}{4} + \dfrac{(y+2)^2}{16} = 1$	Divide each side by 16.

The equation is in standard form with $h = 3$, $k = -2$, $a = \sqrt{16}$ or 4, $b = \sqrt{4}$ or 2, and $c = \sqrt{16 - 4}$ or $2\sqrt{3}$. Use these values to determine the characteristics of the ellipse.

orientation:	vertical	When the equation is in standard form, the y^2-term contains a^2.
center:	$(3, -2)$	(h, k)
foci:	$(3, -2 \pm 2\sqrt{3})$	$(h, k \pm c)$
vertices:	$(3, -6)$ and $(3, 2)$	$(h, k \pm a)$
co-vertices:	$(5, -2)$ and $(1, -2)$	$(h \pm b, k)$
major axis:	$x = 3$	$x = h$
minor axis:	$y = -2$	$y = k$

Graph the center, vertices, foci, and axes. Then make a table of values to sketch the ellipse.

x	y
2	1.46, −5.46
4	1.46, −5.46

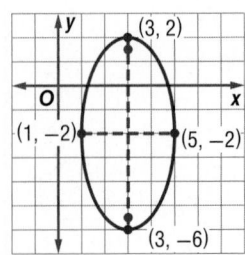

✓ Guided Practice

1A. $\dfrac{(x-6)^2}{9} + \dfrac{(y+3)^2}{16} = 1$

1B. $x^2 + 4y^2 + 4x - 40y + 103 = 0$

EXAMPLE 2 | **Write Equations Given Characteristics**

Write an equation for an ellipse with each set of characteristics.

a. major axis from (−6, 2) to (−6, −8); minor axis from (−3, −3) to (−9, −3)

Use the major and minor axes to determine a and b.

Half the length of major axis	Half the length of minor axis
$a = \dfrac{2 - (-8)}{2}$ or 5	$b = \dfrac{-3 - (-9)}{2}$ or 3

The center of the ellipse is at the midpoint of the major axis.

$(h, k) = \left(\dfrac{-6 + (-6)}{2}, \dfrac{2 + (-8)}{2} \right)$ **Midpoint Formula**

$= (-6, -3)$ **Simplify.**

The x-coordinates are the same for both endpoints of the major axis, so the major axis is vertical and the value of a belongs with the y^2-term. An equation for the ellipse is $\dfrac{(y + 3)^2}{25} + \dfrac{(x + 6)^2}{9} = 1$. The graph of the ellipse is shown in Figure 7.2.1.

b. vertices at (−4, 4) and (6, 4); foci at (−2, 4) and (4, 4)

The length of the major axis, $2a$, is the distance between the vertices.

$2a = \sqrt{(-4 - 6)^2 + (4 - 4)^2}$ **Distance Formula**

$a = 5$ **Solve for a.**

$2c$ represents the distance between the foci.

$2c = \sqrt{(-2 - 4)^2 + (4 - 4)^2}$ **Distance Formula**

$c = 3$ **Solve for c.**

Find the value of b.

$c^2 = a^2 - b^2$ **Equation relating a, b, and c**

$3^2 = 5^2 - b^2$ **$a = 5$ and $c = 3$**

$b = 4$ **Solve for b.**

The vertices are equidistant from the center.

$(h, k) = \left(\dfrac{-4 + 6}{2}, \dfrac{4 + 4}{2} \right)$ **Midpoint Formula**

$= (1, 4)$ **Simplify.**

The y-coordinates are the same for both endpoints of the major axis, so the major axis is horizontal and the value of a belongs with the x^2-term. An equation for the ellipse is $\dfrac{(x - 1)^2}{25} + \dfrac{(y - 4)^2}{16} = 1$. The graph of the ellipse is shown in Figure 7.2.2.

Figure 7.2.1

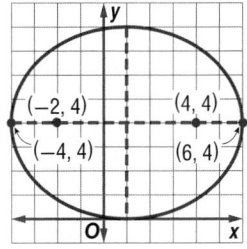

Figure 7.2.2

✓ **Guided Practice**

2A. foci at (19, 3) and (−7, 3); length of major axis equals 30

2B. vertices at (−2, −4) and (−2, 8); length of minor axis equals 10

The **eccentricity** of an ellipse is the ratio of c to a. This value will always be between 0 and 1 and will determine how "circular" or "stretched" the ellipse will be.

Key Concept | **Eccentricity**

For any ellipse, $\dfrac{(x - h)^2}{a^2} + \dfrac{(y - k)^2}{b^2} = 1$ or $\dfrac{(x - h)^2}{b^2} + \dfrac{(y - k)^2}{a^2} = 1$, where $c^2 = a^2 - b^2$, the eccentricity $e = \dfrac{c}{a}$.

The value c represents the distance between one of the foci and the center of the ellipse. As the foci are moved closer together, c and e both approach 0. When the eccentricity reaches 0, the ellipse is a circle and both a and b are equal to the radius of the circle.

$e = 0.94$ $e = 0.71$ $e = 0.32$ $a = b = r$ $c = 0$ $e = 0$

EXAMPLE 3   **Determine the Eccentricity of an Ellipse**

Determine the eccentricity of the ellipse given by $\dfrac{(x-6)^2}{100} + \dfrac{(y+1)^2}{9} = 1$.

First, determine the value of c.

$c^2 = a^2 - b^2$	Equation relating a, b, and c
$c^2 = 100 - 9$	$a^2 = 100$ and $b^2 = 9$
$c = \sqrt{91}$	Solve for c.

Use the values of c and a to find the eccentricity.

$e = \dfrac{c}{a}$	Eccentricity equation
$e = \dfrac{\sqrt{91}}{10}$ or about 0.95	$a = 10$ and $c = \sqrt{91}$

The eccentricity of the ellipse is about 0.95, so the ellipse will appear stretched, as shown in Figure 7.2.3.

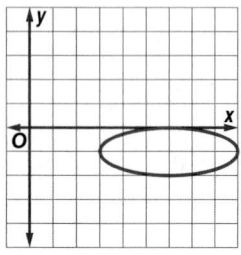

Figure 7.2.3

✓ **Guided Practice**

Determine the eccentricity of the ellipse given by each equation.

3A. $\dfrac{x^2}{18} + \dfrac{(y+8)^2}{48} = 1$ **3B.** $\dfrac{(x-4)^2}{19} + \dfrac{(y+7)^2}{17} = 1$

🌐 **Real-World Career**

Ophthalmic Technician
Ophthalmic technicians work with ophthalmologists to care for patients with eye disease or injury. They perform exams such as visual status and assist in surgical settings. They must complete a one-year training program in addition to a high school diploma or GED.

🌐 **Real-World EXAMPLE 4**   **Use Eccentricity**

OPTICS The shape of an eye can be modeled by a prolate, or three-dimensional, ellipse. The eccentricity of the center cross-section for an eye with normal vision is about 0.28. If a normal eye is about 25 millimeters deep, what is the approximate height of the eye?

Use the eccentricity to determine the value of c.

$e = \dfrac{c}{a}$	Definition of eccentricity
$0.28 = \dfrac{c}{12.5}$	$e = 0.28$ and $a = 12.5$
$c = 3.5$	Solve for c.

Use the values of c and a to determine b.

$c^2 = a^2 - b^2$	Equation relating a, b, and c
$3.5^2 = 12.5^2 - b^2$	$c = 3.5$ and $a = 12.5$
$b = 12$	Solve for b.

Because the value of b is 12, the height of the eye is $2b$ or 24 millimeters.

✓ **Guided Practice**

4. The eccentricity of a nearsighted eye is 0.39. If the depth of the eye is 25 millimeters, what is the height of the eye?

Identify Conic Sections The equation of a circle can be derived using the eccentricity of an ellipse.

$\dfrac{x^2}{a^2} + \dfrac{y^2}{b^2} = 1$ Equation of an ellipse with center at (0, 0)

$\dfrac{x^2}{a^2} + \dfrac{y^2}{a^2} = 1$ $a = b$ when $e = 0$

$x^2 + y^2 = a^2$ Multiply each side by a^2.

$x^2 + y^2 = r^2$ a is the radius of the circle.

Key Concept   Standard Form of Equations for Circles

The standard form of an equation for a circle with center (h, k) and radius r is
$$(x - h)^2 + (y - k)^2 = r^2.$$

If you are given the equation for a conic section, you can determine what type of conic is represented using the characteristics of the equation.

EXAMPLE 5   Determine Types of Conics

Write each equation in standard form. Identify the related conic.

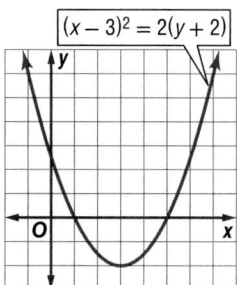

$(x - 3)^2 = 2(y + 2)$

Figure 7.2.4

a. $x^2 - 6x - 2y + 5 = 0$

$x^2 - 6x - 2y + 5 = 0$	Original equation
$(x^2 - 6x) - 2y = -5$	Isolate and group like terms.
$(x^2 - 6x + 9) - 2y = -5 + 9$	Complete the square.
$(x - 3)^2 - 2y = 4$	Factor and simplify.
$(x - 3)^2 = 2y + 4$	Add 2y to each side.
$(x - 3)^2 = 2(y + 2)$	Factor.

Because only one term is squared, the graph is a parabola with vertex $(3, -2)$, as in Figure 7.2.4.

b. $x^2 + y^2 - 12x + 10y + 12 = 0$

$x^2 + y^2 - 12x + 10y + 12 = 0$	Original equation
$(x^2 - 12x) + (y^2 + 10y) = -12$	Isolate and group like terms.
$(x^2 - 12x + 36) + (y^2 + 10y + 25) = -12 + 36 + 25$	Complete the squares.
$(x - 6)^2 + (y + 5)^2 = 49$	Factor and simplify.

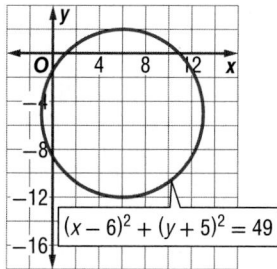

$(x - 6)^2 + (y + 5)^2 = 49$

Figure 7.2.5

Because the equation is of the form $(x - h)^2 + (y - k)^2 = r^2$, the graph is a circle with center $(6, -5)$ and radius 7, as in Figure 7.2.5.

c. $x^2 + 4y^2 - 6x - 7 = 0$

$x^2 + 4y^2 - 6x - 7 = 0$	Original equation
$(x^2 - 6x) + 4y^2 = 7$	Isolate and group like terms.
$(x^2 - 6x + 9) + 4y^2 = 7 + 9$	Complete the square.
$(x - 3)^2 + 4y^2 = 16$	Factor and simplify.
$\dfrac{(x - 3)^2}{16} + \dfrac{y^2}{4} = 1$	Divide each side by 16.

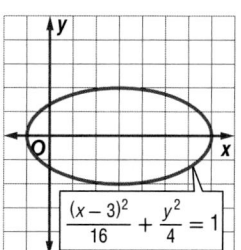

$\dfrac{(x - 3)^2}{16} + \dfrac{y^2}{4} = 1$

Figure 7.2.6

Because the equation is of the form $\dfrac{(x - h)^2}{a^2} + \dfrac{(y - k)^2}{b^2} = 1$, the graph is an ellipse with center $(3, 0)$, as in Figure 7.2.6.

✔ Guided Practice

5A. $y^2 - 3x + 6y + 12 = 0$

5B. $4x^2 + 4y^2 - 24x + 32y + 36 = 0$

5C. $4x^2 + 3y^2 + 36y + 60 = 0$

Graph the ellipse given by each equation. (Example 1)

1. $\dfrac{(x+2)^2}{9} + \dfrac{y^2}{49} = 1$

2. $\dfrac{(x+4)^2}{9} + \dfrac{(y+3)^2}{4} = 1$

3. $x^2 + 9y^2 - 14x + 36y + 49 = 0$

4. $4x^2 + y^2 - 64x - 12y + 276 = 0$

5. $9x^2 + y^2 + 126x + 2y + 433 = 0$

6. $x^2 + 25y^2 - 12x - 100y + 111 = 0$

Write an equation for the ellipse with each set of characteristics. (Example 2)

7. vertices $(-7, -3)$, $(13, -3)$;
   foci $(-5, -3)$, $(11, -3)$

8. vertices $(4, 3)$, $(4, -9)$;
   length of minor axis is 8

9. vertices $(7, 2)$, $(-3, 2)$;
   foci $(6, 2)$, $(-2, 2)$

10. major axis $(-13, 2)$ to $(1, 2)$;
    minor axis $(-6, 4)$ to $(-6, 0)$

11. foci $(-6, 9)$, $(-6, -3)$;
    length of major axis is 20

12. co-vertices $(-13, 7)$, $(-3, 7)$;
    length of major axis is 16

13. foci $(-10, 8)$, $(14, 8)$;
    length of major axis is 30

Determine the eccentricity of the ellipse given by each equation. (Example 3)

14. $\dfrac{(x+5)^2}{72} + \dfrac{(y-3)^2}{54} = 1$

15. $\dfrac{(x+6)^2}{40} + \dfrac{(y-2)^2}{12} = 1$

16. $\dfrac{(x-8)^2}{14} + \dfrac{(y+3)^2}{57} = 1$

17. $\dfrac{(x+8)^2}{27} + \dfrac{(y-7)^2}{33} = 1$

18. $\dfrac{(x-1)^2}{12} + \dfrac{(y+2)^2}{9} = 1$

19. $\dfrac{(x-11)^2}{17} + \dfrac{(y+15)^2}{23} = 1$

20. $\dfrac{x^2}{38} + \dfrac{(y-12)^2}{13} = 1$

21. $\dfrac{(x+9)^2}{10} + \dfrac{(y+11)^2}{8} = 1$

22. **RACING** The design of an elliptical racetrack with an eccentricity of 0.75 is shown. (Example 4)

1000 ft

   a. What is the maximum width w of the track?

   b. Write an equation for the ellipse if the origin x is located at the center of the racetrack.

23. **CARPENTRY** A carpenter has been hired to construct a sign for a pet grooming business. The plans for the sign call for an elliptical shape with an eccentricity of 0.60 and a length of 36 inches. (Example 4)

Sasha's Pet Grooming

36 in.

   a. What is the maximum height of the sign?

   b. Write an equation for the ellipse if the origin is located at the center of the sign.

Write each equation in standard form. Identify the related conic. (Example 5)

24. $x^2 + y^2 + 6x - 4y - 3 = 0$

25. $4x^2 + 8y^2 - 8x + 48y + 44 = 0$

26. $x^2 - 8x - 8y - 40 = 0$

27. $y^2 - 12x + 18y + 153 = 0$

28. $x^2 + y^2 - 8x - 6y - 39 = 0$

29. $3x^2 + y^2 - 42x + 4y + 142 = 0$

30. $5x^2 + 2y^2 + 30x - 16y + 27 = 0$

31 $2x^2 + 7y^2 + 24x + 84y + 310 = 0$

32. **HISTORY** The United States Capitol has a room with an elliptical ceiling. This type of room is called a *whispering gallery* because sound that is projected from one focus of an ellipse reflects off the ceiling and back to the other focus. The room in the Capitol is 96 feet in length, 45 feet wide, and has a ceiling that is 23 feet high.

   a. Write an equation modeling the shape of the room. Assume that it is centered at the origin and that the major axis is horizontal.

   b. Find the location of the two foci.

   c. How far from one focus would one have to stand to be able to hear the sound reflecting from the other focus?

Write an equation for a circle that satisfies each set of conditions. Then graph the circle.

33. center at $(3, 0)$, radius 2

34. center at $(-1, 7)$, diameter 6

35. center at $(-4, -3)$, tangent to $y = 3$

36. center at $(2, 0)$, endpoints of diameter at $(-5, 0)$ and $(9, 0)$

37. **FORMULA** Derive the general form of the equation for an ellipse with a vertical major axis centered at the origin.

38. MEDICAL TECHNOLOGY Indoor Positioning Systems (IPS) use ultrasound waves to detect tags that are linked to digital files containing information regarding a person or item being monitored. Hospitals often use IPS to detect the location of moveable equipment and patients.

   a. If the tracking system receiver must be centrally located for optimal functioning, where should a receiver be situated in a hospital complex that is 800 meters by 942 meters?

   b. Write an equation that models the sonar range of the IPS.

Write an equation for each ellipse.

39.

40.

41.

42.

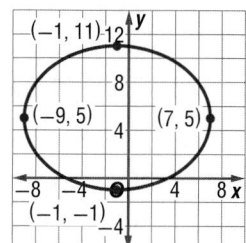

43 PLANETARY MOTION Each of the planets in the solar system move around the Sun in an elliptical orbit, where the Sun is one focus of the ellipse. Mercury is 43.4 million miles from the Sun at its farthest point and 28.6 million miles at its closest, as shown below. The diameter of the Sun is 870,000 miles.

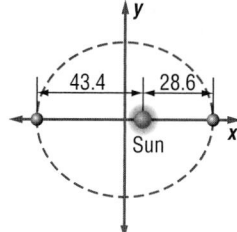

   a. Find the length of the minor axis.

   b. Find the eccentricity of the elliptical orbit.

Find the center, foci, and vertices of each ellipse.

44. $\dfrac{(x+5)^2}{16} + \dfrac{y^2}{7} = 1$

45. $\dfrac{x^2}{100} + \dfrac{(y+6)^2}{25} = 1$

46. $9y^2 - 18y + 25x^2 + 100x - 116 = 0$

47. $65x^2 + 16y^2 + 130x - 975 = 0$

48. TRUCKS Elliptical tanker trucks like the one shown are often used to transport liquids because they are more stable than circular tanks and the movement of the fluid is minimized.

   a. Draw and label the elliptical cross-section of the tank on a coordinate plane.

   b. Write an equation to represent the elliptical shape of the tank.

   c. Find the eccentricity of the ellipse.

Write the standard form of the equation for each ellipse.

49. The vertices are at $(-10, 0)$ and $(10, 0)$, and the eccentricity e is $\dfrac{3}{5}$.

50. The co-vertices are at $(0, 1)$ and $(6, 1)$, and the eccentricity e is $\dfrac{4}{5}$.

51. The center is at $(2, -4)$, one focus is at $(2, -4 + 2\sqrt{5})$, and the eccentricity e is $\dfrac{\sqrt{5}}{3}$.

52. ROLLER COASTERS The shape of a roller coaster loop in an amusement park can be modeled by $\dfrac{y^2}{3306.25} + \dfrac{x^2}{2025} = 1$.

   a. What is the width of the loop along the horizontal axis?

   b. Determine the height of the roller coaster from the ground when it reaches the top of the loop, if the lower rail is 20 feet from ground level.

   c. Find the eccentricity of the ellipse.

53. FOREST FIRES The radius of a forest fire is expanding at a rate of 4 miles per day. The current state of the fire is shown below, where a city is located 20 miles southeast of the fire.

   a. Write the equation of the circle at the current time and the equation of the circle at the time the fire reaches the city.

   b. Graph both circles.

   c. If the fire continues to spread at the same rate, how many days will it take to reach the city?

54. The *latus rectum* of an ellipse is a line segment that passes through a focus, is perpendicular to the major axis of the ellipse, and has endpoints on the ellipse. The length of each latus rectum is $\frac{2b^2}{a}$ units, where a is half the length of the major axis and b is half the length of the minor axis.

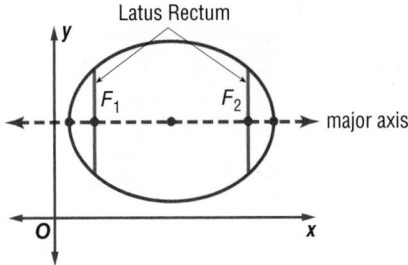

Latus Rectum

Write the equation of a horizontal ellipse with center at $(3, 2)$, major axis is 16 units long, and latus rectum 12 units long.

Find the coordinates of points where a line intersects a circle.

55. $y = x - 8$, $(x - 7)^2 + (y + 5)^2 = 16$

56. $y = x + 9$, $(x - 3)^2 + (y + 5)^2 = 169$

57. $y = -x + 1$, $(x - 5)^2 + (y - 2)^2 = 50$

58. $y = -\frac{1}{3}x - 3$, $(x + 3)^2 + (y - 3)^2 = 25$

59. **REFLECTION** *Silvering* is the process of coating glass with a reflective substance. The interior of an ellipse can be silvered to produce a mirror with rays that originate at the ellipse's focus and then reflect to the other focus as shown.

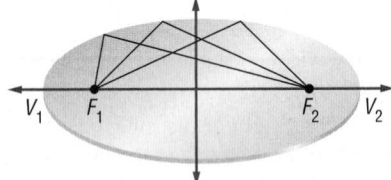

If the segment V_1F_1 is 2 cm long and the eccentricity of the mirror is 0.5, find the equation of the ellipse in standard form.

60. **CHEMISTRY** Distillation columns are used to separate chemical substances based on the differences in their rates of evaporation. The columns may contain plates with bubble caps or small circular openings.

18 in.

a. Write an equation for the plate shown, assuming that the center is at $(-4, -1)$.

b. What is the surface area of the plate not covered by bubble caps if each cap is 2 inches in diameter?

61. **GEOMETRY** The graphs of $x - 5y = -3$, $2x + 3y = 7$, and $4x - 7y = 27$ contain the sides of a triangle. Write the equation of a circle that circumscribes the triangle.

Write the standard form of the equation of a circle that passes through each set of points. Then identify the center and radius of the circle.

62. $(2, 3)$, $(8, 3)$, $(5, 6)$

63. $(1, -11)$, $(-3, -7)$, $(5, -7)$

64. $(0, 9)$, $(0, 3)$, $(-3, 6)$

65. $(7, 4)$, $(-1, 12)$, $(-9, 4)$

H.O.T. Problems   Use **H**igher-**O**rder **T**hinking Skills

66. **ERROR ANALYSIS** Yori and Chandra are graphing an ellipse that has a center at $(-1, 3)$, a major axis of length 8, and a minor axis of length 4. Is either of them correct? Explain your reasoning.

Yori

Chandra

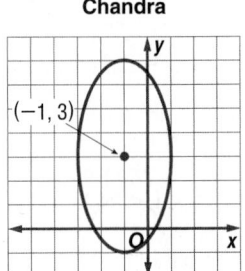

67. **REASONING** Determine whether an ellipse represented by $\frac{x^2}{p} + \frac{y^2}{p + r} = 1$, where $r > 0$, will have the same foci as the ellipse represented by $\frac{x^2}{p + r} + \frac{y^2}{p} = 1$. Explain your reasoning.

CHALLENGE The area A of an ellipse of the form $\frac{x^2}{a^2} + \frac{y^2}{b^2} = 1$ is $A = \pi ab$. Write an equation of an ellipse with each of the following characteristics.

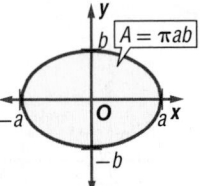

$A = \pi ab$

68. $b + a = 12$, $A = 35\pi$ **69** $a - b = 5$, $A = 24\pi$

70. **WRITING IN MATH** Explain how to find the foci and vertices of an ellipse if you are given the standard form of the equation.

71. **REASONING** Is the ellipse $\frac{x^2}{a^2} + \frac{y^2}{b^2} = 1$ symmetric with respect to the origin? Explain your reasoning.

72. **OPEN ENDED** If the equation of a circle is $(x - h)^2 + (y - k)^2 = r^2$, where $h > 0$ and $k < 0$, what is the domain of the circle? Verify your answer with an example, both algebraically and graphically.

73. **WRITING IN MATH** Explain why an ellipse becomes circular as the value of a approaches the value of c.

For each equation, identify the vertex, focus, axis of symmetry, and directrix. Then graph the parabola. (Lesson 7-1)

74. $y = 3x^2 - 24x + 50$

75. $y = -2x^2 + 5x - 10$

76. $x = 5y^2 - 10y + 9$

77. MANUFACTURING A toy company is introducing two new dolls to its customers: My First Baby, which talks, laughs, and cries, and My Real Baby, which uses a bottle and crawls. In one hour, the company can produce 8 First Babies or 20 Real Babies. Because of the demand, the company must produce at least twice as many First Babies as Real Babies. The company spends no more than 48 hours per week making these two dolls. Find the number and type of dolls that should be produced to maximize the profit. (Lesson 6-5)

Profit per Doll ($)	
First Baby	**Real Baby**
3.00	7.50

Verify each identity. (Lesson 5-4)

78. $\sin(\theta + 30°) + \cos(\theta + 60°) = \cos\theta$

79. $\sin\left(\theta + \frac{\pi}{3}\right) - \cos\left(\theta + \frac{\pi}{6}\right) = \sin\theta$

80. $\sin(3\pi - x) = \sin x$

Find all solutions to each equation in the interval $(0, 2\pi)$. (Lesson 5-3)

81. $\sin\theta = \cos\theta$

82. $\sin\theta = 1 + \cos\theta$

83. $2\sin^2 x + 3\sin x + 1 = 0$

Solve each inequality. (Lesson 2-6)

84. $x^2 - 5x - 24 > 0$

85. $x^2 + 2x - 35 \le 0$

86. $-2y^2 + 7y + 4 < 0$

State the number of possible real zeros and turning points of each function. Then determine all of the real zeros by factoring. (Lesson 2-2)

87. $f(x) = 3x^4 + 18x^3 + 24x^2$

88. $f(x) = 8x^6 + 48x^5 + 40x^4$

89. $f(x) = 5x^5 - 15x^4 - 50x^3$

Simplify. (Lesson 0-2)

90. $(2 + 4i) + (-1 + 5i)$

91. $(-2 - i)^2$

92. $\dfrac{i}{1 + 2i}$

93. SAT/ACT Point B lies 10 units from point A, which is the center of a circle of radius 6. If a tangent line is drawn from B to the circle, what is the distance from B to the point of tangency?

A 6 C 10 E $2\sqrt{41}$
B 8 D $2\sqrt{34}$

94. REVIEW What is the standard form of the equation of the conic given below?

$$2x^2 + 4y^2 - 8x + 24y + 32 = 0$$

F $\dfrac{(x-4)^2}{3} + \dfrac{(y+3)^2}{11} = 1$

G $\dfrac{(x-2)^2}{6} + \dfrac{(y+3)^2}{3} = 1$

H $\dfrac{(x+2)^2}{5} + \dfrac{(y+3)^2}{4} = 1$

J $\dfrac{(x-4)^2}{11} + \dfrac{(y+3)^2}{3} = 1$

95. Ruben is making an elliptical target for throwing darts. He wants the target to be 27 inches wide and 15 inches high. Which equation should Ruben use to draw the target?

A $\dfrac{x^2}{7.5} + \dfrac{y^2}{13.5} = 1$

B $\dfrac{x^2}{56.25} + \dfrac{y^2}{182.25} = 1$

C $\dfrac{x^2}{182.25} + \dfrac{y^2}{56.25} = 1$

D $\dfrac{x^2}{13.5} + \dfrac{y^2}{7.5} = 1$

96. REVIEW If $m = \frac{1}{x}$, $n = 7m$, $p = \frac{1}{n}$, $q = 14p$, and $r = \dfrac{1}{\frac{1}{2^q}}$, find x.

F r H p
G q J $\frac{1}{r}$

Hyperbolas

Why?

Lightning detection systems use multiple sensors to digitize lightning strike waveforms and record details of the strike using extremely accurate GPS timing signals. Two sensors detect a signal at slightly different times and generate a point on a hyperbola where the distance from each sensor is proportional to the difference in the time of arrival. The sensors make it possible to transmit the exact location of a lightning strike in real time.

Then
You analyzed and graphed ellipses and circles. (Lesson 7-2)

Now
- Analyze and graph equations of hyperbolas.
- Use equations to identify types of conic sections.

New Vocabulary
hyperbola
transverse axis
conjugate axis

Math Online
glencoe.com

Analyze and Graph Hyperbolas While an ellipse is the locus of all points in a plane such that the *sum* of the distances from two foci is constant, a **hyperbola** is the locus of all points in a plane such that the *absolute value of the differences* of the distances from two foci is constant.

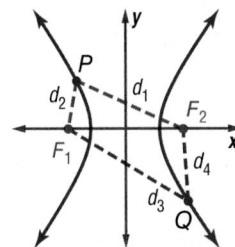

$$|d_1 - d_2| = |d_3 - d_4|$$

The graph of a hyperbola consists of two disconnected branches that approach two asymptotes. The midpoint of the segment with endpoints at the foci is the center. The vertices are at the intersection of this segment and each branch of the curve.

Like an ellipse, a hyperbola has two axes of symmetry. The **transverse axis** has a length of $2a$ units and connects the vertices. The **conjugate axis** is perpendicular to the transverse, passes through the center, and has a length of $2b$ units.

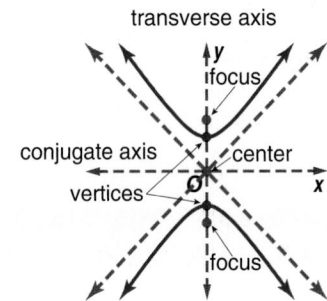

The relationship among the values of a, b, and c is different for a hyperbola than it is for an ellipse. For a hyperbola, the relationship is $c^2 = a^2 + b^2$. In addition, for any point on the hyperbola, the absolute value of the difference between the distances from the point to the foci is $2a$.

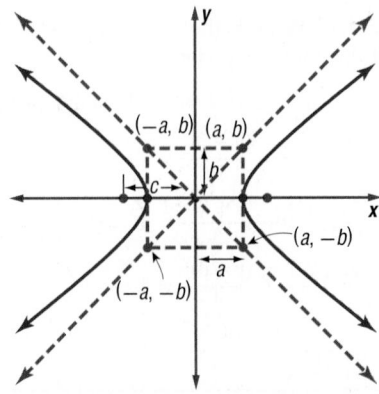

As with other conic sections, the definition of a hyperbola can be used to derive its equation. Let $P(x, y)$ be any point on the hyperbola with center $C(h, k)$. The coordinates of the foci and vertices are shown at the right. By the definition of a hyperbola, the absolute value of the difference of distances from any point on the hyperbola to the foci is constant. Thus, $|PF_1 - PF_2| = 2a$. Therefore, either $PF_1 - PF_2 = 2a$ or $PF_2 - PF_1 = 2a$. For the proof below, we will assume $PF_1 - PF_2 = 2a$.

$$PF_1 - PF_2 = 2a \qquad \text{Definition of hyperbola}$$

$$\sqrt{[x - (h - c)]^2 + (y - k)^2} - \sqrt{[x - (h + c)]^2 + (y - k)^2} = 2a \qquad \text{Distance Formula}$$

$$\sqrt{(x - h + c)^2 + (y - k)^2} = 2a + \sqrt{(x - h - c)^2 + (y - k)^2} \qquad \text{Distributive and Subtraction Properties}$$

$$\sqrt{[(x - h) + c]^2 + (y - k)^2} = 2a + \sqrt{[(x - h) - c]^2 + (y - k)^2} \qquad \text{Associative Property}$$

$$[(x - h) + c]^2 + (y - k)^2 = 4a^2 + 4a\sqrt{[(x - h) - c]^2 + (y - k)^2} + [(x - h) - c]^2 + (y - k)^2$$

$$(x - h)^2 + 2c(x - h) + c^2 + (y - k)^2 = 4a^2 + 4a\sqrt{[(x - h) - c]^2 + (y - k)^2} + (x - h)^2 - 2c(x - h) + c^2 + (y - k)^2$$

$$-4a\sqrt{[(x - h) - c]^2 + (y - k)^2} = 4a^2 - 4c(x - h) \qquad \text{Subtraction Property}$$

$$a\sqrt{[(x - h) - c]^2 + (y - k)^2} = -a^2 + c(x - h) \qquad \text{Divide each side by } -4.$$

$$a^2[(x - h)^2 - 2c(x - h) + c^2 + (y - k)^2] = a^4 - 2a^2c(x - h) + c^2(x - h)^2 \qquad \text{Square each side.}$$

$$a^2(x - h)^2 - 2a^2c(x - h) + a^2c^2 + a^2(y - k)^2 = a^4 - 2a^2c(x - h) + c^2(x - h)^2 \qquad \text{Distributive Property}$$

$$a^2(x - h)^2 - c^2(x - h)^2 + a^2(y - k)^2 = a^4 - a^2c^2 \qquad \text{Addition and Subtraction Properties}$$

$$(a^2 - c^2)(x - h)^2 + a^2(y - k)^2 = a^2(a^2 - c^2) \qquad \text{Distributive Property}$$

$$-b^2(x - h)^2 + a^2(y - k)^2 = a^2(-b^2) \qquad a^2 - c^2 = -b^2$$

$$\frac{(x - h)^2}{a^2} - \frac{(y - k)^2}{b^2} = 1 \qquad \text{Divide each side by } a^2(-b^2).$$

The general equation for a hyperbola centered at (h, k) is given below.

Key Concept   **Standard Forms of Equations for Hyperbolas**

$$\frac{(x - h)^2}{a^2} - \frac{(y - k)^2}{b^2} = 1$$

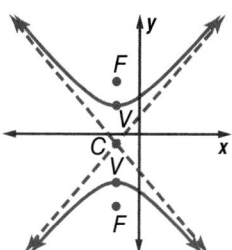

Orientation: horizontal transverse axis

Center: (h, k)

Vertices: $(h \pm a, k)$

Foci: $(h \pm c, k)$

Transverse axis: $y = k$

Conjugate axis: $x = h$

Asymptotes: $y - k = \pm\frac{b}{a}(x - h)$

a, b, c relationship: $c^2 = a^2 + b^2$ or $c = \sqrt{a^2 + b^2}$

$$\frac{(y - k)^2}{a^2} - \frac{(x - h)^2}{b^2} = 1$$

Orientation: vertical transverse axis

Center: (h, k)

Vertices: $(h, k \pm a)$

Foci: $(h, k \pm c)$

Transverse axis: $x = h$

Conjugate axis: $y = k$

Asymptotes: $y - k = \pm\frac{a}{b}(x - h)$

a, b, c relationship: $c^2 = a^2 + b^2$ or $c = \sqrt{a^2 + b^2}$

| EXAMPLE 1 | Graph Hyperbolas in Standard Form |

Graph the hyperbola given by each equation.

a. $\dfrac{y^2}{9} - \dfrac{x^2}{25} = 1$

The equation is in standard form with h and k both equal to zero. Because $a^2 = 9$ and $b^2 = 25$, $a = 3$ and $b = 5$. Use the values of a and b to find c.

$c^2 = a^2 + b^2$ **Equation relating *a, b,* and *c* for a hyperbola**

$c^2 = 3^2 + 5^2$ ***a* = 3 and *b* = 5**

$c = \sqrt{34}$ or about 5.83   **Solve for *c.***

Use these values for $h, k, a, b,$ and c to determine the characteristics of the hyperbola.

orientation:	vertical	**When the equation is in standard form, the x^2-term is subtracted.**
center:	$(0, 0)$	(h, k)
vertices:	$(0, 3)$ and $(0, -3)$	$(h, k \pm a)$
foci:	$(0, \sqrt{34})$ and $(0, -\sqrt{34})$	$(h, k \pm c)$
asymptotes:	$y = \dfrac{3}{5}x$ and $y = -\dfrac{3}{5}x$	$y - k = \pm\dfrac{a}{b}(x - h)$

Graph the center, vertices, foci, and asymptotes. Then make a table of values to sketch the hyperbola.

x	y
−6	−4.69, 4.69
−1	−3.06, 3.06
1	−3.06, 3.06
6	−4.69, 4.69

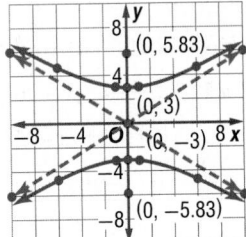

b. $\dfrac{(x + 1)^2}{9} - \dfrac{(y + 2)^2}{16} = 1$

The equation is in standard form with $h = -1, k = -2, a = \sqrt{9}$ or 3, $b = \sqrt{16}$ or 4, and $c = \sqrt{9 + 16}$ or 5. Use these values to determine the characteristics of the hyperbola.

orientation:	horizontal	**When the equation is in standard form, the y^2-term is subtracted.**
center:	$(-1, -2)$	(h, k)
vertices:	$(2, -2)$ and $(-4, -2)$	$(h \pm a, k)$
foci:	$(4, -2)$ and $(-6, -2)$	$(h \pm c, k)$
asymptotes:	$y + 2 = \dfrac{4}{3}(x + 1)$ and $y + 2 = -\dfrac{4}{3}(x + 1)$, or	$y - k = \pm\dfrac{b}{a}(x - h)$
	$y = \dfrac{4}{3}x - \dfrac{2}{3}$ and $y = \dfrac{4}{3}x - \dfrac{10}{3}$	

Graph the center, vertices, foci, and asymptotes. Then make a table of values to sketch the hyperbola.

x	y
−6	−7.33, 3.33
−5	−5.53, 1.53
3	−5.53, 1.53
4	−7.33, 3.33

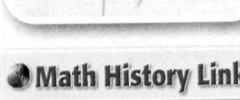

Math History Link

Hypatia
(c. 370 A.D.–415 A.D.)
Hypatia was a mathematician, scientist, and philosopher who worked as a professor at a university in Alexandria, Egypt. Hypatia edited the book *On the Conics of Apollonius,* which developed the ideas of hyperbolas, parabolas, and ellipses.

Source: Agnes Scott College

✓ **Guided Practice**

1A. $\dfrac{x^2}{4} - \dfrac{y^2}{1} = 1$ **1B.** $\dfrac{y^2}{5} - \dfrac{x^2}{3} = 1$

If you know the equation for a hyperbola in standard form, you can use the characteristics to graph the curve. If you are given the equation in another form, you will need to write the equation in standard form to determine the characteristics.

EXAMPLE 2 **Graph a Hyperbola**

Graph the hyperbola given by $25x^2 - 16y^2 + 100x + 96y = 444$.

First, write the equation in standard form.

$25x^2 - 16y^2 + 100x - 96 = 444$	Original equation
$(25x^2 + 100x) - (16y^2 + 96y) = 444$	Group like terms.
$25(x^2 + 4x) - 16(y^2 - 6y) = 444$	Factor.
$25(x^2 + 4x + 4) - 16(y^2 - 6y + 9) = 444 + 25(4) - 16(9)$	Complete the squares.
$25(x + 2)^2 - 16(y - 3)^2 = 400$	Factor and simplify.
$\dfrac{(x + 2)^2}{16} - \dfrac{(y - 3)^2}{25} = 1$	Divide each side by 400.

The equation is now in standard form with $h = -2$, $k = 3$, $a = \sqrt{16}$ or 4, $b = \sqrt{25}$ or 5, and $c = \sqrt{16 + 25}$, which is $\sqrt{41}$ or about 6.4. Use these values to determine the characteristics of the hyperbola.

		When the equation is in standard form, the y^2-term is subtracted.
orientation:	horizontal	
center:	$(-2, 3)$	(h, k)
vertices:	$(-6, 3)$ and $(2, 3)$	$(h \pm a, k)$
foci:	$(-8.4, 3)$ and $(4.4, 3)$	$(h \pm c, k)$
asymptotes:	$y - 3 = \frac{5}{4}(x + 2)$ and $y - 3 = -\frac{5}{4}(x + 2)$, or	$y - k = \pm \frac{b}{a}(x - h)$
	$y = \frac{5}{4}x + \frac{11}{2}$ and $y = -\frac{5}{4}x + \frac{1}{2}$	

Graph the center, vertices, foci, and asymptotes. Then, make a table of values to sketch the hyperbola.

x	y
−9	−4.18, 10.18
−7	−0.75, 6.75
3	−0.75, 6.75
5	−4.18, 10.18

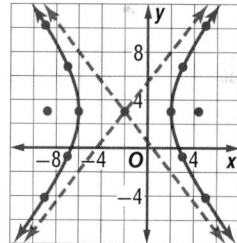

CHECK Solve the equation for y to obtain two functions of x,

$$y = 3 + \sqrt{-25 + \frac{25(x + 2)^2}{16}} \text{ and } 3 - \sqrt{-25 + \frac{25(x + 2)^2}{16}}.$$

Graph the equations in the same window, along with the equations of the asymptote and compare with your graph, by testing a few points. ✔

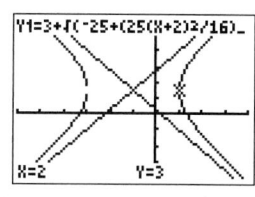

[−12, 8] scl: 1 by [−8, 12] scl: 1

✔ **Guided Practice**

Graph the hyperbola given by each equation.

2A. $\dfrac{(y + 4)^2}{64} - \dfrac{(x + 1)^2}{81} = 1$

2B. $2x^2 - 3y^2 - 12x = 36$

When graphing a hyperbola remember that the graph will approach the asymptotes as it moves away from the vertices. Plot near the vertices to improve the accuracy of your graph.

You can determine the equation for a hyperbola if you are given characteristics that provide sufficient information.

Write Equations Given Characteristics

Write an equation for the hyperbola with the given characteristics.

a. vertices $(-3, -6)$, $(-3, 2)$; foci $(-3, -7)$, $(-3, 3)$

Because the x-coordinates of the vertices are the same, the transverse axis is vertical. Find the center and the values of a, b, and c.

center: $(-3, -2)$	Midpoint of segment between foci
$a = 4$	Distance from each vertex to center
$c = 5$	Distance from each focus to center
$b = 3$	$c^2 = a^2 + b^2$

Because the transverse axis is vertical, the a^2-term goes with the y^2-term. An equation for the hyperbola is $\dfrac{(y+2)^2}{16} - \dfrac{(x+3)^2}{9} = 1$. The graph of the hyperbola is shown in Figure 7.3.1.

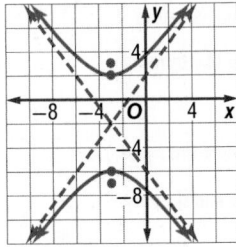

Figure 7.3.1

b. vertices $(-3, 0)$, $(-9, 0)$; asymptotes $y = 2x - 12$, $y = -2x + 12$

Because the y-coordinates of the vertices are the same, the transverse axis is horizontal.

center: $(-6, 0)$	Midpoint of segment between vertices
$a = 3$	Distance from each vertex to center

The slopes of the asymptotes are $\pm\dfrac{b}{a}$. Use the positive slope to find b.

$\dfrac{b}{a} = 2$	Positive slope of asymptote
$\dfrac{b}{3} = 2$	$a = 3$
$b = 6$	Solve for b.

Because the transverse axis is horizontal, the a^2-term goes with the x^2-term. An equation for the hyperbola is $\dfrac{(x+6)^2}{9} - \dfrac{y^2}{36} = 1$. The graph of the hyperbola is shown in Figure 7.3.2.

Figure 7.3.2

☑ Guided Practice

3A. vertices $(3, 2)$, $(3, 6)$; conjugate axis length 10 units

3B. foci $(2, -2)$, $(12, -2)$; asymptotes $y = \dfrac{3}{4}x - \dfrac{29}{4}$, $y = -\dfrac{3}{4}x + \dfrac{13}{4}$

Another characteristic that can be used to describe a hyperbola is the eccentricity. The formula for eccentricity is the same for all conics, $e = \dfrac{c}{a}$. Recall that for an ellipse, the eccentricity is greater than 0 and less than 1. For a hyperbola, the eccentricity will always be greater than 1.

EXAMPLE 4 **Find the Eccentricity of a Hyperbola**

Determine the eccentricity of the hyperbola given by $\dfrac{(y-4)^2}{48} - \dfrac{(x+5)^2}{36} = 1$.

Find c and then determine the eccentricity.

$c^2 = a^2 + b^2$	Equation relating a, b, and c	$e = \dfrac{c}{a}$	Eccentricity equation
$c^2 = 48 + 36$	$a^2 = 48$ and $b^2 = 36$	$= \dfrac{\sqrt{84}}{\sqrt{48}}$	$c = \sqrt{84}$ and $a = \sqrt{48}$
$c = \sqrt{84}$	Solve for c.	≈ 1.32	Simplify.

The eccentricity of the hyperbola is about 1.32.

✓ Guided Practice

Determine the eccentricity of the hyperbola given by each equation.

4A. $\dfrac{(x+8)^2}{64} - \dfrac{(y-4)^2}{80} = 1$ **4B.** $\dfrac{(y-2)^2}{15} - \dfrac{(x+9)^2}{75} = 1$

Identify Conic Sections You can determine the type of conic when the equation for the conic is in general form, $Ax^2 + Bxy + Cy^2 + Dx + Ey + F = 0$. The discriminant, or $B^2 - 4AC$, can be used to identify the conic.

Key Concept   **Classify Conics Using the Discriminant**

The graph of a second degree equation of the form $Ax^2 + Bxy + Cy^2 + Dx + Ey + F = 0$ is

- a circle if $B^2 - 4AC < 0$; $B = 0$ and $A = C$.
- an ellipse if $B^2 - 4AC < 0$; either $B \neq 0$ or $A \neq C$.
- a parabola if $B^2 - 4AC = 0$.
- a hyperbola if $B^2 - 4AC > 0$.

When $B = 0$, the conic will be either vertical or horizontal. When $B \neq 0$, the conic will be neither vertical nor horizontal.

StudyTip

Identifying Conics When a conic has been rotated as in Example 5b, its equation cannot be written in standard form. In this case, only the discriminant can be used to determine the type of conic without graphing. You will learn more about rotated conics in the next lesson.

EXAMPLE 5   **Identify Conic Sections**

Use the discriminant to identify each conic section.

a. $4x^2 + 3y^2 - 2x + 5y - 60 = 0$

A is 4, B is 0, and C is 3.

Find the discriminant.

$B^2 - 4AC = 0^2 - 4(4)(3)$ or -48

The discriminant is less than 0, so the conic must be either a circle or an ellipse. Because $A \neq C$, the conic is an ellipse.

b. $2y^2 + 6x - 3y + 4xy + 2x^2 - 88 = 0$

A is 2, B is 4, and C is 2.

Find the discriminant.

$B^2 - 4AC = 4^2 - 4(2)(2)$ or 0

The discriminant is 0, so the conic is a parabola.

c. $18x - 12y^2 + 4xy + 10x^2 - 6y + 24 = 0$

A is 10, B is 4, and C is -12.

Find the discriminant.

$B^2 - 4AC = 4^2 - 4(10)(-12)$ or 496

The discriminant is greater than 0, so the conic is a hyperbola.

✓ Guided Practice

5A. $3x^2 + 4x - 2y + 3y^2 + 6xy + 64 = 0$

5B. $6x^2 + 2xy - 15x = 3y^2 + 5y + 18$

5C. $4xy + 8x - 3y = 2x^2 + 8y^2$

Researchers can determine the location of a lightning strike on the hyperbolic path formed with the detection sensors located at the foci.

🌐 Real-World EXAMPLE 6 — Apply Hyperbolas

METEOROLOGY Two lightning detection sensors are located 6 kilometers apart, where sensor A is due north of sensor B. As a bolt of lightning strikes, researchers determine the lightning strike occurred east of both sensors and 1.5 kilometers farther from sensor A than sensor B.

a. **Find the equation for the hyperbola on which the lightning strike is located.**

First, place the two sensors on a coordinate grid so that the origin is the midpoint of the segment between sensor A and sensor B. The lightning is east of the sensors and closer to sensor B, so it should be in the 4th quadrant.

The two sensors are located at the foci of the hyperbola, so c is 3. Recall that the absolute value of the difference of the distances from any point on a hyperbola to the foci is $2a$. Because the lightning strike is 1.5 kilometers farther from sensor A than sensor B, $2a = 1.5$ and a is 0.75. Use these values of a and c to find b^2.

$$c^2 = a^2 + b^2 \qquad \text{Equation relating } a, b, \text{ and } c$$
$$3^2 = 0.75^2 + b^2 \qquad c = 3 \text{ and } a = 0.75$$
$$8.4375 = b^2 \qquad \text{Solve for } b^2.$$

The transverse axis is vertical and the center of the hyperbola is located at the origin, so the equation will be of the form $\dfrac{y^2}{a^2} - \dfrac{x^2}{b^2} = 1$. Substituting the values of a^2 and b^2, the equation for the hyperbola is $\dfrac{y^2}{0.5625} - \dfrac{x^2}{8.4375} = 1$.

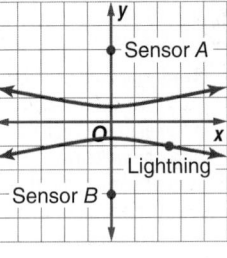

The lightning strike occurred along the hyperbola

$$\dfrac{y^2}{0.5625} - \dfrac{x^2}{8.4375} = 1.$$

b. **Find the coordinates of the lightning strike if it occurred 2.5 kilometers east of the sensors.**

Because the lightning strike occurred 2.5 kilometers east of the sensors, $x = 2.5$. The lightning was closer to sensor B than sensor A, so it lies on the lower branch. Substitute the value of x into the equation and solve for y.

$$\dfrac{y^2}{0.5625} - \dfrac{x^2}{8.4375} = 1 \qquad \text{Original equation}$$
$$\dfrac{y^2}{0.5625} - \dfrac{2.5^2}{8.4375} = 1 \qquad x = 2.5$$
$$y \approx -0.99 \qquad \text{Solve.}$$

The value of y is about -0.99, so the location of the lightning strike is at $(2.5, -0.99)$.

✓ Guided Practice

6. METEOROLOGY Sensor A is located 30 miles due west of sensor B. A lightning strike occurs 9 miles farther from sensor A than sensor B.

   A. Find the equation for the hyperbola on which the lightning strike occurred.

   B. Find the coordinates of the location of the lightning strike if it occurred 8 miles north of the sensors.

Graph the hyperbola given by each equation. (Example 1)

1. $\dfrac{x^2}{16} - \dfrac{y^2}{9} = 1$

2. $\dfrac{y^2}{4} - \dfrac{x^2}{17} = 1$

3. $\dfrac{x^2}{49} - \dfrac{y^2}{30} = 1$

4. $\dfrac{y^2}{34} - \dfrac{x^2}{14} = 1$

5. $\dfrac{x^2}{9} - \dfrac{y^2}{21} = 1$

6. $\dfrac{x^2}{36} - \dfrac{y^2}{4} = 1$

7. $\dfrac{y^2}{81} - \dfrac{x^2}{8} = 1$

8. $\dfrac{y^2}{25} - \dfrac{x^2}{14} = 1$

9. $3x^2 - 2y^2 = 12$

10. $3y^2 - 5x^2 = 15$

11. **LIGHTING** The light projected on a wall by a table lamp can be represented by a hyperbola. The light from a certain table lamp can be modeled by $\dfrac{y^2}{225} - \dfrac{x^2}{81} = 1$. Graph the hyperbola. (Example 1)

Graph the hyperbola given by each equation. (Example 2)

12. $\dfrac{(x+5)^2}{9} - \dfrac{(y+4)^2}{48} = 1$

13. $\dfrac{(y-7)^2}{4} - \dfrac{x^2}{33} = 1$

14. $\dfrac{(x-2)^2}{25} - \dfrac{(y-6)^2}{60} = 1$

15. $\dfrac{(x-5)^2}{49} - \dfrac{(y-1)^2}{17} = 1$

16. $\dfrac{(y-3)^2}{16} - \dfrac{(x-4)^2}{42} = 1$

17. $\dfrac{(x+6)^2}{64} - \dfrac{(y+5)^2}{58} = 1$

18. $x^2 - 4y^2 - 6x - 8y = 27$

19. $-x^2 + 3y^2 - 4x + 6y = 28$

20. $13x^2 - 2y^2 + 208x + 16y = -748$

21. $-5x^2 + 2y^2 - 70x - 8y = 287$

22. **EARTHQUAKES** Shortly after a seismograph detects an earthquake, a second seismograph positioned due north of the first detects the earthquake. The epicenter of the earthquake lies on a branch of the hyperbola represented by $\dfrac{(y-30)^2}{900} - \dfrac{(x-60)^2}{1600} = 1$, where the seismographs are located at the foci. Graph the hyperbola. (Example 2)

Write an equation for the hyperbola with the given characteristics. (Example 3)

23. foci $(-1, 9), (-1, -7)$; conjugate axis length of 14 units

24. vertices $(7, 5), (-5, 5)$; foci $(11, 5), (-9, 5)$

25. foci $(9, -1), (-3, -1)$; conjugate axis length of 6 units

26. vertices $(-1, 9), (-1, 3)$; asymptotes $y = \pm\dfrac{3}{7}x + \dfrac{45}{7}$

27. vertices $(-3, -12), (-3, -4)$; foci $(-3, -15), (-3, -1)$

28. foci $(9, 7), (-17, 7)$; asymptotes $y = \pm\dfrac{5}{12}x + \dfrac{104}{12}$

29. center $(-7, 2)$; asymptotes $y = \pm\dfrac{7}{5}x + \dfrac{59}{5}$, transverse axis length of 10 units

30. center $(0, -5)$; asymptotes $y = \pm\dfrac{\sqrt{19}}{6}x - 5$, conjugate axis length of 12 units

31. vertices $(0, -3), (-4, -3)$; conjugate axis length of 12 units

32. vertices $(2, 10), (2, -2)$; conjugate axis length of 16 units

33. **ARCHITECTURE** The graph below shows the outline of a floor plan for an office building.

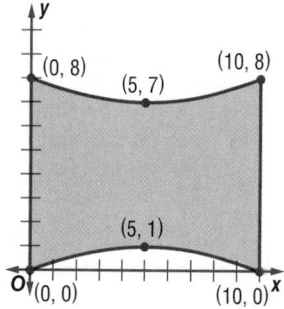

a. Write an equation that could model the curved sides of the building.

b. Each unit on the coordinate plane represents 15 feet. What is the narrowest width of the building? (Example 3)

Determine the eccentricity of the hyperbola given by each equation. (Example 4)

34. $\dfrac{(y-1)^2}{10} - \dfrac{(x-6)^2}{13} = 1$

35. $\dfrac{(x+4)^2}{24} - \dfrac{(y+1)^2}{15} = 1$

36. $\dfrac{(x-3)^2}{38} - \dfrac{(y-2)^2}{5} = 1$

37. $\dfrac{(y+2)^2}{32} - \dfrac{(x+5)^2}{25} = 1$

38. $\dfrac{(y-4)^2}{23} - \dfrac{(x+11)^2}{72} = 1$

39. $\dfrac{(x-1)^2}{16} - \dfrac{(y+4)^2}{29} = 1$

Determine the eccentricity of the hyperbola given by each equation. (Example 4)

40. $11x^2 - 2y^2 - 110x + 24y = -181$

41. $-4x^2 + 3y^2 + 72x - 18y = 321$

42. $3x^2 - 2y^2 + 12x - 12y = 42$

43. $-x^2 + 7y^2 + 24x + 70y = -24$

Use the discriminant to identify each conic section. (Example 5)

44. $14y + y^2 = 4x - 97$

45. $18x - 3x^2 + 4 = -8y^2 + 32y$

46. $14 + 4y + 2x^2 = -12x - y^2$

47. $12y - 76 - x^2 = 16x$

48. $2x + 8y + x^2 + y^2 = 8$

49. $5y^2 - 6x + 3x^2 - 50y = -3x^2 - 113$

50. $x^2 + y^2 + 8x - 6y + 9 = 0$

51. $-56y + 5x^2 = 211 + 4y^2 + 10x$

52. $-8x + 16 = 8y + 24 - x^2$

53. $x^2 - 4x = -y^2 + 12y - 31$

54. PHYSICS A hyperbola occurs naturally when two nearly identical glass plates in contact on one edge and separated by about 5 millimeters at the other edge are dipped in a thick liquid. The liquid will rise by capillarity to form a hyperbola caused by the surface tension. Find a model for the hyperbola if the conjugate axis is 50 centimeters and the transverse axis is 30 centimeters.

55 AVIATION The Federal Aviation Administration performs flight trials to test new technology in aircraft. When one of the test aircraft collected its data, it was 18 kilometers farther from Airport B than Airport A. The two airports are 72 kilometers apart along the same highway, with Airport B due south of Airport A. (Example 6)

a. Write an equation for the hyperbola centered at the origin on which the aircraft was located when the data were collected.

b. Graph the equation, indicating on which branch of the hyperbola the plane was located.

c. When the data were collected, the plane was 40 miles from the highway. Find the coordinates of the plane.

56. ASTRONOMY While each of the planets in our solar system move around the Sun in elliptical orbits, comets may have elliptical, parabolic, or hyperbolic orbits where the center of the sun is a focus. (Example 5)

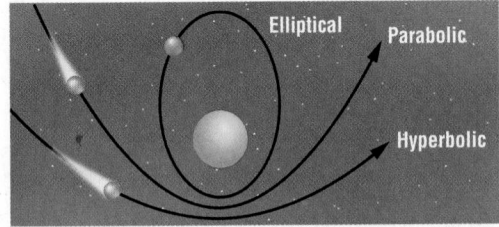

The paths of three comets are modeled below, where the values of x and y are measured in gigameters. Use the discriminant to identify each conic.

a. $3x^2 - 18x - 580850 = 4.84y^2 - 38.72y$

b. $-360x - 8y = -y^2 - 1096$

c. $-24.88y + x^2 = 6x - 3.11y^2 + 412341$

Derive the general form of the equation for a hyperbola with each of the following characteristics.

57. vertical transverse axis centered at the origin

58. horizontal transverse axis centered at the origin

Solve each system of equations. Round to the nearest tenth if necessary.

59. $2y = x - 10$ and $\dfrac{(x-3)^2}{16} - \dfrac{(y+2)^2}{84} = 1$

60. $y = -\dfrac{1}{4}x + 3$ and $\dfrac{x^2}{36} - \dfrac{(y-4)^2}{4} = 1$

61. $y = 2x$ and $\dfrac{(y+2)^2}{64} - \dfrac{(x+5)^2}{49} = 1$

62. $3x - y = 9$ and $\dfrac{(x-5)^2}{36} + \dfrac{y^2}{16} = 1$

63. $\dfrac{y^2}{36} + \dfrac{x^2}{25} = 1$ and $\dfrac{y^2}{36} - \dfrac{x^2}{25} = 1$

64. $\dfrac{x^2}{4} - \dfrac{y^2}{1} = 1$ and $\dfrac{(x+1)^2}{49} + \dfrac{(y+2)^2}{4} = 1$

65. FIREWORKS A fireworks grand finale is heard by Carson and Emmett, who are 3 miles apart talking on their cell phones. Emmett hears the finale about 1 second before Carson. Assume that sound travels at 1100 feet per second.

a. Write an equation for the hyperbola on which the fireworks were located. Place the locations of Carson and Emmett on the x-axis, with Carson on the left and the midpoint between them at the origin.

b. Describe the branch of the hyperbola on which the fireworks display was located.

66. ARCHITECTURE The Kobe Port Tower is a *hyperboloid* structure in Kobe, Japan. This means that the shape is generated by rotating a hyperbola around its conjugate axis. Suppose the hyperbola used to generate the hyperboloid modeling the shape of the tower has an eccentricity of 19.

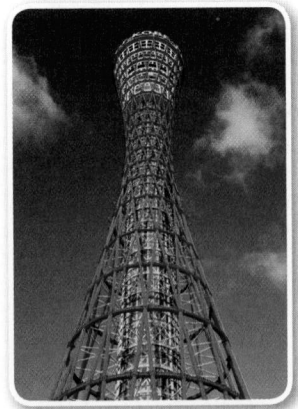

a. If the tower is 8 meters wide at its narrowest point, determine an equation of the hyperbola used to generate the hyperboloid.

b. If the top of the tower is 32 meters above the center of the hyperbola and the base is 76 meters below the center, what is the radius of the top and the radius of the base of the tower?

Write an equation for each hyperbola.

67.

68.

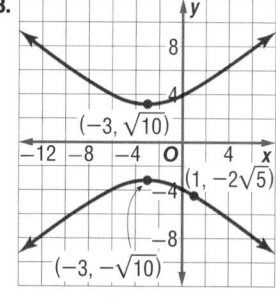

69. SOUND When a tornado siren goes off, three people are located at J, K, and O, as shown on the graph below.

The person at J hears the siren 3 seconds before the person at O. The person at K hears the siren 1 second before the person at O. Find each possible location of the tornado siren. Assume that sound travels at 1100 feet per second. (*Hint*: A location of the siren will be at a point of intersection between the two hyperbolas. One hyperbola has foci at O and J. The other has foci at O and K.)

Write an equation for the hyperbola with the given characteristics.

70. The center is at $(5, 1)$, a vertex is at $(5, 9)$, and an equation of an asymptote is $3y = 4x - 17$.

71. The hyperbola has its center at $(-4, 3)$ and a vertex at $(1, 3)$. The equation of one of its asymptotes is $7x + 5y = -13$.

72. The foci are at $(0, 2\sqrt{6})$ and $(0, -2\sqrt{6})$. The eccentricity is $\dfrac{2\sqrt{6}}{3}$.

73 The eccentricity of the hyperbola is $\dfrac{7}{6}$ and the foci are at $(-1, -2)$ and $(13, -2)$.

74. The hyperbola has foci at $(-1, 9)$ and $(-1, -7)$ and the slopes of the asymptotes are $\pm\dfrac{\sqrt{15}}{7}$.

75. For an *equilateral hyperbola*, $a = b$ when the equation of the hyperbola is written in standard form. The asymptotes of an equilateral hyperbola are perpendicular. Write an equation for the equilateral hyperbola below.

76. ⚙ **MULTIPLE REPRESENTATIONS** In this problem, you will explore a special type of hyperbola called a *conjugate hyperbola*. This occurs when the conjugate axis of one hyperbola is the transverse axis of another.

a. GRAPHICAL Sketch the graphs of $\dfrac{x^2}{36} - \dfrac{y^2}{64} = 1$ and $\dfrac{y^2}{64} - \dfrac{x^2}{36} = 1$ on the same coordinate plane.

b. ANALYTICAL Compare the foci, vertices, and asymptotes of the graphs.

c. ANALYTICAL Write an equation for the conjugate hyperbola for $\dfrac{x^2}{16} - \dfrac{y^2}{9} = 1$.

d. GRAPHICAL Sketch the graphs of the new conjugate hyperbolas.

e. VERBAL Make a conjecture about the similarities of conjugate hyperbolas.

H.O.T. Problems   Use **H**igher-**O**rder **T**hinking Skills

77. OPEN ENDED Write an equation for a hyperbola where the distance between the foci is twice the length of the transverse axis.

78. REASONING Consider $rx^2 = -sy^2 - t$. Describe the type of conic section that is formed for each of the following. Explain your reasoning.

a. $rs = 0$ **b.** $rs > 0$

c. $r = s$ **d.** $rs < 0$

79. WRITING IN MATH Explain why the equation for the asymptotes of a hyperbola changes from $\pm\dfrac{b}{a}$ to $\pm\dfrac{a}{b}$ depending on the location of the transverse axis.

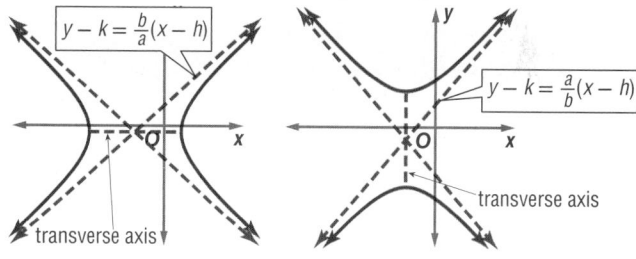

80. REASONING Suppose you are given two of the following characteristics: vertices, foci, transverse axis, conjugate axis, or asymptotes. Is it *sometimes*, *always*, or *never* possible to write the equation for the hyperbola?

81. CHALLENGE A hyperbola has foci at $F_1(0, 9)$ and $F_2(0, -9)$ and contains point P. The distance between P and F_1 is 6 units greater than the distance between P and F_2. Write the equation of the hyperbola in standard form.

82. PROOF An equilateral hyperbola is formed when $a = b$ in the standard form of the equation for a hyperbola. Prove that the eccentricity of every equilateral hyperbola is $\sqrt{2}$.

83. WRITING IN MATH Describe the steps for finding the equation of a hyperbola if the foci and length of the transverse axis are given.

Graph the ellipse given by each equation. (Lesson 7-2)

84. $(x - 8)^2 + \dfrac{(y - 2)^2}{81} = 1$

85. $\dfrac{x^2}{64} + \dfrac{(y + 5)^2}{49} = 1$

86. $\dfrac{(x - 2)^2}{16} + \dfrac{(y + 5)^2}{36} = 1$

87. PROJECTILE MOTION The height of a baseball hit by a batter with an initial speed of 80 feet per second can be modeled by $h = -16t^2 + 80t + 5$, where t is the time in seconds. (Lesson 7-1)

   a. How high above the ground is the vertex located?

   b. If an outfielder's catching height is the same as the initial height of the ball, about how long after the ball is hit will the player catch the ball?

Write each system of equations as a matrix equation, $AX = B$. Then use Gauss-Jordan elimination on the augmented matrix to solve the system. (Lesson 6-2)

88. $3x_1 + 11x_2 - 9x_3 = 25$
$-8x_1 + 5x_2 + x_3 = -31$
$x_1 - 9x_2 + 4x_3 = 13$

89. $x_1 - 7x_2 + 8x_3 = -3$
$6x_1 + 5x_2 - 2x_3 = 2$
$3x_1 - 4x_2 + 9x_3 = 26$

90. $2x_1 - 5x_2 + x_3 = 28$
$3x_1 + 4x_2 + 5x_3 = 17$
$7x_1 - 2x_2 + 3x_3 = 33$

Solve each equation for all values of θ. (Lesson 5-3)

91. $\tan \theta = \sec \theta - 1$

92. $\sin \theta + \cos \theta = 0$

93. $\csc \theta - \cot \theta = 0$

Find the exact values of the six trigonometric functions of θ. (Lesson 4-1)

94.

95.
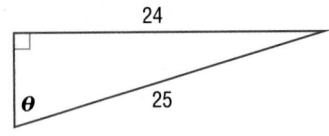

Use the given zero to find all complex zeros of each function. Then write the linear factorization of the function. (Lesson 2-4)

96. $f(x) = 2x^5 - 11x^4 + 69x^3 + 135x^2 - 675x; 3 - 6i$

97. $f(x) = 2x^5 - 9x^4 + 146x^3 + 618x^2 + 752x + 291; 4 + 9i$

Skills Review for Standardized Tests

98. REVIEW What is the equation of the graph?

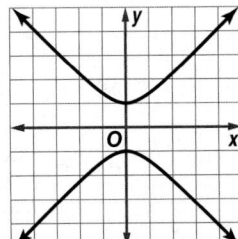

  A $y = x^2 + 1$ **D** $x^2 + y^2 = 1$

  B $y - x = 1$ **E** $xy = 1$

  C $y^2 - x^2 = 1$

99. REVIEW The graph of $\left(\dfrac{x}{4}\right)^2 - \left(\dfrac{y}{5}\right)^2 = 1$ is a hyperbola. Which set of equations represents the asymptotes of the hyperbola's graph?

  F $y = \dfrac{4}{5}x, y = -\dfrac{4}{5}x$ **H** $y = \dfrac{5}{4}x, y = -\dfrac{5}{4}x$

  G $y = \dfrac{1}{4}x, y = -\dfrac{1}{4}x$ **J** $y = \dfrac{1}{5}x, y = -\dfrac{1}{5}x$

100. The foci of the graph are at $(\sqrt{13}, 0)$ and $(-\sqrt{13}, 0)$. Which equation does the graph represent?

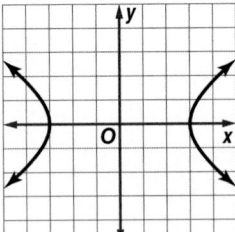

  A $\dfrac{x^2}{9} - \dfrac{y^2}{4} = 1$ **C** $\dfrac{x^2}{3} - \dfrac{y^2}{\sqrt{13}} = 1$

  B $\dfrac{x^2}{3} - \dfrac{y^2}{2} = 1$ **D** $\dfrac{x^2}{9} - \dfrac{y^2}{13} = 1$

101. SAT/ACT If $z = \dfrac{3y}{x^3}$, then what is the effect on the value of z when y is multiplied by 4 and x is doubled?

  F z is unchanged.   **H** z is doubled.

  G z is halved.   **J** z is multiplied by 4.

Write an equation for and graph a parabola with the given focus F and vertex V. (Lesson 7-1)

1. $F(1, 5)$, $V(1, 3)$

2. $F(5, -7)$, $V(1, -7)$

3. MULTIPLE CHOICE In each of the following, a parabola and its directrix are shown. In which parabola is the focus farthest from the vertex? (Lesson 7-1)

A

C

B

D

4. DESIGN The cross-section of the mirror in the flashlight design below is a parabola. (Lesson 7-1)

a. Write an equation that models the parabola.

b. Graph the equation.

Graph the ellipse given by each equation. (Lesson 7-2)

5. $\dfrac{(x+4)^2}{81} + \dfrac{(y+2)^2}{16} = 1$

6. $\dfrac{(x-3)^2}{4} + \dfrac{(y-6)^2}{36} = 1$

Write an equation for the ellipse with each set of characteristics. (Lesson 7-2)

7. vertices $(9, -3)$, $(-3, -3)$; foci $(7, -3)$, $(-1, -3)$

8. foci $(3, 1)$, $(3, 7)$; length of minor axis equals 8

9. major axis $(1, -1)$ to $(1, -13)$; minor axis $(-2, -7)$ to $(4, -7)$

10. vertices $(8, 5)$, $(8, -9)$; length of minor axis equals 6

11. SWIMMING The shape of a swimming pool is designed as an ellipse with a length of 30 feet and an eccentricity of 0.68. (Lesson 7-2)

a. What is the maximum width of the pool?

b. Write an equation for the ellipse if the point of origin is the center of the pool.

12. MULTIPLE CHOICE Which of the following is a possible eccentricity for the graph? (Lesson 7-2)

A 0

B $\dfrac{1}{4}$

C 1

D $\dfrac{9}{5}$

Graph the hyperbola given by each equation. (Lesson 7-3)

13. $\dfrac{x^2}{81} - \dfrac{(y+7)^2}{81} = 1$

14. $\dfrac{(y-3)^2}{4} - \dfrac{(x-3)^2}{16} = 1$

Write an equation for the hyperbola with the given characteristics. (Lesson 7-3)

15. vertices $(0, 5)$, $(0, -5)$; conjugate axis length of 6

16. foci $(10, 0)$, $(-6, 0)$; transverse axis length of 4

17. vertices $(-11, 0)$, $(11, 0)$; foci $(-14, 0)$, $(14, 0)$

18. foci $(5, 7)$, $(5, -9)$; transverse axis length of 10

Use the discriminant to identify each conic section. (Lesson 7-3)

19. $x^2 + 4y^2 - 2x - 24y + 34 = 0$

20. $4x^2 - 25y^2 - 24x - 64 = 0$

21. $2x^2 - y + 5 = 0$

22. $25x^2 + 25y^2 - 100x - 100y + 196 = 0$

Rotations of Conic Sections

Then
You identified and graphed conic sections. (Lessons 7-1 through 7-3)

Now
- Find rotation of axes to write equations of rotated conic sections.
- Graph rotated conic sections.

Math Online
glencoe.com

Why?

Elliptical gears are paired by rotating them about their foci. The driver gear turns at a constant speed, and the driven gear changes its speed continuously during each revolution.

Rotations of Conic Sections In the previous lesson, you learned that when a conic section is vertical or horizontal with its axes parallel to the x- and y-axis, $B = 0$ in its general equation. The equation of such a conic does not contain an xy-term.

$$Ax^2 + Cy^2 + Dx + Ey + F = 0 \qquad \text{Axes of conic are parallel to coordinate axes.}$$

In this lesson, you will examine conics with axes that are rotated and no longer parallel to the coordinate axes. In the general equation for such rotated conics, $B \neq 0$, so there is an xy-term.

$$Ax^2 + Bxy + Cy^2 + Dx + Ey + F = 0 \qquad \text{Axes of conic are rotated from coordinate axes.}$$

If the xy-term were eliminated, the equation of the rotated conic could be written in standard form by completing the square. To eliminate this term, we rotate the coordinate axes until they are parallel to the axes of the conic.

When the coordinate axes are rotated through an angle θ as shown, the origin remains fixed and new axes x' and y' are formed. The equation of the conic in the new $x'y'$-plane has the following general form.

$$A(x')^2 + C(y')^2 + Dx' + Ey' + F = 0 \qquad \text{Equation in } x'y'\text{-plane}$$

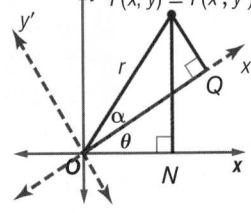

Trigonometry can be used to develop formulas relating a point $P(x, y)$ in the xy-plane and $P(x', y')$ in the $x'y'$ plane.

Consider the figure at the right. Notice that in right triangle PNO, $OP = r$, $ON = x$, $PN = y$, and $m\angle NOP = \alpha + \theta$. Using $\triangle PNO$, you can establish the following relationships.

$x = r \cos(\alpha + \theta)$ **Cosine ratio**
$\quad = r \cos \alpha \cos \theta - r \sin \alpha \sin \theta$ **Cosine Sum Identity**

$y = r \sin(\alpha + \theta)$ **Sine ratio**
$\quad = r \sin \alpha \cos \theta + r \cos \alpha \sin \theta$ **Sine Sum Identity**

Using right triangle POQ, in which $OP = r$, $OQ = x'$, $PQ = y'$, and $m\angle QOP = \alpha$, you can establish the relationships $x' = r \cos \alpha$ and $y' = r \sin \alpha$. Substituting these values into the previous equations, you obtain the following.

$$x = x' \cos \theta - y' \sin \theta \qquad\qquad y = y' \cos \theta + x' \sin \theta$$

Key Concept   **Rotation of Axes of Conics**

An equation $Ax^2 + Bxy + Cy^2 + Dx + Ey + F = 0$ in the xy-plane can be rewritten as $A(x')^2 + C(y')^2 + Dx' + Ey' + F = 0$ in the rotated $x'y'$-plane.

The equation in the $x'y'$-plane can be found using the following equations, where θ is the angle of rotation.

$$x = x' \cos \theta - y' \sin \theta \qquad y = x' \sin \theta + y' \cos \theta$$

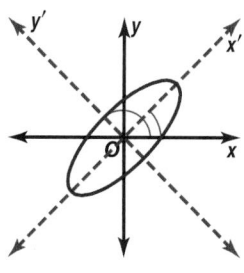
EXAMPLE 1   Write an Equation in the $x'y'$-Plane

Use $\theta = \dfrac{\pi}{4}$ to write $6x^2 + 6xy + 9y^2 = 53$ in the $x'y'$-plane. Then identify the conic.

Find the equations for x and y.

$$x = x' \cos \theta - y' \sin \theta \qquad \textbf{Rotation equations for } x \textbf{ and } y \qquad y = x' \sin \theta + y' \cos \theta$$

$$= \frac{\sqrt{2}}{2}x' - \frac{\sqrt{2}}{2}y' \qquad \sin \frac{\pi}{4} = \frac{\sqrt{2}}{2} \text{ and } \cos \frac{\pi}{4} = \frac{\sqrt{2}}{2} \qquad = \frac{\sqrt{2}}{2}x' + \frac{\sqrt{2}}{2}y'$$

Substitute into the original equation.

$$6x^2 \qquad + \qquad 6xy \qquad + \qquad 9y^2 \qquad = 53$$

$$6\left(\frac{\sqrt{2}x' - \sqrt{2}y'}{2}\right)^2 + 6\left(\frac{\sqrt{2}x' - \sqrt{2}y'}{2}\right)\left(\frac{\sqrt{2}x' + \sqrt{2}y'}{2}\right) + 9\left(\frac{\sqrt{2}x' + \sqrt{2}y'}{2}\right)^2 = 53$$

$$\frac{6[2(x')^2 - 4x'y' + 2(y')^2]}{4} + \frac{6[2(x')^2 - 2(y')^2]}{4} + \frac{9[2(x')^2 + 4x'y' + 2(y')^2]}{4} = 53$$

$$3(x')^2 - 6x'y' + 3(y')^2 + 3(x')^2 - 3(y')^2 + \frac{9}{2}(x')^2 + 9x'y' + \frac{9}{2}(y')^2 - 53 = 0$$

$$6(x')^2 - 12x'y' + 6(y')^2 + 6(x')^2 - 6(y')^2 + 9(x')^2 + 18x'y' + 9(y')^2 - 106 = 0$$

$$21(x')^2 + 6x'y' + 9(y')^2 - 106 = 0$$

The equation in the $x'y'$-plane is $21(x')^2 + 6x'y' + 9(y')^2 - 106 = 0$. For this equation, $B^2 - 4AC = 6^2 - 4(21)(9)$ or -720. Since $-720 < 0$, the conic is an ellipse as shown.

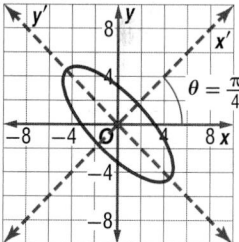

✓ **Guided Practice**

1. Use $\theta = \dfrac{\pi}{6}$ to write $7x^2 + 4\sqrt{3}xy + 3y^2 - 60 = 0$ in the $x'y'$-plane. Then identify the conic.

When the angle of rotation θ is chosen appropriately, the $x'y'$-term is eliminated from the general form equation, and the axes of the conic will be parallel to the axes of the $x'y'$-plane.

After substituting $x = x' \cos \theta - y' \sin \theta$ and $y = x' \sin \theta + y' \cos \theta$ into the general form of a conic, $Ax^2 + Bxy + Cy^2 + Dx + Ey + F = 0$, the coefficient of the $x'y'$-term is $B \cos 2\theta + (C - A) \sin 2\theta$. By setting this equal to 0, the $x'y'$-term can be eliminated.

$$B \cos 2\theta + (C - A) \sin 2\theta = 0 \qquad \textbf{Coefficient of } x'y'\textbf{-term}$$

$$B \cos 2\theta = -(C - A) \sin 2\theta \qquad \textbf{Subtract } (C - A) \sin 2\theta \textbf{ from each side.}$$

$$B \cos 2\theta = (A - C) \sin 2\theta \qquad \textbf{Distributive Property}$$

$$\frac{\cos 2\theta}{\sin 2\theta} = \frac{A - C}{B} \qquad \textbf{Divide each side by } B \sin 2\theta.$$

$$\cot 2\theta = \frac{A - C}{B} \qquad \frac{\cos 2\theta}{\sin 2\theta} = \cot 2\theta$$

Key Concept   Angle of Rotation Used to Eliminate xy-Term

An angle of rotation θ such that $\cot 2\theta = \dfrac{A - C}{B}$, $B \ne 0$, $0 < \theta < \dfrac{\pi}{2}$, will eliminate the xy-term from the equation of the conic section in the rotated $x'y'$-coordinate system.

EXAMPLE 2   Write an Equation in Standard Form

Using a suitable angle of rotation for the conic with equation $8x^2 + 12xy + 3y^2 = 4$, write the equation in standard form.

The conic is a hyperbola because $B^2 - 4AC > 0$. Find θ.

$\cot 2\theta = \dfrac{A-C}{B}$ **Rotation of the axes**

$ = \dfrac{5}{12}$ $A = 8, B = 12,$ and $C = 3$

The figure illustrates a triangle for which $\cot 2\theta = \dfrac{5}{12}$. From this, $\sin 2\theta = \dfrac{12}{13}$ and $\cos 2\theta = \dfrac{5}{13}$.

Use the half-angle identities to determine $\sin \theta$ and $\cos \theta$.

$$\sin \theta = \sqrt{\dfrac{1 - \cos 2\theta}{2}} \qquad \textbf{Half-Angle Identities} \qquad \cos \theta = \sqrt{\dfrac{1 + \cos 2\theta}{2}}$$

$$= \sqrt{\dfrac{1 - \dfrac{5}{13}}{2}} \qquad\qquad \cos 2\theta = \dfrac{5}{13} \qquad\qquad = \sqrt{\dfrac{1 + \dfrac{5}{13}}{2}}$$

$$= \dfrac{2\sqrt{13}}{13} \qquad\qquad\qquad \textbf{Simplify.} \qquad\qquad\qquad = \dfrac{3\sqrt{13}}{13}$$

Next, find the equations for x and y.

$$x = x'\cos\theta - y'\sin\theta \qquad \textbf{Rotation equations for } x \textbf{ and } y \qquad y = x'\sin\theta + y'\cos\theta$$

$$= \dfrac{3\sqrt{13}}{13}x' - \dfrac{2\sqrt{13}}{13}y' \qquad \sin\theta = \dfrac{2\sqrt{13}}{13} \text{ and } \cos\theta = \dfrac{3\sqrt{13}}{13} \qquad = \dfrac{2\sqrt{13}}{13}x' + \dfrac{3\sqrt{13}}{13}y'$$

$$= \dfrac{3\sqrt{13}x' - 2\sqrt{13}y'}{13} \qquad\qquad \textbf{Simplify.} \qquad\qquad = \dfrac{2\sqrt{13}x' + 3\sqrt{13}y'}{13}$$

Substitute these values into the original equation.

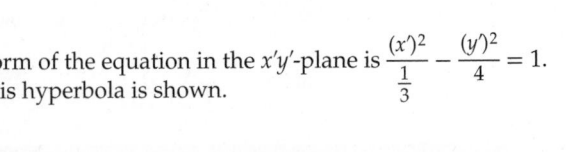

$$8x^2 \qquad\qquad + \qquad\qquad 12xy \qquad\qquad + \qquad\qquad 3y^2 \qquad = 4$$

$$8\left(\dfrac{3\sqrt{13}x' - 2\sqrt{13}y'}{13}\right)^2 + 12\,\dfrac{3\sqrt{13}x' - 2\sqrt{13}y'}{13} \cdot \dfrac{2\sqrt{13}x' + 3\sqrt{13}y'}{13} + 3\left(\dfrac{2\sqrt{13}x' + 3\sqrt{13}y'}{13}\right)^2 = 4$$

$$\dfrac{72(x')^2 - 96x'y' + 32(y')^2}{13} + \dfrac{72(x')^2 + 60x'y' - 72(y')^2}{13} + \dfrac{12(x')^2 + 36x'y' + 27(y')^2}{13} = 4$$

$$\dfrac{156(x')^2 - 13(y')^2}{13} = 4$$

$$3(x')^2 - \dfrac{(y')^2}{4} = 1$$

The standard form of the equation in the $x'y'$-plane is $\dfrac{(x')^2}{\frac{1}{3}} - \dfrac{(y')^2}{4} = 1$. The graph of this hyperbola is shown.

✔ Guided Practice

Using a suitable angle of rotation for the conic with each given equation, write the equation in standard form.

2A. $2x^2 - 12xy + 18y^2 - 4y = 2$

2B. $20x^2 + 20xy + 5y^2 - 12x - 36y - 200 = 0$

Two other formulas relating x' and y' to x and y can be used to find an equation in the xy-plane for a rotated conic.

EXAMPLE 3 Write an Equation in the xy-Plane

PHYSICS Elliptical gears can be used to generate variable output speeds. After a 60° rotation, the equation for the rotated gear in the $x'y'$-plane is $\dfrac{(x')^2}{36} + \dfrac{(y')^2}{18} = 1$. Write an equation for the ellipse formed by the rotated gear in the xy-plane.

Use the rotation formulas for x' and y' to find the equation of the rotated conic in the xy-plane.

$x' = x \cos \theta + y \sin \theta$	**Rotation equations for x' and y'**	$y' = y \cos \theta - x \sin \theta$
$= x \cos 60° + y \sin 60°$	$\theta = 60°$	$= y \cos 60° - x \sin 60°$
$= \dfrac{1}{2}x + \dfrac{\sqrt{3}}{2}y$	$\sin 60° = \dfrac{1}{2}$ and $\cos 60° = \dfrac{\sqrt{3}}{2}$	$= \dfrac{1}{2}y - \dfrac{\sqrt{3}}{2}x$

Substitute these values into the original equation.

$\dfrac{(x')^2}{36} + \dfrac{(y')^2}{18} = 1$	**Original equation**
$(x')^2 + 2(y')^2 = 36$	**Multiply each side by 36.**
$\left(\dfrac{x + \sqrt{3}y}{2}\right)^2 + 2\left(\dfrac{y - \sqrt{3}x}{2}\right)^2 = 36$	**Substitute.**
$\dfrac{x^2 + 2\sqrt{3}xy + 3y^2}{4} + \dfrac{2y^2 - 4\sqrt{3}xy + 6x^2}{4} = 36$	**Simplify.**
$\dfrac{7x^2 - 2\sqrt{3}xy + 5y^2}{4} = 36$	**Combine like terms.**
$7x^2 - 2\sqrt{3}xy + 5y^2 = 144$	**Multiply each side by 4.**
$7x^2 - 2\sqrt{3}xy + 5y^2 - 144 = 0$	**Subtract 144 from each side.**

The equation of the rotated ellipse in the xy-plane is $7x^2 - 2\sqrt{3}xy + 5y^2 - 144 = 0$.

✔ Guided Practice

3. If the equation for the gear after a 30° rotation in the $x'y'$-plane is $(x')^2 + 4(y')^2 - 40 = 0$, find the equation for the gear in the xy-plane.

Graph Rotated Conics When the equations of rotated conics are given for the $x'y'$-plane, they can be graphed by finding points on the graph of the conic and then converting these points to the xy-plane.

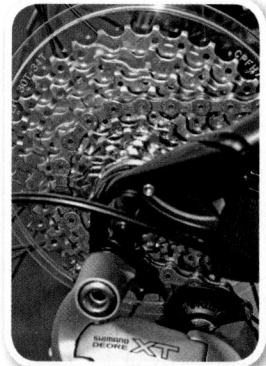

Real-World Link

In a system of gears where both gears spin, such as a bicycle, the speed of the gears in relation to each other is related to their size. If the diameter of one of the gears is $\dfrac{1}{2}$ of the diameter of the second gear, the first gear will rotate twice as fast as the second gear.

Source: *How Stuff Works*

EXAMPLE 4 **Graph a Conic Using Rotations**

Graph $(x' - 2)^2 = 4(y' - 3)$ **if it has been rotated 30° from its position in the** xy-**plane.**

The equation represents a parabola, and it is in standard form. Use the vertex $(2, 3)$ and axis of symmetry $x' = 2$ in the $x'y'$-plane to determine the vertex and axis of symmetry for the parabola in the xy-plane.

Find the equations for x and y for $\theta = 30°$.

$x = x' \cos \theta - y' \sin \theta$	**Rotation equations for x and y**	$y = x' \sin \theta + y' \cos \theta$
$= \dfrac{\sqrt{3}}{2}x' - \dfrac{1}{2}y'$	$\sin 30° = \dfrac{1}{2}$ and $\cos 30° = \dfrac{\sqrt{3}}{2}$	$= \dfrac{1}{2}x' + \dfrac{\sqrt{3}}{2}y'$

Use the equations to convert the $x'y'$-coordinates of the vertex into xy-coordinates.

$x = \dfrac{\sqrt{3}}{2}x' - \dfrac{1}{2}y'$	**Conversion equation**	$y = \dfrac{1}{2}x' + \dfrac{\sqrt{3}}{2}y'$
$= \dfrac{\sqrt{3}}{2}(2) - \dfrac{1}{2}(3)$	$x' = 2$ and $y' = 3$	$= \dfrac{1}{2}(2) + \dfrac{\sqrt{3}}{2}(3)$
$= \sqrt{3} - \dfrac{3}{2}$ or about 0.23	**Multiply.**	$= 1 + \dfrac{3\sqrt{3}}{2}$ or about 3.60

Find the equation for the axis of symmetry.

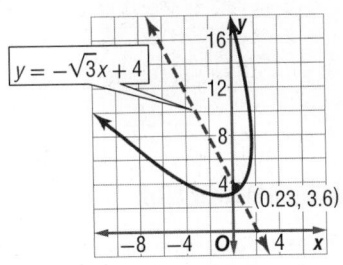

$x' = x \cos \theta + y \sin \theta$	**Conversion equation**
$2 = \dfrac{\sqrt{3}}{2}x + \dfrac{1}{2}y$	$\sin 30° = \dfrac{1}{2}$ and $\cos 30° = \dfrac{\sqrt{3}}{2}$
$y = -\sqrt{3}x + 4$	**Solve for y.**

The new vertex and axis of symmetry can be used to sketch the graph of the parabola in the xy-plane.

$y = -\sqrt{3}x + 4$

$(0.23, 3.6)$

StudyTip

Graphing Convert other points on the conic from $x'y'$-coordinates to xy-coordinates. Then make a table of these values to complete the sketch of the conic.

✓ **Guided Practice**   Graph each equation at the indicated angle.

4A. $\dfrac{(x')^2}{9} - \dfrac{(y')^2}{32} = 1; 60°$ **4B.** $\dfrac{(x')^2}{16} + \dfrac{(y')^2}{25} = 1; 30°$

One method of graphing conic sections with an xy-term is to solve the equation for y and graph with a calculator. Write the equation in quadratic form and then use the Quadratic Formula.

EXAMPLE 5 **Graph a Conic in Standard Form**

Use a graphing calculator to graph the conic given by $4y^2 + 8xy - 60y + 2x^2 - 40x + 155 = 0$.

$4y^2 + 8xy - 60y + 2x^2 - 40x + 155 = 0$	**Original equation**
$4y^2 + (8x - 60)y + (2x^2 - 40x + 155) = 0$	**Quadratic form**
$y = \dfrac{-(8x - 60) \pm \sqrt{(8x - 60)^2 - 4(4)(2x^2 - 40x + 155)}}{2(4)}$	$a = 4, b = 8x - 60,$ and $c = 2x^2 - 40x + 155$
$= \dfrac{-8x + 60 \pm \sqrt{32x^2 - 320x + 1120}}{8}$	**Multiply and combine like terms.**
$= \dfrac{-8x + 60 \pm 4\sqrt{2x^2 - 20x + 70}}{8}$	**Factor out $\sqrt{16}$.**
$= \dfrac{-2x + 15 \pm \sqrt{2x^2 - 20x + 70}}{2}$	**Divide each term by 4.**

Graphing both of these equations on the same screen yields the hyperbola shown.

$[-40, 40]$ scl: 4 by $[-40, 40]$ scl: 4

StudyTip

Arranging Terms Arrange the terms in descending powers of y in order to convert the equation to quadratic form.

✓ **Guided Practice**

5. Use a graphing calculator to graph the conic given by $4x^2 - 6xy + 2y^2 - 60x - 20y + 275 = 0$.

Write each equation in the $x'y'$-plane for the given value of θ. Then identify the conic. (Example 1)

1. $x^2 - y^2 = 9, \theta = \frac{\pi}{3}$

2. $xy = -8, \theta = 45°$

3. $x^2 - 8y = 0, \theta = \frac{\pi}{2}$

4. $2x^2 + 2y^2 = 8, \theta = \frac{\pi}{6}$

5. $y^2 + 8x = 0, \theta = 30°$

6. $4x^2 + 9y^2 = 36, \theta = 30°$

7. $x^2 - 5x + y^2 = 3, \theta = 45°$

8. $49x^2 - 16y^2 = 784, \theta = \frac{\pi}{4}$

9. $4x^2 - 25y^2 = 64, \theta = 90°$

10. $6x^2 + 5y^2 = 30, \theta = 30°$

Using a suitable angle of rotation for the conic with each given equation, write the equation in standard form. (Example 2)

11. $xy = -4$

12. $x^2 - xy + y^2 = 2$

13. $145x^2 + 120xy + 180y^2 = 900$

14. $16x^2 - 24xy + 9y^2 - 5x - 90y + 25 = 0$

15. $2x^2 - 72xy + 23y^2 + 100x - 50y = 0$

16. $x^2 - 3y^2 - 8x + 30y = 60$

17. $5x^2 + 8xy + 3y^2 + 4 = 0$

18. $73x^2 + 72xy + 52y^2 + 30x + 40y - 75 = 0$

Write an equation for each conic in the xy-plane for the given equation in $x'y'$ form and the given value of θ. (Example 3)

19. $(x')^2 + 3(y')^2 = 8, \theta = \frac{\pi}{4}$

20. $\frac{(x')^2}{25} - \frac{(y')^2}{225} = 1, \theta = \frac{\pi}{4}$

21. $\frac{(x')^2}{9} - \frac{(y')^2}{36} = 1, \theta = \frac{\pi}{3}$

22. $(x')^2 = 8y', \theta = 45°$

23. $\frac{(x')^2}{7} + \frac{(y')^2}{28} = 1, \theta = \frac{\pi}{6}$

24. $4x' = (y')^2, \theta = 30°$

25. $\frac{(x')^2}{64} - \frac{(y')^2}{16} = 1, \theta = 45°$

26. $(x')^2 = 5y', \theta = \frac{\pi}{3}$

27. $\frac{(x')^2}{4} - \frac{(y')^2}{9} = 1, \theta = 30°$

28. $\frac{(x')^2}{3} + \frac{(y')^2}{4} = 1, \theta = 60°$

29. ● **ASTRONOMY** Suppose $144(x')^2 + 64(y')^2 = 576$ models the shape in the $x'y'$-plane of a reflecting mirror in a telescope. (Example 4)

   a. If the mirror has been rotated 30°, determine the equation of the mirror in the xy-plane.

   b. Graph the equation.

Graph each equation at the indicated angle.

30. $\frac{(x')^2}{4} + \frac{(y')^2}{9} = 1; 60°$

31. $\frac{(x')^2}{25} - \frac{(y')^2}{36} = 1; 45°$

32. $(x')^2 + 6x' - y' = -9; 30°$

33. $8(x')^2 + 6(y')^2 = 24; 30°$

34. $\frac{(x')^2}{4} - \frac{(y')^2}{16} = 1; 45°$

35. $y' = 3(x')^2 - 2x' + 5; 60°$

36. **COMMUNICATION** A satellite dish tracks a satellite directly overhead. Suppose $y = \frac{1}{6}x^2$ models the shape of the dish when it is oriented in this position. Later in the day, the dish is observed to have rotated approximately 30°. (Example 4)

   a. Write an equation that models the new orientation of the dish.

   b. Use a graphing calculator to graph both equations on the same screen. Sketch this graph on your paper.

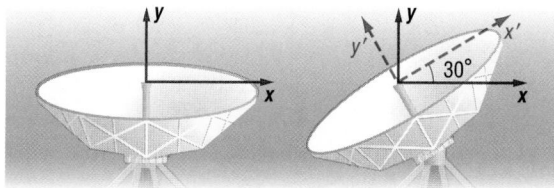

GRAPHING CALCULATOR Graph the conic given by each equation. (Example 5)

37. $x^2 - 2xy + y^2 - 5x - 5y = 0$

38. $2x^2 + 9xy + 14y^2 = 5$

39. $8x^2 + 5xy - 4y^2 = -2$

40. $2x^2 + 4\sqrt{3}xy + 6y^2 + 3x = y$

41. $2x^2 + 4xy + 2y^2 + 2\sqrt{2}x - 2\sqrt{2}y = -12$

42. $9x^2 + 4xy + 6y^2 = 20$

43. $x^2 + 10\sqrt{3}xy + 11y^2 - 64 = 0$

44. $x^2 + y^2 - 4 = 0$

45. $x^2 - 2\sqrt{3}xy - y^2 + 18 = 0$

46. $2x^2 + 9xy + 14y^2 - 5 = 0$

The graph of each equation is a degenerate case. Describe the graph.

47. $y^2 - 16x^2 = 0$

48. $(x + 4)^2 - (x - 1)^2 = y + 8$

49. $(x + 3)^2 + y^2 + 6y + 9 - 6(x + y) = 18$

Match the graph of each conic with its equation.

50.

51.

52.

53.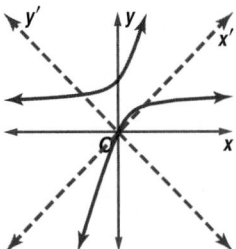

a. $x^2 - xy + y^2 = 2$

b. $145x^2 + 120xy + 180y^2 - 900 = 0$

c. $2x^2 - 72xy + 23y^2 + 100x - 50y = 0$

d. $16x^2 - 24xy + 9y^2 - 5x - 90y + 25 = 0$

54. ROBOTICS A hyperbolic mirror used in robotic systems is attached to the robot so that it is facing to the right. After it is rotated, the shape of its new position is represented by $51.75x^2 - 184.5\sqrt{3}xy - 132.75y^2 = 32{,}400$.

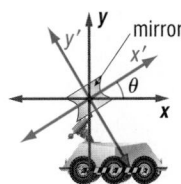

a. Solve the equation for y.

b. Use a graphing calculator to graph the equation.

c. Determine the angle θ through which the mirror has been rotated. Round to the nearest degree.

55. INVARIANTS When a rotation transforms an equation from the xy-plane to the $x'y'$-plane, the new equation is equivalent to the original equation. Some values are invariant under the rotation, meaning their values do not change when the axes are rotated. Use reasoning to explain how each of the following are rotation invariants.

a. $F = F'$

b. $A + C = A' + C'$

c. $B^2 - 4AC = (B')^2 - 4(A'C')$

GRAPHING CALCULATOR Graph each pair of equations and find any points of intersection. If the graphs have no points of intersection, write *no solution*.

56. $x^2 + 2xy + y^2 - 8x - y = 0$
$8x^2 + 3xy - 5y^2 = 15$

57 $9x^2 + 4xy + 5y^2 - 40 = 0$
$x^2 - xy - 2y^2 - x - y + 2 = 0$

58. $x^2 + \sqrt{3}xy - 3 = 0$
$16x^2 - 20xy + 9y^2 = 40$

59. **MULTIPLE REPRESENTATIONS** In this problem, you will investigate angles of rotation that produce the original graphs.

a. TABULAR For each equation in the table, identify the conic and find the minimum angle of rotation needed to transform the equation so that the rotated graph coincides with its original graph.

Equation	Conic	Minimum Angle of Rotation
$x^2 - 5x + 3 - y = 0$		
$6x^2 + 10y^2 = 15$		
$2xy = 9$		

b. VERBAL Describe the relationship between the lines of symmetry of the conics and the minimum angles of rotation needed to produce the original graphs.

c. ANALYTICAL A noncircular ellipse is rotated 50° about the origin. It is then rotated again so that the original graph is produced. What is the second angle of rotation?

H.O.T. Problems · Use Higher-Order Thinking Skills

60. ERROR ANALYSIS Leon and Dario are describing the graph of $x^2 + 4xy + 6y^2 + 3x - 4y = 75$. Leon says that it is an ellipse. Dario thinks it is a parabola. Is either of them correct? Explain your reasoning.

61. CHALLENGE Show that a circle with the equation $x^2 + y^2 = r^2$ remains unchanged under any rotation θ.

62. REASONING *True* or *false*: Every angle of rotation θ can be described as an acute angle. Explain.

63. PROOF Prove $x' = x \cos \theta + y \sin \theta$ and $y' = y \cos \theta - x \sin \theta$. (*Hint:* Solve the system $x = x' \cos \theta - y' \sin \theta$ and $y = x' \sin \theta + y' \cos \theta$ by multiplying one equation by $\sin \theta$ and the other by $\cos \theta$.)

64. REASONING The angle of rotation θ can also be defined as $\tan 2\theta = \dfrac{B}{A - C}$, when $A \neq C$, or $\theta = \dfrac{\pi}{4}$, when $A = C$. Why does defining the angle of rotation in terms of cotangent not require an extra condition with an additional value for θ?

65. WRITING IN MATH The discriminant can be used to classify a conic $A'(x')^2 + C'(y')^2 + D'x' + E'y' + F' = 0$ in the $x'y'$-plane. Explain why the values of A' and C' determine the type of conic. Describe the parameters necessary for an ellipse, a circle, a parabola, and a hyperbola.

66. REASONING *True* or *false*: Whenever the discriminant of an equation of the form $Ax^2 + Bxy + Cy^2 + Dx + Ey + F = 0$ is equal to zero, the graph of the equation is a parabola. Explain.

Graph the hyperbola given by each equation. (Lesson 7-3)

67. $\dfrac{x^2}{9} - \dfrac{y^2}{64} = 1$

68. $\dfrac{y^2}{25} - \dfrac{x^2}{49} = 1$

69. $\dfrac{(x-3)^2}{64} - \dfrac{(y-7)^2}{25} = 1$

Determine the eccentricity of the ellipse given by each equation. (Lesson 7-2)

70. $\dfrac{(x+17)^2}{39} + \dfrac{(y+7)^2}{30} = 1$

71. $\dfrac{(x-6)^2}{12} + \dfrac{(y+4)^2}{15} = 1$

72. $\dfrac{(x-10)^2}{29} + \dfrac{(y+2)^2}{24} = 1$

73. INVESTING Randall has a total of $5000 in his savings account and in a certificate of deposit. His savings account earns 3.5% interest annually. The certificate of deposit pays 5% interest annually if the money is invested for one year. Randall calculates that his interest earnings for the year will be $227.50. (Lesson 6-3)

 a. Write a system of equations for the amount of money in each investment.

 b. Use Cramer's Rule to determine how much money is in Randall's savings account and in the certificate of deposit.

74. OPTICS The amount of light that a source provides to a surface is called the *illuminance*. The illuminance E in foot candles on a surface that is R feet from a source of light with intensity I candelas is $E = \dfrac{I \cos \theta}{R^2}$, where θ is the measure of the angle between the direction of the light and a line perpendicular to the surface being illuminated.

 Verify that $E = \dfrac{I \cot \theta}{R^2 \csc \theta}$ is an equivalent formula. (Lesson 5-2)

Solve each equation. (Lesson 3-4)

75. $\log_4 8n + \log_4 (n-1) = 2$

76. $\log_9 9p + \log_9 (p+8) = 2$

Use the Factor Theorem to determine if the binomials given are factors of $f(x)$. Use the binomials that are factors to write a factored form of $f(x)$. (Lesson 2-3)

77. $f(x) = x^4 - x^3 - 16x^2 + 4x + 48;\ (x-4),\ (x-2)$

78. $f(x) = 2x^4 + 9x^3 - 23x^2 - 81x + 45;\ (x+5),\ (x+3)$

79. SAT/ACT P is the center of the circle and $PQ = QR$. If $\triangle PQR$ has an area of $9\sqrt{3}$ square units, what is the area of the shaded region in square units?

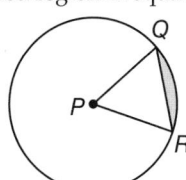

 A $24\pi - 9\sqrt{3}$ **D** $6\pi - 9\sqrt{3}$

 B $9\pi - 9\sqrt{3}$ **E** $12\pi - 9\sqrt{3}$

 C $18\pi - 9\sqrt{3}$

80. REVIEW Which is NOT the equation of a parabola?

 F $y = 2x^2 + 4x - 9$

 G $3x + 2y^2 + y + 1 = 0$

 H $x^2 + 2y^2 + 8y = 8$

 J $x = \tfrac{1}{2}(y-1)^2 + 5$

81. Which is the graph of the conic given by the equation $4x^2 - 2xy + 8y^2 - 7 = 0$?

 A   **C**

 B   **D**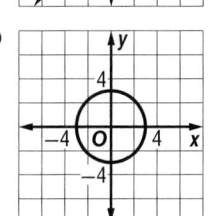

82. REVIEW How many solutions does the system $\dfrac{x^2}{5^2} - \dfrac{y^2}{3^2} = 1$ and $(x-3)^2 + y^2 = 9$ have?

 F 0 **H** 2

 G 1 **J** 4

Objective
Use a graphing calculator to approximate solutions to systems of nonlinear equations and inequalities.

Graphs of conic sections represent a nonlinear system. Solutions of systems of nonlinear equations can be found algebraically. However, approximations can be found by using your graphing calculator.

Recall that graphing calculators can only graph functions. To graph a conic section that is not a function, solve the equation for y.

ACTIVITY 1 | Nonlinear System

Solve the system by graphing.
$x^2 + y^2 = 13$
$xy + 6 = 0$

Step 1 Solve each equation for y.

$$y = \sqrt{13 - x^2} \text{ and } y = -\sqrt{13 - x^2} \quad y = -\frac{6}{x}$$

Step 2 Graph the equations in the appropriate window.

Step 3 Use the **intersect** option from the **CALC** menu to find the four points of intersection.

The solutions are $(-3, 2)$, $(-2, 3)$, $(2, -3)$, and $(3, -2)$.

$[-5, 5]$ scl: 1 by $[-5, 5]$ scl: 1

Exercises

Solve each system of equations by graphing. Round to the nearest tenth.

1. $xy = 2$
$x^2 - y^2 = 3$

2. $49 = y^2 + x^2$
$x = 1$

3. $x = 2 + y$
$x^2 + y^2 = 100$

4. $25 - 4x^2 = y^2$
$2x + y + 1 = 0$

5. $y^2 = 9 - 3x^2$
$x^2 = 10 - 2y^2$

6. $y = -1 - x$
$4 + x = (y - 1)^2$

7. CHALLENGE A house contains two square rooms, the family room and the den. The total area of the two rooms is 468 square feet, and the den is 180 square feet smaller than the family room.

  a. Write a system of second-degree equations that models this situation.

  b. Graph the system found in part **a**, and estimate the length of each room.

Systems of nonlinear inequalities can also be solved using a graphing calculator. Recall from Chapter 1 that inequalities can be graphed by using the *greater than* and *less than* commands from the $\boxed{Y=}$ menu. An inequality symbol is found by scrolling to the left of the equal sign and pressing $\boxed{\text{ENTER}}$ until the shaded triangles are flashing. The triangle above represents *greater than* and the triangle below represents *less than*. The graph of $y \geq x^2$ is shown below.

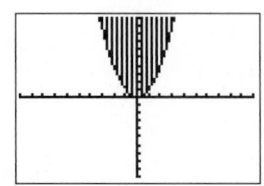

$[-10, 10]$ scl: 1 by $[-10, 10]$ scl: 1

Inequalities with conic sections that are not functions, such as ellipses, circles, and some hyperbolas, can be graphed by using the **Shade(** command from the DRAW menu. The restrictive information required is **Shade(***lowerfunc, upperfunc, Xleft, Xright, 3, 4*).

```
DRAW POINTS STO
1:ClrDraw
2:Line(
3:Horizontal
4:Vertical
5:Tangent(
6:DrawF
7:Shade(
```

This command draws the lower function *lowerfunc* and the upper function *upperfunc* in terms of *x*. It then shades the area that is above *lowerfunc* and below *upperfunc* between the left and right boundaries *Xleft* and *Xright*. The final two entries 3 and 4 specify the type of shading and can remain constant.

TechnologyTip

Clear Screen To clear any drawings from the calculator screen, select ClrDraw from the DRAW menu.

ACTIVITY 2   Nonlinear System of Inequalities

Solve the system of inequalities by graphing.
$$x^2 + y^2 \leq 36$$
$$y - x^2 > 0$$

Step 1   Solve each inequality for *y*.

$$y \leq \sqrt{36 - x^2} \text{ and } y \geq -\sqrt{36 - x^2} \qquad y > x^2$$

Step 2   Graph $y > x^2$, and shade the correct region. Make each inequality symbol by scrolling to the left of the equal sign and selecting ⟦ENTER⟧ until the shaded triangles are flashing.

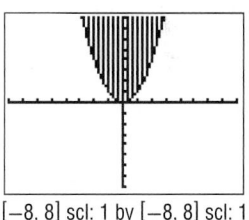

[−8, 8] scl: 1 by [−8, 8] scl: 1

StudyTip

Left and Right Boundaries If the left and right boundaries are not apparent, enter window values that exceed both boundaries. For example, if the boundaries should be $x = -5$ and $x = 5$, entering -10 and 10 will still produce the correct graph.

Step 3   To graph $x^2 + y^2 \leq 36$, the lower boundary is $y = -\sqrt{36 - x^2}$ and the upper boundary is $y = \sqrt{36 - x^2}$. The two halves of the circle meet at $x = -6$ and $x = 6$ as shown.

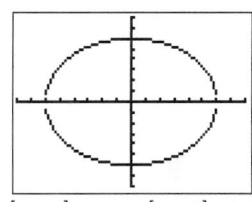

[−8, 8] scl: 1 by [−8, 8] scl: 1

Step 4   From the **DRAW** menu, select **7: Shade**. Enter **Shade(**$-\sqrt{36 - x^2}$, $\sqrt{36 - x}$, $-6, 6, 3, 4$).

```
Shade(-√(36-X²),
√(36-X²),-6,6,3,
4)
```

The solution of the system is represented by the double-shaded area.

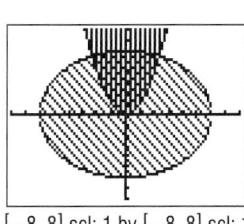

[−8, 8] scl: 1 by [−8, 8] scl: 1

Exercises

Solve each system of inequalities by graphing.

8. $2y^2 \leq 32 - 2x^2$
    $x + 4 \geq y^2$

9. $y + 5 \geq x^2$
    $9y^2 \leq 36 + x^2$

10. $x^2 + 4y^2 \leq 32$
     $4x^2 + y^2 \leq 32$

Parametric Equations

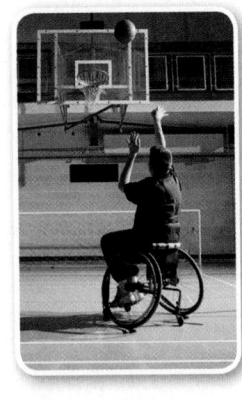

Then
You modeled motion using quadratic functions. (Lesson 1-5)

Now
- Graph parametric equations.
- Solve problems related to the motion of projectiles.

New Vocabulary
parametric equation
parameter
orientation
parametric curve

Math Online
glencoe.com

Why?

You have used quadratic functions to model the paths of projectiles such as a basketball. Parametric equations can also be used to model and evaluate the trajectory and range of projectiles.

Graph Parametric Equations So far in this text, you have represented the graph of a curve in the xy-plane using a single equation involving two variables, x and y. In this lesson you represent some of these same graphs using two equations by introducing a third variable.

Consider the graphs below, each of which models different aspects of what happens when a certain object is thrown into the air. Figure 7.5.1 shows the vertical distance the object travels as a function of time, while Figure 7.5.2 shows the object's horizontal distance as a function of time. Figure 7.5.3 shows the object's vertical distance as a function of its horizontal distance.

Figure 7.5.1

Figure 7.5.2

Figure 7.5.3

Each of these graphs and their equations tells part of what is happening in this situation, but not the whole story. To express the position of the object, both horizontally and vertically, as a function of time we can use **parametric equations**. The equations below both represent the graph shown in Figure 7.5.3.

Rectangular Equation

$$y = -\frac{2}{225}x^2 + x + 40$$

Parametric Equations

$$x = 30\sqrt{2}t \qquad \text{Horizontal component}$$
$$y = -16t^2 + 30\sqrt{2}t + 40 \qquad \text{Vertical component}$$

From the parametric equations, we can now determine where the object was at a given time by evaluating the horizontal and vertical components for t. For example, when $t = 0$, the object was at $(0, 40)$. The variable t is called a **parameter**.

The graph shown is plotted over the time interval $0 \le t \le 4$. Plotting points in the order of increasing values of t traces the curve in a specific direction called the **orientation** of the curve. This orientation is indicated by arrows on the curve as shown.

Key Concept   **Parametric Equations**

If f and g are continuous functions of t on the interval I, then the set of ordered pairs $(f(t), g(t))$ represent a **parametric curve**. The equations

$$x = f(t) \text{ and } y = g(t)$$

are parametric equations for this curve, t is the parameter, and I is the parameter interval.

EXAMPLE 1 | Sketch Curves with Parametric Equations

Sketch the curve given by each pair of parametric equations over the given interval.

a. $x = t^2 + 5$ and $y = \dfrac{t}{2} + 4; -4 \le t \le 4$

Make a table of values for $-4 \le t \le 4$. Then, plot the (x, y) coordinates for each t-value and connect the points to form a smooth curve. The arrows in the graph indicate the orientation of the curve as t moves from -4 to 4.

t	x	y	t	x	y
−4	21	2	1	6	4.5
−3	14	2.5	2	9	5
−2	9	3	3	14	5.5
−1	6	3.5	4	21	6
0	5	4			

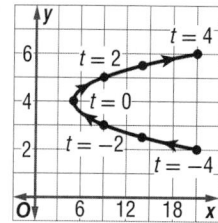

b. $x = \dfrac{t^2}{4} + 5$ and $y = \dfrac{t}{4} + 4; -8 \le t \le 8$

t	x	y	t	x	y
−8	21	2	2	6	4.5
−6	14	2.5	4	9	5
−4	9	3	6	14	5.5
−2	6	3.5	8	21	6
0	5	4			

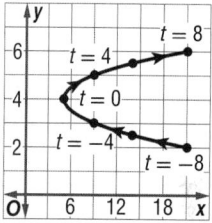

✔ Guided Practice

1A. $x = 3t$ and $y = \sqrt{t} + 6; 0 \le t \le 8$

1B. $x = t^2$ and $y = 2t + 3; -10 \le t \le 10$

Notice that the two different sets of parametric equations in Example 1 trace out the same curve. The graphs differ in their *speeds* or how rapidly each curve is traced out. If t represents time in seconds, then the curve in part **b** is traced in 16 seconds, while the curve in part **a** is traced out in 8 seconds.

Another way to determine the curve represented by a set of parametric equations is to write the set of equations in rectangular form. This can be done using substitution to eliminate the parameter.

EXAMPLE 2 | Write in Rectangular Form

Write $x = -3t$ and $y = t^2 + 2$ in rectangular form.

To eliminate the parameter t, solve $x = -3t$ for t. This yields $t = -\dfrac{1}{3}x$. Then substitute this value for t in the equation for y.

$$y = t^2 + 2 \qquad \text{Equation for } y$$

$$= \left(-\dfrac{1}{3}x\right)^2 + 2 \qquad \text{Substitute } -\dfrac{1}{3}x \text{ for } t.$$

$$= \dfrac{1}{9}x^2 + 2 \qquad \text{Simplify.}$$

This set of parametric equations yields the parabola $y = \dfrac{1}{9}x^2 + 2$.

✔ Guided Practice

2. Write $x = t^2 - 5$ and $y = 4t$ in rectangular form.

In Example 2, notice that a parameter interval for t was not specified. When not specified, the implied parameter interval is all values for t which produce real number values for x and y.

Sometimes the domain must be restricted after converting from parametric to rectangular form.

EXAMPLE 3 | **Rectangular Form with Domain Restrictions**

Write $x = \dfrac{1}{\sqrt{t}}$ and $y = \dfrac{t+1}{t}$ in rectangular form. Then graph the equation. State any restrictions on the domain.

To eliminate t, square each side of $x = \dfrac{1}{\sqrt{t}}$. This yields $x^2 = \dfrac{1}{t}$, so $t = \dfrac{1}{x^2}$. Substitute this value for t in parametric equation for y.

$y = \dfrac{t+1}{t}$ **Parametric equation for y**

$= \dfrac{\dfrac{1}{x^2} + 1}{\dfrac{1}{x^2}}$ **Substitute $\dfrac{1}{x^2}$ for t.**

$= \dfrac{\dfrac{x^2 + 1}{x^2}}{\dfrac{1}{x^2}}$ **Simplify the numerator.**

$= x^2 + 1$ **Simplify.**

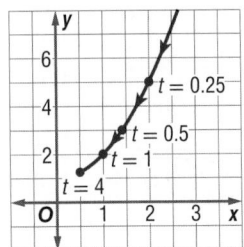

While the rectangular equation is $y = x^2 + 1$, the curve is only defined for $t > 0$. From the parametric equation $x = \dfrac{1}{\sqrt{t}}$, the only possible values for x are values greater than zero. As shown in the graph, the domain of the rectangular equation needs to be restricted to $x > 0$.

✔ Guided Practice

3. Write $x = \sqrt{t+4}$ and $y = \dfrac{1}{t}$ in rectangular form. Graph the equation. State any restrictions on the domain.

The parameter in a parametric equation can also be an angle, θ.

EXAMPLE 4 | **Rectangular Form with θ as Parameter**

Write $x = 2 \cos \theta$ and $y = 4 \sin \theta$ in rectangular form. Then graph the equation.

To eliminate the angular parameter θ, first solve the equations for $\cos \theta$ and $\sin \theta$ to obtain $\cos \theta = \dfrac{x}{2}$ and $\sin \theta = \dfrac{y}{4}$. Then use the Pythagorean Identity to eliminate the parameter θ.

$\cos^2 \theta + \sin^2 \theta = 1$ **Pythagorean Identity**

$\left(\dfrac{x}{2}\right)^2 + \left(\dfrac{y}{4}\right)^2 = 1$ **$\cos \theta = \dfrac{x}{2}$ and $\sin \theta = \dfrac{y}{4}$**

$\dfrac{x^2}{4} + \dfrac{y^2}{16} = 1$ **Simplify.**

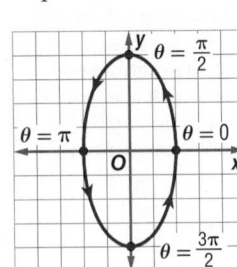

You should recognize this equation as that of an ellipse centered at the origin with vertices at $(0, 4)$ and $(0, -4)$ and covertices at $(2, 0)$ and $(-2, 0)$ as shown. As θ varies from 0 to 2π, the ellipse is traced out counterclockwise.

✔ Guided Practice

4. Write $x = 3 \sin \theta$ and $y = 8 \cos \theta$ in rectangular form. Then sketch the graph.

TechnologyTip

▶ **Parameters** When graphing parametric equations on a calculator, θ and t are interchangeable.

As you saw in Example 1, parametric representations of rectangular graphs are not unique. By varying the definition for the parameter, you can obtain parametric equations that produce graphs that vary only in speed and/or orientation.

StudyTip

Parametric Form The easiest method of converting an equation from rectangular to parametric form is to use $x = t$. When this is done, the other parametric equation is the original equation with t replacing x.

EXAMPLE 5   Write Parametric Equations from Graphs

Use each parameter to write the parametric equations that can represent $y = x^2 - 4$. Then graph the equation, indicating the speed and orientation.

a. $t = x$

$y = x^2 - 4$ **Original equation**

$\quad = t^2 - 4$ **Substitute for x in original equation.**

The parametric equations are $x = t$ and $y = t^2 - 4$. The associated speed and orientation are indicated on the graph.

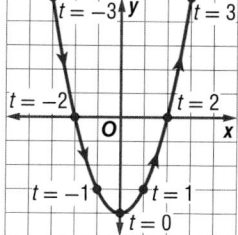

b. $t = 4x + 1$

$x = \dfrac{t - 1}{4}$ **Solve for x.**

$y = \left(\dfrac{t - 1}{4}\right)^2 - 4$ **Substitute for x in original equation.**

$\quad = \dfrac{t^2}{16} - \dfrac{t}{8} - \dfrac{63}{16}$ **Simplify.**

$x = \dfrac{t - 1}{4}$ and $y = \dfrac{t^2}{16} - \dfrac{t}{8} - \dfrac{63}{16}$ are the parametric equations. Notice that the speed is much slower than part **a**.

c. $t = 1 - \dfrac{x}{4}$

$4 - 4t = x$ **Solve for x.**

$y = (4 - 4t)^2 - 4$ **Substitute for x in original equation.**

$\quad = 16t^2 - 32t + 12$ **Simplify.**

The parametric equations are $x = 4 - 4t$ and $y = 16t^2 - 32t + 12$. Notice that the speed is much faster than part **a**. The orientation is also reversed, as indicated by the arrows.

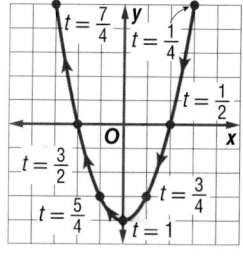

✓ Guided Practice

Use each parameter to determine the parametric equations that can represent $x = 6 - y^2$. Then graph the equation, indicating the speed and orientation.

5A. $t = x + 1$ **5B.** $t = 3x$ **5C.** $t = 4 - 2x$

Projectile Motion Parametric equations are often used to simulate projectile motion. The path of a projectile launched at an angle other than 90° with the horizontal can be modeled by the following parametric equations.

Key Concept   Projectile Motion

For an object launched at an angle θ with the horizontal at an initial velocity v_0, where g is the gravitational constant, t is time, and h_0 is the initial height:

Horizontal Distance $x = tv_0 \cos \theta$

Vertical Position $y = tv_0 \sin \theta - \dfrac{1}{2}gt^2 + h_0$

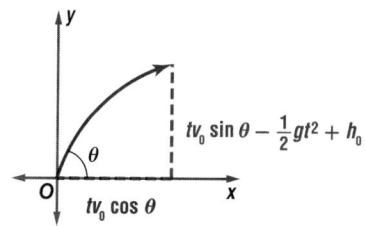

Real-World EXAMPLE 6   Projectile Motion

BASKETBALL Kaylee is practicing free throws for an upcoming basketball game. She releases the ball with an initial velocity of 24 feet per second at an angle of 53° with the horizontal. The horizontal distance from the free throw line to the front rim of the basket is 13 feet. The vertical distance from the floor to the rim is 10 feet. The front of the rim is 2 feet from the backboard. She releases the shot 4.75 feet from the ground. Does Kaylee make the basket?

Make a diagram of the situation.

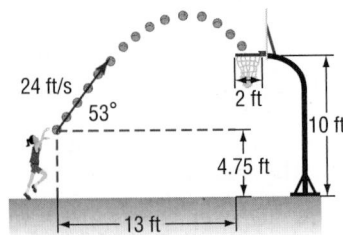

To determine whether she makes the shot, you need the horizontal distance that the ball has traveled when the height of the ball is 10 feet. First, write a parametric equation for the vertical position of the ball.

$y = tv_0 \sin \theta - \frac{1}{2}gt^2 + h_0$ **Parametric equation for vertical position**

$= t(24) \sin 53 - \frac{1}{2}(32)t^2 + 4.75$ $v_0 = 24, \theta = 53°, g = 32,$ and $h_0 = 4.75$

Graph the equation for the vertical position and the line $y = 10$. The curve will intersect the line in two places. The second intersection represents the ball as it is moving down toward the basket. Use **5: intersect** on the **CALC** menu to find the second point of intersection with $y = 10$. The value is about 0.77 second.

[0, 2] scl: 1 by [0, 12] scl: 1

Determine the horizontal position of the ball at 0.77 second.

$x = tv_0 \cos \theta$ **Parametric equation for horizontal position**

$= 0.77(24) \cos 53$ $v_0 = 24, \theta = 53°,$ and $t \approx 0.77$

≈ 11.1 **Use a calculator.**

Because the horizontal position is less than 13 feet when the ball reaches 10 feet for the second time, the shot is short of the basket. Kaylee does not make the free throw.

CHECK You can confirm the results of your calculation by graphing the parametric equations and determining the path of the ball in relation to the basket.

t	x	y	t	x	y
0	0	4.75	0.5	7.22	10.33
0.1	1.44	6.51	0.6	8.67	10.49
0.2	2.89	7.94	0.7	10.11	10.32
0.3	4.33	9.06	0.8	11.55	9.84
0.4	5.78	9.86	0.9	13.00	9.04

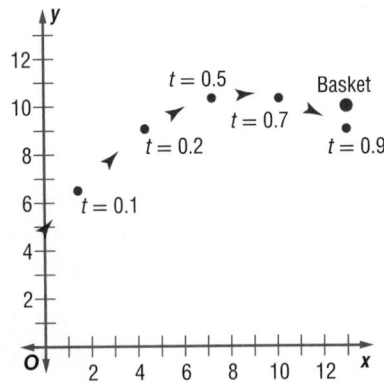

✓ Guided Practice

6. **GOLF** Evan drives a golf ball with an initial velocity of 56 meters per second at an angle of 12° down a flat driving range. How far away will the golf ball land?

Sketch the curve given by each pair of parametric equations over the given interval. (Example 1)

1. $x = t^2 + 3$ and $y = \frac{t}{4} - 5$; $-5 \le t \le 5$

2. $x = \frac{t^2}{2}$ and $y = -4t$; $-4 \le t \le 4$

3. $x = -\frac{5t}{2} + 4$ and $y = t^2 - 8$; $-6 \le t \le 6$

4. $x = 3t + 6$ and $y = \sqrt{t} + 1$; $0 \le t \le 9$

5. $x = 2t - 1$ and $y = -\frac{t^2}{2} + 7$; $-4 \le t \le 4$

6. $x = -2t^2$ and $y = \frac{t}{3} - 6$; $-6 \le t \le 6$

7. $x = \frac{t}{2}$ and $y = -\sqrt{t} + 5$; $0 \le t \le 8$

8. $x = t^2 - 4$ and $y = 3t - 8$; $-5 \le t \le 5$

Write each pair of parametric equations in rectangular form. Then graph the equation and state any restrictions on the domain. (Examples 2 and 3)

9. $x = 2t - 5$, $y = t^2 + 4$

10. $x = 3t + 9$, $y = t^2 - 7$

11. $x = t^2 - 2$, $y = 5t$

12. $x = t^2 + 1$, $y = -4t + 3$

13. $x = -t - 4$, $y = 3t^2$

14. $x = 5t - 1$, $y = 2t^2 + 8$

15. $x = 4t^2$, $y = \frac{6t}{5} + 9$

16. $x = \frac{t}{3} + 2$, $y = \frac{t^2}{6} - 7$

17. MOVIE STUNTS During the filming of a movie, a stunt double leaps off the side of a building. The pulley system connected to the stunt double allows for a vertical fall modeled by $y = -16t^2 + 15t + 100$, and a horizontal movement modeled by $x = 4t$, where x and y are measured in feet and t is measured in seconds. Write and graph an equation in rectangular form to model the stunt double's fall for $0 \le t \le 3$. (Example 3)

Write each pair of parametric equations in rectangular form. Then graph the equation. (Example 4)

18. $x = 3 \cos \theta$ and $y = 5 \sin \theta$

19. $x = 7 \sin \theta$ and $y = 2 \cos \theta$

20. $x = 6 \cos \theta$ and $y = 4 \sin \theta$

21. $x = 3 \cos \theta$ and $y = 3 \sin \theta$

22. $x = 8 \sin \theta$ and $y = \cos \theta$

23. $x = 5 \cos \theta$ and $y = 6 \sin \theta$

24. $x = 10 \sin \theta$ and $y = 9 \cos \theta$

25. $x = \sin \theta$ and $y = 7 \cos \theta$

Use each parameter to write the parametric equations that can represent each equation. Then graph the equations, indicating the speed and orientation. (Example 5)

26. $t = 3x - 2$; $y = x^2 + 9$

27. $t = 8x$; $y^2 = 9 - x^2$

28. $t = 2 - \frac{x}{3}$; $y = \frac{x^2}{12}$

29. $t = \frac{x}{5} + 4$; $y = 10 - x^2$

30. $t = 4x + 7$; $y = \frac{x^2 - 1}{2}$

31. $t = \frac{1 - x}{2}$; $y = \frac{3 - x^2}{4}$

32. BASEBALL A baseball player hits the ball at a 28° angle with an initial speed of 103 feet per second. The bat is 4 feet from the ground at the time of impact. Assuming that the ball is not caught, determine the distance traveled by the ball. (Example 6)

33. FOOTBALL Delmar attempts a 43-yard field goal. He kicks the ball at a 41° angle with an initial speed of 70 feet per second. The goal post is 15 feet high. Is the kick long enough to make the field goal? (Example 6)

Write each pair of parametric equations in rectangular form. Then state the restriction on the domain.

34. $x = \sqrt{t} + 4$
$y = 4t + 3$

35 $x = \log t$
$y = t + 3$

36. $x = \sqrt{t - 7}$
$y = -3t - 8$

37. $x = \log (t - 4)$
$y = t$

38. $x = \frac{1}{\sqrt{t + 3}}$
$y = t$

39. $x = \frac{1}{\log (t + 2)}$
$y = 2t - 4$

40. TENNIS Jill hits a tennis ball 55 centimeters above the ground at an angle of 15° with the horizontal. The ball has an initial speed of 18 meters per second.

a. Use a graphing calculator to graph the path of the tennis ball using parametric equations.

b. How long does the ball stay in the air before hitting the ground?

c. If Jill is 10 meters from the net and the net is 1.5 meters above the ground, will the tennis ball clear the net? If so, by how many meters? If not, by how many meters is the ball short?

Write a set of parametric equations for the line or line segment with the given characteristics.

41. line with a slope of 3 that passes through (4, 7)

42. line with a slope of −0.5 that passes through (3, −2)

43. line segment with endpoints (−2, −6) and (2, 10)

44. line segment with endpoints (7, 13) and (13, 11)

Match each set of parametric equations with its graph.

45. $x = \cos 2t, y = \sin 4t$

46. $x = \cos 3t, y = \sin t$

47. $x = \cos t, y = \sin 3t$

48. $x = \cos 4t, y = \sin 3t$

a.

b.

c.

d.

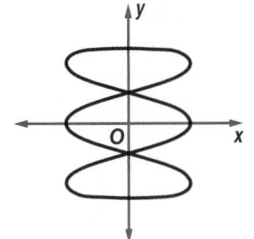

49 BIOLOGY A frog jumps off the bank of a creek with an initial velocity of 0.75 meter per second at an angle of 45° with the horizontal. The surface of the creek is 0.3 meter below the edge of the bank. Let g equal 9.8 meters per second squared.

a. Write the parametric equations to describe the position of the frog at time t. Assume that the surface of the water is located at the line $y = 0$.

b. If the creek is 0.5 meter wide, will the frog reach the other bank, which is level with the surface of the creek? If not, how far from the other bank will it hit the water?

c. If the frog was able to jump on a lily pad resting on the surface of the creek 0.4 meter away and stayed in the air for 0.38 second, what was the initial speed of the frog?

50. RACE Luna and Ruby are competing in a 100-meter dash. When the starter gun fires, Luna runs 8.0 meters per second after a 0.1 second delay from the point $(0, 2)$ and Ruby runs 8.1 meters per second after a 0.3 second delay from the point $(0, 5)$.

a. Using the y-axis as the starting line and assuming that the women run parallel to the x-axis, write parametric equations to describe each runner's position after t seconds.

b. Who wins the race? If the women ran 200 meters instead of 100 meters, who would win? Explain your answer.

51. SOCCER The graph below models the path of a soccer ball kicked by one player and then headed back by another player. The path of the initial kick is shown in blue, and the path of the headed ball is shown in red.

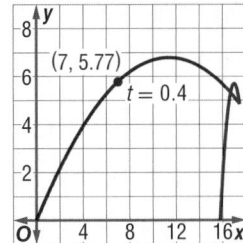

a. If the ball is initially kicked at an angle of 50°, find the initial speed of the ball.

b. At what time does the ball reach the second player if the second player is standing about 17.5 feet away?

c. If the second player heads the ball at an angle of 75°, an initial speed of 8 feet per second, and at a height of 4.75 feet, approximately how long does the ball stay in the air from the time it is first kicked until it lands?

52. ⚙ MULTIPLE REPRESENTATIONS In this problem, you will investigate a *cycloid*, the curve created by the path of a point on a circle with a radius of 1 unit as it is rolled along the x-axis.

a. **GRAPHICAL** Use a graphing calculator to graph the parametric equations $x = t - \sin t$ and $y = 1 - \cos t$, where t is measured in radians.

b. **ANALYTICAL** What is the distance between x-intercepts? Describe what the x-intercepts and the distance between them represent.

c. **ANALYTICAL** What is the maximum value of y? Describe what this value represents and how it would change for circles of differing radii.

H.O.T. Problems   Use Higher-Order Thinking Skills

53. CHALLENGE Consider a line ℓ with parametric equations $x = 2 + 3t$ and $y = -t + 5$. Write a set of parametric equations for the line m perpendicular to ℓ containing the point $(4, 10)$.

54. WRITING IN MATH Explain why there are infinitely many sets of parametric equations to describe one line in the xy-plane.

55. REASONING Determine whether parametric equations for projectile motion can apply to objects thrown at an angle of 90°. Explain your reasoning.

56. CHALLENGE A line in three-dimensional space contains the points $P(2, 3, -8)$ and $Q(-1, 5, -4)$. Find two sets of parametric equations for the line.

57. WRITING IN MATH Explain the advantage of using parametric equations versus rectangular equations when analyzing the horizontal/vertical components of a graph.

Graph each equation at the indicated angle. (Lesson 7-4)

58. $\dfrac{(x')^2}{9} - \dfrac{(y')^2}{4} = 1$ at a 60° rotation from the xy-axis

59. $(x')^2 - (y')^2 = 1$ at a 45° rotation from the xy-axis

Write an equation for the hyperbola with the given characteristics. (Lesson 7-3)

60. vertices $(5, 4)$, $(5, -8)$; conjugate axis length of 4

61. transverse axis length of 4; foci $(3, 5)$, $(3, -1)$

62. WHITE HOUSE There is an open area south of the White House known as The Ellipse. Write an equation to model The Ellipse. Assume that the origin is at its center. (Lesson 7-2)

1057 ft
880 ft
The Ellipse

Simplify each expression. (Lesson 5-1)

63. $\dfrac{\sin x}{\csc x - 1} + \dfrac{\sin x}{\csc x + 1}$

64. $\dfrac{1}{1 - \cos x} + \dfrac{1}{1 + \cos x}$

Use the properties of logarithms to rewrite each logarithm below in the form $a \ln 2 + b \ln 3$, where a and b are constants. Then approximate the value of each logarithm given that $\ln 2 \approx 0.69$ and $\ln 3 \approx 1.10$. (Lesson 3-3)

65. $\ln 54$

66. $\ln 24$

67. $\ln \dfrac{8}{3}$

68. $\ln \dfrac{9}{16}$

For each function, determine any asymptotes and intercepts. Then graph the function and state its domain. (Lesson 2-5)

69. $h(x) = \dfrac{x}{x + 6}$

70. $h(x) = \dfrac{x^2 + 6x + 8}{x^2 - 7x - 8}$

71. $f(x) = \dfrac{x^2 + 8x}{x + 5}$

72. $f(x) = \dfrac{x^2 + 4x + 3}{x^3 + x^2 - 6x}$

Solve each equation. (Lesson 2-1)

73. $\sqrt{3z - 5} - 3 = 1$

74. $\sqrt{5n - 1} = 0$

75. $\sqrt{2c + 3} - 7 = 0$

76. $\sqrt{4a + 8} + 8 = 5$

77. SAT/ACT With the exception of the shaded squares, every square in the figure contains the sum of the number in the square directly above it and the number in the square directly to its left. For example, the number 4 in the unshaded square is the sum of the 2 in the square above it and the 2 in the square directly to its left. What is the value of x?

0	1	2	3	4	5
1	2	4			
2					
3			x		
4					
5					

A 7   B 8   C 15   D 23   E 30

78. Jack and Graham are performing a physics experiment in which they will launch a model rocket. The rocket is supposed to release a parachute 300 feet in the air, 7 seconds after liftoff. They are firing the rocket at a 78° angle from the horizontal. To protect other students from the falling rockets, the teacher needs to place warning signs 50 yards from where the parachute is released. How far should the signs be from the point where the rockets are launched?

F 122 yards

G 127 yards

H 133 yards

J 138 yards

79. FREE RESPONSE An object moves along a curve according to $y = \dfrac{10\sqrt{3t} \pm \sqrt{496 - 2304t}}{62}$, $x = \sqrt{t}$.

a. Convert the parametric equations to rectangular form.

b. Identify the conic section represented by the curve.

c. Write an equation for the curve in the $x'y'$-plane, assuming it was rotated 30°.

d. Determine the eccentricity of the conic.

e. Identify the location of the foci in the $x'y'$-plane, if they exist.

Objective
Use a graphing calculator to model functions parametrically.

As shown in Lesson 7-5, the independent variable t in parametric equations represents time. This parameter reflects the speed with which the graph is drawn. If one graph is completed for $0 \le t \le 5$, while an identical graph is completed for $0 \le t \le 10$, then the first graph is faster.

ACTIVITY   Parametric Graphs of Conic Sections

FOOTBALL Standing side by side, Neva and Owen throw a football at exactly the same time. Neva throws the ball with an initial velocity of 20 meters per second at 60°. Owen throws the ball 15 meters per second at 45°. Assuming that the footballs were thrown from the same initial height, simulate the throws on a graphing calculator.

Step 1 The parametric equations for each throw are as follows.

Neva: $x = 20t \cos 60$ $y = 20t \sin 60 - 4.9t^2$

        $= 10t$ $= 10\sqrt{3}t - 4.9t^2$

Owen: $x = 15t \cos 45$ $y = 15t \sin 45 - 4.9t^2$

        $= 7.5\sqrt{2}t$ $= 7.5\sqrt{2}t - 4.9t^2$

Step 2 Set the mode. In the [MODE] menu, select **degree**, **par**, and **simul**. This allows the equations to be graphed simultaneously. Enter the parametric equations. In parametric form, [X,T,θ,n] uses t instead of x. Set the second set of equations to shade dark to distinguish between the throws.

Step 3 Set the t-values to range from 0 to 8 as an estimate for the duration of the throws. Set **tstep** to 0.1 in order to watch the throws in the graph.

Step 4 Graph the equations.

[0, 50] scl: 5 by [0, 25] scl: 5

[0, 50] scl: 5 by [0, 25] scl: 5

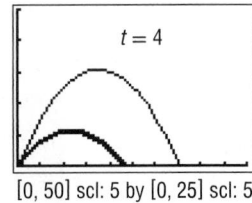

[0, 50] scl: 5 by [0, 25] scl: 5

Neva's throw goes higher and at a greater distance while Owen's lands first.

Exercises

1. **FOOTBALL** Owen's next throw is 21 meters per second at 50°. A second later, Neva throws her football 24 meters per second at 35°. Simulate the throws on a graphing calculator and interpret the results.

2. **BASEBALL** Neva throws a baseball 27 meters per second at 82°. A second later, Owen hits a ball 45 meters per second at 20°. Assuming they are still side by side and the initial height of the hit is one meter lower, simulate the situation on a graphing calculator and interpret the results.

Chapter Summary

Key Concepts

Parabolas (Lesson 7-1)

Equations	Orientation	Vertex	Focus
$(y - k)^2 = 4p(x - h)$	horizontal	(h, k)	$(h + p, k)$
$(x - h)^2 = 4p(y - k)$	vertical	(h, k)	$(h, k + p)$

- p is the distance from the vertex to the focus.

Ellipses and Circles (Lesson 7-2)

Equations	Major Axis	Vertex	Focus
$\dfrac{(x - h)^2}{a^2} + \dfrac{(y - k)^2}{b^2} = 1$	horizontal	$(h \pm a, k)$	$(h \pm c, k)$
$\dfrac{(y - k)^2}{a^2} + \dfrac{(x - h)^2}{b^2} = 1$	vertical	$(h, k \pm a)$	$(h, k \pm c)$

- The eccentricity of an ellipse is given by $e = \frac{c}{a}$, where $a^2 - b^2 = c^2$.
- The standard form of an equation for a circle with center (h, k) and radius r is $(x - h)^2 + (y - k)^2 = r^2$.

Hyperbolas (Lesson 7-3)

Equations	Transverse Axis	Vertex	Focus
$\dfrac{(x - h)^2}{a^2} - \dfrac{(y - k)^2}{b^2} = 1$	horizontal	$(h \pm a, k)$	$(h \pm c, k)$
$\dfrac{(y - k)^2}{a^2} - \dfrac{(x - h)^2}{b^2} = 1$	vertical	$(h, k \pm a)$	$(h, k \pm c)$

- The eccentricity of a hyperbola is given by $e = \frac{c}{a}$, where $a^2 + b^2 = c^2$.

Rotations of Conic Sections (Lesson 7-4)

- An equation in the xy-plane can be transformed to an equation in the $x'y'$-plane using $x = x' \cos \theta - y' \sin \theta$ and $y = x' \sin \theta + y' \cos \theta$.
- An equation in the $x'y'$ plane can be transformed to an equation in the xy plane using $x' = x \cos \theta + y \sin \theta$, and $y' = y \cos \theta - x \sin \theta$.

Parametric Equations (Lesson 7-5)

- Parametric equations are used to describe the horizontal and vertical components of an equation separately, generally in terms of the parameter t.
- For an object launched at an angle θ with the horizontal at an initial velocity of v_0, where g is the gravitational constant, t is time, and h_0 is the initial height, its horizontal distance is $x = tv_0 \cos \theta$, and its vertical distance is $y = tv_0 \sin \theta - \frac{1}{2}gt^2 + h_0$.

Key Vocabulary

axis of symmetry (p. 422)	locus (p. 422)
center (p. 432)	major axis (p. 432)
conic section (p. 422)	minor axis (p. 432)
conjugate axis (p. 442)	orientation (p. 464)
co-vertices (p. 432)	parabola (p. 422)
degenerate conic (p. 422)	parameter (p. 464)
directrix (p. 422)	parametric curve (p. 464)
eccentricity (p. 435)	parametric equation (p. 464)
ellipse (p. 432)	transverse axis (p. 442)
foci (p. 432)	vertex (p. 422)
focus (p. 422)	vertices (p. 432)
hyperbola (p. 442)	

Vocabulary Check

Choose the correct term from the list above to complete each sentence.

1. A _____ is a figure formed when a plane intersects a double-napped right cone.

2. A circle is the _____ of points that fulfill the property that all points be in a given plane and a specified distance from a given point.

3. The _____ of a parabola is perpendicular to its axis of symmetry.

4. The co-vertices of a(n) _____ lie on its minor axis, while the vertices lie on its major axis.

5. From any point on an ellipse, the sum of the distances to the _____ of the ellipse remains constant.

6. The _____ of an ellipse is a ratio that determines how "stretched" or "circular" the ellipse is. It is found using the ratio $\frac{c}{a}$.

7. The _____ of a circle is a single point, and all points on the circle are equidistant from that point.

8. Like an ellipse, a _____ has vertices and foci, but it also has a pair of asymptotes and does not have a connected graph.

9. The equation for a graph can be written using the variables x and y, or using _____ equations, generally using t or the angle θ.

10. The graph of $f(t) = (\sin t, \cos t)$ is a _____ with a shape that is a circle traced clockwise.

Lesson-by-Lesson Review

7-1   Parabolas (pp. 422–431)

For each equation, identify the vertex, focus, axis of symmetry, and directrix. Then graph the parabola.

11. $(x + 3)^2 = 12(y + 2)$

12. $(y - 2)^2 = 8(x - 5)$

13. $(x - 2)^2 = -4(y + 1)$

14. $(x - 5) = \frac{1}{12}(y - 3)^2$

Write an equation for and graph a parabola with the given focus F and vertex V.

15. $F(1, 1), V(1, 5)$

16. $F(-3, 6), V(7, 6)$

17. $F(-2, -3), V(-2, 1)$

18. $F(3, -4), V(3, -2)$

Write an equation for and graph each parabola with focus F and the given characteristics.

19. $F(-4, -4)$; concave left; contains $(-7, 0)$

20. $F(-1, 4)$; concave down; contains $(7, -2)$

21. $F(3, -6)$; concave up; contains $(9, 2)$

EXAMPLE 1

Write an equation for and graph the parabola with focus $(2, 1)$ and vertex $(2, -3)$.

Since the focus and vertex share the same x-coordinate, the graph opens vertically. The focus is $(h, k + p)$, so the value of p is $1 - (-3)$ or 4. Because p is positive, the graph opens up.

Write an equation for the parabola in standard form using the values of h, p, and k.

$4p(y - k) = (x - h)^2$ **Standard form**

$4(4)(y + 3) = (x - 2)^2$ $p = 4, k = -3,$ and $h = 2$

$16(y + 3) = (x - 2)^2$ **Simplify.**

The standard form of the equation is $(x - 2)^2 = 16(y + 3)$. Graph the vertex and focus. Use a table of values to graph the parabola.

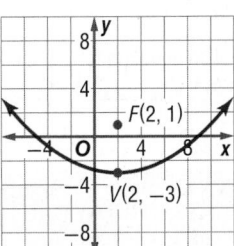

7-2   Ellipses and Circles (pp. 432–441)

Graph the ellipse given by each equation.

22. $\dfrac{x^2}{9} + \dfrac{y^2}{4} = 1$

23. $\dfrac{(x - 3)^2}{16} + \dfrac{(y + 6)^2}{25} = 1$

Write an equation for the ellipse with each set of characteristics.

24. vertices $(7, -3), (3, -3)$; foci $(6, -3), (4, -3)$

25. foci $(1, 2), (9, 2)$; length of minor axis equals 6

26. major axis $(-4, 4)$ to $(6, 4)$; minor axis $(1, 1)$ to $(1, 7)$

Write an equation in standard form. Identify the related conic.

27. $x^2 - 2x + y^2 - 4y - 25 = 0$

28. $4x^2 + 24x + 25y^2 - 300y + 836 = 0$

29. $x^2 - 4x + 4y + 24 = 0$

EXAMPLE 2

Write an equation for the ellipse with a major axis from $(-9, 4)$ to $(11, 4)$ and a minor axis from $(1, 12)$ to $(1, -4)$.

Use the major and minor axes to determine a and b.

Half length of major axis   **Half length of minor axis**

$a = \dfrac{11 - (-9)}{2}$ or 10 $b = \dfrac{12 - (-4)}{2}$ or 8

The center of the ellipse is at the midpoint of the major axis.

$(h, k) = \left(\dfrac{11 + (-9)}{2}, \dfrac{4 + 4}{2} \right)$ **Midpoint Formula**

$= (1, 4)$ **Simplify.**

The y-coordinates are the same for both endpoints of the major axis, so the major axis is horizontal and the value of a belongs with the x^2 term. Therefore, the equation of the ellipse is $\dfrac{(x - 1)^2}{100} + \dfrac{(y - 4)^2}{64} = 1$.

Lesson-by-Lesson Review

7-3 Hyperbolas (pp. 442–452)

Graph the hyperbola given by each equation.

30. $\dfrac{(y+3)^2}{30} - \dfrac{(x-6)^2}{8} = 1$

31. $\dfrac{(x+7)^2}{18} - \dfrac{(y-6)^2}{36} = 1$

32. $\dfrac{(y-1)^2}{4} - (x+1)^2 = 1$

33. $x^2 - y^2 - 2x + 4y - 7 = 0$

Write an equation for the hyperbola with the given characteristics.

34. vertices $(7, 0)$, $(-7, 0)$; conjugate axis length of 8

35. foci $(0, 5)$, $(0, -5)$; vertices $(0, 3)$, $(0, -3)$

36. foci $(1, 15)$, $(1, -5)$; transverse axis length of 16

37. vertices $(2, 0)$, $(-2, 0)$; asymptotes $y = \pm\dfrac{3}{2}x$

Use the discriminant to identify each conic section.

38. $x^2 - 4y^2 - 6x - 16y - 11 = 0$

39. $4y^2 - x - 40y + 107 = 0$

40. $9x^2 + 4y^2 + 162x + 8y + 732 = 0$

EXAMPLE 3

Graph $\dfrac{(y+3)^2}{16} - \dfrac{(x+1)^2}{4} = 1$.

In this equation, $h = -1$, $k = -3$, $a = \sqrt{16}$ or 4, $b = \sqrt{4}$ or 2, and $c = \sqrt{16+4}$ or $2\sqrt{5}$.

Determine the characteristics of the hyperbola.

orientation:	vertical	
center:	$(-1, -3)$	(h, k)
vertices:	$(-1, 1), (-1, -7)$	$(h, k \pm a)$
foci:	$(-1, -3 + 2\sqrt{5})$,	$(h, k \pm c)$
	$(-1, -3 - 2\sqrt{5})$	
asymptotes:	$y + 3 = 2(x + 1)$,	$y - k = \pm\dfrac{a}{b}(x - h)$
	$y + 3 = -2(x + 1)$	

Make a table of values.

x	y
−6	7.77, −13.77
−2	1.47, −7.47
2	4.21, −10.21
6	11.56, −17.56

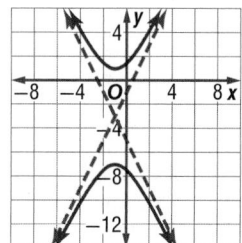

7-4 Rotations of Conic Sections (pp. 454–461)

Use a graphing calculator to graph the conic given by each equation.

41. $x^2 - 4xy + y^2 - 2y - 2x = 0$

42. $x^2 - 3xy + y^2 - 3y - 6x + 5 = 0$

43. $2x^2 + 2y^2 - 8xy + 4 = 0$

44. $3x^2 + 9xy + y^2 = 0$

45. $4x^2 - 2xy + 8y^2 - 7 = 0$

Write each equation in the $x'y'$-plane for the given value of θ. Then identify the conic.

46. $x^2 + y^2 = 4$; $\theta = \dfrac{\pi}{4}$

47. $x^2 - 2x + y = 5$; $\theta = \dfrac{\pi}{3}$

48. $x^2 - 4y^2 = 4$; $\theta = \dfrac{\pi}{2}$

49. $9x^2 + 4y^2 = 36$; $\theta = 90°$

EXAMPLE 4

Use a graphing calculator to graph
$x^2 + 2xy + y^2 + 4x - 2y = 0$.

$x^2 + 2xy + y^2 + 4x - 2y = 0$ **Original equation**

$1y^2 + (2x - 2)y + (x^2 + 4x) = 0$ **Quadratic form**

Use the Quadratic Formula.

$y = \dfrac{-(2x - 2) \pm \sqrt{(2x - 2)^2 - 4(1)(x^2 + 4x)}}{2(1)}$

$= \dfrac{-2x + 2 \pm \sqrt{4x^2 - 8x + 4 - 4x^2 - 16x}}{2}$

$= \dfrac{-2x + 2 \pm 2\sqrt{1 - 6x}}{2}$

$= -x + 1 \pm \sqrt{1 - 6x}$

Graph as

$y_1 = -x + 1 + \sqrt{1 - 6x}$ and

$y_2 = -x + 1 - \sqrt{1 - 6x}$.

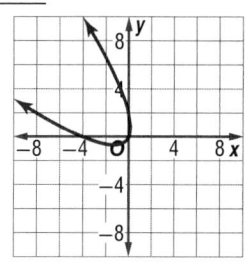

Lesson-by-Lesson Review

7-5 | Parametric Equations (pp. 464–471)

Sketch the curve given by each pair of parametric equations over the given interval.

50. $x = \sqrt{t}, y = 1 - t; 0 \leq t \leq 9$

51. $x = t + 2, y = t^2 - 4; -4 \leq t \leq 4$

Write each pair of parametric equations in rectangular form. Then graph the equation.

52. $x = t + 5$ and $y = 2t - 6$

53. $x = 2t$ and $y = t^2 - 2$

54. $x = t^2 + 3$ and $y = t^2 - 4$

55. $x = t^2 - 1$ and $y = 2t + 1$

EXAMPLE 5

Write $x = 5 \cos t$ and $y = 9 \sin t$ in rectangular form. Then graph the equation.

$$x = 5 \cos t \qquad\qquad y = 9 \sin t$$

$$\cos t = \frac{x}{5} \qquad\qquad \sin t = \frac{y}{9} \qquad \textbf{Solve for sin } t \textbf{ and cos } t.$$

$$\sin^2 t + \cos^2 t = 1$$

$$\left(\frac{x}{5}\right)^2 + \left(\frac{y}{9}\right)^2 = 1$$

$$\frac{x^2}{25} + \frac{y^2}{81} = 1$$

The parametric equations represent the graph of an ellipse.

Applications and Problem Solving

56. MONUMENTS The St. Louis Arch is in the shape of a *catenary*, which resembles a parabola. (Lesson 7-1)

630 ft

630 ft

a. Write an equation for a parabola that would approximate the shape of the arch.

b. Find the location of the focus of this parabola.

57. WATER DYNAMICS A rock dropped in a pond will produce ripples of water made up of concentric expanding circles. Suppose the radii of the circles expand at 3 inches per second. (Lesson 7-2)

a. Write an equation for the circle 10 seconds after the rock is dropped in the pond. Assume that the point where the rock is dropped is the origin.

b. One concentric circle has equation $x^2 + y^2 = 225$. How many seconds after the rock is dropped does it take for the circle to have this equation?

58. ENERGY Cooling towers at a power plant are in the shape of a hyperboloid. The cross section of a hyperboloid is a hyperbola. (Lesson 7-3)

a. Write an equation for the cross section of a tower that is 50 feet tall and 30 feet wide.

b. If the ratio of the height to the width of the tower increases, how is the equation affected?

59. SOLAR DISH Students building a parabolic device to capture solar energy for cooking marshmallows placed at the focus must plan for the device to be easily oriented. Rotating the device directly toward the Sun's rays maximizes the heat potential. (Lesson 7-4)

a. After the parabola is rotated 30° toward the Sun, the equation of the parabola used to create the device in the $x'y'$-plane is $y' = 0.25(x')^2$. Find the equation of the parabola in the xy-plane.

b. Graph the rotated parabola.

60. GEOMETRY Consider $x_1(t) = 4 \cos t, y_1(t) = 4 \sin t,$ $x_2(t) = 4 \cos 2t,$ and $y_2(t) = 4 \sin 2t.$ (Lesson 7-5)

a. Compare the graphs of the two sets of equations: x_1 and y_1; and x_2 and y_2.

b. Write parametric equations for a circle of radius 6 that complete its graph in half the time of $x_1(t)$ and $y_1(t)$.

c. Write the equations from part **b** in rectangular form.

Write an equation for an ellipse with each set of characteristics.

1. vertices $(7, -4)$, $(-3, -4)$; foci $(6, -4)$, $(-2, -4)$

2. foci $(-2, 1)$, $(-2, -9)$; length of major axis is 12

3. **MULTIPLE CHOICE** What value must c be so that the graph of $4x^2 + cy^2 + 2x - 2y - 18 = 0$ is a circle?

   A -8

   B -4

   C 4

   D 8

Write each pair of parametric equations in rectangular form. Then graph the equation.

4. $x = t - 5$ and $y = 3t - 4$

5. $x = t^2 - 1$ and $y = 2t + 1$

6. **BRIDGES** At 1.7 miles long, San Francisco's Golden Gate Bridge was the longest suspension bridge in the world when it was constructed.

   a. Suppose the design of the bridge can be modeled by a parabola and the lowest point of the cable is 15 feet above the road. Write an equation for the design of the bridge.

   b. Where is the focus located in relation to the vertex?

Write an equation for the hyperbola with the given characteristics.

7. vertices $(3, 0)$, $(-3, 0)$; asymptotes $y = \pm\frac{2}{3}x$

8. foci $(8, 0)$, $(8, 8)$; vertices $(8, 2)$, $(8, 6)$

Write an equation for each conic in the xy-plane for the given equation in $x'y'$ form and the given value of θ.

9. $7(x' - 3) = (y')^2$, $\theta = 60°$

10. $\dfrac{(x')^2}{2} + \dfrac{(y')^2}{10} = 1$, $\theta = \dfrac{\pi}{6}$

Graph the hyperbola given by each equation.

11. $\dfrac{x^2}{64} - \dfrac{(y-4)^2}{25} = 1$

12. $\dfrac{(y+3)^2}{4} - \dfrac{(x+6)^2}{36} = 1$

13. **MULTIPLE CHOICE** Which ellipse has the greatest eccentricity?

A

C

B

D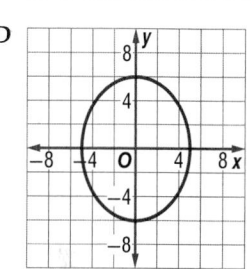

Write an equation for and graph a parabola with the given focus F and vertex V.

14. $F(2, 8)$, $V(2, 10)$

15. $F(2, 5)$, $V(-1, 5)$

Graph the ellipse given by each equation.

16. $\dfrac{(x-5)^2}{49} + \dfrac{(y+3)^2}{9} = 1$

17. $(x + 3)^2 + \dfrac{(y+6)^2}{81} = 1$

18. **CAMPING** In many U.S. parks, campers must secure food and provisions from bears and other animals. One method is to secure food using a bear bag, which is done by tossing a bag tied to a rope over a tall tree branch and securing the rope to the tree. Suppose a tree branch is 30 feet above the ground, and a person 20 feet from the branch throws the bag from 5 feet above the ground.

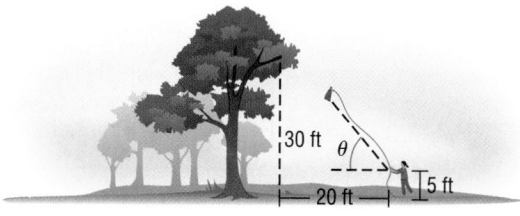

   a. Will a bag thrown at a speed of 40 feet per second at an angle of 60° go over the branch?

   b. Will a bag thrown at a speed of 45 feet per second at an angle of 75° go over the branch?

Use a graphing calculator to graph the conic given by each equation.

19. $x^2 - 6xy + y^2 - 4y - 8x = 0$

20. $x^2 + 4y^2 - 2xy + 3y - 6x + 5 = 0$

In Chapter 2, you learned that integral calculus is a branch of mathematics that focuses on the processes of finding lengths, areas, and volumes. You used rectangles to approximate the areas of irregular shapes, such as those created by a curve and the x-axis. A similar technique can be used to approximate volumes of irregular shapes.

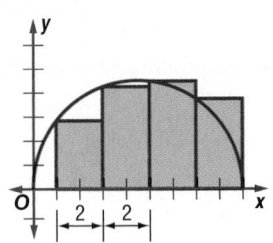

Consider a cone with a height of h and a base with a radius of r. If we did not already know the formula for the volume of a cone, we could approximate the volume by drawing several cylinders of equal height inside the cone. We could then calculate the volume of each cylinder, and find the sum.

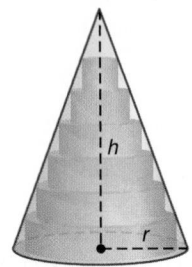

ACTIVITY 1 Sphere

Approximate the volume of the sphere with a radius of 4.5 units and a great circle defined by $f(x) = \pm\sqrt{-x^2 + 9x}$.

Step 1 Sketch a diagram of the sphere.

Step 2 Inscribe a cylinder in the sphere with a base perpendicular to the x-axis and a height of 2 units. Allow for the left edge of the cylinder to begin at $x = 1$ and to extend to the great circle. The radius of the cylinder is $f(1)$.

Step 3 Draw 3 more cylinders all with a height of 2 units. Allow for the left edge of each cylinder to extend to the great circle.

Step 4 Calculate the volume of each cylinder.

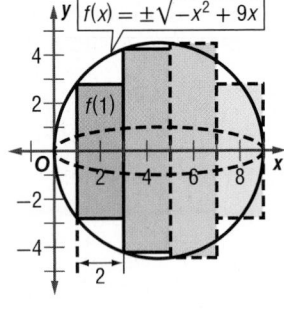

▶ **Analyze the Results**

1. What is the approximation for the volume of the sphere?

2. Calculate the actual volume of the sphere using the radius. How does the approximation compare with the actual volume? What could be done to improve upon the accuracy of the approximation?

When the region between a graph and the x-axis is rotated about the x-axis, a *solid of revolution* is formed. The shape of the graph dictates the shape of the three-dimensional figure formed.

cone

cylinder

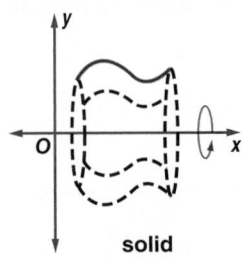

solid

A solid of revolution can be formed by rotating a region in a plane about any fixed line, called the *axis of revolution*. The axis of revolution will dictate the direction and the radii of the cylinders used to approximate the area. If revolving about the x-axis, the cylinders will be parallel to the y-axis and the radii will be given by $f(x)$. If revolving about the y-axis, the cylinders will be parallel to the x-axis and the radii will be given by $f(y)$.

rotating about x-axis   **rotating about y-axis**

ACTIVITY 2   Paraboloid

Approximate the volume of the paraboloid created by revolving the region between $f(x) = -x^2 + 9$, the x-axis, and the y-axis about the y-axis.

Step 1   Sketch a diagram of the paraboloid.

Step 2   Inscribe a cylinder in the paraboloid with a base parallel to the x-axis and a height of 2 units. Allow for the top edge of the cylinder to begin at $y = 8$ and to extend to the edge of the paraboloid.

Step 3   When revolving about the y-axis, the radius is given as $f(y)$. To find $f(y)$, write $f(x)$ as $y = -x^2 + 9$ and solve for y.

Step 4   Draw 3 more cylinders all with a height of 2 units. Allow for the top of each cylinder to extend to the edge of the paraboloid.

Step 5   Calculate the volume of each cylinder.

▶ Analyze the Results

3. What is the approximation for the volume of the paraboloid?

4. Find approximations for the volume of the paraboloid using 8 cylinders with heights of 1 unit and again using 17 cylinders with heights of 0.5 units.

5. As the heights of the cylinders decrease and approach 0, what is happening to the approximations? Explain your reasoning.

6. What shape do the cylinders start to resemble as h approaches 0? Explain your reasoning.

Model and Apply

7. Approximate the volume of the paraboloid created by revolving the region between $f(x) = 2\sqrt{x}$, the x-axis, and the line $x = 6$ about the x-axis. Use 5 cylinders with heights of 1 unit. Let the first cylinder begin at $x = 1$ and the left edge of each cylinder extend to the edge of the paraboloid.

CHAPTER

8

Vectors

Then

In Chapter 4, you used trigonometry to solve triangles.

Now

In Chapter 8, you will:

- Represent and operate with vectors algebraically in the two- and three-dimensional coordinate systems.
- Find the projection of one vector onto another.
- Find cross products of vectors in space, and use triple scalar products to find volumes of parallelepipeds.
- Find the dot products of and angles between vectors in two- and three-dimensional space.

Why?

ROWING Vectors are often used to model changes in direction due to water and air currents. For example, a vector can be used to determine the resultant speed and direction of a kayak that is traveling 8 miles per hour against a 3-mile-per-hour river current.

PREREAD Scan the lesson titles and Key Concept boxes in Chapter 8. Use this information to predict what you will learn in this chapter.

Get Ready for Chapter 8

Diagnose Readiness You have two options for checking Prerequisite Skills.

Text Option

Take the Quick Quiz below.

QuickCheck

Find the distance between each given pair of points and the midpoint of the segment connecting the given points. (Prerequisite Skill)

1. $(1, 4), (-2, 4)$ **2.** $(-5, 3), (-5, 8)$

3. $(2, -9), (-3, -7)$ **4.** $(-4, -1), (-6, -8)$

Find the value of x. Round to the nearest tenth if necessary. (Lesson 4-1)

5.

6.

7.

8.

9. BALLOON A hot air balloon is being held in place by two people holding ropes and standing 35 feet apart. The angle formed between the ground and the rope held by each person is 40°. Determine the length of each rope to the nearest tenth of a foot. (Lesson 4-7)

Find all solutions for the given triangle, if possible. If no solution exists, write *no solution*. Round side lengths to the nearest tenth and angle measures to the nearest degree. (Lesson 4-7)

10. $a = 10, b = 7, A = 128°$ **11.** $a = 15, b = 16, A = 127°$

12. $a = 15, b = 18, A = 52°$ **13.** $a = 30, b = 19, A = 91°$

Online Option

Math Online ▷ Take a self-check Chapter Readiness Quiz at **glencoe.com**.

New Vocabulary

English	• Page •	Español
vector	• 482 •	vector
initial point	• 482 •	punto inicial
terminal point	• 482 •	punto terminal
standard position	• 482 •	posición estándar
direction	• 482 •	dirección
magnitude	• 482 •	magnitud
quadrant bearing	• 483 •	porte de cuadrante
true bearing	• 483 •	porte verdadero
parallel vectors	• 483 •	vectores paralelos
equivalent vectors	• 483 •	vectores equivalentes
opposite vectors	• 483 •	vectores de enfrente
resultant	• 484 •	resultado
zero vector	• 485 •	vector cero
component form	• 492 •	forma componente
unit vector	• 494 •	vector de unidad
dot product	• 500 •	producto de punto
orthogonal	• 500 •	ortogonal
z-axis	• 510 •	z-eje
octants	• 510 •	octants
ordered triple	• 510 •	pedido triple
cross product	• 519 •	producto enfadado
triple scalar product	• 521 •	triplice el producto escalar

Review Vocabulary

scalar • p. P25 • escalar a quantity with magnitude only

dilation • p. 49 • dilatación a nonrigid transformation that has the effect of compressing (shrinking) or expanding (enlarging) the graph of a function vertically or horizontally

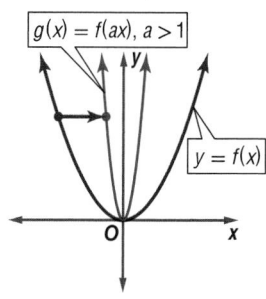

Horizontal dilation

Introduction to Vectors

Then
You used trigonometry to solve triangles. (Lesson 5-4)

Now
- Represent and operate with vectors geometrically.
- Solve vector problems, and resolve vectors into their rectangular components.

New Vocabulary
vector
initial point
terminal point
standard position
direction
magnitude
quadrant bearing
true bearing
parallel vectors
equivalent vectors
opposite vectors
resultant
triangle method
parallelogram method
zero vector
components
rectangular components

Math Online
glencoe.com

Why?

A successful field goal attempt in football depends on several factors. While the speed of the ball after it is kicked is certainly important, the direction the ball takes is as well. We can describe both of these factors using a single quantity called a *vector*.

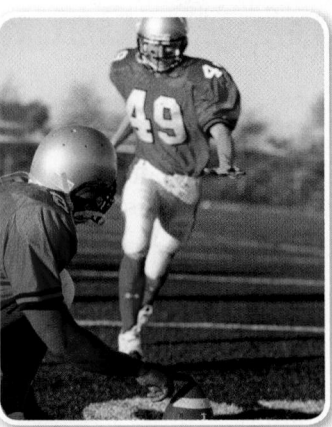

Vectors Many physical quantities, such as speed, can be completely described by a single real number called a *scalar*. This number indicates the *magnitude* or *size* of the quantity. A **vector** is a quantity that has both magnitude and *direction*. The velocity of a football is a vector that describes both the speed and direction of the ball.

EXAMPLE 1 | **Identify Vector Quantities**

State whether each quantity described is a *vector* quantity or a *scalar* quantity.

a. a boat traveling at 15 miles per hour

This quantity has a magnitude of 15 miles per hour, but no direction is given. Speed is a scalar quantity.

b. a hiker walking 25 paces due west

This quantity has a magnitude of 25 paces and a direction of due west. This directed distance is a vector quantity.

c. a person's weight on a bathroom scale

Weight is a vector quantity that is calculated using a person's mass and the downward pull due to gravity. (Acceleration due to gravity is a vector.)

✔ **Guided Practice**

1A. a car traveling 60 miles per hour 15° east of south

1B. a parachutist falling straight down at 12.5 miles per hour

1C. a child pulling a sled with a force of 40 newtons

A vector can be represented geometrically by a directed line segment, or arrow diagram, that shows both magnitude and direction. Consider the directed line segment with an **initial point** *A* (also known as the *tail*) and **terminal point** *B* (also known as the *head* or *tip*) shown. This vector is denoted by $\overrightarrow{AB}$, $\vec{a}$, or **a**.

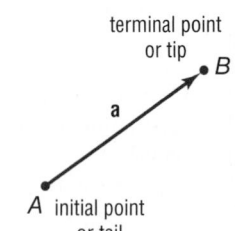

If a vector has its initial point at the origin, it is in **standard position**. The **direction** of a vector is the directed angle between the vector and the horizontal line that could be used to represent the positive *x*-axis. The direction of **a** is 35°.

The length of the line segment represents, and is proportional to, the **magnitude** of the vector. If the scale of the arrow diagram for **a** is 1 cm = 5 ft/s, then the magnitude of **a**, denoted |**a**|, is 2.6×5 or 13 feet per second.

The direction of a vector can also be given as a bearing. A **quadrant bearing** φ, or *phi*, is a directional measurement between 0° and 90° east or west of the north-south line. The quadrant bearing of vector **v** shown is 35° east of south or southeast, written S35°E.

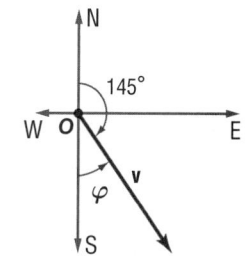

A **true bearing** is a directional measurement where the angle is measured clockwise from north. True bearings are always given using three digits. So, a direction that measures 25° clockwise from north would be written as a true bearing of 025°.

EXAMPLE 2 | **Represent a Vector Geometrically**

Use a ruler and a protractor to draw an arrow diagram for each quantity described. Include a scale on each diagram.

a. **a** = 20 feet per second at a bearing of 030°

Using a scale of 1 in. : 10 ft/sec, draw and label a 20 ÷ 10 or 2-inch arrow at an angle of 30° clockwise from the north.

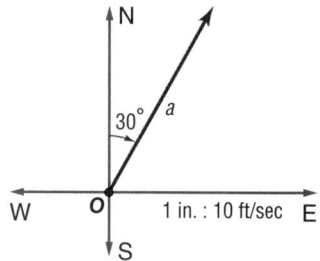

b. **v** = 75 pounds of force at 140° to the horizontal

Using a scale of 1 cm : 25 lb, draw and label a 75 ÷ 25 or 3-centimeter arrow in standard position at a 140° angle to the *x*-axis.

c. **z** = 30 miles per hour at a bearing of S60°W

Using a scale of 1 in. : 20 mi/h, draw and label a 30 ÷ 20 or 1.5-inch arrow 60° west of south.

☑ **Guided Practice**

2A. **t** = 20 feet per second at a bearing of 065°

2B. **u** = 15 miles per hour at a bearing of S25°E

2C. **m** = 60 pounds of force at 80° to the horizontal

In your operations with vectors, you will need to be familiar with the following vector types.

- **Parallel vectors** have the same or opposite direction but not necessarily the same magnitude. In the figure, **a** ‖ **b** ‖ **c** ‖ **e** ‖ **f**.

- **Equivalent vectors** have the same magnitude and direction. In the figure, **a** = **c** because they have the same magnitude and direction. Notice that **a** ≠ **b**, since |**a**| ≠ |**b**|, and **a** ≠ **d**, since **a** and **d** do not have the same direction.

- **Opposite vectors** have the same magnitude but opposite direction. The vector opposite **a** is written −**a**. In the figure, **e** = −**a**.

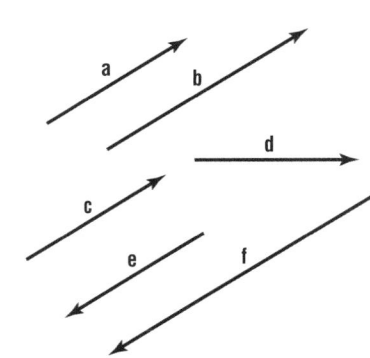

When two or more vectors are added, their sum is a single vector called the **resultant**. The resultant vector has the same effect as applying one vector after the other. Geometrically, the resultant can be found using either the **triangle method** or the **parallelogram method**.

Key Concept — Finding Resultants

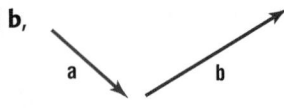

Triangle Method (Tip-to-Tail)

To find the resultant of **a** and **b**, follow these steps.

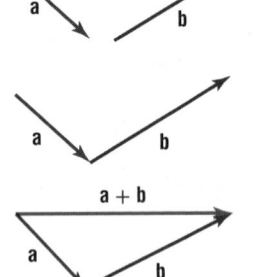

Step 1 Translate **b** so that the tail of **b** touches the tip of **a**.

Step 2 The resultant is the vector from the tail of **a** to the tip of **b**.

Parallelogram Method (Tail-to-Tail)

To find the resultant of **a** and **b**, follow these steps.

Step 1 Translate **b** so that the tail of **b** touches the tail of **a**.

Step 2 Complete the parallelogram that has **a** and **b** as two of its sides.

Step 3 The resultant is the vector that forms the indicated diagonal of the parallelogram.

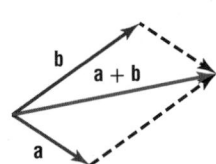

StudyTip

Resultants The parallelogram method must be repeated in order to find the resultant of more than two vectors. The triangle method, however, is easier to use when finding the resultant of three or more vectors. Continue to place the initial point of subsequent vectors at the terminal point of the previous vector.

⊕ Real-World EXAMPLE 3   Find the Resultant of Two Vectors

ORIENTEERING In an orienteering competition, Tia walks N50°E for 120 feet and then walks 80 feet due east. How far and at what quadrant bearing is Tia from her starting position?

Let **p** = walking 120 feet N50°E and **q** = walking 80 feet due east. Draw a diagram to represent **p** and **q** using a scale of 1 cm : 50 ft.

Use a ruler and a protractor to draw a 120 ÷ 50 or 2.4-centimeter arrow 50° east of north to represent **p** and an 80 ÷ 50 or 1.6-centimeter arrow due east to represent **q**.

1 cm : 50 ft

Method 1   Triangle Method

Translate **q** so that its tail touches the tip of **p**. Then draw the resultant vector **p** + **q** as shown.

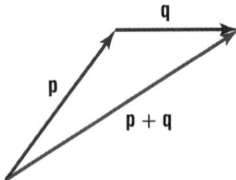

Method 2   Parallelogram Method

Translate **q** so that its tail touches the tail of **p**. Then complete the parallelogram and draw the diagonal, resultant **p** + **q**, as shown.

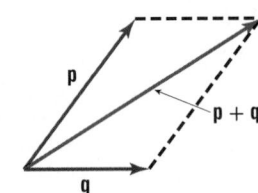

Both methods produce the same resultant vector **p** + **q**. Measure the length of **p** + **q** and then measure the angle this vector makes with the north-south line as shown.

The vector's length of approximately 3.7 centimeters represents 3.7 × 50 or 185 feet. Therefore, Tia is approximately 185 feet at a bearing of 66° east of north or N66°E from her starting position.

1 cm : 50 ft

✓ **Guided Practice**

Find the resultant of each pair of vectors using either the triangle or parallelogram method. State the magnitude of the resultant to the nearest centimeter and its direction relative to the horizontal.

3A.

3B.
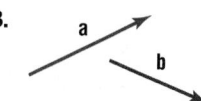

3C. PINBALL A pinball is struck by flipper and is sent 310° at a velocity of 7 inches per second. The ball then bounces off of a bumper and heads 055° at a velocity of 4 inches per second. Find the resulting direction and velocity of the pinball.

When you add two opposite vectors, the resultant is the **zero vector** or *null* vector, denoted by $\vec{0}$ or **0**, which has a magnitude of 0 and no specific direction. Subtracting vectors is similar to subtraction with integers. To find $\mathbf{p} - \mathbf{q}$, add the opposite of $\mathbf{q}$ to $\mathbf{p}$. That is, $\mathbf{p} - \mathbf{q} = \mathbf{p} + (-\mathbf{q})$.

A vector can also be multiplied by a scalar.

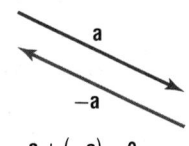

$\mathbf{a} + (-\mathbf{a}) = 0$

Key Concept　　**Multiplying Vectors by a Scalar**

If a vector **v** is multiplied by a real number scalar k, the scalar multiple $k\mathbf{v}$ has a magnitude of $|k|\ |\mathbf{v}|$. Its direction is determined by the sign of k.

• If $k > 0$, $k\mathbf{v}$ has the same direction as **v**.
• If $k < 0$, $k\mathbf{v}$ has the opposite direction as **v**.

EXAMPLE 4　　**Operations with Vectors**

Draw a vector diagram of $3\mathbf{x} - \frac{3}{4}\mathbf{y}$.

Rewrite the expression as the addition of two vectors: $3\mathbf{x} - \frac{3}{4}\mathbf{y} = 3\mathbf{x} + \left(-\frac{3}{4}\mathbf{y}\right)$. To represent $3\mathbf{x}$, draw a vector 3 times as long as **x** in the same direction as **x** (Figure 8.1.1). To represent $-\frac{3}{4}\mathbf{y}$, draw a vector $\frac{3}{4}$ the length of **y** in the opposite direction from **y** (Figure 8.1.2). Then use the triangle method to draw the resultant vector (Figure 8.1.3).

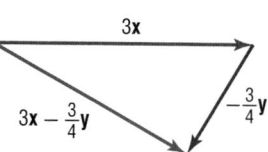

Figure 8.1.1　　　　Figure 8.1.2　　　　Figure 8.1.3

✓ **Guided Practice**

Draw a vector diagram of each expression.

4A. $\mathbf{a} - \mathbf{c} + 2\mathbf{b}$

4B. $\mathbf{m} - \frac{1}{4}\mathbf{p}$

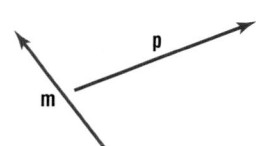

Vector Applications Vector addition and trigonometry can be used to solve vector problems involving triangles which are often oblique.

In navigation, a *heading* is the direction in which a vessel, such as an airplane or boat, is steered to overcome other forces, such as wind or current. The *relative velocity* of the vessel is the resultant when the heading velocity and other forces are combined.

🌐 Real-World EXAMPLE 5 | Use Vectors to Solve Navigation Problems

AVIATION An airplane is flying with an airspeed of 310 knots on a heading of 050°. If a 78-knot wind is blowing from a true heading of 125°, determine the speed and direction of the plane relative to the ground.

Step 1 Draw a diagram to represent the heading and wind velocities (Figure 8.1.4). Translate the wind vector as shown in Figure 8.1.5, and use the triangle method to obtain the resultant vector representing the plane's ground velocity **g**. In the triangle formed by these vectors (Figure 8.1.6), $\gamma = 125° - 50°$ or $75°$.

Figure 8.1.4   **Figure 8.1.5**

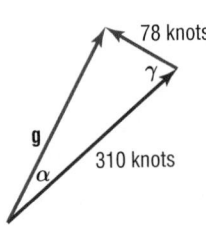

Figure 8.1.6

Step 2 Use the Law of Cosines to find $|\mathbf{g}|$, the plane's speed relative to the ground.

$$c^2 = a^2 + b^2 - 2ab \cos \gamma \qquad \text{Law of Cosines}$$

$$|\mathbf{g}|^2 = 78^2 + 310^2 - 2(78)(310) \cos 75° \qquad c = |\mathbf{g}|, a = 78, b = 310, \text{ and } \gamma = 75°$$

$$|\mathbf{g}| = \sqrt{78^2 + 310^2 - 2(78)(310) \cos 75°} \qquad \text{Take the positive square root of each side.}$$

$$\approx 299.4 \qquad \text{Simplify.}$$

The ground speed of the plane is about 299.4 knots.

Step 3 The heading of the resultant **g** is represented by angle θ, as shown in Figure 8.1.5. To find θ, first calculate α using the Law of Sines.

$$\frac{\sin \alpha}{a} = \frac{\sin \gamma}{c} \qquad \text{Law of Sines}$$

$$\frac{\sin \alpha}{78} = \frac{\sin 75°}{299.4} \qquad c = |\mathbf{g}| \text{ or } 299.4, a = 78, \text{ and } \gamma = 75°$$

$$\sin \alpha = \frac{78 \sin 75°}{299.4} \qquad \text{Solve for } \sin \alpha.$$

$$\alpha = \sin^{-1} \frac{78 \sin 75°}{299.4} \qquad \text{Apply the inverse sine function.}$$

$$\approx 14.6° \qquad \text{Simplify.}$$

The measure of θ is $50° - \alpha$, which is $50° - 14.6°$ or $35.4°$.

Therefore, the speed of the plane relative to the ground is about 299.4 knots at about 035°.

✓ Guided Practice

5. **SWIMMING** Mitchell swims due east at a speed of 3.5 feet per second across a river directly toward the opposite bank. At the same time, the current of the river is carrying him due south at a rate of 2 feet per second. Find Mitchell's speed and direction relative to the shore.

Two or more vectors with a sum that is a vector **r** are called **components** of **r**. While components can have any direction, it often useful to express or *resolve* a vector into two perpendicular components. The **rectangular components** of a vector are horizontal and vertical.

In the diagram, the force **r** exerted to pull the wagon can be thought of as the sum of a horizontal component force **x** that moves the wagon forward and a vertical component force **y** that pulls the wagon upward.

Real-World EXAMPLE 6   Resolve a Force into Rectangular Components

LAWN CARE Heather is pushing the handle of a lawn mower with a force of 450 newtons at an angle of 56° with the ground.

a. Draw a diagram that shows the resolution of the force that Heather exerts into its rectangular components.

Heather's push can be resolved into a horizontal push **x** forward and a vertical push **y** downward as shown.

b. Find the magnitudes of the horizontal and vertical components of the force.

The horizontal and vertical components of the force form a right triangle. Use the sine or cosine ratios to find the magnitude of each force.

$\cos 56° = \dfrac{\lvert \mathbf{x} \rvert}{450}$	**Right triangle definitions of cosine and sine**	$\sin 56° = \dfrac{\lvert \mathbf{y} \rvert}{450}$
$\lvert \mathbf{x} \rvert = 450 \cos 56°$	**Solve for *x* and *y*.**	$\lvert \mathbf{y} \rvert = 450 \sin 56°$
$\lvert \mathbf{x} \rvert \approx 252$	**Use a calculator.**	$\lvert \mathbf{y} \rvert \approx 373$

The magnitude of the horizontal component is about 252 newtons, and the magnitude of the vertical component is about 373 newtons.

✓ Guided Practice

6. **SOCCER** A player kicks a soccer ball so that it leaves the ground with a velocity of 44 feet per second at an angle of 33° with the ground.

A. Draw a diagram that shows the resolution of this force into its rectangular components.

B. Find the magnitude of the horizontal and vertical components of the velocity.

Exercises

State whether each quantity described is a *vector* quantity or a *scalar* quantity. (Example 1)

1. a box being pushed with a force of 125 newtons

2. wind blowing at 20 knots

3. a deer running 15 meters per second due west

4. a baseball thrown with a speed of 85 miles per hour

5. a 15-pound tire hanging from a rope

6. a rock thrown straight up at a velocity of 50 feet per second

Use a ruler and a protractor to draw an arrow diagram for each quantity described. Include a scale on each diagram. (Example 2)

7. h = 13 inches per second at a bearing of 205°

8. g = 6 kilometers per hour at a bearing of N70°W

9. j = 5 feet per minute at 300° to the horizontal

10. k = 28 kilometers at 35° to the horizontal

11. m = 40 meters at a bearing of S55°E

12. n = 32 yards per second at a bearing of 030°

Find the resultant of each pair of vectors using either the triangle or parallelogram method. State the magnitude of the resultant to the nearest tenth of a centimeter and its direction relative to the horizontal. (Example 3)

13.

14.

15.

16.

17.

18.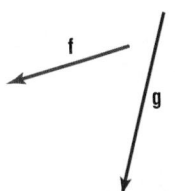

19. **GOLF** While playing a golf video game, Ana hits a ball 35° above the horizontal at a speed of 40 miles per hour with a 5 miles per hour wind blowing, as shown. Find the resulting speed and direction of the ball. (Example 3)

20. **BOATING** A charter boat leaves port on a heading of N60°W for 12 nautical miles. The captain changes course to a bearing of N25°E for the next 15 nautical miles. Determine the ship's distance and direction from port to its current location. (Example 3)

21. **HIKING** Nick and Lauren hiked 3.75 kilometers to a lake 55° east of south from their campsite. Then they hiked 33° west of north to the nature center 5.6 kilometers from the lake. Where is the nature center in relation to their campsite? (Example 3)

Determine the magnitude and direction of the resultant of each vector sum. (Example 3)

22. 18 newtons directly forward and then 20 newtons directly backward

23. 100 meters due north and then 350 meters due south

24. 10 pounds of force at a bearing of 025° and then 15 pounds of force at a bearing of 045°

25. 17 miles east and then 16 miles south

26. 15 meters per second squared at a 60° angle to the horizontal and then 9.8 meters per second squared downward

Use the set of vectors to draw a vector diagram of each expression. (Example 4)

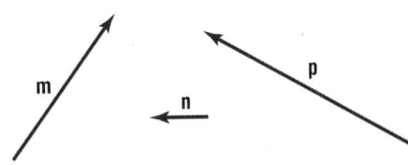

27. $m - 2n$

28. $n - \frac{3}{4}m$

29. $\frac{1}{2}p + 3n$

30. $4n + \frac{4}{5}p$

31. $p + 2n - m$

32. $-\frac{1}{3}m + p - 2n$

33. $3n - \frac{1}{2}p + m$

34. $m - 3n + \frac{1}{4}p$

35. **RUNNING** A runner's resultant velocity is 8 miles per hour due west running with a wind of 3 miles per hour N28°W. What is the runner's speed, to the nearest mile per hour, without the effect of the wind? (Example 5)

36. **GLIDING** A glider is traveling at an air speed of 15 miles per hour due west. If the wind is blowing at 5 miles per hour in the direction N60°E, what is the resulting ground speed of the glider? (Example 5)

37. **CURRENT** Kaya is swimming due west at a rate of 1.5 meters per second. A strong current is flowing S20°E at a rate of 1 meter per second. Find Kaya's resulting speed and direction. (Example 5)

Draw a diagram that shows the resolution of each vector into its rectangular components. Then find the magnitudes of the vector's horizontal and vertical components. (Example 6)

38. $2\frac{1}{8}$ inches at 310° to the horizontal

39. 1.5 centimeters at a bearing of N49°E

40. 3.2 centimeters per hour at a bearing of S78°W

41. $\frac{3}{4}$ inch per minute at a bearing of 255°

42. FOOTBALL For a field goal attempt, a football is kicked with the velocity shown in the diagram below.

90 ft/s
30°

a. Draw a diagram that shows the resolution of this force into its rectangular components.

b. Find the magnitudes of the horizontal and vertical components. (Example 6)

43. CLEANING Aiko is pushing the handle of a push broom with a force of 190 newtons at an angle of 33° with the ground. (Example 6)

190 N
33°

a. Draw a diagram that shows the resolution of this force into its rectangular components.

b. Find the magnitudes of the horizontal and vertical components.

44. GARDENING Carla and Oscar are pulling a wagon full of plants. Each person pulls on the wagon with equal force at an angle of 30° with the axis of the wagon. The resultant force is 120 newtons.

F Carla
30°
30°
F Oscar

a. How much force is each person exerting?

b. If each person exerts a force of 75 newtons, what is the resultant force?

c. How will the resultant force be affected if Carla and Oscar move closer together?

The magnitude and true bearings of three forces acting on an object are given. Find the magnitude and direction of the resultant of these forces.

45. 50 lb at 30°, 80 lb at 125°, and 100 lb at 220°

46. 8 newtons at 300°, 12 newtons at 45°, and 6 newtons at 120°

47. 18 lb at 190°, 3 lb at 20°, and 7 lb at 320°

48. DRIVING Carrie's school is on a direct path three miles from her house. She drives on two different streets on her way to school. She travels at an angle of 20.9° with the path on the first street and then turns 45.4° onto the second street.

Carrie's house
45.4°
Carrie's school
134.6°
20.9°
3 mi

a. How far does Carrie drive on the first street?

b. How far does she drive on the second street?

c. If it takes her 10 minutes to get to school and she averages 25 miles per hour on the first street, what speed does Carrie average after she turns onto the second street?

49 SLEDDING Irwin is pulling his sister on a sled. The direction of his resultant force is 31°, and the horizontal component of the force is 86 newtons.

a. What is the vertical component of the force?

b. What is the magnitude of the resultant force?

50. MULTIPLE REPRESENTATIONS In this problem, you will investigate multiplication of a vector by a scalar.

a. **GRAPHICAL** On a coordinate plane, draw a vector **a** so that the tail is located at the origin. Choose a value for a scalar k. Then draw the vector that results if you multiply the original vector by k on the same coordinate plane. Repeat the process for four additional vectors **b**, **c**, **d**, and **e**. Use the same value for k each time.

b. **TABULAR** Copy and complete the table below for each vector that you drew in part **a**.

Vector	Terminal Point of Vector	Terminal Point of Vector × k
a		
b		
c		
d		
e		

c. **ANALYTICAL** If the terminal point of a vector **a** is located at the point (a, b), what is the location of the terminal point of the vector k**a**?

An *equilibrant* vector is the opposite of a resultant vector. It balances a combination of vectors such that the sum of the vectors and the equilibrant is the zero vector. The equilibrant vector of **a** + **b** is −(**a** + **b**).

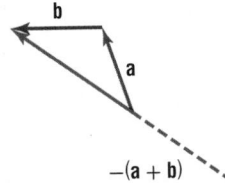

Find the magnitude and direction of the equilibrant vector for each set of vectors.

51 **a** = 15 miles per hour at a bearing of 125°
  b = 12 miles per hour at a bearing of 045°

52. **a** = 4 meters at a bearing of N30W°
  b = 6 meters at a bearing of N20E°

53. **a** = 23 feet per second at a bearing of 205°
  b = 16 feet per second at a bearing of 345°

54. PARTY PLANNING A disco ball is suspended above a dance floor by two wires of equal length as shown.

12 lb

  a. Draw a vector diagram of the situation that indicates that two tension vectors T_1 and T_2 with equal magnitude are keeping the disco ball stationary or at equilibrium.

  b. Redraw the diagram using the triangle method to find $T_1 + T_2$.

  c. Use your diagram from part **b** and the fact that the equilibrant of the resultant $T_1 + T_2$ and the vector representing the weight of the disco ball are equivalent vectors to calculate the magnitudes of T_1 and T_2.

55. CABLE SUPPORT Two cables with tensions T_1 and T_2 are tied together to support a 2500-pound load at equilibrium.

2500 lb

  a. Write expressions to represent the horizontal and vertical components of T_1 and T_2.

  b. Given that the equilibrant of the resultant $T_1 + T_2$ and the vector representing the weight of the load are equivalent vectors, calculate the magnitudes of T_1 and T_2 to the nearest tenth of a pound.

  c. Use your answers from parts **a** and **b** to find the magnitudes of the horizontal and vertical components of T_1 and T_2 to the nearest tenth of a pound.

Find the magnitude and direction of each vector given its vertical and horizontal components and the range of values for the angle of direction θ to the horizontal.

56. horizontal: 0.32 in., vertical: 2.28 in., 90° < θ < 180°

57. horizontal: 3.1 ft, vertical: 4.2 ft, 0° < θ < 90°

58. horizontal: 2.6 cm, vertical: 9.7 cm, 270° < θ < 360°

59. horizontal: 2.9 yd, vertical: 1.8 yd, 180° < θ < 270°

Draw any three vectors a, b, and c. Show geometrically that each of the following vector properties holds using these vectors.

60. Commutative Property: **a** + **b** = **b** + **a**

61. Associative Property: (**a** + **b**) + **c** = **a** + (**b** + **c**)

62. Distributive Property: k(**a** + **b**) = k**a** + k**b**, for k = 2, 0.5, and −2

H.O.T. Problems Use **H**igher-**O**rder **T**hinking Skills

63. OPEN ENDED Consider a vector of 5 units directed along the positive x-axis. Resolve the vector into two perpendicular components in which no component is horizontal or vertical.

64. REASONING Is it *sometimes*, *always*, or *never* possible to find the sum of two parallel vectors using the parallelogram method? Explain your reasoning.

65. REASONING Why is it important to establish a common reference for measuring the direction of a vector, for example, from the positive x-axis?

66. CHALLENGE The resultant of **a** + **b** is equal to the resultant of **a** − **b**. If the magnitude of **a** is 4x, what is the magnitude of **b**?

67. REASONING Consider the statement $|a| + |b| \geq |a + b|$.

  a. Express this statement using words.

  b. Is this statement true or false? Justify your answer.

68. ERROR ANALYSIS Darin and Cris are finding the resultant of vectors **a** and **b**. Is either of them correct? Explain your reasoning.

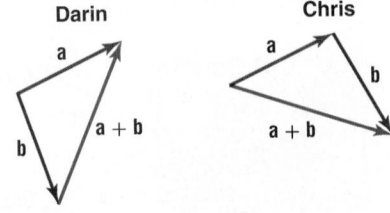

69. REASONING Is it possible for the sum of two vectors to equal one of the vectors? Explain.

70. WRITING IN MATH Compare and contrast the parallelogram and triangle methods of finding the resultant of two or more vectors.

71. KICKBALL Suppose a kickball player kicks a ball at a 32° angle to the horizontal with an initial speed of 20 meters per second. How far away will the ball land? (Lesson 7-5)

72. Graph $(x')^2 + y' - 5 = 1$ if it has been rotated 45° from its position in the xy-plane. (Lesson 7-4)

Write an equation for a circle that satisfies each set of conditions. Then graph the circle. (Lesson 7-2)

73. center at (4, 5), radius 4

74. center at (1, −4), diameter 7

Determine the equation of and graph the parabola with the given focus F and vertex V. (Lesson 7-1)

75. $F(2, 4), V(2, 3)$

76. $F(1, 5), V(−7, 5)$

77. CRAFTS Sanjay is selling wood carvings. He sells large statues for $60, clocks for $40, dollhouse furniture for $25, and chess pieces for $5. He takes the following number of items to the fair: 12 large statues, 25 clocks, 45 pieces of dollhouse furniture, and 50 chess pieces. (Lesson 6-2)

    a. Write an inventory matrix for the number of each item and a cost matrix for the price of each item.

    b. Find Sanjay's total income if he sells all of the items.

Solve each equation for all values of x. (Lesson 5-3)

78. $4 \sin x \cos x - 2 \sin x = 0$

79. $\sin x - 2 \cos^2 x = -1$

80. SAT/ACT If town A is 12 miles from town B and town C is 18 miles from town A, then which of the following *cannot* be the distance from town B to town C?

    A 5 miles   **D** 12 miles

    B 7 miles   **E** 18 miles

    C 10 miles

81. A remote control airplane flew along an initial path of 32° to the horizontal at a velocity of 48 feet per second as shown. Which of the following represent the magnitudes of the horizontal and vertical components of the velocity?

    F 25.4 ft/s, 40.7 ft/s   **H** 56.6 ft/s, 90.6 ft/s

    G 40.7 ft/s, 25.4 ft/s   **J** 90.6 ft/s, 56.6 ft/s

82. REVIEW Triangle ABC has vertices $A(−4, 2)$, $B(−4, −3)$, and $C(3, −3)$. After a dilation, triangle $A'B'C'$ has vertices $A'(−12, 6)$, $B'(−12, −9)$, and $C'(9, −9)$. How many times as great is the area of $\triangle A'B'C'$ than the area of $\triangle ABC$?

    A $\frac{1}{9}$ **C** 3

    B $\frac{1}{3}$ **D** 9

83. REVIEW Holly is drawing a map of her neighborhood. Her house is represented by quadrilateral ABCD with vertices $A(2, 2)$, $B(6, 2)$, $C(6, 6)$, and $D(2, 6)$. She wants to use the same coordinate system to make another map that is one half the size of the original map. What could be the new vertices of Holly's house?

    F $A'(0, 0), B'(2, 1), C'(3, 3), D'(0, 3)$

    G $A'(0, 0), B'(3, 1), C'(2, 3), D'(0, 2)$

    H $A'(1, 1), B'(3, 1), C'(3, 3), D'(1, 3)$

    J $A'(1, 2), B'(3, 0), C'(2, 2), D'(2, 3)$

Vectors in the Coordinate Plane

Then
You performed vector operations using scale drawings. (Lesson 8-1)

Now
- Represent and operate with vectors in the coordinate plane.
- Write a vector as a linear combination of unit vectors.

New Vocabulary
component form
unit vector
linear combination

Math Online

glencoe.com

Why?

Wind can impact the ground speed and direction of an airplane. While pilots can use scale drawings to determine the heading a plane should take to correct for wind, these calculations are more commonly calculated using vectors in the coordinate plane.

Vectors in the Coordinate Plane In Lesson 8-1, you found the magnitude and direction of the resultant of two or more forces geometrically using a scale drawing. Since drawings can be inaccurate, an algebraic approach using a rectangular coordinate system is needed for situations where more accuracy is required or where the system of vectors is complex.

A vector $\overrightarrow{OP}$ in standard position on a rectangular coordinate system (as in Figure 8.2.1) can be uniquely described by the coordinates of its terminal point $P(x, y)$. We denote $\overrightarrow{OP}$ on the coordinate plane by $\langle x, y \rangle$. Notice that x and y are the rectangular components of $\overrightarrow{OP}$. For this reason, $\langle x, y \rangle$ is called the **component form** of a vector.

Figure 8.2.1

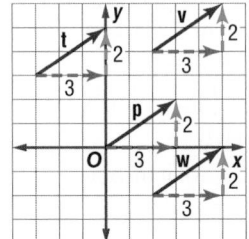

Figure 8.2.2

Since vectors with the same magnitude and direction are equivalent, many vectors can be represented by the same coordinates. For example, vectors **p**, **t**, **v**, and **w** in Figure 8.2.2 are *equivalent* because each can be denoted as $\langle 3, 2 \rangle$. To find the component form of a vector that is not in standard position, you can use the coordinates of its initial and terminal points.

Key Concept   **Component Form of a Vector**

The component form of a vector $\overrightarrow{AB}$ with initial point $A(x_1, y_1)$ and terminal point $B(x_2, y_2)$ is given by

$$\langle x_2 - x_1, y_2 - y_1 \rangle.$$

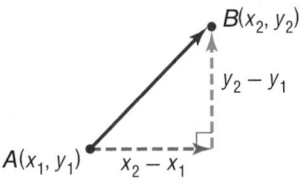

EXAMPLE 1   **Express a Vector in Component Form**

Find the component form of $\overrightarrow{AB}$ with initial point $A(-4, 2)$ and terminal point $B(3, -5)$.

$\overrightarrow{AB} = \langle x_2 - x_1, y_2 - y_1 \rangle$ **Component form**

$\quad = \langle 3 - (-4), -5 - 2 \rangle$ $(x_1, y_1) = (-4, 2)$ and $(x_2, y_2) = (3, -5)$

$\quad = \langle 7, -7 \rangle$ **Subtract.**

✓ **Guided Practice**

Find the component form of $\overrightarrow{AB}$ with the given initial and terminal points.

1A. $A(-2, -7), B(6, 1)$ **1B.** $A(0, 8), B(-9, -3)$

The magnitude of a vector in the coordinate plane is found by using the Distance Formula.

Key Concept **Magnitude of a Vector in the Coordinate Plane**

If **v** is a vector with initial point (x_1, y_1) and terminal point (x_2, y_2), then the magnitude of **v** is given by

$$|\mathbf{v}| = \sqrt{(x_2 - x_1)^2 + (y_2 - y_1)^2}.$$

If **v** has a component form of $\langle a, b \rangle$, then $|\mathbf{v}| = \sqrt{a^2 + b^2}$.

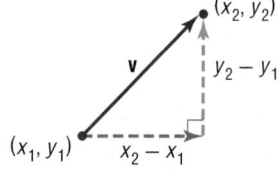

EXAMPLE 2 **Find the Magnitude of a Vector**

Find the magnitude of $\overrightarrow{AB}$ with initial point $A(-4, 2)$ and terminal point $B(3, -5)$.

$$|\overrightarrow{AB}| = \sqrt{(x_2 - x_1)^2 + (y_2 - y_1)^2} \quad \text{Distance Formula}$$
$$= \sqrt{[3 - (-4)]^2 + (-5 - 2)^2} \quad (x_1, y_1) = (-4, 2) \text{ and } (x_2, y_2) = (3, -5)$$
$$= \sqrt{98} \text{ or about } 9.9 \quad \text{Simplify.}$$

CHECK From Example 1, you know that $\overrightarrow{AB} = \langle 7, -7 \rangle$. $|\overrightarrow{AB}| = \sqrt{7^2 + (-7)^2}$ or $\sqrt{98}$. ✔

☑ **Guided Practice**

Find the magnitude of $\overrightarrow{AB}$ with the given initial and terminal points.

2A. $A(-2, -7)$, $B(6, 1)$ **2B.** $A(0, 8)$, $B(-9, -3)$

Addition, subtraction, and scalar multiplication of vectors in the coordinate plane is similar to the same operations with matrices.

Key Concept **Vector Operations**

If $\mathbf{a} = \langle a_1, a_2 \rangle$ and $\mathbf{b} = \langle b_1, b_2 \rangle$ are vectors and k is a scalar, then the following are true.

Vector Addition $\mathbf{a} + \mathbf{b} = \langle a_1 + b_1, a_2 + b_2 \rangle$

Vector Subtraction $\mathbf{a} - \mathbf{b} = \langle a_1 - b_1, a_2 - b_2 \rangle$

Scalar Multiplication $k\mathbf{a} = \langle ka_1, ka_2 \rangle$

EXAMPLE 3 **Operations with Vectors**

Find each of the following for $\mathbf{w} = \langle -4, 1 \rangle$, $\mathbf{y} = \langle 2, 5 \rangle$, and $\mathbf{z} = \langle -3, 0 \rangle$.

a. **w + y**

$$\mathbf{w} + \mathbf{y} = \langle -4, 1 \rangle + \langle 2, 5 \rangle \quad \text{Substitute.}$$
$$= \langle -4 + 2, 1 + 5 \rangle \text{ or } \langle -2, 6 \rangle \quad \text{Vector addition}$$

b. **z − 2y**

$$\mathbf{z} - 2\mathbf{y} = \mathbf{z} + (-2)\mathbf{y} \quad \text{Rewrite subtraction as addition.}$$
$$= \langle -3, 0 \rangle + (-2)\langle 2, 5 \rangle \quad \text{Substitute.}$$
$$= \langle -3, 0 \rangle + \langle -4, -10 \rangle \text{ or } \langle -7, -10 \rangle \quad \text{Scalar multiplication and vector addition}$$

☑ **Guided Practice**

3A. $4\mathbf{w} + \mathbf{z}$ **3B.** $-3\mathbf{w}$ **3C.** $2\mathbf{w} + 4\mathbf{y} - \mathbf{z}$

Unit Vectors A vector that has a magnitude of 1 unit is called a **unit vector**. It is sometimes useful to describe a nonzero vector **v** as a scalar multiple of a unit vector **u** with the same direction as **v**. To find **u**, divide **v** by its magnitude $|\mathbf{v}|$.

$$\mathbf{u} = \frac{\mathbf{v}}{|\mathbf{v}|} \quad \text{or} \quad \frac{1}{|\mathbf{v}|}\mathbf{v}$$

EXAMPLE 4   **Find a Unit Vector with the Same Direction as a Given Vector**

Find a unit vector **u** with the same direction as $\mathbf{v} = \langle -2, 3 \rangle$.

$\mathbf{u} = \dfrac{1}{|\mathbf{v}|}\mathbf{v}$	Unit vector with the same direction as **v**
$= \dfrac{1}{|\langle -2, 3 \rangle|}\langle -2, 3 \rangle$	Substitute.
$= \dfrac{1}{\sqrt{(-2)^2 + 3^2}}\langle -2, 3 \rangle$	$\|\langle a, b \rangle\| = \sqrt{a^2 + b^2}$
$= \dfrac{1}{\sqrt{13}}\langle -2, 3 \rangle$	Simplify.
$= \left\langle -\dfrac{2}{\sqrt{13}}, \dfrac{3}{\sqrt{13}} \right\rangle$	Scalar multiplication
$= \left\langle -\dfrac{2\sqrt{13}}{13}, \dfrac{3\sqrt{13}}{13} \right\rangle$	Rationalize denominators.

CHECK   Since **u** is a scalar multiple of **v**, it has the same direction as **v**. Verify that the magnitude of **u** is 1.

$\|\mathbf{u}\| = \sqrt{\left(-\dfrac{2\sqrt{13}}{13}\right)^2 + \left(\dfrac{3\sqrt{13}}{13}\right)^2}$	Distance Formula
$= \sqrt{\dfrac{52}{169} + \dfrac{117}{169}}$	Simplify.
$= \sqrt{1}$ or 1 ✔	Simplify.

✓ Guided Practice

Find a unit vector with the same direction as the given vector.

4A. $\mathbf{w} = \langle 6, -2 \rangle$ **4B.** $\mathbf{x} = \langle -4, -8 \rangle$

The unit vectors in the direction of the positive *x*-axis and positive *y*-axis are denoted by $\mathbf{i} = \langle 1, 0 \rangle$ and $\mathbf{j} = \langle 0, 1 \rangle$, respectively (Figure 8.2.3). Vectors **i** and **j** are called *standard unit vectors.*

Figure 8.2.3

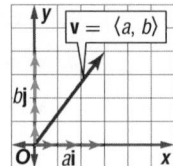

Figure 8.2.4

These vectors can be used to express any vector $\mathbf{v} = \langle a, b \rangle$ as $a\mathbf{i} + b\mathbf{j}$ as shown in Figure 8.2.4.

$\mathbf{v} = \langle a, b \rangle$	Component form of **v**
$= \langle a, 0 \rangle + \langle 0, b \rangle$	Rewrite as the sum of two vectors.
$= a\langle 1, 0 \rangle + b\langle 0, 1 \rangle$	Scalar multiplication
$= a\mathbf{i} + b\mathbf{j}$	$\langle 1, 0 \rangle = \mathbf{i}$ and $\langle 0, 1 \rangle = \mathbf{j}$

The vector sum $a\mathbf{i} + b\mathbf{j}$ is called a **linear combination** of the vectors $\mathbf{i}$ and $\mathbf{j}$.

EXAMPLE 5   Write a Vector as a Linear Combination of Unit Vectors

Let $\overrightarrow{DE}$ be the vector with initial point $D(-2, 3)$ and terminal point $E(4, 5)$. Write $\overrightarrow{DE}$ as a linear combination of the vectors $\mathbf{i}$ and $\mathbf{j}$.

First, find the component form of $\overrightarrow{DE}$.

$\overrightarrow{DE} = \langle x_2 - x_1, y_2 - y_1 \rangle$ **Component form**

$\phantom{\overrightarrow{DE}} = \langle 4 - (-2), 5 - 3 \rangle$ $(x_1, y_1) = (-2, 3)$ and $(x_2, y_2) = (4, 5)$

$\phantom{\overrightarrow{DE}} = \langle 6, 2 \rangle$ **Simplify.**

Then rewrite the vector as a linear combination of the standard unit vectors.

$\overrightarrow{DE} = \langle 6, 2 \rangle$ **Component form**

$\phantom{\overrightarrow{DE}} = 6\mathbf{i} + 2\mathbf{j}$ $\langle a, b \rangle = a\mathbf{i} + b\mathbf{j}$

✔ Guided Practice

Let $\overrightarrow{DE}$ be the vector with the given initial and terminal points. Write $\overrightarrow{DE}$ as a linear combination of the vectors $\mathbf{i}$ and $\mathbf{j}$.

5A. $D(-6, 0), E(2, 5)$ **5B.** $D(-3, -8), E(-7, 1)$

StudyTip

Unit Vector From the statement that $\mathbf{v} = \langle |\mathbf{v}| \cos \theta, |\mathbf{v}| \sin \theta \rangle$, it follows that the unit vector in the direction of $\mathbf{v}$ has the form $\mathbf{v} = \langle 1 \cos \theta, 1 \sin \theta| = \langle \cos \theta, \sin \theta \rangle$.

A way to specify the direction of a vector $\mathbf{v} = \langle a, b \rangle$ is to state the direction angle θ that $\mathbf{v}$ makes with the positive x-axis. From Figure 8.2.5, it follows that $\mathbf{v}$ can be written in component form or as a linear combination of $\mathbf{i}$ and $\mathbf{j}$ using the magnitude and direction angle of the vector.

$\mathbf{v} = \langle a, b \rangle$ **Component form**

$\phantom{\mathbf{v}} = \langle |\mathbf{v}| \cos \theta, |\mathbf{v}| \sin \theta \rangle$ **Substitution**

$\phantom{\mathbf{v}} = |\mathbf{v}| (\cos \theta)\mathbf{i} + |\mathbf{v}| (\sin \theta)\mathbf{j}$ **Linear combination of i and j**

EXAMPLE 6   Find Component Form

Find the component form of the vector $\mathbf{v}$ with magnitude 10 and direction angle 120°.

$\mathbf{v} = \langle |\mathbf{v}| \cos \theta, |\mathbf{v}| \sin \theta \rangle$ **Component form of v in terms of |v| and θ**

$\phantom{\mathbf{v}} = \langle 10 \cos 120°, 10 \sin 120° \rangle$ $|\mathbf{v}| = 10$ and $\theta = 120°$

$\phantom{\mathbf{v}} = \left\langle 10\left(-\dfrac{1}{2}\right), 10\left(\dfrac{\sqrt{3}}{2}\right) \right\rangle$ $\cos 120° = -\dfrac{1}{2}$ and $\sin 120° = \dfrac{\sqrt{3}}{2}$

$\phantom{\mathbf{v}} = \langle -5, 5\sqrt{3} \rangle$ **Simplify.**

CHECK Graph $\mathbf{v} = \langle -5, 5\sqrt{3} \rangle \approx \langle -5, 8.7 \rangle$. The measure of the angle $\mathbf{v}$ makes with the positive x-axis is about 120° as shown, and $|\mathbf{v}| = \sqrt{(-5)^2 + (5\sqrt{3})^2}$ or 10. ✔

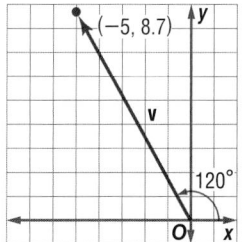

✔ Guided Practice

Find the component form of $\mathbf{v}$ with the given magnitude and direction angle.

6A. $|\mathbf{v}| = 8, \theta = 45°$ **6B.** $|\mathbf{v}| = 24, \theta = 210°$

It also follows from Figure 8.2.5 on the previous page that the direction angle θ of vector $\mathbf{v} = \langle a, b \rangle$ can be found by solving the trigonometric equation $\tan \theta = \dfrac{|\mathbf{v}| \sin \theta}{|\mathbf{v}| \cos \theta}$ or $\tan \theta = \dfrac{b}{a}$.

Figure 8.2.6

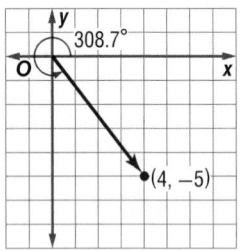

Figure 8.2.7

EXAMPLE 7 Direction Angles of Vectors

Find the direction angle of each vector to the nearest tenth of a degree.

a. $\mathbf{p} = 3\mathbf{i} + 7\mathbf{j}$

$\tan \theta = \dfrac{b}{a}$	**Direction angle equation**
$\tan \theta = \dfrac{7}{3}$	$a = 3$ **and** $b = 7$
$\theta = \tan^{-1} \dfrac{7}{3}$	**Solve for** θ.
$\theta \approx 66.8°$	**Use a calculator.**

So, the direction angle of vector $\mathbf{p}$ is about 67.8° as shown in Figure 8.2.6.

b. $\mathbf{r} = \langle 4, -5 \rangle$

$\tan \theta = \dfrac{b}{a}$	**Direction angle equation**
$\tan \theta = \dfrac{-5}{4}$	$a = 4$ **and** $b = -5$
$\theta = \tan^{-1}\left(-\dfrac{5}{4}\right)$	**Solve for** θ.
$\theta \approx -51.3°$	**Use a calculator.**

Since $\mathbf{r}$ lies in Quadrant IV as shown in Figure 8.2.7, $\theta = 360 + (-51.3)$ or 308.7°.

✓ Guided Practice

7A. $-6\mathbf{i} + 2\mathbf{j}$

7B. $\langle -3, -8 \rangle$

🌐 Real-World EXAMPLE 8 Applied Vector Operations

FOOTBALL A quarterback running forward at 5 meters per second throws a football with a velocity of 25 meters per second at an angle of 40° with the horizontal. What is the resultant speed and direction of the pass?

Since the quarterback moves straight forward, the component form of his velocity $\mathbf{v}_1$ is $\langle 5, 0 \rangle$. Use the magnitude and direction of the football's velocity $\mathbf{v}_2$ to write this vector in component form.

$\mathbf{v}_2 = \langle |\mathbf{v}_2| \cos \theta, |\mathbf{v}_2| \sin \theta \rangle$	**Component form of** $\mathbf{v}_2$
$= \langle 25 \cos 40°, 25 \sin 40° \rangle$	$|\mathbf{v}_2| = 25$ **and** $\theta = 40°$
$\approx \langle 19.2, 16.1 \rangle$	**Simplify.**

Add the algebraic vectors representing $\mathbf{v}_1$ and $\mathbf{v}_2$ to find the resultant velocity, vector $\mathbf{r}$.

$\mathbf{r} = \mathbf{v}_1 + \mathbf{v}_2$	**Resultant vector**
$= \langle 5, 0 \rangle + \langle 19.2, 16.1 \rangle$	**Substitution**
$= \langle 24.2, 16.1 \rangle$	**Vector Addition**

The magnitude of this resultant is $|\mathbf{r}| = \sqrt{24.2^2 + 16.1^2}$ or about 29.1. Next find the resultant direction angle θ.

$\tan \theta = \dfrac{16.1}{24.2}$ $\tan \theta = \dfrac{b}{a}$ **where** $\langle a, b \rangle = \langle 24.2, 16.1 \rangle$

$\theta = \tan^{-1} \dfrac{16.1}{24.2}$ or about 33.6°   **Solve for** θ.

Note: Not drawn to scale.

Therefore, the resultant velocity of the pass is about 29.1 meters per second at an angle of about 33.6° with the horizontal.

✓ Guided Practice

8. FOOTBALL What would the resultant velocity of the football be if the quarterback made the same pass running 5 meters per second backward?

Exercises

Find the component form and magnitude of $\overrightarrow{AB}$ with the given initial and terminal points. (Examples 1 and 2)

1. $A(-3, 1)$, $B(4, 5)$

2. $A(2, -7)$, $B(-6, 9)$

3. $A(10, -2)$, $B(3, -5)$

4. $A(-2, 7)$, $B(-9, -1)$

5. $A(-5, -4)$, $B(8, -2)$

6. $A(-2, 6)$, $B(1, 10)$

7. $A(2.5, -3)$, $B(-4, 1.5)$

8. $A(-4.3, 1.8)$, $B(9.4, -6.2)$

9. $A\left(\frac{1}{2}, -9\right)$, $B\left(6, \frac{5}{2}\right)$

10. $A\left(\frac{3}{5}, -\frac{2}{5}\right)$, $B(-1, 7)$

Find each of the following for $\mathbf{f} = \langle 8, 0 \rangle$, $\mathbf{g} = \langle -3, -5 \rangle$, and $\mathbf{h} = \langle -6, 2 \rangle$. (Example 3)

11. $4\mathbf{h} - \mathbf{g}$

12. $\mathbf{f} + 2\mathbf{h}$

13. $3\mathbf{g} - 5\mathbf{f} + \mathbf{h}$

14. $2\mathbf{f} + \mathbf{g} - 3\mathbf{h}$

15. $\mathbf{f} - 2\mathbf{g} - 2\mathbf{h}$

16. $\mathbf{h} - 4\mathbf{f} + 5\mathbf{g}$

17. $4\mathbf{g} - 3\mathbf{f} + \mathbf{h}$

18. $6\mathbf{h} + 5\mathbf{f} - 10\mathbf{g}$

19. PHYSICS In physics, force diagrams are used to show the effects of all the different forces acting upon an object. The following force diagram could represent the forces acting upon a child sliding down a slide. (Example 3)

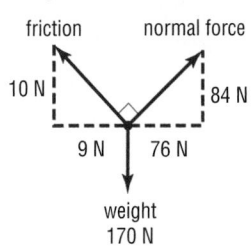

friction   normal force

10 N   84 N

9 N   76 N

weight
170 N

a. Using the blue dot representing the child as the origin, express each force as a vector in component form.

b. Find the component form of the resultant vector representing the force that causes the child to move down the slide.

Find a unit vector $\mathbf{u}$ with the same direction as $\mathbf{v}$. (Example 4)

20. $\mathbf{v} = \langle -2, 7 \rangle$

21. $\mathbf{v} = \langle 9, -3 \rangle$

22. $\mathbf{v} = \langle -8, -5 \rangle$

23. $\mathbf{v} = \langle 6, 3 \rangle$

24. $\mathbf{v} = \langle -2, 9 \rangle$

25. $\mathbf{v} = \langle -1, -5 \rangle$

26. $\mathbf{v} = \langle 1, 7 \rangle$

27. $\mathbf{v} = \langle 3, -4 \rangle$

Let $\overrightarrow{DE}$ be the vector with the given initial and terminal points. Write $\overrightarrow{DE}$ as a linear combination of the vectors $\mathbf{i}$ and $\mathbf{j}$. (Example 5)

28. $D(4, -1)$, $E(5, -7)$

29. $D(9, -6)$, $E(-7, 2)$

30. $D(3, 11)$, $E(-2, -8)$

31. $D(9.5, 1)$, $E(0, -7.3)$

32. $D(-3, -5.7)$, $E(6, -8.1)$

33. $D(-4, -6)$, $E(9, 5)$

34. $D\left(\frac{1}{8}, 3\right)$, $E\left(-4, \frac{2}{7}\right)$

35. $D(-3, 1.5)$, $E(-3, -1.5)$

36. COMMUTE To commute to school, Larisa leaves her house and drives north on Pepper Lane for 2.4 miles. She turns left on Cinnamon Drive for 3.1 miles and then turns right on Maple Street for 5.8 miles. Express Larisa's commute as a linear combination of unit vectors $\mathbf{i}$ and $\mathbf{j}$. (Example 5)

37 ROWING Nadia is rowing across a river at a speed of 5 miles per hour perpendicular to the shore. The river has a current of 3 miles per hour heading downstream. (Example 5)

a. At what speed is she traveling?

b. At what angle is she traveling with respect to the shore?

Find the component form of $\mathbf{v}$ with the given magnitude and direction angle. (Example 6)

38. $|\mathbf{v}| = 12$, $\theta = 60°$

39. $|\mathbf{v}| = 4$, $\theta = 135°$

40. $|\mathbf{v}| = 6$, $\theta = 240°$

41. $|\mathbf{v}| = 16$, $\theta = 330°$

42. $|\mathbf{v}| = 28$, $\theta = 273°$

43. $|\mathbf{v}| = 15$, $\theta = 125°$

Find the direction angle of each vector to the nearest tenth of a degree. (Example 7)

44. $3\mathbf{i} + 6\mathbf{j}$

45. $-2\mathbf{i} + 5\mathbf{j}$

46. $8\mathbf{i} - 2\mathbf{j}$

47. $-4\mathbf{i} - 3\mathbf{j}$

48. $\langle -5, 9 \rangle$

49. $\langle 7, 7 \rangle$

50. $\langle -6, -4 \rangle$

51. $\langle 3, -8 \rangle$

52. SLEDDING Maggie is pulling a sled with a force of 275 newtons by holding its rope at a 58° angle. Her brother is pushing the sled with a force of 320 newtons. Determine the magnitude and direction of the resultant force on the sled. (Example 8)

275 N

58°   320 N

53. NAVIGATION An airplane is traveling due east with a speed of 600 miles per hour. The wind blows at 85 miles per hour at an angle of S59°E. (Example 8)

N

600 mph

59°   85 mph

wind

a. Determine the speed of the airplane's flight.

b. Determine the angle of the airplane's flight.

54. HEADING A pilot needs to plot a course that will result in a velocity of 500 miles per hour in a direction of due west. If the wind is blowing 100 miles per hour from the directed angle of 192°, find the direction and the speed the pilot should set to achieve this resultant.

Determine whether $\overrightarrow{AB}$ and $\overrightarrow{CD}$ with the initial and terminal points given are equivalent. If so, prove that $\overrightarrow{AB} = \overrightarrow{CD}$. If not, explain why not.

55. $A(3, 5), B(6, 9), C(-4, -4), D(-2, 0)$

56. $A(-4, -5), B(-8, 1), C(3, -3), D(1, 0)$

57. $A(1, -3), B(0, -10), C(11, 8), D(10, 1)$

58. RAFTING The Soto family is rafting on the Colorado River. Suppose that they are on a stretch of the river that is 150 meters wide, flowing south at a rate of 1.0 meter per second. In still water, their raft travels 5.0 meters per second.

   a. What is the speed of the raft?

   b. How far downriver will the raft land?

   c. How long does it take them to travel from one bank to the other if they head directly across the river?

59. NAVIGATION A jet is flying with an air speed of 480 miles per hour at a bearing of N82°E. Because of the wind, the ground speed of the plane is 518 miles per hour at a bearing of N79°E.

   a. Draw a diagram to represent the situation.

   b. What are the speed and direction of the wind?

   c. If the pilot increased the air speed of the plane to 500 miles per hour, what would be the resulting ground speed and direction of the plane?

60. TRANSLATIONS You can translate a figure along a translation vector $\langle a, b \rangle$ by adding a to each x-coordinate and b to each y-coordinate. Consider the triangles shown below.

   a. Describe the translation from $\triangle FGH$ to $\triangle F'G'H'$ using a translation vector.

   b. Graph $\triangle F'G'H'$ and its translated image $\triangle F''G''H''$ along $\langle -3, -6 \rangle$.

   c. Describe the translation from $\triangle FGH$ to $\triangle F''G''H''$ using a translation vector.

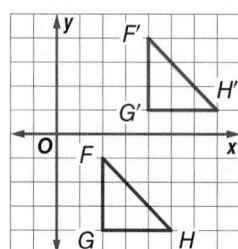

Given the initial point and magnitude of each vector, determine a possible terminal point of the vector.

61. $(-1, 4); \sqrt{37}$ **62.** $(-3, -7); 10$

63. CAMERA A video camera that follows the action at a sporting event is supported by three wires. The tension in each wire can be modeled by a vector.

   a. Find the component form of each vector.

   b. Find the component form of the resultant vector acting on the camera.

   c. Find the magnitude and direction of the resulting force.

64. FORCE A box is stationary on a ramp. Both gravity **g** and friction are exerted on the box. The components of gravity are shown in the diagram. What must be true of the force of friction for this scenario to be possible?

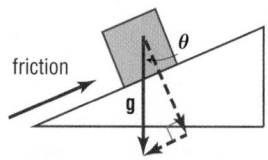

H.O.T. Problems   Use **H**igher-**O**rder **T**hinking Skills

65. REASONING If vectors **a** and **b** are parallel, write a vector equation relating **a** and **b**.

66. CHALLENGE To pull luggage, Greg exerts a force of 150 newtons at an angle of 58° with the horizontal. If the resultant force on the luggage is 72 newtons at an angle of 56.7° with the horizontal, what is the magnitude of the resultant of $\mathbf{F}_{friction}$ and $\mathbf{F}_{weight}$?

67. REASONING If given the initial point of a vector and its magnitude, describe the locus of points that represent possible locations for the terminal point.

68. WRITING IN MATH Explain how to find the direction angle of a vector in the fourth quadrant.

69 CHALLENGE The direction angle of $\langle x, y \rangle$ is $(4y)°$. Find x in terms of y.

PROOF Prove each vector property. Let $\mathbf{a} = \langle x_1, y_1 \rangle$, $\mathbf{b} = \langle x_2, y_2 \rangle$, and $\mathbf{c} = \langle x_3, y_3 \rangle$.

70. $\mathbf{a} + \mathbf{b} = \mathbf{b} + \mathbf{a}$

71. $(\mathbf{a} + \mathbf{b}) + \mathbf{c} = \mathbf{a} + (\mathbf{b} + \mathbf{c})$

72. $k(\mathbf{a} + \mathbf{b}) = k\mathbf{a} + k\mathbf{b}$, where k is a scalar

73. $|k\mathbf{a}| = |k|\,|\mathbf{a}|$, where k is a scalar

74. **TOYS** Roman is pulling a toy by exerting a force of 1.5 newtons on a string attached to the toy. (Lesson 8-1)

   a. The string makes an angle of 52° with the floor. Find the horizontal and vertical components of the force.

   b. If Roman raises the string so that it makes a 78° angle with the floor, what are the magnitudes of the horizontal and vertical components of the force?

Write each pair of parametric equations in rectangular form. (Lesson 7-5)

75. $y = t^2 + 2, x = 3t - 6$

76. $y = t^2 - 5, x = 2t + 8$

77. $y = 7t, x = t^2 - 1$

78. **UMBRELLAS** A beach umbrella has an arch in the shape of a parabola. Write an equation to model the arch, assuming that the origin and the vertex are at the point where the pole and the canopy of the umbrella meet. Express y in terms of x. (Lesson 7-1)

1.5 ft

9 ft

Decompose each expression into partial fractions. (Lesson 6-4)

79. $\dfrac{5z - 11}{2z^2 + z - 6}$

80. $\dfrac{7x^2 + 18x - 1}{(x^2 - 1)(x + 2)}$

81. $\dfrac{9 - 9x}{x^2 - 9}$

Verify each identity. (Lesson 5-4)

82. $\sin(\theta + 180°) = -\sin\theta$

83. $\sin(60° + \theta) + \sin(60° - \theta) = \sqrt{3}\cos\theta$

Express each logarithm in terms of ln 3 and ln 7. (Lesson 3-3)

84. $\ln 189$

85. $\ln 5.\overline{4}$

86. $\ln 441$

87. $\ln\dfrac{9}{343}$

Find each $f(c)$ using synthetic substitution. (Lesson 2-3)

88. $f(x) = 6x^6 - 9x^4 + 12x^3 - 16x^2 + 8x + 24; c = 6$

89. $f(x) = 8x^5 - 12x^4 + 18x^3 - 24x^2 + 36x - 48, c = 4$

90. **SAT/ACT** If $PR = RS$, what is the area of triangle PRS?

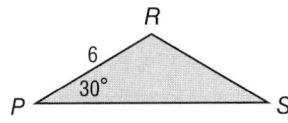

   A $9\sqrt{2}$ **C** $18\sqrt{2}$ **E** $36\sqrt{3}$

   B $9\sqrt{3}$ **D** $18\sqrt{3}$

91. **REVIEW** Dalton has made a game for his younger sister's birthday party. The playing board is a circle divided evenly into 8 sectors. If the circle has a radius of 18 inches, what is the approximate area of one of the sectors?

   F 4 in^2 **H** 127 in^2

   G 32 in^2 **J** 254 in^2

92. Paramedics Lydia Gonzalez and Theo Howard are moving a person on a stretcher. Ms. Gonzalez is pushing the stretcher from behind with a force of 135 newtons at 58° with the horizontal, while Mr. Howard is pulling the stretcher from the front with a force of 214 newtons at 43° with the horizontal. What is the magnitude of the horizontal force exerted on the stretcher?

   A 228 newtons   **C** 299 newtons

   B 260 newtons   **D** 346 newtons

93. **REVIEW** Find the center and radius of the circle with equation $(x - 4)^2 + y^2 - 16 = 0$.

   F $C(-4, 0); r = 4$ units

   G $C(-4, 0); r = 16$ units

   H $C(4, 0); r = 4$ units

   J $C(4, 0); r = 16$ units

Dot Products and Vector Projections

Then
You found the magnitudes of and operated with algebraic vectors.
(Lesson 8-2)

Now
- Find the dot product of two vectors, and use the dot product to find the angle between them.
- Find the projection of one vector onto another.

New Vocabulary
dot product
orthogonal
vector projection
work

Math Online
glencoe.com

Why?

The word *work* can have different meanings in everyday life; but in physics, its definition is very specific. Work is the magnitude of a force applied to an object multiplied by the distance through which the object moves parallel to this applied force. Work, such as that done to push a car a specific distance, can also be calculated using a vector operation called a *dot product*.

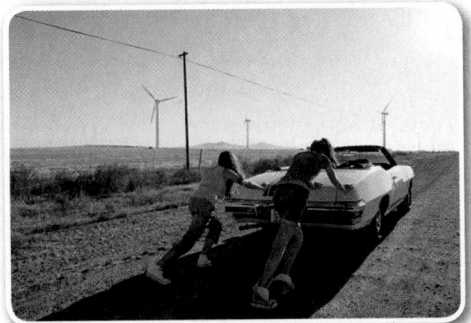

Dot Product In Lesson 8-2, you studied the vector operations of vector addition and scalar multiplication. In this lesson, you will use a third vector operation. Consider two perpendicular vectors in standard position **a** and **b**. Let $\overrightarrow{BA}$ be the vector between their terminal points as shown in the figure. By the Pythagorean Theorem, we know that

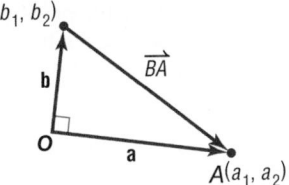

$$\left|\overrightarrow{BA}\right|^2 = |\mathbf{a}|^2 + |\mathbf{b}|^2.$$

Using the definition of the magnitude of a vector, we can find $\left|\overrightarrow{BA}\right|^2$.

$\left\|\overrightarrow{BA}\right\| = \sqrt{(a_1 - b_1)^2 + (a_2 - b_2)^2}$	**Definition of vector magnitude**
$\left\|\overrightarrow{BA}\right\|^2 = (a_1 - b_1)^2 + (a_2 - b_2)^2$	**Square each side.**
$\left\|\overrightarrow{BA}\right\|^2 = a_1{}^2 - 2a_1b_1 + b_1{}^2 + a_2{}^2 - 2a_2b_2 + b_2{}^2$	**Expand each binomial square.**
$\left\|\overrightarrow{BA}\right\|^2 = (a_1{}^2 + a_2{}^2) + (b_1{}^2 + b_2{}^2) - 2(a_1b_1 + a_2b_2)$	**Group the squared terms.**
$\left\|\overrightarrow{BA}\right\|^2 = \|\mathbf{a}\|^2 + \|\mathbf{b}\|^2 - 2(a_1b_1 + a_2b_2)$	$\|\mathbf{a}\| = \sqrt{a_1{}^2 + a_2{}^2}$ so $\|\mathbf{a}\|^2 = a_1{}^2 + a_2{}^2$ and $\|\mathbf{b}\| = \sqrt{b_1{}^2 + b_2{}^2}$, so $\|\mathbf{b}\|^2 = b_1{}^2 + b_2{}^2$.

Notice that the expressions $|\mathbf{a}|^2 + |\mathbf{b}|^2$ and $|\mathbf{a}|^2 + |\mathbf{b}|^2 - 2(a_1b_1 + a_2b_2)$ are equivalent if and only if $a_1b_1 + a_2b_2 = 0$. The expression $a_1b_1 + a_2b_2$ is called the **dot product** of **a** and **b**, denoted **a · b** and read as *a dot b*.

Key Concept **Dot Product of Vectors in a Plane**

The dot product of $\mathbf{a} = \langle a_1, a_2 \rangle$ and $\mathbf{b} = \langle b_1, b_2 \rangle$ is defined as $\mathbf{a} \cdot \mathbf{b} = a_1b_1 + a_2b_2$.

Notice that unlike vector addition and scalar multiplication, the dot product of two vectors yields a scalar, not a vector. As demonstrated above, two nonzero vectors are perpendicular if and only if their dot product is 0. Two vectors with a dot product of 0 are said to be **orthogonal**.

Key Concept **Orthogonal Vectors**

The vectors **a** and **b** are orthogonal if and only if $\mathbf{a} \cdot \mathbf{b} = 0$.

The terms *perpendicular* and *orthogonal* have essentially the same meaning, except when **a** or **b** is the zero vector. The zero vector is orthogonal to any vector **a**, since $\langle 0, 0 \rangle \cdot \langle a_1, a_2 \rangle = 0a_1 + 0a_2$ or 0. However, since the zero vector has no magnitude or direction, it cannot be perpendicular to **a**.

Figure 8.3.1

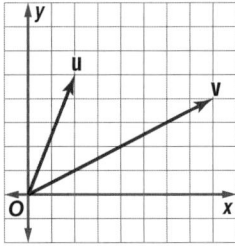

Figure 8.3.2

| EXAMPLE 1 | Find the Dot Product to Determine Orthogonal Vectors |

Find the dot product of **u** and **v**. Then determine if **u** and **v** are orthogonal.

a. $\mathbf{u} = \langle 3, 6 \rangle$, $\mathbf{v} = \langle -4, 2 \rangle$

$\mathbf{u} \cdot \mathbf{v} = 3(-4) + 6(2)$

$\quad\quad = 0$

Since $\mathbf{u} \cdot \mathbf{v} = 0$, **u** and **v** are orthogonal, as illustrated in Figure 8.3.1.

b. $\mathbf{u} = \langle 2, 5 \rangle$, $\mathbf{v} = \langle 8, 4 \rangle$

$\mathbf{u} \cdot \mathbf{v} = 2(8) + 5(4)$

$\quad\quad = 36$

Since $\mathbf{u} \cdot \mathbf{v} \neq 0$, **u** and **v** are not orthogonal, as illustrated in Figure 8.3.2.

✔ **Guided Practice**

1A. $\mathbf{u} = \langle 3, -2 \rangle$, $\mathbf{v} = \langle -5, 1 \rangle$

1B. $\mathbf{u} = \langle -2, -3 \rangle$, $\mathbf{v} = \langle 9, -6 \rangle$

Dot products have the following properties.

Key Concept   **Properties of the Dot Product**

If **u**, **v**, and **w** are vectors and k is a scalar, then the following properties hold.

Commutative Property	$\mathbf{u} \cdot \mathbf{v} = \mathbf{v} \cdot \mathbf{u}$
Distributive Property	$\mathbf{u} \cdot (\mathbf{v} + \mathbf{w}) = \mathbf{u} \cdot \mathbf{v} + \mathbf{u} \cdot \mathbf{w}$
Scalar Multiplication Property	$k(\mathbf{u} \cdot \mathbf{v}) = k\mathbf{u} \cdot \mathbf{v} = \mathbf{u} \cdot k\mathbf{v}$
Zero Vector Dot Product Property	$\mathbf{0} \cdot \mathbf{u} = 0$
Dot Product and Vector Magnitude Relationship	$\mathbf{u} \cdot \mathbf{u} = |\mathbf{u}|^2$

Proof

Proof $\mathbf{u} \cdot \mathbf{u} = |\mathbf{u}|^2$

Let $\mathbf{u} = \langle u_1, u_2 \rangle$.

$\mathbf{u} \cdot \mathbf{u} = u_1{}^2 + u_2{}^2$ Dot product

$\quad\quad = \left(\sqrt{\left(u_1{}^2 + u_2{}^2 \right)} \right)^2$ Write as the square of the square root of $u_1^2 + u_2^2$.

$\quad\quad = |\mathbf{u}|^2$ $\sqrt{u_1^2 + u_2^2} = |\mathbf{u}|$

You will prove the first three properties in Exercises 70–72.

ReadingMath

▶ **Inner and Scalar Products**
The dot product is also called the *inner product* or the *scalar product*.

| EXAMPLE 2 | Use the Dot Product to Find Magnitude |

Use the dot product to find the magnitude of $\mathbf{a} = \langle -5, 12 \rangle$.

Since $|\mathbf{a}|^2 = \mathbf{a} \cdot \mathbf{a}$, then $|\mathbf{a}| = \sqrt{\mathbf{a} \cdot \mathbf{a}}$.

$|\langle -5, 12 \rangle| = \sqrt{\langle -5, 12 \rangle \cdot \langle -5, 12 \rangle}$ $\mathbf{a} = \langle -5, 12 \rangle$

$\quad\quad\quad\quad = \sqrt{(-5)^2 + 12^2}$ or 13   Simplify.

✔ **Guided Practice**   Use the dot product to find the magnitude of the given vector.

2A. $\mathbf{b} = \langle 12, 16 \rangle$

2B. $\mathbf{c} = \langle -1, -7 \rangle$

The angle θ between any two nonzero vectors **a** and **b** is the corresponding angle between these vectors when placed in standard position, as shown. This angle is always measured such that $0 \leq \theta \leq \pi$ or $0° \leq \theta \leq 180°$. The dot product can be used to find the angle between two nonzero vectors.

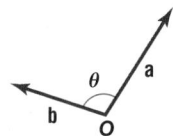

Key Concept   Angle Between Two Vectors

If θ is the angle between nonzero vectors **a** and **b**, then

$$\cos \theta = \frac{\mathbf{a} \cdot \mathbf{b}}{|\mathbf{a}|\,|\mathbf{b}|}.$$

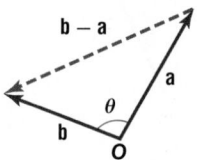

Proof

Consider the triangle determined by **a**, **b**, and **b** − **a** in the figure above.

$|\mathbf{a}|^2 + |\mathbf{b}|^2 - 2\,|\mathbf{a}|\,|\mathbf{b}|\cos\theta = |\mathbf{b} - \mathbf{a}|^2$	**Law of Cosines**
$|\mathbf{a}|^2 + |\mathbf{b}|^2 - 2\,|\mathbf{a}|\,|\mathbf{b}|\cos\theta = (\mathbf{b} - \mathbf{a}) \cdot (\mathbf{b} - \mathbf{a})$	$|\mathbf{u}|^2 = \mathbf{u} \cdot \mathbf{u}$
$|\mathbf{a}|^2 + |\mathbf{b}|^2 - 2\,|\mathbf{a}|\,|\mathbf{b}|\cos\theta = \mathbf{b}\cdot\mathbf{b} - \mathbf{b}\cdot\mathbf{a} - \mathbf{a}\cdot\mathbf{b} + \mathbf{a}\cdot\mathbf{a}$	**Distributive Property for Dot Products**
$|\mathbf{a}|^2 + |\mathbf{b}|^2 - 2\,|\mathbf{a}|\,|\mathbf{b}|\cos\theta = |\mathbf{b}|^2 - 2\mathbf{a}\cdot\mathbf{b} + |\mathbf{a}|^2$	$\mathbf{u}\cdot\mathbf{u} = |\mathbf{u}|^2$
$- 2\,|\mathbf{a}|\,|\mathbf{b}|\cos\theta = -2\mathbf{a}\cdot\mathbf{b}$	**Subtract $|\mathbf{a}|^2 + |\mathbf{b}|^2$ from each side.**
$\cos\theta = \dfrac{\mathbf{a}\cdot\mathbf{b}}{|\mathbf{a}|\,|\mathbf{b}|}$	**Divide each side by $-2|\mathbf{a}|\,|\mathbf{b}|$.**

EXAMPLE 3   Find the Angle Between Two Vectors

Find the angle θ between vectors **u** and **v** to the nearest tenth of a degree.

a. $\mathbf{u} = \langle 6, 2 \rangle$ and $\mathbf{v} = \langle -4, 3 \rangle$

$\cos\theta = \dfrac{\mathbf{u}\cdot\mathbf{v}}{|\mathbf{u}|\,|\mathbf{v}|}$	**Angle between two vectors**
$\cos\theta = \dfrac{\langle 6, 2\rangle \cdot \langle -4, 3\rangle}{|\langle 6, 2\rangle|\,|\langle -4, 3\rangle|}$	$\mathbf{u} = \langle 6, 2\rangle$ and $\mathbf{v} = \langle -4, 3\rangle$
$\cos\theta = \dfrac{-24 + 6}{\sqrt{40}\,\sqrt{25}}$	**Evaluate.**
$\cos\theta = \dfrac{-9}{5\sqrt{10}}$	**Simplify.**
$\theta = \cos^{-1}\dfrac{-9}{5\sqrt{10}}$ or about $124.7°$	**Solve for θ.**

The measure of the angle between **u** and **v** is about 124.7°.

b. $\mathbf{u} = \langle 3, 1 \rangle$ and $\mathbf{v} = \langle 3, -3 \rangle$

$\cos\theta = \dfrac{\mathbf{u}\cdot\mathbf{v}}{|\mathbf{u}|\,|\mathbf{v}|}$	**Angle between two vectors**
$\cos\theta = \dfrac{\langle 3, 1\rangle \cdot \langle 3, -3\rangle}{|\langle 3, 1\rangle|\,|\langle 3, -3\rangle|}$	$\mathbf{u} = \langle 3, 1\rangle$ and $\mathbf{v} = \langle 3, -3\rangle$
$\cos\theta = \dfrac{9 + (-3)}{\sqrt{10}\,\sqrt{18}}$	**Evaluate.**
$\cos\theta = \dfrac{1}{\sqrt{5}}$	**Simplify.**
$\theta = \cos^{-1}\dfrac{1}{\sqrt{5}}$ or about $63.4°$	**Solve for θ.**

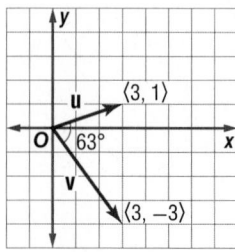

The measure of the angle between **u** and **v** is about 63.4°.

✓ **Guided Practice**

3A. $\mathbf{u} = \langle -5, -2 \rangle$ and $\mathbf{v} = \langle 4, 4 \rangle$ **3B.** $\mathbf{u} = \langle 9, 5 \rangle$ and $\mathbf{v} = \langle -6, 7 \rangle$

Vector Projection In Lesson 8-1, you learned that a vector can be resolved or decomposed into two perpendicular components. While these components are often horizontal and vertical, it is sometimes useful instead for one component to be parallel to another vector.

Key Concept   Projection of u onto v

Let $\mathbf{u}$ and $\mathbf{v}$ be nonzero vectors, and let $\mathbf{w}_1$ and $\mathbf{w}_2$ be vector components of $\mathbf{u}$ such that $\mathbf{w}_1$ is parallel to $\mathbf{v}$ as shown. Then vector $\mathbf{w}_1$ is called the **vector projection** of $\mathbf{u}$ onto $\mathbf{v}$, denoted $\text{proj}_\mathbf{v}\mathbf{u}$, and

$$\text{proj}_\mathbf{v}\mathbf{u} = \left(\frac{\mathbf{u} \cdot \mathbf{v}}{|\mathbf{v}|^2}\right)\mathbf{v}.$$

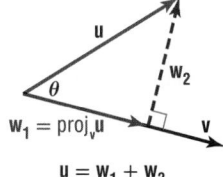

$\mathbf{w}_1 = \text{proj}_\mathbf{v}\mathbf{u}$

$\mathbf{u} = \mathbf{w}_1 + \mathbf{w}_2$

Proof

Since $\text{proj}_\mathbf{v}\mathbf{u}$ is parallel to $\mathbf{v}$, it can be written as a scalar multiple of $\mathbf{v}$. As a scalar multiple of a unit vector $\mathbf{v}_x$ with the same direction as $\mathbf{v}$, $\text{proj}_\mathbf{v}\mathbf{u} = |\mathbf{w}_1|\,\mathbf{v}_x$. Use the right triangle formed by $\mathbf{w}_1$, $\mathbf{w}_2$, and $\mathbf{u}$ and the cosine ratio to find an expression for $|\mathbf{w}_1|$.

$\cos\theta = \dfrac{|\mathbf{w}_1|}{|\mathbf{u}|}$ **Cosine ratio**

$|\mathbf{u}||\mathbf{v}|\cos\theta = \cancel{|\mathbf{u}|}|\mathbf{v}|\dfrac{|\mathbf{w}_1|}{\cancel{|\mathbf{u}|}}$ **Multiply each side by the scalar quantity $|\mathbf{u}|\,|\mathbf{v}|$.**

$\mathbf{u} \cdot \mathbf{v} = |\mathbf{v}|\,|\mathbf{w}_1|$ $\cos\theta = \dfrac{\mathbf{u} \cdot \mathbf{v}}{|\mathbf{u}|\,|\mathbf{v}|}$, so $|\mathbf{u}|\,|\mathbf{v}|\cos\theta = \mathbf{u} \cdot \mathbf{v}$.

$|\mathbf{w}_1| = \dfrac{\mathbf{u} \cdot \mathbf{v}}{|\mathbf{v}|}$ **Solve for $|\mathbf{w}_1|$.**

Now use $\text{proj}_\mathbf{v}\mathbf{u} = |\mathbf{w}_1|\,\mathbf{v}_x$ to find $\text{proj}_\mathbf{v}\mathbf{u}$ as a scalar multiple of $\mathbf{v}$.

$\text{proj}_\mathbf{v}\mathbf{u} = |\mathbf{w}_1|\,\mathbf{v}_x$

$= \dfrac{\mathbf{u} \cdot \mathbf{v}}{|\mathbf{v}|} \cdot \dfrac{\mathbf{v}}{|\mathbf{v}|}$ $|\mathbf{w}_1| = \dfrac{\mathbf{u} \cdot \mathbf{v}}{|\mathbf{v}|}$ and $\mathbf{v}_x = \dfrac{\mathbf{v}}{|\mathbf{v}|}$

$= \left(\dfrac{\mathbf{u} \cdot \mathbf{v}}{|\mathbf{v}|^2}\right)\mathbf{v}$ **Multiply magnitudes.**

EXAMPLE 4   Find the Projection of u onto v

Find the projection of $\mathbf{u} = \langle 3, 2\rangle$ onto $\mathbf{v} = \langle 5, -5\rangle$. Then write $\mathbf{u}$ as the sum of two orthogonal vectors, one of which is the projection of $\mathbf{u}$ onto $\mathbf{v}$.

Step 1 Find the projection of $\mathbf{u}$ onto $\mathbf{v}$.

$\text{proj}_\mathbf{v}\mathbf{u} = \left(\dfrac{\mathbf{u} \cdot \mathbf{v}}{|\mathbf{v}|^2}\right)\mathbf{v}$

$= \dfrac{\langle 3, 2\rangle \cdot \langle 5, -5\rangle}{|\langle 5, -5\rangle|^2}\langle 5, -5\rangle$

$= \dfrac{5}{50}\langle 5, -5\rangle$

$= \left\langle \dfrac{1}{2}, -\dfrac{1}{2}\right\rangle$

Step 2 Find $\mathbf{w}_2$.

Since $\mathbf{u} = \mathbf{w}_1 + \mathbf{w}_2$, $\mathbf{w}_2 = \mathbf{u} - \mathbf{w}_1$.

$\mathbf{w}_2 = \mathbf{u} - \mathbf{w}_1$

$= \mathbf{u} - \text{proj}_\mathbf{v}\mathbf{u}$

$= \langle 3, 2\rangle - \left\langle \dfrac{1}{2}, -\dfrac{1}{2}\right\rangle$

$= \left\langle \dfrac{5}{2}, \dfrac{5}{2}\right\rangle$

Therefore, $\text{proj}_\mathbf{v}\mathbf{u}$ is $\mathbf{w}_1 = \left\langle \dfrac{1}{2}, -\dfrac{1}{2}\right\rangle$ as shown in Figure 8.3.5, and $\mathbf{u} = \left\langle \dfrac{1}{2}, -\dfrac{1}{2}\right\rangle + \left\langle \dfrac{5}{2}, \dfrac{5}{2}\right\rangle$

Figure 8.3.5

✓ Guided Practice

4. Find the projection of $\mathbf{u} = \langle 1, 2\rangle$ onto $\mathbf{v} = \langle 8, 5\rangle$. Then write $\mathbf{u}$ as the sum of two orthogonal vectors, one of which is the projection of $\mathbf{u}$ onto $\mathbf{v}$.

While the projection of **u** onto **v** is a vector parallel to **v**, this vector will not necessarily have the same direction as **v**, as illustrated in the next example.

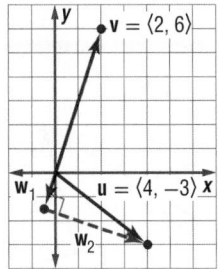

Figure 8.3.6

EXAMPLE 5 — Projection with Direction Opposite v

Find the projection of $\mathbf{u} = \langle 4, -3 \rangle$ onto $\mathbf{v} = \langle 2, 6 \rangle$. Then write u as the sum of two orthogonal vectors, one of which is the projection of u onto v.

Notice that the angle between **u** and **v** is obtuse, so the projection of **u** onto **v** lies on the vector opposite **v** or −**v**, as shown in Figure 8.3.6.

Step 1 Find the projection of **u** onto **v**.

$$\text{proj}_{\mathbf{v}}\mathbf{u} = \left(\frac{\mathbf{u} \cdot \mathbf{v}}{|\mathbf{v}|^2}\right)\mathbf{v}$$

$$= \frac{\langle 4, -3 \rangle \cdot \langle 2, 6 \rangle}{|\langle 2, 6 \rangle|^2}\langle 2, 6 \rangle$$

$$= \frac{-10}{40}\langle 2, 6 \rangle \text{ or } \left\langle -\frac{1}{2}, -\frac{3}{2} \right\rangle$$

Step 2 Find $\mathbf{w}_2$.

Since $\mathbf{u} = \mathbf{w}_1 + \mathbf{w}_2$, $\mathbf{w}_2 = \mathbf{u} - \mathbf{w}_1$ or $\mathbf{u} - \text{proj}_{\mathbf{v}}\mathbf{u}$.

$$\mathbf{u} - \text{proj}_{\mathbf{v}}\mathbf{u} = \langle 4, -3 \rangle - \left\langle -\frac{1}{2}, -\frac{3}{2} \right\rangle$$

$$= \left\langle \frac{9}{2}, -\frac{3}{2} \right\rangle$$

Therefore, $\text{proj}_{\mathbf{v}}\mathbf{u}$, $\mathbf{w}_1 = \left\langle -\frac{1}{2}, -\frac{3}{2} \right\rangle$ and $\mathbf{u} = \left\langle -\frac{1}{2}, -\frac{3}{2} \right\rangle + \left\langle \frac{9}{2}, -\frac{3}{2} \right\rangle$.

✓ Guided Practice

5. Find the projection of $\mathbf{u} = \langle -3, 4 \rangle$ onto $\mathbf{v} = \langle 6, 1 \rangle$. Then write **u** as the sum of two orthogonal vectors, one of which is the projection of **u** onto **v**.

Figure 8.3.7

If the vector **u** represents a force, then $\text{proj}_{\mathbf{v}}\mathbf{u}$ represents the effect of that force acting in the direction of **v**. For example, if you push a box uphill (in the direction **v**) with a force **u** (Figure 8.3.7), the effective force is a component push in the direction of **v**, $\text{proj}_{\mathbf{v}}\mathbf{u}$.

🌐 Real-World EXAMPLE 6 — Use a Vector Projection to Find a Force

CARS A 3000-pound car sits on a hill inclined at 30° as shown. Ignoring the force of friction, what force is required to keep the car from rolling down the hill?

The weight of the car is the force exerted due to gravity, $\mathbf{F} = \langle 0, -3000 \rangle$. To find the force $-\mathbf{w}_1$ required to keep the car from rolling down the hill, project **F** onto a unit vector **v** in the direction of the side of the hill.

Note: Not drawn to scale

Step 1 Find a unit vector **v** in the direction of the hill.

$$\mathbf{v} = \langle |\mathbf{v}|(\cos\theta), |\mathbf{v}|(\sin\theta) \rangle \quad \text{Component form of } \mathbf{v} \text{ in terms of } |\mathbf{v}| \text{ and } \theta$$

$$= \langle 1(\cos 30°), 1(\sin 30°) \rangle \text{ or } \left\langle \frac{\sqrt{3}}{2}, \frac{1}{2} \right\rangle \quad |\mathbf{v}| = 1 \text{ and } \theta = 30°$$

Step 2 Find $\mathbf{w}_1$, the projection of **F** onto unit vector **v**, $\text{proj}_{\mathbf{v}}\mathbf{F}$.

$$\text{proj}_{\mathbf{v}}\mathbf{F} = \left(\frac{\mathbf{F} \cdot \mathbf{v}}{|\mathbf{v}|^2}\right)\mathbf{v} \quad \text{Projection of } \mathbf{F} \text{ onto } \mathbf{v}$$

$$= (\mathbf{F} \cdot \mathbf{v})\mathbf{v} \quad \text{Since } \mathbf{v} \text{ is a unit vector, } |\mathbf{v}| = 1. \text{ Simplify.}$$

$$= \left(\langle 0, -3000 \rangle \cdot \left\langle \frac{\sqrt{3}}{2}, \frac{1}{2} \right\rangle\right)\mathbf{v} \quad \mathbf{F} = \langle 0, -3000 \rangle \text{ and } \mathbf{v} = \left\langle \frac{\sqrt{3}}{2}, \frac{1}{2} \right\rangle$$

$$= -1500\mathbf{v} \quad \text{Find the dot product.}$$

The force required is $-\mathbf{w}_1 = -(-1500\,\mathbf{v})$ or $1500\mathbf{v}$. Since **v** is a unit vector, this means that this force has a magnitude of 1500 pounds and is in the direction of the side of the hill.

✓ Guided Practice

6. SLEDDING Mary sits on a sled on the side of a hill inclined at 60°. What force is required to keep the sled from sliding down the hill if the weight of Mary and the sled is 125 pounds?

Another application of vector projection is the calculation of the work done by a force. Consider a constant force **F** acting on an object to move it from point A to point B as shown in Figure 8.3.8. If **F** is parallel to $\overrightarrow{AB}$, then the **work** W done by **F** is the magnitude of the force times the distance from A to B or $W = |\mathbf{F}||\overrightarrow{AB}|$.

| Figure 8.3.8 | Figure 8.3.9 |

To calculate the work done by a constant force **F** in *any* direction to move an object from point A to B, as shown in Figure 8.3.9 you can use the vector projection of **F** onto $\overrightarrow{AB}$.

$W = |\text{proj}_{\overrightarrow{AB}} \mathbf{F}| \, |\overrightarrow{AB}|$ **Projection formula for work**

$\quad = |\mathbf{F}| (\cos \theta) |\overrightarrow{AB}|$ $\cos \theta = \dfrac{|\text{proj}_{\overrightarrow{AB}} \mathbf{F}|}{|\mathbf{F}|}$, so $|\text{proj}_{\overrightarrow{AB}} \mathbf{F}| = |\mathbf{F}| \cos \theta$.

$\quad = \mathbf{F} \cdot \overrightarrow{AB}$ $\cos \theta = \dfrac{\mathbf{F} \cdot \overrightarrow{AB}}{|\mathbf{F}| \, |\overrightarrow{AB}|}$, so $|\mathbf{F}| \, |\overrightarrow{AB}| \cos \theta = \mathbf{F} \cdot \overrightarrow{AB}$.

Therefore, this work can be calculated by finding the dot product of the constant force **F** and the directed distance $\overrightarrow{AB}$.

🌐 Real-World EXAMPLE 7   Calculate Work

ROADSIDE ASSISTANCE A person pushes a car with a constant force of 120 newtons at a constant angle of 45° as shown. Find the work done in joules moving the car 10 meters.

Method 1   Use the projection formula for work.

The magnitude of the projection of **F** onto $\overrightarrow{AB}$ is $|\mathbf{F}| \cos \theta = 120 \cos 45°$. The magnitude of the directed distance $\overrightarrow{AB}$ is 10.

$W = |\text{proj}_{\overrightarrow{AB}} \mathbf{F}||\overrightarrow{AB}|$ **Projection formula for work**

$\quad = (120 \cos 45°)(10)$ or about 848.5   **Substitution**

Method 2   Use the dot product formula for work.

The component form of the force vector **F** in terms of magnitude and direction angle given is $\langle 120 \cos (-45°), 120 \sin (-45°) \rangle$. The component form of the directed distance the car is moved is $\langle 10, 0 \rangle$.

$W = \mathbf{F} \cdot \overrightarrow{AB}$ **Dot product formula for work**

$\quad = \langle 120 \cos (-45°), 120 \sin (-45°) \rangle \cdot \langle 10, 0 \rangle$ **Substitution**

$\quad = [120 \cos (-45°)](10)$ or about 848.5   **Dot product**

Therefore, the person does about 848.5 joules of work pushing the car.

StudyTip

Units for Work Work is measured in foot-pounds in the customary system of measurement and in newton-meters (N·m) or joules (J) in the metric system.

✔ Guided Practice

7. **CLEANING** Rick is pushing a vacuum cleaner with a force of 85 pounds. The handle of the vacuum cleaner makes a 60° angle with the floor. How much work in foot-pounds does he do if he pushes the vacuum cleaner 6 feet?

Exercises

Find the dot product of u and v. Then determine if u and v are orthogonal. (Example 1)

1. $u = \langle 3, -5 \rangle$, $v = \langle 6, 2 \rangle$ **2.** $u = \langle -10, -16 \rangle$, $v = \langle -8, 5 \rangle$

3. $u = \langle 9, -3 \rangle$, $v = \langle 1, 3 \rangle$ **4.** $u = \langle 4, -4 \rangle$, $v = \langle 7, 5 \rangle$

5. $u = \langle 1, -4 \rangle$, $v = \langle 2, 8 \rangle$ **6.** $u = 11i + 7j$; $v = -7i + 11j$

7. $u = \langle -4, 6 \rangle$, $v = \langle -5, -2 \rangle$ **8.** $u = 8i + 6j$; $v = -i + 2j$

9. SPORTING GOODS The vector $u = \langle 406, 297 \rangle$ gives the numbers of men's basketballs and women's basketballs, respectively, in stock at a sporting goods store. The vector $v = \langle 27.5, 15 \rangle$ gives the prices in dollars of the two types of basketballs, respectively. (Example 1)

  a. Find the dot product $u \cdot v$.

  b. Interpret the result in the context of the problem.

Use the dot product to find the magnitude of the given vector. (Example 2)

10. $m = \langle -3, 11 \rangle$ **11.** $r = \langle -9, -4 \rangle$

12. $n = \langle 6, 12 \rangle$ **13.** $v = \langle 1, -18 \rangle$

14. $p = \langle -7, -2 \rangle$ **15.** $t = \langle 23, -16 \rangle$

Find the angle θ between u and v to the nearest tenth of a degree. (Example 3)

16. $u = \langle 0, -5 \rangle$, $v = \langle 1, -4 \rangle$

17. $u = \langle 7, 10 \rangle$, $v = \langle 4, -4 \rangle$

18. $u = \langle -2, 4 \rangle$, $v = \langle 2, -10 \rangle$

19. $u = -2i + 3j$, $v = -4i - 2j$

20. $u = \langle -9, 0 \rangle$, $v = \langle -1, -1 \rangle$

21. $u = -i - 3j$, $v = -7i - 3j$

22. $u = \langle 6, 0 \rangle$, $v = \langle -10, 8 \rangle$

23. $u = -10i + j$, $v = 10i - 5j$

24. CAMPING Regina and Luis set off from their campsite to search for firewood. The path that Regina takes can be represented by $u = \langle 3, -5 \rangle$. The path that Luis takes can be represented by $v = \langle -7, 6 \rangle$. Find the angle between the pair of vectors. (Example 3)

Find the projection of u onto v. Then write u as the sum of two orthogonal vectors, one of which is the projection of u onto v. (Examples 4 and 5)

25. $u = 3i + 6j$, $v = -5i + 2j$ **26.** $u = \langle 5, 7 \rangle$, $v = \langle -4, 4 \rangle$

27. $u = \langle 8, 2 \rangle$, $v = \langle -4, 1 \rangle$ **28.** $u = 6i + j$, $v = -3i + 9j$

29. $u = \langle 2, 4 \rangle$, $v = \langle -3, 8 \rangle$ **30.** $u = \langle -5, 9 \rangle$, $v = \langle 6, 4 \rangle$

31. $u = 5i - 8j$, $v = 6i - 4j$ **32.** $u = -2i - 5j$, $v = 9i + 7j$

33. WAGON Malcolm is pulling his sister in a wagon up a small slope at an incline of 15°. If the combined weight of Malcolm's sister and the wagon is 78 pounds, what force is required to keep her from rolling down the slope? (Example 6)

34. SLIDE Isabel is going down a slide but stops herself when she notices that another student is lying hurt at the bottom of the slide. What force is required to keep her from sliding down the slide if the incline is 53° and she weighs 62 pounds? (Example 6)

35. PHYSICS Alexa is pushing a construction barrel up a ramp 1.5 meters long into the back of a truck. She is using a force of 534 newtons and the ramp is 25° from the horizontal. How much work in joules is Alexa doing? (Example 7)

36. SHOPPING Sophia is pushing a shopping cart with a force of 125 newtons at a downward angle, or angle of depression, of 52°. How much work in joules would Sophia do if she pushed the shopping cart 200 meters? (Example 7)

Find a vector orthogonal to each vector.

37. $\langle -2, -8 \rangle$ **38.** $\langle 3, 5 \rangle$

39. $\langle 7, -4 \rangle$ **40.** $\langle -1, 6 \rangle$

41. RIDES For a circular amusement park ride, the position vector r is perpendicular to the tangent velocity vector v at any point on the circle, as shown below.

Front View   **Top View**

  a. If the radius of the ride is 20 feet and the speed of the ride is constant at 40 feet per second, write the component forms of the position vector r and the tangent velocity vector v when r is at a directed angle of 35°.

  b. What method can be used to prove that the position vector and the velocity vector that you developed in part a are perpendicular? Show that the two vectors are perpendicular.

Given v and u • v, find u.

42. $\mathbf{v} = \langle 3, -6 \rangle$, $\mathbf{u} \cdot \mathbf{v} = 33$

43. $\mathbf{v} = \langle 4, 6 \rangle$, $\mathbf{u} \cdot \mathbf{v} = 38$

44. $\mathbf{v} = \langle -5, -1 \rangle$, $\mathbf{u} \cdot \mathbf{v} = -8$

45. $\mathbf{v} = \langle -2, 7 \rangle$, $\mathbf{u} \cdot \mathbf{v} = -43$

46. SCHOOL A student rolls her backpack from her Chemistry classroom to her English classroom using a force of 175 newtons.

175 N

θ

a. If she exerts 3060 joules to pull her backpack 31 meters, what is the angle of her force?

b. If she exerts 1315 joules at an angle of 60°, how far did she pull her backpack?

Determine whether each pair of vectors are *parallel*, *perpendicular*, or *neither*. Explain your reasoning.

47. $\mathbf{u} = \left\langle -\frac{2}{3}, \frac{3}{4} \right\rangle$, $\mathbf{v} = \langle 9, 8 \rangle$

48. $\mathbf{u} = \langle -1, -4 \rangle$, $\mathbf{v} = \langle 3, 6 \rangle$

49 $\mathbf{u} = \langle 5, 7 \rangle$, $\mathbf{v} = \langle -15, -21 \rangle$

50. $\mathbf{u} = \langle \sec \theta, \csc \theta \rangle$, $\mathbf{v} = \langle \csc \theta, -\sec \theta \rangle$

Find the angle between the two vectors in radians.

51. $\mathbf{u} = \cos\left(\frac{\pi}{3}\right)\mathbf{i} + \sin\left(\frac{\pi}{3}\right)\mathbf{j}$, $\mathbf{v} = \cos\left(\frac{3\pi}{4}\right)\mathbf{i} + \sin\left(\frac{3\pi}{4}\right)\mathbf{j}$

52. $\mathbf{u} = \cos\left(\frac{7\pi}{6}\right)\mathbf{i} + \sin\left(\frac{7\pi}{6}\right)\mathbf{j}$, $\mathbf{v} = \cos\left(\frac{5\pi}{4}\right)\mathbf{i} + \sin\left(\frac{5\pi}{4}\right)\mathbf{j}$

53. $\mathbf{u} = \cos\left(\frac{\pi}{6}\right)\mathbf{i} + \sin\left(\frac{\pi}{6}\right)\mathbf{j}$, $\mathbf{v} = \cos\left(\frac{2\pi}{3}\right)\mathbf{i} + \sin\left(\frac{2\pi}{3}\right)\mathbf{j}$

54. $\mathbf{u} = \cos\left(\frac{\pi}{4}\right)\mathbf{i} + \sin\left(\frac{\pi}{4}\right)\mathbf{j}$, $\mathbf{v} = \cos\left(\frac{5\pi}{6}\right)\mathbf{i} + \sin\left(\frac{5\pi}{6}\right)\mathbf{j}$

55. WORK Tommy lifts his nephew, who weighs 16 kilograms, a distance of 0.9 meter. The force of weight in newtons can be calculated using $F = mg$, where m is the mass in kilograms and g is 9.8 meters per second squared. How much work did Tommy do to lift his nephew?

The vertices of a triangle on the coordinate plane are given. Find the measures of the angles of each triangle using vectors. Round to the nearest tenth of a degree.

56. $(2, 3), (4, 7), (8, 1)$

57. $(-3, -2), (-3, -7), (3, -7)$

58. $(-4, -3), (-8, -2), (2, 1)$

59. $(1, 5), (4, -3), (-4, 0)$

Given u, |v|, and θ, the angle between u and v, find possible values of v. Round to the nearest hundredth.

60. $\mathbf{u} = \langle 4, -2 \rangle$, $|\mathbf{v}| = 10$, $45°$

61. $\mathbf{u} = \langle 3, 4 \rangle$, $|\mathbf{v}| = \sqrt{29}$, $121°$

62. $\mathbf{u} = \langle -1, -6 \rangle$, $|\mathbf{v}| = 7$, $96°$

63. $\mathbf{u} = \langle -2, 5 \rangle$, $|\mathbf{v}| = 12$, $27°$

64. CARS A car is stationary on a 9° incline. Assuming that the only forces acting on the car are gravity and the 275 newton force applied by the brakes, about how much does the car weigh?

9°

H.O.T. Problems *Use Higher-Order Thinking Skills*

65. REASONING Determine whether the statement below is *true* or *false*. Explain.

If $|\mathbf{d}|$, $|\mathbf{e}|$, and $|\mathbf{f}|$ form a Pythagorean triple, and the angles between $\mathbf{d}$ and $\mathbf{e}$ and between $\mathbf{e}$ and $\mathbf{f}$ are acute, then the angle between $\mathbf{d}$ and $\mathbf{f}$ must be a right angle. Explain your reasoning.

66. ERROR ANALYSIS Beng and Ethan are studying the properties of the dot product. Beng concludes that the dot product is associative because it is commutative; that is, $(\mathbf{u} \cdot \mathbf{v}) \cdot \mathbf{w} = \mathbf{u} \cdot (\mathbf{v} \cdot \mathbf{w})$. Ethan disagrees. Is either of them correct? Explain your reasoning.

67. REASONING Determine whether the statement below is *true* or *false*.
If $\mathbf{a}$ and $\mathbf{b}$ are both orthogonal to $\mathbf{v}$ in the plane, then $\mathbf{a}$ and $\mathbf{b}$ are parallel. Explain your reasoning.

68. CHALLENGE If $\mathbf{u}$ and $\mathbf{v}$ are perpendicular, what is the projection of $\mathbf{u}$ onto $\mathbf{v}$?

69. PROOF Show that if the angle between vectors $\mathbf{u}$ and $\mathbf{v}$ is 90°, $\mathbf{u} \cdot \mathbf{v} = 0$ using the formula for the angle between two nonzero vectors.

PROOF Prove each dot product property. Let $\mathbf{u} = \langle u_1, u_2 \rangle$, $\mathbf{v} = \langle v_1, v_2 \rangle$, and $\mathbf{w} = \langle w_1, w_2 \rangle$.

70. $\mathbf{u} \cdot \mathbf{v} = \mathbf{v} \cdot \mathbf{u}$

71. $\mathbf{u} \cdot (\mathbf{v} + \mathbf{w}) = \mathbf{u} \cdot \mathbf{v} + \mathbf{u} \cdot \mathbf{w}$

72. $k(\mathbf{u} \cdot \mathbf{v}) = k\mathbf{u} \cdot \mathbf{v} = \mathbf{u} \cdot k\mathbf{v}$

73. WRITING IN MATH Explain how to find the dot product of two nonzero vectors.

Find each of the following for $\mathbf{a} = \langle 10, 1 \rangle$, $\mathbf{b} = \langle -5, 2.8 \rangle$, and $\mathbf{c} = \left\langle \frac{3}{4}, -9 \right\rangle$. (Lesson 8-2)

74. $\mathbf{b} - \mathbf{a} + 4\mathbf{c}$

75. $\mathbf{c} - 3\mathbf{a} + \mathbf{b}$

76. $2\mathbf{a} - 4\mathbf{b} + \mathbf{c}$

77. GOLF Jada drives a golf ball with a velocity of 205 feet per second at an angle of 32° with the ground. On the same hole, James drives a golf ball with a velocity of 190 feet per second at an angle of 41°. Find the magnitudes of the horizontal and vertical components for each force. (Lesson 8-1)

Graph the hyperbola given by each equation. (Lesson 7-3)

78. $\dfrac{(x-5)^2}{48} - \dfrac{y^2}{5} = 1$

79. $\dfrac{x^2}{81} - \dfrac{y^2}{49} = 1$

80. $\dfrac{y^2}{36} - \dfrac{x^2}{4} = 1$

Find the exact value of each expression, if it exists. (Lesson 4-6)

81. $\arcsin\left(\sin\dfrac{\pi}{6}\right)$

82. $\arctan\left(\tan\dfrac{1}{2}\right)$

83. $\sin\left(\cos^{-1}\dfrac{3}{4}\right)$

Solve each equation. (Lesson 3-4)

84. $\log_{12}(x^3 + 2) = \log_{12} 127$

85. $\log_2 x = \log_2 6 + \log_2 (x - 5)$

86. $e^{5x-4} = 70$

87. ELECTRICITY A simple electric circuit contains only a power supply and a resistor. When the power supply is off, there is no current in the circuit. When the power supply is turned on, the current almost instantly becomes a constant value. This situation can be modeled by a graph like the one shown at the right. I represents current in amps, and t represents time in seconds. (Lesson 1-3)

   a. At what t-value is this function discontinuous?

   b. When was the power supply turned on?

   c. If the person who turned on the power supply left and came back hours later, what would he or she measure the current in the circuit to be?

Skills Review for Standardized Tests

88. SAT/ACT In the figure below, $\triangle PQR \sim \triangle TRS$. What is the value of x?

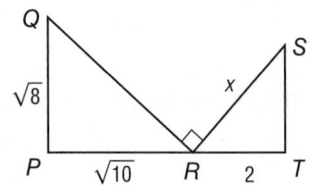

   A $\sqrt{2}$ **C** 3 **E** 6

   B $\sqrt{5}$ **D** $3\sqrt{2}$

89. REVIEW Consider $C(-9, 2)$ and $D(-4, -3)$. Which of the following is the component form and magnitude of $\overrightarrow{CD}$?

   F $\langle 5, -5 \rangle$, $5\sqrt{2}$ **H** $\langle 6, -5 \rangle$, $5\sqrt{2}$

   G $\langle 5, -5 \rangle$, $6\sqrt{2}$ **J** $\langle 6, -6 \rangle$, $6\sqrt{2}$

90. A snow sled is pulled by exerting a force of 25 pounds on a rope that makes a 20° angle with the horizontal, as shown in the figure. What is the approximate work done in pulling the sled 50 feet?

   A 428 foot-pounds   **C** 1175 foot-pounds

   B 1093 foot-pounds   **D** 1250 foot-pounds

91. REVIEW If $\mathbf{s} = \langle 4, -3 \rangle$ $\mathbf{t} = \langle -6, 2 \rangle$, which of the following represents $\mathbf{t} - 2\mathbf{s}$?

   F $\langle 14, 8 \rangle$ **H** $\langle -14, 8 \rangle$

   G $\langle 14, 6 \rangle$ **J** $\langle -14, -8 \rangle$

Find the resultant of each pair of vectors using either the triangle or parallelogram method. State the magnitude of the resultant in centimeters and its direction relative to the horizontal. (Lesson 8-1)

1.

2.

3.

4.
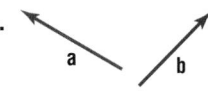

5. **SLEDDING** Alvin pulls a sled through the snow with a force of 50 newtons at an angle of 35° with the horizontal. Find the magnitude of the horizontal and vertical components of the force. (Lesson 8-1)

6. Draw a vector diagram of $\frac{1}{2}\mathbf{c} - 3\mathbf{d}$. (Lesson 8-1)

Let $\overrightarrow{BC}$ be the vector with the given initial and terminal points. Write $\overrightarrow{BC}$ as a linear combination of the vectors i and j. (Lesson 8-2)

7. $B(3, -1), C(4, -7)$

8. $B(10, -6), C(-8, 2)$

9. $B(1, 12), C(-2, -9)$

10. $B(4, -10), C(4, -10)$

11. **MULTIPLE CHOICE** Which of the following is the component form of $\overrightarrow{AB}$ with initial point $A(-5, 3)$ and terminal point $B(2, -1)$? (Lesson 8-2)

   A $\langle 4, -1 \rangle$

   B $\langle 7, -4 \rangle$

   C $\langle 7, 4 \rangle$

   D $\langle -6, 4 \rangle$

12. **BASKETBALL** With time running out in a game, Rachel runs towards the basket at a speed of 2.5 meters per second and from half-court, launches a shot at a speed of 8 meters per second at an angle of 36° to the horizontal. (Lesson 8-2)

   a. Write the component form of the vectors representing Rachel's velocity and the path of the ball.

   b. What is the resultant speed and direction of the shot?

Find the component form and magnitude of the vector with each initial and terminal point. (Lesson 8-2)

13. $A(-4, 2), B(3, 6)$

14. $Q(1, -5), R(-7, 8)$

15. $X(-3, -5), Y(2, 5)$

16. $P(9, -2), S(2, -5)$

Find the angle θ between u and v to the nearest tenth of a degree. (Lesson 8-3)

17. $\mathbf{u} = \langle 9, -4 \rangle, \mathbf{v} = \langle -1, -2 \rangle$

18. $\mathbf{u} = \langle 5, 2 \rangle, \mathbf{v} = \langle -4, 10 \rangle$

19. $\mathbf{u} = \langle 8, 4 \rangle, \mathbf{v} = \langle -2, 4 \rangle$

20. $\mathbf{u} = \langle 2, -2 \rangle, \mathbf{v} = \langle 3, 8 \rangle$

21. **MULTIPLE CHOICE** If $\mathbf{u} = \langle 2, 3 \rangle$, $\mathbf{v} = \langle -1, 4 \rangle$, and $\mathbf{w} = \langle 8, -5 \rangle$, find $(\mathbf{u} \cdot \mathbf{v}) + (\mathbf{w} \cdot \mathbf{v})$. (Lesson 8-3)

   F -18

   G -2

   H 15

   J 38

Find the dot product of u and v. Then determine if u and v are orthogonal. (Lesson 8-3)

22. $\langle 2, -5 \rangle \cdot \langle 4, 2 \rangle$

23. $\langle 4, -3 \rangle \cdot \langle 7, 4 \rangle$

24. $\langle 1, -6 \rangle \cdot \langle 5, 8 \rangle$

25. $\langle 3, -6 \rangle \cdot \langle 10, 5 \rangle$

26. **WAGON** Henry uses a wagon to carry newspapers for his paper route. He is pulling the wagon with a force of 25 newtons at an angle of 30° with the horizontal. (Lesson 8-3)

   a. How much work in joules is Henry doing when he pulls the wagon 150 meters?

   b. If the handle makes an angle of 40° with the ground and he pulls the wagon with the same distance and force, is Henry doing more or less work? Explain your answer.

Find the projection of u onto v. Then write u as the sum of two orthogonal vectors, one of which is the projection of u onto v. (Lesson 8-3)

27. $\mathbf{u} = \langle 7, -3 \rangle, \mathbf{v} = \langle 2, 5 \rangle$

28. $\mathbf{u} = \langle 2, 4 \rangle, \mathbf{v} = \langle 1, 3 \rangle$

29. $\mathbf{u} = \langle 3, 8 \rangle, \mathbf{v} = \langle -9, 2 \rangle$

30. $\mathbf{u} = \langle -1, 4 \rangle, \mathbf{v} = \langle -6, 1 \rangle$

Vectors in Three-Dimensional Space

Why?

The direction of a rocket after takeoff is given in terms of both a two-dimensional bearing and a third-dimensional angle relative to the horizontal. Since directed distance, velocities, and forces are not restricted to the plane, the concept of vectors must extend from two- to three-dimensional space.

Then
You represented vectors both geometrically and algebraically in two-dimensions.
(Lessons 8-2 and 8-3)

Now
- Plot points and vectors in the three-dimensional coordinate system.
- Express algebraically and operate with vectors in space.

New Vocabulary
three-dimensional coordinate system
z-axis
octant
ordered triple

Math Online
glencoe.com

Coordinates in Three Dimensions The Cartesian plane is a two-dimensional coordinate system made up of two perpendicular number lines, the x- and y-axes, that intersect at a point O called the origin. This system allows you to identify and locate points in a plane. We need a **three-dimensional coordinate system** to represent a point in space.

Start with the xy-plane and position it so that it gives the appearance of depth (Figure 8.4.1). Then add a third axis called the **z-axis** that passes through the origin and is perpendicular to both the x- and y-axes (Figure 8.4.2). The additional axis divides space into eight regions called **octants**. To help visualize the first octant, look at the corner of a room (Figure 8.4.3). The floor represents the xy-plane, and the walls represent the xz- and yz-planes.

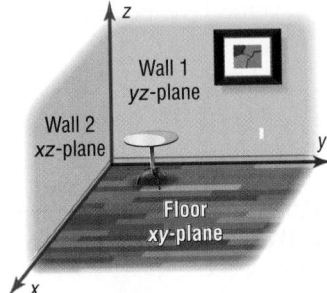

Figure 8.4.1   **Figure 8.4.2**   **Figure 8.4.3**

A point in space is represented by an **ordered triple** of real numbers (x, y, z). To plot such a point, first locate the point (x, y) in the xy-plane and move up or down parallel to the z-axis according to the directed distance given by z.

EXAMPLE 1   Locate a Point in Space

Plot each point in a three-dimensional coordinate system.

a. $(4, 6, 2)$

Locate $(4, 6)$ in the xy-plane and mark it with a cross. Then plot a point **2** units up from this location parallel to the z-axis.

b. $(-2, 4, -5)$

Locate $(-2, 4)$ in the xy-plane and mark it with a cross. Then plot a point **5** units down from this location parallel to the z-axis.

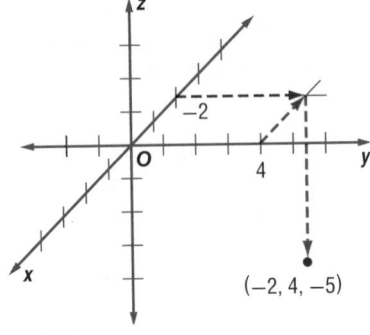

✓ Guided Practice

1A. $(-3, -4, 2)$ **1B.** $(3, 2, -3)$ **1C.** $(5, -4, -1)$

Finding the distance between points and the midpoint of a segment in space is similar to finding distance and a midpoint in the coordinate plane.

Key Concept | **Distance and Midpoint Formulas in Space**

The distance between points $A(x_1, y_1, z_1)$ and $B(x_2, y_2, z_2)$ is given by

$$AB = \sqrt{(x_2 - x_1)^2 + (y_2 - y_1)^2 + (z_2 - z_1)^2}.$$

The midpoint M of AB is given by

$$M\left(\frac{x_1 + x_2}{2}, \frac{y_1 + y_2}{2}, \frac{z_1 + z_2}{2}\right).$$

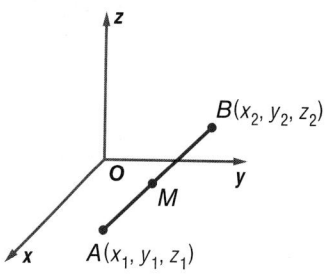

You will prove these formulas in Exercise 66.

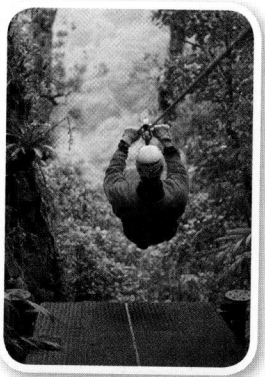

Real-World EXAMPLE 2 Distance and Midpoint of Points in Space

ZIP-LINE A tour of the Sierra Madre Mountains lets guests experience nature by zip-lining from one platform to another over the scenic surroundings. Two platforms that are connected by a zip-line are represented by the coordinates (10, 12, 50) and (70, 92, 30), where the coordinates are given in feet.

a. Find the length of the zip-line needed to connect the two platforms.

Use the Distance Formula for points in space.

$AB = \sqrt{(x_2 - x_1)^2 + (y_2 - y_1)^2 + (z_2 - z_1)^2}$ **Distance Formula**

$\quad = \sqrt{(70 - 10)^2 + (92 - 12)^2 + (30 - 50)^2}$ $(x_1, y_1, z_1) = (10, 12, 50)$ and $(x_2, y_2, z_2) = (70, 92, 30)$

$\quad \approx 101.98$ **Simplify.**

The zip-line needs to be about 102 feet long to connect the two towers.

b. An additional platform is to be built halfway between the existing platforms. Find the coordinates of the new platform.

Use the Midpoint Formula for points in space.

$\left(\dfrac{x_1 + x_2}{2}, \dfrac{y_1 + y_2}{2}, \dfrac{z_1 + z_2}{2}\right)$ **Midpoint Formula**

$= \left(\dfrac{10 + 70}{2}, \dfrac{12 + 92}{2}, \dfrac{50 + 30}{2}\right)$ or (40, 52, 40) $(x_1, y_1, z_1) = (10, 12, 50)$ and $(x_2, y_2, z_2) = (70, 92, 30)$

The coordinates of the new platform will be (40, 52, 40).

✔ Guided Practice

2. AIRPLANES Safety regulations require airplanes to be at least a half a mile apart when in the sky. Two planes are flying above Cleveland with the coordinates (300, 150, 30000) and (450, −250, 28000), where the coordinates are given in feet.

A. Are the two planes in violation of the safety regulations? Explain.

B. If a firework was launched and exploded directly in between the two planes, what are the coordinates of the firework explosion?

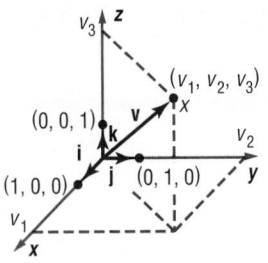

Figure 8.4.4

Vectors in Space In space, a vector **v** in standard position with a terminal point located at (v_1, v_2, v_3) is denoted by $\langle v_1, v_2, v_3 \rangle$. The zero vector is $\mathbf{0} = \langle 0, 0, 0 \rangle$, and the standard unit vectors are $\mathbf{i} = \langle 1, 0, 0 \rangle$, $\mathbf{j} = \langle 0, 1, 0 \rangle$, and $\mathbf{k} = \langle 0, 0, 1 \rangle$ as shown in Figure 8.4.4. The component form of **v** can be expressed as a linear combination of these unit vectors, $\langle v_1, v_2, v_3 \rangle = v_1\mathbf{i} + v_2\mathbf{j} + v_3\mathbf{k}$.

EXAMPLE 3 **Locate a Vector in Space**

Locate and graph $\mathbf{v} = \langle 3, 4, -2 \rangle$.

Plot the point $(3, 4, -2)$.
Draw **v** with terminal point at $(3, 4, -2)$.

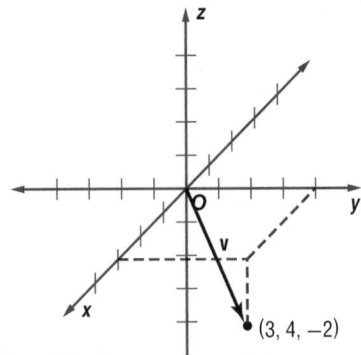

✔ Guided Practice

Locate and graph each vector in space.

3A. $\mathbf{u} = \langle -4, 2, -3 \rangle$

3B. $\mathbf{w} = -\mathbf{i} + -3\mathbf{j} + 4\mathbf{k}$

As with two-dimensional vectors, to find the component form of the directed line segment from $A(x_1, y_1, z_1)$ to $B(x_2, y_2, z_2)$, you subtract the coordinates of its initial point from its terminal point.

$$\overrightarrow{AB} = \langle x_2 - x_1, y_2 - y_1, z_2 - z_1 \rangle$$

Then $|\overrightarrow{AB}| = \sqrt{(x_2 - x_1)^2 + (y_2 - y_1)^2 + (z_2 - z_1)^2}$ or if $\overrightarrow{AB} = \langle a_1, a_2, a_3 \rangle$, then $|\overrightarrow{AB}| = \sqrt{a_1^2 + a_2^2 + a_3^2}$.

A unit vector **u** in the direction of $\overrightarrow{AB}$ is still $\mathbf{u} = \dfrac{\overrightarrow{AB}}{|\overrightarrow{AB}|}$.

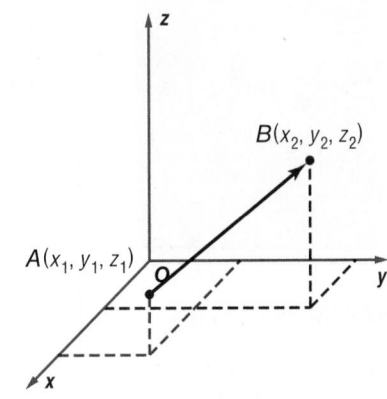

EXAMPLE 4 **Express Vectors in Space Algebraically**

Find the component form and magnitude of $\overrightarrow{AB}$ with initial point $A(-4, -2, 1)$ and terminal point $B(3, 6, -6)$. Then find a unit vector in the direction of $\overrightarrow{AB}$.

$\overrightarrow{AB} = \langle x_2 - x_1, y_2 - y_1, z_2 - z_1 \rangle$ **Component form of vector**

    $= \langle 3 - (-4), 6 - (-2), -6 - 1 \rangle$ or $\langle 7, 8, -7 \rangle$ $(x_1, y_1, z_1) = (-4, -2, 1)$ and $(x_2, y_2, z_2) = (3, 6, -6)$

Using the component form, the magnitude of $\overrightarrow{AB}$ is

$|\overrightarrow{AB}| = \sqrt{7^2 + 8^2 + (-7)^2}$ or $9\sqrt{2}$. $\overrightarrow{AB} = \langle 7, 8, -7 \rangle$

Using this magnitude and component form, find a unit vector **u** in the direction of $\overrightarrow{AB}$.

$\mathbf{u} = \dfrac{\overrightarrow{AB}}{|\overrightarrow{AB}|}$ **Unit vector in the direction of $\overrightarrow{AB}$**

$= \dfrac{\langle 7, 8, -7 \rangle}{9\sqrt{2}}$ or $\left\langle \dfrac{7\sqrt{2}}{18}, \dfrac{4\sqrt{2}}{9}, -\dfrac{7\sqrt{2}}{18} \right\rangle$ $\overrightarrow{AB} = \langle 7, 8, -7 \rangle$ and $|\overrightarrow{AB}| = 9\sqrt{2}$

✔ Guided Practice

Find the component form and magnitude of $\overrightarrow{AB}$ with the given initial and terminal points. Then find a unit vector in the direction of $\overrightarrow{AB}$.

4A. $A(-2, -5, -5)$, $B(-1, 4, -2)$ **4B.** $A(-1, 4, 6)$, $B(3, 3, 8)$

As with vectors in the plane, when vectors in space are in component form or expressed as a linear combination of unit vectors, they can be added, subtracted, or multiplied by a scalar.

Key Concept | **Vector Operations in Space**

If $\mathbf{a} = \langle a_1, a_2, a_3 \rangle$, $\mathbf{b} = \langle b_1, b_2, b_3 \rangle$, and any scalar k, then

Vector Addition $\mathbf{a} + \mathbf{b} = \langle a_1 + b_1, a_2 + b_2, a_3 + b_3 \rangle$

Vector Subtraction $\mathbf{a} - \mathbf{b} = \mathbf{a} + (-\mathbf{b}) = \langle a_1 - b_1, a_2 - b_2, a_3 - b_3 \rangle$

Scalar Multiplication $k\mathbf{a} = \langle ka_1, ka_2, ka_3 \rangle$

StudyTip

Vector Operations
The properties for vector operations in space are the same as those for operations in the plane.

EXAMPLE 5 | **Operations with Vectors in Space**

Find each of the following for $\mathbf{y} = \langle 3, -6, 2 \rangle$, $\mathbf{w} = \langle -1, 4, -4 \rangle$, and $\mathbf{z} = \langle -2, 0, 5 \rangle$.

a. $4\mathbf{y} + 2\mathbf{z}$

$4\mathbf{y} + 2\mathbf{z} = 4\langle 3, -6, 2 \rangle + 2\langle -2, 0, 5 \rangle$ **Substitute.**

$= \langle 12, -24, 8 \rangle + \langle -4, 0, 10 \rangle$ or $\langle 8, -24, 18 \rangle$ **Scalar multiplication and vector addition**

b. $2\mathbf{w} - \mathbf{z} + 3\mathbf{y}$

$2\mathbf{w} - \mathbf{z} + 3\mathbf{y} = 2\langle -1, 4, -4 \rangle - \langle -2, 0, 5 \rangle + 3\langle 3, -6, 2 \rangle$ **Substitute.**

$= \langle -2, 8, -8 \rangle + \langle 2, 0, -5 \rangle + \langle 9, -18, 6 \rangle$ **Scalar multiplication**

$= \langle 9, -10, -7 \rangle$ **Vector addition**

✔ **Guided Practice**

5A. $4\mathbf{w} - 8\mathbf{z}$ **5B.** $3\mathbf{y} + 3\mathbf{z} - 6\mathbf{w}$

🌐 **Real-World EXAMPLE 6** | **Use Vectors in Space**

ROCKETS After liftoff, a model rocket is headed due north and climbing at an angle of 75° relative to the horizontal at 200 miles per hour. If the wind blows from the northwest at 5 miles per hour, find a vector for the resultant velocity of the rocket relative to the point of liftoff.

Let **i** point east, **j** point north, and **k** point up. Vector **v** representing the rocket's velocity and vector **w** representing the wind's velocity are shown. Notice that **w** points toward the southeast, since the wind is blowing *from* the northwest.

Since **v** has a magnitude of 200 and a direction angle of 75°, we can find the component form of **v**, as shown in Figure 8.4.5.

$\mathbf{v} = \langle 0, 200 \cos 75°, 200 \sin 75° \rangle$ or about $\langle 0, 51.8, 193.2 \rangle$

With east as the positive x-axis, **w** has direction angle of 315°. Since $|\mathbf{w}| = 5$, the component form of this vector is $\mathbf{w} = \langle 5 \cos 315°, 5 \sin 315°, 0 \rangle$ or about $\langle 3.5, -3.5, 0 \rangle$, as shown in Figure 8.4.6.

The resultant velocity of the rocket is $\mathbf{v} + \mathbf{w}$.

$\mathbf{v} + \mathbf{w} = \langle 0, 51.8, 193.2 \rangle + \langle 3.5, -3.5, 0 \rangle$

$= \langle 3.5, 48.3, 193.2 \rangle$ or $3.5\mathbf{i} + 48.3\mathbf{j} + 193.2\mathbf{k}$

Figure 8.4.5

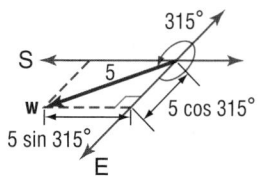

Figure 8.4.6

✔ **Guided Practice**

6. AVIATION After takeoff, an airplane is headed east and is climbing at an angle of 18° relative to the horizontal. Its air speed is 250 miles per hour. If the wind blows from the northeast at 10 miles per hour, find a vector that represents the resultant velocity of the plane relative to the point of takeoff. Let **i** point east, **j** point north, and **k** point up.

Exercises

= **Step-by-Step Solutions** begin on page R29.

Plot each point in a three-dimensional coordinate system. (Example 1)

1. $(1, -2, -4)$ **2.** $(3, 2, 1)$

3. $(-5, -4, -2)$ **4.** $(-2, -5, 3)$

5. $(-5, 3, 1)$ **6.** $(2, -2, 3)$

7. $(4, -10, -2)$ **8.** $(-16, 12, -13)$

Find the length and midpoint of the segment with the given endpoints. (Example 2)

9. $(-4, 10, 4), (1, 0, 9)$ **10.** $(-6, 6, 3), (-9, -2, -2)$

11. $(6, 1, 10), (-9, -10, -4)$ **12.** $(8, 3, 4), (-4, -7, 5)$

13. $(-3, 2, 8), (9, 6, 0)$ **14.** $(-7, 2, -5), (-2, -5, -8)$

15. VACATION A family from Wichita, Kansas, is using a GPS device to plan a vacation to Castle Rock, Colorado. According to the device, the coordinates for the family's home are $(37.7°, 97.2°, 433 \text{ m})$, and the coordinates to Castle Rock are $(39.4°, 104.8°, 1981 \text{ m})$. Determine the longitude, latitude, and altitude of the halfway point between Wichita and Castle Rock. (Example 2)

16. FIGHTER PILOTS During a training session, the location of two F-18 fighter jets are represented by the coordinates $(675, -121, 19,300)$ and $(-289, 715, 16,100)$, where the coordinates are given in feet. (Example 2)

   a. Determine the distance between the two jets.

   b. To what location would one of the fighter pilots have to fly the F-18 in order to reduce the distance between the two jets by half?

Locate and graph each vector in space. (Example 3)

17. $\mathbf{a} = \langle 0, -4, 4 \rangle$ **18.** $\mathbf{b} = \langle -3, -3, -2 \rangle$

19. $\mathbf{c} = \langle -1, 3, -4 \rangle$ **20.** $\mathbf{d} = \langle 4, -2, -3 \rangle$

21. $\mathbf{v} = 6\mathbf{i} + 8\mathbf{j} + -2\mathbf{k}$ **22.** $\mathbf{w} = -10\mathbf{i} + 5\mathbf{k}$

23. $\mathbf{m} = 7\mathbf{i} + -6\mathbf{j} + 6\mathbf{k}$ **24.** $\mathbf{n} = \mathbf{i} + -4\mathbf{j} + -8\mathbf{k}$

Find the component form and magnitude of $\overrightarrow{AB}$ with the given initial and terminal points. Then find a unit vector in the direction of $\overrightarrow{AB}$. (Example 4)

25. $A(-5, -5, -9), B(11, -3, -1)$

26. $A(-4, 0, -3), B(-4, -8, 9)$

27. $A(3, 5, 1), B(0, 0, -9)$

28. $A(-3, -7, -12), B(-7, 1, 8)$

29. $A(2, -5, 4), B(1, 3, -6)$

30. $A(8, 12, 7), B(2, -3, 11)$

31. $A(3, 14, -5), B(7, -1, 0)$

32. $A(1, -18, -13), B(21, 14, 29)$

33. $A(-5, 12, 17), B(6, -11, 4)$

34. $A(9, 3, 7), B(-5, -7, 2)$

35) TETHERBALL In the game of tetherball, a ball is attached to a 10-foot pole by a length of rope. Two players hit the ball in opposing directions in attempt to wind the entire length of rope around the pole. To serve, a certain player holds the ball so that its coordinates are $(5, 3.6, 4.7)$ and the coordinates of the end of the rope connected to the pole are $(0, 0, 9.8)$, where the coordinates are given in feet. Determine the magnitude of the vector representing the length of the rope. (Example 4)

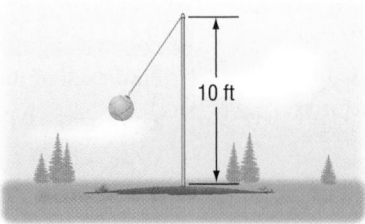

10 ft

Find each of the following for $\mathbf{a} = \langle -5, -4, 3 \rangle$, $\mathbf{b} = \langle 6, -2, -7 \rangle$, and $\mathbf{c} = \langle -2, 2, 4 \rangle$. (Example 5)

36. $6\mathbf{a} - 7\mathbf{b} + 8\mathbf{c}$ **37.** $7\mathbf{a} - 5\mathbf{b}$

38. $2\mathbf{a} + 5\mathbf{b} - 9\mathbf{c}$ **39.** $6\mathbf{b} + 4\mathbf{c} - 4\mathbf{a}$

40. $8\mathbf{a} - 5\mathbf{b} - \mathbf{c}$ **41.** $-6\mathbf{a} + \mathbf{b} + 7\mathbf{c}$

Find each of the following for $\mathbf{x} = -9\mathbf{i} + 4\mathbf{j} + 3\mathbf{k}$, $\mathbf{y} = 6\mathbf{i} - 2\mathbf{j} - 7\mathbf{k}$, and $\mathbf{z} = -2\mathbf{i} + 2\mathbf{j} + 4\mathbf{k}$. (Example 5)

42. $7\mathbf{x} + 6\mathbf{y}$ **43.** $3\mathbf{x} - 5\mathbf{y} + 3\mathbf{z}$

44. $4\mathbf{x} + 3\mathbf{y} + 2\mathbf{z}$ **45.** $-8\mathbf{x} - 2\mathbf{y} + 5\mathbf{z}$

46. $-6\mathbf{y} - 9\mathbf{z}$ **47.** $-\mathbf{x} - 4\mathbf{y} - \mathbf{z}$

48. AIRPLANES An airplane is taking off headed due north with an air speed of 150 miles per hour at an angle of 20° relative to the horizontal. The wind is blowing with a velocity of 8 miles per hour from the southwest. Find a vector that represents the resultant velocity of the plane relative to the point of takeoff. Let $\mathbf{i}$ point east, $\mathbf{j}$ point north, and $\mathbf{k}$ point up. (Example 6)

W   20°   N   150 mph

49. TRACK AND FIELD Lena throws a javelin due south at a speed of 18 miles per hour and at an angle of 48° relative to the horizontal. If the wind is blowing with a velocity of 12 miles per hour at an angle of S15°E, find a vector that represents the resultant velocity of the javelin. Let $\mathbf{i}$ point east, $\mathbf{j}$ point north, and $\mathbf{k}$ point up. (Example 6)

50. SUBMARINE A submarine heading due west dives at a speed of 25 knots and an angle of decline of 55°. The current is moving with a velocity of 4 knots at an angle of S20°W. Find a vector that represents the resultant velocity of the submarine relative to the initial point of the dive. Let **i** point east, **j** point north, and **k** point up. (Example 6)

If N is the midpoint of $\overline{MP}$, find P.

51 $M(3, 4, 5)$; $N\left(\frac{7}{2}, 1, 2\right)$

52. $M(-1, -4, -9)$; $N(-2, 1, -5)$

53. $M(7, 1, 5)$; $N\left(5, -\frac{1}{2}, 6\right)$

54. $M\left(\frac{3}{2}, -5, 9\right)$; $N\left(-2, -\frac{13}{2}, \frac{11}{2}\right)$

55. VOLUNTEERING Jody is volunteering to help guide a balloon in a parade. If the balloon is 35 feet high and she is holding the tether three feet above the ground as shown, how long is the tether to the nearest foot?

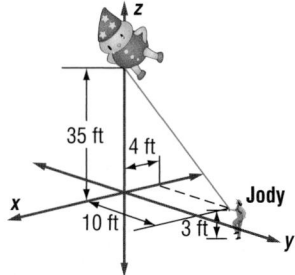

Determine whether the triangle with the given vertices is isosceles or scalene.

56. $A(3, 1, 2)$, $B(5, -1, 1)$, $C(1, 3, 1)$

57. $A(4, 3, 4)$, $B(4, 6, 4)$, $C(4, 3, 6)$

58. $A(-1, 4, 3)$, $B(2, 5, 1)$, $C(0, -6, 6)$

59. $A(-2.2, 4.3, 5.6)$, $B(0.7, 9.3, 15.6)$, $C(3.6, 14.3, 5.6)$

60. TUGBOATS Two tugboats are pulling a disabled supertanker. One of the tow lines makes an angle 23° west of north and the other makes an angle 23° east of north. Each tug exerts a constant force of 2.5×10^6 newtons depressed 15° below the point where the lines attach to the supertanker. They pull the supertanker two miles due north.

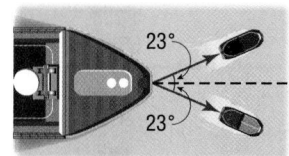

a. Write a three-dimensional vector to describe the force from each tugboat.

b. Find the vector that describes the total force on the supertanker.

c. If each tow line is 300 feet long, about how far apart are the tugboats?

61. SPHERES Use the distance formula for two points in space to prove that the standard form of the equation of a sphere with center (h, k, ℓ) and radius r is $r^2 = (x - h)^2 + (y - k)^2 + (z - \ell)^2$.

Use the formula from Exercise 61 to write an equation for the sphere with the given center and radius.

62. center $= (-4, -2, 3)$; radius $= 4$

63. center $= (6, 0, -1)$; radius $= \frac{1}{2}$

64. center $= (5, -3, 4)$; radius $= \sqrt{3}$

65. center $= (0, 7, -1)$; radius $= 12$

H.O.T. Problems   Use **H**igher-**O**rder **T**hinking Skills

66. REASONING Prove the Distance Formula in Space. (*Hint*: Use the Pythagorean Theorem twice.)

67. CHALLENGE Refer to Example 6.

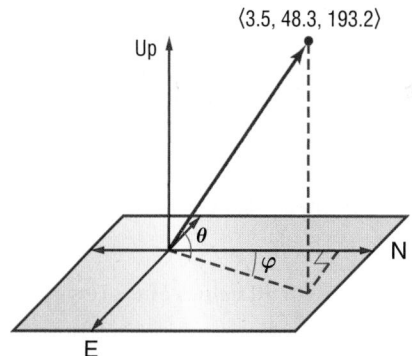

a. Calculate the resultant speed of the rocket.

b. Find the quadrant bearing φ of the rocket.

c. Calculate the resultant angle of incline θ of the rocket relative to the horizontal.

68. CHALLENGE Terri is standing in an open field facing N50°E. She is holding a kite on a 35-foot string that is flying at a 20° angle with the field. Find the components of the vector from Terri to the kite. (*Hint*: Use trigonometric ratios and two right triangles to find **x**, **y**, and **z**.)

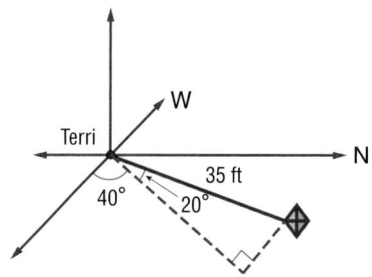

69. WRITING IN MATH Describe a situation where it is more reasonable to use a two-dimensional coordinate system and one where it is more reasonable to use a three-dimensional coordinate system.

Find the projection of u onto v. Then write u as the sum of two orthogonal vectors, one of which is the projection of u onto v. (Lesson 8-3)

70. $\mathbf{u} = \langle 6, 8 \rangle, \mathbf{v} = \langle 2, -1 \rangle$ **71.** $\mathbf{u} = \langle -1, 4 \rangle, \mathbf{v} = \langle 5, 1 \rangle$ **72.** $\mathbf{u} = \langle 5, 4 \rangle, \mathbf{v} = \langle 4, -2 \rangle$

Find the component form and magnitude of $\overrightarrow{AB}$ with the given initial and terminal points.
(Lesson 8-2)

73. $A(6, -4), B(-7, -7)$ **74.** $A(-4, -8), B(1, 6)$ **75.** $A(-5, -12), B(1, 6)$

76. ENTERTAINMENT The Independence Day fireworks at Memorial Park are fired at an angle of 82° with the horizontal. The technician firing the shells expects them to explode about 300 feet in the air 4.8 seconds after they are fired. (Lesson 7-5)

   a. Find the initial speed of a shell fired from ground level.

   b. Safety barriers will be placed around the launch area to protect spectators. If the barriers are placed 100 yards from the point directly below the explosion of the shells, how far should the barriers be from the point where the fireworks are launched?

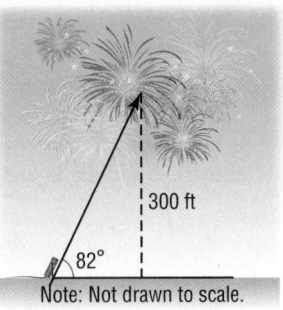

300 ft

82°

Note: Not drawn to scale.

77. CONSTRUCTION A stone fireplace that was designed as an arch in the shape of a semi-ellipse will have an opening with a height of 3 feet at the center and a width of 8 feet along the base. To sketch an outline of the fireplace, a contractor uses a string tied to two thumbtacks. (Lesson 7-2)

   a. At what locations should the thumbtacks be placed?

   b. What length of string needs to be used? Explain your reasoning.

Solve each equation for all values of θ. (Lesson 5-3)

78. $\csc \theta + 2 \cot \theta = 0$ **79.** $\sec^2 \theta - 9 = 0$ **80.** $2 \csc \theta - 3 = 5 \sin \theta$

Sketch the graph of each function. (Lesson 4-6)

81. $y = \cos^{-1}(x - 2)$ **82.** $y = \arccos x + 3$ **83.** $y = \sin^{-1} 3x$

84. SAT/ACT What percent of the area of rectangle *PQRS* is shaded?

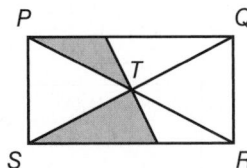

   A 22%   **C** 30%   **E** 35%

   B 25%   **D** $33\frac{1}{3}\%$

85. REVIEW A ship leaving port sails for 75 miles in a direction of 35° north of east. At that point, how far north of its starting point is the ship?

   F 43 miles   **H** 61 miles

   G 55 miles   **J** 72 miles

86. During a storm, the force of the wind blowing against a skyscraper can be expressed by the vector $\langle 132, 3454, -76 \rangle$, where the force of the wind is measured in newtons. What is the approximate magnitude of this force?

   A 3457 N   **C** 3692 N

   B 3568 N   **D** 3717 N

87. REVIEW An airplane is flying due west at a velocity of 100 meters a second. The wind is blowing from the south at 30 meters a second. What is the approximate magnitude of the airplane's resultant velocity?

   F 4 m/s   **H** 100 m/s

   G 95.4 m/s   **J** 104.4 m/s

Objective
Use a graphing calculator to transform vectors using matrices.

In Lesson 8-4, you learned that a vector in space can be transformed when written in component form or when expressed as a linear combination. A vector in space can also be transformed when written as a 3×1 or 1×3 matrix.

$$x\mathbf{i} + y\mathbf{j} + z\mathbf{k} = \begin{bmatrix} x \\ y \\ z \end{bmatrix} \text{ or } [x \ y \ z]$$

Once in matrix form, a vector can be transformed using matrix-vector multiplication.

ACTIVITY **Matrix-Vector Multiplication**

Multiply the vector $\mathbf{B} = 2\mathbf{i} - \mathbf{j} + 2\mathbf{k}$ by the transformation matrix $A = \begin{bmatrix} 3 & 0 & 0 \\ 0 & 3 & 0 \\ 0 & 0 & 3 \end{bmatrix}$, and graph both vectors.

Step 1   Write $\mathbf{B}$ as a matrix.

$$\mathbf{B} = 2\mathbf{i} - \mathbf{j} + 2\mathbf{k} = \begin{bmatrix} 2 \\ -1 \\ 2 \end{bmatrix}$$

Step 2   Enter $\mathbf{B}$ and A in a graphing calculator and find $A\mathbf{B}$. Convert to vector form.

$$A\mathbf{B} = 6\mathbf{i} - 3\mathbf{j} + 6\mathbf{k}$$

Step 3   Graph $\mathbf{B}$ and $A\mathbf{B}$ on a coordinate plane.
$A\mathbf{B}$ is a dilation of $\mathbf{B}$ by a factor of 3.

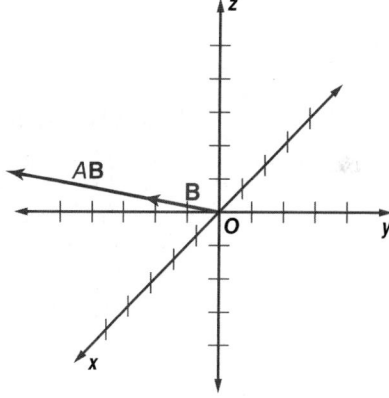

Exercises

Multiply each vector by the transformation matrix. Graph both vectors.

1. $\mathbf{h} = 4\mathbf{i} + \mathbf{j} + 8\mathbf{k}$

$$B = \begin{bmatrix} 0.25 & 0 & 0 \\ 0 & 0.25 & 0 \\ 0 & 0 & 0.25 \end{bmatrix}$$

2. $\mathbf{e} = 5\mathbf{i} + 3\mathbf{j} - 9\mathbf{k}$

$$V = \begin{bmatrix} 2 & 0 & 0 \\ 0 & 2 & 0 \\ 0 & 0 & 2 \end{bmatrix}$$

3. $\mathbf{f} = \mathbf{i} + 7\mathbf{j} - 3\mathbf{k}$

$$W = \begin{bmatrix} 3 & 0 & 0 \\ 0 & -3 & 0 \\ 0 & 0 & 3 \end{bmatrix}$$

4. REASONING Multiply $\mathbf{v} = 3\mathbf{i} - 2\mathbf{j} + 4\mathbf{k}$ by the transformation matrix $B = \begin{bmatrix} 0 & 0 & 1 \\ 0 & 1 & 0 \\ -1 & 0 & 0 \end{bmatrix}$, and graph both vectors. Explain the type of transformation that was performed.

Dot and Cross Products of Vectors in Space

Then
You found the dot product of two vectors in the plane. (Lesson 8-3)

Now
- Find dot products of and angles between vectors in space.
- Find cross products of vectors in space, and use cross products to find area and volume.

New Vocabulary
cross product
torque
parallelepiped
triple scalar product

Math Online

glencoe.com

Why?

The tendency of a hinged door to rotate when pushed is affected by the distance between the location of the push and the hinge, the magnitude of the push, and the direction of the push.

A quantity called *torque* measures how effectively a force applied to a lever causes rotation about an axis.

Dot Products in Space Calculating the dot product of two vectors in space is similar to calculating the dot product of two vectors in a plane. As with vectors in a plane, nonzero vectors in space are perpendicular if and only if their dot product equals zero.

Key Concept — Dot Product and Orthogonal Vectors in Space

The dot product of $\mathbf{a} = \langle a_1, a_2, a_3 \rangle$ and $\mathbf{b} = \langle b_1, b_2, b_3 \rangle$ is defined as

$\mathbf{a} \cdot \mathbf{b} = a_1 b_1 + a_2 b_2 + a_3 b_3$. The vectors $\mathbf{a}$ and $\mathbf{b}$ are orthogonal if and only if $\mathbf{a} \cdot \mathbf{b} = 0$.

EXAMPLE 1 Find the Dot Product to Determine Orthogonal Vectors in Space

Find the dot product of $\mathbf{u}$ and $\mathbf{v}$. Then determine if $\mathbf{u}$ and $\mathbf{v}$ are orthogonal.

a. $\mathbf{u} = \langle -7, 3, -3 \rangle$, $\mathbf{v} = \langle 5, 17, 5 \rangle$

$\mathbf{u} \cdot \mathbf{v} = -7(5) + 3(17) + (-3)(5)$

$= -35 + 51 + (-15)$ or 1

Since $\mathbf{u} \cdot \mathbf{v} \neq 0$, $\mathbf{u}$ and $\mathbf{v}$ are not orthogonal.

b. $\mathbf{u} = \langle 3, -3, 3 \rangle$, $\mathbf{v} = \langle 4, 7, 3 \rangle$

$\mathbf{u} \cdot \mathbf{v} = 3(4) + (-3)(7) + 3(3)$

$= 12 + (-21) + 9$ or 0

Since $\mathbf{u} \cdot \mathbf{v} = 0$, $\mathbf{u}$ and $\mathbf{v}$ are orthogonal.

✓ Guided Practice

1A. $\mathbf{u} = \langle 3, -5, 4 \rangle$, $\mathbf{v} = \langle 5, 7, 5 \rangle$

1B. $\mathbf{u} = \langle 4, -2, -3 \rangle$, $\mathbf{v} = \langle 1, 3, -2 \rangle$

As with vectors in a plane, if θ is the angle between nonzero vectors $\mathbf{a}$ and $\mathbf{b}$, then $\cos \theta = \dfrac{\mathbf{a} \cdot \mathbf{b}}{|\mathbf{a}|\,|\mathbf{b}|}$.

EXAMPLE 2 Angle Between Two Vectors in Space

Find the angle θ between $\mathbf{u}$ and $\mathbf{v}$ to the nearest tenth of a degree if $\mathbf{u} = \langle 3, 2, -1 \rangle$ and $\mathbf{v} = \langle -4, 3, -2 \rangle$.

$\cos \theta = \dfrac{\mathbf{u} \cdot \mathbf{v}}{|\mathbf{u}|\,|\mathbf{v}|}$ Angle between two vectors

$\cos \theta = \dfrac{\langle 3, 2, -1 \rangle \cdot \langle -4, 3, -2 \rangle}{|\langle 3, 2, -1 \rangle|\,|\langle -4, 3, -2 \rangle|}$ $\mathbf{u} = \langle 3, 2, -1 \rangle$ and $\mathbf{v} = \langle -4, 3, -2 \rangle$

$\cos \theta = \dfrac{-4}{\sqrt{14}\,\sqrt{29}}$ Evaluate the dot product and magnitudes.

$\theta = \cos^{-1} \dfrac{-4}{\sqrt{406}}$ or about $101.5°$ Simplify and solve for θ.

The measure of the angle between $\mathbf{u}$ and $\mathbf{v}$ is about $101.5°$.

✓ Guided Practice

2. Find the angle between $\mathbf{u} = -4\mathbf{i} + 2\mathbf{j} + \mathbf{k}$ and $\mathbf{v} = 4\mathbf{i} + 3\mathbf{k}$ to the nearest tenth of a degree.

Cross Products Another important product involving vectors in space is the cross product. Unlike the dot product, the **cross product** of two vectors **a** and **b** in space, denoted **a** × **b** and read *a cross b*, is a vector, not a scalar. The vector **a** × **b** is perpendicular to the plane containing vectors **a** and **b**.

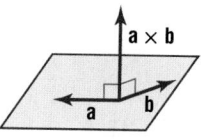

Key Concept   **Cross Product of Vectors in Space**

If $\mathbf{a} = a_1\mathbf{i} + a_2\mathbf{j} + a_3\mathbf{k}$ and $\mathbf{b} = b_1\mathbf{i} + b_2\mathbf{j} + b_3\mathbf{k}$, the cross product of **a** and **b** is the vector

$$\mathbf{a} \times \mathbf{b} = (a_2b_3 - a_3b_2)\mathbf{i} - (a_1b_3 - a_3b_1)\mathbf{j} + (a_1b_2 - a_2b_1)\mathbf{k}.$$

If we apply the formula for calculating the determinant of a 3 × 3 matrix to the following *determinant form* involving **i**, **j**, **k**, and the components of **a** and **b**, we arrive at the same formula for **a** × **b**.

$$\mathbf{a} \times \mathbf{b} = \begin{vmatrix} \mathbf{i} & \mathbf{j} & \mathbf{k} \\ a_1 & a_2 & a_3 \\ b_1 & b_2 & b_3 \end{vmatrix}$$

← Put the unit vectors **i**, **j**, and **k** in Row 1.
← Put the components of **a** in Row 2.
← Put the components of **b** in Row 3.

$$= \begin{vmatrix} a_2 & a_3 \\ b_2 & b_3 \end{vmatrix} \mathbf{i} - \begin{vmatrix} a_1 & a_3 \\ b_1 & b_3 \end{vmatrix} \mathbf{j} + \begin{vmatrix} a_1 & a_2 \\ b_1 & b_2 \end{vmatrix} \mathbf{k}$$

Apply the formula for a 3 × 3 determinant.

$$= (a_2b_3 - a_3b_2)\mathbf{i} - (a_1b_3 - a_3b_1)\mathbf{j} + (a_1b_2 - a_2b_1)\mathbf{k}$$

Compute each 2 × 2 determinant.

EXAMPLE 3   **Find the Cross Product of Two Vectors**

Find the cross product of $\mathbf{u} = \langle 3, -2, 1 \rangle$ and $\mathbf{v} = \langle -3, 3, 1 \rangle$. Then show that **u** × **v** is orthogonal to both **u** and **v**.

$$\mathbf{u} \times \mathbf{v} = \begin{vmatrix} \mathbf{i} & \mathbf{j} & \mathbf{k} \\ 3 & -2 & 1 \\ -3 & 3 & 1 \end{vmatrix}$$

$\mathbf{u} = 3\mathbf{i} - 2\mathbf{j} + \mathbf{k}$ and $\mathbf{v} = -3\mathbf{i} + 3\mathbf{j} + \mathbf{k}$

$$= \begin{vmatrix} -2 & 1 \\ 3 & 1 \end{vmatrix} \mathbf{i} - \begin{vmatrix} 3 & 1 \\ -3 & 1 \end{vmatrix} \mathbf{j} + \begin{vmatrix} 3 & -2 \\ -3 & 3 \end{vmatrix} \mathbf{k}$$

Determinant of a 3 × 3 matrix

$$= (-2 - 3)\mathbf{i} - [3 - (-3)]\mathbf{j} + (9 - 6)\mathbf{k}$$

Determinants of 2 × 2 matrices

$$= -5\mathbf{i} - 6\mathbf{j} + 3\mathbf{k}$$

Simplify.

$$= \langle -5, -6, 3 \rangle$$

Component form

In the graph of **u**, **v**, and **u** × **v**, **u** × **v** is orthogonal to **u** and **u** × **v** is orthogonal to **v**.

To show that **u** × **v** is orthogonal to both **u** and **v**, find the dot product of **u** × **v** with **u** and **u** × **v** with **v**.

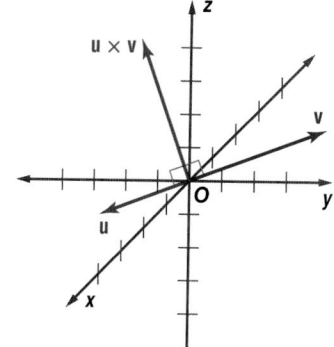

$(\mathbf{u} \times \mathbf{v}) \cdot \mathbf{u}$

$= \langle -5, -6, 3 \rangle \cdot \langle 3, -2, 1 \rangle$

$= -5(3) + (-6)(-2) + 3(1)$

$= -15 + 12 + 3 \text{ or } 0 ✔$

$(\mathbf{u} \times \mathbf{v}) \cdot \mathbf{v}$

$= \langle -5, -6, 3 \rangle \cdot \langle -3, 3, 1 \rangle$

$= -5(-3) + (-6)(3) + 3(1)$

$= 15 + (-18) + 3 \text{ or } 0 ✔$

Because both dot products are zero, the vectors are orthogonal.

☑ **Guided Practice**

Find the cross product of **u** and **v**. Then show that **u** × **v** is orthogonal to both **u** and **v**.

3A. $\mathbf{u} = \langle 4, 2, -1 \rangle$, $\mathbf{v} = \langle 5, 1, 4 \rangle$ **3B.** $\mathbf{u} = \langle -2, -1, -3 \rangle$, $\mathbf{v} = \langle 5, 1, 4 \rangle$

You can use the cross product to find a vector quantity called **torque**. Torque measures how effectively a force applied to a lever causes rotation along the axis of rotation. The torque vector **T** is perpendicular to the plane containing the directed distance **r** from the axis of rotation to the point of the applied force and the applied force **F** as shown. Therefore, the torque vector is **T** = **r** × **F** and is measured in newton-meters (N · m).

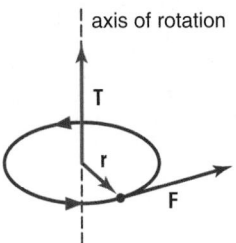
axis of rotation

Real-World EXAMPLE 4   Torque Using Cross Product

AUTO REPAIR Robert uses a lug wrench to tighten a lug nut. The wrench he uses is 50 centimeters or 0.5 meter long. Find the magnitude and direction of the torque about the lug nut if he applies a force of 25 newtons straight down to the end of the handle when it is 40° below the positive x-axis as shown.

Step 1 Graph each vector in standard position (Figure 8.5.1).

Step 2 Determine the component form of each vector.

The component form of the vector representing the directed distance from the axis of rotation to the end of the handle can be found using the triangle in Figure 8.5.2 and trigonometry. Vector **r** is therefore $\langle 0.5 \cos 40°, 0, -0.5 \sin 40° \rangle$ or about $\langle 0.38, 0, -0.32 \rangle$. The vector representing the force applied to the end of the handle is 25 newtons straight down, so **F** = $\langle 0, 0, -25 \rangle$.

Figure 8.5.1   **Figure 8.5.2**

Step 3 Use the cross product of these vectors to find the vector representing the torque about the lug nut.

$$\mathbf{T} = \mathbf{r} \times \mathbf{F}$$ Torque Cross Product Formula

$$= \begin{vmatrix} \mathbf{i} & \mathbf{j} & \mathbf{k} \\ 0.38 & 0 & -0.32 \\ 0 & 0 & -25 \end{vmatrix}$$ Cross product of **r** and **F**

$$= \begin{vmatrix} 0 & -0.32 \\ 0 & -25 \end{vmatrix} \mathbf{i} - \begin{vmatrix} 0.38 & -0.32 \\ 0 & -25 \end{vmatrix} \mathbf{j} + \begin{vmatrix} 0.38 & 0 \\ 0 & 0 \end{vmatrix} \mathbf{k}$$ Determinant of a 3 × 3 matrix

$$= 0\mathbf{i} - (-9.5)\mathbf{j} + 0\mathbf{k}$$ Determinants of 2 × 2 matrices

$$= \langle 0, 9.5, 0 \rangle$$ Component form

Step 4 Find the magnitude and direction of the torque vector.

The component form of the torque vector $\langle 0, 9.5, 0 \rangle$ tells us that the magnitude of the vector is about 9.5 newton-meters parallel to the positive y-axis as shown.

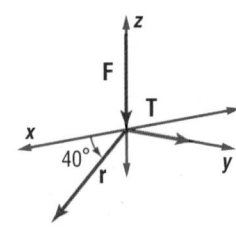

✓ Guided Practice

4. **AUTO REPAIR** Find the magnitude of the torque if Robert applied the same amount of force to the end of the handle straight down when the handle makes a 40° angle above the positive x-axis as shown in Figure 8.5.3.

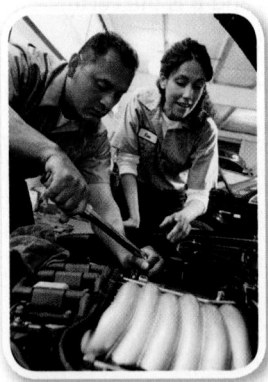

Figure 8.5.3

The cross product of two vectors has several geometric applications. One is that the magnitude of **u** × **v** represents the area of the parallelogram that has **u** and **v** as its adjacent sides (Figure 8.5.4).

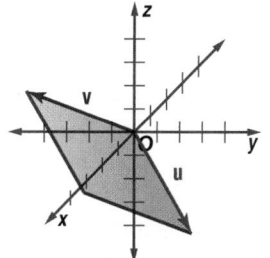

Figure 8.5.4

EXAMPLE 5   Area of a Parallelogram in Space

Find the area of the parallelogram with adjacent sides **u** = 2**i** + 4**j** − 3**k** and **v** = **i** − 5**j** + 3**k**.

Step 1 Find **u** × **v**.

$$\mathbf{u} \times \mathbf{v} = \begin{vmatrix} \mathbf{i} & \mathbf{j} & \mathbf{k} \\ 2 & 4 & -3 \\ 1 & -5 & 3 \end{vmatrix} \qquad \text{**u** = 2**i** + 4**j** − 3**k** and **v** = **i** − 5**j** + 3**k**}$$

$$= \begin{vmatrix} 4 & -3 \\ -5 & 3 \end{vmatrix}\mathbf{i} - \begin{vmatrix} 2 & -3 \\ 1 & 3 \end{vmatrix}\mathbf{j} + \begin{vmatrix} 2 & 4 \\ 1 & -5 \end{vmatrix}\mathbf{k} \qquad \text{Determinant of a 3 × 3 matrix}$$

$$= -3\mathbf{i} - 9\mathbf{j} - 14\mathbf{k} \qquad \text{Determinants of 2 × 2 matrices}$$

Step 2 Find the magnitude of **u** × **v**.

$$|\mathbf{u} \times \mathbf{v}| = \sqrt{(-3)^2 + (-9)^2 + (-14)^2} \qquad \text{Magnitude of a vector in space}$$

$$= \sqrt{286} \text{ or about } 16.9 \qquad \text{Simplify.}$$

The area of the parallelogram shown in Figure 8.5.4 is about 16.9 square units.

✔ Guided Practice

5. Find the area of the parallelogram with adjacent sides **u** = −6**i** −2**j** + 3**k** and **v** = 4**i** +3**j** + **k**.

Three vectors that lie in different planes but share the same initial point determine the adjacent edges of a **parallelepiped** (par-uh-lel-uh-PIE-ped), a polyhedron with faces that are all parallelograms (Figure 8.5.5). The absolute value of the **triple scalar product** of these vectors represents the volume of the parallelepiped.

Key Concept   Triple Scalar Product

If **t** = t_1**i** + t_2**j** + t_3**k**, **u** = u_1**i** + u_2**j** + u_3**k**, **v** = v_1**i** + v_2**j** + v_3**k**, the triple scalar product is

$$\text{given by } \mathbf{t} \cdot (\mathbf{u} \times \mathbf{v}) = \begin{vmatrix} t_1 & t_2 & t_3 \\ u_1 & u_2 & u_3 \\ v_1 & v_2 & v_3 \end{vmatrix}.$$

EXAMPLE 6   Volume of a Parallelepiped

Find the volume of the parallelepiped with adjacent edges **t** = 4**i** − 2**j** − 2**k**, **u** = 2**i** + 4**j** − 3**k**, and **v** = **i** − 5**j** + 3**k**.

$$\mathbf{t} \cdot (\mathbf{u} \times \mathbf{v}) = \begin{vmatrix} 4 & -2 & -2 \\ 2 & 4 & -3 \\ 1 & -5 & 3 \end{vmatrix} \qquad \begin{array}{l} \text{**t** = 4**i** − 2**j** − 2**k**} \\ \text{**u** = 2**i** + 4**j** − 3**k**} \\ \text{and **v** = **i** − 5**j** + 3**k**} \end{array}$$

$$= \begin{vmatrix} 4 & -3 \\ -5 & 3 \end{vmatrix}(4) - \begin{vmatrix} 2 & -3 \\ 1 & 3 \end{vmatrix}(-2) + \begin{vmatrix} 2 & 4 \\ 1 & -5 \end{vmatrix}(-2) \qquad \text{Determinant of a 3 × 3 matrix}$$

$$= -12 + 18 + 28 \text{ or } 34 \qquad \text{Simplify.}$$

The volume of the parallelepiped shown in Figure 8.5.5 is |**t** · (**u** × **v**)| or 34 cubic units.

✔ Guided Practice

6. Find the volume of the parallelepiped with adjacent edges **t** = 2**j** − 5**k**, **u** = −6**i** − 2**j** + 3**k** and **v** = 4**i** + 3**j** + **k**.

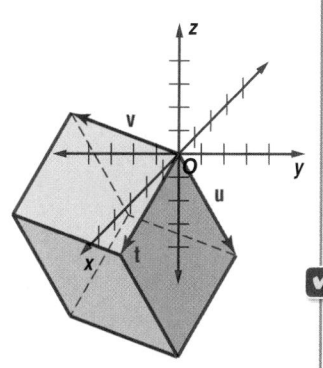

Figure 8.5.5

Exercises

Find the dot product of u and v. Then determine if u and v are orthogonal. (Example 1)

1. $\mathbf{u} = \langle 3, -9, 6 \rangle$, $\mathbf{v} = \langle -8, 2, 7 \rangle$

2. $\mathbf{u} = \langle 5, 0, -4 \rangle$, $\mathbf{v} = \langle 6, -1, 4 \rangle$

3. $\mathbf{u} = \langle 2, -8, -7 \rangle$, $\mathbf{v} = \langle 5, 9, -7 \rangle$

4. $\mathbf{u} = \langle -7, -3, 1 \rangle$, $\mathbf{v} = \langle -4, 5, -13 \rangle$

5. $\mathbf{u} = \langle 11, 4, -2 \rangle$, $\mathbf{v} = \langle -1, 3, 8 \rangle$

6. $\mathbf{u} = 6\mathbf{i} - 2\mathbf{j} - 5\mathbf{k}$, $\mathbf{v} = 3\mathbf{i} - 2\mathbf{j} + 6\mathbf{k}$

7. $\mathbf{u} = 3\mathbf{i} - 10\mathbf{j} + \mathbf{k}$, $\mathbf{v} = 7\mathbf{i} + 2\mathbf{j} - \mathbf{k}$

8. $\mathbf{u} = 9\mathbf{i} - 9\mathbf{j} + 6\mathbf{k}$, $\mathbf{v} = 6\mathbf{i} + 4\mathbf{j} - 3\mathbf{k}$

9. CHEMISTRY A water molecule, in which the oxygen atom is centered at the origin, has one hydrogen atom centered at $\langle 55.5, 55.5, -55.5 \rangle$ and the second hydrogen atom centered at $\langle -55.5, -55.5, -55.5 \rangle$. Determine the bond angle between the vectors formed by the oxygen-hydrogen bonds. (Example 2)

Find the angle θ between vectors u and v to the nearest tenth of a degree. (Example 2)

10. $\mathbf{u} = \langle 3, -2, 2 \rangle$, $\mathbf{v} = \langle 1, 4, -7 \rangle$

11. $\mathbf{u} = \langle 6, -5, 1 \rangle$, $\mathbf{v} = \langle -8, -9, 5 \rangle$

12. $\mathbf{u} = \langle -8, 1, 12 \rangle$, $\mathbf{v} = \langle -6, 4, 2 \rangle$

13. $\mathbf{u} = \langle 10, 0, -8 \rangle$, $\mathbf{v} = \langle 3, -1, -12 \rangle$

14. $\mathbf{u} = -3\mathbf{i} + 2\mathbf{j} + 9\mathbf{k}$, $\mathbf{v} = 4\mathbf{i} + 3\mathbf{j} - 10\mathbf{k}$

15. $\mathbf{u} = -6\mathbf{i} + 3\mathbf{j} + 5\mathbf{k}$, $\mathbf{v} = -4\mathbf{i} + 2\mathbf{j} + 6\mathbf{k}$

Find the cross product of u and v. Then show that u × v is orthogonal to both u and v. (Example 3)

16. $\mathbf{u} = \langle -1, 3, 5 \rangle$, $\mathbf{v} = \langle 2, -6, -3 \rangle$

17. $\mathbf{u} = \langle 4, 7, -2 \rangle$, $\mathbf{v} = \langle -5, 9, 1 \rangle$

18. $\mathbf{u} = \langle 3, -6, 2 \rangle$, $\mathbf{v} = \langle 1, 5, -8 \rangle$

19. $\mathbf{u} = \langle 5, -8, 0 \rangle$, $\mathbf{v} = \langle -4, -2, 7 \rangle$

20. $\mathbf{u} = -2\mathbf{i} - 2\mathbf{j} + 5\mathbf{k}$, $\mathbf{v} = 7\mathbf{i} + \mathbf{j} - 6\mathbf{k}$

21. $\mathbf{u} = -4\mathbf{i} + \mathbf{j} + 8\mathbf{k}$, $\mathbf{v} = 3\mathbf{i} - 4\mathbf{j} - 3\mathbf{k}$

22. RESTAURANTS A restaurant server applies 50 newtons of force to open a door. Find the magnitude and direction of the torque about the door's hinge. (Example 4)

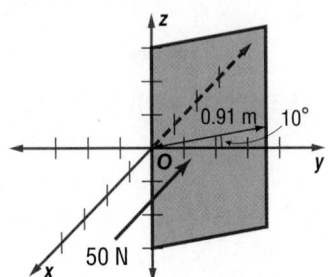

23. WEIGHTLIFTING A weightlifter doing bicep curls applies 212 newtons of force to lift the dumbbell. The weightlifter's forearm is 0.356 meters long and she begins the bicep curl with her elbow bent at a 15° angle below the horizontal in the direction of the positive x-axis. (Example 4)

a. Find the vector representing the torque about the weightlifter's elbow in component form.

b. Find the magnitude and direction of the torque.

Find the area of the parallelogram with adjacent sides u and v. (Example 5)

24. $\mathbf{u} = \langle 2, -5, 3 \rangle$, $\mathbf{v} = \langle 4, 6, -1 \rangle$

25. $\mathbf{u} = \langle -9, 1, 2 \rangle$, $\mathbf{v} = \langle 6, -5, 3 \rangle$

26. $\mathbf{u} = \langle 4, 3, -1 \rangle$, $\mathbf{v} = \langle 7, 2, -2 \rangle$

27. $\mathbf{u} = 6\mathbf{i} - 2\mathbf{j} + 5\mathbf{k}$, $\mathbf{v} = 5\mathbf{i} - 4\mathbf{j} - 8\mathbf{k}$

28. $\mathbf{u} = \mathbf{i} + 4\mathbf{j} - 8\mathbf{k}$, $\mathbf{v} = -2\mathbf{i} + 3\mathbf{j} - 7\mathbf{k}$

29. $\mathbf{u} = -3\mathbf{i} - 5\mathbf{j} + 3\mathbf{k}$, $\mathbf{v} = 4\mathbf{i} - \mathbf{j} + 6\mathbf{k}$

Find the volume of the parallelepiped having t, u, and v as adjacent edges. (Example 6)

30. $\mathbf{t} = \langle -1, -9, 2 \rangle$, $\mathbf{u} = \langle 4, -7, -5 \rangle$, $\mathbf{v} = \langle 3, -2, 6 \rangle$

31. $\mathbf{t} = \langle -6, 4, -8 \rangle$, $\mathbf{u} = \langle -3, -1, 6 \rangle$, $\mathbf{v} = \langle 2, 5, -7 \rangle$

32. $\mathbf{t} = \langle 2, -3, -1 \rangle$, $\mathbf{u} = \langle 4, -6, 3 \rangle$, $\mathbf{v} = \langle -9, 5, -4 \rangle$

33. $\mathbf{t} = -4\mathbf{i} + \mathbf{j} + 3\mathbf{k}$, $\mathbf{u} = 5\mathbf{i} + 7\mathbf{j} - 6\mathbf{k}$, $\mathbf{v} = 3\mathbf{i} - 2\mathbf{j} - 5\mathbf{k}$

34. $\mathbf{t} = \mathbf{i} + \mathbf{j} - 4\mathbf{k}$, $\mathbf{u} = -3\mathbf{i} + 2\mathbf{j} + 7\mathbf{k}$, $\mathbf{v} = 2\mathbf{i} - 6\mathbf{j} + 8\mathbf{k}$

35. $\mathbf{t} = 5\mathbf{i} - 2\mathbf{j} + 6\mathbf{k}$, $\mathbf{u} = 3\mathbf{i} - 5\mathbf{j} + 7\mathbf{k}$, $\mathbf{v} = 8\mathbf{i} - \mathbf{j} + 4\mathbf{k}$

Find a vector that is orthogonal to each vector.

36. $\langle 3, -8, 4 \rangle$

37. $\langle -1, -2, 5 \rangle$

38. $\langle 6, -\frac{1}{3}, -3 \rangle$

39. $\langle 7, 0, 8 \rangle$

Given v and u · v, find u.

40. $\mathbf{v} = \langle 2, -4, -6 \rangle$, $\mathbf{u} \cdot \mathbf{v} = -22$

41. $\mathbf{v} = \langle \frac{1}{2}, 0, 4 \rangle$, $\mathbf{u} \cdot \mathbf{v} = \frac{31}{2}$

42. $\mathbf{v} = \langle -2, -6, -5 \rangle$, $\mathbf{u} \cdot \mathbf{v} = 35$

Determine whether the points are collinear.

43. $(-1, 7, 7)$, $(-3, 9, 11)$, $(-5, 11, 13)$

44. $(11, 8, -1)$, $(17, 5, -7)$, $(8, 11, 5)$

Determine whether each pair of vectors are parallel.

45. $\mathbf{m} = \langle 2, -10, 6 \rangle$, $\mathbf{n} = \langle 3, -15, 9 \rangle$

46. $\mathbf{a} = \langle 6, 3, -7 \rangle$, $\mathbf{b} = \langle -4, -2, 3 \rangle$

47. $\mathbf{w} = \left\langle -\dfrac{3}{2}, \dfrac{3}{4}, -\dfrac{9}{8} \right\rangle$, $\mathbf{z} = \langle -4, 2, -3 \rangle$

Write the component form of each vector.

48. $\mathbf{u}$ lies in the yz-plane, has a magnitude of 8, and makes a 60° angle above the positive y-axis.

49. $\mathbf{v}$ lies in the xy-plane, has a magnitude of 11, and makes a 30° angle to the left of the negative x-axis.

Given the four vertices, determine whether quadrilateral $ABCD$ is a parallelogram. If it is, find its area, and determine whether it is a rectangle.

50. $A(3, 0, -2)$, $B(0, 4, -1)$, $C(0, 2, 5)$, $D(3, 2, 4)$

51. $A(7, 5, 5)$, $B(4, 4, 4)$, $C(4, 6, 2)$, $D(7, 7, 3)$

52. AIR SHOWS In an air show, two airplanes take off simultaneously. The first plane starts at the position $(0, -2, 0)$ and is at the position $(6, -10, 15)$ after three seconds. The second plane starts at the position $(0, 2, 0)$ and is at the position $(6, 10, 15)$ after three seconds. Are the paths of the two planes parallel? Explain.

For $\mathbf{u} = \langle 3, 2, -2 \rangle$ and $\mathbf{v} = \langle -4, 4, 5 \rangle$, find each of the following, if possible.

53. $\mathbf{u} \cdot (\mathbf{u} \times \mathbf{v})$

54. $\mathbf{v} \times (\mathbf{u} \cdot \mathbf{v})$

55. $\mathbf{u} \times \mathbf{u} \times \mathbf{v}$

56. $\mathbf{v} \cdot \mathbf{v} \cdot \mathbf{u}$

57. ELECTRICITY When a wire carrying an electric current is placed in a magnetic field, the force on the wire in newtons is given by $\vec{F} = I\vec{L} \times \vec{B}$, where I represents the current flowing through the wire in amps, $\vec{L}$ represents the vector length of the wire pointing in the direction of the current in meters, and $\vec{B}$ is the force of the magnetic field in teslas. In the figure below, the wire is rotated through an angle θ in the xy-plane.

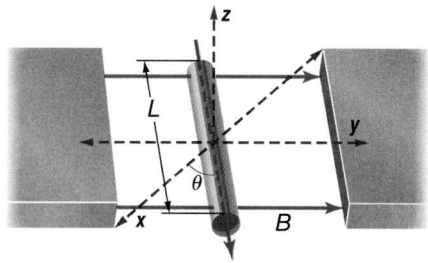

a. If the force of a magnetic field is 1.1 teslas, find the magnitude of the force on a wire in the xy-plane that is 0.15 meter in length carrying a current of 25 amps at an angle of 60°.

b. If the force on the wire is $\vec{F} = \langle 0, 0, -0.63 \rangle$, what is the angle of the wire?

Given v, w, and the volume of the parallelepiped having adjacent edges u, v, and w, find c.

58. $\mathbf{v} = \langle -2, -1, 4 \rangle$, $\mathbf{w} = \langle 1, 0, -2 \rangle$, $\mathbf{u} = \langle c, -3, 1 \rangle$, and $V = 7$ cubic units

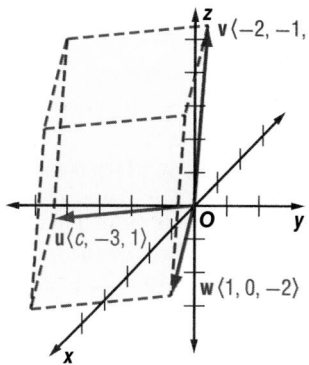

59. $\mathbf{v} = \langle 0, 5, 5 \rangle$, $\mathbf{w} = \langle 5, 0, 5 \rangle$, $\mathbf{u} = \langle 5, c, 0 \rangle$, and $V = 250$ cubic units

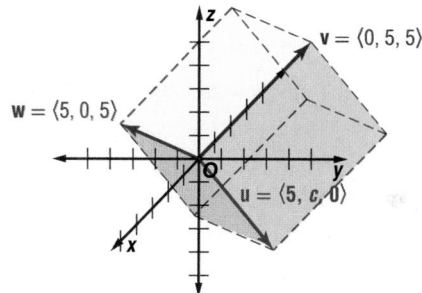

H.O.T. Problems Use Higher-Order Thinking Skills

60. PROOF Verify the formula for the volume of a parallelepiped. (*Hint*: Use the projection of $\mathbf{u}$ onto $\mathbf{v} \times \mathbf{w}$.)

61. REASONING Determine whether the following statement is *sometimes*, *always*, or *never* true. Explain.

For any two nonzero, nonparallel vectors in space, there is a vector that is perpendicular to both.

62. REASONING If $\mathbf{u}$ and $\mathbf{v}$ are parallel in space, then how many vectors are perpendicular to both? Explain.

63. CHALLENGE Given $\mathbf{u} = \langle 4, 6, c \rangle$ and $\mathbf{v} = \langle -3, -2, 5 \rangle$, find the value of c for which $\mathbf{u} \times \mathbf{v} = 34\mathbf{i} - 26\mathbf{j} + 10\mathbf{k}$.

64. REASONING Explain why the cross product is not defined for vectors in the two-dimensional coordinate system.

65. WRITING IN MATH Compare and contrast the methods for determining whether vectors in space are parallel or perpendicular.

Find the length and midpoint of the segment with the given endpoints. (Lesson 8-4)

66. $(1, 10, 13), (-2, 22, -6)$

67. $(12, -1, -14), (21, 19, -23)$

68. $(-22, 24, -9), (10, 10, 2)$

Find the dot product of u and v. Then determine if u and v are orthogonal. (Lesson 8-3)

69. $\langle -8, -7 \rangle \cdot \langle 1, 2 \rangle$

70. $\langle -4, -6 \rangle \cdot \langle 7, 5 \rangle$

71. $\langle 6, -3 \rangle \cdot \langle -3, 5 \rangle$

72. BAKERY Hector's bakery has racks that can hold up to 900 bagels and muffins. Due to costs, the number of bagels produced must be less than or equal to 300 plus twice the number of muffins produced. The demand for bagels is at least three times that of muffins. Hector makes a profit of $3 per muffin sold and $1.25 per bagel sold. How many of each item should he make to maximize profit? (Lesson 6-5)

73. Decompose $\dfrac{2m + 16}{m^2 - 16}$ into partial fractions. (Lesson 6-4)

Verify each identity. (Lesson 5-2)

74. $\tan^2 \theta + \cos^2 \theta + \sin^2 \theta = \sec^2 \theta$

75. $\sec^2 \theta \cot^2 \theta - \cot^2 \theta = 1$

76. $\tan^2 \theta - \sin^2 \theta = \tan^2 \theta \sin^2 \theta$

Solve each triangle. Round side lengths to the nearest tenth and angle measures to the nearest degree. (Lesson 4-7)

77. $a = 20, c = 24, B = 47°$

78. $A = 25°, B = 78°, a = 13.7$

79. $a = 21.5, b = 16.7, c = 10.3$

Write each decimal degree measure in DMS form and each DMS measure in decimal degree form to the nearest thousandth. (Lesson 4-2)

80. $-72.775°$

81. $29° 6' 6''$

82. $132° 18' 31''$

83. SAT/ACT The graph represents the set of all possible solutions to which of the following statements?

A $|x - 1| > 1$

C $|x + 1| < 1$

B $|x - 1| < 1$

D $|x + 1| > 1$

84. What is the cross product of $\mathbf{u} = \langle 3, 8, 0 \rangle$ and $\mathbf{v} = \langle -4, 2, 6 \rangle$?

F $48\mathbf{i} - 18\mathbf{j} + 38\mathbf{k}$

G $48\mathbf{i} - 22\mathbf{j} + 38\mathbf{k}$

H $46\mathbf{i} - 22\mathbf{j} + 38\mathbf{k}$

J $46\mathbf{i} - 18\mathbf{j} + 38\mathbf{k}$

85. FREE RESPONSE A batter hits a ball at a 30° angle with the ground at an initial speed of 90 feet per second.

a. Find the magnitude of the horizontal and vertical components of the velocity.

b. Are the values in part **a** vectors or scalars?

c. Assume that the ball is not caught and the player hit it one yard off the ground. How far will it travel in the air?

d. Assume that home plate is at the origin and second base lies due north. If the ball is hit at a bearing of N20°W and lands at point D, find the component form of $\overrightarrow{CD}$.

e. Determine the unit vector of $\overrightarrow{CD}$.

f. The fielder is standing at $(0, 150)$ when the ball is hit. At what quadrant bearing should the fielder run in order to meet the ball where it will hit the ground?

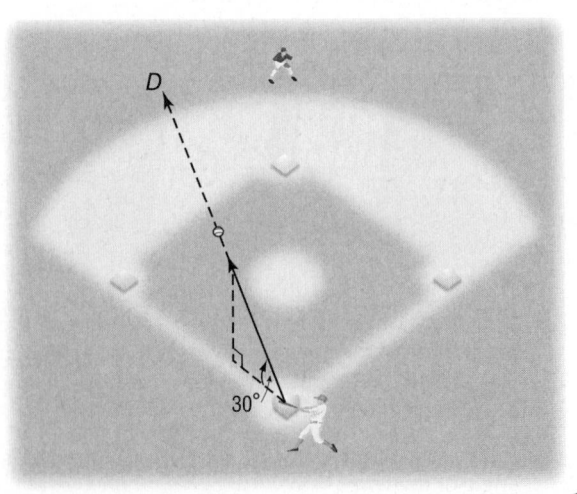

Chapter Summary

Key Concepts

Introduction to Vectors (Lesson 8-1)

- The direction of a vector is the directed angle between the vector and a horizontal line. The magnitude of a vector is its length.

- When two or more vectors are combined, their sum is a single vector called the resultant, which can be found using the triangle or parallelogram method.

Triangle Method

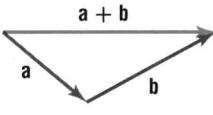

a + b

Parallelogram Method

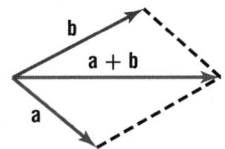

b

a + b

a

Vectors in the Coordinate Plane (Lesson 8-2)

- The component form of a vector with rectangular components x and y is $\langle x, y \rangle$.

- The component form of a vector that is not in standard position, with initial point $A(x_1, y_1)$ and terminal point $B(x_2, y_2)$, is given by $\langle x_2 - x_1, y_2 - y_1 \rangle$.

- The magnitude of a vector $\mathbf{v} = \langle v_1, v_2 \rangle$ is given by $|\mathbf{v}| = \sqrt{(x_2 - x_1)^2 + (y_2 - y_1)^2}$.

- If $\mathbf{a} = \langle a_1, a_2 \rangle$ and $\mathbf{b} = \langle b_1, b_2 \rangle$ are vectors and k is a scalar, then $\mathbf{a} + \mathbf{b} = \langle a_1 + b_1, a_2 + b_2 \rangle$, $\mathbf{a} - \mathbf{b} = \langle a_1 - b_1, a_2 - b_2 \rangle$, and $k\mathbf{a} = \langle ka_1, ka_2 \rangle$.

- The standard unit vectors $\mathbf{i}$ and $\mathbf{j}$ can be used to express any vector $\mathbf{v} = \langle a, b \rangle$ as $a\mathbf{i} + b\mathbf{j}$.

Dot Products (Lesson 8-3)

- The dot product of $\mathbf{a} = \langle a_1, a_2 \rangle$ and $\mathbf{b} = \langle b_1, b_2 \rangle$ is defined as $\mathbf{a} \cdot \mathbf{b} = a_1b_1 + a_2b_2$.

- If θ is the angle between nonzero vectors $\mathbf{a}$ and $\mathbf{b}$, then $\cos \theta = \dfrac{\mathbf{a} \cdot \mathbf{b}}{|\mathbf{a}|\,|\mathbf{b}|}$.

Vectors in Three-Dimensional Space (Lesson 8-4)

- The distance between $A(x_1, y_1, z_1)$ and $B(x_2, y_2, z_2)$ is given by $AB = \sqrt{(x_2 - x_1)^2 + (y_2 - y_1)^2 + (z_2 - z_1)^2}$.

- The midpoint of $\overline{AB}$ is given by $M\left(\dfrac{x_1 + x_2}{2}, \dfrac{y_1 + y_2}{2}, \dfrac{z_1 + z_2}{2}\right)$.

Dot and Cross Products of Vectors in Space (Lesson 8-5)

- The dot product of $\mathbf{a} = \langle a_1, a_2, a_3 \rangle$ and $\mathbf{b} = \langle b_1, b_2, b_3 \rangle$ is defined as $\mathbf{a} \cdot \mathbf{b} = a_1b_1 + a_2b_2 + a_3b_3$.

- If $\mathbf{a} = a_1\mathbf{i} + a_2\mathbf{j} + a_3\mathbf{k}$ and $\mathbf{b} = b_1\mathbf{i} + b_2\mathbf{j} + b_3\mathbf{k}$, the cross product of $\mathbf{a}$ and $\mathbf{b}$ is the vector $\mathbf{a} \times \mathbf{b} = (a_2b_3 - a_3b_2)\mathbf{i} - (a_1b_3 - a_3b_1)\mathbf{j} + (a_1b_2 - a_2b_1)\mathbf{k}$.

Key Vocabulary

component form (p. 492)
components (p. 487)
cross product (p. 519)
direction (p. 482)
dot product (p. 500)
equivalent vectors (p. 483)
initial point (p. 482)
linear combination (p. 495)
magnitude (p. 482)
octants (p. 510)
opposite vectors (p. 483)
ordered triple (p. 510)
orthogonal (p. 500)
parallelepiped (p. 521)
parallelogram method (p. 484)
parallel vectors (p. 483)
quadrant bearing (p. 483)

rectangular components (p. 487)
resultant (p. 484)
standard position (p. 482)
terminal point (p. 482)
three-dimensional coordinate system (p. 510)
torque (p. 520)
triangle method (p. 484)
triple scalar product (p. 521)
true bearing (p. 483)
unit vector (p. 494)
vector (p. 482)
vector projection (p. 503)
work (p. 505)
z-axis (p. 510)
zero vector (p. 485)

Vocabulary Check

Determine whether each statement is *true* or *false*. If false, replace the underlined term or expression to make the statement true.

1. The terminal point of a vector is where the vector <u>begins</u>.

2. If $\mathbf{a} = \langle -4, 1 \rangle$ and $b = \langle 3, 2 \rangle$, the dot product is calculated by <u>$-4(1) + 3(2)$</u>.

3. The midpoint of $\overline{AB}$ with $A(x_1, y_1, z_1)$ and $B(x_2, y_2, z_2)$ is given by <u>$\left(\dfrac{x_1 + x_2}{2}, \dfrac{y_1 + y_2}{2}, \dfrac{z_1 + z_2}{2}\right)$</u>.

4. The <u>magnitude</u> of $\mathbf{r}$ if the initial point is $A(-1, 2)$ and the terminal point is $B(2, -4)$ is $\langle 3, -6 \rangle$.

5. Two vectors are <u>equal</u> only if they have the same direction and magnitude.

6. When two nonzero vectors are orthogonal, the angle between them is <u>180°</u>.

7. The <u>component</u> of $\mathbf{u}$ onto $\mathbf{v}$ is the vector with direction that is parallel to $\mathbf{v}$ *and* with length that is the component of $\mathbf{u}$ along $\mathbf{v}$.

8. To find at least one vector orthogonal to any two vectors in space, calculate the <u>cross product</u> of the two original vectors.

9. When a vector is subtracted, it is equivalent to adding the opposite vector.

10. If $\mathbf{v}$ is a unit vector in the same direction as $\mathbf{u}$, then <u>$\mathbf{v} = \dfrac{|\mathbf{u}|}{\mathbf{u}}$</u>.

Lesson-by-Lesson Review

8-1 Introduction to Vectors (pp. 482–491)

State whether each quantity described is a *vector* quantity or a *scalar* quantity.

11. a car driving 50 miles an hour due east

12. a gust of wind blowing 5 meters per second

Find the resultant of each pair of vectors using either the triangle or parallelogram method. State the magnitude of the resultant to the nearest tenth of a centimeter and its direction relative to the horizontal.

13.

14.

15.

16.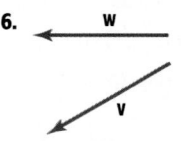

Determine the magnitude and direction of the resultant of each vector sum.

17. 70 meters due west and then 150 meters due east

18. 8 newtons directly backward and then 12 newtons directly backward

EXAMPLE 1

Find the resultant of **r** and **s** using either the triangle or parallelogram method. State the magnitude of the resultant in centimeters and its direction relative to the horizontal.

Triangle Method

Translate **r** so that the tip of **r** touches the tail of **s**. The resultant is the vector from the tail of **r** to the tip of **s**.

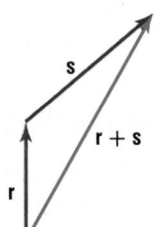

Parallelogram Method

Translate **s** so that the tail of **s** touches the tail of **r**. Complete the parallelogram that has **r** and **s** as two of its sides. The resultant is the vector that forms the indicated diagonal of the parallelogram.

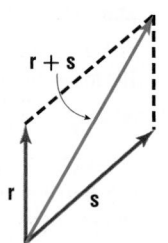

The magnitude of the resultant is 3.4 cm and the direction is 59°.

8-2 Vectors in the Coordinate Plane (pp. 492–499)

Find the component form and magnitude of $\overrightarrow{AB}$ with the given initial and terminal points.

19. $A(-1, 3)$, $B(5, 4)$ 20. $A(7, -2)$, $B(-9, 6)$

21. $A(-8, -4)$, $B(6, 1)$ 22. $A(2, -10)$, $B(3, -5)$

Find each of the following for $\mathbf{p} = \langle 4, 0 \rangle$, $\mathbf{q} = \langle -2, -3 \rangle$, and $\mathbf{t} = \langle -4, 2 \rangle$.

23. $2\mathbf{q} - \mathbf{p}$ 24. $\mathbf{p} + 2\mathbf{t}$

25. $\mathbf{t} - 3\mathbf{p} + \mathbf{q}$ 26. $2\mathbf{p} + \mathbf{t} - 3\mathbf{q}$

Find a unit vector **u** with the same direction as **v**.

27. $\mathbf{v} = \langle -7, 2 \rangle$ 28. $\mathbf{v} = \langle 3, -3 \rangle$

29. $\mathbf{v} = \langle -5, -8 \rangle$ 30. $\mathbf{v} = \langle 9, 3 \rangle$

EXAMPLE 2

Find the component form and magnitude of $\overrightarrow{AB}$ with initial point $A(3, -2)$ and terminal point $B(4, -1)$.

$\overrightarrow{AB} = \langle x_2 - x_1, y_2 - y_1 \rangle$ **Component form**

$= \langle 4 - 3, -1 - (-2) \rangle$ **Substitute.**

$= \langle 1, 1 \rangle$ **Subtract.**

Find the magnitude using the Distance Formula.

$|\overrightarrow{AB}| = \sqrt{(x_2 - x_1)^2 + (y_2 - y_1)^2}$ **Distance Formula**

$= \sqrt{[(4 - 3)]^2 + [-1 - (-2)]^2}$ **Substitute.**

$= \sqrt{2}$ or about 1.4   **Simplify.**

Lesson-by-Lesson Review

8-3   Dot Products and Vector Projections (pp. 500–508)

Find the dot product of **u** and **v**. Then determine if **u** and **v** are orthogonal.

31. $u = \langle -3, 5 \rangle$, $v = \langle 2, 1 \rangle$ **32.** $u = \langle 4, 4 \rangle$, $v = \langle 5, 7 \rangle$

33. $u = \langle -1, 4 \rangle$, $v = \langle 8, 2 \rangle$ **34.** $u = \langle -2, 3 \rangle$, $v = \langle 1, 3 \rangle$

Find the angle θ between **u** and **v** to the nearest tenth of a degree.

35. $u = \langle 5, -1 \rangle$, $v = \langle -2, 3 \rangle$ **36.** $u = \langle -1, 8 \rangle$, $v = \langle 4, 2 \rangle$

EXAMPLE 3

Find the dot product of $x = \langle 2, -5 \rangle$ and $y = \langle -4, 7 \rangle$. Then determine if **x** and **y** are orthogonal.

$$
\begin{aligned}
x \bullet y &= x_1 y_1 + x_2 y_2 && \text{Dot product} \\
&= 2(-4) + -5(7) && \text{Substitute.} \\
&= -8 + (-35) \text{ or } -43 && \text{Simplify.}
\end{aligned}
$$

Since $x \bullet y \neq 0$, **x** and **y** are not orthogonal.

8-4   Vectors in Three-Dimensional Space (pp. 510–516)

Plot each point in a three-dimensional coordinate system.

37. $(1, 2, -4)$ **38.** $(3, 5, 3)$

39. $(5, -3, -2)$ **40.** $(-2, -3, -2)$

Find the length and midpoint of the segment with the given endpoints.

41. $(-4, 10, 4)$, $(2, 0, 8)$ **42.** $(-5, 6, 4)$, $(-9, -2, -2)$

43. $(3, 2, 0)$, $(-9, -10, 4)$ **44.** $(8, 3, 2)$, $(-4, -6, 6)$

Locate and graph each vector in space.

45. $a = \langle 0, -3, 4 \rangle$ **46.** $b = -3i + 3j + 2k$

47. $c = -2i - 3j + 5k$ **48.** $d = \langle -4, -5, -3 \rangle$

EXAMPLE 4

Plot $(-3, 4, -4)$ in a three-dimensional coordinate system.

Locate the point $(-3, 4)$ in the xy-plane and mark it with a cross. Then plot a point 4 units down from this location parallel to the z-axis.

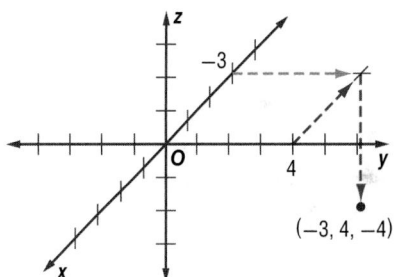

$(-3, 4, -4)$

8-5   Vectors in Three-Dimensional Space (pp. 518–524)

Find the dot product of **u** and **v**. Then determine if **u** and **v** are orthogonal.

49. $u = \langle 2, 5, 2 \rangle$, $v = \langle 8, 2, -13 \rangle$

50. $u = \langle 5, 0, -6 \rangle$, $v = \langle -6, 1, 3 \rangle$

Find the cross product of **u** and **v**. Then show that $u \times v$ is orthogonal to both **u** and **v**.

51. $u = \langle 1, -3, -2 \rangle$, $v = \langle 2, 4, -3 \rangle$

52. $u = \langle 4, 1, -2 \rangle$, $v = \langle 5, -4, -1 \rangle$

EXAMPLE 5

Find the cross product of $u = \langle -4, 2, -3 \rangle$ and $v = \langle 7, 11, 2 \rangle$. Then show that $u \times v$ is orthogonal to both **u** and **v**.

$$
u \times v = \begin{vmatrix} 2 & -3 \\ 11 & 2 \end{vmatrix} i - \begin{vmatrix} -4 & -3 \\ 7 & 2 \end{vmatrix} j + \begin{vmatrix} -4 & 2 \\ 7 & 11 \end{vmatrix} k
$$

$$
= \langle 37, -13, -58 \rangle
$$

$$
(u \times v) \bullet u = \langle 37, -13, -58 \rangle \bullet \langle -4, 2, -3 \rangle
$$
$$
= -148 - 26 + 174 \text{ or } 0 \; \checkmark
$$

$$
(u \times v) \bullet v = \langle 37, -13, -58 \rangle \bullet \langle 7, 11, 2 \rangle
$$
$$
= 259 - 143 - 116 \text{ or } 0 \; \checkmark
$$

Applications and Problem Solving

53. **BASEBALL** A player throws a baseball with an initial velocity of 55 feet per second at an angle of 25° above the horizontal, as shown below. Find the magnitude of the horizontal and vertical components. **(Lesson 8-1)**

54. **STROLLER** Miriam is pushing a stroller with a force of 200 newtons at an angle of 20° below the horizontal. Find the magnitude of the horizontal and vertical components of the force. **(Lesson 8-1)**

55. **LIGHTS** A traffic light at an intersection is hanging from two wires of equal length at 15° below the horizontal as shown. If the traffic light weighs 560 pounds, what is the tension in each wire keeping the light at equilibrium? **(Lesson 8-1)**

56. **AIRPLANE** An airplane is descending at a speed of 110 miles per hour at an angle of 10° below the horizontal. Find the component form of the vector that represents the velocity of the airplane. **(Lesson 8-2)**

57. **LIFEGUARD** A lifeguard at a wave pool swims at a speed of 4 miles per hour at a 60° angle to the side of the pool as shown. **(Lesson 8-2)**

   a. At what speed is the lifeguard traveling if the current in the pool is 2 miles per hour parallel to the side of the pool as shown?

   b. At what angle is the lifeguard traveling with respect to the starting side of the pool?

58. **TRAFFIC** A 1500-pound car is stopped in traffic on a hill that is at an incline of 10°. Determine the force that is required to keep the car from rolling down the hill. **(Lesson 8-3)**

59. **WORK** At a warehouse, Phil pushes a box on sliders with a constant force of 80 newtons up a ramp that has an incline of 15° with the horizontal. Determine the amount of work in joules that Phil does if he pushes the dolly 10 meters. **(Lesson 8-3)**

60. **SATELLITES** The positions of two satellites that are in orbit can be represented by the coordinates (28,625, 32,461, −38,426) and (−31,613, −29,218, 43,015), where (0, 0, 0) represents the center of Earth and the coordinates are given in miles. The radius of Earth is about 3963 miles. **(Lesson 8-4)**

   a. Determine the distance between the two satellites.

   b. If a third satellite were to be placed directly between the two satellites, what would the coordinates be?

   c. Can a third satellite be placed at the coordinates found in part **b**? Explain your reasoning.

61. **BICYCLES** A bicyclist applies 18 newtons of force down on the pedal to put the bicycle in motion. The pedal has an initial position of 47° above the y-axis, and a length of 0.19 meters to the pedal's axle, as shown. **(Lesson 8-5)**

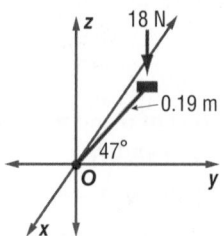

   a. Find the vector representing the torque about the axle of the bicycle pedal in component form.

   b. Find the magnitude and direction of the torque.

Find the resultant of each pair of vectors using either the triangle or parallelogram method. State the magnitude of the resultant to the nearest tenth of a centimeter and its direction relative to the horizontal.

1.
   p
   q

2. c
   d

Find the component form and magnitude of $\overrightarrow{AB}$ with the given initial and terminal points.

3. $A(1, -3), B(-5, 1)$

4. $A\left(\frac{1}{2}, \frac{3}{2}\right), B(-1, 7)$

5. **SOFTBALL** A batter on the opposing softball team hits a ground ball that rolls out to Libby in left field. She runs toward the ball at a velocity of 4 meters per second, scoops it, and proceeds to throw it to the catcher at a speed of 30 meters per second and at an angle of 25° with the horizontal in an attempt to throw out a runner. What is the resultant speed and direction of the throw?

30 m/s
25°
4 m/s

Find a unit vector in the same direction as u.

6. $\mathbf{u} = \langle -1, 4 \rangle$

7. $\mathbf{u} = \langle 6, -3 \rangle$

Find the dot product of u and v. Then determine if u and v are orthogonal.

8. $\mathbf{u} = \langle 2, -5 \rangle, \mathbf{v} = \langle -3, 2 \rangle$

9. $\mathbf{u} = \langle 4, -3 \rangle, \mathbf{v} = \langle 6, 8 \rangle$

10. $\mathbf{u} = 10\mathbf{i} - 3\mathbf{j}, \mathbf{v} = \mathbf{i} + 8\mathbf{j}$

11. **MULTIPLE CHOICE** Write **u** as the sum of two orthogonal vectors, one of which being the projection of **u** onto **v** if $\mathbf{u} = \langle 1, 3 \rangle$ and $\mathbf{v} = \langle -4, 2 \rangle$.

A $\mathbf{u} = \left\langle \frac{2}{5}, -\frac{3}{5} \right\rangle + \left\langle \frac{3}{5}, \frac{18}{5} \right\rangle$

B $\mathbf{u} = \left\langle \frac{2}{5}, \frac{3}{5} \right\rangle + \left\langle \frac{3}{5}, \frac{12}{5} \right\rangle$

C $\mathbf{u} = \left\langle -\frac{4}{5}, \frac{2}{5} \right\rangle + \left\langle \frac{9}{5}, \frac{13}{5} \right\rangle$

D $\mathbf{u} = \left\langle -\frac{2}{5}, \frac{1}{5} \right\rangle + \left\langle \frac{7}{5}, \frac{14}{5} \right\rangle$

12. **MOVING** Tamera is pushing a box along a level floor with a force of 120 pounds at an angle of depression of 20°. Determine how much work is done if the box is moved 25 feet.

Find each of the following for $\mathbf{a} = \langle 2, 4, -3 \rangle, \mathbf{b} = \langle -5, -7, 1 \rangle$, and $\mathbf{c} = \langle 8, 5, -9 \rangle$.

13. $2\mathbf{a} + 5\mathbf{b} - 3\mathbf{c}$

14. $\mathbf{b} - 6\mathbf{a} + 2\mathbf{c}$

15. **HOT AIR BALLOONS** During a festival, twelve hot air balloons take off. A few minutes later, the coordinates of the first two balloons are (20, 25, 30) and (−29, 15, 10) as shown, where the coordinates are given in feet.

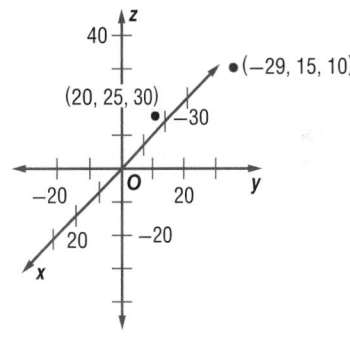

a. Determine the distance between the first two balloons that took off.

b. A third balloon is halfway between the first two balloons. Determine the coordinates of the third balloon.

c. Find a unit vector in the direction of the first balloon if it took off at (0, 0, 0).

Find the angle θ between vectors u and v to the nearest tenth of a degree.

16. $\mathbf{u} = \langle -2, 4, 6 \rangle, \mathbf{v} = \langle 3, 7, 12 \rangle$

17. $\mathbf{u} = -9\mathbf{i} + 5\mathbf{j} + 11\mathbf{k}, \mathbf{v} = -5\mathbf{i} - 7\mathbf{j} - 6\mathbf{k}$

Find the cross product of u and v. Then show that u × v is orthogonal to both u and v.

18. $\mathbf{u} = \langle 1, 7, 3 \rangle, \mathbf{v} = \langle 9, 4, 11 \rangle$

19. $\mathbf{u} = -6\mathbf{i} + 2\mathbf{j} - \mathbf{k}, \mathbf{v} = 5\mathbf{i} - 3\mathbf{j} - 2\mathbf{k}$

20. **BOATING** The tiller is a lever that controls the position of the rudder on a boat. When force is applied to the tiller, the boat will turn. Suppose the tiller on a certain boat is 0.75 meter in length and is currently resting in the xy-plane at a 15° angle from the positive x-axis. Find the magnitude of the torque that is developed about the axle of the tiller if 50 newtons of force is applied in a direction parallel to the positive y-axis.

Vector Fields

Objective
Graph vectors in and identify graphs of vector fields.

In Chapter 8, you examined the effects that wind and water currents have on a moving object. The force produced by the wind and current was represented by a single vector. However, we know that the current in a body of water or the force produced by wind is not necessarily constant; instead it differs from one region to the next. If we want to represent the entire current or air flow in an area, we would need to assign a vector to each point in space, thus creating a *vector field*.

Vector fields are commonly used in engineering and physics to model air resistance, magnetic and gravitational forces, and the motion of liquids. While these applications of vector fields require multiple dimensions, we will analyze vector fields in only two dimensions.

A vector field $\mathbf{F}(x, y)$ is a function that converts two-dimensional coordinates into sets of two-dimensional vectors.

$$\mathbf{F}(x, y) = \langle f_1(x, y), f_2(x, y) \rangle, \text{ where } f_1(x, y) \text{ and } f_2(x, y) \text{ are scalar functions.}$$

To graph a vector field, evaluate $\mathbf{F}(x, y)$ at (x, y) and graph the vector using (x, y) as the initial point. This is done for several points.

ACTIVITY 1 | **Vector Fields**

Evaluate **F(2, 1)**, **F(−1, −1)**, **F(1.5, −2)**, and **F(−3, 2)** for the vector field $\mathbf{F}(x, y) = \langle y^2, x - 1 \rangle$. Graph each vector using (x, y) as the initial point.

Step 1   To evaluate $\mathbf{F}(2, 1)$, let $x = 2$ and $y = 1$.

$$\langle y^2, x - 1 \rangle = \langle 1^2, 2 - 1 \rangle$$
$$= \langle 1, 1 \rangle$$

Step 2   To graph, let $(2, 1)$ represent the initial point of the vector.
This makes $(2 + 1, 1 + 1)$ or $(3, 2)$ the terminal point.

Step 3   Repeat Steps 1–2 for $\mathbf{F}(-1, -1)$, $\mathbf{F}(1.5, -2)$ and $\mathbf{F}(-3, 2)$.

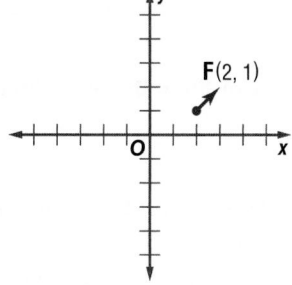

▶ **Analyze the Results**

1. Are the magnitudes and directions of the vectors the *same* or *different*?

2. Make a conjecture as to why a vector field can be defined as a function.

3. Is it possible to graph every vector in a vector field? Explain your reasoning.

A graph of a vector field $\mathbf{F}(x, y)$ should include a variety of vectors all with initial points at (x, y). Graphing devices are typically used to graph vector fields because sketching vector fields by hand is often too difficult.

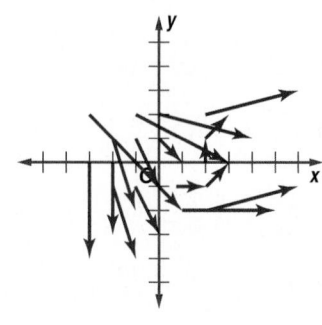

To keep vectors from overlapping each other and to prevent the graph from looking too jumbled, the graphing devices proportionally reduce the lengths of the vectors and only construct vectors at certain intervals. For example, if we continue to graph more vectors for the vector field from Activity 1, the result would be the graph on the right.

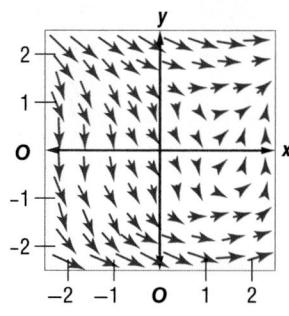

Analyze the component functions of a vector field to identify the type of graph it will produce.

ACTIVITY 2   Vector Fields

Match each vector field to its graph.

$$\mathbf{F}(x, y) = \langle 2, 1 + 2xy \rangle \qquad \mathbf{G}(x, y) = \langle e^y, x \rangle \qquad \mathbf{H}(x, y) = \langle e^y, y \rangle$$

Figure 1

Figure 2

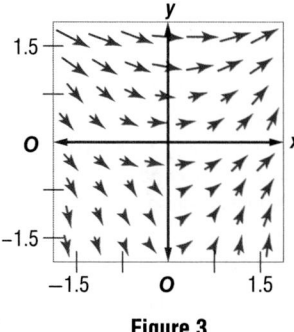

Figure 3

Step 1 Start by analyzing the components that make up $\mathbf{F}(x, y)$. The second component $(1 + 2xy)$ will produce a positive outcome when x and y have the same sign. This occurs in Quadrants I and III. Thus, the vertical component of the vectors in Quadrants I and III is positive, which makes the vectors in these quadrants point upward.

Step 2 Find the graph that has vectors pointing upward in Quadrants I and III. The only graph with this characteristic is Figure 2.

Step 3 Repeat Steps 1–2 for the remaining vector fields.

▶ Analyze the Results

4. Suppose the vectors in a vector field represent a force. What is the relationship between the force, the magnitude, and the length of a vector?

5. Representing wind by a single vector assumed that the force created remained constant for an entire area. If the force created by wind is represented by multiple vectors in a vector field, what assumption would have to be made about the third dimension?

Model and Apply

6. Complete the table for the vector field $\mathbf{F}(x, y) = \langle -y, x \rangle$. Then graph each vector.

(x, y)	⟨−y, x⟩	(x, y)	⟨−y, x⟩
(2, 0)		(−2, 1)	
(1, 2)		(−2, 0)	
(2, 1)		(−1, −2)	
(0, 2)		(0, −2)	
(−1, 2)		(1, −2)	
(−2, −1)		(2, −1)	

Then

In Chapter 7, you identified and graphed rectangular equations of conic sections.

Now

In Chapter 9, you will:

- Graph polar coordinates and equations.
- Convert between polar and rectangular coordinates and equations.
- Identify polar equations of conic sections.
- Convert complex numbers between polar and rectangular form.

Why?

CONCERTS Polar equations can be used to model sound patterns to help determine stage arrangement, speaker and microphone placement, and volume and recording levels. Polar equations can also be used with lighting and camera angles when concerts are filmed.

PREREAD Use the Lesson Openers in Chapter 9 to make two or three predictions about what you will learn in this chapter.

Get Ready for Chapter 9

Diagnose Readiness You have two options for checking Prerequisite Skills.

Text Option

Take the Quick Quiz below.

*Quick*Check

Graph each function using a graphing calculator. Analyze the graph to determine whether each function is *even*, *odd*, or *neither*. Confirm your answer algebraically. If odd or even, describe the symmetry of the graph of the function. (Lesson 1-2)

1. $f(x) = x^2 + 10$ **2.** $f(x) = -2x^3 + 5x$

3. $g(x) = \sqrt{x + 9}$ **4.** $h(x) = \sqrt{x^2 - 3}$

5. $g(x) = 3x^5 - 7x$ **6.** $h(x) = \sqrt{x^2 - 5}$

7. BALLOON The distance in meters between a balloon and a person can be represented by $d(t) = \sqrt{t^2 + 3000}$, where t represents time in seconds. Analyze the graph to determine whether the function is *even*, *odd*, or *neither*. (Lesson 1-2)

Approximate to the nearest hundredth the relative or absolute extrema of each function. State the *x*-values where they occur. (Lesson 1-4)

8. $f(x) = 4x^2 - 20x + 24$ **9.** $g(x) = -2x^2 + 9x - 1$

10. $f(x) = -x^3 + 3x - 2$ **11.** $f(x) = x^3 + x^2 - 5x$

12. ROCKET A rocket is fired into the air. The function $h(t) = -16t^2 + 35t + 15$ represents the height h of the rocket in feet after t seconds. Find the extrema of this function. (Lesson 1-4)

Identify all angles that are coterminal with the given angle. Then find and draw one positive and one negative angle coterminal with the given angle. (Lesson 4-2)

13. $165°$ **14.** $205°$ **15.** $-10°$

16. $\dfrac{\pi}{6}$ **17.** $\dfrac{4\pi}{3}$ **18.** $-\dfrac{\pi}{4}$

Online Option

Math Online Take a self-check Chapter Readiness Quiz at **glencoe.com**.

New Vocabulary

English	• Page •	Español
polar coordinate system	• 534 •	sistema de coordenadas polares
pole	• 534 •	polo
polar axis	• 534 •	eje polares
polar coordinates	• 534 •	coordenadas polares
polar equation	• 536 •	ecuación polar
polar graph	• 536 •	gráfico polar
limaçon	• 543 •	limaçon
cardioid	• 544 •	cardioide
rose	• 545 •	rosa
lemniscate	• 546 •	lemniscate
spiral of Archimedes	• 546 •	espiral de Arquímedes
complex plane	• 569 •	plano complejo
real axis	• 569 •	eje real
imaginary axis	• 569 •	eje imaginario
Argand plane	• 569 •	avión de Argand
absolute value of a complex number	• 569 •	valor absoluto de un número complejo
polar form	• 570 •	forma polar
trigonometric form	• 570 •	forma trigonométrica
modulus	• 570 •	módulo
argument	• 570 •	argumento

Review Vocabulary

initial side of an angle • p. 231 • lado inicial de un ángulo the starting position of the ray

terminal side of an angle • p. 231 • lado terminal de un ángulo the ray's position after rotation

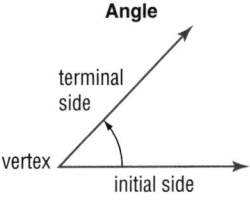

measure of an angle • p. 231 • medida de un ángulo the amount and direction of rotation necessary to move from the initial side to the terminal side of the angle

Polar Coordinates

Then
You drew positive and negative angles given in degrees and radians in standard position. (Lesson 4-2)

Now
- Graph points with polar coordinates.
- Graph simple polar equations.

New Vocabulary
polar coordinate system
pole
polar axis
polar coordinates
polar equation
polar graph

Math Online
glencoe.com

Why?

To provide safe routes and travel, air traffic controllers use advanced radar systems to direct the flow of airplane traffic. This ensures that airplanes keep a sufficient distance from other aircraft and landmarks. The radar uses angle measure and directional distance to plot the positions of aircraft. Controllers can then relay this information to the pilots.

Graph Polar Coordinates To this point, you have graphed equations in a rectangular coordinate system. When air traffic controllers record the locations of airplanes using distances and angles, they are using a **polar coordinate system** or polar plane.

In a rectangular coordinate system, the x- and y-axes are horizontal and vertical lines, respectively, and their point of intersection O is called the origin. The location of a point P is identified by rectangular coordinates of the form (x, y), where x and y are the horizontal and vertical *directed distances*, respectively, to the point. For example, the point $(3, -4)$ is 3 units to the right of the y-axis and 4 units below the x-axis.

Rectangular Coordinate System

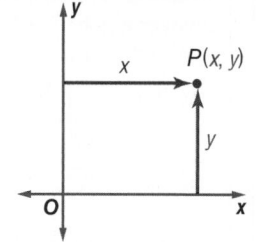

In a polar coordinate system, the origin is a fixed point O called the **pole**. The **polar axis** is an initial ray from the pole, usually horizontal and directed toward the right. The location of a point P in the polar coordinate system can be identified by **polar coordinates** of the form (r, θ), where r is the directed distance from the pole to the point and θ is the *directed angle* from the polar axis to $\overrightarrow{OP}$.

Polar Coordinate System

To graph a point given in polar coordinates, remember that a positive value of θ indicates a counterclockwise rotation from the polar axis, while a negative value indicates a clockwise rotation. If r is positive, then P lies on the terminal side of θ. If r is negative, P lies on the ray opposite the terminal side of θ.

EXAMPLE 1 **Graph Polar Coordinates**

Graph each point.

a. $A(2, 45°)$

Because $\theta = 45°$, sketch the terminal side of a $45°$ angle with the polar axis as its initial side. Because $r = 2$, plot a point 2 units from the pole along the terminal side of this angle.

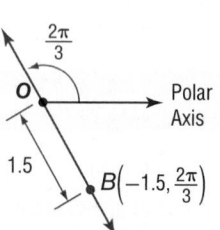

b. $B\left(-1.5, \dfrac{2\pi}{3}\right)$

Because $\theta = \dfrac{2\pi}{3}$, sketch the terminal side of a $\dfrac{2\pi}{3}$ angle with the polar axis as its initial side. Because r is negative, extend the terminal side of the angle in the *opposite* direction and plot a point 1.5 units from the pole along this extended ray.

c. *C*(3, −30°)

Because θ = −30°, sketch the terminal side of a −30° angle with the polar axis as its initial side. Because *r* = 3, plot a point 3 units from the pole along the terminal side of this angle.

 Guided Practice

Graph each point.

1A. $D\left(-1, \frac{\pi}{2}\right)$ **1B.** $E(2.5, 240°)$ **1C.** $F\left(4, -\frac{5\pi}{6}\right)$

Just as rectangular coordinates are graphed on a rectangular grid, polar coordinates are graphed on a circular or *polar* grid representing the polar plane.

EXAMPLE 2   **Graph Points on a Polar Grid**

Graph each point on a polar grid.

a. $P\left(3, \frac{4\pi}{3}\right)$

Because $θ = \frac{4\pi}{3}$, sketch the terminal side of a $\frac{4\pi}{3}$ angle with the polar axis as its initial side. Because *r* = 3, plot a point 3 units from the pole along the terminal side of the angle.

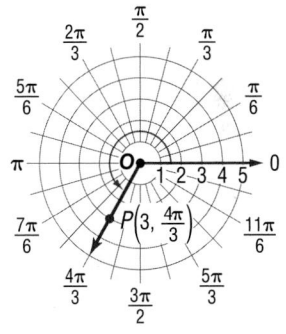

b. *Q*(−3.5, 150°)

Because θ = 150°, sketch the terminal side of a 150° angle with the polar axis as its initial side. Because *r* is negative, extend the terminal side of the angle in the *opposite* direction and plot a point 3.5 units from the pole along this extended ray.

 Guided Practice

2A. $R\left(1.5, -\frac{7\pi}{6}\right)$ **2B.** $S(-2, -135°)$

In a rectangular coordinate system, each point has a unique set of coordinates. This is *not* true in a polar coordinate system. In Lesson 4-2, you learned that a given angle has infinitely many coterminal angles. As a result, if a point has polar coordinates (*r*, θ), then it also has polar coordinates (*r*, θ ± 360°) or (*r*, θ ± 2π) as shown.

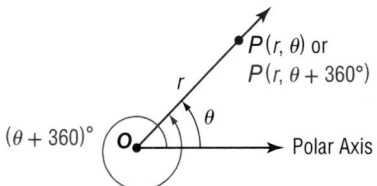

Additionally, because r is a directed distance, (r, θ) and $(-r, \theta \pm 180°)$ or $(-r, \theta \pm \pi)$ represent the same point as shown.

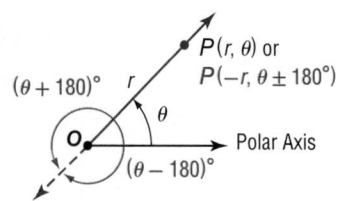

In general, if n is any integer, the point with polar coordinates (r, θ) can also be represented by polar coordinates of the form $(r, \theta + 360n°)$ or $(-r, \theta + (2n + 1)180°)$. Likewise, if θ is given in radians and n is any integer, the other representations of (r, θ) are of the form $(r, \theta + 2n\pi)$ or $(-r, \theta + (2n + 1)\pi)$.

EXAMPLE 3   Multiple Representations of Polar Coordinates

Find four different pairs of polar coordinates that name point T if $-360° \le \theta \le 360°$.

One pair of polar coordinates that name point T is $(4, 135°)$. The other three representations are as follows.

$(4, 135°) = (4, 135° - 360°)$ **Subtract 360° from θ.**

$= (4, -225°)$

$(4, 135°) = (-4, 135° + 180°)$ **Replace r with $-r$ and**

$= (-4, 315°)$ **add 180° to θ.**

$(4, 135°) = (-4, 135° - 180°)$ **Replace r with $-r$ and**

$= (-4, -45°)$ **subtract 180° from θ.**

☑ Guided Practice

Find three additional pairs of polar coordinates that name the given point if $-360° \le \theta \le 360°$ or $-2\pi \le \theta \le \pi$.

3A. $(5, 240°)$ **3B.** $\left(2, \dfrac{\pi}{6}\right)$

Graphs of Polar Equations An equation expressed in terms of polar coordinates is called a **polar equation**. For example, $r = 2 \sin \theta$ is a polar equation. A **polar graph** is the set of all points with coordinates (r, θ) that satisfy a given polar equation.

You already know how to graph equations in the Cartesian, or *rectangular*, coordinate system. Graphs of equations involving constants like $x = 2$ and $y = -3$ are considered basic in the Cartesian coordinate system. Similarly, the graphs of the polar equations $r = k$ and $\theta = k$, where k is a constant, are considered basic in the polar coordinate system.

EXAMPLE 4   Graph Polar Equations

Graph each polar equation.

a. $r = 2$

The solutions of $r = 2$ are ordered pairs of the form $(2, \theta)$, where θ is any real number.

The graph consists of all points that are 2 units from the pole, so the graph is a circle centered at the origin with radius 2.

b. $\theta = \dfrac{\pi}{6}$

The solutions of $\theta = \dfrac{\pi}{6}$ are ordered pairs of the form $\left(r, \dfrac{\pi}{6}\right)$, where r is any real number. The graph consists of all points on the line that makes an angle of $\dfrac{\pi}{6}$ with the positive polar axis.

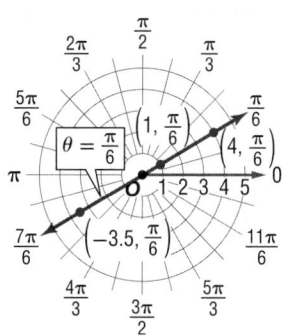

✔ Guided Practice

Graph each polar equation.

4A. $r = 3$

4B. $\theta = \dfrac{2\pi}{3}$

The distance between two points in the polar plane can be found using the following formula.

Key Concept   **Polar Distance Formula**

If $P_1(r_1, \theta_1)$ and $P_2(r_2, \theta_2)$ are two points in the polar plane, then the distance P_1P_2 is given by

$$\sqrt{{r_1}^2 + {r_2}^2 - 2r_1r_2 \cos(\theta_2 - \theta_1)}.$$

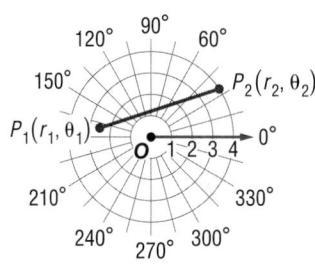

You will prove this formula in Exercise 63.

🌐 **Real-World EXAMPLE 5**   **Find the Distance Between Polar Coordinates**

AIR TRAFFIC An air traffic controller is tracking two airplanes that are flying at the same altitude. The coordinates of the planes are $A(5, 310°)$ and $B(6, 345°)$, where the directed distance is measured in miles.

a. Sketch a graph of this situation.

Airplane A is located 5 miles from the pole on the terminal side of the angle 310°, and airplane B is located 6 miles from the pole on the terminal side of the angle 345°, as shown.

b. If regulations prohibit airplanes from passing within three miles of each other, are these airplanes in violation? Explain.

Use the Polar Distance Formula.

$AB = \sqrt{{r_1}^2 + {r_2}^2 - 2r_1r_2 \cos(\theta_2 - \theta_1)}$ Polar Distance Formula

$= \sqrt{5^2 + 6^2 - 2(5)(6) \cos(345° - 310°)}$ or about 3.44 $(r_2, \theta_2) = (6, 345°)$ and $(r_1, \theta_1) = (5, 310°)$

The planes are about 3.44 miles apart, so they are not in violation of this regulation.

✔ Guided Practice

5. BOATS A naval radar is tracking two aircraft carriers. The coordinates of the two carriers are $(8, 150°)$ and $(3, 65°)$, with r measured in miles.

  A. Sketch a graph of this situation.

  B. What is the distance between the two aircraft carriers?

Exercises

Graph each point on a polar grid. (Examples 1 and 2)

1. $R(1, 120°)$

2. $T(-2.5, 330°)$

3. $F\left(-2, \frac{2\pi}{3}\right)$

4. $A\left(3, \frac{\pi}{6}\right)$

5. $Q\left(4, -\frac{5\pi}{6}\right)$

6. $B(5, -60°)$

7. $D\left(-1, -\frac{5\pi}{3}\right)$

8. $G\left(3.5, -\frac{11\pi}{6}\right)$

9. $C(-4, \pi)$

10. $M(0.5, 270°)$

11. $P(4.5, -210°)$

12. $W(-1.5, 150°)$

13. **ARCHERY** The target in competitive target archery consists of 10 evenly spaced concentric circles with score values from 1 to 10 points from the outer circle to the center. Suppose an archer using a target with a 60-centimeter radius shoots arrows at $(57, 45°)$, $(41, 315°)$, and $(15, 240°)$. (Examples 1 and 2)

1 pt   10 pts
60 cm

a. Plot the points where the archer's arrows hit the target on a polar grid.

b. How many points did the archer earn?

Find three different pairs of polar coordinates that name the given point if $-360° \le \theta \le 360°$ or $-2\pi \le \theta \le 2\pi$. (Example 3)

14. $(1, 150°)$

15. $(-2, 300°)$

16. $\left(4, -\frac{7\pi}{6}\right)$

17. $\left(-3, \frac{2\pi}{3}\right)$

18. $\left(5, \frac{11\pi}{6}\right)$

19. $\left(-5, -\frac{4\pi}{3}\right)$

20. $(2, -30°)$

21. $(-1, -240°)$

22. **SKYDIVING** In competitive accuracy landing, skydivers attempt to land as near as possible to "dead center," the center of a target marked by a disk 2 meters in diameter. (Example 4)

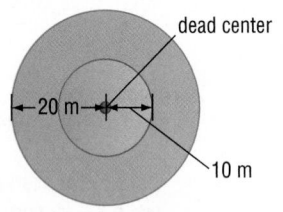
dead center
20 m
10 m

a. Write polar equations representing the three target boundaries.

b. Graph the equations on a polar grid.

Graph each polar equation. (Example 4)

23. $r = 4$

24. $\theta = 225°$

25. $r = 1.5$

26. $\theta = -\frac{7\pi}{6}$

27. $\theta = -15°$

28. $r = -3.5$

29. **DARTBOARD** A certain dartboard has a radius of 225 millimeters. The bull's-eye has two sections. The 50-point section has a radius of 6.3 millimeters. The 25-point section surrounds the 50-point section for an additional 9.7 millimeters. (Example 4)

a. Write and graph polar equations representing the boundaries of the dartboard and these sections.

b. What percentage of the dartboard's area does the bull's-eye comprise?

Find the distance between each pair of points. (Example 5)

30. $(2, 30°), (5, 120°)$

31. $\left(3, \frac{\pi}{2}\right), \left(8, \frac{4\pi}{3}\right)$

32. $(6, 45°), (-3, 300°)$

33. $\left(7, -\frac{\pi}{3}\right), \left(1, \frac{2\pi}{3}\right)$

34. $\left(-5, \frac{7\pi}{6}\right), \left(4, \frac{\pi}{6}\right)$

35. $(4, -315°), (1, 60°)$

36. $(-2, -30°), (8, 210°)$

37. $\left(-3, \frac{11\pi}{6}\right), \left(-2, \frac{5\pi}{6}\right)$

38. $\left(1, -\frac{\pi}{4}\right), \left(-5, \frac{7\pi}{6}\right)$

39. $(7, -90°), (-4, -330°)$

40. $\left(8, -\frac{2\pi}{3}\right), \left(4, -\frac{3\pi}{4}\right)$

41. $(-5, 135°), (-1, 240°)$

42. **SURVEYING** A surveyor mapping out the land where a new housing development will be built identifies a landmark 223 feet away and 45° left of center. A second landmark is 418 feet away and 67° right of center. Determine the distance between the two landmarks. (Example 5)

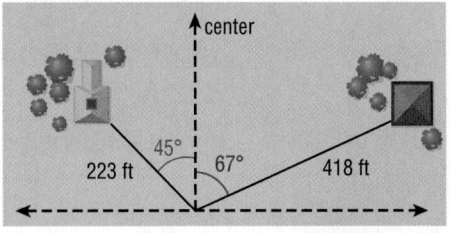
center
45°   67°
223 ft   418 ft

43 **SURVEILLANCE** A mounted surveillance camera oscillates and views part of a circular region determined by $-60° \le \theta \le 150°$ and $0 \le r \le 40$, where r is in meters.

a. Sketch a graph of the region that the security camera can view on a polar grid.

b. Find the area of the region.

Find a different pair of polar coordinates for each point such that $0 \le \theta \le 180°$ or $0 \le \theta \le \pi$.

44. $(5, 960°)$

45. $\left(-2.5, \frac{5\pi}{2}\right)$

46. $\left(4, \frac{11\pi}{4}\right)$

47. $(1.25, -920°)$

48. $\left(-1, -\frac{21\pi}{8}\right)$

49. $(-6, -1460°)$

50. AMPHITHEATER Suppose a singer is performing at an amphitheater. We can model this situation with polar coordinates by assuming that the singer is standing at the pole and is facing the direction of the polar axis. The seats can then be described as occupying the area defined by $-45° \leq \theta \leq 45°$ and $30 \leq r \leq 240$, where r is measured in feet.

    a. Sketch a graph of this region on a polar grid.

    b. If each person needs 5 square feet of space, how many seats can fit in the amphitheater?

51. SECURITY A security light mounted above a house illuminates part of a circular region defined by $\frac{\pi}{6} \leq \theta \leq \frac{5\pi}{6}$ and $x \leq r \leq 20$, where r is measured in feet. If the total area of the region is approximately 314.16 square feet, find x.

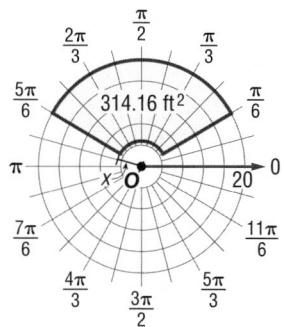

Find a value for the missing coordinate that satisfies the following condition.

52. $P_1 = (3, 35°); P_2 = (r, 75°); P_1P_2 = 4.174$

53. $P_1 = (5, 125°); P_2 = (2, \theta); P_1P_2 = 4; 0 \leq \theta \leq 180°$

54. $P_1 = (3, \theta); P_2 = \left(4, \frac{7\pi}{9}\right); P_1P_2 = 5; 0 \leq \theta \leq \pi$

55 $P_1 = (r, 120°); P_2 = (4, 160°); P_1P_2 = 3.297$

56. ⬛ **MULTIPLE REPRESENTATIONS** In this problem, you will investigate the relationship between polar coordinates and rectangular coordinates.

    a. GRAPHICAL Plot points $A\left(2, \frac{\pi}{3}\right)$ and $B\left(4, \frac{5\pi}{6}\right)$ on a polar grid. Sketch a rectangular coordinate system on top of the polar grid so that the origins coincide and the x-axis aligns with the polar axis. The y-axis should align with the line $\theta = \frac{\pi}{2}$. Form one right triangle by connecting point A to the origin and perpendicular to the x-axis. Form another right triangle by connecting point B to the origin and perpendicular to the x-axis.

    b. NUMERICAL Calculate the lengths of the legs of each triangle.

    c. ANALYTICAL How do the lengths of the legs relate to rectangular coordinates for each point?

    d. ANALYTICAL Explain the relationship between the polar coordinates (r, θ) and the rectangular coordinates (x, y).

Write an equation for each polar graph.

57.

58.

59.
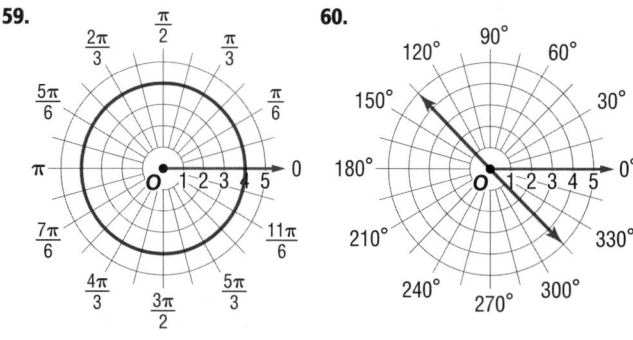

60.

H.O.T. Problems   Use **H**igher-**O**rder **T**hinking Skills

61. REASONING Explain why the order of the points used in the Polar Distance Formula is not important. That is, why can you choose either point to be P_1 and the other to be P_2?

62. CHALLENGE Find an ordered pair of polar coordinates to represent the point with rectangular coordinates $(-3, -4)$. Round the angle measure to the nearest degree.

63. PROOF Prove that the distance between two points with polar coordinates $P_1(r_1, \theta_1)$ and $P_2(r_2, \theta_2)$ is

$$P_1P_2 = \sqrt{r_1^2 + r_2^2 - 2r_1r_2 \cos(\theta_2 - \theta_1)}.$$ (*Hint:* Use the Law of Cosines.)

64. REASONING Describe what happens to the Polar Distance Formula when $\theta_2 - \theta_1 = \frac{\pi}{2}$. Explain this change.

65. ERROR ANALYSIS Sona and Erina both graphed the polar coordinates $(5, 45°)$. Is either of them correct? Explain your reasoning.

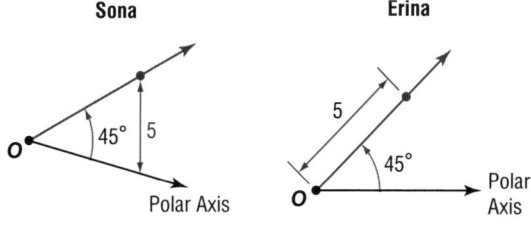

66. WRITING IN MATH Make a conjecture as to why having the polar coordinates for an aircraft is not enough to determine its exact position.

Find the dot product of u and v. Then determine if u and v are orthogonal. (Lesson 8-5)

67. $\mathbf{u} = \langle 4, 10, 1 \rangle$, $\mathbf{v} = \langle -5, 1, 7 \rangle$
68. $\mathbf{u} = \langle -5, 4, 2 \rangle$, $\mathbf{v} = \langle -4, -9, 8 \rangle$
69. $\mathbf{u} = \langle -8, -3, 12 \rangle$, $\mathbf{v} = \langle 4, -6, 0 \rangle$

Find each of the following for a = $\langle -4, 3, -2 \rangle$, b = $\langle 2, 5, 1 \rangle$, and c = $\langle 3, -6, 5 \rangle$. (Lesson 8-4)

70. $3\mathbf{a} + 2\mathbf{b} + 8\mathbf{c}$
71. $-2\mathbf{a} + 4\mathbf{b} - 5\mathbf{c}$
72. $5\mathbf{a} - 9\mathbf{b} + \mathbf{c}$

For each equation, identify the vertex, focus, axis of symmetry, and directrix. Then graph the parabola. (Lesson 7-1)

73. $-14(x - 2) = (y - 7)^2$
74. $(x - 7)^2 = -32(y - 12)$
75. $y = \frac{1}{2}x^2 - 3x + \frac{19}{2}$

76. STATE FAIR If Curtis and Drew each purchased the number of game and ride tickets shown below, what was the price for each type of ticket? (Lesson 6-3)

Person	Ticket Type	Tickets	Total ($)
Curtis	game	6	93
	ride	15	
Drew	game	7	81
	ride	12	

Write the augmented matrix for the system of linear equations. (Lesson 6-1)

77. $12w + 14x - 10y = 23$
$4w - 5y + 6z = 33$
$11w - 13x + 2z = -19$
$19x - 6y + 7z = -25$

78. $-6x + 2y + 5z = 18$
$5x - 7y + 3z = -8$
$y - 12z = -22$
$8x - 3y + 2z = 9$

79. $x + 8y - 3z = 25$
$2x - 5y + 11z = 13$
$-5x + 8z = 26$
$y - 4z = 17$

Solve each equation for all values of x. (Lesson 5-3)

80. $2 \cos^2 x + 5 \sin x - 5 = 0$
81. $\tan^2 x + 2 \tan x + 1 = 0$
82. $\cos^2 x + 3 \cos x = -2$

Skills Review for Standardized Tests

83. SAT/ACT If the figure shows part of the graph of $f(x)$, then which of the following could be the range of $f(x)$?

A $\{y \mid y \geq -2\}$
B $\{y \mid y \leq -2\}$
C $\{y \mid -2 < y < 1\}$
D $\{y \mid -2 \leq y \leq 1\}$
E $\{y \mid y > -2\}$

84. REVIEW Which of the following is the component form of $\overrightarrow{RS}$ with initial point $R(-5, 3)$ and terminal point $S(2, -7)$?

F $\langle 7, -10 \rangle$
G $\langle -3, 10 \rangle$
H $\langle -7, 10 \rangle$
J $\langle -3, -10 \rangle$

85. The lawn sprinkler shown can cover the part of a circular region determined by the polar inequalities $-30° \leq \theta \leq 210°$ and $0 \leq r \leq 20$, where r is measured in feet. What is the approximate area of this region?

A 821 square feet
B 838 square feet
C 852 square feet
D 866 square feet

86. REVIEW What type of conic is represented by $25y^2 = 400 + 16x^2$?

F circle
G ellipse
H hyperbola
J parabola

Objective
Use a graphing calculator to explore the shape and symmetry of graphs of polar equations.

In Lesson 9-1, you graphed polar coordinates and simple polar equations on the polar coordinate system. Now you will explore the shape and symmetry of more complex graphs of polar equations by using a graphing calculator.

ACTIVITY **Graph Polar Equations**

Graph each equation. Then describe the shape and symmetry of the graph.

a. $r = 3 \cos \theta$

First, change the graph mode to polar and the angle mode to radians. Next, enter $r = 3 \cos \theta$ for r_1 in the **Y=** list. Use the window shown.

The graph of $r = 3 \cos \theta$ is a circle with a center at $(1.5, 0)$ and radius 1.5 units. The graph is symmetric with respect to the polar axis.

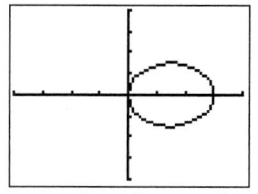

$[0, 2\pi]$ scl: $\frac{\pi}{24}$ by $[-4, 4]$ scl: 1 by $[-4, 4]$ scl: 1

b. $r = 2 \cos 3\theta$

Clear the equation from part **a** in the **Y=** list and insert $r = 2 \cos 3\theta$. Use the window shown.

The graph of $r = 2 \cos 3\theta$ is a classic polar curve called a *rose*, which will be covered in Lesson 9-2. The graph has 3 leaves and is symmetric with respect to the polar axis.

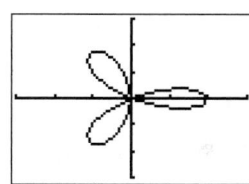

$[0, 2\pi]$ scl: $\frac{\pi}{24}$ by $[-3, 3]$ scl: 1 by $[-3, 3]$ scl: 1

c. $r = 1 + 2 \sin \theta$

Clear the equation from part **b** in the **Y=** list, and enter $r = 1 + 2 \sin \theta$. Adjust the window to display the entire graph.

The graph of $r = 1 + 2 \sin \theta$ is a classic polar curve called a *limaçon*, which will be covered in Lesson 9-2. The graph has a curve with an inner loop and is symmetric with respect to the line $\theta = \frac{\pi}{2}$.

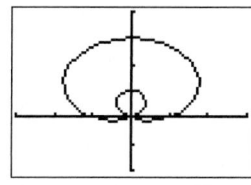

$[0, 2\pi]$ scl: $\frac{\pi}{24}$ by $[-3, 3]$ scl: 1 by $[-2, 4]$ scl: 1

Exercises

Graph each equation. Then describe the shape and symmetry of the graph.

1. $r = -3 \cos \theta$ **2.** $r = 3 \sin \theta$ **3.** $r = -3 \sin \theta$

4. $r = 2 \cos 4\theta$ **5.** $r = 2 \cos 5\theta$ **6.** $r = 2 \cos 6\theta$

7. $r = 2 + 4 \sin \theta$ **8.** $r = 1 - 3 \sin \theta$ **9.** $r = 1 + 2 \sin (-\theta)$

Analyze the Results

10. ANALYTICAL Explain how each value affects the graph of the given equation.

    a. the value of n in $r = a \cos n\theta$

    b. the value of $|a|$ in $r = b \pm a \sin n\theta$

11. MAKE A CONJECTURE Without graphing $r = 10 \cos 24\theta$, describe the shape and symmetry of the graph. Explain your reasoning.

Graphs of Polar Equations

Then
You graphed functions in the rectangular coordinate system. (Lesson 1-2)

Now
- Graph polar equations.
- Identify and graph classical curves.

New Vocabulary
limaçon
cardioid
rose
lemniscate
spiral of Archimedes

Math Online
glencoe.com

Why?

To reduce background noise, networks that broadcast sporting events use directional microphones to capture the sounds of the game. Directional microphones have the ability to pick up sound primarily from one direction or region. The sounds that these microphones can detect can be expressed as polar functions.

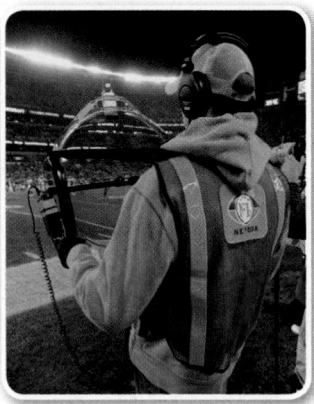

Graphs of Polar Equations When you graphed equations on a rectangular coordinate system, you began by using an equation to obtain a set of ordered pairs. You then plotted these coordinates as points and connected them with a smooth curve. In this lesson, you will approach the graphing of polar equations in a similar manner.

EXAMPLE 1 Graph Polar Equations by Plotting Points

Graph each equation.

a. $r = \cos \theta$

Make a table of values to find the r-values corresponding to various values of θ on the interval $[0, 2\pi]$. Round each r-value to the nearest tenth.

θ	0	$\frac{\pi}{6}$	$\frac{\pi}{3}$	$\frac{\pi}{2}$	$\frac{2\pi}{3}$	$\frac{5\pi}{6}$	π	$\frac{7\pi}{6}$	$\frac{4\pi}{3}$	$\frac{3\pi}{2}$	$\frac{5\pi}{3}$	$\frac{11\pi}{6}$	2π
$r = \cos \theta$	1	0.9	0.5	0	−0.5	−0.9	−1	−0.9	−0.5	0	0.5	0.9	1

Graph the ordered pairs (r, θ) and connect them with a smooth curve. It appears that the graph shown in Figure 9.2.1 is a circle with center at $(0.5, 0)$ and radius 0.5 unit.

b. $r = \sin \theta$

θ	0	$\frac{\pi}{6}$	$\frac{\pi}{3}$	$\frac{\pi}{2}$	$\frac{2\pi}{3}$	$\frac{5\pi}{6}$	π	$\frac{7\pi}{6}$	$\frac{4\pi}{3}$	$\frac{3\pi}{2}$	$\frac{5\pi}{3}$	$\frac{11\pi}{6}$	2π
$r = \sin \theta$	0	0.5	0.9	1	0.9	0.5	0	−0.5	−0.9	−1	−0.9	−0.5	0

Graph the ordered pairs and connect them with a smooth curve. It appears that the graph shown in Figure 9.2.2 is a circle with center at $\left(0.5, \frac{\pi}{2}\right)$ and radius 0.5 unit.

Figure 9.2.1

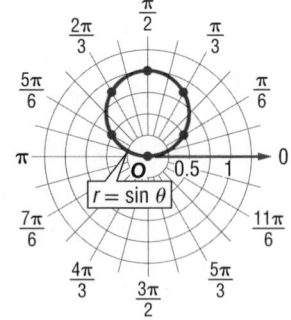

Figure 9.2.2

✓ Guided Practice

1A. $r = -\sin \theta$

1B. $r = 2 \cos \theta$

1C. $r = \sec \theta$

Notice that as θ increases on $[0, 2\pi]$, each graph above is traced twice. This is because the polar coordinates obtained on $[0, \pi]$ represent the same points as those obtained on $[\pi, 2\pi]$.

Like knowing whether a graph in the rectangular coordinate system has symmetry with respect to the x-axis, y-axis, or origin, knowing whether the graph of a polar equation is symmetric can help reduce the number of points needed to sketch its graph. Graphs of polar equations can be symmetric with respect to the line $\theta = \frac{\pi}{2}$, the polar axis, or the pole, as shown below.

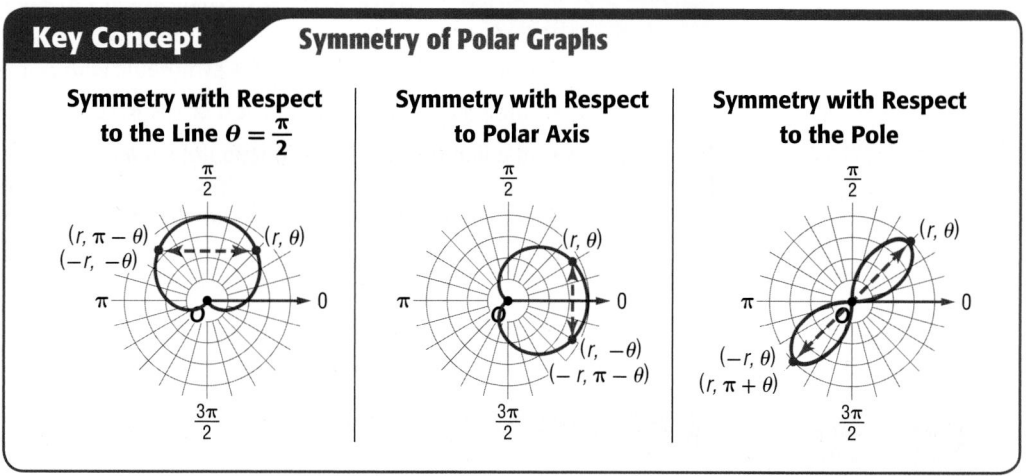

Key Concept   Symmetry of Polar Graphs

| Symmetry with Respect to the Line $\theta = \frac{\pi}{2}$ | Symmetry with Respect to Polar Axis | Symmetry with Respect to the Pole |

The graphical definitions above provide a way of testing a polar equation for symmetry. For example, if replacing (r, θ) in a polar equation with $(r, -\theta)$ or $(-r, \pi - \theta)$ produces an equivalent equation, then its graph is symmetric with respect to the polar axis. If an equation passes one of the symmetry tests, this is sufficient to guarantee that the equation has that type of symmetry. The converse, however, is *not* true. If a polar equation fails all of these tests, the graph may still have symmetry.

EXAMPLE 2   Polar Axis Symmetry

Use symmetry to graph $r = 1 - 2 \cos \theta$.

Replacing (r, θ) with $(r, -\theta)$ yields $r = 1 - 2 \cos (-\theta)$. Because cosine is an even function, $\cos (-\theta) = \cos \theta$, so this equation simplifies to $r = 1 - 2 \cos \theta$. Because the replacement produced an equation equivalent to the original equation, the graph of this equation is symmetric with respect to the polar axis.

Because of this symmetry, you need only make a table of values to find the r-values corresponding to θ on the interval $[0, \pi]$.

θ	0	$\frac{\pi}{6}$	$\frac{\pi}{4}$	$\frac{\pi}{3}$	$\frac{\pi}{2}$	$\frac{2\pi}{3}$	$\frac{3\pi}{4}$	$\frac{5\pi}{6}$	π
$r = 1 - 2 \cos \theta$	-1	-0.7	-0.4	0	1	2	2.4	2.7	3

StudyTip

Graphing Polar Equations
It is customary to graph polar functions in radians, rather than in degrees.

Plotting these points and using polar axis symmetry, you obtain the graph shown.

The type of curve is called a **limaçon** (LIM-uh-son). Some limaçons have inner loops like this one. Other limaçons come to a point, have a dimple, or just curve outward.

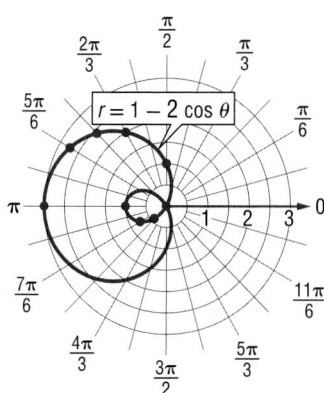

✓ **Guided Practice**

Use symmetry to graph each equation.

2A. $r = 1 - \cos \theta$ **2B.** $r = 2 + \cos \theta$

In Examples 1 and 2, notice that the graphs of $r = \cos\theta$ and $r = 1 - 2\cos\theta$ are symmetric with respect to the polar axis, while the graph of $r = \sin\theta$ is symmetric with respect to the line $\theta = \frac{\pi}{2}$. These observations can be generalized as follows.

Key Concept   **Quick Tests for Symmetry in Polar Graphs**

Words   The graph of a polar equation is symmetric with respect to

- the polar axis if it is a function of $\cos\theta$, and
- the line $\theta = \frac{\pi}{2}$ if it is a function of $\sin\theta$.

Example   The graph of $r = 3 + \sin\theta$ is symmetric with respect to the line $\theta = \frac{\pi}{2}$.

You will justify these tests for specific cases in Exercises 65–66.

Symmetry can be used to graph polar functions that model real-world situations.

🌐 **Real-World EXAMPLE 3**   **Symmetry with Respect to the Line $\theta = \frac{\pi}{2}$**

AUDIO TECHNOLOGY During a concert, a directional microphone was placed facing the audience from the center of stage to capture the crowd noise for a live recording. The area of sound the microphone captures can be represented by $r = 3.5 + 3.5\sin\theta$. Suppose the front of the stage faces due north.

a. Graph the polar pattern of the microphone.

Because this polar equation is a function of the sine function, it is symmetric with respect to the line $\theta = \frac{\pi}{2}$. Therefore, make a table and calculate the values of r on $\left[-\frac{\pi}{2}, \frac{\pi}{2}\right]$.

θ	$-\frac{\pi}{2}$	$-\frac{\pi}{3}$	$-\frac{\pi}{4}$	$-\frac{\pi}{6}$	0	$\frac{\pi}{6}$	$\frac{\pi}{4}$	$\frac{\pi}{3}$	$\frac{\pi}{2}$
$r = 3.5 + 3.5\sin\theta$	0	0.5	1.0	1.8	3.5	5.25	6.0	6.5	7

Plotting these points and using symmetry with respect to the line $\theta = \frac{\pi}{2}$, you obtain the graph shown. This type of curve is called a **cardioid** (CAR-dee-oid). A cardioid is a special limaçon that has a heart shape.

b. Describe what the polar pattern tells you about the microphone.

The polar pattern indicates that the microphone will pick up sounds up to 7 units away directly in front of the microphone and up to 3.5 units away directly to the left or right of the microphone.

✓ **Guided Practice**

3. VIDEOTAPING A high school teacher is videotaping presentations performed by her students using a stationary video camera positioned in the back of the room. The area of sound captured by the camera's microphone can be represented by $r = 5 + 2\sin\theta$. Suppose the front of the classroom is due north of the camera.

    A. Graph the polar pattern of the microphone.

    B. Describe what the polar pattern tells you about the microphone.

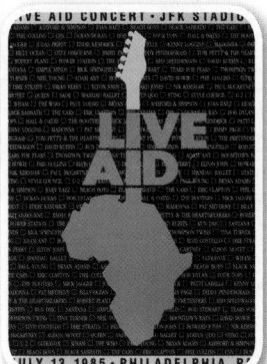

🌐 Real-World Link

Live Aid was a 1985 rock concert held in an effort to raise $1 million for Ethiopian aid. Concerts in London, Philadelphia, and other cities were televised and viewed by 1.9 billion people in 150 countries. The event raised $140 million.

Source: CNN

Watch Out!

▸ **Graphing over the Period** Usually the period of the trigonometric function used in a polar equation is sufficient to trace the entire graph, but sometimes it is not. The best way to know if you have graphed enough to discern a pattern is to plot more points.

In Lesson 4-4, you used maximum and minimum points along with zeros to aid in graphing trigonometric functions. On the graph of a polar function, r is at its maximum for a value of θ when the distance between that point (r, θ) and the pole is maximized. To find the maximum point(s) on the graph of a polar equation, find the θ-values for which $|r|$ is maximized. Additionally, if $r = 0$ for some value of θ, you know that the graph intersects the pole.

EXAMPLE 4   Symmetry, Zeros, and Maximum *r*-Values

Use symmetry, zeros, and maximum *r*-values to graph $r = 2 \cos 3\theta$.

Determine the symmetry of the graph.

This function is symmetric with respect to the polar axis, so you can find points on the interval $[0, \pi]$ and then use polar axis symmetry to complete the graph.

Find the zeros and the maximum *r*-value.

Sketch the graph of the rectangular function $y = 2 \cos 3x$ on the interval $[0, \pi]$.

From the graph, you can see that $|y| = 2$ when $x = 0, \frac{\pi}{3}, \frac{2\pi}{3}$, and π and $y = 0$ when $x = \frac{\pi}{6}, \frac{\pi}{2}$, and $\frac{5\pi}{6}$.

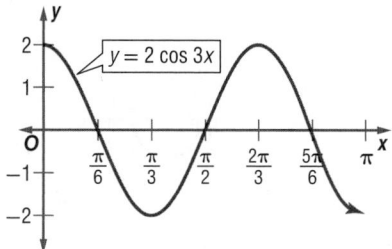

Interpreting these results in terms of the polar equation $r = 2 \cos 3\theta$, we can say that $|r|$ has a maximum value of 2 when $\theta = 0, \frac{\pi}{3}, \frac{2\pi}{3}$, or π and $r = 0$ when $\theta = \frac{\pi}{6}, \frac{\pi}{2}$, or $\frac{5\pi}{6}$.

Graph the function.

Use these and a few additional points to sketch the graph of the function.

θ	0	$\frac{\pi}{12}$	$\frac{\pi}{6}$	$\frac{\pi}{4}$	$\frac{\pi}{3}$	$\frac{5\pi}{12}$	$\frac{\pi}{2}$	$\frac{7\pi}{12}$	$\frac{2\pi}{3}$	$\frac{3\pi}{4}$	$\frac{5\pi}{6}$	$\frac{11\pi}{12}$	π
$r = 2 \cos 3\theta$	2	1.4	0	-1	-2	-1.4	0	1.4	2	1.4	0	-1.4	-2

Notice that polar axis symmetry can be used to complete the graph after plotting points on $\left[0, \frac{\pi}{2}\right]$.

This type of curve is called a **rose**. Roses can have three or more equal loops.

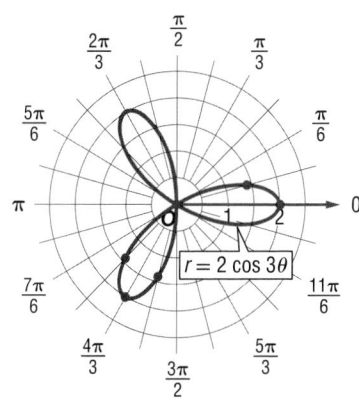

StudyTip

Alternative Method
Solving the rectangular function $y = 2 \cos 3x$, we find that the function has extrema when $x = 0, \frac{\pi}{3}, \frac{2\pi}{3}$, or π. Similarly, the function has zeros when $x = \frac{\pi}{6}, \frac{\pi}{2}$, or $\frac{5\pi}{6}$.

✔ Guided Practice

Use symmetry, zeros, and maximum *r*-values to graph each function.

4A. $r = 3 \sin 2\theta$

4B. $r = \cos 5\theta$

Classic Polar Curves Circles, limaçons, cardioids, and roses are examples of classic curves. The forms and model graphs of these and other classic curves are summarized below.

Concept Summary | **Special Types of Polar Graphs**

Circles

$$r = a \cos \theta \text{ or } r = a \sin \theta$$

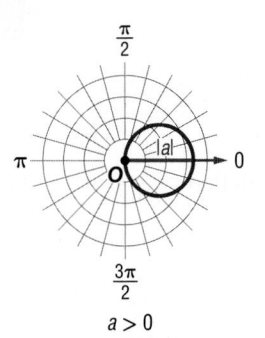

$a > 0$
$r = a \cos \theta$

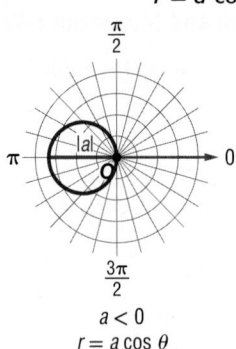

$a < 0$
$r = a \cos \theta$

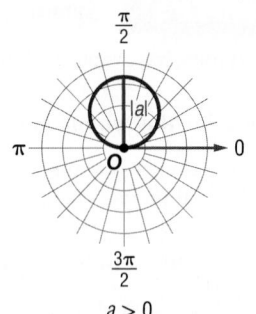

$a > 0$
$r = a \sin \theta$

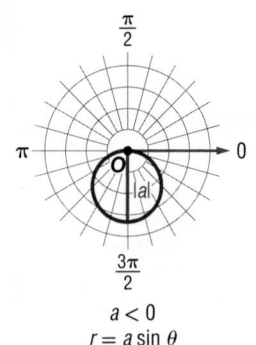

$a < 0$
$r = a \sin \theta$

Limaçons

$$r = a \pm b \cos \theta \text{ or } r = a \pm b \sin \theta, \text{ where } a \text{ and } b \text{ are both positive}$$

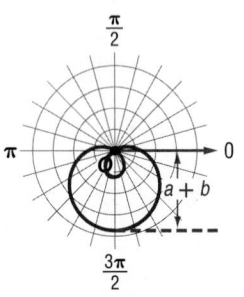

Limaçon with inner loop
$a < b$
$r = a - b \sin \theta$

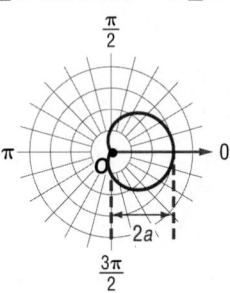

Cardioid
$a = b$
$r = a + b \cos \theta$

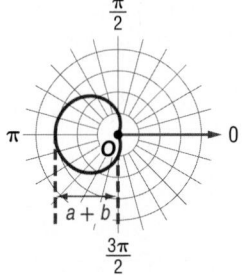

Dimpled limaçon
$b < a < 2b$
$r = a - b \cos \theta$

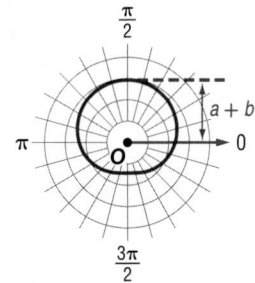

Convex limaçon
$a \geq 2b$
$r = a + b \sin \theta$

Roses

$$r = a \cos n\theta \text{ or } r = a \sin n\theta, \text{ where } n \geq 2 \text{ is an integer}$$
The rose has n petals if n is odd and $2n$ petals if n is even.

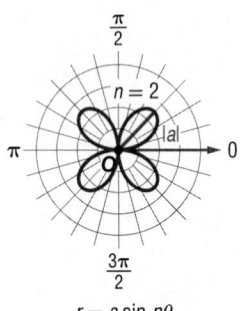

$n = 2$
$r = a \sin n\theta$

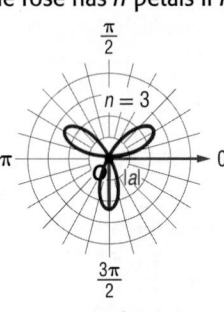

$n = 3$
$r = a \sin n\theta$

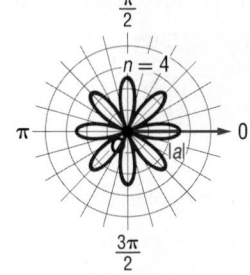

$n = 4$
$r = a \cos n\theta$

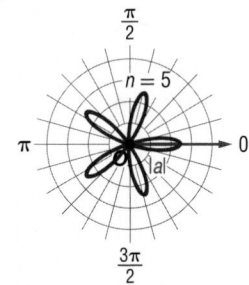

$n = 5$
$r = a \cos n\theta$

Lemniscates (LEM-nis-keyts)

$$r^2 = a^2 \cos 2\theta \text{ or } r^2 = a^2 \sin 2\theta$$

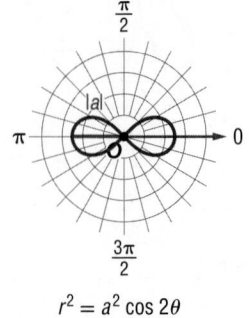

$r^2 = a^2 \cos 2\theta$

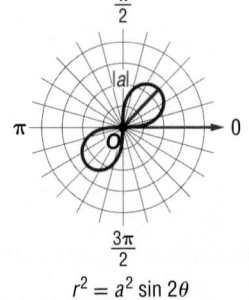

$r^2 = a^2 \sin 2\theta$

Spirals of Archimedes (ahr-kuh-MEE-deez)

$$r = a\theta + b$$

$\theta > 0, r = a\theta + b$

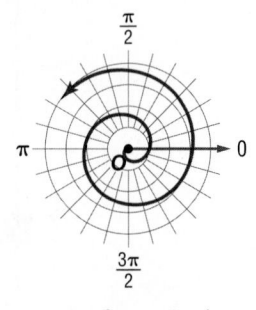

$\theta < 0, r = a\theta + b$

| EXAMPLE 5 | Identify and Graph Classic Curves |

Identify the type of curve given by each equation. Then use symmetry, zeros, and maximum *r*-values to graph the function.

a. $r^2 = 16 \sin 2\theta$

Type of Curve and Symmetry

The equation is of the form $r^2 = a^2 \sin 2\theta$, so its graph is a lemniscate. Replacing (r, θ) with $(-r, \theta)$ yields $(-r)^2 = 16 \sin 2\theta$ or $r^2 = 16 \sin 2\theta$. Therefore, the function has symmetry with respect to the pole.

Maximum *r*-Value and Zeros

The equation $r^2 = 16 \sin 2\theta$ is equivalent to $r = \pm 4\sqrt{\sin 2\theta}$, which is undefined when $\sin 2\theta < 0$. Therefore, the domain of the function is restricted to the intervals $\left[0, \frac{\pi}{2}\right]$ or $\left[\pi, \frac{3\pi}{2}\right]$.

Because you can use pole symmetry, you need only graph points in the interval $\left[0, \frac{\pi}{2}\right]$. The function attains a maximum *r*-value of $|a|$ or 4 when $\theta = \frac{\pi}{4}$ and zero *r*-value when $x = 0$ and $\frac{\pi}{2}$.

Graph

Use these points and the indicated symmetry to sketch the graph of the function.

θ	0	$\frac{\pi}{12}$	$\frac{\pi}{6}$	$\frac{\pi}{4}$	$\frac{\pi}{3}$	$\frac{5\pi}{12}$	$\frac{\pi}{2}$
r	0	± 2.8	± 3.7	± 4	± 3.7	± 2.8	0

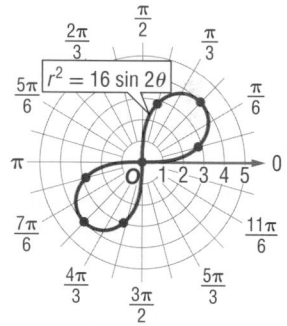

b. $r = 3\theta$

Type of Curve and Symmetry

The equation is of the form $r = a\theta + b$, so its graph is a spiral of Archimedes. Replacing (r, θ) with $(-r, -\theta)$ yields $(-r) = 3(-\theta)$ or $r = 3\theta$. Therefore, the function has symmetry with respect to the line $\theta = \frac{\pi}{2}$.

Maximum *r*-Value and Zeros

Spirals are unbounded. Therefore, the function has no maximum *r*-values and only one zero when $\theta = 0$.

Graph

Use points on the interval $[0, 4\pi]$ to sketch the graph of the function. To show symmetry, points on the interval $[-4\pi, 0]$ should also be graphed.

θ	0	$\frac{\pi}{4}$	$\frac{\pi}{2}$	π	$\frac{3\pi}{2}$	2π	3π	4π
r	0	2.4	4.7	9.4	14.1	18.8	28.3	37.7

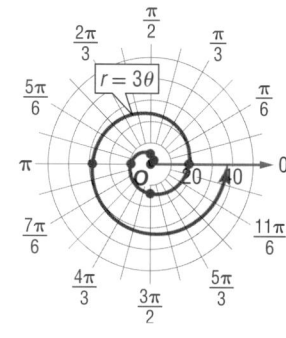

TechnologyTip

▶ **Window Settings** θmin and θmax determine the values of θ that will be graphed. Normal settings for these are θmin=0 and θmax=2π, although it may be necessary to change these values to obtain a complete graph. θstep determines the interval for plotting points. The smaller this value is, the smoother the look of the graph.

✔ **Guided Practice**

5A. $r^2 = 9 \cos 2\theta$

5B. $r = 3 \sin 5\theta$

Exercises

= **Step-by-Step Solutions** begin on page R29.

Graph each equation by plotting points. (Example 1)

1. $r = -\cos\theta$ **2.** $r = \csc\theta$

3. $r = \frac{1}{2}\cos\theta$ **4.** $r = 3\sin\theta$

5. $r = -\sec\theta$ **6.** $r = \frac{1}{3}\sin\theta$

7. $r = -4\cos\theta$ **8.** $r = -\csc\theta$

Use symmetry to graph each equation. (Examples 2 and 3)

9. $r = 3 + 3\cos\theta$ **10.** $r = 1 + 2\sin\theta$

11. $r = 4 - 3\cos\theta$ **12.** $r = 2 + 4\cos\theta$

13. $r = 2 - 2\sin\theta$ **14.** $r = 3 - 5\cos\theta$

15. $r = 5 + 4\sin\theta$ **16.** $r = 6 - 2\sin\theta$

Use symmetry, zeros, and maximum r-values to graph each function. (Example 4)

17. $r = \sin 4\theta$ **18.** $r = 2\cos 2\theta$

19. $r = 5\cos 3\theta$ **20.** $r = 3\sin 2\theta$

21. $r = \frac{1}{2}\sin 3\theta$ **22.** $r = 4\cos 5\theta$

23. $r = 2\sin 5\theta$ **24.** $r = 3\cos 4\theta$

25. MARINE BIOLOGY Rose curves can be observed in marine wildlife. Determine the symmetry, zeros, and maximum r-values of each function modeling a marine species for $0 \le \theta \le \pi$. Then use the information to graph the function. (Example 4)

   a. The pores forming the petal pattern of a sand dollar (Figure 9.2.3) can be modeled by $r = 3\cos 5\theta$.

   b. The outline of the body of a crown-of-thorns sea star (Figure 9.2.4) can be modeled by $r = 20\cos 8\theta$.

     Figure 9.2.3   **Figure 9.2.4**

Identify the type of curve given by each equation. Then use symmetry, zeros, and maximum r-values to graph the function. (Example 5)

26. $r = \frac{1}{3}\cos\theta$ **27.** $r = 4\theta + 1;\ \theta > 0$

28. $r = 2\sin 4\theta$ **29.** $r = 6 + 6\cos\theta$

30. $r^2 = 4\cos 2\theta$ **31.** $r = 5\theta + 2;\ \theta > 0$

32. $r = 3 - 2\sin\theta$ **33.** $r^2 = 9\sin 2\theta$

34. FIGURE SKATING The original focus of figure skating was to carve figures, known as *compulsory figures,* into the ice. The shape of one of these figures can be modeled by $r^2 = 25\cos 2\theta$. (Example 5)

   a. Which classic curve does the figure model?

   b. Graph the model.

Write an equation for each graph.

35.

36.

37.

38.

39.

40.

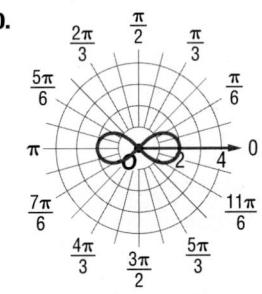

41 FAN A ceiling fan has a central motor with five blades that each extend 4 units from the center. The shape of the fan can be represented by a rose curve.

   a. Write two polar equations that can be used to represent the fan.

   b. Sketch two graphs of the fan using the equations that you wrote.

Use one of the three tests to prove the specified symmetry.

42. $r = 3 + \sin\theta$, symmetric about the line $\theta = \frac{\pi}{2}$

43. $r^2 = 4\sin 2\theta$, symmetric about the pole

44. $r = 3\sin 2\theta$, symmetric about the polar axis

45. $r = 5\cos 8\theta$, symmetric about the line $\theta = \frac{\pi}{2}$

46. $r = 2\sin 4\theta$, symmetric about the pole

47. FOUR-LEAF CLOVER The shape of a certain type of clover can be represented using a rose curve. Write a polar equation for the clover if it has:

   a. 5 petals with a length of 2 units each.

   b. 4 petals with a length of 7 units each.

   c. 8 petals with a length of 6 units each.

48. CONCERT For a concert, a circular stage is constructed and placed in the center so fans can completely surround the musicians. To record the sound of the crowd, two directional microphones are placed next to each other on the stage, one facing due east and the other facing due west. The patterns of the microphones can be represented by the polar equations $r = 2.5 + 2.5 \cos \theta$ and $r = -2.5 - 2.5 \cos \theta$.

a. Identify the type of curve given by each polar equation.

b. Sketch the graph of each microphone pattern on the same polar grid.

c. Describe what the graph tells you about the area covered by the microphones.

49. CANDY Write an equation that can model this lollipop in the shape of a limaçon if it is symmetric with respect to the line $\theta = \frac{\pi}{2}$ and measures 4 inches from the top of the lollipop to where the candy meets the stick.

4 in.

Match each equation with its graph.

50. $r = 1 + 4 \cos 3\theta$

51. $r = 1 - 4 \sin 4\theta$

52. $r = 1 - 3 \sin 3\theta$

53. $r = 1 + 3 \cos 4\theta$

a.

b.

c.

d.
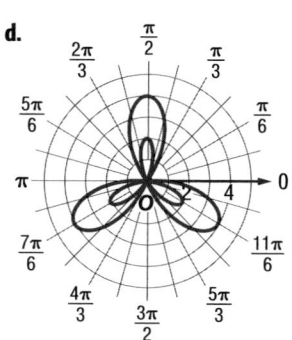

Find x for the interval $0 \le \theta \le x$ so that x is a minimum and the graph is complete.

54. $r = 3 + 2 \cos \theta$

55 $r = 2 - \sin 2\theta$

56. $r = 1 + \cos \frac{\theta}{3}$

Match each equation with an equation that produces an equivalent graph.

57. $r = 5 + 4 \cos \theta$

58. $r = -5 + 4 \sin \theta$

59. $r = 5 - 4 \sin \theta$

60. $r = -5 - 4 \cos \theta$

a. $r = 5 + 4 \sin \theta$

b. $r = -5 + 4 \cos \theta$

c. $r = 5 - 4 \cos \theta$

d. $r = -5 - 4 \sin \theta$

61. **MULTIPLE REPRESENTATIONS** In this problem, you will investigate a spiral of Archimedes.

a. **GRAPHICAL** Sketch separate graphs of $r = \theta$ for the intervals $0 \le \theta \le 3\pi$, $-3\pi \le \theta \le 0$, and $-3\pi \le \theta \le 3\pi$.

b. **VERBAL** Make a conjecture as to the symmetry of $r = \theta$. Explain your reasoning.

c. **ANALYTICAL** Prove your conjecture from part **b** by using one of the symmetry tests discussed in this lesson.

d. **VERBAL** How does changing the interval for θ affect the other classic curves? How does this differ from how the interval affects a spiral of Archimedes? Explain your reasoning.

H.O.T. Problems Use **Higher-Order Thinking Skills**

62. ERROR ANALYSIS Haley and Ella are graphing polar equations. Ella says that $r = 7 \sin 2\theta$ is not a function because it does not pass the vertical line test. Haley says the vertical line test does not apply in a polar grid. Is either of them correct? Explain your reasoning.

63. REASONING Sketch the graphs of $r_1 = \cos \theta$, $r_2 = \cos \left(\theta - \frac{\pi}{2} \right)$, and $r_3 = \cos (\theta - \pi)$ on the same polar grid. Describe the relationship between the three graphs. Make a conjecture as to the change in a graph when a value d is subtracted from θ.

64. CHALLENGE Solve the following system of polar equations algebraically on $[0, 2\pi]$. Graph the system and compare the points of intersection with the solutions that you found. Explain any discrepancies.

$r = 1 + 2 \sin \theta$

$r = 4 \sin \theta$

65. PROOF Prove that the graph of $r = a + b \cos 2\theta$ is symmetric with respect to the line $\theta = \frac{\pi}{2}$.

66. PROOF Prove that the graph of $r = a \sin 2\theta$ is symmetric with respect to the polar axis.

67. WRITING IN MATH Describe the effect of a in the graph of $r = a \cos \theta$.

68. OPEN ENDED Sketch the graph of a rose with 8 petals. Then write the equation for your graph.

Graph each polar equation. (Lesson 9-1)

69. $r = 3.5$

70. $\theta = -\dfrac{\pi}{3}$

71. $\theta = 225°$

Find the angle θ between vectors u and v to the nearest tenth of a degree. (Lesson 8-5)

72. $\mathbf{u} = \langle 4, -3, 5 \rangle$, $\mathbf{v} = \langle 2, 6, -8 \rangle$

73. $\mathbf{u} = 2\mathbf{i} - 4\mathbf{j} + 7\mathbf{k}$, $\mathbf{v} = 5\mathbf{i} + 6\mathbf{j} - 11\mathbf{k}$

74. $\mathbf{u} = \langle -1, 1, 5 \rangle$, $\mathbf{v} = \langle 7, -6, 9 \rangle$

Let $\overrightarrow{DE}$ be the vector with the given initial and terminal points. Write $\overrightarrow{DE}$ as a linear combination of the vectors i and j. (Lesson 8-2)

75. $D\left(-5, \dfrac{2}{3}\right)$, $E\left(-\dfrac{4}{5}, 0\right)$

76. $D\left(-\dfrac{1}{2}, \dfrac{4}{7}\right)$, $E\left(-\dfrac{3}{4}, \dfrac{5}{7}\right)$

77. $D(9.7, -2.4)$, $E(-6.1, -8.5)$

78. YARDWORK Kyle is pushing a wheelbarrow full of leaves with a force of 525 newtons at a 48° angle with the ground. (Lesson 8-1)

525 N

48°

  a. Draw a diagram that shows the resolution of the force that Kyle is exerting into its rectangular components.

  b. Find the magnitudes of the horizontal and vertical components of the force.

Graph the hyperbola given by each equation. (Lesson 7-3)

79. $\dfrac{x^2}{9} - \dfrac{y^2}{25} = 1$

80. $\dfrac{(y-4)^2}{16} - \dfrac{(x+2)^2}{9} = 1$

81. $\dfrac{(x+1)^2}{4} - \dfrac{(y+3)^2}{9} = 1$

Write an equation for and graph each parabola with focus F and the given characteristics. (Lesson 7-1)

82. $F(-5, 8)$; opens right; contains $(-5, 12)$

83. $F(-1, -5)$; opens left; contains $(-1, 5)$

Skills Review for Standardized Tests

84. SAT/ACT In the figure, C is the center of the circle, $AC = 12$, and $m\angle BAD = 60°$. What is the perimeter of the shaded region?

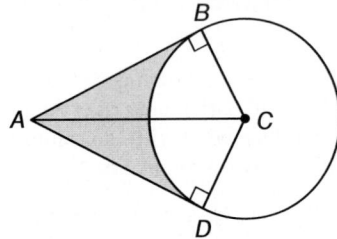

A $12 + 3\pi$

B $6\sqrt{3} + 4\pi$

C $6\sqrt{3} + 3\pi$

D $12\sqrt{3} + 3\pi$

E $12\sqrt{3} + 4\pi$

85. REVIEW While mapping a level site, a surveyor identifies a landmark 450 feet away and 30° left of center and another landmark 600 feet away and 50° right of center. What is the approximate distance between the two landmarks?

F 672 feet

G 685 feet

H 691 feet

J 703 feet

86. Which type of curve does the figure represent?

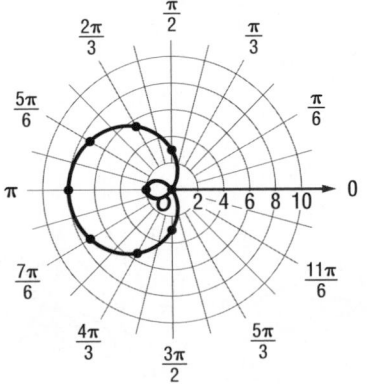

A lemniscate

B limaçon

C rose

D cardioid

87. REVIEW An air traffic controller is tracking two jets at the same altitude. The coordinates of the jets are $(5, 310°)$ and $(6, 345°)$, with r measured in miles. What is the approximate distance between the jets?

F 2.97 miles

G 3.25 miles

H 3.44 miles

J 3.71 miles

Polar and Rectangular Forms of Equations

Then
You used a polar coordinate system to graph points and equations. (Lessons 9-1 and 9-2)

Now
- Convert between polar and rectangular coordinates.
- Convert between polar and rectangular equations.

Math Online

glencoe.com

Why?

An ultrasonic sensor attached to a robot emits an outward beam that rotates through a full circle. The sensor receives a return signal when the beam intercepts an object, and it calculates the position of the object in terms of its distance r and the angle measure θ relative to the front of the robot. The sensor relays these polar coordinates to the robot, which converts them to rectangular coordinates so it can plot the object on an internal map.

Polar and Rectangular Coordinates Recall from Chapter 4 that the coordinates of a point $P(x, y)$ corresponding to an angle θ on a unit circle with radius 1 can be written in terms of θ as $P(\cos \theta, \sin \theta)$ because

$$\cos \theta = \frac{x}{r} = \frac{x}{1} \text{ or } x \quad \text{and} \quad \sin \theta = \frac{y}{r} = \frac{y}{1} \text{ or } y.$$

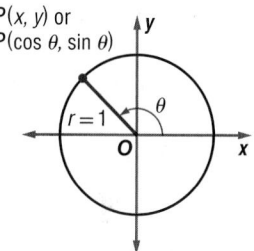

If we let r take on any real value, we can write a point $P(x, y)$ in terms of both r and θ.

$$\cos \theta = \frac{x}{r} \quad \text{and} \quad \sin \theta = \frac{y}{r}$$

$$r \cos \theta = x \qquad r \sin \theta = y \qquad \text{Multiply each side by } r.$$

If we let the polar axis and pole in the polar coordinate system coincide with the positive x-axis and origin in the rectangular coordinate system, respectively, we now have a means of converting polar coordinates to rectangular coordinates.

Key Concept   Convert Polar to Rectangular Coordinates

If a point P has polar coordinates (r, θ), then the rectangular coordinates (x, y) of P are given by

$$x = r \cos \theta \quad \text{and} \quad y = r \sin \theta.$$

That is, $(x, y) = (r \cos \theta, r \sin \theta)$.

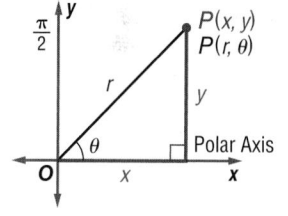

EXAMPLE 1   Polar Coordinates to Rectangular Coordinates

Find the rectangular coordinates for each point with the given polar coordinates.

a. $P\left(4, \frac{\pi}{6}\right)$

For $P\left(4, \frac{\pi}{6}\right)$, $r = 4$ and $\theta = \frac{\pi}{6}$.

$x = r \cos \theta$	**Conversion formula**	$y = r \sin \theta$
$= 4 \cos \frac{\pi}{6}$	$r = 4$ and $\theta = \frac{\pi}{6}$	$= 4 \sin \frac{\pi}{6}$
$= 4\left(\frac{\sqrt{3}}{2}\right)$	**Simplify.**	$= 4\left(\frac{1}{2}\right)$
$= 2\sqrt{3}$		$= 2$

The rectangular coordinates of P are $(2\sqrt{3}, 2)$ or approximately $(3.46, 2)$ as shown.

b. $Q(-2, 135°)$

For $Q(-2, 135°)$, $r = -2$ and $\theta = 135°$.

$x = r \cos \theta$	**Conversion formula**	$y = r \sin \theta$
$= -2 \cos 135°$	$r = -2$ and $\theta = 135°$	$= -2 \sin 135°$
$= -2\left(-\dfrac{\sqrt{2}}{2}\right)$	**Simplify.**	$= -2\left(\dfrac{\sqrt{2}}{2}\right)$
$= \sqrt{2}$		$= -\sqrt{2}$

The rectangular coordinates of Q are $(\sqrt{2}, -\sqrt{2})$ or approximately $(1.41, -1.41)$ as shown.

c. $V(3, -120°)$

For $V(3, -120°)$, $r = 3$ and $\theta = -120°$.

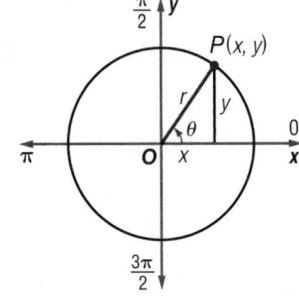

$x = r \cos \theta$	**Conversion formula**	$y = r \sin \theta$
$= 3 \cos -120°$	$r = 3$ and $\theta = -120°$	$= 3 \sin -120°$
$= 3\left(-\dfrac{1}{2}\right)$	**Simplify.**	$= 3\left(-\dfrac{\sqrt{3}}{2}\right)$
$= -\dfrac{3}{2}$		$= -\dfrac{3\sqrt{3}}{2}$

The rectangular coordinates of V are $\left(-\dfrac{3}{2}, -\dfrac{3\sqrt{3}}{2}\right)$ or approximately $(-1.5, -2.6)$ as shown.

✓ Guided Practice

1A. $R(-6, -120°)$ **1B.** $S\left(5, \dfrac{\pi}{3}\right)$ **1C.** $T(-3, 45°)$

To write a pair of rectangular coordinates in polar form, you need to find the distance r a point (x, y) is from the origin or pole and the angle measure θ that point is from the x- or polar axis.

To find the distance r from the point (x, y) to the origin, use the Pythagorean Theorem.

$r^2 = x^2 + y^2$	**Pythagorean Theorem**
$r = \sqrt{x^2 + y^2}$	**Take the positive square root of each side.**

The angle θ is related to x and y by the tangent function.

$\tan \theta = \dfrac{y}{x}$	**Tangent Ratio**
$\theta = \tan^{-1} \dfrac{y}{x}$	**Definition of inverse tangent function**

Recall that the inverse tangent function is only defined on the interval $\left[-\dfrac{\pi}{2}, \dfrac{\pi}{2}\right]$ or $[-90°, 90°]$. In the rectangular coordinate system, this refers to θ-values in Quadrants I and IV or when $x > 0$, as shown in Figure 9.3.1. If a point is located in Quadrant II or III, which is when $x < 0$, you must add π or $180°$ to the angle measure given by the inverse tangent function, as shown in Figure 9.3.2.

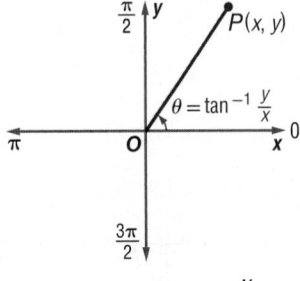

When $x > 0$, $\theta = \tan^{-1} \dfrac{y}{x}$.

Figure 9.3.1

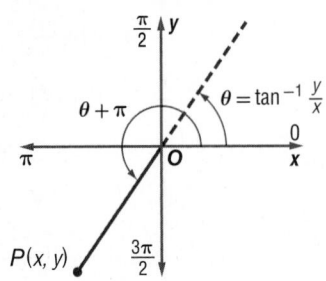

When $x < 0$, $\theta = \tan^{-1} \dfrac{y}{x} + \pi$ or $\theta = \tan^{-1} \dfrac{y}{x} + 180°$.

Figure 9.3.2

If a point P has rectangular coordinates (x, y) then the polar coordinates (r, θ) of P are given by

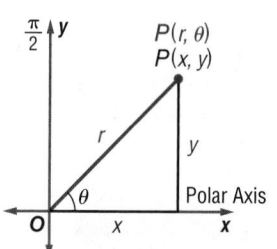

$$r = \sqrt{x^2 + y^2} \quad \text{and} \quad \theta = \tan^{-1}\frac{y}{x}, \text{ when } x > 0$$

$$\theta = \tan^{-1}\frac{y}{x} + \pi \text{ or}$$

$$\theta = \tan^{-1}\frac{y}{x} + 180°, \text{ when } x < 0.$$

Recall that polar coordinates are not unique. The conversion from rectangular coordinates to polar coordinates results in just *one* representation of the polar coordinates. There are, however, infinitely many polar representations for a point given in rectangular form.

EXAMPLE 2 **Rectangular Coordinates to Polar Coordinates**

Find two pairs of polar coordinates for each point with the given rectangular coordinates.

a. $S\left(1, -\sqrt{3}\right)$

For $S(x, y) = (1, -\sqrt{3})$, $x = 1$ and $y = -\sqrt{3}$. Because $x > 0$, use $\tan^{-1}\frac{y}{x}$ to find θ.

$r = \sqrt{x^2 + y^2}$	**Conversion formula**	$\theta = \tan^{-1}\frac{y}{x}$
$= \sqrt{1^2 + (-\sqrt{3})^2}$	$x = 1$ **and** $y = -\sqrt{3}$	$= \tan^{-1}\frac{-\sqrt{3}}{1}$
$= \sqrt{4}$ or 2	**Simplify.**	$= -\frac{\pi}{3}, \frac{5\pi}{3}$

One set of polar coordinate for S is $\left(2, -\frac{\pi}{3}\right)$. Another representation that uses a positive θ-value is $\left(2, -\frac{\pi}{3} + 2\pi\right)$ or $\left(2, \frac{5\pi}{3}\right)$, as shown.

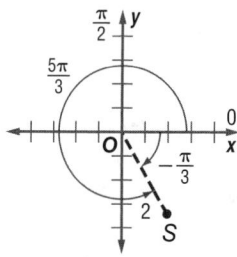

b. $T(-3, 6)$

For $T(x, y) = (-3, 6)$, $x = -3$ and $y = 6$.

Because $x < 0$, use $\tan^{-1}\frac{y}{x} + \pi$ to find θ.

$r = \sqrt{x^2 + y^2}$	**Conversion formula**	$\theta = \tan^{-1}\frac{y}{x} + \pi$
$= \sqrt{(-3)^2 + 6^2}$	$x = -3$ **and** $y = 6$	$= \tan^{-1}\left(-\frac{6}{3}\right) + \pi$
$= \sqrt{45}$ or about 6.71	**Simplify.**	$= \tan^{-1}(-2) + \pi$ or about 2.03

One set of polar coordinates for T is approximately $(6.71, 2.03)$. Another representation that uses a negative r-value is $(-6.71, 2.03 + \pi)$ or $(-6.71, 5.17)$, as shown.

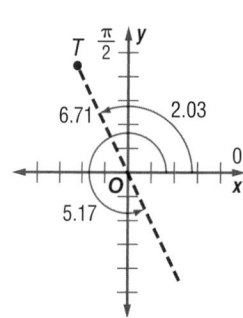

✔ **Guided Practice**

Find two pairs of polar coordinates for each point with the given rectangular coordinates. Round to the nearest hundredth, if necessary.

2A. $V(8, 10)$ **2B.** $W(-9, -4)$

For some real-world phenomena, it is useful to be able to convert between polar coordinates and rectangular coordinates.

Real-World EXAMPLE 3   Conversion of Coordinates

ROBOTICS Refer to the beginning of the lesson. Suppose the robot is facing due east and its sensor detects an object at (5, 295°).

a. What are the rectangular coordinates that the robot will need to calculate?

$x = r \cos \theta$	**Conversion formula**	$y = r \sin \theta$
$= 5 \cos 295°$	$r = 5$ **and** $\theta = 295°$	$= 5 \sin 295°$
≈ 2.11	**Simplify.**	≈ -4.53

The object is located at the rectangular coordinates $(2.11, -4.53)$.

b. If a previously detected object has rectangular coordinates of (3, 7), what are the distance and angle measure of the object relative to the front of the robot?

$r = \sqrt{x^2 + y^2}$	**Conversion formula**	$\theta = \tan^{-1} \dfrac{y}{x}$
$= \sqrt{3^2 + 7^2}$	$x = 3$ **and** $y = 7$	$= \tan^{-1} \dfrac{7}{3}$
≈ 7.62	**Simplify.**	$\approx 66.8°$

The object is located at the polar coordinates $(7.62, 66.8°)$.

✔ Guided Practice

3. **FISHING** A fish finder is a type of radar that is used to locate fish under water. Suppose a boat is facing due east, and a fish finder gives the polar coordinates of a school of fish as (6, 125°).

   A. What are the rectangular coordinates for the school of fish?

   B. If a previously detected school of fish had rectangular coordinates of (−2, 6), what are the distance and angle measure of the school relative to the front of the boat?

Polar and Rectangular Equations In calculus, you will sometimes need to convert from the rectangular form of an equation to its polar form and vice versa to facilitate some calculations. Some complicated rectangular equations have much simpler polar equations. Consider the rectangular and polar equations of the circle graphed below.

Rectangular Equation

$x^2 + y^2 = 9$

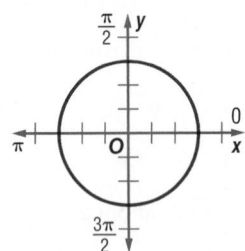

Polar Equation

$r = 3$

Likewise, some polar equations have much simpler rectangular equations, such as the line graphed below.

Polar Equation

$r = \dfrac{6}{2 \cos \theta - 3 \sin \theta}$

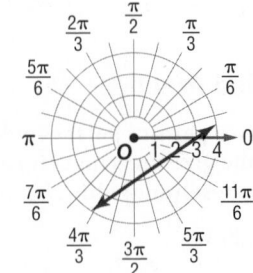

Rectangular Equation

$2x - 3y = 6$

The conversion of a rectangular equation to a polar equation is fairly straightforward. Replace x with $r \cos \theta$ and y with $r \sin \theta$, and then simplify the resulting equation using algebraic manipulations and trigonometric identities.

EXAMPLE 4 | Rectangular Equations to Polar Equations

Identify the graph of each rectangular equation. Then write the equation in polar form. Support your answer by graphing the polar form of the equation.

a. $(x - 4)^2 + y^2 = 16$

The graph of $(x - 4)^2 + y^2 = 16$ is a circle with radius 4 centered at $(4, 0)$. To find the polar form of this equation, replace x with $r \cos \theta$ and y with $r \sin \theta$. Then simplify.

$(x - 4)^2 + y^2 = 16$	**Original equation**
$(r \cos \theta - 4)^2 + (r \sin \theta)^2 = 16$	$x = r \cos \theta$ and $y = r \sin \theta$
$r^2 \cos^2 \theta - 8r \cos \theta + 16 + r^2 \sin^2 \theta = 16$	**Multiply.**
$r^2 \cos^2 \theta - 8r \cos \theta + r^2 \sin^2 \theta = 0$	**Subtract 16 from each side.**
$r^2 \cos^2 \theta + r^2 \sin^2 \theta = 8r \cos \theta$	**Isolate the squared terms.**
$r^2(\cos^2 \theta + \sin^2 \theta) = 8r \cos \theta$	**Factor.**
$r^2(1) = 8r \cos \theta$	**Pythagorean Identity**
$r = 8 \cos \theta$	**Divide each side by r.**

The graph of this polar equation (Figure 9.3.3) is a circle with radius 4 centered at $(4, 0)$.

b. $y = x^2$

The graph of $y = x^2$ is a parabola with vertex at the origin that opens up.

$y = x^2$	**Original equation**
$r \sin \theta = (r \cos \theta)^2$	$x = r \cos \theta$ and $y = r \sin \theta$
$r \sin \theta = r^2 \cos^2 \theta$	**Multiply.**
$\dfrac{\sin \theta}{\cos^2 \theta} = r$	**Divide each side by $r \cos^2 \theta$.**
$\dfrac{\sin \theta}{\cos \theta} \cdot \dfrac{1}{\cos \theta} = r$	**Rewrite.**
$\tan \theta \sec \theta = r$	**Quotient and Reciprocal Identities**

The graph of the polar equation $r = \tan \theta \sec \theta$ (Figure 9.3.4) is a parabola with vertex at the pole that opens up.

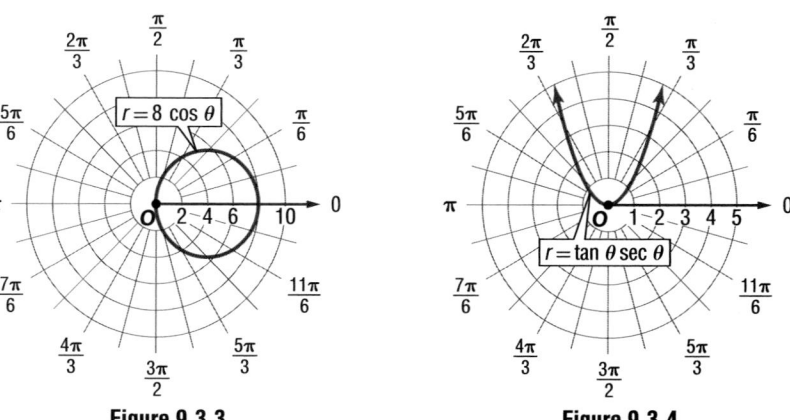

Figure 9.3.3   Figure 9.3.4

✔ Guided Practice

4A. $x^2 + (y - 3)^2 = 9$

4B. $x^2 - y^2 = 1$

To write a polar equation in rectangular form, you also make use of the relationships $r^2 = x^2 + y^2$, $x = r \cos \theta$, and $y = r \sin \theta$, as well as the relationship $\tan \theta = \frac{y}{x}$. The process, however, is not as straightforward as converting from rectangular to polar form.

StudyTip

Alternative Method Two points on the line $\theta = \frac{\pi}{6}$ are $\left(2, \frac{\pi}{6}\right)$ and $\left(4, \frac{\pi}{6}\right)$. In rectangular form, these points are $(\sqrt{3}, 1)$ and $(2\sqrt{3}, 2)$. The equation of the line through these points is $y = \frac{\sqrt{3}}{3}x$.

EXAMPLE 5 **Polar Equations to Rectangular Equations**

Write each equation in rectangular form, and then identify its graph. Support your answer by graphing the polar form of the equation.

a. $\theta = \frac{\pi}{6}$

$\theta = \dfrac{\pi}{6}$	Original equation
$\tan \theta = \dfrac{\sqrt{3}}{3}$	Find the tangent of each side.
$\dfrac{y}{x} = \dfrac{\sqrt{3}}{3}$	$\tan \theta = \dfrac{y}{x}$
$y = \dfrac{\sqrt{3}}{3}x$	Multiply each side by x.

The graph of this equation is a line through the origin with slope $\frac{\sqrt{3}}{3}$ or about $\frac{2}{3}$, as supported by the graph of $\theta = \frac{\pi}{6}$ shown.

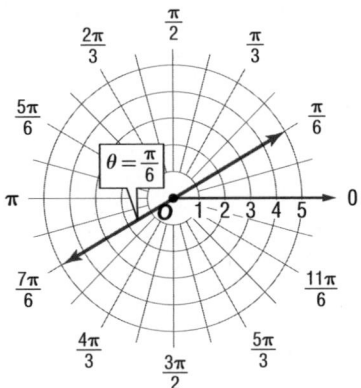

StudyTip

Converting to Rectangular Form Other useful substitutions are variations of the equations $x = r \cos \theta$ and $y = r \sin \theta$, such as $r = \dfrac{x}{\cos \theta}$ and $r = \dfrac{y}{\sin \theta}$.

b. $r = 7$

$r = 7$	Original equation
$r^2 = 49$	Square each side.
$x^2 + y^2 = 49$	$r^2 = x^2 + y^2$

The graph of this rectangular equation is a circle with center at the origin and radius 7, supported by the graph of $r = 7$ shown.

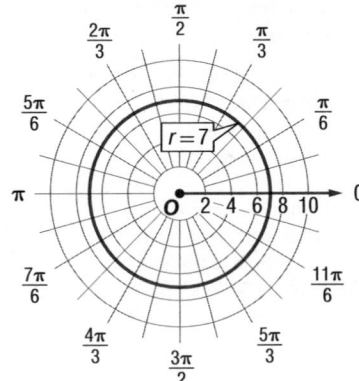

c. $r = -5 \sin \theta$

$r = -5 \sin \theta$	Original equation
$r^2 = -5r \sin \theta$	Multiply each side by r.
$x^2 + y^2 = -5y$	$r^2 = x^2 + y^2$ and $y = r \sin \theta$
$x^2 + y^2 + 5y = 0$	Add $5y$ to each side.

Because in standard form, $x^2 + (y + 2.5)^2 = 6.25$, you can identify the graph of this equation as a circle centered at $(0, -2.5)$ with radius 2.5, as supported by the graph of $r = -5 \sin \theta$.

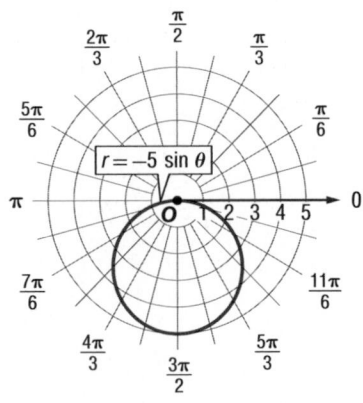

✅ **Guided Practice**

5A. $r = -3$ **5B.** $\theta = \dfrac{\pi}{3}$ **5C.** $r = 3 \cos \theta$

Find the rectangular coordinates for each point with the given polar coordinates. Round to the nearest hundredth, if necessary. (Example 1)

1. $\left(2, \frac{\pi}{4}\right)$

2. $\left(\frac{1}{4}, \frac{\pi}{2}\right)$

3. $(5, 240°)$

4. $(2.5, 250°)$

5. $\left(-2, \frac{4\pi}{3}\right)$

6. $(-13, -70°)$

7. $\left(3, \frac{\pi}{2}\right)$

8. $\left(\frac{1}{2}, \frac{3\pi}{4}\right)$

9. $(-2, 270°)$

10. $(4, 210°)$

11. $\left(-1, -\frac{\pi}{6}\right)$

12. $\left(5, \frac{\pi}{3}\right)$

Find two pairs of polar coordinates for each point with the given rectangular coordinates if $0 \leq \theta \leq 2\pi$. Round to the nearest hundredth, if necessary. (Example 2)

13. $(7, 10)$

14. $(-13, 4)$

15. $(-6, -12)$

16. $(4, -12)$

17. $(2, -3)$

18. $(0, -173)$

19. $(a, 3a), a > 0$

20. $(-14, 14)$

21. $(52, -31)$

22. $(3b, -4b), b > 0$ **23.** $(1, -1)$

24. $(2, \sqrt{2})$

25. DISTANCE Standing on top of his apartment building, Nicolas determines that a concert arena is 53° east of north. Suppose the arena is exactly 1.5 miles from Nicolas' apartment. (Example 3)

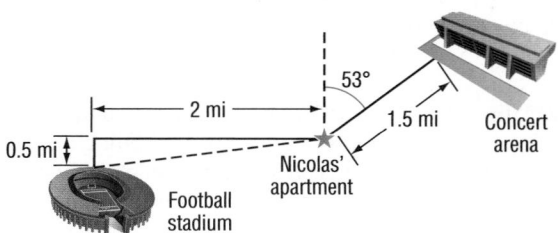

a. How many miles north and east will Nicolas have to travel to reach the arena?

b. If a football stadium is 2 miles west and 0.5 mile south of Nicolas' apartment, what are the polar coordinates of the stadium if Nicolas' apartment is at the pole?

Identify the graph of each rectangular equation. Then write the equation in polar form. Support your answer by graphing the polar form of the equation. (Example 4)

26. $x = -2$

27. $(x + 5)^2 + y^2 = 25$

28. $y = -3$

29. $x = y^2$

30. $(x - 2)^2 + y^2 = 4$

31. $(x - 1)^2 - y^2 = 1$

32. $x^2 + (y + 3)^2 = 9$

33. $y = \sqrt{3}x$

34. $x^2 + (y + 1)^2 = 1$

35. $x^2 + (y - 8)^2 = 64$

Write each equation in rectangular form, and then identify its graph. Support your answer by graphing the polar form of the equation. (Example 5)

36. $r = 3 \sin \theta$

37. $\theta = -\frac{\pi}{3}$

38. $r = 10$

39. $r = 4 \cos \theta$

40. $\tan \theta = 4$

41. $r = 8 \csc \theta$

42. $r = -4$

43. $\cot \theta = -7$

44. $\theta = \frac{3\pi}{4}$

45. $r = \sec \theta$

46. EARTHQUAKE An equation to model the seismic waves of an earthquake is $r = 12.6 \sin \theta$, where r is measured in miles. (Example 5)

a. Graph the polar pattern of the earthquake.

b. Write an equation in rectangular form to model the seismic waves.

c. Find the rectangular coordinates of the epicenter of the earthquake, and describe the area that is affected by the earthquake.

47 MICROPHONE The polar pattern for a directional microphone at a football game is given by $r = 2 + 2 \cos \theta$. (Example 5)

a. Graph the polar pattern.

b. Will the microphone detect a sound that originates from the point with rectangular coordinates $(-2, 0)$? Explain.

Write each equation in rectangular form, and then identify its graph. Support your answer by graphing the polar form of the equation.

48. $r = \frac{1}{\cos \theta + \sin \theta}$

49. $r = 10 \csc\left(\theta + \frac{7\pi}{4}\right)$

50. $r = 3 \csc\left(\theta - \frac{\pi}{2}\right)$

51. $r = -2 \sec\left(\theta - \frac{11\pi}{6}\right)$

52. $r = 4 \sec\left(\theta - \frac{4\pi}{3}\right)$

53. $r = \frac{5 \cos \theta + 5 \sin \theta}{\cos^2 \theta - \sin^2 \theta}$

54. $r = 2 \sin\left(\theta + \frac{\pi}{3}\right)$

55. $r = 4 \cos\left(\theta + \frac{\pi}{2}\right)$

56. ASTRONOMY Polar equations are used to model the paths of satellites or other orbiting bodies in space. Suppose the path of a satellite is modeled by $r = \frac{4}{4 + 3 \sin \theta}$, where r is measured in tens of thousands of miles, with Earth at the pole.

a. Sketch a graph of the path of the satellite.

b. Determine the minimum and maximum distances the satellite is from Earth at any time.

c. Suppose a second satellite passes through a point with rectangular coordinates $(1.5, -3)$. Are the two satellites at risk of ever colliding at this point? Explain.

Identify the graph of each rectangular equation. Then write the equation in polar form. Support your answer by graphing the polar form of the equation.

57. $6x - 3y = 4$

58. $2x + 5y = 12$

59. $(x - 6)^2 + (y - 8)^2 = 100$

60. $(x + 3)^2 + (y - 2)^2 = 13$

Write rectangular and polar equations for each graph.

61.

62.

63.

64.
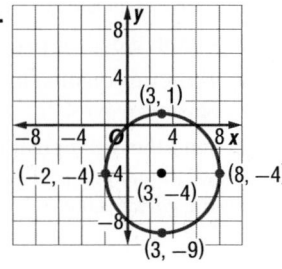

65. GOLF On the 18th hole at Hilly Pines Golf Course, the circular green is surrounded by a ring of sand as shown in the figure. Find the area of the region covered by sand assuming the hole acts as the pole for both equations and units are given in yards.

$r = 6 \cos \theta + 2 \sin \theta$

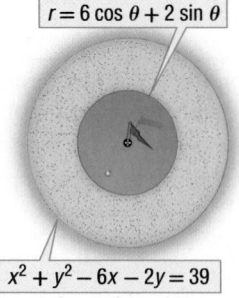

$x^2 + y^2 - 6x - 2y = 39$

66. CONSTRUCTION Boom cranes operate on three-dimensional counterparts of polar coordinates called *spherical coordinates*. A point in space has spherical coordinates (r, θ, ϕ), where r represents the distance from the pole, θ represents the angle of rotation about the vertical axis, and ϕ represents the polar angle from the positive vertical axis. Given a point in spherical coordinates (r, θ, ϕ) find the rectangular coordinates (x, y, z) in terms of r, θ, and ϕ.

67. **MULTIPLE REPRESENTATIONS** In this problem, you will investigate the relationship between complex numbers and polar coordinates.

a. GRAPHICAL The complex number $a + bi$ can be plotted on a complex plane using the ordered pair (a, b), where the x-axis is the real axis R and the y-axis is the imaginary axis i. Graph the complex number $6 + 8i$.

b. NUMERICAL Find polar coordinates for the complex number using the rectangular coordinates plotted in part **a** if $0 < \theta < 360°$. Graph the coordinates on a polar grid.

c. GRAPHICAL Graph the complex number $-3 + 3i$ on a rectangular coordinate system.

d. GRAPHICAL Find polar coordinates for the complex number using the rectangular coordinates plotted in part **c** if $0 < \theta < 360°$. Graph the coordinates on a polar grid.

e. ANALYTICAL For a complex number $a + bi$, find an expression for converting to polar coordinates.

H.O.T. Problems   Use Higher-Order Thinking Skills

68. ERROR ANALYSIS Becky and Terrell are writing the polar equation $r = \sin \theta$ in rectangular form. Terrell believes that the answer is $x^2 + \left(y - \frac{1}{2}\right)^2 = \frac{1}{4}$. Becky believes that the answer is simply $y = \sin x$. Is either of them correct? Explain your reasoning.

69. CHALLENGE The equation for a circle is $r = 2a \cos \theta$. Write this equation in rectangular form. Find the center and radius of the circle.

70. REASONING Given a set of rectangular coordinates (x, y) and a value for r, write expressions for finding θ in terms of sine and in terms of cosine. (*Hint:* You may have to write multiple expressions for each function, similar to the expressions given in this lesson using tangent.)

71. WRITING IN MATH Make a conjecture about when graphing an equation is made easier by representing the equation in polar form rather than rectangular form and vice versa.

72. PROOF Use $x = r \cos \theta$ and $y = r \sin \theta$ to prove that $r = x \sec \theta$ and $r = y \csc \theta$.

73 CHALLENGE Write $r^2(4 \cos^2 \theta + 3 \sin^2 \theta) + r(-8a \cos \theta + 6b \sin \theta) = 12 - 4a^2 - 3b^2$ in rectangular form. (*Hint:* Distribute before substituting values for r^2 and r. The rectangular equation should be a conic.)

74. WRITING IN MATH Use the definition of a polar axis given in Lesson 9-1 to explain why it was necessary to state that the robot in Example 3 was facing due east. How can the use of quadrant bearings help to eliminate this?

Use symmetry to graph each equation. (Lesson 9-2)

75. $r = 1 - 2 \sin \theta$

76. $r = -2 - 2 \sin \theta$

77. $r = 2 \sin 3\theta$

Find three different pairs of polar coordinates that name the given point if $-360° < \theta \leq 360°$ or $-2\pi < \theta \leq 2\pi$. (Lesson 9-1)

78. $T(1.5, 180°)$

79. $U\left(-1, \frac{\pi}{3}\right)$

80. $V(4, 315°)$

Find the angle θ between **u** and **v**. (Lesson 8-3)

81. $\mathbf{u} = \langle 6, -4 \rangle, \mathbf{v} = \langle -5, -7 \rangle$

82. $\mathbf{u} = \langle 2, 3 \rangle, \mathbf{v} = \langle -9, 6 \rangle$

83. $\mathbf{u} = \langle 1, 10 \rangle, \mathbf{v} = \langle 8, -2 \rangle$

Write each pair of parametric equations in rectangular form. Then graph and state any restrictions on the domain. (Lesson 7-5)

84. $y = t + 6$ and $x = \sqrt{t}$

85. $y = \frac{t}{2} + 1$ and $x = \frac{t^2}{4}$

86. $y = -3 \sin t$ and $x = 3 \cos t$

87. NAVIGATION Two LORAN broadcasting stations are located 460 miles apart. A ship receives signals from both stations and determines that it is 108 miles farther from Station 2 than Station 1. (Lesson 7-3)

   a. Determine the equation of the hyperbola centered at the origin on which the ship is located.

   b. Graph the equation, indicating on which branch of the hyperbola the ship is located.

   c. Find the coordinates of the location of the ship on the coordinate grid if it is 110 miles from the x-axis.

88. BICYCLES Woodland Bicycles makes two models of off-road bicycles: the Adventure, which sells for $250, and the Grande Venture, which sells for $350. Both models use the same frame. The painting and assembly time required for the Adventure is 2 hours, while the time is 3 hours for the Grande Venture. If there are 375 frames and 450 hours of labor available for production, how many of each model should be produced to maximize revenue? What is the maximum revenue? (Lesson 6-5)

Solve each system of equations using Gauss-Jordan elimination. (Lesson 6-1)

89. $3x + 9y + 6z = 21$
$4x - 10y + 3z = 15$
$-5x + 12y - 2z = -6$

90. $x + 5y - 3z = -14$
$2x - 4y + 5z = 18$
$-7x - 6y - 2z = 1$

91. $2x - 4y + z = 20$
$5x + 2y - 2z = -4$
$6x + 3y + 5z = 23$

92. SAT/ACT A square is inscribed in circle B. If the circumference of the circle is 50π, what is the length of the diagonal of the square?

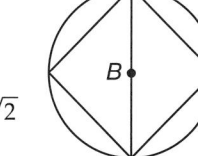

   A $10\sqrt{2}$ **D** 50

   B 25 **E** $50\sqrt{2}$

   C $25\sqrt{2}$

93. REVIEW Which of the following could be an equation for a rose with three petals?

   F $r = 3 \sin \theta$ **H** $r = 6 \sin \theta$

   G $r = \sin 3\theta$ **J** $r = \sin 6\theta$

94. What is the polar form of $x^2 + (y - 2)^2 = 4$?

   A $r = \sin \theta$

   B $r = 2 \sin \theta$

   C $r = 4 \sin \theta$

   D $r = 8 \sin \theta$

95. REVIEW Which of the following could be an equation for a spiral of Archimedes that passes through $A\left(\frac{\pi}{4}, \frac{\pi}{2}\right)$?

   F $r = \frac{\sqrt{2}\pi}{2} \cos \theta$ **H** $r = \frac{3}{4}$

   G $r = \theta$ **J** $r = \frac{\theta}{2}$

Graph each point on a polar grid. (Lesson 9-1)

1. $A(-2, 45°)$

2. $D(1, 315°)$

3. $C\left(-1.5, -\frac{4\pi}{3}\right)$

4. $B\left(3, -\frac{5\pi}{6}\right)$

Graph each polar equation. (Lesson 9-1)

5. $r = 3$

6. $\theta = -\frac{3\pi}{4}$

7. $\theta = 60°$

8. $r = -1.5$

9. HELICOPTERS A helicopter rotor consists of five equally spaced blades. Each blade is 11.5 feet long. (Lesson 9-1)

a. If the angle blade A makes with the polar axis is 3°, write an ordered pair to represent the tip of each blade on a polar grid. Assume that the rotor is centered at the pole.

b. What is the distance d between the tips of the helicopter blades to the nearest tenth of a foot?

Graph each equation. (Lesson 9-2)

10. $r = \frac{1}{4} \sec \theta$

11. $r = \frac{1}{3} \cos \theta$

12. $r = 3 \csc \theta$

13. $r = 4 \sin \theta$

14. STAINED GLASS A rose window is a circular window seen in gothic architecture. The pattern of the window radiates from the center. The window shown can be approximated by the equation $r = 3 \sin 6\theta$. Use symmetry, zeros, and maximum r-values of the function to graph the function. (Lesson 9-2)

Identify and graph each classic curve. (Lesson 9-2)

15. $r = \frac{1}{2} \sin \theta$

16. $r = \frac{1}{3}\theta + 3, \theta \geq 0$

17. $r = 1 + 2 \cos \theta$

18. $r = 5 \sin 3\theta$

19. MULTIPLE CHOICE Identify the polar graph of $y^2 = \frac{1}{2}x$. (Lesson 9-3)

A

C

B

D
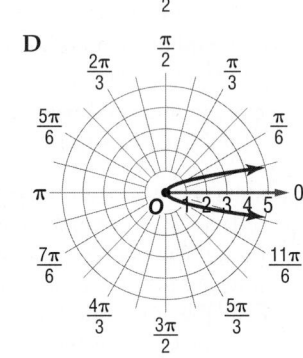

Find the rectangular coordinates for each point with the given polar coordinates. (Lesson 9-3)

20. $\left(4, \frac{2\pi}{3}\right)$

21. $\left(-2, -\frac{\pi}{4}\right)$

22. $(-1, 210°)$

23. $(3, 30°)$

Find two pairs of polar coordinates for each point with the given rectangular coordinates if $0 \leq \theta \leq 2\pi$. Round to the nearest hundredth. (Lesson 9-3)

24. $(-3, 5)$

25. $(8, 1)$

26. $(7, -6)$

27. $(-4, -10)$

Write a rectangular equation for each graph. (Lesson 9-3)

28.
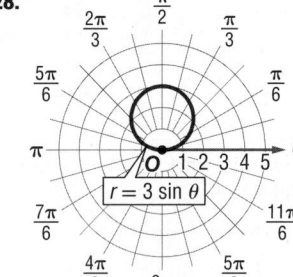
$r = 3 \sin \theta$

29.
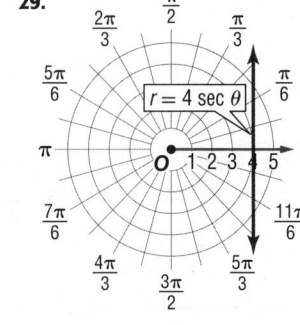
$r = 4 \sec \theta$

Polar Forms of Conic Sections

Why?

Polar equations of conic sections can be used to model orbital motion, such as the orbit of a planet around the Sun or the orbit of a satellite around a planet.

Then
You defined conic sections.
(Lessons 7-1 through 7-3)

Now
- Identify polar equations of conics.
- Write and graph the polar equation of a conic given its eccentricity and the equation of its directrix.

Math Online

glencoe.com

Use Polar Equations of Conics In Chapter 7, you defined conic sections in terms of the distance between a focus and directrix (parabola) or between two foci (ellipse and hyperbola). Alternatively, we can define all of these curves using the focus-directrix definition of a parabola.

In general, a conic section can be defined as the locus of points such that the distance from a point P to the focus and the distance from the point to a fixed line not containing P (the directrix) is a constant ratio. This constant ratio $\frac{PF}{PQ}$ represents the eccentricity of a conic and is denoted e.

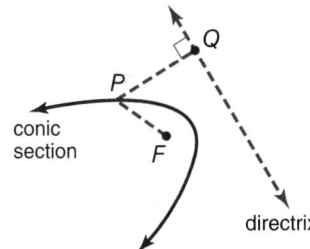

e as Constant Ratio		e as Constant Multiplier
$e = \dfrac{PF}{PQ}$	or	$PF = e \cdot PQ$

Recall that for a parabola, $PF = PQ$. Therefore, a parabola has eccentricity $\frac{PQ}{PQ}$ or 1. Other values of e give us other conics. These eccentricities are summarized below.

Concept Summary   **Eccentricities of Conics**

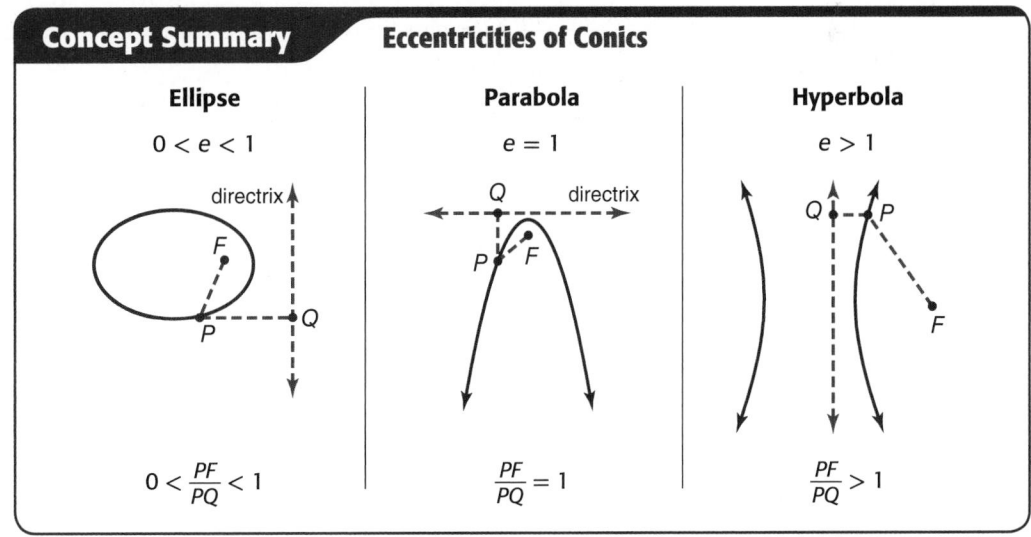

Ellipse	Parabola	Hyperbola
$0 < e < 1$	$e = 1$	$e > 1$
$0 < \dfrac{PF}{PQ} < 1$	$\dfrac{PF}{PQ} = 1$	$\dfrac{PF}{PQ} > 1$

Recall too that when the center of a conic section lies at the origin, the rectangular equations of conics take on a simpler form.

Ellipses	Parabolas	Hyperbolas
$\dfrac{x^2}{a^2} + \dfrac{y^2}{b^2} = 1$ or $\dfrac{x^2}{b^2} + \dfrac{y^2}{a^2} = 1$	$x^2 = 4pv$ or $y^2 = 4px$	$\dfrac{x^2}{a^2} - \dfrac{y^2}{b^2} = 1$ or $\dfrac{y^2}{a^2} - \dfrac{x^2}{b^2} = 1$

Using the focus-directrix definition, the equation of a conic in polar form is simplified if a *focus* of the conic lies at the origin.

Other Conics When defining conics in terms of their eccentricity, *e* is a strictly positive constant. There are no circles, lines, or other degenerate conics.

Consider a conic with its focus located at the origin and its directrix to the right at $x = d$. For any point $P(x, y)$ on the curve, the distance PF is given by $\sqrt{x^2 + y^2}$, and the distance PQ is given by $d - x$. We can substitute these expressions in the definition of a conic section.

$$PF = e \cdot PQ \qquad \text{Definition of a conic section}$$

$$\sqrt{x^2 + y^2} = e(d - x) \qquad PF = \sqrt{x^2 + y^2} \text{ and } PQ = d - x$$

The expression $\sqrt{x^2 + y^2}$ should make you think of polar coordinates. In fact, the equation above has a simpler form in the polar coordinate system.

$$\sqrt{x^2 + y^2} = e(d - x) \qquad \text{Rectangular form of conic defined in terms of its eccentricity } e$$

$$r = e(d - r \cos \theta) \qquad r = \sqrt{x^2 + y^2} \text{ and } x = r \cos \theta$$

$$r = ed - er \cos \theta \qquad \text{Distributive Property}$$

$$r + er \cos \theta = ed \qquad \text{Isolate } r\text{-terms.}$$

$$r(1 + e \cos \theta) = ed \qquad \text{Factor.}$$

$$r = \frac{ed}{1 + e \cos \theta} \qquad \text{Solve for } r.$$

This last equation is the polar form of an equation for the conic sections with focus at the pole and vertical directrix and center or vertex to the right of the pole. Different orientations of the focus and directrix can produce different forms of this polar equation as summarized below.

ReadingMath

Eccentricity In each of these polar equations, the letter *e* is a variable that represents the eccentricity of the conic. It should *not* be confused with the transcendental number *e*, which is a constant.

Key Concept — **Polar Equations of Conics**

The conic section with eccentricity $e > 0$, $d > 0$, and focus at the pole has the polar equation:

- $r = \dfrac{ed}{1 + e \cos \theta}$ if the directrix is the vertical line $x = d$ (Figure 9.4.1),

- $r = \dfrac{ed}{1 - e \cos \theta}$ if the directrix is the vertical line $x = -d$ (Figure 9.4.2),

- $r = \dfrac{ed}{1 + e \sin \theta}$ if the directrix is the horizontal line $y = d$ (Figure 9.4.3), and

- $r = \dfrac{ed}{1 - e \sin \theta}$ if the directrix is the horizontal line $y = -d$ (Figure 9.4.4).

In each of the examples below, $e = 1$, so the conic takes the form of a parabola.

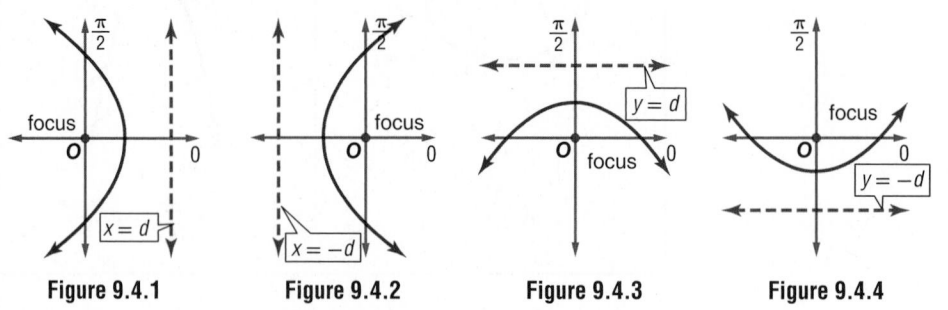

| Figure 9.4.1 | Figure 9.4.2 | Figure 9.4.3 | Figure 9.4.4 |

You will derive the last three of these equations in Exercises 50–52.

Notice that for $r = \dfrac{ed}{1 - e \cos \theta}$, the directrix of the conic is to the left of the pole. For $r = \dfrac{ed}{1 + e \sin \theta}$, the directrix is above the pole. For $r = \dfrac{ed}{1 - e \sin \theta}$, the directrix is below the pole.

To analyze the polar equation of a conic, begin by writing the equation in standard form, $r = \dfrac{ed}{1 \pm e \cos \theta}$ or $r = \dfrac{ed}{1 \pm e \sin \theta}$. In this form, determine the eccentricity and use this value to identify the type of conic the equation represents. Then determine the equation of the directrix, and use it to describe the orientation of the conic.

EXAMPLE 1 Identify Conics from Polar Equations

Determine the eccentricity, type of conic, and equation of the directrix for each polar equation.

a. $r = \dfrac{9}{3 + 2.25 \cos \theta}$

Write the equation in standard form, $r = \dfrac{ed}{1 + e \cos \theta}$.

$r = \dfrac{9}{3 + 2.25 \cos \theta}$ **Original equation**

$r = \dfrac{3(3)}{3(1 + 0.75 \cos \theta)}$ **Factor the numerator and denominator.**

$r = \dfrac{3}{1 + 0.75 \cos \theta}$ **Divide the numerator and denominator by 3.**

In this form, you can see from the denominator that $e = 0.75$. Therefore, the conic is an ellipse. For polar equations of this form, the equation of the directrix is $x = d$. From the numerator, we know that $ed = 3$, so $d = 3 \div 0.75$ or 4. Therefore, the equation of the directrix is $x = 4$.

CHECK Sketch the graph of $r = \dfrac{9}{3 + 2.25 \cos \theta}$ and its directrix $x = 4$ using either the techniques shown in Lesson 9-2 or a graphing calculator. The graph is an ellipse with its directrix to the right of the pole. ✔

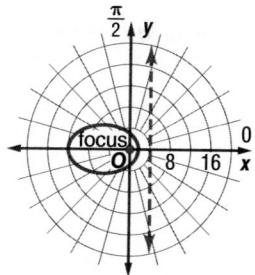

StudyTip

Focus-Directrix Pairs
While a parabola has one focus and one directrix, ellipses and hyperbolas have two foci-directrix pairs. Either focus-directrix pair can be used to generate the conic.

b. $r = \dfrac{-16}{4 \sin \theta - 2}$

Write the equation in standard form.

$r = \dfrac{-16}{4 \sin \theta - 2}$ **Original equation**

$r = \dfrac{-2(8)}{-2(1 - 2 \sin \theta)}$ **Factor the numerator and denominator.**

$r = \dfrac{8}{1 - 2 \sin \theta}$ **Divide the numerator and denominator by −2.**

The equation is of the form $r = \dfrac{ed}{1 - e \sin \theta}$, so $e = 2$. Therefore, the conic is a hyperbola. For polar equations of this form, the equation of the directrix is $y = -d$. Because $ed = 8$, $d = 8 \div 2$ or 4. Therefore, the equation of the directrix is $y = -4$.

CHECK Sketch the graph of $r = \dfrac{-16}{4 \sin \theta - 2}$ and its directrix $y = -4$. The graph is a hyperbola with one focus at the origin, above the directrix. ✔

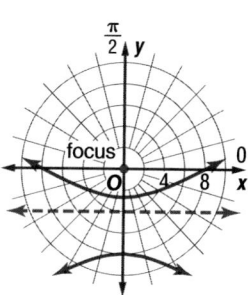

✔ Guided Practice

1A. $r = \dfrac{-6}{3 \cos \theta - 1}$

1B. $r = \dfrac{9}{3 + 3 \sin \theta}$

1C. $r = \dfrac{1}{6 + 1.2 \cos \theta}$

Write Polar Equations of Conics You can write the polar equation of a conic given its eccentricity and the equation of the directrix or its eccentricity and some other characteristics.

EXAMPLE 2 Write Polar Equations of Conics

Write and graph a polar equation and directrix for the conic with the given characteristics.

a. $e = 2$; directrix: $y = 4$

Because $e = 2$, the conic is a hyperbola. The directrix $y = 4$ is above the pole, so the equation is of the form $r = \dfrac{ed}{1 + e \sin \theta}$. Use the values for e and d to write the equation.

$r = \dfrac{ed}{1 + e \sin \theta}$ **Polar form of conic with directrix $y = d$**

$r = \dfrac{2(4)}{1 + 2 \sin \theta}$ or $\dfrac{8}{1 + 2 \sin \theta}$ **$e = 2$ and $d = 4$**

Sketch the graph of this polar equation and its directrix. The graph is a hyperbola with its directrix above the pole.

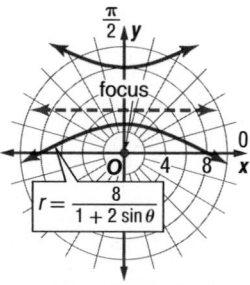

b. $e = 0.5$; vertices at $(-4, 0)$ and $(12, 0)$

Because $e = 0.5$, the conic is an ellipse. The center of the ellipse is at $(4, 0)$, the midpoint of the segment between the given vertices. This point is to the right of the pole. Therefore, the directrix will be to the left of the pole at $x = -d$. The polar equation of a conic with this directrix is $r = \dfrac{ed}{1 - e \cos \theta}$.

Use the value of e and the polar form of a point on the conic to find the value of d. The vertex point $(12, 0)$ has polar coordinates $(r, \theta) = \left(\sqrt{12^2 + 0^2}, \tan^{-1} \dfrac{0}{12} \right)$ or $(12, 0)$.

$r = \dfrac{ed}{1 - e \cos \theta}$ **Polar form of conic with directrix $x = -d$**

$12 = \dfrac{0.5d}{1 - 0.5 \cos 0}$ **$e = 0.5$, $r = 12$, and $\theta = 0$**

$12 = \dfrac{0.5d}{0.5}$ **$\cos 0 = 1$**

$12 = d$ **Simplify.**

Therefore, the equation for the ellipse is $r = \dfrac{0.5 \cdot 12}{1 - 0.5 \cos \theta}$ or $r = \dfrac{6}{1 - 0.5 \cos \theta}$. Because $d = 12$, the equation of the directrix is $x = -12$. The graph is an ellipse with vertices at $(-4, 0)$ and $(12, 0)$.

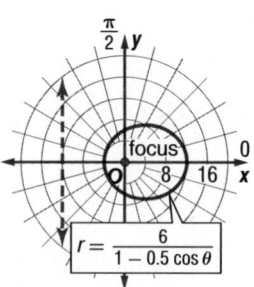

✓ **Guided Practice**

2A. $e = 1$; directrix: $x = 2$

2B. $e = 2.5$; vertices at $(0, -3)$ and $(0, -7)$

In Lessons 7-1 through 7-3, you analyzed the rectangular equations of conics in standard form to describe the geometric properties of parabolas, ellipses, and hyperbolas. You can use the geometric analysis of the graph of a conic given in polar form to write the equation in rectangular form.

EXAMPLE 3 **Write the Polar Form of Conics in Rectangular Form**

Write each polar equation in rectangular form.

a. $r = \dfrac{4}{1 - \sin \theta}$

Step 1 Analyze the polar equation.

For this equation, $e = 1$ and $d = 4$. The eccentricity and form of the equation determine that this is a parabola that opens vertically with focus at the pole and a directrix $y = -4$. The general equation of such a parabola in rectangular form is $(x - h)^2 = 4p(y - k)$.

Step 2 Determine values for h, k, and p.

The vertex lies between the focus F and directrix of the parabola, occurring when $\theta = \dfrac{3\pi}{2}$, as shown in Figure 9.4.5. Evaluating the function at this value, we find that the vertex lies at polar coordinates $\left(2, \dfrac{3\pi}{2}\right)$, which correspond to rectangular coordinates $(0, -2)$. So, $(h, k) = (0, -2)$. The distance p from the vertex at $(0, -2)$ to the focus at $(0, 0)$ is 2.

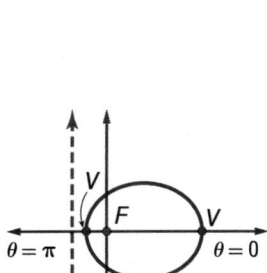

Figure 9.4.5

Step 3 Substitute the values for h, k, and p into the standard form of an equation for a parabola.

$(x - h)^2 = 4p(y - k)$ **Standard form of a parabola**

$(x - 0)^2 = 4(2)[y - (-2)]$ **$h = 0$, $k = -2$, and $p = 2$**

$x^2 = 8y + 16$ **Simplify.**

b. $r = \dfrac{3.2}{1 - 0.6 \cos \theta}$

Step 1 Analyze the polar equation.

For this equation, $e = 0.6$ and $d \approx 5.3$. The eccentricity and form of the equation determine that this is an ellipse with directrix $x = -5.3$. Therefore, the major axis of the ellipse lies along the polar or x-axis. The general equation of such an ellipse in rectangular form is $\dfrac{(x - h)^2}{a^2} + \dfrac{(y - k)^2}{b^2} = 1$.

Step 2 Determine values for h, k, a, and b.

The vertices are the endpoints of the major axis and occur when $\theta = 0$ and π as shown in Figure 9.4.6. Evaluating the function at these values, we find that the vertices have polar coordinates $(8, 0)$ and $(2, \pi)$, which correspond to rectangular coordinates $(8, 0)$ and $(-2, 0)$. The ellipse's center is the midpoint of the segment between the vertices, so $(h, k) = (3, 0)$.

The distance a between the center and each vertex is 5. The distance c from the center to the focus at $(0, 0)$ is 3. By the Pythagorean relation $b = \sqrt{a^2 - c^2}$, $b = \sqrt{5^2 - 3^2}$ or 4.

Step 3 Substitute the values for h, k, a, and b into the standard form of an equation for an ellipse.

$\dfrac{(x - h)^2}{a^2} + \dfrac{(y - k)^2}{b^2} = 1$ **Standard form of an ellipse**

$\dfrac{(x - 3)^2}{5^2} + \dfrac{(y - 0)^2}{4^2} = 1$ **$h = 3$, $k = 0$, $a = 5$, and $b = 4$**

$\dfrac{(x - 3)^2}{25} + \dfrac{y^2}{16} = 1$ **Simplify.**

Figure 9.4.6

☑ **Guided Practice**

3A. $r = \dfrac{2.5}{1 - 1.5 \cos \theta}$

3B. $r = \dfrac{5}{1 + \sin \theta}$

Determine the eccentricity, type of conic, and equation of the directrix for each polar equation. (Example 1)

1. $r = \dfrac{20}{4 + 4 \sin \theta}$

2. $r = \dfrac{18}{2 - 6 \cos \theta}$

3. $r = \dfrac{21}{3 \cos \theta + 1}$

4. $r = \dfrac{24}{4 \sin \theta + 8}$

5. $r = \dfrac{-12}{6 \cos \theta - 6}$

6. $r = \dfrac{9}{4 - 3 \sin \theta}$

7. $r = \dfrac{-8}{\sin \theta - 0.25}$

8. $r = \dfrac{10}{2.5 + 2.5 \cos \theta}$

9 **TELESCOPES** The Cassegrain Telescope, invented in 1692, produces an image by reflecting light off of parabolic and hyperbolic mirrors. Determine the eccentricity, type of conic, and the equation of the directrix for each equation modeling a mirror in the telescope. (Example 1)

light path

Hyperbolic Mirror   Parabolic Mirror

a. $r = \dfrac{7}{2 \sin \theta + 2}$

b. $r = \dfrac{28}{12.5 \cos \theta + 5}$

Write and graph a polar equation and directrix for the conic with the given characteristics. (Example 2)

10. $e = 1$: directrix: $y = 6$

11. $e = 0.75$; directrix: $x = -8$

12. $e = 5$; directrix: $x = 2$

13. $e = 0.1$; directrix: $y = 8$

14. $e = 6$; directrix: $y = -7$

15. $e = 1$; directrix: $x = -1.5$

16. $e = 0.8$; vertices at $(-36, 0)$ and $(4, 0)$

17. $e = 1.5$; vertices at $(-3, 0)$ and $(-15, 0)$

Write each polar equation in rectangular form. (Example 3)

18. $r = \dfrac{4.8}{1 + \sin \theta}$

19. $r = \dfrac{30}{4 + \cos \theta}$

20. $r = \dfrac{5}{1 - 1.5 \cos \theta}$

21. $r = \dfrac{5.1}{1 + 0.7 \sin \theta}$

22. $r = \dfrac{12}{1 - \cos \theta}$

23. $r = \dfrac{6}{0.25 - 0.75 \sin \theta}$

24. $r = \dfrac{4.5}{1 + 1.25 \sin \theta}$

25. $r = \dfrac{8.4}{1 - 0.4 \cos \theta}$

GRAPHING CALCULATOR Determine the type of conic for each polar equation. Then graph each equation.

26. $r = \dfrac{2}{2 + \sin \left(\theta + \dfrac{\pi}{3}\right)}$

27. $r = \dfrac{3}{1 + \cos \left(\theta - \dfrac{\pi}{4}\right)}$

28. $r = \dfrac{2}{1 - \cos \left(\theta + \dfrac{\pi}{6}\right)}$

29. $r = \dfrac{4}{1 + 2 \sin \left(\theta + \dfrac{3\pi}{4}\right)}$

Match each polar equation with its graph.

a.

b.

c.

d.
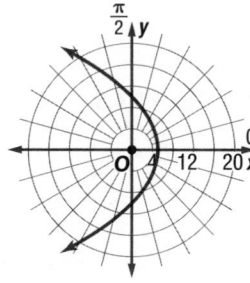

30. $r = \dfrac{10}{1 + \cos \theta}$

31. $r = \dfrac{4}{1 - \sin \theta}$

32. $r = \dfrac{5}{2 - \cos \theta}$

33. $r = \dfrac{12}{1 + 3 \sin \theta}$

Determine the eccentricity, type of conic, and equation of the directrix for each polar equation. Then sketch the graph of the equation, and label the directrix.

34. $r = \dfrac{12}{2 - 0.75 \cos \theta}$

35. $r = \dfrac{1}{0.2 - 0.2 \sin \theta}$

36. $r = \dfrac{6}{1.2 \sin \theta + 0.3}$

37. $r = \dfrac{8}{\cos \theta + 5}$

38. **ASTRONOMY** The comet Borrelly travels in an elliptical orbit around the Sun with eccentricity $e = 0.624$. The point in a comet's orbit nearest to the Sun is defined as the *perihelion*, while the farthest point from the Sun is defined as the *aphelion*. The aphelion occurs at a distance of 5.83 AU (astronomical units, based on the distance between Earth and the Sun) from the Sun and the perihelion occurs at a distance of 1.35 AU. The diameter of the Sun is about 0.0093 AU.

a. Write a polar equation for the elliptical orbit of the comet Borrelly, and graph the equation.

b. Determine the distance in miles between the comet Borrelly and the Sun at the aphelion and perihelion if 1 AU ≈ 93 million miles.

PROOF Prove each of the following.

39. $b = a\sqrt{1 - e^2}$ for an ellipse

40. $b = a\sqrt{e^2 - 1}$ for a hyperbola

41. PROOF Use the definition for the eccentricity of a conic, $PF = ePQ$, and the drawing of the hyperbola shown below, to verify that $d = \dfrac{a(e^2 - 1)}{e}$ for any hyperbola.

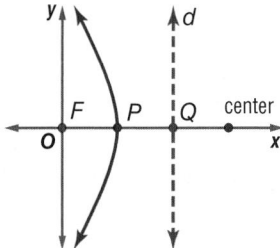

Write each rectangular equation in polar form.

42. $x^2 = 4y + 4$

43. $-10y + 25 = x^2$

44. $\dfrac{(x-2)^2}{16} + \dfrac{y^2}{12} = 1$

45. $\dfrac{(x+4)^2}{64} + \dfrac{y^2}{48} = 1$

46. ASTRONOMY The planets travel around the Sun in approximately elliptical orbits with the Sun at one focus, as shown below.

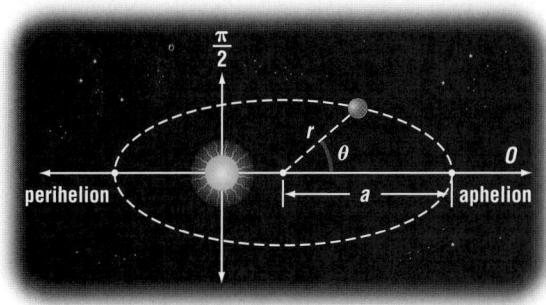

perihelion   a   aphelion

a. Show that the polar equation of the planets' orbit can be written as $r = \dfrac{a(1 - e^2)}{(1 - e \cos \theta)}$.

b. Prove that the perihelion distance of any planet is $a(1 - e)$, and the aphelion distance is $a(1 + e)$.

c. Use the formulas from part **a** to find the perihelion and aphelion distances for each of the planets.

Planet	a	e	Planet	a	e
Earth	1.000	0.017	Neptune	30.06	0.009
Jupiter	5.203	0.048	Saturn	9.539	0.056
Mars	1.524	0.093	Uranus	19.18	0.047
Mercury	0.387	0.206	Venus	0.723	0.007

d. For which planet is the distance between the perihelion and aphelion the smallest? the greatest?

Write each equation in polar form. (*Hint*: Translate each conic so that a focus lies on the pole.)

47. $\dfrac{(x-2)^2}{64} - \dfrac{y^2}{36} = 1$

48. $3(x+5)^2 + 4y^2 = 192$

49. **MULTIPLE REPRESENTATIONS** In this problem, you will investigate the effects of varying the eccentricity and the directrix on graphs of conic sections.

a. NUMERICAL Write an equation for a conic section with focus $(0, 0)$ and directrix $x = 3$ for $e = 0.4, 0.6, 1, 1.6,$ and 2. Then identify the type of conic that each equation represents.

b. GRAPHICAL Graph and label the eccentricity for each of the equations that you found in part **a** on the same coordinate plane.

c. VERBAL Describe the changes in the graphs from part **b** as e approaches 2.

d. NUMERICAL Write an equation for a conic section with focus $(0, 0)$ and eccentricity $e = 0.5$ for $d = 0.25, 1,$ and 4.

e. GRAPHICAL Graph each of the equations on the same coordinate plane.

f. VERBAL Describe the relationship between the value of d and the distances between the vertices and the foci of the graphs from part **e**.

Derive each of the following polar equations of conics as shown on page 562 for the equation $r = \dfrac{ed}{1 + e \cos \theta}$. **Include a diagram with each derivation.**

50. $r = \dfrac{ed}{1 - e \cos \theta}$

51. $r = \dfrac{ed}{1 + e \sin \theta}$

52. $r = \dfrac{ed}{1 - e \sin \theta}$

H.O.T. Problems Use **Higher-Order** Thinking Skills

53. WRITING IN MATH Describe two definitions that can be used to define a conic section.

54. REASONING Explain why $r = \dfrac{ed}{1 + e \sin \theta}$ does not produce a true circle for any value of e.

CHALLENGE Determine a polar equation for the ellipse with the given vertices if one focus is at the pole.

55 and **56.**

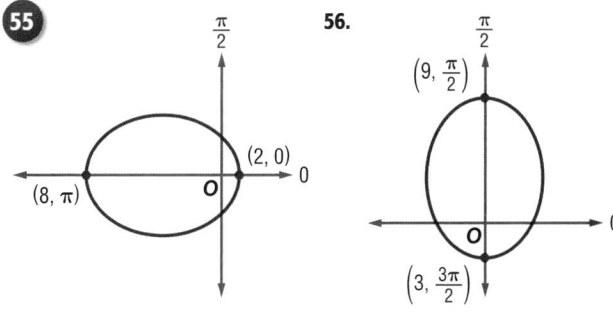

57. WRITING IN MATH Explain how a polar equation with focus $(0, 0)$ can be used to find the distance from the focus to any point on the conic.

Find two pairs of polar coordinates for each point with the given rectangular coordinates if $0 \leq \theta \leq 2\pi$. If necessary, round to the nearest hundredth. (Lesson 9-3)

58. $\left(-\sqrt{2}, \sqrt{2}\right)$

59. $(-2, -5)$

60. $(8, -12)$

Identify and graph each classic curve. (Lesson 9-2)

61. $r = 3 + 3\cos\theta$

62. $r = -2\sin 3\theta$

63. $r = \frac{5}{2}\theta, \theta \geq 0$

Determine an equation of an ellipse with each set of characteristics. (Lesson 7-2)

64. co-vertices (5,8), (5, 0); foci (8, 4), (2, 4)

65. major axis $(-2, 4)$ to $(8, 4)$; minor axis $(3, 1)$ to $(3, 7)$

66. foci $(1, -1), (9, -1)$; length of minor axis equals 6

67. OLYMPICS In the Olympic Games, team standings are determined according to each team's total points. Each type of Olympic medal earns a team a given number of points. Use the information to determine the Olympics in which the United States earned the most points. (Lesson 6-2)

Olympics	Gold	Silver	Bronze
1996	44	32	25
2000	37	24	31
2004	35	39	29
2008	36	38	36

Medal	Points
gold	3
silver	2
bronze	1

Find the values of $\sin 2\theta$, $\cos 2\theta$, and $\tan 2\theta$ for the given value and interval. (Lesson 5-5)

68. $\sin\theta = \frac{2}{3}, (0°, 90°)$

69. $\tan\theta = -\frac{24}{7}, \left(\frac{\pi}{2}, \pi\right)$

70. $\sin\theta = -\frac{4}{5}, \left(\pi, \frac{3\pi}{2}\right)$

Locate the vertical asymptotes, and sketch the graph of each function. (Lesson 4-5)

71. $y = \sec\left(x + \frac{\pi}{3}\right)$

72. $y = 4\cot\frac{x}{2}$

73. $y = 2\cot\left[\frac{2}{3}\left(x - \frac{\pi}{2}\right)\right] + 0.75$

Find the exact values of the five remaining trigonometric functions of θ. (Lesson 4-3)

74. $\sec\theta = 2$, where $\sin\theta > 0$ and $\cos\theta > 0$

75. $\csc\theta = \sqrt{5}$, where $\sin\theta > 0$ and $\cos\theta > 0$

76. SAT/ACT A pulley with a 9-inch diameter is belted to a pulley with a 6-inch diameter, as shown in the figure. If the larger pulley runs at 120 rpm (revolutions per minute), how fast does the smaller pulley run?

A 80 rpm   C 160 rpm   E 200 rpm

B 120 rpm   D 180 rpm

77. What type of conic is given by $r = \dfrac{3}{2 - 0.5\cos\theta}$?

F circle   H parabola

G ellipse   J hyperbola

78. REVIEW Which of the following includes the component form and magnitude of $\overrightarrow{AB}$ with initial point $A(3, 4, -2)$ and terminal point $B(-5, 2, 1)$?

A $\langle -8, -2, 3\rangle, \sqrt{77}$

B $\langle 8, -2, 3\rangle, \sqrt{77}$

C $\langle -8, -2, 3\rangle, \sqrt{109}$

D $\langle 8, -2, 3\rangle, \sqrt{109}$

79. REVIEW What is the eccentricity of the ellipse described by $\dfrac{y^2}{47} + \dfrac{(x - 12)^2}{34} = 1$?

F 0.38   H 0.53

G 0.41   J 0.62

Complex Numbers and De Moivre's Theorem

Then
You performed operations with complex numbers written in rectangular form. (Lesson 0-6)

Now
- Convert complex numbers from rectangular to polar form and vice versa.
- Find products, quotients, powers, and roots of complex numbers in polar form.

New Vocabulary
complex plane
real axis
imaginary axis
Argand plane
absolute value of a complex number
polar form
trigonometric form
modulus
argument
*p*th roots of unity

Math Online
glencoe.com

Why?

Electrical engineers use complex numbers to describe certain relationships of electricity. Voltage E, impedance Z, and current I are the three quantities related by the equation $E = I \cdot Z$ used to describe alternating current. Each variable can be written as a complex number in the form $a + bj$, where j is an imaginary number (engineers use j to not be confused with current I). For impedance, the real part a represents the opposition to current flow due to resistors, and the imaginary part b is related to the opposition due to inductors and capacitors.

Polar Forms of Complex Numbers A complex number given in rectangular form, $a + bi$, has a real component a and an imaginary component bi. You can graph a complex number on the **complex plane** by representing it with the point (a, b). Similar to a coordinate plane, we need two axes to graph a complex number. The real component is plotted on the horizontal axis called the **real axis**, and the imaginary component is plotted on the vertical axis called the **imaginary axis**. The complex plane may also be referred to as the **Argand Plane** (ar GON).

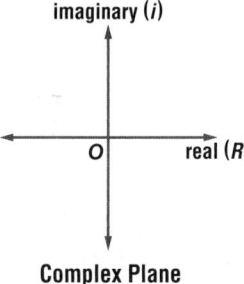

Complex Plane

Consider a complex number where $b = 0$, $a + 0i$. The result is a real number a that can be graphed using just a real number line or the real axis. When $b \neq 0$, the imaginary axis is needed to represent the imaginary component.

Recall that the absolute value of a real number is its distance from zero on the number line. Similarly, the **absolute value of a complex number** is its distance from zero in the complex plane. When $a + bi$ is graphed in the complex plane, the distance from zero can be calculated using the Pythagorean Theorem.

Key Concept   **Absolute Value of a Complex Number**

The absolute value of the complex number $z = a + bi$ is

$$|z| = |a + bi| = \sqrt{a^2 + b^2}.$$

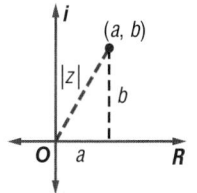

EXAMPLE 1 **Graphs and Absolute Values of Complex Numbers**

Graph each number in the complex plane, and find its absolute value.

a. $z = 4 + 3i$

$(a, b) = (4, 3)$

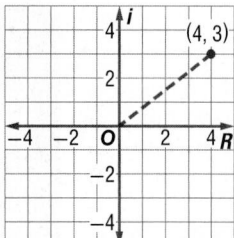

$|z| = \sqrt{a^2 + b^2}$ **Absolute value formula**

    $= \sqrt{4^2 + 3^2}$ **a = 4 and b = 3**

    $= \sqrt{25}$ or 5   **Simplify.**

The absolute value of $4 + 3i$ is 5.

b. $z = -2 - i$

$(a, b) = (-2, -1)$

$|z| = \sqrt{a^2 + b^2}$ **Absolute value formula**

    $= \sqrt{(-2)^2 + (-1)^2}$ **a = −2 and b = −1**

    $= \sqrt{5}$ or 2.24   **Simplify.**

The absolute value of $-2 - i \approx 2.24$.

✔️ **Guided Practice**

1A. $5 + 2i$ **1B.** $-3 + 4i$

Watch Out!

Polar Form The *polar form of a complex number* should not be confused with *polar coordinates of a complex number*. The polar form of a complex number is another way to represent a complex number. Polar coordinates of a complex number will be discussed later in this lesson.

Just as rectangular coordinates (x, y) can be written in polar form, so can the coordinates that represent the graph of a complex number in the complex plane. The same trigonometric ratios that were used to find values of x and y can be applied to represent values for a and b.

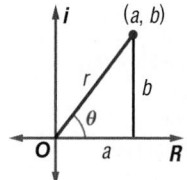

$\cos \theta = \frac{a}{r}$ and $\sin \theta = \frac{b}{r}$

$r \cos \theta = a$ $r \sin \theta = b$ **Multiply each side by r.**

Substituting the polar representations for a and b, we can calculate the **polar form** or **trigonometric form** of a complex number.

$z = a + bi$ **Original complex number**

   $= r \cos \theta + (r \sin \theta)i$ **a = r cos θ and b = r sin θ**

   $= r(\cos \theta + i \sin \theta)$ **Factor.**

In the case of a complex number, r represents the absolute value, or **modulus**, of the complex number and can be found using the same process you used when finding the absolute value, $r = |z| = \sqrt{a^2 + b^2}$. The angle θ is called the **argument** of the complex number. Similar to finding θ with rectangular coordinates (x, y), when using a complex number, $\theta = \tan^{-1} \frac{b}{a}$ or $\theta = \tan^{-1} \frac{b}{a} + \pi$ if $a < 0$.

StudyTip

Argument The argument of a complex number is also called the *amplitude*. Just as in polar coordinates, θ is not unique, although it is normally given in the interval $-2\pi < \theta < 2\pi$.

Key Concept   **Polar Form of a Complex Number**

The polar or trigonometric form of the complex number $z = a + bi$ is

$$z = r(\cos \theta + i \sin \theta), \text{ where}$$

$r = |z| = \sqrt{a^2 + b^2}$, $a = r \cos \theta$, $b = r \sin \theta$, and $\theta = \tan^{-1} \frac{b}{a}$ for $a > 0$ or $\theta = \tan^{-1} \frac{b}{a} + \pi$ for $a < 0$.

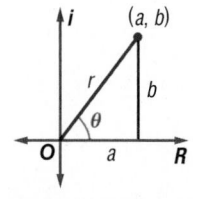

EXAMPLE 2   Complex Numbers in Polar Form

Express each complex number in polar form.

a. $-6 + 8i$

Find the modulus r and argument θ.

$$r = \sqrt{a^2 + b^2} \qquad\qquad \text{Conversion formula} \qquad \theta = \tan^{-1}\frac{b}{a} + \pi$$

$$= \sqrt{(-6)^2 + 8^2} \text{ or } 10 \qquad a = -6 \text{ and } b = 8 \qquad = \tan^{-1}\frac{8}{6} + \pi \text{ or about } 2.21$$

The polar form of $-6 + 8i$ is about $10(\cos 2.21 + i \sin 2.21)$.

b. $4 + \sqrt{3}i$

Find the modulus r and argument θ.

$$r = \sqrt{a^2 + b^2} \qquad\qquad \text{Conversion formula} \qquad \theta = \tan^{-1}\frac{b}{a}$$

$$= \sqrt{4^2 + (\sqrt{3})^2} \qquad a = 4 \text{ and } b = \sqrt{3} \qquad = \tan^{-1}\frac{\sqrt{3}}{4}$$

$$= \sqrt{19} \text{ or about } 4.36 \qquad\qquad \textbf{Simplify.} \qquad\qquad \approx 0.41$$

The polar form of $4 + \sqrt{3}i$ is about $4.36(\cos 0.41 + i \sin 0.41)$.

☑ Guided Practice

2A. $9 + 7i$ **2B.** $-2 - 2i$

> **ReadingMath**
>
> **Polar Form** $r(\cos \theta + i \sin \theta)$ is often abbreviated as r cis θ. In Example 2a, $-6 + 8i$ can also be expressed as 10 cis 2.21, where $10 = \sqrt{(-6)^2 + 8^2}$ and $2.21 = \tan^{-1}\frac{8}{6}$.

You can use the polar form of a complex number to graph the number on a polar grid by using the r and θ values as your polar coordinates (r, θ). You can also take a complex number written in polar form and convert it to rectangular form by evaluating.

EXAMPLE 3   Graph and Convert the Polar Form of a Complex Number

Graph $z = 3\left(\cos \dfrac{\pi}{6} + i \sin \dfrac{\pi}{6}\right)$ on a polar grid. Then express it in rectangular form.

The value of r is 3, and the value of θ is $\dfrac{\pi}{6}$.

Plot the polar coordinates $\left(3, \dfrac{\pi}{6}\right)$.

To express the number in rectangular form, evaluate the trigonometric values and simplify.

$$3\left(\cos \frac{\pi}{6} + i \sin \frac{\pi}{6}\right) \qquad \textbf{Polar form}$$

$$= 3\left[\frac{\sqrt{3}}{2} + i\left(\frac{1}{2}\right)\right] \qquad \textbf{Evaluate for cosine and sine.}$$

$$= \frac{3\sqrt{3}}{2} + \frac{3}{2}i \qquad \textbf{Distributive Property}$$

The rectangular form of $z = 3\left(\cos \dfrac{\pi}{6} + i \sin \dfrac{\pi}{6}\right)$ is $z = \dfrac{3\sqrt{3}}{2} + \dfrac{3}{2}i$.

> **TechnologyTip**
>
> **Complex Number Conversions** You can convert a complex number in polar form to rectangular form by entering the expression in polar form, then selecting ENTER. To be in polar mode, select MODE then a + bi.
>
> ```
> 3(cos(π/6)+isin(
> π/6))
>        2.598076211+1.5i
> ```

☑ Guided Practice

Graph each complex number on a polar grid. Then express it in rectangular form.

3A. $5\left(\cos \dfrac{3\pi}{4} + i \sin \dfrac{3\pi}{4}\right)$ **3B.** $4\left(\cos \dfrac{5\pi}{3} + i \sin \dfrac{5\pi}{3}\right)$

Products, Quotients, Powers, and Roots of Complex Numbers The polar form of complex numbers, along with the sum and difference formulas for cosine and sine, greatly aid in the multiplication and division of complex numbers. The formula for the product of two complex numbers in polar form can be derived by performing the multiplication.

$z_1z_2 = r_1(\cos\theta_1 + i\sin\theta_1) \cdot r_2(\cos\theta_2 + i\sin\theta_2)$	Original equation
$= r_1r_2(\cos\theta_1\cos\theta_2 + i\cos\theta_1\sin\theta_2 + i\sin\theta_1\cos\theta_2 + i^2\sin\theta_1\sin\theta_2)$	FOIL
$= r_1r_2[(\cos\theta_1\cos\theta_2 - \sin\theta_1\sin\theta_2) + (i\cos\theta_1\sin\theta_2 + i\sin\theta_1\cos\theta_2)]$	$i^2 = -1$ and group imaginary terms.
$= r_1r_2[(\cos\theta_1\cos\theta_2 - \sin\theta_1\sin\theta_2) + i(\cos\theta_1\sin\theta_2 + \sin\theta_1\cos\theta_2)]$	Factor out i.
$= r_1r_2[\cos(\theta_1 + \theta_2) + i\sin(\theta_1 + \theta_2)]$	Sum identities for cosine and sine

---

Key Concept   **Product and Quotient of Complex Numbers in Polar Form**

Given the complex numbers $z_1 = r_1(\cos\theta_1 + i\sin\theta_1)$ and $z_2 = r_2(\cos\theta_2 + i\sin\theta_2)$:

Product Formula $z_1z_2 = r_1r_2[\cos(\theta_1 + \theta_2) + i\sin(\theta_1 + \theta_2)]$

Quotient Formula $\dfrac{z_1}{z_2} = \dfrac{r_1}{r_2}[\cos(\theta_1 - \theta_2) + i\sin(\theta_1 - \theta_2)]$, where z_2 and $r_2 \neq 0$

You will prove the Quotient Formula in Exercise 77.

Notice that when multiplying complex numbers, you multiply the moduli and add the arguments. When dividing, you divide the moduli and subtract the arguments.

EXAMPLE 4   **Product of Complex Numbers in Polar Form**

Find $2\left(\cos\dfrac{5\pi}{3} + i\sin\dfrac{5\pi}{3}\right) \cdot 4\left(\cos\dfrac{\pi}{6} + i\sin\dfrac{\pi}{6}\right)$ in polar form. Then express the product in rectangular form.

$2\left(\cos\dfrac{5\pi}{3} + i\sin\dfrac{5\pi}{3}\right) \cdot 4\left(\cos\dfrac{\pi}{6} + i\sin\dfrac{\pi}{6}\right)$	Original expression
$= 2(4)\left[\cos\left(\dfrac{5\pi}{3} + \dfrac{\pi}{6}\right) + i\sin\left(\dfrac{5\pi}{3} + \dfrac{\pi}{6}\right)\right]$	Product Formula
$= 8\left(\cos\dfrac{11\pi}{6} + i\sin\dfrac{11\pi}{6}\right)$	Simplify.

Now find the rectangular form of the product.

$8\left(\cos\dfrac{11\pi}{6} + i\sin\dfrac{11\pi}{6}\right)$	Polar form
$= 8\left(\dfrac{\sqrt{3}}{2} - i\dfrac{1}{2}\right)$	Evaluate.
$= 4\sqrt{3} - 4i$	Distributive Property

The polar form of the product is $8\left(\cos\dfrac{11\pi}{6} + i\sin\dfrac{11\pi}{6}\right)$. The rectangular form of the product is $4\sqrt{3} - 4i$.

✔ **Guided Practice**

Find each product in polar form. Then express the product in rectangular form.

4A. $3\left(\cos\dfrac{\pi}{3} + i\sin\dfrac{\pi}{3}\right) \cdot 5\left(\cos\dfrac{\pi}{4} + i\sin\dfrac{\pi}{4}\right)$

4B. $-6\left(\cos\dfrac{3\pi}{4} + i\sin\dfrac{3\pi}{4}\right) \cdot 2\left(\cos\dfrac{2\pi}{3} + i\sin\dfrac{2\pi}{3}\right)$

As mentioned at the beginning of this lesson, quotients of complex numbers can be used to show relationships in electricity.

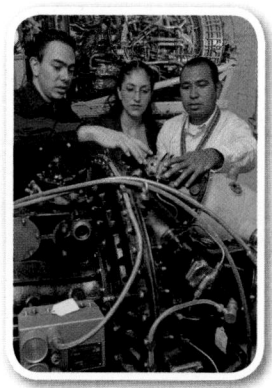

Real-World Career

Electrical Engineers Electrical engineers design and create new technology used to manufacture global positioning systems, giant generators that power entire cities, turbine engines used in jet aircrafts, and radar and navigation systems. They also work on improving various products such as cell phones, cars, and robots.

🌐 **Real-World EXAMPLE 5** / **Quotient of Complex Numbers in Polar Form**

ELECTRICITY If a circuit has a voltage E of 150 volts and an impedance Z of $6 - 3j$ ohms, find the current I amps in the circuit in rectangular form. Use $E = I \cdot Z$.

Express each number in polar form.

$150 = 150(\cos 0 + j \sin 0)$ $r = \sqrt{150^2 + 0^2}$ or 150; $\theta = \tan^{-1}\frac{0}{150}$ or 0

$6 - 3j = 3\sqrt{5}[\cos(-0.46) + j \sin(-0.46)]$ $r = \sqrt{6^2 + (-3)^2}$ or $3\sqrt{5}$; $\theta = \tan^{-1}-\frac{3}{6}$ or about -0.46

Solve for the current I in $I \cdot Z = E$.

$I \cdot Z = E$	**Original equation**
$I = \dfrac{E}{Z}$	**Divide each side by Z.**
$I = \dfrac{150(\cos 0 + j \sin 0)}{3\sqrt{5}[\cos(-0.46) + j \sin(-0.46)]}$	$E = 150(\cos 0 + j \sin 0)$ and $Z = 3\sqrt{5}[\cos(-0.46) + j \sin(-0.46)]$
$I = \dfrac{150}{3\sqrt{5}}\{\cos[0 - (-0.46)] + j \sin[0 - (-0.46)]\}$	**Quotient Formula**
$I = 10\sqrt{5}(\cos 0.46 + j \sin 0.46)$	**Simplify.**

Now convert the current to rectangular form.

$I = 10\sqrt{5}(\cos 0.46 + j \sin 0.46)$	**Original equation**
$= 10\sqrt{5}(0.90 + 0.44j)$	**Evaluate.**
$= 20.12 + 9.84j$	**Distributive Property**

The current is about $20.12 + 9.84j$ amps.

✅ **Guided Practice**

5. ELECTRICITY If a circuit has a voltage of 120 volts and a current of $8 + 6j$ amps, find the impedance of the circuit in rectangular form.

Before calculating the powers and roots of complex numbers, it may be helpful to express the complex numbers in polar form. Abraham De Moivre (duh MWAHV ruh) is credited with discovering a useful pattern for evaluating powers of complex numbers.

We can use the formula for the product of complex numbers to help visualize the pattern that De Moivre discovered.

First, find z^2 by taking the product of $z \cdot z$.

$z \cdot z = r(\cos \theta + i \sin \theta) \cdot r(\cos \theta + i \sin \theta)$	**Multiply.**
$z^2 = r^2[\cos(\theta + \theta) + i \sin(\theta + \theta)]$	**Product Formula**
$z^2 = r^2(\cos 2\theta + i \sin 2\theta)$	**Simplify.**

Now find z^3 by calculating $z^2 \cdot z$.

$z^2 \cdot z = r^2(\cos 2\theta + i \sin 2\theta) \cdot r(\cos \theta + i \sin \theta)$	**Multiply.**
$z^3 = r^3[\cos(2\theta + \theta) + i \sin(2\theta + \theta)]$	**Product Formula**
$z^3 = r^3(\cos 3\theta + i \sin 3\theta)$	**Simplify.**

Notice that when calculating these powers of a complex number, you take the nth power of the modulus and multiply the argument by n.

This pattern is summarized below.

Key Concept   **De Moivre's Theorem**

If the polar form of a complex number is $z = r(\cos \theta + i \sin \theta)$, then for positive integers n

$$z^n = [r(\cos \theta + i \sin \theta)]^n \text{ or } r^n(\cos n\theta + i \sin n\theta).$$

You will prove De Moivre's Theorem in Lesson 10-4.

EXAMPLE 6   **De Moivre's Theorem**

Find $\left(4 + 4\sqrt{3}i\right)^6$, and express it in rectangular form.

First, write $4 + 4\sqrt{3}i$ in polar form.

$r = \sqrt{a^2 + b^2}$	**Conversion formula**	$\theta = \tan^{-1} \dfrac{b}{a}$
$= \sqrt{4^2 + (4\sqrt{3})^2}$	$a = 4$ and $b = 4\sqrt{3}$	$= \tan^{-1} \dfrac{4\sqrt{3}}{4}$
$= \sqrt{16 + 48}$	**Simplify.**	$= \tan^{-1} \sqrt{3}$
$= 8$	**Simplify.**	$= \dfrac{\pi}{3}$

The polar form of $4 + 4\sqrt{3}i$ is $8\left(\cos \dfrac{\pi}{3} + i \sin \dfrac{\pi}{3}\right)$.

Now use De Moivre's Theorem to find the sixth power.

$(4 + 4\sqrt{3}i)^6 = \left[8\left(\cos \dfrac{\pi}{3} + i \sin \dfrac{\pi}{3}\right)\right]^6$	**Original equation**
$= 8^6\left[\cos 6\left(\dfrac{\pi}{3}\right) + i \sin 6\left(\dfrac{\pi}{3}\right)\right]$	**De Moivre's Theorem**
$= 262{,}144(\cos 2\pi + i \sin 2\pi)$	**Simplify.**
$= 262{,}144(1 + 0i)$	**Evaluate.**
$= 262{,}144$	**Simplify.**

Therefore, $(4 + 4\sqrt{3}i)^6 = 262{,}144$.

✓ Guided Practice

Find each power, and express it in rectangular form.

6A. $(1 + \sqrt{3}i)^4$ **6B.** $(2\sqrt{3} - 2i)^8$

In the real number system, $x^4 = 256$ has two solutions, 4 and −4. The graph of $y = x^4 - 256$ shows that there are two real zeros at $x = 4$ and −4. In the complex number system, however, there are two real solutions and two complex solutions.

In Lesson 2-4, you learned through the Fundamental Theorem of Algebra that polynomials of degree n have exactly n zeros in the complex number system. Therefore, the equation $x^4 = 256$, rewritten as $x^4 - 256 = 0$, has exactly four solutions, or roots: 4, −4, $4i$, and $-4i$. In general, all nonzero complex numbers have p distinct pth roots. That is, they each have two square roots, three cube roots, four fourth roots, and so on.

To find all of the roots of a polynomial, we can use De Moivre's Theorem to arrive at the following expression.

Key Concept   **Distinct Roots**

For a positive integer p, the complex number $r(\cos \theta + i \sin \theta)$ has p distinct pth roots. They are found by

$$r^{\frac{1}{p}}\left(\cos \frac{\theta + 2n\pi}{p} + i \sin \frac{\theta + 2n\pi}{p}\right),$$

where $n = 0, 1, 2, \ldots, p - 1$.

We can use this formula for the different values of n, but we can stop when $n = p - 1$. When n equals or exceeds p, the roots repeat as the following shows.

$$\frac{\theta + 2\pi p}{p} = \frac{\theta}{p} + 2\pi \qquad \text{Coterminal with } \frac{\theta}{p}, \text{ when } n = 0$$

EXAMPLE 7   ***p*th Roots of a Complex Number**

Find the fourth roots of $-4 - 4i$.

First, write $-4 - 4i$ in polar form.

$$-4 - 4i = 4\sqrt{2}\left(\cos \frac{5\pi}{4} + i \sin \frac{5\pi}{4}\right) \qquad\qquad r = \sqrt{(-4)^2 + (-4)^2} \text{ or } 4\sqrt{2}; \ \theta = \tan^{-1}\frac{-4}{-4} + \pi \text{ or } \frac{5\pi}{4}$$

Now write an expression for the fourth roots.

$$(4\sqrt{2})^{\frac{1}{4}}\left(\cos \frac{\frac{5\pi}{4} + 2n\pi}{4} + i \sin \frac{\frac{5\pi}{4} + 2n\pi}{4}\right) \qquad\qquad \theta = \frac{5\pi}{4}, p = 4, \text{ and } r^{\frac{1}{p}} = (4\sqrt{2})^{\frac{1}{4}}$$

$$= \sqrt[8]{32}\left[\cos\left(\frac{5\pi}{16} + \frac{n\pi}{2}\right) + i \sin\left(\frac{5\pi}{16} + \frac{n\pi}{2}\right)\right] \qquad\qquad \text{Simplify.}$$

Let $n = 0, 1, 2,$ and 3 successively to find the fourth roots.

Let $n = 0$. $\sqrt[8]{32}\left[\cos\left(\frac{5\pi}{16} + \frac{(0)\pi}{2}\right) + i \sin\left(\frac{5\pi}{16} + \frac{(0)\pi}{2}\right)\right]$ **Distinct Roots**

$$= \sqrt[8]{32}\left(\cos \frac{5\pi}{16} + i \sin \frac{5\pi}{16}\right) \text{ or } 0.86 + 1.28i \qquad \text{First fourth root}$$

Let $n = 1$. $\sqrt[8]{32}\left[\cos\left(\frac{5\pi}{16} + \frac{(1)\pi}{2}\right) + i \sin\left(\frac{5\pi}{16} + \frac{(1)\pi}{2}\right)\right]$

$$= \sqrt[8]{32}\left(\cos \frac{13\pi}{16} + i \sin \frac{13\pi}{16}\right) \text{ or } -1.28 + 0.86i \qquad \text{Second fourth root}$$

Let $n = 2$. $\sqrt[8]{32}\left[\cos\left(\frac{5\pi}{16} + \frac{(2)\pi}{2}\right) + i \sin\left(\frac{5\pi}{16} + \frac{(2)\pi}{2}\right)\right]$

$$= \sqrt[8]{32}\left(\cos \frac{21\pi}{16} + i \sin \frac{21\pi}{16}\right) \text{ or } -0.86 - 1.28i \qquad \text{Third fourth root}$$

Let $n = 3$. $\sqrt[8]{32}\left[\cos\left(\frac{5\pi}{16} + \frac{(3)\pi}{2}\right) + i \sin\left(\frac{5\pi}{16} + \frac{(3)\pi}{2}\right)\right]$

$$= \sqrt[8]{32}\left(\cos \frac{29\pi}{16} + i \sin \frac{29\pi}{16}\right) \text{ or } 1.28 - 0.86i \qquad \text{Fourth fourth root}$$

The fourth roots of $-4 - 4i$ are approximately $0.86 + 1.28i$, $-1.28 + 0.86i$, $-0.86 - 1.28i$, and $1.28 - 0.86i$.

✓ **Guided Practice**

7A. Find the cube roots of $2 + 2i$.   **7B.** Find the fifth roots of $4\sqrt{3} - 4i$.

We can make observations about the distinct roots of a number by graphing the roots on a coordinate plane. As shown at the right, the four fourth roots found in Example 7 lie on a circle. If we look at the polar form of each complex number, each has the same modulus of $\sqrt[8]{32}$, which acts as the radius of the circle. The roots are also equally spaced around the circle as a result of the arguments differing by $\frac{\pi}{2}$.

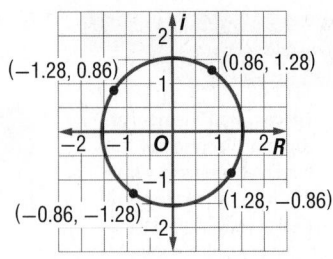

A special case of finding roots occurs when finding the pth roots of 1. When 1 is written in polar form, $r = 1$. As mentioned in the previous paragraph, the modulus of our roots is the radius of the circle that is formed from plotting the roots on a coordinate plane. Thus, the pth roots of 1 lie on the unit circle. Finding the pth roots of 1 is referred to as finding the **pth roots of unity**.

StudyTip

The pth Roots of a Complex Number Each root will have the same modulus of $r^{\frac{1}{p}}$. The argument of the first root is $\frac{\theta}{p}$, and each successive root is found by repeatedly adding $\frac{2\pi}{p}$ to the argument.

EXAMPLE 8 The pth Roots of Unity

Find the eighth roots of unity.

First, write 1 in polar form.

$$1 = 1 \cdot (\cos 0 + i \sin 0) \qquad r = \sqrt{1^2 + 0^2} \text{ or } 1 \text{ and } \theta = \tan^{-1}\frac{0}{1} \text{ or } 0$$

Now write an expression for the eighth roots.

$$1\left(\cos \frac{0 + 2n\pi}{8} + i \sin \frac{0 + 2n\pi}{8}\right) \qquad \theta = 0, p = 8, \text{ and } r^{\frac{1}{p}} = 1^{\frac{1}{8}} \text{ or } 1$$

$$= \cos \frac{n\pi}{4} + i \sin \frac{n\pi}{4} \qquad \textbf{Simplify.}$$

Let $n = 0$ to find the first root of 1.

$$n = 0 \qquad \cos \frac{(0)\pi}{4} + i \sin \frac{(0)\pi}{4} \qquad \textbf{Distinct Roots}$$

$$= \cos 0 + i \sin 0 \text{ or } 1 \qquad \textbf{First root}$$

Notice that the modulus of each complex number is 1. The arguments are found by $\frac{n\pi}{4}$, resulting in θ increasing by $\frac{\pi}{4}$ for each successive root. Therefore, we can calculate the remaining roots by adding $\frac{\pi}{4}$ to each previous θ.

$\cos 0 + i \sin 0$ or 1   **1ˢᵗ root**

$\cos \frac{\pi}{4} + i \sin \frac{\pi}{4}$ or $\frac{\sqrt{2}}{2} + \frac{\sqrt{2}}{2}i$ **2ⁿᵈ root**

$\cos \frac{\pi}{2} + i \sin \frac{\pi}{2}$ or i **3ʳᵈ root**

$\cos \frac{3\pi}{4} + i \sin \frac{3\pi}{4}$ or $-\frac{\sqrt{2}}{2} + \frac{\sqrt{2}}{2}i$ **4ᵗʰ root**

$\cos \pi + i \sin \pi$ or -1 **5ᵗʰ root**

$\cos \frac{5\pi}{4} + i \sin \frac{5\pi}{4}$ or $-\frac{\sqrt{2}}{2} - \frac{\sqrt{2}}{2}i$ **6ᵗʰ root**

$\cos \frac{3\pi}{2} + i \sin \frac{3\pi}{2}$ or $-i$ **7ᵗʰ root**

$\cos \frac{7\pi}{4} + i \sin \frac{7\pi}{4}$ or $\frac{\sqrt{2}}{2} - \frac{\sqrt{2}}{2}i$ **8ᵗʰ root**

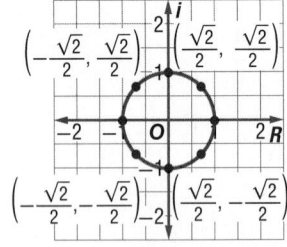

Figure 9.5.1

The eighth roots of 1 are $1, \frac{\sqrt{2}}{2} + \frac{\sqrt{2}}{2}i, i, -\frac{\sqrt{2}}{2} + \frac{\sqrt{2}}{2}i, -1, -\frac{\sqrt{2}}{2} - \frac{\sqrt{2}}{2}i, -i,$ and $\frac{\sqrt{2}}{2} - \frac{\sqrt{2}}{2}i$ as shown in Figure 9.5.1.

✓ Guided Practice

8A. Find the cube roots of unity.

8B. Find the seventh roots of unity.

Exercises

● = **Step-by-Step Solutions** begin on page R29.

Graph each number in the complex plane, and find its absolute value. (Example 1)

1. $z = 4 + 4i$ **2.** $z = -3 + i$

3. $z = -4 - 6i$ **4.** $z = 2 - 5i$

5. $z = 3 + 4i$ **6.** $z = -7 + 5i$

7. $z = -3 - 7i$ **8.** $z = 8 - 2i$

9. VECTORS The force on an object is given by $z = 10 + 15i$, where the components are measured in newtons (N). (Example 1)

   a. Represent z as a vector in the complex plane.

   b. Find the magnitude and direction angle of the vector.

Express each complex number in polar form. (Example 2)

10. $4 + 4i$ **11.** $-2 + i$

12. $4 - \sqrt{2}i$ **13.** $2 - 2i$

14. $4 + 5i$ **15.** $-2 + 4i$

16. $-1 - \sqrt{3}i$ **17.** $3 + 3i$

Graph each complex number on a polar grid. Then express it in rectangular form. (Example 3)

18. $10(\cos 6 + i \sin 6)$ **19.** $2(\cos 3 + i \sin 3)$

20. $4\left(\cos \frac{\pi}{3} + i \sin \frac{\pi}{3}\right)$ **21.** $3\left(\cos \frac{\pi}{4} + i \sin \frac{\pi}{4}\right)$

22. $\left(\cos \frac{11\pi}{6} + i \sin \frac{11\pi}{6}\right)$ **23.** $2\left(\cos \frac{4\pi}{3} + i \sin \frac{4\pi}{3}\right)$

24. $-3(\cos 180° + i \sin 180°)$ **25.** $\frac{3}{2}(\cos 360° + i \sin 360°)$

Find each product or quotient, and express it in rectangular form. (Examples 4 and 5)

26. $6\left(\cos \frac{\pi}{2} + i \sin \frac{\pi}{2}\right) \cdot 4\left(\cos \frac{\pi}{4} + i \sin \frac{\pi}{4}\right)$

27. $5(\cos 135° + i \sin 135°) \cdot 2\,(\cos 45° + i \sin 45°)$

28. $3\left(\cos \frac{3\pi}{4} + i \sin \frac{3\pi}{4}\right) \div \frac{1}{2}(\cos \pi + i \sin \pi)$

29. $2(\cos 90° + i \sin 90°) \cdot 2(\cos 270° + i \sin 270°)$

30. $3\left(\cos \frac{\pi}{6} + i \sin \frac{\pi}{6}\right) \div 4\left(\cos \frac{2\pi}{3} + i \sin \frac{2\pi}{3}\right)$

31. $4\left(\cos \frac{9\pi}{4} + i \sin \frac{9\pi}{4}\right) \div 2\left(\cos \frac{3\pi}{2} + i \sin \frac{3\pi}{2}\right)$

32. $\frac{1}{2}(\cos 60° + i \sin 60°) \cdot 6(\cos 150° + i \sin 150°)$

33. $6\left(\cos \frac{3\pi}{4} + i \sin \frac{3\pi}{4}\right) \div 2\left(\cos \frac{\pi}{4} + i \sin \frac{\pi}{4}\right)$

34. $5(\cos 180° + i \sin 180°) \cdot 2(\cos 135° + i \sin 135°)$

35. $\frac{1}{2}\left(\cos \frac{\pi}{3} + i \sin \frac{\pi}{3}\right) \div 3\left(\cos \frac{\pi}{6} + i \sin \frac{\pi}{6}\right)$

Find each power, and express it in rectangular form. (Example 6)

36. $(2 + 2\sqrt{3}i)^6$ **37.** $(12i - 5)^3$

38. $\left[4\left(\cos \frac{\pi}{2} + i \sin \frac{\pi}{2}\right)\right]^4$ **39.** $(\sqrt{3} - i)^3$

40. $(3 - 5i)^4$ **41.** $(2 + 4i)^4$

42. $(3 - 6i)^4$ **43.** $(2 + 3i)^2$

44. $\left[3\left(\cos \frac{\pi}{6} + i \sin \frac{\pi}{6}\right)\right]^3$ **45.** $\left[2\left(\cos \frac{\pi}{4} + i \sin \frac{\pi}{4}\right)\right]^4$

46. DESIGN Stella works for an advertising agency. She wants to incorporate a design comprised of regular hexagons as the artwork for one of her proposals. Stella can locate the vertices of a regular hexagon by graphing the solutions to $x^6 - 1 = 0$ in the complex plane. Find the vertices of one of the hexagons. (Example 7)

Find all of the distinct pth roots of the complex number. (Examples 7 and 8)

47. sixth roots of i

48. fifth roots of $-i$

49. fourth roots of $4\sqrt{3} - 4i$

50. cube roots of $-117 + 44i$

51. fifth roots of $-1 + 11\sqrt{2}i$

52. square root of $-3 - 4i$

53. find the square roots of unity

54. find the fourth roots of unity

55. ELECTRICITY The impedance in one part of a series circuit is $5(\cos 0.9 + j \sin 0.9)$ ohms. In the second part of the circuit, it is $8(\cos 0.4 + j \sin 0.4)$ ohms.

   a. Convert each expression to rectangular form.

   b. Add your answers from part **a** to find the total impedance in the circuit.

   c. Convert the total impedance back to polar form.

Find each product. Then repeat the process by multiplying the polar forms of each pair of complex numbers using the Product Formula.

56. $(1 - i)(4 + 4i)$ **57.** $(3 + i)(3 - i)$

58. $(4 + i)(3 - i)$ **59.** $(-6 + 5i)(2 - 3i)$

60. $(\sqrt{2} + 2i)(1 + i)$ **61.** $(3 - 2i)(1 + \sqrt{3}i)$

62. FRACTALS A *fractal* is a geometric figure that is made up of a pattern that is repeated indefinitely on successively smaller scales, as shown below.

In this problem, you will generate a fractal through iterations of $f(z) = z^2$. Consider $z_0 = 0.8 + 0.5i$.

a. Calculate $z_1, z_2, z_3, z_4, z_5, z_6,$ and z_7 where $z_1 = f(z_0)$, $z_2 = f(z_1)$, and so on.

b. Graph each of the numbers on the complex plane.

c. Predict the location of z_{100}. Explain.

63. TRANSFORMATIONS There are certain operations with complex numbers that correspond to geometric transformations in the complex plane. Describe the transformation applied to point z to obtain point w in the complex plane for each of the following operations.

a. $w = z + (3 - 4i)$

b. w is the complex conjugate of z.

c. $w = i \cdot z$

d. $w = 0.25z$

Find z and the pth roots of z given each of the following.

64. $p = 3$, one cube root is $\frac{5}{2} - \frac{5\sqrt{3}}{2}i$

65. $p = 4$, one fourth root is $-1 - i$

66. GRAPHICS By representing each vertex by a complex number in polar form, a programmer dilates and then rotates the square below 45° counterclockwise so that the new vertices lie at the midpoints of the sides of the original square.

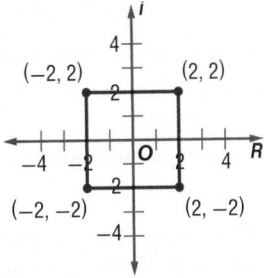

a. By what complex number should the programmer multiply each number to produce this transformation?

b. What happens if the numbers representing the original vertices are multiplied by the square of your answer to part a?

Use the Distinct Roots Formula to find all of the solutions of each equation. Express the solutions in rectangular form.

67. $x^3 = i$

68. $x^3 + 3 = 128$

69. $x^4 = 81i$

70. $x^5 - 1 = 1023$

71. $x^3 + 1 = i$

72. $x^4 - 2 + i = -1$

H.O.T. Problems   Use Higher-Order Thinking Skills

73. ERROR ANALYSIS Alma and Blake are evaluating $\left(-\frac{\sqrt{3}}{2} + \frac{1}{2}i\right)^5$. Alma uses De Moivre's Theorem and gets an answer of $\cos\frac{5\pi}{6} + i\sin\frac{5\pi}{6}$. Blake tells her that she has only completed part of the problem. Is either of them correct? Explain your reasoning.

74. REASONING Suppose $z = a + bi$ is one of the 29th roots of 1.

a. What is the maximum value of a?

b. What is the maximum value of b?

CHALLENGE Find the roots shown on each graph and write them in polar form. Then identify the complex number with the given roots.

75   **76.**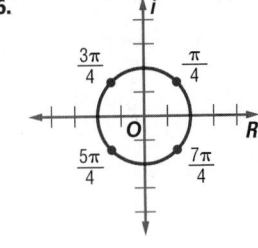

77. PROOF Given $z_1 = r_1(\cos\theta_1 + i\sin\theta_1)$ and $z_2 = r_2(\cos\theta_2 + i\sin\theta_2)$, where $r_2 \neq 0$, prove that $\frac{z_1}{z_2} = \frac{r_1}{r_2}[\cos(\theta_1 - \theta_2) + i\sin(\theta_1 - \theta_2)]$.

REASONING Determine whether each statement is *sometimes*, *always*, or *never* true. Explain your reasoning.

78. The pth roots of a complex number z are equally spaced around the circle centered at the origin with radius $r^{\frac{1}{p}}$.

79. The complex conjugate of $z = a + bi$ is $\bar{z} = a - bi$. For any z, $z + \bar{z}$ and $z \cdot \bar{z}$ are real numbers.

80. OPEN ENDED Find two complex numbers $a + bi$ in which $a \neq 0$ and $b \neq 0$ with an absolute value of $\sqrt{17}$.

81. WRITING IN MATH Explain why the sum of the imaginary parts of the p distinct pth roots of any positive real number must be zero. (*Hint*: The roots are the vertices of a regular polygon.)

Write each polar equation in rectangular form. (Lesson 9-4)

82. $r = \dfrac{15}{1 + 4\cos\theta}$

83. $r = \dfrac{14}{2\cos\theta + 2}$

84. $r = \dfrac{-6}{\sin\theta - 2}$

Identify the graph of each rectangular equation. Then write the equation in polar form. Support your answer by graphing the polar form of the equation. (Lesson 9-3)

85. $(x - 3)^2 + y^2 = 9$

86. $x^2 - y^2 = 1$

87. $x^2 + y^2 = 2y$

Graph the conic given by each equation. (Lesson 7-4)

88. $y = x^2 + 3x + 1$

89. $y^2 - 2x^2 - 16 = 0$

90. $x^2 + 4y^2 + 2x - 24y + 33 = 0$

Find the center, foci, and vertices of each ellipse. (Lesson 7-2)

91. $\dfrac{(x + 8)^2}{9} + \dfrac{(y - 7)^2}{81} = 1$

92. $25x^2 + 4y^2 + 150x + 24y = -161$

93. $4x^2 + 9y^2 - 56x + 108y = -484$

Solve each system of equations using Gauss-Jordan elimination. (Lesson 6-1)

94. $x + y + z = 12$
$6x - 2y - z = 16$
$3x + 4y + 2z = 28$

95. $9g + 7h = -30$
$8h + 5j = 11$
$-3g + 10j = 73$

96. $2k - n = 2$
$3p = 21$
$4k + p = 19$

97. POPULATION In the beginning of 2008, the world's population was about 6.7 billion. If the world's population grows continuously at a rate of 2%, the future population P, in billions, can be predicted by $P = 6.5e^{0.02t}$, where t is the time in years since 2008. (Lesson 3-4)

   a. According to this model, what will be the world's population in 2018?

   b. Some experts have estimated that the world's food supply can support a population of at most 18 billion people. According to this model, for how many more years will the food supply be able to support the trend in world population growth?

98. SAT/ACT The graph on the xy-plane of the quadratic function g is a parabola with vertex at $(3, -2)$. If $g(0) = 0$, then which of the following must also equal 0?

   A $g(2)$

   B $g(3)$

   C $g(4)$

   D $g(6)$

   E $g(7)$

99. Which of the following expresses the complex number $20 - 21i$ in polar form?

   F $29(\cos 5.47 + i\sin 5.47)$

   G $29(\cos 5.52 + i\sin 5.52)$

   H $32(\cos 5.47 + i\sin 5.47)$

   J $32(\cos 5.52 + i\sin 5.52)$

100. FREE RESPONSE Consider the graph at the right.

   a. Give a possible equation for the function.

   b. Describe the symmetries of the graph.

   c. Give the zeroes of the function over the domain $0 \le \theta \le 2\pi$.

   d. What is the minimum value of r over the domain $0 \le \theta \le 2\pi$?

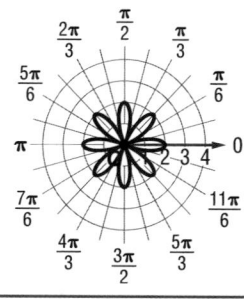

Chapter Summary

Key Concepts

Polar Coordinates (Lesson 9-1)

• In the polar coordinate system, a point (r, θ) is located using its directed distance r and directed angle θ.

• The distance between $P_1(r_1, \theta_1)$ and $P_2(r_2, \theta_2)$ in the polar plane is $P_1P_2 = \sqrt{r_1^2 + r_2^2 - 2r_1r_2 \cos(\theta_2 - \theta_1)}$.

Graphs of Polar Equations (Lesson 9-2)

• Circle: $r = a \cos \theta$ or $r = a \sin \theta$
• Limaçon: $r = a \pm b \cos \theta$ or $r = a \pm b \sin \theta$, $a > 0, b > 0$
• Rose: $r = a \cos n\theta$ or $r = a \sin n\theta$, $n \geq 2, n \in \mathbb{Z}$
• Lemniscate: $r^2 = a^2 \cos 2\theta$ or $r^2 = a^2 \sin 2\theta$
• Spirals of Archimedes: $r = a\theta + b$, $\theta \geq 0$

Polar and Rectangular Forms of Equations (Lesson 9-3)

• A point $P(r, \theta)$ has rectangular coordinates $(r \cos \theta, r \sin \theta)$.

• To convert a point $P(x, y)$ from rectangular to polar coordinates, use the equations $r = \sqrt{x^2 + y^2}$ and $\theta = \tan^{-1} \frac{y}{x}$, when $x > 0$ or $\theta = \tan^{-1} \frac{y}{x} + \pi$, when $x < 0$.

Polar Forms of Conic Sections (Lesson 9-4)

• The polar equation of a conic section is of the form $r = \dfrac{ed}{1 \pm e \cos \theta}$ or $r = \dfrac{ed}{1 \pm e \sin \theta}$, depending on the location and orientation of the directrix.

Complex Numbers and De Moivre's Theorem (Lesson 9-5)

• The polar or trigonometric form of the complex number $a + bi$ is $r (\cos \theta + i \sin \theta)$.

• The product formula for two complex numbers z_1 and z_2 is $z_1z_2 = r_1r_2 [\cos(\theta_1 + \theta_2) + i \sin(\theta_1 + \theta_2)]$.

• The quotient formula for two complex numbers z_1 and z_2 is $\dfrac{z_1}{z_2} = \dfrac{r_1}{r_2} [\cos(\theta_1 - \theta_2) + i \sin(\theta_1 - \theta_2)]$, where z_2 and $r_2 \neq 0$.

• De Moivre's Theorem states that if the polar form of a complex number is $z = r (\cos \theta + i \sin \theta)$, then $z^n = r^n (\cos n\theta + i \sin n\theta)$ for positive integers n.

Key Vocabulary

absolute value of a complex number (p. 569)

Argand plane (p. 569)

argument (p. 570)

cardioid (p.544)

complex plane (p. 569)

imaginary axis (p. 569)

lemniscate (p. 546)

limaçon (p. 543)

modulus (p. 570)

polar axis (p. 534)

polar coordinate system (p. 534)

polar coordinates (p. 534)

polar equation (p. 536)

polar form (p. 570)

polar graph (p. 536)

pole (p. 534)

*p*th roots of unity (p. 576)

real axis (p. 569)

rose (p. 545)

spiral of Archimedes (p. 546)

trigonometric form (p. 570)

Vocabulary Check

Choose the correct term from the list above to complete each sentence.

1. A(n) _____ is the set of all points with coordinates (r, θ) that satisfy a given polar equation.

2. A plane that has an axis for the real component and an axis for the imaginary component is a(n) _____.

3. The location of a point in the _____ is identified using the directed distance from a fixed point and the angle from a fixed axis.

4. A special type of limaçon with equation of the form $r = a + b \cos \theta$ where $a = b$ is called a(n) _____.

5. The _____ is the angle θ of a complex number written in the form $r(\cos \theta + i \sin \theta)$.

6. The origin of a polar coordinate system is called the _____.

7. The absolute value of a complex number is also called the _____.

8. The _____ is another name for the complex plane.

9. The graph of a polar equation of the form $r^2 = a^2 \cos 2\theta$ or $r^2 = a^2 \sin 2\theta$ is called a(n) _____.

10. The _____ is an initial ray from the pole, usually horizontal and directed toward the right.

Lesson-by-Lesson Review

9-1 **Polar Coordinates** (pp. 534–540)

Graph each point on a polar grid.

11. $W(-0.5, 210°)$

12. $X\left(1.5, \dfrac{7\pi}{4}\right)$

13. $Y(4, -120°)$

14. $Z\left(-3, \dfrac{5\pi}{6}\right)$

Graph each polar equation.

15. $\theta = -60°$

16. $r = \dfrac{9}{2}$

17. $r = 7$

18. $\theta = \dfrac{11\pi}{6}$

Find the distance between each pair of points.

19. $\left(5, \dfrac{\pi}{2}\right), \left(2, -\dfrac{7\pi}{6}\right)$

20. $(-3, 60°), (4, 240°)$

21. $(-1, -45°), (6, 270°)$

22. $\left(7, \dfrac{5\pi}{6}\right), \left(2, \dfrac{4\pi}{3}\right)$

EXAMPLE 1

Graph $r = 5$.

The solutions of $r = 5$ are ordered pairs of the form $(5, \theta)$ where θ is any real number. The graph consists of all points that are 5 units from the pole, so the graph is a circle centered at the pole with radius 5.

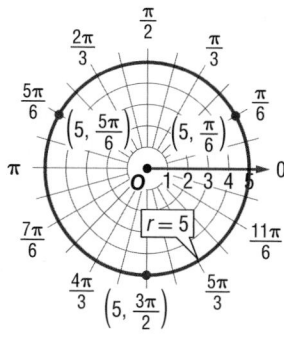

9-2 **Graphs of Polar Equations** (pp. 542–550)

Use symmetry, zeros, and maximum r-values to graph each function.

23. $r = \sin 3\theta$

24. $r = 2 \cos \theta$

25. $r = 5 \cos 2\theta$

26. $r = 4 \sin 4\theta$

27. $r = 2 + 2 \cos \theta$

28. $r = 1.5\theta, \theta \geq 0$

Use symmetry to graph each equation.

29. $r = 2 - \sin \theta$

30. $r = 1 + 5 \cos \theta$

31. $r = 3 - 2 \cos \theta$

32. $r = 4 + 4 \sin \theta$

33. $r = -3 \sin \theta$

34. $r = -5 + 3 \cos \theta$

EXAMPLE 2

Use symmetry to graph $r = 4 + 3 \cos \theta$.

Replacing (r, θ) with $(r, -\theta)$ yields $r = 4 + 3 \cos(-\theta)$, which simplifies to $r = 4 + 3 \cos \theta$ because cosine is even. The equations are equivalent, so the graph of this equation is symmetric with respect to the polar axis. Therefore, you can make a table of values to find the r-values corresponding to θ on the interval $[0, \pi]$.

θ	r
0	7
$\dfrac{\pi}{4}$	$\dfrac{8 + 3\sqrt{2}}{2}$
$\dfrac{\pi}{2}$	4
$\dfrac{3\pi}{4}$	$\dfrac{8 - 3\sqrt{2}}{2}$
π	1

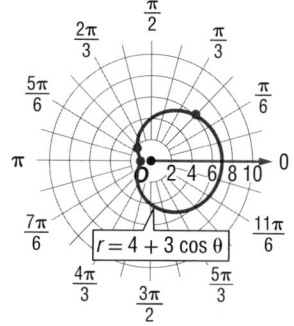

By plotting these points and using polar axis symmetry, you obtain the graph shown.

Lesson-by-Lesson Review

9-3 Polar and Rectangular Forms (pp. 551–559)

Find two pairs of polar coordinates for each point with the given rectangular coordinates if $0 \leq \theta \leq 2\pi$. Round to the nearest hundredth.

35. $(-1, 5)$

36. $(3, 7)$

37. $(2a, 0), a > 0$

38. $(4b, -6b), b > 0$

Write each equation in rectangular form, and then identify its graph. Support your answer by graphing the polar form of the equation.

39. $r = 5$

40. $r = -4 \sin \theta$

41. $r = 6 \sec \theta$

42. $r = \frac{1}{3} \csc \theta$

EXAMPLE 3

Write $r = 2 \cos \theta$ in rectangular form, and then identify its graph. Support your answer by graphing the polar form of the equation.

$$\begin{array}{ll} r = 2 \cos \theta & \text{Original equation} \\ r^2 = 2r \cos \theta & \text{Multiply each side by } r. \\ x^2 + y^2 = 2x & r^2 = x^2 + y^2 \text{ and } x = r \cos \theta \\ x^2 + y^2 - 2x = 0 & \text{Subtract } 2x \text{ from each side.} \end{array}$$

In standard form, $(x - 1)^2 + y^2 = 1$, you can identify the graph of this equation as a circle centered at $(1, 0)$ with radius 1, as supported by the graph of $r = 2 \cos \theta$.

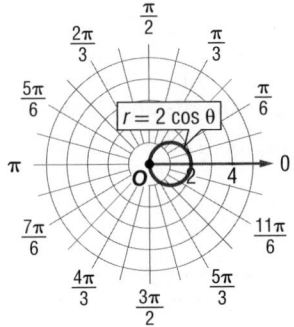

9-4 Polar Forms of Conic Sections (pp. 561–568)

Determine the eccentricity, type of conic, and equation of the directrix for each polar equation.

43. $r = \dfrac{3.5}{1 + \sin \theta}$

44. $r = \dfrac{1.2}{1 + 0.3 \cos \theta}$

45. $r = \dfrac{14}{1 - 2 \sin \theta}$

46. $r = \dfrac{6}{1 - \cos \theta}$

Write and graph a polar equation and directrix for the conic with the given characteristics.

47. $e = 0.5$; vertices at $(0, -2)$ and $(0, 6)$

48. $e = 1.5$; directrix: $x = 5$

Write each polar equation in rectangular form.

49. $r = \dfrac{1.6}{1 - 0.2 \sin \theta}$

50. $r = \dfrac{5}{1 + \cos \theta}$

EXAMPLE 4

Determine the eccentricity, type of conic, and equation of the directrix for $r = \dfrac{7}{3.5 - 3.5 \cos \theta}$.

Write the equation in standard form, $r = \dfrac{ed}{1 + e \cos \theta}$.

$$r = \frac{7}{3.5 - 3.5 \cos \theta} \quad \text{Original equation}$$

$$r = \frac{3.5(2)}{3.5(1 - \cos \theta)} \quad \text{Factor the numerator and denominator.}$$

$$r = \frac{2}{1 - \cos \theta} \quad \text{Divide the numerator and denominator by 3.5.}$$

In this form, you can see from the denominator that $e = 1$; therefore, the conic is a parabola. For polar equations of this form, the equation of the directrix is $x = -d$. From the numerator, we know that $ed = 2$, so $d = 2 \div 1$ or 2. Therefore, the equation of the directrix is $x = -2$.

Lesson-by-Lesson Review

9-5 **Complex Numbers and De Moivre's Theorem** (pp. 569–579)

Graph each number in the complex plane, and find its absolute value.

51. $z = 3 - i$ **52.** $z = 4i$

53. $z = -4 + 2i$ **54.** $z = 6 - 3i$

Express each complex number in polar form.

55. $3 + \sqrt{2}i$ **56.** $-5 + 8i$

57. $-4 - \sqrt{3}i$ **58.** $\sqrt{2} + \sqrt{2}i$

Graph each complex number on a polar grid. Then express it in rectangular form.

59. $z = 3\left(\cos\frac{\pi}{2} + i\sin\frac{\pi}{2}\right)$

60. $z = 5\left(\cos\frac{\pi}{3} + i\sin\frac{\pi}{3}\right)$

61. $z = -2\left(\cos\frac{\pi}{4} + i\sin\frac{\pi}{4}\right)$

62. $z = 4\left(\cos\frac{5\pi}{6} + i\sin\frac{5\pi}{6}\right)$

Find each product or quotient, and express it in rectangular form.

63. $-2\left(\cos\frac{5\pi}{6} + i\sin\frac{5\pi}{6}\right) \cdot -4\left(\cos\frac{\pi}{3} + i\sin\frac{\pi}{3}\right)$

64. $8(\cos 225° + i\sin 225°) \cdot \frac{1}{2}(\cos 120° + i\sin 120°)$

65. $5\left(\cos\frac{\pi}{2} + i\sin\frac{\pi}{2}\right) \div \frac{1}{3}\left(\cos\frac{\pi}{6} + i\sin\frac{\pi}{6}\right)$

66. $6(\cos 210° + i\sin 210°) \div 3(\cos 150° + i\sin 150°)$

Find each power, and express it in rectangular form.

67. $(4 - i)^5$

68. $(\sqrt{2} + 3i)^4$

Find all of the distinct pth roots of the complex number.

69. cube roots of $6 - 4i$

70. fourth roots of $1 + i$

EXAMPLE 5

Graph $4 - 6i$ in the complex plane and express in polar form.

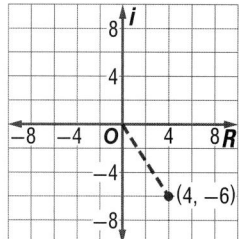

Find the modulus.

$r = \sqrt{a^2 + b^2}$ **Conversion formula**

$\quad = \sqrt{4^2 + (-6)^2}$ or $2\sqrt{13}$ $a = 4$ and $b = -6$

Find the argument.

$\theta = \tan^{-1}\frac{b}{a}$ **Conversion formula**

$\quad = \tan^{-1}\left(-\frac{6}{4}\right)$ $a = 4$ and $b = -6$

$\quad = -0.98$ **Simplify.**

The polar form of $4 - 6i$ is approximately $2\sqrt{13}\,[(\cos(-0.98) + i\sin(-0.98)]$.

EXAMPLE 6

Find $-3\left(\cos\frac{\pi}{4} + i\sin\frac{\pi}{4}\right) \cdot 5\left(\cos\frac{7\pi}{6} + i\sin\frac{7\pi}{6}\right)$ in polar form. Then express the product in rectangular form.

$-3\left(\cos\frac{\pi}{4} + i\sin\frac{\pi}{4}\right) \cdot 5\left(\cos\frac{7\pi}{6} + i\sin\frac{7\pi}{6}\right)$ **Original expression**

$= (-3 \cdot 5)\left[\cos\left(\frac{\pi}{4} + \frac{7\pi}{6}\right) + i\sin\left(\frac{\pi}{4} + \frac{7\pi}{6}\right)\right]$ **Product Formula**

$= -15\left[\cos\left(\frac{17\pi}{12}\right) + i\sin\left(\frac{17\pi}{12}\right)\right]$ **Simplify.**

Now find the rectangular form of the product.

$-15\left[\cos\left(\frac{17\pi}{12}\right) + i\sin\left(\frac{17\pi}{12}\right)\right]$ **Polar form**

$\quad = -15[-0.26 + i(-0.97)$ **Evaluate.**

$\quad = 3.9 + 14.5i$ **Distributive Property**

The polar form of the product is $-15\left[\cos\left(\frac{17\pi}{12}\right) + i\sin\left(\frac{17\pi}{12}\right)\right]$

The rectangular form of the product is $3.9 + 14.5i$.

Applications and Problem Solving

71. GAMES An arcade game consists of rolling a ball up an incline at a target. The region in which the ball lands determines the number of points earned. The model shows the point value for each region. (Lesson 9-1)

a. If, on a turn, a player rolls the ball to the point (3.5, 165°), how many points does he get?

b. Give two possible locations that a player will receive 50 points.

72. LANDSCAPING A landscaping company uses an adjustable lawn sprinkler that can rotate 360° and can cover a circular region with radius of 20 feet. (Lesson 9-1)

a. Graph the dimensions of the region that the sprinkler can cover on a polar grid if it is set to rotate 360°.

b. Find the area of the region that the sprinkler covers if the rotation is adjusted to −30° ≤ θ ≤ 210°.

73. BIOLOGY The pattern on the shell of a snail can be modeled using $r = \frac{1}{3}\theta + \frac{1}{2}$, $\theta \geq 0$. Identify and graph the classic curve that models this pattern. (Lesson 9-2)

74. RIDES The path of a Ferris wheel can be modeled by $r = 50 \sin \theta$, where r is given in feet. (Lesson 9-3)

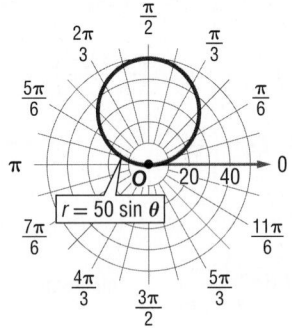

a. What are the polar coordinates of a rider located at $\theta = \frac{\pi}{12}$? Round to the nearest tenth, if necessary.

b. What are the rectangular coordinates of the rider's location? Round to the nearest tenth, if necessary.

c. What is the rider's distance above the ground if the polar axis represents the ground?

75. ORIENTEERING Orienteering requires participants to make their way through an area using a topographic map. One orienteer starts at Checkpoint A and walks 5000 feet at an angle of 35° measured clockwise from due east. A second orienteer starts at Checkpoint A and walks 3000 feet due west and then 2000 feet due north. How far, to the nearest foot, are the two orienteers from each other? (Lesson 9-3)

76. SATELLITE The orbit of a satellite around Earth has eccentricity of 0.05, and the distance from a vertex of the path to the center of Earth is 32,082 miles. Write a polar equation that can be used to model the path of the satellite if Earth is located at the focus closest to the given vertex. (Lesson 9-4)

77. ELECTRICITY Most circuits in Europe are designed to accommodate 220 volts. For parts **a** and **b**, use $E = I \cdot Z$, where voltage E is measured in volts, impedance Z is measured in ohms, and current I is measured in amps. Round to the nearest tenth. (Lesson 9-5)

a. If the circuit has a current of $2 + 5j$ amps, what is the impedance?

b. If a circuit has an impedance of $1 - 3j$ ohms, what is the current?

78. COMPUTER GRAPHICS Geometric transformation of figures can be performed using complex numbers. If a programmer starts with square $ABCD$, as shown below, each of the vertices can be represented by a complex number in polar form. Multiplication can then be used to rotate and dilate the square, producing the square $A'B'C'D'$. By what complex number should the programmer multiply each number to produce this transformation? (Lesson 9-5)

Find four different pairs of polar coordinates that name point P if $-2\pi \le \theta \le 2\pi$.

1.

2.

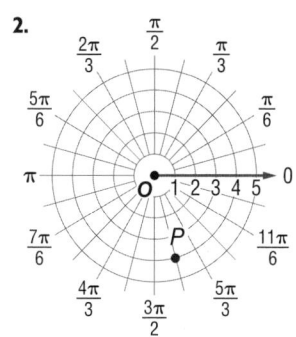

Graph each polar equation.

3. $\theta = 30°$

4. $r = 1$

5. $r = 2.5$

6. $\theta = \dfrac{5\pi}{3}$

7. $r = \dfrac{2}{3}\sin \theta$

8. $r = -\dfrac{1}{2}\sec \theta$

9. $r = -4 \csc \theta$

10. $r = 2 \cos \theta$

Identify and graph each classic curve.

11. $r = 1.5 + 1.5 \cos \theta$

12. $r^2 = 6.25 \sin 2\theta$

13. RADAR An air traffic controller is tracking an airplane with a current location of $(66, 115°)$. The value of r is given in miles.

   a. What are the rectangular coordinates of the airplane? Round to the nearest tenth mile.

   b. If a second plane is located at the point $(50, -75)$, what are the polar coordinates of the plane if $r > 0$ and $0 \le \theta \le 360°$? Round to the nearest mile and the nearest tenth of a degree, if necessary.

   c. What is the distance between the two planes? Round to the nearest mile.

Identify the graph of each rectangular equation. Then write the equation in polar form. Support your answer by graphing the polar form of the equation.

14. $(x - 7)^2 + y^2 = 49$

15. $y = 3x^2$

Determine the eccentricity, type of conic, and equation of the directrix for each polar equation.

16. $r = \dfrac{2}{1 - 0.4 \sin \theta}$

17. $r = \dfrac{6}{2 \cos \theta + 1}$

Write the equation for each polar graph in rectangular form.

18.

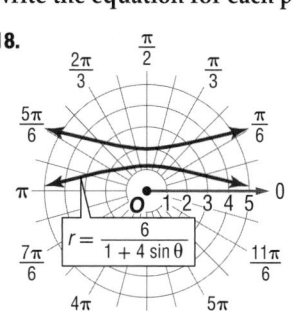

$$r = \frac{6}{1 + 4 \sin \theta}$$

19.

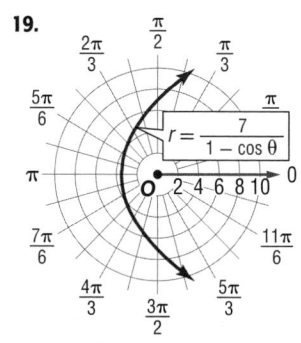

$$r = \frac{7}{1 - \cos \theta}$$

20. ELECTRICITY If a circuit has a voltage E of 135 volts and a current I of $3 - 4j$ amps, find the impedance Z of the circuit in ohms in rectangular form. Use the equation $E = I \cdot Z$.

21. MULTIPLE CHOICE Identify the graph of point P with complex coordinates $(-\sqrt{3}, -1)$ on the polar coordinate plane.

A

C

B

D

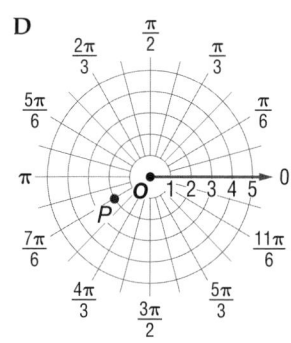

Find each power, and express it in rectangular form. Round to the nearest tenth.

22. $(-1 + 4i)^3$

23. $(-7 - 3i)^5$

24. $(6 + i)^4$

25. $(2 - 5i)^6$

Objective
Approximate the arc length of a curve.

You can find the length of a line segment by using the Distance Formula. You can find the length of an arc by using proportions. In calculus, you will need to calculate many lengths that are not represented by line segments or sections of a circle.

$$\frac{\ell}{2\pi r} = \frac{x}{360}$$

As mentioned in Chapter 2, *integral calculus* focuses on areas, volumes, and lengths. It can be used to find the length of a curve for which we do not have a standard equation, such as a curve defined by a quadratic, cubic, or polar function. *Riemann sums* and *definite integrals*, two concepts that you will learn more about in the following chapters, are needed to calculate the exact length of a curve, or *arc length*, denoted s.

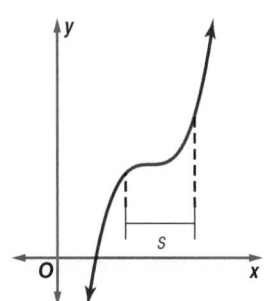

In this lesson, we will approximate the arc length of a curve using a process similar to the method that you applied to approximate the rate of change at a point. Recall that in Chapter 1, you calculated the slopes of secant lines to approximate the rates of change for graphs at specific points. Decreasing the distance between the two points on the secant lines increased the accuracy of the approximations as shown in the graph at the right.

$$m = \frac{f(x_2) - f(x_1)}{x_2 - x_1}$$

ACTIVITY 1 | Approximate Arc Length

Approximate the arc length of the graph of $y = x^2$ for $0 \le x \le 3$.

Step 1 Graph $y = x^2$ for $0 \le x \le 3$ as shown.

Step 2 Graph points on the curve at $x = 1, 2,$ and 3. Connect the points using line segments as shown.

Step 3 Use the Distance Formula to find the length of each line segment.

Step 4 Approximate the length of the arc by finding the sum of the lengths of the line segments.

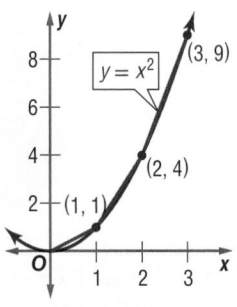

Analyze the Results

1. Is your approximation *greater* or *less* than the actual length? Explain your reasoning.

2. Approximate the arc length a second time using 6 line segments that are formed by the points $x = 0, 0.5, 1.0, 1.5, 2.0, 2.5,$ and 3.0. Include a sketch of the graph with your approximation.

3. Describe what happens to the approximation for the arc length as shorter line segments are used.

4. For the two approximations, the endpoints of the line segments were equally spaced along the x-axis. Do you think this will always produce the most accurate approximation? Explain your reasoning.

Notice that for the first activity, the endpoints of the line segments were equally spaced along the *x*-axis. For example, dividing the curve into 6 segments on the interval [0, 3] resulted in line segments with endpoints that were equally spaced 0.5 units apart along the *x*-axis. When using advanced methods of calculus to find *exact* arc length, a constant difference between a pair of endpoints along the *x*-axis is essential. This difference is denoted Δx.

For the purpose of accurately *approximating* arc length, using a constant Δx to create the line segments may not always be the most efficient method. The shape of the arc will dictate the spacing of the endpoints, thus creating different values for Δx. For example, if a graph shows an increase or decrease over a large interval for *x*, a large line segment may be used for the approximation. If a graph includes a turning point, it is better to use small line segments to account for the curve in the graph.

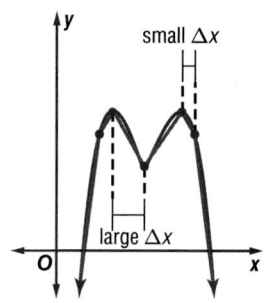

In Lesson 9-1, you learned how to calculate the distance between polar coordinates. This formula can be used to approximate the arc length of a curve represented by a polar equation.

StudyTip

Polar Graphs Create a table of values for *r* and θ when calculating the arc length for a polar graph. This will help to reduce errors created by functions that produce negative values for *r*.

ACTIVITY 2 | **Approximate Arc Length**

Approximate the arc length of the graph of $r = 4 + 4 \sin \theta$ for $0 \le x \le 2\pi$.

Step 1 Graph $r = 4 + 4 \sin \theta$ for $0 \le x \le 2\pi$ as shown.

Step 2 Draw 6 points on the curve at $\theta = 0, \frac{\pi}{3}, \frac{\pi}{2}, \frac{2\pi}{3}, \pi$, and $\frac{3\pi}{2}$. Connect the points using line segments as shown.

Step 3 Use the Polar Distance Formula to find the length of each line segment.

Step 4 Approximate the length of the arc by finding the sum of the lengths of the line segments.

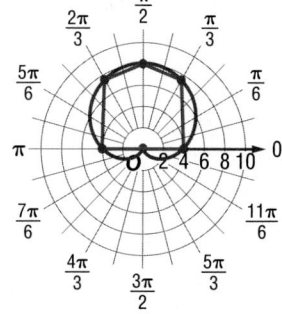

▶ **Analyze the Results**

5. Explain how symmetry can be used to reduce the number of calculations in Step 3.

6. Approximate the arc length using at least 10 segments. Include a sketch of the graph.

7. Let *n* be the number of line segments used in an approximation and $\Delta\theta$ be a constant difference in θ between the endpoints of a line segment. Make a conjecture regarding the relationship between *n*, θ, and the approximation for an arc length.

Model and Apply

Approximate the arc length for each graph. Include a sketch of your graph.

8.

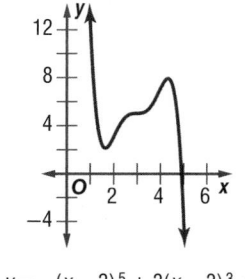

$$y = -(x-3)^5 + 3(x-3)^3 + 5$$
$$\text{for } 1 \le x \le 5$$

9.

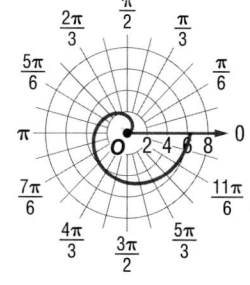

$$r = \theta \text{ for } 0 \le \theta \le 2\pi$$

CHAPTER 10

Sequences and Series

Then

In Chapters 1–4, you modeled data using various types of functions.

Now

In Chapter 10, you will:

- Relate sequences and functions.
- Use sigma notation to represent and calculate sums of series.
- Use arithmetic and geometric sequences and series.
- Prove statements by using mathematical induction.
- Expand powers by using the Binomial Theorem.

Why?

MARCHING BAND Sequences and series can be used to predict patterns. For example, arithmetic sequences can be used to determine the number of band members in a specified row of a pyramid formation.

PREREAD Use the text on this page to predict the organization of Chapter 10.

Get Ready for Chapter 10

Diagnose Readiness You have two options for checking Prerequisite Skills.

Text Option

Take the Quick Check below.

*Quick*Check

Expand each binomial. (Prerequisite Skill)

1. $(x + 3)^3$ **2.** $(2x - 1)^4$

Use the graph of each function to describe its end behavior. Support the conjecture numerically. (Lesson 1-3)

3. $f(x) = \dfrac{4}{x}$

4. 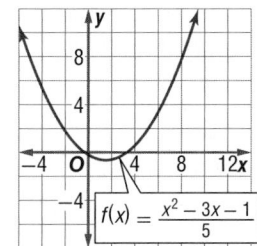 $f(x) = \dfrac{x^2 - 3x - 1}{5}$

Sketch and analyze the graph of each function. Describe its domain, range, intercepts, asymptotes, end behavior, and where the function is increasing or decreasing. (Lesson 3-1)

5. $f(x) = 3^{-x}$ **6.** $r(x) = 5^{-x}$

7. $h(x) = 0.1^{x+2}$ **8.** $k(x) = -2^x$

Evaluate each expression. (Lesson 3-2)

9. $\log_2 16$ **10.** $\log_{10} 10$ **11.** $\log_6 \dfrac{1}{216}$

12. MUSIC The table shows the type and number of CDs that Adam and Lindsay bought. Write and solve a system of equations to determine the price of each type of CD. (Lesson 6-1)

Buyer	New CD	Used CD	Price ($)
Adam	2	5	49
Lindsay	3	4	56

Online Option

Math Online Take a self-check Chapter Readiness Quiz at **glencoe.com**.

New Vocabulary

English	• Page •	Español
sequence	• 590 •	sucesión
term	• 590 •	término
finite sequence	• 590 •	sucesión finita
infinite sequence	• 590 •	sucesión infinita
recursive sequence	• 591 •	sucesión recursiva
explicit sequence	• 591 •	sucesión explícita
Fibonacci sequence	• 591 •	sucesión de Fibonacci
converge	• 592 •	converge
diverge	• 592 •	diverge
series	• 593 •	serie
finite series	• 593 •	serie finita
nth partial sum	• 593 •	suma parcial enésima
infinite series	• 593 •	serie infinita
sigma notation	• 594 •	notación de suma
arithmetic sequence	• 599 •	sucesión aritmética
common difference	• 599 •	diferencia común
arithmetic series	• 602 •	serie aritmética
common ratio	• 608 •	razón común
geometric means	• 611 •	medios geométricos
geometric series	• 611 •	serie geométrica
binomial coefficients	• 628 •	coeficientes binomiales
power series	• 636 •	serie de potencias

Review Vocabulary

exponential function • p. 158 • funciones exponenciales a function in which the base is a constant and the exponent is a variable

Exponential Growth Function

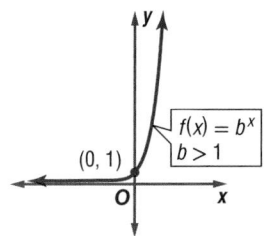

$f(x) = b^x$
$b > 1$

$(0, 1)$

Sequences, Series, and Sigma Notation

Then
You used functions to generate ordered pairs and used graphs to analyze end behavior. (Lesson 1-1 and 1-3)

Now
- Investigate several different types of sequences.
- Use sigma notation to represent and calculate sums of series.

New Vocabulary
sequence
term
finite sequence
infinite sequence
recursive sequence
explicit sequence
Fibonacci sequence
converge
diverge
series
finite series
*n*th partial sum
infinite series
sigma notation

Math Online

glencoe.com

Why?
Khari developed a Web site where students at her high school can post their own social networking Web pages. A student at the high school is given a free page if he or she refers the Web site to five friends. The site starts with one page created by Khari, who in turn, refers five friends that each create a page. Those five friends refer five more people each, all of whom develop pages, and so on.

In mathematics, a **sequence** is an ordered list of numbers. Each number in the sequence is known as a **term**. A **finite sequence**, such as 1, 3, 5, 7, 9, 11, contains a finite number of terms. An **infinite sequence**, such as 1, 3, 5, 7, …, contains an infinite number of terms.

Each term of a sequence is a function of its position. Therefore, an infinite sequence is a function whose domain is the set of natural numbers and can be written as $f(1) = a_1, f(2) = a_2, f(3) = a_3, …,$ $f(n) = a_n, …,$ where a_n denotes the nth term. If the domain of the function is only the first n natural numbers, the sequence is finite.

Infinitely many sequences exist with the same first few terms. To sufficiently define a *unique* sequence, a formula for the nth term or other information *must* be given. When defined *explicitly*, an **explicit formula** gives the nth term a_n as a function of n.

EXAMPLE 1 | Find Terms of Sequences

a. Find the next four terms of the sequence 2, 7, 12, 17, … .

The nth term of this sequence is not given. One possible pattern is that each term is 5 greater than the previous term. Therefore, a sample answer for the next four terms is 22, 27, 32, and 37.

b. Find the next four terms of the sequence 2, 5, 10, 17, … .

The nth term of this sequence is not given. If we subtract each term from the term that follows, we start to see a possible pattern.

$$a_2 - a_1 = 5 - 2 \text{ or } 3 \qquad a_3 - a_2 = 10 - 5 \text{ or } 5 \qquad a_4 - a_3 = 17 - 10 \text{ or } 7$$

It appears that each term is generated by adding the next successive odd number. However, looking at the pattern, it may also be determined that each term is 1 more than each perfect square, or $a_n = n^2 + 1$. Using either pattern, a sample answer for the next four terms is 26, 37, 50, and 65.

c. Find the first four terms of the sequence given by $a_n = 2n(-1)^n$.

Use the explicit formula given to find a_n for $n = 1, 2, 3,$ and 4.

$$a_1 = 2 \cdot 1 \cdot (-1)^1 \text{ or } -2 \qquad n = 1 \qquad a_2 = 2 \cdot 2 \cdot (-1)^2 \text{ or } 4 \qquad n = 2$$
$$a_3 = 2 \cdot 3 \cdot (-1)^3 \text{ or } -6 \qquad n = 3 \qquad a_4 = 2 \cdot 4 \cdot (-1)^4 \text{ or } 8 \qquad n = 4$$

The first four terms in the sequence are $-2, 4, -6,$ and 8.

✓ Guided Practice

Find the next four terms of each sequence.

1A. 32, 16, 8, 4, …

1B. 1, 2, 4, 7, 11, 16, 22, …

1C. Find the first four terms of the sequence given by $a_n = n^3 - 10$.

Sequences can also be defined *recursively*. Recursively defined sequences give one or more of the first few terms and then define the terms that follow using those previous terms. The formula defining the *n*th term of the sequence is called a **recursive formula** or a *recurrence relation*.

StudyTip

Notation The term denoted a_n represents the *n*th term of a sequence. The term denoted a_{n-1} represents the term immediately before a_n. The term a_{n-2} represents the term two terms before a_n.

EXAMPLE 2   **Recursively Defined Sequences**

Find the fifth term of the recursively defined sequence $a_1 = 1$, $a_n = a_{n-1} + 2n - 1$, where $n \geq 2$.

Since the sequence is defined recursively, all the terms before the fifth term must be found first. Use the given first term, $a_1 = 2$, and the recursive formula for a_n.

$a_2 = a_{2-1} + 2(2) - 1$ **$n = 2$**

   $= a_1 + 3$ **Simplify.**

   $= 2 + 3$ or 5   **$a_1 = 2$**

$a_3 = a_{3-1} + 2(3) - 1$ **$n = 3$**

   $= a_2 + 5$ or 10   **$a_2 = 5$**

$a_4 = a_{4-1} + 2(4) - 1$ **$n = 4$**

   $= a_3 + 7$ or 17   **$a_3 = 10$**

$a_5 = a_{5-1} + 2(5) - 1$ **$n = 5$**

   $= a_4 + 9$ or 26   **$a_4 = 17$**

☑ **Guided Practice**

Find the sixth term of each sequence.

2A. $a_1 = 3$, $a_n = (-2)a_{n-1}$, $n \geq 2$ **2B.** $a_1 = 8$, $a_n = 2a_{n-1} - 7$, $n \geq 2$

The **Fibonacci sequence** describes many patterns found in nature. This sequence is often defined recursively.

🌐 **Real-World EXAMPLE 3**   **Fibonacci Sequence**

NATURE Suppose that when a plant first starts to grow, the stem has to grow for two months before it is strong enough to support branches. At the end of the second month, it sprouts a new branch and will continue to sprout one new branch each month. The new branches also each grow for two months and then start to sprout one new branch each month. If this pattern continues, how many branches will the plant have after 10 months?

During the first two months, there will only be one branch, the stem. At the end of the second month, the stem will produce a new branch, making the total for the third month two branches. The new branch will grow and develop two months before producing a new branch of its own, but the original branch will now produce a new branch each month.

🌐 **Real-World Link**

Along with being found in flower petals, sea shells, and the bones in a human hand, Fibonacci sequences can also be found in pieces of art, music, poetry, and architecture.

Source: *Universal Principles of Design*

(continued on the next page)

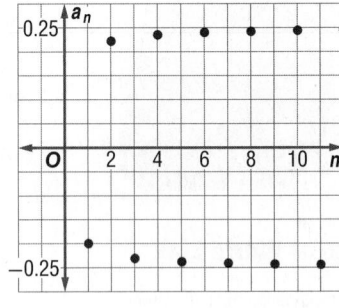

Watch Out!

Notation The first term of a sequence is occasionally denoted as a_0. When this occurs, the domain of the function describing the sequence is the set of whole numbers.

The following table shows the pattern.

Month	1	2	3	4	5	6	7	8	9	10
Branches	1	1	2	3	5	8	13	21	34	55

Each term is the sum of the previous two terms. This pattern can be written as the recursive formula $a_0 = 1, a_1 = 1, a_n = a_{n-2} + a_{n-1}$, where $n \geq 2$.

✓ Guided Practice

3. NATURE How many branches will a plant like the one described in Example 3 have after 15 months if no branches are removed?

In Lesson 1-3, you examined the end behavior of the graphs of functions. You learned that as the domains of some functions approach ∞, the ranges approach a unique number called a limit. As a function, an infinite sequence may also have a limit. If a sequence has a limit such that the terms approach a unique number, then it is said to **converge**. If not, the sequence is said to **diverge**.

EXAMPLE 4   Convergent and Divergent Sequences

Determine whether each sequence is *convergent* or *divergent*.

a. $a_n = -3n + 12$

The first eight terms of this sequence are 12, 9, 6, 3, 0, −3, −6, and −9. From the graph at the right, you can see that a_n does not approach a finite number. Therefore, this sequence is divergent.

b. $a_1 = 36, a_n = -\frac{1}{2}a_{n-1}, a \geq 2$

The first eight terms of this sequence are 36, −18, 9, −4.5, 2.25, −1.125, 0.5625, −0.28125, and 0.140625. From the graph at the right, you can see that a_n approaches 0 as n increases. This sequence has a limit and is therefore convergent.

TechnologyTip

Convergent or Divergent Sequences If an explicit formula for a sequence is known, you can enter the formula in the **Y=** menu of a graphing calculator and graph the related function. Analyzing the end behavior of the graph can help you to determine whether the sequence is convergent or divergent.

c. $a_n = \frac{(-1)^n \cdot n}{4n + 1}$

The first twelve terms of this sequence are given or approximated below.

$a_1 = -0.2$ $a_2 \approx 0.222$
$a_3 \approx -0.231$ $a_4 \approx 0.235$
$a_5 \approx -0.238$ $a_6 = 0.24$
$a_7 \approx -0.241$ $a_8 \approx 0.242$
$a_9 \approx -0.243$ $a_{10} \approx 0.244$
$a_{11} \approx -0.244$ $a_{12} \approx 0.245$

It appears that when n is odd, a_n approaches $-\frac{1}{4}$, and when n is even, a_n approaches $\frac{1}{4}$. Since a_n does not approach one particular value, the sequence has no limit. Therefore, the sequence is divergent.

✓ Guided Practice

4A. $a_n = \frac{64}{2n}$ **4B.** $a_1 = 9, a_n = a_{n-1} + 4$ **4C.** $a_n = 3(-1)^n$

Series and Sigma Notation A **series** is the indicated sum of all of the terms of a sequence. Like sequences, series can be finite or infinite. A **finite series** is the indicated sum of all the terms of a finite sequence, and an **infinite series** is the indicated sum of all the terms of an infinite sequence.

	Sequence	Series
Finite	1, 3, 5, 7, 9	$1 + 3 + 5 + 7 + 9$
Infinite	1, 3, 5, 7, 9, ...	$1 + 3 + 5 + 7 + 9 + \ldots$

The sum of the first n terms of a series is called the **nth partial sum** and is denoted S_n. The nth partial sum of any series can be found by calculating each term up to the nth term and then finding the sum of those terms.

EXAMPLE 5 | **The nth Partial Sum**

a. Find the fourth partial sum of $a_n = (-2)^n + 3$.

Find the first four terms.

$a_1 = (-2)^1 + 3$ or 1 $n = 1$
$a_2 = (-2)^2 + 3$ or 7 $n = 2$
$a_3 = (-2)^3 + 3$ or -5 $n = 3$
$a_4 = (-2)^4 + 3$ or 19 $n = 4$

The fourth partial sum is $S_4 = 1 + 7 + (-5) + 19$ or 22.

b. Find S_3 of $a_n = \dfrac{4}{10^n}$.

Find the first three terms.

$a_1 = \dfrac{4}{10^1}$ or 0.4 $n = 1$

$a_2 = \dfrac{4}{10^2}$ or 0.04 $n = 2$

$a_3 = \dfrac{4}{10^3}$ or 0.004 $n = 3$

The third partial sum is $S_3 = 0.4 + 0.04 + 0.004$ or 0.444.

✔ Guided Practice

5A. Find the sixth partial sum of $a_1 = 8$, $a_n = 0.5(a_{n-1})$, $n \geq 2$.

5B. Find the seventh partial sum of $a_n = 3\left(\dfrac{1}{10}\right)^n$.

Since an infinite series does not have a finite number of terms, you might assume that an infinite series has no sum S. This is true for the series below.

Infinite Sequence	**Infinite Series**	**Sequence of First Four Partial Sums**
1, 4, 7, 10, ...	$1 + 4 + 7 + 10 + \ldots$	1, 5, 12, 22, ...

However, some infinite series do have sums. For an infinite series to have a fixed sum S, the infinite sequence associated with this series must converge to 0. Notice the sequence of partial sums in the infinite series below appears to approach a sum of $0.\overline{1}$ or $\dfrac{1}{9}$.

Infinite Sequence	**Infinite Series**	**Sequence of First Three Partial Sums**
0.1, 0.01, 0.001, ...	$0.1 + 0.01 + 0.001 + \ldots$	0.1, 0.11, 0.111, ...

We will take a closer look at sums of infinite sequences in Lesson 10-3.

Series are often more conveniently notated using the uppercase Greek letter sigma Σ. A series written using this letter is said to be expressed using *summation notation* or **sigma notation**.

Key Concept | **Sigma Notation**

For any sequence $a_1, a_2, a_3, a_4, \ldots$, the sum of the first k terms is denoted

$$\sum_{n=1}^{k} a_n = a_1 + a_2 + a_3 + \ldots + a_k,$$

where n is the index of summation, k is the upper bound of summation, and 1 is the lower bound of summation.

ReadingMath

Sigma Notation $\sum_{n=1}^{k} a_n$ is read *the summation from $n = 1$ to k of a sub n.*

In this notation, the lower bound indicates where to begin summing the terms of the sequence and the upper bound indicates where to end the sum. If the upper bound is given as ∞, the sigma notation represents an infinite series.

$$\sum_{n=1}^{\infty} a_n = a_1 + a_2 + a_3 + \ldots$$

Watch Out!

Variations in Sigma Notation The index of summation does not have to be the letter n. It can be represented by any variable. For example, the summation in Example 6a could also be written as

$$\sum_{i=1}^{5} (4i - 3).$$

EXAMPLE 6 | **Sums in Sigma Notation**

Find each sum.

a. $\displaystyle\sum_{n=1}^{5} (4n - 3)$

$$\sum_{n=1}^{5} (4n - 3) = [4(1) - 3] + [4(2) - 3] + [4(3) - 3] + [4(4) - 3] + [4(5) - 3]$$
$$= 1 + 5 + 9 + 13 + 17 \text{ or } 45$$

b. $\displaystyle\sum_{n=3}^{7} \frac{6n - 3}{2}$

$$\sum_{n=3}^{7} \frac{6n - 3}{2} = \frac{6(3) - 3}{2} + \frac{6(4) - 3}{2} + \frac{6(5) - 3}{2} + \frac{6(6) - 3}{2} + \frac{6(7) - 3}{2}$$
$$= 7.5 + 10.5 + 13.5 + 16.5 + 19.5 \text{ or } 67.5$$

c. $\displaystyle\sum_{n=1}^{\infty} \frac{7}{10^n}$

$$\sum_{n=1}^{\infty} \frac{7}{10^n} = \frac{7}{10^1} + \frac{7}{10^2} + \frac{7}{10^3} + \frac{7}{10^4} + \frac{7}{10^5} + \ldots$$
$$= 0.7 + 0.07 + 0.007 + 0.0007 + 0.00007 + \ldots$$
$$= 0.77777\ldots \text{ or } \frac{7}{9}$$

 Guided Practice

6A. $\displaystyle\sum_{n=1}^{5} \frac{n^2 - 1}{2}$ **6B.** $\displaystyle\sum_{n=7}^{13} (n^3 - n^2)$ **6C.** $\displaystyle\sum_{n=1}^{\infty} \frac{6}{10^n}$

Note that while the lower bound of a summation is often 1, a sum can start with any term p in a sequence as long as $p < k$. In Example 6b, the summation started with the 3rd term of the sequence and ended with the 7th term.

Find the next four terms of each sequence. (Example 1)

1. $1, 8, 15, 22, \ldots$

2. $3, -6, 12, -24, \ldots$

3. $81, 27, 9, 3, \ldots$

4. $1, 3, 7, 13, \ldots$

5. $-2, -15, -28, -41 \ldots$

6. $1, 4, 10, 19, \ldots$

Find the first four terms of each sequence. (Example 1)

7. $a_n = n^2 - 1$

8. $a_n = -2^n + 7$

9. $a_n = \dfrac{n + 7}{9 - n}$

10. $a_n = (-1)^{n + 1} + n$

11. **AUTOMOBILE LEASES** Lease agreements often contain clauses that limit the number of miles driven per year by charging a per-mile fee over that limit. For the car shown below, the lease requires that the number of miles driven each year must be no more than 15,000. (Example 2)

a. Write the sequence describing the maximum number of allowed miles on the car at the end of every 12 months of the lease if the car has 1350 miles at the beginning of the lease.

b. Write the first 4 terms of the sequence that gives the cumulative cost of the lease for a given month.

c. Write an explicit formula to represent the sequence in part **b**.

d. Determine the total amount of money paid by the end of the lease.

Find the specified term of each sequence. (Example 2)

12. 4th term, $a_1 = 5, a_n = -3a_{n-1} + 10, n \geq 2$

13. 7th term, $a_1 = 14, a_n = 0.5a_{n-1} + 3, n \geq 2$

14. 4th term, $a_1 = 0, a_n = 3^{a_{n-1}}, n \geq 2$

15. 3rd term, $a_1 = 3, a_n = (a_{n-1})^2 - 5a_{n-1} + 4, n \geq 2$

16. **WEB SITE** Khari, the student from the beginning of the lesson, had great success expanding her Web site. Each student who received a referral developed a Web page and referred five more students to Khari's site. (Example 3)

a. List the first five terms of a sequence modeling the number of new Web pages created through Khari's site.

b. Suppose the school has 1576 students. After how many rounds of referrals did the entire student body have a Web page?

17. **BEES** Female honeybees come from fertilized eggs (male and female parent), while male honeybees come from unfertilized eggs (only one female parent). (Example 3)

a. Draw a family tree showing the 3 previous generations of a male honeybee (parents only).

b. Determine the number of parent bees in the 11th previous generation of a male honeybee.

Determine whether each sequence is *convergent* or *divergent*. (Example 4)

18. $a_1 = 4, 1.5a_{n-1}, n \geq 2$

19. $a_n = \dfrac{5}{10^n}$

20. $a_n = -n^2 - 8n + 106$

21. $a_1 = -64, \dfrac{3}{4}a_{n-1}, n \geq 2$

22. $a_1 = 1, a_n = 4 - a_{n-1}, n \geq 2$

23. $a_n = n^2 - 3n + 1$

24. $a_n = \dfrac{n^2 + 4}{3 + n}$

25. $a_1 = 9, a_n = \dfrac{a_{n-1} + 3}{2}, n \geq 2$

26. $a_n = \dfrac{5n + 6}{n}$

27. $a_n = \dfrac{5n}{5^n} + 1$

Find the indicated sum for each sequence. (Example 5)

28. 5th partial sum of $a_n = n(n - 4)(n - 3)$

29. 6th partial sum of $a_n = \dfrac{-5n + 3}{n}$

30. S_8 of $a_1 = 1, a_n = a_{n-1} + (18 - n), n \geq 2$

31. S_4 of $a_1 = 64, a_n = -\dfrac{3}{4}a_{n-1}, n \geq 2$

32. 11th partial sum of $a_1 = 4, a_n = (-1)^{n-1}(|a_{n-1}| + 3), n \geq 2$

33. S_9 of $a_1 = -35, a_n = a_{n-1} + 8, n \geq 2$

34. 4th partial sum of $a_1 = 3, a_n = (a_{n-1} - 2)^3, n \geq 2$

35. S_4 of $a_n = \dfrac{(-3)^n}{10}$

Find each sum. (Example 6)

36. $\displaystyle\sum_{n=1}^{8} (6n - 11)$

37. $\displaystyle\sum_{n=4}^{11} (30 - 4n)$

38. $\displaystyle\sum_{n=1}^{7} [n^2(n - 5)]$

39. $\displaystyle\sum_{n=2}^{7} (n^2 - 6n + 1)$

40. $\displaystyle\sum_{n=8}^{15} \left(\dfrac{n}{4} - 7\right)$

41. $\displaystyle\sum_{n=1}^{10} [(n - 4)^2(n - 5)]$

42. $\displaystyle\sum_{n=0}^{6} [(-2)^n - 9]$

43. $\displaystyle\sum_{n=1}^{3} 7\left(\dfrac{1}{10}\right)^{2n}$

44. $\displaystyle\sum_{n=1}^{\infty} 5\left(\dfrac{1}{10^n}\right)$

45. $\displaystyle\sum_{n=1}^{\infty} \dfrac{8}{10^n}$

46. **FINANCIAL LITERACY** Jim's bank account had an initial deposit of $380, earning 3.5% interest per year compounded annually.

a. Find the balance each year for the first five years.

b. Write a recursive and an explicit formula defining his account balance.

c. For very large values of n, which formula gives a more accurate balance? Explain.

47. INVESTING Melissa invests $200 every 3 months. The investment pays an annual percentage rate of 8%, and the interest is compounded quarterly. If Melissa makes each payment at the beginning of the quarter and the interest is posted at the end of the quarter, what will the total value of the investment be after 2 years?

48. RIDES The table shows the number of riders of the Mean Streak each year from 1998 to 2007. This ridership data can be approximated by $a_n = -\frac{1}{20}n + 1.3$, where $n = 1$ represents 1998, $n = 2$ represents 1999, and so on.

Mean Streak Roller Coaster			
Year	Number of Riders (millions)	Year	Number of Riders (millions)
1998	1.31	2003	0.99
1999	1.15	2004	0.95
2000	1.14	2005	0.89
2001	1.09	2006	0.81
2002	1.05	2007	0.82

Source: Cedar Fair Entertainment Company

a. Sketch a graph of the number of riders from 1998 to 2007. Then determine whether the sequence appears to be *convergent* or *divergent*. Does this make sense in the context of the situation? Explain your reasoning.

b. Use the table to find the total number of riders from 1998 to 2005. Then use the explicit sequence to find the 8th partial sum of a_n. Compare the results.

c. If the sequence continues, find a_{14}. What does this number represent?

Copy and complete the table.

	Recursive Formula	Sequence	Explicit Formula
49.		6, 8, 10, 12, …	
50.	$a_1 = 15, a_n = a_{n-1} - 1, n \geq 2$		
51.		7, 21, 63, 189, …	
52.			$a_n = 10(-2)^n$
53.			$a_n = 8n - 3$
54.	$a_1 = 2, a_n = 4a_{n-1}, n \geq 2$		
55.	$a_1 = 3, a_n = a_{n-1} + 2n - 1, n \geq 2$		
56.			$a_n = n^2 + 1$

Write each series in sigma notation. The lower bound is given.

57. $-2 - 1 + 0 + 1 + 2 + 3 + 4 + 5; n = 1$

58. $\frac{1}{20} + \frac{1}{25} + \frac{1}{30} + \frac{1}{35} + \frac{1}{40} + \frac{1}{45}; n = 4$

59 $8 + 27 + 64 + \cdots + 1000; n = 2$

60. $\frac{1}{2} + \frac{1}{4} + \frac{1}{8} + \cdots + \frac{1}{128}; n = 1$

61. $-8 + 16 - 32 + 64 - 128 + 256 - 512; n = 3$

62. $8\left(-\frac{1}{3}\right) + 8\left(\frac{1}{9}\right) + 8\left(-\frac{1}{27}\right) + \cdots + 8\left(-\frac{1}{243}\right); n = 1$

Determine whether each sequence is *convergent* or *divergent*. Then find the fifth partial sum of the sequence.

63. $a_n = \sin \frac{n\pi}{2}$ **64.** $a_n = n \cos \pi$ **65.** $a_n = e^{-\frac{n}{2}} \cos \pi n$

66. WATER PRESSURE The pressure exerted on the human body at sea level is 14.7 pounds per square inch (psi). For each additional foot below sea level, the pressure is about 0.445 psi greater, as shown.

sea level = 14.7 psi
+0.445 psi/ft
n ft

a. Write a recursive formula to represent a_n, the pressure at n feet below sea level. (*Hint:* Let $a_0 = 14.7$.)

b. Write the first three terms of the sequence and describe what they represent.

c. Scuba divers cannot safely dive deeper than 100 feet. Write an explicit formula to represent a_n. Then use the formula to find the water pressure at 100 feet below sea level.

Match each sequence with its graph.

67.

68.

69.

70.

71.

72.
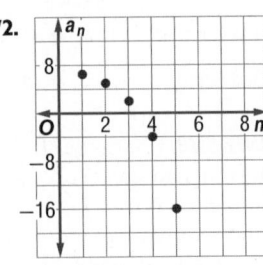

a. $a_n = \frac{4}{3}n$

b. $a_n = -\frac{3}{4}n + 9$

c. $a_n = \left(-\frac{4}{3}\right)^n$

d. $a_n = 8 - \frac{3}{4}(2^n)$

e. $a_n = 9 - 2n$

f. $a_n = \left(\frac{4}{3}\right)^n + 8$

73. GOLDEN RATIO Consider the Fibonacci sequence 1, 1, 2, 3, ..., $a_{n-2} + a_{n-1}$.

   a. Find $\dfrac{a_n}{a_{n-1}}$ for the second through eleventh terms of the Fibonacci sequence.

   b. Sketch a graph of the terms found in part **a**. Let $n - 1$ be the x-coordinate and $\dfrac{a_n}{a_{n-1}}$ be the y-coordinate.

   c. Based on the graph found in part **b**, does this sequence appear to be convergent? If so, describe the limit to three decimal places. If not, explain why not.

   d. In a *golden rectangle*, the ratio of the length to the width is about 1.61803399. This is called the *golden ratio*. How does the limit of the sequence $\dfrac{a_n}{a_{n-1}}$ compare to the golden ratio?

   e. Golden rectangles are common in art and architecture. The Parthenon, in Greece, is an example of how golden rectangles are used in architecture.

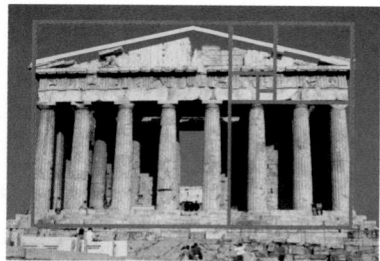

   Research golden rectangles and find two more examples of golden rectangles in art or architecture.

Determine whether each sequence is *convergent* or *divergent*.

74.

75.

76.

77.
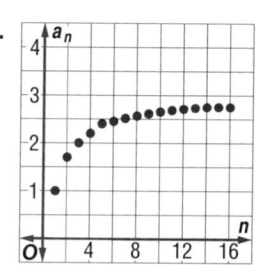

Write an explicit formula for each recursively defined sequence.

78. $a_1 = 10; a_n = a_{n-1} + 5$

79. $a_1 = 1.25; a_n = a_{n-1} - 0.5$

80. $a_1 = 128; a_n = 0.5a_{n-1}$

81. ⚙ **MULTIPLE REPRESENTATIONS** In this problem, you will investigate sums of infinite series.

   a. **NUMERICAL** Calculate the first five terms of the infinite sequence $a_n = \dfrac{4}{10^n}$.

   b. **GRAPHICAL** Use a graphing calculator to sketch $y = \dfrac{4}{10^x}$.

   c. **VERBAL** Describe what is happening to the terms of the sequence as $n \to \infty$.

   d. **NUMERICAL** Find the sum of the first 5 terms, 7 terms, and 9 terms of the series.

   e. **VERBAL** Describe what is happening to the partial sums S_n as n increases.

   f. **VERBAL** Predict the sum of the first n terms of the series. Explain your reasoning.

H.O.T. Problems   Use Higher-Order Thinking Skills

82. CHALLENGE Consider the recursive sequence below.
$$a_n = a_{n-a_{n-1}} + a_{n-a_{n-2}} \text{ for } a_1 = 1, a_2 = 1, n \geq 3$$

   a. Find the first eight terms of the sequence.

   b. Describe the similarities and differences between this sequence and the other recursive sequences in this lesson.

83. OPEN ENDED Write a sequence either recursively or explicitly that has the following characteristics.

   a. converges to 0

   b. converges to 3

   c. diverges

84. WRITING IN MATH Describe why an infinite sequence must not only converge, but converge to 0, in order for there to be a sum.

REASONING Determine whether each statement is *true* or *false*. Explain your reasoning.

85. $\displaystyle\sum_{n=1}^{5} (n^2 + 3n) = \sum_{n=1}^{5} n^2 + 3\sum_{n=1}^{5} n$

86. $\displaystyle\sum_{n=1}^{5} 3^n = \sum_{n=3}^{7} 3^{n-2}$

87 CHALLENGE Find the sum of the first 60 terms of the sequence below. Explain how you determined your answer.
$$15, 17, 2, -15, -17, ...,$$
$$\text{where } a_n = a_{n-1} - a_{n-2} \text{ for } n \geq 3$$

88. WRITING IN MATH Make an outline that could be used to describe the steps involved in finding the 300th partial sum of the infinite sequence $a_n = 2n - 3$. Then explain how to express the same sum using sigma notation.

Graph each complex number on a polar grid. Then express it in rectangular form. (Lesson 9-5)

89. $2\left(\cos\frac{5\pi}{4} + i\sin\frac{5\pi}{4}\right)$

90. $2.5(\cos 1 + i\sin 1)$

91. $5(\cos 0 + i\sin 0)$

Determine the eccentricity, type of conic, and equation of the directrix given by each polar equation. (Lesson 9-4)

92. $r = \dfrac{3}{2 - 0.5\cos\theta}$

93. $r = \dfrac{6}{1.2\sin\theta + 0.3}$

94. $r = \dfrac{1}{0.2 - 0.2\sin\theta}$

Determine whether the points are collinear. Write *yes* or *no*. (Lesson 8-5)

95. $(-3, -1, 4)$, $(3, 8, 1)$, $(5, 12, 0)$

96. $(4, 8, 6)$, $(0, 6, 12)$, $(8, 10, 0)$

97. $(0, -4, 3)$, $(8, -10, 5)$, $(12, -13, 2)$

98. $(-7, 2, -1)$, $(-9, 3, -4)$, $(-5, 1, 2)$

Find the length and the midpoint of the segment with the given endpoints. (Lesson 8-4)

99. $(2, -15, 12)$, $(1, -11, 15)$

100. $(-4, 2, 8)$, $(9, 6, 0)$

101. $(7, 1, 5)$, $(-2, -5, -11)$

102. TIMING The path traced by the tip of the hour-hand of a clock can be modeled by a circle with parametric equations $x = 6\sin t$ and $y = 6\cos t$. (Lesson 7-5)

a. Find an interval for t in radians that can be used to describe the motion of the tip as it moves from 12 o'clock noon to 12 o'clock noon the next day.

b. Simulate the motion described in part **a** by graphing the equation in parametric mode on a graphing calculator.

c. Write an equation in rectangular form that models the motion of the hour-hand. Find the radius of the circle traced out by the hour-hand if x and y are given in inches.

Find the exact value of each expression. (Lesson 5-4)

103. $\tan\dfrac{\pi}{12}$

104. $\sin 75°$

105. $\cos 165°$

Find the partial fraction decomposition of each rational expression. (Lesson 6-4)

106. $\dfrac{10x^2 - 11x + 4}{2x^2 - 3x + 1}$

107. $\dfrac{1}{2x^2 + x}$

108. $\dfrac{x + 1}{x^3 + x}$

Skills Review for Standardized Tests

109. SAT/ACT The first term in a sequence is -5, and each subsequent term is 6 more than the term that immediately precedes it. What is the value of the 104th term?

  A 607

  B 613

  C 618

  D 619

  E 615

110. REVIEW Find the exact value of $\cos 2\theta$ if $\sin\theta = -\dfrac{\sqrt{5}}{3}$ and $180° < \theta < 270°$.

  F $-\dfrac{\sqrt{6}}{6}$ **H** $-\dfrac{\sqrt{30}}{6}$

  G $-\dfrac{4\sqrt{5}}{9}$ **J** $-\dfrac{1}{9}$

111. The first four terms of a sequence are 144, 72, 36, and 18. What is the tenth term in the sequence?

  A 0   **C** $\dfrac{9}{32}$

  B $\dfrac{9}{64}$ **D** $\dfrac{9}{16}$

112. REVIEW How many 5-inch cubes can be stacked inside a box that is 10 inches long, 15 inches wide, and 5 inches tall?

  F 5   **H** 15

  G 6   **J** 20

10-2 Arithmetic Sequences and Series

Then
You found terms of sequences and sums of series. (Lesson 10-1)

Now
- Find *n*th terms and arithmetic means of arithmetic sequences.
- Find sums of *n* terms of arithmetic series.

New Vocabulary
arithmetic sequence
common difference
arithmetic means
first difference
second difference
arithmetic series

Math Online
glencoe.com

Why?

With cross country season approaching, Meg decides to train every day until the first day of practice. She plans to run 1 mile the first day, 1.25 miles the second day, 1.5 miles the third day, and so on. Her goal is to run a total of 100 miles before the first day of practice.

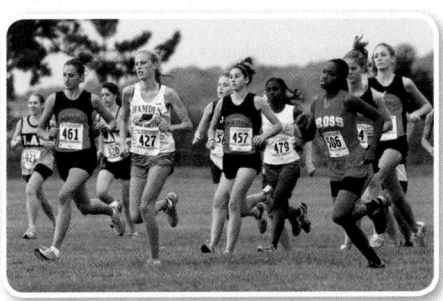

Arithmetic Sequences A sequence in which the difference between successive terms is a constant is called an **arithmetic sequence**. The constant is referred to as the **common difference**, denoted d. To find the common difference of an arithmetic sequence, subtract any term from its succeeding term. To find the next term in the sequence, add the common difference to the given term.

EXAMPLE 1   Arithmetic Sequences

Determine the common difference, and find the next four terms of the arithmetic sequence $17, 12, 7, \ldots$.

First, find the common difference.

$a_2 - a_1 = 12 - 17$ or -5 **Find the difference between two pairs of consecutive terms to verify the common difference.**

$a_3 - a_2 = 7 - 12$ or -5

The common difference is -5. Add -5 to the third term to find the fourth term, and so on.

$a_4 = 7 + (-5)$ or 2 $a_5 = 2 + (-5)$ or -3 $a_6 = -3 + (-5)$ or -8 $a_7 = -8 + (-5)$ or -13

The next four terms are $2, -3, -8,$ and -13.

✔ Guided Practice

Determine the common difference, and find the next four terms of each arithmetic sequence.

1A. $-129, -98, -67, \ldots$ **1B.** $244, 187, 130, \ldots$

Each term in an arithmetic sequence is found by adding the common difference to the preceding term. Therefore, $a_n = a_{n-1} + d$. You can use this recursive formula to develop an explicit formula for generating an arithmetic sequence. Consider the arithmetic sequence in which $a_1 = 6$ and $d = 3$.

first term	a_1	a_1	6
second term	a_2	$a_1 + d$	$6 + 1(3) = 9$
third term	a_3	$a_1 + 2d$	$6 + 2(3) = 12$
fourth term	a_4	$a_1 + 3d$	$6 + 3(3) = 15$
fifth term	a_5	$a_1 + 4d$	$6 + 4(3) = 18$
*n*th term	a_n	$a_1 + (n-1)d$	$6 + (n-1)3$

The pattern formed leads to the following formula for finding the *n*th term of an arithmetic sequence.

Key Concept   The *n*th Term of an Arithmetic Sequence

Words   The *n*th term of an arithmetic sequence with first term a_1 and common difference d is given by $a_n = a_1 + (n-1)d$.

Example   The 16th term of $2, 5, 8, \ldots$ is $a_{16} = 2 + (16-1) \cdot 3$ or 47.

EXAMPLE 2   **Explicit and Recursive Formulas**

Find both an explicit formula and a recursive formula for the nth term of the arithmetic sequence 12, 21, 30, ...

First, find the common difference.

$a_2 - a_1 = 21 - 12$ or 9   **Find the difference between two pairs of consecutive terms to verify the common difference.**

$a_3 - a_2 = 30 - 21$ or 9

For an explicit formula, substitute $a_1 = 12$ and $d = 9$ in the formula for the nth term of an arithmetic sequence.

$a_n = a_1 + (n - 1)d$ ***n*th term of an arithmetic sequence**

$\quad = 12 + (n - 1)9$ **$a_1 = 12$ and $d = 9$**

$\quad = 12 + 9(n - 1)$ or $9n + 3$ **Simplify.**

For a recursive formula, state the first term a_1 and then indicate that the next term is the sum of the previous term a_{n-1} and d.

$a_1 = 12, a_n = a_{n-1} + 9$

✓ Guided Practice

2. Find both an explicit formula and a recursive formula for the nth term of the arithmetic sequence 35, 23, 11, ...

The formula for the nth term of an arithmetic sequence can be used to find a specific term in a sequence.

EXAMPLE 3   ***n*th Terms**

a. Find the 68th term of the arithmetic sequence 25, 17, 9, ...

First, find the common difference.

$a_2 - a_1 = 17 - 25$ or -8 **Find the difference between two pairs of consecutive terms to verify the common difference.**

$a_3 - a_2 = 9 - 17$ or -8

Use the formula for the nth term of an arithmetic sequence to find a_{68}.

$a_n = a_1 + (n - 1)d$ ***n*th term of an arithmetic sequence**

$a_{68} = 25 + (68 - 1)(-8)$ **$n = 68, a_1 = 25,$ and $d = -8$**

$a_{68} = -511$ **Simplify.**

b. Find the first term of the arithmetic sequence for which $a_{25} = 139$ and $d = \frac{3}{4}$.

Substitute $a_{25} = 139$, $n = 25$, and $d = \frac{3}{4}$ in the formula for the nth term of an arithmetic sequence to find a_1.

$a_n = a_1 + (n - 1)d$ ***n*th term of an arithmetic sequence**

$139 = a_1 + (25 - 1)\frac{3}{4}$ **$n = 25, a_n = 139,$ and $d = \frac{3}{4}$**

$a_1 = 121$ **Simplify.**

✓ Guided Practice

3A. Find the 38th term of the arithmetic sequence $-29, -2, 25, \ldots$.

3B. Find d of the arithmetic sequence for which $a_1 = 75$ and $a_{38} = 56.5$.

If two nonconsecutive terms of an arithmetic sequence are known, the terms between them can be calculated. These terms are called **arithmetic means**. In the sequence below, 17, 27, and 37 are the arithmetic means between 7 and 47.

$$-3, 7, 17, 27, 37, 47, 57$$

StudyTip

Alternative Method An alternative method to find d would be to subtract the first term from the last term and divide by the total number of terms minus 1.

EXAMPLE 4 **Arithmetic Means**

Write an arithmetic sequence that has four arithmetic means between 4.3 and 12.8.

The sequence will resemble 4.3, __?__, __?__, __?__, __?__, 12.8. Note that 12.8 is the sixth term of the sequence or a_6.

First, find the common difference using $a_6 = 12.8$, $a_1 = 4.3$, and $n = 6$.

$a_n = a_1 + (n - 1)d$	**nth term of an arithmetic sequence**
$12.8 = 4.3 + (6 - 1)d$	**$a_n = 12.8$, $a_1 = 4.3$, and $n = 6$**
$12.8 = 4.3 + 5d$	**Simplify.**
$d = 1.7$	**Solve for d.**

Then determine the arithmetic means by using $d = 1.7$.

$a_2 = 4.3 + \mathbf{1.7}$ or 6

$a_3 = 6 + \mathbf{1.7}$ or 7.7

$a_4 = 7.7 + \mathbf{1.7}$ or 9.4

$a_5 = 9.4 + \mathbf{1.7}$ or 11.1

The sequence is 4.3, 6, 7.7, 9.4, 11.1, 12.8.

✓ Guided Practice

4. Write a sequence that has six arithmetic means between 12.4 and −24.7.

The **first differences** of a sequence are found by subtracting each term from its successive term.

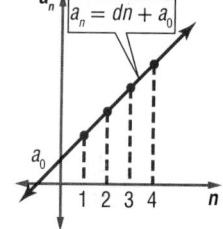

When the first differences are all the same, the sequence is arithmetic and the nth term can be modeled by a linear function of the form $a_n = dn + a_0$, as shown.

If the first differences are not the same, the sequence is not arithmetic. However, the differences may still help to identify the type of function that can be used to model the sequence. Consecutive first differences may be subtracted from one another, thus producing **second differences**.

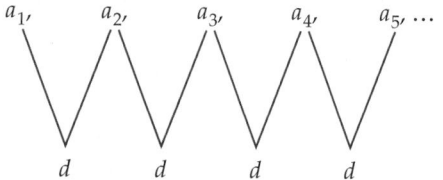

If the second differences are constant, then the nth term of the sequence can be modeled by a quadratic function. This function can be found by solving a system of equations, as demonstrated in Example 5.

EXAMPLE 5	Second Differences

Find a quadratic model for the sequence 12, 20, 30, 42, 56, 72, … .

The nth term can be represented by a quadratic equation of the form $a_n = an^2 + bn + c$. Substitute values for a_n and n into the equation.

$12 = a(1)^2 + b(1) + c$ $a_n = 12$ and $n = 1$

$20 = a(2)^2 + b(2) + c$ $a_n = 20$ and $n = 2$

$30 = a(3)^2 + b(3) + c$ $a_n = 30$ and $n = 3$

This yields a system of linear equations in three variables.

$12 = a + b + c$ **Simplified first equation**

$20 = 4a + 2b + c$ **Simplified second equation**

$30 = 9a + 3b + c$ **Simplified third equation**

Solving for a, b, and c gives $a = 1$, $b = 5$, and $c = 6$. Substituting these values in the equation for a_n, the model for the sequence is $a_n = n^2 + 5n + 6$, as shown in Figure 10.2.1.

Figure 10.2.1

✅ Guided Practice

5. Find a quadratic model for the sequence $-14, -8, 0, 10, 22, 36, …$.

If calculating second differences does not result in a constant difference, higher differences may be found. This process is similar to the process needed for finding a quadratic equation. The function that will model a sequence is dependent upon how many computed differences are necessary before finding a constant difference.

Higher differences may never result in constant differences. In this case, there may not be a polynomial model that can be used to describe the sequence.

Differences	Model
first	linear
second	quadratic
third	cubic
fourth	quartic
fifth	quintic

Arithmetic Series An **arithmetic series** is the indicated sum of the terms of an arithmetic sequence.

Arithmetic Sequence	Arithmetic Series
$-6, -3, 0, 3, 6$	$-6 + (-3) + 0 + 3 + 6$
$4.25, 4, 3.75, 3.5, 3.25$	$4.25 + 4 + 3.75 + 3.5 + 3.25$
$a_1, a_2, a_3, a_4, …, a_n$	$a_1 + a_2 + a_3 + a_4 + \cdots + a_n$

To develop a formula for finding the sum of a finite arithmetic series, start by looking at the series S_n that has terms created by adding multiples of d to a_1. If we combine this with the same series written in reverse order, we can find a formula for calculating the sum of a finite arithmetic series.

$$S_n = a_1 + (a_1 + d) + (a_1 + 2d) + \cdots + (a_n - 2d) + (a_n - d) + a_n$$
$$(+)\ S_n = a_n + (a_n - d) + (a_n - 2d) + \cdots + (a_1 + 2d) + (a_1 + d) + a_1$$
$$\overline{2S_n = (a_1 + a_n) + (a_1 + a_n) + (a_1 + a_n) + \cdots + (a_1 + a_n) + (a_1 + a_n) + (a_1 + a_n)}$$
$$2S_n = n(a_1 + a_n) \qquad \text{There are } \textit{n} \text{ terms in the series, all of which are } (a_1 + a_n).$$

Therefore, $S_n = \frac{n}{2}(a_1 + a_n)$. When the value of the last term is unknown, you can still determine the nth partial sum of the series by combining the nth term of an arithmetic sequence formula and the sum of a finite arithmetic series formula.

$S_n = \frac{n}{2}(a_1 + a_n)$ **Sum of a finite arithmetic series formula**

$S_n = \frac{n}{2}\{a_1 + [a_1 + (n-1)d]\}$ $a_n = a_1 + (n-1)d$, **nth term of an arithmetic sequence formula**

$S_n = \frac{n}{2}[2a_1 + (n-1)d]$ **Simplify.**

Key Concept — Sum of a Finite Arithmetic Series

The sum of a finite arithmetic series with n terms or the nth partial sum of an arithmetic series can be found using one of two related formulas.

Formula 1 $S_n = \frac{n}{2}(a_1 + a_n)$

Formula 2 $S_n = \frac{n}{2}[2a_1 + (n-1)d]$

EXAMPLE 6   Sums of Arithmetic Series

Find the indicated sum of each arithmetic series.

a. $-5 + 2 + 9 + \cdots + 317$

In this sequence, $a_1 = -5$, $a_n = 317$, and $d = 2 - (-5)$ or 7. Use the nth term formula to find the number of terms in the sequence n.

$a_n = a_1 + (n-1)d$ **nth term of an arithmetic sequence**

$317 = -5 + (n-1)7$ **$a_n = 317$, $a_1 = -5$, and $d = 7$**

$47 = n$ **Simplify.**

Now use Formula 1 to find the sum of the series.

$S_n = \frac{n}{2}(a_1 + a_n)$ **Formula 1**

$S_{47} = \frac{47}{2}(-5 + 317)$ **$n = 47$, $a_1 = -5$, and $a_n = 317$**

$= 7332$ **Simplify.**

b. the 28th partial sum of $27 + 14 + 1 + \cdots$

In this sequence, $a_1 = -5$ and $d = 14 - 27$ or -13. Use Formula 2 to find the 28th partial sum.

$S_n = \frac{n}{2}[2a_1 + (n-1)d]$ **Formula 2**

$S_{28} = \frac{28}{2}[2(27) + (28-1)(-13)]$ **$n = 28$, $a_1 = 27$, and $d = -13$**

$= -4158$ **Simplify.**

c. $\displaystyle\sum_{n=6}^{28}(5n - 17)$

$\displaystyle\sum_{n=6}^{28}(5n - 17) = [5(6) - 17] + [5(7) - 17] + \cdots + [5(28) - 17]$

$= 13 + 18 + \cdots + 123$

The first term of this series is 13 and the last term is 123. The number of terms is equal to the upper bound minus the lower bound plus one, which is $28 - 6 + 1$ or 23. Therefore $a_1 = 13$, $a_n = 123$, and $n = 23$. Use Formula 1 to find the sum of the series.

$S_n = \frac{n}{2}(a_1 + a_n)$ **Formula 1**

$S_{23} = \frac{23}{2}(13 + 123)$ **$n = 23$, $a_1 = 13$, and $a_n = 123$**

$= 1564$ **Simplify.**

✓ Guided Practice

6A. $211 + 193 + 175 + \cdots + (-455)$

6B. the 19th partial sum of $-19 + 23 + 65 + \cdots$

6C. $\displaystyle\sum_{n=23}^{37}(2n + 3)$

6D. $\displaystyle\sum_{n=12}^{18}(-2n + 57)$

Arithmetic series have many useful real-life applications.

🌐 Real-World EXAMPLE 7   Sum of an Arithmetic Series

VIDEO GAMES A video game tournament, in which gamers compete in multiple games and accumulate an overall score, pays the top 20 finishers. First place receives $5000, second place receives $4800, third place receives $4600, and so on. How much total prize money is awarded?

$S_n = \frac{n}{2}[2a_1 + (n-1)d]$ **Formula 2**

$S_{20} = \frac{20}{2}[2(5000) + (20-1)(-200)]$ $n = 20, a_1 = 5000, \text{ and } d = -200$

$= 62{,}000$ **Simplify.**

The total prize money awarded is $62,000.

✔ Guided Practice

7. VIDEO GAMES Selma is playing a video game. She scores 50 points if she clears the first level. Each following level is worth 50 more points than the previous level. Thus, she scores 100 points for clearing the second level, 150 for the third, and so on. What is the total amount of points Selma will score after she clears the ninth level?

The formula for the sum of a finite arithmetic series can also be used to solve for values of n.

🌐 Real-World EXAMPLE 8   Sum of an Arithmetic Series

BASEBALL Carter has been collecting baseball cards since his father gave him a 20-card collection. During each month, Carter's father gives him 5 more cards than the previous month. In how many months will Carter reach 1000 cards?

$S_n = \frac{n}{2}[2a_1 + (n-1)d]$ **Formula 2**

$1000 = \frac{n}{2}[2(20) + (n-1)5]$ $S_n = 1000, a_1 = 20, \text{ and } d = 5$

$2000 = n(5n + 35)$ **Multiply each side by 2 and simplify.**

$0 = 5n^2 + 35n - 2000$ **Distribute and subtract 2000 from each side.**

$0 = n^2 + 7n - 400$ **Divide each side by 5.**

$n = \dfrac{-7 \pm \sqrt{7^2 - 4(1)(-400)}}{2(1)}$ **Use the Quadratic Formula.**

$n \approx 16.8 \text{ and } -23.8$ **Simplify.**

Because time cannot be negative, Carter will reach 1000 cards in 17 months.

CHECK $20 + 25 + \cdots + 100 = \dfrac{17}{2}(20 + 100)$
$= 1020 \checkmark$

In seventeen months, Carter will have 1020 baseball cards, which is more than 1000.

✔ Guided Practice

8. LAWN SERVICE Kevin runs a lawn mowing service. He currently has 14 clients. He has gained 2 new clients at the beginning of each of the past three years. Each year, he mows each client's lawn an average of 15 times. Starting now, if Kevin continues to gain 2 clients each year and if he charges $30 per lawn, after how many years will he earn a total of $51,300?

Determine the common difference, and find the next four terms of each arithmetic sequence. (Example 1)

1. $20, 17, 14, \ldots$

2. $3, 16, 29, \ldots$

3. $117, 108, 99, \ldots$

4. $-83, -61, -39, \ldots$

5. $-3, 1, 5, \ldots$

6. $4, 21, 38, \ldots$

7. $-4.5, -9.5, -14.5, \ldots$

8. $-97, -29, 39, \ldots$

9. **MARCHING BAND** A marching band begins its performance in a pyramid formation. The first row has 1 band member, the second row has 3 band members, the third row has 5 band members, and so on. (Examples 1 and 2)

   a. Find the number of band members in the 8th row.

   b. Write an explicit formula and a recursive formula for finding the number of band members in the nth row.

Find both an explicit formula and a recursive formula for the nth term of each arithmetic sequence. (Example 2)

10. $2, 5, 8, \ldots$

11. $-6, 5, 16, \ldots$

12. $-9, -16, -23, \ldots$

13. $4, 19, 34, \ldots$

14. $25, 11, -3, \ldots$

15. $7, -3.5, -14, \ldots$

16. $-18, 4, 26, \ldots$

17. $1, 37, 73, \ldots$

Find the specified value for the arithmetic sequence with the given characteristics. (Example 3)

18. If $a_{14} = 85$ and $d = 9$, find a_1.

19. Find d for $-24, -31, -38, \ldots$.

20. If $a_n = 14$, $a_1 = -36$, and $d = 5$, find n.

21. If $a_1 = 47$ and $d = -5$, find a_{12}.

22. If $a_{22} = 95$ and $a_1 = 11$, find d.

23. Find a_6 for $84, 5, -74, \ldots$.

24. If $a_n = -20$, $a_1 = 46$, and $d = -11$, find n.

25. If $a_{35} = -63$ and $a_1 = 39$, find d.

26. **CONSTRUCTION** Each 8-foot section of a wooden fence contains 14 pickets. Let a_n represent the number of pickets in n sections. (Example 3)

   a. Find the first 5 terms of the sequence.

   b. Write a recursive formula for the sequence in part **a**.

   c. If 448 pickets were used to fence in the customer's backyard, how many feet of fencing was used?

Find the indicated arithmetic means for each set of nonconsecutive terms. (Example 4)

27. 3 means; 19 and -5

28. 5 means; -62 and -8

29. 4 means; 3 and 88

30. 8 means; -5.5 and 23.75

31. 7 means; -4.5 and 7.5

32. 10 means; 6 and 259

Find a quadratic model for each sequence. (Example 5)

33. $12, 19, 28, 39, 52, 67, \ldots$

34. $-11, -9, -5, 1, 9, 19, \ldots$

35. $8, 3, -6, -19, -36, -57, \ldots$

36. $-7, -2, 9, 26, 49, 78, \ldots$

37. $6, -2, -12, -24, -38, -54, \ldots$

38. $-3, 1, 13, 33, 61, 97, \ldots$

Find the indicated sum of each arithmetic series. (Example 6)

39. 26th partial sum of $3 + 15 + 27 + \cdots + 303$

40. $-28 + (-19) + (-10) + \cdots + 242$

41. 42nd partial sum of $120 + 114 + 108 + \cdots$

42. 54th partial sum of $213 + 205 + 197 + \cdots$

43. $-17 + 1 + 19 + \cdots + 649$

44. $89 + 58 + 27 + \cdots + (-562)$

45. **RUNNING** Refer to the beginning of the lesson. (Example 6)

   a. Determine the number of miles Meg will run on her 12th day of training.

   b. During which day of training will Meg reach her goal of 100 total miles?

Find the indicated sum of each arithmetic series. (Example 6)

46. $\displaystyle\sum_{n=1}^{20} (3 + 2n)$

47. $\displaystyle\sum_{n=1}^{28} (100 - 4n)$

48. $\displaystyle\sum_{n=7}^{18} (-9n - 26)$

49. $\displaystyle\sum_{n=6}^{52} (7n + 1)$

50. $\displaystyle\sum_{n=7}^{42} (84 - 3n)$

51. $\displaystyle\sum_{n=1}^{13} [32 + 4(n - 1)]$

52. $\displaystyle\sum_{n=20}^{24} \left(\frac{n}{2} - 9\right)$

53. $\displaystyle\sum_{n=2}^{9} (-15n - 12)$

54. **CONSTRUCTION** A crew is tiling a hotel lobby with a trapezoidal mosaic pattern. The shorter base of the trapezoid begins with a row of 8 tiles. Each row has two additional tiles until the 20th row. Determine the number of tiles needed to create the mosaic design. (Example 7)

55. **SNOWMOBILING** A snowmobiling competitor travels 12 feet in the first second of a race. If the competitor travels 1.5 additional feet each subsequent second, how many feet did the competitor travel in 64 seconds? (Example 7)

56. FUNDRAISING Lalana organized a charity walk. In the first year, the walk generated $3000. She hopes to increase this amount by $900 each year for the next several years. If her goal is met, in how many years will the walk have generated a total of at least $65,000? (Example 8)

57. Find a_n if $S_n = 490$, $a_1 = -5$, and $n = 100$.

58. If $S_n = 51.7$, $n = 22$, $a_n = -11.3$, find a_1.

59. Find n for $-7 + (-5.5) + (-4) + \cdots$ if $S_n = -14$ and $a_n = 3.5$.

60. Find a_1 if $S_n = 1287$, $n = 22$, and $d = 5$.

61. If $S_{26} = 1456$, and $a_1 = -19$, find d.

62. If $S_{12} = 174$, $a_{12} = 39$, find d.

Write each arithmetic series in sigma notation. The lower bound is given.

63. $6 + 12 + 18 + \cdots + 66$; $n = 1$

64. $-1 + 0 + 1 + \cdots + 7$; $n = 1$

65. $17 + 21 + 25 + \cdots + 61$; $n = 4$

66. $1 + 0 + (-1) + (-2) + \cdots + (-13)$; $n = 6$

67. $-\dfrac{13}{5} + \left(-\dfrac{12}{5}\right) + \left(-\dfrac{11}{5}\right) + \cdots + \left(-\dfrac{3}{5}\right)$; $n = 2$

68. $9.25 + 8.5 + 7.75 + \cdots - 2$; $n = 1$

69. CONCERTS The seating in a concert auditorium is arranged as shown below.

Row 35
34 seats — Row 3
29 seats — Row 2
24 seats — Row 1
Stage

a. Write a series in sigma notation to represent the number of seats in the auditorium, if the seating pattern shown in the first 3 rows continues for each successive row.

b. Find the total number of seats in the auditorium.

c. Another auditorium has 32 rows with 18 seats in the first row and 4 more seats in each of the successive rows. How many seats are there in this auditorium?

Write a function that can be used to model the nth term of each sequence.

70. $2, 5, 8, 11, 14, 17, \ldots$

71. $8, 13, 20, 29, 40, 53, \ldots$

72. $2, 2, 4, 8, 14, 22, \ldots$

73. $5, 31, 97, 221, 421, 715, \ldots$

74. $-6, -8, -6, 6, 34, 84, \ldots$

75. $0, 23, 134, 447, 1124, 2375, \ldots$

Find each common difference.

76. $\displaystyle\sum_{n=1}^{100} (6n + 2)$

77. $\displaystyle\sum_{n=21}^{65} \left(8 - \dfrac{2n}{3}\right)$

78. $a_{12} = 63$, $a_{19} = 7$

79. $a_8 = -4$, $a_{27} = \dfrac{7}{3}$

80. CALCULUS The area between the graph of a continuous function and the x-axis can be approximated using sequences. Consider $f(x) = x^2$ on the interval $[1, 3]$.

a. Write the sequence x_n formed when there are 5 arithmetic means between 1 and 3.

b. Write the sequence y_n formed when $y_n = f(x_n)$.

c. Write the sequence p_n defined by $d \cdot y_n$.

d. The *left-hand* approximation of the area is given by $L_n = \displaystyle\sum_{k=1}^{n} p_k$. Find L_6.

e. The *right-hand* approximation of the area is given by $R_n = \displaystyle\sum_{k=2}^{n+1} p_k$. Find R_6.

H.O.T. Problems   Use Higher-Order Thinking Skills

81. ERROR ANALYSIS Peter and Candace are given the arithmetic sequence 2, 9, 16, … . Peter wrote the explicit formula $a_n = 2 + 7(n - 1)$ for the sequence. Candace's formula is $a_n = 7n - 5$. Is either of them correct? Explain.

82. OPEN ENDED You have learned that the nth term of an arithmetic sequence can be modeled by a linear function. Can the sequence of partial sums of an arithmetic series also be modeled by a linear function? If yes, provide an example. If no, how can the sequence be modeled? Explain.

83. CHALLENGE Prove that for an arithmetic sequence, $a_n = a_k + (n - k)d$ for integers k in the domain of the sequence.

REASONING Determine whether each statement is *true* or *false* for finite arithmetic series. Explain.

84. If you know the sum and d, you can solve for a_1.

85. If you only know the first and last terms, then you can find the sum.

86. If the first three terms of a sequence are positive, then all of the terms of the sequence are positive or the sum of the series is positive.

87 CHALLENGE Consider the arithmetic sequence of odd natural numbers.

a. Find S_7 and S_9.

b. Make a conjecture about the pattern that results from the sums of the corresponding arithmetic series.

c. Write an algebraic proof verifying the conjecture that you made in part **b.**

88. WRITING IN MATH Explain why the arithmetic series 25 + 20 + 15 + … does not have a sum.

Find the next four terms of each sequence. (Lesson 10-1)

89. 12, 16, 20, …

90. 3, 1, −1, …

91. 31, 24, 17, …

Find each product or quotient and express it in rectangular form. (Lesson 9-5)

92. $6\left[\cos\left(-\frac{\pi}{3}\right) + i\sin\left(-\frac{\pi}{3}\right)\right] \cdot 3\left[\cos\frac{5\pi}{6} + i\sin\frac{5\pi}{6}\right]$

93. $3\left(\cos\frac{7\pi}{3} + i\sin\frac{7\pi}{3}\right) \div \left(\cos\frac{\pi}{2} + i\sin\frac{\pi}{2}\right)$

Find the dot product of u and v. Then determine if u and v are orthogonal. (Lesson 8-3)

94. $\mathbf{u} = \langle 4, -1\rangle, \mathbf{v} = \langle 1, 5\rangle$

95. $\mathbf{u} = \langle 8, -3\rangle, \mathbf{v} = \langle 4, 2\rangle$

96. $\mathbf{u} = \langle 4, 6\rangle, \mathbf{v} = \langle 9, -5\rangle$

Find the direction angle of each vector to the nearest tenth of a degree. (Lesson 8-2)

97. $-\mathbf{i} - 3\mathbf{j}$

98. $\langle -9, 5\rangle$

99. $\langle -7, 7\rangle$

100. MANUFACTURING A cam in a punch press is shaped like an ellipse with the equation $\frac{x^2}{81} + \frac{y^2}{36} = 1$. The camshaft goes through the focus on the positive axis. (Lesson 7-4)

a. Graph a model of the cam.

b. Find an equation that translates the model so that the camshaft is at the origin.

c. Find the equation of the model in part **b** when the cam is rotated to an upright position.

101. Use the graph of $f(x) = \ln x$ to describe the transformation that results in the graph of $g(x) = 3\ln(x - 1)$. Then sketch the graphs of f and g. (Lesson 3-2)

102. SAT/ACT What is the units digit of 3^{36}?

A 0

B 1

C 3

D 7

E 9

103. Using the table, which formula can be used to determine the nth term of the sequence?

F $a_n = 6n$

G $a_n = n + 5$

H $a_n = 2n + 1$

J $a_n = 4n + 2$

n	a_n
1	6
2	10
3	14
4	18

104. REVIEW If $a_1 = 3$, $a_2 = 5$, and $a_n = a_{n-2} + 3n$, find a_{10}.

A 59

C 89

B 75

D 125

105. REVIEW Which of the sequences shown below is convergent?

F

H

G

J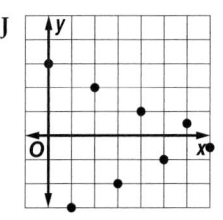

10-3

Geometric Sequences and Series

Why?

The first summer X Games took place in Rhode Island in 1995 and included 27 events. Due to their growing popularity, the winter X Games were introduced at Big Bear Lake, California, in 1997. With an immense fan base, the annual X Games now receive live 24-hour network coverage. Since its inaugural year, the event has seen an average growth of 13% in revenue each year.

Then
You found terms and means of arithmetic sequences and sums of arithmetic series. (Lesson 10-2)

Now
- Find *n*th terms and geometric means of geometric sequences.
- Find sums of *n* terms of geometric series and the sums of infinite geometric series.

New Vocabulary
geometric sequence
common ratio
geometric means
geometric series

Math Online

glencoe.com

Geometric Sequences A sequence in which the ratio between successive terms is a constant is called a **geometric sequence**. The constant is referred to as the **common ratio**, denoted r. To find the common ratio of a geometric sequence, divide any term following the first term by the preceding term. Given a term of the sequence, to find the next term of the sequence, multiply the given term by the common ratio. While the rate of change of an arithmetic sequence is constant, the rate of change of a geometric sequence can either increase or decrease.

EXAMPLE 1   Geometric Sequences

Determine the common ratio, and find the next three terms of each geometric sequence.

a. $8, -2, \frac{1}{2}, \ldots$

First, find the common ratio.

$a_2 \div a_1 = -2 \div 8 \text{ or } -\frac{1}{4}$ **Find the ratio between two pairs of consecutive terms to verify the common ratio.**

$a_3 \div a_2 = \frac{1}{2} \div -2 \text{ or } -\frac{1}{4}$

The common ratio is $-\frac{1}{4}$. Multiply the third term by $-\frac{1}{4}$ to find the fourth term, and so on.

$a_4 = \frac{1}{2}\left(-\frac{1}{4}\right) \text{ or } -\frac{1}{8}$ $a_5 = -\frac{1}{8}\left(-\frac{1}{4}\right) \text{ or } \frac{1}{32}$ $a_6 = \frac{1}{32}\left(-\frac{1}{4}\right) \text{ or } -\frac{1}{128}$

The next three terms are $-\frac{1}{8}, \frac{1}{32},$ and $-\frac{1}{128}$.

b. $w + 3, 2w + 6, 4w + 12, \ldots$

First, find the common ratio.

$a_2 \div a_1 = \frac{2w + 6}{w + 3}$ $\begin{aligned} a_2 &= 2w + 6 \text{ and} \\ a_1 &= w + 3 \end{aligned}$ $a_3 \div a_2 = \frac{4w + 12}{2w + 6}$ $\begin{aligned} a_3 &= 4w + 12 \text{ and} \\ a_2 &= 2w + 6 \end{aligned}$

$\quad\quad = \frac{2(w + 3)}{w + 3}$ **Factor.** $\quad\quad = \frac{4(w + 3)}{2(w + 3)}$ **Factor.**

$\quad\quad = 2$ **Simplify.** $\quad\quad = 2$ **Simplify.**

The common ratio is 2. Multiply the third term by 2 to find the fourth term, and so on.

$a_4 = 2(4w + 12) \text{ or } 8w + 24$

$a_5 = 2(8w + 24) \text{ or } 16w + 48$

$a_6 = 2(16w + 48) \text{ or } 32w + 96$

The next three terms are $8w + 24, 16w + 48,$ and $32w + 96$.

✓ Guided Practice

1A. $4, 11, 30.25, \ldots$ **1B.** $64r - 128, -16r + 32, 4r - 8, \ldots$

Watch Out!

Type of Sequence
Remember that if a sequence is not arithmetic, it does not necessarily mean that the sequence is geometric. Test several terms for a common ratio before determining that the sequence is indeed geometric.

In Lesson 10-2, you learned that arithmetic sequences can be defined both recursively and explicitly. This also applies to geometric sequences. A geometric sequence can be expressed recursively, where a term a_n is found by taking the product of the previous term a_{n-1} and r, or $a_n = a_{n-1}r$, as illustrated by the previous example. To develop an explicit formula for a geometric sequence, consider the pattern created by the geometric sequence for which $a_1 = 3$ and $r = 4$.

Term		Expanded Form	Exponential Form	Example
first term	a_1	a_1	a_1	3
second term	a_2	$a_1 \cdot r$	$a_1 r^1$	$3 \cdot 4 = 12$
third term	a_3	$a_1 \cdot r \cdot r$	$a_1 r^2$	$3 \cdot 4^2 = 48$
fourth term	a_4	$a_1 \cdot r \cdot r \cdot r$	$a_1 r^3$	$3 \cdot 4^3 = 192$
fifth term	a_5	$a_1 \cdot r \cdot r \cdot r \cdot r$	$a_1 r^4$	$3 \cdot 4^4 = 768$
nth term	a_n	$\underbrace{a_1 \cdot r \cdot r \cdot r \cdot \ldots \cdot r}_{n-1 \text{ factors}}$	$a_1 r^{n-1}$	$3 \cdot 4^{n-1}$

Key Concept **The nth Term of a Geometric Sequence**

Words The nth term of a geometric sequence with first term a_1 and common ratio r is given by $a_n = a_1 r^{n-1}$.

Example The 9th term of 2, 10, 50, ... is $a_9 = 2 \cdot 5^{9-1}$ or 781,250.

EXAMPLE 2 **Explicit and Recursive Formulas**

Write an explicit formula and a recursive formula for finding the nth term of the geometric sequence given in Example 1a.

For an explicit formula, substitute $a_1 = 8$ and $r = -0.25$ in the nth term formula.

$a_n = a_1 r^{n-1}$ *nth term of a geometric sequence*

$\quad = 8(-0.25)^{n-1}$ *$a_1 = 8$ and $r = -0.25$*

For a recursive formula, state the first term a_1. Then indicate that the next term is the product of the previous term a_{n-1} and r.

$a_1 = 8, a_n = (-0.25)a_{n-1}$

✓ **Guided Practice**

2. Write an explicit formula and a recursive formula for finding the nth term in the sequence 2, 25, 312.5, ...

Finding the nth term of a geometric sequence is simplified by explicit formulas.

EXAMPLE 3 **nth Terms**

Find the 27th term of the geometric sequence 189, 151.2, 120.96, ...

First, find the common ratio.

$a_2 \div a_1 = 151.2 \div 189$ or 0.8   *Find the ratio between two pairs of consecutive terms to verify the common ratio.*

$a_3 \div a_2 = 120.96 \div 151.2$ or 0.8

Use the formula for the nth term of a geometric sequence.

$a_n = a_1 r^{n-1}$ *nth term of a geometric sequence*

$a_{27} = 189(0.8)^{27-1}$ *$n = 27$, $a_1 = 189$, and $r = 0.8$*

$a_{27} \approx 0.57$ *Simplify.*

Find the specified term of each geometric sequence or sequence with the given characteristics.

3A. a_9 for 4, 14, 49, ...

3B. a_{12} if $a_3 = 32$ and $r = -4$

Just as arithmetic sequences are linear functions with restricted domains, geometric sequences are also functions. Consider the exponential function $f(x) = 2(2)^x$ and the explicit formula for the geometric sequence $a_n = 2(2)^n$.

Notice that the graphs of the terms of the geometric sequence lie on a curve, as shown. A geometric sequence can be modeled by an exponential function in which the domain is restricted to the natural numbers.

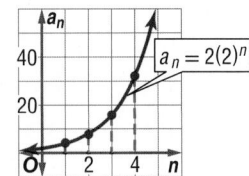

🌐 **Real-World EXAMPLE 4** | ***n*th Term of a Geometric Sequence**

AUTOMOBILE Damian purchased a late-model car for $15,000. At the end of each year, the value of the car depreciates 11%.

a. Write an explicit formula for the value of Damian's car after n years.

If the car's value depreciates at a rate of 11% per year, it retains 100% − 11% or 89% of its original value. Note that the original value given represents the a_0 and not the a_1 term, so we need to use an adjusted formula for the *n*th term of this geometric sequence.

first term $a_1 = a_0 r$

second term $a_2 = a_0 r^2$

⋮   ⋮

*n*th term $a_n = a_0 r^n$

Use this adjusted formula to find an explicit formula for the value of the car after n years.

$a_n = a_0 r^n$ **Adjusted *n*th term of a geometric sequence**

$a_n = 15{,}000(0.89)^n$ $a_0 = 15{,}000, r = 0.89$

b. What is the value of Damian's car at the end of the seventh year?

Evaluate the formula found in part **a** for $n = 7$.

$a_n = 15{,}000(0.89)^n$ **Original equation**

$= 15{,}000(0.89)^7$ ***n = 7***

≈ 6634.70 **Simplify.**

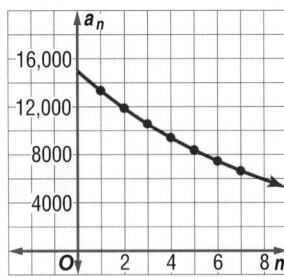

The value of the car at the beginning of the seventh year is about $6634.70.

The value of the car at each year is shown as a point on the graph. The function connecting the points represents exponential decay.

4. WATERCRAFT Rohan purchased a personal watercraft for $9000. Assume that by the end of each year, the value of the watercraft depreciates 30%.

  A. Write an explicit formula for finding the value of Rohan's watercraft after n years.

  B. What is the value of Rohan's watercraft after 5 years?

🌐 **Real-World Link**

The value of a newly purchased vehicle can depreciate by as much as 30–35% in its first year. Each year after, the value continues to depreciate by 7–12%, depending on the make and model. After a five-year period, on average, cars are worth 35% of the original sticker price, making car buying costly.

Source: *Kelly Blue Book*

Similar to arithmetic sequences, if two nonconsecutive terms of a geometric sequence are known, the terms between them can be calculated. These terms are called **geometric means**.

EXAMPLE 5 Geometric Means

Write a sequence that has two geometric means between 480 and −7.5.

The sequence will resemble 480, __?__, __?__, −7.5.

Note that $a_1 = 480$, $n = 4$, and $a_4 = -7.5$. Find the common ratio using the nth term for a geometric sequence formula.

$a_4 = a_1 r^{n-1}$ **nth term of a geometric sequence**

$-7.5 = 480r^{4-1}$ **$a_4 = -7.5$, $a_1 = 480$, and $n = 4$**

$-\dfrac{1}{64} = r^3$ **Simplify and divide each side by 480.**

$-\dfrac{1}{4} = r$ **Take the cube root of each side.**

The common ratio is $-\dfrac{1}{4}$. Use r to find the geometric means.

$a_2 = 480(-0.25)$ or -120 $a_3 = -120(-0.25)$ or 30

Therefore, a sequence with two geometric means between 480 and −7.5, is 480, −120, 30, −7.5.

StudyTip

Geometric Means
Sometimes, more than one set of geometric means are possible. For example, the three geometric means between 3 and 48 can be 6, 12, and 24 or −6, 12, and −24.

✔ Guided Practice

Find the indicated geometric means for each pair of nonconsecutive terms.

5A. −4 and 13.5; 2 means   **5B.** 10 and 0.016; 3 means

Geometric Series A **geometric series** is the sum of the terms of a geometric sequence.

Geometric Sequence	Geometric Series
2, 4, 8, 16, 32	2 + 4 + 8 + 16 + 32
27, 9, 3, 1, $\dfrac{1}{3}$	27 + 9 + 3 + 1 + $\dfrac{1}{3}$
$a_1, a_2, a_3, a_4, \ldots, a_n$	$a_1 + a_2 + a_3 + a_4 + \cdots + a_n$

A formula for the sum S_n of the first n terms of a finite geometric series can be developed by looking at the series S_n and rS_n. To create the terms for rS_n, each term in S_n is multiplied by r. These series are then aligned so that similar terms are grouped together and then rS_n is subtracted from S_n.

$$S_n = \quad a_1 + a_1 r + a_1 r^2 + \cdots + a_1 r^{n-2} + a_1 r^{n-1}$$

$$(-)\, rS_n = \qquad\quad a_1 r + a_1 r^2 + \cdots + a_1 r^{n-2} + a_1 r^{n-1} + a_1 r^n$$

$S_n - rS_n = a_1 - a_1 r^n$ **Subtract.**

$S_n(1 - r) = a_1 - a_1 r^n$ **Factor.**

$S_n = \dfrac{a_1 - a_1 r^n}{1 - r}$ **Divide each side by 1 − r.**

$S_n = \dfrac{a_1(1 - r^n)}{1 - r}$ **Factor.**

Therefore, $S_n = a_1\left(\dfrac{1 - r^n}{1 - r}\right)$.

If the value of n is not provided, the sum of a finite geometric series can still be found. If we take a look at the next-to-last step of the proof, we can substitute for $a_1 r^n$.

$$S_n = a_1\left(\frac{1 - r^n}{1 - r}\right) \qquad \text{Sum of a finite geometric series formula}$$

$$S_n = \frac{a_1 - a_1 r^n}{1 - r} \qquad \text{Multiply.}$$

$$= \frac{a_1 - a_1 r^{n-1} \cdot r}{1 - r} \qquad \text{Factor one } r \text{ from } a_1 r^n.$$

$$= \frac{a_1 - a_n r}{1 - r} \qquad a_1 r^{n-1} = a_n, \text{ } n\text{th term of a geometric sequence formula}$$

Key Concept — Sum of a Finite Geometric Series

The sum of a finite geometric series with n terms or the nth partial sum of a geometric series can be found using one of two related formulas.

Formula 1 $\quad S_n = a_1\left(\dfrac{1 - r^n}{1 - r}\right)$

Formula 2 $\quad S_n = \dfrac{a_1 - a_n r}{1 - r}$

StudyTip

Infinite vs. Finite Geometric Series Notice that the series given in Example 6a is an infinite geometric series. Because you are asked to find the sum of the *first six terms* of the series, you are actually finding the sum of a finite series.

EXAMPLE 6 Sums of Geometric Series

a. Find the sum of the first six terms of the geometric series $8 + 14 + 24.5 + \dots$.

First, find the common ratio.

$a_2 \div a_1 = 14 \div 8$ or 1.75 **Find the ratio between two pairs of consecutive terms to**

$a_3 \div a_2 = 24.5 \div 14$ or 1.75 **verify the common ratio.**

The common ratio is 1.75. Use Formula 1 to find the sum of the series.

$$S_n = a_1\left(\frac{1 - r^n}{1 - r}\right) \qquad \text{Formula 1}$$

$$S_6 = 8\left(\frac{1 - 1.75^6}{1 - 1.75}\right) \qquad n = 6, a_1 = 8, \text{ and } r = 1.75$$

$$S_6 \approx 295.71 \qquad \text{Simplify.}$$

The sum of the first six terms of the geometric series is about 295.71.

CHECK The next three terms of the related sequence are 42.875, 75.03125, and 131.3046875.

$$8 + 14 + 24.5 + 42.875 + 75.03125 + 131.3046875 \approx 295.71 \checkmark$$

b. Find the sum of the first n terms of a geometric series with $a_1 = 3$, $a_n = 768$, and $r = -2$.

Use Formula 2 for the sum of a finite geometric series.

$$S_n = \frac{a_1 - a_n r}{1 - r} \qquad \text{Formula 2}$$

$$= \frac{3 - 768(-2)}{1 - (-2)} \qquad a_1 = 3, a_n = 768, \text{ and } r = -2$$

$$= 513 \qquad \text{Simplify.}$$

The sum of the first n terms of the geometric series is 513.

✓ Guided Practice

6A. Find the sum of the first 11 terms of the geometric series $7 + (-24.5) + 85.75 + \dots$.

6B. Find the sum of the first n terms of a geometric series with $a_1 = -8$, $a_n = 131{,}072$, and $r = -4$.

Geometric series may also be represented in sigma notation.

EXAMPLE 7 **Geometric Sum in Sigma Notation**

Find $\displaystyle\sum_{n=2}^{7} 3(5)^{n-1}$.

Find n, a_1, and r.

$n = 7 - 2 + 1$ or 6 **Upper bound minus lower bound plus 1**

$a_1 = 3(5)^2 - 1$ or 15 **$n = 2$**

$r = 5$ **r is the base of the exponential function.**

Method 1 Substitute $n = 6$, $a_1 = 15$, and $r = 5$ into Formula 1.

$S_n = a_1 \left(\dfrac{1 - r^n}{1 - r} \right)$ **Formula 1**

$S_6 = 15 \left(\dfrac{1 - 5^6}{1 - 5} \right)$ **$a_1 = 15$, $r = 5$, and $n = 6$**

$S_6 = 58,590$ **Simplify.**

Method 2 Find a_n.

$a_n = a_1 r^{n-1}$ **nth term of a geometric sequence**

$= 15(5)^{6-1}$ **$a_1 = 15$, $r = 5$, and $n = 6$**

$= 46,875$ **Simplify.**

Substitute $a_1 = 15$, $a_n = 46,875$, and $r = 5$ into Formula 2.

$S_n = \dfrac{a_1 - a_n r}{1 - r}$ **Formula 2**

$S_6 = \dfrac{15 - (46,875)(5)}{1 - 5}$ **$a_1 = 15$, $a_n = 46,875$, and $r = 5$**

$S_6 = 58,590$ **Simplify.**

Therefore, $\displaystyle\sum_{n=2}^{7} 3(5)^{n-1} = 58,590$.

☑ **Guided Practice**

7A. $\displaystyle\sum_{n=16}^{31} 0.5(2)^{n-1}$

7B. $\displaystyle\sum_{n=4}^{11} 120(0.5)^{n-1}$

In Lesson 10-1, you learned that calculating the sums of infinite series may be possible if the sequence of terms converges to 0. For this reason, the sums of infinite arithmetic series cannot be found.

The formula for the sum of a finite geometric series can be used to develop a formula for the sum of an infinite geometric series. If $|r| > 1$, then $|r^n|$ increases without limit as $n \to \infty$. However, when $|r| < 1$, r^n approaches 0 as $n \to \infty$. Thus,

$S = a_1 \left(\dfrac{1 - r^n}{1 - r} \right)$ **Sum of a finite geometric series formula**

$= a_1 \left(\dfrac{1 - 0}{1 - r} \right)$ **r^n approaches 0 as $n \to \infty$.**

$= \dfrac{a_1}{1 - r}$ **Simplify and multiply.**

Key Concept **The Sum of an Infinite Geometric Series**

The sum S of an infinite geometric series for which $|r| < 1$ is given by

$$S = \dfrac{a_1}{1 - r}.$$

The formula for the sum of an infinite geometric series involves three variables: S, a_1, and r. If any two of the three variables are known, you can solve for the third.

EXAMPLE 8 **Sums of Infinite Geometric Series**

If possible, find the sum of each infinite geometric series.

a. $9 + 3 + 1 + \ldots$

First, find the common ratio.

$a_2 \div a_1 = 3 \div 9$ or $\frac{1}{3}$ Find the ratio between two pairs of consecutive terms to verify the common ratio.

$a_3 \div a_2 = 1 \div 3$ or $\frac{1}{3}$

The common ratio r is $\frac{1}{3}$, and $\left|\frac{1}{3}\right| < 1$. This infinite geometric series has a sum. Use the formula for the sum of an infinite geometric series.

$S = \dfrac{a_1}{1 - r}$ Sum of an infinite geometric series formula

$= \dfrac{9}{1 - \frac{1}{3}}$ $a_1 = 9$ and $r = \frac{1}{3}$

$= 13.5$ Simplify.

The sum of the infinite series is 13.5.

StudyTip

Common Ratio Recall that $|r| < 1$ is equivalent to $-1 < r < 1$.

b. $0.25 + (-1.25) + 6.25 + \ldots$

First, find the common ratio.

$a_2 \div a_1 = -1.25 \div 0.25$ or -5 Find the ratio between two pairs of consecutive terms to verify the common ratio.

$a_3 \div a_2 = 6.25 \div (-1.25)$ or -5

The common ratio r is -5, and $|-5| > 1$. Therefore, this infinite geometric series has no sum.

c. $\displaystyle\sum_{n=4}^{\infty} 4(0.2)^{n-1}$

The common ratio r is 0.2, and $|0.2| < 1$. Therefore, this infinite geometric series has a sum. Find a_1.

$a_1 = 4(0.2)^{4-1}$ Lower bound = 4

$= 0.032$ Simplify.

Use the formula for the sum of an infinite geometric series to find the sum.

$S = \dfrac{a_1}{1 - r}$ Sum of an infinite geometric series formula

$= \dfrac{0.032}{1 - 0.2}$ $a_1 = 0.032$ and $r = 0.2$

$= 0.04$ Simplify.

The sum of the infinite series is 0.04.

✓ **Guided Practice**

8A. $10 + (-5) + 2.5 + \ldots$ **8B.** $20 + 15 + 10 + \ldots$ **8C.** $\displaystyle\sum_{n=1}^{\infty} 120(0.8)^{n-1}$

Exercises

● = **Step-by-Step Solutions** begin on page R29.

Determine the common ratio, and find the next three terms of each geometric sequence. (Example 1)

1. $-\frac{1}{4}, \frac{1}{2}, -1, \dots$

2. $\frac{1}{2}, -\frac{3}{8}, \frac{9}{32}, \dots$

3. $0.5, 0.75, 1.125, \dots$

4. $8, 20, 50, \dots$

5. $2x, 10x, 50x, \dots$

6. $64x, 16x, 4x, \dots$

7. $x + 5, 3x + 15, 9x + 45, \dots$

8. $-9 - y, 27 + 3y, -81 - 9y, \dots$

9. **GEOMETRY** Consider a sequence of circles with diameters that form a geometric sequence: d_1, d_2, d_3, d_4, d_5.

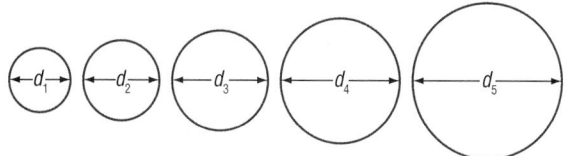

   a. Show that the sequence of circumferences of the circles is also geometric. Identify r.

   b. Show that the sequence of areas of the circles is also geometric. Identify the common ratio.

Write an explicit formula and a recursive formula for finding the nth term of each geometric sequence. (Example 2)

10. $36, 12, 4, \dots$

11. $64, 16, 4, \dots$

12. $-2, 10, -50, \dots$

13. $4, -12, 36, \dots$

14. $4, 8, 16, \dots$

15. $20, 30, 45, \dots$

16. $15, 5, \frac{5}{3}, \dots$

17. $\frac{1}{32}, \frac{1}{16}, \frac{1}{8}, \dots$

18. **CHAIN E-MAIL** Melina receives a chain e-mail that she forwards to 7 of her friends. Each of her friends forwards it to 7 of their friends. (Example 2)

   a. Write an explicit formula for the pattern.

   b. How many will receive the e-mail after 6 forwards?

19. **BIOLOGY** A certain bacteria divides every 15 minutes to produce two complete bacteria. (Example 2)

   a. If an initial colony contains a population of b_0 bacteria, write an equation that will determine the number of bacteria b_t present after t hours.

   b. Suppose a Petri dish contains 12 bacteria. Use the equation found in part **a** to determine the number of bacteria present 4 hours later.

Find the specified term for each geometric sequence or sequence with the given characteristics. (Example 3)

20. a_9 for $60, 30, 15, \dots$

21. a_4 for $7, 14, 28, \dots$

22. a_5 for $3, 1, \frac{1}{3}, \dots$

23. a_6 for $540, 90, 15, \dots$

24. a_7 if $a_3 = 24$ and $r = 0.5$

25. a_6 if $a_3 = 32$ and $r = -0.5$

26. a_6 if $a_1 = 16{,}807$ and $r = \frac{3}{7}$

27. a_8 if $a_1 = 4096$ and $r = \frac{1}{4}$

28. **ACCOUNTING** Julian Rockman is an accountant for a small company. On January 1, 2009, the company purchased $50,000 worth of computers, printers, scanners, and hardware. Because this equipment is a company asset, Mr. Rockman needs to determine how much the computer equipment is presently worth. He estimates that the computer equipment depreciates at a rate of 45% per year. What value should Mr. Rockman assign the equipment in his 2014 year-end accounting report? (Example 4)

29. Find the sixth term of a geometric sequence with a first term of 9 and a common ratio of 2. (Example 4)

30. If $r = 4$ and $a_8 = 100$, what is the first term of the geometric sequence? (Example 4)

31. **X GAMES** Refer to the beginning of the lesson. The X Games netted approximately $40 million in revenue in 2002. If the X Games continue to generate 13% more revenue each year, how much revenue will the X Games generate in 2020? (Example 4)

Find the indicated geometric means for each pair of nonconsecutive terms. (Example 5)

32. 4 and 256; 2 means

33. 256 and 81; 3 means

34. $\frac{4}{7}$ and 7; 1 mean

35. -2 and 54; 2 means

36. 1 and 27; 2 means

37. 48 and -750; 2 means

38. i and -1; 4 means

39. t^8 and t^{-7}; 4 means

Find the sum of each geometric series described. (Example 6)

40. first six terms of $3 + 9 + 27 + \cdots$

41. first nine terms of $0.5 + (-1) + 2 + \cdots$

42. first eight terms of $2 + 2\sqrt{3} + 6 + \cdots$

43. first n terms of $a_1 = 4, a_n = 2000, r = -3$

44. first n terms of $a_1 = 5, a_n = 1{,}747{,}625, r = 4$

45. first n terms of $a_1 = 3, a_n = 46{,}875, r = -5$

46. first n terms of $a_1 = -8, a_n = -256, r = 2$

47. first n terms of $a_1 = -36, a_n = 972, r = 7$

Find each sum. (Example 7)

48. $\displaystyle\sum_{n=1}^{6} 5(2)^{n-1}$

49. $\displaystyle\sum_{n=1}^{5} -4(3)^{n-1}$

50. $\displaystyle\sum_{n=1}^{5} (-3)^{n-1}$

51. $\displaystyle\sum_{n=1}^{6} 2(1.4)^{n-1}$

52. $\displaystyle\sum_{n=1}^{6} 100\left(\frac{1}{2}\right)^{n-1}$

53. $\displaystyle\sum_{n=1}^{9} \frac{1}{27}(-3)^{n-1}$

54. $\displaystyle\sum_{n=1}^{7} 144\left(-\frac{1}{2}\right)^{n-1}$

55. $\displaystyle\sum_{n=1}^{20} 3(2)^{n-1}$

If possible, find the sum of each infinite geometric series. (Example 8)

56. $\dfrac{1}{20} + \dfrac{1}{40} + \dfrac{1}{80} + \cdots$

57. $\dfrac{2}{7} + \dfrac{4}{7} + \dfrac{8}{7} + \cdots$

58. $18 + (-27) + 40.5 + \cdots$

59. $12 + (-7.2) + 4.32 + \cdots$

60. $\displaystyle\sum_{n=1}^{\infty} 6(-0.4)^{n-1}$

61. $\displaystyle\sum_{n=1}^{\infty} 40\left(\frac{3}{5}\right)^{n-1}$

62. $\displaystyle\sum_{n=1}^{\infty} \frac{1}{2}\left(\frac{3}{8}\right)^{n-1}$

63 $\displaystyle\sum_{n=1}^{\infty} 35\left(-\frac{3}{4}\right)^{n-1}$

64. BUNGEE JUMPING A bungee jumper falls 35 meters before his cord causes him to spring back up. He rebounds $\frac{2}{5}$ of the distance after each fall. (Example 8)

   a. Find the first five terms of the infinite sequence representing the vertical distance traveled by the bungee jumper. Include each drop and rebound distance as separate terms.

   b. What is the total vertical distance the jumper travels before coming to rest? (*Hint*: Rewrite the infinite sequence suggested by part **a** as two infinite geometric sequences.)

Find the missing quantity for the geometric sequence with the given characteristics.

65. Find a_1 if $S_{12} = 1365$ and $r = 2$.

66. If $S_6 = 196.875$, $a_1 = 100$, $r = 0.5$, find a_6.

67. Find r if $a_1 = 0.12$, $S_n = 590.52$, and $a_n = 787.32$.

68. Find n for $4.1 + 8.2 + 16.4 + \cdots$ if $S_n = 61.5$.

69. If $15 - 18 + 21.6 - \cdots$, $S_n = 23.784$, find a_n.

70. If $r = -0.4$, $S_5 = 144.32$, and $a_1 = 200$, find a_5.

71. Find a_1 if $S_n = 468$, $a_n = 375$, and $r = 5$.

72. If $S_n = \dfrac{61}{40}$, $\dfrac{5}{8} + \dfrac{1}{2} + \dfrac{2}{5} + \cdots$, find n.

73. LOANS Marc is making monthly payments on a loan. Suppose instead of the same monthly payment, the bank requires a low initial payment that grows at an exponential rate each month. The total cost of the loan is represented by $\displaystyle\sum_{n=1}^{k} 5(1.1)^{n-1}$.

   a. What is Marc's initial payment and at what rate is this payment increasing?

   b. If the sum of Marc's payments at the end of the loan is $7052, how many payments did Marc make?

Find the common ratio for the geometric sequence with the given terms.

74. $a_3 = 12$, $a_6 = 187.5$

75. $a_2 = -6$, $a_7 = -192$

76. $a_4 = -28$, $a_6 = -1372$

77. $a_5 = 6$, $a_8 = -0.048$

78. ADVERTISING Word-of-mouth advertising can be an effective form of marketing, or it can be very harmful. Consider a new restaurant that serves 27 customers on its opening night.

   a. Of the 27 customers, 25 found the experience enjoyable and each told 3 friends over the next month. This group each told 3 friends over the next month, and so on, for 6 months. Assuming that no one heard twice, how many people have had a positive experience or heard positive reviews of the restaurant?

   b. Suppose the 2 unhappy customers each told 6 friends over the next month about the experience. This group then each told 6 friends, and so on, for 6 months. Assuming that no one heard a review twice, how many people have had a negative experience or heard a negative review?

Write the first 3 terms of the infinite geometric series with the given characteristics.

79. $S = 12$, $r = \dfrac{1}{2}$

80. $S = -25$, $r = 0.2$

81. $S = 44.8$, $a_1 = 56$

82. $S = \dfrac{2}{3}$, $a_1 = \dfrac{8}{9}$

83. $S = -60$, $r = 0.4$

84. $S = -126.25$, $a_1 = -50.5$

85. $S = -115$, $a_1 = -138$

86. $S = \dfrac{891}{20}$, $r = -\dfrac{1}{9}$

87. $\displaystyle\sum_{n=1}^{\infty} 12\left(-\frac{1}{4}\right)^{n-1}$

88. $\displaystyle\sum_{n=1}^{\infty} \frac{1}{6}\left(\frac{2}{3}\right)^{n-1}$

Determine whether each sequence is *arithmetic*, *geometric*, or *neither*. Then find the next three terms of the sequence.

89. $\dfrac{1}{4}, \dfrac{2}{6}, \dfrac{3}{8}, \dfrac{4}{10}, \cdots$

90. $\dfrac{9}{2}, \dfrac{17}{4}, 4, \dfrac{15}{4}, \cdots$

91. $12, 24, 36, 48, \ldots$

92. $128, 96, 72, 54, \ldots$

93. $36k, 49k, 64k, 81k, \ldots$

94. $7.2y, 9.1y, 11y, 12.9y, \ldots$

95. $3\sqrt{5}, 15, 15\sqrt{5}, 75, \ldots$

96. $2\sqrt{3}, 2\sqrt{6}, 2\sqrt{9}, 2\sqrt{12}, \ldots$

Write each geometric series in sigma notation.

97. $3 + 12 + 48 + \cdots + 3072$

98. $9 + 18 + 36 + \cdots + 1152$

99. $50 + 85 + 144.5 + \cdots + 417.605$

100. $\dfrac{1}{8} - \dfrac{1}{4} + \dfrac{1}{2} - \cdots + 8$

101. $0.2 - 1 + 5 - \cdots - 625$

102. HORSES For each of the first few months after a horse is born, the amount of weight that it gains is about 120% of the previous month's weight gain. In the first month, a horse has gained 30 pounds.

   a. Write a geometric series in sigma notation that can be used to model the horse's weight gain for the first five months.

   b. About how much weight did the horse gain in the fourth month?

   c. If the horse weighed 150 pounds at birth, about how much did it weigh after 5 months?

   d. Will the horse continue to grow at this rate indefinitely? Explain.

103. MEDICINE A newly developed and researched medicine has a half-life of about 1.5 hours after it is administered. The medicine is given to patients in doses of d milligrams every 6 hours.

   a. What fraction of the first dose will be left in the patient's system when the second dose is taken?

   b. Find the first four terms of the sequence that represents the amount of medicine in the patient's system after the first 4 doses.

   c. Write a recursive formula that can be used to determine the amount of medicine in the patient's system after the nth dose.

104. 🔀 **MULTIPLE REPRESENTATIONS** In this problem, you will investigate the limits of $\frac{1 - r^n}{1 - r}$.

   a. GRAPHICAL Graph $S_n = \frac{1 - r^n}{1 - r}$ for $r = 0.2, 0.5$, and 0.9 on the same graph.

   b. TABULAR Copy and complete the table shown below.

n	$S_n, r = 0.2$	$S_n, r = 0.5$	$S_n, r = 0.9$
0			
4			
8			
12			
16			
20			
24			

   c. ANALYTICAL For each graph in part **a**, describe the values of S_n as $n \to \infty$.

   d. GRAPHICAL Graph $S_n = \frac{1 - r^n}{1 - r}$ for $r = 1.2, 2.5$, and 4 on the same graph.

   e. ANALYTICAL For each graph in part **d**, describe the values of S_n as $n \to \infty$.

   f. ANALYTICAL Make a conjecture about what happens to S_n as $n \to \infty$ for $S_n = \frac{1 - 8.6^n}{1 - 8.6}$.

H.O.T. Problems   Use **H**igher-**O**rder **T**hinking Skills

105 **ERROR ANALYSIS** Emilio believes that the sum of the infinite geometric series $16 + 4 + 1 + 0.25 + \ldots$ can be calculated. Annie disagrees. Is either of them correct? Explain your reasoning.

106. CHALLENGE A ball is dropped from a height of 5 meters. On each bounce, the ball rises to 65% of the height it reached on the previous bounce.

   a. Approximate the total vertical distance the ball travels, until it stops bouncing.

   b. The ball makes its first complete bounce in 2 seconds, that is, from the moment it first touches the ground until it next touches the ground. Each complete bounce that follows takes 0.8 times as long as the preceding bounce. Estimate the total amount of time that the ball bounces.

107. WRITING IN MATH Explain why an infinite geometric series will not have a sum if $|r| > 1$.

REASONING Determine whether each statement is *true* or *false*. Explain your reasoning.

108. If the first two terms of a geometric sequence are positive, then the third term is positive.

109. If you know r and the sum of a finite geometric series, you can find the last term.

110. If r is negative, then the geometric sequence converges.

111. REASONING Determine whether the following statement is *sometimes*, *always*, or *never* true. Explain your reasoning.

   If all of the terms of an infinite geometric series are negative, then the series has a sum that is a negative number.

112. CHALLENGE The midpoints of the sides of a square are connected so that a new square is formed. Suppose this process is repeated indefinitely.

   a. What is the perimeter of the square with side lengths of x inches?

   b. What is the sum of the perimeters of all the squares?

   c. What is the sum of the areas of all the squares?

Find each sum. (Lesson 10-2)

113. $\displaystyle\sum_{n=1}^{7} (2n + 1)$

114. $\displaystyle\sum_{n=3}^{7} (3n + 4)$

115. $\displaystyle\sum_{n=1}^{150} (11 + 2n)$

116. TOURIST ATTRACTIONS To prove that objects of different weights fall at the same rate, Marlene dropped two objects with different weights from the Leaning Tower of Pisa in Italy. The objects hit the ground at the same time. When an object is dropped from a tall building, it falls about 16 feet in the first second, 48 feet in the second second, and 80 feet in the third second, regardless of its weight. If this pattern continues, how many feet would an object fall in the sixth second? (Lesson 10-1)

117. TEXTILES Patterns in fabric can often be created by modifying a mathematical graph. The pattern at the right can be modeled by a lemniscate. (Lesson 9-2)

   a. Suppose the designer wanted to begin with a lemniscate that was 6 units from end to end. What polar equation could have been used?

   b. What polar equation could have been used to generate a lemniscate that was 8 units from end to end?

Graph each polar equation on a polar grid. (Lesson 9-1)

118. $\theta = -\dfrac{\pi}{4}$

119. $r = 1.5$

120. $\theta = -150°$

Find the cross product of u and v. Then show that u × v is orthogonal to both u and v.
(Lesson 8-5)

121. $\mathbf{u} = \langle 1, 9, -1 \rangle, \mathbf{v} = \langle -2, 6, -4 \rangle$

122. $\mathbf{u} = \langle -3, 8, 2 \rangle, \mathbf{v} = \langle 1, -5, -7 \rangle$

123. $\mathbf{u} = \langle 9, 0, -4 \rangle, \mathbf{v} = \langle -6, 2, 5 \rangle$

Find the component form and magnitude of $\overline{AB}$ with the given initial and terminal points. Then find a unit vector in the direction of $\overline{AB}$. (Lesson 8-4)

124. $A(6, 7, 9), B(18, 21, 18)$

125. $A(24, -6, 16), B(8, 12, -4)$

126. $A(3, -5, 9), B(-1, 15, -7)$

Skills Review for Standardized Tests

127. SAT/ACT In the geometric sequence $\dfrac{1}{8}, \dfrac{1}{4}, \dfrac{1}{2}, \dots$, each term after the first is equal to the previous term times a constant. What is the value of the 13th term?

   A 2^7

   B 2^8

   C 2^9

   D 2^{10}

   E 2^{11}

128. REVIEW The pattern of dots shown below continues infinitely, with more dots being added at each step.

      First   Second   Third
      Step   Step   Step

   Which expression can be used to determine the number of dots at the nth step?

   F $2n$ **H** $n(n + 1)$

   G $n(n + 2)$ **J** $2(n + 1)$

129. The first term of a geometric series is −1, and the common ratio is −3. How many terms are in the series if its sum is 182?

   A 6

   B 7

   C 8

   D 9

130. REVIEW Cora begins a phone tree to notify her friends about a party. In step 1, she calls 3 friends. In step 2, each of those friends calls 3 new friends. In step 3, each of those friends calls 3 more new friends. After step 3, how many people know about the party, including Cora?

   F 12

   G 13

   H 39

   J 40

Objective
Use a graphing calculator to represent continued fractions.

An expression of the following form is called a *continued fraction*. Continued fractions can be used to write sequences that approach limits.

$$a + \cfrac{b}{a + \cfrac{b}{a + \cfrac{b}{a + \cdots}}}$$

TechnologyTip

▶ **Memory** You may need to clear the calculator's memory to eliminate any previously stored values.

ACTIVITY 1 $a = b = 1$

Evaluate $1 + \cfrac{1}{1 + \cfrac{1}{1 + \cfrac{1}{1 + \cdots}}}$.

Step 1 On the home screen, press 1 and $\boxed{\text{ENTER}}$.

Step 2 Then type in 1 $\boxed{+}$ 1 $\boxed{\div}$ $\boxed{\text{2nd}}$ [ANS] $\boxed{\text{ENTER}}$.

Step 3 Press $\boxed{\text{ENTER}}$ four more times.

```
              1
1+1/Ans
              2
            1.5
    1.666666667
            1.6
          1.625
```

Analyze the Results

1A. Write the expression that was evaluated to generate each of the values obtained in Steps 2–3.

1B. Press $\boxed{\text{ENTER}}$ 20 more times and record the result to approximate the limit of this sequence. This number is an approximation for the *golden ratio*.

1C. Solve $x = 1 + \frac{1}{x}$. How do you think this relates to the golden ratio?

ACTIVITY 2 $a = 3$ and $b = 2$

Evaluate $3 + \cfrac{2}{3 + \cfrac{2}{3 + \cfrac{2}{3 + \cdots}}}$.

Step 1 On the home screen, press 3 and $\boxed{\text{ENTER}}$.

Step 2 Then type in 3 $\boxed{+}$ 2 $\boxed{\div}$ $\boxed{\text{2nd}}$ [ANS] $\boxed{\text{ENTER}}$.

Step 3 Press $\boxed{\text{ENTER}}$ four more times.

```
              3
3+2/Ans
    3.666666667
    3.545454545
    3.564102564
    3.561151079
    3.561616162
■
```

StudyTip

▶ **Finding Expressions** When developing the expression in 2C, think of different ways in which b can make the expression equivalent to the limit of the sequence. Use different values of a and b to confirm your answer.

Analyze the Results

2A. Write the expression that was evaluated to generate each of the values obtained in Steps 2–3.

2B. Press $\boxed{\text{ENTER}}$ 20 more times. What is the approximation for the limit of this sequence?

2C. **MAKE A CONJECTURE** Use $a = 3$ and the approximation found in Exercise 2B to develop an expression for both a and b. (*Hint:* Solve for the radicand, and determine how b can be used to make it equal.) This is the general expression for the limit of the continued fraction sequence for any values of a and b.

Exercises

Approximate the value of the continued fraction with the given values for a and b.

1. $a = 4$ and $b = 3$ **2.** $a = 5$ and $b = 2$ **3.** $a = 3$ and $b = 1$

Find the next four terms of each sequence. (Lesson 10-1)

1. 109, 97, 85, 73, …

2. 2, 6, 14, 30, …

3. 0, 1, 5, 14, …

4. −2187, 729, −243, 81, …

5. **NATURE** A petting zoo starts a population of rabbits with one newborn male and one newborn female. Assuming that each adult pair will produce one male and one female offspring per month starting at two months, how many rabbits will there be after 6 months? (Lesson 10-1)

Determine whether each sequence is *convergent* or *divergent*. (Lesson 10-1)

6. 3, 5, 8, 12, …

7. $a_1 = 15, a_n = \frac{a_{n-1} - 1}{3}$

8. 48, 24, 12, 6, …

9. $a_n = n^2 + 5n$

Find each sum. (Lesson 10-1)

10. $\sum\limits_{n=0}^{9} \frac{n^2}{4}$

11. $\sum\limits_{n=-5}^{0} (n^3 + 7)$

12. $\sum\limits_{n=1}^{6} (2^n - 4)$

13. $\sum\limits_{n=8}^{13} (4n - 10)$

14. **GOLF** In a charity golf tournament, each of the top ten finishers wins a donation to the charity of his or her choice. The amount of the donation follows the arithmetic sequence shown below. What is the total amount of money donated to charity as a result of the tournament? (Lesson 10-2)

Annual Tri-Cities
Charity Golf Tournament

1st Place: $2500
2nd Place: $2250
3rd Place: $2000

Write an explicit formula and a recursive formula for finding the *n*th term of each arithmetic sequence. (Lesson 10-2)

15. −11, −15, −19, −23, …

16. −96, −84, −72, −60, …

17. 7, 10, 13, 16, …

18. 32, 30, 28, 26, …

19. **JEWELRY** Mary Anne is hosting a jewelry party. For each guest who buys an item of jewelry, she gets a hostess bonus in the amount shown. She receives a larger amount for each guest making a purchase. (Lesson 10-2)

Guests Purchasing Jewelry	Amount Mary Anne Receives ($)
first	10
second	15
third	20

a. How much will Mary Anne receive for the 12th guest who makes a purchase?

b. If she wants a total hostess bonus of $100, how many guests need to make a purchase?

Write an explicit formula and a recursive formula for finding the *n*th term of each geometric sequence. (Lesson 10-3)

20. $\frac{1}{2}, \frac{3}{2}, \frac{9}{2}, \frac{27}{2}, …$

21. $9, -3, 1, -\frac{1}{3}, …$

22. 3, 18, 108, 648, …

23. −4, 20, −100, 500, …

24. **POPULATION** The population of Sandy Shores is currently 55,000 and is decreasing at a rate of 3% annually. (Lesson 10-3)

a. Write an explicit formula for finding the population of Sandy Shores during the *n*th year.

b. What do you predict will be the population of Sandy Shores after 10 years?

c. After how many years do you predict the population of Sandy shores will reach 37,000?

25. **MULTIPLE CHOICE** If possible, find the sum of the geometric series $12 + 3 + \frac{3}{4} + \frac{3}{16} + …$. (Lesson 10-3)

A 13.5

B 16

C 18

D not possible

10-4 Mathematical Induction

Then
You found the next term in a sequence or series.
(Lesson 10-1)

Now
- Use mathematical induction to prove summation formulas and properties of divisibility involving a positive integer n.
- Use extended mathematical induction.

New Vocabulary
principle of mathematical induction
anchor step
inductive hypothesis
inductive step
extended principle of mathematical induction

Math Online

glencoe.com

Why?

Raini draws points on a circle and connects every pair of points with a chord, dividing the circle into regions. After drawing circles with 2, 3, and 4 points, Raini conjectures that if there are n points on a circle, then connecting each pair of points will divide the circle into 2^{n-1} regions. While his conjecture holds for $n = 2, 3,$ and 4, are these three examples sufficient to prove that his conjecture is true?

2 points
2 regions = 2^1

3 points
4 regions = 2^2

4 points
8 regions = 2^3

For n points, 2^{n-1} regions?

Mathematical Induction When looking for patterns and making conjectures, it is often tempting to assume that if a conjecture holds for several cases, then it is true in all cases. In the situation above, Raini may be convinced that his conjecture is true once he shows that it holds for $n = 5$ because connecting 5 points does form 16 or 2^4 regions. "Proof by example," however, is *not* a logically valid method of proof because it does not show that a conjecture is true for *all* cases. In fact, you can show that Raini's conjecture fails when $n = 6$.

While all that is required to prove mathematical conjectures false is a counterexample, proving one true requires a more formal method. One such method uses the **principle of mathematical induction.** The essential idea behind the principle of mathematical induction is that a conjecture can be proven true if you can:

1. show that something works for the first case (base or **anchor step**),

2. assume that it works for any particular case (**inductive hypothesis**), and then

3. show that it works for the next case (**inductive step**).

This principle, described more formally below, is a powerful tool for proving many conjectures about positive integers.

Key Concept — **The Principle of Mathematical Induction**

Let P_n be a statement about a positive integer n. Then P_n is true for all positive integers n if and only if

- P_1 is true, and
- for every positive integer k, if P_k is true, then P_{k+1} is true.

To understand why the principle of mathematical induction works, imagine a ladder with an infinite number of rungs (Figure 10.4.1). If you can get on the ladder (anchor step) and then move from one rung to the next (inductive hypothesis and step), you can climb the whole ladder. Similarly, imagine an unending line of dominos (Figure 10.4.2) arranged so that if any kth domino falls, the $(k + 1)$th domino will also fall (inductive hypothesis and step). By pushing over the first domino (anchor step), you start a chain reaction that knocks down the whole line.

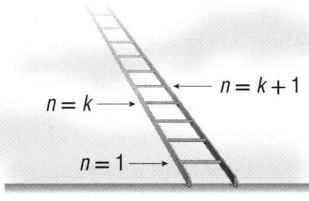

$n = k + 1$
$n = k$
$n = 1$

Figure 10.4.1

$n = k$ $n = k + 1$
$n = 1$

Figure 10.4.2

To apply the principle of mathematical induction, follow these steps.

Step 1   Verify that a conjecture P_n is true for $n = 1$. (Anchor Step)

Step 2   Assume that P_n is true for $n = k$. (Inductive Hypothesis)

Step 3   Use this assumption to prove that P_n is also true for $n = k + 1$. (Inductive Step)

EXAMPLE 1   Prove a Summation Formula

Use mathematical induction to prove that the sum of the first n even positive integers is $n^2 + n$. That is, prove that $2 + 4 + 6 + \cdots + 2n = n^2 + n$ is true for all positive integers n.

Conjecture   Let P_n be the statement that $2 + 4 + 6 + \cdots + 2n = n^2 + n$.

Anchor Step   Verify that P_n is true for $n = 1$.

$P_n\!: 2 + 4 + 6 + \cdots + 2n = n^2 + n$ **Original statement P_n**

$P_1\!:$ $2 = 1^2 + 1$ **P_n for $n = 1$, the first partial sum**

Because $2 = 1^2 + 1$ is a true statement, P_n is true for $n = 1$.

Inductive Hypothesis   Assume that P_n is true for $n = k$.

To write the inductive hypothesis, replace n with k in P_n. That is, assume that
$P_k\!: 2 + 4 + 6 + \cdots + 2k = k^2 + k$ is true.

Inductive Step   Use the inductive hypothesis to prove that P_n is true for $n = k + 1$.

To prove that P_n is true for $n = k + 1$, we need to show that P_{k+1} must be true. Start with your inductive hypothesis and then add the next term, the $(k + 1)$th term, to each side.

$2 + 4 + 6 + \cdots + 2k = k^2 + k$	**Inductive hypothesis**
$2 + 4 + 6 + \cdots + 2k + 2(k + 1) = k^2 + k + 2(k + 1)$	**Add the $(k + 1)$th term to each side.**
$2 + 4 + 6 + \cdots + 2k + 2(k + 1) = k^2 + k + 2k + 2$	**Simplify the right-hand side.**
$2 + 4 + 6 + \cdots + 2k + 2(k + 1) = (k^2 + 2k + 1) + (k + 1)$	**Rewrite 2 as $1 + 1$ and regroup.**
$2 + 4 + 6 + \cdots + 2k + 2(k + 1) = (k + 1)^2 + (k + 1)$	**Factor $k^2 + 2k + 1$.**

This final statement is exactly the statement for P_{k+1}, so P_{k+1} is true. It follows that if P_n is true for $n = k$, then P_n is also true for $n = k + 1$.

Conclusion   Because P_n is true for $n = 1$ and P_k implies P_{k+1}, P_n is true for $n = 2$, $n = 3$, and so on. That is, by the principle of mathematical induction, $P_n\!: 2 + 4 + 6 + \cdots + 2n = n^2 + n$ is true for all positive integers n.

✔ Guided Practice

1. Use mathematical induction to prove that the sum of the first n even positive integers is n^2. That is, prove that $1 + 3 + 5 + \cdots + (2n - 1) = n^2$ is true for all positive integers n.

StudyTip

Representing the Next Term To determine the $(k + 1)$th term, substitute the quantity $k + 1$ for k in the expression for the general form of the next term in the series. In Example 1, because $2k$ represents the kth term, $2(k + 1)$ represents the $(k + 1)$th term.

Watch Out!

Use the Inductive Hypothesis To show that P_n is true for $n = k + 1$, you do not substitute $k + 1$ for n on each side of the equation for P_n. Doing so would give you nothing to prove. To complete the inductive step, you *must* use the inductive hypothesis.

In Example 1, notice that you are *not* trying to prove that P_n is true for $n = k$. Instead, you *assume* that P_n is true for $n = k$ and use that assumption to show that P_n is true for the next number, $n = k + 1$. If while trying to complete either the anchor step or the inductive step you arrive at a contradiction, then the assumption you made in your inductive hypothesis *may be* false. For example, if there is a contradiction in the anchor step, then you know it does not work for *that value only*. The inductive hypothesis may still be true.

Mathematical induction can be used to prove divisibility. Recall that an integer p is *divisible* by an integer q if $p = qr$ for some integer r.

StudyTip

Inductive Step There is no "fixed" way of completing the inductive step for a proof by mathematical induction. Each problem has its own special characteristics that require a different technique to complete the proof.

EXAMPLE 2 **Prove Divisibility**

Prove that $3^n - 1$ is divisible by 2 for all positive integers n.

Conjecture and Anchor Step Let P_n be the statement that $3^n - 1$ is divisible by 2. P_1 is the statement that $3^1 - 1$ is divisible by 2. P_1 is true because $3^1 - 1$ is 2, which is divisible by 2.

Inductive Hypothesis and Step Assume that $3^k - 1$ is divisible by 2. That is, assume that $3^k - 1 = 2r$ for some integer r. Use this hypothesis to show that $3^{k+1} - 1$ is divisible by 2.

$3^k - 1 = 2r$	**Inductive hypothesis**
$3^k = 2r + 1$	**Add 1 to each side.**
$3 \cdot 3^k = 3(2r + 1)$	**Multiply each side by 3.**
$3^{k+1} = 6r + 3$	**Simplify.**
$3^{k+1} - 1 = 6r + 2$	**Subtract 1 from each side.**
$3^{k+1} - 1 = 2(3r + 1)$	**Factor.**

Because r is an integer, $3r + 1$ is an integer and $2(3r + 1)$ is divisible by 2. Therefore, $3^{k+1} - 1$ is divisible by 2.

Conclusion Because P_n is true for $n = 1$ and P_k implies P_{k+1}, P_n is true for $n = 2$, $n = 3$, and so on. By the principle of mathematical induction, $3^n - 1$ is divisible by 2 for all positive integers n.

✔ **Guided Practice**

2. Prove that $4^n - 1$ is divisible by 3 for all positive integers n.

You can also prove statements of inequality using mathematical induction.

StudyTip

Proofs of Inequalities The approach of showing that the difference between some quantity and $k + 1$ is greater than or less than zero, along with the Transitive Property of Inequality, is used in many inequality proofs by mathematical induction.

EXAMPLE 3 **Prove Statements of Inequality**

Prove that $n < 2^n$ for all positive integers n.

Conjecture and Anchor Step Let P_n be the statement $n < 2^n$. P_1 and P_2 are true, since $1 < 2^1$ and $2 < 2^2$ are true inequalities. Showing P_2 to be true provides the anchor for the second part of our inductive hypothesis below.

Inductive Hypothesis and Step Assume that $k < 2^k$ is true for a positive $k > 1$. Use both parts of this inductive hypothesis to show that $k + 1 < 2^{k+1}$ is true.

$k < 2^k$	**Inductive hypothesis**	$k > 1$
$2 \cdot k < 2 \cdot 2^k$		$k - 1 > 0$
$2k < 2^{k+1}$		$2k - k - 1 > 0$
		$2k - (k + 1) > 0$
		$2k > k + 1$
		$k + 1 < 2k$

By the Transitive Property of Inequality, if $k + 1 < 2k$ and $2k < 2^{k+1}$, then $k + 1 < 2^{k+1}$.

Conclusion Because P_n is true for $n = 1$ and 2 and P_k implies P_{k+1} for $k \geq 2$, P_n is true for $n = 3$, $n = 4$, and so on. By the principle of mathematical induction, $n < 2^n$ is true for all positive integers n.

✔ **Guided Practice**

3. Prove that $2n < 3^n$ for all positive integers n.

Extended Mathematical Induction Sometimes you will be asked to prove a statement that is true for an arbitrary value greater than 1. In situations like this, you can use a variation on the principle of mathematical induction called the **extended principle of mathematical induction.** Instead of verifying that P_n is true for $n = 1$, you can instead verify that P_n is true for the first possible case.

EXAMPLE 4 **Use Extended Mathematical Induction**

Prove that $n! > 2^n$ for integer values of $n \geq 4$.

Conjecture and Anchor Step Let P_n be the statement $n! > 2^n$. P_4 is true since $4! > 2^4$ or $24 > 16$ is a true statement.

Inductive Hypothesis and Step Assume that $k! > 2^k$ is true for a positive integer $k \geq 4$. Show that $(k + 1)! > 2^{k+1}$ is true. Use this inductive hypothesis and its restriction that $k \geq 4$.

$k! > 2^k$	**Inductive hypothesis**
$(k + 1) \cdot k! > (k + 1) \cdot 2^k$	**Multiply each side by $k + 1$.**
$(k + 1)! > (k + 1) \cdot 2^k$	$(k + 1) \cdot k! = (k + 1)!$
$(k + 1)! > (k + 1) \cdot 2^k > 2 \cdot 2^k$	$k + 1 > 2$ is true for $k \geq 4$; therefore by the Multiplication Property of Inequality $(k + 1) \cdot 2^k > 2 \cdot 2^k$.
$(k + 1)! > 2 \cdot 2^k$	**Transitive Property of Inequality**
$(k + 1)! > 2^{k+1}$	**Simplify.**

Therefore, $(k + 1)! > 2^{k+1}$ is true.

Conclusion Because P_n is true for $n = 4$ and P_k implies P_{k+1} for $k \geq 4$, P_n is true for $n = 5$, $n = 6$, and so on. That is, by the extended principle of mathematical induction, $n! > 2^n$ is true for integer values of $n \geq 4$.

☑ **Guided Practice**

4. Prove that $n! > 3^n$ for integer values of $n \geq 7$.

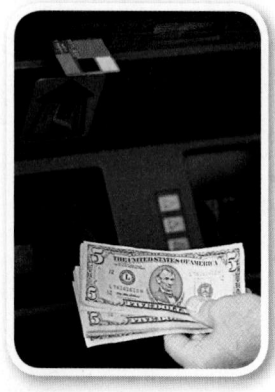

Real-World Link

Most automatic teller machines dispense withdrawals in $10 or $20 increments, but a few can go as low as $5.

Source: *The Economist*

🌐**Real-World EXAMPLE 5** **Apply Extended Mathematical Induction**

MONEY Prove that all multiples of $10 greater than $40 can be formed using just $20 and $50 bills.

Conjecture and Anchor Step P_n: There exists a set of $20 and $50 bills that adds to 10n$ for $n > 4$. For $n = 5$, the first possible case, the conjecture is true because $10(5) = \$20(0) + \$50(1)$.

Inductive Hypothesis and Step Assume that there exists a set of $20 and/or $50 bills that adds to 10k$. Show that this implies the existence of a set of $20 and/or $50 bills that adds to $10($k + 1$)$.

Case 1 The set contains at least one $50 bill. Replace one $50 bill in the set with three $20 bills and the value of the set is increased by $10 to 10k$ + 10 or $10($k + 1$)$, which is exactly P_{k+1}.

Case 2 The set contains no $50 bills. The set must contain at least three $20 bills because the value of the set must be greater than $40. Replace two of the $20 bills with a $50 bill and the value of the set is increased by $10, to 10k$ + 10 or $10($k + 1$)$, which is exactly P_{k+1}.

Conclusion In both cases, P_n is true for $n = k + 1$. Because P_n is true for $n = 5$ and P_k implies P_{k+1} for $k \geq 5$, P_n is true for $n = 6$, $n = 7$, and so on. That is, by the extended principle of mathematical induction, all multiples of $10 greater than $40 can be formed using just $20 and $50 bills.

☑ **Guided Practice**

5. AMUSEMENT Prove that all rides at the fair requiring more than 7 tickets can be paid for using just 3-ticket and 5-ticket vouchers offered by the school for donations of canned goods.

Use mathematical induction to prove that each conjecture is true for all positive integers n. (Example 1)

1. $3 + 5 + 7 + \cdots + (2n + 1) = n(n + 2)$

2. $1 + 2 + 3 + \cdots + n = \dfrac{n(n + 1)}{2}$

3. $2 + 2^2 + 2^3 + \cdots + 2^n = 2(2^n - 1)$

4. $3 + 7 + 11 + \cdots + (4n - 1) = 2n^2 + n$

5. $1 + 4 + 7 + \cdots + (3n - 2) = \dfrac{n(3n - 1)}{2}$

6. $1 + 8 + 27 + \cdots + n^3 = \dfrac{n^2(n + 1)^2}{4}$

7. $1 + 2 + 4 + \cdots + 2^{n-1} = 2^n - 1$

8. $1^2 + 3^2 + 5^2 + \cdots + (2n - 1)^2 = \dfrac{n(2n - 1)(2n + 1)}{3}$

9. $\dfrac{1}{2} + \dfrac{1}{2^2} + \dfrac{1}{2^3} + \cdots + \dfrac{1}{2^n} = 1 - \dfrac{1}{2^n}$

10. $\dfrac{1}{1 \cdot 2} + \dfrac{1}{2 \cdot 3} + \dfrac{1}{3 \cdot 4} + \cdots + \dfrac{1}{n(n + 1)} = \dfrac{n}{n + 1}$

11 **TRIANGULAR NUMBERS** Triangular numbers are numbers that can be represented by a triangular array of dots, with n dots on each side. The first three triangular numbers are 1, 3, and 6. (Example 1)

a. Find the next five triangular numbers.

b. Write a general formula for the nth term of this sequence.

c. Prove that the sum of first n triangular numbers can be found using $S_n = \dfrac{n(n + 1)(n + 2)}{6}$.

12. ICEBREAKER At freshman orientation, students are separated into groups to play an icebreaker game. The game requires each student in a group to have one individual interaction with every other student in the group. (Example 1)

a. Develop a formula to calculate the total number of interactions taking place during the icebreaker for a group of n students.

b. Prove that the formula is true for all positive integer values of n.

c. Determine the length of the icebreaker game in minutes for a group of 12 students if each interaction is allotted 30 seconds.

Prove that each conjecture is true for all positive integers n. (Examples 2)

13. $9^n - 1$ is divisible by 8.

14. $6^n + 4$ is divisible by 5.

15. $2^{3n} - 1$ is divisible by 7.

16. $5^n - 2^n$ is divisible by 3.

Prove that each inequality is true for the indicated values of n. (Examples 3 and 4)

17. $3^n \geq 3n, n \geq 1$

18. $n! > 4^n, n \geq 9$

19. $2^n > 2n, n \geq 3$

20. $9n < 3^n, n \geq 4$

21. $3n < 4^n, n \geq 1$

22. $2^n > 10n + 7, n \geq 10$

23. $2n < 1.5^n, n \geq 7$

24. $1.5^n > 10 + 0.5n, n \geq 7$

25. POSTAGE Prove that all postage greater than 20¢ can be formed using just 5¢ and 6¢ stamps. (Example 5)

26. ENTERTAINMENT All of the activities at the Family Fun Entertainment Center, such as video games, paintball, and go-kart racing, require tokens worth 25¢, 50¢, 75¢, and so on. Prove that all of the activities costing more than \$1.50 can be paid for using just 50¢ and 75¢ tokens. (Example 5)

27. OBLONG NUMBERS Oblong numbers are numbers that can be represented by a rectangular array having one more column than rows.

Prove that the sum of the first n oblong numbers is given by $S_n = \dfrac{n^3 + 3n^2 + 2n}{3}$.

Use mathematical induction to prove that each conjecture is true for all positive integers n.

28. $\displaystyle\sum_{a=1}^{n} (4a - 3) = n(2n - 1)$

29. $\displaystyle\sum_{a=1}^{n} (3a - 2) = \dfrac{n}{2}(3n - 1)$

30. $\displaystyle\sum_{a=1}^{n} (a^2 + a) = \dfrac{n(n + 1)(n + 2)}{3}$

31. $\displaystyle\sum_{a=1}^{n} \dfrac{1}{4a^2 - 1} = \dfrac{n}{2n + 1}$

32. $\displaystyle\sum_{a=1}^{n} \dfrac{1}{2a(a + 1)} = \dfrac{n}{2(n + 1)}$

33. $\displaystyle\sum_{a=1}^{n} \dfrac{1}{(a + 1)(a + 2)} = \dfrac{n}{2(n + 2)}$

Prove that each statement is true for all positive integers n or find a counterexample.

34. $1 + 6 + 11 + \cdots + (5n - 4) = n(2n - 1)$

35. $\dfrac{1}{2 \cdot 4} + \dfrac{1}{4 \cdot 6} + \dfrac{1}{6 \cdot 8} + \cdots + \dfrac{1}{2n(2n + 2)} = \dfrac{n}{2(2n + 2)}$

36. $n^2 - n + 5$ is prime.

37. $3^n + 4n + 1$ is divisible by 4.

38. $4^n + 6n - 1$ is divisible by 9.

39. $2^{2n + 1} + 3^{2n + 1}$ is divisible by 5.

Prove the inequality for the indicated integer values of n and indicated values of a, b, and x.

40. $\left(\dfrac{a}{b}\right)^n > \left(\dfrac{a}{b}\right)^{n + 1}$, $n \geq 1$ and $0 < a < b$

41. $(x + 1)^n \geq nx$, $n \geq 1$ and $x \geq 1$

42. $(1 + a)^n > 1 + na$, $n \geq 2$ and $a > 0$

Find a formula for the sum S_n of the first n terms of each sequence. Then prove that your formula is true using mathematical induction.

43. $2, 6, 10, 14, \ldots, (4n - 2)$

44. $2, 7, 12, 17, \ldots, (5n - 3)$

45. $-\dfrac{1}{2}, -\dfrac{1}{4}, -\dfrac{1}{8}, -\dfrac{1}{16}, \ldots, -\dfrac{1}{2^n}$

46. $\dfrac{1}{6}, \dfrac{1}{18}, \dfrac{1}{36}, \dfrac{1}{60}, \ldots, \dfrac{1}{3n(n + 1)}$

47. FIBONACCI NUMBERS In the Fibonacci sequence, 1, 1, 2, 3, 5, 8, ..., each element after the first two is found by adding the previous two terms. Numbers in the Fibonacci sequence occur throughout nature. For example, the number of scales in the clockwise and counterclockwise spirals on a pinecone are Fibonacci numbers. If f_n represents the nth Fibonacci number, prove that $f_1 + f_2 + \cdots + f_n = f_{n + 2} - 1$.

48. COMPLEX NUMBERS Prove that DeMoivre's Theorem for finding the power of a complex number written in polar form is true for any positive integer n.

$$[r(\cos \theta + i \sin \theta)]^n = r^n(\cos n\theta + i \sin n\theta)$$

49. GEOMETRY According to the Interior Angle Sum Formula, the sum of the measures of the interior angles of a convex polygon with n sides is $180(n - 2)$ degrees. Use extended mathematical induction to prove this formula for $n \geq 3$.

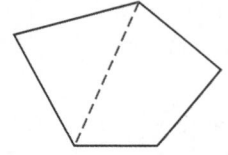

Use mathematical induction to prove that each conjecture is true for all positive integers n.

50. $(xy)^n = x^n y^n$

51. $\left(\dfrac{x}{y}\right)^n = \dfrac{x^n}{y^n}$

52. $x^{-n} = \dfrac{1}{x^n}$

53. $\cos n\pi = (-1)^n$

Use mathematical induction to prove each formula for the sum of the first n terms in a series.

54. $S_n = a_1\left(\dfrac{1 - r^n}{1 - r}\right)$

55. $S_n = \dfrac{n}{2}[2a_1 + (n - 1)d]$

56. CHALLENGE Prove that $n! < n^n$ when $n > 1$.

57 OPEN ENDED Consider the following statement.

$$a^n + b \text{ is divisible by } c.$$

   a. Make a divisibility conjecture by replacing a, b, and c with positive integers.

   b. Use mathematical induction to prove that the conjecture you found in part **a** is true. If it is not true, find a counterexample.

58. REASONING Describe the error in the proof by mathematical induction shown below.

Conjecture and Anchor Step
Let P_n be the statement that in a room with n students, all of the students have the same birthday. P_1 is true since one student has only one birthday.

Inductive Hypothesis and Step
Assume that in a room with k students, all of the students have the same birthday. Suppose $k + 1$ students are in a room. If one student leaves, then the remaining k students must have the same birthday, according to the inductive hypothesis. Then if the first student returns and another student leaves, then that group (one) of k students must have the same birthday. So, P_n is true for $n = k + 1$. Therefore, P_n is true for all positive integers n. That is, in a room with n students, all of the students have the same birthday.

59. CHALLENGE If a_n is represented by the sequence

$$\sqrt{6}, \sqrt{6 + \sqrt{6}}, \sqrt{6 + \sqrt{6 + \sqrt{6}}}, \sqrt{6 + \sqrt{6 + \sqrt{6 + \sqrt{6}}}}, \ldots,$$

prove that the nth term in this sequence is always less than 3.

60. WRITING IN MATH In the inductive hypothesis step of mathematical induction, you assume that P_n is true for $n = k$. Explain why you cannot simply assume that P_n is true for n.

Find the specified nth term of each geometric sequence. (Lesson 10-3)

61. a_9 for $a_1 = \frac{1}{5}, 1, 5, \ldots$

62. $a_4 = 16, r = 0.5, n = 8$

63. $a_6 = 3, r = 2, n = 12$

64. GAMES In a game, the first person in a group states his or her name and an interesting fact about himself or herself. The next person must repeat the first person's name and fact and then say his or her own information. Each person must repeat the information for all those who preceded him or her. If there are 20 people in a group, what is the total number of times the names and facts will be stated? (Lesson 10-2)

Graph each number in the complex plane, and find its absolute value. (Lesson 9-5)

65. $z = 5 - 3i$

66. $z = -9 - 8i$

67. $z = 2 + 6i$

Find rectangular coordinates for each point with the given polar coordinates. (Lesson 9-3)

68. $\left(3, -\frac{5\pi}{4}\right)$

69. $\left(2, \frac{7\pi}{6}\right)$

70. $(-4, 1.4)$

Write the component form of each vector. (Lesson 8-5)

71. $\mathbf{w}$ lies in the xz plane, has a magnitude of 2, and makes a $45°$ angle to the left of the positive z-axis.

72. $\mathbf{u}$ lies in the yz plane, has a magnitude of 9, and makes a $30°$ angle to the right of the negative z-axis.

For each equation, identify the vertex, focus, axis of symmetry, and directrix. Then graph the parabola. (Lesson 7-1)

73. $(x - 4)^2 = -(y + 7)$

74. $6(x + 6) = (y - 4)^2$

75. $(y - 6)^2 = 4x$

76. BUSINESS A factory is making skirts and dresses from the same fabric. Each skirt requires 1 hour of cutting and 1 hour of sewing. Each dress requires 2 hours of cutting and 3 hours of sewing. The cutting and sewing departments can work up to 120 and 150 hours each week, respectively. If profits are \$12 for each skirt and \$18 for each dress, how many of each item should the factory make for maximum profit? (Lesson 6-5)

77. SAT/ACT Triangle ABC is inscribed in circle O. $\overleftrightarrow{CD}$ is tangent to circle O at point C. If $m\angle BCD = 40°$, find $m\angle A$.

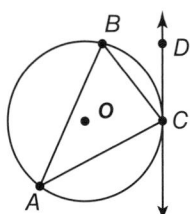

A $60°$ **C** $40°$ **E** $25°$

B $50°$ **D** $30°$

78. Which of the following is divisible by 2 for all positive integers n?

F $1^n - 1$ **H** $3^n - 1$

G $2^n - 1$ **J** $4^n - 1$

79. REVIEW What is the first term in the arithmetic sequence?

$$\underline{\quad}, 8\frac{1}{3}, 7, 5\frac{2}{3}, 4\frac{1}{3}, \ldots$$

A 3

B $9\frac{2}{3}$

C $10\frac{1}{3}$

D 11

80. REVIEW What is the tenth term in the arithmetic sequence that begins 10, 5.6, 1.2, -3.2, $\ldots$?

F -39.6

G -29.6

H 29.6

J 39.6

The Binomial Theorem

Then
You represented infinite series using sigma notation. (Lesson 10-1)

Now
- Use Pascal's triangle to write binomial expansions.
- Use the Binomial Theorem to write and find the coefficients of specified terms in binomial expansions.

New Vocabulary
binomial coefficients
Pascal's triangle
Binomial Theorem

Math Online

glencoe.com

Why?

Suppose a biologist studying an endangered species of gibbon has found that on average 80% of gibbon offspring are female and 20% are male. Zoo workers anticipate that their captive gibbons will produce n offspring and want to know the probability that none of these offspring will be male. The biologist can use a term from the binomial expansion of $(0.8 + 0.2)^n$ to solve this problem.

Pascal's Triangle Recall that a *binomial* is an algebraic expression involving the sum of two unlike terms. An important series is generated by the expansion of a binomial raised to an integer power. Examine this series generated by the expansion of $(a + b)^n$ for several nonnegative integer values of n.

$$(a + b)^0 = 1a^0b^0$$
$$(a + b)^1 = 1a^1b^0 + 1a^0b^1$$
$$(a + b)^2 = 1a^2b^0 + 2a^1b^1 + 1a^0b^2$$
$$(a + b)^3 = 1a^3b^0 + 3a^2b^1 + 3a^1b^2 + 1a^0b^3$$
$$(a + b)^4 = 1a^4b^0 + 4a^3b^1 + 6a^2b^2 + 4a^1b^3 + 1a^0b^4$$
$$(a + b)^5 = 1a^5b^0 + 5a^4b^1 + 10a^3b^2 + 10a^2b^3 + 5a^1b^4 + 1a^0b^5$$

Observe the following patterns in the expansions of $(a + b)^n$ above.

- Each expansion has $n + 1$ terms.
- The first term is a^n, and the last term is b^n.
- In successive terms, the powers of a decrease by 1 and the powers of b increase by 1.
- The sum of the exponents in each term is n.
- The coefficients, in red above, increase and then decrease in a symmetric pattern.

If just the coefficients of these expansions, called the **binomial coefficients**, are extracted and arranged in a triangular array, they form a pattern called **Pascal's triangle**, named after the French mathematician Blaise Pascal. The top row in this triangle is called the *zeroth row* because it corresponds to the binomial expansion of $(a + b)^0$.

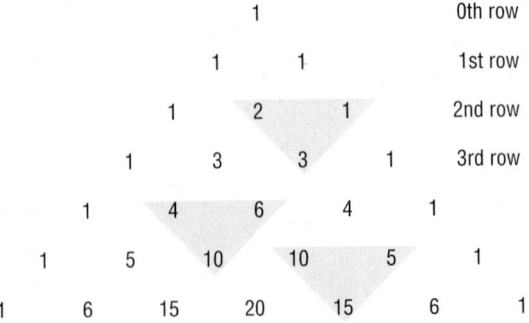

				1					0th row
			1		1				1st row
		1		2		1			2nd row
	1		3		3		1		3rd row
1		4		6		4		1	
1	5		10		10		5	1	
1	6	15	20		15	6	1		

Notice that the first and last numbers in each row are 1 and every other number is formed by adding the two numbers immediately above that number in the previous row. Pascal's triangle can be extended indefinitely using the recursive relationship that the coefficients in the $(n - 1)$th row can be used to determine the coefficients in the nth row.

By extending Pascal's triangle and using the patterns observed in the first 5 expansions of $(a + b)^n$, you can expand a binomial raised to any whole number power.

EXAMPLE 1   **Power of a Binomial Sum**

Use Pascal's triangle to expand each binomial.

a. $(a + b)^7$

Step 1   Write the series for $(a + b)^7$, omitting the coefficients. Because the power is 7, this series should have $7 + 1$ or 8 terms. Use the pattern of increasing and decreasing exponents to complete the series.

$a^7b^0 + a^6b^1 + a^5b^2 + a^4b^3 + a^3b^4 + a^2b^5 + a^1b^6 + a^0b^7$

> **Exponents of a decrease from 7 to 0.**
> **Exponents of b increase from 0 to 7.**

Step 2   Use the numbers in the seventh row of Pascal's triangle as the coefficients of the terms. To find these numbers, extend Pascal's triangle to the 7th row.

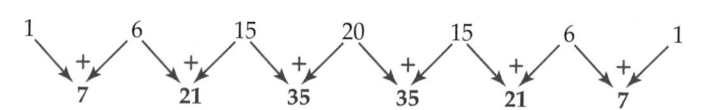

$(a + b)^7 = 1a^7b^0 + 7a^6b^1 + 21a^5b^2 + 35a^4b^3 + 35a^3b^4 + 21a^2b^5 + 7a^1b^6 + 1a^0b^7$

$= a^7 + 7a^6b + 21a^5b^2 + 35a^4b^3 + 35a^3b^4 + 21a^2b^5 + 7ab^6 + b^7$ **Simplify.**

b. $(3x + 2)^4$

Step 1   Write the series for $(a + b)^4$, omitting the coefficients and replacing a with $3x$ and b with 2. The series has $4 + 1$ or 5 terms.

$(3x)^4(2)^0 + (3x)^3(2)^1 + (3x)^2(2)^2 + (3x)^1(2)^3 + (3x)^0(2)^4$

> **Exponents of $3x$ decrease from 4 to 0.**
> **Exponents of 2 increase from 0 to 4.**

Step 2   The numbers in the 4th row of Pascal's triangle are 1, 4, 6, 4, and 1. Use these numbers as the coefficients of the terms in the series. Then simplify.

$(3x + 2)^4 = 1(3x)^4(2)^0 + 4(3x)^3(2)^1 + 6(3x)^2(2)^2 + 4(3x)^1(2)^3 + 1(3x)^0(2)^4$

$= 81x^4 + 216x^3 + 216x^2 + 96x + 16$

✔ **Guided Practice**

1A. $(a + b)^8$ **1B.** $(2x + 3y)^5$

To expand a binomial difference, first rewrite the expression as a binomial sum.

EXAMPLE 2   **Power of a Binomial Difference**

Use Pascal's triangle to expand $(x - 4y)^5$.

Because $(x - 4y)^5 = [x + (-4y)]^5$, write the series for $(a + b)^5$, replacing a with x and b with $-4y$. Use the numbers in the 5th row of Pascal's triangle, 1, 5, 10, 10, 5, and 1, as the binomial coefficients. Then simplify.

$(x - 4y)^5 = 1x^5(-4y)^0 + 5x^4(-4y)^1 + 10x^3(-4y)^2 + 10x^2(-4y)^3 + 5x^1(-4y)^4 + 1x^0(-4y)^5$

$= x^5 - 20x^4y + 160x^3y^2 - 640x^2y^3 + 1280xy^4 - 1024y^5$

✔ **Guided Practice**

Use Pascal's triangle to expand each binomial.

2A. $(2x - 7)^3$ **2B.** $(2x - 3y)^4$

The Binomial Theorem While it is possible to expand any binomial using Pascal's triangle, the recursive method of computing the binomial coefficients makes expansions of $(a + b)^n$ for large values of n time consuming. An explicit formula for computing each binomial coefficient is developed by considering $(a + b)^n$ as the product of n factors in which each factor contributes an a or a b to each product in the expansion.

$$(a + b)^n = \underbrace{(a + b)(a + b)(a + b)\,(a + b)\cdots(a + b)}_{n\text{ factors}} = \ldots + \underbrace{a^{n-r}b^r}_{\text{If there are } r \text{ letter } b\text{'s, then there are } (n-r) \text{ letter } a\text{'s.}} + \ldots$$

Consider $(a + b)^3$. There are three ways to choose 1 a and 2 b's from each of its three factors to form the product ab^2, and 3 is the binomial coefficient of the ab^2 term in the expansion.

$$\left. \begin{array}{l} (a + b)(a + b)(a + b) = \ldots + ab^2 + \ldots \\ (a + b)(a + b)(a + b) = \ldots + ab^2 + \ldots \\ (a + b)(a + b)(a + b) = \ldots + ab^2 + \ldots \end{array} \right\} \text{—— 3 ways ——}$$

$$(a + b)^3 = a^3 + 3a^2b + 3ab^2 + b^3$$

ReadingMath

Combinations The notations $_nC_r$ and $\binom{n}{r}$ are both read as *the combination of n things taken r at a time*. To review computing combinations, see Lesson 0-7.

Because those factors that do not contribute a b will by default contribute an a, the number of ways to form the product $a^{n-r}b^r$ can be more simply thought of as the number of ways to choose r factors to contribute a b to the product from the n factors available. This is the combination given by $_nC_r = \dfrac{n!}{(n-r)!\,r!}$, also written $\binom{n}{r}$.

Key Concept   **Formula for the Binomial Coefficients of $(a + b)^n$**

Words   The binomial coefficient of the $a^{n-r}b^r$ term in the expansion of $(a + b)^n$ is given by $_nC_r = \dfrac{n!}{(n-r)!\,r!}$.

Example $(a + b)^3 = {_3}C_0a^3b^0 + {_3}C_1a^2b^1 + {_3}C_2a^1b^2 + {_3}C_3a^0b^3$

$$= \frac{3!}{(3-0)!\,0!}a^3 + \frac{3!}{(3-1)!\,1!}a^2b + \frac{3!}{(3-2)!\,2!}ab^2 + \frac{3!}{(3-3)!\,3!}b^3$$

$$= 1a^3 + 3a^2b + 3ab^2 + 1b^3$$

In the example above, notice that for the first term $r = 0$, for the second term $r = 1$, for the third term $r = 2$, and so on. In general, to find the binomial coefficient of the kth term in an expansion of the form $(a + b)^n$, use the formula $_nC_r$ and let $r = k - 1$.

EXAMPLE 3   **Find Binomial Coefficients**

Find the coefficient of the 5th term in the expansion of $(a + b)^7$.

To find the coefficient of the 5th term, evaluate $_nC_r$ for $n = 7$ and $r = 5 - 1$ or 4.

$$_7C_4 = \frac{7!}{(7-4)!\,4!} \qquad\qquad _nC_r = \frac{n!}{(n-r)!\,r!}$$

$$= \frac{7!}{3!\,4!} \qquad\qquad \text{Subtract.}$$

$$= \frac{7 \cdot 6 \cdot 5 \cdot \cancel{4!}}{3!\,\cancel{4!}} \qquad \text{Rewrite 7! as } 7 \cdot 6 \cdot 5 \cdot 4! \text{ and divide out common factorials.}$$

$$= \frac{7 \cdot 6 \cdot 5}{3 \cdot 2 \cdot 1} \text{ or } 35 \qquad \text{Simplify.}$$

The coefficient of the 5th term in the expansion of $(a + b)^7$ is 35.

CHECK From Example 1a, you know the 5th term in the expansion of $(a + b)^7$ is $35a^3b^4$. ✔

✓ **Guided Practice**

Find the coefficient of the indicated term in each expansion.

3A. $(x + y)^9$, 6th term   **3B.** $(a - b)^{13}$, 3rd term

EXAMPLE 4 **Binomials with Coefficients Other than 1**

Find the coefficient of the x^7y^2 term in the expansion of $(4x - 3y)^9$.

For $(4x - 3y)^9$ to have the form $(a + b)^n$, let $a = 4x$ and $b = -3y$. The coefficient of the term containing $a^{n-r}b^r$ in the expansion of $(a + b)^n$ is given by $_nC_r$. So, to find the binomial coefficient of the term containing a^7b^2 in the expansion of $(a + b)^9$, evaluate $_nC_r$ for $n = 9$ and $r = 2$.

$$_9C_2 = \frac{9!}{(9 - 2)!\, 2!} \qquad\qquad _nC_r = \frac{n!}{(n - r)!\, r!}$$

$$= \frac{9!}{7!\, 2!} \qquad\qquad\qquad \text{Subtract.}$$

$$= \frac{9 \cdot 8 \cdot 7!}{7!\, 2!} \qquad\qquad \text{Rewrite 9! as } 9 \cdot 8 \cdot 7! \text{ and divide out common factorials.}$$

$$= \frac{9 \cdot 8}{2 \cdot 1} \text{ or } 36 \qquad\qquad \text{Simplify.}$$

Thus, the binomial coefficient of the a^7b^2 term in $(a + b)^9$ is 36. Substitute $4x$ for a and $-3y$ for b to find the coefficient of the x^7y^2 term in the original binomial expansion.

$$36a^7b^2 = 36(4x)^7(-3y)^2 \qquad a = 4x \text{ and } b = -3y$$

$$= 5{,}308{,}416x^7y^2 \qquad\qquad \text{Simplify.}$$

Therefore, the coefficient of the x^7y^2 term in the expansion of $(4x - 3y)^9$ is 5,308,416.

✔️ **Guided Practice**

Find the coefficient of the indicated term in each binomial expansion.

4A. $(2x - 3y)^8$, x^3y^5 term

4B. $(2p + 1)^{15}$, 11th term

You can use the coefficients of binomial expansions to solve real-world problems in which there are only two outcomes for an event. Problems that can be solved using a binomial expansion are called *binomial experiments*. Such experiments occur if and only if: (1) the experiment consists of n identical trials, (2) each trial results in *one* of two outcomes, and (3) the trials are independent.

For n independent trials of an experiment, if the probability of a success is p and the probability of a failure is $q = 1 - p$, then the term $_nC_x\, p^x q^{n-x}$ in the expansion of $(p + q)^n$ gives the probability of x successes for those n trials.

🌐 **Real-World EXAMPLE 5** **Use Binomial Coefficients**

BASEBALL The probability that Andres gets a hit when at bat is $\frac{1}{5}$. What is the probability that Andres gets exactly 4 hits during his next 10 at bats?

A success in this situation is Andres getting a hit, so $p = \frac{1}{5}$ and $q = 1 - \frac{1}{5}$ or $\frac{4}{5}$. Each at bat represents a trial, so $n = 10$. You want to find the probability that Andres succeeds 4 times out of those 10 trials, so let $x = 4$. To find this probability, find the value of the term $_nC_x\, p^x q^{n-x}$ in the expansion of $(p + q)^n$.

$$_nC_x\, p^x q^{n-x} = {_{10}C_4}\left(\frac{1}{5}\right)^4\left(\frac{4}{5}\right)^{10-4} \qquad p = \frac{1}{5},\, q = \frac{4}{5},\, n = 10, \text{ and } x = 4$$

$$= \frac{10!}{(10 - 4)!\, 4!}\left(\frac{1}{5}\right)^4\left(\frac{4}{5}\right)^6 \qquad _nC_r = \frac{n!}{(n - r)!\, r!}$$

$$\approx 0.088 \qquad\qquad\qquad \text{Use a calculator.}$$

So, the probability that Andres gets 4 hits during his next 10 at bats is about 0.088 or 8.8%.

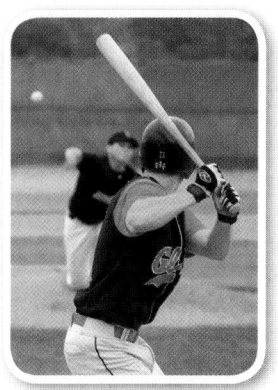

🌎 Real-World Link

An *at bat* in baseball is any time that a batter faces a pitcher except when the player "(i) hits a sacrifice bunt or sacrifice fly; (ii) is awarded first base on four called balls; (iii) is hit by a pitched ball; or (iv) is awarded first base because of interference or obstruction."

Source: Major League Baseball

✔️ **Guided Practice**

5. COIN TOSS A fair coin is flipped 8 times. Find the probability of each outcome.

A. exactly 3 heads

B. exactly 6 tails

The formula for finding the coefficients of a binomial expansion leads us to a theorem about expanding powers of binomials called the **Binomial Theorem**.

Key Concept   **Binomial Theorem**

For any positive integer n, the expansion of $(a + b)^n$ is given by

$$(a + b)^n = {}_nC_0\, a^n b^0 + {}_nC_1\, a^{n-1}b^1 + {}_nC_2\, a^{n-2}b^2 + \cdots + {}_nC_r\, a^{n-r}b^r + \cdots + {}_nC_n\, a^0 b^n,$$

where $r = 0, 1, 2, \ldots, n$.

You will prove the Binomial Theorem in Exercise 75.

EXAMPLE 6   **Expand a Binomial Using the Binomial Theorem**

Use the Binomial Theorem to expand each binomial.

a. $(3x - y)^4$

Apply the Binomial Theorem to expand $(a + b)^4$, where $a = 3x$ and $b = -y$.

$$(3x - y)^4 = {}_4C_0\, (3x)^4(-y)^0 + {}_4C_1\, (3x)^3(-y)^1 + {}_4C_2\, (3x)^2(-y)^2$$
$$+ {}_4C_3\, (3x)^1(-y)^3 + {}_4C_4\, (3x)^0(-y)^4$$
$$= 1(81x^4)(1) + 4(27x^3)(-y) + 6(9x^2)(y^2) + 4(3x)(-y^3) + 1(1)(y^4)$$
$$= 81x^4 - 108x^3y + 54x^2y^2 - 12xy^3 + y^4$$

TechnologyTip

Combinations To evaluate $_{10}C_4$ using a calculator, enter 10, select nCr from the MATH ▶ PRB menu, and then enter 4.

b. $(2p + q^2)^5$

Apply the Binomial Theorem to expand $(a + b)^5$, where $a = 2p$ and $b = q^2$.

$$(2p + q^2)^5 = {}_5C_0\, (2p)^5(q^2)^0 + {}_5C_1\, (2p)^4(q^2)^1 + {}_5C_2\, (2p)^3(q^2)^2 + {}_5C_3\, (2p)^2(q^2)^3 + {}_5C_4\, (2p)^1(q^2)^4$$
$$+ {}_5C_5\, (2p)^0(q^2)^5$$
$$= 1(32p^5)(1) + 5(16p^4)(q^2) + 10(8p^3)(q^4) + 10(4p^2)(q^6) + 5(2p)(q^8) + 1(1)(q^{10})$$
$$= 32p^5 + 80p^4q^2 + 80p^3q^4 + 40p^2q^6 + 10pq^8 + q^{10}$$

✓ **Guided Practice**

6A. $(5m + 4)^3$ **6B.** $(8x^2 - 2y)^6$

Because a binomial expansion is a sum, the Binomial Theorem is often written using sigma notation. In addition, the notation $_nC_r$ is usually replaced by $\binom{n}{r}$.

$$(a + b)^n = \sum_{r=0}^{n} \binom{n}{r} a^{n-r} b^r$$

EXAMPLE 7   **Write a Binomial Expansion Using Sigma Notation**

Represent the expansion of $(5x - 7y)^{20}$ using sigma notation.

Apply the Binomial Theorem to represent the expansion of $(a + b)^{20}$ using sigma notation, where $a = 5x$ and $b = -7y$.

$$(5x - 7y)^{20} = \sum_{r=0}^{20} \binom{20}{r} (5x)^{20-r}(-7y)^r$$

✓ **Guided Practice**

7. Represent the expansion of $(3a + 12b)^{30}$ using sigma notation.

Exercises

Use Pascal's triangle to expand each binomial.
(Examples 1 and 2)

1. $(2 + x)^4$

2. $(n + m)^5$

3. $(4a - b)^3$

4. $(x + y)^6$

5. $(3x + 2y)^7$

6. $(n - 4)^6$

7. $(3c - d)^4$

8. $(m - a)^5$

9. $(a - b)^3$

10. $(3p - 2q)^4$

Find the coefficient of the indicated term in each expansion.
(Examples 3 and 4)

11. $(x - 2)^{10}$, 5th term

12. $(4m + 1)^8$, 3rd term

13. $(x + 3y)^{10}$, 8th term

14. $(2c - d)^{12}$, 6th term

15. $(a + b)^8$, 4th term

16. $(2a + 3b)^{10}$, 5th term

17. $(x - y)^9$, 6th term

18. $(x + y)^{12}$, 7th term

19. $(x + 2)^7$, 4th term

20. $(a - 3)^8$, 5th term

21. $(2a + 3b)^{10}$, a^6b^4 term

22. $(2x + 3y)^9$, x^6y^3 term

23. $\left(x + \frac{1}{3}\right)^7$, 4th term

24. $\left(x - \frac{1}{2}\right)^{10}$, 6th term

25. $(x + 4y)^7$, x^2y^5 term

26. $(3x + 5y)^{10}$, x^6y^4 term

27. TESTING Alfonso is taking a test that contains a section of 16 true-false questions. (Example 3)

   a. How many of the possible sets of answers to these questions have exactly 12 correct answers of false?

   b. How many of the possible sets of answers to these questions have exactly 8 correct answers of true?

28. BUSINESS The probability of a certain sales representative successfully making a sale is $\frac{1}{5}$. The sales representative has 12 appointments this week. (Example 5)

   a. Find the probability that the sales representative makes no sales this week.

   b. What is the probability that the sales representative makes exactly 3 sales this week?

   c. Find the probability that the sales representative will make 10 sales this week.

29. BIOLOGY Refer to the beginning of the lesson. Assume that the zoo workers expect 30 gibbon offspring this year. (Example 5)

   a. What is the probability that there will be no male gibbon offspring this year?

   b. Find the probability that there will be exactly 2 male gibbon offspring this year.

   c. What is the probability that there will be 23 female gibbon offspring this year?

30. BOWLING Carol averages 2 strikes every 10 frames. What is the probability that Carol will get exactly 4 strikes in the next 10 frames? (Example 5)

Bowling Castle Scorecard

Carol	1	2	3	4	5	6	7	8	9	10	Total
	7/	X	5 4	9/	8/	X	X	7 0	X	8 1	165
	20	39	48	66	86	113	130	137	156	165	
	1	2	3	4	5	6	7	8	9	10	Total

Use the Binomial Theorem to expand each binomial.
(Example 6)

31. $(4t + 3)^5$

32. $(8 - 5y)^3$

33. $(2m - n)^6$

34. $(9h + 2j)^4$

35. $(3p + q)^7$

36. $(a^2 - 2b)^8$

37. $(7c^2 + 3d)^5$

38. $(2w - 4x^3)^7$

Represent the expansion of each expression using sigma notation. (Example 7)

39. $(2q + 3)^{15}$

40. $(m - 8n)^{25}$

41. $(11x + y)^{31}$

42. $(4a + 7b)^{19}$

43. $\left(3f - \frac{3}{4}g\right)^{22}$

44. $\left(\frac{1}{2}s - 5t\right)^{36}$

45 COMPUTER GAMES In a computer game, when a treasure chest is opened, it contains either gold coins or rocks. The probability that it contains gold coins is $\frac{5}{6}$.

   a. The treasure chest is opened 15 times per game. In one game, how many different ways is it possible to open the chest and find gold coins exactly 9 times?

   b. What is the probability that a person playing the game will find gold in the chest more than 12 times?

46. COMMUNITY OUTREACH At a food bank, canned goods are received and distributed to people in the community who are in need. Volunteers check the quality of the food before distribution. The probability that a canned good received at a food bank is distributed to the needy is $\frac{4}{5}$.

   a. A volunteer checks 30 canned goods per hour. In one hour, how many different ways is it possible to check a canned good and donate it exactly 23 times?

   b. What is the probability that a volunteer checking canned goods will find themselves throwing out items less than 4 times?

Use the Binomial Theorem to expand and simplify each expression.

47. $(2d + \sqrt{5})^4$

48. $(\sqrt{a} - \sqrt{b})^5$

49. $\left(4s + \frac{1}{2}t\right)^5$

50. $\left(\frac{1}{y} - 3z\right)^6$

Find the coefficient of the indicated term in each expansion.

51. $(k - \sqrt{5})^9$, 5th term

52. $(\sqrt{2} + 2c)^{10}$, middle term

53. $\left(\frac{1}{4}p + q\right)^{11}$, 7th term

54. $\left(\sqrt{h} - 3\sqrt{j}\right)^{11}$, 6th term

Use the Binomial Theorem to expand and simplify each power of a complex number.

55. $(i + 2)^4$

56. $(i - 3)^3$

57. $(1 - 4i)^5$

58. $\left(2 + \sqrt{7}i\right)^4$

59. $\left(\frac{\sqrt{2}}{2}i - \frac{1}{2}\right)^3$

60. $\left(\sqrt{-16}i + 3\right)^5$

The graph of $g(x)$ is a translation of the graph of $f(x)$. Use the Binomial Theorem to find the polynomial function for $g(x)$ in standard form.

61. $f(x) = x^4 + 5x$
    $g(x) = f(x + 3)$

62. $f(x) = x^5 + 1$
    $g(x) = f(x - 4)$

63. $f(x) = x^6 + 2x^3$
    $g(x) = f(x - 1)$

64. $f(x) = x^7 - 3x^4 + 2x$
    $g(x) = f(x + 2)$

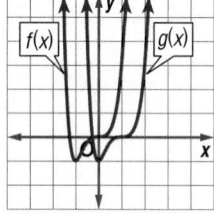

65. ⚙ **MULTIPLE REPRESENTATIONS** In this problem, you will use the Binomial Theorem to investigate the difference quotient $\dfrac{f(x + h) - f(x)}{h}$ for power functions.

   a. **ANALYTICAL** Use the Binomial Theorem to expand and simplify the difference quotient for $f(x) = x^3$, $f(x) = x^4$, $f(x) = x^5$, $f(x) = x^6$, and $f(x) = x^7$. Use the pattern to simplify the difference quotient for $f(x) = x^n$.

   b. **TABULAR** Evaluate each expression in part **a** for $h = 0.1, 0.01, 0.001$, and 0.0001 and record your results in a table. What do you observe?

   c. **GRAPHICAL** Graph the set of resulting functions from part **b** for $f(x) = x^3$ on the same coordinate plane. What do you observe?

   d. **ANALYTICAL** As h approaches 0, write an expression for the difference quotient when $f(x) = x^n$, where n is a positive integer.

66. In the expansion of $(ax + b)^5$ the numerical coefficient of the second term is 400 and the numerical coefficient of the third term is 2000. Find the values of a and b.

67. **RESEARCH** Although Pascal's triangle is named for Blaise Pascal, other mathematicians applied their knowledge of the triangle hundreds of years before Pascal. Use the Internet or another source to research at least one other person who used the properties of the triangle before Pascal was born. Then describe other patterns found in Pascal's triangle that are not described in this lesson.

H.O.T. Problems   Use **H**igher-**O**rder **T**hinking Skills

68. **ERROR ANALYSIS** Jena and Gil are finding the 6th term of the expansion of $(x + y)^{14}$. Jena says that the coefficient of the term is 3003. Gil thinks that it is 2002. Is either of them correct? Explain your reasoning.

69. **CHALLENGE** Describe a strategy that uses the Binomial Theorem to expand $(x + y + z)^n$. Then write and simplify an expansion for the expression.

70. **PROOF** The sums of the coefficients in the first five rows of Pascal's triangle are shown below.

row 0	1	$= 2^0$
row 1	$1 + 1$	$= 2^1$
row 2	$1 + 2 + 1$	$= 2^2$
row 3	$1 + 3 + 3 + 1$	$= 2^3$
row 4	$1 + 4 + 6 + 4 + 1$	$= 2^4$
row 5	$1 + 5 + 10 + 10 + 5 + 1$	$= 2^5$

Prove that the sum of the coefficients in the nth row of Pascal's triangle is 2^n. (*Hint*: Write 2^n as $(1 + 1)^n$. Then use the Binomial Theorem to expand.)

71. **WRITING IN MATH** Describe how to find the numbers in each row of Pascal's triangle. Then write a few sentences to describe how the expansions of $(a + b)^{n-1}$ and $(a - b)^n$ are different from the expansion of $(a + b)^n$.

72. **REASONING** Determine whether the following statement is *sometimes*, *always*, or *never* true. Justify your reasoning.

   If a binomial is raised to the power 5, the two middle terms of the expansion have the same coefficients.

73. **CHALLENGE** Explain how you could find a term in the expansion of $\left(\frac{1}{2v} + 6v^7\right)^8$ that does not contain the variable v. Then find the term.

74. **PROOF** Use the principle of mathematical induction to prove the Binomial Theorem.

Prove that each statement is true for all positive integers *n* or find a counterexample. (Lesson 10-4)

75. $1^2 + 2^2 + 3^2 + \ldots + n^2 = \dfrac{n(3n - 1)}{2}$

76. $10^{2n-1} + 1$ is divisible by 11.

77. GENEALOGY In the book *Roots*, author Alex Haley traced his family history back many generations. If you could trace your family back for 15 generations, starting with your parents, how many ancestors would there be? (Lesson 10-3)

78. Let $\overrightarrow{DE}$ be the vector with initial point $D(5, -12)$ and terminal point $E(8, -17)$. Write $\overrightarrow{DE}$ as a linear combination of the vectors **i** and **j**. (Lesson 8-2)

Write each equation in standard form. Identify the related conic. (Lesson 7-2)

79. $x^2 + y^2 - 16x + 10y + 64 = 0$

80. $y^2 + 16x - 10y + 57 = 0$

81. $x^2 + y^2 + 2x + 24y + 141 = 0$

Find the partial fraction decomposition of each rational expression. (Lesson 6-4)

82. $\dfrac{5x^2 - 14}{(x^2 - 2)^2}$

83. $\dfrac{x^3 - 8x^2 + 21x - 22}{x^2 - 8x + 15}$

84. $\dfrac{3x}{(x - 3)^2}$

Determine whether each matrix is in row-echelon form. (Lesson 6-1)

85. $\begin{bmatrix} 1 & 10 & -5 & | & 3 \\ 0 & 1 & 14 & | & -2 \\ 0 & 1 & 9 & | & 6 \end{bmatrix}$

86. $\begin{bmatrix} 1 & 3 & -7 & | & 11 \\ 0 & 1 & -13 & | & 18 \\ 0 & 0 & 0 & | & 1 \end{bmatrix}$

87. $\begin{bmatrix} 1 & 0 & | & 8 \\ 0 & 1 & | & -4 \\ 0 & 0 & | & 1 \end{bmatrix}$

Find the exact value of each trigonometric expression. (Lesson 5-4)

88. $\tan 195°$

89. $\csc \dfrac{5\pi}{12}$

90. $\sin \dfrac{\pi}{12}$

91. SAVINGS Janet's father deposited \$30 into a bank account for her. They forgot about the money and made no further deposits or withdrawals. The table shows the account balance for several years. (Lesson 3-5)

a. Make a scatter plot of the data.

b. Find an exponential function to model the data.

c. Use the function to predict the balance of the account in 41 years.

Elapsed Time (yr)	Balance (\$)
0	30.00
5	41.10
10	56.31
15	77.16
20	105.71
25	144.83
30	198.43

92. SAT/ACT In the figure below, *ABCD* is a parallelogram. What are the coordinates of point *C*?

   A $(d + a, y)$

   B $(d - a, b)$

   C $(d + x, b)$

   D $(d + a, b)$

   E $(d + b, a)$

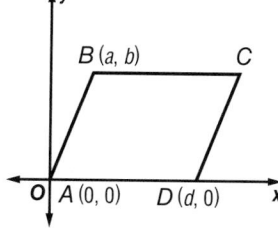

93. REVIEW What is the value of $\dfrac{12!}{8!\,4!}$?

   F 495

   G 500

   H 660

   J 710

94. Mrs. Thomas is giving a four-question multiple-choice quiz. Each question can be answered A, B, C, or D. How many ways could a student answer the questions using each answer A, B, C, or D once?

   A 20   **C** 24

   B 22   **D** 26

95. REVIEW Which expression is equivalent to $(2x - 2)^4$?

   F $16x^4 + 64x^3 - 96x^2 - 64x + 16$

   G $16x^4 - 32x^3 - 192x^2 - 64x + 16$

   H $16x^4 - 64x^3 + 96x^2 - 64x + 16$

   J $16x^4 + 32x^3 - 192x^2 - 64x + 16$

Functions as Infinite Series

Then
You found the nth term of an infinite series expressed using sigma notation. (Lesson 10-1)

Now
- Use a power series to represent a rational function.
- Use power series representations to approximate values of transcendental functions.

New Vocabulary
power series
exponential series
Euler's Formula

| Math Online >

glencoe.com

Why?

The music to which you listen on a digital audio player is first performed by an artist. The waveform of each sound in that performance is then broken down into its component parts and stored digitally. These parts are then retrieved and combined to reproduce each original sound of the performance. The analysis of a special series is an essential ingredient in this process.

Power Series Earlier in this chapter, you saw how some series of numbers can be expressed as functions. In this lesson, you will see that some functions can be broken down into infinite series of component functions.

In Lesson 10-3, you learned that the sum of an infinite geometric series,

$$1 + r + r^2 + \cdots + r^n + \cdots, a_1 = 1$$

with common ratio r, converges to a sum of $\frac{a_1}{1-r}$ if $|r| < 1$. Replacing r with x,

$$\sum_{n=0}^{\infty} x^n = 1 + x + x^2 + \cdots + x^n + \cdots = \frac{1}{1-x}, \text{ for } |x| < 1.$$

It follows that $f(x) = \frac{1}{1-x}$ can be expressed as an infinite series. That is,

$$f(x) = \frac{1}{1-x} = \sum_{n=0}^{\infty} x^n \text{ or } 1 + x + x^2 + \cdots + x^n + \cdots \text{ for } |x| < 1.$$

The figures below show the graph of $f(x) = \frac{1}{1-x}$ and the second through fifth partial sums $S_n(x)$ of the series: $S_2(x) = 1 + x$, $S_3(x) = 1 + x + x^2$, $S_4(x) = 1 + x + x^2 + x^3$, and $S_5(x) = 1 + x + x^2 + x^3 + x^4$.

   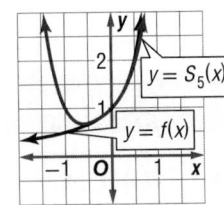

Notice that as n increases, the graph of $S_n(x)$ appears to come closer and closer to the graph of $f(x)$ on the interval $(-1, 1)$ or $|x| < 1$. Notice too that each of the partial sums of the series is a polynomial function, so the series can be thought of as an "infinite" polynomial. An infinite series of this type is called a **power series**.

| **Key Concept** | **Power Series** |

An infinite series of the form

$$\sum_{n=0}^{\infty} a_n x^n = a_0 + a_1 x + a_2 x^2 + a_3 x^3 + \cdots,$$

where x and a_n can take on any values for $n = 0, 1, 2, \ldots$, is called a power series in x.

If you know the power series representation of one function, you can use it to find the power series representations of other related functions.

EXAMPLE 1 **Power Series Representation of a Rational Function**

Use $\displaystyle\sum_{n=0}^{\infty} x^n$ to find a power series representation of $g(x) = \dfrac{1}{3 - x}$. Indicate the interval on which the series converges. Use a graphing calculator to graph $g(x)$ and the sixth partial sum of its power series.

To find the transformation that relates $f(x)$ to $g(x)$, use u-substitution. Substitute u for x in f, equate the two functions, and solve for u as shown.

$$g(x) = f(u)$$

$$\frac{1}{3 - x} = \frac{1}{1 - u}$$

$$1 - u = 3 - x$$

$$-u = 2 - x$$

$$u = x - 2$$

Therefore, $g(x) = f(x - 2)$. Replacing x with $x - 2$ in $f(x) = \dfrac{1}{1 - x} = \displaystyle\sum_{n=0}^{\infty} x^n$ for $|x| < 1$ yields

$$f(x - 2) = \sum_{n=0}^{\infty} (x - 2)^n \text{ for } |x - 2| < 1.$$

Therefore, $g(x) = \dfrac{1}{3 - x}$ can be represented by the power series $\displaystyle\sum_{n=0}^{\infty} (x - 2)^n$.

This series converges for $|x - 2| < 1$, which is equivalent to $-1 < x - 2 < 1$ or $1 < x < 3$.

The sixth partial sum $S_6(x)$ of this series is

$$\sum_{n=0}^{5} (x - 2)^n \text{ or } 1 + (x - 2) + (x - 2)^2 + (x - 2)^3 + (x - 2)^4 + (x - 2)^5.$$

The graphs of $g(x) = \dfrac{1}{3 - x}$ and $S_6(x) = 1 + (x - 2) + (x - 2)^2 + (x - 2)^3 + (x - 2)^4 + (x - 2)^5$ are shown. Notice that on the interval $(1, 3)$, the graph of $S_6(x)$ comes close to the graph of $g(x)$.

[0.5, 3.5] scl: 0.5 by [0, 4] scl: 0.5

[0.5, 3.5] scl: 0.5 by [0, 4] scl: 0.5

[0.5, 3.5] scl: 0.5 by [0, 4] scl: 0.5

Watch Out!

When finding the kth partial sum of a series where the lower bound starts at 0 use the series $\displaystyle\sum_{n=0}^{k-1}$. For instance in Example 1, the sixth partial sum is called for, but since the lower bound is 0, the upper bound is $6 - 1$ or 5, not 6.

StudyTip

Graphs of Series Notice that the graphs of $f(x)$ and $S_n(x)$ only converge on an interval. The graphs may differ greatly outside of that interval.

✓ **Guided Practice**

Use $\displaystyle\sum_{n=0}^{\infty} x^n$ to find a power series representation of $g(x)$. Indicate the interval on which the series converges. Use a graphing calculator to graph $g(x)$ and the sixth partial sum of its power series.

1A. $g(x) = \dfrac{1}{1 - 2x}$

1B. $g(x) = \dfrac{2}{1 - x}$

In calculus, power series representations are often easier to use in calculations than other representations of functions when determining functions called *derivatives* and *integrals*. A more immediate application can be seen by looking at the power series representations of transcendental functions such as $f(x) = e^x$, $f(x) = \sin x$, and $f(x) = \cos x$.

ReadingMath

Euler Number The Swiss mathematician Leonhard Euler (pronounced OY ler), published a work in which he developed this irrational number, called *e*, the Euler number.

Transcendental Functions as Power Series In Lesson 3-1, you learned that the transcendental number e is given by $\lim_{n \to \infty} \left(1 + \frac{1}{n}\right)^n$. Thus, $e^x = \lim_{x \to \infty} \left(1 + \frac{1}{n}\right)^{nx}$. We can use this definition along with the Binomial Theorem to derive a power series representation for $f(x) = e^x$.

If we let $u = \frac{1}{n}$ and $k = nx$, then $\left(1 + \frac{1}{n}\right)^{nx}$ becomes $(1 + u)^k$. Applying the Binomial Theorem,

$$(1 + u)^k = {_kC_0}(1)^k u^0 + {_kC_1}(1)^{k-1}u + {_kC_2}(1)^{k-2}u^2 + {_kC_3}(1)^{k-3}u^3 + \cdots$$

$$= \frac{k!}{(k-0)!\,0!}(1) + \frac{k!}{(k-1)!\,1!}(1)u + \frac{k!}{(k-2)!\,2!}(1)u^2 + \frac{k!}{(k-3)!\,3!}(1)u^3 + \cdots$$

$$= 1 + \frac{k(k-1)!}{(k-1)!}u + \frac{k(k-1)(k-2)!}{(k-2)!\,2!}u^2 + \frac{k(k-1)(k-2)(k-3)!}{(k-3)!\,3!}u^3 + \cdots$$

$$= 1 + ku + \frac{k(k-1)}{2!}u^2 + \frac{k(k-1)(k-2)}{3!}u^3 + \cdots$$

Now replace u with $\frac{1}{n}$ and k with nx and find the limit as n approaches infinity. Use the fact that as n approaches infinity, the fraction $\frac{1}{n}$ gets increasingly smaller, so $\lim_{n \to \infty} \frac{1}{n} = 0$.

$$\lim_{x \to \infty} \left(1 + \frac{1}{n}\right)^{nx} = 1 + (nx)\frac{1}{n} + \frac{nx(nx-1)}{2!}\left(\frac{1}{n}\right)^2 + \frac{nx(nx-1)(nx-2)}{3!}\left(\frac{1}{n}\right)^3 + \cdots$$

$$= 1 + x + \frac{x\left(x - \frac{1}{n}\right)}{2!} + \frac{x\left(x - \frac{1}{n}\right)\left(x - \frac{2}{n}\right)}{3!} + \cdots$$

$$= 1 + x + \frac{x^2}{2!} + \frac{x^3}{3!} + \cdots$$

This series is often called the **exponential series**.

StudyTip

Defining *e* The exponential series provides yet another way to define *e*. When $x = 1$, $e^1 = \sum_{n=0}^{\infty} \frac{1^n}{n!}$ or $\sum_{n=0}^{\infty} \frac{1}{n!}$.

Key Concept   **Exponential Series**

The power series representing e^x is called the exponential series and is given by

$$e^x = \sum_{n=0}^{\infty} \frac{x^n}{n!} = 1 + x + \frac{x^2}{2!} + \frac{x^3}{3!} + \frac{x^4}{4!} + \frac{x^5}{5!} + \cdots,$$

which is convergent for all *x*.

The graph of $f(x) = e^x$ and the partial sums $S_3(x)$, $S_4(x)$, and $S_5(x)$ of the exponential series are shown below.

  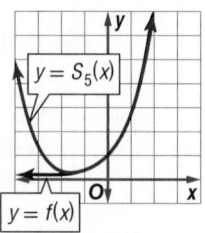

You can see from the graphs that the partial sums of the exponential series approximate the graph of $f(x) = e^x$ on increasingly wider intervals of the domain for increasingly greater values of n.

Notice that the calculations involved in the exponential series are relatively simple: multiplications (for powers and factorials), divisions, and additions. Because of this, calculators and computer programs use partial sums of the exponential series to evaluate e^x to desired degrees of accuracy.

EXAMPLE 2   Exponential Series

Use the fifth partial sum of the exponential series to approximate the value of $e^{1.5}$. Round to three decimal places.

$$e^x \approx 1 + x + \frac{x^2}{2!} + \frac{x^3}{3!} + \frac{x^4}{4!} \qquad\qquad e^x \approx \sum_{n=0}^{4} \frac{x^n}{n!}$$

$$e^{1.5} \approx 1 + 1.5 + \frac{1.5^2}{2!} + \frac{1.5^3}{3!} + \frac{1.5^4}{4!} \qquad x = 1.5$$

$$\approx 4.398 \qquad\qquad\qquad\qquad \textbf{Simplify.}$$

CHECK A calculator, using a partial sum of the exponential series with many more terms, returns an approximation of 4.48 for $e^{1.5}$. Therefore, an approximation of 4.398 is reasonable. ✔

✔ Guided Practice

Use the fifth partial sum of the exponential series to approximate each value. Round to three decimal places.

2A. $e^{-0.75}$ **2B.** $e^{0.25}$

Other transcendental functions have power series representations as well. Calculators and computers use **power series** to approximate the values of cosine and sine functions.

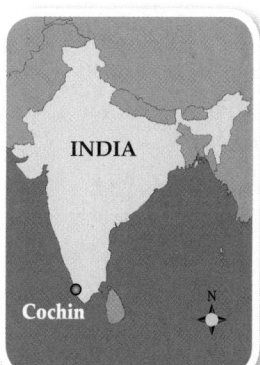

INDIA

Cochin

Key Concept   Power Series for Cosine and Sine

The power series representations for cos x and sin x are given by

$$\cos x = \sum_{n=0}^{\infty} \frac{(-1)^n x^{2n}}{(2n)!} = 1 - \frac{x^2}{2!} + \frac{x^4}{4!} - \frac{x^6}{6!} + \frac{x^8}{8!} - \cdots, \text{ and}$$

$$\sin x = \sum_{n=0}^{\infty} \frac{(-1)^n x^{2n+1}}{(2n+1)!} = x - \frac{x^3}{3!} + \frac{x^5}{5!} - \frac{x^7}{7!} + \frac{x^9}{9!} - \cdots,$$

which are convergent for all x.

By replacing x with any angle measure expressed in radians and carrying out the computations, approximate values of the cosine and sine functions can be found to any desired degree of accuracy.

EXAMPLE 3   Trigonometric Series

a. Use the fifth partial sum of the power series for cosine to approximate the value of $\cos \frac{\pi}{7}$. Round to three decimal places.

$$\cos x \approx 1 - \frac{x^2}{2!} + \frac{x^4}{4!} - \frac{x^6}{6!} + \frac{x^8}{8!} \qquad\qquad \cos x \approx \sum_{n=0}^{4} \frac{(-1)^n x^{2n}}{(2n)!}$$

$$\cos \frac{\pi}{7} \approx 1 - \frac{(0.449)^2}{2!} + \frac{(0.449)^4}{4!} - \frac{(0.449)^6}{6!} + \frac{(0.449)^8}{8!} \qquad x = \frac{\pi}{7} \text{ or about } 0.449$$

$$\approx 0.901 \qquad\qquad\qquad\qquad\qquad\qquad\qquad\qquad \textbf{Simplify.}$$

CHECK A calculator, using a partial sum of the power series for cosine with many more terms, returns an approximation of 0.901, to three decimal places, for $\cos \frac{\pi}{7}$. Therefore, an approximation of 0.901 is reasonable. ✔

b. Use the fifth partial sum of the power series for sine to approximate the value of $\sin \frac{\pi}{5}$. Round to three decimal places.

$$\sin x \approx x - \frac{x^3}{3!} + \frac{x^5}{5!} - \frac{x^7}{7!} + \frac{x^9}{9!} \qquad\qquad \sin x \approx \sum_{n=0}^{5} \frac{(-1)^n x^{2n+1}}{(2n+1)!}$$

$$\sin \frac{\pi}{5} \approx 0.628 - \frac{(0.628)^3}{3!} + \frac{(0.628)^5}{5!} - \frac{(0.628)^7}{7!} + \frac{(0.628)^9}{9!} \qquad x = \frac{\pi}{5} \text{ or about } 0.628$$

$$\approx 0.588 \qquad\qquad\qquad\qquad \textbf{Simplify.}$$

> **CHECK** Using a calculator, $\sin \frac{\pi}{5} \approx 0.588$. Therefore, an approximation of 0.588 is reasonable. ✔

☑ **Guided Practice**

Use the fifth partial sum of the power series for cosine or sine to approximate each value. Round to three decimal places.

3A. $\sin \frac{\pi}{11}$ **3B.** $\cos \frac{2\pi}{17}$

You may have noticed similarities in the power series representations of $f(x) = e^x$ and the power series representations of $f(x) = \sin x$ and $f(x) = \cos x$. A relationship is derived by replacing x by $i\theta$ in the exponential series, where i is the imaginary unit and θ is the measure of an angle in radians.

$$e^{i\theta} = 1 + i\theta + \frac{(i\theta)^2}{2!} + \frac{(i\theta)^3}{3!} + \frac{(i\theta)^4}{4!} + \frac{(i\theta)^5}{5!} + \frac{(i\theta)^6}{6!} + \frac{(i\theta)^7}{7!} + \cdots \qquad e^{i\theta} = \sum_{n=0}^{\infty} \frac{(i\theta)^n}{n!}$$

$$= 1 + i\theta - \frac{\theta^2}{2!} - i\frac{\theta^3}{3!} + \frac{\theta^4}{4!} + i\frac{\theta^5}{5!} - \frac{\theta^6}{6!} - i\frac{\theta^7}{7!} + \cdots \qquad \begin{array}{l} i^2 = -1, i^3 = -i, i^4 = 1, \\ i^5 = i, i^6 = -1, i^7 = -i \end{array}$$

$$= \left(1 - \frac{\theta^2}{2!} + \frac{\theta^4}{4!} - \frac{\theta^6}{6!} + \cdots\right) + \left(i\theta - i\frac{\theta^3}{3!} + i\frac{\theta^5}{5!} - i\frac{\theta^7}{7!} + \cdots\right) \qquad \textbf{Group real and imaginary terms.}$$

$$= \left(1 - \frac{\theta^2}{2!} + \frac{\theta^4}{4!} - \frac{\theta^6}{6!} + \cdots\right) + i\left(\theta - \frac{\theta^3}{3!} + \frac{\theta^5}{5!} - \frac{\theta^7}{7!} + \cdots\right) \qquad \textbf{Distributive Property}$$

$$= \cos\theta + i\left(\theta - \frac{\theta^3}{3!} + \frac{\theta^5}{5!} - \frac{\theta^7}{7!} + \cdots\right) \qquad \cos\theta = 1 - \frac{\theta^2}{2!} + \frac{\theta^4}{4!} - \frac{\theta^6}{6!} + \cdots$$

$$= \cos\theta + i\sin\theta \qquad \sin\theta = \theta - \frac{\theta^3}{3!} + \frac{\theta^5}{5!} - \frac{\theta^7}{7!} + \frac{\theta^9}{9!} - \cdots$$

This relationship is called **Euler's Formula**.

Key Concept   **Euler's Formula**

For any real number θ, $e^{i\theta} = \cos\theta + i\sin\theta$.

From your work in Lesson 9-5, you should recognize the right-hand side of this equation as being part of the polar form of a complex number. Applying Euler's Formula to the polar form of a complex number yields the following result.

$$a + bi = r(\cos\theta + i\sin\theta) \qquad \textbf{Polar form of a complex number}$$

$$= re^{i\theta} \qquad\qquad\qquad \textbf{Euler's Formula}$$

Therefore, Euler's Formula gives us a way of expressing a complex number in exponential form.

Key Concept   **Exponential Form of a Complex Number**

The exponential form of a complex number $a + bi$ is given by

$$a + bi = re^{i\theta},$$

where $r = \sqrt{a^2 + b^2}$ and $\theta = \tan^{-1}\frac{b}{a}$ for $a > 0$ and $\theta = \tan^{-1}\frac{b}{a} + \pi$ for $a < 0$.

EXAMPLE 4 **Write a Complex Number in Exponential Form**

Write $-\sqrt{3} + i$ in exponential form.

Write the polar form of $-\sqrt{3} + i$. In this expression, $a = -\sqrt{3}$, $b = 1$, and $a < 0$. Find r.

$$r = \sqrt{\left(-\sqrt{3}\right)^2 + 1^2} \qquad r = \sqrt{a^2 + b^2}$$

$$= \sqrt{4} \text{ or } 2 \qquad\qquad \text{Simplify.}$$

Now find θ.

$$\theta = \tan^{-1} \frac{1}{-\sqrt{3}} + \pi \qquad \theta = \tan^{-1} \frac{b}{a} + \pi \text{ for } a < 0$$

$$= -\frac{\pi}{6} + \pi \text{ or } \frac{5\pi}{6} \qquad \text{Simplify.}$$

Therefore, because $a + bi = re^{i\theta}$, the exponential form of $-\sqrt{3} + i$ is $2e^{i\frac{5\pi}{6}}$.

✔ Guided Practice

Write each complex number in exponential form.

4A. $1 + \sqrt{3}i$

4B. $\sqrt{2} + \sqrt{2}i$

From your study of logarithms in Chapter 3, you know that no *real* number can be the logarithm of a negative number. We can use Euler's Formula to show that the natural logarithm of a negative number does exist in the *complex* number system.

$$e^{i\theta} = \cos\theta + i\sin\theta \qquad \text{Euler's Formula}$$

$$e^{i\pi} = \cos\pi + i\sin\pi \qquad \text{Let } \theta = \pi.$$

$$e^{i\pi} = -1 + i(0) \qquad \cos\pi = -1 \text{ and } \sin\pi = 0$$

$$e^{i\pi} = -1 \qquad\qquad \text{Simplify.}$$

$$\ln e^{i\pi} = \ln(-1) \qquad \text{Take the natural logarithm of each side.}$$

$$i\pi = \ln(-1) \qquad \text{Power Property of Logarithms}$$

This result indicates that the natural logarithm of -1 exists and is the complex number $i\pi$. You can use this result to find the natural logarithm of any negative number $-k$, for $k > 0$.

$$\ln(-k) = \ln[(-1)k] \qquad -k = (-1)k$$

$$= \ln(-1) + \ln k \qquad \text{Product Property of Logarithms}$$

$$= i\pi + \ln k \qquad \ln(-1) = i\pi$$

$$= \ln k + i\pi \qquad \text{Write in the form } a + bi.$$

TechnologyTip

Complex Numbers You can use your calculator to evaluate the natural logarithm of a negative number by changing from REAL to $a + bi$ under MODE.

EXAMPLE 5 **Natural Logarithm of a Negative Number**

Find the value of $\ln(-5)$ in the complex number system.

$$\ln(-5) = \ln 5 + i\pi \qquad \ln(-k) = \ln k + i\pi$$

$$\approx 1.609 + i\pi \qquad \text{Use a calculator to compute ln 5.}$$

✔ Guided Practice

Find the value of each natural logarithm in the complex number system.

5A. $\ln(-8)$

5B. $\ln(-6.24)$

Use $\displaystyle\sum_{n=0}^{\infty} x^n$ to find a power series representation of $g(x)$.

Indicate the interval on which the series converges. Use a graphing calculator to graph $g(x)$ and the sixth partial sum of its power series. (Example 1)

1. $g(x) = \dfrac{4}{1-x}$

2. $g(x) = \dfrac{3}{1-2x}$

3. $g(x) = \dfrac{2}{1-x^2}$

4. $g(x) = \dfrac{3}{2-x}$

5. $g(x) = \dfrac{2}{5-3x}$

6. $g(x) = \dfrac{4}{3-2x^2}$

Use the fifth partial sum of the exponential series to approximate each value. Round to three decimal places. (Example 2)

7. $e^{0.5}$

8. $e^{-0.25}$

9. $e^{-2.5}$

10. $e^{0.8}$

11. $e^{-0.3}$

12. $e^{3.5}$

13 **ECOLOGY** The population density P per square meter of zebra mussels in the Upper Mississippi River can be modeled by $P = 3.5e^{0.08t}$, where t is measured in weeks. Use the fifth partial sum of the exponential series to estimate the zebra mussel population density after 4 weeks, 12 weeks, and 1 year. (Example 2)

Use the fifth partial sum of the power series for cosine or sine to approximate each value. Round to three decimal places. (Example 3)

14. $\sin\dfrac{\pi}{9}$

15. $\cos\dfrac{2\pi}{13}$

16. $\sin\dfrac{5\pi}{13}$

17. $\cos\dfrac{3\pi}{10}$

18. $\cos\dfrac{2\pi}{9}$

19. $\sin\dfrac{3\pi}{19}$

20. **AMUSEMENT PARK** A ride at an amusement park is in the shape of a giant pendulum that swings riders back and forth in a 240° arc to a maximum height of 137 feet. The pendulum is supported by a tower that is 85 feet tall and dips below ground-level into a pit when swinging below the tower. Use the fifth partial sum of the power series for cosine or sine to approximate the length of the pendulum. (Example 3)

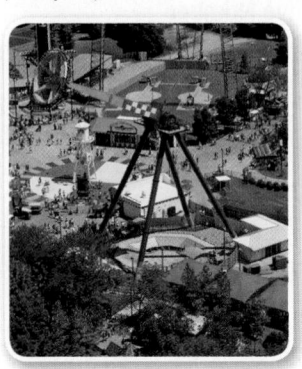

Write each complex number in exponential form. (Example 4)

21. $\sqrt{3} + i$

22. $\sqrt{3} - i$

23. $\sqrt{2} - \sqrt{2}i$

24. $-\sqrt{3} - i$

25. $1 - \sqrt{3}i$

26. $-1 + \sqrt{3}i$

27. $-\sqrt{2} + \sqrt{2}i$

28. $-1 - \sqrt{3}i$

Find the value of each natural logarithm in the complex number system. (Example 5)

29. $\ln(-6)$

30. $\ln(-3.5)$

31. $\ln(-2.45)$

32. $\ln(-7)$

33. $\ln(-4.36)$

34. $\ln(-9.12)$

35. **POWER SERIES** Use the power series representations of $\sin x$ and $\cos x$ to answer each of the following questions.

   a. Graph $f(x) = \sin x$ and the third partial sum of the power series representing $\sin x$. Repeat for the fourth and fifth partial sums. Describe the interval of convergence for each.

   b. Repeat part **a** for $f(x) = \cos x$ and the third, fourth, and fifth partial sums of the power series representing $\cos x$. Describe the interval of convergence for each.

   c. Describe how the interval of convergence changes as n increases. Then make a conjecture as to the relationship between each trigonometric function and its related power series as $n \rightarrow \infty$.

Solve for z over the imaginary numbers. Round to three decimal places.

36. $2e^z + 5 = 0$

37. $e^{2z} + 12 = 0$

38. $4e^{2z} + 7 = 6$

39. $3(e^z - 1) + 5 = -2$

40. $e^{2z} - e^z = 2$

41. $10e^{2z} + 17e^z = -3$

42. **ECONOMICS** The total value of an investment of P dollars compounded continuously at an annual interest rate of r over t years is Pe^{rt}. Use the first five terms of the exponential series to approximate the value of an investment of \$10,000 compounded continuously at 5.25% for 5 years.

43. **RELATIVE ERROR** *Relative error* is the absolute error in estimating a quantity divided by its true value. The relative error of an approximation a of a quantity b is given by $\dfrac{|b-a|}{b}$. Find the relative error in approximating $e^{2.1}$ using two, three, and six terms of the exponential series.

Approximate the value of each expression using the first four terms of the power series for sine and cosine. Then find the expected value of each.

44. $\sin^2 \frac{1}{2} + \cos^2 \frac{1}{2}$

45. $\sec^2 1 - \tan^2 1$

46. RAINBOWS *Airy's equation*, which is used in physics to model the diffraction of light, can also be used to explain how a light wave front is converted into a curved wave front in forming rainbows.

Raindrop

Light wave front

Cubic wave front

Rainbow ray

Reflected wave fronts

This equation can be represented by the power series shown below.

$$f(x) = 1 + \sum_{k=1}^{\infty} \frac{x^{3k}}{(2 \cdot 3)(5 \cdot 6) \cdots [(3k-1) \cdot (3k)]}$$

Use the fifth partial sum of the series to find $f(3)$. Round to the nearest hundredth.

47. ELECTRICITY When an electric charge is accompanied by an equal and opposite charge nearby, such an object is called an *electric dipole*. It consists of charge q at the point $x = d$ and charge $-q$ at $x = -d$, as shown below.

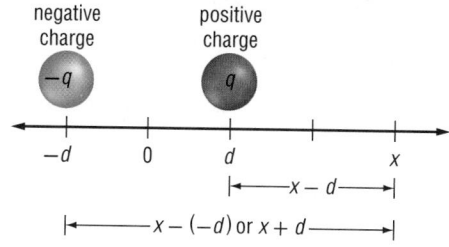

negative charge

positive charge

$-q$

q

$-d$ 0 d x

$\longmapsto x - d \longmapsto$

$\longmapsto x - (-d)$ or $x + d \longmapsto$

Along the x-axis, the electric field strength at x is the sum of the electric fields from each of the two charges. This is given by $E(x) = \frac{kq}{(x-d)^2} - \frac{kq}{(x+d)^2}$. Find a power series representing $E(x)$ if k is a constant and $d = 1$.

48. SOUND The *Fourier Series* represents a periodic function of time $f(t)$ as a summation of sine waves and cosine waves with frequencies that start at 0 and increase by integer multiples. The series below represents a sound wave from the digital data fed from a CD into a CD player.

$$f(t) = 0.7 + \sum_{n=1}^{\infty} \left(\frac{(-1)^n}{n} \cos 270.6nt + \frac{1}{2n-1} \sin 270.6nt \right)$$

Graph the series for $n = 4$. Then analyze the graph.

IDENTITIES Use power series representations from this lesson to verify each trigonometric identity.

49 $\sin(-x) = -\sin x$

50. $\cos(-x) = \cos x$

51. APPROXIMATIONS The infinite series for the inverse tangent function $f(x) = \tan^{-1} x$, is given by $\sum_{k=0}^{\infty} \frac{(-1)^k}{2k+1} x^{2k+1}$. However, this series is only valid for values of x on the interval $(-1, 1)$.

a. Write the first five terms of the infinite series representation for $f(x) = \tan^{-1} x$.

b. Use the first five terms of the series to approximate $\tan^{-1} 0.1$.

c. On the same coordinate plane, graph $f(x) = \tan^{-1} x$ and the third partial sum of the power series representing $f(x) = \tan^{-1} x$. On another coordinate plane, graph $f(x)$ and the fourth partial sum. Then graph $f(x)$ and the fifth partial sum.

d. Describe what happens on the interval $(-1, 1)$ and in the regions $x \geq 1$ or $x \leq -1$.

H.O.T. Problems Use **Higher-Order Thinking Skills**

52. WRITING IN MATH Describe how using additional terms in the approximating series for e^x affects the outcome.

53. REASONING Use the power series for sine to explain why, for x-values on the interval $[-0.1, 0.1]$, a close approximation of $\sin x$ is x.

54. CHALLENGE Prove that $2 \sin \theta \cos \theta = \dfrac{e^{2\theta i} - e^{-2\theta i}}{2i}$

55. REASONING For what values of α and β does $e^{i\alpha} = e^{i\beta}$? Explain.

PROOF Show that for all real numbers x, the following are true.

56. $\sin x = \dfrac{e^{ix} - e^{-ix}}{2i}$

57. $\cos x = \dfrac{e^{ix} + e^{-ix}}{2}$

58. CHALLENGE The hyperbolic sine and hyperbolic cosine functions are analogs of the trigonometric functions that you studied in Chapters 4 and 5. Just as the points $(\cos x, \sin x)$ form a unit circle, the points $(\cosh t, \sinh t)$ form the right half of an equilateral hyperbola. An equilateral hyperbola has perpendicular asymptotes. The hyperbolic sine (sinh) and hyperbolic cosine (cosh) functions are defined below. Find the power series representations for these functions.

$$\sinh x = \frac{e^x - e^{-x}}{2} \qquad\qquad \cosh x = \frac{e^x + e^{-x}}{2}$$

Use Pascal's triangle to expand each binomial. (Lesson 10-5)

59. $\left(3m + \sqrt{2}\right)^4$

60. $\left(\frac{1}{2}n + 2\right)^5$

61. $(p^2 + q)^8$

62. Prove that $4 + 7 + 10 + \cdots + (3n + 1) = \dfrac{n(3n + 5)}{2}$ for all positive integers n. (Lesson 10-4)

Find each power, and express it in rectangular form. (Lesson 9-5)

63. $(-2 + 2i)^3$

64. $\left(1 + \sqrt{3}i\right)^4$

65. $\left(\sqrt{2} + \sqrt{2}i\right)^{-2}$

66. Given $\mathbf{t} = \langle -9, -3, c \rangle$, $\mathbf{u} = \langle 8, -4, 3 \rangle$, $\mathbf{v} = \langle 2, 5, -6 \rangle$, and that the volume of the parallelepiped having adjacent edges $\mathbf{t}$, $\mathbf{u}$, and $\mathbf{v}$ is 93 cubic units, find c. (Lesson 8-5)

Use an inverse matrix to solve each system of equations, if possible. (Lesson 6-3)

67. $x - 8y = -7$
$2x + 5y = 28$

68. $4x + 7y = 22$
$-9x + 11y = 4$

69. $w + 2x + 3y = 18$
$4w - 8x + 7y = 41$
$-w + 9x - 2y = -4$

Determine whether A and B are inverse matrices. (Lesson 6-2)

70. $A = \begin{bmatrix} 1 & -2 \\ 7 & -6 \end{bmatrix}$, $B = \begin{bmatrix} -6 & 2 \\ -7 & 1 \end{bmatrix}$

71. $A = \begin{bmatrix} -11 & -5 \\ 9 & 4 \end{bmatrix}$, $B = \begin{bmatrix} 4 & 5 \\ -9 & -11 \end{bmatrix}$

72. $A = \begin{bmatrix} 6 & 2 \\ -2 & 8 \end{bmatrix}$, $B = \begin{bmatrix} -1 & 1 \\ -3 & -5 \end{bmatrix}$

73. CONFERENCE A university sponsored a conference for 680 women. The Venn diagram shows the number of participants in three of the activities offered. Suppose women who attended the conference were randomly selected for a survey. (Lesson 0-1)

a. What is the probability that a woman selected participated in hiking or sculpting?

b. Describe a set of women such that the probability of being selected is about 0.39.

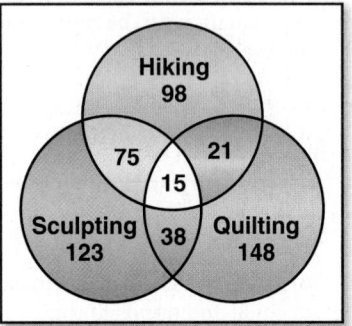

74. SAT/ACT $PQRS$ is a square. What is the ratio of the length of diagonal $\overline{QS}$ to the length of side $\overline{RS}$?

A 2

D $\dfrac{\sqrt{2}}{2}$

B $\sqrt{2}$

E $\dfrac{\sqrt{3}}{2}$

C 1

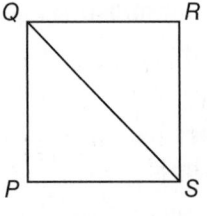

75. REVIEW What is the sum of the infinite geometric series $\dfrac{1}{3} + \dfrac{1}{6} + \dfrac{1}{12} + \dfrac{1}{24} + \cdots$?

F $\dfrac{2}{3}$

H $1\dfrac{1}{3}$

G 1

J $1\dfrac{2}{3}$

76. FREE RESPONSE Consider the pattern of dots shown.

a. Draw the next figure in this sequence.

b. Write the sequence, starting with 1, that represents the number of dots that must be added to each figure in the sequence to get the number of dots in the next figure.

c. Find the expression for the nth term of the sequence found in part **b**.

d. Find the expression for the number of dots in the nth figure in the original sequence.

e. Prove, through mathematical induction, that the sum of the sequence found in part **b** is equal to the expression found in part **d**.

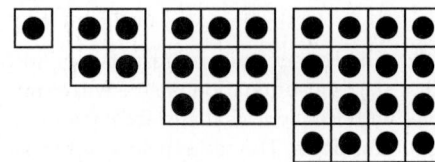

Chapter Summary

Key Concepts

Sequences, Series, and Sigma Notation (Lesson 10-1)

- A *finite sequence* is a sequence with a set number of terms. An *infinite sequence* has infinitely many terms.

- A sequence with a limit is said to *converge*. A sequence with no limit is said to *diverge*.

Divergent Sequence

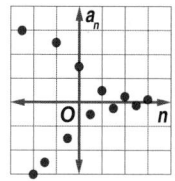
Convergent Series

- A *series* is the sum of all of the terms of a sequence.

$$\sum_{n=1}^{k} a_n = a_1 + a_2 + a_3 + a_4 + \cdots + a_k$$

Arithmetic Sequences and Series (Lesson 10-2)

- The *n*th term of an arithmetic sequence with first term a_1 and common difference *d* is given by $a_n = a_1 + (n-1)d$.

- The sum of a finite arithmetic series is given by $S_n = \frac{n}{2}(a_1 + a_n)$ or $S_n = \frac{n}{2}[2a_1 + (n-1)d]$.

Geometric Sequences and Series (Lesson 10-3)

- The *n*th term of a geometric sequence with first term a_1 and common ratio *r* is given by $a_n = a_1 r^{n-1}$.

- The sum of the first *n* terms of a geometric series is given by
$S_n = a_1\left(\frac{1-r^n}{1-r}\right)$ or $S_n = \frac{a_1 - a_n r}{1-r}$.

- The sum of an infinite geometric series is given by $S = \frac{a_1}{1-r}$, for $|r| < 1$.

Mathematical Induction (Lesson 10-4)

- If P_n is a statement about a positive integer *n*, then P_n is true for all positive integers *n* if and only if P_1 is true, and for every positive integer *k*, if P_k is true, then P_{k+1} is true.

The Binomial Theorem (Lesson 10-5)

- The expression $(a+b)^n$ can be expanded using the nth row of Pascal's triangle to determine the coefficients of each term.

- The binomial coefficient of the $a^{n-r}b^r$ term in the expansion of $(a+b)^n$ is given by $_nC_r = \frac{n!}{(n-r)!\,r!}$.

Functions as Infinite Series (Lesson 10-6)

- A power series in *x* is an infinite series of the form $\sum_{n=0}^{\infty} a_n x^n$.

- Euler's Formula states that for any real number θ, $e^{i\theta} = \cos\theta + i\sin\theta$.

Key Vocabulary

anchor step (p. 621)	geometric means (p. 611)
arithmetic means (p. 601)	geometric sequence (p. 608)
arithmetic sequence (p. 599)	geometric series (p. 611)
arithmetic series (p. 602)	inductive hypothesis (p. 621)
binomial coefficients (p. 628)	inductive step (p. 621)
Binomial Theorem (p. 632)	infinite sequence (p. 590)
common difference (p. 599)	infinite series (p. 593)
common ratio (p. 608)	*n*th partial sum (p. 593)
converge (p. 592)	power series (p. 636)
diverge (p. 592)	recursive formula (p. 591)
Euler's Formula (p. 640)	second difference (p. 601)
exponential series (p. 638)	sequence (p. 590)
Fibonacci sequence (p. 591)	series (p. 593)
finite sequence (p. 590)	sigma notation (p. 594)
finite series (p. 593)	term (p. 590)
first difference (p. 601)	

Vocabulary Check

State whether each sentence is *true* or *false*. If *false*, replace the underlined term to make a true sentence.

1. In mathematical induction, the assumption that a conjecture works for any particular case is called the <u>inductive hypothesis</u>.

2. A sequence that has a set number of terms is called an <u>infinite sequence</u>.

3. A sequence a_n defined as a function of *n* is defined <u>recursively</u>.

4. The step in which you show that something works for the first case is called the <u>inductive step</u> in mathematical induction.

5. <u>Explicitly</u> defined sequences give one or more of the first few terms and then define the terms that follow using those previous terms.

6. The sum of the first *n* terms of a finite or infinite sequence is called a <u>finite series</u>.

7. The difference between the terms of an arithmetic sequence is called the <u>common ratio</u>.

8. A <u>geometric series</u> is the sum of the terms of a geometric sequence.

9. If a sequence does not have a limit, it is said to <u>converge</u>.

10. The <u>Binomial Theorem</u> is a recursive sequence that describes many patterns found in nature.

Lesson-by-Lesson Review

10-1 Sequences, Series, and Sigma Notation (pp. 590–598)

Find the next four terms of each sequence.

11. $1, 9, 17, 25, \ldots$ **12.** $-1, 2, 7, 14, \ldots$

Find the indicated sum for each sequence.

13. fourth partial sum of $a_n = 2n - 10$

14. S_7 of $a_n = -n^3$

Find each sum.

15. $\displaystyle\sum_{n=2}^{9} \frac{4n - 6}{3}$ **16.** $\displaystyle\sum_{n=1}^{6} 7n - 4$

EXAMPLE 1

Find the sum of $\displaystyle\sum_{n=1}^{4} n^2 - 5$.

Find the values for a_n for $n = 1, 2, 3,$ and 4.

$a_1 = (1)^2 - 5$ or -4	$n = 1$
$a_2 = (2)^2 - 5$ or -1	$n = 2$
$a_3 = (3)^2 - 5$ or 4	$n = 3$
$a_4 = (4)^2 - 5$ or 11	$n = 4$

Therefore, $\displaystyle\sum_{n=1}^{4} n^2 - 5 = -4 + (-1) + 4 + 11$ or 10.

10-2 Arithmetic Sequences and Series (pp. 599–607)

Write an explicit formula and a recursive formula for finding the nth term of each arithmetic sequence.

17. $-6, -1, 4, \ldots$ **18.** $23, 15, 7, \ldots$

Find each sum.

19. 50th partial sum of $55 + 62 + 69 + \ldots + 398$

20. 37th partial sum of $9 + 3 + (-3) + \ldots$

Find each sum.

21. $\displaystyle\sum_{n=24}^{35} -3n - 11$ **22.** $\displaystyle\sum_{n=8}^{27} 4n + 14$

EXAMPLE 2

Find the 43rd partial sum of the arithmetic series $104 + 100 + 96 + \ldots$.

Use the second sum of a finite arithmetic series formula.

$S_n = \frac{n}{2}[2a_1 + (n - 1)d]$ **Sum of a finite arithmetic series formula**

$S_{43} = \frac{43}{2}[2(104) + (43 - 1)(-4)]$ **$n = 43$, $a_1 = 104$, and $d = -4$**

$= 860$ **Simplify.**

The 43rd partial sum is $S_{43} = 860$.

10-3 Geometric Sequences and Series (pp. 608–618)

Determine the common ratio, and find the next three terms of each geometric sequence.

23. $5, 7.5, 11.25, \ldots$

24. $3 + a, -12 - 4a, 48 + 16a, \ldots$

Write an explicit formula and a recursive formula for finding the nth term of each geometric sequence.

25. $10, -20, 40, \ldots$ **26.** $162, 54, 18, \ldots$

Find each sum.

27. first seven terms of $80 + 32 + \frac{64}{5} + \ldots$

28. $a_1 = 11, a_n = 360,448, r = 8$

EXAMPLE 3

Determine the common ratio, and find the next three terms of $27, -9, 3, \ldots$.

First, find the common ratio.

$a_2 \div a_1 = -9 \div 27$ or $-\frac{1}{3}$ **Find the ratio between two pairs of consecutive terms to verify the common ratio.**

$a_3 \div a_2 = 3 \div -9$ or $-\frac{1}{3}$

The common ratio is $-\frac{1}{3}$. Multiply the third term by $-\frac{1}{3}$ to find the fourth term, and so on.

$a_4 = 3 \cdot -\frac{1}{3}$ or -1 $a_5 = -1 \cdot -\frac{1}{3}$ or $\frac{1}{3}$

$a_6 = \frac{1}{3} \cdot -\frac{1}{3}$ or $-\frac{1}{9}$

The next three terms are $-1, \frac{1}{3},$ and $-\frac{1}{9}$.

Lesson-by-Lesson Review

10-4 Mathematical Induction (pp. 621–627)

Use mathematical induction to prove that each conjecture is true for all positive integers n.

29. $6^n - 9$ is divisible by 3.

30. $7^n - 5$ is divisible by 2.

31. $5^n + 3$ is divisible by 4.

32. $2 \cdot 3 + 4 \cdot 5 + 6 \cdot 7 + \cdots + 2n\,(2n + 1) = \dfrac{n(n + 1)(4n + 5)}{3}$

33. $\dfrac{1}{1 \cdot 2 \cdot 3} + \dfrac{1}{2 \cdot 3 \cdot 4} + \cdots + \dfrac{1}{n(n + 1)(n + 2)} = \dfrac{n(n + 3)}{4(n + 1)(n + 2)}$

Prove each inequality for the indicated values of n.

34. $4^n \geq 4n$, for all positive integers n

35. $5n < 6^n$, for all positive integers n

EXAMPLE 4

Prove that $5^n - 1$ is divisible by 4 for all positive integers n.

Conjecture and Anchor Step   Let P_n be the statement that $5^n - 1$ is divisible by 4. When $n = 1$, $5^n - 1 = 5^1 - 1$ or 4. Since 4 is divisible by 4, the statement P_1 is true.

Inductive Hypothesis and Step   Assume that $5^k - 1$ is divisible by 4. That is, assume that $5^k - 1 = 4r$ for some integer r. Use this inductive hypothesis to show that $5^{k+1} - 1$ is divisible by 4.

$5^k - 1 = 4r$	Inductive hypothesis
$5^k = 4r + 1$	Add 1 to each side.
$5 \cdot 5^k = 5(4r + 1)$	Multiply each side by 5.
$5^{k+1} = 20r + 5$	Simplify.
$5^{k+1} - 1 = 20r + 4$	Subtract 1 from each side.
$5^{k+1} - 1 = 4(5r + 1)$	Factor.

Since r is an integer, $5r + 1$ is an integer and $4(5r + 1)$ is divisible by 4. Therefore, $5^{k+1} - 1$ is divisible by 4.

Conclusion   Since P_n is true for $n = 1$ and P_k implies P_{k+1}, P_n is true for $n = 2$, $n = 3$, and so on. By the principle of mathematical induction, $5^n - 1$ is divisible by 4 for all positive integers n.

10-5 The Binomial Theorem (pp. 628–635)

Use Pascal's triangle to expand each binomial.

36. $(4x + 6)^5$

37. $(m - 5n)^6$

Find the coefficient of the indicated term in each expansion.

38. $(6x - 3y)^{10}$, x^4y^6 term

39. $(2y + 3)^{13}$, 8th term

Use the Binomial Theorem to expand each binomial.

40. $(2p^2 - 7)^4$

41. $(4m + 3n)^7$

EXAMPLE 5

Use the Binomial Theorem to expand $(3x + 10)^5$.

Apply the Binomial Theorem to expand $(a + b)^5$, where $a = 3x$ and $b = 10$.

$(3x + 10)^5 = {}_5C_0(3x)^5(10)^0 + {}_5C_1(3x)^4(10)^1 + {}_5C_2(3x)^3(10)^2 +$
$\qquad {}_5C_3(3x)^2(10)^3 + {}_5C_4(3x)^1(10)^4 + {}_5C_5(3x)^0(10)^5$

$= 1(243x^5)(1) + 5(81x^4)(10) + 10(27x^3)(100) +$
$\qquad 10(9x^2)(1000) + 5(3x)(10{,}000) + 1(1)(100{,}000)$

$= 243x^5 + 4050x^4 + 27{,}000x^3 + 90{,}000x^2 +$
$\qquad 150{,}000x + 100{,}000$

Lesson-by-Lesson Review

10-6 Functions as Infinite Series (pp. 636–644)

Use $\sum_{n=0}^{\infty} x^n$ to find a power series representation of $g(x)$. Indicatetnt the interval on which the series converges. Use a graphing calculator to graph $g(x)$ and the 6th partial sum of its power series.

42. $g(x) = \dfrac{1}{1 - 5x}$

43. $g(x) = \dfrac{3}{1 - 2x}$

Use the fifth partial sum of the exponential series to approximate each value. Round to three decimal places.

44. $e^{\frac{1}{4}}$

45. $e^{-1.5}$

Find the value of each natural logarithm in the complex number system.

46. $\ln(-4)$

47. $\ln(-7.15)$

EXAMPLE 6

Use $\sum_{n=0}^{\infty} x^n$ to find a power series representation of $g(x) = \dfrac{4}{1 - x}$. Indicate the interval on which the series converges.

A geometric series converges to $f(x) = \dfrac{1}{1 - x} = \sum_{n=0}^{\infty} x^n$ for $|x| < 1$. Replace x with $\dfrac{x + 3}{4}$ since $g(x)$ is a transformation of $f(x)$ and $g(x) = f\left(\dfrac{x + 3}{4}\right)$. The result is

$$f\left(\dfrac{x + 3}{4}\right) = \sum_{n=0}^{\infty} \left(\dfrac{x + 3}{4}\right)^n \text{ for } \left|\dfrac{x + 3}{4}\right| < 1.$$

Therefore, $g(x) = \dfrac{4}{1 - x}$ can be represented by

$\sum_{n=0}^{\infty} \left(\dfrac{x + 3}{4}\right)^n$. This series converges for $\left|\dfrac{x + 3}{4}\right| < 1$, which is equivalent to $-1 < \dfrac{x + 3}{4} < 1$ or $-7 < x < 1$.

Applications and Problem Solving

48. RETAIL A chain of retail coffee stores has a business plan that includes opening 6 new stores nationwide annually. If there were 480 stores on January 1, 2005, how many stores will there be at the end of the year in 2012? (Lesson 10-1)

49. DANCE Mara's dance team is performing at a recital. In one formation, there are three dancers in the front row and two additional dancers in each row behind the first row. (Lesson 10-2)

   a. How many dancers are in the fourth row?

   b. If there are eight rows, how many members does the dance team have?

50. AGRICULTURE The population of a herd of cows increases as shown over the course of four years. (Lesson 10-3)

Year	1	2	3	4
Population	47	51	56	61

   a. Write an explicit formula for finding the population of the herd after n years. Assume that the sequence shown above is geometric.

   b. What will the population of the herd be after 7 years?

   c. About how many years will it take the population of the herd to exceed 85?

51. NUMBER THEORY Consider the statement $0.\overline{9} = 1$. (Lesson 10-4)

   a. Prove that $0.9 + 0.09 + 0.009 + \cdots + \dfrac{9}{10^n} = \dfrac{10^n - 1}{10^n}$ for any positive integer n.

   b. Use your understanding of limits and the statement you proved in part **a** to explain why $0.\overline{9} = 1$ is true.

52. BASKETBALL Julie usually makes 4 out of every 6 free throws that she attempts. What is the probability that Julie will make 5 out of 6 of the next free throws that she attempts? (Lesson 10-5)

53. HEIGHT Lina is estimating the height of a tree. She stands 30 feet from the base and estimates that her angle of sight to the top of the tree is 40°. If she uses the fifth partial sum of the trigonometric series for cosine and sine approximated to three decimal places to calculate the height of the tree, what is Lina's estimate? (Lesson 10-6)

Find the specified term of each sequence.

1. ninth term, $a_n = \dfrac{n^3}{n+3}$

2. sixth term, $a_1 = 156$, $a_n = \dfrac{a_{n-1} - 4}{2}$

Find the indicated sum for each series.

3. fifth partial sum of $a_n = 3^n + 4$

4. S_8 of $-2, 3, 8, 13, \ldots$

Find the indicated arithmetic means for each set of nonconsecutive terms.

5. 3 means; -10 and -2

6. 4 means; -4 and 56

7. **EXERCISE** While training for a 5-mile fun run, Kirsten ran 1 mile each workout during the first week of training and added 0.5 mile per week to each run so that she ran 5 miles each workout the week before the run. If Kirsten had training three times per week, how many miles did she run during her training?

8. **FIREWORKS** If the year to year increase in attendance at the fireworks show is constant, what was the attendance each year from 2003 to 2007?

Wednesday   July 2008

Show a success

Attendance at annual fireworks show increases from 12,000 in 2002 to 13,121 in 2008.

If possible, find the sum of each infinite geometric series.

9. $\dfrac{4}{10}, \dfrac{4}{5}, \dfrac{8}{5}, \ldots$

10. $\displaystyle\sum_{n=3}^{\infty} -2(0.6)^{n-1}$

11. **GEOMETRY** The measure of each interior angle a of a regular polygon with n sides is $a_n = \dfrac{180(n-2)}{n}$, where $n \geq 3$.

$n = 3$ $n = 4$ $n = 5$

  a. Find the measure of an exterior angle of a regular triangle, square, pentagon, and hexagon.

  b. Write a general formula for the measure of an exterior angle of a regular polygon with n sides.

  c. Determine whether the sequence is convergent or divergent. Does this make sense in the context of the situation? Explain your reasoning.

Prove each inequality for the indicated values of n.

12. $n! > 5^n$, $n \geq 12$

13. $4n < 7^n$, $n \geq 1$

14. $3^n > n + 8$, $n \geq 3$

15. $3n < \left(\dfrac{5}{2}\right)^n$, $n \geq 2$

Use Pascal's triangle to expand each binomial.

16. $(2x + 3y)^4$

17. $(x - 6)^7$

18. **MULTIPLE CHOICE** Ellis' basketball scoring statistics are shown below. Based on his statistics, what is the probability that he will score a 2- or 3-point field goal on 3 of his next 7 shots?

Shots on Goal	2-point Shots Scored	3-point Shots Scored
20	11	3

A 0.10 %

B 0.24 %

C 9.72 %

D 23.88 %

Use the Binomial Theorem to expand and simplify each expression.

19. $(x - 4y)^4$

20. $(3a + b^3)^5$

Use $\displaystyle\sum_{n=0}^{\infty} x^n$ to find a power series representation of $g(x)$. Indicate the interval on which the series converges. Use a graphing calculator to graph $g(x)$ and the 6th partial sum of its power series.

21. $g(x) = \dfrac{3}{1-x}$

22. $g(x) = \dfrac{2}{1-4x}$

23. **SKATEBOARDING** A skateboarding ramp has an incline of 24°. Use the 5th partial sum of the trigonometric series for cosine or sine approximated to three decimal places to find the length of the ramp.

ℓ 3.74 ft   24°   8.4 ft

Write each complex number in exponential form.

24. $-\sqrt{3} - i$

25. $1 - \sqrt{3}i$

Objective
Develop notation for approximating the area of a region bound by a curve and the x-axis.

In Chapter 2, you learned to approximate the area between a curve and the x-axis. You divided the area into rectangles, found the area of each individual rectangle, and then calculated the sum of the areas. In calculus, this process is assigned special notation and is studied further in an effort to calculate exact areas. We will analyze the components of this process to better understand the notation.

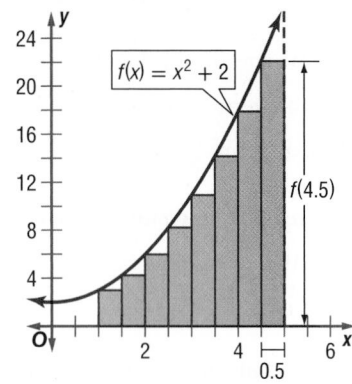

The area A of the region shown above can be approximated as follows.

$$A = 0.5 \cdot f(1) + 0.5 \cdot f(1.5) + 0.5 \cdot f(2.0) + 0.5 \cdot f(2.5) + 0.5 \cdot f(3.0) + 0.5 \cdot f(3.5) + 0.5 \cdot f(4.0) + 0.5 \cdot f(4.5)$$

Notice that we can factor out the width.

$$A = \underbrace{0.5}_{\text{width}} \cdot \underbrace{[f(1) + f(1.5) + f(2.0) + f(2.5) + f(3.0) + f(3.5) + f(4.0) + f(4.5)]}_{\text{sum of the heights}}$$

The approximation is equal to the product of the width of the rectangles and the sum of their heights. We will examine both of these components separately.

The first component used to approximate the area of a region is the width of the rectangles. The width of the rectangles, denoted Δx, is the difference between the left endpoint and the right endpoint of a rectangle, such as $2.5 - 2$ or 0.5. Generally, we are not given any of the x-coordinates of our rectangles. Instead, we get the *lower bound a* and the *upper bound b* of the interval $[a, b]$ and the number of rectangles n.

ACTIVITY 1 — **Find Δx**

Find Δx if we want to approximate the area between the graph of $f(x) = -x^2 + 5x$ and the x-axis on the interval [1, 4] using 6, 12, and 24 rectangles.

Step 1 Find the total length of the interval by calculating $b - a$.

Step 2 Divide the answer from Step 1 by 6.

Step 3 Repeat Step 2 for 12 and 24 rectangles.

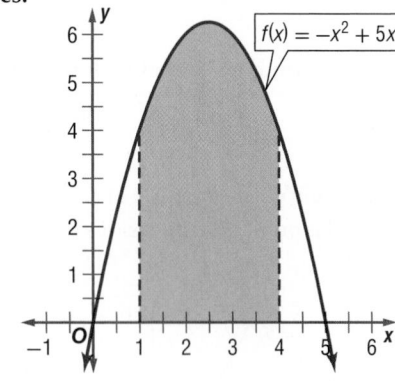

▶ **Analyze the Results**

1. As the number of rectangles increases, what is happening to Δx? How would this affect your approximation for the area?

2. Find Δx if we want to approximate the area between a curve and the x-axis on the interval $[a, b]$ using n rectangles.

3. As n approaches ∞, what is happening to Δx?

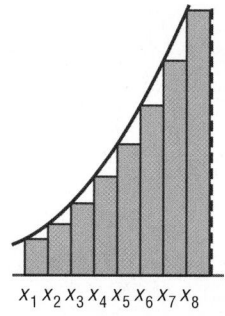

$x_1\,x_2\,x_3\,x_4\,x_5\,x_6\,x_7\,x_8$

Figure 10.CAP.1

The second component needed to approximate the area of a region is the sum of the heights of the rectangles. The sum of the heights resembles the sum of a series. For the example presented at the beginning of this lesson, this sum is

$$f(1) + f(1.5) + f(2.0) + f(2.5) + f(3.0) + f(3.5) + f(4.0) + f(4.5).$$

Since there were 8 rectangles, $f(x)$ is evaluated for 8 values of x. We can write this series as

$$f(x_1) + f(x_2) + f(x_3) + f(x_4) + f(x_5) + f(x_6) + f(x_7) + f(x_8),$$

where $x_1, x_2, \ldots, x_8$ are the x-coordinates used to find the heights of the rectangles, as shown in the figure. We can represent this series using sigma notation as $\sum_{i=1}^{8} f(x_i)$.

For example, the sum of the heights for $f(x)$ can be written in *expanded form* as

$$\sum_{i=1}^{8} f(x_i) = \overset{x_1}{f(1)} + \overset{x_2}{f(1.5)} + \overset{x_3}{f(2.0)} + \overset{x_4}{f(2.5)} + \overset{x_5}{f(3.0)} + \overset{x_6}{f(3.5)} + \overset{x_7}{f(4.0)} + \overset{x_8}{f(4.5)}.$$

In general, the sum of the heights for n rectangles can be described as $\sum_{i=1}^{n} f(x_i)$.

You now have the two components for approximating the area of a region using n rectangles. We can multiply our width by our expression for the sum of the heights to develop the notation $\sum_{i=1}^{n} f(x_i) \cdot \Delta x$. This expression is called a *Riemann sum*.

StudyTip

Notation When having to evaluate the sum, it may be easier to represent the approximation for the area of a region as

$$\Delta x \sum_{i=1}^{n} f(x_i).$$

ACTIVITY 2   Approximate Area Under a Curve

Approximate the area between the graph of $f(x) = -x^2 + 5x$ and the x-axis on the interval [1, 4] using 6 rectangles. Let the left endpoint of each rectangle represent the height.

Step 1 Let $a = 1$, $b = 4$, and $n = 6$. Calculate Δx.

Step 2 Write the approximation in sigma notation. Substitute the value found in Step 1 for Δx and let $n = 6$.

$$\sum_{i=1}^{6} f(x_i) \cdot 0.5$$

Step 3 Write the expression in expanded form.
$$0.5\,[f(x_1) + f(x_2) + f(x_3) + f(x_4) + f(x_5) + f(x_6)]$$

Step 4 Find each value for x. x_1 will start at 1. Each successive value for x can be found by adding Δx to each previous value. For example, $x_2 = 1 + 0.5$, $x_3 = x_2 + 0.5$, and so on.

Step 5 Calculate the area of the rectangles.

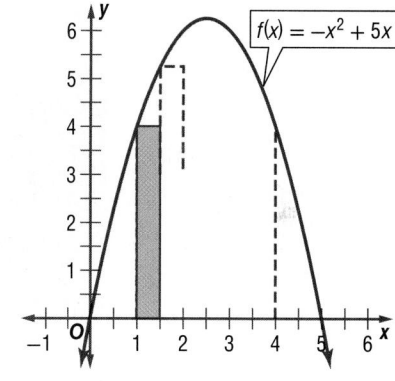

▶ **Analyze the Results**

4. If n increased to 12, how would it change the expression in expanded form? What would happen to the x-values found in Step 4?

5. As n approaches ∞, what happens to the calculation?

Model and Apply

Given n and an interval $[a, b]$, find Δx. Then write the approximation for finding the area between the graph of $f(x) = -x^2 + 10x$ and the x-axis in sigma notation. Calculate the area. Let the left endpoint of each rectangle represent the height.

6. $n = 4$; [1, 2]   7. $n = 10$; [6, 10]   8. $n = 24$; [3, 9]

CHAPTER
11
Inferential Statistics

Then

In Chapter 0, you found measures of center and spread and organized statistical data.

Now

In Chapter 11, you will:

- Use the shape of a distribution to select appropriate descriptive statistics.
- Construct and use probability distributions.
- Use the Central Limit Theorem.
- Find and use confidence intervals, and perform hypothesis testing.
- Analyze and predict using bivariate data.

Why?

🌐 **ENVIRONMENTAL ENGINEERING** Statistics are extremely important in engineering. In environmental engineering, hypothesis testing can be used to determine if a change in an emission level for a chemical has a significant impact on overall pollution. Also, confidence intervals can be used to help suggest restrictions on by-product wastes in ground water.

PREREAD Scan the Study Guide and Review and use it to make two or three predictions about what you will learn in Chapter 11.

Get Ready for Chapter 11

Diagnose Readiness You have two options for checking Prerequisite Skills.

Text Option

Take the Quick Quiz below.

*Quick*Check

Find each value. (Lesson 0-7)

1. $_5P_2$ **2.** $_9P_4$ **3.** $_8C_3$

4. INTERNET The table shows the survey results of 18 high school students who were asked how many hours they spent on the Internet the previous week. (Lesson 0-8)

Hours Spent on the Internet

2	3.5	1	8	2.5	7.5
10	4	5.5	3.5	7.5	1.5
4.5	11	3.5	5	8	6.5

a. Make a histogram of the data.

b. Were there more students on the Internet for fewer than 3 hours or more than 6 hours?

For Exercises 5 and 6, complete each step.

a. Linearize the data according to the given model.

b. Graph the linearized data, and find the linear regression equation.

c. Use the linear model to find a model for the original data. (Lesson 3-5)

5. exponential

x	y
0	11.1
1	40.7
2	149.5
3	548.4
4	2012.1
5	7383.1

6. quadratic

x	y
0	2.0
1	0.9
2	6.0
3	17.3
4	34.8
5	58.5

Online Option

Math Online Take a self-check Chapter Readiness Quiz at **glencoe.com**.

New Vocabulary

English	• Page •	Español
percentiles	• 658 •	percentiles
random variable	• 664 •	variable aleatoria
probability distribution	• 665 •	distribución probabilística
binomial distribution	• 669 •	distribución binomial
normal distribution	• 674 •	distribución normal
z-value	• 676 •	valor de z
standard error of the mean	• 685 •	error estándar del media
inferential statistics	• 696 •	inferencia estadística
confidence level	• 696 •	nivel de confianza
critical values	• 697 •	valores críticos
confidence interval	• 697 •	intervalo de confianza
t-distribution	• 699 •	distribución t
hypothesis test	• 705 •	prueba de hipótesis
level of significance	• 706 •	nivel de significancia
p-value	• 708 •	valor p
correlation coefficient	• 713 •	coeficiente de correlación
regression line	• 716 •	recta de regresión
residual	• 716 •	residual

Review Vocabulary

statistics • p. P33 • estadística   the science of collecting, analyzing, interpreting, and presenting data

histogram • p. P35 • histograma   numerical data organized into equal intervals and displayed using bars

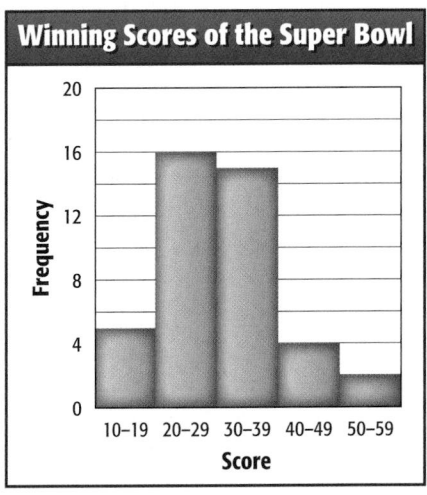

Descriptive Statistics

Why?

A high school newspaper reports that according to a random survey of students, the mean and median number of unexcused tardies received by students last year were 7 and 5, respectively. While both of these values can be used to describe the center of the survey data, only a graph of the data can reveal which measure best represents the typical number of student tardies.

Describing Distributions In Lesson 0-8, you described distributions of **univariate** or one-variable data numerically. You did this by calculating and reporting a distribution's

- center using either the mean or median and
- spread or variability using either the standard deviation or five-number summary (quartiles).

To determine which summary statistics you should choose to best describe the center and spread of a data set, you must identify the shape of the distribution. Three common shapes are given below.

Key Concept Symmetric and Skewed Distributions

Negative or Left-Skewed Distribution	Symmetrical Distribution	Positive or Right-Skewed Distribution
		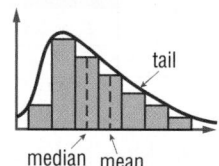
In a **negatively skewed distribution**, the mean is less than the median, the majority of the data is on the right, and the tail extends to the left.	In a **symmetrical distribution**, the data are evenly distributed on both sides of the mean. The mean and median are approximately equal.	In a **positively skewed distribution**, the mean is greater than the median, the majority of the data is on the left, and the tail extends to the right.

When a distribution is reasonably symmetrical, the mean and median are close together. In skewed distributions, however, the mean is located closer to the tail than the median. Outliers, which are extremely high or low values in a data set, will cause the mean to drift even farther toward the tail. The median is less affected by the presence of outliers. For these reasons, the median is called a **resistant statistic** and the mean a *nonresistant statistic*.

Since standard deviation measures the spread of a distribution by how far data values are from the mean, this statistic is also nonresistant to the effects of outliers. This leads to the following guidelines about choosing summary statistics to describe a distribution.

Key Concept Choosing Summary Statistics

When choosing measures of center and spread to describe a distribution, first examine the shape of the distribution.

- If the distribution is reasonably symmetrical and free of outliers, use the mean and standard deviation.
- If the distribution is skewed or has strong outliers, the five-number summary (minimum, quartile 1, median, quartile 3, maximum) usually provides a better summary of the overall pattern in the data.

When identifying the shape of a distribution, focus on major peaks in the graph instead of minor ups and downs.

EXAMPLE 1 **Skewed Distribution**

REAL ESTATE The table shows the selling prices for a sample of new homes in a community.

New Home Selling Prices (thousands of dollars)				
248	219	234	250	225
299	205	212	215	245
257	228	221	233	212
220	213	231	212	266
238	249	292	223	235
218	227	209	242	217

a. Construct a histogram and use it to describe the shape of the distribution.

On a graphing calculator, press [STAT] EDIT and input the data into L1. Then turn on **Plot1** under the **STAT PLOT** menu and choose ▟█. Graph the histogram by pressing **ZoomStat** or by pressing **Graph** and adjusting the window manually.

The graph shown has a single peak. Using the **TRACE** feature, you can determine that this peak represents selling prices from $210 to $220 thousand.

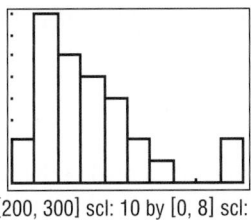
[200, 300] scl: 10 by [0, 8] scl: 1

[200, 300] scl: 10 by [0, 8] scl: 1

The graph is positively skewed. Most of the selling prices appear to fall between $210 and $250 thousand, but a few were much higher, so the tail of the distribution trails off to the right.

b. Summarize the center and spread of the data using either the mean and standard deviation or the five-number summary. Justify your choice.

Since the distribution is skewed, use the five-number summary instead of the mean and standard deviation to summarize the center and spread of the data. To display this summary, press [STAT], select 1-Var Stats under the **Calc** submenu, and scroll down.

```
1-Var Stats
↑n=30
 minX=205
 Q1=217
 Med=227.5
 Q3=245
 maxX=299
```

The five-number summary (minX, Q1, Med, Q3, and maxX) indicates that while the prices range from $205 to $299 thousand, the median selling price was $227.5 thousand and half of the prices were between $217 and $245 thousand.

☑ **Guided Practice**

1. **LAB GRADES** The laboratory grades of all of the students in a biology class are shown.

   A. Construct a histogram, and use it to describe the shape of the distribution.

   B. Summarize the center and spread of the data using either the mean and standard deviation or the five-number summary. Justify your choice.

Lab Grades (percent)					
72	84	67	80	75	87
86	76	89	91	96	74
68	83	80	76	63	98
92	73	80	88	94	78

Watch Out!

Skew Direction The tail of the distribution indicates in which direction a distribution is skewed, not the peak.

TechnologyTip

Bin Width On a graphing calculator, each bar is called a *bin*. The bin widths are chosen by the calculator when you use the ZoomStat feature. The bin width can be adjusted by changing the Xscl parameter under Window. A bin width that is too narrow or too wide will affect the apparent shape of a distribution.

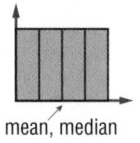
Distributions of data are not always symmetrical or skewed. Sometimes data will fall into subgroups or **clusters**. If a distribution has a gap in the middle, two separate clusters of data may result. A distribution of data that has two modes, and therefore two peaks, is known as a **bimodal distribution**.

In data that represent a reported preference about a topic, a bimodal distribution can indicate a polarization of opinions. Often, however, a bimodal distribution indicates that the sample data comes from two or more overlapping distributions.

EXAMPLE 2 | Bimodal Distribution

TUITION The annual cost of tuition for a sample of 20 colleges at a college fair are shown.

College Tuition Costs ($)				
32,000	10,100	31,000	11,000	31,500
5500	35,000	10,800	3600	11,500
7400	15,100	18,200	25,600	33,100
36,200	32,000	30,400	14,300	12,400

a. **Construct a histogram, and use it to describe the shape of the distribution.**

The histogram of the data has not one but two major peaks. Therefore, the distribution is neither symmetrical nor skewed but bimodal. The two separate clusters suggest that two types of colleges are mixed in the data set. It is likely that the 11 colleges with less expensive tuitions are public colleges, and the 9 colleges with more expensive tuitions are private colleges.

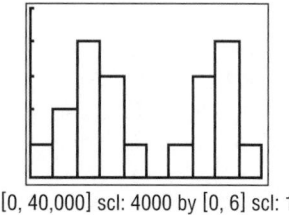

[0, 40,000] scl: 4000 by [0, 6] scl: 1

b. **Summarize the center and spread of the data using either the mean and standard deviation or the five-number summary. Justify your choice.**

Since the distribution is bimodal, an overall summary of center and spread would give an inaccurate depiction of the data. Instead, summarize the center and spread of each cluster. Since each cluster appears fairly symmetrical, enter each cluster separately and summarize the data using the mean and standard deviation of each cluster.

```
1-Var Stats
x̄=10900
Σx=119900
Σx²=1487370000
Sx=4248.05838
σx=4050.364742
↓n=11
```

```
1-Var Stats
x̄=31866.66667
Σx=286800
Σx²=9211820000
Sx=3009.568075
σx=2837.447993
↓n=9
```

The mean cost of Cluster 1 is $10,900 with a standard deviation of about $4050, while the mean cost of the Cluster 2 is $31,866 with a standard deviation of about $2837.

✔ Guided Practice

2. **TRACK** The numbers of minutes that 30 members of a high school cross-country team ran during a practice session are shown.

Practice Session Times (min)									
26	36	31	58	51	29	56	23	61	46
30	50	45	22	64	49	34	42	53	55
41	37	28	54	32	50	59	48	62	39

A. Construct a histogram, and use it to describe the shape of the distribution.

B. Summarize the center and spread of the data using either the mean and standard deviation or the five-number summary. Justify your choice.

You can also examine a box-and-whisker plot or *box plot* of a set of data to identify the shape of a distribution. To determine symmetry or skewness from a box plot, you must consider both the position of the line representing the median and the length of each "whisker."

Key Concept | **Symmetric and Skewed Box Plots**

Negatively Skewed	Symmetrical	Positively Skewed
The left whisker is longer than right whisker, and the line representing the median is closer to Q_3 than to Q_1.	The whiskers are the same length, and the line representing the median is exactly between Q_1 and Q_3.	The right whisker is longer than left whisker, and the line representing the median is closer to Q_1 than to Q_3.

EXAMPLE 3 | **Describe a Distribution Using a Box Plot**

POPULATION The table shows the populations, in thousands of people, during a recent year for fifteen cities in Florida.

Population (Thousands)				
151	95	303	89	186
362	137	109	152	118
102	226	139	736	248

a. **Construct a box plot, and use it to describe the shape of the distribution.**

Input the data into L1 on a graphing calculator. Then turn on **Plot1** under the **STAT PLOT** menu and choose ⊡. Graph the box plot by pressing **ZoomStat** or by pressing **Window** and adjusting the window manually.

Since the right whisker is longer than the left whisker and the line representing the median is closer to Q_1 than to Q_3, the distribution is positively skewed. Notice that the distribution has an outlier at 736.

[0, 800] scl: 50 by [0, 1] scl: 0.5

Review Vocabulary

interquartile range the difference between the upper quartile and lower quartile of a data set (Lesson 0-8)

b. **Summarize the center and spread of the data using either the mean and standard deviation or the five-number summary. Justify your choice.**

Since the distribution is skewed, use the five-number summary. This summary indicates that while the populations ranged from 89,000 to 736,000, the median population was 151,000. Populations in the middle half of the data varied by 248,000 – 109,000 or 139,000 people, which is the interquartile range.

```
1-Var Stats
↑n=15
 minX=89
 Q₁=109
 Med=151
 Q₃=248
 maxX=736
```

✓ **Guided Practice**

3. **TRUCKS** The costs on a used car web site for twelve trucks that are the same make, model, and year are shown.

   A. Construct a box plot, and use it to describe the shape of the distribution.

   B. Summarize the center and spread of the data using either the mean and standard deviation or the five-number summary. Justify your choice.

Used Truck Costs ($)		
9000	8200	9200
7800	8900	8500
6500	7500	7800
8000	6400	5500

Measures of Position The quartiles given by the five-number summary specify the positions of data values within a distribution. For this reason, box plots are most useful for side-by-side comparisons of two or more distributions.

EXAMPLE 4 | **Compare Position Using Box Plots**

BASKETBALL The number of games won by the Boston Celtics during two different 15-year periods are shown. Construct side-by-side box plots of the data sets. Then use this display to compare the distributions.

1st 15-Year Period				
49	52	59	57	60
60	54	62	59	58
54	48	34	44	56

2nd 15-Year Period				
48	32	35	33	15
36	19	36	35	49
44	36	45	33	24

Input the data into L1 and L2. Then turn on Plot1 and Plot2 under the STAT PLOT menu, choose ⊡, and graph the box plots by pressing ZoomStat or by pressing Graph and adjusting the window manually.

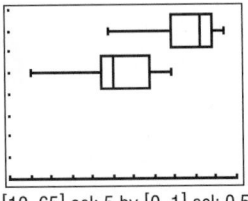

[10, 65] scl: 5 by [0, 1] scl: 0.5

Compare Measures of Position

The median number of games won each season in the first 15-year period is greater than those won every season in the second 15-year period. The first quartile for the first 15-year period is approximately equal to the maximum value for the second 15-year period. This means that 75% of the data values for the first 15-year period are greater than any of the values in the second 15-year period. Therefore, we can conclude that the Celtics had significantly more successful seasons during the first 15-year period than during the second 15-year period.

Compare Spreads

The spread of the middle half of the data, represented by the box, is roughly the same in each distribution. Therefore, the variability in the number of games won each season in those 15-year periods was about the same.

✓ Guided Practice

4. **BASEBALL** The number of home runs hit in Major League Baseball in 1927 and 2007 by the top 20 home run hitters is shown. Construct side-by-side box plots of the data sets. Then use this display to compare the distributions.

1927				
19	10	13	30	16
14	18	14	12	60
30	14	47	14	15
12	20	18	26	17

2007				
40	32	47	31	34
34	33	35	30	46
32	54	31	33	32
36	31	34	50	35

StudyTip

Fractiles Quartiles and percentiles are two types of *fractiles*—numbers that divide an ordered set of data into equal groups. *Deciles* divide a data set into ten equal groups.

In addition to quartiles, you can also use *percentiles* to indicate the relative position of an individual value within a data set. **Percentiles** divide a distribution into 100 equal groups and are symbolized by $P_1, P_2, P_3, ..., P_{99}$. The nth percentile or P_n, is the value such that n% of the data are less than P_n and $(100 - n)$% of the data are equal to or greater than P_n. The highest percentile that a data value can be is the 99th percentile.

A **percentile graph** uses the same values as a cumulative relative frequency graph, except that the proportions are instead expressed as percents.

You can use a percentile graph to approximate the percentile rank of a given value for a variable.

EXAMPLE 5 | **Construct and Use a Percentile Graph**

GPA The table gives the frequency distribution of the GPAs of the 200 students at Ashlyn's high school.

Class Boundaries	f	Class Boundaries	f
2.00–2.25	10	3.00–3.25	36
2.25–2.50	28	3.25–3.50	32
2.50–2.75	30	3.50–3.75	26
2.75–3.00	32	3.75–4.00	6

a. Construct a percentile graph of the data.

First, find the cumulative frequencies. Then find the cumulative percentages by expressing the cumulative frequencies as percents. The calculations for the first two classes are shown.

Class Boundaries	f	Cumulative Frequency	Cumulative Percentages
2.00–2.25	10	10	$\frac{10}{200}$ or 5%
2.25–2.50	28	10 + 28 or 38	$\frac{38}{200}$ or 19%
2.50–2.75	30	68	34%
2.75–3.00	32	100	50%
3.00–3.25	36	136	68%
3.25–3.50	32	168	84%
3.50–3.75	26	194	97%
3.75–4.00	6	200	100%

Finally, graph the data with the class boundaries along the *x*-axis and the cumulative percentages along the *y*-axis, as shown.

b. Estimate the percentile rank a GPA of 3.4 would have in this distribution, and interpret its meaning.

Find 3.4 on the *x*-axis and draw a vertical line to the graph. This point on the graph corresponds to approximately the 78th percentile. Therefore, a student with a GPA of 3.4 has a better grade-point average than about 78% of the students at Ashlyn's school.

✔ Guided Practice

5. HEIGHT The table gives the frequency distribution of the heights of girls in Mr. Lee's precalculus classes.

  A. Construct a percentile graph of the data.

  B. Estimate the percentile rank a girl with a height of 68 inches would have in this distribution, and interpret its meaning.

Class Boundaries	Frequency (f)
58.5–61.5	11
61.5–64.5	15
64.5–67.5	15
67.5–70.5	12
70.5–73.5	7

Watch Out!

Percentage Versus Percentile *Percentages* are not the same as *percentiles*. If a student gets 85 problems correct out of a possible 100, he obtains a percentage score of 85. This does not indicate whether the grade was high or low compared to the rest of the class.

Watch Out!

Understanding Percentiles Saying that a girl's height is at the 75th percentile does *not* mean that her height is 75% of some ideal height. Instead, her height is greater than 75% of all girls in the precalculus class.

For Exercises 1–4, complete each step.

a. Construct a histogram, and use it to describe the shape of the distribution.

b. Summarize the center and spread of the data using either the mean and standard deviation or the five-number summary. Justify your choice. (Examples 1 and 2)

1. **AVIATION** The landing speeds in miles per hour of 20 commercial airplane flights at a certain airport are shown.

Landing Speeds (mph)			
150	157	153	145
155	158	158	162
149	142	138	154
156	161	146	148
158	144	151	152

2. **COMPUTERS** The retail prices of laptop and desktop computers at a certain electronics store are shown.

Computer Prices ($)		
950	1000	975
1150	450	1075
675	1250	540
1025	1180	925
580	950	890

3. **BOWLING** Bowling scores range between 0 and 300. The scores for randomly selected players at a certain bowling alley are shown.

Bowling Scores				
116	81	234	173	75
61	205	92	219	156
134	259	273	53	241
105	190	94	127	235
228	248	271	46	112
99	223	142	217	68

4. **SALARIES** The starting salary for an employee at a certain new company ranges between $20,000 and $90,000. Starting salary depends in part on the employee's years of previous experience and the level of the position for which they were hired. The starting salaries for all the company's new-hires last year are shown.

Salaries (thousands of dollars)			
24	40	34	59
48	52	65	54
68	26	85	32
36	42	33	45
38	89		

For Exercises 5–6, complete each step.

a. Construct a box plot, and use it to describe the shape of the distribution.

b. Summarize the center and spread of the data using either the mean and standard deviation or the five-number summary. Justify your choice. (Example 3)

5. **VIDEO GAMES** The amount of time that a sample of students at East High School spends playing video games each week is shown.

Time Spent Playing Video Games (hours)					
1.5	2.5	0	4.5	12.5	1
2.5	4	2	8.5	1.5	9
1	0	2	1.5	5.5	2

6. **ACT** The students in Mrs. Calhoun's homeroom class recently took the ACT. The score for each student is shown.

ACT Scores							
32	21	24	35	28	29	28	30
28	25	29	19	24	23	25	22
23	29	27	24	27	29	21	18

For Exercises 7–8, complete each step.

a. Construct side-by-side box plots of the data sets.

b. Use this display to compare the two distributions. (Example 4)

7. **HYBRID CARS** The fuel efficiency in miles per gallon for 18 hybrid cars manufactured during two recent years are shown.

Year 1								
23	48	31	27	28	35	27	28	24
15	16	28	33	22	16	28	40	24

Year 2								
29	34	25	33	26	35	27	40	27
22	48	29	34	21	24	29	21	34

8. **EARTHQUAKES** The Richter scale magnitudes of 18 earthquakes that occurred in recent years in Alaska and California are shown.

Alaska								
6.6	6.6	6.4	7.2	6.5	6.7	4.8	6.8	6.8
7.8	6.9	7.1	6.6	7.9	6.7	5.3	7.9	7.7

California								
5.4	5.4	5.6	4.4	4.2	4.3	5.2	4.5	4.7
6.6	4.9	7.2	5.2	4.1	6.0	3.0	6.6	3.5

9. MARINE BIOLOGY The table gives the frequency distribution of the weights, in pounds, of 40 adult female sea otters in Washington. (Example 5)

a. Construct a percentile graph of the data.

b. Estimate the percentile rank a weight of 55 pounds would have in this distribution, and interpret its meaning.

Class Boundaries	f
40.5–45.5	4
45.5–50.5	5
50.5–55.5	7
55.5–60.5	12
60.5–65.5	9
65.5–70.5	3

10. RAINFALL The table gives the frequency distribution of the average annual rainfall in inches for all 50 U.S. states. (Example 5)

Class Boundaries	f
0–9.5	3
9.5–19.5	8
19.5–29.5	4
29.5–39.5	14
39.5–49.5	16
49.5–59.5	5

a. Construct a percentile graph of the data.

b. Estimate the percentile rank an average rainfall of 50 inches would have in this distribution, and interpret its meaning.

Write the letter of the box plot that corresponds to each of the following histograms.

11.

12.

13.

14.

15. ATTENDANCE The average number of New York Yankee home game attendees in thousands of people, per season, from 1979 to 2008 is shown.

Home Game Attendance					
31.7	32.4	30.2	25.2	27.9	22.5
27.5	28.0	30.0	32.7	27.0	24.8
23.0	21.6	29.8	29.7	23.5	27.8
31.9	36.5	40.7	38.0	40.8	42.7
42.8	47.8	50.5	51.9	52.7	53.1

a. Construct a histogram and box plot, and use the graphs to describe the shape of the distribution.

b. Find the average number of people who attended home games during the past 30 years.

c. Which of the graphs would be best to use when estimating the average? Explain your reasoning.

d. Can either of the graphs from part **a** be used to describe any trends in home game attendance during that period? Explain your reasoning.

16. VACATION The percentile graph represents the ages of people who went on three different two-week vacations.

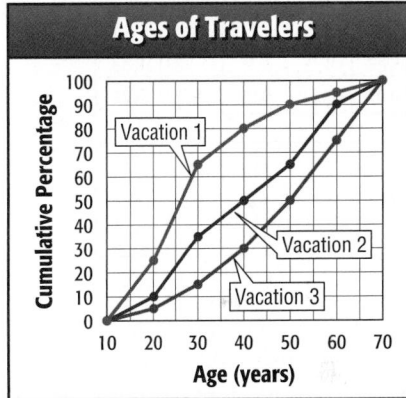

a. Describe the shape of each of the distributions.

b. Which vacation had younger travelers? older travelers? Explain your reasoning.

17. MANUFACTURING The lifetimes, measured in number of charging cycles, for two brands of rechargeable batteries are shown.

Brand A				
998	950	1020	1003	990
942	1115	973	1018	981
1047	1002	997	1110	1003
Brand B				
892	1044	1001	999	903
950	998	993	1002	995
990	1000	1005	997	1004

a. Construct a histogram of each data set.

b. Which of the brands has a greater variation in lifetime?

18. BASKETBALL The heights in inches for the players on the U.S. men's and women's national basketball teams during the 2008 Olympics are shown.

Men's Heights					
81	76	80	75	78	76
78	83	82	72	81	80
Women's Heights					
69	73	69	68	73	77
72	74	72	78	71	76

a. Construct a percentile graph of the data.

b. Estimate the percentile ranks that a male and a female player with a height of 75 inches would have in each distribution. Interpret their meaning.

c. Suppose the 78-inch-tall women's player is replaced with a 74-inch-tall player. What percentile rank would the new player have in the corresponding distribution?

Another measure of center known as the *midquartile* is given by $\frac{Q_1 + Q_3}{2}$. Find Q_1, Q_2, Q_3, and the midquartile for each set of data.

19.

0.12	0.25	0.19	0.38	0.28	0.16
0.41	0.29	0.32	0.11	0.04	0.25
0.29	0.07	0.26	0.09	0.31	0.23

20.

112	101	138	200	176	199
105	127	146	128	116	154
167	202	191	143	205	130

21. ENERGY Petroleum consumption from 1988 to 2007 for the United States and North America is shown.

United States (thousands of barrels/day)				
16,700	17,300	17,300	17,000	16,700
17,000	17,200	17,700	17,700	18,300
18,600	18,900	19,500	19,700	19,600
19,800	20,000	20,700	20,800	20,700
North America (thousands of barrels/day)				
19,900	20,600	20,800	20,000	20,200
20,600	20,800	21,400	21,300	22,000
22,400	22,800	23,500	23,800	23,700
23,800	24,200	25,000	25,200	25,000

a. Construct side-by-side box plots and histograms.

b. Compare the average petroleum consumption for the U.S. and North America.

c. Which of the graphs is easier to use when comparing measures of center and spread?

d. On average, what percent of petroleum consumption in North America can be attributed to the U.S.? Round to the nearest percent.

22. 🖘 **MULTIPLE REPRESENTATIONS** In this problem, you will investigate how a linear transformation affects the shape, center, and spread of a distribution of data. Consider the table shown.

52	37	59	31	45
23	48	42	65	39
40	53	14	49	56
68	32	77	44	28

a. GRAPHICAL Construct a histogram and use it to describe the shape of the distribution.

b. NUMERICAL Find the mean and standard deviation of the data set.

c. TABULAR Perform each of the following linear transformations of the form $X' = a + bX$, where X is the initial data value and X' is the transformed data value. Record each set of transformed data values (i–iii) in a separate table.
  i. $a = 3, b = 5$ **ii.** $a = 10, b = 1$ **iii.** $a = 0, b = 5$

d. GRAPHICAL Repeat parts **a** and **b** for each set of transformed data values that you found in part **c**. Adjust the bin width for each appropriately.

e. VERBAL Describe how a linear transformation affects the shape, center, and spread of a distribution of data.

f. ANALYTICAL If every value in a data set is multiplied by a constant c, what will happen to the mean and standard deviation of the distribution?

H.O.T. Problems   Use Higher-Order Thinking Skills

23. WRITING IN MATH Explain why using the range can be an ineffective method for measuring the spread of a distribution of data.

24. CHALLENGE Suppose 20% of a data set lies between 35 and 55. If 10 is added to each value in the set and then each result is doubled, what values will 20% of the resulting data lie between?

REASONING The gas prices, in dollars, at four gas stations over a period of one month are shown.

25. Which of the stations has the greatest variation in gas prices? the least variation? Explain your reasoning.

26. Which of the distributions is positively skewed? negatively skewed? symmetrical? Explain your reasoning.

27. WRITING IN MATH Why is the median less affected by outliers than the mean? Justify your answer.

Write each complex number in exponential form. (Lesson 10-6)

28. $\sqrt{3} + \sqrt{3}i$

29. $\sqrt{5} - \sqrt{5}i$

30. $\sqrt{2} - \sqrt{6}i$

Use the Binomial Theorem to expand each binomial. (Lesson 10-5)

31. $(3a + 4b)^5$

32. $(5c - 2d)^4$

33. $(-2x + 4y)^6$

Find the sum of each geometric series described. (Lesson 10-3)

34. first five terms of $\frac{5}{3} + 5 + 15 + \cdots$

35. first six terms of $65 + 13 + 2.6 + \cdots$

36. first ten terms of $1 - \frac{3}{2} + \frac{9}{4} - \cdots$

Find the angle θ between vectors **u** and **v**. (Lesson 8-5)

37. $\mathbf{u} = 4\mathbf{i} - 2\mathbf{j} + 9\mathbf{k}, \mathbf{v} = 3\mathbf{i} + 7\mathbf{j} - 10\mathbf{k}$

38. $\mathbf{u} = \langle -7, 4, 2 \rangle, \mathbf{v} = \langle 9, -5, 1 \rangle$

39. $\mathbf{u} = \langle 4, 4, -6 \rangle, \mathbf{v} = \langle 8, -5, 2 \rangle$

Graph the ellipse given by each equation. (Lesson 7-2)

40. $\dfrac{x^2}{4} + \dfrac{(y-3)^2}{25} = 1$

41. $\dfrac{(x+6)^2}{16} + \dfrac{(y-5)^2}{9} = 1$

42. $\dfrac{(x-2)^2}{28} + \dfrac{y^2}{8} = 1$

43. SURVEYING To determine the new height of a volcano after an eruption, a surveyor measured the angle of elevation to the top of the volcano to be 37° 45′. The surveyor then moved 1000 feet closer to the volcano and measured the angle of elevation to be 40° 30′. Determine the new height of the volcano. (Lesson 4-1)

37° 45′

1000 ft   40° 30′

44. REVIEW An amusement park ride operates like the bob of a pendulum. On its longest swing, the ship travels through an arc 75 meters long. Each successive swing is two-fifths the length of the preceding swing. What is the total distance the ship will travel from the beginning of its longest swing if the ride is allowed to continue without intervention?

A 75 m

B 125 m

C 150 m

D 187.5 m

75 m

45. REVIEW The value of a certain car depreciated at a constant rate. If the initial value was $25,000 and the car was worth $8192 after five years, find the annual rate of depreciation.

F 10%   H 30%

G 20%   J 40%

46. SAT/ACT The values of each house in a city are collected and analyzed. Which descriptive statistic will best describe the data?

A mean   D range

B median   E standard deviation

C mode

47. The table shows the frequency distribution of scores on the state driving test at a particular center on a given day. Estimate the percentile rank of someone who scored a 72 that day.

F 27%

G 30%

H 34%

J 72%

Class Boundaries	Frequency f
0–65.5	12
65.5–70.5	3
70.5–75.5	4
75.5–80.5	1
80.5–85.5	9
85.5–90.5	13
90.5–95.5	8
95.5–100	6

Probability Distributions

Then
You found probabilities of events involving combinations. (Lesson 0-7)

Now
- Construct a probability distribution, and calculate its summary statistics.
- Construct and use a binomial distribution, and calculate its summary statistics.

New Vocabulary
random variable
discrete random variable
continuous random variable
probability distribution
expected value
binomial experiment
binomial distribution
binomial probability
   distribution function

Math Online
glencoe.com

Why?
Car insurance companies use statistics to measure the risk associated with particular events, such as collisions. Using data about what has happened in the past, they assign probabilities to all possible outcomes relating to the event and calculate statistics based on how these probabilities are distributed. With these statistics, they can predict the likelihood of certain outcomes and make decisions accordingly.

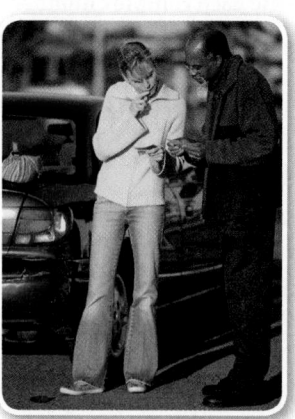

Probability Distributions In the previous lesson, you used descriptive statistics to analyze a *variable*, a characteristic of a population. In that lesson, the values the variable could take on were determined by collecting data. In this lesson, you will consider variables with values that are determined by chance.

A **random variable** X represents a numerical value assigned to an outcome of a probability experiment. There are two types of random variables: discrete and continuous.

Key Concept   Discrete and Continuous Random Variables

A **discrete random variable** can take on a finite or countable number of possible values.

Example

Number of Car Accidents Per Year, X

0 1 2 3 4 5 6 7 8

A **continuous random variable** can take on an infinite number of possible values within a specified interval.

Example

Number of Miles Driven, X

0 2 4 6 8 10

Since different statistical techniques are used to analyze these two types of random variables, it is important to be able to distinguish between them. To correctly classify a random variable, consider whether X represents counted or measured data.

EXAMPLE 1   Classify Random Variables as Discrete or Continuous

Classify each random variable X as *discrete* or *continuous*. Explain your reasoning.

a. X represents the weight of the cereal in a 15-ounce box of cereal chosen at random from those on an assembly line.

The weight of the cereal could be any weight between 0 and 15 ounces. Therefore, X is a continuous random variable.

b. X represents the number of cars in a school parking lot chosen at a random time during the school day.

The number of cars in the parking lot is countable. There could be 0, 1, 2, 3, or some other whole number of cars. Therefore, X is a discrete random variable.

✓ Guided Practice

1A. X represents the time it takes to serve a fast-food restaurant customer chosen at random.

1B. X represents the attendance at a randomly selected monthly school board meeting.

The sample space for the familiar theoretical probability experiment of tossing two coins is {TT, TH, HT, HH}. If X is the random variable for the number of heads, then X can assume the value 0, 1, or 2. From the sample space, you can find the theoretical probability of getting no, one, or two heads.

$$P(0) = \frac{1}{4} \qquad P(1) = \frac{1}{2} \qquad P(2) = \frac{1}{4}$$

The table below and the graph at the right show the *probability distribution* of X.

Number of heads, X	0	1	2
Probability, $P(X)$	$\frac{1}{4}$	$\frac{1}{2}$	$\frac{1}{4}$

ReadingMath

Probabilities of Random Variables The notation $P(1)$ is read the *probability that the random variable X is equal to 1*.

Key Concept — Probability Distribution

A **probability distribution** of a random variable X is a table, equation, or graph that links each possible value of X with its probability of occurring. These probabilities are determined theoretically or by observation.

A probability distribution must satisfy the following conditions.

• The probability of each value of X must be between 0 and 1. That is, $0 \leq P(X) \leq 1$.

• The sum of all the probabilities for all the values of X must equal 1. That is, $\sum P(X) = 1$.

To construct a discrete probability distribution using observed instead of theoretical data, use the frequency of each observed value to compute its probability.

StudyTip

Continuous Distributions This lesson focuses on discrete random variables. You will study continuous probability distributions in Lesson 11-3.

EXAMPLE 2 — Construct a Probability Distribution

TEACHER EVALUATION On a teacher evaluation form, students were asked to rate the teacher's explanations of the subject matter using a score from 1 to 5, where 1 was too simplified and 5 was too technical. Use the frequency distribution shown to construct and graph a probability distribution for the random variable X.

Score, X	Frequency
1	1
2	8
3	20
4	16
5	5

To find the probability that X takes on each value, divide the frequency of each value by the total number of students rating this teacher, which is $1 + 8 + 20 + 16 + 5$ or 50.

$$P(1) = \frac{1}{50} \text{ or } 0.02 \qquad P(2) = \frac{8}{50} \text{ or } 0.16 \qquad P(3) = \frac{20}{50} \text{ or } 0.40$$

$$P(4) = \frac{16}{50} \text{ or } 0.32 \qquad P(5) = \frac{5}{50} \text{ or } 0.10$$

The probability distribution of X is shown below, and its graph is shown at the right.

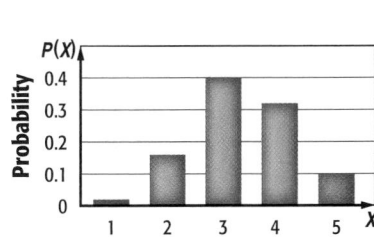

Score, X	1	2	3	4	5
$P(X)$	0.02	0.16	0.40	0.32	0.10

CHECK Note that all of the probabilities in the table are between 0 and 1 and that
$$\sum P(X) = 0.02 + 0.16 + 0.4 + 0.32 + 0.1 \text{ or } 1. ✔$$

☑ Guided Practice

2. **CAR SALES** A car salesperson tracked the number of cars she sold each day during a 30-day period. Use the frequency distribution of the results to construct and graph a probability distribution for the random variable X, rounding each probability to the nearest hundredth.

Cars Sold, X	0	1	2	3
Frequency	20	7	2	1

To compute the mean of a probability distribution, we must use a formula different from that used to compute the mean of a population. To understand why, consider computing the mean of the number of heads X resulting from an infinite number of two-coin tosses. We cannot compute the mean using $\mu = \frac{\sum X}{N}$, since N would be infinite. However, the probability distribution of X tells us what fraction of those tosses we would *expect* to have a value of 0, 1, or 2.

Number of Heads after Two Coin Tosses

{TT, TT, ..., TT, TT,   HT, HT, ..., HT, HT, TH, TH, ..., TH, TH,   HH, HH, ..., HH, HH}
{0,   0 ..., 0,   0,   1,   1, ..., 1,   1,   1,   1, ..., 1,   1,   2,   2, ..., 2,   2}

$\frac{1}{4}$ $\frac{1}{2}$ $\frac{1}{4}$

Therefore, we would expect that on average the number of heads for many or an infinite number of tosses would be $\frac{1}{4} \cdot 0 + \frac{1}{2} \cdot 1 + \frac{1}{4} \cdot 2$ or 1. This method for finding the mean of a probability distribution is summarized below.

Key Concept   **Mean of a Probability Distribution**

Words   To find the mean of a probability distribution of X, multiply each value of X by its probability and find the sum of the products.

Symbols   The mean of a random variable X is given by $\mu = \sum[X \cdot P(X)]$, where $X_1, X_2, ..., X_n$ are the values of X and $P(X_1), P(X_2), ..., P(X_n)$ are the corresponding probabilities.

EXAMPLE 3   **Mean of a Probability Distribution**

TEACHER EVALUATION The table shows the probability distribution for the teacher evaluation question from Example 2. Find the mean score to the nearest hundredth, and interpret its meaning in the context of the problem situation.

Score, X	$P(X)$
1	0.02
2	0.16
3	0.40
4	0.32
5	0.10

Multiply each score by its probability, and find the sum of these products. Organize your calculations by extending the table.

StudyTip

Rounding Rule The mean, as well as the variance and standard deviation, discussed on the next page, should be rounded to one decimal place more than that of an actual value that X can assume.

Score, X	$P(X)$	$X \cdot P(X)$
1	0.02	$1 \cdot 0.02 = 0.02$
2	0.16	$2 \cdot 0.16 = 0.32$
3	0.40	$3 \cdot 0.40 = 1.20$
4	0.32	$4 \cdot 0.32 = 1.28$
5	0.10	$5 \cdot 0.10 = 0.50$
		$\sum[X \cdot P(X)]$ or 3.3

Therefore, the mean μ of this probability distribution is about 3.3.

Since a score of 3 indicates that the teacher's explanations were neither too simplified nor too technical, a mean of 3.3 indicates that on average, students felt that this teacher's explanations were appropriate but leaned slightly towards being too technical.

☑ **Guided Practice**

3. CAR SALES Find the mean of the probability distribution that you constructed in Guided Practice 2 and interpret its meaning in the context of the problem situation.

The variance formula used for population distributions can also not be used to calculate the variance or standard deviation of a probability distribution, because the value of N would be infinite. Instead, the following formulas are used to find the spread of a probability distribution.

Key Concept **Variance and Standard Deviation of Probability Distribution**

Words To find the variance of a probability distribution of X, subtract the mean of the probability distribution from each value of X and square the difference. Then multiply each difference by its corresponding probability and find the sum of the products. The standard deviation is the square root of the variance.

Symbols The variance of a random variable X is given by $\sigma^2 = \sum[(X - \mu)^2 \cdot P(X)]$, and the standard deviation is given by $\sigma = \sqrt{\sigma^2}$.

EXAMPLE 4 **Variance and Standard Deviation of a Probability Distribution**

TEACHER EVALUATION Find the variance and standard deviation of the probability distribution for the teacher evaluation question from Example 2 to the nearest hundredth.

Score, X	$P(X)$
1	0.02
2	0.16
3	0.40
4	0.32
5	0.10

Subtract each value of X from the mean found in Example 3, 3.32 and square the difference. Then multiply each difference by its corresponding probability and find the sum of the products.

Score, X	$P(X)$	$(X - \mu)^2$	$(X - \mu)^2 \cdot P(X)$
1	0.02	$(1 - 3.32)^2 \approx 5.38$	$5.38 \cdot 0.02 \approx 0.1076$
2	0.16	$(2 - 3.32)^2 \approx 1.74$	$1.74 \cdot 0.16 \approx 0.2788$
3	0.40	$(3 - 3.32)^2 \approx 0.10$	$0.10 \cdot 0.40 \approx 0.0410$
4	0.32	$(4 - 3.32)^2 \approx 0.46$	$0.46 \cdot 0.32 \approx 0.1480$
5	0.10	$(5 - 3.32)^2 \approx 2.82$	$2.82 \cdot 0.10 \approx 0.2822$
			$\sum[(X - \mu)^2 \cdot P(X)] = 0.8576$

The variance σ^2 is about 0.86, and the standard deviation is $\sqrt{0.8576}$ or about 0.93.

✅ **Guided Practice**

4. **CAR SALES** Find the variance and standard deviation of the probability distribution that you constructed in Guided Practice 2 to the nearest hundredth.

The **expected value** $E(X)$ of a random variable for a probability distribution is equal to the mean of the random variable. That is, $E(X) = \mu = \sum[X \cdot P(X)]$.

EXAMPLE 5 **Find an Expected Value**

FUNDRAISERS At a raffle, 500 tickets are sold at $1 each for three prizes of $100, $50, and $10. What is the expected value of your net gain if you buy a ticket?

Construct a probability distribution for the possible net gains. Then find the expected value. The net gain for each prize is the value of the prize minus the cost of the tickets purchased.

Gain, X	$100 - 1$ or $99	$50 - 1$ or $49	$10 - 1$ or $9	$0 - 1$ or $-1
Probability, $P(X)$	$\frac{1}{500}$ or 0.002	$\frac{1}{500}$ or 0.002	$\frac{1}{500}$ or 0.002	$\frac{497}{500}$ or 0.994

$E(X) = \sum[X \cdot P(X)]$
$= (99 \cdot 0.002) + (49 \cdot 0.002) + (9 \cdot 0.002) + (-1 \cdot 0.994)$ or about $-\$0.68$

This expected value means that the average loss for someone purchasing a ticket is $0.68.

✓ **Guided Practice**

5. WATER PARK A water park makes $350,000 when the weather is normal and loses $80,000 per season when there are more bad weather days than normal. If the probability of having more bad weather days than normal this season is 35%, find the park's expected profit.

Binomial Distribution Many probability experiments can be reduced to one involving only two outcomes: success or failure. For example, a multiple-choice question with five answer choices can be classified as simply correct or incorrect, or a medical treatment can be classified as effective or ineffective. Such experiments have been reduced to *binomial* experiments.

Key Concept **Binomial Experiment**

A **binomial experiment** is a probability experiment that satisfies the following conditions.

- The experiment is repeated for a fixed number of independent trials n.
- Each trial has only two possible outcomes, success S or failure F.
- The probability of success $P(S)$ or p is the same in every trial. The probability of failure $P(F)$ or q is $1 - p$.
- The random variable X represents the number of successes in n trials.

EXAMPLE 6 **Identify a Binomial Experiment**

Determine whether each experiment is a binomial experiment or can be reduced to a binomial experiment. If it can be presented as a binomial experiment, state the values of n, p, and q. Then list all possible values of the random variable. If it is not, explain why not.

a. The results of a school survey indicate that 68% of students own an MP3 player. Six students are randomly selected and asked if they own an MP3 player. The random variable represents the number of students who say that they do own an MP3 player.

The experiment satisfies the conditions of a binomial experiment.

- Each student selected represents one trial, and the selection of each of the six students is independent of the others.
- There are only two possible outcomes: either the student owns an MP3 player S or does not own an MP3 player F.
- The probability of success is the same for each student selected, $P(S) = 0.68$.

In this experiment, $n = 6$ and $p = P(S)$ or 0.68. The probability of failure is $q = 1 - p$, so $q = 1 - 0.68$ or 0.32. X represents the number of students who own an MP3 player out of those selected, so $X = 0, 1, 2, 3, 4, 5,$ or 6.

b. Five cards are drawn at random from a deck to make a hand for a card game. The random variable represents the number of spades.

In this experiment, each card selected represents one trial. The probability of drawing a spade for the first card is $\frac{13}{52}$ or $\frac{1}{4}$. However, since this card is kept to make a player's hand, the trials are not independent, and the probability of success for each draw will not be the same. Therefore, this experiment cannot be reduced to a binomial experiment.

✓ **Guided Practice**

6A. The results of a survey indicate that 61% of students like the new school uniforms and 24% do not. Twenty students are randomly selected and asked if they like the uniforms. The random variable represents the number who say that they do like the uniforms.

6B. You complete a test by randomly guessing the answers to 20 multiple-choice questions that each have 4 answer choices, only one of which is correct. The random variable represents the number of correct answers.

Real-World Link

One in five American teens ages 12 years and older owns a portable MP3 player. More than one in twenty teens own more than one player.

Source: Digital Trends

The distribution of the outcomes of a binomial experiment and their corresponding probabilities is called a **binomial distribution.** The probabilities in this distribution can be calculated using the following formula, which represents the $p^x q^{n-x}$ term in the binomial expansion of $(p + q)^n$.

Key Concept   **Binomial Probability Formula**

The probability of X successes in n independent trials of a binomial experiment is

$$P(X) = {}_nC_x\, p^x q^{n-x} = \frac{n!}{(n-x)!x!}\, p^x q^{n-x},$$

where p is the probability of success and q is the probability of failure for an individual trial.

Notice that this formula represents a discrete function of the random variable X, known as the **binomial probability distribution function**.

EXAMPLE 7   **Binomial Distributions**

EXERCISE In a recent poll, 35% of teenagers said they exercise regularly. Five teenagers chosen at random are asked if they exercise regularly. Construct and graph a binomial distribution for the random variable X, which represents the number of teenagers who said yes. Then find the probability that at least three of these teenagers said yes.

This is a binomial experiment in which $n = 5$, $p = 0.35$, $q = 1 - 0.35$ or 0.65. Use a calculator to compute the probability of each possible value for X using the Binomial Probability Formula.

$P(0) = {}_5C_0 \cdot 0.35^0 \cdot 0.65^5 \approx 0.116$

$P(1) = {}_5C_1 \cdot 0.35^1 \cdot 0.65^4 \approx 0.312$

$P(2) = {}_5C_2 \cdot 0.35^2 \cdot 0.65^3 \approx 0.336$

$P(3) = {}_5C_3 \cdot 0.35^3 \cdot 0.65^2 \approx 0.181$

$P(4) = {}_5C_4 \cdot 0.35^4 \cdot 0.65^1 \approx 0.049$

$P(5) = {}_5C_5 \cdot 0.35^5 \cdot 0.65^0 \approx 0.005$

The probability distribution of X and its graph are shown below.

X	P(X)
0	0.116
1	0.312
2	0.336
3	0.181
4	0.049
5	0.005

To find the probability that *at least* three of the teenagers exercise regularly, find the sum of $P(3)$, $P(4)$, and $P(5)$.

$P(X \geq 3) = P(3) + P(4) + P(5)$ **P (at least three)**

$= 0.181 + 0.049 + 0.005$ **P(3) = 0.181, P(4) = 0.049, and P(5) = 0.005**

$= 0.235$ or 23.5%   **Simplify.**

☑ **Guided Practice**

7. **CLASSES** In a certain high school graduating class, 48% of the students took a world language during their senior year. Seven students chosen at random are asked if they took a world language during their final year. Construct and graph a probability distribution for the random variable X, which represents the number of students who said yes. Then find the probability that fewer than 4 of these students said yes.

Use the following formulas to find the mean, variance, and standard deviation of a binomial distribution.

Key Concept   **Mean and Standard Deviation of a Binomial Distribution**

The mean, variance, and standard deviation of a random variable X that has a binomial distribution are given by the following formulas.

Mean $\mu = np$

Variance $\sigma^2 = npq$

Standard Deviation $\sigma = \sqrt{\sigma^2}$ or $\sqrt{npq}$

These formulas are simpler than, but algebraically equivalent to, the formulas that you used to find the mean, variance, and standard deviation of probability distributions.

EXAMPLE 8   **Mean and Standard Deviation of a Binomial Distribution**

EXERCISE The table shows the binomial distribution in Example 7. Find the mean, variance, and standard deviation of this distribution. Interpret the mean in the context of the problem situation.

X	0	1	2	3	4	5
P(X)	0.116	0.312	0.336	0.181	0.049	0.005

Method 1 Use the formulas for the mean, variance, and standard deviation of a probability distribution.

$$\mu = \sum[X \cdot P(X)]$$
$$= 0(0.116) + 1(0.312) + 2(0.336) + 3(0.181) + 4(0.049) + 5(0.005)$$
$$= 1.748$$

$$\sigma^2 = \sum[(X - \mu)^2 \cdot P(X)]$$
$$= (0 - 1.748)^2 \cdot 0.116 + (1 - 1.748)^2 \cdot 0.312 + (2 - 1.748)^2 \cdot 0.336 +$$
$$\quad (3 - 1.748)^2 \cdot 0.181 + (4 - 1.748)^2 \cdot 0.049 + (5 - 1.748)^2 \cdot 0.005$$
$$\approx 1.1354$$

$$\sigma = \sqrt{\sigma^2}$$
$$= \sqrt{1.1354} \text{ or about } 1.0656$$

Method 2 Use the formulas for the mean, variance, and standard deviation of a binomial probability distribution. In this binomial experiment, $n = 5$, $p = 0.35$, and $q = 0.65$.

$$\mu = np$$
$$= 5(0.35) \text{ or } 1.75$$

$$\sigma^2 = npq$$
$$= 5(0.35)(0.65) \text{ or } 1.1375$$

$$\sigma = \sqrt{\sigma^2}$$
$$= \sqrt{1.1375} \text{ or about } 1.0665$$

Both methods give approximately the same results. Therefore, the mean of the distribution is about 1.8 or 2, which means that on average about 2 out of the 5 students would say that they exercise regularly. The variance and standard deviation of the distribution are both about 1.1.

✓ **Guided Practice**

8. **CLASSES** Find the mean, variance, and standard deviation of the distribution that you constructed in Guided Practice 7. Interpret the mean in the context of the problem situation.

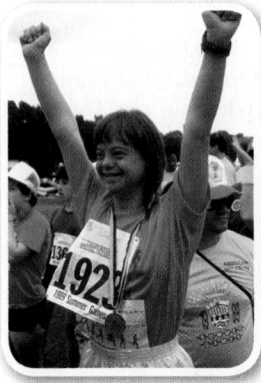

🌐 Real-World Link

According to a recent Gallup poll, approximately 58% of U.S. teens are classified as highly active. Activities in which the teens said that they regularly participate include basketball, running, jogging, biking, and swimming.

Source: The Gallup Poll

Classify each random variable *X* as *discrete* or *continuous*. Explain your reasoning. (Example 1)

1. *X* represents the number of text messages sent by a randomly chosen student during a given day.

2. *X* represents the time it takes a randomly selected student to complete a physics test.

3. *X* represents the weight of a chocolate chip cookie selected at random in the school cafeteria.

4. *X* represents the number of CDs owned by a student chosen at random during a given day.

5. *X* represents the number of votes received by a candidate selected at random for a particular election.

6. *X* represents the weight of a wrestler selected at random on a given day.

Construct and graph a probability distribution for each random variable *X*. Find and interpret the mean in the context of the given situation. Then find the variance and standard deviation. (Examples 2–4)

7. **MUSIC** Students were asked how many MP3 players they own.

Players, *X*	Frequency
0	9
1	17
2	9
3	5
4	2

8. **AMUSEMENT** There were 20 participants in a pie eating contest at a county fair.

Pies Eaten, *X*	Frequency
1	1
2	5
3	9
4	3
5	2

9. **BREAKFAST** A sample of high school students was asked how many days they ate breakfast last week.

Days, *X*	Frequency
0	5
1	3
2	17
3	27
4	6
5	19
6	18
7	65

10. **HEALTH** Patients at a dentist's office were asked how many times a week they floss their teeth.

Flosses, *X*	Frequency
1	9
2	15
3	5
4	2
5	1
6	0
7	1

11. **CAR INSURANCE** A car insurance policy that costs $300 will pay $25,000 if the car is stolen and not recovered. If the probability of a car being stolen is $p = 0.0002$, what is the expected value of the profit (or loss) to the insurance company for this policy? (Example 5)

12. **FUNDRAISERS** A school hosts an annual fundraiser where raffle tickets are sold for baked goods, the values of which are indicated below. Suppose 100 tickets were sold for a drawing for each of the four cakes.

$5 $10 $15 $20

What is the expected value of a participant's net gain if he or she buys a ticket for $1? (Example 5)

Determine whether each experiment is a binomial experiment or can be reduced to a binomial experiment. If it can be presented as a binomial experiment, state the values of *n*, *p*, and *q*. Then list all possible values of the random variable. If it is not, explain why not. (Example 6)

13. You survey 25 students to find out how many are left-handed. The random variable represents the number of left-handed people.

14. You survey 200 people to see if they watch Monday Night Football. The random variable represents the number who watched Monday Night Football.

15. You roll a die 10 times to see if a 5 appears. The random variable represents the number of 5s.

16. You toss a coin 20 times to see how many tails occur. The random variable represents the number of tails.

17. You ask 15 people how old they are. The random variable represents their age.

18. You survey 40 students to find out whether they passed their driving test. The random variable represents their scores.

19. You select 10 cards from a deck without replacement. The random variable represents the number of hearts.

Construct and graph a binomial distribution for each random variable. Find and interpret the mean in the context of the given situation. Then find the variance and standard deviation. (Examples 7 and 8)

20. In a recent poll, 89% of Americans order toppings on their pizza. Five teenagers chosen at random are asked if they order toppings.

21. In Eureka, California, 21% of the days are sunny. Consider the number of sunny days in February.

22. According to a survey, 26% of a company's employees have surfed the Internet at work. Ten co-workers were selected at random and asked if they have surfed the Internet at work.

23. A high school newspaper reported that 65% of students wear their seatbelts while driving. Eight students chosen at random are asked if they wear seatbelts.

24. According to a recent survey, 41% of high school students own a car. Seven students chosen at random are asked if they own a car.

25. **GAME SHOWS** The prize wheel on a game show has 16 numbers. During one turn, a bet is made and the wheel is spun.

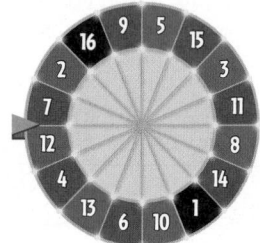

The payoffs for a $5 bet are shown. If a player bets $5, find the expected value for each.

Bet	Payoff	Bet	Payoff
even or odd	$7	1	$50
red or green	$10	16	$50
1–4	$15	even and red	$25
5–8	$15	odd and green	$25
9–12	$15	1 or 16	$30

   a. green
   b. even and red
   c. odd
   d. 1 or 16
   e. 1

26. **VOLUNTEERING** In a recent poll, 62% of Americans said that they had donated their time volunteering for a charity in the past year. If a random sample of 10 Americans is selected, find each of the following probabilities.

   a. Exactly 6 people donated their time to a charity.
   b. At least 5 people donated their time to a charity.
   c. At most 3 people donated their time to a charity.
   d. More than 8 people donated their time to a charity.

27. 🔁 **MULTIPLE REPRESENTATIONS** In this problem, you will investigate the shape of a binomial distribution.

   a. **GRAPHICAL** Construct and graph the binomial distribution that corresponds to each of the following experiments.

      i. $n = 6, p = 0.5$ ii. $n = 6, p = 0.3$
      iii. $n = 6, p = 0.7$ iv. $n = 8, p = 0.5$
      v. $n = 10, p = 0.5$

   b. **VERBAL** Describe the shape of each of the distributions you found in part **a**.

   c. **ANALYTICAL** Make a conjecture regarding the shape of a distribution with each of the following probabilities of success: $p < 0.5$, $p = 0.5$, and $p > 0.5$.

   d. **ANALYTICAL** What happens to the spread of a binomial distribution as n increases?

H.O.T. Problems Use **H**igher-**O**rder **T**hinking Skills

28. **PROOF** Use the distribution below to prove that $\mu = np$ and $\sigma^2 = npq$ for a binomial distribution, given $\mu = \sum[X \cdot P(X)]$ and $\sigma^2 = \sum[(X - \mu)^2 \cdot P(X)]$ for a probability distribution.

X	P(X)
0	$1 - p$
1	p

29. **REASONING** Suppose a coin is tossed ten times and lands on heads each time. Will the probability of the coin landing on tails increase during the next toss? Explain your reasoning.

30. **OPEN ENDED** A probability distribution in which all of the values of the random variable occur with equal probability is called a *uniform probability distribution*. Describe an example of an experiment that would produce a uniform distribution. Then find the theoretical probabilities that would result from this experiment. Include a table and graph of the distribution.

REASONING Determine whether each of the following statements is *true* or *false*. Explain your reasoning.

31. The probabilities associated with rolling two dice are determined theoretically.

32. The mean of a random variable is always a possible outcome of the experiment.

33 **CHALLENGE** Consider a binomial distribution in which $n = 50$ and $\sigma = 1.54$. What is the mean of the distribution? (*Hint*: p is closer to 0 than 1.)

34. **WRITING IN MATH** Describe another way you could find the probability that at least three of the teenagers exercise regularly or $P(X \geq 3)$ from Example 7. Give an example of when this method would be faster to use.

35. ART The prices in dollars of paintings sold at an art auction are shown. (Lesson 11-1)

a. Construct a histogram, and use it to describe the shape of the distribution.

b. Summarize the center and spread of the data using either the mean and standard deviation or the five-number summary. Justify your choice.

Art Prices ($)					
1800	600	750	600	600	1800
1350	450	300	1200	750	600
750	450	2700	600	750	300
750	2300	600	450	2100	1200

Use the fifth partial sum of the exponential series to approximate each value to three decimal places. (Lesson 10-6)

36. $e^{0.2}$

37. $e^{-0.4}$

38. $e^{-0.75}$

Find the indicated geometric means for each pair of nonconsecutive terms. (Lesson 10-3)

39. 8 and 312.5; 3 means

40. $\frac{2}{9}$ and 54; 4 means

41. $\frac{3}{4}$ and $\frac{24}{3125}$; 4 means

Find the next four terms of each sequence. (Lesson 10-1)

42. $a_1 = -12, a_n = a_{n-1} + 3, n \geq 2$

43. $a_1 = 19, a_n = a_{n-1} - 13, n \geq 2$

44. $a_1 = 81, a_n = a_{n-1} - 72, n \geq 2$

Find the dot product of u and v. Then determine if u and v are orthogonal. (Lesson 8-5)

45. $\mathbf{u} = \langle 2, 9, -2 \rangle, \mathbf{v} = \langle -4, 7, 6 \rangle$

46. $\mathbf{u} = 3\mathbf{i} - 5\mathbf{j} + 6\mathbf{k}$ and $\mathbf{v} = -7\mathbf{i} + 8\mathbf{j} + 9\mathbf{k}$

47. $\mathbf{u} = \langle 8, -2, -2 \rangle, \mathbf{v} = \langle -6, 6, -10 \rangle$

Graph the hyperbola given by each equation. (Lesson 7-3)

48. $\dfrac{(y+6)^2}{36} - \dfrac{(x-1)^2}{24} = 1$

49. $\dfrac{(y+5)^2}{49} - \dfrac{(x-6)^2}{20} = 1$

50. $\dfrac{(y+3)^2}{9} - \dfrac{(x+5)^2}{4} = 1$

Find AB and BA, if possible. (Lesson 6-2)

51. $A = [2, -1], B = \begin{bmatrix} 5 \\ 4 \end{bmatrix}$

52. $A = \begin{bmatrix} 3 & -2 \\ 5 & 1 \end{bmatrix}, B = \begin{bmatrix} 4 & 1 \\ 2 & 7 \end{bmatrix}$

53. $A = \begin{bmatrix} 4 & -1 \\ 6 & 1 \\ 5 & -8 \end{bmatrix}, B = \begin{bmatrix} 2 & 9 & -3 \\ 4 & -1 & 0 \end{bmatrix}$

Skills Review for Standardized Tests

54. REVIEW Find the sum of $16 + 8 + 4 + \ldots$

A 28

B 32

C 48

D 64

55. In a recent poll, 48% of Americans said that they shopped online for at least one holiday gift. If a random sample of 10 Americans is selected, what is the probability that at least 7 shopped online for a gift?

F 3.4%

G 4.8%

H 10.0%

J 14.1%

56. SAT/ACT Find the area of the shaded region.

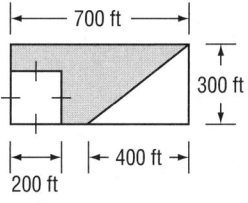

A 90,000 ft^2 C 130,000 ft^2 E 210,000 ft^2

B 110,000 ft^2 D 150,000 ft^2

57. REVIEW Which of the following distributions best describes the data?

{14, 15, 11, 13, 13, 14, 15, 14, 12, 13, 14, 15}

F positively skewed   H normal

G negatively skewed   J binomial

The Normal Distribution

Why?

In a recent year, approximately 107 million Americans 20 years and older had a total blood cholesterol level of 200 milligrams per deciliter or higher. Physicians use variables of this type to compare patients' cholesterol levels to *normal* cholesterol ranges. In this lesson, you will determine the probability of a randomly selected person having a specific cholesterol level.

Then
You analyzed probability distributions for discrete random variables.
(Lesson 11-2)

Now
- Find area under normal distribution curves.
- Find probabilities for normal distributions, and find data values given probabilities.

New Vocabulary
normal distribution
empirical rule
z-value
standard normal distribution

Math Online

glencoe.com

The Normal Distribution The probability distribution for a continuous variable is called a *continuous probability distribution*. The most widely used continuous probability distribution is called the **normal distribution**. The characteristics of the normal distribution are as follows.

Key Concept — **Characteristics of the Normal Distribution**

- The graph of the curve is bell-shaped and symmetric with respect to the mean.
- The mean, median, and mode are equal and located at the center.
- The curve is continuous.
- The curve approaches, but never touches, the *x*-axis.
- The total area under the curve is equal to 1 or 100%.

Consider a continuous probability distribution of times for a 400-meter run in a random sample of 100 athletes. By increasing sample size and decreasing class width, the distribution becomes more and more symmetrical. If it were possible to sample the entire population, the distribution would approach the normal distribution, as shown.

$n = 100$

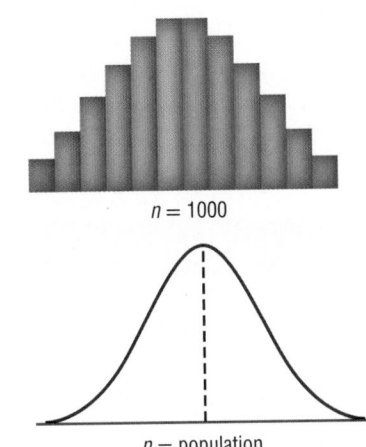

$n = 1000$

$n = 10,000$

$n = $ population

For every normally distributed random variable, the shape and position of the normal distribution curve are dependent on the mean and standard deviation. For example, in Figure 11.3.1, you can see that a larger standard deviation results in a flatter curve. A change in the mean, as shown in Figure 11.3.2, results in a horizontal translation of the curve.

Figure 11.3.1

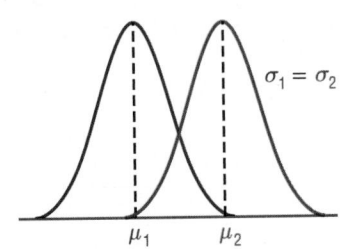

Figure 11.3.2

The area under the normal distribution curve between two data values represents the percent of data values that fall within that interval. The **empirical rule** can be used to describe areas under the normal curve over intervals that are one, two, or three standard deviations from the mean.

Key Concept   **The Empirical Rule**

In a normal distribution with mean μ and standard deviation σ:

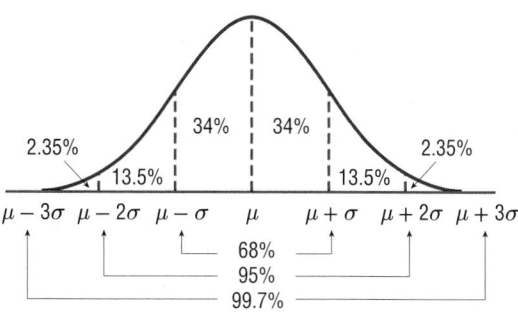

- approximately 68% of the data values fall between $\mu - \sigma$ and $\mu + \sigma$.
- approximately 95% of the data values fall between $\mu - 2\sigma$ and $\mu + 2\sigma$.
- approximately 99.7% of the data values fall between $\mu - 3\sigma$ and $\mu + 3\sigma$.

You can solve problems involving approximately normal distributions using the empirical rule.

EXAMPLE 1   **Use the Empirical Rule**

HEIGHT The heights of the 880 students at East High School are normally distributed with a mean of 67 inches and a standard deviation of 2.5 inches.

a. Approximately how many students are more than 72 inches tall?

To determine the number of students that are more than 72 inches tall, find the corresponding area under the curve.

In the figure shown, you can see that 72 is 2σ from the mean. Because 95% of the data values fall within two standard deviations from the mean, each tail represents 2.5% of the data. The area to the right of 72 is 2.5% of 880 or 22.

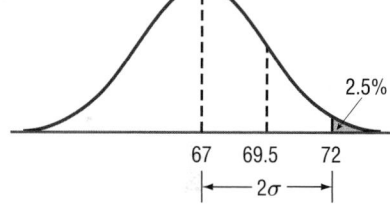

Thus, about 22 students are more than 72 inches tall.

b. What percent of the students are between 59.5 and 69.5 inches tall?

The percent of students between 59.5 and 69.5 inches tall is represented by the shaded area in the figure at the right, which is between $\mu - 3\sigma$ and $\mu + \sigma$. The total area under the curve between 59.5 and 69.5 is equal to the sum of the areas of each region.

$$2.35\% + 13.5\% + 68\% = 83.85\%$$

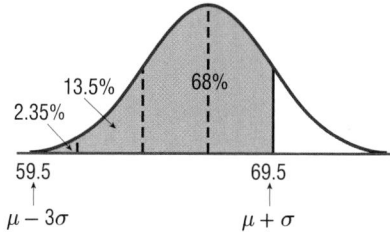

Therefore, about 84% of the students are between 59.5 and 69.5 inches tall.

✓ Guided Practice

1. **MANUFACTURING** A machine used to fill water bottles dispenses slightly different amounts into each bottle. Suppose the volume of water in 120 bottles is normally distributed with a mean of 1.1 liters and a standard deviation of 0.02 liter.

   A. Approximately how many bottles of water are filled with less than 1.06 liters?

   B. What percent of the bottles have between 1.08 and 1.14 liters?

While the empirical rule can be used to analyze a normal distribution, it is only useful when evaluating specific values, such as $\mu + \sigma$. A normally distributed variable can be transformed into a standard value or z-value, which can be used to analyze any range of values in the normal distribution. This transformation is known as *standardizing*. The **z-value**, also known as the z-score and z test statistic, represents the number of standard deviations that a given data value is from the mean.

Key Concept   **Formula for z-Values**

The z-value for a data value in a set of data is given by $z = \dfrac{X - \mu}{\sigma}$, where X is the data value, μ is the mean, and σ is the standard deviation.

You can use z-values to determine the position of *any* data value within a set of data. For example, consider a distribution with $\mu = 40$ and $\sigma = 6$. A data value of 57.5 is located about 2.92 standard deviations away from the mean, as shown. Therefore, in this distribution, $X = 57.5$ correlates to a z-value of 2.92.

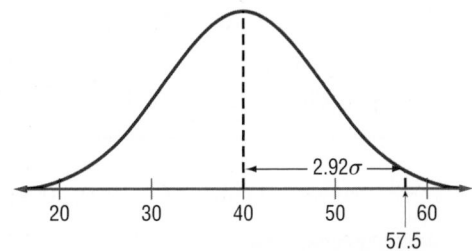

EXAMPLE 2   **Find z-Values**

Find each of the following.

a. z if $X = 24$, $\mu = 29$, and $\sigma = 4.2$

$$z = \dfrac{X - \mu}{\sigma} \qquad \text{Formula for z-values}$$

$$= \dfrac{24 - 29}{4.2} \qquad X = 24, \mu = 29, \text{ and } \sigma = 4.2$$

$$\approx -1.19 \qquad \text{Simplify.}$$

The z-value that corresponds to $X = 24$ is -1.19. Therefore, 24 is 1.19 standard deviations less than the mean in the distribution.

b. X if $z = -1.73$, $\mu = 48$, and $\sigma = 2.3$

$$z = \dfrac{X - \mu}{\sigma} \qquad \text{Formula for z-values}$$

$$-1.73 = \dfrac{X - 48}{2.3} \qquad \mu = 48, \sigma = 2.3, \text{ and } z = -1.73$$

$$-3.979 = X - 48 \qquad \text{Multiply each side by 2.3.}$$

$$44.021 = X \qquad \text{Add 48 to each side.}$$

A z-value of -1.73 corresponds to a data value of approximately 44 in the distribution.

☑ **Guided Practice**

2A. z if $X = 32$, $\mu = 28$, and $\sigma = 1.7$ **2B.** X if $z = 2.15$, $\mu = 39$, and $\sigma = 0.4$

Every normally distributed random variable has a unique mean and standard deviation, which affect the position and shape of the curve. As a result, there are infinitely many normal probability distributions. Fortunately, they can all be related to one distribution known as the *standard* normal distribution. The **standard normal distribution** is a normal distribution of z-values with a mean of 0 and a standard deviation of 1.

The characteristics of the standard normal distribution are summarized below.

Key Concept — Characteristics of the Standard Normal Distribution

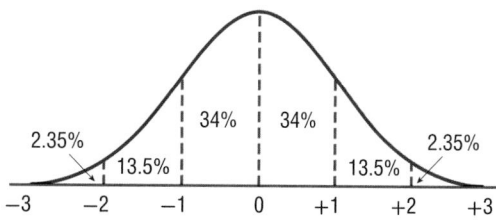

- The total area under the curve is equal to 1 or 100%.
- Almost all of the area is between $z = -3$ and $z = 3$.
- The distribution is symmetric.
- The mean is 0, and the standard deviation is 1.
- The curve approaches, but never touches, the x-axis.

You can solve normal distribution problems by finding the z-value that corresponds to a given X-value, and then finding the approximate area under the standard normal curve. The corresponding area can be found by using a table of z-values that shows the area *to the left* of a given z-value. For example, the area under the curve to the left of a z-value of 1.42 is 0.9222, as shown.

You can also find the area under the curve that corresponds to any z-value with a graphing calculator. This method will be used for the remainder of this chapter.

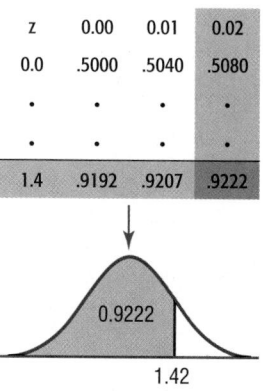

z	0.00	0.01	0.02
0.0	.5000	.5040	.5080
.	.	.	.
.	.	.	.
1.4	.9192	.9207	.9222

EXAMPLE 3 — Use the Standard Normal Distribution

COMMUNICATION The average number of phone calls received by a customer service representative each day during a 30-day month was 105 with a standard deviation of 12. Find the number of days with fewer than 110 phone calls. Assume that the number of calls is normally distributed.

$$z = \frac{X - \mu}{\sigma} \qquad \text{Formula for } z\text{-values}$$

$$= \frac{110 - 105}{12} \text{ or about } 0.42 \qquad X = 110, \mu = 105, \text{ and } \sigma = 12$$

Although the standard normal distribution extends to negative and positive infinity, when you are finding the area less than or greater than a given value, you can use a lower value of -4 and an upper value of 4.

In this case, enter a lower z-value of -4 and an upper z-value of 0.42. The resulting area is 0.66. Since there were 30 days in the month, there were fewer than 110 calls on $30 \cdot 0.66$ or 19.8 days.

```
normalcdf(-4,0.4
2)
         .6627255515
```

Therefore, there were approximately 20 days with fewer than 110 calls.

TechnologyTip

Area Under the Normal Curve You can use a graphing calculator to find the area under a standard normal curve that corresponds to any pair of z-values by selecting [2nd] [DISTR] and normalcdf(*lower z value, upper z value*).

✓ Guided Practice

3. **BASKETBALL** The average number of points that a basketball team scored during a single season was 63 with a standard deviation of 18. If there were 15 games during the season, find the percentage of games in which the team scored more than 70 points. Assume that the number of points is normally distributed.

In Example 3, you found the area under the normal curve that corresponds to a z-value. You can also find z-values that correspond to specific areas. For example, you can find the z-value that corresponds to a cumulative area of 1%, 20%, or 99%. You can also find intervals of z-values that contain or are between a certain percentage of data.

middle 20%   outside 1%

EXAMPLE 4 **Find z-Values Corresponding to a Given Area**

Find the interval of z-values associated with each area.

a. middle 50% of the data

The middle 50% of the data corresponds to the data between 25% and 75% of the distribution, or 0.25 and 0.75, as shown.

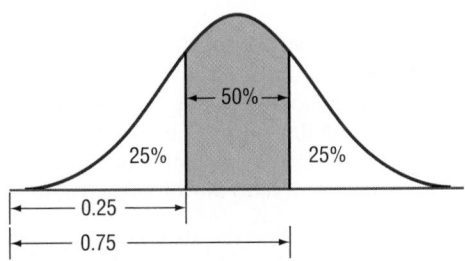

To find the z-scores that correspond to 0.25 and 0.75, select 2nd [DISTR] to display the **DISTR** menu on a graphing calculator. Select **invNorm(** and enter 0.25. Repeat to find the value corresponding to 0.75. As shown at the right, the z-value corresponding to 0.25 is −0.67 and the z-value corresponding to 0.75 is 0.67.

```
invNorm(0.25)
        -.6744897495
invNorm(0.75)
         .6744897495
```

Therefore, the interval that represents the middle 50% of the data is −0.67 < z < 0.67.

b. the outside 20% of the data

The outside 20% of the data represents the top 10% and the bottom 10% of the distribution or 0.1 and 0.9, as shown.

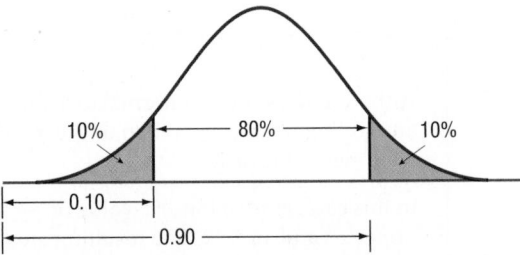

To find the z-value corresponding to 0.10, enter 0.10 into a graphing calculator under **invNorm(** and repeat for 0.90. As shown, the z-value corresponding to 0.10 is −1.28 and the z-value corresponding to 0.90 is 1.28.

```
invNorm(0.10)
       -1.281551567
invNorm(0.90)
        1.281551567
```

Therefore, the interval that represents the outside 20% of the data is −1.28 > z or z > 1.28.

✓ **Guided Practice**

4A. the middle 25% of the data   **4B.** the outside 60% of the data

Probability and the Normal Distribution You have seen how the area under the normal curve corresponds to the proportion of data values in an interval. The area also corresponds to the probability of data values falling within a given interval. If a z-value is chosen randomly, the probability of choosing a value between 0 and 1 would be equivalent to the area under the curve between 0 and 1.00, which is 0.3413. Therefore, the probability of choosing a value between 0 and 1 would be approximately 34%.

EXAMPLE 5 **Find Probabilities**

METEOROLOGY The temperatures for one month for a city in California are normally distributed with $\mu = 81°$ and $\sigma = 6°$. Find each probability, and use a graphing calculator to sketch the corresponding area under the curve.

a. $P(70° < X < 90°)$

The question is asking for the percentage of temperatures that were between 70° and 90°. First, find the corresponding z-values for $X = 70$ and $X = 90$.

$z = \dfrac{X - \mu}{\sigma}$ **Formula for z-values**

$= \dfrac{70 - 81}{6}$ **X = 70, μ = 81, and σ = 6**

≈ -1.83 **Simplify.**

Use 90 to find the other z-value.

$z = \dfrac{X - \mu}{\sigma}$ **Formula for z-values**

$= \dfrac{90 - 81}{6}$ **X = 90, μ = 81, and σ = 6**

≈ 1.5 **Simplify.**

You can use a graphing calculator to display the area that corresponds to any z-value by selecting [2nd] [DISTR]. Then, under the **DRAW** menu, select **ShadeNorm(** *lower z value, upper z value*). The area between $z = -1.83$ and $z = 1.5$ is 0.899568, as shown.

[−4, 4] scl: 1 by [0, 0.5] scl: 0.125

Therefore, approximately 90% of the temperatures were between 70 and 90.

b. $P(X \geq 95°)$

$z = \dfrac{X - \mu}{\sigma}$ **Formula for z-values**

$= \dfrac{95 - 81}{6}$ **X = 95, μ = 81, and σ = 6**

≈ 2.33 **Simplify.**

Using a graphing calculator, you can find the area between $z = 2.33$ and $z = 4$ to be 0.0099.

Therefore, the probability that a randomly selected temperature is at least 95° is about 0.1%.

[−4, 4] scl: 1 by [0, 0.5] scl: 0.125

✔ **Guided Practice**

5. TESTING The scores on a standardized test are normally distributed with $\mu = 72$ and $\sigma = 11$. Find each probability and use a graphing calculator to sketch the corresponding area under the curve.

A. $P(X < 89)$

B. $P(65 < X < 85)$

You can find specific intervals of data for given probabilities or percentages by using the standard normal distribution.

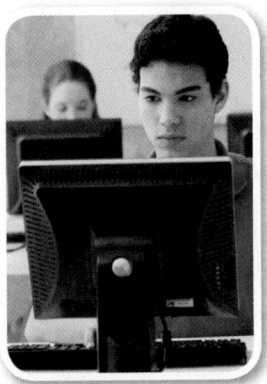

EXAMPLE 6   Find Intervals of Data

COLLEGE The scores for the entrance exam for a college's mathematics department is normally distributed with $\mu = 65$ and $\sigma = 8$.

a. If Ramona wants to be in the top 20%, what score must she get?

To find the top 20% of the exam scores, you must find the exam score X that separates the upper 20% of the area under the normal curve, as shown. The top 20% correlates with $1 - 0.2$ or 0.8. Using a graphing calculator, you can find the corresponding z-value to be 0.84.

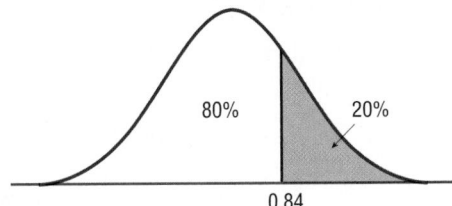

Now, use the formula for the z-value for a population to find the corresponding exam score.

$$z = \frac{X - \mu}{\sigma} \quad \text{Formula for } z\text{-values}$$

$$0.84 = \frac{X - 65}{8} \quad \mu = 65, \sigma = 8, \text{ and } z = 0.84$$

$$6.72 = X - 65 \quad \text{Multiply each side by 8.}$$

$$71.72 = X \quad \text{Add 65 to each side.}$$

Ramona needs a score of at least 72 to be in the top 20%.

b. Ramona expects to earn a grade in the middle 90% of the distribution. What range of scores fall in this category?

The middle 90% of the exam scores represents 45% on each side of the mean and therefore corresponds to the interval of area from 0.05 to 0.95. Using a graphing calculator, the z-values that correspond to 0.05 and 0.95 are -1.645 and 1.645, respectively.

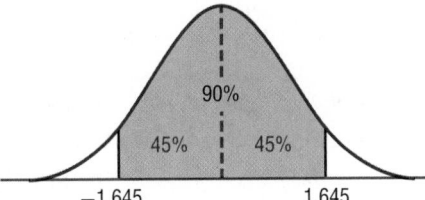

Use the z-values to find each value of X.

$$z = \frac{X - \mu}{\sigma} \quad \text{Formula for } z\text{-values} \qquad z = \frac{X - \mu}{\sigma}$$

$$-1.645 = \frac{X - 65}{8} \quad \mu = 65 \text{ and } \sigma = 8 \qquad 1.645 = \frac{X - 65}{8}$$

$$-13.16 = X - 65 \quad \text{Multiply.} \qquad 13.16 = X - 65$$

$$51.84 = X \quad \text{Simplify.} \qquad 78.16 = X$$

Therefore, Ramona expects to score between 52 and 78.

✔ Guided Practice

6. RESEARCH As part of a medical study, a researcher selects a study group with a mean weight of 190 pounds and a standard deviation of 12 pounds. Assume that the weights are normally distributed.

A. If the study will mainly focus on participants whose weights are in the middle 80% of the data set, what range of weights will this include?

B. If participants whose weights fall in the outside 5% of the distribution are contacted 2 weeks after the study, people in what weight range will be contacted?

1. **NOISE POLLUTION** As part of a noise pollution study, researchers measured the sound level in decibels of a busy city street for 30 days. According to the study, the average noise was 82 decibels with a standard deviation of 6 decibels. Assume that the data are normally distributed. (Example 1)

   a. If a normal conversation is held at about 64 decibels, determine the number of hours during the study that the noise level was this low.

   b. Determine the percent of the study during which the noise was between 76 decibels and 88 decibels.

2. **GAS MILEAGE** Dion commutes 290 miles each week for work. His car averages 29.6 miles per gallon with a standard deviation of 5.4 miles per gallon. Assume that the data are normally distributed. (Example 1)

   a. Approximate the number of miles that Dion's car gets a gas mileage of 35 miles per gallon or better.

   b. For what percentage of Dion's commute does his car have a gas mileage between 24.2 miles per gallon and 40.4 miles per gallon?

Find each of the following. (Example 2)

3. z if $X = 19$, $\mu = 22$, and $\sigma = 2.6$

4. X if $z = 2.3$, $\mu = 64$, and $\sigma = 1.3$

5. z if $X = 52$, $\mu = 43$, and $\sigma = 3.7$

6. X if $z = 2.5$, $\mu = 27$, and $\sigma = 0.4$

7. z if $X = 32$, $\mu = 38$, and $\sigma = 2.8$

8. X if $z = 1.7$, $\mu = 49$, and $\sigma = 4.1$

9. **ICHTHYOLOGY** As part of a science project, José studied the growth rate of 797 green gold catfish and found the following information. Assume that the data are normally distributed. (Example 3)

The green gold catfish reaches its maximum length within its first 3 months of life.

• Average length at birth 4.69 millimeters
• Standard deviation 0.258 millimeters

   a. Determine the number of fish with a length less than 4.5 millimeters at birth.

   b. Determine the number of fish with a length greater than 5 millimeters at birth.

10. **ROLLER COASTER** The average wait in line for the 16,000 daily passengers of a roller coaster is 72 minutes with a standard deviation of 15 minutes. Assume that the data are normally distributed. (Example 3)

    a. Determine the number of passengers who wait less than 60 minutes to ride the roller coaster.

    b. Determine the number of passengers who wait more than 90 minutes to ride the roller coaster.

Find the interval of z-values associated with each area. (Example 4).

11. middle 30%

12. outside 15%

13. outside 40%

14. middle 10%

15. outside 25%

16. middle 84%

17. **BATTERY** The life of a certain brand of AA battery is normally distributed with $\mu = 8$ hours and $\sigma = 1.5$ hours. Find each probability. (Example 5)

    a. The battery will last less than 6 hours.

    b. The battery will last more than 12 hours.

    c. The battery will last between 8 and 9 hours.

18. **HEALTH** The average blood cholesterol level in adult Americans is 203 mg/dL (milligrams per deciliter) with a standard deviation of 38.8 mg/dL. Find each probability. Assume that the data are normally distributed. (Example 5)

    a. a blood cholesterol level below 160 mg/dL, which is considered low and can lead to a higher risk of stroke

    b. a blood cholesterol level above 240 mg/dL, which is considered high and can lead to higher risk of heart disease

    c. a blood cholesterol level between 180 and 200 mg/dL, which is considered normal

19. **SNOWFALL** The average annual snowfall in centimeters for the U.S. and Canada region from 45°N to 55°N is normally distributed with $\mu = 260$ and $\sigma = 27$. (Example 6)

    a. Determine the minimum amount of snowfall occurring in the top 15% of the distribution.

    b. Determine the maximum amount of snowfall occurring in the bottom 30%.

    c. What range of snowfall occurs in the middle 60%?

20. **TRAFFIC SPEED** The average speed in miles per hour of traffic on North Street is normally distributed with $\mu = 37.5$ and $\sigma = 5.5$. (Example 6)

    a. Determine the maximum speed of the slowest 10% of cars driving on North Street.

    b. Determine the minimum speed of the fastest 5% of cars driving on North Street.

    c. At what range of speed do the middle 25% of cars on North Street drive?

21. **TESTS** Miki took the ACT and SAT and earned the scores shown. Which of the scores has a higher relative position? Explain your reasoning.

Test	Miki's Score	National Average	Standard Deviation
ACT	27	21	4.7
SAT	620	508	111

22. EXAMS Bena scored 76 on a physics test that had a mean of 72 and a standard deviation of 10. She also scored 81 on a sociology test that had a mean of 78 and a standard deviation of 9. Compare her relative scores on each test. Assume that the data are normally distributed.

Find the area that corresponds to each shaded region.

23.

−0.96

24.

−1.35

25.

0   2.45

26.

−0.88   1.65

27. FRACTILES Recall from Lesson 11-1 that quartiles, percentiles, and deciles are three types of fractiles, which divide an ordered set of data into equal groups. Find the z-values that correspond to each of the following fractiles.

a. D_{20}, D_{40}, and D_{80}

b. Q_1, Q_2, and Q_3

c. P_{10}, P_{40}, and P_{90}

28. METEOROLOGY The humidity observed in the morning during the same day in Chicago, Orlando, and Phoenix is shown. Assume that the data are normally distributed.

City	Humidity	Average Humidity	Standard Deviation
Chicago	85%	82%	12%
Orlando	94%	91%	15%
Phoenix	46%	43%	10%

a. Which city has the highest relative humidity? the lowest relative humidity? Explain your reasoning.

b. How would a fourth city compare that has a relative humidity of 81% and an average humidity of 78% with a standard deviation of 8%?

29. JOBS The salaries of employees in the sales department of an advertising agency are normally distributed with a standard deviation of $8000. During the holiday season, employees who earn less than $35,000 receive a gift basket.

a. Suppose 10% of the employees receive a gift basket. What was the mean salary of the sales department?

b. Suppose employees who make $10,000 greater than the mean salary receive an incentive bonus. If 200 employees work in the sales department, how many employees will receive a bonus?

30. MULTIPLE REPRESENTATIONS In this problem, you will investigate the shape of a normal distribution. Consider a population of 4, 10, 6, 8.

a. GRAPHICAL Construct a bar graph, and use it to describe the shape of the distribution. Then find the mean and standard deviation of the data set.

b. GRAPHICAL Select eight random samples of size 2, with replacement, from the data set. Construct a bar graph, and use it to describe the shape of the distribution. Find the mean and standard deviation of the sample means.

c. TABULAR The table includes all samples of size 2 that can be taken, with replacement, from the data set. Find the mean of each sample and the mean and standard deviation of all of the sample means.

Sample	Mean	Sample	Mean
4, 4		8, 4	
4, 6		8, 6	
4, 8		8, 8	
4, 10		8, 10	
6, 4		10, 4	
6, 6		10, 6	
6, 8		10, 8	
6, 10		10, 10	

d. GRAPHICAL Construct a bar graph of the sample means from part **c** and use it to describe the shape of the distribution. What happens to the shape of a distribution of data as the sample size increases?

e. ANALYTICAL Divide the standard deviation of the population that you found in part **a** by the square root of the sample size. What do you think happens to the mean and standard deviation of a distribution of data as the sample size increases?

H.O.T. Problems Use **H**igher-**O**rder **T**hinking Skills

31. ERROR ANALYSIS Chad and Lucy are finding the z interval associated with the outside 35% of a distribution of data. Chad thinks it is the interval $z < -0.39$ or $z > 0.39$, while Lucy thinks it is the interval $z < -0.93$ or $z > 0.93$. Is either of them correct? Explain your reasoning.

32. REASONING In real-life applications, z-values usually fall between −3 and +3 in the standard normal distribution. Why do you think this is the case? Explain your reasoning.

33. CHALLENGE Find two z-values, one positive and one negative, so that the combined area of the two equivalent tails is equal to each of the following.

a. 1%   **b.** 5%   **c.** 10%

34. REASONING Continuous variables *sometimes*, *always*, or *never* have normal distributions. Explain your reasoning.

35. WRITING IN MATH Compare and contrast the characteristics of a normal distribution and the standard normal distribution.

36. BASEBALL The number of hits by each Wildcats player during a doubleheader is shown in the frequency distribution. (Lesson 11-2)

    a. Construct and graph a probability distribution for the random variable X.

    b. Find and interpret the mean in the context of the situation.

    c. Find the variance and standard deviation.

Hits, X	Frequency
0	3
1	1
2	8
3	2
4	3

37. FOOTBALL The number of penalties a professional football team received for each game during two recent seasons is shown. Construct side-by-side box plots of the data sets. Then use this display to compare the distributions. (Lesson 11-1)

Season 1				Season 2			
8	11	6	13	9	1	3	5
9	18	16	11	8	3	6	4
15	14	14	9	10	6	3	1
8	5	10	5	5	5	3	2

Find the sum of each arithmetic series. (Lesson 10-2)

38. S_{51} of $-92 + (-88) + (-84) + \ldots$ **39.** 24th partial sum of $-13 + 2 + 17 + \ldots$ **40.** S_{46} of $295 + 281 + 267 + \ldots$

Find rectangular coordinates for each point with the given polar coordinates. (Lesson 9-3)

41. $\left(\frac{1}{4}, \frac{\pi}{2}\right)$ **42.** $\left(3, \frac{\pi}{3}\right)$ **43.** $(-2, \pi)$

Given v and u · v, find u. There may be more than one answer. (Lesson 8-5)

44. $\mathbf{v} = \langle -4, 2, -7 \rangle$, $\mathbf{u} \cdot \mathbf{v} = 17$ **45.** $\mathbf{v} = \langle 2, 8, 5 \rangle$, $\mathbf{u} \cdot \mathbf{v} = -6$ **46.** $\mathbf{v} = \left\langle \frac{2}{3}, -3, \frac{1}{3} \right\rangle$, $\mathbf{u} \cdot \mathbf{v} = 10$

Find the direction angle of each vector. (Lesson 8-2)

47. $6\mathbf{i} + 3\mathbf{j}$ **48.** $-3\mathbf{i} + 4\mathbf{j}$ **49.** $2\mathbf{i} - 8\mathbf{j}$

Write an equation of an ellipse with each set of characteristics. (Lesson 7-2)

50. vertices $(-3, 11)$, $(-3, -9)$; foci $(-3, 7)$, $(-3, -5)$ **51.** co-vertices $(-1, -6)$, $(-3, -6)$; length of major axis equals 10   **52.** vertices $(-4, 2)$, $(8, 2)$; length of minor axis equals 8

Skills Review for Standardized Tests

53. SAT/ACT If X is the sum of the first 1000 positive even integers and Y is the sum of the first 500 positive odd integers, about what percent greater is X than Y?

    A 100%   **C** 300%   **E** 500%

    B 200%   **D** 400%

54. REVIEW In a recent year, the mean and standard deviation for scores on the ACT was 21.0 and 4.7. Assume that the scores were normally distributed. What is the approximate probability that a test taker scored higher than 30.4?

    F 1%   **H** 2%

    G 1.5%   **J** 2.5%

55. The length of each song in a music collection is normally distributed with $\mu = 4.12$ minutes and $\sigma = 0.68$ minutes. Find the probability that a song selected from the collection at random is longer than 5 minutes.

    A 10%   **C** 39%

    B 19%   **D** 89%

56. REVIEW Find $\mathbf{u} \cdot \mathbf{v}$.

    F -47 **H** -6

    G -24 **J** 47

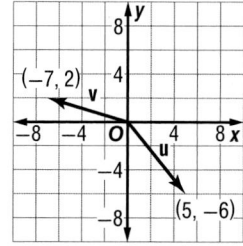

Objective
Use a graphing calculator to transform skewed data into data that resemble a normal distribution.

It is common for biological, medical, and other data to be positively skewed. It can sometimes be helpful to *transform* the original data so that it better resembles a normal distribution. This allows for the data to be spread out as opposed to being bunched at one end of a display.

ACTIVITY   **Transform Data Using Natural Logarithms**

Use the data to construct a histogram, and describe the shape of the distribution. Then transform the data by calculating the common logarithm of each entry. Graph the new data, and describe the shape of the distribution.

Data									
15	7	2	5	8	17	15	8	3	4
9	18	13	10	9	8	10	23	26	10
7	14	25	7	6	13	35	48	14	6

Step 1 Input the data into L1. Construct a histogram for the data using the intervals and scales shown.

The data appear to be positively skewed.

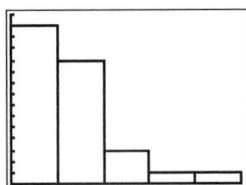

[0, 50] scl: 10 by [0, 15] scl: 1

Step 2 Input the common logarithm for each value into L2. Place the cursor on L2. Press **LOG** and enter L1. Press ENTER.

L1	L2	L3 2
15	1.1761	------
7	.8451	
2	.30103	
5	.69897	
8	.90309	
17	1.2304	
15	1.1761	

L2 =log(L1)█

Step 3 Construct a histogram for the new data using the intervals and scales shown.

The data appear to have a normal distribution.

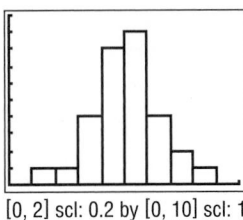

[0, 2] scl: 0.2 by [0, 10] scl: 1

Data may also be transformed by calculating the square roots or powers of the entries. When data are transformed, the type of operation performed should always be specified. A transformation will not always result in the new data being normally distributed.

Exercise

Use the data to construct a histogram, and describe the shape of the distribution. Then transform the data by calculating the square root of each entry. Graph the new data, and describe the shape of the distribution. Explain how the transformation affected the summary statistics.

Data									
23	30	36	39	36	24	31	33	42	36
26	32	46	45	27	34	52	41	28	33
43	20	24	34	30	40	29	35	61	35

11-4

The Central Limit Theorem

Then
You used the normal distribution to find probabilities for intervals of data values in distributions. (Lesson 11-3)

Now
- Use the Central Limit Theorem to find probabilities.
- Find normal approximations of binomial distributions.

New Vocabulary
sampling distribution
standard error of the mean
sampling error
continuity correction factor

Math Online >
glencoe.com

Why?

In manufacturing processes, quality control systems are used to determine when a process is outside of upper and lower control limits or "out of control." The mean of the process is controlled; therefore, successive sample means should be normally distributed around the actual mean. You can determine whether a process is out of control by finding the mean and standard deviation of the sample means.

The Central Limit Theorem Sampling is an important statistical tool in which subgroups of a population are selected so that inferences can be made about the entire population. The means of these subgroups, or sample means, can be compared to the mean of the population by using a sampling distribution. A **sampling distribution** is a distribution of the means of random samples of a certain size that are taken from a population.

Consider a population consisting of 16, 18, 20, 20, 22, and 24, with $\mu = 20$ and $\sigma = 2.582$. Suppose 12 random samples of size 2 are taken, with replacement. The mean $\bar{x}$ of each sample is shown.

Sample	$\bar{x}$	Sample	$\bar{x}$	Sample	$\bar{x}$
20,22	21	20,18	19	22,22	22
22,18	20	16,22	19	18,18	18
20,24	22	24,16	20	20,16	18
20,20	20	20,24	22	24,22	23

The distribution of the means of the 12 random samples, shown in Figure 11.4.1, does not appear to be normal. However, if all 36 samples of size of 2 from the population are found, the distribution of sample means will approach the normal distribution, as shown in Figure 11.4.2.

Figure 11.4.1

Figure 11.4.2

The mean of the means of every possible sample of size 2 from the population is

$$\mu_{\bar{x}} = \frac{16 + 17 + \dots + 24}{36} = \frac{720}{36} \text{ or } 20.$$

Notice that this value is equal to the population mean $\mu = 20$. So, when the mean of the means of every possible sample of size 2 are found, $\mu_{\bar{x}} = \mu$. The standard deviation of the sample means $\sigma_{\bar{x}}$ and the standard deviation of the population σ when divided by the square root of the sample of size n are

$$\sigma_{\bar{x}} = \frac{\sqrt{(16 - 20)^2 + (17 - 20)^2 + \dots + (24 - 20)^2}}{36} \approx 1.826 \qquad \text{and} \qquad \frac{\sigma}{\sqrt{n}} = \frac{2.582}{\sqrt{2}} \approx 1.826.$$

Since these two values are equal, the standard deviation of the sample means, also known as the **standard error of the mean**, can be found by using the formula $\sigma_{\bar{x}} = \frac{\sigma}{\sqrt{n}}$.

In general, randomly selected samples will have sample means that differ from the population mean. These differences are caused by **sampling error**, which occurs because the sample is not a complete representation of the population. However, if *all* possible samples of size n are taken from a population with mean μ and a standard deviation σ, the distribution of sample means will have:

- a mean $\mu_{\bar{x}}$ that is equal to μ and
- a standard deviation $\sigma_{\bar{x}}$ that is equal to $\frac{\sigma}{\sqrt{n}}$.

When the sample size n is large, regardless of the shape of the original distribution, the Central Limit Theorem states that the shape of the distribution of the sample means will approach a normal distribution.

Key Concept　　　**Central Limit Theorem**

As the sampling size n increases:

- the shape of the distribution of the sample means of a population with mean μ and standard deviation σ will approach a normal distribution and
- the distribution will have a mean μ and standard deviation $\sigma_{\bar{x}} = \frac{\sigma}{\sqrt{n}}$.

The Central Limit Theorem can be used to answer questions about sample means in the same way that the normal distribution was used to answer questions about individual values. In this case, we can use a formula for the z-value of a sample mean.

Key Concept　　　***z*-Value of a Sample Mean**

The z-value for a sample mean in a population is given by $z = \frac{\bar{x} - \mu}{\sigma_{\bar{x}}}$, where $\bar{x}$ is the sample mean, μ is the mean of the population, and $\sigma_{\bar{x}}$ is the standard error.

EXAMPLE 1　　　**Use the Central Limit Theorem**

AGE According to a recent study, the average age that an American adult leaves home is 26 years old. Assume that this variable is normally distributed with a standard deviation of 2.4 years. If a random sample of 20 adults is selected, find the probability that the mean age the participants left home is greater than 25 years old.

Since the variable is normally distributed, the distribution of the sample means will be approximately normal with $\mu = 26$ and $\sigma_{\bar{x}} = \frac{2.4}{\sqrt{20}}$ or about 0.537. Find the z-value.

$z = \dfrac{\bar{x} - \mu}{\sigma_{\bar{x}}}$　　　***z*-value for a sample mean**

$= \dfrac{25 - 26}{0.537}$　　　$\bar{x} = 25, \mu = 26$, and $\sigma_{\bar{x}} = 0.537$

≈ -1.86　　　**Simplify.**

The area to the right of a z-value of -1.86 is 0.9685. Therefore, the probability that the mean age of the sample is greater than 25 or $P(\bar{x} > 25)$ is about 96.85%.

✓ Guided Practice

1. **TORNADOES** The average number of tornadoes in Kansas is 47 per year, with a standard deviation of approximately 14.2 tornadoes. If a random sample of 15 years is selected, find the probability that the mean number of tornadoes is less than 50.

You can also determine the probability that a sample mean will fall within a given interval of the sampling distribution.

EXAMPLE 2 / **Find the Area Between Two Sample Means**

BATTERY LIFE A company that produces rechargeable batteries is designing a battery that will need to be recharged after an average of 19.3 hours of use. Assume that the distribution is normal with a standard deviation of 2.4 hours. If a random sample of 20 batteries is selected, find the probability that the mean life of the batteries before recharging is between 18 and 20 hours.

The area that corresponds to an interval of 18 to 20 hours is shown at the right.

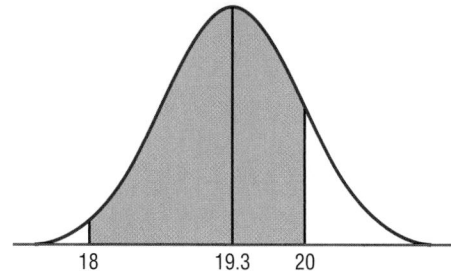

First, find the standard deviation of the sample means.

$$\sigma_{\bar{x}} = \frac{\sigma}{\sqrt{n}} \qquad \text{Standard deviation of a sample mean}$$

$$= \frac{2.4}{\sqrt{20}} \qquad \sigma = 2.4 \text{ and } n = 20$$

$$\approx 0.536 \qquad \text{Simplify.}$$

Use the z-value formula for a sample mean to find the corresponding z-values for 18 and 20.

z-value for $\bar{x} = 18$:

$$z = \frac{\bar{x} - \mu}{\sigma_{\bar{x}}} \qquad \text{z-value formula for a sample mean}$$

$$= \frac{18 - 19.3}{0.536} \qquad \bar{x} = 18, \mu = 19.3, \text{ and } \sigma_{\bar{x}} = 0.536$$

$$\approx -2.42 \qquad \text{Simplify.}$$

z-value for $\bar{x} = 20$:

$$z = \frac{\bar{x} - \mu}{\sigma_{\bar{x}}} \qquad \text{z-value formula for a sample mean}$$

$$= \frac{20 - 19.3}{0.536} \qquad \bar{x} = 20, \mu = 19.3, \text{ and } \sigma_{\bar{x}} = 0.536$$

$$\approx 1.30 \qquad \text{Simplify.}$$

Using a graphing calculator, select **normalcdf(** to find the area between $z = -2.42$ and $z = 1.30$.

The area between z-values of -2.42 and 1.30 is 0.8954. Therefore, $P(18 < \mu < 20)$ is 89.54%. So, the probability that the mean life of the batteries is between 18 and 20 hours is 89.54%.

✔ Guided Practice

2. **DAIRY** The average cost of a gallon of milk in a U.S. city is $3.49 with a standard deviation of $0.24. If a random sample of 40 1-gallon containers of milk is selected, find the probability that the mean of the sample will be between $3.40 and $3.60.

EXAMPLE 3 **Analyze Individual Values and Sample Means**

CLASS SIZE According to a recent study, the average class size in high schools nationwide is 24.7 students per class. Assume that the distribution is normal with a standard deviation of 3.6 students.

a. Find the probability that a randomly selected class will have fewer than 23 students.

The question is asking for an individual value in which $P(x < 23)$. Use the z-value formula for an individual data value to find the corresponding z-value.

$z = \dfrac{X - \mu}{\sigma}$ **z-value formula for an individual value**

$= \dfrac{23 - 24.7}{3.6}$ or about -0.47 **X = 23, μ = 24.7, and σ = 3.6**

The area associated with $z < -0.47$, or $P(z < -0.47)$, is 0.3192. Therefore, the probability that a randomly selected class has fewer than 23 students is 31.9%.

b. If a sample of 15 classes is selected, find the probability that the mean of the sample will be fewer than 23 students per class.

This question deals with a sample mean, so use the z-value formula for a sample mean to find the corresponding z-value. First, find the standard error of the mean.

$\sigma_{\bar{x}} = \dfrac{\sigma}{\sqrt{n}}$ **Standard error of the mean**

$= \dfrac{3.6}{\sqrt{15}}$ or about 0.93 **σ = 3.6 and n = 15**

Next, find the z-value using the z-value formula for a sample mean.

$z = \dfrac{\bar{x} - \mu}{\sigma_{\bar{x}}}$ **z-value formula for a sample mean**

$= \dfrac{23 - 24.7}{0.93}$ or about -1.83 **$\bar{x}$ = 23, μ = 24.7, and $\sigma_{\bar{x}}$ = 0.93**

The area associated with $z < -1.83$, or $P(z < -1.83)$, is 0.0336. Therefore, the probability that a sample of 15 classes will have a mean class size of fewer than 23 students is 3.36%.

> **StudyTip**
>
> **z-Value Formulas** Notice that the difference between the z-value formula for an individual data value and the z-value formula for a sample mean is that $\bar{x}$ is substituted for X and $\sigma_{\bar{x}}$ is substituted for σ in the formula for an individual value.

✓ **Guided Practice**

3. APPLES Consumers in the U.S. eat an average of 19 pounds of apples per year. Assume that the standard deviation is 4 pounds and the distribution is approximately normal.

   A. Find the probability that a randomly selected person consumes more than 21 pounds of apples per year.

   B. If a sample of 30 people is selected, find the probability that the mean of the sample would be more than 21 pounds of apples per year.

Notice in Figure 11.4.3 that the probability that an individual class has fewer than 23 students is much greater than the probability associated with the mean of a sample being fewer than 23 shown in Figure 11.4.4. This means that as the sample size increases, the distribution becomes narrower and the variability decreases.

Figure 11.4.3

Figure 11.4.4

The Normal Approximation According to the Central Limit Theorem, any sampling distribution can approach the normal distribution as n increases. As a result, other distributions such as the binomial distribution can be approximated with the normal distribution. Recall from Lesson 11-2 that the binomial distribution can be determined by using the equation

$$P(X) = {}_nC_x \, p^x q^{n-x},$$

where n is the number of trials, p is the probability of success, and q is the probability of failure.

If the number of trials increases or the probability of success gets close to 0.5, the shape of the binomial distribution begins to resemble the normal distribution. For example, consider the binomial distribution at the right. When $p = 0.2$ and $n = 6$, the distribution is positively skewed.

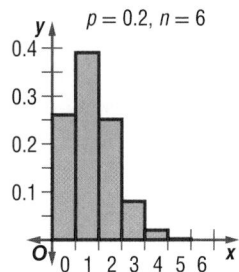

However, when $p = 0.5$ and $n = 6$ or when $p = 0.2$ and $n = 50$, as shown below, the distribution is approximately normal.

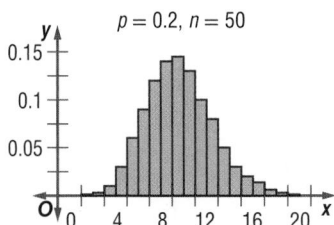

When the probability of success is close to 0 or 1 and the number of trials is relatively small, the normal approximation is not accurate. Therefore, as a rule, the normal approximation is typically used only when $np \geq 5$ and $nq \geq 5$.

Key Concept **Approximation Rule for Binomial Distributions**

Words The normal distribution can be used to approximate a binomial distribution when $np \geq 5$ and $nq \geq 5$.

Example If p is 0.4 and n is 5, then $np = 5(0.4)$ or 2. Since $2 < 5$, the normal distribution should not be used to approximate the binomial distribution.

It also is important to remember that the normal distribution should only be used to approximate a binomial distribution if the original variable is normally distributed or $n \geq 30$.

Since binomial distributions are *discrete* and normal distributions are *continuous*, a correction for continuity called the **continuity correction factor** must be used when approximating a binomial distribution. To use the correction factor, 0.5 unit is added to or subtracted from a given discrete boundary. For example, to find $P(X = 6)$ in a discrete distribution, the correction would be to find $P(5.5 < X < 6.5)$ for a continuous distribution, as shown below.

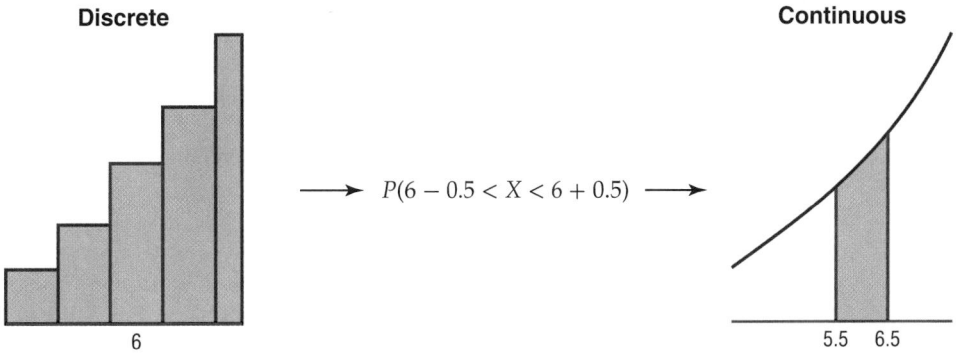

Use the following steps to approximate a binomial distribution with the normal distribution.

StudyTip

Binomial Formulas Recall from Lesson 11-2 that the mean μ and standard deviation σ of a binomial distribution are found using $\mu = np$ and $\sigma = \sqrt{npq}$, respectively.

Key Concept **Normal Approximation of a Binomial Distribution**

The procedure for the normal approximation of a binomial distribution is as follows.

Step 1 Find the mean μ and standard deviation σ.

Step 2 Write the problem in probability notation using X.

Step 3 Find the continuity correction factor, and rewrite the problem to show the corresponding area under the normal distribution.

Step 4 Find any corresponding z-values for X.

Step 5 Use a graphing calculator to find the corresponding area.

EXAMPLE 4 **Normal Approximation of a Binomial Distribution**

COLLEGE A school newspaper reported that 20% of the current senior class would be attending an out-of-state college. If 35 seniors are selected at random, find the probability that fewer than 5 of the seniors will be attending an out-of-state college.

In this binomial experiment, $n = 35$, $p = 0.2$, and $q = 0.8$.

Step 1 Find the mean μ and standard deviation σ.

$\mu = np$	**Mean and standard deviation of a binomial distribution**	$\sigma = \sqrt{npq}$
$= 35 \cdot 0.2$	$n = 35$, $p = 0.2$, and $q = 0.8$	$= \sqrt{35 \cdot 0.2 \cdot 0.8}$
$= 7$	**Simplify.**	≈ 2.37

Since $np = 35(0.2)$ or 7 and $nq = 35(0.8)$ or 28, which are both greater than 5, the normal distribution can be used to approximate the binomial distribution.

Step 2 Write the problem in probability notation using X.

The probability that fewer than 5 of the seniors will be attending an out-of-state college is $P(X < 5)$.

Step 3 Rewrite the problem with the continuity factor included.

Since the question is asking for the probability that *fewer than* 5 will be attending, subtract 0.5 unit from 5.

$P(X < 5) = P(X < 5 - 0.5)$ or $P(X < 4.5)$

Watch Out!

z-Value Formula When approximating the binomial distribution using the normal distribution, remember to use the z-value formula for an individual data value, not the formula for a sample mean.

Step 4 Find the corresponding z-value for X.

$z = \dfrac{X - \mu}{\sigma}$ **z-value formula**

$= \dfrac{4.5 - 7}{2.37}$ $X = 4.5$, $\mu = 7$, and $\sigma = 2.37$

≈ -1.05 **Simplify.**

Step 5 Use a graphing calculator to find the area to the left of z.

The approximate area to the left of $z = -1.05$ is 0.147, as shown at the right. Therefore, the probability that fewer than 5 seniors will be attending an out-of-state college in a random sample of 35 seniors is about 14.7%.

```
normalcdf(-4,-1.
05)
         .1468273946
```

✓ **Guided Practice**

4. **ADVERTISING** According to the results of an advertising survey sent to customers selected at random, 65% of the customers had not seen a recent television advertisement. Find the probability that from a sample of 50 customer responses, 15 or more did not see the advertisement.

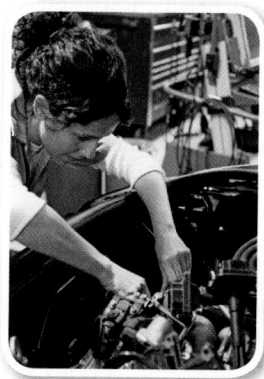

EXAMPLE 5 Normal Approximation of a Binomial Distribution

MANUFACTURING An automaker has discovered a defect in a new model. The defect is expected to affect 30% of the cars that were produced. What is the probability that there are at least 10 and at most 15 cars with the defect in a random sample of 40 cars?

In this binomial experiment, $n = 40$, $p = 0.3$, and $q = 0.7$.

Step 1 Begin by finding the mean μ and standard deviation σ.

$\mu = np$	**Mean and standard deviation of a binomial distribution**	$\sigma = \sqrt{npq}$
$= 40 \cdot 0.3$	$n = 40$, $p = 0.3$, and $q = 0.7$	$= \sqrt{40 \cdot 0.3 \cdot 0.7}$
$= 12$	**Simplify.**	≈ 2.9

Since $np = 40(0.3)$ or 12 and $nq = 40(0.7)$ or 28, which are both greater than 5, the normal distribution can be used to approximate the binomial distribution.

Step 2 Write the problem in probability notation: $P(10 \leq X \leq 15)$.

Step 3 Rewrite the problem with the continuity factor included.
$P(10 \leq X \leq 15) = P(10 - 0.5 < X < 15 + 0.5)$ or $P(9.5 \leq X \leq 15.5)$

Step 4 Find the corresponding z-values for $X = 9.5$ and $X = 15.5$.

$z = \dfrac{X - \mu}{\sigma}$	**z-value formula**	$z = \dfrac{X - \mu}{\sigma}$
$= \dfrac{9.5 - 12}{2.9}$	**Substitute**	$= \dfrac{15.5 - 12}{2.9}$
≈ -0.86	**Simplify.**	≈ 1.21

Step 5 Use a graphing calculator to find the area between $z = -0.86$ and $z = 1.21$.

The approximate area that corresponds to $-0.86 < z < 1.21$ is 0.692, as shown at the right. Therefore, the probability of there being at least 10 and at most 15 cars with the defect in a random sample of 40 cars is about 69.2%.

```
normalcdf(-0.86,
1.21)
        .6919660179
```

✔ **Guided Practice**

5. **MANUFACTURING** Suppose a defect in a second model by the same automaker is expected to affect 20% of the cars that were produced. What is the probability that there are at least 8 and at most 10 defects in a random sample of 30 cars?

It may seem difficult to know whether to add or subtract 0.5 unit from a discrete data value to find the continuity correction factor. The table below shows each case.

Concept Summary Binomial Distribution Correction Factors

When using the normal distribution to approximate a binomial distribution, the following correction factors should be used, where c is a given data value in the binomial distribution.

Binomial	Normal
$P(X = c)$	$P(c - 0.5 < X < c + 0.5)$
$P(X > c)$	$P(X > c + 0.5)$
$P(X \geq c)$	$P(X > c - 0.5)$
$P(X < c)$	$P(X < c - 0.5)$
$P(X \leq c)$	$P(X < c + 0.5)$

Exercises

= **Step-by-Step Solutions** begin on page R29.

1. **VIDEO GAMES** The average prices for three video games at an online auction site are shown. Assume that the variable is normally distributed. (Examples 1 and 2)

Game	Average Price ($)
Column Craze	35
Dungeon Attack!	45
Space Race	52

   a. For a sample of 35 online prices for Column Craze, find the probability that the mean price is more than $38, if the standard deviation is $9.

   b. For a random sample of 40 online prices for Space Race, find the probability that the mean price will be between $50 and $55 if the standard deviation is $12.

2. **CHEWING GUM** Americans chew an average of 182 sticks of gum per year. Assume a standard deviation of 13 sticks for each question. Assume that the variable is normally distributed. (Examples 1 and 2)

   a. Find the probability that 50 randomly selected people chew an average of 175 sticks or more per year.

   b. If a random sample of 45 people is selected, find the probability that the mean number of sticks of gum they chew per year is between 180 and 185.

3. **EXERCISE** The average number of days per week that Americans from four different age groups spent exercising during a recent year is shown. Assume that the variable is normally distributed. (Examples 1 and 2)

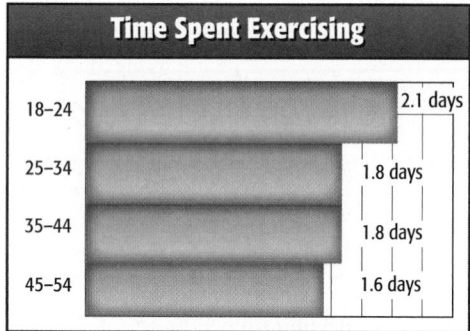

   a. Find the probability that a random sample of 30 Americans ages 45 to 54 spent more than 1.5 days a week exercising, if the standard deviation is 0.5 day.

   b. Assuming a standard deviation of 1.2 days, in a random sample of 30 Americans ages 18 to 24, find the probability that the average time spent exercising is between 2 to 2.5 days per week.

4. **MEDICINE** The mean recovery time for patients with a certain virus is 4.5 days with a standard deviation of 2 days. Assume that the variable is normally distributed. (Examples 1 and 2)

   a. Find the probability of an average recovery time of less than 4 days for a random sample of 75 people.

   b. In a random sample of 80 people, find the probability that average recovery time is between 4.4 and 4.8 days.

5. **TOURISM** The average number of tourists that visit a national monument every month is 55,000, with a standard deviation of 8000. Assume that the variable is normally distributed. (Example 3)

   a. If a random month is selected, find the probability that there would be fewer than 50,000 visiting tourists.

   b. If a sample of 10 months is selected, find the probability that there would be fewer than 50,000 visiting tourists.

6. **NUTRITION** The average protein content of a certain brand of energy bar is 12 grams with a standard deviation of 2 grams. Assume that the variable is normally distributed. (Example 3)

   a. Find the probability that a randomly selected bar will have more than 10 grams of protein.

   b. In a sample of 15 bars, find the probability that the average protein content will be greater than 10 grams.

7. **WORLD CUP** In a recent year, 33% of Americans said that they were planning to watch the World Cup soccer tournament. What is the probability that in a random sample of 45 people, fewer than 14 people plan to watch the World Cup? Assume that the variable is normally distributed. (Example 4)

8. **MOVIES** According to a national poll, in a recent year, 27% of Americans saw 5 or more movies in theaters. What is the probability that in a random sample of 40 people, between 6 and 11 people saw more than 5 movies in a movie theater that year? Assume that the variable is normally distributed. (Example 5)

9. **LIBRARY** A poll was conducted at a library to approximate the percent of books, CDs, magazines, and movies that were checked out during one month. The results are shown. Assume that the variable is normally distributed. (Examples 4 and 5)

Resources	Percent
books	45
CDs	20
magazines	3
movies	32

   a. What is the probability that of 65 randomly selected resources, exactly 35 were books?

   b. Find the probability that of 85 randomly selected resources, at least 15 and at most 18 were CDs.

10. **DRIVING** A driving instructor has found that 12% of students cancel or forget about lessons. Assume that the variable is normally distributed. (Examples 4 and 5)

   a. If the instructor has 60 lessons what is the probability that more than 10 of the students will miss a lesson?

   b. What is the probability that of 80 lessons, exactly 7 students will miss a lesson?

11. TESTS A multiple-choice test consists of 50 questions, with possible answers A, B, C, and D. Find the probability that, with random guessing, the number of correct answers will be each of the following.

a. less than 18 **b.** exactly 12

c. at least 14 **d.** between 10 and 15

Find the minimum sample size needed for each probability so that the normal distribution can be used to approximate the binomial distribution.

12. $p = 0.1$ **13.** $p = 0.4$

14. $p = 0.5$ **15.** $p = 0.8$

16. BASKETBALL The average points per game scored by four different basketball players are shown.

Player	A	B	C	D
Average	8.1	6.3	4.9	10.3

a. Find the mean and standard deviation of the averages.

b. Identify each possible combination of 3 players' averages, and find the mean of each combination.

c. Find the mean of each of the means that you found in part **b**. How does this compare to the mean that you found in part **a**?

17. BICYCLES Consider the bicycle rim shown, where $\mu = 25$ inches and $\sigma = 0.125$ inch.

25 in.

The diameters for 10 random samples of 3 bicycle rims from a company's assembly line are shown.

Sample	Diameter	Sample	Diameter
1	25.2, 24.9, 25	6	24.9, 25.1, 24.8
2	25.1, 25, 24.8	7	25.3, 24.9, 25.1
3	25.3, 24.9, 24.8	8	25.4, 24.8, 25.3
4	24.9, 25.3, 25.2	9	24.8, 24.9, 25.2
5	25, 25.2, 24.7	10	25, 25.3, 24.7

a. Find $\bar{x}$ and s for each sample.

b. Construct a scatter plot with the sample number on the *x*-axis and the sample means on the *y*-axis.

c. In this process, the upper control limit is $\bar{x} + \dfrac{3\sigma}{\sqrt{n}}$ and the lower control limit is $\bar{x} - \dfrac{3\sigma}{\sqrt{n}}$. If the process is in control, all values should fall within the control limits. Use the graph from part **b** to determine whether the process is in control. Explain your reasoning.

18. BLOOD TYPES The distributions of blood types of U.S. and Canadian citizens are shown.

U.S.		Canada	
Type	Distribution	Type	Distribution
O	44%	O	46%
A	42%	A	42%
B	10%	B	9%
AB	4%	AB	3%

a. If 50 U.S. citizens are selected at random, find the probability that fewer than 20 of those chosen will have type O blood.

b. Find the probability that out of 100 randomly selected Canadian citizens, between 80 and 90 of those chosen will have types O or A blood.

c. What is the probability that two randomly chosen people from the U.S. or Canada will have the same blood type?

H.O.T. Problems   Use Higher-Order Thinking Skills

19. ERROR ANALYSIS Weston and Hannah are calculating results for a survey that they are taking as part of a summer internship. They found that of the residents they surveyed, 65% do not recycle. Weston found the probability that fewer than 30 out of 50 random residents do not recycle is 18.7%, while Hannah found that it would be 27.7%. Is either of them correct? Explain your reasoning.

20. WRITING IN MATH Explain how the Central Limit Theorem can be used to describe the shape, center, and spread of a distribution of sample means.

21 CHALLENGE In the United States, 7% of the male population and 0.4% of the female population are color-blind. Suppose random samples of 100 men and 1500 women are selected. Is there a greater probability that the men's or women's sample will include at least 10 people who are color-blind? Explain your reasoning.

22. OPEN ENDED Give an example of a population and a sample of the population. Explain what is meant by the corresponding sampling distribution.

REASONING Determine whether each statement is *true* or *false*. Explain your reasoning.

23. As the sample size increases, a sampling distribution of sample means will approach the normal distribution.

24. In a binomial distribution, $P(X \geq c) \neq P(X > c)$.

25. WRITING IN MATH Explain why the normal distribution can be used to approximate a binomial distribution, what conditions are necessary to do so, and why a correction for continuity is needed.

26. **COMMUNITY SERVICE** A recent study of 1286 high school seniors revealed that the students completed an average of 38 hours of volunteer work over the summer with a standard deviation of 6.7 hours. Determine the number of seniors who completed more than 42 hours of community service. Assume that the variable is normally distributed. (Lesson 11-3)

27. **GAMES** Managers of a fitness club randomly surveyed 56 members and recorded the number of days that each member attended the club in a given week. Use the frequency distribution shown to construct a probability distribution for the random variable X. Then find the mean, variation, and standard deviation of the probability distribution. (Lesson 11-2)

Days, X	Frequency	Days, X	Frequency
0	3	4	11
1	5	5	9
2	10	6	3
3	14	7	1

Find each sum. (Lesson 10-2)

28. $\displaystyle\sum_{n=1}^{19} -50 + 5n$

29. $\displaystyle\sum_{n=12}^{68} 5 - \frac{n}{4}$

30. $\displaystyle\sum_{n=10}^{16} 24n - 90$

Find the specified term of each sequence. (Lesson 10-1)

31. 7th term, $a_n = (a_{n-1} - 6)^2$, $a_1 = 4$

32. 6th term, $a_n = 3n^2 - 4n$

33. 4th term, $a_n = (a_{n-1})^2 - 11$, $a_1 = 3$

Find rectangular coordinates for each point with the given polar coordinates. (Lesson 9-3)

34. $\left(2, \frac{\pi}{2}\right)$

35. $\left(\frac{1}{4}, \frac{\pi}{4}\right)$

36. $(6, 210°)$

Find each of the following for p $= \langle 4, 0 \rangle$, q $= \langle -2, -3 \rangle$, and t $= \langle -4, 2 \rangle$. (Lesson 8-2)

37. p $-$ t $-$ 2q

38. q $-$ 4p $+$ 3t

39. 4p $+$ 3q $-$ 6t

Write an equation for and graph each parabola with focus F and the given characteristics. (Lesson 7-1)

40. $F(-6, 8)$; opens up; contains $(0, 16)$

41. $F(2, -5)$; opens down; contains $(10, -11)$

42. **YOGURT** The Frozen Yogurt Shack sells cones in three sizes: small, \$2.89; medium, \$3.19; and large, \$3.39. On Friday, 78 cones were sold totaling \$246.42. The Shack sold six more medium cones than small cones that day. Use Cramer's Rule to determine the number of each type of cone sold on Friday. (Lesson 6-3)

43. **SAT/ACT** What is the value of x?

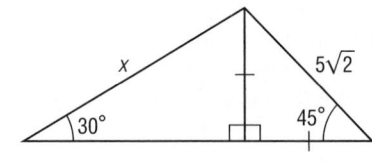

   A $2\sqrt{2}$　　　C $5\sqrt{3}$　　　E $5\sqrt{6}$

   B 5　　　D 10

44. **REVIEW** In a study, 62% of registered voters said they voted in the 2008 presidential election. If 6 registered voters are chosen at random, what is the probability that at least 4 of them voted?

   F 32%　　　H 58.6%

   G 41.2%　　　J 73.2%

45. The average number of patients who are seen every week at a certain hospital is normally distributed. The average per week is 12,423, with a standard deviation of 3269. If a week is selected at random, find the probability that there would be fewer than 4000 patients.

   A 0.50%　　　C 32.20%

   B 2.37%　　　D 36.73%

46. **REVIEW** Find the area that corresponds to the shaded region of this standard normal distribution.

   F 0.02

   G 0.04

   H 0.96

   J 0.98

1. **AUDITIONS** The ages of 20 students who auditioned for roles in a high school production of *Gone with the Wind* are shown. (Lesson 11-1)

Ages of Students				
14	15	17	16	14
16	17	16	18	16
15	16	18	15	17
14	18	15	17	16

   a. Construct a histogram, and use it to describe the shape of the distribution.

   b. Summarize the center and spread of the data using either the mean and standard deviation or the five-number summary. Justify your choice.

2. **VACATION** Angie is planning a trip for spring vacation. She has narrowed her choices down to two locations. The temperatures during twelve days around the time of spring vacation are shown below for each location. (Lesson 11-1)

Cape Hatteras, North Carolina					
52	60	62	57	55	63
64	59	54	52	54	60
Orlando, Florida					
77	77	76	76	72	71
72	74	74	72	73	73

   a. Construct side-by-side box plots of the data sets, and use this display to compare the center and spread of the distributions.

   b. Which location has the greater variation in temperature?

3. **MULTIPLE CHOICE** Which of the following displays a data set that is positively skewed? (Lesson 11-1)

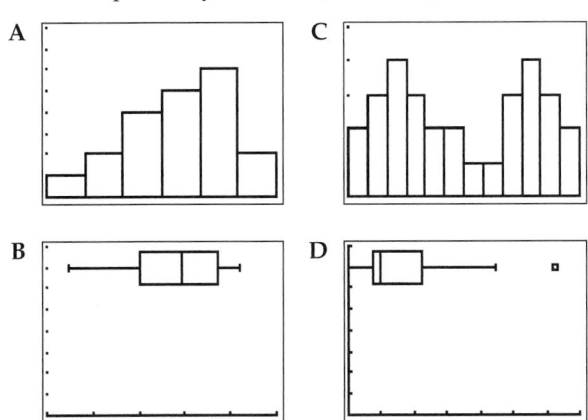

Classify each random variable X as *discrete* or *continuous*. Explain your reasoning. (Lesson 11-2)

4. X represents the number of times a coin lands on heads if flipped a random number of times.

5. X represents the time it takes for a randomly chosen marathon runner to complete a race.

6. **TRAVEL** In a recent poll, 20% of teenagers said that they have visited Washington, D.C. Find the probability that out of 6 randomly chosen teenagers, at least 3 have visited the nation's capital. (Lesson 11-2)

7. **SHAMPOO** The amount of water in milliliters in a particular shampoo is normally distributed with $\mu = 125$ and $\sigma = 7$. Find each of the following. (Lesson 11-3)

   a. $P(X < 105)$

   b. $P(X > 140)$

   c. $P(115 < X < 130)$

8. **GOLF** A random sample of 130 golfers resulted in an average score of 78 with a standard deviation of 6.3. Find the number of golfers with an average of 70 or lower. (Lesson 11-3)

Find the area that corresponds with the shaded region. (Lesson 11-3)

9.
   -0.62

10.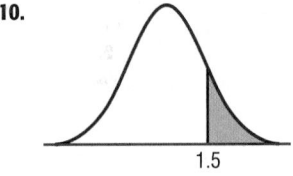
   1.5

11. **PROJECTS** The scores on a science project for one class are normally distributed with $\mu = 78$ and $\sigma = 8$. Find each probability. (Lesson 11-3)

   a. $P(X \geq 96)$

   b. $P(60 < X < 85)$

Find the probability of each sample mean. (Lesson 11-4)

12. $P(\bar{x} < 38); \mu = 40, \sigma = 5.5, n = 25$

13. $P(\bar{x} > 82.2); \mu = 82.5, \sigma = 4.1, n = 50$

14. **EMPLOYMENT** According to a recent study, the average age that a person starts his or her first job is 16.8 years old. Assume that this variable is normally distributed with a standard deviation of 1.7 years. If a random sample of 25 people is selected, find the probability that the mean age the participants started their first jobs is greater than 17 years old. (Lesson 11-4)

Confidence Intervals

Why?

Executives at a film studio want to know the average age of people seeing a movie. A survey of 200 people who saw the movie finds that the average age was 20.4 years. The studio executives decide to estimate the mean age a for all customers as between 18.1 and 22.7 or $18.1 < a < 22.7$.

Then
You analyzed sample means and the effect of the Central Limit Theorem on a sampling distribution. (Lesson 11-4)

Now
- Use the normal distributions to find confidence intervals for the mean.
- Use t-distributions to find confidence intervals for the mean.

New Vocabulary
inferential statistics
parameter
point estimate
interval estimate
confidence level
maximum error of estimate
critical value
confidence interval
t-distribution
degrees of freedom

Math Online
glencoe.com

Normal Distribution In **inferential statistics**, a sample of data is analyzed and conclusions are made about the entire population. This procedure is used because it is usually too challenging to get information from every member of a population. A measure that describes a characteristic of a population, such as the mean or standard deviation, is called a **parameter**. Many different parameters may be used to analyze data; but in this lesson, we will concentrate on the mean.

The average age of 20.4 years is an example of a **point estimate**, a single value estimate of an unknown population parameter. Due to sampling error and relatively small sampling size, the point estimate will most likely not match the population mean. For this reason, the studio executives used an **interval estimate** of $18.1 < a < 22.7$. An interval estimate is a range of values used to estimate an unknown population parameter. To form an interval estimate, a point estimate is used as the center of the interval and a margin of error is added to and subtracted from the point estimate. For this study, the studio executives let the margin of error be 2.3 years.

Before an interval estimate is created, it is helpful to know just how reliable you want it to be. The probability that the interval estimate will include the actual population parameter is known as the **confidence level**, and is denoted as c. We can illustrate a confidence level using the normal distribution if the standard deviation of the population σ is known and the population is normally distributed or if $n \geq 30$. Recall that the Central Limit Theorem states that when $n \geq 30$, the sampling distribution of sample means resembles a normal distribution.

The confidence level for a normal distribution is equal to the area under the standard normal curve between $-z$ and z, as shown. The remaining area in the two tails is then $1 - c$, or $\frac{1}{2}(1 - c)$ for each tail.

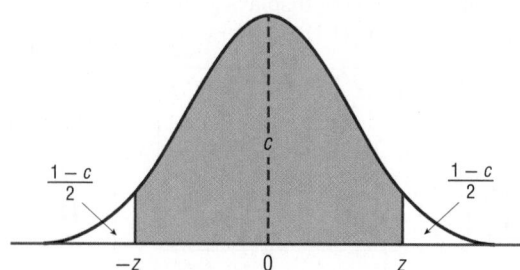

Suppose you are conducting an experiment in which you want a confidence level of 95%. When $c = 95\%$, 2.5% of the area lies to the left of $-z$ and 2.5% lies to the right of z. Using a graphing calculator, you can find the corresponding value for $-z$ to be -1.96 and z to be 1.96. By calculating the product of the z-values and the standard deviation of the sample means $\sigma_{\bar{x}}$, the **maximum error of estimate** E, the maximum difference between the point estimate and the actual value of the parameter, can be determined.

Key Concept — Maximum Error of Estimate

The maximum error of estimate E for a population mean μ is given by

$$E = z \cdot \sigma_{\bar{x}} \text{ or } z \cdot \frac{\sigma}{\sqrt{n}},$$

where z is a critical value that corresponds to a particular confidence level, and $\sigma_{\bar{x}}$ or $\frac{\sigma}{\sqrt{n}}$ is the standard deviation of the sample means. When $n \geq 30$, s, the sample standard deviation, may be substituted for σ.

StudyTip

Maximum Error of Estimate The maximum error of estimate E will be a positive value since it is the maximum difference between the point estimate and the actual value of the parameter.

The z-values that correspond to a particular confidence level are known as **critical values**. The three most widely used confidence levels and their corresponding critical values are shown below.

Confidence Level	z-Value
90%	1.645
95%	1.960
99%	2.576

EXAMPLE 1 — Find Maximum Error of Estimate

TEXTBOOKS A poll of 75 randomly selected college students showed that the students spent an average of $230 on textbooks per session. Assume from past studies that the standard deviation is $55. Use a 99% confidence level to find the maximum error of estimate for the amount of money spent by students on textbooks.

In a 99% confidence interval, 0.5% of the area lies in each tail. You can find the corresponding z-value to be 2.576 by using a graphing calculator or the table above.

$$E = z \cdot \frac{\sigma}{\sqrt{n}} \qquad \textbf{Maximum Error of Estimate}$$

$$= 2.576 \cdot \frac{55}{\sqrt{75}} \qquad \textbf{\textit{z} = 2.576, } \sigma \textbf{ = 55, and } \textit{n} \textbf{ = 75}$$

$$\approx 16.36 \qquad \textbf{Simplify.}$$

This means that you can be 99% confident that the population mean of money spent on textbooks will be no more than $16.36 from the sample mean of $230.

✓ Guided Practice

1. **MUSIC** Executives at a music label surveyed 125 people who actively download music and found that the listeners have an average of 740 MP3s downloaded to their computers. Assume a standard deviation of 86 MP3s. Use a 94% confidence level to find the maximum error of estimate for the amount of MP3s on the computer of someone who actively downloads music.

Once a confidence level is established and a maximum error of estimate is calculated, it can be used to determine a **confidence interval**. A confidence interval, denoted CI, is a specific interval estimate of a parameter and can be found when the maximum error of estimate is added to and subtracted from the sample mean.

Key Concept — Confidence Interval for the Mean

A confidence interval CI for a population mean μ is given by

$$CI = \bar{x} \pm E \text{ or } \bar{x} \pm z \cdot \frac{\sigma}{\sqrt{n}},$$

where $\bar{x}$ is the sample mean and E is the maximum error of estimate.

EXAMPLE 2 **Find Confidence Intervals When σ is Known**

HOMEWORK A poll of 20 randomly selected high school students showed that the students spent a mean time of 35 minutes per weeknight on homework. Assume a normal distribution with a standard deviation of 12 minutes. Find a 90% confidence interval for the mean of all of the students.

Substitute 1.645, the corresponding z-value for a 90% confidence interval, into the confidence interval formula.

$$CI = \bar{x} \pm z \cdot \frac{\sigma}{\sqrt{n}}$$ **Confidence Interval for the Mean**

$$= 35 \pm 1.645 \cdot \frac{12}{\sqrt{20}}$$ $\bar{x} = 35, z = 1.645, \sigma = 12$, and $n = 20$

$$\approx 35 \pm 4.41$$ **Simplify.**

Add and subtract the margin of error.

Left Boundary	**Right Boundary**
$35 - 4.41 = 30.59$	$35 + 4.41 = 39.41$

A 90% confidence interval is $30.59 < \mu < 39.41$. Therefore, with 90% confidence, the mean time spent on homework by students is between 30.6 and 39.4 minutes.

✔ Guided Practice

2. SHOPPING A sample of 65 randomly selected mall patrons showed that they spent an average of $70 that day. Assume a standard deviation of $12. Find a 95% confidence interval for the average amount spent by a mall patron that day.

In many real-world situations, the population standard deviation is unknown. When this is the case, the standard deviation s of the sample can be used in place of the population standard deviation, as long as the variable is normally distributed and $n \geq 30$.

EXAMPLE 3 **Find Confidence Intervals When σ is Unknown**

ENGINEERING Tensile strength is the stress at which a material breaks or deforms. A company wants to estimate the mean tensile strength of a new material. A random sample of 40 units is normally distributed with an average tensile strength of 36.3 ksi, or 36,300 pounds per square inch, and a standard deviation of 2.9 ksi. Find a 98% confidence interval for the mean tensile strength of the material.

In a 98% confidence interval, 1% of the area lies in each tail. You can find the corresponding z-value to be 2.33 by using a graphing calculator.

```
invNorm(0.01)
          -2.326347877
invNorm(1-0.01)
           2.326347877
```

Since the distribution is normal and $n \geq 30$, the sample standard deviation can be used to find a confidence interval.

$$CI = \bar{x} \pm z \cdot \frac{s}{\sqrt{n}}$$ **Confidence Interval for the Mean**

$$= 36.3 \pm 2.33 \cdot \frac{2.9}{\sqrt{40}}$$ $\bar{x} = 36.3, z = 2.33, s = 2.9$, and $n = 40$

$$\approx 36.3 \pm 1.06$$ **Simplify.**

Therefore, a 98% confidence interval is $35.2 < \mu < 37.4$.

✔ Guided Practice

3. ENGINEERING Suppose a random sample of 50 units of the same material is normally distributed with an average tensile strength of 39.2 ksi and a standard deviation of 3.1 ksi. Estimate the mean tensile strength with 99% confidence.

🌀 Real-World Career

Engineering Engineers use science and math to find economical solutions to technical problems. A bachelor's degree in engineering is usually required for most entry-level jobs.

TechnologyTip

▶ **Calculate Confidence Intervals** You can check your answer by using a graphing calculator. Press [STAT] and select ZInterval under the TESTS menu. For Input: select Stats and then enter each of the values. Then select Calculate.

StudyTip

Distributions That Are
Not Normal You cannot
use a normal distribution or
t-distribution to construct a
confidence interval if the
population is not normally
or approximately normally
distributed.

t-Distribution In many cases, the population standard deviation is not known and, due to constraints such as time and cost, sample sizes exceeding 30 are not realistic. In these cases, another distribution called the *t*-distribution can be used, as long as the variable is approximately normally distributed.

The **t-distribution** is a family of curves that are dependent on a parameter known as the *degrees of freedom*. The **degrees of freedom (d.f.)** are equal to $n - 1$ and represent the number of values that are free to vary after a sample statistic is determined.

For example, if $\bar{x} = 4$ in a sample of 10 values, 9 of the 10 values are free to vary. Once the first 9 values are selected, the tenth value must be a specific number in order for $\bar{x} = \frac{40}{10}$. So, the degrees of freedom are $10 - 1$ or 9, which corresponds to a specific curve.

Notice in the figure below that as the degrees of freedom increase, or as d.f. approaches 30, the *t*-distribution approaches the standard normal distribution.

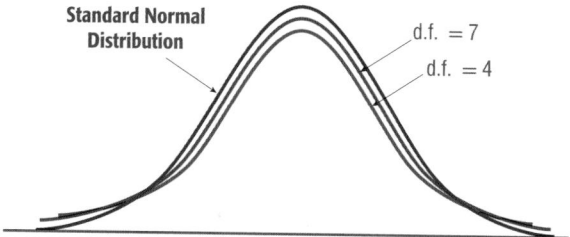

The characteristics of the *t*-distribution are summarized below.

Key Concept — **Characteristics of the *t*-Distribution**

- The distribution is bell-shaped and symmetric about the mean.
- The mean, median, and mode equal 0 and are at the center of the distribution.
- The curve never touches the *x*-axis.
- The standard deviation is greater than 1.
- The distribution is a family of curves based on the sample size *n*.
- As *n* increases, the distribution approaches the standard normal distribution.

Similar to the normal distribution, the *t*-distribution can be used to construct a confidence interval by using a *t*-value rather than a *z*-value to calculate the maximum error of the estimate *E*. A *t*-value can be found by

$$t = \frac{\bar{x} - \mu}{\frac{s}{\sqrt{n}}} \text{ or } \frac{\bar{x} - \mu}{\sigma_{\bar{x}}}, \text{ where } \mu \text{ is the population mean.}$$

You will use a graphing calculator to find values for *t* since the population mean μ is the parameter that you are attempting to estimate. You can find a confidence interval when using the *t*-distribution by using the formula shown.

StudyTip

Distributions When
$n \geq 30$, it is standard
procedure to use the
normal distribution.
However, t-distributions
can still be used.

Key Concept — **Confidence Interval Using *t*-Distribution**

A confidence interval *CI* for the *t*-distribution is given by

$$CI = \bar{x} \pm t \cdot \frac{s}{\sqrt{n}},$$

where $\bar{x}$ is the sample mean, *t* is a critical value with $n - 1$ degrees of freedom, *s* is the sample standard deviation, and *n* is the sample size.

EXAMPLE 4 **Find Confidence Intervals with the *t*-Distribution**

CAPACITY The capacities of eight randomly selected tanks are measured. The mean capacity is 143 liters and the standard deviation is 3.0. Find the 95% confidence interval of the mean capacity of the tanks. Assume that the variable is normally distributed.

The population standard deviation is unknown and $n < 30$, so the *t*-distribution must be used. Since $n = 8$, there are $8 - 1$ or 7 degrees of freedom. You can use a graphing calculator to find the corresponding *t*-value.

In the **DISTR** Menu, select **InvT(*a*, df)**. The *a* value represents the area of one tail of the distribution and **df** represents degrees of freedom. So, for a 95% confidence interval, the area in either tail of the *t*-distribution is half of 5% or 0.025.

$CI = \bar{x} \pm t \cdot \dfrac{s}{\sqrt{n}}$ **Confidence Interval Using *t*-Distribution**

$= 143 \pm 2.365 \cdot \dfrac{3}{\sqrt{8}}$ $\bar{x} = 143$, $t = 2.365$, $s = 3$, and $n = 8$

$= 143 \pm 2.5$ **Simplify**.

Therefore, the 95% confidence interval is $140.5 < \mu < 145.5$.

CHECK You can check your answer by using a graphing calculator. In the **STAT** menu, select **TESTS** and **TInterval**. Under the **TInterval** menu, select **Stats** and enter each of the values. Then select **Calculate**.

 Guided Practice

4. RESTAURANTS The waiting time of ten randomly selected customers at a restaurant was measured with a mean of 25 minutes and a standard deviation of 4 minutes. Find the 99% confidence interval of the mean waiting time for all customers, assuming that the variable is normally distributed.

StudyTip

Using the *t*-Distribution Most real-world inferences about the population mean will be completed using the *t*-value because σ is rarely known.

It may be difficult to determine whether to use a normal distribution or a *t*-distribution for a given problem. The chart shown below summarizes when to use each, assuming that the population is normally or approximately normally distributed.

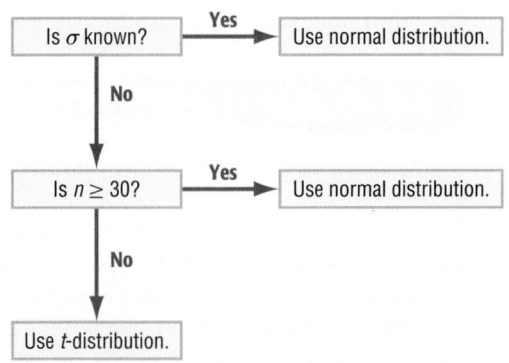

In all statistical experiments, the user determines the confidence level, which directly affects the confidence interval. With all other variables held constant, increasing the confidence level will expand the confidence interval. Expanding a confidence interval reduces the accuracy of the estimate. For example, observe the confidence interval from Example 2 when the confidence level is raised to 99%.

	90% Confidence Level	99% Confidence Level
z-value	1.645	2.576
E	4.41	6.91
CI	$30.59 < \mu < 39.41$	$28.09 < \mu < 41.91$

Generally, a high confidence level and a small confidence interval are desired. This can be achieved by increasing the sample size n. You can find the minimum sample size needed for a specific maximum error of estimate by starting with the formula for E and solving for n.

$$E = z \cdot \frac{\sigma}{\sqrt{n}} \qquad \text{Maximum Error of Estimate}$$

$$\sqrt{n} \cdot E = z \cdot \sigma \qquad \text{Multiply each side by } \sqrt{n}.$$

$$\sqrt{n} = \frac{z \cdot \sigma}{E} \qquad \text{Divide each side by } E.$$

$$n = \left(\frac{z\sigma}{E}\right)^2 \qquad \text{Square each side.}$$

Key Concept   Minimum Sample Size Formula

The minimum sample size needed when finding a confidence interval for the mean is given by $n = \left(\frac{z\sigma}{E}\right)^2$, where n is the sample size and E is the maximum error of estimate.

StudyTip

Rounding It is not possible to have a sample size that is a fraction. Therefore, when finding a minimum sample size, always round answers in the form of a fraction or a decimal to the next greater whole number.

EXAMPLE 5   Find a Minimum Sample Size

PRODUCT DEVELOPMENT You are testing the reliability of a thermometer. Your manager asks you to conduct an experiment with results that are accurate to ± 0.05 degree with 95% confidence. If $\sigma = 0.8$, how many measurements are needed?

$$n = \left(\frac{z\sigma}{E}\right)^2 \qquad \text{Minimum Sample Size Formula}$$

$$= \left(\frac{1.96(0.8)}{0.05}\right)^2 \qquad z = 1.96, \sigma = 0.8, \text{ and } E = 0.05$$

$$= 983.45 \qquad \text{Simplify.}$$

At least 984 observations are needed to have a margin of error of ± 0.05 with 95% confidence.

✓ Guided Practice

5. **MARKETING** Executives at a car dealership want to estimate the average age of their customers before making a television commercial. They want to be 90% confident that the mean age is ± 2 years of the sample mean. If the standard deviation from a previous study is 12 years, how large should the sample be?

Exercises

1. **TRANSPORTATION** A random sample of 85 New York City residents showed that the average commuting time to work was 36.5 minutes. Assume that the standard deviation from previous studies was 11.3 minutes. Find the maximum error of estimate for a 99% confidence level. Then create a confidence interval for the mean commuting time of all New York City residents. (Examples 1–3)

2. **ORANGES** The owner of an orange grove randomly selects 50 oranges of the same type and weighs them with a resulting mean weight of 7.45 ounces and a standard deviation of 0.8 ounce. Find the maximum error of estimate for a 98% confidence level. Then estimate the mean weight of the oranges using a confidence interval. (Examples 1–3)

3. **TEMPERATURE** The average body temperature for 15 randomly selected polar bears was 97.5°F. Assume that the standard deviation from a recent study was 2.8°F. Find the maximum error of estimate for a 95% confidence level. Then estimate the mean body temperature for all polar bears in that region using a confidence interval. (Examples 1–3)

4. **TYPING SPEED** In a random sample of 20 students in a computer class, the average keyboarding speed was 40 words per minute (WPM) with a standard deviation of 8 WPM. Estimate the mean keyboarding speed for all students taking the class using a 90% confidence level. (Example 4)

5. **TEXT MESSAGES** A random sample of 25 students with cell phones found on average, the students send or receive 68 text messages a day with a standard deviation of 13 messages. Estimate the mean number of text messages for all students with cell phones using a 96% confidence level. (Example 4)

6. **COLLEGE VISITS** A random sample of 20 college-bound juniors found on average they visited 6.4 colleges with a standard deviation of 1.9. Estimate the mean number of college visits for all college-bound juniors using a 95% confidence level. (Example 4)

Determine whether the normal distribution or *t*-distribution should be used for each question. Then find each confidence interval given the following information. (Examples 2–4)

7. 90%; $\bar{x} = 128$, $s = 7$, $n = 20$.

8. 95%; $\bar{x} = 65$, $s = 15.9$, $n = 300$.

9. 95%; $\bar{x} = 39.4$, $s = 1.2$, $n = 15$

10. 98%; $\bar{x} = 122.3$, $\sigma = 2.2$, $n = 2000$

11. 99%; $\bar{x} = 28.3$, $\sigma = 4.5$, $n = 75$

12. 99%; $\bar{x} = 2489$, $\sigma = 18.3$, $n = 160$

13. **COFFEE** The owner of a coffee shop wants to determine the average price for a small cup of coffee in his city. How large should the sample be if he wishes to be accurate to within $0.015 at 90% confidence? A previous study showed that the standard deviation of the price was $0.10. (Example 5)

14. **TESTS** A teacher wants to estimate the average amount of time it takes students to finish a 25-question test. How large should the sample be if the teacher wishes to be 99% accurate within 8 minutes? A previous study showed that the standard deviation of the time was 11.3 minutes. (Example 5)

15. **SCHOOL** A survey was taken by 26 randomly selected students, recording the amount of time each student participated in after-school activities for a given week. Assume that the time is normally distributed.

Time (hours)						
11	7	2	7	6	12	9
10	8	6	4	8	8	7
4	7	8	8	6	5	
9	9	10	15	12	13	

a. Decide the type of distribution that can be used to estimate the population mean. Explain your reasoning.

b. Calculate the mean and the standard deviation to the nearest tenth.

c. Construct a 95% confidence interval for the average amount of time students participate in after-school activities.

d. Interpret the confidence interval in the context of the problem.

16. **WAGES** In a previous study, the standard deviation for starting wages among employed high school students was $0.50. A survey of 20 randomly selected employed high school students was conducted and their starting wages were recorded. Assume that the wages are normally distributed.

Wages ($)				
6.75	6.50	6.50	5.50	6.75
5.75	6.50	7.50	7.25	6.00
6.50	7.25	6.75	6.00	5.75
6.00	6.50	6.75	7.00	6.25

a. Decide the type of distribution that can be used to estimate the population mean. Explain your reasoning.

b. Calculate the mean to the nearest hundredth.

c. Construct a 90% confidence interval for the average starting wage for an employed high school student.

d. Interpret the confidence interval in the context of the problem.

17. AGE Julian wants to estimate the average age of teachers with a 95% confidence level. He knows that the standard deviation from past studies is 9 years. If Julian has only 50 teachers at his school to survey, how accurate can he make his estimate?

18. TELEVISION Felix and Tanya want to compare the average amount of time per day in minutes that boys and girls watch television. They surveyed 16 female students and 16 male students chosen at random and recorded the viewing times.

Female		Male	
115	120	90	140
125	130	120	110
120	120	105	115
125	105	125	120
110	115	105	130
105	110	150	125
120	125	120	110
110	115	115	90

  a. Calculate the mean and sample standard deviation for each data set.

  b. Construct two 99% confidence intervals for the average amount of time spent watching television for both boys and girls.

  c. Make a statement comparing the effectiveness of the two intervals.

19. RESTAURANT The owner of a restaurant wants a mean preparation time of 20 minutes for each order that is placed. To help ensure that the goal is achieved, the owner timed 24 randomly selected orders and found an average preparation time of 22 minutes with a standard deviation of 4 minutes. The owner will be satisfied if the goal preparation time falls within the 99% confidence interval at which the restaurant is currently operating. Is the owner satisfied? Explain your reasoning.

20. INCOME Diego is being transferred by his employer and has his choice of three cities. Before making a decision, he wants to compare the average incomes of his fellow employees in the three cities. With help from the human resource department, he records the following information. The sample size for each city was 35 employees.

City	$\bar{x}$ ($)	σ ($)
1	46,700	6300
2	47,800	3000
3	45,000	8000

  a. Construct a 95% confidence interval for the average income of employees in each city.

  b. If salary is all that is considered, to what city should Diego choose to be transferred? Explain your reasoning.

21. CELL PHONES A cell phone manufacturer wants the mean talk time, the time a phone is engaged in sending a message or transmitting a conversation, for its long-life batteries to be 62 hours. To ensure the quality of its batteries, the manufacturer randomly samples 14 phones and records the talk times in hours.

Talk Time (hours)						
61.0	63.1	63.3	59.1	63.4	61.5	60.0
62.6	62.3	60.3	62.9	61.3	62.4	63.6

The manufacturer will be satisfied if the mean talk time falls within the 99% confidence interval at which the batteries are currently operating. Is the manufacturer satisfied? Explain your reasoning.

22. READING Rama is conducting a study on the average amount of time that people between the ages of 17–25 spend reading each day. She surveys 20 people chosen at random. She currently has a confidence interval with a maximum error of estimate of 10 minutes. What will Rama's sample size need to be if she wants to reduce the error to 5 minutes? 2.5 minutes?

H.O.T. Problems *Use Higher-Order Thinking Skills*

23 CHALLENGE A confidence interval of $40.872 < \mu < 49.128$ is created by a random study. If the sample standard deviation was 10 and the *t*-value used was 2.064, find the degrees of freedom for the *t*-distribution.

24. WRITING IN MATH Most studies desire results with high confidence levels. Explain why a 99% confidence level is not used for every study.

REASONING Determine whether each statement is *true* or *false*. Explain your reasoning.

25. Increasing the sample size will expand a confidence interval.

26. Increasing the confidence level will expand a confidence interval.

27. Increasing the standard deviation will expand a confidence interval.

28. Increasing the mean will expand a confidence interval.

29. REASONING If a person conducting a study wants to reduce the maximum error of estimate by $\frac{1}{x}$, what would the person have to do to the sample size?

30. CHALLENGE A study conducted with a random sample of size $n = 64$ results in a confidence interval of $3.19 < \mu < 4.01$. If the interval was created using a 90% confidence level, find the sample standard deviation.

31. WRITING IN MATH Explain why parameters are needed in statistics.

32. EDUCATION According to a recent survey, 35% of adult Americans had obtained a bachelor's degree. What is the probability that in a random sample of 50 people, between 12 and 16 people had earned a bachelor's degree? (Lesson 11-4)

33. CAR BATTERY The useful life of a certain car battery is normally distributed with a mean of 100,000 miles and a standard deviation of 10,000 miles. The company makes 20,000 batteries a month. What is the probability that if you select a car battery at random, it will last between 80,000 and 110,000 miles? (Lesson 11-3)

Use the fifth partial sum of the trigonometric series for cosine or sine to approximate each value to three decimal places. (Lesson 10-6)

34. $\sin \frac{\pi}{7}$

35. $\cos \frac{2\pi}{11}$

36. $\sin \frac{4\pi}{17}$

Determine the eccentricity, type of conic, and equation of the directrix given by each polar equation. (Lesson 9-4)

37. $r = \dfrac{8}{\cos \theta + 5}$

38. $r = \dfrac{4}{7 \cos \theta + 4}$

39. $r = \dfrac{2}{\sin \theta + 3}$

Use the dot product to find the magnitude of the given vector. (Lesson 8-3)

40. $\mathbf{u} = \langle -8, 0 \rangle$

41. $\mathbf{v} = \langle 7, 2 \rangle$

42. $\mathbf{u} = \langle 4, 8 \rangle$

43. THRILL RIDES At an amusement park, there is an additional cost per person to ride the Cloud Coaster as well as the Danger Coaster as a 1-person, 2-person, and 3-person ride. The table shows how many people paid for the rides during the first four hours that the park was open. Write and solve a system of equations to determine the cost of each ride per person. Interpret your solution. (Lesson 6-1)

Hour	Cloud Coaster	Danger Coaster			Total Paid ($)
		1 Person	2 Person	3 Person	
1	8	5	10	3	575
2	10	8	2	6	574
3	16	4	8	3	661
4	13	11	6	0	722

44. SAT/ACT Which line best fits the data in the graph?

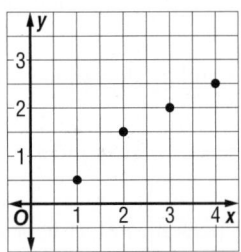

A $y = x$

B $y = -0.5x + 4$

C $y = -0.5x - 4$

D $y = 0.5 + 0.5x$

E $y = 0.75x$

45. REVIEW People were chosen at random and asked how many times they went out to eat per week. If $\sigma = 0.6$, the results had 95% confidence, and they were accurate to ± 0.05, about how many people were asked?

F 6   **G** 23   **H** 144   **J** 554

46. In a random sample of 28 college-educated adults aged 25 to 35, the average amount of student-loan debt was $5566 with a standard deviation of $1831. Estimate the mean student-loan debt for all college-educated adults ages 25 to 35 using a 90% confidence interval.

A $4188 < \mu < 6944$

B $4319 < \mu < 6813$

C $4507 < \mu < 6625$

D $4997 < \mu < 6135$

47. REVIEW A school has two independent backup generators having probabilities of 0.9 and 0.95, respectively, of successful operation in case of a power outage. What is the probability that at least one backup generator operates during a power outage?

F 0.855   **H** 0.95

G 0.89   **J** 0.995

Hypothesis Testing

Why?

Luther and Jimmy are shooting baskets when Jimmy proudly boasts, "I can make at least 90% of my free throws." Luther is curious about Jimmy's remark and wants to test the accuracy of his claim.

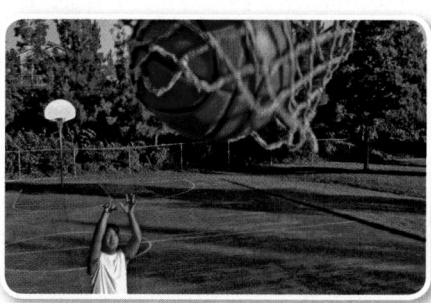

Then
You found confidence intervals for the means of distributions. (Lesson 11-5)

Now
- Write null and alternative hypotheses, and identify which represents the claims.
- Perform hypothesis testing using test statistics and *p*-values.

New Vocabulary
hypothesis test
null hypothesis
alternative hypothesis
level of significance
left-tailed test
two-tailed test
right-tailed test
p-value

Math Online
glencoe.com

Hypotheses A **hypothesis test** assesses evidence provided by data about a claim concerning a population parameter. A claim of this type is called a *statistical hypothesis* and may or may not be true. Jimmy's claim at the beginning of the lesson is an example of a statistical hypothesis.

To test the validity of a claim, write it as a mathematical statement. Jimmy's claim can be written as $\mu \geq 90\%$, where μ is his average shooting percentage. The statement $\mu < 90\%$ is the complement of the original statement, which represents Jimmy not meeting his claim. These two statements represent the pair of hypotheses that need to be stated to test a claim.

- The **null hypothesis** H_0: There *is not* a significant difference between the sample value and the population parameter. It will contain a statement of *equality*, such as $\geq$, $=$, or $\leq$. In this example, $\mu \geq 90\%$ is the null hypothesis.

- The **alternative hypothesis** H_a: There is a difference between the sample value and the population parameter. It will contain a statement of *inequality*, such as $>$, $\neq$, or $<$. In this example, $\mu < 90\%$ is the alternative hypothesis.

If a claim k is made about a population mean μ, the possible combinations for the hypotheses are:

H_0: $\mu = k$ and H_a: $\mu \neq k$ H_0: $\mu \geq k$ and H_a: $\mu < k$ H_0: $\mu \leq k$ and H_a: $\mu > k$

EXAMPLE 1 | **Null and Alternative Hypotheses**

For each statement, write the null and alternative hypotheses. State which hypothesis represents the claim.

a. Makers of a gum brand claim that their gum will keep its flavor for at least 5 hours.

This claim becomes $\mu \geq 5$ and is the null hypothesis since it includes an equality symbol. The complement is $\mu < 5$.

H_0: $\mu \geq 5$ **(Claim)** H_a: $\mu < 5$

b. Technicians of an automotive company claim that they will perform an oil change on a car in less than 15 minutes.

This claim becomes $\mu < 15$ and is the alternative hypothesis since it includes an inequality symbol. The complement is $\mu \geq 15$.

H_0: $\mu \geq 15$ H_a: $\mu < 15$ **(Claim)**

c. A teacher claims that the average amount of time that his students are spending on homework every night is 35 minutes.

This claim becomes $\mu = 35$ and is the null hypothesis since it includes an equality symbol. The complement is $\mu \neq 35$.

H_0: $\mu = 35$ **(Claim)** H_a: $\mu \neq 35$

✓ Guided Practice

1A. A football player claims that he can achieve more than 100 rushing yards per game.

1B. A cross country star claims that it will take her no more than 4 minutes to run a mile.

1C. A salesperson claims that she averages 12 sales per month.

Significance and Tests To validate a claim, the null hypothesis is always tested. In the example at the beginning of the lesson, $\mu \geq 90\%$ would be tested. After a sample of data is analyzed, one of two decisions is made.

- Reject the null hypothesis.
- Do not reject the null hypothesis.

Every shot that Jimmy could ever take cannot be recorded. Luther can only analyze a sample of data, such as having Jimmy take 100 shots. Thus, there is always a chance that Luther can make the wrong decision. When the decision is incorrect, it is either a *Type I* or a *Type II* error.

	H_0 is True	H_0 is False
H_0 is rejected	**Type I Error** The null hypothesis is rejected, when it is actually true. Luther rejects the statement $\mu \geq 90\%$ when Jimmy actually shoots 90% or better.	**Correct decision** Luther rejects the statement $\mu \geq 90\%$ when Jimmy actually shoots less than 90%.
H_0 is not rejected	**Correct decision** Luther does not reject the statement $\mu \geq 90\%$ when Jimmy actually shoots 90% or better.	**Type II Error** The null hypothesis is not rejected, when it is actually false. Luther does not reject the statement $\mu \geq 90\%$ when Jimmy actually shoots less than 90%.

This suggests that there are actually four possible outcomes when a decision about the null hypothesis is made. The only way to guarantee complete accuracy is to test the entire population.

The **level of significance**, denoted α, is the maximum allowable probability of committing a Type I error. For example, if $\alpha = 0.10$, there is a 10% chance that H_0 was rejected when it was actually true, or there is a 90% chance that a correct decision was made. Any level of significance can be chosen. The three most commonly used levels are $\alpha = 0.10$, $\alpha = 0.05$, and $\alpha = 0.01$.

After a level of significance is chosen, a critical value can be found using either a z- or t-value. Similar to confidence intervals, the decision to use a z- or t-value is based on the characteristics of the study.

- If σ is known or $n \geq 30$, use a z-value.
- If σ is unknown and $n < 30$, use a t-value.

The z- or t-value and the alternative hypothesis will determine the *critical region*, the range of values that suggest a significant enough difference to reject the null hypothesis. The location of a critical region is determined by the inequality sign of the alternate hypothesis, which indicates whether the test is left-tailed, right-tailed, or two-tailed.

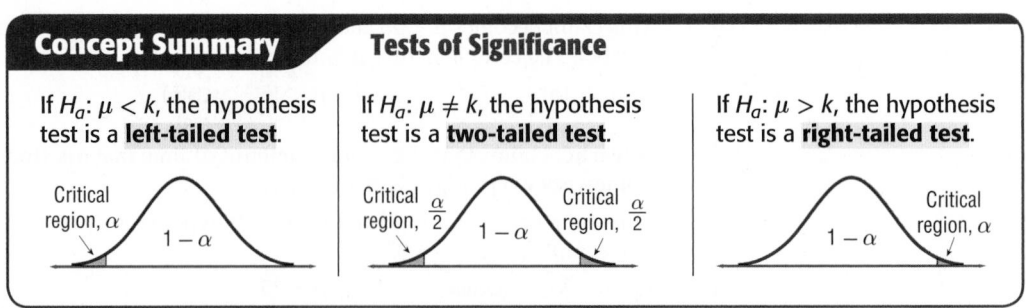

Concept Summary   **Tests of Significance**

If H_a: $\mu < k$, the hypothesis test is a **left-tailed test**.

If H_a: $\mu \neq k$, the hypothesis test is a **two-tailed test**.

If H_a: $\mu > k$, the hypothesis test is a **right-tailed test**.

Once the area corresponding to the level of significance is determined, a *test statistic* for the sample mean is calculated. The test statistic is the z- or t-value for the sample and will be referred to as the *z statistic* or *t statistic*. If the z or t statistic for the sample:

- is in the critical region, H_0 should be rejected.
- is *not* in the critical region, H_0 should not be rejected.

The steps to conduct a hypothesis test are summarized below.

Key Concept — Steps for Hypothesis Testing

Step 1 State the hypotheses, and identify the claim.

Step 2 Determine the critical value(s) and region.

Step 3 Calculate the test statistic.

Step 4 Reject or fail to reject the null hypothesis.

EXAMPLE 2 — One-Sided Hypothesis Test

NUTRITION Representatives of a company report that their product contains no more than 5 grams of fat. A researcher tests a random sample of 50 products and finds that $\bar{x} = 5.03$ grams. If the standard deviation of the population is 0.14 gram, use a 5% level of significance to determine whether there is sufficient evidence to reject the company's claim.

Step 1 State the null and alternative hypotheses, and identify the claim.

The claim written as a mathematical statement is $\mu \leq 5$. This is the null hypothesis. The alternative hypothesis is $\mu > 5$.

$H_0: \mu \leq 5$ **(claim)** $\qquad H_a: \mu > 5$

Step 2 Determine the critical value(s) and region.

The population standard deviation is known and $n \geq 30$, so you can use the z-value. The test is right-tailed since $\mu > 5$. Because a 5% level of significance is called for, $\alpha = 0.05$. Use a graphing calculator to find the z-value.

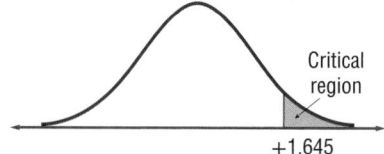

Step 3 Calculate the test statistic.

Find the z statistic. Since $\sigma = 0.14$ and $n = 50$, $\sigma_{\bar{x}} = \dfrac{0.14}{\sqrt{50}}$ or about 0.02.

$z = \dfrac{\bar{x} - \mu}{\sigma_{\bar{x}}}$ **Formula for z statistic**

$= \dfrac{5.03 - 5}{0.02}$ $\bar{x} = 5.03, \mu = 5,$ **and** $\sigma_{\bar{x}} = 0.02$

$= 1.5$ **Simplify.**

Step 4 Reject or fail to reject the null hypothesis.

H_0 is not rejected since the test statistic does not fall within the critical region.

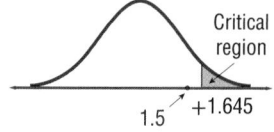

Therefore, there is not enough evidence to reject the claim that there is no more than 5 grams of fat per product.

✓ Guided Practice

2. JOBS Employees at a bookstore claim that the mean wage per hour is less than the competitor's mean wage of $10.50. If a random sample of 20 employees shows a mean wage of $10.05 with a standard deviation of $0.75, test the employees' claim at $\alpha = 0.01$.

For a two-sided test, the level of significance α must be divided by 2 in order to determine the critical value at each tail.

EXAMPLE 3 | Two-Sided Hypothesis Test

FRUIT SNACKS Representatives of a company have stated that each box of fruit snacks contains 80 pieces. A researcher wants to determine if this is true. A random sample of 25 boxes is selected, with a sample mean of 84.1 pieces and a standard deviation of 7 pieces. Is this statistically significant at $\alpha = 0.01$?

Step 1 State the null and alternative hypotheses, and identify the claim.

The claim written as a mathematical statement is $\mu = 80$. This is the null hypothesis. The alternative hypothesis is $\mu \neq 80$.

$H_0: \mu = 80$ **(claim)** $H_a: \mu \neq 80$

Step 2 Determine the critical value(s) and region.

The t-value should be used since $n < 30$ and σ is unknown. The test is two-tailed since $\mu \neq 80$, so the critical values are determined by $\frac{\alpha}{2}$ or 0.005. Using a graphing calculator, the critical values for $\alpha = 0.005$ with $25 - 1$ or 24 degrees of freedom are $t = -2.8$ and $t = 2.8$.

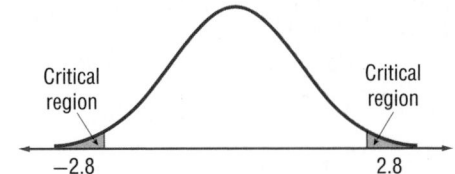

> **Watch Out!**
>
> **Determining Critical Values** Remember to use the InvT(function to find t-values for the t-distribution and InvNorm(to find z-values for the normal distribution.

Step 3 Calculate the test statistic.

$$t = \frac{\bar{x} - \mu}{\sigma_{\bar{x}}}$$ **Formula for t statistic**

$$= \frac{84.1 - 80}{1.4}$$ $\bar{x} = 84.1$, $\mu = 80$, and $\sigma_{\bar{x}} = \frac{7}{\sqrt{25}}$ or 1.4

$$\approx 2.93$$ **Simplify.**

Step 4 Reject or fail to reject the null hypothesis.

H_0 is rejected since the test statistic falls within the critical region.

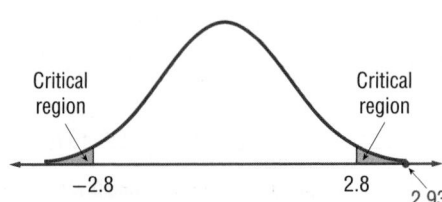

There is enough evidence to reject the claim that there are 80 pieces in each box.

☑ Guided Practice

3. **TRAVEL** Representatives of a travel bureau in a U.S. city claim that in a recent year, an average of 110 people visited the city every day. In a sample of 90 days, there was an average of 115 visitors per day, with a standard deviation of 18 visitors. At $\alpha = 0.05$, is there enough evidence to reject the claim?

The **p-value** can also be used to determine whether H_0 should be rejected. The p-value is the lowest level of significance at which H_0 can be rejected for a given set of data. After calculating the z or t statistic for a hypothesis test, it can be converted into a p-value using a graphing calculator. To use the p-value to evaluate H_0, compare the p-value to α.

- If $p \leq \alpha$, then reject H_0.
- If $p > \alpha$, then do not reject H_0.

StudyTip

Law of Large Numbers
The value of $\bar{x}$ is rarely identical to the true μ and often varies from sample to sample. However, due to the Law of Large Numbers, if we take larger and larger samples, $\bar{x}$ is guaranteed to get closer and closer to the true μ. This is true for *any* distribution.

The *p*-value corresponds to the area found under the normal curve to the left or right of the *z* or *t* statistic calculated for the sample data. The location of the area is determined by the type of test being preformed.

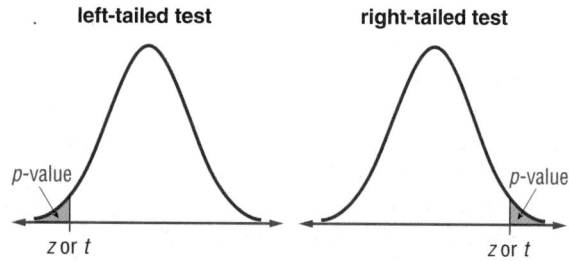

left-tailed test   right-tailed test

The α value is chosen by the researcher before the statistical test is performed, while the *p*-value is calculated after the sample mean is determined.

EXAMPLE 4   Hypothesis Testing and *p*-Values

HORTICULTURE A biologist treated 40 plants with a chemical and then compared the amount of growth with untreated plants. For the untreated plants, the mean height is 21.6 centimeters. The treated plants have a mean height of 22.4 centimeters and $s = 1.8$ centimeters. The biologist claims that the chemical increased plant growth. Determine whether this result is significant at $\alpha = 0.01$.

The claim written as a mathematical statement is $\mu > 21.6$. This is the alternative hypothesis. The null hypothesis is $\mu \leq 21.6$.

$H_0: \mu \leq 21.6$ $\qquad\qquad$ $H_a: \mu > 21.6$ $\qquad$ **(claim)**

Since $n \geq 30$, the *z* statistic is used.

$z = \dfrac{\bar{x} - \mu}{\sigma_{\bar{x}}}$ $\qquad$ **Formula for z statistic**

$\approx \dfrac{22.4 - 21.6}{0.285}$ $\qquad$ $\bar{x} = 22.4$, $\mu = 21.6$, and $\sigma_{\bar{x}} = \dfrac{1.8}{\sqrt{40}}$ or about 0.285

≈ 2.807 $\qquad$ **Simplify.**

TechnologyTip

Calculating the Area Under a *t*-Curve You can use a graphing calculator to find the area under the *t*-curve that corresponds to any *t*-value by selecting 2nd [DISTR] and tcdf(*lower t value, upper t value, d.f.*).

This is a right-tailed test since the alternative hypothesis is $\mu > 21.6$. The area associated with $z = 2.807$ is 0.0025.

right-tailed test

The *p*-value 0.0025 is less than 0.01. Therefore, the null hypothesis is rejected and there is significant evidence that the chemical increased plant growth.

☑ Guided Practice

4. **DRUGS** Makers of a sleep-aid claim that their product provides more than 8 hours of uninterrupted sleep. In a test of 50 patients, the mean amount of uninterrupted sleep was 8.07 hours with a standard deviation of 0.3 hour. Find the *p*-value and determine if there is enough evidence to reject the claim at $\alpha = 0.03$.

It is important to remember that statistical tests do not prove that a claim is true or false. These types of tests simply state that there is or is not enough evidence to say that a claim is likely to be true.

Write the null and alternative hypotheses for each statement, and state which hypothesis represents the claim. (Example 1)

1. Makers of a cereal brand claim their product contains 4 grams of fiber.

2. A student claims that he has received at least an 85% on his math tests.

3. Darcy claims that she can drive to school from her house in less than 10 minutes.

4. Vanessa claims that her bowling average is 183.

5. Ian claims that he can recite the names of more than 38 former presidents of the United States.

Calculate the test statistic, and determine whether there is enough evidence to reject the null hypothesis. Then make a statement regarding the original claim.

6. **ADVERTISING** Company representatives claim that they will ship a product in less than four days. If a random selection of 60 delivery times has a sample mean of 3.9 days and a standard deviation of 0.6 day, is there enough evidence to reject the claim at $\alpha = 0.05$? Explain. (Examples 2 and 3)

7. **HEALTH** A researcher claims that a supplement does not increase bone density by at least 0.05 gram per square centimeter. If a study shows that the supplement increased bone density in a random sample of 35 people by 0.048 gram per square centimeter with a standard deviation of 0.004, is there enough evidence to reject the claim at $\alpha = 0.01$? Explain. (Examples 2 and 3)

8. **HOTELS** Owners of a hotel chain claim that the average cost of one of their hotel rooms in the U.S. is $82. Sample data for 25 hotel rooms is collected. The average cost was found to be $85 with a standard deviation of $8. Is there enough evidence to reject the owners' claim at $\alpha = 0.02$? Explain. (Examples 2 and 3)

9. **CALCULATORS** A teacher claims that the average cost of a certain brand of graphing calculator is at least $90. A random sample of 40 stores shows that the mean cost is $89.25 with a standard deviation of $4.95. Is there enough evidence to reject the teacher's claim at $\alpha = 0.05$? Explain. (Examples 2 and 3)

For each claim k, use the specified information to calculate the test statistic and determine whether there is enough evidence to reject the null hypothesis. Then make a statement regarding the original claim. (Examples 2 and 3)

10. $k: \mu = 1240, \alpha = 0.05, \bar{x} = 1245, s = 32, n = 50$

11. $k: \mu > 88, \alpha = 0.05, \bar{x} = 91.2, s = 3.9, n = 22$

12. $k: \mu < 500, \alpha = 0.01, \bar{x} = 490, s = 27, n = 35$

13. $k: \mu \neq 5500, \alpha = 0.01, \bar{x} = 5430, s = 236, n = 200$

14. $k: \mu \leq 10,000, \alpha = 0.01, \bar{x} = 10,015, s = 85, n = 18$

15. **REAL ESTATE** A researcher wants to test a claim that the average home sale price in the U.S. is less than $260,000. She selects a sample of 40 homes and finds the mean sale price of the sample to be $254,500 with a standard deviation of $12,500. Find the p-value, and determine whether there is enough evidence to support the claim at $\alpha = 0.05$. (Example 4)

16. **MUSIC** Representatives of an electronics company claim that the average lifetime of an MP3 player is at least 5 years. A random sample of 100 MP3s shows a mean life span of 5.2 years with a standard deviation of 1.2 years. Find the p-value, and determine whether there is enough evidence to support the claim at $\alpha = 0.01$. (Example 4)

17. **BASEBALL** Shelby believes that the cost of attending a baseball game for a family of two adults and two children is under $125. She surveys 18 families selected at random and finds that the average cost is $122.88 with a standard deviation of $13.21. Find the p-value, and determine whether there is enough evidence to support the claim at $\alpha = 0.10$.

18. **CROSS COUNTRY** Pablo claims that the average mile time for the students in his school is less than 7 minutes. He records the mile times of 20 randomly selected students. Determine whether Pablo's claim is supported at $\alpha = 0.05$.

Mile Times (minutes)									
5.25	7.27	5.46	7.63	7.75	5.42	6.00	8.17	9.45	6.20
6.63	7.38	6.97	7.85	7.03	6.53	6.87	7.22	7.16	6.92

a. Write the null and alternative hypotheses, and state which hypothesis represents the claim.

b. Determine whether there is enough evidence to reject the null hypothesis using critical values.

c. Make a statement regarding the original claim.

19. **HOMEWORK** Ms. Taylor claims that her math students spend 25 minutes each night on homework. Ava asks her classmates to record the average amount of time that they spend on homework each night over the course of a week. Determine whether Ms. Taylor's claim is supported at $\alpha = 0.10$.

Times (minutes)											
45	40	10	15	18	20	34	36	20	25	28	25
26	30	22	25	24	29	26	28	23	28	25	26
29	30	22	20	22	24	23	24	25	29	25	

a. Write the null and alternative hypotheses, and state which hypothesis represents the claim.

b. Determine whether there is enough evidence to reject the null hypothesis using critical values.

c. Make a statement regarding the original claim.

Describe the outcome if a type I or a type II error is committed when the null hypothesis is tested.

20. The accused person is not guilty.

21. The *X*-ray came back positive for an ankle sprain.

22. Students use study time efficiently.

23. The majority of students do not have jobs.

24. The average lifespan of a goldfish is 2 years.

25. The venom from the snake is not poisonous.

26. SLEEP Mr. King believes that college students get less than 6 hours of sleep each night. He randomly selected a group of students and recorded the average amount of sleep each student gets each night.

Average Sleep (hours/night)								
5.4	6.7	6.5	5.5	5.5	6.0	5.8	6.7	6.8
4.5	5.7	7.5	5.4	5.3	8.0	4.5	4.5	5.0

a. Write the null and alternative hypotheses, and state which hypothesis represents the claim.

b. Find the *p*-value.

c. Determine whether there is enough evidence to reject the null hypothesis at $\alpha = 0.05$.

d. Make a statement regarding the original claim.

27. ACT The average composite score on the ACT is a 21. Instructors of an ACT preparation class claim that they can raise test takers' scores. The scores of randomly selected attendees were recorded.

ACT Scores											
24	23	27	23	19	16	33	30	22	25	23	26
21	30	22	18	28	21	26	32	20	17	23	24
25	28	19	22	21	19	18	20	25	22	24	23

a. Write the null and alternative hypotheses, and state which hypothesis represents the claim.

b. Find the *p*-value.

c. Determine whether there is enough evidence to reject the null hypothesis at $\alpha = 0.01$.

d. Make a statement regarding the original claim.

28. QUANTITY A chocolatier claims that his candy averages at least 82 pieces per bag. Conrad randomly selects bags of the candies and counts the pieces. Assume that $\sigma = 1.2$.

Number of Pieces								
81	80	82	82	83	82	84	81	81
80	83	83	82	81	80	84	81	81

a. Write the null and alternative hypotheses, and state which hypothesis represents the claim.

b. Determine whether there is enough evidence to reject the null hypothesis at $\alpha = 0.05$ using critical values.

c. Find the *p*-value. Determine whether there is enough evidence to reject the null hypothesis at $\alpha = 0.05$.

29. GRADES Mr. Lewis claims that the average grade for his students is an 85%. Two of his students, Victor and Malina, collect the following samples of grades from students in their classes.

Victor's Scores								
64	84	86	99	76	90	79	94	85
84	85	88	91	80	85	76	86	96

Malina's Scores							
95	86	95	83	86	85	84	88
88	86	87	88	95	86	85	95

a. Write the null and alternative hypotheses, and state which hypothesis represents the claim.

b. Suppose $\alpha = 0.10$. For each class, determine whether there is enough evidence to reject the null hypothesis and make a statement regarding the original claim.

c. Determine whether there is enough evidence to reject the null hypothesis if the two samples of data are combined. Use the result to make a statement regarding the original claim.

d. Make a conjecture in regard to the result found in part **c** and the Law of Large Numbers.

30. ERROR ANALYSIS Trish and Molly are completing their statistics homework. Trish claims that it is always better to have a type I error rather than a type II error. Molly disagrees. Is either of them correct? Explain your reasoning.

31. WRITING IN MATH Describe the difference between conducting a hypothesis test using test statistics and critical values and conducting a hypothesis test using *p*-values.

REASONING Determine whether each statement is *true* or *false*. Explain your reasoning.

32. If the null hypothesis is rejected, then the claim is always rejected.

33. The alternative hypothesis can include an equality symbol if it represents the claim.

34. The *p*-value is always going to be a positive value.

35 **CHALLENGE** For a random set of data, $p = 0.0104$, $s = 0.3$, and $\bar{x} = 14.9$. If the study was conducted to test a claim of $\mu < 15$, find *n*.

36. WRITING IN MATH Explain why it may not always be in the researcher's best interest to have the lowest possible significance level in order to reduce the possibility of a type I error.

37. SHOES A sample of 35 pairs of running shoes found the average cost to be $45.25 with a standard deviation of $7.60. Estimate the mean cost for running shoes using a confidence interval given a 90% confidence level. (Lesson 11-5)

38. RARE BOOKS The average prices for three antique books are shown. The prices vary due to the age and condition of each book. (Lesson 11-4)

Book	Average Price ($)
Don Quixote	155
Moby Dick	98
Oliver Twist	118

   a. For a sample of 25 copies of *Don Quixote*, find the probability that the mean price is more than $160, if the standard deviation is $18.

   b. For a random sample of 40 copies of *Oliver Twist*, find the probability that the mean price will be between $115 and $120, if the standard deviation is $15.

Determine whether each sequence is *convergent* or *divergent*. (Lesson 10-1)

39. $a_n = (10 - n)^2$ **40.** $a_n = n^2 + 10n - 9$ **41.** $a_n = \dfrac{3n + 2}{n}$

Find the angle θ between vectors u and v. (Lesson 8-5)

42. $\mathbf{u} = 2\mathbf{i} + 8\mathbf{j} + 7\mathbf{k},\ \mathbf{v} = -7\mathbf{i} - 3\mathbf{j} - 9\mathbf{k}$ **43.** $\mathbf{u} = -2\mathbf{i} - 8\mathbf{j} + 4\mathbf{k},\ \mathbf{v} = 3\mathbf{i} - 4\mathbf{j} - 7\mathbf{k}$ **44.** $\mathbf{u} = 5\mathbf{i} - 9\mathbf{k},\ \mathbf{v} = 2\mathbf{i} - 3\mathbf{j} - 8\mathbf{k}$

Find a unit vector in the same direction as v. (Lesson 8-2)

45. $\mathbf{v} = \langle -9, 2 \rangle$ **46.** $\mathbf{v} = \langle -5, -1 \rangle$ **47.** $\mathbf{v} = \langle 4, 3 \rangle$

Write an equation for the hyperbola with the given characteristics. (Lesson 7-3)

48. vertices $(-6, 3)$, $(4, 3)$; conjugate axis is 8 units   **49.** vertices $(-2, 6)$, $(-2, -4)$; conjugate axis is 6 units

---

Skills Review for Standardized Tests

50. REVIEW Estimate the median and spread of the data represented by the box plot.

[0, 80] scl: 10 by [0, 10] scl: 1

   A median ≈ 30, spread ≈ 50

   B median ≈ 30, spread ≈ 65

   C median ≈ 50, spread ≈ 50

   D median ≈ 50; spread ≈ 65

51. REVIEW Find the solutions of $x^2 = i$.

   F i and $-i$

   G $\cos \dfrac{\pi}{4} + i \sin \dfrac{\pi}{4}$ and $\cos \dfrac{5\pi}{4} + i \sin \dfrac{5\pi}{4}$

   H 1 and -1

   J $\cos \dfrac{\pi}{8} + i \sin \dfrac{\pi}{8}$ and $\cos \dfrac{9\pi}{8} + i \sin \dfrac{9\pi}{8}$

52. SAT/ACT Which of the following statements are true if n is an integer?

     I. $3n + 6$ is divisible by 3.

     II. $10n + 8$ is divisible by 2.

     III. $4n - 2$ is divisible by 4.

   A I only   **D** I and III only.

   B II only   **E** I, II, and III are true.

   C I and II only

53. Marcel believes that the average price of gasoline is still under $2.50 per gallon. He randomly calls 40 different service stations and finds that the average price is $2.51 with a standard deviation of $0.06. Find the p-value, and determine whether there is enough evidence to support the claim at $\alpha = 0.10$.

   F 0.85; not enough evidence

   G 0.95; enough evidence

   H 0.15; not enough evidence

   J 0.05; enough evidence

Correlation and Linear Regression

Why?

A feature writer for a school newspaper is interested in determining whether the number of hours of sleep a student gets each night is related to his or her overall grade point average. In statistical terms, the writer would like to know if there is a *correlation* between sleep and grades.

Correlation Thus far in this chapter, you have graphed, characterized, and used summary statistics to describe distributions of one-variable data sets. You have used sample statistics of such univariate data to make inferences about a population by developing confidence intervals and performing hypothesis tests. **Correlation** is another area of inferential statistics that involves determining whether a relationship exists between two variables in a set of **bivariate** data.

Bivariate data can be represented as ordered pairs (x, y), where x is the independent or **explanatory variable** and y is the dependent or **response variable**. To determine whether there may be a linear, a nonlinear, or no correlation between the variables, you can use a scatter plot.

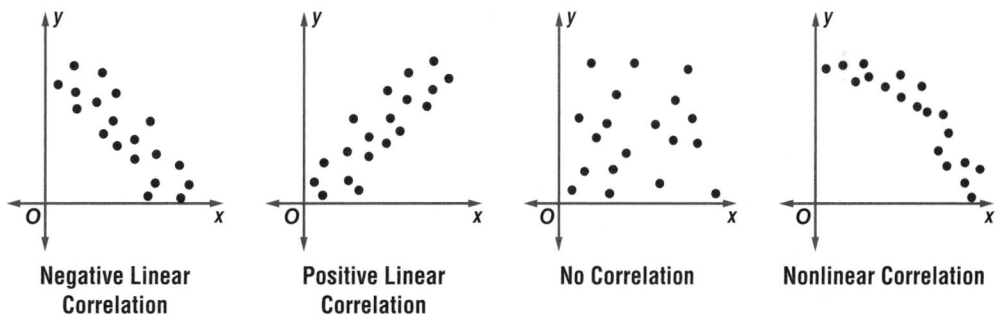

| Negative Linear Correlation | Positive Linear Correlation | No Correlation | Nonlinear Correlation |

We say that the data have a strong linear relationship if the points lie close to a straight line and weak if they are widely scattered about the line, but interpreting correlation using a scatter plot tends to be subjective. A more precise way to determine the type and strength of the linear relationship between two variables is to calculate the **correlation coefficient**. A formula for this measure is given below.

Key Concept   Correlation Coefficient

For n pairs of sample data for the variables x and y, the correlation coefficient r between x and y is given by

$$r = \frac{1}{n-1} \sum \left(\frac{x_i - \bar{x}}{s_x}\right)\left(\frac{y_i - \bar{y}}{s_y}\right),$$

where x_i and y_i represent the values for the ith pair of data, $\bar{x}$ and $\bar{y}$ represent the means of the two variables, and s_x and s_y represent the standard deviations of the two variables.

The correlation coefficient can take on values from -1 to 1. This value indicates the strength and type of linear correlation between x and y as shown in the diagram below.

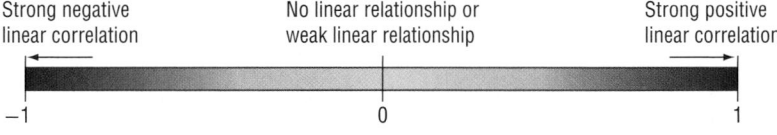

| Strong negative linear correlation | No linear relationship or weak linear relationship | Strong positive linear correlation |

-1   0   1

Notice from the formula that the correlation coefficient is the average of the products of the standardized values for x and the standardized values for y. The correlation coefficient can be tedious to calculate by hand, so we most often rely on computer software or a graphing calculator.

EXAMPLE 1   Calculate and Interpret the Correlation Coefficient

SLEEP/GPA STUDY A feature writer for a student newspaper conducts a study to determine whether there is a linear relationship between the average number of hours a student sleeps each night and his or her overall grade point average. The table shows the data that the writer collected. Make a scatter plot of the data, and identify the relationship. Then calculate and interpret the correlation coefficient.

Hours of Sleep	GPA	Hours of Sleep	GPA
6.6	2.2	8.0	2.9
6.6	2.4	8.0	3.1
6.7	2.3	8.1	3.3
6.8	2.3	8.2	3.3
6.8	2.2	8.2	3.2
7.0	2.6	8.3	2.8
7.0	2.7	8.4	3.1
7.2	2.8	8.6	3.3
7.4	2.6	8.7	3.4
7.4	3.0	8.8	3.1
7.4	2.9	8.8	3.2
7.5	2.7	8.8	3.4
7.7	2.8	9.1	3.3
7.9	2.9	9.2	3.8
7.9	3.0	9.2	3.5

StudyTip

Resistance of Correlation Coefficient Like the mean and standard deviation, r is a nonresistant statistic. It can be affected by outliers.

Step 1 Graph a scatter plot of the data.

Enter the data into L1 and L2 on your calculator. Then turn on **Plot1** under the **STAT PLOT** menu and choose ⌐ using L1 for the **Xlist** and L2 for the **Ylist**. Graph the scatterplot by pressing **ZoomStat** or by pressing **Graph** and adjusting the window manually (Figure 11.7.1). From the graph, it appears that the data have a positive linear correlation.

Step 2 Calculate and interpret the correlation coefficient.

Press ⌐STAT⌐ and select **LinReg(ax+b)** under the **CALC** menu (Figure 11.7.2). The correlation coefficient r is about 0.9148. Because r is close to 1, this suggests that the data may have a strong positive linear correlation. This numerical assessment of the data is consistent with our graphical assessment.

[6, 9.5] scl: 1 by [2, 4] scl: 1

Figure 11.7.1

Figure 11.7.2

✔ Guided Practice

Rain (in.)	Temp. (°F)
5.35	41.3
4.03	44.3
3.77	46.6
2.51	50.4
1.84	56.1
1.59	61.4
0.85	65.3
1.22	65.7
1.94	60.8
3.25	53.5
5.65	46.3
6.00	41.6

Figure 11.7.3

1. METEOROLOGY A weather program is featuring a special on a city where a study was conducted to determine whether there is a linear relationship between the average monthly rainfall and temperature. The table in Figure 11.7.3 shows the data collected. Make a scatter plot of the data. Then calculate and interpret the correlation coefficient for the data.

In Example 1, the data collected represent just a sample of the entire school population; therefore, r represents a *sample correlation coefficient*. In order for r to be a valid estimate of the *population correlation coefficient* ρ, the following assumptions must be valid.

- The variables x and y are *linearly* related.

- The variables are *random* variables.

- The two variables have a *bivariate normal distribution*. That is, x and y each come from a normally distributed population.

We would like to use the value of r to make an inference about the relationship between the variables x and y for the entire population. In order to do that, we need to determine whether the value of $|r|$ is great enough to conclude that there is a significant relationship between x and y.

To make this determination, you can perform a hypothesis test. The null and alternative hypotheses for a two-tailed test of the population correlation coefficient ρ are as follows.

H_0: $\rho = 0$ There is no correlation between the x and y variables in the population.

H_a: $\rho \neq 0$ There is a correlation between the x and y variables in the population.

We can use a t-test as described below to test the significance of the correlation coefficient.

Key Concept   **Formula for the *t*-Test for the Correlation Coefficient**

For a t-test of the correlation between two variables, the test statistic for ρ is the sample correlation coefficient r and the standardized test statistic t is given by

$$t = r\sqrt{\frac{n-2}{1-r^2}}, \text{ where } n-2 \text{ is the degrees of freedom.}$$

EXAMPLE 2   **Test for Significance**

SLEEP/GPA STUDY In Example 1, you calculated the correlation coefficient r for the 30 pairs of student sleep and GPA data to be about 0.9148. Test the significance of this correlation coefficient at the 5% level.

Step 1   State the hypotheses.

       H_0: $\rho = 0$ H_a: $\rho \neq 0$

Step 2   Determine the critical values.

Testing for significance at the 5% level means that $\alpha = 0.05$. Since this is a two-tailed test, the critical values are determined by $\frac{\alpha}{2}$ or 0.025. Using a graphing calculator, the critical values for $\alpha = 0.025$ with $30 - 2$ or 28 degrees of freedom are $t = \pm 2.0$.

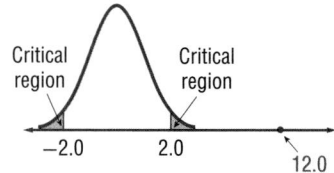

```
invT(0.025,28)
         -2.048407113
invT(1-0.025,28)
          2.048407113
```

Step 3   Calculate the test statistic.

$$t = r\sqrt{\frac{n-2}{1-r^2}}$$ **Calculate the test statistic for ρ.**

$$= 0.9148\sqrt{\frac{30-2}{1-(0.9148)^2}} \text{ or about 12.0}$$ **$r = 0.9148$ and $n = 30$**

Step 4   Reject or fail to reject the null hypothesis.

Since $12.0 > 2.0$, the statistic falls within the critical region and the null hypothesis is rejected.

Critical region   Critical region

-2.0 2.0 12.0

At the 5% level, there is enough evidence to conclude that there is a significant correlation between the average amount of sleep a student gets each night and his or her overall GPA.

☑ **Guided Practice**

2. METEOROLOGY In the Guided Practice for Example 1, you calculated the correlation coefficient r for the 12 pairs of rainfall and temperature data. Test the significance of this correlation coefficient at the 10% level.

Linear Regression Once the correlation between two variables is determined to be significant, the next step is to determine the equation of the **regression line**, also called a **line of best fit**. The regression line describes how the response variable y changes as the explanatory variable x changes.

While many lines of best fit can be drawn through a set of points, the one used most often is determined by specific criteria. Consider the scatter plot and regression line shown. The difference d between an observed y-value and its predicted y-value on the regression line is called a **residual**.

residual d_i = observed y_i − predicted y_i

Residuals are positive when the observed value is above the line, negative when the observed value is below the line, and zero when it is on the line. The **least-squares regression line** is the line for which the sum of the squares of these residuals is a minimum.

Key Concept   **Equation of the Least-Squares Regression Line**

The equation of the least-squares regression line for an explanatory variable x and response variable y is $\hat{y} = ax + b$.

The slope a and y-intercept b in this equation are given by

$$a = r\frac{s_y}{s_x} \text{ and } b = \bar{y} - a\bar{x},$$

where r represents the correlation coefficient between the two variables, $\bar{x}$ and $\bar{y}$ represent their means, and s_x and s_y represent their standard deviations.

As with the correlation coefficient, it is not necessary to calculate the least-squares regression equation by hand. Computer software or a graphing calculator will provide the slope a and the y-intercept b of the least-squares regression line for keyed-in values of the variables.

EXAMPLE 3   **Find the Least-Squares Regression Line**

SLEEP/GPA STUDY Find the equation of the regression line for the data used in Example 1. Interpret the slope and intercept in context. Then assess the fit of the modeling equation by graphing it, along with the scatter plot of the data, in the same window.

Using the same screen you used to obtain the correlation coefficient (Figure 11.7.4), the least-squares regression equation is approximately $\hat{y} = 0.457x - 0.667$. The slope $a = 0.457$ indicates that for each additional hour of sleep, a student will raise his or her GPA by 0.457 point. The y-intercept $b = -0.667$ indicates that when a student averages no sleep, his or her GPA will be less than 0, which is not possible.

Since the data appear to be randomly scattered about the line $\hat{y} = 0.457x - 0.667$, this regression line appears to be a good fit for the data (Figure 11.7.5).

[6, 9.5] scl: 1 by [2, 4] scl: 1

Figure 11.7.4   **Figure 11.7.5**

☑ **Guided Practice**

3. **METEOROLOGY** Find the equation of the regression line for the rainfall and temperature data used in the Guided Practice for Example 1. Interpret the slope and intercept in context. Then assess the fit of the modeling equation by graphing it and the scatter plot of the data in the same window.

A least-squares regression line describes the overall pattern in a set of bivariate data. As with univariate data analysis, you should always be on the lookout for striking deviations, or outliers, from this pattern. Remember that the residuals measure how much the data deviate from the regression line.

Scatterplot with Regression Line

Examining a scatter plot of the residuals, called a **residual plot**, can help you assess how well the regression line describes the data. In a residual plot, the horizontal line at zero corresponds to the regression line. You can create a residual plot using your graphing calculator. If the plot of the residuals appears to be randomly scattered and centered about $y = 0$, the use of a linear model for the data is supported. If the plot displays a curved pattern, the use of a linear model would not be supported.

Residual Plot

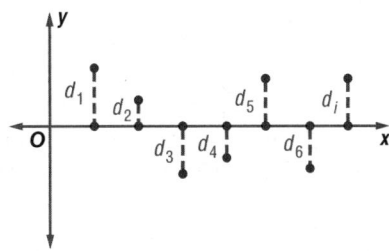

EXAMPLE 4   **Graph and Analyze a Residual Plot**

SLEEP/GPA STUDY Graph and analyze the residual plot for the average sleep hours and GPA data in Example 1 to determine whether the linear model found in Example 3 is appropriate.

After calculating the least-squares regression line in Example 3, you can obtain the residual plot of the data by turning on **Plot2** under the **STAT PLOT** menu and choosing ⊾, using **L1** for the **Xlist** and **RESID** for the **Ylist**. You can obtain **RESID** by pressing [2nd] [STAT] and selecting **RESID** from the list of names. Graph the scatter plot of the residuals by pressing **ZoomStat**.

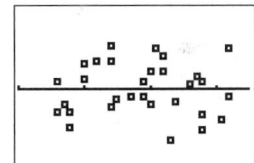

[6, 9.5] scl: 1 by [−0.5, 0.5] scl: 1

The residuals appear to be randomly scattered and centered about the regression line at $y = 0$. This supports the claim that the use of a linear model is appropriate.

✔ **Guided Practice**

4. **METEOROLOGY** Graph and analyze the residual plot for the rainfall and temperature data to determine if the linear model found in the Guided Practice for Example 3 is appropriate.

The residual plot magnifies deviations of the data points from the regression line, making it easier to see outliers in the data that lie in the y-direction. Outliers in the y-direction can indicate errors in data recording or unique cases, especially when describing societal trends or behavioral traits.

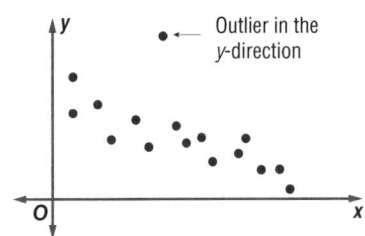

Outlier in the y-direction

Outliers in the *x*-direction can have a strong influence on the position of a regression line. In the figure, two least-squares regression lines are shown. The solid line is calculated using all the data, while the dashed line is calculated leaving out the outlier in the *x*-direction. Notice that leaving out this point noticeably moves the regression line.

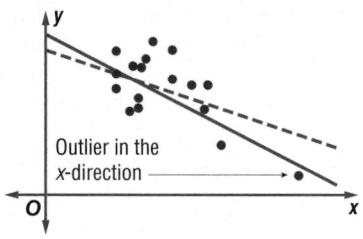

An individual data point that substantially changes a regression line is said to be **influential**. Outliers in the *x*-direction are often influential to the least-squares regression line. To determine if a point is an influential outlier, calculate and graph regression lines with and without this point. The point is influential if there is a substantial difference in the positions of the regression lines when the point is removed.

StudyTip

Influence The influence of an outlier is not a yes or no question. It is a matter of degree and is therefore subjective.

EXAMPLE 5   Identify an Influential Outlier

SLEEP/GPA STUDY Suppose the feature writer in Example 1 conducting the sleep/GPA study later received the additional piece of data listed in the table, which is an outlier.

Hours of Sleep	GPA
10.7	3.6

a. **Make a new scatter plot of the sleep/GPA data that includes the additional data point.**

Add the data point to the end of **L1** and **L2** and then graph the data, adjusting your window as necessary. From the graph you can see that this point is an outlier in the *x*-direction.

[6, 11] scl: 1 by [2, 4] scl: 1

b. **Calculate the correlation coefficient and least squares regression line with this outlier. Describe the effect this outlier has on the strength of the correlation and on the slope and intercept of the regression line.**

Original data: $r \approx 0.9148$ $\hat{y} = 0.457x - 0.667$

Data with outlier: $r \approx 0.8934$ $\hat{y} = 0.394x - 0.181$

The outlier has reduced the strength of the correlation. The change in the slope of the regression equation has caused the rate at which a student's GPA is raised due to additional sleep to drop from 0.457 points per hour to 0.394 per hour. At the same time, this outlier has raised the *y*-intercept, indicating that a student who gets no sleep will have a GPA close to 0.

c. **Plot both regression lines in the same window. Then state whether the outlier is influential. Explain your reasoning.**

The graph of the regression lines shows that the regression line moves more than a small amount when the outlier is added.
Therefore, the outlier (10.7, 3.6) is influential.

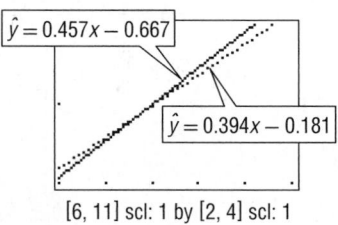

$\hat{y} = 0.457x - 0.667$

$\hat{y} = 0.394x - 0.181$

[6, 11] scl: 1 by [2, 4] scl: 1

✓ Guided Practice

5. **METEOROLOGY** Suppose the value (2.51, 50.4) for the rainfall and temperature data from Guided Practice 1 was replaced with (0.5, 50.4).

   A. Make a scatter plot of the original temperature/rainfall data that includes this outlier.

   B. Calculate the correlation coefficient and least squares regression line with this outlier. Describe the effect this outlier has on the strength of the correlation and on the slope and intercept of the regression line.

   C. Plot both regression lines in the same window. Then state whether the outlier is influential. Explain your reasoning.

Once you determine that the linear correlation coefficient for a set of data is significant and you find the least-squares regression line, you can then use the equation to make predictions over the range of the data. Making such predictions is called **interpolation**. Using the equation to make predictions far outside the range of the x-values you used to obtain the regression line is called **extrapolation**. Extrapolation should be avoided, since few real-world relationships are linear for all values of the explanatory variable.

EXAMPLE 6 Predictions with Regression

SLEEP/GPA STUDY The regression equation for the average hours of sleep x and GPA y from Example 3 was $\hat{y} = 0.457x - 0.667$. Use this equation to predict the expected GPA (to the nearest tenth) for a student who averages the following hours of sleep and state whether this prediction is reasonable. Explain.

a. 8 hours

Evaluate the regression equation for $x = 8$ to calculate $\hat{y}$.

$\hat{y} = 0.457x - 0.667$	**Regression equation**
$= 0.457(8) - 0.667$	$x = 8$
$= 3.656 - 0.667$	**Multiply.**
$= 2.989$	**Subtract.**

Using this model, we would expect that a student averaging 8 hours of sleep would have GPA of about 3.0. This GPA is reasonable since 8 is an x-value in the range of the original data.

b. 10.5 hours

$\hat{y} = 0.457x - 0.667$	**Regression equation**
$= 0.457(10.5) - 0.667$	$x = 10.5$
$= 4.7985 - 0.667$	**Multiply.**
$= 4.1315$	**Subtract.**

Using this model, we would expect that a student averaging 10.5 hours of sleep would have a GPA of about 4.1. This value is not reasonable, since we are extrapolating a y-value for an x-value that falls far outside the range of the original data. It is also not meaningful, since a student cannot earn a GPA higher than a 4.0 in this model.

✓ Guided Practice

6. METEOROLOGY Use the regression equation for the rainfall and temperature data from Guided Practice 3 to predict the expected temperature (to the nearest tenth of a degree) for months with each average rainfall. State whether this prediction is reasonable. Explain.

A. 3 in.   **B.** 8 in.

When analyzing bivariate data, follow the steps summarized below.

Concept Summary Analyzing Bivariate Data

Step 1 Make a scatter plot, and decide whether the variables appear to be linearly related.

Step 2 If they appear to be linearly related, calculate the strength of the relationship by calculating the correlation coefficient.

Step 3 Use a t-test to determine if the correlation is significant.

Step 4 If significant, find the least-squares regression equation that models the data.

Exercises

● = **Step-by-Step Solutions** begin on page R29.

For Exercises 1–6, analyze the bivariate data. (Examples 1–6)

a. Make a scatter plot of the data, and identify the relationship. Then calculate and interpret the correlation coefficient.

b. Determine if the correlation coefficient is significant at the 1%, 5%, and 10% levels. Explain your reasoning.

c. If the correlation is significant at the 10% level, state the least-squares regression equation and interpret the slope and intercept in context.

d. Graph and analyze the residual plot.

e. Identify any influential outliers. Describe the effect the outlier has on the strength of the original correlation and on the slope and intercept of the original regression line.

f. If any data were removed, reassess the significance of the correlation at the 10% level and, if still appropriate, recalculate the regression equation.

g. Use the regression equation to make the specified predictions. Interpret your results, and state whether the prediction is reasonable. Explain your reasoning.

1 **FAT GRAMS AND PROTEIN** An athlete wondered if there is a significant linear correlation between grams of fat and grams of protein in various foods. If appropriate, use the data below to predict the amount of protein (per serving) of an item with 1, 5, or 13 grams of fat.

Fat (g)	Protein (g)	Fat (g)	Protein (g)
12	14	9	13
57	30	18	24
9	15	30	25
20	25	18	25
12	15	32	24
39	28		

2. FIBER AND CALORIES The following data shows the caloric counts and amount of fiber in a variety of breakfast cereals. Use the data to predict the Calories in a serving of cereal that has 4.5, 5.5, or 7 grams of fiber.

Fiber (g)	Calories	Fiber (g)	Calories
1.5	133.5	1	149
0.5	115.5	1.5	114.5
1	143	0.5	85.5
2.5	109.5	1	116
0	119	1.5	110
0.5	113.5	0	53.5
0.5	102	8	196.5
0.5	117.5	0.5	99.5
6	186.5	6.5	114.5
1	154	3.5	140.5
11	389	0.5	122.5
4	114.5	2	110

3. EDUCATION AND HEALTH CARE The following data lists the performance rankings of education and health care in 14 states. If appropriate, use the data to predict the health care ranking if the education ranking is 15, 28, or 42.

Education	Health Care	Education	Health Care
1	45	8	35
2	48	9	18
3	50	10	13
4	37	11	20
5	39	12	28
6	26	13	15
7	21	14	29

4. WEIGHT AND MPG A shopper wants to determine if there is a significant linear correlation between the weight of cars and their highway fuel efficiency. If appropriate, use the data below to predict the gas mileage of automobiles that weigh 2900, 3300, and 4000 pounds.

Weight (lb)	MPG	Weight (lb)	MPG
3450	32	3460	28
3216	32	2897	36
2636	34	2805	32
2690	40	3067	28
2875	51	2716	31
2403	36	2595	38
2972	35	2326	39
2811	34	2911	30

5. GRADUATION AND UNEMPLOYMENT An economist took a sample of the graduation rates and unemployment rates of various states in a given year. If appropriate, use the data below to predict the unemployment rate if the graduation rates are 70%, 80%, or 90%.

Graduated	73	85	64	79	68	82
Unemployed	6.9	4.1	3.2	2.9	4.3	5.1
Graduated	71	81	76	64	77	82
Unemployed	4.1	5.5	5	6.8	4.8	5.2

6. POPULATION AND CRIME The following data lists the performance rankings of population and crime in 14 states. If appropriate, use the data to predict the crime ranking if the population ranking is 15.

Population	1	2	3	4	5	6	7
Crime	14	15	13	4	5	9	7
Population	8	9	10	11	12	13	14
Crime	11	3	12	10	8	1	6

Match each graph to the corresponding correlation coefficient.

a.

b.

c.

d.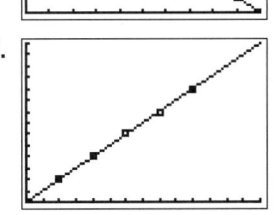

7. $r = -0.90$

8. $r = 0.50$

9. $r = 1.00$

10. $r = -1.00$

11 **INCOME AND DINING OUT** A restaurant is conducting a study to determine the relationship between a person's monthly income and the number of times that person dines out each month.

Income	500	1125	300	750	1250	950
Meals	4	10	3	6	12	8

a. Make a scatter plot of the data, and linearize the data by finding $(x, \ln y)$.

b. Make a scatter plot of the linearized data, and calculate and interpret r.

c. Determine if r is significant at the 5% level.

d. If r is significant, find the least-squares regression equation by using the model for the linearized data to find a model for the original data. (*Hint*: You can review this in Lesson 3-5.)

e. If appropriate, use the regression equation to predict the number of times that a person with a monthly income of $2000 will dine out. Is the prediction reasonable? Explain your reasoning.

12. **ADS AND SALES** An advertising firm wants to determine the strength of the relationship between the number of television ads aired each week and the amount of sales (in thousands of dollars) of the product.

Ads	2	3	5	7	7
Sales ($)	3	4	6	8	9
Ads	8	9	10	10	12
Sales ($)	10	12	12	13	15

a. Make a scatter plot of the data, and identify the relationship. Then find the correlation coefficient.

b. Determine if the correlation coefficient is significant at the 10% level. If so, find the least-squares regression equation.

c. Suppose the firm airs 15 ads during one week and 18 ads during the following week, and each ad spot costs $500. Make a prediction about the increase in profit from the first week to the second week.

13. 🔄 **MULTIPLE REPRESENTATIONS** In this problem, you will investigate the *coefficient of determination*.

a. **GRAPHICAL** Make a scatter plot of the data below. Then calculate the correlation coefficient r.

x	1	2	3	4	5	6
y	4	9	12	15	20	24

b. **NUMERICAL** Find the mean $\bar{y}$ of the y-values.

c. **NUMERICAL** Determine the least squares regression equation, and find the predicted $\hat{y}$-values by substituting each value of x into the equation.

d. **NUMERICAL** Use the following formulas to find the total variation $\Sigma(y - \bar{y})^2$, explained variation $\Sigma(\hat{y} - \bar{y})^2$ and unexplained variation $\Sigma(y - \hat{y})^2$.

e. **NUMERICAL** The coefficient of determination is given by $r^2 = \dfrac{\text{explained variation}}{\text{total variation}}$. Use the formula and your answers from part **d** to find r^2.

f. **ANALYTICAL** If the explained variation is the variation that can be explained by the relationship between x and y, what do you think the value of the coefficient of determination that you found means?

H.O.T. Problems *Use Higher-Order Thinking Skills*

REASONING Determine whether each statement is *true* or *false*. Explain your reasoning.

14. An r value of -0.85 indicates a stronger linear correlation than an r value of 0.75.

15. If the null hypothesis is rejected, it means that the value of ρ is not significantly different from 0.

16. **CHALLENGE** Consider two sets of bivariate data, C and D, which represent exponential relationships. With an exponential regression, the value of the base b in C is the reciprocal of the value of b in D. The correlation coefficients for each are equal to 0.99. What is the relationship of the linearized regression lines of C and D?

17. **REASONING** Consider the data set below where row A represents the explanatory variable and row B represents the response variable.

A	21	30	44	49	52	59
B	114	127	148	154	169	179

a. Make a scatter plot of the data. Then determine the equation for the least squares regression line and graph it in the same window as the scatter plot.

b. Interchange A and B and repeat part **a**.

c. What effect does switching the explanatory and response variables have on the regression line?

18. **WRITING IN MATH** Describe the strengths and weaknesses of the correlation coefficient as a measure of linear correlation for a set of bivariate data.

19. FOOTBALL Alexi claims that she can throw a football at least 55 yards. After 37 throws, her average distance is 57.7 yards with a standard deviation of 3.6 yards. Is there enough evidence to reject Alexi's claim at $\alpha = 0.05$? Explain your reasoning. (Lesson 11-6)

20. BOWLING Sonia and Pearl want to compare their bowling scores. They recorded their scores for 16 games as shown. (Lesson 11-5)

   a. Calculate the mean and sample standard deviation for each data set.

   b. Construct two 99% confidence intervals for the average score for both Sonia and Pearl.

   c. Make a statement comparing the effectiveness of the two intervals.

Sonia		Pearl	
112	109	88	169
98	116	129	190
143	131	146	99
109	98	170	108
121	122	95	181
84	128	111	183
106	121	108	122
100	107	181	99

If possible, find the sum of each infinite geometric series. (Lesson 10-3)

21. $a_1 = 4, r = \dfrac{5}{7}$ **22.** $a_1 = 14, r = \dfrac{7}{3}$ **23.** $16 + 12 + 9 + \ldots$

Write an explicit formula and a recursive formula for finding the nth term of each arithmetic sequence. (Lesson 10-2)

24. $10, 26.5, 43, \ldots$ **25.** $15, -9, -33, \ldots$ **26.** $3, \dfrac{11}{3}, \dfrac{13}{3}, \ldots$

Express each complex number in polar form. (Lesson 9-5)

27. $6 - 8i$ **28.** $-4 + i$ **29.** $3 + 2i$

Determine whether each pair of vectors are parallel. (Lesson 8-5)

30. $\mathbf{g} = \langle 3, 4, -6 \rangle, \mathbf{h} = \langle 9, 12, -18 \rangle$ **31.** $\mathbf{j} = \langle 9, -15, 11 \rangle, \mathbf{k} = \langle -14, 10, 7 \rangle$ **32.** $\mathbf{n} = \langle -16, -8, -13 \rangle, \mathbf{p} = \langle -15, 9, 5 \rangle$

33. SAT/ACT Which of the following must be true about the graph?

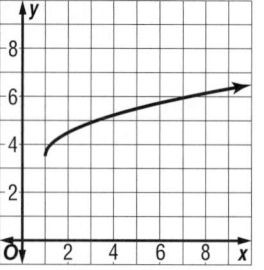

   I. The domain is all real numbers.

   II. The function is $y = \sqrt{x} + 3.5$.

   III. The range is about $\{y \mid y \geq 3.5\}$.

   A I only   **D** II only

   B II and III   **E** III only

   C I, II, and III

34. The table shows the total attendance for minor league baseball in some recent years. Which of the following is a regression equation for the data?

Year	Attendance (millions)
1990	18.4
1995	25.2
2000	33.1
2005	37.6

   F $y = 1.31x - 2588.15$ **H** $y = 1.31x - 18.4$

   G $y = 1.46x - 2588.15$ **J** $y = 1.46x - 18.4$

35. FREE RESPONSE For the following problem, consider a real-life situation that exhibits the characteristics of exponential or logistic growth or decay.

   a. Identify the situation and the type of growth or decay that it represents.

   b. Pose a question or make a claim about the situation.

   c. Make a hypothesis to the answer of the question.

   d. Develop, justify, and implement a method to collect, organize, and analyze the related data.

   e. Extend the nature of collected, discrete data to that of a continuous function that describes the known data set.

   f. Generalize the results and make a conclusion.

   g. Compare the hypothesis and the conclusion.

Objective
Use a TI-Nspire to find a median-fit line to model a relationship shown in a scatter plot.

In previous lessons, you have used regression equations to represent a set of data. Another type of regression used to model data is a *median-fit line*.

A median-fit line is found by dividing a set of data into three equal-sized groups and using the medians of those groups to determine a regression equation for the data.

ACTIVITY 1   Draw a Median-Fit Line

Use the data in the table to draw a median-fit line.

U.S. Energy Related Carbon Dioxide Emissions (million metric tons)			
Year	Emissions	Year	Emissions
1995	5301	2002	5820
1996	5489	2003	5872
1997	5570	2004	5966
1998	5607	2005	5974
1999	5669	2006	5888
2000	5848	2007	5984
2001	5754		

Source: Energy Information Administration

Step 1   Enter the data in a spreadsheet. Then make a scatter plot of the data. Let the *x*-axis represent the number of years where 0 represents the year 2000 and the *y*-axis the metric tons of carbon dioxide.

TechnologyTip

Icons Use different icons for the median points to easily distinguish them from the regular data points. Grab each point and select Attributes to change the icon type.

Step 2   Divide the data into three relatively equal and symmetric groups. The second group will have 5 data values, and the other groups will have 4. Then find the medians for the *x*- and *y*-values of each group.

Group 1 median: $(-3.5, 5529.5)$

Group 2 median: $(1, 5820)$

Group 3 median: $(5.5, 5970)$

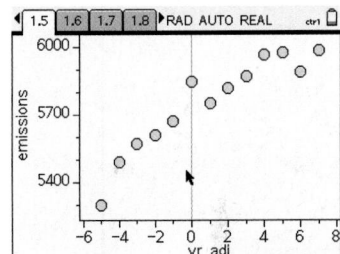

Step 3   The median-fit line uses the median points from the 1st and 3rd groups to determine slope and the average of the three median points as a point on the line. By using the point-slope form, $y = m(x - a) + b$, where m = slope and (a, b) is the average, you can form the median-fit line.

(continued on the next page)

Step 4 Define the average of y as $avey = \dfrac{medy1 + medy2 + medy3}{3}$. Define the median-fit line as $med\_med(x) = 48.944(x - 1) + 5773.17$. Then graph the median-fit line.

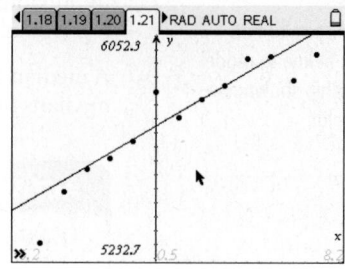

ACTIVITY 2 — Calculate a Median-Fit Line

Use the data in Activity 1 to calculate the median-fit line.

Step 1 Remove the three ordered pairs that represent the medians. Then make a scatter plot of the data.

Step 2 Calculate the equation of the median-fit line. Then graph the line.

Open a new **Calculator** screen. Under the **Statistics: Stat Calculations** menu, select **Median-Median Line**. Enter the lists for the x- and y-values.

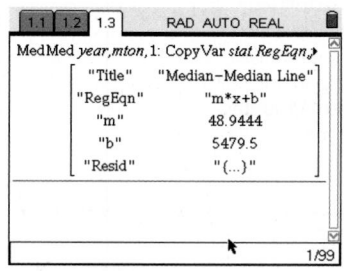

Notice that the equation of the median-fit line found in Activity 1 is identical to the calculator regression equation.

Analyze the Results

1. Explain the meaning of the slope of median-fit-line in this situation.

2. Is it reasonable to expect this line to represent the data indefinitely? Explain why or why not.

3. How many metric tons of carbon dioxide emissions can be expected in 2015?

Chapter Summary

Key Concepts

Descriptive Statistics (Lesson 11-1)

• The three most common shapes for distributions of data are negatively skewed, symmetrical, and positively skewed.

Probability Distributions (Lesson 11-2)

• The probability distribution of a random variable X links each possible value for X with its probability of occurring.

The Normal Distribution (Lesson 11-3)

• The z-value represents the number of standard deviations that a given data value is from the mean, and is given by $z = \frac{X - \mu}{\sigma}$.

• The standard normal distribution is a distribution of z-values with mean 0 and standard deviation 1.

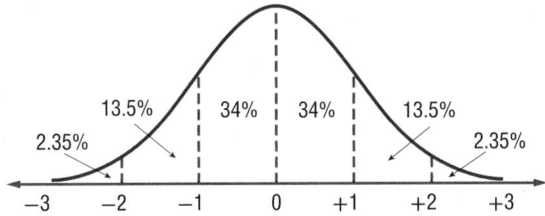

The Central Limit Theorem (Lesson 11-4)

• As the sampling size n increases, the shape of the distribution of the sample means approaches a normal distribution.

Confidence Intervals (Lesson 11-5)

• When $n \geq 30$, $CI = \bar{x} \pm z \cdot \frac{\sigma}{\sqrt{n}}$; when $n < 30$ and σ is unknown, $CI = \sigma \pm t \cdot \frac{s}{\sqrt{n}}$.

Hypothesis Testing (Lesson 11-6)

• The steps to conduct a hypothesis test are as follows.

Step 1 State the hypotheses, and identify the claim.

Step 2 Determine the critical value(s) and region.

Step 3 Calculate the test statistic.

Step 4 Accept or reject the null hypothesis.

Correlation and Linear Regression (Lesson 11-7)

• To analyze bivariate data:

Step 1 Make a scatter plot, and decide whether the variables appear to be linearly related.

Step 2 Calculate the correlation coefficient.

Step 3 Use a t-test to determine if the correlation is significant.

Step 4 Find the least-squares regression equation.

Key Vocabulary

alternative hypothesis (p. 705)
binomial distribution (p. 669)
confidence interval (p. 697)
continuous random variable (p. 664)
correlation (p. 713)
correlation coefficient (p. 713)
critical values (p. 697)
discrete random variable (p. 664)
empirical rule (p. 675)
explanatory variable (p. 713)
extrapolation (p. 719)
hypothesis test (p. 705)
inferential statistics (p. 696)
interpolation (p. 719)
least squares regression line (p. 716)

negatively skewed distribution (p. 654)
normal distribution (p. 674)
null hypothesis (p. 705)
percentiles (p. 658)
positively skewed distribution (p. 654)
probability distribution (p. 665)
random variable (p. 664)
regression line (p. 716)
response variable (p. 713)
sampling distribution (p. 685)
sampling error (p. 686)
standard normal distribution (p. 676)
symmetrical distribution (p. 654)
t-distribution (p. 699)
z-value (p. 676)

Vocabulary Check

Identify the word or phrase that best completes each sentence.

1. The mean is less than the median and the majority of the data are on the right in a (negatively skewed, positively skewed) distribution.

2. A (continuous, discrete) random variable can take on an infinite number of possible values within a specified interval.

3. A distribution of z-values with a mean of 0 and a standard deviation of 1 is called a (binomial, standard normal) distribution.

4. The standard deviation of the sample means is called the (sampling error, standard error of the mean).

5. The (Central Limit Theorem, empirical rule) states that as n increases, the shape of the distribution of the sample means will approach a normal distribution.

6. A single value estimate of an unknown population parameter is called a(n) (point, interval) estimate.

7. The (alternative, null) hypothesis states that there is not a significant difference between a sample value and a population parameter.

8. Using an equation to make predictions far outside the range of the x-values you used to obtain the regression line is called (extrapolation, interpolation).

Lesson-by-Lesson Review

11-1 Descriptive Statistics (pp. 654–663)

9. SAT SCORES The table gives the math SAT scores for 24 precalculus students.

SAT Math Scores					
373	437	477	491	503	516
392	454	479	491	508	519
405	463	485	498	508	522
417	470	485	499	513	533

a. Construct a histogram, and use it to describe the shape of the distribution.

b. Summarize the center and spread of the data using either the mean and standard deviation or the five-number summary.

10. RADON GAS The table shows the amount in picocuries per liter of radon gas in a sample of homes.

Amount of Radon (pCi/L)				
0.5	1.1	1.9	2.4	4.0
0.7	1.4	2.2	2.5	4.2
1.0	1.5	2.2	2.9	5.4
1.0	1.7	2.2	2.9	6.3
1.1	1.8	2.3	3.1	7.0

a. Construct a box plot, and use it to describe the shape of the distribution.

b. Summarize the center and spread of the data using either the mean and standard deviation or the five-number summary. Justify your choice.

11. MARATHON The table gives the frequency distribution of the completion times for the Boston Marathon for the first 322 women finishers. Construct a percentile graph. Estimate the percentile rank for those finishing below 3 hours, and interpret its meaning.

Time (hours)	Runners
2:45–2:49:59	3
2:50–2:54:59	4
2:55–2:59:59	28
3:00–3:04:59	35
3:05–3:09:59	54
3:10–3:14:59	80
3:15+	118

EXAMPLE 1

BACKPACKS The table shows the weight of school backpacks for a sample of high school students.

Average Backpack Weight (lb)					
11.5	15.0	16.0	17.0	19.0	24.5
12.5	15.5	16.0	17.5	21.0	25.0
14.5	15.5	16.5	18.0	21.0	25.0
14.5	15.5	17.0	18.0	21.5	27.0
15.0	16.0	17.0	18.5	23.5	30.0

a. Construct a histogram, and use it to describe the shape of the distribution.

[10, 34] scl: 4 by [0, 16] scl: 4

The graph is positively skewed. Most of the backpacks appear to weigh between 14 and 22 pounds, with a few that are heavier, so the tail of the distribution trails off to the right.

b. Summarize the center and spread of the data using either the mean and standard deviation or the five-number summary.

```
1-Var Stats
↑n=30
 minX=11.5
 Q₁=15.5
 Med=17
 Q₃=21
 maxX=30
```

The distribution of data is skewed; therefore, the five-number summary can be used to describe the distribution. The five-number summary indicates that while the weights range from 11.5 to 30 pounds, the median weight is 17 pounds and half of the weights are between 15.5 and 21 pounds.

Lesson-by-Lesson Review

11-2 Probability Distributions (pp. 664–673)

Classify each random variable X as *discrete* or *continuous*. Explain your reasoning.

12. X represents the number of people attending an opening show of a new movie on a given day.

13. X represents the amount of blood donated per person at a recent blood drive.

14. FAMOUS PEOPLE In a survey, 63% of adults said they recognized a certain famous athlete. Five adults are selected at random and asked if they recognize the athlete.

   a. Construct and graph a binomial distribution for the random variable X representing the number of adults who recognized the athlete.

   b. Find the probability that more than 2 adults recognized the athlete.

15. DOGS Find the variance and standard deviation of the probability distribution for the number of dogs per household in Greenville, South Carolina.

Dogs	Frequency
0	17,519
1	2720
2	1614
3	774
4	333

EXAMPLE 2

GRAPHING In a school survey, 45% of the students said that they knew how to graph a conic. Five students chosen at random are asked if they can graph a conic.

a. Construct and graph a binomial distribution for the random variable X representing the number of students who said they could graph a conic.

Here $n = 5$, $p = 0.45$, and $q = 1 - 0.45$ or 0.55.

$P(0) = {}_5C_0 \cdot 0.45^0 \cdot 0.55^5 \approx 0.050$
$P(1) = {}_5C_1 \cdot 0.45^1 \cdot 0.55^4 \approx 0.206$
$P(2) = {}_5C_2 \cdot 0.45^2 \cdot 0.55^3 \approx 0.337$
$P(3) = {}_5C_3 \cdot 0.45^3 \cdot 0.55^2 \approx 0.276$
$P(4) = {}_5C_4 \cdot 0.45^4 \cdot 0.55^1 \approx 0.113$
$P(5) = {}_5C_5 \cdot 0.45^5 \cdot 0.55^0 \approx 0.018$

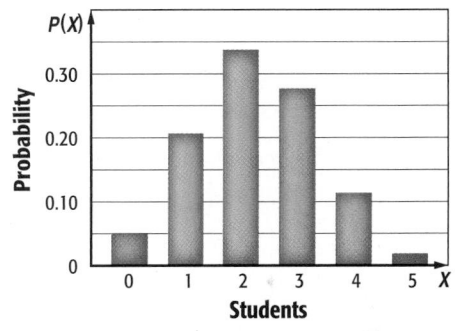

b. Find the probability that fewer than three of the students interviewed could graph a conic.

$P(X < 3) = P(0) + P(1) + P(2)$
$\quad = 0.05 + 0.21 + 0.34$ or 0.60 or 60%

11-3 The Normal Distribution (pp. 674–683)

Find each of the following.

16. z if $X = 1.5$, $\mu = 1.1$, and $\sigma = 0.3$

17. X if $z = 2.34$, $\mu = 105$, and $\sigma = 18$

18. z if $X = 125$, $\mu = 100$, and $\sigma = 15$

19. X if $z = -1.12$, $\mu = 35$, and $\sigma = 3.4$

Find the interval of z-values associated with each area.

20. outside 55%   **21.** middle 24%

22. middle 96%   **23.** outside 49%

EXAMPLE 3

Find z if $X = 36$, $\mu = 31$, and $\sigma = 1.3$.

$z = \dfrac{X - \mu}{\sigma}$ **Formula for z-values**

$\quad = \dfrac{36 - 31}{1.3}$ $X = 36$, $\mu = 31$, and $\sigma = 1.3$

$\quad \approx 3.85$ **Simplify.**

Lesson-by-Lesson Review

11-4 The Central Limit Theorem (pp. 685–694)

24. GRADES The average grade-point average or GPA in a particular school is 2.88 with a standard deviation of approximately 0.67. Find each probability for a random sample of 50 students from that school.

   a. the probability that the mean GPA will be less than 2.75

   b. the probability that the mean GPA will be greater than 3.05

   c. the probability that the mean GPA will be greater than 3.0 but less than 3.75

25. PHOTOGRAPHY A local photographer reported that 55% of seniors had their senior photos taken outdoors. If 15 seniors are selected at random, find the probability that fewer than 5 of the seniors will get their pictures taken outdoors.

Find each of the following if z is the z-value, $\bar{x}$ is the sample mean, μ is the mean of the population, n is the sample size, and σ is the standard deviation.

26. z if $\bar{x} = 5.8$, $\mu = 5.5$, $n = 18$, and $\sigma = 0.2$

27. μ if $\bar{x} = 14.8$, $z = 4.49$, $n = 14$, and $\sigma = 1.5$

28. n if $z = 1.5$, $\bar{x} = 227$, $\mu = 224$, and $\sigma = 10$

29. σ if $z = -2.67$, $\bar{x} = 38.2$, $\mu = 40$, and $n = 16$

EXAMPLE 4

WEATHER The average annual snowfall for Albany, New York, is 62 inches with a standard deviation of approximately 20 inches. Find the probability that the mean snowfall will be between 60 and 70 inches using a random sample of data for 7 years.

z-value for $\bar{x} = 60$:

$$= \frac{60 - 62}{7.56} \qquad \bar{x} = 60, \mu = 62, \text{ and } \sigma_{\bar{x}} = \frac{20}{\sqrt{7}} \approx 7.56$$

$$\approx -0.26 \qquad \textbf{Simplify.}$$

z-value for $\bar{x} = 70$:

$$= \frac{70 - 62}{7.56} \qquad \bar{x} = 70, \mu = 62, \text{ and } \sigma_{\bar{x}} = \frac{20}{\sqrt{7}} \approx 7.56$$

$$\approx 1.06 \qquad \textbf{Simplify.}$$

There is a 45.8% probability that the snowfall will be between 60 and 70 inches.

```
normalcdf(-0.26,
1.06)
        .4579957305
■
```

11-5 Confidence Intervals (pp. 696–704)

Determine whether the normal distribution or t-distribution should be used for each question. Then find each confidence interval given the following information.

30. $c = 90\%$, $\bar{x} = 73$, $s = 4.8$, $n = 12$

31. $c = 96\%$, $\bar{x} = 34$, $\sigma = 2.3$, $n = 38$

32. $c = 99\%$, $\bar{x} = 16$, $s = 1.6$, $n = 55$

33. $c = 90\%$, $\bar{x} = 5.8$, $\sigma = 1.1$, $n = 47$;

Determine the minimum sample size needed to conduct an experiment that has the given requirements.

34. $c = 90\%$, $\sigma = 3.9$, $E = 0.8$

35. $c = 95\%$, $\sigma = 1.6$, $E = 0.6$

36. $c = 98\%$, $\sigma = 6.8$, $E = 1.2$

37. $c = 92\%$, $\sigma = 10.2$, $E = 3.5$

EXAMPLE 5

Determine whether the normal distribution or t-distribution should be used to find a 95% confidence interval in which $\bar{x} = 12.8$, $s = 3.8$, and $n = 50$. Then find the confidence interval.

Since $n \geq 30$, the normal distribution should be used.

In a 95% confidence interval, 2.5% of the area lies in each tail. Use a graphing calculator to find z.

$$CI = \bar{x} \pm z \cdot \frac{s}{\sqrt{n}} \qquad \textbf{Confidence Interval for the Mean}$$

$$= 12.8 \pm 1.96 \cdot \frac{3.8}{\sqrt{50}} \qquad \bar{x} = 12.8, z = 1.96, s = 3.8, \text{ and } n = 50$$

$$= 12.8 \pm 1.05 \qquad \textbf{Simplify.}$$

The 95% confidence interval is $11.75 < \mu < 13.85$.

Lesson-by-Lesson Review

11-6 **Hypothesis Testing** (pp. 705–712)

For each statement, write the null and alternative hypotheses, and state which hypothesis represents the claim.

38. Jenna claims that she did not drive above 50 miles per hour during the entire trip.

39. Rafael claims that he can type faster than 60 words per minute.

40. Natasha claims that on average, it takes her less than 3 days to read a short novel.

41. Grant claims that he can bake at least 6 dozen cookies per hour.

For each claim k, use the specified information to calculate the test statistic and determine whether there is enough evidence to reject the null hypothesis. Then make a statement regarding the original claim.

42. $k: \mu \leq 26.5, \alpha = 0.10, \bar{x} = 27.8, s = 1.0, n = 46$

43. $k: \mu = 56, \alpha = 0.05, \bar{x} = 58.9, s = 6.7, n = 98$

44. $k: \mu < 18, \alpha = 0.01, \bar{x} = 17.6, s = 0.8, n = 26$

45. $k: \mu \geq 39, \alpha = 0.10, \bar{x} = 38.6, s = 2.6, n = 42$

EXAMPLE 6

For claim k, use the specified information to calculate the test statistic and determine whether there is enough evidence to reject the null hypothesis. Then make a statement regarding the original claim.

$k: \mu \geq 62, \alpha = 0.05, \bar{x} = 61.5, s = 4.3, n = 70$

State the null and alternative hypotheses, and identify the claim.

$H_0: \mu \geq 62$ **(claim)** $H_a: \mu < 62$

Determine the critical value(s) and region.

Use the z-value since $n \geq 30$ and a left-tailed test since $\mu < 62$. Since $\alpha = 0.05$, you can use a graphing calculator to find $z = -1.645$.

```
invNorm(0.05)
          -1.644853626
```

Calculate the test statistic.

$z = \dfrac{61.5 - 62}{0.51} \approx -0.98$ $\bar{x} = 61.5, \mu = 62,$ and $\sigma_{\bar{x}} = \dfrac{4.3}{\sqrt{70}}$

Reject or fail to reject the null hypothesis.

H_0 is not rejected since the test statistic does not fall within the critical region. Therefore, there is not enough evidence to reject the claim.

11-7 **Correlation and Linear Regression** (pp. 713–722)

46. GRADES The table shows the pre-test and final grades for a high school college prep class. (x = pre-test, y = final)

Scores for a College Prep Class

x	y	x	y	x	y	x	y
86	3.5	77	2.5	85	3.0	62	1.9
70	3.0	97	3.9	85	3.8	92	3.6
100	4.0	79	3.0	68	2.2	84	3.0
87	3.8	69	2.4	73	2.4	84	3.6
99	4.0	67	2.1	91	3.7	74	2.8

a. Make a scatter plot of the data, and identify the relationship. Then calculate and interpret the correlation coefficient.

b. Test the significance of this correlation coefficient at the 10% level.

EXAMPLE 7

HEIGHT The table shows the heights of brothers and sisters. Make a scatter plot of the data and identify the relationship. Then calculate and interpret the correlation coefficient.

Brother	71	68	66	67	70
Sister	69	64	63	63	68
Brother	71	70	73	72	65
Sister	69	67	70	71	61

The correlation coefficient r is about 0.9773. Since r is close to 1, this suggests that the data have a strong positive linear correlation. This numerical assessment of the data agrees with our graphical assessment.

[60, 75] scl: 1 by [60, 75] scl: 1

Applications and Problem Solving

47. SPORTS The body-fat levels of 20 professional basketball players are shown. (Lesson 11-1)

Body-Fat Levels (%)			
3.4	5.5	6.1	4.8
8.3	7.7	6.5	6.5
4.9	3.7	3.9	4.0
7.3	8.9	9.5	9.8
3.9	7.1	6.3	6.1

a. Construct a histogram, and use it to describe the shape of the distribution.

b. Summarize the center and spread of the data using either the mean and standard deviation or the five-number summary. Justify your choice.

48. EXERCISE The number of hours that a sample of students exercises each week is shown. (Lesson 11-1)

Time Spent Exercising (hours)		
3	2.5	0
1.5	3	2
3.5	2	0
1.5	9.5	0
8	0.5	1.5
1	10	4

a. Construct a box plot, and use it to describe the shape of the distribution.

b. Summarize the center and spread of the data using either the mean and standard deviation or the five-number summary. Justify your choice.

49. AP CLASSES The table shows the number of AP classes per senior. Find the mean, variance, and standard deviation of this distribution. (Lesson 11-2)

X	0	1	2	3	4
Frequency	12	18	25	19	11

50. IQ IQs for a group of people are normally distributed with a mean of 105 and a standard deviation of 22. Find the probability that a randomly chosen person will have an IQ that corresponds to each of the following. (Lesson 11-3)

a. above 101

b. below 94

c. between 110 and 120

51. COOKIES The number of chocolate chips in a cookie is normally distributed with $\mu = 25$ and $\sigma = 3$. Find each of the following. (Lesson 11-3)

a. $P(X < 35)$

b. $P(21 < X < 29)$

c. $P(X > 15)$

52. WRESTLING The average number of fans attending East High School's wrestling meets is normally distributed with $\mu = 88$ and $\sigma = 16$. If 6 random meets are selected, find the probability that the mean of the sample would be more than 90 fans. (Lesson 11-4)

53. EXERCISE A sample of 58 students found that on average, the students spend 185 minutes engaged in physical activity each week. Assume that the standard deviation from a recent study was 28 minutes. Estimate the mean time students spend engaged in physical activity each week using a confidence interval given a 95% confidence level. (Lesson 11-5)

54. FLIGHT An airline claims that its flights from Cleveland to Texas take less than 3.0 hours. A random sample of 30 flights found an average time of 2.9 hours and a standard deviation of 0.25 hour. Determine whether the airline's claim is supported at $\alpha = 0.05$. (Lesson 11-6)

a. Write the null and alternative hypotheses, and state which hypothesis represents the claim.

b. Calculate the test statistic.

c. Determine whether there is enough evidence to reject the null hypothesis.

d. Make a statement regarding the original claim.

55. SCORES The table shows the aptitude and writing test scores for a class over the same material. (Lesson 11-7)

Aptitude					Writing				
135	146	153	154	139	26	33	55	50	32
131	149	137	133	149	25	44	31	31	34
141	164	146	149	147	32	47	37	46	36
152	143	146	141	136	47	36	35	28	28
154	151	155	140	143	36	48	36	33	42
148	149	141	137	135	32	32	29	34	30

a. Make a scatter plot of the data, and identify the relationship. Then calculate and interpret the correlation coefficient.

b. Test the significance of this correlation coefficient at the 5% level.

c. Find the equation of the regression line.

d. Use this equation to predict writing score for a student who scored a 142 on the aptitude test.

1. **RACING** The ages of the last 20 winners of the Indianapolis 500 are shown.

Age (years)									
24	26	28	33	40	25	27	30	36	42
26	27	32	35	43	26	27	33	38	46

   a. Construct a histogram, and use it to describe the shape of the distribution.

   b. Summarize the center and spread of the data using either the mean and standard deviation or the five-number summary.

2. **TELEVISION** The number of televisions per household for 100 students is shown.

Televisions	0	1	2	3	4	5
Frequency	1	3	21	53	16	6

   a. Use the frequency distribution to construct and graph a probability distribution for the random variable X.

   b. Find the mean score, and interpret its meaning in the context of the problem situation.

   c. Find the variance and standard deviation of the probability distribution.

3. **FOOD** Ms. Martinez's Spanish class took a poll to find how many drive-through trips students made in a week.

X	0	1	2	3	4	5
Frequency	10	16	12	22	8	2

   a. Use the frequency distribution of the results to construct and graph a probability distribution for the random variable X, rounding each probability to the nearest hundredth.

   b. Find the mean of the probability distribution.

   c. Find the variance and standard deviation.

4. **VACATION** In the summer, the average temperature at a Caribbean vacation resort is 89°F with a standard deviation of 4.8°F. For a randomly selected day, find the probability that the temperature will be as follows.

   a. above 72°F

   b. below 68°F

   c. between 85°F and 93°F

PACKAGING The mean weight of a box of cereal is 362 grams and standard deviation of 5. If a random selection of 5 boxes is sampled, find the following.

5. probability that the mean weight is less than 355

6. probability that the mean weight is greater than 370

7. **CONCESSIONS** A survey of 97 movie patrons found that customers spent an average of $12.50 at the concessions counter. Assume that the standard deviation from a recent study was $2.25. Estimate the mean amount of money customers spend given a 95% confidence level.

8. **RENT** Phil claims that the average college student spends less than $400 a month on rent. A sample of 48 students found that students spent an average of $385 on rent each month and a standard deviation of $30. At $\alpha = 0.10$, determine whether there is enough evidence to reject the null hypothesis, and make a statement regarding the original claim.

9. **MULTIPLE CHOICE** Identify the graph that could have a correlation coefficient of -0.96 in a linear regression.

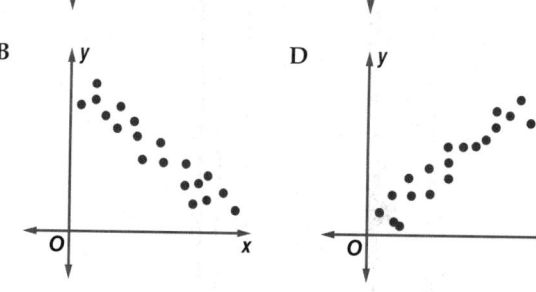

10. **DRIVING** The table lists the average number of accidents per month for sections of highway with the given speed limits.

   a. Make a scatter plot of the data, and identify the relationship.

   b. Calculate and interpret the correlation coefficient.

   c. Determine if the correlation coefficient is significant at the 5% level. Explain your reasoning.

Speed (mph)	Accidents
25	2.6
30	3.5
35	6.9
40	10.3
45	15.2
50	18.3
55	22.3
60	24.8
65	26.0
70	29.2

Objective
Create confidence intervals for population proportions.

In Lesson 11-2, you learned that the probability of a success in a single trial of a binomial experiment is p and it can be expressed as a fraction, decimal, or percentage. For example, the probability of tossing a fair coin and recording a tail is $\frac{1}{2}$, 0.5, or 50%. This probability is a *population proportion* because for the fair coin, the entire population, both heads and tails, is accounted for.

It is not always plausible to calculate population proportions. For example, calculating the percentage of high school students that own their own car would require surveying every high school student. Thus, population proportions can be estimated using *sample proportions* in the same way that sample means were used to estimate population means in this chapter.

The sample proportion $\hat{p}$ is the proportion of successes in a sample and is given by $\hat{p} = \frac{x}{n}$, where x is the number of successes in the sample and n is the sample size. The probability of failure is then given by $\hat{q} = 1 - \hat{p}$.

ACTIVITY 1   Sample Proportion

A sample of 2582 high school students found that 362 students own their own car. Estimate the population proportion of high school students that own their own car by calculating the sample proportion $\hat{p}$.

Step 1 Substitute $x = 362$ and $n = 2582$ into the formula for $\hat{p}$ and simplify.

Step 2 Interpret the result.

The percent of all high school students that own their own car is approximately 14%.

▶ Analyze the Results

1. Is the sample proportion an accurate estimate for the population proportion? Explain your reasoning.

2. If the sample is conducted with a larger n, what can be said about the relationship between the sample proportion and the population proportion?

3. Will the sample proportion ever equal the population proportion? If not, what can be done to the sample proportion in addition to increasing n to give a better estimate for the population proportion? Explain your reasoning.

We know from Lesson 11-5 that the $\hat{p}$ found in Activity 1 is a *point estimate*. If we wanted to create a better estimate, we would want to construct an interval. The behavior of the distribution of sample proportions is similar to the distribution of sample means. As the sample size increases, the distribution becomes approximately normal and the average of the sample proportions approaches the population proportion p.

Just as a confidence interval can be calculated for a population mean by adding and subtracting a maximum error of estimate E to and from a sample mean $\bar{x}$, a maximum error of estimate can be added to and subtracted from a sample proportion $\hat{p}$ to create a confidence interval for a population proportion.

StudyTip

Normal Distribution and z-Values Recall from Lesson 11-4, that the normal distribution is used for binomial distribution when $np \geq 5$ and $nq \geq 5$. Thus we can find and use z-values to calculate E in the same way as we did in Lesson 11-5.

Key Concept   Confidence Interval For a Population Proportion

The confidence interval *CI* for a population proportion is given by

$$CI = \hat{p} \pm E,$$

where $\hat{p}$ is the sample proportion and E is the maximum error of estimate represented by $z\sqrt{\frac{\hat{p}\hat{q}}{n}}$.

Confidence Interval for a Proportion

A random survey of 825 college applicants recorded the students' high school grade point average a. Find the 90% confidence interval for the proportion of all college applicants with a grade point average of 3.0 or higher.

GPA a	Applicants
$4.0 \leq a$	33
$3.0 \leq a < 4.0$	600
$2.0 \leq a < 3.0$	175
$a < 2.0$	17

Step 1 Find $\hat{p}$ and $\hat{q}$.

$\hat{p} = \dfrac{x}{n}$ **Sample Proportion Formula**

    $= \dfrac{633}{825}$ or about 0.77 $x = 633$ and $n = 825$

Therefore, $\hat{q} = 1 - 0.77$ or about 0.23.

Step 2 Verify that $n\hat{p} \geq 5$ and $n\hat{q} \geq 5$.

$n\hat{p} \approx 825(0.77)$ or 635.25 $n\hat{q} \approx (825)(0.23)$ or 189.75

Since $n\hat{p} \geq 5$ and $n\hat{q} \geq 5$, the sampling distribution of $\hat{p}$ can be approximated by the normal distribution.

Step 3 Find the z-value.

For a 90% confidence level, $z = 1.645$.

Step 4 Find the maximum error of estimate.

$E = z\sqrt{\dfrac{\hat{p}\hat{q}}{n}}$ **Maximum Error of Estimate Formula**

    $\approx 1.645\sqrt{\dfrac{0.77(0.23)}{825}}$ or about 0.0241 $z = 1.645, \hat{p} \approx 0.77, \hat{q} \approx 0.23,$ and $n = 825$

Step 5 Find the left and right endpoints of the confidence interval.

$CI = \hat{p} \pm E$ **Confidence Interval for a Proportion**

    $= 0.77 \pm 0.0241$ $\hat{p} = 0.77$ and $E = 0.0241$

Left Boundary	**Right Boundary**
$0.77 - 0.0241 = 0.7459$	$0.77 + 0.0241 = 0.7941$

The 90% confidence interval is then $0.74 < p < 0.79$. Therefore, we are 90% confident that the proportion of applicants with a G.P.A. of 3.0 or higher is between 74.6% and 79.4%.

▶ **Analyze the Results**

4. Describe two ways that the confidence interval found in Step 5 can be narrowed.

5. If the confidence level is held constant, what would n need to be to reduce the maximum error of estimate by $\dfrac{1}{2}$?

Model and Apply

6. In a 2006 Gallup Poll of 1000 adults, 480 felt that the money the government spent on the space shuttle should have been spent on something else. Find the 95% confidence interval for the proportion of all adults who felt this way.

7. A random sample of 279 households found that 58% had at least one sport-utility vehicle (SUV). Find the 99% confidence interval for the proportion of all households that own a SUV.

CHAPTER 12
Limits and Derivatives

Then
In Chapter 1, you learned about limits and rates of change.

Now
In Chapter 12, you will:

- Evaluate limits of polynomial and rational functions.
- Find instantaneous rates of change.
- Find and evaluate derivatives of polynomial functions.
- Approximate the area under a curve using definite integrals and integration.
- Find antiderivatives, and use the Fundamental Theorem of Calculus.

Why?

BUNGEE JUMPING The basic tools of calculus, derivatives and integrals, are very useful when working with rates that are not constant. A bungee jumper experiences varying rates of descent and ascent, as well as changing acceleration, depending on her position during the jump.

PREREAD Use the Mid-Chapter Quiz to write two or three questions about the first three lessons that will help you to predict the organization of the first half of Chapter 12.

Get Ready for Chapter 12

Diagnose Readiness You have two options for checking Prerequisite Skills.

Text Option

Take the Quick Quiz below.

QuickCheck

Use the graph of each function to describe its end behavior. (Lesson 1-3)

1. $q(x) = -\frac{2}{x}$

2. $f(x) = \frac{7}{x}$

3. $p(x) = \frac{x+5}{x-4}$

4. $m(x) = \frac{7-10x}{2x+7}$

5. MUSIC The average cost to produce x CDs can be represented by $A(x) = \frac{1700}{x} + 1200$. Find the limit as x approaches positive infinity. (Lesson 1-3)

Find the average rate of change of each function on the given interval. (Lesson 1-4)

6. $g(x) = 2x^2 + 4x - 1; [-2, 1]$

7. $f(x) = -2x^3 - 5x^2 + 6; [-4, -1]$

8. $f(x) = 4x^3 - x^2 + 9x - 1; [-2, 4]$

9. BOOKS The cost of producing x books can be represented by $C(x) = 2x^2 + 140x + 25$. Find the average rate of change of the cost if 50 books are produced instead of 25 books. (Lesson 1-4)

Find the domain of each function and the equations of the vertical or horizontal asymptotes, if any. (Lesson 2-5)

10. $f(x) = \frac{4x^2}{2x^2 + 1}$

11. $h(x) = \frac{2x^2 - 8}{x - 10}$

12. $f(x) = \frac{(x-1)(x+5)}{(x+2)(x-4)}$

13. $g(x) = \frac{x^2 - 16}{(x-2)(x+4)}$

Find the next four terms of each arithmetic or geometric sequence. (Lesson 10-2 and 10-3)

14. $3, 7, 11, 15, \ldots$

15. $8, 3, -2, -7, \ldots$

16. $5, -1, -7, -13, \ldots$

17. $-4, 12, -36, 108, \ldots$

18. $5, -10, 20, -40, \ldots$

19. $-28, -21, -14, -7, \ldots$

Online Option

Math Online > Take a self-check Chapter Readiness Quiz at glencoe.com.

New Vocabulary

English	• Page •	Español
one-sided limit	• 738 •	límite unilateral
two-sided limit	• 738 •	límite bilateral
direct substitution	• 748 •	sustitución directa
indeterminate form	• 749 •	forma indeterminada
tangent line	• 758 •	tangente
instantaneous rate of change	• 758 •	tasa instantánea de cambio
instantaneous velocity	• 760 •	velocidad instantánea
derivative	• 766 •	derivada
differentiation	• 766 •	diferenciación
differential equation	• 766 •	ecuación diferencial
differential operator	• 766 •	operador diferencial
regular partition	• 777 •	partición regular
definite integral	• 777 •	integral definida
lower limit	• 777 •	límite inferior
upper limit	• 777 •	límite superior
right Riemann sum	• 777 •	suma de Riemann por la derecha
integration	• 777 •	integración
antiderivative	• 784 •	antiderivada
indefinite integral	• 785 •	integral indefinida
Fundamental Theorem of Calculus	• 786 •	teorema fundamental del cálculo

Review Vocabulary

limit • p. 24 • límite the concept of approaching a value without necessarily ever reaching it

asymptotes • p. 130 • asíntota a line or curve that a graph approaches

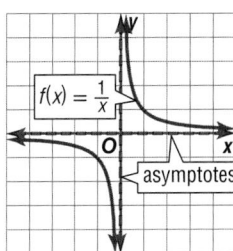

holes • p. 135 • agujeros removable discontinuities on the graph of a function that occur when the numerator and denominator of the function have common factors

Estimating Limits Graphically

Why?

Are there limits to world records set by athletes? At the 2008 Bejing Olympics, Elena Isinbaeva of Russia won the pole vault, setting a new world record by vaulting 5.05 meters. The logistic function $f(x) = \dfrac{5.334}{1 + 62548.213e^{-0.129x}}$, where x is the number of years since 1900, models the world record heights in women's pole vaulting from 1996 to 2008. You can use the limit of this function as x goes to infinity to predict a limiting height for this track and field event.

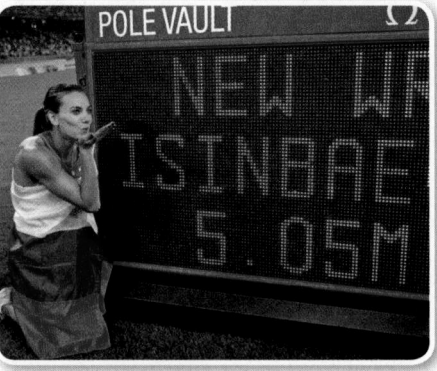

Estimate Limits at a Point Calculus centers around two fundamental problems:
- finding the equation of a line tangent to the graph of a function at a point, and
- finding the area between the graph of a function and the x-axis.

Fundamental to the solutions of each of these problems is an understanding of the concept of a limit. Recall from Lesson 1-3 that if $f(x)$ approaches a unique value L as x approaches c from either side, then the limit of $f(x)$ as x approaches c is L, written $\lim\limits_{x\to c} f(x) = L$.

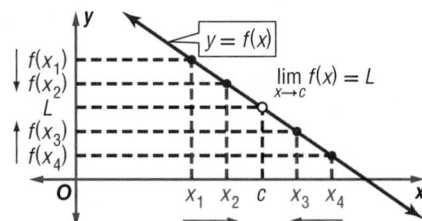

You can apply this description to estimate the limit of a function $f(x)$ as x approaches a fixed value c or $\lim\limits_{x\to c} f(x)$ by using a graph or making a table of values.

EXAMPLE 1   Estimate a Limit = $f(c)$

Estimate $\lim\limits_{x\to 2} (-3x + 1)$ using a graph. Support your conjecture using a table of values.

Analyze Graphically

The graph of $f(x) = -3x + 1$ suggests that as x gets closer to 2, the corresponding function value gets closer to -5. Therefore, we can estimate that $\lim\limits_{x\to 2} (-3x + 1)$ is -5.

Support Numerically

Make a table of values for f, choosing x-values that approach 2 by using some values slightly less than 2 and some values slightly greater than 2.

	⟶ *x* approaches 2 ⟶				⟵ *x* approaches 2 ⟵		
x	1.9	1.99	1.999	2	2.001	2.01	2.1
f(x)	−4.7	−4.97	−4.997		−5.003	−5.03	−5.3

The pattern of outputs suggests that as x gets closer to 2 from the left or from the right, $f(x)$ gets closer to -5. This supports our graphical analysis.

✔ Guided Practice

Estimate each limit using a graph. Support your conjecture using a table of values.

1A. $\lim\limits_{x\to -3} (1 - 5x)$ **1B.** $\lim\limits_{x\to 1} (x^2 - 1)$

In Example 1, $\lim\limits_{x\to2}(-3x+1)$ is the same as the value of $f(2)$. However, the limit of a function is not always equal to a function value.

TechnologyTip

▶ **Tables** To help create a table using a graphing calculator, enter the function using the [Y=] menu. Then use the Table function by pressing [2nd] [TABLE]. To approach a specific value, change the starting point and interval for x in your table by pressing [2nd] [TBLSET] and adjusting the TBLSET options.

EXAMPLE 2 | **Estimate a Limit $\neq$ f(c)**

Estimate $\lim\limits_{x\to3}\dfrac{x^2-9}{x-3}$ using a graph. Support your conjecture using a table of values.

Analyze Graphically

The graph of $f(x)=\dfrac{x^2-9}{x-3}$ suggests that as x gets closer to 3, the corresponding function value approaches 6. Therefore, we can estimate that $\lim\limits_{x\to2}\dfrac{x^2-9}{x-3}$ is 6.

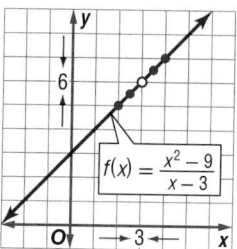

Support Numerically

Make a table of values, choosing x-values that approach 3 from either side.

	← x approaches 3 →				← x approaches 3 →		
x	2.9	2.99	2.999	3	3.001	3.01	3.1
f(x)	5.9	5.99	5.999		6.001	6.01	6.1

The pattern of outputs suggests that as x gets closer to 3, $f(x)$ gets closer to 6. This supports our graphical analysis.

✓ **Guided Practice**

Estimate each limit using a graph. Support your conjecture using a table of values.

2A. $\lim\limits_{x\to-2}\dfrac{x+2}{x^2-4}$

2B. $\lim\limits_{x\to5}\dfrac{x^2-4x-5}{x-5}$

In Example 2, notice that the limit as x approaches 3 is 6, even though $f(3)\neq6$. In fact, $f(3)$ does not even exist, because the expression $\dfrac{x^2-9}{x-3}$ is not defined when $x=3$. This illustrates an important point about limits.

Key Concept | **Independence of Limit from Function Value at a Point**

Words The limit of a function $f(x)$ as x approaches c does *not* depend on the value of the function *at* point c.

Example

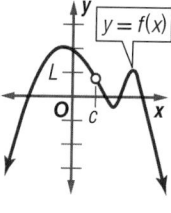

$\lim\limits_{x\to c}f(x)=L$

$f(c)$ is undefined.

$\lim\limits_{x\to c}g(x)=L$

$g(c)=n$

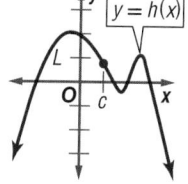

$\lim\limits_{x\to c}h(x)=L$

$h(c)=L$

It is important to understand that a limit is not about what happens at the number that x is approaching. Instead, a limit is about what happens *near* or *close to* that number.

In finding limits using a table or a graph, we have looked at the value of $f(x)$ as x approaches c from *each* side. We can describe the behavior of a graph from the left and right of x more concisely in terms of **one-sided limits**.

Key Concept — One-Sided Limits

Left-Hand Limit	**Right-Hand Limit**
If the value of $f(x)$ approaches a unique number L_1 as x approaches c from the left, then	If the value of $f(x)$ approaches a unique number L_2 as x approaches c from the right, then
$$\lim_{x \to c^-} f(x) = L_1,\ \text{which is read}$$	$$\lim_{x \to c^+} f(x) = L_2,\ \text{which is read}$$
The limit of f(x) as x approaches c from the left is L_1.	*The limit of f(x) as x approaches c from the right is L_2.*

Using these definitions, we can state more concisely what it means for a **two-sided limit** to exist.

Key Concept — Existence of a Limit at a Point

The limit of a function $f(x)$ as x approaches c exists if and only if both one-sided limits exist *and* are equal. That is, if

$$\lim_{x \to c^-} f(x) = \lim_{x \to c^+} f(x) = L, \text{ then } \lim_{x \to c} f(x) = L.$$

EXAMPLE 3   Estimate One-Sided and Two-Sided Limits

Estimate each one-sided or two-sided limit, if it exists.

a. $\lim\limits_{x \to 0^-} \dfrac{|2x|}{x}$, $\lim\limits_{x \to 0^+} \dfrac{|2x|}{x}$, and $\lim\limits_{x \to 0} \dfrac{|2x|}{x}$

The graph of $f(x) = \dfrac{|2x|}{x}$ suggests that

$$\lim_{x \to 0^-} \frac{|2x|}{x} = -2 \quad \text{and} \quad \lim_{x \to 0^+} \frac{|2x|}{x} = 2.$$

Because the left- and right-hand limits of $f(x)$ as x approaches 0 are not the same, $\lim\limits_{x \to 0} \dfrac{|2x|}{x}$ does not exist.

b. $\lim\limits_{x \to -3^-} g(x)$, $\lim\limits_{x \to -3^+} g(x)$, and $\lim\limits_{x \to -3} g(x)$, where $g(x) = \begin{cases} 4 & \text{if } x \neq -3 \\ -2 & \text{if } x = -3 \end{cases}$

The graph of $g(x)$ suggests that

$$\lim_{x \to -3^-} g(x) = 4 \quad \text{and} \quad \lim_{x \to -3^+} g(x) = 4.$$

Because the left- and right-hand limits of $g(x)$ as x approaches 3 are the same, $\lim\limits_{x \to -3} g(x)$ exists and is 4.

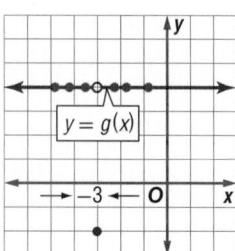

✓ Guided Practice

3A. $\lim\limits_{x \to 1^-} f(x)$, $\lim\limits_{x \to 1^+} f(x)$, and $\lim\limits_{x \to 1} f(x)$,

where $f(x) = \begin{cases} x^3 + 2 & \text{if } x < 1 \\ 2x + 1 & \text{if } x \geq 1 \end{cases}$

3B. $\lim\limits_{x \to -2^-} g(x)$, $\lim\limits_{x \to -2^+} g(x)$, and $\lim\limits_{x \to -2} g(x)$,

where $g(x) = \begin{cases} -0.5x + 2 & \text{if } x < -2 \\ -x^2 & \text{if } x \geq -2 \end{cases}$

Another way a limit can fail to exist is when the value of $f(x)$ as x approaches c does not approach a fixed finite value. Instead, the value of $f(x)$ increases without bound, indicated by ∞, or decreases without bound, indicated by $-\infty$.

ReadingMath

Without Bound For $f(x)$ to increase or decrease *without bound* as $x \to c$ means that by choosing an x-value arbitrarily close to c, you can obtain a function value with an absolute value that is as great as you want. The closer to c this x-value is chosen, the greater $|f(x)|$ is.

EXAMPLE 4   Limits and Unbounded Behavior

Estimate each limit, if it exists.

a. $\displaystyle\lim_{x \to 4} \frac{1}{(x-4)^2}$

Analyze Graphically The graph of $f(x) = \dfrac{1}{(x-4)^2}$ suggests that

$$\lim_{x \to 4^-} \frac{1}{(x-4)^2} = \infty \qquad \text{and} \qquad \lim_{x \to 4^+} \frac{1}{(x-4)^2} = \infty,$$

because as x gets closer to 4, the function values of the graph increase.

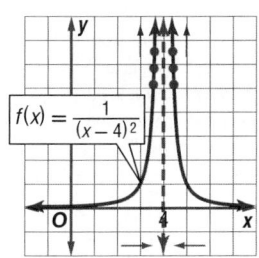

Neither one-sided limit at $x = 4$ exists; therefore, we can conclude that $\displaystyle\lim_{x \to 4} \frac{1}{(x-4)^2}$ does not exist. However, because both sides agree (both tend to ∞), we describe the behavior of $f(x)$ at 4 by writing $\displaystyle\lim_{x \to 4} \frac{1}{(x-4)^2} = \infty$.

Support Numerically

	← x approaches 4 →				← x approaches 4 →		
x	3.9	3.99	3.999	4	4.001	4.01	4.1
f(x)	100	10,000	1,000,000		1,000,000	10,000	100

The pattern of outputs suggests that as x gets closer to 4 from the left and the right, $f(x)$ grows without bound. This supports our graphical analysis.

b. $\displaystyle\lim_{x \to 0} \frac{1}{x}$

Watch Out!

Infinite Limits It is important to understand that the expressions $\displaystyle\lim_{x \to 0^-} f(x) = -\infty$ and $\displaystyle\lim_{x \to 0} f(x) = \infty$ are descriptions of why these limits fail to exist. The symbols ∞ and $-\infty$ do not represent real numbers.

Analyze Graphically The graph of $f(x) = \dfrac{1}{x}$ suggests that

$$\lim_{x \to 0^-} \frac{1}{x} = -\infty \qquad \text{and} \qquad \lim_{x \to 0^+} \frac{1}{x} = \infty,$$

because as x gets closer to 0, the function values from the left decrease and the function values from the right increase.

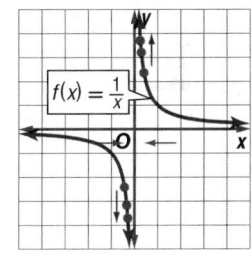

Neither one-sided limit at $x = 0$ exists; therefore, $\displaystyle\lim_{x \to 0} \frac{1}{x}$ does not exist. In this case, we cannot describe the behavior of $f(x)$ at 0 using a single expression because the unbounded behaviors from the left and right differ.

Support Numerically

	← x approaches 0 →				← x approaches 0 →		
x	−0.1	−0.01	−0.001	0	0.001	0.01	0.1
f(x)	−10	−100	−1000		1000	100	10

The pattern of outputs suggests that as x gets closer to 0 from the left and the right, $f(x)$ decreases and increases without bound, respectively. This supports our graphical analysis.

✔ Guided Practice

4A. $\displaystyle\lim_{x \to 3} \frac{x^2 - 4}{x - 3}$ **4B.** $\displaystyle\lim_{x \to 0} -\frac{2}{x^4}$

[−0.25, 0.25] scl: 0.05 by
[−1.5, 1.5] scl: 1

A limit can also fail to exist if instead of approaching a fixed value, $f(x)$ *oscillates* or bounces back and forth between two values.

EXAMPLE 5   Limits and Oscillating Behavior

Estimate $\lim\limits_{x \to 0} \cos \dfrac{1}{x}$, if it exists.

The graph of $f(x) = \cos \dfrac{1}{x}$ suggests that as x gets closer to 0, the corresponding function values oscillate between −1 and 1. This means that for an x_1-value close to 0 such that $f(x_1) = 1$, you can always find an x_2-value closer to 0 such that $f(x_2) = -1$. Likewise, for an x_3-value close to 0 such that $f(x_3) = -1$, you can always find an x_4-value closer to 0 such that $f(x_4) = 1$.

Therefore, $\lim\limits_{x \to 0} \cos \dfrac{1}{x}$ does not exist.

✅ Guided Practice

Estimate each limit, if it exists.

5A. $\lim\limits_{x \to 0} \sin \dfrac{1}{x}$ **5B.** $\lim\limits_{x \to 0} (x^2 \sin x)$

The three most common reasons why the limit of a function fails to exist at a point are summarized below.

Concept Summary   Why Limits at a Point Do Not Exist

The limit of $f(x)$ as x approaches c does not exist if:

- $f(x)$ approaches a different value from the left of c than from the right,
- $f(x)$ increases or decreases without bound from the left and/or the right of c, or
- $f(x)$ oscillates between two fixed values.

Estimate Limits at Infinity To this point, limits have been used to describe how a function $f(x)$ behaves as x approaches a fixed value c. In Lesson 1-3, you learned that limits can also be used to describe the end behavior of a function, that is, how a function behaves as x increases or decreases without bound. The notation for such limits is summarized below.

Key Concept   Limits at Infinity

- If the value of $f(x)$ approaches a unique number L_1 as x increases, then $\lim\limits_{x \to \infty} f(x) = L_1$, which is read *the limit of $f(x)$ as x approaches infinity is L_1.*

- If the value of $f(x)$ approaches a unique number L_2 as x decreases, then $\lim\limits_{x \to -\infty} f(x) = L_2$, which is read *the limit of $f(x)$ as x approaches negative infinity is L_2.*

In Lesson 2-4, you learned that unbounded behavior that can be described by ∞ or $-\infty$ indicates the location of a vertical asymptote. You also learned that the existence of a limit at infinity indicates the location of a horizontal asymptote. That is,

- $x = c$ is a *vertical asymptote* of the graph of $f(x)$ if $\lim\limits_{x \to c^-} f(x) = \pm\infty$ or $\lim\limits_{x \to c^+} f(x) = \pm\infty$, and

- $y = c$ is a *horizontal asymptote* of the graph of $f(x)$ if $\lim\limits_{x \to -\infty} f(x) = c$ or $\lim\limits_{x \to \infty} f(x) = c$.

EXAMPLE 6 **Estimate Limits at Infinity**

Estimate each limit, if it exists.

a. $\lim\limits_{x \to \infty} \frac{1}{x}$

Analyze Graphically The graph of $f(x) = \frac{1}{x}$ suggests that $\lim\limits_{x \to \infty} \frac{1}{x} = 0$. As x increases, $f(x)$ gets closer to 0.

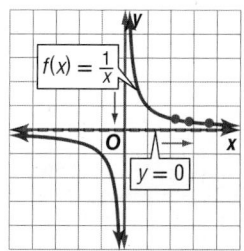

Support Numerically

— *x* approaches ∞ —→

x	10	100	1000	10,000	100,000
f(x)	0.1	0.01	0.001	0.0001	0.00001

The pattern of outputs suggests that as x grows increasingly larger, $f(x)$ approaches 0. ✔

StudyTip

Asymptotes The limit in Example 6a indicates that there is an asymptote at $y = 0$, while the limit in Example 6b indicates that there is an asymptote at $y = 2$.

b. $\lim\limits_{x \to -\infty} \left(-\frac{3}{x^2} + 2 \right)$

Analyze Graphically The graph of $f(x) = -\frac{3}{x^2} + 2$ suggests that $\lim\limits_{x \to -\infty} \left(-\frac{3}{x^2} + 2 \right) = 2$. As x decreases, $f(x)$ gets closer to 2.

Support Numerically

←— *x* approaches −∞ —

x	−100,000	−10,000	−1000	−100	−10
f(x)	1.99999	1.99999	1.99999	1.9997	1.97

The pattern of outputs suggests that as x decreases, $f(x)$ approaches 2.

c. $\lim\limits_{x \to -\infty} e^x \sin 3\pi x$ and $\lim\limits_{x \to \infty} e^x \sin 3\pi x$

Analyze Graphically The graph of $f(x) = e^x \sin 3x\pi$ suggests that $\lim\limits_{x \to -\infty} e^x \sin 3\pi x = 0$. As x decreases, $f(x)$ oscillates but tends toward 0.

The graph suggests that $\lim\limits_{x \to \infty} e^x \sin 3\pi x$ does not exist. As x increases, $f(x)$ oscillates between ever increasing values.

Support Numerically

←— *x* approaches −∞ —   — *x* approaches ∞ —→

x	−100	−50	−10	0	10	50	100
f(x)	3×10^{-44}	-2.0×10^{-22}	−0.00005	0	21966	4.8×10^{21}	-2.0×10^{43}

The pattern of outputs suggests that as x decreases, $f(x)$ approaches 0 and as x increases, $f(x)$ oscillates.

Watch Out!

Oscillating Behavior Do not assume that because a function $f(x)$ exhibits oscillating behavior, it has no limit as x approaches either ∞ or −∞. If the oscillations fluctuate between two fixed values or are unbounded, then the limit does not exist. If, however, the oscillations diminish and approach a fixed value, then the limit does exist.

☑ **Guided Practice**

6A. $\lim\limits_{x \to \infty} \left(\frac{1}{x^4} - 3 \right)$

6B. $\lim\limits_{x \to -\infty} e^x$

6C. $\lim\limits_{x \to \infty} \sin x$

You can use graphical and numerical approaches to estimate limits at infinity in many real-world situations.

🌐 Real-World EXAMPLE 7 — Estimate Limits at Infinity

a. HYDRAULICS A low-energy swing door uses a spring to close the door and a hydraulic mechanism to dampen or slow down the door's movement. If the door is opened to an angle of $\frac{\pi}{4}$ and then released from this resting position, the angle θ of the door t seconds after it is released is given by $\theta(t) = \frac{\pi}{4}(1 + 2t)e^{-2t}$. Estimate $\lim\limits_{t \to \infty} \theta(t)$, if it exists, and interpret your result.

Low-Energy Swing Door

Estimate the Limit

Graph $\theta(t) = \frac{\pi}{4}(1 + 2t)e^{-2t}$ using a graphing calculator. The graph shows that when $t = 0$, $\theta(t) \approx 0.785$ or about $\frac{\pi}{4}$. Notice that as t increases, the function values of the graph tend toward 0. So, we can estimate that $\lim\limits_{t \to \infty} \theta(t)$ is 0.

[0, 5] scl: 0.5 by [−0.5, 3.5] scl: 0.5

Interpret the Result

A limit of 0 in this situation indicates that the angle the door makes with its closed position tends toward a measure of 0 radians. That is, as the number of seconds after the door has been released increases, the door comes closer and closer to being completely closed.

b. MEDICINE The concentration in milligrams per milliliter of a certain medicine in a patient's bloodstream t hours after the medicine has been administered is given by $C(t) = Ate^{-0.18t}$, where A is a positive constant. Estimate $\lim\limits_{t \to \infty} C(t)$, if it exists, and interpret your result.

Estimate the Limit

Graph the related function $C_1(t) = te^{-0.18t}$ using a graphing calculator. The graph shows that as t increases, the function values of the graph tend toward 0. So, we can estimate that $\lim\limits_{t \to \infty} C_1(t)$ is 0.

[−5, 60] scl: 5 by [−1, 2.5] scl: 0.5

Because A is a positive constant, the graph of $C(t) = Ate^{-0.18t}$ will be the graph of $C_1(t) = te^{-0.18t}$ expanded vertically by a factor of A but will still tend to 0 as t increases without bound. So, we can estimate that $\lim\limits_{t \to \infty} C(t)$ is also 0.

Interpret the Result

A limit of 0 in this situation indicates that all of the medicine will eventually be eliminated from the patient's bloodstream.

✓ Guided Practice

7A. ELECTRICITY The typical voltage V supplied by an electrical outlet in the United States can be modeled by the function $V(t) = 165 \sin 120\pi t$, where t is the time in seconds. Estimate $\lim\limits_{t \to \infty} V(t)$, if it exists, and interpret your result.

7B. BIOLOGY Fruit flies are placed in a half-pint milk bottle with a piece of fruit and a yeast plant. The fruit fly population after t days is given by $P(t) = \dfrac{230}{1 + 56.5e^{-0.37t}}$. Estimate $\lim\limits_{t \to \infty} P(t)$, if it exists, and interpret your result.

Exercises

= **Step-by-Step Solutions** begin on page R29.

Estimate each limit using a graph. Support your conjecture using a table of values. (Examples 1 and 2)

1. $\lim\limits_{x \to 5} (4x - 10)$

2. $\lim\limits_{x \to 2} \left(\frac{1}{2} x^5 - 2x^3 + 3x^2 \right)$

3. $\lim\limits_{x \to -2} (x^2 + 2x - 15)$

4. $\lim\limits_{x \to -2} \frac{x^3 + 8}{x^2 - 4}$

5. $\lim\limits_{x \to 3} (2x^3 - 10x + 1)$

6. $\lim\limits_{x \to 0} \frac{x \cos x}{x^2 + x}$

7. $\lim\limits_{x \to 0} [5(\cos^2 x - \cos x)]$

8. $\lim\limits_{x \to 4} \frac{x - 4}{\sqrt{x} - 2}$

9. $\lim\limits_{x \to 6} (x + \sin x)$

10. $\lim\limits_{x \to -5} \frac{x^2 + x - 20}{x + 5}$

Estimate each one-sided or two-sided limit, if it exists. (Example 3)

11. $\lim\limits_{x \to 0^+} \frac{\sin x - x}{x}$

12. $\lim\limits_{x \to 0^-} \frac{|4x|}{x}$

13. $\lim\limits_{x \to 0} \frac{2x^2}{|x|}$

14. $\lim\limits_{x \to 9^+} \frac{3 - \sqrt{x}}{x - 9}$

15. $\lim\limits_{x \to 3^-} \frac{x^2 - 5x + 6}{x - 3}$

16. $\lim\limits_{x \to -\frac{1}{2}^-} \frac{|2x + 1|}{x}$

17. $\lim\limits_{x \to -2} \frac{x^2 + 5x + 6}{|x + 2|}$

18. $\lim\limits_{x \to -7} \frac{x^2 - x - 56}{x + 7}$

19. $\lim\limits_{x \to 0^-} (\sqrt{-x} - 7)$

20. $\lim\limits_{x \to 5} \frac{x^2 - 25}{x - 5}$

21. $\lim\limits_{x \to 4^+} \frac{x^2 - x - 12}{|x - 4|}$

22. $\lim\limits_{x \to 0^+} (\sqrt{x} + 2x + 3)$

23. $\lim\limits_{x \to 0} \frac{|3x|}{2x}$

24. $\lim\limits_{x \to -1} \frac{|x + 1|}{x^2 - 1}$

25. $\lim\limits_{x \to 0} f(x)$ where $f(x) = \begin{cases} \sqrt{-x} & \text{if } x < 0 \\ \sqrt{x} & \text{if } x \geq 0 \end{cases}$

26. $\lim\limits_{x \to 3} f(x)$ where $f(x) = \begin{cases} 3x & \text{if } x < 3 \\ x^2 & \text{if } x \geq 3 \end{cases}$

27. $\lim\limits_{x \to 0} f(x)$ where $f(x) = \begin{cases} x - 5 & \text{if } x < 0 \\ x^2 + 5 & \text{if } x \geq 0 \end{cases}$

28. $\lim\limits_{x \to 0} f(x)$ where $f(x) = \begin{cases} -x^2 + 2 & \text{if } x < 0 \\ \frac{2x}{x} & \text{if } x \geq 0 \end{cases}$

For each function below, estimate each limit if it exists. (Examples 1–4)

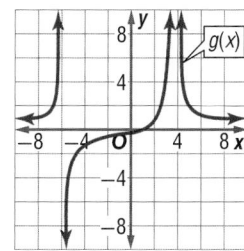

29. $\lim\limits_{x \to -6} f(x)$

30. $\lim\limits_{x \to 4} f(x)$

31. $\lim\limits_{x \to 4} g(x)$

32. $\lim\limits_{x \to -6} g(x)$

Estimate each limit, if it exists. (Examples 4–6)

33. $\lim\limits_{x \to -4} \frac{-17}{x^2 + 8x + 16}$

34. $\lim\limits_{x \to 5} \frac{x^2}{x^2 - 10x + 25}$

35. $\lim\limits_{x \to 4} \frac{|x|}{x - 4}$

36. $\lim\limits_{x \to \infty} e^{2x - 5}$

37. $\lim\limits_{x \to 6} \frac{5}{(x - 6)^2}$

38. $\lim\limits_{x \to -\infty} (x^5 - 7x^4 - 4x + 1)$

39. $\lim\limits_{x \to \infty} \frac{x^2 + x - 22}{4x^3 - 13}$

40. $\lim\limits_{x \to -3} \frac{x^2 + 9x + 20}{x + 3}$

41. $\lim\limits_{x \to \infty} \frac{3x - 4}{9x + 3}$

42. $\lim\limits_{x \to \infty} x \cos x$

43. $\lim\limits_{x \to -\infty} \frac{3^x + 3^{-x}}{3^x - 3^{-x}}$

44. $\lim\limits_{x \to 0} \frac{\sin |x|}{x}$

45. $\lim\limits_{x \to 0} x^2 \cos \frac{1}{x}$

46. $\lim\limits_{x \to -\infty} \frac{4x - 13}{2x + 8}$

47. MEDICINE A vaccine was quickly administered to combat a small outbreak of a minor infection. The number of reported cases of the infection w weeks since the vaccine was administered is shown. (Example 7)

a. Use the graph to estimate $\lim\limits_{w \to 1} f(w)$ and $\lim\limits_{w \to 3} f(w)$.

b. Use the graph to estimate $\lim\limits_{w \to \infty} f(w)$ if it exists, and interpret your results.

48. TRACK AND FIELD The logistic function $f(x) = \dfrac{5.334}{1 + 62548.213e^{-0.129x}}$, where x is the number of years since 1900, models the world record heights in meters for women's pole vaulting from 1996 to 2008. (Example 7)

a. Graph the function for $96 \leq x \leq 196$.

b. Estimate $\lim\limits_{x \to \infty} \dfrac{5.334}{1 + 62548.213e^{-0.129x}}$, if it exists.

c. Explain the relationship between the limit of the function and the world record heights.

49 INTERNET VIDEO A group of friends created a video parody of several popular songs and posted it online. As word of the video spread, interest grew. A model that can be used to estimate the number of people p that viewed the video is $p(d) = 12(1.25012)^d - 12$, where d is days since the video was originally posted. (Example 7)

a. Graph the function for $0 \leq d \leq 20$.

b. Estimate the number of people who viewed the video by the ends of the 5th, 10th, and 20th days. How many people will have viewed it after 2 months? (Use $d = 60$.)

c. Estimate $\lim\limits_{d \to \infty} p(d)$, if it exists, and interpret your results.

50. TECHNOLOGY The number of cell phone owners between the ages of 18 and 25 has steadily increased since the 1990s. A sequence model that can estimate the number of people ages 18–25 per cell phone is $a_n = 64.39(0.82605)^n + 1$, where n represents years since 1993. (Example 7)

   a. Graph the function for the years 1993 to 2011.

   b. Use the graph to estimate the amount of people per cell phone for 1998, 2007, and 2011.

   c. Use your graph to estimate $\lim\limits_{n \to \infty} a_n$.

   d. Explain the relationship between the limit of the function and the number of people per cell phone.

51 CHEMICALS An underground pipeline is leaking a toxic chemical. After the leak began, it spread as shown below. The distance the chemical spreads each year can be defined as $d(t) = 2000(0.7)^{t-1}$, for $t \geq 1$, where t is years since the leak began. (Example 7)

   a. Graph the function for $1 \leq t \leq 15$.

   b. Use your graph to find values of d for $t = 5, 10$, and 15 years.

   c. Use your graph to estimate $\lim\limits_{t \to \infty} d(t)$.

   d. Will the chemical ever spread to a hospital that is located 7000 meters away from the leak? Recall that the sum of an infinite geometric series can be found by $\frac{a_1}{1 - r}$.

52. DEPRECIATION Chuck purchases a motorcycle for $11,000, and it depreciates each year that he owns it. The value v of the motorcycle after t years can be estimated by the model $v(t) = 11,000(0.76)^t$. (Example 7)

   a. Graph the function for $0 \leq t \leq 10$.

   b. Use your graph to estimate the value of the motorcycle for $t = 3, 7$, and 10 years.

   c. Use your graph to estimate $\lim\limits_{t \to \infty} v(t)$.

   d. Explain the relationship between the limit of the function and the value of Chuck's motorcycle.

For the function below, estimate each limit if it exists.

53. $\lim\limits_{x \to 0^-} f(x)$

54. $\lim\limits_{x \to 0^+} f(x)$

55. $\lim\limits_{x \to 0} f(x)$

56. $\lim\limits_{x \to 2^-} f(x)$

57. $\lim\limits_{x \to 2^+} f(x)$

58. $\lim\limits_{x \to 1} f(x)$

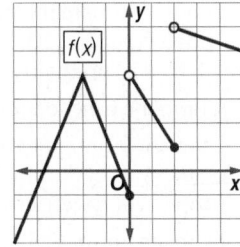

GRAPHING CALCULATOR Determine whether each limit exists. If not, describe what is happening graphically at the limit.

59. $\lim\limits_{x \to 1} \dfrac{x^2 - 1}{x^2 - 2x + 1}$

60. $\lim\limits_{x \to 2} \dfrac{x^2 + x}{x^2 - x - 2}$

61. $\lim\limits_{x \to 0} 3 \cos \dfrac{\pi}{x}$

62. $\lim\limits_{x \to -5} \dfrac{|x + 5|}{x + 5}$

63. $\lim\limits_{x \to 0} \dfrac{7}{1 + e^{\frac{1}{x}}}$

64. $\lim\limits_{x \to -4} \dfrac{\sqrt{2 - x} - 3}{x + 4}$

H.O.T. Problems Use **H**igher-**O**rder **T**hinking Skills

65. ERROR ANALYSIS Will and Kenyi are finding the limit of the function below as x approaches -6. Will says that the limit is -4. Kenyi disagrees, arguing that the limit is 3. Is either of them correct? Explain your reasoning.

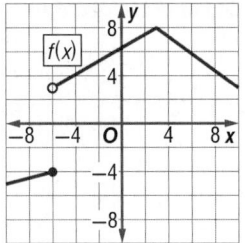

66. OPEN ENDED Give an example of a function f such that $\lim\limits_{x \to 0} f(x)$ exists but $f(0)$ does not exist. Give an example of a function g such that $g(0)$ exists but $\lim\limits_{x \to 0} g(x)$ does not exist.

67. CHALLENGE Suppose $f(x) = \dfrac{x^2 + 1}{x - 1}$ and $g(x) = \dfrac{x + 1}{x^2 - 4}$. Estimate $\lim\limits_{x \to 1} f(x)$ and $\lim\limits_{x \to 2} g(x)$. If $g(a) = 0$ and $f(a) \neq 0$, what can you assume about $\lim\limits_{x \to a} \dfrac{f(x)}{g(x)}$? Explain your reasoning.

68. REASONING Determine whether the following statement is *always*, *sometimes*, or *never* true. Explain your reasoning.

$$\text{If } f(c) = L, \text{ then } \lim\limits_{x \to c} f(x) = L.$$

69. OPEN ENDED Sketch the graph of a function such that $\lim\limits_{x \to 0} f(x) = -3$, $f(0) = 2$, $f(2) = 5$, and $\lim\limits_{x \to 2} f(x)$ does not exist.

70. CHALLENGE For the function below, estimate each limit if it exists.

$$f(x) = \begin{cases} 2x + 4 & \text{if } x < -1 \\ -1 & \text{if } -1 \leq x \leq 0 \\ x^2 & \text{if } 0 < x \leq 2 \\ x - 3 & \text{if } x > 2 \end{cases}$$

   a. $\lim\limits_{x \to -1} f(x)$ **b.** $\lim\limits_{x \to 0^-} f(x)$ **c.** $\lim\limits_{x \to 2^+} f(x)$

71. WRITING IN MATH Explain what method you could use to estimate limits if a function is continuous. Explain how this differs from methods used to estimate functions that are not continuous.

72. FUEL ECONOMY The table shows various engine sizes available from an auto manufacturer and their respective fuel economies. (Lesson 11-7)

a. Make a scatter plot of the data, and identify the relationship.

b. Calculate and interpret the correlation coefficient. Determine whether it is significant at the 10% level.

c. If the correlation is significant at the 10% level, find the least-squares regression equation and interpret the slope and intercept in context.

d. Use the regression equation that you found in part **c** to predict the expected miles per gallon that a car would get for an engine size of 8.0 liters. State whether this prediction is reasonable. Explain.

Engine Size (liters)	Highway Mileage (MPG)
1.6	34
2.2	37
2.0	30
6.2	26
7.0	24
3.5	29
5.3	24
2.4	33
3.6	26
6.0	24
4.4	23
4.6	24

For each statement, write the null and alternative hypotheses and state which hypothesis represents the claim. (Lesson 11-6)

73. A brand of dill pickles claims to contain 4 Calories.

74. A student claims that he exercises 85 minutes a day.

75. A student claims that she can get ready for school in less than 10 minutes.

76. Use Pascal's triangle to expand $\left(3a + \frac{2}{3}b\right)^4$. (Lesson 10-5)

Write and graph a polar equation and directrix for the conic with the given characteristics. (Lesson 9-4)

77. $e = 1$; vertex at $(0, -2)$

78. $e = 3$; vertices at $(0, 3)$ and $(0, 6)$

Find the angle between each pair of vectors to the nearest tenth of a degree. (Lesson 8-5)

79. $\mathbf{u} = \langle 2, 9, -2 \rangle$, $\mathbf{v} = \langle -4, 7, 6 \rangle$

80. $\mathbf{m} = 3\mathbf{i} - 5\mathbf{j} + 6\mathbf{k}$ and $\mathbf{n} = -7\mathbf{i} + 8\mathbf{j} + 9\mathbf{k}$

Use a graphing calculator to graph the conic given by each equation. (Lesson 7-4)

81. $7x^2 - 50xy + 7y^2 = -288$

82. $x^2 - 2\sqrt{3}xy + 3y^2 + 16\sqrt{3}x + 16y = 0$

83. SAT/ACT What is the area of the shaded region?

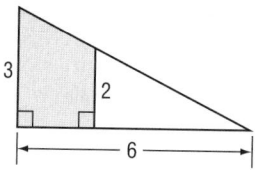

A 5   C 7   E 9
B 6   D 8

84. REVIEW Which of the following best describes the end behavior of $f(x) = x^{10} - x^9 + 5x^8$?

F $f(x) \to \infty$ as $x \to \infty$, $f(x) \to -\infty$ as $x \to \infty$

G $f(x) \to \infty$ as $x \to \infty$, $f(x) \to \infty$ as $x \to -\infty$

H $f(x) \to -\infty$ as $x \to \infty$, $f(x) \to \infty$ as $x \to -\infty$

J $f(x) \to -\infty$ as $x \to -\infty$, $f(x) \to \infty$ as $x \to \infty$

85. According to the graph of $y = f(x)$, $\lim\limits_{x \to 0} f(x) =$

A 0   C 3
B 1   D The limit does not exist.

86. REVIEW Which of the following describes the graph of $g(x) = \frac{1}{x^2}$?

I It has an infinite discontinuity.

II It has a jump discontinuity.

III It has a point discontinuity.

F I only   G II only   H I and II only

J I and III only   K I, II and III

Evaluating Limits Algebraically

Why?

Suppose the width in millimeters of an animal's pupil is given by $d(x) = \dfrac{152x^{-0.45} + 85}{4x^{-0.45} + 10}$, where x is the illuminance of the light shining on the pupils measured in lux. You can evaluate limits to find the width of the animal's pupils when the light is at its minimum and maximum intensity.

Compute Limits at a Point In Lesson 12-1, you learned how to estimate limits by using a graph or by making a table of values. In this lesson, we will explore computational techniques for evaluating limits.

Key Concepts   **Limits of Functions**

Limits of Constant Functions

Words	The limit of a constant function at any point c is the constant value of the function.
Symbols	$\lim\limits_{x \to c} k = k$

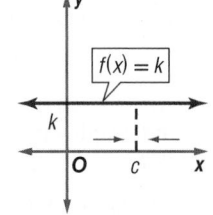

Limits of the Identity Function

Words	The limit of the identity function at any point c is c.
Symbols	$\lim\limits_{x \to c} x = c$

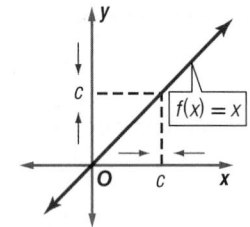

When combined with the following properties of limits, these constant and identity function limits become very useful.

Key Concepts   **Properties of Limits**

If k and c are real numbers, n is a positive integer, and $\lim\limits_{x \to c} f(x)$ and $\lim\limits_{x \to c} g(x)$ both exist, then the following are true.

Sum Property	$\lim\limits_{x \to c} [f(x) + g(x)] = \lim\limits_{x \to c} f(x) + \lim\limits_{x \to c} g(x)$
Difference Property	$\lim\limits_{x \to c} [f(x) - g(x)] = \lim\limits_{x \to c} f(x) - \lim\limits_{x \to c} g(x)$
Scalar Multiple Property	$\lim\limits_{x \to c} [kf(x)] = k \lim\limits_{x \to c} f(x)$
Product Property	$\lim\limits_{x \to c} [f(x) \cdot g(x)] = \lim\limits_{x \to c} f(x) \cdot \lim\limits_{x \to c} g(x)$
Quotient Property	$\lim\limits_{x \to c} \dfrac{f(x)}{g(x)} = \dfrac{\lim\limits_{x \to c} f(x)}{\lim\limits_{x \to c} g(x)}$, if $\lim\limits_{x \to c} g(x) \neq 0$
Power Property	$\lim\limits_{x \to c} [f(x)^n] = \left[\lim\limits_{x \to c} f(x) \right]^n$
nth Root Property	$\lim\limits_{x \to c} \sqrt[n]{f(x)} = \sqrt[n]{\lim\limits_{x \to c} f(x)}$, if $\lim\limits_{x \to c} f(x) > 0$ when n is even.

StudyTip

Limit Properties Each of the limit properties given on the previous page also hold for one-sided limits and limits at infinity, so long as each limit exists.

EXAMPLE 1 Use Limit Properties

Use the properties of limits to evaluate each limit.

a. $\lim\limits_{x \to 4} (x^2 - 6x + 3)$

$\lim\limits_{x \to 4} (x^2 - 6x + 3) = \lim\limits_{x \to 4} x^2 - \lim\limits_{x \to 4} 6x + \lim\limits_{x \to 4} 3$ Sum and Difference Properties

$= \left(\lim\limits_{x \to 4} x \right)^2 - 6 \cdot \lim\limits_{x \to 4} x + \lim\limits_{x \to 4} 3$ Power and Scalar Multiple Properties

$= 4^2 - 6 \cdot 4 + 3$ Limits of Constant and Identity Functions

$= -5$ Simplify.

CHECK The graph of $f(x) = x^2 - 6x + 3$ supports this result. ✔

b. $\lim\limits_{x \to -2} \dfrac{4x^3 + 1}{x - 5}$

$\lim\limits_{x \to -2} \dfrac{4x^3 + 1}{x - 5} = \dfrac{\lim\limits_{x \to -2} (4x^3 + 1)}{\lim\limits_{x \to -2} (x - 5)}$ Quotient Property

$= \dfrac{\lim\limits_{x \to -2} 4x^3 + \lim\limits_{x \to -2} 1}{\lim\limits_{x \to -2} x - \lim\limits_{x \to -2} 5}$ Sum and Difference Properties

$= \dfrac{4\left(\lim\limits_{x \to -2} x\right)^3 + \lim\limits_{x \to -2} 1}{\lim\limits_{x \to -2} x - \lim\limits_{x \to -2} 5}$ Scalar Multiple and Power Properties

$= \dfrac{4(-2)^3 + 1}{-2 - 5}$ Limits of Constant and Identity Functions

$= \dfrac{31}{7}$ Simplify.

CHECK Make a table of values, choosing x-values that approach -2 from either side. ✔

	← x approaches −2 →				→ x approaches −2 ←		
x	−2.1	−2.01	−2.001	−2	−1.999	−1.99	−1.9
f(x)	5.08	4.49	4.43		4.42	4.37	3.83

c. $\lim\limits_{x \to 3} \sqrt{8 - x}$

$\lim\limits_{x \to 3} \sqrt{8 - x} = \sqrt{\lim\limits_{x \to 3} (8 - x)}$ nth Root Property

$= \sqrt{\lim\limits_{x \to 3} 8 - \lim\limits_{x \to 3} x}$ Difference Property

$= \sqrt{8 - 3}$ Limits of Constant and Identity Functions

$= \sqrt{5}$ Simplify.

✔ **Guided Practice**

1A. $\lim\limits_{x \to 2} (-x^3 + 4)$ **1B.** $\lim\limits_{x \to 2} \dfrac{x - 3}{2x^2 - x - 15}$ **1C.** $\lim\limits_{x \to -1} \sqrt{x + 3}$

Notice that for each of the functions in Example 1, the limit of $f(x)$ as x approaches c is the same value that you would get if you calculated $f(c)$. While this is not true of every function, it is true of polynomial functions and of rational functions as described at the top of the next page.

Key Concepts   Limits of Functions

Limits of Polynomial Functions

If $p(x)$ is a polynomial function and c is a real number, then $\lim\limits_{x \to c} p(x) = p(c)$.

Limits of Rational Functions

If $r(x) = \dfrac{p(x)}{q(x)}$ is a rational function and c is a real number, then $\lim\limits_{x \to c} r(x) = r(c) = \dfrac{p(c)}{q(c)}$ if $q(c) \neq 0$.

More simply stated, limits of polynomial and rational functions can be found by **direct substitution**, so long as the denominator of the rational function evaluated at c is not 0.

EXAMPLE 2   Use Direct Substitution

Use direct substitution, if possible, to evaluate each limit. If not possible, explain why not.

a. $\lim\limits_{x \to -1} (-3x^4 + 5x^3 - 2x^2 + x + 4)$

Since this is the limit of a polynomial function, we can apply the method of direct substitution to find the limit.

$$\lim\limits_{x \to -1} (-3x^4 + 5x^3 - 2x^2 + x + 4) = -3(-1)^4 + 5(-1)^3 - 2(-1)^2 + (-1) + 4$$
$$= -3 - 5 - 2 - 1 + 4 \text{ or } -7$$

CHECK The graph of $f(x) = -3x^4 + 5x^3 - 2x^2 + x + 4$ supports this result. ✔

$f(x) = -3x^4 + 5x^3 - 2x^2 + x + 4$

b. $\lim\limits_{x \to 3} \dfrac{2x^3 - 6}{x - x^2}$

This is the limit of a rational function, the denominator of which is nonzero when $x = 3$. Therefore, we can apply the method of direct substitution to find the limit.

$$\lim\limits_{x \to 3} \dfrac{2x^3 - 6}{x - x^2} = \dfrac{2(3)^3 - 6}{3 - 3^2}$$
$$= \dfrac{48}{-6} \text{ or } -8$$

c. $\lim\limits_{x \to 1} \dfrac{x^2 - 1}{x - 1}$

This is the limit of a rational function. Since the denominator of this function is 0 when $x = 1$, the limit cannot be found by direct substitution.

✔ Guided Practice

2A. $\lim\limits_{x \to 4} (x^3 - 3x^2 - 5x + 7)$ **2B.** $\lim\limits_{x \to -5} \dfrac{x + 1}{x^2 + 3}$ **2C.** $\lim\limits_{x \to -8} \sqrt{x + 6}$

Suppose you incorrectly applied the Quotient Property of Limits or direct substitution to evaluate $\lim\limits_{x \to 1} \dfrac{x^2 - 1}{x - 1}$.

$$\lim\limits_{x \to 1} \dfrac{x^2 - 1}{x - 1} = \dfrac{\lim\limits_{x \to 1} (x^2 - 1)}{\lim\limits_{x \to 1} (x - 1)} = \dfrac{1^2 - 1}{1 - 1} \text{ or } \dfrac{0}{0}$$ **This is <u>incorrect</u> because the limit of the denominator is 0.**

It is customary to describe the resulting fraction $\frac{0}{0}$ as having an **indeterminate form** because you cannot determine the limit of the function with 0 in the denominator. A limit of this type may exist and have a real number value, or it may not exist, possibly diverging to ∞ or $-\infty$. In this case, from the graph of $f(x) = \frac{x^2 - 1}{x - 1}$, it appears that $\lim\limits_{x \to 1} \frac{x^2 - 1}{x - 1}$ does exist and has a value of 2.

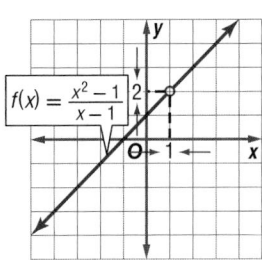

While an indeterminate form limit results from an incorrect application of limit properties or theorems, an analysis of this form can provide a clue as to what technique *should* be applied to find a limit.

If you evaluate the limit of a rational function and reach the indeterminate form $\frac{0}{0}$, you should try to simplify the expression algebraically by factoring and dividing out a common factor.

EXAMPLE 3 **Use Factoring**

Evaluate each limit.

a. $\lim\limits_{x \to -4} \dfrac{x^2 - x - 20}{x + 4}$

By direct substitution, you obtain $\dfrac{(-4)^2 - (-4) - 20}{-4 + 4}$ or $\dfrac{0}{0}$. Since this is an indeterminate form, try factoring and dividing out any common factors.

$$\lim\limits_{x \to -4} \dfrac{x^2 - x - 20}{x + 4} = \lim\limits_{x \to -4} \dfrac{(x - 5)(x + 4)}{x + 4} \qquad \text{Factor the numerator.}$$

$$= \lim\limits_{x \to -4} \dfrac{(x - 5)\cancel{(x + 4)}}{\cancel{x + 4}} \qquad \text{Divide out the common factor.}$$

$$= \lim\limits_{x \to -4} (x - 5) \qquad \text{Simplify.}$$

$$= (-4) - 5 \text{ or } -9 \qquad \text{Apply direct substitution and simplify.}$$

CHECK The graph of $f(x) = \dfrac{x^2 - x - 20}{x + 4}$ supports this result. ✔

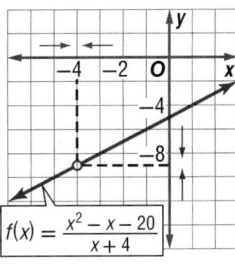

b. $\lim\limits_{x \to 3} \dfrac{x - 3}{x^3 - 3x^2 - 7x + 21}$

Watch Out!

Factoring If the entire expression in the numerator is divided out, the result is a 1, not a 0.

By direct substitution, you obtain $\dfrac{3 - 3}{3^3 - 3(3)^2 - 7(3) + 21}$ or $\dfrac{0}{0}$.

$$\lim\limits_{x \to 3} \dfrac{x - 3}{x^3 - 3x^2 - 7x + 21} = \lim\limits_{x \to 3} \dfrac{x - 3}{(x^2 - 7)(x - 3)} \qquad \text{Factor the denominator.}$$

$$= \lim\limits_{x \to 3} \dfrac{\cancel{x - 3}}{(x^2 - 7)\cancel{(x - 3)}} \qquad \text{Divide out the common factor.}$$

$$= \lim\limits_{x \to 3} \dfrac{1}{x^2 - 7} \qquad \text{Simplify.}$$

$$= \dfrac{1}{(3)^2 - 7} \text{ or } \dfrac{1}{2} \qquad \text{Apply direct substitution and simplify.}$$

☑ **Guided Practice**

3A. $\lim\limits_{x \to -2} \dfrac{x^3 - 3x^2 - 4x + 12}{x + 2}$

3B. $\lim\limits_{x \to 6} \dfrac{x^2 - 7x + 6}{3x^2 - 11x - 42}$

While this method of dividing out a common factor is valid, it requires some justification. In Example 3a, dividing out a common factor from $f(x)$ resulted in a new function, $g(x)$, where

$$f(x) = \frac{x^2 - x - 20}{x + 4} \quad \text{and} \quad g(x) = x - 5.$$

These two functions yield the same function values for every x except when $x = -4$. If two functions differ only at a value c in their domain, their limits as x approaches c are the same. This is because the value of a limit at a point is not dependent on the value of the *function* at that point.

Therefore, $\displaystyle\lim_{x \to -4} \frac{x^2 - x - 20}{x + 4} = \lim_{x \to -4} (x - 5)$.

Another technique to find limits that have indeterminate form is to rationalize the numerator or denominator of a function and then divide out any common factors.

EXAMPLE 4 Use Rationalizing

Evaluate $\displaystyle\lim_{x \to 9} \frac{\sqrt{x} - 3}{x - 9}$.

By direct substitution, you obtain $\dfrac{\sqrt{9} - 3}{9 - 9}$ or $\dfrac{0}{0}$. Rationalize the numerator of the fraction, and then divide common factors.

$$\lim_{x \to 9} \frac{\sqrt{x} - 3}{x - 9} = \lim_{x \to 9} \frac{\sqrt{x} - 3}{x - 9} \cdot \frac{\sqrt{x} + 3}{\sqrt{x} + 3}$$ Multiply the numerator and denominator by $\sqrt{x} + 3$, the conjugate of $\sqrt{x} - 3$.

$$= \lim_{x \to 9} \frac{x - 9}{(x - 9)(\sqrt{x} + 3)}$$ Simplify.

$$= \lim_{x \to 9} \frac{\cancel{x - 9}}{\cancel{(x - 9)}(\sqrt{x} + 3)}$$ Divide out the common factor.

$$= \lim_{x \to 9} \frac{1}{\sqrt{x} + 3}$$ Simplify.

$$= \frac{1}{\sqrt{9} + 3}$$ Apply direct substitution.

$$= \frac{1}{6}$$ Simplify.

Figure 12.2.1

CHECK The graph of $f(x) = \dfrac{\sqrt{x} - 3}{x - 9}$ in Figure 12.2.1 supports this result. ✔

☑ Guided Practice

Evaluate each limit.

4A. $\displaystyle\lim_{x \to 25} \frac{x - 25}{\sqrt{x} - 5}$

4B. $\displaystyle\lim_{x \to 0} \frac{2 - \sqrt{x + 4}}{x}$

Compute Limits at Infinity In Lesson 2-1, you learned that all even-degree power functions have the same end behavior, and all odd-degree power functions have the same end behavior. This can be described in terms of limits as shown below.

Key Concept Limits of Power Functions at Infinity

For any positive integer n,

- $\displaystyle\lim_{x \to \infty} x^n = \infty$.

- $\displaystyle\lim_{x \to -\infty} x^n = \infty$ if n is even.

- $\displaystyle\lim_{x \to -\infty} x^n = -\infty$ if n is odd.

$f(x) = x^n$, n is even

$f(x) = x^n$, n is odd

In Lesson 2-2, you also learned that the end behavior of a polynomial function is determined by the end behavior of the power function related to its highest-powered term. This can also be described using limits.

Key Concept   **Limits of Polynomial Functions at Infinity**

Let p be a polynomial function $p(x) = a_n x^n + \ldots + a_1 x + a_0$. Then $\lim\limits_{x \to \infty} p(x) = \lim\limits_{x \to \infty} a_n x^n$ and $\lim\limits_{x \to -\infty} p(x) = \lim\limits_{x \to -\infty} a_n x^n$.

You can use these properties to evaluate limits of polynomial functions at infinity. Remember, notating that the limit of a function is ∞ or $-\infty$ does not indicate that the limit exists but instead describes the behavior of the function as increasing or decreasing without bound, respectively.

EXAMPLE 5   **Limits of Polynomial Functions at Infinity**

Evaluate each limit.

a. $\lim\limits_{x \to -\infty} (x^3 - 2x^2 + 5x - 1)$

$\lim\limits_{x \to -\infty} (x^3 - 2x^2 + 5x - 1) = \lim\limits_{x \to -\infty} x^3$ **Limits of Polynomial Functions at Infinity**

$\qquad\qquad\qquad\qquad\qquad\quad = -\infty$ **Limits of Power Functions at Infinity**

b. $\lim\limits_{x \to \infty} (4 + 3x - x^2)$

$\lim\limits_{x \to \infty} (4 + 3x - x^2) = \lim\limits_{x \to \infty} -x^2$ **Limits of Polynomial Functions at Infinity**

$\qquad\qquad\qquad\quad = -\lim\limits_{x \to \infty} x^2$ **Scalar Multiple Property**

$\qquad\qquad\qquad\quad = -\infty$ **Limits of Power Functions at Infinity**

c. $\lim\limits_{x \to -\infty} (5x^4 - 3x)$

$\lim\limits_{x \to -\infty} (5x^4 - 3x) = \lim\limits_{x \to -\infty} 5x^4$ **Limits of Polynomial Functions at Infinity**

$\qquad\qquad\qquad\quad = 5 \lim\limits_{x \to -\infty} x^4$ **Scalar Multiple Property**

$\qquad\qquad\qquad\quad = 5 \cdot \infty \text{ or } \infty$ **Limits of Power Functions at Infinity**

✓ **Guided Practice**   Evaluate each limit.

5A. $\lim\limits_{x \to \infty} (-x^3 - 4x^2 + 9)$ **5B.** $\lim\limits_{x \to -\infty} (4x^6 + 3x^5 - x)$ **5C.** $\lim\limits_{x \to -\infty} (2x - 6x^2 + 4x^5)$

To evaluate limits of rational functions at infinity, we need another limit property.

Key Concept   **Limits of Reciprocal Function at Infinity**

Words   The limit of a reciprocal function at positive or negative infinity is 0.

Symbols $\lim\limits_{x \to \infty} \dfrac{1}{x} = \lim\limits_{x \to -\infty} \dfrac{1}{x} = 0$

   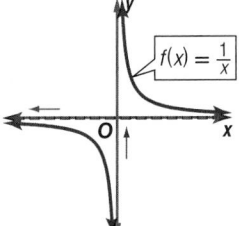

Corollary   For any positive integer n, $\lim\limits_{x \to \pm\infty} \dfrac{1}{x^n} = 0$.

If we divide the numerator and denominator of a rational function by the highest power of x that occurs in the function, we can use this property to find limits of rational functions at infinity.

EXAMPLE 6 **Limits of Rational Functions at Infinity**

Evaluate each limit.

a. $\displaystyle\lim_{x\to\infty} \frac{4x+5}{8x-3}$

$$\lim_{x\to\infty} \frac{4x+5}{8x-3} = \lim_{x\to\infty} \frac{\dfrac{4x}{x}+\dfrac{5}{x}}{\dfrac{8x}{x}-\dfrac{3}{x}}$$
Divide each term by highest-powered term, x.

$$= \lim_{x\to\infty} \frac{4+\dfrac{5}{x}}{8-\dfrac{3}{x}}$$
Simplify.

$$= \frac{\displaystyle\lim_{x\to\infty} 4 + 5\lim_{x\to\infty}\dfrac{1}{x}}{\displaystyle\lim_{x\to\infty} 8 - 3\lim_{x\to\infty}\dfrac{1}{x}}$$
Quotient, Sum, Difference and Scalar Multiple Properties

$$= \frac{4+5\cdot 0}{8-3\cdot 0} \text{ or } \frac{1}{2}$$
Limit of a Constant Function and
Limits of Reciprocal Function at Infinity

CHECK The graph of $f(x) = \dfrac{4x+5}{8x-3}$ in supports this result. ✔

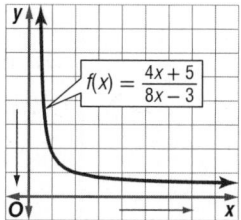

$f(x) = \dfrac{4x+5}{8x-3}$

b. $\displaystyle\lim_{x\to-\infty} \frac{6x^2-x}{3x^3+1}$

$$\lim_{x\to-\infty} \frac{6x^2-x}{3x^3+1} = \lim_{x\to-\infty} \frac{\dfrac{6x^2}{x^3}-\dfrac{x}{x^3}}{\dfrac{3x^3}{x^3}+\dfrac{1}{x^3}}$$
Divide each term by highest-powered term, x^3.

$$= \lim_{x\to-\infty} \frac{\dfrac{6}{x}-\dfrac{1}{x^2}}{3+\dfrac{1}{x^3}}$$
Simplify.

$$= \frac{6\displaystyle\lim_{x\to-\infty}\dfrac{1}{x} - \lim_{x\to-\infty}\dfrac{1}{x^2}}{\displaystyle\lim_{x\to-\infty} 3 + \lim_{x\to-\infty}\dfrac{1}{x^3}}$$
Quotient, Sum, Difference and Scalar Multiple Properties

$$= \frac{6\cdot 0 - 0}{3+0} \text{ or } 0$$
Limit of a Constant Function and
Limits of Reciprocal Function at Infinity

c. $\displaystyle\lim_{x\to\infty} \frac{5x^4}{9x^3+2x}$

$$\lim_{x\to\infty} \frac{5x^4}{9x^3+2x} = \lim_{x\to\infty} \frac{5}{\dfrac{9}{x}+\dfrac{2}{x^3}}$$
Divide each term by highest-powered term, x^4. Then simplify.

$$= \frac{\displaystyle\lim_{x\to\infty} 5}{9\displaystyle\lim_{x\to\infty}\dfrac{1}{x} + 2\lim_{x\to\infty}\dfrac{1}{x^3}}$$
Quotient, Sum, and Scalar Multiple Properties

$$= \frac{5}{9\cdot 0 + 2\cdot 0} \text{ or } \frac{5}{0}$$
Limit of a Constant Function and
Limits of Reciprocal Function at Infinity

Because the limit of the denominator is 0, we know that we have not correctly applied the Quotient Property of Limits. However, we can argue that as 5 is divided by increasingly smaller values approaching 0, the value of the fraction gets increasingly larger. Therefore, the limit can be described as approaching ∞.

✔ **Guided Practice**

6A. $\displaystyle\lim_{x\to-\infty} \frac{5}{x-10}$

6B. $\displaystyle\lim_{x\to\infty} \frac{-3x^2+7}{5x+1}$

6C. $\displaystyle\lim_{x\to\infty} \frac{7x^3-3x^2+1}{2x^3+4x}$

TechnologyTip

Evaluating Limits Using a calculator is not a foolproof way of evaluating
$\displaystyle\lim_{x\to c} f(x)$ or $\displaystyle\lim_{x\to\pm\infty} f(x)$.
You may only analyze the values of $f(x)$ for a few values of x near c or for a few large values of x. However, the function may do something unexpected as x gets even closer to c or as x gets even larger or smaller. You should use algebraic methods whenever possible to find limits.

In Lesson 10-1, you learned that since a sequence is a function of the natural numbers, the limit of a sequence is the limit of a function as $n \to \infty$. If this limit exists, then its value is the number to which the sequence converges. For example, the sequence $a_n = 1, \frac{1}{2}, \frac{1}{3}, \frac{1}{4}, \ldots$ can be described as $f(n) = \frac{1}{n}$, where n is a positive integer. Because $\lim\limits_{n \to \infty} \frac{1}{n} = 0$, the sequence converges to 0.

EXAMPLE 7 Limits of Sequences

Write the first five terms of each sequence. Then find the limit of the sequence, if it exists.

a. $a_n = \dfrac{3n + 1}{n + 5}$

The first five terms of this sequence are $\dfrac{3(1) + 1}{1 + 5}, \dfrac{3(2) + 1}{2 + 5}, \dfrac{3(3) + 1}{3 + 5}, \dfrac{3(4) + 1}{4 + 5},$ and $\dfrac{3(5) + 1}{5 + 5}$ or approximately 0.667, 1, 1.25, 1.444, and 1.6. To find the limit of the sequence, find $\lim\limits_{n \to \infty} \dfrac{3n + 1}{n + 5}$.

$$\lim_{n \to \infty} \frac{3n + 1}{n + 5} = \lim_{n \to \infty} \frac{3 + \frac{1}{n}}{1 + \frac{5}{n}} \qquad \text{Divide each term by highest-powered term, } n, \text{ and simplify.}$$

$$= \frac{\lim\limits_{n \to \infty} 3 + \lim\limits_{n \to \infty} \frac{1}{n}}{\lim\limits_{n \to \infty} 1 + 5 \lim\limits_{n \to \infty} \frac{1}{n}} \qquad \text{Quotient, Sum, and Scalar Multiple Properties}$$

$$= \frac{3 + 0}{1 + 5 \cdot 0} \text{ or } 3 \qquad \text{Limit of a Constant Function and Limits of Reciprocal Function at Infinity}$$

So, the limit of the sequence is 3. That is, the sequence converges to 3.

CHECK The graph of $a_n = \dfrac{3n + 1}{n + 5}$ supports this result. ✔

StudyTip

Check for Reasonableness
To check the reasonableness of the results in Example 7, find the 100th, 1000th, and 10,000th terms in each sequence. In Example 7a, these terms are 2.867, 2.986, and 2.999, respectively. Since these values appear to be approaching 3, a limit of 3 is reasonable.

b. $b_n = \dfrac{5}{n^4}\left[\dfrac{n^2 (n + 1)^2}{4}\right]$

The first five terms of this sequence are approximately 5, 2.813, 2.222, 1.953, and 1.8. Now find the limit of the sequence.

$$\lim_{n \to \infty} \frac{5}{n^4}\left[\frac{n^2(n + 1)^2}{4}\right] = \lim_{n \to \infty} \frac{5}{n^4}\left[\frac{n^2(n^2 + 2n + 1)}{4}\right] \qquad \text{Square the binomial.}$$

$$= \lim_{n \to \infty} \frac{5n^4 + 10n^3 + 5n^2}{4n^4} \qquad \text{Multiply.}$$

$$= \frac{\lim\limits_{n \to \infty} 5 + 10 \lim\limits_{n \to \infty} \frac{1}{n} + 5 \lim\limits_{n \to \infty} \frac{1}{n^2}}{\lim\limits_{n \to \infty} 4} \qquad \text{Divide each term by highest-powered term. Then use Quotient, Sum, and Scalar Multiple Properties.}$$

$$= \frac{5}{4} \text{ or } 1.25 \qquad \text{Limit of a Constant Function and Limits of Reciprocal Function at Infinity}$$

So, the limit of b_n is 1.25. That is, the sequence converges to 1.25.

CHECK Make a table of values, choosing large n-values that grow increasingly larger. ✔

⎯⎯⎯ *n* approaches ∞ ⎯⎯→

n	10	100	1000	10,000	100,000
a_n	1.51	1.28	1.25	1.25	1.25

✔ Guided Practice

7A. $a_n = \dfrac{4}{n^2 + 1}$ **7B.** $b_n = \dfrac{2n^3}{3n + 8}$ **7C.** $c_n = \dfrac{9}{n^3}\left[\dfrac{n(n + 1)(2n + 1)}{6}\right]$

Use the properties of limits to evaluate each limit. (Example 1)

1. $\lim\limits_{x \to -3} (5x - 10)$

2. $\lim\limits_{x \to 5} \dfrac{x^2 + 4x + 13}{x - 3}$

3. $\lim\limits_{x \to -1} (7x^2 - 6x - 3)$

4. $\lim\limits_{x \to -2} \dfrac{2x^5 - 4x^3 - 2x - 12}{x^3 + 5x^2}$

5. $\lim\limits_{x \to 9} \left(\dfrac{1}{x} + 2x + \sqrt{x} \right)$

6. $\lim\limits_{x \to -4} [x^2(x + 1) + 2]$

7. $\lim\limits_{x \to 12} \dfrac{x^2 - 10x}{\sqrt{x + 4}}$

8. $\lim\limits_{x \to 1} \dfrac{x^3 + 2x - 11}{x + 3}$

9. $\lim\limits_{x \to 2} (26 - 3x)$

10. $\lim\limits_{x \to -6} \dfrac{x^4 - x^3}{x^2}$

Use direct substitution, if possible, to evaluate each limit. If not possible, explain why not. (Example 2)

11. $\lim\limits_{x \to 16} \dfrac{x^2 + 9}{\sqrt{x} - 4}$

12. $\lim\limits_{x \to 2} 4x^3 - 3x^2 + 10$

13. $\lim\limits_{x \to 3} \dfrac{x^3 + 9x + 6}{x^2 + 5x + 6}$

14. $\lim\limits_{x \to 3} \sqrt{2 - x}$

15. $\lim\limits_{x \to -4} \dfrac{5x^5 - 16x^4}{x + 5}$

16. $\lim\limits_{x \to 4} \dfrac{x + 4}{x - 4}$

17. $\lim\limits_{x \to 5} \dfrac{x^3}{\sqrt{x + 4} - 5}$

18. $\lim\limits_{x \to 9} (3x^2 - 10x + 35)$

19. $\lim\limits_{x \to 5} \dfrac{2x + 11}{x^2 - x - 20}$

20. $\lim\limits_{x \to 1} (-x^2 + 3x + \sqrt{x})$

21. PHYSICS According to the special theory of relativity developed by Albert Einstein, the mass of an object traveling at speed v is given by $m = \dfrac{m_0}{\sqrt{1 - \dfrac{v^2}{c^2}}}$, where c is the speed of light and m_0 is the initial, or *rest mass*, of the object. (Example 2)

  a. Find $\lim\limits_{v \to 0} m$. Explain the relationship between this limit and m_0.

  b. What happens to the mass of an object if its speed were able to approach the speed of light?

22. GEOMETRY The area of an ellipse is defined as $A = \pi a \sqrt{a^2 - c^2}$, where a is the distance from the vertices to the center and c is the distance from the foci to the center. (Example 2)

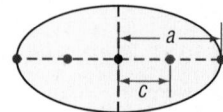

  a. What is the area of an ellipse for $a = 5$ and $c = 3$?

  b. What happens to the eccentricity of an ellipse as the foci move closer to the center of the ellipse?

  c. What is the limit of the area of the ellipse as c approaches 0 in terms of a?

Evaluate each limit. (Examples 3 and 4)

23. $\lim\limits_{x \to 4} \dfrac{2x^2 - 5x - 12}{x - 4}$

24. $\lim\limits_{x \to 0} \dfrac{4x}{\sqrt{x + 1} - 1}$

25. $\lim\limits_{x \to 1} \dfrac{x^2 + 4x - 5}{x^2 - 1}$

26. $\lim\limits_{x \to 9} \dfrac{\sqrt{x} - 3}{x - 9}$

27. $\lim\limits_{x \to -5} \dfrac{4x^2 + 21x + 5}{3x^2 + 17x + 10}$

28. $\lim\limits_{x \to 7} \dfrac{5 - \sqrt{18 + x}}{x - 7}$

29. $\lim\limits_{x \to -2} \dfrac{x + 2}{\sqrt{6 + x} - 2}$

30. $\lim\limits_{x \to \frac{1}{2}} \dfrac{8x^2 + 2x - 3}{12x^2 + 8x - 7}$

31. $\lim\limits_{x \to 0} \dfrac{2x}{3 - \sqrt{x + 9}}$

32. $\lim\limits_{x \to -3} \dfrac{x^2 - 2x - 15}{x + 3}$

33. $\lim\limits_{x \to 6} \dfrac{\sqrt{x + 3} - 3}{x - 6}$

34. $\lim\limits_{x \to 0} \dfrac{\sqrt{16 + x} - 4}{x}$

Evaluate each limit. (Examples 5 and 6)

35. $\lim\limits_{x \to \infty} (5 - 2x^2 + 7x^3)$

36. $\lim\limits_{x \to \infty} \dfrac{3x^3 - 10x + 2}{4x^3 + 20x^2}$

37. $\lim\limits_{x \to \infty} \dfrac{2x^2 + 7x - 17}{3x^5 + 4x^2 + 2}$

38. $\lim\limits_{x \to \infty} (10x + 14 + 6x^2 - x^4)$

39. $\lim\limits_{x \to \infty} \dfrac{x^6 + 12x}{3x^6 + 2x^2 + 11x}$

40. $\lim\limits_{x \to \infty} \dfrac{14x^3 - 12x}{4x^2 + 13x - 8}$

41. $\lim\limits_{x \to \infty} (7x^3 + 4x^4 + x)$

42. $\lim\limits_{x \to \infty} \dfrac{6x^5 - 12x^2 + 14x}{2x^5 + 13x^3}$

43. $\lim\limits_{x \to \infty} (x^3 - 6x^7 + 2x^6)$

44. $\lim\limits_{x \to \infty} \dfrac{6x^3 + 2x - 11}{-x^5 + 17x^3 + 4x}$

45. $\lim\limits_{x \to \infty} \dfrac{10x^4 - 2}{5x^4 + 3x^3 - 2x}$

46. $\lim\limits_{x \to -\infty} (2x^5 - 4x^2 + 10x - 8)$

47 **SPONGE** A gel capsule contains a sponge animal. When the capsule is submerged in water, it immediately dissolves, allowing the sponge to absorb water and quickly grow in size. The length ℓ in millimeters of the sponge animal after being submerged in water for t seconds can be defined as $\ell(t) = \dfrac{105t^2}{10 + t^2} + 25$. (Example 6)

  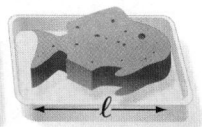

$t = 0$ $t = t_1$ $t = t_n$

  a. What is the length of the capsule before it is submerged in water?

  b. What is the limit of this function as $t \to \infty$?

  c. Explain how the limit of this function relates to the length of the sponge animal.

48. PUPPIES Suppose the weight w in pounds of a puppy d days after birth can be estimated by $w(d) = \dfrac{50}{2 + 98(0.85)^d}$. (Example 6)

  a. What is the weight of the puppy at birth?

  b. How much will the puppy eventually weigh (that is, the weight as $d \to \infty$)?

Find the limit of each sequence, if it exists. (Example 7)

49. $a_n = \dfrac{n^3 - 2}{n^2}$

50. $a_n = \dfrac{8n + 1}{n^2 - 3}$

51. $a_n = \dfrac{-4n^2 + 6n - 1}{n^2 + 3n}$

52. $a_n = \dfrac{4 - 3n}{2n^3 + 5}$

53. $a_n = \dfrac{12n^2 + 2}{6n^2 - 1}$

54. $a_n = \dfrac{8n^2 + 5n + 2}{3 + 2n}$

55. $a_n = \dfrac{5}{n^2}\left[\dfrac{n(n+1)}{2}\right]$

56. $a_n = \dfrac{3}{n^3}\left[\dfrac{n(2n+1)(n+1)}{6}\right]$

57. $a_n = \dfrac{1}{n^4}\left[\dfrac{n^2(n+1)^2}{4}\right]$

58. $a_n = \dfrac{12}{n^2}\left[\dfrac{n(2n+1)(n+1)}{6}\right]$

59. POPULATION After being ranked one of the best cities in which to live by a national publication, Northfield Heights experienced a rise in population that could be modeled by $p(t) = \dfrac{36t^3 - 12t + 13}{3t^3 + 90}$, where p is the total rise in population in thousands and t is the number of years after 2006. (Example 7)

Years Since 2006	Rise in Population
1	?
2	?
3	?

  a. Complete the table for 2007–2009.

  b. What was the total rise in population by 2011?

  c. What is the limit of the population growth?

  d. Explain why a city's population growth may have a limit.

Find each limit, if it exists, by using direct substitution to evaluate the corresponding one-sided limits.

60. $\lim\limits_{x \to -2} \begin{cases} x - 3 & \text{if } x \le -2 \\ 2x - 1 & \text{if } x > -2 \end{cases}$

61. $\lim\limits_{x \to 0} \begin{cases} 4x + 2 & \text{if } x \le 0 \\ 2 - x^2 & \text{if } x > 0 \end{cases}$

62. $\lim\limits_{x \to 0} \begin{cases} 5 - x^2 & \text{if } x \le 0 \\ 5 - x & \text{if } x > 0 \end{cases}$

63. $\lim\limits_{x \to 2} \begin{cases} (x - 2)^2 + 1 & \text{if } x \le 2 \\ x - 6 & \text{if } x > 2 \end{cases}$

Find each limit, if it exists, using any method.

64. $\lim\limits_{x \to \pi} \dfrac{\sin x}{x}$

65. $\lim\limits_{x \to 0} (1 + x + 2^x - \cos x)$

66. $\lim\limits_{x \to 0} \dfrac{\tan 2x}{x}$

67. $\lim\limits_{x \to 0} \dfrac{3x - \sin 3x}{x^2 \sin x}$

68. $\lim\limits_{x \to 1} \dfrac{\ln x}{\ln (2x - 1)}$

69. $\lim\limits_{x \to 1} \dfrac{1 - \sqrt{x}}{x - 1}$

70. BIOLOGY Suppose the width in millimeters of an animal's pupil is given by $d(x) = \dfrac{152x^{-0.45} + 85}{4x^{-0.45} + 10}$, where x is the illuminance of the light shining on the pupils measured in lux.

  a. Write a limit to describe the width of the animal's pupils when the light is at its minimum illuminance. Then find the limit, and interpret your results.

  b. Write a limit to describe the width of the animal's pupils when the light is at its maximum illuminance. Then find the limit, and interpret your results.

Find $\lim\limits_{h \to 0} \dfrac{f(x + h) - f(x)}{h}$ **for each function.**

71. $f(x) = 2x - 1$

72. $f(x) = 7 - 9x$

73 $f(x) = \sqrt{x}$

74. $f(x) = \sqrt{x + 1}$

75. $f(x) = x^2$

76. $f(x) = x^2 + 8x + 4$

77. PHYSICS An object that is in motion possesses an energy of motion called *kinetic energy* because it can do work when it impacts another object. The kinetic energy of an object with mass m is given by $k(t) = \frac{1}{2}m \cdot [v(t)]^2$, where $v(t)$ is the velocity of the object at time t and mass is given in kilograms. Suppose $v(t) = \dfrac{50}{1 + t^2}$ for all $t \ge 0$. What value does the kinetic energy of an object that has a mass of one kilogram approach as time approaches 100?

H.O.T. Problems   Use Higher-Order Thinking Skills

78. PROOF Use the properties of limits to show that for any $p(x) = a_n x^n + a_{n-1}x^{n-1} + \cdots + a_2 x^2 + a_1 x + a_0$ and a real number c, $\lim\limits_{x \to c} p(x) = p(c)$.

79. PROOF Use mathematical induction to show that if $\lim\limits_{x \to c} f(x) = L$, then $\lim\limits_{x \to c} [f(x)]^n = [\lim\limits_{x \to c} f(x)]^n$ or L^n for any integer n.

80. CHALLENGE Find

$$\lim\limits_{x \to \infty} \dfrac{a_n x^n + a_{n-1}x^{n-1} + \cdots + a_2 x^2 + a_1 x + a_0}{b_n x^m + b_{n-1}x^{m-1} + \cdots + b_2 x^2 + b_1 x + b_0},$$

where $a_n \ne 0$ and $b_m \ne 0$. (*Hint:* Consider the cases where $m < n$, $n = m$, and $m > n$.)

81. REASONING If $r(x)$ is a rational function, is it *sometimes*, *always*, or *never* true that $\lim\limits_{x \to c} r(x) = r(c)$? Explain your reasoning.

82. WRITING IN MATH Use a spreadsheet or a table to summarize the properties of limits. Give an example of each.

83. WRITING IN MATH Consider $\lim\limits_{x \to a} \dfrac{p(x)}{q(x)} = \dfrac{\infty}{\infty}$. Susan says that this answer means the limit is 1. Why is she incorrect? What further analysis could be used to determine the limit, if it exists?

Use the graph of $y = f(x)$ to find each value. (Lesson 12-1)

84. $\displaystyle\lim_{x \to -2} f(x)$ and $f(-2)$

85. $\displaystyle\lim_{x \to 0} f(x)$ and $f(0)$

86. $\displaystyle\lim_{x \to 3} f(x)$ and $f(3)$

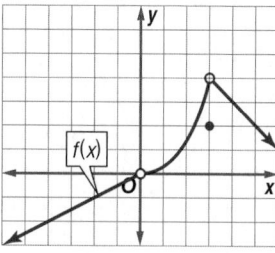

87. HEALTH The table shows the average life expectancy for people born in various years in the United States. (Lesson 11-7)

a. Make a scatter plot of the data, and identify the relationship.

b. Calculate and interpret the correlation coefficient. Determine whether it is significant at the 5% level.

c. If the correlation is significant at the 5% level, find the least-squares regression equation and interpret the slope and intercept in context.

d. Use the regression equation that you found in part **c** to predict the average life expectancy for 2080. State whether this prediction is reasonable. Explain.

Years Since 1900	Life Expectancy
10	50
20	54.1
30	59.7
40	62.9
50	68.2
60	69.7
70	70.8
80	73.7
90	75.4
100	76.9

88. ACOUSTICS Polar coordinates can be used to model the shape of a concert amphitheater. Suppose the performer is placed at the pole and faces the direction of the polar axis. The seats have been built to occupy the region with $-\frac{\pi}{3} \le \theta \le \frac{\pi}{3}$ and $0.25 \le r \le 3$, where r is measured in hundreds of feet. (Lesson 9-1)

a. Sketch this region in the polar plane.

b. How many seats are there if each person has 6 square feet of space?

89. Write the pair of parametric equations, $x = 2 \sin t$ and $y = 5 \cos t$, in rectangular form. Then sketch the graph. (Lesson 7-5)

90. SAT/ACT According to the data in the table, by what percent did the number of applicants to Green College increase from 1995 to 2000?

Number of Applicants to Green College	
Year	Applicants
1990	18,000
1995	20,000
2000	24,000
2005	25,000

A 15%   C 25%   E 29%

B 20%   D 27%

91. REVIEW What is $\displaystyle\lim_{h \to 0} \frac{2h^3 - h^2 + 5h}{h}$?

F 3   H 5

G 4   J The limit does not exist.

92. What value does $g(x) = \dfrac{x + \pi}{\cos(x + \pi)}$ approach as x approaches 0?

A $-\pi$ C $-\frac{1}{2}\pi$

B $-\frac{3}{4}$ D 0

93. REVIEW Consider the graph of $y = f(x)$ shown. What is the $\displaystyle\lim_{x \to 2^+} f(x)$?

F 0   H 5

G 1   J The limit does not exist.

Objective
Use a TI-Nspire to estimate the slope of a curve.

The slope of a line as a constant rate of change is a familiar concept. General curves do not have a constant rate of change because the slope is different at every point on the graph.

However, the graphs of most functions are *locally linear*. That is, if you examine the graph of a function on a very small interval, it will appear linear.

By looking at successive secant lines, it is possible to apply slope to curves.

Estimate the slope of the graph of $y = (x - 2)^3 + 1$ **at (3, 2).**

Step 1 Enter $y = (x - 2)^3 + 1$ in **f1**. Then calculate the slope of the line secant to the graph of $y = (x - 2)^3 + 1$ through $x = 2$ and $x = 4$.

The slope of the secant line is 4.

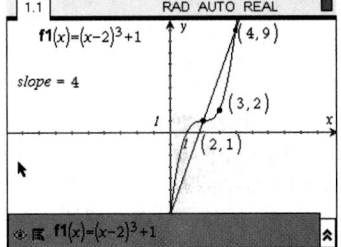

Step 2 Find the slope of the line secant to the graph of $y = (x - 2)^3 + 1$ through $x = 2.5$ and $x = 3.5$.

The slope of the secant line is 3.25.

Step 3 Find the slope of the line secant to the graph of $y = (x - 2)^3 + 1$ through $x = 2.8$ and $x = 3.2$.

The slope of the secant line is 3.04.

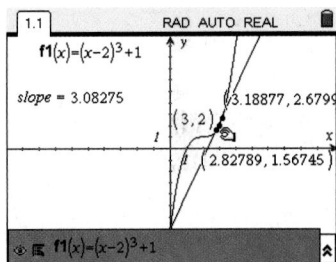

Step 4 Find the slope of 3 more secant lines on decreasing intervals around (3, 2).

As the interval around (3, 2) decreases, the slope of the secant line approaches 3. So, the slope of $y = (x - 2)^3 + 1$ at (3, 2) is about 3.

Exercises

Estimate the slope of each function at the given point.

1. $y = (x + 1)^2$; $(-4, 9)$

2. $y = x^3 - 5$; $(2, 3)$

3. $y = 4x^4 - x^2$; $(0.5, 0)$

4. $y = \sqrt{x}$; $(1, 1)$

Analyze the Results

5. ANALYZE Describe the change to a line secant to the graph of a function as the points of intersection approach a given point (a, b).

6. MAKE A CONJECTURE Describe how you could determine the exact slope of a curve at a given point.

Tangent Lines and Velocity

Why?

When a skydiver exits a plane, gravity causes the speed of his or her fall to increase. For this reason, the velocity of the sky diver at each instant before terminal velocity is achieved or the parachute is opened varies.

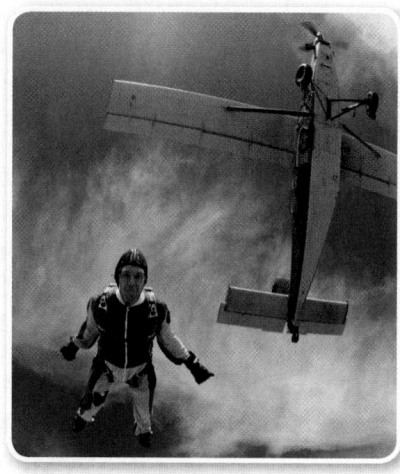

Tangent Lines In Lesson 1-4, you calculated the average rate of change between two points on the graph of a nonlinear function by finding the slope of the secant line through these points. In this lesson, we develop a way to find the slope of such functions at one instant or point on the graph.

The graphs below show successively better approximations of the slope of $y = x^2$ at (1, 1) using secant lines.

Figure 12.3.1

Figure 12.3.2

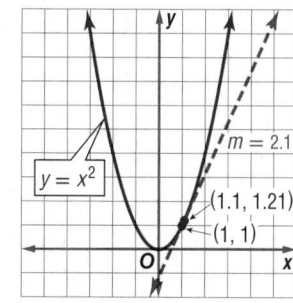

Figure 12.3.3

Notice as the rightmost point moves closer and closer to (1, 1), the secant line provides a better linear approximation of the curve near that point. We call the best of these linear approximations the **tangent line** to the graph at (1, 1). The slope of this line represents the rate of change in the slope of the curve at that instant. To define each of these terms more precisely, we use limits.

To define the slope of the tangent line to $y = f(x)$ at the point $(x, f(x))$, find the slope of the secant line through this point and one other point on the curve. Let the x-coordinate of the second point be $x + h$ for some small value of h. The corresponding y-coordinate for this point is then $f(x + h)$, as shown in Figure 12.3.4. The slope of the secant line through these two point is given by

$$m = \frac{f(x + h) - f(x)}{(x + h) - x} \text{ or } \frac{f(x + h) - f(x)}{h}.$$

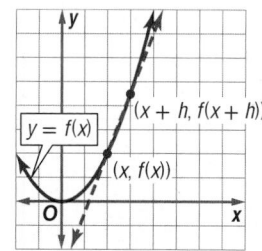

Figure 12.3.4

This expression is called the **difference quotient**.

As the second point approaches the first, or as $h \to 0$, the secant line approaches the tangent line at $(x, f(x))$. We define the slope of the tangent line at x, which represents the instantaneous rate of change of the function at that point, by finding the limits of the slopes of the secant lines as $h \to 0$.

| **Key Concept** | **Instantaneous Rate of Change** | |

The instantaneous rate of change of the graph of $f(x)$ at the point $(x, f(x))$ is the slope m of the tangent line at $(x, f(x))$ given by $m = \lim\limits_{h \to 0} \dfrac{f(x + h) - f(x)}{h}$, provided the limit exists.

You can use this expression to find the slope of the tangent line to a graph for a specified point.

StudyTip

Instantaneous Rate of Change When calculating the limit of the slopes of the secant lines as $h \to 0$, any term containing a value of h that has not been divided out will become 0.

EXAMPLE 1 **Slope of a Graph at a Point**

Find the slope of the line tangent to the graph of $y = x^2$ at (1, 1).

$m = \lim\limits_{h \to 0} \dfrac{f(x + h) - f(x)}{h}$ **Instantaneous Rate of Change Formula**

$= \lim\limits_{h \to 0} \dfrac{f(1 + h) - f(1)}{h}$ **$x = 1$**

$= \lim\limits_{h \to 0} \dfrac{(1 + h)^2 - 1^2}{h}$ **$f(1 + h) = (1 + h)^2$ and $f(1) = 1^2$**

$= \lim\limits_{h \to 0} \dfrac{1 + 2h + h^2 - 1}{h}$ **Multiply.**

$= \lim\limits_{h \to 0} \dfrac{h(2 + h)}{h}$ **Simplify and factor.**

$= \lim\limits_{h \to 0} (2 + h)$ **Divide by h.**

$= 2 + 0 \text{ or } 2$ **Sum Property of Limits and Limits of Constant and Identity Functions**

The slope of the graph at (1, 1) is 2, as shown.

✓ **Guided Practice**

Find the slope of the line tangent to the graph of each function at the given point.

1A. $y = x^2$; (3, 9)   **1B.** $y = x^2 + 4$; (−2, 8)

The expression for instantaneous rate of change can also be used to find an equation for the slope of the tangent line to a graph at any point x.

EXAMPLE 2 **Slope of a Graph at Any Point**

Find an equation for the slope of the graph of $y = \dfrac{4}{x}$ at any point.

$m = \lim\limits_{h \to 0} \dfrac{f(x + h) - f(x)}{h}$ **Instantaneous Rate of Change Formula**

$= \lim\limits_{h \to 0} \dfrac{\frac{4}{x + h} - \frac{4}{x}}{h}$ **$f(x + h) = \dfrac{4}{x + h}$ and $f(x) = \dfrac{4}{x}$**

$= \lim\limits_{h \to 0} \dfrac{\frac{-4h}{x(x + h)}}{h}$ **Add fractions in the numerator and simplify.**

$= \lim\limits_{h \to 0} \dfrac{-4h}{xh(x + h)}$ **Simplify.**

$= \lim\limits_{h \to 0} \dfrac{-4}{x^2 + xh}$ **Divide by h and multiply.**

$= \dfrac{-4}{x^2 + x(0)}$ **Quotient and Sum Properties of Limits and Limits of Constant and Identity Functions**

$= \dfrac{-4}{x^2}$ **Simplify.**

An equation for the slope of the graph at any point is $m = -\dfrac{4}{x^2}$, as shown.

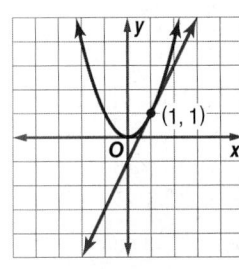

✓ **Guided Practice**

Find an equation for the slope of the graph m of each function at any point.

2A. $y = x^2 - 4x + 2$ **2B.** $y = x^3$

Instantaneous Velocity In Lesson 1-4, you calculated the average speed of a dropped object by dividing the distance traveled by the time it took for the object to cover that distance. Velocity is speed with the added dimension of direction. You can calculate average velocity using the same approach that you used when calculating average speed.

> ### Key Concept   Average Velocity
>
> If position is given as a function of time $f(t)$, for any two points in time a and b, the average velocity v is given by
>
> $$v_{avg} = \frac{\text{change in distance}}{\text{change in time}} = \frac{f(b) - f(a)}{b - a}.$$

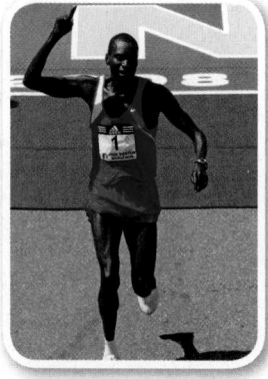

🌐 Real-World EXAMPLE 3   Average Velocity of an Object

MARATHON The distance in miles that a runner competing in the Boston Marathon has traveled after a certain time t in hours can be found by $f(t) = -1.3t^2 + 12t$. What was the runner's average velocity between the second and third hour of the race?

First, find the total distance traveled by the runner for $a = 2$ and $b = 3$.

$f(t) = -1.3t^2 + 12t$	**Original equation**	$f(t) = -1.3t^2 + 12t$
$f(2) = -1.3(2)^2 + 12(2)$	**$a = 2$ and $b = 3$**	$f(3) = -1.3(3)^2 + 12(3)$
$f(2) = 18.8$	**Simplify.**	$f(3) = 24.3$

Now use the formula for average velocity.

$v_{avg} = \dfrac{f(b) - f(a)}{b - a}$	**Average Velocity Formula**
$= \dfrac{24.3 - 18.8}{3 - 2}$	**$f(b) = 24.3$, $f(a) = 18.8$, $b = 3$, and $a = 2$**
$= 5.5$	**Simplify.**

The average velocity of the runner during the third hour was 5.5 miles per hour forward.

✔️ Guided Practice

3. **WATER BALLOON** A water balloon is propelled straight up using a launcher. The height of the balloon in feet t seconds after it is launched can be defined by $d(t) = 5 + 65t - 16t^2$. What was the balloon's average velocity between $t = 1$ and 2?

Looking more closely at Example 3, we can see that the velocity was found by calculating the slope of the secant line that connects the two points (2, 18.8) and (3, 24.3), as shown in the graph. The velocity that was calculated is the average velocity traveled by the runner over a period of time and does not represent the **instantaneous velocity**, the velocity or speed the runner achieved at a specific point in time.

To find the actual velocity of the runner at a specific time t, we find the instantaneous rate of change of the graph of $f(t)$ at t.

> ### Key Concept   Instantaneous Velocity
>
> If the distance an object travels is given as a function of time $f(t)$, then the instantaneous velocity $v(t)$ at a time t is given by
>
> $$v(t) = \lim_{h \to 0} \frac{f(t + h) - f(t)}{h},$$
>
> provided the limit exists.

EXAMPLE 4   **Instantaneous Velocity at a Point**

A baseball is dropped from the top of a building 2000 feet above the ground. The height of the baseball in feet after t seconds is given by $f(t) = 2000 - 16t^2$. Find the instantaneous velocity $v(t)$ of the baseball at 5 seconds.

To find the instantaneous velocity, let $t = 5$ and apply the formula for instantaneous velocity.

$v(t) = \lim\limits_{h \to 0} \dfrac{f(t+h) - f(t)}{h}$	**Instantaneous Velocity Formula**
$v(5) = \lim\limits_{h \to 0} \dfrac{2000 - 16(5 + h)^2 - [2000 - 16(5)^2]}{h}$	$f(t + h) = 2000 - 16(5 + h)^2$ and $f(t) = 2000 - 16(5)^2$
$= \lim\limits_{h \to 0} \dfrac{-160h - 16h^2}{h}$	**Multiply and simplify.**
$= \lim\limits_{h \to 0} \dfrac{h(-160 - 16h)}{h}$	**Factor.**
$= \lim\limits_{h \to 0} (-160 - 16h)$	**Divide by h.**
$= -160 - 16(0)$ or -160	**Difference Property of Limits and Limits of Constant and Identity Functions**

The instantaneous velocity of the baseball at 5 seconds is 160 feet per second. The negative sign indicates that the height of the ball is decreasing.

✔ Guided Practice

4. A window washer accidentally drops his lunch off his scaffold 1400 feet above the ground. The position of the lunch in relation to the ground is given as $d(t) = 1400 - 16t^2$, where time t is given in seconds and the position of the lunch is given in feet. Find the instantaneous velocity $v(t)$ of the lunch at 7 seconds.

Equations for finding the instantaneous velocity of an object at any time t can also be determined.

EXAMPLE 5   **Instantaneous Velocity at Any Point**

The distance a particle moves along a path is given by $s(t) = 18t - 3t^3 - 1$, where t is given in seconds and the distance of the particle from its starting point is given in centimeters. Find the equation for the instantaneous velocity $v(t)$ of the particle at any point in time.

Apply the formula for instantaneous velocity.

$v(t) = \lim\limits_{h \to 0} \dfrac{s(t+h) - s(t)}{h}$	**Instantaneous Velocity Formula**
$= \lim\limits_{h \to 0} \dfrac{18(t + h) - 3(t + h)^3 - 1 - [18t - 3t^3 - 1]}{h}$	$s(t + h) = 18(t + h) - 3(t + h)^3 - 1$ and $s(t) = 18t - 3t^3 - 1$
$= \lim\limits_{h \to 0} \dfrac{18h - 9t^2h - 9th^2 - 3h^3}{h}$	**Multiply and simplify.**
$= \lim\limits_{h \to 0} \dfrac{h(18 - 9t^2 - 9th - 3h^2)}{h}$	**Factor.**
$= \lim\limits_{h \to 0} (18 - 9t^2 - 9th - 3h^2)$	**Divide by h.**
$= 18 - 9t^2 - 9t(0) - 3(0)^2$	**Difference Property of Limits and Limits of Constant and Identity Functions**
$= 18 - 9t^2$	**Simplify.**

The instantaneous velocity of the particle at time t is $v(t) = 18 - 9t^2$.

✔ Guided Practice

5. The distance in feet of a water rocket from the ground after t seconds is given by $s(t) = 90t - 16t^2$. Find the expression for the instantaneous velocity $v(t)$ of the rocket at any time t.

Exercises

● = **Step-by-Step Solutions** begin on page R29.

Find the slope of the lines tangent to the graph of each function at the given points. (Example 1)

1. $y = x^2 - 5x$; $(1, -4)$ and $(5, 0)$

2. $y = 6 - 3x$; $(-2, 12)$ and $(6, -12)$

3. $y = x^2 + 7$; $(3, 16)$ and $(6, 43)$

4. $y = \frac{3}{x}$; $(1, 3)$ and $(3, 1)$

5. $y = x^3 + 8$; $(-2, 0)$ and $(1, 9)$

6. $y = \frac{1}{x + 2}$; $(2, 0.25)$ and $(-1, 1)$

Find an equation for the slope of the graph of each function at any point. (Example 2)

7. $y = 4 - 2x$

8. $y = -x^2 + 4x$

9. $y = x^2 + 3$

10. $y = x^3$

11. $y = 8 - x^2$

12. $y = 2x^2$

13. $y = -2x^3$

14. $y = x^2 + 2x - 3$

15. $y = \frac{1}{\sqrt{x}}$

16. $y = \frac{1}{x^2}$

17. SLEDDING A person's vertical position on a sledding hill after traveling a horizontal distance x units away from the top of the hill is given by $y = 0.06x^3 - 1.08x^2 + 51.84$. (Example 2)

a. Find an equation for the hill's slope m at any distance x.

b. Find the hill's slope for $x = 2, 5$, and 7.

The position of an object in miles after t minutes is given by $s(t)$. Find the average velocity of the object in miles per hour for the given interval of time. Remember to convert from minutes to hours. (Example 3)

18. $s(t) = 0.4t^2 - \frac{1}{20}t^3$ for $3 \le t \le 5$

19. $s(t) = 1.08t - 30$ for $4 \le t \le 8$

20. $s(t) = 0.2t^2$ for $2 \le t \le 4$

21. $s(t) = 0.01t^3 - 0.01t^2$ for $4 \le t \le 7$

22. $s(t) = -0.5(t - 5)^2 + 3$ for $4 \le t \le 4.5$

23. $s(t) = 0.6t + 20$ for $3.8 \le t \le 5.7$

24. TYPING The number of words w a person has typed after t minutes is given by $w(t) = 10t^2 - \frac{1}{2}t^3$. (Example 3)

a. What was the average number of words per minute the person typed between the 2nd and 4th minutes?

b. What was the average number of words per minute the person typed between the 3rd and 7th minutes?

The distance d an object is above the ground t seconds after it is dropped is given by $d(t)$. Find the instantaneous velocity of the object at the given value for t. (Example 4)

25. $d(t) = 100 - 16t^2$; $t = 3$

26. $d(t) = 38t - 16t^2$; $t = 0.8$

27. $d(t) = -16t^2 - 47t + 300$; $t = 1.5$

28. $d(t) = 500 - 30t - 16t^2$; $t = 4$

29. $d(t) = -16t^2 - 400t + 1700$; $t = 3.5$

30. $d(t) = 150t - 16t^2$; $t = 2.7$

31. $d(t) = 1275 - 16t^2$; $t = 3.8$

32. $d(t) = 853 - 48t - 16t^2$; $t = 1.3$

Find an equation for the instantaneous velocity $v(t)$ if the path of an object is defined as $s(t)$ for any point in time t. (Example 5)

33. $s(t) = 14t^2 - 7$

34. $s(t) = t - 3t^2$

35. $s(t) = 5t + 8$

36. $s(t) = 18 - t^2 + 4t$

37. $s(t) = t^3 - t^2 + t$

38. $s(t) = 11t^2 - t$

39. $s(t) = \sqrt{t} - 3t^2$

40. $s(t) = 12t^2 - 2t^3$

41 SKYDIVER Refer to the beginning of the lesson. The position d of the skydiver in feet relative to the ground can be defined by $d(t) = 15{,}000 - 16t^2$, where t is seconds passed after the skydiver exited the plane. (Example 5)

a. What is the average velocity of the skydiver between the 2nd and 5th seconds of the jump?

b. What was the instantaneous velocity of the sky diver at 2 and 5 seconds?

c. Find an equation for the instantaneous velocity $v(t)$ of the skydiver.

42. DIVING A cliff diver's distance d in meters above the surface of the water after t seconds is given.

t	0.5	0.75	1.0	1.5	2.0	2.5	3.0
d	43.7	42.1	40.6	33.8	25.3	14.2	0.85

a. Calculate the diver's average velocity for the interval $0.5 \le t \le 1.0$.

b. Use quadratic regression to find an equation to model $d(t)$. Graph $d(t)$ and the data on the same coordinate plane.

c. Find an expression for the instantaneous velocity $v(t)$ of the diver and use it to estimate the velocity of the diver at 3 seconds.

43. SOCCER A goal keeper can kick a ball at an upward velocity of 75 feet per second. Suppose the height d of the ball in feet t seconds after it is kicked is given by $d(t) = -16t^2 + 75t + 2.5$.

a. Find an equation for the instantaneous velocity $v(t)$ of the soccer ball.

b. How fast is the ball traveling 0.5 second after it is kicked?

c. If the instantaneous velocity of the ball is 0 when the ball reaches its maximum height, at what time will the ball reach its maximum height?

d. What is the maximum height of the ball?

Find an equation for a line that is tangent to the graph of the function and perpendicular to the given line. Then use a graphing calculator to graph the function and both lines on the same coordinate plane.

44. $f(x) = x^2 + 2x; y = -\frac{1}{2}x + 3$

45. $g(x) = -4x^2; y = \frac{1}{4}x + 5$

46. $f(x) = -\frac{1}{6}x^2; y - x = 2$

47 $g(x) = \frac{1}{2}x^2 + 4x; y = -\frac{1}{6}x + 9$

48. PHYSICS The distance s of a particle moving in a straight line is given by $s(t) = 3t^3 + 8t + 4$, where t is given in seconds and s is measured in meters.

a. Find an equation for the instantaneous velocity $v(t)$ of the particle at any given point in time.

b. Find the velocity of the particle for $t = 2, 4,$ and 6 seconds.

Each graph represents an equation for the slope of a function at any point. Match each graph with its original function.

49. $f(x) = \frac{a}{x}$

50. $g(x) = ax^5$

51. $h(x) = ax^4$

52. $j(x) = a\sqrt{x}$

a.

b.

c.

d.

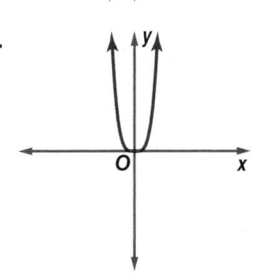

53. PROJECTILE When an object is thrown straight down, the total distance y the object falls can be modeled by $y = 16t^2 + v_0 t$, where time t is measured in seconds and the initial velocity v_0 is measured in feet per second.

a. If an object thrown straight down from a height of 816 feet takes 6 seconds to hit the ground, what was the initial velocity of the object?

b. What was the average velocity of the object?

c. What was the object's velocity when it hit the ground?

54. ✂ **MULTIPLE REPRESENTATIONS** In this problem, you will investigate the *Mean Value Theorem*. The theorem states that if a function f is continuous and differentiable on (a, b), then there exists a point c in (a, b) such that the tangent line is parallel to the line passing through $(a, f(a))$ and $(b, f(b))$.

a. ANALYTICAL Find the average rate of change for $f(x) = -x^2 + 8x$ on the interval [1, 6], and find an equation for the related secant line through $(1, f(1))$ and $(6, f(6))$.

b. ANALYTICAL Find an equation for the slope of $f(x)$ at any point.

c. ANALYTICAL Find a point on the interval (1, 6) at which the slope of the tangent line to $f(x)$ is equal to the slope of the secant line found in part **a**. Find an equation for the line tangent to $f(x)$ at this point.

d. VERBAL How are the secant line in part **a** and the tangent line in part **b** related? Explain.

e. GRAPHICAL Using a graphing calculator, graph $f(x)$, the secant line, and the tangent line on the same screen. Does the graph verify your answer in part **d**? Explain.

H.O.T. Problems   Use **H**igher-**O**rder **T**hinking Skills

55. ERROR ANALYSIS Chase and Jillian were asked to find an equation for the slope at any point for $f(x) = |x|$. Chase thinks the graph of the slope equation will be continuous because the original function is continuous. Jillian disagrees. Is either of them correct? Explain your reasoning.

56. CHALLENGE Find an equation for the slope of $f(x) = 2x^4 + 3x^3 - 2x$ at any point.

57. REASONING *True* or *false*: The instantaneous velocity of an object modeled by $s(t) = at + b$ is always a.

58. REASONING Show that $\lim_{h \to 0} \dfrac{f(a + h) - f(a)}{h} = \lim_{x \to a} \dfrac{f(x) - f(a)}{x - a}$ for $f(x) = x^2 + 1$.

59. WRITING IN MATH Suppose that $f(t)$ represents the balance in dollars in a bank account t years after the initial deposit. Interpret each of the following.

a. $\dfrac{f(4) - f(0)}{4} \approx 41.2$

b. $\lim_{h \to 0} \dfrac{f(4 + h) - f(4)}{h} \approx 42.9$

Evaluate each limit. (Lesson 12-2)

60. $\lim\limits_{x\to 4} (x^2 + 2x - 2)$

61. $\lim\limits_{x\to -1} (-x^4 + x^3 - 2x + 1)$

62. $\lim\limits_{x\to 0} (x + \sin x)$

63. HYDRAULICS The velocity of a molecule of liquid flowing through a pipe depends on the distance of the molecule from the center of the pipe. The velocity, in inches per second, of a molecule is given by $v(r) = k(R^2 - r^2)$, where r is the distance of the molecule from the center of the pipe in inches and k is a constant. Suppose for a particular liquid and a particular pipe that $k = 0.65$ and $R = 0.5$. (Lesson 12-1)

  a. Graph $v(r)$.

  b. Determine the limiting velocity of molecules closer and closer to the wall of the pipe.

64. HEIGHT The mean height of a sample of 100 high school seniors is 68 inches with a standard deviation of 4 inches. Determine the interval of heights such that the probability is 90% that the mean height of the entire population lies within that interval. (Lesson 11-5)

65. EDUCATION A college professor plans to grade a test on a curve. The mean score on the test is 65, and the standard deviation is 7. The professor wants the grades distributed as shown in the table. Assume that the grades are normally distributed. (Lesson 11-3)

  a. What is the lowest score for an A?

  b. If a D is the lowest passing letter grade, find the lowest passing score.

  c. What is the interval for the Bs?

Grade	Percent of Class
A	15
B	20
C	30
D	20
F	15

Find the specified nth term of each geometric sequence. (Lesson 10-3)

66. $a_4 = 50, r = 2, n = 8$

67. $a_4 = 1, r = 3, n = 10$

68. a_6 for $a_n = \frac{1}{5}a_{n-1}, a_1 = -2$

69. a_5 for $a_n = (-3)a_{n-1}, a_1 = 11$

Find the indicated arithmetic means for each set of nonconsecutive terms. (Lesson 10-2)

70. 7 means; 62 and -2

71. 4 means; 17.2 and 47.7

72. 3 means; -5.6 and 8

73. 9 means; -45 and 115

74. SAT/ACT If the radius of the circle with center O is 4, what is the length of arc RST?

  A 2π

  B 4π

  C 8π

  D 12π

  E 16π

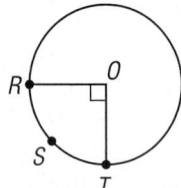

75. REVIEW Which of the following best describes the point at $(0, 0)$ on $f(x) = 2x^5 - 5x^4$?

  F absolute maximum

  G relative maximum

  H relative minimum

  J absolute minimum

76. When a bowling ball is dropped, the distance $d(t)$ that it falls in t seconds is given by $d(t) = 16t^2$. Its velocity after 2 seconds is given by $\lim\limits_{h\to 0} \frac{d(2+h) - d(2)}{h}$. What is the velocity of the bowling ball after 2 seconds?

  A 46 feet per second   **C** 64 feet per second

  B 58 feet per second   **D** 72 feet per second

77. REVIEW The monthly profit P of a manufacturing company depends on the number of units x manufactured and can be described by $P(x) = \frac{1}{3}x^3 - 34x^2 + 1012x, 0 \le x \le 50$. How many units should be manufactured monthly in order to maximize profits?

  F 15   **H** 37

  G 22   **J** 46

Estimate each one-sided or two-sided limit, if it exists. (Lesson 12-1)

1. $\lim\limits_{x \to 0^+} \dfrac{\sin x}{x}$

2. $\lim\limits_{x \to 0} \dfrac{|x|}{x}$

3. $\lim\limits_{x \to 3^-} \dfrac{2x^2 - 18}{x - 3}$

4. $\lim\limits_{x \to 0^-} \dfrac{\cos x - 1}{x}$

Estimate each limit, if it exists. (Lesson 12-1)

5. $\lim\limits_{x \to 3} \dfrac{2x}{x^2 + 1}$

6. $\lim\limits_{x \to 1} \sqrt{x^3 + 3}$

7. $\lim\limits_{x \to -2} e^{2x + 3}$

8. $\lim\limits_{x \to -4} \dfrac{\sqrt{x + 20}}{x}$

9. **COLLECTIBLES** The value of Jorge's baseball card has been increasing every year. The value v of the baseball card after t years can be represented by the model $v(t) = \dfrac{400t - 2}{2t + 15}$. (Lesson 12-1)

   a. Graph the function for $0 \le t \le 10$.

   b. Use your graph to estimate the value of the baseball card for $t = 2, 5,$ and 10 years.

   c. Use your graph to evaluate $\lim\limits_{t \to \infty} v(t)$.

   d. Explain the relationship between the limit of the function and the value of Jorge's baseball card.

Use direct substitution, if possible, to evaluate each limit. If not possible, explain why not. (Lesson 12-2)

10. $\lim\limits_{x \to 9} \dfrac{x^2 + 1}{\sqrt{x} - 3}$

11. $\lim\limits_{x \to -2} 2x^3 + x^2 - 8$

12. **WILDLIFE** A deer population P in hundreds at a national park after t years can be estimated by the model $P(t) = \dfrac{10t^3 - 40t + 2}{2t^3 + 14t + 12}$, where $t \ge 3$. The population for 5 years is shown below. What is the maximum number of deer that can live at the national park? (Lesson 12-2)

Deer Population

Evaluate each limit. (Lesson 12-2)

13. $\lim\limits_{x \to \infty} 15 - x^2 + 8x^3$

14. $\lim\limits_{x \to \infty} \dfrac{2x^3 - x - 2}{4x^3 + 5x^2}$

15. $\lim\limits_{x \to \infty} \dfrac{2x^2 + 5x - 1}{2x^4 - 14x^2 + 2}$

16. $\lim\limits_{x \to \infty} 10x^3 - 4 + x^2 - 7x^4$

17. **MULTIPLE CHOICE** Evaluate $\lim\limits_{x \to 0} \dfrac{2x^2 + 5}{10 - e^{\frac{16}{x}}}$. (Lesson 12-1)

   A does not exist

   B $\dfrac{1}{2}$

   C $\dfrac{1}{5}$

   D $\dfrac{1}{10}$

Find the slope of the line tangent to the graph of each function at the given points. (Lesson 12-3)

18. $y = x^2 - 3x$; $(2, -2)$ and $(-1, 4)$

19. $y = 2 - 5x$; $(-2, 12)$ and $(3, -13)$

20. $y = x^3 - 4x^2$; $(1, -3)$ and $(3, -9)$

21. **FIREWORKS** Fireworks are launched with an upward velocity of 90 feet per second. Suppose the height d of the fireworks in feet t seconds after they are launched is defined as $d(t) = -16t^2 + 90t + 3.2$. (Lesson 12-3)

   a. Find an equation for the instantaneous velocity $v(t)$ of the fireworks.

   b. How fast are the fireworks traveling 0.5 seconds after they are lit?

   c. What is the maximum height of the fireworks?

22. **MULTIPLE CHOICE** Find an equation for the slope of the graph of $y = 7x^2 - 2$ at any point. (Lesson 12-3)

   F $m = 7x$

   G $m = 7x - 2$

   H $m = 14x$

   J $m = 14x - 2$

The position of an object in miles after t minutes is given by $s(t)$. Find the average velocity of the object in miles per hour given two values of time t. Remember to convert from minutes to hours. (Lesson 12-3)

23. $s(t) = 12 + 0.7t$ for $t = 2$ and 5

24. $s(t) = 2.05t - 11$ for $t = 1$ and 7

25. $s(t) = 0.9t - 25$ for $3 \le t \le 6$

26. $s(t) = 0.5t^2 - 4t$ for $4 \le t \le 8$

Find an equation for the instantaneous velocity $v(t)$ if the position of an object is defined as $s(t)$ for any point in time t. (Lesson 12-3)

27. $s(t) = 4t^2 - 9t$

28. $s(t) = 2t - 13t^2$

29. $s(t) = 2t - 5t^2$

30. $s(t) = 6t^2 - t^3$

Derivatives

Why?

Zach, who lives on the sixth floor of an apartment building, accidentally drops a ball out of his window. Gabe, standing on the ground outside of Zach's building, retrieves the ball and attempts to throw it back up to Zach. If Gabe can throw the ball at a speed of 65 feet per second, can he get it to Zach's window 70 feet above the ground?

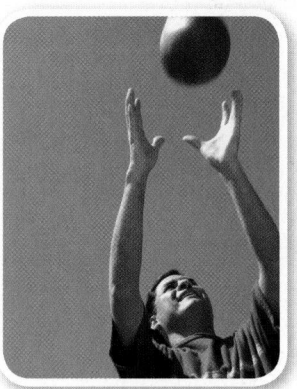

Then

You calculated the slope of tangent lines to find the instantaneous rate of change. (Lesson 12-3)

Now

- Find instantaneous rates of change by calculating derivatives.
- Use the Product and Quotient Rules to calculate derivatives.

New Vocabulary

derivative
differentiation
differential equation
differential operator

Math Online

glencoe.com

Basic Rules In Lesson 12-3, you used limits to determine the slope of a line tangent to the graph of a function at any point. This limit is called the derivative of a function. The **derivative** of $f(x)$ is $f'(x)$, which is given by

$$f'(x) = \lim_{h \to 0} \frac{f(x+h) - f(x)}{h},$$

provided the limit exists. The process of finding a derivative is called **differentiation** and the result is called a **differential equation**.

EXAMPLE 1 | Derivative of a Function at Any Point

Find the derivative of $f(x) = 4x^2 - 5x + 8$. Then evaluate the derivative at $x = 1$ and 5.

$$f'(x) = \lim_{h \to 0} \frac{f(x+h) - f(x)}{h}$$ **Definition of a derivative**

$$= \lim_{h \to 0} \frac{4(x+h)^2 - 5(x+h) + 8 - (4x^2 - 5x + 8)}{h}$$ $f(x+h) = 4(x+h)^2 - 5(x+h) + 8$ and $f(x) = 4x^2 - 5x + 8$

$$= \lim_{h \to 0} \frac{8xh + 4h^2 - 5h}{h}$$ **Expand and simplify.**

$$= \lim_{h \to 0} \frac{h(8x + 4h - 5)}{h}$$ **Factor.**

$$= \lim_{h \to 0} (8x + 4h - 5)$$ **Divide by h.**

$$= 8x + 4(0) - 5 \text{ or } 8x - 5$$ **Sum and Difference Properties of Limits and Limits of Constant and Identity Functions**

The derivative of $f(x)$ is $f'(x) = 8x - 5$. Evaluate $f'(x)$ for $x = 1$ and 5.

$f'(x) = 8x - 5$	**Original equation**	$f'(x) = 8x - 5$
$f'(1) = 8(1) - 5$	**x = 1 and 5**	$f'(5) = 8(5) - 5$
$f'(1) = 3$	**Simplify.**	$f'(5) = 35$

✓ Guided Practice

Find the derivative of $f(x)$. Then evaluate the derivative for the given values of x.

1A. $f(x) = 6x^2 + 7$; $x = 2$ and 5   **1B.** $f(x) = -5x^2 + 2x - 12$; $x = 1$ and 4

The derivative of the function $y = f(x)$ may also be denoted y', $\frac{df}{dx}$, or $\frac{dy}{dx}$. If a function is preceded by a **differential operator** $\frac{d}{dx}$, then you are to take the derivative of the function.

To this point, you have had to evaluate limits as they approach 0 in order to calculate derivatives, slopes of tangent lines, and instantaneous velocity. A very helpful rule for simplifying this process and for reducing calculation errors is the Power Rule. It allows for the evaluation of derivatives without the need to calculate limits.

ReadingMath

Derivatives The notation for a derivative $f'(x)$ is read *f prime of x* or *the derivative of f with respect to x.*

Key Concept   **Power Rule for Derivatives**

Words   The power of x in the derivative is one less than the power of x in the original function, and the coefficient of the power of x in the derivative is the same as the power of x in the original function.

Symbols   If $f(x) = x^n$ and n is a real number, then $f'(x) = nx^{n-1}$.

EXAMPLE 2   **Power Rule for Derivatives**

Find the derivative of each function.

a. $f(x) = x^9$

$f(x) = x^9$ Original equation

$f'(x) = 9x^{9-1}$ Power Rule

$\quad = 9x^8$ Simplify.

b. $g(x) = \sqrt[5]{x^7}$

$g(x) = \sqrt[5]{x^7}$ Original equation

$g(x) = x^{\frac{7}{5}}$ Rewrite using a rational exponent.

$g'(x) = \frac{7}{5}x^{\frac{7}{5}-1}$ Power Rule

$\quad = \frac{7}{5}x^{\frac{2}{5}}$ or $\frac{7}{5}\sqrt[5]{x^2}$ Simplify.

c. $h(x) = \dfrac{1}{x^8}$

$h(x) = \dfrac{1}{x^8}$ Original equation

$h(x) = x^{-8}$ Rewrite using a negative exponent.

$h'(x) = -8x^{-8-1}$ Power Rule

$\quad = -8x^{-9}$ or $-\dfrac{8}{x^9}$ Simplify.

Watch Out!

Negative Derivatives The derivative of $f(x) = x^{-4}$ is not $f'(x) = -4x^{-3}$. Remember that 1 must be *subtracted* from the exponent and that $-4 - 1 = -4 + (-1)$ or -5. Therefore $f'(x) = -4x^{-5}$.

✅ **Guided Practice**

2A. $j(x) = x^4$ **2B.** $k(x) = \sqrt{x^3}$ **2C.** $m(x) = \dfrac{1}{x^5}$

There are several other rules of derivatives that are useful when finding the derivatives of functions that contain several terms.

Key Concept   **Other Derivative Rules**

Constant	The derivative of a constant function is zero. That is, if $f(x) = c$, then $f'(x) = 0$.
Constant Multiple of a Power	If $f(x) = cx^n$, where c is a constant and n is a real number, then $f'(x) = cnx^{n-1}$.
Sum or Difference	If $f(x) = g(x) \pm h(x)$, then $f'(x) = g'(x) \pm h'(x)$.

EXAMPLE 3 Derivative Rules

Find the derivative of each function.

a. $f(x) = 5x^3 + 4$

$f(x) = 5x^3 + 4$	Original equation
$f'(x) = 5 \cdot 3x^{3-1} + 0$	Constant, Constant Multiple of a Power, and Sum Rules
$= 15x^2$	Simplify.

b. $g(x) = x^5(2x^3 + 4)$

$g(x) = x^5(2x^3 + 4)$	Original equation
$g(x) = 2x^8 + 4x^5$	Distributive Property
$g'(x) = 2 \cdot 8x^{8-1} + 4 \cdot 5x^{5-1}$	Constant Multiple of a Power and Sum Rules
$= 16x^7 + 20x^4$	Simplify.

c. $h(x) = \dfrac{5x^3 - 12x + 6\sqrt{x^5}}{x}$

$h(x) = \dfrac{5x^3 - 12x + 6\sqrt{x^5}}{x}$	Original equation
$h(x) = \dfrac{5x^3}{x} - \dfrac{12x}{x} + \dfrac{6\sqrt{x^5}}{x}$	Divide each term in the numerator by x.
$h(x) = 5x^2 - 12 + 6x^{\frac{3}{2}}$	$x^{\frac{5}{2}} \cdot x^{-1} = x^{\frac{3}{2}}$
$h'(x) = 5 \cdot 2x^{2-1} + 0 + 6 \cdot \dfrac{3}{2}x^{\frac{3}{2}-1}$	Constant, Constant Multiple of a Power, and Sum and Difference Rules
$= 10x + 9x^{\frac{1}{2}}$ or $10x + 9\sqrt{x}$	Simplify.

✔ Guided Practice

3A. $f(x) = 2x^5 - x^3 - 102$ **3B.** $g(x) = 3x^4(x + 2)$ **3C.** $h(x) = \dfrac{4x^4 - 3x^2 + 5x}{x}$

Now that you are familiar with the basic rules of derivatives, problems involving the slopes of tangent lines and instantaneous velocity can be calculated in just a few steps. Example 5 in Lesson 12-3 involved finding an expression for the instantaneous velocity of a particle. Notice the simplification of the problem as a result of the derivative rules.

EXAMPLE 4 Instantaneous Velocity

The distance a particle moves along a path is defined by $s(t) = 18t - 3t^3 - 1$, where t is given in seconds and the distance of the particle is given in centimeters. Find the expression for the instantaneous velocity $v(t)$ of the particle.

The instantaneous velocity $v(t)$ is equivalent to $s'(t)$.

$s(t) = 18t - 3t^3 - 1$	Original equation
$s'(t) = 18 \cdot 1t^{1-1} - 3 \cdot 3t^{3-1} - 0$	Constant, Constant Multiple of a Power, and Difference Rules
$= 18 - 9t^2$	Simplify.

The instantaneous velocity is $v(t) = 18 - 9t^2$. Notice that this is the same result as found in Example 5 of Lesson 12-3.

✔ Guided Practice

4. A soccer ball is kicked straight up. The height of the ball is defined by $h(t) = 55t - 16t^2$, where time t is given in seconds and the height of the ball is given in feet. Find the expression for the instantaneous velocity $v(t)$ of the ball at any point in time.

In Lesson 1-4, you found relative and absolute extrema of functions graphically and numerically. On a closed interval, these values can be found using the derivative and the following theorem.

Key Concept   Extreme Value Theorem

If a function f is continuous on a closed interval $[a, b]$, then $f(x)$ attains both a maximum and a minimum on $[a, b]$.

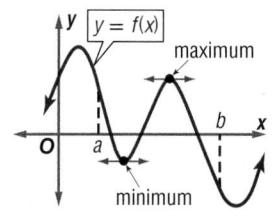

Relative extrema occur only at *critical points* where the slope of the tangent line, the derivative of the function, is 0 or undefined. To locate the maximum and minimum of a polynomial function $f(x)$ on $[a, b]$, evaluate the function at a, b, and at any values x in $[a, b]$ for which $f'(x) = 0$.

●Real-World EXAMPLE 5   Maximum and Minimum

ROLLER COASTER The height h in feet of a car travelling along the track of a roller coaster can be modeled by $h(t) = -\frac{1}{3}t^3 + 4t^2 + \frac{11}{3}$ on the interval $[1, 12]$, where time t is given in seconds. Find the maximum and minimum heights of the car.

Find the derivative of $h(t)$.

$h(t) = -\frac{1}{3}t^3 + 4t^2 + \frac{11}{3}$ **Original equation**

$h'(t) = -\frac{1}{3} \cdot 3t^{3-1} + 4 \cdot 2t^{2-1} + 0$ **Constant, Constant Multiple of a Power, and Sum and Difference Rules**

$\quad\quad = -t^2 + 8t$ **Simplify.**

Solve $h'(t) = 0$ to find where the critical points of $h(x)$ occur.

$-t^2 + 8t = 0$ $h'(t) = -t^2 + 8t$

$-t(t - 8) = 0$ **Factor.**

Critical points for this function occur when $t = 0$ and 8. Note that although $t = 0$ is a critical point of the function $h(t)$, it does not lie on the interval $[1, 12]$. To find the maximum and minimum of the function on $[1, 12]$ evaluate $h(t)$ for 1, 8, and 12.

$h(1) = -\frac{1}{3}(1)^3 + 4(1)^2 + \frac{11}{3}$ or 7.33

$h(8) = -\frac{1}{3}(8)^3 + 4(8)^2 + \frac{11}{3}$ or 89   **maximum**

$h(12) = -\frac{1}{3}(12)^3 + 4(12)^2 + \frac{11}{3}$ or 3.67   **minimum**

The car will achieve a maximum height of 89 feet 8 seconds into the ride and a minimum height of about 3.7 feet 12 seconds into the ride.

CHECK The graph of $h(t) = -\frac{1}{3}t^3 + 4t^2 + \frac{11}{3}$ shows that $h(t)$ has a maximum of 89 at $x = 8$ and a minimum of about 3.7 at $x = 12$ on the interval $[1, 12]$. ✔

✔ Guided Practice

5. BUNGEE JUMPING A bungee jumper's height h in feet relative to the ground can be modeled by $h(t) = 20t^2 - 160t + 330$ on the interval $[0, 6]$, where time t is given in seconds. Find the maximum and minimum heights of the jumper.

Product and Quotient Rules Earlier, you learned that the derivative of the sum of functions is equal to the sum of the individual derivatives. Is the derivative of a product of functions equal to the product of the derivatives? Consider the functions $f(x) = x$ and $g(x) = 3x^3$.

Derivative of Product

$$\frac{d}{dx}[f(x) \cdot g(x)] = \frac{d}{dx}[x \cdot 3x^3]$$

$$= \frac{d}{dx}(3x^4)$$

$$= 12x^3$$

Product of Derivatives

$$\frac{d}{dx}f(x) \cdot \frac{d}{dx}g(x) = \frac{d}{dx}(x) \cdot \frac{d}{dx}(3x^3)$$

$$= 1 \cdot 9x^2$$

$$= 9x^2$$

It is clear that the derivative of the product is not necessarily the product of the derivatives. The following rule can be applied when calculating the derivative of products.

Key Concept | **Product Rule for Derivatives**

If f and g are differentiable at x, then $\frac{d}{dx}[f(x)g(x)] = f'(x)g(x) + f(x)g'(x)$.

You will prove the Product Rule for Derivatives in Exercise 64.

EXAMPLE 6 | **Product Rule**

Find the derivative of each product.

a. $h(x) = (x^3 - 2x + 7)(3x^2 - 5)$

Let $f(x) = x^3 - 2x + 7$ and $g(x) = 3x^2 - 5$. So, $h(x) = f(x)g(x)$.

$f(x) = x^3 - 2x + 7$	Original equation
$f'(x) = 3x^2 - 2$	Power, Constant Multiple of a Power, Constant, and Sum and Difference Rules
$g(x) = 3x^2 - 5$	Original equation
$g'(x) = 6x$	Constant Multiple of a Power, Constant, and Difference Rules

Use $f(x)$, $f'(x)$, $g(x)$, and $g'(x)$ to find the derivative of $h(x)$.

$h'(x) = f'(x)g(x) + f(x)g'(x)$	Product Rule
$= (3x^2 - 2)(3x^2 - 5) + (x^3 - 2x + 7)(6x)$	Substitution
$= 15x^4 - 33x^2 + 42x + 10$	Distribute and simplify.

b. $h(x) = (x^3 - 4x^2 + 48x - 64)(6x^2 - x - 2)$

Let $f(x) = x^3 - 4x^2 + 48x - 64$ and $g(x) = 6x^2 - x - 2$.

$f(x) = x^3 - 4x^2 + 48x - 64$	Original equation
$f'(x) = 3x^2 - 8x + 48$	Power, Constant Multiple of a Power, Constant, and Sum and Difference Rules
$g(x) = 6x^2 - x - 2$	Original equation
$g'(x) = 12x - 1$	Constant Multiple of a Power, Power, Constant, and Difference Rules

Use $f(x)$, $f'(x)$, $g(x)$, and $g'(x)$ to find the derivative of $h(x)$.

$h'(x) = f'(x)g(x) + f(x)g'(x)$	Product Rule
$= (3x^2 - 8x + 48)(6x^2 - x - 2) + (x^3 - 4x^2 + 48x - 64)(12x - 1)$	Substitution
$= 30x^4 - 100x^3 + 870x^2 - 848x - 32$	Distribute and simplify.

StudyTip

Product Rule The Product Rule results in an answer that can still be simplified. Unless there is an easy simplification to make or a reason to do so, you may leave the answer as is.

✓ **Guided Practice**

6A. $h(x) = (x^5 + 13x^2)(7x^3 - 5x^2 + 18)$ **6B.** $h(x) = (x^2 + x^3 + x)(8x^2 + 3)$

The same reasoning used for the derivatives of products can be applied to quotients. The following rule can be applied when calculating the derivative of quotients.

Key Concept   Quotient Rule for Derivatives

If f and g are differentiable at x and $g(x) \neq 0$, then $\dfrac{d}{dx}\left[\dfrac{f(x)}{g(x)}\right] = \dfrac{f'(x)g(x) - f(x)g'(x)}{[g(x)]^2}$.

You will prove the Quotient Rule for Derivatives in Exercise 67.

EXAMPLE 7   Quotient Rule

Find the derivative of each quotient.

a. $h(x) = \dfrac{5x^2 - 3}{x^2 - 6}$

Let $f(x) = 5x^2 - 3$ and $g(x) = x^2 - 6$. So, $h(x) = \dfrac{f(x)}{g(x)}$.

$f(x) = 5x^2 - 3$	Original equation
$f'(x) = 10x$	Constant Multiple of a Power, Constant, and Difference Rules
$g(x) = x^2 - 6$	Original equation
$g'(x) = 2x$	Power, Constant, and Difference Rules

Use $f(x), f'(x), g(x)$, and $g'(x)$ to find the derivative of $h(x)$.

$h'(x) = \dfrac{f'(x)g(x) - f(x)g'(x)}{[g(x)]^2}$	Quotient Rule
$\quad = \dfrac{10x(x^2 - 6) - (5x^2 - 3)(2x)}{(x^2 - 6)^2}$	Substitution
$\quad = \dfrac{10x^3 - 60x - 10x^3 + 6x}{(x^2 - 6)^2}$	Distributive Property
$\quad = \dfrac{-54x}{(x^2 - 6)^2}$	Simplify.

b. $h(x) = \dfrac{x^2 + 8}{x^3 - 2}$

Let $f(x) = x^2 + 8$ and $g(x) = x^3 - 2$.

$f(x) = x^2 + 8$	Original equation
$f'(x) = 2x$	Power, Constant, and Sum Rules
$g(x) = x^3 - 2$	Original equation
$g'(x) = 3x^2$	Power, Constant, and Difference Rules

Use $f(x), f'(x), g(x)$, and $g'(x)$ to find the derivative of $h(x)$.

$h'(x) = \dfrac{f'(x)g(x) - f(x)g'(x)}{[g(x)]^2}$	Quotient Rule
$\quad = \dfrac{2x(x^3 - 2) - (x^2 + 8)3x^2}{(x^3 - 2)^2}$	Substitution
$\quad = \dfrac{-x^4 - 24x^2 - 4x}{(x^3 - 2)^2}$	Expand and simplify.

StudyTip

Quotient Rule
Simplification with the Quotient Rule tends to be more significant and useful. However, it is not necessary to expand the denominator if doing so does not result in further simplification.

✓ Guided Practice

7A. $j(x) = \dfrac{7x - 10}{12x + 5}$

7B. $k(x) = \dfrac{6x}{2x^2 + 4}$

Evaluate limits to find the derivative of each function. Then evaluate the derivative of each function for the given values of each variable. (Example 1)

1. $f(x) = 4x^2 - 3; x = 2$ and -1

2. $g(t) = -t^2 + 2t + 11; t = 5$ and 3

3. $m(j) = 14j - 13; j = -7$ and -4

4. $v(n) = 5n^2 + 9n - 17; n = 7$ and 2

5. $h(c) = c^3 + 2c^2 - c + 5; c = -2$ and 1

6. $r(b) = 2b^3 - 10b; b = -4$ and -3

Find the derivative of each function. (Examples 2 and 3)

7. $y(f) = -11f$

8. $z(n) = 2n^2 + 7n$

9. $p(v) = 7v + 4$

10. $g(h) = 2h^{\frac{1}{2}} + 6h^{\frac{1}{3}} - 2h^{\frac{3}{2}}$

11. $b(m) = 3m^{\frac{2}{3}} - 2m^{\frac{3}{2}}$

12. $n(t) = \frac{1}{t} + \frac{3}{t^2} + \frac{2}{t^3} + 4$

13. $f(x) = 3x^{\frac{1}{2}} - x^{\frac{3}{2}} + 2x^{-\frac{1}{2}}$

14. $q(c) = c^9 - 3c^5 + 5c^2 - 3c$

15. $p(k) = k^{5.2} - 8k^{4.8} + 3k$

16. $f(x) = -5x^3 - 9x^4 + 8x^5$

17 **TEMPERATURE** The temperature in degrees Fahrenheit over a 24-hour period in a certain city can be defined as $f(h) = -0.0036h^3 - 0.01h^2 + 2.04h + 52$, where h is the number of hours since midnight. (Example 4)

a. Find an equation for the instantaneous rate of change for the temperature.

b. Find the instantaneous rate of change for $h = 2, 14$, and 20.

c. Find the maximum temperature for $0 \le h \le 24$.

Use the derivative to find any critical points of the function. Then find the maximum and minimum points of each graph on the given interval. (Example 5)

18. $f(x) = 2x^2 + 8x; [-5, 0]$

19. $g(m) = m^3 - 4m + 10; [-3, 3]$

20. $r(t) = t^4 + 6t^2 - 2; [1, 4]$

21. $t(u) = u^3 + 15u^2 + 75u + 115; [-6, -3]$

22. $k(p) = p^4 - 8p^2 + 2; [0, 3]$

23. $f(x) = -5x^2 - 90x; [-11, -8]$

24. $z(k) = k^3 - 3k^2 + 3k; [0, 3]$

25. $a(d) = d^4 - 3d^3 + 2; [-1, 4]$

26. $c(n) = \frac{1}{3}n^3 + \frac{1}{2}n^2 - 6n + 8; [-5, 5]$

27. **THROWN OBJECT** Refer to the application at the beginning of the lesson. The height h of the ball in feet after t seconds can be defined as $h(t) = 65t - 16t^2 + 5$ for $0 \le t \le 4$. (Example 5)

a. Find $h'(t)$.

b. Find the maximum and minimum points of $h(t)$ on the interval.

c. Can Gabe throw the ball up to Zach's window?

Find the derivative of each function. (Example 6)

28. $f(x) = (4x + 3)(x^2 + 9)$

29. $g(x) = (3x^4 + 2x)(5 - 3x)$

30. $h(x) = (-7x^2 + 4)(2 - x)$

31. $s(t) = \left(t^{\frac{1}{2}} + 2\right)(3t^{11} - 4t)$

32. $g(x) = \left(x^{\frac{3}{2}} + 2x\right)(0.5x^4 - 3x)$

33. $c(t) = (t^3 + 2t - t^7)(t^6 + 3t^4 - 22t)$

34. $p(r) = (r^{2.5} + 8r)(r - 7r^2 + 108)$

35. $q(a) = \left(a^{\frac{9}{8}} + a^{-\frac{1}{4}}\right)\left(a^{\frac{5}{4}} - 13a\right)$

36. $f(x) = (1.4x^5 + 2.7x)(7.3x^9 - 0.8x^5)$

37. $h(x) = \left(\frac{1}{8}x^{\frac{2}{3}} + \frac{2}{5}x^{-\frac{1}{6}}\right)\left(x^{\frac{5}{2}} + x^{\frac{7}{8}}\right)$

38. **BASEBALL** A pitch is struck by a bat with a mass of m kilograms. Suppose the initial speed of the ball after being struck is given by $s(m) = \frac{185m - 15}{m + 0.15}$. (Example 7)

a. Find an equation for the instantaneous rate of change for the initial speed of the ball.

b. Use a calculator to graph the equation you found in part **a** on $0 \le m \le 2$. What is happening to the instantaneous rate of change for the initial speed of the ball as the mass of the bat increases?

c. If the mass of the bat varies inversely with the hitter's control of the swing, is it wise to use a bat that weighs 1.05 kilograms over a bat that weighs 0.80 kilogram? Explain your reasoning.

Use the Quotient Rule to find the derivative of each function. (Example 7)

39. $f(m) = \frac{3 - 2m}{3 + 2m}$

40. $g(n) = \frac{3n + 2}{2n + 3}$

41. $r(t) = \frac{t^2 + 2}{3 - t^2}$

42. $m(q) = \frac{q^4 + 2q^2 + 3}{q^3 - 2}$

43. $v(t) = \frac{t^2 - 5t + 3}{t^3 - 4t}$

44. $c(m) = \frac{m^4 + 1}{-m^3 + 2m}$

45. $f(x) = \frac{x^3 + 2x}{-x^2 + 3}$

46. $q(r) = \frac{1.5r^3 + 5 - r^2}{r^3}$

47. $t(w) = \frac{w + w^4}{w^2}$

48. $m(x) = \frac{x^5 + 3x}{-x^4 - 2x^3 - 2x - 3}$

49. ECONOMICS Nayla and Deirdre are selling sweatshirts to raise money for the junior class. Their weekly revenue is given by $r(x) = 0.125x^3 - 11.25x^2 + 250x$, where x is the cost of one sweatshirt.

    a. Find $r'(x)$.

    b. Find the solutions of $r'(x) = 0$.

    c. What do the solutions you found in part **b** represent in terms of the given situation?

Find the equation of the line tangent to $f(x)$ at the given point. Verify your answer graphically.

50. $f(x) = 3x^2 + 2x - 7$; $(1, -2)$

51. $f(x) = -5x^2 - 10x + 25$; $(-2, 25)$

52. $f(x) = -0.2x^2 + 1.5x - 0.75$; $(5, 1.75)$

53. $f(x) = 4x^2 - 12x - 35$; $(-1.2, -14.84)$

54. $f(x) = 0.8x^2 + 0.64x - 12$; $(10, 74.4)$

55. DERIVATIVES Let $f'(x)$ be the derivative of a function $f(x)$. If it exists, we can calculate the derivative of $f'(x)$, which is called the second derivative, and is denoted $f''(x)$ or $f^{(2)}(x)$. We can continue and find the derivative of $f''(x)$, which is called the third derivative and is denoted $f'''(x)$ or $f^{(3)}(x)$. These are examples of higher-order derivatives. Find the indicated derivative of each function.

    a. second derivative of $f(x) = 4x^5 - 2x^3 + 6$

    b. third derivative of $g(x) = -2x^7 + 4x^4 - 7x^3 + 10x$

    c. fourth derivative of $h(x) = 3x^{-3} + 2x^{-2} + 4x^2$

Sketch a graph for a function that has the given characteristics.

56. The derivative is 0 at $x = -1$ and $x = 1$.

57. The derivative is -2 at $x = -1$, $x = 0$, and $x = 2$.

58. The derivative is 0 at $x = -1$, $x = 2$, and $x = 4$.

59. The derivative is undefined at $x = 4$.

60. STUDYING Sayra kept track of the amount of time t in minutes that she studied the night before an exam and the percentage p that she earned on the exam.

t	30	60	90	120	180	210	240
p	39	68	86	96	90	76	56

    a. Find a quadratic equation $p(t)$ that can be used to model the data. Round the coefficients to the nearest ten thousandth. Graph the data and $p(t)$ on the same screen.

    b. Use $p'(t)$ to find the maximum test score that Sayra can earn and the amount of time that she would need to study to achieve this score.

    c. Explain why more time spent studying does not necessarily result in a higher exam score.

61. ⚏ **MULTIPLE REPRESENTATIONS** In this problem, you will investigate how derivatives relate to some geometric properties.

    a. ANALYTICAL Find the derivatives of the formulas for the area A of a circle and the volume V of a sphere in terms of r.

    b. VERBAL Explain the relationship between each formula and its derivative.

    c. GEOMETRIC Draw a square and its apothem a. Draw a cube and the apothem a for three faces that are joined at a shared vertex.

    d. ANALYTICAL Write formulas for the area A of the square and the volume V of the cube in terms of the apothem a. Find the derivative of each formula in terms of a.

    e. VERBAL Explain the relationship between each formula and its derivative.

H.O.T. Problems   Use **H**igher-**O**rder **T**hinking Skills

62. ERROR ANALYSIS Alicia and Ryan are finding $[f'(x)]^2$, where $f(x) = 6x^2 + 4x$. Ryan thinks that the answer is $144x^2 + 96x + 16$, but Alicia thinks that the answer is $144x^3 + 144x^2 + 32x$. Is either of them correct? Explain your reasoning.

63 CHALLENGE Find $f'(y)$ if
$f(y) = 10x^2y^3 + 5xz^2 - 6xy^2 + 8x^5 - 11x^8yz^7$.

64. PROOF Prove the Product Rule for Derivatives by showing that

$$f'(x)g(x) + f(x)g'(x) = \lim_{h \to 0} \frac{f(x+h)g(x+h) - f(x)g(x)}{h}.$$

(*Hint:* Work with the right side. Add and subtract $f(x)g(x+h)$ in the numerator.)

65. REASONING Determine whether the following statement is *true* or *false*. Explain your reasoning. If $f(x) = x^{5n+3}$, then $f'(x) = (5n+3)x^{5n+2}$.

66. PREWRITE Use a plot pyramid to map out the process of finding the derivative of $f(x) = 4x^2 - 2x + 5$ evaluated at $x = 1$.

67. PROOF Prove the Quotient Rule for Derivatives by showing that

$$\frac{f'(x)g(x) - f(x)g'(x)}{[g(x)]^2} = \lim_{h \to 0} \frac{f(x+h)g(x) - f(x)g(x+h)}{hg(x+h)g(x)}.$$

(*Hint:* Work with the right side. Add and subtract $f(x)g(x)$ in the numerator.)

68. WRITING IN MATH Can two different functions have the same derivative? Explain why this is or is not possible, and provide examples to support your answer.

Find the slope of the lines tangent to the graph of each function at the given points. (Lesson 12-3)

69. $y = x^2 - 3x$; $(0, 0)$ and $(3, 0)$

70. $y = 4 - 2x$; $(-2, 8)$ and $(6, -8)$

71. $y = x^2 + 9$; $(3, 18)$ and $(6, 45)$

Evaluate each limit. (Lesson 12-2)

72. $\lim\limits_{x \to -4} \dfrac{x^2 - 16}{x + 4}$

73. $\lim\limits_{x \to -2} \dfrac{x^2 + 5x + 6}{x^2 + x - 2}$

74. $\lim\limits_{x \to 2} \dfrac{3x + 9}{x^2 - 5x - 24}$

75. EXERCISE A gym teacher asked his students to track how many days they exercised each week. Use the frequency distribution shown to construct and graph a probability distribution for the random variable X, rounding each probability to the nearest hundredth. (Lesson 11-2)

Days, X	Frequency
0	3
1	6
2	7
3	8
4	4
5	2

76. SPORTS The number of hours per week members of the North High School basketball team spent practicing, either as a team or individually, are listed below. (Lesson 11-1)

$$15, 18, 16, 20, 22, 18, 19, 20, 24, 18, 16, 18$$

a. Construct a histogram and use it to describe the shape of the distribution.

b. Summarize the center and spread of the data using either the mean and standard deviation or the five-number summary. Justify your choice.

Use the fifth partial sum of the trigonometric series for cosine or sine to approximate each value to three decimal places. (Lesson 10-6)

77. $\cos \dfrac{2\pi}{11}$

78. $\sin \dfrac{3\pi}{14}$

79. $\sin \dfrac{\pi}{13}$

Write an explicit formula and a recursive formula for finding the nth term of each geometric sequence. (Lesson 10-3)

80. $1.25, -1.5, 1.8, \ldots$

81. $1.4, -3.5, 8.75, \ldots$

82. $\dfrac{1}{8}, \dfrac{1}{4}, \dfrac{1}{2}, \ldots$

83. SAT/ACT The figure shows the dimensions, in feet, of a stone slab. How many slabs are required to construct a rectangular patio 24 feet long and 12 feet wide?

A 18 **C** 24 **E** 40

B 20 **D** 36

84. REVIEW What is the slope of the line tangent to the graph of $y = 2x^2$ at the point $(1, 2)$?

F 1 **H** 4

G 2 **J** 8

85. The Better Book Company finds that the cost in dollars to print x copies of a book is given by $C(x) = 1000 + 10x - 0.001x^2$. The derivative $C'(x)$ is called the *marginal cost function*. The marginal cost is the approximate cost of printing one more book after x copies have been printed. What is the marginal cost when 1000 books have been printed?

A $7 **C** $9

B $8 **D** $10

86. REVIEW Find the derivative of $f(x) = 5\sqrt[3]{x^8}$.

F $f'(x) = \dfrac{40}{3}x^{\frac{5}{3}}$ **H** $f'(x) = 225x^{\frac{5}{3}}$

G $f'(x) = \dfrac{40}{3}x^{\frac{8}{3}}$ **J** $f'(x) = 225x^{\frac{8}{3}}$

Area Under a Curve and Integration

Why?
Marginal cost is the approximate cost that a company incurs to produce an additional unit of a product. The equation for marginal cost is found by taking the derivative of the equation representing the *actual cost* of producing a product. The marginal cost function for a particular publisher is $f(x) = 10 - 0.002x$, where x is the number of books manufactured and $f(x)$ is in dollars.

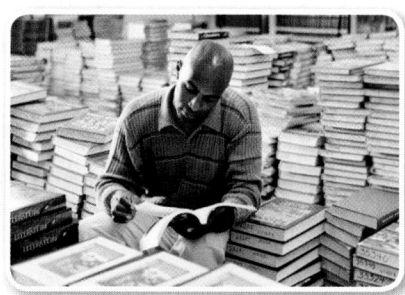

Area Under a Curve In geometry, you learned how to calculate the area of a basic figure, such as a triangle, a rectangle, or a regular polygon. You also learned how to calculate the area of a composite figure, that is, a region comprised of basic shapes. However, many regions are not a collection of the basic shapes. As a result, you need a general approach for calculating the area of any two-dimensional figure.

We can approximate the area of an irregular shape by using a basic figure with a known formula for area, the rectangle. For example, consider the graph of $f(x) = -x^2 + 12x$ on the interval [0, 12]. We can *approximate* the area between the curve and the x-axis using rectangles of equal width.

EXAMPLE 1   **Area Under a Curve Using Rectangles**

Approximate the area between the curve $f(x) = -x^2 + 12x$ and the x-axis on the interval [0, 12] using 4, 6, and 12 rectangles. Use the right endpoint of each rectangle to determine the height.

Using the figures from below for reference, notice that the rectangles were drawn with a height equal to $f(x)$ at each right endpoint. For example, the heights of the rectangles in the first figure are $f(3)$, $f(6)$, $f(9)$, and $f(12)$. We can use these heights and the length of the base of each rectangle to approximate the area under the curve.

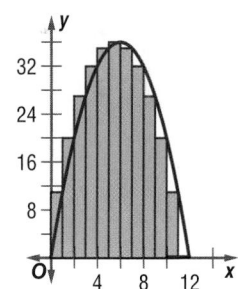

Area using 4 rectangles

$R_1 = 3 \cdot f(3)$ or   81
$R_2 = 3 \cdot f(6)$ or   108
$R_3 = 3 \cdot f(9)$ or   81
$R_4 = 3 \cdot f(12)$ or   0
     total area = 270

Area using 6 rectangles

$R_1 = 2 \cdot f(2)$ or   40
$R_2 = 2 \cdot f(4)$ or   64
$R_3 = 2 \cdot f(6)$ or   72
$R_4 = 2 \cdot f(8)$ or   64
$R_5 = 2 \cdot f(10)$ or   40
$R_6 = 2 \cdot f(12)$ or   0
     total area = 280

Area using 12 rectangles

$R_1 = 1 \cdot f(1)$ or   11
$R_2 = 1 \cdot f(2)$ or   20
$R_3 = 1 \cdot f(3)$ or   27
$R_4 = 1 \cdot f(4)$ or   32
$R_5 = 1 \cdot f(5)$ or   35
$R_6 = 1 \cdot f(6)$ or   36
$R_7 = 1 \cdot f(7)$ or   35
$R_8 = 1 \cdot f(8)$ or   32
$R_9 = 1 \cdot f(9)$ or   27
$R_{10} = 1 \cdot f(10)$ or   20
$R_{11} = 1 \cdot f(11)$ or   11
$R_{12} = 1 \cdot f(12)$ or   0
     total area = 286

The approximation for the area under the curve using 4, 6, and 12 rectangles is 270 square units, 280 square units, and 286 square units, respectively.

TechnologyTip

Tables To help generate multiple heights of rectangles with your graphing calculator, enter the function using the [Y=] menu. Then use the Table function by pressing [2nd] [TABLE]. This will generate a list of heights for different values of *x*. You can also change the interval for the *x* values in your table by pressing [2nd] [TBLSET] and adjusting the TBLSET options.

✔ **Guided Practice**

  1. Approximate the area between the curve $f(x) = -x^2 + 24x$ and the *x*-axis on the interval [0, 24] using 6, 8, and 12 rectangles. Use the right endpoint of each rectangle to determine the height.

Notice that the thinner the rectangles are, the better they fit the region, and the better their total area approximates the area of the region. Also, the rectangles were drawn so that the right endpoint of each one evaluated at $f(x)$ represents the height. The left endpoints may also be used to determine the height of each rectangle and can produce a different result for the approximated area.

Using right or left endpoints may result in adding or excluding area that does or does not lie between the curve and the *x*-axis. In some cases, better approximations may be obtained by calculating the area using both left and right endpoints and then averaging the results.

EXAMPLE 2   **Area Under a Curve Using Left and Right Endpoints**

Approximate the area between the curve $f(x) = x^2$ and the *x*-axis on the interval [0, 4] by first using the right endpoints and then by using the left endpoints of the rectangles. Use rectangles with a width of 1.

Using right endpoints for the height of each rectangle produces four rectangles with a width of 1 unit (Figure 12.5.1). Using left endpoints for the height of each rectangle produces four rectangles with a width of 1 unit (Figure 12.5.2).

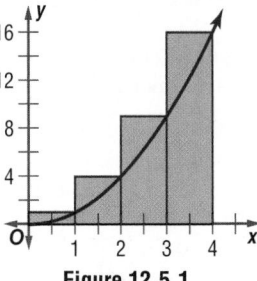

Figure 12.5.1

Area using right endpoints

$R_1 = 1 \cdot f(1)$ or   1

$R_2 = 1 \cdot f(2)$ or   4

$R_3 = 1 \cdot f(3)$ or   9

$R_4 = 1 \cdot f(4)$ or 16

total area $= 30$

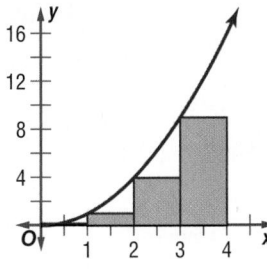

Figure 12.5.2

Area using left endpoints

$R_1 = 1 \cdot f(0)$ or 0

$R_2 = 1 \cdot f(1)$ or 1

$R_3 = 1 \cdot f(2)$ or 4

$R_4 = 1 \cdot f(3)$ or 9

total area $= 14$

The area using the right and left endpoints is 30 and 14 square units, respectively. We now have lower and upper estimates for the area of the region, $14 <$ area < 30. Averaging the two areas would give a better approximation of 22 square units.

✔ **Guided Practice**

  2. Approximate the area between the curve $f(x) = \frac{12}{x}$ and the *x*-axis on the interval [1, 5] by first using the right endpoints and then by using the left endpoints. Use rectangles with a width of 1 unit. Then find the average of the two approximations.

Any point within the width of the rectangles may be used as the heights when approximating the area between the graph of a curve and the *x*-axis. The most commonly used are the left endpoints, the right endpoints, and the midpoints.

Integration As we saw in Example 1, as the rectangles get thinner, their total area approaches the exact area of the region under the curve. We can conclude that the area of the region under a curve is the limit of the total area of the rectangles as the widths of the rectangles approach 0.

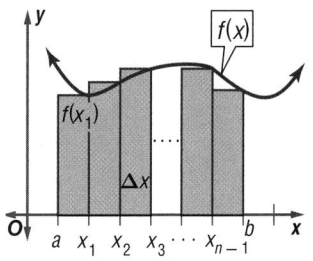

ReadingMath

Sigma Notation

$$\sum_{i=1}^{n} f(x_i) \, \Delta x$$ is read *the summation of the product of f of x sub i from 1 to n and the change in x.*

In the figure, the interval from a to b has been subdivided into n equal subintervals. This is called a **regular partition**. The length of the entire interval from a to b is $b - a$, so the width of each of the n rectangles is $\frac{b-a}{n}$ and is denoted Δx. The height of each rectangle at the right endpoint corresponds with the value of the function at that point. Thus, the height of the first rectangle is $f(x_1)$, the height of the second is $f(x_2)$, and so on, with the height of the last rectangle being $f(x_n)$.

The area of each rectangle can now be calculated by taking the product of Δx and the corresponding height. The area of the first rectangle is $f(x_1)\Delta x$, the area of the second rectangle is $f(x_2)\Delta x$, and so on. The total area A of the n rectangles is given by the sum of the areas and can be written in sigma notation.

$A = f(x_1)\Delta x + f(x_2)\Delta x + \cdots + f(x_n)\Delta x$ **Add the areas.**

$A = \Delta x[f(x_1) + f(x_2) + \cdots + f(x_n)]$ **Factor out Δx.**

$A = \Delta x \sum\limits_{i=1}^{n} f(x_i)$ **Write the sum of the heights in sigma notation.**

$A = \sum\limits_{i=1}^{n} f(x_i) \, \Delta x$ **Commutative Property of Multiplication**

To assist in future calculations, we can derive a formula to find any x_i. The width Δx of each rectangle is the distance between successive x_i-values. Consider the x-axis.

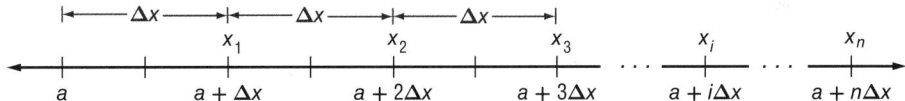

We can see that $x_i = a + i\Delta x$. This formula will be useful when finding the area under the graph of any function.

To make the width of the rectangles approach 0, we let the number of rectangles approach infinity. This limit is called a **definite integral** and is given special notation.

Key Concept   Definite Integral

The area of a region under the graph of a function is

$$\int_a^b f(x)\,dx = \lim_{n \to \infty} \sum_{i=1}^{n} f(x_i)\Delta x,$$

where a and b are the **lower limit** and **upper limit** respectively, $\Delta x = \frac{b-a}{n}$, and $x_i = a + i\Delta x$. This method is referred to as the **right Riemann sum**.

Watch Out!

Summations The sum of a constant c is cn, not 0 or ∞. For instance,

$$\sum_{i=1}^{n} 5 = 5n.$$

The Riemann sum is named for the German mathematician Bernhard Riemann (1826–1866). He is credited with formulating the expression for approximating the area under a curve using limits. The expression may be altered to use left endpoints or midpoints.

The process of evaluating an integral is called **integration**. The following summation formulas will assist in evaluating definite integrals.

$$\sum_{i=1}^{n} c = cn,\ c \text{ is a constant}$$

$$\sum_{i=1}^{n} i = \frac{n(n+1)}{2}$$

$$\sum_{i=1}^{n} i^2 = \frac{n(n+1)(2n+1)}{6}$$

$$\sum_{i=1}^{n} i^3 = \frac{n^2(n+1)^2}{4}$$

$$\sum_{i=1}^{n} i^4 = \frac{6n^5 + 15n^4 + 10n^3 - n}{30}$$

$$\sum_{i=1}^{n} i^5 = \frac{2n^6 + 6n^5 + 5n^4 - n^2}{12}$$

There are two summation properties that are needed to evaluate some integrals.

$$\sum_{i=1}^{n}(a_i + b_i) = \sum_{i=1}^{n} a_i \pm \sum_{i=1}^{n} b_i \qquad\qquad \sum_{i=1}^{n} ci = c\sum_{i=1}^{n} i, c \text{ is a constant}$$

EXAMPLE 3 Area Under a Curve Using Integration

Use limits to find the area of the region between the graph of $y = x^2$ and the x-axis on the interval [0, 4], or $\int_0^4 x^2\, dx$.

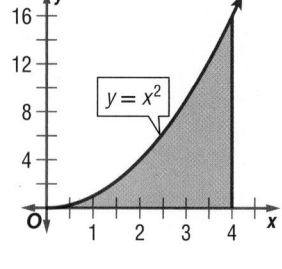

First, find Δx and x_i.

$$\Delta x = \frac{b-a}{n} \qquad\qquad \text{Formula for } \Delta x$$

$$= \frac{4-0}{n} \text{ or } \frac{4}{n} \qquad b = 4 \text{ and } a = 0$$

$$x_i = a + i\Delta x \qquad\qquad \text{Formula for } x_i$$

$$= 0 + i\frac{4}{n} \text{ or } \frac{4i}{n} \qquad a = 0 \text{ and } \Delta x = \frac{4}{n}$$

Calculate the definite integral that gives the area.

StudyTip

Limits Factor each summation until the expression remaining involves only a constant or i. Then apply the necessary summation formula.

$$\int_0^4 x^2\, dx = \lim_{n\to\infty}\sum_{i=1}^{n} f(x_i)\,\Delta x \qquad\qquad \text{Definition of a definite integral}$$

$$= \lim_{n\to\infty}\sum_{i=1}^{n}(x_i)^2\,\Delta x \qquad\qquad f(x_i) = x_i^2$$

$$= \lim_{n\to\infty}\sum_{i=1}^{n}\left(\frac{4i}{n}\right)^2\left(\frac{4}{n}\right) \qquad\qquad x_i = \frac{4i}{n} \text{ and } \Delta x = \frac{4}{n}$$

$$= \lim_{n\to\infty}\frac{4}{n}\sum_{i=1}^{n}\left(\frac{4i}{n}\right)^2 \qquad\qquad \text{Factor.}$$

$$= \lim_{n\to\infty}\frac{4}{n}\sum_{i=1}^{n}\frac{16i^2}{n^2} \qquad\qquad \text{Expand.}$$

$$= \lim_{n\to\infty}\frac{4}{n}\left(\frac{16}{n^2}\sum_{i=1}^{n}i^2\right) \qquad\qquad \text{Factor.}$$

$$= \lim_{n\to\infty}\frac{4}{n}\left[\frac{16}{n^2}\cdot\frac{n(n+1)(2n+1)}{6}\right] \qquad \sum_{i=1}^{n}i^2 = \frac{n(n+1)(2n+1)}{6}$$

$$= \lim_{n\to\infty}\frac{4}{n}\left[\frac{16n(2n^2+3n+1)}{6n^2}\right] \qquad\qquad \text{Multiply and expand.}$$

$$= \lim_{n\to\infty}\frac{64n(2n^2+3n+1)}{6n^3} \qquad\qquad \text{Multiply.}$$

$$= \lim_{n\to\infty}\frac{64(2n^2+3n+1)}{6n^2} \qquad\qquad \text{Divide by } n.$$

$$= \lim_{n\to\infty}\frac{64}{6}\left[\frac{(2n^2+3n+1)}{n^2}\right] \qquad\qquad \text{Factor.}$$

$$= \lim_{n\to\infty}\frac{64}{6}\left(2 + \frac{3}{n} + \frac{1}{n^2}\right) \qquad\qquad \text{Divide each term by } n^2.$$

$$= \left(\lim_{n\to\infty}\frac{64}{6}\right)\left[\lim_{n\to\infty}2 + \left(\lim_{n\to\infty}3\right)\left(\lim_{n\to\infty}\frac{1}{n}\right) + \lim_{n\to\infty}\frac{1}{n^2}\right] \qquad \text{Limit theorems}$$

$$= \frac{64}{6}[2 + 3(0) + 0] \text{ or } \frac{64}{3} \qquad\qquad \text{The area is } \frac{64}{3} \text{ or } 21\frac{1}{3} \text{ square units.}$$

✓ Guided Practice

Use limits to find the area between the graph of each function and the x-axis given by the definite integral.

3A. $\int_0^1 3x^2\, dx$ 　　　　　　　　　　　**3B.** $\int_0^3 x\, dx$

The areas of regions that do not have a lower limit at the origin can also be found using limits.

EXAMPLE 4   **Area Under a Curve Using Integration**

Use limits to find the area of the region between the graph of $y = 4x^3$ and the x-axis on the interval [1, 3], or $\int_1^3 4x^3\,dx$.

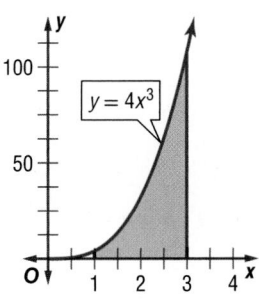

First, find Δx and x_i.

$$\Delta x = \frac{b - a}{n} \qquad\qquad \text{Formula for } \Delta x$$

$$= \frac{3 - 1}{n} \text{ or } \frac{2}{n} \qquad\qquad b = 3 \text{ and } a = 1$$

$$x_i = a + i\Delta x \qquad\qquad \text{Formula for } x_i$$

$$= 1 + i\frac{2}{n} \text{ or } 1 + \frac{2i}{n} \qquad a = 1 \text{ and } \Delta x = \frac{2}{n}$$

Calculate the definite integral that gives the area.

$$\int_1^3 4x^3\,dx = \lim_{n\to\infty} \sum_{i=1}^{n} f(x_i)\Delta x \qquad\qquad \text{Definition of a definite integral}$$

$$= \lim_{n\to\infty} \sum_{i=1}^{n} 4(x_i)^3 \Delta x \qquad\qquad f(x_i) = 4(x_i)^3$$

$$= \lim_{n\to\infty} \sum_{i=1}^{n} 4\left(1 + \frac{2i}{n}\right)^3 \left(\frac{2}{n}\right) \qquad\qquad x_i = 1 + \frac{2i}{n} \text{ and } \Delta x = \frac{2}{n}$$

$$= \lim_{n\to\infty} \frac{8}{n} \sum_{i=1}^{n} \left(1 + \frac{2i}{n}\right)^3 \qquad\qquad \text{Factor}$$

$$= \lim_{n\to\infty} \frac{8}{n} \sum_{i=1}^{n} \left[1 + 3\left(\frac{2i}{n}\right) + 3\left(\frac{2i}{n}\right)^2 + \left(\frac{2i}{n}\right)^3\right] \qquad \text{Expand.}$$

$$= \lim_{n\to\infty} \frac{8}{n} \sum_{i=1}^{n} \left(1 + \frac{6i}{n} + \frac{12i^2}{n^2} + \frac{8i^3}{n^3}\right) \qquad \text{Simplify.}$$

$$= \lim_{n\to\infty} \frac{8}{n} \left(\sum_{i=1}^{n} 1 + \sum_{i=1}^{n} \frac{6i}{n} + \sum_{i=1}^{n} \frac{12i^2}{n^2} + \sum_{i=1}^{n} \frac{8i^3}{n^3}\right) \qquad \text{Apply summations.}$$

$$= \lim_{n\to\infty} \frac{8}{n} \left(\sum_{i=1}^{n} 1 + \frac{6}{n}\sum_{i=1}^{n} i + \frac{12}{n^2}\sum_{i=1}^{n} i^2 + \frac{8}{n^3}\sum_{i=1}^{n} i^3\right) \qquad \text{Factor constants.}$$

Watch Out!

Limits When finding the area under a curve by using limits, evaluate the summation expressions for the given values of i before distributing the width Δx or any other constants.

$$= \lim_{n\to\infty} \frac{8}{n} \left[n + \frac{6}{n} \cdot \frac{n(n + 1)}{2} + \frac{12}{n^2} \cdot \frac{n(n + 1)(2n + 1)}{6} + \frac{8}{n^3} \cdot \frac{n^2(n + 1)^2}{4}\right] \qquad \begin{array}{l}\text{Summation}\\ \text{formulas}\end{array}$$

$$= \lim_{n\to\infty} \left[\frac{8n}{n} + \frac{48n(n + 1)}{2n^2} + \frac{96n(2n^2 + 3n + 1)}{6n^3} + \frac{64n^2(n^2 + 2n + 1)}{4n^4}\right] \qquad \text{Distribute } \frac{8}{n}.$$

$$= \lim_{n\to\infty} \left[8 + \frac{24(n + 1)}{n} + \frac{16(2n^2 + 3n + 1)}{n^2} + \frac{16(n^2 + 2n + 1)}{n^2}\right] \qquad \text{Simplify.}$$

$$= \lim_{n\to\infty} \left[8 + 24\left(1 + \frac{1}{n}\right) + 16\left(2 + \frac{3}{n} + \frac{1}{n^2}\right) + 16\left(1 + \frac{2}{n} + \frac{1}{n^2}\right)\right] \qquad \begin{array}{l}\text{Factor and}\\ \text{perform}\\ \text{division.}\end{array}$$

$$= \lim_{n\to\infty} 8 + 24\lim_{n\to\infty}\left(1 + \frac{1}{n}\right) + 16\lim_{n\to\infty}\left(2 + \frac{3}{n} + \frac{1}{n^2}\right) + 16\lim_{n\to\infty}\left(1 + \frac{2}{n} + \frac{1}{n^2}\right) \qquad \text{Limit Theorems}$$

$$= 8 + 24(1 + 0) + 16(2 + 0 + 0) + 16(1 + 0 + 0) \text{ or } 80 \qquad \text{Simplify.}$$

The area of the region is 80 square units.

✅ **Guided Practice**

Use limits to find the area between the graph of each function and the x-axis given by the definite integral.

4A. $\int_1^3 x^2\,dx$ **4B.** $\int_2^4 x^3\,dx$

Definite integrals can be used to find the areas of other irregular shapes.

Real-World EXAMPLE 5 — Area Under a Curve

LANDSCAPING Foster charges $2.40 per square foot of mulch to deliver and install it. He is hired to create two identical flower beds in the back corners of a residential lot. If the area of each flower bed can be found by $\int_0^{10} (10 - 0.1x^2)\,dx$, how much will Foster charge for the flower beds if x is given in terms of feet?

First, find Δx and x_i.

$$\Delta x = \frac{b-a}{n} \qquad \text{Formula for } x$$

$$= \frac{10-0}{n} \text{ or } \frac{10}{n} \qquad b = 10 \text{ and } a = 0$$

$$x_i = a + i\Delta x \qquad \text{Formula for } x_i$$

$$= 0 + i\frac{10}{n} \text{ or } \frac{10i}{n} \qquad a = 0 \text{ and } \Delta x = \frac{10}{n}$$

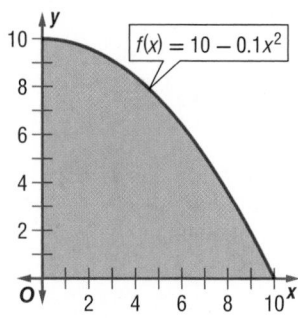

$f(x) = 10 - 0.1x^2$

Calculate the definite integral that gives the area.

$$\int_0^{10} (10 - 0.1x^2)\,dx = \lim_{n\to\infty} \sum_{i=1}^{n} f(x_i)\Delta x \qquad \text{Definition of a definite integral}$$

$$= \lim_{n\to\infty} \sum_{i=1}^{n} (10 - 0.1x_i^2)\Delta x \qquad f(x_i) = 10 - 0.1x_i^2$$

$$= \lim_{n\to\infty} \sum_{i=1}^{n} \left[10 - 0.1\left(\frac{10i}{n}\right)^2 \right] \cdot \frac{10}{n} \qquad x_i = \frac{10i}{n} \text{ and } \Delta x = \frac{10}{n}$$

$$= \lim_{n\to\infty} \frac{10}{n} \sum_{i=1}^{n} \left(10 - \frac{10i^2}{n^2} \right) \qquad \text{Expand and simplify.}$$

$$= \lim_{n\to\infty} \frac{10}{n} \left(\sum_{i=1}^{n} 10 - \sum_{i=1}^{n} \frac{10i^2}{n^2} \right) \qquad \text{Apply summations.}$$

$$= \lim_{n\to\infty} \frac{10}{n} \left(\sum_{i=1}^{n} 10 - \frac{10}{n^2} \sum_{i=1}^{n} i^2 \right) \qquad \text{Factor.}$$

$$= \lim_{n\to\infty} \frac{10}{n} \left(10n - \frac{10}{n^2} \cdot \frac{n(n+1)(2n+1)}{6} \right) \qquad \text{Summation formulas}$$

$$= \lim_{n\to\infty} \left(\frac{100n}{n} - \frac{100n(2n^2+3n+1)}{3n^2} \right) \qquad \text{Distribute } \frac{10}{n}.$$

$$= \lim_{n\to\infty} \left(100 - \frac{50(2n^2+3n+1)}{3n^2} \right) \qquad \text{Divide by } n.$$

$$= \lim_{n\to\infty} \left[100 - \frac{50}{3}\left(2 + \frac{3}{n} + \frac{1}{n^2} \right) \right] \qquad \text{Factor and perform division.}$$

$$= \lim_{n\to\infty} 100 - \frac{50}{3} \lim_{n\to\infty} \left(2 + \frac{3}{n} + \frac{1}{n^2} \right) \qquad \text{Limit Theorems}$$

$$= 100 - \frac{50}{3}(2 + 0 + 0) \text{ or } 66\frac{2}{3} \qquad \text{Simplify.}$$

The area of one flower bed is about 66.67 square feet. For Foster to install both flower beds, he would have to charge $2.40 \cdot \left(66\frac{2}{3} \cdot 2 \right)$ or $320.

✓ Guided Practice

5. **PAINTING** Mrs. Sung's art class is painting a large mural depicting a winter ski scene. The students want to paint a ski hill at both the beginning and the end of the mural, but they only have enough paint to cover 30 square feet. If the area of each ski hill can be found by $\int_0^5 (5 - 0.2x^2)\,dx$, will the students have enough paint for both hills? Explain.

Approximate the area of the shaded region for each function using the indicated number of rectangles. Use the specified endpoints to determine the heights of the rectangles. (Example 1)

1. 5 rectangles
right endpoints

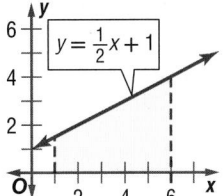

2. 4 rectangles
left endpoints

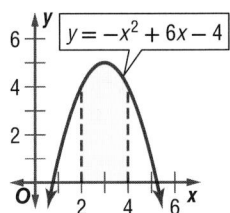

3. 8 rectangles
right endpoints

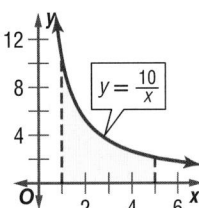

4. 8 rectangles
left endpoints

5. 4 rectangles
left endpoints

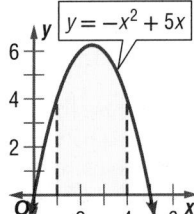

6. 5 rectangles
right endpoints

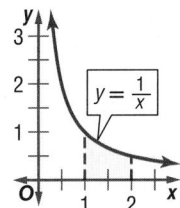

7 **FLOORING** Len is installing a hardwood floor and has to cover a semicircular section defined by $f(x) = (-x^2 + 10x)^{0.5}$. (Example 1)

a. Approximate the area of the semicircular region using left endpoints and rectangles one unit in width.

b. Len decides that using both left and right endpoints may give a better estimate because it would eliminate any area outside the semicircular region. Approximate the area of the semicircular region as shown in the figure.

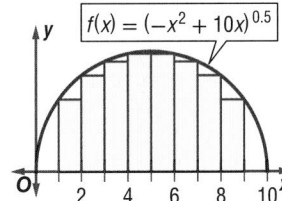

c. Find the area of the region by using the formula for the area of a semicircle. Which approximation is closer to the actual area of the region? Explain why this estimate gives a better approximation.

Approximate the area of the shaded region for each function by first using the right endpoints and then by using the left endpoints. Then find the average for both approximations. Use the specified width for the rectangles. (Example 2)

8. width = 0.5

9. width = 1.0

10. width = 0.75

11. width = 0.5

12. width = 0.75

13. width = 0.5

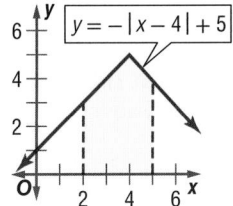

Use limits to find the area between the graph of each function and the x-axis given by the definite integral. (Examples 3 and 4)

14. $\int_1^4 4x^2 \, dx$

15. $\int_2^6 (2x + 5) \, dx$

16. $\int_0^2 6x \, dx$

17. $\int_1^3 (2x^2 + 3) \, dx$

18. $\int_2^5 (x^2 + 4x - 2) \, dx$

19. $\int_1^2 8x^3 \, dx$

20. $\int_0^4 (4x - x^2) \, dx$

21. $\int_3^4 (-x^2 + 6x) \, dx$

22. $\int_0^3 (x^3 + x) \, dx$

23. $\int_2^4 (-3x + 15) \, dx$

24. $\int_1^5 (x^2 - x + 1) \, dx$

25. $\int_1^3 12x \, dx$

26. $\int_0^3 (8 - 0.6x^2) \, dx$

27. $\int_3^8 0.5x^2 \, dx$

28. PUBLISHING Refer to the beginning of the lesson. The publisher would like to increase daily production from 1000 books to 1500 books. Find the cost of the increase if it is defined as $\int_{1000}^{1500} (10 - 0.002x)\, dx$. (Example 5)

29. ARCHWAY The homecoming committee decided the entrance into the dance would be a balloon archway. In addition, the committee wanted to hang streamers that would extend from the top of the arch down to the floor, completely covering the entrance. Find the area beneath the balloon archway if it can be defined as $\int_{1}^{13} (-0.2x^2 + 2.8x - 1.8)\, dx$, where x is given in feet. (Example 5)

30. LOGO Part of a company's logo is in the shape of the region shown. If this part of the logo is to be sewn on a flag, how much material is required if x is given in feet? (Example 5)

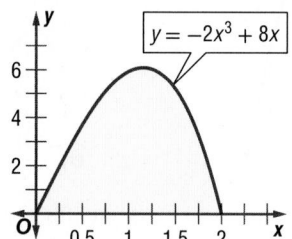

31. NEGATIVE LIMITS Definite integrals can be calculated for both positive and negative limits.

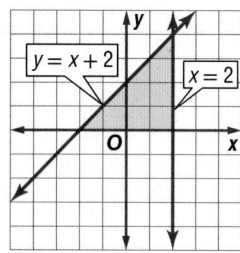

a. Find the height and length of the base of the triangle. Then calculate the area of the triangle using its height and base.

b. Calculate the area of the triangle by evaluating $\int_{-2}^{2} (x + 2)\, dx$.

Use limits to find the area between the graph of each function and the x-axis given by the definite integral.

32. $\int_{-1}^{1} x^2\, dx$

33 $\int_{-1}^{0} (x^3 + 2)\, dx$

34. $\int_{-4}^{-2} (-x^2 - 6x)\, dx$

35. $\int_{-3}^{-2} -5x\, dx$

36. $\int_{-2}^{0} (2x + 6)\, dx$

37. $\int_{-1}^{0} (x^3 - 2x)\, dx$

Use limits to find the area between the graph of each function and the x-axis given by the definite integral.

38. $\int_{-3}^{-1} (-2x^2 - 7x)\, dx$

39. $\int_{-2}^{0} (-x^3)\, dx$

40. $\int_{-4}^{3} 2\, dx$

41. $\int_{-2}^{-1} \left(-\frac{1}{2}x + 3\right) dx$

42. ⚙ **MULTIPLE REPRESENTATIONS** In this problem, you will investigate the process of finding the area between two curves.

a. GRAPHICAL Graph $f(x) = -x^2 + 4$ and $g(x) = x^2$ on the same coordinate plane and shade the areas represented by $\int_{0}^{1} (-x^2 + 4)\, dx$ and $\int_{0}^{1} x^2\, dx$.

b. ANALYTICAL Evaluate $\int_{0}^{1} (-x^2 + 4)\, dx$ and $\int_{0}^{1} x^2\, dx$.

c. VERBAL Explain why the area between the two curves is equal to $\int_{0}^{1} (-x^2 + 4)\, dx - \int_{0}^{1} x^2\, dx$. Then calculate this value using the values found in part **b**.

d. ANALYTICAL Find $f(x) - g(x)$. Then evaluate $\int_{0}^{1} [f(x) - g(x)]\, dx$.

e. VERBAL Make a conjecture about the process of finding the area between two curves.

H.O.T. Problems   Use Higher-Order Thinking Skills

43. ERROR ANALYSIS Pete says that when you use the right endpoints of rectangles to estimate the area between a curve and the x-axis, the total area of the rectangles will always be greater than the actual area. Lara says that the area of the rectangles is always greater when you use the left endpoints. Is either of them correct? Explain.

44. CHALLENGE Evaluate $\int_{0}^{1} (5x^4 + 3x^2 - 2x + 1)\, dx$.

45. REASONING Suppose every vertical cross section of a tunnel can be modeled by $f(x)$ on $[a, b]$. Explain how the volume of the tunnel can be calculated by using $\ell \cdot \int_{a}^{b} f(x)\, dx$, where ℓ is the length of the tunnel.

46. PREWRITE Write an outline that could be used to describe the steps involved in estimating the area between the x-axis and the curve of a function on a given interval.

47. CHALLENGE Evaluate $\int_{0}^{t} (x^2 + 2)\, dx$.

48. WRITING IN MATH Explain the effectiveness of using triangles and circles to approximate the area between a curve and the x-axis. Which shape do you think gives the best approximation?

Find the derivative of each function. (Lesson 12-4)

49. $j(x) = (2x^3 + 11x)(2x^8 - 12x^2)$ **50.** $f(k) = (k^{15} + k^2 + 2k)(k - 7k^2)$ **51.** $s(t) = (\sqrt{t} - 7)(3t^8 - 5t)$

Find the slope of the line tangent to the graph of each function at $x = 1$. (Lesson 12-3)

52. $y = x^3$ **53.** $y = x^3 - 7x^2 + 4x + 9$ **54.** $y = (x + 1)(x - 2)$

Evaluate each limit. (Lesson 12-2)

55. $\displaystyle\lim_{x \to 0} \frac{x^2 + 3x}{x}$ **56.** $\displaystyle\lim_{x \to 1} \frac{x^2 - 3x + 2}{x - 1}$ **57.** $\displaystyle\lim_{x \to 3} \frac{x^2 - 9}{x^3 - 27}$

58. MOVIES Veronica believes that the price of a ticket for a movie is still under $7.00. She randomly visits 14 movie theatres and records the prices of a ticket. Find the p-value and determine whether there is enough evidence to support the claim at $\alpha = 0.10$. (Lesson 11-6)

Ticket Prices ($)						
5.25	7.27	5.46	7.63	7.75	5.42	6.00
6.63	7.38	6.97	7.85	7.03	6.53	6.87

59. VIDEO GAMES A random sample of 85 consumers of video games showed that the average price of a video game was $36.50. Assume that the standard deviation from previous studies was $11.30. Find the maximum error of estimate given a 99% confidence level. Then create a confidence interval for the mean price of a video game. (Lesson 11-5)

60. SHOPPING In a recent year, 33% of Americans said that they were planning to go shopping on Black Friday, the day after Thanksgiving. What is the probability that in a random sample of 45 people, fewer than 14 people plan to go shopping on Black Friday? (Lesson 11-4)

Classify each random variable X as *discrete* or *continuous*. Explain your reasoning. (Lesson 11-2)

61. X represents the number of mobile phone calls made by a randomly chosen student during a given day.

62. X represents the time it takes a randomly selected student to run a mile.

63. SAT/ACT If the statement below is true, then which of the following must also be true?
If at least 1 bear is sleepy,
then some ponies are happy.

  A If all bears are sleepy, then all ponies are happy.

  B If all ponies are happy, then all bears are sleepy.

  C If no bear is sleepy, then no pony is happy.

  D If no pony is happy, then no bear is sleepy.

  E If some ponies are happy, then at least 1 bear is sleepy.

64. REVIEW What is $\displaystyle\lim_{x \to 3} \frac{x^2 + 3x - 10}{x^2 + 5x + 6}$?

  F $\dfrac{1}{15}$ **H** $\dfrac{3}{15}$

  G $\dfrac{2}{15}$ **J** $\dfrac{4}{15}$

65. Find the area of the region between the graph of $y = -x^2 + 3x$ and x-axis on the interval $[0, 3]$ or $\displaystyle\int_0^3 (-x^2 + 3x)\, dx$.

  A $3\dfrac{3}{4}$ units^2 **C** $21\dfrac{1}{4}$ units^2

  B $4\dfrac{1}{2}$ units^2 **D** $22\dfrac{1}{2}$ units^2

66. REVIEW Find the derivative of $n(a) = \dfrac{4}{a} - \dfrac{5}{a^2} + \dfrac{3}{a^4} + 4a$.

  F $n'(a) = 8a - 5a^2 + 3a^4$

  H $n'(a) = 4a^2 - 5a^3 + 3a^4 + 4$

  G $n'(a) = -\dfrac{4}{a^2} + \dfrac{5}{a^3} - \dfrac{3}{a^5} + 4$

  J $n'(a) = -\dfrac{4}{a^2} + \dfrac{10}{a^3} - \dfrac{12}{a^5} + 4$

The Fundamental Theorem of Calculus

Then
You used limits to approximate the area under a curve. (Lesson 12-5)

Now
- Find antiderivatives.
- Use the Fundamental Theorem of Calculus.

New Vocabulary
antiderivative
indefinite integral
Fundamental Theorem of Calculus

Math Online
glencoe.com

Why?

During the initial ascent of a hot-air balloon ride, Kristina realized that she had her brother's cell phone in her pocket. Before the balloon rose too high, Kristina dropped the phone over the edge to her brother waiting on the ground. Knowing that the velocity of the phone could be described as $v(t) = -32t$, where t is given in seconds and velocity is given in feet per second, Kristina was able to determine how high she was from the ground when she dropped the phone over the edge.

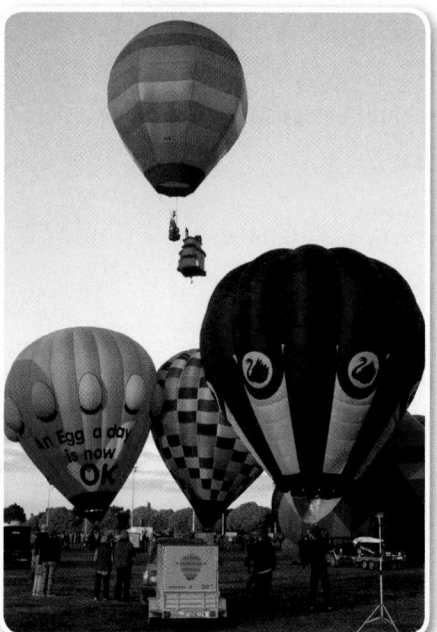

Antiderivatives and Indefinite Integrals In Lessons 12-3 and 12-4, you learned that if the position of an object is defined as $f(x) = x^2 + 2x$, then an expression for the velocity of the object is the derivative of $f(x)$ or $f'(x) = 2x + 2$. However, if you are given an expression for velocity, but need to know the formula from which it was obtained, we need a way to work backward or undo the differentiation.

In other words, if given $f(x)$, we want to find an equation $F(x)$ such that $F'(x) = f(x)$. The function $F(x)$ is an **antiderivative** of $f(x)$.

EXAMPLE 1   Find Antiderivatives

Find an antiderivative for each function.

a. $f(x) = 3x^2$

We need to find a function that has a derivative of $3x^2$. Recall that the derivative has an exponent that is one less than the exponent in the original function. Therefore, $F(x)$ will be raised to a power of three. Also, the coefficient of a derivative is determined in part by the exponent of the original function. $F(x) = x^3$ fits this description. The derivative of $x^3 = 3x^{3-1}$ or $3x^2$.

However, x^3 is not the only function that works. The function $G(x) = x^3 + 10$ is another because its derivative is $G'(x) = 3x^{3-1} + 0$ or $3x^2$. Another answer could be $H(x) = x^3 - 37$.

b. $f(x) = -\dfrac{8}{x^9}$

Rewrite $f(x)$ with a negative exponent, $-8x^{-9}$. Again, the derivative has an exponent that is one less than the exponent in the original function, so $F(x)$ will be raised to the negative eighth power. We can try $F(x) = x^{-8}$. The derivative of x^{-8} is $-8x^{-8-1}$ or $-8x^{-9}$.

$G(x) = x^{-8} + 3$ and $H(x) = x^{-8} - 12$ are other antiderivatives.

✓ Guided Practice

Find two different antiderivatives for each function.

1A. $2x$

1B. $-3x^{-4}$

In Example 1, notice that adding or subtracting constants to the original antiderivative produced other antiderivatives. In fact, because the derivative of any constant is 0, adding or subtracting a constant C to an antiderivative will not affect its derivative. Therefore, there are an infinite number of antiderivatives for a given function. Antiderivatives that include a constant C are said to be in *general form*.

As with derivatives, there are rules for finding antiderivatives.

Key Concept — Antiderivative Rules

Power Rule

If $f(x) = x^n$, where n is a rational number other than -1,
$$F(x) = \frac{x^{n+1}}{n+1} + C.$$

Constant Multiple of a Power

If $f(x) = kx^n$, where n is a rational number other than -1
and k is a constant, then $F(x) = \frac{kx^{n+1}}{n+1} + C.$

Sum and Difference

If the antiderivatives of $f(x)$ and $g(x)$ are $F(x)$ and $G(x)$
respectively, then an antiderivative of $f(x) \pm g(x)$ is
$F(x) \pm G(x).$

StudyTip

Antiderivatives An antiderivative of a constant k is kx. For instance, if $f(x) = 3$, then $F(x) = 3x$.

EXAMPLE 2 Antiderivative Rules

Find all antiderivatives for each function.

a. $f(x) = 4x^7$

$f(x) = 4x^7$ Original equation

$F(x) = \frac{4x^{7+1}}{7+1} + C$ Constant Multiple of a Power

$= \frac{1}{2}x^8 + C$ Simplify.

b. $f(x) = \frac{2}{x^4}$

$f(x) = \frac{2}{x^4}$ Original equation

$= 2x^{-4}$ Rewrite with a negative exponent.

$F(x) = \frac{2x^{-4+1}}{-4+1} + C$ Constant Multiple of a Power

$= -\frac{2}{3}x^{-3} + C$ or $-\frac{2}{3x^3} + C$ Simplify.

c. $f(x) = x^2 - 8x + 5$

$f(x) = x^2 - 8x + 5$ Original equation

$= x^2 - 8x^1 + 5x^0$ Rewrite the function so each term has a power of x.

$F(x) = \frac{x^{2+1}}{2+1} - \frac{8x^{1+1}}{1+1} + \frac{5x^{0+1}}{0+1} + C$ Antiderivative Rule

$= \frac{1}{3}x^3 - 4x^2 + 5x + C$ Simplify.

✓ Guided Practice

2A. $f(x) = 6x^4$ **2B.** $f(x) = \frac{10}{x^3}$ **2C.** $f(x) = 8x^7 + 6x + 2$

The general form of an antiderivative is given a special name and notation.

Key Concept — Indefinite Integral

The **indefinite integral** of $f(x)$ is defined by $\int f(x)\, dx = F(x) + C$, where $F(x)$ is an antiderivative of $f(x)$ and C is any constant.

Real-World EXAMPLE 3 Indefinite Integral

EGG DROP Students of Ms. Nair's technology class are participating in an egg drop competition. Each team of students must build a protective device that will keep an egg from cracking after a 30-foot drop. The instantaneous velocity of an egg can be defined as $v(t) = -32t$, where t is given in seconds and velocity is measured in feet per second.

a. Find the position function $s(t)$ of a dropped egg.

To find the function for the position of the egg, find the antiderivative of $v(t)$.

$$s(t) = \int v(t)\, dt \qquad \text{Relationship between position and velocity}$$

$$= \int -32t\, dt \qquad v(t) = -32t$$

$$= -\frac{32t^{1+1}}{1+1} + C \qquad \text{Constant Multiple of a Power}$$

$$= -16t^2 + C \qquad \text{Simplify.}$$

Find C by substituting 30 feet for the initial height and 0 for the initial time.

$$s(t) = -16t^2 + C \qquad \text{Antiderivative of } v(t)$$

$$30 = -16(0)^2 + C \qquad s(t) = 30 \text{ and } t = 0$$

$$30 = C \qquad \text{Simplify.}$$

The position function for the egg is $s(t) = -16t^2 + 30$.

b. Find how long it will take for the egg to hit the ground.

Solve for t when $s(t) = 0$.

$$s(t) = -16t^2 + 30 \qquad \text{Position function for the egg}$$

$$0 = -16t^2 + 30 \qquad s(t) = 0$$

$$-30 = -16t^2 \qquad \text{Subtract 30 from each side.}$$

$$1.875 = t^2 \qquad \text{Divide each side by } -16.$$

$$1.369 = t \qquad \text{Take the positive square root of each side.}$$

The egg will hit the ground in about 1.37 seconds.

✔ Guided Practice

3. FALLING OBJECT A maintenance worker, securely standing on a catwalk in a gymnasium, is fixing a light positioned 120 feet above the floor when his wallet accidentally falls out of his pocket. The instantaneous velocity of his wallet can be defined as $v(t) = -32t$, where t is given in seconds and velocity is measured in feet per second.

A. Find the position function $s(t)$ of the dropped wallet.

B. Find how long it will take for the wallet to hit the floor.

Fundamental Theorem of Calculus Notice that the notation used for indefinite integrals looks very similar to the notation used in Lesson 12-5 for definite integrals. The only difference appears to be the lack of upper and lower limits in indefinite integrals. In fact, finding an antiderivative of a function is a shortcut to calculating a definite integral by evaluating a Riemann sum. This connection between definite integrals and antiderivatives is so important that it is called the **Fundamental Theorem of Calculus**.

Key Concept Fundamental Theorem of Calculus

If f is continuous on $[a, b]$ and $F(x)$ is any antiderivative of $f(x)$, then

$$\int_a^b f(x)\, dx = F(b) - F(a).$$

The difference $F(b) - F(a)$ is commonly denoted by $F(x)\Big|_a^b =$.

Math History Link

Maria Gaetana Agnesi (1718–1799)
An Italian linguist, mathematician, and philosopher, Maria Agnesi authored *Analytical Institutions*, the first book to discuss both differential and integral calculus. She is also known for an equation she developed for a curve called the "Witch of Agnesi."

A by-product of the Fundamental Theorem of Calculus is that it makes a connection between integrals and derivatives. Integration is the process of calculating antiderivatives, while differentiation is the process of calculating derivatives. Thus, integration and differentiation are inverse processes. We can use the Fundamental Theorem of Calculus to evaluate definite integrals without having to use limits.

EXAMPLE 4 **Area Under a Curve**

Use the Fundamental Theorem of Calculus to find the area of the region between the graph of each function and the *x*-axis on the given interval.

a. $y = 4x^3$ on the interval [1, 3], or $\int_1^3 4x^3\, dx$.

First, find the antiderivative.

$$\int 4x^3\, dx = \frac{4x^{3+1}}{3+1} + C \qquad \text{Constant Multiple of a Power}$$

$$= x^4 + C \qquad \text{Simplify.}$$

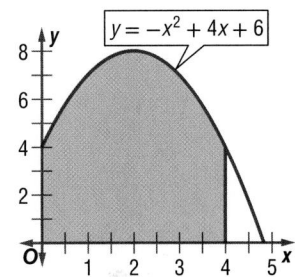

Now evaluate the antiderivative at the upper and lower limits, and find the difference.

$$\int_1^3 4x^3\, dx = x^4 + C \Big|_1^3 \qquad \text{Fundamental Theorem of Calculus}$$

$$= (3^4 + C) - (1^4 + C) \qquad b = 3 \text{ and } a = 1$$

$$= 81 - 1 \text{ or } 80 \qquad \text{Simplify.}$$

The area under the graph on the interval [1, 3] is 80 square units.

b. $y = -x^2 + 4x + 6$ on the interval [0, 4], or $\int_0^4 (-x^2 + 4x + 6)\, dx$

First, find the antiderivative.

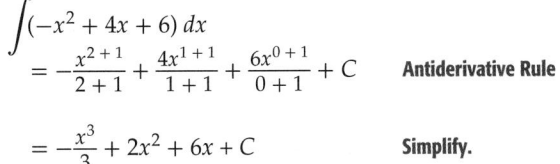

$$\int (-x^2 + 4x + 6)\, dx$$

$$= -\frac{x^{2+1}}{2+1} + \frac{4x^{1+1}}{1+1} + \frac{6x^{0+1}}{0+1} + C \qquad \text{Antiderivative Rule}$$

$$= -\frac{x^3}{3} + 2x^2 + 6x + C \qquad \text{Simplify.}$$

Now evaluate the antiderivative at the upper and lower limits, and find the difference.

$$\int_0^4 (-x^2 + 4x + 6)\, dx = -\frac{x^3}{3} + 2x^2 + 6x + C \Big|_0^4 \qquad \text{Fundamental Theorem of Calculus}$$

$$= \left(-\frac{(4)^3}{3} + 2(4)^2 + 6(4) + C\right) - \left(-\frac{(0)^3}{3} + 2(0)^2 + 6(0) + C\right) \qquad \begin{array}{l} b = 4 \text{ and} \\ a = 0 \end{array}$$

$$= 34.67 - 0 \text{ or } 34.67 \qquad \text{Simplify.}$$

The area under the graph on the interval [0, 4] is 34.67 square units.

☑ **Guided Practice**

Evaluate each definite integral.

4A. $\int_2^5 3x^2\, dx$

4B. $\int_1^2 (16x^3 - 6x^2)\, dx$

Notice that when the antiderivative is evaluated at the upper and lower limits and the difference is found, we do not have an exact value for *C*. However, regardless of the value of the constant, because it is present in each antiderivative, the difference between the constants is 0. Therefore, when evaluating definite integrals using the Fundamental Theorem of Calculus, you can disregard the constant term when writing the antiderivative.

Before evaluating an integral, determine whether it is indefinite or definite.

Watch Out!

Integrals Although the constant *C* can be excluded when evaluating definite integrals, it must be included when evaluating indefinite integrals because it is part of the antiderivative.

EXAMPLE 5 Indefinite and Definite Integrals

Evaluate each integral.

a. $\int (9x - x^3)\, dx$

This is an indefinite integral. Use the antiderivative rules to evaluate.

$\int (9x - x^3)\, dx = \dfrac{9x^{1+1}}{1+1} - \dfrac{x^{3+1}}{3+1} + C$ **Constant Multiple of a Power**

$= \dfrac{9}{2}x^2 - \dfrac{x^4}{4} + C$ **Simplify.**

b. $\int_2^3 (9x - x^3)\, dx$

This is a definite integral. Evaluate the integral using the given upper and lower limits.

$\int_2^3 (9x - x^3)\, dx = \left(\dfrac{9}{2}x^2 - \dfrac{x^4}{4}\right)\Big|_2^3$ **Fundamental Theorem of Calculus**

$= \left(\dfrac{9}{2}(3)^2 - \dfrac{3^4}{4}\right) - \left[\dfrac{9}{2}(2)^2 - \dfrac{2^4}{4}\right]$ **b = 3 and a = 2**

$= 20.25 - 14 \text{ or } 6.25$ **Simplify.**

The area under the graph on the interval [2, 3] is 6.25 square units.

✔ Guided Practice

5A. $\int (6x^2 + 8x - 3)\, dx$ **5B.** $\int_1^3 (-x^4 + 8x^3 - 24x^2 + 30x - 4)\, dx$

Notice that indefinite integrals give the antiderivative of a function while definite integrals not only give the antiderivative but also require it to be evaluated for given upper and lower limits. Thus, an indefinite integral gives a function, the antiderivative, for finding the area under a curve for *any* set of limits. The integral becomes definite when a set of limits is provided and the antiderivative can be evaluated.

EXAMPLE 6 Definite Integrals

The work, in joules, required to stretch a certain spring a distance of 0.5 meter beyond its natural length is given by $\int_0^{0.5} 360x\, dx$. How much work is required?

Evaluate the definite integral for the given upper and lower limits.

$\int_0^{0.5} 360x\, dx = 180x^2 \Big|_0^{0.5}$ **Constant Multiple of a Power and the Fundamental Theorem of Calculus**

$= 180(0.5)^2 - 180(0)^2$ **Let b = 0.5 and a = 0 and subtract.**

$= 45 - 0 \text{ or } 45$ **Simplify.**

The work required is 45 joules.

✔ Guided Practice

Find the work required to stretch a spring if it is defined by the following integrals.

6A. $\int_0^{0.7} 476x\, dx$ **6B.** $\int_0^{1.4} 512x\, dx$

Exercises

Find all antiderivatives for each function. (Examples 1 and 2)

1. $f(x) = x^5$

2. $h(b) = -5b - 3$

3. $f(z) = \sqrt[3]{z}$

4. $n(t) = \frac{1}{4}t^4 - \frac{2}{3}t^2 + \frac{3}{4}$

5. $q(r) = \frac{3}{4}r^{\frac{2}{5}} + \frac{5}{8}r^{\frac{1}{3}} + r^{\frac{1}{2}}$

6. $w(u) = \frac{2}{3}u^5 + \frac{1}{6}u^3 - \frac{2}{5}u$

7. $g(a) = 8a^3 + 5a^2 - 9a + 3$

8. $u(d) = \frac{12}{d^5} + \frac{5}{d^3} - 6d^2 + 3.5$

9. $m(t) = 16t^3 - 12t^2 + 20t - 11$

10. $p(h) = 72h^8 + 24h^5 - 12h^2 + 14$

11. CELL PHONE Refer to the beginning of the lesson. Suppose the phone took exactly two seconds to drop from the balloon to the ground. (Example 3)

   a. Evaluate $s(t) = \int -32t\, dt$.

   b. Solve for C in the position function $s(t)$ by substituting two seconds for t and 0 for $s(t)$.

   c. How far from the ground is the phone after 1.5 seconds of falling?

Evaluate each integral. (Examples 4 and 5)

12. $\int (6m + 12m^3)\, dm$

13. $\int (20n^3 - 9n^2 - 18n + 4)\, dn$

14. $\int_1^4 2x^3\, dx$

15. $\int_2^5 (a^2 - a + 6)\, da$

16. $\int_1^2 (4g + 6g^2)\, dg$

17. $\int_2^{10} \left(\frac{2}{5}p^{\frac{1}{8}} + \frac{5}{4}p^{\frac{2}{7}} + \frac{1}{4} \right) dp$

18. $\int_1^3 \left(\frac{1}{2}h^2 + \frac{2}{3}h^3 - \frac{1}{5}h^4 \right) dh$

19. $\int_0^2 (-v^4 + 2v^3 + 2v^2 + 6)\, dv$

20. $\int (3.4t^4 - 1.2t^3 + 2.3t - 5.7)\, dt$

21. $\int (14.2w^{6.1} - 20.1w^{5.7} + 13.2w^{2.3} + 3)\, dw$

22. SURVEYOR A plot of land has two perpendicular fences and a stream for borders as shown.

Suppose the edge of the stream that borders the plot can be modeled by $f(x) = -0.00005x^3 + 0.004x^2 - 1.04x + 40$, where the fences are the x and y-axes, and x is given in miles. Evaluate $\int_0^{40} f(x)\, dx$ to find the area of the land. (Example 6)

23 INSECTS The velocity of a flea's jump can be defined as $v(t) = -32t + 34$, where t is given in seconds and velocity is given in feet per second. (Example 6)

   a. Find the position function $s(t)$ for the flea's jump. Assume that for $t = 0$, $s(t) = 0$.

   b. After a flea jumps, how long does it take before the flea lands on the ground?

24. NATIONAL MONUMENT A magician wants to make the Gateway Arch in St. Louis disappear. To attempt the illusion, he needs to cover the arch with a large canopy. Before constructing the canopy, the magician wants an approximation of the area under the arch. One equation that can be used to model the shape of the arch is $y = -\frac{x^2}{157.5} + 4x$, where x is given in feet. Find the area under the arch. (Example 6)

25. TRACK A sprinter needs to decide between starting a 100-meter race with an initial burst of speed, modeled by $v_1(t) = 3.25t - 0.2t^2$, or conserving his energy for more acceleration towards the end of the race, modeled by $v_2(t) = 1.2t + 0.03t^2$, where velocity is measured in meters per second after t seconds.

   a. Use a graphing calculator to graph both velocity functions on the same screen for $0 \le t \le 12$.

   b. Find the position function $s(t)$ for $v_1(t)$ and $v_2(t)$.

   c. How long does it take for the sprinter to finish a 100-meter race using each strategy?

Evaluate each integral.

26. $\int_{-3}^1 3\, dx$

27. $\int_{-1}^2 (-x^2 + 10)\, dx$

28. $\int_{-6}^{-3} (-x^2 - 9x - 10)\, dx$

29. $\int_{-3}^{-1} (x^3 + 8x^2 + 21x + 20)\, dx$

30. $\int_{-2}^{-1} \left(\frac{x^5}{2} + \frac{5x^4}{4} \right) dx$

31. $\int_{-1}^1 (x^4 - 2x^3 - 4x + 8)\, dx$

32. CATAPULT Pumpkins are launched by catapults at The Pumpkin Chunking World Championships in Delaware. A pumpkin launched by a catapult has a velocity of $v(t) = -32t + 120$ feet per second after t seconds. After 3 seconds, the pumpkin has a height of 228 feet.

    a. Find the maximum height of the pumpkin.

    b. Find the pumpkin's velocity when it hits the ground.

Evaluate each integral.

33. $\displaystyle\int_x^2 (3t^2 + 8t)\, dt$ **34.** $\displaystyle\int_5^x (10t^4 - 12t^2 + 5)\, dt$

35. $\displaystyle\int_3^{2x} (4t^3 + 10t + 2)\, dt$ **36.** $\displaystyle\int_{-x}^6 (-9t^2 + 4t)\, dt$

37 $\displaystyle\int_x^{x^2} (16t^3 - 15t^2 + 7)\, dt$ **38.** $\displaystyle\int_{2x}^{x+3} (3t^2 + 6t + 1)\, dt$

39. SPHERE The volume of a sphere with radius R can be found by slicing the sphere vertically into circular cross sections and then integrating the areas.

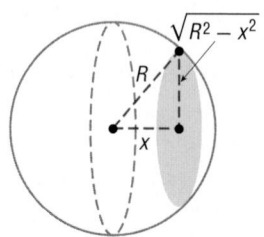

The radius of each cross section is $\sqrt{R^2 - x^2}$. Thus, the area of a cross-section is equal to $\pi\left(\sqrt{R^2 - x^2}\right)^2$.

Evaluate $\displaystyle\int_{-R}^R (\pi R^2 - \pi x^2)\, dx$ to find the volume of a sphere.

40. AREA Calculate the area bound by $f(x)$, $g(x)$, and the x-axis on the interval $1 \leq x \leq 3$.

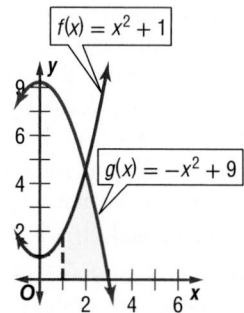

The integral $\displaystyle\int_0^{n + 0.5} x^k\, dx$ gives a reasonable estimate of the sum of the series $\displaystyle\sum_{i=1}^n i^k$. Use the integral to estimate each sum and then find the actual sum.

41. $\displaystyle\sum_{i=1}^{20} i^3$ **42.** $\displaystyle\sum_{i=1}^{100} i^2$

43. $\displaystyle\sum_{i=1}^{25} i^4$ **44.** $\displaystyle\sum_{i=1}^{30} i^5$

45. MULTIPLE REPRESENTATIONS In this problem, you will investigate the relationship between the *total area* and the *signed area* of a region bound by a curve and the x-axis.

    a. GEOMETRIC Graph $f(x) = x^3 - 6x^2 + 8x$, and shade the region bound by $f(x)$ and the x-axis for $0 \leq x \leq 4$.

    b. ANALYTICAL Evaluate $\displaystyle\int_0^2 (x^3 - 6x^2 + 8x)\, dx$ and $\displaystyle\int_2^4 (x^3 - 6x^2 + 8x)\, dx$.

    c. VERBAL Make a conjecture for the area found above and below the x-axis.

    d. ANALYTICAL Calculate the *signed area* by evaluating $\displaystyle\int_0^4 (x^3 - 6x^2 + 8x)\, dx$. Then calculate the *total area* by evaluating

$$\left|\int_0^2 (x^3 - 6x^2 + 8x)\, dx\right| + \left|\int_2^4 (x^3 - 6x^2 + 8x)\, dx\right|.$$

    e. VERBAL Make a conjecture for the difference between the *signed area* and the *total area*.

46. CHALLENGE Evaluate $\displaystyle\int_{-r}^r \sqrt{r^2 - x^2}\, dx$, where r is a constant. (*Hint*: Find the area between the graph of $y = \sqrt{r^2 - x^2}$ and the x-axis)

REASONING Determine whether each statement is *always*, *sometimes*, or *never* true. Explain your reasoning.

47. $\displaystyle\int_a^b f(x)\, dx = \int_b^a f(x)\, dx$

48. $\displaystyle\int_a^b f(x)\, dx = \int_{-b}^{-a} f(x)\, dx$

49. $\displaystyle\int_a^b f(x)\, dx = \int_{|a|}^{|b|} f(x)\, dx$

50. PROOF Prove that for constants n and m
$$\int_a^b (n + m)\, dx = \int_a^b n\, dx + \int_a^b m\, dx.$$

51. REASONING Describe the values of $f(x)$, $\displaystyle\sum_{i=1}^n f(x_i)\Delta x$, and $\displaystyle\int_a^b f(x)\, dx$, where the graph of $f(x)$ lies below the x-axis for $a \leq x \leq b$.

52. WRITING IN MATH Explain why the constant term C in an antiderivative can be disregarded when evaluating a definite integral.

53. WRITING IN MATH Write an outline that could be used to describe the steps involved in finding the area of the region between the graph of $y = 6x^2$ and the x-axis on the interval $[0, 2]$.

Use limits to approximate the area between the graph of each function and the x-axis given by the definite integral. (Lesson 12-5)

54. $\displaystyle\int_{-2}^{2} 14x^2\,dx$

55. $\displaystyle\int_{0}^{6} (x+2)\,dx$

Use the Quotient Rule to find the derivative of each function. (Lesson 12-4)

56. $j(k) = \dfrac{k^8 - 7k}{2k^4 + 11k^3}$

57. $g(n) = \dfrac{2n^3 + 4n}{n^2 + 1}$

58. **FASHION** The average prices for three designer handbags on an online auction site are shown. (Lesson 11-4)

  a. If a random sample of 35 A-style handbags is selected, find the probability that the mean price is more than \$138 if the standard deviation of the population is \$9.

  b. If a random sample of 40 C-style handbags is selected, find the probability that the mean price is between \$150 and \$155 if the standard deviation of the population is \$12.

Handbag Style	Average Price ($)
A	135
B	145
C	152

59. **BASEBALL** The average age of a major league baseball player is normally distributed with a mean of 28 and a standard deviation of 4 years. (Lesson 11-3)

  a. About what percent of major league baseball players are younger than 24?

  b. If a team has 35 players, about how many are between the ages of 24 and 32?

60. Find two pairs of polar coordinates for the point with the given rectangular coordinates $(3, 8)$, if $-2\pi \le \theta \le 2\pi$. (Lesson 9-3)

61. **SAT/ACT** In the figure, C and E are midpoints of the edges of the cube. A triangle is to be drawn with R and S as two of the vertices. Which of the following points should be the third vertex of the triangle if it is to have the largest possible perimeter?

  A A
  B B
  C C
  D D
  E E

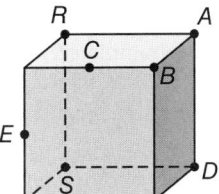

62. The work in joules required to pump all of the water out of a 10-meter by 5-meter by 2-meter swimming pool is given by $\displaystyle\int_{0}^{2} 490,000x\,dx$. If you evaluate this integral, what is the required work?

  F 980,000 J
  G 985,000 J
  H 990,000 J
  J 995,000 J

63. **FREE RESPONSE** A particle moves along a line so that its position at any time $t \ge 0$ is given by the function $s(t) = t^2 - 3t + 10$ where s is measured in meters and t is measured in seconds.

  a. Find the displacement of the particle during the first 4 seconds. That is, how far has the particle moved from its original starting position after 4 seconds?

  b. Find the average velocity of the particle during the first 4 seconds.

  c. Write an equation for the instantaneous velocity of the particle at any time t.

  d. Find the instantaneous velocity of the particle when $t = 1$ and $t = 4$.

  e. At what value for t does $s(t)$ attain a minimum value?

  f. What does the value of t you found in part **d** represent in terms of the motion of the particle?

Chapter Summary

Key Concepts

Estimating Limits Graphically (Lesson 12-1)

• The limit of a function $f(x)$ as x approaches c exists if and only if both one-sided limits exist and are equal.

• A limit of $f(x)$ as x approaches c does not exist if $f(x)$ approaches a different value from the left of c than from the right of c, increases or decreases without bound from the left and/or right of c, or *oscillates* back and forth between two values.

Evaluating Limits Algebraically (Lesson 12-2)

• Limits of polynomial and rational functions can often be found by using direct substitution.

• If you evaluate a limit and reach the indeterminate form $\frac{0}{0}$, simplify the expression algebraically by factoring and dividing out a common factor or by rationalizing the numerator or denominator and then dividing out any common factors.

Tangent Lines and Velocity (Lesson 12-3)

• The instantaneous rate of change of the graph of $f(x)$ at the point $(x, f(x))$ is the slope m of the tangent line given by $m = \lim\limits_{h \to 0} \dfrac{f(x+h) - f(x)}{h}$.

Derivatives (Lesson 12-4)

• The derivative of $f(x) = x^n$ is $f'(x)$ given by $f'(x) = nx^{n-1}$, where n is a real number.

Area Under a Curve and Integration (Lesson 12-5)

• The area of a region under the graph of a function is $\int_a^b f(x)\, dx = \lim\limits_{n \to \infty} \sum\limits_{i=1}^{n} f(x_i)\, \Delta x$, where a and b are the lower and upper limits, respectively, $\Delta x = \dfrac{b-a}{n}$, and $x_i = a + i\Delta x$.

The Fundamental Theorem of Calculus (Lesson 12-6)

• The antiderivative of $f(x) = x^n$ is $F(x)$ given by $F(x) = \dfrac{x^{n+1}}{n+1} + C$, for some constant C.

• If $F(x)$ is the antiderivative of the continuous function $f(x)$, then $\int_a^b f(x)\, dx = F(b) - F(a)$.

Key Vocabulary

antiderivative (p. 784)

definite integral (p. 777)

derivative (p. 766)

difference quotient (p. 758)

differential equation (p. 766)

differential operator (p. 766)

differentiation (p. 766)

direct substitution (p. 748)

indefinite integral (p. 785)

indeterminate form (p. 749)

instantaneous rate of change (p. 758)

instantaneous velocity (p. 760)

integration (p. 777)

lower limit (p. 777)

one-sided limit (p. 738)

regular partition (p. 777)

right Riemann sum (p. 777)

tangent line (p. 758)

two-sided limit (p. 738)

upper limit (p. 777)

Vocabulary Check

Choose the correct term to complete each sentence.

1. The slope of a nonlinear graph at a specific point is the _____ and can be represented by the slope of the tangent line to the graph at that point.

2. The process of evaluating an integral is called _____.

3. Limits of polynomial and rational functions can be found by _____, so long as the denominator of the rational function evaluated at c is not 0.

4. The function $F(x)$ is the _____ of $f(x)$.

5. Since it is not possible to determine the limit of a function with 0 in the denominator, it is customary to describe the resulting fraction $\frac{0}{0}$ as having a(n) _____.

6. To find the limits of rational functions at infinity, divide the numerator and denominator by the _____ power of x that occurs in the function.

7. The process of finding a derivative is called _____.

8. If a function is preceded by a(n) _____ $\frac{d}{dx}$, then you are to take the derivative of the function.

9. The velocity or speed achieved at any specific point in time is the _____.

10. The indefinite integral of $f(x)$ is defined by $\int f(x)\, dx = $ _____.

Lesson-by-Lesson Review

12-1 Estimating Limits Graphically (pp. 736–745)

Estimate each limit using a graph. Support your conjecture using a table of values.

11. $\lim\limits_{x \to 3} (2x - 7)$

12. $\lim\limits_{x \to 1} (0.5x^4 + 3x^2 - 5)$

Estimate each one-sided or two-sided limit, if it exists.

13. $\lim\limits_{x \to 2^+} \dfrac{x^2 + x - 6}{x - 2}$

14. $\lim\limits_{x \to 4} \dfrac{x^2 + x + 20}{x - 4}$

Estimate each limit, if it exists.

15. $\lim\limits_{x \to 4} \dfrac{9}{x^2 - 8x + 16}$

16. $\lim\limits_{x \to 2} \dfrac{x^2 - 7x - 10}{x - 2}$

EXAMPLE 1

Estimate $\lim\limits_{x \to 2} \dfrac{x^2 - 4}{x - 2}$ using a graph. Support your conjecture using a table of values.

Analyze Graphically The graph of $f(x) = \dfrac{x^2 - 4}{x - 2}$ suggests that as x gets closer to 2, the corresponding function value approaches 4. Therefore, we can estimate that $\lim\limits_{x \to 2} \dfrac{x^2 - 4}{x - 2}$ is 4.

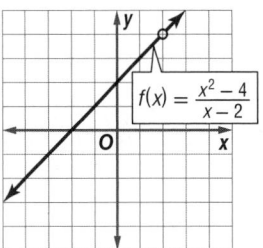

$f(x) = \dfrac{x^2 - 4}{x - 2}$

Support Numerically Make a table of values, choosing x-values that approach 2 from either side. ✔

	—— x approaches 2 ——→				←—— x approaches 2 ——		
x	1.9	1.99	1.999	2	2.001	2.01	2.1
f(x)	3.9	3.99	3.999		4.001	4.01	4.1

12-2 Evaluating Limits Algebraically (pp. 746–756)

Use the properties of limits to evaluate each limit.

17. $\lim\limits_{x \to 5} \dfrac{x^2 + 2x + 10}{x}$

18. $\lim\limits_{x \to -1} (5x^2 - 2x + 12)$

Use direct substitution, if possible, to evaluate each limit. If not possible, explain why not.

19. $\lim\limits_{x \to 25} \dfrac{x^2 + 1}{\sqrt{x} - 5}$

20. $\lim\limits_{x \to 2} (-3x^3 - 2x^2 + 15)$

Evaluate each limit.

21. $\lim\limits_{x \to -2} \dfrac{x + 2}{x^2 - 2x - 8}$

22. $\lim\limits_{x \to \infty} \left(2 - 4x^3 + x^2\right)$

EXAMPLE 2

Use direct substitution, if possible, to evaluate each limit. If not possible, explain why not.

a. $\lim\limits_{x \to 2} (2x^3 - x^2 + 4x + 1)$

This is the limit of a polynomial function. Therefore, direct substitution can be used to find the limit.

$$\lim\limits_{x \to 2} (2x^3 - x^2 + 4x + 1) = 2(2)^3 - 2^2 + 4(2) + 1$$
$$= 16 - 4 + 8 + 1 \text{ or } 21$$

b. $\lim\limits_{x \to -4} \dfrac{2x - 7}{2 - x^2}$

This is the limit of a rational function, the denominator of which is nonzero when $x = -4$. Therefore, direct substitution can be used to find the limit.

$$\lim\limits_{x \to -4} \dfrac{2x - 7}{2 - x^2} = \dfrac{2(-4) - 7}{2 - (-4)^2} = \dfrac{-8 - 7}{2 - 16} \text{ or } \dfrac{15}{14}$$

Lesson-by-Lesson Review

12-3 Tangent Lines and Velocity (pp. 758–764)

Find the slope of the lines tangent to the graph of each function at the given points.

23. $y = 6 - x$; $(-1, 7)$ and $(3, 3)$

24. $y = x^2 + 2$; $(0, 2)$ and $(-1, 3)$

The distance d an object is above the ground t seconds after it is dropped is given by $d(t)$. Find the instantaneous velocity of the object at the given value for t.

25. $y = -x^2 + 3x$

26. $y = x^3 + 4x$

Find the instantaneous velocity if the position of an object in feet is defined as $h(t)$ for given values of time t given in seconds.

27. $h(t) = 15t + 16t^2$; $t = 0.5$

28. $h(t) = -16t^2 - 35t + 400$; $t = 3.5$

Find an equation for the instantaneous velocity $v(t)$ if the path of an object is defined as $h(t)$ for any point in time t.

29. $h(t) = 12t^2 - 5$ **30.** $h(t) = 8 - 2t^2 + 3t$

EXAMPLE 3

Find the slope of the line tangent to the graph of $y = x^2$ at $(2, 4)$.

$$m = \lim_{h \to 0} \frac{f(x + h) - f(x)}{h} \qquad \text{Instantaneous Rate of Change Formula}$$

$$= \lim_{h \to 0} \frac{f(2 + h) - f(2)}{h} \qquad x = 2$$

$$= \lim_{h \to 0} \frac{(2 + h)^2 - 2^2}{h} \qquad f(2 + h) = (2 + h)^2 \text{ and } f(2) = 2^2$$

$$= \lim_{h \to 0} \frac{4 + 4h + h^2 - 4}{h} \qquad \text{Multiply.}$$

$$= \lim_{h \to 0} \frac{h(4 + h)}{h} \qquad \text{Simplify and factor.}$$

$$= \lim_{h \to 0} (4 + h) \qquad \text{Divide by } h.$$

$$= 4 + 0 \text{ or } 4 \qquad \text{Sum Property of Limits and Limits of Constant and Identity Functions}$$

Therefore, the slope of the line tangent to the graph of $y = x^2$ at $(2, 4)$ is 4.

12-4 Derivatives (pp. 766–774)

Evaluate limits to find the derivative of each function. Then evaluate the derivative of each function for the given values of each variable.

31. $g(t) = -t^2 + 5t + 11$; $t = -4$ and 1

32. $m(j) = 10j - 3$; $j = 5$ and -3

Find the derivative of each function.

33. $p(v) = -9v + 14$ **34.** $z(n) = 4n^2 + 9n$

35. $t(x) = -3 \sqrt[5]{x^6}$ **36.** $g(h) = 4h^{\frac{3}{4}} - 8h^{\frac{1}{2}} + 5$

Use the quotient rule to find the derivative of each function.

37. $f(m) = \dfrac{5 - 3m}{5 + 2m}$ **38.** $m(q) = \dfrac{2q^4 - q^2 + 9}{q^2 - 12}$

EXAMPLE 4

Find the derivative of $h(x) = \dfrac{x^2 - 5}{x^3 + 2}$.

Let $f(x) = x^2 - 5$, and $g(x) = x^3 + 2$. So, $h(x) = \dfrac{f(x)}{g(x)}$.

Find the derivatives of $f(x)$ and $g(x)$

$f(x) = x^2 - 5$ **Original equation**

$f'(x) = 2x$ **Power and Constant Rules**

$g(x) = x^3 + 2$ **Original equation**

$g'(x) = 3x^2$ **Power and Constant Rules**

Use $f(x)$, $f'(x)$, $g(x)$, and $g'(x)$ to find the derivative of $h(x)$.

$$h'(x) = \frac{f'(x)g(x) - f(x)g'(x)}{[g(x)]^2} \qquad \text{Quotient Rule}$$

$$= \frac{2x(x^3 + 2) - (x^2 - 5)\,3x^2}{(x^3 + 2)^2} \qquad \text{Substitution}$$

$$= \frac{-x^4 + 15x^2 + 4x}{(x^3 + 2)^2} \qquad \text{Simplify.}$$

Lesson-by-Lesson Review

12-5 **Area Under a Curve and Integration** (pp. 775–783)

Approximate the area of the shaded region for each function using right endpoints and 5 rectangles.

39.

$f(x) = \dfrac{8}{x}$

40.
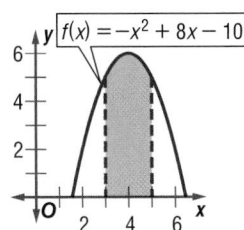
$f(x) = -x^2 + 8x - 10$

Use limits to find the area between the graph of each function and the x-axis given by the definite integral.

41. $\displaystyle\int_1^2 2x^2\, dx$

42. $\displaystyle\int_0^3 (2x^3 - 1)\, dx$

43. $\displaystyle\int_0^2 (x^2 + x)\, dx$

44. $\displaystyle\int_1^4 (3x^2 - x)\, dx$

EXAMPLE 5

Use limits to find the area of the region between the graph of $y = 2x^2$ and the x-axis on the interval $[0, 2]$, or $\displaystyle\int_0^2 2x^2\, dx$.

First, find Δx and x_i.

$\Delta x = \dfrac{b - a}{n}$ **Formula for Δx**

$\Delta x = \dfrac{2 - 0}{n}$ or $\dfrac{2}{n}$ $b = 2$ and $a = 0$

$x_i = 0 + i\dfrac{2}{n}$ or $\dfrac{2i}{n}$ $a = 0$ and $\Delta x = \dfrac{2}{n}$

$\displaystyle\int_0^2 2x^2\, dx = \lim_{n\to\infty} \sum_{i=1}^n 2\left(\dfrac{2i}{n}\right)^2\left(\dfrac{2}{n}\right)$ $x_i = \dfrac{2i}{n}$ and $\Delta x = \dfrac{2}{n}$

$= \displaystyle\lim_{n\to\infty} \dfrac{4}{n}\left(\sum_{i=1}^n \dfrac{4i^2}{n^2}\right)$ **Simplify.**

$= \displaystyle\lim_{n\to\infty} \dfrac{4}{n}\left(\dfrac{4}{n^2} \cdot \dfrac{n(n+1)(2n+1)}{6}\right)$ $\displaystyle\sum_{i=1}^n i^2 = \dfrac{n(n+1)(2n+1)}{6}$

$= \displaystyle\lim_{n\to\infty} \left(\dfrac{8(2n^2 + 3n + 1)}{3n^2}\right)$ **Multiply and divide by n.**

$= \displaystyle\lim_{n\to\infty} \left[\dfrac{8}{3} \cdot \left(2 + \dfrac{3}{n} + \dfrac{1}{n^2}\right)\right]$ **Factor and divide each term by n^2.**

$= \dfrac{16}{3}$ or $5\dfrac{1}{3}$ **Limit theorems and simplify.**

12-6 **The Fundamental Theorem of Calculus** (pp. 784–791)

Find all antiderivatives for each function.

45. $g(n) = 5n - 2$

46. $r(q) = -3q^2 + 9q - 2$

47. $m(t) = 6t^3 - 12t^2 + 2t - 11$

48. $p(h) = 7h^6 + 4h^5 - 12h^3 - 4$

Evaluate each integral.

49. $\displaystyle\int 8x^2\, dx$

50. $\displaystyle\int (2x^2 - 4)\, dx$

51. $\displaystyle\int_3^5 (2x^2 - 4 + 5x^3 + 3x^4)\, dx$

52. $\displaystyle\int_1^4 (-x^2 + 4x - 2x^3 + 5x^5)\, dx$

EXAMPLE 6

Find all antiderivatives for each function.

a. $f(x) = \dfrac{4}{x^5}$

$f(x) = 4x^{-5}$ **Rewrite with a negative exponent.**

$F(x) = \dfrac{4x^{-5+1}}{-5+1} + C$ **Constant Multiple of a Power**

$= -1x^{-4} + C$ or $-\dfrac{1}{x^4} + C$ **Simplify.**

b. $f(x) = x^2 - 7$

$f(x) = x^2 - 7$ **Original equation**

$= x^2 - 7x^0$ **Rewrite the function so each term has a power of x.**

$F(x) = \dfrac{x^{2+1}}{2+1} - \dfrac{7x^{0+1}}{0+1} + C$ **Antiderivative Rule**

$= \dfrac{1}{3}x^3 - 7x + C$ **Simplify.**

Applications and Problem Solving

53. STAMPS Suppose the value v of a stamp in dollars after t years can be represented as $v(t) = \dfrac{450}{5 + 25(0.4)^t}$. (Lesson 12-1)

a. Complete the following table.

Years	0	1	2	3
Value				

b. Graph the function for $0 \le t \le 10$.

c. Use the graph to estimate $\lim\limits_{t \to \infty} v(t)$, if it exists.

d. Explain the relationship between the limit of the function and the value of the stamp.

54. ANIMALS A wildlife conservation's population P in hundreds after t years can be estimated by

$P(t) = \dfrac{120}{1 + 24e^{-0.25t}}$. (Lesson 12-1)

a. Use a graphing calculator to graph the function for $0 \le t \le 50$.

b. Estimate $\lim\limits_{t \to \infty} \dfrac{120}{1 + 24e^{-0.25t}}$, if it exists.

c. Interpret your results from part **b**.

55. COLLECTORS The value of Eldin's coin collection has been increasing every year. The value v of the coins after t years can be represented by $v(t) = \dfrac{800t - 21}{4t + 19}$. (Lesson 12-2)

a. Find $\lim\limits_{t \to \infty} v(t)$.

b. Explain the relationship between the limit of the function and the value of Eldin's coin collection.

c. After 10 years, a coin dealer offers Eldin $300 for his collection. Should Eldin sell his collection? Explain your answer.

56. ROCKETS A rocket is launched into the sky with an upward velocity of 150 feet per second. Suppose the height d of the rocket in feet t seconds after it is launched is modeled by $d(t) = -16t^2 + 150t + 8.2$. (Lesson 12-3)

a. Find an equation for the instantaneous velocity $v(t)$ of the rocket.

b. How fast is the rocket traveling 1.5 seconds after it is launched?

c. For what value of t will the rocket reach its maximum height?

d. What is the maximum height that the rocket will reach?

57. ARCHERY An archer shoots an arrow with a velocity of 35 feet per second towards the target. Suppose the height s of the arrow in feet t seconds after it is shot is defined as $s(t) = -16t^2 + 35t + 1.5$. (Lesson 12-3)

a. Find an equation for the instantaneous velocity $v(t)$ of the arrow.

b. How fast is the arrow traveling 0.5 second after it is shot?

c. For what value of t will the arrow reach its maximum height?

d. What is the maximum height of the arrow?

58. DESIGN The owner of a ski resort is designing a new logo to put on his employees' uniforms. The design is in the shape of the region shown in the figure. If this part of the design is to be sewn on to the uniforms, how much material is required if x is given in inches? (Lesson 12-6)

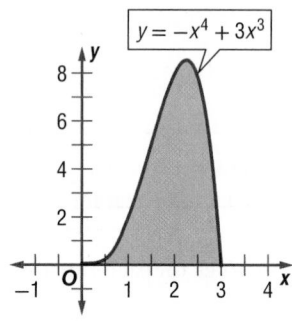

59. FROGS A frog can jump with a velocity modeled by $v(t) = -32t + 26$, where t is given in seconds and velocity is given in feet per second. (Lesson 12-6)

a. Find the position function $s(t)$ for the frog's jump. Assume that for $t = 0$, $s(t) = 0$.

b. How long will the frog stay in the air when it jumps?

60. BIRDS A cardinal in a tree 20 feet above the ground drops some food. The instantaneous velocity of his food can be defined as $v(t) = -32t$, where t is given in seconds and velocity is measured in feet per second. (Lesson 12-6)

a. Find the position function $s(t)$ of the dropped food.

b. Find how long it will take for the food to hit the ground.

Estimate each one-sided or two-sided limit, if it exists.

1. $\lim\limits_{x \to 0^+} \sqrt{x+4} - 8$

2. $\lim\limits_{x \to 4} \dfrac{x^2 - 16}{x - 4}$

Estimate each limit, if it exists.

3. $\lim\limits_{x \to 7} \dfrac{6}{x - 7}$

4. $\lim\limits_{x \to \infty} x^3 + 5x^2 - 2x + 21$

5. **ELECTRONICS** The average cost C in dollars of x number of personal digital assistants can be modeled by $C(x) = \dfrac{100x + 7105}{x}$.

   a. Determine the limit of the function as x approaches infinity.

   b. Interpret the results from part **a**.

Use direct substitution, if possible, to evaluate each limit. If not possible, explain why not.

6. $\lim\limits_{x \to 5} \dfrac{x^2}{\sqrt{x - 4} - 2}$

7. $\lim\limits_{x \to 9} (2x^3 - 12x + 3)$

8. **SCHOOL** The number of students S absent with the flu after t days at a school can be modeled by $S(t) = \dfrac{2000t^2 + 4}{1 + 50t^2}$.

   a. How many students are initially sick?

   b. How many students will eventually be sick?

Evaluate each limit.

9. $\lim\limits_{x \to \infty} (x^2 - 7x + 2)$

10. $\lim\limits_{x \to \infty} (2x^3 - 8x^2 - 5)$

11. $\lim\limits_{x \to \infty} \dfrac{2x^3 - x - 1}{-x^4 + 7x^3 + 4}$

12. $\lim\limits_{x \to \infty} \dfrac{\sqrt{25 + x} - 4}{x}$

13. **MULTIPLE CHOICE** Evaluate $\lim\limits_{x \to 0} \dfrac{\frac{1}{x+3} - \frac{1}{3}}{x}$.

   A $-\dfrac{1}{9}$ **C** $\dfrac{1}{9}$

   B 0 **D** does not exist

Find the slope of the lines tangent to the graph of each function at the given points.

14. $y = x^2 + 2x - 8$; $(-5, 7)$ and $(-2, -8)$

15. $y = \dfrac{4}{x^3} + 2$; $(-1, -2)$ and $\left(2, \dfrac{5}{2}\right)$

16. $y = (2x + 1)^2$; $(-3, 25)$ and $(0, 1)$

Find an equation for the instantaneous velocity $v(t)$ if the path of an object is defined as $h(t)$ for any point in time t.

17. $h(t) = 9t + 3t^2$

18. $h(t) = 10t^2 - 7t^3$

19. $h(t) = 3t^3 - 2 + 4t$

Find the derivative of each function.

20. $f(x) = -3x - 7$

21. $b(c) = 4c^{\frac{1}{2}} - 8c^{\frac{2}{3}} + 5c^{\frac{4}{5}}$

22. $w(y) = 3y^{\frac{4}{3}} + 6y^{\frac{1}{2}}$

23. $g(x) = (x^2 - 4)(2x - 5)$

24. $h(t) = \dfrac{t^3 + 4t^2 + t}{t^2}$

25. **FOOTBALL** The marginal cost c of producing x footballs is represented by $c(x) = 15 - 0.005x$.

   a. Determine the function that represents the actual cost function.

   b. Determine the cost of increasing the daily production from 1500 footballs to 2000 footballs.

Use limits to find the area between the graph of each function and the x-axis given by the definite integral.

26. $\displaystyle\int_1^4 (x^2 - 3x + 4)\, dx$

27. $\displaystyle\int_3^8 10x^4\, dx$

28. $\displaystyle\int_2^5 (7 - 2x + 4x^2)\, dx$

Find all antiderivatives for each function.

29. $d(a) = 4a^3 + 9a^2 - 2a + 8$

30. $w(z) = \dfrac{3}{4}z^4 + \dfrac{1}{6}z^2 - \dfrac{2}{5}$

Evaluate each integral.

31. $\displaystyle\int (5x^3 - 6x^2 + 4x - 3)\, dx$

32. $\displaystyle\int_1^4 (x^2 + 4x - 2)\, dx$

33. **AREA** Calculate the area bound by $f(x)$ and $g(x)$ on the interval $2 \le x \le 4$.

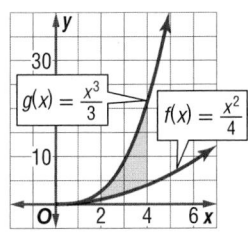

   F $17\dfrac{5}{12}$ **H** $15\dfrac{1}{3}$

   G $17\dfrac{1}{3}$ **J** 16

Objective
Differentiate composite functions using the Chain Rule.

Learning to differentiate polynomials by using the Power Rule led to the evaluation of definite integrals. This allowed for the calculation of the area found between a curve and the x-axis. However, our integration work was limited to basic polynomial functions. So, how can we calculate the areas found between the x-axis and the curves defined by composite functions such as $h(x) = (x^4 - 3)^2$ or $t(x) = \sqrt{x^3 - 4}$?

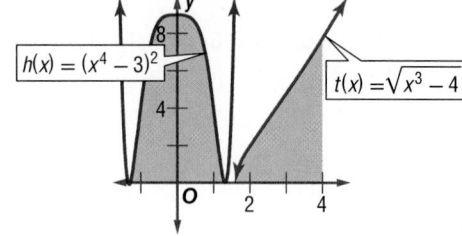

Just as in Lesson 12-4, you must learn how to differentiate these functions before you can integrate. Begin with $h(x)$. You can use the Product Rule to develop the derivative.

$h(x) = (x^4 - 3)^2$	**Original equation**
$= (x^4 - 3)(x^4 - 3)$	**Rewrite without the power.**
$h'(x) = 4x^3(x^4 - 3) + (x^4 - 3)4x^3$	**Product Rule**
$= 2(x^4 - 3)4x^3$	**Simplify.**

Although the derivative of $h(x)$ can be simplified further, leave it as shown in order to derive a rule for composite functions.

ACTIVITY 1 | **Derivative of a Composite Function**

Find the derivative of $k(x) = (x^4 - 3)^3$.

Step 1 Rewrite $k(x)$ to include the factor $(x^4 - 3)^2$.
$k(x) = (x^4 - 3)(x^4 - 3)^2$

Step 2 Let $m(x) = (x^4 - 3)$ and $n(x) = (x^4 - 3)^2$. Calculate the derivative of each function.
$m'(x) = 4x^3$ **Power Rule** $n'(x) = 2(x^4 - 3)4x^3$ **Derived above**

Step 3 Use the Product Rule to find $k'(x)$.
$k'(x) = m'(x)n(x) + m(x)n'(x)$
$\qquad = 4x^3(x^4 - 3)^2 + (x^4 - 3)2(x^4 - 3)4x^3$ **Substitution**
$\qquad = (x^4 - 3)^2 4x^3 + 2(x^4 - 3)^2 4x^3$ **Simplify.**
$\qquad = 3(x^4 - 3)^2 4x^3$ **Add.**

Therefore, $k'(x) = 3(x^4 - 3)^2 4x^3$.

CHECK You can use a graphing calculator to evaluate the derivative of a function at a point. Graph $k(x)$, and from the **CALC** menu, select **6:dy/dx**. After the screen returns to the graphing window, press 1 and then ENTER. The value of $k'(1) = 48$. Now verify the answer found in Step 3 by substituting $x = 1$ into $k'(x)$.
$k'(1) = 3(1^4 - 3)^2 4(1)^3$ or 48 ✔

[−5, 5] scl: 1 by [−20, 20] scl: 2

▶ Analyze the Results

1. Make a conjecture as to why $h'(x)$ and $k'(x)$ have $4x^3$ as a factor.

2. Make a conjecture as to why $h'(x)$ has 2 as a factor and $k'(x)$ has 3 as a factor.

3. Without rewriting $p(x)$ as a product, find the derivative of $p(x) = (x^4 - 3)^4$.

Notice that for the derivatives of $h(x)$ and $k(x)$, each expression was multiplied by its exponent, 1 was subtracted from its exponent, and then each expression was multiplied by the derivative of the expression inside the parentheses.

Multiply by exponent.
Subtract 1 from exponent.
$$h(x) = (x^4 - 3)^2$$
Multiply by derivative.

Multiply by exponent.
Subtract 1 from exponent.
$$k(x) = (x^4 - 3)^3$$
Multiply by derivative.

$$h'(x) = 2(x^4 - 3)^{2-1}4x^3$$
$$= 2(x^4 - 3)4x^3$$

$$k'(x) = 3(x^4 - 3)^{3-1}4x^3$$
$$= 3(x^4 - 3)^2 4x^3$$

Recall from Lesson 1-6 that we can decompose a function. For example, $h(x) = (x^4 - 3)^2$ can be written as $h(x) = f[g(x)]$, where $f(x) = x^2$ and $g(x) = (x^4 - 3)$. This can be used to develop the Chain Rule.

Key Concept   Chain Rule for Derivatives

Words   The derivative of a composite function $f[g(x)]$ is the derivative of the outer function f evaluated at the inner function g multiplied by the derivative of the inner function g.

Symbols   If g is differentiable at x and f is differentiable at $g(x)$, then
$$\frac{d}{dx} f[g(x)] = f'[g(x)]g'(x).$$

ACTIVITY 2   Derivative of a Composite Function

Find the derivative of $t(x) = \sqrt{x^3 - 4}$.

Step 1   Decompose $t(x)$ into two functions f and g.
$$f(x) = \sqrt{x} \qquad\qquad g(x) = x^3 - 4$$

Step 2   Find $f'(x)$ and $g'(x)$.

$f(x) = \sqrt{x}$ or $x^{\frac{1}{2}}$	**Original equation**	$g(x) = x^3 - 4$
$f'(x) = \frac{1}{2}x^{-\frac{1}{2}}$	**Power Rule**	$g'(x) = 3x^2$

Step 3   Substitute into the Chain Rule.

$$t'(x) = f'[g(x)]\, g'(x) \qquad \text{Chain Rule}$$

$$= \frac{1}{2}(x^3 - 4)^{-\frac{1}{2}} 3x^2 \qquad f'(x) = \frac{1}{2}x^{-\frac{1}{2}}, g(x) = x^3 - 4, \text{ and } g'(x) = 3x^2$$

$$= \frac{3x^2}{2\sqrt{x^3 - 4}} \qquad \text{Simplify.}$$

$$= \frac{3x^2\sqrt{x^3 - 4}}{2(x^3 - 4)} \qquad \text{Simplify.}$$

▶ Analyze the Results

4. Can $t'(x)$ be found using the same method that was used at the beginning of the lesson to find $h'(x)$? Explain your reasoning.

Model and Apply   Find the derivative of each function.

5. $s(t) = (t^2 - 1)^4$ **6.** $b(x) = (1 - 5x)^6$ **7.** $c(r) = (3r - 2r^2)^3$

8. $h(x) = (x^3 + x - 1)^3$ **9.** $f(x) = \sqrt{100 + 8x}$ **10.** $g(m) = \sqrt{m^2 + 4}$

How to Use the Student Handbook

The Student Handbook is the additional skill and reference material found at the end of the text. This Handbook can help you answer these questions.

The **Key Concepts** section lists all of the important concepts that were highlighted in your text along with the pages where they appear.

The answers to odd-numbered problems are included in **Selected Answers and Solutions**. Check your answers to make sure you understand how to solve all of the assigned problems.

The **English-Spanish Glossary** provides a list of important or difficult words used throughout the textbook. It provides a definition in English and Spanish as well as the page number(s) where the word can be found.

The **Index** alphabetically lists the subjects covered throughout the entire textbook and the pages on which each subject can be found.

Inside the back cover of your math book are several lists of **Formulas, Identities, and Symbols** that are used in the book.

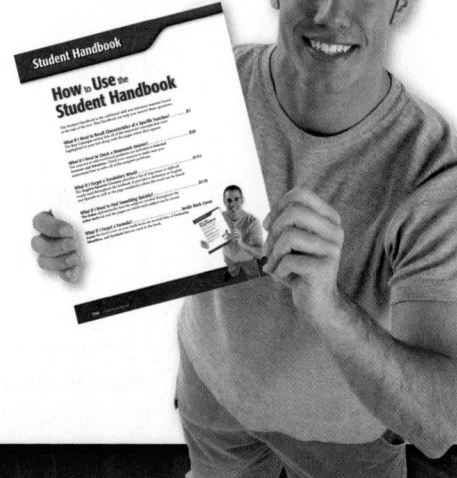

Key Concepts

Chapter 0

Powers of i (p. P6) To find the value of i^n, let R be the remainder when n is divided by 4.

$n < 0$	$n > 0$
$R = -3 \rightarrow i^n = i$	$R = 1 \rightarrow i^n = i$
$R = -2 \rightarrow i^n = -1$	$R = 2 \rightarrow i^n = -1$
$R = -1 \rightarrow i^n = -i$	$R = 3 \rightarrow i^n = -i$
$R = 0 \rightarrow i^n = 1$	$R = 0 \rightarrow i^n = 1$

Graph of a Quadratic Function (p. P9) Consider the graph of $y = ax^2 + bx + c$, where $a \neq 0$.

- The y-intercept is $a(0)^2 + b(0) + c$ or c.
- The equation of the axis of symmetry is $x = -\dfrac{b}{2a}$.
- The x-coordinate of the vertex is $-\dfrac{b}{2a}$.

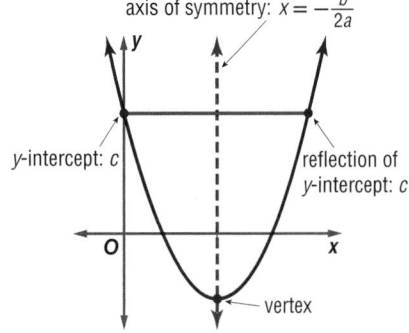

axis of symmetry: $x = -\dfrac{b}{2a}$

y-intercept: c

reflection of y-intercept: c

vertex

Maximum and Minimum Values (p. P10) The graph of $f(x) = ax^2 + bx + c$, where $a \neq 0$,
- opens up and has a minimum value when $a > 0$, and
- opens down and has a maximum value when $a < 0$.

Zero Product Property (p. P11) For any real numbers a and b, if $ab = 0$, then either $a = 0$, $b = 0$, or both a and b equal zero.

Completing the Square (p. P12) To complete the square for any quadratic expression of the form $x^2 + bx$, follow the steps below.

Step 1 Find one half of b, the coefficient of x.

Step 2 Square the result in Step 1.

Step 3 Add the result of Step 2 to $x^2 + bx$.

$$x^2 + bx + \left(\frac{b}{2}\right)^2 = \left(x + \frac{b}{2}\right)^2$$

Quadratic Formula (p. P12) The solutions of a quadratic equation of the form $ax^2 + bx + c = 0$, where $a \neq 0$, are given by the following formula.

$$x = \frac{-b \pm \sqrt{b^2 - 4ac}}{2a}$$

nth Root of a Number (p. P14) Let a and b be real numbers and let n be any positive integer greater than 1.

- If $a = b^n$, then b is an nth root of a.
- If a has an nth root, the principal nth root of a is the root having the same sign as a.

The principal nth root of a is denoted by the radical expression $\sqrt[n]{a}$, where n is the index of the radical and a is the radicand.

Basic Properties of Radicals (p. P15) Let a and b be real numbers, variables, or algebraic expressions, and m and n be positive integers greater than 1, where all of the roots are real numbers and all of the denominators are greater than 0. Then the following properties are true.

Product Property $\sqrt[n]{a} \cdot \sqrt[n]{b} = \sqrt[n]{ab}$

Quotient Property $\dfrac{\sqrt[n]{a}}{\sqrt[n]{b}} = \sqrt[n]{\dfrac{a}{b}}$

Rational Exponents (p. P16) If b is a real number, variable, or algebraic expression and m and n are positive integers greater than 1, then

- $b^{\frac{1}{n}} = \sqrt[n]{b}$, if the principal nth root of b exists, and
- $b^{\frac{m}{n}} = \left(\sqrt[n]{b}\right)^m$ or $\sqrt[n]{b^m}$, if $\frac{m}{n}$ is in reduced form.

Substitution Method (p. P18)
Step 1 Solve one equation for one of the variables in terms of the other.
Step 2 Substitute the expression found in Step 1 into the other equation to obtain an equation in one variable. Then solve the equation.
Step 3 Substitute the value found in Step 2 into the expression obtained in Step 1 to find the value of the remaining variable.

Elimination Method (p. P19)
Step 1 Multiply one or both equations by a number to result in two equations that contain opposite terms.
Step 2 Add the equations, eliminating one variable. Then solve the equation.
Step 3 Substitute to solve for the other variable.

Solving Systems of Inequalities (p. P21)
Step 1 Graph each inequality, shading the correct area. Use a solid line to graph inequalities that contain $\geq$ or $\leq$. Use a dashed line to graph inequalities that contain $>$ or $<$.
Step 2 Identify the region that is shaded for all of the inequalities. This is the solution of the system.
Step 3 Check the solution using a test point within the solution region.

Adding and Subtracting Matrices (p. P24) To add or subtract two matrices with the same dimensions, add or subtract their corresponding elements.

$$\begin{matrix} A & + & B & = & A + B \end{matrix}$$
$$\begin{bmatrix} a & b \\ c & d \end{bmatrix} + \begin{bmatrix} e & f \\ g & h \end{bmatrix} = \begin{bmatrix} a+e & b+f \\ c+g & d+h \end{bmatrix}$$

$$\begin{matrix} A & - & B & = & A - B \end{matrix}$$
$$\begin{bmatrix} a & b \\ c & d \end{bmatrix} - \begin{bmatrix} e & f \\ g & h \end{bmatrix} = \begin{bmatrix} a-e & b-f \\ c-g & d-h \end{bmatrix}$$

Properties of Matrix Operations (p. P25)
For any matrices A, B, and C for which the matrix sum and product are defined and any scalar k, the following properties are true.

Commutative Property of Addition $A + B = B + A$
Associative Property of Addition $(A + B) + C = A + (B + C)$
Left Scalar Distributive Property $k(A + B) = kA + kB$
Right Scalar Distributive Property $(A + B)k = kA + kB$

Fundamental Counting Principle (p. P28) Let A and B be two events. If event A has n_1 possible outcomes and is followed by event B that has n_2 possible outcomes, then event A followed by event B has $n_1 \cdot n_2$ possible outcomes.

Permutations (p. P29)

The number of permutations of n objects taken n at a time is

$$_nP_n = n!.$$

The number of permutations of n objects taken r at a time is

$$_nP_r = \frac{n!}{(n-r)!}.$$

Combinations (p. P30) The number of combinations of n objects taken r at a time is

$$_nC_r = \frac{n!}{(n-r)!\, r!}.$$

Measures of Central Tendency (p. P32)
Mean   The sum of the numbers in a set of data divided by the number of items.

Population

$$\mu = \frac{\Sigma x}{n}$$

Sample

$$\bar{x} = \frac{\Sigma x}{n}$$

Median The middle number in a set of data when the data are arranged in numerical order.
Mode   The number or numbers that appear most often in a set of data.

Measures of Spread (p. P32)

Range   The difference between the greatest and least values in a set of data.

Variance   The mean of the squares of the deviations from the mean.

Population Variance	**Sample Variance**
$\sigma^2 = \dfrac{\Sigma(x - \mu)^2}{n}$	$s^2 = \dfrac{\Sigma(x - \bar{x})^2}{n - 1}$

Standard Deviation   The average amount by which individual items deviate from the mean of all the data.

Population Standard Deviation	**Sample Standard Deviation**
$\sigma = \sqrt{\dfrac{\Sigma(x - \mu)^2}{n}}$	$s = \sqrt{\dfrac{\Sigma(x - \bar{x})^2}{n - 1}}$

Chapter 1

Functions from a Calculus Perspective

Real Numbers (p. 4)

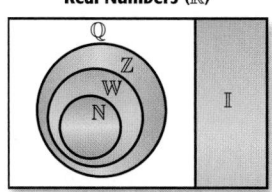

Real Numbers ($\mathbb{R}$)

Letter	Set	Examples
$\mathbb{Q}$	rationals	$0.125, -\frac{7}{8}, \frac{2}{3} = 0.666\ldots$
$\mathbb{I}$	irrationals	$\sqrt{3} = 1.73205\ldots$
$\mathbb{Z}$	integers	$-5, 17, -23, 8$
$\mathbb{W}$	wholes	$0, 1, 2, 3\ldots$
$\mathbb{N}$	naturals	$1, 2, 3, 4\ldots$

Function (p. 5) A function f from set A to set B is a relation that assigns to each element x in set A *exactly one* element y in set B.

Vertical Line Test (p. 6) A set of points in the coordinate plane is the graph of a function if each possible vertical line intersects the graph in at most one point.

Tests for Symmetry (p. 16)

Graphical Test	Model	Algebraic Test
The graph of a relation is *symmetric with respect to the x-axis* if and only if for every point (x, y) on the graph, the point $(x, -y)$ is also on the graph.		Replacing y with $-y$ produces an equivalent equation.
The graph of a relation is *symmetric with respect to the y-axis* if and only if for every point (x, y) on the graph, the point $(-x, y)$ is also on the graph.		Replacing x with $-x$ produces an equivalent equation.
The graph of a relation is *symmetric with respect to the origin* if and only if for every point (x, y) on the graph, the point $(-x, -y)$ is also on the graph.		Replacing x with $-x$ and y with $-y$ produces an equivalent equation.

Even and Odd Functions (p. 18)

Type of Function	Algebraic Test
Functions that are symmetric with respect to the y-axis are called even functions.	For every x in the domain of f, $f(-x) = f(x)$.
Functions that are symmetric with respect to the origin are called odd functions.	For every x in the domain of f, $f(-x) = -f(x)$.

Limits (p. 24) If the value of $f(x)$ approaches a unique value L as x approaches c from each side, then the limit of $f(x)$ as x approaches c is L.

Types of Discontinuity (p. 24)

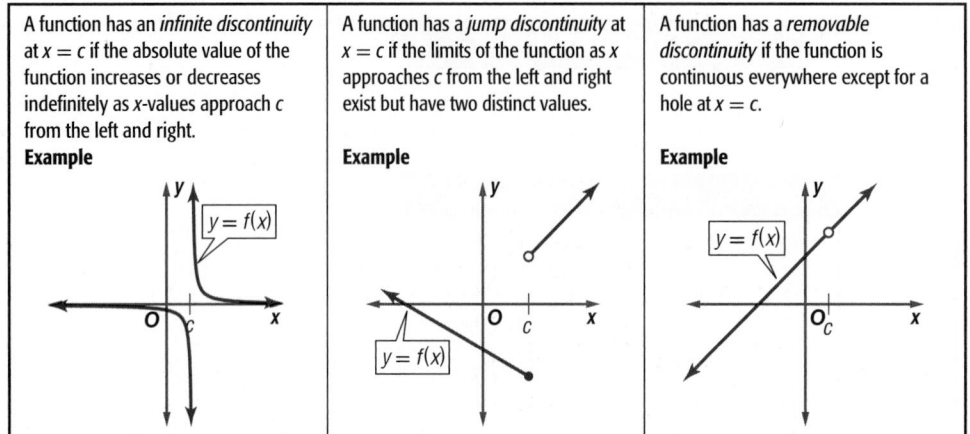

A function has an *infinite discontinuity* at $x = c$ if the absolute value of the function increases or decreases indefinitely as x-values approach c from the left and right. **Example**	A function has a *jump discontinuity* at $x = c$ if the limits of the function as x approaches c from the left and right exist but have two distinct values. **Example**	A function has a *removable discontinuity* if the function is continuous everywhere except for a hole at $x = c$. **Example**

Continuity Test (p. 25) A function $f(x)$ is continuous at $x = c$ if it satisfies the following conditions.

- $f(x)$ is defined at c. That is, $f(c)$ exists.
- $f(x)$ approaches the same value from either side of c. That is, $\lim_{x \to c} f(x)$ exists.
- The value that $f(x)$ approaches from each side of c is $f(c)$. That is, $\lim_{x \to c} f(x) = f(c)$.

Intermediate Value Theorem (p. 27) If $f(x)$ is a continuous function and $a < b$ and there is a value n such that n is between $f(a)$ and $f(b)$, then there is a number c, such that $a < c < b$ and $f(c) = n$.

Increasing, Decreasing, and Constant Functions (p. 34)
- A function f is *increasing* on an interval I if and only if for any two points in I, a positive change in x results in a positive change in $f(x)$.
- A function f is *decreasing* on an interval I if and only if for any two points in I, a positive change in x results in a negative change in $f(x)$.
- A function f is *constant* on an interval I if and only if for any two points in I, a positive change in x results in a zero change in $f(x)$.

Relative and Absolute Extrema (p. 36)
- A *relative maximum* of a function f is the greatest value $f(x)$ can attain on some interval of the domain. If a relative maximum is the greatest value a function f can attain over its entire domain, then it is the *absolute maximum*.
- A *relative minimum* of a function f is the least value $f(x)$ can attain on some interval of the domain. If a relative minimum is the least value a function f can attain over its entire domain, then it is the *absolute minimum*.

Average Rate of Change (p. 38) The average rate of change between any two points on the graph of f is the slope of the line through those points. The line through two points on a curve is called a secant line. The slope of the secant line is denoted m_{sec}. The average rate of change on the interval $[x_1, x_2]$ is

$$m_{\text{sec}} = \frac{f(x_2) - f(x_1)}{x_2 - x_1}.$$

Linear and Polynomial Parent Functions (p. 45)

A *constant function* has the form $f(x) = c$, where c is any real number. Its graph is a horizontal line. When $c = 0$, $f(x)$ is the *zero function*. 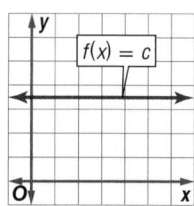	The *identity function* $f(x) = x$ passes through all points with coordinates (a, a).
The *quadratic function* $f(x) = x^2$ has a U-shaped graph. 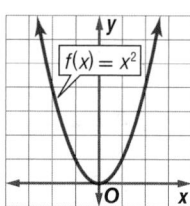	The *cubic function* $f(x) = x^3$ is symmetric about the origin.

Square Root and Reciprocal Parent Functions (p. 45)

The *square root function* has the form $f(x) = \sqrt{x}$. 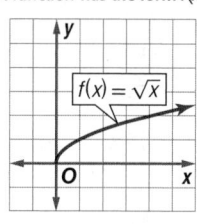	The *reciprocal function* has the form $f(x) = \frac{1}{x}$. 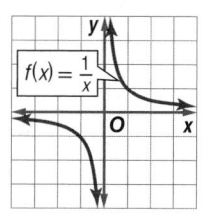

Absolute Value and Greatest Integer Functions (p. 46)

The *absolute value function*, denoted $f(x) = \lvert x \rvert$, is a V-shaped function defined as $$f(x) = \begin{cases} -x & \text{if } x < 0 \\ x & \text{if } x \geq 0 \end{cases}$$ 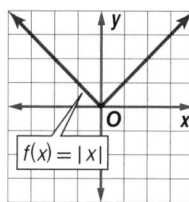	The *greatest integer function*, denoted $f(x) = [\![x]\!]$, is defined as the greatest integer less than or equal to x. 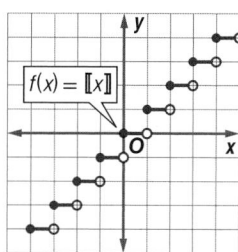

Vertical and Horizontal Translations (p. 47)

- **Vertical Translations:** The graph of $g(x) = f(x) + k$ is the graph of $f(x)$ translated k units up when $k > 0$, and k units down when $k < 0$.
- **Horizontal Translations:** The graph of $g(x) = f(x - h)$ is the graph of $f(x)$ translated h units right when $h > 0$, and h units left when $h < 0$.

Reflections in the Coordinate Axes (p. 48)

- **Reflection in x-axis:** $g(x) = -f(x)$ is the graph of $f(x)$ reflected in the x-axis.
- **Reflection in y-axis:** $g(x) = f(-x)$ is the graph of $f(x)$ reflected in the y-axis.

Vertical and Horizontal Dilations (p. 49)
- **Vertical Dilations:** If a is a positive real number, then $g(x) = a \cdot f(x)$, is the graph of $f(x)$ expanded vertically, if $a > 1$; the graph of $f(x)$ compressed vertically, if $0 < a < 1$.
- **Horizontal Dilations:** If a is a positive real number, then $g(x) = f(ax)$, is the graph of $f(x)$ compressed horizontally, if $a > 1$; the graph of $f(x)$ expanded horizontally, if $0 < a < 1$.

Transformations with Absolute Value (p. 51)

$g(x) = \lvert f(x) \rvert$	$g(x) = f(\lvert x \rvert)$
This transformation reflects any portion of the graph of $f(x)$ that is below the x-axis so that it is above the x-axis.	This transformation results in the portion of the graph of $f(x)$ that is to the left of the y-axis being replaced by a reflection of the portion to the right of the y-axis.
	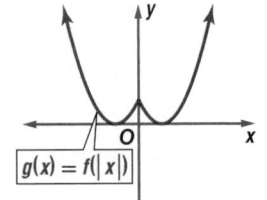

Operations with Functions (p. 57) Let f and g be two functions with intersecting domains. Then for all x-values in the intersection, the sum, product, difference, and quotient of f and g are new functions defined as follows.

Sum $\qquad (f + g)(x) = f(x) + g(x)$

Difference $\quad (f - g)(x) = f(x) - g(x)$

Product $\qquad (f \cdot g)(x) = f(x) \cdot g(x)$

Quotient $\qquad \left(\dfrac{f}{g}\right)(x) = \dfrac{f(x)}{g(x)}, \; g(x) \neq 0$

Composition of Functions (p. 58) The composition of function f with function g is defined by $[f \circ g](x) = f[g(x)]$. The domain of $f \circ g$ includes all x-values in the domain of g that map to $g(x)$-values in the domain of f.

Horizontal Line Test (p. 65) A function f has an inverse function f^{-1} if and only if each horizontal line intersects the graph of the function in at most one point.

Finding an Inverse Function (p. 66)
- **Step 1** Determine whether the function has an inverse by checking to see if it is one-to-one using the horizontal line test.
- **Step 2** In the equation for $f(x)$, replace $f(x)$ with y and then interchange x and y.
- **Step 3** Solve for y and then replace y with $f^{-1}(x)$ in the new equation.
- **Step 4** State any restrictions on the domain of f^{-1}. Then show that the domain of f is equal to the range of f^{-1} and the range of f is equal the domain of f^{-1}.

Compositions of Inverse Functions (p. 68) Two functions, f and g, are inverse functions if and only if
- $f[g(x)] = x$ for every x in the domain of $g(x)$, and
- $g[f(x)] = x$ for every x in the domain of $f(x)$.

Power Function (p. 86) A power function is any function of the form $f(x) = ax^n$ where a and n are nonzero constant real numbers.

Monomial Functions (p. 86) Let f be the power function $f(x) = ax^n$, where n is a positive integer.

n Even, a Positive	**n Even, a Negative**
	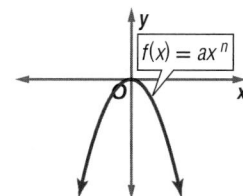
Domain: $(-\infty, \infty)$ **Range:** $[0, \infty)$	**Domain:** $(-\infty, \infty)$ **Range:** $(-\infty, 0]$
x- and y-intercept: 0	**x- and y-intercept:** 0
Continuity: continuous for $x \in \mathbb{R}$	**Continuity:** continuous for $x \in \mathbb{R}$
Symmetry: y-axis **Minimum:** (0, 0)	**Symmetry:** y-axis **Maximum:** (0, 0)
Decreasing: $(-\infty, 0)$ **Increasing:** $(0, \infty)$	**Decreasing:** $(0, \infty)$ **Increasing:** $(-\infty, 0)$
End behavior: $\lim\limits_{x \to -\infty} f(x) = \infty$ and $\lim\limits_{x \to \infty} f(x) = \infty$	**End behavior:** $\lim\limits_{x \to -\infty} f(x) = -\infty$ and $\lim\limits_{x \to \infty} f(x) = -\infty$
n Odd, a Positive	**n Odd, a Negative**
	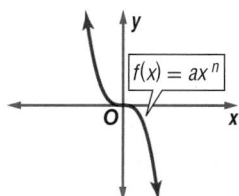
Domain and Range: $(-\infty, \infty)$	**Domain and Range:** $(-\infty, \infty)$
x- and y-intercept: 0	**x- and y-intercept:** 0
Continuity: continuous on $(-\infty, \infty)$	**Continuity:** continuous for $x \in \mathbb{R}$
Symmetry: origin	**Symmetry:** origin
Extrema: none **Increasing:** $(-\infty, \infty)$	**Extrema:** none **Decreasing:** $(-\infty, \infty)$
End Behavior: $\lim\limits_{x \to -\infty} f(x) = -\infty$ and $\lim\limits_{x \to \infty} f(x) = \infty$	**End Behavior:** $\lim\limits_{x \to -\infty} f(x) = \infty$ and $\lim\limits_{x \to \infty} f(x) = -\infty$

Radical Functions (p. 90) Let f be the radical function $f(x) = \sqrt[n]{x}$ where n is a positive integer.

n Even	**n Odd**
	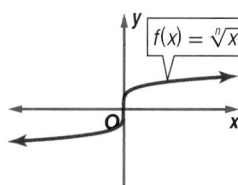
Domain and Range: $[0, \infty)$	**Domain and Range:** $(-\infty, \infty)$
x- and y-intercept: 0	**x- and y-intercept:** 0
Continuity: continuous on $[0, \infty)$	**Continuity:** continuous on $(-\infty, \infty)$
Symmetry: none **Increasing:** $(0, \infty)$	**Symmetry:** origin **Increasing:** $(-\infty, \infty)$
Extrema: absolute minimum at (0, 0)	**Extrema:** none
End Behavior: $\lim\limits_{x \to \infty} f(x) = \infty$	**End Behavior:** $\lim\limits_{x \to -\infty} f(x) = -\infty$ and $\lim\limits_{x \to \infty} f(x) = \infty$

Leading Term Test for Polynomial End Behavior (p. 98)

The end behavior of any polynomial function $f(x) = a_nx^n + \dots + a_1x + a_0$ can be described in one of the following four ways.

Zeros and Turning Points of Polynomial Functions (p. 100) A polynomial function f of degree $n \geq 1$ has at most n distinct real zeros and at most $n - 1$ turning points.

Quadratic Form (p. 100) A polynomial expression in x is in quadratic form if it is written as $au^2 + bu + c$ for any numbers a, b, and c, $a \neq 0$, where u is some expression in x.

Repeated Zeros of Polynomial Functions (p. 102) If $(x - c)^m$ is the highest power of $(x - c)$ that is a factor of polynomial function f, then c is a zero of multiplicity m of f, where m is a natural number.

- If a zero c has odd multiplicity, then the graph of f crosses the x-axis at $x = c$ and the value of $f(x)$ changes signs at $x = c$.
- If a zero c has even multiplicity, then the graph of f is tangent to the x-axis at $x = c$ and the value of $f(x)$ does not change signs at $x = c$.

Polynomial Division (p. 110) Let $f(x)$ and $d(x)$ be polynomials such that the degree of $d(x)$ is less than or equal to the degree of $f(x)$ and $d(x) \neq 0$. Then there exist unique polynomials $q(x)$ and $r(x)$ such that $\dfrac{f(x)}{d(x)} = q(x) + \dfrac{r(x)}{d(x)}$ or $f(x) = d(x) \cdot q(x) + r(x)$, where $r(x) = 0$ or the degree of $r(x)$ is less than the degree of $d(x)$. If $r(x) = 0$, then $d(x)$ *divides evenly* into $f(x)$.

Synthetic Division Algorithm (p. 111) To divide a polynomial by the factor $x - c$, complete each step.

Step 1 Write the coefficients of the dividend in standard form. Write the related zero c of the divisor $x - c$ in the box. Bring down the first coefficient.

Step 2 Multiply the first coefficient by c. Write the product under the second coefficient.

Step 3 Add the product and the second coefficient.

Step 4 Repeat Steps 2 and 3 until you reach a sum in the last column. The numbers along the bottom row are the coefficients of the quotient. The power of the first term is one less than the degree of the dividend. The final number is the remainder.

Example

Divide $6x^3 - 25x^2 + 18x + 9$ by $x - 3$.

Remainder Theorem (p. 112) If a polynomial $f(x)$ is divided by $x - c$, the remainder is $r = f(c)$.

Factor Theorem (p. 113) A polynomial $f(x)$ has a factor $(x - c)$ if and only if $f(c) = 0$.

Rational Zero Theorem (p. 119) If f is a polynomial function of the form $f(x) = a_n x^n + a_{n-1} x^{n-1} + \ldots + a_2 x^2 + a_1 x + a_0$, with degree $n \geq 1$, integer coefficients, and $a_0 \neq 0$, then every rational zero of f has the form $\frac{p}{q}$, where
- p and q have no common factors other than ± 1,
- p is an integer factor of the constant term a_0, and
- q is an integer factor of the leading coefficient a_n.

Upper and Lower Bound Tests (p. 121) Let f be a polynomial function of degree $n \geq 1$, real coefficients, and a positive leading coefficient. Suppose $f(x)$ is divided by $x - c$ using synthetic division.
- If $c \leq 0$ and every number in the last line of the division is alternately nonnegative and nonpositive, then c is a *lower bound* for the real zeros of f.
- If $c \geq 0$ and every number in the last line of the division is nonnegative, then c is an *upper bound* for the real zeros of f.

Descartes' Rule of Signs (p. 123) If $f(x) = a_n x^n + a_{n-1} x^{n-1} + \ldots + a_1 x + a_0$ is a polynomial function with real coefficients, then
- the number of *positive* real zeros of f is equal to the number of variations in sign of $f(x)$ or less than that number by some even number, and
- the number of *negative* real zeros of f is the same as the number of variations in sign of $f(-x)$ or less than that number by some even number.

Fundamental Theorem of Algebra (p. 123) A polynomial function of degree n, where $n > 0$, has at least one zero (real or imaginary) in the complex number system.

Linear Factorization Theorem (p. 124) If $f(x)$ is a polynomial function of degree $n > 0$, then f has exactly n linear factors and

$$f(x) = a_n(x - c_1)(x - c_2) \cdots (x - c_n)$$

where a_n is some nonzero real number and $c_1, c_2, \ldots, c_n$ are the complex zeros (including repeated zeros) of f.

Factoring Polynomial Functions Over the Reals (p. 124) Every polynomial function of degree $n > 0$ with real coefficients can be written as the product of linear factors and irreducible quadratic factors, each with real coefficients.

Vertical and Horizontal Asymptotes (p. 131)

The line $x = c$ is a *vertical asymptote* of the graph of f if $\lim\limits_{x \to c^-} f(x) = \pm\infty$ or $\lim\limits_{x \to c^+} f(x) = \pm\infty$.

The line $y = c$ is a *horizontal asymptote* of the graph of f if $\lim\limits_{x \to -\infty} f(x) = c$ or $\lim\limits_{x \to \infty} f(x) = c$.

Graphs of Rational Functions (p. 132) If f is the rational function given by

$$f(x) = \frac{a(x)}{b(x)} = \frac{a_n x^n + a_{n-1} x^{n-1} + \ldots + a_1 x + a_0}{b_m x^m + b_{m-1} x^{m-1} + \ldots + b_1 x + b_0},$$

where $b(x) \neq 0$ and $a(x)$ and $b(x)$ have no common factors other than ± 1, then the graph of f has the following characteristics.

Vertical Asymptotes Vertical asymptotes may occur at the real zeros of $b(x)$.
Horizontal Asymptote The graph has either one or no horizontal asymptotes as determined by comparing the degree n of $a(x)$ to the degree m of $b(x)$.
- If $n < m$, the horizontal asymptote is $y = 0$.
- If $n = m$, the horizontal asymptote is $y = \frac{a_n}{b_m}$.
- If $n > m$, there is no horizontal asymptote.

Intercepts The x-intercepts, if any, occur at the real zeros of $a(x)$. The y-intercept, if it exists, is the value of f when $x = 0$.

Oblique Asymptotes (p. 134) If f is the rational function given by

$$f(x) = \frac{a(x)}{b(x)} = \frac{a_n x^n + a_{n-1} x^{n-1} + \ldots + a_1 x + a_0}{b_m x^m + b_{m-1} x^{m-1} + \ldots + b_1 x + b_0},$$

where $b(x)$ has a degree greater than 0 and $a(x)$ and $b(x)$ have no common factors other than 1, then the graph of f has an oblique asymptote if $n = m + 1$. The function for the oblique asymptote is the quotient polynomial $q(x)$ resulting from the division of $a(x)$ by $b(x)$.

Exponential and Logarithmic Functions

Exponential Function (p. 158) An exponential function with base b has the form $f(x) = ab^x$, where x is any real number and a and b are real number constants such that $a \neq 0$, b is positive, and $b \neq 1$.

Properties of Exponential Functions (p. 159)

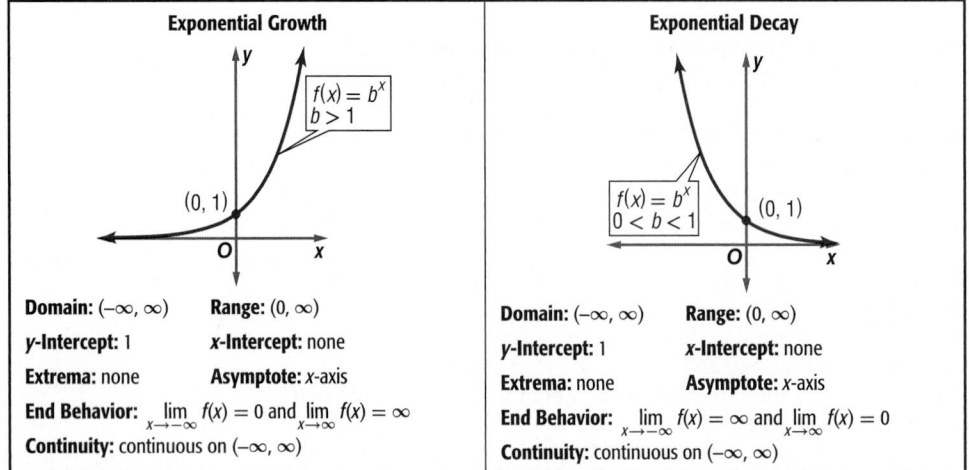

Exponential Growth

$f(x) = b^x$
$b > 1$

$(0, 1)$

Domain: $(-\infty, \infty)$ **Range:** $(0, \infty)$

y-Intercept: 1 **x-Intercept:** none

Extrema: none **Asymptote:** x-axis

End Behavior: $\lim\limits_{x \to -\infty} f(x) = 0$ and $\lim\limits_{x \to \infty} f(x) = \infty$

Continuity: continuous on $(-\infty, \infty)$

Exponential Decay

$f(x) = b^x$
$0 < b < 1$

$(0, 1)$

Domain: $(-\infty, \infty)$ **Range:** $(0, \infty)$

y-Intercept: 1 **x-Intercept:** none

Extrema: none **Asymptote:** x-axis

End Behavior: $\lim\limits_{x \to -\infty} f(x) = \infty$ and $\lim\limits_{x \to \infty} f(x) = 0$

Continuity: continuous on $(-\infty, \infty)$

Compound Interest Formula (p. 162) If a principal P is invested at an annual interest rate r (in decimal form) compounded n times a year, then the balance A in the account after t years is given by

$$A = P\left(1 + \frac{r}{n}\right)^{nt}.$$

Continuous Compound Interest Formula (p. 163) If a principal P is invested at an annual interest rate r (in decimal form) compounded continuously, then the balance A in the account after t years is given by

$$A = Pe^{rt}.$$

Exponential Growth or Decay Formulas (p. 164)

If an initial quantity N_0 grows or decays at an exponential rate r or k (as a decimal), then the final amount N after a time t is given by the following formulas.

Exponential Growth or Decay	**Continuous Exponential Growth or Decay**
$N = N_0(1 + r)^t$	$N = N_0 e^{kt}$
If r is a *growth rate*, then $r > 0$.	If k is a *continuous growth rate*, then $k > 0$.
If r is a *decay rate*, then $r < 0$.	If k is a *continuous decay rate*, then $k < 0$.

Relating Logarithmic and Exponential Forms (p. 172) If $b > 0$, $b \neq 1$, and $x > 0$, then

Logarithmic Form		**Exponential Form**
$\log_b x = y$	if and only if	$b^y = x$.
base exponent		base exponent

Basic Properties of Logarithms (p. 173) If $b > 0$, $b \neq 1$, and x is a real number, then the following statements are true.

- $\log_b 1 = 0$
- $\log_b b = 1$

- $\log_b b^x = x$ ⎫
- $b^{\log_b x} = x, x > 0$ ⎬ **Inverse Properties**

Basic Properties of Common Logarithms (p. 173) If x is a real number, then the following statements are true.

- $\log 1 = 0$
- $\log 10 = 1$

- $\log 10^x = x$ ⎫
- $\log 10^{\log x} = x, x > 0$ ⎬ **Inverse Properties**

Basic Properties of Natural Logarithms (p. 174) If x is a real number, then the following statements are true.

- $\ln 1 = 0$
- $\ln e = 1$

- $\ln e^x = x$
- $e^{\ln x} = x, x > 0$ } **Inverse Properties**

Properties of Logarithmic Functions (p. 176)

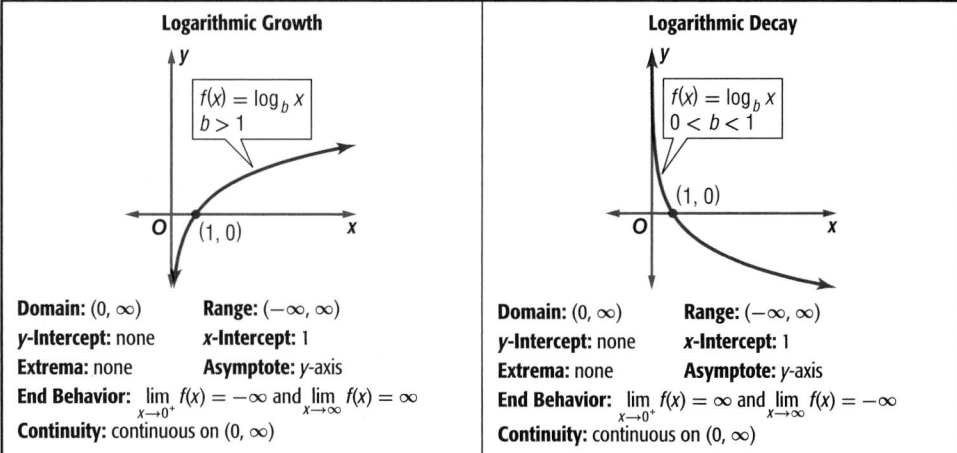

Logarithmic Growth

$f(x) = \log_b x$
$b > 1$

$(1, 0)$

Domain: $(0, \infty)$ **Range:** $(-\infty, \infty)$
y-Intercept: none **x-Intercept:** 1
Extrema: none **Asymptote:** y-axis
End Behavior: $\lim\limits_{x \to 0^+} f(x) = -\infty$ and $\lim\limits_{x \to \infty} f(x) = \infty$
Continuity: continuous on $(0, \infty)$

Logarithmic Decay

$f(x) = \log_b x$
$0 < b < 1$

$(1, 0)$

Domain: $(0, \infty)$ **Range:** $(-\infty, \infty)$
y-Intercept: none **x-Intercept:** 1
Extrema: none **Asymptote:** y-axis
End Behavior: $\lim\limits_{x \to 0^+} f(x) = \infty$ and $\lim\limits_{x \to \infty} f(x) = -\infty$
Continuity: continuous on $(0, \infty)$

Properties of Logarithms (p. 181) If b, x, and y are positive real numbers, $b \neq 1$, and p is a real number, then the following statements are true.

Product Property $\qquad \log_b xy = \log_b x + \log_b y$
Quotient Property $\qquad \log_b \frac{x}{y} = \log_b x - \log_b y$
Power Property $\qquad \log_b x^p = p \log_b x$

Change of Base Formula (p. 183) For any positive real numbers a, b, and x, $a \neq 1$, $b \neq 1$,

$$\log_b x = \frac{\log_a x}{\log_a b}.$$

One-to-One Property of Exponential Functions (p. 190) For $b > 0$ and $b \neq 1$, $b^x = b^y$ if and only if $x = y$.

One-to-One Property of Logarithmic Functions (p. 191) For $b > 0$ and $b \neq 1$, $\log_b x = \log_b y$ if and only if $x = y$.

Logistic Growth Function (p. 202) A logistic growth function has the form $f(t) = \dfrac{c}{1 + ae^{-bt}}$,

where t is any real number, a, b, and c are positive constants, and c is the limit to growth.

Transformations for Linearizing Data (p. 204) To linearize data modeled by:

- a quadratic function $y = ax^2 + bx + c$, graph $\left(x, \sqrt{y}\right)$.
- an exponential function $y = ab^x$, graph $(x, \ln y)$.
- a logarithmic function $y = a \ln x + b$, graph $(\ln x, y)$.
- a power function $y = ax^b$, graph $(\ln x, \ln y)$.

Trigonometric Functions **Chapter 4**

Trigonometric Functions (p. 220)

Let θ be an acute angle in a right triangle and the abbreviations opp, adj, and hyp refer to the length of the side opposite θ, the length of the side adjacent to θ, and the length of the hypotenuse, respectively. Then the six trigonometric functions of θ are defined as follows.

$\text{sine } (\theta) = \sin \theta = \dfrac{\text{opp}}{\text{hyp}}$ $\qquad$ $\text{cosecant } (\theta) = \csc \theta = \dfrac{\text{hyp}}{\text{opp}}$

$\text{cosine } (\theta) = \cos \theta = \dfrac{\text{adj}}{\text{hyp}}$ $\qquad$ $\text{secant } (\theta) = \sec \theta = \dfrac{\text{hyp}}{\text{adj}}$

$\text{tangent } (\theta) = \tan \theta = \dfrac{\text{opp}}{\text{adj}}$ $\qquad$ $\text{cotangent } (\theta) = \cot \theta = \dfrac{\text{adj}}{\text{opp}}$

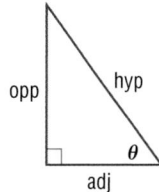

opp / hyp / adj / θ

Trigonometric Values of Special Angles (p. 222)

30°-60°-90° Triangle

45°-45°-90° Triangle

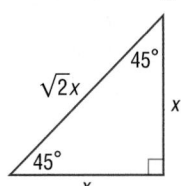

θ	30°	45°	60°
$\sin \theta$	$\frac{1}{2}$	$\frac{\sqrt{2}}{2}$	$\frac{\sqrt{3}}{2}$
$\cos \theta$	$\frac{\sqrt{3}}{2}$	$\frac{\sqrt{2}}{2}$	$\frac{1}{2}$
$\tan \theta$	$\frac{\sqrt{3}}{3}$	1	$\sqrt{3}$
$\csc \theta$	2	$\sqrt{2}$	$\frac{2\sqrt{3}}{3}$
$\sec \theta$	$\frac{2\sqrt{3}}{3}$	$\sqrt{2}$	2
$\cot \theta$	$\sqrt{3}$	1	$\frac{\sqrt{3}}{3}$

Inverse Trigonometric Functions (p. 223)

Inverse Sine If θ is an acute angle and the sine of θ is x, then the *inverse sine* of x is the measure of angle θ. That is, if $\sin \theta = x$, then $\sin^{-1} x = \theta$.

Inverse Cosine If θ is an acute angle and the cosine of θ is x, then the *inverse cosine* of x is the measure of angle θ. That is, if $\cos \theta = x$, then $\cos^{-1} x = \theta$.

Inverse Tangent If θ is an acute angle and the tangent of θ is x, then the *inverse tangent* of x is the measure of angle θ. That is, if $\tan \theta = x$, then $\tan^{-1} x = \theta$.

Radian Measure (p. 232) The measure θ in radians of a central angle of a circle is equal to the ratio of the length of the intercepted arc s to the radius r of the circle. In symbols, $\theta = \frac{s}{r}$, where θ is measured in radians.

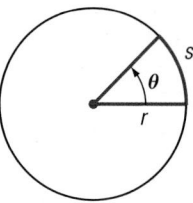

Degree/Radian Conversion Rules (p. 233)

- To convert a degree measure to radians, multiply by $\frac{\pi \text{ radians}}{180°}$.

- To convert a radian measure to degrees, multiply by $\frac{180°}{\pi \text{ radians}}$.

Coterminal Angles (p. 234)

Degrees If α is the degree measure of an angle, then all angles measuring $\alpha + 360n°$, where n is an integer, are coterminal with α.

Radians If α is the radian measure of an angle, then all angles measuring $\alpha + 2n\pi$, where n is an integer, are coterminal with α.

Arc Length (p. 235) If θ is a central angle in a circle of radius r, then the length of the intercepted arc s is given by $s = r\theta$, where θ is measured in radians.

Linear and Angular Speed (p. 236) Suppose an object moves at a constant speed along a circular path of radius r.

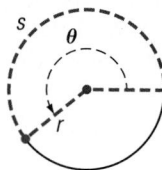

- If s is the arc length traveled by the object during time t, then the object's *linear speed* v is given by $v = \frac{s}{t}$.

- If θ is the angle of rotation (in radians) through which the object moves during time t, then the *angular speed* ω of the object is given by $\omega = \frac{\theta}{t}$.

Area of a Sector (p. 237) The area A of a sector of a circle with radius r and central angle θ is $A = \frac{1}{2}r^2\theta$, where θ is measured in radians.

Trigonometric Functions of Any Angle (p. 242) Let θ be any angle in standard position and point $P(x, y)$ be a point on the terminal side of θ. Let r represent the nonzero distance from P to the origin. That is, let $r = \sqrt{x^2 + y^2} \neq 0$. Then the trigonometric functions of θ are as follows.

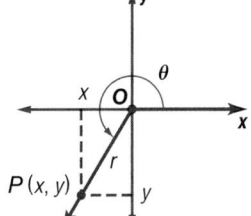

$$\sin\theta = \frac{y}{r} \qquad\qquad \csc\theta = \frac{r}{y},\, y \neq 0$$

$$\cos\theta = \frac{x}{r} \qquad\qquad \sec\theta = \frac{r}{x},\, x \neq 0$$

$$\tan\theta = \frac{y}{x},\, x \neq 0 \qquad \cot\theta = \frac{x}{y},\, y \neq 0$$

Common Quadrantal Angles (p. 243)

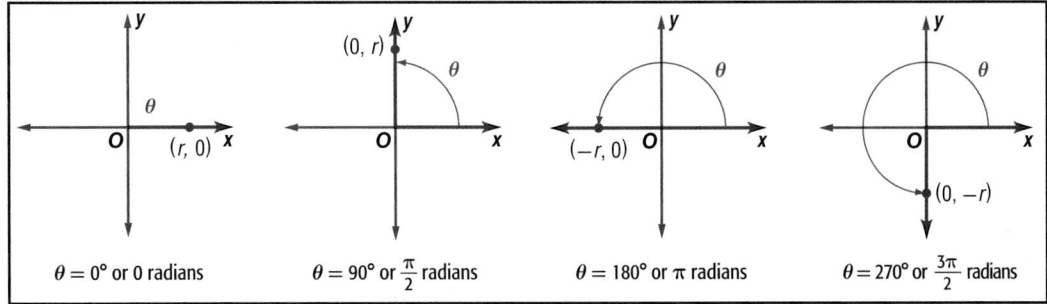

Reference Angle Rules (p. 244)

If θ is an angle in standard position, its reference angle θ' is the acute angle formed by the terminal side of θ and the x-axis. The reference angle θ' for any angle θ, $0° < \theta < 360°$ or $0 < \theta < 2\pi$, is defined as follows.

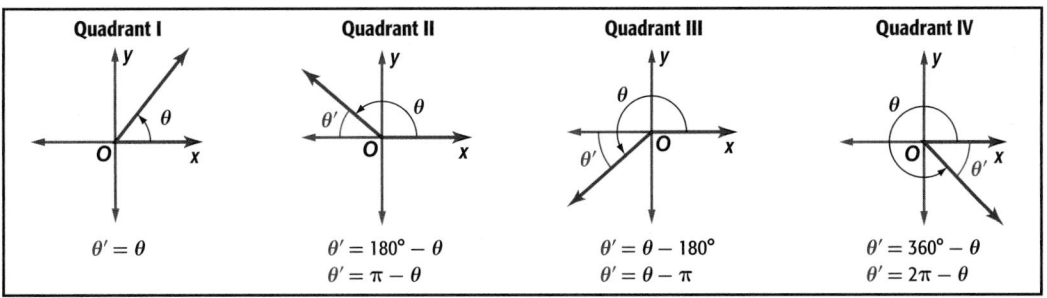

Evaluating Trigonometric Functions of Any Angle (p. 245)

Step 1 Find the reference angle θ'.

Step 2 Find the value of the trigonometric function for θ'.

Step 3 Using the quadrant in which the terminal side of θ lies, determine the sign of the trigonometric function value of θ.

Trigonometric Functions on the Unit Circle (p. 248) Let t be any real number on a number line and let $P(x, y)$ be the point on t when the number line is wrapped onto the unit circle. Then the trigonometric functions of t are as follows.

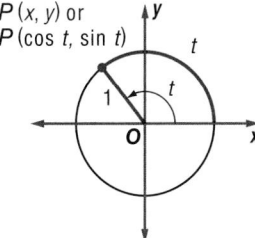

$$\sin t = y \qquad \cos t = x \qquad \tan t = \frac{y}{x},\, x \neq 0$$

$$\csc t = \frac{1}{y},\, y \neq 0 \qquad \sec t = \frac{1}{x},\, x \neq 0 \qquad \cot t = \frac{x}{y},\, y \neq 0$$

Therefore, the coordinates of P corresponding to the angle t can be written as $P(\cos t, \sin t)$.

Periodic Functions (p. 250) A function $y = f(t)$ is periodic if there exists a positive real number c such that $f(t + c) = f(t)$ for all values of t in the domain of f. The smallest number c for which f is periodic is called the period of f.

Properties of the Sine and Cosine Functions (p. 256)

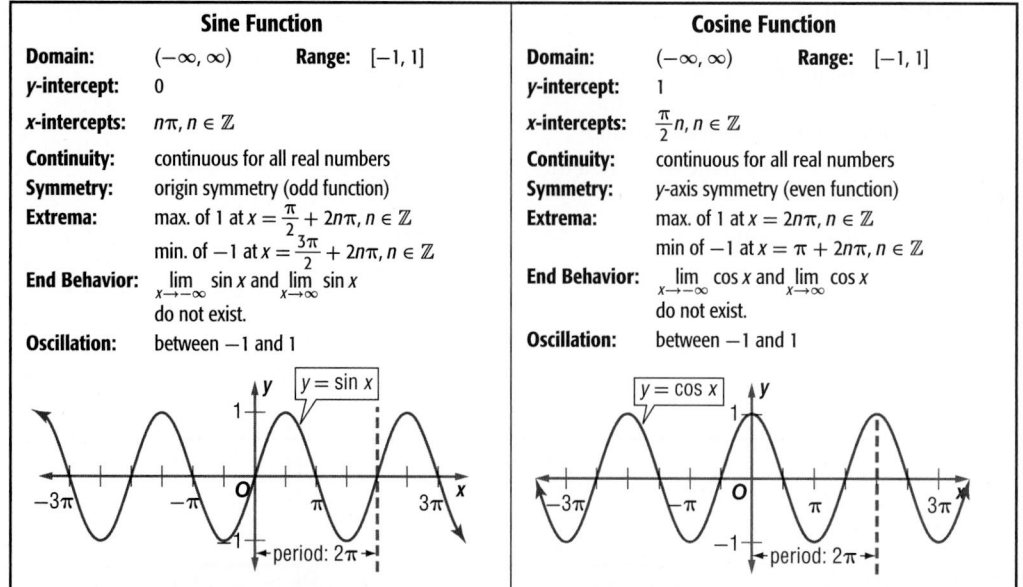

	Sine Function		**Cosine Function**	
Domain:	$(-\infty, \infty)$ **Range:** $[-1, 1]$	**Domain:**	$(-\infty, \infty)$ **Range:** $[-1, 1]$
y-intercept:	0	**y-intercept:**	1
x-intercepts:	$n\pi, n \in \mathbb{Z}$	**x-intercepts:**	$\frac{\pi}{2}n, n \in \mathbb{Z}$
Continuity:	continuous for all real numbers	**Continuity:**	continuous for all real numbers
Symmetry:	origin symmetry (odd function)	**Symmetry:**	y-axis symmetry (even function)
Extrema:	max. of 1 at $x = \frac{\pi}{2} + 2n\pi, n \in \mathbb{Z}$ min. of -1 at $x = \frac{3\pi}{2} + 2n\pi, n \in \mathbb{Z}$	**Extrema:**	max. of 1 at $x = 2n\pi, n \in \mathbb{Z}$ min of -1 at $x = \pi + 2n\pi, n \in \mathbb{Z}$
End Behavior:	$\lim\limits_{x \to -\infty} \sin x$ and $\lim\limits_{x \to \infty} \sin x$ do not exist.	**End Behavior:**	$\lim\limits_{x \to -\infty} \cos x$ and $\lim\limits_{x \to \infty} \cos x$ do not exist.
Oscillation:	between -1 and 1	**Oscillation:**	between -1 and 1

Amplitudes of Sine and Cosine Functions (p. 257) The amplitude of a sinusoidal function is half the distance between the maximum and minimum values of the function or half the height of the wave. For $y = a \sin (bx + c) + d$ and $y = a \cos (bx + c) + d$, amplitude $= |a|$.

Periods of Sine and Cosine Functions (p. 259) The period of a sinusoidal function is the distance between any two sets of repeating points on the graph of the function. So, period $= \frac{2\pi}{|b|}$.

Frequency of Sine and Cosine Functions (p. 260) The frequency of a sinusoidal function is the number of cycles the function completes in a one unit interval. The frequency is the reciprocal of the period. So, frequency $= \frac{|b|}{2\pi}$.

Phase Shift of Sine and Cosine Functions (p. 261) The phase shift of a sinusoidal function is the difference between the horizontal position of the function and that of an otherwise similar sinusoidal function. So, phase shift $= -\frac{c}{|b|}$.

Properties of the Tangent Function (p. 269)

Domain:	$x \in \mathbb{R}, x \neq \frac{\pi}{2} + n\pi, n \in \mathbb{R}$
Range:	$(-\infty, \infty)$
x-intercepts:	$n\pi, n \in \mathbb{R}$
y-intercept:	0
Continuity:	infinite discontinuity at $x = \frac{\pi}{2} + n\pi$, $n \in \mathbb{R}$
Asymptotes:	$x = \frac{\pi}{2} + n\pi, n \in \mathbb{R}$
Symmetry:	origin symmetry (odd function)
Extrema:	none
End Behavior:	$\lim\limits_{x \to -\infty} \tan x$ and $\lim\limits_{x \to \infty} \tan x$ do not exist. The function oscillates between $-\infty$ and ∞.

Period of the Tangent Function (p. 270) The *period* of a tangent function is the distance between any two consecutive vertical asymptotes. So, period $= \frac{\pi}{|b|}$.

Properties of the Cotangent Function (p. 272)

Domain:	$x \in \mathbb{R}, x \neq n\pi, n \in \mathbb{R}$
Range:	$(-\infty, \infty)$
x-intercepts:	$\frac{\pi}{2} + n\pi, n \in \mathbb{R}$
y-intercept:	none
Continuity:	infinite discontinuity at $x = n\pi, n \in \mathbb{R}$
Asymptotes:	$x = n\pi, n \in \mathbb{R}$
Symmetry:	origin (odd function)
Extrema:	none
End Behavior:	$\lim\limits_{x \to -\infty} \cot x$ and $\lim\limits_{x \to \infty} \cot x$ do not exist. The function oscillates between $-\infty$ and ∞.

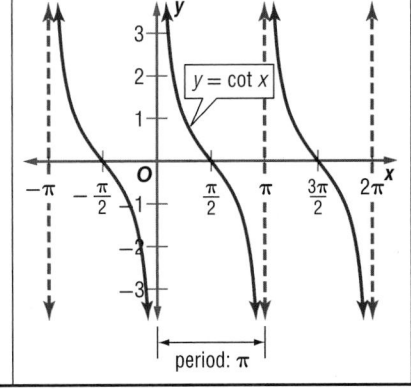

Properties of the Cosecant and Secant Functions (p. 273)

Cosecant Function		**Secant Function**	
Domain:	$x \in \mathbb{R}, x \neq n\pi, n \in \mathbb{R}$	**Domain:**	$x \in \mathbb{R}, x \neq \frac{\pi}{2} + n\pi, n \in \mathbb{R}$
Range:	$(-\infty, -1]$ and $[1, \infty)$	**Range:**	$(-\infty, -1]$ and $[1, \infty)$
x-intercepts:	none	**x-intercepts:**	none
y-intercept:	none	**y-intercept:**	1
Continuity:	infinite discontinuity at $x = n\pi, n \in \mathbb{R}$	**Continuity:**	infinite discontinuity at $x = \frac{\pi}{2} + n\pi, n \in \mathbb{R}$
Asymptotes:	$x = n\pi, n \in \mathbb{R}$	**Asymptotes:**	$x = \frac{\pi}{2} + n\pi, n \in \mathbb{R}$
Symmetry:	origin (odd function)	**Symmetry:**	y-axis (even function)
End Behavior:	$\lim\limits_{x \to -\infty} \csc x$ and $\lim\limits_{x \to \infty} \csc x$ do not exist. The function oscillates between $-\infty$ and ∞.	**Behavior:**	$\lim\limits_{x \to -\infty} \sec x$ and $\lim\limits_{x \to \infty} \sec x$ do not exist. The function oscillates between $-\infty$ and ∞.

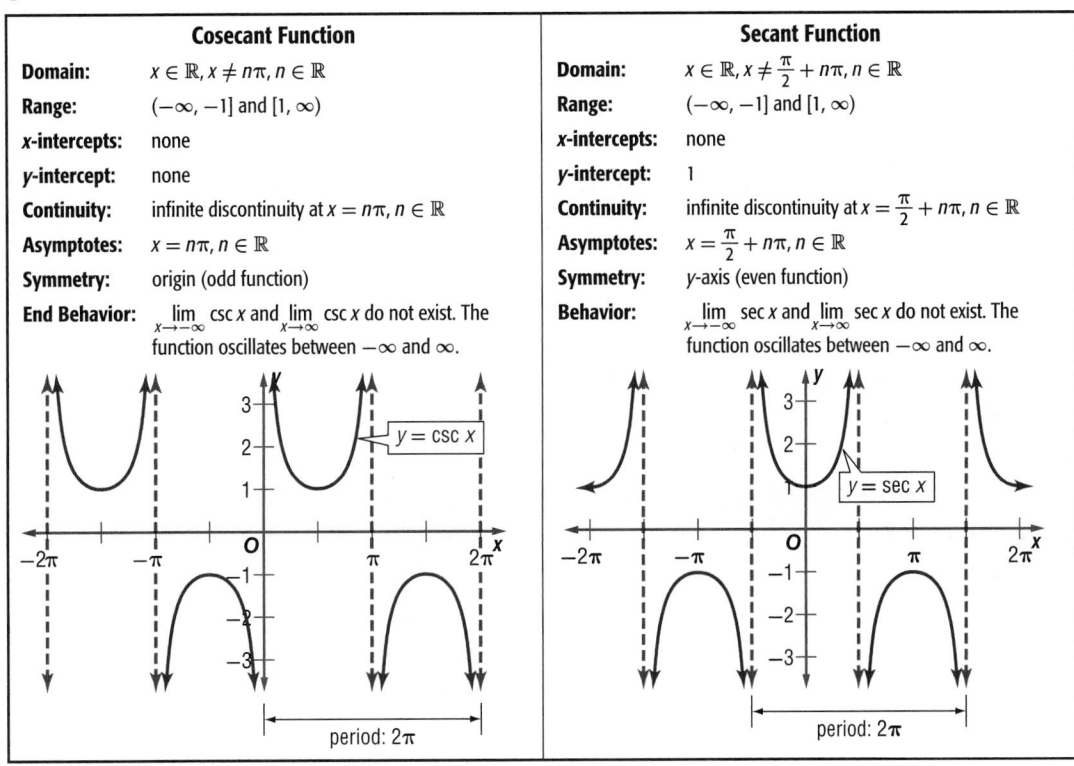

Damped Harmonic Motion (p. 276) An object is in damped harmonic motion when the amplitude is determined by the function $a(t) = ke^{-ct}$.

Domain of Compositions of Trigonometric Functions (p. 286)

$f(f^{-1}(x)) = x$	$f^{-1}(f(x)) = x$
If $-1 \leq x \leq 1$, then $\sin(\sin^{-1} x) = x$.	If $-\frac{\pi}{2} \leq x \leq \frac{\pi}{2}$, then $\sin^{-1}(\sin x) = x$.
If $-1 \leq x \leq 1$, then $\cos(\cos^{-1} x) = x$.	If $0 \leq x \leq \pi$, then $\cos^{-1}(\cos x) = x$.
If $-\infty < x < \infty$, then $\tan(\tan^{-1} x) = x$.	If $-\frac{\pi}{2} < x < \frac{\pi}{2}$, then $\tan^{-1}(\tan x) = x$.

Law of Sines (p. 291) If $\triangle ABC$ has side lengths a, b, and c representing the lengths of the sides opposite the angles with measures A, B, and C, then $\frac{\sin A}{a} = \frac{\sin B}{b} = \frac{\sin C}{c}$.

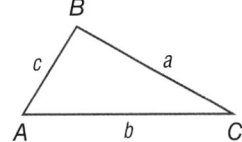

The Ambiguous Case (SSA) (p. 293)

Consider a triangle in which a, b, and A are given. For the acute case, $\sin A = \frac{h}{b}$, so $h = b \sin A$.

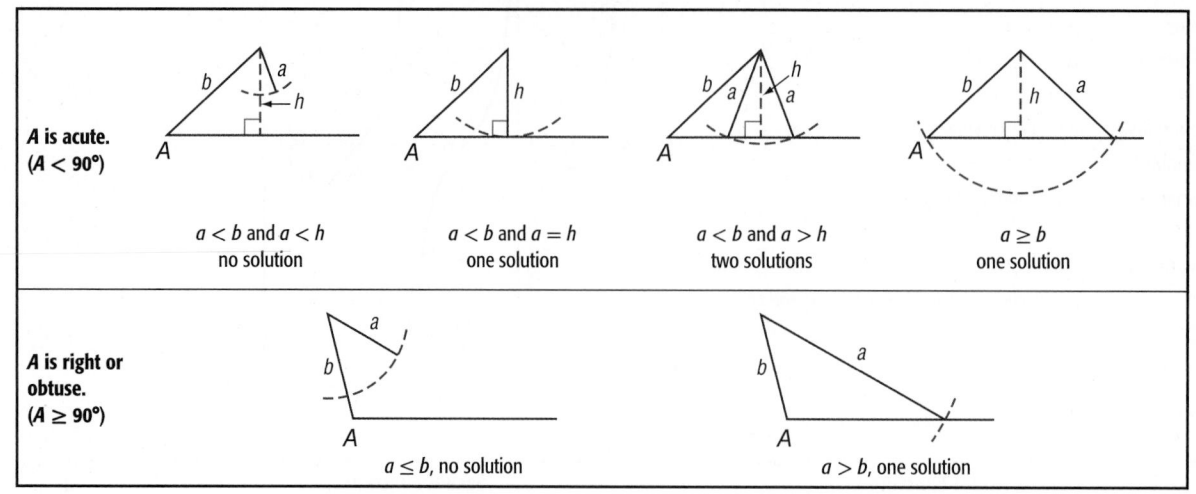

A is acute. $(A < 90°)$	$a < b$ and $a < h$ no solution	$a < b$ and $a = h$ one solution	$a < b$ and $a > h$ two solutions	$a \geq b$ one solution
A is right or obtuse. $(A \geq 90°)$	$a \leq b$, no solution		$a > b$, one solution	

Law of Cosines (p. 295) In $\triangle ABC$, if sides with lengths a, b, and c are opposite angles with measures A, B, and C, respectively, then the following are true.

- $a^2 = b^2 + c^2 - 2bc \cos A$
- $b^2 = a^2 + c^2 - 2ac \cos B$
- $c^2 = a^2 + b^2 - 2ab \cos C$

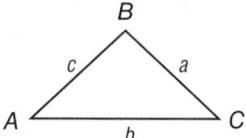

Heron's Formula (p. 296) If the measures of the sides of $\triangle ABC$ are a, b, and c, then the area of the triangle is given by Area $= \sqrt{s(s-a)(s-b)(s-c)}$, where $s = \frac{1}{2}(a+b+c)$.

Area of a Triangle Given SAS (p. 297) The area of a triangle is one half the product of the lengths of two sides and the sine of their included angle.

Chapter 5

Trigonometric Identities and Equations

Reciprocal and Quotient Identities (p. 312)

Reciprocal Identities

$$\sin \theta = \frac{1}{\csc \theta} \qquad \cos \theta = \frac{1}{\sec \theta} \qquad \tan \theta = \frac{1}{\cot \theta}$$

$$\csc \theta = \frac{1}{\sin \theta} \qquad \sec \theta = \frac{1}{\cos \theta} \qquad \cot \theta = \frac{1}{\tan \theta}$$

Quotient Identities

$$\tan \theta = \frac{\sin \theta}{\cos \theta}$$

$$\cot \theta = \frac{\cos \theta}{\sin \theta}$$

Pythagorean Identities (p. 313)

$$\sin^2 \theta + \cos^2 \theta = 1 \qquad \tan^2 \theta + 1 = \sec^2 \theta \qquad \cot^2 \theta + 1 = \csc^2 \theta$$

Cofunction Identities (p. 314)

$$\sin \theta = \cos \left(\frac{\pi}{2} - \theta \right) \qquad \tan \theta = \cot \left(\frac{\pi}{2} - \theta \right) \qquad \sec \theta = \csc \left(\frac{\pi}{2} - \theta \right)$$

$$\cos \theta = \sin \left(\frac{\pi}{2} - \theta \right) \qquad \cot \theta = \tan \left(\frac{\pi}{2} - \theta \right) \qquad \csc \theta = \sec \left(\frac{\pi}{2} - \theta \right)$$

Odd-Even Identities (p. 314)

$$\sin(-\theta) = -\sin \theta \qquad \cos(-\theta) = \cos \theta \qquad \tan(-\theta) = -\tan \theta$$

$$\csc(-\theta) = -\csc \theta \qquad \sec(-\theta) = \sec \theta \qquad \cot(-\theta) = -\cot \theta$$

Sum and Difference Identities (p. 337)

Sum Identities

$$\cos(\alpha + \beta) = \cos \alpha \cos \beta - \sin \alpha \sin \beta$$

$$\sin(\alpha + \beta) = \sin \alpha \cos \beta + \cos \alpha \sin \beta$$

$$\tan(\alpha + \beta) = \frac{\tan \alpha + \tan \beta}{1 - \tan \alpha \tan \beta}$$

Difference Identities

$$\cos(\alpha - \beta) = \cos \alpha \cos \beta + \sin \alpha \sin \beta$$

$$\sin(\alpha - \beta) = \sin \alpha \cos \beta - \cos \alpha \sin \beta$$

$$\tan(\alpha - \beta) = \frac{\tan \alpha - \tan \beta}{1 + \tan \alpha \tan \beta}$$

Double-Angle Identities (p. 346)

$$\sin 2\theta = 2\sin\theta\cos\theta \qquad \cos 2\theta = \cos^2\theta - \sin^2\theta$$

$$\tan 2\theta = \frac{2\tan\theta}{1 - \tan^2\theta} \qquad \cos 2\theta = 2\cos^2\theta - 1$$

$$\cos 2\theta = 1 - 2\sin^2\theta$$

Power-Reducing Identities (p. 347)

$$\sin^2\theta = \frac{1 - \cos 2\theta}{2} \qquad \cos^2\theta = \frac{1 + \cos 2\theta}{2} \qquad \tan^2\theta = \frac{1 - \cos 2\theta}{1 + \cos 2\theta}$$

Half-Angle Identities (p. 348)

$$\sin\frac{\theta}{2} = \pm\sqrt{\frac{1 - \cos\theta}{2}} \qquad \tan\frac{\theta}{2} = \pm\sqrt{\frac{1 - \cos\theta}{1 + \cos\theta}} \qquad \tan\frac{\theta}{2} = \frac{\sin\theta}{1 + \cos\theta}$$

$$\cos\frac{\theta}{2} = \pm\sqrt{\frac{1 + \cos\theta}{2}} \qquad \tan\frac{\theta}{2} = \frac{1 - \cos\theta}{\sin\theta}$$

Product-to-Sum Identities (p. 350)

$$\sin\alpha\sin\beta = \frac{1}{2}[\cos(\alpha - \beta) - \cos(\alpha + \beta)] \qquad \sin\alpha\cos\beta = \frac{1}{2}[\sin(\alpha + \beta) + \sin(\alpha - \beta)]$$

$$\cos\alpha\cos\beta = \frac{1}{2}[\cos(\alpha - \beta) + \cos(\alpha + \beta)] \qquad \cos\alpha\sin\beta = \frac{1}{2}[\sin(\alpha + \beta) - \sin(\alpha - \beta)]$$

Sum-to-Product Identities (p. 350)

$$\sin\alpha + \sin\beta = 2\sin\left(\frac{\alpha + \beta}{2}\right)\cos\left(\frac{\alpha - \beta}{2}\right) \qquad \cos\alpha + \cos\beta = 2\cos\left(\frac{\alpha + \beta}{2}\right)\cos\left(\frac{\alpha - \beta}{2}\right)$$

$$\sin\alpha - \sin\beta = 2\cos\left(\frac{\alpha + \beta}{2}\right)\sin\left(\frac{\alpha - \beta}{2}\right) \qquad \cos\alpha - \cos\beta = -2\sin\left(\frac{\alpha + \beta}{2}\right)\sin\left(\frac{\alpha - \beta}{2}\right)$$

Chapter 6

Systems of Equations and Matrices

Operations that Produce Equivalent Systems (p. 364) Each of the following operations produces an equivalent system of linear equations.
- Interchange any two equations.
- Multiply one of the equations by a nonzero real number.
- Add a multiple of one equation to another equation.

Elementary Row Operations (p. 366) Each of the following row operations produces an equivalent augmented matrix.
- Interchange any two rows.
- Multiply one row by a nonzero real number.
- Add a multiple of one row to another row.

Row-Echelon Form (p. 367) A matrix is in row-echelon form if the following conditions are met.
- Rows consisting entirely of zeros (if any) appear at the bottom of the matrix.
- The first nonzero entry in a row is 1, called a *leading 1*.
- For two successive rows with nonzero entries, the leading 1 in the higher row is farther to the left than the leading 1 in the lower row.

Matrix Multiplication (p. 375) If A is an $m \times r$ matrix and B is an $r \times n$ matrix, then the product AB is an $m \times n$ matrix obtained by adding the products of the entries of a row in A to the corresponding entries of a column in B.

Properties of Matrix Multiplication (p. 377) For any matrices A, B, and C for which the matrix product is defined and any scalar k, the following properties are true.

Associative Property of Matrix Multiplication	$(AB)C = A(BC)$
Associative Property of Scalar Multiplication	$k(AB) = (kA)B = A(kB)$
Left Distributive Property	$C(A + B) = CA + CB$
Right Distributive Property	$(A + B)C = AC + BC$

Identity Matrix (p. 378) The identity matrix of order n, denoted I_n, is an $n \times n$ matrix consisting of all 1s on its main diagonal, from upper left to lower right, and 0s for all other elements.

Inverse of a Square Matrix (p. 379) Let A be an $n \times n$ matrix. If there exists a matrix B such that $AB = BA = I_n$, then B is called the inverse of A and is written as A^{-1}. So, $AA^{-1} = A^{-1}A = I_n$.

Inverse and Determinant of a 2 × 2 Matrix (p. 381) Let $A = \begin{bmatrix} a & b \\ c & d \end{bmatrix}$. A is invertible if and only if $ad - cb \neq 0$. If A is invertible, then $A^{-1} = \dfrac{1}{ad - cb} \begin{bmatrix} d & -b \\ -c & a \end{bmatrix}$. The number $ad - cb$ is called the determinant of the 2 × 2 matrix and is denoted $\det(A) = |A| = \begin{vmatrix} a & b \\ c & d \end{vmatrix} = ad - cb$.

Invertible Square Linear Systems (p. 388) Let A be the coefficient matrix of a system of n linear equations in n variables given by $AX = B$, where X is the matrix of variables and B is the matrix of constants. If A is invertible, then the system of equations has a unique solution given by $X = A^{-1}B$.

Cramer's Rule (p. 390) Let A be the coefficient matrix of a system of n linear equations in n variables given by $AX = B$. If $\det(A) \neq 0$, then the unique solution of the system is given by

$$x_1 = \frac{|A_1|}{|A|}, x_2 = \frac{|A_2|}{|A|}, x_3 = \frac{|A_3|}{|A|}, \dots, x_n = \frac{|A_n|}{|A|},$$

where A_i is obtained by replacing the ith column of A with the column of constant terms B. If $\det(A) = 0$, then $AX = B$ has either no solution or infinitely many solutions.

Vertex Theorem for Optimization (p. 405) If a linear programming problem can be optimized, an optimal value will occur at one of the vertices of the region representing the set of feasible solutions.

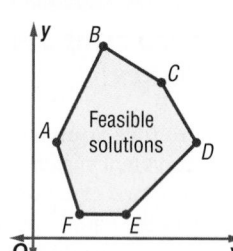

Linear Programming (p. 406) To solve a linear programming problem, follow these steps.
Step 1 Graph the region corresponding to solution of the system of constraints.
Step 2 Find the coordinates of the vertices of the region formed.
Step 3 Evaluate the objective function at each vertex to determine which x- and y-values, if any, maximize or minimize the function.

Partial Fraction Decomposition of $f(x)/d(x)$ (p. 401)
1. If the degree of $f(x) \geq$ the degree of $d(x)$, use polynomial long division and the division algorithm to write $\dfrac{f(x)}{d(x)} = q(x) + \dfrac{r(x)}{d(x)}$. Then apply partial fraction decomposition to $\dfrac{r(x)}{d(x)}$.

2. If $\dfrac{f(x)}{d(x)}$ is proper, factor $d(x)$ into a product of linear and/or prime quadratic factors.

3. For each factor of the form $(ax + b)^n$ in the denominator, the partial fraction decomposition must include the sum of n fractions

$$\frac{A_1}{ax + b} + \frac{A_2}{(ax + b)^2} + \frac{A_3}{(ax + b)^3} + \dots + \frac{A_n}{(ax + b)^n},$$

where $A_1, A_2, A_3, \dots, A_n$ are real numbers.

4. For each prime quadratic factor that occurs n times in the denominator, the partial fraction decomposition must include the sum of n fractions

$$\frac{B_1 x + C_1}{ax^2 + bx + c} + \frac{B_2 x + C_2}{(ax^2 + bx + c)^2} + \frac{B_3 x + C_3}{(ax^2 + bx + c)^3} + \dots + \frac{B_n x + C_n}{(ax^2 + bx + c)^n},$$

where $B_1, B_2, B_3, \dots, B_n$ and $C_1, C_2, C_3, \dots, C_n$ are real numbers.

5. The partial fraction decomposition of the original function is the sum of $q(x)$ from part 1 and the fractions in parts 3 and 4.

Chapter 7

Conic Sections and Parametric Equations

Standard Form of Equations for Parabolas (p. 423)

$(x - h)^2 = 4p(y - k)$

$p > 0$ $p < 0$

Orientation: opens vertically

Vertex: (h, k) **Focus:** $(h, k + p)$ **Directrix d:** $y = k - p$

Axis of Symmetry a: $x = h$

$(y - k)^2 = 4p(x - h)$

$p > 0$ $p < 0$

Orientation: opens horizontally

Vertex: (h, k) **Focus:** $(h + p, k)$ **Directrix d:** $x = h - p$

Axis of Symmetry a: $y = k$

Standard Form of Equations for Ellipses (p. 433)

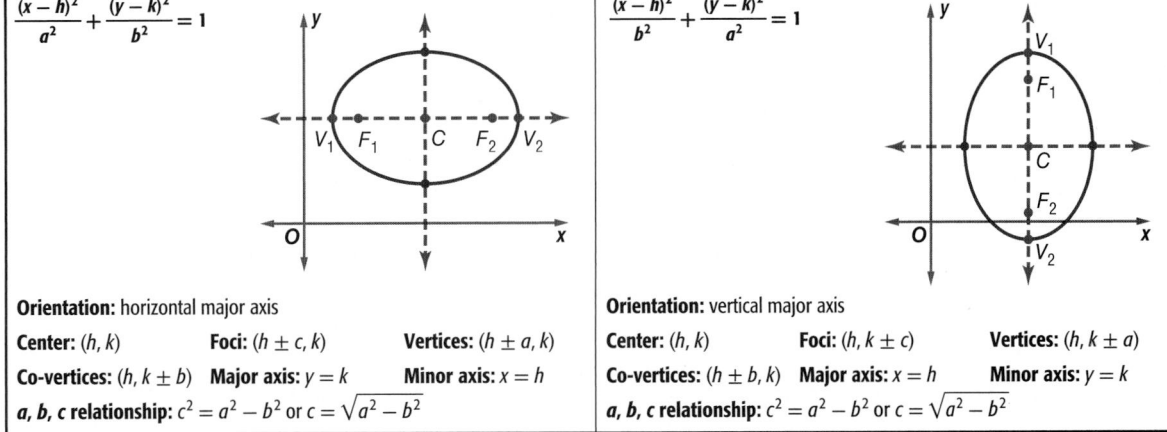

$\dfrac{(x - h)^2}{a^2} + \dfrac{(y - k)^2}{b^2} = 1$

Orientation: horizontal major axis

Center: (h, k) **Foci:** $(h \pm c, k)$ **Vertices:** $(h \pm a, k)$

Co-vertices: $(h, k \pm b)$ **Major axis:** $y = k$ **Minor axis:** $x = h$

a, b, c relationship: $c^2 = a^2 - b^2$ or $c = \sqrt{a^2 - b^2}$

$\dfrac{(x - h)^2}{b^2} + \dfrac{(y - k)^2}{a^2} = 1$

Orientation: vertical major axis

Center: (h, k) **Foci:** $(h, k \pm c)$ **Vertices:** $(h, k \pm a)$

Co-vertices: $(h \pm b, k)$ **Major axis:** $x = h$ **Minor axis:** $y = k$

a, b, c relationship: $c^2 = a^2 - b^2$ or $c = \sqrt{a^2 - b^2}$

Eccentricity (p. 435) For any ellipse, $\dfrac{(x - h)^2}{a^2} + \dfrac{(y - k)^2}{b^2} = 1$ or $\dfrac{(x - h)^2}{b^2} + \dfrac{(y - k)^2}{a^2} = 1$, where $c^2 = a^2 - b^2$, the eccentricity $e = \dfrac{c}{a}$.

Standard Form of Equations for Circles (p. 437) The standard form of an equation for a circle with center (h, k) and radius r is $(x - h)^2 + (y - k)^2 = r^2$.

Standard Forms of Equations for Hyperbolas (p. 443)

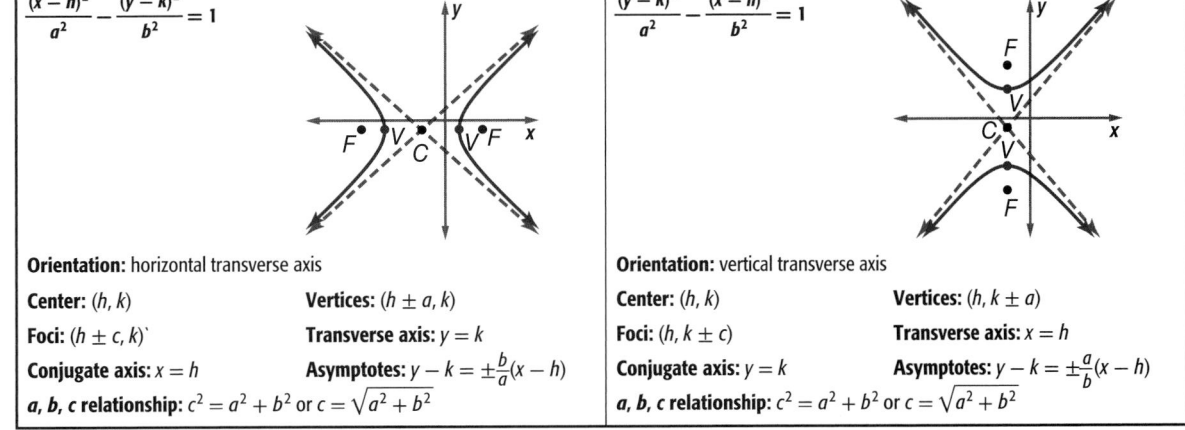

$\dfrac{(x - h)^2}{a^2} - \dfrac{(y - k)^2}{b^2} = 1$

Orientation: horizontal transverse axis

Center: (h, k) **Vertices:** $(h \pm a, k)$

Foci: $(h \pm c, k)$ **Transverse axis:** $y = k$

Conjugate axis: $x = h$ **Asymptotes:** $y - k = \pm \dfrac{b}{a}(x - h)$

a, b, c relationship: $c^2 = a^2 + b^2$ or $c = \sqrt{a^2 + b^2}$

$\dfrac{(y - k)^2}{a^2} - \dfrac{(x - h)^2}{b^2} = 1$

Orientation: vertical transverse axis

Center: (h, k) **Vertices:** $(h, k \pm a)$

Foci: $(h, k \pm c)$ **Transverse axis:** $x = h$

Conjugate axis: $y = k$ **Asymptotes:** $y - k = \pm \dfrac{a}{b}(x - h)$

a, b, c relationship: $c^2 = a^2 + b^2$ or $c = \sqrt{a^2 + b^2}$

Classify Conics Using the Discriminant (p. 447) The graph of a second degree equation of the form $Ax^2 + Bxy + Cy^2 + Dx + Ey + F = 0$ is
- a circle if $B^2 - 4AC < 0$; $B = 0$ and $A = C$.
- an ellipse if $B^2 - 4AC < 0$; either $B \neq 0$ or $A \neq C$.
- a parabola if $B^2 - 4AC = 0$.
- a hyperbola if $B^2 - 4AC > 0$.

Rotation of Axes of Conics (p. 454) An equation $Ax^2 + Bxy + Cy^2 + Dx + Ey + F = 0$ in the xy-plane can be rewritten as $A(x')^2 + C(y')^2 + Dx' + Ey' + F = 0$ in the $x'y'$-plane by rotating the coordinate axes through an angle θ. The equation in the $x'y'$-plane can be found using $x = x' \cos \theta - y' \sin \theta$, and $y = x' \sin \theta + y' \cos \theta$.

Angle of Rotation Used to Eliminate xy-Term (p. 455) An angle of rotation θ such that $\cot 2\theta = \frac{A - C}{B}$, $B \neq 0$, $0 < \theta < \frac{\pi}{2}$, will eliminate the xy-term from the equation of the conic section in the rotated $x'y'$-coordinate system.

Rotation of Axes of Conics (p. 457) When an equation of a conic section is rewritten in the $x'y'$-plane by rotating the coordinate axes through θ, the equation in the xy-plane can be found using $x' = x \cos \theta + y \sin \theta$, and $y' = y \cos \theta - x \sin \theta$.

Parametric Equations (p. 464) If f and g are continuous functions of t on the interval I, then the set of ordered pairs $(f(t), g(t))$ represent a *parametric curve*. The equations $x = f(t)$ and $y = g(t)$ are parametric equations for this curve, t is the parameter, and I is the parameter interval.

Projectile Motion (p. 467) For an object launched at an angle θ with the horizontal at an initial velocity v_0, where g is the gravitational constant, t is time, and h_0 is the initial height:

Horizontal Distance $\quad x = tv_0 \cos \theta$ $\qquad$ **Vertical Position** $\quad y = tv_0 \sin \theta - \frac{1}{2}gt^2 + h_0$

Chapter 8

Vectors

Finding Resultants (p. 484)

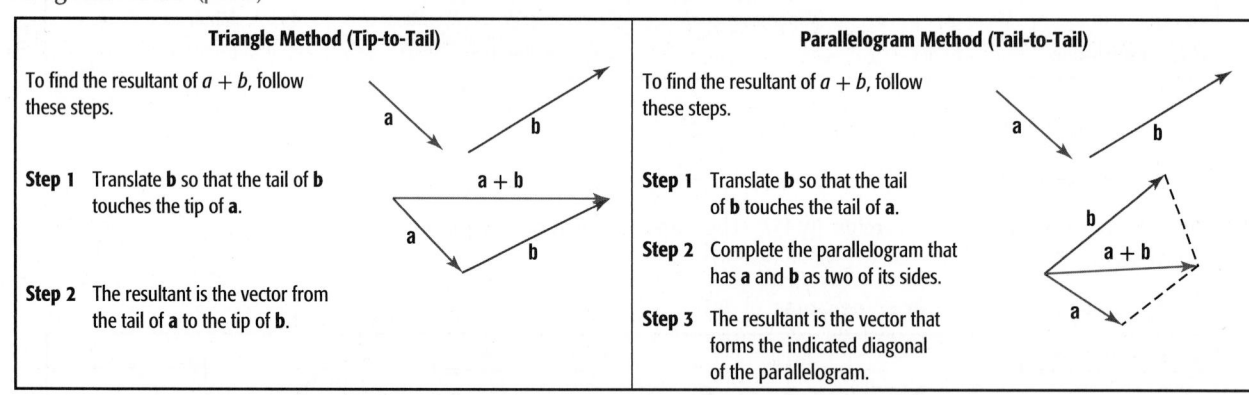

Multiplying Vectors by a Scalar (p. 485) If a vector $\mathbf{v}$ is multiplied by a real number scalar k, the scalar multiple $k\mathbf{v}$ has a magnitude of $|k| |\mathbf{v}|$. If $k > 0$, $k\mathbf{v}$ has the same direction as $\mathbf{v}$. If $k < 0$, $k\mathbf{v}$ has the opposite direction as $\mathbf{v}$.

Component Form of a Vector (p. 492) The component form of a vector $\overrightarrow{AB}$ with initial point $A(x_1, y_1)$ and terminal point $B(x_2, y_2)$ is given by $\langle x_2 - x_1, y_2 - y_1 \rangle$.

Magnitude of a Vector in the Coordinate Plane (p. 493) If $\mathbf{v}$ is a vector with initial point (x_1, y_1) and terminal point (x_2, y_2), then the magnitude of $\mathbf{v}$ is given by $|\mathbf{v}| = \sqrt{(x_2 - x_1)^2 + (y_2 - y_1)^2}$. If $\mathbf{v}$ has a component form of $\langle a, b \rangle$, then $|\mathbf{v}| = \sqrt{a^2 + b^2}$.

Vector Operations (p. 493) If $\mathbf{a} = \langle a_1, a_2 \rangle$ and $\mathbf{b} = \langle b_1, b_2 \rangle$ are vectors and k is a scalar, then the following are true.

Vector Addition	$\mathbf{a} + \mathbf{b} = \langle a_1 + b_1, a_2 + b_2 \rangle$
Vector Subtraction	$\mathbf{a} - \mathbf{b} = \langle a_1 - b_1, a_2 - b_2 \rangle$
Scalar Multiplication	$k\mathbf{a} = \langle ka_1, ka_2 \rangle$

Dot Product of Vectors in a Plane (p. 500) The dot product of $\mathbf{a} = \langle a_1, a_2 \rangle$ and $\mathbf{b} = \langle b_1, b_2 \rangle$ is defined as $\mathbf{a} \cdot \mathbf{b} = a_1 b_1 + a_2 b_2$.

Orthogonal Vectors (p. 500) The vectors $\mathbf{a}$ and $\mathbf{b}$ are orthogonal if and only if $\mathbf{a} \cdot \mathbf{b} = 0$.

Properties of the Dot Product (p. 501) If $\mathbf{u}$, $\mathbf{v}$, and $\mathbf{w}$ are vectors and k is a scalar, then the following properties hold.

Commutative Property	$\mathbf{u} \cdot \mathbf{v} = \mathbf{v} \cdot \mathbf{u}$
Distributive Property	$\mathbf{u} \cdot (\mathbf{v} + \mathbf{w}) = \mathbf{u} \cdot \mathbf{v} + \mathbf{u} \cdot \mathbf{w}$
Scalar Multiplication Property	$k(\mathbf{u} \cdot \mathbf{v}) = k\mathbf{u} \cdot \mathbf{v} = \mathbf{u} \cdot k\mathbf{v}$
Zero Vector Dot Product Property	$\mathbf{0} \cdot \mathbf{u} = 0$
Dot Product and Vector Magnitude Relationship	$\mathbf{u} \cdot \mathbf{u} = |\mathbf{u}|^2$

Angle Between Two Vectors (p. 502) If θ is the angle between nonzero vectors $\mathbf{a}$ and $\mathbf{b}$, then

$$\cos \theta = \frac{\mathbf{a} \cdot \mathbf{b}}{|\mathbf{a}||\mathbf{b}|}.$$

Projection of u onto v (p. 503) Let $\mathbf{u}$ and $\mathbf{v}$ be nonzero vectors, and let $\mathbf{w}_1$ and $\mathbf{w}_2$ be vector components of $\mathbf{u}$ such that $\mathbf{w}_1$ is parallel to $\mathbf{v}$ as shown. Then vector $\mathbf{w}_1$ is called the *vector projection* of $\mathbf{u}$ onto $\mathbf{v}$, denoted $\text{proj}_\mathbf{v}\mathbf{u}$, and $\text{proj}_\mathbf{v}\mathbf{u} = \left(\dfrac{\mathbf{u} \cdot \mathbf{v}}{|\mathbf{v}|^2}\right)\mathbf{v}$.

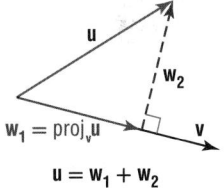

Distance Formula in Space (p. 511) The distance between points $A(x_1, y_1, z_1)$ and $B(x_2, y_2, z_2)$ is given by $AB = \sqrt{(x_2 - x_1)^2 + (y_2 - y_1)^2 + (z_2 - z_1)^2}$.

Midpoint Formula in Space (p. 511) The midpoint M of AB is given by $M\left(\dfrac{x_1 + x_2}{2}, \dfrac{y_1 + y_2}{2}, \dfrac{z_1 + z_2}{2}\right)$.

Vector Operations in Space (p. 513) If $\mathbf{a} = \langle a_1, a_2, a_3 \rangle$, $\mathbf{b} = \langle b_1, b_2, b_3 \rangle$, and any scalar k, then

Vector Addition	$\mathbf{a} + \mathbf{b} = \langle a_1 + b_1, a_2 + b_2, a_3 + b_3 \rangle$
Vector Subtraction	$\mathbf{a} - \mathbf{b} = \mathbf{a} + (-\mathbf{b}) = \langle a_1 - b_1, a_2 - b_2, a_3 - b_3 \rangle$
Scalar Multiplication	$k\mathbf{a} = \langle ka_1, ka_2, ka_3 \rangle$

Dot Product and Orthogonal Vectors in Space (p. 518) The dot product of $\mathbf{a} = \langle a_1, a_2, a_3 \rangle$ and $\mathbf{b} = \langle b_1, b_2, b_3 \rangle$ is defined as $\mathbf{a} \cdot \mathbf{b} = a_1 b_1 + a_2 b_2 + a_3 b_3$. The vectors $\mathbf{a}$ and $\mathbf{b}$ are orthogonal if and only if $\mathbf{a} \cdot \mathbf{b} = 0$.

Cross Product of Vectors in Space (p. 519) If $\mathbf{a} = a_1\mathbf{i} + a_2\mathbf{j} + a_3\mathbf{k}$ and $\mathbf{b} = b_1\mathbf{i} + b_2\mathbf{j} + b_3\mathbf{k}$, the cross product of $\mathbf{a}$ and $\mathbf{b}$ is the vector $\mathbf{a} \times \mathbf{b} = (a_2 b_3 - a_3 b_2)\mathbf{i} - (a_1 b_3 - a_3 b_1)\mathbf{j} + (a_1 b_2 - a_2 b_1)\mathbf{k}$.

Triple Scalar Product (p. 521) If $\mathbf{t} = t_1\mathbf{i} + t_2\mathbf{j} + t_3\mathbf{k}$, $\mathbf{u} = u_1\mathbf{i} + u_2\mathbf{j} + u_3\mathbf{k}$, $\mathbf{v} = v_1\mathbf{i} + v_2\mathbf{j} + v_3\mathbf{k}$, the triple scalar product is given by $\mathbf{t} \cdot (\mathbf{u} \times \mathbf{v}) = \begin{vmatrix} t_1 & t_2 & t_3 \\ u_1 & u_2 & u_3 \\ v_1 & v_2 & v_3 \end{vmatrix}$.

Chapter 9

Polar Coordinates and Complex Numbers

Polar Distance Formula (p. 537) If $P_1(r_1, \theta_1)$ and $P_2(r_2, \theta_2)$ are two points in the polar plane, then the distance $P_1 P_2$ is given by

$$\sqrt{r_1{}^2 + r_2{}^2 - 2r_1 r_2 \cos(\theta_2 - \theta_1)}.$$

Symmetry of Polar Graphs (p. 543)

Symmetry with Respect to the Line $\theta = \dfrac{\pi}{2}$	Symmetry with Respect to Polar Axis	Symmetry with Respect to the Pole

Quick Tests for Symmetry in Polar Graphs (p. 544)
The graph of a polar equation is symmetric with respect to the polar axis if it is a function of $\cos\theta$, and the line $\theta = \dfrac{\pi}{2}$ if it is a function of $\sin\theta$.

Special Types of Polar Graphs (p. 546)

Circles	$r = a\cos\theta$ or $r = a\sin\theta$
Limaçons	$r = a \pm b\cos\theta$ or $r = a \pm b\sin\theta$, where a and b are both positive
Roses	$r = a\cos n\theta$ or $r = a\sin n\theta$, where $n \geq 2$ is an integer. The rose has n petals if n is odd and $2n$ petals if n is even.
Lemniscates	$r^2 = a^2\cos 2\theta$ or $r^2 = a^2\sin 2\theta$
Spirals of Archimedes	$r = a\theta + b$

Convert Polar to Rectangular Coordinates (p. 551)
If a point P has polar coordinates (r, θ), then the rectangular coordinates (x, y) of P are given by $x = r\cos\theta$ and $y = r\sin\theta$. That is, $(x, y) = (r\cos\theta, r\sin\theta)$.

Convert Rectangular to Polar Coordinates (p. 553)
If a point P has rectangular coordinates (x, y) then the polar coordinates (r, θ) of P are given by

$$r = \sqrt{x^2 + y^2} \qquad \text{and} \qquad \theta = \tan^{-1}\frac{y}{x}, \text{ when } x > 0$$

$$\theta = \tan^{-1}\frac{y}{x} + \pi \text{ or } \theta = \tan^{-1}\frac{y}{x} + 180°, \text{ when } x < 0.$$

Polar Equations of Conics (p. 562)
The conic section with eccentricity $e > 0$, $d > 0$, and focus at the pole has the polar equation:

- $r = \dfrac{ed}{1 + e\cos\theta}$ if the directrix is the vertical line $x = d$,

- $r = \dfrac{ed}{1 - e\cos\theta}$ if the directrix is the vertical line $x = -d$,

- $r = \dfrac{ed}{1 + e\sin\theta}$ if the directrix is the horizontal line $y = d$, and

- $r = \dfrac{ed}{1 - e\sin\theta}$ if the directrix is the horizontal line $y = -d$.

Absolute Value of a Complex Number (p. 569)
The absolute value of the complex number $z = a + bi$ is $|z| = |a + bi| = \sqrt{a^2 + b^2}$.

Polar Form of a Complex Number (p. 570)
The polar or trigonometric form of the complex number $z = a + bi$ is $z = r(\cos\theta + i\sin\theta)$, where $r = |z| = \sqrt{a^2 + b^2}$, $a = r\cos\theta$, $b = r\sin\theta$, and $\theta = \tan^{-1}\frac{b}{a}$ for $a > 0$ or $\theta = \tan^{-1}\frac{b}{a} + \pi$ for $a < 0$.

Product and Quotient of Complex Numbers in Polar Form (p. 572)
Given the complex numbers $z_1 = r_1(\cos\theta_1 + i\sin\theta_1)$ and $z_2 = r_2(\cos\theta_2 + i\sin\theta_2)$:

Product Formula	$z_1 z_2 = r_1 r_2[\cos(\theta_1 + \theta_2) + i\sin(\theta_1 + \theta_2)]$
Quotient Formula	$\dfrac{z_1}{z_2} = \dfrac{r_1}{r_2}[\cos(\theta_1 - \theta_2) + i\sin(\theta_1 - \theta_2)]$, where z_2 and $r_2 \neq 0$

De Moivre's Theorem (p. 574) If the polar form of a complex number is $z = r(\cos \theta + i \sin \theta)$, then for positive integers n, $z^n = [r(\cos \theta + i \sin \theta)]^n$ or $r^n(\cos n\theta + i \sin n\theta)$.

Distinct Roots (p. 575) For a positive integer p, the complex number $r(\cos \theta + i \sin \theta)$ has p distinct pth roots. They are found by

$$r^{\frac{1}{p}}\left(\cos \frac{\theta + 2n\pi}{p} + i \sin \frac{\theta + 2n\pi}{p}\right), \text{ where } n = 0, 1, 2, \ldots, p - 1.$$

Sequences and Series — Chapter 10

Sigma Notation (p. 594) For any sequence $a_1, a_2, a_3, a_4, \ldots$, the sum of the first k terms is denoted $\sum\limits_{n=1}^{k} a_n = a_1 + a_2 + a_3 + a_4 + \cdots + a_k$, where n is the index of summation, k is the upper bound of summation, and 1 is the lower bound of summation.

The nth Term of an Arithmetic Sequence (p. 599) The nth term of an arithmetic sequence with first term a_1 and common difference d is given by $a_n = a_1 + (n - 1)d$.

Sum of a Finite Arithmetic Series (p. 603) The sum of a finite arithmetic series with n terms or the nth partial sum of an arithmetic series can be found using one of two related formulas.

Formula 1 $S_n = \frac{n}{2}(a_1 + a_n)$ **Formula 2** $S_n = \frac{n}{2}[2a_1 + (n - 1)d]$

The nth Term of a Geometric Sequence (p. 609) The nth term of a geometric sequence with first term a_1 and common ratio r is given by $a_n = a_1 r^{n - 1}$.

Sum of a Finite Geometric Series (p. 612) The sum of a finite geometric series with n terms or the nth partial sum of a geometric series can be found using one of two related formulas.

Formula 1 $S_n = a_1\left(\dfrac{1 - r^n}{1 - r}\right)$ **Formula 2** $S_n = \dfrac{a_1 - a_n r}{1 - r}$

The Sum of an Infinite Geometric Series (p. 613) The sum S of an infinite geometric series for which $|r| < 1$ is given by $S = \dfrac{a_1}{1 - r}$.

The Principle of Mathematical Induction (p. 621) Let P_n be a statement about a positive integer n. Then P_n is true for all positive integers n if and only if P_1 is true, and for every positive integer k, if P_k is true, then $P_{k + 1}$ is true.

Formula for the Binomial Coefficients of $(a + b)^n$ (p. 630) The binomial coefficient of the $a^{n - r} b^r$ term in the expansion of $(a + b)^n$ is given by $_nC_r = \dfrac{n!}{(n - r)!\, r!}$.

Binomial Theorem (p. 632) For any positive integer n, the expansion of $(a + b)^n$ is given by
$$(a + b)^n = {_nC_0}\, a^n b^0 + {_nC_1}\, a^{n - 1} b^1 + {_nC_2}\, a^{n - 2} b^2 + \cdots + {_nC_r}\, a^{n - r} b^r + \cdots + {_nC_n}\, a^0 b^n,$$
where $r = 0, 1, 2, \ldots, n$.

Power Series (p. 636) An infinite series of the form $\sum\limits_{n=0}^{\infty} a_n x^n = a_0 + a_1 x + a_2 x^2 + a_3 x^3 + \cdots$, where x and a can take on any values and $n = 0, 1, 2, \ldots$, is called a power series in x.

Exponential Series (p. 638) The power series representing e^x is called the *exponential series* and is given by $e^x = \sum\limits_{n=0}^{\infty} \dfrac{x^n}{n!} = 1 + x + \dfrac{x^2}{2!} + \dfrac{x^3}{3!} + \dfrac{x^4}{4!} + \dfrac{x^5}{5!} + \cdots$, which is convergent for all x.

Power Series for Cosine and Sine (p. 639) The power series representations for $\cos x$ and $\sin x$ are given by

$$\cos x = \sum_{n=0}^{\infty} \frac{(-1)^n x^{2n}}{(2n)!} = 1 - \frac{x^2}{2!} + \frac{x^4}{4!} - \frac{x^6}{6!} + \frac{x^8}{8!} - \cdots, \text{ and}$$

$$\sin x = \sum_{n=0}^{\infty} \frac{(-1)^n x^{2n+1}}{(2n+1)} = x - \frac{x^3}{3!} + \frac{x^5}{5!} - \frac{x^7}{7!} + \frac{x^9}{9!} - \cdots, \text{ which are convergent for all } x.$$

Euler's Formula (p. 640) For any real number θ, $e^{i\theta} = \cos\theta + i\sin\theta$.

Exponential Form of a Complex Number (p. 640) The exponential form of a complex number $a + bi$ is given by $a + bi = re^{i\theta}$, where $r = \sqrt{a^2 + b^2}$ and $\theta = \tan^{-1}\frac{b}{a}$ for $a > 0$ and $\theta = \tan^{-1}\frac{b}{a} + \pi$ for $a < 0$.

Inferential Statistics | Chapter 11

Symmetric and Skewed Distributions (p. 654)

Negative or Left-Skewed Distribution	Symmetrical Distribution	Positive or Right-Skewed Distribution
In a *negatively skewed distribution*, the mean is less than the median, the majority of the data is on the right, and the tail extends to the left.	In a *symmetrical distribution*, the data are evenly distributed on both sides of the mean. The mean and median are approximately equal.	In a *positively skewed distribution*, the mean is greater than the median, the majority of the data is on the left, and the tail extends to the right.

Choosing Summary Statistics (p. 654) When choosing measures of center and spread to describe a distribution, first examine the shape of the distribution.
- If the distribution is reasonably symmetrical and free of outliers, use the mean and standard deviation.
- If the distribution is skewed or has strong outliers, the five-number summary (minimum, quartile 1, median, quartile 3, maximum) usually provides a better summary of the overall pattern in the data.

Symmetric and Skewed Box Plots (p. 657)

Negatively Skewed	Symmetrical	Positively Skewed
		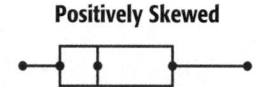
The left whisker is longer than right whisker and the line representing the median is closer to Q_3 than to Q_1.	The whiskers are the same length and the line representing the median is exactly between Q_1 and Q_3.	The right whisker is longer than left whisker and the line representing the median is closer to Q_1 than to Q_3.

Discrete and Continuous Random Variables (p. 664)
- A *discrete random variable* can take on a finite or countable number of possible values.
- A *continuous random variable* can take on an infinite number of possible values within a specified interval.

Probability Distribution (p. 665) A probability distribution of a random variable X is a table, equation, or graph that links each possible value of X with its probability of occurring. These probabilities are determined theoretically or by observation. A probability distribution must satisfy the following conditions.
- The probability of each value of X must be between 0 and 1. That is, $0 \le P(X) \le 1$.
- The sum of all the probabilities for all the values of X must equal 1. That is, $\Sigma P(X) = 1$.

Mean of a Probability Distribution (p. 666) To find the mean of a probability distribution of X, multiply each value of X by its probability and find the sum of the products.

Variance and Standard Deviation of a Probability Distribution (p. 667) To find the variance of a probability distribution, subtract the mean of the probability distribution from each value and square the difference. Then multiply each difference by its corresponding probability and find the sum of the products. The standard deviation is the square root of the variance.

Binomial Experiment (p. 668) A *binomial experiment* is a probability experiment that satisfies the following conditions.
- The experiment is repeated for a fixed number of independent trials n.
- Each trial has only two possible outcomes, success S or failure F.
- The probability of success $P(S)$ or p is the same in every trial. The probability of failure $P(F)$ or q is $1 - p$.
- The random variable X represents the number of successes in n trials.

Binomial Probability Formula (p. 669) The probability of X successes in n independent trials of a binomial experiment is $P(X) = {}_nC_x\, p^x q^{n-x} = \dfrac{n!}{(n-x)!\, x!}\, p^x q^{n-x}$, where p is the probability of success and q is the probability of failure for an individual trial.

Mean and Standard Deviation of a Binomial Distribution (p. 670) The mean, variance, and standard deviation of a random variable X that has a binomial distribution are given by the following formulas.

mean:	$\mu = np$
variance:	$\sigma^2 = npq$
standard deviation:	$\sigma = \sqrt{\sigma^2}$ or $\sqrt{npq}$

Characteristics of the Normal Distribution (p. 674)
- The graph of the curve is bell-shaped and symmetric with respect to the mean.
- The mean, median, and mode are equal and located at the center.
- The curve is continuous.
- The curve approaches, but never touches, the x-axis.
- The total area under the curve is equal to 1 or 100%.

The Empirical Rule (p. 675) In a normal distribution with mean μ and standard deviation σ:

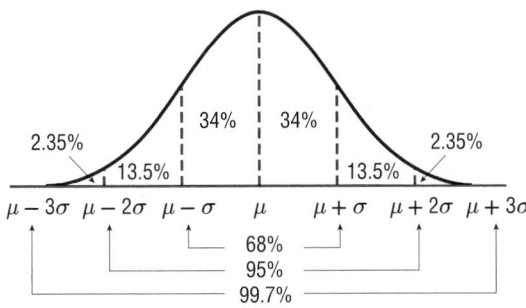

- approximately 68% of the data values fall between $\mu - \sigma$ and $\mu + \sigma$.
- approximately 95% of the data values fall between $\mu - 2\sigma$ and $\mu + 2\sigma$.
- approximately 99.7% of the data values fall between $\mu - 3\sigma$ and $\mu + 3\sigma$.

Formula for z-Values (p. 676) The z-value for a data value in a set of data is given by $z = \dfrac{X - \mu}{\sigma}$, where X is the data value, μ is the mean, and σ is the standard deviation.

Characteristics of the Standard Normal Distribution (p. 677)
- The total area under the curve is equal to 1 or 100%.
- Almost all of the area is between $z = -3$ and $z = 3$.
- The distribution is symmetric.
- The mean is 0, and the standard deviation is 1.
- The curve approaches, but never touches, the x-axis.

Central Limit Theorem (p. 686) As the sampling size n increases:
- the shape of the distribution of the sample means of a population with mean μ and standard deviation σ will approach a normal distribution, and
- the distribution will have a mean μ and standard deviation $\sigma_{\bar{x}} = \dfrac{\sigma}{\sqrt{n}}$.

z-Value of a Sample Mean (p. 686) The z-value for a sample mean in a population is given by $z = \dfrac{\bar{x} - \mu}{\sigma_{\bar{x}}}$, where $\bar{x}$ is the sample mean, μ is the mean of the population, and $\sigma_{\bar{x}}$ is the standard error.

Approximation Rule for Binomial Distributions (p. 689) The normal distribution can be used to approximate a binomial distribution when $np \geq 5$ and $nq \geq 5$.

Normal Approximation of a Binomial Distribution (p. 690) The procedure for the normal approximation of a binomial distribution is as follows.

Step 1 Find the mean μ and standard deviation σ.
Step 2 Write the problem in probability notation using X.
Step 3 Find the continuity correction factor, and rewrite the problem to show the corresponding area under the normal distribution.
Step 4 Find any corresponding z-values for X.
Step 5 Use a graphing calculator to find the corresponding area.

Maximum Error of Estimate (p. 697) The maximum error of estimate E for a population mean μ is given by $E = z \cdot \sigma_{\bar{x}}$ or $z \cdot \dfrac{\sigma}{\sqrt{n}}$, where z is a critical value that corresponds to a particular confidence level, and $\sigma_{\bar{x}}$ or $\dfrac{\sigma}{\sqrt{n}}$ is the standard deviation of the sample means. When $n \geq 30$, s, the sample standard deviation, may be substituted for σ.

Confidence Interval for the Mean (p. 697) A confidence interval CI for a population mean μ is given by $CI = \bar{x} \pm E$ or $\bar{x} \pm z \cdot \dfrac{\sigma}{\sqrt{n}}$, where $\bar{x}$ is the sample mean and E is the maximum error of estimate.

Characteristics of the t-Distribution (p. 699)
- The distribution is bell-shaped and symmetric about the mean.
- The mean, median, and mode equal 0 and are at the center of the distribution.
- The curve never touches the x-axis.
- The standard deviation is greater than 1.
- The distribution is a family of curves based on the sample size n.
- As n increases, the distribution approaches the standard normal distribution.

Confidence Interval Using t-Distribution (p. 699) A confidence interval CI for the t-distribution is given by $CI = \bar{x} \pm t \cdot \dfrac{s}{\sqrt{n}}$, where $\bar{x}$ is the sample mean, t is a critical value with $n - 1$ degrees of freedom, s is the sample standard deviation, and n is the sample size.

Minimum Sample Size Formula (p. 701) The minimum sample size needed when finding a confidence interval for the mean is given by $n = \left(\dfrac{z\sigma}{E}\right)^2$, where n is the sample size and E is the maximum error of estimate.

Steps for Hypothesis Testing (p. 707)
Step 1 State the hypotheses, and identify the claim.
Step 2 Determine the critical value(s) and region.
Step 3 Calculate the test statistic.
Step 4 Reject or fail to reject the null hypothesis.

Correlation Coefficient (p. 713) For n pairs of sample data for the variables x and y, the correlation coefficient r between x and y is given by $r = \dfrac{1}{n-1}\sum\left(\dfrac{x_i - \bar{x}}{s_x}\right)\left(\dfrac{y_i - \bar{y}}{s_y}\right)$ where x_i and y_i represent the values for the ith pair of data, $\bar{x}$ and $\bar{y}$ represent the means of the two variables, and s_x and s_y represent the standard deviations of the two variables.

Formula for the t-Test for the Correlation Coefficient (p. 715) For a t-test of the correlation between two variables, the test statistic for ρ is the sample correlation coefficient r and the standardized test statistic t is given by $t = r\sqrt{\dfrac{n-2}{1-r^2}}$, where $n - 2$ is the degrees of freedom.

Equation of the Least-Squares Regression Line (p. 716) The equation of the least-squares regression line for an explanatory variable x and response variable y is $\hat{y} = ax + b$. The slope a and y-intercept b in this equation are given by $a = r\dfrac{s_y}{s_x}$ and $b = \bar{y} - a\bar{x}$, where r represents the correlation coefficient between the two variables, $\bar{x}$ and $\bar{y}$ represent their means, and s_x and s_y represent their standard deviations.

Limits and Derivatives — Chapter 12

Independence of Limit from Function Value at a Point (p. 737) The limit of a function $f(x)$ as x approaches c *does not* depend on the value of the function *at* point c.

One-Sided Limits (p. 738)

Left-Hand Limit If the value of $f(x)$ approaches a unique number L_1 as x approaches c from the left, then $\lim\limits_{x \to c^-} f(x) = L_1$.

Right-Hand Limit If the value of $f(x)$ approaches a unique number L_2 as x approaches c from the right, then $\lim\limits_{x \to c^+} f(x) = L_2$.

Existence of a Limit at a Point (p. 738) The limit of a function $f(x)$ as x approaches c exists if and only if both one-sided limits exist *and* are equal. That is, if

$$\lim_{x \to c^-} f(x) = \lim_{x \to c^+} f(x) = L, \text{ then } \lim_{x \to c} f(x) = L.$$

Limits at Infinity (p. 740)
- If the value of $f(x)$ approaches a unique number L_1 as x increases, then $\lim\limits_{x \to \infty} f(x) = L_1$, which is read *the limit of $f(x)$ as x approaches infinity is L_1*.
- If the value of $f(x)$ approaches a unique number L_2 as x decreases, then $\lim\limits_{x \to -\infty} f(x) = L_2$, which is read *the limit of $f(x)$ as x approaches negative infinity is L_2*.

Limits of Constant Functions (p. 746) The limit of a constant function at any point is the constant value of the function or $\lim\limits_{x \to c} k = k$.

Limits of the Identity Function (p. 746) The limit of the identity function at any point c is c or $\lim\limits_{x \to c} x = c$.

Properties of Limits (p. 746) If k and c are real numbers, n is a positive integer, and $\lim\limits_{x \to c} f(x)$ and $\lim\limits_{x \to c} g(x)$ both exist, then the following are true.

Sum Property $\lim\limits_{x \to c} [f(x) + g(x)] = \lim\limits_{x \to c} f(x) + \lim\limits_{x \to c} g(x)$

Difference Property $\lim\limits_{x \to c} [f(x) - g(x)] = \lim\limits_{x \to c} f(x) - \lim\limits_{x \to c} g(x)$

Scalar Multiple Property $\lim\limits_{x \to c} [kf(x)] = k \lim\limits_{x \to c} f(x)$

Product Property $\lim\limits_{x \to c} [f(x) \cdot g(x)] = \lim\limits_{x \to c} f(x) \cdot \lim\limits_{x \to c} g(x)$

Quotient Property $\lim\limits_{x \to c} \dfrac{f(x)}{g(x)} = \dfrac{\lim\limits_{x \to c} f(x)}{\lim\limits_{x \to c} g(x)}$, if $\lim\limits_{x \to c} g(x) \neq 0$

Power Property $\lim\limits_{x \to c} [f(x)^n] = \left[\lim\limits_{x \to c} f(x)\right]^n$

nth Root Property $\lim\limits_{x \to c} \sqrt[n]{f(x)} = \sqrt[n]{\lim\limits_{x \to c} f(x)}$, if $\lim\limits_{x \to c} f(x) > 0$ when n is even.

Limits of Polynomial Functions (p. 748) If $p(x)$ is a polynomial function and c is a real number, then $\lim\limits_{x \to c} p(x) = p(c)$.

Limits of Rational Functions (p. 748) If $r(x) = \dfrac{p(x)}{q(x)}$ is a rational function and c is a real number, then $\lim\limits_{x \to c} r(x) = r(c) = \dfrac{p(c)}{q(c)}$ if $q(c) \neq 0$.

Limits of Power Functions at Infinity (p. 750) For any positive integer n,
- $\lim\limits_{x \to \infty} x^n = \infty$.
- $\lim\limits_{x \to -\infty} x^n = \infty$ if n is even.
- $\lim\limits_{x \to -\infty} x^n = -\infty$ if n is odd.

Limits of Polynomial Functions at Infinity (p. 751) Let p be a polynomial function $p(x) = a_n x^n + \ldots + a_1 x + a_0$. Then $\lim_{x \to \infty} p(x) = \lim_{x \to \infty} a_n x^n$ and $\lim_{x \to -\infty} p(x) = \lim_{x \to -\infty} a_n x^n$.

Limits of Reciprocal Function at Infinity (p. 751) The limit of a reciprocal function at positive or negative infinity is 0 or $\lim_{x \to \infty} \frac{1}{x} = \lim_{x \to -\infty} \frac{1}{x} = 0$.

Instantaneous Rate of Change (p. 758) The instantaneous rate of change of the graph of $f(x)$ at the point $(x, f(x))$ is the slope m of the tangent line given by $m = \lim_{h \to 0} \frac{f(x+h) - f(x)}{h}$, provided the limit exists.

Average Velocity (p. 760) If position is given as a function of time $f(t)$, for any two points in time a and b, the average velocity v is given by $v_{\text{avg}} = \frac{\text{change in distance}}{\text{change in time}} = \frac{f(b) - f(a)}{b - a}$.

Instantaneous Velocity (p. 760) If the distance an object travels is given as a function of time $f(t)$, then the instantaneous velocity $v(t)$ at a time t is given by $v(t) = \lim_{h \to 0} \frac{f(a+h) - f(a)}{h}$, provided the limit exists.

Power Rule for Derivatives (p. 767) The power of x in the derivative is one less than the power of x in the original function, and the coefficient of the power of x in the derivative is the same as the power of x in the original function.

Rules for Derivatives (p. 767)

Constant The derivative of a constant function is zero. If $f(x) = c$, then $f'(x) = 0$.

Constant Multiple of a Power If $f(x) = cx^n$, where c is a constant and n is a real number, then $f'(x) = cnx^{n-1}$.

Sum and Difference If $f(x) = g(x) \pm h(x)$, then $f'(x) = g'(x) \pm h'(x)$.

Extreme Value Theorem (p. 769) If a function f is continuous on a closed interval $[a, b]$, then $f(x)$ attains both an absolute maximum and an absolute minimum on $[a, b]$.

Product Rule for Derivatives (p. 770) If f and g are differentiable at x, then
$$\frac{d}{dx}[f(x)g(x)] = f'(x)g(x) + f(x)g'(x).$$

Quotient Rule for Derivatives (p. 771) If f and g are differentiable at x and $g(x) \neq 0$, then
$$\frac{d}{dx}\left[\frac{f(x)}{g(x)}\right] = \frac{f'(x)g(x) - f(x)g'(x)}{[g(x)]^2}.$$

Definite Integral (p. 777) The area of a region under the graph of a function is
$$\int_a^b f(x)dx = \lim_{n \to \infty} \sum_{i=1}^{n} f(x_i)\Delta x,$$
where a and b are the lower limits and upper limits respectively, $\Delta x = \frac{b-a}{n}$, and $x_i = a + i\Delta x$. This method is referred to as the *right Riemann sum*.

Rules for Antiderivatives (p. 785)

Power Rule If $f(x) = x^n$, where n is a rational number other than -1, $F(x) = \frac{x^{n+1}}{n+1} + C$.

Constant Multiple of a Power If $f(x) = kx^n$, where n is a rational number other than -1 and k is a constant, then $F(x) = \frac{kx^{n+1}}{n+1} + C$.

Sum and Difference If the antiderivatives of $f(x)$ and $g(x)$ are $F(x)$ and $G(x)$ respectively, then the antiderivative of $f(x) \pm g(x)$ is $F(x) \pm G(x)$.

Indefinite Integral (p. 785) The indefinite integral of $f(x)$ is defined by $\int f(x)dx = F(x) + C$, where $F(x)$ is an antiderivative of $f(x)$ and C is any constant.

Fundamental Theorem of Calculus (p. 786) If f is continuous on $[a, b]$ and $F(x)$ is any antiderivative of $f(x)$, then $\int_a^b f(x)dx = F(b) - F(a)$. The difference $F(b) - F(a)$ is commonly denoted by $F(x)\Big|_a^b$.

Selected Answers and Solutions

For Homework Help, go to **Hotmath.com**
Complete, step-by-step solutions of most odd-numbered exercises are provided free of charge.

Chapter 0 Preparing for Precalculus

Page P5   Lesson 0-1

1. $J = \{0, 3, 6, 9, 12\}$; false   **3.** $L = \{2, 3, 5, 7, 11, 13\}$; false
5. $N = \{1, 2, 3, 4, 5, 6, 7, 8, 9, 10, 11\}$; false   **7.** $A = \{$all girls
names starting with A$\}$; true   **9.** False; $D = \{4, 5, 10\}$; 3 is not
an element of D.   **11.** True; all of the elements of B are also
elements of A.   **13.** False; 5 is an element of D.   **15.** False;
0 is not an element of F.   **17.** $\{2, 3, 4, 5, 7, 8, 10, 12\}$
19. $\{0, 3, 4, 5, 7, 8, 11\}$ **21.** $\{6\}$ **23.** $\{1, 2, 6, 9, 10, 12\}$
25. $\{1, 2, 3, 6, 7, 9, 10\}$ **27.** $\{1, 2, 6, 11, 12, 14, 15\}$ **29.** $\{6\}$

31. **a.**

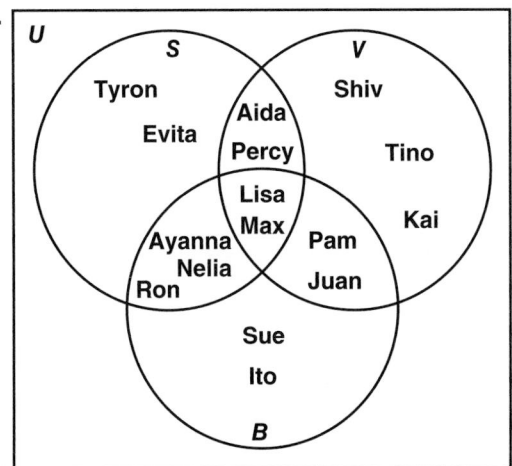

b. $S \cap V$ represents the students in Mr. Frank's gym class
who participate in both soccer and volleyball. The
students who participate in both soccer and volleyball
are Percy, Aida, Max, and Lisa. Therefore, $S \cap V =$
$\{$Percy, Aida, Max, Lisa$\}$.

c. S' represents the students in Mr. Frank's gym class
who participate in sports other than soccer. The
students who participate in sports other than soccer
are are Shiv, Tino, Kai, Juan, Pam, Ito, and Sue.
Therefore, $S' = \{$Shiv, Tino, Kai, Juan, Pam, Ito, Sue$\}$.

d. $B \cup V$ represents the students in Mr. Frank's gym class
who participate in basketball, volleyball, or both. The
students who participate in basketball, volleyball, or
both are Ito, Kai, Nelia, Percy, Juan, Tino, Aida, Ron,
Shiv, Max, Ayanna, Pam, Sue, Lisa. Therefore, $B \cup V =$
$\{$Ito, Kai, Nelia, Percy, Juan, Tino, Aida, Ron, Shiv,
Max, Ayanna, Pam, Sue, Lisa$\}$.

33. Sample answer: Let set $J = \{$apple juice, orange juice,
grape juice$\}$. Let set $S = \{$Brand A, Brand B$\}$. The word *or* in
"juice or soda" indicates the union of these sets. Since the
sets have no elements in common, $J \cup S = \{$apple juice,
orange juice, grape juice, Brand A soda, Brand B soda$\}$,
which includes the total number of elements from each set.
Thus, the sum of the number of elements in each set would
be the total number of elements in the union of the sets. This
principle could not be applied if the sets had elements in
common.   **35.** $\overline{AC}$ **37.** $\{D\}$ **39.** $\overrightarrow{EB}$

Page P8   Lesson 0-2

1. -1 **3.** $1 - i$ **5.** i **7.** $2i$ **9.** $-1 + 8i$ **11.** $-1.5 + 2i$

13. $3.5 + 10.4i$ **15.** 3 **17.** $9 + 2j$ ohms   **19.** $-15 + 8i$
21. $8 + 6i$ **23.** 34 **25.** -13 **27.** $\frac{2}{5} + \frac{1}{5}i$
29. Rationalize the denominator by multiplying the
numerator and denominator by the conjugate $-4 + i$.

$$\frac{3 - 2i}{-4 - i} \cdot \frac{-4 + i}{-4 + i} = \frac{(3 - 2i)(-4 + i)}{(-4 - i)(-4 + i)}$$
$$= \frac{-12 + 3i + 8i - 2i^2}{16 - i^2}$$
$$= \frac{-12 + 11i - 2(-1)}{16 - (-1)}$$
$$= \frac{-10 + 11i}{17}$$
$$= -\frac{10}{17} + \frac{11}{17}i$$

31. $\frac{23}{26} - \frac{11}{26}i$ **33.** $\left(\frac{1}{3} - \frac{\sqrt{6}}{3}\right) + \left(\frac{\sqrt{2}}{3} + \frac{\sqrt{3}}{3}\right)i$
35. $-13 - 34j$ volts   **37.** $28 + 57j$ volts   **39.** $\pm 4i$ **41.** $\pm 2\sqrt{3}i$
43. $\pm 13i$ **45.** $2, 5$ **47.** $3, -4$ **49.** $6, 4$ **51.** $12 - 31i$
53. $20 + 15i$

Page P13   Lesson 0-3

1.

$f(x) = x^2 + 5x + 6$

3.

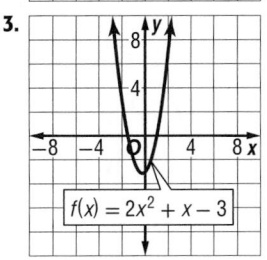

$f(x) = 2x^2 + x - 3$

5.

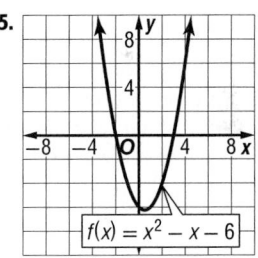

$f(x) = x^2 - x - 6$

7.

$d(t) = 80t - 16t^2 + 3.5$

9.

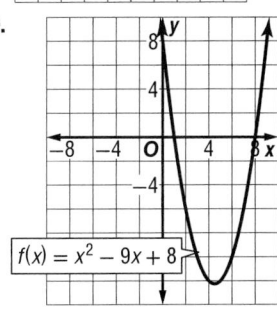

$f(x) = x^2 - 9x + 8$

11.

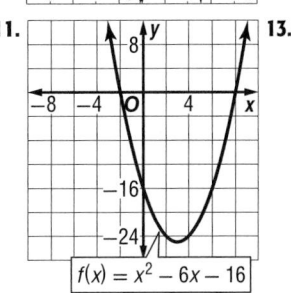

$f(x) = x^2 - 6x - 16$

13.

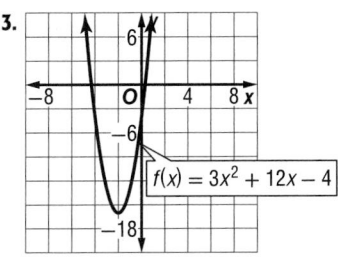

$f(x) = 3x^2 + 12x - 4$

15. maximum; $\left(\frac{3}{4}, -3\frac{7}{8}\right)$; domain: $\mathbb{R}$; range: $y \le -3\frac{7}{8}$, for $y \in \mathbb{R}$ **17.** maximum; $\left(\frac{3}{2}, -2\frac{3}{4}\right)$; domain: $\mathbb{R}$; range: $y \le -2\frac{3}{4}$, for $y \in \mathbb{R}$ **19.** minimum; $(-1, 5)$; domain: $\mathbb{R}$; range: $y \ge 5$, for $y \in \mathbb{R}$ **21.** maximum; $\left(-\frac{1}{3}, -\frac{2}{3}\right)$; domain: $\mathbb{R}$; range: $y \le -\frac{2}{3}$, for $y \in \mathbb{R}$ **23.** $3, 7$ **25.** $-4, 7$ **27.** $\frac{5}{4}, -1$ **29.** $-10, 2$ **31.** $-4, 6$

33.
$t^2 - 3t - 7 = 0$	Rewrite so that the left side is of the form $x^2 + bx$.
$t^2 - 3t + \frac{9}{4} = 7 + \frac{9}{4}$	Because $\left(\frac{3}{2}\right)^2 = \frac{9}{4}$, add $\frac{9}{4}$ to each side.
$\left(t - \frac{3}{2}\right)^2 = \frac{37}{4}$	Write the left side as a perfect square.
$t - \frac{3}{2} = \pm\sqrt{\frac{37}{4}}$	Take the square root of each side.
$t = \frac{3 \pm \sqrt{37}}{2}$	Add $\frac{3}{2}$ to each side.

35. -6 **37.** $-\frac{3}{2}, \frac{1}{3}$ **39.** $\frac{1 \pm i\sqrt{35}}{4}$ **41a.** $(8 - 2x)(10 - 2x) = 20$

41b.

41c. about 2.21 in. **43.** $5, 8$
45. 6 **47.** $\pm\frac{\sqrt{21}}{7}i$ **49.** $\frac{-3 \pm \sqrt{15}}{2}$
51. $-8, -9$

53. Let x be the first number and $-9 - x$ be the other number.
$$x(-9 - x) = 24$$
$$-x^2 - 9x - 24 = 0$$
Because the graph of the related function does not intersect the x-axis, this equation has no real solutions. Therefore, no such numbers exist.

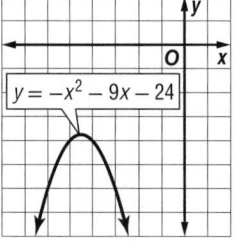

Page P17 Lesson 0-4

1. -13 **3.** $\frac{6}{5}$ **5.** not a real number **7.** 3 **9.** $-3x^3$
11. $\sqrt[8]{8y^2}$ **13.** $2x^8\sqrt{5}$ **15.** $|a^3|b^2\sqrt[4]{b}$ **17.** $b^{\frac{7}{4}}$ or $b\sqrt[4]{b^3}$
19. $2\sqrt[6]{10a^3}$ **21.** $\sqrt[15]{2^{14}}$ **23a.** ≈ 36.8 **23b.** ≈ 12.0 ft **25.** 6
27. $\frac{1}{7}$ **29a.** about 1.26 **29b.** 2; sample answer: The perfect 8th is 12 half tones above the perfect 1st. Evaluating $r = \left(\sqrt[12]{2}\right)^n$ for $n = 12$, yields $r = \left(\sqrt[12]{2}\right)^{12}$ or $2^{\frac{12}{12}}$, which simplifies to 2^1 or 2. **31.** $4a^3b^2c\sqrt[3]{2bc}$
33. $\sqrt[6]{64x^7y^6z^{18}} = \sqrt[6]{2^6x^6xy^6z^{18}} = \sqrt[6]{(2xyz^3)^6}\ x = 2x|yz^3|\sqrt[6]{x}$ Because you are taking an even root of an even power and the result is an odd power, you must use the absolute value of y and z^3. It is not necessary to use absolute value on x because $\sqrt[6]{x^7}$ is an even root of an odd power.
35. $wxy^2z^2\sqrt[5]{wx^3z^3}$

---

1. $(1, 3)$

3. $(2, -4)$

5. $(-3, -2)$

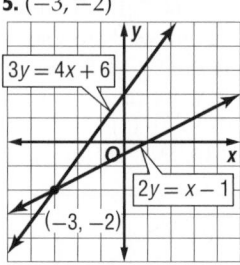

7. $(3, -1)$ **9.** $(5.25, 0.75)$ **11.** $(9, 6)$ **13.** movie rental: 10 hours, walking dog: 3 hours **15.** $(8, 5)$ **17.** $\left(\frac{1}{2}, \frac{2}{3}\right)$
19. $(-4, -5)$ **21.** $(7, -1, 1)$ **23.** $(-6, -4, 7)$ **25.** $(10, -7, 1)$

27. Eliminate the y term first.
$4x - 2y + z = -5$	Equation 1
$(+)\ \underline{10x + 2y + 6z = 12}$	$2 \times$ Equation 2
$14x + 7z = 7$	Add.

and
$-15x - 3y - 9z = -18$	$-3 \times$ Equation 2
$(+)\ \underline{-2x + 3y + 2z = -4}$	Equation 3
$-17x - 7z = -22$	Add.

Solve the resulting system of two equations.
$$14x + 7z = 7$$
$$-17x - 7z = -22$$
$14x + 7z = 7$	Equation 1
$(+)\ \underline{-17x - 7z = -22}$	Equation 2
$-3x = -15$	Add.
$x = 5$	Divide.

Substitute $x = 5$ into Equation 1 and solve for z.
$14(5) + 7z = 7$	$x = 5$
$70 + 7z = 7$	Multiply.
$7z = -63$	Subtract 70 from each side.
$z = -9$	Divide each side by 7.

Substitute the values for x and z into Equation 1 and solve for y.
$4(5) - 2y - 9 = -5$	$x = 5$ and $z = -9$
$-2y + 11 = -5$	Simplify.
$-2y = -16$	Subtract 11 from each side.
$y = 8$	Divide each side by -2.

The solution of the system is equations is $(5, 8, -9)$.
29. no solution; inconsistent **31.** infinitely many solutions; consistent and dependent **33.** $\left(\frac{1}{2}, 3\right)$; consistent and independent **35.** $(4, -8, 5)$; consistent and independent

37.

39.

41.

43.

no solution

45.

47.

49.

---

35. $\begin{bmatrix} 8 & -7 & -6 \\ 45 & -35 & 24 \end{bmatrix}$ **37.** $\begin{bmatrix} 8 & 12 \\ -7 & 9 \end{bmatrix}$ **39.** $\begin{bmatrix} 4 & -3 & 1 \\ 7 & -3 & -1 \\ 5 & -4 & 3 \end{bmatrix}$

41. $\begin{bmatrix} 0 & 4 & 8 \\ -8 & 12 & 0 \\ 16 & 16 & -8 \end{bmatrix}$ **43.** impossible   **45a.** $\begin{bmatrix} 0.11 \text{ s} \\ 0.51 \text{ s} \\ 1.31 \text{ s} \\ 2.12 \text{ s} \end{bmatrix}$

45b. $\frac{1}{60}A$

Page P31   Lesson 0-7

1. 30   **3.** 48   **5.** 300   **7.** 720   **9.** 720   **11.** 35   **13.** 8
15. 120   **17.** 15,120   **19.** 495   **21a.** 3003   **21b.** 360,360
21c. $\frac{56}{143}$ **23a.** 8,000,000   **23b.** $\frac{1}{8,000,000}$ **23c.** 62,500   **23d.** $\frac{1}{128}$
23e. 10,000; $\frac{1}{800}$ **25.** $\frac{1}{204}$

27 There are 7 red and 8 orange gumballs to choose from
and a total of $7 + 8 + 9 + 7 + 5$ or 36 gumballs in the
bag. Since order does not matter, use combinations to
calculate this probability. The number of favorable
outcomes is the combination of 7 red taken 1 at a time
times the number of combinations of 8 orange taken 2 at
a time. The total number of possible outcomes is the
combination of 36 gumballs taken 3 at a time.

$$P(1 \text{ R and } 2 \text{ N}) = \frac{{}_7C_1 \cdot {}_8C_2}{{}_{36}C_3}$$

$$= \frac{\frac{7!}{(7-1)!1!} \cdot \frac{8!}{(8-2)!2!}}{\frac{36!}{(36-3)!3!}}$$

$$= \frac{\frac{7 \cdot 6!}{6!1} \cdot \frac{8 \cdot 7 \cdot 6!}{6!2 \cdot 1}}{\frac{36 \cdot 35 \cdot 34 \cdot 33!}{33!3 \cdot 2 \cdot 1}}$$

$$= \frac{7 \cdot 4 \cdot 7}{6 \cdot 35 \cdot 34}$$

$$= \frac{7}{255}$$

29. $\frac{1}{68}$ **31.** $\frac{1}{285}$ or about 0.4%   **33.** $\frac{880}{2691}$ or about 32.7%

Page P37   Lesson 0-8

1. mean: 28, median: 28.5, mode: 31   **3.** mean: 8.5, median:
8.5, modes: 6 and 11   **5.** range: \$12.75, variance: $\approx$\$16.86,
standard deviation: $\approx$\$4.11   **7.** range: 1.9, variance: $\approx$0.54,
standard deviation: $\approx$0.73

9 **a.** range = maximum − minimum, which is $74 - 60$ or
14. To calculate the population variance σ^2, first
calculate the population mean μ. Then use μ to
calculate σ^2.

$$\mu = \frac{66 + 72 + 70 + \dots + 62 + 62 + 61}{20} \text{ or } 66.65$$

$$\sigma^2 = \frac{\Sigma(x - \mu)^2}{n}$$

$$= \frac{(66 - 66.65)^2 + (72 - 66.65)^2 + \dots + (62 - 66.65)^2 + (61 - 66.65)^2}{20}$$

$$= 18.1275$$

The population variance is about 18.1. The population
standard deviation σ is $\sqrt{\sigma^2} = \sqrt{18.1275}$ or about 4.3.
b. The sample range is $73 - 61$ or 12. To calculate the
sample variance, first calculate the sample mean $\bar{x}$.
Then use $\bar{x}$ to calculate s^2.

Page P27   Lesson 0-6

1. 2×2 **3.** 2×4 **5.** 36   **7.** 29   **9.** −11

11. $\begin{bmatrix} 14 & -9 \\ -3 & -1 \\ 8 & 15 \end{bmatrix}$ **13.** $\begin{bmatrix} -16 & 5 & 6 \\ 10 & -23 & 8 \end{bmatrix}$ **15.** $\begin{bmatrix} 12 & -3 \\ 7 & -19 \\ -16 & 1 \end{bmatrix}$

17. $\begin{bmatrix} 93 & 101 & 164 \\ 94 & 105 & 170 \\ 101 & 99 & 167 \end{bmatrix}$ **19.** $\begin{bmatrix} -9 & -45 \\ 72 & 36 \end{bmatrix}$

21. $\begin{bmatrix} -18 & 60 & -30 & 54 \end{bmatrix}$ **23.** $\begin{bmatrix} -16 & 24 \\ -48 & 20 \\ 12 & 16 \end{bmatrix}$ **25.** $\begin{bmatrix} 4 & 20 \\ 13 & -17 \\ 9 & -26 \end{bmatrix}$

27. $\begin{bmatrix} \frac{3}{2} & 2 \\ \frac{5}{2} & -\frac{9}{2} \\ 5 & \frac{3}{2} \end{bmatrix}$ **29.** $\begin{bmatrix} 1 & 16 \\ 8 & -8 \\ -1 & -29 \end{bmatrix}$ **31.** $\begin{bmatrix} 5 & -\frac{5}{2} & -3 \\ -5 & 4 & \frac{9}{2} \end{bmatrix}$

33 $3X = 2J - L$

$$3X = \begin{bmatrix} 16 & -20 & 6 \\ -8 & 2 & 24 \end{bmatrix} - \begin{bmatrix} 4 & 1 & -8 \\ 11 & -7 & 6 \end{bmatrix}$$

$$3X = \begin{bmatrix} 12 & -21 & 14 \\ -19 & 9 & 18 \end{bmatrix}$$

$$X = \frac{1}{3}\begin{bmatrix} 12 & -21 & 14 \\ -19 & 9 & 18 \end{bmatrix}$$

$$X = \begin{bmatrix} 4 & -7 & \frac{14}{3} \\ -\frac{19}{3} & 3 & 6 \end{bmatrix}$$

$$\bar{x} = \frac{68 + 71 + 70 + 72 + 73 + 65 + 63 + 62 + 62 + 61}{10} \text{ or } 66.7$$

$$s^2 = \frac{\Sigma(x - \bar{x})^2}{n - 1}$$

$$= \frac{(68 - 66.7)^2 + (71 - 66.7)^2 + \cdots + (62 - 66.65)^2 + (61 - 66.65)^2}{10 - 1}$$

$$\approx 21.34444444$$

The sample variance is about 21.3. The sample standard deviation s is $\sqrt{s^2} \approx \sqrt{21.34444444}$ or about 4.6.

11. The sample standard deviations for Stores 1 and 2 are $8.94 and $6.61, respectively. So, there is a greater variation in prices of graphic novels at Store 1.

13.

Chapter 1 Functions from a Calculus Perspective

Page 3   Chapter 1   Get Ready

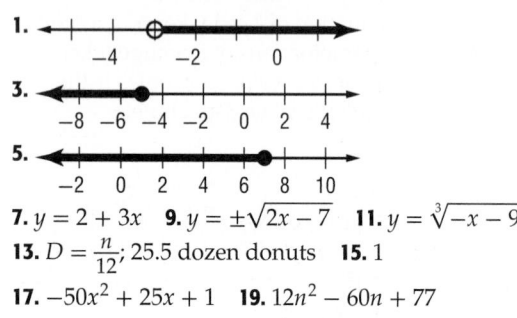

7. $y = 2 + 3x$ **9.** $y = \pm\sqrt{2x - 7}$ **11.** $y = \sqrt[3]{-x - 9}$
13. $D = \frac{n}{12}$; 25.5 dozen donuts   **15.** 1
17. $-50x^2 + 25x + 1$ **19.** $12n^2 - 60n + 77$

Pages 9–12   Lesson 1-1

1. $\{x \mid x > 50, x \in \mathbb{R}\}$; $(50, \infty)$ **3.** $\{x \mid x \leq -4, x \in \mathbb{R}\}$; $(-\infty, -4]$
5. $\{x \mid 8 < x < 99, x \in \mathbb{R}\}$; $(8, 99)$ **7.** $\{x \mid x < -19 \text{ or } x > 21,$
$x \in \mathbb{R}\}$; $(-\infty, -19) \cup (21, \infty)$ **9.** $\{x \mid 0.25n = x, n \geq -1,$
$n \in \mathbb{Z}\}$ **11.** $\{x \mid x \leq -45 \text{ or } x > 86, x \in \mathbb{R}\}$; $(-\infty, -45] \cup$
$(86, \infty)$ **13.** $\{x \mid x = 5n, n \in \mathbb{Z}\}$ **15.** function   **17.** function
19. function   **21.** function   **23.** function   **25.** function
27. not a function   **29a.** $\{(1, 70), (2, 75), (3, 70), (4, 62),$
$(5, 65)\}$ **29b.** Yes; there is exactly one estimated high temperature each day. Yes; there is exactly one low temperature each day.   **31a.** -207 **31b.** $24y^3 + 12y + 9$
31c. $-375b^3 - 675b^2 - 435b - 90$

33 **a.** $g(-2) = \dfrac{3(-2)^3}{(-2)^2 + (-2) - 4}$

$$= \frac{3(-8)}{4 + (-2) - 4}$$

$$= \frac{-24}{-2}$$

$$= 12$$

b. $g(5x) = \dfrac{3(5x)^3}{(5x)^2 + (5x) - 4}$

$$= \frac{3(125x^3)}{25x^2 + 5x - 4}$$

$$= \frac{375x^3}{25x^2 + 54 - 4}$$

c. $g(8 - 4b) = \dfrac{3(8 - 4b)^3}{(8 - 4b)^2 + (8 - 4b) - 4}$

$$= \frac{3(512 - 768b + 384b^2 - 64b^3)}{64 - 64b + 16b^2 + 8 - 4b - 4}$$

$$= \frac{-192b^3 + 1152b^2 - 2304b + 1536}{16b^2 - 68b + 68}$$

$$= \frac{-48b^3 + 288b^2 - 576b + 384}{4b^2 - 17b + 17}$$

35a. -0.8 **35b.** $-1 - \dfrac{1}{8x}$ **35c.** $\dfrac{-48b^3 + 288b^2 - 576b + 384}{4b^2 - 17b + 17}$
37a. $20\sqrt{6}$ **37b.** $10|x|\sqrt{6}$ **37c.** $5\,|7 + n|\sqrt{6}$
39. $(-\infty, -4) \cup (-4, -1) \cup (-1, \infty)$ **41.** $(-\infty, \infty)$
43. $(0.25, \infty)$ **45.** $(-\infty, -1) \cup (-1, 0) \cup (0, \infty)$
47. Yes; sample answer: Because length must be positive, the domain of the function is $(0, \infty)$.   **49.** 21; 157   **51.** 1; $8\frac{1}{6}$
53a. 8.65 million; 9.36 million   **53b.** integral values inside the interval $[0, 10]$
55. No; a vertical line would pass through infinitely many points.   **57.** No; the y-axis is a vertical line that passes through two points on the graph, $(0, 0)$ and $(0, -4)$.

59 Sample answer: Because presidential elections are held every 4 years, and do not have a finite end, it is impractical to display the set in interval notation. If set-builder notation is used, the interval can be taken into account and a finite interval is not necessary. Therefore, the set of presidential election years beginning in 1792 can be described in set-builder notation as $\{x \mid x = 4n + 1792, n \in \mathbb{W}\}$.

61. $D = \{x \mid x \geq 1874, x \in \mathbb{W}\}$ **63.** $-5; -5; 0$
65. $\dfrac{1}{a + 4}; \dfrac{1}{a + h + 4}; \dfrac{-1}{a^2 + ah + 8a + 4h + 16}$ **67.** $a^2 - 6a + 8;$
$a^2 + 2ah + h^2 - 6a - 6h + 8; 2a - 6 + h$ **69.** $-a^5; -a^5 - 5a^4h - 10a^3h^2 - 10a^2h^3 - 5ah^4 - h^5; -5a^4h - 10a^3h - 10a^2h^2 - 5ah^3 - h^4$ **71.** $7a - 3; 7a + 7h - 3; 7$ **73.** $a^3; a^3 + 3a^2h + 3ah^2 + h^3; 3a^2 + 3ah + h^2$ **75a.** $A(\ell) = \dfrac{\ell^2}{1.8}$; $[5, 11.5]$
75b. $A(h) = 2.1h^2$; $[2.4, 5.5]$ **75c.** 52.9 in^2 **77.** No; sample answer: Most nonnegative x-values are paired with two y-values because it is necessary to take both the positive and negative values of the absolute value of x when solving the equation for y.

79a.

[−10, 10] scl: 1 by [−10, 10] scl: 1   [−10, 10] scl: 1 by [−10, 10] scl: 1

[−10, 10] scl: 1 by [−10, 10] scl: 1   [−10, 10] scl: 1 by [−10, 10] scl: 1

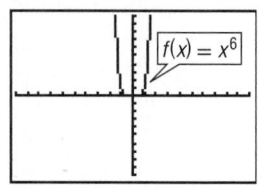

[−10, 10] scl: 1 by [−10, 10] scl: 1   [−10, 10] scl: 1 by [−10, 10] scl: 1

79b.

n	Range
1	$(-\infty, \infty)$
2	$[0, \infty)$
3	$(-\infty, \infty)$
4	$[0, \infty)$
5	$(-\infty, \infty)$
6	$[0, \infty)$

79c. Sample answer: When n is even in $f(x) = x^n$, the range is $[0, \infty)$.
79d. Sample answer: When n is odd in $f(x) = x^n$, the range is $(-\infty, \infty)$.

81 When the denominator of $\dfrac{1}{(x+3)(x+1)(x-5)}$ is zero, the expression is undefined. Therefore, $f(x)$ is undefined for $x = -3, x = -1$, and $x = 5$. In interval notation the domain can be described as $(-\infty, -3) \cup (-3, -1) \cup (-1, 5) \cup (5, \infty)$, and in set-builder notation the domain can be described as $\{x \mid x \neq -3, x \neq -1, x \neq -5, x \in \mathbb{R}\}$. I prefer set-builder notation because instead of listing the four intervals in which x can be an element, it lists the three real numbers that x cannot be equal to.

83. true **85.** False; sample answer: Two or more elements in X may be matched with the same element in Y. **87.** Sample answer: If each of the possible inputs is assigned to exactly one output in the verbal description, the relation is a function. **89.** Sample answer: If each input value in the table is paired with a unique output value, the relation is a function. **91.** Sample answer: If each x-value can be paired with exactly one y-value after the equation is solved for y, then the relation is a function. **93.** 0.73 **95.** 216,216
97. $(-3.5, 2)$ **99.** $(9, 6)$; consistent and independent
101. $(2, 2)$; consistent and independent **103.** $\{1, 2, 4, 5, 8, 9, 11, 12\}$ **105.** $\varnothing$ **107.** B **109.** A

Pages 19–23   Lesson 1-2

1a. 37.75 **1b.** 32.68 **1c.** 28.21

3 The function value at $x = -8$ appears to be about 10. Find $f(-8)$.
$$f(x) = |x| + 2$$
$$f(-8) = |-8| + 2$$
$$= 8 + 2 \text{ or } 10$$
The function value at $x = -3$ appears to be about 5. Find $f(-3)$.
$$f(-3) = |-3| + 2$$
$$= 3 + 2 \text{ or } 5$$
The function value at $x = 0$ appears to be about 2. Find $f(0)$.
$$f(0) = |0| + 2$$
$$= 2$$

5a. $\dfrac{4}{3}$ **5b.** -1 **5c.** undefined **7a.** 36,000 tons, 46,000 tons, 54,000 tons; 35,750 tons, 46,082 tons, 53,113 tons
7b. about 2004 or $x = 11$; $f(10) = -0.0013(10)^4 + 0.0513(10)^3 - 0.662(10)^2 + 4.128(10) + 35.75 \approx 49$; $f(12) = -0.0013(12)^4 + 0.0513(12)^3 - 0.662(12)^2 + 4.128(12) + 35.75 \approx 52$
9. $D = (-\infty, \infty)$, $R = [2, \infty)$ **11.** $D = (-4, 4]$, $R = [-1, 6]$
13. $D = [-5, \infty)$, $R = [-2, \infty)$ **15a.** Sample answer: copper: $D = [-150, 150]$, $R = [1.75]$; aluminum: $D = [-150, 150]$, $R = [0.6, 1.5]$; zinc: $D = [-150, 150]$, $R = [0.5, 1.3]$; steel: $D = [-150, 150]$, $R = [0.2, 1.75]$ **15b.** Sample answer: copper ≈ 1.75 J, aluminum ≈ 1.2 J, zinc ≈ 0.5 J, steel ≈ 1.5 J

17 From the graph, it appears that $f(x)$ intersects the y-axis at approximately $(0, 0)$, so the y-intercept is 0. Find $f(0)$.
$$f(0) = 2(0)^3 - (0)^2 - 3(0)$$
$$f(0) = 0$$
Therefore, the y-intercept is 0.
From the graph, the x-intercepts appear to be at about -1 and $\dfrac{3}{2}$. Let $f(x) = 0$ and solve for x.
$$2x^3 - x^2 - 3x = 0$$
$$x(2x - 3)(x + 1) = 0$$
$$x = 0 \text{ or } 2x - 3 = 0 \text{ or } x + 1 = 0$$
$$x = \frac{3}{2} \qquad x = -1$$
Therefore, the zeros of f are 0, $\dfrac{3}{2}$, and -1.

19. y-intercept: 3; no zeros;
$$\sqrt{x} + 3 = 0$$
$$\sqrt{x} \neq -3$$
21. y-intercept: -2; zeros: $-\dfrac{1}{2}, \dfrac{2}{3}$;
$$6x^2 - x - 2 = 0$$
$$(2x + 1)(3x - 2) = 0$$
$$2x + 1 = 0 \text{ or } 3x - 2 = 0$$
$$x = -\frac{1}{2} \qquad x = \frac{2}{3}$$
23. y-intercept: 6; zeros: $-2, -3$;
$$x^2 + 5x + 6 = 0$$
$$(x + 2)(x + 3) = 0$$
$$x + 2 = 0 \text{ or } x + 3 = 0$$
$$x = -2 \qquad x = -3$$
25. x-axis;

x	y	(x, y)
1	2	$(1, 2)$
1	-2	$(1, -2)$
2	$\sqrt{5}$	$(2, \sqrt{5})$
2	$-\sqrt{5}$	$(2, -\sqrt{5})$
3	$\sqrt{6}$	$(3, \sqrt{6})$
3	$-\sqrt{6}$	$(3, -\sqrt{6})$

Because $x = (-y)^2 - 3$ is equivalent to $x = y^2 - 3$, the graph is symmetric with respect to the x-axis.

27. x-axis, y-axis, and origin;

x	y	(x, y)
1	$\dfrac{2\sqrt{2}}{5}$	$\left(1, \dfrac{2\sqrt{2}}{5}\right)$
1	$-\dfrac{2\sqrt{2}}{5}$	$\left(1, -\dfrac{2\sqrt{2}}{5}\right)$
2	$\dfrac{\sqrt{35}}{5}$	$\left(2, \dfrac{\sqrt{35}}{5}\right)$
2	$-\dfrac{\sqrt{35}}{5}$	$\left(2, -\dfrac{\sqrt{35}}{5}\right)$
3	$\dfrac{4\sqrt{5}}{5}$	$\left(3, \dfrac{4\sqrt{5}}{5}\right)$
3	$-\dfrac{4\sqrt{5}}{5}$	$\left(3, -\dfrac{4\sqrt{5}}{5}\right)$

x	y	(x, y)
-3	$\dfrac{4\sqrt{5}}{5}$	$\left(-3, \dfrac{4\sqrt{5}}{5}\right)$
-2	$\dfrac{\sqrt{35}}{5}$	$\left(-2, \dfrac{\sqrt{35}}{5}\right)$
-1	$\dfrac{2\sqrt{2}}{5}$	$\left(-1, \dfrac{2\sqrt{2}}{5}\right)$
1	$\dfrac{2\sqrt{2}}{5}$	$\left(1, \dfrac{2\sqrt{2}}{5}\right)$
2	$\dfrac{\sqrt{35}}{5}$	$\left(2, \dfrac{\sqrt{35}}{5}\right)$
3	$\dfrac{4\sqrt{5}}{5}$	$\left(3, \dfrac{4\sqrt{5}}{5}\right)$

Because $9x^2 - 25(-y)^2 = 1$ is equivalent to $9x^2 - 25y^2 = 1$, the graph is symmetric with respect to the x-axis.

Because $9(-x)^2 - 25y^2 = 1$ is equivalent to $9x^2 - 25y^2 = 1$, the graph is symmetric with respect to the y-axis.

(Continued on the next page)

x	y	(x, y)
-3	$-\frac{4\sqrt{5}}{5}$	$\left(-3, -\frac{4\sqrt{5}}{5}\right)$
-2	$-\frac{\sqrt{35}}{5}$	$\left(-2, -\frac{\sqrt{35}}{5}\right)$
-1	$-\frac{2\sqrt{2}}{5}$	$\left(-1, -\frac{2\sqrt{2}}{5}\right)$
1	$\frac{2\sqrt{2}}{5}$	$\left(1, \frac{2\sqrt{2}}{5}\right)$
2	$\frac{\sqrt{35}}{5}$	$\left(2, \frac{\sqrt{35}}{5}\right)$
3	$\frac{4\sqrt{5}}{5}$	$\left(3, \frac{4\sqrt{5}}{5}\right)$

Because $9(-x^2) - 25(-y)^2 = 1$ is equivalent to $9x^2 - 25y^2 = 1$, the graph is symmetric with respect to the origin.

29. origin;

x	y	(x, y)
-10	1	(-10, 1)
-5	2	(-5, 2)
-1	10	(-1, 10)
1	-10	(1, -10)
5	-2	(5, -2)
10	-1	(10, -1)

Because $-y = -\frac{10}{(-x)}$ is equivalent to $y = -\frac{10}{x}$, the graph is symmetric with respect to the origin.

31. y-axis

x	y	(x, y)
-3	9	(-3, 9)
-2	-16	(-2, -16)
-1	-7	(-1, -7)
1	-7	(1, -7)
2	-16	(2, 16)
3	9	(3, 9)

Because $y = (-x)^4 - 8(-x)^2$ is equivalent to $y = x^4 - 8x^2$, the graph is symmetric with respect to the y-axis.

33. y-axis;

x	y	(x, y)
$-2\sqrt{2}$	6	$(-2\sqrt{2}, 6)$
$-\frac{\sqrt{14}}{2}$	12	$\left(-\frac{\sqrt{14}}{2}, 12\right)$
-1	$2\sqrt{14} + 6$	$(-1, 2\sqrt{14} + 6)$
$\frac{\sqrt{14}}{2}$	12	$\left(\frac{\sqrt{14}}{2}, 12\right)$
1	$2\sqrt{14} + 6$	$(1, 2\sqrt{14} + 6)$
$2\sqrt{2}$	6	$(2\sqrt{2}, 6)$

Because $(y - 6)^2 + 8(-x)^2 = 64$ is equivalent to $(y - 6)^2 + 8x^2 = 64$, the graph is symmetric with respect to the y-axis.

35.

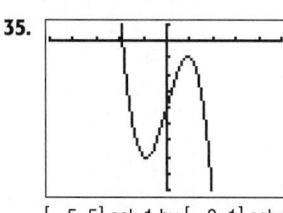

neither;
$$f(-x) = -2(-x)^3 + 5(-x) - 4 = -2x^3 - 5x - 4$$

[-5, 5] scl: 1 by [-9, 1] scl: 1

37.

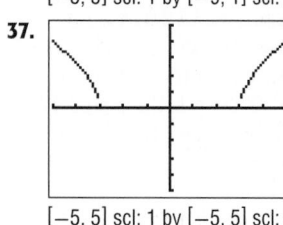

Even; the graph of h(x) is symmetric with respect to the y-axis.
$$h(-x) = \sqrt{(-x)^2 - 9} = \sqrt{x^2 - 9} = h(x)$$

[-5, 5] scl: 1 by [-5, 5] scl: 1

39.

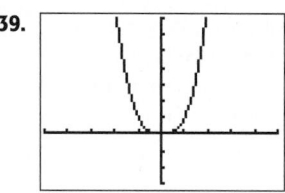

Even; the graph of f(x) is symmetric with respect to the y-axis.
$$f(-x) = |(-x)^3| = |-x^3| = f(x)$$

[-5, 5] scl: 1 by [-3, 7] scl: 1

41.

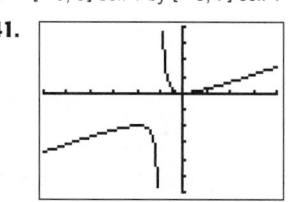

neither;
$$g(-x) = \frac{(-x)^2}{-x + 1} = \frac{x^2}{-x + 1}$$

[-6, 4] scl: 1 by [-12, 8] scl: 2

43a. undefined **43b.** 0 **43c.** 1.5

45 a. Because x represents years since 2001 and the model is only valid from 2001 to 2005, the relevant domain of the function is $\{x \mid 0 \leq x \leq 4, x \in \mathbb{R}\}$. The graph does not appear to extend below $f(0) = 1.2$ or above $f(4) \approx 11.2$. Therefore, the range of the function is approximately R = {y | 1.2 million ≤ y ≤ 11.2 million, $y \in \mathbb{R}$}.

b. Because 2003 is 2 years after 2001, $x = 2$. Sample answer: From the graph, it appears that the function value when $x = 2$ is about 4.1. Find h(2).
$$h(2) = 0.5(2)^2 + 0.5(2) + 1.2 = 2 + 1 + 1.2 \text{ or } 4.2$$

c. Sample answer: From the graph, it appears that the y-intercept is about 1.1. Find h(0).
$$h(0) = 0.5(0)^2 + 0.5(0) + 1.2 = 1.2$$
The y-intercept represents the number of millions of households with only wireless service in 2001.

d. There are no zeros because there were more than zero households with only wireless phone service for all of the years in the domain.

47a.

[-5, 15] scl: 2 by [-100, 400] scl: 50

47b. [0, 6]; The relevant domain represents the interval of time beginning when the dose was first taken and ending when the pain reliever left the bloodstream. Because time cannot be negative, $x \geq 0$. The amount of pain reliever in the bloodstream is zero when the dose is first taken at $x = 0$, and is zero again at $x = 6$. Therefore, $x \leq 6$. Combining these restrictions, the relevant domain is $\{x \mid 0 \leq x \leq 6, x \in \mathbb{R}\}$ or [0, 6]. **47c.** about 346 milligrams

49.

no zeros

[-30, 30] scl: 6 by [-36, 24] scl: 6

51.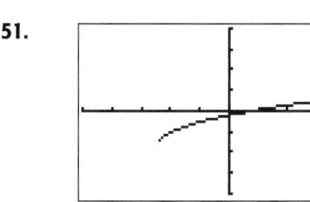

[−25, 15] scl: 5 by [−20, 20] scl: 5

$2\sqrt{x + 12} - 8 = 0$
$2\sqrt{x + 12} = 8$
$\sqrt{x + 12} = 4$
$x + 12 = 16$
$x = 4$

53.

[−20, 20] scl: 5 by [−20, 20] scl: 5

$\dfrac{6}{x} + 3 = 0$
$6 + 3x = 0$
$3x = -6$
$x = -2$

55. $D = (-8, -4] \cup (-2, \infty)$, $R = (-6, \infty)$
57. $D = (-\infty, -6] \cup (0, 4) \cup [7, \infty)$, $R = [-8, \infty)$
59a. $D = [0, 70]$, $R \approx [-19.19, 12.58]$
59b. Sample answers: 12, 12.58; The y-intercept represents the percent population change in 1930.
59c. 15.4; The zero represents the time at which the percent population change was 0.
59d. 203.83%; Sample answer: This value does not seem realistic because it would mean that the population increased by about 204%, which does not seem likely in comparison to the population trends from 1930 to 2000.

61a.

x	1.99	1.999	2	2.001	2.01
$f(x)$	−100	−1000	undefined	1000	100

61b. Sample answer: As x approaches 2 from the left, the value of $f(x)$ becomes greater and greater. As x approaches 2 from the right, the value of $f(x)$ becomes more and more negative.

61c.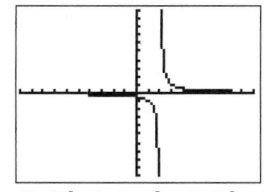

[−10, 10] scl: 1 by [−10, 10] scl: 1

Sample answer: As x approaches 2 from the left, the graph decreases without bound. As x approaches 2 from the right, the graph increases without bound.

61d. Sample answer: As the value of x becomes increasingly large, or when $x > 3$, the denominator of the fraction will also become increasingly large. This creates smaller fractions that will continue to decrease but will never reach zero or cross the x-axis. This also applies as x decreases at an increasing rate, or when $x < -1$. However, since x is negative, the value will also be negative but it too will never reach 0. As the value of x approaches 2, the difference of x and 2 becomes smaller and smaller. When the difference d is $-1 < d < 1$, the denominator is smaller than the numerator, thus producing a larger number. If the difference is positive, the fraction will approach infinity. If the difference is negative, the fraction will approach negative infinity. Only at 2 will the fraction fail to exist because the difference in the denominator cannot be 0.

63.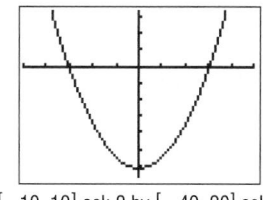

[−10, 10] scl: 2 by [−40, 20] scl: 6

Even; the graph of $g(n)$ is symmetric with respect to the y-axis.
$g(-n) = n^2 - 37$
$= (-n)^2 - 37$
$= n^2 - 37$
$= g(n)$

65.

[−5, 5] scl: 1 by [−5, 5] scl: 1

Odd; the graph of $f(g)$ is symmetric with respect to the origin.
$f(-g) = (-g)^9$
$= -g^9$
$= -f(g)$

67.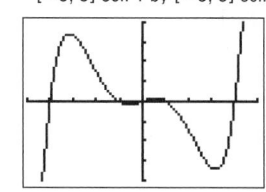

[−5, 5] scl: 1 by [−200, 200] scl: 50

Odd; the graph of $h(y)$ is symmetric with respect to the origin.
$h(-y) = (-y)^5 - 17(-y)^3 + 16(-y)$
$= -y^5 + 17y^3 - 16y$
$= -h(y)$

69. Sample answer:

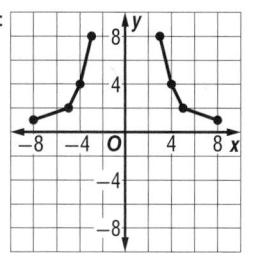

71 The graph of a relation is symmetric with respect to the origin if, for every point (x, y) on the graph, the point $(-x, -y)$ is also on the graph. Therefore, the points $(3, 18)$, $(2, 9)$, and $(1, 3)$ are also on the graph.

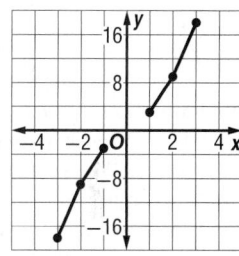

73. Sample answer: For a relation to be a function, each x-value must map to exactly one y-value. A relation with two y-intercepts located at $(0, y_1)$ and $(0, y_2)$ cannot be a function because two different y-values cannot map to the same x-value. A relation with two x-intercepts located at $(x_1, 0)$ and $(x_2, 0)$ can be a function because two different x-values can map to the same y-value. **75.** False; sample answer: If n is 0, the range is $\{y \mid y = 0\}$. If n is negative, the range is $\{y \mid y \leq 0, y \in \mathbb{R}\}$. The range will only be $\{y \mid y \geq 0, y \in \mathbb{R}\}$ if n is positive. **77.** False; sample answer: Consider the odd function $y = x^3$ and the point $(2, 8)$ on its graph. Reflecting the point $(2, 8)$ in the line $y = -x$ produces the point $(-8, -2)$ which is not on the graph of $y = x^3$.

79. odd; Sample answer: **Proof:**
   Given: a is an odd function and $b(x) = a(-x)$
   Prove: b is an odd function
   1. $b(x) = a(-x)$ (Given)
   2. a is an odd function   (Given)
   3. $a(-x) = -a(x)$ (Def. odd function)
   4. $b(x) = -a(x)$ (Trans. Prop. of Equality using 1 and 3)
   5. $-b(x) = a(x)$ (Div. Prop. of Equality)
   6. $a(x) = -b(x)$ (Reflexive Prop. of Equality)

7. $b(-x) = a(-(-x))$ (Substitute $-x$ for x in 1)
8. $b(-x) = a(x)$ (Substitute x for $-(-x)$.)
9. $b(-x) = -b(x)$ (Trans. Prop. of Equality using 8 and 6)
10. b is an odd function.   (Def. odd function)

81. even; Sample answer: **Proof:**
Given: a is an odd function and $b(x) = [a(x)]^2$
Prove: b is an even function
1. $b(x) = [a(x)]^2$ (Given)
2. a is an odd function   (Given)
3. $a(-x) = -a(x)$ (Def. odd function)
4. $b(-x) = [a(-x)]^2$ (Substitute $-x$ for x in 1)
5. $b(-x) = [-a(x)]^2$ (Trans. Prop. of Equality using 4 and 3)
6. $b(-x) = [a(x)]^2$ (Mult. Prop.)
7. $b(-x) = b(x)$ (Trans. Prop. of Equality using 6 and 1)
8. b is an even function.   (Def. even function)

83. odd; Sample answer: **Proof:**
Given: a is an odd function and $b(x) = [a(x)]^3$
Prove: b is an odd function
1. $b(x) = [a(x)]^3$ (Given)
2. a is an odd function   (Given)
3. $a(-x) = -a(x)$ (Def. odd function)
4. $b(-x) = [a(-x)]^3$ (Substitute $-x$ for x in 1)
5. $b(-x) = [-a(x)]^3$ (Trans. Prop. of Equality using 4 and 3)
6. $b(-x) = -[a(x)]^3$ (Mult. Prop.)
7. $-b(-x) = [a(x)]^3$ (Div. Prop. of Equality)
8. $-b(-x) = b(x)$ (Trans. Prop. of Equality using 7 and 1)
9. $b(-x) = -b(x)$ (Div. Prop. of Equality)
10. b is an odd function.   (Def. odd function)

85. Sometimes; sample answer: The relation described by $x^2 + (y - 2)^2 = 4$ is symmetric with respect to the line $y = 2$, but is not a function. A relation whose points all lie on the line $y = 2$ is symmetric with respect to the line $y = 2$ and is a function. **87.** Sometimes; sample answer: The relation described by $x^2 + y^2 = 25$ is symmetric with respect to the x- and y-axes but is not a function. The relation $\{(x, 0) \mid -1 \le x \le 1, x \in \mathbb{R}\}$ is symmetric with respect to the x- and y-axes and is a function. **89a.** -13 **89b.** $16x^2 + 40x + 3$ **89c.** $9n^2 - 24n - 6$ **91a.** 8 **91b.** $\dfrac{2x^6 + 2}{x^4 - 2}$ **91c.** $\dfrac{2x^3 + 6x^2 + 6x + 4}{x^2 + 2x - 1}$ **93a.** 22,308 **93b.** 144 **93c.** 792 **95.** no solution **97.** 3 **99.** $\dfrac{1}{7}$ **101.** 125

103. $R^2 + 2RW + W^2$ **105.** $-15 + 8i$ **107.** B **109.** B

Pages 30–33   Lesson 1-3

1. Continuous; $f(-5) = \sqrt{21}$ or about 4.58, $\lim\limits_{x \to -5} \approx 4.58$, and $\lim\limits_{x \to -5} = f(-5)$. **3.** Discontinuous at $x = -6$; $h(-6)$ is undefined and $\lim\limits_{x \to -6} = -12$, so $h(x)$ has a removable discontinuity at $x = -6$. Continuous at $x = 6$. $h(6) = 0$, $\lim\limits_{x \to 6} h(x) = 0$, and $\lim\limits_{x \to 6} h(x) = h(6)$. **5.** Discontinuous; $g(1)$ is undefined and $g(x)$ approaches $-\infty$ as x approaches 1 from the left and ∞ as x approaches 1 from the right, so $g(x)$ has an infinite discontinuity at $x = 1$. **7.** Discontinuous at $x = 1$; $h(1)$ is undefined and $h(x)$ approaches $-\infty$ as x approaches 1 from the left and ∞ as x approaches 1 from the right, so $h(x)$ has an infinite discontinuity at $x = 1$. Discontinuous at $x = 4$; $h(4)$ is undefined and $\lim\limits_{x \to 4} h(x) = \dfrac{1}{3}$, so $h(x)$ has a removable

discontinuity at $x = 4$.   **9.** Discontinuous at $x = -6$; $f(x)$ approaches -25 as x approaches -6 from the left and 8 as x approaches -6 from the right, so $f(x)$ has a jump discontinuity at $x = -6$.

11 a. Find $f(0.4)$.
$$f(w) = \frac{7.4}{w}$$
$$f(0.4) = \frac{7.4}{(0.4)}$$
$$= 18.5$$

The function is defined at $w = 0.4$.
Find $\lim\limits_{w \to 0.4} f(w)$. Construct a table that shows values of $f(w)$ for w-values approaching 0.4 from the left and from the right.

w	0.39	0.399	0.3999	0.4	0.4001	0.401	0.41
f(w)	18.974	18.546	18.505		18.495	18.454	18.049

$\lim\limits_{w \to 0.4} f(w) = 18.5$.
Because $\lim\limits_{w \to 0.4} = f(0.4)$, $f(w)$ is continuous as $w = 0.4$.

b. The function is discontinuous at $w = 0$. Construct a table that shows values of $f(w)$ for w-values approaching 0 from the right. The function does not have to be examined for negative values of w since a wall cannot have a negative thickness.

w	0	0.001	0.01	0.1
f(w)		7400	740	74

Because $f(0)$ is undefined and $f(w)$ approaches ∞ as w approaches 0 from the right, $f(w)$ is discontinuous at $w = 0$ and has an infinite discontinuity at $w = 0$.

c. Evaluate the function for several w-values in its domain. Use these points to construct a graph.

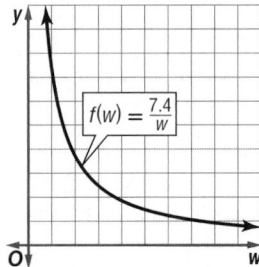

13. 1 and 2 **15.** -1 and 0, 1 and 2 **17.** -1 and 0, 0 and 1, 2 and 3 **19.** no zeros on the interval **21.** 2 and 3 **23.** From the graph, it appears that $f(x) \to \infty$ as $x \to -\infty$ and $f(x) \to -\infty$ as $x \to \infty$.

x	−10,000	−1000	0	1000	10,000
f(x)	$5 \cdot 10^{12}$	$5 \cdot 10^9$	-1	$-5 \cdot 10^9$	$-5 \cdot 10^{12}$

25. From the graph, it appears that $f(x) \to -4$ as $x \to -\infty$ and $f(x) \to -4$ as $x \to \infty$.

x	−10,000	−1000	0	1000	10,000
f(x)	-3.9981	-3.9811	-0.8333	-4.0191	-4.0019

27. From the graph, it appears that $f(x) \to 0$ as $x \to -\infty$ and $f(x) \to 0$ as $x \to \infty$.

x	−10,000	−1000	0	1000	10,000
f(x)	$-8 \cdot 10^{-4}$	$-8 \cdot 10^{-3}$	0	$8 \cdot 10^{-3}$	$8 \cdot 10^{-4}$

29. From the graph, it appears that $f(x) \to -2$ as $x \to -\infty$ and $f(x) \to -2$ as $x \to \infty$.

x	−1000	−100	0	100	1000
f(x)	−1.99999	−1.9999	−5	−1.9999	−1.99999

31a.

[0, 200] scl: 20 by [0, 0.5] scl: 0.05

31b. Sample answer: The end behavior of the graph indicates that as the concentration of the catalyst solution is increased, the chemical reaction rate approaches 0.5.

x	0	10	100	1000	10,000
R(x)	0	0.2273	0.4464	0.4941	0.4994

33. Sample answer: As $x \to \infty$, the fraction will decrease, and $q(x)$ will approach 0. **35.** Sample answer: As $x \to \infty$, the fraction will approach $\frac{x}{x}$, so $p(x)$ will approach 1. **37.** Sample answer: As $x \to \infty$, the fraction will decrease, and $c(x)$ will approach 0. **39.** Sample answer: As $x \to \infty$, $h(x)$ will increase without bound and approach ∞. **41.** Sample answer: As the mass of the railway car continues to increase, the railway car's kinetic energy will approach 0.

43 $h(x)$ has an infinite discontinuity at $x = 0$ because $h(0)$ is undefined and $h(x)$ approaches ∞ as x approaches 0 from the left and the right.

w	−0.1	−0.01	−0.001	0	0.001	0.01	0.1
h(x)	136.4	13,636	1.36×10^6		1.36×10^6	13,636	136.4

From the graph, it appears that as $x \to -\infty$ and ∞, $h(x) \to 0$.

x	h(x)
−10,000	1.36×10^{-8}
−1000	1.36×10^{-6}
0	undefined
1000	1.36×10^{-6}
10,000	1.36×10^{-8}

The table supports this conjecture.

45.

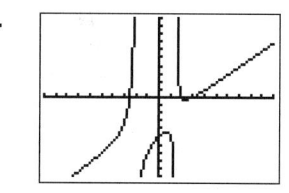

[−5, 5] scl: 1 by [−10, 10] scl: 1

discontinuity: infinite at $x = -1$, $x = 2$, and $x = 3$; end behavior: $\lim_{x \to -\infty} f(x) = 0$, $\lim_{x \to \infty} f(x) = 0$; zeros: $x = 0$

47.

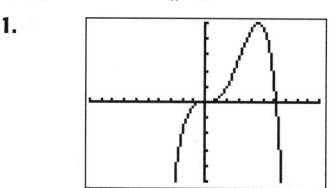

[−20, 20] scl: 2 by [−20, 20] scl: 2

discontinuity: infinite at $x = 3$ and $x = -6$; end behavior: $\lim_{x \to -\infty} h(x) = 4$, $\lim_{x \to \infty} h(x) = 4$; zeros: $x = -3$ and $\frac{1}{4}$

49.

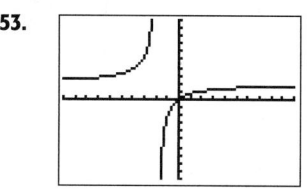

[−20, 20] scl: 2 by [−20, 20] scl: 2

discontinuity: infinite at $x = -4$ and $x = 3$; end behavior: $\lim_{x \to -\infty} h(x) = -\infty$, $\lim_{x \to \infty} h(x) = \infty$; zeros: $x = -5$, 4, and 6

51.

end behavior:
$\lim_{x \to -\infty} f(x) = -\infty$,
$\lim_{x \to \infty} f(x) = -\infty$

[−20, 20] scl: 2 by [−2500, 2500] scl: 500

x	−10,000	−1000	−100	0	100	1000	10,000
f(x)	$-1 \cdot 10^{16}$	$-1 \cdot 10^{12}$	$-1 \cdot 10^8$	−4	$-9 \cdot 10^7$	$-1 \cdot 10^{12}$	$-1 \cdot 10^{16}$

53.

end behavior:
$\lim_{x \to -\infty} f(x) = 16$,
$\lim_{x \to \infty} f(x) = 16$

[−80, 80] scl: 8 by [−80, 80] scl: 8

x	−10,000	−1000	−100	0	100	1000	10,000
f(x)	16.024	16.244	18.824	−	13.913	15.764	15.976

55a. $f(n) = \dfrac{3n + 4000}{n}$ **55b.**

[0, 1000] scl: 100 by [0, 400] scl: 50

55c. \$3; Sample answer: As n approaches ∞, $f(n)$ approaches 3. Therefore, the average cost approaches \$3 as the number of shirts sold increases.

57a. Sample answer: If n is even, $f(x)$ approaches ∞ as x approaches $-\infty$ and ∞.

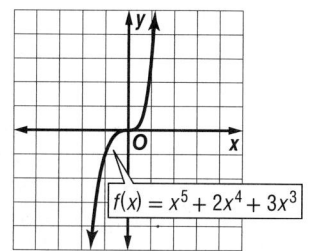

$f(x) = x^4 + 3x^3 + 2x^2$

57b. Sample answer: If n is odd, $f(x)$ approaches $-\infty$ as x approaches $-\infty$ and ∞ as x approaches ∞.

$f(x) = x^5 + 2x^4 + 3x^3$

59. Infinite; $f(0)$ is undefined, and $f(x)$ approaches $-\infty$ as x approaches 0 from the left and ∞ as x approaches 0 from the right.

61 The graph is continuous at $x = 3$, then by substitution:
$$(3)^2 + a = (3)b + a \qquad x = 3$$
$$9 + a = 3b + a \qquad \text{Simplify.}$$
$$9 = 3b \qquad \text{Subtract.}$$
$$b = 3 \qquad \text{Divide.}$$
The graph is continuous at $x = -3$, then by substitution:
$$(-3)b + a = \sqrt{-b + 3} \qquad x = -3$$
$$(-3)3 + a = \sqrt{-3 + 3} \qquad b = 3$$
$$-9 + a = 0 \qquad \text{Simplify.}$$
$$a = 9 \qquad \text{Subtract.}$$

63. $\lim\limits_{x \to -\infty} f(x) = \infty$; because f is odd, $f(-x) = -f(x)$. **65.** $\lim\limits_{x \to -\infty}$ $f(x) = \infty$; with y-axis symmetry, $f(x) = f(-x)$.

67.

Even; the graph of $h(x)$ is symmetric with respect to the y-axis.
$$h(-x) = \sqrt{(-x)^2 - 16}$$
$$= \sqrt{x^2 - 16}$$
$$= h(x)$$

[−10, 10] scl: 2 by [−10, 10] scl: 2

69. no zeros

71. $D = \left(-\infty, 1 - \sqrt{11}\right) \cup$ $\left(1 - \sqrt{11}, 1 + \sqrt{11}\right) \cup$ $\left(1 + \sqrt{11}, \infty\right)$

[−10, 10] scl: 2 by [−10, 10] scl: 2

73a. 100,000   **73b.** 7290   **73c.** 999,900,000

75. $\begin{bmatrix} -32 & 50 & -16 \\ 4 & -30 & 10 \end{bmatrix}$ **77.** infinite solutions   **79.** $(2, 3, -1)$

81. J   **83.** D

Pages 40–43   Lesson 1-4

1. f is increasing on $(-\infty, -0.5)$, decreasing on $(-0.5, 1)$, and increasing on $(1, \infty)$.   **3.** f is decreasing on $(-\infty, 2.5)$ and increasing on $(2.5, \infty)$.   **5.** f is increasing on $(-\infty, 0)$ and increasing on $(0, \infty)$.   **7.** f is increasing on $(-\infty, -2)$, decreasing on $(0, 4)$, and increasing on $(4, \infty)$.   **9.** f is constant on $(-\infty, -5)$, increasing on $(-5, -3.5)$, and decreasing on $(-3.5, \infty)$.

11a.

$f(t) = -16t^2 + 23.8t + 5$

11b. about 13.9 ft

13–21. Your extrema values should be close to the actual extrema values given.

13 It appears that $f(x)$ has an absolute maximum at $x = -1.5$, a relative minimum at $x = 0$, and an absolute maximum at $x = 1.5$. It also appears that $\lim\limits_{x \to -\infty} f(x) = -\infty$ and $\lim\limits_{x \to \infty} f(x) = -\infty$, so we conjecture that this function has no absolute minimum.

x	−100	−2	−1.5	−1
f(x)	$-1 \cdot 10^9$	−1	2.94	2

Because $f(-1.5) > f(-2)$ and $f(-1.5) > f(-1)$, there is an absolute maximum in the interval $(-2, -1)$. The approximate value of the absolute maximum is 2.94.

x	−0.5	0	0.5
f(x)	−0.063	−1	−0.063

Because $f(0) < f(-0.5)$ and $f(0) < f(0.5)$, there is a relative maximum in the interval $(-0.5, 0.5)$. The approximate value of this relative minimum is –1.

x	1	1.5	2	100
f(x)	2	2.94	−1	$-1 \cdot 10^9$

Because $f(1.5) > f(1)$ and $f(1.5) > f(2)$, there is also an absolute maximum in the interval $(1, 2)$. The approximate value of the absolute maximum is 2.94.

15. abs. min: $(-3.71, -1334.6)$; rel. max: $(0.11, 0.001)$; rel. min: $(3.59, -1042.52)$ **17.** rel. min: $(-1, -4)$; rel. max: $(1.08, 4.18)$ **19.** rel. min: $(1.2, -1.11)$; rel. max: $(2, 0)$ **21.** rel. max: $(-1.14, -1.58)$; rel. min: $(1.14, -8.42)$ **23.** rel. max: $(1.08, 0.04)$; rel. min: $(-1.08, -10.04)$ **25.** abs. min: $(-1.38, -7.08)$ **27.** rel. max: $(1.11, 2.12)$; rel. min: $(-0.17, -1.08)$ **29.** rel. max: $(-1.11, 1.32)$; rel. min: $(0, -4)$ **31.** rel. max: $(-1.66, 3.43)$; rel. min: $(0.93, -3.82)$ **33.** $r = 1.85$ in.; $h = 3.70$ in.   **35.** 28   **37.** -16 **39.** -309 **41.** 472   **43.** -0.45 **45.** ≈ 0.219

47 **a.** $f(0) = 12$ and $f(10) = 15.6$
$$m = \frac{15.6 - 12}{10 - 0} \qquad \text{Rate of change}$$
$$= \frac{3.6}{10} \text{ or } 0.36 \text{ million pounds more per year}$$

b. $f(5) = 11.925$ and $f(10) = 15.6$
$$m = \frac{15.6 - 11.925}{10 - 5} \qquad \text{Rate of change}$$
$$= \frac{3.675}{5} \text{ or } 0.735 \text{ million pounds more per year}$$

49a. [5, 15]: 5; [15, 20]: 3; [25, 45]: 0.5   **49b.** The object is increasing in speed, or accelerating, along all three intervals. It is accelerating at the fastest rate for the interval [5, 15]. While it is very slowly accelerating for the interval [25, 45], it is *still* increasing its speed.   **49c.** Steep graph = high magnitude rate of change = rapidly increasing or decreasing; flat graph = low magnitude rate of change = minimal increasing or decreasing.

51a.

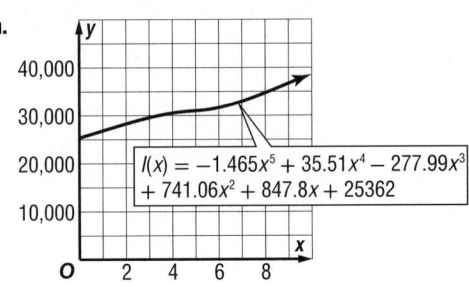

$l(x) = -1.465x^5 + 35.51x^4 - 277.99x^3 + 741.06x^2 + 847.8x + 25362$

51b. about $1280.93; Sample answer: This value represents the average change in net personal income in the U.S. from 2000 to 2007. **51c.** The rate of change was lowest from 2000 to 2004 at about $826.43 and highest from 2003 to 2007 at about $1711.44.

53. 14.46 in. × 14.46 in. × 14.46 in.; Sample answer: The function for the surface area of the box in terms of the length of one side of the base is $f(w) = 2w^2 + \dfrac{12,096}{w}$. Graph the function. The absolute minimum occurred when $w = 14.46$ in.

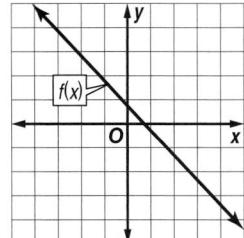

[0, 100] scl: 10 by [0, 5000] scl: 500

55. Sample answer:

57. Sample answer:

59. Sample answer:

61. $(-5, -1)$; maximum

63. $(2.3, 8)$; minimum

65. $(0, -5)$; minimum

67. Sample answer: One reason for the variation in average rate of change may be that Simeon's family encountered traffic later in their trip, thus slowing them down. Another may be that Simeon's family traveled on residential roads for the first hour before entering a highway for the next three hours. The two instances on the graph where Simeon's family appeared to not travel any distance may be as a result of the family needing to stop to eat or take a break. They may have also encountered an accident where traffic was completely stopped.

69. Sample answer:

(71) 0; Sample answer: When a function is constant on an interval, the y-coordinates of each point in the interval are the same. Because $f(x)$ is constant on the interval $[a, b]$, $f(a) = f(b)$.

$$m = \frac{f(a) - f(b)}{a - b}$$
$$= \frac{0}{a - b} \text{ or } 0$$

So, the slope of the secant line is zero.

73.

[−1080, 1080] scl: 90 by [−1, 1] scl: 0.1

There are an infinite number of relative maxima and minima. The relative maximum is 1 and occurs at $\{x \mid x = 90 + 360n, n \in \mathbb{Z}\}$. The relative minimum is −1 and occurs at $\{x \mid x = 270 + 360n, n \in \mathbb{Z}\}$.

75. Sample answer: When a function is increasing on an interval, the average rate of change is positive. When a function is decreasing on an interval, the average rate of change is negative. When a function is constant on an interval, the average rate of change is zero.

77. Continuous; $f(3) = 2$, $\lim\limits_{x \to 3^-} f(x) = 2$, and $\lim\limits_{x \to 3} f(x) = f(3)$.

79.

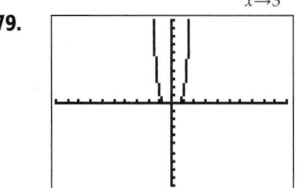

[−10, 10] scl: 1 by [−10, 10] scl: 1

Even; the graph of $f(x)$ is symmetric with respect to the y-axis.
$$f(-x) = |(-x)^5|$$
$$= |x^5|$$
$$= f(x)$$

81.

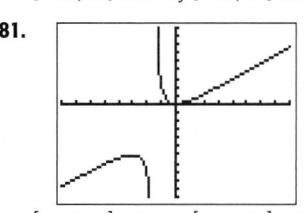

[−16, 16] scl: 2 by [−22, 18] scl: 2

neither; $g(-x) = \dfrac{(-x)^2}{-x + 3}$
$$= \dfrac{x^2}{-x + 3}$$

83. $(-\infty, -3] \cup [3, \infty)$ **85.** $(5, 3, 2)$ **87.** $-6, 3$ **89.** $7, -3$
91. $9 - 7i$ **93.** 200 or 400 amps **95.** G **97.** G

Pages 52–55 Lesson 1-5

1. D = $\{x \mid x \in \mathbb{R}\}$, R = $\{y \mid y \in \mathbb{Z}\}$. The graph has a y-intercept at $(0, 0)$ and x-intercepts for $\{x \mid 0 \le x \le 1, x \in \mathbb{R}\}$. The graph has no symmetry. The graph has a jump discontinuity for $\{x \mid x \in \mathbb{Z}\}$. $\lim\limits_{x \to -\infty} f(x) = -\infty$ and $\lim\limits_{x \to \infty} f(x) = \infty$. The graph is constant for $\{x \mid x \notin \mathbb{Z}\}$. The graph increases for $\{x \mid x \in \mathbb{Z}\}$.

3. D = $\{x \mid x \in \mathbb{R}\}$, R = $\{y \mid y \in \mathbb{R}\}$. The graph has an intercept at $(0, 0)$. The graph is symmetric with respect to the origin. The graph is continuous. $\lim\limits_{x \to -\infty} f(x) = -\infty$ and $\lim\limits_{x \to \infty} f(x) = \infty$. The graph is increasing on $(-\infty, \infty)$.

5. D = $\{x \mid x \in \mathbb{R}\}$, R = $\{y \mid y = c, c \in \mathbb{R}\}$. If $c = 0$, all real numbers are x-intercepts. If $c \ne 0$, there are no x-intercepts. The graph has a y-intercept at $(0, c)$. If $c \ne 0$, the graph is symmetric with respect to the y-axis. If $c = 0$, the graph is symmetric with respect to the x-axis, y-axis, and origin. The graph is continuous. $\lim\limits_{x \to -\infty} f(x) = c$ and $\lim\limits_{x \to \infty} f(x) = c$. The graph is constant on $(-\infty, \infty)$.

7.

9.

11.

13.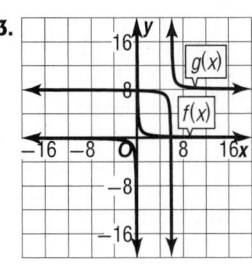

15. The graph of $g(x)$ is the graph of $f(x)$ translated 5 units to the right and $g(x) = [\![x - 5]\!]$, or translated 5 units down and $g(x) = [\![x]\!] - 5$. **17.** The graph of $g(x)$ is the graph of $f(x)$ reflected in the y-axis and translated 5 units right when $g(x) = [\![5 - x]\!]$, or reflected in the y-axis and translated 5 units up when $g(x) = [\![-x]\!] + 5$.

19 Because there is a two-month delay, $d(x)$ is the graph of $p(x)$ translated two units to the right. Therefore, $d(x) = 10(x - 2)^3 - 70(x - 2)^2 + 150(x - 2) + 2$

21. The graph of $g(x)$ is the graph of $f(x)$ translated 5 units down; $g(x) = |x| - 5$.

23. The graph of $g(x)$ is the graph of $f(x)$ translated 1 unit to the right and 2 units down; $g(x) = |x - 1| - 2$.

25. $f(x) = \sqrt{x}$; the graph of $g(x)$ is the graph of $f(x)$ translated 8 units to the left and expanded vertically.

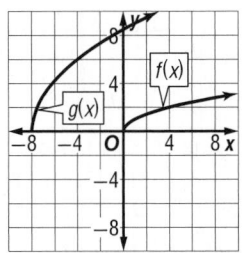

27. $f(x) = [\![x]\!]$; the graph of $g(x)$ is the graph of $f(x)$ translated 6 units to the right and expanded vertically.

29. $f(x) = |x|$; $g(x)$ is the graph of $f(x)$ translated 5 units to the left, expanded vertically, and reflected in the x-axis.

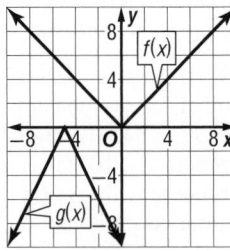

31. $f(x) = \sqrt{x}$; $g(x)$ is the graph of $f(x)$ translated 3 units to the left and compressed vertically.

33.

35.

37.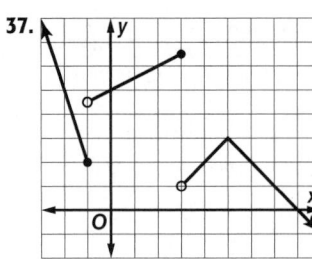

39a. The graph of $c(x)$ is the graph of $f(x)$ compressed vertically and translated 1.99 units up.
39b. $c(x) = 2.49 + 0.05[\![x]\!]$

39c.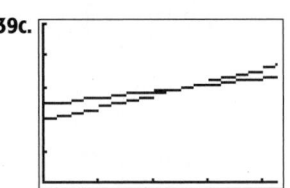

[0, 17] scl: 4 by [0, 5] scl: 1

39d. Yes; the plans will equal each other at 10 minutes.

41.

43.

45.

47a. $f(x) = \begin{cases} 0.4[\![x]\!] + 2.50 & \text{if } [\![x]\!] = x \\ 0.4[\![x+1]\!] + 2.50 & \text{if } [\![x]\!] < x \end{cases}$

47b.

Taxi Cost

47c. The graph of $f(x)$ is translated 0.5 unit up.

Taxi Cost

49. A vertical shift of 5 is $f(x) + 5$ or $\frac{1}{x} + 5$.

A horizontal shift of -7 is $f(x - (-7))$ or $\frac{1}{x+7}$.

A vertical expansion by a factor of 2 is $2f(x)$ or $2\left(\frac{1}{x}\right) = \frac{2}{x}$.

Combining the attributes, $g(x) = 2f(x - (-7)) + 5$ or $\frac{2}{x+7} + 5$.

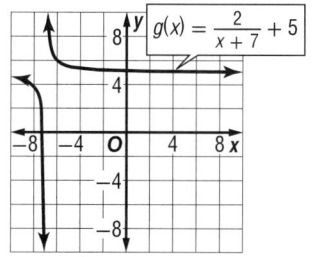

51. translated one unit to the left; translated one unit down **53.** translated 2 units to the left; expanded vertically; translated 7 units down **55.** $g(x) = \frac{2}{x-3} + 4$ **57.** $g(x) = 4\sqrt{x+4} - 6$ **59a.** $g(x) = 1.12f(x)$ **59b.** There is no affect on $f(x)$. While the opening is delayed, the number of days after the opening, which determines the domain of the function, is unaffected. **59c.** $g(x) = f(x) - 0.45$ **61.** $f(x) = x^3$; Sample answer: The graph of $g(x)$ is the graph of $f(x)$ translated 4 units to the right, expanded vertically, reflected in the x-axis, and translated 2 units up. **63.** $f(x) = \sqrt{x}$; Sample answer: The graph of $g(x)$ is the graph of $f(x)$ translated 3 units to the right, reflected in the x-axis, and translated 5 units up.

65.

67.

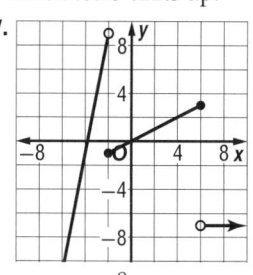

69. $g(x) = -\frac{24}{\sqrt{x+6}} + 18$ **71.** $g(x) = \frac{8}{\sqrt{2x+7}} + 4$

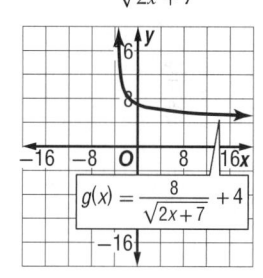

73. Sample answer: Both; for the greatest integer function, a shift of a units left is identical to a shift of a units up. **75.** Sample answer: Order is important because different graphs can be obtained depending on the order the transformations are performed. For example, if (a, b) is on the original graph and there is a translation 6 units up and then a reflection in the x-axis, the resulting point on the transformation will be $(a, -b - 6)$. However, if (a, b) is reflected in the x-axis first and then translated 6 units up, the resulting point will be $(a, -b + 6)$. **77.** Sometimes; sample answer: $f(x) = x^3$ is an odd function and $f(-x) \neq -|f(x)|$ when $x = -1$. However, $f(x) = 0$ is an odd function and $f(-x) = -|f(x)|$ for all x.

79 Sample answer: There are an infinite number of functions that meet these criteria. Any values of a, b, and c for $g(x) = a\sqrt{x+b} + c$ are valid such that $f(-2) = -6$. To simplify this process, let $a = 1$.

$$f(x) = a\sqrt{x+b} - c$$
$$-6 = \sqrt{-2+b} - c$$
$$c - 6 = \sqrt{-2+b}$$
$$(c-6)^2 = -2 + b$$
$$(c-6)^2 + 2 = b$$

Any values of b and c that satisfy this equation will be valid. If $c = 8$, $b = 6$ and $g(x) = \sqrt{x+6} - 8$.

83. 6 **85.** 0; Sample answer: As $x \to \infty$, the denominator of the fraction will increase and the value of the fraction will approach 0, so $g(x)$ will approach 0. **87.** 1; Sample answer: As $x \to \infty$, the fraction will get closer and closer to $\frac{x}{x}$, so $p(x)$ will approach 1.

89. y-intercept $= 0$; zeros: $-1, 0, 2$;

$$x^3 - x^2 - 2x = 0$$
$$x(x^2 - x - 2) = 0$$
$$x(x - 2)(x + 1) = 0$$
$$x = 0 \quad \text{or} \quad x - 2 = 0 \quad \text{or} \quad x + 1 = 0$$
$$x = 2 \qquad\qquad x = -1$$

91. about 118.60 **93.** B **95.** G

Pages 61–64 Lesson 1-6

1. $(f + g)(x) = x^2 + \sqrt{x} + 4$; $D = [0, \infty)$; $(f - g)(x) = x^2 - \sqrt{x} + 4$; $D = [0, \infty)$; $(f \cdot g)(x) = x^{\frac{5}{2}} + 4x^{\frac{1}{2}}$; $D = [0, \infty)$; $\left(\dfrac{f}{g}\right)(x) = \dfrac{x^2 + 4}{\sqrt{x}}$; $D = (0, \infty)$ **3.** $(f + g)(x) = x^2 + 6x + 8$; $D = (-\infty, \infty)$; $(f - g)(x) = x^2 + 4x + 4$; $D = (-\infty, \infty)$; $(f \cdot g)(x) = x^3 + 7x^2 + 16x + 12$; $D = (-\infty, \infty)$; $\left(\dfrac{f}{g}\right)(x) = x + 3$; $D (-\infty, -2) \cup (-2, \infty)$ **5.** $(f + g)(x) = x^2 + 10x$; $D = (-\infty, \infty)$; $(f - g)(x) = x^2 - 8x$; $D = (-\infty, \infty)$; $(f \cdot g)(x) = 9x^3 + 9x^2$; $D = (-\infty, \infty)$; $\left(\dfrac{f}{g}\right)(x) = \dfrac{x + 1}{9}$; $D = (-\infty, 0) \cup (0, \infty)$ **7.** $(f + g)(x) = x^3 + x + \dfrac{6}{x}$; $D = (-\infty, 0) \cup (0, \infty)$; $(f - g)(x) = -x^3 - x + \dfrac{6}{x}$; $D = (-\infty, 0) \cup (0, \infty)$; $(f \cdot g)(x) = 6x^2 + 6$; $D = (-\infty, 0) \cup (0, \infty)$; $\left(\dfrac{f}{g}\right)(x) = \dfrac{6}{x^4 + x^2}$; $D = (-\infty, 0) \cup (0, \infty)$ **9.** $(f + g)(x) = \dfrac{1}{\sqrt{x}} - 4\sqrt{x}$; $D = (0, \infty)$; $(f - g)(x) = \dfrac{1}{\sqrt{x}} - 4\sqrt{x}$; $D = (0, \infty)$; $(f \cdot g)(x) = 4$; $D = (0, \infty)$; $\left(\dfrac{f}{g}\right)(x) = \dfrac{1}{4x}$; $D = (0, \infty)$ **11.** $(f + g)(x) = \sqrt{x + 8} + \sqrt{x + 5} - 3$; $D = [-5, \infty)$; $(f - g)(x) = \sqrt{x + 8} - \sqrt{x + 5} + 3$; $D = [-5, \infty)$; $(f \cdot g)(x) = \sqrt{x^2 + 13x + 40} - 3\sqrt{x + 8}$; $D = [-5, \infty)$; $\left(\dfrac{f}{g}\right)(x) = \dfrac{\sqrt{x^2 + 13x + 40} + 3\sqrt{x + 8}}{x - 4}$; $D = [-5, 4) \cup (4, \infty)$ **13a.** $(f + g)(x) = 40x + 550$; $x \geq 0$ **13b.** $(f + g)(x)$ represents all factors that influence the budget. **13c.** \$710; The budget for 4 weeks.

15. $[f \circ g](x) = 8x - 19$; $[g \circ f](x) = 8x - 20$; $[f \circ g](6) = 29$ **17.** $[f \circ g](x) = -x^4 - 2x^3 - 3x^2 - 2x + 7$; $[g \circ f](x) = x^4 - 17x^2 + 73$; $[f \circ g](6) = -1841$ **19.** $[f \circ g](x) = -x^6 - 2x^3 + 2$; $[g \circ f](x) = -x^6 + 9x^4 - 27x^2 + 28$; $[f \circ g](6) = -47{,}086$ **21.** $[f \circ g](x) = \dfrac{1}{x^2 - 3}$ for $x \neq \pm\sqrt{3}$ **23.** $[f \circ g](x) = |x|$ **25.** $[f \circ g](x) = \dfrac{5\sqrt{6 - x}}{6 - x}$ for $x < 6$ **27.** $[f \circ g](x) = |x + 2|$ **29. a.** Since the denominator cannot be zero, $1 - \dfrac{c^2}{v^2} \neq 0$. Thus, $v \neq c$. Because velocity cannot be negative, $D = \{v \mid 0 \leq v < c, v \in \mathbb{R}\}$.

b.
$$m(10) = \dfrac{100}{\sqrt{1 - \dfrac{10^2}{(3 \times 10^8)^2}}} = 100$$

$$m(10{,}000) = \dfrac{100}{\sqrt{1 - \dfrac{10{,}000^2}{(3 \times 10^8)^2}}} = 100.000000056$$

$$m(1{,}000{,}000) = \dfrac{100}{\sqrt{1 - \dfrac{(1 \times 10^6)^2}{(3 \times 10^8)^2}}} = 100.00055556$$

c. $\displaystyle\lim_{v \to c} m(v) = \infty$, since as m approaches c, $\dfrac{v^2}{c^2}$ approaches 1, making $\sqrt{1 - \dfrac{v^2}{c^2}}$ very small. 100 divided by a very small number results in a very large quotient.

d. Sample answer: Let $g(v) = 1 - \dfrac{v^2}{c^2}$ and $f(v) = \dfrac{100}{\sqrt{v}}$.
$$m(v) = f[g(v)]$$
$$= \dfrac{100}{\sqrt{1 - \dfrac{v^2}{c^2}}}$$

31. Sample answer: $f(x) = \dfrac{6}{x} - 8$; $g(x) = x + 5$ **33.** Sample answer: $f(x) = [\![-3x]\!]$; $g(x) = x - 9$ **35.** Sample answer: $f(x) = x^3$; $g(x) = \sqrt{x} + 4$ **37.** Sample answer: $f(x) = \dfrac{8}{x^2}$; $g(x) = x - 5$ **39.** Sample answer: $f(x) = \dfrac{x + 6}{\sqrt{x}}$; $g(x) = x - 1$ **41a.** $h[f(x)]$; The bonus is found after subtracting 300,000 from the total sales. **41b.** \$6000 **43.** Sample answer: $f(x) = x - \dfrac{4}{x^2 + 1}$; $g(x) = \sqrt{x - 1}$ **45.** Sample answer: $f(x) = \dfrac{8}{x + 2} + 5\sqrt{x}$; $g(x) = x^2$ **47.** Sample answer: $f(x) = \dfrac{x - 4}{2x - 9} + \sqrt{\dfrac{4}{x - 4}}$; $g(x) = x + 4$ **49.** $f(0.5) = -0.75$, $f(-6) = 22$, $f(x + 1) = x^2 + 4x + 1$ **51.** $f(0.5) = -3.6$, $f(-6) = -300$, $f(x + 1) = -9x^2 - 22x + \dfrac{x + 1}{10} - \dfrac{62}{5}$

53 The domain of $h(x) = \sqrt{x} + 3$ is $\{x \mid x \geq 0, x \in \mathbb{R}\}$ and of $g(x) = x^2 - 6$ is all real numbers. Therefore the domain of $g \circ h$ is $\{x \mid x \geq 0, x \in \mathbb{R}\}$.

$$[g \circ h](x) = g[(h(x)]$$
$$= g(\sqrt{x} + 3)$$
$$= [(\sqrt{x} + 3)^2 - 6]$$
$$= [x + 6\sqrt{x} + 9 - 6]$$
$$= x + 6\sqrt{x} + 3$$

Since the domain of $f(x) = x + 8$ is all real numbers, the domain of $f \circ g \circ h$ is $\{x \mid x \geq 0, x \in \mathbb{R}\}$.

$$[f \circ g \circ h](x) = f(g[(h(x)])$$
$$= f(x + 6\sqrt{x} + 3)$$
$$= (x + 6\sqrt{x} + 3) + 8$$
$$= x + 6\sqrt{x} + 11$$

Therefore, $[f \circ g \circ h](x) = x + 6\sqrt{x} + 11$ for $x \geq 0$.

55. $[f \circ g \circ h](x) = \sqrt{\dfrac{1}{x^2} + 2}$ for $x \neq 0$ **57a.** $g(x) = x^2 + 4$ **57b.** $g(x) = 4x + 8$ **59a.** $g(x) = \dfrac{1}{4x}$ **59b.** $g(x) = \dfrac{1}{32}x^{\frac{1}{3}}$ **61.** 1 **63.** 0 **65.** 0 **67a.** $\{m \mid m > 0, m \in \mathbb{R}\}$; The molar mass of the gas cannot be negative or zero. **67b.** about 7.22 m/s **67c.** The velocity will decrease. **67d.** Sample answer: $v(m) = f[g(x)]$; $f(m) = \sqrt{m}$; $g(m) = \dfrac{(24.9435)(303)}{m}$ **69.** Sample answer: $f(x) = \sqrt{x}$; $g(x) = x^2 + 8$; $h(x) = x - 5$ **71.** Sample answer: $f(x) = \dfrac{4}{x}$; $g(x) = x^2 + 1$; $h(x) = \sqrt{x} + 3$ **73.** $[f \circ g](x) = x$ for $x \geq -19$; $[g \circ f](x) = |x + 4| - 4$ **75.** $[f \circ g](x) = \sqrt{\sqrt{9 - x^2}}$ for $-3 \leq x \leq 3$; $[g \circ f](x) = \sqrt{9 - x}$ for $0 \leq x \leq 9$ **77.** $[f \circ g](x) = \dfrac{24 - 6x}{12 - x}$ for $x \neq 4$ and $x \neq 12$; $[g \circ f](x) = \dfrac{4x + 2}{4x - 1}$ for $x \neq -\dfrac{1}{2}$ and $x \neq \dfrac{1}{4}$

79.

81.

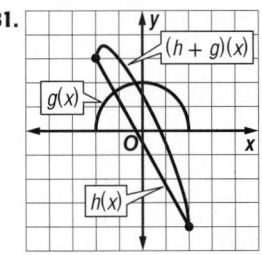

83. $\{x \mid -10 \leq x \leq -4 \text{ or } 2 \leq x \leq 8, x \in \mathbb{R}\}$ **85.** $\{x \mid 8 \leq x \leq 10, x \in \mathbb{R}\}$ **87.** odd **89.** even

91. Sample answer: $f(x) = \frac{1}{x}, x \neq 0$

93. Sample answer: This can occur when the range of $g(x)$ is a subset of the domain of $f(x)$. For example, when $f(x) = \sqrt{x}$ and $g(x) = x^2 + 4$, the range of $g(x)$ is all real numbers greater than or equal to 4. This range is a subset of the domain of $f(x)$, which is all real numbers greater than or equal to 0. Thus, $f[g(x)]$ exists for all real numbers.

95 The domain of $h(x) = \frac{4}{x}$ is $\{x \mid x \neq 0, x \in \mathbb{R}\}$. To evaluate $g \circ h$, you must be able to evaluate $g(x) = \sqrt{x + 1}$ for each of these $h(x)$-values, which is only possible when $h(x) \geq -1$. That is, when $\frac{4}{x} \geq -1$. Since we do not know if x is positive or negative, we cannot multiply both sides of this inequality by x to solve for x. Instead, rewrite the inequality as $\frac{4}{x} - 1 \geq 0$. When $x = -4$, $\frac{4}{x} + 1 = 0$ and when $x = 0$, $\frac{4}{x} + 1$ is undefined. Test intervals to determine the solutions of $\frac{4}{x} + 1 \geq 0$ are therefore $x \leq -4$, $-4 \leq x < 0$, and $x > 0$. Test points in these intervals to determine the solutions of the inequality. Since $\frac{4}{x} + 1 \geq 0$ is true for $x = -5$ and $x = 1$, but not for $x = -1$, the solution is $x \leq -4$ or $x > 0$. Therefore the domain of $g \circ h$ is $\{x \mid x \leq -4 \text{ or } x \geq 0, x \in \mathbb{R}\}$. To find $f \circ g \circ h$, you must be able to evaluate $f(x) = \frac{1}{x - 2}$, for each of these $g[h(x)]$-values, which can only be done for $g[h(x)] \neq -2$. This means that we must exclude from the domain those values for which $\sqrt{\frac{4 + x}{x}} = -2$.

$$\sqrt{\frac{4+x}{x}} = -2$$
$$\frac{4+x}{x} = 4$$
$$4x = 4 + x$$
$$3x = 4$$
$$x = \frac{4}{3}$$

Therefore, the domain of $f \circ g \circ h$ is $\{x \mid x \leq -4 \text{ or } 0 < x < \frac{4}{3} \text{ or } x > \frac{4}{3} \in \mathbb{R}\}$. In interval notation, the domain is $(-\infty, -4] \cup \left(0, \frac{4}{3}\right) \cup \left(\frac{4}{3}, \infty\right)$.

97. Sample answer: Order is important because there usually will be different outputs for $f[g(x)]$ and $g[f(x)]$. For example, if $f(x) = x^2$ and $g(x) = x + 2$, $f[g(x)] = x^2 + 4x + 4$ and $g[f(x)] = x^2 + 2$.

99.

[−10, 10] scl: 1 by [−10, 10] scl: 1
rel. max: (0, 4); rel. min: (1, 3)

101.

[−10, 10] scl: 1 by [−10, 10] scl: 1
abs. min: (−0.75, −2.11)

103. −1.73, 1.73

105a.

105b. D = {43, 48, 54}, R = {137, 146, 148, 150, 156} **105c.** No; each of the domain values 48 and 54 is paired with two different range values. **107.** G

Pages 70–73 Lesson 1-7

1. no **3.** no **5.** yes **7.** yes **9.** yes **11.** yes **13.** no **15.** no **17.** no **19.** yes; $f^{-1}(x) = \frac{4}{x + 1}; x \neq -1$ **21.** yes; $f^{-1}(x) = 8 - \frac{36}{x^2}; x > 0$ **23.** yes; $f^{-1}(x) = \frac{8x + 3}{x - 6}; x \neq 6$ **25.** no

27. $f[g(x)] = \frac{-6(3 - x)}{6} + 3 = \frac{-18 + 6x}{6} + 3 = -3 + x + 3 = x$; $g[f(x)] = \frac{3 - (-6x + 3)}{6} = \frac{6x}{6} = x$ **29.** $f[g(x)] = -3\left(\sqrt{\frac{5 - x}{3}}\right)^2 + 5 = \frac{-3(5 - x)}{3} + 5 = x - 5 + 5 = x$; $g[f(x)] = \sqrt{\frac{5 - (-3x^2 + 5)}{3}} = \sqrt{\frac{3x^2}{3}} = \sqrt{x^2} = x$ **31.** $f[g(x)] = 2\left(\sqrt[3]{\frac{x + 6}{2}}\right)^3 - 6 = \frac{2(x + 6)}{2} - 6 = x + 6 - 6 = x$; $g[f(x)] = \sqrt[3]{\frac{2x^3 - 6 + 6}{2}} = \sqrt[3]{\frac{2x^3}{2}} = \sqrt[3]{x^3} = x$

33. $f[g(x)] = \left(\sqrt{x + 8} - 4\right)^2 + 8\sqrt{x + 8} - 32 + 8 = x + 8 - 8\sqrt{x + 8} + 16 + 8\sqrt{x + 8} - 24 = x + 24 - 24 = x$; $g[f(x)] = \sqrt{x^2 + 8x + 8 + 8} - 4 = \sqrt{x^2 + 8x + 16} - 4 = \sqrt{(x + 4)^2} - 4 = x + 4 - 4 = x$ **35.** $f[g(x)] = \frac{\frac{4}{x - 1} + 4}{\frac{4}{x - 1}} = \frac{4 + 4x - 4}{x - 1} \div \frac{4}{x - 1} = \frac{4 + 4x - 4}{4} = \frac{4x}{4} = x$; $g[f(x)] = \frac{4}{\frac{x + 4}{x} - 1} = \frac{4}{\frac{x + 4 - x}{x}} = 4 \div \frac{4}{x} = x$

37 a. If $y = 0.5mx^2$, the inverses comes from $x = \frac{1}{2}my^2$.

$2x = my^2$ Original equation.
$y^2 = \frac{2x}{m}$ Divide.
$y = \sqrt{\frac{2x}{m}}$ Take the square root and reject the negative root.
$g(x) = \sqrt{\frac{2x}{m}}$ Replace y with $g(x)$.

$g(x) = $ velocity in m/s, $x = $ kinetic energy in J and $m = $ mass in kg.

b. $f[g(x)] = f\left(\sqrt{\frac{2x}{m}}\right) = \frac{1}{2}m\left(\sqrt{\frac{2x}{m}}\right)^2 = \frac{1}{2}m\left(\frac{2x}{m}\right) = x$; $g[f(x)] = g\left(\frac{1}{2}mx^2\right) = \left(\sqrt{\frac{2\left(\frac{1}{2}mx^2\right)}{m}}\right) = \sqrt{x^2} = x$; because $f[g(x)] = g[f(x)] = x$, the functions are inverses. However they are not inverse functions until the domain of $f(x)$ is restricted to D = $[0, \infty)$.

(*Continued on the next page*)

c.

[0, 10] scl: 1 by [0, 10] scl: 1

39.

41.

43.

45a. Sample answer: The graph of the function is linear, so it passes the horizontal line test. Therefore, it is a one-to-one function and it has an inverse; $f^{-1}(x) = \frac{x}{0.66}$.

45b. x represents the value of the currency in U.S. dollars, and $f^{-1}(x)$ represents the value of the currency in Euros.
45c. $x \geq 0$; You cannot exchange negative money.
45d. 151.52　**47.** f^{-1} exists.　**49.** f^{-1} exists.　**51.** f^{-1} does not exist.　**53.** f^{-1} exists.

x	8	7	6	5	4	3
$f^{-1}(x)$	−10	−9	−8	−7	−6	−5

55 Since the vertex of the parabola is at $(5, 0)$, either $x \leq 5$ or $x \geq 5$ will make the inverse one-to-one. Choose $x \geq 5$.

$$x = (y - 5)^2 \quad \text{Original function.}$$
$$\sqrt{x} = y - 5 \quad \text{Interchange } x \text{ and } y.$$
$$\sqrt{x} + 5 = y \quad \text{Add 5 to each side.}$$

Thus, $f^{-1}(x) = \sqrt{x} + 5$.

57. Sample answer: $x \geq 0, f^{-1}(x) = \sqrt{x + 6}$　**59.** f: D = $\{x \,|\, x \geq 6, x \in \mathbb{R}\}$, R = $\{y \,|\, y \geq 0, y \in \mathbb{R}\}$; f^{-1}: D = $\{x \,|\, x \geq 0, x \in \mathbb{R}\}$, R = $\{y \,|\, y \geq 6, y \in \mathbb{R}\}$　**61.** f: D = $\{x \,|\, x \neq 4, x \in \mathbb{R}\}$, R = $\{y \,|\, y \neq 3, y \in \mathbb{R}\}$; f^{-1}: D = $\{x \,|\, x \neq 3, x \in \mathbb{R}\}$, R = $\{y \,|\, y \neq 4, y \in \mathbb{R}\}$　**63a.** Sample answer: $f(x) = 272.6x + 1710.4$

63b. $f^{-1}(x) = \frac{x - 1710.4}{272.6}$; x represents the number of nesting pairs and $f^{-1}(x)$ represents the number of years after 1984.　**63c.** 1996

65. $f^{-1}(x) = \begin{cases} -\sqrt{x} & \text{if } 16 \leq x \\ 2.5 - 0.5x & \text{if } 13 > x \end{cases}$

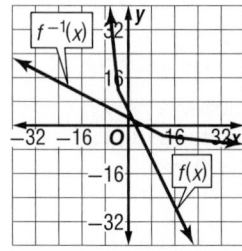

67a. $v^{-1}(x) = 3x$; represents the formula for the flow rate of the gas.　**67b.** $5000 \frac{\text{ft}}{\text{s}}$　**67c.** $\approx 35.3 \frac{\text{ft}^3}{\text{s}}$　**69.** $[f^{-1} \circ g^{-1}](x) = \frac{x + 2}{16}$　**71.** $[f \circ g]^{-1}(x) = \frac{x - 44}{16}$　**73.** $(f \cdot g)^{-1}(x) \frac{\sqrt{x + 49} - 5}{4}$, $x \geq -1.25$　**75.** $[f^{-1} \circ g^{-1}](x) = \sqrt{x^2 + 3}$ for $x \geq 0$
77. $[f \circ g]^{-1}(x) = x + 3$ for $x \geq 1$
79. $(f \cdot g^{-1})(x) = x^4 + 5x^2 + 4$ for $x \geq 0$

81a.

81b. D = $\{x \,|\, x \in \mathbb{R}\}$, R = $\{y \,|\, y \in$ positive multiples of 0.4$\}$

81c.

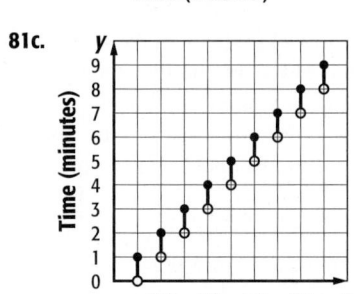

81d. D = $\{x \,|\, x \in$ positive multiples of 0.4$\}$; R = $\{y \,|\, y \in \mathbb{R}\}$
81e. The inverse gives the number of possible minutes spent using the scanner that costs x dollars.

83 The inverse of (x, y) is (y, x). Thus, if f contains a zero at 6, $(6, 0)$ is a point on the graph. By definition of an inverse, the graph of the inverse must contain $(0, 6)$, which is its y-intercept.

85. Sample answer: False; constant functions are linear, but they do not pass the horizontal line test. Therefore, constant functions are not one-to-one functions and do not have inverse functions.　**87.** Sample answer: Yes; one function that does this is $f(x) = \frac{1}{x}$. Even though both limits approach 0, they do it from opposite sides of 0 and no x-values ever share a corresponding y-value. Therefore, the function passes the horizontal line test.　**89.** Sample answer: If $f(x) = x^2$, f does not have an inverse because it is not one-to-one. If the domain is restricted to $x \geq 0$, then the function is now one-to-one and f^{-1} exists; $f^{-1}(x) = \sqrt{x}$.　**91.** $[f \circ g](x) = \frac{1}{2}x - 4$ for $\{x \,|\, x \in \mathbb{R}\}$; $[g \circ f](x) = \frac{1}{2}x - 1$ for $\{x \,|\, x \in \mathbb{R}\}$　**93a.** expanded horizontally　**93b.** translated 5 units to the right and 2 units down　**93c.** expanded vertically, translated 6 units up
95a. compressed horizontally　**95b.** translated 5 units to the right　**95c.** compressed horizontally, translated 4 units down
97. $(1, -4, 4)$　**99.** $(10, -7, 1)$　**101.** D　**103.** A

Pages 76-80　Chapter 1　Study Guide and Review

1. true　**3.** true　**5.** false; end behavior　**7.** true　**9.** true
11. function　**13.** function　**15.** 14　**17.** D = $\{x \,|\, x \in \mathbb{R}\}$
19. D = $\{a \,|\, a \neq -5, a \in \mathbb{R}\}$　**21.** D = $[-8, 8]$, R = $[0, 8]$
23. $-9; \frac{9}{4}$　**25.** 0; 0, 4, −4　**27.** continuous at $x = 4$; The function is defined when $x = 4$. The function approaches

Selected Answers and Solutions

4 when x approaches 4 from both sides, and $f(4) = 4$.
29. continuous at $x = 0$; The function is defined when $x = 0$. The function approaches 0 when x approaches 0 from both sides, and $f(0) = 0$. continuous at $x = 7$; The function is defined when $x = 7$. The function approaches 0.5 when x approaches 7 from both sides, and $f(7) = 0.5$. **31.** continuous at $x = 1$; The function is defined when $x = 1$. The function approaches 2 when x approaches 1 from both sides, and $f(1) = 2$. **33.** From the graph, it appears that as $x \to \infty$, $f(x) \to 0$; as $x \to -\infty$, $f(x) \to 0$. **35.** f is decreasing on $(-\infty, -3)$, increasing on $(-3, -1.5)$, decreasing on $(-1.5, 0.5)$, and increasing on $(0.5, \infty)$; rel. min. at $(-3, 3)$, rel. max. at $(-1.5, 6)$ and rel. min. at $(0.5, -7)$. **37.** 0

39. $f(x) = x^2$; $g(x)$ is the graph of $f(x)$ reflected in the x-axis and translated 6 units to the right and 5 units down.

41. $f(x) = [\![x]\!]$; $g(x)$ is the graph of $f(x)$ compressed vertically by a factor of $\frac{1}{4}$ and translated 3 units up.

43. The graph is reflected in the x-axis and translated 4 units right and 1 unit up; $g(x) = -\sqrt{x - 4} + 1$. **45.** $(f + g)(x) = 4x^2 + 5x - 2$; $D = (-\infty, \infty)$; $(f - g)(x) = 4x^2 - 5x$; $D = (-\infty, \infty)$; $(f \cdot g)(x) = 20x^3 - 4x^2 - 5x + 1$; $D = (-\infty, \infty)$; $\left(\dfrac{f}{g}\right)(x) = \dfrac{4x^2 - 1}{5x - 1}$; $D = \left(-\infty, \frac{1}{5}\right) \cup \left(\frac{1}{5}, \infty\right)$ **47.** $(f + g)(x) = \dfrac{x + 1}{x^2}$; $D = (-\infty, 0) \cup (0, \infty)$; $(f - g)(x) = \dfrac{x - 1}{x^2}$; $D = (-\infty, 0) \cup (0, \infty)$; $(f \cdot g)(x) = \dfrac{1}{x^3}$; $D = (-\infty, 0) \cup (0, \infty)$; $\left(\dfrac{f}{g}\right)(x) = x$; $D = (-\infty, 0) \cup (0, \infty)$ **49.** $x^2 - 8x + 23$; $x^2 + 2x + 3$; 11

51. $[f \circ g](x) = \dfrac{1}{2x - 9}$ for $x \neq \dfrac{9}{2}$ **53.** no **55.** yes **57.** $f^{-1}(x) = \sqrt[3]{x + 2}$ **59.** $h^{-1}(x) = \frac{1}{4}x^2 - 3, x \geq 0$

61a. $f(x) = \begin{cases} 39.99 & \text{if } 0 \leq x \leq 500 \\ 39.99 + 0.2(x - 500) & \text{if } x > 500 \end{cases}$

61b. \$39.99; \$49.99

61c.

63. No; sample answer: At the time of her promotion, her income had a jump discontinuity.

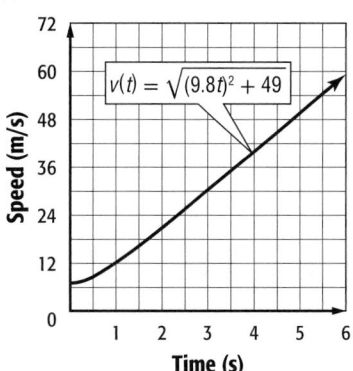

65.

67a. $A(x) = 6.4516x \text{ cm}^2$
67b. $A^{-1}(x) = \dfrac{1}{6.4516}x \text{ in}^2$

Pages 82–83   Chapter 1   Connect to AP Calculus
1. 5 **3.** Sample answer: 5 **5.** Sample answer: The secant line appears to become a line tangent to the graph of f. **7.** Sample answer: The exact rate of change is 5 which is equivalent to what the other values appeared to be approaching. **9.** Sample answer: To find the rate of change of a function $f(x)$ at the point x, find the difference quotient, and simplify the expression. Substitute $h = 0$ into the simplified expression, and the result is the rate of change of the function at x.

Chapter 2 Polynomial and Rational Functions

Page 85   Chapter 2   Get Ready
1. $(x - 4)(x + 5)$ **3.** $(2x - 3)(x - 7)$ **5.** $(4x - 5)(3x + 7)$
7. $4x + 7$

9.

11.

13.

15. $\{x \mid x \leq 6, x \in \mathbb{R}\}$; $(-\infty, 6]$
17. $\{x \mid -2 < x < 9, x \in \mathbb{R}\}$; $(-2, 9)$ **19.** $\{x \mid x < -4 \text{ or } x > 5, x \in \mathbb{R}\}$; $(-\infty, -4) \cup (5, \infty)$
21. $\{x \mid 9.99 < x < 19.99, x \in \mathbb{R}\}$; $(9.99, 19.99)$

Pages 92–95   Lesson 2-1
1.
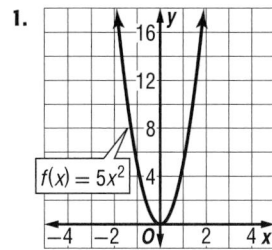

$D = (-\infty, \infty)$, $R = [0, \infty)$; intercept: 0; $\lim\limits_{x \to -\infty} f(x) = \infty$ and $\lim\limits_{x \to \infty} f(x) = \infty$; continuous for all real numbers; decreasing: $(-\infty, 0)$; increasing: $(0, \infty)$

3.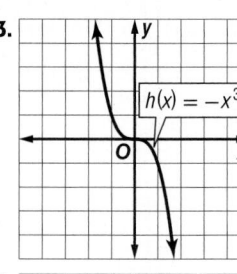

$D = (-\infty, \infty), R = (-\infty, \infty)$; intercept: 0; $\lim\limits_{x \to -\infty} f(x) = \infty$ and $\lim\limits_{x \to \infty} f(x) = -\infty$; continuous for all real numbers; decreasing: $(-\infty, \infty)$

5.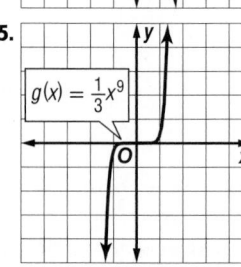

$D = (-\infty, \infty), R = (-\infty, \infty)$; intercept: 0; $\lim\limits_{x \to -\infty} f(x) = -\infty$ and $\lim\limits_{x \to \infty} f(x) = \infty$; continuous for all real numbers; increasing: $(-\infty, \infty)$

7.

$D = (-\infty, \infty), R = (-\infty, \infty)$; intercept: 0; $\lim\limits_{x \to -\infty} f(x) = \infty$ and $\lim\limits_{x \to \infty} f(x) = -\infty$; continuous for all real numbers; decreasing: $(-\infty, \infty)$

9.

$D = (-\infty, 0) \cup (0, \infty), R = (0, \infty)$; no intercepts; $\lim\limits_{x \to -\infty} f(x) = 0$ and $\lim\limits_{x \to \infty} f(x) = 0$; infinite discontinuity at $x = 0$; increasing: $(-\infty, 0)$; decreasing: $(0, \infty)$

11.

$D = (-\infty, 0) \cup (0, \infty)$, $R = (-\infty, 0) \cup (0, \infty)$; no intercepts; $\lim\limits_{x \to -\infty} f(x) = 0$ and $\lim\limits_{x \to \infty} f(x) = 0$; infinite discontinuity at $x = 0$; increasing: $(-\infty, 0)$ and $(0, \infty)$

13.

$D = (-\infty, 0) \cup (0, \infty)$, $R = (-\infty, 0) \cup (0, \infty)$; no intercepts; $\lim\limits_{x \to -\infty} f(x) = 0$ and $\lim\limits_{x \to \infty} f(x) = 0$; infinite discontinuity at $x = 0$; increasing: $(-\infty, 0)$ and $(0, \infty)$

15.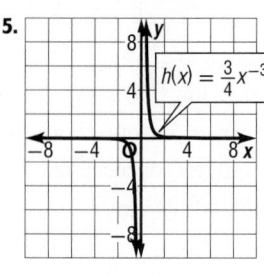

$D = (-\infty, 0) \cup (0, \infty)$, $R = (-\infty, 0) \cup (0, \infty)$; no intercepts; $\lim\limits_{x \to -\infty} f(x) = 0$ and $\lim\limits_{x \to \infty} f(x) = 0$; infinite discontinuity at $x = 0$; decreasing: $(-\infty, 0)$ and $(0, \infty)$

17a. $D = (0, \infty), R = (0, \infty)$

17b.

19.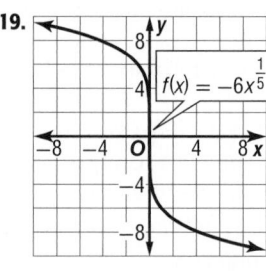

$D = (-\infty, \infty), R = (-\infty, \infty)$; intercept: 0; $\lim\limits_{x \to -\infty} f(x) = \infty$ and $\lim\limits_{x \to \infty} f(x) = -\infty$; continuous for all real numbers; decreasing: $(-\infty, \infty)$

21.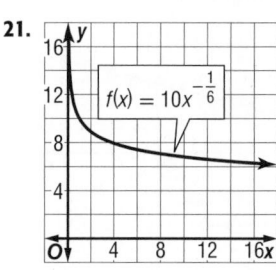

$D = (0, \infty), R = (0, \infty)$; no intercepts; $\lim\limits_{x \to \infty} f(x) = 0$; continuous on $(0, \infty)$; decreasing: $(0, \infty)$

23.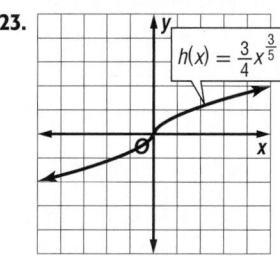

$D = (-\infty, \infty), R = (-\infty, \infty)$; intercept: 0; $\lim\limits_{x \to -\infty} f(x) = -\infty$ and $\lim\limits_{x \to \infty} f(x) = \infty$; continuous for all real numbers; increasing: $(-\infty, \infty)$

25.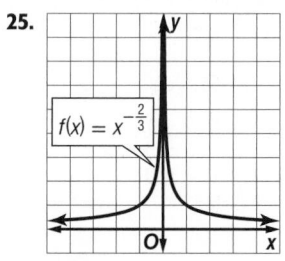

$D = (-\infty, 0) \cup (0, \infty)$, $R = (0, \infty)$; no intercepts; $\lim\limits_{x \to -\infty} f(x) = 0$ and $\lim\limits_{x \to \infty} f(x) = 0$; infinite discontinuity at $x = 0$; increasing: $(-\infty, 0)$; decreasing: $(0, \infty)$

27.

$D = [0, \infty), R = (-\infty, 0]$; intercept: 0; $\lim\limits_{x \to \infty} f(x) = -\infty$; continuous on $[0, \infty)$; decreasing: $(0, \infty)$

29.

$D = (-\infty, 0) \cup (0, \infty)$, $R = (0, \infty)$; no intercepts; $\lim\limits_{x \to -\infty} f(x) = 0$ and $\lim\limits_{x \to \infty} f(x) = 0$; infinite discontinuity at $x = 0$; increasing: $(-\infty, 0)$; decreasing: $(0, \infty)$

31a.

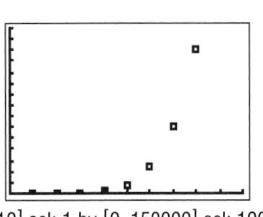

[0, 10] scl: 1 by [0, 150000] scl: 10000

31b. $y = 0.77x^{5.75}$
31c. about 235,906,039

33a.

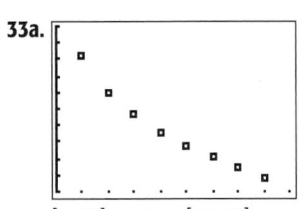

[0, 45] scl: 5 by [40, 50] scl: 1

33b. $f(x) = 55.14x^{-0.0797}$
33c. about 39.54°F

35.

$D = (-\infty, \infty)$, $R = (-\infty, \infty)$; x-intercept: -128, y-intercept: -8; $\lim\limits_{x \to -\infty} f(x) = \infty$ and $\lim\limits_{x \to \infty} f(x) = -\infty$; continuous for all real numbers; decreasing: $(-\infty, \infty)$

37 Evaluate the function for several x-values in its domain.

x	$\frac{12}{7}$	2	3	4	5	6	7
f(x)	4	5.4	7	8	8.8	9.5	10.1

Use these points to construct a graph.

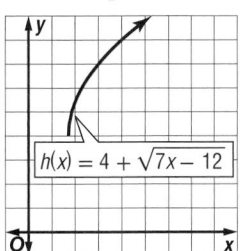

Since it is an even-degree radical function, the domain is restricted to nonnegative values for the radicand, $7x = 12$. Solve for x when the radicand is 0 to find the restriction on the domain.

$0 = 7x - 12$

$12 = 7x$

$\frac{12}{7} = x$

$D = \left[\frac{12}{7}, \infty\right)$, $R = [4, \infty)$; no intercepts; $\lim\limits_{x \to \infty} f(x) = \infty$; continuous on $\left[\frac{12}{7}, \infty\right)$; increasing: $\left(\frac{12}{7}, \infty\right)$

39.

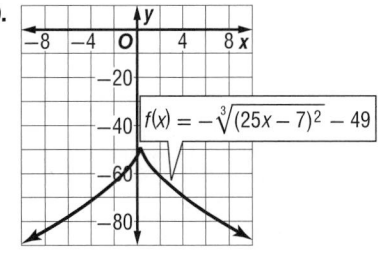

$D = (-\infty, \infty)$, $R = (-\infty, -49.00]$; y-intercept: -52.66; $\lim\limits_{x \to -\infty} f(x) = -\infty$ and $\lim\limits_{x \to \infty} f(x) = -\infty$; continuous for all real numbers; increasing: $(-\infty, 0.28)$; decreasing: $(0.28, \infty)$

41.

$D = [1, 22]$, $R = [-\sqrt{63}, \sqrt{21}]$; x-intercept: 6.25; continuous on $[1, 22]$; decreasing: $(1, 22)$

43a. about 6.89 Mcal **43b.** about 7.18 lb **45.** no solution
47. 7 **49.** 2, 4 **51.** 8 **53.** no solution **55.** 10 **57.** Yes; sample answer: The function follows the form $f(x) = ax^n$, where n is a positive integer. In this case, $a = -2a$ and $n = 4$. **59.** Yes; sample answer: The function follows the form $f(x) = ax^n$, where n is a positive integer. In this case, $a = \frac{7}{3}$ and $n = ab$. **61.** No; sample answer: The function is not a monomial function because the exponent for x is negative. **63.** $x \geq -2$ **65.** $x \leq -6$ **67.** $x \geq 5$

69 a. **Volume vs. Pressure**

b. Use the power regression function on the graphing calculator to find values for a and n.
$P(v) = 3.62v^{-1}$

```
PwrReg
y=a*x^b
a=3.624708022
b=-.9999645702
```

c. Sample answer: Yes; the problem states that the volume and pressure are inversely proportional, and in the power function, the exponent of the volume variable is -1.

d. Graph the regression equation using a graphing calculator. To predict the pressure of the gas if the volume is 3.25 liters, use the **CALC** function on the graphing calculator. Let $v = 3.25$.

[0, 5] scl: 0.5 by [0, 5] scl: 0.5

The pressure of 3.25 liters of the gas is about 1.12 atmospheres.

e. Graph the regression equation using a graphing calculator. To predict the pressure of the gas if the volume is 6 liters, use the **CALC** function on the graphing calculator. Let $v = 6$.

[0, 6] scl: 0.5 by [0, 5] scl: 0.5

The pressure of 6 liters of the gas is about 0.60 atmospheres.

71. c **73.** d

75 The function is undefined at $x = 0$, therefore; it is a power function of the form $f(x) = ax^{-n}$ or $f(x) = a\dfrac{1}{x^n}$, where a is a real constant and n is a natural number. Start by assuming $n = 1$ and solve for a by substituting a set of points $(2, 2)$ for x and y.

$$f(x) = a\frac{1}{x^n}$$
$$2 = a\frac{1}{2^1}$$
$$2 = a\frac{1}{2}$$
$$4 = a$$

When $n = 1$, $f(x) = 4 \cdot \dfrac{1}{x}$ or $\dfrac{4}{x}$ for the point $(2, 2)$. If this power function is true for the second point $(-1, -4)$, then $f(x)$ can represent the graph.

$$f(x) = \frac{4}{x}$$
$$-4 = \frac{4}{-1}$$
$$-4 = -4$$

Since $f(x)$ is true for $(-1, -4)$, $f(x) = \dfrac{4}{x}$ or $4x^{-1}$ is a power function for the graph.

77. $f(x) = 3x^{\frac{1}{2}}$ **79a.** 7.98 mm **79b.** 7.35 mm **79c.** increase

81. $\sqrt{\dfrac{8^n \cdot 2^7}{4^{-n}}} = \sqrt{\dfrac{(2^3)^n \cdot 2^7}{(2^2)^{-n}}}$

$$= \sqrt{\frac{2^{3n} \cdot 2^7}{2^{-2n}}}$$
$$= \sqrt{2^{(3n+7)-(-2n)}}$$
$$= \sqrt{2^{5n+7}}$$
$$= \sqrt{2^{4n+6} \cdot 2^{n+1}}$$
$$= \sqrt{2^{4n+6}} \cdot \sqrt{2^{n+1}}$$
$$= 2^{2n+3}\sqrt{2^{n+1}}$$

85. Sample answer: As n increases, the value of $\dfrac{1}{n}$ approaches 0. This means that the value of $\dfrac{1}{x^n}$ will approach 1 when x is positive and -1 when x is negative. Therefore, for positive values of x, $f(x)$ will approach $1 + 5$ or 6 and will resemble the line $y = 6$. For negative values of x, $f(x)$ will approach $-1 + 5$ or 4 and will resemble the line $y = 4$. **87a.** $r = -1 + \dfrac{\sqrt[3]{B}}{10}$

87b. 3.23% **89.** $(f + g)(x) = \dfrac{x^3 + x^2 - 1}{x + 1}$, $D = (-\infty, -1) \cup (-1, \infty)$; $(f - g)(x) = \dfrac{-x^3 - x^2 + 2x + 1}{x + 1}$, $D = (-\infty, -1) \cup (-1, \infty)$; $(f \cdot g)(x) = x^2 - x$; $D = (-\infty, -1) \cup (-1, \infty)$; $\left(\dfrac{f}{g}\right)(x) = \dfrac{x}{x^3 + x^2 - x - 1}$, $D = (-\infty, -1) \cup (-1, 1) \cup (1, \infty)$

91.

93.

95. f is increasing on $(-\infty, -4)$ and increasing on $(-4, \infty)$.

97. $\left(\dfrac{1}{6} - \dfrac{\sqrt{6}}{3}\right) + \left(\dfrac{\sqrt{3}}{3} + \dfrac{\sqrt{2}}{6}\right)i$ **99.** $-\dfrac{24}{169} + \dfrac{10}{169}i$ **101.** J

103. H

Pages 104–107   Lesson 2-2

1.

3.

5.

7.

$f(x) = (x - 3)^4 + 6$

9.

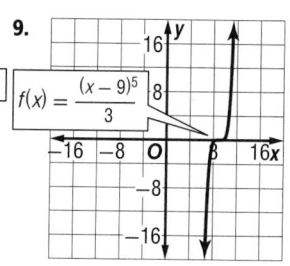

$f(x) = \dfrac{(x - 9)^5}{3}$

11.

Draining Water

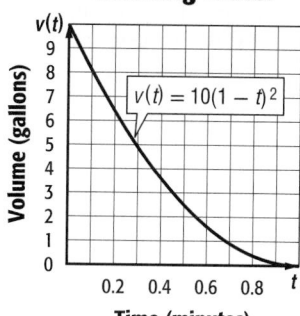

$v(t) = 10(1 - t)^2$

Time (minutes)

13. The degree is 6, and the leading coefficient is 2. Because the degree is even and the leading coefficient is positive, $\lim\limits_{x \to -\infty} f(x) = \infty$ and $\lim\limits_{x \to \infty} f(x) = \infty$. **15.** The degree is 6, and the leading coefficient is −6. Because the degree is even and the leading coefficient is negative, $\lim\limits_{x \to -\infty} f(x) = -\infty$ and $\lim\limits_{x \to \infty} f(x) = -\infty$. **17.** The degree is 5, and the leading coefficient is −2. Because the degree is odd and the leading coefficient is negative, $\lim\limits_{x \to -\infty} f(x) = \infty$ and $\lim\limits_{x \to \infty} f(x) = -\infty$. **19.** The degree is 4, and the leading coefficient is −2. Because the degree is even and the leading coefficient is negative, $\lim\limits_{x \to -\infty} f(x) = -\infty$ and $\lim\limits_{x \to \infty} f(x) = -\infty$. **21.** The degree is 6, and the leading coefficient is 1. Because the degree is even and the leading coefficient is positive, $\lim\limits_{x \to -\infty} f(x) = \infty$ and $\lim\limits_{x \to \infty} f(x) = \infty$.

23 The degree of $f(x)$ is 5, so it will have at most five real zeros and four turning points.

$0 = x^5 + 3x^4 + 2x^3$

$0 = x^3(x^2 + 3x + 2)$

$0 = x^3(x + 2)(x + 1)$

So, the zeros are −1, −2, and 0.

25. 4 real zeros and 3 turning points; $\pm\sqrt{3}$ **27.** 6 real zeros and 5 turning points; 2 and $\sqrt[3]{-2}$ **29.** 6 real zeros and 5 turning points; 0, −2, and 2 **31.** 4 real zeros and 3 turning points; 0, $-\dfrac{1}{2}$, and $\dfrac{3}{2}$ **33a.** The degree is 4, and the leading coefficient is 1. Because the degree is even and the leading coefficient is positive, $\lim\limits_{x \to -\infty} f(x) = \infty$ and $\lim\limits_{x \to \infty} f(x) = \infty$.

33b. 0, −4, 1 (multiplicity: 2)

33c. Sample answer: (−5, 180), (−2, −36), (0.5, 0.5625), (2, 12)

33d.

$f(x) = x(x + 4)(x - 1)^2$

35a. The degree is 4, and the leading coefficient is −1. Because the degree is even and the leading coefficient is negative, $\lim\limits_{x \to -\infty} f(x) = -\infty$ and $\lim\limits_{x \to \infty} f(x) = -\infty$.

35b. 0, −3 (multiplicity: −2), 5

35c. Sample answer: (−4, −36), (−1, −24), (2, 150), (6, −486)

35d.

$f(x) = -x(x + 3)^2(x - 5)$

37a. The degree is 5, and the leading coefficient is −1. Because the degree is odd and the leading coefficient is negative, $\lim\limits_{x \to -\infty} f(x) = \infty$ and $\lim\limits_{x \to \infty} f(x) = -\infty$.

37b. 0, 3, −2 (multiplicity: 3) **37d.**

37c. Sample answer: (−3, 18), (−1, −4), (1, 54), (4, −864)

$f(x) = -x(x - 3)(x + 2)^3$

39a. The degree is 3, and the leading coefficient is 3. Because the degree is odd and the leading coefficient is positive, $\lim\limits_{x \to -\infty} f(x) = -\infty$ and $\lim\limits_{x \to \infty} f(x) = \infty$. **39b.** 0, 4, −3

39c. Sample answer: (−4, −96), (−2, 36), (2, −60), (5, 120)

39d.

$f(x) = 3x^3 - 3x^2 - 36x$

41a. The degree is 4, and the leading coefficient is 1. Because the degree is even and the leading coefficient is positive, $\lim\limits_{x \to -\infty} f(x) = \infty$ and $\lim\limits_{x \to \infty} f(x) = \infty$. **41b.** 0 (multiplicity: 2), 4, −5 **41c.** Sample answer: (−6, 360), (−2, −72), (2, −56), (5, 250)

41d.

$f(x) = x^4 + x^3 - 20x^2$

43a. Sample answer: $f(x) = -0.0175x^4 + 0.355x^3 - 2.276x^2 + 5.72x + 3.5$ **43b.** 14.8 ft **45.** Sample answer: $f(x) = 0.09x^3 - 2.70x^2 + 24.63x - 65.21$ **47.** Sample answer: $f(x) = 4.05x^4 - 0.09x^3 + 6.69x^2 - 222.03x + 2697.74$ **49a.** According to the data, the values are increasing as x increases, therefore, $\lim\limits_{x \to \infty} f(x) = \infty$.

49b.

Sample answer: $f(x) = 0.146x^4 - 3.526x^3 + 32.406x^2 - 63.374x + 473.255$; The line is not a good fit. There are many outlying points.

[0, 12] scl: 1 by [400, 1400] scl: 200

49c. The leading coefficient is 0.146, so $\lim\limits_{x \to \infty} f(x) = \infty$. Sample answer: The prediction was accurate because the leading coefficient is positive, as $x \to \infty$, $f(x) \to \infty$. **51.** yes **53.** No; undefined at $x = 0$ **55.** Sample answer: $f(x) = x^3 - 9x^2 + 27x - 27$ **57.** Sample answer: $f(x) = x^4 + 11x^3 + 15x^2 - 175x - 500$ **59.** Sample answer: $f(x) = x^5 + 4x^4$ **61.** Sample answer: $f(x) = x^5 - x^4 - 6x^3$ **63.** Sample answer: $f(x) = x^6 + 3x^4 + 3x^2 + 1$ **65.** n is odd; a_n is positive. **67.** n is even; a_n is negative. **69.** $f(x) = x^3 - 8x^2 - 3x + 90$ **71.** $f(x) = x^5 - 9x^4 + 23x^3 - 3x^2 - 36x$ **73.** $f(x) = 12x^4 + 83x^3 + 131x^2 - 54x - 72$

75. a. Sample answer: Enter (0, 56.1), (10, 63), (20, 68.6), (30, 74.8), (40, 74.9), (50, 79.2) into STAT EDIT; the linear regression ($r = 0.97$) is $f(x) = 0.45x + 58.19$

```
LinReg
y=ax+b
a=.4497142857
b=58.19047619
r²=.9464926266
r=.9728785262
■
```

b. Find $f(x)$ for $x = 65$ since 2015 is 65 years since 1950. Graph the regression equation using a graphing calculator. Use the CALC feature on the calculator to find $f(65)$.

```
Y1=.4488571428571X+58.1
X=65        Y=87.370952
```

[0, 70] scl: 5 by [0, 100] scl: 10

$f(65) = 87.4\%$.

c. Sample answer: If $f(x) = 85$, then

$$85 = 0.45x + 58.2$$
$$26.8 = 0.45x$$
$$59.\overline{5} = x.$$

So 59.5 years after 1950, about 2010, 85% of the population will live in metropolitan areas.

77. Sample answer: $f(x) = -(x + 2)(x + 3)(x - 4)(x - 1) \cdot (x^2 + 6)$

$f(x) = -(x + 2)(x + 3)(x - 4)(x - 1)(x^2 + 6)$

[graph with y-axis values 640, 320, -320 and x-axis values -4, -2, 2, 4]

79. Sample answer: $f(x) = -(x - 1)(x - 2)(x + 1)^2(x^2 + 2)$

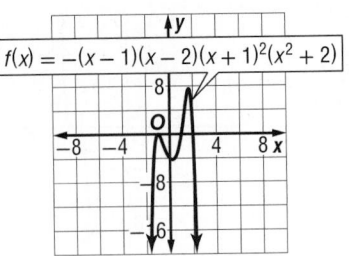

$f(x) = -(x - 1)(x - 2)(x + 1)^2(x^2 + 2)$

81a. degree = 4 **81b.** $-6, -2$ (multiplicity: 2), 4 **81c.** $f(x) = 0.5(x + 2)^2(x + 6)(x - 4)$ **83a.** degree = 4 **83b.** $-4, 3$ (multiplicity: 2), -1 **83c.** $f(x) = \frac{1}{8}(x - 3)^2(x + 1)(x + 4)$

85. 4 real zeros and 3 turning points; no real zeros

87 The degree of the function is 4, so f has at most 4 distinct real zeros and at most $4 - 1$ or 3 turning points. To find the real zeros, solve the related equation $f(x) = 0$ by factoring.

$$-24x^4 + 24x^3 - 6x^2 = 0$$
$$-6x^2(4x^2 - 4x + 1) = 0$$
$$-6x^2(2x - 1)(2x - 1) = 0$$
$$-6x^2(2x - 1)^2 = 0$$

The expression above has 4 factors, but solving for x yields only 2 distinct real zeros: $x = 0$ and $x = \frac{1}{2}$. Both of the zeros have multiplicity 2.

89a.

[−10, 10] scl: 1 by [−10, 10] scl: 1

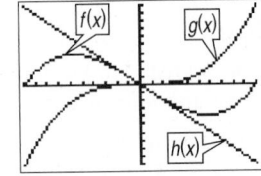

[−10, 10] scl: 1 by [−10, 10] scl: 1

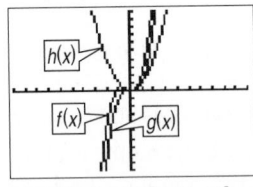

[−10, 10] scl: 1 by [−10, 10] scl: 1

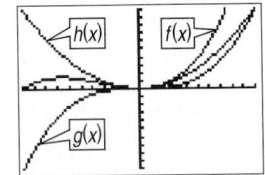

[−1, 1] scl: 0.1 by [−1, 1] scl: 0.1

[−10, 10] scl: 1 by [−10, 10] scl: 1

[−1, 1] scl: 0.1 by [−1, 1] scl: 0.1

89b. Sample answer: Very close to the origin, the function approximates the behavior of the term of lower degree, or $h(x)$. **89c.** Sample answer: As $x \to \infty$ and $-\infty$, the function approximates the behavior of the term of higher degree, or $g(x)$. **89d.** Sample answer: As $x \to \infty$ and $-\infty$, the function approximates the behavior of function a and, very close to the origin, the function approximates the behavior of function b. **91.** Sample answer: No; a polynomial function $f(x)$ cannot have both an absolute maximum and absolute minimum because as $x \to \infty$, $f(x)$ will approach either ∞ or $-\infty$. If $f(x) \to \infty$ as $x \to \infty$, then an absolute maximum is impossible. If $f(x) \to -\infty$ as $x \to \infty$, then an absolute minimum is impossible. **93.** Rearranging the terms gives

$f(x) = x^3 - x^2 - 12x + 5x^2 - 5x - 60$. Notice how the first set of three terms has a common factor of x and the second set of three terms has a common factor of 5. After factoring using the Distributive Property, $f(x) = x(x^2 - x - 12) + 5(x^2 - x - 12)$. Now notice how the factors inside the parentheses are identical. Using the Commutative Property, $f(x) = (x + 5) \cdot (x^2 - x - 12)$. After factoring the second factor, $f(x) = (x + 5) \cdot (x - 4)(x + 3)$. The function is now completely factored and the zeros of the function are -5, 4, and -3. These were determined by setting each factor equal to zero and solving for x. **95.** Sample answer: There is one turning point at the absolute maximum, one at the relative minimum, and one at the relative maximum. Therefore, the minimum degree is 3 + 1 or 4. **97.** 46 **99.** no solution **101.** $2x^2 + x + 7$ **103.** $12x^2 - 30x + 22$ **105.** The graph of $g(x)$ is the graph of $f(x)$ reflected in the x-axis and translated 2 units left and 4 units down; $g(x) = -(x + 2)^2 - 4$. **107a.** $50 **107b.** Sample answer: The company's competition might offer a similar product at a lower cost. **109.** ≈ 12.54 **111.** F **113.** H

Pages 115–117   Lesson 2-3

1. $(x + 3)(x - 5)(x + 4)$ **3.** $(x + 5)(x + 2)(x - 4)$ **5.** $-3(x - 5)(x + 6)(x - 6)$ **7.** $(x + 7)(x - 1)(x + 3)^2$ **9.** $5x^3 + 17x^2 + 74x + 295 + \frac{1192}{x - 4}$ **11.** $2x^3 - 8x^2 + 22x - 47 + \frac{100}{x + 2}$ **13.** $3x^5 + 3x^3 - 6x^2 - 2x + 4 - \frac{2}{2x - 1}$ **15.** $x + 4$ **17.** $2x^2 - 4x + 2 + \frac{2}{3x^3 + 2x + 3}$ **19.** $x^3 + x^2 + 5x + 4 + \frac{2}{x - 2}$

21 Because $x - 4$, $c = 4$. Set up the synthetic division as follows, using a zero placeholder for the missing x^2-term in the dividend. Then follow the synthetic division procedure.

$$
\begin{array}{r|rrrrr}
4 & 3 & -9 & 0 & -24 & -48 \\
  &   & 12 & 12 & 48 & 96 \\
\hline
  & 3 & 3 & 12 & 24 & 48
\end{array}
$$

The quotient is $3x^3 + 3x^2 + 12x + 24 + \frac{48}{x - 4}$.

23. $6x^4 + 14x^3 + 12x^2 + 12x + 18 + \frac{46}{2x - 3}$ **25.** $15x^4 + 12x^3 + 9x^2 + 6x + \frac{20}{3} + \frac{76}{3(3x - 2)}$ **27.** $12x^5 + 6x^4 - 3x^3 - 5$ **29.** 2,151,000 students **31.** 711 **33.** 45,536 **35.** $-67,978$ **37.** $-125,184$ **39.** no, no **41.** yes, yes; $f(x) = (3x - 1)(x - 5) \cdot (x - 4)(x + 2)$ **43.** no, no **45.** yes, no; $f(x) = (4x + 3)(x^4 - 3x^3 + 12x^2 - 3x + 21)$ **47.** 5 seconds **49.** $(x^{2n} + 2x^n - 10) \cdot (x^n - 1)$ **51.** $3(x^n - 3)(x^n + 6)(3x^n - 1)$ **53.** 2 **55.** -4

57a. $v(x) = -x^3 + 12x^2 - 47x + 60$

57b.

Volume of Clay

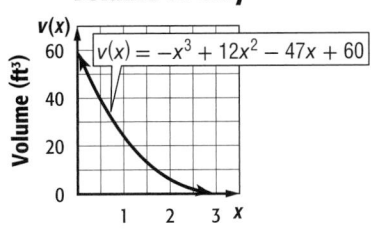

Removed Amount (ft)

57c. $36 = -x^3 + 12x^2 - 47x + 60$ **57d.** about 0.60 ft **59.** $f(x) = (x + 6)(x - 2)^2(3x - 7)(2x + 5)$

61 Using the Factor Theorem, $(x - 1)$ is a factor if $f(1) = 0$.

$f(1) = 18(1)^{165} - 15(1)^{135} + 8(1)^{105} - 15(1)^{55} + 4$
$= 18(1) - 15(1) + 8(1) - 15(1) + 4$
$= 18 - 15 + 8 - 15 + 4$
$= 0$

Since $f(1) = 0$, $(x - 1)$ is a factor of the polynomial. **63.** True; sample answer: The Remainder Theorem states that if $h(y)$ is divided by $y - (-2)$, then the remainder is $r = h(-2)$, which is -1. **65.** -30 **67.** -5 **69.** even; negative **71.** odd; negative **73a.** $h = \frac{v^2}{64}$ **73b.** Yes; the pump can propel water to a height of about 88 ft. **75.** $(-1, 1)$ **77.** $\left(\frac{1}{3}, \frac{2}{3}\right)$ **79.** $\left(-\frac{6}{43}, -\frac{64}{43}\right)$ **81.** K **83.** H

Pages 127–129   Lesson 2-4

1. $\pm 1, \pm 2, \pm 3, \pm 4, \pm 5, \pm 6, \pm 9, \pm 10, \pm 12, \pm 15, \pm 18, \pm 20, \pm 30, \pm 36, \pm 45, \pm 60, \pm 90, \pm 180$; 6, 5, 1, -6 **3.** $\pm 1, \pm 2, \pm 3, \pm 5, \pm 6, \pm 10, \pm 15, \pm 30$; -5, 6, -1, 1 **5.** $\pm 1, \pm 2, \pm 3, \pm 4, \pm 5, \pm 6, \pm 10, \pm 12, \pm 15, \pm 20, \pm 30, \pm 60, \pm \frac{1}{2}, \pm \frac{3}{2}, \pm \frac{5}{2}, \pm \frac{15}{2}, \pm \frac{1}{3}, \pm \frac{2}{3}, \pm \frac{4}{3}, \pm \frac{5}{3}, \pm \frac{10}{3}, \pm \frac{20}{3}, \pm \frac{1}{6}, \pm \frac{5}{6}$; $-\frac{5}{3}, -\frac{1}{2}$ **7.** $\pm 1, \pm 2, \pm 3, \pm 4, \pm 6, \pm 8, \pm 9, \pm 12, \pm 16, \pm 18, \pm 24, \pm 27, \pm 36, \pm 48, \pm 54, \pm 72, \pm 108, \pm 144, \pm 216, \pm 432$; 4 (multiplicity: 2), 3 **9.** 1 in., 5 in., 9 in. **11.** -4, 1, -3 (multiplicity: 2) **13.** 4, -5, 2 (multiplicity: 2) **15.** $\frac{1}{2}, \frac{2}{3}, -4$ (multiplicity: 2) **17.** 2 days **19.** Sample answer: $[-6, 5]$; 3, 2, -4, $-\frac{1}{2}$ **21.** Sample answer: $[-4, 7]$; -2, 5, 1, $\frac{3}{2}$ **23.** Sample answer: $[-2, 7]$; 3 (multiplicity: 2), -1, $\frac{3}{2}$ **25.** Sample answer: $[-3, 5]$; $\frac{1}{2}$ (multiplicity: 2), -2, 3 (multiplicity: 2) **27.** 3 or 1 positive zeros, 1 negative zero **29.** 2 or 0 positive zeros, 2 or 0 negative zeros **31.** 4, 2 or 0 positive zeros, 1 negative zero **33.** Sample answer: $f(x) = x^4 + 4x^3 - 19x^2 - 106x - 120$ **35.** Sample answer: $f(x) = x^4 - 19x^3 + 113x^2 - 163x - 296$ **37.** Sample answer: $f(x) = x^4 - 5x^3 - 21x^2 + 119x - 130$ **39.** Sample answer: $f(x) = x^4 - 6x^3 + 19x^2 + 36x - 150$ **41.** Sample answer: $f(x) = x^4 - 28x^3 + 296x^2 - 1372x + 2263$

43 **a.** $g(x)$ has possible rational zeros of $\pm 1, \pm 2, \pm 4, \pm 8$. By using synthetic division, it can be determined that $x = -2$ is a rational zero.

$$
\begin{array}{r|rrrrr}
-2 & 1 & -3 & -12 & 0 & 8 \\
   &   & -2 & 10 & 4 & -8 \\
\hline
   & 1 & -5 & -2 & 4 & 0
\end{array}
$$

By using synthetic division on the depressed polynomial, it can be determined that $x = -1$ is a rational zero.

$$
\begin{array}{r|rrrr}
-1 & 1 & -5 & -2 & 4 \\
   &   & -1 & 6 & -4 \\
\hline
   & 1 & -6 & 4 & 0
\end{array}
$$

The remaining quadratic factor $(x^2 - 6x + 4)$ yields no rational zeros. Use the quadratic formula to find the zeros.

$$
\frac{-b \pm \sqrt{b^2 - 4ac}}{2a} = \frac{-(-6) \pm \sqrt{(-6)^2 - 4(1)(4)}}{2(1)}
$$
$$
= \frac{6 \pm \sqrt{20}}{2}
$$
$$
= 3 \pm \sqrt{5}
$$

(Continued on the next page)

So, $g(x)$ written as a product of linear and irreducible quadratic factors is $g(x) = (x + 2)(x + 1)(x - 3 + \sqrt{5})(x - 3 - \sqrt{5})$.
b. $g(x)$ written as a product of linear factors is $g(x) = (x + 2)(x + 1)(x - 3 + \sqrt{5})(x - 3 - \sqrt{5})$.
c. The zeros are $-2, -1, 3 - \sqrt{5}, 3 + \sqrt{5}$.

45a. $f(x) = (x^2 - 4x + 13)(4x - 3)(x - 4)$ **45b.** $f(x) = (4x - 3) \cdot (x - 4)(x - 2 + 3i)(x - 2 - 3i)$ **45c.** $\frac{3}{4}, 4, 2 - 3i, 2 + 3i$

47a. $h(x) = (x + 3)(x - 5)(x + \sqrt{2})(x - \sqrt{2})$ **47b.** $h(x) = (x + 3)(x - 5)(x + \sqrt{2})(x - \sqrt{2})$ **47c.** $-3, 5, \sqrt{2}, -\sqrt{2}$

49. $3, -4, \frac{1}{2}, 3i, -3i; h(x) = (x - 3) \cdot (x + 4)(2x - 1)(x + 3i) \cdot (x - 3i)$ **51.** $-3, -2, 1, 3 + i, 3 - i; g(x) = (x + 3)(x + 2) \cdot (x - 1)(x - 3 + i)(x - 3 - i)$ **53.** $3, 2\sqrt{2}, -2\sqrt{2}, 2i, -2i; f(x) = (x - 3)(x + 2\sqrt{2})(x - 2\sqrt{2})(x + 2i)(x - 2i)$ **55a.** $V(\ell) = \frac{1}{3} \cdot \ell^3 - 3\ell^2$ **55b.** $6300 = \frac{1}{3}\ell^3 - 3\ell^2$ **55c.** base: 30 in. by 30 in.; height: 21 in. **57.** Sample answer: $f(x) = x^3 - 6$

59. Sample answer: $f(x) = x^3 + 2$ **61.** $g(x) = 3(x - 4)(x - 1) \cdot (x + 5i)(x - 5i); 4, 1, \pm 5i$ **63.** -1 **65.** no rational zeros

67 Solve $d = 0.0000008x^2(200 - x)$ for $d = 0.8$.
$0.8 = 0.0000008x^2(200 - x)$
$\quad = 0.0000008x^2(200 - x) - 0.8$

Use a graphing calculator to graph $y = 0.0000008x^2 \cdot (200 - x) - 0.8$ and solve for $y = 0$. Using the **CALC** menu, find the zeros of the function.

[0, 200] scl: 10 by [−1, 1] scl: 0.5

The zeros of the function occur at $x = 100$ and $x = 161.8$. A vertical deflection of 0.8 feet will occur when a weight is placed 100 feet or about 161.8 feet from the left piling. **69.** $(x + 2\sqrt[3]{2})(x^2 - 2x\sqrt[3]{2} + 4\sqrt[3]{4})$ **71.** $(3x^2 + \sqrt[3]{4}) \cdot (9x^4 - 3x^2\sqrt[3]{4} + 2\sqrt[3]{2})$ **73.** Julius; sample answer: Angie did not divide by the factors of the leading coefficient.
75a. The graphs of $-f(x)$ are the graphs of $f(x)$ reflected in the x-axis. The zeros are the same. **75b.** The graphs of $f(-x)$ are the graphs of $f(x)$ reflected in the y-axis. The zeros are opposites. **77.** False; sample answer: The third-degree polynomial $x^3 + x^2 - 17x + 15$ has three real zeros and no nonreal zeros. **79.** $\frac{x_1}{3}, \frac{x_2}{3}, \frac{x_3}{3}$ **81.** $-x_1, -x_2, -x_3$
83. Sample answer: If a polynomial has an imaginary zero, then its complex conjugate is also a zero of the polynomial. Therefore, any polynomial that has one imaginary zero has at least two imaginary zeros. **85.** $x^3 + 3x^2 + 6x + 12 + \frac{23}{x - 2}$ **87.** $2x^2 + 2x - \frac{3}{x - 1}$ **89.** The degree is 6, and the leading coefficient is 4. Because the degree is even and the leading coefficient is positive, $\lim_{x \to -\infty} f(x) = \infty$ and $\lim_{x \to \infty} f(x) = \infty$. **91.** It appears that $f(x)$ has a relative minimum of -3 at $x = 0$ and a relative maximum of 3 at $x = -2$. **93.** It appears that $f(x)$ has a relative maximum of 0 at $x = 0$ and $x = -4$ and a relative minimum of -80 at $x = -2$ and -150 at $x = 2$. **95.** E **97.** A

1. $D = \{x \mid x \neq 2, -2, x \in \mathbb{R}\}; x = 2, x = -2, y = 1$ **3.** $D = \{x \mid x \neq 4, -3, x \in \mathbb{R}\}; x = 4, x = -3$ **5.** $D = \{x \mid x \neq 0, -2, x \in \mathbb{R}\}; x = 0, x = -2, y = 2$ **7.** $D = \{x \mid x \neq 2, -4, x \in \mathbb{R}\}; x = 2, x = -4, y = 0$ **9.** asymptotes: $x = -4, x = 5, y = 1$; x-intercepts: $-2, 3$; y-intercept: $\frac{3}{10}$; $D = \{x \mid x \neq -4, 5, x \in \mathbb{R}\}$

11. asymptotes: $x = -2$, $x = 2, y = 0$; y-intercept: -2; $D = \{x \mid x \neq -2, 2, x \in \mathbb{R}\}$

13. asymptotes: $x = -5, x = 6, y = 0$; x-intercept: -2; y-intercept: $-\frac{1}{15}$; $D = \{x \mid x \neq -5, 6, x \in \mathbb{R}\}$

15. asymptotes: $x = -3$, $x = -1$; x-intercepts: -5, $2, 0$; y-intercept: 0; $D = \{x \mid x \neq -3, -1, x \in \mathbb{R}\}$

17. asymptote: $y = 0$; x-intercept: 8; y-intercept: $-\frac{8}{5}$; $D = \{x \mid x \in \mathbb{R}\}$

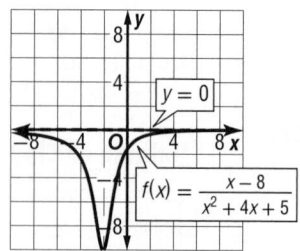

19a. $D = \{z \mid z \geq 0, z \in \mathbb{R}\}$ **19b.** horizontal asymptote: $y = \frac{3}{2}$; x-intercept: 1; y-intercept: $-\frac{3}{5}$

19c.

Car Wash

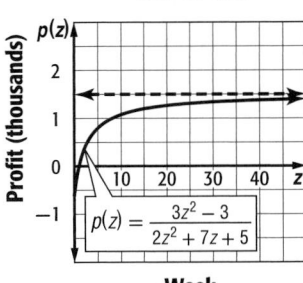

$$p(z) = \frac{3z^2 - 3}{2z^2 + 7z + 5}$$

Week

21. asymptotes: $x = \sqrt[3]{-\frac{4}{3}}$, $y = 0$; y-intercept: $\frac{1}{4}$;

$D = \left\{x \,|\, x \neq \sqrt[3]{-\frac{4}{3}}, x \in \mathbb{R}\right\}$

$$h(x) = \frac{4x^2 - 2x + 1}{3x^3 + 4}$$

23. asymptotes: $x = 4$, $y = 1$; x-intercept: -7; y-intercept: $-\frac{7}{4}$;

$D = \{x \,|\, x \neq 4, x \in \mathbb{R}\}$

$$g(x) = \frac{x+7}{x-4}$$

25. asymptote: $x = 4$; x-intercepts: $-2, -1, 0$; y-intercept: 0;

$D = \{x \,|\, x \neq 4, x \in \mathbb{R}\}$

$$g(x) = \frac{x^3 + 3x^2 + 2x}{x - 4}$$

27 $g(x)$ can be written as $g(x) = \dfrac{(x-2)(x+2)}{(x-2)(x+2)(x+1)}$ or $\dfrac{1}{x+1}$.

The function is undefined at $b(x) = 0$, so $D = \{x \,|\, x \neq -2, -1, 2, x \in \mathbb{R}\}$.

There is a vertical asymptote at $x = -1$, the real zero of the simplified denominator.

There is a horizontal asymptote at $y = 0$, because the degree of the denominator is greater than the degree of the numerator.

Since the simplified numerator has no real zeros, there are no x-intercepts. The y-intercept is 1 because $g(0) = 1$.

There are holes at $(-2, -1)$ and $\left(2, \frac{1}{3}\right)$ because the original function is undefined when $x = -2$ and $x = 2$. Graph the asymptotes and intercept. Then find and plot points.

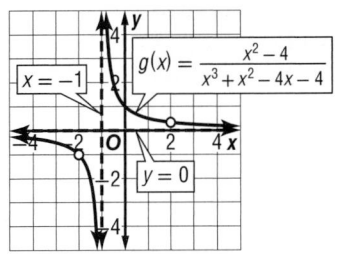

$$g(x) = \frac{x^2 - 4}{x^3 + x^2 - 4x - 4}$$

29. asymptotes: $x = -4$; $y = 0$; hole: $\left(5, \frac{11}{81}\right)$; x-intercept: $-\frac{1}{2}$;

y-intercept: $\frac{1}{16}$; $D = \{x \,|\, x \neq -4, 5, x \in \mathbb{R}\}$

$$g(x) = \frac{(2x+1)(x-5)}{(x-5)(x+4)^2}$$

31a. $\frac{1}{8} = \frac{1}{d_i} + \frac{1}{32}$ **31b.** $10\frac{2}{3}$ cm   **33.** $\pm 2\sqrt{3} - 2$ **35.** no

solution   **37.** 5   **39.** -6 **41.** -1 **43.** Sample answer:

$f(x) = \dfrac{x(x-4)}{(x-1)(x-6)^2}$

45 **a.** There is a vertical asymptote at the real zero of the denominator $r_1 = 30$. There is a horizontal asymptote at $r_2 = \dfrac{30}{1}$ or $r_2 = 30$, the ratio of the leading coefficients of the numerator and denominator, because the degrees of the polynomials are equal.

[0, 240] scl: 30 by [0, 100] scl: 10

b. Evaluate the function for each value of r_1 to determine r_2.

$r_2 = \dfrac{30(45)}{45 - 30}$ $r_2 = \dfrac{30(60)}{60 - 30}$

$= 90$ $= 60$

$r_2 = \dfrac{30(50)}{50 - 30}$ $r_2 = \dfrac{30(65)}{65 - 30}$

$= 75$ $= 55.7$

$r_2 = \dfrac{30(55)}{55 - 30}$ $r_2 = \dfrac{30(70)}{70 - 30}$

$= 66$ $= 52.5$

r_1	45	50	55	60	65	70
r_2	90	75	66	60	55.7	52.5

c. No; sample answer: As r_1 approaches infinity, r_2 approaches 30. This suggests that the average speeds reached during the first leg of the trip have no bounds. Being able to reach an infinite speed is not reasonable. The same holds true about r_2 as r_1 approaches 30 from the right.

47. Sample answer: $f(x) = \dfrac{2x^2 - 14x + 20}{x^2 + x - 12}$ **49.** $\approx -2.65, \approx 2.65$

51. $\approx -3.87, \approx 1.20, \approx 3.70$

53a.

Function	Horizontal Asymptote
$f(x) = \dfrac{x^2 - 5x + 4}{x^3 + 2}$	$y = 0$
$h(x) = \dfrac{x^3 - 3x^2 + 4x - 12}{x^4 - 4}$	$y = 0$
$g(x) = \dfrac{x^4 - 1}{x^5 + 3}$	$y = 0$

53b.

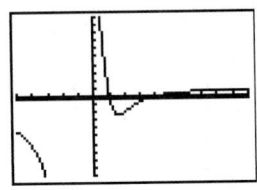

[−5, 10] scl: 1 by [−1, 1] scl: 0.1

[−5, 10] scl: 1 by [−1, 1] scl: 0.1

[−5, 10] scl: 1 by [−1, 1] scl: 0.1

53c.

Function	Real Zeros of Numerator
$f(x) = \dfrac{x^2 - 5x + 4}{x^3 + 2}$	1, 4
$h(x) = \dfrac{x^3 - 3x^2 + 4x - 12}{x^4 - 4}$	3
$g(x) = \dfrac{x^4 - 1}{x^5 + 3}$	−1, 1

53d. Sample answer: When the degree of the numerator is less than the degree of the denominator and the numerator has at least one real zero, the graph of the function will have $y = 0$ as an asymptote and will intersect the asymptote at the real zeros of the numerator. **59.** Sample answer: The test intervals are used to determine the location of points on the graph. Because many rational functions are not continuous, one interval may include y-values that are vastly different than the next interval. Therefore, at least one, and preferably more than one, point is needed for every interval in order to sketch a reasonably accurate graph of the function. **61.** ±1, ±2, ±3, ±6, ±9, ±18; −2 **63.** no; yes; $f(x) = (x - 2) \cdot (x^3 - 13x - 12)$ **65.** yes; no; $f(x) = (3x - 2) \cdot (2x^3 + 21x^2 + 60x + 25)$ or $f(x) = (3x - 2)(2x + 1)(x + 5)^2$ **67.** yes; no; $f(x) = (x + 2)(4x^4 + 7x^3 - 2x^2)$ or $f(x) = x^2(x + 2)^2(4x - 1)$

69.

$f(x) = (x + 7)^2$

71.

$f(x) = x^4 - 5$

73a. $(12 - 2x)(16 - 2x) = 96$

73b.

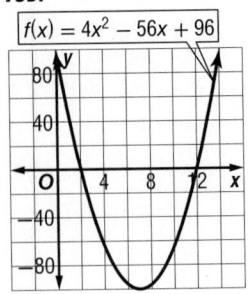

$f(x) = 4x^2 - 56x + 96$

73c. 8 ft by 12 ft **75.** $-4 + 4i$
77. C **79.** C

Pages 145–147 Lesson 2-6

1. $[-4, 2]$ **3.** $\left(-\infty, -\dfrac{1}{3}\right] \cup [8, \infty)$ **5.** $\left(-\infty, -\dfrac{1}{2}\right) \cup \left(\dfrac{2}{3}, \infty\right)$

7. $\left(-3, -\dfrac{3}{4}\right) \cup (0, \infty)$ **9.** $(-\infty, \infty)$ **11.** ∅ **13.** {4} **15.** {−2}

17. $-0.0004x^2 + 80x - 1,000,000 \geq 2,000,000$; [50,000, 150,000] or $50,000 \leq x \leq 150,000$ **19.** $(-\infty, 5)$

21.
$$\frac{3x - 2}{x + 3} < 6$$
$$\frac{3x - 2}{x + 3} - 6 < 0$$
$$\frac{3x - 2}{x + 3} - 6\left(\frac{x + 3}{x + 3}\right) < 0$$
$$\frac{3x - 2}{x + 3} - \frac{6x + 18}{x + 3} < 0$$
$$\frac{-3x - 20}{x + 3} < 0$$

Let $f(x) = \dfrac{-3x - 20}{x + 3}$. The zeros and undefined points of the inequality are the zeros of the numerator, $x = -\dfrac{20}{3}$, and denominator, $x = -3$. Create a sign chart. Then choose and test x-values in each interval to determine if $f(x)$ is positive or negative.

(−) (+) (−)

$-\dfrac{20}{3}$ −3 x

The solutions of $\dfrac{-3x - 20}{x + 3} < 0$ are x-values such that $f(x)$ is negative. From the sign chart, you can see that the solution set is $\left(-\infty, -\dfrac{20}{3}\right) \cup (-3, \infty)$.

23. $\left(-\infty, \dfrac{14}{13}\right] \cup \left(\dfrac{5}{3}, \infty\right)$ **25.** $\left(-3, \dfrac{2}{3}\right] \cup (1, \infty)$ **27.** $\left(-\infty, \dfrac{13}{6}\right) \cup (4, \infty)$ **29.** $\dfrac{750 + 25x}{x} < 120$; 8 to 14 people **31.** $(-\infty, -5] \cup [8, \infty)$ **33.** $(-\infty, -3] \cup [3, \infty)$ **35.** $(-\infty, -6) \cup (-6, 6) \cup (6, \infty)$ **37.** $[-4, 3]$ **39a.** $\ell(56 - \ell) \geq 588$ or $-\ell^2 + 56\ell - 588 \geq 0$ **39b.** [14, 42]; The length of the playing field is at least 14 and at most 42 feet. **39c.** $0 < \ell(56 - \ell) \leq 588$; (0, 14] ∪ [42, 56); Sample answer: The area of the playing field must be greater than 0 square feet but at most 588 square feet, so $0 < \ell \leq 14$ or $42 \leq \ell < 56$. **41.** $\left[\dfrac{40}{9}, 8\right]$ **43.** $\left(-12, \dfrac{25}{12}\right]$ **45.** $x^2 + 2x - 8 < 0$ **47.** $\left(-5, -\dfrac{4}{3}\right) \cup (0, 4)$ **49.** $(-\infty, -8] \cup \left[-3, -\dfrac{1}{3}\right] \cup [1, 6]$ **51.** $(-\infty, -3) \cup \left(-3, -\dfrac{1}{2}\right) \cup \left(-\dfrac{1}{2}, 4\right)$ **53.** $(-\infty, -2) \cup (3, 6) \cup (6, \infty)$ **55a.** $75 + \left\lceil \dfrac{t}{5} \right\rceil \geq 89.5$

55b. 75; Sample answer: Since $t \geq 72.5$, Jarrick will have to spend 75 minutes studying for the test. **57.** 36 units² **59.** $(-k, k)$ **61.** $(-\infty, \infty)$ **63.** Neither; sample answer: The expression is undefined at 3, but it is positive everywhere that it is defined, so the solution set is $(-\infty, 3) \cup (3, \infty)$. **65.** $a < 0$ and $|a + b| > |c + d|$

67 Because $(x - a)^{2n}$ is always an even degree, by definition, the end behavior is to approach ∞ as $x \to \pm \infty$. The absolute minimum occurs at $x - a = 0$ or $x = a$. Since any other value of $(x - a)$ to an even power must be positive in the reals, $(x - a)^{2n} > 0$ for all values of x except when $x = a$, since $(0)^{2n} > 0$ is false. Thus, the solution set is $(-\infty, a) \cup (a, \infty)$.

69. Sample answer: There will be instances when $x - 2$ is a negative value. When this occurs, we are multiplying an inequality by a negative and the inequality symbol will need to be reversed. The problem occurs here because $x - 2$ could be positive or negative. **71.** $D = \{x \mid x \neq -6, x \in \mathbb{R}\}; x = -6$ **73.** about 0.37 cm or 9 cm **75.** $3a^3 - 9a^2 + 7a - 6$ **77.** $x^2 - xy + y^2$ **79a.** 100,000 **79b.** 30,240 **79c.** 8000; 2352 **81.** F

Pages 148–152   Chapter 2   Study Guide and Review

1. leading coefficient **3.** repeated zeros **5.** Factor Theorem **7.** Vertical **9.** Oblique

11.

$D = (-\infty, \infty), R = [0, \infty)$; intercept: 0; $\lim\limits_{x \to -\infty} f(x) = \infty$ and $\lim\limits_{x \to \infty} f(x) = \infty$; continuous for all real numbers; decreasing: $(-\infty, 0)$; increasing: $(0, \infty)$

13.

$D = (-\infty, 0) \cup (0, \infty)$, $R = (-\infty, 0) \cup (0, \infty)$; no intercepts; $\lim\limits_{x \to -\infty} f(x) = 0$ and $\lim\limits_{x \to \infty} f(x) = 0$; infinite discontinuity at $x = 0$; decreasing: $(-\infty, 0)$ and $(0, \infty)$

15.

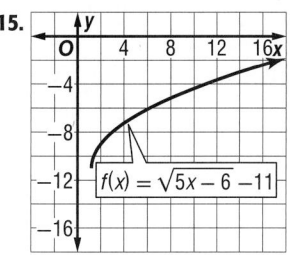

$D = [1.2, \infty), R = [-11, \infty)$; x-intercept: 25.4; $\lim\limits_{x \to \infty} f(x) = \infty$; continuous on $[1.2, \infty)$; increasing: $(1.2, \infty)$

17. 4 **19.** 4 **21.** $\lim\limits_{x \to \infty} f(x) = -\infty$; $\lim\limits_{x \to -\infty} f(x) = -\infty$; The degree is even and the leading coefficient is negative. **23.** $\lim\limits_{x \to \infty} f(x) = \infty$ and $\lim\limits_{x \to -\infty} f(x) = \infty$; The degree is even and the leading coefficient is positive. **25.** 3 real zeros and 2 turning points; 0, 3, and 4 **27.** 4 real zeros and 3 turning points; $-3, -1, 1$, and 3 **29a.** $\lim\limits_{x \to \infty} f(x) = \infty$; $\lim\limits_{x \to -\infty} f(x) = \infty$ **29b.** 0 (multiplicity 3), 3, -4 (multiplicity 2) **29c.** Sample answer: $(-1, 36), (0, 0), (1, -50), (2, -288)$

29d.

31. $x^2 + 10x + 20 + \dfrac{35}{x - 2}$ **33.** $x^4 + 3x^3 - x^2 - 9 + \dfrac{1}{2x - 1}$ **35.** $x^3 + x^2 + 9x + 9$ **37.** yes; $f(x) = (x + 3)(x^2 - 8)$ **39.** yes; yes; $f(x) = (x - 2)^2(x + 1)^2$ **41.** $\pm 1, \pm 3, \pm 5, \pm 15; -3$ **43.** $\pm 1, \pm 2, \pm 5, \pm 10, \pm\frac{1}{3}, \pm\frac{2}{3}, \pm\frac{5}{3}, \pm\frac{10}{3}; 2, -\frac{1}{3}$ **45.** $2, \frac{1}{2}, \frac{4}{3}$ **47.** $\frac{3}{2}, -2, 2, 4$ **49.** $1, 3, 2i, -2i; f(x) = (x - 3)(x - 1)(x - 2i)(x + 2i)$

51. $D = \{x \mid x \neq 5, -5, x \in \mathbb{R}\}; x = 5, x = -5, y = 1$

53. $D = \{x \mid x \neq -3, -9, x \in \mathbb{R}\}; x = -3, x = -9, y = 1$

55. asymptotes: $x = -4, y = 1$; x-intercept: 2; y-intercept: $-\frac{1}{2}$; $D = \{x \mid x \neq -4, x \in \mathbb{R}\}$

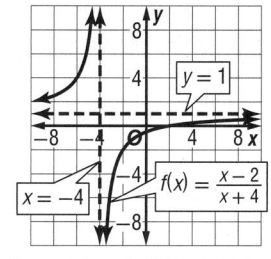

57. asymptotes: $x = -6, x = 3, y = 1$; x-intercepts: 0 and -7; y-intercept: 0; $D = \{x \mid x \neq -6, 3, x \in \mathbb{R}\}$

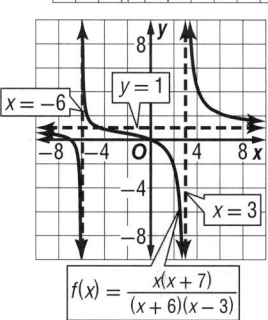

59. $x = 0, x = 1, x = 5, y = 0$; x-intercepts: 4 and -4; $D = \{x \mid x \neq 0, 1, 5, x \in \mathbb{R}\}$

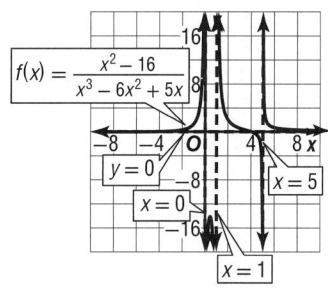

61. -3 **63.** $\dfrac{7}{3}$ **65.** $(-\infty, -2) \cup (8, \infty)$ **67.** $\left(-5, -\dfrac{3}{2}\right)$ **69.** $\varnothing$

71. $(0, 5)$ **73.** $\left(3, \dfrac{26}{7}\right) \cup (4, \infty)$

75a.

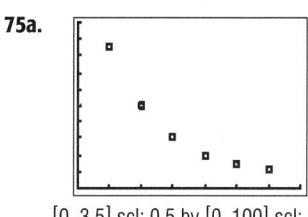

[0, 3.5] scl: 0.5 by [0, 100] scl: 10

75b. $v(t) = 43.63t^{-1.12}$
75c. about 35.6 mi/h
75d. about 0.94 second
77. 6.13 ft^2 **79.** 15 m by 50 m or 25 m by 30 m
81. 16 cakes

Pages 154–155   Chapter 2   Connect to AP Calculus

1. 21.86 units^2 **3.** 25.13 units^2; Sample answer: The approximation is less than the actual area. As more rectangles are used, the approximation approaches the actual area. **5.** Sample answer: 48 units^2 **7.** Yes; sample answer: For this example, using right endpoints for the rectangles would produce rectangles that overlap the curve and include area that lies above the curve. This would produce a higher approximation. However, this is not always the case. If the curve is symmetrical, like the previous example, the approximations will be the same regardless of the endpoint used.   **9.** Sample answer: The more rectangles that are used, the better the approximation of the area. Smaller rectangles fit the desired region better than larger rectangles, thus producing more accurate approximations.   **11a.** 4 rectangles: 14 units^2; 8 rectangles: 13.75   **11b.** Sample answer: Approximating the area using 4 or 8 rectangles probably does not give a good representation of the actual area. The nature of the curve prevents rectangles of widths of 1 and 0.5 units from really fitting well beneath it.   **11c.** Yes; sample answer: The graph is not symmetrical. Using right endpoints to determine the height of the rectangles would produce different rectangles with different areas.

Chapter 3 Exponential and Logarithmic Functions

Page 157   Chapter 3   Get Ready

1. $162x^{11}$ **3.** y^3 **5.** c^3 **7.** $10a^5$ ft^2

9. yes

[−5, 5] scl: 1 by [−5, 5] scl: 1

11. yes

[−5, 5] scl: 1 by [−5, 5] scl: 1

13. yes

[−10, 10] scl: 1 by [−10, 10] scl: 1

15.

$D = (-\infty, \infty), R = [0, \infty)$; x-intercept: 0, y-intercept: 0; $\lim\limits_{x\to-\infty} f(x) = \infty$ and $\lim\limits_{x\to\infty} f(x) = \infty$; continuous for all real numbers; decreasing: $(-\infty, 0)$, increasing: $(0, \infty)$

$f(x) = 2x^2$

17.

$D = (-\infty, \infty), R = (-\infty, \infty)$; x-intercept: 0, y-intercept: 0; $\lim\limits_{x\to-\infty} f(x) = \infty$ and $\lim\limits_{x\to\infty} f(x) = -\infty$; continuous for all real numbers; decreasing: $(-\infty, \infty)$

$h(x) = -3x^3$

Pages 166–169   Lesson 3-1

1.

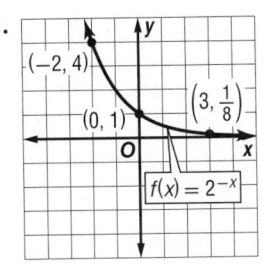

$f(x) = 2^{-x}$

$D = (-\infty, \infty)$; $R = (0, \infty)$; y-intercept: 1; asymptote: x-axis; $\lim\limits_{x\to-\infty} f(x) = \infty$, $\lim\limits_{x\to\infty} f(x) = 0$; decreasing on $(-\infty, \infty)$

3.

$h(x) = 0.2^{x+2}$

$D = (-\infty, \infty)$; $R = (0, \infty)$; y-intercept: 0.04; asymptote: x-axis; $\lim\limits_{x\to-\infty} h(x) = \infty$, $\lim\limits_{x\to\infty} h(x) = 0$; decreasing on $(-\infty, \infty)$

5.

$m(x) = -0.25^x$

$D = (-\infty, \infty)$; $R = (-\infty, 0)$; y-intercept: -1; asymptote: x-axis; $\lim\limits_{x\to-\infty} m(x) = -\infty$, $\lim\limits_{x\to\infty} m(x) = 0$; increasing on $(-\infty, \infty)$

7.

$q(x) = \left(\frac{1}{6}\right)^x$

$D = (-\infty, \infty)$; $R = (0, \infty)$; y-intercept: 1; asymptote: x-axis; $\lim\limits_{x\to-\infty} q(x) = \infty$, $\lim\limits_{x\to\infty} q(x) = 0$; decreasing on $(-\infty, \infty)$

9.

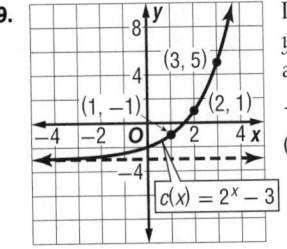

$c(x) = 2^x - 3$

$D = (-\infty, \infty)$; $R = (-3, \infty)$; y-intercept: -2; x-intercept: 1.6; asymptote: $y = -3$; $\lim\limits_{x\to-\infty} c(x) = -3$, $\lim\limits_{x\to\infty} c(x) = \infty$; increasing on $(-\infty, \infty)$

11.

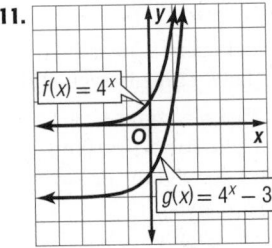

$f(x) = 4^x$

$g(x) = 4^x - 3$

$g(x)$ is the graph of $f(x)$ translated 3 units down.

13.

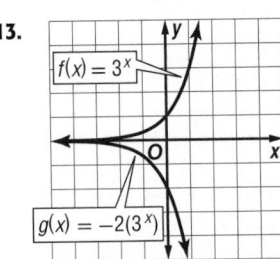

$f(x) = 3^x$

$g(x) = -2(3^x)$

$g(x)$ is the graph of $f(x)$ reflected in the x-axis and expanded vertically by a factor of 2.

15.

g(x) is the graph of f(x) reflected in the y-axis and translated 3 units to the right.

17.

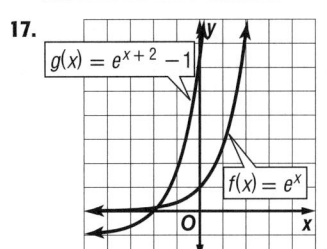

g(x) is the graph of f(x) translated 2 units to the left and 1 unit down.

19.

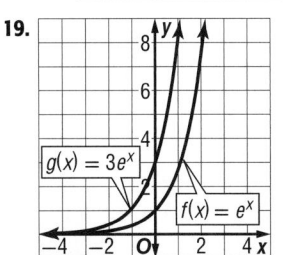

g(x) is the graph of f(x) expanded vertically by a factor of 3.

21.

n	1	4	12	365	continuously
A	$579.64	$580.59	$580.81	$580.91	$580.92

23.

n	1	4	12	365	continuously
A	$2653.30	$2701.48	$2712.64	$2718.10	$2718.28

25. a. The inheritance is $20,000. The interest rate r is 4.6% or 0.046. The interest is compounded monthly, so $n = 12$. There are 10 years from when Brady is 8 to when he is 18. Therefore, $t = 10$. Use the compound interest formula.

$$A = P\left(1 + \frac{r}{n}\right)^{nt}$$
$$= 20{,}000\left(1 + \frac{0.046}{12}\right)^{12(10)}$$
$$\approx \$31{,}653.63$$

Brady will have $31,653.63 when he turns 18.

b. Use the continuous compound interest formula. The interest rate r is 4.6% or 0.046. The time t is 10.

$$A = Pe^{rt}$$
$$= 20{,}000e^{0.042(10)}$$
$$\approx \$30{,}439.23$$

Brady will have $30,439.23 when he turns 18.

27.

t	5	10	15	20	50
N	12,774	10,308	8317	6711	1853

29.

t	5	10	15	20	50
N	19,572	21,652	23,953	26,499	48,575

31a. about 1289.75 million or 1.29 billion gallons; about 5654.12 million or 5.65 billion gallons **31b.** about 1351.89 million or 1.35 billion gallons; about 6212.13 million or 6.21 billion gallons **33.** No; Exterm-inc has to spray 5 times in order to guarantee that they have eliminated all but 1% of a termite colony.

35. a. $t = 17$, $N_0 = 1.19$; $N(17) = 3.86$

$$3.86 = 1.19(1 + r)^{17}$$
$$\frac{3.86}{1.19} = (1 + r)^{17}$$
$$\sqrt[17]{\frac{3.86}{1.19}} = 1 + r$$
$$1 + r = 1.0717$$

So, $N(t) = 1.19(1.0717)^t$, where t is the years since 1990.

b. For 2011, $t = 21$, $N(21) = 1.19(1.0717)^{21} = \$5.09$.
For 2013, $t = 23$, $N(23) = 1.19(1.0717)^{23} = \$5.85$.

c. 2015 − 2016, since $N(25) = \$6.72$ and $N(26) = \$7.20$.

d. Sample answer: The cost of gasoline usually does not increase or decrease at a constant rate. It fluctuates in response to many factors such as the economy, the availability of oil, etc.

37a. about 199.8 g **37b.** about 2.29 g

39a. $N(t) = N_0 e^{-0.1209681t}$ **39b.** about 1.48 g **39c.** in about 16.09 thousand years **41a.** 16,198 articles; 113%
41b. 2006 **41c.** 66,318,857 articles
43a. $f(x) = 0.7378 \cdot 1.131^x$

43b. 26.8 months
43c. 271.7 million

[0, 35] scl: 5 by [0, 35] scl: 5

45 The interest rate r is 9%. Interest is compounded daily, so n is 365. They are investing from age 0 to age 18, so t is 18. Their goal for the end of the investment is $N = \$60{,}000$. They want to determine N_0, the initial amount.

Solve $60{,}000 = N_0\left(1 + \frac{0.09}{365}\right)^{365(18)}$. N_0 is about $11,876.29.

47. $x = 0.55$ **49.** $x \geq -0.75$
51. D $= (-\infty, \infty)$; R $= (3, \infty)$; continuous; decreasing for $(-\infty, \infty)$; Sample answer:
53. $f(x) = 5^{x+3} - 4$
55. $f(x) = -4^{x+1} + 6$
57. reflection in the x-axis and translation 4 units up

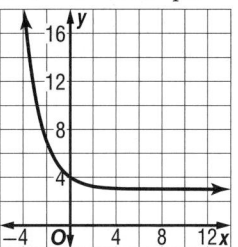

59 Sample answer: Sonja; she used the equation

$$A = P\left(1 + \frac{r}{n}\right)^{nt} = 550 \cdot \left(1 + \frac{0.035}{12}\right)^{(12)(12)} = \$836.57.$$ Eric

determined the worth if the interest was compounded continuously. $A = Pe^{rt} = 550 \cdot e^{(0.035)(12)} = \$837.08$.

61. True; sample answer: The range of an exponential function will always approach a value of the range that it will never reach, so there will always be a restriction on the range. **63.** Sample answer: $f(x) = 2^x - 4$ **65.** Always; sample answer: All functions of the form $f(x) = b^x$ have the point $(0, 1)$ in common because any number b raised to the 0 power is 1. **67.** $[-2, 3]$ **69.** $(-\infty, 1.1882623] \cup [6.3117377, \infty)$ **71.** D $= \{x \mid x \neq -5 \text{ or } 1\}$; asymptote: $x = -5$, $y = 0$; hole: $x = 1$ **73a.** $F^{-1}(x) = \frac{5}{9}(x - 32)$; $F[F^{-1}(x)] = F^{-1}[F(x)] = x$

73b. It can be used to convert Fahrenheit to Celsius. **75a.** {(13,264, 4184), (27,954, 4412), (21,484, 6366), (13,423, 4851), (16,849, 2415), (19,563, 5982), (17,284, 6949)} **75b.** {13,264, 16,849, 17,284, 19,563, 21,484, 13,423, 27,954}; {2415, 4184, 4412, 4851, 5982, 6366, 6949} **75c.** Yes; no member of the domain is paired with more than one member of the range. **75d.** Sample answer: No; it is not reasonable to make a prediction equation for this situation. You cannot determine the number of attendees at a school based on the number of applications received.
77. H **79.** G

Pages 178–180 Lesson 3-2

1. 3 **3.** −2 **5.** 2 **7.** 4 **9.** 1.623 **11.** 3.722 **13.** 12 **15.** $\frac{1}{10}$

17. 2.803 **19.** ∅ **21.** 2.079 **23.** 7 **25a.** 0.14 **25b.** 67%
27. 76

29.

D = $(0, \infty)$; R = $(-\infty, \infty)$; x-intercept: 1; asymptote: y-axis; $\lim_{x \to 0} f(x) = -\infty$ and $\lim_{x \to \infty} f(x) = \infty$; increasing on $(0, \infty)$

31.

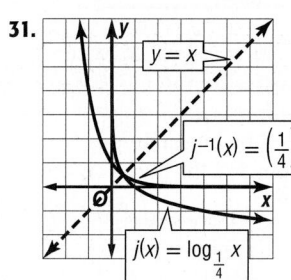

D = $(0, \infty)$; R = $(-\infty, \infty)$; x-intercept; 1; asymptote: y-axis; $\lim_{x \to 0} f(x) = \infty$ and $\lim_{x \to \infty} f(x) = -\infty$; decreasing on $(0, \infty)$

33.

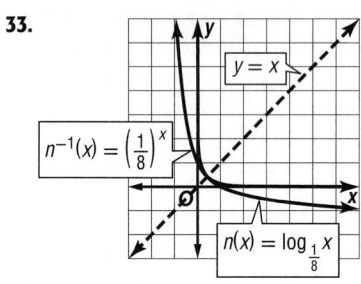

D = $(0, \infty)$; R = $(-\infty, \infty)$; x-intercept: 1; asymptote: y-axis; $\lim_{x \to 0} f(x) = \infty$ and $\lim_{x \to \infty} f(x) = -\infty$; decreasing on $(0, \infty)$

35. The graph of $g(x)$ is the graph of $f(x)$ translated 1 unit to the right.

37. The graph of $g(x)$ is the graph of $f(x)$ compressed vertically by a factor of 0.5.

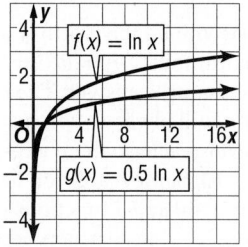

39. The graph of $g(x)$ is the graph of $f(x)$ expanded vertically by a factor of 3 and translated 1 unit up.

41. The graph of $g(x)$ is the graph of $f(x)$ reflected across the y-axis.

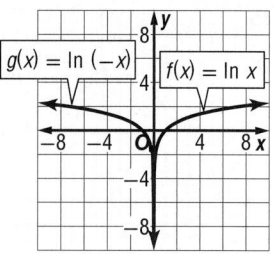

43. D = $\{x \mid x > -7, x \in \mathbb{R}\}$; R = $\{y \mid y \in \mathbb{R}\}$; x-intercept: −6; asymptote: $x = -7$ **45.** D = $\{x \mid x > 3, x \in \mathbb{R}\}$; R = $\{y \mid y \in \mathbb{R}\}$; x-intercept: 4; asymptote: $x = 3$ **47.** $y = \frac{\ln x}{3}$

49 If $y = 4e^{2x}$, the inverse is $x = 4e^{2y}$. Solve the inverse equation for y.
$$x = 4e^{2y}$$
$$\frac{x}{4} = e^{2y}$$
$$\ln\left(\frac{x}{4}\right) = \ln e^{2y}$$
$$\ln\left(\frac{x}{4}\right) = 2y \ln e$$
$$\ln\left(\frac{x}{4}\right) = 2y$$
$$y = \frac{1}{2} \ln\left(\frac{x}{4}\right)$$

51. $y = \log_{20} x$ **53.** D = $\{x \mid x \in \mathbb{R}\}$; R = $\{y \mid y > 0, y \in \mathbb{R}\}$

55a. $P = I \cdot 2^{\frac{t}{2}}$ **55b.** 8

57. D = $(-3, \infty)$; R = $(-\infty, \infty)$; no symmetry; continuous; increasing for $(-3, \infty)$; sample answer:

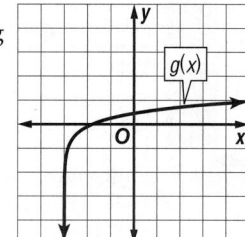

59. D = $(0, \infty)$; R = $(-\infty, \infty)$; no symmetry; continuous; decreasing for $(0, \infty)$; sample answer:

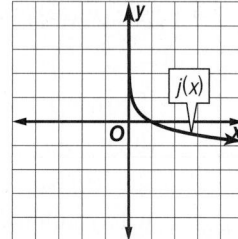

61. $h(x) = \log x - 3$

63 The graph $k(x)$ is the graph of $f(x)$ shifted 4 units left, so x needs to be changed to $(x + 4)$. The graph of $k(x)$ is the graph of $f(x)$ shifted 3 units down, so we must subtract 3 from $f(x)$. Therefore, $k(x) = \log (x + 4) - 3$. **65.** true
67. false

69a.

69b. $f(x) = 24$, $g(x) = 10$ **69c.** The growth rate for an exponential function is greater than the growth rate of a power function.

69d.

69e. $f(x) = 0.203$, $g(x) = 0.225$ **69f.** The growth rate for a root function is greater than the growth rate of the natural log function.   **71.** Sample answer: The value of b cannot be negative because even roots of negative numbers are undefined in the set of real numbers. For example, $\frac{1}{2} = \log_{-2} x$ can be rewritten as $(-2)^{\frac{1}{2}} = x$ by using the definition of a logarithm. The value of $(-2)^{\frac{1}{2}}$ is undefined.
73. Sample answer: Exponential functions with larger bases increase more rapidly than with smaller bases. For example, $10^9 > 6^{10} > 2^{25}$. As the exponent increases, the difference between the y-values for each base will increase. Also, logarithmic functions with larger bases increase less rapidly than with smaller bases. For example, $\log_2 36 > \log_6 216 > \log_{10} 500$.   **75.** False; sample answer: The range of $y = \ln (x^2 + 1)$ is $[0, \infty)$.

79. $(-\infty, -2)$ or $(5, \infty)$ **81.** $(-\infty, 0)$ or $(9, \infty)$ **83.** $\left[-\frac{3}{2}, 39\right]$
85. $-\frac{13}{3}, 3$ **87.** $\frac{1 \pm \sqrt{145}}{4}$

89.

$D = (-\infty, 0) \cup (0, \infty)$; $R = (0, \infty)$; no intercepts; $\lim\limits_{x \to -\infty} f(x) = 0$ and $\lim\limits_{x \to \infty} f(x) = 0$; infinite discontinuity at $x = 0$; increasing: $(-\infty, 0)$, decreasing: $(0, \infty)$

91a. polynomial   **91b.** $0.5 < t < 39.5$ **91c.** $0 < t < 0.5$ and $t > 39.5$
93. F   **95.** G

Pages 185–188   Lesson 3-3

1. $2 \ln 2 - \ln 5$ **3.** $4 \ln 2 + \ln 5$ **5.** $\ln 2 - \ln 5$ **7.** $4 \ln 2 + 3 \ln 5$ **9.** $2 \ln 3 + \ln 7$ **11.** $\ln 7 - 2 \ln 3$ **13.** $3 \ln 3 + 2 \ln 7$
15. $4 \ln 7 - 4 \ln 3$ **17a.** $\log K_w = \log [\mathrm{H}^+] + \log [\mathrm{OH}^-]$
17b. $-14 = \log [\mathrm{H}^+] + \log [\mathrm{OH}^-]$ **17c.** 1×10^{-5} moles per liter   **19.** $\frac{1}{2}$ **21.** 47   **23.** 3   **25.** 6   **27.** $\frac{5}{6}$ **29.** $\log_9 6 + 3 \log_9 x + 5 \log_9 y + \log_9 z$ **31.** $2 \log_3 p + \log_3 q - \frac{1}{5} \log_3 (3q - 1)$
33. $\log_{11} a - 4 \log_{11} b + 12 \log_{11} c + 7 \log_{11} d$ **35.** $\log_4 10 + 2 \log_4 t + \log_4 u - 3 \log_4 v$ **37.** $\ln 3 + 4 \ln a + 7 \ln b + \ln c - \frac{1}{4} \ln (b - 9)$ **39.** $\log_5 \frac{x^3}{\sqrt{6 - x}}$ or $\log_5 \frac{x^3 \sqrt{6 - x}}{6 - x}$ **41.** $\log_3 \frac{a^7 b}{64 c^2}$
43. $\log_8 \left(\frac{81x^2}{2x - 5}\right)$ **45.** $\log_6 25 a^2 b c^7$ **47.** $\ln \frac{\sqrt[4]{2a - b}}{\sqrt[5]{3b + c}}$ or $\ln \frac{\sqrt[4]{2a - b} \cdot \sqrt[5]{(3b + c)^4}}{3b + c}$ **49.** 1.473   **51.** 0.827   **53.** 2.003
55. 1.002   **57.** no real solution

59 **a.** If $R = \log_2 n$, then for $n = 240$,
$$R = \log_2 240 = \frac{\log 240}{\log 2} \approx 7.9 \text{ sec.}$$

b. If $R = 8.45$, then
$$8.45 = \log_2 n$$
$$2^{8.45} = n$$
$$n = 3.49.71 \text{ or about 350 steps.}$$
61. 1   **63.** 2   **65.** -3 **67.** 1   **69.** $\frac{3}{5} \ln x + \frac{1}{5} \ln (x + 3)$

71
$$\log_{14} \frac{11}{\sqrt[4]{x^5 (8x - 1)}} = \log_{14} 11 - \log_{14} \sqrt[4]{x^5 (8x - 1)}$$
$$= \log_{14} 11 - \log_{14} [x^5 (8x - 1)]^{\frac{1}{4}}$$
$$= \log_{14} 11 - \frac{1}{4} \log_{14} [x^5 (8x - 1)]$$
$$= \log_{14} 11 - \frac{1}{4} [\log_{14} x^5 + \log_{14} (8x - 1)]$$
$$= \log_{14} 11 - \frac{1}{4} [5 \log_{14} x + \log_{14} (8x - 1)]$$
$$= \log_{14} 11 - \frac{5}{4} \log_{14} x - \frac{1}{4} \log_{14} (8x - 1)$$

73. $\frac{3}{7} \log_8 x + \frac{2}{7} \log_8 y + \frac{1}{7} \log_8 (z - 1)$ **75a.** $M = \frac{2}{3} (\log E - \log 10^{4.4})$ **75b.** 5   **75c.** 1.67   **75d.** 7.94×10^8 joules
77. $\log_2 \frac{15x^5}{\sqrt[3]{x(x + 3)}}$ or $\log_2 \frac{15x^5 \sqrt[3]{(x^2 + 3x)^2}}{x^2 + 3x}$ **79.** $\log_6 \frac{8x^3 y^8}{\sqrt[5]{x^4 y^3 z}}$ or $\log_6 \frac{8x^2 y^7 \sqrt[5]{xy^2 z^4}}{z}$ **81.** $\ln \frac{\sqrt{x^5(y + 8)}}{y^3(10 - x)}$ **83.** $4 \ln 2 + \ln 3$; 3.86
85. $3 \ln 2 + 3 \ln 3$; 5.37   **87.** $2 \ln 2 - 2 \ln 3$; -0.82 **89.** $5 \ln 2 - 2 \ln 3$; 1.25   **91.** b   **93.** c   **95.** e   **97.** $\log_5 \sqrt{100}$, $\log_5 55$, $3 \log_5 \sqrt[3]{75}$ **99.** $f(x) = \log_{10} x$ **101.** $h(x) = \log_2 x$
103a. $pK_a = -(\log [\mathrm{H}^+] + \log [\mathrm{F}^-] - \log [\mathrm{HF}])$ **103b.** 4.30
103c. 1×10^{-25} **103d.** 1×10^{-17} **105.** 4   **107.** 2

109 Use the change of base formula to solve.
$$(\log_2 7)(\log_5 2) = \frac{\log 7}{\log 2} \cdot \frac{\log 2}{\log 5} = \frac{\log 7}{\log 5} = \log_5 7$$

111. $\log_5 8$

113.

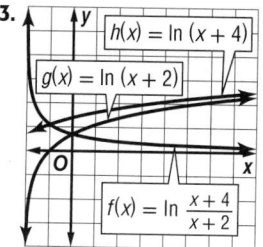

The graph of $f(x)$ represents the difference between the graphs of $h(x)$ and $g(x)$.

Let $x = \log_b m$. Let $y = \log_b n$.
Then $b^x = m$; and $b^y = n$.

$\log_b \left(\frac{m}{n}\right) = \log_b \left(\frac{b^x}{b^y}\right)$	Substitution
$= \log_b(b^{x - y})$	Quotient Property
$= x - y$	Definition of logarithm
$= \log_b m - \log_b n$	Substitution

115. Let $\log_a n = x$.

$a^x = n$	Definition of logarithm
$\log_b a^x = \log_b n$	One-to-One Property of logarithms
$x \log_b a = \log_b n$	Power Property of Logarithms
$x = \frac{\log_b n}{\log_b a}$	Divide each side by $\log_b a$.
$\log_a n = \frac{\log_b n}{\log_b a}$	Replace x with $\log_a n$.

117. $\ln x$ cannot be simplified for values of x that are prime numbers.

119. $\dfrac{\log_5 (nt)^2}{\log_4 \left(\frac{t}{r}\right)} = \dfrac{2\log_5 nt}{\log_4 \left(\frac{t}{r}\right)}$

$= \dfrac{2\log_5 n + \log_5 t}{\log_4 t - \log_4 r}$

$= \dfrac{2\left(\dfrac{\log n}{\log 5} + \dfrac{\log t}{\log 5}\right)}{\dfrac{\log t}{\log 4} - \dfrac{\log r}{\log 4}}$

$= \dfrac{2\left(\dfrac{\log n + \log t}{\log 5}\right)}{\dfrac{\log t - \log r}{\log 4}}$

$= \dfrac{2\log 4(\log n + \log t)}{\log 5(\log t - \log r)}$

$= \dfrac{2\log n \log 4 + 2\log t \log 4}{\log 5 \log t - \log 5 \log r}$

121.

D = $(0, \infty)$; R = $(-\infty, \infty)$; x-intercept: 1; asymptote: y-axis; $\lim\limits_{x\to 0^+} f(x) = -\infty$ and $\lim\limits_{x\to\infty} f(x) = \infty$; increasing on $(0, \infty)$

123.

D = $(0, \infty)$; R = $(-\infty, \infty)$; x-intercept: 25; asymptote: y-axis; $\lim\limits_{x\to 0^+} f(x) = -\infty$ and $\lim\limits_{x\to\infty} f(x) = \infty$; increasing on $(0, \infty)$

125. $g(x)$ is the graph of $f(x)$ translated 3 units left.

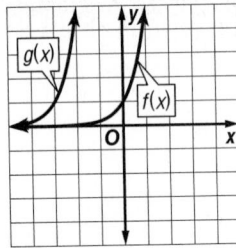

127a. $A(x) = \dfrac{480}{x} + 2x^2$

127b.

127c. The surface area approaches infinity.

129. $x^2 - 4x + 3$

131. $f(g(x))$

$= -\dfrac{2}{3}\left(-\dfrac{3}{2}x + \dfrac{1}{4}\right) + \dfrac{1}{6}$

$= x - \dfrac{1}{6} + \dfrac{1}{6} = x$;

$g(f(x))$

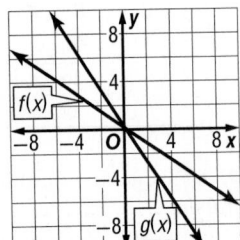

$= -\dfrac{3}{2}\left(-\dfrac{2}{3}x + \dfrac{1}{6}\right) + \dfrac{1}{4}$

$= x - \dfrac{1}{4} + \dfrac{1}{4} = x$

133. $f(g(x))$

$= [(\sqrt[3]{x-4} + 3) - 3]^3 + 4$

$= (\sqrt[3]{x-4})^3 + 4$

$= x - 4 + 4 = x$;

$g(f(x))$

$= \sqrt[3]{[(x-3)^3 + 4] - 4} + 3$

$= \sqrt[3]{(x-3)^3} + 3$

$= x - 3 + 3 = x$

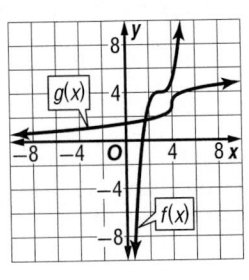

135. D **137.** A

Pages 196–199 Lesson 3-4

1. 5 **3.** $\dfrac{10}{3}$ **5.** 8 **7.** 12 **9.** 11 **11.** $e^4 \approx 54.60$ **13.** no real solution **15.** $\dfrac{e^2}{7} \approx 1.06$ **17.** $e^{-3} \approx 0.05$ **19.** 65,536

21 If $H = 200$, then
$200 = \log_{1.003} \dfrac{D}{1.394}$
$1.003^{200} = \dfrac{D}{1.394}$
$D = 1.394(1.003^{200})$
$D = 2.54$ liters.

23. ± 9 **25.** 12 **27.** 1 **29.** 3.85 **31.** 0.98 **33.** 0.53 **35.** 1.20

37. 1.38 **39.** $\dfrac{\ln \frac{27}{7}}{\ln \frac{49}{3}} \approx 0.48$ **41.** $\dfrac{4\ln 6}{\ln \frac{32}{9}} \approx 5.65$ **43.** $\dfrac{3\ln \frac{4}{3}}{\ln 648} \approx 0.13$

45. $\dfrac{\ln 4500}{\ln \frac{6}{25}} \approx -5.89$ **47.** $\dfrac{\ln \frac{1}{16}}{\ln 2} = -4$ **49a.** 449,032 **49b.** -1.9

49c. -2.8 **51.** $\ln 7 \approx 1.95$ and $\ln 8 \approx 2.08$ **53.** $\ln \dfrac{3}{2} \approx 0.41$

55. $0, \dfrac{\ln \frac{7}{8}}{2} \approx -0.067$ **57.** $\dfrac{\ln 5}{3} \approx 0.54$ **59.** $\ln 5 \approx 1.61$

61. 2 **63.** no solution **65.** no solution **67.** $\pm\sqrt[4]{2} \approx \pm 1.19$

69. no solution

71
$$\log x = 1 - \log (x - 3)$$
$$\log x + \log (x - 3) = 1$$
$$\log (x(x - 3)) = 1$$
$$x(x - 3) = 10^1$$
$$x^2 - 3x - 10 = 0$$
$$(x - 5)(x + 2) = 0$$
So, $x = 5$ or $x = -2$. Since $\log (-2)$ is undefined, $x = 5$ is the only solution.

73. 3 **75.** 45 **77.** 1 **79.** 6 **81a.** $P = 0.93e^{0.3328(t-1)}$

81b. about $13.33 **83.** $e^{24} \approx 2.65$ **85.** $3^8 = 6561$

87. cannot be solved algebraically; $x \approx 1.37$

89. cannot be solved algebraically; $x \approx 0.7$

91a. phosphorous-32: $14.282 \dfrac{\ln R}{-0.693}$; sulfur-35: $87.51 \dfrac{\ln R}{-0.693}$

91b. $R = 0.954$ **93.** $2 < y < 3$ **95.** $-3 < f \le -1$

97. $5.15 > q > 3$ **99.** $4 < a < 5.44$ **101.** $-0.69 \ge c \ge -4.61$

103a. about 23 **103b.** $\dfrac{4\ln \left(\frac{5276}{3p} - \frac{100}{3}\right)}{3}$ **103c.** about 10

105. 22 **107.** 1.10 **109.** $\ln \sqrt{\dfrac{7}{5}} \approx 0.17$ **111.** $\ln \sqrt{3} \approx 0.549$

113. A logarithmic equation can have infinitely many extraneous solutions. For example, an infinite number of terms, $\ln x_1, \ln x_2, \ln x_3$, etc., can be combined to form an equation like $\ln x_1 + \ln x_2 + \ln x_3 + \ldots = n$. This equation can be simplified to $(x - a)(x - b)(x - c) \ldots = 0$, where $a, b, c,$ ... appear to be solutions but are extraneous because they cause $\ln x_1, \ln x_2, \ln x_3, \ldots$ to have no real solution.

115 Since we are tripling the principal P, then $A = 3P$.

$$A = P\left(1 + \frac{r}{n}\right)^{nt}$$

$$3P = P\left(1 + \frac{r}{12}\right)^{12t}$$

$$3 = \left(1 + \frac{r}{12}\right)^{12t}$$

$$\ln 3 = \ln\left(1 + \frac{r}{12}\right)^{12t}$$

$$\ln 3 = 12t\ln\left(1 + \frac{r}{12}\right)$$

$$\frac{\ln 3}{12\ln\left(1 + \frac{r}{12}\right)} = t$$

$$\frac{1}{12}\ln\left(3 - \left[1 + \frac{r}{12}\right]\right) = t$$

$$\frac{1}{12}\ln\left(2 - \frac{r}{12}\right) = t$$

117. $f(x)$ is defined for $x < -1$ or $x > -1$, while $g(x)$ is defined only for $x > 1$. **119.** 1.3023 **121.** 4

123a. The degree is 3, and the leading coefficient is 1. Because the degree is odd and the leading coefficient is positive, $\lim_{x \to -\infty} f(x) = -\infty$ and $\lim_{x \to \infty} f(x) = \infty$.
123b. 0, 7, 1 **123c.** $(-1, -16)$, $(0.5, 1.625)$, $(3, -24)$, $(8, 56)$

123d.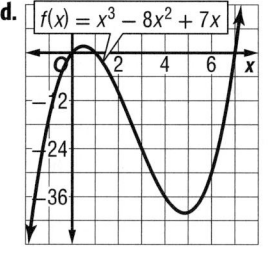

125a. The degree is 4, and the leading coefficient is -1. Because the degree is even and the leading coefficient is negative, $\lim_{x \to -\infty} f(x) = -\infty$ and $\lim_{x \to \infty} f(x) = -\infty$. **125b.** $-2, 0, 4$
125c. $(-2.5, -52.8)$, $(-1, 25)$, $(2, -32)$, $(5, -35)$

125d.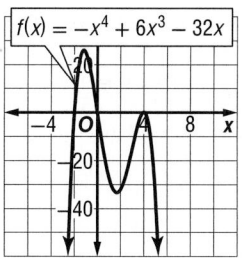

127. 31 **129.** ∞; Sample answer: As $x \to \infty$, the leading term, x^{10} approaches infinity, so $f(x)$ will approach infinity.
131. ∞; Sample answer: As $x \to \infty$, the value inside the absolute value symbols approaches x^2. The absolute value of x^2 is x^2, so $h(x)$ will approach infinity. **133.** 8.2, 2.9
135. E **137.** C

Pages 207–210   Lesson 3-5

1a. $y = 3.19(1.98)^x$ **1b.** 2,735,863.90 **3a.** $y = 24.98(0.25)^x$
3b. 2.27×10^{-11}

5 a. Enter the table in STAT EDIT, then choose STAT CALC, ExpReg. The regression equation is $y = 4476(0.810)^x$.
b. Graph the full regression line and locate the intersection of this line and $y = 12$.

[15, 30] scl: 3 by [0, 80] scl: 8

15 a. Enter the table in STAT EDIT, then Choose STAT CALC, Logistic. The rounded regression equation is
$$f(x) = \frac{1.43}{1 + 0.125e^{-0.089x}}$$

b. The numerator of the logistic function is 1.43, so, according to the function, Maine's growth will level off at about 1.43 million.

c. Sample answer: The model only remains effective as long as the birth and death rates remain consistent and the available resources do not change.

17a. Both logarithmic and logistic regressions can be used to model the data, but logistic is a better model because the king snake eel does not grow indefinitely long. Rounded logarithmic model: $y \approx -1.19 + 6.83 \ln x$; Rounded logistic model: $y \approx \dfrac{18.99}{1 + 3.56e^{-0.25x}}$

17b. logarithmic: 18.91 in.   logistic: 18.47 in.

[0, 25] scl: 5 by [0, 25] sc: 5   [0, 25] scl: 5 by [0, 25] sc: 5

19a.

x	0	1	2	3	4	5	6	7
$\sqrt{y}$	1	2.57	4.12	5.67	7.22	8.77	10.32	11.87

19b. $\hat{y} = 1.55x + 1.01$ **19c.** $y = 2.4x^2 + 3.1x + 1$

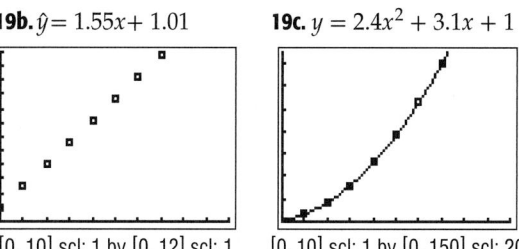

[0, 10] scl: 1 by [0, 12] scl: 1   [0, 10] scl: 1 by [0, 150] scl: 20

21a.

In x	0	0.69	1.10	1.39	1.61	1.79	1.95	2.08
In y	1.61	3.04	3.78	4.37	4.79	5.19	5.52	5.77

21b. $\hat{y} = 1.99\hat{x} + 1.62$ **21c.** $y = 5.05x^{1.99}$

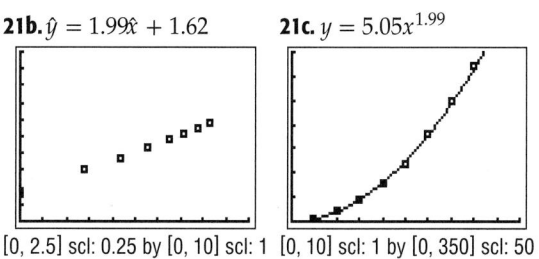

[0, 2.5] scl: 0.25 by [0, 10] scl: 1   [0, 10] scl: 1 by [0, 350] scl: 50

23a.

x	0	3	6	9	12	15
ln y	11.26	11.31	11.36	11.42	11.47	11.52

23b. Sample answer: $\hat{y} = 0.0175x + 11.26$

23c. $y = 77,653e^{0.0175x}$; about $118,185 **25.** Power; the graph increases at a constant percent change.

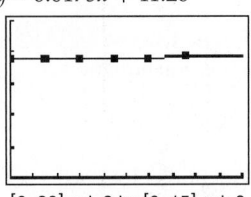

[0, 20] scl: 2 by [0, 15] scl: 3

27.

Exponential		Power		Logarithmic	
x	ln y	ln x	ln y	ln x	y
2	0.92	0.69	0.92	0.69	2.5
4	1.99	1.39	1.99	1.39	7.3
6	2.62	1.79	2.62	1.79	13.7
8	3.06	2.08	3.06	2.08	21.3
10	3.41	2.30	3.41	2.30	30.2
12	3.69	2.48	3.69	2.48	40.0
14	3.93	2.64	3.93	2.64	50.8
16	4.14	2.77	4.14	2.77	62.5

Exponential: Power:

[0, 20] scl: 2 by [0, 5] scl: 0.5   [0, 3] scl: 0.3 by [0, 5] scl: 0.5

Logarithmic:

[0, 3] scl: 0.3 by [0, 70] scl: 7

The graph of the linearized power regression most closely resembles a line.

29.

Exponential		Power		Logarithmic	
x	ln y	ln x	ln y	ln x	y
1	3.63	0	3.63	0	37.8
2	2.83	0.69	2.83	0.69	17.0
3	2.04	1.10	2.04	1.10	7.7
4	1.22	1.39	1.22	1.39	3.4
5	0.47	1.61	0.47	1.61	1.6
6	−0.36	1.79	−0.36	1.79	0.7
7	−1.20	1.95	−1.20	1.95	0.3
8	−2.30	2.08	−2.30	2.08	0.1

Exponential: Power:

[0, 10] scl: 1 by [−5, 5] scl: 1   [0, 2.5] scl: 0.25 by [−5, 5] scl: 1

Logarithmic:

[0, 2.5] scl: 0.25 by [0, 50] scl: 5

The graph of the linearized exponential regression most closely resembles a line.

31. Logarithmic regression is invalid when the domain is 0 because logarithmic functions are undefined at that value and regression does not work with undefined values.

33. Sample answer: Yes; a logistic function has a y-intercept at $y = \dfrac{c}{1+a}$, but cannot have an x-intercept because a constant can never be zero.

35
$$y = ax^b$$
$$\ln y = \ln(ax^b)$$
$$\ln y = \ln a + b \ln x$$
$$\ln y = b \ln x + \ln a$$
Since ln a and ln b are constants, this equation is linear when $y = \ln y$ and $x = \ln x$.

37. Sample answer: Data sets modeled by exponential regression have a restricted range. Data sets modeled by logarithmic regression have a restricted domain. **39.** 10
41. 3 **43.** ±4 **45a.** $y = 1,000,000(1.044)^x$ **45b.** $2,578,760.27; Sample answer: Yes, the money is continuing to grow at a faster rate each year. In the first 10 years it grew by $678,000, and in the next ten years it grew about $900,000.
47. $-75i$ **49.** $6 - 7i$ **51.** $\dfrac{10}{17} - \dfrac{6}{17}i$ **53.** F

Pages 211–214 Chapter 3 Study Guide and Review

1. common logarithm **3.** transcendental functions
5. logistic growth function **7.** linearize **9.** continuously compounded interest
11. D: $(-\infty, \infty)$; R: $(0, \infty)$; y-intercept: 1; asymptote: x-axis; $\lim\limits_{x \to -\infty} f(x) = 0$, $\lim\limits_{x \to \infty} f(x) = \infty$; increasing for $(-\infty, \infty)$

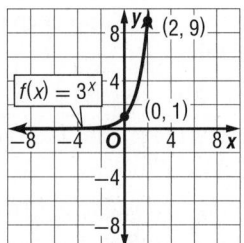

13. D $= (-\infty, \infty)$; R $= (0, \infty)$; y-intercept: 1; asymptote: x-axis; $\lim\limits_{x \to -\infty} f(x) = 0$, $\lim\limits_{x \to \infty} f(x) = \infty$; increasing on $(-\infty, \infty)$

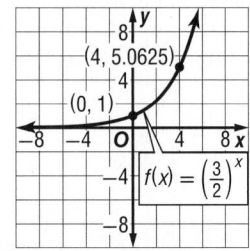

15. The graph of g is the graph of f translated 2 units up.

17. The graph of g is the graph of f expanded vertically by a factor of 2 and translated 5 units down.

19.

n	1	4	12	365	continuously
A	$375.18	$379.11	$380.03	$380.48	$380.49

21. 5 **23.** $\frac{1}{2}$ **25.** 11 **27.** 1.90

29. The graph of g is the graph of f reflected in the x-axis and translated 4 units to the left.

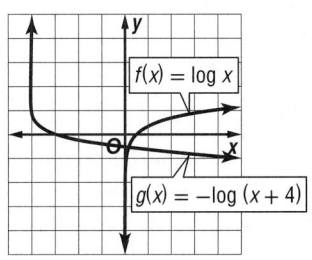

31. The graph of g is the graph of f compressed vertically by a factor of $\frac{1}{4}$ and then translated 2 units down.

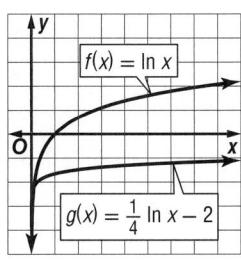

33. $2 \log_5 x + 7 \log_5 a + \frac{1}{2} \log_5 b$ **35.** $\frac{1}{2}(\log g + 5 \log j + \log k) - 2$ **37.** $\log_2 [\sqrt[3]{a}(b+1)]$ **39.** $\frac{9}{2}$ **41.** $\ln 3; \ln 5$ **43.** 6

45a.

x	0	1	2	3	4	5	6
ln y	0.69	1.61	2.83	3.97	5.11	6.25	7.39

45b. $y = 1.13x + 0.59$ **45c.** $y = 1.80e^{1.13x}$

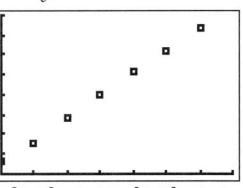

[0, 7] scl: 1 by [0, 8] scl: 1   [0, 7] scl: 1 by [0, 1700] scl: 200

47a. $2.75; $3.43 **47b.** $5.32 **47c.** 2008 **49a.** 28 years **49b.** 18% **51.** 75 years **53a.** about 6.76 years **53b.** about 8.1%

---

1. 3 **3.** For $x = 1$, the average rates of change approach 2.7184. For $x = 2$, the average rates of change approach 7.3894. For $x = 3$, the average rates of change approach 20.0865. **5.** Sample answer: 8.15; The rate of change at $x = 1$ is multiplied by 3. Instead of being 2.72, it is 3(2.72) or 8.15. **7.** For $x = 2$, the average rates of change approach $\frac{1}{2}$. For $x = 3$, the average rates of change approach $\frac{1}{3}$. For $x = 4$, the average rates of change approach $\frac{1}{4}$. **9a.** For $x = 2$, the average rates of change approach -1.5 or $-\frac{3}{2}$. For $x = 3$, the average rates of change approach -1. **9b.** Sample answer: The average rates of change for $g(x)$ are the average rates of change for $f(x)$ multiplied by -3. **9c.** The rate of change for $g(x) = a \ln x$ at x is $\frac{a}{x}$.

Chapter 4 Trigonometric Functions

1. $6\sqrt{2}$ **3.** 3.52 **5.** no **7.** 7 **9.** vertical asymptote at $x = -5$ **11.** $x = 3, x = 5, y = 0$ **13.** no asymptotes

1. $\sin \theta = \frac{4\sqrt{2}}{9}$, $\cos \theta = \frac{7}{9}$, $\tan \theta = \frac{4\sqrt{2}}{7}$, $\csc \theta = \frac{9\sqrt{2}}{8}$, $\sec \theta = \frac{9}{7}$, $\cot \theta = \frac{7\sqrt{2}}{8}$ **3.** $\sin \theta = \frac{9\sqrt{97}}{97}$, $\cos \theta = \frac{4\sqrt{97}}{97}$, $\tan \theta = \frac{9}{4}$, $\csc \theta = \frac{\sqrt{97}}{9}$, $\sec \theta = \frac{\sqrt{97}}{4}$, $\cot \theta = \frac{4}{9}$ **5.** $\sin \theta = \frac{\sqrt{165}}{29}$, $\cos \theta = \frac{26}{29}$, $\tan \theta = \frac{\sqrt{165}}{26}$, $\csc \theta = \frac{29\sqrt{165}}{165}$, $\sec \theta = \frac{29}{26}$, $\cot \theta = \frac{26\sqrt{165}}{165}$ **7.** $\sin \theta = \frac{3}{5}$, $\cos \theta = \frac{4}{5}$, $\tan \theta = \frac{3}{4}$, $\csc \theta = \frac{5}{3}$, $\sec \theta = \frac{5}{4}$, $\cot \theta = \frac{4}{3}$ **9.** $\cos \theta = \frac{3}{5}$, $\tan \theta = \frac{4}{3}$, $\csc \theta = \frac{5}{4}$, $\sec \theta = \frac{5}{3}$, $\cot \theta = \frac{3}{4}$ **11.** $\sin \theta = \frac{3\sqrt{10}}{10}$, $\cos \theta = \frac{\sqrt{10}}{10}$, $\csc \theta = \frac{\sqrt{10}}{3}$, $\sec \theta = \sqrt{10}$, $\cot \theta = \frac{1}{3}$ **13.** $\sin \theta = \frac{2\sqrt{14}}{9}$, $\tan \theta = \frac{2\sqrt{14}}{5}$, $\csc \theta = \frac{9\sqrt{14}}{28}$, $\sec \theta = \frac{9}{5}$, $\cot \theta = \frac{5\sqrt{14}}{28}$ **15.** $\sin \theta = \frac{\sqrt{26}}{26}$, $\cos \theta = \frac{5\sqrt{26}}{26}$, $\tan \theta = \frac{1}{5}$, $\csc \theta = \sqrt{26}$, $\sec \theta = \frac{\sqrt{26}}{5}$ **17.** $\sin \theta = \frac{\sqrt{77}}{9}$, $\cos \theta = \frac{2}{9}$, $\tan \theta = \frac{\sqrt{77}}{2}$, $\csc \theta = \frac{9\sqrt{77}}{77}$, $\cot \theta = \frac{2\sqrt{77}}{77}$ **19.** 3.2 **21.** 4.1 **23.** 43.0 **25.** 7.5

(27)

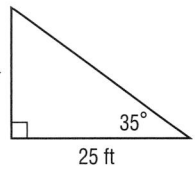

An acute angle measure and the adjacent side length are given, so the tangent function can be used to find the length of the opposite side.

$\tan \theta = \dfrac{\text{opp}}{\text{adj}}$ Tangent function

$\tan 35° = \dfrac{x}{25}$ $\theta = 35°$, opp = x, and adj = 25

$25 \tan 35° = x$ Multiply each side by 25.

$17.5 = x$ Use a calculator.

So, the ravine is about 17.5 feet wide.

29a.

29b. 1.3 mi **31.** 14°
33. 35° **35.** 59°
37. 53° **39.** 424 ft

41a.

41b. 307 ft **43.** about 3.1 ft

45 a. Because the angle of depression from the lighthouse to a ship and the angle of elevation from that ship to the lighthouse are congruent, you can label the angles of elevation at each ship as shown. Label the distance from the lighthouse to the first ship y and the distance between the two ships $x + y$.

b. From the smaller right triangle, you can use the tangent function to find the value of y.

$\tan\theta = \dfrac{\text{opp}}{\text{adj}}$ Tangent function

$\tan 27° = \dfrac{156}{y}$ $\theta = 27°$, opp = 156, and adj = y

$y\tan 27° = 156$ Multiply each side by y.

$y = \dfrac{156}{\tan 27°}$ Divide each side by tan 27°.

$y \approx 306.167$ Use a calculator.

From the larger right triangle, you can use the tangent function to find x.

$\tan 7° = \dfrac{156}{x+y}$ $\theta = 7°$, opp = 156, and adj = $x + y$

$(x + y)\tan 7° = 156$ Multiply each side by $x + y$.

$x + y = \dfrac{156}{\tan 7°}$ Divide each side by tan 7°.

$x = \dfrac{156}{\tan 7°} - y$ Subtract y from each side.

$x = \dfrac{156}{\tan 7°} - 306.167$ $y = 306.167$

$x \approx 964.35$ Use a calculator.

Therefore, the ships are about 964 feet apart.

47. $B = 70°$, $b \approx 16.5$, $c \approx 17.5$ **49.** $P \approx 43°$, $Q \approx 47°$, $r \approx 34.0$
51. $K \approx 19°$, $j \approx 18.0$, $k \approx 6.2$ **53.** $H = 41°$, $f \approx 19.6$, $h \approx 17.1$

55a.

55b. 9°

57. $\dfrac{\sqrt{3}}{2}$ **59.** $\dfrac{2\sqrt{3}}{3}$ **61.** $\sqrt{3}$ **63.** 45° **65.** 60° **67.** 30° **69.** 2

71.

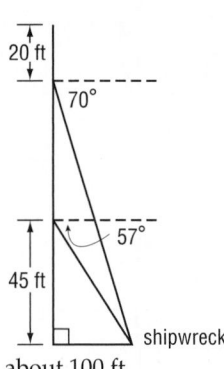

about 100 ft

73. about 0.92 **75.** > **77.** <
79. > **81.** about 17.9 in.

83. Sample answer: For an acute angle θ of a right triangle, $\sin\theta = \dfrac{\text{opp}}{\text{hyp}}$, $\cos\theta = \dfrac{\text{adj}}{\text{hyp}}$, $\tan\theta = \dfrac{\text{opp}}{\text{adj}}$, and $\cot\theta = \dfrac{\text{adj}}{\text{opp}}$.

Using these definitions, $\dfrac{\sin\theta}{\cos\theta} = \dfrac{\frac{\text{opp}}{\text{hyp}}}{\frac{\text{adj}}{\text{hyp}}} = \dfrac{\text{opp}}{\text{adj}} = \tan\theta$.

Similarly, $\dfrac{\cos\theta}{\sin\theta} = \dfrac{\frac{\text{adj}}{\text{hyp}}}{\frac{\text{opp}}{\text{hyp}}} = \dfrac{\text{adj}}{\text{opp}} = \cot\theta$.

85. The trigonometric functions are transcendental functions because they cannot be expressed in terms of algebraic operations. For example, there is no way to find the value of θ in $y = \cos\theta$ by adding, subtracting, multiplying, or dividing a constant and θ or raising θ to a rational power.

87 Sample answer: Given a triangle with hypotenuse c and legs a and b, let $\sin\theta = \dfrac{a}{c}$ and $\cos\theta = \dfrac{b}{c}$. By the Pythagorean Theorem, $c^2 = a^2 + b^2$.

$(\sin\theta)^2 + (\cos\theta)^2 = \left(\dfrac{a}{c}\right)^2 + \left(\dfrac{b}{c}\right)^2$

$= \dfrac{a^2}{c^2} + \dfrac{b^2}{c^2}$

$= \dfrac{a^2 + b^2}{c^2}$

$= \dfrac{c^2}{c^2}$ or 1

Therefore, $(\sin\theta)^2 + (\cos\theta)^2 = 1$.

89. False; sample answer: In $\triangle ABC$, the measure of angle B is less than the measure of angle A, $\cos B \approx 0.7986$, and $\cos A \approx 0.6018$. Therefore, $\cos B > \cos A$, and thus the statement is false.

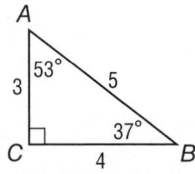

91. Sample answer: Since the cosine function is the reciprocal of the secant function, the sine function is the reciprocal of the cosecant function, and the tangent function is the reciprocal of the cotangent function, you can find the value of a secant, cosecant, or cotangent function on a graphing calculator by finding one divided by the reciprocal of the function. **93.** 0.64

95.

$D = (-\infty, \infty)$, $R = (-\infty, 0)$; no x-intercept, y-intercept: $-\dfrac{1}{9}$; horizontal asymptote at $y = 0$; $\lim\limits_{x\to-\infty} = 0$, $\lim\limits_{x\to\infty} = -\infty$; decreasing on $(-\infty, \infty)$

97.

$D = (-\infty, \infty)$, $R = (-\infty, 0)$; no x-intercept, y-intercept: -4096; horizontal asymptote at $y = 0$; $\lim\limits_{x \to -\infty} = -\infty$, $\lim\limits_{x \to \infty} = 0$; increasing on $(-\infty, \infty)$

99. 4

101a.

[0, 10] scl: 1 by [0, 1000] scl: 100

101b. $y = 904.254x^{-0.149}$
101c. 611,068
103. G   **105.** G

Pages 238–241   Lesson 4-2

1. $11° \, 46' \, 23''$ **3.** $141° \, 32' \, 56''$ **5.** $87.886°$ **7.** $45.357°$ **9.** $17.63°$
11. $\frac{5\pi}{4}$ **13.** $-\frac{\pi}{4}$ **15.** $450°$ **17.** $-210°$ **19–25.** n is an integer.
19. $-75° + 360n°$; Sample answer: $285°$, $-435°$

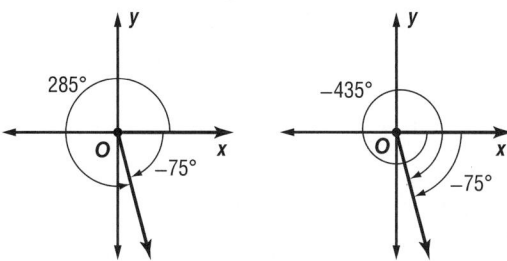

21. $-150° + 360n°$; Sample answer: $210°$, $-510°$

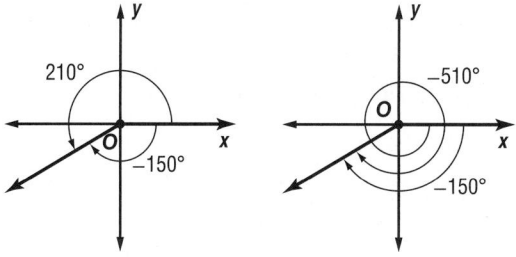

23. $-\frac{3\pi}{4} + 2n\pi$; Sample answer: $\frac{5\pi}{4}$, $-\frac{11\pi}{4}$

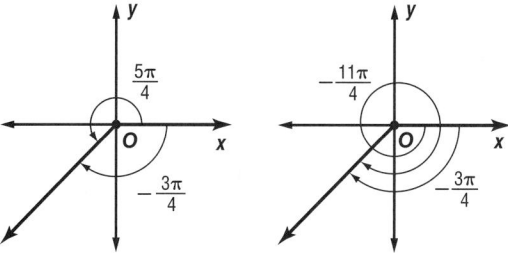

25. $\frac{3\pi}{2} + 2n\pi$; Sample Answer: $\frac{7\pi}{2}$, $-\frac{\pi}{2}$

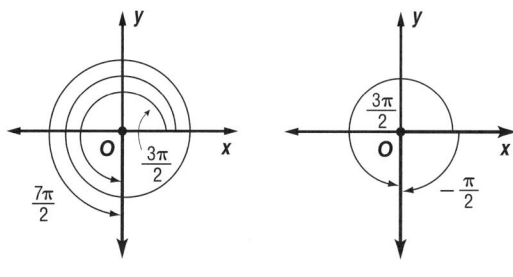

27. 1.3 m   **29.** 5.2 yd   **31.** 3.9 mi

33 **a.** Convert $3024°$ to radian measure, and then use the arc length formula $s = r\theta$.
$$3024° = 3024° \left(\frac{\pi \text{ radians}}{180°} \right)$$
$$= 16.8\pi \text{ radians}$$
Substitute $r = 13$ and $\theta = 16.8\pi$.
$$s = r\theta$$
$$= 13(16.8\pi)$$
$$\approx 686.124 \text{ or about } 686 \text{ feet}$$

b. Let r_1 represent the distance the first rider is from the center of the carousel and r_2 represent the distance the second rider is from the center.
$$s = r_2\theta - r_1\theta$$
$$= 18(16.8\pi) - 13(16.8\pi)$$
$$\approx 263.894 \text{ or about } 264 \text{ feet}$$

35. $1.125 \, \frac{\text{rev}}{\text{min}}$ **37.** 6 m   **39.** 16.3 in.   **41a.** $1292\pi \, \frac{\text{rad}}{\text{min}}$ to $1680\pi \, \frac{\text{rad}}{\text{min}}$ **41b.** 50 mi/h to 65 mi/h   **43.** 2.0 in^2 **45.** 90.5 yd^2
47. 500.5 ft^2 **49.** 12.7 in^2

51 First, convert $68°$ to radians.
$$\theta = 68° \left(\frac{\pi}{180} \right)$$
$$= \frac{17}{45}\pi \text{ radians}$$
Use the area formula for a sector to find the radius.
$$A = \frac{1}{2}r^2\theta$$
$$29 = \frac{1}{2}r^2\left(\frac{17}{45}\pi\right) \qquad A = 29 \text{ and } \theta = \frac{17}{45}\pi$$
$$29 = \frac{17}{90}\pi r^2 \qquad \text{Multiply.}$$
$$r^2 = \frac{2610}{17\pi} \qquad \text{Solve for } r^2.$$
$$r = \sqrt{\frac{2610}{17\pi}} \qquad \text{Take the positive square root of each side.}$$
$$r \approx 6.99 \text{ or about 7 ft}$$

53. 12 in.   **55a.** Quadrant I: $0 < \theta < \frac{\pi}{2}$ **55b.** Quadrant II: $\frac{\pi}{2} < \theta < \pi$ **55c.** Quadrant III: $\pi < \theta < \frac{3\pi}{2}$
55d. Quadrant IV: $\frac{3\pi}{2} < \theta < 2\pi$ **57.** about 537 mi

59. 3.3 radians or about $191°$ **61.** 1.5 radians or about $85.9°$
63a. about 18.8 in.   **63b.** about 5.2 ft

65 Let θ_1 represent the sector with a central angle of $53°$ and θ_2 represent the sector with a central angle of $72°$. First, convert each angle to radian measure.
$$53° = 53°\left(\frac{\pi \text{ radians}}{180°}\right) \text{ or } 0.29\pi$$
$$72° = 72°\left(\frac{\pi \text{ radians}}{180°}\right) \text{ or } 0.40\pi$$
Then use the area formula for a sector to find the area of the entire shaded region.
$$A = \frac{1}{2}r^2\theta_1 + \frac{1}{2}r^2\theta_2$$
$$= \frac{1}{2}(11)^2(0.29\pi) + \frac{1}{2}(11)^2(0.40\pi)$$
$$\approx 132 \text{ in}^2$$

67a. about 0.43 mi/h   **67b.** about 41 mi/h   **69.** The angle does not have a complement since it is greater than $\frac{\pi}{2}$ or $90°$; supplement: $\frac{\pi}{6}$ **71.** Complements and supplements are not defined for negative angles.   **73.** Sarah; sample answer: The formula for the length of an intercepted arc is $s = r\theta$.

Therefore, the perimeter of the sector is the sum of the length of the intercepted arc and twice the radius, or $P = r\theta + 2r$. Because $P = 10r$, using substitution, $10r = r\theta + 2r$. Upon simplifying, $\theta = 8$ radians.   **75.** Decrease; sample answer: If the radius decreased, then the linear speed would also decrease because the linear speed is directly proportional to the radius.   **77.** Increase; sample answer: The equation for linear speed can also be written as $v = r\omega$. If the angular speed increased, then the linear speed would also increase because the linear speed is directly proportional to the angular speed.   **79a.** The perimeter would double; sample answer: The perimeter of a sector of a circle P is equal to the sum of the arc length s and two times the radius r, so $P = s + 2r$. Because $s = r\theta$, if the radius doubled, the arc length would become a new arc length $s' = (2r)\theta$, which is equal to $s' = 2(r\theta)$ or $s' = 2s$. So, the perimeter would be $P = 2s + 2(2r)$ or $P = 2(s + 2r)$. Therefore, the perimeter would double.   **79b.** The area would quadruple; sample answer:

Because $A = \frac{1}{2}r^2\theta$, if the radius doubled, the area would become a new area $A' = \frac{1}{2}(2r)^2\theta$, which is equal to $A' = \frac{1}{2}4r^2\theta$ or $A' = 4\left(\frac{1}{2}r^2\theta\right)$. Therefore, the area would quadruple.

81. $\cos \theta = \frac{\sqrt{161}}{15}$, $\tan \theta = \frac{8\sqrt{161}}{161}$, $\csc \theta = \frac{15}{8}$, $\sec \theta = \frac{15\sqrt{161}}{161}$, $\cot \theta = \frac{\sqrt{161}}{8}$ **83.** $\sin \theta = \frac{19\sqrt{26}}{130}$, $\cos \theta = \frac{17\sqrt{26}}{130}$, $\tan \theta = \frac{19}{17}$, $\csc \theta = \frac{5\sqrt{26}}{19}$, $\sec \theta = \frac{5\sqrt{26}}{17}$ **85.** $2 \ln 5 - 4 \ln 2$ **87.** $\ln 2 - \ln 5$
89. $\pm 1, \pm 2, \pm 4, \pm 5, \pm 10, \pm 20; -2, 2, 5$ **91.** $f(x) \to -\infty$ as $x \to -\infty$, $f(x) \to \infty$ as $x \to \infty$ **93.** $h(x) \to 2$ as $x \to -\infty$, $h(x) \to 2$ as $x \to \infty$ **95.** $\{x \mid -4 \le x < 10, x \in \mathbb{R}\}$; $[-4, 10)$ **97.** B   **99.** C

Pages 251–253   Lesson 4-3

1. $\sin \theta = \frac{4}{5}$, $\cos \theta = \frac{3}{5}$, $\tan \theta = \frac{4}{3}$, $\csc \theta = \frac{5}{4}$, $\sec \theta = \frac{5}{3}$, $\cot \theta = \frac{3}{4}$ **3.** $\sin \theta = -\frac{3}{5}$, $\cos \theta = -\frac{4}{5}$, $\tan \theta = \frac{3}{4}$, $\csc \theta = -\frac{5}{3}$, $\sec \theta = -\frac{5}{4}$, $\cot \theta = \frac{4}{3}$ **5.** $\sin \theta = -\frac{8\sqrt{65}}{65}$, $\cos \theta = \frac{\sqrt{65}}{65}$, $\tan \theta = -8$, $\csc \theta = -\frac{\sqrt{65}}{8}$, $\sec \theta = \sqrt{65}$, $\cot \theta = -\frac{1}{8}$

7. $\sin \theta = \frac{15}{17}$, $\cos \theta = -\frac{8}{17}$, $\tan \theta = -\frac{15}{8}$, $\csc \theta = \frac{17}{15}$, $\sec \theta = -\frac{17}{8}$, $\cot \theta = -\frac{8}{15}$ **9.** 1   **11.** undefined   **13.** 0   **15.** 0
17. 45°
19. $\frac{5\pi}{12}$

21. 45°
23. $\frac{\pi}{6}$

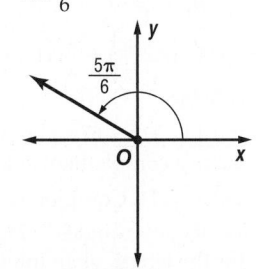

25. $-\frac{1}{2}$ **27.** $\frac{\sqrt{2}}{2}$ **29.** 2   **31.** $-\frac{\sqrt{3}}{3}$ **33.** $\sin \theta = \frac{2\sqrt{5}}{5}$, $\cos \theta = \frac{\sqrt{5}}{5}$, $\csc \theta = \frac{\sqrt{5}}{2}$, $\sec \theta = \sqrt{5}$, $\cot \theta = \frac{1}{2}$
35. $\cos \theta = \frac{2\sqrt{6}}{5}$, $\tan \theta = -\frac{\sqrt{6}}{12}$, $\csc \theta = -5$, $\sec \theta = \frac{5\sqrt{6}}{12}$, $\cot \theta = -2\sqrt{6}$ **37.** $\sin \theta = -\frac{\sqrt{6}}{3}$, $\cos \theta = \frac{\sqrt{3}}{3}$, $\tan \theta = -\sqrt{2}$, $\csc \theta = -\frac{\sqrt{6}}{2}$, $\cot \theta = -\frac{\sqrt{2}}{2}$ **39.** $\sin \theta = -\frac{\sqrt{2}}{2}$, $\cos \theta = \frac{\sqrt{2}}{2}$, $\csc \theta = -\sqrt{2}$, $\sec \theta = \sqrt{2}$, $\cot \theta = -1$

41 Let the center of the carousel represent the origin on the coordinate plane and Zoe's position after the 210° rotation have coordinates (x, y). The seat rotates 210°, so the reference angle is $210° - 180°$ or 30°. Because the final position of the seat corresponds to Quadrant III, the sine and cosine of 210° are negative.

$$\cos \theta = \frac{x}{r} \qquad \sin \theta = \frac{y}{r}$$
$$\cos 210° = \frac{x}{40} \qquad \sin 210° = \frac{y}{40}$$
$$-\cos 30° = \frac{x}{12} \qquad \sin 30° = \frac{y}{12}$$
$$-\frac{\sqrt{3}}{2} = \frac{x}{40} \qquad -\frac{1}{2} = \frac{y}{40}$$
$$-20\sqrt{3} = x \qquad -20 = y$$

Therefore, the coordinates of Zoe's seat after a rotation of 210° are $(-20\sqrt{3}, -20)$ or $(-34.6, -20)$.

43. -2 **45.** $\frac{1}{2}$ **47.** 2   **49.** undefined   **51.** $\sqrt{3}$ **53.** $-\sqrt{3}$
55. $\frac{\sqrt{3}}{2}$ **57.** $-\sqrt{3}$

59.

t	0	0.5	1	1.5	2	2.5
θ	22	0	−22	0	22	0

61. $\frac{\pi}{2}$ **63.** $\frac{4\pi}{3}$ or $\frac{5\pi}{3}$ **65.** $\frac{\pi}{6}$ **67.** $\frac{8}{11}, \frac{11}{8}$ **69.** $\frac{12}{13}, \frac{12}{13}$
71. $\sin \theta = -\frac{2\sqrt{5}}{5}$, $\cos \theta = -\frac{\sqrt{5}}{5}$, $\tan \theta = 2$, $\csc \theta = -\frac{\sqrt{5}}{2}$, $\sec \theta = -\sqrt{5}$, $\cot \theta = \frac{1}{2}$ **73.** $\left(-\frac{5}{2}, \frac{5\sqrt{3}}{2}\right)$ **75.** $(-4, -4\sqrt{3})$
77a. 11 m   **77b.** 7:00 A.M. and 7:00 P.M.

79 **a.** On the unit circle, $\cos \theta = 0$ at $\frac{\pi}{2}$ and $\frac{3\pi}{2}$. Since the cosine function has a period of 2π, $\cos \theta$ will also equal 0 at 2π multiples of $\frac{\pi}{2}$ and $\frac{3\pi}{2}$. Therefore, $\cos\left(n \cdot \frac{\pi}{2}\right) = 0$ when n is an odd integer.

b. The cosecant function is the reciprocal of the sine function. So, $\csc \theta$ will be undefined when $\sin \theta = 0$. On the unit circle, $\sin \theta = 0$ at 0, π, and 2π multiples of those angles. Therefore, $\csc\left(n \cdot \frac{\pi}{2}\right)$ is undefined when n is an even integer.
81. False; sample answer: The expression $\tan(-t) = -\tan t$ means that tangent is an odd function. The tangent of an angle depends on what quadrant the terminal side of the angle lies in.   **83.** Sample answer: The sine function is

represented by the y-coordinate on the unit circle. Comparing $\sin t$ and $\sin(-t)$ for different values of t, notice that the y-coordinate is positive for $\sin t$ and is negative for $\sin(-t)$. For instance, on the first unit circle, $\sin t = b$ and $\sin(-t) = -b$. Now find $-(\sin t)$ to verify the relationship. $-(\sin t) = -(b)$ or $-b$, which is equivalent to $\sin(-t)$. Thus, $-\sin t = \sin(-t)$.

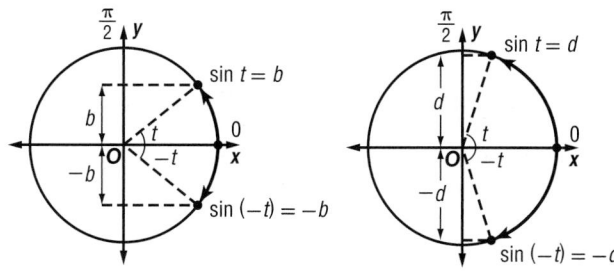

85. Sample answer: Since $\tan t = \dfrac{\sin t}{\cos t}$, we can analyze $-\tan t$ and $\tan(-t)$ by first looking at $\tan t$ and $\tan(-t)$ on the unit circle for a given value of t. Regardless of the sign of t, the value of cosine remains the same. However, the value of sine is positive for t but negative for $-t$. This results in $\tan t = \dfrac{a}{b}$, but $\tan(-t) = -\dfrac{a}{b}$. Now find $-\tan t$ to verify the relationship. $-\tan t = -\left(\dfrac{a}{b}\right)$ or $-\dfrac{a}{b}$ which is equivalent to $\tan(-t)$. Thus, $-\tan t = \tan(-t)$.

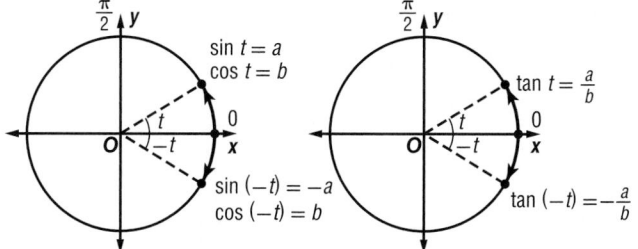

87. $168°\ 21'$ **89.** $14.089°$ **91a.** $6°$ **91b.** about 3.2 in. **93.** $\dfrac{1}{3}$
95. 3.5 **97.** $\pm 1, \pm 3; -3$ **99.** $\pm 1, \pm 3, \pm\dfrac{1}{2}, \pm\dfrac{3}{2}; -3, \dfrac{1}{2}, 1$ **101.**
$\pm 1, \pm 2, \pm\dfrac{1}{4}, \pm\dfrac{1}{2}; -\dfrac{1}{4}$ **103.** $45°$ **105.** J

Pages 264–266   Lesson 4-4

1. The graph of $g(x)$ is the graph of $f(x)$ compressed vertically. The amplitude of $g(x)$ is $\dfrac{1}{2}$.

3. The graph of $g(x)$ is the graph of $f(x)$ expanded vertically. The amplitude of $g(x)$ is 6.

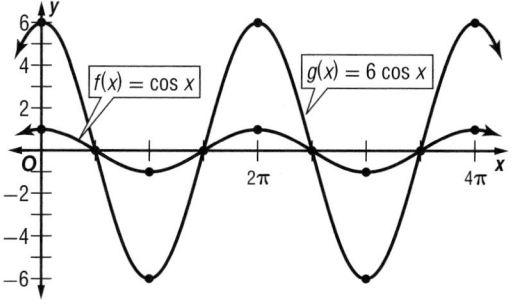

5. The graph of $g(x)$ is the graph of $f(x)$ compressed horizontally. The period of $g(x)$ is $\dfrac{\pi}{2}$.

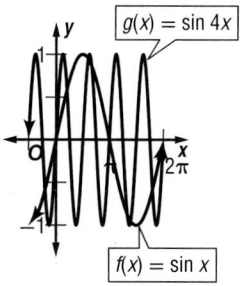

7. The graph of $g(x)$ is the graph of $f(x)$ expanded horizontally. The period of $g(x)$ is 10π.

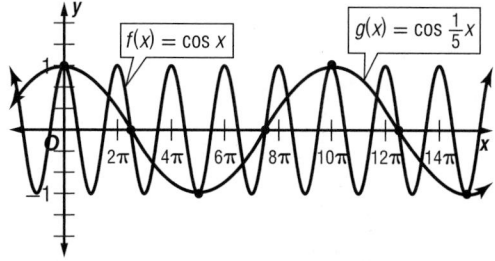

9. Sample answer: $y = 0.15 \sin 330\pi t$ **11.** Sample answer: $y = 0.25 \sin 1864\pi t$ **13.** Sample answer: $y = 0.2 \sin 1246\pi t$
15. amplitude $= 1$; period $= 6\pi$; frequency $= \dfrac{1}{6\pi}$; phase shift $= -\dfrac{3\pi}{2}$; vertical shift $= 0$

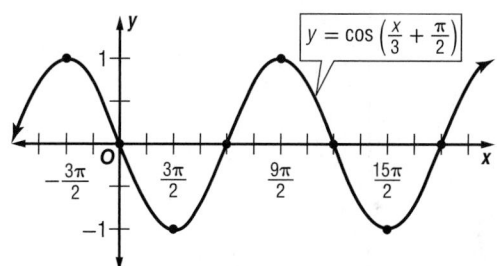

17. amplitude $= 1$; period $= \dfrac{2\pi}{3}$; frequency $= \dfrac{3}{2\pi}$; phase shift $= 0$; vertical shift $= -2$

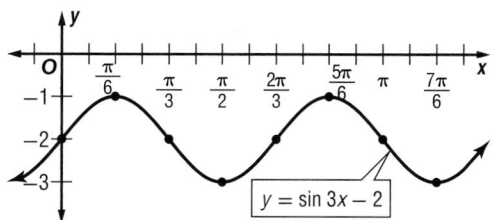

19. amplitude $= 1$; period $= 2\pi$; frequency $= \dfrac{1}{2\pi}$; phase shift $= -\dfrac{5\pi}{6}$; vertical shift $= 4$

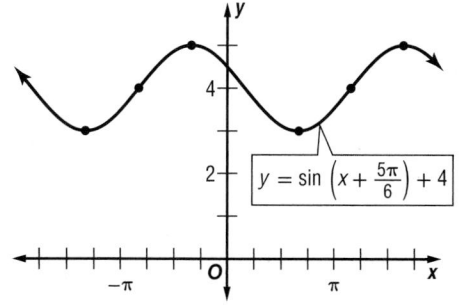

21a. amplitude = 5.465; period = 13; phase shift = 4.417;
vertical shift = 7.485   **21b.** Sample answer:
$y = 5.465 \cos\left(\frac{\pi}{6.5}x - \frac{\pi}{1.47}\right) + 7.485$ **21c.** about 7.28 ft

23. $x = -\frac{\pi}{4}$ and $\frac{3\pi}{4}$ **25.** $x = -\frac{\pi}{4}$ and $\frac{3\pi}{4}$ **27.** none

29 First, find the period of $y = 1.5 \sin(2t + c)$.

Period = $\frac{2\pi}{|b|}$ Period formula

   = $\frac{2\pi}{|2|}$ or π $b = 2$

Since the period is π, the function will reach a maximum
height at $\frac{\pi}{4}$ radians. The phase shift is the difference
between the horizontal position of the function at $\frac{\pi}{4}$
and $\frac{7\pi}{12}$ radians, which is $\frac{\pi}{3}$ radians. Substitute $\frac{\pi}{3}$ and b
into the phase shift formula to find c.

Phase shift = $-\frac{c}{|b|}$ Phase shift formula

   $\frac{\pi}{3} = -\frac{c}{|2|}$ phase shift = $\frac{\pi}{3}$ and $b = 2$

   $-\frac{2\pi}{3} = c$ Multiply each side by −2.

Therefore, the equation is $y = 1.5 \sin\left(2t - \frac{2\pi}{3}\right)$.

31. Sample answer: $y = 3 \sin 2x$ **33.** Sample answer:
$y = 2 \cos 4x + 1$ **35.** Sample answer: $y = 5 \cos 2x$

37. Sample answer: $y = \frac{3}{2}\cos 4x$

39a.

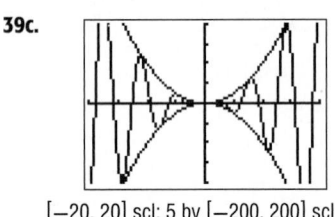

[−20, 20] scl: 5 by [−40, 40] scl: 5

39b. The graph of $y = 2x$
cos x oscillates between
the graphs of $y = 2x$ and
$y = -2x$.

39c.

[−20, 20] scl: 5 by [−200, 200] scl: 50

39d. The graph of
$y = x^2 \sin x$ oscillates
between the graphs of
$y = x^2$ and $y = -x^2$.
39e. The graph of $y = f(x)$
sin x or $y = f(x) \cos x$
will oscillate between
the graphs of $y = f(x)$
and $y = -f(x)$.

41. True; sample answer: The graph of $y = \cos x$ is a
horizontal translation of graph of $y = \sin x$. Therefore, a
cosine function can be written from any sine function using
the same amplitude and period by applying the necessary
phase shift.

43 The graph of $y = \cos x$ completes one cycle on the
interval $[0, 2\pi]$. Therefore, the graph of $y = \cos 1500x$
will complete 1500 cycles on $[0, 2\pi]$. Since cos $1500x$ has
two zeros per cycle and there are 1500 cycles, there will
be 1500(2) or 3000 zeros.

45. Sample answer: Although the pulse of light can be
represented as a function with a period, it is not a sinusoidal
function because the distance the pulse of light is from the
ground changes at a constant rate. As a result, the graph of
this function would resemble the graph below.

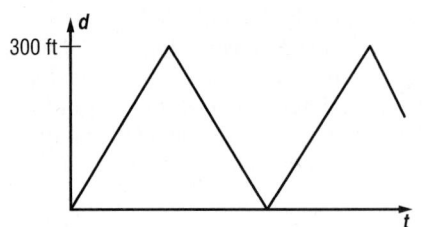

47. $\sin \theta = -\frac{\sqrt{17}}{17}$, $\cos \theta = \frac{4\sqrt{17}}{17}$, $\tan \theta = -\frac{1}{4}$, $\csc \theta = -\sqrt{17}$,
$\sec \theta = \frac{\sqrt{17}}{4}$, $\cot \theta = -4$ **49.** $\sin \theta = \frac{5\sqrt{41}}{41}$, $\cos \theta = \frac{4\sqrt{41}}{41}$,
$\tan \theta = \frac{5}{4}$, $\csc \theta = \frac{\sqrt{41}}{5}$, $\sec \theta = \frac{\sqrt{41}}{4}$, $\cot \theta = \frac{4}{5}$ **51.** $-\frac{7\pi}{3}$
53. 480°   **55.** 3 real zeros and 2 turning points; −4, 0, and 2
57. 5 real zeros and 4 turning points; −2, 0, and 2   **59.** yes;
$f^{-1}(x) = -x - 2$ **61.** no   **63.** C   **65.** C

Pages 277–279   Lesson 4-5

R68   Selected Answers and Solutions

17. The damping factor is $\frac{3}{5}x$; the amplitude of the function is decreasing as x approaches 0 from both directions.

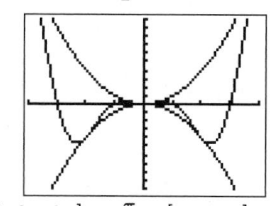

$[-5\pi, 5\pi]$ scl: $\frac{\pi}{2}$ by $[-10, 10]$ scl: 1

19. The damping factor is $2x^2$; the amplitude of the function is decreasing as x approaches 0 from both directions.

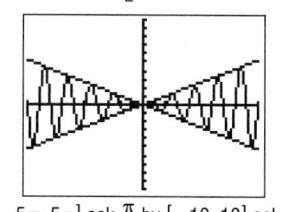

$[-2\pi, 2\pi]$ scl: $\frac{\pi}{2}$ by $[-50, 50]$ scl: 5

21. The damping factor is $\frac{1}{3}x$; the amplitude of the function is decreasing as x approaches 0 from both directions.

$[-5\pi, 5\pi]$ scl: $\frac{\pi}{2}$ by $[-10, 10]$ scl: 1

23. The damping factor is $e^{0.5x}$; the amplitude of the function is decreasing as x approaches $-\infty$.

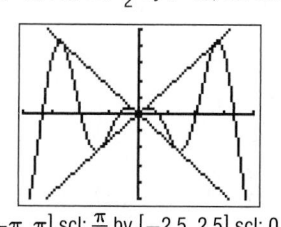

$[-\pi, 3\pi]$ scl: $\frac{\pi}{2}$ by $[-50, 50]$ scl: 1

25. The damping factor is $|x|$; the amplitude of the function is decreasing as x approaches 0 from both directions.

$[-\pi, \pi]$ scl: $\frac{\pi}{4}$ by $[-2.5, 2.5]$ scl: 0.5

27a. Sample answer: $y = 8e^{-3t}\cos 5\pi t$ **27b.** about 0.06 second

29.

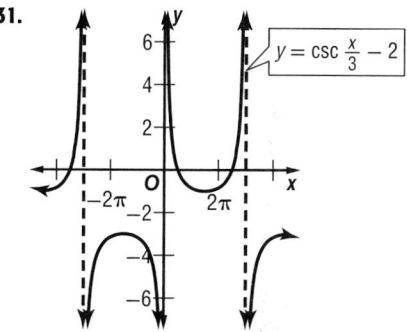

$y = \sec x + 3$

31.

$y = \csc \frac{x}{3} - 2$

33.

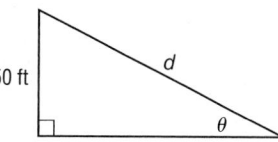

$y = \cot(2x + \pi) - 3$

35 a.

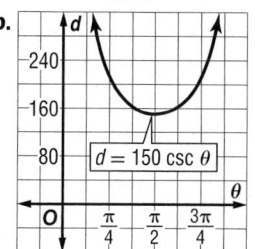

150 ft, d, θ

Since d is the length of the hypotenuse and you are given the length of the side opposite θ, you can use the sine function to write d as a function of θ.

$$\sin \theta = \frac{\text{opp}}{\text{hyp}}$$

$$\sin \theta = \frac{150}{d}$$

$$d \sin \theta = 150$$

$$d = \frac{150}{\sin \theta} \text{ or } d = 150 \csc \theta$$

b.

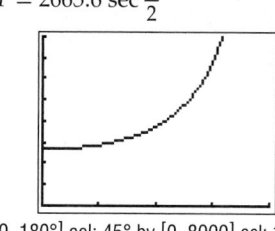

$d = 150 \csc \theta$

c. Substitute $\theta = 45°$ into the function you found in part **a**.

$d = 150 \csc \theta$
$= 150 \csc 45°$
$= 150\sqrt{2}$
≈ 212.1 ft

37. 0.427 and 2.715   **39.** -2.880, -1.833, 0.262, and 1.309
41. -2.356 and 0.785   **43a.** 5331.2 N
43b. $T = 2665.6 \sec \frac{\theta}{2}$

43c.

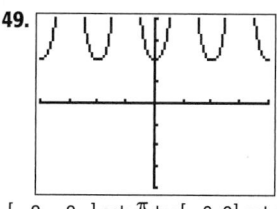

$[0, 180°]$ scl: 45° by $[0, 8000]$ scl: 1000

43d. total rope: about 12.9 m; tension: about 3769.7 N   **43e.** $\theta \approx 97.0°$; about 4023 N

45. b   **47.** a

49.

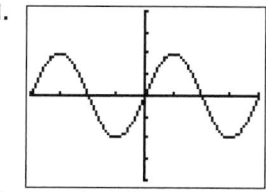

51.

$[-2\pi, 2\pi]$ scl: $\frac{\pi}{2}$ by $[-2, 2]$ scl: 0.5 $[-2\pi, 2\pi]$ scl: $\frac{\pi}{2}$ by $[-2, 2]$ scl: 0.5

The expressions are equivalent.

The expressions are not equivalent for all real numbers.

53 The general form of the tangent function is $y = a \tan(bx + c) + d$, where the amplitude is equal to $|a|$, the period is equal to $\frac{\pi}{b}$, the phase shift is equal to $-\frac{c}{|b|}$, and the vertical shift is equal to d. Since we are given the period, phase shift, and vertical shift, we can find the values of b, c, and d and substitute them into the general form of the tangent function.

$$\text{period} = \frac{\pi}{2} \qquad\qquad \text{phase shift} = -\frac{c}{|b|}$$

$$\frac{\pi}{|b|} = \frac{\pi}{2} \qquad\qquad \frac{\pi}{4} = -\frac{c}{\pm 2}$$

$$\pi = \frac{\pi|b|}{2} \qquad\qquad \frac{\pi}{4} = -\frac{c}{2}$$

$$2 = |b| \qquad\qquad -\frac{\pi}{2} = c$$

$$\pm 2 = b$$

vertical shift $= -1$

Therefore, a tangent function with a period $\frac{\pi}{2}$, a phase shift of $\frac{\pi}{4}$ units to the right, and a vertical shift of 1 unit down is given by $y = \tan\left(2x - \frac{\pi}{2}\right) - 1$ or $y = \tan\left(-2x - \frac{\pi}{2}\right) - 1$.

55. $y = \cot\left(\frac{x}{3} - \frac{\pi}{6}\right) + 4$, $y = \cot\left(-\frac{x}{3} - \frac{\pi}{6}\right) + 4$

57. To find the y-intercept, let $t = 0$.

$y = ke^{-ct} \cos wt$	Original equation
$= ke^{-c(0)} \cos (w \cdot 0)$	Substitute $t = 0$.
$= ke^0 \cos 0$	Multiply.
$= k(1)(1)$	$e^0 = 1$, $\cos 0 = 1$
$= k$	Simplify.

59. True; sample answer: Since $y = \csc x = \frac{1}{\sin x}$, asymptotes will occur for values of x when $\sin x = 0$. Since $y = \cot x = \frac{\cos x}{\sin x}$, asymptotes will also occur for values of x when $\sin x = 0$. **61.** Sample answer: $y = \csc\left(x + \frac{\pi}{2}\right)$; $y = -\cot\left(x - \frac{\pi}{2}\right)$

63. amplitude $= 3$; period $= \pi$; frequency $= \frac{1}{\pi}$; phase shift $= \frac{\pi}{6}$; vertical shift $= 10$

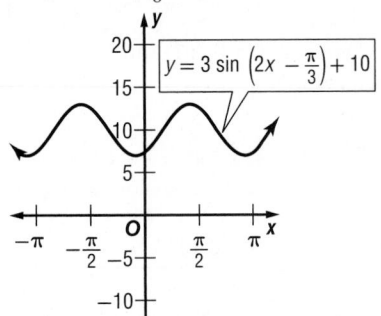

65. amplitude $= \frac{1}{2}$; period $= \frac{\pi}{2}$; frequency $= \frac{2}{\pi}$; phase shift $= \frac{\pi}{4}$; vertical shift $= 1$

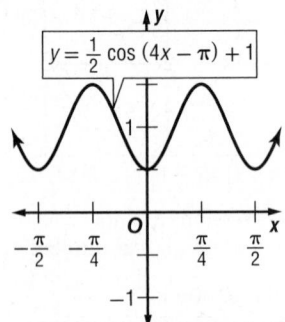

67. $\sin \theta = \frac{\sqrt{37}}{37}$, $\tan \theta = \frac{1}{6}$, $\csc \theta = \sqrt{37}$, $\sec \theta = \frac{\sqrt{37}}{6}$, $\cot \theta = 6$ **69.** about 3.6% **71.** $(x + 1)(x + 2)(x - 1)$ **73.** $(x - 4)(x + 2)(x + 1)$ **75.** C **77.** B

1. 0 **3.** $\frac{\pi}{4}$ **5.** $-\frac{\pi}{4}$ **7.** $\frac{\pi}{4}$ **9.** $\frac{\pi}{6}$ **11.** $\frac{\pi}{4}$ **13.** $\frac{\pi}{6}$ **15.** $\frac{\pi}{6}$

17.

19.

21.

23.

25.

27 a.

Since x is the length of the side opposite θ and you are given the length of the side adjacent to θ, you can use the tangent function to write θ as a function of x.

$$\tan \theta = \frac{\text{opp}}{\text{adj}}$$

$$\tan \theta = \frac{x}{30}$$

$$\theta = \tan^{-1} \frac{x}{30} \text{ or } \arctan \frac{x}{30}$$

b. Substitute $x = 6$ and $x = 14$ into the function you found in part **a.**

$$\theta = \tan^{-1} \frac{x}{30} \qquad \theta = \tan^{-1} \frac{x}{30}$$

$$\theta = \tan^{-1} \frac{6}{30} \qquad \theta = \tan^{-1} \frac{14}{30}$$

$$\approx 11.3° \qquad\qquad \approx 25.0°$$

29. $\frac{3}{4}$ **31.** $\frac{2}{9}$ **33.** $\frac{\pi}{4}$ **35.** $\frac{\sqrt{2}}{2}$ **37.** 1 **39.** $\frac{\sqrt{2}}{2}$

41 Let $u = \arccos x$, so $\cos u = x$. You can draw a right triangle, where the side adjacent to u is equal to x and the hypotenuse is equal to 1.

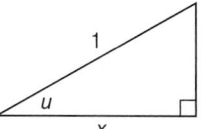

The domain of the inverse cosine function is restricted to Quadrants I and II, so u must lie in one of those quadrants. Using the Pythagorean Theorem, you can find the length of the side opposite u to be $\sqrt{1^2 - x^2}$. Now, find $\tan u$.

$$\tan u = \frac{\text{opp}}{\text{adj}}$$

$$= \frac{\sqrt{1^2 - x^2}}{x} \text{ or } \frac{\sqrt{1 - x^2}}{x}$$

So, $\tan (\arccos x) = \dfrac{\sqrt{1 - x^2}}{x}$.

43. $\sqrt{1 - x^2}$ **45.** $\frac{1}{x}$ **47.** $\dfrac{x\sqrt{1 - x^2}}{1 - x^2}$ **49.** The graph of $g(x)$ is the graph of $f(x)$ translated 1 unit to the right and 2 units down. **51.** The graph of $g(x)$ is the graph of $f(x)$ expanded vertically and translated 6 units down **53.** The graph of $g(x)$ is the graph of $f(x)$ compressed horizontally and translated 5 units up. **55a.** about 36° **55b.** about 11 ft

57. D = $\{x \mid -1 \le x \le 1\}$;
R = $\{y \mid 0 \le y \le 1\}$

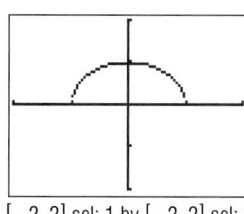

[−2, 2] scl: 1 by [−2, 2] scl: 1

59. D = $\{x \mid x \in \mathbb{R}\}$;
R = $\left\{y \mid -\dfrac{\pi}{2} \le y \le \dfrac{\pi}{2}\right\}$

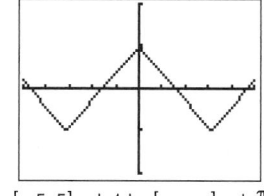

[−5, 5] scl: 1 by [−π, π] scl: $\frac{\pi}{2}$

61. D = $\{x \mid -1 \le x \le 1, x \ne 0\}$; R = $\{y \mid y \ne 0\}$

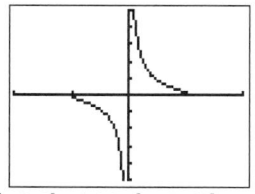

[−2, 2] scl: 1 by [−10, 10] scl: 2

63. $\tan (\sin^{-1} x)$ or $\cot (\cos^{-1} x)$ **65a.** Sample answer: For $f \circ f^{-1}$, the domain is $[-1, 1]$ and the range is $[-1, 1]$. For $f^{-1} \circ f$, the domain is $(-\infty, \infty)$ and the range is $\left[-\dfrac{\pi}{2}, \dfrac{\pi}{2}\right]$.

65b. Sample answer:

x	−2.0	−1.5	−1.0	−0.75	−0.5	−0.25	0
$f \circ f^{-1}$			−1.0	−0.75	−0.5	−0.25	0

x	0.25	0.5	0.75	1.0	1.5	2.0	
$f \circ f^{-1}$	0.25	0.5	0.75	1.0			

x	−2.0	−1.5	−1.0	−0.75	−0.5	−0.25	0
$f^{-1} \circ f$	−1.142	−1.5	−1.0	−0.75	−0.5	−0.25	0

x	0.25	0.50	0.75	1.0	1.5	2.0	
$f^{-1} \circ f$	0.25	0.50	0.75	1.0	1.5	1.142	

$f \circ f^{-1}$

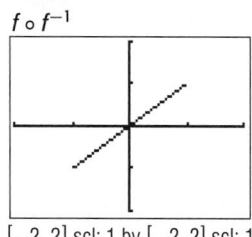

[−2, 2] scl: 1 by [−2, 2] scl: 1

$f^{-1} \circ f$

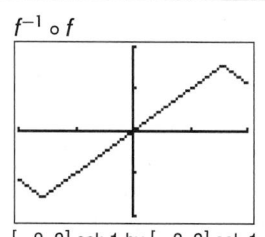

[−2, 2] scl: 1 by [−2, 2] scl: 1

65c. Sample answer: For $g \circ g^{-1}$, the domain is $[-1, 1]$ and the range is $[-1, 1]$. For $g^{-1} \circ g$, the domain is $(-\infty, \infty)$ and the range is $[0, \pi]$. The graph of $g \circ g^{-1}$ should be the line $y = x$ for $-1 \le x \le 1$. The inverse property of trigonometric functions states that on the closed interval $[-1, 1]$, $\cos (\cos^{-1} x) = x$. The graph of $g^{-1} \circ g$ should be the line $y = x$ for $0 \le x \le \pi$. Once the graph reaches π, it will turn and decrease until it reaches the x-axis at the same rate. When it reaches the x-axis, it will turn again and increase until it reaches π. It will continue to do this as x approaches infinity.

65d. $g \circ g^{-1}$

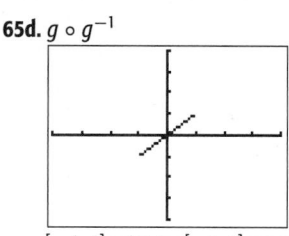

[−4, 4] scl: 1 by [−4, 4] scl: 1

$g^{-1} \circ g$

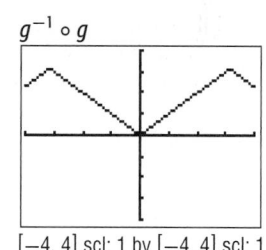

[−4, 4] scl: 1 by [−4, 4] scl: 1

65e. Sample answer: For $f(x) = \tan (\tan^{-1} x)$, due to the inverse property of trigonometric functions, for all values of x, $f(x) = x$. This should result in the line $y = x$ for all real numbers. The graph of $g(x) = \tan^{-1} (\tan x)$ will be different because $\tan x$ is undefined for multiples of π. As a result, asymptotes for multiples of π can be expected. We can also expect a range of $\left[-\dfrac{\pi}{2}, \dfrac{\pi}{2}\right]$ due to the definition of arctan.

$f(x) = \tan(\tan^{-1} x)$

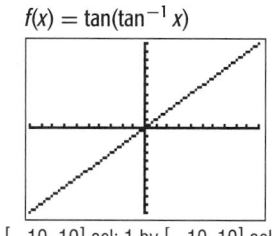

[−10, 10] scl: 1 by [−10, 10] scl: 1

$g(x) = \tan^{-1}(\tan x)$

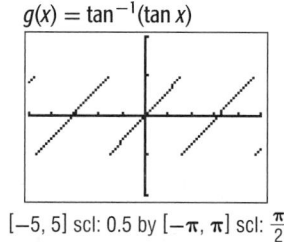

[−5, 5] scl: 0.5 by [−π, π] scl: $\frac{\pi}{2}$

67.

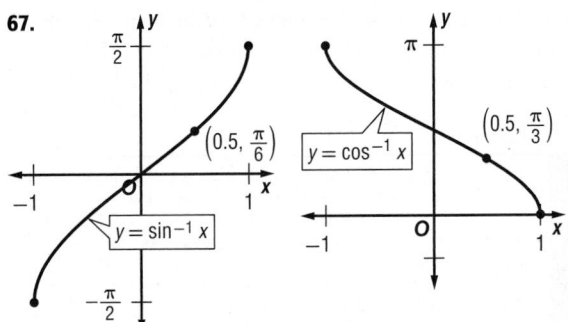

Notice from the graphs of $y = \sin^{-1} x$ and $y = \cos^{-1} x$ that when $x = 0.5$, $\sin^{-1} x = \frac{\pi}{6}$ and $\cos^{-1} x = \frac{\pi}{3}$. So, $\sin^{-1} x + \cos^{-1} x$ is $\frac{\pi}{6} + \frac{\pi}{3}$ or $\frac{\pi}{2}$. When $x = 1$, $\sin^{-1} x = \frac{\pi}{2}$ and $\cos^{-1} x = 0$, so $\sin^{-1} x + \cos^{-1} x$ is $\frac{\pi}{2} + 0$ or $\frac{\pi}{2}$. When $x = -1$, $\sin^{-1} x = -\frac{\pi}{2}$ and $\cos^{-1} x = \pi$, so $\sin^{-1} x + \cos^{-1} x$ is $-\frac{\pi}{2} + \pi$ or $\frac{\pi}{2}$. Therefore, on the interval $[-1, 1]$ it appears that $\sin^{-1} x + \cos^{-1} x = \frac{\pi}{2}$. The graph of $y = \sin^{-1} x + \cos^{-1} x$ supports this conjecture.

69. Sample answer: The function is odd. The definition of an odd function states for every x in the domain of f, $f(-x) = -f(x)$. If we let $\sin^{-1} x = u$, we have $x = \sin u$. From Lesson 4-3, we know that the sine function is odd, so $-x = \sin (-u)$. From here, we can get $\sin^{-1} (-x) = -u$. Graphically, it can be seen that for every x in the domain of f, $f(-x) = -f(x)$.
71. Sample answer: The function is odd. The definition of an odd function states for every x in the domain of f, $f(-x) = -f(x)$. If we let $\tan^{-1} x = u$, we have $x = \tan u$. From Lesson 4-3, we know that the tangent function is odd, so $-x = \tan (-u)$. From here, we can get $\tan^{-1} (-x) = -u$. Graphically, it can be seen that for every x in the domain of f, $f(-x) = -f(x)$.

73.

75.

77. 5.8   **79.** 6   **81.** $[f \circ g](x) = 6 - \frac{5}{x}$; $[g \circ f](x) = \frac{1}{6 - 5x}$; $[f \circ g](4) = 4.75$ **83.** 10 questions   **85.** H   **87.** F

1. $A = 32°$, $a = 11.2$, $b = 19.8$ **3.** $K = 82°$, $j = 16.2$, $\ell = 21.4$
5. $T = 20°$, $t = 29.0$, $u = 17.7$ **7.** about 0.85 ft

9   a. Let $B = 8°$, $C = 157°$, and $a = 90$.

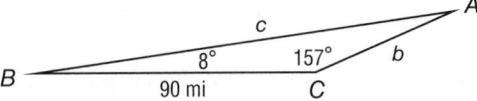

First, find A.
$m \angle A = 180° - 8° - 157°$ or $15°$
Since the measures of A, B, and a are known, you can use the Law of Sines to find b.
$$\frac{\sin 15°}{90} = \frac{\sin 8°}{b}$$
$$b = \frac{90 \sin 8°}{\sin 15°} \text{ or } 48.4 \text{ mi}$$
So, the entire trip is $90 + 48.4$ or 138.4 miles.

b. If the direct flight distance is c,
$$\frac{\sin 15°}{90} = \frac{\sin 157°}{c}$$
$$c = \frac{90 \sin 157°}{\sin 15°}$$
$$= 135.9$$
Therefore, the distance of a direct flight to the destination is about 135.9 miles.

11. no solution   **13.** $B = 43°$, $C = 45°$, $c = 24.7$ **15.** $B = 90°$, $C = 60°$, $c = 32.9$ **17.** $B = 90°$, $C = 45°$, $c = 10$ **19.** $B = 63°$, $C = 78°$, $c = 18.7$ and $B = 117°$, $C = 24°$, $c = 7.8$ **21.** $B = 70°$, $C = 49°$, $c = 12.2$ and $B = 110°$, $C = 9°$, $c = 2.5$ **23.** $B = 70°$, $C = 56°$, $c = 31.8$ and $B = 110°$, $C = 16°$, $c = 10.6$ **25.** 33.4 mi
27. $B = 39°$, $C = 99°$, $a = 12.9$ **29.** $Q = 27°$, $R = 80°$, $p = 14.6$
31. $R = 96°$, $S = 39°$, $T = 45°$ **33.** $C = 162°$, $D = 2°$, $b = 24.1$
35. $15.6°$; $71.5°$; $92.9°$ **37.** 47.6 cm²   **39.** 1133.0 ft²
41. 20.9 yd²   **43a.** 4511.5 steps² **43b.** 14,617 ft²   **45.** 51.5 mm²
47. 313.2 ft²   **49.** 259.8 in²   **51.** about 22.3 ft or about 26.1 ft

53 To find the total area of the figure, find the sum of the areas of $\triangle FJE$ and $\triangle FGH$.
For $\triangle FJE$: $s = \frac{1}{2}(a + b + c)$
$$= \frac{1}{2}(12 + 20 + 27)$$
$$= 29.5$$
$$\text{Area} = \sqrt{s(s - a)(s - b)(s - c)}$$
$$= \sqrt{29.5(29.5 - 12)(29.5 - 20)(29.5 - 27)}$$
$$\approx 110.7 \text{ mm}^2$$
For $\triangle FGH$: $s = \frac{1}{2}(17 + 21 + 28)$
$$= 33$$
$$\text{Area} = \sqrt{33(33 - 17)(33 - 21)(33 - 28)}$$
$$\approx 178.0 \text{ mm}^2$$
So, the total area is $110.7 + 178.0$ or 288.7 mm².

55. 1062.6 ft²   **57.** South Bay is about 25.72 mi from the boat and Steep Rock is about 31.92 mi from the boat.
59. 994.2 mm²   **61.** 884.2 in²   **63.** ≈ 40.9 m   **65.** Neither; for the acute case, $h = 21 \sin 34°$ or 11.7. Because $a < b$ and $h < a$, there are two solutions.

67. Sample answer: The domain of inverse cosine includes angles measured 0 to 180 degrees. The domain of inverse sine includes angles measured −90 to 90 degrees.

69.

A is acute.

A is obtuse.

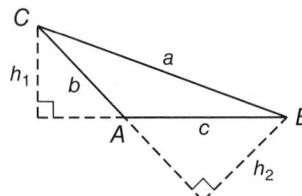

Sample answer: Let h_1 be an altitude of either triangle shown above. From the definition of the sine function, $\sin A = \dfrac{h_1}{b}$ or $h_1 = b \sin A$, and $\sin B = \dfrac{h_1}{a}$ or $h_1 = a \sin B$. Equating the two values of h_1, $b \sin A = a \sin B$, or $\dfrac{\sin A}{a} = \dfrac{\sin B}{b}$, where $\sin A \neq 0$ and $\sin B \neq 0$. When an altitude h_2 is drawn from vertex B to side AC (extended in the obtuse triangle), $\sin A = \dfrac{h_2}{c}$ or $h_2 = c \sin A$, and $\sin C = \dfrac{h_2}{a}$ or $h_2 = a \sin C$. Equating the two values of h_2, $c \sin A = a \sin C$, or $\dfrac{\sin A}{a} = \dfrac{\sin C}{c}$. By the Transitive Property of equality, $\dfrac{\sin A}{a} = \dfrac{\sin B}{b} = \dfrac{\sin C}{c}$.

71 Draw a diagram to represent the situation.

To find the amount of time that it takes to complete one full orbit, we need to find the measure of the arc intercepted by points B and X. From the information given, $m\angle XAC = 90° + 35°$ or 125° and the length of XC is $2110 + 850$ or 2960 miles. Now that we have two side lengths and an angle measure, we can find $m\angle C$ in $\triangle XAC$.

First, use the Law of Sines to find $\angle X$.

$$\frac{\sin 125°}{2960} = \frac{\sin X}{2110}$$
$$2110 \sin 125° = 2960 \sin X$$
$$\frac{2110 \sin 125°}{2960} = \sin X$$
$$\sin^{-1}\left(\frac{2110 \sin 125°}{2960}\right) = X$$
$$35.73° \approx X$$

So, $\angle ACX$ is $180° - (125° + 35.73°)$ or about 19.27°. Because point B is directly above point A, $\angle ACX \cong \angle BCX$. From geometry, $m\angle BCX = m\overset{\frown}{XB}$. Therefore, $m\overset{\frown}{XB} \approx 19.27°$. Find the amount of time that it takes to complete one full orbit or 360°.

$$\frac{19.27°}{14} = \frac{360°}{x}$$
$$19.27°x = (360°)(14)$$
$$x \approx 261.55$$

Convert 261.55 minutes to hours.

$$261.55 \text{ min} \cdot \frac{1 \text{ h}}{60 \text{ min}} \approx 4.36 \text{ h}$$

Therefore, one full orbit takes approximately 4.36 hours or 4 hours and 22 minutes.

73. $\dfrac{2\pi}{3}$ **75.** $\dfrac{\pi}{4}$

77. $f(x) = -2x$; the amplitude of the function is decreasing as x approaches 0 from both sides.

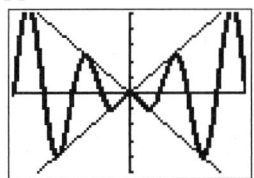

[−4π, 4π] scl: π by [−20, 20] scl: 4

79. $f(x) = (x - 1)^2$; the amplitude of the function is decreasing as x approaches 0 from both sides.

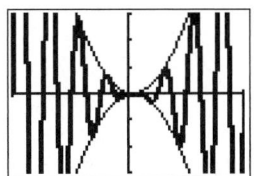

[−8π, 8π] scl: 2π by [−150, 150] scl: 50

81.

Latitude	Distance
0°	24,881.4
30°	21,547.9
45°	17,593.8
60°	12,440.7
90°	0

The distances range from about 24,881 mi to 0 mi.

83. Sample answer: $f(x) = x^3 - 5x^2 - x + 5$ **85.** Sample answer: $f(x) = x^3 + 3x^2 + 4x + 12$ **87.** E

89a.

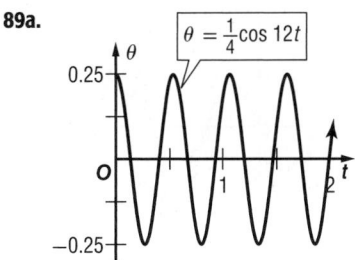

89b. The period is $\dfrac{\pi}{6}$ or about 0.524 seconds. This represents the amount of time it takes for the pendulum to complete one full swing or cycle. The amplitude is $\dfrac{1}{4}$. This represents the maximum angular displacement of the pendulum. The frequency is $\dfrac{6}{\pi}$ or about 1.91. This represents the number of swings the pendulum completes per second. **89c.** The maximum angular displacement is 14.3°. **89d.** The midline represents when the pendulum is vertical and there is no angular displacement. **89e.** about $0.101 + 0.524n$ seconds and $0.161 + 0.524n$ seconds

Pages 302–306 Chapter 4 Study Guide and Review

1. true **3.** false, angle of depression **5.** true **7.** false, amplitude **9.** false, arccosine **11.** $\sin\theta = \dfrac{12}{13}$, $\cos\theta = \dfrac{5}{13}$, $\tan\theta = \dfrac{12}{5}$, $\csc\theta = \dfrac{13}{12}$, $\sec\theta = \dfrac{13}{5}$, $\cot\theta = \dfrac{5}{12}$ **13.** 5.0 **15.** 19° **17.** $\dfrac{3\pi}{4}$ **19.** 315°

21. $342° + 360n°$, n is an integer; Sample answer: $702°$, $-18°$

 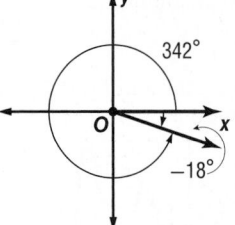

23. 53.0 in^2

25. $60°$

27. $\frac{\pi}{4}$

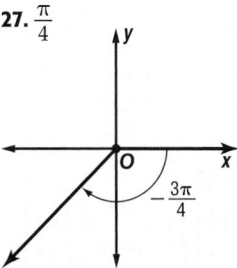

29. $\sin \theta = \frac{\sqrt{21}}{5}$, $\tan \theta = \frac{\sqrt{21}}{2}$, $\csc \theta = \frac{5\sqrt{21}}{21}$, $\sec \theta = \frac{5}{2}$, $\cot \theta = \frac{2\sqrt{21}}{21}$ **31.** $\cos \theta = \frac{12}{13}$, $\csc \theta = \frac{13}{5}$, $\sec \theta = \frac{13}{12}$, $\tan \theta = -\frac{5}{12}$, $\cot \theta = -\frac{12}{5}$ **33.** 0 **35.** undefined

37. The graph of $g(x)$ is the graph of $f(x)$ stretched vertically. The amplitude of $g(x)$ is 5, and the period is 2π.

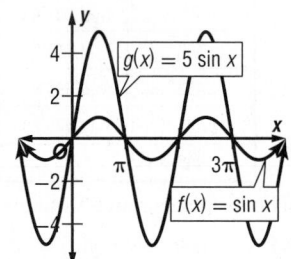

39. The graph of $g(x)$ is the graph of $f(x)$ compressed vertically. The amplitude of $g(x)$ is $\frac{1}{2}$, and the period is 2π.

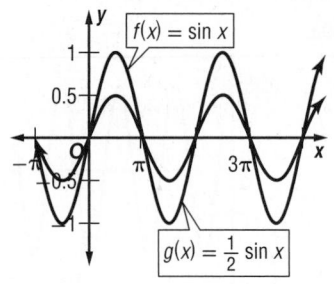

41. amplitude $= 2$; period $= 2\pi$; frequency $= \frac{1}{2\pi}$; phase shift $= \pi$; vertical shift $= 0$

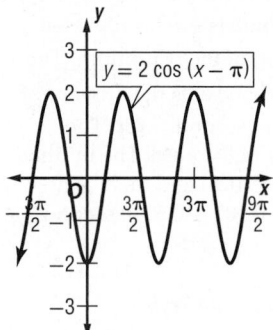

43. amplitude $= \frac{1}{2}$; period $= 2\pi$; frequency $= \frac{1}{2\pi}$; phase shift $= -\frac{\pi}{2}$; vertical shift $= 0$

45.

47.

49.

51.

53. $-\frac{\pi}{2}$ **55.** $-\frac{\pi}{6}$ **57.** $-\frac{\pi}{4}$ **59.** $-\frac{\pi}{3}$
61. $B = 12°$, $C = 146°$, $c = 16.4$ **63.** $B = 29°$, $C = 73°$, $c = 19.5$
65. $A = 78°$, $B = 65°$, $C = 37°$ **67a.** 43 ft
67b. 0.07
69a. 150π/min or 2.5π/s
69b. 27,000°/min
71a. $60°$ **71b.** 550 ft
71c. 2.5 min

73a. 12 h 30 min   **73b.** 11:15 P.M.   **75.** 6.3 ft   **77a.** $63°$
77b. 84 units^2

1.

3. Sample answer: If one of the cars slowed down, the distance between the two cars would grow at a slower rate. Therefore, the average rate of change of the distance between the two cars would be smaller. If one of the cars sped up, the distance between the two cars would grow at a faster rate. Therefore, the average rate of change of the distance between the two cars would be larger.   **5.** For $t = 1$, $A \approx 79$ cm^2. For $t = 2$, $A \approx 314$ cm^2. For $t = 3$, $A \approx 707$ cm^2. For $t = 4$, $A \approx 1257$ cm^2. For $t = 5$, $A \approx 1963$ cm^2.   **7.** For $t = 2$, m approaches 314 cm^2/sec. For $t = 3$, m approaches 471 cm^2/sec. For $t = 4$, m approaches 628 cm^2/sec.   **9.** Sample answer: In the first activity, the graph resembled a linear function. Therefore, the average rate of change could be found by using any two points since the rate of change remained constant. In the second activity, the graph resembled a quadratic function. The difference quotient had to be used to approximate the rate of change at each individual value of t.

Chapter 5 Trigonometric Identities and Equations

Page 311   Chapter 1   Get Ready

1. $-8, 3$ **3.** $-\frac{1}{2}, 5$ **5.** $-2, 0, 3$ **7.** 12 s   **9.** $a \approx 13.62$, $b \approx 23.30$, $A = 32°$ **11.** $k \approx 9.13$, $J \approx 42.63°$, $L \approx 75.37°$
13. $\frac{\sqrt{2}}{2}$ **15.** $-\sqrt{3}$ **17.** $\frac{1}{2}$

Pages 317–319   Lesson 5-1

1. $\frac{7}{5}$ **3.** 5   **5.** $\frac{\sqrt{35}}{35}$ **7.** $\frac{7\sqrt{10}}{20}$ **9.** $\sec \theta = \sqrt{26}$, $\cos \theta = \frac{1}{\sqrt{26}}$ or $\frac{\sqrt{26}}{26}$ **11.** $\tan \theta = \sqrt{15}$, $\sin \theta = \frac{\sqrt{15}}{4}$ **13.** $\cos \theta = \frac{\sqrt{55}}{8}$, $\tan \theta = \frac{3}{\sqrt{55}}$ or $\frac{3\sqrt{55}}{55}$ **15.** $\cot \theta = -\frac{2}{\sqrt{77}}$ or $-\frac{2\sqrt{77}}{77}$, $\sin \theta = \frac{\sqrt{77}}{9}$ **17.** -1.24 **19.** 1.52   **21.** -1.35 **23.** $\sin x$
25. $2 \sin x$ **27.** $2 \cos x$ **29.** $\tan^2 x$ **31.** $\sin x \cos x$ **33.** $\sin x$
35. $\cot x - \cos x + \tan x + \sin x \tan x$

37 **a.** $I = I_0 - \dfrac{I_0}{\csc^2 \theta}$
$= I_0 \left(1 - \dfrac{1}{\csc^2 \theta}\right)$
$= I_0 (1 - \sin^2 \theta)$
$= I_0 \cos^2 \theta$
b. $I = I_0 \cos^2 \theta$
$= I_0 \left(\dfrac{\sqrt{3}}{2}\right)^2$
$= I_0 \left(\dfrac{3}{4}\right)$
$= \dfrac{3I_0}{4}$
Thus, $\frac{3}{4}$ of the original intensity emerges.

39. $\sec^2 x (\csc x + 1)$ **41.** $\sec x (\csc x - 1)$ **43.** $2 - 2 \cos x$
45. $\cos x (\csc x + 1)$ **47.** $\sec x - 1$ **49.** Even; the graph shown is $f(x) = \sec x$. $f(-x) = \sec(-x) = \sec x = f(x)$. Since $f(-x) = f(x)$, $f(x) = \sec x$ is an even function.   **51.** $-\cot x$
53. $-\csc x$
55a.

x	-2π	$-\pi$	0	π	2π
y_1	1	1	1	1	1
y_2	1	1	1	1	1
y_3	0	0	0	0	0
y_4	0	0	0	0	0

55b.

$[-2\pi, 2\pi]$ scl: $\frac{\pi}{2}$ by $[-4, 4]$ scl: 1

$y_2 = \sec x \cos x - \sin x \sec x$

$[-2\pi, 2\pi]$ scl: $\frac{\pi}{2}$ by $[-4, 4]$ scl: 1

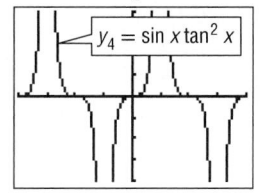

$[-2\pi, 2\pi]$ scl: $\frac{\pi}{2}$ by $[-4, 4]$ scl: 1

$y_4 = \sin x \tan^2 x$

$[-2\pi, 2\pi]$ scl: $\frac{\pi}{2}$ by $[-4, 4]$ scl: 1

55c. $y_1 \neq y_2$, $y_3 = y_4$ **55d.** Sample answer: The graphs of y_1 and y_2 support the conjecture that $y_1 \neq y_2$. Although the table from part **a** and the graphs of y_3 and y_4 support the conjecture that $y_3 = y_4$, it is impossible to determine whether the two equations are equal over the entire domain unless the equations are verified algebraically.   **57.** $-2 \ln |\cos x|$
59. $\ln |\csc^2 x|$ **61a.** $\csc \theta = \dfrac{d}{m\lambda}$ **61b.** $\theta = 0.11$ radian
63. Jenelle; $\dfrac{1 - \sin^2 x}{\sin^2 x - \cos^2 x} = \dfrac{1 - \sin^2 x}{(1 - \cos^2 x) - \cos^2 x}$
$= \dfrac{\cos^2 x}{1 - 2\cos^2 x}$ **65.** $\sin x = \pm\sqrt{1 - \cos^2 x}$,
$\csc x = \pm\dfrac{\sqrt{1 - \cos^2 x}}{1 - \cos^2 x}$, $\sec x = \dfrac{1}{\cos x}$, $\tan x = \pm\dfrac{\sqrt{1 - \cos^2 x}}{\cos x}$,
$\cot x = \pm\dfrac{\cos x \sqrt{1 - \cos^2 x}}{1 - \cos^2 x}$ **67.** False; sample answer: The domain does not include values of x where $\csc x$ is undefined, such as $n\pi$.

69 According to the Pythagorean Theorem, $x^2 + y^2 = r^2$.

When each side of the equation is divided by x^2, $1 + \dfrac{y^2}{x^2} = \dfrac{r^2}{x^2}$. Since $\tan = \dfrac{y}{x}$ and $\sec = \dfrac{r}{x}$, $1 + \dfrac{y^2}{x^2} = \dfrac{r^2}{x^2}$ is equivalent to $1 + \tan^2 x = \sec^2 x$ or $\tan^2 x + 1 = \sec^2 x$.

73. $C = 73°, a \approx 55.6, b \approx 48.2$ **75.** $B \approx 46°, C \approx 69°, c \approx 5.1$

77. $A \approx 40°, B \approx 65°, b \approx 2.8$ **79.** 3 **81.** $\dfrac{\sqrt{2}}{2}$ **83.** $\dfrac{4}{5}$

85. True; all of the elements of A are also elements of B.

87. D **89.** B

Pages 324–326 Lesson 5-2

1. $(\sec^2\theta - 1)\cos^2\theta = (\tan^2\theta)\cos^2\theta$
$= \left(\dfrac{\sin^2\theta}{\cos^2\theta}\right)\cos^2\theta$
$= \sin^2\theta$

3. $\sin\theta - \sin\theta\cos^2\theta = \sin\theta(1 - \cos^2\theta)$
$= \sin\theta\sin^2\theta$
$= \sin^3\theta$

5. $\cot^2\theta\csc^2\theta - \cot^2\theta = \cot^2\theta(\csc^2\theta - 1)$
$= \cot^2\theta\cot^2\theta$
$= \cot^4\theta$

7. $\dfrac{\sec\theta}{\sin\theta} - \dfrac{\sin\theta}{\cos\theta} = \dfrac{\frac{1}{\cos\theta}}{\sin\theta} - \dfrac{\sin\theta}{\cos\theta}$
$= \dfrac{1}{\sin\theta\cos\theta} - \dfrac{\sin^2\theta}{\sin\theta\cos\theta}$
$= \dfrac{1 - \sin^2\theta}{\sin\theta\cos\theta}$
$= \dfrac{\cos^2\theta}{\sin\theta\cos\theta}$
$= \dfrac{\cos\theta}{\sin\theta}$
$= \cot\theta$

9. $\dfrac{\cos\theta}{1 + \sin\theta} + \tan\theta$
$= \dfrac{\cos\theta}{1 + \sin\theta} + \dfrac{\sin\theta}{\cos\theta}$
$= \dfrac{\cos\theta}{\cos\theta} \cdot \dfrac{\cos\theta}{1 + \sin\theta} + \dfrac{1 + \sin\theta}{1 + \sin\theta} \cdot \dfrac{\sin\theta}{\cos\theta}$
$= \dfrac{\cos^2\theta + \sin\theta + \sin^2\theta}{\cos\theta(1 + \sin\theta)}$
$= \dfrac{1 + \sin\theta}{\cos\theta(1 + \sin\theta)}$
$= \dfrac{1}{\cos\theta}$
$= \sec\theta$

11. $\dfrac{1}{1 - \tan^2\theta} + \dfrac{1}{1 - \cot^2\theta}$
$= \dfrac{1}{1 - \frac{\sin^2\theta}{\cos^2\theta}} + \dfrac{1}{1 - \frac{\cos^2\theta}{\sin^2\theta}}$
$= \dfrac{1}{\frac{\cos^2\theta}{\cos^2\theta} - \frac{\sin^2\theta}{\cos^2\theta}} + \dfrac{1}{\frac{\sin^2\theta}{\sin^2\theta} - \frac{\cos^2\theta}{\sin^2\theta}}$
$= \dfrac{1}{\frac{\cos^2\theta - \sin^2\theta}{\cos^2\theta}} + \dfrac{1}{\frac{\sin^2\theta - \cos^2\theta}{\sin^2\theta}}$
$= \dfrac{\cos^2\theta}{\cos^2\theta - \sin^2\theta} + \dfrac{\sin^2\theta}{\sin^2\theta - \cos^2\theta}$
$= \dfrac{\cos^2\theta}{\cos^2\theta - \sin^2\theta} + \dfrac{-1}{-1} \cdot \dfrac{\sin^2\theta}{\sin^2\theta - \cos^2\theta}$
$= \dfrac{\cos^2\theta}{\cos^2\theta - \sin^2\theta} + \dfrac{-\sin^2\theta}{\cos^2\theta - \sin^2\theta}$

$= \dfrac{\cos^2\theta - \sin^2\theta}{\cos^2\theta - \sin^2\theta}$
$= 1$

13. $(\csc\theta - \cot\theta)(\csc\theta + \cot\theta) = \csc^2\theta - \cot^2\theta = 1$

15. $\dfrac{1}{1 - \sin\theta} + \dfrac{1}{1 + \sin\theta}$
$= \dfrac{1 + \sin\theta}{1 + \sin\theta} \cdot \dfrac{1}{1 - \sin\theta} + \dfrac{1 - \sin\theta}{1 - \sin\theta} \cdot \dfrac{1}{1 + \sin\theta}$
$= \dfrac{1 + \sin\theta}{1 - \sin^2\theta} + \dfrac{1 - \sin\theta}{1 - \sin^2\theta}$
$= \dfrac{2}{1 - \sin^2\theta}$
$= \dfrac{2}{\cos^2\theta}$
$= 2\sec^2\theta$

17. $\csc^4\theta - \cot^4\theta$
$= (\csc^2\theta - \cot^2\theta)(\csc^2\theta + \cot^2\theta)$
$= [\csc^2\theta - (\csc^2\theta - 1)][\csc^2\theta + (\csc^2\theta - 1)]$
$= 2\csc^2\theta - 1$
$= 2(\cot^2\theta + 1) - 1$
$= 2\cot^2\theta + 2 - 1$
$= 2\cot^2\theta + 1$

19 a. $\dfrac{v^2\tan^2\theta}{2g\sec^2\theta} = \dfrac{v^2\left(\frac{\sin^2\theta}{\cos^2\theta}\right)}{2g\left(\frac{1}{\cos^2\theta}\right)}$
$= \dfrac{v^2\sin^2\theta}{2g}$

b. $h = \dfrac{v^2\sin^2\theta}{2g}$
$= \dfrac{(110)^2(\sin 80)^2}{2(9.8)} \approx 598.7 \text{ m}$

21. $\tan^2\theta - \sin^2\theta = \dfrac{\sin^2\theta}{\cos^2\theta} - \sin^2\theta$
$= \dfrac{\sin^2\theta}{\cos^2\theta} - \sin^2\theta\left(\dfrac{\cos^2\theta}{\cos^2\theta}\right)$
$= \dfrac{\sin^2\theta}{\cos^2\theta} - \dfrac{\sin^2\theta\cos^2\theta}{\cos^2\theta}$
$= \dfrac{\sin^2\theta - \sin^2\theta\cos^2\theta}{\cos^2\theta}$
$= \dfrac{\sin^2\theta(1 - \cos^2\theta)}{\cos^2\theta}$
$= \dfrac{\sin^2\theta\sin^2\theta}{\cos^2\theta}$
$= \sin^2\theta\left(\dfrac{\sin^2\theta}{\cos^2\theta}\right)$
$= \sin^2\theta\tan^2\theta$

23. $\dfrac{1 + \csc\theta}{\sec\theta} = \dfrac{1 + \frac{1}{\sin\theta}}{\frac{1}{\cos\theta}}$
$= \dfrac{\frac{\sin\theta}{\sin\theta} + \frac{1}{\sin\theta}}{\frac{1}{\cos\theta}}$
$= \dfrac{\sin\theta + 1}{\sin\theta} \cdot \dfrac{\cos\theta}{1}$
$= \dfrac{\sin\theta\cos\theta + \cos\theta}{\sin\theta}$
$= \dfrac{\sin\theta\cos\theta}{\sin\theta} + \dfrac{\cos\theta}{\sin\theta}$
$= \cos\theta + \cot\theta$

25. $\dfrac{1 + \tan^2 \theta}{1 - \tan^2 \theta} = \dfrac{1 + \dfrac{\sin^2 \theta}{\cos^2 \theta}}{1 - \dfrac{\sin^2 \theta}{\cos^2 \theta}}$

$= \dfrac{\dfrac{\cos^2 \theta}{\cos^2 \theta} + \dfrac{\sin^2 \theta}{\cos^2 \theta}}{\dfrac{\cos^2 \theta}{\cos^2 \theta} - \dfrac{\sin^2 \theta}{\cos^2 \theta}}$

$= \dfrac{\dfrac{\cos^2 \theta + \sin^2 \theta}{\cos^2 \theta}}{\dfrac{\cos^2 \theta - \sin^2 \theta}{\cos^2 \theta}}$

$= \dfrac{\cos^2 \theta + \sin^2 \theta}{\cos^2 \theta} \cdot \dfrac{\cos^2 \theta}{\cos^2 \theta - \sin^2 \theta}$

$= \dfrac{\cos^2 \theta + \sin^2 \theta}{\cos^2 \theta - \sin^2 \theta}$

$= \dfrac{1}{\cos^2 \theta - \sin^2 \theta}$

$= \dfrac{1}{\cos^2 \theta - (1 - \cos^2 \theta)}$

$= \dfrac{1}{2\cos^2 \theta - 1}$

27. $\sec \theta - \cos \theta = \dfrac{1}{\cos \theta} - \cos \theta$

$= \dfrac{1}{\cos \theta} - \dfrac{\cos^2 \theta}{\cos \theta}$

$= \dfrac{1 - \cos^2 \theta}{\cos \theta}$

$= \dfrac{\sin^2 \theta}{\cos \theta}$

$= \tan \theta \sin \theta$

29. $(\csc \theta - \cot \theta)^2 = \left(\dfrac{1}{\sin \theta} - \dfrac{\cos \theta}{\sin \theta}\right)^2$

$= \left(\dfrac{1 - \cos \theta}{\sin \theta}\right)^2$

$= \dfrac{(1 - \cos \theta)^2}{\sin^2 \theta}$

$= \dfrac{(1 - \cos \theta)^2}{1 - \cos^2 \theta}$

$= \dfrac{(1 - \cos \theta)(1 - \cos \theta)}{(1 - \cos \theta)(1 + \cos \theta)}$

$= \dfrac{1 - \cos \theta}{1 + \cos \theta}$

31. $\dfrac{2 + \csc \theta \sec \theta}{\csc \theta \sec \theta} = \dfrac{2}{\csc \theta \sec \theta} + 1$

$= 2 \cdot \dfrac{1}{\csc \theta} \cdot \dfrac{1}{\sec \theta} + 1$

$= 2 \sin \theta \cos \theta + (\sin^2 + \cos^2 \theta)$

$= \sin^2 \theta + 2\sin \theta \cos \theta + \cos^2 \theta$

$= (\sin \theta + \cos \theta)^2$

33. $I_m - \dfrac{I_m}{\cot^2 \theta + 1} = I_m\left(1 - \dfrac{1}{\cot^2 \theta + 1}\right)$

$= I_m\left(1 - \dfrac{1}{\csc^2 \theta}\right)$

$= I_m(1 - \sin^2 \theta)$

$= I_m \cos^2 \theta$

35.

$[-2\pi, 2\pi]$ scl: $\dfrac{\pi}{2}$ by $[-4, 4]$ scl: 1

$[-2\pi, 2\pi]$ scl: $\dfrac{\pi}{2}$ by $[-4, 4]$ scl: 1

$\dfrac{1}{\sec x - \tan x} = \dfrac{1}{\dfrac{1}{\cos x} - \dfrac{\sin x}{\cos x}}$

$= \dfrac{1}{\dfrac{1 - \sin x}{\cos x}}$

$= \dfrac{\cos x}{1 - \sin x} \cdot \dfrac{1 + \sin x}{1 + \sin x}$

$= \dfrac{\cos x + \sin x \cos x}{1 - \sin^2 x}$

$= \dfrac{\cos x + \sin x \cos x}{\cos^2 x}$

$= \dfrac{1}{\cos x} + \dfrac{\sin x}{\cos x}$

$= \sec x + \tan x$

37.

 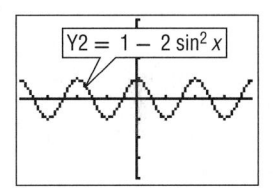

$[-2\pi, 2\pi]$ scl: $\dfrac{\pi}{2}$ by $[-4, 4]$ scl: 1 $[-2\pi, 2\pi]$ scl: $\dfrac{\pi}{2}$ by $[-4, 4]$ scl: 1

$\dfrac{\cot^2 x - 1}{1 + \cot^2 x} = \dfrac{\cot^2 x - 1}{\csc^2 x}$

$= \dfrac{\dfrac{\cos^2 x}{\sin^2 x} - 1}{\dfrac{1}{\sin^2 x}}$

$= \cos^2 x - \sin^2 x$

$= (1 - \sin^2 x) - \sin^2 x$

$= 1 - 2 \sin^2 x$

39.

$[-2\pi, 2\pi]$ scl: $\dfrac{\pi}{2}$ by $[-4, 4]$ scl: 1

$[-2\pi, 2\pi]$ scl: $\dfrac{\pi}{2}$ by $[-4, 4]$ scl: 1

$\dfrac{\cot x - \tan x}{\tan x + \cot x} = \dfrac{\dfrac{\cos x}{\sin x} - \dfrac{\sin x}{\cos x}}{\dfrac{\sin x}{\cos x} + \dfrac{\cos x}{\sin x}}$

$= \dfrac{\dfrac{\cos^2 x}{\sin x \cos x} - \dfrac{\sin^2 x}{\sin x \cos x}}{\dfrac{\sin^2 x}{\sin x \cos x} + \dfrac{\cos^2 x}{\sin x \cos x}}$

$$= \dfrac{\dfrac{\cos^2 x - \sin^2 x}{\sin x \cos x}}{\dfrac{\sin^2 x + \cos^2 x}{\sin x \cos x}}$$

$$= \dfrac{\cos^2 x - \sin^2 x}{\sin^2 x + \cos^2 x}$$

$$= \cos^2 x - \sin^2 x$$

41. $\sqrt{\dfrac{\sec x - 1}{\sec x + 1}} = \sqrt{\dfrac{\sec x - 1}{\sec x + 1}} \cdot \sqrt{\dfrac{\sec x - 1}{\sec x - 1}}$

$$= \sqrt{\dfrac{(\sec x - 1)^2}{\sec^2 x - 1}}$$

$$= \sqrt{\dfrac{(\sec x - 1)^2}{\tan^2 x}}$$

$$= \left| \dfrac{\sec x - 1}{\tan x} \right|$$

43. $\ln |\cot x| + \ln |\tan x \cos x|$

$$= \ln \left| \dfrac{\cos x}{\sin x} \right| + \ln \left| \dfrac{\sin x}{\cos x} \cdot \cos x \right|$$

$$= \ln \left| \dfrac{\cos x}{\sin x} \cdot \sin x \right|$$

$$= \ln |\cos x|$$

45. $\sin^4 \theta - \cos^4 \theta - 1$

$$= (\sin^2 \theta + \cos^2 \theta)(\sin^2 \theta - \cos^2 \theta) - 1$$

$$= (\sin^2 \theta - \cos^2 \theta) - 1$$

$$= 1 - \cos^2 \theta - \cos^2 \theta - 1$$

$$= -2 \cos^2 \theta$$

47. $\sec^6 \theta - \tan^6 \theta$

$$= (\sec^3 \theta - \tan^3 \theta)(\sec^3 \theta + \tan^3 \theta)$$

$$= (\sec \theta - \tan \theta)(\sec^2 \theta + \sec \theta \tan \theta + \tan^2 \theta) \cdot$$
$$\quad (\sec \theta + \tan \theta)(\sec^2 \theta - \sec \theta \tan \theta + \tan^2 \theta)$$

$$= (\sec^2 \theta - \tan^2 \theta)[(1 + \tan^2 \theta) + \sec \theta \tan \theta + \tan^2 \theta] \cdot$$
$$\quad [(1 + \tan^2 \theta) - \sec \theta \tan \theta + \tan^2 \theta]$$

$$= (1 + 2 \tan^2 \theta + \sec \theta \tan \theta)(1 + 2 \tan^2 \theta - \sec \theta \tan \theta)$$

$$= (1 + 2 \tan^2 \theta)^2 - (\sec \theta \tan \theta)^2$$

$$= 1 + 4 \tan^2 \theta + 4 \tan^4 \theta - \sec^2 \theta \tan^2 \theta$$

$$= 1 + \tan^2 \theta (4 + 4 \tan^2 \theta - \sec^2 \theta)$$

$$= 1 + \tan^2 \theta [4 + 4(\sec^2 \theta - 1) - \sec^2 \theta]$$

$$= 1 + \tan^2 \theta (4 + 4 \sec^2 \theta - 4 - \sec^2 \theta)$$

$$= 1 + \tan^2 \theta (3 \sec^2 \theta)$$

$$= 1 + 3 \tan^2 \theta \sec^2 \theta$$

$$= 3 \sec^2 \theta \tan^2 \theta + 1$$

49. $\sec^2 x \csc^2 x = (\tan^2 x + 1)\csc^2 x$

$$= \left(\dfrac{\sin^2 x}{\cos^2 x} + 1 \right)\left(\dfrac{1}{\sin^2 x} \right)$$

$$= \dfrac{1}{\cos^2 x} + \dfrac{1}{\sin^2 x}$$

$$= \sec^2 x + \csc^2 x$$

51. $\cosh^2 x - \sinh^2 x$

$$= \tfrac{1}{4}(e^x + e^{-x})^2 - \tfrac{1}{4}(e^x - e^{-x})^2$$

$$= \tfrac{1}{4}[e^{2x} + 2 + e^{-2x} - (e^{2x} - 2 + e^{-2x})]$$

$$= \tfrac{1}{4}(4)$$

$$= 1$$

53. $1 - \tanh^2 x = 1 - \dfrac{\sinh^2 x}{\cosh^2 x}$

$$= \dfrac{\cosh^2 x}{\cosh^2 x} - \dfrac{\sinh^2 x}{\cosh^2 x}$$

$$= \dfrac{\cosh^2 x - \sinh^2 x}{\cosh^2 x}$$

$$= \dfrac{1}{\cosh^2 x}$$

$$= \text{sech}^2 x$$

55.

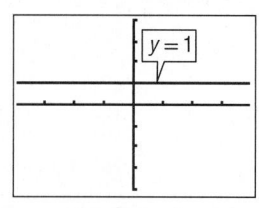

$[-2\pi, 2\pi]$ scl: $\frac{\pi}{2}$ by $[-4, 4]$ scl: $[-2\pi, 2\pi]$ scl: $\frac{\pi}{2}$ by $[-4, 4]$ scl: 1

$$\dfrac{\sec x}{\cos x} - \dfrac{\tan x \sec x}{\csc x} = \dfrac{1}{\cos x} \cdot \dfrac{1}{\cos x} - \dfrac{\dfrac{\sin x}{\cos x} \cdot \dfrac{1}{\cos x}}{\dfrac{1}{\sin x}}$$

$$= \dfrac{1}{\cos^2 x} - \dfrac{\sin^2 x}{\cos^2 x}$$

$$= \sec^2 x - \tan^2 x$$

$$= 1$$

57.

$[-2\pi, 2\pi]$ scl: $\frac{\pi}{2}$ by $[-4, 4]$ scl: 1 $[-2\pi, 2\pi]$ scl: $\frac{\pi}{2}$ by $[-4, 4]$ scl: 1

$(\tan x + \sec x)(1 - \sin x)$

$$= \tan x - \tan x \sin x + \sec x - \sec x \sin x$$

$$= \dfrac{\sin x}{\cos x} - \dfrac{\sin x}{\cos x} \cdot \sin x + \dfrac{1}{\cos x} - \dfrac{1}{\cos x} \cdot \sin x$$

$$= \dfrac{\sin x}{\cos x} - \dfrac{\sin x}{\cos x} \cdot \sin x + \dfrac{1}{\cos x} - \dfrac{1}{\cos x} \cdot \sin x$$

$$= \dfrac{\sin x}{\cos x} - \dfrac{\sin^2 x}{\cos x} + \dfrac{1}{\cos x} - \dfrac{\sin x}{\cos x}$$

$$= \dfrac{\sin^2 x}{\cos x} + \dfrac{1}{\cos x}$$

$$= \dfrac{1 - \sin^2 x}{\cos x}$$

$$= \dfrac{\cos^2 x}{\cos x}$$

$$= \cos x$$

59a. $\sin x = \dfrac{1}{2}$

59b. The graphs of $y = \sin x$ and $y = \frac{1}{2}$ intersect at $\frac{\pi}{6}$ and $\frac{5\pi}{6}$ over $[0, 2\pi)$.

59c.

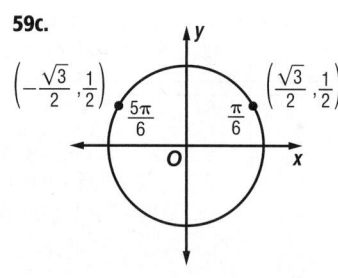

$[0, 2\pi]$ scl: $\frac{\pi}{2}$ by $[-2, 2]$ scl: 1

59d. The graphs of $y = \sin x$ and $y = \frac{1}{2}$ intersect at $-\frac{11\pi}{6}$, $-\frac{7\pi}{6}$, $\frac{\pi}{6}$, and $\frac{5\pi}{6}$ over $(-2\pi, 2\pi)$.

$[-2\pi, 2\pi]$ scl: $\frac{\pi}{2}$ by $[-2, 2]$ scl: 1

59e. Sample answer: Since sine is a periodic function, the solutions of $\sin x = \frac{1}{2}$ are $x = \frac{\pi}{6} + 2n\pi$ and $x = \frac{5\pi}{6} + 2n\pi$, where n is an integer.

61 Using the Law of Sines, $\dfrac{\sin \beta}{b} = \dfrac{\sin \alpha}{a}$, so $b = \dfrac{a \sin \beta}{\sin \alpha}$.

$$A = \dfrac{1}{2}ab \sin \gamma$$
$$A = \dfrac{1}{2}a\left(\dfrac{a \sin \beta}{\sin \alpha}\right)\sin \gamma$$
$$A = \dfrac{a^2 \sin \beta \sin \gamma}{2 \sin \alpha}$$
$$A = \dfrac{a^2 \sin \beta \sin \gamma}{2 \sin (180° - (\beta + \gamma))}$$
$$A = \dfrac{a^2 \sin \beta \sin \gamma}{2 \sin (\beta + \gamma)}$$

63. Sample answers: $\tan x \sin x + \cos x = \sec x$ and $\sin x + \cot x \cos x = \csc x$;

$$\tan x \sin x + \cos x = \dfrac{\sin x}{\cos x} \cdot \sin x + \cos x$$
$$= \dfrac{\sin^2 x}{\cos x} + \cos x$$
$$= \dfrac{1 - \cos^2 x}{\cos x} + \cos x$$
$$= \dfrac{1}{\cos x} - \cos x + \cos x$$
$$= \dfrac{1}{\cos x}$$
$$= \sec x$$

$$\sin x + \cot x \cos x = \sin x + \dfrac{\cos x}{\sin x} \cdot \cos x$$
$$= \sin x + \dfrac{\cos^2 x}{\sin x}$$
$$= \sin x + \dfrac{1 - \sin^2 x}{\sin x}$$
$$= \sin x + \dfrac{1}{\sin x} - \sin x$$
$$= \dfrac{1}{\sin x}$$
$$= \csc x$$

65. Sample answer: You could start on the left side of the identity and simplify it as much as possible. Then, you could move to the right side and simplify until it matches the left side. **67.** 1 **69.** $\cot^2 \theta$ **71.** 1

73.

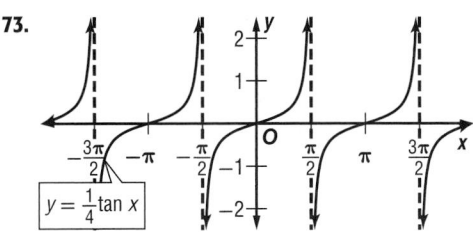

$y = \dfrac{1}{4}\tan x$

75.

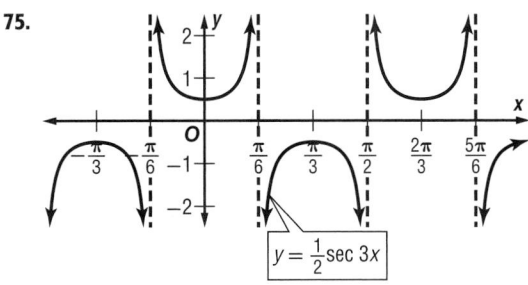

$y = \dfrac{1}{2}\sec 3x$

77. $\dfrac{19\pi}{6}$ **79.** $1305°$ **81.** $\dfrac{1620}{\pi} \approx 515.7°$ **83.** $(-7, 4)$
85. $(-\infty, -4] \cup [4, \infty)$ **87.** $(-\infty, -7] \cup [1, \infty)$ **89.** B **91.** D

---

1. $\dfrac{7\pi}{6} + 2n\pi, \dfrac{11\pi}{6} + 2n\pi, n \in \mathbb{Z}$ **3.** $\dfrac{\pi}{3} + 2n\pi, \dfrac{2\pi}{3} + 2n\pi,$
$\dfrac{4\pi}{3} + 2n\pi, \dfrac{5\pi}{3} + 2n\pi, n \in \mathbb{Z}$ **5.** $\dfrac{\pi}{6} + n\pi, \dfrac{5\pi}{6} + n\pi, n \in \mathbb{Z}$
7. $\dfrac{\pi}{4} + 2n\pi, \dfrac{3\pi}{4} + 2n\pi, n \in \mathbb{Z}$ **9.** $\dfrac{\pi}{4} + n\pi, \dfrac{3\pi}{4} + n\pi, n \in \mathbb{Z}$
11. $\dfrac{\pi}{3} + n\pi, n \in \mathbb{Z}$ **13.** $\dfrac{\pi}{2}, \dfrac{3\pi}{2}$ **15.** $\dfrac{\pi}{2}, \dfrac{3\pi}{2}$ **17.** $0, \pi$

19 **a.** $d = \dfrac{1}{32}v_0^2 \sin 2\theta$

$$40 = \dfrac{1}{32}(50)^2 \sin 2\theta$$
$$\dfrac{1280}{2500} = \sin 2\theta$$
$$2\theta = 30.8°$$
$$\theta = 15.4°$$

Sine is also positive in Quadrant II.
So, $180 - 30.8 = 149.2°$.
So if $2\theta = 149.2°$, $\theta = 74.6°$.
Hence, the possible interval is $[15.4°, 74.6°]$.

b. $50 = \dfrac{1}{32}(50)^2 \sin 2\theta$
$$\dfrac{1600}{2500} = \sin 2\theta$$
$$2\theta = 39.8 \quad\text{or}\quad 2\theta = 140.2$$
$$\theta = 19.9° \qquad\qquad \theta = 70.1°$$

21. $\dfrac{\pi}{2}, \dfrac{7\pi}{6}, \dfrac{11\pi}{6}$ **23.** $0, \dfrac{2\pi}{3}, \dfrac{4\pi}{3}$ **25.** $\dfrac{\pi}{2}, \dfrac{7\pi}{6}, \dfrac{11\pi}{6}$ **27.** $\dfrac{\pi}{2}, 0$
29. $\dfrac{\pi}{3}, \dfrac{7\pi}{6}$ **31.** $\dfrac{\pi}{4} + 2n\pi, \dfrac{5\pi}{4} + 2n\pi, n \in \mathbb{Z}$ **33.** $\dfrac{2\pi}{3}, \dfrac{4\pi}{3}$ **35.** $\dfrac{\pi}{3},$
$\dfrac{5\pi}{3}$ **37.** 0.41 **39.** $1.84, 4.49$ **41.** $\dfrac{\pi}{2}, \dfrac{3\pi}{2}$ **43.** $\dfrac{\pi}{4}, \dfrac{5\pi}{4}$ **45.** $\dfrac{\pi}{4}, \dfrac{5\pi}{4},$
$\dfrac{9\pi}{4}, \dfrac{13\pi}{4}$ **47.** $\dfrac{\pi}{4}, \dfrac{3\pi}{4}, \dfrac{5\pi}{4}, \dfrac{7\pi}{4}, \dfrac{9\pi}{4}, \dfrac{11\pi}{4}, \dfrac{13\pi}{4}, \dfrac{15\pi}{4}$ **49a.** 2.39
radians **49b.** 3.01 radians **51.** $0 \le x < \dfrac{\pi}{4}$ or $\dfrac{7\pi}{4} < x < 2\pi$

53. $0 < x \le 2\pi$ **55.** $0 \le x < \dfrac{\pi}{4}$ or $\dfrac{3\pi}{4} < x < 2\pi$ **57.** Both;
sample answer: Vijay's solutions are correct; however, they are not stated in the simplest form. For example, his solutions of $x = \dfrac{\pi}{4} + n\pi$ and $x = \dfrac{5\pi}{4} + n\pi$ could simply be stated as $x = \dfrac{\pi}{4} + n\pi$ because when $n = 1$, $\dfrac{\pi}{4} + n\pi$ is
equivalent to $\dfrac{5\pi}{4}$.

59 $4\cos^2 x - 4\sin^2 \cos^2 x + 3\sin^2 x - 3 = 0$
$$4\cos^2 x (1 - \sin^2 x) + 3(\sin^2 x - 1) = 0$$
$$4\cos^2 x (1 - \sin^2 x) - 3(1 - \sin^2 x) = 0$$
$$(1 - \sin^2 x)(4\cos^2 x - 3) = 0$$
$1 - \sin^2 x = 0 \quad\text{or}\quad 4\cos^2 x - 3 = 0$
$$\sin^2 x = 1 \qquad\qquad \cos^2 x = \dfrac{3}{4}$$
$$\sin x = \pm 1 \qquad\qquad \cos x = \pm\dfrac{\sqrt{3}}{2}$$
$$x = \dfrac{\pi}{2}, \dfrac{3\pi}{2} \qquad\qquad x = \dfrac{\pi}{6}, \dfrac{5\pi}{6}, \dfrac{7\pi}{6}, \dfrac{11\pi}{6}$$

61. Sample answer: $2\sin^2 x = \sqrt{2}\sin x$ **63.** Sample answer: When solving an equation, you use properties of equality to manipulate each side of the equation to isolate a variable. When verifying an identity, you transform an expression on one side of the identity into the expression on the other side through a series of algebraic steps.

65. $\dfrac{1 + \tan \theta}{1 + \cot \theta} = \dfrac{1 + \dfrac{\sin \theta}{\cos \theta}}{1 + \dfrac{\cos \theta}{\sin \theta}}$

$$= \dfrac{\dfrac{\sin \theta + \cos \theta}{\cos \theta}}{\dfrac{\sin \theta + \cos \theta}{\sin \theta}}$$

$$= \frac{\sin \theta + \cos \theta}{\cos \theta} \cdot \frac{\sin \theta}{\sin \theta + \cos \theta}$$

$$= \frac{\sin \theta}{\cos \theta}$$

67. $\frac{\sqrt{3}}{3}$ **69.** $\sqrt{2}$ **71.** $(f - g)(x) = 2x^2 - 11x - 1$

73. $\left(\frac{f}{g}\right)(x) = \frac{2x^2 - 5x + 3}{6x + 4}, x \neq -\frac{2}{3}$ **75.** D **77.** D

Pages 341–343   Lesson 5-4

1. $\frac{\sqrt{6} - \sqrt{2}}{4}$ **3.** $\frac{\sqrt{6} - \sqrt{2}}{4}$ **5.** $-2 + \sqrt{3}$ **7.** $-\cos nwt$

9a. $y = 31.65 \sin \left(\frac{\pi}{6}x - 2.09\right) + 42.65$ **9b.** The new function $h(x)$ represents the average of the high and low temperatures for each month. **11.** $\frac{\sqrt{3}}{3}$ **13.** 1 **15.** $\frac{1}{2}$ **17.** $\tan \theta$

19. $\sin 4y$ **21.** $\cos x$

㉓ $V_L = IwL \cos \left(wt + \frac{\pi}{2}\right)$

$$= IwL \left[\cos (wt) \cos \frac{\pi}{2} - \sin (wt) \sin \frac{\pi}{2}\right]$$

$$= IwL \left[\cos (wt) \cdot 0 - \sin (wt) \cdot 1\right]$$

$$= IwL \left[-\sin (wt)\right]$$

$$= -IwL \sin wt$$

25. $2x\left(\sqrt{1 - x^2}\right) - x\left(\sqrt{1 - 4x^2}\right)$ **27.** $\frac{(\sqrt{3} - x)\sqrt{x^2 + 1}}{2x^2 + 2}$

29. $\frac{-x^4 + x^3 + x + (x^3 + x)\sqrt{1 - x^2}}{x^4}$

31. $\frac{-x^2 + 1 + (2x - 2x^3)\sqrt{1 - x^2}}{2x^4 - 3x^2 + 1}$

33. $\sec \left(\frac{\pi}{2} - x\right) = \frac{1}{\cos \left(\frac{\pi}{2} - x\right)}$

$$= \frac{1}{\cos \frac{\pi}{2} \cos x + \sin \frac{\pi}{2} \sin x}$$

$$= \frac{1}{0(\cos x) + 1(\sin x)}$$

$$= \frac{1}{\sin x}$$

$$= \csc x$$

35. $\cos (\pi - \theta) = \cos \pi \cos \theta + \sin \pi \sin \theta$

$$= -1(\cos \theta) + 0(\sin \theta)$$

$$= -\cos \theta$$

37. $\sin (\pi - \theta) = \sin \pi \cos \theta - \cos \pi \sin \theta$

$$= 0(\cos \theta) - (-1)(\sin \theta)$$

$$= \sin \theta$$

39. $\cos (270° - \theta) = \cos 270° \cos \theta + \sin 270° \sin \theta$

$$= 0(\cos \theta) + (-1)(\sin \theta)$$

$$= -\sin \theta$$

41. $\frac{2\pi}{3}, \frac{4\pi}{3}$ **43.** $\frac{\pi}{3}, \frac{5\pi}{3}$ **45.** $\frac{\pi}{4}, \frac{5\pi}{4}$

47. $\frac{\cos (\alpha + \beta)}{\sin \alpha \sin \beta} = \frac{\cos \alpha \cos \beta - \sin \alpha \sin \beta}{\sin \alpha \cos \beta}$

$$= \frac{\cos \alpha \cos \beta}{\sin \alpha \cos \beta} - \frac{\sin \alpha \sin \beta}{\sin \alpha \cos \beta}$$

$$= \frac{\cos \alpha}{\sin \alpha} - \frac{\sin \beta}{\cos \beta}$$

$$= \cot \alpha - \tan \beta$$

49. $\sin (a + b) + \sin (a - b) = \sin a \cos b + \cos a \sin b + \sin a \cos b - \cos a \sin b = 2 \sin a \cos b$

51. The function is equivalent to $y = 1$.

$$\cos^2 \left(x + \frac{\pi}{4}\right) + \cos^2 \left(x - \frac{\pi}{4}\right)$$

$$= \left(\cos x \cos \frac{\pi}{4} - \sin x \sin \frac{\pi}{4}\right)^2 + \left(\cos x \cos \frac{\pi}{4} + \sin x \sin \frac{\pi}{4}\right)^2$$

$$= \left(\frac{\sqrt{2}}{2} \cos x - \frac{\sqrt{2}}{2} \sin x\right)^2 + \left(\frac{\sqrt{2}}{2} \cos x + \frac{\sqrt{2}}{2} \sin x\right)^2$$

$$= \frac{1}{2} \cos^2 x - \sin x \cos x + \frac{1}{2} \sin^2 x + \frac{1}{2} \cos^2 x +$$
$$\sin x \cos x + \frac{1}{2} \sin^2 x$$

$$= \cos^2 x + \sin^2 x$$

$$= 1$$

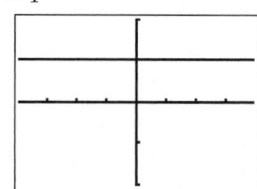

$[-2\pi, 2\pi]$ scl: $\frac{\pi}{2}$ by $[-2, 2]$ scl: 1

53. $\sin z = \sin [\pi - (x + y)]$

$$= \sin \pi \cos (x + y) - \cos \pi \sin (x + y)$$

$$= 0 \cdot \cos (x + y) - [(-1) \sin (x + y)]$$

$$= 0 - [(-1)(\sin x \cos y + \cos x \sin y)]$$

$$= \sin x \cos y + \cos x \sin y$$

55a. $\frac{\sin x \cos h + \cos x \sin h - \sin x}{h}$

55b.

  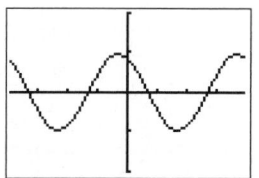

$[-2\pi, 2\pi]$ scl: $\frac{\pi}{2}$ by $[-2, 2]$ scl: 1 $[-2\pi, 2\pi]$ scl: $\frac{\pi}{2}$ by $[-2, 2]$ scl: 1

$[-2\pi, 2\pi]$ scl: $\frac{\pi}{2}$ by $[-2, 2]$ scl: 1 $[-2\pi, 2\pi]$ scl: $\frac{\pi}{2}$ by $[-2, 2]$ scl: 1

55c. $\cos x$

57. $\tan (\alpha + \beta) = \frac{\sin (\alpha + \beta)}{\cos (\alpha + \beta)}$

$$= \frac{\sin \alpha \cos \beta + \cos \alpha \sin \beta}{\cos \alpha \cos \beta - \sin \alpha \sin \beta}$$

$$= \frac{\frac{\sin \alpha \cos \beta}{\cos \alpha \cos \beta} + \frac{\cos \alpha \sin \beta}{\cos \alpha \cos \beta}}{\frac{\cos \alpha \cos \beta}{\cos \alpha \cos \beta} - \frac{\sin \alpha \sin \beta}{\cos \alpha \cos \beta}}$$

$$= \frac{\frac{\sin \alpha}{\cos \alpha} + \frac{\sin \beta}{\cos \beta}}{1 - \frac{\sin \alpha \sin \beta}{\cos \alpha \cos \beta}}$$

$$= \frac{\tan \alpha + \tan \beta}{1 - \tan \alpha \tan \beta}$$

59. $\sin (\alpha + \beta) = \cos [90° - (\alpha + \beta)]$

$$= \cos [(90° - \alpha) - \beta)]$$

$$= \cos (90° - \alpha) \cos \beta + \sin (90° - \alpha) \sin \beta$$

$$= \sin \alpha \cos \beta + \cos \alpha \sin \beta$$

61. $\sin (x + y + z)$

$$= \sin [(x + y) + z]$$

$$= \sin (x + y) \cos z + \cos (x + y) \sin z$$

$$= (\sin x \cos y + \cos x \sin y) \cos z +$$
$$(\cos x \cos y - \sin x \sin y) \sin z$$

$$= \sin x \cos y \cos z + \cos x \sin y \cos z +$$
$$\cos x \cos y \sin z - \sin x \sin y \sin z$$

63 $\sin(x - y) = \sin x \cos y - \cos x \sin y$
$$= \left(-\frac{2}{3}\right)\left(\frac{1}{3}\right) - \left(\frac{\sqrt{5}}{3}\right)\left(\frac{2\sqrt{2}}{3}\right)$$
$$= -\frac{2}{9} - \frac{2\sqrt{10}}{9}$$
$$= \frac{-2 - 2\sqrt{10}}{9}$$

65a.
$$\sin 3x \cos 2x = \cos 3x \sin 2x$$
$$\sin 3x \cos 2x - \cos 3x \sin 2x = 0$$
$$\sin(3x - 2x) = 0$$
$$\sin x = 0$$
$$x = 0 \text{ and } x = \pi$$

65b.

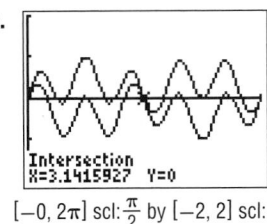

Intersection
X=3.1415927 Y=0

$[-0, 2\pi]$ scl: $\frac{\pi}{2}$ by $[-2, 2]$ scl: 1

67. $\dfrac{\cos(x + h) - \cos x}{h}$
$$= \frac{\cos x \cos h - \sin x \sin h - \cos x}{h}$$
$$= \frac{\cos x (\cos h - 1) - \sin x \sin h}{h}$$
$$= \cos x \left(\frac{\cos h - 1}{h}\right) - \sin x \left(\frac{\sin h}{h}\right)$$

69. $\approx 31.3°$

71. $\dfrac{\sec \theta}{\sin \theta} - \dfrac{\sin \theta}{\cos \theta} = \dfrac{1}{\sin \theta \cos \theta} - \dfrac{\sin \theta}{\cos \theta}$
$$= \frac{1}{\sin \theta \cos \theta} - \frac{\sin^2 \theta}{\sin \theta \cos \theta}$$
$$= \frac{\cos^2 \theta}{\sin \theta \cos \theta}$$
$$= \frac{\cos \theta}{\sin \theta}$$
$$= \cot \theta$$

73. $\frac{\pi}{3}$ **75a.** Exponential; the base, $1 + \frac{r}{n}$, is fixed, but the exponent, nt, is variable since the time t can vary.
75b. $A(t) = 1000(1.01)^{4t}$ **75c.** \$2216.72 **77.** $\pm 1, \pm 2, \pm 3, \pm 5,$ $\pm 6, \pm 10, \pm 15$ and $\pm 30; -3, 2$ **79.** D **81.** A

Pages 352–354   Lesson 5-5

1. $-\frac{24}{25}; -\frac{7}{25}; \frac{24}{7}$ **3.** $-\frac{720}{1681}; -\frac{1519}{1681}; \frac{720}{1519}$ **5.** $-\frac{4}{5}; \frac{3}{5}; -\frac{4}{3}$

7. $-\frac{24}{25}; -\frac{7}{25}; \frac{24}{7}$ **9.** $\frac{\pi}{6}, \frac{\pi}{2}, \frac{5\pi}{6}, \frac{3\pi}{2}$ **11.** $\frac{\pi}{6}, \frac{5\pi}{6}, \frac{3\pi}{2}$ **13.** $\frac{\pi}{3}, \frac{5\pi}{3}$

15 **a.**
$$d = \frac{v_0{}^2 \sin 2\theta}{32} \qquad \text{Distance equation}$$
$$242 = \frac{88^2 \sin 2\theta}{32} \qquad d = 242 \text{ and } v_0^2 = 88$$
$$\frac{(32)(242)}{88^2} = \sin 2\theta \qquad \text{Solve for } \sin 2\theta.$$
$$1 = \sin 2\theta \qquad \text{Simplify.}$$
$$90° = 2\theta \qquad \sin^{-1} 1 = 90°$$
$$45° = \theta \qquad \text{Divide each side by 2.}$$

b. $d = \dfrac{v_0{}^2 \sin 2\theta}{32}$
$$= \frac{2v_0{}^2 \sin \theta \cos \theta}{32} \qquad \sin 2\theta = 2 \sin \theta \cos \theta$$
$$= \frac{v_0{}^2 \sin \theta \cos \theta}{16} \qquad \text{Simplify.}$$

17. $\dfrac{\tan \theta - \tan \theta \cos 2\theta}{1 + \cos 2\theta}$ **19.** $\dfrac{\cos \theta + \cos \theta \cos 2\theta}{1 - \cos 2\theta}$

21. $\dfrac{\cos \theta - \cos \theta \cos 4\theta}{8}$ **23.** $\dfrac{3 - 4 \cos 2\theta + \cos 4\theta}{4 + 4 \cos 2\theta}$

25. $\frac{\pi}{6} + n\pi, \frac{5\pi}{6} + n\pi, n \in \mathbb{Z}$ **27.** $n\pi, \frac{3\pi}{4} + 2n\pi, n \in \mathbb{Z}$

29. $\dfrac{\sqrt{2 + \sqrt{2}}}{2}$ **31.** $1 - \sqrt{2}$ **33.** $0, \frac{\pi}{3}, \frac{5\pi}{3}$ **35.** 0

37. $\frac{1}{2} \cos 2\theta + \frac{1}{2} \cos 4\theta$ **39.** $\frac{1}{2} \sin 5x + \frac{1}{2} \sin x$ **41.** $\dfrac{1 + \sqrt{3}}{2}$

43. $\dfrac{\sqrt{2} - 1}{6}$ **45.** $\dfrac{\sqrt{2}}{2}$ **47.** $-\dfrac{3\sqrt{6}}{2}$ **49.** $n\pi, \frac{\pi}{2} + n\pi, n \in \mathbb{Z}$

51. $\frac{\pi}{2}n, \frac{\pi}{4} + \frac{\pi}{2}n, n \in \mathbb{Z}$ **53.** $\frac{2\pi n}{5}, \frac{\pi}{5} + \frac{2\pi n}{5}, n\pi, n \in \mathbb{Z}$

55. $\pm\cos 3x$ **57.** $\frac{1}{2}(\cos 2a + \cos 2b)$ **59.** $\frac{1}{2}[\sin(2b + \theta + \pi) +$
$\sin(\theta - \pi)]$ **61a.** $\dfrac{1 + \cos \ell + \sin \ell}{1 + \cos \ell - \sin \ell}$ **61b.** $2 + \sqrt{3}$

63. $\cos 2\theta = \cos(\theta + \theta)$
$$= \cos \theta \cos \theta - \sin \theta \sin \theta$$
$$= \cos^2 \theta - \sin^2 \theta$$

65. $\tan 2\theta = \tan(\theta + \theta)$
$$= \frac{\tan \theta + \tan \theta}{1 - \tan \theta \tan \theta}$$
$$= \frac{2 \tan \theta}{1 - \tan^2 \theta}$$

67. $\tan^2 \theta = \dfrac{1 - \cos 2\theta}{1 + \cos 2\theta}$
$$\tan \theta = \pm\sqrt{\frac{1 - \cos 2\theta}{1 + \cos 2\theta}}$$
$$\tan \frac{\theta}{2} = \pm\sqrt{\frac{1 - \cos \theta}{1 + \cos \theta}}$$

69. i. $2 \cos^2 5\theta - 1 = 2\left[\frac{1}{2}(1 + \cos 10\theta)\right] - 1$
$$= 1 + \cos 10\theta - 1$$
$$= \cos 10\theta$$

ii. $2 \cos^2 5\theta - 1 = 2 \cos 5\theta \cos 5\theta - 1$
$$= 2\left\{\frac{1}{2}[\cos(5\theta - 5\theta) + \cos(5\theta + 5\theta)]\right\} - 1$$
$$= (\cos 0 + \cos 10\theta) - 1$$
$$= 1 + \cos 10\theta - 1$$
$$= \cos 10\theta$$

71. $\frac{1}{16}[5 - 7 \cos 2\theta + 3 \cos 4\theta - \cos 2\theta \cos 4\theta]$

73. $\frac{1}{16}(5 \cos \theta + 7 \cos \theta \cos 2\theta + 3 \cos \theta \cos 4\theta +$
$\cos \theta \cos 2\theta \cos 4\theta)$

75a.

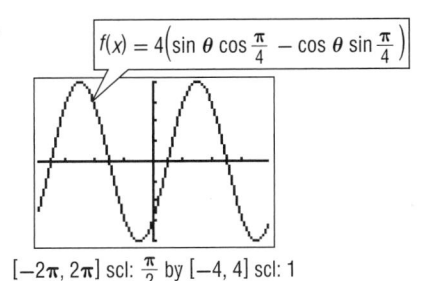

$f(x) = 4\left(\sin \theta \cos \frac{\pi}{4} - \cos \theta \sin \frac{\pi}{4}\right)$

$[-2\pi, 2\pi]$ scl: $\frac{\pi}{2}$ by $[-4, 4]$ scl: 1

75b. $4 \sin \left(\theta - \frac{\pi}{4}\right);$

$4 \sin \left(\theta - \frac{\pi}{4}\right) = 4 \left(\sin \theta \cos \frac{\pi}{4} - \cos \theta \sin \frac{\pi}{4}\right)$, difference of two angles identity

75c.

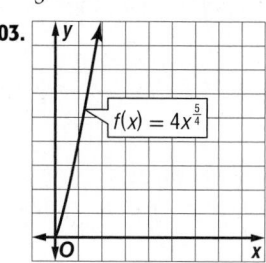

$g(x) = \cos^2 \left(\theta - \frac{\pi}{3}\right) - \sin^2 \left(\theta - \frac{\pi}{3}\right)$

$[-2\pi, 2\pi]$ scl: $\frac{\pi}{2}$ by $[-2, 2]$ scl: 1

75d. $\cos \left(2\theta - \frac{2\pi}{3}\right);$

$\cos \left(2\theta - \frac{2\pi}{3}\right) = \cos \left[2 \left(\theta - \frac{\pi}{3}\right)\right]$

$= \cos^2 \left(\theta - \frac{\pi}{3}\right) - \sin^2 \left(\theta - \frac{\pi}{3}\right)$

77. I, II, or between I and II; I   **79.** I, II, or between I and II; II

81 $\cos 4\theta = \cos 2(2\theta)$

$= 1 - 2 \sin^2 2\theta$

$= 1 - 2 (\sin 2\theta)(\sin 2\theta)$

$= 1 - 2 (2 \sin \theta \cos \theta)(2 \sin \theta \cos \theta)$

$= 1 - 8 \sin^2 \theta \cos^2 \theta$

83. $\tan^2 \theta = \frac{\sin^2 \theta}{\cos^2 \theta}$

$= \frac{\frac{1 - \cos 2\theta}{2}}{\frac{1 + \cos 2\theta}{2}}$

$= \frac{1 - \cos 2\theta}{1 + \cos 2\theta}$

85. $\frac{1}{2}[\sin (\alpha + \beta) + \sin (\alpha - \beta)]$

$= \frac{1}{2}(\sin \alpha \cos \beta + \cos \alpha \sin \beta + \sin \alpha \cos \beta - \cos \alpha \sin \beta)$

$= \frac{1}{2}(2 \sin \alpha \cos \beta)$

$= \sin \alpha \cos \beta$

87. Let $x = \frac{\alpha + \beta}{2}$ and let $y = \frac{\alpha - \beta}{2}$.

$2 \cos \left(\frac{\alpha + \beta}{2}\right) \cos \left(\frac{\alpha - \beta}{2}\right)$

$= 2 \cos x \cos y$

$= 2 \left\{\frac{1}{2}[\cos (x + y) + \cos (x - y)]\right\}$

$= \cos (x + y) + \cos (x - y)$

$= \cos \left(\frac{\alpha + \beta}{2} + \frac{\alpha - \beta}{2}\right) + \cos \left(\frac{\alpha + \beta}{2} - \frac{\alpha - \beta}{2}\right)$

$= \cos \alpha + \cos \beta$

89. Let $x = \frac{\alpha + \beta}{2}$ and let $y = \frac{\alpha - \beta}{2}$.

$-2 \sin \left(\frac{\alpha + \beta}{2}\right) \sin \left(\frac{\alpha - \beta}{2}\right)$

$= -2 \sin x \sin y$

$= -2 \left\{\frac{1}{2}[\cos (x - y) - \cos (x + y)]\right\}$

$= -\cos (x - y) + \cos (x + y)$

$= -\cos \left(\frac{\alpha + \beta}{2} - \frac{\alpha - \beta}{2}\right) + \cos \left(\frac{\alpha + \beta}{2} + \frac{\alpha - \beta}{2}\right)$

$= -\cos \beta + \cos \alpha$

$= \cos \alpha - \cos \beta$

91. $\frac{\sqrt{6} + \sqrt{2}}{4}$ **93.** $\frac{1}{2}$ **95.** $-\frac{\sqrt{3}}{2}$ **97.** the 104th day of the year, or around April 14

99. $\frac{\sqrt{3}}{3}$ **101.** -1

103.

$f(x) = 4x^{\frac{5}{4}}$

$D = [0, \infty), R = [0, \infty);$
intercept: $(0, 0)$; $\lim\limits_{x \to \infty} f(x) = \infty;$
continuous on $[0, \infty);$
increasing on $(0, \infty)$

105.

$f(x) = 4x^5$

$D = (-\infty, \infty), R = (-\infty, \infty);$
intercept: $(0, 0)$; $\lim\limits_{x \to -\infty} f(x) = -\infty$ and $\lim\limits_{x \to \infty} f(x) = \infty;$
continuous for all real numbers;
increasing on $(-\infty, \infty)$

107. J

Pages 355–358   Chapter 5   Study Guide and Review

1. $\frac{1}{\cos \theta}$; reciprocal identity   **3.** $\tan^2 \theta$; pythagorean identity
5. $-\tan \theta$; odd-even identity   **7.** $\cos 2\alpha$; double-angle identity   **9.** $\sin^2 \theta$; power-reducing identity

11. $\sec \theta = \sqrt{10}, \cos \theta = \frac{1}{\sqrt{10}}$ or $\frac{\sqrt{10}}{10}$ **13.** $\csc \theta = -\frac{5}{4},$
$\tan \theta = -\frac{4}{3}$ **15.** $\sec \theta = \frac{5}{2\sqrt{5}}$ or $\frac{\sqrt{5}}{2}, \sin \theta = -\frac{1}{\sqrt{5}}$ or
$-\frac{\sqrt{5}}{5}$ **17.** 1   **19.** $\sec x$ **21.** $\sec^2 x + \tan x \sec x$

23. $\frac{\sin \theta}{1 - \cos \theta} + \frac{\sin \theta}{1 + \cos \theta}$

$= \frac{\sin \theta (1 + \cos \theta)}{1 - \cos^2 \theta} + \frac{\sin \theta (1 - \cos \theta)}{1 - \cos^2 \theta}$

$= \frac{\sin \theta (1 + \cos \theta)}{\sin^2 \theta} + \frac{\sin \theta (1 - \cos \theta)}{\sin^2 \theta}$

$= \frac{1 + \cos \theta}{\sin \theta} + \frac{1 - \cos \theta}{\sin \theta}$

$= \frac{2}{\sin \theta}$

$= 2 \csc \theta$

25. $\frac{\cot \theta}{1 + \csc \theta} + \frac{1 + \csc \theta}{\cot \theta}$

$= \frac{\cot^2 \theta}{\cot \theta (1 + \csc \theta)} + \frac{(1 + \csc \theta)^2}{\cot \theta (1 + \csc \theta)}$

$= \frac{\cot^2 \theta + (1 + \csc \theta)^2}{\cot \theta (1 + \csc \theta)}$

$= \frac{\csc^2 \theta - 1 + 1 + 2 \csc \theta + \csc^2 \theta}{\cot \theta (1 + \csc \theta)}$

$= \frac{2 \csc^2 \theta + 2 \csc \theta}{\cot \theta (1 + \csc \theta)}$

$= \frac{2 \csc \theta (\csc \theta + 1)}{\cot \theta (1 + \csc \theta)}$

$= \frac{2}{\sin \theta} \cdot \frac{\sin \theta}{\cos \theta}$

$= \frac{2}{\cos \theta}$

$= 2 \sec \theta$

27. $\dfrac{\cot^2\theta}{1+\csc\theta} = \dfrac{\cot^2\theta\,(1-\csc\theta)}{(1+\csc\theta)(1-\csc\theta)}$

$\qquad = \dfrac{\cot^2\theta\,(1-\csc\theta)}{1-\csc^2\theta}$

$\qquad = \dfrac{\cot^2\theta\,(1-\csc\theta)}{-\cot^2\theta}$

$\qquad = -(1-\csc\theta)$

$\qquad = \csc\theta - 1$

29. $\dfrac{\sec\theta+\csc\theta}{1+\tan\theta} = \dfrac{\dfrac{1}{\cos\theta}+\dfrac{1}{\sin\theta}}{\dfrac{\cos\theta}{\cos\theta}+\dfrac{\sin\theta}{\cos\theta}}$

$\qquad = \dfrac{\dfrac{\sin\theta+\cos\theta}{\sin\theta\cos\theta}}{\dfrac{\sin\theta+\cos\theta}{\cos\theta}}$

$\qquad = \dfrac{\sin\theta+\cos\theta}{\sin\theta\cos\theta}\cdot\dfrac{\cos\theta}{\sin\theta+\cos\theta}$

$\qquad = \dfrac{1}{\sin\theta}$

$\qquad = \csc\theta$

31. $\dfrac{\sin\theta}{\sin\theta+\cos\theta} = \dfrac{\dfrac{\sin\theta}{\cos\theta}}{\dfrac{\sin\theta}{\cos\theta}+\dfrac{\cos\theta}{\cos\theta}}$

$\qquad = \dfrac{\tan\theta}{1+\tan\theta}$

33. $\dfrac{\pi}{4},\dfrac{3\pi}{4}$ **35.** $\dfrac{\pi}{3},\dfrac{2\pi}{3},\dfrac{4\pi}{3},\dfrac{5\pi}{3}$ **37.** $0,\pi,\dfrac{\pi}{6},\dfrac{5\pi}{6}$

39. $n\pi,\dfrac{\pi}{2}+2n\pi,n\in\mathbb{Z}$ **41.** $\dfrac{2\pi}{3}+2n\pi,\dfrac{4\pi}{3}+2n\pi,n\in\mathbb{Z}$

43. $2n\pi,\dfrac{\pi}{2}+n\pi,n\in\mathbb{Z}$ **45.** $\dfrac{\sqrt6+\sqrt2}{4}$ **47.** $2-\sqrt3$

49. $\dfrac{-\sqrt6-\sqrt2}{4}$ **51.** 0 **53.** $\dfrac{\sqrt2}{2}$

55. $\cos(\theta+30°)-\sin(\theta+60°)$

$\qquad = \cos\theta\cos30° - \sin\theta\sin30° -$
$\qquad\quad (\sin\theta\cos60° + \cos\theta\sin60°)$

$\qquad = \dfrac{\sqrt3}{2}\cos\theta - \dfrac{1}{2}\sin\theta - \dfrac{1}{2}\sin\theta - \dfrac{\sqrt3}{2}\cos\theta$

$\qquad = -\sin\theta$

57. $\cos\left(\theta-\dfrac{\pi}{3}\right)+\cos\left(\theta+\dfrac{\pi}{3}\right)$

$\qquad = \cos\theta\cos\dfrac{\pi}{3} + \sin\theta\sin\dfrac{\pi}{3} + \cos\theta\cos\dfrac{\pi}{3} - \sin\theta\sin\dfrac{\pi}{3}$

$\qquad = \dfrac{1}{2}\cos\theta + \dfrac{\sqrt3}{2}\sin\theta + \dfrac{1}{2}\cos\theta - \dfrac{\sqrt3}{2}\sin\theta$

$\qquad = \cos\theta$

59. $\dfrac{4\sqrt2}{9},-\dfrac{7}{9},-\dfrac{4\sqrt2}{7}$ **61.** $-\dfrac{24}{25},-\dfrac{7}{25},\dfrac{24}{7}$ **63.** $\dfrac{\sqrt{2+\sqrt3}}{2}$

65. $\sqrt2+1$ **67.** $-\dfrac{\sqrt{2-\sqrt2}}{2}$ **69.** $\dfrac{\sqrt{145}}{145},\dfrac{12\sqrt{145}}{145}$

71. $\dfrac{\sin\alpha}{1-\cos\alpha} = \dfrac{\sin\alpha}{1-\cos\alpha}\cdot\dfrac{1+\cos\alpha}{1+\cos\alpha}$ **73.** yes

$\qquad = \dfrac{\sin\alpha\,(1+\cos\alpha)}{1-\cos^2\alpha}$

$\qquad = \dfrac{\sin\alpha\,(1+\cos\alpha)}{\sin^2\alpha}$

$\qquad = \dfrac{1+\cos\alpha}{\sin\alpha}$

Pages 360–361 **Chapter 5** **Connect to AP Calculus**

1. Sample answer: The values state that the rate of change for $f(x)$ at $\dfrac{\pi}{2}$ is 0, at 0 is 1, and at π is -1. This should be the slope

of the lines tangent to $f(x)$ at those values of x. The graph of $f(x)$ and the tangent lines verify this conclusion.

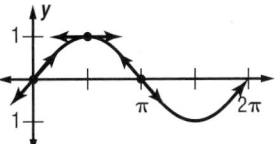

3. For $x=\dfrac{3\pi}{2}$, the rate of change is 0. For $x=2\pi$, the rate of change is 1. For $x=\dfrac{5\pi}{2}$, the rate of change is 0.

5a. $m = \dfrac{\cos(x+h)-\cos x}{h}$

5b. $m = \cos x\left(\dfrac{\cos h - 1}{h}\right) - \sin x\left(\dfrac{\sin h}{h}\right)$

5c. $\lim\limits_{h\to o}\dfrac{\cos h-1}{h}=0;\ \lim\limits_{h\to 0}\dfrac{\sin h}{h}=1$

5d. $m = \cos x\,(0) - \sin x\,(1)$ **5e.** $m = -\sin x$

5f. For $x=0$, the rate of change is 0. For $x=\dfrac{\pi}{2}$, the rate of change is -1. For $x=\pi$, the rate of change is 0. For $x=\dfrac{3\pi}{2}$, the rate of change is 1.

Chapter 6 System of Equations and Matrices

Page 363 **Chapter 6** **Get Ready**

1. $(4,1)$ **3.** $(-2,4)$ **5.** $(-3,-2)$ **7.** dog: \$15, cat: \$12

9. $\begin{bmatrix} -10 & -8 & 0 \\ 34 & 4 & -10 \end{bmatrix}$ **11.** $\begin{bmatrix} 6 & 23 & -7 \\ 13 & 3 & 23 \end{bmatrix}$ **13.** $\begin{bmatrix} -7 & 9 & -5 \\ 45 & 4 & 4 \end{bmatrix}$

15. $2x^3 - x + 4 + \dfrac{-12}{x+2}$

17. $2x^4 + 4x^3 + 7x^2 + 16x + 41 + \dfrac{88}{x-2}$

Pages 372–374 **Lesson 6-1**

1. $(-2,-7)$ **3.** $\left(\dfrac{1}{2},5\right)$ **5.** $(8,9,5)$ **7.** $\left(-6,4,\dfrac{1}{6}\right)$

9. $\left[\begin{array}{cc|c} 12 & -5 & -9 \\ -3 & 8 & 10 \end{array}\right]$ **11.** $\left[\begin{array}{ccc|c} 3 & -5 & 7 & 9 \\ -10 & 1 & 8 & 6 \\ 4 & 0 & -15 & -8 \end{array}\right]$

13. $\left[\begin{array}{cccc|c} 1 & -8 & 5 & 0 & 11 \\ 7 & 2 & -3 & 9 & -5 \\ 6 & 0 & 12 & -15 & 4 \\ 0 & 3 & 4 & -8 & -13 \end{array}\right]$

15 **a.** The total amount of money raised can be represented by $30c + 40p + 200g = 684.50$. A linear equation that relates the cost of a pie to the cost of a cake is $p - c = -2$, and a linear equation that relates the cost of a cake to a giant cookie is $c - 5g = 0$.

b. Let the first column of the augmented matrix represent c, the second column represent p, the third column represent g, and the fourth column represent the cost in dollars.

$\left[\begin{array}{ccc|c} 30 & 40 & 200 & 684.50 \\ -1 & 1 & 0 & -2 \\ 1 & 0 & -5 & 0 \end{array}\right]$

c. Perform elementary row operations to obtain a row-echelon form of the matrix. Begin by switching the third and first rows so that the first row has a 1 in the first entry.

$$\begin{bmatrix} 30 & 10 & 200 & | & 684.5 \\ -1 & 1 & 0 & | & -2 \\ 1 & 0 & -5 & | & 0 \end{bmatrix}$$

$$\text{interchange rows} \rightarrow \begin{bmatrix} 1 & 0 & -5 & | & 0 \\ -1 & 1 & 0 & | & -2 \\ 30 & 40 & 200 & | & 684.5 \end{bmatrix}$$

$$R_1 + R_2 \rightarrow \begin{bmatrix} 1 & 0 & -5 & | & 0 \\ 0 & 1 & -5 & | & -2 \\ 30 & 40 & 200 & | & 684.5 \end{bmatrix}$$

$$-30R_1 + R_3 \searrow \begin{bmatrix} 1 & 0 & -5 & | & 0 \\ 0 & 1 & -5 & | & -2 \\ 0 & 40 & 350 & | & 684.5 \end{bmatrix}$$

$$-40R_2 + R_3 \searrow \begin{bmatrix} 1 & 0 & -5 & | & 0 \\ 0 & 1 & -5 & | & -2 \\ 0 & 0 & 550 & | & 764.5 \end{bmatrix}$$

$$\frac{1}{550}R_3 \searrow \begin{bmatrix} 1 & 0 & -5 & | & 0 \\ 0 & 1 & -5 & | & -2 \\ 0 & 0 & 1 & | & 1.39 \end{bmatrix}$$

You can use substitution to find that $c = 6.95$ and $p = 4.95$. Therefore, the solution is (6.95, 4.95, 1.39). So, a cake costs $6.95, a pie costs $4.95, and a giant cookie costs $1.39.
17. yes **19.** yes **21.** yes **23.** $(-3, 1)$ **25.** $(-1, 2)$ **27.** $(10, -2, 1)$ **29.** $(52, 10, 24)$ **31.** Sample answer: There are an infinite number of solutions to the system. One of the rows in the matrix is eliminated, so the solution set is $(-2.75s + 8, 0.5s + 0.25, s)$. More information is needed to determine which of these solutions is valid.
33. $(z + 5, 9 - z, z)$ **35.** $(3z + 1, 2z + 9, z)$
37. $(5z - 3, 4z + 1, z)$ **39.** $(w - 2, 3w - 5, 2w - 4, w)$

41. $\begin{bmatrix} 1 & \frac{5}{6} & | & \frac{10}{3} \\ 0 & 1 & | & -8 \end{bmatrix}; \begin{bmatrix} 1 & 0 & | & 10 \\ 0 & 1 & | & -8 \end{bmatrix}$

43. $\begin{bmatrix} 1 & -\frac{10}{159} & \frac{43}{159} & | & \frac{148}{159} \\ 0 & 1 & -0.16\ldots & | & \frac{3535}{2144} \\ 0 & 0 & 1 & | & -7 \end{bmatrix}; \begin{bmatrix} 1 & 0 & 0 & | & 3 \\ 0 & 1 & 0 & | & 2.8 \\ 0 & 0 & 1 & | & -7 \end{bmatrix}$ **45.** $R_2 + R_3$

47 The fourth row of the second matrix is different from the fourth row in the first matrix. Therefore, an operation involving the fourth row was performed. The first entry in the fourth row was -3 in the first matrix but is 0 in the second matrix. Thus, 3 was added to this entry. Since there is no 3 in the first column for any of the three other rows, one of the rows must have been multiplied by a factor. Starting with the first row, if the row is multiplied by 3, the entries will become 3, 45, 6, 8, and 42. Adding this new row to the fourth row results in the fourth row for the second matrix. Thus, the row operation performed is $3R_1 + R_4$.
49a. $x + y = 10; 0.20x + 0.40y = 2.5$ **49b.** (7.5, 2.5); The nurses will need to mix 7.5 L of 20% and 2.5 L of 40%

saline. **51.** 3 or $\frac{27}{34}$ **53.** False; Sample answer: Because the equation in the last row is $0 = 0$, the value of one of the variables cannot be determined. Therefore, the corresponding system of equations may have infinitely many solutions. **55.** Sample answer: In Gaussian elimination, matrices are in row-echelon form. In Gauss-Jordan elimination, matrices are in reduced row-echelon form. Matrices in row-echelon form and reduced row-echelon form may have rows consisting entirely of 0s. If they do, these rows appear at the bottom of the matrices. In both forms of matrices, the first entry in any row with nonzero entries is 1. This is called a leading 1. In both forms of matrices, if there are two successive rows with nonzero entries, the leading 1 in the higher row is farther to the left than the leading 1 in the lower row. For a system with a unique solution, when the matrix is in reduced row-echelon form, every column that has a leading 1 has 0s above and/or below that 1.

57. $\tan^2\frac{x}{2} = \frac{\sin^2\frac{x}{2}}{\cos^2\frac{x}{2}}$

$$= \frac{1 - \cos\left[2\left(\frac{x}{2}\right)\right]}{1 + \cos\left[2\left(\frac{x}{2}\right)\right]}$$

$$= \frac{1 - \cos x}{1 + \cos x}$$

59. $\frac{\sqrt{2} - \sqrt{6}}{4}$ **61.** $\frac{\sqrt{2} - \sqrt{6}}{4}$ **63.** $2 - \sqrt{3}$ **65.** about 46.1 ft
67a. $y = 65{,}000(6.2)^x$ **67b.** about 22,890,495,000 **69.** 2, 9
71. B **73.** C

Pages 383–386 Lesson 6-2

1. $AB = [19 \quad -54]; BA$ is undefined. **3.** $AB = [7 \quad 15 \quad -16];$

BA is undefined. **5.** AB is undefined; $BA = \begin{bmatrix} 18 \\ -11 \end{bmatrix}$.

7. $AB = \begin{bmatrix} -9 & 6 & 12 \\ -41 & -14 & 65 \end{bmatrix}; BA$ is undefined.

9 Let a 3×3 matrix represent the number of free throws, 2-pointers, and 3-pointers that each player made, and let a 3×1 matrix represent the point value of each shot.

	FT	2-pt	3-pt
Rey	44	32	25
Chris	37	24	31
Jerry	35	39	29

FT	1
2-pt	2
3-pt	3

Perform matrix multiplication to determine the total amount of points scored by each player.

$$\begin{bmatrix} 44 & 32 & 25 \\ 37 & 24 & 31 \\ 35 & 39 & 29 \end{bmatrix} \cdot \begin{bmatrix} 1 \\ 2 \\ 3 \end{bmatrix} = \begin{bmatrix} 183 \\ 178 \\ 200 \end{bmatrix}$$

So, Rey scored 183 points, Chris scored 178 points, and Jerry scored 200 points.

11. $\begin{bmatrix} 2 & -5 & 3 \\ 4 & 1 & -6 \\ -3 & 9 & -7 \end{bmatrix} \cdot \begin{bmatrix} x_1 \\ x_2 \\ x_3 \end{bmatrix} = \begin{bmatrix} 9 \\ 35 \\ -6 \end{bmatrix}; (1, -5, -6)$

13. $\begin{bmatrix} 2 & -10 & 7 \\ 6 & -1 & 5 \\ -4 & 8 & -3 \end{bmatrix} \cdot \begin{bmatrix} x_1 \\ x_2 \\ x_3 \end{bmatrix} = \begin{bmatrix} 7 \\ -2 \\ -22 \end{bmatrix}$; $(2.5, -3, -4)$

15. $\begin{bmatrix} 2 & 6 & -5 \\ 8 & -12 & 7 \\ -4 & 10 & -1 \end{bmatrix} \cdot \begin{bmatrix} x_1 \\ x_2 \\ x_3 \end{bmatrix} = \begin{bmatrix} -20 \\ 28 \\ 7 \end{bmatrix}$; $(0.5, 1.5, 6)$

17. $\begin{bmatrix} -1 & -3 & 9 \\ -5 & 11 & 8 \\ 2 & 1 & -13 \end{bmatrix} \cdot \begin{bmatrix} x_1 \\ x_2 \\ x_3 \end{bmatrix} = \begin{bmatrix} 25 \\ 33 \\ -45 \end{bmatrix}$; $(14, 5, 6)$ **19.** yes

21. no   **23.** yes   **25.** yes   **27.** singular   **29.** $A^{-1} = \begin{bmatrix} -3 & -5 \\ 2 & 3 \end{bmatrix}$

31. $A^{-1} = \begin{bmatrix} -44 & -5 & -14 \\ 16 & 2 & 5 \\ 9 & 1 & 3 \end{bmatrix}$ **33.** singular   **35.** 3; $\begin{bmatrix} -\frac{2}{3} & \frac{5}{3} \\ -1 & 2 \end{bmatrix}$

37. 6; $\begin{bmatrix} \frac{3}{2} & \frac{7}{6} \\ -1 & -\frac{2}{3} \end{bmatrix}$ **39.** -11; $\begin{bmatrix} \frac{1}{11} & \frac{5}{11} & \frac{8}{11} \\ 0 & 1 & 2 \\ -\frac{4}{11} & \frac{13}{11} & \frac{23}{11} \end{bmatrix}$

41. 0; singular

43. -14; $\begin{bmatrix} \frac{12}{7} & 1 & -\frac{1}{14} \\ \frac{15}{7} & 1 & -\frac{3}{14} \\ -\frac{13}{7} & -1 & \frac{2}{7} \end{bmatrix}$

45 Find the determinate of the matrix formed by the three vertices.

$\begin{vmatrix} 1 & 1 & 1 \\ 4 & -1 & 1 \\ 4 & 3 & 1 \end{vmatrix} = 1\begin{vmatrix} -1 & 1 \\ 3 & 1 \end{vmatrix} - 1\begin{vmatrix} 4 & 1 \\ 4 & 1 \end{vmatrix} + 1\begin{vmatrix} 4 & -1 \\ 4 & 3 \end{vmatrix}$

$= -4 - 0 + 16$ or 12

Since the area of triangle is $\frac{1}{2}$ of the determinate of the coordinates, $A = \frac{1}{2}(12)$ or 6 units^2.

47. $13\frac{1}{2}$ units^2 **49.** $\begin{bmatrix} 2 & 8 \\ -5 & 4 \end{bmatrix}$ **51.** $x = 0.5, y = -3$ **53.** c^3

55a. $A^2 = \begin{bmatrix} 1 & 2 \\ 0 & 1 \end{bmatrix}$, $A^3 = \begin{bmatrix} 1 & 3 \\ 0 & 1 \end{bmatrix}$, $A^4 = \begin{bmatrix} 1 & 4 \\ 0 & 1 \end{bmatrix}$, $A^n = \begin{bmatrix} 1 & n \\ 0 & 1 \end{bmatrix}$

55b. $B^2 = \begin{bmatrix} 2 & 2 \\ 2 & 2 \end{bmatrix}$, $B^3 = \begin{bmatrix} 4 & 4 \\ 4 & 4 \end{bmatrix} = \begin{bmatrix} 2^2 & 2^2 \\ 2^2 & 2^2 \end{bmatrix}$, $B^4 = \begin{bmatrix} 8 & 8 \\ 8 & 8 \end{bmatrix} = $

$\begin{bmatrix} 2^3 & 2^3 \\ 2^3 & 2^3 \end{bmatrix}$, $B^5 = \begin{bmatrix} 16 & 16 \\ 16 & 16 \end{bmatrix} = \begin{bmatrix} 2^4 & 2^4 \\ 2^4 & 2^4 \end{bmatrix}$, $B^n = \begin{bmatrix} 2^{n-1} & 2^{n-1} \\ 2^{n-1} & 2^{n-1} \end{bmatrix}$

or $B^n = 2^{n-1}\begin{bmatrix} 1 & 1 \\ 1 & 1 \end{bmatrix}$ **55c.** $C^2 = \begin{bmatrix} 4 & 0 \\ 8 & 4 \end{bmatrix} = \begin{bmatrix} 2^2 & 0 \\ 2^3 & 2^2 \end{bmatrix}$,

$C^3 = \begin{bmatrix} 8 & 0 \\ 24 & 8 \end{bmatrix} = \begin{bmatrix} 2^3 & 0 \\ 3 \cdot 2^3 & 2^3 \end{bmatrix}$, $C^4 = \begin{bmatrix} 16 & 0 \\ 64 & 16 \end{bmatrix} = \begin{bmatrix} 2^4 & 0 \\ 4 \cdot 2^4 & 2^4 \end{bmatrix}$,

$C^5 = \begin{bmatrix} 32 & 0 \\ 160 & 32 \end{bmatrix} = \begin{bmatrix} 2^5 & 0 \\ 5 \cdot 2^5 & 2^5 \end{bmatrix}$, $C^n = \begin{bmatrix} 2^n & 0 \\ n \cdot 2^n & 2^n \end{bmatrix}$

55d. $\begin{bmatrix} 128 & 0 \\ 896 & 128 \end{bmatrix}$ **57.** $\begin{bmatrix} 12 & 4 \\ -4 & 0 \\ 32 & 6 \end{bmatrix}$ **59.** $\begin{bmatrix} 13 & 27 & -10 \\ 7 & 7 & 8 \\ -5 & 16 & -43 \end{bmatrix}$

61. $\begin{bmatrix} 3 & 5 \\ -2 & -1 \end{bmatrix}$ **63.** $\begin{bmatrix} 2 & 1 \\ \frac{1}{3} & -\frac{4}{3} \\ 1 & 3 \end{bmatrix}$ **65a.** 4   **65b.** 4

65c. They are the same.

65d. $\begin{vmatrix} a & b & c \\ d & e & f \\ g & h & i \end{vmatrix} = \begin{vmatrix} a & b & c & a & b \\ d & e & f & d & e \\ g & h & i & g & h \end{vmatrix}$

$= (aei + bfg + cdh) - (gec + hfa + idb)$
$= aei + bfg + cdh - gec - hfa - idb$
$= a(ei - hf) + b(fg - id) + c(dh - ge)$
$= a(ei - hf) - b(id - fg) + c(dh - ge)$
$= a(ei - hf) - b(di - gf) + c(dh - ge)$

$= a\begin{vmatrix} e & f \\ h & i \end{vmatrix} - b\begin{vmatrix} d & f \\ g & i \end{vmatrix} + c\begin{vmatrix} d & e \\ g & h \end{vmatrix}$ **65e.** No; sample answer:

If $A = \begin{bmatrix} 1 & 2 & 3 & 0 \\ 0 & 1 & 2 & 3 \\ 2 & 3 & 1 & 1 \\ 3 & 1 & 2 & 1 \end{bmatrix}$, $\det(A) = -60$.

$\begin{vmatrix} 1 & 2 & 3 & 0 \\ 0 & 1 & 2 & 3 \\ 2 & 3 & 1 & 1 \\ 3 & 1 & 2 & 1 \end{vmatrix} = \begin{vmatrix} 1 & 2 & 3 & 0 & 1 & 2 & 3 \\ 0 & 1 & 2 & 3 & 0 & 1 & 2 \\ 2 & 3 & 1 & 1 & 2 & 3 & 1 \\ 3 & 1 & 2 & 1 & 3 & 1 & 2 \end{vmatrix}$

$= 1(1)(1)(1) + 2(2)(1)(3) + 3(3)(2)(1) + 0(0)(3)(2) -$
$[3(3)(2)(0) + 1(1)(3)(1) + 2(1)(0)(2) + 1(2)(1)(3)]$
$\overset{?}{=} 1 + 12 + 18 + 0 - (0 + 3 + 0 + 6)$
$= 22$

Since $22 \neq -60$, this method does not work for a 4×4 matrix.   **67.** Suppose B and C are both inverses of square matrix A. Then by definition, $AB = BA = I$ and $AC = CA = I$. Then $B(AC) = BI = B$ and $(BA)C = IC = C$. But $B(AC) = (BA)C$ by the Associative Property of Matrix Multiplication. Therefore, $B = C$.

69 Let $B = \begin{bmatrix} r & s & t \\ u & v & w \\ x & y & z \end{bmatrix}$.

Set up three systems of equations.
$AB = A \cdot B$

$\begin{bmatrix} 14 & 6 & 33 \\ 4 & 4 & 13 \\ 1 & 18 & 12 \end{bmatrix} = \begin{bmatrix} 3 & -1 & 5 \\ 1 & 0 & 2 \\ 6 & 4 & 1 \end{bmatrix} \cdot \begin{bmatrix} r & s & t \\ u & v & w \\ x & y & z \end{bmatrix}$

$\begin{array}{lll} 3r - u + 5x = 14 & 3s - v + 5y = 6 & 3t - w + 5z = 33 \\ r + 0u + 2x = 4 & s + 0v + 2y = 4 & t + 0w + 2z = 13 \\ 6r + 4u + x = 1 & 6s + 4v + y = 18 & 6t + 4w + z = 12 \end{array}$

Solve the first system.
$r + 2x = 4$
$r = 4 - 2x$

$(4)(3r - u + 5x = 14)$
$(+)6r + 4u + x = 1$
_____
$12r - 4u + 20x = 56$
$(+)6r + 4u + x = 1$
_____
$18r + 21x = 57$

$18(4 - 2x) + 21x = 57$
$72 - 36x + 21x = 57$
$-15x = -15$
$x = 1$
$r = 4 - 2(1)$
$r = 2$

$$3r - u + 5x = 14$$
$$3(2) - u + 5(1) = 14$$
$$6 - u + 5 = 14$$
$$-u = 3$$
$$u = -3$$

Solve the next system.
$$s + 2y = 4$$
$$s = 4 - 2y$$

$$(4)(3s - v + 5y = 6)$$
$$(+)6s + 4v + y = 18$$

$$12s - 4v + 20y = 24$$
$$(+)6s + 4v + y\ \ \ = 18$$
$$\overline{18s + 21y = 42}$$

$$18s + 21y = 42$$
$$18(4 - 2y) + 21y = 42$$
$$72 - 36y + 21y = 42$$
$$-15y = -30$$
$$y = 2$$
$$s = 4 - 2(2)$$
$$s = 0$$

$$3s - v + 5y = 6$$
$$3(0) - v + 5(2) = 6$$
$$-v + 10 = 6$$
$$-v = -4$$
$$v = 4$$

Solve the last system.
$$t + 2z = 13$$
$$t = 13 - 2z$$

$$4(3t - w + 5z = 33)$$
$$(+)6t + 4w + z = 12$$

$$12t - 4w + 20z = 132$$
$$(+)6t + 4w + z\ \ \ = 12$$
$$\overline{18t + 21z = 144}$$

$$18t + 21z = 144$$
$$18(13 - 2z) + 21z = 144$$
$$234 - 36z + 21z = 144$$
$$-15z = -90$$
$$z = 6$$
$$t = 13 - 2(6)$$
$$t = 1$$

$$3t - w + 5z = 33$$
$$3(1) - w + 5(6) = 33$$
$$3 - w + 30 = 33$$
$$-w = 0$$
$$w = 0$$

$$B = \begin{bmatrix} r & s & t \\ u & v & w \\ x & y & z \end{bmatrix} = \begin{bmatrix} 2 & 0 & 1 \\ -3 & 4 & 0 \\ 1 & 2 & 6 \end{bmatrix}$$

71. Sample answer: $A = \begin{bmatrix} 2 & 2 \\ 2 & 2 \end{bmatrix}$, $B = \begin{bmatrix} 3 & 3 \\ 3 & 3 \end{bmatrix}$

73. $C(A + B) = \begin{bmatrix} c_{11} & c_{12} \\ c_{21} & c_{22} \end{bmatrix}\left(\begin{bmatrix} a_{11} & a_{12} \\ a_{21} & a_{22} \end{bmatrix} + \begin{bmatrix} b_{11} & b_{12} \\ b_{21} & b_{22} \end{bmatrix}\right)$

$$= \begin{bmatrix} c_{11} & c_{12} \\ c_{21} & c_{22} \end{bmatrix}\begin{bmatrix} a_{11} + b_{11} & a_{12} + b_{12} \\ a_{21} + b_{21} & a_{22} + b_{22} \end{bmatrix}$$

$$= \begin{bmatrix} c_{11}(a_{11} + b_{11}) + c_{12}(a_{21} + b_{21}) \\ c_{21}(a_{11} + b_{11}) + c_{22}(a_{21} + b_{21}) \end{bmatrix}$$
$$\begin{bmatrix} c_{11}(a_{12} + b_{12}) + c_{12}(a_{22} + b_{22}) \\ c_{21}(a_{12} + b_{12}) + c_{22}(a_{22} + b_{22}) \end{bmatrix}$$

$$= \begin{bmatrix} c_{11}a_{11} + c_{11}b_{11} + c_{12}a_{21} + c_{12}b_{21} \\ c_{21}a_{11} + c_{21}b_{11} + c_{22}a_{21} + c_{22}b_{21} \end{bmatrix}$$
$$\begin{bmatrix} c_{11}a_{12} + c_{11}b_{12} + c_{12}a_{22} + c_{12}b_{22} \\ c_{21}a_{12} + c_{21}b_{12} + c_{22}a_{22} + c_{22}b_{22} \end{bmatrix}$$

$$= \begin{bmatrix} c_{11}a_{11} + c_{12}a_{21} + c_{11}b_{11} + c_{12}b_{21} \\ c_{21}a_{11} + c_{22}a_{21} + c_{21}b_{11} + c_{22}b_{21} \end{bmatrix}$$
$$\begin{bmatrix} c_{11}a_{12} + c_{12}a_{22} + c_{11}b_{12} + c_{12}b_{22} \\ c_{21}a_{12} + c_{22}a_{22} + c_{21}b_{12} + c_{22}b_{22} \end{bmatrix}$$

$$= \begin{bmatrix} c_{11}a_{11} + c_{12}a_{21} & c_{11}a_{12} + c_{12}a_{22} \\ c_{21}a_{11} + a_{22}a_{21} & c_{21}a_{12} + c_{22}a_{22} \end{bmatrix} +$$
$$\begin{bmatrix} c_{11}b_{11} + c_{12}b_{21} & c_{11}b_{12} + c_{12}b_{22} \\ c_{21}b_{11} + c_{22}b_{21} & c_{21}b_{12} + c_{22}b_{22} \end{bmatrix}$$

$$= CA + CB$$

75. $c(AB) = c\left(\begin{bmatrix} a_{11} & a_{12} \\ a_{21} & a_{22} \end{bmatrix}\begin{bmatrix} b_{11} & b_{12} \\ b_{21} & b_{22} \end{bmatrix}\right)$ Subst.

$$= c\begin{bmatrix} a_{11}b_{11} + a_{12}b_{21} & a_{11}b_{12} + a_{12}b_{22} \\ a_{21}b_{11} + a_{22}b_{21} & a_{21}b_{12} + a_{22}b_{22} \end{bmatrix}$$ Def. Matrix Mult.

$$= \begin{bmatrix} c(a_{11}b_{11} + a_{12}b_{21}) & c(a_{11}b_{12} + a_{12}b_{22}) \\ c(a_{21}b_{11} + a_{22}b_{21}) & c(a_{21}b_{12} + a_{22}b_{22}) \end{bmatrix}$$ Def. Scalar Mult.

$$= \begin{bmatrix} ca_{11}b_{11} + ca_{12}b_{21} & ca_{11}b_{12} + ca_{12}b_{22} \\ ca_{21}b_{11} + ca_{22}b_{21} & ca_{21}b_{12} + ca_{22}b_{22} \end{bmatrix}$$ Dist. Prop.

$$= \begin{bmatrix} ca_{11} & ca_{12} \\ ca_{21} & ca_{22} \end{bmatrix}\begin{bmatrix} b_{11} & b_{12} \\ b_{21} & b_{22} \end{bmatrix}$$ Def. Matrix Mult.

$$= \left(c\begin{bmatrix} a_{11} & a_{12} \\ a_{21} & a_{22} \end{bmatrix}\right)\begin{bmatrix} b_{11} & b_{12} \\ b_{21} & b_{22} \end{bmatrix}$$ Def. Scalar Mult.

$$= (cA)B$$ Subst.

$$= c(AB) = c\left(\begin{bmatrix} a_{11} & a_{12} \\ a_{21} & a_{22} \end{bmatrix}\begin{bmatrix} b_{11} & b_{12} \\ b_{21} & b_{22} \end{bmatrix}\right)$$ Subst.

$$= c\begin{bmatrix} a_{11}b_{11} + a_{12}b_{21} & a_{11}b_{12} + a_{12}b_{22} \\ a_{21}b_{11} + a_{22}b_{21} & a_{21}b_{12} + a_{22}b_{22} \end{bmatrix}$$ Def. Matrix Mult.

$$= \begin{bmatrix} c(a_{11}b_{11} + a_{12}b_{21}) & c(a_{11}b_{12} + a_{12}b_{22}) \\ c(a_{21}b_{11} + a_{22}b_{21}) & c(a_{21}b_{12} + a_{22}b_{22}) \end{bmatrix}$$ Def. Scalar Mult.

$$= \begin{bmatrix} ca_{11}b_{11} + ca_{12}b_{21} & ca_{11}b_{12} + ca_{12}b_{22} \\ ca_{21}b_{11} + ca_{22}b_{21} & ca_{21}b_{12} + ca_{22}b_{22} \end{bmatrix}$$ Dist. Prop.

$$= \begin{bmatrix} a_{11}cb_{11} + a_{12}cb_{21} & a_{11}cb_{12} + a_{12}cb_{22} \\ a_{21}cb_{11} + a_{22}cb_{21} & a_{21}cb_{12} + a_{22}cb_{22} \end{bmatrix}$$ Comm. Prop.

$$= \begin{bmatrix} a_{11} & a_{12} \\ a_{21} & a_{22} \end{bmatrix}\begin{bmatrix} cb_{11} & cb_{12} \\ cb_{21} & cb_{22} \end{bmatrix}$$ Def. Matrix Mult.

$$= A(cB)$$ Subst.

77. 6×8; sample answer: Since the number of columns of matrix A must be equal to the number of rows of matrix B, matrix B must have 6 rows. The product matrix AB has the same number of columns as matrix B. So, matrix B must have 8 columns.

79. $\begin{bmatrix} 10 & -3 & \vdots & -12 \\ -6 & 4 & \vdots & 20 \end{bmatrix}$ **81.** $\begin{bmatrix} 1 & -6 & 14 & 0 & \vdots & 19 \\ 3 & 2 & -4 & 8 & \vdots & -2 \\ 9 & 0 & 18 & -12 & \vdots & 3 \\ 0 & 5 & 10 & -16 & \vdots & -26 \end{bmatrix}$

83. $-\dfrac{\pi}{18}$ **85.** $135°$ **87.** 8 **89.** 5 **91.** 9 **93.** C **95.** G

Pages 392–394 Lesson 6-3

1. $(3, 2)$ **3.** $(-1, 6)$ **5.** $(-6, 7, 8)$ **7.** no unique solution
9. 4 sitcoms, 3 talk shows, and 2 movies **11.** $(-2, -2)$
13. $\left(1, \dfrac{1}{2}\right)$ **15.** no unique solution **17.** $(6, 3, -4)$
19. 15.5 gallons

21 Let $x =$ the cost of each extra gaming minute, $y =$ the cost of each extra call minute, and $z =$ the cost of each text message. The corresponding system of equations is shown below.
$30x + 12y + 40z = 52.90$
$18x + 15y + 55z = 48.07$
$6x + 7y = 13.64$

The coefficient matrix is $\begin{bmatrix} 30 & 12 & 40 \\ 18 & 15 & 55 \\ 6 & 7 & 0 \end{bmatrix}$. Calculate the

determinant of the matrix.

$\begin{vmatrix} 30 & 12 & 40 \\ 18 & 15 & 55 \\ 6 & 7 & 0 \end{vmatrix} = 30\begin{vmatrix} 15 & 55 \\ 7 & 0 \end{vmatrix} - 12\begin{vmatrix} 18 & 55 \\ 6 & 0 \end{vmatrix} + 40\begin{vmatrix} 18 & 15 \\ 6 & 7 \end{vmatrix}$

$= 30(-385) - 12(-330) + 40(36) \text{ or } -6150$

Since the determinant does not equal 0, you can apply Cramer's Rule to find x, y, and z.

$x = \dfrac{|A_x|}{|A|} = \dfrac{\begin{vmatrix} 52.90 & 12 & 40 \\ 48.07 & 15 & 55 \\ 13.64 & 7 & 0 \end{vmatrix}}{-6150} = 0.99$

$y = \dfrac{|A_y|}{|A|} = \dfrac{\begin{vmatrix} 30 & 52.90 & 40 \\ 18 & 48.07 & 55 \\ 6 & 13.64 & 0 \end{vmatrix}}{-6150} = 1.10$

$z = \dfrac{|A_z|}{|A|} = \dfrac{\begin{vmatrix} 30 & 12 & 52.90 \\ 18 & 15 & 48.07 \\ 6 & 7 & 13.64 \end{vmatrix}}{-6150} = 0.6$

Each extra gaming minute is $0.99, each extra call minute is $1.10, and each text message is $0.25.

23. $\begin{bmatrix} 1 & -2 \\ 2 & 1 \end{bmatrix}$ **25.** $\begin{bmatrix} -1 & 2 \\ -4 & 3 \end{bmatrix}$ **27.** $(-6, 2, 5)$ **29.** $(-2, 4, 1)$

31 When the determinant of a 2×2 matrix is 0, the matrix

is singular and has no inverse. Let $A = \begin{bmatrix} n & -8 \\ 1 & 2 \end{bmatrix}$.

Find an equation for the determinant of A and solve for n when the determinant is 0.
$\begin{vmatrix} n & -8 \\ 1 & 2 \end{vmatrix} = n(2) - 1(-8) \text{ or } 2n + 8$
$2n + 8 = 0$
$2n = -8$
$n = -4$

Therefore, when $n = -4$, the system cannot be solved using an inverse matrix.

33. 0 **35.** 0.4 kg of 35% alloy, 0.8 kg of 55% alloy, and 1.3 kg of 60% alloy **37.** $AB = 32$ mm, $BC = 36$ mm, $AC = 21$ mm

39. $\begin{bmatrix} \dfrac{2x}{2-3x} & -\dfrac{3}{2-3x} \\ \dfrac{-x^2}{2-3x} & \dfrac{1}{2-3x} \end{bmatrix}$ **41.** $\begin{bmatrix} -2i & -3 \\ -1 & i \end{bmatrix}$ **43.** $X = (A + B)^{-1}(D)$

45. $X = (-C - D)(I - A)^{-1}$ **47.** $X = (B - A)^{-1}(AD)$
49. Neither; if the determinant of the coefficient matrix is 0, then there is no unique solution. The system may have no solution, or it may have an infinite number of solutions. The system of equations shown has an infinite number of solutions. **51.** Yes; sample answer:

$A^2 = \begin{bmatrix} a^2 + bc & ab + bd \\ ac + cd & bc + d^2 \end{bmatrix}$. Therefore,

$(A^2)^{-1} = \dfrac{1}{a^2 d^2 - 2abcd + b^2 c^2} \cdot \begin{bmatrix} bc + d^2 & -ab - bd \\ -ac - cd & a^2 + bc \end{bmatrix}$ and

$A^{-1} = \dfrac{1}{ad - bc} \cdot \begin{bmatrix} d & -b \\ -c & a \end{bmatrix} = \begin{bmatrix} \dfrac{d}{ad - bc} & \dfrac{-b}{ad - bc} \\ \dfrac{-c}{ad - bc} & \dfrac{a}{ad - bc} \end{bmatrix}$, so

$(A^{-1})^2 = \dfrac{1}{a^2 d^2 - 2abcd + b^2 c^2} \cdot \begin{bmatrix} bc + d^2 & -ab - bd \\ -ac - cd & a^2 + bc \end{bmatrix}$.

Thus, $(A^2)^{-1} = (A^{-1})^2$. **53a.** Sample answer: Gauss-Jordan elimination can be used to solve any system of linear equations. It is possible to perform row operations on any matrix. **53b.** Sample answer: Inverse matrices can only be used to solve systems with square coefficient matrices because matrix multiplication can only be done if the number of columns of the first matrix is equal to the number of rows of the second matrix. **53c.** Sample answer: Cramer's Rule uses determinants to solve systems, and because it is only possible to find the determinant of a square matrix, this method can only be used to solve square systems.

55. $AB = \begin{bmatrix} 54 & 26 & -76 \\ 202 & -58 & -4 \\ 136 & 67 & -66 \end{bmatrix}$, $BA = \begin{bmatrix} 38 & 2 & 112 \\ 43 & -122 & 27 \\ -84 & 38 & 14 \end{bmatrix}$

57. yes **59a.** Sirius **59b.** Sirius: 1.45, Vega: 0.58 **59c.** Vega
61. F **63.** H

Pages 402–404 Lesson 6-4

1. $\dfrac{2}{x+3} + \dfrac{-1}{x+2}$ **3.** $\dfrac{10}{x+3} + \dfrac{-9}{x+4}$ **5.** $\dfrac{3}{-2x-9} + \dfrac{1}{x+5}$
7. $3 + \dfrac{2}{x} + \dfrac{5}{x-2}$ **9.** $-2 + \dfrac{4}{x} + \dfrac{1}{x-2} + \dfrac{3}{x+4}$
11. $x + 4 + \dfrac{-8}{x+3} + \dfrac{-6}{x+5}$ **13.** $\dfrac{-3}{x} + \dfrac{4}{x+1} + \dfrac{2}{(x+1)^2}$
15. $\dfrac{-2}{x} + \dfrac{1}{x+5} + \dfrac{-7}{(x+5)^2}$ **17.** $\dfrac{4}{x} + \dfrac{-4}{x-8} + \dfrac{49}{(x-8)^2}$

19 **a.** Decompose $s(x) = \dfrac{20x + 10\pi x + 20}{\pi x^3 + \pi x^2}$. First, divide the numerator by π.

$s(x) = \dfrac{20x + 10\pi x + 20}{\pi x^3 + \pi x^2}$

$= \dfrac{\frac{20}{\pi}x + 10x + \frac{20}{\pi}}{x^3 + x^2}$

The fraction is proper, so we can rewrite the expression as partial fractions with constant numerators, A, B, and C, and denominators that are the linear factors of the original denominator. This time, however, we have repeated factors in the denominator.

$x^3 + x^2 = (x^2)(x + 1)$

Because the factor x^2 has a multiplicity of 2, include partial fractions with denominators of x, x^2, and $x + 1$.

$$\frac{\frac{20}{\pi}x + 10x + \frac{20}{\pi}}{x^3 + x^2} = \frac{A}{x} + \frac{B}{x^2} + \frac{C}{x + 1}$$

Multiply each side by the LCD, $x^3 + x^2$.

$$\frac{\frac{20}{\pi}x + 10x + \frac{20}{\pi}}{x^3 + x^2} = \frac{A}{x} + \frac{B}{x^2} + \frac{C}{x + 1}$$

$$\frac{\frac{20}{\pi}x + 10x + \frac{20}{\pi}}{x^3 + x^2} = \frac{A(x)(x + 1)}{x^3 + x^2} + \frac{B(x + 1)}{x^3 + x^2} + \frac{Cx^2}{x^3 + x^2}$$

$$\frac{20}{\pi}x + 10x + \frac{20}{\pi} = A(x)(x + 1) + B(x + 1) + Cx^2$$

$$\left(\frac{20}{\pi} + 10\right)x + \frac{20}{\pi} = Ax^2 + Ax + Bx + B + Cx^2$$

$$\left(\frac{20}{\pi} + 10\right)x + \frac{20}{\pi} = Ax^2 + Cx^2 + Ax + Bx + B$$

$$\left(\frac{20}{\pi} + 10\right)x + \frac{20}{\pi} = (A + C)x^2 + (A + B)x + B$$

Equate the coefficients on the left and right side of the equation to form a system of equations. In other words, the coefficients of the x-terms on the left side of the equation must equal the coefficients of the x-terms on the right side.

$A + C = 0$

$A + B = \frac{20}{\pi} + 10$

$B = \frac{20}{\pi}$

Use any method to solve the system.

$B = \frac{20}{\pi}$

$A + B = \frac{20}{\pi} + 10$

$A + \frac{20}{\pi} = \frac{20}{\pi} + 10$

$A = 10$

$A + C = 0$

$10 + C = 0$

$C = -10$

Replace A, B, and C with 10, $\frac{20}{\pi}$, and -10 in the partial fraction decomposition.

$$\frac{\frac{20}{\pi}x + 10x + \frac{20}{\pi}}{x^3 + x^2} = \frac{A}{x} + \frac{B}{x^2} + \frac{C}{x + 1}$$

$$= \frac{10}{x} + \frac{\pi}{x^2} + \frac{-10}{x + 1}$$

b.

[−10, 10] scl: 1 by [−100, 100] scl: 10

21. $\frac{2}{x} + \frac{x}{x^2 + 3} + \frac{-11x + 8}{(x^2 + 3)^2}$ **23.** $\frac{8x}{x^2 - 6} + \frac{7}{(x^2 - 6)^2}$

25. $\frac{4x}{x^2 - 3x + 3} + \frac{-17x + 20}{(x^2 - 3x + 3)^2}$ **27.** $\frac{3.4x + 0.6}{4x^2 + 1} + \frac{-0.6}{x + 1}$

29. $\frac{-2}{x} + \frac{3}{x - 2} + \frac{4}{x + 2}$ **31.** $\frac{1}{x^2 + 3} + \frac{x + 2}{(x^2 + 3)^2}$ **33.** $2x + \frac{-17}{x + 5}$

$+ \frac{4}{x + 1}$ **35.** $\frac{6}{x}, \frac{-6}{x + 1}, \frac{-7}{(x + 1)^2}$ **37.** $A = \frac{2}{3}t, B = r - \frac{2}{3}t - 12$

39. $A = -t, B = 2, C = t + 3, D = 5r - 2$

41. Rewrite the expression as partial fractions with constant numerators A, B, C, and D and denominators that are linear factors of the original denominator.

$$\frac{x^3 + 4}{(x^2 - 1)(x^2 + 3x + 2)} = \frac{A}{(x - 1)} + \frac{B}{(x + 1)} + \frac{C}{(x + 1)^2}$$

$$+ \frac{D}{(x + 2)}$$

$x^3 + 4 = A(x + 1)^2(x + 2) + B(x + 1)(x + 2)(x - 1) + C(x - 1)(x + 2) + D(x + 1)^2(x - 1)$

$x^3 + 4 = A(x^3 + 4x^2 + 5x + 2) + B(x^3 + 2x^2 - x - 2) + C(x^2 + x - 2) + D(x^3 + x^2 - x - 1)$

$x^3 + 4 = x^3(A + B + D) + x^2(4A + 2B + C + D) + x(5A - B + C - D) + 2A - 2B - 2C - D$

Equate the coefficients on the left and right sides of the equation.

$A + B + D = 1$

$4A + 2B + C + D = 0$

$5A - B + C - D = 0$

$2A - 2B - 2C - D = 0$

To solve the system, you can write it in matrix form $EX = F$ and solve for X.

$$\begin{array}{ccc} E & X & F \end{array}$$

$$\begin{bmatrix} 1 & 1 & 0 & 1 \\ 4 & 2 & 1 & 1 \\ 5 & -1 & 1 & -1 \\ 2 & -2 & -2 & -1 \end{bmatrix} \cdot \begin{bmatrix} A \\ B \\ C \\ D \end{bmatrix} = \begin{bmatrix} 1 \\ 0 \\ 0 \\ 4 \end{bmatrix}$$

You can use a graphing calculator to find $X = E^{-1}F$.

So, $A = \frac{5}{12}$, $B = -\frac{3}{4}$, $C = -\frac{3}{2}$, and $D = \frac{4}{3}$. Therefore,

$$\frac{x^3 + 4}{(x^2 - 1)(x^2 + 3x + 2)} = \frac{-\frac{3}{4}}{x + 1} + \frac{-\frac{3}{2}}{(x + 1)^2} + \frac{\frac{5}{12}}{x - 1} + \frac{\frac{4}{3}}{x + 2}$$

43. $7x + 2 + \frac{1}{x} + \frac{8}{x - 1} + \frac{2}{(x - 1)^2} + \frac{-4}{x + 2} + \frac{-4x + 4}{x^2 + 1}$

45. I   **47.** 0; The partial fraction decomposition of $f(x)$ is $\frac{1}{x + 1} + \frac{-3}{(x + 1)^2} + \frac{3}{(x + 1)^3}$. As x approaches infinity, the values of the three terms approach 0. So, the limit of the expression is 0.   **49.** b   **51.** d

53. True; $\frac{-5}{x} + \frac{4 + x}{x^2 - 3} + \frac{x^2 + 1}{(x^2 - 3)^2}$

$$= \frac{-5(x^2 - 3)^2 + (4 + x)(x)(x^2 - 3) + x(x^2 + 1)}{x(x^2 - 3)^2}$$

$$= \frac{-5x^4 + 30x^2 - 45 + 4x^3 - 12x + x^4 - 3x^2 + x^3 + x}{x(x^2 - 3)^2}$$

$$= \frac{-4x^4 + 5x^3 + 27x^2 - 11x - 45}{x(x^2 - 3)^2}$$

55. Sample answer: First, factor the denominator, if necessary, and rewrite the equation with constant numerators, A, B, C, ..., and denominators that are the linear or prime quadratic factors of the original denominator. Write the original expression equal to the new expression and multiply each side by the LCD. Use the Distributive Property to group like terms. Equate the coefficients on both sides of the equation to get a system of two equations. Then solve the system by writing it in matrix form $CX = D$ and solving for X. Finally, use substitution to find the partial fraction decomposition.   **57.** (2, −1, 3)   **59a.** $24,900   **59b.** $26,360   **59c.** She made $1460.   **61.** $\tan^2 \theta$ **63a.** $y = 0.25 \sin 128 \pi t$, $y = 0.25 \sin 512 \pi t$, $y = 0.25 \sin 1024 \pi t$

63b. Sample answer: As the frequency increases, the period decreases. **65.** F **67.** H

Pages 410–412 Lesson 6-5

1. 31 at (10, 1); 1 at (0, 1) **3.** 5 at (2, −3); −6 at (−2, 4)
5. 11 at (5, 2); −5 at (1, 4) **7.** 0 at (4, 1); −12 at (4, 4)

9 **a.** The objective function is $f(x, y) = 125x + 55y$, where x is the number of physicals and y is the number of checkups. The constraints are $0 \le x \le 6$, $y \ge 0$, and $20y + 40x \le 420$ (7 hours = 420 minutes).

b.

Physicals

c. The polygonal region of feasible solutions has four vertices located at (0, 21), (6, 9), (6, 0), and (0, 0). Find the value of the objective function at each of the vertices.

$f(0, 0) = 0$ $f(6, 9) = 1245$
$f(6, 0) = 750$ $f(0, 21) = 1155$

So, the maximum income is $1245 when Olivia schedules 6 checkups and 9 physicals.

11a. $f(w, a) = 600w + 700a$, $w \ge 0$, $a \ge 0$, $10w + 15a \le 70$, $4.5w + 9a \le 36$

11b.

Web Sites

11c. The company should create 7 Web sites and no E-albums for a maximum profit of $4200.

13. The problem has multiple maxima of 15 from (1, 2) to (3, 1), or along the line $y = -0.5x + 2.5$ in [−1, 3]. There is a minimum of 0 at (0, 0). **15.** The problem has multiple minima of −8 from (0, 2) to (2, 5), or along the line $y = 1.5x + 2$ in [0, 2]. There is a maximum of 18 at (3, 0).
17. The problem has multiple maxima of 40 from (0, 4) to (5, 0), or along the line $y = -0.8x + 4$ in [0, 5]. There is a minimum of 0 at (0, 0). **19.** Sample answer: 1 **21.** Sample answer: −3 **23.** Sample answer: 5 **25.** Sample answer: $f(x, y) = 4x - y$ **27.** Sample answer: $f(x, y) = -2x + 5y$
29a. $f(x, y) = 16x + 12y$, $3x + 2y \le 40$, $x + y \le 15$, $x \ge 0$, $y \ge 0$
29b. 10 batches of Mango Sunrise and 5 batches of Oasis Dream; $220 **31.** 48 at (1, 10); −24 at (−3, −6) **33.** 40 units^2
35. 46.5 units^2 **37.** No; sample answer: If a and b are negative, then the maximum value of P, if it exists, will be negative. **39a.** Sample answer: A bakery makes up to 100 apple and lemon meringue pies each week. Due to labor costs, the number of lemon meringue pies that are made needs to be less than or equal to 25 plus two times the number of apple pies. The customer demand for lemon meringue pies is at least twice that of apple pies. The bakery makes a profit of $8 per apple pie and $10 per lemon meringue pie. How many of each type of pie should be made to maximize profit? **39b.** The constraints are $y \le 25 + 2x$, $y \ge 2x$, $x + y \le 100$, $x \ge 0$, and $y \ge 0$. The objective function is given by $f(x, y) = 8x + 10y$.

39c.

Apple Pies

The graph of the system of inequalities forms a polygonal region with vertices at (0, 0), (0, 25), (25, 75), and (33.3, 66.7). **39d.** Since f is greatest at (25, 75), the bakery should make 25 apple pies and 75 lemon meringue pies to earn a maximum profit of $950. **41.** $\dfrac{5}{y - 1} + \dfrac{3}{y + 2}$
43. $\dfrac{3.5}{m + 2} + \dfrac{1.5}{m - 2}$ **45.** race car, 6.5 points; snowboard, 5.25 points
47. $\tan \theta \sin \theta \cos \theta \csc^2 \theta$

$= \dfrac{\sin \theta}{\cos \theta} \cdot \sin \theta \cdot \cos \theta \cdot \dfrac{1}{\sin^2 \theta}$

$= 1$
49. 19.5 yd^2 **51.** 6.4 cm^2 **53.** G

Pages 413–416 Chapter 6 Study Guide and Review

1. augmented matrix **3.** reduced row-echelon **5.** inverse matrix **7.** determinant **9.** constraints **11.** (1, 1)
13. (2, 3, −1) **15.** (33, 15, −10) **17.** (1, 3) **19.** (2, 2, 3)
21. (1, 0, 5) **23.** AB is undefined; $BA = \begin{bmatrix} -13 & 15 & -39 \\ 15 & -3 & 14 \end{bmatrix}$.
25. $AB = [-11 \ 23]$; BA is undefined. **27.** $\begin{bmatrix} \dfrac{8}{19} & -\dfrac{1}{19} \\ \dfrac{3}{19} & \dfrac{2}{19} \end{bmatrix}$
29. singular **31.** (−4, 5) **33.** (0, 1, −1) **35.** (2, −5, 8)
37. (9, −3) **39.** (3, −3, −4) **41.** (0, −2, 7)
43. $\dfrac{1}{x - 2} + \dfrac{1}{x + 2}$ **45.** $\dfrac{1}{x - 5} - \dfrac{3}{x - 6}$ **47.** $2x - \dfrac{1}{x} + \dfrac{3}{x - 7}$
49. $2 - \dfrac{2}{x} + \dfrac{6}{x - 2}$ **51.** $\dfrac{1}{2} + \dfrac{2}{x} - \dfrac{1.5}{2x - 3}$ **53.** max at (3, 4) = 2, min at (0, 7) = −7 **55.** max at (10, 8) = 18, min at (0, 4) = 4
57. max at (−4, 20) = 44, min at (2, 2) = 14 **59.** plain hamburgers = $3.00, cheeseburgers = $4.50, veggie burgers = $5.50 **61.** 11 strawberry, 13 pineapple, 24 cherry
63a. $f(p, g) = 25p + 20g$, $p + g \ge 60$, $0 \le p \le 50$, $0 \le g \le 30$
63b. 50 lb of paper and 10 lb of glass **63c.** $1450

Pages 418–419 Chapter 6 Connect to AP Calculus

1. $0 < x < 8$; Sample answer: x must be greater than 0 for the box to have any height, but it must also be less than 8 or the

value for the width of the box would be negative or zero.
3. height ≈ 2.94 in., length ≈ 14.12 in., width ≈ 10.12 in.,
maximum volume ≈ 420.11 in³   **5.** radius = 1.55 in.,
height = 3.09 in. minimum surface area = 45.19 in²
7. (6.67, 1.83), Sample answer: A minimum time of 1.83 hours
will be obtained if the racer takes a path that will have
him/her at the sidewalk 3.33 miles before the pier.
9. $m = 0$; Sample answer: The slope of the tangent line is 0.
The tangent line is a horizontal line.

Chapter 7 Conic Sections and Parametric Equations

Page 421   Chapter 7   Get Ready

1. $x = 1$; (0, −12); (1, −13)   **3.** $x = -1$; (0, −8); (−1, −10)
5. $x = 2$; (0, −4); (2, −16)   **7.** $x = 25$; (0, 550); (25, 543.75)
9. 108   **11.** 100   **13.** −172   **15.** vertical asymptote at $x = -5$
17. vertical asymptotes at $x = -5$ and $x = 1$; horizontal
asymptote at $y = 0$ **19.** vertical asymptote at $x = 4$.

Pages 428–431   Lesson 7-1

1. vertex: (3, 7);
focus: (3, 10);
directrix: $y = 4$;
axis of symmetry: $x = 3$

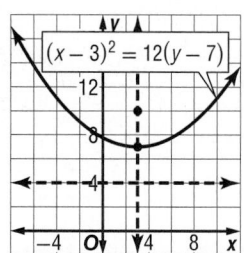

3. vertex: (−2, 4);
focus: (3, 4);
directrix: $x = -7$;
axis of symmetry: $y = 4$

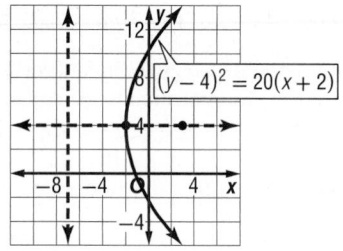

5. vertex: (−8, 3);
focus: (−8, 5);
directrix: $y = 1$;
axis of symmetry: $x = -8$

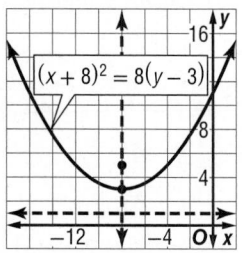

7. vertex: (1, −5);
focus: (7, −5);
directrix: $x = -5$;
axis of symmetry: $y = -5$

9. vertex: (−8, −2);
focus: (−8, −3);
directrix: $y = -1$;
axis of symmetry: $x = -8$

11. Sample answer: The focus is 4 feet above the ground.

13 **a.** $y^2 - 180x + 10y + 565 = 0$
$$y^2 + 10y = 180x - 565$$
$$y^2 + 10y + 25 = 180x - 565 + 25$$
$$y^2 + 10y + 25 = 180x - 540$$
$$(y + 5)^2 = 180(x - 3)$$

b. The equation in standard form has y as the squared
term, which means that the parabola opens
horizontally. Because $4p = 180$, $p = 45$ and the graph
opens to the right. The equation is in the form
$(y - k)^2 = 4p(x - h)$, so $h = 3$ and $k = -5$. Since the
stern is located at the vertex of the parabola formed, it
is at the point (h, k) or (3, −5). The swimmer is at the
focus, located at $(h + p, k)$, which is (3 + 45, −5) or
(48, −5). The distance the swimmer is from the stern
of the boat represents the length of rope needed to
attach the swimmer to the stern. Using the distance
formula, the distance between these two points is
$$\sqrt{(48 - 3)^2 + (-5 - (-5))^2}$$ or 45. Therefore, the length
of rope attaching the swimmer to the stern of the boat
is 45 feet.

15. $x^2 = 8(y + 7)$;
vertex: (0, −7);
focus: (0, −5);
directrix: $y = -9$;
axis of symmetry: $x = 0$

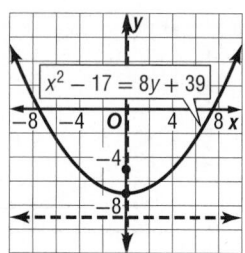

17. $x^2 = -24(y + 1)$;
vertex: (0, −1);
focus: (0, −7);
directrix: $y = 5$;
axis of symmetry: $x = 0$

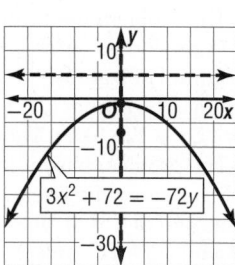

19. $y^2 = 20(x - 3)$;
vertex: (3, 0);
focus: (8, 0);
directrix: $x = -2$;
axis of symmetry: $y = 0$

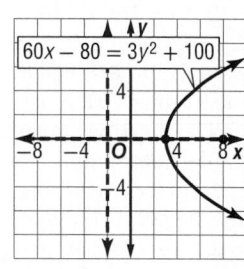

21. $(y - 4)^2 = 10(x - 2)$;
vertex: $(2, 4)$;
focus: $(4.5, 4)$;
directrix: $x = -0.5$;
axis of symmetry: $y = 4$

$-72 = 2y^2 - 16y - 20x$

23. $(x + 6)^2 = 18(y + 9)$;
vertex: $(-6, -9)$;
focus: $(-6, -4.5)$;
directrix: $y = -13.5$;
axis of symmetry: $x = -6$

$x^2 - 18y + 12x = 126$

25. 9 in.
27. $(y + 1)^2 = 24(x + 4)$

$(y + 1)^2 = 24(x + 4)$

29. $(x + 3)^2 = 8(y - 2)$

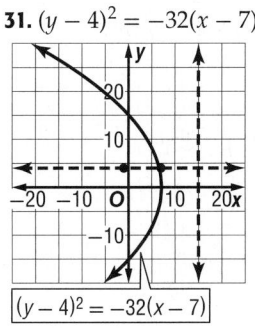

$(x + 3)^2 = 8(y - 2)$

31. $(y - 4)^2 = -32(x - 7)$

$(y - 4)^2 = -32(x - 7)$

33. $(x - 1)^2 = -12(y - 6)$

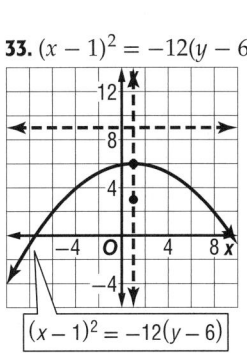

$(x - 1)^2 = -12(y - 6)$

35. $(x - 8)^2 = 16(y + 7)$

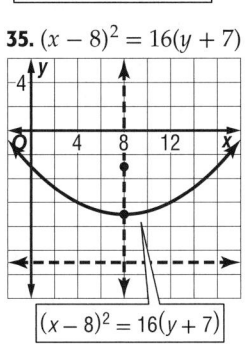

$(x - 8)^2 = 16(y + 7)$

37. $(x - 1)^2 = -12(y - 5)$

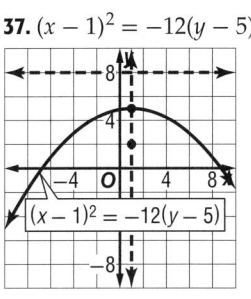

$(x - 1)^2 = -12(y - 5)$

39. $(x + 4)^2 = -4(y - 1)$

$(x + 4)^2 = -4(y - 1)$

41. $(y + 9)^2 = 4(x + 6)$

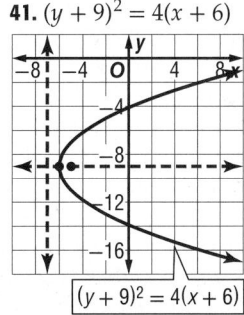

$(y + 9)^2 = 4(x + 6)$

43. $(x + 5)^2 = 16(y + 6)$

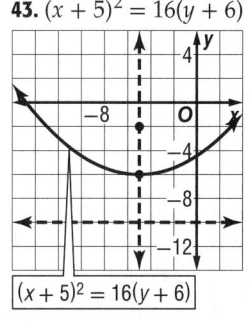

$(x + 5)^2 = 16(y + 6)$

45. $y = -8x - 45$ **47.** $y = 4x + 14$

49 The graph opens vertically. Determine the vertex and focus.

$$-0.25(x - 6)^2 = y - 9$$
$$(x - 6)^2 = -4(y - 9)$$

Because $4p = -4$, $p = -1$. The vertex is $(6, 9)$ and the focus is $(6, 8)$. We need to determine d, the distance between the focus and the point of tangency, C. This is one leg of the isosceles triangle.

$$d = \sqrt{(x_2 - x_1)^2 + (y_2 - y_1)^2}$$
$$= \sqrt{(10 - 6)^2 + (5 - 8)^2}$$
$$= 5$$

Use d to find A, the endpoint of the other leg of the isosceles triangle. Since p is negative, the parabola opens down and A will be above the focus.

$$A = (6, 8 + 5) \text{ or } (6, 13)$$

Points A and C both lie on the line tangent to the parabola. To find an equation for this line, first use two points to find the slope m.

$$m = \frac{5 - 13}{10 - 6} \text{ or } -2$$

Then use a point on the line and the slope to write an equation for the line.

$y - y_1 = m(x - x_1)$	Point-slope formula
$y - 5 = -2(x - 10)$	$(x_1, y_1) = (10, 5)$ and $m = -2$
$y - 5 = -2x + 20$	Simplify.
$y = -2x + 25$	Solve for y.

51. opens downward   **53.** opens upward   **55.** $(x - 3)^2 = -4(y - 5)$ **57.** $(y - 1)^2 = -16(x + 5)$ **59a.** Sample answer: $x^2 = -180.27(y + 20)$ **59b.** about 30.35 m

61. $(y - 4)^2 = -4(x + 6)$

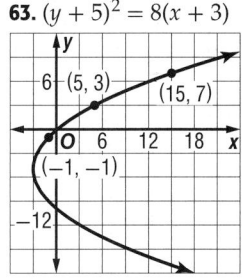

63. $(y + 5)^2 = 8(x + 3)$

65. $(2, 8)$ **67.** $(0, -1)$ **69a.** $y^2 = \frac{2}{3}x$ **69b.** 1.53 ft

71a. $(y - 2)^2 = 8(x + 3)$ or $(y - 2)^2 = -8(x + 3)$

71b.

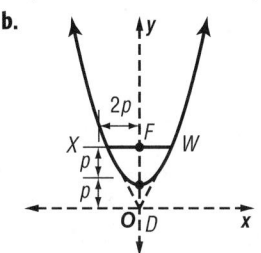

Sample answer: To prove that the endpoints of the latus rectum X and W and the point of intersection of the axis and directrix D are the vertices of a right isosceles triangle $\triangle XDW$, we need to show that $\angle XDW$ is a right angle and that $\overline{XD} \cong \overline{WD}$. Since $\overline{XF} \cong \overline{FW}$, $\overline{FD} \cong \overline{FD}$, and $\angle XFD \cong \angle WFD$, $\triangle XFD \cong \triangle FWD$ by SAS. Thus, $\overline{XD} \cong \overline{WD}$. To prove that $\angle XDW$ is a right angle, we can first prove that $\angle XDF$ and $\angle FDW$ are both 45° angles. Since $\overline{XF} \cong \overline{FD}$ and $\angle XFD$ is a right angle, $\triangle XFD$ is an isosceles right triangle. Therefore, $\angle XDF$ is 45°. This also means that $\angle FDW$ is 45°. Likewise, $\angle XDW$ is a right angle. Thus, $\triangle XDW$ is an isosceles right triangle. **73.** Sample answer: $x^2 = 4(y - 2)$ **75.** Sample answer: $y^2 = 12x$ **77a. i.** 1 unit; **ii.** 2 units; **iii.** 4 units

77b.

77c. As the focus is moved farther away from the vertex, the parabolas become wider. **77d.** Sample answer: $(x + 1)^2 = 4(y + 7)$

77e. Sample answer: The parabolas all have a vertex of $(0, -1)$ and open downward. The first equation produces the narrowest parabola and the second equation produces the widest parabola.

79 The width of the sector is 3 units, so $y = 1.5$. $A = \frac{4}{3}xy$ and the area of the sector is 2.4 square units. Substitute the values into the equation and solve for x.

$2.4 = \frac{4}{3}x(1.5)$

$2.4 = 2x$

$1.2 = x$

Therefore, the vertex of the parabola is $(0, 0)$, and two other points on the parabola are $(1.2, 1.5)$ and $(1.2, -1.5)$. Set up and solve a system of equations using y as the independent variable.

$ay^2 + by + c = x$

$\quad a(0)^2 + b(0) + c = 0$

$\quad a(1.5)^2 + b(1.5) + c = 1.2$

$a(-1.5)^2 + b(-1.5) + c = 1.2$

$\qquad\qquad\qquad c = 0$

$\quad 2.25a + 1.5b = 1.2$

$\quad 2.25a - 1.5b = 1.2$

$\qquad\qquad 4.5a = 2.4$

$\qquad\qquad\quad a = \frac{8}{15}$

$2.25 \cdot \frac{8}{15} + 1.5b = 1.2$

$\quad 1.2 + 1.5b = 1.2$

$\qquad\quad 1.5b = 0$

$\qquad\qquad b = 0$

The equation of the parabola is $\frac{8}{15}y^2 = x$ or $y^2 = \frac{15}{8}x$.

81. Quadrants I and IV; the vertex is $(-2, 5)$ and $p = -2$. Since the vertex is to the left of the y-axis, and the parabola opens to the left, no points will be to the right of the y-axis, or in Quadrants I and IV.

85. max at $(7, 8.5) = 81.5$, min at $(2, 0) = 16$ **87.** $\frac{-1}{y + 2} + \frac{3}{y + 1}$

89. $\cot \theta = \frac{7}{6}$, $\csc \theta = \frac{\sqrt{85}}{6}$

91.

93.

95. $2x$

97.

99.

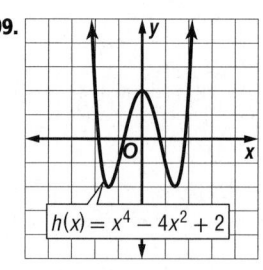

101. G **103.** G

Pages 438–441 Lesson 7-2

1.

3.

5.

7. $\frac{(x - 3)^2}{100} + \frac{(y + 3)^2}{36} = 1$ **9.** $\frac{(x - 2)^2}{25} + \frac{(y - 2)^2}{9} = 1$

11. $\frac{(x + 6)^2}{64} + \frac{(y - 3)^2}{100} = 1$ **13.** $\frac{(x - 2)^2}{225} + \frac{(y - 8)^2}{81} = 1$

15. 0.837 **17.** 0.426 **19.** 0.511 **21.** 0.447 **23a.** 28.8 in.

23b. $\frac{x^2}{324} + \frac{y^2}{207} = 1$ **25.** $\frac{(x - 1)^2}{8} + \frac{(y + 3)^2}{4} = 1$; ellipse

27. $(y + 9)^2 = 12(x - 6)$; parabola **29.** $\frac{(x - 7)^2}{3} + \frac{(y + 2)^2}{9} = 1$; ellipse

31

$2x^2 + 7y^2 + 24x + 84y + 310 = 0$

$2x^2 + 24x + 7y^2 + 84y + 310 = 0$

$$2(x^2 + 12x + 36) + 7(y^2 + 12y + 36) + 310 = 0 + 72 + 252$$
$$2(x + 6)^2 + 7(y + 6)^2 = 14$$
$$\frac{(x + 6)^2}{7} + \frac{(y + 6)^2}{2} = 1$$

The related conic is an ellipse because $a \neq b$ and the graph is of the form $\frac{(x - h)^2}{a^2} + \frac{(y - k)^2}{b^2} = 1$.

33. $(x - 3)^2 + y^2 = 4$ **35.** $(x + 4)^2 + (y + 3)^2 = 36$

37.

$$PF_1 + PF_2 = 2a$$
$$\sqrt{(x - 0)^2 + (y - c)^2} + \sqrt{(x - 0)^2 + (y - (-c))^2} = 2a$$
$$\sqrt{x^2 + (y - c)^2} + \sqrt{x^2 + (y + c)^2} = 2a$$
$$\sqrt{x^2 + (y - c)^2} = 2a - \sqrt{x^2 + (y + c)^2}$$
$$x^2 + y^2 - 2cy + c^2 = 4a^2 - 4a\sqrt{x^2 + (y + c)^2} +$$
$$x^2 + y^2 + 2cy + c^2$$
$$4a\sqrt{x^2 + (y + c)^2} = 4a^2 + 4cy$$
$$a\sqrt{x^2 + (y + c)^2} = a^2 + cy$$
$$a^2(x^2 + y^2 + 2cy + c^2) = a^4 + 2a^2cy + c^2y^2$$
$$a^2x^2 + a^2y^2 + 2a^2cy + a^2c^2 = a^4 + 2a^2cy + c^2y^2$$
$$a^2y^2 - c^2y^2 + a^2x^2 = a^4 - a^2c^2$$
$$y^2(a^2 - c^2) + a^2x^2 = a^2(a^2 - c^2)$$
$$y^2b^2 + a^2x^2 = a^2b^2$$
$$\frac{y^2}{a^2} + \frac{x^2}{b^2} = 1$$

39. $\frac{(x - 1)^2}{16} + \frac{(y - 3)^2}{1} = 1$ **41.** $\frac{(x - 2)^2}{16} + \frac{(y - 2)^2}{49} = 1$

43 a. The length of the major axis is $43.4 + 28.6 + 0.87$ or 72.87. The value of a is $\frac{72.87}{2}$ or 36.435. The distance from the focus (the sun) to the vertex is $28.6 + \frac{0.87}{2}$ or 29.035. Therefore, the value of c, the focus to the center, is $36.435 - 29.035$ or 7.4. In an ellipse, $c^2 = a^2 - b^2$. The value of b is $\sqrt{a^2 - c^2} = \sqrt{36.435^2 - 7.4^2} \approx 35.676$. The value of $2b$, the length of the minor axis, is about $35.676 \cdot 2$ or about 71.35 million mi.
b. The eccentricity e is equal to $\frac{c}{a}$. $e = \frac{7.4}{36.435} \approx 0.203$

45. center: $(0, -6)$; foci: $(\pm 5\sqrt{3}, -6)$; vertices: $(\pm 10, -6)$
47. center: $(-1, 0)$; foci: $(-1, \pm 7)$; vertices: $(-1, \pm\sqrt{65})$
49. $\frac{x^2}{100} + \frac{y^2}{64} = 1$ **51.** $\frac{(y + 4)^2}{36} + \frac{(x - 2)^2}{16} = 1$
53a. $x^2 + y^2 = 64$, $x^2 + y^2 = 784$

53b.

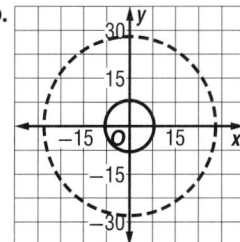

53c. 5 days

55. $(7, -1)$, $(3, -5)$ **57.** $(-2, 3)$, $(6, -5)$ **59.** $\frac{x^2}{16} + \frac{y^2}{12} = 1$
61. $(x - 6.5)^2 + (y - 4.5)^2 = 32.5$ **63.** $(x - 1)^2 + (y + 7)^2 = 16$; center: $(1, -7)$, radius: 4 **65.** $(x + 1)^2 + (y - 4)^2 = 64$; center: $(-1, 4)$, radius: 8 **67.** Sample answer: No; if $a^2 = p$ and $b^2 = p + r$, then $c = \sqrt{-r}$. If $a^2 = p + r$ and $b^2 = p$, then $c = \sqrt{r}$ and the foci are $(0, \pm\sqrt{r})$.

69 The area of the ellipse $A = 24\pi$. Substitute into the area formula to obtain an equation with a and b.
$$24\pi = \pi ab$$
$$24 = ab$$
Substitute $a = b + 5$ into the equation and solve for b.
$$24 = (b + 5)b$$
$$24 = b^2 + 5b$$
$$0 = b^2 + 5b - 24$$
$$0 = (b + 8)(b - 3)$$
Since b cannot be negative, $b = 3$ and $a = 3 + 5$ or 8. Therefore, an equation for the ellipse is
$$\frac{x^2}{8^2} + \frac{y^2}{3^2} = 1 \text{ or } \frac{x^2}{64} + \frac{y^2}{9} = 1.$$
71. Yes; sample answer: If (x, y) is a point on the ellipse, then $(-x, -y)$ must also be on the ellipse.
$$\frac{x^2}{a^2} + \frac{y^2}{b^2} = 1$$
$$\frac{(-x)^2}{a^2} + \frac{(-y)^2}{b^2} = 1$$
$$\frac{x^2}{a^2} + \frac{y^2}{b^2} = 1$$
Thus, $(-x, -y)$ is also a point on the ellipse and the ellipse is symmetric with respect to the origin.

73. Sample answer: When a is much greater than c, $\frac{c}{a}$ is close to 0. Since $e = \frac{c}{a}$, the value of e is close to zero and the foci are near the center of the ellipse. So, the ellipse is nearly circular.

75. vertex: $\left(\frac{5}{4}, -\frac{55}{8}\right)$; focus: $\left(\frac{5}{4}, -7\right)$; directrix: $y = -\frac{27}{4}$; axis of symmetry: $x = \frac{5}{4}$

77. 160 My Real Babies and 320 My First Babies

79. $\sin\left(\theta + \frac{\pi}{3}\right) - \cos\left(\theta + \frac{\pi}{6}\right)$
$$= \sin\theta\cos\frac{\pi}{3} + \cos\theta\sin\frac{\pi}{3} - \cos\theta\cos\frac{\pi}{6} + \sin\theta\sin\frac{\pi}{6}$$
$$= \frac{1}{2}\sin\theta + \frac{\sqrt{3}}{2}\cos\theta - \frac{\sqrt{3}}{2}\cos\theta + \frac{1}{2}\sin\theta$$

Selected Answers and Solutions

$$= \frac{1}{2}\sin\theta + \frac{1}{2}\sin\theta$$
$$= \sin\theta$$

81. $\frac{\pi}{4}, \frac{5\pi}{4}$ **83.** $\frac{3\pi}{2}, \frac{7\pi}{6}, \frac{11\pi}{6}$ **85.** $[-7, 5]$ **87.** 4 real zeros and 3 turning points; 0, −4, and −2 **89.** 5 real zeros and 4 turning points; 0, 5, and −2 **91.** $3 + 4i$ **93.** B **95.** C

Pages 449–452   Lesson 7-3

1.

3.

5.

7.

9.

11.

13.

15.

17.

19.

21.
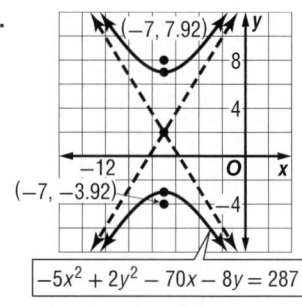

23. $\frac{(y-1)^2}{15} - \frac{(x+1)^2}{49} = 1$ **25.** $\frac{(x-3)^2}{27} - \frac{(y+1)^2}{9} = 1$

27. $\frac{(y+8)^2}{16} - \frac{(x+3)^2}{33} = 1$ **29.** $\frac{(x+7)^2}{25} - \frac{(y-2)^2}{49} = 1$

31 The vertices are 4 units apart, so $2a = 4$, $a = 2$, and $a^2 = 4$. The center is the midpoint of the vertices, or $(-2, -3)$. Therefore, $h = -2$ and $k = -3$. The conjugate axis is of length 12, so $2b = 12$, $b = 6$, and $b^2 = 36$. The x-coordinates of the vertices are different, so the transverse axis is horizontal and the a^2-term goes with the x^2-term. The equation for the hyperbola is
$$\frac{(x+2)^2}{4} - \frac{(y+3)^2}{36} = 1.$$

33a. $\frac{(y-4)^2}{9} - \frac{(x-5)^2}{\frac{225}{7}} = 1$ **33b.** 90 ft **35.** 1.27 **37.** 1.33

39. 1.68 **41.** 1.32 **43.** 2.83 **45.** hyperbola **47.** parabola **49.** ellipse **51.** hyperbola **53.** circle

55 a. The common difference is 18 kilometers and the absolute value of the difference of the distances from any point on a hyperbola to the foci is $2a$, so $2a = 18$, $a = 9$, and $a^2 = 81$. The two airports are the foci of the hyperbola and are 72 kilometers apart, so $2c = 72$, $c = 36$, and $c^2 = 1296$. $c^2 = a^2 + b^2$, so $b^2 = 1296 - 81$ or 1215. Airport B is due south of airport A, so the transverse axis is vertical and a^2-term goes with the y^2-term. The equation for the hyperbola is
$$\frac{y^2}{81} - \frac{x^2}{1215} = 1.$$

b.

The plane is on the top branch because it is closer to airport A.

c. Because the highway is the transverse axis and the plane is 40 kilometers from the highway, $x = 40$. Substitute 40 for x in the equation.

R94   Selected Answers and Solutions

$$\frac{y^2}{81} - \frac{40^2}{1215} = 1$$

$$\frac{y^2}{81} - 1.317 = 1$$

$$\frac{y^2}{81} = 2.317$$

$$y^2 = 187.67$$

$$y \approx 13.7$$

The coordinates of the plane are (40, 13.7).

57.

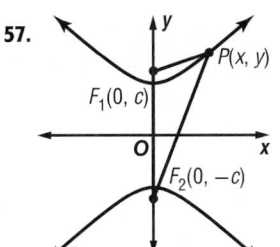

$$|PF_1 - PF_2| = 2a$$

$$\left| \sqrt{(x - 0)^2 + (y - c)^2} - \sqrt{(x - 0)^2 + [y - (-c)]^2} \right| = 2a$$

$$\sqrt{x^2 + (y - c)^2} - \sqrt{x^2 + (y + c)^2} = 2a$$

$$\sqrt{x^2 + (y - c)^2} = 2a + \sqrt{x^2 + (y + c)^2}$$

$$x^2 + (y - c)^2 = 4a^2 + 4a\sqrt{x^2 + (y + c)^2} + x^2 + (y + c)^2$$

$$-4cy - 4a^2 = 4a\sqrt{x^2 + (y + c)^2}$$

$$-cy - a^2 = a\sqrt{x^2 + (y + c)^2}$$

$$c^2y^2 + 2a^2cy + a^4 = a^2x^2 + 2a^2cy + a^2c^2 + a^2y^2$$

$$c^2y^2 - a^2y^2 - a^2x^2 = a^2c^2 - a^4$$

$$(c^2 - a^2)y^2 - a^2x^2 = a^2(c^2 - a^2)$$

$$\frac{y^2}{a^2} - \frac{x^2}{c^2 - a^2} = 1$$

$$\frac{y^2}{a^2} - \frac{x^2}{b^2} = 1$$

By the Pythagorean Theorem, $c^2 - a^2 = b^2$.

59. (−1.3, −5.7), (7, −1.5)   **61.** (−5, −10), (6.9, 13.7)

63. (0, 6), (0, −6)   **65a.** $\frac{x^2}{550^2} - \frac{y^2}{7901^2} = 1$ **65b.** on the right

branch   **67.** $\frac{x^2}{25} - \frac{y^2}{9} = 1$ **69.** (641.2, 3178.3), (1212.8, 640.6),

(331.6, 4454.4), (2351.0, 100.8)   **71.** $\frac{(x + 4)^2}{25} - \frac{(y - 3)^2}{49} = 1$

73 The foci are 14 units apart, so $c = 7$ and $c^2 = 49$.
The center is the midpoint of the foci, (6, −2), so $h = 6$
and $k = -2$. The eccentricity $\frac{c}{a}$ is $\frac{7}{6}$, so $a = 6$ and $a^2 = 36$.
$b^2 = c^2 - a^2 = 49 - 36$ or 13.
The x-coordinates of the vertices are different, so the
transverse axis is horizontal and the a^2-term goes with
the x^2-term. The equation for the hyperbola is

$$\frac{(x - 6)^2}{36} - \frac{(y + 2)^2}{13} = 1.$$

75. $\frac{2x^2}{121} - \frac{2y^2}{121} = 1$ **77.** Sample answer: $\frac{x^2}{16} - \frac{y^2}{48} = 1$

79. Sample answer: If the transverse is horizontal, then a is a
horizontal distance and b is a vertical distance. So, the slopes
of the asymptotes are $\pm\frac{b}{a}$. If the transverse is vertical, then a
is a vertical distance and b is a horizontal distance. So, the
slopes of the asymptotes are $\pm\frac{a}{b}$.

81. $\frac{y^2}{9} - \frac{x^2}{72} = 1$ **83.** Sample answer: First, determine
whether the orientation of the hyperbola is vertical or
horizontal. Then use the foci to locate the center of the
hyperbola and determine the values of h and k. Use the
transverse axis length to find a^2. Find c, the distance from the
center to a focus. Use the equation $b^2 = c^2 - a^2$ to find b^2.
Finally, use the correct standard form to write the equation,
depending on whether the transverse axis is parallel to the
x-axis or to the y-axis.

85.

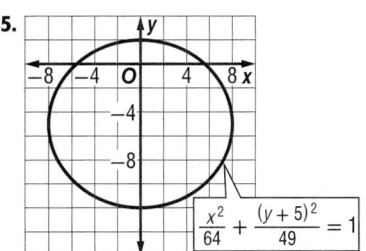

87a. 105 ft
87b. 5 seconds

89. $\begin{bmatrix} 1 & -7 & 8 \\ 6 & 5 & -2 \\ 3 & -4 & 9 \end{bmatrix} \cdot \begin{bmatrix} x_1 \\ x_2 \\ x_3 \end{bmatrix} = \begin{bmatrix} -3 \\ 2 \\ 26 \end{bmatrix}$; (−5, 10, 9)   **91.** $2n\pi; n \in \mathbb{Z}$

93. no solution   **95.** $\sin\theta = \frac{24}{25}, \cos\theta = \frac{7}{25}, \tan\theta = \frac{24}{7}$,

$\csc\theta = \frac{25}{24}, \sec\theta = \frac{25}{7}, \cot\theta = \frac{7}{24}$ **97.** $-\frac{3}{2}, -1$ (multiplicity: 2),

$4 + 9i, 4 - 9i; (2x + 3)(x + 1)^2(x - 4 + 9i)(x - 4 - 9i)$

99. H   **101.** G

Pages 459–461   Lesson 7-4

1. hyperbola; $(x')^2 + 2\sqrt{3}x'y' - (y')^2 + 18 = 0$ **3.** parabola;
$(y')^2 - 8x' = 0$ **5.** parabola; $(x')^2 + 2\sqrt{3}x'y' + 3(y')^2 + 16\sqrt{3}x' - 16y' = 0$ **7.** ellipse; $2(x')^2 + 2(y')^2 - 5\sqrt{2}x + 5\sqrt{2}y - 6 = 0$ **9.** hyperbola; $25(x')^2 - 4(y')^2 + 64 = 0$
11. $\frac{y'^2}{8} - \frac{x'^2}{8} = 1$ **13.** $\frac{x'^2}{4} + \frac{y'^2}{9} = 1$ **15.** $\frac{(y' - 1)^2}{\frac{1}{2}} - \frac{(x' - 1)}{1} = 1$

17. $\frac{(x')^2}{\frac{1}{6}} - \frac{(y')^2}{2} = 1$ **19.** $x^2 - xy + y^2 - 4 = 0$

21. $x^2 + 10\sqrt{3}xy + 11y^2 - 144 = 0$ **23.** $13x^2 + 6\sqrt{3}xy + 7y^2 - 112 = 0$ **25.** $3x^2 - 10xy + 3y^2 + 128 = 0$
27. $23x^2 + 26\sqrt{3}xy - 3y^2 - 144 = 0$

29 **a.** Find the equations for x and y for $\theta = 30°$.
$x' = x\cos\theta + y\sin\theta$

$\quad = \frac{\sqrt{3}}{2}x + \frac{1}{2}y$

$y' = y\cos\theta - x\sin\theta$

$\quad = \frac{\sqrt{3}}{2}y - \frac{1}{2}x$

Substitute into $144(x')^2 + 64(y')^2 = 576$

$144\left(\frac{\sqrt{3}}{2}x + \frac{1}{2}y\right)^2 + 64\left(\frac{\sqrt{3}}{2}y - \frac{1}{2}x\right)^2 = 576$

$144\left(\frac{3}{4}x^2 + \frac{1}{4}y^2 + \frac{\sqrt{3}}{2}xy\right) + 64\left(\frac{3}{4}y^2 + \frac{1}{4}x^2 - \frac{\sqrt{3}}{2}xy\right) = 576$

$108x^2 + 36y^2 + 72\sqrt{3}xy + 48y^2 + 16x^2 - 32\sqrt{3}xy = 576$

$124x^2 + 40\sqrt{3}xy + 84y^2 - 576 = 0$

$31x^2 + 10\sqrt{3}xy + 21y^2 - 144 = 0$

b. Graph the equation by solving for y.

$21y^2 + \left(10\sqrt{3}x\right)y + \left(31x^2 - 144\right) = 0$
Use the quadratic formula.

$$y = \frac{-10\sqrt{3}x \pm \sqrt{(10\sqrt{3}x)^2 - 4(21)(31x^2 - 144)}}{2(21)}$$

$$y = \frac{-10\sqrt{3}x \pm \sqrt{12096 - 2304x^2}}{42}$$

Graph the conic.

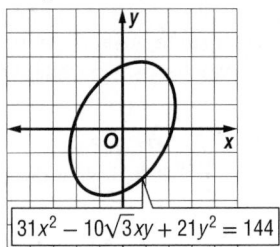

$31x^2 - 10\sqrt{3}xy + 21y^2 = 144$

31.

$\dfrac{(x')^2}{25} - \dfrac{(y')^2}{36} = 1$

33.

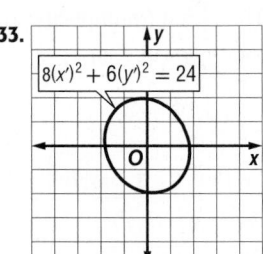

$8(x')^2 + 6(y')^2 = 24$

35.

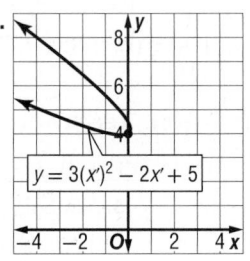

$y = 3(x')^2 - 2x' + 5$

37. $x^2 - 2xy + y^2 - 5x - 5y = 0$

[−6.61, 14.6] scl: 1
by [−2, 12] scl: 1

39. $8x^2 + 5xy - 4y^2 = -2$

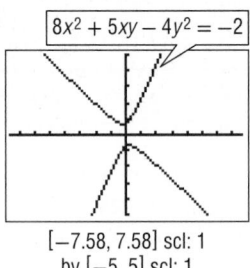

[−7.58, 7.58] scl: 1
by [−5, 5] scl: 1

41.

$2x^2 + 4xy + 2y^2 + 2\sqrt{2}x - 2\sqrt{2}y = -12$

[−10.58, 4.58] scl: 1 by
[−2, 8] scl: 1

43. $x^2 + 10\sqrt{3}xy + 11y^2 - 64 = 0$

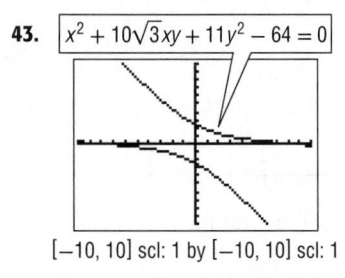

[−10, 10] scl: 1 by [−10, 10] scl: 1

45.

$x^2 - 2\sqrt{3}xy - y^2 + 18 = 0$

[−10, 10] scl: 1 by [−10, 10] scl: 1

47. intersecting lines
$y = 4x$ and $y = -4x$
49. point at $(0, 0)$ **51.** b
53. c

55a. $Ax^2 + Bxy + Cy^2 + Dx + Ey + F = 0$ is equivalent to $A'(x')^2 + B'x'y' + C'(y')^2 + D'x' + E'y' + F' = 0$ by rotating the conic section through θ such that $\cot 2\theta = \dfrac{A - C}{B}$. F is unaffected by this rotation, so $F = F'$.

55b. $Ax^2 + Bxy + Cy^2 + Dx + Ey + F = A'(x')^2 + B'x'y' + C'(y')^2 + D'x' + E'y' + F'$; by subtracting out the xy, x, y, and F terms, the remaining statement is true. $Ax^2 + Cy^2 = A'(x')^2 + C'(y')^2$; A must equal A' and B must equal B' in order for the statement to be true.   **55c.** $Ax^2 + Bxy + Cy^2 + Dx + Ey + F = A'(x')^2 + B'x'y' + C'(y')^2 + D'x' + E'y' + F'$; by subtracting out the x, y, and F terms, the remaining statement is true. $Ax^2 + Bxy + Cy^2 = A'(x')^2 + B'x'y' + C'(y')^2$; A must equal A', B must equal B', and C must equal C' in order for the statement to be true. Therefore, $B^2 - 4AC = (B')^2 - 4(A'C')$; $A(x \cos \theta + y \sin \theta)^2 + B(x \cos \theta + y \sin \theta)(y \cos \theta - x \sin \theta)$.

57 For each equation, solve for y using the quadratic formula. Then graph using a graphing calculator.

$$9x^2 + 4xy + 5y^2 - 40 = 0$$
$$5y^2 + 4x(y) + (9x^2 - 40) = 0$$

$$y = \frac{-4x \pm \sqrt{(4x)^2 - 4(5)(9x^2 - 40)}}{2(5)}$$

$$= \frac{-4x \pm \sqrt{800 - 164x^2}}{10}$$

$$x^2 - xy - 2y^2 - x - y + 2 = 0$$
$$-2y^2 + (-x - 1)y + (x^2 - x + 2) = 0$$

$$y = \frac{x + 1 \pm \sqrt{(-x - 1)^2 - 4(-2)(x^2 - x + 2)}}{2(-2)}$$

$$= \frac{x + 1 \pm \sqrt{9x^2 - 6x + 17}}{-4}$$

[−10, 10] scl: 1 by [−10, 10] scl: 1

The graphs intersect at four points: $(-1.9, 2.2)$, $(-1.5, -1.5)$, $(1.9, -2.2)$, $(1.9, 0.8)$.

59a.

Equation	Graph	Minimum Angle of Rotation
$x^2 - 5x + 3 - y = 0$	parabola	360°
$6x^2 + 10y^2 = 15$	ellipse	180°
$2xy = 9$	hyperbola	180°

59b. Sample answer: A parabola has 1 line of symmetry and the minimum angle of rotation is a complete circle. An ellipse and a hyperbola have 2 lines of symmetry and the minimum angle of rotation is a half circle.   **59c.** 130°

61. Let $x = x' \cos \theta + y' \sin \theta$ and $y = -x' \sin \theta + y' \cos \theta$.
$$r^2 = x^2 + y^2$$
$$= (x' \cos \theta + y' \sin \theta)^2 + (-x' \sin \theta + y' \cos \theta)^2$$
$$= (x')^2 \cos^2 \theta + 2x'y' \cos \theta \sin \theta + (y')^2 \sin^2 \theta + (x') \sin^2 \theta$$
$$\quad - 2x'y' \cos \theta \sin \theta + (y')^2 \cos^2 \theta$$
$$= [(x')^2 + (y')^2] \cos^2 \theta + [(x')^2 + (y')^2] \sin^2 \theta$$
$$= [(x')^2 + (y')^2](\cos^2 \theta + \sin^2 \theta)$$
$$= [(x')^2 + (y')^2](1)$$
$$= (x')^2 + (y')^2$$

63.
$$\cos \theta \,(x = x' \cos \theta - y' \sin \theta)$$
$$\sin \theta \,(y = x' \sin \theta + y' \cos \theta)$$
$$x \cos \theta = x' \cos^2 \theta - y' \sin \theta \cos \theta$$
$$\underline{+ \; y \sin \theta = x' \sin^2 \theta + y' \sin \theta \cos \theta}$$
$$x \cos \theta + y \sin \theta = x' \cos^2 \theta + x' \sin^2 \theta$$
$$x \cos \theta + y \sin \theta = x' (\cos^2 \theta + \sin^2 \theta)$$
$$x \cos \theta + y \sin \theta = x'$$

$$\sin \theta \,(x = x' \cos \theta - y' \sin \theta)$$
$$\cos \theta \,(y = x' \sin \theta + y' \cos \theta)$$
$$x \sin \theta = x' \cos \theta \sin \theta - y' \sin^2 \theta$$
$$\underline{- \; y \cos \theta = x' \cos \theta \sin \theta + y' \cos^2 \theta}$$
$$x \sin \theta - y \cos \theta = - y' \sin^2 \theta - y' \cos^2 \theta$$
$$x \sin \theta - y \cos \theta = -y' (\sin^2 \theta + \cos^2 \theta)$$
$$y \cos \theta - x \sin \theta = y'$$

65. Sample answer: The discriminant is defined as $B^2 - 4AC$, or in this instance, $(B')^2 - 4A'C'$. Since a conic that is rotated has no B' term, the discriminant reduces to $-4A'C'$. Thus, only the A' and C' terms determine the type of conic. Therefore, $-4A'C' < 0$ would be an ellipse or a circle, $-4A'C' = 0$ would be a parabola, and $-4A'C' > 0$ would be a hyperbola. For a circle or an ellipse, A' and C' need to share the same sign. For a parabola, either A' or C' has to be equal to 0. For a hyperbola, A' and C' need to have opposite signs.

67.

69.
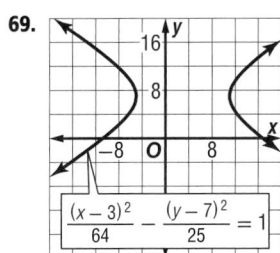

71. 0.447 **73a.** $s + d = 5000$, $0.035s + 0.05d = 227.50$
73b. savings account: $1500; certificate of deposit: $3500
75. 2 **77.** yes, yes; $(x - 4)$ $(x - 2)(x + 2)(x + 3)$ **79.** D
81. A

Pages 469–471   Lesson 7-5

1.

3.

5.

7.
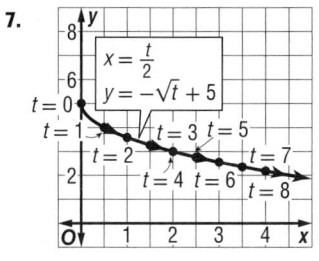

9. $y = 0.25x^2 + 2.5x + 10.25$
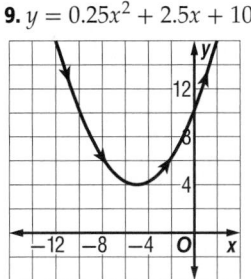

11. $y = 5\sqrt{x + 2}$
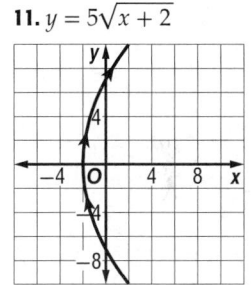

13. $y = 3x^2 + 24x + 48$

15. $y = \dfrac{3\sqrt{x}}{5} + 9$
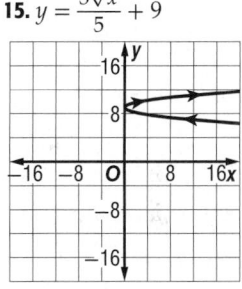

17. $y = -x^2 + \dfrac{15}{4}x + 100$

19. $\dfrac{x^2}{49} + \dfrac{y^2}{4} = 1$
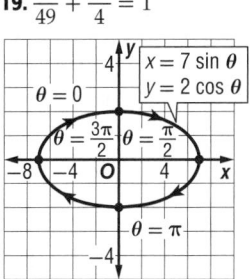

21. $\dfrac{x^2}{9} + \dfrac{y^2}{9} = 1$
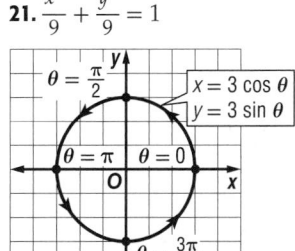

23. $\dfrac{x^2}{25} + \dfrac{y^2}{36} = 1$

25. $x^2 + \dfrac{y^2}{49} = 1$

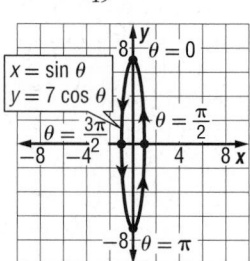

27. $x = \dfrac{t}{8}$ and $y = \sqrt{9 - \dfrac{t^2}{64}}$

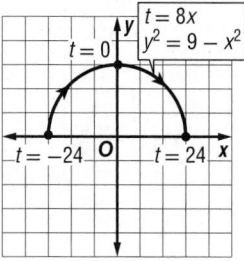

29. $x = 5t - 20$ and
$y = -25t^2 + 200t - 390$

31. $x = 1 - 2t$ and
$y = -t^2 + t + \dfrac{1}{2}$

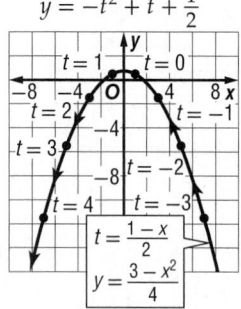

33. yes

35 Solve the first equation for t.
$$x = \log t$$
$$10^x = t$$

Substitute into second equation.
$$y = t + 3$$
$$y = 10^x + 3$$

$\ln x = \log t$, x must be > 0, so the domain restriction is $x > 0$.

37. $y = 10^x + 4$ **39.** $y = 2 \cdot 10^{\frac{1}{x}} - 8, x \neq 0$ **41.** $x = t + 4$ and $y = 7 + 3t$ **43.** $x = t - 2$ and $y = -6 + 4t, 0 \leq t \leq 4$ **45.** b
47. a

49 **a.** The position equations are $x = tv_0 \cos \theta$ and
$y = tv_0 \sin \theta - \dfrac{1}{2}gt^2 + h_0$. The initial velocity v_0 is
0.75 and θ is 45°. The surface of the water is at
$y = 0$, so the value of the initial height h_0 is 0.3. By
substitution, the position equations are $x = t \cdot$
$0.75 \cos 45°$ and $y = t \cdot 0.75 \sin 45° - 4.9t^2 + 0.3$.

b. Find the value of t for which $y = 0$ in the vertical
position equation.
$$0 = t \cdot 0.75 \sin 45° - 4.9t^2 + 0.3$$
$$0 = \dfrac{3\sqrt{2}}{8}t - 4.9t^2 + 0.3.$$

The positive zero in the graph of the equation is about
0.3074, so $t \approx 0.3074$ seconds. Substitute this for t in
$x = t \cdot 0.75 \cos 45°$, the horizontal position equation.
$$x = 0.3074(0.75) \cdot \dfrac{\sqrt{2}}{2} \approx 0.16.$$

This means that when the frog reaches the surface
of the water, he is only about 0.16 m from where
he jumped. Therefore, he is about $0.5 - 0.16$ or
0.34 meters from the other bank.

c. The 0.4 meter distance is a horizontal distance, so
substitute the values into the horizontal position
equation and solve for v_0.

$$x = tv_0 \cos \theta$$
$$0.4 = 0.38 \cdot v_0 \cdot \dfrac{\sqrt{2}}{2}$$
$$1.49 = v_0$$

51a. about 27.2 ft/s   **51b.** about 1 second   **51c.** about
1.84 seconds   **53.** Sample answer: $x = 4 + t, y = 10 + 3t$
55. Sample answer: The horizontal distance is modeled by
the cosine function, which is 0 at 90°. This would imply that
the projectile has no horizontal movement. The
corresponding parametric equation would be $x = 0$.
57. Sample answer: Parametric equations show both
horizontal and vertical positions of an object over time,
while rectangular equations can only show one or the other.

59.

61. $\dfrac{(y-2)^2}{4} - \dfrac{(x-3)^2}{5} = 1$
63. $2\tan^2 x$
65. $3\ln 3 + \ln 2; 3.99$
67. $3\ln 2 - \ln 3; 0.97$

69. horizontal asymptote at $y = 1$;
vertical asymptote at $x = -6$;
$D = \{x \mid x \neq -6, x \in \mathbb{R}\}$

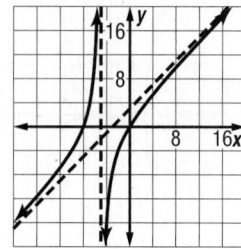

71. vertical asymptotes at $x = -5$;
oblique asymptote at $y = x + 3$;
x-intercepts 0 and -8; y-intercept 0;
$D = (-\infty, -5) \cup (-5, \infty)$

73. 7   **75.** 23   **77.** E   **79a.** $21x^2 + 31y^2 - 10\sqrt{3}xy = 4$
79b. ellipse   **79c.** $4(x')^2 + 9(y')^2 = 1$ **79d.** ≈ 0.745

79e. $\left(\dfrac{\sqrt{5}}{6}, 0\right), \left(-\dfrac{\sqrt{5}}{6}, 0\right)$

Pages 473–476   Chapter 7   Study Guide and Review

1. conic section   **3.** directrix   **5.** foci   **7.** center
9. parametric

11. vertex: $(-3, -2)$,
focus: $(-3, 1)$;
axis of symmetry: $x = -3$;
directrix: $y = -5$

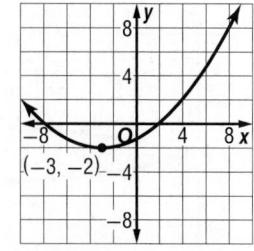

13. vertex: $(2, -1)$, focus: $(2, -2)$; axis of symmetry: $x = 2$; directrix: $y = 0$

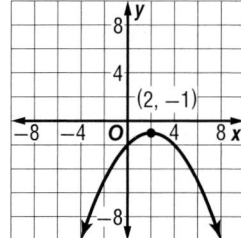

15. $(x - 1)^2 = -16(y - 5)$

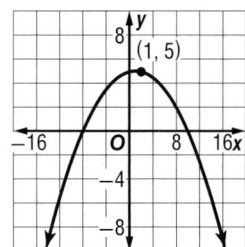

17. $(x + 2)^2 = -16(y - 1)$

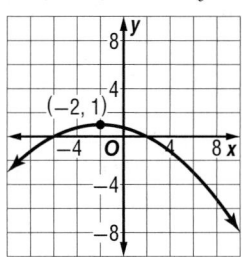

19. $(y + 4)^2 = -4(x + 3)$

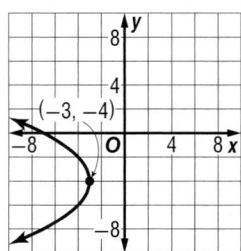

21. $(x - 3)^2 = 4(y + 7)$

23.

$\dfrac{(x - 3)^2}{16} + \dfrac{(y + 6)^2}{25} = 1$

25. $\dfrac{(x - 5)^2}{25} + \dfrac{(y - 2)^2}{9} = 1$

27. $(x - 1)^2 + (y - 2)^2 = 30$; circle

29. $(x - 2)^2 = -4(y + 5)$; parabola

31.

$\dfrac{(x + 7)^2}{18} - \dfrac{(y - 6)^2}{36} = 1$

33.

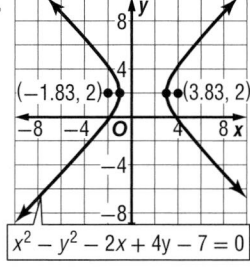

$x^2 - y^2 - 2x + 4y - 7 = 0$

35. $\dfrac{y^2}{9} - \dfrac{x^2}{16} = 1$

37. $\dfrac{x^2}{4} - \dfrac{y^2}{9} = 1$ **39.** parabola

41.

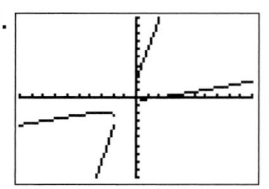

$[-10, 10]$ scl: 1 by $[-10, 10]$ scl: 1

43.

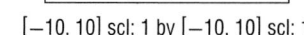

$[-10, 10]$ scl: 1 by $[-10, 10]$ scl: 1

45.

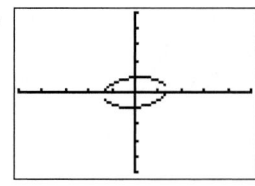

$[-5, 5]$ scl: 1 by $[-5, 5]$ scl: 1

47. $(x')^2 - 2\sqrt{3}x'y' + 3(y')^2 + (2\sqrt{3} - 4)x' + (4\sqrt{3} + 2)y' = 20$; parabola   **49.** $9(y')^2 + 4(x')^2 = 36$; ellipse

51.

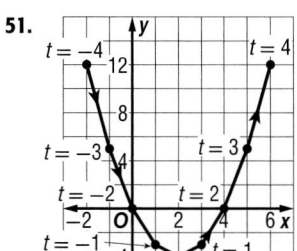

53. $y = \dfrac{x^2}{4} - 2$

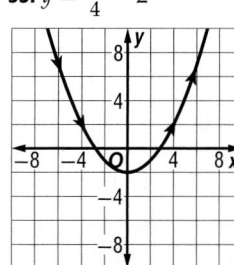

55. $(y - 1)^2 = 4(x + 1)$

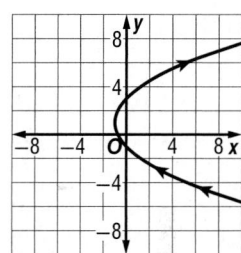

57a. $x^2 + y^2 = 900$
57b. 5 seconds
59a. $3x^2 + 2\sqrt{3}xy + y^2 + 8x - 8\sqrt{3}y = 0$

59b.

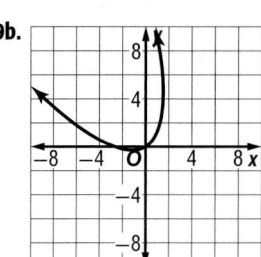

Pages 478–479 Chapter 7 Connect to AP Calculus

1. 376.99 units^3 **3.** 100.53 units^3 **5.** Sample answer: The approximation will get closer to the actual volume of the solid because as the cylinders decrease in height, they will better fill the volume of the solid.   **7.** 188.5 units^3

Chapter 8 Vectors

Page 481 Chapter 8 Get Ready

1. 3; $\left(-\dfrac{1}{2}, 4\right)$ **3.** $\sqrt{29}$; $\left(-\dfrac{1}{2}, -8\right)$ **5.** 5.4   **7.** 4.0   **9.** 22.8 ft
11. no solution   **13.** $B \approx 39°$, $C \approx 50°$, $c \approx 23.0$

1. scalar **3.** vector **5.** vector

7. Sample answer: **9.** Sample answer:

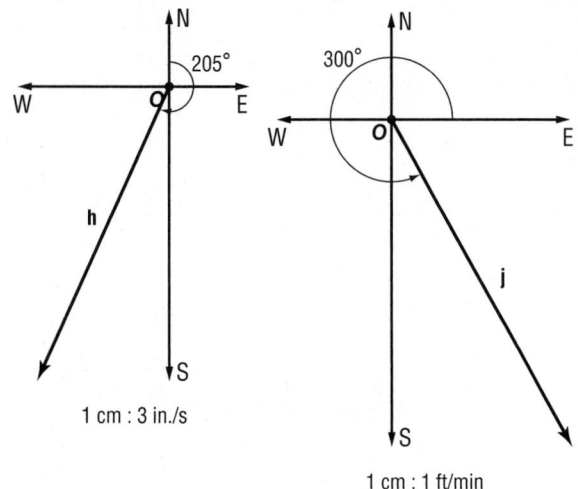

1 cm : 3 in./s

1 cm : 1 ft/min

11. Sample answer:

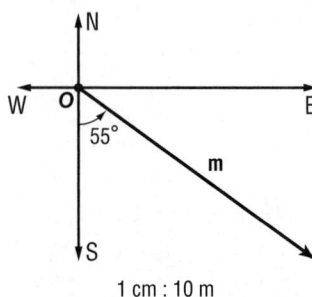

1 cm : 10 m

13. 1.4 cm, 50° **15.** 1.0 cm, 46° **17.** 2.3 cm, 188° **19.** 45 mph, 31° **21.** 2.6 km due north **23.** 250 m due south

25 Let **a** = 17 miles east and **b** = 16 miles south. Draw a diagram to represent **a** and **b** using a scale of 1 cm : 4 mi.

1 cm : 4 mi

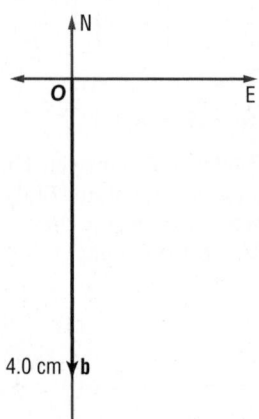

Translate **b** so that its tail touches the tip of **a**. Then draw the resultant vector **a** + **b**.

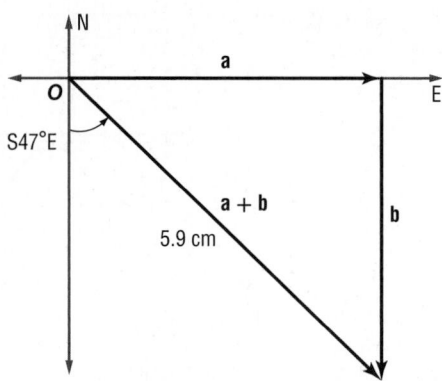

Measure the length of **a** + **b** and then measure the angle this vector makes with the horizontal. The length of the vector is approximately 5.9 centimeters, which is 5.9 × 4 or 23.6 miles. Therefore, the resultant is about 23.6 miles at a bearing of S47°E.

27.   **29.**

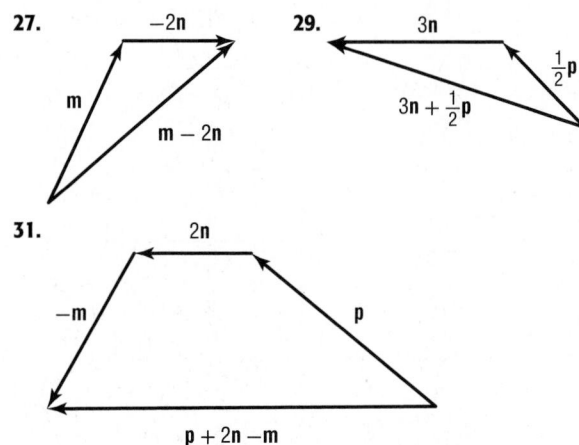

31.

33.   **35.** 7 mi/h
37. about 1.49 m/s at a bearing of S51°W

39. about 1.13 cm;   **41.** about 0.72 in./min;
about 0.98 cm   about 0.19 in./min

43a.   **43b.** about 159.3 N; about 103.5 N
45. 84 lb at 162° **47.** 11.7 lb at 215°

1 cm : 50 N

49 **a.** Let **v** represent the vertical component of the force and **r** represent the magnitude of the resultant force.

Use the tangent ratio to find **v**.

$\tan 31° = \dfrac{|v|}{86}$

$|\mathbf{x}| = 86 \tan 31°$

≈ 51.674

The vertical component of the force is about 52 newtons.

b. Use the cosine ratio to find **r**.

$\cos 31° = \dfrac{86}{|\mathbf{r}|}$

$|\mathbf{r}| \cos 31° = 86$

$|\mathbf{r}| = \dfrac{86}{\cos 31°}$

≈ 100.33

The magnitude of the resultant force is about 100 newtons.

51 Draw a diagram to represent **a** and **b** using a scale of 1 cm : 4 mi/h.

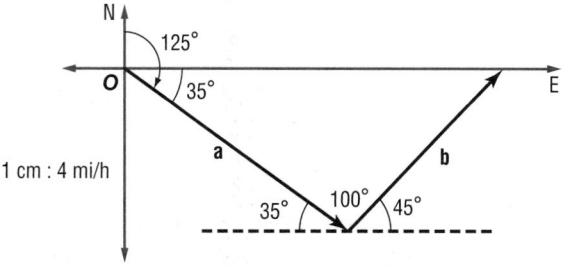

The angle created by **a** and the *x*-axis is 125° − 90° or 35°. Draw a horizontal where the tip of **a** and the tail of **b** meet, as shown. **b** makes a 45° angle and **a** makes a 35° angle with the horizontal. Thus, the angle created by **a** and **b** is 180° − 35° − 45° or 100°. Draw the resultant **a** + **b**. The three vectors form a triangle.

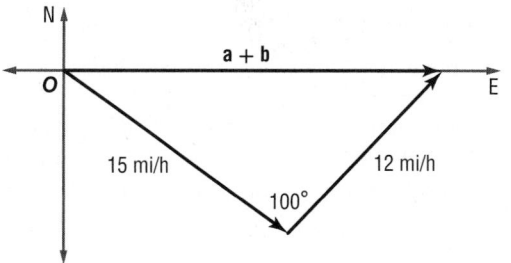

Use the Law of Cosines to find the magnitude of **a** + **b**.

$c^2 = a^2 + 2b^2 - 2ab \cos \theta$

$|\mathbf{a} + \mathbf{b}|^2 = 15^2 + 12^2 - 2(15)(12) \cos 100°$

$|\mathbf{a} + \mathbf{b}| = \sqrt{15^2 + 12^2 - 2(15)(12) \cos 100°}$

≈ 20.77

Use the Law of Sines to find the angle opposite of **b**.

$\dfrac{\sin 100°}{20.77} = \dfrac{\sin B}{12}$

$\dfrac{12 \sin 100°}{20.77} = \sin B$

$\sin^{-1} \dfrac{12 \sin 100°}{20.77} = B$

$34.68 \approx B$

The angle opposite **b** is about 35°. To find the bearing of **a** + **b**, subtract 35° from 125°. Thus, the direction of **a** + **b** is a bearing of 90°. Since the equilibrant vector is the opposite of the resultant vector, it will have a magnitude of about 20.77 mi/h at a bearing of about 270°.

53. about 14.87 ft/s at a bearing of 69° **55a.** Sample answers: $\mathbf{T}_{1x} = \mathbf{T}_1 \cos 65°$; $\mathbf{T}_{1y} = \mathbf{T}_1 \sin 65°$; $\mathbf{T}_{2x} = \mathbf{T}_2 \cos 35°$; $\mathbf{T}_{2y} = \mathbf{T}_2 \sin 35°$ **55b.** $\mathbf{T}_1 \approx 2079.5$ lb; $\mathbf{T}_2 \approx 1072.8$ lb **55c.** $\mathbf{T}_{1x} \approx 878.8$ lb; $\mathbf{T}_{1y} \approx 1884.7$ lb; $\mathbf{T}_{2x} \approx 878.8$ lb; $\mathbf{T}_{2y} \approx 615.3$lb **57.** about 5.2 ft at 54° **59.** about 3.4 yd at 212°

61. Sample answer:

63. Sample answer:

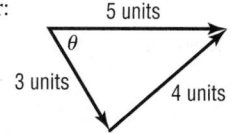

65. Sample answer: In order for the direction to have a consistent meaning, it must be measured using a common reference. Lack of a common reference would cause ambiguity in the reporting of the direction of the vector. **67a.** The magnitude of **a** added to the magnitude of **b** is greater than or equal to the magnitude of the vector created by **a** + **b**. **67b.** True; sample answer: The vector created by **a** + **b** has to account for the direction of both vectors. This can result in a very small magnitude, $|\mathbf{a} + \mathbf{b}|$, if **a** and **b** have opposite directions. Calculating the sum of the magnitudes, $|\mathbf{a}| + |\mathbf{b}|$, will result in the greatest possible value because direction is not being considered. This value can only be achieved by $|\mathbf{a} + \mathbf{b}|$ if **a** and **b** are parallel vectors with the same direction. **69.** Yes; sample answer: It is possible for the sum of two vectors to be equal to one of the components only when one of the vectors is the zero vector. **71.** 36.7 m

73. $(x - 4)^2 + (y - 5)^2 = 16$ **75.** $(x - 2)^2 = 4(y - 3)$

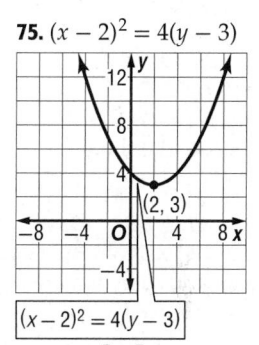

$(x - 4)^2 + (y - 5)^2 = 16$

$(x - 2)^2 = 4(y - 3)$

77a. Sample answer: $[12 \quad 25 \quad 45 \quad 50]$, $\begin{bmatrix} 60 \\ 40 \\ 25 \\ 5 \end{bmatrix}$

77b. \$3095 **79.** $\dfrac{\pi}{6} + 2n\pi, \dfrac{5\pi}{6} + 2n\pi,$ $\dfrac{3\pi}{2} + 2n\pi; n \in \mathbb{Z}$ **81.** G **83.** H

Pages 497–499 Lesson 8-2

1. $\langle 7, 4 \rangle$; $\sqrt{65} \approx 8.1$ **3.** $\langle -7, -3 \rangle$; $\sqrt{58} \approx 7.6$ **5.** $\langle 13, 2 \rangle$; $\sqrt{173} \approx 13.2$ **7.** $\langle -6.5, 4.5 \rangle$; $\sqrt{62.5} \approx 7.9$ **9.** $\left\langle \dfrac{11}{2}, \dfrac{23}{2} \right\rangle$; $\sqrt{\dfrac{325}{2}} \approx 12.7$ **11.** $\langle -21, 13 \rangle$ **13.** $\langle -55, -13 \rangle$ **15.** $\langle 26, 6 \rangle$ **17.** $\langle -42, -18 \rangle$ **19a.** $\mathbf{f} = \langle -9, 10 \rangle$, $\mathbf{n} = \langle 76, 84 \rangle$, $\mathbf{w} = \langle 0, -170 \rangle$ **19b.** $\langle 67, -76 \rangle$

21. $\mathbf{u} = \left\langle \dfrac{3\sqrt{10}}{10}, -\dfrac{\sqrt{10}}{10} \right\rangle$ **23.** $\mathbf{u} = \left\langle \dfrac{2\sqrt{5}}{5}, \dfrac{\sqrt{5}}{5} \right\rangle$ **25.** $\mathbf{u} = \left\langle -\dfrac{\sqrt{26}}{26}, -\dfrac{5\sqrt{26}}{26} \right\rangle$ **27.** $\mathbf{u} = \left\langle \dfrac{3}{5}, -\dfrac{4}{5} \right\rangle$ **29.** $-16\mathbf{i} + 8\mathbf{j}$ **31.** $-9.5\mathbf{i} - 8.3\mathbf{j}$ **33.** $13\mathbf{i} + 11\mathbf{j}$ **35.** 0

37 **a.** Nadia's rowing can be represented by the vector $\mathbf{r} = \langle 0, 5 \rangle$ and the current can be represented by the vector $\mathbf{c} = \langle 3, 0 \rangle$.

Add the vectors representing $\mathbf{r}$ and $\mathbf{c}$ to find the resultant vector, $\mathbf{v}$.

$\mathbf{v} = \mathbf{r} + \mathbf{c}$
$\phantom{\mathbf{v}} = \langle 0, 5 \rangle + \langle 3, 0 \rangle \text{ or } \langle 3, 5 \rangle$

The speed at which Nadia is traveling is the magnitude of $\mathbf{v}$.

$|\mathbf{v}| = \sqrt{3^2 + 5^2}$
$\phantom{|\mathbf{v}|} = \sqrt{34} \text{ or about } 5.8$

Thus, she is traveling at about 5.8 miles per hour.

b. Find θ by constructing a right triangle with the given vectors and using the tangent ratio.

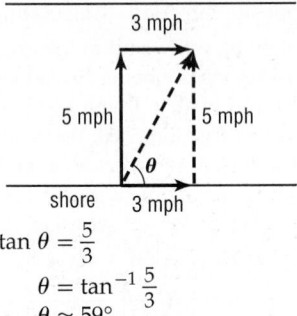

$\tan \theta = \dfrac{5}{3}$

$\theta = \tan^{-1} \dfrac{5}{3}$

$\theta \approx 59°$

Thus, Nadia is traveling at an angle of about 59° with respect to the shore.

39. $\langle -2\sqrt{2}, 2\sqrt{2} \rangle$ **41.** $\langle 8\sqrt{3}, -8 \rangle$ **43.** $\langle -8.6, 12.29 \rangle$
45. 111.8°   **47.** 216.9°   **49.** 45°   **51.** 290.6°   **53a.** about 674.3 mph   **53b.** about S86°E   **55.** No; sample answer: The magnitude and direction are not the same for both vectors, so they are not equivalent.   **57.** Yes; sample answer: The magnitude and direction are the same for both vectors, so they are equivalent.

59a. Sample answer:

Groundspeed: 518 mph   3°   wind
Air speed: 480 mph

59b. about 46.1 mph; about N46°E   **59c.** about 538 mph; about N79°E

61. Sample answer: $(0, -2)$ **63a.** wire 1: $\langle -755.2, 932.6 \rangle$; wire 2: $\langle 898.8, 438.4 \rangle$; wire 3: $\langle 544.0, -440.5 \rangle$
63b. $\langle 687.6, 930.5 \rangle$ **63c.** about 1157 N; about 53.5°
65. Sample answer: $\mathbf{a} = k\mathbf{b}$, where k is any real number scalar.
67. Sample answer: If the initial point of a vector is (a, b) and the vector has magnitude m, then the locus of terminal points (x, y) are the points that satisfy the equation $\sqrt{(x-a)^2 + (y-b)^2} = m$.

69 $\tan \theta = \dfrac{b}{a}$

$\tan 4y = \dfrac{y}{x}$

$x \tan 4y = y$

$x = \dfrac{y}{\tan 4y}$

71. $(\mathbf{a} + \mathbf{b}) + \mathbf{c} = (\langle x_1, y_1 \rangle + \langle x_2, y_2 \rangle) + \langle x_3, y_3 \rangle$
$\phantom{(\mathbf{a} + \mathbf{b}) + \mathbf{c}} = \langle x_1 + x_2, y_1 + y_2 \rangle + \langle x_3, y_3 \rangle$
$\phantom{(\mathbf{a} + \mathbf{b}) + \mathbf{c}} = \langle x_1 + x_2 + x_3, y_1 + y_2 + y_3 \rangle$
$\phantom{(\mathbf{a} + \mathbf{b}) + \mathbf{c}} = \langle x_1, y_1 \rangle + \langle x_2 + x_3, y_2 + y_3 \rangle$
$\phantom{(\mathbf{a} + \mathbf{b}) + \mathbf{c}} = \langle x_1, y_1 \rangle + (\langle x_2, y_2 \rangle + \langle x_3, y_3 \rangle)$
$\phantom{(\mathbf{a} + \mathbf{b}) + \mathbf{c}} = \mathbf{a} + (\mathbf{b} + \mathbf{c})$

73. $|k\mathbf{a}| = |k\langle x_1, y_1 \rangle|$
$\phantom{|k\mathbf{a}|} = |\langle kx_1, ky_1 \rangle|$
$\phantom{|k\mathbf{a}|} = \sqrt{(kx_1)^2 + (ky_1)^2}$
$\phantom{|k\mathbf{a}|} = \sqrt{k^2 x_1{}^2 + k^2 y_1{}^2}$
$\phantom{|k\mathbf{a}|} = \sqrt{k^2(x_1{}^2 + y_1{}^2)}$
$\phantom{|k\mathbf{a}|} = \sqrt{k^2}\sqrt{x_1{}^2 + y_1{}^2}$
$\phantom{|k\mathbf{a}|} = |k| |\langle x_1, y_1 \rangle|$
$\phantom{|k\mathbf{a}|} = |k| |\mathbf{a}|$

75. $y = \dfrac{x^2}{9} + \dfrac{4x}{3} + 6$ **77.** $y = 7\sqrt{x+1}$ **79.** $\dfrac{3}{z+2} - \dfrac{1}{2z-3}$
81. $\dfrac{-6}{x+3} + \dfrac{-3}{x-3}$
83. $\sin(60° + \theta) + \sin(60° - \theta)$
$= \sin 60° \cos \theta + \cos 60° \sin \theta + \sin 60° \cos \theta - \cos 60° \sin \theta$
$= \dfrac{\sqrt{3}}{2}\cos \theta + \dfrac{1}{2}\sin \theta + \dfrac{\sqrt{3}}{2}\cos \theta - \dfrac{1}{2}\sin \theta$
$= \dfrac{2\sqrt{3}}{2}\cos \theta$
$= \sqrt{3}\cos \theta$

85. $2\ln 7 - 2\ln 3$ **87.** $2\ln 3 - 3\ln 7$ **89.** 5984   **91.** H   **93.** H

Pages 506–508   Lesson 8-3

1. 8; not orthogonal   **3.** 0; orthogonal   **5.** −30; not orthogonal   **7.** 8; not orthogonal   **9a.** 15,620
9b. The total revenue that can be made by selling all of the basketballs is \$15,620.   **11.** $\sqrt{97} \approx 9.8$ **13.** $5\sqrt{13} \approx 18.0$
15. $\sqrt{785} \approx 28.0$ **17.** 100.0°   **19.** 82.9°   **21.** 48.4°   **23.** 159.1°

25. $\text{proj}_\mathbf{v} \, \mathbf{u} = \left\langle \dfrac{15}{29}, -\dfrac{6}{29} \right\rangle$; $\mathbf{u} = \left\langle \dfrac{15}{29}, -\dfrac{6}{29} \right\rangle + \left\langle \dfrac{72}{29}, \dfrac{180}{29} \right\rangle$

27. $\text{proj}_\mathbf{v} \, \mathbf{u} = \left\langle \dfrac{120}{17}, -\dfrac{30}{17} \right\rangle$; $\mathbf{u} = \left\langle \dfrac{120}{17}, -\dfrac{30}{17} \right\rangle + \left\langle \dfrac{16}{17}, \dfrac{64}{17} \right\rangle$

29. $\text{proj}_\mathbf{v} \, \mathbf{u} = \left\langle -\dfrac{78}{73}, \dfrac{208}{73} \right\rangle$; $\mathbf{u} = \left\langle -\dfrac{78}{73}, \dfrac{208}{73} \right\rangle + \left\langle \dfrac{224}{73}, \dfrac{84}{73} \right\rangle$

31. $\text{proj}_\mathbf{v} \, \mathbf{u} = \left\langle \dfrac{93}{13}, -\dfrac{62}{13} \right\rangle$; $\mathbf{u} = \left\langle \dfrac{93}{13}, -\dfrac{62}{13} \right\rangle + \left\langle -\dfrac{28}{13}, -\dfrac{42}{13} \right\rangle$

33 The combined weight of Malcolm's sister and the wagon is the force exerted due to gravity, $\mathbf{F} = \langle 0, -78 \rangle$. To find the force $-\mathbf{w}_1$ required to keep her from sliding down the slope, project $\mathbf{F}$ onto a unit vector $\mathbf{v}$ in the direction of the side of the hill.

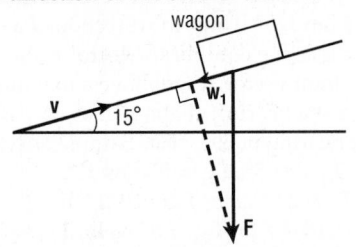

Step 1 Find a unit vector $\mathbf{v}$ in the direction of the side of the hill.

$$\mathbf{v} = \langle |\mathbf{v}| \cos \theta, |\mathbf{v}| \sin \theta \rangle$$
$$= \langle 1 \cdot \cos 15°, 1 \cdot \sin 15° \rangle$$
$$= \langle 0.966, 0.259 \rangle$$

Step 2 Find $\mathbf{w}_1$, the projection of $\mathbf{F}$ onto the unit vector $\mathbf{v}$, $\text{proj}_{\mathbf{v}}\mathbf{F}$.

$$\text{proj}_{\mathbf{v}}\mathbf{F} = \left(\frac{\mathbf{F} \cdot \mathbf{v}}{|\mathbf{v}|^2}\right)\mathbf{v}$$
$$= (\mathbf{F} \cdot \mathbf{v})\mathbf{v}$$
$$= (\langle 0, -78 \rangle \cdot \langle 0.966, 0.259 \rangle)\mathbf{v}$$
$$\approx -20.2\mathbf{v}$$

Since $\mathbf{w}_1$ points down the hill, the force required is $-\mathbf{w}_1 = -(-20.2\mathbf{v})$ or $20.2\mathbf{v}$. Since $\mathbf{v}$ is a unit vector, 20.2 pounds represents the force required to keep Malcolm's sister from sliding down the hill.

35. 801 J   **37.** Sample answer: $\langle -12, 3 \rangle$ **39.** Sample answer: $\langle 8, 14 \rangle$ **41a.** $\langle 16.38, 11.47 \rangle$, $\langle 22.94, -32.77 \rangle$
41b. dot product; $(20 \cos 35°)(40 \cos 55°) + (20 \sin 35°) \cdot (-40 \sin 55°) = 0$ **43.** Sample answer: $\mathbf{u} = \langle -1, 7 \rangle$
45. Sample answer: $\mathbf{u} = \langle 4, -5 \rangle$ **47.** Perpendicular; sample answer: Since $\mathbf{u} \cdot \mathbf{v} = 0$, the vectors are perpendicular.

49 Find the angle between $\mathbf{u}$ and $\mathbf{v}$.

$$\cos \theta = \frac{\mathbf{u} \cdot \mathbf{v}}{|\mathbf{u}||\mathbf{v}|}$$

$$\cos \theta = \frac{\langle 5, 7 \rangle \cdot \langle -15, -21 \rangle}{|\langle 5, 7 \rangle||\langle -15, -21 \rangle|}$$

$$\cos \theta = \frac{-75 + (-147)}{\sqrt{74}\,\sqrt{666}}$$

$$\cos \theta = \frac{-222}{222}$$

$$\cos \theta = -1$$
$$\theta = \cos^{-1}(-1) \text{ or } 180°$$

Parallel; sample answer: The angle between the two vectors is 180°. They are headed in opposite directions.

51. $\frac{5\pi}{12}$ **53.** $\frac{\pi}{2}$ **55.** about 141.1 J   **57.** 39.8°, 50.2°, 90°
59. 48.9°, 65.6°, 65.6°   **61.** $\langle -5.36, 0.55 \rangle$ or $\langle 2.03, -4.99 \rangle$
63. $\langle -9.03, 7.90 \rangle$ or $\langle 1.09, 11.95 \rangle$ **65.** False; sample answer: All three vectors may originate at the same point and not form a triangle at all. If so, the angle between $\mathbf{d}$ and $\mathbf{f}$ may be acute, right, or obtuse.   **67.** False; sample answer: If either $\mathbf{a}$ or $\mathbf{b}$ is a zero vector, it would be orthogonal to $\mathbf{v}$, but could not be parallel to the other vector.

69. $\cos 90° = \dfrac{\mathbf{u} \cdot \mathbf{v}}{|\mathbf{u}| \cdot |\mathbf{v}|}$ Angle between $\mathbf{u}$ and $\mathbf{v}$ where $\theta = 90°$.

$0 = \dfrac{\mathbf{u} \cdot \mathbf{v}}{|\mathbf{u}| \cdot |\mathbf{v}|}$ $\cos 90° = 0$

$0 = \mathbf{u} \cdot \mathbf{v}$ Multiply each side by $|\mathbf{u}| \cdot |\mathbf{v}|$.

71. $\mathbf{u} \cdot (\mathbf{v} + \mathbf{w}) = \mathbf{u} \cdot \mathbf{v} + \mathbf{u} \cdot \mathbf{w}$
$\langle u_1, u_2 \rangle \cdot (\langle v_1, v_2 \rangle + \langle w_1, w_2 \rangle) \overset{?}{=} \langle u_1, u_2 \rangle \cdot \langle v_1, v_2 \rangle + \langle u_1, u_2 \rangle \cdot \langle w_1, w_2 \rangle$
$\langle u_1, u_2 \rangle \cdot \langle v_1 + w_1, v_2 + w_2 \rangle \overset{?}{=} (u_1 v_1 + u_2 v_2) + (u_1 w_1 + u_2 w_2)$
$u_1(v_1 + w_1) + u_2(v_2 + w_2) \overset{?}{=} u_1 v_1 + u_1 w_1 + u_2 v_2 + u_2 w_2$
$u_1 v_1 + u_1 w_1 + u_2 v_2 + u_2 w_2 = u_1 v_1 + u_1 w_1 + u_2 v_2 + u_2 w_2$

73. Sample answer: For two nonzero vectors $\langle a, b \rangle$ and $\langle c, d \rangle$, the dot product $\langle a, b \rangle \cdot \langle c, d \rangle$ is the sum of the products of the x-coordinates and the y-coordinates, or $ac + bd$.

---

75. $\left\langle -\dfrac{137}{4}, -9.2 \right\rangle$ **77.** Jada: about 173.8 ft/s and about 108.6 ft/s; James: about 143.4 ft/s and about 124.7 ft/s

79.

$\dfrac{x^2}{81} - \dfrac{y^2}{49} = 1$

81. $\dfrac{\pi}{6}$ **83.** $\dfrac{\sqrt{7}}{4}$ **85.** 6
87a. $t = 4$ **87b.** when $t = 4$
87c. 10 amps   **89.** F   **91.** H

Pages 514–516   Lesson 8-4

1.

$(1, -2, -4)$

3.

$(-5, -4, -2)$

5.

$(-5, 3, 1)$

7.

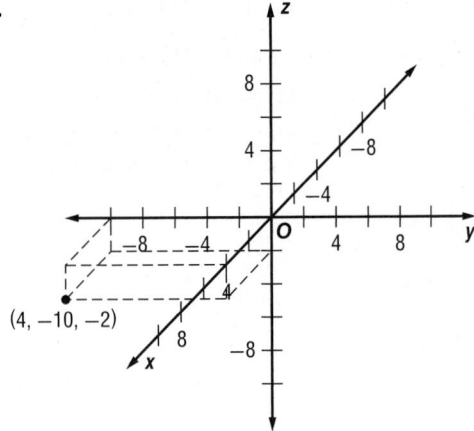

$(4, -10, -2)$

9. $5\sqrt{6} \approx 12.2; \left(-\frac{3}{2}, 5, \frac{13}{2}\right)$ **11.** $\sqrt{542} \approx 23.3; \left(-\frac{3}{2}, -\frac{9}{2}, 3\right)$

13. $4\sqrt{14} \approx 15.0; (3, 4, 4)$ **15.** longitude: 38.55°, latitude: 101°, altitude: 1207 m

17.

$(0, -4, 4)$

19.

$(-1, 3, -4)$

21.

$(6, 8, -2)$

23.

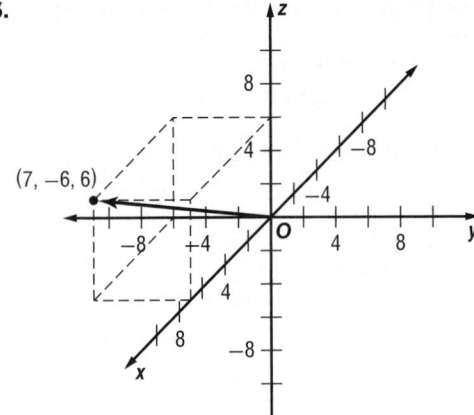

$(7, -6, 6)$

25. $\langle 16, 2, 8 \rangle; 18; \left\langle \frac{8}{9}, \frac{1}{9}, \frac{4}{9} \right\rangle$ **27.** $\langle -3, -5, -10 \rangle; \sqrt{134};$

$\left\langle -\frac{3\sqrt{134}}{134}, -\frac{5\sqrt{134}}{134}, -\frac{5\sqrt{134}}{67} \right\rangle$

29. $\langle -1, 8, -10 \rangle; \sqrt{165}; \left\langle -\frac{\sqrt{165}}{165}, \frac{8\sqrt{165}}{165}, -\frac{2\sqrt{165}}{33} \right\rangle$

31. $\langle 4, -15, 5 \rangle; \sqrt{266}; \left\langle \frac{2\sqrt{266}}{133}, -\frac{15\sqrt{266}}{266}, \frac{5\sqrt{266}}{266} \right\rangle$

33. $\langle 11, -23, -13 \rangle; 3\sqrt{91}; \left\langle \frac{11\sqrt{91}}{273}, -\frac{23\sqrt{91}}{273}, -\frac{\sqrt{91}}{21} \right\rangle$

35 Let P represent the point at which the rope and the pole are connected and let B represent the position of the ball as it is being served. Then $\overrightarrow{PB}$ represents the vector from the pole to the ball.

$\overrightarrow{PB} = \langle 5 - 0, 3.6 - 0, 4.7 - 9.8 \rangle$
$= \langle 5, 3.6, -5.1 \rangle$

The magnitude of $\overrightarrow{PB}$ is

$\sqrt{5^2 + 3.6^2 + (-5.1)^2} \approx 8.0.$

Thus, the rope is about 8 feet long.

37. $\langle -65, -18, 56 \rangle$ **39.** $\langle 48, 12, -38 \rangle$ **41.** $\langle 22, 36, 3 \rangle$
43. $-63\mathbf{i} + 28\mathbf{j} + 56\mathbf{k}$ **45.** $50\mathbf{i} - 18\mathbf{j} + 10\mathbf{k}$ **47.** $-13\mathbf{i} + 2\mathbf{j} + 21\mathbf{k}$ **49.** $3.1\mathbf{i} - 23.6\mathbf{j} + 13.4\mathbf{k}$

51 Let $P = (x_2, y_2, z_2)$. Use the Midpoint Formula and the points M and N to solve for P.

$N = \left(\frac{x_1 + x_2}{2}, \frac{y_1 + y_2}{2}, \frac{z_1 + z_2}{2} \right)$

$\left(\frac{7}{2}, 1, 2 \right) = \left(\frac{3 + x_2}{2}, \frac{4 + y_2}{2}, \frac{5 + z_2}{2} \right)$

Solve for x_2, y_2, and z_2.

$\frac{3 + x_2}{2} = \frac{7}{2}$ $\frac{4 + y_2}{2} = 1$ $\frac{5 + z_2}{2} = 2$
$3 + x_2 = 7$ $4 + y_2 = 2$ $5 + z_2 = 4$
$x_2 = 4$ $y_2 = -2$ $z_2 = -1$

So, $P = (4, -2, -1)$.

53. $(3, -2, 7)$ **55.** 34 ft **57.** scalene **59.** isosceles

61. $d = \sqrt{(x_2 - x_1)^2 + (y_2 - y_1)^2 + (z_2 - z_1)^2}$

$r = \sqrt{(x - h)^2 + (y - k)^2 + (z - \ell)^2}$

$r^2 = (x - h)^2 + (y - k)^2 + (z - \ell)^2$

63. $(x - 6)^2 + y^2 + (z + 1)^2 = \frac{1}{4}$ **65.** $x^2 + (y - 7)^2 + (z + 1)^2 = 144$ **67a.** about 199.2 mi/h **67b.** N4.1°E

67c. about 75.9° **69.** Sample answer: It is more reasonable to use two dimensions when describing position on a map since the map itself is two-dimensional. When the description of the position of an object requires the added dimension of altitude, such as the position of an airplane, it is more reasonable to use three dimensions.
71. $\langle -0.19, -0.04 \rangle$; $\mathbf{u} = \langle -0.19, -0.04 \rangle + \langle -0.81, 4.04 \rangle$
73. $\langle -13, -3 \rangle$, $\sqrt{178} \approx 13.3$ **75.** $\langle 6, 18 \rangle$, $\sqrt{360} \approx 19.0$
77a. about 2.6 feet to the left and to the right of the center of the arch **77b.** 8 ft; Sample answer: With the string anchored by thumbtacks at the foci of the arch and held taut by a pencil, the sum of the distances from each thumbtack to the pencil will remain constant. When the pencil is at one of the bottom corners of the arch, exactly 4 feet from the center, the distance from the far thumbtack to the bottom corner will be $4 + 2.6$ or about 6.6 feet and the distance from the other thumbtack to the bottom corner will be $4 - 2.6$ or about 1.4 feet. Thus, the total length of string needed is $6.6 + 1.4$ or 8 feet.
79. $\cos^{-1}\left(\frac{1}{3}\right) + n\pi, \cos^{-1}\left(-\frac{1}{3}\right) + n\pi, n \in \mathbb{Z}$

81.

83.

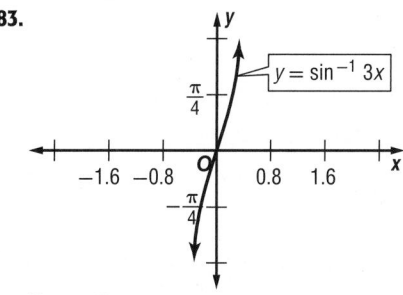

85. F **87.** J

Pages 522–524 Lesson 8-5

1. 0; orthogonal **3.** −13; not orthogonal **5.** −15; not orthogonal **7.** 0; orthogonal

9 Let $\mathbf{u} = \langle 55.5, 55.5, -55.5 \rangle$ and $\mathbf{v} = \langle -55.5, -55.5, -55.5 \rangle$. Use the formula for the angle between two vectors.

$$\cos \theta = \frac{\mathbf{u} \cdot \mathbf{v}}{|\mathbf{u}||\mathbf{v}|}$$

$$\cos \theta = \frac{\langle 55.5, 55.5, -55.5 \rangle \cdot \langle -55.5, -55.5, -55.5 \rangle}{|\langle 55.5, 55.5, -55.5 \rangle||\langle -55.5, -55.5, -55.5 \rangle|}$$

$$\cos \theta = \frac{-3080.25 + (-3080.25) + 3080.25}{\sqrt{9240.75} \sqrt{9240.75}}$$

$$\cos \theta = \frac{-3080.25}{9240.75}$$

$$\theta = \cos^{-1}\left(-\frac{1}{3}\right) \text{ or about } 109.5°$$

The angle between the vectors formed by the oxygen-hydrogen bonds is about 109.5°.

11. 88.9° **13.** 37.5° **15.** 16.6°

---

17. $\langle 25, 6, 71 \rangle$

$(\mathbf{u} \times \mathbf{v}) \cdot \mathbf{u}$
$= \langle 25, 6, 71 \rangle \cdot \langle 4, 7, -2 \rangle$
$= 25(4) + 6(7) + 71(-2)$
$= 0$

$(\mathbf{u} \times \mathbf{v}) \cdot \mathbf{v}$
$= \langle 25, 6, 71 \rangle \cdot \langle -5, 9, 1 \rangle$
$= 25(-5) + 6(9) + 71(1)$
$= 0$

19. $\langle -56, -35, -42 \rangle$

$(\mathbf{u} \times \mathbf{v}) \cdot \mathbf{u}$
$= \langle -56, -35, -42 \rangle \cdot \langle 5, -8, 0 \rangle$
$= (-56)(5) + (-35)(-8) + (-42)(0)$
$= 0$

$(\mathbf{u} \times \mathbf{v}) \cdot \mathbf{v}$
$= \langle -56, -35, -42 \rangle \cdot \langle -4, -2, 7 \rangle$
$= (-56)(-4) + (-35)(-2) + (-42)(7)$
$= 0$

21. $\langle 29, 12, 13 \rangle$

$(\mathbf{u} \times \mathbf{v}) \cdot \mathbf{u}$
$= \langle 29, 12, 13 \rangle \cdot \langle -4, 1, 8 \rangle$
$= 29(-4) + 12(1) + 13(8)$
$= 0$

$(\mathbf{u} \times \mathbf{v}) \cdot \mathbf{v}$
$= \langle 29, 12, 13 \rangle \cdot \langle 3, -4, -3 \rangle$
$= 29(3) + 12(-4) + 13(-3)$
$= 0$

23a. $\langle 0, -72.08, 0 \rangle$ **23b.** about 72.08 N · m parallel to the negative y-axis **25.** $13\sqrt{19} \approx 56.7$ units^2 **27.** $\sqrt{6821} \approx$ 82.6 units^2 **29.** $\sqrt{2158} \approx 46.5$ units^2 **31.** 206 units^3 **33.** 102 units^3 **35.** 69 units^3 **37.** Sample answer: $\langle 5, 5, 3 \rangle$ **39.** Sample answer: $\langle -8, 0, 7 \rangle$ **41.** Sample answer: $\langle -1, 0, 4 \rangle$ **43.** not collinear **45.** parallel **47.** parallel
49. $\left\langle -\frac{11\sqrt{3}}{2}, -\frac{11}{2}, 0 \right\rangle$ **51.** parallelogram; about 9.4 units^2; rectangle **53.** 0 **55.** $\langle 0, 0, 0 \rangle$ **57a.** about 2.1 N **57b.** about 98.8° **59.** 5 **61.** Always; sample answer: The cross product of two three-dimensional vectors results in a vector that is perpendicular to both original vectors.

63 Calculate $\mathbf{u} \times \mathbf{v}$.

$$\mathbf{u} \times \mathbf{v} = \begin{vmatrix} \mathbf{i} & \mathbf{j} & \mathbf{k} \\ 4 & 6 & c \\ -3 & -2 & 5 \end{vmatrix}$$

$$= \begin{vmatrix} 6 & c \\ -2 & 5 \end{vmatrix} \mathbf{i} - \begin{vmatrix} 4 & c \\ -3 & 5 \end{vmatrix} \mathbf{j} + \begin{vmatrix} 4 & 6 \\ -3 & -2 \end{vmatrix} \mathbf{k}$$

$$= (30 + 2c)\mathbf{i} - (20 + 3c)\mathbf{j} + 10\mathbf{k}$$

For $\mathbf{u} \times \mathbf{v}$ to equal $34\mathbf{i} - 26\mathbf{j} + 10\mathbf{k}$, set the components equal and solve for c.
$30 + 2c = 34$, so $c = 2$.
$-(20 + 3c) = -26$, so $c = 2$.
This implies the system is consistent. Thus, $c = 2$.

65. Sample answer: To determine if two vectors are parallel or perpendicular, you can use the formula for the angle between two vectors. If the angle measures 0 or 180, they are parallel. If the angle measures 90, they are perpendicular. You can also find the component form of the two vectors and use the ratios of the coordinates to determine if two vectors are parallel. If the ratios of all three coordinates of the component vectors are the same, then the vectors are parallel. To determine if two vectors are perpendicular, you can find the dot product of the vectors. If the dot product is 0, then the vectors are perpendicular. **67.** $\sqrt{562} \approx 23.7$; $\left(\frac{33}{2}, 9, -\frac{37}{2}\right)$ **69.** −22; not orthogonal **71.** −33; not orthogonal **73.** $\frac{3}{m - 4} + \frac{-1}{m + 4}$
75. $\sec^2 \theta \cot^2 \theta - \cot^2 \theta = \cot^2 \theta (\sec^2 \theta - 1)$
$= \cot^2 \theta \tan^2 \theta$
$= 1$

77. $A \approx 55°, C \approx 78°, b \approx 17.9$ **79.** $A \approx 103°, B \approx 49°, C \approx 28°$
81. $29.102°$ **83.** D **85a.** horizontal: $45\sqrt{3}$ ft/s; vertical:
45 ft/s **85b.** scalars **85c.** about 226.0 ft **85d.** $\langle -77.3, 212.4 \rangle$
85e. $\langle -0.342, 0.940 \rangle$ **85f.** about N51.1°W

Pages 525–528 Chapter 8 Study Guide and Review

1. false; ends **3.** true **5.** true **7.** false; projection **9.** true
11. vector **13.** 4.0 cm, 11° **15.** 2.8 cm, 297°
17. 80 m due east **19.** $\langle 6, 1 \rangle$; $\sqrt{37} \approx 6.1$ **21.** $\langle 14, 5 \rangle$;
$\sqrt{221} \approx 14.9$ **23.** $\langle -8, -6 \rangle$ **25.** $\langle -18, -1 \rangle$
27. $\left\langle -\dfrac{7\sqrt{53}}{53}, \dfrac{2\sqrt{53}}{53} \right\rangle$ **29.** $\left\langle -\dfrac{5\sqrt{89}}{89}, -\dfrac{8\sqrt{89}}{89} \right\rangle$
31. -1; not orthogonal **33.** 0; orthogonal **35.** 135.0°

37.

(1, 2, −4)

39.

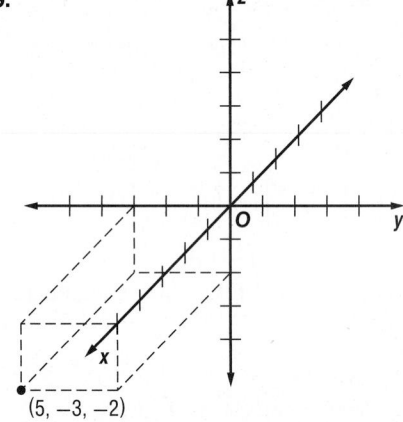

(5, −3, −2)

41. $2\sqrt{38} \approx 12.3$; $(-1, 5, 6)$ **43.** $4\sqrt{19} \approx 17.4$; $(-3, -4, 2)$

45.

(0, −3, 4)

47.

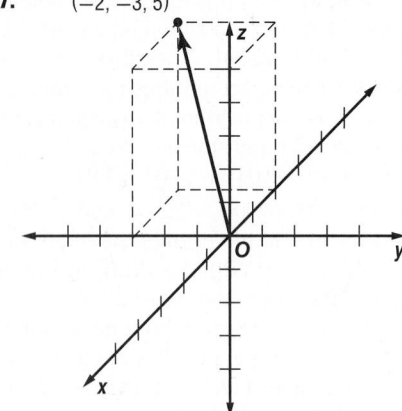

(−2, −3, 5)

49. 0; orthogonal **51.** $\langle 17, -1, 10 \rangle$; $\langle 17, -1, 10 \rangle \cdot$
$\langle 1, -3, -2 \rangle = 0$ and $\langle 17, -1, 10 \rangle \cdot \langle 2, 4, -3 \rangle = 0$ **53.** about
49.8 ft/s; about 23.2 ft/s **55.** about 1081.8 lb **57a.** about
5.3 mph **57b.** about 40.9° **59.** 800 J **61a.** $\langle -2.3, 0, 0 \rangle$
61b. about 2.3 N · m parallel to the negative x-axis

Pages 530–531 Chapter 8 Connect to AP Calculus

1. different **3.** No; sample answer: Every point (x, y) in a
plane has a vector associated to it, and there are infinitely
many points in a given plane. **5.** Sample answer:
Representing a vector field in only two dimensions ignores
the third dimension and assumes that the force created by
wind would remain constant along the z-plane.

Chapter 9 Polar Coordinates and Complex Numbers

Page 533 Chapter 9 Get Ready

1.

[−10, 10] scl: 2 by [−2, 18] scl: 2

even; symmetric with
respect to the y-axis
$f(-x) = (-x)^2 + 10$
$= x^2 + 10$
$= f(x)$

3.

[−10, 10] scl: 1 by [−10, 10] scl: 1

neither
$g(-x) = \sqrt{-x + 9}$ which
is not equal to $g(x)$ or
$-g(x)$.

5.

[−10, 10] scl: 1 by [−10, 10] scl: 1

odd; symmetric with
respect to the origin
$g(-x) = 3(-x)^5 - 7(-x)$
$= 3(-x^5) + 7x$
$= -3x^5 + 7x$
$= -g(x)$

7. even **9.** abs. max.: (2.25, 9.13) **11.** real. max.: (−1.67,
6.48); rel. min.: (1, −3)

13. $165° + 360k°$; $525°$; $-195°$

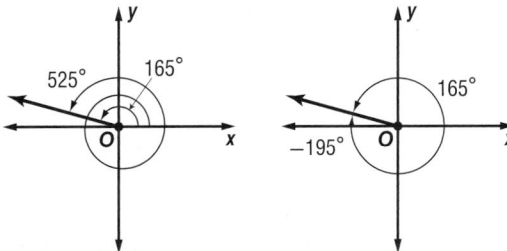

15. $-10° + 360k°$; $350°$; $-370°$

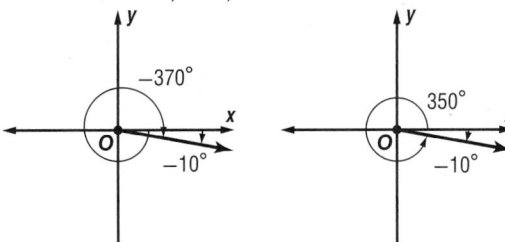

17. $\frac{4\pi}{3} + 2k\pi$; $\frac{10\pi}{3}$; $-\frac{2\pi}{3}$

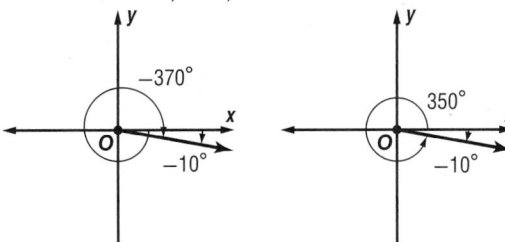

Pages 538–540   Lesson 9-1

1.

3.   **5.**

7.   **9.**

11.   **13a.**

13b. 13 points   **15.** $(2, 120°)$, $(2, -240°)$, $(-2, -60°)$

17. $\left(3, \frac{5\pi}{3}\right)$, $\left(3, -\frac{\pi}{3}\right)$, $\left(-3, -\frac{4\pi}{3}\right)$ **19.** $\left(5, \frac{5\pi}{3}\right)$, $\left(5, -\frac{\pi}{3}\right)$, $\left(-5, \frac{2\pi}{3}\right)$ **21.** $(1, 300°)$, $(1, -60°)$, $(-1, 120°)$

23.   **25.**

27.   **29a.** $r = 6.3$, $r = 16$, $r = 225$

29b. $\approx 0.5\%$ **31.** ≈ 10.70 **33.** 8   **35.** ≈ 3.05 **37.** 5
39. ≈ 6.08 **41.** ≈ 5.35

43 **a.** The parameters $-60° \leq \theta \leq 150°$ and $0 \leq r \leq 40$ describe a sector of a circle with radius 40 units from $-60°$ to $150°$.

b. The area of a sector of a circle with radius r and central angle θ is $A = \frac{1}{2}r^2\theta$, where θ is measured in radians. The angle of the sector is $150° - (-60°)$ or $210°$, which is $210° \times \frac{\pi}{180°}$ or $\frac{7\pi}{6}$ radians. Therefore, the area of the sector is $\frac{1}{2}(40)^2\left(\frac{7\pi}{6}\right)$ or about 2932.2 m².

45. $\left(-2.5, \frac{\pi}{2}\right)$ **47.** $(1.25, 160°)$ **49.** $(6, 160°)$ **51.** ≈ 10 ft
53. $\approx 174.46°$

55 Substitute the given information into the Polar Distance Formula.

$$P_1P_2 = \sqrt{r_1{}^2 + r_2{}^2 - 2r_1r_2 \cos(\theta_2 - \theta_1)}$$

$$3.297 = \sqrt{r^2 + 4^2 - 2r(4)\cos(160° - 120°)}$$

$$3.297^2 = r^2 + 16 - 8r \cos 40°$$

$$= r^2 - 8r \cos 40°r + (16 - 3.297^2)$$

Use the Quadratic Formula to solve for r.

$$r = \frac{8\cos 40° \pm \sqrt{(-8\cos 40°)^2 - 4(1)(16 - 3.297^2)}}{2(1)}$$

$$\approx 1 \text{ or } 5.13$$

57. Sample answer: $\theta = \frac{\pi}{12}$ **59.** $r = 4$ or $r = -4$ **61.** Sample answer: The sum and product of the r-values are calculated. Both of these operations are commutative. Due to the odd-even identities of trigonometric functions, $\cos(-\theta) = \cos\theta$. Therefore, the order of the angles does not matter.

63.

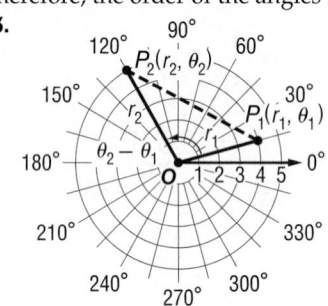

The triangle created by the two points and the origin results in a triangle with two known sides and an included angle. Therefore, by the Law of Cosines,

$(P_1P_2)^2 = r_1{}^2 + r_2{}^2 - 2r_1r_2 \cos(\theta_2 - \theta_1)$ or

$P_1P_2 = \sqrt{r_1{}^2 + r_2{}^2 - 2r_1r_2 \cos(\theta_2 - \theta_1)}$.

65. Erina; sample answer: Sona plotted a point where the distance between the polar axis and the ray equals 5 units. She should have measured 5 units along the terminal side of the angle.   **67.** -3; not orthogonal   **69.** -14; not orthogonal   **71.** $\langle 1, 44, -17 \rangle$

73. vertex: $(2, 7)$;
focus: $(-1.5, 7)$;
directrix: $x = 5.5$;
axis of symmetry: $y = 7$

75. vertex: $(3, 5)$;
focus: $(3, 5.5)$;
directrix: $y = 4.5$;
axis of symmetry: $x = 3$

77. $\begin{bmatrix} 12 & 14 & -10 & 0 & | & 23 \\ 4 & 0 & -5 & 6 & | & 33 \\ 11 & -13 & 0 & 2 & | & -19 \\ 0 & 19 & -6 & 7 & | & -25 \end{bmatrix}$ **79.** $\begin{bmatrix} 1 & 8 & -3 & | & 25 \\ 2 & -5 & 11 & | & 13 \\ -5 & 0 & 8 & | & 26 \\ 0 & 1 & -4 & | & 17 \end{bmatrix}$

81. $\frac{3\pi}{4} + \pi n; n \in \mathbb{Z}$ **83.** A   **85.** B

R108   Selected Answers and Solutions

Pages 548–550   Lesson 9-2

1.

3.

5.

7.

9.

11.

13.

15.

17.

19.

21.

23.

25a.

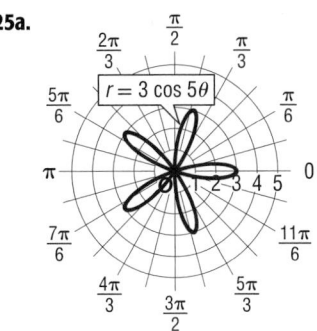

symmetry with respect to the polar axis; $|r| = 3$ when $\theta = 0, \dfrac{\pi}{5}, \dfrac{2\pi}{5}, \dfrac{3\pi}{5}, \dfrac{4\pi}{5}$, and π; zeros for r when $\theta = \dfrac{\pi}{10}, \dfrac{3\pi}{10}, \dfrac{\pi}{2}, \dfrac{7\pi}{10}$, and $\dfrac{9\pi}{10}$

25b.

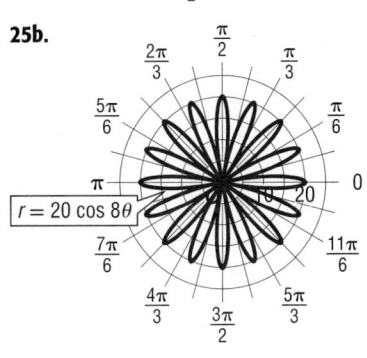

symmetry with respect to the polar axis, the line $\theta = \dfrac{\pi}{2}$, and the pole; $|r| = 20$ when $\theta = 0, \dfrac{\pi}{8}, \dfrac{\pi}{4}, \dfrac{3\pi}{8}, \dfrac{\pi}{2}, \dfrac{5\pi}{8}, \dfrac{3\pi}{4}, \dfrac{7\pi}{8}$, and π; zeros for r when $\theta = \dfrac{\pi}{16}, \dfrac{3\pi}{16}, \dfrac{5\pi}{16}, \dfrac{7\pi}{16}, \dfrac{9\pi}{16}, \dfrac{11\pi}{16}, \dfrac{13\pi}{16}$, and $\dfrac{15\pi}{16}$

27. Spiral of Archimedes;

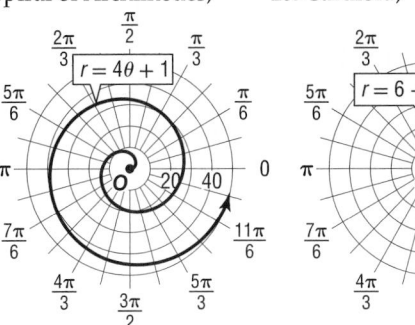

29. Cardioid;

31. Spiral of Archimedes;

32. Lemniscate;

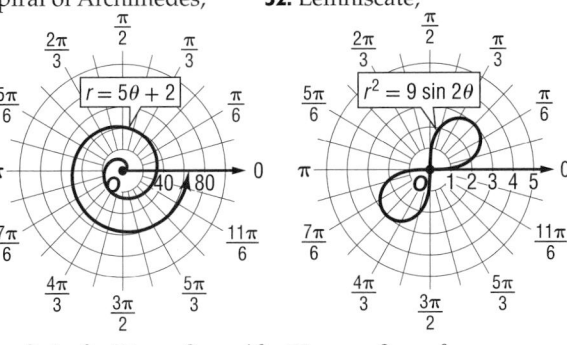

35. $r = 3 \sin \theta$ **37.** $r = 3 \cos 4\theta$ **39.** $r = -2 \cos \theta$

41 **a.** Since the fan is in the shape of a rose curve, the equation is of the form $r = a \cos n\theta$ or $r = a \sin n\theta$. Since the fan has 5 blades and 5 is odd, $n = 5$. The length of

each blade is 4 ft, so $a = 4$. Thus, two equations that can represent the blades are $r = 4 \cos 5\theta$ and $r = 4 \sin 5\theta$.

b.

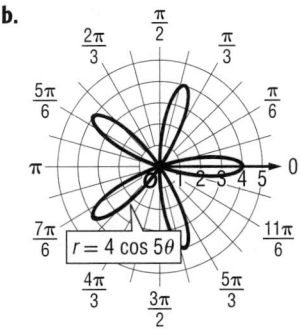

$r = 4 \cos 5\theta$ is symmetric with respect to the polar axis, so you can find points on the interval $[0, \pi]$ and then use polar axis symmetry to complete the graph.

θ	0	$\dfrac{\pi}{6}$	$\dfrac{\pi}{4}$	$\dfrac{\pi}{3}$	$\dfrac{\pi}{2}$	$\dfrac{2\pi}{3}$	$\dfrac{3\pi}{4}$	$\dfrac{5\pi}{6}$	π
r	4	-3.5	-2.8	2	0	-2	2.8	3.5	-4

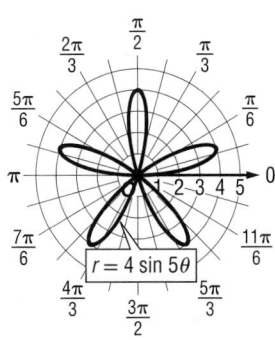

$r = 4 \sin 5\theta$ is symmetric with respect to the line $\theta = \dfrac{\pi}{2}$, so you can find points on the interval $\left[-\dfrac{\pi}{2}, \dfrac{\pi}{2} \right]$ and then use symmetry to complete the graph.

θ	$-\dfrac{\pi}{2}$	$-\dfrac{\pi}{3}$	$-\dfrac{\pi}{4}$	$-\dfrac{\pi}{6}$	0	$\dfrac{\pi}{6}$	$\dfrac{\pi}{4}$	$\dfrac{\pi}{3}$	$\dfrac{\pi}{2}$
r	-4	3.5	2.8	-2	0	2	-2.8	-3.5	4

43. Substitute $(-r, \theta)$ for (r, θ).
$$r^2 = 4 \sin 2\theta$$
$$(-r)^2 = 4 \sin 2\theta$$
$$r^2 = 4 \sin 2\theta$$
Since this substitution produces an equivalent equation, $r^2 = 4 \sin 2\theta$ is symmetric with respect to the pole.

45. Substitute $(r, \pi - \theta)$ for (r, θ).
$$r = 5 \cos 8\theta$$
$$r = 5 \cos 8(\pi - \theta)$$
$$r = 5 \cos (8\pi - 8\theta)$$
$$r = 5(\cos 8\pi \cos 8\theta + \sin 8\pi \sin 8\theta)$$
$$r = 5[(1) \cos 8\theta - (0) \sin 8\theta]$$
$$r = 5 \cos 8\theta$$
Since this substitution produces an equivalent equation, $r = 5 \cos 8\theta$ is symmetric with respect to the line $\theta = \dfrac{\pi}{2}$. Substitute $\pi - \theta$ for θ.

47a. $r = 2 \sin 5\theta$ or $r = 2 \cos 5\theta$ **47b.** $r = 7 \sin 2\theta$ or $r = 7 \cos 2\theta$ **47c.** $r = 6 \sin 4\theta$ or $r = 6 \cos 4\theta$ **49.** $r = 2 + 2 \sin \theta$ **51.** a   **53.** b

55 Make a table of values for $r = 2 - \sin 2\theta$.

θ	0	$\dfrac{\pi}{6}$	$\dfrac{\pi}{4}$	$\dfrac{\pi}{3}$	$\dfrac{\pi}{2}$	$\dfrac{2\pi}{3}$	$\dfrac{3\pi}{4}$	$\dfrac{5\pi}{6}$
r	2	1.13	1	1.13	2	2.86	3	2.86

θ	π	$\dfrac{7\pi}{6}$	$\dfrac{5\pi}{4}$	$\dfrac{4\pi}{3}$	$\dfrac{3\pi}{2}$	$\dfrac{5\pi}{3}$	$\dfrac{7\pi}{4}$	$\dfrac{11\pi}{6}$
r	2	1.13	1	1.13	2	2.86	3	2.86

θ	2π	$\dfrac{13\pi}{6}$	$\dfrac{9\pi}{4}$	$\dfrac{7\pi}{3}$	$\dfrac{5\pi}{2}$	$\dfrac{8\pi}{3}$	$\dfrac{11\pi}{4}$	$\dfrac{17\pi}{6}$
r	2	1.13	1	1.13	2	2.86	3	2.86

Notice that the points in the first and third tables are equivalent polar coordinates. Therefore, for a complete polar graph, the function must be graphed on the interval $0 \le \theta \le 2\pi$. So $x = 2\pi$.

57. b **59.** d

61a.

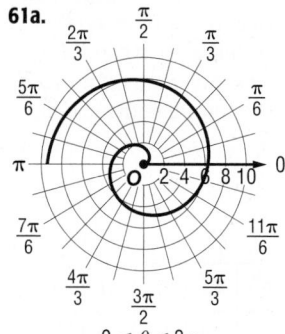

$0 \le \theta \le 3\pi$

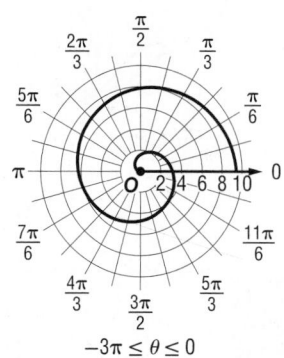

$-3\pi \le \theta \le 0$

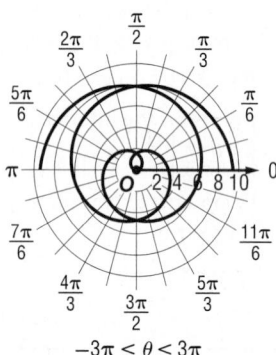

$-3\pi \le \theta \le 3\pi$

61b. Sample answer: The equation $r = \theta$ is symmetric with respect to the line $\theta = \frac{\pi}{2}$ when the interval for θ is $-a \le \theta \le a$.

61c. Substitute $(-r, -\theta)$ for (r, θ).

$r = \theta$
$-r = -\theta$
$r = \theta$

This substitution produces an equivalent equation, therefore $r = \theta$ is symmetric with respect to the line $\theta = \frac{\pi}{2}$.

61d. Sample answer: It does not affect the other classic curves. The classic curves all consist of either a sine or cosine function. Therefore, to achieve a complete graph, they just need to be graphed for all values of θ within their period. Extending the interval of θ to include additional values outside of the period will result in the graph repeating itself. Because the spiral of Archimedes does not contain a trigonometric function, additional values of θ result in different values of r.

63. Sample answer: r_2 and r_3 are the graphs of r_1 after a rotation about the pole of $\frac{\pi}{2}$ and π, respectively. The graph of $r = \cos(\theta - d)$ will be the graph of $r = \cos\theta$ after a rotation of d about the pole.

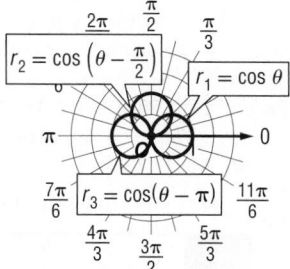

65. To test for symmetry with respect to the line $\theta = \frac{\pi}{2}$, substitute $(r, \pi - \theta)$ for (r, θ).

$r = a + b\cos 2(\pi - \theta)$
$r = a + b\cos(2\pi - 2\theta)$
$r = a + b(\cos 2\pi \cos 2\theta + \sin 2\pi \sin 2\theta)$
$r = a + b[(1)\cos 2\theta + (0)\sin 2\theta]$
$r = a + b\cos 2\theta$

Since this substitution produces an equivalent equation, the graph of $r = a + b\cos 2\theta$ is symmetric with respect to the line $\theta = \frac{\pi}{2}$.

67. Sample answer: The value of a determines the diameter of the circle. If $a > 0$, the graph is located in Quadrants I and IV of the plane with the polar axis containing a diameter. If $a < 0$, the graph is located in Quadrants II and III of the plane with the polar axis containing a diameter.

69.

71.

73. $144.3°$ **75.** $\frac{21}{5}\mathbf{i} - \frac{2}{3}\mathbf{j}$ **77.** $-15.8\mathbf{i} - 6.1\mathbf{j}$

79.

81.

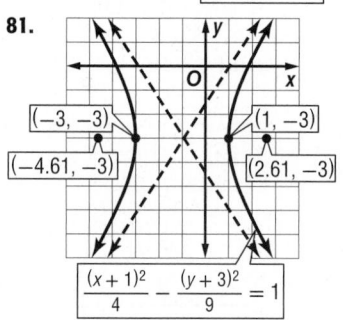

83. $(y + 5)^2 = -20(x - 4)$

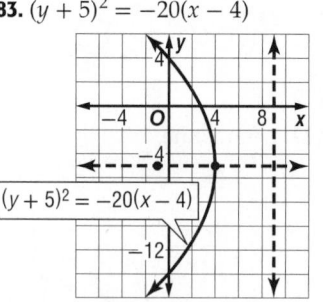

85. G
87. H

Pages 557–559 Lesson 9-3

1. $(\sqrt{2}, \sqrt{2})$ **3.** $\left(-\frac{5}{2}, -\frac{5\sqrt{3}}{2}\right)$ **5.** $(1, \sqrt{3})$ **7.** $(0, 3)$ **9.** $(0, 2)$

11. $\left(-\frac{\sqrt{3}}{2}, \frac{1}{2}\right)$ **13.** $(12.21, 0.96)$ and $(-12.2, 4.10)$ **15.** $(13.42, 4.25)$ and $(-13.42, 1.11)$ **17.** $(3.61, 5.30)$ and $(-3.61, 2.16)$ **19.** $(3.16a, 1.25)$ and $(-3.16a, 4.39)$ **21.** $(60.54, 5.74)$ and $(-60.54, 2.61)$ **23.** $\left(\sqrt{2}, \frac{7\pi}{4}\right)$ and $\left(-\sqrt{2}, \frac{3\pi}{4}\right)$

25a. about 0.90 mi north and about 1.20 mi east

25b. Sample answer: (2.06, 194.04°)

27. circle; $r = -10 \cos \theta$

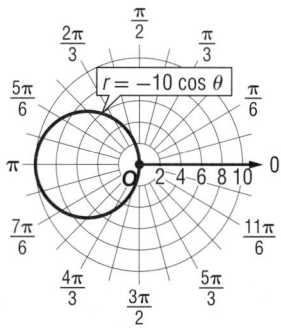

29. parabola; $r = \cot \theta \csc \theta$

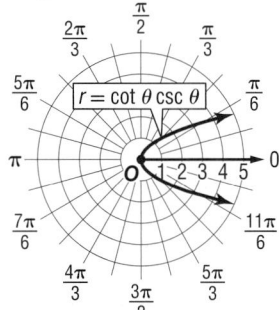

31. hyperbola; $r = 2 \sec 2\theta \cos \theta$

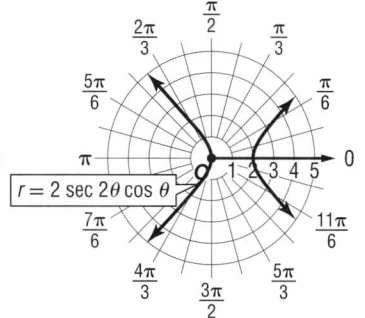

33. line; $\theta = \frac{\pi}{3}$

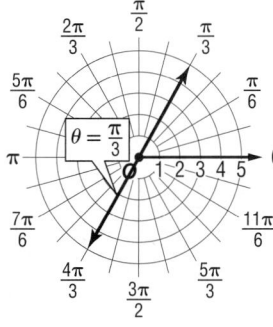

35. circle; $r = 16 \sin \theta$

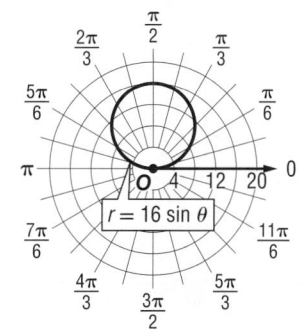

37. $y = -\sqrt{3}x$; line

39. $x^2 - 4x + y^2 = 0$; circle

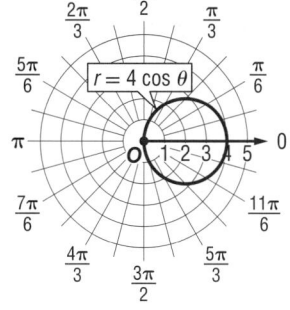

41. $y = 8$; line

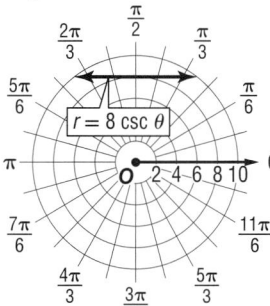

43. $x = -7y$ or $-\frac{1}{7}x = y$; line

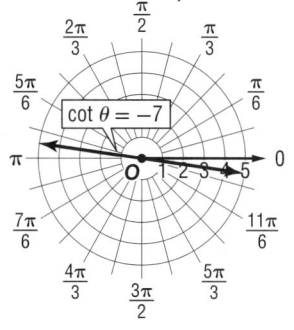

45. $x = 1$; line

47 **a.**

This graph is symmetric with respect to the polar axis, so you can find points on the interval $[0, \pi]$ and then use polar axis symmetry to complete the graph.

θ	0	$\frac{\pi}{6}$	$\frac{\pi}{4}$	$\frac{\pi}{3}$	$\frac{\pi}{2}$	$\frac{2\pi}{3}$	$\frac{3\pi}{4}$	$\frac{5\pi}{6}$	π
r	4	3.7	3.4	3	2	1	0.6	0.3	0

b. No; sample answer: The given point is on the negative x-axis, directly behind the microphone. The polar pattern indicates that the microphone does not pick up any sound from this direction.

49.

$\frac{\sqrt{2}}{2}y - \frac{\sqrt{2}}{2}x = 10$ or $y = x + 10\sqrt{2}$; line

51.

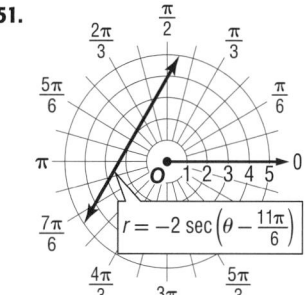

$\frac{\sqrt{3}}{2}x - \frac{1}{2}y = -2$ or $y = \sqrt{3}x + 4$; line

53.

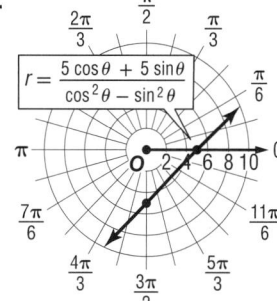

$x - y = 5$ or $y = x - 5$; line

55.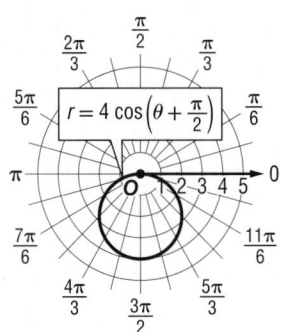
$x^2 + y^2 + 4y = 0$ or $x^2 + (y + 2)^2 = 4$; circle

57.
line; $r = \dfrac{4}{6 \cos \theta - 3 \sin \theta}$

59.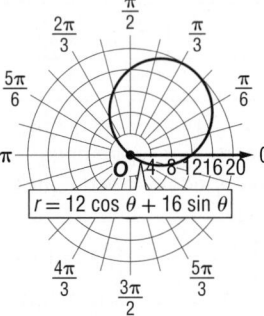
circle; $r = 12 \cos \theta + 16 \sin \theta$

61. Sample answer: $y = -4$; $r = -4 \csc \theta$ **63.** Sample answer: $x^2 + (y - 2)^2 = 4$; $r = 4 \sin \theta$ **65.** 39π yd² or about 122.52 yd²

67a.

67b.
(10, 53.13°)

67c.

67d.
(4.24, 135°)

67e. $r = \sqrt{a^2 + b^2}$, $\theta = \tan^{-1} \frac{b}{a}$, when a is positive; $\theta = \tan^{-1} \frac{b}{a} + 180°$ when a is negative. **69.** $(x - a)^2 + y^2 = a^2$, radius $= a$, center $= (a, 0)$ **71.** Sample answer: Rectangular equations that are not functions, such as equations representing ellipses or circles, are easier to graph in polar form. Equations that represent functions, such as linear functions, are easier to graph in rectangular form.

73 $r^2(4 \cos^2 \theta + 3 \sin^2 \theta) + r(-8a \cos \theta + 6b \sin \theta) = 12 - 4a^2 - 3b^2$
$4r^2 \cos^2 \theta + 3r^2 \sin^2 \theta - 8ar \cos \theta + 6br \sin \theta = 12 - 4a^2 - 3b^2$
$4(r \cos \theta)^2 + 3(r \sin \theta)^2 - 8a(r \cos \theta) + 6b(r \sin \theta) = 12 - 4a^2 - 3b^2$
$4x^2 + 3y^2 - 8ax + 6by = 12 - 4a^2 - 3b^2$
$4x^2 - 8ax + 4a^2 + 3y^2 + 6by + 3b^2 = 12$
$4(x^2 - 2ax + a^2) + 3(y^2 + 2by + b^2) = 12$
$4(x - a)^2 + 3(y + b)^2 = 12$
$\dfrac{(x - a)^2}{3} + \dfrac{(y + b)^2}{4} = 1$

75.

77.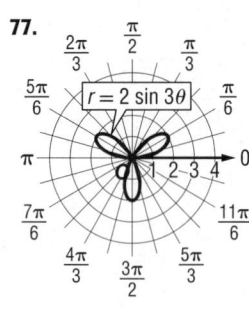

79. $\left(1, -\dfrac{2\pi}{3}\right), \left(1, \dfrac{4\pi}{3}\right), \left(-1, -\dfrac{5\pi}{3}\right)$

81. 91.8°, not orthogonal **83.** 98.3°, not orthogonal

85. $(y - 1)^2 = x$ **87a.** $\dfrac{x^2}{2916} - \dfrac{y^2}{49,984} = 1$

87b.

87c. (−60.2, 110)

89. $\begin{bmatrix} 1 & 0 & 0 & | & -4 \\ 0 & 1 & 0 & | & -1 \\ 0 & 0 & 1 & | & 7 \end{bmatrix}$ **91.** $\begin{bmatrix} 1 & 0 & 0 & | & 2 \\ 0 & 1 & 0 & | & -3 \\ 0 & 0 & 1 & | & 4 \end{bmatrix}$ **93.** G **95.** J

Pages 566–568 Lesson 9-4

1. $e = 1$; parabola; $y = 5$ **3.** $e = 3$; hyperbola; $x = 7$ **5.** $e = 1$; parabola; $x = -2$ **7.** $e = 4$; hyperbola; $y = -8$

9 a. Factor 2 from the denominator, to rewrite the equation as $r = \dfrac{\frac{7}{2}}{1 + \sin \theta}$. Since the coefficient of $\sin \theta$ is 1, $e = 1$, which means the conic is a parabola. Then $\frac{7}{2} = ed$, but since $e = 1$, $d = \frac{7}{2}$, so the directrix is $y = \frac{7}{2}$ or $y = 3.5$.

b. Factor 5 from the denominator, to rewrite the equation as $r = \dfrac{\frac{28}{5}}{1 + 2.5 \cos \theta}$. The coefficient of $\cos \theta$ is 2.5, so $e = 2.5 > 1$, which means the conic is a hyperbola. Then $\frac{28}{5} = ed = 2.5d$. Dividing by 2.5, $d = \frac{28}{12.5} = 2.24$, so the directrix is $x = 2.24$.

11.

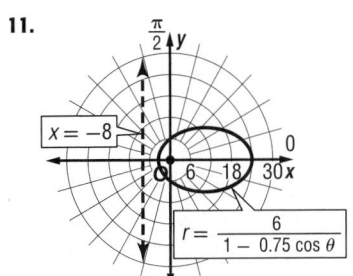

$r = \dfrac{6}{1 - 0.75 \cos \theta}$

13.

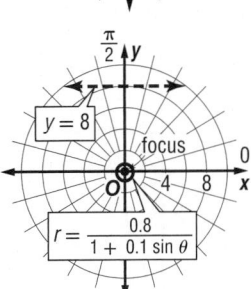

$r = \dfrac{0.8}{1 + 0.1 \sin \theta}$

15.

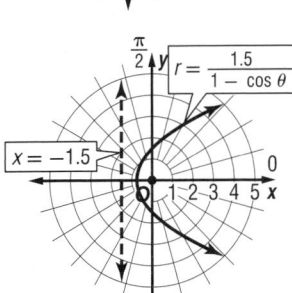

$r = \dfrac{1.5}{1 - \cos \theta}$

17.

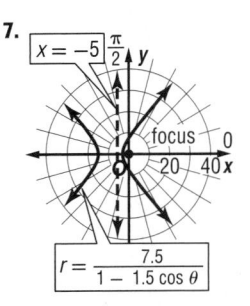

$r = \dfrac{7.5}{1 - 1.5 \cos \theta}$

19. $\dfrac{(x+2)^2}{64} + \dfrac{y^2}{60} = 1$ **21.** $\dfrac{x^2}{51} + \dfrac{(y+7)^2}{100} = 1$

23. $\dfrac{(y+9)^2}{9} - \dfrac{x^2}{72} = 1$ **25.** $\dfrac{(x-4)^2}{100} + \dfrac{y^2}{84} = 1$

27. parabola **29.** hyperbola

$[0, 2\pi]$ scl: $\frac{\pi}{24}$ by $[-14, 6]$ scl: 2 by $[-14, 6]$ scl: 2 $[0, 2\pi]$ scl: $\frac{\pi}{24}$ by $[-3, 7]$ scl: 1 by $[-7, 3]$ scl: 1

31. a **33.** c

35. $e = 1$; parabola; $y = -5$ **37.** $e = \frac{1}{5}$; ellipse; $x = 8$

39. $b^2 = a^2 - c^2$
$b^2 = a^2 - (ae)^2$
$b^2 = a^2 - a^2 e^2$
$b^2 = a^2(1 - e^2)$
$b = a\sqrt{1 - e^2}$

41. $PF = ePQ$
$c - a = e[a - (c - d)]$
$c - a = e(a - c + d)$
$ae - a = e(a - ae + d)$
$ae - a = ae - ae^2 + de$
$-a = -ae^2 + de$
$ae^2 - a = de$
$\dfrac{ae^2 - a}{e} = d$
$\dfrac{a(e^2 - 1)}{e} = d$

43. $r = \dfrac{5}{1 + \sin \theta}$ **45.** $r = \dfrac{6}{1 + 0.5 \cos \theta}$ **47.** $r' = \dfrac{4.5}{1 + 1.25 \cos \theta'}$

49a. Sample answer: $r = \dfrac{1.2}{1 + 0.4 \cos \theta'}$, ellipse; $r = \dfrac{1.8}{1 + 0.6 \cos \theta'}$ ellipse; $r = \dfrac{3}{1 + \cos \theta'}$, parabola; $r = \dfrac{4.8}{1 + 1.6 \cos \theta'}$, hyperbola; $r = \dfrac{6}{1 + 2 \cos \theta'}$, hyperbola

49b.

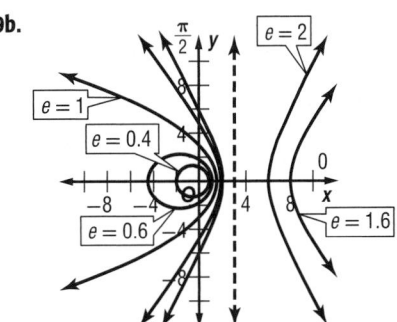

49c. Sample answer: The ellipse with the smaller eccentricity is much more circular than the ellipse with greater

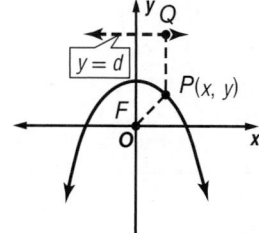

eccentricity. As e approaches 1, the ellipses open farther to the left, approaching the parabola with $e = 1$. For values where $e > 1$, the hyperbola with the greater eccentricity approaches its asymptotes much more rapidly than the hyperbola with the smaller eccentricity. Also, as e approaches 1, one branch of the hyperbola moves farther and farther away from its asymptotes, approaching the parabola with $e = 1$.

49d. Sample answer: $r = \dfrac{0.125}{1 + 0.5 \cos \theta}$; $r = \dfrac{0.5}{1 + 0.5 \cos \theta}$; $r = \dfrac{2}{1 + 0.5 \cos \theta}$

49e.

49f. Sample answer: As the value of the directrix increases from 0.25 to 4, the distances between the vertices and the foci of the ellipses increase.

51.

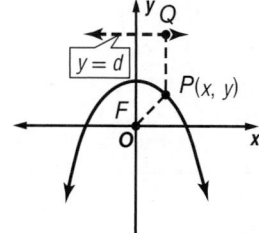

Statements	(Reasons)
1. $PF = ePQ$	(Definition of a conic section)
2. $\sqrt{x^2 + y^2} = e(d - y)$	$\left(PF = \sqrt{x^2 + y^2} \text{ and } PQ = d - y\right)$
3. $r = e(d - r \sin \theta)$	$\left(r = \sqrt{x^2 + y^2} \text{ and } y = r \sin \theta\right)$
4. $r = ed - er \sin \theta$	(Distributive Property)
5. $r + er \sin \theta = ed$	(Isolate the r-terms.)
6. $r(1 + e \sin \theta) = ed$	(Factor.)
7. $r = \dfrac{ed}{1 + e \sin \theta}$	(Solve for r.)

53. Sample answer: A conic section can be defined as a figure that is formed when a plane intersects a double-napped right cone or as the locus of points such that the distance from a point to the focus and the distance from the point to the directrix is a constant ratio.

55 Since one focus is at the pole and the other focus appears to be to the left of the pole, the equation is of the form $r = \dfrac{ed}{1 + e \cos \theta}$. Substitute the points $(2, 0)$ and $(8, \pi)$ and solve for e in the resulting system of equations.

$$2 = \frac{ed}{1 + e \cos 0} \qquad\qquad 8 = \frac{ed}{1 + e \cos \pi}$$

$$2 = \frac{ed}{1 + e} \qquad\qquad\quad 8 = \frac{ed}{1 - e}$$

$$2(1 + e) = ed \qquad\qquad 8(1 - e) = ed$$

$$2(1 + e) = 8(1 - e) \qquad ed = 8(1 - e)$$

$$2 + 2e = 8 - 8e \qquad \text{Multiply.}$$

$$10e = 6 \qquad\qquad\quad \text{Add } 8e - 2 \text{ to each side.}$$

$$e = \frac{6}{10} \text{ or } \frac{3}{5} \qquad \text{Solve for } e.$$

Substitute $e = \dfrac{3}{5}$ into $2(1 + e) = ed$ and solve for d.

$$2\left(1 + \frac{3}{5}\right) = \frac{3}{5}d \qquad e = \frac{3}{5}$$

$$\frac{16}{5} = \frac{3}{5}d \qquad \text{Simplify.}$$

$$d = \frac{16}{3} \qquad \text{Solve for } d.$$

Therefore, $r = \dfrac{\frac{3}{5} \cdot \frac{16}{3}}{1 + \frac{3}{5} \cos \theta}$ or $r = \dfrac{16}{5 + 3 \cos \theta}$.

57. Sample answer: Let $P = (r, \theta)$ be a point on the conic. Then the distance from P to the focus located at $(0, 0)$ is $|r|$. In terms of θ, depending on the conic, $|r| = \left|\dfrac{ed}{1 + e \cos \theta}\right|$, $|r| = \left|\dfrac{ed}{1 - e \cos \theta}\right|$, $|r| = \left|\dfrac{ed}{1 + e \sin \theta}\right|$, or $|r| = \left|\dfrac{ed}{1 - e \sin \theta}\right|$.

59. $(5.39, 4.33), (-5.39, 1.19)$

61. cardioid

63. spiral of Archimedes

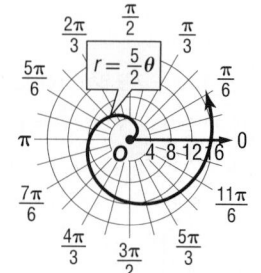

65. $\dfrac{(x - 3)^2}{25} + \dfrac{(y - 4)^2}{9} = 1$ **67.** 1996 Olympics

69. $-\dfrac{336}{625}; -\dfrac{527}{625}; \dfrac{336}{527}$ **71.** asymptotes at $x = \dfrac{\pi}{6} + n\pi$

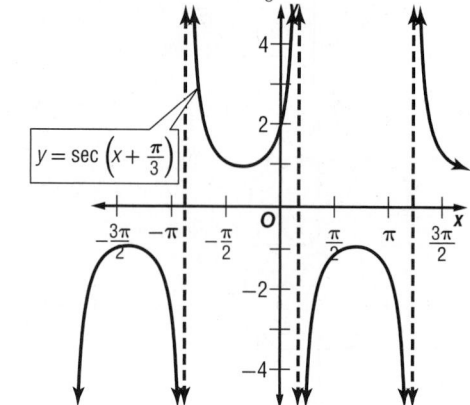

73. asymptotes at $x = \dfrac{\pi}{2} + \dfrac{3\pi}{2}n$

75. $\sin \theta = \frac{\sqrt{5}}{5}$, $\cos \theta = \frac{2\sqrt{5}}{5}$, $\tan \theta = \frac{1}{2}$, $\sec \theta = \frac{\sqrt{5}}{2}$, $\cot \theta = 2$ **77.** G   **79.** H

Pages 577–579   Lesson 9-5

1. $4\sqrt{2} \approx 5.66$

3. $2\sqrt{13} \approx 7.21$

5. 5

7. $\sqrt{58} \approx 7.62$

9a.

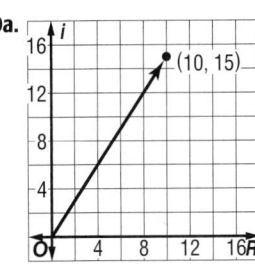

9b. 18.03 N; about 56.31°
11. $\sqrt{5}(\cos 2.68 + i \sin 2.68)$
13. $2\sqrt{2}\left(\cos \frac{7\pi}{4} + i \sin \frac{7\pi}{4}\right)$
15. $2\sqrt{5}(\cos 2.03 + i \sin 2.03)$
17. $3\sqrt{2}\left(\cos \frac{\pi}{4} + i \sin \frac{\pi}{4}\right)$

19. $-1.98 + 0.28i$

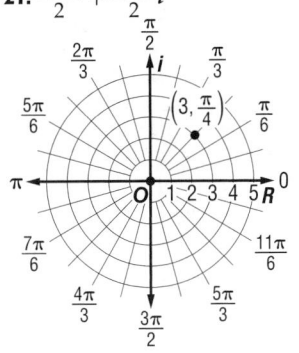

21. $\frac{3\sqrt{2}}{2} + \frac{3\sqrt{2}}{2}i$

23. $-1 - \sqrt{3}i$

25. $\frac{3}{2}$

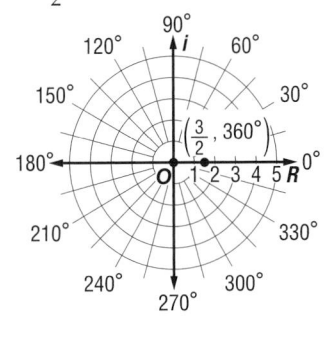

27. $10(\cos 180° + i \sin 180°)$; -10 **29.** $4(\cos 360° + i \sin 360°)$; 4   **31.** $2\left(\cos \frac{3\pi}{4} + i \sin \frac{3\pi}{4}\right)$; $-\sqrt{2} + \sqrt{2}i$

33. $3\left(\cos \frac{\pi}{2} + i \sin \frac{\pi}{2}\right)$; $3i$ **35.** $\frac{1}{6}\left(\cos \frac{\pi}{6} + i \sin \frac{\pi}{6}\right)$; $\frac{\sqrt{3}}{12} + \frac{1}{12}i$
37. $2035 - 828i$ **39.** $-8i$ **41.** $-112 - 384i$ **43.** $-5 + 12i$
45. -16 **47.** $0.97 + 0.26i, 0.26 + 0.97i, -0.71 + 0.71i, -0.97 - 0.26i, -0.26 - 0.97i, 0.71 - 0.71i$ **49.** $0.22 + 1.67i, -1.67 + 0.22i, -0.22 - 1.67i, 1.67 - 0.22i$ **51.** $1.64 + 0.55i, -0.02 + 1.73i, -1.65 + 0.52i, -1.00 - 1.41i, 1.03 - 1.39i$ **53.** $1, -1$

55 **a.** Evaluate the trigonometric values and simplify.
$5(\cos 0.9 + j \sin 0.9) = 5\cos 0.9 + 5j \sin 0.9$
$\approx 3.11 + 3.92\,j$ ohms
$8(\cos 0.4 + j \sin 0.4) = 8\cos 0.4 + 8j \sin 0.4$
$\approx 7.37 + 3.12\,j$ ohms

b. Find the sum.
$(3.11 + 3.92j) + (7.37 + 3.12j) = 10.48 + 7.04j$ ohms
c. Find the modulus r and argument θ.

$r = \sqrt{a^2 + b^2}$ $\theta = \tan^{-1}\frac{b}{a}$

$= \sqrt{10.48^2 + 7.04^2}$ $= \tan^{-1}\frac{7.04}{10.48}$

≈ 12.63 ≈ 0.59

The polar form of $10.48 + 7.04j$ is $12.63(\cos 0.59 + j \sin 0.59)$ ohms.

57. $(3 + i)(3 - i) = 10$; $\sqrt{10}(\cos 0.3218 + i \sin 0.3218) \cdot \sqrt{10}[\cos(-0.3218) + i \sin(-0.3218)] = 10$ **59.** $(-6 + 5i) \cdot (2 - 3i) = 3 + 28i$; $\sqrt{61}(\cos 2.4469 + i \sin 2.4469) \cdot \sqrt{13}[\cos(-0.9828) + i \sin(-0.9828)] \approx 3 + 28i$
61. $(3 - 2i)(1 + \sqrt{3}i) \approx 6.464 + 3.196i$; $\sqrt{13}[\cos(-0.5880) + i \sin(-0.5880)] \cdot 2(\cos 1.0472 + i \sin 1.0472) \approx 6.464 + 3.196i$
63a. a translation 3 units to the right and 4 units down
63b. a reflection in the real axis   **63c.** a rotation of 90° counterclockwise about the origin   **63d.** a dilation by a factor of 0.25   **65.** $-4; 1 + i, -1 + i, -1 - i, 1 - i$
67. $\frac{\sqrt{3}}{2} + \frac{1}{2}i, -\frac{\sqrt{3}}{2} + \frac{1}{2}i, i$
69. $2.77 + 1.15i, -1.15 + 2.77i, -2.77 - 1.15i, 1.15 - 2.77i$
71. $0.79 + 0.79i, -1.08 + 0.29i, 0.29 - 1.08i$ **73.** Blake; sample answer: Alma only converted the expression into polar form. She needed to use De Moivre's Theorem to find the fifth power.

75 Let r be any of the roots depicted. Since r lies on a circle of radius 3, $|r| = 3$. The root r_1 at $\frac{\pi}{6}$ can be represented by $r_1 = 3\left(\cos \frac{\pi}{6} + i \sin \frac{\pi}{6}\right)$. The root r_2 at $\frac{5\pi}{6}$ can be represented by $r_2 = 3\left(\cos \frac{5\pi}{6} + i \sin \frac{5\pi}{6}\right)$. The root r_3 at $\frac{3\pi}{2}$ can be represented by $r_3 = 3\left(\cos \frac{3\pi}{2} + i \sin \frac{3\pi}{2}\right)$.

To determine the number whose roots are r_1, r_2, and r_3, use De Moivre's Theorem to cube any one of them.

$r_1{}^3 = \left[3\left(\cos \frac{\pi}{6} + i \sin \frac{\pi}{6}\right)\right]^3 = 27\left(\cos \frac{\pi}{2} + i \sin \frac{\pi}{2}\right) = 27i$

77. Given: $z_1 = r_1(\cos \theta_1 + i \sin \theta_1)$ and $z_2 = r_2(\cos \theta_2 + i \sin \theta_2)$; Prove: $\frac{z_1}{z_2} = \frac{r_1}{r_2}[\cos(\theta_1 - \theta_2) + i \sin(\theta_1 - \theta_2)]$

$\frac{z_1}{z_2} = \frac{r_1(\cos \theta_1 + i \sin \theta_1)}{r_2(\cos \theta_2 + i \sin \theta_2)}$

$= \frac{r_1}{r_2}\left(\frac{\cos \theta_1 + i \sin \theta_1}{\cos \theta_2 + i \sin \theta_2}\right) \cdot \left(\frac{\cos \theta_2 - i \sin \theta_2}{\cos \theta_2 - i \sin \theta_2}\right)$

$$= \frac{r_1}{r_2} \cdot$$

$$\left(\frac{\cos \theta_1 \cos \theta_2 - i \sin \theta_2 \cos \theta_1 + i \sin \theta_1 \cos \theta_2 - i^2 \sin \theta_1 \sin \theta_2}{\cos^2 \theta_2 - i \sin \theta_2 \cos \theta_2 + i \sin \theta_2 \cos \theta_2 - i^2 \sin^2 \theta_2} \right)$$

$$= \frac{r_1}{r_2} \cdot$$

$$\left(\frac{\cos \theta_1 \cos \theta_2 - i \sin \theta_2 \cos \theta_1 + i \sin \theta_1 \cos \theta_2 - i^2 \sin \theta_1 \sin \theta_2}{\cos^2 \theta_2 + \sin^2 \theta_2} \right)$$

$$= \frac{r_1}{r_2} (\cos \theta_1 \cos \theta_2 - i \sin \theta_2 \cos \theta_1 + i \sin \theta_1 \cos \theta_2 + \sin \theta_1 \sin \theta_2)$$

$$= \frac{r_1}{r_2} [(\cos \theta_1 \cos \theta_2 + \sin \theta_1 \sin \theta_2) + i(\sin \theta_1 \cos \theta_2 - \sin \theta_2 \cos \theta_1)]$$

$$= \frac{r_1}{r_2} [\cos (\theta_1 - \theta_2) + i \sin (\theta_1 - \theta_2)]$$

79. Always; sample answer: If $z = a + bi$ and $\bar{z} = a - bi$, $z + \bar{z} = a + bi + a - bi$ or $2a$ and $z \cdot \bar{z} = (a + bi)(a - bi)$ or $a^2 + b^2$.
81. Sample answer: Since one of the roots must be a positive real number, a vertex of the polygon lies on the positive real axis and the polygon is symmetric about the real axis. This means that the non-real complex roots occur in conjugate pairs. Since the imaginary part of the sum of two complex conjugates is 0, the imaginary part of the sum of all of the roots must be 0.   **83.** $y^2 = -14(x - 3.5)$

85. circle; $r = 6 \cos \theta$

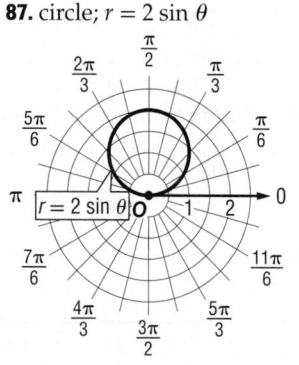

87. circle; $r = 2 \sin \theta$

89.

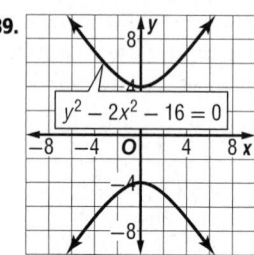

$y^2 - 2x^2 - 16 = 0$

91. center: $(-8, 7)$; foci: $(-8, 7 \pm 6\sqrt{2})$; vertices: $(-8, 16)$, $(-8, -2)$ **93.** center: $(7, -6)$; foci: $(7 \pm \sqrt{5}, -6)$; vertices: $(10, -6)$, $(4, -6)$
95. $(-1, -3, 7)$ **97a.** about 7.94 billion   **97b.** about 51 years
99. F

Pages 580–584   Chapter 9   Study Guide and Review
1. polar graph   **3.** polar coordinate system   **5.** argument
7. modulus   **9.** lemniscate

11.

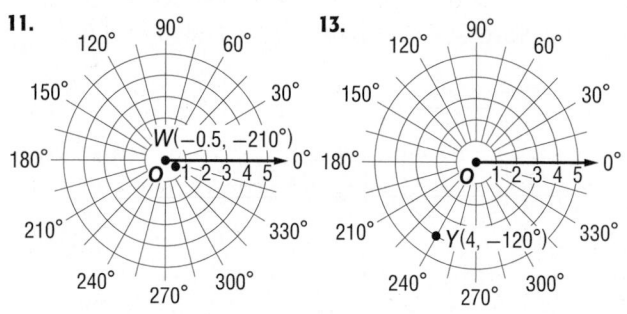

$W(-0.5, -210°)$

13.

$Y(4, -120°)$

15.

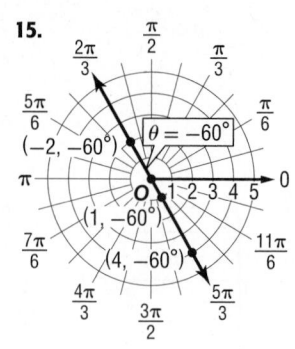

$\theta = -60°$
$(-2, -60°)$
$(1, -60°)$
$(4, -60°)$

17.

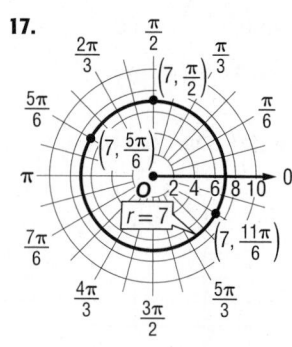

$(7, \frac{\pi}{2})$
$(7, \frac{5\pi}{6})$
$r = 7$
$(7, \frac{11\pi}{6})$

19. ≈ 4.36 **21.** ≈ 6.74

23.

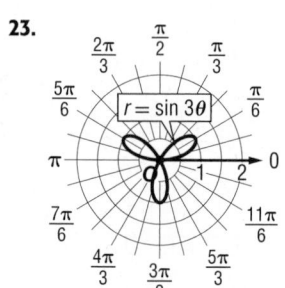

$r = \sin 3\theta$

25.

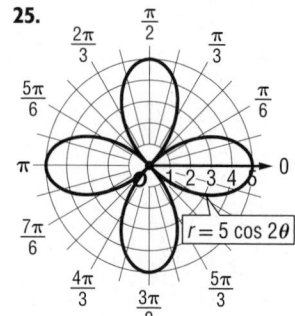

$r = 5 \cos 2\theta$

27.

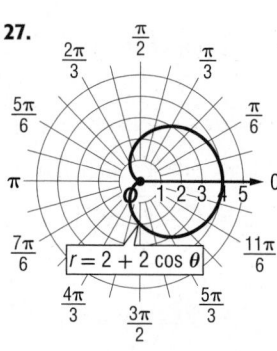

$r = 2 + 2 \cos \theta$

29.

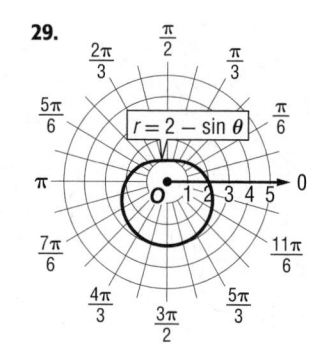

$r = 2 - \sin \theta$

31.

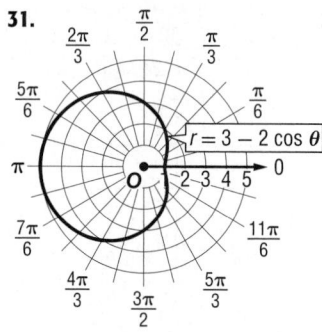

$r = 3 - 2 \cos \theta$

33.

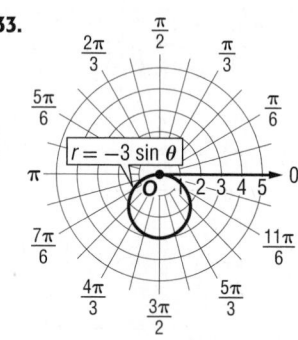

$r = -3 \sin \theta$

35. $(-5.10, 4.91)$ and $(5.10, 1.77)$ **37.** $(2a, 0)$ and $(-2a, \pi)$

39. $x^2 + y^2 = 25$; circle

$r = 5$

41. $x = 6$; line

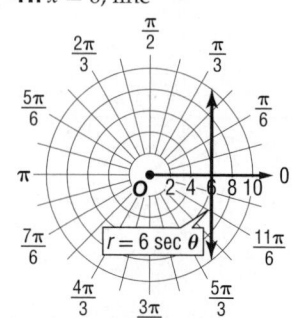

$r = 6 \sec \theta$

43. $e = 1$; parabola; $y = 3.5$ **45.** $e = 2$; hyperbola; $y = -7$

47. $r = \dfrac{3}{1 - 0.5 \sin \theta}; y = -6$ **49.** $\dfrac{\left(y - \frac{1}{3}\right)^2}{\frac{25}{9}} + \dfrac{x^2}{\frac{8}{3}} = 1$

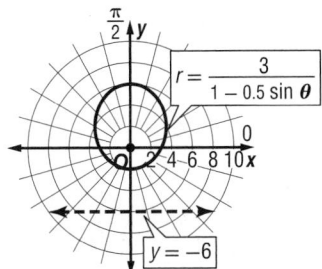

51. $\sqrt{10} \approx 3.16$ **53.** $2\sqrt{5} \approx 4.47$

55. $3.317(\cos 0.441 + i \sin 0.441)$
57. $4.359(\cos 3.550 + i \sin 3.550)$
59. $3i$ **61.** $-\sqrt{2} - \sqrt{2}i$

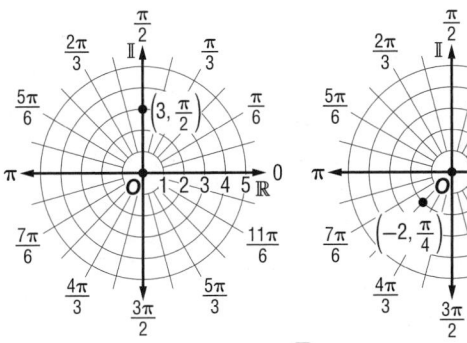

63. $-4\sqrt{3} - 4i$ **65.** $\dfrac{15}{2} + \dfrac{15\sqrt{3}}{2}i$ **67.** $404 - 1121i$ **69.** $1.895 - 0.376i, -0.622 + 1.829i, -1.273 - 1.453i$ **71a.** 20
71b. Sample answer: $(2, 0°)$ or $(2, 180°)$
73. spiral of Archimedes

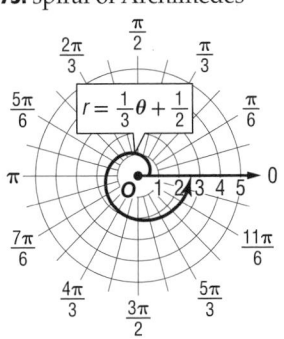

75. 8605 ft   **77a.** $15.2 - 37.9j$ ohms   **77b.** $21.9 + 66.0j$ amps

Pages 586–587   Chapter 9   Connect to AP Calculus

1. Less; sample answer: The shortest distance between two points is a straight line, which is given by our line segments. Since the graph connects the points with a curve, it is going to be longer than the measurements given by the line segments.   **3.** Sample answer: The approximation approaches the actual arc length.   **5.** Sample answer: The graph is symmetric with respect to the y-axis. If the lengths

---

of three line segments that are on either side of the y-axis are found, the result can be doubled to account for the line segments on the other side.
7. Sample answer: As n increases, $\Delta\theta$ decreases. The approximation approaches the actual arc length.

9. Sample answer: 20.6 units

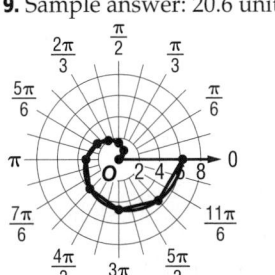

Chapter 10 Sequences and Series

Page 589   Chapter 10   Get Ready

1. $x^3 + 9x^2 + 27x + 27$ **3.** As $x \to \infty, f(x) \to 0$; as $x \to -\infty, f(x) \to 0$.

x	$-10{,}000$	-1000	0	1000	$10{,}000$
$f(x)$	$-4 \cdot 10^{-4}$	-0.004	undefined	0.004	$4 \cdot 10^{-4}$

5. $D = (-\infty, \infty)$; $R = (0, \infty)$;
y-intercept: 1;
asymptote: x-axis;
$\lim\limits_{x \to -\infty} f(x) = \infty$, $\lim\limits_{x \to \infty} f(x) = 0$;
decreasing on $(-\infty, \infty)$

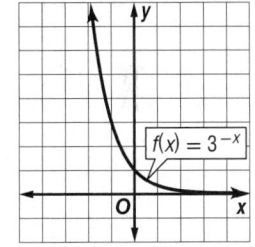

7. $D = (-\infty, \infty)$; $R = (0, \infty)$;
y-intercept: 0.01;
asymptote: x-axis;
$\lim\limits_{x \to -\infty} h(x) = \infty$, $\lim\limits_{x \to \infty} h(x) = 0$;
decreasing on $(-\infty, \infty)$

9. 4   **11.** -3

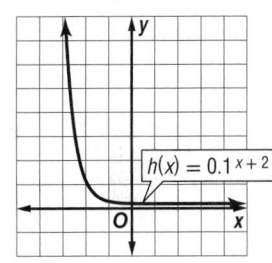

Pages 595–598   Lesson 10-1

1. Sample answer: 29, 36, 43, 50   **3.** Sample answer: $1, \dfrac{1}{3}, \dfrac{1}{9}, \dfrac{1}{27}$ **5.** Sample answer: $-54, -67, -80, -93$ **7.** 0, 3, 8, 15
9. $1, \dfrac{9}{7}, \dfrac{5}{3}, \dfrac{11}{5}$ **11a.** 16,350; 31,350; 46,350   **11b.** 2098, 2497, 2896, 3295   **11c.** $a_n = 1699 + 399n$ **11d.** \$16,063   **13.** 6.125
15. 18

17 **a.** In the family tree, each male will have only a female parent, and each female will have one of each parent. Starting from the left, the male has a mother. His mother has both a mother and father, who have two and one parent, respectively.

$$M - F \underset{\displaystyle F}{\overset{\displaystyle M}{<}} \begin{matrix} M - F \\ M \\ F \end{matrix}$$

b. Let B_n be the number of bees in the nth previous level of the tree, with level 0 corresponding to the original male bee. Each of the B_{n-1} bees at level $n - 1$ will have a mother, so there are B_{n-1} female bees at level n. Each female at level $n - 1$ will also have a father and each of these is a mother to exactly one of the bees at level $n - 2$, so there are B_{n-2} male bees at level n. Thus, $B_n = B_{n-1} + B_{n-2}$ with $B_0 = 1$ and $B_1 = 1$. This is the Fibonacci sequence, in which the previous two terms are added to get the next term. The 11th term in this sequence is 144, so there will be 144 parent bees in the 11th previous generation.

19. convergent **21.** convergent **23.** divergent
25. convergent **27.** convergent **29.** -22.65 **31.** 25 **33.** -27

35. 6 **37.** 0 **39.** -17 **41.** 300 **43.** $\dfrac{70,707}{1,000,000}$ **45.** $\dfrac{8}{9}$

47. $1750.93 **49.** Sample answer: $a_1 = 6$, $a_n = a_{n-1} + 2$, $n \geq 2$; $a_n = 2n + 4$ **51.** Sample answer: $a_1 = 7$, $a_n = 3a_{n-1}$, $n \geq 2$; $a_n = 7(3)^{n-1}$ **53.** $a_1 = 5$, $a_n = a_{n-1} + 8$, $n \geq 2$; 5, 13, 21, 29, ... **55.** 3, 6, 11, 18, ... ; $a_n = n^2 + 2$ **57.** $\displaystyle\sum_{n=1}^{8}(n - 3)$

59 In the sum $8 + 27 + 64 + \cdots + 1000$, each term is a perfect cube. The first term is $8 = 2^3$ and the last term is $1000 = 10^3$. Thus, $8 + 27 + 64 + \cdots + 1000 = \displaystyle\sum_{n=2}^{10} n^3$.

61. $\displaystyle\sum_{n=3}^{9}(-2)^n$ **63.** divergent; 1 **65.** convergent; ≈ -0.4

67. b **69.** c **71.** e **73a.** $\dfrac{1}{1}, \dfrac{2}{1}, \dfrac{3}{2}, \dfrac{5}{3}, \dfrac{8}{5}, \dfrac{13}{8}, \dfrac{21}{13}, \dfrac{34}{21}, \dfrac{55}{34}, \dfrac{89}{55}$

73b.

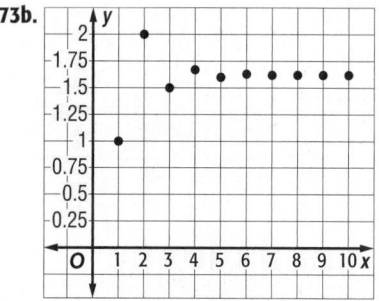

73c. yes; 1.618 **73d.** The two ratios are equivalent to three decimal places. **73e.** Sample answer: the Mona Lisa and the Taj Mahal **75.** divergent **77.** convergent
79. $a_n = -0.5n + 1.75$ **81a.** 0.4, 0.04, 0.004, 0.0004, 0.00004

81b.

81c. As $n \to \infty$, the terms approach a value of 0.
81d. 0.44444; 0.4444444; 0.444444444 **81e.** Sample answer: The sum approaches $\dfrac{4}{9}$.

81f. Sample answer: The sum will be the decimal 0.44..., with n number of 4s. The sum of the first five terms has 5 fours, the sum of the first seven terms has 7 fours, and sum of the first nine terms has 9 fours. **83a.** Sample answer: $a_n = \dfrac{1}{n}$ **83b.** Sample answer: $a_n = \dfrac{1}{n} + 3$ **83c.** Sample answer: $a_n = 2n$ **85.** True; the left side is equal to $4 + 10 + 18 + 28 + 40$ or 100. The right side is equal to $(1 + 4 + 9 + 16 + 25) + 3(1 + 2 + 3 + 4 + 5)$ or 100.

87 Write out the next several terms of the sequence.
$a_6 = a_5 - a_4 = -17 - (-15) = -2$
$a_7 = a_6 - a_5 = -2 - (-17) = 15$
$a_8 = a_7 - a_6 = 15 - (-2) = 17$
Since $a_7 = a_1$ and $a_8 = a_2$, the sequence will repeat the first six terms. The sum of these six terms is 0, so the sum of the first 60 terms will also be zero.

89. $-\sqrt{2} - \sqrt{2}i$ **91.** 5

93. $e = 4$; hyperbola; $y = 5$ **95.** no **97.** no **99.** $\sqrt{26} \approx 5.10$; $(1.5, -13, 13.5)$ **101.** $\sqrt{373} \approx 19.31$; $(2.5, -2, -3)$
103. $2 - \sqrt{3}$ **105.** $-\dfrac{\sqrt{2} + \sqrt{6}}{4}$ **107.** $\dfrac{1}{x} - \dfrac{2}{2x + 1}$ **109.** B
111. C

Pages 605–607   Lesson 10-2

1. -3; 11, 8, 5, 2 **3.** -9; 90, 81, 72, 63 **5.** 4; 9, 13, 17, 21
7. -5; $-19.5, -24.5, -29.5, -34.5$

9 **a.** In this problem, $a_1 = 1$, $a_2 = 3$, and $a_3 = 5$. Find the common difference.
$a_2 - a_1 = 3 - 1$ or 2
$a_3 - a_2 = 5 - 3$ or 2
Add 2 to the third term to find the fourth term, and so on to find the eighth term.
$a_4 = 5 + 2$ or 7
$a_5 = 7 + 2$ or 9
$a_6 = 9 + 2$ or 11
$a_7 = 11 + 2$ or 13
$a_8 = 13 + 2$ or 15
So, there are 15 band members in the 8th row.
b. For an explicit formula, substitute $a_1 = 1$ and $d = 2$ in the formula for the nth term of an arithmetic sequence.
$a_n = a_1 + (n - 1)d$
$a_n = 1 + (n - 1)2$
$a_n = 1 + 2(n - 1)$ or $2n - 1$
For the recursive formula, state the first term a_1 and then indicate that the next term is the sum of the previous term a_{n-1} and d.
$a_1 = 1$, $a_n = a_{n-1} + 2$

11. $a_n = -6 + 11(n - 1)$; $a_1 = -6$, $a_n = a_{n-1} + 11$ **13.** $a_n = 4 + 15(n - 1)$; $a_1 = 4$, $a_n = a_{n-1} + 15$ **15.** $a_n = 7 + (-10.5) \cdot (n - 1)$; $a_1 = 7$, $a_n = a_{n-1} - 10.5$ **17.** $a_n = 1 + 36(n - 1)$; $a_1 = 1$, $a_n = a_{n-1} + 36$ **19.** $d = -7$ **21.** $a_{12} = -8$
23. $a_6 = -311$ **25.** $d = -3$ **27.** 13, 7, 1 **29.** 20, 37, 54, 71
31. $-3, -1.5, 0, 1.5, 3, 4.5, 6$ **33.** $a_n = n^2 + 4n + 7$
35. $a_n = -2n^2 + n + 9$ **37.** $a_n = -n^2 - 5n + 12$ **39.** 3978
41. -126 **43.** 12,008 **45a.** 3.75 mi **45b.** day 25 **47.** 1176
49. 9588 **51.** 728 **53.** -756 **55.** 3792 ft **57.** 14.8 **59.** 8
61. 6 **63.** $\displaystyle\sum_{n=1}^{11} 6n$ **65.** $\displaystyle\sum_{n=4}^{15}(4n + 1)$ **67.** $\displaystyle\sum_{n=2}^{12}\left(\dfrac{n}{5} - 3\right)$
69a. Sample answer: $\displaystyle\sum_{n=1}^{35}(5n + 19)$ **69b.** 3815 seats
69c. 2560 seats **71.** quadratic; $a_n = n^2 + 2n + 5$ **73.** cubic; $a_n = 3n^3 + 2n^2 - n + 1$ **75.** quartic; $a_n = 2n^4 - n^3 - 1$
77. $-\dfrac{2}{3}$ **79.** $\dfrac{1}{3}$ **81.** Both; sample answer: Candace's formula could be simplified to Peter's formula.
83. Sample answer: We know that $a_n = a_1 + (n - 1)d$. Similarly, $a_k = a_1 + (k - 1)d$. Solving the second equation for a_1 and substituting into the first yields:

$$a_n = a_k - (k-1)d + (n-1)d$$
$$= a_k - kd + d + nd - d$$
$$= a_k + nd - kd$$
$$= a_k + (n-k)d$$

85. False; sample answer: you must also know n.

87. a. The arithmetic sequence of odd natural numbers is
$a_1 = 1, a_2 = 3, a_3 = 5, a_5 = 7, a_6 = 9, a_7 = 11, a_8 = 13,$
$a_9 = 15\ldots$. Find the common difference.
$a_2 - a_1 = 3 - 1$ or 2
$a_3 - a_2 = 5 - 3$ or 2

Find S_7.
$S_n = \frac{n}{2}[2a_1 + (n-1)d]$
$S_7 = \frac{7}{2}[2(1) + (7-1)(2)]$
$= \frac{7}{2}(14)$ or 49

Find S_9.
$S_n = \frac{n}{2}[2a_1 + (n-1)d]$
$S_9 = \frac{9}{2}[2(1) + (9-1)(2)]$
$= \frac{9}{2}(18)$ or 81

b. Sample answer: Since the sum of the first 7 odd natural numbers is 49 and the sum of the first 9 is 81, the sum of the first n terms of the sequence of odd natural numbers is n^2.

c. Write an explicit formula for the arithmetic sequence of odd natural numbers.
$a_n = a_1 + (n-1)d$
$= 1 + (n-1)(2)$
$= 1 + 2n - 2$ or $2n - 1$
Substitute $a_n = 2n - 1$ and $a_1 = 1$ into the formula for the sum of a finite arithmetic series.
$S_n = \frac{n}{2}(a_1 + a_n)$
$= \frac{n}{2}[1 + (2n - 1)]$
$= \frac{n}{2}(2n)$ or n^2

89. Sample answer: 24, 28, 32, 36 **91.** Sample answer: 10, 3, $-4, -11$ **93.** $3\left(\cos\frac{11\pi}{6} + i\sin\frac{11\pi}{6}\right); \frac{3\sqrt{3}}{2} - \frac{3}{2}i$ **95.** 26; not orthogonal **97.** 251.6° **99.** 135°

101.
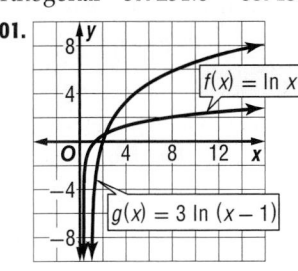

The graph of $g(x)$ is the graph of $f(x)$ shifted 1 unit to the right and expanded vertically.

103. J **105.** J

Pages 615–618 Lesson 10-3

1. $-2; 2, -4, 8$ **3.** $1.5; 1.6875, 2.53125, 3.796875$ **5.** $5; 250x,$ $1250x, 6250x$ **7.** $3; 27x + 135, 81x + 405, 243x + 1215$

9a. Sample answer: The sequence of circumferences is $\pi d_1,$ $\pi d_2, \pi d_3, \pi d_4, \pi d_5$. The common ratio is $\frac{\pi d_2}{\pi d_1}$ or $\frac{d_2}{d_1}$.

9b. Sample answer: The sequence of areas is $\pi\left(\frac{d_1}{2}\right)^2, \pi\left(\frac{d_2}{2}\right)^2,$ $\pi\left(\frac{d_3}{2}\right)^2, \pi\left(\frac{d_4}{2}\right)^2, \pi\left(\frac{d_5}{2}\right)^2$. The common ratio is $\frac{\pi\left(\frac{d_2}{2}\right)^2}{\pi\left(\frac{d_1}{2}\right)^2} = \frac{d_2^2}{d_1^2}$ or $\left(\frac{d_2}{d_1}\right)^2$. **11.** $a_n = 64\left(\frac{1}{4}\right)^{n-1}; a_1 = 64, a_n = \frac{1}{4}a_{n-1}$ **13.** $a_n = 4$

$(-3)^{n-1}; a_1 = 4, a_n = -3a_{n-1}$ **15.** $a_n = 20(1.5)^{n-1}; a_1 = 20,$ $a_n = 1.5a_{n-1}$ **17.** $a_n = \frac{1}{32}(2)^{n-1}; a_1 = \frac{1}{32}, a_n = 2a_{n-1}$

19. a. The number of bacteria doubles every 15 minutes, so in the course of an hour, it increases by a factor of 2^4 or 16, since there are 4 15-minute intervals. Thus, the number of bacteria in the dish after t hours is $b_t = b_0 \cdot 16^t$.

b. Substituting $b_0 = 12$ and $t = 4$, the number of bacteria after 4 hours is $b_4 = 12 \cdot 16^4$ or 786,432.

21. 56 **23.** $\frac{5}{72}$ **25.** -4 **27.** $\frac{1}{4}$ **29.** 288 **31.** $\approx$\$360.97 million **33.** $-192, 144, -108$ or $192, 144, 108$ **35.** $6, -18$ **37.** $-120,$ 300 **39.** t^5, t^2, t^{-1}, t^{-4} **41.** 85.5 **43.** 1501 **45.** 39,063 **47.** 1140 **49.** -484 **51.** 32.64768 **53.** $\frac{4921}{27}$ **55.** 3,145,725 **57.** does not exist **59.** 7.5 **61.** 100

63. The common ratio r is $\left|-\frac{3}{4}\right| < 1$. Therefore, the infinite geometric series has a sum. Find a_1.
$a_1 = 35\left(-\frac{3}{4}\right)^{1-1}$ or 35
Use the formula for the sum of an infinite geometric series to find the sum.
$S = \frac{a_1}{1-r}$
$= \frac{35}{1-\left(-\frac{3}{4}\right)}$ or 20

65. $\frac{1}{3}$ **67.** -3 **69.** 31.104 **71.** 3 **73a.** \$5, 1.1 **73b.** 52 **75.** 2 **77.** -0.2 **79.** $6, 3, \frac{3}{2}$ **81.** $56, -14, 3.5$ **83.** $-36, -14.4,$ -5.76 **85.** $-138, 27.6, -5.52$ **87.** $12, -3, \frac{3}{4}$ **89.** neither; $\frac{5}{12},$ $\frac{6}{14}, \frac{7}{16}$ **91.** arithmetic; 60, 72, 84 **93.** neither; $100k, 121k, 144k$ **95.** geometric; $75\sqrt{5}, 375, 375\sqrt{5}$ **97.** $\sum_{n=1}^{6} 3(4)^{n-1}$ **99.** $\sum_{n=1}^{5} 50(1.7)^{n-1}$ **101.** $\sum_{n=1}^{6} 0.2(-5)a_{n-1}$ **103a.** $\frac{1}{16}$ **103b.** $d,$ $\frac{17}{16}d, \frac{273}{256}d, \frac{4369}{4096}d$ **103c.** $a_1 = d, a_n = a_{n-1} + \left(\frac{1}{16}\right)^{n-1}d$

105. Find the common ratio.
$a_2 \div a_1 = 4 \div 16$ or 0.25
$a_3 \div a_2 = 1 \div 4$ or 0.25
Sample answer: The common ratio is less than 1, so the sequence is converging to 0, and the sum of the series can be calculated. Therefore, Emilio is correct.

107. If $|r| > 1$, then $|S_n|$ increases without limit. Therefore, the corresponding sequence will be divergent, and the sum of the series cannot be calculated. **109.** False; you must also know the value of the first term. **111.** Sometimes; if $|r| < 1$, then the series has a sum, and the sum is a negative number. If $|r| > 1$, then the corresponding sequence is divergent, and the series has no sum. **113.** 63 **115.** 24,300 **117a.** $r^2 = 9\cos 2\theta$ or $r^2 = 9\sin 2\theta$ **117b.** $r^2 = 16\cos 2\theta$ or $r^2 = 16\sin 2\theta$

119.

121. $\langle -30, 6, 24\rangle$

$\langle -30, 6, 24\rangle \cdot \langle 1, 9, -1\rangle = -30(1) + 6(9) + 24(-1)$
$= 0$

$\langle -30, 6, 24\rangle \cdot \langle -2, 6, -4\rangle = -30(-2) + 6(6) + 24(-4)$
$= 0$

123. $\langle 8, -21, 18\rangle$

$\langle 8, -21, 18\rangle \cdot \langle 9, 0, -4\rangle = 8(9) + (-21)0 + 18(-4)$
$= 0$

$\langle 8, -21, 18\rangle \cdot \langle -6, 2, 5\rangle = 8(-6) + (-21)2 + 18(5)$
$= 0$

125. $\langle -16, 18, -20\rangle$; $14\sqrt{5}$; $\left\langle -\dfrac{8\sqrt{5}}{35}, \dfrac{9\sqrt{5}}{35}, -\dfrac{2\sqrt{5}}{7}\right\rangle$ **127.** C
129. A

Pages 625–627   Lesson 10-4

1. Let P_n be the statement $3 + 5 + 7 + \cdots + (2n + 1) = n(n + 2)$. Because $3 = 1(1 + 2)$ is a true statement, P_n is true for $n = 1$. Assume that $P_k\colon 3 + 5 + 7 + \cdots + (2k + 1) = k(k + 2)$ is true for a positive integer k. Show that P_{k+1} must be true.

$$3 + 5 + 7 + \cdots + (2k + 1) = k(k + 2)$$
$$3 + 5 + 7 + \cdots + (2k + 1) + [2(k + 1) + 1] = k(k + 2) + [2(k + 1) + 1]$$
$$3 + 5 + 7 + \cdots + (2k + 1) + [2(k + 1) + 1] = k^2 + 4k + 3$$
$$3 + 5 + 7 + \cdots + (2k + 1) + [2(k + 1) + 1] = (k + 1)(k + 3)$$
$$3 + 5 + 7 + \cdots + (2k + 1) + [2(k + 1) + 1] = (k + 1) \cdot [(k + 1) + 2]$$

This final statement is exactly P_{k+1}, so P_{k+1} is true. Because P_n is true for $n = 1$ and P_k implies P_{k+1}, P_n is true for $n = 2$, $n = 3$, and so on. That is, by the principle of mathematical induction, $3 + 5 + 7 + \cdots + (2n + 1) = n(n + 2)$ is true for all positive integers n.

3. Let P_n be the statement $2 + 2^2 + 2^3 + \cdots + 2^n = 2(2^n - 1)$. Because $2 = 2(2^1 - 1)$ is a true statement, P_n is true for $n = 1$. Assume that $2 + 2^2 + 2^3 + \cdots + 2^k = 2(2^k - 1)$ is true for a positive integer k. Show that P_{k+1} must be true.

$$2 + 2^2 + 2^3 + \cdots + 2^k = 2(2^k - 1)$$
$$2 + 2^2 + 2^3 + \cdots + 2^k + 2^{k+1} = 2(2^k - 1) + 2^{k+1}$$
$$2 + 2^2 + 2^3 + \cdots + 2^k + 2^{k+1} = 2 \cdot 2^{k+1} - 2$$
$$2 + 2^2 + 2^3 + \cdots + 2^k + 2^{k+1} = 2(2^{k+1} - 1)$$

This final statement is exactly P_{k+1}, so P_{k+1} is true. Because P_n is true for $n = 1$ and P_k implies P_{k+1}, P_n is true for $n = 2$, $n = 3$, and so on. That is, by the principle of mathematical induction, $2 + 2^2 + 2^3 + \cdots + 2^n = 2(2^n - 1)$ is true for all positive integers n.

5. Let P_n be the statement $1 + 4 + 7 + \cdots + (3n - 2) = \dfrac{n(3n - 1)}{2}$. Because $1 = \dfrac{1 \cdot [3(1) - 1]}{2}$ is a true statement, P_n is true for $n = 1$. Assume that $1 + 4 + 7 + \cdots + (3k - 2) = \dfrac{k(3k - 1)}{2}$ is true for a positive integer k. Show that P_{k+1} must be true.

$$1 + 4 + 7 + \cdots + (3k - 2) = \frac{k(3k - 1)}{2}$$
$$1 + 4 + 7 + \cdots + (3k - 2) + [3(k + 1) - 2] = \frac{k(3k - 1)}{2} + [3(k + 1) - 2]$$
$$1 + 4 + 7 + \cdots + (3k - 2) + [3(k + 1) - 2] = \frac{k(3k - 1) + 2(3k + 1)}{2}$$
$$1 + 4 + 7 + \cdots + (3k - 2) + [3(k + 1) - 2] = \frac{3k^2 + 5k + 2}{2}$$
$$1 + 4 + 7 + \cdots + (3k - 2) + [3(k + 1) - 2] = \frac{(k + 1)(3k + 2)}{2}$$
$$1 + 4 + 7 + \cdots + (3k - 2) + [3(k + 1) - 2] = \frac{(k + 1)[3(k + 1) - 1]}{2}$$

This final statement is exactly P_{k+1}, so P_{k+1} is true. Because P_n is true for $n = 1$ and P_k implies P_{k+1}, P_n is true for $n = 2$, $n = 3$, and so on. That is, by the principle of mathematical induction, $1 + 4 + 7 + \cdots + (3n - 2) = \dfrac{n(3n - 1)}{2}$ is true for all positive integers n.

7. Let P_n be the statement $1 + 2 + 4 + \cdots + 2^{n-1} = 2^n - 1$. Because $1 = 2^1 - 1$ is a true statement, P_n is true for $n = 1$. Assume that $1 + 2 + 4 + \cdots + 2^{k-1} = 2^k - 1$ is true for a positive integer k. Show that P_{k+1} must be true.

$$1 + 2 + 4 + \cdots + 2^{k-1} = 2^k - 1$$
$$1 + 2 + 4 + \cdots + 2^{k-1} + 2^{(k+1)-1} = 2^k - 1 + 2^{(k+1)-1}$$
$$1 + 2 + 4 + \cdots + 2^{k-1} + 2^{(k+1)-1} = 2(2^k) - 1$$
$$1 + 2 + 4 + \cdots + 2^{k-1} + 2^{(k+1)-1} = 2^{k+1} - 1$$

This final statement is exactly P_{k+1}, so P_{k+1} is true. Because P_n is true for $n = 1$ and P_k implies P_{k+1}, P_n is true for $n = 2$, $n = 3$, and so on. That is, by the principle of mathematical induction, $1 + 2 + 4 + \cdots + 2^{n-1} = 2^n - 1$ is true for all positive integers n.

9. Let P_n be the statement $\dfrac{1}{2} + \dfrac{1}{2^2} + \dfrac{1}{2^3} + \cdots + \dfrac{1}{2^n} = 1 - \dfrac{1}{2^n}$. Because $\dfrac{1}{2} = 1 - \dfrac{1}{2^1}$ is a true statement, P_n is true for $n = 1$. Assume that $\dfrac{1}{2} + \dfrac{1}{2^2} + \dfrac{1}{2^3} + \cdots + \dfrac{1}{2^k} = 1 - \dfrac{1}{2^k}$ is true for a positive integer k. Show that P_{k+1} must be true.

$$\frac{1}{2} + \frac{1}{2^2} + \frac{1}{2^3} + \cdots + \frac{1}{2^k} = 1 - \frac{1}{2^k}$$
$$\frac{1}{2} + \frac{1}{2^2} + \frac{1}{2^3} + \cdots + \frac{1}{2^k} + \left(\frac{1}{2^{k+1}}\right) = 1 - \frac{1}{2^k} + \left(\frac{1}{2^{k+1}}\right)$$
$$\frac{1}{2} + \frac{1}{2^2} + \frac{1}{2^3} + \cdots + \frac{1}{2^k} + \left(\frac{1}{2^{k+1}}\right) = 1 - \frac{2}{2 \cdot 2^k} + \frac{1}{2^{k+1}}$$
$$\frac{1}{2} + \frac{1}{2^2} + \frac{1}{2^3} + \cdots + \frac{1}{2^k} + \left(\frac{1}{2^{k+1}}\right) = 1 - \frac{2}{2^{k+1}} + \frac{1}{2^{k+1}}$$
$$\frac{1}{2} + \frac{1}{2^2} + \frac{1}{2^3} + \cdots + \frac{1}{2^k} + \left(\frac{1}{2^{k+1}}\right) = 1 - \frac{1}{2^{k+1}}$$

This final statement is exactly P_{k+1}, so P_{k+1} is true. Because P_n is true for $n = 1$ and P_k implies P_{k+1}, P_n is true for $n = 2$, $n = 3$, and so on. That is, by the principle of mathematical induction, $\dfrac{1}{2} + \dfrac{1}{2^2} + \dfrac{1}{2^3} + \cdots \dfrac{1}{2^n} = 1 - \dfrac{1}{2^n}$ is true for all positive integers n.

11 a. From the pattern of dots, we can see that each triangle contains one greater number of dots than the previous row.

4th row: $1 + 2 + 3 + 4 = 10$
5th row: $1 + 2 + 3 + 4 + 5 = 15$
6th row: $1 + 2 + 3 + 4 + 5 + 6 = 21$
7th row: $1 + 2 + 3 + 4 + 5 + 6 + 7 = 28$
8th row: $1 + 2 + 3 + 4 + 5 + 6 + 7 + 8 = 36$

b. The sequence of numbers shows a pattern of common second differences.

$$\begin{array}{ccccccccccccccc}
1 & & 3 & & 6 & & 10 & & 15 & & 21 & & 28 & & 36 \\
 & 2 & & 3 & & 4 & & 5 & & 6 & & 7 & & 8 & \\
 & & 1 & & 1 & & 1 & & 1 & & 1 & & 1 & &
\end{array}$$

This means there exists a quadratic expression of the form $an^2 + bn + c$ to model this sequence. Write and solve a system of three equations to find this expression.

Equation 1: $a(1)^2 + b(1) + c = 1$
Equation 2: $a(2)^2 + b(2) + c = 3$
Equation 3: $a(3)^2 + b(3) + c = 6$

Simplify.

Equation 1: $a + b + c = 1$
Equation 2: $4a + 2b + c = 3$
Equation 3: $9a + 3b + c = 6$

Add (-1)Equation 1 to Equation 2 and (-1)Equation 1 to Equation 3 to get the following system and solve for a.

$$3a + b = 2 \qquad\qquad -6a - 2b = -4$$
$$\underline{8a + 2b = 5} \qquad (+) \ \underline{8a + 2b = 5}$$
$$\qquad\qquad\qquad\qquad 2a = 1$$
$$\qquad\qquad\qquad\qquad a = \tfrac{1}{2}$$

Then solve for b and then c.

$$3\left(\tfrac{1}{2}\right) + b = 2$$

$$\tfrac{3}{2} + b = 2 \qquad\qquad \tfrac{1}{2} + \tfrac{1}{2} + c = 1$$
$$b = 2 - \tfrac{3}{2} \qquad\qquad 1 + c = 1$$
$$b = \tfrac{1}{2} \qquad\qquad\qquad c = 0$$

Therefore, the expression that models this sequence is $\tfrac{1}{2}n^2 + \tfrac{1}{2}n$, which is $\dfrac{n^2 + n}{2}$ or $\dfrac{n(n+1)}{2}$. So, $a_n = \dfrac{n(n+1)}{2}$.

c. Let P_n be the statement $1 + 3 + 6 + \cdots + \dfrac{n(n+1)}{2} = \dfrac{n(n+1)(n+2)}{6}$. Because $1 = \dfrac{(1)[(1)+1][(1)+2]}{6}$ is a true statement, P_n is true for $n = 1$. Assume that $1 + 3 + 6 + \cdots + \dfrac{k(k+1)}{2} = \dfrac{k(k+1)(k+2)}{6}$ is true for a positive integer k. Show that P_{k+1} must be true.

$$1 + 3 + 6 + \cdots + \frac{k(k+1)}{2} = \frac{k(k+1)(k+2)}{6}$$

$$1 + 3 + 6 + \cdots + \frac{k(k+1)}{2} + \frac{(k+1)[(k+1)+1]}{2} = \frac{k(k+1)(k+2)}{6} + \frac{(k+1)[(k+1)+1]}{2}$$

$$1 + 3 + 6 + \cdots + \frac{k(k+1)}{2} + \frac{(k+1)[(k+1)+1]}{2} = \frac{k(k+1)(k+2) + 3(k+1)(k+2)}{6}$$

$$1 + 3 + 6 + \cdots + \frac{k(k+1)}{2} + \frac{(k+1)[(k+1)+1]}{2} = \frac{(k+1)(k+2)(k+3)}{6}$$

$$1 + 3 + 6 + \cdots + \frac{k(k+1)}{2} + \frac{(k+1)[(k+1)+1]}{2} = \frac{(k+1)[(k+1)+1][(k+1)+2]}{6}$$

This final statement is exactly P_{k+1}, so P_{k+1} is true. Because P_n is true for $n = 1$ and P_k implies P_{k+1}, P_n is true for $n = 2$, $n = 3$, and so on. That is, by the principle of mathematical induction, $1 + 3 + 6 + \cdots + \dfrac{n(n+1)}{2} = \dfrac{n(n+1)(n+2)}{6}$ is true for all positive integers n.

13. Let P_n be the statement $9^n - 1$ is divisible by 8. P_1 is the statement that $9^1 - 1$ is divisible by 8. P_1 is true because $9^1 - 1$ is 8, which is divisible by 8. Assume P_k is true, where k is a positive integer, and show that P_{k+1} must be true. That is, show that $9^k - 1 = 8r$ for some integer r implies that $9^{k+1} - 1$ is divisible by 8.

$$9^k - 1 = 8r$$
$$9^k = 8r + 1$$
$$9 \cdot 9^k = 9(8r + 1)$$
$$9^{k+1} = 72r + 9$$
$$9^{k+1} - 1 = 72r + 8$$
$$9^{k+1} - 1 = 8(9r + 1)$$

Because r is an integer, $9r + 1$ is an integer and $8(9r + 1)$ is divisible by 8. Therefore, $9^{k+1} - 1$ is divisible by 8. Because P_n is true for $n = 1$ and P_k implies P_{k+1}, P_n is true for $n = 2$, $n = 3$, and so on. By the principle of mathematical induction, $9^n - 1$ is divisible by 8 for all positive integers n.

15. Let P_n be the statement $2^{3n} - 1$ is divisible by 7. P_1 is true because $2^{3(1)} - 1$ is 7, which is divisible by 7. Assume P_k is true, where k is a positive integer, and show that P_{k+1} must be true. That is, show that $2^{3k} - 1 = 7r$ for some integer r implies that $2^{3(k+1)} - 1$ is divisible by 7.

$$2^{3k} - 1 = 7r$$
$$2^{3k} = 7r + 1$$
$$2^3 \cdot 2^{3k} = 2^3(7r + 1)$$
$$2^{3(k+1)} = 56r + 8$$
$$2^{3(k+1)} - 1 = 56r + 7$$
$$2^{3(k+1)} - 1 = 7(8r + 1)$$

Because r is an integer, $8r + 1$ is an integer and $7(8r + 1)$ is divisible by 7. Therefore, $2^{3(k+1)} - 1$ is divisible by 7. Because P_n is true for $n = 1$ and P_k implies P_{k+1}, P_n is true for $n = 2$, $n = 3$, and so on. By the principle of mathematical induction, $2^{3n} - 1$ is divisible by 7 for all positive integers n.

17. Let P_n be the statement $3^n \geq 3n$. P_1 is true, since $3^1 \geq 3(1)$ is a true inequality. Assume P_k is true, that $3^k \geq 3k$ if k is a positive integer $k \geq 1$, and show that P_{k+1} must be true. That is, show that $3^k \geq 3k$ implies $3^{k+1} \geq 3(k+1)$. Use both parts of the inductive hypothesis.

$3^k \geq 3k$	$k \geq 1$
$3 \cdot 3^k \geq 3 \cdot 3k$	$6k \geq 6$
$3^{k+1} \geq 9k$	If $6k \geq 6$ and $6 \geq 3$, then by the Transitive Property of Inequality, $6k \geq 3$.
	$6k \geq 3$
	$3k + 6k \geq 3k + 3$
	$9k \geq 3(k+1)$

By the Transitive Property of Inequality, if $3^{k+1} \geq 9k$ and $9k \geq 3(k+1)$, then $3^{k+1} \geq 3(k+1)$. Since P_n is true for $n = 1$ and P_k implies P_{k+1} for $k \geq 1$, P_n is true for $n = 2$, $n = 3$, and so on. By the principle of mathematical induction, $3^n \geq 3n$ is true for integer values $n \geq 1$.

19. Let P_n be the statement $2^n > 2n$. P_3 is true, since $2^3 \geq 2(3)$ is a true statement. Assume P_k is true, that $2^k > 2k$ if k is a positive integer $k \geq 3$, and show that P_{k+1} must be true. That is, show that $2^k > 2k$ implies $2^{k+1} > 2(k+1)$. Use both parts of the inductive hypothesis.

$2^k > 2k$	$k \geq 3$
$2 \cdot 2^k > 2 \cdot 2k$	$2k \geq 6$
$2^{k+1} > 4k$	If $2k \geq 6$ and $6 \geq 2$, then by the Transitive Property of Inequality, $2k \geq 2$.
	$2k \geq 2$
	$2k + 2k \geq 2k + 2$
	$4k \geq 2(k+1)$

By the Transitive Property of Inequality, if $2^{k+1} > 4k$ and $4k \geq 2(k+1)$, then $2^{k+1} > 2(k+1)$. Since P_n is true for $n = 3$ and P_k implies P_{k+1} for $k \geq 3$, P_n is true for $n = 4$, $n = 5$, and so on. By the extended principle of mathematical induction, $2^n > 2n$ is true for integer values $n \geq 3$.

21. Let P_n be the statement $3n < 4^n$. P_1 is true, since $3(1) < 4^1$ is a true statement. Assume P_k is true, that $3k < 4^k$ if k is a positive integer $k \geq 1$, and show that P_{k+1} must be true. That is, show that $3k < 4^k$ implies $3(k+1) < 4^{k+1}$. Use both parts of the inductive hypothesis.

$3k < 4^k$	$k \geq 1$
$4 \cdot 3k < 4 \cdot 4^k$	$3k \geq 3$
$12k < 4^{k+1}$	$3k - 3 \geq 0$
	$6k - 3k - 3 \geq 0$
	$6k - 3(k+1) \geq 0$
	$6k \geq 3(k+1)$
	$3(k+1) \leq 6k$

Since k is a positive integer, $6k \leq 12k$. If $3(k+1) \leq 6k$ and $6k \leq 12k$ then $3(k+1) \leq 12k$.

By the Transitive Property of Inequality, if $3(k + 1) \leq 12k$ and $12k < 4^{k+1}$, then $3(k + 1) < 4^{k+1}$. Since P_n is true for $n = 1$ and P_k implies P_{k+1} for $k \geq 1$, P_n is true for $n = 2$, $n = 3$, and so on. By the principle of mathematical induction, $3n < 4^n$ is true for integer values $n \geq 1$.

23. Let P_n be the statement $2n < 1.5^n$. P_7 is true, since $2(7) < 1.5^7$ is a true statement. Assume P_k is true, that $2k < 1.5^k$ if k is a positive integer $k \geq 7$, and show that P_{k+1} must be true. That is, show that $2k < 1.5^k$ implies $2(k + 1) < 1.5^{k+1}$. Use both parts of the inductive hypothesis.

$2k < 1.5^k$	$k \geq 7$
$1.5 \cdot 2k < 1.5 \cdot 1.5^k$	If $k \geq 7$ and $7 \geq 2$, then by the
$3k < 1.5^{k+1}$	Transitive Property of Inequality, $k \geq 2$.
	$k \geq 2$
	$2 \leq k$
	$2k + 2 \leq 2k + k$
	$2(k + 1) \leq 3k$

By the Transitive Property of Inequality, if $2(k + 1) \leq 3k$ and $3k < 1.5^{k+1}$, then $2(k + 1) < 1.5^{k+1}$. Since P_n is true for $n = 7$ and P_k implies P_{k+1} for $k \geq 7$, P_n is true for $n = 8$, $n = 9$, and so on. By the extended principle of mathematical induction, $2n < 1.5^n$ is true for all integers $n \geq 7$.

25. Let P_n be the statement that there exists a set of \$0.05 and \$0.06 stamps that adds to \$0.01n for $n > 20$. For $n = 21$, the conjecture is true, because \$0.01(21) = \$0.05(3) + \$0.06(1). Assume that there exists a set of \$0.05 and/or \$0.06 stamps that adds to \$0.01k for $k > 20$. Show that this implies the existence of a set of \$0.05 and/or \$0.06 stamps that adds to \$0.01(k + 1).

Case 1 The set contains at least one \$0.05 stamp. Replace it with a \$0.06 stamp and the set's value is increased to \$0.01k + 0.01 or \$0.01(k + 1), which is exactly P_{k+1}.

Case 2 The set contains no \$0.05 stamps. The set must contain at least four \$0.06 stamps, because the set's value must be greater than \$0.20. Replace four of the \$0.06 stamps with five \$0.05 stamps and the set's value is increased by \$0.01, to \$0.01k + \$0.01 or \$0.01(k + 1), which is exactly P_{k+1}.

In both cases, P_n is true for $n = k + 1$. Because P_n is true for $n = 21$ and P_k implies P_{k+1} for $k > 20$, P_n is true for $n = 22$, $n = 23$, and so on. That is, by the extended principle of mathematical induction, all stamps greater than \$0.20 can be formed using just \$0.05 and \$0.06 stamps.

27. Since each oblong number has one more column than row, the sequence of oblong numbers is defined explicitly by the formula $a_n = n(n + 1)$. To prove that the sum of the first n oblong numbers is given by $S_n = \dfrac{n^3 + 3n^2 + 2n}{3}$, you must prove that $2 + 6 + \cdots + n(n + 1) = \dfrac{n^3 + 3n^2 + 2n}{3}$ is true for all positive integers n. Let this statement be P_n. Because $2 = \dfrac{1^3 + 3(1)^2 + 2(1)}{3}$ is a true statement, P_n is true for $n = 1$. Assume that $2 + 6 + \cdots + k(k + 1) = \dfrac{k^3 + 3k^2 + 2k}{3}$ is true for a positive integer k. Show that P_{k+1} must be true.

$$2 + 6 + \cdots + k(k + 1) = \frac{k^3 + 3k^2 + 2k}{3}$$

$$2 + 6 + \cdots + k(k + 1) + (k + 1)[(k + 1) + 1] = \frac{k^3 + 3k^2 + 2k}{3} + (k + 1)[(k + 1) + 1]$$

$$2 + 6 + \cdots + k(k + 1) + (k + 1)[(k + 1) + 1] = \frac{k^3 + 3k^2 + 2k}{3} + (k^2 + 3k + 2)$$

$$2 + 6 + \cdots + k(k + 1) + (k + 1)[(k + 1) + 1] = \frac{k^3 + 3k^2 + 2k + 3k^2 + 9k + 6}{3}$$

$$2 + 6 + \cdots + k(k + 1) + (k + 1)[(k + 1) + 1] = \frac{k^3 + 3k^2 + 2k + 3k^2 + (3k + 6k) + (3 + 1 + 2)}{3}$$

$$2 + 6 + \cdots + k(k + 1) + (k + 1)[(k + 1) + 1] = \frac{(k^3 + 3k^2 + 3k + 1) + (3k^2 + 6k + 3) + (2k + 2)}{3}$$

$$2 + 6 + \cdots + k(k + 1) + (k + 1)[(k + 1) + 1] = \frac{(k + 1)^3 + 3(k + 1)^2 + 2(k + 1)}{3}$$

This final statement is exactly P_{k+1}, so P_{k+1} is true. Because P_n is true for $n = 1$ and P_k implies P_{k+1}, P_n is true for $n = 2$, $n = 3$, and so on. That is, by the principle of mathematical induction, $2 + 6 + \cdots + n(n + 1) = \dfrac{n^3 + 3n^2 + 2n}{3}$ is true for all positive integers n.

29. Let P_n be $1 + 4 + \cdots + (3n - 2) = \dfrac{n}{2}(3n - 1)$. Because $1 = \dfrac{1}{2}[3(1) - 1]$ is a true statement, P_n is true for $n = 1$. Assume that $1 + 4 + \cdots + (3k - 2) = \dfrac{k}{2}(3k - 1)$ is true for a positive integer k. Show that P_{k+1} must be true.

$$1 + 4 + \cdots + (3k - 2) = \frac{k}{2}(3k - 1)$$

$$1 + 4 + \cdots + (3k - 2) + (3k + 1) = \frac{k}{2}(3k - 1) + (3k + 1)$$

$$1 + 4 + \cdots + (3k - 2) + (3k + 1) = \frac{k(3k - 1)}{2} + \frac{2(3k + 1)}{2}$$

$$1 + 4 + \cdots + (3k - 2) + (3k + 1) = \frac{3k^2 + 5k + 2}{2}$$

$$1 + 4 + \cdots + (3k - 2) + (3k + 1) = \frac{(k + 1)(3k + 2)}{2}$$

$$1 + 4 + \cdots + (3k - 2) + (3k + 1) = \frac{(k + 1)}{2}[3(k + 1) - 1]$$

This final statement is exactly P_{k+1}, so P_{k+1} is true. Because P_n is true for $n = 1$ and P_k implies P_{k+1}, P_n is true for $n = 2$, $n = 3$, and so on. That is, by the principle of mathematical induction, $1 + 4 + \cdots + (3n - 2) = \dfrac{n}{2}(3n - 1)$ or $\displaystyle\sum_{a=1}^{n}(3a - 2) = \dfrac{n}{2}(3n - 1)$ is true for all positive integers n.

31. Let P_n be $\dfrac{1}{3} + \dfrac{1}{15} + \cdots + \dfrac{1}{4n^2 - 1} = \dfrac{n}{2n + 1}$. Because $\dfrac{1}{3} = \dfrac{1}{2(1) + 1}$ is a true statement, P_n is true for $n = 1$. Assume that $\dfrac{1}{3} + \dfrac{1}{15} + \cdots + \dfrac{1}{4n^2 - 1} = \dfrac{k}{2k + 1}$ is true for a positive integer k. Show that P_{k+1} must be true.

$$\frac{1}{3} + \frac{1}{15} + \cdots + \frac{1}{4k^2 - 1} = \frac{k}{2k + 1}$$

$$\frac{1}{3} + \frac{1}{15} + \cdots + \frac{1}{4k^2 - 1} + \frac{1}{4(k + 1)^2 - 1} = \frac{k}{2k + 1} + \frac{1}{4(k + 1)^2 - 1}$$

$$\frac{1}{3} + \frac{1}{15} + \cdots + \frac{1}{4k^2 - 1} + \frac{1}{4k^2 + 8k + 3} = \frac{k}{2k + 1} + \frac{1}{4k^2 + 8k + 3}$$

$$\frac{1}{3} + \frac{1}{15} + \cdots + \frac{1}{4k^2 - 1} + \frac{1}{4k^2 + 8k + 3} = \frac{k}{2k + 1} + \frac{1}{(2k + 1)(2k + 3)}$$

$$\frac{1}{3} + \frac{1}{15} + \cdots + \frac{1}{4k^2 - 1} + \frac{1}{4k^2 + 8k + 3} = \frac{k(2k + 3) + 1}{(2k + 1)(2k + 3)}$$

$$\frac{1}{3} + \frac{1}{15} + \cdots + \frac{1}{4k^2 - 1} + \frac{1}{4k^2 + 8k + 3} = \frac{2k^2 + 3k + 1}{(2k + 1)(2k + 3)}$$

$$\frac{1}{3} + \frac{1}{15} + \cdots + \frac{1}{4k^2 - 1} + \frac{1}{4k^2 + 8k + 3} = \frac{(2k + 1)(k + 1)}{(2k + 1)(2k + 3)}$$

$$\frac{1}{3} + \frac{1}{15} + \cdots + \frac{1}{4k^2 - 1} + \frac{1}{4k^2 + 8k + 3} = \frac{k + 1}{2k + 3}$$

$$\frac{1}{3} + \frac{1}{15} + \cdots + \frac{1}{4k^2 - 1} + \frac{1}{4k^2 + 8k + 3} = \frac{k + 1}{2(k + 1) + 1}$$

This final statement is exactly P_{k+1}, so P_{k+1} is true. Because P_n is true for $n = 1$ and P_k implies P_{k+1}, P_n is true for $n = 2$, $n = 3$, and so on. That is, by the principle of mathematical induction, $\frac{1}{3} + \frac{1}{15} + \cdots + \frac{1}{4n^2 - 1} = \frac{n}{2n+1}$ or $\sum_{a=1}^{n} \frac{1}{4a^2 - 1} = \frac{n}{2n+1}$ is true for all positive integers n.

33. Let P_n be $\frac{1}{6} + \frac{1}{12} + \cdots + \frac{1}{(n+1)(n+2)} = \frac{n}{2(n+2)}$.
Because $\frac{1}{6} = \frac{1}{(1+1)(1+2)}$ is a true statement, P_n is true for $n = 1$. Assume that $\frac{1}{6} + \frac{1}{12} + \cdots + \frac{1}{(k+1)(k+2)} = \frac{k}{2(k+2)}$ is true for a positive integer k. Show that P_{k+1} must be true.

$$\frac{1}{6} + \frac{1}{12} + \cdots + \frac{1}{(k+1)(k+2)} = \frac{k}{2(k+2)}$$

$$\frac{1}{6} + \frac{1}{12} + \cdots + \frac{1}{(k+1)(k+2)} + \frac{1}{[(k+1)+1][(k+1)+2]} = \frac{k}{2(k+2)} + \frac{1}{[(k+1)+1][(k+1)+2]}$$

$$\frac{1}{6} + \frac{1}{12} + \cdots + \frac{1}{(k+1)(k+2)} + \frac{1}{(k+2)(k+3)} = \frac{k}{2(k+2)} + \frac{1}{(k+2)(k+3)}$$

$$\frac{1}{6} + \frac{1}{12} + \cdots + \frac{1}{(k+1)(k+2)} + \frac{1}{(k+2)(k+3)} = \frac{k(k+3) + 2}{2(k+2)(k+3)}$$

$$\frac{1}{6} + \frac{1}{12} + \cdots + \frac{1}{(k+1)(k+2)} + \frac{1}{(k+2)(k+3)} = \frac{k^2 + 3k + 2}{2(k+2)(k+3)}$$

$$\frac{1}{6} + \frac{1}{12} + \cdots + \frac{1}{(k+1)(k+2)} + \frac{1}{(k+2)(k+3)} = \frac{(k+2)(k+1)}{2(k+2)(k+3)}$$

$$\frac{1}{6} + \frac{1}{12} + \cdots + \frac{1}{(k+1)(k+2)} + \frac{1}{(k+2)(k+3)} = \frac{k+1}{2(k+3)}$$

$$\frac{1}{6} + \frac{1}{12} + \cdots + \frac{1}{(k+1)(k+2)} + \frac{1}{(k+2)(k+3)} = \frac{k+1}{2[(k+1)+2]}$$

This final statement is exactly P_{k+1}, so P_{k+1} is true. Because P_n is true for $n = 1$ and P_k implies P_{k+1}, P_n is true for $n = 2$, $n = 3$, and so on. That is, by the principle of mathematical induction, $\frac{1}{6} + \frac{1}{12} + \cdots + \frac{1}{(n+1)(n+2)} = \frac{n}{2(n+2)}$ or $\sum_{a=1}^{n} \frac{1}{(a+1)(a+2)} = \frac{n}{2(n+2)}$ is true for all positive integers n.

35. Let P_n be $\frac{1}{2 \cdot 4} + \frac{1}{4 \cdot 6} + \frac{1}{6 \cdot 8} + \cdots + \frac{1}{(2n)(2n+2)} = \frac{n}{2(2n+2)}$. Because $\frac{1}{2 \cdot 4} = \frac{1}{2(2+2)}$ is a true statement, P_n is true for $n = 1$. Assume that $\frac{1}{2 \cdot 4} + \frac{1}{4 \cdot 6} + \frac{1}{6 \cdot 8} + \cdots + \frac{1}{(2k)(2k+2)} = \frac{k}{2(2k+2)}$ is true for a positive integer k. Show that P_{k+1} must be true.

$$\frac{1}{2 \cdot 4} + \frac{1}{4 \cdot 6} + \frac{1}{6 \cdot 8} + \cdots + \frac{1}{(2k)(2k+2)} = \frac{k}{2(2k+2)}$$

$$\frac{1}{2 \cdot 4} + \frac{1}{4 \cdot 6} + \frac{1}{6 \cdot 8} + \cdots + \frac{1}{(2k)(2k+2)} + \frac{1}{[2(k+1)][2(k+1)+2]}$$
$$= \frac{k}{2(2k+2)} + \frac{1}{[2(k+1)][2(k+1)+2]}$$

$$\frac{1}{2 \cdot 4} + \frac{1}{4 \cdot 6} + \frac{1}{6 \cdot 8} + \cdots + \frac{1}{(2k)(2k+2)} + \frac{1}{(2k+2)(2k+4)} =$$
$$\frac{k}{2(2k+2)} + \frac{1}{(2k+2)(2k+4)}$$

$$\frac{1}{2 \cdot 4} + \frac{1}{4 \cdot 6} + \frac{1}{6 \cdot 8} + \cdots + \frac{1}{(2k)(2k+2)} + \frac{1}{(2k+2)(2k+4)} =$$
$$\frac{k(2k+4) + 1(2)}{2(2k+2)(2k+4)}$$

$$\frac{1}{2 \cdot 4} + \frac{1}{4 \cdot 6} + \frac{1}{6 \cdot 8} + \cdots + \frac{1}{(2k)(2k+2)} + \frac{1}{(2k+2)(2k+4)} =$$
$$\frac{2k^2 + 4k + 2}{2(2k+2)(2k+4)}$$

$$\frac{1}{2 \cdot 4} + \frac{1}{4 \cdot 6} + \frac{1}{6 \cdot 8} + \cdots + \frac{1}{(2k)(2k+2)} + \frac{1}{(2k+2)(2k+4)} =$$
$$\frac{2(k+1)^2}{4(k+1)(2k+4)}$$

$$\frac{1}{2 \cdot 4} + \frac{1}{4 \cdot 6} + \frac{1}{6 \cdot 8} + \cdots + \frac{1}{(2k)(2k+2)} + \frac{1}{(2k+2)(2k+4)} =$$
$$\frac{k+1}{4k+8}$$

$$\frac{1}{2 \cdot 4} + \frac{1}{4 \cdot 6} + \frac{1}{6 \cdot 8} + \cdots + \frac{1}{(2k)(2k+2)} + \frac{1}{(2k+2)(2k+4)} =$$
$$\frac{k+1}{2[2(k+1)+2]}$$

This final statement is exactly P_{k+1}, so P_{k+1} is true. Because P_n is true for $n = 1$ and P_k implies P_{k+1}, P_n is true for $n = 2$, $n = 3$, and so on. That is, by the principle of mathematical induction, $\frac{1}{2 \cdot 4} + \frac{1}{4 \cdot 6} + \frac{1}{6 \cdot 8} + \cdots + \frac{1}{(2n)(2n+2)} = \frac{n}{2(2n+2)}$ is true for all positive integers n.

37. false; $n = 2$

39. Let P_n be the statement that $2^{2n+1} + 3^{2n+1}$ is divisible by 5. P_1 is the statement that $2^{2(1)+1} + 3^{2(1)+1}$ is divisible by 5. P_1 is true, since $2^{2(1)+1} + 3^{2(1)+1} = 35$, which is divisible by 5. Assume P_k is true, where k is a positive integer, and show that P_{k+1} must be true. That is, show that $2^{2k+1} + 3^{2k+1} = 5r$ for some integer r implies that $2^{2(k+1)+1} + 3^{2(k+1)+1}$ is divisible by 5.

$$2^{2k+1} + 3^{2k+1} = 5r$$
$$2^{2k+1} = 5r - 3^{2k+1}$$
$$2^{2k+1} \cdot 2^2 = 5r - 3^{2k+1} \cdot 2^2$$
$$2^{2k+1} \cdot 2^2 = (5r - 3^{2k+1})(3^2 - 5)$$
$$2^{2k+3} = 45r - 25r - 3^{2k+3} + 5(3^{2k+1})$$
$$2^{2k+3} + 3^{2k+3} = 45r - 25r - 3^{2k+3} + 5(3^{2k+1}) + 3^{2k+3}$$
$$2^{2k+3} + 3^{2k+3} = 20r + 5(3^{2k+1})$$
$$2^{2(k+1)+1} + 3^{2(k+1)+1} = 5(4r + 3^{2k+1})$$

Because r and k are integers, $4r + 3^{2k+1}$ is an integer and $5(4r + 3^{2k+1})$ is divisible by 5. Therefore, $2^{2k+3} + 3^{2k+3}$ is divisible by 5. Because P_n is true for $n = 1$ and P_k implies P_{k+1}, P_n is true for $n = 2$, $n = 3$, and so on. By the principle of mathematical induction, $2^{2n+1} + 3^{2n+1}$ is divisible by 5 for all positive integers n.

41. Let P_n be the statement $(x + 1)^n \geq nx$, for integer values $n \geq 1$ and $x \geq 1$. P_1 is true, since when $n = 1$ and $x \geq 1$, $(x + 1)^1 \geq 1x$ is a true statement. Assume that $(x + 1)^k \geq kx$ is true for an integer $k \geq 1$. Show that $(x + 1)^{k+1} \geq (k+1)x$ must be true.

$$(x + 1)^k \geq kx$$
$$(x + 1)(x + 1)^k \geq (x + 1)kx$$
$$(x + 1)^{k+1} \geq kx^2 + kx$$

Now we need to prove that $kx^2 + kx \geq (k+1)x$. First note that since $k \geq 1$ and $x \geq 1$, both kx and $x - 1$ are nonnegative. Hence their product $kx(x - 1)$ is nonnegative. Thus $kx^2 - kx \geq 0$, which implies $kx^2 \geq kx$. But $kx \geq x$, so by the Transitive Property of Inequality, $kx^2 \geq x$. Add kx to each side of $kx^2 \geq x$ and factor to obtain $kx^2 + kx \geq (k+1)x$. Since $(x + 1)^{k+1} \geq kx^2 + kx$ and $kx^2 + kx \geq (k+1)x$, then $(x + 1)^{k+1} \geq (k+1)x$. Because P_n is true for $n = 1$ and P_k implies P_{k+1}, P_n is true for $n = 2$, $n = 3$, and so on. By the principle of mathematical induction, $(x + 1)^n \geq nx$ is true for all integers $n \geq 1$ and for $x \geq 1$.

43. $S_n = 2n^2$; Let P_n be the statement $2 + 6 + 10 + 14 + \cdots + (4n - 2) = 2n^2$. Because $2 = 2(1)^2$ is a true statement, P_n is true for $n = 1$. Assume that $2 + 6 + 10 + 14 + \cdots + (4k - 2)$ $= 2k^2$ is true for a positive integer k. Show that P_{k+1} must be true.

$$2 + 6 + 10 + 14 + \cdots + (4k - 2) = 2k^2$$
$$2 + 6 + 10 + 14 + \cdots + (4k - 2) + [4(k + 1) - 2] = 2k^2 + [4(k + 1) - 2]$$
$$2 + 6 + 10 + 14 + \cdots + (4k - 2) + (4k + 2) = 2k^2 + (4k + 2)$$
$$2 + 6 + 10 + 14 + \cdots + (4k - 2) + (4k + 2) = 2(k^2 + 2k + 1)$$
$$2 + 6 + 10 + 14 + \cdots + (4k - 2) + (4k + 2) = 2(k + 1)^2$$

This final statement is exactly P_{k+1}, so P_{k+1} is true. Because P_n is true for $n = 1$ and P_k implies P_{k+1}, P_n is true for $n = 2$, $n = 3$, and so on. That is, by the principle of mathematical induction, $2 + 6 + 10 + 14 + \cdots + (4n - 2) = 2n^2$ is true for all positive integers n.

45. $S_n = \frac{1}{2^n} - 1$; Let P_n be $-\frac{1}{2} + \left(-\frac{1}{4}\right) + \left(-\frac{1}{8}\right) + \left(-\frac{1}{16}\right) + \cdots$ $+ \left(-\frac{1}{2^n}\right) = \frac{1}{2^n} - 1$. Because $-\frac{1}{2} = \frac{1}{2^1} - 1$ is a true statement, P_n is true for $n = 1$. Assume that $-\frac{1}{2} + \left(-\frac{1}{4}\right) + \left(-\frac{1}{8}\right) + \left(-\frac{1}{16}\right)$ $+ \cdots + \left(-\frac{1}{2^k}\right) = \frac{1}{2^k} - 1$ is true for a positive integer k. Show that P_{k+1} must be true.

$$-\frac{1}{2} + \left(-\frac{1}{4}\right) + \left(-\frac{1}{8}\right) + \left(-\frac{1}{16}\right) + \cdots + \left(-\frac{1}{2^k}\right) = \frac{1}{2^k} - 1$$

$$-\frac{1}{2} + \left(-\frac{1}{4}\right) + \left(-\frac{1}{8}\right) + \left(-\frac{1}{16}\right) + \cdots + \left(-\frac{1}{2^k}\right) + \left(-\frac{1}{2^{k+1}}\right) = \frac{1}{2^k} - 1 - \frac{1}{2^{k+1}}$$

$$-\frac{1}{2} + \left(-\frac{1}{4}\right) + \left(-\frac{1}{8}\right) + \left(-\frac{1}{16}\right) + \cdots + \left(-\frac{1}{2^k}\right) + \left(-\frac{1}{2^{k+1}}\right) = \frac{2\left(\frac{1}{2^k}\right)}{2} - 1 - \frac{1}{2^{k+1}}$$

$$-\frac{1}{2} + \left(-\frac{1}{4}\right) + \left(-\frac{1}{8}\right) + \left(-\frac{1}{16}\right) + \cdots + \left(-\frac{1}{2^k}\right) + \left(-\frac{1}{2^{k+1}}\right) = \frac{2}{2^{k+1}} - 1 - \frac{1}{2^{k+1}}$$

$$-\frac{1}{2} + \left(-\frac{1}{4}\right) + \left(-\frac{1}{8}\right) + \left(-\frac{1}{16}\right) + \cdots + \left(-\frac{1}{2^k}\right) + \left(-\frac{1}{2^{k+1}}\right) = \frac{1}{2^{k+1}} - 1$$

This final statement is exactly P_{k+1}, so P_{k+1} is true. Because P_n is true for $n = 1$ and P_k implies P_{k+1}, P_n is true for $n = 2$, $n = 3$, and so on. That is, by the principle of mathematical induction, $-\frac{1}{2} + \left(-\frac{1}{4}\right) + \left(-\frac{1}{8}\right) + \left(-\frac{1}{16}\right) + \cdots + \left(-\frac{1}{2^n}\right) = \frac{1}{2^n} - 1$ is true for all positive integers n.

47. Let P_n be the statement that $f_1 + f_2 + \cdots + f_n = f_{n+2} - 1$. Because $f_1 = 1$ and $f_{1+2} - 1 = f_3 - 1 = 2 - 1$ or 1, then $f_1 = f_{1+2} - 1$ is true. So, P_n is true for $n = 1$. Assume that $f_1 + f_2 + \cdots + f_k = f_{k+2} - 1$ is true for a positive integer k. Show that P_{k+1} must be true.

$$f_1 + f_2 + \cdots + f_k = f_{k+2} - 1$$
$$f_1 + f_2 + \cdots + f_k + f_{k+1} = f_{k+2} - 1 + f_{k+1}$$
$$f_1 + f_2 + \cdots + f_k + f_{k+1} = f_{k+3} - 1$$

This final statement is exactly P_{k+1}, so P_{k+1} is true. Because P_n is true for $n = 1$ and P_k implies P_{k+1}, P_n is true for $n = 2$, $n = 3$, and so on. So, $f_1 + f_2 + \cdots + f_n = f_{n+2} - 1$ is true for all positive integers n.

49. Let P_n be the statement that if a polygon has n sides, the sum of the interior angles is $180(n - 2)$. P_3 is true, since when $n = 3$, the figure is a triangle and the sum of its interior angles is 180 and $180(3 - 2) = 180$. Assume that if a polygon

has k sides, the sum of the interior angles is $180(k - 2)$ for a positive integer $k \geq 3$. Show that P_{k+1} must be true. That is, show that if a polygon has $k + 1$ sides, the sum of the interior angles is $180(k + 1 - 2)$. Use the induction hypothesis and its restriction that $k \geq 3$. Consider a convex polygon with $k + 1$ vertices. Because $k \geq 3$, it follows that $k + 1 \geq 3$. If we take one vertex x, there is another vertex y such that there is one vertex between x and y in one direction, and $k - 2$ vertices in the other direction. Join x and y by a new edge, dividing the original polygon into two polygons. The sum of the new polygons' interior angle sums is the sum of the original polygon's interior angles. One of the new polygons is a triangle and the other a polygon has k vertices (all but the one isolated between x and y). The triangle has interior angle sum of 180, and by the inductive hypothesis, the other polygon has an interior angle sum of $(k - 2)180$. Summing these, we get $(k + 1 - 2)180$, and the theorem is proved. Because P_n is true for $n = 1$ and P_k implies P_{k+1}, P_n is true for $n = 2$, $n = 3$, and so on. That is, by the extended principle of mathematical induction, if a polygon has n sides, the sum of the interior angles is $180(n - 2)$ is true for integer values of $n \geq 3$.

51. Let P_n be the statement $\left(\frac{x}{y}\right)^n = \frac{x^n}{y^n}$. Because $\left(\frac{x}{y}\right)^1 = \frac{x^1}{y^1}$ is a true statement, P_n is true for $n = 1$. Assume that $\left(\frac{x}{y}\right)^k = \frac{x^k}{y^k}$ is true for a positive integer k. Show that P_{k+1} must be true.

$$\left(\frac{x}{y}\right)^k = \frac{x^k}{y^k}$$
$$\frac{x}{y} \cdot \left(\frac{x}{y}\right)^k = \frac{x}{y} \cdot \frac{x^k}{y^k}$$
$$\left(\frac{x}{y}\right)^{k+1} = \frac{x^{k+1}}{y^{k+1}}$$

This final statement is exactly P_{k+1}, so P_{k+1} is true. Because P_n is true for $n = 1$ and P_k implies P_{k+1}, P_n is true for $n = 2$, $n = 3$, and so on. That is, by the principle of mathematical induction, $\left(\frac{x}{y}\right)^n = \frac{x^n}{y^n}$ is true for all positive integers n.

53. Let P_n be the statement $\cos n\pi = (-1)^n$ for any positive integer n. Because $\cos \pi = (-1)^1$ is a true statement, P_n is true for $n = 1$. Assume that $\cos k\pi = (-1)^k$ is true for a positive integers k. Show that P_{k+1} must be true.

$$\cos k\pi = (-1)^k$$
$$\cos k\pi \cdot \cos \pi = (-1)^k \cdot \cos \pi$$
$$\cos k\pi \cos \pi + (-\sin k\pi \sin \pi + \sin k\pi \sin \pi) = (-1)^k \cdot (-1)$$
$$(\cos k\pi \cos \pi - \sin k\pi \sin \pi) + \sin k\pi \sin \pi = (-1)^{k+1}$$
$$\cos (k\pi + \pi) + \sin k\pi \sin \pi = (-1)^{k+1}$$
$$\cos (k\pi + \pi) + \sin k\pi \cdot 0 = (-1)^{k+1}$$
$$\cos (k\pi + \pi) + 0 = (-1)^{k+1}$$
$$\cos [(k + 1)\pi] = (-1)^{k+1}$$

This final statement is exactly P_{k+1}, so P_{k+1} is true. Because P_n is true for $n = 1$ and P_k implies P_{k+1}, P_n is true for $n = 2$, $n = 3$, and so on. That is, by the principle of mathematical induction, $\cos n\pi = (-1)^n$ is true for all positive integers n.

55. Let $S_n = a_1 + (a_1 + d) + (a_1 + 2d) + \cdots + [a_1 + (n - 1)d]$ and P_n be the statement $a_1 + (a_1 + d) + (a_1 + 2d) + \cdots + [a_1 + (n - 1)d] = \frac{n}{2}[2a_1 + (n - 1)d]$. Because $a_1 = \frac{1}{2}[2a_1 + (1 - 1)d]$ is a true statement, P_n is true for $n = 1$. Assume $a_1 + (a_1 + d) + (a_1 + 2d) + \cdots + [a_1 + (k - 1)d] = \frac{k}{2}[2a_1 + (k - 1)d]$ is true for a positive integer k. Show that P_{k+1} must be true.

$$a_1 + (a_1 + d) + (a_1 + 2d) + \cdots + [a_1 + (k - 1)d] = \frac{k}{2}[2a_1 + (k - 1)d]$$

$$a_1 + (a_1 + d) + (a_1 + 2d) + \cdots + [a_1 + (k-1)d] + (a_1 + kd) =$$
$$\frac{k}{2}[2a_1 + (k-1)d] + (a_1 + kd)$$

$$a_1 + (a_1 + d) + (a_1 + 2d) + \cdots + [a_1 + (k-1)d] + (a_1 + kd) =$$
$$\frac{k[2a_1 + (k-1)d] + 2(a_1 + kd)}{2}$$

$$a_1 + (a_1 + d) + (a_1 + 2d) + \cdots + [a_1 + (k-1)d] + (a_1 + kd) =$$
$$\frac{2a_1 k + k(k-1)d + 2a_1 + 2kd}{2}$$

$$a_1 + (a_1 + d) + (a_1 + 2d) + \cdots + [a_1 + (k-1)d] + (a_1 + kd) =$$
$$\frac{2a_1 k + 2a_1 + [k(k-1) + 2k]d}{2}$$

$$a_1 + (a_1 + d) + (a_1 + 2d) + \cdots + [a_1 + (k-1)d] + (a_1 + kd) =$$
$$\frac{2a_1(k+1) + (k^2 + k)d}{2}$$

$$a_1 + (a_1 + d) + (a_1 + 2d) + \cdots + [a_1 + (k-1)d] + (a_1 + kd) =$$
$$\frac{2a_1(k+1) + (k+1)kd}{2}$$

$$a_1 + (a_1 + d) + (a_1 + 2d) + \cdots + [a_1 + (k-1)d] + (a_1 + kd) =$$
$$\frac{k+1}{2}(2a_1 + kd)$$

$$a_1 + (a_1 + d) + \cdots + [a_1 + (k-1)d] + \{a_1 + [(k+1)-1]d\} =$$
$$\frac{k+1}{2}\{2a_1 + [(k+1)-1]d\}$$

This final statement is exactly P_{k+1}, so P_{k+1} is true. Because P_n is true for $n = 1$ and P_k implies P_{k+1}, P_n is true for $n = 2$, $n = 3$, and so on. That is, by the principle of mathematical induction, $a_1 + (a_1 + d) + (a_1 + 2d) + \cdots + [a_1 + (n-1)d] = \frac{n}{2}[2a_1 + (n-1)d]$ is true for all positive integers n.

57. a. Sample answer: The conjecture: $2^n + 1$ is divisible by 3 is false.
b. counterexample: $n = 2$
59. Let P_n be the statement $a_n < 3$. Because $\sqrt{6} < 3$ is a true statement, P_n is true for $n = 1$. Assume that $a_k < 3$ is true for a positive integer k. Show that P_{k+1} must be true.

$a_{k+1} = \sqrt{6 + a_k}$	Each successive term is the square root of 6 more than the previous term.
$(a_{k+1})^2 = 6 + a_k$	Square both sides.
$(a_{k+1})^2 - 6 = a_k$	Solve for a_k.
$a_k < 3$	Inductive hypothesis
$(a_{k+1})^2 - 6 < 3$	Substitution
$(a_{k+1})^2 < 9$	Add 6 to each side.
$a_{k+1} < 3$	Take the square root of each side.

This final statement is exactly P_{k+1}, so P_{k+1} is true. Because P_n is true for $n = 1$ and P_k implies P_{k+1}, P_n is true for $n = 2$, $n = 3$, and so on. That is, by the principle of mathematical induction, P_n is true for all positive integers n.
61. 78,125   **63.** 192
65. 5.83

67. 6.32

69. $(-\sqrt{3}, -1)$ **71.** $\langle -\sqrt{2}, 0, \sqrt{2} \rangle$

73. vertex: $(4, -7)$; focus: $\left(4, -7\frac{1}{4}\right)$; axis of symmetry: $x = 4$; directrix: $y = -6\frac{3}{4}$

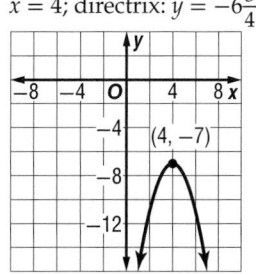

75. vertex: $(0, 6)$; focus: $(1, 6)$; axis of symmetry: $y = 6$; directrix: $x = -1$

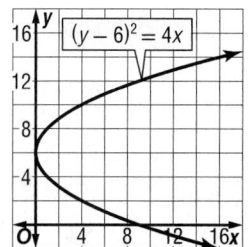

77. C   **79.** B

Pages 633–635   Lesson 10-5

1. $16 + 32x + 24x^2 + 8x^3 + x^4$ **3.** $64a^3 - 48a^2 b + 12ab^2 - b^3$
5. $2187x^7 + 10{,}206x^6 y + 20{,}412x^5 y^2 + 22{,}680x^4 y^3 + 15{,}120x^3 y^4 + 6048x^2 y^5 + 1344xy^6 + 128y^7$ **7.** $81c^4 - 108c^3 d + 54c^2 d^2 - 12cd^3 + d^4$ **9.** $a^3 - 3a^2 b + 3ab^2 - b^3$ **11.** 3360
13. 262,440   **15.** 56   **17.** -126 **19.** 280   **21.** 1,088,640   **23.** $\frac{35}{27}$
25. 21,504   **27a.** 1820   **27b.** 12,870   **29a.** 0.12%   **29b.** 3.37%
29c. 15.38%   **31.** $1024t^5 + 3840t^4 + 5760t^3 + 4320t^2 + 1620t + 243$ **33.** $64m^6 - 192m^5 n + 240m^4 n^2 - 160m^3 n^3 + 60m^2 n^4 - 12mn^5 + n^6$ **35.** $2187p^7 + 5103p^6 q + 5103p^5 q^2 + 2835p^4 q^3 + 945p^3 q^4 + 189p^2 q^5 + 21pq^6 + q^7$ **37.** $16{,}807c^{10} + 36{,}015c^8 d + 30{,}870c^6 d^2 + 13{,}230c^4 d^3 + 2835c^2 d^4 + 243d^5$

39. $\displaystyle\sum_{r=0}^{15} \binom{15}{r}(2q)^{15-r}(3)^r$ **41.** $\displaystyle\sum_{r=0}^{31} \binom{31}{r}(11x)^{31-r}(y)^r$

43. $\displaystyle\sum_{r=0}^{22} \binom{22}{r}(3f)^{22-r}\left(-\frac{3}{4}g\right)^r$

45. a. Each opening of a treasure chest represents a trial, so $n = 15$. You want to find how many different ways is it possible to open the chest and find gold coins exactly 9 times, so let $x = 9$. Find the value of the term $_nC_x$.
$$_nC_x = {_{15}C_9}$$
$$= \frac{15!}{(15-9)!9!}$$
$$= 5005$$
b. A success in this situation is finding gold coins, so $p = \frac{5}{6}$ and $q = 1 - \frac{5}{6}$ or $\frac{1}{6}$. Each opening of a treasure chest represents a trial, so $n = 15$. You want to find the probability of finding gold more than 12 times out of those 15 trials, so find the sum of $x = 13$, $x = 14$, and $x = 15$. To find these probabilities, find the value of the terms $_nC_x\, p^x q^{n-x}$ in the expansions of $(p + q)^n$.

$$_nC_x\, p^x q^{n-x} = {_{15}C_{13}}\left(\frac{5}{6}\right)^{13}\left(\frac{1}{6}\right)^{15-13}$$
$$= \frac{15!}{(15-13)!13!}\left(\frac{5}{6}\right)^{13}\left(\frac{1}{6}\right)^2$$
$$= 0.2726$$

$$_nC_x\, p^x q^{n-x} = {_{15}C_{14}}\left(\frac{5}{6}\right)^{14}\left(\frac{1}{6}\right)$$
$$= \frac{15!}{(15-14)!14!}\left(\frac{5}{6}\right)^{14}\left(\frac{1}{6}\right)^1$$
$$= 0.1947$$

$$C_x\, p^x\, q^{n-x} = {}_{15}C_{15}\left(\frac{5}{6}\right)^{15}\left(\frac{1}{6}\right)^{15-15}$$

$$= \frac{15!}{(15-15)!15!}\left(\frac{5}{6}\right)^{15}\left(\frac{1}{6}\right)^{0}$$

$$= 0.0649$$

The probability of finding gold more than 12 times is $0.2726 + 0.1947 + 0.0649$ or 0.5322 or 53.22%.

47. $16d^4 + 32\sqrt{5}d^3 + 120d^2 + 40\sqrt{5}d + 25$ **49.** $1024s^5 +$ $640s^4t + 160s^3t^2 + 20s^2t^3 + \frac{5}{4}st^4 + \frac{1}{32}t^5$ **51.** 3150 **53.** $\frac{231}{512}$
55. $-7 + 24i$ **57.** $1121 - 404i$ **59.** $\frac{5+\sqrt{2}i}{8}$ **61.** $g(x) = x^4 +$
$12x^3 + 54x^2 + 113x + 96$ **63.** $g(x) = x^6 - 6x^5 + 15x^4 - 18x^3$
$+ 9x^2 - 1$ **65a.** $f(x) = x^3$: $3x^2 + 3xh + h^2,\ h \neq 0$ $f(x) = x^4$: $4x^3$
$+ 6x^2h + 4xh^2 + h^3,\ h \neq 0$ $f(x) = x^5$: $5x^4 + 10x^3h + 10x^2h^2 +$
$5xh^3 + h^4,\ h \neq 0$ $f(x) = x^6$: $6x^5 + 15x^4h + 20x^3h^2 + 15x^2h^3 +$
$6xh^4 + h^5,\ h \neq 0$ $f(x) = x^7$: $7x^6 + 21x^5h + 35x^4h^2 + 35x^3h^3 +$
$21x^2h^4 + 7xh^5 + h^6,\ h \neq 0$ $f(x) = x^n$: $nx^{n-1} + {}_nC_2x^{n-2}h +$
${}_nC_3x^{n-3}h^2 + \cdots + {}_nC_rx^{n-r}h^{r-1} + \cdots + h^{n-1},\ h \neq 0$

65b.

	$f(x) = x^3$	$f(x) = x^4$	$f(x) = x^5$
$h = 0.1$	$3x^2 + 0.3x + 0.01$	$4x^3 + 0.6x^2 +$ $0.04x + 0.001$	$5x^4 + 1x^3 + 0.1x^2$ $+ 0.005x +$ 0.0001
$h = 0.01$	$3x^2 + 0.03x +$ 0.0001	$4x^3 + 0.06x^2 +$ $0.0004x +$ 0.000001	$5x^4 + 0.1x^3 +$ $0.001x^2 +$ $0.000005x +$ (1×10^{-8})
$h = 0.001$	$3x^2 + 0.003x +$ 0.000001	$4x^3 + 0.006x^2 +$ $0.000004x +$ (1×10^{-9})	$5x^4 + 0.01x^3 +$ $0.00001x^2 +$ $(5 \times 10^{-9})x +$ (1×10^{-12})
$h = 0.0001$	$3x^2 + 0.0003x +$ 0.00000001	$4x^3 + 0.0006x^2 +$ $0.00000004x +$ (1×10^{-12})	$5x^4 + 0.001x^3 +$ $(1 \times 10^{-7})x^2 +$ $(5 \times 10^{-12})x +$ (1×10^{-16})

	$f(x) = x^6$	$f(x) = x^7$
$h = 0.1$	$6x^5 + 1.5x^4 + 0.2x^3 +$ $0.015x^2 + 0.0006x +$ 0.00001	$7x^6 + 2.1x^5 + 0.35x^4 +$ $0.035x^3 + 0.0021x^2 +$ $(7 \times 10^{-5})x + (1 \times 10^{-6})$
$h = 0.01$	$6x^5 + 0.15x^4 + 0.002x^3 +$ $0.000015x^2 + (6 \times 10^{-8})x +$ (1×10^{-10})	$7x^6 + 0.21x^5 + 0.0035x^4 +$ $(3.5 \times 10^{-5})x^3 +$ $(2.1 \times 10^{-7})x^2 +$ $(7 \times 10^{-10})x +$ (1×10^{-12})
$h = 0.001$	$6x^5 + 0.015x^4 + 0.00002x^3$ $+ (1.5 \times 10^{-8})x^2 +$ $(6 \times 10^{-12})x +$ (1×10^{-15})	$7x^6 + 0.021x^5 + 0.000035x^4 +$ $(3.5 \times 10^{-8})x^3 +$ $(2.1 \times 10^{-11})x^2 +$ $(7 \times 10^{-15})x +$ (1×10^{-18})
$h = 0.0001$	$6x^5 + 0.0015x^4 +$ $0.0000002x^3 +$ $(1.5 \times 10^{-11})x^2 +$ $(6 \times 10^{-16})x +$ (1×10^{-20})	$7x^6 + 0.0021x^5 +$ $0.00000035x^4 +$ $(3.5 \times 10^{-11})x^3 +$ $(2.1 \times 10^{-15})x^2 +$ $(7 \times 10^{-20})x +$ (1×10^{-24})

Sample answer: As h decreases, all the terms except the first in each expression get closer to 0.

65c.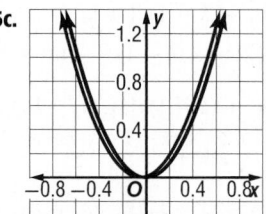
Sample answer: The graphs of the functions are almost exactly the same.
65d. nx^{n-1}

67. Sample answer: Chinese mathematician Chu Shih-chieh used the triangle in 1303. If the second number of any row is prime, all other numbers in the row (excluding the 1s) are divisible by the prime. The second diagonal of the triangle is the set of natural numbers. The third diagonal is the set of triangular numbers. **69.** Sample answer: Treat $x + y$ as a single term and expand $[(x + y) + z]^n$; $(x + y)^n +$ $n(x + y)^{n-1}z + {}_nC_2(x + y)^{n-2}z^2 + {}_nC_3(x + y)^{n-3}z^3 + \cdots +$ ${}_nC_r(x + y)^{n-r}z^r + \cdots + z^n$. **71.** Sample answer: The first and last numbers in each row are 1. Every other number is formed by adding the two numbers just above that number in the previous row. The expansion of $(a + b)^{n-1}$ has n terms and the expansion of $(a + b)^n$ has $n + 1$ terms. The coefficients of the terms of the expansion of $(a + b)^n$ are all positive, whereas every other coefficient in the expansion of $(a - b)^n$ is negative.

73 Each term in the expansion of $\left(\frac{1}{2v} + 6v^7\right)^8$ will be of the form ${}_8C_r\left(\frac{1}{2v}\right)^{8-r}(6v^7)^r$. The term ${}_8C_r\left(\frac{1}{2v}\right)^{8-r}(6v^7)^r$ can be written as ${}_8C_r\left(\frac{1}{2^{8-r}v^{8-r}}\right)(6^rv^{7r})$ or ${}_8C_r\left(\frac{6^rv^{7r}}{2^{8-r}v^{8-r}}\right)$. For v to be completely divided out of the term, the exponent of the v in the numerator must be equal to the exponent of the v in the denominator. So, $7r = 8 - r$ or $r = 1$. Since $r = 1$, this will be the second term of the expansion.

$${}_8C_r\left(\frac{1}{2v}\right)^{8-r}(6v^7)^r = {}_8C_1\left(\frac{1}{2v}\right)^{8-1}(6v^7)^1$$

$$= {}_8C_1\left(\frac{1}{128v^7}\right)(6v^7)$$

$$= 8\left(\frac{6v^7}{128v^7}\right)$$

$$= \frac{3}{8}$$

75. Sample answer: $n = 3$ **77.** $65{,}534$ **79.** circle: $(x - 8)^2 +$ $(y + 5)^2 = 25$ **81.** circle: $(x + 1)^2 + (y + 12)^2 = 4$
83. $x + \frac{2}{x-3} + \frac{4}{x-5}$ **85.** no **87.** yes **89.** $\sqrt{6} - \sqrt{2}$

91a.

91b. $y = 30(1.065)^x$
91c. $\approx \$396.70$ **93.** F
95. H

1. $g(x) = \dfrac{4}{1-x} =$

$\displaystyle\sum_{n=0}^{\infty} \left(\dfrac{x+3}{4}\right)^n$

for $-7 < x < 1$

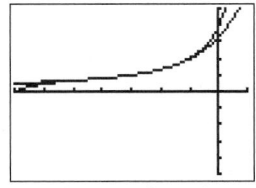

[−7, 1] scl: 1 by [−5, 5] scl: 1

3. $g(x) = \dfrac{2}{1-x^2} =$

$\displaystyle\sum_{n=0}^{\infty} \left(\dfrac{x^2+1}{2}\right)^n$

for $-1 < x < 1$

[−1, 1] scl: 0.1 by [−5, 5] scl: 1

5. $g(x) = \dfrac{2}{5-3x} =$

$\displaystyle\sum_{n=0}^{\infty} \left(\dfrac{3x-3}{2}\right)^n$

for $\dfrac{1}{3} < x < \dfrac{5}{3}$

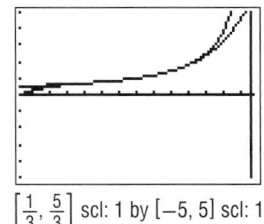

$\left[\dfrac{1}{3}, \dfrac{5}{3}\right]$ scl: 1 by [−5, 5] scl: 1

7. 1.648 **9.** 0.648 **11.** 0.741

13 Use the power series representation of e^x to find the fifth partial sum of $P = 3.5e^{0.08t}$ and evaluate the expression for $t = 4$, $t = 12$, and $t = 52$.

$3.5e^{0.08t} \approx 3.5\left(1 + 0.08t + \dfrac{(0.08t)^2}{2!} + \dfrac{(0.08t)^3}{3!} + \dfrac{(0.08t)^4}{4!}\right)$

$3.5e^{0.08(4)}$

$\approx 3.5\left(1 + 0.08(4) + \dfrac{[0.08(4)]^2}{2!} + \dfrac{[0.08(4)]^3}{3!} + \dfrac{[0.08(4)]^4}{4!}\right)$

≈ 4.8 mussels/m^2

$3.5e^{0.08(12)}$

$\approx 3.5\left(1 + 0.08(12) + \dfrac{[0.08(12)]^2}{2!} + \dfrac{[0.08(12)]^3}{3!} + \dfrac{[0.08(12)]^4}{4!}\right)$

≈ 9.1 mussels/m^2

$3.5e^{0.08(52)}$

$\approx 3.5\left(1 + 0.08(52) + \dfrac{[0.08(52)]^2}{2!} + \dfrac{[0.08(52)]^3}{3!} + \dfrac{[0.08(52)]^4}{4!}\right)$

≈ 134.0 mussels/m^2

15. 0.886 **17.** 0.588 **19.** 0.476 **21.** $2e^{i\frac{\pi}{6}}$ **23.** $2e^{i\frac{7\pi}{4}}$ **25.** $2e^{i\frac{5\pi}{3}}$

27. $2e^{i\frac{3\pi}{4}}$ **29.** $\approx 1.792 + i\pi$ **31.** $\approx 0.896 + i\pi$ **33.** $\approx 1.472 + i\pi$

35a.

Sample answer:
convergence, (−1.5, 1.5)

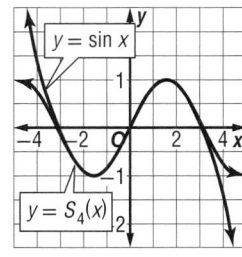

Sample answer:
convergence, (−2.5, 2.5)

Sample answer:
convergence, (−3.5, 3.5)

35b.

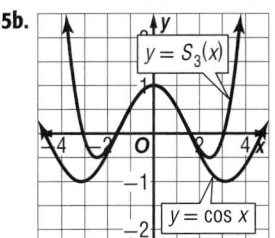

Sample answer:
convergence, (−1.5, 1.5)

Sample answer:
convergence, (−2.5, 2.5)

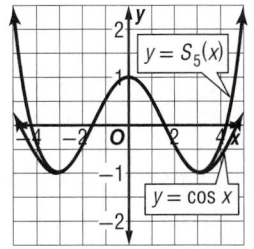

Sample answer:
convergence, (−3, 3)

35c. Sample answer: The interval of convergence widens as n increases. As n approaches infinity, the power series equals the trigonometric function that it represents.

37. $\approx 1.243 + \dfrac{i\pi}{2}$ **39.** $\approx 0.288 + i\pi$ **41.** $\approx 0.406 + i\pi$,

$\approx -1.609 + i\pi$

43. two terms: $\dfrac{|e^{2.1} - (1 + 2.1)|}{e^{2.1}} = 0.62$ or 62%;

three terms: $\dfrac{\left|e^{2.1} - \left(1 + 2.1 + \dfrac{2.1^2}{2}\right)\right|}{e^{2.1}} = 0.35$ or 35%;

six terms: $\dfrac{\left|e^{2.1} - \left(1 + 2.1 + \dfrac{2.1^2}{2} + \dfrac{2.1^3}{6} + \dfrac{2.1^4}{24} + \dfrac{2.1^5}{120}\right)\right|}{e^{2.1}} = 0.02$ or 2%

45. $\sin 1 \approx 1 - \dfrac{1^3}{3}! + \dfrac{1^5}{5}! - \dfrac{1^7}{7!}$ or 0.8415

$\cos 1 \approx 1 - \dfrac{1^2}{2!} + \dfrac{1^4}{4!} - \dfrac{1^6}{6!}$ or 0.5403

$\tan 1 = \dfrac{\sin 1}{\cos 1} \approx \dfrac{0.8415}{0.5403}$ or 1.5575

$\sec 1 = \dfrac{1}{\cos 1} = \dfrac{1}{0.5403}$ or 1.8508

$1.8508^2 - 1.5575^2 \approx 0.9997$

The expected value is exactly 1 because $\sec^2 x - \tan^2 x = 1$.

47. $\displaystyle\sum_{n=0}^{\infty} \left(\dfrac{4kqx - x^4 + 2x^2 - 1}{4kqx}\right)^n$

49 $\sin x = x - \dfrac{x^3}{3!} + \dfrac{x^5}{5!} - \dfrac{x^7}{7!} + \cdots$

$\sin(-x) = -x - \dfrac{(-x)^3}{3!} + \dfrac{(-x)^5}{5!} - \dfrac{(-x)^7}{7!} + \cdots$

$= -x + \dfrac{x^3}{3!} - \dfrac{x^5}{5!} + \dfrac{x^7}{7!} - \cdots$

$= -\left(x - \dfrac{x^3}{3!} + \dfrac{x^5}{5!} - \dfrac{x^7}{7!} + \cdots\right)$

$= -\sin x$

51a. $\tan^{-1} x \approx x - \dfrac{x^3}{3} + \dfrac{x^5}{5} - \dfrac{x^7}{7} + \dfrac{x^9}{9}$ **51b.** 0.0997

51c.

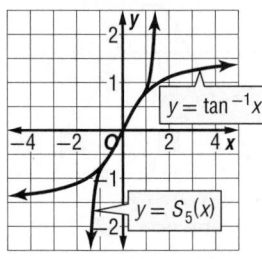

51d. As n increases, the graphs of the partial sums more closely resemble the graph of $f(x) = \tan^{-1} x$ on the interval $(-1, 1)$. Outside of the interval $(-1, 1)$, the end behavior of the polynomial approximations causes the graphs of the partial sums to diverge from the graph of $f(x) = \tan^{-1} x$.

53. Sample answer: The power series representation for sine is given by $\sin x \approx x - \dfrac{x^3}{3!} + \dfrac{x^5}{5!} - \dfrac{x^7}{7!} + \dfrac{x^9}{9!} - \cdots$. For x-values on the interval $[-0.1, 0.1]$, the cubic term and those of higher degree have values less than or equal to $\left|\dfrac{1}{6000}\right|$. This represents less than one-tenth of 1 percent of the value. Therefore, for the values of x on the interval $[-0.1, 0.1]$, the first term of the series, x, is a close approximation of $\sin x$.

55. If α and β differ by an integer multiple of 2π, then the complex numbers represented by $e^{i\alpha}$ and $e^{i\beta}$ will be the same.

57. $e^{ix} + e^{-ix} = (\cos x + i \sin x) + (\cos x - i \sin x)$
$e^{ix} + e^{-ix} = 2 \cos x$
$\cos x = \dfrac{e^{ix} + e^{-ix}}{2}$

59. $81m^4 + 108\sqrt{2}m^3 + 108m^2 + 24\sqrt{2}m + 4$ **61.** $p^{16} + 8p^{14}q + 28p^{12}q^2 + 56p^{10}q^3 + 70p^8q^4 + 56p^6q^5 + 28p^4q^6 + 8p^2q^7 + q^8$

63. $16 + 16i$ **65.** $-\dfrac{1}{4}i$ **67.** $(9, 2)$ **69.** $(7, 1, 3)$ **71.** yes

73a. 54.41%   **73b.** Sample answer: women who participated in hiking or quilting but not sculpting   **75.** F

Pages 645–648   Chapter 10   Study Guide and Review

1. true   **3.** false; explicitly   **5.** false; recursively   **7.** false; common difference   **9.** false; diverge   **11.** Sample answer: 33, 41, 49, 57   **13.** −20   **15.** $42\dfrac{2}{3}$ **17.** $a_n = -6 + 5(n - 1)$; $a_1 = -6$, $a_n = a_{n-1} + 5$ **19.** 11,325   **21.** −1194   **23.** 1.5; 16.875, 25.3125, 37.96875   **25.** $a_n = 10(-2)^{n-1}$; $a_1 = 10$, $a_n = (-2)a_{n-1}$
27. 133.11

29. Let P_n be the statement $6^n - 9$ is divisible by 3. P_1 is the statement that $6^1 - 9$ is divisible by 3. P_1 is true because $6^1 - 9$ is −3, which is divisible by 3. Assume P_k is true, where k is a positive integer, and show that P_{k+1} must be true. That is, show that $6^k - 9 = 3r$ for some integer r implies that $6^{k+1} - 9$ is divisible by 3.

$6^k - 9 = 3r$
$6^k = 3r + 9$
$6 \cdot 6^k = 6(3r + 9)$
$6^{k+1} = 18r + 54$
$6^{k+1} - 9 = 18r + 45$
$6^{k+1} - 9 = 3(6r + 15)$

Because r is an integer, $6r + 15$ is an integer, and $3(6r + 15)$ is divisible by 3. Therefore, $6^{k+1} - 9$ is divisible by 3. Because P_n is true for $n = 1$ and P_k implies P_{k+1}, P_n is true for $n = 2$, $n = 3$, and so on. By the principle of mathematical induction, $6^n - 9$ is divisible by 3 for all positive integers n.

31. Let P_n be the statement $5^n + 3$ is divisible by 4. P_1 is the statement that $5^1 + 3$ is divisible by 4. P_1 is true because $5^1 + 3$ is 8, which is divisible by 4. Assume P_k is true, where k is a positive integer, and show that P_{k+1} must be true. That is, show that $5^n + 3 = 4r$ for some integer r implies that $5^{k+1} + 3$ is divisible by 4.

$5^k + 3 = 4r$
$5^k = 4r - 3$
$5 \cdot 5^k = 5(4r - 3)$
$5^{k+1} = 20r - 15$
$5^{k+1} + 3 = 20r - 12$
$5^{k+1} + 3 = 4(5r - 3)$

Because r is an integer, $5r - 3$ is an integer, and $4(5r - 3)$ is divisible by 4. Therefore, $5^{k+1} + 3$ is divisible by 4. Because P_n is true for $n = 1$ and P_k implies P_{k+1}, P_n is true for $n = 2$, $n = 3$, and so on. By the principle of mathematical induction, $5^n + 3$ is divisible by 4 for all positive integers n.

33. Let P_n be the statement $\dfrac{1}{1 \cdot 2 \cdot 3} + \dfrac{1}{2 \cdot 3 \cdot 4} + \dfrac{1}{3 \cdot 4 \cdot 5} + \cdots + \dfrac{1}{n(n+1)(n+2)} = \dfrac{n(n+3)}{4(n+1)(n+2)}$. Because $\dfrac{1}{(1+1)(1+2)} = \dfrac{(1+3)}{4(1+1)(1+2)}$ is a true statement, P_n is true for $n = 1$.

Assume that $\dfrac{1}{1 \cdot 2 \cdot 3} + \dfrac{1}{2 \cdot 3 \cdot 4} + \dfrac{1}{3 \cdot 4 \cdot 5} + \cdots + \dfrac{1}{k(k+1)(k+2)} = \dfrac{k(k+3)}{4(k+1)(k+2)}$ is true for a positive integer k. Show that P_{k+1} must be true.

$\dfrac{1}{1 \cdot 2 \cdot 3} + \dfrac{1}{2 \cdot 3 \cdot 4} + \cdots + \dfrac{1}{k(k+1)(k+2)} = \dfrac{k(k+3)}{4(k+1)(k+2)}$

$\dfrac{1}{1 \cdot 2 \cdot 3} + \dfrac{1}{2 \cdot 3 \cdot 4} + \cdots + \dfrac{1}{k(k+1)(k+2)} + \dfrac{1}{(k+1)[(k+1)+1][(k+1)+2]} = \dfrac{k(k+3)}{4(k+1)(k+2)} + \dfrac{1}{(k+1)[(k+1)+1][(k+1)+2]}$

$\dfrac{1}{1 \cdot 2 \cdot 3} + \dfrac{1}{2 \cdot 3 \cdot 4} + \cdots + \dfrac{1}{k(k+1)(k+2)} + \dfrac{1}{(k+1)[(k+1)+1][(k+1)+2]} = \dfrac{k(k+3)}{4(k+1)(k+2)} + \dfrac{1}{(k+1)(k+2)(k+3)}$

$\dfrac{1}{1 \cdot 2 \cdot 3} + \dfrac{1}{2 \cdot 3 \cdot 4} + \cdots + \dfrac{1}{k(k+1)(k+2)} + \dfrac{1}{(k+1)[(k+1)+1][(k+1)+2]} = \dfrac{k(k+3)^2 + 4}{4(k+1)(k+2)(k+3)}$

$\dfrac{1}{1 \cdot 2 \cdot 3} + \dfrac{1}{2 \cdot 3 \cdot 4} + \cdots + \dfrac{1}{k(k+1)(k+2)} + \dfrac{1}{(k+1)[(k+1)+1][(k+1)+2]} = \dfrac{k^3 + 6k^2 + 9k + 4}{4(k+1)(k+2)(k+3)}$

$\dfrac{1}{1 \cdot 2 \cdot 3} + \dfrac{1}{2 \cdot 3 \cdot 4} + \cdots + \dfrac{1}{k(k+1)(k+2)} + \dfrac{1}{(k+1)[(k+1)+1][(k+1)+2]} = \dfrac{(k+1)^2 + (k+4)}{4(k+1)(k+2)(k+3)}$

$$\frac{1}{1 \cdot 2 \cdot 3} + \frac{1}{2 \cdot 3 \cdot 4} + \cdots + \frac{1}{k(k+1)(k+2)} +$$
$$\frac{1}{(k+1)[(k+1)+1][(k+1)+2]} = \frac{(k+1)(k+4)}{4(k+2)(k+3)}$$
$$\frac{1}{1 \cdot 2 \cdot 3} + \frac{1}{2 \cdot 3 \cdot 4} + \cdots + \frac{1}{k(k+1)(k+2)} +$$
$$\frac{1}{(k+1)[(k+1)+1][(k+1)+2]} = \frac{(k+1)[(k+1)+3]}{4[(k+1)+1][(k+1)+2]}$$

This final statement is exactly P_{k+1}, so P_{k+1} is true. Because P_n is true for $n = 1$ and P_k implies P_{k+1}, P_n is true for $n = 2$, $n = 3$, and so on. That is, by the principle of mathematical induction, $\frac{1}{1 \cdot 2 \cdot 3} + \frac{1}{2 \cdot 3 \cdot 4} + \frac{1}{3 \cdot 4 \cdot 5} + \cdots + \frac{1}{n(n+1)(n+2)}$ $= \frac{n(n+3)}{4(n+1)(n+2)}$ is true for all positive integers n.

35. Let P_n be the statement $5n < 6^n$. P_1 is true, since $5(1) < 6^1$ is a true statement. Assume P_k is true, that $5k < 6^k$ if k is a positive integer, and show that P_{k+1} must be true. That is, show that $5k < 6^k$ implies $5(k+1) < 6^{k+1}$. Use both parts of the inductive hypothesis.

$$
\begin{array}{c|c}
5k < 6^k & k \geq 1 \\
6 \cdot 5k < 6 \cdot 6^k & 1 \leq k \\
30k < 6^{k+1} & 25 \cdot 1 \leq 25 \cdot k \\
 & 25 \leq 25k
\end{array}
$$

If $5 \leq 25$ and $25 \leq 25k$, then by the Transitive Property of Inequality, $5 \leq 25k$.
$$5 \leq 25k$$
$$5k + 5 \leq 25k + 5k$$
$$5(k+1) \leq 30k$$

By the Transitive Property of Inequality, if $5(k+1) \leq 30k$ and $30k < 6^{k+1}$, then $5(k+1) < 6^{k+1}$. Since P_n is true for $k \geq 1$ and P_k implies P_{k+1} for $k \geq 1$, P_n is true for $n = 2$, $n = 3$, and so on. By the principle of mathematical induction, $5n < 6^n$ is true for all positive integers n.

37. $m^6 - 30m^5n + 375m^4n^2 - 2500m^3n^3 + 9375m^2n^4 - 18{,}750mn^5 + 15{,}625n^6$ **39.** 240,185,088 **41.** $16{,}384m^7 + 86{,}016m^6n + 193{,}536m^5n^2 + 241{,}920m^4n^3 + 181{,}440m^3n^4 + 81{,}648m^2n^5 + 20{,}412mn^6 + 2187n^7$

43. $\displaystyle\sum_{n=0}^{\infty} \left(\frac{2x+2}{3}\right)^n$; $-\frac{5}{2} < x < \frac{1}{2}$

$y = g(x)$
$y = S_5(x)$
[−5, 5] scl: 1 by [−2, 8] scl: 1

45. 0.273 **47.** $1.967 + i\pi$ **49a.** 9 **49b.** 80

51. Let P_n be the set of \$10 and \$25 bills that adds to \$10n for $n \geq 6$. For $n = 6$, the first possible case, the conjecture is valid, since \$10(6) = \$10(1) + \$25(2). Assume that for $n = k$, there exists a set of \$10 and/or \$25 bills that adds to \$10k. Show that P_n is valid for $n = k + 1$. There are two cases to consider.

Case 1: P_k contains at least three \$25 bills. Replace two \$25 bills in P_k with six \$10 bills and the value of P_k is increased by \$10, to \$10k + 10 or \$10(k + 1), which is P_{k+1}.

Case 2: P_k contains no \$25 bills. P_k must contain at least six \$10 bills, since the value of P_k must be greater than or equal to \$60. Replace four of the \$10 bills with two \$25 bills and the value of P_k is increased by \$10, to \$10k + 10 or \$10(k + 1), which is P_{k+1}.

In both cases, P_n is valid for $n = k + 1$. Since P_n is valid for $n = 6$ and for $n = k + 1$, it is valid for $n = 7$, $n = 8$, and so on. That is, by the extended principle of mathematical induction, all multiples of \$5 greater than \$60 can be formed using just \$10 and \$25 bills. **53.** 31.173 ft

Pages 650–651 Chapter 10 Connect to AP Calculus

1. Sample answer: Δx is decreasing. As Δx becomes smaller, the rectangles will better approximate the area of the region. **3.** Sample answer: Δx approaches 0. **5.** Sample answer: Δx approaches 0, the number of terms approaches ∞, and the area approaches 16.5. **7.** $\displaystyle\sum_{i=1}^{10} f(x_i) \cdot 0.4$; 63.36 units^2

Chapter 11 Inferential Statistics

Page 653 Chapter 11 Get Ready

1. 20 **3.** 56

5a.

x	0	1	2	3	4	5
ln y	2.41	3.71	5.01	6.31	7.61	8.91

5b. $\hat{y} = 1.3\hat{x} + 2.41$ **5c.** $y = 11.1e^{1.3x}$

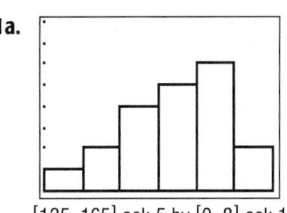

[0, 6] scl: 1 by [0, 10] scl: 2

Pages 660–663 Lesson 11-1

1a.

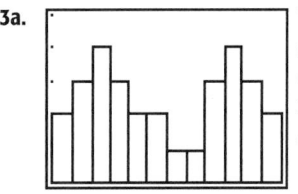

[135, 165] scl: 5 by [0, 8] scl: 1

The graph is negatively skewed. The majority of the landings occurred between 155 and 160 mph, with few that were greater than 160 mph or less than 140 mph.

1b. Sample answer: Since the distribution is skewed, the five-number summary can be used to describe the distribution of data; the landing speeds ranged from 138 to 162 mph, the median landing speed was 152.5 mph, and half of the speeds were between 147 and 157.5 mph.

3a.

The graph is bimodal, which suggests that there are two different types of bowlers mixed in the data set: average and experienced.

[40, 280] scl: 20 by [0, 5] scl: 1

3b. Sample answer: Since the lower cluster of data is approximately symmetrical, a mean bowling score of 97.6 and standard deviation of 31.5 can be used to describe the center and spread of the data, respectively. The upper cluster is positively skewed. Therefore, the five-number summary indicates that the scores ranged from 173 to 273, the median score was 231, and half of the scores were between 217 and 248.

5 **a.** Input the data into L1 on a graphing calculator. Then turn on **Plot1** under the **STAT PLOT** menu and choose "mini box plot." Graph the box plot and select **ZoomStat**. Because the right whisker is longer than the left whisker, and the median is closer to Q_1 than to Q_3, the data is positively skewed.

[0, 14] scl: 2 by [0, 1] scl: 0.125

b. Sample answer: Since the data is skewed, the five-number summary is a better representation. Press **STAT**, select the **Calc** submenu, and choose **1-Var Stats**. The summary shows that the number of hours played ranged from 0 to 12.5 hours and the median number of hours played was 2. Half the students spent between 1.5 and 4.5 hours playing.

7.

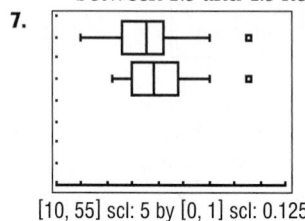

[10, 55] scl: 5 by [0, 1] scl: 0.125

The median fuel efficiency for cars from Year 1 is slightly lower than those from Year 2. The minimum values from Year 1 are lower than Year 2, but the maximum values are approximately equal. Therefore, there was not much of an improvement in fuel efficiency for hybrid cars from the first year to the second year. The spread of the middle half of the data is slightly larger for the cars from Year 2, but roughly the same as those from Year 1.

9a.

Adult Female Sea Otter Weights

9b. Sample answer: An adult female sea otter that weighs 55 pounds would be in the 40th percentile, which means that 40% of the otters have a lower weight. **11.** B **13.** A

15 **a.** Input the values in L1. To obtain the histogram, enter the **STAT PLOT** menu, turn on **Plot1** and change to the histogram setting. To view the graph, select **ZoomStat** or press **Graph** and adjust the window manually. To obtain the box plot, use the box plot setting instead of the histogram setting. Because more of the data appears to the left, the data is positively skewed.

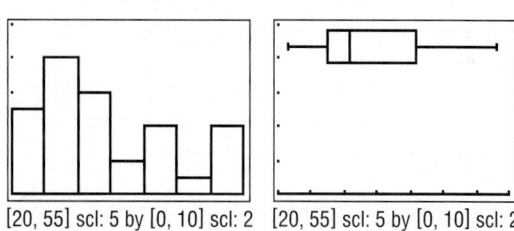

[20, 55] scl: 5 by [0, 10] scl: 2   [20, 55] scl: 5 by [0, 10] scl: 2

b. Enter the **Calc** submenu of the **STAT** menu and select **1-Var Stats**. The mean is 34,157 people.

c. The box plot; sample answer: Since the distribution is skewed, the median is the best measure of center. The median can be easily read off of the graph as slightly higher than 30,000.

d. No; sample answer: Since the histogram only displays the frequency at which the values occur in a given interval, and the box plot only displays the percentage of values that fall within a given interval, there is no way to associate any one attendance with a specific year using these types of graphs.

17a.

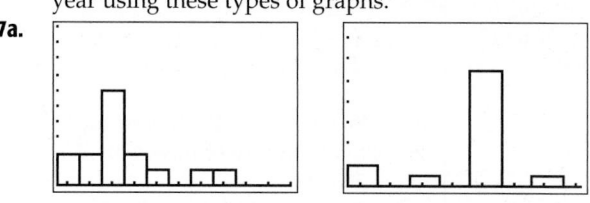

[940, 1200] scl: 25 by [0, 10] scl: 1 [890, 1080] scl: 25 by [0, 15] scl: 2

17b. Brand A

19 Input the data in L1. Select **1-Var Stats** from the **Calc** submenu of the **Stats** menu. $Q_1 = 0.12$, $Q_2 = \text{Med} = 0.25$, $Q_3 = 0.29$. Compute $\dfrac{Q_1 + Q_3}{2} = \dfrac{0.12 + 0.29}{2} = 0.205$.

21a.

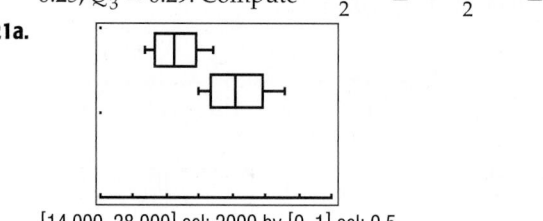

[14,000, 28,000] scl: 2000 by [0, 1] scl: 0.5

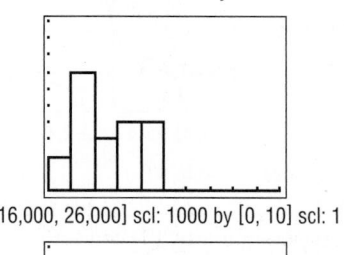

[16,000, 26,000] scl: 1000 by [0, 10] scl: 1

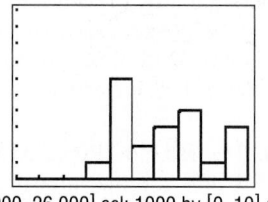

[16,000, 26,000] scl: 1000 by [0, 10] scl: 1

21b. Sample answer: The median consumption of petroleum in the U.S. from 1987 to 2007 was 18,450, while the mean was 18,560. In North America, the median was 22,200 and the mean was 22,350. **21c.** Sample answer: It is easier to determine the median of each distribution from the side-by-side box plots than the means of the histograms. While you can determine the spread of the entire range of data from either graph, you can also determine the range of each quartile from the box plots. **21d.** 83% **23.** Sample answer: The range only takes into account the maximum and minimum values of the distribution. Outlying data values could have a dramatic effect on the range. **25.** Sample answer: The variation is measured by the spread of the prices. Station C has the greatest variation in gas prices from $3.05 to $3.45, and Station B has the least variation in gas prices from $3.20 to $3.35. **27.** Sample answer: The mean formula includes every data value, while the median formula only includes the middle value if the number of data values is odd, or the middle two values if the number of values is even. When there are outlying data values, the mean will include these values and may be pushed to the left or right of the distribution, while the median will not be strongly affected by the values. For example, the data set {30, 40, 50, 60, 70} has a mean and median of 50. If data values of 98 and 99 are added to the data set, the median increases slightly to 55, while the mean increases to about 63.9. **29.** $\sqrt{10}e^{i\frac{7\pi}{4}}$ **31.** $243a^5 + 1620a^4b + 4320a^3b^2 + 5760a^2b^3 + 3840ab^4 + 1024b^5$ **33.** $64x^6 - 768x^5y + 3840x^4y^2 - 10{,}240x^3y^3 + 15{,}360x^2y^4 - 12{,}288xy^5 + 4096y^6$ **35.** $\frac{50{,}778}{625} \approx 81.2448$ **37.** 136.7° **39.** 90°

41.

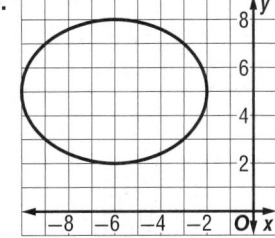

43. 8287 ft **45.** G **47.** G

Pages 671–673   Lesson 11-2

1. Discrete; the number of text messages is countable and therefore discrete.
3. Continuous; the weight can be any number.
5. Discrete; the number of votes is countable.

7.

Players, X	0	1	2	3	4
P(X)	0.21	0.41	0.21	0.12	0.05

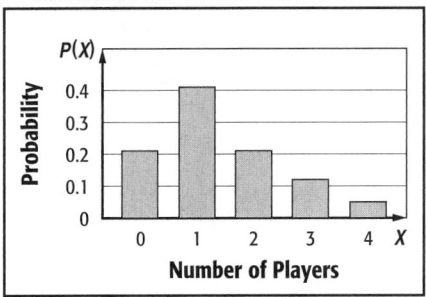

1.39; Sample answer: On average, the students have had 1 or 2 MP3 players; 1.20; 1.09

9.

Days, X	P(X)
0	0.03
1	0.02
2	0.11
3	0.17
4	0.04
5	0.12
6	0.11
7	0.41

5; Sample answer: The students ate breakfast an average of 5 days a week; 4.5, 2.1.

11 Use the formula $E(X) = \Sigma\,[X \cdot P(X)]$ with $X = \$25{,}000$ and $P(X) = 0.0002$. The expected value of the loss for the insurance company is $\$25{,}000 \cdot 0.0002 = \$5$, so the insurance company should expect to gain $\$300 - \$5 = \$295$.

13. Binomial experiment; $n = 25$, $P(S) =$ the student is left-handed, $P(F) =$ the student is right-handed; sample answer: The student will either be left- or right-handed. **15.** Binomial experiment; $n = 10$, $P(S) = \frac{1}{6}$ each roll, $P(F) = \frac{5}{6}$; sample answer: You will either roll a 5 or another number. **17.** Not a binomial experiment; sample answer: This is not a binomial experiment because there are more than two possible outcomes. A person's age can be any reasonable number. **19.** Not a binomial experiment; sample answer: Since you are drawing cards without replacement, the probability changes each time because there are fewer cards.

21.

X	P(X)	X	P(X)	X	P(X)
0	0.00	10	0.03	20	0.00
1	0.01	11	0.01	21	0.00
2	0.04	12	0.01	22	0.00
3	0.08	13	0.00	23	0.00
4	0.14	14	0.00	24	0.00
5	0.18	15	0.00	25	0.00
6	0.18	16	0.00	26	0.00
7	0.15	17	0.00	27	0.00
8	0.11	18	0.00	28	0.00
9	0.06	19	0.00		

Number of Sunny Days in February

5.88; Sample answer: Out of 28 days in February, there will be 5.88 sunny days; 4.65; 2.16.

23.

X	0	1	2	3	4	5	6	7	8
P(X)	0.00	0.003	0.02	0.08	0.19	0.28	0.26	0.14	0.03

High School Students Wearing Seatbelts

5.22; Sample answer: Out of 8 students, 5.22 will wear their seatbelts; 1.82; 1.35.

25a. −$0.63   **25b.** −$0.31   **25c.** −$1.50   **25d.** −$1.25
25e. −$1.88
27a.

$n = 6, p = 0.5$	
X	P(X)
0	0.016
1	0.094
2	0.234
3	0.312
4	0.234
5	0.094
6	0.016

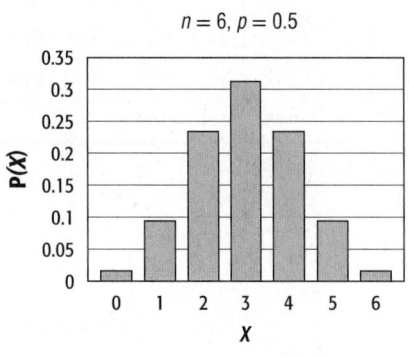

$n = 6, p = 0.5$

$n = 6, p = 0.3$	
X	P(X)
0	0.118
1	0.303
2	0.324
3	0.185
4	0.059
5	0.010
6	0.001

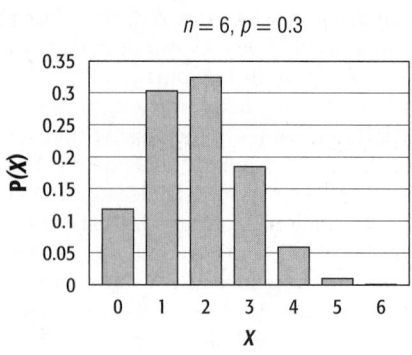

$n = 6, p = 0.3$

$n = 6, p = 0.7$	
X	P(X)
0	0.001
1	0.010
2	0.059
3	0.185
4	0.324
5	0.303
6	0.118

$n = 6, p = 0.7$

$n = 8, p = 0.5$	
X	P(X)
0	0.004
1	0.031
2	0.109
3	0.219
4	0.273
5	0.219
6	0.109
7	0.031
8	0.004

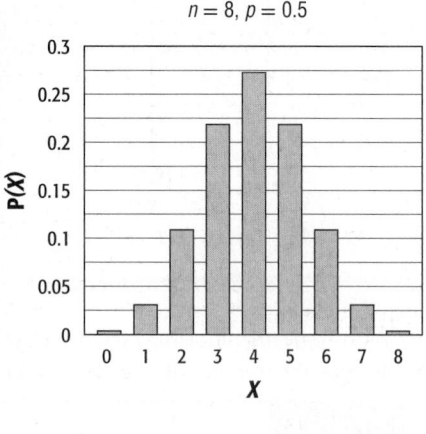

$n = 8, p = 0.5$

$n = 10, p = 0.5$	
X	P(X)
0	0.001
1	0.010
2	0.044
3	0.117
4	0.205
5	0.246
6	0.205
7	0.117
8	0.044
9	0.010
10	0.001

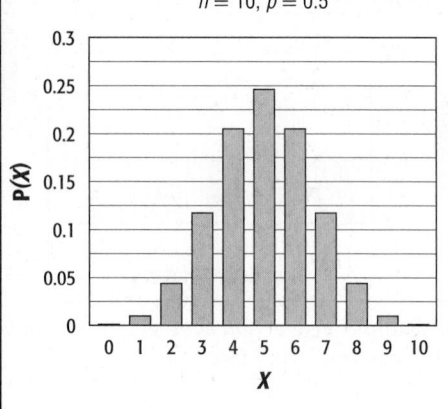

$n = 10, p = 0.5$

27b. Sample answer: The distribution corresponding to $n = 6$ and $p = 0.5$ is symmetrical. The distribution corresponding to $n = 6$ and $p = 0.3$ is positively skewed and the distribution corresponding to $n = 6$ and $p = 0.7$ is negatively skewed. The distributions corresponding to $n = 8$, $p = 0.5$ and $n = 10$, $p = 0.5$ are both symmetrical.   **27c.** Sample answer: When $p = 0.5$ the shape of the binomial distribution will be symmetrical, when $p < 0.5$ the distribution will be positively skewed, and when $p > 0.5$ the distribution will be negatively skewed.   **27d.** Sample answer: As n increases, the probability distribution widens and the standard deviation increases.   **29.** No; sample answer: Tossing a coin is a binomial experiment, in which each trial is independent. Therefore, the probability of the coin landing on tails is the same in every trial and the outcome of each of the previous trials is independent of the current trial.   **31.** True; sample answer: Rolling dice is an example of a theoretical probability experiment.

33 Use $\mu = np$ and $\sigma^2 = npq$ with $n = 50$, $q = 1 - p$ and $\sigma = 1.54$. Solve for p. Then use p and n to find μ.
$$1.54^2 = 50p(1 - p)$$
$$\frac{1.54^2}{50} = p(1 - p)$$
$$0.047432 = p - p^2$$
$$p^2 - p + 0.047432 = 0$$

Use the quadratic formula to solve for p.
$$p = \frac{-(-1) \pm \sqrt{(-1)^2 - 4 \cdot 1 \cdot 0.047432}}{2 \cdot 1}$$
So $p = 0.95$ or $p = 0.05$. From the hint, $p = 0.05$, so $\mu = np = 50 \cdot 0.05 = 2.5$.

35a.

[300, 3100] scl: 400 by [0, 15] scl: 1

The graph is positively skewed, which suggests that most of the prices are on the low-end, at about $700 or less.

35b. The graph is positively skewed, so we can use the five-number summary. The prices range from $300 to $2700 and half of the prices are between $600 and $1275.   **37.** 0.670
39. ± 20, 50, ± 125 **41.** $\frac{3}{10}, \frac{3}{25}, \frac{6}{125}, \frac{12}{625}$ **43.** 16, -7, -20, -33
45. 43; not orthogonal   **47.** -40; not orthogonal

49.
(graph showing hyperbola with points (6, 3.31) and (6, −13.31))

51. $AB = [6]$, $BA = \begin{bmatrix} 10 & -5 \\ 8 & -4 \end{bmatrix}$

53. $AB = \begin{bmatrix} 4 & 37 & -12 \\ 16 & 53 & -18 \\ -22 & 53 & -15 \end{bmatrix}$,
$BA = \begin{bmatrix} 47 & 31 \\ 10 & -5 \end{bmatrix}$

55. J   **57.** G

Pages 681–683   Lesson 11-3

1a. 1.08 hours   **1b.** 68%   **3.** -1.15 **5.** 2.43   **7.** -2.14
9a. 184   **9b.** 92   **11.** $-0.39 < z < 0.39$ **13.** $z < -0.84$ or $z > 0.84$ **15.** $z < -1.15$ or $z > 1.15$ **17a.** 9%   **17b.** 0.4%
17c. 25%

---

19 a. To find the top 15% of snowfall amounts, find the amount X that separates the upper 15% of the area under the normal curve. The top 15% correlates with $1 - 0.15 = 0.85$. Using a graphing calculator, the corresponding z-value is 1.036. Use the formula for the z-value for a population to find the corresponding snowfall.
$$z = \frac{X - \mu}{\sigma}$$
$$1.036 = \frac{X - 260}{27}$$
$$28.0 = X - 260$$
$$288.0 = X$$

b. To find the bottom 30% of snowfall amounts, find the amount X that separates the bottom 30% of the area under the normal curve. The bottom 30% correlates with 0.30. Using a graphing calculator, the corresponding z-value is -0.5244. Use the formula for the z-value for a population to find the corresponding snowfall.
$$z = \frac{X - \mu}{\sigma}$$
$$-0.5244 = \frac{X - 260}{27}$$
$$-14.2 = X - 260$$
$$245.8 = X$$

c. The middle 60% represents 30% on each side of the mean and therefore corresponds to the interval from 0.2 to 0.8. Using a graphing calculator, the z-values that correspond to 0.2 and 0.8 are -0.8416 and 0.8416, respectively.
$$z = \frac{X - \mu}{\sigma} \qquad\qquad z = \frac{X - \mu}{\sigma}$$
$$-0.8416 = \frac{X - 260}{27} \qquad 0.8416 = \frac{X - 260}{27}$$
$$-22.7 = X - 260 \qquad\quad 22.7 = X - 260$$
$$237.3 = X \qquad\qquad\quad 282.7 = X$$
The range of snowfall is 237.3 cm $-$ 282.7 cm.
21. Sample answer: The ACT score has a z-value of 1.28 and the SAT score has a z-value of 1. Therefore, the ACT score of 27 has a higher relative position than the SAT score of 620.
23. 0.168   **25.** 0.493   **27a.** -0.84, -0.25, 0.84   **27b.** -0.67, 0, 0.67   **27c.** -1.28, -0.25, 1.28   **29a.** about $45,252   **29b.** 22 employees   **31.** Lucy; the outside 35% corresponds to 0.175 and 0.825, which is associated with z-values $z < -0.93$ or $z > 0.93$. Chad found the z-values associated with the outside 70%, which corresponds to 0.35 and 0.65.

33 a. If the combined area of the two tails is 1%, then 0.5% lies in each, so the z-values correlate to 0.005 and 0.995. Using a graphing calculator, these are -2.58 and 2.58, respectively.

b. If the combined area of the two tails is 5%, then 2.5% lies in each, so the z-values correlate to 0.025 and 0.975. Using a graphing calculator, these are -1.96 and 1.96, respectively.

c. If the combined area of the two tails is 10%, then 5% lies in each, so the z-values correlate to 0.05 and 0.95. Using a graphing calculator, these are -1.64 and 1.64, respectively.

35. Sample answer: The total area under both the normal and standard normal distributions is equal to 1. Both curves are also continuous and symmetric with respect to the mean. However, the standard distribution has a mean of 1 and a standard deviation of 0, while the mean and standard deviation of the normal distribution could be any value.

37.

[0, 20] scl: 1 by [−10, 10] scl: 1

Sample answer: The median number of penalties for Season 1 is much larger than the median for Season 2. The maximum number of penalties in Season 2 is less than the median for Season 1, which means that more than 50% of the data values for Season 1 are higher than for Season 2. Therefore, the number of penalties received by the team during Season 2 is generally lower than those in Season 1.

39. 3828   **41.** $\left(0, \frac{1}{4}\right)$ **43.** (2, 0)   **45.** Sample answer: ⟨0, 3, −6⟩   **47.** 26.6°   **49.** 284.0°   **51.** $(x + 2)^2 + \dfrac{(y + 6)^2}{25} = 1$ **53.** C   **55.** A

Pages 692–694   Lesson 11-4

1a. 2.4%   **1b.** 79.7%   **3a.** 86.3%   **3b.** 64.2%   **5a.** 26.6%   **5b.** 2.4%

7 Find μ and σ.

$$\mu = np \qquad\qquad \sigma = \sqrt{npq}$$
$$= 45 \cdot 0.33 \qquad = \sqrt{45(0.33)(0.67)}$$
$$= 14.85 \qquad\qquad \approx 3.15$$

Since $np = 45 \cdot 0.33 = 14.85$ and $nq = 45 \cdot 0.67 = 30.15$ are both greater than 5, the normal distribution can be used to approximate the binomial distribution. Rewrite the problem with the continuity factor included. Since the question is asking for the probability that fewer than 14 people plan to watch the World Cup, subtract 0.5 from 14: $P(X < 14) = P(X < 14 − 0.5) = P(X < 13.5)$. Find the corresponding z-value for 13.5.

$$z = \frac{X - \mu}{\sigma} = \frac{13.5 - 14.85}{3.15} \approx -0.43$$

Use a graphing calculator to find the area to the left of −0.43. The approximate area to the left of $z = -0.43$ is 0.334. Therefore, the probability that fewer than 14 people plan to watch the World Cup is 33.4%.

9a. 3.6%   **9b.** 40.9%   **11a.** 94.9%   **11b.** 12.8%   **11c.** 37.2%   **11d.** 48.6%   **13.** 12.5   **15.** 25

17a.

Sample	$\overline{X}$	s
1	25.03	0.153
2	24.97	0.153
3	25	0.265
4	25.13	0.208
5	24.97	0.252
6	24.93	0.153
7	25.1	0.200
8	25.17	0.321
9	24.97	0.208
10	25	0.300

17b.

Control Chart of Sample Means

17c. Sample answer: The upper control limit is 25.22 and the lower control limit is 24.78. Since all of the sample means fall within these limits, the process is in control.   **19.** Weston; sample answer: Hannah found an incorrect continuity correction of $P(X > 30.5)$.

21 Sample answer: Find μ and σ for each population.

men:
$$\mu = np \qquad\qquad \sigma = \sqrt{npq}$$
$$= 100 \cdot 0.07 \qquad = \sqrt{100(0.07)(0.93)}$$
$$= 7 \qquad\qquad \approx 2.55$$

women:
$$\mu = np \qquad\qquad \sigma = \sqrt{npq}$$
$$= 1500 \cdot 0.004 \qquad = \sqrt{1500(0.004)(0.996)}$$
$$= 6 \qquad\qquad \approx 2.44$$

Since np and nq are greater than 5 for each population, the normal distribution can be used to approximate the binomial distribution. Rewrite the problem with the continuity factor included. Since the question asks for the probability that either population will have at least 10 people who are colorblind, instead find the probability that each population does not have at least 10: $P(X < 10) = P(X < 10 − 0.5) = P(X < 9.5)$. Find the corresponding z-value for 9.5 for each population.

men:
$$z = \frac{X - \mu}{\sigma}$$
$$= \frac{9.5 - 7}{2.55}$$
$$\approx 0.98$$

women:
$$z = \frac{X - \mu}{\sigma}$$
$$= \frac{9.5 - 6}{2.44}$$
$$\approx 1.43$$

Use a graphing calculator to find the areas to the left of 0.98 and 1.43. The approximate areas to the left of these values are 0.84 and 0.92, respectively. To find the probabilities that each population has at least 10 people who are colorblind, subtract each of these from 1 to obtain $1 − 0.84 = 0.16$ and $1 − 0.92 = 0.08$, respectively. Thus, there is a 16% probability that at least 10 men are colorblind and an 8% probability that at least 10 women are, so it more likely to have 10 colorblind men.

23. False; sample answer: A distribution will approach the normal distribution as the sample size increases, not the total number of samples.   **25.** Sample answer: According to the Central Limit Theorem, any distribution will approach the normal distribution as the sample size increases. Therefore, as long as $np \geq 5$, $nq \geq 5$, and the original variable is normally distributed or $n \geq 30$, the normal distribution can be used to approximate a binomial distribution. A correction for continuity is needed because the binomial distribution is discrete, whereas the normal distribution is continuous.

27.

Days, X	0	1	2	3	4	5	6	7
P(X)	0.05	0.09	0.18	0.25	0.20	0.16	0.05	0.02

3.24; 2.54; 1.59
29. -285 **31.** 4 **33.** 38 **35.** $\left(\frac{\sqrt{2}}{8}, \frac{\sqrt{2}}{8}\right)$ **37.** $\langle 12, 4\rangle$
39. $\langle 34, -21\rangle$
41. $(x-2)^2 = -8(y+3)$ **43.** D **45.** A

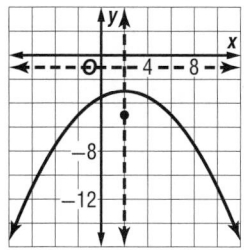

Pages 702–704   Lesson 11-5

1. 3.2 min; $33.3 < \mu < 39.7$ **3.** 1.4°F; $96.1 < \mu < 98.9$
5. $62.4 < \mu < 73.6$ **7.** t-distribution; $125.3 < \mu < 130.7$
9. t-distribution; $38.7 < \mu < 40.1$ **11.** normal distribution;
$27 < \mu < 29.6$

13 Use $z = 1.645$, $\sigma = 0.1$, and $E = 0.015$.
$$n = \left(\frac{z\sigma}{E}\right)^2$$
$$= \left(\frac{1.645 \cdot 0.1}{0.015}\right)^2$$
$$= 120.27$$
He should use 121 samples to be accurate within 90%.
15a. t-distribution; sample answer: The sample size is less than 30 and the standard deviation is unknown.
15b. mean = 8.1; sample standard deviation = 3.0
15c. $6.9 < \mu < 9.3$ **15d.** Sample answer: Based on the survey, we are 95% confident that students average between 6.9 and 9.3 hours of after-school activities per week. **17.** within 2.5 years **19.** Yes; sample answer: Calculating a 99% confidence interval using t-distribution and letting $n = 24$, $s = 4$, and $\bar{x} = 22$ results in $19.7 < \mu < 24.3$. The desired preparation time of 20 minutes falls within this interval. **21.** Yes; sample answer: Calculating a 99% confidence interval using t-distribution and letting $n = 14$, $s = 1.41$, and $\bar{x} = 61.91$ results in $60.77 < \mu < 63.05$. The desired talk time of 62 hours falls within this interval.

23 Work backward to determine the degrees of freedom. The error of estimate E is equal to one half of the range of the confidence interval CI.
$$E = \frac{49.128 - 40.872}{2}$$
$$E = 4.128$$
$$E = t \cdot \frac{s}{\sqrt{n}} = 4.128$$
The sample standard deviation s and the t-value are known. Solve for n.
$$t \cdot \frac{s}{\sqrt{n}} = 4.128$$
$$2.064 \cdot \frac{10}{\sqrt{n}} = 4.128$$
$$\frac{10}{\sqrt{n}} = 2$$
$$10 = 2\sqrt{n}$$
$$5 = \sqrt{n}$$
$$25 = n$$
The degrees of freedom are $n - 1$ or 24.
25. False; increasing the sample size will give a better depiction of the actual parameter and will narrow the confidence interval, making it more accurate. **27.** True;

an increase in the standard deviation will increase the maximum error of estimate and will expand the confidence interval. **29.** The person will have to multiply the sample size by x^2. **31.** Sample answer: By using a parameter, a characteristic of an entire population can be represented with a single number or range of numbers instead of a large number of individual data points. **33.** 81.86% **35.** 0.841
37. $e = 0.2$; ellipse; $x = 8$ **39.** $e = \frac{1}{3}$; ellipse; $y = 2$
41. $\sqrt{53} \approx 7.28$

43. $8w + 5x + 10y + 3z = 575$
$10w + 8x + 2y + 6z = 574$
$16w + 4x + 8y + 3z = 661$
$13w + 11x + 6y = 722$
(20, 30, 22, 15); Cloud Coaster: costs \$20/person; Danger Coaster, 1 person: costs \$30/person; Danger Coaster, 2-person: costs \$22/person; Danger Coaster, 3 person: costs \$15/person **45.** J **47.** J

Pages 710–712   Lesson 11-6

1. $H_0: \mu = 4$ (claim), $H_a: \mu \neq 4$ **3.** $H_0: \mu \geq 10$, $H_a: \mu < 10$ (claim) **5.** $H_0: \mu \leq 38$, $H_a: \mu > 38$ (claim) **7.** $z \approx -2.5$; The null hypothesis is rejected. Therefore, the evidence does not reject the claim of $\mu < 0.5$. **9.** $z \approx -0.96$; The null hypothesis is not rejected. Therefore, the evidence does not reject the claim of $\mu \geq 90$. **11.** $z \approx 3.85$; The null hypothesis is rejected; the claim is not rejected. **13.** $z \approx -4.19$; The null hypothesis is rejected; the claim is not rejected.

15 The claim can be written as $\mu < 260,000$. This is the alternative hypothesis. The null hypothesis is $\mu \geq 260,000$.
$H_0: \mu \geq 260,000$ $H_a: \mu < 260,000$
Since $n \geq 30$, the z statistic is used.
$$\sigma_{\bar{x}} = \frac{12,500}{\sqrt{40}} \approx 1976.42$$
$$z = \frac{\bar{x} - \mu}{\sigma_{\bar{x}}}$$
$$\approx \frac{254,500 - 260,000}{1976.42}$$
$$\approx -2.783$$
This is a left-tailed test since $\mu < 260,000$. The area associated with $z = -2.783$ is about 0.003. The p-value 0.003 is less than 0.05. Therefore, the null hypothesis is rejected and there is significant evidence that the average home sale price is less than \$260,000.
17. ≈ 0.194; The null hypothesis of $\mu \geq 7.50$ is not rejected. This does not support the claim of $\mu < 7.50$.
19a. $H_0: \mu = 25$ (claim); $H_a: \mu \neq 25$ **19b.** $z \approx 0.6766$; The null hypothesis is not rejected. **19c.** Sample answer: The claim of $\mu = 25$ is supported by the evidence. **21.** Type I: The X-ray determines no ankle sprain when there is actually an ankle sprain. Type II: The X-ray determines an ankle sprain where there is actually no ankle sprain. **23.** Type I: It is determined that the majority of students do have jobs when they actually do not. Type II: It is determined that the majority of students do not have jobs when they actually do. **25.** Type I: The venom is determined poisonous when it actually is not. Type II: The venom is determined not poisonous when it actually is. **27a.** $H_0: \mu \leq 21$, $H_a: \mu > 21$ (claim) **27b.** ≈ 0.0004
27c. The null hypothesis is rejected. **27d.** There is enough evidence to support the claim of $H_a: \mu > 21$. **29a.** $H_0: \mu = 85$ (claim), $H_a: \mu \neq 85$ **29b.** Victor's class: $t \approx -0.58$; the null hypothesis is not rejected. The claim of $\mu = 85$ is supported

by the evidence. Malina's class: $t \approx 3.06$; The null hypothesis is rejected. The claim of $\mu = 85$ is not supported by the evidence. **29c.** $z \approx 1.27$; The null hypothesis is not rejected. The claim of $\mu = 85$ is supported by the evidence. **29d.** The Law of Large Numbers states that the larger the sample, the closer $\bar{x}$ will get to μ. Combining the two samples increases n and produces a result that is more representative of the population mean. It also happens to correspond with the claim. **31.** Sample answer: To use test statistics and critical values, the significance level is first converted into a critical value, which is then used to locate a critical region. The test statistic is then compared to the critical region. To use p-values, the test statistic is converted to a p-value. This number is then compared to the level of significance. **33.** False; sample answer: The alternative hypothesis will only include inequality symbols, even if it represents the claim.

35 To determine z, calculate $\text{InvNorm}(P) = \text{InvNorm}(0.0104) = -2.3166$. Since $s = 0.3$, $\sigma_{\bar{x}} = \dfrac{0.3}{\sqrt{n}}$. Use the z-statistic and solve for n.

$$z = \frac{\bar{x} - \mu}{\sigma_{\bar{x}}}$$

$$-2.3166 = \frac{14.9 - 15}{\frac{0.3}{\sqrt{n}}}$$

$$-2.3166 = \frac{-0.1}{\frac{0.3}{\sqrt{n}}}$$

$$-2.3166 = \frac{-0.1\sqrt{n}}{0.3}$$

$$-0.69498 = -0.1\sqrt{n}$$

$$6.9498 = \sqrt{n}$$

$$48 \approx n$$

Therefore, $n = 48$.

37. $43.14 < \mu < 47.36$ **39.** divergent **41.** convergent

43. $91.5°$ **45.** $\left\langle -\dfrac{9\sqrt{85}}{85}, \dfrac{2\sqrt{85}}{85} \right\rangle$ **47.** $\left\langle \dfrac{4}{5}, \dfrac{3}{5} \right\rangle$

49. $\dfrac{(y-1)^2}{25} - \dfrac{(x+2)^2}{9} = 1$ **51.** G **53.** F

Pages 720–722 Lesson 11-7

1 **a.** Enter the data into L₁ and L₂ on your calculator. Then turn on **Plot1** under the **STAT PLOT** menu. Graph the scatter plot by pressing **ZoomStat** or by pressing **Graph** and adjusting the window manually. Press **STAT** and select **LinReg(ax+b)** under the **CALC** menu. The correlation coefficient is about 0.8320. Because r is close to 1, this suggests that the data has a positive correlation.

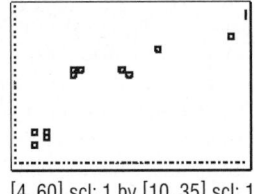

[4, 60] scl: 1 by [10, 35] scl: 1

b. Calculate the test statistic.

$$t = r\sqrt{\frac{(n-2)}{(1-r)^2}}$$

$$= 0.8320\sqrt{\frac{11-2}{1-0.8320^2}}$$

$$\approx 4.499$$

To determine if the correlation coefficient is significant at the 1% level, the critical values are determined by $\dfrac{\alpha}{2}$, or 0.005 since this is a two-tailed test. Use the **invT** function on graphing calculator to determine the critical values with $11 - 2 = 9$ degrees of freedom are $t = \pm 3.250$. Since $4.499 > 3.25$, the statistic falls within the critical region and the null hypothesis of $\rho \neq 0$ is rejected. At the 1% level, there is enough evidence to conclude that there is a significant correlation between the fat and protein content of various foods. Thus, there is sufficient evidence at the 5% and 10% level as well.

c. On the same screen from which the correlation coefficient was obtained, the least-squares regression line is approximately $\hat{y} = 0.34155x + 13.6875$; the slope $a = 0.34155$ indicates that each gram of fat provides roughly 0.34155 grams of protein. The y-intercept $b = 13.6875$ indicates that for a serving of food with zero grams of fat, there will be about 13.6875 grams of protein, which will likely not always be the case.

d. Turn on **Plot2** under the **STAT PLOT** menu and choose L₁ for the Xlist and **RESID** for the Ylist. Graph the residual plot by pressing **ZoomStat**. The residuals appear to be randomly scattered and centered about the regression line at $y = 0$. This supports the claim that the use of a linear model is appropriate.

[4.2, 61.8] scl: 1 by [−5.3, 6.6] scl: 1

e. The point (57, 30) appears to be an outlier in the x-direction. Remove the point from the data and recalculate the regression line to obtain $\hat{y} = 0.461x + 11.633$. Since there is a significant change in the regression line, the outlier is influential.

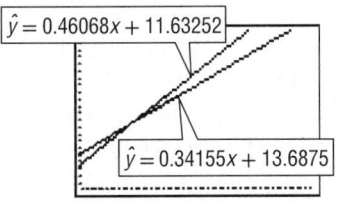

$\hat{y} = 0.46068x + 11.63252$

$\hat{y} = 0.34155x + 13.6875$

[4.2, 61.8] scl: 1 by [−5.3, 6.6] scl: 1

f. The new correlation coefficient is approximately $r = 0.8344$. Calculate t with $n = 10$.

$$t = r\sqrt{\frac{n-2}{1-r^2}}$$

$$= 0.8344\sqrt{\frac{10-2}{1-0.8344^2}}$$

$$\approx 4.282$$

To determine if the correlation coefficient is significant at the 10% level, the critical values are determined by $\dfrac{\alpha}{2}$, or 0.05 since this is a two-tailed test. Use the **invT** function on a graphing calculator to determine the critical values with $10 - 2 = 8$ degrees of freedom are $t = \pm 1.86$. Since $4.282 > 1.860$, the correlation is significant at the 10% level.

g. Use the regression line to predict the protein content at $x = 1$, $x = 5$, and $x = 13$.
$\hat{y} = 0.461 \cdot 1 + 11.633 = 12.09$
$\hat{y} = 0.461 \cdot 5 + 11.633 = 13.94$
$\hat{y} = 0.461 \cdot 13 + 11.633 = 17.63$
This means that for items with 1, 5, and 13 grams of fat, there would be 12.09, 13.94, and 17.63 grams of protein, respectively. The predictions for 5 and 13 grams of fat are reasonable, since both values are within or fall only slightly beyond the range of the original data on which the revised regression equation is based. The prediction for 1 gram of fat may not be as meaningful since it falls well beyond the range of the original data.

3a. The data appear to have a weak negative linear correlation; $r \approx -0.769$; the correlation coefficient indicates that the data have a negative linear correlation.

[0, 15] scl: 1 by [0, 60] scl: 10

3b. Since $t \approx -4.167$ and $-4.167 < -3.055$, $-4.167 < -2.179$, and $-4.167 < -1.782$, the statistic falls within the critical region and the null hypothesis is rejected. Therefore, the correlation is significant at the 1%, 5%, and 10% levels.

3c. $\hat{y} = -2.259x + 47.231$; The slope $a = -2.259$ indicates that the health care ranking drops by 2 for each increase in education ranking. The y-intercept $b = 47.23$ indicates that for an education ranking of 0, there will be a health care ranking of 47.23, although an education ranking of 0 is not possible.

3d.

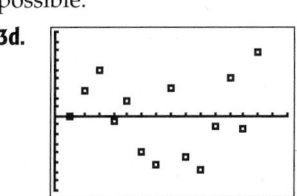

[0, 16] scl: 1 by [−16, 18] scl: 2

The residuals appear to be randomly scattered and centered about the regression line at $y = 0$. This supports our claim that the use of a linear model is appropriate.

3e. no real outliers   **3f.** no real outliers   **3g.** A state with an education ranking of 15 would have a health care ranking of 13, which is reasonable. A state with an education ranking of 28 would have a health care ranking of −16, while a state with an education ranking of 42 would have a health care ranking of −47, which are both unreasonable. Rankings cannot be negative.   **5a.** The data appear to have a negative linear correlation; $r \approx -0.079$; the correlation coefficient indicates that the data have an extremely weak negative linear correlation.

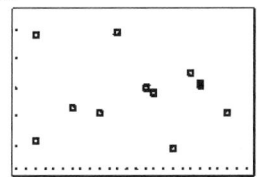

[61.9, 87.1] scl: 1 by [2.22, 7.58] scl: 1

5b. Since $t \approx -0.251$ and $-0.251 > -3.169$, $-0.251 > -2.228$, and $-0.251 > -1.812$, the correlation is not significant at the 1%, 5%, or 10% levels.   **7.** b   **9.** d

11 **a.** Input the original data in L1 and L2 and create a scatter plot of the data by turning on **Plot1** under the **STAT PLOT** menu. Graph the scatter plot by pressing **ZoomStat** or by pressing **Graph** and adjusting the window manually.

[0, 1600] scl: 200 by [0, 14] scl: 2

To linearize the data, take the natural logarithm of each meal amount.

x	500	1125	300	750	1250	950
ln y	1.39	2.30	1.10	1.79	2.48	2.08

b. Create a scatter plot of the linearized data by replacing the data in L2 with the ln y values from the linearized data. Press **STAT** and select **LinReg(ax+b)** under the **CALC** menu. The correlation coefficient is about 0.999, which shows there is an extremely strong correlation between income and the natural logarithm of number of meals a person dines out each month.

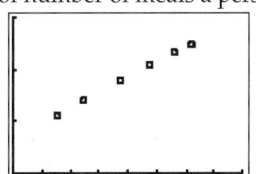

[0, 1600] scl: 200 by [0, 3] scl: 1

c. Sample answer: Calculate the test statistic.
$$t = r\sqrt{\frac{n-2}{1-r^2}}$$
$$= 0.999\sqrt{\frac{6-2}{1-0.999^2}}$$
$$\approx 44.69.$$
To determine if the correlation coefficient is significant at the 5% level, the critical values are determined by $\frac{\alpha}{2}$, or 0.025 since this is a two-tailed test. Use the **invT** function on graphing calculator to determine the critical values with $6 - 2 = 4$ degrees of freedom are $t = \pm 2.78$. Since $44.69 > 2.78$, the statistic falls within the critical region and the null hypothesis of $\rho = 0$ is rejected. At the 5% level, there is enough evidence to conclude that there is a significant correlation.

d. Press **STAT** and select **LinReg(ax+b)** under the **CALC** menu to obtain $\ln \hat{y} = 0.00146x + 0.6724$. Solve for $\hat{y}$.
$e^{\ln \hat{y}} = e^{0.00146x + 0.6724}$
$\hat{y} = e^{0.6724} e^{0.00146x}$
$\hat{y} = 1.96 \, e^{0.00146x}$

e. Substitute $x = 2000$.
$\hat{y} = 1.96 \, e^{0.00146 \cdot 2000} \approx 36$. This is not a reasonable answer since $2000 is more than other monthly incomes and there are more meals out than days of the month.

13a.

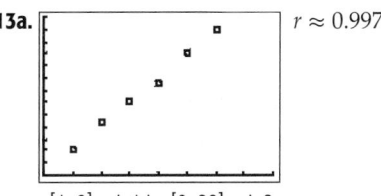

$r \approx 0.997$

[1, 8] scl: 1 by [0, 26] scl: 2

13b. $\bar{y} = 14$
13c. $\hat{y} = 3.886x + 0.4$

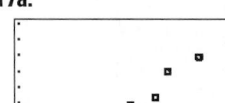

x	1	2	3	4	5	6
$\hat{y}$	4.286	8.172	12.058	15.944	19.83	23.716

13d. total variation = 266, explained variation ≈ 264.27, unexplained variation ≈ 1.77 **13e.** $r^2 \approx 0.993$ **13f.** Sample answer: A value of 0.993 means that 99.3% of the total variation of y can be explained by the relationship between x and y. **15.** False; sample answer: When the null hypothesis is rejected, it means that there is a significant difference between the value of ρ and 0.

17a.

[20, 70] scl: 5 by [100, 200] scl: 10   [20, 70] scl: 5 by [100, 200] scl: 10

$\hat{y} = 1.700x + 76.243$

17b.

[100, 200] scl: 10 by [20, 70] scl: 5   [100, 200] scl: 10 by [20, 70] scl: 5

$\hat{y} = 0.575x - 42.850$

17c. Sample answer: The slope remains positive, but the slope of one regression equation is nearly the reciprocal of the slope of the other regression equation. The correlation coefficient was unchanged. **19.** No; $z = 4.562$, so the null hypothesis is not rejected. Therefore, Alexi's claim that $\mu \geq 55$ is not rejected. **21.** 14 **23.** 64 **25.** $a_n = 15 + (-24)(n - 1)$; $a_1 = 15$, $a_n = a_{n-1} - 24$ **27.** $10(\cos 5.36 + i \sin 5.36)$ **29.** $\sqrt{13}(\cos 0.59 + i \sin 0.59)$ **31.** not parallel **33.** E

35a. Sample answer: With continued practice, an individual's bowling average increases slowly and theoretically can reach a maximum of 300. This represents a logistic growth function because with continued practice, an individual's average will slowly increase and a person cannot average higher than 300. **35b.** Sample answer: Jimmy says that, with practice, he can average 180, but George says he will average less than that. **35c.** Sample answer: H_0: $\mu = 180$; H_a: $\mu \neq 180$
35d. Sample answer: Jimmy practices for a few months. Then he and George join a bowling league and keep track of Jimmy's average. The league will simulate competitive bowling and accurately measure Jimmy's average. When the league is over, they will review Jimmy's average over time.
35e. Sample answer: George tracked Jimmy's average over each week of the bowling league and used the data to develop a logistic regression. The logistic regression equation was $y = \dfrac{155.32}{1 + 0.867e^{-0.2356x}}$. The logarithmic regression equation was $y = 144.95 + 3.93 \ln x$. **35f.** Sample answer: The logarithmic regression is a more accurate representation of the data because Jimmy may eventually have a better average than 155. While the equation does appear to increase infinitely, the rate of change is so minimal that Jimmy would not hit a 300 average until he bowled over a million games.
35g. Sample answer: The null hypothesis is rejected and George's hypothesis that Jimmy cannot average 180 is accepted.

1. negatively skewed **3.** standard normal **5.** Central Limit Theorem **7.** null

9a.

The distribution is slightly negatively skewed.

[370, 550] scl: 25 by [0, 8] scl: 1

9b. The distribution of data is skewed; therefore, the five-number summary of 373, 458.5, 488, 508, and 533 can be used to describe the distribution. The scores ranged from 373 to 533, the median landing score was 488, and half of the scores were between 458.5 and 508.

11.

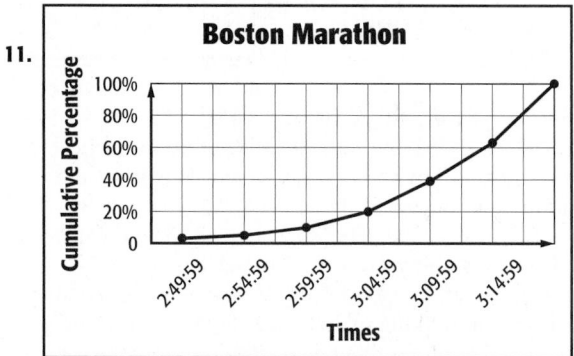

11% of the women ran the race in less than 3 hours.
13. Continuous; the amount of blood can be any value.
15. $\sigma^2 \approx 0.77$, $\sigma \approx 0.88$ **17.** 147.12 **19.** 31.192
21. $-0.31 < z < 0.31$ **23.** $z < -0.69$ and $z > 0.69$ **25.** 2.6%
27. 13 **29.** 2.7 **31.** normal; $33.23 < \mu < 34.77$ **33.** normal; $5.54 < \mu < 6.06$ **35.** 28 **37.** 27 **39.** H_0: $\mu \leq 60$, H_a: $\mu > 60$ (claim) **41.** H_0: $\mu \geq 6$ (claim), H_a: $\mu < 6$ **43.** $z \approx 4.28$; The null hypothesis is rejected; the claim is rejected. **45.** $z \approx -1.0$; The null hypothesis is not rejected; the claim is not rejected.

47a.

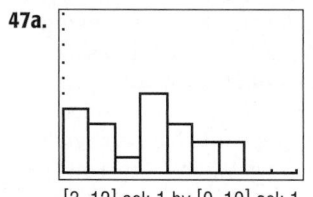

The graph appears clustered. The body-fat levels seem to fluctuate between 3% and 10%.

[3, 12] scl: 1 by [0, 10] scl: 1

47b. The mean body-fat level is 6.2% with a standard deviation of 1.9%. The distribution is not skewed, so the five-number summary is not needed.

49. 1.97; 1.51; 1.23 **51a.** 99.95% **51b.** 81.65% **51c.** 99.95%
53. $177.8 < \mu < 192.2$

55a.

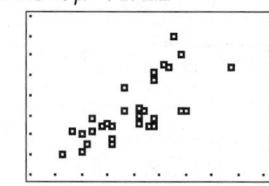

The data appear to have a positive correlation. $r \approx 0.7424$, which shows a positive linear relationship between the aptitude and writing tests.

[125, 170] scl: 5 by [20, 60] scl: 5

55b. $t = 5.864$, critical values are ± 2.048. Since $5.864 > 2.048$, the statistic falls within the critical region and the null hypothesis is rejected. Thus there is a correlation between the aptitude and writing tests.

Pages 732–733   Chapter 11   Connect to AP Calculus

1. Sample answer: The $\hat{p}$ found in Step 1 may be an accurate estimate. However, depending on the size of the population, $\hat{p}$ could be too low or too high. Increasing the sample size would help to obtain a better estimate.   **3.** Sample answer: No; establishing an interval for the population proportion gives a better estimate.   **5.** 3301   **7.** $50.4\% < p < 65.6\%$

Chapter 12 Limits and Derivatives

Page 735   Chapter 12   Get Ready

1. From the graph, it appears that $f(x) \to 0$ as $x \to -\infty$ and $f(x) \to 0$ as $x \to \infty$.   **3.** From the graph, it appears that $f(x) \to 1$ as $x \to -\infty$ and $f(x) \to 1$ as $x \to \infty$.   **5.** 1200   **7.** -17 **9.** \$290   **11.** $D = \{x \mid x \neq 10, x \in \mathbb{R}\}; x = 10$ **13.** $D = \{x \mid x \neq 2, x \neq -4, x \in \mathbb{R}\}; x = 2, y = 1$ **15.** $-12, -17, -22, -27$ **17.** $-324, 972, -2916, 8748$ **19.** $0, 7, 14, 21$

Pages 743–745   Lesson 12-1

1. 10

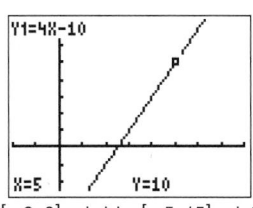
[−2, 8] scl: 1 by [−5, 15] scl: 2

x	4.99	4.999	5	5.001	5.01
f(x)	9.96	9.996		10.004	10.04

3. -15

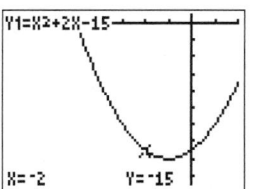
[−8, 2] scl: 1 by [−19, 1] scl: 2

x	−2.01	−2.001	−2	−1.999	−1.99
f(x)	−14.98	−14.998		−15.002	−15.02

5. 25

[−5, 5] scl: 1 by [−10, 30] scl: 5

x	2.99	2.999	3	3.001	3.01
f(x)	24.56	24.956		25.044	25.44

7. 0

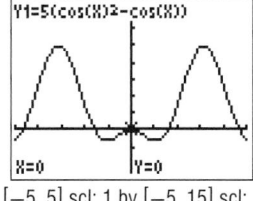
[−5, 5] scl: 1 by [−5, 15] scl: 2

x	−0.01	−0.001	0	0.001	0.01
f(x)	−0.0002	−0.000002		−0.000002	−0.0002

9. 5.72

[−2, 8] scl: 1 by [−2, 8] scl: 1

x	5.99	5.999	6	6.001	6.01
f(x)	5.70	5.719		5.723	5.74

11. 0   **13.** 0   **15.** 1   **17.** does not exist   **19.** -7 **21.** 7   **23.** does not exist   **25.** 0   **27.** does not exist   **29.** 4   **31.** ∞ **33.** $-\infty$ **35.** does not exist   **37.** ∞ **39.** 0   **41.** $\frac{1}{3}$ **43.** -1 **45.** 0   **47a.** $\lim\limits_{w \to 1} f(w) = 250$; $\lim\limits_{w \to 3} f(w) = 100$ **47b.** 0; Sample answer: Eventually, the vaccine will eliminate all cases of the infection.

49 **a.** Using a graphing calculator, graph $Y_1 = 12(1.25012)^X - 12$ in the window $[0, 20] \times [0, 1100]$.

[0, 20] scl: 2 by [0, 1100] scl: 100

b. Use the TRACE feature. Move the cursor until X is close to 5, 10, and 20. The y-values near these values are about 25, 100, and 1031. Therefore, the number of people who have viewed the video after 60 days is $p(60) = 12(1.25012)^{60}$ or about 7,880,000.
c. ∞; Sample answer: The graph appears to increase without bound, which suggests that an infinite amount of people may view the video.

51 **a.** Using a graphing calculator, graph $Y_1 = 2000(0.7)^{(X-1)}$ in the window $[1, 15] \times [0, 2000]$.

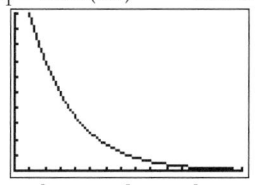
[1, 15] scl: 1 by [0, 2000] scl: 200

b. Use the TRACE feature. Move the cursor until X is close to 5, 10, and 15. The y-values near these values are about 480.2, 80.71, and 13.56.
c. The graph approaches the x-axis, so $\lim\limits_{t \to \infty} = 0$.
d. No; the sum of the infinite series is approximately 6666.67 meters, which is less than the 7000 meters required to reach the hospital.

53. -1 **55.** does not exist   **57.** 6   **59.** no; vertical asymptote   **61.** no; oscillations   **63.** no; vertical asymptote   **65.** Neither; sample answer: If $f(x)$ approaches a different value from the left than from the right, then the limit at that point does not exist.

67. $\lim_{x\to 1} f(x)$ does not exist; $\lim_{x\to 1} g(x)$ does not exist; $\lim_{x\to a} \dfrac{f(x)}{g(x)}$ does not exist; sample answer: If the denominator of a rational function is equal to zero for a given point, the limit at this point does not exist.

69. Sample answer:

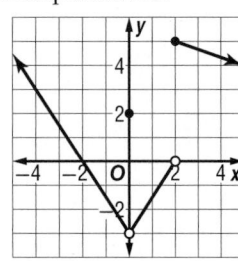

71. Sample answer: If $f(x)$ is continuous at $x = a$, you can substitute a into the function. If the function is not continuous, you may be able to simplify it and then substitute in a. If neither of these methods works, you must estimate the limit graphically.

73. H_0: $\mu = 4$ (claim), H_a: $\mu \neq 4$ **75.** H_0: $\mu \geq 10$, H_a: $\mu < 10$ (claim)

77. $r = \dfrac{4}{1 - \sin\theta}$

79. 63.0°

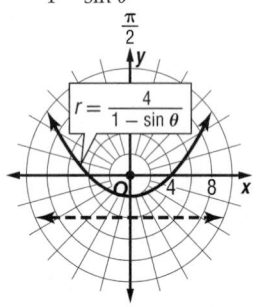

81.

[−10, 10] scl: 1 by [−10, 10] scl: 1

83. A   **85.** C

Pages 754–756   Lesson 12-2

1. −25   **3.** 10   **5.** $21\frac{1}{9}$ **7.** 6   **9.** 20   **11.** Not possible; when $x = 16$, the denominator is 0.   **13.** 2   **15.** −9216   **17.** $-62\frac{1}{2}$
19. Not possible; when $x = 5$, the denominator is 0.
21a. $\lim_{v\to 0} m = m_0$. As the velocity of the object approaches 0, the mass of the object will approach its initial, or rest mass.   **21b.** The mass of the object would begin to increase without bound.   **23.** 11   **25.** 3   **27.** $1\frac{6}{13}$
29. 4   **31.** −12   **33.** $\frac{1}{6}$ **35.** ∞ **37.** 0   **39.** $\frac{1}{3}$ **41.** ∞
43. $-\infty$ **45.** 2

47. a. The length of the capsule before it is submerged is its length at $t = 0$.
$$\ell(0) = \frac{105(0)^2}{10 + 0^2} + 25 = 25 \text{ mm}$$
b. $\lim_{t\to\infty} \dfrac{105t^2}{10 + t^2} + 25 = \lim_{t\to\infty} \dfrac{\frac{105t^2}{t^2}}{\frac{10}{t^2} + \frac{t^2}{t^2}} + 25$

$= \lim_{t\to\infty} \dfrac{105}{\frac{10}{t^2} + 1} + 25$

$= \dfrac{\lim_{t\to\infty} 105}{\lim_{t\to\infty} \frac{10}{t^2} + \lim_{t\to\infty} 1} + \lim_{t\to\infty} 25$

$= \dfrac{105}{0 + 1} + 25$

$= 130$

c. The sponge animal will not grow more than 130 mm in length.

49. ∞ **51.** −4   **53.** 2   **55.** $\frac{5}{2}$ **57.** $\frac{1}{4}$

59a.

Years Since 2006	Rise in Population
1	398
2	2430
3	5550

59b. approximately 9576 people   **59c.** 12,000 people

59d. Sample answer: A city's borders may limit the amount of possible growth and building opportunities.   **61.** 2
63. No limit exists.   **65.** 1   **67.** 4.5   **69.** $-\dfrac{1}{2}$ **71.** 2

73. $\lim_{h\to 0} \dfrac{f(x + h) - f(x)}{h} = \lim_{h\to 0} \dfrac{\sqrt{x + h} - \sqrt{x}}{h}$

$= \lim_{h\to 0} \dfrac{\sqrt{x + h} - \sqrt{x}}{h} \cdot \dfrac{\sqrt{x + h} + \sqrt{x}}{\sqrt{x + h} + \sqrt{x}}$

$= \lim_{h\to 0} \dfrac{(x + h) - x}{h(\sqrt{x + h} + \sqrt{x})}$

$= \lim_{h\to 0} \dfrac{h}{h(\sqrt{x + h} + \sqrt{x})}$

$= \lim_{h\to 0} \dfrac{1}{\sqrt{x + h} + \sqrt{x}}$

$= \dfrac{1}{\lim_{h\to 0}\sqrt{x + h} + \lim_{h\to 0}\sqrt{x}}$

$= \dfrac{1}{2\sqrt{x}}$ or $\dfrac{\sqrt{x}}{2x}$

75. $2x$ **77.** 0.0000125

79. Let P_n be the statement if $\lim_{x\to c} f(x) = L$, then $\lim_{x\to c} [f(x)]^n = [\lim_{x\to c} f(x)]^n$ or L^n for any integer n.
Because $\lim_{x\to c} [f(x)]^1 = L^1$ is a true statement, P_1 is true.
Assume that $\lim_{x\to c} [f(x)]^k = L^k$ is true for an integer k. Show that P_{k+1} must be true.
$$\lim_{x\to c} [f(x)]^k = L^k$$
$$\lim_{x\to c} [f(x)]^k \cdot \lim_{x\to c} [f(x)]^1 = L^k \cdot \lim_{x\to c} [f(x)]^1$$
$$\lim_{x\to c} [f(x)]^{k+1} = L^k \cdot L$$
$$\lim_{x\to c} [f(x)]^{k+1} = L^{k+1}$$
This final statement is exactly P_{k+1}, so P_{k+1} is true. Because P_n is true for $n = 1$ and P_k implies P_{k+1}, P_n is true for $n = 2$, $n = 3$, and so on. That is, by the principle of mathematical induction, if $\lim_{x\to c} f(x) = L$, then $\lim_{x\to c}[f(x)]^n = [\lim_{x\to c} f(x)]^n$ or L^n for any integer n.

81. Sample answer: Sometimes; if the limit in question is not at a vertical asymptote, then it is true; if the limit in question is at a vertical asymptote, then it is not true.   **83.** Sample answer: A limit of $\frac{\infty}{\infty}$ does not translate to 1 because infinity is not a real number; it is more of a concept. Further analyze this problem by graphing the original rational function and observing the behavior of the graph around the limit.   **85.** 0; undefined

87a.

The data appear to have a positive linear correlation.

[0, 105] scl: 10 by [40, 100] scl: 5

87b. $r \approx 0.975$; The correlation coefficient indicates that the data have a strong positive linear correlation. Since $t \approx 2.306$ and $12.41 > 2.306$, the statistic falls within the critical region and the null hypothesis is rejected. Therefore, the correlation is significant at the 5% level.

87c. $\hat{y} = 0.295x + 49.927$; A slope $a = 0.295$ indicates that for each additional year, the life expectancy will increase by a rate of 0.295 per year. The y-intercept $b = 49.927$ indicates that in 1900, the average life expectancy was approximately 50 years.

87d. Using this model, the average life expectancy in 2080 will be about 103 years, which is probably not reasonable.

89. $\dfrac{x^2}{4} = \dfrac{y^2}{25} = 1$

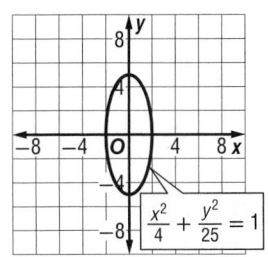

$$\dfrac{x^2}{4} + \dfrac{y^2}{25} = 1$$

91. H   **93.** G

Pages 762–764   Lesson 12-3

1. $-3; 5$ **3.** $6; 12$ **5.** $12; 3$ **7.** $m = -2$ **9.** $m = 2x$

11. $m = -2x$ **13.** $m = -6x^2$ **15.** $m = -\dfrac{\sqrt{x}}{2x^2}$

17a. $m = 0.18x^2 - 2.16x$ **17b.** $-3.6, -6.3, -6.3$ **19.** 64.8 mi/h
21. 49.2 mi/h   **23.** 36 mi/h   **25.** -96 ft/s   **27.** -95 ft/s
29. -512 ft/s   **31.** -121.6 ft/s   **33.** $v(t) = 28t$ **35.** $v(t) = 5$

37. $v(t) = 3t^2 - 2t + 1$ **39.** $v(t) = \dfrac{\sqrt{t}}{2t} - 6t$

41 **a.** Determine the height of the skydiver at times $t = 2$ and $t = 5$.
$h(2) = 15{,}000 - 16 \cdot 2^2 = 14{,}936$
$h(5) = 15{,}000 - 16 \cdot 5^2 = 14{,}600$
The average velocity between the 2nd and 5th seconds is

$$\dfrac{h(5) - h(2)}{5 - 2} = \dfrac{14600 - 14936}{3} = -112 \text{ ft/s}$$

b. Calculate $v(2)$ and $v(5)$.

$$v(2) = \lim_{h \to 0} \dfrac{h(2 + h) - h(2)}{h}$$
$$= \lim_{h \to 0} \dfrac{15000 - 16(2 + h)^2 - \left[15000 - 16 \cdot 2^2\right]}{h}$$
$$= \lim_{h \to 0} \dfrac{15000 - 16(4 + 4h + h^2) - 15000 + 64}{h}$$
$$= \lim_{h \to 0} \dfrac{-64 - 64h - 16h^2 + 64}{h}$$
$$= \lim_{h \to 0} \dfrac{h(-64 - 16h)}{h}$$
$$= \lim_{h \to 0} (-64 - 16h)$$
$$= -64$$
$$v(5) = \lim_{h \to 0} \dfrac{h(5 + h) - h(5)}{h}$$
$$= \lim_{h \to 0} \dfrac{15000 - 16(5 + h)^2 - \left[15000 - 16 \cdot 5^2\right]}{h}$$
$$= \lim_{h \to 0} \dfrac{15000 - 16(25 + 10h + h^2) - 15000 + 400}{h}$$
$$= \lim_{h \to 0} \dfrac{-400 - 160h - 16h^2 + 400}{h}$$

$$= \lim_{h \to 0} \dfrac{h(-160 - 16h)}{h}$$
$$= \lim_{h \to 0} (-160 - 16h)$$
$$= -160$$
The velocities at times $t = 2$ and $t = 5$ are -64 ft/s and -160 ft/s.

c. $v(t) = \lim_{h \to 0} \dfrac{h(t + h) - h(t)}{h}$
$$= \lim_{h \to 0} \dfrac{15000 - 16(t + h)^2 - \left[15000 - 16t^2\right]}{h}$$
$$= \lim_{h \to 0} \dfrac{15000 - 16(t^2 + 2th + h^2) - 15000 + 16t^2}{h}$$
$$= \lim_{h \to 0} \dfrac{-16t^2 - 32th - 16h^2 + 16t^2}{h}$$
$$= \lim_{h \to 0} \dfrac{h(-32t - 16h)}{h}$$
$$= \lim_{h \to 0}(-32t - 16h)$$

$$= -32t$$

43a. $v(t) = -32t + 75$ **43b.** 59 ft/s
43c. $t \approx 2.344$ seconds   **43d.** ≈ 90.39 ft   **45.** $y = -4x + 1$

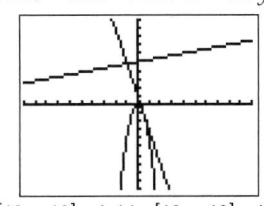

[10, −10] scl: 1 by [10, −10] scl: 1

47 Find an equation for the slope of $g(x)$ and determine the x-value for which it equals the slope of the line perpendicular to $y = -\dfrac{1}{6}x + 9$, namely $m = 6$.

$$m = \lim_{h \to 0} \dfrac{g(x + h) - g(x)}{h}$$
$$= \lim_{h \to 0} \dfrac{\frac{1}{2}(x + h)^2 + 4(x + h) - \left(\frac{1}{2}x^2 + 4x\right)}{h}$$
$$= \lim_{h \to 0} \dfrac{xh + \frac{1}{2}h^2 + 4h}{h}$$
$$= \lim_{h \to 0} \left(x + \frac{1}{2}h + 4\right)$$
$$= x + \frac{1}{2}(0) + 4 \text{ or } x + 4$$

Set this expression equal to the desired slope of 6 and solve.
$x + 4 = 6$
$\quad x = 2$
The y-value for $g(x)$ at $x = 2$ is $g(2) = \frac{1}{2} \cdot 2^2 + 4 \cdot 2$ or 10. Use the point-slope formula to find the equation of the line with slope $m = 6$ that passes through $(2, 10)$.
$y - y_1 = m(x - x_1)$
$y - 10 = 6(x - 2)$
$y - 10 = 6x - 12$
$\quad\quad y = 6x - 2$

[10, −10] scl: 1 by [10, −10] scl: 1

49. c **51.** b **53a.** −40 ft/s **53b.** −136 ft/s **53c.** −232 ft/s
55. Jillian; sample answer: The graph of $f(x)$ has a slope of −1 when $x < 0$ and a slope of 1 when $x > 0$. Therefore, the graph of this equation would be two horizontal lines

$$y = \begin{cases} -1 & \text{if} & x < 0 \\ 1 & \text{if} & x > 0 \end{cases}$$ and would not be continuous.

57. False; sample answer: If a graph is not circular, a tangent line can intersect a graph at more than one point. For example, the graph of $y = \sin x$. **63a.** The average growth of the investment for the first 4 years is about $41.20 per year. **63b.** After exactly 4 years, the investment is growing at a rate of about $42.90 per year. **61.** 1

63a.
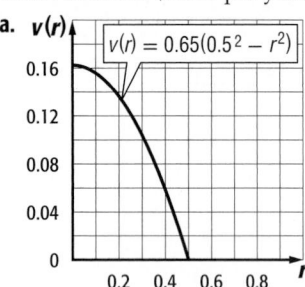

63b. 0 in./s **65a.** 72
65b. 58 **65c.** 68–71
67. 729 **69.** 891
71. 23.3, 29.4, 35.5, 41.6
73. −29, −13, 3, 19, 35, 51, 67, 83, 99 **75.** G
77. G

Pages 772–774 Lesson 12-4

1. $f'(x) = 8x; f'(2) = 16, f'(-1) = -8$ **3.** $m'(j) = 14; m'(-7) = 14, m'(-4) = 14$ **5.** $f'(c) = 3c^2 + 4c - 1; f'(-2) = 3, f'(1) = 6$

7. $y'(f) = -11$ **9.** $p'(v) = 7$ **11.** $b'(m) = 2m^{-\frac{1}{3}} - 3m^{\frac{1}{2}}$

13. $f'(x) = \frac{3}{2}x^{-\frac{1}{2}} - \frac{3}{2}x^{\frac{1}{2}} - x^{-\frac{3}{2}}$ **15.** $p'(k) = 5.2k^{4.2} - 38.4k^{3.8} + 3$

17 **a.** Find $f'(h)$.

$f'(h) = -0.0036 \cdot 3h^2 - 0.01 \cdot 2h + 2.04$
$= -0.0108h^2 - 0.02h + 2.04$

b. Evaluate $f'(h)$ at $h = 2$, $h = 14$, and $h = 20$.

$f'(2) = -0.0108 \cdot 2^2 - 0.02 \cdot 2 + 2.04$
$\approx 1.96 \text{ deg/hr}$
$f'(14) = -0.0108 \cdot 14^2 - 0.02 \cdot 14 + 2.04$
$\approx -0.36 \text{ deg/hr}$
$f'(20) = -0.0108 \cdot 20^2 - 0.02 \cdot 20 + 2.04$
$\approx -2.68 \text{ deg/hr}$

c. Determine when $f'(h) = 0$.

$-0.0108h^2 - 0.02h + 2.04 = 0$

$h = \dfrac{-(-0.02) \pm \sqrt{(-0.02)^2 - 4(-0.0108)(2.04)}}{2(-0.0108)}$

$h \approx -14.70$ or $h \approx 12.85$

Since $0 \le h \le 24$, choose only the positive value. Evaluate f at $h = 0$, $h = 12.85$ and $h = 24$.
$f(0) = -0.0036 \cdot 0^3 - 0.01 \cdot 0^2 + 2.04 \cdot 0 + 52 = 52$
$f(12.85) = -0.0036 \cdot 12.85^3 - 0.01 \cdot 12.85^2 + 2.04 \cdot$
$12.85 + 52 \approx 68.92$
$f(24) = -0.0036 \cdot 24^3 - 0.01 \cdot 24^2 + 2.04 \cdot 24 +$
$52 \approx 43.43$
Thus, the maximum temperature is 68.92°F.

19. critical points: $\left(\frac{2\sqrt{3}}{3}, 6.92\right)$ and $\left(-\frac{2\sqrt{3}}{3}, 13.08\right)$; max: 25, min: −5 **21.** critical point: (−5, −10); max: −2, min: −11 **23.** critical point: (−9, 405); max: 405, min: 385 **25.** critical points: (0, 2) and (2.25, −6.54); max: 66, min: −6.54

27a. $h'(t) = 65 - 32t$ **27b.** (2.03, 71.02); (0, 5) **27c.** Yes; the maximum height that Gabe can throw the ball is approximately 71 ft. This is more than the 70 ft needed to reach Zach's window. **29.** $g'(x) = -45x^4 + 60x^3 - 12x + 10$

31. $s'(t) = \frac{69}{2}t^{\frac{21}{2}} + 66t^{10} - 6t^{\frac{1}{2}} - 8$ **33.** $c'(t) = 13t^{12} - 33t^{10} +$
$9t^8 - 132t^7 + 35t^6 + 30t^4 - 88t$ **35.** $q'(a) = \frac{19}{8}a^{\frac{11}{8}} - \frac{221}{8}a^{\frac{9}{8}} +$
$a - \frac{39}{4}a^{-\frac{1}{4}}$ **37.** $h'(x) = \frac{19}{48}x^{\frac{13}{6}} + \frac{37}{192}x^{\frac{13}{24}} + \frac{14}{15}x^{\frac{4}{3}} + \frac{17}{60}x^{-\frac{7}{24}}$

39. $f'(m) = \frac{-12}{(3 + 2m)^2}$ **41.** $r'(t) = \frac{10t}{(3 - t^2)^2}$

43. $v'(t) = \frac{-t^4 + 10t^3 - 13t^2 + 12}{(t^3 - 4t)^2}$

45. $f'(x) = \frac{-x^4 + 11x^2 + 6}{(-x^2 + 3)^2}$ **47.** $t'(w) = \frac{2w^3 - 1}{w^2}$

49a. $r'(x) = 0.375x^2 - 22.5x + 250$ **49b.** 14.72, 45.28 **49c.** Sample answer: The solution of 14.72 represents the price that Nayla and Deirdre should charge per sweatshirt to make the maximum possible revenue. The solution of 45.28 is not relevant because the revenue is 0 when $x = 40$.

51. $y = 10x + 45$

53. $y = -21.6x - 40.76$

55a. $f''(x) = 80x^3 - 12x$ **55b.** $g'''(x) = -420x^4 + 96x - 42$
55c. $h^{(4)}(x) = 1080x^{-7} + 240x^{-6}$
57. Sample answer: **59.** Sample answer:

 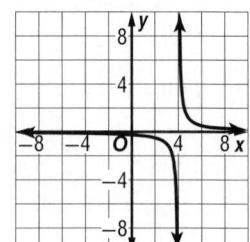

61a. $A' = 2\pi r$; $V' = 4\pi r^2$ **61b.** Sample answer: The derivative of the formula for the area of a circle is the formula for the circumference of a circle. The derivative of the formula for the volume of a sphere is the formula for the surface area of a sphere.

61c.

61d. $A = 4a^2$; $A' = 8a$; $V = 8a^3$; $V' = 24a^2$ **61e.** Sample answer: When the area of a square is written in terms of its apothem, the derivative is the formula for the perimeter of the square. When the volume of a cube is written in terms of the apothems of its faces, the derivative is the formula for the surface area of the cube.

63 Treat x and z as constants.
$$f'(y) = 10x^2 \cdot 3y^2 + 0 - 6x \cdot 2y + 0 - 11x^8 \cdot 1 \cdot z^7$$
$$= 30x^2y^2 - 12xy - 11x^8z^7$$

65. True; sample answer: The power for $f(x)$ is $5n + 3$. Due to the Power Rule, this will become the coefficient of the derivative. The power of the derivative will be one less than the original power. It will then be $(5n + 3) - 1$ or $5n + 2$.

67. Sample answer:
$$\frac{f'(x)g(x) - f(x)g'(x)}{[g(x)]^2}$$
$$= \lim_{h \to 0} \frac{\frac{f(x+h)g(x) - f(x)g(x+h)}{hg(x+h)g(x)}}{}$$
$$= \lim_{h \to 0} \frac{f(x+h)g(x) - f(x)g(x) + f(x)g(x) - f(x)g(x+h)}{hg(x+h)g(x)}$$
$$= \lim_{h \to 0} \frac{[f(x+h) - f(x)]g(x) - [g(x+h) - g(x)]f(x)}{hg(x+h)g(x)}$$
$$= \lim_{h \to 0} \frac{\frac{f(x+h) - f(x)}{h}g(x) - \frac{g(x+h) - g(x)}{h}f(x)}{g(x+h)g(x)}$$
$$= \frac{g(x)\lim\limits_{h\to 0}\frac{f(x+h)-f(x)}{h} - f(x)\lim\limits_{h\to 0}\frac{g(x+h)-g(x)}{h}}{\lim\limits_{h\to 0}g(x+h)g(x)}$$
$$= \frac{f'(x)g(x) - f(x)g'(x)}{[g(x)]^2}$$

69. $-3; 3$ **71.** $6; 12$ **73.** $-\frac{1}{3}$

75.

Gym Class Exercise Amount for a Week

Days, X	0	1	2	3	4	5
P(x)	0.10	0.20	0.23	0.27	0.13	0.07

77. 0.841 **79.** 0.239 **81.** $a_n = 1.4(-2.5)^{n-1}$; $a_1 = 1.4$, $a_n = -2.5a_{n-1}$ **83.** D **85.** B

Pages 781–783 Lesson 12-5

1. 15 units^2 **3.** 18.29 units^2 **5.** 16.23 units^2

7 a. The graph of the semicircle is on the interval [1, 10]. Form rectangles with left endpoints and a width of 1.
$$R_1 = 1 \cdot [-(1)^2 + 10(1)]^{0.5} = 3$$
$$R_2 = 1 \cdot [-(2)^2 + 10(2)]^{0.5} = 4$$
$$R_3 = 1 \cdot [-(3)^2 + 10(3)]^{0.5} \approx 4.58$$
$$R_4 = 1 \cdot [-(4)^2 + 10(4)]^{0.5} \approx 4.90$$

$$R_5 = 1 \cdot [-(5)^2 + 10(5)]^{0.5} = 5$$
$$R_6 = 1 \cdot [-(6)^2 + 10(6)]^{0.5} \approx 4.90$$
$$R_7 = 1 \cdot [-(7)^2 + 10(7)]^{0.5} \approx 4.58$$
$$R_8 = 1 \cdot [-(8)^2 + 10(8)]^{0.5} = 4$$
$$R_9 = 1 \cdot [-(9)^2 + 10(9)]^{0.5} = 3$$
Total area ≈ 37.96

b. For this section, some rectangles use right endpoints and some use left endpoints.
$$R_1 = 1 \cdot [-(1)^2 + 10(1)]^{0.5} = 3$$
$$R_2 = 1 \cdot [-(2)^2 + 10(2)]^{0.5} = 4$$
$$R_3 = 1 \cdot [-(3)^2 + 10(3)]^{0.5} \approx 4.58$$
$$R_4 = 1 \cdot [-(4)^2 + 10(4)]^{0.5} \approx 4.90$$
$$R_5 = 1 \cdot [-(4)^2 + 10(4)]^{0.5} \approx 4.90$$
$$R_6 = 1 \cdot [-(7)^2 + 10(7)]^{0.5} \approx 4.58$$
$$R_7 = 1 \cdot [-(8)^2 + 10(8)]^{0.5} = 4$$
$$R_8 = 1 \cdot [-(9)^2 + 10(9)]^{0.5} = 3$$
Total area ≈ 32.96

c. The radius is 5.
$$\text{Area} = \frac{1}{2}\pi r^2$$
$$= \frac{1}{2}\pi 5^2$$
$$= 12.5\pi$$
$$\approx 39.27$$

The first estimate is closer. Sample answer: The additional area outside of the semicircle that is included in the first estimate helps account for the area in the region that is not being included by the rectangles

9. right endpoints: 12.6 units^2; left endpoints: 9.4 units^2; average: 11 units^2 **11.** right endpoints: 18.91 units^2; left endpoints: 19.66 units^2; average: 19.285 units^2 **13.** right endpoints: 12.75 units^2; left endpoints: 12.25 units^2; average: 12.5 units^2 **15.** 52 units^2 **17.** $23\frac{1}{3}$ units^2 **19.** 30 units^2 **21.** $8\frac{2}{3}$ units^2 **23.** 12 units^2 **25.** 48 units^2 **27.** $80\frac{5}{6}$ units^2 **29.** 67.2 ft^2 **31a.** height: 4 units, base: 4 units; 8 units^2 **31b.** 8 units^2

33 Find Δx and x_i.
$$\Delta x = \frac{b-a}{n} = \frac{0-(-1)}{n} = \frac{1}{n}$$
$$x_i = a + i\Delta x = -1 + \frac{i}{n}.$$
Calculate the definite integral that gives the area.
$$\int_{-1}^{0} (x^3 + 2)dx = \lim_{n \to \infty} \sum_{i=1}^{n} f(x_i)\Delta x$$
$$= \lim_{n \to \infty} \sum_{i=1}^{n} \left[\left(-1 + \frac{i}{n}\right)^3 + 2\right]\frac{1}{n}$$
$$= \lim_{n \to \infty} \frac{1}{n} \sum_{i=1}^{n} \left[-1 + 3\left(\frac{i}{n}\right) - 3\left(\frac{i}{n}\right)^2 + \left(\frac{i}{n}\right)^3 + 2\right]$$
$$= \lim_{n \to \infty} \frac{1}{n} \sum_{i=1}^{n} \left[1 + \frac{3i}{n} - \frac{3i^2}{n^2} + \frac{i^3}{n^3}\right]$$
$$= \lim_{n \to \infty} \frac{1}{n} \left[\sum_{i=1}^{n} 1 + \sum_{i=1}^{n} \frac{3i}{n} - \sum_{i=1}^{n} \frac{3i^2}{n^2} + \sum_{i=1}^{n} \frac{i^3}{n^3}\right]$$
$$= \lim_{n \to \infty} \frac{1}{n} \left[\sum_{i=1}^{n} 1 + \frac{3}{n}\sum_{i=1}^{n} i - \frac{3}{n^2}\sum_{i=1}^{n} i^2 + \frac{1}{n^3}\sum_{i=1}^{n} i^3\right]$$

$$= \lim_{n \to \infty} \frac{1}{n}\left(n + \frac{3}{n}\frac{n(n+1)}{2} - \frac{3}{n^2}\frac{n(n+1)(2n+1)}{6} + \frac{1}{n^3}\frac{n^2(n+1)^2}{4}\right)$$

$$= \lim_{n \to \infty}\left(\frac{n}{n} + \frac{3n(n+1)}{2n^2} - \frac{3n(2n^2+3n+1)}{6n^3} + \frac{n^2(n^2+2n+1)}{4n^4}\right)$$

$$= \lim_{n \to \infty}\left(1 + \frac{3(n+1)}{2n} - \frac{2n^2+3n+1}{2n^2} + \frac{n^2+2n+1}{4n^2}\right)$$

$$= \lim_{n \to \infty}\left(1 + 3\left(\frac{1}{2} + \frac{1}{2n}\right) - \left(1 + \frac{3}{2n} + \frac{1}{2n^2}\right) + \left(\frac{1}{4} + \frac{1}{2n} + \frac{1}{4n^2}\right)\right)$$

$$= \lim_{n \to \infty}(1) + 3\lim_{n \to \infty}\left(\frac{1}{2} + \frac{1}{2n}\right) - \lim_{n \to \infty}\left(1 + \frac{3}{2n} + \frac{1}{2n^2}\right) +$$

$$\lim_{n \to \infty}\left(\frac{1}{4} + \frac{1}{2n} + \frac{1}{4n^2}\right)$$

$$= 1 + 3\left(\frac{1}{2} + 0\right) - (1 + 0 + 0) + \left(\frac{1}{4} + 0 + 0\right) = 1\frac{3}{4} \text{ or } 1.75$$

35. $12\frac{1}{2}$ units^2 **37.** $\frac{3}{4}$ units^2 **39.** 4 units^2 **41.** $3\frac{3}{4}$ units^2

43. Neither; sample answer: If the function is increasing, then using the right endpoints will cause the area of the rectangles to be greater than the actual area while using the left endpoints causes the area of the rectangles to be less. But, if the function is decreasing, then using the left endpoints will cause the area of the rectangles to be greater than the actual area while using the right endpoints causes the area of the rectangles to be less. **45.** Sample answer: The integral gives the area of each cross section. Multiplying this area by the total length of the tunnel would compute the volume of the tunnel.

47. $\frac{t^3}{3} + 2t$ units^2 **49.** $j'(x) = 44x^{10} + 198x^8 - 120x^4 - 396x^2$
51. $s'(t) = \frac{51}{2}t^{\frac{15}{2}} - 168t^7 - \frac{5}{2}t^{\frac{1}{2}} + 35$ **53.** -7 **55.** 3

57. $\frac{2}{9}$ **59.** $E = 3.16$; $33.34 < \mu < 39.66$ **61.** Discrete; the number of mobile phone calls is countable and therefore discrete. **63.** D **65.** B

Pages 789–791 Lesson 12-6

1. $F(x) = \frac{1}{6}x^6 + C$ **3.** $F(z) = \frac{3}{4}z^{\frac{4}{3}} + C$ **5.** $Q(r) = \frac{15}{28}r^{\frac{7}{5}} +$
$\frac{15}{32}r^{\frac{4}{3}} + \frac{2}{3}r^{\frac{3}{2}} + C$ **7.** $G(a) = 2a^4 + \frac{5}{3}a^3 - \frac{9}{2}a^2 + 3a + C$
9. $M(t) = 4t^4 - 4t^3 + 10t^2 - 11t + C$ **11a.** $s(t) = -16t^2 + C$
11b. 64 **11c.** 28 ft **13.** $5n^4 - 3n^3 - 9n^2 + 4n + C$ **15.** 46.5
17. 22.37 **19.** 18.93 **21.** $2w^{7.1} - 3w^{6.7} + 4w^{3.3} + 3w + C$

23 **a.** Find the antiderivative of $v(t)$ to find $s(t)$.

$$s(t) = -32\frac{t^{1+1}}{1+1} + 34t + C = -16t^2 + 34t + C$$

Substitute $t = 0$ and $s(t) = 0$ and solve for C.

$$0 = -16 \cdot 0^2 + 34 \cdot 0 + C$$
$$0 = C$$

So, $s(t) = -16t^2 + 34t$.

b. Solve $s(t) = 0$ to determine when the flea lands.

$$-16t^2 + 34t = 0$$
$$(-16t + 34)t = 0$$
$$-16t + 34 = 0 \text{ or } t = 0$$

Since $t = 0$ is when the flea jumps, solve $-16t + 34 = 0$.
$-16t = -34$. So, $t = \frac{-34}{-16}$ or $\frac{17}{8}$ or 2.125 seconds.

25a.
[0, 12] scl: 1 by [0, 24] scl: 2

25b. $s_1(t) = \frac{3.25t^2}{2} - \frac{0.2t^3}{3}$;
$s_2(t) = \frac{1.2t^2}{2} + \frac{0.03t^3}{3}$

25c. 10.34 seconds for strategy one and 11.80 seconds for strategy two **27.** 27 **29.** 5.33 **31.** 16.4 **33.** $-x^3 - 4x^2 + 24$
35. $16x^4 + 20x^2 + 4x - 132$

37 $\displaystyle\int_{x}^{x^2}(16t^3 - 15t^2 + 7)dt$

$$= 16\frac{t^{3+1}}{3+1} - 15\frac{t^{2+1}}{2+1} + 7t\Big|_{x}^{x^2}$$

$$= 4t^4 - 5t^3 + 7t\Big|_{x}^{x^2}$$

$$= \left[4(x^2)^4 - 5(x^2)^3 + 7x^2\right] - \left[4x^4 - 5x^3 + 7x\right]$$

$$= 4x^8 - 5x^6 - 4x^4 + 5x^3 + 7x^2 - 7x$$

39. $\frac{4}{3}\pi R^3$ **41.** 44,152.52; 44,100 **43.** 2,156,407.8; 2,153,645

45a.

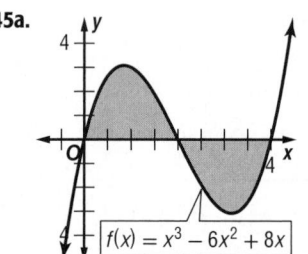

$f(x) = x^3 - 6x^2 + 8x$

45b. 4; -4 **45c.** The two areas are equivalent; however, the value of the integral of $f(x)$ that corresponds to the area above the x-axis is positive, and the value of the integral of $f(x)$ that corresponds to the area below the x-axis is negative.
45d. 0; 8 **45e.** Signed area is the difference between the absolute values of the areas found above and below the x-axis. Total area is the sum of the absolute values of the areas found above and below the x-axis. **47.** Sometimes; sample answer: Changing the order of the limits will change the sign of the original answer unless the answer is 0. **49.** Sometimes; sample answer: If $f(x)$ is an even function, $a \geq 0$, and $b \geq 0$, the identity will hold true. If $f(x)$ is an odd function, $a \geq 0$ and $b \geq 0$, or $a \leq 0$ and $b \leq 0$, the identity will hold true.
51. Because the graph is below the x-axis, $f(x)$ is negative. Each $f(x)$ is negative and Δx is positive, so each term in the

sum $\displaystyle\sum_{i=1}^{n}f(x_i)\Delta x$ is negative. Therefore, the sum is negative.

Because $\displaystyle\int_{a}^{b}f(x)dx$ is a limit of negative sums, it is also

negative.
53. Sample answer:
  1) Decide which rule to apply to find the antiderivative of the function: $2x^3 + C$.
     a) Power rule.
     b) Constant Multiple of a Power rule (applicable here).
     c) Sum and Difference rules.
  2) Remove/ignore the constant C since this is a definite integral.
  3) Evaluate the antiderivative at the upper and lower limits and find the difference.

$$2x^3 \Big|_0^2 = 2(2)^3 - 2(0)^3 = 16 - 0 = 16$$

4) The area under the graph on the interval [0, 2] is 16 square units.

55. 30 **57.** $g'(n) = \dfrac{2n^4 + 2n^2 + 4}{(n^2 + 1)^2}$ **59a.** 16% **59b.** 24 **61.** B

63a. 4 m **63b.** 1 m/s **63c.** $s'(t) = 2t - 3$ **63d.** -1 m/s; 5 m/s

63e. 1.5 s **63f.** Sample answer: After 1.5 s, the particle changes direction, moving to the right instead of to the left.

Pages 792–796 Chapter 12 Study Guide and Review

1. instantaneous rate of change **3.** direct substitution
5. indeterminate form **7.** differentiation **9.** instantaneous velocity **11.** -1

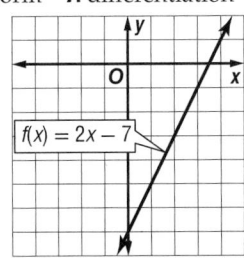

x	2.99	2.999	3	3.001	3.01
$f(x)$	-1.02	-1.002		-0.998	-0.98

13. 5 **15.** ∞ **17.** 9 **19.** Not possible; when $x = 25$, the denominator is 0. **21.** $-\dfrac{1}{6}$ **23.** $-1; -1$ **25.** $m = -2x + 3$
27. 31 ft/s **29.** $v(t) = 24t^6$ **31.** $g'(t) = -2t + 5$;
$g'(-4) = 13$ and $g'(1) = 3$ **33.** $p'(v) = -9$ **35.** $t'(x) = -\dfrac{18}{5}x^{\frac{1}{5}}$

37. $f'(m) = \dfrac{-25}{(5 + 2m)^2}$ **39.** 5.16 units^2 **41.** $4\dfrac{2}{3}$ units^2

43. $4\dfrac{2}{3}$ units^2 **45.** $G(n) = \dfrac{5}{2}n^2 - 2n + C$

47. $M(t) = \dfrac{3}{2}t^4 - 4t^3 + t^2 - 11t + C$ **49.** $\dfrac{8}{3}x^3 + C$

51. 2466.53 units^2

53a.

t	0	1	2	3
V	15	30	50	68.2

53b.

53c. 90 **53d.** Sample answer: The stamp will peak in value at about \$90. **55a.** 200 **55b.** The maximum value of Eldin's coin collection is equal to $\lim\limits_{t \to \infty} v(t)$. **55c.** Yes; the offer is more than the collection is worth. **57a.** $v(t) = -32t + 35$
57b. 19 ft/s **57c.** ≈ 1.09 s **57d.** ≈ 20.64 ft **59a.** $s(t) = -16t^2 + 26t$ **59b.** 1.63 s

Pages 798–799 Chapter 12 Connect to AP Calculus

1. Sample answer: $4x^3$ is the derivative of the expression within the parentheses. **3.** Sample answers: $4(x^4 - 3)^3 4x^3$ or $16x^3(x^4 - 3)^3$ **5.** $s'(t) = 8t(t^2 - 1)^3$

7. $c'(r) = 3(3r - 2r^2)^2(3 - 4r)$ **9.** $f'(x) = \dfrac{\sqrt{100 + 8x}}{25 + 2x}$ or $\dfrac{2\sqrt{25 + 2x}}{25 + 2x}$

Photo Credits

Glossary/Glosario

Cómo usar el glosario en español:

1. Busca el término en inglés que desees encontrar.

2. El término en español, junto con la definición, se encuentran en la columna de la derecha.

English / Español

A

absolute value function (p. 46) A function that contains an absolute value of the independent variable, with parent function $f(x) = |x|$.

función valor absoluto (pág. 46) Función que contiene un valor absoluto de la variable independiente, cuya función generatriz es $f(x) = |x|$.

absolute value of a complex number (p. 569) A complex number's distance from zero in the complex plane.

valor absoluto de un número complejo (pág. 569) Distancia entre un número complejo y el cero sobre un plano complejo.

algebraic function (p. 158) A function with values that are obtained by adding, subtracting, multiplying, or dividing constants and the independent variable or raising the independent variable to a rational power.

función algebraica (pág. 158) Función cuyos valores se obtienen al sumar, restar, multiplicar o dividir constantes y la variable independiente o al elevar la variable independiente a una potencia racional.

alternative hypothesis (p. 705) One of two hypotheses that need to be stated to test a claim; states that there is a difference between the sample value and the population parameter. The alternative hypothesis contains a statement of inequality such as $>$, $\neq$, or $<$.

hipótesis alternativa (pág. 705) Una de las dos hipótesis que se requieren para probar una afirmación. Establece que hay una diferencia entre el valor de la muestra y el parámetro demográfico correspondiente. La hipótesis alternativa contiene un enunciado de desigualdad como $>$, $\neq$ o $<$.

ambiguous case (p. 292) Given the measures of two sides and a nonincluded angle, either no triangle exists, exactly one triangle exists, or two triangles exist.

caso ambiguo (pág. 292) Dadas las medidas de dos lados y de un ángulo distinto al incluido, se puede obtener: ningún triángulo, un solo triángulo o dos triángulos.

amplitude (p. 257) Half the distance between the maximum and minimum values of a sinusoidal function. For $y = a \sin(bx + c) + d$ and $y = a \cos(bx + c) + d$, amplitude $= |a|$.

amplitud (pág. 257) Equivale a la mitad de la distancia entre los valores máximo y mínimo de una función sinusoidal. Para $y = a \operatorname{sen}(bx + c) + d$ y $y = a \cos(bx + c) + d$, la amplitud $= |a|$.

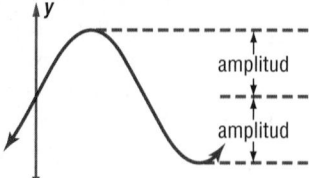

anchor step (p. 621) In mathematical induction, showing that something works for the first case, or that P_1 is true.

paso o caso base (pág. 621) Demostrar en una inducción matemática que una condición se cumple para el primer caso, es decir que P_1 es verdadera.

angle of depression (p. 224) The angle formed by a horizontal line and an observer's line of sight to an object below.

angle of elevation (p. 224) The angle formed by a horizontal line and an observer's line of sight to an object above.

angular speed (p. 236) The rate at which the object rotates about a fixed point.

antiderivative (p. 784) $F(x)$ is an antiderivative of $f(x)$ if $F'(x) = f(x)$.

arccosine function (p. 280) The inverse cosine function, written as $y = \cos^{-1} x$ or $y = \arccos x$, that has a domain of $[-1, 1]$ and a range of $[0, \pi]$.

arcsine function (p. 282) The inverse sine function, written as $y = \sin^{-1} x$ or $y = \arcsin x$, that has a domain of $[-1, 1]$ and a range of $\left[-\dfrac{\pi}{2}, \dfrac{\pi}{2}\right]$.

arctangent function (p. 283) The inverse tangent function, written as $y = \tan^{-1} x$ or $y = \arctan x$, that has a domain of $(-\infty, \infty)$ and a range of $\left(-\dfrac{\pi}{2}, \dfrac{\pi}{2}\right)$.

Argand Plane (p. 569) The complex plane.

argument (p. 570) The angle θ of a complex number written in the form $r(\cos \theta + i \sin \theta)$.

arithmetic means (p. 600) The terms between two nonconsecutive terms of an arithmetic sequence.

arithmetic series (p. 602) The sum of the terms of an arithmetic sequence.

arithmetic sequence (p. 599) A sequence in which the difference between successive terms is a constant.

asymptote (p. 130) A line or curve that a graph approaches.

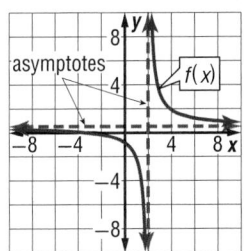

augmented matrix (p. 366) A matrix that contains the coefficients and constant terms of a system of linear equations, each written in standard form with the constant terms to the right of the equals sign.

average rate of change (p. 38) The slope of the line through any two points on the graph of a nonlinear function f.

ángulo de depresión (pág. 224) Ángulo formado por la horizontal y la línea visual dirigida hacia un objeto ubicado por debajo de la horizontal.

ángulo de elevación (pág. 224) Ángulo formado por la horizontal y la línea visual dirigida hacia un objeto ubicado por arriba de la horizontal.

rapidez angular (pág. 236) Tasa a la cual un objeto rota alrededor de un punto fijo.

antiderivada (pág. 784) $F(x)$ es la antiderivada de $f(x)$ si $F'(x) = f(x)$.

función arcocoseno (pág. 280) Inversa de la función coseno. Se escribe como $y = \cos^{-1} x$, o $y = \arccos x$. Su dominio es $[-1, 1]$ y su rango es $[0, \pi]$.

función arcoseno (pág. 282) Inversa de la función seno. Se escribe como $y = \operatorname{sen}^{-1} x$, o $y = \operatorname{arcsen} x$. Su dominio es $[-1, 1]$ y su rango es $\left[-\dfrac{\pi}{2}, \dfrac{\pi}{2}\right]$.

función arcotangente (pág. 283) Inversa de la función tangente. Se escribe como $y = \tan^{-1} x$, o $y = \arctan x$. Su dominio es $(-\infty, \infty)$ y su rango es $\left(-\dfrac{\pi}{2}, \dfrac{\pi}{2}\right)$.

plano de Argand (pág. 569) El plano complejo.

argumento (pág. 570) Ángulo θ de un número complejo expresado en la forma $r(\cos \theta + i \operatorname{sen} \theta)$.

medios aritméticos (pág. 600) Términos entre dos términos no consecutivos de una sucesión aritmética.

serie aritmética (pág. 602) Suma de los términos de una sucesión aritmética.

sucesión aritmética (pág. 599) Sucesión en la cual la diferencia entre términos sucesivos es una constante.

asíntota (pág. 130) Recta o curva a la cual se aproxima una gráfica.

matriz aumentada (pág. 366) Matriz que contiene los coeficientes y los términos constantes de un sistema de ecuaciones lineales, con las ecuaciones escritas en forma estándar y las constantes a la derecha del signo de igualdad.

tasa promedio de cambio (pág. 38) Pendiente de la recta entre dos puntos cualesquiera de la gráfica de una función no lineal f.

axis of symmetry (pp. P9, 422) A line about which a figure is symmetric. In a parabola, the axis of symmetry is perpendicular to the directrix and passes through the focus.

eje de simetría (págs. P9 y 422) Recta que divide una figura en partes simétricas. El eje de simetría de una parábola es perpendicular a la directriz y pasa a través del foco.

B

bimodal distribution (p. 656) A graph of a distribution of data that has two modes.

distribución bimodal (pág. 656) Gráfica de una distribución de datos que tiene dos modas.

binomial coefficients (p. 628) The coefficients of the terms of an expanded binomial $(a + b)^n$.

coeficientes binomiales (pág. 628) Coeficientes de los términos del binomio desarrollado $(a + b)^n$.

binomial distribution (p. 669) The distribution of the outcomes of a binomial experiment and their corresponding probabilities.

distribución binomial (pág. 669) Distribución de los resultados de un experimento binomial y sus probabilidades correspondientes.

binomial experiment (p. 668) A probability experiment in which there are a fixed number of independent trials, there are exactly two possible outcomes for each trial, and the probability of success is the same for each trial.

experimento binomial (pág. 668) Experimento probabilístico en el cual hay un número fijo de pruebas independientes, con dos posibles resultados para cada prueba y en el cual la probabilidad de éxito es igual en cada prueba.

binomial probability distribution function (p. 669) A discrete function of the random variable X, represented in the binomial probability formula.

función de distribución de la probabilidad binomial (pág. 669) Función discreta de la variable aleatoria X, representada por la fórmula de probabilidad binomial.

Binomial Theorem (p. 632) For any positive integer n, $(a + b)^n = {_nC_0}\,a^n b^0 + {_nC_1}\,a^{n-1} b^1 + {_nC_2}\,a^{n-2}\,b^2 + \ldots + {_nC_r}\,a^{n-r} b^r + \ldots + {_nC_n}\,a^0 b^n$, where $r = 0, 1, 2, \ldots, n$.

teorema del binomio (pág. 632) Para todo entero positivo $n, (a + b)^n = {_nC_0}\,a^n b^0 + {_nC_1}\,a^{n-1} b^1 + {_nC_2}\,a^{n-2}\,b^2 + \ldots + {_nC_r}\,a^{n-r} b^r + \ldots + {_nC_n}\,a^0 b^n$, donde $r = 0, 1, 2, \ldots, n$.

bivariate data (p. 713) Data with two variables.

datos bivariados (pág. 713) Datos con dos variables.

C

cardioid (p. 544) The graph of a polar equation of the form $r = a \pm a \cos \theta$ or $r = a \pm a \sin \theta$, where a is positive.

cardioide (pág. 544) Gráfica de una ecuación polar de la forma $r = a \pm a \cos \theta$ o $r = a \pm a$ sen θ, donde a es positiva.

center of an ellipse (p. 432) The midpoint of the major and minor axes of an ellipse.

centro de una elipse (pág. 432) Punto medio de los ejes mayor y menor de una elipse.

circular function (p. 248) A trigonometric function defined as a function of the real number system using the unit circle.

función circular (pág. 248) Función trigonométrica definida como una función del sistema de los números reales usando el círculo unitario.

class (p. P34) A data value or group of data values.

clase (pág. P34) Dato o grupo de datos.

class width (p. P34) The range of values for each class of data.

amplitud de una clase (pág. P34) Rango de valores de cada clase de datos.

clusters (p. 656) Subgroups of data.

conglomerados (pág. 656) Subconjuntos de datos.

coefficient matrix (p. 366) A matrix that contains only the coefficients of a system of linear equations.

matriz de coeficientes (pág. 366) Matriz que contiene solamente los coeficientes de un sistema de ecuaciones lineales.

column matrix (p. P23) A matrix that has only one column.

matriz columna (pág. P23) Matriz que sólo tiene una columna.

combination (p. P30) An arrangement of objects in which order is not important.

combinación (pág. P30) Arreglo de elementos en que el orden no es importante.

common difference (p. 599) The difference between successive terms of an arithmetic sequence.

common logarithm (p. 173) A logarithm with base 10, usually written log x.

common ratio (p. 608) The ratio of successive terms of a geometric sequence.

complement (p. P3) The complement of an event A consists of all the outcomes in the sample space that are not included as outcomes of event A.

completing the square (p. P12) A process used to make a quadratic expression into a perfect square trinomial.

complex conjugates (pp. P7, 124) Two complex numbers of the form $a + bi$ and $a - bi$, where $b \neq 0$

complex plane (p. 569) A plane used to graph complex numbers. The real component of a complex number is graphed on the horizontal and the imaginary component is graphed on the vertical axis.

complex number (p. P6) Any number that can be written in the form $a + bi$, where a and b are real numbers and i is the imaginary unit.

component form (p. 492) A vector represented by its rectangular components. If the initial point of a vector is $A(x_1, y_1)$ and the terminal point is at $B(x_2, y_2)$, the component form is given by $\langle x_2 - x_1, y_2 - y_1 \rangle$.

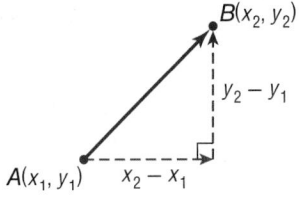

components (p. 487) Two or more vectors with a sum that is a given vector.

composition (p. 58) The combining of functions by using the result of one function to evaluate a second function. The composition of function f with function g is defined by $[f \circ g](x) = f[g(x)]$.

confidence interval (p. 697) A specific interval estimate of a parameter in an experiment that can be found when the maximum error of estimate is added to and subtracted from the sample mean.

confidence level (p. 696) The probability that the interval estimate will include the actual population parameter.

conic section (p. 422) A figure that is formed when a plane intersects a double-napped right cone, also called a *conic*.

conjugate axis (p. 442) The segment that is perpendicular to the transverse axis of a hyperbola, passes through the center, and has a length of $2b$ units.

diferencia común (pág. 599) Diferencia entre términos consecutivos de una sucesión aritmética.

logaritmo común (pág. 173) Logaritmo de base 10. Generalmente se escribe como log x.

razón común (pág. 608) Razón entre términos consecutivos de una sucesión geométrica.

complemento (pág. P3) El complemento de un evento A consiste en todos los resultados en el espacio muestral que no se incluyen como resultados del evento A.

completar el cuadrado (pág. P12) Proceso mediante el cual una expresión cuadrática se transforma en un trinomio cuadrado perfecto.

conjugados complejos (págs. P7 y 124) Dos números complejos de la forma $a + bi$ y $a - bi$, donde b no es 0.

plano complejo (pág. 569) Plano que se usa para graficar números complejos. El componente real del número complejo se grafica en el eje horizontal y el componente imaginario se grafica en el eje vertical.

número complejo (pág. P6) Cualquier número que puede escribirse de la forma $a + bi$, donde a y b son números reales e i es la unidad imaginaria.

forma componentes rectangulares (pág. 492) Vector representado por sus componentes rectangulares. Si el punto inicial de un vector es $A(x_1, y_1)$ y el punto final está en $B(x_2, y_2)$, la forma rectangular está dada por $\langle x_2 - x_1, y_2 - y_1 \rangle$.

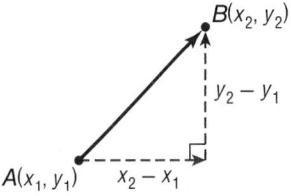

componentes (pág. 487) Dos o más vectores cuya suma es un vector dado.

composición (pág. 58) Combinación de funciones que se obtiene al usar el resultado de una función para evaluar una segunda función. La compuesta de la función f con la función g se define por $[f \circ g](x) = f[g(x)]$.

intervalo de confianza (pág. 697) Intervalo específico de la estimación de un parámetro resultado de un experimento, que se obtiene al sumar y restar de la media de la muestra al error máximo de la estimación.

nivel de confianza (pág. 696) Probabilidad de que el intervalo de la estimación incluya el parámetro.

sección cónica (pág. 422) Figura que se forma cuando un plano interseca un cono doble recto. También se conoce como *cónica*.

eje conjugado (pág. 442) Segmento de recta que es perpendicular al eje transversal de una hipérbola, la atraviesa por el centro y tiene una longitud de $2b$ unidades.

Conjugate Root Theorem (p. 124) When a polynomial equation in one variable has a zero of the form $a + bi$, where $b \neq 0$, then its complex conjugate, $a - bi$, is also a root.

consistent (p. P20) A system of equations that has at least one solution.

constant (p. 34) Describes a function f or an interval of a function in which for any two points, a positive change in x results in a zero change in $f(x)$.

constant function (p. 45) A function of the form $f(x) = c$, where c is any real number.

constraints (p. 405) Conditions given to variables in a two-dimensional linear programming problem, often expressed as a system of linear inequalities.

continuity correction factor (p. 689) A correction for continuity that must be used when approximating a binomial distribution.

continuous compound interest (p. 163) Interest that is reinvested continuously so that there is no waiting period between interest payments.

continuous function (p. 24) A function that can be graphed with no breaks, holes, or gaps.

continuous random variable (p. 664) A random variable that can take on an infinite number of possible values within a specified interval in a probability experiment.

converge (p. 592) If a sequence has a limit such that the terms approach a unique number, it is said to converge.

correlation (p. 713) An area of inferential statistics that involves determining whether a relationship exists between two variables.

correlation coefficient (p. 713) A measure that determines the type and strength of the linear relationship between the variables in bivariate data.

cosecant (p. 220) In a right triangle with acute angle θ, the ratio comparing the length of the hypotenuse to the side opposite of θ. It is the reciprocal of the sine ratio, or $\csc \theta = \dfrac{1}{\sin \theta}$.

cosine (p. 220) In a right triangle with acute angle θ, the ratio comparing the length of the side adjacent to θ and the hypotenuse.

cotangent (p. 220) In a right triangle with acute angle θ, the ratio comparing the length of the side adjacent to θ and the side opposite θ. It is the reciprocal of the tangent ratio, or $\cot \theta = \dfrac{1}{\tan \theta}$.

coterminal angles (p. 234) Angles in standard position that have the same initial and terminal sides, but different measures.

teorema sobre las raíces conjugadas complejas (pág. 124) Cuando una ecuación polinómica en una variable tiene un cero de la forma $a + bi$, donde $b \neq 0$, entonces su complejo conjugado, $a - bi$, también es una raíz del polinomio.

consistente (pág. P20) Sistema de ecuaciones que tiene por lo menos una solución.

constante (pág. 34) Describe una función f, o el intervalo de una función, en la cual un cambio positivo en x para dos puntos cualesquiera no produce ningún cambio en $f(x)$.

función constante (pág. 45) Función de la forma $f(x) = c$, donde c es cualquier número real.

restricciones (pág. 405) Condiciones aplicadas a las variables en un problema de programación lineal bidimensional. A menudo se expresan como un sistema de desigualdades lineales.

factor de corrección por continuidad (pág. 689) Corrección por continuidad que se aplica al aproximar una distribución binomial.

interés continuo compuesto (pág. 163) Interés que es reinvertido continuamente para evitar el período de espera entre los pagos de intereses.

función continua (pág. 24) Función que al ser graficada no presenta espacios o huecos.

variable aleatoria continua (pág. 664) Variable aleatoria que puede tomar un número infinito de valores posibles dentro de un intervalo dado, en un experimento probabilístico.

convergir (pág. 529) Si una sucesión tiene un límite tal que sus términos se aproximan a un valor único, se dice que converge.

correlación (pág. 713) Área de la inferencia estadística que se encarga de determinar si existe una relación entre dos variables.

coeficiente de correlación (pág. 713) Medida que determina el tipo y la magnitud de la relación lineal entre dos variables de datos bivariados.

cosecante (pág. 220) Razón que compara la longitud de la hipotenusa con la longitud del lado opuesto a θ, en un triángulo rectángulo con ángulo agudo θ. Es el recíproco de la razón seno, o $\csc \theta = \dfrac{1}{\operatorname{sen} \theta}$.

coseno (pág. 220) Razón que compara la longitud del lado adyacente a θ con la longitud de la hipotenusa, en un triángulo rectángulo con ángulo agudo θ.

cotangente (pág. 220) Razón que compara la longitud del lado adyacente a θ con el lado opuesto en un triángulo rectángulo, con un ángulo agudo θ. Es el recíproco de la razón tangente, $\cot \theta = \dfrac{1}{\tan \theta}$.

ángulos coterminales (pág. 234) Ángulos en posición normal que tienen el mismo lado inicial y terminal, pero distinta medida.

Glossary/Glosario

co-vertices (p. 432) The endpoints of the minor axis of an ellipse.

Cramer's Rule (p. 390) A method that uses determinants to solve square systems of linear equations.

critical values (p. 697) The z-values that correspond to a particular confidence level.

cross product (p. 519) If $\mathbf{a} = a_1\mathbf{i} + a_2\mathbf{j} + a_3\mathbf{k}$ and $\mathbf{b} = b_1\mathbf{i} + b_2\mathbf{j} + b_3\mathbf{k}$, the cross product of $\mathbf{a}$ and $\mathbf{b}$ is the vector $\mathbf{a} \times \mathbf{b} = (a_2b_3 - a_3b_2)\mathbf{i} - (a_1b_3 - a_3b_1)\mathbf{j} + (a_1b_2 - a_2b_1)\mathbf{k}$.

cubic function (p. 45) A function of the form $f(x) = ax^3 + bx^2 + cx + d$, where $a \neq 0$, with parent function $f(x) = x^3$.

cumulative frequency (p. P34) The sum of a frequency and all frequencies of previous classes.

cumulative relative frequency (p. P34) The ratio of the cumulative frequency of the class to all the data

covértices (pág. 432) Extremos del eje menor de una elipse.

regla de Cramer (pág. 390) Método en que se usan determinantes para resolver sistemas cuadrados de ecuaciones lineales.

valores críticos (pág. 697) Valores de z que corresponden a cierto nivel de confianza.

productos cruzados (pág. 519) Si $\mathbf{a} = a_1\mathbf{i} + a_2\mathbf{j} + a_3\mathbf{k}$ y $\mathbf{b} = b_1\mathbf{i} + b_2\mathbf{j} + b_3\mathbf{k}$, entonces los productos cruzados de $\mathbf{a}$ y $\mathbf{b}$ son el vector $\mathbf{a} \times \mathbf{b} = (a_2b_3 - a_3b_2)\mathbf{i} - (a_1b_3 - a_3b_1)\mathbf{j} + (a_1b_2 - a_2b_1)\mathbf{k}$.

función cúbica (pág. 45) Función de la forma $f(x) = ax^3 + bx^2 + cx + d$, donde $a \neq 0$, cuya función generatriz es $f(x) = x^3$.

frecuencia acumulada (pág. P34) Suma de una frecuencia y todas las frecuencias de clases anteriores.

frecuencia acumulada relativa (pág. P34) Razón de las frecuencias acumuladas de las clases a todos los datos.

D

damped harmonic motion (p. 276) The motion of an object whose amplitude decreases with time due to friction.

damped oscillation (p. 275) The reduction in amplitude of a sinusoidal wave of a damped trigonometric function.

damped trigonometric function (p. 275) The function formed when a sinusoidal function $y = \sin bx$ or $y = \cos bx$ is multiplied by another function $y = f(x)$. A function of the form $y = f(x) \sin bx$ or $y = f(x) \cos bx$.

damped wave (p. 275) A wave whose amplitude decreases, such as the graph of a damped trigonometric function.

damping factor (p. 275) In a damped trigonometric function of the form $y = f(x) \sin bx$ or $y = f(x) \cos bx$, $f(x)$ is the damping factor.

decreasing (p. 34) Describes a function f or an interval of a function in which for any two points, a positive change in x results in a negative change in $f(x)$.

definite integral (p. 777) An integral that has lower and upper bounds, given by $\displaystyle\int_a^b f(x)dx = \lim_{n\to\infty}\sum_{i=1}^{n} f(x_i)\,\Delta x$.

movimiento armónico amortiguado (pág. 276) Movimiento de un objeto cuya amplitud decrece con el tiempo debido a la fricción.

oscilación amortiguada (pág. 275) Reducción en amplitud de la onda sinusoidal de una función trigonométrica amortiguada.

función trigonométrica amortiguada (pág. 275) Función que se obtiene cuando una función sinusoidal $y = \operatorname{sen} bx$ o $y = \cos bx$ se multiplica por una función $y = f(x)$. Función de la forma $y = f(x) \operatorname{sen} bx$ o $y = f(x) \cos bx$.

onda amortiguada (pág. 275) Onda cuya amplitud disminuye; por ejemplo, la gráfica de una función trigonométrica amortiguada.

factor de amortiguamiento (pág. 275) En una función trigonométrica amortiguada de la forma $y = f(x) \operatorname{sen} bx$ o $y = f(x) \cos bx$, $f(x)$ es el factor de amortiguamiento.

decreciente (pág. 34) Describe una función f, o el intervalo de una función, en la cual un cambio positivo en x para dos puntos cualesquiera produce un cambio negativo en $f(x)$.

integral definida (pág. 777) Integral cuyos límites superior e inferior están dados por $\displaystyle\int_a^b f(x)dx = \lim_{n\to\infty}\sum_{i=1}^{n} f(x_i)\,\Delta x$.

degenerate conic (p. 422) A point, a line, or two intersecting lines that are formed when a plane intersects the vertex of a double-napped right cone.

degrees of freedom (d.f.) (p. 699) Represent the number of values that are free to vary after a sample statistic is determined, and are equal to $n - 1$ in a sample of n values.

dependent (p. P20) When a system of linear equations has an infinite number of solutions.

dependent events (p. P28) Two or more events in which the outcome of one event affects the outcome of the other events.

dependent variable (p. 7) In a function, the variable, usually y, that represents any value in the range.

depressed polynomial (p. 112) The quotient when a polynomial is divided by one of its binomial factors $x - c$.

derivative (p. 766) The derivative of the function $f(x)$ is the function $f'(x)$ given by $f'(x) = \lim_{h \to 0} \dfrac{f(x + h) - f(x)}{h}$.

Descartes' Rule of Signs (p. 123) A rule that gives information about the number of positive and negative real zeros of a polynomial function by looking at a polynomial's variations in sign.

determinant (p. 381) If $A = \begin{bmatrix} a & b \\ c & d \end{bmatrix}$, the determinant of A, written det(A) or $|A|$, is the difference of the product of the two diagonals of the matrix, or $ad - cb$. If $B = \begin{bmatrix} a & b & c \\ d & e & f \\ g & h & i \end{bmatrix}$, then $|B| = a\begin{vmatrix} e & f \\ h & i \end{vmatrix} - b\begin{vmatrix} d & f \\ g & i \end{vmatrix} + c\begin{vmatrix} d & e \\ g & h \end{vmatrix}$.

difference quotient (p. 758) Given a secant line through the points $(x, f(x))$ and $(x + h, f(x + h))$, the difference quotient is the slope of the line, $\dfrac{f(x + h) - f(x)}{h}$.

differential equation (p. 766) The result when finding the derivative of a function.

differential operator (p. 766) An operator such as $\dfrac{d}{dx}$, which specifies the action of taking the derivative of a function.

differentiation (p. 766) The process of finding the derivative of a function.

dilation (p. 49) A transformation in which the graph of a function is compressed or expanded vertically or horizontally.

dimensions (p. P23) A description of the number of rows and columns of a matrix.

cónica degenerada (pág. 422) Punto, recta o intersección de rectas que se forma cuando un plano interseca el vértice de un cono recto doble.

grados de libertad (gl) (pág. 699) Representa el número de valores que pueden variar libremente luego de definir una estadística muestral. Es igual a $n - 1$ en una muestra de n valores.

dependiente (pág. P20) Cuando un sistema de ecuaciones tiene un número infinito de soluciones.

eventos dependientes (pág. P28) Dos o más eventos en los cuales el resultado de un evento afecta el resultado de los demás eventos.

variable dependiente (pág. 7) Variable de una función que representa los valores en el rango; generalmente es y.

polinomio deprimido (pág. 112) Cociente que se obtiene cuando un polinomio se divide entre uno de sus factores binomiales $x - c$.

derivada (pág. 766) La derivada de una función $f(x)$ es la función $f'(x)$ dada por $f'(x) = \lim_{h \to 0} \dfrac{f(x + h) - f(x)}{h}$.

regla de los signos de Descartes (pág. 123) Regla que proporciona información sobre el número de ceros reales positivos y negativos de una función polinomial, mediante el examen de las variaciones de signo del polinomio.

determinante (pág. 381) Si $A = \begin{bmatrix} a & b \\ c & d \end{bmatrix}$, entonces el determinante de A, det(A) o $|A|$, es la diferencia del producto de las dos diagonales de la matriz, o $ad - cb$. Si $B = \begin{bmatrix} a & b & c \\ d & e & f \\ g & h & i \end{bmatrix}$, entonces $|B| = a\begin{vmatrix} e & f \\ h & i \end{vmatrix} - b\begin{vmatrix} d & f \\ g & i \end{vmatrix} + c\begin{vmatrix} d & e \\ g & h \end{vmatrix}$.

cociente de diferencias (pág. 758) Dada una secante que atraviesa los puntos $(x, f(x))$ y $(x + h, f(x + h))$, el cociente de diferencias es la pendiente de la recta, $\dfrac{f(x + h) - f(x)}{h}$.

ecuación diferencial (pág. 766) Resultado obtenido al calcular la derivada de una función.

operador diferencial (pág. 766) Operador como $\dfrac{d}{dx}$ que indica la acción de obtener la derivada de una función.

diferenciación (pág. 766) Proceso de cálculo de la derivada de una función.

homotecia (pág. 49) Transformación en la cual la gráfica de una función se comprime o expande vertical u horizontalmente.

dimensión (pág. P23) Una descripción del número de filas y de columnas de una matriz.

direction (p. 482) The directed angle between the vector and the horizontal line that could be used to represent the positive x-axis.

directrix (p. 422) A specific line from which all points on a parabola are equidistant.

direct substitution (p. 748) A method of evaluating the limit of a polynomial or rational function $f(x)$ as x approaches c by finding $f(c)$.

discontinuous function (p. 24) A function that is not continuous.

discrete random variable (p. 664) A random variable that can take on a finite number of possible values in a probability experiment.

diverge (p. 592) If a sequence does not have a limit, it is said to diverge.

dot product (p. 500) The dot product of vectors $\mathbf{a} = \langle a_1, a_2 \rangle$ and $\mathbf{b} = \langle b_1, b_2 \rangle$ is defined as $\mathbf{a} \cdot \mathbf{b} = a_1 b_1 + a_2 b_2$.

dirección (pág. 482) Ángulo dirigido entre el vector y la recta horizontal que se usa para representar el eje x positivo.

directriz (pág. 422) Recta específica a partir de la cual todos los puntos de una parábola son equidistantes.

sustitución directa (pág. 748) Método de evaluación del límite de un polinomio, o función racional $f(x)$, a medida que x se aproxima a c, mediante el cálculo de $f(c)$.

función discontinua (pág. 24) Función que no es continua.

variable aleatoria discreta (pág. 664) Variable aleatoria que puede tomar un número finito de valores posibles en un experimento probabilístico.

divergir (pág. 592) Si una sucesión no tiene límite, se dice que diverge.

producto punto o producto vector (pág. 500) Producto punto de los vectores $\mathbf{a} = \langle a_1, a_2 \rangle$ y $\mathbf{b} = \langle b_1, b_2 \rangle$ se define como $\mathbf{a} \cdot \mathbf{b} = a_1 b_1 + a_2 b_2$.

E

eccentricity (p. 435) A measure that determines how "circular" or "stretched" an ellipse will be. For any ellipse, $\frac{(x - h)^2}{a^2} + \frac{(y - k)^2}{b^2} = 1$ or $\frac{(x - h)^2}{b^2} + \frac{(y - k)^2}{a^2} = 1$, where $c^2 = a^2 - b^2$, the eccentricity $e = \frac{c}{a}$.

element (p. P3, P23) **1.** Each object or number in a set. **2.** Each entry in a matrix.

elementary row operations (p. 366) The operations shown below are used to transform an augmented matrix into an equivalent matrix.
• Interchange any two rows.
• Multiply one row by a nonzero real number.
• Add a multiple of one row to another row.

elimination method (p. P19) Eliminate one of the variables in a system of equations by adding or subtracting the equations.

ellipse (p. 432) The locus of points in a plane such that the sum of the distances from two fixed points, called foci, is constant.

excentricidad (pág. 435) Medida que determina el alargamiento o achatamiento de una elipse. Para toda elipse, $\frac{(x - h)^2}{a^2} + \frac{(y - k)^2}{b^2} = 1$ ó $\frac{(x - h)^2}{b^2} + \frac{(y - k)^2}{a^2} = 1$, donde $c^2 = a^2 - b^2$, la excentricidad $e = \frac{c}{a}$.

elemento (págs. P3 y P23) **1.** Cada objeto o número de un conjunto. **2.** Cada entrada de una matriz.

operaciones elementales de fila (pág. 366) Las siguientes operaciones se utilizan para transformar una matriz aumentada en una matriz equivalente.
• Intercambia dos filas cualesquiera.
• Multiplica una de las filas por un número real distinto de cero.
• Suma el múltiplo de una fila a otra fila.

método de eliminación (pág. P19) Eliminar una de las variables en un sistema de ecuaciones sumando o restando las ecuaciones.

elipse (pág. 432) Locus de los puntos en un plano tal que la suma de las distancias desde dos puntos fijos, llamados focos, es constante.

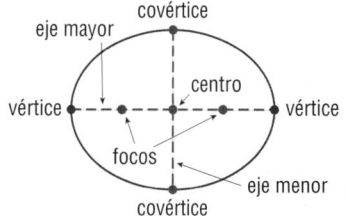

empirical rule (p. 675) Describes areas under the normal curve over intervals that are one, two, and three standard deviations from either side of the mean. About 68% of the values are within one standard deviation of the mean, 95% are within two standard deviations from the mean, and 99.7% are within three standard deviations of the mean.

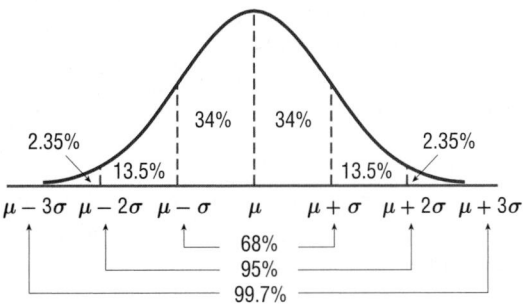

empty set (p. P4) A set with no elements, symbolized by { } or ø.

end behavior (p. 28) Describes what happens to the value of $f(x)$ as x increases or decreases without bound.

equal matrices (p. P23) Two matrices that have the same dimensions and each element of one matrix is equal to the corresponding element of the other matrix.

equivalent vectors (p. 483) Vectors that have the same magnitude and direction.

Euler's Formula (p. 640) For any real number θ, $e^{i\theta} = \cos\theta + i\sin\theta$.

even function (p. 18) A function that is symmetric with respect to the y-axis.

expected value (p. 667) The mean of the random variable in a probability distribution.

experiment (p. P28) A situation involving chance or probability that leads to specific outcomes.

explanatory variable (p. 713) The independent variable x in bivariate data.

exponential function (p. 158) A function of the form $f(x) = ab^x$, where x is any real number and a and b are real number constants such that $a \neq 0$, b is positive, and $b \neq 1$.

exponential series (p. 638) The power series that approximates e^x as $e^x = 1 + x + \dfrac{x^2}{2!} + \dfrac{x^3}{3!} + \dfrac{x^4}{4!} + \ldots$.

extended principle of mathematical induction (p. 624) Instead of verifying that P_n is true for $n = 1$, as in the principle of mathematical induction, instead verify that P_n is true for the first possible case.

regla empírica (pág. 675) Describe áreas bajo la curva normal que están a una, dos y tres desviaciones estándar hacia ambos lados de la media. Cerca del 68% de los valores se hallan dentro del intervalo de una desviación estándar de la media; cerca del 95 por ciento se encuentran dentro del intervalo de dos desviaciones estándar y alrededor del 99.7% de los datos están dentro de un intervalo de tres desviaciones estándar.

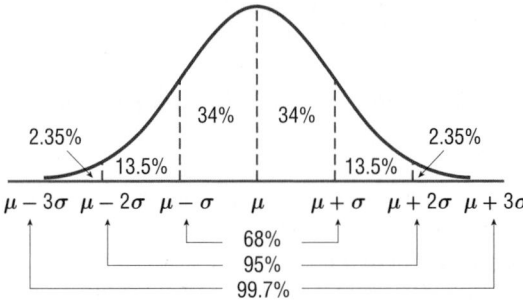

conjunto vacío (pág. P4) Conjunto sin elementos, denotado por { } o ø.

comportamiento extremo (pág. 28) Describe lo que le sucede al valor de $f(x)$ a medida que x aumenta o disminuye sin cotas.

matrices iguales (pág. P23) Dos matrices que tienen las mismas dimensiones y en las cuales cada elemento de una de ellas es igual al elemento correspondiente de la otra matriz.

vectores equivalentes (pág. 483) Vectores que tienen la misma magnitud y dirección.

fórmula de Euler (pág. 640) Para todo número real θ, $e^{i\theta} = \cos\theta + i\,\text{sen}\,\theta$.

función par (pág. 18) Función que es simétrica respecto al eje y.

valor esperado (pág. 667) Media de la variable aleatoria en una distribución de probabilidad.

experimento (pág. P28) Situación aleatoria o probabilística que conduce a resultados específicos.

variable explicatoria (pág. 713) Variable independiente x de datos bivariados.

función exponencial (pág. 158) Función de la forma $f(x) = ab^x$, donde x es cualquier número real y a y b son constantes reales tales que $a \neq 0$, b es positiva y $b \neq 1$.

serie exponencial (pág. 638) Serie de potencias que aproxima el valor de e^x como $e^x = 1 + x + \dfrac{x^2}{2!} + \dfrac{x^3}{3!} + \dfrac{x^4}{4!} + \ldots$.

principio extendido de inducción matemática (pág. 624) En lugar de verificar que P_n se cumple para $n = 1$, como se hace al aplicar el principio de inducción matemática, se verifica que P_n se cumple para el primer caso posible.

extraneous solution (p. 91) A solution that does not satisfy the original equation.

extrapolation (p. 719) To use the equation of the least-squares regression line to make predictions far outside the range of the x-values that were used to obtain the regression line.

extrema (p. 36) The maximum and minimum values of a function.

solución extraña (pág. 91) Solución que no satisface la ecuación original.

extrapolación (pág. 719) Uso de la ecuación de regresión lineal obtenida por mínimos cuadrados para predecir valores más allá del rango de los valores de x que se usaron para obtener la recta de regresión.

extremos (pág. 36) Valores máximo y mínimo de una función.

F

factorial (p. P28) If n is a positive integer, then $n! = n(n-1)(n-2) \cdot \ldots \cdot 2 \cdot 1$.

feasible solutions (p. 405) The set of possible solutions of a system of inequalities in a linear programming problem, which are points of the form (x, y).

Fibonacci sequence (p. 591) A sequence in which the first two terms are 1 and each of the additional terms is the sum of the two previous terms.

finite sequence (p. 590) A sequence that has a finite number of terms.

finite series (p. 593) The sum of the first n terms of a finite or infinite sequence.

first differences (p. 601) Values obtained by subtracting each term in a sequence from its successive term.

five-number summary (p. P36) A statistic that includes the minimum value, lower quartile, median, upper quartile, and the maximum value of a data set.

foci (pp. 432, 442) Two fixed points used to define an ellipse or hyperbola. *See ellipse and hyperbola.*

focus (pp. 422, 432, 442) See *parabola, ellipse, hyperbola.*

frequency (p. 260) For a sinusoidal function, the number of cycles the function completes in a one unit interval. The frequency is the reciprocal of the period. For $y = a \sin(bx + c) + d$ and $y = a \cos(bx + c) + d$, frequency $= \dfrac{1}{\text{period}}$ or $\dfrac{|b|}{2\pi}$.

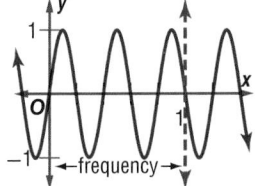

frequency distribution (p. P34) A table used to organize data by groups, classes or intervals.

function (p. 5) A relation that assigns to each element in the domain exactly one element in the range.

factorial (pág. P28) Si n es un entero positivo, entonces $n! = n(n-1)(n-2) \cdot \ldots \cdot 2 \cdot 1$.

soluciones factibles (pág. 405) Conjunto de soluciones posibles a un sistema de desigualdades en un problema de programación lineal. Son puntos de la forma (x, y).

sucesión de Fibonacci (pág. 591) Sucesión en la cual los dos primeros términos son 1 y cada uno de los siguientes términos es la suma de los dos términos anteriores.

sucesión finita (pág. 590) Sucesión que tiene un número finito de términos.

serie finita (pág. 593) Suma de los primeros n términos de una sucesión finita o infinita.

primeras diferencias (pág. 601) Valores que se obtienen restando cada término en una sucesión de su término sucesivo.

resumen de cinco números (pág. P36) Estadística que incluye el valor mínimo, el cuartil inferior, la mediana, el cuartil superior y el valor máximo de un conjunto de datos.

focos (págs. 432 y 442) Dos puntos fijos que se usan para definir una elipse o una hipérbola. *Ver elipse e hipérbola.*

foco (págs. 422, 432 y 442) Ver: *parábola, elipse* e *hipérbola.*

frecuencia (pág. 260) Para una función sinusoidal, el número de ciclos que completa la función en un intervalo de una unidad. La frecuencia es el recíproco del período. Para $y = a \operatorname{sen}(bx + c) + d$ y $y = a \cos(bx + c) + d$, frecuencia $= \dfrac{1}{\text{período}}$ ó $\dfrac{|b|}{2\pi}$.

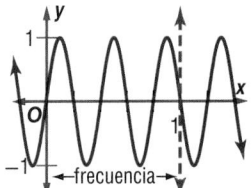

distribución de frecuencias (pág. P34) Tabla que se usa para organizar datos en grupos, clases o intervalos.

función (pág. 5) Relación que asigna a cada elemento del dominio uno y sólo un elemento del rango.

function notation (p. 7) An equation of y in terms of x can be rewritten so that $y = f(x)$. For example, $y = 4x$ can be written as $f(x) = 4x$.

Fundamental Theorem of Algebra (p. 123) A polynomial function of degree n, where $n > 0$, has at least one zero (real or imaginary) in the complex number system.

Fundamental Theorem of Calculus (p. 786) If f is continuous on $[a, b]$ and $F(x)$ is any antiderivative of $f(x)$, then $\int_a^b f(x)\, dx = F(b) - F(a)$.

notación funcional (pág. 7) Una ecuación de y en términos de x se puede replantear de manera que $y = f(x)$. Por ejemplo, $y = 4x$ se puede escribir como $f(x) = 4x$.

teorema fundamental del álgebra (pág. 123) Función polinomial de grado n, donde $n > 0$, tiene por lo menos un cero (real o imaginario) en el sistema de números complejos.

teorema fundamental del cálculo (pág. 786) Si f es continua en $[a, b]$ y $F(x)$ es cualquier antiderivada de $f(x)$, entonces, $\int_a^b f(x)\, dx = F(b) - F(a)$.

G

Gaussian elimination (p. 364) Using the operations below to transform a system of linear equations into an equivalent system.
- Interchange any two equations.
- Multiply one of the equations by a nonzero real number.
- Add a multiple of one equation to another equation.

Gauss-Jordan Elimination (p. 369) Solving a system of linear equations by transforming an augmented matrix so that it is in reduced row-echelon form.

geometric means (p. 611) The terms between two nonconsecutive terms of a geometric sequence.

geometric sequence (p. 608) A sequence in which the ratio between successive terms is a constant.

geometric series (p. 611) The sum of the terms of a geometric sequence.

greatest integer function (p. 46) Has the parent function $f(x) = [\![x]\!]$, which is defined as the greatest integer less than or equal to x.

eliminación de Gauss (pág. 364) Uso de las siguientes operaciones para transformar un sistema de ecuaciones lineales en un sistema equivalente.
- Intercambia dos ecuaciones cualesquiera.
- Multiplica una de las ecuaciones por un número real distinto de cero.
- Suma un múltiplo de una ecuación a la otra ecuación.

eliminación de Gauss-Jordan (pág. 369) Solución de un sistema de ecuaciones lineales mediante la transformación de una matriz aumentada, de modo que quede en forma escalón reducida.

medios geométricos (pág. 611) Términos entre dos términos no consecutivos de una sucesión geométrica.

sucesión geométrica (pág. 608) Sucesión en la cual la razón entre términos consecutivos es una constante.

serie geométrica (pág. 611) Suma de los términos de una sucesión geométrica.

función parte entera o mayor entero (pág. 46) Su función principal es $f(x) = [\![x]\!]$, que se define como el mayor entero menor que o igual a x.

H

Heron's Formula (p. 296) If $\triangle ABC$ has side lengths a, b, and c, then the area of the triangle is $\sqrt{s(s-a)(s-b)(s-c)}$, where $s = \frac{1}{2}(a + b + c)$.

holes (p. 135) Removable discontinuities on the graph of a function that occur when the numerator and denominator of the function have common factors. The holes occur at the zeros of the common factors.

horizontal asymptote (p. 131) The line $y = c$ is a horizontal asymptote of the graph of f if $\lim_{x \to -\infty} f(x) = c$ or $\lim_{x \to \infty} f(x) = c$.

fórmula de Herón (pág. 296) Si la longitud de los lados de $\triangle ABC$ es a, b y c, entonces el área del triángulo es igual a $\sqrt{s(s-a)(s-b)(s-c)}$, donde $s = \frac{1}{2}(a + b + c)$.

huecos (pág. 135) Discontinuidades evitables en la gráfica de una función que suceden cuando el numerador y el denominador de la función tienen factores comunes. Los huecos se presentan en los ceros de los factores comunes.

asíntota horizontal (pág. 131) La recta $y = c$ es la asíntota horizontal de la gráfica de f, si $\lim_{x \to -\infty} f(x) = c$ o $\lim_{x \to \infty} f(x) = c$.

Glossary/Glosario

hyperbola (p. 442) The set of all points in a plane such that the absolute value of the differences of the distances from two foci is constant.

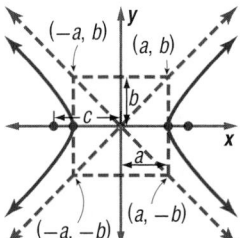

hipérbola (pág. 442) Conjunto de todos los puntos en un plano tales que el valor absoluto de la diferencia de las distancias desde dos focos es constante.

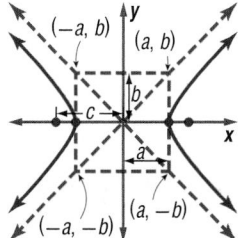

hypothesis test (p. 705) Assesses evidence provided by data about a claim concerning a population parameter.

prueba de hipótesis (pág. 705) Evalúa la evidencia proporcionada por los datos acerca de una conjetura relacionada con un parámetro demográfico.

I

---

identity (p. 312) An equation in which the left side is equal to the right side for all values of the variable for which both sides are defined.

identidad (pág. 312) Ecuación en la cual su lado izquierdo es igual a su lado derecho para todos los valores de la variable para los cuales ambos lados están definidos.

identity function (p. 45) The function $f(x) = x$, which passes through all points with coordinates (a, a).

función identidad (pág. 45) Función $f(x) = x$, la cual pasa a través de todos los puntos con coordenadas (a, a).

identity matrix (p. 378) The identity matrix of order n, I_n, is an $n \times n$ matrix consisting of all 1s on its main diagonal, from upper left to lower right, and 0s for all other elements. If A is an $n \times n$ matrix, then $AI_n = I_nA = A$.

matriz identidad (pág. 378) La matriz identidad de orden n, I_n es una matriz $n \times n$ que contiene solamente números 1 en su diagonal principal, desde la esquina superior izquierda a la esquina inferior derecha y ceros en los otros elementos. Si A es una matriz $n \times n$, entonces $AI_n = I_nA = A$.

imaginary axis (p. 569) The vertical axis of a complex plane on which the imaginary component of a complex number is graphed.

eje imaginario (pág. 569) Eje vertical de un plano complejo, en el cual se grafica el componente imaginario de un número complejo.

imaginary number (p. P6) Another name for a complex number of the form $a + bi$, when $b \neq 0$.

número imaginario (pág. P6) Otro nombre para un número complejo de la forma $a + bi$, cuando $b \neq 0$.

imaginary part (p. P6) In an imaginary number $a + bi$, b is the imaginary part.

parte imaginaria (pág. P6) En un número imaginario $a + bi$, b es la parte imaginaria.

imaginary unit (p. P6) i, or the principal square root of -1.

unidad imaginaria (pág. P6) i, o la raíz cuadrada principal de -1.

implied domain (p. 7) In a function with an unspecified domain, the set of all real numbers for which the expression used to define the function is real.

dominio implícito (pág. 7) En una función de dominio no especificado, es el conjunto de todos los números reales para los que la expresión usada para definir la función es real.

inconsistent (p. P20) A system of equations that has no solutions.

inconsistente (pág. P20) Sistema de ecuaciones que no tiene solución alguna.

increasing (p. 34) Describes a function f or an interval of a function in which for any two points, a positive change in x results in a positive change in $f(x)$.

creciente (pág. 34) Describe una función f, o el intervalo de una función, en la cual un cambio positivo en x para dos puntos cualesquiera produce un cambio positivo en $f(x)$.

indefinite integral (p. 785) The indefinite integral of $f(x)$ is defined by $\int f(x)dx = F(x) + C$, where $F(x)$ is an antiderivative of $f(x)$ and C is any constant.

integral indefinida (pág. 785) La integral indefinida de $f(x)$ está definida por $\int f(x)dx = F(x) + C$, donde $F(x)$ es la antiderivada de $f(x)$ y C es una constante cualquiera.

independent (p. P20) When a system of linear equations has exactly one solution.

independent events (p. P28) Events that do not affect each other.

independent variable (p. 7) In a function, the variable, usually x, that represents any value in the domain.

indeterminate form (p. 749) An expression obtained when evaluating a limit that does not give enough information to determine the original limit.

inductive hypothesis (p. 621) In mathematical induction, assuming that something works for any particular case, or that assuming that P_k is true.

inductive step (p. 621) In mathematical induction, showing that something works for the case after P_k, or showing that P_{k+1} is true.

inferential statistics (p. 696) A sample of data is analyzed and conclusions are made about the entire population.

infinite discontinuity (p. 24) A characteristic of a function in which the absolute value of the function increases or decreases indefinitely as x-values approach c from the left and right.

infinite sequence (p. 590) A sequence that has infinitely many terms.

infinite series (p. 593) The sum of the terms of an infinite sequence.

influential (p. 718) An individual data point that substantially changes a regression line.

initial point (p. 482) The starting point of a vector that is represented by a directed line segment. Also known as the *tail* of the vector.

initial side (p. 231) The starting position of a ray when forming an angle.

instantaneous rate of change (p. 758) For the graph of $f(x)$, the slope m of the line tangent at the point $(x, f(x))$ given by $m = \lim\limits_{h \to 0} \dfrac{f(x+h) - f(x)}{h}$, provided the limit exists.

instantaneous velocity (p. 760) The velocity achieved at a specific point in time.

integration (p. 777) The process of evaluating an integral.

interpolation (p. 719) To use the equation of the least-squares regression line to make predictions over the range of the data.

interquartile range (p. P36) The range of the middle half of a set of data. It is the difference between the upper quartile and the lower quartile.

independiente (pág. P20) Cuando un sistema de ecuaciones lineales tiene exactamente una solución.

eventos independientes (pág. P28) Eventos que no se afectan mutuamente.

variable independiente (pág. 7) Variable de una función que representa cualquier valor del dominio, generalmente es x.

forma indeterminada (pág. 749) Expresión que se obtiene cuando se evalúa un límite que no provee suficiente información para determinar el límite original.

hipótesis inductiva (pág. 621) En el proceso de inducción matemática, suponer que algo se cumple para cualquier caso particular o suponer que se cumple P_k.

paso inductivo (pág. 621) En el proceso de inducción matemática, se trata de demostrar que algo se cumple para el caso que sigue a P_k o demostrar que P_{k+1} es verdadera.

inferencia estadística (pág. 696) Análisis y obtención de conclusiones sobre una población completa a partir de una muestra de datos.

discontinuidad infinita (pág. 24) Característica de una función en la cual el valor absoluto de la función crece o disminuye indefinidamente a medida que los valores de x se aproximan a c por la derecha o por la izquierda.

sucesión infinita (pág. 590) Sucesión con un número infinito de términos.

serie infinita (pág. 593) Suma de los términos de una sucesión infinita.

dato influyente (pág. 718) Dato individual que afecta significativamente una recta de regresión.

punto inicial (pág. 482) Punto donde comienza un vector, representado por un segmento de recta dirigido. Algunas veces se le llama la *cola* del vector.

lado inicial (pág. 231) Posición inicial de un rayo cuando forma un ángulo.

tasa de cambio instantánea (pág. 758) En la gráfica de $f(x)$, la pendiente m de la recta tangente al punto $(x, f(x))$ dada por $m = \lim\limits_{h \to 0} \dfrac{f(x+h) - f(x)}{h}$, dado que existe el límite.

velocidad instantánea (pág. 760) Velocidad alcanzada en un punto determinado del tiempo.

integración (pág. 777) Proceso de evaluación de una integral.

interpolación (pág. 719) Uso de la ecuación de regresión lineal obtenida por mínimos cuadrados para predecir valores dentro del rango de los datos.

rango intercuartílico (pág. P36) Rango de la mitad central de un conjunto de datos. Es la diferencia entre el cuartil superior y el inferior.

intersection (p. P4) The intersection of sets A and B is all elements found in both A and B, written as $A \cap B$.

interval (pág. P34) A data value or group of data values.

interval estimate (p. 696) A range of values used to estimate an unknown population parameter.

interval notation (p. 5) An expression that uses inequalities to describe subsets of real numbers.

inverse (p. 379) Let A be an $n \times n$ matrix. If there exists a matrix B such that $AB = BA = I_n$, then B is called the inverse of A and is written as A^{-1}.

inverse cosine (p. 223) If θ is an acute angle and $\cos \theta = x$, then the inverse cosine of x, or $\cos^{-1} x$, is the measure of angle θ.

inverse function (p. 65) Two functions f and f^{-1} are inverse functions if and only if $f[f^{-1}(x)] = x$ for every x in the domain of $f^{-1}(x)$, and $f^{-1}[f(x)] = x$ for every x in the domain of $f(x)$.

inverse matrix (p. 379) The multiplicative inverse of a square matrix. The product of a matrix A and its inverse A^{-1} must equal the identity matrix I_n.

inverse relation (p. 65) Two relations are inverse relations if and only if one relation contains the element (b, a) whenever the other relation contains the element (a, b).

inverse sine (p. 223) If θ is an acute angle and $\sin \theta = x$, then the inverse sine of x, or $\sin^{-1} x$, is the measure of angle θ.

inverse tangent (p. 223) If θ is an acute angle and $\tan \theta = x$, then the inverse tangent of x, or $\tan^{-1} x$, is the measure of angle θ.

inverse trigonometric function (p. 223) The inverse sine of x or $\sin^{-1} x$, the inverse cosine of x or $\cos^{-1} x$, and the inverse tangent of x or $\tan^{-1} x$.

invertible matrix (p. 379) A matrix that has an inverse.

irreducible over the reals (p. 124) A quadratic expression that has real coefficients but no real zeros associated with it.

intersección (pág. P4) La intersección de los conjuntos A y B son todos los elementos que se encuentran tanto en A como en B; se escribe como $A \cap B$.

intervalo (pág. P34) Dato o grupo de datos.

intervalo de la estimación (pág. 696) Rango de valores que sirven para estimar un parámetro demográfico desconocido.

notación de intervalo (pág. 5) Expresión que utiliza desigualdades para describir subconjuntos de números reales.

inversa (pág. 379) Sea A la matriz $n \times n$. Si existe una matriz B tal que $AB = BA = I_n$, entonces B es la inversa de A y se escribe como A^{-1}.

coseno inverso (pág. 223) Si θ es un ángulo agudo y $\cos \theta = x$, entonces el coseno inverso de x, o $\cos^{-1} x$, es la medida del ángulo θ.

función inversa (pág. 65) Dos funciones f y f^{-1} son inversas si y sólo si $f[f^{-1}(x)] = x$ para toda x en el dominio de $f^{-1}(x)$, y $f^{-1}[f(x)] = x$ para toda x en el dominio de $f(x)$.

inversa de la matriz (pág. 379) El inverso multiplicativo de una matriz cuadrada. El producto de una matriz A por su inversa A^{-1} es igual a la matriz identidad I_n.

relación inversa (pág. 65) Dos relaciones son inversas si y sólo si una de las relaciones contiene el elemento (b, a) siempre que la otra contenga el elemento (a, b).

seno inverso (pág. 223) Si θ es un ángulo agudo y $\operatorname{sen} \theta = x$, entonces el seno inverso de x, o $\operatorname{sen}^{-1} x$, es la medida del ángulo θ.

tangente inversa (pág. 223) Si θ es un ángulo agudo y $\tan \theta = x$, entonces la tangente inversa de x, o $\tan^{-1} x$, es la medida del ángulo θ.

función trigonométrica inversa (pág. 223) El seno inverso de x, o $\operatorname{sen}^{-1} x$; el coseno inverso de x, o $\cos^{-1} x$; y la tangente inversa de x, o $\tan^{-1} x$.

matriz invertible (pág. 379) Matriz que tiene inversa.

irreducible sobre los reales (pág. 124) Expresión cuadrática que tiene coeficientes reales, pero que no tiene ceros reales asociados con ella.

J

jump discontinuity (p. 24) A characteristic of a function in which the function has two distinct limit values as x-values approach c from the left and right.

discontinuidad de salto (pág. 24) Característica de una función que tiene dos valores de límite distintos, dependiendo de si los valores de x se aproximan a c por la derecha o por la izquierda.

latus rectum (p. 430) The line segment that passes through the focus of a parabola, is perpendicular to the axis of symmetry, and has endpoints on the parabola.

Law of Cosines (p. 295) If $\triangle ABC$ has side lengths a, b, and c representing the lengths of the sides opposite the angles with measures A, B, and C, respectively. Then the following are true.
$$a^2 = b^2 + c^2 - 2bc \cos A$$
$$b^2 = a^2 + c^2 - 2ac \cos B$$
$$c^2 = a^2 + b^2 - 2ab \cos C$$

Law of Sines (p. 291) If $\triangle ABC$ has side lengths a, b, and c representing the lengths of the sides opposite the angles with measures A, B, and C, respectively, then
$$\frac{\sin A}{a} = \frac{\sin B}{b} = \frac{\sin C}{c}.$$

leading coefficient (p. 97) In a polynomial function, the coefficient of the variable with the greatest exponent.

leading term test (p. 98) Uses the power and coefficient of the leading term of a polynomial to determine the end behavior of a polynomial function.

least-squares regression line (p. 716) The line for which the sum of the squares of the residuals is at a minimum.

left-tailed test (p. 706) The hypothesis test if $H_a: \mu < k$.

level of significance (p. 706) The maximum allowable probability of committing a Type I error, denoted α.

lemniscates (p. 546) The graph of a polar equation of the form $r^2 = a^2 \cos 2\theta$ or $r^2 = a^2 \sin 2\theta$.

limaçon (p. 543) The graph of a polar equation of the form $r = a \pm b \cos \theta$ or $r = a \pm b \sin \theta$, where a and b are both positive.

limit (p. 24) The unique value that a function approaches as x-values of the function approach c from the left and right sides.

linear combination (p. 495) The sum of two vectors, each multiplied by a scalar, that is used to represent a vector with a given initial point and terminal point.

Linear Factorization Theorem (p. 124) If $f(x)$ is a polynomial function of degree $n > 0$, then f has exactly n linear factors and $f(x) = a_n(x - c_1)(x - c_2) \ldots (x - c_n)$, where a_n is some nonzero real number and $c_1, c_2, \ldots, c_n$ are the complex zeros (including repeated zeros) of f.

linearize (p. 204) Transform data so that they appear to cluster about a line by applying a function to one or both of the variables in the data set.

latus rectum o lado recto (pág. 430) Segmento de recta que atraviesa el foco de una parábola, es perpendicular al eje de simetría y tiene sus extremos en la parábola.

ley de los cosenos (pág. 295) Si un $\triangle ABC$ tiene lados de longitud a, b y c, que representan la longitud de los lados opuestos a los ángulos de medida A, B y C, respectivamente, entonces los siguientes enunciados son verdaderos.
$$a^2 = b^2 + c^2 - 2bc \cos A$$
$$b^2 = a^2 + c^2 - 2ac \cos B$$
$$c^2 = a^2 + b^2 - 2ab \cos C$$

ley de los senos (pág. 291) Si un $\triangle ABC$ tiene lados con longitud a, b y c, que representan la longitud de los lados opuestos a los ángulos de medida A, B y C, respectivamente, entonces $\frac{\operatorname{sen} A}{a} = \frac{\operatorname{sen} B}{b} = \frac{\operatorname{sen} C}{c}$.

coeficiente principal (pág. 97) El coeficiente de la variable con el mayor exponente en una función polinomial.

prueba del término principal (pág. 98) Uso de la potencia y el coeficiente del término principal de un polinomio para determinar el comportamiento extremo de una función polinomial.

recta de regresión de mínimos cuadrados (pág. 716) Recta para la cual la suma de los cuadrados de los residuales es mínima.

prueba de cola izquierda (pág. 706) Prueba de hipótesis si $H_a: \mu < k$.

nivel de significancia (pág. 706) Máxima probabilidad permisible de cometer un error Tipo I; es llamado error α.

lemniscata (pág. 546) Gráfica de una ecuación polar de la forma $r^2 = a^2 \cos 2\theta$ o $r^2 = a^2 \operatorname{sen} 2\theta$.

caracol (pág. 543) Gráfica de una ecuación polar de la forma $r = a \pm b \cos \theta$ o $r = a \pm b \operatorname{sen} \theta$, donde a y b son positivos.

límite (pág. 24) Valor único al que una función se aproxima a medida que el valor de x se aproxima a c por el lado derecho y por el izquierdo.

combinación lineal (pág. 495) Suma de dos vectores, ambos multiplicados por un escalar, que sirve para representar un vector con los puntos inicial y final dados.

teorema de la factorización lineal (pág. 124) Si $f(x)$ es una función polinomial de grado $n > 0$, entonces f tiene exactamente n factores lineales y $f(x) = a_n(x - c_1)(x - c_2)$ $\ldots (x - c_n)$, donde a_n es un número real distinto de cero y $c_1, c_2, \ldots, c_n$ son los ceros complejos de f (incluyendo los ceros repetidos).

linealizar (pág. 204) Transformación de datos para que se distribuyan alrededor de una recta, aplicando una función a una o ambas variables de un conjunto de datos.

Glossary/Glosario

linear programming (p. 405) The process of finding a minimum or maximum value of a linear function for a region defined by linear inequalities.

linear speed (p. 236) The rate at which an object moves along a circular path.

line of best fit (p. 716) A line drawn through a set of data points that describes how the response variable y changes as the explanatory variable x changes. Also called a *regression line*.

line symmetry (p. 16) Describes graphs that can be folded along a line so that the two halves match exactly.

locus (p. 422) A set of all points that fulfill a geometric property.

logarithm (p. 172) In the function $x = b^y$, y is called the logarithm, base b, of x. Usually written as $y = \log_b x$ and is read *log base b of x*.

logarithmic function with base b (p. 172) A function of the form $y = \log_b x$, where $b > 0$, $b \neq 1$, and $x > 0$, which is the inverse of the exponential function of the form $b^y = x$.

logistic growth function (p. 202) A function that models exponential growth with limiting factors. Logistic growth functions are bounded by horizontal asymptotes $y = 0$ and $y = c$, where c is the limit to growth.

lower bound (p. 121) A real number a that is less than or equal to the least real zero of a polynomial function.

lower limit (p. 777) The lower bound of a definite integral.

programación lineal (pág. 405) Proceso de cálculo de un valor máximo o mínimo de una función lineal, para una región definida por desigualdades lineales.

rapidez lineal (pág. 236) Tasa a la cual un cuerpo recorre un trayecto circular.

recta de mejor ajuste (pág. 716) Recta trazada a través de un conjunto de puntos que describe la manera en que cambia la variable de respuesta y a medida que cambia la variable explicativa x. También se llama *recta de regresión*.

simetría lineal (pág. 16) Describe gráficas que se pueden doblar por la mitad, en dos secciones que se corresponden exactamente.

locus (pág. 422) Conjunto de todos los puntos que satisfacen una propiedad geométrica.

logaritmo (pág. 172) En la función $x = b^y$, y es el logaritmo base b de x. Generalmente se escribe como $y = \log_b x$ y se lee como el *logaritmo base b de x*.

función logarítmica de base b (pág. 172) Función de la forma $y = \log_b x$, donde $b > 0$, $b \neq 1$ y $x > 0$. Es la inversa de la función exponencial de la forma $b^y = x$.

función de crecimiento logístico (pág. 202) Función que modela el crecimiento exponencial con factores limitantes. Las funciones de crecimiento logístico están limitadas por asíntotas horizontales $y = 0$ y $y = c$, donde c es el límite de crecimiento.

cota inferior (pág. 121) Número real a que es menor que o igual al mínimo cero real de una función polinomial.

límite inferior (pág. 777) Cota inferior de una integral definida.

M

magnitude (p. 482) The length of the directed line segment that represents the vector.

major axis (p. 432) The segment that contains the foci of an ellipse and has endpoints on the ellipse.

matrix (p. P23) Any rectangular array of variables or constants in horizontal rows and vertical columns.

maximum (p. 36) For a function f, the greatest value of $f(x)$. A critical point on the graph of a function where the curve changes from increasing to decreasing.

maximum error of estimate (p. 696) The maximum difference between the point estimate and the actual value of the parameter in an experiment.

mean (p. P32) The sum of numbers in a set of data divided by the number of items in the data set.

measure of central tendency (p. P32) A number that represents the center or middle of a set of data.

magnitud (pág. 482) Longitud del segmento de recta dirigido que representa al vector.

eje mayor (pág. 432) Segmento que contiene los focos de una elipse y cuyos extremos se encuentran en la elipse.

matriz (pág. P23) Cualquier arreglo rectangular de variables o constantes en filas y columnas.

máximo (pág. 36) Para una función, f es el valor mayor de $f(x)$. Punto crítico de la gráfica de una función donde una curva creciente se convierte en decreciente.

error máximo de la estimación (pág. 696) Diferencia máxima entre la estimación puntual y el valor real del parámetro en un experimento.

media (pág. P32) La suma de los números en un conjunto de datos dividida entre el numero total de artículos en el conjunto.

medida de tendencia central (pág. P32) Número que representa el centro o medio de un conjunto de datos.

measures of spread (or variation) (p. P32) A representation of how spread out or scattered a set of data is.

median (p. P32) The middle number in a set of data when the data are arranged in numerical order. If the data set has an even number, the median is the mean of the two middle numbers.

midline (p. 262) A horizontal axis that is the reference line about which the graph of a sinusoidal function oscillates.

minimum (p. 36) For a function f, the least value of $f(x)$. A critical point on the graph of a function where the curve changes from decreasing to increasing.

minor axis (p. 432) The segment through the center of an ellipse that is perpendicular to the major axis and has endpoints on the ellipse.

mode (p. P32) The number(s) that appear most often in a set of data.

modulus (p. 570) The absolute value of a complex number, the number r when a complex number is written in the form $r(\cos \theta + i \sin \theta)$.

monomial function (p. 86) A function that can be written as $f(x) = a$ or $f(x) = ax^n$, where a and n are nonzero constant real numbers.

multiple optimal solutions (p. 408) Solutions that occur when the graph of the equation related to the objective function f to be optimized is coincident with one side of the region of feasible solutions.

multiplicity (p. 102) If $(x - c)^m$ is the highest power of $(x - c)$ that is a factor of polynomial function f, then c is a zero of multiplicity m of f, where m is a natural number.

multivariable linear system (p. 364) A system of linear equations in two or more variables, also called a *multivariate* linear system.

medidas de dispersión (o variación) (pág. P32) Representación de cómo están dispersos o esparcidos los datos en un conjunto.

mediana (pág. P32) El número central en un conjunto de datos, una vez que los datos han sido ordenados numéricamente. Si hay un número par de datos, la mediana es la media de los dos números del medio.

línea media (pág. 262) Eje horizontal que sirve como línea de referencia alrededor de la cual oscila la gráfica de una función sinusoidal.

mínimo (pág. 36) Para una función, f es el valor menor de $f(x)$. Punto crítico de la gráfica de una función donde una curva decreciente se convierte en creciente.

eje menor (pág. 432) Segmento que atraviesa el centro de una elipse, es perpendicular al eje mayor de elipse y tiene sus extremos en la misma.

moda (pág. P32) Número o números que aparecen con más frecuencia en un conjunto de datos.

módulo (pág. 570) Valor absoluto de un número complejo que corresponde al número r cuando un número complejo se expresa en la forma $r(\cos \theta + i \operatorname{sen} \theta)$.

función monomio (pág. 86) Función que se puede escribir de la forma $f(x) = a$ o $f(x) = ax^n$, donde a y n son constantes de números reales distintos de cero.

soluciones óptimas múltiples (pág. 408) Soluciones que se obtienen cuando la gráfica de la ecuación relacionada con la función objetivo f a ser optimizada coincide con uno de los lados de la región de soluciones factibles.

multiplicidad (pág. 102) Si $(x - c)^m$ es la potencia mayor de $(x - c)$ que es un factor de la función polinómica f, entonces c es un cero de multiplicidad m de f, donde m es un número natural.

sistema lineal multivariable (pág. 364) Sistema de ecuaciones lineales con dos o más variables. También se llaman sistemas lineales *multivariados*.

N

natural base (p. 160) The irrational number e, which is approximately equal to 2.718281828....

natural logarithm (p. 174) A logarithm with base e, written ln x.

base natural (pág. 160) El número irracional e, el cual es aproximadamente igual a 2.718281828....

logaritmo natural (pág. 174) Logaritmo de base e. Se escribe ln x.

negatively skewed distribution (p. 654) In a data distribution, the mean is less than the median, the majority of the data are on the right, and the tail extends to the left.

nonremovable discontinuity (p. 25) Describes infinite and jump discontinuities because they cannot be eliminated by redefining the function at that point.

nonsingular matrix (p. 379) A matrix that has an inverse.

normal distribution (p. 674) A continuous probability distribution in which the graph of the curve is bell-shaped and symmetric with respect to the mean; the mean, median, and mode are equal and located at the center. The curve is continuous and approaches, but never touches, the x-axis; the total area under the curve is equal to 1 or 100%.

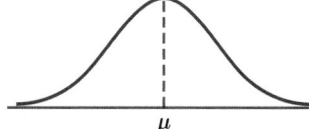

nth partial sum (p. 593) The sum of the first n terms of a finite or infinite series.

nth root (p. P14) For any real numbers a and b, and any positive integer n, if $a^n = b$, then a is an nth root of b.

null hypothesis (p. 705) One of two hypotheses that need to be stated to test a claim; states that there is not a significant difference between the sample value and the population parameter. The null hypothesis contains a statement of equality such as $\geq$, $=$, or $\leq$.

distribución asimétrica negativa (pág. 654) Sucede cuando en una distribución de datos la media es menor que la mediana, la mayoría de los datos están a la derecha de la distribución y la cola de la distribución se extiende a la izquierda.

discontinuidad no evitable (pág. 25) Describe discontinuidades infinitas y de salto porque no se pueden eliminar al redefinir la función en dicho punto.

matriz no singular (pág. 379) Matriz que tiene inversa.

distribución normal (pág. 674) Distribución probabilística continua en la cual la gráfica de la curva tiene forma de campana y es simétrica con respecto a la media; además de que la media, la mediana y la moda son iguales y están ubicadas en el centro de la distribución. La curva es continua y se aproxima, pero nunca interseca, el eje x. El área total bajo la curva es igual a 1 ó 100%.

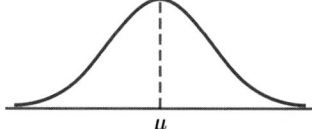

suma parcial enésima (pág. 593) Suma de los primeros n términos de una serie finita o infinita.

raíz enésima (pág. P14) Para cualquier número real a y b y cualquier entero positivo n, si $a^n = b$, entonces a es una raíz enésima de b.

hipótesis nula (pág. 705) Una de las dos hipótesis que se necesitan establecer para probar una afirmación. Establece que no hay una diferencia significativa entre el valor de la muestra y el parámetro demográfico. La hipótesis nula contiene algún enunciado de igualdad como $\geq$, $=$ o $\leq$.

O

objective function (p. 405) A linear function of the form $f(x, y) = ax + by + c$ to be optimized in a two-dimensional linear programming problem.

oblique asymptote (p. 134) An asymptote that is neither horizontal nor vertical that occurs when the degree of the numerator of a rational function is exactly one more than the degree of the denominator. Also called a *slant asymptote*.

oblique triangle (p. 291) A triangle that is not a right triangle.

octants (p. 510) Eight regions into which the three axes of a three-dimensional coordinate system divide space.

odd function (p. 18) A function that is symmetric with respect to the origin.

one-sided limit (p. 738) The limit L_1 of $f(x)$ as x approaches c from the left or the limit L_2 of $f(x)$ as x approaches c from the right.

función objetivo (pág. 405) Función lineal de la forma $f(x, y) = ax + by + c$ que se optimiza durante un problema de programación lineal bidimensional.

asíntota oblicua (pág. 134) Asíntota que no es vertical ni horizontal y que ocurre cuando el grado del numerador de una función racional es exactamente un grado mayor que el grado del denominador. También se conoce como *asíntota inclinada*.

triángulo oblicuo (pág. 291) Triángulo que no es rectángulo.

octantes (pág. 510) Las ocho regiones en las que dividen el espacio los tres ejes de un sistema tridimensional de coordenadas.

función impar (pág. 18) Función que es simétrica con respecto al origen.

límite unilateral (pág. 738) El límite L_1 de $f(x)$ a medida que x se aproxima a c por la izquierda o el límite L_2 de $f(x)$ a medida que x se aproxima a c por la derecha.

one-to-one (p. 66) **1.** A function in which no *x*-value is matched with more than one *y*-value and no *y*-value is matched with more than one *x*-value. **2.** A function whose inverse is a function.

opposite vectors (p. 483) Vectors that have the same magnitude but opposite direction.

optimization (p. 405) The process of finding a minimum or maximum value for a specific quantity, usually to minimize costs in order to maximize profits in business.

ordered triple (p. 510) Coordinates of the location of a point in space given by real numbers (x, y, z).

orientation (p. 464) Plotting points of parametric equations in the order of increasing values of *t* traces the curve in a specific direction of the curve.

orthogonal (p. 500) Two vectors with a dot product of 0.

outliers (p. P36) Data that are more than 1.5 times the interquartile range beyond the upper and lower quartiles.

uno a uno o inyectiva (pág. 66) **1.** Función en la cual ningún valor de *x* se corresponde con más de un valor de *y*, y en la cual ningún valor de *y* se corresponde con más de un valor de *x*. **2.** Función cuya inversa es una función.

vectores opuestos (pág. 483) Vectores que tienen la misma magnitud, pero dirección opuesta.

optimización (pág. 405) Proceso de obtención del valor máximo o mínimo para una cantidad específica; generalmente se aplica en negocios para minimizar costos y maximizar ganancias.

triple ordenado (pág. 510) Coordenadas de la ubicación de un punto en el espacio, dadas por los números reales (x, y, z).

orientación (pág. 464) Trazado de la curva en una dirección específica que sucede al graficar los puntos de ecuaciones paramétricas en orden creciente de los valores de *t*.

ortogonal (pág. 500) Dos vectores con un producto escalar de 0.

valores atípicos (pág. P36) Datos que están a más de 1.5 veces el rango intercuartílico del cuartil superior o el inferior.

P

parabola (pp. P9, 422) The locus of points in a plane that are equidistant from a fixed point, called the focus, and a specific line, called the directrix.

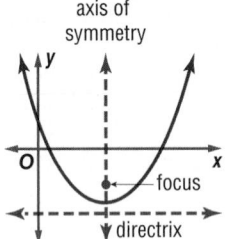
axis of symmetry
focus
directrix

parallelepiped (p. 521) A polyhedron with faces that are all parallelograms.

parallelogram method (p. 484) A method of finding the resultant vector by translating one vector so that its tail touches the tail of another. A parallelogram is drawn and the diagonal is the resultant vector.

parallel vectors (p. 483) Vectors that have the same or opposite direction, but not necessarily the same magnitude.

parameter (pp. 464, 696) **1.** Arbitrary values, usually time or angle measurement, used in parametric equations. **2.** A measure that describes a characteristic of a population.

parametric curve (p. 464) If *f* and *g* are continuous functions, then the set of ordered pairs $(f(t), g(t))$ is a plane curve with $x = f(t)$ and $y = g(t)$ as the parametric equations and *t* as the parameter.

parametric equation (pp. 74, 464) An equation that can express the position of an object as a function of time.

parábola (págs. P9, 422) Locus de los puntos en un plano que son equidistantes de un punto fijo llamado foco y una recta específica llamada directriz.

eje de simetría
foco
directriz

paralelepípedo (pág. 521) Poliedro cuyas caras son todas paralelogramos.

método del paralelogramo (pág. 484) Método de cálculo del vector resultante en que se traslada un vector de manera que su cola toca la cola de otro vector. Luego, se traza un paralelogramo y la diagonal es el vector resultante.

vectores paralelos (pág. 483) Vectores que tienen una misma dirección o una dirección opuesta, pero no necesariamente la misma magnitud.

parámetro (págs. 464, 696) **1.** Valores arbitrarios, generalmente el tiempo o la medida de un ángulo, que se utilizan en ecuaciones paramétricas. **2.** Medida que describe una característica de una población.

curva paramétrica (pág. 464) Si *f* y *g* son funciones continuas, entonces el conjunto de pares ordenados $(f(t), g(t))$ es un plano curvo con $x = f(t)$ y $y = g(t)$ como ecuaciones paramétricas y *t* como parámetro.

ecuación paramétrica (págs. 74, 464) Ecuación que expresa la posición de un objeto como función del tiempo.

parent function (p. 45) The simplest function in a family of functions. A function that is transformed to create other members in a family of functions.

partial fraction (p. 398) When a rational function is written as the sum of two fractions with denominators that are linear factors of the original denominator, each fraction in the sum is a partial fraction.

partial fraction decomposition (p. 398) A rational expression rewritten as the sum of two simpler rational expressions.

Pascal's triangle (p. 628) A triangular array of numbers such that the first and last numbers in each row are 1 and every other number is formed by adding the two numbers immediately above that number in the previous row. The $(n + 1)$th row contains the coefficients of the terms of the expansion $(a + b)^n$ for $n = 0, 1, 2, \ldots$.

percentile graph (p. 658) Uses the same values as a cumulative relative frequency graph, except that the proportions are instead expressed as percents.

percentiles (p. 658) Divide a distribution into 100 equal groups and are symbolized by $P_1, P_2, P_3, \ldots, P_{99}$. The nth percentile or P_n, is the value such that $n\%$ of the data are lower than P_n.

period (p. 250) For a function $y = f(t)$, the smallest positive number c for which $f(t + c) = f(t)$.

periodic function (p. 250) A function with values that repeat at regular intervals. There exists a positive real number c such that $f(t + c) = f(t)$ for all values of t in the domain of f.

permutation (p. P29) An arrangement of objects in which order is important.

phase shift (p. 261) For a sinusoidal function, the difference between the horizontal position of a function and that of an otherwise similar sinusoidal function. For $y = a \sin (bx + c) + d$ and $y = a \cos (bx + c) + d$, phase shift $= -\frac{c}{|b|}$.

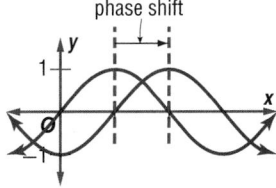
phase shift

piecewise-defined function (p. 8) A function that is defined using two or more expressions for different intervals of the domain.

point estimate (p. 696) A single value estimate of an unknown population parameter.

point symmetry (p. 16) Describes graphs that can be rotated 180° with respect to a point and appear unchanged.

función principal (pág. 45) La función más simple de una familia de funciones. Función que es transformada para obtener los otros miembros de una familia de funciones.

fracción parcial (pág. 398) Cuando una función racional se escribe como la suma de dos fracciones cuyos denominadores son factores lineales del denominador original, cada fracción de la suma es una fracción parcial.

descomposición por fracciones parciales (pág. 398) Expresión racional expresada como la suma de dos expresiones racionales más simples.

triángulo de Pascal (pág. 628) Arreglo triangular de números ordenados de manera que el primero y el último número de cada fila es un 1 y cada número siguiente está formado por la suma de los dos números ubicados en la fila previa, inmediatamente arriba de dicho número. La fila $(n + 1)$ contiene los coeficientes de los términos de la expansión $(a + b)^n$, para $n = 0, 1, 2, \ldots$.

gráfica de percentiles (pág. 658) Utiliza los mismos valores que una gráfica de frecuencias relativas acumuladas, excepto que las proporciones se expresan como porcentajes.

percentiles (pág. 658) Dividen una distribución en 100 grupos iguales y se representan como $P_1, P_2, P_3, \ldots, P_{99}$. El $n$ésimo percentil P_n tiene un valor tal que $n\%$ de los datos son menores que P_n.

período (pág. 250) Para una función $y = f(t)$, el menor número positivo c para el cual $f(t + c) = f(t)$.

función periódica (pág. 250) Función cuyos valores se repiten a intervalos regulares. Existe un número positivo real c tal que $f(t + c) = f(t)$ para todo valor de t en el dominio de f.

permutación (pág. P29) Arreglo de elementos en que el orden es importante.

desplazamiento de fase (pág. 261) En una función sinusoidal, se refiere a la diferencia entre la posición horizontal de una función y la posición horizontal de otra función sinusoidal semejante. Para $y = a \operatorname{sen} (bx + c) + d$ y $y = a \cos (bx + c) + d$, el desplazamiento de fase $= -\frac{c}{|b|}$.

desplazamiento de fase

función por partes (pág. 8) Función que se define usando dos o más expresiones para distintos intervalos del dominio.

estimación puntual (pág. 696) Estimación individual de un parámetro demográfico desconocido.

simetría respecto a un punto (pág. 16) Describe gráficas que se pueden rotar 180° con respecto a un punto, sin que sufran cambios aparentes.

polar axis (p. 534) An initial ray from the pole in the polar coordinate system, usually horizontal and directed toward the right.

polar coordinates (p. 534) Describes the location of a point $P(r, \theta)$ in the polar coordinate system, where r is the directed distance from the pole O to the point and θ is the directed angle from the polar axis to $\overrightarrow{OP}$.

polar coordinate system (p. 534) A coordinate system in which the location of a point is identified by polar coordinates of the form (r, θ), where r is the distance from the center, or the pole, to the given point and θ is the measure of the angle formed by the polar axis and a line from the pole through the point.

polar equation (p. 536) An equation expressed in terms of polar coordinates.

polar form (p. 570) The complex number $z = a + bi$ written as $z = r(\cos \theta + i \sin \theta)$, where $r = |z| = \sqrt{a^2 + b^2}$, $a = r \cos \theta$, $b = r \sin \theta$, and $\theta = \tan^{-1} \frac{b}{a}$ for $a > 0$, $\theta = \tan^{-1} \frac{b}{a} + \pi$ for $a < 0$.

polar graph (p. 536) The set of all points with coordinates (r, θ) that satisfy a given polar equation.

pole (p. 534) The origin of the polar coordinate system, O.

polynomial function (p. 97) A function of the form $f(x) = a_n x^n + a_{n-1} x^{n-1} + \ldots + a_2 x^2 + a_1 x + a_0$, where $a_0, a_1, a_2, \ldots, a_{n-1}, a_n$ are real numbers.

polynomial function of degree n (p. 97) $f(x) = a_n x^n + a_{n-1} x^{n-1} + \ldots + a_2 x^2 + a_1 x + a_0$, where n is a nonnegative integer and $a_0, a_1, a_2, \ldots, a_{n-1}, a_n$, are real numbers with $a_n \neq 0$.

polynomial inequality (p. 141) An inequality of the form $f(x) \leq 0$, $f(x) < 0$, $f(x) \neq 0$, $f(x) > 0$, or $f(x) \geq 0$, where $f(x)$ is a polynomial function.

population (p. P32) An entire group of living things or objects.

positively skewed distribution (p. 654) In a data distribution, the mean is greater than the median, the majority of the data are on the left, and the tail extends to the right.

power function (p. 86) A function of the form $f(x) = ax^n$, where a and n are nonzero real numbers.

eje polar (pág. 534) Rayo inicial que parte del polo en el sistema de coordenadas polares; generalmente es horizontal y está dirigido hacia la derecha.

coordenadas polares (pág. 534) Describe la ubicación del punto $P(r, \theta)$ en el sistema de coordenadas polares, donde r es la distancia dirigida desde el polo O hacia el punto y θ es el ángulo dirigido desde el eje polar hacia $\overrightarrow{OP}$.

sistema de coordenadas polares (pág. 534) Sistema de coordenadas en el cual la ubicación de un punto está determinada por coordenadas polares de la forma (r, θ), donde r es la distancia desde el centro (o polo) hasta el punto dado y θ es la medida del ángulo formado por el eje polar y la recta que va desde el polo hasta dicho punto.

ecuación polar (pág. 536) Ecuación expresada en términos de coordenadas polares.

forma polar (pág. 570) Número complejo $z = a + bi$ expresado como $z = r(\cos \theta + i \operatorname{sen} \theta)$, donde $r = |z| = \sqrt{a^2 + b^2}$, $a = r \cos \theta$, $b = r \operatorname{sen} \theta$ y $\theta = \tan^{-1} \frac{b}{a}$ para $a > 0$ y $\theta = \tan^{-1} \frac{b}{a} + \pi$ para $a < 0$.

gráfica polar (pág. 536) Conjunto de todos los puntos con coordenadas (r, θ) que satisfacen una ecuación polar dada.

polo (pág. 534) Origen en el sistema de coordenadas polares; se denota como O.

función polinomial (pág. 97) Función de la forma $f(x) = a_n x^n + a_{n-1} x^{n-1} + \ldots + a_2 x^2 + a_1 x + a_0$, donde $a_0, a_1, a_2, \ldots, a_{n-1}, a_n$ son números reales.

función polinomial de grado n (pág. 97) $f(x) = a_n x^n + a_{n-1} x^{n-1} + \ldots + a_2 x^2 + a_1 x + a_0$, donde n es un entero no negativo y $a_0, a_1, a_2, \ldots, a_{n-1}, a_n$, son números reales donde $a_n \neq 0$.

desigualdad polinomial (pág. 141) Desigualdad de la forma $f(x) \leq 0$, $f(x) < 0$, $f(x) \neq 0$, $f(x) > 0$ o $f(x) \geq 0$, donde $f(x)$ es una función polinomial.

población (pág. P32) Un grupo entero de cosas o de objetos vivos.

distribución asimétrica positiva (pág. 654) Sucede cuando en una distribución de datos la media es mayor que la mediana, la mayoría de los datos están a la izquierda de la distribución y la cola de la distribución se extiende a la derecha.

función potencia (pág. 86) Función de la forma $f(x) = ax^n$, donde a y n son números reales distintos de cero.

power series (p. 636) An infinite series of the form $\sum_{n=0}^{\infty} a_n x^n = a_0 + a_1 x + a_2 x^2 + a_3 x^3 + ...$, where x and a_n can take on any values for $n = 0, 1, 2, ...$.

principal *n*th root (p. P14) The nonnegative *n*th root.

principle of mathematical induction (p. 621) Let P_n be a statement about a positive integer n. Then P_n is true for all positive integers n if and only if
- P_1 is true, and
- for every positive integer k, if P_k is true, then P_{k+1} is true.

probability distribution (p. 665) A table, equation, or graph that links each possible value for a random variable with its probability of occurring.

***p*th roots of unity** (p. 576) Finding the *p*th roots of 1.

***p*-value** (p. 708) The lowest level of significance at which H_0 can be rejected for a given set of data.

pure imaginary number (p. P6) An imaginary number $(a + bi)$, where $a = 0$.

serie de potencias (pág. 636) Serie infinita de la forma $\sum_{n=0}^{\infty} a_n x^n = a_0 + a_1 x + a_2 x^2 + a_3 x^3 + ...$, donde x y a pueden tomar cualquier valor y $n = 0, 1, 2, ...$.

raíz principal *enésima* (pág. P14) La raíz *enésima* no negativa.

principio de inducción matemática (pág. 621) Sea P_n un enunciado sobre un entero positivo n. Entonces P_n es verdadera para todos los enteros positivos n si y sólo si
- P_1 es verdadera y
- para todo entero positivo k, si P_k es verdadera entonces P_{k+1} es verdadera.

distribución probabilística (pág. 665) Tabla, ecuación o gráfica que relaciona todo valor posible de una variable aleatoria con la probabilidad de que dicho valor suceda.

raíces *p-enésimas* de la unidad (pág. 576) Cálculo de las raíces primitivas *enésimas* de 1.

valor *p* (pág. 708) El menor nivel de significancia al cual se puede rechazar H_0 para un conjunto de datos dado.

número imaginario puro (pág. P6) Número imaginario $(a + bi)$, donde $a = 0$.

Q

quadrant bearing (p. 483) A directional measurement of a vector between 0° and 90° east or west of the north-south line.

quadrantal angle (p. 243) An angle in standard position that has a terminal side that lies on one of the coordinate axes.

quadratic equation (p. P11) A polynomial equation of degree two, in the form $ax^2 + bx + c$, where $a \neq 0$.

quadratic form (p. 100) A polynomial expression that is written in the form $au^2 + bu + c$ for any numbers a, b, and c, where $a \neq 0$ and u is some expression in x.

Quadratic Formula (p. P12) The solutions of a quadratic equation of the form $ax^2 + bx + c$, where $a \neq 0$, are given by the Quadratic Formula, which is $x = \dfrac{-b \pm \sqrt{b^2 - 4ac}}{2a}$.

quadratic function (p. 45) A function of the form $f(x) = ax^2 + bx + c$, where $a \neq 0$, with parent function $f(x) = x^2$.

quartic function (p. 99) A function that contains a fourth-degree polynomial.

quartiles (p. P35) The values that divide a set of data into four equal parts.

rumbo cuadrantal (pág. 483) Medición de la dirección de un vector entre 0° y 90° hacia el este o el oeste, con respecto a una recta con dirección norte sur.

ángulo cuadrantal (pág. 243) Ángulo en posición estándar cuyo lado terminal yace en uno de los ejes de coordenadas.

ecuación cuadrática (pág. P11) Ecuación polinómica de segundo grado, de la forma $ax^2 + bx + c$, donde $a \neq 0$.

forma cuadrática (pág. 100) Expresión polinomial escrita en la forma $au^2 + bu + c$, para todo número a, b y c, donde $a \neq 0$ y u es una expresión en x.

fórmula cuadrática (pág. P12) Las soluciones de una ecuación cuadrática de la forma $ax^2 + bx + c$, donde $a \neq 0$, vienen dadas por la fórmula cuadrática, $x = \dfrac{-b \pm \sqrt{b^2 - 4ac}}{2a}$.

función cuadrática (pág. 45) Función de la forma $f(x) = ax^2 + bx + c$, donde $a \neq 0$, cuya función generatriz es $f(x) = x^2$.

función cuártica (pág. 99) Función que contiene un polinomio de cuarto grado.

cuartiles (pág. P35) Valores que dividen un conjunto de datos en cuatro partes iguales.

R

radians (p. 232) A unit of angular measurement equal to $\dfrac{180°}{\pi}$ or about 57.296°.

radianes (pág. 232) Unidad de medida angular igual a $\dfrac{180°}{\pi}$ ó aproximadamente 57.296°.

radical function (p. 89) A function that can be written as $f(x) = \sqrt[n]{x^p}$, where n and p are positive integers greater than 1 that have no common factors.

random variable (p. 664) Represents a numerical value assigned to an outcome of a probability experiment.

range (p. P32) The difference between the greatest and least values in a set of data.

rational function (p. 130) A function of the form $f(x) = \frac{a(x)}{b(x)}$, where $a(x)$ and $b(x)$ are polynomial functions, and $b(x) \neq 0$.

rational inequality (p. 143) An inequality that contains one or more rational expressions.

Rational Zero Theorem (p. 119) Describes how the leading coefficient and constant term of a polynomial function with integer coefficients can be used to determine a list of all possible rational zeros.

real axis (p. 569) The horizontal axis of a complex plane on which the real component of a complex number is graphed.

real part (p. P6) In an imaginary number $a + bi$, a is the real part.

reciprocal function (p. 45, 200) **1.** A function of the form $f(x) = \frac{1}{a(x)}$, where $a(x)$ is a linear function and $a(x) \neq 0$, with parent function $f(x) = \frac{1}{x}$.
2. Trigonometric functions that are reciprocals of each other.

rectangular components (p. 487) Horizontal and vertical components of a vector.

recursive formula (p. 591) A formula used to determine the nth term of a sequence using one or more of the preceding terms.

reduced row-echelon form (p. 369) An augmented matrix in which the first nonzero element of each row of the coefficient portion of the matrix is 1 and the rest of the elements in the same column as this element are 0.

reduction identity (p. 340) An identity that results when a sum or difference identity is used to rewrite a trigonometric expression in which one of the angles is a multiple of $90°$ or $\frac{\pi}{2}$ radians.

reference angle (p. 244) The acute angle formed by the terminal side of an angle in standard position and the x-axis.

reflection (p. 48) A transformation in which a mirror image of the graph of a function is produced with respect to a specific line.

regression line (p. 716) A line drawn through a set of data points that describes how the response variable y changes as the explanatory variable x changes. Also called a *line of best fit*.

función radical (pág. 89) Función que se puede escribir de la forma $f(x) = \sqrt[n]{x^p}$, donde n y p son enteros positivos mayores que 1, que no tienen factores comunes.

variable aleatoria (pág. 664) Representa el valor numérico asignado a un resultado de un experimento probabilístico.

rango (pág. P32) Diferencia entre el valor mayor y el menor en un conjunto de datos.

función racional (pág. 130) Función de la forma $f(x) = \frac{a(x)}{b(x)}$, donde $a(x)$ y $b(x)$ son funciones polinómicas y $b(x) \neq 0$.

desigualdad racional (pág. 143) Desigualdad que contiene una o más expresiones racionales.

teorema de los ceros racionales (pág. 119) Describe cómo el coeficiente principal y las constantes de una función polinómica se pueden usar para determinar una lista de todos los ceros racionales posibles.

eje real (pág. 569) Eje horizontal de un plano complejo en el cual se grafica el componente real de un número complejo.

parte real (pág. P6) En un número imaginario $a + bi$, a es la parte real.

función inversa (págs. 45, 200) **1.** Función de la forma $f(x) = \frac{1}{a(x)}$, donde $a(x)$ es una función lineal y $a(x) \neq 0$, cuya función generatriz es $f(x) = \frac{1}{x}$.
2. Funciones trigonométricas recíprocas entre sí.

componentes rectangulares (pág. 487) Los componentes horizontal y vertical de un vector.

fórmula recursiva (pág. 591) Fórmula que se usa para determinar el enésimo término en una sucesión mediante el uso de uno o más de los términos anteriores.

forma escalón reducida por filas (pág. 369) Matriz aumentada en la cual el primer elemento distinto de cero en cada fila de la porción del coeficiente de la matriz es 1 y el resto de los elementos en la misma columna del elemento son iguales a 0.

identidad de reducción (pág. 340) Identidad que se obtiene cuando una identidad suma o diferencia se utiliza para replantear una expresión trigonométrica en la cual uno de los ángulos es un múltiplo de $90°$ ó $\frac{\pi}{2}$ radianes.

ángulo de referencia (pág. 244) Ángulo agudo formado por el lado terminal de un ángulo en posición normal y el eje x.

reflexión (pág. 48) Transformación en la cual se produce una imagen especular con respecto a una línea específica de la gráfica de una función.

recta de regresión (pág. 716) Recta trazada a través de un conjunto de puntos que describe cómo cambia la variable de respuesta y a medida que cambia la variable explicativa x. También se conoce como *recta de mejor ajuste*.

regular partition (p. 777) In the area under the graph of a function, an interval that is subdivided into equal subintervals.

relative frequency (p. P34) In a frequency table, the frequency of occurrence for each data value.

relevant domain (p. 8) In a function, the part of the domain that is relevant to a model.

removable discontinuity (p. 24) A characteristic of a function in which the function is continuous everywhere except for a hole at $x = c$.

repeated zero (p. 101) The related zero c of a function when a factor $(x - c)$ occurs more than once in the completely factored form of $f(x)$.

residual (p. 716) The difference between an observed y-value of a data point and its predicted y-value on a regression line.

residual plot (p. 717) A scatter plot of the residuals in which the horizontal line at zero corresponds to the regression line.

resistant statistic (p. 654) A statistic that is not highly affected by the presence of outlying data values.

response variable (p. 713) The dependent variable y in bivariate data.

resultant (p. 484) A single vector that results when two or more vectors are added.

right Riemann sum (p. 777) A method for approximating the area under a curve by using the values at the right endpoints.

right-tailed test (p. 706) The hypothesis test if $H_a: \mu > k$.

root (p. 15) For a function $f(x)$, a solution of the equation $f(x) = 0$.

rose (p. 545) The graph of a polar equation of the form $r = a \cos n\theta$ or $r = a \sin n\theta$, where $n \geq 2$ is an integer.

row-echelon form (p. 367) A matrix is in row-echelon form if the following conditions are met.
- Rows of all zeros (if any) appear at the bottom of the matrix.
- The first nonzero entry in any row is 1.
- For two successive rows with nonzero entries, the leading 1 in the higher row is farther to the left than the leading 1 in the lower row.

$$\begin{bmatrix} 1 & a & b & | & c \\ 0 & 1 & d & | & e \\ 0 & 0 & 1 & | & f \\ 0 & 0 & 0 & | & 0 \end{bmatrix}$$

partición regular (pág. 777) En el área bajo la gráfica de una función, es un intervalo que se divide en subintervalos iguales.

frecuencia relativa (pág. P34) En una tabla de frecuencias, la frecuencia con que ocurre cada dato.

dominio relevante o de interés (pág. 8) Parte del dominio de una función que es relevante para un modelo.

discontinuidad evitable (pág. 24) Característica de una función en la cual la función es continua en todos lados, excepto por una brecha en $x = c$.

ceros múltiples (pág. 101) El cero relacionado c de una función cuando el factor $(x - c)$ ocurre más de una vez en $f(x)$ factorizada a su mínima expresión.

residual (pág. 716) Diferencia entre el valor observado de y en un punto y el valor predicho de y en la recta de regresión.

gráfica de residuales (pág. 717) Diagrama de dispersión de los residuales en el cual la recta horizontal sobre el cero corresponde a la recta de regresión.

estadístico resistente (pág. 654) La media, porque es más afectada por la presencia de valores atípicos que la mediana.

variable de respuesta (pág. 713) Variable dependiente y en un conjunto de datos bivariados.

resultante (pág. 484) Vector único que se obtiene al sumar dos o más vectores.

suma de Riemann por la derecha (pág. 777) Método para aproximar el área bajo una curva, utilizando los valores en los extremos derechos.

prueba de cola derecha (pág. 706) La prueba de hipótesis si $H_a: \mu > k$.

raíz (pág. 15) Es una solución de la ecuación $f(x) = 0$ para una función $f(x)$.

rosa (pág. 545) Gráfica de una ecuación polar de la forma $r = a \cos n\theta$ o $r = a \sin n\theta$, donde $n \geq 2$ es un entero.

forma escalón por filas (pág. 367) Una matriz están en forma escalón por filas si cumple las siguientes condiciones.
- Las filas que contienen sólo ceros (si las hay) están en las filas inferiores de la matriz.
- La primera entrada distinta de cero en cualquier fila es 1.
- Por cada dos filas sucesivas con entradas distintas de cero, el 1 principal en la fila más alta está más a la izquierda que el 1 principal en la fila inferior.

$$\begin{bmatrix} 1 & a & b & | & c \\ 0 & 1 & d & | & e \\ 0 & 0 & 1 & | & f \\ 0 & 0 & 0 & | & 0 \end{bmatrix}$$

row matrix (p. P23) A matrix that has only one row.

matriz fila (pág. P23) Matriz que sólo tiene una fila.

S

sample (p. P32) A part of a population.

sample correlation coefficient (p. 714) A measure that determines the type and strength of the linear relationship between the variables in bivariate data that represent a sample of the population.

sampling distribution (p. 685) A distribution of the means of random samples of a certain size that are taken from a population.

sampling error (p. 686) Occurs when a sample is not a complete representation of the population and causes differences between sample means and the population mean.

sample space (p. P28) The set of all possible outcomes of an experiment.

scalar (p. P24) A constant.

secant (p. 220) In a right triangle with acute angle θ, the ratio comparing the length of the hypotenuse to the side adjacent to θ. It is the reciprocal of the cosine ratio, or $\sec \theta = \frac{1}{\cos \theta}$.

secant line (p. 38) The line through two points on a curve.

second differences (p. 601) Differences that are found by subtracting consecutive first differences from one another.

sector (p. 237) In a circle, the region bounded by a central angle and its intercepted arc.

sequence (p. 590) An ordered list of numbers.

series (p. 593) The sum of all the terms of a finite or infinite sequence.

set (p. P3) A collection of objects or numbers, often shown using braces { } and usually named by a capital letter.

set-builder notation (p. 4) An expression that describes a set of numbers by using the properties of numbers in the set to define the set, for example $\{x \mid x \geq 8, x \in \mathbb{W}\}$.

sigma notation (pp. P32, 594) For any sequence $a_1, a_2, a_3,$ ..., the sum of the first k terms is denoted $\sum_{n=1}^{k} a_n$, which is read *the summation from n = 1 to k of a_n*. Thus $\sum_{n=1}^{k} a_n = a_1 + a_2 + a_3 + \ldots + a_k$ where k is an integer value.

sign chart (p. 141) Used to determine on which intervals a polynomial function is positive or negative.

muestra (pág. P32) Parte de una población.

coeficiente de correlación de la muestra (pág. 714) Medida que determina el tipo y la magnitud de la relación lineal entre las variables de datos bivariados que representan una muestra de la población.

distribución de muestreo (pág. 685) Distribución de las medias de muestras aleatorias de cierto tamaño que se toman de una población.

error muestral (pág. 686) Sucede cuando una muestra no es una representación completa de la población y se producen diferencias entre las medias de la muestra y la media de la población.

espacio muestral (pág. P28) Conjunto de todos los resultados posibles de un experimento probabilístico.

escalar (pág. P24) Una constante.

secante (pág. 220) Razón que compara la longitud de la hipotenusa con la longitud del lado adyacente a θ, en un triángulo rectángulo con ángulo agudo θ. Es el recíproco de la razón coseno, o $\sec \theta = \frac{1}{\cos \theta}$.

recta secante (pág. 38) Recta que atraviesa dos puntos en una curva.

segundas diferencias (pág. 601) Diferencias que se obtienen al restar primeras diferencias consecutivas entre sí.

sector (pág. 237) Región de un círculo limitada por un ángulo central y el arco que interseca.

sucesión (pág. 590) Función cuyo dominio es el conjunto de los números naturales.

serie (pág. 593) Suma de todos los términos de una sucesión finita o infinita.

conjunto (pág. P3) Colección de objetos o números, que a menudo se representan con corchetes { } y que por lo general se denotan con una letra mayúscula.

notación de construcción de conjuntos (pág. 4) Expresión que describe un conjunto de números aplicando las propiedades de los números en el conjunto para definir el conjunto, por ejemplo $\{x \mid x \geq 8, x \in \mathbb{W}\}$.

notación de sumatoria (págs. P32 y 594) Para cualquier sucesión $a_1, a_2, a_3,$ ..., la suma de los primeros k términos se denota como $\sum_{n=1}^{k} a_n$, lo cual se lee *la suma de n = 1 a k de a_n*. Así, $\sum_{n=1}^{k} a_n = a_1 + a_2 + a_3 + \ldots + a_k$ donde k es un valor entero.

tabla de signos (pág. 141) Sirve para determinar en qué intervalos una función polinomial es positiva o negativa.

Glossary/Glosario

sine (p. 220) In a right triangle with acute angle θ, the ratio comparing the length of the side opposite θ and the hypotenuse.

singular matrix (p. 379) A matrix that does not have an inverse.

sinusoid (p. 256) Any transformation of a sine function. The general forms of sinusoidal functions are $y = a \sin(bx + c) + d$ and $y = a \cos(bx + c) + d$, where a, b, c, and d are constants and neither a nor b is 0.

solve a right triangle (p. 226) To find the measures of all of the sides and angles of a right triangle.

spiral of Archimedes (p. 546) The graph of a polar equation of the form $r = a\theta + b$.

square matrix (p. P23) A matrix with the same number of rows and columns.

square root function (p. 45) A function that contains a square root of the independent variable, with parent function $f(x) = \sqrt{x}$.

square system (p. 388) A system of linear equations that has the same number of equations as variables.

standard deviation (p. P32) The average amount by which individual items deviate from the mean of all the data found by taking the square root of the variance and represented by σ.

standard error of the mean (p. 685) The standard deviation of the sample means, given by $\sigma_x = \dfrac{\sigma}{\sqrt{n}}$.

standard form (p. P6) A complex number written in the form of $a + bi$.

standard normal distribution (p. 676) A normal distribution of z-values with a mean of 0 and a standard deviation of 1.

standard position (pp. 231, 482) **1.** In the coordinate plane, an angle positioned so that its vertex is at the origin and its initial side is along the positive x-axis. **2.** A vector that has its initial point at the origin.

statistics (p. P32) The science of collecting, analyzing, interpreting and presenting data.

step function (p. 46) A piecewise-defined function in which the graph is a series of line segments that resemble a set of stairs.

subset (p. P3) If every element of set B a set is contained in set A, then B is a subset of A.

substitution method (p. P18) A method of solving a system of equations in which one equation is solved for one variable in terms of the other.

seno (pág. 220) Razón que compara la longitud del lado opuesto a θ con la longitud de la hipotenusa, en un triángulo rectángulo con ángulo agudo θ.

matriz singular (pág. 379) Matriz que no tiene inversa.

sinusoide (pág. 256) Toda transformación de una función seno. Las formas generales de una función sinusoidal son: $y = a \operatorname{sen}(bx + c) + d$ y $y = a \cos(bx + c) + d$, donde a, b, c y d son constantes y a y b son distintas de cero.

resolver un triángulo rectángulo (pág. 226) Hallar las medidas de todos los lados y los ángulos de un triángulo rectángulo.

espiral de Arquímedes (pág. 546) Gráfica de una ecuación polar de la forma $r = a\theta + b$.

matriz cuadrada (pág. P23) Matriz con el mismo número de filas y columnas.

función raíz cuadrada (pág. 45) Función que contiene un raíz cuadrada de la variable independiente, cuya función generatriz es $f(x) = \sqrt{x}$.

sistema cuadrado (pág. 388) Sistema de ecuaciones lineales que tiene el mismo número de ecuaciones que de variables.

desviación estándar (pág. P32) Cantidad promedio por la cual se desvían los datos individuales de la media de todos los datos. Se obtiene calculando la raíz cuadrada de la varianza y representándola por σ.

error estándar de la media (pág. 685) Desviación estándar de las medias de la muestra; está dada por $\sigma_x = \dfrac{\sigma}{\sqrt{n}}$.

forma estándar (pág. P6) Número complejo escrito en la forma $a + bi$.

distribución normal estándar (pág. 676) Distribución normal de valores de z con una media de 0 y una desviación estándar de 1.

posición normal (págs. 231, 482) **1.** En un plano de coordenadas se refiere al ángulo colocado de manera que su vértice se halla en el origen y su lado inicial se sobrepone al eje x positivo. **2.** Vector que tiene su punto inicial en el origen.

estadística (pág. P32) Ciencia de reunir, analizar, interpretar y representar datos.

función escalón (pág. 46) Función por partes en la cual la gráfica es una serie de segmentos de recta que semejan una serie de escalones.

subconjunto (pág. P3) Si cada elemento del conjunto B está contenido en el conjunto A, entonces B es un subconjunto de A.

método de sustitución (pág. P18) Método para resolver un sistema de ecuaciones en el cual se despeja una variable en una de las ecuaciones y luego se despeja la segunda variable en términos de la primera.

symmetrical distribution (p. 654) In a data distribution, the data are evenly distributed on both sides of the mean.

synthetic division (p. 111) A shortcut for dividing a polynomial by a linear factor of the form $x - c$.

synthetic substitution (p. 113) The use of synthetic division to evaluate a function.

system of equations (p. P18) A set of equations with the same variables.

system of inequalities (p. P21) A set of inequalities with the same variables.

distribución simétrica (pág. 654) Distribución de datos en las que los datos están distribuidos uniformemente a ambos lados de la media.

división sintética (pág. 111) Método que facilita la división de un polinomio entre un factor lineal de la forma $x - c$.

sustitución sintética (pág. 113) El uso de la división sintética para evaluar una función.

sistema de ecuaciones (pág. P18) Conjunto de ecuaciones con las mismas variables.

sistema de desigualdades (pág. P21) Conjunto de desigualdades con las mismas variables.

T

tangent (p. 220) **1.** A line that intersects a circle at exactly one point. **2.** In a right triangle with acute angle θ, the ratio comparing the length of the side opposite θ and the side adjacent to θ.

tangent line (p. 758) The tangent line to $f(x)$ at x is the line passing through the point $(x, f(x))$ with slope m, where $m = \lim\limits_{h \to 0} \frac{f(x + h) - f(x)}{h}$.

t-distribution (p. 699) A family of curves that are dependent on a parameter known as the *degrees of freedom*.

term (p. 590) **1.** The monomials that make up a polynomial. **2.** Each number in a sequence or series.

terminal point (p. 482) The ending point of a vector that is represented by a directed line segment. Also known as the *head* or *tip* of the vector.

terminal side (p. 231) The final position of a ray after rotation when forming an angle.

three-dimensional coordinate system (p. 510) A coordinate system formed by three perpendicular number lines, the x-, y-, and z-axes, that intersect at the origin O. Each point is represented by an ordered triple of real numbers (x, y, z).

torque (p. 520) A vector quantity that measures how effectively a force applied to a lever causes rotation along the axis of rotation.

transcendental function (p. 158) A function that cannot be expressed in terms of algebraic operations, such as an exponential or logarithmic function.

transformation (p. 46) A change in the position or shape of the graph of a parent function.

translation (p. 47) A rigid transformation that has the effect of shifting the graph of a function

transverse axis (p. 442) The segment that has a length of $2a$ units and connects the vertices of a hyperbola.

tangente (pág. 220) **1.** Recta que interseca un círculo en exactamente un punto. **2.** Razón que compara la longitud del lado opuesto a θ con la longitud del lado adyacente a θ, en un triángulo rectángulo con ángulo agudo θ.

recta tangente (pág. 758) La recta tangente a $f(x)$ en x es la recta que pasa por el punto $(x, f(x))$ con pendiente m, donde $m = \lim\limits_{h \to 0} \frac{f(x + h) - f(x)}{h}$.

distribución t (pág. 699) Familia de curvas que son dependientes de un parámetro conocido como *grados de libertad*.

término (pág. 590) **1.** Monomios que componen un polinomio. **2.** Cada número en una sucesión o serie.

punto final (pág. 482) Punto donde termina un vector representado por un segmento de recta dirigido. También se conoce como *punta* del vector.

lado terminal (pág. 231) Posición final de un rayo después de una rotación para formar un ángulo.

sistema tridimensional de coordenadas (pág. 510) Sistema de coordenadas formado por tres rectas numéricas perpendiculares, los ejes x, y y z, que se intersecan en el origen O. Cada punto se representa mediante un triple ordenado de números reales (x, y, z).

par de fuerzas o par de torsión (pág. 520) Vector cantidad que mide la efectividad con que la fuerza aplicada a una palanca produce rotación alrededor del eje de rotación.

función trascendental (pág. 158) Función que no se puede expresar en términos de operaciones algebraicas; por ejemplo una función exponencial o logarítmica.

transformación (pág. 46) Cambio en la posición o la forma de la gráfica de una función principal.

traslación (pág. 47) Transformación rígida cuyo efecto es desplazar la gráfica de una función.

eje transversal (pág. 442) Segmento de recta con una longitud de $2a$ unidades y que une los vértices de una hipérbola.

triangle method (p. 484) A method of finding the resultant vector by translating one vector so that its tail touches the tip of another. The resultant vector is drawn to form a triangle.

trigonometric form (p. 570) See *polar form*.

trigonometric function (p. 220) Let θ be an acute angle in a right triangle and opp, adj, and hyp are the lengths of the side opposite θ, the side adjacent to θ, and the hypotenuse, respectively. Then the trigonometric functions of θ are defined below.

$$\sin\theta = \frac{\text{opp}}{\text{hyp}} \qquad \cos\theta = \frac{\text{adj}}{\text{hyp}} \qquad \tan\theta = \frac{\text{opp}}{\text{adj}}$$

$$\csc\theta = \frac{\text{hyp}}{\text{opp}} \qquad \sec\theta = \frac{\text{hyp}}{\text{adj}} \qquad \cot\theta = \frac{\text{adj}}{\text{opp}}$$

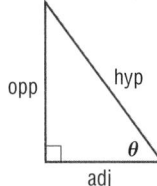

trigonometric identity (p. 312) An equation that involves trigonometric functions that is true for all values of the variable.

trigonometric ratios (p. 220) Ratios that are formed using the side measures of a right triangle and a reference angle θ.

triple scalar product (p. 521) If $\mathbf{t} = t_1\mathbf{i} + t_2\mathbf{j} + t_3\mathbf{k}$, $\mathbf{u} = u_1\mathbf{i} + u_2\mathbf{j} + u_3\mathbf{k}$, and $\mathbf{v} = v_1\mathbf{i} + v_2\mathbf{j} + v_3\mathbf{k}$, the triple scalar product is given by $\mathbf{t} \cdot (\mathbf{u} \times \mathbf{v}) = \begin{vmatrix} t_1 & t_2 & t_3 \\ u_1 & u_2 & u_3 \\ v_1 & v_2 & v_3 \end{vmatrix}$.

A triple scalar product of vectors represents the volume of a parallelepiped.

true bearing (p. 483) A directional measurement of a vector where the angle is measured clockwise from north.

turning point (p. 99) A point on the graph of a function that indicates where the graph changes from increasing to decreasing, or vice versa. The location of a relative maximum or minimum.

two-sided limit (p. 738) The limit of $f(x)$ as x approaches c from the left and from the right, which exists only when both one-sided limits exist and are equal.

two-tailed test (p. 706) The hypothesis test if H_a: $\mu \neq k$.

método del triángulo (pág. 484) Método para el cálculo del vector resultante mediante la traslación de un vector, de modo que su cola toque la punta de otro vector. Luego, se traza el vector resultante y se obtiene un triángulo.

forma trigonométrica (pág. 570) Ver *forma polar*.

función trigonométrica (pág. 220) Sea θ un ángulo agudo de un triángulo rectángulo y op, ady e hip las longitudes del lado opuesto a θ, el lado adyacente a θ y la hipotenusa, respectivamente. Entonces, las funciones trigonométricas de θ son las siguientes.

$$\text{sen}\,\theta = \frac{\text{op}}{\text{hip}} \qquad \cos\theta = \frac{\text{ady}}{\text{hip}} \qquad \tan\theta = \frac{\text{op}}{\text{ady}}$$

$$\csc\theta = \frac{\text{hip}}{\text{op}} \qquad \sec\theta = \frac{\text{hip}}{\text{ady}} \qquad \cot\theta = \frac{\text{ady}}{\text{op}}$$

identidad trigonométrica (pág. 312) Ecuación que incluye funciones trigonométricas y que se cumple para todos los valores de la variable.

razones trigonométricas (pág. 220) Razones que se obtienen al relacionar las medidas de los lados de un triángulo rectángulo en referencia a un ángulo θ.

producto escalar triple (pág. 521) Si $\mathbf{t} = t_1\mathbf{i} + t_2\mathbf{j} + t_3\mathbf{k}$, $\mathbf{u} = u_1\mathbf{i} + u_2\mathbf{j} + u_3\mathbf{k}$, y $\mathbf{v} = v_1\mathbf{i} + v_2\mathbf{j} + v_3\mathbf{k}$, el triple producto escalar está dado por

$$\mathbf{t} \cdot (\mathbf{u} \times \mathbf{v}) = \begin{vmatrix} t_1 & t_2 & t_3 \\ u_1 & u_2 & u_3 \\ v_1 & v_2 & v_3 \end{vmatrix}.$$

Un triple producto escalar de vectores representa el volumen de un paralelepípedo.

rumbo verdadero (pág. 483) Medición de la dirección de un vector en la cual los ángulos se miden en el sentido de las manecillas del reloj, a partir del norte.

punto de inflexión (pág. 99) Punto de la gráfica de una función en donde la gráfica cambia de ser creciente a decreciente, o viceversa. Es el punto donde se halla un máximo o mínimo relativo.

límite bilateral (pág. 738) El límite de $f(x)$ a medida que x se aproxima a c por la izquierda es igual al límite de $f(x)$ a medida que x se aproxima a c por la derecha.

prueba de dos colas (pág. 706) Prueba de hipótesis si H_a: $\mu \neq k$.

U

unbounded (p. 408) A region formed by a system of linear inequalities in a linear programming problem that is not a polygon.

ilimitada (pág. 408) Región que no forma un polígono y que está definida por un sistema de desigualdades lineales de un problema de programación lineal.

union (p. P4) The union of sets A and B is all elements in both A and B, written as $A \cup B$.

unit circle (p. 247) A circle of radius 1 centered at the origin of a coordinate system.

unit vector (p. 494) A vector that has a magnitude of 1 unit.

univariate data (pp. P32, 654) Data with one variable.

universal set (p. P3) The set of all possible elements for a situation.

upper bound (p. 121) A real number b that is greater than or equal to the greatest real zero of a polynomial function.

upper limit (p. 777) The upper bound of a definite integral.

unión (pág. P4) La unión de los conjuntos A y B son todos los elementos tanto en A como en B; se escribe como $A \cup B$.

círculo unitario (pág. 247) Círculo de radio 1 y cuyo centro está en el origen de un sistema de coordenadas.

vector unitario (pág. 494) Vector que tiene una magnitud de 1 unidad.

datos univariados (págs. P32 y 654) Datos con una variable.

conjunto universal (pág. P3) Conjunto de todos los elementos posibles de una situación.

cota superior (pág. 121) Número real b que es mayor que o igual al mayor cero real de una función polinomial.

límite superior (pág. 777) La cota superior de una integral definida.

V

variance (p. P32) The mean of the squares of the deviations from the arithmetic mean.

vector (p. 482) A quantity that has both magnitude and direction.

vector projection (p. 503) Let $\mathbf{u}$ and $\mathbf{v}$ be nonzero vectors, and let $\mathbf{w}_1$ and $\mathbf{w}_2$ be vector components of $\mathbf{u}$ such that $\mathbf{w}_1$ is parallel to $\mathbf{v}$. Then vector $\mathbf{w}_1$ is called the vector projection of $\mathbf{u}$ onto $\mathbf{v}$, denoted $\text{proj}_\mathbf{v}\mathbf{u}$, and $\text{proj}_\mathbf{v}\mathbf{u} = \left(\dfrac{\mathbf{u} \cdot \mathbf{v}}{|\mathbf{v}|^2}\right)\mathbf{v}$.

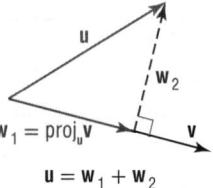

varianza (pág. P32) Media de los cuadrados de las desviaciones de la media aritmética.

vector (pág. 482) Cantidad que tiene magnitud y dirección.

vector proyección (pág. 503) Sean $\mathbf{u}$ y $\mathbf{v}$ vectores distintos de cero; sean $\mathbf{w}_1$ y $\mathbf{w}_2$ vectores componentes de $\mathbf{u}$ tales que $\mathbf{w}_1$ es paralelo a $\mathbf{v}$. Entonces el vector $\mathbf{w}_1$ se conoce como el vector proyección de $\mathbf{u}$ sobre $\mathbf{v}$, que se denota como $\text{proy}_\mathbf{v}\mathbf{u}$, donde $\text{proy}_\mathbf{v}\mathbf{u} = \left(\dfrac{\mathbf{u} \cdot \mathbf{v}}{|\mathbf{v}|^2}\right)\mathbf{v}$.

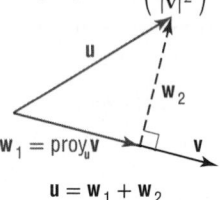

verify an identity (p. 320) To prove that both sides of the equation are equal for all values of the variable for which both sides are defined.

vertex (pp. P9, 231, 422, 432) **1.** The common endpoint of two or more noncollinear rays. **2.** A point at which a parabola intersects its axis of symmetry. **3.** The two endpoints of the major axis of an ellipse.

vertical asymptote (p. 131) The line $x = c$ is a vertical asymptote of the graph of f if $\lim\limits_{x \to c^-} f(x) = \pm\infty$ or $\lim\limits_{x \to c^+} f(x) = \pm\infty$.

vertical shift (p. 262) For a sinusoidal function, a vertical translation that is the average of the maximum and minimum values of the function.

vertices (p. 432) The endpoints of the major axis of an ellipse.

verificar una identidad (pág. 320) Demostrar que ambos lados de una ecuación son iguales para todos los valores de la variable para los cuales ambos lados están definidos.

vértice (págs. P9, 231, 422, 432) **1.** Extremo común de dos o más rayos no colineales. **2.** Punto en que una parábola interseca su eje de simetría. **3.** Los dos extremos del eje mayor de una elipse.

asíntota vertical (pág. 131) La recta $x = c$ es la asíntota vertical de la gráfica de f, si $\lim\limits_{x \to c^-} f(x) = \pm\infty$ o $\lim\limits_{x \to c^+} f(x) = \pm\infty$.

desplazamiento vertical (pág. 262) Para una función sinusoidal, es una traslación vertical que es igual al promedio de los valores máximo y mínimo de la función.

vértices (pág. 432) Los extremos del eje mayor de una elipse.

work (p. 505) If a constant force **F** acts on an object to move it from point A to point B, then the work done equals the dot product of the constant force **F** and the directed distance $\overrightarrow{AB}$, or $\mathbf{F} \cdot \overrightarrow{AB}$.

trabajo (pág. 505) Si una fuerza constante **F** actúa sobre un cuerpo para moverlo del punto A al punto B, el trabajo realizado es igual al producto vector de la fuerza constante **F** y la distancia dirigida $\overrightarrow{AB}$, o $\mathbf{F} \cdot \overrightarrow{AB}$.

zeros (p. 15) The x-intercepts of the graph of a function.

ceros (pág. 15) Intersecciones con el eje x de la gráfica de una función.

zero function (p. 45) The function sometimes known as the zero function is the constant function with constant $c = 0$. In other words, $f(x) = 0$.

función cero (pág. 45) La función a veces conocida como la función cero es la función constante con la constante $c = 0$. En otras palabras, $f(x) = 0$.

zero matrix (p. P23) A matrix in which every element is zero.

matriz nula (pág. P23) Matriz cuyos elementos son todos iguales a cero.

zero vector (p. 485) The resultant when two opposite vectors are added, has a magnitude of 0 and no specific direction. Also called the *null* vector, denoted by $\vec{0}$ or **0**.

vector nulo (pág. 485) La resultante que se obtiene cuando se suman dos vectores opuestos; su magnitud es 0 y no tiene dirección específica. También se llama vector *cero*. Se denota como $\vec{0}$ ó **0**.

z-axis (p. 510) a third axis in a three-dimensional coordinate system that passes through the origin and is perpendicular to both the x- and y-axes.

eje z (pág. 510) Tercer eje en un sistema tridimensional de coordenadas. Pasa por el origen y es perpendicular al eje x y al eje y.

z-value (p. 676) Represents the number of standard deviations that a given data value is from the mean. Also known as the *z-score* and *z test statistic*.

valor de z (pág. 676) Representa el número de desviaciones estándar que separan un valor dado de los datos de la media. También se conoce como *estadístico z* o *prueba de z*.

Index

Index

Index

Index

Index

H

Index

Index

Index

Index

Index

Trigonometric inequalities, 334

Trigonometric ratios, 220–221

Triple scalar product, 521

True bearing, 483

t statistic, 707–709

t-test for the correlation coefficient, 715

Two-sided limits, 738–739

Two-tailed test, 706–709

U

Unbounded behavior, 739

Unbounded interval, 5

Unbounded regions, 408

Uniform distribution, 656

Union, P4

Unit circle, 344
    16-point, 248
    trigonometric functions on, 242–253
    using
        to evaluate trigonometric functions, 336
        to solve trigonometric equations, 328

Unit vectors, 494–495
    linear combination of, 495
    notation of, 494
    standard, 494

Univariate data, P32, 654

Universal set, P3

Upper bound test, 122

Upper class limit, P34

Upper limits, 777

Upper quartile, P35

V

Values
    convenient, 399
    critical, 697, 707
    expected, 667–668
    of functions
        estimating, 13–14
        finding, 7
    principal, 282

    P-values, 708–709
    y-values, 6

Variables, P32
    continuous random, 664
    dependent, 7, 713
    discrete random, 664
    explanatory, 713
    independent, 7, 15, 713
    normally distributed, 685
    random, 664–670, 714

response, 713

Variance, P33
    of a probability distribution, 667

Variation, in sign, 123

Vector fields, 530–531

Vector projections, 500–509
    to find force, 504

Vectors, 480–531
    adding, 485, 493, 513
    angles between, 502
        in space, 518
    applied operations of, 496
    component form of, 492–499
    components of, 487
    in the coordinate plane, 492–499
    of cross products, 518–524
    cross products of
        in space, 518–524
    direction angles of, 496
    directions of, 482–483, 485, 492
    dot product operations, 500–509
    dot products of
        in space, 518–524
    equivalent, 483, 492
    expressing in space algebraically, 512
    initial points of, 482
    magnitudes of, 482–483, 492–493, 500, 501
    multiplying
        by a matrix, 517
        by a scalar, 485, 493
        by a scalar in space, 513
    operations in space, 513
    operations with, 485
    opposite, 483
    orthogonal, 500–501, 518
    parallel, 483, 485, 502
    perpendicular, 500, 502, 503
    rectangular components of, 487
    resultants of, 484
    to solve navigation problems, 486
    standard position of, 482
    subtracting, 485, 493, 513
    terminal points of, 482
    in three-dimensional space, 510–516
    transformations of, 517
    unit, 494–495
    zero, 485

Velocity, 758–764. *See also* speed
    average, 760
    instantaneous, 760–761, 769

Venn diagrams, P3

Verbal relations, 5

Verify identities, 320–326, 339, 340
    by combining fractions, 321
    by factoring, 322
    by multiplying, 321
    strategies for, 323
    by working each side separately, 322

Vertex Theorem for Optimization, 405

Vertical asymptotes, 130–134, 269, 274, 740

Vertical dilations, 49

Vertical line test, 6

Vertical shift, 262

Vertical translation, 47

Vertices, 231. *See also* Co-vertices
    of ellipses, 432–433
    of hyperbolas, 442–443
    of parabolas, P9–P10, 422–427
    of polygons, 387

Vocabulary, P1, 3, 85, 157, 219, 311, 363, 421, 481, 533, 589, 653, 735. *See also* Reading Math; Review Vocabulary; Vocabulary Check

Vocabulary Check, 76, 148, 211, 302, 355, 413, 473, 525, 580, 645, 725, 792

Volume
    maximum, 418
    of paraboloid, 479
    of parallelepiped, 521
    of spheres, 478

W

Watch Out!
    amplitude, 258
    combinations, P30
    common error, P5, P16
    common misconception, 221
    completing the square, P12
    complex conjugates, P7
    complex numbers, 126
    compositions and inverses, 286
    correlation vs. causation, 719
    cross product, 519
    determining critical values, 708
    determining period, 259
    determining signs, 348
    disjoint sets, P4
    dividing by trigonometric factors, 28
    division by zero, 390
    elimination, P19
    evaluating e^x, 639
    factoring, 749
    graphing, 315
    graphing over the period, 544
    horizontal line test, 66
    infinite limits, 739
    integrals, 788
    interpreting graphs, 769
    intervals, 35
    limits, 779
    logarithm of a sum, 183
    magnitude, 483
    making predictions, 719
    matrix equations, P25
    misinterpreting expected value, 667
    mode, 537
    negative derivatives, 767

Formulas

Complex Numbers

Product Formula $z_1 z_2 = r_1 r_2 [\cos(\theta_1 + \theta_2) + i \sin(\theta_1 + \theta_2)]$

Quotient Formula $\dfrac{z_1}{z_2} = \dfrac{r_1}{r_2}[\cos(\theta_1 - \theta_2) + i \sin(\theta_1 - \theta_2)]$

Distinct Roots Formula $r^{\frac{1}{p}}\left(\cos \dfrac{\theta + 2n\pi}{p} + i \sin \dfrac{\theta + 2n\pi}{p}\right)$

De Moivre's Theorem $z^n = [r(\cos\theta + i\sin\theta)]^n$ or $r^n(\cos n\theta + i\sin n\theta)$

Sequences and Series

Sum of Finite Arithmetic Series $S_n = \dfrac{n}{2}(a_1 + a_n)$ or $S_n = \dfrac{n}{2}[2a_1 + (n-1)d]$

Sum of Finite Geometric Series $S_n = a_1\left(\dfrac{1 - r^n}{1 - r}\right)$ or $S_n = \dfrac{a_1 - a_n r}{1 - r}$

Sum of Infinite Geometric Series $S = \dfrac{a_1}{1 - r}, \, |r| < 1$

Euler's Formula $e^{i\theta} = \cos\theta + i\sin\theta$

Power Series $\displaystyle\sum_{n=0}^{\infty} a_n x^n = a_0 + a_1 x + a_2 x^2 + a_3 x^3 + \cdots$

Exponential Series $e^x = \displaystyle\sum_{n=0}^{\infty} \dfrac{x^n}{n!} = 1 + x + \dfrac{x^2}{2!} + \dfrac{x^3}{3!} + \dfrac{x^4}{4!} + \dfrac{x^5}{5!} + \cdots$

Binomial Theorem $(a + b)^n = {}_nC_0\, a^n b^0 + {}_nC_1\, a^{n-1} b^1 + {}_nC_2\, a^{n-2} b^2 + \cdots + {}_nC_r\, a^{n-r} b^r + \cdots + {}_nC_n\, a^0 b^n$

Cosine and Sine Power Series $\cos x = \displaystyle\sum_{n=0}^{\infty} \dfrac{(-1)^n x^{2n}}{(2n)!} = 1 - \dfrac{x^2}{2!} + \dfrac{x^4}{4!} - \dfrac{x^6}{6!} + \dfrac{x^8}{8!} - \cdots$ $\sin x = \displaystyle\sum_{n=0}^{\infty} \dfrac{(-1)^n x^{2n+1}}{(2n+1)!} = x - \dfrac{x^3}{3!} + \dfrac{x^5}{5!} - \dfrac{x^7}{7!} + \dfrac{x^9}{9!} - \cdots$

Statistics

z-Values $z = \dfrac{X - \mu}{\sigma}$

z-Value of a Sample Mean $z = \dfrac{\bar{x} - \mu}{\sigma_{\bar{x}}}$

Binomial Probability $P(X) = {}_nC_x\, p^x q^{n-x} = \dfrac{n!}{(n-x)!\,x!}\, p^x q^{n-x}$

Maximum Error of Estimate $E = z \cdot \sigma_{\bar{x}}$ or $z \cdot \dfrac{\sigma}{\sqrt{n}}$

Confidence Interval, Normal Distribution $CI = \bar{x} \pm E$ or $\bar{x} \pm z \cdot \dfrac{\sigma}{\sqrt{n}}$

Confidence Interval, t-Distribution $CI = \bar{x} \pm t \cdot \dfrac{s}{\sqrt{n}}$

Correlation Coefficient $r = \dfrac{1}{n-1}\displaystyle\sum\left(\dfrac{x_i - \bar{x}}{s_x}\right)\left(\dfrac{y_i - \bar{y}}{s_y}\right)$

t-Test for the Correlation Coefficient $t = r\sqrt{\dfrac{n-2}{1-r^2}}$, degrees of freedom: $n - 2$

Limits

Sum $\displaystyle\lim_{x\to c}[f(x) + g(x)] = \lim_{x\to c}f(x) + \lim_{x\to c}g(x)$

Difference $\displaystyle\lim_{x\to c}[f(x) - g(x)] = \lim_{x\to c}f(x) - \lim_{x\to c}g(x)$

Scalar Multiple $\displaystyle\lim_{x\to c}[k\,f(x)] = k\lim_{x\to c}f(x)$

Product $\displaystyle\lim_{x\to c}[f(x)\cdot g(x)] = \lim_{x\to c}f(x) \cdot \lim_{x\to c}g(x)$

Quotient $\displaystyle\lim_{x\to c}\dfrac{f(x)}{g(x)} = \dfrac{\lim_{x\to c}f(x)}{\lim_{x\to c}g(x)}$, if $\displaystyle\lim_{x\to c}g(x) \neq 0$

Power $\displaystyle\lim_{x\to c}[f(x)^n] = \left[\lim_{x\to c}f(x)\right]^n$

nth Root $\displaystyle\lim_{x\to c}\sqrt[n]{f(x)} = \sqrt[n]{\lim_{x\to c}f(x)}$, if $\displaystyle\lim_{x\to c}f(x) > 0$ when n is even

Velocity   Average   Instantaneous $v_{\text{avg}} = \dfrac{f(b) - f(a)}{b - a}$ $v(t) = \displaystyle\lim_{h\to 0}\dfrac{f(t+h) - f(t)}{h}$

Derivatives

Power Rule If $f(x) = x^n$, $f'(x) = nx^{n-1}$

Sum or Difference If $f(x) = g(x) \pm h(x)$, then $f'(x) = g'(x) \pm h'(x)$

Product Rule $\dfrac{d}{dx}[f(x)g(x)] = f'(x)g(x) + f(x)g'(x)$

Quotient Rule $\dfrac{d}{dx}\left[\dfrac{f(x)}{g(x)}\right] = \dfrac{f'(x)g(x) - f(x)g'(x)}{[g(x)]^2}$

Integrals

Indefinite Integral $\displaystyle\int f(x)\, dx = F(x) + C$

Fundamental Theorem of Calculus $\displaystyle\int_a^b f(x)\, dx = F(b) - F(a)$

Formulas

Function Operations

Addition	$(f + g)(x) = f(x) + g(x)$	Multiplication	$(f \cdot g)(x) = f(x) \cdot g(x)$
Subtraction	$(f - g)(x) = f(x) - g(x)$	Division	$\left(\dfrac{f}{g}\right)(x) = \dfrac{f(x)}{g(x)}, g(x) \neq 0$

Exponential and Logarithmic Functions

Compound Interest	$A = P\left(1 + \dfrac{r}{n}\right)^{nt}$	Exponential Growth or Decay	$N = N_0(1 + r)^t$
Continuous Compound Interest	$A = Pe^{rt}$	Continuous Exponential Growth or Decay	$N = N_0 e^{kt}$
Product Property	$\log_b xy = \log_b x + \log_b y$	Power Property	$\log_b x^p = p \log_b x$
Quotient Property	$\log_b \dfrac{x}{y} = \log_b x - \log_b y$	Change of Base	$\log_b x = \dfrac{\log_a x}{\log_a b}$
Logistic Growth	$f(t) = \dfrac{c}{1 + ae^{-bt}}$		

Conic Sections

Parabola	$(x - h)^2 = 4p(y - k)$ or $(y - k)^2 = 4p(x - h)$	Circle	$x^2 + y^2 = r^2$ or $(x - h)^2 + (y - k)^2 = r^2$
Ellipse	$\dfrac{(x - h)^2}{a^2} + \dfrac{(y - k)^2}{b^2} = 1$ or $\dfrac{(x - h)^2}{b^2} + \dfrac{(y - k)^2}{a^2} = 1$	Hyperbola	$\dfrac{(x - h)^2}{a^2} - \dfrac{(y - k)^2}{b^2} = 1$ or $\dfrac{(y - k)^2}{a^2} - \dfrac{(x - h)^2}{b^2} = 1$
Rotation of Conics	$x' = x \cos \theta + y \sin \theta$ and $y' = y \cos \theta - x \sin \theta$		

Parametric Equations

Vertical Position	$y = tv_0 \sin \theta - \dfrac{1}{2}gt^2 + h_0$	Horizontal Distance	$x = tv_0 \cos \theta$

Vectors

Addition in Plane	$\mathbf{a} + \mathbf{b} = \langle a_1 + b_1, a_2 + b_2 \rangle$	Addition in Space	$\mathbf{a} + \mathbf{b} = \langle a_1 + b_1, a_2 + b_2, a_3 + b_3 \rangle$
Subtraction in Plane	$\mathbf{a} - \mathbf{b} = \langle a_1 - b_1, a_2 - b_2 \rangle$	Subtraction in Space	$\mathbf{a} - \mathbf{b} = \mathbf{a} + (-\mathbf{b})$ $= \langle a_1 - b_1, a_2 - b_2, a_3 - b_3 \rangle$
Scalar Multiplication in Plane	$k\mathbf{a} = \langle ka_1, ka_2 \rangle$	Scalar Multiplication in Space	$k\mathbf{a} = \langle ka_1, ka_2, ka_3 \rangle$
Dot Product in Plane	$\mathbf{a} \cdot \mathbf{b} = a_1 b_1 + a_2 b_2$	Dot Product in Space	$\mathbf{a} \cdot \mathbf{b} = a_1 b_1 + a_2 b_2 + a_3 b_3$
Angle Between Two Vectors	$\cos \theta = \dfrac{\mathbf{a} \cdot \mathbf{b}}{|\mathbf{a}||\mathbf{b}|}$	Projection of u onto v	$\text{proj}_\mathbf{v}\mathbf{u} = \left(\dfrac{\mathbf{u} \cdot \mathbf{v}}{|\mathbf{v}|^2}\right)\mathbf{v}$
Magnitude of a Vector	$|\mathbf{v}| = \sqrt{(x_2 - x_1)^2 + (y_2 - y_1)^2}$	Triple Scalar Product	$\mathbf{t} \cdot (\mathbf{u} \times \mathbf{v}) = \begin{vmatrix} t_1 & t_2 & t_3 \\ u_1 & u_2 & u_3 \\ v_1 & v_2 & v_3 \end{vmatrix}$
Cross Product	$\mathbf{a} \times \mathbf{b} = (a_2 b_3 - a_3 b_2)\mathbf{i} - (a_1 b_3 - a_3 b_1)\mathbf{j} + (a_1 b_2 - a_2 b_1)\mathbf{k}$		

Trigonometric Functions and Identities

Trigonometric Functions

Trigonometric Functions

$$\sin \theta = \frac{\text{opp}}{\text{hyp}} \qquad \cos \theta = \frac{\text{adj}}{\text{hyp}} \qquad \tan \theta = \frac{\text{opp}}{\text{adj}} = \frac{\sin \theta}{\cos \theta}$$

$$\csc \theta = \frac{\text{hyp}}{\text{opp}} = \frac{1}{\sin \theta} \qquad \sec \theta = \frac{\text{hyp}}{\text{adj}} = \frac{1}{\cos \theta} \qquad \cot \theta = \frac{\text{adj}}{\text{opp}} = \frac{\cos \theta}{\sin \theta}$$

Law of Cosines $\quad a^2 = b^2 + c^2 - 2bc \cos A \qquad b^2 = a^2 + c^2 - 2ac \cos B \qquad c^2 = a^2 + b^2 - 2ab \cos C$

Law of Sines $\quad \dfrac{\sin A}{a} = \dfrac{\sin B}{b} = \dfrac{\sin C}{c} \qquad\qquad$ **Heron's Formula** $\text{Area} = \sqrt{s(s-a)(s-b)(s-c)}$

Linear Speed $\quad v = \dfrac{s}{t} \qquad\qquad$ **Angular Speed** $\quad \omega = \dfrac{\theta}{t}$

Trigonometric Identities

Reciprocal

$$\sin \theta = \frac{1}{\csc \theta} \qquad \cos \theta = \frac{1}{\sec \theta} \qquad \tan \theta = \frac{1}{\cot \theta}$$

$$\csc \theta = \frac{1}{\sin \theta} \qquad \sec \theta = \frac{1}{\cos \theta} \qquad \cot \theta = \frac{1}{\tan \theta}$$

Pythagorean

$$\sin^2 \theta + \cos^2 \theta = 1 \qquad \tan^2 \theta + 1 = \sec^2 \theta \qquad \cot^2 \theta + 1 = \csc^2 \theta$$

Cofunction

$$\sin \theta = \cos\left(\frac{\pi}{2} - \theta\right) \qquad \tan \theta = \cot\left(\frac{\pi}{2} - \theta\right) \qquad \sec \theta = \csc\left(\frac{\pi}{2} - \theta\right)$$

$$\cos \theta = \sin\left(\frac{\pi}{2} - \theta\right) \qquad \cot \theta = \tan\left(\frac{\pi}{2} - \theta\right) \qquad \csc \theta = \sec\left(\frac{\pi}{2} - \theta\right)$$

Odd-Even

$$\sin(-\theta) = -\sin \theta \qquad \cos(-\theta) = \cos \theta \qquad \tan(-\theta) = -\tan \theta$$

$$\csc(-\theta) = -\csc \theta \qquad \sec(-\theta) = \sec \theta \qquad \cot(-\theta) = -\cot \theta$$

Sum & Difference

$$\cos(\alpha + \beta) = \cos \alpha \cos \beta - \sin \alpha \sin \beta \qquad \cos(\alpha - \beta) = \cos \alpha \cos \beta + \sin \alpha \sin \beta$$

$$\sin(\alpha + \beta) = \sin \alpha \cos \beta + \cos \alpha \sin \beta \qquad \sin(\alpha - \beta) = \sin \alpha \cos \beta - \cos \alpha \sin \beta$$

$$\tan(\alpha + \beta) = \frac{\tan \alpha + \tan \beta}{1 - \tan \alpha \tan \beta} \qquad \tan(\alpha - \beta) = \frac{\tan \alpha - \tan \beta}{1 + \tan \alpha \tan \beta}$$

Double-Angle

$$\cos 2\theta = \cos^2 \theta - \sin^2 \theta \qquad \cos 2\theta = 2\cos^2 \theta - 1 \qquad \cos 2\theta = 1 - 2\sin^2 \theta$$

$$\sin 2\theta = 2 \sin \theta \cos \theta \qquad \tan 2\theta = \frac{2 \tan \theta}{1 - \tan^2 \theta}$$

Power-Reducing

$$\sin^2 \theta = \frac{1 - \cos 2\theta}{2} \qquad \cos^2 \theta = \frac{1 + \cos 2\theta}{2} \qquad \tan^2 \theta = \frac{1 - \cos 2\theta}{1 + \cos 2\theta}$$

Half-Angle

$$\sin \frac{\theta}{2} = \pm\sqrt{\frac{1 - \cos \theta}{2}} \qquad \cos \frac{\theta}{2} = \pm\sqrt{\frac{1 + \cos \theta}{2}}$$

$$\tan \frac{\theta}{2} = \pm\sqrt{\frac{1 - \cos \theta}{1 + \cos \theta}} \qquad \tan \frac{\theta}{2} = \frac{1 - \cos \theta}{\sin \theta} \qquad \tan \frac{\theta}{2} = \frac{\sin \theta}{1 + \cos \theta}$$

Product-to-Sum

$$\sin \alpha \sin \beta = \frac{1}{2}[\cos(\alpha - \beta) - \cos(\alpha + \beta)] \qquad \sin \alpha \cos \beta = \frac{1}{2}[\sin(\alpha + \beta) + \sin(\alpha - \beta)]$$

$$\cos \alpha \cos \beta = \frac{1}{2}[\cos(\alpha - \beta) + \cos(\alpha + \beta)] \qquad \cos \alpha \sin \beta = \frac{1}{2}[\sin(\alpha + \beta) - \sin(\alpha - \beta)]$$

Sum-to-Product

$$\sin \alpha + \sin \beta = 2 \sin\left(\frac{\alpha + \beta}{2}\right) \cos\left(\frac{\alpha - \beta}{2}\right) \qquad \cos \alpha + \cos \beta = 2 \cos\left(\frac{\alpha + \beta}{2}\right) \cos\left(\frac{\alpha - \beta}{2}\right)$$

$$\sin \alpha - \sin \beta = 2 \cos\left(\frac{\alpha + \beta}{2}\right) \sin\left(\frac{\alpha - \beta}{2}\right) \qquad \cos \alpha - \cos \beta = -2 \sin\left(\frac{\alpha + \beta}{2}\right) \sin\left(\frac{\alpha - \beta}{2}\right)$$